*Freshwater
Algae of
North America*

Ecology and Classification

AQUATIC ECOLOGY Series

Series Editor
James H. Thorp
Kansas Biological Survey
University of Kansas
Lawrence, Kansas

Editorial Advisory Board
Alan P. Covich, Jack A. Stanford, Roy Stein and Robert G. Wetzel

Other titles in the series:

Groundwater Ecology
Janine Gilbert, Dan L. Danielopol, Jack A. Stanford

Algal Ecology
R. Jan Stevenson, Max L. Bothwell, Rex. L. Lowe

Streams and Ground Waters
Jeremy B. Jones, Patrick J. Mulholland

Freshwater Ecology
Walter K. Dodds

Freshwater Algae of North America

Ecology and Classification

Edited by

JOHN D. WEHR
Louis Calder Center—Biological Station
and Department of Biological Sciences
Fordham University
Armonk, New York

and

ROBERT G. SHEATH
Office of Provost and Vice President for Academic Affairs
California State University, San Marcos
San Marcos, California

ACADEMIC PRESS
An Imprint of Elsevier

Amsterdam Boston London New York Oxford Paris San Diego
San Francisco Singapore Sydney Tokyo

Front cover image: Kandis Elliot
Institute of Applied Science.

This book is printed on acid-free paper.

Copyright 2003, Elsevier

All Rights Reserved.
No part of this publication may be reproduced or trasmitted in any form or by any means, electronic or mechanical, including photocopy, recording, or any information storage and retrieval system, without permission in writing from the publisher.

The appearance of the code at the bottom of the first page of a chapter in this book indicates the Publisher's consent that copies of the chapter may be made for personal or internal use of specific clients. This consent is given on the condition, however, that the copier pay the stated per copy fee through the Copyright Clearance Center, Inc. (www.copyright.com), for copying beyond that permitted by Sections 107 or 108 of the U.S. Copyright Law. This consent does not extend to other kinds of copying, such as copying for general distribution, for advertising or promotional purposes, for creating new collective works, or for resale.
Copy fees for pre-2003 chapters are as shown on the title pages. If no fee code appears on the title page, the copy fee is the same as for current chapters.
$35.00

Permissions may be sought directly from Elsevier's Science and Technology Rights Department in Oxford, UK. Phone: (44) 1865 843830, Fax: (44) 1865 853333, e-mail: permissions@elsevier.co.uk. You may also complete your request on-line via the Elsevier homepage: http://www.elsevier.com by selecting "Customer Support" and then "Obtaining Permissions".

Academic Press
An Imprint of Elsevier
525 B Street, Suite 1900, San Diego, California 92101-4495, USA
http://www.academicpress.com

Academic Press
An Imprint of Elsevier
84 Theobald's Road, London WC1X 8RR, UK
http://www.academicpress.com

Academic Press
200 Wheeler Road Burlington, Massachusetts 01803, USA
http://www.acadmicpressbooks.com

Library of Congress Control Number: 2002107708

International Standard Book Number: 0-12-741550-5

PRINTED IN THE UNITED STATES OF AMERICA
05 06 07   9 8 7 6 5 4 3

We acknowledge the numerous phycological pioneers who blazed many trails for those of us who followed in the study of North American freshwater algae, especially Gilbert M. Smith, Gerald W. Prescott, George J. Schumacher, Larry A. Whitford, Francis Wolle and Hoaratio C. Wood.

John D. Wehr
Robert G. Sheath

My heartfelt thanks and love go to Deborah Donaldson for her faith and support.

John D. Wehr

My love and appreciation to Mary Koske who supported me through those many collecting trips to far-flung places and the myriad of career and location changes.

Robert G. Sheath

Contents

Contributors xiii
Preface xv

1
INTRODUCTION TO FRESHWATER ALGAE
Robert G. Sheath and John D. Wehr

    I. Introduction 1
    II. Classification 5
    III. Taxonomic Chapters in This Book 8
    Literature Cited 9

2
FRESHWATER HABITATS OF ALGAE
John D. Wehr and Robert G. Sheath

    I. What is Fresh Water? 11
    II. Lentic Environments 12
    III. Lotic Environments 28
    IV. Wetlands 38
    V. Thermal and Acidic Environments 40
    VI. Unusual Environments 42
    Literature Cited 45

3
COCCOID AND COLONIAL CYANOBACTERIA
Jiří Komárek

    I. Introduction 59
    II. Morphology and Diversity 60
    III. Ecology and Distribution 63
    IV. Collection, Preparation, and Culture 67
    V. Key and Descriptions of Genera 68
    VI. Guide to Literature for Species Identification 110
    Literature Cited 110

4
FILAMENTOUS CYANOBACTERIA
Jiří Komárek, Hedy Kling, and Jaroslava Komárková

    I. Introduction 117
   II. Morphology 118
  III. Ecology 120
  IV. Methods 121
   V. Key and Descriptions of Genera 121
      Note Added in Proof 189
  VI. Guide to Literature for Species Identification 191
      Literature Cited 191

5
RED ALGAE
Robert G. Sheath

    I. Introduction 197
   II. Diversity and Morphology 197
  III. Ecology and Distribution 202
  IV. Collection and Preparation for Identification 206
   V. Key and Descriptions of Genera 207
  VI. Guide to Literature for Species Identification 221
      Literature Cited 221

6
FLAGELLATED GREEN ALGAE
Hisayoshi Nozaki

    I. Introduction 225
   II. Diversity and Morphology 225
  III. Ecology and Distribution 226
  IV. Collection and Preparation for Identification 227
   V. Key and Descriptions of Genera 227
  VI. Guide to Literature for Species Identification 247
      Literature Cited 248

7
NONMOTILE COCCOID AND COLONIAL GREEN ALGAE
L. Elliot Shubert

    I. Introduction 253
   II. Diversity and Morphology 254
  III. Ecology and Distribution 257
  IV. Collection and Preparation for Identification 258
   V. Key and Descriptions of Genera 259
  VI. Guide to Literature for Species Identification 307
      Literature Cited 307

8
FILAMENTOUS AND PLANTLIKE GREEN ALGAE
David M. John

     I. Introduction 311
    II. Diversity and Morphology 311
   III. Classification of Green Algae 312
   IV. Ecology and Distribution 313
    V. Collection and Preparation of Samples 315
   VI. Key and Descriptions of Genera 316
  VII. Guide to Literature for Species Identification 347
       Literature Cited 349

9
CONJUGATING GREEN ALGAE AND DESMIDS
Joseph F. Gerrath

    I. Introduction 353
   II. Diversity and Morphology 354
  III. Ecology and Distribution 363
  IV. Collection and Preparation for Identification 365
   V. Key and Descriptions of Genera 365
  VI. Guide to Literature for Species Identification 379
      Literature Cited 379

10
PHOTOSYNTHETIC EUGLENOIDS
James R. Rosowski

    I. Introduction 383
   II. Diversity and Morphology 387
  III. Ecology and Distribution 405
  IV. Collection, Culturing, and Preparation for Identification 408

V. Key and Descriptions of North American Genera   410
VI. Guide to Literature for Species Identification   415
    Literature Cited   416

11
EUSTIGMATOPHYTE, RAPHIDOPHYTE, AND TRIBOPHYTE ALGAE
Donald W. Ott and Carla K. Oldham-Ott

I. General Introduction   423
II. Eustigmatophytes   424
III. Raphidophytes   427
IV. Tribophytes   429
V. Collection and Preparation for Identification   463
VI. Guide to Literature for Species Identification   465
    Literature Cited   466

12
CHRYSOPHYCEAN ALGAE
Kenneth H. Nicholls and Daniel E. Wujek

I. Introduction   471
II. Diversity and Morphology   473
III. Ecology   485
IV. Collection and Preparation for Identification   490
V. Key and Descriptions of Genera   491
VI. Guide to Literature for Species Identification   503
    Literature Cited   503

13
HAPTOPHYTE ALGAE
Kenneth H. Nicholls

I. Introduction   511
II. Diversity and Morphology   512
III. Ecology and Distribution   513
IV. Collection and Preparation for Identification   514
V. Key and Descriptions of Genera   515
VI. Guide to Literature for Species Identification   519
    Literature Cited   519

14
SYNUROPHYTE ALGAE
Peter A. Siver

I. Introduction   523
II. Diversity and Morphology   524
III. Ecology and Distribution   534
IV. Collection and Preparation for Identification   539
V. Keys to Genera and Common Species from North America   541
VI. Guide to Literature for Species Identification   551
    Literature Cited   552

15
CENTRIC DIATOMS
Eugene F. Stoermer and Matthew L. Julius
J. P. Kociolek and S. A. Spaulding (Introduction)

I. General Introduction to the Diatoms   559
II. Introduction to Centric Diatoms   562
III. Classification   563
IV. Morphology and Physiology   565
V. Ecology and Evolution   568
VI. Collection and Study Methods   570
VII. Key and Descriptions of Genera   571
VIII. Guide to Literature for Species Identification   587
    Literature Cited   588

16
ARAPHID AND MONORAPHID DIATOMS
John C. Kingston

I. Introduction   595
II. Diversity and Morphology   596
III. Ecology and Distribution   604
IV. Collection and Preparation for Identification   605
V. Key and Descriptions of Genera   605
VI. Guide to Literature for Species Identification   628
    Literature Cited   631

17
SYMMETRICAL NAVICULOID DIATOMS
J. P. Kociolek and S. A. Spaulding

    I. Introduction 637
   II. Ecology and Distribution 638
  III. Key and Descriptions of Genera 639
  IV. Guide to Literature for Species Identification 651
      Literature Cited 651

18
EUNOTIOID AND ASYMMETRICAL NAVICULOID DIATOMS
J. P. Kociolek and S. A. Spaulding

    I. Introduction 655
   II. Diversity and Morphology 656
  III. Ecology and Distribution 661
  IV. Key and Descriptions of North American Genera 662
   V. Guide to Literature for Species Identification 666
      Literature Cited 666

19
KEELED AND CANALLED RAPHID DIATOMS
Rex L. Lowe

    I. Introduction 669
   II. Diversity and Morphology 670
  III. Ecology and Distribution 671
  IV. Collection and Preparation for Identification 674
   V. Keys and Descriptions of Genera 675
  VI. Guide to Literature for Species Identification 682
      Literature Cited 682

20
DINOFLAGELLATES
Susan Carty

    I. Introduction 685

   II. Morphology and Diversity 687
  III. Ecology and Distribution 699
  IV. Collection and Preparation for Identification 702
   V. Key and Descriptions of Genera 703
  VI. Guide to Literature for Species Identification 709
      Literature Cited 710

21
CRYPTOMONADS
Paul Kugrens and Brec L. Clay

     I. Introduction 715
    II. Unique Features of Cryptomonads 716
   III. Origin of Cryptomonads 736
   IV. Ecology 738
    V. Collection, Preparation for Isolation, and Culturing 740
   VI. Classification, Key, and Descriptions 740
  VII. Availability of Cryptomonads 749
 VIII. Family Katablepharidaceae 749
      Literature Cited 751

22
BROWN ALGAE
John D. Wehr

    I. Introduction 757
   II. Diversity and Morphology 758
  III. Ecology and Distribution 763
  IV. Methods for Collection and Identification 766
   V. Key and Descriptions of Genera 767
  VI. Guide to Literature for Species Identification 771
      Literature Cited 772

23
USE OF ALGAE IN ENVIRONMENTAL ASSESSMENTS
R. Jan Stevenson and John P. Smol

    I. Introduction 775

II. Goals of Environmental Assessment with Algae 776
III. Sampling and Assessing Algal Assemblages for Environmental Assessment 778
IV. Developing Metrics for Hazard Assessment 786
V. Exposure Assessment: What Are Environmental Conditions? 790
VI. Stressor–Response Relations 794
VII. Risk Characterization and Management Decisions 795
VIII. Conclusions 796
Literature Cited 797

24
CONTROL OF NUISANCE ALGAE
Carole A. Lembi

I. Introduction 805
II. Problems Associated with Algae 805
III. Control Methods for Nuisance Algae 812
Literature Cited 826

Glossary 835

Author Index 849

Subject Index 885

Taxonomic Index 897

Contributors

Number in parentheses indicate the pages on which the authors' contributions begin.

Susan Carty (685) Department of Biology, Heidelberg College, Tiffin, Ohio 44883.

Brec L. Clay (715) CH Diagnostic and Consulting Service, Loveland, Colorado 80538.

Joseph F. Gerrath (353) Department of Botany, University of Guelph, Guelph, Ontario, Canada N1G 2W1.

David M. John (311) Department of Botany, The Natural History Museum, London SW7 5BD, United Kingdom.

Matthew L. Julius (559) Department of Biological Sciences, St. Cloud State University, St. Cloud, Minnesota 56301.

John C. Kingston (595) Center for Water and the Environment, Natural Resources Research Institute, University of Minnesota Duluth, Ely, Minnesota 55731.

Hedy Kling (117) Freshwater Institute, Winnipeg, Manitoba, Canada, R3T 2N6.

J. P. Kociolek (559, 637, 655) Diatom Collection, California Academy of Sciences, Golden Gate Park, San Francisco, California 94118.

Jaroslava Komárková (117) Hydrobiological Institute, Academy of Sciences of the Czech Republic, Faculty of Biological Sciences, University of South Bohemia, CZ-37005 České Budějovice, Czech Republic.

Jiří Komárek (59, 117) Institute of Botany, Academy of Sciences of the Czech Republic, Faculty of Biological Sciences, University of South Bohemia, CZ-37982 Třeboň, Czech Republic.

Paul Kugrens (715) Department of Biology, Colorado State University, Fort Collins, Colorado 80523.

Carole A. Lembi (805) Department of Botany and Plant Pathology, Purdue University, West Lafayette, Indiana 47907.

Rex L. Lowe (669) Biological Sciences, Bowling Green State University, Bowling Green, Ohio 43403 and University of Michigan Biological Station, Pellston, Michigan 49769.

Kenneth H. Nicholls (471, 511) S-15 Concession 1, RR #1 Sunderland, Ontario, Canada L0C 1H0.

Hisayoshi Nozaki (225) Department of Biological Sciences, Graduate School of Science, University of Tokyo, Hongo, Bunkyo-ku, Tokyo 113-0033, Japan.

Carla K. Oldham-Ott (423) Department of Biology, University of Akron, Akron, Ohio 44325.

Donald W. Ott (423) Department of Biology, University of Akron, Akron, Ohio 44325.

James R. Rosowski (383) School of Biological Sciences, College of Arts and Sciences, University of Nebraska–Lincoln, Lincoln, Nebraska 68588.

Robert G. Sheath (1, 11, 197) Office of Provost and Vice President for Academic Affairs, California State University, San Marcos, San Marcos, California 92096.

L. Elliot Shubert (253) Department of Botany, The Natural History Museum, London SW7 5BD, United Kingdom.

Peter A. Siver (523) Botany Department, Connecticut College, New London, Connecticut, 06320.

John P. Smol (775) Department of Biology, Paleoecological Environmental Assessment and Research Laboratory (PEARL), Queen's University, Kingston, Ontario, Canada K7L 3N6.

S. A. Spaulding (559, 637, 655) Diatom Collection, California Academy of Sciences, Golden Gate Park, San Francisco, California 94118.

R. Jan Stevenson (775) Department of Zoology, Michigan State University, East Lansing, Michigan, 48824.

Eugene F. Stoermer (559) Michigan Herbarium University of Michigan, Ann Arbor, Michigan 48109.

John D. Wehr (1, 11, 757) Louis Calder Center—Biological Station and Department of Biological Sciences, Fordham University, Armonk, New York 10504.

Daniel E. Wujek (471) Department of Biology, Central Michigan University, Mt. Pleasant, Michigan 48859.

Preface

The study of freshwater algae in North America has a long and rich history, with some of the early monographic works dating back to the late 1800's. In recent years, there has been an enormous and remarkable level of research on this very diverse and heterogeneous collection of organisms, making any definitive taxonomic or ecological treatise always out of date. Nonetheless, it is our goal with this book to synthesize and update much of this vast knowledge, and to provide a practical and comprehensive guide to all of the genera of freshwater algae known from throughout the continent, in one volume. Chapters also provide guides to other publications and specialized works for the identification and ecological information at the species level. Our intent is to combine the necessary ecological and taxonomic information in a practical book that can be used by all scientists working in aquatic environments, whether their specialty is in environmental monitoring, ecology, evolution, systematics, biodiversity, or molecular biology. This is the first book of its sort covering the entire continent. We also hope that this book will serve to encourage new generations of aquatic biologists to explore freshwater algae carefully, rather than regarding phytoplankton or benthic algae as simply quantities of chlorophyll or carbon. The enormous variety of algae in lakes, rivers and other aquatic habitats is part of the ecological content of aquatic communities, and their ecosystem function varies with the species that occur there.

Many of the previous monographs dealing with a broad geographic region are still useful, such as Smith's (1950) *Freshwater Algae of the United States,* but most are decades old and do not contain recent taxonomic changes. Our approach is to include chapters authored by experts who have specialized in the study of specific groups of freshwater algae. Given the great quantity of research that has been produced on all of the major algal taxa, it is no longer possible for one or two authors to produce an authoritative book of this kind, and one which will span the entire range of taxonomic and ecological detail that is now known about all the organisms termed algae. This volume is modeled closely after the book by Thorp and Covich on freshwater invertebrates (*Ecology and Classification of North American Freshwater Invertebrates*), also published by Academic Press.

The organization of this book includes an introduction to the freshwater algae (with a guide to the taxonomic chapters that follow), an overview of freshwater habitats, 20 taxonomic chapters, and finally chapters on the use of algae in environmental assessments and control of nuisance algae. More than 770 genera are described and illustrated in this book, and each taxonomic chapter includes an introduction to the key terms and characteristics of the group, ecological distribution, and a guide to the taxonomic literature to distinguish species within each genus. While we have undoubtedly omitted some less common or yet unrecorded freshwater genera, this compilation represents an increase in the taxonomic scope and geographic coverage of the freshwater algae of North America. This compares with roughly 490 genera recorded from the United States by Smith (1950), about 335 in Prescott's (1962) coverage of the Western Great Lakes region, and nearly 380 genera from the southeastern U.S. reported by Whitford and Schumacher (1984). Since not all algal groups are equally well studied, coverage in the present volume varies among taxa and chapters. We hope that students, scientists working in water management agencies, and experienced phycologists will use this book thoroughly and provide us with feedback, such as missing taxa or incomplete geographic information. We will endeavor to incorporate this information into a future edition.

We are extremely grateful to the contributors of this volume who took much time and effort to research and prepare the chapters and follow through with the reviewer's and editorial suggestions. We extend our sincere thanks to the reviewers whose helpful comments enhanced the quality of the final presentation. The reviewers were as follows: Robert Andersen, J. Craig Bailey, Barry Biggs, Alan Brook, Alain Couté, Eileen Cox, David Czarnecki, Gary Dillard, Gary Floyd, Paul Hamilton, Kyle Hoagland, Ronald Hoham, Jeffrey Johansen, John Kingston, Dag Klaveness, Hedy Kling, Lothar Krienitz, Jorgen Kristiansen, Elsadore Kusel-Fetzmann, Carole Lembi, Rex Lowe, David Mann, Richard McCourt, Øjvind Moestrup, Orlando Necchi Jr., Kenneth Nichols, Gianfranco Novarino, Hans Paerl, Russell Rhodes, Frank Round, Robert Sheath, Alan Steinman, Eugene Stoermer, Francis Trainor, Richard Treimer, Herb Vandermeulen, Morgan Vis, James Wee, John Wehr, Robert Wetzel, Ruth Willey, and David Williams. We wish to thank our editor at Academic Press, Dr. Charles R. Crumly, and his assistant, Ms. Christine Vogelei, for their continued support throughout the creation and completion of this book. We also thank Ms. Geri Mattson at Mattson Publishing Services for providing professional assistance in the production of the final copy of the book. We also wish to acknowledge the many colleagues who generously permitted reproduction of published and unpublished material; these are acknowledged in the individual chapters. Kandis Elliot designed the beautifully illustrated cover.

The production of this book was partially supported by the Routh Endowment of the Louis Calder Center for JDW and NSERC grant number RGP 0105629 to RGS, as well as general support from Fordham University, University of Guelph and California State University, San Marcos. Help in manuscript production from Pam Anderson, Petra DelValle (Fordham), Marcy Boyle (California State University, San Marcos), and Toni Pellizzari (Guelph) is appreciated. Lastly, we would like to thank our spouses, Deb Donaldson (Wehr) and Mary Koske (Sheath) for their understanding and support through this long process.

John D. Wehr and Robert G. Sheath

LITERATURE CITED

Prescott, G.W. 1962. *Algae of the Western Great Lakes Area*, 2nd Edn. W.C. Brown, Dubuque, Iowa.

Smith, G. M. 1950. *The Freshwater Algae of the United States*, 2nd Edn. McGraw-Hill, New York.

Whitford, L.A. & Schumacher, G.J. 1984. *A Manual of Fresh-Water Algae*, Revised Edn., Sparks Press, Raleigh, NC.

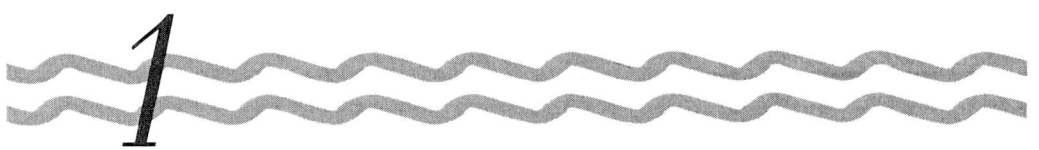

INTRODUCTION TO FRESHWATER ALGAE

Robert G. Sheath
Office of Provost and Vice President for Academic Affairs
California State University, San Marcos
San Marcos, California 92096

John D. Wehr
Louis Calder Center—Biological Station and Department of Biological Sciences
Fordham University,
Armonk, New York 10504

I. Introduction
II. Classification
  A. Cyanobacteria
  B. Red Algae
  C. Green Algae
  D. Euglenoids
  E. Eustigmathophyte, Raphidiophyte, and Tribophyte Algae
  F. Chrysophycean Algae
  G. Haptophyte Algae
  H. Synurophyte Algae
  I. Diatoms
  J. Dinoflagellates
  K. Cryptomonads
  L. Brown Algae
III. Taxonomic Chapters in This Book
  A. Key
Literature Cited

I. INTRODUCTION

Algae are treated in this book in the same sense as they are in many introductory phycology texts (e.g., Van den Hoek *et al.*, 1995; Sze 1998; Graham and Wilcox, 2000); that is, they are considered to be a loose group of organisms that have all or most of the following characteristics: aquatic, photosynthetic, simple vegetative structures without a vascular system, and reproductive bodies that lack a sterile layer of protecting cells. As such, algae are no longer regarded as a phylogenetic concept, but still represent an ecologically meaningful and important collection of organisms. Both prokaryotic (cells that have no membrane-bound organelles) and eukaryotic taxa (cells with organelles) are included. In addition, there is a wide range of vegetative morphologies, including the following:

1. **Unicells**: species that occur as solitary cells that may be nonmotile or motile. Motile cells may have one or more flagella or they may glide. A wide variety of forms exists among unicells, including those contained within a gelatinous sheath (Fig. 1A), with intricate cell walls (Fig. 1B), having flexible cell shapes (Fig 1C), with two flagella of unequal length (Fig. 1D) or two equal flagella (Fig. 1E), with cells drawn out into hornlike extensions (Fig. 1F), and having cells contained in a hardened case or lorica (Fig. 1G).

2. **Colonies**: an aggregation of cells that are held together either in a loose (Fig. 1H and I) or tight, well organized fashion (Fig. 2B, D, and E). Depending on the algal taxon, colonies may contain a variable number of cells or they may be constant throughout their development (Fig. 2B). Colonies may contain flagellated or nonflagellated cells. The basis for cellular connection varies among colonies, including a surrounding gelatinous matrix (Fig. 1H and I), gelatinous stalks (Fig. 2A), common parental wall (Fig. 2B), and direct attachment at the cellular edges (Fig. 2C) or at the middle portion of each cell (Fig. 2C). Alternately, cells may be connected via their loricae (Fig. 2E).

3. **Pseudofilaments**: an aggregation of cells in an end-to-end fashion. The cells are not directly connected to each other; rather, they are spaced

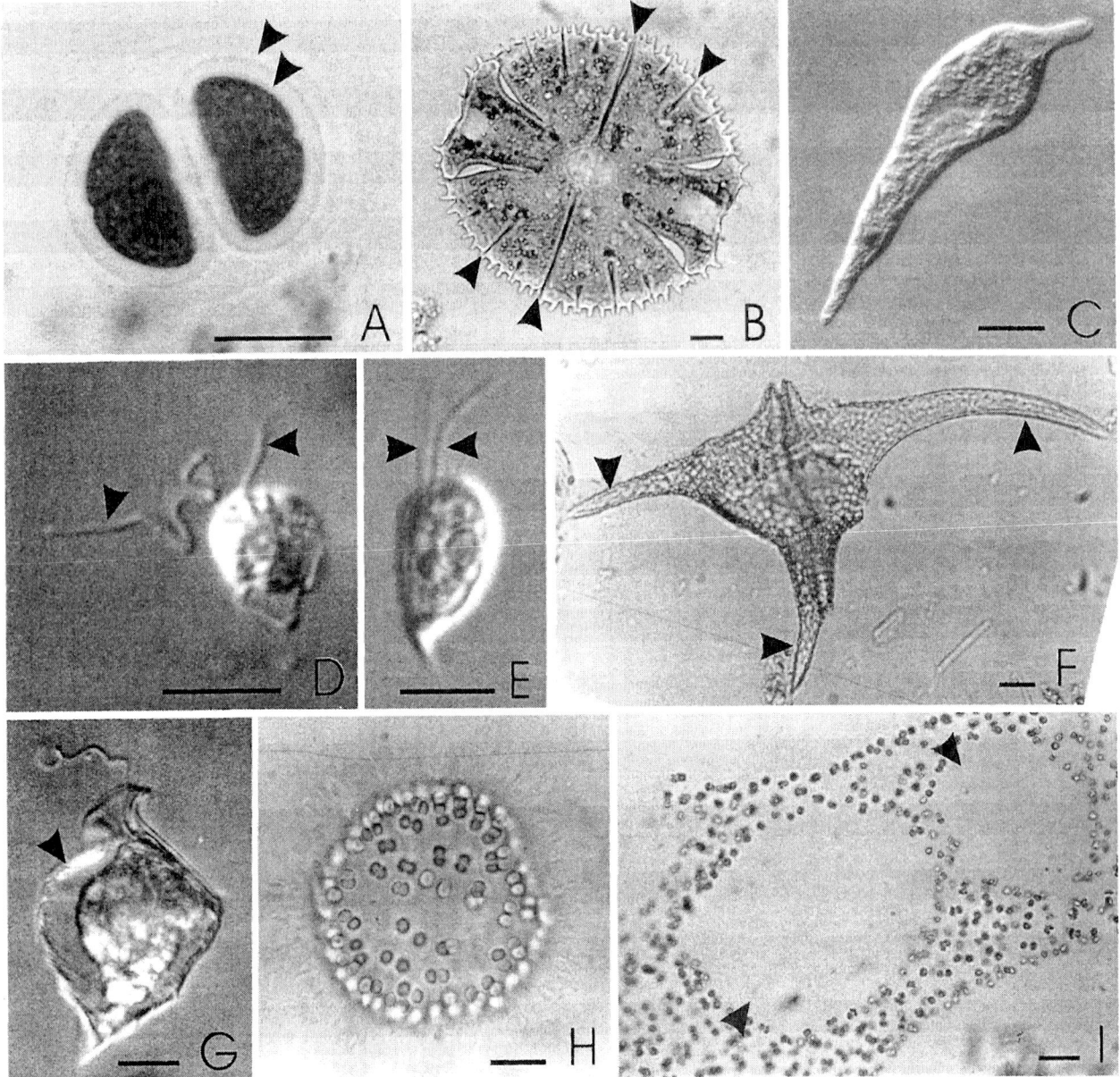

FIGURE 1 Unicellular and colonial forms of freshwater algae. A. *Gloeocapsa* (cyanobacterium), a unicell to small grouping of cells contained within concentrically layered gelatinous sheaths (arrows). B. *Micrasterias* (green alga, desmid), a unicell with many regular cell wall incisions (arrows) that form a series of lobes and lobules. C. *Euglena* (euglenoid), a unicell that does not produce walls and can readily change shape. D. *Ochromonas* (chrysophycean alga), a unicell with one long and one short apically inserted flagellum (arrows). E. *Pyrenomonas* (cryptomonad), a unicell with two equal subapically inserted flagella. F. *Ceratium* (dinoflagellate), a unicell with a theca composed of cellulose plates and cellular extensions or horns (arrows). G. *Strombomonas* (euglenoid), a flagellated unicell within a rigid lorica (arrow). H. *Coelosphaerium* (cyanobacterium), a colony with spherical cells loosely arranged at the periphery of a gelatinous matrix. I. *Dermatochrysis* (chrysophycean alga), a colony with spherical cells in a single layer scattered in a gelatinous matrix that has distinct perforations (arrows). Scale bars = 10 μm.

apart and contained in a common gelatinous matrix (Fig. 2F).
4. **Filaments:** a chain or series of cells in which the cells are arranged in an end-to-end manner, where adjacent cells share a common cross wall (Figs. 2H–J, and 3B and C). Linear colonies can be distinguished from true filaments by the fact that abutting colonial cells each possess their own entire walls (Fig. 2D). Filaments may be arranged in a single series (uniseriate or

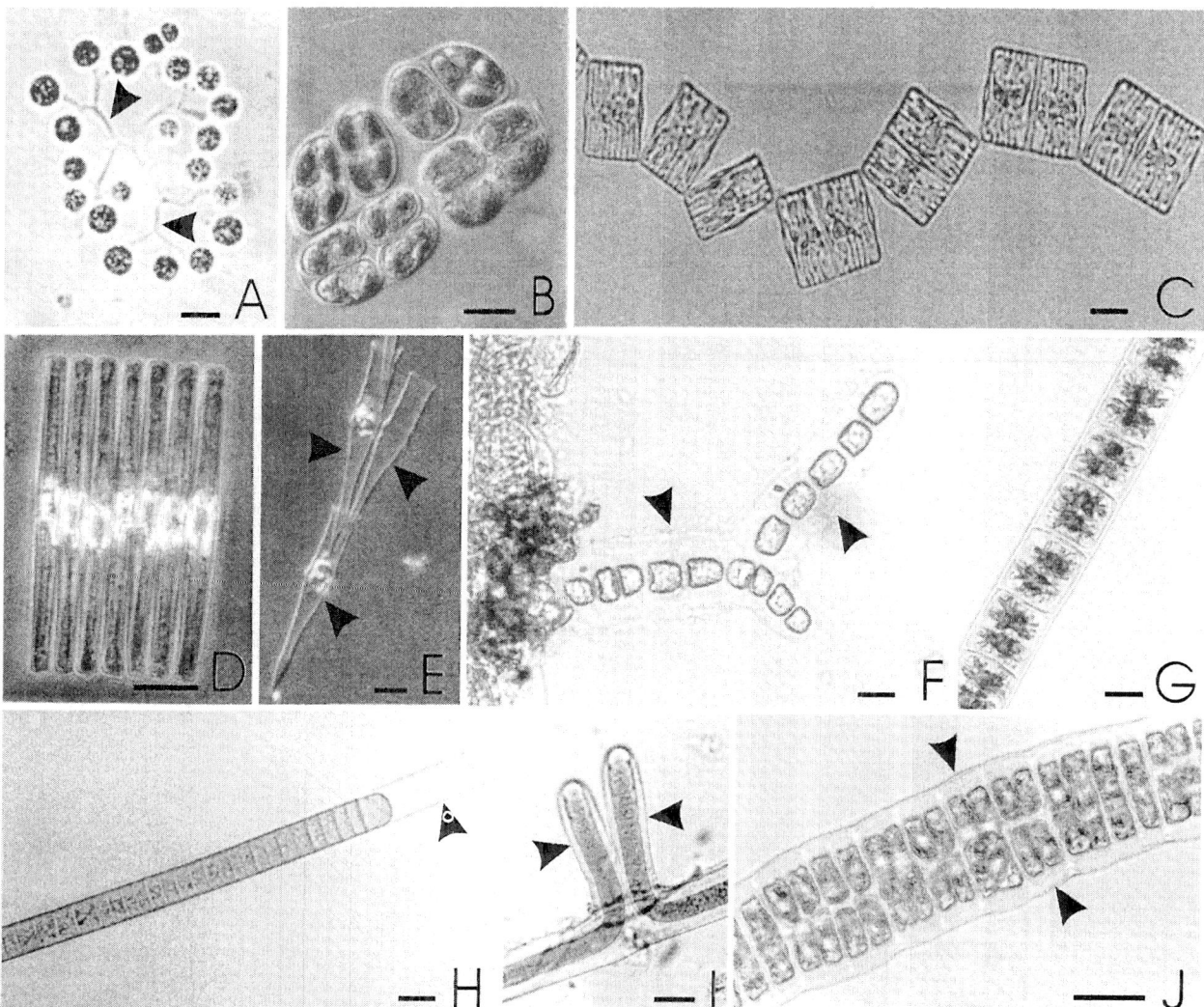

FIGURE 2 Colonial, pseudofilamentous, and filamentous forms of freshwater algae. A. *Porphyridium* (red alga), a colony with spherical cells attached together by gelatinous strands (arrows). B. *Crucigenia* (green alga), a colony with consistent groups of four cells produced inside the walls of the parent cells. C. *Tabellaria* (diatom), a colony with cells attached at their edges in a zig-zag fashion. D. *Asterionella* (diatom), a linear colony with cells attached only at the central region. E. *Dinobryon* (chrysophycean alga), a colony with cells attached by their loricae (arrows). F. *Chroodactylon* (red alga), a pseudofilament with cells arranged in an end-to-end pattern in a common gelatinous matrix (arrows), but not directly connected to each other. G. *Zygnema* (green alga), an unbranched filament without a gelatinous matrix. H. *Lyngbya* (cyanobacterium), an unbranched filament that is contained in a gelatinous sheath that is evident at the filament tip (arrow). I. *Scytonema* (cyanobacterium), a filament that produces double false branches (arrows) that result from breakage and further growth of each fragment. J. *Bangia* (red alga), a multiseriate filament in part with at least two cells across (arrows). Scale bars = 10 μm.

uniaxial) (Fig. 2G–I) or they may be in more than one series of cells (multiseriate or multiaxial) (Fig. 2J). Filaments may be unbranched (Figs. 2G and H) or they can produce branches in a new plane that are morphologically similar to the main axis (Fig. 3B) or that are quite distinct (Fig. 3C). Branching may be dichotomous or forked (Fig. 3B), alternate (Fig. 3C), opposite, or whorled (Fig. 3D). False branches are formed in some cyanobacteria, such as *Scytonema* (Fig. 2I), by fragmentation and continued growth of one or both fragments. Other types of filaments include those that are heterotrichous, that is, they have a distinct prostrate system with attached erect branches. Differentiated filaments have specialized cells within the chain. The main axis may have a surrounding layer of small cells termed cortication (Fig. 3A).

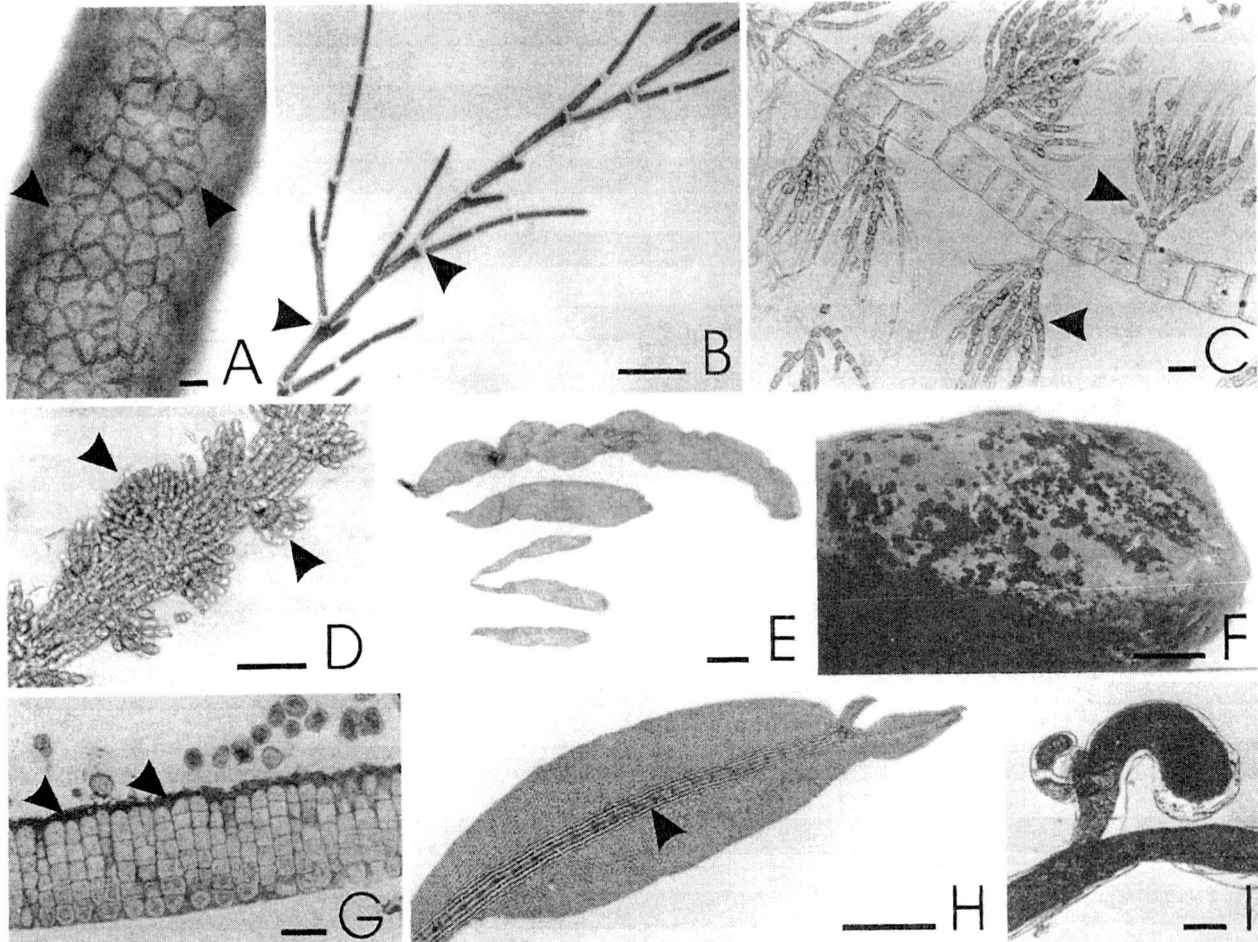

FIGURE 3 Filamentous, saclike, crustose, pseudoparenchymatous, and siphonous forms of freshwater algae. A. *Compsopogon* (red alga), a filamentous form with small cortical cells (arrows) covering the main filament. B. *Cladophora* (green alga), a filament that has dichotomous (forked) branches (arrows). C. *Draparnaldia* (green alga), a filament that has tuftlike lateral branches with cells that are considerably smaller than those of the main axis. D. *Batrachospermum* (red alga), a filament with whorllike lateral branches (arrows). E. *Boldia* (red alga), a saclike thallus that consists of a single layer of cells. F. *Heribaudiella* (brown alga), a crust that is tightly adherent to the rock substratum. G. *Hildenbrandia* (red alga), a cross section of a crust that shows vertical files of cells (arrows). H. *Caloglossa* (red alga), a pseudoparenchymatous thallus composed of a main filamentous axis (arrow) with tightly compacted lateral branches. I. *Vaucheria* (yellow–green alga), a siphonous thallus without cross walls separating the nuclei. Scale bars = 10 μm, except B = 250 μm, E = 1 cm, and F = 2 cm. Figure A courtesy of Tara Rintoul; Figure B from Vis *et al.* (1994) reprinted with permission of University of Hawaii Press; Figure E from Sheath (1984) with permission; Figure G courtesy of Alison Sherwood.

5. **Pseudoparenchymatous structures:** tissue-like thalli that consist of closely appressed branches of a uniseriate or multiseriate filament (Fig. 3H). Crustose forms may be composed of short, compacted filaments, such as the brown alga *Heribaudiella* (Fig. 3F) and the rhodophyte *Hildenbrandia* (Fig. 3G).
6. **Parenchymatous forms:** true tissues composed of a solid mass of cells that is three dimensional, variously shaped, and not filamentous in construction. The cells may be differentiated into an outer photosynthetic layer (the cortex) and an inner non-photosynthetic region (the medulla). Most tissue-like forms in freshwater habitats are simple, such as the saccate red alga *Boldia*, which consists of a single layer of cells (Fig. 3E).
7. **Coenocytic or siphonous forms:** large multinucleate forms of various shapes without cross walls to separate the nuclei or other organelles. An example is the yellow–green alga *Vaucheria* (Fig. 3I).

Freshwater algae exhibit all of these morphologies, but the macroscopic pseudoparenchymatous and parenchymatous forms tend to be smaller than those found in marine systems (Sheath and Hambrook, 1990). In addition, planktonic (floating) forms are typically small and microscopic, and mostly consist of the simpler forms. In contrast, benthic (attached) algae include the entire range of morphologies, although flagellated taxa are less common than in plankton.

II. CLASSIFICATION

Algae do not represent a formal taxonomic group of organisms, but rather constitute a loose collection of divisions or phyla with representatives that have the characteristics noted previously. The divisions are distinguished from each other based on a combination of characteristics, including photosynthetic pigments, starchlike reserve products, cell covering, and other aspects of cellular organization (e.g., Van den Hoek et al., 1995; Sze, 1998; Graham and Wilcox, 2000). There is little consensus among phycologists as to the exact number of algal divisions; 8–11 have been recognized in recent texts (Van den Hoek et al., 1995; Sze, 1998; Graham and Wilcox, 2000).

The 12 major algal groups (divisions and classes) recognized in this book are distinguished from each other in Table I. Each of the major groupings is briefly presented in the following sections, but the reader should refer to the relevant chapter(s) for more details. The number of freshwater genera now reported (>800) from North America, as discussed in Chapters 3–22, has greatly increased from earlier treatises (e.g., Smith,

TABLE I Major Distinguishing Features of the Major Algal Groups Presented Herein

Algal group (chapter number)	Photosynthetic pigments[a]	Chloroplast outer membranes	Thylakoid associations	Starch-like reserve[b]	External covering[c]	Flagella
Cyanobacteria (3 & 4)	chl^a, PE, PC, APC	0	0	Cyanophycean	Pepitoglycan matrices or walls	0
Red algae (5)	chl^a, PE, PC, APC	2	0	Floridean	Walls with a galactose polymer matrix	0
Green algae (6–9)	chl$^{a, b}$	2	2–6	True	Cellulosic walls, scales	0 – many
Euglenoid Algae (10)	chl$^{a, b}$	3	3	Paramylon	Pellicle	1–2 emergent
Yellow–green and related algae (11)	chl$^{a, c}$	4	3	Chrysolaminarin	Mostly cellulosic walls	2 unequal if present
Chrysophyte algae (12)	chl$^{a, c}$ fucoxanthin	4	3	Chrysolaminarin	None, scales, lorica	2 unequal
Haptophyte algae (13)	chl$^{a, c}$ fucoxanthin	4	3	Chrysolaminarin	Nonsiliceous scales	2 equal + haptonema
Synurophyte algae (14)	chl$^{a, c}$ fucoxanthin	4	3	Chrysolaminarin	Siliceous scales	2 unequal
Diatoms (15–19)	chl$^{a, c}$ fucoxanthin	4	4	Chrysolaminarin	Siliceous frustule	1, reproductive cells only
Dinoflagellates (20)	chl$^{a, c}$ peridinin	3	3	True	Theca	2 unequal
Cryptomonads (21)	chl$^{a, c}$ PC or PE	4	2	True	Periplast	2 equal
Brown algae (22)	chl$^{a, c}$ fucoxanthin	4	3	Laminarin	Walls with alginate matrices	2 unequal

Source: Various phycology textbooks (e.g., Sze, 1998; and Graham and Wilcox, 2000).

[a] chl = chlorophyll (green); PE = phycoerythrin (red); PC = phycocyanin (blue); APC = allophycocyanin (blue); fucoxanthin and peridinin (golden to brown).

[b] All of the reserves are polymers of glucose. They differ by their linkages: cyanophycean and floridean α1, 4 and α1, 6 branches; true starch with amylose α1, 4 and amylopectin α1, 4 and α1, 6 branches; paramylon β1, 3; chrysolamin and laminarin β1, 3 and β1, 6 branches. Only true starch stains positively with iodine (purple to black).

[c] Pellicle and periplast within plasma membrane; the rest are external to it.

1950; Prescott, 1962), but is still highly tentative and likely to be an underestimate of the region's biodiversity.

A. Cyanobacteria

Cyanobacteria or blue–green algae are prokaryotes, that is, cells that have no membrane-bound organelles, including chloroplasts (Table I; Chap. 3). Other characteristics of this division include unstacked thylakoids, phycobiliprotein pigments, cyanophycean starch, and peptidoglycan matrices or walls. There are 124 genera reported from inland habitats in North America, of which 53 are unicellular or colonial (Chap. 3) and 71 are filamentous (Chap. 4). However, the taxonomy of this division is currently in a state of flux, as noted in Chapter 3, and the number of genera should be considered to be tentative.

Cyanobacteria inhabit the widest variety of freshwater habitats on Earth and can become important in surface blooms in nutrient-rich standing waters (Chaps. 3 and 4). Some of these blooms can be toxic to zooplankton and fish, as well as livestock that drink water containing these organisms. Inland cyanobacteria also occur in extreme environments, such as hot springs, saline lakes, and endolithic desert soils and rocks.

B. Red Algae

Rhodophyta or red algae represent a division that is characterized by chloroplasts that have no external endoplasmic reticulum and unstacked thylakoids, phycobiliprotein pigments, floridean starch, and lack of flagella (Table I; Chap. 5). They are predominantly marine in distribution; only approximately 3% of over 5000 species occur in truly freshwater habitats. In North America, 25 genera are recognized in inland habitats (Chap. 5).

Freshwater red algae are largely restricted to streams and rivers, but also can occur in other inland habitats, such as lakes, hot springs, soils, caves, and even sloth hair (Chap. 5).

C. Green Algae

Chlorophyta or green algae constitute a division that has the following set of attributes: chloroplasts with no external endoplasmic reticulum, thylakoids typically in stacks of two to six, chlorophyll-*a* and -*b* as photosynthetic pigments, true starch, and cellulosic walls or scales (Table I). This is a diverse group in inland habitats of North America that includes 44 flagellated genera (Chap. 6), at least 129 coccoid and nonmotile colonies (Chap. 7), 81 filamentous and plantlike genera (Chap. 8), and 48 conjugating genera and desmids (Chap. 9). Some members of the green algae (Charophyceae) are part of a lineage that is thought to be ancestral to higher plants.

Green algae are widespread in inland habitats, but certain groups may have specific ecological requirements. For example, flagellated chlorophytes tend to be more abundant in standing waters that are nutrient rich (Chap. 6). Coccoid unicells and colonies are common in the plankton of standing waters and slowly moving rivers when nutrients, light and temperature are reasonably high (Chap. 7). The majority of filamentous and plantlike Chlorophyta are attached to hard surfaces in standing or flowing waters, but some can exist in the floating state or on soils or other subaerial habitats (Chap. 8). Filamentous conjugating green algae are most frequent in stagnant waters of roadside ditches and ponds, and in the littoral zones of lakes, where they can form free-floating mats or intermingle with other algae in attached or floating masses (Chap. 9). Desmids are more common in ponds and streams that have low conductance and moderate nutrient levels, and often intermingle with macrophytes.

D. Euglenoids

Photosynthetic Euglenophyta or euglenoids have chloroplasts surrounded by three membranes, thylakoids in stacks of three, chlorophyll-*a* and -*b* as photosynthetic pigments, paramylon, and a pellicle (Table I; Chap. 10). Ten genera are reported from North American freshwater habitats (Chap. 10).

Euglenoids are particularly abundant in the plankton of standing waters rich in nutrients and organic matter, and they can be associated with sediments, fringing higher plants, and leaf litter, although some may dominate in highly acidic environments (Chap. 10).

E. Eustigmatophyte, Raphidiophyte, and Tribophyte Algae

Eustigmatophyte, raphidiophyte, and tribophyte algae comprise a loose group of algae that share the following characteristics: chloroplasts with four surrounding membranes, thylakoids in stacks of three, chlorophyll-*a* and -*c* as the typical photosynthetic pigments, and chrysolaminarin as the photosynthetic reserve product (where known) (Table I; Chap. 11). The yellow–green algae are quite diverse in freshwater habitats of North America: at least 90 genera have been reported, whereas the eustigmatophytes and raphdiophytes are relatively small groups that comprise eight and three genera, respectively (Chap. 11). Many of these genera seldom have been reported.

Members of this group of algae have been collected from a wide variety of habitats, but many are collected primarily in northern habitats (Chap. 11). They are both planktonic and associated with a variety of substrata.

F. Chrysophycean Algae

Chrysophyceae or chrysomonads are distinguished by chloroplasts that have four surrounding membranes, thylakoids in stacks of three, fucoxanthin that typically masks chlorophyll-*a* and -*c*, and chrysolaminarin as the photosynthetic reserve. At least 72 genera are reported from inland habitats of North America (Chap. 12).

Chrysophycean algae are typically associated with standing bodies of water that have low or moderate nutrients, alkalinity, and conductances, and a pH that is slightly acidic to neutral (Chap. 12). In addition, the majority of genera tend to be planktonic; attached forms occur to a lesser extent.

G. Haptophyte Algae

Haptophyceae are characterized by chloroplasts that have four surrounding membranes, thylakoids in stacks of three, fucoxanthin that masks chlorophyll-*a* and -*c*, chrysolaminarin as the photosynthetic reserve, and a unique appendage associated with the flagellar apparatus, the haptonema (Table I; Chap. 13). Only three freshwater genera are found in North America (Chap. 13).

The two common genera are planktonic in lakes and ponds, and occasionally form predominant blooms, particularly in areas with low conductance (Chap. 13). *Chrysochromulina breviturrita* has been used as an indicator of moderately acidic water.

H. Synurophyte Algae

Synurophyceae is characterized by chloroplasts that have four surrounding membranes, thylakoids in stacks of three, fucoxonthin that masks chlorophyll-*a* and -*c*, chrysolaminarin as the photosynthetic reserve product, and siliceous scales (Table I; Chap. 14). Three genera are found in North American freshwater habitats (a fourth is known only from Australia), but the genera are species-rich, such as *Mallomonus* and *Synura* (Chap. 14).

Synurophytes are exclusively freshwater phytoplankters in lakes, ponds, and slowly flowing rivers (Chap. 14). Habitats that support the largest flora are slightly acidic, low in conductance, alkalinity, and nutrients, and have moderate amounts of humic substances.

I. Diatoms

Bacillariophyceae or diatoms are distinguished by chloroplasts that have four surrounding membranes, thylakoids in stacks of three, fucoxanthin that masks chlorophyll-*a* and -*c*, chrysolaminarin as the photosynthetic reserve product, and a siliceous frustule that makes up the external covering (Table I; Chap. 15). The diatoms are a complex and diverse group in terms of frustule morphology. The North American freshwater genera consist of 25 centrics (Chap. 15), 28 araphid and monoraphid diatoms (Chap. 16), 37 symmetrical naviculoid taxa (Chap. 17) 14 eunotioid and asymmetrical naviculoid diatoms (Chap. 18), and 14 keeled and canalled forms (Chap. 19).

Diatoms are found in all freshwater habitats, including standing and flowing waters, and planktonic and benthic habitats, and they can often dominate the microscopic flora. Because diatoms inhabit a broad array of habitats but many have specific habitat requirements, they have been used in freshwater environment assessment and to monitor long-term changes in ecological characteristics (Chap. 23).

J. Dinoflagellates

Pyrrhophyta or dinoflagellates are characterized by chloroplasts that have three surrounding membranes, thylakoids in stacks of three, peridinin that masks chlorophyll-*a* and -*c*, true starch, a nucleus that has condensed chromosomes in cell cycle phases, a theca covering, and frequently a transverse and posterior flagellum. (Table I; Chap. 20). There are 37 genera in North American freshwater habitats (Chap. 20).

The dinoflagellates are typically minor components of the phytoplankton of lakes and ponds, but sometimes form dense blooms, particularly in the presence of high levels of nitrates and phosphates (Chap. 20).

K. Cryptomonads

Cryptophyta, cryptomonads or cryptophyte algae, have chloroplasts that have four surrounding membranes in which a nucleomorph occurs between the outer and inner two membranes, thylakoids in loose pairs, phycocyanin or phycoerythrin that masks chlorophyll-*a* and -*c*, true starch as the photosynthetic reserve, a periplast, and two subapical flagella (Table I; Chap. 21). There are 12 genera reported from the inland waters of North America (Chap. 21).

Cryptomonads are typically planktonic in lakes and ponds, and are particularly diverse in temperate regions (Chap. 21).

L. Brown Algae

Phaeophyceae or brown algae are distinguished by chloroplasts that have four surrounding membranes, thylakoids in stacks of three, fucoxanthin that masks chlorophyll-*a* and -*c*, laminarin as the photosynthetic reserve, and alginates commonly as the wall matrix component. There are six genera of freshwater brown algae, four of which have been collected in North America (Chap. 22).

Brown algae are predominantly marine in distribution; less than 1% of the species are from fresh water. The inland species are benthic, either in lakes or streams, and distribution is quite scattered (Chap. 22).

III. TAXONOMIC CHAPTERS IN THIS BOOK

The approach of this book is to break the major algal groups into manageable taxonomic units, resulting in multiple chapters for those divisions that have many freshwater representatives. The following key gives major characteristics to allow the reader to immediately proceed to the appropriate chapter to determine an unknown algal sample.

A. Key

1a.	Cells with no chloroplasts; typically blue–green colored throughout (occasionally black, purple, brown, or reddish)	2
1b.	Cells with variously colored pigments localized in one or more chloroplasts	3
2a.	Organisms unicellular or colonial (**coccoid cyanobacteria**)	Chapter 3
2b.	Organisms filamentous (**filamentous cyanobacteria**)	Chapter 4
3a.	Cells stain positively (purple to black) with iodine for true starch	4
3b.	Cells do not stain positively (orange to reddish brown) with iodine for starch	9
4a.	Green-colored chloroplasts with chlorophyll-*a* and -*b* as predominant photosynthetic pigments	5
4b.	Chloroplasts with other colors and predominant photosynthetic pigments	8
5a.	Organisms flagellated in the vegetative state (**flagellated green algae**)	Chapter 6
5b.	Organisms nonflagellated in the vegetative stage	6
6a.	Organisms coccoid (nonmotile unicells of various shapes) or colonial in forms without conjugation (**coccoid and colonial nonmotile green algae**)	Chapter 7
6b.	Organisms filamentous, plantlike, or with sexual reproduction by conjugation	7
7a.	Organisms with filamentous, bladelike, or plantlike morphologies without conjugation (**filamentous and plantlike green algae**)	Chapter 8
7b.	Organisms with coccoid or filamentous morphologies with conjugation (**conjugating green algae filaments and desmids**)	Chapter 9
8a.	Cells with golden-colored chloroplasts with peridinin as the predominant photosynthetic pigment; two separate flagella typically with transverse and posterior insertions (**dinoflagellates**)	Chapter 20
8b.	Cells with blue-, brown-, or red-colored chloroplasts with either phycocyanin or phycoerythrin as the predominant photosynthetic pigment; flagella subapical (**cryptophyte algae**)	Chapter 21
9a.	Cells with blue- or red-colored chloroplasts with phycocyanin or phycoerythrin as the predominant photosynthetic pigment (**red algae**)	Chapter 5
9b.	Cells green or golden colored with other predominant photosynthetic pigments	10
10a.	Motile green-colored cells with a pellicle (layer below the plasma membrane that often appears as spiral strips on the cell) (**euglenoids**)	Chapter 10
10b.	Nonmotile or motile, yellow–green- or golden-colored cells; naked or walled cells without a pellicle	11
11a.	Yellow–green-colored chloroplasts with chlorophyll-*a* and -*c* as the predominant photosynthetic pigments (**eustigmatophyte, raphidiophyte, and tribophyte algae**)	Chapter 11
11b.	Golden-colored chloroplasts with fucoxanthin as the predominant photosynthetic pigment	12
12a.	Cells with a silica frustule covering (**diatoms**)	13

12b.	Cells with no covering or with one that is not a siliceous frustule (may be siliceous scales)	17
13a.	Frustules in the valve view are radially symmetrical or symmetrical in more than two planes (**centric diatoms**)	Chapter 15
13b.	Frustules symmetrical in the valve view in one or two planes	14
14a.	Frustules without a raphe or with one-on-one valve only (**araphid and monoraphid diatoms**)	Chapter 16
14b.	Frustules with two raphes	15
15a.	Raphe in an elevated keel or a canal (**keeled and canalled raphid diatoms**)	Chapter 19
15b.	Raphe not in a keel or canal	16
16a.	Frustules bilaterally symmetrical in valve view (**biraphid symmetrical diatoms**)	Chapter 17
16b.	Frustules not bilaterally symmetrical in valve view (**eunotioid and asymmetrical biraphid diatoms**)	Chapter 18
17a.	Cells with a specialized appendage, the haptonema, in vegetative and/or reproductive stages (**haptophyte algae**)	Chapter 13
17b.	Cells without a haptonema in any stage	18
18a.	Vegetative cells with siliceous scales (**synurophyte algae**)	Chapter 14
18b.	Vegetative cells without siliceous scales	19
19a.	Exclusively benthic; often macroscopic thalli with no unicellular or colonial representatives; cell walls with alginates (**brown algae**)	Chapter 22
19b.	Mostly planktonic with some attached representatives; numerous unicellular or colonial representatives; mostly microscopic thalli; cell walls without alginates (**chrysophyte algae**)	Chapter 12

LITERATURE CITED

Graham, L. E., Wilcox, L. W. 2000. Algae. Prentice–Hall, Upper Saddle River, NJ, 640 pp., glossary, literature cited, and index.

Prescott, G. W. 1962. Algae of the Western Great Lakes area. W. B. Brown, Dubuque, IA, 977 p.

Sheath, R. G. 1984. The biology of freshwater red algae. Progress in Phycological Research 3:89–157.

Sheath, R. G., Hambrook, J. A. 1990. Freshwater ecology, *in*: Cole, K. M., Sheath, R. G., Eds. Biology of the red algae. Cambridge University Press, Cambridge, UK, pp. 423–453.

Smith, G. M. 1950. Freshwater algae of the United States, 2nd ed. McGraw–Hill, New York, 719 p.

Sze, P. 1998. A biology of the algae 3rd ed. McGraw–Hill, Boston, 278 p.

Van den Hoek, C., Mann, D. G., Jahns, H. M. 1995. Algae. An introduction to phycology. Cambridge University Press, Cambridge, UK, 623 p.

Vis, M. L., Sheath, R. G., Hambrook, J. A., Cole, K. M. 1994. Stream macroalgae of the Hawaiian Islands: A preliminary study. Pacific Science 48:175–187.

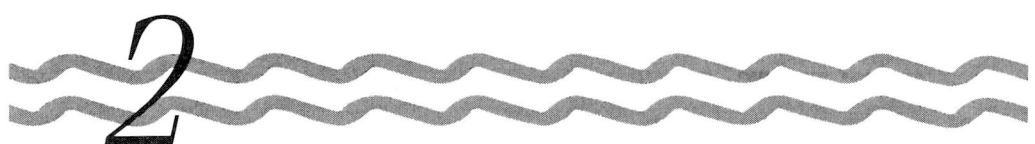

FRESHWATER HABITATS OF ALGAE

John D. Wehr

Louis Calder Center — Biological Station and Department of Biological Sciences Fordham University, Armonk, New York 10504

Robert G. Sheath

Office of Provost and Vice President for Academic Affairs California State University, San Marcos San Marcos, California 92096

I. What is Fresh Water?
II. Lentic Environments
    A. Major Lakes of North America
    B. Lake Basins
    C. Lake Community Structure and Productivity
    D. Ponds, Temporary Pools, and Bogs
    E. Phytoplankton of Lakes and Ponds
    F. Benthic Algal Assemblages of Lakes
III. Lotic Environments
    A. Major Rivers of North America
    B. Geomorphology of Rivers
    C. The River Continuum and Other Models
    D. Benthic Algal Communities of Rivers
    E. Phytoplankton Communities of Rivers
IV. Wetlands
    A. Functional Importance of Algae in Wetlands
    B. Algal Diversity in Freshwater Wetlands
    C. Algal Communities of Bogs
V. Thermal and Acidic Environments
    A. Thermal Springs
    B. Acid Environments
VI. Unusual Environments
    A. Saline Lakes and Streams
    B. Snow and Ice
    C. Other Unusual Habitats
Literature Cited

I. WHAT IS FRESH WATER?

The study of freshwater algae is really the study of organisms from many diverse habitats, some of which are not entirely "fresh." Although the oceans are clearly saline (≈ 35 g salts L^{-1}) and most lakes are relatively dilute (world average < 0.1 g L^{-1}; Wetzel, 1983a), there is enormous variation in the chemical composition of the nonmarine habitats that algae occupy. Conditions in lakes and rivers vary not only in salinity, but also in size, depth, transparency, nutrient conditions, pH, pollution, and many other important factors. Aquatic ecologists also use the term "inland" waters to encompass a greater range of aquatic ecosystems. Even this term may be unsatisfactory, because algae occupy many other habitats, such as snow, soils, cave walls, and symbiotic associations (Round, 1981).

Organisms grouped together in this volume as freshwater algae fall into a large, but ecologically meaningful collection of environments: all habitats that are at least slightly wet, other than oceans and estuaries. One reason for such a broad scope is that inland saline lakes, snow and ice, damp soils, and wetlands are studied by phycologists and ecologists who also examine more traditional freshwater environments. Some genera with terrestrial species, such as *Vaucheria*, *Nostoc*, *Chlorella*, and *Prasiola*, also have species found principally in streams or lakes (Smith, 1950; Whitton, 1975). In North America, the variety of freshwater habitats colonized by algae is very rich, and offers an enormous and fascinating range of environments for their study.

The distinction between marine and freshwater habitats is revealed in the variety of algae that occur in these environments. There are no exclusively freshwater divisions of algae, but certain groups exhibit greater abundance and diversity within fresh waters, especially Cyanobacteria, Chlorophyta, and Charophyta (Smith, 1950). Within the green algae, conjugating greens and desmids (Zygnematales, Chap. 9) comprise a very rich collection of species that almost exclusively occupy fresh water. Other groups, such as the diatoms and chrysophytes, are well represented in both spheres. Other groups, particularly the Phaeophyta, Pyrrophyta, and Rhodophyta, exhibit greater diversity in marine waters (Smith, 1950; Bourrelly, 1985). Most freshwater algae are best described as cosmopolitan, although there are reports of endemic chrysophytes, green algae, rhodophytes, and diatoms (Tyler, 1996; Kociolek et al. 1998), and at least some species of cyanobacteria (Hoffmann, 1996). Many algal taxa have particular environmental tolerances or requirements, and are ecologically restricted, but still geographically widespread. The euglenophyte *Colacium* is almost exclusively epizooic on aquatic invertebrates, but is widely distributed throughout North America (Smith, 1950; Chap. 10). The chrysophyte *Hydrurus foetidus* is an exclusive inhabitant of cold mountain streams, but is distributed worldwide (Smith, 1950; Whitton, 1975, Chap. 12). Even specialized taxa such as *Basicladia chelonum* (Chlorophyceae), which is restricted mainly to the shells of turtles, has been collected from many habitats throughout eastern North America (Smith, 1950; Prescott, 1962; Colt et al., 1995). The actual distribution of apparently disjunct freshwater species must therefore be viewed with some caution until detailed surveys have been conducted (see, for example, Linne von Berg and Kowallik, 1996; Müller et al., 1998).

Inland waters represent only about 0.02% of all water in the biosphere, and nearly 90% of this total is contained within only about 250 of the world's largest lakes (Wetzel, 1983a). Nonetheless, it is fresh water that is most important for human consumption and is most threatened by human activities. Algal ecologists play an important role in the understanding of aquatic ecosystems, their productivity, and water quality issues (Round, 1981; Brock, 1985a; Hoffmann, 1998; Dow and Swoboda, 2000; Oliver and Ganf, 2000, Chaps. 23 and 24). This chapter examines the habitats of freshwater algae and how differences in these systems affect algal communities.

II. LENTIC ENVIRONMENTS

Lentic environments include standing waters from the smallest ponds (a few square meters) to enormous bodies of water (e.g., Laurentian Great Lakes: 245,000 km²). Their formation, geography, limnology, and conservation have been covered in several texts (Hutchinson, 1957, 1967, 1975; Frey, 1963; Wetzel, 1983a; Cole, 1994; Abel et al., 2000). This section summarizes some features of lentic environments as they pertain to the ecology and distribution of freshwater algae.

A. Major Lakes of North America

Worldwide, the single largest volume of freshwater — nearly 20% of the world's total — is located in Lake Baikal, Siberia (23,000 km³), but the North American Great Lakes (Fig. 1A) collectively represent the largest total volume of nonsaline water on Earth, approximately 24,600 km³ (Wetzel, 1983a). North America is home to many spectacular large and deep freshwater systems, nearly half of all the world's lakes greater than 500 km² (Hutchinson, 1957). Two of the most impressive lakes are subarctic: Great Slave Lake (28,200 km²; 614 m deep; deepest in North America) and Great Bear Lake (30,200 km²; >300 m deep) in the Northwest Territories (Hutchinson, 1957). Crater Lake in Oregon (Fig. 1B) is much smaller (64 km² in area), but is the deepest lake in the United States (608 m) and seventh deepest in the world (Edmondson, 1963). The largest lakes on the continent are located in northern and temperate regions, although Great Salt Lake (Utah) is a massive remnant lake (>6000 km²) that has a mean depth (≈ 9 m) and very high salinity (130–280 g L^{-1}) that fluctuate with available moisture, and occupies a portion of the Pleistocene Lake Bonneville, which had an area >51,000 km² and a depth of 320 m (Hutchinson, 1957).

B. Lake Basins

Sizes and shapes of lake basins (their morphometry) have profound effects on the physics, chemistry, and biology of lake ecosystems, and influence the composition of algal communities and their productivity. Lake basins differ in morphometry as a result of the forces that created them, many of which were catastrophic events from the past, principally glacial, seismic, and volcanic activity. Hutchinson (1957) distinguished 76 different lake types based on their origins; these were classified into a simpler scheme by Wetzel (1983a) that is summarized in Table I.

Glacial activity is the most important agent in North America. It created millions of small and large basins from the arctic south to the southern extent of the Wisconsin ice sheet. In this period (15,000–5000 years BP), many basins became closed by morainal deposits, including the Laurentian Great Lakes. Some morainal lakes occur at the ends of long valleys after glaciers have receded, including the Finger Lakes of

FIGURE 1 Examples of different types of lakes in North America: A, Laurentian Great Lakes; B, Crater Lake, Oregon, a deep caldera lake; C, New York Finger Lakes; D, kettle lakes in Becher's Prairie, central British Columbia; E, pothole lakes in Qu'appelle Valley, Saskatchewan; F, Louise Lake, a cirque in Mt. Rainier National Park, Washington. Photos A and C courtesy of U.S. Geological Survey EROS Data Center, reproduced with permission; photo B by R. G. Sheath; photo D by R. J. Cannings; photo E by P. R. Leavitt, reproduced with permission; photo F by J. D. Wehr.

New York (Fig. 1C), which are elongate, radially arranged basins that range from small ponds to large lakes, such as Seneca (175 km² area, 188 m depth; Hutchinson, 1957; Berg, 1963). However, most glacially formed lakes are small kettles scattered across the continent (Fig. 1D and E). Glacial scouring in mountainous terrain may form deep amphitheater-like cirques (Fig. 1F), which are common from Alaska through the western mountain ranges south to tropical locations in Costa Rica (Haberyan et al., 1995). Glacial basins within narrow valleys may form deep fjord lakes (Fig. 2A), or a chain of smaller lakes known as paternosters. Several forces, including glacial scour and lava flow, combined to form the small (9.9 km²) but deep (259 m) fjordlike Garibaldi Lake (Northcote and Larkin, 1963; Fig. 2B). Ice-formed (thermokarst) lakes, which result from freezing and thawing action in ice and soil, are common in the Arctic. All are shallow but vary from large elliptical basins (up 70 km² area) to small (10–50 m diameter), "polygon" ponds (Fig. 2C),

TABLE I Major Lake Basin Types, Grouped According to the Principal Forces That Shaped Them

Type of forces	Basins	Principal force	North American examples
Catastrophic	1. Glacial	Glacial scouring	Great Slave Lake, NWT
		Moraine deposits	Finger Lakes, NY; Moraine Lake, AK
		Kettles	Linsley Pond, MA; Cedar Bog Lake, MN
		Cryogenic	Many polygon lakes, AK
	2. Tectonic	Graben	Lake Tahoe, CA-NV
		Uplift	Lake Okeechobee, FL
		Landslide	Mountain Lake, VA
	3. Volcanic	Caldera	Crater Lake, OR
		Maar	Zuni Salt Lake, Mexico
	4. Meteor	Meteor	New Quebec Lake, PQ
Noncatastrophic	5. Solution	Doline	Deep Lake, FL; limestone areas of KY-IN-TN
		Salt collapse	Montezuma Well, AZ; Bottomless Lakes, NM
	6. Rivers	Oxbows	Lake Providence, LA
		Plunge	Fayetteville Green Lake, NY
	7. Coastal	Shoreline	Along Laurentian Great Lakes; Cape Cod area
	8. Wind	Deflation	Moses Lake, WA; Sandhills region, NB
	9. Organic	Beaver	Many locations in northern regions
		Human	Lake Mead, AZ-NV; Cherokee Reservoir, TN

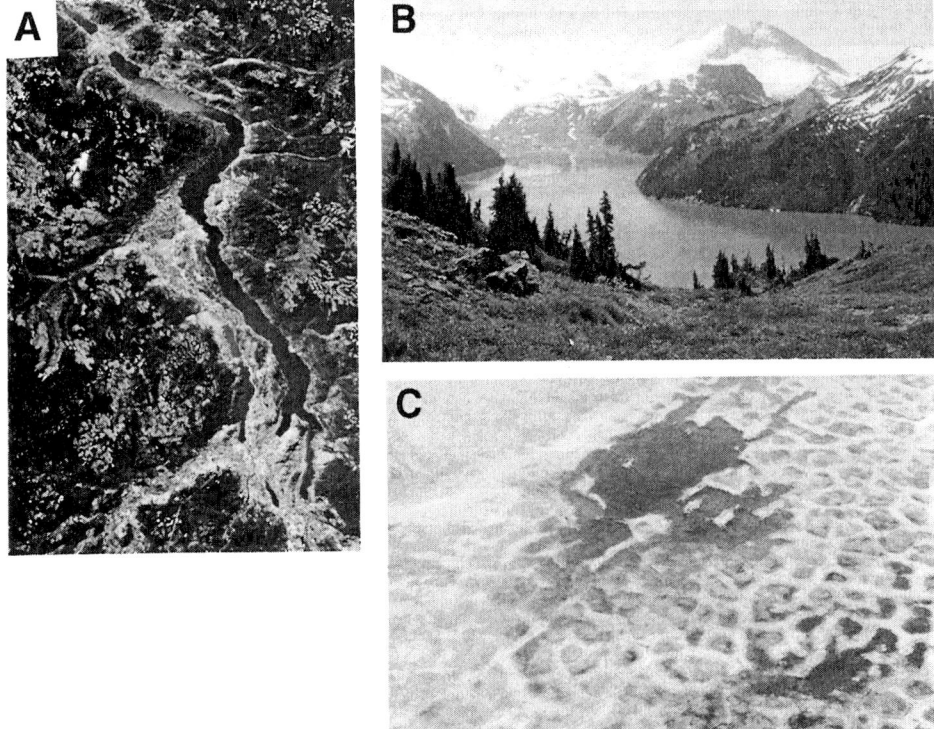

FIGURE 2 Other glacial and ice-formed lakes: A, Okanagan Lake (BC), a fjord lake; B, Garibaldi Lake, formed by glacial scour and lava damming; C, arctic polygon ponds. Photos A courtesy of NASA, reproduced with permission; photos B and C by R. J. Cannings, reproduced with permission.

which are estimated at more than a million in number (Livingstone, 1963; Sheath, 1986).

Tectonic basins are formed by movements of the Earth's crust. Among these, grabens form when fault lines create often enormous depressions, such as Lake Tahoe, a symmetrical, deep (505 m), and steep-sided lake (Fig. 3A). Tahoe (double fault lines) and Lakes Baikal and Tanganyika (single faults) include the deepest lakes in the world, although less spectacular examples also occur. Lago de Peten (Guatemala) is

FIGURE 3 Tectonic lakes: A, Lake Tahoe, a graben, viewed from the northeast; B, Spirit Lake, Washington, a landslide lake near Mt. St. Helens; C, Lake Okeechobee, Florida (larger lake on lower right), and surrounding lakes. Photos A and B by J. D. Wehr; photo C reproduced with permission of the South Florida Water Management District.

the largest (567 km^2) and deepest (> 32 m) lake on the Yucatan Peninsula (Covich, 1976), and the largest lake in Mexico, Lago Chapala (1109 km^2), is also a tectonic trench (Serruya and Pollingher, 1983). Landslide lakes form when water flows through an existing depression and is blocked by rock or other material, as in as Spirit Lake, near Mount St. Helens (WA; Fig. 3B) and Mountain Lake (VA; Parker *et al.*, 1975). In the Pliocene, shallow marine areas were raised above sea level and existing depressions filled with freshwater, as with Lake Okeechobee (Fig. 3C), a large (1840 km^2), shallow (4 m) subtropical lake in Florida.

Volcanic lakes are among the deepest and most steep-sided lakes on the continent. The exceptionally clear water in Crater Lake (OR), a collapsed caldera (Fig. 1B), has Secchi depths between 20 and 30 m, with 1% surface light down to 100 m (Larson *et al.*, 1996). Lake Atitlán in Guatemala is an alpine tropical caldera (8.2 km^2; 1550 m elevation) that reaches a depth of 341 m, making its ratio of depth to surface area four times greater than better known Crater Lake. Lake Nicaragua (Fig. 4A; 7700 km^2 area; depth ≈ 60 m) was formed by volcanic lava damming an existing valley (Cole, 1963). Some volcanic lakes, such as Yellowstone Lake (WY) and Surprise (AK) Lake, have hydrothermal vents that influence temperature, pH, and O_2 conditions, and may contribute trace metals (Pierce, 1987; Larson, 1989; Cameron and Larson, 1993). Other volcanic lakes formed from violent explosions of cinder cones (maars) and are often nearly circular in outline, such as Big Soda Lake in Nevada, Lago Chamico in El Salvador (Cole, 1963), and Laguna Hule in Costa Rica (Umana-Villalobos, 1993). Volcanic lakes also occur in Mexico, Guatemala, Nicaragua, and El Salvador (but not all are deep), and some were formed as recently as 500 year ago (Hutchinson, 1957; Cole, 1963).

In limestone regions, solution or sinkhole lakes form from the dissolution of bedrock by surface and underground waters charged with CO_2 (Cole, 1994). Sinks may be circular, elliptical, or irregular in outline, and occur throughout Kentucky, Indiana, Tennessee, Florida, Mexico, and Guatemala (Fig. 4C). Florida is especially rich in sinkholes, with several hundred lakes and ponds ranging from less than 1 ha to several square kilometers (Fig. 3C). In north Florida, some lakes are relatively dilute and colored with organic matter from pine litter, whereas others are clear,

FIGURE 4 Volcanic and solution lakes: A, Lakes Managua (near) and Nicaragua, two large calderas; B, Volcan Maderas (Nicaragua), a small caldera located on an island in Lake Nicaragua; C, unnamed, deep limestone sink in the Yucatan Peninsula; D, Montezuma Well (AZ), a collapsed travertine limestone sink. Photo A courtesy of NASA, reproduced with permission; photo B by A. Merola; photo C by L. P. Burney; photo D by D. W. Blinn, reproduced with permission.

hardwater systems (Shannon and Brezonik, 1972). Spring-fed sinkholes may become isolated or thermally constant, such as Montezuma Well (AZ; Fig. 4D), creating an environment with high algal productivity (600 g C m^{-2} y^{-1}; Boucher *et al.*, 1984) and unusual communities. Montezuma is a collapsed travertine system with several endemic invertebrates, but no fish, rotifers, or cladocerans (Cole, 1994).

Lakes may form through wind action, whereby deposited sand blocks existing valleys (e.g., Moses Lake, WA) or forms depressions in dunes, as in Nebraska and Texas (Cole, 1963; Edmondson, 1963). River-formed lakes, including oxbows, occur across North America where rivers traverse level terrain, enabling siltation of meandering valleys (Fig. 5A). Other small basins form in the plunge pools of waterfalls (Fig. 5B). Fayetteville Green Lake is a relatively deep (59 m; 0.3 km^2 area) plunge-pool lake in central New York that apparently has never fully mixed; it was formed during the Pleistocene when melting glaciers formed a vast waterfall (Berg, 1963; Brunskill and Ludlam, 1969).

For many centuries the principal biological agent responsible for creating lakes in North America was the beaver (*Castor canadensis*), which dams smaller rivers to form lakes and ponds (Hutchinson, 1957; Fig. 5C). Today, reservoirs are a more important group of lentic ecosystems, the size and number of which are increasing worldwide (Fig. 5D). The physical and chemical properties of reservoirs differ from natural lakes with respect to dendritic or eccentric morphometry (deepest near the dam), shorter flushing period, irregular water level, greater dissolved and suspended solids, and less stable littoral zone (Wetzel, 1990). Because reservoirs serve hydroelectric, flood control, or drinking water uses, they occur in many biomes.

A few lakes may have been formed by meteor impact, such as New Quebec Lake, a nearly perfectly circular basin (3.4 km diameter) in a region of irregular, glacially formed lakes in northern Quebec (Cole, 1994). Carolina Bays, which are not truly bays, are a series of roughly 150,000 small, shallow, elliptical basins, with a distinctive NW–SE orientation, and concentrated along the Atlantic coast from New Jersey south to Florida. It is their directional orientation that has caused some to speculate that their origin may be from meteor showers, whereas others have suggested wind action or artesian springs (Hutchinson, 1967;

FIGURE 5 Lakes formed by river action and other agents: A, an oxbow lake in Texas; B, a plunge pool below Waimea Falls, Hawaii; C, a beaver-dam lake (southern British Columbia); D, reservoir (eastern Colorado). Photo A by J. Cotner; photo B by D. Burney, reproduced with permission; photos C and D by J. D. Wehr.

Cole, 1994). Many are now filled, but those with aquatic habitats are shallow and have extensive macrophyte beds and low algal production (Schalles and Shure, 1989).

C. Lake Community Structure and Productivity

1. Lake Zones and Thermal Patterns

Regions within lakes exhibit physical and chemical differences that affect algal communities. The open water region of lakes is termed the pelagic (or limnetic) zone, whereas close to shore is the littoral zone, where the greatest exchange between nutrient-rich sediments and the water occurs (Fig. 6). The littoral zone is colonized by submersed (e.g., *Ceratophyllum*, *Potamogeton*, and *Vallisneria*) and emergent (e.g., *Scirpus* and *Typha*) flowering plants, although some macroalgae (e.g., *Chara*, *Nitella*, and *Batrachospermum*) and nonflowering plants (mosses, liverworts) also occur (Hutchinson, 1975). Vertical zones also develop in temperate regions. In early spring, most temperate lakes are well mixed, with similar temperatures and chemical conditions from top to bottom. As temperatures increase, upper mixed waters become thermally isolated from deeper and colder waters, a process which is termed stratification. The upper epilimnion continues to become warmer, receives greater irradiance, and is well mixed and oxygenated; deeper waters remain cool (ca. 4°C in deep lakes) and dissolved gases are consumed by microbial activity. At an intermediate depth (the metalimnion), temperature declines, often sharply, with reduced heat penetration and reduced mixing (= thermocline if $\geq \Delta 1° \, m^{-1}$). Density gradients formed by this thermal barrier may be sufficient to support dense algal populations (Pick *et al.*, 1984); here light is adequate for photosynthesis coupled with a greater supply of nutrients. Algal production may create metal-

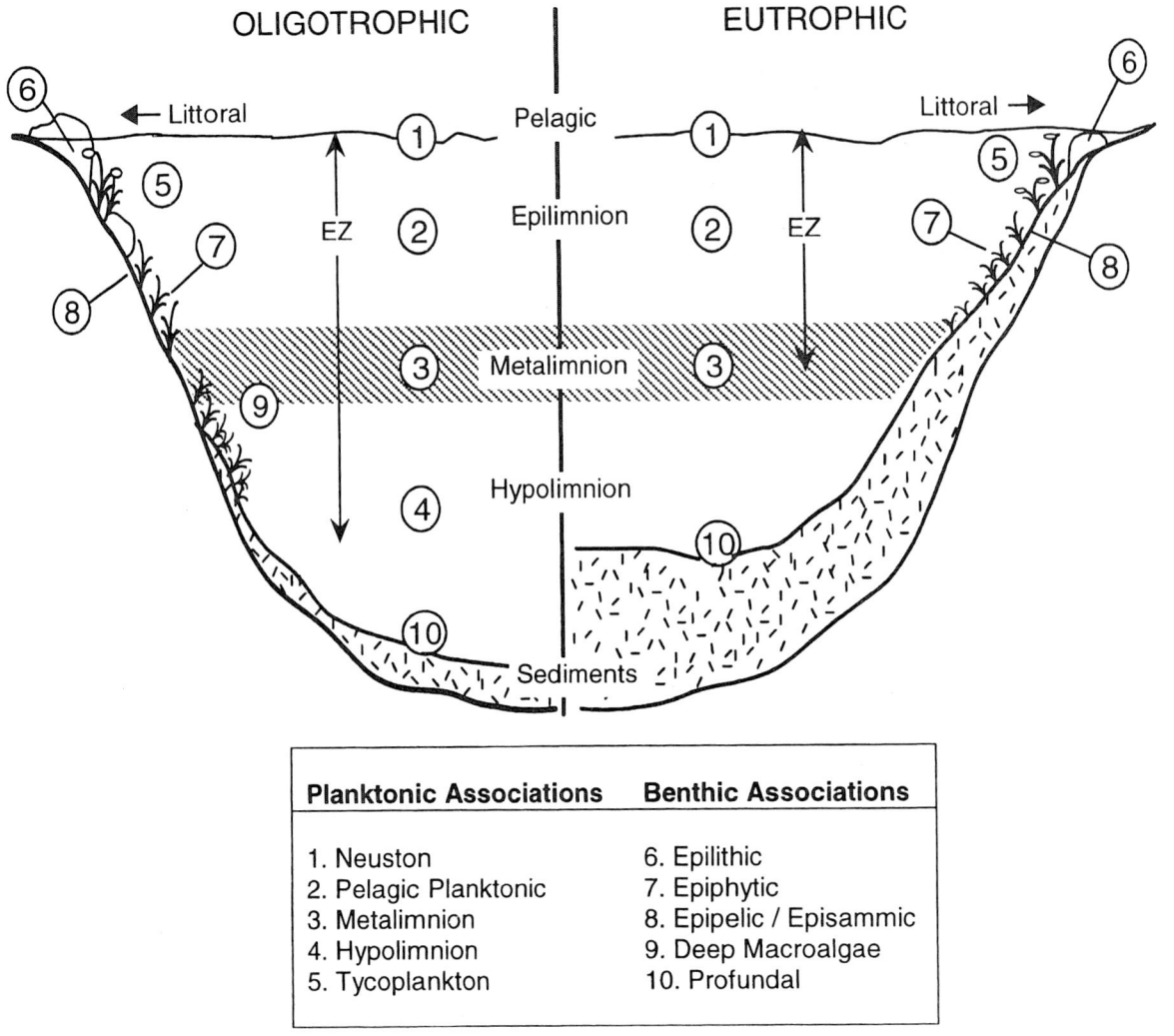

FIGURE 6 Diagram that represents the zones and algal habitats within typical oligotrophic and eutrophic lakes (EZ = euphotic zone).

imnetic oxygen maxima in clear oligotrophic lakes (Wetzel, 1983a; Parker et al., 1991). The hypolimnion is a deeper, cooler region with greater nutrient supply, but reduced (approaching zero) oxygen levels; light may be too low for photosynthetic algal growth except in very clear lakes.

Patterns of thermal stratification and mixing differ with altitude and across biomes. A dimictic pattern, in which lakes stratify in the summer, mix in the autumn, stratify in the winter after ice cover, and mix in the spring after ice-out, is most common in temperate climates. Warm monomictic lakes (stratification and one mixing period; summer epilimnion > 4°C) occur in warmer climates or in large basins without ice cover (Wetzel, 1983a). Examples include the Great Lakes, larger Finger Lakes, Lake Tahoe, lakes in warm or coastal climates, and subtropical, high altitude lakes. Cold monomictic lakes, with a single turnover in summer or late spring, occur mainly in alpine and arctic areas (temperatures ≤ 4°C). Oligomictic lakes have rare mixing periods (less than once per year), where temperature strata (summer epilimnion > 4°C) may remain for some years; this pattern is most common in deep tropical lakes (Wetzel, 1983a). Polymictic lakes are shallow systems with frequent or continuous mixing, and occur in tropical and equatorial areas such as Lake Managua (Xolotlán), Nicaragua (Erikson et al., 1997). Amictic lakes, uncommon in North America (some in Greenland; common in the Antarctic), are perennially ice covered and do not turn over. A special class of lakes in which upper waters (mixolimnion) mix, but deeper waters (monimolimnion) never circulate, are termed meromictic (Wetzel, 1983a). These lakes have a very stable chemical and temperature density gradient, known as a chemocline, that results in anoxic conditions, H_2S, and purple sulfur bacteria in the water column. A strong depth to surface area ratio is usually necessary to maintain meromixis; Fayetteville Green Lake in New York and Hot Lake in Washington are examples (Hutchinson, 1957).

2. Lake Productivity

Limnologists distinguish lakes according to a gradient of primary production (^{14}C uptake, algal growth) or biomass, from oligotrophic (annual average < 50–300 mg C m^{-2} d^{-1}; < 0.05–1 μg chlorophyll-a L^{-1}) to eutrophic (> 1000 mg C m^{-2} d^{-1}; 15–100 μg chlorophyll-a L^{-1}), and these levels are influenced, at least in part, by the properties of lake basins (Wetzel, 1983a; Likens, 1985). Oligotrophic lakes are poor in nutrients, usually deep and steep-sided, and have high transparency, a narrow littoral zone, abundant dissolved oxygen with depth, and larger relative hypolimnion volume. Lake Tahoe is an ultra-oligotrophic lake that has exceptional water clarity, although average Secchi depths have declined from about 30 to 23 m and primary productivity levels have more than doubled (40–100 mg C m^{-2} d^{-1}) since the late 1950s, following increased nutrient loading from regional development (Goldman, 1988). Eutrophic lakes are nutrient-rich, often shallower with a broad littoral zone, and have depleted summer hypolimnetic oxygen and reduced transparency. Eutrophication of lakes often result when nutrients are added as sewage, detergents, or fertilizers (Wetzel, 1983a). Lake Mendota, Wisconsin, is an example of a larger lake (39 km^2) that has an average depth of only 12 m and receives substantial input from agricultural and urban sources. The lake has experienced algal blooms and high nutrient levels for more than a century, and water transparency has been consistently low for more than 80 years (Brock, 1985a).

D. Ponds, Temporary Pools, and Bogs

Smaller lentic environments, often called ponds, may seem very similar to lakes except for their size, but they have distinct properties. Ponds are shallow enough either to support rooted aquatic vegetation across the entire basin or to fail to stratify during the summer. The term "pond," however, is not a precise concept, despite its frequent usage. In New England, the word is often applied to fairly substantial lakes, such as Linsley Pond (9.4 ha, max depth 14.8 m; Brooks and Deevey, 1963) and Long Pond (40 ha, 22 m; Canavan and Siver, 1995), both of which stratify in summer. Small limnetic systems can be divided into those that contain water year round, often called permanent, and temporary waters that become dry each year and are termed vernal ponds (Wetzel, 1983a).

Temporary ponds are located in low-lying areas that fill with snow melt or spring runoff, but dry up in the summer (Wetzel, 1983a); resident algal populations may be quite substantial during the growing season and rely on resistant resting stages (e.g., dinoflagellate cysts [Chap. 20], cyanobacterial akinetes [Chap. 4], and zygospores of Zygnematales [Chap. 9]) capable of surviving extended adverse conditions (Reynolds, 1984a, b). Arctic tundra ponds typically freeze solid for many months, resulting in a growing season of only 60–100 days (Wetzel, 1983a; Sheath, 1986). Algal communities often consist of small flagellates (particularly Cryptophyceae and Chrysophyceae; see Chaps. 21 and 12–14) that bloom during the brief summer of long daylight, although many species apparently do not form resting stages and survive freezing conditions in their vegetative condition (Sheath, 1986).

E. Phytoplankton of Lakes and Ponds

Assemblages of planktonic algae vary greatly among lake basins and biogeographic regions. They include members of all algal divisions except the Rhodophyta and Phaeophyta. Their ecology has been the subject of many reviews, including Hutchinson (1967), Kalff and Knoechel (1978), Round (1981), Reynolds (1984a), Munawar and Talling (1986), Sandgren (1988), Munawar and Munawar (1996, 2000), Stoermer and Smol (1999), and Whitton and Potts (2000).

1. Phytoplankton Diversity, Composition, and Seasonal Succession

Every collection of freshwater algae is characterized by a fascinating and perplexing diversity of species, many of which are potential competitors for common resources. Up to several hundred algal species may comprise the phytoplankton community of a typical north-temperate lake (Kalff and Knoechel, 1978). This paradox of many potentially competing phytoplankton species coexisting in a relatively uniform habitat was proposed by Hutchinson (1961) to be possible because of the numerous niches within a lake, as well as variation within the environment over time. Many factors contribute to phytoplankton diversity and production, including temporal and spatial variations in nutrient supply, grazing, temperature, and parasitism (Turpin and Harrison, 1979; Crumpton and Wetzel, 1982; Sommer, 1984; Bergquist and Carpenter, 1986). The biomass of phytoplankton is thought to be driven mainly by nutrient supply and herbivory in all lakes, but their temporal dynamics differ in eutrophic and oligotrophic systems (Walters *et al.*, 1987; Carpenter *et al.*, 1993).

Ecosystem-level studies of lake food webs in some cases have ignored phytoplankton species composition, and instead treat this component as a black box. This approach is incomplete because the functional properties of algal assemblages vary strongly with species composition. Taxonomic information is important for ecological studies because many of the features used to classify algae, such as photosynthetic pigments, storage products, motility, reproduction, cell ultrastructure, and even DNA sequence information, have functional importance. For example, among freshwater phytoplankton, only cyanobacteria and some cryptomonads possess the red accessory pigment phycoerythrin, which has an absorption maximum (540–560 nm) that broadens the photosynthetic capacity of cells and may facilitate growth at greater depths (Goodwin, 1974, Chap. 21). Similarly, only certain species of cyanobacteria are able to fix N_2, mainly those that possess heterocysts (e.g. *Anabaena*, *Aphanizomenon*, and *Nostoc*), although some non-heterocystous forms with thick sheaths or that form dense aggregations (*Gloeocapsa*, *Oscillatoria*, *Microcoleus*, and *Plectonema*; see Chap. 4) also have this ability (Bothe, 1982; Paerl et al., 1989). Algal flagellates are a polyphyletic collection of different protists, but they form an important ecological group because of their tendency toward mixotrophy (Porter, 1988). This fact is not coincidental, because the flagellar apparatus is directly involved in the capture of bacterial prey (Andersen and Wetherbee, 1992).

Diversity in size is also an important property of phytoplankton communities. One scheme (Sieburth et al., 1978) categorizes sizes into groups that differ over orders of magnitude: picoplankton (> 0.2–2 µm), nanoplankton (> 2–20 µm), microplankton (> 20–200 µm), and mesoplankton (> 200–2000 µm) include most algal cells and colonies in freshwater. Expressed in terms of volume, the sizes of freshwater phytoplankton span at least 8 orders of magnitude (Reynolds, 1984b). The bacteria-sized picoplankton have attracted recent interest because they have been found to dominate (at least numerically) many phytoplankton communities in lakes and marine systems (Stockner et al., 2000). They occur in great numbers (10^5–10^6 mL^{-1}), possess rapid growth rates, and are highly productive; most often reported are cyanobacteria (e.g., *Cyanobium*, *Cyanothece*, *Synechococcus*, and *Synechocystis*; see Chap. 3) and some green algae (e.g., *Nannochloris*). Their importance (percentage of biomass or primary production) seems to be greatest in oligotrophic and least in eutrophic lakes (Burns and Stockner, 1991; Hawley and Whitton, 1991; Wehr, 1991), although there are exceptions (Wehr, 1990; Weisse, 1993). Autotrophic picoplankton have a strong competitive ability in P-limited conditions (Suttle et al., 1987, 1988; Wehr, 1989) and are grazed mainly by micro-zooplankton (ciliates, flagellates), rather than cladocerans or copepods (Pernthaler et al., 1996; Hadas et al., 1998), making them important links between microbial and classical food webs (Christoffersen et al., 1990; Sommaruga, 1995).

Size affects sinking rate and thus the ability of cells to remain in the euphotic zone. Smaller cells tend to be spherical or ellipsoid and thus sink more slowly, whereas larger forms have more elongate or complex shapes to reduce sinking. The dinoflagellate *Ceratium hirundinella* is a large planktonic alga (up to 400 µm) common in mesotrophic and eutrophic lakes with stable stratification (Chap. 20); the alga regulates its position in the water column by active migration and perhaps by changes in the shape and size of hornlike projections (Heaney and Furnass, 1980; Pollingher, 1987; Heaney et al., 1988). The silica walls of diatoms result in heavier cellular densities, making them susceptible to sinking. Diatoms are estimated to have 3–16 times faster sinking rates than nonsiliceous algae of equivalent sizes (Sommer, 1988). Larger colonial diatoms, such as *Asterionella formosa*, *Fragilaria crotonensis*, and *Tabellaria fenestrata* (see Chap. 16), have more elongate or elaborate morphologies, which may reduce sinking rates or cause some cells to rotate (Smayda, 1970; Barber and Haworth, 1981). Planktonic cyanobacteria and desmids produce extracellular mucilage, which may aid in buoyancy (Round, 1981). Some cyanobacteria, such as *Anabaena flos-aquae* and *Microcystis aeruginosa*, also maintian their position in the water column using gas vacuoles (Chaps. 3 and 4). Observed changes in the population size may be the result of differences in vertical migration and sinking to lower strata, rather than actual changes in numbers.

Regular seasonal changes (seasonal succession) in phytoplankton populations over many years have been observed widely and reported in long-term limnological records. Few species have been documented as thoroughly as *Asterionella formosa*, a diatom found in mesotrophic, temperate lakes (Round, 1981; Reynolds, 1984a, Chap. 16). In Lake Windermere, populations increase after turnover and peak in the spring when light levels and temperatures are increasing and the Si:P ratio is near maximum (Fig. 7). Populations decline over the summer and have a second (usually smaller) peak in the autumn (Lund, 1964; Reynolds, 1984a; Neale et al., 1991). *Asterionella* was the spring dominant in Lake Erie from at least 1931 until about 1950, when nutrient enrichment selected for diatom genera such as *Stephanodiscus*, *Aulacoseira* (*Melosira*; see Chap. 15), and *Fragilaria*, and filamentous cyanobacteria such as *Anabaena* and *Aphanizomenon* (Davis,

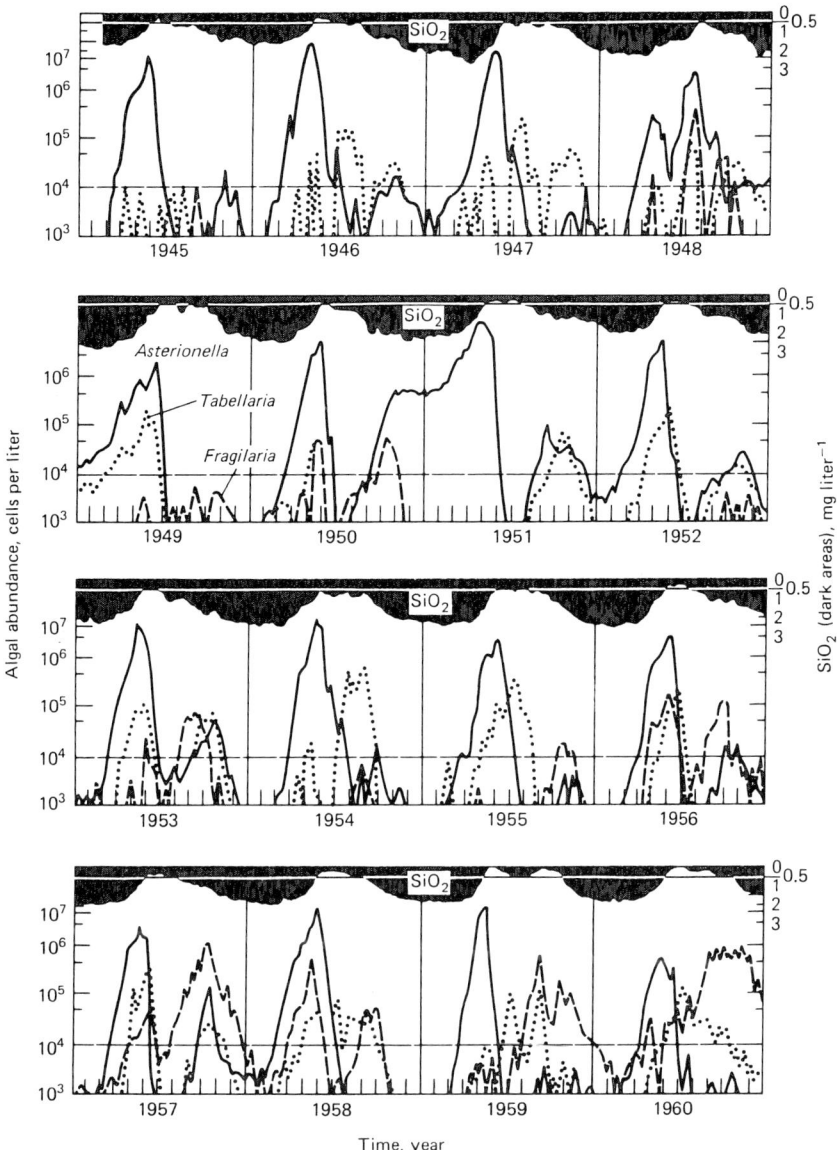

FIGURE 7 Seasonal periodicity in the abundances (cells per liter) of *Asterionella formosa* (solid line), *Fragilaria crotonensis* (dashed line), and *Tabellaria flocculosa* (dotted line), and concentrations of dissolved silica (upper, black; milligrams per liter) in Windermere, English Lake District 1945–1960. Reproduced with permission from A. Horne and C. R. Goldman, Limnology, 2nd ed. Copyright © 1994, McGraw–Hill, New York.

1964). As eutrophication receded, there was a 70–98% reduction in numbers of *Stephanodiscus* spp. and *Aphanizomenon flos-aquae*, and the reappearance of *Asterionella* (Makarewicz, 1993). Further details are given in Munawar and Munawar (1996).

Population dynamics generally follow predictable changes in temperature, sunlight, nutrients, and other factors. However, algal population dynamics exhibit more abrupt changes than these gradual trends predict, suggesting that other factors may drive seasonal succession. Round (1971) described these as "shock" periods: times in the lake cycle, such as turnover, that lead to sharp changes in chemical or physical conditions. Lakes with less predictable conditions or more shock periods exhibit frequent changes in species composition and shorter growth peaks (Fig. 8A), but when conditions are more stable, populations may persist over longer time periods (Fig. 8B). Furthermore, coexistence (temporal overlap) among species will result if a greater variety of habitats are available, perhaps through stratification, basin complexity, or multiple inflows (Fig. 8C). These temporal patterns repeat only if nutrient

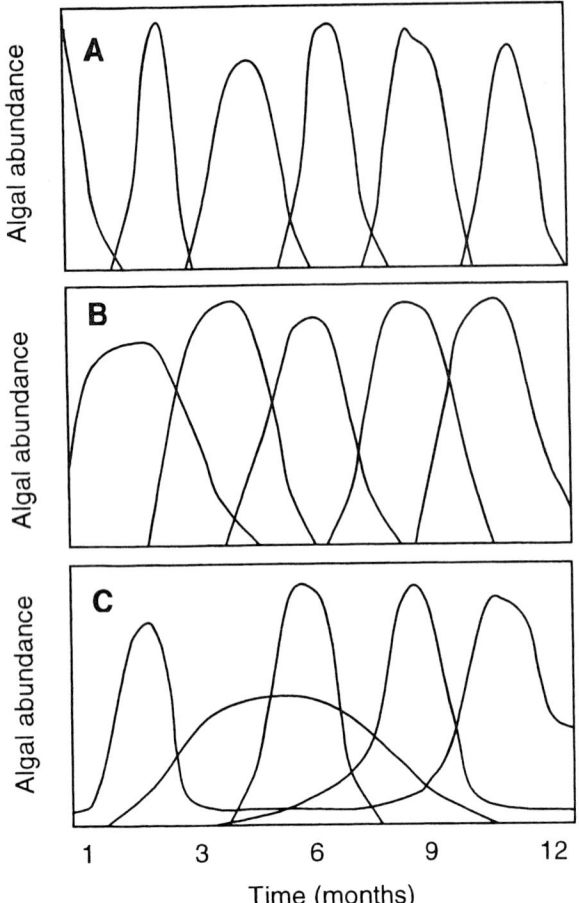

FIGURE 8 Diagrammatic representation of seasonal growth curves of freshwater phytoplankton species in lakes with different ecological conditions and habitat complexity: A, variable ecological conditions or frequent shock periods, with short peaks and low overlap; B, more stable conditions or longer stratification periods, with greater temporal overlap; C, complex lake basins with more habitats and peaks of different duration, overlap, and frequency. Redrawn and adapted from Round (1972).

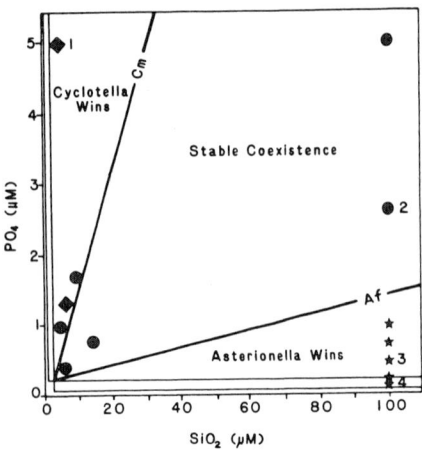

FIGURE 9 Predicted (lines) and observed (points) outcomes of competition between *Asterionella formosa* (stars = dominant) and *Cyclotella* (*Stephanocyclus*) *meneghiniana* (diamonds = dominant) under varying levels of Si and P, indicating that coexistence (solid circles) is possible in an intermediate range of Si:P ratios (From Tilman, D.; Resource Competition and Community Structure. Copyright © 1982 by Princeton University Press. Reprinted by permission of Princeton University Press.

or other conditions are stable and other disturbances are kept to a minimum. For example, in Lake Michigan *Tabellaria* sp. and *Asterionella formosa* exhibited regular peaks in the spring phytoplankton community for several decades, but *Asterionella* numbers have declined in more recent years, while *Stephanodiscus* and filamentous cyanobacteria have increased following increases in P supply (Makarewicz and Baybutt, 1981).

Laboratory experiments with *Stephanocyclus* (*Cyclotella*) *meneghiniana* and *Asterionella formosa*, which have different Si and P requirements and consumption rates, have indentified the levels of these nutrients at which the two species may coexist (Fig. 9, Tilman, 1977, 1982). Studies have quantified resource competition among other species and with other resources, such as C:P and N:P (Kilham and Kilham, 1978; Rhee and Gotham, 1980; Sommer and Kilham, 1985; Olsen et al., 1989). *In situ* manipulations of phosphorus and light have demonstrated clear differences among species: some respond positively to P addition alone (e.g., *Synedra radians*), whereas others, such as chrysophytes (e.g., *Dinobryon sertularia* and *Synura uvella*), increase under greater light (de Noyelles et al., 1980; Wehr, 1993). Species composition and size structure each can be influenced by nutrient supply. With higher N:P supply ratios, assemblages in an oligotrophic lake were dominated by pico-cyanobacteria (*Synechococcus* sp.; see Chap. 3), but at lower supply ratios, larger diatoms (*Nitzschia* and *Synedra*) dominated (Suttle et al., 1987). Whole-lake N and P additions to oligotrophic Kennedy Lake (BC), used to enhance production of sockeye salmon, also affected competitive interactions among phytoplankton (Stockner and Shortreed, 1988). Loadings at N:P ratios between 10 and 25 increased algal biomass with a summer community dominated by N_2-fixing *Anabaena circinalis*; increasing the N:P ratio to 35 retained the higher biomass, but shifted the community dominance to small-celled *Synechococcus* spp.

Manipulations of fish densities in a small Wisconsin lake varied predation pressure on zooplankton, which in turn created different levels of grazing pressure on phytoplankton; nested nutrient-permeable chambers separated effects of recycled nutrients from relaxed herbivory (Vanni and Layne, 1997). Phytoplankton biomass and the abundance of many algal taxa (e.g., *Peridinium inconspicuum*, *Chrysochromulina* sp., and *Staurastrum dejectum*) increased with greater fish biomass, although a similar effect was seen in grazer-free diffusion chambers for some species, which suggested that the positive effect of fish may have been mediated through recycled nutrients, rather than lower grazing pressure.

Some algae occupy both planktonic and benthic habitats at different periods of the year or even daily. This strategy may involve overwintering as inactive stages (e.g., algal cysts) or active recruitment of benthic forms into the pelagic zone. In many systems, flagellates such as *Ceratium*, *Cryptomonas*, and *Euglena*, exhibit diurnal vertical migrations toward the surface by day and into deeper strata at night (Palmer and Round, 1965; Heaney and Talling, 1980; Hansson et al., 1994), responding to patterns in light availability, temperature cues, mixing conditions, or nutrient supply. In spring and early summer, the cyanobacterium *Gloeotrichia echinulata* colonizes shallow sediments or submersed plants in eutrophic lakes, but in late summer, gas-vacuolate colonies migrate into the pelagic zone, representing as much as 40% of the planktonic assemblage (Barbiero and Welch, 1992). In subtropical Lake Apopka, Florida, benthic and settled planktonic diatoms (and their resting stages) are regularly resuspended into the water column by wind action, making a major contribution to the chlorophyll budget of the lake (Schelske et al., 1995). In these shallow lakes, buoyant cyanobacteria contribute to recycling phosphorus (internal loading) from sediments back into the water column (Salonen et al., 1984; Pettersson et al., 1993; Moss et al. 1997).

2. Factors That Regulate Phytoplankton Production in Lakes

A large body of data clearly indicates that phytoplankton production and biomass in most lentic systems is controlled by P supply (Schindler, 1978; Wetzel, 1983a; Hecky and Kilham, 1988). Although other factors (e.g., grazing, light availability, temperature) are clearly involved in regulating phytoplankton production, only nutrients are amenable to regulation (Schindler, 1978). This concept was demonstrated in Lake Washington, which received secondary sewage (a source of P) from the city of Seattle. Blooms of planktonic cyanobacteria, especially *Planktothrix* (*Oscillatoria*) *rubescens*, were common in past decades (Edmondson, 1977). Today eutrophication in Lake Washington has been largely reversed through monitoring and sewage diversion, which reduced P inputs to near zero, summertime chlorophyll-*a* from about 45 to 5 µg L^{-1} and the percentage of phytoplankton as cyanobacteria from ≈100 to 10% or less (Edmondson and Lehman, 1981).

Eutrophic lakes with low N:P ratios favor blooms of N$_2$-fixing cyanobacteria; hence, nutrient cycling within these lakes is closely coupled these organisms (Schindler, 1977, 1985). Algal species have different micronutrient requirements, such as Si, Mg, Ca, Fe, Mo, and Se. Diatom dominance in phytoplankton assemblages is dependent on recycled Si following spring turnover; lakes with lower Si levels may lack a springtime diatom pulse altogether. The haptophyte *Chrysochromulina breviturrita* develops large populations in dilute lakes undergoing the early stages of acidification (Nicholls et al., 1982, Chap. 13). Its requirement for Se and NH$_4^+$, and its inability to use NO$_3^-$ favor its success in oligotrophic lakes within the pH range 5.5–6.5 (Wehr and Brown, 1985; Wehr et al., 1987). Nutrient requirements may have important interactions, for example, Mo, which is an essential micronutrient for some cyanobacteria as a co-factor for N$_2$ fixation; its assimilation may be inhibited by elevated SO$_4^{2-}$, which is common in saline lakes (Howarth and Cole, 1985; Marino et al., 1990).

Light supply within the water column is a critical factor that also affects phytoplankton production and species composition in lakes. Although the abundance of many species is greatest in the epilimnion where irradiance is greatest, other species, including several algal flagellates are adapted to deeper waters (Lund and Reynolds, 1982). Under relatively warm and calm conditions in Blelham Tarn, more than 90% of *Uroglena* sp. colonies and *Eudorina elegans* aggregated in the upper 2 m, but in October, *Trachelomonas* occupied a narrow band near the thermocline (6–8 m depth). In eutrophic or highly turbid waters, some algae remain buoyant using gas vesicles (e.g., *Anabaena*, *Aphanizomenon*, *Coelosphaerium*, and *Microcystis*; see Chaps. 3 and 4) or flagella (e.g., *Ceratium*, *Chlamydomonas*, and *Euglena*). Species also exhibit different photosynthesis–irradiance and temperature optima. Some species, such as *Asterionella formosa*, have especially high photosynthetic efficiency at low light (Reynolds, 1984a), whereas others, including species of *Oscillatoria*, may vary their chlorophyll-*a* levels and employ accessory pigments that saturate at lower and spectrally altered light levels (Mur and Bejsdorf, 1978). Flagellates such as *Dinobryon*, *Poterioochromonas*, *Cryptomonas*, and *Ceratium* may ingest bacteria under low light conditions or under ice; these mixotrophic species switch between photosynthetic and bacterivorous metabolism (Bird and Kalff, 1987; Porter, 1988; Berninger et al., 1992; Caron et al., 1993). This strategy may also serve to supplement inorganic nutrients as well as organic-C needs.

F. Benthic Algal Assemblages of Lakes

Benthic algae—those attached to or closely associated with various substrata or bottom surfaces—occupy an enormous variety of microhabitats, including stones, macrophytes, sediments, sand grains, and logs, as well as a variety of artificial substrata. All divisions of algae have benthic representatives, although many more freshwater species in Chrysophyta (Chaps. 12–14), Xanthophyta (Chap. 11), Cryptophyta (Chap. 21), and Pyrrophyta (Chap. 20) are planktonic, whereas all species of freshwater Phaeophyta and Rhodophyta are benthic (but rare in lakes). Perhaps the most widely used term for benthic algae is periphyton, a word that has obscure etymology and debatable usage (Sládeckova, 1962; Round, 1981; Wetzel, 1983a; Aloi, 1990; Stevenson, 1996a). The term may have been coined in the 1920s by Russian limnologists (Sládeckova, 1962) to refer to a collection of organisms (bacteria, fungi, protozoa) and detritus (Wetzel, 1983c). The word is analogous to the German *aufwuchs*, which means "to grow upon," but some authors (e.g., Round, 1981) argue against using it because it is often used incorrectly to describe only the algal community and is imprecise with regard to habitat. Its use probably will not be abandoned, however. We recommend the use of the most precise and descriptive terminology for particular algal communities (e.g., epilithic diatom) that includes the nature of the substratum. Otherwise the term "benthic" is perhaps most suitable for general uses or when the substratum is not defined.

Modes of algal attachment are diverse. Some are firmly attached or encrusting, such as *Chamaesiphon*, *Coleochaete*, and *Cocconeis*, making them resistant to wave scour or other disturbances, but susceptible to competition from canopy-forming morphologies, such as *Stigeoclonium* or *Ulothrix* (Hoagland and Peterson, 1990; Maltais and Vincent, 1997; Graham and Vinebrooke, 1998). Diatoms, such as *Frustulia*, *Navicula*, *Nitzschia*, *Pinnularia*, and *Stauroneis*, maintain contact with various surfaces (and glide among these microhabitats) by means of a slit in the wall that is termed a raphe (Chaps. 16–19). Other diatoms (e.g., *Cymbella* and *Gomphonema*) attach by means of gelatinous pads or stalks. Some filamentous cyanobacteria, such as *Oscillatoria*, *Hapalosiphon*, *Lyngbya*, and *Microcoleus*, exhibit motility by gliding, although the mechanisms are not clear (Castenholz, 1982). Filamentous green algae, such as *Cladophora*, *Oedogonium* (see Chap. 8), *Spirogyra*, and *Zygnema* (Chap. 9) produce holdfast-like rhizoids that enable them to remain attached in turbulent conditions. Other benthic forms are loosely associated with plants or sediments, and include filamentous species like *Mougeotia*, flagellates such as *Cryptomonas*, *Euglena*, and *Eudorina*, and chains of cells such as *Tabellaria* (Hutchinson, 1975; Graham and Vinebrooke, 1998).

The range of sizes in freshwater benthic algae exceeds that of planktonic forms. The smallest include unicells like *Nannochloris* or *Synechococcus* (0.8–2 μm diameter) to actual macrophytes, such as *Chara*, *Cladophora*, and *Hydrodictyon*, which range from a few centimeters to more than a meter in length. In terms of length, this range is more than 6 orders of magnitude, and in biovolume, perhaps as great as a factor of 10^{10}. The tremendous variety of microhabitats, morphologies, sizes, and architectures found in benthic algal associations has led to the suggestion that these organisms may represent a more diverse community and greater trophic complexity than phytoplankton (Havens *et al.*, 1996; Stevenson, 1996a). In shallow lakes, production of epiphytic algae often equals or exceeds that of phytoplankton per unit area (Wetzel, 1983a). Much of the literature on freshwater benthic algae has been reviewed in several important works, including Hutchinson (1975), Round (1981), Wetzel (1983b), and Stevenson *et al.* (1996).

1. Epiphytic Communities

Epiphytic algae colonize submersed and emergent plants. These are the most widely studied group of benthic algae in lakes, perhaps because of their obvious accumulation in the littoral zone. Larger forms, such as *Cladophora*, *Chara*, *Hydrodictyon*, and *Oedogonium*, serve as additional substrata for microalgae. Epiphytic algae are important in macrophyte communities, because greater densities may cover and shade their hosts (Losee and Wetzel, 1983). Evidence includes negative relationships between epiphyte and macrophyte biomass (Sand-Jensen and Søndergaard, 1981; Cattaneo *et al.*, 1998) and more rapid host senescence with greater epiphyte cover (Neely, 1994).

There are often differences in species composition and biomass of epiphytic algae among different macrophyte host species. Prowse (1959) recognized that densities of three common epiphytes, *Gomphonema gracile*, *Eunotia pectinalis*, and *Oedogonium* sp., differed among three macrophyte species in one small pond. Many subsequent studies have reported differences in epiphyte biomass or species composition on different plant hosts (Gough and Woelkerling, 1976; Eminson and Moss, 1980; Lodge, 1986; Blindow, 1987; Douglas and Smol, 1995; Hawes and Schwarz, 1996), although not in all cases (Siver, 1977). In one shallow lake, epiphyte biomass on submersed macrophytes (*Myriophyllum spicatum*, *Ceratophyllum demersum*, and *Najas marina*) was 10–40 times greater than on floating-leaved plants (*Trapa natans*), but species diversity was less (Cattaneo et al., 1998). In the Great Lakes, *Cladophora glomerata* is a host to many microalgal epiphytes, but the red alga *Chroodactylon ornatum* (as *C. ramosum*) is attached only to this species (Sheath and Morrison, 1982). The three-dimensional architecture of epiphyte assemblages also varies with the type of substratum. Colonization by epiphytic algae has been compared to terrestrial plant succession, which comprises temporal changes in vertical structure and diversity, an increase in the dominance of larger organisms, and possible facilitative effects of earlier colonizers (Hoagland et al., 1982).

The reasons for differences in epiphytic communities among host plant species can be attributed to features of the macrophyte, such as leaf orientation, texture, or chemical properties. One survey revealed a correspondence between epiphytic communities and species of submersed macrophytes in less productive lakes, but little pattern was observed in eutrophic lakes where nutrient macrophyte interactions might be less (Eminson and Moss, 1980). However, plants may inhabit different zones within lakes that indirectly offer different ecological conditions for algal colonization. Nonetheless, direct evidence shows that living macrophytes translocate and release small quantities of P (about 3.5 µg P g^{-1} macrophyte shoot), which can be taken up by algal epiphytes, and that algal species differ in their ability to sequester released P (Moeller et al., 1988). A *Synedra–Fragilaria* complex obtained more than 50% of released P, but erect forms such as *Mougeotia* and *Lyngbya*, and stalked *Gomphonema* obtained most of their P from the surrounding water. Alkaline phosphatase activity of epiphytic algae on artificial (plastic) plants was shown to be greater than on natural plants under similar conditions (Burkholder and Wetzel, 1990).

Epiphytic communities are important and complex components of lake food webs. In mesotrophic Lake Mann (WI), herbivorous snails consume and regulate benthic algal biomass, but pumpkinseed sunfish also exert predatory control on snails (Brönmark et al., 1992). Algal-feeding snails, benthic insects, and other invertebrates also have a qualitative impact on epiphytic communities, because many consumers graze more effectively on erect or filamentous forms, thereby shifting the community toward more compact or adherent forms like *Cocconeis placentula* and *Coleochaete* spp. (Kesler, 1981; Lodge, 1986; Marks and Lowe, 1993). In eutrophic lakes, snails similarly avoid larger colonies of epiphytic *Gloeotrichia* (Cattaneo, 1983; Brönmark et al., 1992). In contrast, the limpet *Ferrissia fragilis* grazes mainly understory species, such as *Epithemia* spp., *Cocconeis placentula*, and *Achnanthidium minutissimum*, and avoids upright forms such as *Synedra ulna* and *Fragilaria vaucheriae* (Blinn et al., 1989). Grazers of epiphytic algae may have indirect effects on host plants by reducing shade and enhancing plant growth (Lodge et al., 1994).

Because of the difficulties of sampling epiphytic algae, artificial substrata, such as glass slides, plastic flagging, styrofoam floats, plexiglas plates, and plastic aquarium plants, are employed. Advantages include reduced variability, known surface area, standardized conditions, and no nutritional or chlorophyll artifacts from the host. An implicit assumption in their use is that the community sampled is representative of the "true" epiphyte community on aquatic plants, but studies suggest this is rarely true (Tippet, 1970; Robinson, 1983; Aloi, 1990; Cattaneo and Amireault, 1992). Glass microscope slides were among the first materials used (Sládecková, 1962), but differences in biomass, seasonal patterns, and community structure (different species proportions) suggest this approach may provide unreliable estimates (Tippet, 1970). Evidence suggests that biomass of most epiphytic algae is overestimated when some types of artificial substrata are used, although green algae and cyanobacteria may be undersampled (Aloi, 1990; Cattaneo and Amireault, 1992). Synthetic materials are much simpler in surface texture and chemistry than natural substrata, and this is likely to affect the grazing, production, and community structure of epiphytes. Although artificial substrata should not be assumed to mimic natural habitats fully, they can be useful in comparative analyses or replicated studies on the effects of disturbances on benthic algal communities (Robinson, 1983; Aloi, 1990).

2. Epilithic Communities

Epilithic algae colonize stones, boulders, and bedrock in lakes, and may dominate wave-swept littoral zones and oligotrophic lakes that have minimal macrophyte cover (Loeb et al., 1983). Species composition differs more strongly from phytoplankton than do epiphytic communities, but do exhibit pronounced seasonality in response to changes in nutrients, temperature, and other factors (Hutchinson, 1975; Lowe, 1996).

Epilithic communities in turbulent littoral habitats are distinct from epiphytic communities within the same lake and comprise species known mainly from streams, such as *Chamaesiphon* spp., *Gongrosira incrustans*, *Hildenbrandia rivularis*, *Tolypothrix distorta*, or *Heribaudiella fluviatilis* (Kann, 1941, 1978; Auer et al., 1983). Vertical zonation is often observed. Many epilithic species are restricted to the upper littoral zone, whereas others occur in deeper waters where wave action is less severe (Hoagland and Peterson, 1990; Lowe, 1996). *Bangia atropurpurea* and *Ulothrix zonata* occur on rocks in the upper splash zone of the Laurentian Great Lakes, just above a mat of *Phormidium* sp., whereas a *Cladophora glomerata* zone is found in deeper water (Sheath and Cole, 1980). In Lake Traunsee (Austria), Kann (1959) documented that zonation patterns can differ among regions in a lake according to differences in slope. Epilithic algae form distinct communities in different regions of oligotrophic Lac à l'Eau Claire (subarctic Quebec), that are influenced by ice scour and wave action (Maltais and Vincent, 1997). A *Gloeocapsa*-dominated community colonized shallow areas, while a filamentous community dominated by *Ulothrix zonata* occurred in open, south-facing shores. In a study of 35 arctic ponds, freezing and other habitat factors were found to be of greater importance to benthic communities than were chemical variables (Douglas and Smol, 1995). Light or other factors may interact with the effects of wave action. An epilithic population of *U. zonata* exhibited a greater photosynthesis irradiance optimum (1200 µmol photons m^{-2} s^{-1}) than *C. glomerata* (300 µmol photons m^{-2} s^{-1}) isolated from the same region of Lake Huron (Auer *et al.*, 1983). Temperature tolerances and nutrient requirements may further interact with irradiance optima, and affect local and regional distributions (Graham *et al.*, 1985). *Bangia*, recent invader to the Great Lakes, has displaced *Ulothrix zonata* in many locations, perhaps because of its ability to produce more durable holdfasts (Lin and Blum, 1977) or an ability to resist epiphyte cover by sloughing its cell wall (Lowe *et al.*, 1982). Interestingly, Kann (1959) similarly observed *Bangia* occupying rocks in the splash zones of the Traunsee, but co-occurring with another filamentous green alga, *Mougeotia*. In the calmer epilithic community of Montezuma Well (AZ), only 8 of the 83 benthic diatom taxa identified were restricted to this habitat (Czarnecki, 1979). The upper littoral zone can be a harsh habitat, where algal cells experience abrasive turbulence or desiccation during an annual cycle.

Controls on epilithic production vary among different lake types. The epilithon of softwater, oligotrophic Lakes in the Experimental lakes Area (ELA; ON) was dominated by diatoms and filamentous green algae; production levels tended to be low but quite variable (Stockner and Armstrong, 1971; Schindler *et al.*, 1973). Comparing nutrient-amended, pH-manipulated, and reference lakes in the ELA, Turner *et al.* (1994) concluded that epilithon production is unrelated to N or P supply (despite positive effects on phytoplankton), but is limited by dissolved inorganic carbon (DIC). Experiments using nutrient-diffusing substrata determined that DIC and P supply were the most important influences on biomass and species composition in another oligotrophic, softwater lake in Pennsylvania (Fairchild *et al.*, 1989). The epilithon of meso-oligotrophic Flathead Lake (Marks and Lowe, 1993) was limited principally by N and P, but individual species differed in their response to nutrient and shading manipulations. In sublittoral Lake Tahoe, epilithic populations of *Calothrix*, *Tolypothrix*, and *Nostoc* were strongly N-limited and exhibited N-fixation activity, in contrast to resident phytoplankton (Reuter *et al.*, 1986). Recent increases in atmospheric N deposition may change the nutrient economy toward P limitation (Jassby *et al.*, 1995).

Grazing by benthic invertebrates is also important to epilithic algae. Snails (*Planorbis contortus*) and limpets (*Ancylus fluviatilis*) that inhabit the stony littoral zone of a small calcareous lake consumed substantial quantities of algae and detritus, and each preferentially grazed certain algal species (Calow, 1973a, b). Selectivity and more intense grazing activity by limpets exerted greater effects on algal community structure than did snails. Light availability and grazing pressure are factors that logically would be expected to be more important than nutrients for epilithic algae in eutrophic lakes, but grazing had minor impacts on biomass and seasonal patterns of epilithic algae in Crosmere, a eutrophic lake in the English Midlands, although caddisfly larvae may have contributed to spatial patchiness of *Cladophora* (Harrison and Hildrew, 1998). Studies on epilithic food webs that consisted of crayfish, invertebrates, and macrophytes in one Swedish lake found little top-down control by crayfish on epilithic algae (Nyström *et al.*, 1996). The general importance of benthic algae in lake food webs is not well established, in part because of difficulties in their quantification. However, a study of arctic, temperate, and tropical lakes using stable isotopes suggests that previous efforts may have underestimated the importance of algae in benthic food webs (Hecky and Hesslein, 1995). Prior studies were based on net production of phytoplankton, benthic algae, and macrophytes, instead of ease of grazing, edibility, or nutritional quality; all of these qualities were predicted to be greatest in benthic algae. The importance of epilithic algae in some lakes may be increasing, due to the expansion of filamentous algae in many acidifying lakes (Stokes, 1986; Turner *et al.*, 1995). In one neutral lake, tadpoles suppressed the growth of filamentous algae on tiles, and favored communities of adherent and encrusting species (*Coleochaete scutata*, *Achnanidium minutissimum*); grazers had no such effects on epilithic communities in acidified lakes (Graham and Vinebrooke, 1998). Transplants of epilithic algae (on natural quartz tiles) across lakes of varying acidity, coupled with grazer exclosures confirmed that pH was the key factor that regulated algal communities, but grazer control was important in neutral lakes (Vinebrooke, 1996).

3. Epipelic and Epipsammic Communities

Algal communities that colonize sediments (epipelic) and sand (epipsammic) are among the least studied benthic associations. This knowledge gap is surprising, considering the large area of many lakes not covered by stones or aquatic plants. It is difficult to separate epipelic from epipsammic habitats because these substrata are often mixed and, due to wave action, are especially unstable (Hickman, 1978; Kingston et al., 1983). Methods used to quantify living algal cells from lake sediments require considerable care and may still result in fairly high relative error (30% or greater), which may not be consistent among taxa (Eaton and Moss, 1966).

Epipelic and epipsammic communities occur in all lake types, but their relative importance is greatest in small, shallow systems. Diatoms are the most common algal group in most systems (numbers and biomass), although cyanobacteria, cryptomonads, desmids, euglenoids, and colonial and filamentous green algae often are observed (Gruendling, 1971; Round, 1972; Hickman, 1978; Roberts and Boylen, 1988). Epipelic algae may live on or within the first few millimeters of sediment and so must be able to tolerate conditions of very low light or oxygen, making motility a distinct advantage. Nonmotile epipelic and epipsammic forms may be capable of heterotrophic growth, which allows them to tolerate dark conditions and utilize greater levels of dissolved nutrients. Despite these constraints, epipelic and epipsammic communities are often very diverse. In a study of arctic ponds, Moore (1974) identified 357 algal taxa from sediments, of which 226 were benthic diatoms. Very high standing crops were also measured, some exceeding 10^8 cells cm^{-2}, which perhaps was influenced by long photoperiods during arctic summers. A diverse assemblage of 255 taxa was observed along a depth gradient in Lake Michigan, where benthic forms predominated in shallow and mid-depth communities (6–15 m) and settled, living planktonic forms were more common in deep (23–27 m), low light conditions (Stevenson and Stoermer, 1981).

Patterns of seasonal succession seem to differ from those observed for lake phytoplankton. Over a three-year period, Round (1972) observed that populations of epipelic algae in two shallow ponds had distinct and reproducible periods of abundance peaks, but their duration was longer (months rather than weeks) and varied markedly among species. *Stauroneis anceps* and *Oscillatoria* sp. persisted for several winter months, whereas other species like *Navicula hungarica* and *N. cryptocephala* maintained sizeable populations for more than nine months. In acidified Woods Lake (NY), several epipelic species were abundant from May through October, such as *Navicula tenuicephala* at 1 m and *Hapalosiphon pumilis* at 7 m (Roberts and Boylen, 1988). Following liming, several acidophilic species were replaced by other taxa, but diatom species still dominated the epipelon (Roberts and Boylen, 1989). Round (1972) suggested that epipelic habitats provide greater habitat diversity than the water column, enabling more species to coexist.

4. Benthic Macroalgae

Several species of macroalgae, those that form macroscopic or plantlike morphologies, can be important in benthic lake communities. Some investigators (e.g., Hutchinson, 1975) restrict this category to taxa that are attached by means of rhizoids, which include only charophytes (*Chara*, *Lamprothamnium*, *Nitella*, *Tolypella*, and *Nitellopsis*). From an ecological perspective it makes sense to include other algae, such as *Cladophora*, *Enteromorpha*, and *Nostoc*, which also may function as structuring elements within the benthic zone and as hosts for epiphytic microalgae. Astounding colonies of *Nostoc* greater than 30 cm in diameter ("mare's eggs") have been observed (Dodds et al., 1995), and tubes of *Enteromorpha* may exceed 1 m in length (Wehr, unpublished). *Chara* may form large underwater meadows in the littoral zone of calcareous, nutrient-poor lakes and frequently is encrusted with marl ($CaCO_3$).

In clear, oligotrophic lakes, charophytes colonize lake bottoms down to depths of 30 m or more. In nonturbid systems, wave action is suggested to be the primary limiting factor, rather than light (Schwarz and Hawes, 1997). Various species of *Chara* and *Nitella* differ in their depth distributions, presumably as a function of individual light requirements (Wood, 1950). The lower depth boundary for *Nitella* meadows in more productive lakes is influenced by total irradiance and the supply of red light; in deep, clear lakes, their distribution is limited by the availability of blue light (Stross et al., 1995). Reduced charophyte abundance in lakes with eutrophication has been attributed to excessive or even toxic levels of P (Hutchinson, 1975; Phillips et al., 1978). However, the macrophyte community of eutrophic Lake Luknajno (Poland) is dominated (90% of total dry mass) by seven charophyte species (codominants: *C. aculeolata* and *C. tomentosa*), where the mean biomass is greater than 1 kg dry mass m^{-2} (Krolikowska, 1997). *Nitella hookeri* was found to grow best at very high P concentrations, about 20 mg L^{-1} (Starling et al., 1974). Reduced charophyte abundance in eutrophic lakes may be the result of light limitation, given that some species do grow at greater depths but only in clear lakes (Blindow, 1992; Steinman *et al.*, 1997). Recovery of a previously eutrophic lake resulted in a 15-fold increase in the benthic cover of charophytes in just a two-year period (Meijer and Hosper, 1997).

The epiphytic algae that colonize charophytes can be dense and may differ in composition among specific charophyte taxa. In one deep oligotrophic lake, epiphytes on charophytes differed strongly with depth; filamentous green algae and stalked diatoms predominated in shallow water (≤ 5 m), while adnate forms, such as *Cocconeis placentula*, predominated at greater depths (Hawes and Schwarz, 1996). Littoral epiphyte densities in one eutrophic Swedish lake were greater on *C. tomentosa*, *C. globularis*, and *Nitellopsis obtusa* than on *Potamogeton pectinatus* (Blindow, 1987). *Nitellopsis* harbored the highest total densities, but differences in epiphyte species composition were most pronounced between the two *Chara* species, probably a result of differences in marl encrustation. Benthic invertebrates and zebra mussel larvae also colonize species of *Chara* more readily than other macrophyte hosts (Lewandowski and Ozimek, 1997; Van Den Berg *et al.*, 1997).

III. LOTIC ENVIRONMENTS

Running water ecosystems, from headwater streams to the largest rivers, are termed lotic environments and differ from standing waters in several important respects. Lotic systems are turbulent and generally well mixed; hence stratification is uncommon except for brief periods in slowly flowing, lowland rivers (Hynes, 1970). Waters with greater current velocity tend to have abundant dissolved oxygen; even large rivers with substantial current speed, such as the Ohio, are generally well mixed (Thorp *et al.*, 1994). In shallow rivers or littoral regions of large rivers, organisms experience less temperature fluctuation than organisms in lakes of comparable depths. Many organisms adopt a benthic habit and are attached to a variety of substrata, the sizes of which are a function of current velocity and discharge. Rapid rivers are typified by large stones and boulders, whereas the bottom material in more slowly flowing systems consists of sand and silt.

Rivers are intimately connected with the surrounding watershed, which is responsible for regulating water and chemical balances within the aquatic environment (Likens *et al.*, 1977). Rivers are open systems that transport materials and energy from one part of the watershed to downstream areas. Descriptions of the physical and geological attributes of river systems are given in the treatises by Leopold *et al.*, (1964) and Morisawa (1968), and in syntheses by Hynes (1970) and Beaumont (1975). Several important reviews discuss ecological feature of rivers, in particular, those by Hynes (1970), Whitton (1975, 1984), Lock and Williams (1981), Allan (1995), and Petts and Callow (1996).

A. Major Rivers of North America

The Mississippi–Missouri River system is longest in the world (6970 km), has the third largest drainage area (3270×10^3 km^2), and is the sixth largest in terms of discharge (18,390 m^3 s^{-1}; Hynes, 1970; Milliman and Meade, 1983). At least 10 other rivers in North America exceed 1000 km in length. Most of them have been altered substantially by navigation or hydroelectric dams, channelization, wetland removal, and pollutants (Sparks, 1995; Wehr and Descy, 1998). The Missouri River (3770 km) has six major impoundments over 1230 km (33%) of its length; another 1200 km (32%) have been channelized. Only 35% (1330 km) of all river sections remain free-flowing, although discharge still is influenced by reservoir conditions upriver (Hesse *et al.*, 1989). No major impoundments are located on the Ohio or Mississippi Rivers, but navigation dams and channelization have been built throughout their lengths to facilitate ship traffic.

B. Geomorphology of Rivers

Streams and rivers are part of a network of connected, increasing tributaries that have hydrological features that vary in predictable ways. Many features, such as discharge, substratum size, stream width, and depth, affect the species composition and productivity of lotic algae and their consumers. For example, sizes of substrata available for colonization vary with differences in current velocity (Table II). Physical features of rivers may be described in a system of stream orders, which assign increasing numbers to streams when two tributaries of equal order join. The most widely adopted system (Strahler, 1957) defines a headwater stream or spring with no (permanent) tributaries as first order and the junction of two such streams a second order (Fig. 10). A second-order stream increases only when it is joined by another second-order stream and so on. Larger order streams are wider and longer segments, drain larger areas, and have a more gradual slope than smaller streams. The network of these stream segments forms a treelike structure that is used in hydrological models to predict average discharge, behavior of flood events, and quantity of suspended matter (Beaumont, 1975). A simpler scheme divides rivers into three zones

TABLE II Relationship between Minimum Current Velocity and Mean Size of Stone Substrata That Can Be Moved along a Streambed (based on Hynes, 1970; Reid and Wood 1976)

Current velocity (m s-1)	Particle size (cm)	Stream bed characteristics	Habitat
3.0	180	Bedrock	Torrential
2.0	80	Boulders	Rapids
1.0	20	Large stones	Riffles
0.8	10	Stones and gravel	Riffles
0.5	5	Gravel & coarse sand	Run
0.2	1	Sand	Run
0.1	0.2	Silt	Pool

(Schumm, 1977). Zone 1 (erosional) includes headwater and small order streams that function as the source of water and sediments for downstream reaches. In Zone 2 (conveyance), water and sediments are transported along the mainstem with no net gain or loss of materials. Zone 3 (deposition) includes lower reaches that receive sediments from upriver, including river deltas and estuaries. Some studies suggest that benthic algal communities differ similarly along these zones (Rott and Pfister, 1988).

Although many aspects of drainage basins follow an ordered structure within an individual watershed (catchment), physical conditions and biological communities in streams of equal order at two locations may be quite different. Current velocity, stream width, and substrata in a second- or third-order mountain stream in New England differ from a similar order cool-desert stream in the western Great Basin (Minshall, 1978; Minshall et al., 1983; Benke et al., 1988). Landscape factors such as climate, terrestrial vegetation, and external nutrient supplies, may exert a substantial effect on the biological properties of rivers. For example, rivers that flow through a limestone region will possess greater concentrations of certain ions (Ca, Mg) than would be found in streams flowing through a region composed of granite or basalt.

River basins also differ in the amount of interaction between river channels and their watershed, which is mainly influenced by geological features of the region and amount of interface between groundwater and surface waters (Dahm et al., 1998). Some rivers are geologically constricted, such as the Hudson River and large sections of the Ohio and Columbia. Floodplain rivers have substantial watershed interaction, such as the lower Mississippi. The floodplain includes portions of the watershed (tributaries, adjacent wetlands, flood-

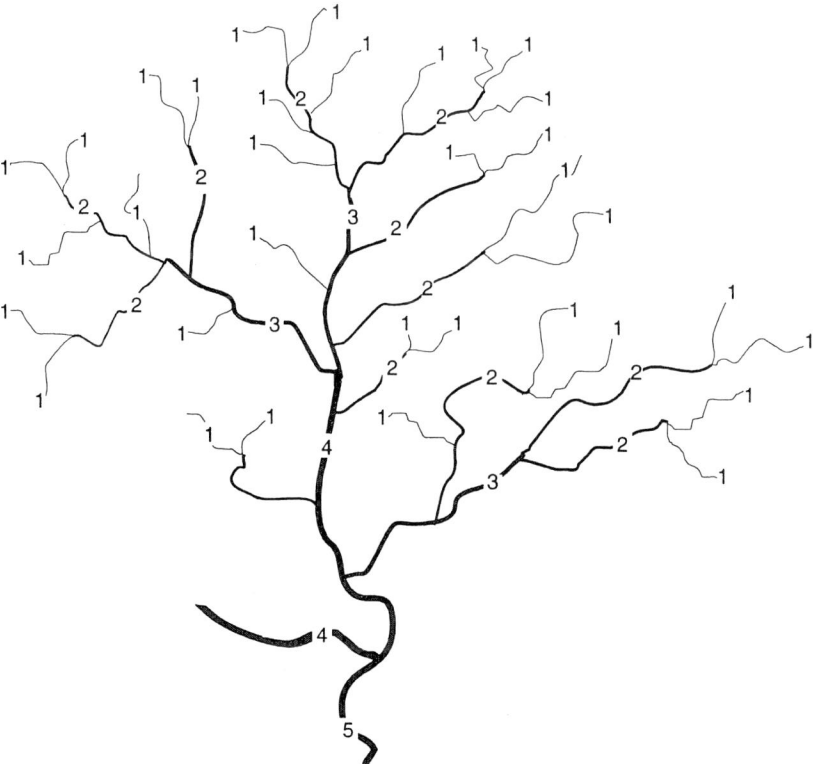

FIGURE 10 Structure of tributaries in a watershed, indicating numbering of stream orders.

plain lakes, riparian zones) that are seasonally inundated during periods of high flow. Meandering rivers and those with more islands have greater littoral and floodplain interaction, and more complex and varied current regimes than rivers with less sinuous courses (Fig. 11A and B). This increased habitat complexity may offer refuge to larger, more slowly growing algae. Small "islands" also form in lowland rivers from large stands of submersed angiosperms (Butcher, 1933; Holmes and Whitton, 1977).

Within a given reach, there are alternating regions of erosion and deposition. Bottom materials are eroded from river margins and deposited downstream in point bars. Differences in current velocity and substratum also create regular, alternating patterns of riffles and pools (Fig. 11C and D). Riffles are shallow sections with larger substrata and greater current velocity, and are spaced at fairly regular intervals, about five to seven stream-widths apart (Hynes, 1970). These regions have greater turbulence and concentrations of dissolved gases, which may provide a physiologically richer habitat for benthic algae. Deeper pools form downstream of riffles, where organisms experience reduced current velocity, possibly depleted dissolved gases, and perhaps light limitation. Bottom materials consist of smaller particles, primarily sand and silt, and current velocity is reduced during average flow periods. However, during flooding, pools lack stable substrata, which makes them susceptible to scouring and erosion.

C. The River Continuum and Other Models

For many decades ecologists lacked broad conceptual models for describing and testing patterns in the structure and function of river communities. Early concepts, some of which were developed by algal biolo-

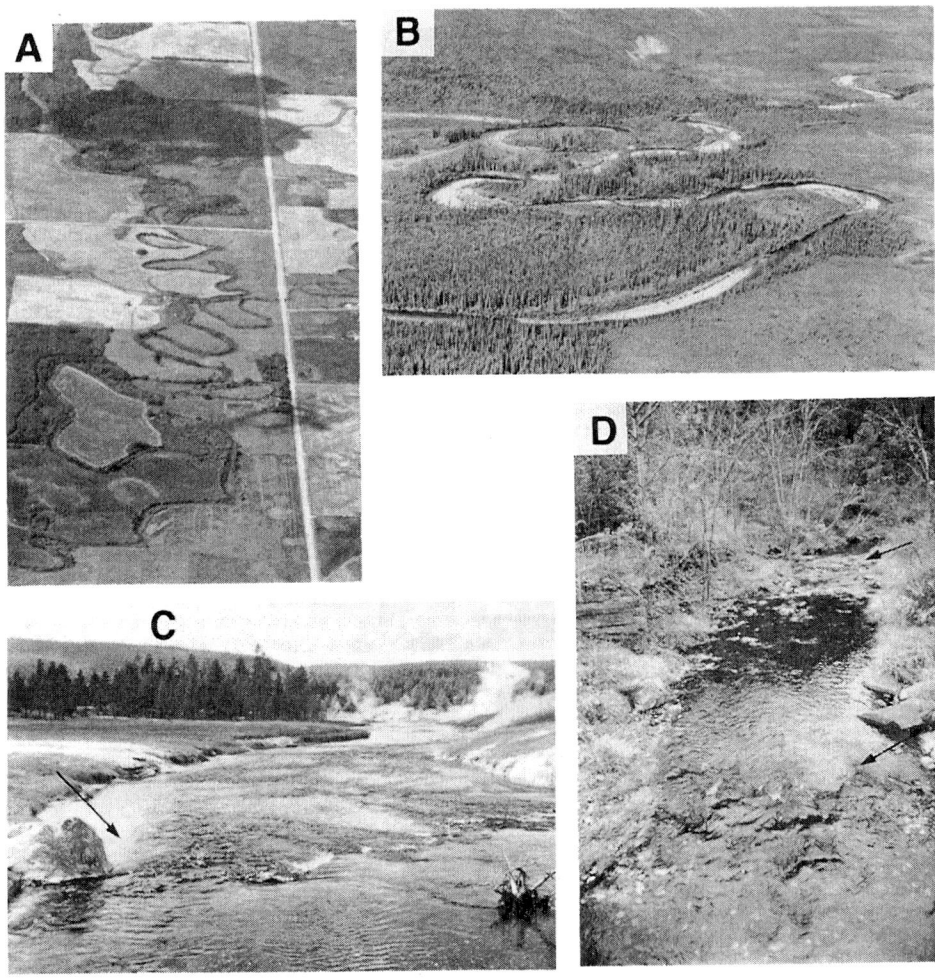

FIGURE 11 Examples of river features: A, B, meanders in lowland rivers; C, riffle (arrows) and pool sequences in a larger (Firehole River, WY) and smaller (Little Beaverkill, NY) streams. All photos by J. D. Wehr.

gists, borrowed ideas from lake systems, such as the concept of oligotrophic–eutrophic gradients and climax communities (Blum, 1956; Hynes, 1970). These ideas do not adapt easily to lotic ecosystems, and a recent synthesis of nutrient and chlorophyll data from more than 200 temperate streams suggest that nutrient–algal biomass relationships are weaker than in lakes, perhaps because of the effects of nonalgal turbidity in running waters (Dodds *et al.*, 1998). Some ecologists recognize different zones along the length of a river that possess different physical conditions and habitats for riverine organisms (Hawkes, 1975). A river's features, however, do not fall into discrete zones, but rather vary continuously along a river's course. The combination of hydrological principles with changes in biological and chemical processes along river gradients led to the development of the river continuum concept (RCC; Fig. 12; Vannote *et al.*, 1980).

The RCC characterizes lotic ecosystems as a network of streams with a continuum of longitudinally linked environmental (e.g., width, depth, flow) and resource (nutrients, light) gradients. Biological communities respond to longitudinal changes in geomorphology, water chemistry, and energy sources in several ways. In addition, organisms, by their own activities, influence conditions and communities downstream (Fig. 12). This is evidenced by changes in the types of invertebrate consumers that are localized in rivers of different sizes. The model was based on data largely from smaller, temperate forest streams, and broadened into a theory for rivers as a whole. Many other ideas and studies concerning nutrient spiraling, benthic invertebrates, fish production, and the influence of dams have emerged from the general framework laid out in the RCC (Newbold *et al.*, 1981; Welcomme *et al.*, 1989; Minshall *et al.*, 1983; Ward and Stanford, 1983; Thorp and Delong, 1994).

The role of algae and other primary producers was also considered in the RCC. The metabolism of first- to third-order streams was viewed to be largely dependent on external or allochthonous sources of terrestrial carbon. Hence, consumers in smaller streams were mainly shredders and collectors of coarse particulate matter. Algae were viewed as a minor component of food webs in headwater communities because of light limitation due to heavy riparian shading and subsidies of terrestrial organic matter. The overall effect is a net heterotrophic community, in which system-level respiration (R) exceeds *in situ* primary production (P; $P:R < 1$). Long-term studies confirm that the mass of allochthonous litter (leaves, wood) forms the dominant organic carbon source for small streams (Benke *et al.*, 1988; Findlay *et al.*, 1997), although food web studies suggest that benthic algae still may be an important or even dominant food source for some benthic animals, perhaps because of greater assimilation efficiency with algae than detritus (Fuller *et al.*, 1986; Mayer and Likens, 1987). Algal production is predicted to increase in mid-sized (fourth to sixth order) rivers, in response to greater sunlight and a reduction in subsidies of allochthonous organic matter, resulting in $P:R \geq 1$. In such systems, consumers are likely to be dominated by grazers of (mainly benthic) algal material and collectors of finer organic matter transported downstream from upper reaches. The model suggests that algal (plus macrophyte) primary production still may not come to dominate river metabolism in mid-sized rivers if the reach receives substantial supplies of allochthonous organic matter from smaller tributaries, which would increase system-level respiration and reduce light penetration. In larger rivers (greater than sixth order), the RCC predicts that even though the river basin is open to full sun, higher levels of fine particulate matter from upstream, greater depth, and resuspended sediments

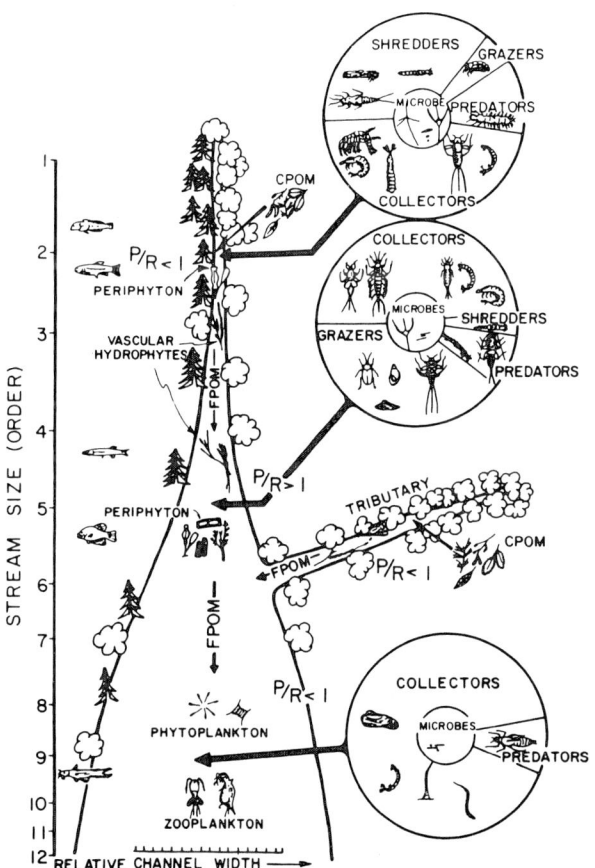

FIGURE 12 Diagrammatic representation of the river continuum concept, predicting changes in P/R ratios, consumer groups, and sources of organic matter along the length of a river. Reprinted from R. L. Vannote *et al.*, Canadian Journal of Fisheries and Aquatic Science 55:668–681, 1980, with permission from NRC Research Press.

limit algal production. Although phytoplankton usually dominate the algal community in large rivers, these systems are predicted to be too turbid and too deep to support high levels of algal production. This forecast appears to be the case in the Hudson River, where depth and tidally driven mixing create a light-limited environment for phytoplankton (Cole *et al.*, 1992), but studies on other large rivers have found that substantial phytoplankton populations can develop (e.g., Descy and Gosselain, 1994; Lair and Reyes-Marchant, 1997; Wehr and Descy, 1998).

Ideas from the RCC have stimulated a great increase in research on and discussion of lotic ecosystems, but several exceptions have been raised (Wetzel, 1975; Lock and Williams, 1981; Benke *et al.*, 1988). Rivers in different biomes have different climate, lithology, and amounts of riparian vegetation, which may shift the relative importance of autochthonous versus allochthonous production for river metabolism (Minshall, 1978; Minshall *et al.*, 1983; Wetzel and Ward, 1992). Not all rivers (small or mid-sized) in North America are shaded heavily (e.g., high altitudes, deserts, agricultural areas) nor do they receive large subsidies of terrestrial organic matter from their watershed. In more open rivers, algal production may be substantial. A metanalysis of studies on 30 streams from five biomes showed that gross primary production (GPP) levels varied by more than 4 orders of magnitude; the greatest rates were in desert areas (Lamberti and Steinman, 1997). Variations in GPP were related to variables such as watershed area, inorganic P, temperature, discharge, and canopy cover, but not to stream order or latitude, further suggesting that variation among rivers may be greater than predicted by the RCC. River systems are also very patchy, from differences in sunlight within a local reach to watershed-level differences in sources of carbon or inorganic nutrients (Pringle *et al.*, 1988). Subsequent syntheses have emphasized that lotic ecosystems also can be regulated by biotic interactions, such as predator–prey dynamics, competition, and food-chain length, rather than solely controlled by physical (e.g., hydraulic) factors (Power *et al.*, 1988). In turn, physical factors, such as the frequency of floods, have significant effects on food web interactions at several levels, including algal biomass (Wootton *et al.*, 1996).

Other syntheses have focused on large rivers (greather than sixth order). Sedell *et al.* (1989) and Junk *et al.* (1989) pointed out that many large rivers receive materials, energy, and organisms from the adjacent floodplain, often in greater quantity than may have been transported from upriver. The influence of algal production was regarded as minimal because large rivers were generally turbid and light-limited. In their flood-pulse model, Junk *et al.* (1989) suggested that large river metabolism may be driven more by batch processes—the pulses of resources from flood events—than by the continuous processes emphasized in the RCC. This process is undoubtedly true for rivers such as the Amazon and the lower Mississippi. However, not all large rivers occur in extensive floodplains. River basins also flow through constricted channels that have minimal floodplain influence. Thorp and Delong (1994) proposed a river productivity model that suggests that local autochthonous production may have been underestimated in large, constricted-channel rivers, which have firm substrata and less turbid water. In such systems, for example, a 350 km stretch of the Ohio River, substantial phytoplankton biomass and production have been observed, and during some seasons may be the principal carbon source for planktonic consumers, such as small-bodied cladocerans and rotifers (Thorp *et al.*, 1994; Wehr and Thorp, 1997). Also, many, if not most large rivers worldwide have been altered substantially due to industrialization, which has had profound influence on ecological conditions. Flow regulation and nutrient inputs favor greater phytoplankton productivity in large rivers, such that $P:R$ ratios may exceed 1.0, at least during the spring and the summer (Admiraal *et al.*, 1994; Descy and Gosselain, 1994; Wehr and Descy, 1998).

D. Benthic Algal Communities of Rivers

Most studies of river algae concern benthic species. The necessity to remain in a stable position while water flows downstream is an important selective force for all benthic organisms in lotic environments. Earlier reviews pointed out the paucity of studies on lotic algae relative to those on lakes (Blum, 1956; Hynes, 1970; Whitton, 1975), yet the pace of work on benthic algae in streams and rivers has increased substantially. Much of what has been learned in the interim has been summarized in several reviews (Lock *et al.*, 1984; Reynolds, 1996; Biggs, 1996; Steinman, 1996), which reveal an evolution from mainly descriptive studies to structural and functional analyses of benthic algae and their importance in lotic food webs.

1. Benthic Algal Diversity, Composition, and Biogeography

In streams, diatoms often comprise the dominant algal group in terms of species number and biomass (Blum, 1956; Douglas, 1958; Round, 1981; Kawecka, 1981). Their diversity in species and growth form (upright frustules, short- and long-stalked, rosettes, tube-dwelling, filamentous, mucilaginous matrix, and prostrate cells) enables them to colonize a variety of microhabitats. In more slowly flowing or less flood-prone systems, filamentous species like *Melosira varians* and upright, stalked forms such as *Gomphoneis herculeana* may predominate, along with filamentous nondiatom species (Stevenson, 1996a; Biggs, 1996). In very rapid water, firmly attached diatoms, such as *Cocconeis placentula*, *Achnanthidium minutissimum*, and *Hannaea arcus*, may occur with encrusting nondiatoms, such as *Hildenbrandia rivularis*, *Gongrosira* spp., and *Chamaesiphon* spp., and corticated forms like *Lemanea* spp. (Fritsch, 1929; Whitton, 1975; Kawecka, 1980; Kann, 1978).

One of the fascinations of studying benthic algae in rivers is that many species (although not the majority) are macroscopic and recognizable in the field (Holmes and Whitton, 1977; Kann, 1978; Entwisle, 1989; Sheath and Cole, 1992). These include cyanobacteria, and green and red algae, as well as chrysophytes, xanthophytes, and brown algae (Whitton, 1975; Sheath and Cole, 1992). Morphologies are diverse and include encrusting, turflike, filamentous, cartilaginous, mucilaginous, tubular, and bladelike thalli (Fig. 13). In some instances, diatoms, including species of *Eunotia*, *Fragilaria*, and *Melosira*, and stalked forms, such as *Didymosphenia geminata* and *Cymbella* spp., may be recognized by their gross appearance, but are identified only using microscopy (Holmes and Whitton, 1981; Steinman and Sheath, 1984; Sheath and Cole, 1992).

Some benthic stream algae have limited distributions. *Prasiola mexicana*, a seaweed-like green alga (Fig. 13B), thus far has been recorded only in streams located in the arctic tundra and western coniferous biomes, whereas the encrusting brown alga *Heribaudiella fluviatilis* has been confirmed in North America from streams in western Canada and the United States only (Sheath and Cole, 1992; Wehr and Stein, 1985). Some taxa have been regarded to be limited in geography or habitat, only to be found later in other locations. *Thorea violacea* is a large red alga that was thought to be restricted to streams in warmer biomes or in temperate areas only during the summer (Smith, 1950; Sheath and Hambrook, 1990, Chap. 5), but it was discovered growing profusely in the upper Hudson River in cool (15°C), rapidly flowing water (Pueschel *et al.*, 1995). Few long-term studies of benthic stream macroalgae exist that may help to explain the dispersal patterns of these organisms. In one of the few cases of

FIGURE 13 Examples of different forms of macroalgae from rivers: A, *Zygnema* sp., simple, flexible filaments (arrow = direction of flow); B, *Prasiola mexicana*, a flat, seaweed-like thallus; C, *Lemanea fluviatilis*, corticated tubes; D, *Nostoc verrucosum*, mucilaginous colonies. Photos A and D by J. D. Wehr; photos B and C by R. G. Sheath.

a long-term (more than 40 years) database, Holmes and Whitton (1977, 1981) were able to characterize species as previously overlooked, currently increasing, or recently extirpated from the River Tees (UK). No such studies are known from North America.

Ecological endemism in some species of river algae may be caused by the spread of marine taxa into estuarine environments, followed by adaptation to lower salinity. Such may be the case with genera such as *Audouinella*, *Hildenbrandia*, *Prasiola*, and *Enteromorpha*: genera that each has marine and freshwater species and are found in rivers but only rarely in lakes (Flint, 1955; Whitton, 1975; Sheath and Cole, 1980; Sheath *et al.*, 1985; Hamilton and Edlund, 1994). However, other species appear to be human-accelerated invaders from other freshwater systems. Populations of the red alga *Bangia*, which invaded North American freshwaters in the 1960s through the St. Lawrence River, are not closely related to any marine populations (based on *rbc*-L, RuBisCo spacer, and 18s rDNA sequences) and show strong affinities with freshwater European populations (Müller *et al.*, 1998). These data suggest that some vector, such as a ship's ballast water, enabled invasion, rather than gradual spread and adaptation from a marine environment.

Biogeographic data suggest that green algae are the most common group of stream macroalgae across all biomes in North America. Based on 1000 stream segments studied, the lowest species diversity was found in streams in arctic tundra and the greatest in boreal forests (Sheath and Cole, 1992). There was no increase in macroalgal species diversity from the arctic to the tropics (contrary to marine species), which may be the result of periodic flooding common in streams across all regions. Arctic streams tend to have more species of macroalgal cyanobacteria, whereas tropical streams have more species of Rhodophyta (Sheath and Cole, 1992; Sheath *et al.*, 1996; Sheath and Müller, 1997).

2. Factors That Regulate Benthic Algae in Rivers

Several reviews (Blum, 1956; Hynes, 1970; Whitton, 1975; Biggs, 1996; Stevenson, 1997) discuss many ecological factors, including current velocity, substratum, geology, nutrient conditions, grazers, temperature, pollutants, and light availability, and their effects on benthic algae in rivers. Factors often interact to affect algal growth and survival, and multivariate analyses have been useful in determining the key environmental factors that affect species composition (e.g., Hufford and Collins, 1976; Wehr, 1981; Lowe and Pan, 1996). The following discussion provides a brief overview of current perspectives on the influence of habitat variables on benthic stream algae and, in particular, efforts to integrate structural aspects (species composition, diversity, architecture) with functional studies (production, food web dynamics, nutrient cycling).

In all river systems, there are proximate variables (those we measure) that affect organisms, such as light availability or nutrient supply, and larger scale factors such as climate, watershed features, and land use practices, that drive proximate variables (Stevenson, 1997). Biggs (1996) and Biggs *et al.* (1998) proposed a conceptual, disturbance–resource supply–grazer model that categorizes controls on benthic algal production in streams in terms of two processes: factors that affect (1) biomass accrual and (2) biomass loss. The model recognizes how ecosystem-level changes, such as floodplain modifications, can affect proximate controls, such as current velocity. An energetic balance sheet is constructed for each side of the ledger that can be used to predict to algal production and species composition (Fig. 14). In this scheme, biomass accrual increases as a function of resource supply, whereas biomass losses are a function of disturbance and grazing. In rivers with infrequent, low-intensity floods (= disturbance) and modest grazing intensity, the model predicts that biomass accrual dominates to a level dictated by resources. Under low resource supplies, growth continues at lower rates, favoring an adnate and turflike community dominated by filamentous cyanobacteria (e.g., *Schizothrix*, *Phormidium*, and *Tolypothrix*; Chap. 4), red algae (e.g., *Audouinella*; Chapt. 5), and many benthic diatoms (e.g., *Epithemia* and *Navicula*; Chaps. 17–19). Under similar hydraulic and grazing conditions but greater nutrient supply, a greater biomass of filamentous taxa, such as *Cladophora*, *Ulothrix*, or *Melosira*, is expected. For each combination of plus and minus factors, specific predictions can be made about the most important variables that drive algal production and community composition.

Empirical evidence supports these predictions. In the Colorado River, where nutrients and sunlight are generally nonlimiting, release of water from Glen Canyon Dam (greater disturbance) decreased the biomass and relative importance of *Cladophora glomerata*, and reduced total benthic primary productivity (Blinn *et al.*, 1998). The physiognomy of epiphytic algae changed from upright assemblage (*Diatoma vulgare*, *Rhoicosphenia curvata*) to more closely adherent forms (*Achnanthes* spp. and *Cocconeis pediculus*; Hardwick *et al.*, 1992). Herbivorous invertebrates have been widely shown to reduce the density and biomass of benthic algae, but losses due to grazer activity come from scouring as well as consumption (Allan, 1995). Furthermore, ingestion and assimilation of benthic algae vary widely (30–70%), depending on the species of both the alga and the consumer (Lamberti *et al.*,

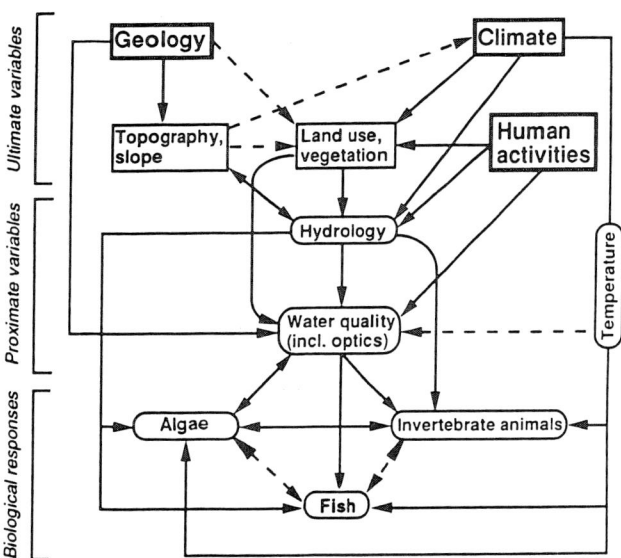

FIGURE 14 Conceptual model of ultimate (landscape) and proximate (stream) variables that control benthic algal communities, their consumers, and interactions in streams (solid lines = strong effects; dashed lines = weaker effects; double arrows indicate feedback interactions. Reproduced with permission from Biggs, B. J., in Stevenson, R. J. et al., Eds., Algal Ecology: Freshwater Benthic Ecosystems. Copyright © 1996 by Academic Press.

1989; Pandian and Marian, 1986). Resultant biomass and compositional changes, in turn, have effects on benthic food webs, because many invertebrates find an adherent community less easily grazed (Colletti et al., 1987). The Riviere de L'Achigan (Quebec) is an unshaded stony stream interspersed with a chain of small lakes that alter flow conditions immediately downstream (lower disturbance; Cattaneo, 1996). Biomass and species composition of benthic algae on gravel varied inversely with distance from lake outlets, yet this impoundment effect was unimportant for algae that colonize boulders. Only boulders supported communities of large filamentous and plumose forms such as *Draparnaldia* (Chlorophyceae), *Stigonema* (Cyanobacteria), and *Batrachospermum* (Rhodophyta). An experimental study in Big Sulphur Creek (CA) demonstrated the interactive effects of shading (resource supply) and invertebrate grazing on epilithic algal communities (Feminella et al., 1989). Algal biomass in low-grazer conditions declined by 75% with greater (15–95%) canopy cover, but was unaffected by light availability at normal grazer densities. The food web—resource supply interaction in Big Sulphur Creek is complicated by the fact that densities of trichopteran grazers (e.g., *Gumaga nigricula*) declined significantly in shaded conditions, while other species were unaffected. Abundances of algal grazers are often driven by food supply, but differences in irradiance also affect the chemical composition of algal assemblages, including changes in protein, total lipids, and fatty acid composition (Steinman et al., 1988). In general, cyanobacteria have different complements of fatty acids, including greater oleic, linoleic, and linolenic acids than diatoms (McIntire et al., 1969). Reductions in light availability to Kingsley Creek (NY) caused significant reductions in epilithic algal biomass and densities of a herbivorous mayfly (*Baetis tricaudatus*), but did not affect densities of a filter-feeding blackfly larva (*Simulium* spp.), which mainly consumed detritus (Fuller et al., 1986). However, most field studies on grazing are conducted under summer base-flow conditions when disturbance is low and resource factors are less important (Feminella and Hawkins, 1995).

Current velocity is of great importance to benthic algae. Species that colonize areas of rapid current velocity are firmly attached to substrata using rhizoidal or holdfast-like structures (Israelson, 1949; Whitton, 1975). Greater current velocity provides a continuous replenishment of nutrients from upstream and a steeper diffusion gradient near the cell surface (Whitford, 1960; Horner et al., 1990). Species restricted to riffles with strong current velocity (e.g., *Hydrurus foetidus* and *Lemanea fluviatilis*) may have a higher metabolic demand for nutrients, which must be met by a greater physical supply. Studies suggest that there is no simple positive relationship between current velocity and algal metabolism or growth rate; greater current speed may also decrease algal biomass through scouring (Borchardt et al., 1994; Stevenson, 1996b). Few correlative studies are able to attribute differences in algal communities to current velocity effects because many variables change along a river's course. A more clear-cut study of parallel streams draining the same reservoir found that the regulated stream had greater diatom cover and large populations of *Prasiola fluviatilis*, while the stream that lacked flood control had populations of *Hydrurus foetidus* and *Ulothrix zonata*, and greater species number, but a lower biomass of diatoms (Kawecka, 1990).

Current velocity affects the growth form of individual algae, such as in *Cladophora*, which develops compact tufts with narrow branching in greater current velocity and develops plumose forms with widely branched filaments in calmer flow (Whitton, 1975; Dodds and Gudder, 1992, Chap. 8). The architecture of algal communities affects their response to current speed. A mucilaginous, stalked diatom community (e.g., *Gomphoneis herculeana* and *Navicula avenacea*) exhibited increased production with greater near-bed current velocity, whereas a community consisting of long filamentous algae (*Oedogonium* sp. and *Phormidium* sp.) experienced decreaseed biomass accrual (Biggs *et al.*, 1998). The morphology of *Nostoc parmelioides* (a filamentous colonial cyanobacterium) in stony streams is altered from spherical to ear-shaped colonies by the presence of an endosymbiotic midge larva (Ward *et al.*, 1985), and only ear-shaped colonies exhibit greater photosynthesis and N_2-fixation rates with greater current velocity (Dodds, 1989). Such studies that aim to link metabolic patterns with differences in community composition or form are needed to improve our understanding of the factors that regulate lotic communities.

Several studies of benthic stream algae also have considered the influence of substratum. Several decades ago, Douglas (1958) recognized that different sizes of stones supported different densities and species of epilithic algae, which was likely the result of differences in their susceptibility to flood disturbance. A comparison of epilithic communities on different substratum sizes and similar in nutrient conditions found that most of the variation in algal biomass was explained by total-P and seston levels, but size of the substratum also exerted a significant effect (Cattaneo *et al.*, 1997). Stones that are disturbed or scoured during floods may have algal crusts or propagules that still remain on their surfaces (Power and Stewart, 1987). The degree to which algae can recolonize disturbed substrata in a stream is a function of (1) their resilience, through immigration and greater growth rates, or (2) their resistance, as influenced by the species' morphology and community physiognomy (Peterson, 1996). Benthic algal biomass and species composition are also influenced by substratum–current interactions. Diatom immigration onto bare substrata may increase with either reduced current speed or greater surface complexity (Stevenson, 1983). Substrata conditioned with a simulated mucilage (agar coating) were colonized twice as rapidly as clean surfaces, but responses were species-specific; some increased (*Navicula gregaria* and *Synedra ulna*), while others declined (*Achnanthidium minutissima* and *Diatoma vulgare*) or were unaffected (*Diatoma tenue* and *Gomphonema olivaceum*). A matrix of organic matter and bacteria probably facilitates colonization by most benthic algae in rivers (Karlström, 1978; Korte and Blinn, 1983; Sheldon and Wellnitz, 1998). A comparison of benthic diatom assemblages that colonize natural rocks, sterilized rocks, and clay tiles in Fleming Creek (MI) found greater total densities and species diversity on natural rocks (Tuchman and Stevenson, 1980). Otherwise identical natural substrata constructed from three rock types (basalt, sandstone, and limestone) that occur in Oak Creek (AZ) were compared for their effects on diatom colonization of riffles (Blinn *et al.*, 1980). Densities on sandstone substrata were 60–80% greater than on basalt and limestone, but species composition and diversity were similar. In Mack Creek (OR), greater biovolume (but not chlorophyll-*a*) and diversity of benthic algae were observed colonizing pieces of wood than clay tiles (Sabater *et al.*, 1998). Certain taxa, such as *Cymbella minuta*, *Hannaea arcus*, and *Zygnema* sp., were more abundant on wood, while some closely adherent forms, such as *A. minutissima* and *A. lanceolata*, were more abundant on clay tiles. The physical structure of the stream bed also influences benthic communities. Experimental manipulations of rocks and bricks with different density and surface texture resulted in significant changes in the diversity and the abundance of benthic invertebrates and epilithic algae in the Steavenson River, Australia (Downes *et al.*, 1998). Greater densities of *Audouinella hermannii* were observed on substrata without large crevices, but total biomass was greatest on surfaces that were roughened, independent of the presence of crevices.

E. Phytoplankton Communities of Rivers

Although benthic algae typically dominate rocky streams and smaller rivers, phytoplankton become important in larger rivers and lowland streams (Rosemarin, 1975; Reynolds and Descy, 1996). A long history of studies on river phytoplankton dates back to at least the 1890s, when Zacharias (1898) coined the term "potamoplankton," to refer to the suspended organisms in flowing waters. In North America, phytoplankton have been studied since the early years of limnological research in rivers including the Illinois (Kofoid, 1903, 1908), Mississippi (Reinhard, 1931), Ohio (Eddy, 1934), San Joaquin (Allen, 1921), and Sacramento (Greenberg, 1964). Much of the early research focused on whether a true phytoplankton community (populations that survive and reproduce within rivers) actually existed, as opposed to dislodged benthic forms or plankton washed in from lakes within the watershed. Indeed, plankton in most rivers consists of all three components in varying proportions (Reynolds, 1988), but in a single river sample, it is difficult to distinguish these sources, although certain algal taxa may be considered typical of each.

Many benthic algae become suspended. These meroplanktonic forms can be washed out from sediments, plants, or other substrata. A metanalysis of 67 studies suggests that about 50% of suspended algal taxa in rivers are either benthic or meroplanktonic (Rojo *et al.*, 1994). Among diatoms, most raphe-bearing species are likely nonplanktonic, but it is difficult to distinguish true potamoplankton from those that originated from lakes (Reynolds, 1988). One distinction may be found in species' responses to flow regime and other physical factors, as has been attempted for phytoplankton species in the Ohio River (Peterson and Stevenson, 1989; Wehr and Thorp, 1997). The abundances of most species were negatively related to discharge, but tributary rivers were found to have little effect on species composition. However, during low flow (summer to early autumn), there were significantly greater amounts of colonial cyanobacteria (e.g., *Aphanocapsa saxicola*) and certain diatoms (e.g., *Stephanocyclus* [= *Cyclotella*] *meneghiniana*) in the river downstream of these tributaries, suggesting that some populations may have originated from outside the main river.

The fact that chlorophyll-*a* concentrations in large rivers usually vary inversely with discharge (Schmidt, 1994) strongly suggests that potamoplankton is not derived primarily from benthic habitats. This pattern was observed for individual species in the River Thames, although smaller cells were less affected or even increased with greater discharge (Ruse and Love, 1997). In general, smaller forms appear to be more successful members of the potamoplankton, perhaps due to greater growth rates and surface area to volume ratios (Reynolds, 1988; Rae and Vincent, 1998). Diatoms are clearly the most diverse and abundant group, with *Cyclotella* and smaller species of *Stephanodiscus* especially common in larger rivers worldwide (Chap. 15). Algal flagellates rarely achieve large numbers in river plankton, except some cryptomonads, chrysophytes, and members of the Volvocales (see Chap. 6). In some instances, poor preservation techniques may cause underreporting (see Chaps. 12 and 21).

Two major limitations to survival and growth of river phytoplankton are the continuous removal of organisms by downstream flow (so-called washout) and mixing within the water column, which places cells in variable and often aphotic light fields. Hence, most studies conclude that riverine phytoplankton production is controlled by discharge (Baker and Baker, 1979; Soballe and Kimmel, 1987; Cole et al., 1992; Reynolds and Descy, 1996). Assuming no other limiting resources, rivers must be sufficiently long and/or the flow rate sufficiently low for net positive algal growth rates. This principle was demonstrated clearly in early studies on the Sacramento River in which peaks of abundance in potamoplankton became progressively more pronounced further downstream, a region that provided reduced current velocity and more time for populations to develop (Greenberg, 1964). A similar increase was seen along the Rhine (Germany–Netherlands; de Ruyter van Steveninck et al., 1992). In the lowland River Spree (Germany; Köhler, 1993, 1995), phytoplankton biomass declined in mid-river in response to increased turbidity and Fe precipitation of P, but then increased further downstream as a result of impoundments and flow regulation. A rather different longitudinal pattern is seen in the St. Lawrence River, in which a gradient of increased P and reduced current velocity is counterbalanced by greater suspended matter, causing a net decrease in river plankton densities downstream (Hudon et al., 1996). The contrary and complex influences of discharge and algal growth rates have been discussed in detail by Reynolds (1988, 1995). Despite these limits on algal growth, large accumulations of phytoplankton frequently develop in summer and other low-flow periods. With greater nutrient supplies, surface blooms of cyanobacteria (*Microcystis*, *Anabaena* and *Aphanizomenon*) may occur, although their prevalence appears to be greater in warmer climates or in temperate zones under low-flow conditions (Baker and Baker, 1979; Krogmann et al., 1986; Paerl and Bowles, 1987). Unlike many lakes, nutrient limitation is uncommon in larger rivers (Reynolds, 1988; Reynolds and Descy, 1996; Wehr and Descy, 1998). Therefore, the principal factor that regulates phytoplankton production in rivers is discharge, which regulates dilution rates, turbidity, and mixing of cells within the water column (Reynolds and Descy, 1996).

Algal productivity in larger and lowland rivers can be substantial despite frequent turbidity and continuous mixing of algal cells within the water column (Descy et al., 1987, 1994; Reynolds and Descy, 1996). A delicate balance exists between phytoplankton production and respiratory losses during periods of higher turbidity (Descy et al., 1994; Reynolds and Descy, 1996). In freshwater tidal sections of the Hudson River, turbidity is further complicated by tidally driven mixing, resulting in a net heterotrophic balance for most of the year (Cole et al., 1992). Following invasion of the Hudson by zebra mussels (*Dreissena polymorpha*), phytoplankton biomass declined from a summertime mean of about 30 to about 5 µg chlorophyll-*a* L^{-1} and species composition shifted from colonial cyanobacteria to diatoms (Caraco et al., 1997; Smith et al., 1998). In the River Spree, a positive autotrophic balance is established during the spring (mainly diatoms), but in the summer, cyanobacteria dominate (Köhler, 1995). How large river systems maintain large phytoplankton populations throughout the year is still something of a mystery, but the main channel may receive subsidies of algae and nutrients from tributaries, wetlands, or backwaters (Owens and Crumpton, 1995; Reynolds, 1996). Species composition may provide a clue: slower growth rates of larger colonial species (e.g., *Aphanizomenon* and *Planktothrix*) may have higher respiratory costs ($P:R < 1$) for maintaining populations than smaller centric diatoms (e.g., *Cyclotella* spp. and

Stephanodiscus hantzschii; see Chap. 15), but may be stable if they are ineffectively grazed by small-bodied zooplankton (Gosselain *et al.*, 1998). Only zebra mussels, which utilize a wider particle size range, may crop these larger forms, as in the Hudson River. In the Meuse, Rhine, Danube, and upper Mississippi Rivers, higher levels of nutrients coupled with less turbid conditions enable high levels of phytoplankton production to be sustained for several months of the year (Baker and Baker, 1979; Descy *et al.*, 1987; Lange and Rada, 1993; Admiraal *et al.*, 1994; Kiss, 1994). With greater primary production, potamoplankton influence the biogeochemical properties of large rivers, including dissolved O_2 (Köhler, 1995; Reynolds and Descy, 1996), dissolved Si (Admiraal *et al.*, 1993), and dissolved organic matter (Wehr *et al.*, 1997).

Phytoplankton are an important food source for zooplankton in rivers, even if grazers do not regulate algal biomass or production as effectively as in lakes. Grazing pressure is less important because the zooplankton community is usually dominated by small-bodied cladocerans and rotifers (Winner, 1975, Köhler, 1995). Rivers select for small-bodied zooplankton because of their ability to grow rapidly enough to compensate for downstream losses (Viroux, 1997). Biomass and density of zooplankton in larger rivers also may be less than in lakes (Pace *et al.*, 1992; Thorp *et al.*, 1994). Although discharge and turbidity typically drive phytoplankton production in rivers on an annual basis, zooplankton grazing may still be an important loss factor during summer low-flow periods (Gosselain *et al.*, 1994). Grazing activity in the River Meuse (mainly *Bosmina* and rotifers) appears to exert strong control over summertime phytoplankton numbers and to cause a shift in phytoplankton size structure toward larger celled forms (Gosselain *et al.*, 1998).

Models designed to predict phytoplankton production in large rivers primarily have been devised for specific conditions, such as the influence of temperature and irradiance on phytoplankton production in one reach of the Great Whale River, Quebec (Rae and Vincent, 1998). Light and temperature explained between 74 and 98% of the variation in photosynthetic activity in this subarctic river. A larger model for the Rhine, which included the effects of irradiance, light attenuation, flow, nutrients, phytoplankton biomass, and grazing, was successful in predicting the fate of algal production, although some parameters (e.g., zooplankton grazing) were based on data from lakes (Admiraal *et al.*, 1993). Efforts to develop nutrient–algal biomass models have been less effective than in lakes, owing to weaker relationships between N or P and chlorophyll-*a*, and the complicating effects of discharge and turbidity (Van Niewenhuyse and Jones, 1996; Dodds *et al.*, 1998). Even more difficult to predict are changes in species composition of phytoplankton communities in rivers. This information is important for water management agencies, because certain algae affect taste and odor conditions of water that may ultimately be used for domestic consumption. Despite the inclusion of many physical and chemical variables, only about 20% of the total variation in phytoplankton species composition for the River Thames could be explained using a canonical correspondence model (Ruse and Hutchings, 1996). One model, which incorporates several physical factors and river order, found that hydrological conditions exert an overriding effect on potamoplankton development, but biological controls (mainly grazers) are important during low-flow conditions (Billen *et al.*, 1994). Reynolds (1988, 1995) has suggested that more fundamental work is still needed to acquire the necessary data to build meaningful predictive models for plankton communities in rivers.

IV. WETLANDS

Wetlands regulate nutrient fluxes between terrestrial and aquatic systems, serve as nurseries for many of the world's fisheries, and are among the most productive and threatened ecosystems worldwide (Whittaker, 1975; Mitsch and Gosselink, 1993). Freshwater wetlands occur from arctic to tropical biomes across North America. Many are situated in the upper littoral zone of lakes and rivers, and also include marshes, peat bogs, fens, wet alpine meadows, cypress swamps, and forested lowlands. Unifying features of wetland include more or less continuously saturated soils, shallow water depth (≤ 2 m), fluctuating water levels, an accumulation of plant detritus and organic matter, and vegetation adapted to wet conditions (Mitsch and Gosselink, 1993; Goldsborough and Robinson, 1996).

A. Functional Importance of Algae in Wetlands

Most wetland algae are benthic, loosely associated with emergent plants (metaphyton), attached to plants, or colonized in sediments; most suspended forms have been dislodged from various surfaces. A five-year study of epipelic, epiphytic, metaphytic, and planktonic primary production in Delta Marsh, Manitoba, determined that metaphyton contributed roughly 70% of the total algal productivity, compared with only 6% for phytoplankton (Robinson *et al.*, 1997). Values for algal productivity (400–1100 g C m^{-2} y^{-1}) over the year are comparable to or exceed that of the emergent macrophytes present (aboveground: 100–1700 g C m^{-2} y^{-1}). Among several constructed wetlands in Illinois, benthic algae are estimated to contribute between 1 and 65% of the total system primary production (Cronk and Mitsch, 1994). However, not all wetland algal communities are highly productive, presumably a result of nutrient limitation (Murkin et al., 1991; Goldsborough and Robinson, 1996).

Algal production is important for many invertebrate consumers that preferentially consume algal material over either live or detrital macrophyte tissues (Campeau et al., 1994; Goldsborough and Robinson, 1996). Attached algal material may be especially important in winter when emergent macrophytes are dead or senescent (Meulemans and Hienis, 1983). In a wetland along western Lake Superior, δ^{13}C data suggest that entrained algae are an important primary food source for the grazing food web in addition to macrophyte detritus (Keough et al., 1998). Given the greater proportion of structural tissues in emergent plants and the greater turnover rates among algal cells, the importance of algae in wetland food webs is often substantial. Nutrient additions designed to enhance algal biomass in wetland enclosures also resulted in greater densities of cladocerans and copepods in nearshore water, as well as benthic invertebrates such as snails and chironomids (Gabor et al., 1994). Using pigment tracers, Bianchi et al. (1993) estimated that benthic diatoms comprise a major food source for invertebrates in Hudson River wetlands and, combined with lower C:N ratios, may be a better resource for benthic consumers than detritus. Algal biomass may, however, be spatially variable. In wetlands of Lake Gooimeer (Netherlands), stable isotope data indicate that within *Phragmites* beds, macrophyte detritus is the major carbon source for benthic invertebrates, but algal material dominates littoral food webs outside the reed bed (Boschker et al., 1995). The littoral zone of lakes is an important region of nutrient exchange between nearshore and pelagic zones, and attached algal communities are important components of this exchange (Mickle and Wetzel, 1978; Aziz and Whitton, 1988; Moeller et al., 1988). During decomposition of wetland plants, attached algae may enhance breakdown (Neely, 1994), although other data suggest that dissolved organic carbon (DOC) released during this process may inhibit algal growth (Cooksey and Cooksey, 1978).

B. Algal Diversity in Freshwater Wetlands

The communities of algae in freshwater wetlands are nearly as diverse as those found in lakes. As in other systems, nutrient conditions, climate, and geology influence species composition, but in wetlands, water level, macrophyte plant composition, and degree of mixing with other water bodies are also important (Goldsborough and Robinson, 1996). In the Everglades, benthic algae colonize many macrophyte and sediment surfaces; species composition and production vary with species of macrophyte, water level, nutrient inputs, and degree of CaCO$_3$ incrustation (Browder et al., 1994). Filamentous cyanobacteria, including *Scytonema*, *Schizothrix*, *Oscillatoria*, and *Microcoleus*, are often abundant. Filamentous green algae (*Spirogyra*, *Bulbochaete*, and *Oedogonium*), desmids, and diatoms (*Cymbella*, *Gomphonema*, and *Mastogloia*) are common in less calcareous conditions. Algae represent between 30 and 50% of primary producer biomass in these systems, and their activity is apparent in the large diurnal changes in dissolved O$_2$ and CO$_2$. Because of their close connection with water chemistry, benthic algae help regulate water quality in wetlands, especially P loading from agricultural and urban runoff (McCormick and Stevenson, 1998).

In nutrient-rich wetlands, the algal flora is typified by filamentous green algae such as *Stigeoclonium*, *Oedogonium*, and *Cladophora*, cyanobacteria such as *Lyngbya*, *Oscillatoria*, and *Nostoc*, and many diatoms, including species of *Amphora*, *Epithemia*, *Navicula*, *Nitzschia*, and *Surirella* (Chaps. 17–19). Filamentous cyanobacteria *Rivularia*, *Calothrix*, *Microcoleus*, and *Gloeotrichia* (Chap. 4), and meadows of *Chara* (Chap. 8) typically dominate systems that have greater Ca^{2+} levels. In more oligotrophic systems, filamentous members of the Zygnematales (Chap. 9), including *Mougeotia*, *Zygnema*, and *Spirogyra*, commonly occur, along with epiphytic diatoms such as *Tabellaria*, *Eunotia*, and *Fragilaria* (see Chaps. 17 and 18), sediment-dwelling species such as *Frustulia* and *Pinnularia* (Chap. 17), and many desmids (Hooper-Reid and Robinson, 1978; Livingstone and Whitton, 1984; Goldsborough and Robinson, 1996; Pan and Stevenson, 1996, Chap. 9). Similar to patterns in softwater lakes, wetlands too may be impacted by acidic precipitation, leading to increases in *Mougeotia*, *Zygnema*, and, in severely impacted systems, *Zygogonium* (Stokes, 1986; Turner et al., 1995). Invertebrate grazers mediate species composition. Experimental removal of cladocerans and copepods from marsh enclosures resulted in a shift from a simple community dominated by *Stigeoclonium* to a more diverse and structurally complex assemblage of diatoms, filamentous green algae, and cyanobacteria (Hann, 1991).

An important reason for the success of certain algal species in wetland habitats is their ability to tolerate variations in water level and desiccation. One model predicted specific wetland algal communities that depend on varying water level: dry, sheltered, or lakelike (Goldsborough and Robinson, 1996). Water levels may fluctuate several times in a few months or persist for several years. Algae that occupy a variable moisture regime must have adaptations to tolerate extremes of conditions. Some epipelic desmids (*Closterium* and *Micrasterias* species) are capable of surviving extended periods of drying and darkness (Brook and Williamson, 1988). Other filamentous forms (e.g., *Oscillatoria*, *Lyngbya*, and *Oedogonium*) may form thick mats during the open (flooded) state that protect algal cells during a later dry phase.

C. Algal Communities of Bogs

Bogs are a special class of algal habitats, and include wetlands, streams, and ponds where water retention and chemistry are usually influenced by *Sphagnum*. Most bogs have lower pH (4.0–5.5), low Ca^{2+}, are poor in nutrients, and have high levels of dissolved organic matter, which casts a yellow or brown stain to the water (Gorham *et al.*, 1985; Cole, 1994). These dystrophic systems, along with lakes high in humic materials, accumulate poorly decomposed organic matter as peat. Ombrotrophic bogs are hydrologically isolated and depend on precipitation as their water source. Bogs are scattered throughout North America, especially in cool or cold regions that have an excess of moisture most of the year (Wetzel, 1983a). They are common across the subarctic, New England–Maritime region, northern Great Lakes, Pacific Northwest, and scattered alpine areas (Brooks and Deevey, 1963; Northcote and Larkin, 1963; Yung *et al.*, 1986). Bogs are not restricted to cool climates, however. Many of the Carolina Bays are dystrophic ponds or bogs and exhibit low algal production (Schalles and Shure, 1989). Bogs also occur in semidesert, desert, and tropical areas, such as in east-central Texas, Arizona, and Costa Rica (Cole, 1963). Some bogs are islands of acidic waters and soils surrounded by an alkaline "sea" (Glime *et al.*, 1982).

In bog lakes, mats of vegetation (bryophytes, angiosperms, algae) may float out over the littoral zone and grow toward the center for many years as the bottom of the lake fills in with peat (Whittaker, 1975). Bogs are a stage in the long-term succession of some lake basins that are in the gradual process of filling in. A pioneering limnological study of Cedar Bog Lake (MN; Lindeman, 1941a, b, 1942) documented the vegetational history and aquatic food webs, and was one of the first estimates of freshwater algal productivity. Lindeman's study helped launch the trophic–dynamic concept in ecology and served as the basis for subsequent energetic studies of freshwater food webs and development of the ecosystem concept (Cole, 1994).

Algal communities of bogs are typically species-poor, although diversity, especially that of the desmids, may be much greater in systems connected to other lakes or streams (Woelkerling, 1976; Hooper, 1981; Mataloni and Tell, 1996). A tangle of filamentous green algae (Zygnematales and Ulotrichales) and desmids (see Chap. 9) are common, but contrary to common wisdom, although desmids are numerous and diverse, they are rarely important in terms of algal biomass (Yung *et al.*, 1986). Further details on their distribution and ecology are given in Chapter 9. At least three major algal habitats are found in bogs: (1) pools and open water, (2) habitats associated with *Sphagnum*, and (3) epiphytic habitats of *Nuphar* or other macrophytes. Data for 31 eastern bog systems demonstrated that algal species richness increases (especially desmids) with proximity to the Atlantic coast, and is less in systems with greater color and lower pH (Yung *et al.*, 1986). The flagellate *Gonyostomum* (Raphidophyceae; Chap. 11, Sect. III) and the red alga *Batrachospermum turfosum* (as *B. keratophytum*; Chap. 5) are two unusual species that are characteristic of boggy systems (Prescott, 1962; Bourelly, 1985; Yung *et al.*, 1986; Sheath *et al.*, 1994).

Diatoms are generally less diverse than desmids, but certain species are frequently observed, including *Anomoeoneis brachysira*, *Frustulia rhomboides* var. *saxonica*, *Eunotia elegans*, *E. exigua*, *Navicula subtilissima*, *Pinnularia viridis*, and several types of *Stauroneis* spp. (Kingston, 1982; Cochrane-Stafira and Andersen, 1984; Mataloni and Tell, 1996). Diatom species composition may change with successional stage of the surrounding vascular plant community, and apparently are responsive to many of the same variables, such as Ca, pH, and specific conductance (Cochrane-Stafira and Andersen, 1984). Although it has been argued that cyanobacteria may be unable to tolerate lower pH, especially values less than 4.0 (Brock, 1973), there are several species that are common in these waters, including species of *Aphanocapsa*, *Chroococcus*, *Dacytlococcopsis*, *Hapalosiphon*, *Microchaete*, *Nostoc*, and *Stigonema*. Because many of these species are heterocystous, their N-cycling role in these nutrient-poor environments deserves attention. Some species of dinoflagellates, chrysophytes and synurophytes also may be found, although their abundance is usually low.

V. THERMAL AND ACIDIC ENVIRONMENTS

A. Thermal Springs

Thermal springs and streams (hot springs) are extreme environments in geologically active regions where temperatures are influenced by geothermal sources and can range from 35 to 110°C. In North America, thermal springs are common from Alaska south to Costa Rica, and in scattered locations in Arkansas, Florida, Georgia, Virginia, and north to Massachusetts. Among the best known for aquatic organisms is the spectacular thermal area of springs, geysers, and fumaroles in Yellowstone National Park. Some thermal springs are less obvious, exerting their chemical and thermal effects on larger lakes and rivers, as in Yellowstone Lake (and adjacent lakes), and several streams and lakes in eastern Costa Rica (Pringle et al., 1993; Theriot et al., 1997).

The geology, chemistry, and organisms of hot springs have been reviewed by Castenholz and Wickstrom (1975), Brock (1986), and Ward and Castenholz (2000). Temperatures can be fairly constant near the source, but can range from about 110°C (with a high concentrations of salts) to just above ambient, depending on the temperature and volume of thermal water, distance from the source, and volumes of nonthermal surface water entering a system. Temperature is not the only extreme condition for thermal organisms; most springs have elevated concentrations (50–150 mg L^{-1}) of inorganic ions (Ca^{2+}, Mg^{2+}, Na^+, HCO_3^-, SO_4^{2-}, Cl^-, Si, and H_2S) and elevated pH (8–10). These conditions select for highly adapted organisms, especially chemoautotrophic and heterotrophic bacteria in very hot (>70°C to ≈94°C) conditions (Brock, 1985b). Among photosynthetic organisms, cyanobacteria are most common, with an upper limit between 70 and 73°C; eukaryotic algae are restricted to a maximum of about 55°C. Along a thermal stream, distinct zones of bright colors and morphologies that correspond to different species along the thermal gradient can be observed.

A diversity of cyanobacteria dominate thermal waters. They include masses of coccoid species of *Aphanocapsa*, *Chroococcus*, *Cyanobacterium*, and *Synechococcus*, and filamentous species of *Mastigocladus*, *Oscillatoria*, and *Phormidium* (Ward and Castenholz, 2000, Chaps. 3 and 4). In cooler waters further from the source (35–50°C), diatoms (*Achnanthes* and *Pinnularia*) and green algae (*Spirogyra* and *Mougeotia*) proliferate (Stockner, 1967; Castenholz and Wickstrom, 1975). Cyanobacterial assemblages may form mats several centimeters thick and have extremely high rates of primary production (>10 g C m^{-2} d^{-1}; Castenholz and Wickstrom, 1975). In one hot spring in Costa Rica (62°C, pH 7.0), a species of *Oscillatoria* dominated, while nearby streams with less extreme temperatures (35–36°C; pH 7.8–8.0) had a greater diversity of algal species, including cyanobacteria (*Oscillatoria*, *Phormidium* and *Lyngbya*) and diatoms (*Pinnularia*; Pringle et al., 1993). Unusual consumers are associated with hot springs, because few metazoa tolerate temperatures greater than about 50°C. Invertebrates include ostracods, water mites, and rotifers, but little is known of their dynamics or food webs. Adult beetles and flies are successful in some systems. Brine flies (*Paracoenia* and *Epihydra*) lay eggs in microbial mats found in springs in Yellowstone Park within the 30–40°C range. Both adult and larval stages consume algal and bacterial material, which may in turn enhance primary productivity (Brock, 1967; Brock et al., 1969).

Some thermal springs are highly acidic, which further limits their species diversity. A notable eukaryote, the red alga *Cyanidium caldarium*, is often the sole photosynthetic organism in very acid (pH 2–4) hot springs up to about 55°C. This somewhat enigmatic organism was variously classified as a cyanobacterium, green alga, cryptomonad, and an evolutionary link between red and green algae (Seckbach, 1991); today it is placed in the division Rhodophyta based on pigments, chloroplast structure, and molecular features (Steinmüller et al., 1983; Pueschel, 1990). In acid hot springs, other enigmatic rhodophytes including *Cyanidioschyzon merolae* and *Galdiera sulphuraria* (DeLuca and Moretti, 1983), also have been reported, also have. It seems to be unlikely that either high temperature or low pH is solely responsible for this peculiar flora, because alkaline hot springs and nonthermal acid springs have very different algal communities.

B. Acid Environments

Most highly acid springs and streams are nonthermal and support a characteristic algal flora that is unlike those in other aquatic environments. Although bogs may exhibit relatively low pH (4.0–5.0), highly acidic environments typically are regarded as systems with H^+ concentrations at least an order of magnitude greater, that is, pH values ≤3.0, and they usually receive acidic inputs from either geological or anthropogenic sources (Hargreaves et al., 1975). Nearly all have elevated concentrations of metals, including Al, Fe, Mn, Pb, Co, Cu, and Zn, which may be near saturation levels even for very low pH, resulting in the formation of metal salt precipitates along stream margins or on algal colonies. Acid springs with very high Fe concentrations also may have very low dissolved O_2, due to $Fe(OH)_2$ and $FeO(OH)_2$ precipitates (Van Everdingen, 1970). Laguna de Alegría, a crater lake in El Salvador, is influenced by sulfur-rich fumaroles and exhibits pH values as low as 2.0 (Cole, 1963).

The earliest detailed studies of algae in highly acidic systems in North America were conducted in acid mine drainages in Indiana, Kentucky, Ohio,

Pennsylvania, and West Virginia (Lackey, 1938; Bennett, 1969; Warner, 1971). These studies, as well as those in the United Kingdom (Hargreaves *et al.*, 1975), all reveal low species diversity and a remarkable similarity in composition. *Euglena mutabilis* is the most widespread and often most abundant species, occurring in systems as acidic as pH 1.5. *E. mutabilis* is also common in naturally acidic streams, such as the Rio Agrio (pH 2.3) in Costa Rica (Pringle *et al.*, 1993), and in acidic ponds in the Smoking Hills region of the Northwest Territories (Sheath *et al.*, 1982). The latter site also supports populations of *Chlamydomonas acidophila*. *E. mutabilis* is not apparent in natural acid springs in Kootenay Paint Pots (BC) although a few diatoms and green algae are present (Wehr and Whitton, 1983). Other common elements in many highly acidic environments include *Klebsormidium* (previously *Hormidium*) *rivulare*, *Eunotia tenella*, *Pinnularia microstauron*, *P. braunii*, and *Gloeochrysis turfosa*. Acid sites in West Virginia contain many of the same species found in the United Kingdom. (Bennett, 1969), although an apparent absence of *G. turfosa* from North American sites may be the result of the alga being overlooked. No studies report cyanobacteria in these highly acidic environments, in agreement with Brock's (1973) recommendation for a lower pH limit of less than 4.0. Isolates of *E. mutabilis*, *Chlamydomonas acidophila*, *Klebsormidium rivulare*, *Gloeochrysis turfosa*, and *Stichococcus bacillaris* from one acid stream (pH 2.6–3.1) were able to tolerate and grow at pH levels less than the lowest measured in their collecting site (Hargreaves and Whitton, 1976a). In addition, an acid strain of *Klebsormidium rivulare* tolerated greater Zn and Cu concentrations in the pH range 3.0–4.0, than at pH ≥ 6.0, suggesting a H$^+$–metal interaction (Hargreaves and Whitton, 1976b).

VI. UNUSUAL ENVIRONMENTS

A. Saline Lakes and Streams

Saline lakes and streams make up a large and heterogeneous collection of water bodies that have elevated total dissolved salts (> 500 mg L^{-1}; Williams, 1996). Many are closed basins or desert playas that gain salinity over the year as they lose water (Hammer, 1986; Evans and Prepas, 1996). The term "saline" does not mean simply greater concentrations of NaCl. The ion content of inland saline lakes is influenced by Na$^+$, K$^+$, Ca^{2+}, and Mg^{2+}, and the major anions are typically Cl$^-$, SO$_4^{2-}$, HCO$_3^-$, and CO$_3^{2-}$ (Wetzel, 1983a; Hammer, 1986; Cole, 1994). Systems are usually well buffered (high in HCO$_3^-$ and/or CO$_3^{2-}$) and neutral to alkaline in pH (7.5–10.0). Among 47 saline lakes in the western United States and Canada, anion and cation chemistries vary significantly with latitude: lakes north of 47° latitude are dominated by SO$_4^{2-}$ and either Na$^+$ or Mg^{2+}, whereas lakes in more southern locations are dominated by CO$_3^{2-}$ or Cl$^-$ in conjunction with Na$^+$ ions, reflecting climatic as well as geological characteristics of each region (Blinn, 1993).

Saline lakes of athalassic (nonmarine) origin differ from oceanic systems in several important ways. Total dissolved salts vary considerably more (0.5–600 g L^{-1}) than in oceans (35–40 g L^{-1}), both among systems and over time. Most athalassic lakes are shallow, which makes seasonal and longer term (climatic, anthropogenic) changes in salinity important for algal survival, and results in low biotic diversity (Cole, 1994). A few are large and relatively deep, such as Pyramid Lake (532 km^2; 102 m deep) and Big Soda Lake (1.5 km^2; 64 m deep) in Nevada (Hutchinson, 1957), Mono Lake (150 km^2; 40–50 m deep) in California (National Academy of Sciences 1987), and Soap Lake (3.6 km^2; 27 m deep) in Washington (Castenholz, 1960). Because of differences in concentrations of salts with depth, many of the deeper saline lakes, such as Soap and Mono Lakes, are meromictic. Water level in Mono Lake varies substantially as a function of water diversion for human usage. In Redberry Lake, Saskatchewan, mean depth has decreased by about 37%, while salinity has increased by roughly 41% since the 1940s, due to changes in land use (Evans *et al.*, 1996).

Saline lakes are concentrated in arid environments, especially in the U.S. Southwest, Mexico, and interior regions of California, Oregon, Washington, northern prairies, and British Columbia. Some of the most extensive surveys of the chemistry and biology of saline lakes were conducted in Saskatchewan (Rawson and Moore, 1944; Hammer *et al.*, 1983). Few inland saline lakes are found in eastern North America, although Onondaga Lake (12 km^2, 20.5 m depth) is a saline lake in the New York Finger Lakes region. Levels of [Na$^+$ + K$^+$] and Cl^{-1} exceeded 500 and 1400 mg L^{-1}, respectively, in part from salt springs, plus pollution from an adjacent soda ash facility (Berg, 1963; Sze and Kingsbury, 1972). Controls on salt waste have resulted in reduced total salinity (450 mg Cl$^-$ L^{-1}), although levels are still greater than pre-industrial times (ca. 230 mg Cl$^-$ L^{-1}; Effler and Owens, 1996; Rowell, 1996).

Algal communities of saline lakes differ among systems and their diversity varies inversely with salinity (Blinn, 1993; Cole, 1994). Because many saline lakes are shallow and subject to wind-driven mixing, it is often difficult to distinguish between benthic and planktonic forms. Several studies have evaluated the influences of salinity and ion composition on algal communities (Castenholz, 1960; Hammer *et al.*, 1983;

Blinn, 1993; Fritz et al., 1993; Evans and Prepas, 1996; Wilson et al., 1996). In mildly saline (total salts 500–2000 mg L^{-1}) lakes, the algal flora is fairly rich and composed of a variety of diatoms (e.g., species of *Amphora*, *Campylodiscus*, *Cyclotella*, *Epithemia*, *Fragilaria*, *Navicula*, *Nitzschia*, and *Rhopalodia*), green algae (*Crucigenia*, *Pediastrum*, *Oocystis*, and *Sphaerocystis*), and cyanobacteria (e.g., species of *Anabaena*, *Aphanizomenon*, *Chroococcus*, *Lyngbya*, *Merismopedia*, *Microcystis*, and *Oscillatoria*), especially if N and P are high. In more strongly saline conditions (2–20 g L^{-1}), many species are eliminated, but the community still includes taxa found in nonsaline waters, like *Cladophora glomerata*, *Botryococcus braunii*, *Cocconeis placentula*, *Mastogloia* spp., *Nitzschia palea*, *Plagioselmis* (as *Rhodomonas*) *minuta*, and several cyanobacteria (*Aphanizomenon flos-aquae*, *Anabaena* spp., *Microcystis aeruginosa*, and *Oscillatoria* spp.). Species typical of higher salinities also co-occur within this range, probably because concentrations can vary by an order of magnitude or more over a year in many lakes. Under hypersaline conditions (20–600 g L^{-1}), diversity is very low, and includes some species that are restricted to higher salt levels, such as the diatoms *Amphora coffeiformis*, *Anomoneis sphaerophora*, *Navicula subinflatoides*, *Nitzschia communis*, and *N. frustulum*, cyanobacteria *Nodularia spumigena* and *Aphanothece halophytica*, and the filamentous green alga *Ctenocladus circinnatus* (Blinn, 1971; Herbst and Bradley, 1989; Wurtsbaugh and Berry, 1990; Kociolek and Herbst, 1992; Reuter et al., 1993). A few taxa that have marine distributions, such as *Enteromorpha intestinalis*, diatoms *Dunaliella salina*, *D. viridus*, *Chaetoceros muelleri*, and *Thalassiosira pseudonana*, and the coccolithophorid *Pleurochrysis carterae*, have been observed (Sze and Kingsbury, 1972; Stephens and Gillespie, 1976; National Academy of Sciences, 1987; Johansen et al., 1988). Experiments in which salinity was varied within Mono Lake mesocosms determined that diatom dominance, algal diversity, chlorophyll-*a*, and photosynthesis in benthic communities all declined at higher salinities, while *Ctenocladus circinnatus* and an *Oscillatoria* spp. dominated, although the diatom *Nitzschia monoensis* increased (Herbst and Blinn, 1998).

Primary production by algae is often high (300–1000 g C m^{-2} y^{-1}), along with very high chlorophyll-*a* levels (50 to >500 mg L^{-1}), because many saline lakes are located in sunny locations and have surplus P levels (total P: 50 to >1000 mg L^{-1}; Hammer, 1981; National Academy of Sciences, 1987; Robarts et al., 1992). Much of this production comes from epipelic and epilithic assemblages (e.g., *Ctenocladus*) rather than phytoplankton (Wetzel, 1964). Surface blooms of cyanobacteria may at times represent a significant proportion of total production (Sze and Kingsbury, 1972; Robarts et al., 1992). Many lakes in the Canadian prairies have lower algal biomass and primary productivity than would be predicted using standard nutrient models (Campbell and Prepas, 1986), apparently because of greater densities of macrozooplankton that flourish in saline waters and perhaps due to fewer zooplanktivorous fish (Evans et al., 1996). In Great Salt Lake, weather-induced decreases in salinity (250–50 g L^{-1}) resulted in an increase in grazers from one species (*Artemia*) to four, where rotifers and copepods were dominant (Wurtsbaugh and Berry, 1990).

Paleoecological studies have considered whether these systems were saline in the past (Fritz et al., 1993; Rowell, 1996). A study of the diatoms from 219 lakes in British Columbia indicated that some species are limited by salinities as dilute as 0.02 g L^{-1} (e.g., *Achnanthes pusilla*), whereas others tolerate more than 500 g L^{-1} (*Amphora coffeiformis* and *Nitzschia frustulum*; Wilson et al., 1996). Models from these data have been used to infer temporal and spatial differences in precipitation, which in turn influences salinity in these closed basins. Long-term changes in fossil algal pigments suggest that as prairie lakes became more saline, phytoplankton communities have shifted from a diatom–chrysophyte–dinoflagellate community to one dominated by greens, cyanobacteria, and diatoms (Vinebrooke et al., 1998). Due to possible effects of global warming, semiarid regions of North America may experience an increase in the number of saline waters in the next few decades (Evans and Prepas, 1996).

B. Snow and Ice

People who have hiked in alpine regions are familiar with red snow, especially where snowfields accumulate for most of the year. Aristotle also apparently observed red snow many centuries ago (Kol, 1968). This phenomenon is most often caused by the green alga *Chloromonas* (previously recorded as *Chlamydomonas*) *nivalis* (see Chap. 6). The red color is the result of an accumulation of secondary carotenoids, mainly astaxanthin, in resting cells (Round, 1981; Bidigare et al., 1993). Not all snow communities are colored red: orange, brown, and green patches are also seen, depending on the species present, dominant pigments, exposure to sunlight, and perhaps pH (Stein and Amundsen, 1967; Kol, 1968; Hoham and Blinn, 1979). Cells aggregate near the surface, and as successive snowfalls accumulate, layers or bands of pigmented algal communities can be seen in vertical cuts through a snow bank (Hoham and Mullet, 1977). Although motile stages can be found in snow, most cells occur either as thick-walled resting zygotes or as asexual hypnospores (Stein and Amundsen, 1967). Cryophilic flora include other algal flagellates, such as *Chloromonas nivalis*, *C. brevispina*, *Carteria nivale* (= planozygotes of *Chloromonas* sp.?), *Scotiella cryophila*, and *Chromulina chionophila*, and nonflagellated green algae, such as *Raphidonema nivale* and *Stichococcus* spp. (Stein and Amundsen, 1967; Kol, 1968; Hoham, 1975; Hoham and Blinn, 1979). The life cycles of many snow algae are incompletely known; thus, some cells previously identified as new taxa have been shown to be zygotes of other known species (Hoham and Blinn, 1979; R.W. Hoham, personal communication).

Snow algae exhibit measurable but low photosynthesis rates, many of which reach a maximum near 10°C, although some peak near freezing and decline at higher temperatures (Mosser *et al.*, 1977; Round, 1981). A strain of *Chloromonas pichinchae* from snowfields in Washington state grew best at 1°C and pH 6.0, whereas the optimum for an isolate of *Raphidonema nivale* from the same location was 4–5°C, with a pH optimum ≥7.0, despite lower environmental pH (ca. 5.0) and temperature (near 0°C; Hoham, 1975). Strains of *Chloromonas* isolated from snowfields in the Adirondack Mountains (exposed to decades of acid deposition) have significantly greater (1.5- to 2.2-fold greater) growth yields at pH 4.0 than *Chloromonas* isolated from the (less acidic) White Mountains of Arizona (Hoham and Mohn, 1985). Algal cells that colonize snow surfaces at high altitude experience extreme levels of solar (including UV) radiation. Carotenoid pigments provide photoprotection by reducing total radiation and filtering certain shorter wavelengths from photosynthetic pigments (Bidigare *et al.*, 1993). Experiments with snow algae from Tioga Pass in the Sierra Nevada found that UV radiation inhibited photosynthesis in green snow by 85%, but only 25% in red snow (Thomas and Duval, 1995). Snow algal communities also are influenced by the amount of wind-blown soil (plus factor: nutrient source), exposure (plus or minus factor: direction and quantity of sunlight), snow albedo (usually minus factor: reflective property), and water content (plus factor) of the snow (Stein and Amundsen, 1967; Hoham and Blinn, 1979; Thomas and Duval, 1995). These communities have cryophilic food webs with protozoan consumers, and bacterial and fungal decomposers that are active at very low temperatures (Felip *et al.*, 1995; Thomas and Duval, 1995).

C. Other Unusual Habitats

Algae form thick mats on tank walls, outflow weirs, pipes, and other substrata in sewage treatment plants, where concentrations of nutrients greatly exceed requirements (N: 1–20 mg L^{-1}, P: 0.1–2 mg L^{-1}). The most common taxon appears to be *Stigeoclonium tenue*; lesser abundances of *Chlorella* spp., *Nitzschia palea*, *Oedogonium* sp., *Oscillatoria* spp., *Pleurocapsa minor*, *Pseudanabaena catenata*, *Scenedesmus quadricauda*, and *Tribonema* spp. also are observed (Palmer, 1962; Sládecková *et al.*, 1983; Davis *et al.*, 1990a). Communities exhibit seasonal changes in composition: *Tribonema* dominates in the spring, and *Oscillatoria*, *Scenedesmus*, and *Stigeoclonium* reach their maxima in warmer months. These shifts are largely driven by changes in light and temperature, not nutrient supply (Davis *et al.*, 1990a). Substantial biomass and rapid growth rates have made algae candidates for nutrient removal plans (Sládecková *et al.*, 1983; Davis *et al.*, 1990b). Treatment ponds have been devised in which >95% of P may be removed by algal assemblages alone (Hoffmann, 1998).

Algae colonize caves in many limestone or dolomite regions where fractures or underground streams form underground pockets or large caverns. Some cave-dwelling algae, including species of *Gloeocapsa* (Cyanobacteria), survive very low light levels by having densely packed thylakoids in their cells and very slow growth rates (Pentecost and Whitton, 2000). By some observers caves may be considered to be dark, unproductive environments, but substantial algal floras develop where surface are open to the sunlight (fissures, mouths) or in show caverns where artificial lighting has been added. Cyanobacteria (e.g., *Aphanocapsa*, *Calothrix*, *Chroococcus*, *Gloeocapsa*, *Hapalosiphon*, and *Schizothrix*) may be especially abundant near light sources (Claus, 1962; Round, 1981). Species diversity is typically low: Timpanogos Cave (UT) and Seneca Cavern (OH) each support an algal flora of fewer than 30 species (St. Clair and Rushforth, 1976; Dayner and Johansen,1991). Species include diatoms (*Navicula tantula* and *N. contenta*), unicellular green algae (*Chlorella miniata*), and the xanthophyte *Pleurochloris commutata*. CaCO$_3$ from cave walls may become incorporated into algal colonies and thalli, which is why artificial lighting and the resultant algal colonization has been cited as a prime cause of cave wall destruction in show caves (Gurnee, 1994).

The calcareous cave flora are similar to wet limestone seepages above ground, and exhibit an abundance of calcified cyanobacteria and diatoms (Pentecost, 1982; Pentecost and Whitton, 2000). Algae from pools adjacent to caves may be an important carbon source for food webs inside caves (Pohlman *et al.*, 1997). Noncalcareous, temporary rock pools that form in weathered bedrock along lake and river margins also become algal habitats. These small systems (a few liters or less) experience great variations in water level and nutrients, and extremes in solar radiation. The most common inhabitant is the algal flagellate *Haematococcus lacustris*, which accumulates red secondary carotenoids much like snow algae. This adaptation serves a similar photoprotective role and is not influenced by nutrient levels (Yong and Lee, 1991; Lee and Soh, 1991). Humans have created a widespread habitat for *H. lacustris*, namely birdbaths (Canter-Lund and Lund, 1995); hence, many people report red water in their birdbaths after a year or so of use.

Endophytic algae live in cavities of higher plants or colonize plant tissues intracellularly. Algae colonize microhabitats created by bromeliad cups, and colonized by algae, and differ in pH and dissolved O_2 levels among different plants (Laessle, 1961). Perhaps the most ecologically important endophytic algae in freshwaters are species of cyanobacteria that colonize various aquatic plants and bryophytes. The best known is *Anabaena azollae*, a symbiont within the water fern *Azolla*, which is used as a nitrogen source or "green fertilizer" for rice crops worldwide (Bothe, 1982). Roots of some cycads (*Cyas*, *Encephalartos*, and *Macrozamia*) are colonized by several cyanobacteria, especially *Anabaena*, *Calothrix*, and *Nostoc* (Huang and Grobbelaar, 1989); when sectioned, cyanobacteria appear as a healthy blue–green color. Green algae (e.g., *Chlorella*, *Chlamydomonas*) and some xanthophytes (see Frost *et al.*, 1997, Chaps. 6 and 7) are symbionts in a variety of aquatic organisms, including freshwater sponges (e.g., *Spongilla lacustris* and *Corvomeyenia everetti*), cnidarians (e.g., *Hydra* spp.), ciliates (e.g., *Euplotes* and *Ophrydium*), and other organisms (Slobodkin, 1964; Frost and Williamson, 1980; Berninger *et al.*, 1986; Sand-Jensen *et al.*, 1997). In the spring, green, baseball-sized (up to 20 cm) green masses can be observed in ponds and lake outflows that are actually mucilaginous masses of amphibian eggs (*Amblystoma* and *Rana*) colonized by the green alga *Chlamydomonas* (syn. *Oöphila*) *amblystomatis*. This association appears to provide N for the alga and added O_2 for developing amphibian eggs (Goff and Stein, 1978; Bachmann *et al.*, 1986; Pinder and Friet, 1994). Much more intimate symbioses are found among algae and cyanobacteria with fungi, as lichens. The treatise by Ahmadjian (1993) can be consulted for further information.

Many algae colonize terrestrial habitats such as soils, trees, and other surfaces, and serve important ecological functions in soil and moisture retention, seed germination, nutrient dynamics, and succession of terrestrial vegetation (Carson and Brown, 1978; Bell, 1993; Vazquez *et al.*, 1998). Many of the more common terrestrial species, appear as a green "felt" on stone walls, tree bark, and wooden fences such as *Apatococcus* and *Trentepohlia* (Chlorophyta), that may be mistaken for moss (Canter-Lund and Lund, 1995). Epiphytic green algae have been used as biological indicators of air quality (Hanninen *et al.*, 1993). A number of genera that colonize plant leaves, walls, stones, or soils are also common in fresh waters, including many cyanobacteria (*Nostoc*, *Oscillatoria*, *Lyngbya*, and *Plectonema*), green algae (*Chlamydomonas*, *Chlorella*, *Chlorococcum*, and *Klebsormidium*), and diatoms (*Achnanthes*, *Hantzschia*, and *Navicula*; Segal, 1969; Cox and Hightower, 1972; King and Ward, 1977; Hunt *et al.*, 1979). Diatoms from these habitats are often desiccation-resistant and many are regarded as obligate aerial taxa (Johansen, 1999). Other genera, such as *Prasiola* and *Zygogonium*, have very distinct ecological requirements. *Prasiola* colonizes N-rich (e.g., guano) rocks and walls in aerial (often shaded) and even urban environments (Jackson, 1997; Rindi *et al.*, 1999), whereas aquatic species colonize cool, nutrient-poor streams (Sheath and Cole, 1992; Kawecka, 1990; Hamilton and Edlund, 1994). A few terrestrial species colonize the interstitial spaces within crystalline rocks in arid or semiarid regions, where water supply may be limited (Bell, 1993; Johansen, 1993). Further details on the algal flora of soils and terrestrial habitats can be found in reviews by Fritsch (1922), Starks *et al.* (1981), and Johansen (1993). Some algae are parasitic within plant tissues (leaves and twigs), such as *Cephaleuros*, which is the cause of red rust disease in higher plants (Thompson and Wujek, 1997). Perhaps one of the most unusual of all algal habits is that of the red alga *Rufusia*, which lives within the hairs of two- and three-toed sloths, but apparently does not colonize nearby vegetation (Chap. 5).

ACKNOWLEDGMENTS

Thanks are due Dr. R. Jan Stevenson (Michigan State University) and Dr. Robert G. Wetzel (University of Alabama) for helpful reviews and advice on relevant literature. Dr. Ronald W. Hoham (Colgate University) provided advice on species of snow algae. We also thank D. W. Blinn, D. Burney, R. J. Cannings, S. J. Cannings, J. Cotner, P. R. Leavitt, A. Merola, NASA, and the South Florida Water District for permission to reproduce photos.

LITERATURE CITED

Abel, R. A., Olson, D. M., Dinerstein, E., Hurley, P. Eds. 2000. Freshwater ecoregions of North America: A conservation assessment. World Wildlife Fund, Washington, DC.

Admiraal, W., Mylius, S. D., de Ruijter van Steveninck, E. D., Tubbing, D. M. J. 1993. A model of phytoplankton production in the lower River Rhine verified by observed changes in silicate concentration. Journal of Plankton Research 15:659–682.

Admiraal, W., Breebaart, L., Tubbing, G. M. J., Van Zanten, B., de Ruijter van Steveninck, E. D., Bijerk, R. 1994. Seasonal variation in composition and production of planktonic communities in the lower River Rhine. Freshwater Biology 32:519–531.

Ahmadjian, V. 1993. The lichen symbiosis. J Wiley, New York.

Allan, J. D., Ed. 1995. Stream ecology: Structure and function of running waters. Chapman & Hall, New York.

Allen, W. E. 1921. A quantitative and statistical study of the plankton of the San Joaquin River and its tributaries in and near Stockton, California, in 1913. University California Publications Zoology 22:1–292.

Aloi, J. E. 1990. A critical review of recent freshwater periphyton field methods. Canadian Journal of Fisheries and Aquatic Sciences 47:656–670.

Andersen, R. A., Wetherbee, R. 1992. Microtubules of the flagellar apparatus are active during prey capture in the chrysophycean alga *Epipyxis pulchra*. Protoplasma 166:8–20.

Auer, M. T., Graham, J. M., Graham, L. E., Kranzfelder, J. A. 1983. Factors relating the spatial and temporal distribution of *Cladophora* and *Ulothrix* in the Laurentian Great Lakes, *in*: Wetzel, R. G. Ed., Periphyton of freshwater ecosystems. Junk, The Hague, pp. 135–145.

Aziz, A., Whitton, B. A. 1988. Influence of light flux on nitrogenase activity of the deepwater rice-field cyanobacterium (blue–green alga) *Gloeotrichia pisum* in field and laboratory. *Microbios* 53:7–19.

Bachmann, M. D., Carlton, R. G., Burkholder, J. M., Wetzel, R. G. 1986. Symbiosis between salamander eggs and green algae: Microelectrode measurements inside eggs demonstrate effect of photosynthesis on oxygen concentration. Canadian Journal of Zoology 64:1586–1588.

Baker, A. L., Baker, K. K. 1979. Effects of temperature and current discharge on the concentration and photosynthetic activity of the phytoplankton in the upper Mississippi River. Freshwater Biology 9:191–198.

Barber, H. G., Haworth, E. Y. 1981. A guide to the morphology of the diatom frustule. Freshwater Biology Association Science Publication 41, 112 p.

Barbiero, R. P., Welch E. B. 1992. Contribution of benthic blue–green algal recruitment to lake populations and phosphorus translocation. Freshwater Biology 27:249–620.

Beaumont, P. 1975. Hydrology, *in*: Whitton, B. A., Ed., River ecology. Blackwell Science, Oxford, UK, pp. 1–38.

Bell, R. A. 1993. Cryptoendolithic algae of hot semiarid lands and deserts. Journal of Phycology 29:133–139.

Benke, A. C., Hall, C. A. S., Hawkins, C. P., Lowe-McConnell, R. H., Stanford, J. A., Suberkrop, K., Ward, J. V. 1988. Bioenergetic considerations in the analysis of stream ecosystems. Journal of the North American Benthological Society 7:480–502.

Bennett, H. D. 1969. Algae in relation to mine water. Castanea 34:306–328.

Berg, C. O. 1963. Middle Atlantic states, *in*: Frey, D.G., Ed., Limnology in North America. University of Wisconsin Press, Madison, pp. 191–237.

Bergquist, A. M., Carpenter, S. R. 1986. Limnetic herbivory: Effects on phytoplankton populations and primary production. Ecology 67:1351–1360.

Berninger, U. G., Finlay, B. J., Canter, H. M. 1986. The distribution and ecology of zoochlorellae-bearing ciliates in a productive pond. Journal of Protozoology 33:557–563.

Berninger, U. G., Caron, D. A., Sanders, R. W. 1992. Mixotrophic algae in three ice-covered lakes of the Pocono Mountains, U.S.A. Freshwater Biology 28:263–272.

Bianchi, T. S., Findlay, S., Dawson, R. 1993. Organic matter sources in the water column and sediments of the Hudson River estuary: The use of plant pigments as tracers. Estuarine, Coastal, and Shelf Science 36:359–376.

Bidigare, R. R., Ondrisek, M. E., Kennicutt, M. C., Iturriaga, R., Harvey, H. R., Hoham, R. W., Macko, S. A. 1993. Evidence for a photoprotective function for secondary carotenoids of snow algae. Journal of Phycology 29:427–434.

Biggs, B. J. F. 1996. Patterns in benthic algae of streams, *in*: Stevenson, R. J., Bothwell, M. L., Lowe, R. L., Eds. Algal ecology: Freshwater benthic ecosystems. Academic Press, San Diego, pp. 31–56.

Biggs, B. F., Goring, D. G., Nikora, V. I. 1998. Subsidy and stress responses of stream periphyton to gradients in water velocity as a function of community growth form. Journal of Phycology 34:598–607.

Billen, G., Garnier, J., Hanset, P. 1994. Modelling phytoplankton development in whole drainage networks: the RIVER-STRAHLER model applied to the Seine river system. Hydrobiologia 289:119–137.

Bird, D. F., Kalff, J. 1987. Algal phagotrophy: Regulating factors and importance relative to photosynthesis in *Dinobryon* (Chrysophyceae). Limnology and Oceanography 32:277–284.

Blindow, I. 1987. The composition and density of epiphyton on several species of submersed macrophytes — The neutral substrate hypothesis tested. Aquatic Botany 29:157–168.

Blindow, I. 1992. Decline of charophytes during eutrophication: Comparison with angiosperms. Freshwater Biology 28:9–14.

Blinn, D. W. 1971. Autecology of a filamentous alga, *Ctenocladus circinnatus* (Chlorophyceae), in saline environments. Canadian Journal of Botany 49:735–743.

Blinn, D. W. 1993. Diatom community structure along physicochemical gradients in saline lakes. Ecology 74:1246–1263.

Blinn, D. W., Fredericksen, A., Korte, V. 1980. Colonization rates and community structure of diatoms on three different rock substrata in a lotic system. British Phycological Journal 15:303–310.

Blinn, D. W., Truitt, R. T., Pickart, A. 1989. Feeding ecology and radular morphology of the freshwater limpet *Ferrissia fragilis*. Journal of the North American Benthological Society 8:237–242.

Blinn, D. W., Shannon, J. P., Benenati, P. L., Wilson, K. P. 1998. Algal ecology in tailwater stream communities: The Colorado River below Glen Canyon Dam, Arizona. Journal of Phycology 34:734–740.

Blum, J. L. 1956. The ecology of river algae. Botanical Review 22:291–341.

Borchardt, M. A., Hoffmann, J. P., Cook, P. W. 1994. Phosphorus uptake kinetics of *Spirogyra fluviatilis* (Charophyceae) in flowing water. Journal of Phycology 30:403–417.

Boschker, H. T. S., Dekkers, E. M. J., Pel, R., Cappenberg, T. E. 1995. Sources of organic carbon in the littoral of Lake Gooimeer as indicated by stable isotope and carbohydrate compositions. Biogeochemistry 29:89–105.

Bothe, H. 1982. Nitrogen fixation, *in*: Carr, N. G., Whitton, B. A., Eds., The biology of cyanobacteria. Blackwell Science, Oxford, UK, pp. 87–104.

Boucher, P., Blinn, D. W., Johnson, D. B. 1984. Phytoplankton ecology in an unusually stable environment (Montezuma Well, Arizona, U.S.A.). Hydrobiologia 119:149–160.

Bourelly, P. 1985. Les algues d'eau douce. III. Les algues bleues et rouges, les Eugléniens, Peridiniens et Cryptomonadines. Soc. Nouvelle Éditions Boubée, Paris.

Brock, M. L., Wiegert, R. G., Brock, T. D. 1969. Feeding by *Paracoenia* and *Ephydra* (Diptera: Ephydridae) on the microorganisms of hot springs. Ecology 50:192–199.

Brock, T. D. 1967. Relationship between primary productivity and standing crop along a hot spring thermal gradient. Ecology 48:566–571.

Brock, T. D. 1973. Lower pH limit for the existence of blue–green algae: Evolutionary and ecological implications. Science 179:480–483.

Brock, T. D. 1985a. A eutrophic lake: Lake Mendota, Wisconsin. Springer-Verlag, New York.

Brock, T. D. 1985b. Life at high temperatures. Science 230:132–138.

Brock, T. D., Ed. 1986. Thermophiles: General, molecular, and applied microbiology. Wiley, New York.

Brönmark, C., Klosiewski, S. P., Stein, R. A. 1992. Indirect effects of predation in a freshwater, benthic food chain. Ecology 73:1662–1674.

Brook, A. J., Williamson, D. B. 1988. The survival of desmids on the drying mud of a small lake, in: Round, F. E., Ed., Algae and the aquatic environment. Biopress Ltd., Bristol, UK, pp. 185–196.

Brooks, J. L., Deevey, E. S. 1963. New England, in: Frey, D. G., Ed., Limnology in North America. University of Wisconsin Press, Madison, pp. 117–162.

Browder, J. A., Gleason, P. J., Swift, D. R. 1994. Periphyton in the Everglades: Spatial variation, environmental correlates, and ecological implications, in: Davis, S. M., Ogden, J. C., Eds., Everglades. The ecosystem and its restoration. St. Lucie Press, Delray Beach, FL, pp. 379–418.

Brunskill, G. J., Ludlam, S. D. 1969. Fayetteville Green Lake, New York. 1. Physical and chemical limnology. Limnology and Oceanography 14:817–829.

Burkholder, J. M., Wetzel, R. G. 1990. Epiphytic alkaline phosphatase on natural and artificial plants in an oligotrophic lake: Re-evaluation of the role of macrophytes as a phosphorus source for epiphytes. Limnology and Oceanography 35:736–747.

Burns, C. W., Stockner, J. G. 1991. Picoplankton in size New Zealand lakes: Abundance in relation to season and trophic state. Internationale Revue der Gesamten Hydrobiologie 76:523–536.

Butcher, R. W. 1933. Studies on the ecology of rivers. 1. On the distribution of macrophytic vegetation in the rivers of Britain. Journal of Ecology 21:58–91.

Calow, P. 1973a. Field observations and laboratory experiments on the general food requirements of two species of freshwater snail, *Planorbis contortus* (Linn.) and *Ancylus fluviatilis* Müll. Proceedings of Malacological Society of London 40:483–498.

Calow, P. 1973b. The food of *Ancylus fluviatilis* (Müll.), a littoral stone-dwelling herbivore. Oecologia 13:113–133.

Cameron, W. A., Larson, G. L. 1993. Limnology of a caldera lake influenced by hydrothermal processes. Archiv für Hydrobiologie 128:13–38.

Campbell, C. E., Prepas, E. E. 1986. Evaluation of factors related to the unusually low chlorophyll levels in prairie saline lakes. Canadian Journal of Fisheries and Aquatic Sciences 43:846–854.

Campeau, S., Murkin, H. R., Titman, R. D. 1994. Relative importance of algae and emergent plant litter to freshwater marsh invertebrates. Canadian Journal of Fisheries and Aquatic Sciences 51:681–692.

Canavan, R. W., Siver, P. A. 1995. Connecticut lakes. A study of the chemical and physical properties of fifty-six Connecticut lakes. Connecticut College Arboretum, New London, CT.

Canter-Lund, H., Lund, J. W. G. 1995. Freshwater algae: Their microscopic world explored. Biopress Ltd., Bristol, UK.

Caraco, N. F., Cole, J. J., Raymond, P. A., Strayer, D. L., Pace, M. L., Findlay, S. E. G., Fischer, D. T. 1997. Zebra mussel invasion in a large, turbid river: Phytoplankton response to increased grazing. Ecology 78:588–602.

Caron, D. A., Sanders, R. W., Lim, E. L., Marrase, C., Amaral, L. A., Whitney, S., Aoki, R. B., Porter, K. G. 1993. Light-dependent phagotrophy in the freshwater mixotrophic chrysophyte *Dinobryon cylindricum*. Microbe Ecology 25:93–111.

Carpenter, S. R., Lathrop, R. C., Muñoz-del-Rio, A. 1993. Comparison of dynamic models for edible phytoplankton. Canadian Journal of Fisheries and Aquatic Sciences 50:1757–1767.

Carson, J. L., Brown, R. M. 1978. Studies of Hawaiian freshwater and soil algae II. Algal colonization and succession on a dates volcanic substrate. Journal of Phycology 14:171–178.

Castenholz, R. W. 1960. Seasonal changes in the attached algae of freshwater and saline lakes in the Lower Coulee, Washington. Limnology and Oceanography 5:1–28.

Castenholz, R. W. 1982. Motility and taxes, in: Carr, N. G., Whitton, B. A., Eds., The biology of the cyanobacteria. Blackwell Science, Oxford, UK, pp. 413–439.

Castenholz, R. W, Wickstrom, C. E. 1975. Thermal streams, in: Whitton, B. A., Ed., River ecology. Blackwell Science, Oxford, UK, pp. 264–285.

Cattaneo, A. 1983. Grazing on epiphytes. Limnology and Oceanography 28:124–132.

Cattaneo, A. 1996. Algal seston and periphyton distribution along a stream linking a chain of lakes on the Canadian Shield. Hydrobiologia 325:183–192.

Cattaneo, A., Amireault, M. C. 1992. How artificial are artificial substrata for periphyton? Journal of the North American Benthological Society 11:244–256.

Cattaneo, A., Kerimian, T., Roberge, M., Marty, J. 1997. Periphyton distribution and abundance on substrata of different size along a gradient of stream trophy. Hydrobiologia 354:101–110.

Cattaneo, A., Galanti, G., Gentinetta, S., Romo, S. 1998. Epiphytic algae and macroinvertebrates on submerged and floating-leaved macrophytes in an Italian lake. Freshwater Biology 39:725–740.

Christoffersen, K., Riemann, B., Hansen, L. R., Klysner, A., Sorensen, H.B. 1990. Qualitative importance of the microbial loop and plankton community structure in a eutrophic lake during a bloom of cyanobacteria. Microbial Ecology 20:253–272.

Claus, G. 1962. Beitrage zur Kenntnis der Algenflora der Abaligeter Hohle. Hydrobiologia 19:192–222.

Cochran-Stafira, D. L., Andersen, R. A. 1984. Diatom flora of a kettle-hole bog in relation to hydrarch succession zones. Hydrobiologia 109:265–273.

Cole, G. A. 1963. The American Southwest and Middle America, in: Frey, D. G., Ed., Limnology in North America. University of Wisconsin Press, Madison, pp. 393–434.

Cole, G. A. 1994. Textbook of Limnology, 4th ed. Waveland Press, Prospect Heights, IL.

Cole, J. J., Caraco, N. F., Peierls, B. 1992. Can phytoplankton maintain a positive carbon balance in a turbid, freshwater, tidal estuary? Limnology and Oceanography 37:1608–1617.

Colletti, P. J., Blinn, D. W., Pickart, A., Wagner, V. T. 1987. Influence of different densities of the mayfly grazer *Heptagenia criddlei* on lotic diatom communities. Journal of the North American Benthological Society 6:270–280.

Colt, L. C., Jr., Saumure, R. A., Jr., Baskinger, S. 1995. First record of the algal genus *Basicladia* (Chlorophyta, Cladophorales) in Canada. Canadian Field-Naturalist 109:454–455.

Cooksey, K. E., Cooksey, B. 1978. Growth-influencing substances in sediment extracts from a subtropical wetland: Investigation using a diatom bioassay. Journal of Phycology 14:347–352.

Covich, A. P. 1976. Recent changes in molluscan species diversity of a large tropical lake (Lago de Peten, Guatemala). Limnology and Oceanography 21:51–59.

Cox, E. R., Hightower, J. 1972. Some corticolous algae of McMinn County, Tennessee, U.S.A. Journal of Phycology 8:203–205.

Cronk, J. K., Mitsch, W. J. 1994. Periphyton productivity on artificial and natural surfaces in constructed freshwater wetlands under different hydrologic regimes. Aquatic Botany 48:325–341.

Crumpton, W. G., Wetzel, R. G. 1982. Effects of differential growth and mortality in the seasonal succession of phytoplankton populations in Lawrence Lake, Michigan. Ecology 63:1729–1739.

Czarnecki, D. 1979. Epipelic and epilithic diatom assemblages in Montezuma Well National Monument, Arizona. Journal of Phycology 15:346–352.

Dahm, C. N., Grimm, N. B., Marmonier, P., Valett, H. M., Vervier, P. 1998. Nutrient dynamics at the interface between surface waters and groundwaters. Freshwater Biology 40:641–654.

Davis, C. C. 1964. Evidence for the eutrophication of Lake Erie from phytoplankton records. Limnology and Oceanography 9:275–283.

Davis, L. S., Hoffmann, J. P., Cook, P. W. 1990a. Seasonal succession of algal periphyton from a wastewater treatment facility. Journal of Phycology 26:611–617.

Davis, L. S., Hoffmann, J. P., Cook, P. W. 1990b. Production and nutrient accumulation by periphyton in a wastewater treatment facility. Journal of Phycology 26:617–623.

Dayner, D. M., Johansen, J. R. 1991. Observations on the algal flora of Seneca Cavern, Seneca County, Ohio. Ohio Journal of Science 91:118–21.

DeLuca, P., Moretti, A. 1983. Floridosides in *Cyanidium caldarium*, *Cyanidioschyzon marolae* and *Galdiera sulphuraria* (Rhodophyta, Cyanidiophyceae). Journal of Phycology 19:368–369.

de Noyelles, F., Knoechel, R., Reinke, D., Treanor, D., Altenhofen, C. 1980. Continuous culturing of natural phytoplankton communities in the Experimental Lakes Area: Effects of enclosure, *in situ* incubation, light, phosphorus, and cadmium. Canadian Journal of Fisheries and Aquatic Sciences 37:424–433.

de Ruyter van Stevenink, E. D., Admiraal, W., Breebaart, L., Tubbing, G. M. J., van Zanten, B. 1992. Plankton in the River Rhine: structural and functional changes observed during downstream transport. Journal of Plankton Research 14:1351–1368.

Descy, J.-P., Gosselain, V. 1994. Development and ecological importance of phytoplankton in a large lowland river (River Meuse, Belgium). Hydrobiologia 289:139–155.

Descy, J.-P., Servais, P., Smitz, J. S., Billen, G., Everbecq, E. 1987. Phytoplankton biomass and production in the River Meuse (Belgium). Water Research 21:1557–1566.

Descy, J.-P., Gosselain, V., Evrard, F. 1994. Respiration and photosynthesis of river phytoplankton. Internationale Vereinigung für Theoretische und Angewandte Limnologie Verhandlung 25:1555–1560.

Dodds, W. K. 1989. Photosynthesis of two morphologies of *Nostoc parmelioides* (Cyanobacteria) as related to current velocities and diffusion patterns. Journal of Phycology 25:258–262.

Dodds, W. K., Gudder, D. A. 1992. The ecology of *Cladophora*. Journal of Phycology 28:415–427.

Dodds, W. K., Gudder, D. A., Mollenhauer, D. 1995. The ecology of *Nostoc*. Journal of Phycology 31:2–18.

Dodds, W. K., Jones, J. R., Welch, E. B. 1998. Suggested classification of stream trophic state: distributions of temperate stream types by chlorophyll, total nitrogen, and phosphorus. Water Research 32:1455–1462.

Douglas, B. 1958. The ecology of the attached diatoms and other algae in a small stony stream. Journal of Ecology 45:295–322.

Douglas, M. S. V., Smol, J. P. 1995. Periphytic diatom assemblages from high arctic ponds. Journal of Phycology 31:60–69.

Dow, C. S., Swoboda, U. K. 2000. Cyanotoxins, *In*: Whitton, B. A., Potts, M., Eds., The ecology of cyanobacteria: Their diversity in time and space. Kluwer, Dordrecht, pp. 613–632.

Downes, B. J., Lake, P. S., Schreiber, E. S. G., Glaister, A. 1998. Habitat structure and regulation of local species diversity in a stony, upland stream. Ecology Monographs 68:237–257.

Eaton, J. W., Moss, B. 1966. The estimation of numbers and pigment content in epipelic algal populations. Limnology and Oceanography 11:584–595.

Eddy, S. 1934. A study of the fresh-water plankton communities. Illinois Biological Monographs 12(4), 93 pp.

Edmondson, W. T. 1963. Pacific coast and Great Basin, *in*: Frey, D. G., Ed., Limnology in North America. University of Wisconsin Press, Madison, pp. 371–392.

Edmondson, W. T. 1977. Recovery of Lake Washington from eutrophication, *in*: Cairns, J., Dickson, K. L., Herricks, E. E., Eds., Recovery and restoration of damaged ecosystems. University Press Virginia, Charlottesville, pp. 102–109.

Edmondson, W. T., Lehman, J. T. 1981. The effect of changes in the nutrient income on the condition of Lake Washington. Limnology and Oceanography 26:1–29.

Effler, S. W., Owens, E. M. 1996. Density stratification in Onondaga Lake: 1968–1994. Lake Reservoir Management 12:25–33.

Eminson, D., Moss, B. 1980. The composition and ecology of periphyton communities in freshwaters. 1. The influence of host type and external environment on community composition. British Phycological Journal 15:429–446.

Entwisle, T. J. 1989. Macroalgae in the Yarra River basin: Flora and distribution. Proceedings of the Royal Society of Victoria 101:1–76.

Erikson, R., Pum, M., Vammen, K., Cruz, A., Ruiz, M., Zamora, H. 1997. Nutrient availability and the stability of phytoplankton biomass and production in Lake Xolotlán (Lake Managua, Nicaragua). Limnologica 27:157–164.

Evans, J. C., Arts, M. T., Robarts, R. D. 1996. Algal productivity, algal biomass, and zooplankton biomass in a phosphorus-rich, saline lake: deviations from regression model predictions. Canadian Journal of Fisheries and Aquatic Sciences 53:1048–1060.

Evans, J. C., Prepas, E. E. 1996. Potential effects of climate change on ion chemistry and phytoplankton communities in prairie saline lakes. Limnology and Oceanography 41:1063–1076.

Fairchild, G. W., Sherman, J. W., Acker, F. W. 1989. Effects of nutrient (N, P, C) enrichment, grazing and depth upon littoral periphyton of a softwater lake. Hydrobiologia 173:69–83.

Felip, M., Sattler, B., Psenner, R., Catalan, J. 1995. Highly active microbial communities in the ice and snow cover of high mountain lakes. Applied Environmental Microbiology 61:2394–2401.

Feminella, J. W., Hawkins, C. P. 1995. Interactions between stream herbivores and periphyton: A quantitative analysis of past experiments. Journal of the North American Benthological Society 14:465–509.

Feminella, J. W., Power, M. E., Resh, V. H. 1989. Periphyton responses to invertebrate grazing and riparian canopy in three northern California coastal streams. Freshwater Biology 22:445–457.

Findlay, S., Likens, G. E., Hedin, L., Fisher, S. G., McDowell, W. H. 1997. Organic matter dynamics in Bear Brook, Hubbard Brook Experimental Forest, New Hampshire, USA. Journal of the North American Benthological Society 16:43–46.

Flint, L. H. 1955. *Hildenbrandia* in America. Phytomorphology 5:185–189.

Frey, D. G., Ed. 1963. Limnology in North America. University of Wisconsin Press, Madison.

Fritsch, F. E. 1922. The terrestrial algae. Journal of Ecology 10:220–236.
Fritsch, F. E. 1929. The encrusting algal communities of certain fast-flowing streams. New Phytologist 28:165–196.
Fritz, S. C., Juggins, S., Batterbee, R. W. 1993. Diatom assemblages and ionic characterization of freshwater and saline lakes of the Northern Great Plains, North America: A tool for reconstructing past salinity and climate fluctuations. Canadian Journal of Fisheries and Aquatic Sciences 50:1844–1856.
Frost, T. M., Williamson, C. E. 1980. In situ determination of the effect of symbiotic algae on the growth of the freshwater sponge *Spongilla lacustris*. Ecology 61:1361–1370.
Frost, T. M., Graham, L. E., Elias, J. E., Haase, M. J., Kretchmer, D. W., Kranzfelder, J. A. 1997. A yellow–green algal symbiont in the freshwater sponge, *Corvomeyenia everetti*:convergent evolution of symbiotic associations. Freshwater Biology 38:395–399.
Fuller, R. L., Roelofs, J. L., Fry, T. J. 1986. The importance of algae to stream invertebrates. Journal of the North American Benthological Society 5:290–296.
Gabor, T. S., Murkin, H. R., Stainton, M. P., Boughen, J. A., Titman, R. D. 1994. Nutrient additions to wetlands in the interlake region of Manitoba, Canada: Effects of a single pulse addition in spring. Hydrobiologia 279/280:497–510.
Glime, J. M., Wetzel, R. G., Kennedy, B. J. 1982. The effects of bryophytes on succession from alkaline marsh to *Sphagnum* bog. American Midland Naturalist 108:209–223.
Goff, L. J., Stein, J. R. 1978. Ammonia: basis for algal symbiosis in salamander egg masses. Life Science 22:1463–1468.
Goldman, C. R. 1988. Primary productivity, nutrients, and transparency during the early onset of eutrophication in ultra-oligotrophic Lake Tahoe, California–Nevada. Limnology and Oceanography 33:1321–1333.
Goldsborough, L. G., Robinson, G. G. C. 1996. Pattern in wetlands, *in*: Stevenson, R. J., Bothwell, M. L., Lowe, R. L., Eds., Algal ecology: Freshwater benthic ecosystems. Academic Press, San Diego, pp. 77–117.
Goodwin, T. W. 1974. Carotenoids and biliproteins, *in*: Stewart, W. D. P., Ed., Algal physiology and biochemistry. University of California Press, Berkeley, pp. 176–205.
Gorham, E., Eisenreich, S. J. Ford, J., Santelmann, M. V. 1985. The chemistry of bog waters, *in*: Stumm, W., Ed., Chemical processes in lakes, Wiley, New York, pp. 339–363.
Gosselain, V., Descy, J.-P., Everbecq, E. 1994. The phytoplankton community of the River Meuse, Belgium: Seasonal dynamics (year 1992) and the possible incidence of zooplankton grazing. Hydrobiologia 289:179–191.
Gosselain, V., Viroux, L., Descy, J.-P. 1998. Can a community of small-bodied grazers control phytoplankton in rivers? Freshwater Biology 39:9–24.
Gough, S. B., Woelkerling, W. J. 1976. Wisconsin desmids. 2. Aufwuchs and plankton communities of selected soft water lakes, hard water lakes, and calcareous spring ponds. Hydrobiologia 49:3–25.
Graham, J. M., Kranzfelder, J. A., Auer, M. T. 1985. Light and temperature as factors regulating seasonal growth and distribution of *Ulothrix zonata* (Ulvophyceae). Journal of Phycology 21:228–234.
Graham, M. D., Vinebrooke, R. D. 1998. Trade-offs between herbivore resistance and competitiveness in periphyton of acidified lakes. Canadian Journal of Fisheries and Aquatic Sciences 55:806–814.
Greenberg, A. E. 1964. Plankton of the Sacramento River. Ecology 45:40–49.
Gruendling, G. K. 1971. Ecology of the epipelic algal communities in Marion Lake, British Columbia. Journal of Phycology 7:239–249.
Gurnee, J. 1994. Management of some unusual features in the show caves of the United States. International Journal of Speleology 23:13–17.
Haberyan, K. A., Umana, G., Collado, C., Horn, S. P. 1995. Observations on the plankton of some Costa Rican lakes. Hydrobiologia 312:75–85.
Hadas, O., Malinsky-Rushansky, N., Pinkas, R., Cappenberg, T. E. 1998. Grazing on autotrophic and heterotrophic picoplankton by ciliates isolated from Lake Kinneret, Israel. Journal of Plankton Research 20:1435–1448.
Hamilton, P. B., Edlund, S. A. 1994. Occurrence of *Prasiola fluviatilis* (Chlorophyta) on Ellesmere Island in the Canadian Arctic. Journal of Phycology 30:217–221.
Hammer, U. T. 1981. Primary production in saline lakes. Hydrobiologia 81:47–57.
Hammer, U. T. 1986. Saline lake ecosystems of the world. Kluwer, The Hague.
Hammer, U. T., Shamess, J., Haynes, R. C. 1983. The distribution and abundance of algae in saline lakes of Saskatchewan, Canada. Hydrobiologia 105:1–26.
Hann, B. J. 1991. Invertebrate grazer–periphyton interactions in a eutrophic marsh pond. Freshwater Biology 26:87–96.
Hanninen, O., Ruuskanen, J., Oksanen, J. 1993. A method for facilitating the use of algae growing on tree trunks as bioindicators of air quality. Environmental Monitoring and Assessment 28:215–220.
Hansson, L.-A., Rudstam, L. G., Johnson, T. B., Soranno, P., Allen, Y. 1994. Patterns in recruitment from sediment to water in a dimictic, eutrophic lake. Canadian Journal of Fisheries and Aquatic Sciences 51:2825–2833.
Hardwick, G. G., Blinn, D. W., Usher, H. D. 1992. Epiphytic diatoms on *Cladophora glomerata* in the Colorado Rover, Arizona: Longitudinal and vertical distribution in a regulated river. Southwestern Naturalist 37:148–156.
Hargreaves, J. W., Whitton, B. A. 1976a. Effects of pH on growth of acid stream algae. British Phycological Journal 11:215–223.
Hargreaves, J. W., Whitton, B. A. 1976b. Effects of pH on tolerance of *Hormidium rivulare* to zinc and copper. Oecologia 26:235–243.
Hargreaves, J. W., Lloyd, E. J. H., Whitton, B. A. 1975. Chemistry and vegetation of highly acidic streams. Freshwater Biology 5:563–576.
Harrison, S. S. C., Hildrew, A. G. 1998. Patterns in the epilithic community of a lake littoral. Freshwater Biology 39:477–492.
Havens, K. E., Bull, L. A., Warren, G. L., Crisman, T. L., Phlips, E. J., Smith, J. P. 1996. Food web structure in a subtropical lake ecosystem. Oikos 75:20–32.
Hawes, I., Schwarz, A.-M. 1996. Epiphytes from a deep-water characean meadow in an oligotrophic New Zealand lake: Species composition, biomass and photosynthesis. Freshwater Biology 36:297–313.
Hawkes, H. A. 1975. River zonation and classification, *in*: Whitton, B. A., Ed., River ecology. Blackwell Science, Oxford, UK, pp. 312–374.
Hawley, G. R. W., Whitton, B. A. 1991. Seasonal changes in chlorophyll-containing picoplankton populations in ten lakes in Northern England. Internationale Revue der Gesamten Hydrobiologie 76:545–554.
Heaney, S. I., Furnass, T. I. 1980. Laboratory models of diel vertical migration in the dinoflagellate *Ceratium hirundinella*. Freshwater Biology 10:163–170.
Heaney, S. I., Talling, J. F. 1980. Dynamics aspects of dinoflagellate distribution patterns in a small productive lake. Journal of Ecology 68:75–94.

Heaney, S. I., Lund, J. W. G., Canter, H., Gray, K. 1988. Population dynamics of *Ceratium* spp. in three English lakes, 1945–1985. Hydrobiologia 161:133–148.

Hecky, R. E., Hesslein, R. H. 1995. Contribution of benthic algae to lake food webs as revealed by stable isotope analysis. Journal of the North American Benthological Society 14:631–653.

Hecky, R. E., Kilham, P. 1988. Nutrient limitation of phytoplankton in freshwater and marine environments: A review of recent evidence on the effects of enrichment. Limnology and Oceanography 33:796–822.

Herbst, D. B., Blinn, D. W. 1998. Experimental mesocosm studies of salinity effects on the benthic algal community of a saline lake. Journal of Phycology 34:772–778.

Herbst, D. B., Bradley, T. J. 1989. Salinity and nutrient limitations on growth of benthic algae from two alkaline salt lakes of the western Great Basin (USA). Journal of Phycology 25:673–678.

Hesse, L. W., Schmulbach, J. C., Carr, J. M., Keenlyne, K. D., Unkenholz, D. G., Robinson, J. W., Mestl, G. E. 1989. Missouri River fishery resources in relation to past, present, and future stresses, in: Dodge, D. P., Ed., Proceedings of the International large river symposium, Canadian Special Publication in Fisheries and Aquatic Science 106:352–371.

Hickman, M. 1978. Ecological studies on the epipelic algal community in five prairie-parkland lakes in central Alberta. Canadian Journal of Botany 56:991–1009.

Hoagland, K. D., Peterson, C. G. 1990. Effects of light and wave disturbance on vertical zonation of attached microalgae in a large reservoir. Journal of Phycology 26:450–457.

Hoagland, K. D., Roemer, S. C., Rosowski, J. R. 1982. Colonization and community structure of two periphyton assemblages, with emphasis on the diatoms (Bacillariophyceae). American Journal of Botany 69:188–213.

Hoffmann, J. P. 1998. Wastewater treatment with suspended and nonsuspended algae. Journal of Phycology 34:757–763.

Hoffmann, L. 1996. Geographic distribution of freshwater bluegreen alage. Hydrobiologia 336:33–40.

Hoham, R. W. 1975. Optimum temperatures and temperature ranges for growth of snow algae. Arctic and Alpine Research 7:13–24.

Hoham, R. W., Blinn, D. W. 1979. Distribution of cryophilic algae in an arid region, the American Southwest. Phycologia 18:133–145.

Hoham, R. W., Mohn, W. W. 1985. The optimum pH of four strains of acidophilic snow algae in the genus *Chloromonas* (Chlorophyta) and possible effects of acid precipitation. Journal of Phycology 21:603–609.

Hoham, R. W., Mullet, J. E. 1977. The life history and ecology of the snow alga *Chloromonas cryophila* sp. nov. (Chlorophyta, Volvocales). Phycologia 16:53–68.

Holmes, N. T. H., Whitton, B. A. 1977. The macrophytic vegetation of the River Tees in 1975: observed and predicted changes. Freshwater Biology 7:43–60.

Holmes, N. T. H., Whitton, B. A. 1981. Phytobenthos of the River Tees and its tributaries. Freshwater Biology 11:139–163.

Hooper, C.A. 1981. Microcommunities of algae on a *Sphagnum* mat. Holarctic Ecology 4:201–207.

Hooper-Reid, N. M., Robinson, G. G. C. 1978. Seasonal dynamics of epiphytic algal growth in a marsh pond: productivity, standing crop, and community composition. Canadian Journal of Botany 56:2434–2440.

Horne, A., Goldman, C. R. 1994. Limnology, 2nd ed. McGraw–Hill, New York.

Horner, R. R., Welch, E. B., Veenstra, R. B. 1990. Responses of periphyton to changes in current velocity, suspended sediment, and phosphorus concentration. Freshwater Biology 24:215–232.

Howarth, R. W., Cole, J. J. 1985. Molybdenum availability, nitrogen limitation, and phytoplankton growth in natural waters. Science 229:653–655.

Huang, T. C., Grobbelaar, N. 1989. Isolation and characterization of endosymbiotic *Calothrix* (Cyanophyceae) in *Encephalartos hildenbrandtii* (Cycadales). Phycologia 28:464–468.

Hudon, C., Paquet, S., Jarry, V. 1996. Downstream variations of phytoplankton in the St. Lawrence River (Québec, Canada). Hydrobiologia 337:11–26.

Hufford, T. L., Collins, G. B. 1976. Distribution patterns of diatoms in Cedar Run. Ohio Journal of Science 76:172–184.

Hunt, M. E., Floyd, G. L., Stout, B. B. 1979. Soil algae in field and forest environments. Ecology 60:362–375.

Hutchinson, G. E. 1957. A treatise on limnology, Vol. 1. Geography, Physics, and chemistry. Wiley, London.

Hutchinson, G. E. 1961. The paradox of the plankton. American Naturalist 95:137–146.

Hutchinson, G. E. 1967. A Treatise on limnology, Vol. 2. Introduction to lake biology and the limnoplankton. Wiley, London.

Hutchinson, G. E. 1975. A treatise on limnology, Vol. 3. Limnological botany. Wiley, London.

Hynes, H. B. N. 1970. The ecology of running waters. University Toronto Press.

Israelson, G. 1949. On some attached Zygnemales and their significance in classifying streams. Botaniska Notiser 21:313–358.

Jackson, A. E. 1997. Physiological adaptations to freezing and UV radiation exposure in *Prasiola crispa*, an Antarctic terrestrial alga, in: Battaglia, B., Valencia, J., Walton D. W. H., Eds., Antarctic communities: Species, structure and survival. Cambridge University Press, Cambridge, UK, pp. 226–233.

Jassby, A. D., Goldman, C. R., Reuter, J. E. 1995. Long-term change in Lake Tahoe (California–Nevada, U.S.A.) and its relation to atmospheric deposition of algal nutrients. Archiv für Hydrobiologie 135:1–21.

Johansen, J. R. 1993. Cryptogamic crusts of semiarid and arid lands of North America. Journal of Phycology 29:140–147.

Johansen, J. R. 1999. Diatoms of aerial habitats, in: Stoermer, E. F., Smol, J. P., Eds., The diatoms: Applications for the environmental and earth sciences. Cambridge University Press, Cambridge, UK, pp. 264–273.

Johansen, J. R., Doucette, G. J., Barclay, W. R., Bull, J. D. 1988. The morphology and ecology of *Pleurochrysis carterae* var. *dentata* var. nov. (Prymnesiophyceae), a new coccolithophorid from an inland saline pond in New Mexico, USA. Phycologia 27:78–88.

Junk, W. J., Bayley, P. B., Sparks, R. E. 1989. The flood-pulse concept in river–floodplain systems, in: Dodge, D. P., Ed. Proceedings of the International large river symposium, Canadian Special Publication in Fisheries and Aquatic Science 106:110–127.

Kalff, J., Knoechel, R. 1978. Phytoplankton and their dynamics in oligotrophic and eutrophic lakes. Annual Review of Ecology and Systematics 9:475–495.

Kann, E. 1941. Krustensteine in Seen. Archiv für Hydrobiologie 37:504–532.

Kann, E. 1959. Die eulittorale Algenzone im Traunsee (Oberösterreich). Archiv für Hydrobiologie 55:129–192.

Kann, E. 1978. Systematik und Ökologie der Algen österreichischer Bergbäche. Archiv für Hydrobiologie Supplement 53:405–643.

Karlström, U. 1978. Role of the organic layer on stones in detrital metabolism in streams. Internationale Vereinigung für Theoretische und Angewandte Limnologie Verhandlung 20:1463–1470.

Kawecka, B. 1980. Sessile algae in European mountain streams. 1. The ecological characteristics of communities. Acta Hydrobiologica 22:361–420.

Kawecka, B. 1981. Sessile algae in European mountain streams. 2. Taxonomy and autecology. Acta Hydrobiologica 23:17–46.

Kawecka, B. 1990. The effect of flood-control regulation of a montane stream on the communities of sessile algae. Acta Hydrobiology 32:345–354.

Keough, J. R., Hagley, C. A., Ruzycki, E., Sierszen, M. 1998. $\delta^{13}C$ composition of primary producers and role of detritus in a freshwater coastal ecosystem. Limnology and Oceanography 43:734–740.

Kesler, D. H. 1981. Periphyton grazing by *Amnicola limnosa*: An enclosure–exclosure experiment. Journal of Freshwater Ecology 1:51–59.

Kilham, S. S., Kilham, P. 1978. Natural community bioassays: Predictions of results based nutrient physiology and competition. Internationale Vereinigung für Theoretische und Angewandte Limnologie Verhandlung 20:68–74.

King, J. M., Ward, C. H. 1977. Distribution of edaphic algae as related to land usage. Phycologia 16:23–30.

Kingston, J. C. 1982. Association and distribution of common diatoms in surface samples from northern Minnesota peatlands. Nova Hedwigia 73:333–346.

Kingston, J. C., Lowe, R. L., Stoermer, E. F., Ladewski, T. B. 1983. Spatial and temporal distribution of benthic diatoms in northern Lake Michigan. Ecology 64:1566–1580.

Kiss, K. T. 1994. Trophic level and eutrophication of the River Danube in Hungary. Internationale Vereinigung für Theoretische und Angewandte Limnologie Verhandlung 25:1688–1691.

Kociolek, J. P., Herbst, D. B. 1992. Taxonomy and distribution of benthic diatoms from Mono Lake, California, USA. Transactions of the American Microscopical Society 111:338–355.

Kociolek, J. P., Spaulding, S. A., Kingston, J. C. 1998. Valve morphology and systematic position of *Navicula walkeri* (Bacillariophyceae), a diatom endemic to Oregon and California (USA). Nova Hedwigia 67:235–245.

Kofoid, C. A. 1903. The plankton of the Illinois River, 1894–1899, with introductory notes upon the hydrography of the Illinois River and its basin. Part I. Quantitative investigations and general results. Bulletin of the Illinois State Laboratory of Natural History 6:95–629.

Kofoid, C. A. 1908. The plankton of the Illinois River, 1894–1899, with introductory notes upon the hydrography of the Illinois River and its basin. Part II. Constituent organisms and their seasonal distribution. Bulletin of the Illinois State Laboratory of Natural History 8:1–361.

Köhler, J. 1993. Growth, production and losses of phytoplankton in the lowland River Spree. I. Population dynamics. Journal of Plankton Research 15:335–349.

Köhler, J. 1995. Growth, production and losses of phytoplankton in the lowland River Spree: Carbon balance. Freshwater Biology 34:501–512.

Kol, E. 1968. Kryobiologie. Biologie und Limnologie des Schnees und Eises. I. Kryovegetation, in: Elster, H.-J., Ohle, W., Eds., Die Binnengewasser, Vol. 24, E. Schweizerbart'sche Verlag, Stuttgart.

Korte, V. L., Blinn, D. W. 1983. Diatom colonization on artificial substrata in pool and riffle zones studied by light and scanning electron microscopy. Journal of Phycology 19:332–341.

Krogmann, D. W., Buttala, R., Sprinkle, J. 1986. Blooms of Cyanobacteria in the Potomac River. Plant Physiology 80:667–671.

Krolikowska, J. 1997. Eutrophication processes in a shallow, macrophyte dominated lake — Species differentiation, biomass and the distribution of submerged macrophytes in Lake Luknajno (Poland). Hydrobiologia 342/343:411–416.

Lackey, J. B. 1938. The fauna and flora of surface waters polluted by acid mine drainage. U.S. Public Health Reports 53:1499–1507.

Laessle, A. M. 1961. A micro-limnological study of Jamaican bromeliads. Ecology 42:499–517.

Lair, N., Reyes-Marchant, P. 1997. The potamoplankton of the Middle Loire and the role of the "moving littoral" in downstream transfer of algae and rotifers. Hydrobiologia 356:33–52.

Lamberti, G. A., Steinman, A. D. 1997. A comparison of primary production in stream ecosystems. Journal of the North American Benthological Society 16:95–104.

Lamberti, G. A., Gregory, S. V., Ashkenas, L. R., Steinman, A. D., McIntire, C. D. 1989. Productive capacity of periphyton as a determinant of plant–herbivore interactions in streams. Ecology 70:1840–1856.

Lange, T. R., Rada, R. G. 1993. Quantitative studies on the phytoplankton in a typical navigation pool in the upper Mississippi River. Journal of the Iowa Academy of Science 100:21–7.

Larson, G. K. 1989. Geographical distribution, morphology and water quality of caldera lakes: A review. Hydrobiologia 171:24–32.

Larson, G. L., McIntire, C. D., Hurley, M., Buktenica, M. W. 1996. Temperature, water chemistry, and optical properties of Crater Lake. Journal of Lake and Reservoir Management 12:230–247.

Lee, Y.-K., Soh, C.-W. 1991. Accumulation of astaxanthin in *Haematococcus lacustris* (Chlorophyta). Journal of Phycology 27:575–577.

Leopold, L. B., Wolman, M. G., Miller, J. P. 1964. Fluvial processes in geomorphology. W. H. Freeman, San Francisco.

Lewandowski, K., Ozimek, T. 1997. Relationship of *Dreissena polymorpha* (Pall.) to various species of submerged macrophytes. Polskie Archiwum Hydrobiologii 44:457–466.

Likens, G. E., Ed. 1985. An ecosystem approach to aquatic ecology. Mirror Lake and its environment. Springer-Verlag, New York.

Likens, G. E., Bormann, F. H., Pierce, R. S., Eaton, J. S., Johnson, N. M. 1977. Biogeochemistry of a forested ecosystem. Springer-Verlag, New York.

Lin, C. K., Blum, J. L. 1977. Recent invasion of a red alga (*Bangia atropurpurea*) in Lake Michigan. Journal of the Fisheries Research Board of Canada 34:2413–2416.

Lindeman, R. L. 1941a. The developmental history of Cedar Creek Bog. American Midland Naturalist 25:101–112.

Lindeman, R. L. 1941b. Seasonal food-cycle dynamics in a senescent lake. American Midland Naturalist 25:636–673.

Lindeman, R. L. 1942. The trophic–dynamic concept aspect of ecology. Ecology 23:399–418.

Linne von Berg, K.-H., Kowallik, K. V. 1996. Biogeography of *Vaucheria* species from European freshwater/soil habitats: Implications from chloroplast genomes. Hydrobiologia 336:83–91.

Livingstone, D., Whitton, B. A. 1984. Water chemistry and phosphatase activity of the blue–green alga *Rivularia* in upper Teesdale streams. Journal of Ecology 72:405–421.

Livingstone, D. A. 1963. Alaska, Yukon, Northwest Territories, and Greenland, in: Frey, D. G., Ed., Limnology in North America. University of Wisconsin Press, Madison, pp. 559–574.

Lock, M. A., Williams, D. D., Eds. 1981. Perspectives in running water ecology. Plenum Press, New York.

Lock, M. A., Wallace, R. R., Costerton, J. W., Ventulloa, R. M., Charlton, S. E. 1984. River epilithon: Toward a structural functional model. Oikos 42:10–22.

Lodge, D. M. 1986. Selective grazing on periphyton: a determinant of freshwater gastropod microdistributions. Freshwater Biology 16:831–841.

Lodge, D. L., Kershner, M. W., Aloi, J. E., Covich, A. P. 1994. Effects of an omnivorous crayfish (*Orconectes rusticus*) on a freshwater littoral food web. Ecology 75:1265–1281.

Loeb, S. L., Reuter, J. E., Goldman, C. R. 1983. Littoral zone production of oligotrophic lakes, *in*: Wetzel, R. G., Ed., Periphyton of freshwater ecosystems. Junk, The Hague, pp. 161–7.

Losee, R. F., Wetzel, R. G. 1983. Selective light attenuation by the periphyton complex, *in*: Wetzel, R. G., Ed., Periphyton of freshwater ecosystems. The Hague, pp. 89–96.

Lowe, R. L. 1996. Periphyton patterns in lakes, *in*: Stevenson, R. J., Bothwell, M. L., Lowe, R. L., Eds., Algal ecology: Freshwater benthic ecosystems. Academic Press, San Diego, pp. 57–76.

Lowe, R. L., Pan, Y. 1996. Benthic algal communities as biological monitors, *in*: Stevenson, R. J., Bothwell, M. L., Lowe, R. L., Eds., Algal ecology: Freshwater benthic ecosystems. Academic Press, San Diego, pp. 705–739.

Lowe, R. L., Rosen, B. H., Kingston, J. C. 1982. A comparison of epiphytes on *Bangia atropurpurea* (Rhodophyta) and *Cladophora glomerata* (Chlorophyta) from northern Lake Michigan. Journal of Great Lakes Research 8:164–168.

Lund, J. W. G. 1964. Primary production and periodicity of phytoplankton. Internationale Vereinigung für Theoretische und Angewandte Limnologie Verhandlung 15:37–56.

Lund, J. W. G., Reynolds, C. S. 1982. The development and operation of large limnetic enclosures in Blelham Tarn, English Lake District, and their contribution to phytoplankton ecology. Progress in Phycological Research 1:1–65.

Makarewicz, J. C. 1993. Phytoplankton biomass and species composition in Lake Erie 1970 to 1987. Journal of Great Lakes Research 19:258–274.

Makarewicz, J. C., Baybutt, R. I. 1981. Long-term (1927–1978) changes in the phytoplankton community of Lake Michigan at Chicago. Bulletin of the Torrey Botanical Club 108:240–254.

Maltais, M.-J., Vincent, W. F. 1997. Periphyton community structure and dynamics in a subarctic lake. Canadian Journal of Botany 75:1556–1569.

Marino, R., Howarth, R. W., Shamess, J., Prepas, E. 1990. Molybdenum and sulfate as controls on the abundance of nitrogen-fixing cyanobacteria in saline lakes in Alberta. Limnology and Oceanography 35:245–259.

Marks, J. C., Lowe, R. L. 1993. Interactive effects of nutrient availability and light levels on the periphyton composition of a large oligotrophic lake. Canadian Journal of Fisheries and Aquatic Sciences 50:1270–1278.

Mataloni, G., Tell, G. 1996. Comparative analysis of the phytoplankton communities of a peat bog from Tierra del Fuego. Hydrobiologia 325:101–112.

Mayer, M. S., Likens, G. E. 1987. The importance of algae in a shaded headwater stream as food for an abundant caddisfly (Trichoptera). Journal of the North American Benthological Society 6:262–269.

McCormick, P. V., Stevenson, R. J. 1998. Periphyton as a tool for ecological assessment and management in the Florida Everglades. Journal of Phycology 34:726–733.

McIntire, C. D., Tinsley, I. J., Lowry, R. R. 1969. Fatty acids in lotic periphyton: Another measure of community structure. Journal of Phycology 5:26–32.

Meijer, M.-L., Hosper, H. 1997. Effects of biomanipulation in the large and shallow Lake Wolderwijd, The Netherlands. Hydrobiologia 342/343:335–349.

Meulemans, J. T., Heinis, F. 1983. Biomass and production of periphyton attached to dead reed stems in Lake Maarseveen, *in*: Wetzel, R. G., Ed., Periphyton of freshwater ecosystems. Junk, The Hague, pp. 169–173.

Mickle A. M., Wetzel, R. G. 1978. Effectiveness of submersed angiosperm – epiphyte complexes on exchange of nutrients and organic carbon in littoral systems. Aquatic Botany 4:303–316.

Milliman, J. D., Meade, R. H. 1983. World-wide delivery of river sediment to the oceans. Journal of Geology 91:1–21.

Minshall, G. W. 1978. Autotrophy in stream ecosystems. BioScience 28:767–771.

Minshall, G. W., Petersen, R. C., Cummins, K. W., Bott, T. L., Sedell, J. R., Cushing, C. E., Vannote, R. L. 1983. Interbiome comparison of stream ecosystem dynamics. Ecological Monographs 53:1–25.

Mitsch W. J., Gosselink, J. G. 1993. Wetlands. Van Nostrand Reinhold, New York.

Moeller, R. E., Burkholder, J. M., Wetzel, R. G. 1988. Significance of sedimentary phosphorus to a rooted submersed macrophyte (*Najas flexilis*) and its algal epiphytes. Aquatic Botany 32:261–281.

Moore, J. W. 1974. Benthic algae of southern Baffin Island. II. The epipelic communities in temporary ponds. Journal of Ecology 62:809–819.

Morisawa, M. 1968. Streams: Their dynamics and morphology. McGraw-Hill, New York.

Moss, B., Beklioglu, M., Carvalho, L., Kilinc, S., McGowan, S., Stephen, D. 1997. Vertically-challenged limnology: Contrasts between deep and shallow lakes. Hydrobiologia 342/343: 257–267.

Mosser, J. L., Mosser, A. G., Brock, T. D. 1977. Photosynthesis in the snow: The alga *Chlamydomonas nivalis* (Chlorophyceae). Journal of Phycology 13:22–27.

Müller, K. M., Sheath, R. G., Vis, M. L., Crease, T. J., Cole, K. M. 1998. Biogeography and systematics of *Bangia* (Bangiales, Rhodophyta) based on the Rubisco spacer, rbcL gene and 18s rDNA gene sequences and morphometric analyses. 1. North America. Phycologia 37:195–207.

Munawar, M., Munawar, I. F. 1996. Phytoplankton dynamics in the North American Great Lakes, Vol. 1. Lakes Ontario, Eirie and St. Clair. SPB, Amsterdam.

Munawar, M., Munawar, I. F. 2000. Phytoplankton dynamics in the North American Great Lakes, Vol. 2. Lakes Superior and Michigan, North Channel, Georgian Bay and Lake Huron. Backhuys, Leiden.

Munawar, M., Talling, J. F. 1986. Seasonality of freshwater phytoplankton: A global perspective. Junk, The Hague.

Mur, L., Bejsdorf, R. O. 1978. A model of the succession from green to blue-green algae based on light limitation. Internationale Vereinigung für Theoretische und Angewandte Limnologie Verhandlung 20:259–262.

Murkin, H. R., Stainton, M. P., Boughen, J. A., Pollard, J. B., Titman, R. D. 1991. Nutrient status of wetlands in the Interlake region of Manitoba, Canada. Wetlands 11:105–122.

National Academy of Sciences. 1987. The Mono Basin ecosystem. Effects of changing lake level. National Academy Press, Washington, DC.

Neale, P. J., Talling, J. F., Heaney, S. I., Reynolds, C. S., Lund, J. W. G. 1991. Long time series from the English Lake District: Irradiance-dependent phytoplankton dynamics during the spring maximum. Limnology and Oceanography 36:751–760.

Neely, R. K. 1994. Evidence for positive interactions between epiphytic algae and heterotrophic decomposers during the decomposition of *Typha latifolia*. Archiv für Hydrobiologie 129:443–457.

Newbold, J. D., Elwood, J. W., O'Neill, R. V., Van Winkle, W. 1981. Measuring nutrient spiralling in streams. Canadian Journal of Fisheries and Aquatic Sciences 38:860–863.

Nicholls, K. H., Beaver, J. R., Estabrook, R. H. 1982. Lakewide odours in Ontario and New Hampshire caused by *Chrysochromulina brevitturita* Nich. (Prymnesiophyceae). Hydrobiologia 96:91–95.

Northcote, T. G., Larkin, P. A. 1963. Western Canada, *in*: Frey, D. G., Ed., Limnology in North America. University of Wisconsin Press, Madison, pp. 451–485.

Nyström, P., Brönmark, C., Granéli, W. 1996. Patterns in benthic food webs: A role for omnivorous crayfish? Freshwater Biology 36:631–646.

Oliver, R. L., Ganf, G. G. 2000. Freshwater blooms, in: Whitton, B. A., Potts, M., Eds., The ecology of cyanobacteria: Their diversity in time and space. Kluwer, Dordrecht, pp. 149–194.

Olsen, Y., Vadstein, O., Andersen, T., Jensen, A. 1989. Competition between *Staurastrum luetkemuellerii* (Chlorophyceae) and *Microcystis aeruginosa* (Cyanophyceae) under varying modes of phosphate supply. Journal of Phycology 25:499–508.

Owens, J. L., Crumpton, W. G. 1995. Primary production and light dynamics in an upper Mississippi River backwater. Regulated of Rivers and Reservoir Management 11:185–192.

Pace, M. L., Findlay, S. E. G., Links, D. 1992. Zooplankton in advective environments: The Hudson River community and a comparative analysis. Canadian Journal of Fisheries and Aquatic Sciences 49:1060–1069.

Paerl, H. W., Bowles, N. D. 1987. Dilution bioassays: their application to assessments of nutrient limitation in hypereutrophic waters. Hydrobiologia 146:265–273.

Paerl, H. W., Priscu, J. C., Brawner, D. L. 1989. Immunochemical localization of nitrogenase in marine *Trichodesmium* aggregates: Relationship to N_2 fixation potential. Applied and Environmental Microbiology 55:2965–2975.

Palmer, C. M. 1962. Algae in water supplies. U.S. Public Health Service publication 657, Washington, DC.

Palmer, J. D., Round, F. E. 1965. Persistent, vertical-migration rhythms in benthic microflora. I. The effect of light and temperature on the rhythmic behavior of *Euglena obtusa*. Journal of the Marine Biology Association of the United Kingdom 45:567–582.

Pan, Y., Stevenson, R. J. 1996. Gradient analysis of diatom assemblages in western Kentucky wetlands. Journal of Phycology 32:222–232.

Pandian, T. J., Marian, M. P. 1986. An indirect procedure for the estimation of assimilation efficiency of aquatic insects. Freshwater Biology 16:93–98.

Parker, B. C., Wolfe, H. E., Howard, R. V. 1975. On the origin and history of Mountain Lake, Virginia. Southeastern Geology 16:213–226.

Parker, B. C., Wenkert, L. J., Parson, M. J. 1991. Cause of the metalimnetic oxygen maximum in Mountain Lake, Virginia. Journal of Freshwater Ecology 6:293–303.

Pentecost, A. 1982. A quantitative study of calcareous stream and tintenstriche algae from the Malham District, northern England. British Phycological Journal 17:443–456.

Pentecost, A., Whitton, B. A. 2000. Limestones, in: Whitton, B. A., Potts, M., Eds., The ecology of cyanobacteria: Their diversity in time and space. Kluwer, Dordrecht, pp. 257–279.

Pernthaler, J., Simek, K., Sattler, B., Schwarzenbacher, A., Bobkova, J., Psenner, R. 1996. Short-term changes of protozoan control on autotrophic picoplankton in an oligo-mesotrophic lake. Journal of Plankton Research 18:443–462.

Peterson, C. G. 1996. Response of benthic algal communities to natural physical disturbance, in: Stevenson, R. J., Bothwell, M. L., Lowe, R. L., Eds., Algal ecology: Freshwater benthic ecosystems. Academic Press, San Diego, pp. 375–402.

Peterson, C. G., Stevenson, R. J. 1989. Seasonality in river phytoplankton: Multivariate analyses of data from the Ohio River and six Kentucky tributaries. Hydrobiologia 182:99–114.

Pettersson, K., Herlitz, E., Istvanovics, V. 1993. The role of *Gloeotrichia echinulata* in the transfer of phosphorus from sediment to water in Lake Erken. Hydrobiologia 253:123–129.

Petts, G., Calow, P. 1996. River biota: Diversity and dynamics. Blackwell Science, Oxford, UK.

Phillips. G. L., Eminson, D., Moss, B. 1978. A mechanism to account for macrophyte decline in progressively eutrophicated freshwater. Aquatic Botany 4:103–126.

Pick, F. R., Nalewajko, C., Lean, D. R. S. 1984. The origin of a metalimnetic chrysophyte peak. Limnology and Oceanography 29:125–134.

Pierce, S. 1987. The lakes of Yellowstone. The Mountaineers, Seattle, WA.

Pinder, A. W., Friet, S. C. 1994. Oxygen transport in egg masses of the amphibians *Rana sylvatica* and *Amblystoma maculatum*: convection, diffusion and oxygen production by algae. Journal of Experimental Biology 197:17–30.

Pohlman, J. W., Iliffe, T. M., Cifuentes, L. A. 1997. A stable isotope study of organic cycling and the ecology of an anchialine cave ecosystem. Marine Ecology Progress Series 155:17–27.

Pollingher, U. 1987. Freshwater ecosystems, in: Taylor, F. J. R., Ed., The biology of dinoflagellates. Blackwell Science, Palo Alto, CA, pp. 502–529.

Porter, K. G. 1988. Phagotrophic phytoflagellates in microbial food webs. Hydrobiologia 159:89–97.

Power, M. E., Stewart, A. J. 1987. Disturbance and recovery of an algal assemblage following flooding in an Oklahoma stream. American Midland Naturalist 117:333–345.

Power, M. E., Stout, R. J., Cushing, C. E., Harper, P. P., Hauer, F. R., Matthews, W. J., Moyle, P. B., Statzner, B., Wais De Badgen, I. R. 1988. Biotic and abiotic controls in river and stream communities. Journal of the North American Benthological Society 7:456–479.

Prescott, G. W. 1962. Algae of the western Great Lakes Area, 2nd ed. WC. Brown, Dubuque, IA.

Pringle, C. M., Naiman, R. J., Bretschko, G., Karr, J. R., Oswood, M. W., Webster, J. R. Welcomme, R. L., Winterbourn, M. J. 1988. Patch dynamics in lotic systems: The stream as a mosaic. Journal of the North American Benthological Society 7:503–524.

Pringle, C. M., Rowe, G. L., Triska, F. J., Fernandez, J. F., West, J. 1993. Landscape linkages between geothermal activity and solute composition and ecological response in surface waters draining the Atlantic slope of Costa Rica. Limnology and Oceanography 38:753–774.

Prowse, G. A. 1959. Relationship between epiphytic algal species and their macrophyte hosts. Nature 183:1204–1205.

Pueschel, C. M. 1990. Cell structure, in: Cole, K. M., Sheath, R. G., Eds., Biology of the red algae. Cambridge University Press, Cambridge, UK, pp. 7–41.

Pueschel, C. M., Sullivan P. G., Titus, J. E. 1995. Occurrence of the red algae *Thorea violacea* (Batrachospermales: Thoreaceae) in the Hudson River, New York state. Rhodora 97:328–338.

Rae, R., Vincent, W. F. 1998. Phytoplankton development in subarctic lake and river ecosystems: Development of a photosynthesis–temperature–irradiance model. Journal of Plankton Research 20:1293–1312.

Rawson, D. S., Moore, A. J. 1944. The saline lakes of Saskatchewan. Canadian Journal of Research Series D 22:141–201.

Reinhard, E. G. 1931. The plankton ecology of the upper Mississippi, Minneapolis to Winona. Ecological Monographs 1:395–464.

Reuter, J. E., Loeb, S. L., Goldman, C. R. 1986. Inorganic nitrogen uptake by epilithic periphyton in a N-deficient lake. Limnology and Oceanography 31:149–160.

Reuter, J. E., Rhodes, C. L., Lebo, M. E., Klotzman, M., Goldman, C. R. 1993. The importance of nitrogen in Pyramid Lake (Nevada, USA), a saline, desert lake. Hydrobiologia 267:179–189.

Reynolds, C. S. 1984a. The ecology of freshwater phytoplankton. Cambridge University Press, Cambridge, UK.

Reynolds, C. S. 1984b. Phytoplankton periodicity: The interactions of form, function and environmental variability. Freshwater Biology 14:111–142.

Reynolds, C. S. 1988. Potamoplankton: Paradigms, paradoxes, prognoses, in: Round, F. E., Ed., Algae and the aquatic environment. Biopress, Bristol, UK, pp. 285–311.

Reynolds, C. S. 1995. River plankton: The paradigm regained, in: Harper, D. M., Ferguson, A. J. D., Eds. The ecological basis for river management. Wiley, New York, pp. 161–174.

Reynolds, C. S. 1996. Algae, in: Petts, G., Calow, P., Eds., River biota: Diversity and Dynamics. Blackwell Science, Oxford, UK, pp. 6–26.

Reynolds, C. S., Descy, J.-P. 1996. The production, biomass and structure of phytoplankton in large rivers. Archiv für Hydrobiologie Supplement 113:161–187.

Rhee, G. Y., Gotham, I. J. 1980. Optimum N:P ratios and coexistence of planktonic algae. Journal of Phycology 16:486–489.

Rindi, F., Guiry, M. D., Barbiero, R. P., Cinelli, F. 1999. The marine and terrestrial Prasiolales (Chlorophyta) of Galway City, Ireland: A morphological and ecological study. Journal of Phycology 35:469–482.

Robarts, R. D., Evans, M. S., Arts, M. T. 1992. Light, nutrients, and water temperature as determinants of phytoplankton production in two saline, prairies lakes with high sulphate concentrations. Canadian Journal of Fisheries and Aquatic Sciences 49:2281–2290.

Roberts, D. A., Boylen, C. W. 1988. Patterns of epipelic algal distribution in an acidic Adirondack lake. Journal of Phycology 24:146–152.

Roberts, D. A., Boylen, C. W. 1989. Effects of liming on the epipelic algal community of Woods Lake, New York. Canadian Journal of Fisheries and Aquatic Sciences 46:287–294.

Robinson, G. G. C. 1983. Methodology: the key to understanding periphyton, in: Wetzel, R. G., Ed., Periphyton of freshwater ecosystems. Junk, The Hague, pp. 245–251.

Robinson, G. G. C., Gurney, S. E., Goldsborough, L. G. 1997. The primary productivity of benthic and planktonic algae in a prairie wetlands under controlled water-level regimes. Wetlands 17:182–194.

Rojo, C., Cobelas, M. A., Arauzo, M. 1994. An elementary, structural analysis of river phytoplankton. Hydrobiologia 289:43–55.

Rosemarin, A. S. 1975. Comparison of primary productivity (^{14}C) per unit biomass between phytoplankton and periphyton in the Ottawa River new Ottawa, Canada. Internationale Vereinigung für Theoretische und Angewandte Limnologie Verhandlung 19:1584–1592.

Rott, E., Pfister, P. 1988. Natural epilithic algal communities in fast-flowing mountain streams and rivers and some man-induced changes. Internationale Vereinigung für Theoretische und Angewandte Limnologie Verhandlung 23:1320–1324.

Round, F. E. 1971. The growth and succession of algal populations in freshwaters. Internationale Vereinigung Limnologie Mitteilungen 19:70–99.

Round, F. E. 1972. Patterns of seasonal succession of freshwater epipelic algae. British Phycological Journal 7:213–220.

Round, F. E. 1981. The ecology of algae. Cambridge University Press, Cambridge, UK.

Rowell, H. C. 1996. Paleolimnology of Onondaga Lake: The history of anthropogenic impacts on water quality. Lake and Reservoir Management 12:35–45.

Ruse, L. P., Hutchings, A. J. 1996. Phytoplankton composition of the River Thames in relation to certain environmental variables. Archiv für Hydrobiologie Supplement 113:189–201.

Ruse, L., Love, A. 1997. Predicting phytoplankton composition in the River Thames, England. Regulated Rivers and Reservoir Management 13:171–183.

Sabater, S., Gregory, S. V., Sedell, J. R. 1998. Community dynamics and metabolism of benthic algae colonizing wood and rock substrata in a forest stream. Journal of Phycology 34:561–567.

Salonen, K., Jones, R., Arvola, L. 1984. Hypolimnetic phosphorus retrieval by diel vertical migration of lake phytoplankton. Freshwater Biology 14:431–438.

Sandgren, C. D., Ed. 1988. Growth and reproductive strategies of freshwater phytoplankton. Cambridge University Press, Cambridge, UK.

Sand-Jensen, K., Søndergaard, M. 1981. Phytoplankton and epiphyte development and their shading effect on submerged macrophytes in lakes of different nutrient status. Internationale Revue der Gesamten Hydrobiologie 66:529–552.

Sand-Jensen, K., Pedersen, O., Geertz-Hansen, O. 1997. Regulation and role of photosynthesis in the colonial ciliate Ophyridium versatile. Limnology and Oceanography 42:866–873.

Schalles, J. F., Shure, D. F. 1989. Hydrology, community structure, and productivity patterns of a dystrophic Carolina Bay wetland. Ecology Monographs 59:365–385.

Schelske, C. L., Carrick, H. J., Aldridge, F. J. 1995. Can wind-induced resuspension of meroplankton affect phytoplankton dynamics? Journal of the North American Benthological Society 14:616–630.

Schindler, D. W. 1977. Evolution of phosphorus limitation in lakes. Science 195:260–262.

Schindler, D. W. 1978. Factors regulating phytoplankton production and standing crop in the world's freshwaters. Limnology and Oceanography 23:478–486.

Schindler, D. W. 1985. The coupling of elemental cycles by organisms: Evidence from whole-lake chemical perturbations, in: Stumm, W., Ed., Chemical processes in lakes. Wiley, New York, pp. 225–250.

Schindler, D. W., Frost, V. E., Schmidt, R. V. 1973. Production of epilithiphyton in two lakes of the Experimental Lakes Area, northwestern Ontario. Journal of the Fisheries Research Board of Canada 30:1511–1524.

Schmidt, A. 1994. Main characteristics of the phytoplankton of the southern Hungarian section of the River Danube. Hydrobiologia 289:97–108.

Schumm, S. A. 1977. The fluvial system. Wiley, New York.

Schwarz, A.-M., Hawes, I. 1997. Effects of changing water clarity on characean biomass and species composition in a large oligotrophic lake. Aquatic Botany 56:169–181.

Seckbach, J. 1991. Systematic problems with Cyanidium caldarium and Galdiera sulphuraria and their implications for molecular biology studies. Journal of Phycology 27:794–796.

Sedell, J. R., Richey, J. E., Swanson, F. J. 1989. The river continuum concept: a basis for the expected ecosystem behavior of very large rivers?, in: Dodge, D. P., Ed., Proceedings of the International Large River Symposium Canadian Special Publication on Fisheries and Aquatic Science 106:49–55.

Segal, S. 1969. Ecological notes on wall vegetation. Junk, The Hague.

Serruya, C., Pollingher, U. 1983. Lakes of the warm belt. Cambridge University Press, Cambridge, UK.

Shannon, E. E., Brezonik, P.L. 1972. Limnological characteristics of north and central Florida lakes. Limnology and Oceanography 17:97–110.

Sheath, R. G. 1986. Seasonality of phytoplankton in northern tundra ponds. Hydrobiologia 138:75–83.

Sheath, R. G., Cole, K. M. 1980. Distribution and salinity adaptations of Bangia atropurpurea (Rhodophyta), a putative migrant into the Laurentian Great Lakes. Journal of Phycology 16:412–420.

Sheath, R. G., Cole, K. M. 1992. Biogeography of stream macroalgae in North America. Journal of Phycology 28:448–460.

Sheath, R. G., Hambrook, J. A. 1990. Freshwater ecology, *in*: Cole, K. M., Sheath, R. G., Eds., Biology of the red algae. Cambridge University Press, Cambridge, UK, pp. 423–453.

Sheath, R. G., Morrison, M. O. 1982. Epiphytes on *Cladophora glomerata* in the Great Lakes and St. Lawrence seaway with particular reference to the red alga *Chroodactylon ramosum* (= *Asterocytis smargdina*). Journal of Phycology 18:385–391.

Sheath, R. G., Müller, K. M. 1997. Distribution of stream macroalgae in four high arctic drainage basins. Arctic 50:355–364.

Sheath, R. G., Havas, M., Hellebust, J. A., Hutchinson, T. C. 1982. Effects of long-term natural acidification on the algal communities of tundra ponds at the Smoking Hills, N.W.T., Canada. Canadian Journal of Botany 60:58–72.

Sheath, R. G., Van Alstyne, K. L., Cole, K. M. 1985. Distribution, seasonality and reproductive phenology of *Bangia atropurpurea* (Rhodophyta) in Rhode Island, U.S.A. Journal of Phycology 21:297–303.

Sheath, R. G., Vis, M. L., Cole, K. M. 1994. Distribution and systematics of *Batrachospermum* (Batrachospermales, Rhodophyta) in North America. 6. Section *turfosa*. Journal of Phycology 30:872–884.

Sheath, R. G., Vis, M. L., Hambrook, J. A., Cole, K. M. 1996. Tundra stream macroalgae of North America: Composition, distribution and physiological adaptations. Hydrobiologia 336:67–82.

Sheldon, S. P., Wellnitz, T. A. 1998. Do bacteria mediate algal colonization in iron-enriched streams? Oikos 83:85–92.

Sieburth, J. M., Smetacek, V., Lenz, J. 1978. Pelagic ecosystem structure: Heterotrophic compartments of the plankton and their relationship to plankton size fractions. Limnology and Oceanography 23:1256–1263.

Siver, P. A. 1977. Comparison of attached diatoms communities of natural and artificial substrates. Journal of Phycology 13:402–406.

Sládecková, A. 1962. Limnological investigation methods for the periphyton ("aufwuchs") community. Botany Review 28:286–350.

Sládecková, A., Marvan, P., Vymazal, J. 1983. The utilization of periphyton in waterworks pre-treatment for nutrient removal from enriched influents, *in*: Wetzel, R. G., Ed., Periphyton of freshwater ecosystems. Junk, The Hague, pp. 299–303.

Slobodkin, L. B. 1964. Experimental populations of the hydrida. Journal of Animal Ecology Supplement 33:131–148.

Smayda, T. J. 1970. The suspension and sinking of phytoplankton in the sea. Oceanography and Marine Biology Annual Review 8:353–414.

Smith, G. M. 1950. The freshwater algae of the United States, 2nd ed., McGraw-Hill, New York.

Smith, T. E., Stevenson, R. J., Caraco, N. F., Cole, J. J. 1998. Changes in phytoplankton community structure during the zebra mussel (*Dreissena polymorpha*) invasion of the Hudson River. Journal of Plankton Research 20:1567–1579.

Soballe, D. M., Kimmel, B. L. 1987. A large-scale comparison of factors influencing phytoplankton abundance in rivers, lakes, and impoundments. Ecology 68:1943–1954.

Sommaruga, R. 1995. Microbial and classical food webs: A visit to a hypertrophic lake. FEMS Microbiology Ecology 17:257–270.

Sommer, U. 1984. The paradox of plankton: Fluctuations in phosphorus availability maintain diversity in flow-through cultures. Limnology and Oceanography 29:633–636.

Sommer, U. 1988. Growth and survival strategies of planktonic diatoms, *in*: Sandgren, C. D., Ed., Growth and reproductive strategies of freshwater phytoplankton. Cambridge University Press, Cambridge, UK, pp. 227–260.

Sommer, U., Kilham, S. S. 1985. Phytoplankton natural community experiments: A reinterpretation. Limnology and Oceanography 30:438–442.

Sparks, R. E. 1995. Need for ecosystem management of large rivers and their floodplains. BioScience 45:168–181.

Starks, T. L., Shubert, L. E., Trainor, F. R. 1981. Ecology of soil algae: A review. Phycologia 20:65–80.

Starling, M. B., Chapman, V. J., Brown, J. M. A. 1974. A contribution to the biology of *Nitella hookeri* A. Br. I. Inorganic nutritional requirements. Hydrobiologia 45:91–113.

St. Clair, L. L. Rushforth, S. R. 1976. The diatom flora of Timpanogos Cave National Monument, Utah. American Journal of Botany 63:49–59.

Stein, J. R., Amundsen, C. C. 1967. Studies on snow algae and fungi from the front range of Colorado. Canadian Journal of Botany 45:2033–2045.

Steinman, A. D. 1996. Effects of grazers on freshwater benthic algae, *in*: Stevenson, R. J., Bothwell, M. L., Lowe, R. L., Eds., Algal ecology: Freshwater benthic ecosystems. Academic Press, San Diego, pp. 341–373.

Steinman, A. D., Sheath, R. G. 1984. Morphological variability of *Eunotia pectinalis* (Bacillariophyceae) in a softwater Rhode Island stream. Journal of Phycology 20:266–276.

Steinman, A. D., McIntire, C. D., Lowry, R. R. 1988. Effects of irradiance and age on chemical constituents of algal assemblages in laboratory streams. Archiv für Hydrobiologie 114:45–61.

Steinman, A. D., Meeker, R. H., Rodusky, A. J., Davis, W. P., Hwang, S.-J. 1997. Ecological properties of charophytes in a large subtropical lake. Journal of the North American Benthological Society 16:781–793.

Steinmüller, K., Kaling, M., Zetsche, K. 1983. In-vitro synthesis of phycobiliproteins and ribulose-1,5-biphosphate carboxylase by non-poly-adenylated-RNA of *Cyanidium calderium* and *Porphyridium aerugineum*. Planta 159:308–313.

Stephens, D. W., Gillespie, D. M. 1976. Phytoplankton production in the Great Salt Lake, Utah, and a laboratory study of algal response to enrichment. Limnology and Oceanography 21:74–87.

Stevenson, R. J. 1983. Effects of current and conditions simulating autogenically changing microhabitats on benthic diatom immigration. Ecology 64:1514–1524.

Stevenson, R. J. 1996a. An introduction to algal ecology in freshwater benthic habitats, *in*: Stevenson, R. J., Bothwell, M. L., Lowe, R. L., Eds., Algal ecology: Freshwater benthic ecosystems. Academic Press, San Diego, pp. 3–30.

Stevenson, R. J., 1996b. The stimulation and drag of current, *in*: Stevenson, R. J., Bothwell, M. L., Lowe, R. L., Eds., Algal ecology: Freshwater benthic ecosystems. Academic Press, San Diego, pp. 321–340.

Stevenson, R. J. 1997. Scale-dependent determinants and consequences of benthic algal heterogeneity. Journal of the North American Benthological Society 16:248–262.

Stevenson, R. J., Stoermer, E. F. 1981. Quantitative differences between benthic algal communities along a depth gradient in Lake Michigan. Journal of Phycology 17:29–36.

Stevenson, R. J., Bothwell, M. L., Lowe, R. L., Eds. 1996. Algal ecology: Freshwater benthic ecosystems. Academic Press, San Diego.

Stockner, J. G. 1967. Observations of thermophilic algal communities in Mount Rainier and Yellowstone National Parks. Limnology and Oceanography 12:13–17.

Stockner, J. G., Armstrong, F. A. J. 1971. Periphyton studies in selected lakes in the Experimental Lakes Area, northwestern Ontario. Journal of the Fisheries Research Board of Canada 28:215–230.

Stockner J. G., Shortreed, K. S. 1988. Response of *Anabaena* and *Synechococcus* to manipulation of nitrogen : phosphorus ratios in a lake fertilization experiment. Limnology and Oceanography 33:1348–1361.

Stockner, J. G., Callieri, C., Cronberg, G. 2000. Picoplankton and other non-blooming cyanobacteria in lakes, in: Whitton, B. A., Potts, M., Eds., The ecology of cyanobacteria: Their diversity in time and space. Kluwer, Dordrecht, pp. 195–231.

Stoermer, E. F., Smol, J. P. Eds. 1999. The diatoms: Applications for the environmental and earth sciences. Cambridge University Press, Cambridge, UK.

Stokes, P. M. 1986. Ecological effects of acidification on primary producers in aquatic systems. Water, Air and Soil Pollution 30:421–438.

Strahler, A. N. 1957. Quantitative analysis of watershed geomorphology. Transactions of the American Geophysical Union 38:913–920.

Stross, R. G., Sokol, R. C., Schwarz, A.-M., Howard-Williams, C. 1995. Lake optics and depth limits photogenesis and photosynthesis in charophyte meadows. Hydrobiologia 302:11–19.

Suttle, C. A., Stockner, J. G., Harrison, P. J. 1987. Effects of nutrient pulses on community structure and cell size of a freshwater phytoplankton assemblage in culture. Canadian Journal of Fisheries and Aquatic Sciences 44:1768–1774.

Suttle, C. A., Stockner, J. G., Shortreed, K. S., Harrison, P. J. 1988. Time-courses of size-fractionated phosphate uptake: Are larger cells better competitors for pulses of phosphate than smaller cells? Oecologia 74:571–576.

Sze, P., Kingsbury J. M. 1972. Distribution of phytoplankton in a polluted saline lake, Onondaga Lake, New York. Journal of Phycology 8:25–37.

Theriot, E. C., Fritz, S. C., Gresswell, R. E. 1997. Long-term limnological data from the larger lakes of Yellowstone National Park, Wyoming, U.S.A. Arctic and Alpine Research 29:304–313.

Thomas, W. H., Duval, B. 1995. Sierra Nevada, California, U.S.A., snow algae: snow albedo changes, algal–bacterial interrelationships, and ultraviolet radiation effects. Arctic and Alpine Research 27:389–399.

Thompson, R. H., Wujek, D. E. 1997. *Trentepohliales: Cephaleuros, Phycopeltis*, and *Stomatochroon*: Morphology, taxonomy, and ecology. Science Publishers, Enfield, NH.

Thorp, J. H., Delong, M. D. 1994. The riverine productivity model: An heuristic view of carbon sources and organic processing in large river ecosystems. Oikos 70:305–308.

Thorp, J. H., Black, A. R., Haag, K. H., Wehr, J. D. 1994. Zooplankton assemblages in the Ohio River: Seasonal, tributary, and navigation dam effects. Canadian Journal of Fisheries and Aquatic Sciences 51:1634–1643.

Tilman, D. 1977. Resource competition between planktonic algae: An experimental and theoretical approach. Ecology 58:338–348.

Tilman, D. 1982. Resource competition and community structure. Princeton University Press.

Tippet, R. 1970. Artificial surfaces as a method of studying populations of benthic micro-algae in fresh water. British Phycology Journal 5:187–199.

Tuchman, M. L., Stevenson, R. J. 1980. Comparison of clay tile, sterilized rock, and natural substrate diatom communities in a small stream in southeastern Michigan, USA. Hydrobiologia 75:73–79.

Turner, M. A., Howell, E. T., Robinson, G. G. C., Campbell, P., Hecky, R. E., Schindler, E. U. 1994. Roles of nutrients in controlling growth of epilithon in oligotrophic lakes of low alkalinity. Canadian Journal of Fisheries and Aquatic Sciences 51:2784–2793.

Turner, M. A., Schindler, D. W., Findlay, D. L., Jackson, M. B., Robinson, G. G. C. 1995. Disruption of littoral algal associations by experimental lake acidification. Canadian Journal of Fisheries and Aquatic Sciences 52:2238–2250.

Turpin, D. H., Harrison, P. J. 1979. Limiting nutrient patchiness and its role in phytoplankton ecology. Journal of Experimental Marine Biology and Ecology 39:151–166.

Tyler, P. A. 1996. Endemism in freshwater algae. Hydrobiologia 336:127–135.

Umana-Villalobos, G. 1993. The planktonic community of Laguna Hule, Costa Rica. Revista de Biologia Tropical 41:499–507.

Van Den Berg, M. S., Coops, H., Noordhuis, R., van Schie, J., Simons, J. 1997. Macroinvertebrate communities in relation to submerged vegetation in two *Chara*-dominated lakes. Hydrobiologia 343/343:143–150.

Van Everdingen, R. P. 1970. The Paint Pots, Kootenay National Park, British Columbia — acid spring water with extreme heavy metal content. Canadian Journal of Earth Science 7:831–852.

Vanni, M., Layne, C. D. 1997. Nutrient recycling and herbivory as mechanisms in the "top-down" effect of fish on algae in lakes. Ecology 78:21–40.

Van Niewenhuyse, E. E., Jones, J. R. 1996. Phosphorus–chlorophyll relationships in temperate streams and its variation with stream catchment area. Canadian Journal of Fisheries and Aquatic Sciences 53:99–105.

Vannote, R. L., Minshall, G. W., Cummins, K. W., Sedell, J. R., Cushing, C. E. 1980. The river continuum concept. Canadian Journal of Fisheries and Aquatic Sciences 37:130–137.

Vazquez, G., Moreno-Casasola, P., Barrera, O. 1998. Interaction between algae and seed germination in tropical dune slack species: A facilitation process. Aquatic Botany 60:409–416.

Vinebrooke, R. D. 1996. Abiotic and biotic regulation of periphyton in recovering acidified lakes. Journal of the North American Benthological Society 15:318–331.

Vinebrooke, R. D., Hall, R. I., Leavitt, P. R., Cumming, B. F. 1998. Fossil pigments as indicators of phototrophic response to salinity and climate in lakes of western Canada. Canadian Journal of Fisheries and Aquatic Sciences 55:668–681.

Viroux, L. 1997. Zooplankton development in two large lowland rivers, the Moselle (France) and the Meuse (Belgium), in 1993. Journal of Plankton Research 19:1743–1762.

Walters, C. J., Krause, E., Neill, W. E., Northcote, T. G. 1987. Equilibrium models for seasonal dynamics of plankton biomass in four oligotrophic lakes. Canadian Journal of Fisheries and Aquatic Sciences 44:1002–1017.

Ward, A. K., Dahm, C. N., Cummins, K. W. 1985. *Nostoc* (Cyanophyta) productivity in Oregon stream ecosystems: Invertebrate influences and differences between morphological types. Journal of Phycology 21:223–227.

Ward, D. M., Castenholz, R. W. 2000. Cyanobacteria in geothermal habitats, in: Whitton, B. A., Potts, M., Eds., The ecology of cyanobacteria: Their diversity in time and space. Kluwer, Dordrecht, pp. 37–59.

Ward, J. V., Stanford, J. A. 1983. The serial discontinuity concept of lotic ecosystems, in: Fontaine, T. D., Bartells, S. M., Eds., Dynamics of lotic ecosystems. Ann Arbor Science, Ann Arbor, MI, pp. 29–42.

Warner, R. W. 1971. Distribution of biota in a stream polluted by acid mine-drainage. Ohio Journal of Science 71:202–215.

Wehr, J. D. 1981. Analysis of seasonal succession of attached algae in a mountain stream, the North Alouette River, British Columbia. Canadian Journal of Botany 59:1465–1474.

Wehr, J. D. 1989. Experimental tests of nutrient limitation in freshwater picoplankton. Applied and Environmental Microbiology 55:1605–1611.

Wehr, J. D. 1990. Predominance of picoplankton and nanoplankton in eutrophic Calder Lake. Hydrobiologia 203:35–44.

Wehr, J. D. 1991. Nutrient- and grazer-mediated effects on picoplankton and size structure in phytoplankton communities. Internationale Revue der Gesamten Hydrobiologie 76:643–656.

Wehr, J. D. 1993. Effects of experimental manipulations of light and phosphorus supply on competition among picoplankton and nanoplankton in an oligotrophic lake. Canadian Journal of Fisheries and Aquatic Sciences 50:936–945.

Wehr, J. D., Brown, L. M. 1985. Selenium requirement of a bloom-forming planktonic alga from softwater and acidified lakes. Canadian Journal of Fisheries and Aquatic Sciences 42:1783–1788.

Wehr, J. D., Descy, J.-P. 1998. Use of phytoplankton in large river management. Journal of Phycology 34:741–749.

Wehr, J. D., Stein, J. R. 1985. Studies on the autecology and biogeography of the freshwater brown alga *Heribaudiella fluviatilis* (Aresch.) Sved. Journal of Phycology 21:81–93.

Wehr, J. D., Thorp, J. H. 1997. Impacts of navigation dams, tributaries, and littoral zones on phytoplankton communities in the Ohio River. Canadian Journal of Fisheries and Aquatic Sciences 54:378–395.

Wehr, J. D., Whitton, B. A. 1983. Aquatic cryptogams of natural acid springs enriched with heavy metals: The Kootenay Paint Pots, British Columbia. Hydrobiologia 98:97–105.

Wehr, J. D., Brown, L. M., O'Grady, K. 1987. Highly specialized nitrogen metabolism in a freshwater phytoplankter, *Chrysochromulina breviturrita*. Canadian Journal of Fisheries and Aquatic Sciences 44:736–742.

Wehr, J. D., Lonergan, S. P., Thorp, J. H. 1997. Concentrations and controls of dissolved organic matter in a constricted-channel region of the Ohio River. Biogeochemistry 38:41–65.

Weisse, T. 1993. Dynamics of autotrophic picoplankton in marine and freshwater ecosystems. Microbial Ecology 13:327–370.

Welcomme, R. L., Ryder, R. A., Sedell, J. A. 1989. Dynamics of fish assemblages in river systems — A synthesis, *in*: Dodge, D. P., Ed., Proceedings of the international large river symposium, Canadian Special Publication on Fisheries and Aquatic Sciences 106:569–577.

Wetzel, R. G. 1964. A comparative study of the primary productivity of higher aquatic plants, periphyton, and phytoplankton in a large shallow lake. Internationale Revue der Gesamten Hydrobiologie 49:1–61.

Wetzel, R. G. 1975. Primary production, *in*: Whitton, B. A., Ed., River ecology. Blackwell, Oxford, UK, pp. 230–247.

Wetzel, R. G. 1983a. Limnology, 2nd ed. Saunders, Philadelphia.

Wetzel, R. G., Ed. 1983b. Periphyton of freshwater ecosystems. Junk, The Hague.

Wetzel, R. G. 1983c. Opening remarks, *in*: Wetzel, R. G., Ed., Periphyton of freshwater ecosystems. Junk, The Hague, pp. 3–4.

Wetzel, R. G. 1990. Reservoir ecosystems: conclusions and speculations, *in*: Thornton, K. W., Kimmel, B. L., Payne, F. E., Eds., Reservoir limnology: Ecological perspectives. Wiley, New York, pp. 227–238.

Wetzel, R. G., Ward, A. K. 1992. Primary production, *in*: Calow, P., Petts, G. E., Eds., Rivers handbook. Blackwell, Oxford, UK, pp. 354–369.

Whitford, L. A. 1960. The current effect and growth of freshwater algae. Transactions of the American Microscopical Society 79:302–309.

Whittaker, R. H. 1975. Communities and ecosystems. 2nd ed. MacMillan, New York.

Whitton, B. A. 1975. Algae, *in*: Whitton, B. A., Ed., River ecology. Blackwell Science, Oxford, UK, pp. 81–105.

Whitton, B. A., Ed. 1984. Ecology of European rivers. Blackwell Science, Oxford, UK.

Whitton, B. A., Potts, M., Eds. 2000. The ecology of cyanobacteria. Their diversity in time and space, Kluwer, Boston.

Williams, W. D. 1996. The largest, highest and lowest lakes of the world: saline lakes. Internationale Vereinigung für Theoretische und Angewandte Limnologie Verhandlungen 26:61–79.

Wilson, S. E., Cumming, B. F., Smol, J. P. 1996. Assessing the reliability of salinity inference models from diatom assemblages: An examination of a 219-lake data set from western North America. Canadian Journal of Fisheries and Aquatic Sciences 53:1580–1515.

Winner, J. M. 1975. Zooplankton, *in*: Whitton, B. A., Ed., River ecology. Blackwell, Oxford, UK, pp. 155–169.

Woelkerling, W. J. 1976. Wisconsin desmids. I. Aufwuchs and plankton communities of selected acid bogs, alkaline bogs, and closed bogs. Hydrobiologia 48:209–232.

Wood, R. D. 1950. Stability and zonation of Characeae. Ecology 31:641–647.

Wootton, J. T., Parker, M. S., Power, M. E. 1996. Effects of disturbance on river food webs. Science 273:1558–1561.

Wurtsbaugh, W. A., Berry, T. S. 1990. Cascading effects of decreased salinity on the plankton, chemistry, and physics of the Great Salt Lake (Utah). Canadian Journal of Fisheries and Aquatic Sciences 47:100–109.

Yong, Y. Y. R., Lee, Y.-K. 1991. Do carotenoids play a photoprotective role in the cytoplasm of *Haematococcus lacustris* (Chlorophyta)? Phycologia 30:257–261.

Yung, Y.-K. Stokes, P., Gorham, E. 1986. Algae of selected continental and maritime bogs in North America. Canadian Journal of Botany 64:1825–1833.

Zacharias, O. 1898. Das Potamoplankton. Zoologischer Anzeiger 21:41–48.

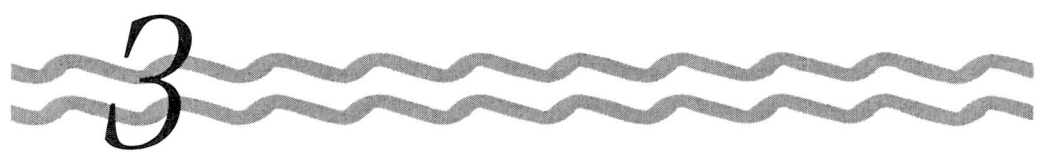

COCCOID AND COLONIAL CYANOBACTERIA

Jiří Komárek

Institute of Botany
Academy of Sciences of the Czech Republic
Faculty of Biological Sciences
University of South Bohemia
CZ-37982
Třeboň, Czech Republic

I. Introduction
II. Morphology and Diversity
    A. Cellular and Colonial Morphology
    B. Cytology
    C. Cell Division and Reproduction
    D. Diversity and Classification
III. Ecology and Distribution
    A. Soils
    B. Subaerial Environments
    C. Aquatic Environments
    D. Extreme Environments
    E. Geographic Distribution
IV. Collection, Preparation, and Culture
V. Key and Descriptions of Genera
    A. Key
    B. Descriptions of Genera
VI. Guide to Literature for Species Identification
Literature Cited

I. INTRODUCTION

Cyanobacteria (cyanoprokaryotes, cyanophytes, and blue–green algae) represent an ancient, but diverse and abundant group of microorganisms that possess prokaryotic (bacterial) cell structure and predominantly CO_2-dependent oxygen-evolving photosynthesis. Unlike all other algae, cyanobacteria are classified within the Eubacteria due to their simple, prokaryotic cells and gram-negative (peptidoglycan) cell walls (Stanier and Cohen-Bazire, 1977). Cells contain chlorophyll-*a* and several phycobilin–protein complexes that produce a variety of pigmentations, including the characteristic blue–green color. Many species possess nitrogenase and fix atmospheric N_2; several are capable of precipitating calcium carbonate, and they form stromatolites and travertine deposits (Fogg et al., 1973; Carr and Whitton, 1973, 1982; Whitton and Carr, 1982). Therefore, they belong to very important photosynthetic microscopic organisms in natural environments. Cyanobacteria are thought to have evolved during the early Precambrian period, and during their long existence, they have colonized nearly all freshwater, marine, and terrestrial habitats on Earth. Many cyanobacteria occur in extreme habitats, such as hot springs (up to 70°C), hypersaline lakes, and hot and cold deserts. Some planktonic[1] species form massive surface blooms, have the ability to produce toxins, and play an important role in water quality (Reynolds and Walsby, 1975).

Cyanobacteria occur in a wide variety of morphologies and ecological forms. Hence, historically their classification has been based on simple morphological characteristics (e.g., Gomont, 1892; Geitler, 1932; Desikachary, 1959), whereas many bacteriologists have applied physiological and biochemical properties for those species that exist in culture (e.g., Castenholz and Waterbury, 1989; Castenholz, 2001). During the last four decades, ecological characteristics, ultrastruc-

---

[1] The adjective "planktonic," derived from the Greek term "plankton," is incorrect according to ancient and modern *Greek grammars* (it should be "planktic," similarly to "benthic," "metaphytic," etc.), but it is the common term in English speaking countries.

tural features, and molecular evidence have substantially influenced our knowledge of this group. As a result, species concepts and classification are undergoing radical changes (Anagnostidis and Komárek, 1985; Castenholz, 1992; Komárek, 1994; Castenholz, 2001). However, the cellular–morphological approach to classification (including ecospecies) cannot yet satisfactorily be replaced by any other criteria. The molecular and ultrastructural studies explain relationships well, but thus far encompass only a small portion of the cyanobacterial diversity and variability that occur in natural habitats. Many well-known taxa (e.g., from the genera *Gloeocapsa*, *Gomphosphaeria*, *Woronichinia*, *Chroococcus*, *Hyella*, and many others) have not yet been studied in culture, which is a first step to experimental revisions of the taxonomic status of species (Rippka et al., 1979; Komárek, 1994; Castenholz, 2001). There are also difficulties in identifying cyanobacteria from different geographic areas. Misinterpretations result with both field populations and isolated strains, due to the fact that so few collections and identifications have been made from certain habitats, especially tropical and subarctic regions, and from extreme environments (Frémy, 1930b; DiCastri and Younéz, 1994; Watanabe, 1999). Changes in the classification and identification of cyanobacteria can be expected in the future, but presently the traditional approach is still necessary, especially for field studies.

The distribution and taxonomy of cyanobacteria in North America has been studied by a number of phycologists over many decades (e.g., Wood, 1872; Tilden, 1910; Smith, 1920, 1925, 1933, 1950; Gardner, 1927; Copeland, 1936; Drouet, 1936a, b, 1938, 1942; Daily, 1942, 1946; Drouet and Daily, 1948, 1952; Prescott, 1962; Golubić, 1965, 1967a, b, 1980; Friedmann, 1971; Croasdale, 1973; Komárek and Anagnostidis, 1995). However, recent studies of natural populations are few and much needed to understand the level of cyanobacterial diversity in North America. Specialists are interested mainly in revisions and characterization of individual species and genera (Jaag, 1941; Komárek, 1970; Parker, 1982; Lukas and Golubić, 1983; Wilmotte and Stam, 1984; Komárek and Hindák, 1989; Gold-Morgan et al., 1994), or cyanobacterial diversity in particular habitats, some of which may employ molecular data (Komárek, 1999). Based on these historical and present-day studies, plus some preliminary studies by me and co-workers (Komárek and Montejano, 1994; Tavera and Komárek, 1996; Komárek and Komárková-Legnerová, 2002) on habitats in central America (e.g., volcanic regions, marine and brackish waters, subaerial habitats from wet and desert rocks, and soil communities), it is estimated that no more than 10% of all natural forms of cyanobacteria (species and varieties) from North America have been described. Chapters 3 and 4 represent an effort to update the general knowledge of the North American cyanobacterial flora and to stimulate greater study of these organisms across the continent.

Cyanobacteria have been traditionally classified into several orders. This chapter concerns coccoid unicelled and colonial cyanobacteria [order Chroococcales Wettstein 1924; see Geitler (1932, 1942), including Pleurocapsales Geitler 1925; for details see Geitler (1932), Waterbury and Stanier (1978), and Komárek and Anagnostidis (1986, 1998)]. This is a heterogeneous group, as concerns ultrastructural and molecular criteria, and will be most likely reclassified along several distinct evolutionary lines. However, the data available do not yet allow a full revision and therefore all coccoid cyanobacteria are considered and reviewed as one group in this chapter.

II. MORPHOLOGY AND DIVERSITY

A. Cellular and Colonial Morphology

The simplest cyanobacteria grow as solitary cells, which may be enveloped by a thin, diffuse or firm gelatinous layer (sheath), or may be clustered into colonies; very thin sheaths may not be apparent in the light microscope (Drews and Weckesser, 1982). Cell size varies from less than 1 µm in diameter (picoplankton; Platt and Li, 1986; Komárek, 1999; Šmajs and Šmarda, 1999) to cells greater than 20 µm in length. Solitary cells and/or specialized cells in a cluster or colony may reach a diameter up to 100 µm (Geitler, 1932; Komárek, 1976). Some colonial forms (e.g., species of *Microcystis*) produce massive accumulations in surface blooms that are recognizable with the naked eye. Cells vary from spherical, oval, fusiform, and rod-like, to irregular in shape; in some colonial genera, the shape may appear polygonal or elongate. An important evolutionary trend within chroococcal cyanobacteria is the polarization of cells and thalli. For example, in some forms the cells are heteropolar, morphologically differentiated in basal and apical parts, and possess specific patterns of cell division (Chamaesiphonaceae and Dermocarpellaceae; Figs. 16B–18). A heteropolar thallus occurs in more complicated forms, where a collection of heterogeneous cells forms a colony (Xenococcaceae and Hyellaceae; Figs. 20–23). Both amorphous and polarized colonies occur in this group; each develops from solitary cells and forms small clusters, which may grow into distinct colonies or macroscopic mats. Mucilaginous colonies of simple genera can be spherical, oval, or irregular. The structure of sheaths and envelopes around cells is diverse: the mucilage can form hyaline, amorphous mass, diffused and marginal envelopes, or structured and layered sheaths. Gelatinous envelopes (fine mucilage) and sheaths (firm or structured external layer) appear to have a protective function in habitats that are exposed to intense solar radiation; often they are colored by carotenoid pigments (scytonemin and gloeocapsin; Garcia-Pichel and Castenholz, 1991; Whitton, 1992).

B. Cytology

The prokaryotic structure of cyanobacterial cells corresponds with other members of the Eubacteria. The distribution of DNA appears fairly uniform by electron microscopy (EM), but is actually concentrated in the central region, forming nucleoids that have shapes that are diagnostic among different genera (Cepák, 1993). The position and pattern of photosynthetic thylakoids seen with EM differ among various genera. In numerous species, three to six or more thylakoids are peripherally arranged in parallel and cause a dark peripheral region within cells that is recognizable with the light microscope (LM), and is sometimes termed chromatoplasm in older literature. The internal part without thylakoids is called the centroplasm (or nucleoplasm; Geitler, 1932, 1942). Genera with different thylakoid patterns probably belong to different evolutionary lines. Regularly widened thylakoids have been observed in several species (Komárek et al., 1975; Roussomoustakaki and Anagnostidis, 1991). Their occurrence and intensity depend on light intensities, but the function and metabolic processes connected with this variability are still unclear. Under the LM, cells appear to be pale or bright blue–green, olive green, greyish, pinkish, or violet, depending on the ratio of photosynthetic pigments, particularly chlorophyll-a and phycobilins, allophycocyanin, phycocyanins (blue), and phycoerythrin (red). Phycobilins are localized in phycobilisomes and their ratio can change in several species and within the same population (called photoacclimation or chromatic adaptation in older literature). In species or strains capable of photoacclimation, green light stimulates synthesis of phycoerythrin, which more efficiently captures these wavelengths than chlorophyll-a; conversely, increased red light stimulates phycocyanin synthesis (Cohen-Bazire and Bryant, 1982; Bryant, 1994). Different carotenoids have been found in several species from individual genera.

Granules of various types occur throughout most cells. Phycobilisomes associated with thylakoids contain both phycobilins (Cohen-Bazire and Bryant, 1982). Assimilatory materials, polyphosphates, and carboxysomes (site of ribulose bisphosphate carboxylase/oxygenase RUBISCO) are the most common inclusions (Healy, 1982; Smith, 1982; Whitton, 1992). The character of most granular inclusions can be confirmed only with EM, although polyphosphate bodies can be recognized under the LM and their production can be induced under a variety of nutritional conditions (Sinclair and Whitton, 1977; Lawry and Simon, 1982). Physiological and evolutionary implications of various granules in cyanobacteria were reviewed by Jensen (1984, 1985). When observed with EM, ribosomes are also recognizable.

Important intracellular inclusions, which are distinctive in many cyanobacteria, are gas vesicles, which are present in many planktonic species (e.g., *Microcystis* and several species of *Woronichinia*); their presence may be characteristic for certain genera and species. The gas vesicles (Walsby 1972, 1978) are gathered in clusters (hundreds or thousands per cell) and are visible in vegetative cells as dark brownish, irregular bodies of various sizes (aerotopes; the older term is gas vacuoles). When aerotopes are abundant, they may render the observed color of vegetative cells as brownish (due to refraction of light), which can be mistaken as the actual pigmentation. In such cases, slight or sometimes substantial pressure applied to the specimens or samples will collapse the vesicles to reveal the actual pigmentation (Walsby, 1972). This process is reversible and dependent on environmental conditions, including nutrients, light, temperature, and atmospheric pressure (Walsby, 1972, 1981; Fay, 1983).

Cyanobacterial walls are mainly proteinous, with a peptidoglycan layer (electron opaque) that overlays the cytoplasmic membrane; an external sheath, as well as lipoproteins and lipopolysaccharides also may be present (Drews and Weckesser, 1982). This basic structure of cell walls is largely uniform in cyanoprokaryotes. A facultative crystalline S-layer is unique in a few unicellular and simple filamentous types; its structure differs among genera (Šmarda, 1991; Šmarda et al., 1996, 2002; Komárek, 1999). The cell surface is smooth and without sculpturing. Almost all cyanobacteria produce mucilaginous compounds that form variably structured slimy masses around cells. The form of characteristic colonies is dependent on these mucilaginous formations, the genetic stability of which is not known well. Gelatinous envelopes may be colored by various pigments (red, blue, yellow, or brown) and may be characteristic for species separation. However, the intensity and character depend on environmental conditions, particularly light and pH (Jaag, 1945). Details on the development and ultrastructure of cyanobacterial walls were reviewed by Drews and Weckesser (1982).

C. Cell Division and Reproduction

Cells divide mainly by simple binary fission, in which the cell wall projects into the protoplast and the cell splits into two isomorphic or (less frequently) asymmetrical daughter cells. This process proceeds by one of two methods (Drews and Weckesser, 1982). The simplest (in gram-negative bacteria) is the constrictive type via simultaneous invagination (pinching) of all cell layers of the cell wall. In the other type, a septum forms from invagination of the cytoplasmic membrane and the peptidoglycan layer (septum type) with later invagination of the outer layer (cleavage). In general, cell division may occur in one, two, or three planes, which are more or less perpendicular to one another in successive generations. This process is regularly repeated and characteristic of different families. A special type of this cell division is a facultative asymmetrical binary fission in simple genera under suboptimal conditions, as seen in some members of the Synechococcaceae or simple Chamaesiphonaceae (Waterbury and Stanier, 1977; Rippka et al., 1979; Komárek and Anagnostidis, 1986; Figs. 4Ab and 17Ba), or regularly asymmetrical in the upper part of polarized cells (*Stichosiphon*; Montejano et al., 1997). Daughter cells usually grow to the original size (in several genera also to the original shape) before the next division. In more diversified genera, cell division is irregular, resulting in packet-like colonies (e.g., *Gloeocapsopsis* and *Cyanosarcina*), or division in one direction prevails in genera with polarized thalli (e.g., *Chlorogloea*, *Entophysalis*, and *Hydrococcus*) and pseudofilamentous forms (e.g., *Pleurocapsa*, *Hyella*).

A more complicated type of cell division is an asymmetrical cell differentiation, in which the mother cell divides in rapid sequence in the upper part of heteropolar cells (*Chamaesiphon* and *Chamaecalyx*). Daughter cells of such distinctly and obligately polarized cells are called exocytes (exospores in early studies), which separate singly and/or successively in rows, and rarely in three-dimensional clusters from the apical cell end (Chamaesiphonaceae; Figs. 16B–18A). Exocytes have the capability for repeated divisions in a few genera (*Stichococcus*; Montejano et al., 1997). Another specialized type of cell division is multiple fission, in which the whole cell or a part of it (apical) divides in rapid sequence or almost simultaneously into numerous small daughter cells, which liberate from gelatinized or split envelopes and contribute to the reproduction. If the daughter cells develop from cells localized in fine, diffuse mucilage, they are called nanocytes (Komárek and Anagnostidis, 1998; Fig. 19B). Other daughter cells, termed baeocytes, originate from mother cells enclosed in firm sheaths (Waterbury and Stanier, 1978; Rippka et al., 1979; Figs. 18B, 19A, and 20). Both cell division patterns, binary and multiple fission, may be combined in several genera, where the cell division differs during growth of colonies, and nanocytes/baeocytes predominantly contribute to the reproduction (Komárek and Anagnostidis, 1998, Table 4).

In some genera, morphologically different developmental stages appear that seem to depend on environmental conditions (Microcystaceae, Chroococcaceae, and Entophysalidaceae; Jaag, 1945; Komárek, 1993). Several developmental (and dormant) stages of complicated forms are easily confused with different taxa (e.g., certain species of *Asterocapsa*, *Gloeocapsa*, and *Gloeocapsopsis*; Figs. 12A, and 13A and B). In morphologically complex families, especially in Hyellaceae, the thallus develops from initial cells into morphologically diverse cells in different parts of the thallus (Waterbury and Stanier, 1978; Lukas and Golubić, 1981; Golubić et al., 1985; LeCampion-Alsumard and Golubić, 1985).

Reproduction in coccoid cyanobacteria proceeds by ordinary cell division in the unicellular types, by solitary cells or their small clusters liberated from colonies, by fragmentation of thalli, or by the production of exocytes, nanocytes, and baeocytes. Sexual processes have never been observed. Motility has been recognized in baeocytes and in solitary cells of several genera (Waterbury and Stanier, 1978; Waterbury and Rippka, 1989). This motility is probably facultative and its mechanism is not yet satisfactorily explained. No motility organelles have been found in any cyanobacteria.

D. Diversity and Classification

The taxonomy and classification of cyanobacteria has been under study since about the middle of the 19th century using morphological and cellular criteria, similar to other microalgae (e.g., Kützing, 1849; Nägeli, 1849). However, more modern approaches in the last four decades have emphasized important structural and molecular characteristics of these bacterial organisms with plantlike metabolism (Stanier and Cohen-Bazire, 1977; Rippka et al., 1979). Diversification through ecological acclimation, adaptation and stabilization of diverse morpho- and ecotypes, as well as changes caused by mutation and possibly also genome transfers (Rudi et al., 1998), give rise to new cyanobacterial types in many habitats.

Knowledge of the full diversity and evolutionary hierarchy of cyanobacterial taxonomic entities is still unclear. Thus far, DNA-based genotypes more or less correspond to traditional (morphologically based) genera, or to phenotypically distinct and definable clusters of species within known genera (Castenholz, 1992). However, the infrageneric diversity is heterogeneous, and stable traditional species that are morphologically and ecologically well defined (e.g., *Aphanothece*, *Chroococcus*, *Chamaesiphon*, and *Hyella*) occur repeatedly in similar but distant habitats and over extended time intervals (Smith, 1950; Prescott, 1962; Whitford and Schumacher, 1969; Bourrelly, 1985). In contrast, variable taxa also exist and occur in numerous morphological forms, some of which are described as separate taxonomic species (e.g., *Microcystis* and *Gloeocapsa*). Nearly all populations of cyanobacteria from different locations differ to some degree from each other, and these deviations may stabilize in cultures (Kondrateva, 1968; Waterbury and Rippka, 1989; Kato and Watanabe, 1993). This process indicates that new forms continually develop and are stabilized under new constant conditions.

Diversification within the cyanobacteria is a continuing process in which new types develop from continually modified cyanobacterial genotypes under various environmental situations. Recent data also reveal variation within cyanobacterial genomes, such as changes in the toxicity of certain strains (Neilan *et al.*, 1997) and possible interchange of genetic material (e.g., Rudi *et al.*, 1998). Other well-adapted forms, so-called traditional species, appear to persist over the long term and contain a wide spectrum of stable types for long periods (Komárek and Anagnostidis, 1998). Therefore, it is believed that a substantially greater diversity of cyanobacteria exists in nature than has been recognized up until now, making any current synopsis a low estimate, even with traditionally defined taxa. Arbitrary use of taxonomic names in strain designations, and in ecological and floristic studies also complicates this estimate. The present account of cyanobacteria (Chaps. 3 and 4) recognizes 53 coccoid genera (in 9 families) and 71 filamentous genera (in 16 families) from North America. Future studies will undoubtedly reveal more, especially as classification of known entities into genera, species, and other taxonomic or nontaxonomic units (e.g., strains) receives further research.

III. ECOLOGY AND DISTRIBUTION

Cyanobacteria colonize nearly all habitats on Earth and have attracted the attention of ecologists for centuries. They can occur in great masses, and are important in many aquatic and terrestrial communities for their substantial biomass and primary production, N_2 fixation, production of toxic compounds, creation of stromatolites, boring in limestone substrates, and importance in symbioses. Coccoid cyanobacteria contribute significantly to almost all the metabolic activities mentioned, as well as toward global carbon and oxygen budgets. The present review considers both coccoid and filamentous forms. For further information on the habitats and ecology of cyanobacteria, consult Fogg *et al.* (1973), Carr and Whitton (1973, 1982), Fay and VanBaalen (1987), Mann and Carr (1992), and Whitton and Potts (2000). Available data indicate at least some ecological specialization in all cyanobacterial taxa; no species is known to be ubiquitous across all or most freshwater environments.

A. Soils

Species of cyanobacteria are widely distributed in nearly all types of soil. The species present may be specialized to particular habitats, but may still be quite broadly distributed geographically. Examples include species from the genera *Synechococcus*, *Aphanothece*, and the filamentous genera *Leptolyngbya*, *Pseudophormidium*, *Phormidium*, *Mastigocoleus*, *Nostoc*, and *Schizothrix* (Starks *et al.*, 1981). Differences among plant communities, soil types, and local physical–chemical conditions affect the composition and abundance of algal taxa in soils. The most commonly reported factor that favors cyanobacteria in soils appears to be higher pH (Fairchild and Wilson, 1967; MacEntree *et al.*, 1972; King and Ward, 1977). Some studies suggest that cyanobacteria are important in the early stages of primary succession processes in soils (Starks *et al.*, 1981; Lukešová, 1993), especially because many species are N_2 fixers, aid in stabilizing soils against erosion, or facilitate production of organic matter in new soils (Bailey *et al.*, 1973; Jürgensen, 1973; Kubečkova *et al.*, in press). Their importance appears to be greater in regions with sparse vegetation, such as deserts, semi-deserts, and polar regions (Cameron, 1963; Cameron *et al.*, 1965; Forest and Weston, 1966), although more recent data suggest their abundance and productivity can also be substantial in agricultural and forest soils (Hunt *et al.*, 1979; Shimmel and Darley, 1985). The majority of soil cyanobacteria are filamentous species, but several coccoid forms (particularly from the genus *Synechococcus*) are also common. Cyanobacterial assemblages are recognizable with the naked eye on wet soils, where they develop in sufficient quantity to form mats on the soil surface, or as crusts (microzones) under the surface, especially in sandy and periodically flooded soils (e.g., Eskew and Ting, 1978; Garcia-Pichel and Belnap, 1996), as well as on metal-contaminated soils (e.g., Maxwell, 1991).

B. Subaerial Environments

Cyanobacteria are among the most important autotrophs that colonize periodically or continually moist soils, rocks, and walls. They are found together with lichens and mosses, which sometimes represent a later successional stages (Jaag, 1945; Golubić, 1967a). Some cyanobacteria grow only under conditions of high humidity (aerophytic), in sites with periodic water supply (subaerial), or sites with a continual supply of water vapor (atmophytic). Their development depends on the periodicity of moisture, chemical composition of the substratum, intensity of illumination, temperature, and so forth. The composition of tropical algal communities of this kind is also distinctly different from those in temperate to polar zones (personal observation). Often, assemblages form dark stripes on rocky walls, in dripping water, and near waterfalls in mountainous areas. Tree bark may also serve as a habitat for cyanobacteria, especially in humid tropical forests (Frémy, 1930a, b; Desikachary, 1959). Our recent investigations indicate that specialized types of cyanobacteria occur in these habitats, but their taxonomy is still almost unknown.

Various species from the genera *Gloeocapsa, Gloeothece, Aphanocapsa, Aphanothece, Gloeocapsopsis, Asterocapsa, Chroococcus,* and *Chroococcidiopsis* are common components of communities on wet rocks (Jaag, 1945; Komárek, 1993). The morphological variability and importance of various (possibly dormant) stages substantially complicate their taxonomic position and identification. Distinct differences in taxa from various geographically distant regions remain unclear, due to both lack of study and many taxonomic misinterpretations.

C. Aquatic Environments

1. Standing Waters

A wide spectrum of cyanobacterial types occur in planktonic, benthic, metaphytic, and periphytic habitats, and the literature on phytoplankton, including cyanobacteria, based on these habitats is enormous (Bourrelly, 1966; Baker and Baker, 1979; Gibson and Smith, 1982; Whitton and Potts, 2000) and beyond the scope of this review (but see Chapters 2 and 24). Aquatic species are the best known examples of cyanobacteria, particularly as a result of their abundance and diversity in different lakes and reservoirs. Their importance typically increases with eutrophication (Pick and Lean, 1987), with accompanying surface blooms and massive planktonic biomass (Reynolds and Walsby, 1975; Paerl, 1996). Several species have worldwide distributions (e.g., from the genera *Anabaena* and *Nodularia*); others have distinctly limited distributions (several species of *Aphanizomenon, Cylindrospermopsis,* and others). A long list of specialized bloom-forming nanoplanktonic and picoplanktonic species are known. Their abundance depends on a variety of ecological factors, particularly nutrient status, illumination, turbulence, and temperature. Distinct differences in environmental requirements exist between species; some occur in strictly saline habitats or oligotrophic freshwater habitats. Several species are characteristic of particular regions (e.g., temperate *Gomphosphaeria aponina,* tropical *G. multiplex,* and subpolar *G. natans*), but more species are known to be cosmopolitan. Some species (e.g., *Aphanothece stagnina,* or several picoplanktonic or water-bloom forming species) may have a specialized developmental cycle, where initial growth is in the benthos and they later become planktonic (Barbiero and Welch, 1992). Attached species may be restricted to one type of the substratum, such as limestone, submerged plants, mosses, or filamentous algae (e.g., different species of the genus *Chamaesiphon*). The following brief overview summarizes a few of the differences in cyanobacterial assemblages from the major habitats in which they are commonly found; other ecological details are discussed in Chapter 2.

2. Bloom-Forming Species

Cyanobacteria that contain gas vesicles in their cells are well adapted to the planktonic habit and commonly occur in moderately productive to hypertrophic lakes throughout temperate and tropical regions. Their importance as biomass and toxin producers has been recognized for many years (Gorham *et al.,* 1964; Paerl and Millie, 1996). Other cyanobacteria are responsible for taste and odor problems in surface waters (Jüttner, 1987). Specialized bloom-forming taxa also occur in coastal brackish and marine (also oceanic) waters (species of *Nodularia* and *Trichodesmium*). Of the coccoid forms, the genus *Microcystis* is one of the most widely reported (although taxonomically difficult), and typically forms blooms in the spring in temperate zones and may persist throughout the summer in eutrophic lakes (Reynolds *et al.,* 1981; Doers and Park, 1988). Several species occur in tropical regions over the entire year. These species establish massive surface blooms ("hyperscums") under eutrophic conditions, low turbulence (calm winds), and high irradiance (Zohary and Breen, 1989). *Microcystis* blooms can represent more than 50% of the total algal biomass in some lakes (Fallon and Brock, 1981; Zohary and Robarts, 1990), and some strains produce hepatotoxins (Carmichael, 1992, 1997; Codd, 1995; Chorus and Bartram, 1999) that can inhibit herbivorous zooplankton (Fulton and Paerl, 1987; Hietala *et al.,* 1997). Other planktonic coccoid genera contribute to water blooms, such as species of *Coelosphaerium, Snowella,* and *Woronichinia* (Komárek and Komárková-Legnerová,

1992). There is a large literature on bloom-forming cyanobacteria (reviewed in Reynolds and Walsby, 1975; Paerl, 1996), but in many individual studies, the characteristics of species and populations are often incompletely documented. Further details concerning filamentous forms are discussed in Chapter 4.

3. Picoplanktonic and Nanoplanktonic Species

Many other species of coccoid cyanobacteria that lack gas vesicles are well adapted to a planktonic mode of life. Several live as solitary cells and are very small (ca. 1–2 μm diameter); these members of the picoplankton are important or even dominant primary producers in oligotrophic waters (Fahnenstiel et al., 1986; Fogg, 1986; Stockner and Shortreed, 1991; Burns and Stockner, 1991; Corpe and Jensen, 1992) and occasionally in more productive systems (Platt and Li, 1986; Wehr, 1990; Weisse, 1993; Komárková, 2001). Their importance is typically greater in large oligotrophic or mesotrophic lakes and the open ocean, and at moderate or greater depths (Eguchi et al., 1996; Waterbury, 1979; Komárková and Šimek, 2003). Some of these tiny species are superior competitors for P (Suttle and Harrison, 1988; Wehr, 1989), and some populations may also be adapted to low light regimes or altered spectra (Hauschild et al., 1991; Wehr, 1992; Callieri et al., 1995). Picoplanktonic (unicellular) species have most commonly been assigned to the genus *Synechococcus* (some of them perhaps erroneously), but the genera *Cyanobium* and *Synechocystis* may be equally or more important (Stockner, 1988; Albertano and Capucci, 1997; Šmajs and Šmarda, 1999; Komárek, 1999). Other coccoid species form microscopic mucilaginous colonies that passively float in the water (nanoplankton). Nanoplanktonic species are common in mesotrophic and eutrophic waters, but rarely produce a substantial biomass. Common coccoid nanoplanktonic species include the genera *Chroococcus*, *Cyanodictyon*, *Aphanothece*, *Rhabdoderma*, *Aphanocapsa*, *Merismopedia*, *Coelosphaerium*, *Coelomoron*, *Snowella*, and *Gomphosphaeria* (Komárek and Komárková-Legnerová, 1992; Komárek and Anagnostidis, 1998). A few species are more common in waters with higher salinity (*Pannus spumosus* and several *Aphanothece* species), and appear to be halotolerant by means of osmoregulatory compounds (Reed et al., 1986).

4. Benthic and Epiphytic Species

Benthic cyanobacteria occur widely in lentic and lotic ecosystems, in epipelic, epipsammic, and epilithic habits or are epiphytic on filamentous algae, mosses, and vascular plants. Benthic mats develop on sediments in standing waters and later may form floating mats (*Aphanothece*, *Oscillatoria*, and *Phormidium*). Many are specialized with respect to substratum and habitat (see Chap. 2). For example, some benthic cyanobacteria colonize specific types of rock (limestone, sandstone, granite). Characteristic communities with a dominant cyanobacterial flora develop in mountain streams, particularly in limestone areas, where mainly filamentous species form characteristic microzonation on stones (Geitler, 1932). Similar communities occur in splash zones of larger lakes, and include the coccoid genera *Chamaesiphon* subg. *Godlewskia*, *Chlorogloea*, and *Hydrococcus*, plus filamentous species of *Homoeothrix*, *Schizothrix*, *Phormidium*, *Calothrix*, and *Rivularia* (Golubić, 1967a). Many epilithic species form distinctly colored macroscopic patches on stones, particularly species of *Chamaesiphon* (Geitler, 1932; Kann, 1973). Epiphytic communities typically include other species of cyanobacteria, such as members of the coccoid genera *Stichosiphon*, *Chamaesiphon*, *Chamaecalyx*, *Cyanocystis*, and *Xenococcus*, plus many filamentous species in the genera *Leptolyngbya*, *Heteroleibleinia*, *Hapalosiphon*, *Cylindrospermum*, *Trichormus*, and *Oscillatoria*, among others.

5. Transported Plankton in Rivers and Streams

A number of cyanobacterial species occur in the potamoplankton of large streams and rivers, although their relative importance (and diversity) appears to be less than centric diatoms or coccoid green algae (Wehr and Descy, 1998). Among those commonly reported are species of *Synechococcus*, *Chroococcus*, *Pseudanabaena*, *Planktothrix*, and *Oscillatoria* (Reynolds and Descy, 1996). Current data suggest that, in general, their importance is greater in slower flowing and warm–temperate to tropical rivers (especially those with backwaters), where longer residence times permit their slower growth rates to attain greater population sizes (Reynolds and Descy, 1996). Well-studied examples include the Neuse River, North Carolina (Paerl and Bowles, 1987) and the Darling River, Australia (Hötzel and Croome, 1994), both of which receive large quantities of nutrients. Summertime cyanobacterial blooms (e.g., *Microcystis*, *Anabaena*, and *Aphanizomenon*) have also been reported periodically in the upper Mississippi and Potomac Rivers over at least five decades (Reinhard, 1931; Baker and Baker, 1981; Krogman et al., 1986). Some data suggest that picoplanktonic species may be quite numerous in large rivers, but apparently are not dominant in terms of biomass (Wehr and Thorp, 1997). The many studies over the last 100 years suggest that few, if any, species reported are strictly potamoplanktonic forms (Wehr and Descy, 1998); indeed many genera and their species (e.g., species of *Anabaena*, *Chroococcus*, and *Oscillatoria*) are also commonly observed in lakes. Species composition often depends on tributaries and especially on plankton from adjacent lakes or reservoirs (Reynolds and Descy, 1996), and is more dependent on the planktonic cyanobacterial flora typical for that region.

6. Wetlands

Cyanobacteria are well represented in the metaphyton and periphyton of swamps, temporary pools, and bogs. Many such systems possess a rich cyanobacterial flora, especially alkaline systems, where species of *Chroococcus, Aphanothece, Leptolyngbya, Phormidium, Lyngbya, Microcoleus, Scytonema,* and *Schizothrix* are frequently reported, many of which become encrusted with $CaCO_3$ (Browder et al., 1994; Goldsborough and Robinson, 1996). This flora has been described from alkaline wetlands in the Caribbean and Cuba (Gardner, 1927; Schiller, 1956), the Florida Everglades (McCormick and Stevenson, 1998), and coastal wetlands in Belize and Mexico (Rejmánková et al., 1996). Some of these communities contain cyanobacterial species that have been recognized as possibly endemic to the region. In contrast, a number of species from the genera *Cyanothece, Aphanothece, Rhabdoderma, Merismopedia, Eucapsis,* and *Chroococcus* are known mainly from acidic peat or salt swamps. Among the best studied wetland cyanobacteria may those that colonize rice fields, because many are important N_2 fixers (Whitton, 1992). Species from the littoral regions of mesotrophic or eutrophic ponds and lakes are less specific, and often represent dislodged benthic forms or accumulations of entrained algae from pelagic regions.

D. Extreme Environments

Cyanobacteria are the main and often sole autotrophic organisms in many extreme environments, which include thermal, saline, arid, and endolithic habitats.

1. Thermal Waters

The average upper temperature limit for plant existence is slightly above 40°C, but species adapted to about 70°C exist among cyanobacteria (Castenholz, 1969a, b, 1977; see also Chap. 2, Sect. V). Although modern taxonomic studies are sorely needed for thermal cyanobacterial species, North American hot springs belong to the best known thermal sites in the world. Several coccoid forms in Yellowstone National Park were described more than five decades ago by Copeland (1936), who described many of the major representatives of high-temperature cyanobacteria, including species of *Synechococcus* (*S. lividus* and *S. vulcanus*) and *Cyanobacterium* (*C. minervae*). A number of important filamentous cyanobacteria in hot springs include species of *Leptolyngbya* (several species), *Mastigocladus* (*M. laminosus*), and *Phormidium*. These assemblages often develop an extremely large biomass, which often forms layered mats of various colors (Castenholz and Wickstrom, 1975). Many species experience not only extreme temperatures, but also high levels of sulfide and UV exposure (Castenholz, 1976; Miller et al., 1998).

2. Saline Inland and Coastal Habitats

Cyanobacteria adapted to high salinity can often be the main autotrophs in hypersaline environments (Setchell and Gardner, 1905; Gardner, 1906; Taylor, 1928; Frémy, 1933; Feldman, 1958; Golubić, 1980; Humm and Wicks, 1980; Komárek and Lukavský, 1988; Montoya and Golubić, 1991; Anagnostidis and Pantazidou, 1991; Roussomoustakaki and Anagnostidis, 1991). Pools, lakes, and swamps with high salt concentrations that periodically dry and often contain unusual combinations of salts (sulfates, Mg salts, and sulfides; see also Chap. 2, Sect. VI.A), represent very specialized habitats for coccoid and filamentous cyanobacterial species. Lakes in desert regions, some volcanic lakes, and coastal pools (with periodic marine influences) belong to these environments. Some of the exclusively halophilic species are members of coccoid genera *Aphanothece, Cyanothece, Merismopedia, Stanieria,* and *Synechocystis* (Golubić, 1980; Anagnostidis and Pantazidou, 1991; Roussomoustakaki and Anagnostidis, 1991; Palinska et al., 1996) or species of filamentous genera *Arthrospira* and *Nodularia*. Many of the species regarded as halophiles have been found to be salt-tolerant, and their occurrence in saltwater environments is often influenced by their acclimation to these conditions and less by competition with other phototrophic organisms (Komárek and Lukavský, 1988). Mechanisms of acclimation, and development of halophilic species of cyanobacteria has been reviewed by Golubić (1980).

3. Arid Environments

Several cyanobacteria live as endolithic species inside rocks (within the crystal lattice) in very arid, hot or cold deserts. Three types are recognized: (1) euendoliths, which bore into rocks (see the next section), (2) chasmoendoliths, which occur in fissures and cavities, and (3) cryptoendoliths, which colonize fractured or porous rocks, usually forming layers parallel to the rock surface (Whitton and Potts, 1982; Friedmann and Ocampo-Friedmann, 1984). The crypto- and chasmoendoliths are unique phototrophic organisms that occur mainly in very arid, hot or cold deserts. Their presence is dependent at least on traces of periodic humidity, but they represent some of the most interesting forms of autotrophic, oxygen-evolving life on Earth (Friedman, 1971, 1980; Friedman and Ocampo,

1985; Friedman and Ocampo-Friedman, 1984; Golubić *et al.*, 1981; Büdel and Wessels, 1991; Büdel *et al.*, 1994; Johansen and Rushford, 1985; Johansen, 1993; Johansen *et al.*, 1993; Kubečková *et al.*, 2003). Species of the genus *Chroococcidiopsis* are among the most commonly reported organisms in this unusual habitat.

4. Endolithic and Lithogenic Aquatic Habitats

A specialized and unique ecological group of cyanobacteria occurs in both marine and freshwater regions that contain euendolithic and lithogenic species (Golubić *et al.*, 1975). The boring types are known mainly from marine rocky littoral zones (*Cyanosaccus* sp. div., *Hyella balani*, *H. tenuior*, *Scytonema endolithicum*, and others), but a few species are known also from freshwaters. In contrast, under subaerophytic and submersed freshwater conditions and in submersed marine habitats, many species participate in travertine and stromatolite formation (Golubić *et al.*, 1975, 1981). A few species of the coccoid genera (*Chroococcidium*, *Bacularia*, *Chlorogloea*, and *Entophysalis*) were described from such habitats (Geitler and Ruttner, 1935; Copeland, 1936; Komárek and Montejano, 1994).

E. Geographic Distribution

In accordance with their considerable morphological diversity, many of the genera and species of cyanobacteria that have been described are considered to be distributed in many locations worldwide (Whitton, 1992). However, the geographic distribution of cyanbacterial species is nonetheless dependent on their ecological requirements. Some species, which are almost cosmopolitan in their distribution, occur in distinctly specialized habitats; hence, even very specialized taxa (e.g., species from hot springs) can still be cosmopolitan among suitable habitats. Other species are known to tolerate a wide range of special ecological factors (e.g., temperature and salinity), but are quite sensitive to other environmental factors (e.g., halophilic *Nodularia harveyana* and *Arthronema africanum*; Komárek and Lukavský, 1988; personal observation). A consistent pattern of tolerances to and requirements of ecological factors does not exist for cyanobacteria; therefore, the geographic distribution of various species and forms often varies. Data suggest there are many very specialized types, the occurrence of which is restricted to small defined habitats and localities; several of these distributions are described in Section V.B. Notable examples can be found in thermal springs, certain tropical habitats, volcanic lakes, and near marine coastal regions. For example *Chlorogloea lithogenes* is known from central Mexico, and *Bacularia indurata* from Yellowstone National Park, *Mantellum rubrum* from volcanic lakes in Mexico, *Aphanothece bacilloidea* and *Gomphosphaeria semen-vitis* from alkaline swamps in Central America. Each appears to occupy restricted habitats (Gardner, 1927; Copeland, 1936; Komárek and Montejano, 1994; Tavera and Komárek, 1996).

There are, however, forms with broader ecological requirements, but restricted geographic distribution. For example, a number of freshwater planktonic (or metaphytic) species have been confirmed only from certain locations in North America, such as *Coelomoron regulare*, *Merismopedia smithii*, *Woronichinia klingae*, *Gomphosphaeria natans*, *Snowella rosea*, *Chroococcus multicoloratus*, and *Stanieria cyanosphaera* (see the ecological data in Komárek and Anagnostidis, 1998). Some species occur more widely, but with morphologies that are phenotypically uniform in one area and variable in another (including more morphotypes). Examples of variable populations (in comparison with other regions) are *Woronichinia naegeliana* in Canadian lakes (Komárek and Komárková-Legnerová, 1992) and *Microcystis wesenbergii* in Northern Europe (Cronberg and Komárek, 1994). For these reasons, use of identification keys based on microflora from other regions (especially Europe) for North American (and especially tropical) cyanobacteria, should be used with utmost care.

IV. COLLECTION, PREPARATION, AND CULTURE

Samples of natural cyanobacterial populations should be collected by methods widely used for other microalgae (e.g., Prescott, 1962; Whitford and Schumacher, 1969). For example, plankton nets (5, 10, 20, or rarely 45 μm) are well suited for most planktonic and metaphytic species. However, small nano- and picoplanktonic forms would be undersampled using nets and generally require collection of whole water samples, which can be concentrated by sedimentation or centrifugation. Epiphytic, epilithic, and endolithic species can be collected with parts of the substrata whereas parts of massive macroscopic colonies and mats can be collected directly into tubes of different sizes. The exact documentation and description of samples is an important requirement. The study of living material is preferred for species identification.

Fixation with 2% formalin (final concentration) is preferred for most samples, because higher concentrations produce artifacts in the mucilaginous sheaths, and distort cell size and shape (Komárek, 1958). Lugol's solution, commonly used in limnological studies for quantitative estimates (e.g., Utermöhl, 1958), is not recommended for taxonomic purposes, due to disintegration of colonies, changes in color, and distortions in sheaths or mucilage. Glutaraldehyde (to 2% concentration) is recommended for EM examination. In most cases, drying cyanobacterial samples is not recommended (especially for delicate forms), although some old herbarium specimens could be usable for comparisons based on general morphology and DNA-based methods. Periodically dried preparations (i.e., slides usable for microscopic treatment) are not commonly used, but may be used over longer periods if specimens are first preserved with formalin or glutaraldehyde.

Isolation and culture is important for cyanobacterial taxonomy. Methods used to culture freshwater (and terrestrial) algae and cyanobacteria are described in several general treatises (Pringsheim, 1946; Venkataraman, 1969; Carr and Whitton, 1973). However, most species have yet to be isolated into pure culture. Obtaining cultures is particularly difficult for ecologically specialized forms, such as thermal, endolithic, and even some bloom-forming species (see the lists of strains in world collections). However, problems also exist for many common planktonic (e.g., *Aphanothece*, *Lemmermanniella*, *Cyanodictyon*, *Woronichinia*, and *Microcystis*) and benthic species (e.g., *Chamaesiphon* species from mountain streams) that have not yet been studied in culture. Another complication is that in culture, many species lose the characteristic colony shape or change cell size, making it difficult to link strains with known field populations (Van Baalen, 1965; Komárek, 1976). Much useful information could be gained in future studies by carefully recording morphologies and phenotypic variation of field-collected cyanobacteria (under natural conditions) prior to isolation. Currently cultures that contain similar, small, rodlike cells may belong to species or forms with different colony forms in nature; hence, standardization and simplification of media and growth conditions are not able to clarify different accounts of cyanobacterial morphology.

There is a special problem with identification and designation of experimental strains that have unknown field populations. Strains should be identified before isolation or during the course of the isolation process. Based on the collective experience in our laboratory, in more than 50% of the cyanobacterial strains from the world collections, the taxonomic name on the label was a different (nomenclatural or taxonomic) type (according to both botanical and bacteriological codes) than what we found inside the tube. Unfortunately, many experimental scientists (e.g., physiologists using a strain for particular enzyme assays or systematists measuring phylogenetic affinities) do not revise the phenotype identification. This can lead to numerous discrepancies in morphological and molecular data in modern cladograms, simply due to incorrectly identified or labeled strains. Specialized collections with well-defined strains (e.g., National Institute for Environmental Studies – NIES in Japan), are rare. The designation of strains may also lead to confusion. Several proposals (Komárek, 1969; Lhotský and Komárek, 1981) aimed to unify methods for strain designations, but a single unified method was never accepted and each collection has its own code. Many strains thus occur in the literature under different numbers, for example, *Nostoc* sp. (strain NIVA-CYA 246) and *Anabaena* sp. (strain PCC 7120), which are derived from the same original strain (their identity was recently proved by molecular sequencing). It is recommended that researchers use at least the citation of the acronym of the collection (prescribed by the International Committee of Strain Collections; Garcia Reina, 1997; Sugawara and Miyazaki, 1999) and the full strain number of the corresponding collection.

Despite these problems, the use of cultures is quite necessary for modern cyanobacterial taxonomy. Molecular analyses and almost all ultrastructural studies are impossible without defined strains, and several cyanobacterial types (picoplanktonic species) are not recognizable or identifiable without culturing. However, new isolates should always be added to collections with documentation and characterization of the original (natural) material, its phenotype features, and ecological parameters.

V. KEY AND DESCRIPTIONS OF GENERA

A. Key[2]

1a.	Cells solitary (rarely aggregated in groups), heteropolar, with distinct basal and apical ends, usually attached by one (basal) end to the substratum (plants or stones); family Chamaesiphonaceae	31
1b.	Cells of different shape, solitary or in various colonies, but not distinctly diversified into basal (attached) and apical ends (cells may be polarized within colonies)	2

[2]Note: Cyanobacterial genera are characterized mainly by molecular sequencing and cell ultrastructure in modern taxonomy. It is sometimes difficult to find a single morphological or phenotypic feature that is diagnostic. Therefore, it is necessary to compare the entire set of characteristics from the key for identification.

2a.	Cells widely oval to cylindrical (rodlike), which divide exclusively perpendicular to one (usually longer) cell axis; family Synechococcaceae	3
2b.	Cells spherical, hemispherical, ovoid, irregularly elongated, polygonal–rounded, or of irregular outline, dividing in two or more planes	16
3a.	Cells solitary (or agglomerations of solitary cells)	4
3b.	Cells united regularly within mucilaginous colonies of various kinds	7
4a.	Cells cylindrical to long rodlike; cell content more or less homogeneous with only solitary granules; chromatoplasm sometimes barely visible (Fig. 4A)	*Synechococcus*
4b.	Cells widely oval to wide cylindrical	5
5a.	Cells with differentiated centro- and chromatoplasm, 0.4–4.5 (up to 10) μm long (Fig. 1A)	*Cyanobium*
5b.	Cells usually with lengthwise striated or reticulate content, longer than (2)3.7 μm (up to 30–100 μm long)	6
6a.	Cells with lengthwise striated content (special thylakoid arrangement), 2–30 μm long (Fig. 3B)	*Cyanobacterium*
6b.	Cells with reticulate content (radial arrangement of thylakoids), 6.2–40 (100) μm long (Fig. 5B)	*Cyanothece*
7a.	Cell length/width ratio = 1 to 2(3):1; colonies spherical or formless	8
7b.	Cells more than (2)3 times longer than wide; colonies irregular or distinctly elongated	13
8a.	Colonies $CaCO_3$ encrusted; component of travertine formations (not pictured)	*Lithomyxa*
8b.	Colonies not $CaCO_3$ encrusted; not in travertine formations	9
9a.	Cells arranged only in microscopic, mucilaginous, spherical or irregular netlike colonies; at least partly in rows	10
9b.	Cells within or on the surface of irregularly spherical colonies or in amorphous gelatinous colonies; not in rows	11
10a.	Colonies more or less spherical or slightly elongated, with cells clustered in the center and in short, radiating rows at the colonial periphery (Fig. 1C)	*Radiocystis*
10b.	Colonies irregular–reticulate, composed of mucilaginous strands; cells more or less in not radiating rows (Fig. 1B)	*Cyanodictyon*
11a.	Oval cells situated on the surface of mucilaginous spheres (Fig. 2A)	*Epigloeosphaera*
11b.	Cells inside of micro- to macroscopic gelatinous colonies, sometimes enveloped by their individual, sometimes lamellated envelopes	12
12a.	Cells entirely (or at least central cells) without individual envelopes (Fig. 2B)	*Aphanothece*
12b.	All cells with individual, occasionally lamellated envelopes (Fig. 3A)	*Gloeothece*
13a.	Cells cylindrical or fusiform, always longer than wide, randomly distributed within colonies (occasionally regularly oriented along longer axes in one direction); colony usually elongated with cells not in pseudofilamentous rows	14
13b.	Cells shorter than long (cells divide perpendicular to shorter axis of cells), arranged in uniseriate pseudofilamentous rows, forming elongated mucilaginous colonies (Fig. 6A)	*Johannesbaptistia*
14a.	Colonial mucilage indistinct, colorless, diffuse; cells distant from one another	15
14b.	Colonial mucilage distinctly limited, sometimes encrusted; cells arranged clearly and usually ± densely in one direction (Fig. 5A)	*Bacularia*
15a.	Cells cylindrical with rounded ends (Fig. 4B)	*Rhabdoderma*
15b.	Cells fusiform or with tapering cell ends (Fig. 4C)	*Rhabdogloea*
16a.	Cells divide exclusively in two directions in successive generations; cells solitary or in flat to nearly spherical colonies; family Merismopediaceae	17
16b.	Cells divide in three or more directions (or irregularly) in successive generations, cells in colonies of various forms	27
17a.	Solitary cells, always spherical (Fig. 6B)	*Synechocystis*
17b.	Cells in mucilaginous colonies; cells spherical or slightly elongated (oval, ovoid)	18
18a.	Colonies distinctly platelike (cells arranged more or less in one flat layer)	19
18b.	Colonies three dimensional, irregular, or more or less spherical	22
19a.	Cells spherical, in perpendicular rows in colonies	20

19b.	Cells slightly but distinctly elongated, in short rows or irregularly arranged	21
20a.	Cells in free-living, platelike colonies (Fig. 7A)	*Merismopedia*
20b.	Flat colonies attached (monolayer), lying flat on substrata (Fig. 7B)	*Mantellum*
21a.	Colonies always only few-celled, longer axis in the plane of colony; cells oval to ovoid (Fig. 7C)	*Cyanotetras*
21b.	Old colonies many-celled, longer axis perpendicular to plane of the colony; cells oval to almost rodlike (Fig. 8A)	*Microcrocis*
22a.	Colonies amorphous, irregular; cells irregularly (sometimes densely) arranged (Fig. 6C)	*Aphanocapsa*
22b.	Colonies more or less spherical; cells arranged peripherally and sometimes radially near the surface of mucilaginous spheres	23
23a.	Colonial mucilage of older (larger) colonies more or less homogeneous; without visible gelatinous stalks (not revealed by any stain)	24
23b.	Within colonies develops a gelatinous system of stalks; cells or cell groups joined to the ends of stalks (revealed sometimes only by stain)	25
24a.	Cells spherical, arranged peripherally; colonies spherical (rarely irregular), mucilaginous (Fig. 8B)	*Coelosphaerium*
24b.	Cells slightly elongated, arranged radially in the marginal parts in more or less spherical colonies, or several spherical colonies joined together (Fig. 8C)	*Coelomoron*
25a.	Stalks thin, stringlike, pseudodichotomously divided, joined in the center to form a widened gelatinous cluster; cells more or less distant from one another (Fig. 9A)	*Snowella*
25b.	Stalks thick, not stringlike	26
26a.	Stalk system composed of fine, gelatinous, parallel tubes radiating from center (phase contrast or staining); each stalk bears radially-arranged spherical or oval cells; cells may form dense peripheral layer in old colonies (Fig. 9B)	*Woronichinia*
26b.	Stalks thick, divided and diffuse; widened at the ends and enveloping peripherally arranged cells; cells more or less obovoid and remain connected after division, often forming pairs of cordiform cells (Fig. 10A)	*Gomphosphaeria*
27a.	Cells spherical, dividing regularly in three perpendicular planes in successive generations; family Microcystaceae	28
27b.	Cells hemispherical, polygonal, elongate, or irregular, dividing usually in three or more variously oriented planes	37
28a.	Cells arranged distinctly in three-dimensional, cubical colonies, in more or less regular, perpendicular rows (Fig. 10B)	*Eucapsis*
28b.	Cells not arranged in a cubical manner; colonies are irregular, sometimes forming amorphous masses	29
29a.	Cells without distinct individual sheaths, with aerotopes (groups of gas vesicles); planktonic colonies irregular in shape with irregularly (sometimes densely) scattered cells (Fig. 11)	*Microcystis*
29b.	Cells (or small groups) with individual sheaths (may be colored), without aerotopes; subaerial, metaphytic, or benthic irregularly shaped colonies with irregularly arranged cells	30
30a.	Cell groupings within colonies usually spherical, with more or less widened, spheroidal gelatinous envelopes; mainly subaerial species (Fig. 12A)	*Gloeocapsa*
30b.	Cell groupings within colonies in dense polyhedral clusters; mainly submersed species (Fig. 12B)	*Chondrocystis*
31a.	Sessile cells reproduce by exocytes (daughter cells separate asymmetrically and individually, or remain in rows or clusters at the apex of the mother cell); family Chamaesiphonaceae	32
31b.	Sessile cells reproduce by baeocytes (formed from successive or almost spontaneous divisions of the whole mother cell; baeocytes disperse from gelatinized or ruptured sheaths); family Dermocarpellaceae	36
32a.	Sessile cells rodlike, without gelatinous sheaths; attached to the substratum by a small mucilaginous pad (revealed with staining or EM sections) (Fig. 16B)	*Geitleribactron*
32b.	Sessile cells always with firm gelatinous sheath (sometimes very thin and indistinct, or visible only during exocyte liberation)	33
33a.	Thin sheath (may be obscure) of vegetative cells narrowed at apex into conspicuous, short or long, hairlike projection (Fig. 16C)	*Clastidium*
33b.	Sheaths of vegetative cells without apical hairlike formations	34
34a.	Cells (more or less cylindrical) produce long rows of exocytes that are retained within the sheath of the mother cell, forming pseudofilamentous formations (Fig. 17A)	*Stichosiphon*
34b.	Cells (usually oval, ovoid to cylindrical) do not form long, pseudofilamentous rows of exocytes	35
35a.	Exocytes separate individually or arranged at the apex of the mother cell in a ± short single series (Fig. 17B)	*Chamaesiphon* subg. *Chamaesiphon*
35b.	Exocytes arranged apically in three-dimensional clusters (Fig. 18A)	*Chamaecalyx*

36a.	The first division plane is vertical during successive divisions of the mother cell into baeocytes (Fig. 18B)	*Cyanocystis*
36b.	The first division plane into baeocytes is horizontal (Fig. 18C)	*Dermocarpella*
37a.	Cells spherical to subspherical, gathered in small agglomerations without slime production; cell division solely into baeocytes: entire whole mother cell divides into numerous small daughter cells, which liberate from a ruptured sheath; family Dermocarpellaceae (Fig. 19A)	*Stanieria*
37b.	Cells of variable shape, forming irregular colonies with slime production or a differentiated thallus; cell division irregular in various planes in exocytes or partially in one direction, sometimes combined with baeocyte production; several types of division are combined in some cases	38
38a.	Cells within colonies divide only by binary fission, never by exocytes or baeocytes	39
38b.	Cells within colonies divide by binary fission and via specialized cells, by exocytes or baeocytes	46
39a.	Colonies irregular, never polarized; all cells similar in character; family Chroococcaceae	40
39b.	Colonies often polarized and slightly differentiated, cells often with different shape in basal, central, and apical regions of colony	44
40a.	Cells with narrow, firm sheaths; gathered into dense packets with numerous cells	41
40b.	Sheaths more widened, forming distinct gelatinous layers around cells; colonies usually few-celled (but sometimes clustered together)	42
41a.	Packets of cells with more or less cubical (or sarcinoid) arrangement; sheaths colorless (Fig. 14B)	*Cyanosarcina*
41b.	Packets of clustered and ensheathed cells irregularly arranged; sheaths often colored (Fig. 13B)	*Gloeocapsopsis*
42a.	Sheaths (fine or firm) around cells often spheroidal and colored, and often with surface structure (granular or warty) (Fig. 13A)	*Asterocapsa*
42b.	Sheaths usually fine, colorless, delimited or diffuse, in vegetative stages without surface structures	43
43a.	Cells usually arranged in three-dimensional, spheroidal clusters (Fig. 14A)	*Chroococcus*
43b.	Cells arranged in very short, parallel rows, slightly distant from one another (Fig. 13C)	*Cyanokybus*
44a.	Colonies slightly polarized, three-dimensional, forming large mucilaginous clusters on various substrata; family Entophysalidaceae	45
44b.	Colonies more or less flat, forming discoidal, often one- or few-layered biofilms on substrata, growing radially, with morphologically different central and marginal (apical) cells; family Hydrococcaceae (Fig. 16A)	*Hydrococcus*
45a.	Central and apical cells usually spherical, in marginal or old colonial parts with individual gelatinous envelopes; mucilage mainly colorless, rarely colored (Fig. 15A)	*Chlorogloea*
45b.	All cells ± irregular, most (or small groups) with individual gelatinous sheaths, which are usually colored (Fig. 15B)	*Entophysalis*
46a.	Colonies multilayered, ± polarized, attached to substratum; most cells divide in apical areas as exocytes (remain attached to open sheaths of the mother cell), also by binary fission; family Chamaesiphonaceae (Fig. 17B)	*Chamaesiphon* subg. *Godlewskia*
46b.	Cells divide by binary fission and by baeocytes (rarely by nanocytes), never by exocytes	47
47a.	Thallus not distinctly polarized, usually three-dimensional, free-living or attached to substrata; baeocytes arise from specialized cells developing in parts of a colony; family Xenococcaceae	48
47b.	Thallus polarized (sometimes very short with few cells) with pseudofilamentous structure (epiphytic, epilithic, sometimes boring forms); baeocytes arise from specialized basal (old) or apical cells; family Hyellaceae	53
48a.	Cells of different size in one colony; cell groupings enveloped by fine, diffuse, gelatinous envelopes (Fig. 19B)	*Chroococcidium*
48b.	Cells more or less uniform in size in one colony; cells enveloped by thin, but firm, distinct sheaths	49
49a.	Colonies free-living, not directly attached to substrata but sometimes occurring in epiphytic or endolithic assemblages; three-dimensional but without any polarized cells	50
49b.	Colonies attached to substrata; at least portions of cells in colonies have distinct polarity	52
50a.	Packet-like or sarcinoid colonies, sometimes gathered together, with irregular cells (Fig. 20B)	*Myxosarcina*
50b.	Cells in irregular agglomerations, not forming packet-like colonies; cells subspherical	51
51a.	Most cells of similar, more or less uniform spheroidal shape; only young and old cells differ in size; cells never in rows (Fig. 20A)	*Chroococcidiopsis*

51b.	Cells in colonies are different in size, sometimes arranged in short, irregular rows (Fig. 21B).................Chroococcopsis
52a.	Cells form a monolayer on substrata; a few may reproduce by baeocytes (Fig. 21A).................Xenococcus
52b.	Cells in two or more layers on substrata, may form irregular hemispherical colonies; baeocytes produced from randomly arranged mother cells at periphery of colony (Fig. 21C).................Xenotholos
53a.	Pseudofilaments erect, arising parallel from basal cells that creep along substrata; baeocytes occur in apical, enlarged cells (Fig. 22A).................Radaisia
53b.	Pseudofilaments creeping along substrata, some are endophytic or endolithic; baeocytes do not arise from apical cells.................54
54a.	Pseudofilaments only creeping along substrata, never boring; baeocyte-forming cells arise in various parts of a thallus (Fig. 22B).................Pleurocapsa
54b.	Pseudofilaments distinctly polarized, creeping on substrata and/or boring into limestone, or growing into the intercellular spaces of host plants (marine seaweeds); baeocytes arise from cells in oldest parts of thallus (Fig. 23).................Hyella

B. Descriptions of Genera[3]

Synechococcaeae (Figs. 1–6A)

Aphanothece Nägeli (Fig. 2B)

Colonies are microscopic or macroscopic (up to several centimeters in diameter), multicellular, more or less spherical or mostly amorphous, greyish, greenish, blue-green, or brownish colored, with irregularly, sometimes densely arranged cells, which are embedded by colonial mucilage. Two subgenera are defined: (i) *Anathece*—cells in diffuse mucilage, without individual envelopes—and (ii) *Aphanothece*—mucilage usually distinctly delimited, and cells (especially marginal) enveloped facultatively within individual sheaths or envelopes. Cells are widely oval to cylindrical, straight or slightly arcuate, pale greyish blue–green, green to bright blue–green (rarely reddish), usually with apparent chromatoplasm in larger cells. Gas vesicles and aerotopes are present in some planktonic species. Reproduction by transversal binary fission of cells, which is perpendicular to the longer cell axis.

A common genus with about 60 species (based on morphology) from diverse environments, but numerous morpho- and ecotypes are not identifiable. Ecological information is important for species identification, yet distribution data for various species is uncertain due to difficulties with identification and numerous misinterpretations in the literature. The genus is common in many North American environments (Prescott, 1962; Whitford and Schumacher, 1969; Duthie and Socha, 1976; Stein and Borden, 1979), with numerous species from wet rocks (*A. castagnei*, and *A. saxicola*), soils (*A. pallida*), benthos of ponds and pools (*A. microscopica*, *A. nidulans*, *A. stagnina*, and *A. uliginosa*), and thermal springs (*A. bullosa*, and *A. thermalis*), but many records need revision. Several species (possibly endemic and ecologically restricted) occur in alkaline swamps in subtropical and tropical areas, including Florida (Everglades), Cuba (Zapata peninsula, Ciénaga de Lanier), Belize, and Mexico (Yucatán). These species include *A. bacilloidea*, *A. cylindracea*, and *A. opalescens* (Gardner, 1927; Komárek, 1995). One tropical species with aerotopes, *A. conglomerata*, occasionally is present in surface blooms in reservoirs in Florida (original unpublished data). Some species colonize coastal and inland brackish environments, for example, *A. utahensis* and *A. karukerae* (Caribbean), and a few other species that require revision. Many species from the subgenus *Anathece* are common components of lake phytoplankton; *A. minutissima*, *A. bachmannii*, *A. clathrata*, and *A. smithii* are known from mesotrophic Canadian lakes (original unpublished data, H. Kling, personal communication).

Bacularia Borzì (Fig. 5A)

Cylindrical cells are arranged more or less in parallel (lengthwise, in one direction) in narrow, filamentous or tubular, elongated mucilaginous colonies, tapering and pointed or open at both ends. Cells are slightly distant, but not in rows or in pseudofilaments. The mucilage is fine, homogeneous, colorless, and usually distinctly delimited at the margin. Cells are always cylindrical, straight, with rounded ends, solitary or in pairs after division, pale blue–green, (2.2)3–15(20) × 0.5–3 μm. Cells divide perpendicularly to the long cell axis in two, more or less isomorphic daughter cells.

All five described species are from metaphytic and periphytic habitats in freshwater. Two species are known from thermal springs (Copeland, 1936; Frémy, 1949). Among them, *B. indurata* was described from thermal sites in Yellowstone National Park (62–70°C)

---

[3]Figures are largely based on North American material and literature. However, the documentation of several species had to be selected from non-North American papers. Localities are cited only in figures from this continent. Schemes of cell division, life cycles, and colony formations are selected according to various authors from Komárek and Anagnostidis (1998).

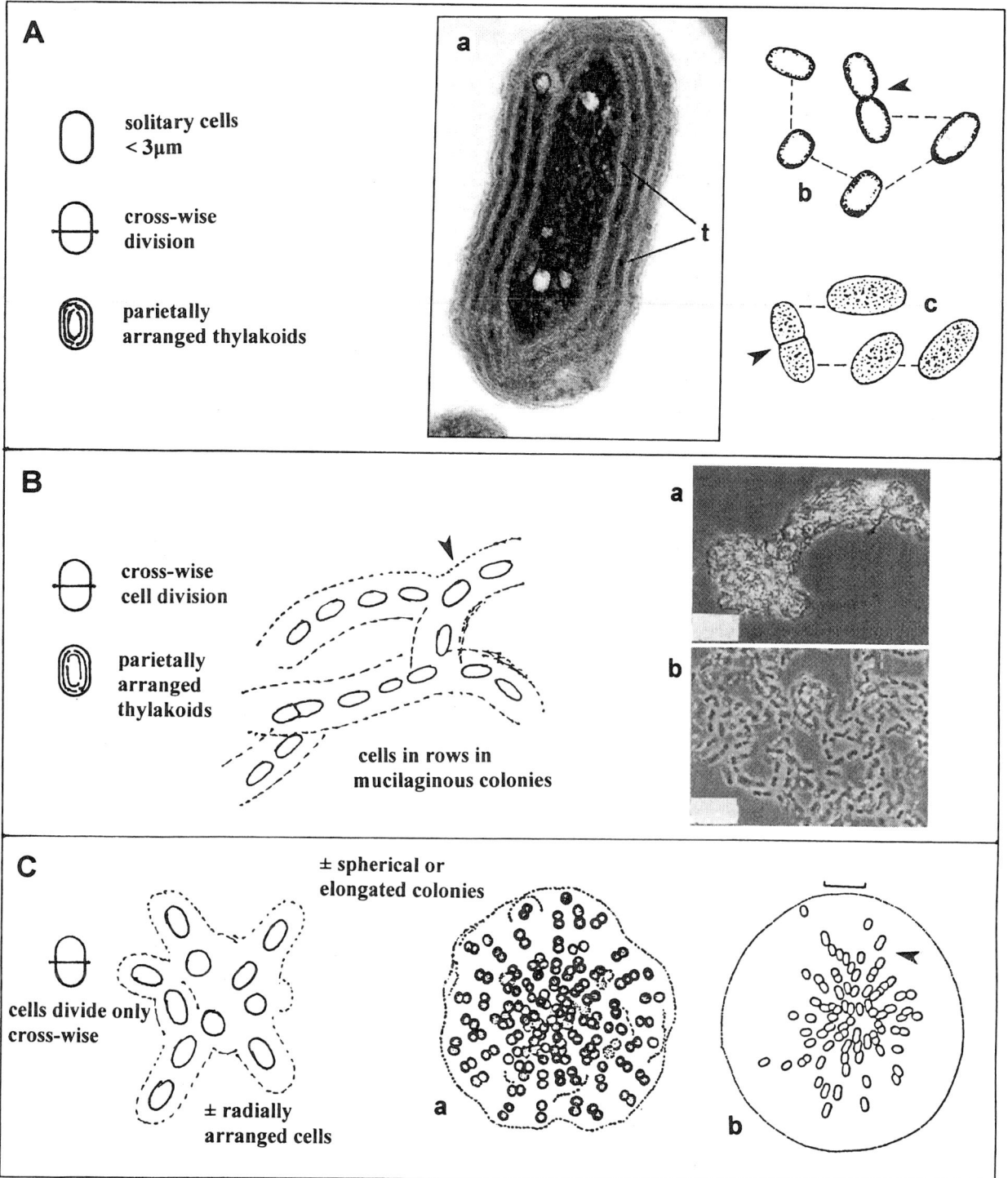

FIGURE 1 (A) *Cyanobium*; a. *C. gracile*, cross section through the cell with parietally localized thylakoids (t = thylakoids; bar = 0.3 μm; after Komárek, 1999); b. *C. eximium* and c. *C. roseum* (both described from thermal springs in Yellowstone National Park under *Synechococcus*; after Copeland, 1936). (B) *Cyanodictyon*; a, b. *C. planctonicum* (bars = 10 μm; photo after Hickel, 1981). (C) *Radiocystis*: a. *R. geminata* (after Skuja, 1948); b. *R. elongata* (bar = 10 μm; after Hindák, 1996, from lakes in central Canada).

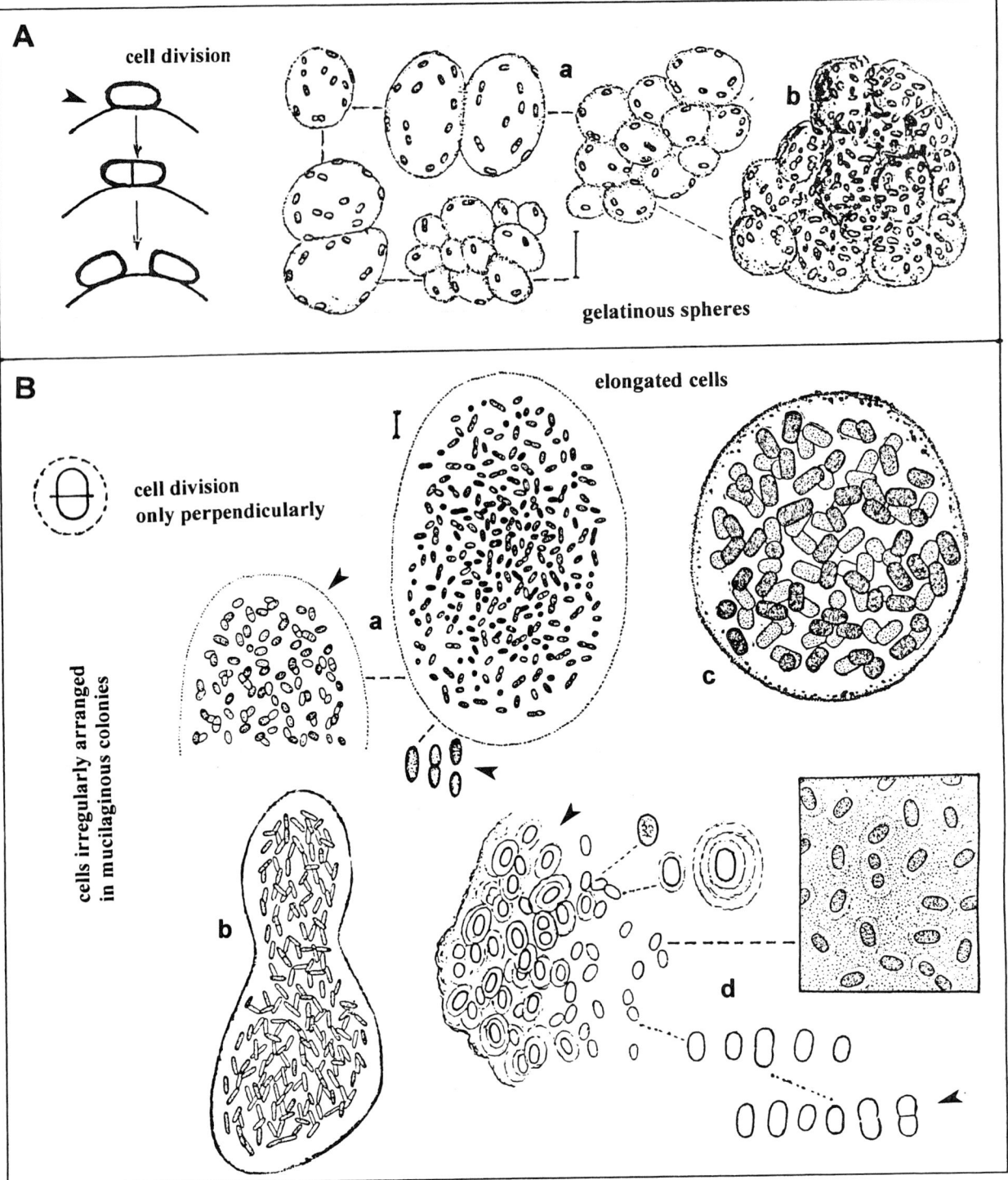

FIGURE 2 (A) *Epigloeosphaera*: **a, b.** *E. glebulenta* (bar = 10 μm; after Komárková-Legnerová, 1991, and Zalessky, 1926). (B) *Aphanothece*: **a.** *A. (Anathece) smithii* (bar = 10 μm; after Smith, 1920, from Wisconsin lakes, and Komárková-Legnerová and Cronberg, 1994); **b.** *A. (Anathece) clathrata* (after Smith, 1920, from Wisconsin lakes); **c.** *A. (Aphanothece) stagnina* (after Smith, 1920, from Wisconsin lakes); **d.** *A. (Aphanothece) castagnei* (after Komárek and Anagnostidis, 1998, and Smith, 1950, from North America; sub *Anacystis rupestris*).

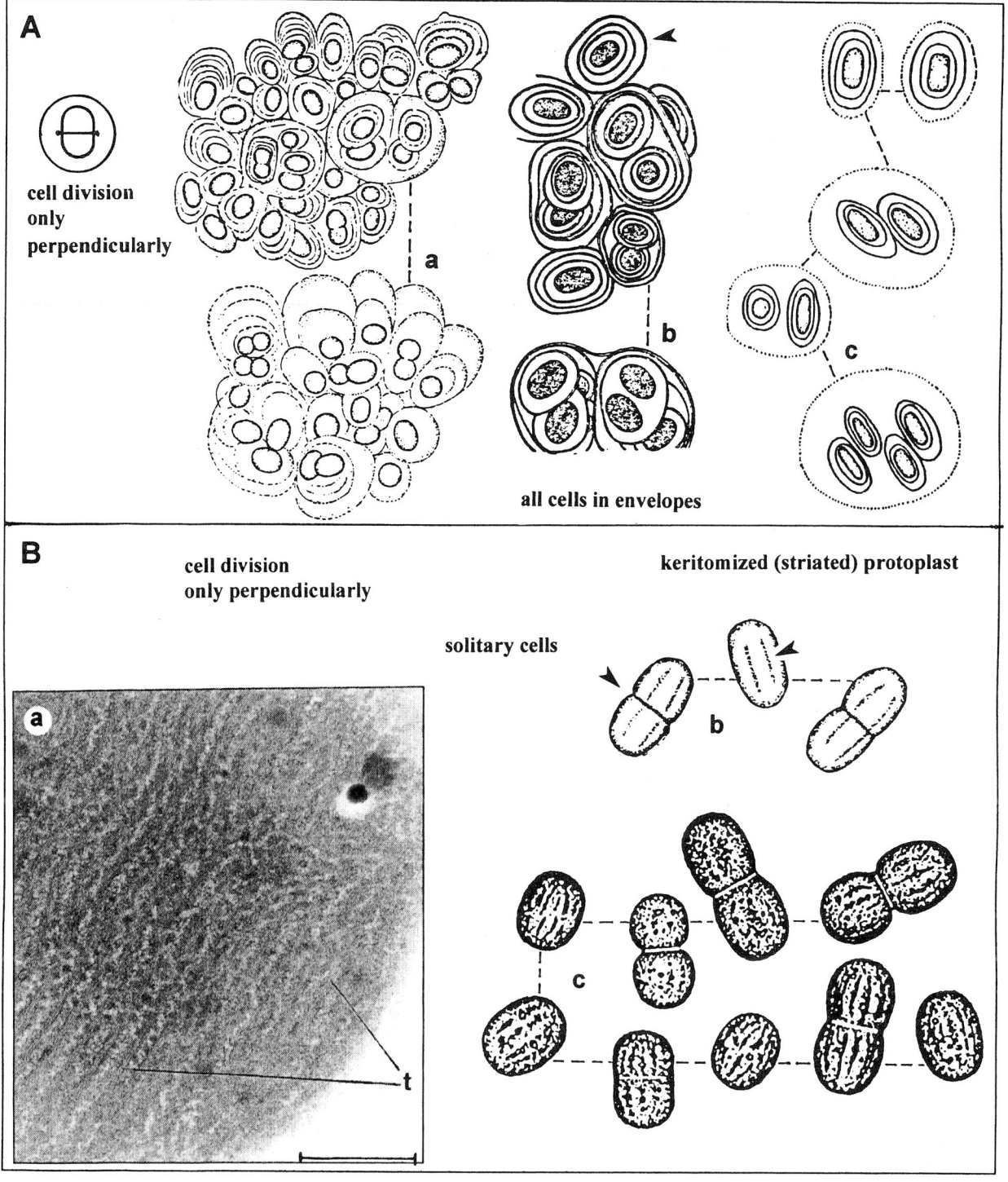

FIGURE 3 (A) *Gloeothece*: a. *G. heufleri* (after Prichod'kova from Kondrateva *et al.*, 1984); b. *G. rupestris* (after Bourrelly and Manguin, 1952, from Guadeloupe); c. *G. interspersa* (after Gardner, 1927, from Puerto Rico). (B) *Cyanobacterium*: a. *C.* cf. *cedrorum*, lengthwise section with characteristic position of thylakoids (t); b. *C. minervae* (after Copeland, 1936, from thermal springs of Yellowstone National Park); c. *C. diachloros* (after Skuja, 1939).

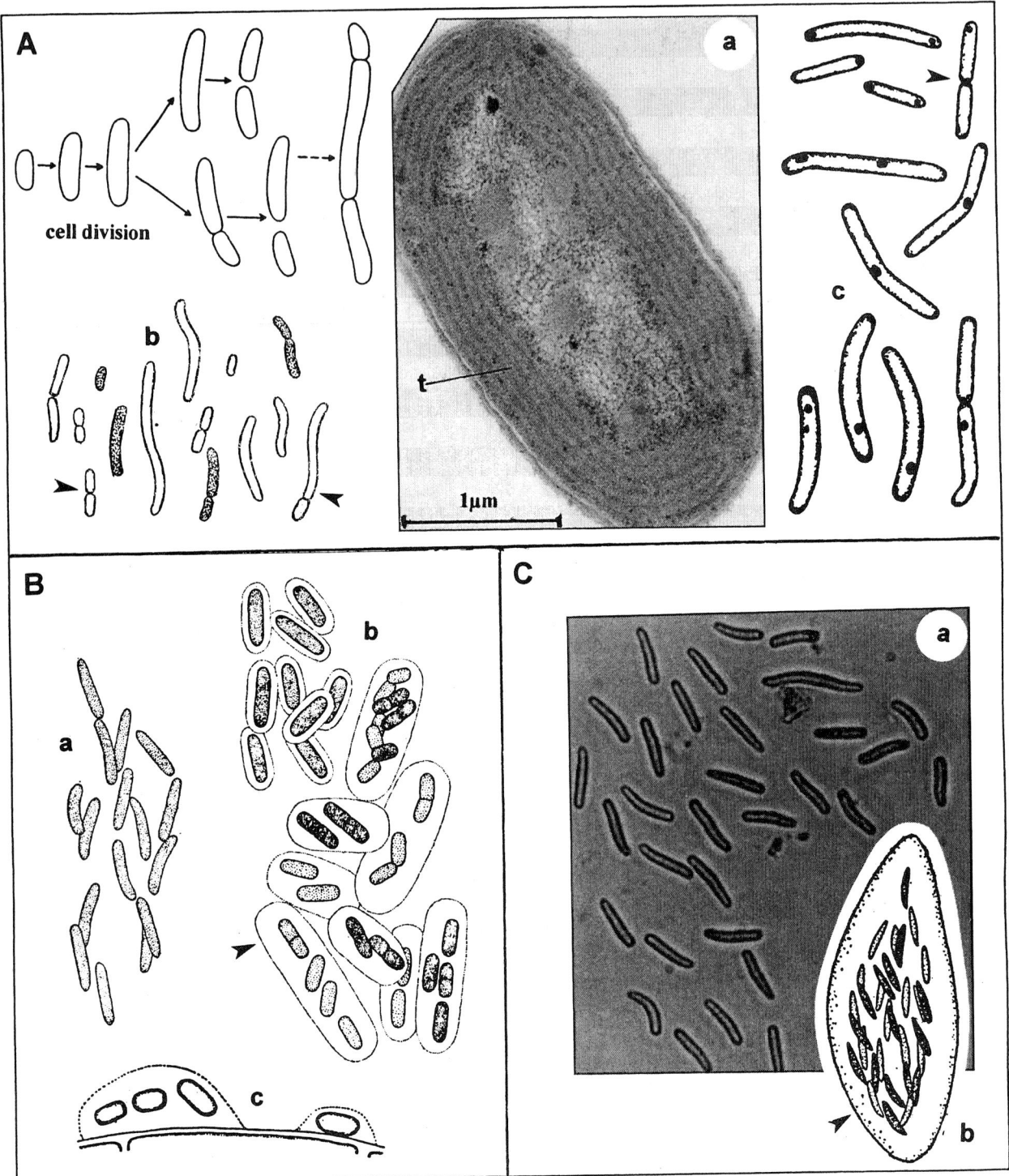

FIGURE 4 (A) *Synechococcus*: **a.** *Synechococcus* sp., lengthwise section through a cell (t = thylakoids, bar = 1 μm); **b.** *S. nidulans* (after Komárek, 1989, from Cuba); **c.** *S. lividus*, diversity of cells in various populations (after Copeland, 1936, from Yellowstone National Park). (B) *Rhabdoderma*: **a.** *R. lineare* (after Smith, 1920; planktonic species from Wisconsin lakes); **b.** *R. compositum* (after Smith, 1920; planktonic species from Wisconsin lakes; sub *Gloeothece linearis* var. *composita*); **c.** *R. zygnemicolum* (after Copeland, 1936, epiphytic species from Yellowstone National Park). (C) *Rhabdogloea*: **a.** *R. smithii* (after Komárek, 1969), **b.** *R. smithii* (after Smith, 1920, from Wisconsin lakes; sub *Dactylococcopsis raphidioides*).

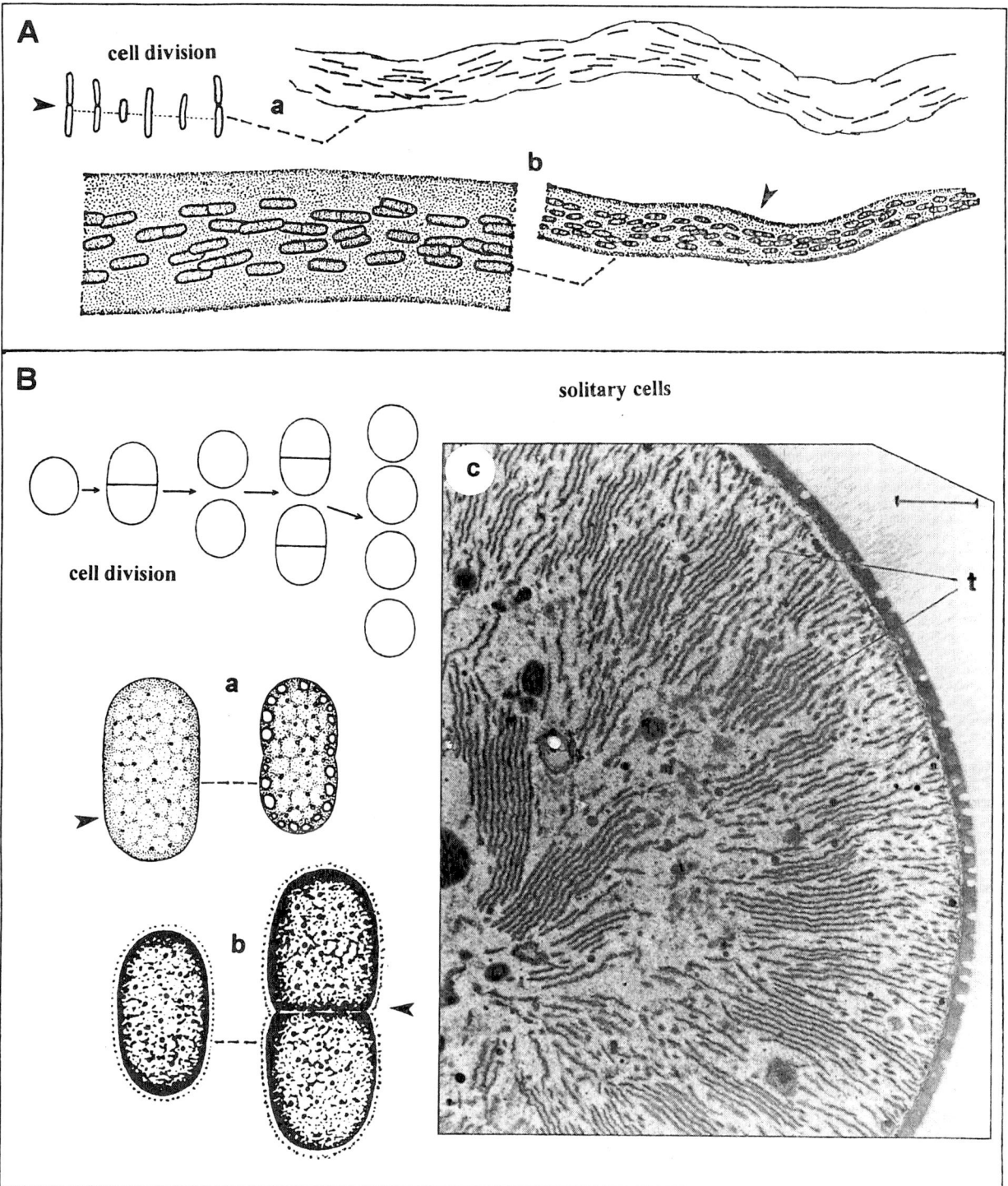

FIGURE 5 (A) *Bacularia*: **a.** *B. gracilis* (after Komárek, 1995, from Cuba); **b.** *B. indurata* (after Copeland, 1936, from hot springs in Yellowstone National Park under *Bacillosiphon induratus*). (B) *Cyanothece*: **a.** *C. aeruginosa* (after Geitler, 1925, under *Synechococcus aeruginosus*); **b.** *C. maior* (after Starmach, 1973, under *Synechococcus maior*); **c.** *C. aeruginosa* (t = thylakoids; bar = 2 μm; after Komárek and Cepák, 1998; cross section through the vegetative cell).

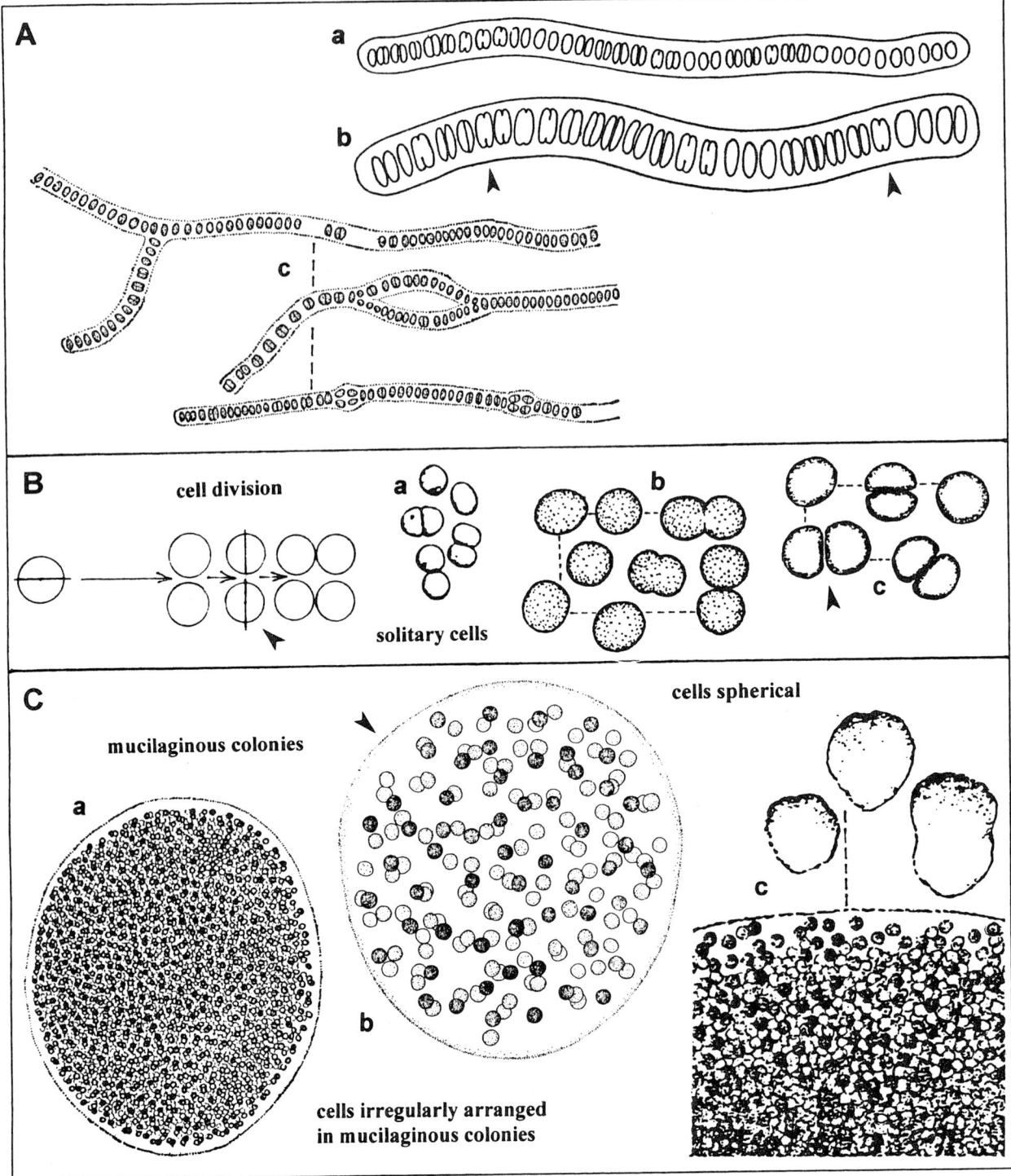

FIGURE 6 (A) *Johannesbaptistia*: **a.** *J. primaria* (after Gardner, 1927, from Puerto Rico; sub *Cyanothrix primaria*); **b.** *J. pellucida* (after Gardner, 1927, from Puerto Rico; sub *Cyanothrix willei*); **c.** *J. schizodichotoma* (after Copeland, 1936, and Smith, 1950, from Yellowstone National Park; sub *Heterohormogonium schizodichotomum*). (B) *Synechocystis*: **a.** *S. aquatilis* (after Setchell and Gardner, 1919, from Pacific coast in California); **b.** *S. willei* (after Gardner, 1927, from Puerto Rico); **c.** *S. thermalis* (after Copeland, 1936, from Yellowstone National Park). (C) *Aphanocapsa*: **a.** *A. incerta* (after Smith, 1920, from Wisconsin lakes; sub *Polycystis incerta*); **b.** *A. grevillei* (after Smith, 1920, from Wisconsin lakes); **c.** *A. farlowiana* (after Drouet, 1942, from northern United States).

with lime-encrusted sheaths as *Bacillosiphon*. The metaphytic species *B. gracilis* was discovered in littoral zones of alkaline lakes in Cuba (Komárek, 1995), and occurs more in tropical America.

Cyanobacterium Rippka *et* Cohen-Bazire (Fig. 3B)

Cells are solitary or in groups and/or aggregates, do not form mucilaginous colonies, have no enveloping mucilage or only a very fine, indistinct, and narrow gelatinous layer around the cell surface. Cells are widely oval to cylindrical and rod-shaped, rarely slightly arcuate, after division in pairs. The cell content is homogeneous, sometimes slightly keritomized or lengthwise striated (parallel position of thylakoids, sometimes in entire cell volume), pale to bright blue-green or olive green, sometimes with distinct granulation, without obvious gas vesicles, (2)3.7–17(30) × (1)3–12(20) μm. Cell division only by perpendicular binary fission (mainly pinching); daughter cells grow into the original size and shape before the next division.

This recently defined genus had a few species previously described under *Synechococcus*. *Cyanobacterium* now comprises eight well-known species, but further descriptions are expected. Species are not widely reported and most are from extreme environments. *C. minervae* (originally *Synechococcus minervae*) was described from hot springs in Yellowstone National Park, but is now known from thermal waters worldwide. *C. cedrorum* was recorded from soils in North Carolina (as *Synechococcus cedrorum*) by Whitford and Schumacher (1969), but this species occurs in various concepts in the literature. Other species occur in swamps and rice fields in subtropical zones (Florida and Cuba).

Cyanobium Rippka *et* Cohen-Bazire (Fig. 1A)

Cells are solitary or in pairs after division, without mucilage or with very fine, narrow, usually diffuse slimy envelopes around cells; never in colonies. Cells are oval to short rod-shaped, up to 4 μm × 1–2 μm, without gas vesicles, pale blue-green to olive green, greyish blue-green, or reddish, usually with distinctly visible chromatoplasm (= parietal thylakoids). Cell division by binary fission (usually by pinching), always transversely to the longer cell axis; daughter cells are more or less isodiametric, and grow to the original shape and size before the next division.

This widely distributed but little-known genus occurs in plankton (may be a major component of freshwater and marine picoplankton), in subaerial and submersed benthic mats, and as an epiphyte (Drews *et al.*, 1961; Waterbury *et al.*, 1979; Komárek and Anagnostidis, 1998; Komárek *et al.*, 1999). Only 12 species currently have been described, but *Cyanobium* is likely present in many environments in North America. In older literature it is often identified as a "small *Synechococcus* species." Earlier identifications from North America may require revision. A few types have been recorded from oligotrophic lakes in New York (Corpe and Jensen, 1992), but without species identification. Three thermophilic species, *C. amethystinum*, *C. eximium*, and *C. roseum*, are known from Yellowstone Park and other thermal springs in North America (Copeland, 1936; Komárek *et al.*, 1999).

Cyanodictyon Pascher (Fig. 1B)

Cells occur in spherical to irregularly reticulate, slimy, microscopic flat or three-dimensional colonies, which are sometimes elongate, composed from irregularly branching and anastomosing mucilaginous strands, later forming an amorphous mass with holes. The mucilage is fine, colorless, and sometimes diffuse. Cells are arranged in strands in one or more rows, and later irregularly. Cells are more or less spherical, slightly elongated to rod-shaped, small, up to 4.5 μm long (or diameter), without obvious gas vesicles, and pale blue-green, greyish, or olive green. Cell division is by binary fission, always perpendicularly to one (longer) cell axis. Daughter cells are isomorphic and grow to the original shape and size before the next division.

Cyanodictyon species are mainly planktonic in oligotrophic to mesotrophic (rarely dystrophic) water bodies; one species is endgloeic within the mucilage of *Anabaena* (Geitler, 1932). Probably widely distributed. Of eight described species, four were recorded from Canadian lakes: *C. reticulatum*, *C. tubiforme*, *C. planctonicum*, and *C. filiforme* (H. Kling, personal communication).

Cyanothece Komárek (Fig. 5B)

Cells are solitary or in pairs after division, rarely aggregated in groups, but never forming distinct colonies. They have no individual envelopes, but are surrounded by narrow, very fine, diffuse and structured colonial mucilage. Cells are widely oval to almost cylindrical with widely rounded ends, (6.2)7–45(100) × (6)7–30(76) μm, and a length/width ratio less than 3:1. Cells appear pale to bright blue-green, olive green, greyish, or pinkish. Higher magnifications show distinct netlike keritomy (= irregular to radial arrangement of thylakoidal fascicles, sometimes with intra-thylakoidas spaces). Involution cells are irregular. Cells divide by binary fission (cleavage), transversely to one (longer) cell axis, into two isomorphic hemispherical daughter cells, which grow to original size and shape before the next division. The ultrastructure of large cells is unlike most other morphologically similar genera (Komárek and Cepák, 1998).

Six well-known and several uncertain species that occur mostly in metaphyton in various habitats have been identified. Two species are known from peaty, cold stenotherm swamps and lakes: *C. aeruginosa* (in previous literature *Synechococcus aeruginosus*) has cosmopolitan distribution and is well known from North America (Daily, 1942; Prescott, 1962; Whitford and Schumacher, 1969); *C. major* occurs in subarctic and alpine regions in Alaska and Canada (Smith, 1950; Croasdale, 1973). *C. lineata* was collected from small mesotrophic lakes in Mexico (Komárek and Komárková-Legnerová, 2002).

Epigloeosphaera Komárková-Legnerová (Fig. 2A)

Colonies are microscopic to (rarely, when old) macroscopic, composed of clusters of irregular mucilaginous spheres or elongated formations, with smooth surfaces, on which scarce and irregularly situated solitary cells are observed (from above). Mucilage is firm, homogeneous, colorless, and distinctly delimited. Cells are widely oval to cylindrical, up to 4.2 µm long, distant (after division in pairs), with no obvious gas vesicles, and pale blue–green. Cell division is only in one plane in successive generations, perpendicular to the longer cell axis.

Populations typically develop in benthic habitats (epipelic, among plants) in oligotrophic and mesotrophic pools, ponds, and lakes; sometimes they float in metaphyton or in plankton. Of two species, *E. glebulenta* occurs rarely in clear, cold lakes in the northern temperate areas. In North America it was observed in small lakes in central Canada (our data). (The second species is known only from South Africa.)

Gloeothece Nägeli (Fig. 3A)

Colonies are composed of numerous groups of cells that have their own, usually lamellated and delimited mucilaginous sheaths or envelopes. They form slimy macroscopic layers on various substrata, sometimes surrounded by fine common mucilage. The mucilaginous envelopes are colorless or colored bluish, violet, reddish, or yellow–brown. Cells are widely oval to rod-shaped, pale or bright blue–green, olive green, or violet, with no obvious gas vesicles. Cell division is by binary fission, only transversely to the longer cell axis; reproduction is by disintegration of clusters of cells. Variability and taxonomic relationships of various populations are poorly known, and it has not yet in culture. Nanocytic cell division was described in several species (Geitler, 1942).

The genus contains mainly subaerial (over 30 known) species growing on wet rocks; a few species are found in aquatic habitats, and one species is nanoplanktonic (Skuja, 1964). Several species are recorded from North America (*G. confluens*, *G. palea*, *G. linearis*, *G. fusco-lutea*, *G. rupestris*, and others; Prescott, 1962; Whitford and Schumacher, 1969; Duthie and Socha, 1976; Stein and Borden, 1979; Sheath and Steinman, 1982), but because their taxonomy requires revision, other species are likely. Species cited in North American literature in various concepts (e.g., *G. distans* and *G. membranacea*) are taxonomically unclear. A few species are known from (possibly restricted to) subtropical regions (Puerto Rico, Cuba, and Mexico), including *G. endochromatica*, *G. interspersa*, *G. opalothecata*, and *G. prototypa* (Gardner, 1927). Taxonomic revision of the genus is needed.

Johannesbaptistia De Toni (Fig. 6A)

Short discoid cells are arranged in uniseriate pseudofilaments within wide, mucilaginous, tubelike strands. Strands may be simple, unbranched, or (rarely) pseudodichotomously divided and anastomozing, and straight or slightly wavy. Mucilaginous envelopes are homogeneous, colorless, usually distinctly delimited at the margin, rarely diffuse, and rounded at the ends. Cells are short discoid, but in a lateral view are narrow oval, slightly to distinctly distant, pale grey–blue with fine granular cell content and no gas vesicles, 1–5.5(6.5) × (1.2)6.6–8.3(10) µm. Apical cells are usually rounded. Numerous necridic cells are often in pseudofilaments. Cells divide transversely to their shorter axis, that is, perpendicularly to the pseudofilaments.

Numerous species have been described (some under *Cyanothrix*; Gardner, 1927; Kiselev, 1947), but interspecific features are presently unclear. A common but variable species, *J. pellucida*, requires taxonomic attention. It occurs mainly in metaphyton of coastal swamps with elevated salinity (in Massachusetts; Drouet and Daily, 1956), in freshwaters in warmer regions (Florida, Louisiana, California, Caribbean islands, Mexico, and Belize), and in some freshwater ponds and small lakes in Connecticut, Iowa, and Indiana (Drouet and Daily, 1956). A second well-defined species, *J. schizodichotoma*, was described from slightly acidic mineral springs (as *Heterohormogonium*; Copeland, 1936) in Yellowstone National Park (28–42°C), and possibly is endemic.

Lithomyxa Howe (Not pictured)

Cells irregularly arranged in macroscopic, flat, originally slimy and later hard, crustose colonies, usually heavily encrusted by lime. Cells are enveloped by their own gelatinous envelopes. Cells are spherical to oval and short cylindrical, slightly distant from one another, rarely in groups, only slightly over 1 µm in diameter, with no obvious gas vesicles, and pale blue–green. Cell division probably is in one plane in successive generations.

Specimens of this poorly known monotypic genus (needing revision) were described (with the species *L. calcigena*) from the United States (Howe, 1932) as an important lithogenic and travertine-forming species. Not collected since the description, the herbarium type specimen does not contain similar cyanobacterial type. According to Drouet and Daily (1956), it belongs to the Eubacteria.

Radiocystis Skuja (Fig. 1C)

Colonies are microscopic, more or less spherical to slightly elongated, and have radially oriented rows of cells, which are later (particularly in the center of colonies) irregularly arranged. Colonial mucilage is fine, colorless, and diffuse (apparent in phase contrast or after any staining). Cells are slightly distant or in pairs, nearly spherical to oval, to 5 (in tropical species up to 8) µm in length (or diameter), sometimes with visible aerotopes, and pale greenish in color. Cells divide only in one plane in successive generations (perpendicularly to radiating rows); however, their radial position in colonies may be indistinct.

Species are mainly nanoplanktonic and may be part of surface blooms in mesotrophic and eutrophic waters (Skuja, 1948; Hindák and Moustaka, 1988; Hindák, 1996). Of five described species, two are tropical (*R. fernandoi* occurs in tropical regions of central America; Komárek and Komárková-Legnerová, 1993), and two (*R. geminata* and *R. elongata*) are known from plankton of mesotrophic water bodies in north-temperate regions of North America, including Ontario (Duthie and Socha, 1976; Hindák, 1996; H. Kling, personal communication).

Rhabdoderma Schmidle et Lauterborn (Fig. 4B)

Cylindrical cells are arranged more or less in the same direction, but usually not chainlike, irregularly distant, in microscopic, irregularly oval or elongated mucilaginous colonies. The mucilage is fine, homogeneous, colorless, diffuse or delimited at the margin, and sometimes indistinct. Cells always are cylindrical, rod-shaped, or slightly curved, rounded at the ends, sometimes several times longer than wide, (2)4–12(33) × (0.5)1–3.5 µm, with pale blue–green, greyish, or olive-green cell content and no obvious gas vesicles. Cell division is perpendicular to the longer axis, and sometimes asymmetrical (particularly under suboptimal conditions). Daughter cells sometimes remain joined together in short pseudofilaments; long filamentous involution cells are known.

This species never occurs in masses, but several species commonly occur in the plankton of large oligotrophic and mesotrophic lakes, particularly in northern regions of the temperate zone (Prescott, 1962; Stein and Borden, 1979; Komárek and Anagnostidis, 1998). *R. lineare* and *R. compositum* are common species in the plankton of various lakes and water bodies in the northern United States and Canada. *R. curtum* was described from saline coastal swamps and pools in California, and epiphytic *R. zygnemicolum* is known only from Yellowstone National Park in temperatures > 30°C (Copeland, 1936).

Rhabdogloea Schröder (Fig. 4C)

Fusiform cells are arranged irregularly, distant, rarely more or less in one direction, in microscopic mucilaginous colonies. Mucilage is fine, homogeneous, colorless, and usually diffuse or indistinct at the margin. Cells are fusiform or cylindrical with conical cell ends, sometimes pointed, often slightly curved, (1.8)3–12(22) × (0.5)1–3.5(6) µm, with pale blue–green or olive-green content. Involution filamentous cells are up to more than 25 µm long (known only from culture). Cell division is by binary fission, transverse to the long axis into two more or less isomorphic daughter cells. Reproduction is by disintegration of colonies.

Ten to fifteen species are known, mostly planktonic in lakes and reservoirs, but also endogloeic in *Microcystis* colonies (Komárek and Anagnostidis, 1998). One metaphytic species occurs in acidic peaty swamps. Planktonic *R. smithii* commonly is found in northern lakes in North America (our results) and *R. hungarica* was collected from snowfields in the Rocky Mountains (as *Gloeothece transsylvanica*; Garric, 1965). *R. subtropica* was described from the plankton of small reservoirs and swamps in Cuba (Hindák, 1984).

Synechococcus Nägeli (Fig. 4A)

Cells are solitary, or in irregular groups or agglomerations. They do not form distinct colonies, have no slimy envelopes or only a very fine, diffuse, and narrow gelatinous layer. Cells are cylindrical to long rod-shaped, sometimes slightly arcuate, after division in pairs. Cell content is homogeneous, occasionally has a slightly recognizable centro- and chromatoplasm (= few to several parietal thylakoids), have pale blue–green, olive green, or reddish color, no gas vesicles, and sometimes prominent granules. Cells are (1.5)3–15(40) × 0.4–3(6) µm. Cell division is by perpendicular binary fission (usually cleavage), sometimes asymmetrically. Genus specific filamentous involution cells occur under stress conditions.

This important, but possibly uncommon genus comprises several species. Cultures are used widely as experimental model organisms. Over 20 species are known after recent taxonomic revisions. *S. nidulans* is distributed in small water bodies, mainly in temperate zones, and the strain, isolated from the United States and known in the literature as *Anacystis nidulans* (Kratz and Allen's strain; Allen, 1968) is an important experimental strain. A group of thermophilic types (*S. bigranulatus*, *S. koidzumii*, and particularly *S. lividus* and *S. vulcanus*) are adapted to temperatures of about 70°C (Copeland, 1936; Castenholz, 1969a, b, 1970) and have been recorded from hot springs in Yellowstone National Park. Morphologically similar species are known from other hot springs in North America, Japan, Africa, and Indonesia (Geitler and Ruttner, 1935; Hirose and Hirano, 1981). Several species are known from littoral zones of lakes, soils, and swamps, and could be expected in similar environments elsewhere on the continent, but their taxonomy and distribution are poorly known. *S. sigmoideus* is a characteristic species from nanoplankton of lakes, including those in North Dakota (Moore and Carter, 1923).

Merismopediaceae (Figs. 6B–10A)

Aphanocapsa Nägeli (Fig. 6C)

Aphanocapsa comprises microscopic to macroscopic, mucilaginous colonies with irregularly arranged cells of various densities that are spherical or irregular. Colonial mucilage is mainly homogeneous and colorless, and usually has a diffuse, rarely delimited margin. Cells have no individual envelopes or, rarely, have a fine surrounding slimy layer that is distinct from the common mucilage. Cells are spherical, but after division, they are hemispherical, pale or bright blue–green or olive green, rarely (in marine species) red or pinkish in color, with no gas vesicles, and 0.4–6(12) μm in diameter. Cells divide by binary fission in two perpendicular planes in successive generations (cleavage). Reproduction is by disintegration of colonies.

Among many described species, almost 50 are well-defined. They colonize mainly subaerial, soil, and aquatic habitats (Komárek and Anagnostidis, 1998). *A. delicatissima*, *A. incerta*, *A. holsatica*, *A. conferta*, and *A. planctonica*, which are planktonic and grow in microscopic colonies in mainly mesotrophic lakes, are perhaps the most important species of this genus in North America (Duthie and Socha, 1976; Sheath and Steinman, 1982). *A. farlowiana* produces macroscopic colonies, is possibly endemic in the northern United States and Canada, and sometimes occurs in large masses in the littoral zone of ponds and smaller lakes (Drouet, 1942). *A. thermalis*, *A. botryoides*, *A. protea*, and *A. tolliana* are known from thermal springs (Copeland, 1936). Several species grow on wet rocks in mountains (*A. muscicola*) or are benthic in streams (sometimes identified as *Polycystis montana* f. *minor*); several of these need revision. *A. arctica* was described from Canadian arctic regions (Whelden, 1947), and *A. intertexta* was obtained from Puerto Rico (Gardner, 1927). Other species occur in coastal marine, brackish, and saline swamps and pools (e.g., *A. marina*). This genus has a very simple morphology, leading to many misinterpretations in the literature.

Coelomoron Buell (Fig. 8C)

Cells are arranged more or less peripherally in one or a few layers (may be variable) near the surface of the microscopic, spherical, free-living colonies, which are sometimes composed of subcolonies. Colonial mucilage is fine, homogeneous, colorless, and usually diffuse at the margin, but usually more densely clustered in the center. Cells in young colonies are randomly and sparsely distributed, distant and variably arranged, and usually slightly shifted from one layer; later cells are arranged distinctly peripherally, forming one to three irregular layers. Cells are slightly elongated, widely oval to almost spherical, and radially situated in the colony (generic feature), pale blue–green or olive green, and 1–6.5 × 0.8–4 μm. One species (*C. minimum*) has distinct aerotopes (gas vacuoles). Cell division occurs in two planes in successive generations, perpendicular to one another and more or less to the colony surface. Reproduction is by disintegration of colonies.

Seven described species (Komárek and Anagnostidis, 1998) exist, but further species may be discovered. The type species *C. regulare* was described from freshwater metaphyton in the United States (Buell, 1938; Geitler, 1942), but has not been seen again. Nanoplanktonic *C. pusillum* occurs in mesotrophic waters in temperate and subtropical zones (syn. = *Coelosphaerium collinsii* Drouet et Daily 1942, described from the United States). *C. microcystoides* and *C. vestitum* were described from tropical and subtropical regions (Komárek, 1989). The aerotopated, pantropical *C. minimum* was recorded from the plankton of tropical Lake Catemaco in Mexico (Komárková-Legnerová and Tavera, 1996).

Coelosphaerium Nägeli (Fig. 8B)

Coelosphaerium comprises free-living, spherical or oval mucilaginous colonies, in which the cells are arranged irregularly in one marginal layer (but sometimes slightly shifted to one another), or near the surface of the sphere. Colonies sometimes are composed of subcolonies. Mucilage is colorless and homogeneous. Cells are spherical, pale or bright blue–green, and 1–7 μm in diameter. One species has visible aerotopes. Cell division occurs in two planes in successive generations, perpendicular to one another and to the colony surface. Reproduction is by disintegration of a colony and by liberation of subcolonies.

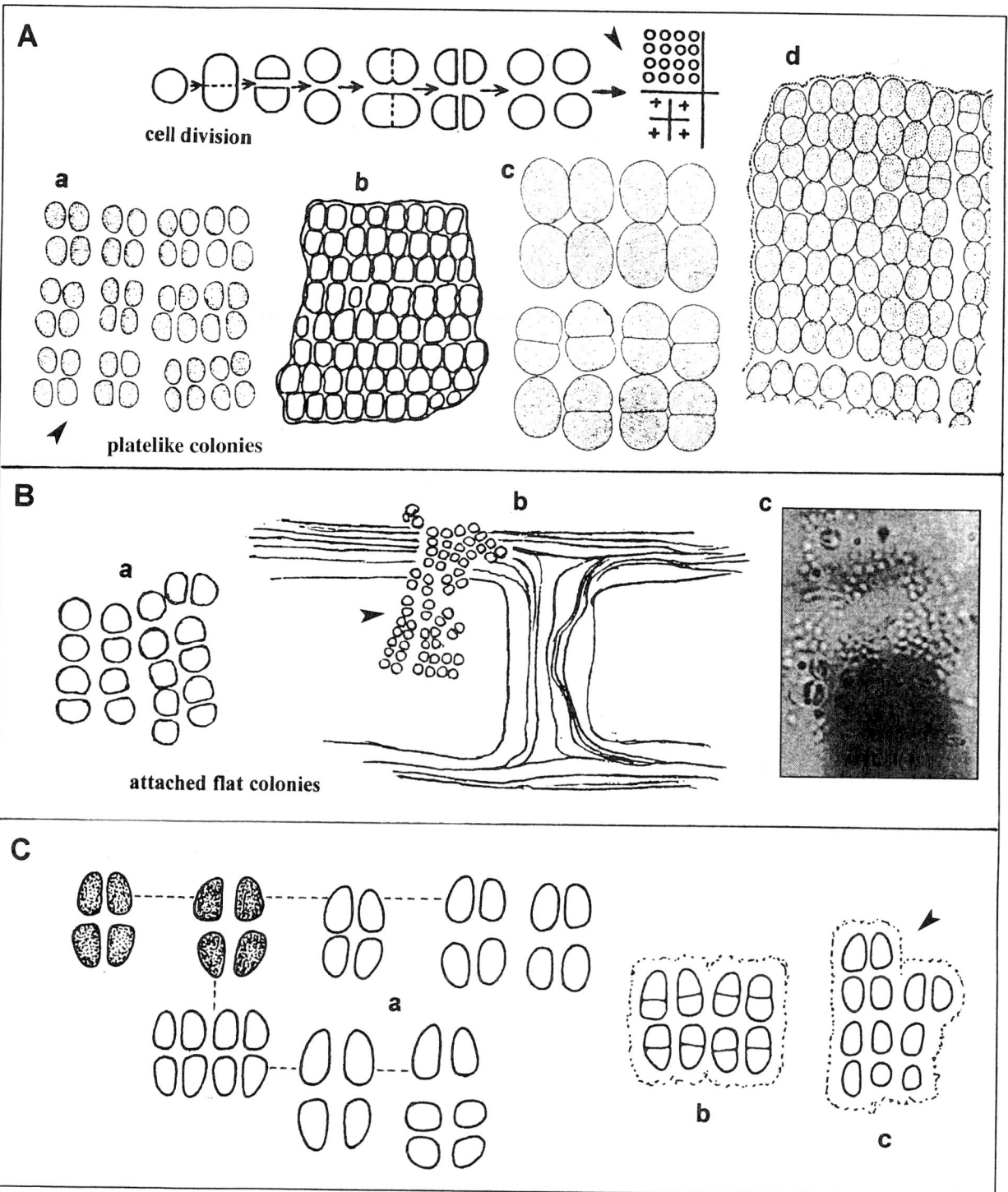

FIGURE 7 (A) *Merismopedia*: **a.** *M. punctata* (after Smith, 1920, from Wisconsin lakes); **b.** *M. angularis* (after Thompson, 1938, from eastern Kansas); **c.** *M. smithii* (after Smith, 1920, from Wisconsin lakes; sub *M. elegans* var. *maior*); **d.** *M. elegans* (after Smith, 1920, from Wisconsin lakes). (B) *Mantellum*: **a–c.** *M. rubrum* (after Tavera and Komárek, 1996, from a volcanic lake in Mexico). (C) *Cyanotetras*: **a–c.** *C. crucigenielloides* (after Komárek, 1995; variation in form of colonies from swamps in Cuba).

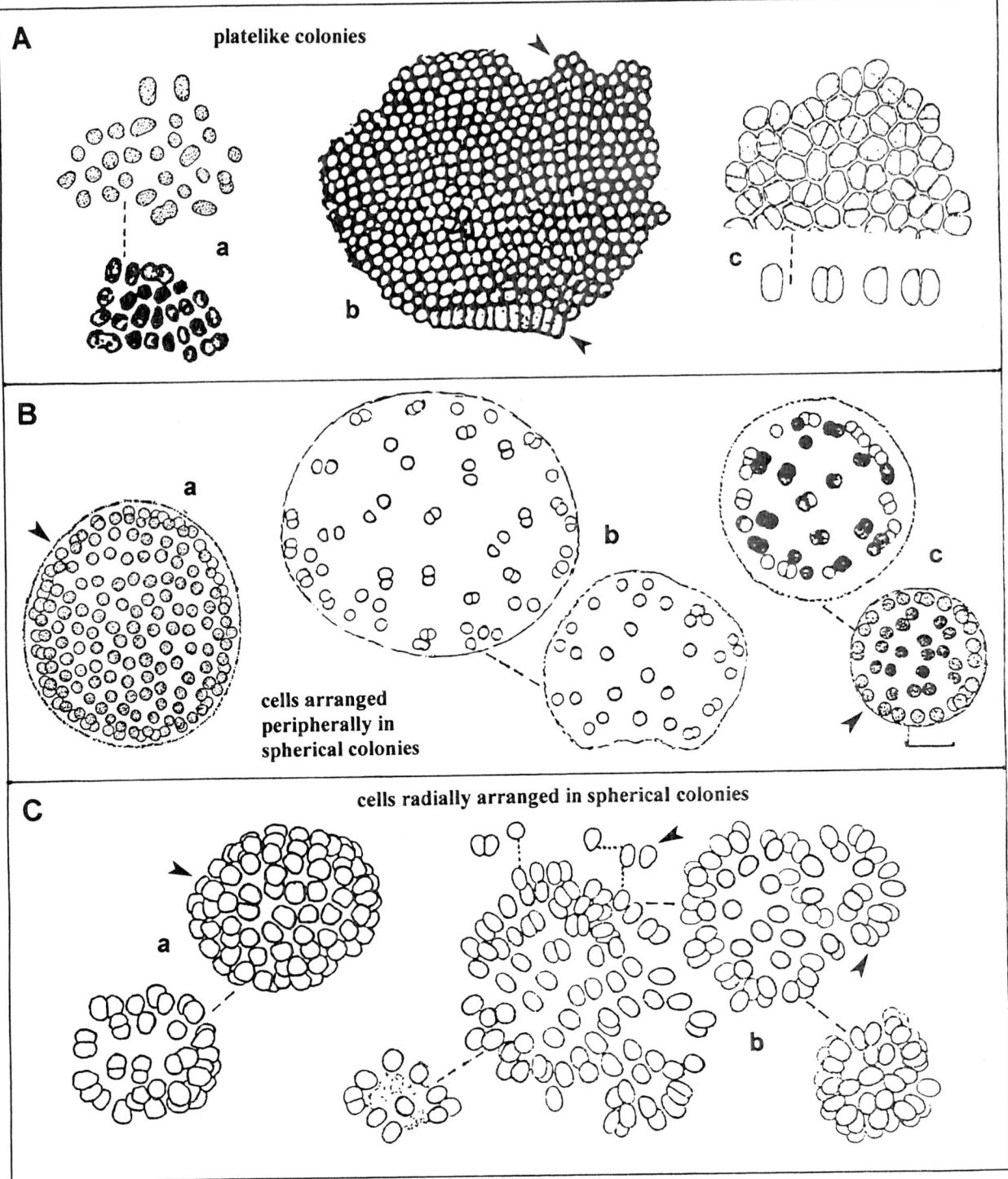

FIGURE 8 (A) *Microcrocis*: a. *M. irregulare* (after Lagerheim, from Smith, 1950); b. *M. obvoluta* (after Tiffany, 1934, from North America, Lake Erie); c. *M. pulchella* (after Buell, 1938, from North America). (B) *Coelosphaerium*: a. *C. kuetzingianum* (after Smith, 1950, from North America); b. *C. subarcticum* (after Komárek and Komárková-Legnerová, 1992, from Canadian lakes); c. *C. aerugineum* (bar = 10 μm; after Smith, 1920, from Wisconsin lakes, and after Komárek, 1958). (C) *Coelomoron*: a. *C. tropicalis* (after Senna et al., 1998, from Brazil); b. *C. microcystoides* (after Komárek, 1989, populations from Cuba).

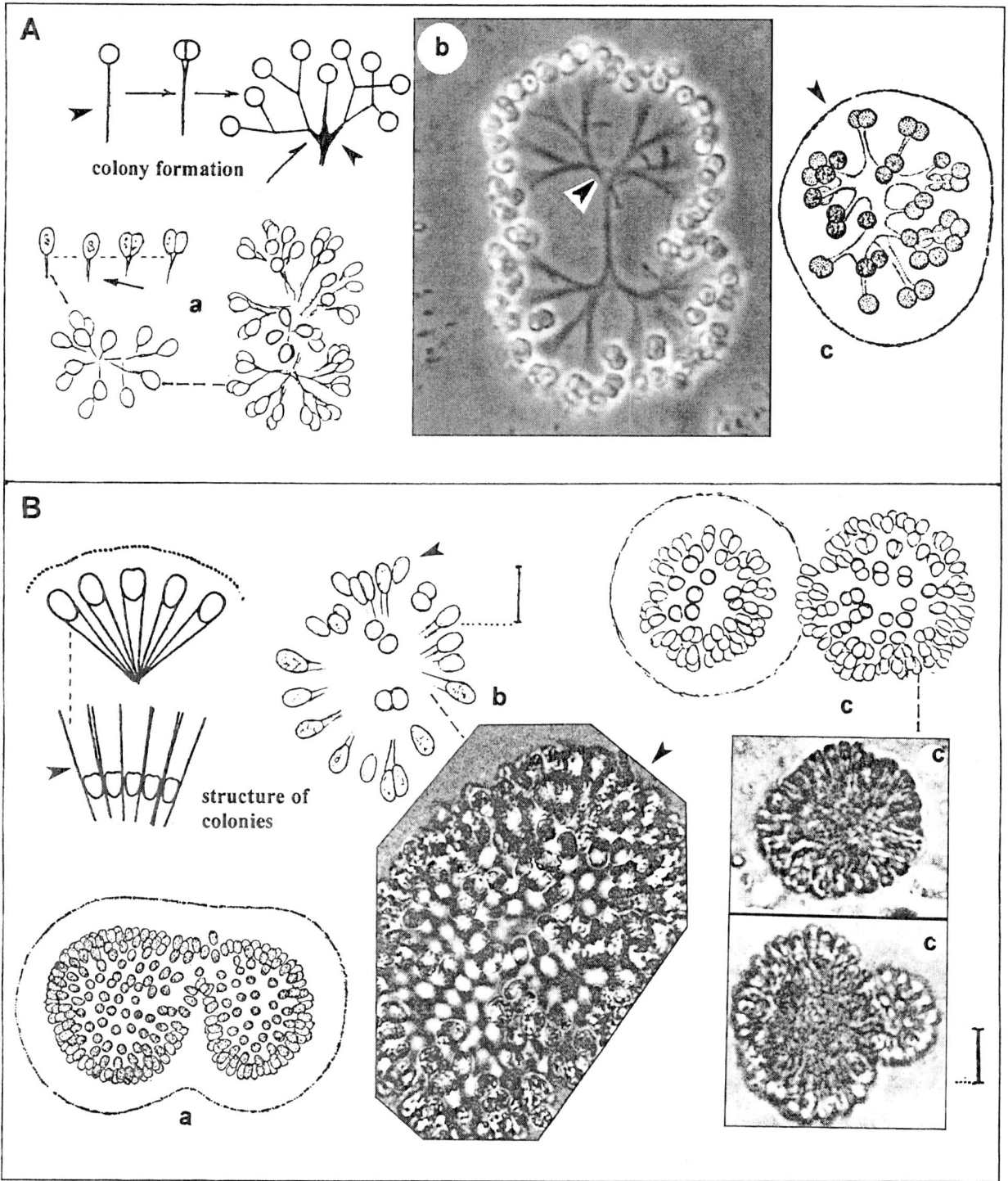

FIGURE 9 (A) *Snowella*: a. *S. fennica* (after Komárek and Komárková-Legnerová, 1992, from Canadian lakes); b. *S. litoralis* (original photo by H. Kling, from lakes in Central Canada); c. *S. litoralis* (after Smith, 1950, from Wisconsin lakes; sub "*Gomphosphaeria lacustris*"). (B) *Woronichinia*: a. *W. naegeliana* (after Smith, 1950, from North America; sub *Coelosphaerium naegelianum*); b. *W. naegeliana* (bar = 10 μm; after Komárek and Komárková-Legnerová, 1992, from central Canadian lakes); c. *W. klingae* (bar = 10 μm; after Komárek and Komárková-Legnerová, 1992, from Manitoba, Canada).

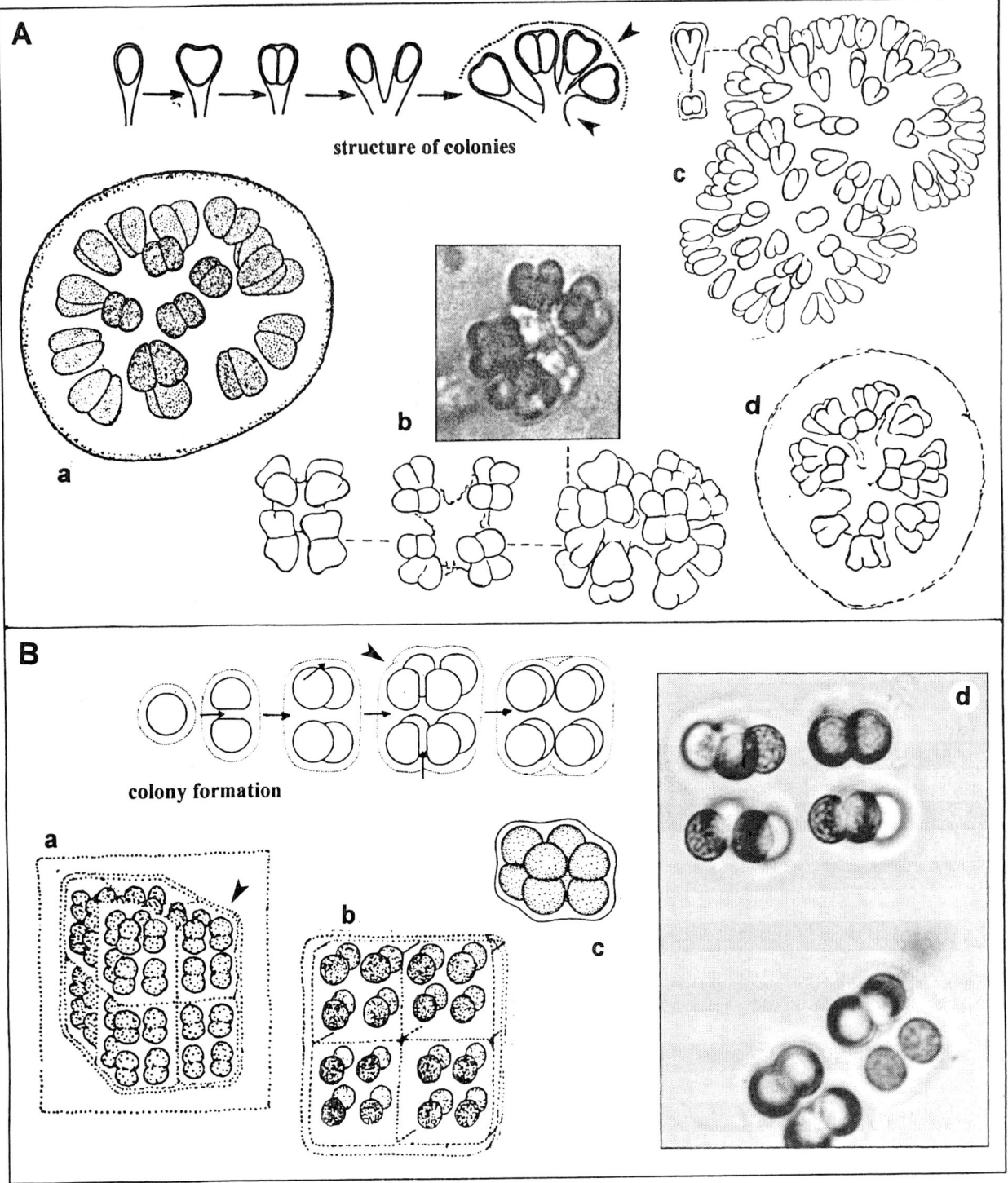

FIGURE 10 (A) *Gomphosphaeria*: a. *G. aponina* (after Smith, 1920, from Wisconsin lakes); b. *G. natans* (after Komárek and Komárková-Legnerová, 1992, from central Canadian lakes); c. *G. semen-vitis* (after Komárek, 1989, from alkaline swamps in Cuba and in Florida); d. *G. virieuxii* (after Komárek and Komárková-Legnerová, 1992, from central Canadian lakes). (B) *Eucapsis*: a. *E. alpina* (after Clements and Shantz, 1909, original drawing from Colorado); b. *E. alpina* (after Prescott and Croasdale, 1937, from Minnesota); c. *E. alpina* forma (after Thompson, 1938, from eastern Kansas); d. *E. alpina* (original photo from central Europe).

Ten species are known from plankton of lakes and reservoirs across North America and worldwide (Prescott, 1962; Whitford and Schumacher, 1969; Duthie and Socha, 1976; Stein and Borden, 1979; Sheath and Steinman, 1982; some = *Woronichinia* spp.); they rarely occur in metaphyton (Komárek and Anagnostidis, 1998). Nanoplanktonic *C. minutissimum*, *C. aerugineum*, and *C. kuetzingianum* are known from lakes in the temperate zone of North America. *C. subarcticum* occurs in clear lakes of central Canada, but is probably more widely distributed in similar environments. The occurrence of *C. dubium* and *C. confertum* in North America is unconfirmed.

Cyanotetras Hindák (Fig. 7C)

Cells occur in microscopic, free-floating colonies that are flat or platelike and composed from slightly elongated cells, the longer axis of which lies in the plane of the colony. Colonies typically are few-celled. The cells are arranged in short perpendicular rows, but have irregularities. The colony is surrounded by very fine, colorless, and diffuse mucilage. Cells are oval or ovoid, sometimes in twos or flat tetrads, have pale blue–green color and, homogeneous content. One species has gas vesicles and visible aerotopes; another species has iron precipitates within the mucilage. Cells are 1.5–5 μm long and 1–4.5 μm wide. Cells divide by binary fission in two planes in successive generations, perpendicularly to one another and to the plane of the colony.

Three species have been described. Two are from the subtropical regions of North America (Komárek, 1995; Komárek and Komárková-Legnerová, 2002): *C. crucigenielloides* is common in the metaphyton of alkaline swamps, ponds, and lakes in the Caribbean region; *C. aerotopa* was discovered recently in plankton of mesotrophic and eutrophic waters in Mexico.

Gomphosphaeria Kützing (Fig. 10A)

Colonies are free-living, spherical or irregularly oval, and sometimes composed of subcolonies. They have a central system of thick mucilaginous stalks that are nearly pseudodichotomously divided and may be diffuse within the colony. Stalks widen at the ends and envelope individual cells with a thin mucilage layer. Cells are slightly elongate, obovate, or club-shaped, and radially oriented more or less at the colony periphery, which is sometimes enveloped by a fine, colorless, and diffuse mucilage. Cells have a homogeneous pale or bright blue–green, olive-green, or red content and are (4.2)6–12(15) × 2–8(13.2) μm. After division, the cells may remain joined together and form a characteristic cordiform shape. Cells in colonies are slightly distant and sometimes slightly radially displaced from one another. Cell division occurs in two planes in successive generations, perpendicular to one another and to the colony surface. Reproduction is by colony disintegration.

Six species have been recorded from North America (Prescott, 1962; Duthie and Socha, 1976; Stein and Borden, 1979; Sheath and Steinman, 1982). *G. natans* is planktonic in northern mesotrophic lakes (from Canada: Komárek and Hindák, 1988; Komárek and Komárková-Legnerová, 1992) and possibly endemic to the continent. *G. virieuxii* (planktonic in lakes), *G. aponina* (metaphytic in clear, slightly acidic swamps) and *G. salina* (metaphytic in saline pools and swamps), are known from across the temperate zones. *G. multiplex* and *G. semen-vitis* were recorded from alkaline swamps in subtropical North America (Caribbean; Komárek, 1989). *G. cordiformis* (*sensu* Smith, 1920) was recorded from Wisconsin lakes, but needs taxonomic revision and confirmation.

Mantellum Dangeard (Fig. 7B)

Cells occur in flat, microscopic formations attached to substrata (as epiphytes on other algae) in one layer and arranged more or less in perpendicular rows. Cells are spherical, pale blue–green or reddish in color, have homogeneous content or distinguishable centro- and chromatoplasm, and are 0.8–4 μm in diameter. Cells usually are enveloped by thin, very fine, colorless, and diffuse mucilage. Division is by binary fission, in two planes perpendicular to one another and to the substrate; from this process, flat formations arise on the substrate, but sometimes have slightly shifted cells.

Only three epiphytic species, one marine and two freshwater, have been described to date. One freshwater species with reddish cells (*M. rubrum*) is known from deep volcanic lakes in central Mexico (Tavera and Komárek, 1996).

Merismopedia Meyen (Fig. 7A)

Merismopedia comprises flattened, free-living, platelike (rectangular), more or less rectangular colonies that have one layer of cells, arranged loosely or densely in perpendicular rows and enveloped by fine, colorless, usually indistinct, and marginally diffuse mucilage. Colonies are flat or slightly wavy, usually microscopic (except for a few species that are macroscopically visible), and sometimes composed of subcolonies. Cells are spherical or widely oval before the division, pale or bright blue–green, (rarely reddish), and sometimes have visible centro- and chromatoplasm (parietal thylakoids). Several planktonic species have gas vesicles (few or solitary in cell centers). Occasionally the cells have slimy envelopes. After division, the cells are hemispherical and (0.4)1.2–6.5(17) μm in diameter.

When elongated, the longer axis is situated in the plane of the colony. Cell division (binary fission) occurs regularly in two planes perpendicular to the plane of colony; the daughter cells do not move from their position after division. Reproduction is by fragmentation of colonies.

Over 30 species are known from freshwater and saline environments (Komárek and Anagnostidis, 1998). Several common species occur across temperate regions in the plankton and metaphyton of eutrophic and mesotrophic waters in North America, especially *M. punctata* and *M. tenuissima*; two common species, *M. glauca* and *M. elegans* occur in the metaphyton (rarely in plankton) of mesotrophic lakes, ponds, and swamps (Prescott, 1962; Whitford and Schumacher, 1969; Duthie and Socha, 1976; Stein and Borden, 1979; Sheath and Steinman, 1982). *M. smithii* (syn. = *M. major*), is a species that has large cells and is typical of temperate swamps only in North America (not confirmed from other regions). *M. angularis* occurs in acidic swamps in cold temperate to subarctic regions. *M. gardneri* grows in pools on the American Pacific coast. *M. convoluta* is a conspicuous metaphytic species from warmer regions of the continent (North Carolina, Florida, and the Greater Antilles). Records of a species designated *Agmenellum quadruplicatum* in floras and ecological studies (according to Drouet and Daily, 1956) may be referred to either *M. punctata* or *M. glauca*. Strains designated by the same name in world collections and experimental papers (e.g., Van Baalen and O'Donnell, 1972) contain the species *Synechococcus nidulans*.

Microcrocis Richter (Fig. 8A)

Cells are arranged in free-living, flat, microscopic to macroscopic colonies, in one layer, and form perpendicular rows in young colonies; later they are irregularly aggregated. Colonies are enveloped by fine, colorless, homogeneous, and diffuse mucilage. Cells have no individual sheaths. A critical feature is the presence of elongated cells situated with the longer axes perpendicular to the colony plane. Cells are widely oval to rodlike with rounded ends, homogeneous content, and pale to bright blue–green color. The cell structure has been studied only in one species, in which the thylakoids were flexuous throughout (different from other related genera that have parietal thylakoids). Cell dimensions are (2)5–16(19) × 1.5–6(8) μm. Cells divide longitudinally in two perpendicular planes in successive generations. Reproduction is by fragmentation of colonies.

This rare and poorly known genus has 10–15 species, of which several need revision (Komárek and Anagnostidis, 1998). Metaphytic, epipelic, and epipsammic species exist in freshwater and marine environments. *M. irregularis*, *M. obvoluta*, *M. gigas*, and *M. pulchella* were described from North America (Komárek and Anagnostidis, 1998), but their taxonomy and ecology require confirmation.

Snowella Elenkin (Fig. 9A)

Snowella comprises free-living, spherical or oval microscopic colonies, sometimes composed of subcolonies, that have a central system of thin, nearly pseudodichotomously divided gelatinous stalks (may be confluent) that bear cells at their ends (periphery of colony). Colonies are enveloped by very fine, unstructured, and diffuse mucilage. Cells are spherical or slightly radially elongated, and pale blue–green, olive green, or yellowish in color; one species may be pinkish. Two species have one or a few central aerotopes. The cells are 0.6–4.2 μm in diameter (or long). Cell division is by binary fission in two planes, perpendicular to one another and to the colonial surface. Reproduction is by disintegration of colonies.

Seven species are known from the plankton of temperate fresh and brackish waters, particularly from cold, northern, mesotrophic lakes (Komárek and Anagnostidis, 1998). *S. septentrionalis* and *S. fennica* were recorded from Canada (original unpublished data). The tychoplanktonic *S. rosea* has reddish cells and the nanoplanktonic *S. litoralis* and *S. lacustris* occur sporadically in the temperate zones of North America, although *S. lacustris* may also occur in subtropical regions (Komárek and Hindák, 1988).

Synechocystis Sauvageau (Fig. 6B)

Cells are solitary or in pairs short time after division, have no mucilage or a fine, narrow, colorless, indistinct, mucilage layer around the cells that is diffuse at the margins. Cells are spherical, rarely to widely oval before division, usually pale or rarely, bright blue–green or olive green, in few species, reddish violet or red. Cells are 0.7–15(30) μm in diameter. Cell content is more or less homogeneous, sometimes has solitary granules and a distinguishable centro- and chromatoplasm (parietal position of thylakoids). The species with irregularly distributed thylakoids and larger dimensions probably belong to another taxonomic type. Cell division is by binary fission (cleavage), regularly in two perpendicular planes in successive generations. Cells grow to the original size and shape before the next division.

Over 20 species have been described. Most are planktonic and metaphytic, but some occur in thermal springs, mineral and saline environments, and from marine Ascidians (epizoic on the surface) and are symbiotic in dinoflagellates (Norris, 1967; Schulz-Baldes and Lewin, 1976; Komárek and Anagnostidis, 1998).

In North America, *S. aquatilis* and *S. salina* are planktonic in higher conductivity pools and lakes (Stein and Borden, 1979), *S. sallensis* occurs in northern peat bogs, and *S. thermalis* and *S. minuscula* are known from hot springs (Copeland, 1936). *S. fuscopigmentosa*, *S. primigenia*, and *S. willei* were recorded from swamps in the Caribbean region (Gardner, 1927, Kováčik, 1988). An interesting *Synechocystis* species was found in hollow hairs from polar bears in zoological gardens in California (Lewin and Robinson, 1979) and given the preliminary name *Aphanocapsa montana*, but its taxonomy is not resolved. Picoplanktonic forms that have spherical cells were recorded from clear lakes in New York (Corpe and Jensen, 1992) and probably belong to this genus.

Woronichinia Elenkin (Fig. 9B)

Woronichinia comprises spherical or irregularly oval, free-living colonies, commonly composed of subcolonies, that have a central system of simple, radially arranged, colorless, mucilaginous (sometimes tubelike) stalks, joined in the center of the colony. The colony is surrounded by a fine, diffuse, colorless mucilage. Cells are at (or within) the ends of mucilaginous stalks and radially arranged, forming a layer of cells at the colony periphery. Stalks are densely packed, causing radial lamellation, but sometimes are diffuse near the center. Cells may be nearly spherical, but are usually slightly elongate, widely oval, or obovoid; in old colonies the cells are densely arranged. The cells are pale blue–green, olive green, or slightly reddish in color, sometimes have gas vesicles and visible aerotopes, and are (1)2.5–7 × 1–4(5) µm. Cell division occurs in two planes, perpendicular in successive generations and to the colony surface. Reproduction is by disintegration of colonies and by solitary cells ("expulsion cells"), which sometimes liberate spontaneously from a colony.

Of 15 described species, 3 contain gas vesicles and may form surface blooms in mesotrophic and eutrophic waters (Komárek and Anagnostidis, 1998). In temperate regions, the aerotopated (with gas vacuoles) and cosmopolitan *W. naegeliana* (syn. *Coelosphaerium naegelianum*) is widespread in the plankton of many lakes and may form numerous morphotypes, of which few are regarded as endemic in North America (Cronberg and Komárek 1994). A similar species named *Gomphosphaeria wichurae* has been recorded in North America, but probably falls within the range of variation in *W. naegeliana*. *W. fremyi* was described from the Caribbean region (Komárek, 1984). *W. karelica* and *W. elorantae*, originally described from the Baltic region in Europe, are nanoplanktonic species that occur in northern lakes in Canada (original unpublished data). *W. klingae* is an interesting planktonic species that has reddish cells and solitary gas vesicles, and possibly is endemic to stagnant, swampy waters in central Canada (Komárek and Komárková-Legnerová, 1992).

Microcystaceae (Figs. 10B–12)

Chondrocystis Lemmermann (Fig. 12B)

Colonies occur in microscopic or macroscopic masses, are free-living, irregular in outline, packet-like, gelatinous, in granular agglomerations, and composed of numerous subcolonies, sometimes with inner $CaCO_3$ precipitates. The mucilage is firm, colorless to yellow–brown, and distinctly limited. Cells and their groups are surrounded by individual firm sheaths and envelopes, which are tightly and irregularly aggregated together. Cells are spherical, pale blue–green, 1.5–6 µm in diameter, have no gas vesicles, and divide by binary fission in three perpendicular planes in successive generations. Reproduction is by disintegration of colonies.

Three halophilic (metaphytic) and one benthic species from limestone mountainous streams (*C. dermochroa*) are known (Komárek and Anagnostidis, 1998), but not yet confirmed from North America. *C. schauinslandii* from Hawaii and *C. bracei* from the Bahamas and Bermuda were described (Lemmermann, 1899; Howe, 1924).

Eucapsis Clements *et* Shantz (Fig. 10B)

Colonies are microscopic, mucilaginous, free-living, more or less cubic in form, and sometimes composed of subcolonies with cells arranged three dimensionally (cubelike) in more or less regular perpendicular rows; rows may be disturbed in a few species. The mucilage is colorless, hyaline, and usually diffuse at the margin. Cells are spherical or slightly oval before division, pale or bright blue–green or olive green, 1–6(11) µm in diameter, and have no obvious gas vesicles. Cell division occurs regularly in three perpendicular planes in successive generations. Daughter cells more or less keep their position in a colony. Reproduction is by disintegration of colonies.

Eight species and several varieties have been described. The often are metaphytic in swamps and bogs, but also are found in volcanic soils (Komárek and Hindák, 1989; Komárek and Anagnostidis, 1998). *E. minor* and *E. alpina* are found in peaty habitats in North America. The multicellular *E. alpina* var. *maior* (known only from Alaska; Prescott and Vinyard, 1965) probably represents a separate type of this genus (but the name is a later homonym and must be changed).

Gloeocapsa Kützing (Fig. 12A)

Colonies are microscopic, usually spherical, usually aggregated in a macroscopic mucilaginous, amorphous mass that colonizes wet stony substrates and, less

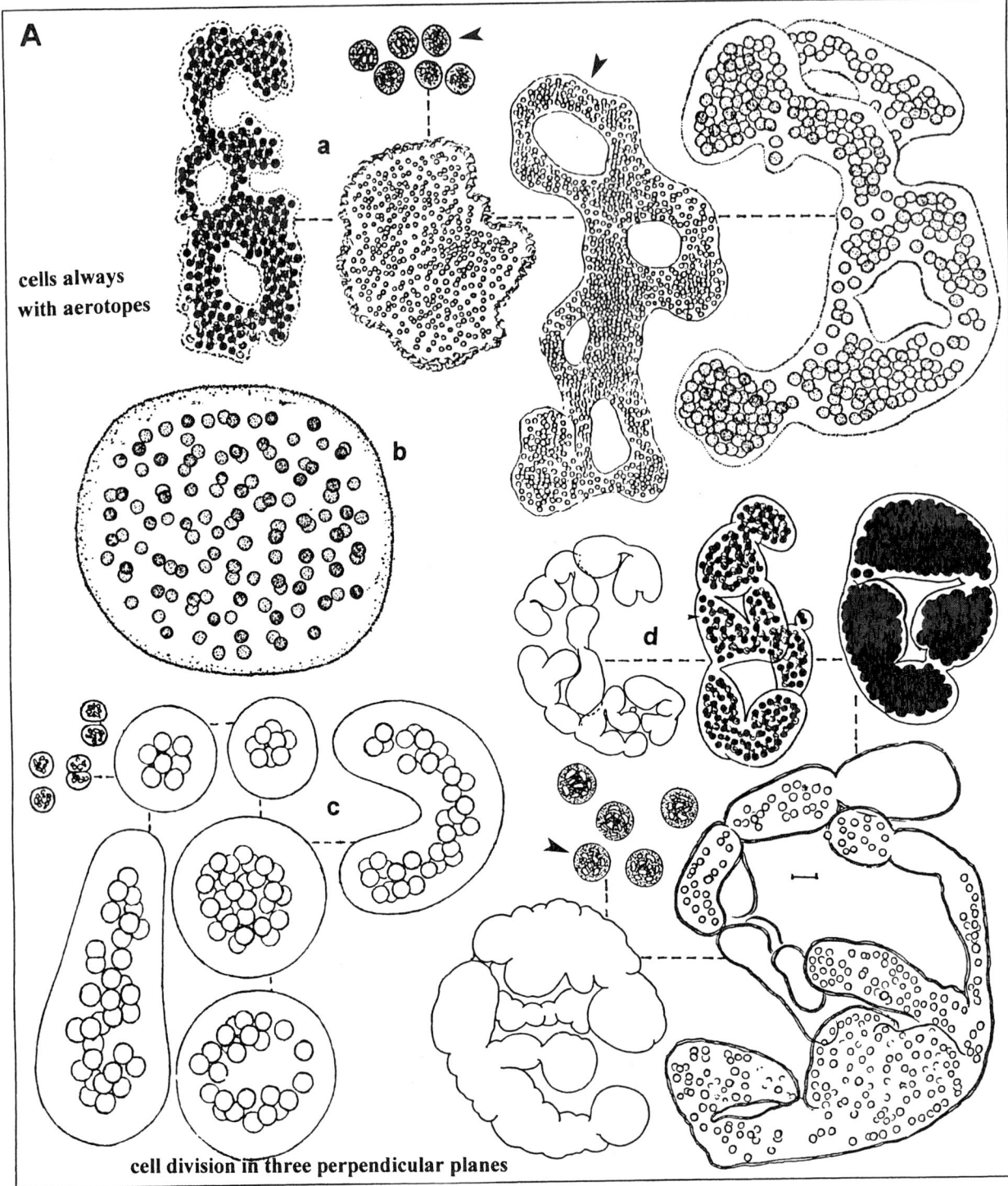

FIGURE 11 (A) *Microcystis*: **a.** *M. aeruginosa* (after Teiling, 1941, Komárek, 1958, and Smith, 1950); **b.** *M. pulchra* (after Smith, 1920, from Wisconsin lakes); **c.** *M. comperei* (after Komárek, 1964, tropical species, from Central America); **d.** *M. wesenbergii* (after Teiling, 1941, and Wojciechowski 1971).

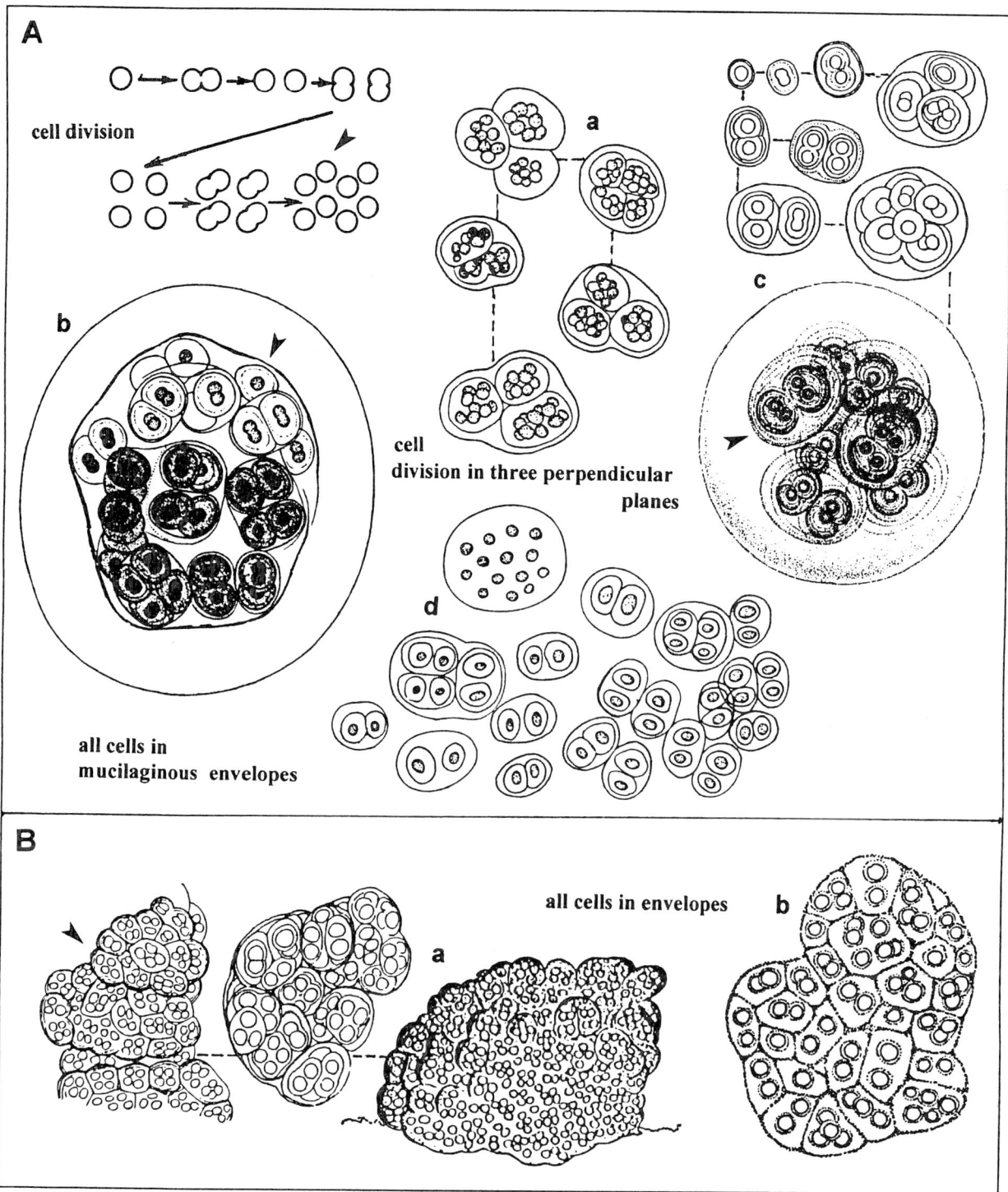

FIGURE 12 (A) *Gloeocapsa*: **a.** *G. conglomerata* (after Kützing from Geitler, 1925); **b.** *G. alpina* (after Geitler, 1932); **c.** *G. sanguinea* (from Anagnostidis and Komárek, 1998); **d.** *G. gelatinosa* (after Kützing from Tilden, 1910). (B) *Chondrocystis;* **a.** *C. dermochroa* (after Geitler, 1925, and Starmach, 1966; sub *Gloeocapsa dermochroa*); **b.** *C. schauinslandii* (after Lemmermann, 1905, from Hawaii Islands).

frequently tree bark or aquatic habitats (metaphyton or plankton). Cells and their groups are surrounded by wide, spheroidal, concentrically layered gelatinous sheaths (with distinct or indistinct lamellation) and sharply delimited margins, which may be intensely colored (some species) yellow, yellow–brown, orange, red, blue, or violet. Sheath colors may change according to environmental pH (Jaag, 1945). Cells are spherical, but after division, are hemispherical; only in dormant stages are they irregular (not dividing). Cells have firm, gelatinous, usually rounded envelopes, with pale to bright blue–green or olive green protoplast, are 0.7–6(11) µm in diameter (without envelopes), and have no obvious gas vesicles. Cell division occurs in three perpendicular planes in successive generations. Daughter cells grow to the original size and shape before next division (Golubić, 1965, 1967a, b). Reproduction is by disintegration of colonies.

Of the more than 100 species described, perhaps 50 are clearly defined (Komárek and Anagnostidis, 1998). Most forms grow on wet rocks and walls, commonly in mountain areas. Several species are restricted to calcareous or acidic rocky substrates (Jaag, 1945; Golubić, 1967). *G. conglomerata*, *G. caldariorum*, *G. decorticans*, *G. gelatinosa*, *G. granosa*, *G. arenaria*, and *G. atrata* (= *G. alpicola*) have colorless sheaths, *G. nigrescens* and *G. alpina* have blue or violet sheaths, *G. sanguinea* has red sheaths, and *G. fusco-lutea*, *G. kuetzingiana*, and *G. rupestris* have yellow or yellow-brown sheaths; all have been recorded in North America (Prescott, 1962; Duthie and Socha, 1976; Stein and Borden, 1979; Sheath and Steinman, 1982). *G. sparsa* is known only from the eastern United States (Wood, 1869). *G. thermophila* was described from thermal springs in California, whereas *G. acervata*, *G. calcicola*, *G. sphaerica*, and several other spesies are known from limestone substrates in Puerto Rico (Gardner, 1927). Taxonomic revision of numerous species is needed; typical *Gloeocapsa* species are not yet in culture.

Microcystis Kützing ex Lemmermann (Fig. 11)

This genus has irregular micro- or macroscopic colonies that are free-floating, compact or clathrate, may be composed of clustered subcolonies, and has sparsely or densely, irregularly arranged cells. The mucilage is fine, colorless, and diffuse or distinctly delimited, sometimes forming a wide margin around the cells (rarely with indistinct structures), or delimited along cell agglomerations. Cells are spherical or hemispherical after division and pale blue–green, but they appear brownish due to aerotopes that mask the blue–green color of the protoplast. Cells are 0.8–6(9.4) µm in diameter and have no individual mucilaginous envelopes. Cell division is by binary fission in three perpendicular planes in successive generations. The daughter cells grow to the original shape and size before the next division.

Based on current revisions, about 25 planktonic species are known worldwide, and many form dense blooms in eutrophic waters (Reynolds and Walsby, 1975). Because many strains (species) are toxic, *Microcystis* is one of the most important cyanobacteria in limnological studies (Gorham and Carmichael, 1988; Chorus and Bartram, 1999). Several species have been recorded from North America, including toxic forms of *M. viridis*, *M. aeruginosa*, and *M. ichthyoblabe* (Smith, 1920; Prescott, 1950, 1962; Whitford and Schumacher, 1969; Duthie and Socha, 1976; Stein and Borden, 1979; Fallon and Brock, 1981; Doers and Park, 1988; our results). In a few North American species, including *M. natans*, *M. smithii* (as *Aphanocapsa pulchra*), *M. flos-aquae*, and *M. wesenbergii*, toxicity has not yet been recorded or definitively proved. Identification of species is difficult and further taxonomic investigation is needed. At least 50% of the species in the genus are restricted to tropical and subtropical regions (Huber-Pestalozzi, 1938; Desikachary, 1959; Komárek et al., 2002). Of these, *M. comperei* was described from Cuba and other species probably occur in southern locations. *M. glauca*, recorded from the United States by Smith (1950), and other species (denoted *Polycystis firma*, *P. pulverea*, and *P. marginata*) were recorded from North Carolina by Whitford and Schumacher (1969), but they are taxonomically uncertain.

Chroococcaceae (Figs. 13 and 14)

Asterocapsa Chu (Fig. 13A)

Cells are solitary or occur in spherical, microscopic colonies, sometimes agglomerated in granular, macroscopic, gelatinous masses. Cells and colonies are enveloped by distinctly delimited, usually spherical, firm mucilaginous sheaths, which are colorless or colored (blue, violet, orange, reddish), sometimes slightly concentrically lamellated and have a smooth or warty surface. Cells are subspherical, oval, irregular, or polygonal–rounded in outline, pale or bright blue–green or olive green, often have fine granulations in the protoplast, are 2–8 µm in diameter, and have no obvious gas vesicles. Cells divide by binary fission in various planes, irregularly. Reproduction is by cells that are liberated from split firm, gelatinous envelopes. Unicellular and colonial types are included in this genus, but may become classified as different genera.

About 15 species have been described, most of which are subaerial on wet rocks (Komárek and Anagnostidis, 1998). Several interesting species are known from Mexico (Komárek, 1993, *A. divina*)

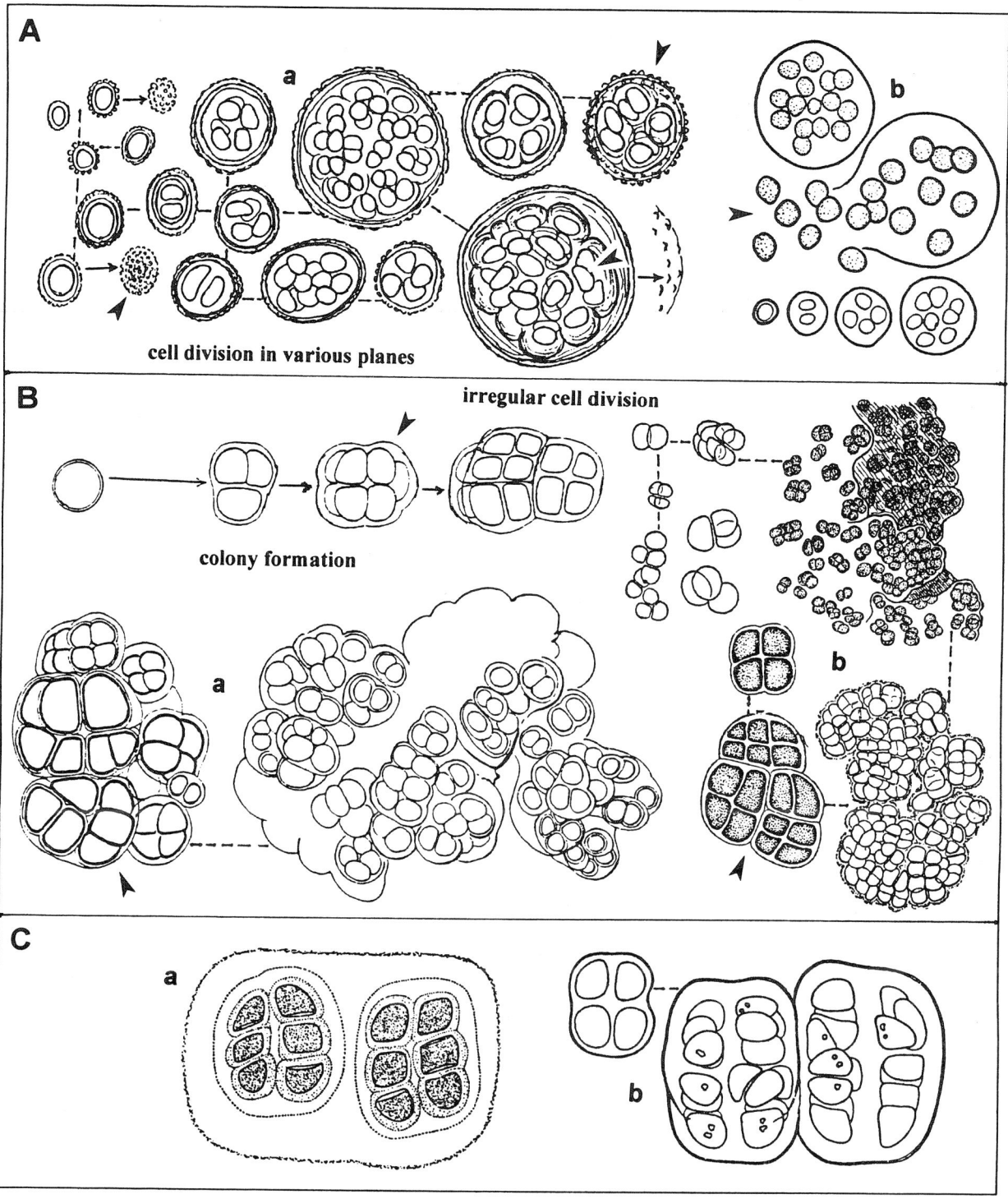

FIGURE 13 (A) *Asterocapsa*: **a.** *A. divina* (after Komárek, 1993, from San Luis Potosí, Mexico); **b.** *A.* sp. (after Gardner, 1927, from Puerto Rico; sub *Anacystis nigroviolacea*). (B) *Gloeocapsopsis*: **a.** *G. magma* (after Geitler, 1932); **b.** *G. crepidinum* (after Bornet and Thuret, and Hollerbach from Kosinskaja, 1948, sub *Gloeocapsa crepidinum*, marine species). (C) *Cyanokybus*: **a.** *C. venezuelae* (after Schiller, 1956, from Los Aves Islands, Venezuela); **b.** *C. venezuelae* (after Komárek, 1994, from Cuba).

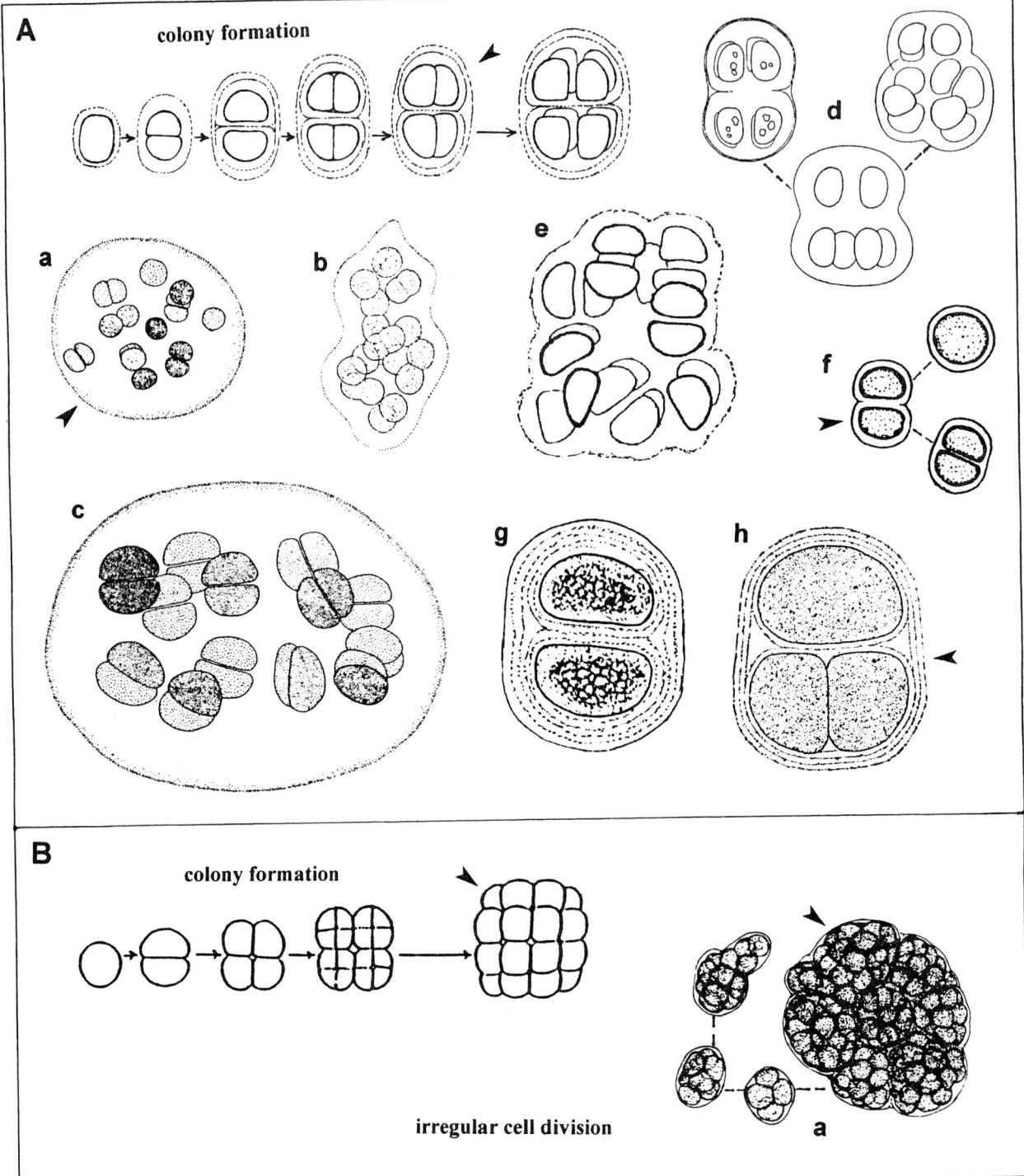

FIGURE 14 (A) *Chroococcus*: a. *C. (Limnococcus) dispersus* (after Smith, 1920, from Wisconsin lakes; sub *C. limneticus* var. *subsalsus*); b. *C. (Limnococcus) sonorensis* (after Drouet, 1942, from Baja California, Mexico); c. *C. (Limnococcus) limneticus* (after Smith, 1920, from Wisconsin lakes); d. *C. (Chroococcus) mipitanensis* (after Komárek and Novelo, 1994, from central America); e. *C. (Chroococcus) polymorphus* (after Komárek and Novelo, 1994, from central America); f. *C. (Chroococcus) minutus* var. *thermalis* (after Copeland, 1936, from Yellowstone National Park); g. *C. (Chroococcus) yellowstonensis* (after Copeland, 1936, from Yellowstone National Park); h. *C. (Chroococcus) turgidus* (after Smith, 1920, from Wisconsin lakes). (B) *Cyanosarcina*; a. *C.* sp. (after Hollenberg, 1939, from California; sub *Microcystis splendens*; marine species).

and Puerto Rico (Gardner, 1927, *A. magnifica* and *A. pulchra*).

Chroococcus Nägeli (Fig. 14A)

Cells or groups of cells (mainly two to four cells), are surrounded by mucilaginous envelopes, and usually occur in microscopic, spherical or composed colonies; rarely form agglomerations. Mucilaginous envelopes are colorless or yellowish, usually copying the cell form, sometimes lamellated, distinct or diffuse at the margin (subg. *Chroococcus*) or fine, homogeneous, and diffuse, in which the cells or cell groups are irregularly arranged (subg. *Limnococcus*). Cells at first are spherical or oval, and later are hemispherical or in the form of a segment of the sphere. The cells are 0.7–50 µm in diameter. The cell content is grey, blue–green, olive green, orange, or reddish violet and granular. Only in few planktonic species are there gas vesicles. Cell division is by binary fission in three or more planes, or irregular (in old colonies). Reproduction is by fragmentation of colonies.

Many species have been described, and over 60 are well defined (Komárek and Anagnostidis, 1998); many occur in North America. Planktonic species from the subgenus *Limnococcus* are common in temperate and northern water bodies and include *C. microscopicus*, *C. minimus*, *C. dispersus*, *C. distans*, and *C. limneticus*. *C. prescottii*, *C. refractus*, and *C. sonorensis* were described from the United States, but especially *C. prescottii* is more common in cold, usually slightly acidic waters (our data). Most species from the subgenus *Chroococcus* are metaphytic, although the type species *C. rufescens* colonizes soils (Daily, 1942). From this group, *C. varius*, *C. minutus*, *C. pallidus*, *C. schizodermaticus*, *C. multicoloratus*, *C. submarinus* (halophilic), and *C. turgidus* are the most common. *C. thermalis*, *C. tenacoides*, *C. yellowstonensis*, and *C. endophyticus* are known from thermal springs (Copeland, 1936). Several species are typical of tropical and subtropical regions, including *C. aeruginosus*, *C. cubicus*, *C. deltoides*, *C. heanogloios*, *C. mipitanensis*, and *C. polyedriformis* (Komárek and Novelo, 1994). This genus has been widely reported across North America, including North Carolina (Whitford and Schumacher, 1969), the western Great Lakes region (Prescott, 1962), British Columbia (Stein and Borden, 1979), Ontario (Duthie and Socha, 1976), and the Northwest Territories (Sheath and Steinman, 1982).

Cyanokybus Schiller (Fig. 13C)

Colonies are few-celled, microscopic, and usually oval in outline. The cells are arranged in short perpendicular, irregular rows, more or less distant one from another, within a colorless, homogeneous, distinctly delimited mucilaginous envelope. The cells are hemispherical, rectangular, or polyhedral–rounded, have or do not have individual fine envelopes, are pale to bright blue–green, up to 15–20 µm in diameter, and have no obvious gas vesicles. Cell division is by binary fission in three planes, later repeatedly in one plane (cells in rows). Reproduction is by colony disintegration.

This is a monotypic genus. The type species, *C. venezuelae*, was described from an island near the Venezuelan coast, but it is distributed sporadically in alkaline swamps throughout the Caribbean region (Schiller, 1956; Komárek, 1994).

Cyanosarcina Kováčik (Fig. 14B)

Colonies are microscopic, packet-like, sarcinoid or irregular, and have densely aggregated cells, enveloped by thin, colorless, firm envelopes that tightly enclose clusters of cells and later occasionally from macroscopic agglomerations. Cells are more or less spherical, subspherical, or irregularly rounded, enveloped by thin individual gelatinous layers, usually tightly packed, colored pale or bright blue–green, olive green, or reddish, finely granular, and 2–10 µm in diameter. Cells have no obvious gas vesicles. Cell division is by binary fission regularly in three, later in more planes; cells enlarge prior to divisions. Reproduction is by disintegration of colonies.

Twelve species, usually metaphytic or periphytic, are described from various countries (Kováčik, 1988; Komárek and Anagnostidis, 1998). None is explicitly recorded from North America, but certain species have been described by Gardner (1927) from Puerto Rico (as *Endospora rubra*) and by Hollenberg (1939) from California (as *Microcystis splendens*), which probably belong to this genus. Further observations are needed.

Gloeocapsopsis Geitler *ex* Komárek (Fig. 13B)

Cells are usually densely aggregated in irregular, packet-like, microscopic colonies (rarely solitary), later forming a macroscopic flat or granular, gelatinous mass on substrata. Cells are subspherical or irregular-rounded in outline, sometimes slightly elongated, 2.5–14(20) µm in diameter, and usually have individual gelatinous envelopes. Colonies are surrounded by thin, firm, narrow, distinctly limited, and often colored (blue, red, or yellow–brown) sheaths that usually follow the irregular outline of cell aggregates by their shape. Cell content is pale or bright blue–green, finely granular, and has no obvious gas vesicles. Cell division is by irregular binary fission in various planes. Reproduction is by disintegration of colonies.

Often confused with *Gloeocapsa* in older literature (because they resemble morphologically dormant stages of *Gloeocapsa* spp.), the geographic distribution of most of the 10 known species (Komárek and Anagnostidis, 1998) is still unclear. *G. magma* (which has reddish envelopes) may be widely distributed on wet rocks in high mountains (Tilden, 1910; Stein and Borden, 1979).

Entophysalidaceae (Fig. 15)

Chlorogloea Wille (Fig. 15A)

Colonies are multicellular, mucilaginous, spherical or irregular, and have a rough surface. Colonies usually are composed of subcolonies, that later form macroscopic gelatinous masses attached to various substrata; they are rarely epiphytic or metaphytic. Cells are arranged irregularly, but in the margin are more or less in radial rows (the result of terminal or heteropolar growth). Cells are spherical or slightly irregular–rounded in outline to polygonal–rounded, usually enveloped by individual envelopes, but later in homogeneous, colorless, and delimited mucilage that has a more or less homogeneous, pale grey–green, blue–green, or reddish content. The cells are 1–6 µm in diameter and have no gas vesicles. Cell division usually occurs in three perpendicular planes, less frequently irregularly, and sometimes with repeated division in one direction. Reproduction is by disintegration of colonies.

This a poorly known genus comprises ecologically distinct types, most of which require revision. About 20 species have been described (Komárek and Anagnostidis, 1998), including three aquatic species (*C. cuauhtemocii*, *C. epiphytica*, and *C. lithogenes*) from alkaline environments in Mexico (Komárek and Montejano, 1994), and *C. tuberculosa* and *C. regularis* from saline environments in Pacific locations (Setchell and Gardner, 1924).

Entophysalis Kützing (Fig. 15B)

Colonies are multicellular, mucilaginous, microscopic to macroscopic, have polarized growth, are attached to the substratum, and are composed of groups of cells that are enveloped by their own distinct gelatinous and usually delimited envelopes. The gelatinous envelopes are firm, delimited, often layered, colorless or colored (red, dark violet, yellow–brown), and sometimes enveloped by distinct common mucilage. Cells (and their groups) are often arranged in irregular radial rows, spherical to irregular, often of variable size in the same colony, have pale blue–green, olive green, or yellowish content, are finely granular, (0.8)2–9 µm in diameter, and have no gas vesicles. Cells divide by binary fission, in various planes, but sometimes (in marginal parts) the perpendicular cell division repeats. Reproduction is by disintegration of colonies.

More than 20 species are accepted in modern literature (Komárek and Anagnostidis, 1998), but few are well known in terms of variability, ecology, and distribution. *E. cornuana* was observed in U.S. mountain streams by Silva (Whitford and Schumacher, 1969). *E. atrata* and *E. lithophila* were described from stromatolites in a saline volcanic lake in central Mexico (Tavera and Komárek, 1996), and *E. willei* was found in wet rocks in Puerto Rico (Gardner, 1927). Probably more forms belong to this genus, but taxonomic revision is needed. The very widely conceived *E. lemaniae* (recorded from North America by Drouet and Daily, 1956) contains a cluster of various types; the original *E. lemaniae* was described from the Baltic Sea and in this (taxonomically unclear) concept probably has not been collected from North America. Several marine species are known from Pacific coasts, North Carolina, and the Bahamas.

Hydrococcaceae (Fig. 16A)

Several species from the genera *Hormathonema* Ercegović and *Placoma* Schousboe *ex* Bornet and Thuret occur in North America, but are known only from marine environments.

Hydrococcus Kützing (Fig. 16A)

Colonies of cells are initially flat, in single layers, and pseudoparenchymatous (nematoparenchymatous at the margin) with more or less radially arranged cells and more or less circular outline. Aggregates of cells and erect pseudofilaments grow from a colonial center and form clusters of cells in older colonies. Pseudofilaments and groups of cells are embedded in thin, colorless, and confluent sheaths. Old colonies are flattened to hemispherical and are blackish green, brownish, or violet in color. In the center are groups of cells enveloped by individual mucilaginous sheaths, distant from one another, sometimes even forming sarcinoid packets. The cells are spherical, oval, irregular–rounded, or elongate at the ends of pseudofilaments, pale blue–green or olive green, and have more or less homogeneous content. They are 1.2–7.5 µm in diameter and have no obvious gas vesicles. Cell division occurs in various planes in successive cell cycles, but crosswise binary fission prevails in marginal parts of the colonies. Reproduction is by liberated solitary cells and clusters of cells.

Several species have been described (partly under the synonymous genus *Oncobyrsa* Meneghini), of which *H. cesatii* and *H. rivularis* are known from North America (Smith, 1950); one morphotype was registered in Canadian Arctic (Whelden, 1947). Species have been reported on a variety of subtrata (e.g., rocks and wood) from streams in eastern and central Canada and the United States, Puerto Rico, and Hawaii (as *Entophysalis rivularis*; Drouet and Daily, 1956). However, the whole genus is little known.

Chamaesiphonaceae (Figs. 16A–18A)

Chamaecalyx Komárek *et* Anagnostidis (Fig. 18A)

Cells are heteropolar, solitary or in groups,

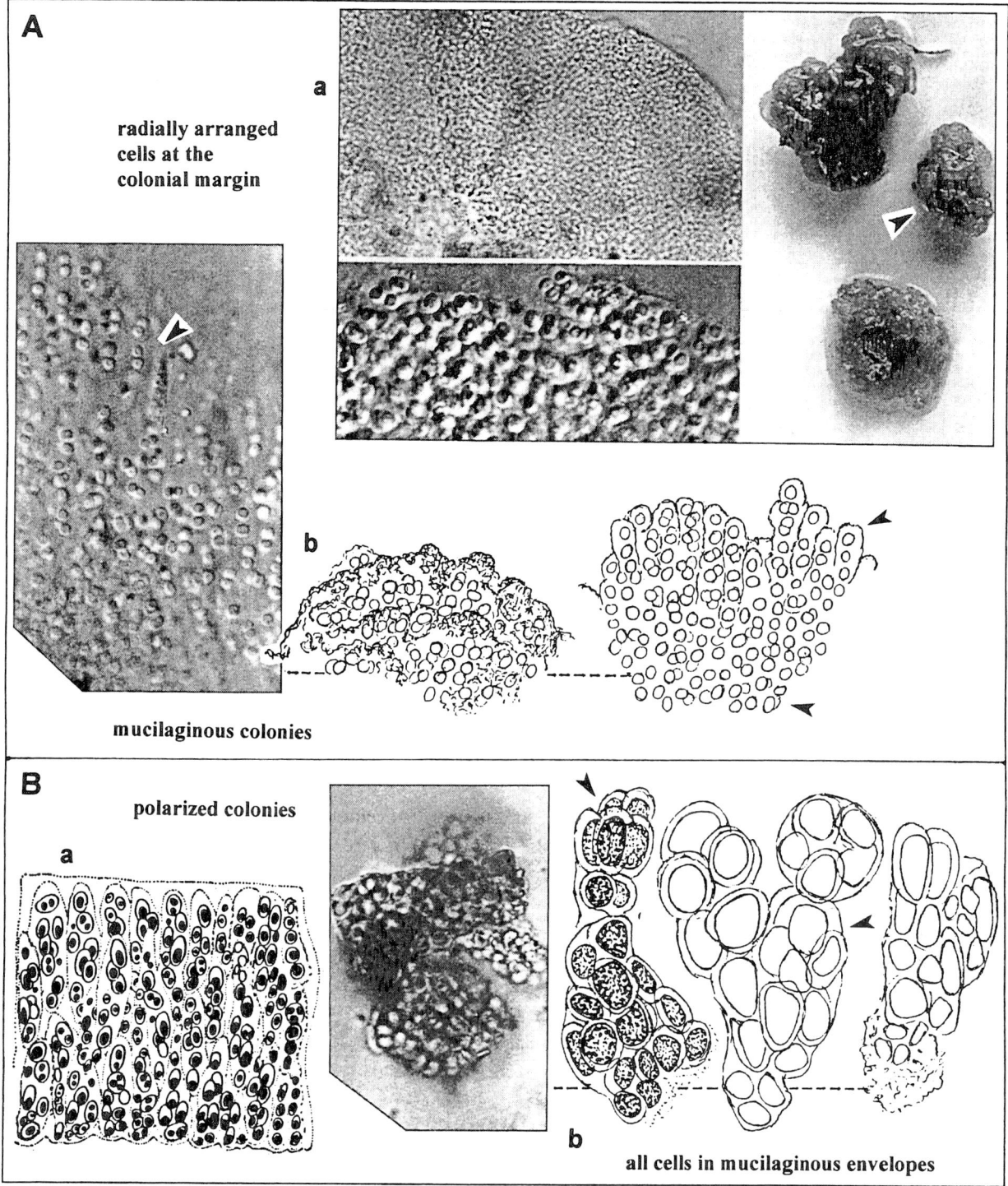

FIGURE 15 (A) *Chlorogloea*: a. *C. epiphytica* (after Komárek and Montejano, 1994, from San Luis Potosí, Mexico); b. *C. lithogenes* (after Komárek and Montejano, 1994, from San Luis Potosí, Mexico). (B) *Entophysalis*; a. *E. willei* (after Gardner, 1927, from Puerto Rico); b. *E. lithophila* (after Tavera and Komárek, 1996, from volcanic lakes in central Mexico).

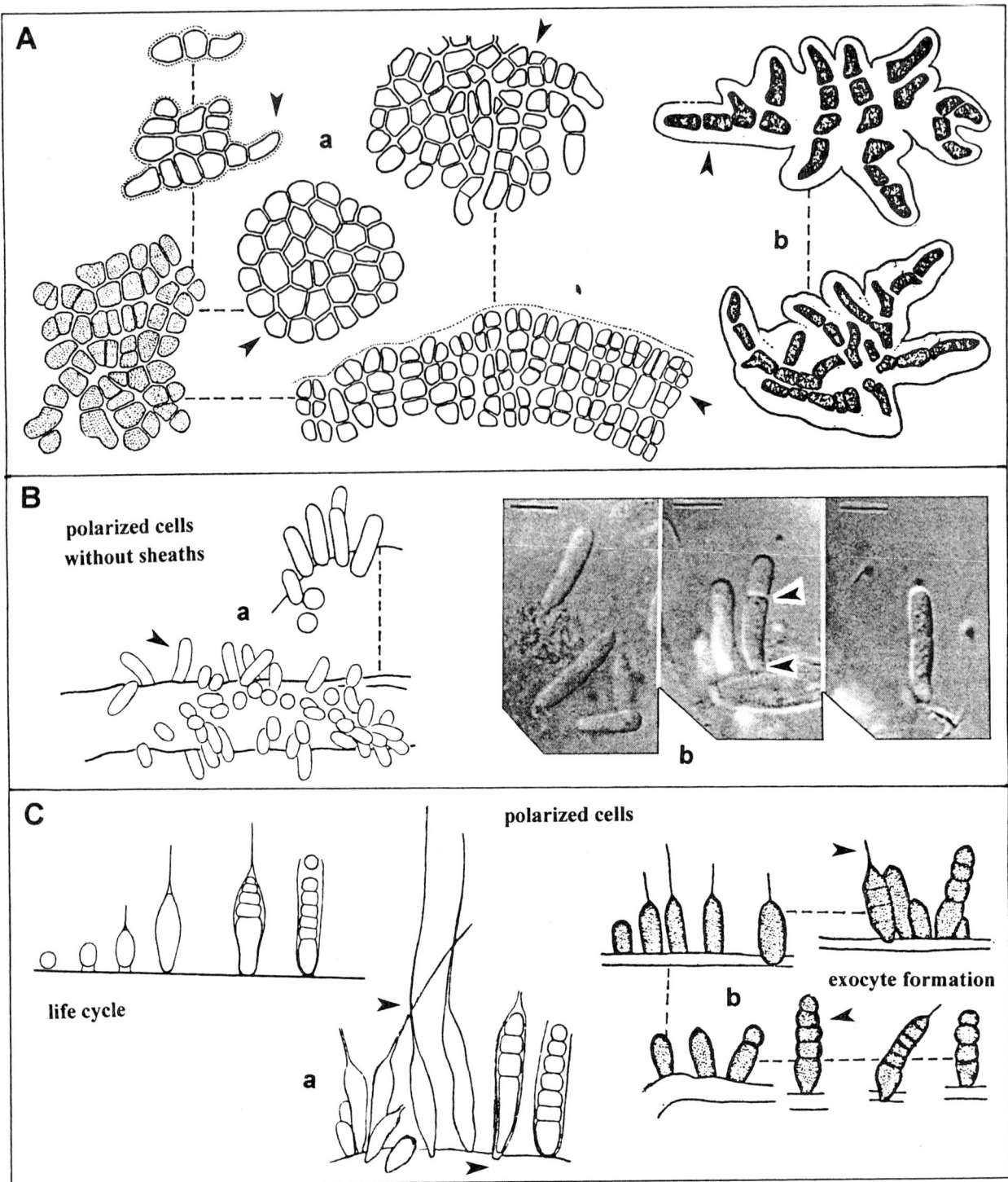

FIGURE 16 (A) *Hydrococcus*: **a.** *H. cesatii* (after Geitler, 1960); **b.** *H.* sp. (after Whelden, 1947, from Canadian Arctic). (B) *Geitleribactron*: **a.** *G. periphyticum* (after Komárek, 1975, and Hällfors and Munsterhjelm, 1982); **b.** *G. crassum* (bar = 10 μm; after Gold-Morgan et al., 1996, from central Mexico). (C) *Clastidium*: **a.** *C. setigerum* (after Starmach, 1966); **b.** *C. cylindricum* (after Whelden, 1947, from Canadian Arctic).

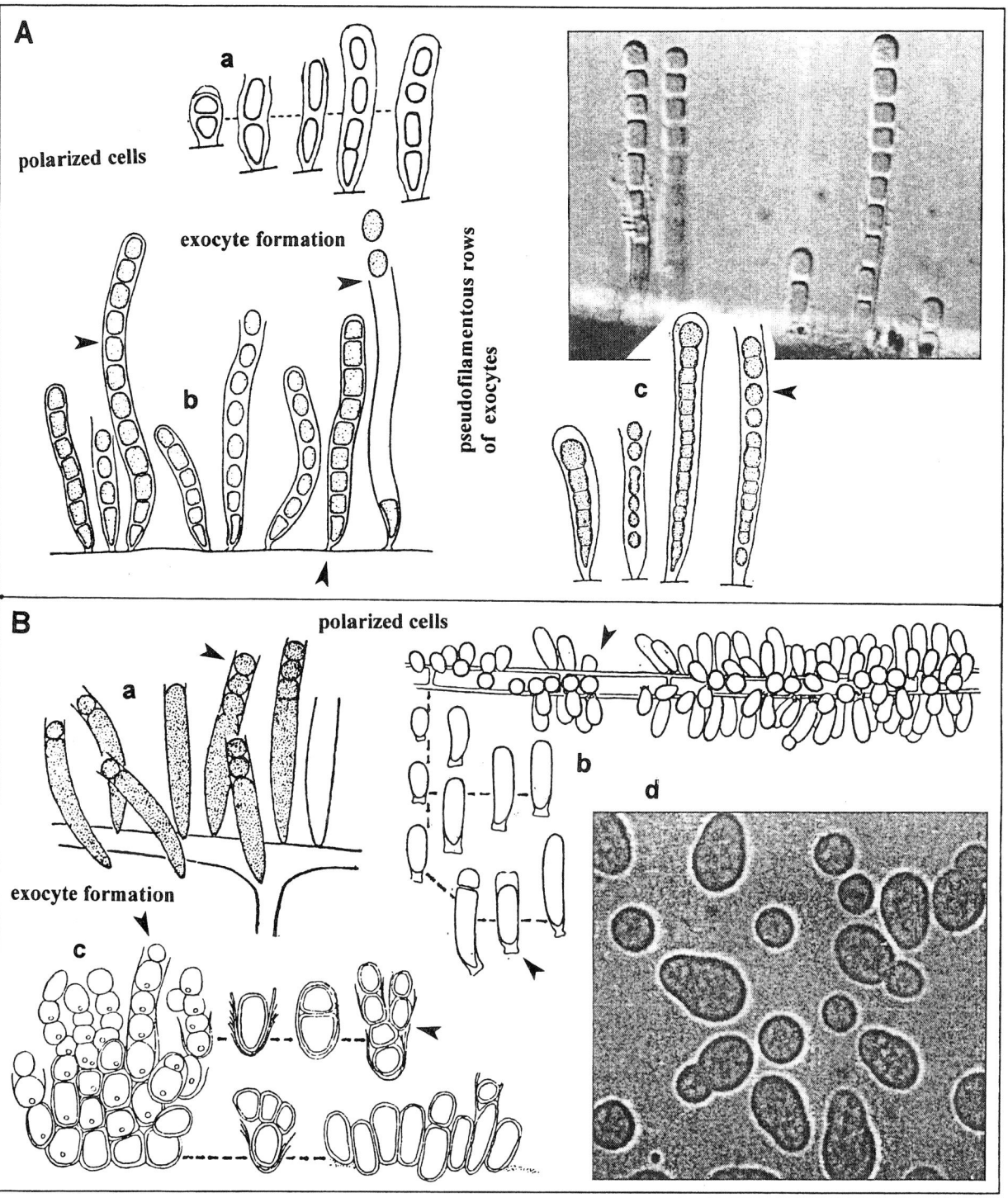

FIGURE 17 (A) *Stichosiphon*: **a.** *S. exiguus* (after Montejano *et al.*, 1997, from San Luis Potosí, Mexico); **b.** *S. willei* (after Gardner, 1927, from Puerto Rico, sub *Chamaesiphon willei*); **c.** *S. sansibaricus* (after Whelden, 1941, from Florida, and, microphoto after Montejano *et al.*, 1997, from central Mexico). (B) *Chamaesiphon*: **a.** *C. (Chamaesiphon) incrustans* (after Smith, 1950, from the United States); **b.** *C. (Chamaesiphon) amethystinus* (after Komárek, 1989, from Cuba); **c.** *C. (Godlewskia) polonicus* (from Geitler, 1932); **d.** *C. (Godlewskia) subglobosus* (after Waterbury and Stanier, 1977; sub *Chamaesiphon* sp., from culture).

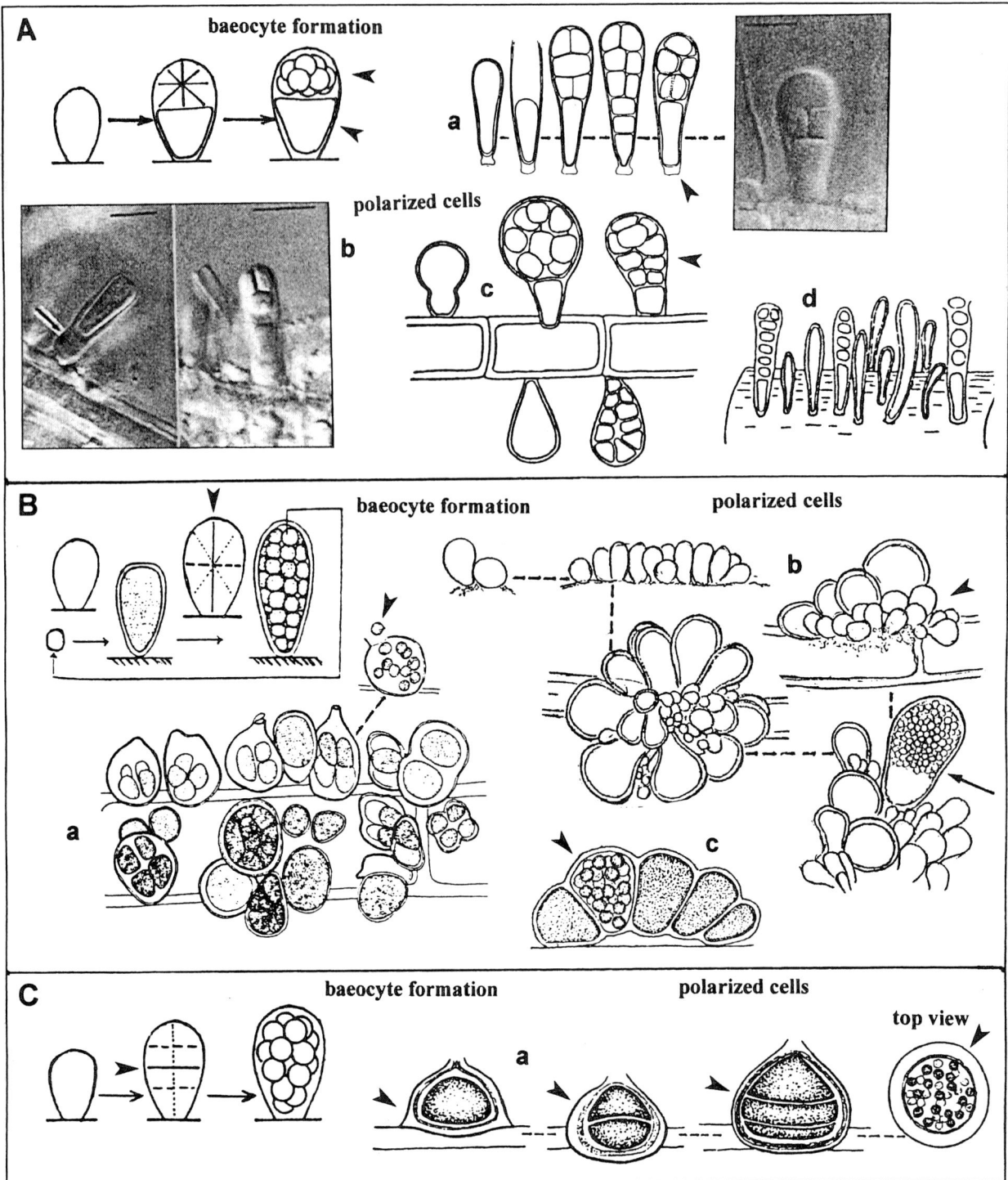

FIGURE 18 (A) *Chamaecalyx*: a. *C. swirenkoi* (after Geitler, 1932; microphoto from Gold-Morgan *et al.*, 1996, from central Mexico); b. *C. calyculatus* (after Gold-Morgan *et al.*, 1996, from San Luis Potosí, Mexico); c. *C. suffultus* (after Setchell and Gardner from Geitler, 1932, from California, marine species); d. *C. clavatus* (after Setchell and Gardner from Kosinskaja, 1948, from Baja California, Mexico; marine species). (B) *Cyanocystis*: a. *C. valiae-allorgei* (after Bourrelly, 1985, from thermal waters in Guadeloupe); b. *C. mexicana* (after Montejano *et al.*, 1993, from central America); c. *C. pacifica* (after Setchell and Gardner from Smith, 1950; a marine species from North America; sub *Dermocarpa pacifica*). (C) *Dermocarpella*: a. *D. hemisphaerica* (after Lemmermann from Geitler, 1932; freshwater species from Chatham Islands, New Zealand).

obovoid to club-shaped, attached to substrata by narrowing basal ends (sometimes combined with a sheath pad), enveloped by a thin or slightly thickened, firm, colorless sheath (pseudovagina). The dimensions of older cells are 8–30(55) × 3.5–10.5(20) µm. The cell content usually is homogeneous (thylakoids are regularly distributed throughout the cells) and is pale blue–green, olive green, or reddish. Reproduction is by exocytes, which differentiate three dimensionally from the upper part of the cells after the first crosswise (horizontal) and the second vertical cell division, and which liberate from the sheaths of the mother cell by the apical opening. Differentiation of baeocytes is successive or simultaneous, while the basal part remains undivided; basal parts may grow into new mother cells, but rarely does the entire cell divide.

Of the 12 known species, the freshwater *C. swirenkoi* and (possibly endemic) *C. calyculatus* were observed growing on filamentous algae in running waters in central Mexico (Gold-Morgan *et al.*, 1996). Four species are known from marine coastal environments of North America (Setchell and Gardner, 1930; Komárek and Anagnostidis, 1986).

Chamaesiphon A. Braun *et* Grunow in Rabenhorst (Fig. 17B)

Cells are heteropolar, solitary or in groups (subg. *Chamaesiphon*), or form multilayered colonies with numerous cells (more or less parallely arranged) that form radial or shrublike, microscopic to macroscopic colonies (subg. *Godlewskia*) attached by their bases to plants, algae, or stony substrata. Cells are slightly or distinctly elongated, subspherical, oval, cylindrical up to slightly club-shaped, always enveloped by a sheath (pseudovagina) or (subg. *Godlewskia*) surrounded by sheaths and a common gelatinous, colorless or slightly yellowish to brownish matrix, (1)2.5–9(–70–200) × 1–8.5(13) µm in size, and have pale blue–green, olive green, greyish, pinkish, or reddish violet content, and no obvious gas vesicles, but sometimes a few prominent granules. Cells are sometimes slightly withdrawn from the narrowed basal sheaths. Cell division is asymmetric, crosswise near the apex; the exocytes separate solitarily or form short vertical rows before release from opened sheaths. In the subgenus *Godlewskia*, the exocytes attach to the margin of the opened sheaths or pseudovaginae (the origin of shrublike or multilayered colonies).

This widespread and occasionally important genus in running waters has about 30 known species, mainly from alpine waters or from clear, cold streams worldwide (Geitler, 1925; Kann, 1972, 1973). *C. britannicus* (= *C. regularis*), *C. incrustans*, *C. rostafinskii*, *C. polonicus*, *C. confervicolus*, and possibly also *C. geitleri* are known from mountainous streams in North American temperate to subarctic zones (Smith, 1950; Stein and Borden, 1979; Sheath and Steinman, 1982); also a common epiphyte on *Cladophora* in hardwater lakes (e.g., Great Lakes). *C. minutus* occurs in mountains in Mexico (Gold-Morgan *et al.*, 1996). *C. amethystinus*, *C. fallax*, and *C. portoricensis* occur in warmer regions (Cuba, Mexico, and Puerto Rico; Gardner, 1927; Komárek, 1989; Gold-Morgan *et al.*, 1996). *C. halophilus*, a red-pigmented species, was described from a volcanic lake in Mexico and possibly is endemic (Tavera and Komárek, 1996).

Clastidium Kirchner (Fig. 16C)

Cells are heteropolar, solitary or in groups, elongated, and attached to the substrate by morphologically differentiated basal ends (usually narrowed and rounded with gelatinous pad). The apical part is conically tapered. The cells are enveloped by very thin, fine, colorless mucilaginous sheaths, which may be elongated into hairlike processes at the apex. The resulting cell form is oval, ovoid, ellipsoidal, cylindrical to slightly club-shaped, or pear-shaped, the cells are 2–15(38) × 1–4(6) µm, and the hairs are up to 52(75) µm long (different lengths among species). The cell content is pale blue–green, homogeneous, and has no obvious gas vesicles or prominent granules. Cell division is transverse, more or less simultaneously near the apex (rarely along the whole cell length) into a row of nearly spherical or oval and motile exocytes, which separate successively. The remnant of the mother cell may form the new vegetative cell.

Five epilithic and epiphytic species are known from clear (Komárek and Anagnostidis, 1998), usually cold mountain streams and, rarely, lakes. Two species are known in this region: *C. setigerum* from the Northwest Territories, British Columbia, the northern United States, Alaska, Louisiana, and Mississippi (Drouet and Daily, 1956; Stein and Borden, 1979; Sheath and Steinman, 1982) and *C. cylindricum* from arctic Canada (Whelden, 1947).

Geitleribactron Komárek (Fig. 16B)

Cells are heteropolar, solitary or in groups, elongated, attached by one end to substrata (usually by indistinct, hyaline, gelatinous pad), and in groups or in stellate clusters. Cells are oval to cylindrical and rod-shaped, rounded at the apex, slightly and shortly attenuate or rounded at the base, have no sheaths, are pale blue–green or olive green, 4–25(50) × 1.3–6 µm in size, have finely granular content with no obvious gas vesicles, and sometimes have visible centro- and chromatoplasm (= parietal thylakoids). Cell division is by transverse, symmetrical or asymmetrical binary fision;

rarely by two simultaneous divisions. Released daughter cells (exocytes) attach to substrata by either end.

Three freshwater periphytic and epilithic species have been described (Komárek and Anagnostidis, 1998). *G. crassum* was described from running waters in central Mexico and is epiphytic on filamentous algae (Gold-Morgan *et al.*, 1996).

Stichosiphon Geitler (Fig. 17A)

Cells are heteropolar, solitary or in groups, attached by the slightly narrowed base to substrata (usually also by means of a mucilaginous pad) and later form a long, uniseriate, straight or curved row of transversally differentiating exocytes at the apical end. The exocytes remain joined and form moniliform or trichome-like pseudofilaments, enveloped by thin, colorless sheaths that open at the apex. Exocytes may be irregularly clustered in the terminal part of a sheath. The pseudofilamentous rows of exocytes differ in the lengths among species, and measure (4)8–200(450) × (2.5)3–9 µm. Collectively, the complete cells with exocytes have the appearance of heteropolar, filamentous forms. Cells are shortly cylindrical or club-shaped when young, pale blue–green, olive green, or greyish in color, finely granular, have no gas vesicles, rarely have solitary granules, and usually have homogeneous or keritomized cell contents (thylakoids are located over the whole cell volume). Cell division is transverse with numerous exocytes in a close sequence; exocytes remain for extended periods in a filamentous formation within sheaths and can divide repeatedly. Exocytes are rarely rounded, usually slightly cylindrical, and after liberation attach to the substrate by either end.

More than 10 species are known, mainly from warmer regions (Komárek and Anagnostidis, 1998). *S. willei* and *S. gardneri* have been described from swamps in the Greater Antilles; *S. regularis* and *S. filamentosus* were recorded in Mexico, where *S. exiguus* was also described from a limestone region in San Luis Potosí (Gardner, 1927; Komárek, 1989; Montejano *et al.*, 1997). *S. sansibaricus* was described from wetlands in Florida (Whelden, 1941; Smith, 1950) and Mexico (Montejano *et al.*, 1997).

Dermocarpellaceae (Figs. 18B and 19A)

Cyanocystis Borzì (Fig. 18B)

Cells are typically heteropolar, solitary or agglomerated in flat groups, usually slightly to distinctly elongated, widely oval, obovoid, club-shaped, or pear-shaped, rarely nearly spherical, and rounded apically. The cells are attached to substrata by means of a sheath, which may be slightly widened at the base; sheaths (pseudovaginae) are thin, firm, and colorless. Cells vary in size and have homogeneous, pale blue–green, olive green, or violet content, never have obvious gas vesicles, rarely have scattered prominent granules, and have very different sizes in various species, (5)6.2–30(90) × (0.7)3.6–21(30) µm. Cell division is by successive multiple fission into numerous spherical, nonmotile baeocytes, which liberate from the sheath by the rupture at the cell apex. The first division plane is always vertical from the apex to the base.

About 17 species have been revised, mainly from marine, rarely freshwater environments (Komárek and Anagnostidis, 1998). Freshwater species *C. pseudoxenococcoides* (from Guadeloupe) and *C. mexicana* (from Mexico) were described from running waters of Central America (Bourrelly, 1985; Montejano *et al.*, 1993). *C. valiae-allorgei* is known from thermal waters in Guadeloupe (Bourrelly and Manguin, 1952; Bourrelly, 1985). Several marine species have been recorded, particularly from the California coast (*C. hemisphaerica*, *C. pacifica*, and *C. sphaeroidea*); *C. olivacea* and *C. violacea* are probably distributed in marine coastal waters up to temperate zone (Hua *et al.*, 1989). Many other similar species need revision.

Dermocarpella Lemmermann (Fig. 18C)

Cells are heteropolar, solitary or in groups, hemispherical, oval or club-shaped, rounded at the apex, and attached to substrata at the widened base or by narrowed ends. Sheaths are thin, firm, and colorless. Cells are more or less homogeneous or finely granular, pale or bright blue–green, olive green, or brownish, have no gas vesicles, and are (6)12–27(100) × (3)4–27(40) µm in size. Cell division is only by complete, successive multiple fission into baeocytes, which liberate from the sheaths by the rupture at the apex. The first (and usually first several) division plane is always horizontal to the substrate.

Only five species are confirmed, two of them from freshwaters (Komárek and Anagnostidis, 1998). Marine epiphytic species *D. prasina* and *D. protea* occur along both the Atlantic and Pacific coasts of North America.

Stanieria Komárek *et* Anagnostidis (Fig. 19A)

Cells are solitary or irregularly clustered in groups, more or less spherical, and attached to substrata by means of mucilaginous sheaths, but have no distinct cell polarity (attachment by any side). Sheaths are thin or slightly thickened, firm, and colorless. Cells are of variable size, pale blue–green, blue–green, yellowish, olive green, or pinkish-reddish, and have more or less homogeneous content and no prominent granules or

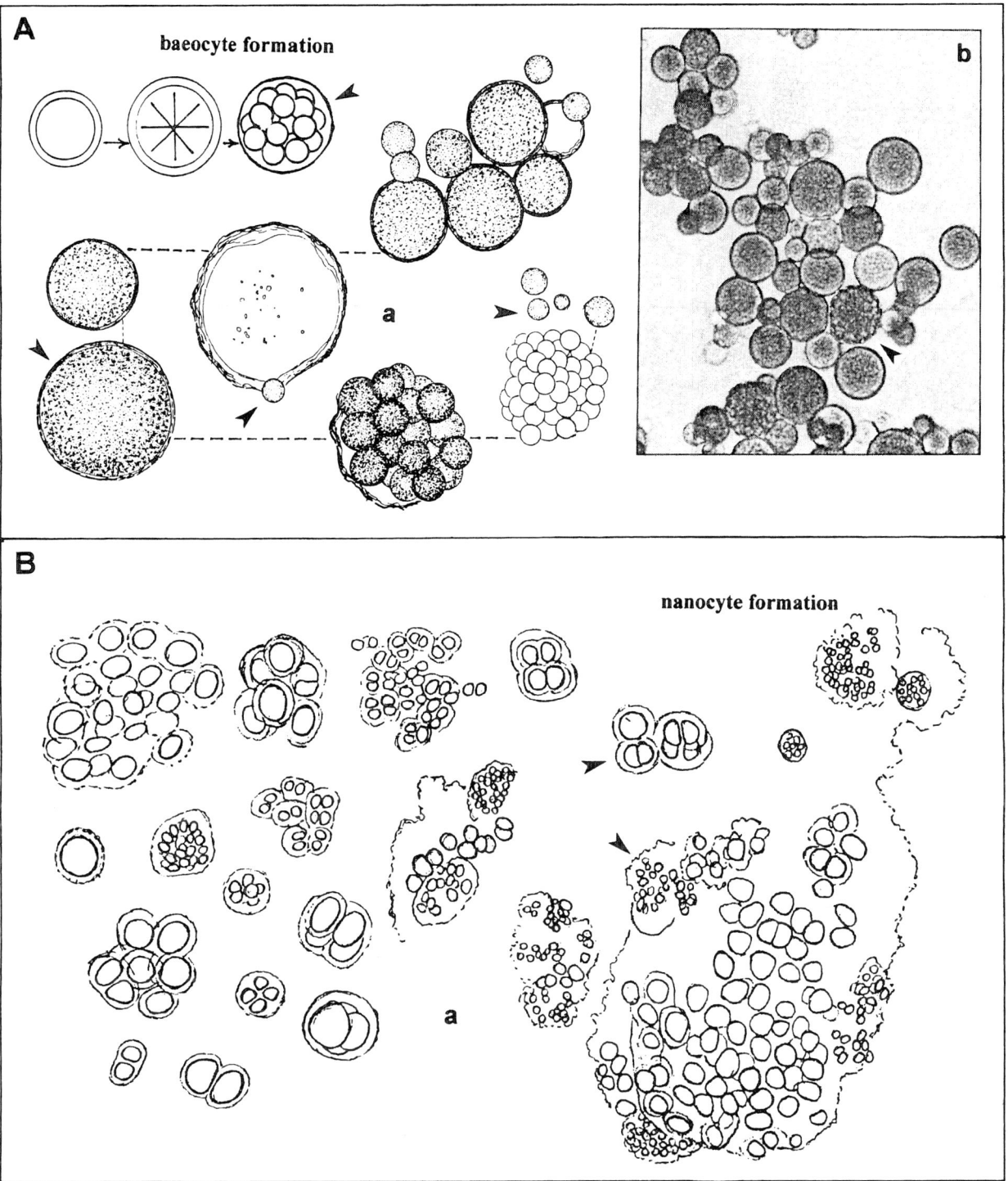

FIGURE 19 (A) *Stanieria*: a, b. *S. cyanosphaera* (after Komárek and Hindák, 1975, from Cuba; sub *Chroococcidiopsis cyanosphaera*). (B) *Chroococcidium*: a. *C. gelatinosum* (after Tavera and Komárek, 1996, from a volcanic lake in central Mexico).

gas vesicles. Cell division (total) is by multiple fission, which proceeds in rapid sequence in various directions or almost simultaneously. The resulting numerous, motile spherical baeocytes liberate after splitting of the sheath. The difference in size of the small baeocytes and the mother cells is obvious. Baeocytes grow to the original size before the next division.

Several marine species have been recognized. *S. sphaerica* was originally described from California and a few of them are not yet well defined taxonomically. *S. cyanosphaera* (one of three freshwater species) is known from alkaline swamps and littoral of lakes in Cuba (Komárek and Anagnostidis, 1998).

Xenococcaceae (Figs. 19B–21)

Chroococcidiopsis Geitler (Fig. 20A)

Cells are solitary or aggregated in irregular groups, enveloped by thin, firm, colorless sheaths, and usually attached to various, mostly stony substrates in a variety of subaerial and aquatic environments, sometimes under extreme conditions; endolithic species are also known. Cells are spherical, oval to irregular–rounded, of varying sizes, have more or less homogeneous pigmentation (thylakoids are distributed irregularly), pale or bright blue–green, rarely greyish or violet, and 1.5–20 (rarely more) μm diameter. Cells divide irregularly, successively, or in rapid sequence in cells of different size, or in numerous baeocytes, which liberate from ruptured sheaths; sheaths sometimes gelatinize.

More than 20 morpho- and ecotypes (species?) are mentioned in the literature (Komárek and Anagnostidis, 1998), but sometimes are not clearly defined. *C. thermalis* has been recorded from hot springs and *C. cubana* has been noted from wetlands and littoral regions of standing waters in Florida, Cuba, and Mexico (Komárek and Hindák, 1975; Komárek and Anagnostidis, 1998). Taxonomically not well described forms occur in deserts of southwestern regions of North America, forming lichens (in Mexico; Büdel, 1985; Büdel and Henssen, 1983), or in various habitats in Puerto Rico (but assigned to *Anacystis* species; Gardner, 1927).

Chroococcidium Geitler (Fig. 19B)

Cells are clustered in irregular groups within thin, amorphous, colorless, and diffuse mucilage, usually attached to stony substrata; less frequently they are epiphytic. Cells are of different size, mainly spherical or irregularly rounded, sometimes with diffuse individual envelopes around cells and around cell groups, and have 3–18 μm diameter, with homogeneous cell content, yellow–green or blue–green color. Cell division is irregular in various planes, sometimes rapidly into numerous nanocytes, which liberate by gelatinization of envelopes.

One species, *C. gelatinosum*, originally described from Indonesia, occurs rarely in volcanic lakes in central Mexico (Geitler and Ruttner, 1935; Tavera and Komárek, 1996). Probably more species will be recognized in the future.

Chroococcopsis Geitler (Fig. 21B)

Cells are solitary, in few-celled clusters, or compact groups ensheathed by firm, thin, or slightly thickened envelopes; rarely they are organized in short, indistinct and irregular rows. Distinctly larger, terminal cells develop near the margins of older colonies. Sheaths are colorless and sometimes slightly layered. Cells are spherical, subspherical, oval, or irregular–rounded; marginal cells are club-shaped or pyriform (radially arranged). Cell size varies widely (2.5–36 μm in the whole genus). Cell content is homogeneous or finely granular and blue–green, olive green to slightly violet in color. Cell division is by irregular binary fission and, in older colonies (usually marginal), by multiple fission in numerous baeocytes, which liberate from the ruptured and/or gelatinized sheaths.

Of five species, *C. fluviatilis* (probably a special morphotype) has been recorded from the United States (Smith, 1950; sub *Pleurocapsa fluviatilis*). An undescribed species was observed in Cuban streams on limestone substrata (Komárek, 1985).

Myxosarcina Printz (Fig. 20B)

Cells are densely agglomerated in packet-like, "sarcinoid" groups, irregular or polygonal–rounded, and sometimes in slightly flattened colonies. Subcolonies are aggregated sometimes in granular mats, attached to substrata or free-living among other algae or in detritus. Firm, mucilaginous envelopes are thin or distinct, and colorless, rarely yellowish brownish. Cells are homogeneous, dark or pale olive green or blue–green in color, rarely violet. Reproduction is by irregular binary fission in three or more planes in successive generations, sometimes obliquely, forming packet-like colonies; several cells in colonies divide rapidly into motile baeocytes, which usually liberate from the split sheaths. Colonies occasionally disintegrate.

More than 10 species have been recognized (Komárek and Anagnostidis, 1998). At least two have been described from North America: *M. amethystina* from thermal springs in Yellowstone Park (Copeland, 1936) and *M. gloeocapsoides* from salt marshes in California (Gardner, 1918). *M. rubra* has been recorded from moist aerial environments (on rocks and wood near springs) from Puerto Rico (Gardner, 1927; Bourrelly, 1985), but its taxonomic identity needs

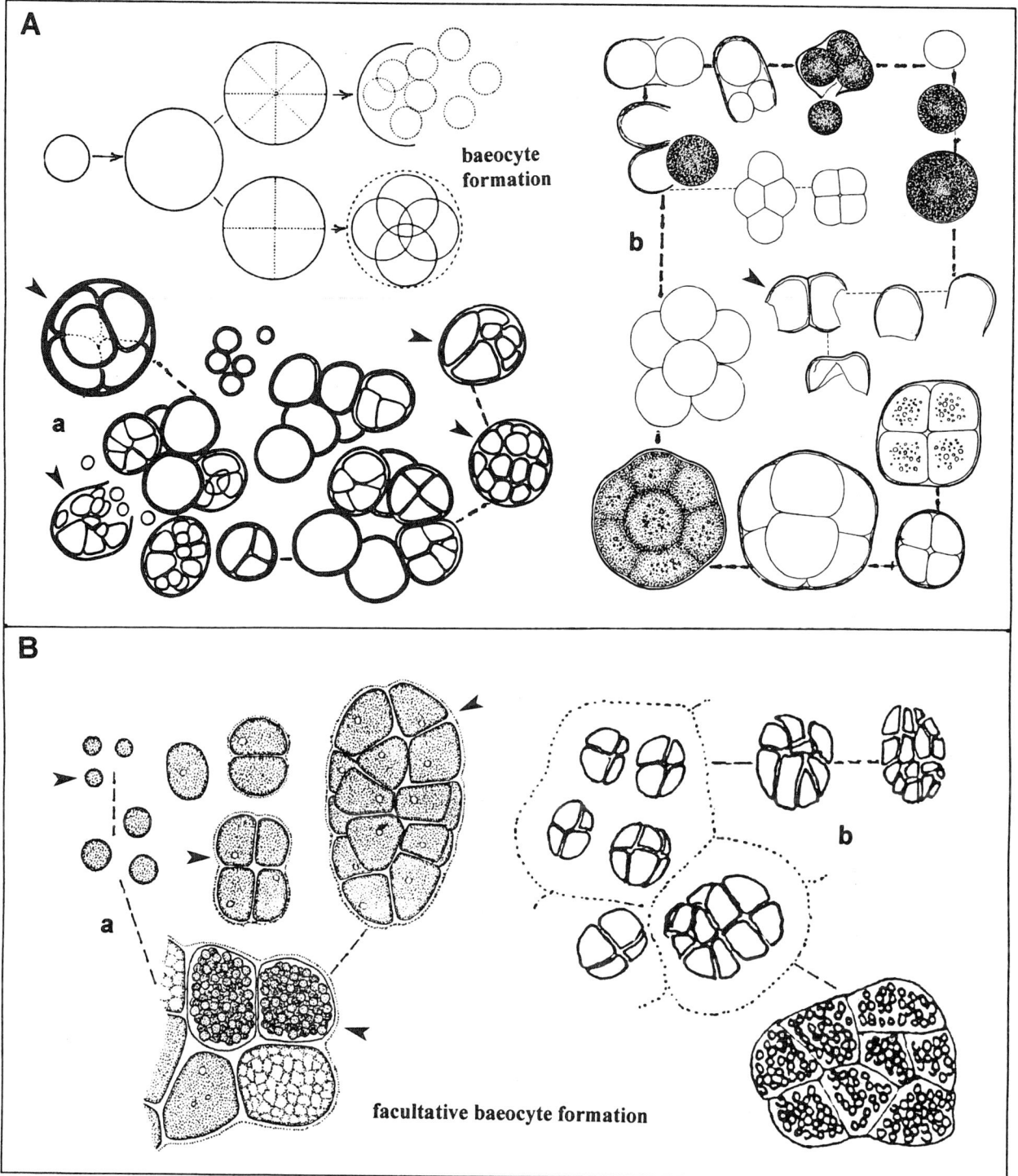

FIGURE 20 (A) *Chroococcidiopsis*: a. *C. thermalis* (after Geitler from Geitler and Ruttner, 1935); b. *C. cubana* (after Komárek and Hindák, 1975, from Cuba). (B) *Myxosarcina*: a. *M. amethystina* (after Copeland, 1936, from Yellowstone National Park); b. *M. gloeocapsoides* (after Setchell and Gardner in Gardner, 1918, from California; marine species; sub *Pleurocapsa gloeocapsoides*).

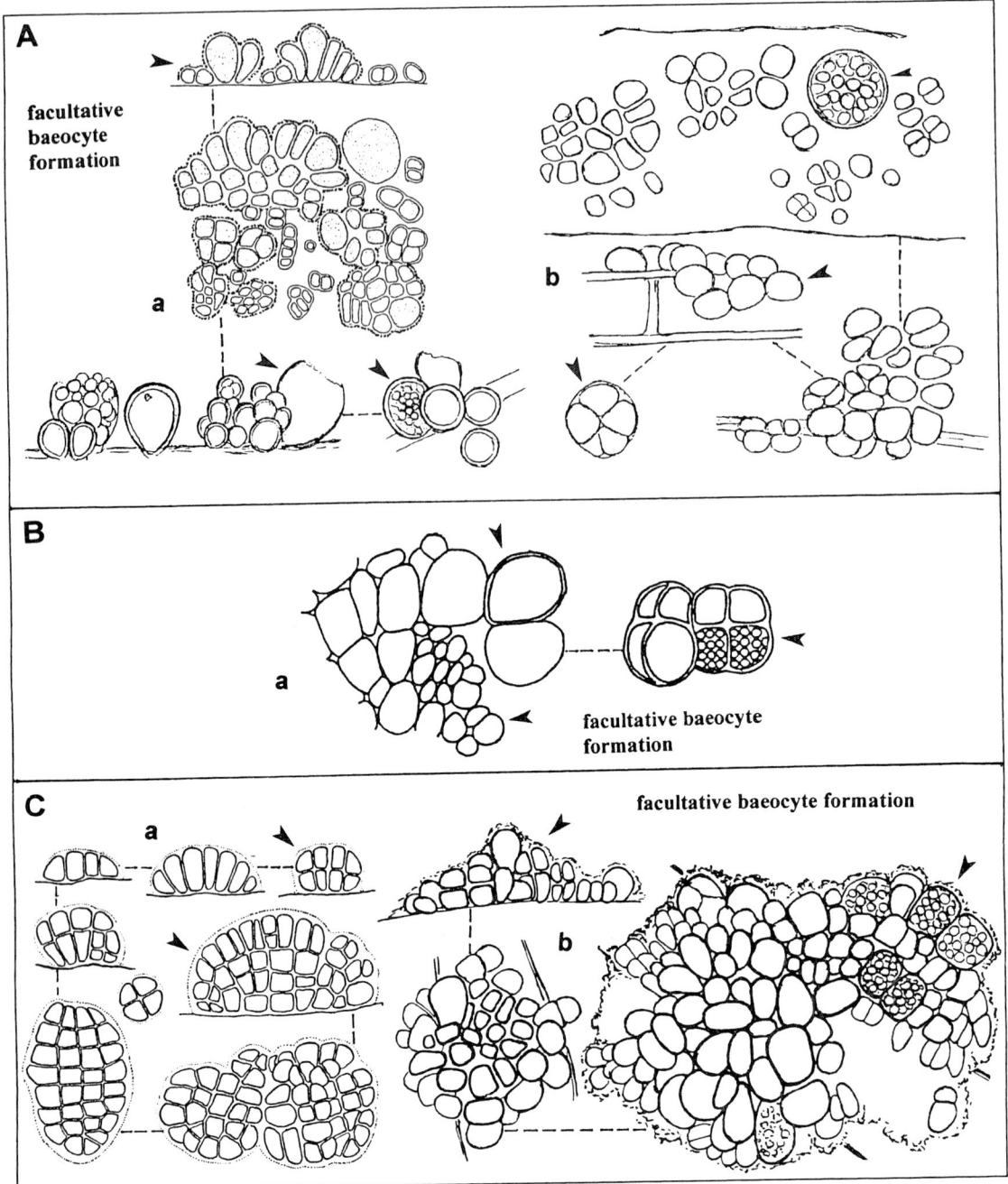

FIGURE 21 (A) *Xenococcus*: a. *X. willei* (after Gardner, 1927, from Puerto Rico, and Montejano *et al.*, 1993, from Mexico); b. *X. bicudoi* (after Montejano *et al.*, 1993, from Mexico). (B) *Chroococcopsis*: a. *C. fluviatilis* (after Geitler, 1932, and Lagerheim from Geitler, 1942). (C) *Xenotholos*: a. *X. kerneri* (after Geitler, 1932); b. *X. huastecanus* (after Gold-Morgan *et al.*, 1994, from San Luis Potosí, Mexico).

clarification. More species may be described from Caribbean habitats.

Xenococcus Thuret in Bornet *et* Thuret (Fig. 21A)

Cells are attached to substrata (usually filamentous algae) in clusters (sometimes densely arranged), enveloped by thin, firm, rarely gelatinizing, colorless or yellowish–brownish sheaths. Cells in clusters are always in one layer, typically heteropolar, spherical to oval, pear-shaped, or club-shaped, and rounded at the apex. Cell content is homogeneous, pale or bright blue–green, yellow–green, or pinkish violet. Cells are 1.5–12(23) μm in diameter. Binary fission occurs in two or more planes, usually perpendicular to the substratum. Occasionally, cells divide by multiple fission in numerous small baeocytes, which liberate from split sheaths. Baeocytes are formed from whole cells or upper portions of mother cells.

Twenty-six well described species, plus a similar number not yet revised (Komárek and Anagnostidis, 1998), are known from freshwater and saline environments worldwide. *X. yellowstonensis* was described from thermal waters (> 50°C) in Yellowstone (Copeland, 1936). *X. bicudoi*, *X. lamellosus*, and *X. willei* were described from freshwater habitats in Mexico and Puerto Rico, mainly as epiphytes on algae in streams (Gardner, 1927; Montejano et al., 1993; Gold-Morgan et al., 1994). *X. candelariae*, which has reddish cells, is a deep-water species from a volcanic lake in central Mexico that is epiphytic on *Cladophora* (Tavera and Komárek, 1996). Marine species *X. pallidus*, *X. schousboei*, *X. pyriformis*, *X. angulatus*, *X. chaetomorphae*, *X. cladophorae*, *X. deformans*, and *X. gilkeyae* have been recorded from various localities along the Atlantic and Pacific coasts (Gardner, 1918; Setchell and Gardner, 1924; Smith, 1950).

Xenotholos Gold-Morgan, Montejano *et* Komárek
(Fig. 21C)

Cells are arranged in thalli (colonies) attached to the substrata. Colonies develop from solitary cells that are first disklike and later layered, usually forming a slightly globose (irregular–hemispherical) colony, in which the cells are organized more or less in radial rows, and finally in two or more layers. The colony is enveloped by a thin, firm, colorless sheath. Cells are hemispherical, polygonal–rounded to slightly elongated, sometimes with individual sheaths. Cells divide in various planes, occasionally (marginal parts) into baeocytes. Reproduction is by solitary cells and baeocytes.

X. kerneri is not common, but is a widespread species in cold mountain streams, probably across North America (Smith, 1950, under *Xenococcus kerneri*). Three other species were described from mountains in central Mexico (Gold-Morgan et al., 1994).

Hyellaceae (Figs. 22 and 23)

Hyella Bornet *et* Flahault (Fig. 23)

Cells are organized in differentiated thalli composed of irregular clusters of cells and pseudofilaments, sometimes ramified. The thalli creep on calcareous substrata or are epiphytic creeping into intercellular spaces of host plants, or endolithic actively growing into substrata. Pseudofilaments are uni- to multiseriate, sometimes pseudodichotomously divided; on the surface (older thalli) they remain in nematoparenchymatous cell clusters. The developed thallus is usually composed from an epilithic (or epiphytic) component with irregularly arranged cells, and endolithic or endophytic component of pseudofilaments boring into substrata (or growing in intercellular spaces of seaweeds). Mucilaginous sheaths are firm, thick, and rarely gelatinizing. Cell morphology varies in each part of the thallus. Cells of very different sizes (2.5–55 μm in diameter) are more or less spherical, subspherical, polygonal–rounded, or elongate, pale or bright blue–green, olive green, or pinkish violet in color, and oval or club-shaped at the ends of pseudofilaments (= apical with respect to growth). Cells divide by binary fission in various planes, in pseudofilaments perpendicularly to the long axis, before branching longitudinally or obliquely. The oldest cells (basal, i.e., near the surface of stony substrata or of host plants) may divide into baeocytes—reproductive cells.

About 30 species have been described (Golubić et al., 1975, 1981; Al-Thukair et al., 1994; Lukas and Golubić, 1983; Lukas and Hoffman, 1984), usually from marine coastal environments (warm seas). *H. kalligrammos* has been collected from freshwater habitats in limestone areas of Mexico, but other forms occur in central and North America (e.g., *Hyella* cf. *fontana* in creeks in limestone areas in Cuba; Komárek, 1985). From marine locations, endolithic *H. balani*, *H. gigas*, *H. pyxis*, *H. caespitosa*, *H. tenuior*, *H. linearis*, *H. littorinae*, and *H. vacans*, and the epiphytic *H. seriata* are known, particularly from warmer seas (including the Bahamas and Bermuda; Gardner, 1918; Hollenberg, 1939; Lukas and Golubić, 1983; Gektidis and Golubić, 1996).

Pleurocapsa Thuret in Hauck (Fig. 22B)

Cells are arranged in irregular groups or rows and pseudofilaments, creep on substrata (mostly stones), and sometimes pseudodichotomously divide. Pseudofilaments are partly endolithic in several species or form crustose layers. Rows of cells are uni- to multiseriate and enveloped more or less by thin, firm, sometimes lamellate and yellow brownish, confluent sheaths. Cells are irregular, variable in size, sometimes slightly elongate, have homogeneous or slightly granular content, blue–green, pale blue–green, or pinkish color, and are (2.4)3–15(20) μm in diameter. Cells divide irregularly by binary fission in various planes; in pseudofilaments predominantly crosswise. Enlarged cells, which arise in different parts of a thallus, divide into baeocytes, which escape from the gelatinized and divided cells.

Over 20 species have been described (Komárek and Anagnostidis, 1998), but require revision. The genus is poorly known and perhaps has two distinct groups (Waterbury and Stanier, 1978). *P. minor* occurs in

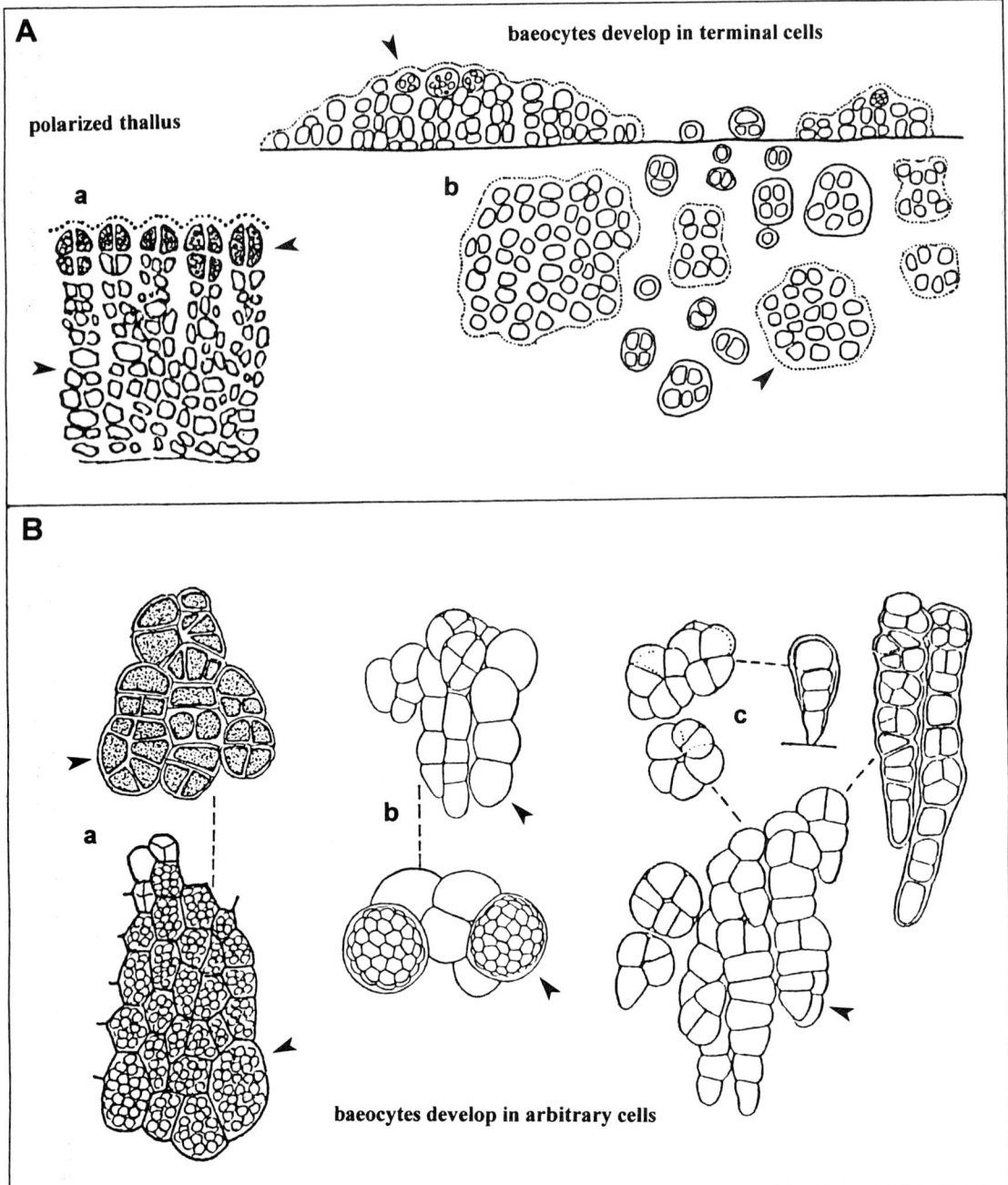

FIGURE 22 (A) *Radaisia*: **a.** *R. epiphytica* (after Setchell and Gardner in Gardner, 1918, from California, marine species); **b.** *R. gardneri* (after Gardner, 1927, from Puerto Rico; sub *Pleurocapsa epiphytica*). (B) *Pleurocapsa*: **a.** *P. minor* (after Geitler, 1932); **b.** *P. crepidinum* (after Geitler, 1932, from coasts of North America; marine species); **c.** *P. minuta* (after Geitler, 1932; marine species).

calcareous streams in North America (Smith, 1950), but other freshwater species occur (e.g., taxonomically uncertain *P. varia* from numerous localities in the United States; Daily, 1942). Four marine species have been recorded from North American coasts (Setchell and Gardner, 1919; Setchell, 1924; Weber van Bosse, 1925).

Radaisia Sauvageau (Fig. 22A)

Cells are arranged initially in flat, discoid, or irregular, nemato- or blastoparenchymatous layers on substrata, from which grow erect, more or less parallel pseudofilamentous rows of cells, which may be straight or slightly curved and sometimes divided. Old colonies form flat, gelatinous and crustose layers of pseudo-

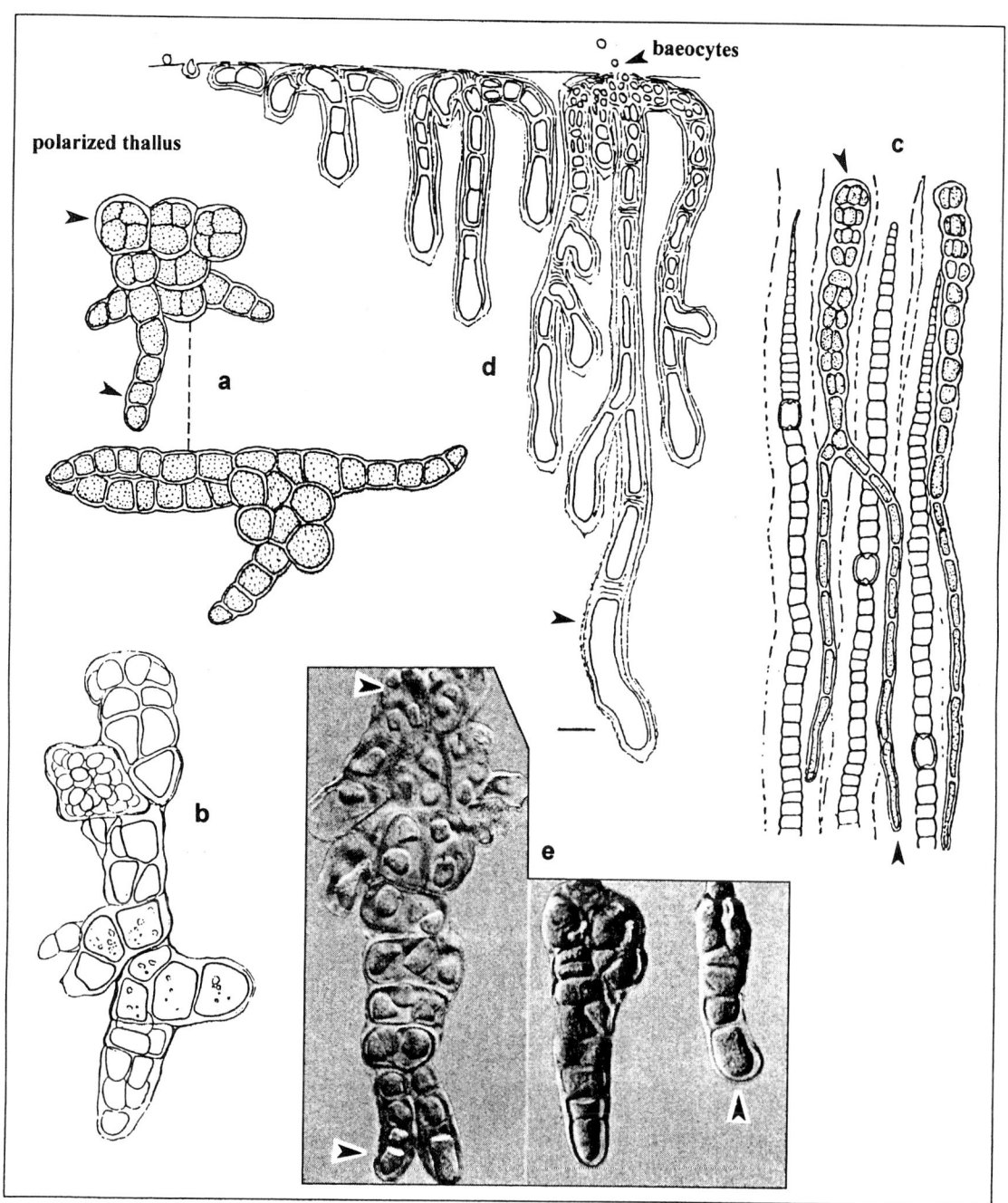

FIGURE 23 *Hyella*: a. *H. fontana* (after Smith, 1950, from North America); b. *H.* cf. *fontana* (after Komárek, 1985, from Cuba); c. *H. seriata* (after Hollenberg, 1939, from California; epiphytic marine species); d. *H. cae-spitosa* (after LeCampion-Alsumard and Golubić, 1985; endolithic marine species); e. *H. balani* (after LeCampion-Alsumard and Golubić, 1985; endolithic marine species; bar = 10 μm).

filaments in parallel arrangement, perpendicular to the substratum. Mucilaginous sheaths present around the pseudofilaments, are confluent later in a homogeneous mass. Cells are irregular, polyhedral–rounded, usually elongated toward the ends of the pseudofilaments, pale blue–green or reddish violet in color, and have 2.5–9(10) μm diameter. Cell division is irregular, usually crosswise to the pseudofilament. Apical, usually larger cells divide into baeocytes, which arise by simultaneous or successive cell division and escape from the ruptured or gelatinized envelopes.

About 10 species have been described (Komárek and Anagnostidis, 1998): three freshwater species are known from standing waters in Puerto Rico (*R. conflu-*

ens, *R. gardneri*, and *R. willei*; Gardner, 1927; Komárek and Anagnostidis, 1998); and three marine species were described from coastal areas of California; one from Florida (Gardner, 1918, 1927; Weber van Bosse, 1926).

VI. GUIDE TO LITERATURE FOR SPECIES IDENTIFICATION

This guide applies to cyanobacterial taxa covered in chapters 3 and 4. The taxonomy of cyanobacteria has changed drastically as a consequence of current molecular, biochemical, ecological, and ultrastructural data, but few of these changes are reflected even in newer identification keys. The reader must still rely on the older literature until these changes are incorporated into general keys. For North America, Tilden (1910) is still useful. Several cyanobacterial species in the Great Lakes region can be identified using Prescott (1962). Two other taxonomic works written in English are from India (Desikachary, 1959) and the British Isles (Whitton, 2002). Most other important works were not written in English, but should be consulted, especially Geitler (1932), as well as monographs covering continental Europe (Huber-Pestalozzi, 1938; Starmach, 1966), and Russia (Elenkin, 1936, 1938, 1949; Kosinskaja, 1948; Kondrateva, 1968).

Some modern results are incorporated into in recent monographs (in English) on the Chroococcales (Komárek and Anagnostidis, 1986, 1998), and the filamentous orders Oscillatoriales, Stigonematales, and Nostocales (Anagnostidis and Komárek, 1985, 1988, 1990; Komárek and Anagnostidis, 1989). Castenholz (2001) has recently written (in English) an overview of cyanobacterial classification.

LITERATURE CITED

Albertano, P., Capucci, E. 1997. Cyanobacterial picoplankton from the Central Baltic Sea: Cell size classification by image-analyzed fluorescence microscopy. Journal of Plankton Research 19:1405–1416.

Allen, M. M. 1968. Ultrastructure of the cell wall and cell division of unicellular blue–green algae. Journal of Bacteriology 96:842–852.

Al-Thukair, A. A., Golubić, S., Rosen, G. 1994. New endolithic cyanobacteria from the Bahama bank and the Arabian Gulf: *Hyella racemus* sp. nov. Journal of Phycology 30:764–769.

Anagnostidis, K., Komárek, J. 1988. Modern approach to the classification system of cyanophytes. 3. Oscillatoriales. Hydrobiologie/Algological Studies. 50/53:327–472.

Anagnostidis, K., Komárek, J. 1990. Modern approach to the classification system of cyanophytes. 5. Stigonematales. Archüv für Hydrobiologie/Algological Studies 59:1–73.

Anagnostidis, K., Komárek, J. 1990. Modern approach to the classification system of cyanophytes, 1—Introduction. Archiv für Hydrobiologie/Algological Studies 38/39:291–302.

Anagnostidis, K., Pantazidou, A. 1991. Marine and aerophytic *Cyanosarcina*, *Stanieria* and *Pseudocapsa* (Chroococcales) species from Hellas (Greece). Archiv für Hydrobiologie/Algological Studies 65:141–157.

Bailey, D., Mazurak, A. P., Rosowski, J. R. 1973. Aggregation of soil particles by algae. Journal of Phycology 9:99–101.

Baker, A. L., Baker, K. K. 1979. Effects of temperature and current discharge on the concentration and photosynthetic activity of the phytoplankton in the upper Mississippi River. Freshwater Biology 9:191–198.

Baker, A. L., Baker, K. K. 1981. Seasonal succession of the phytoplankton in the upper Mississippi River. Hydrobiologia 83:295–301.

Barbiero, R. P., Welch, H. 1992. Contribution of benthic blue–green algal recruitment to lake populations and phosphorus translocation. Freshwater Biology 27:249–260.

Bourrelly, P. 1966. Quelques algues d'eau douce du Canada. Internationale Revue der Gesamten Hydrobiologie 51:45–126.

Bourrelly, P. 1985. Les algues d'eau douce III. Les algues bleues et rouges, les Eugleniens, Peridiniens et Cryptomonadines, 2nd. ed. Boubée, Paris, 606 p.

Bourrelly, P., Manguin, E. 1952. Algues d'eau douce de la Guadeloupe. Paris, 281 pp.

Browder, J. A., Gleason, P. J., Swift, D. R. 1994. Periphyton in the Everglades: Spatial variation, environmental correlates, and ecological implications, *in*: Davis, S. M., Ogden, J. C., Eds., Everglades. The ecosystem and its restoration. St. Lucie Press, Delray Beach, FL, pp. 379–418.

Bryant, D. A. 1994. The molecular biology of cyanobacteria. Kluwer, Dordrecht, 908 p.

Büdel, B. 1985. Blue–green phycobionts in the lichen family Lichinaceae. Archiv für Hydrobiologie/Algological Studies 38/39:355–357.

Büdel, B., Henssen, A. 1983. *Chroococcidiopsis* (Cyanophyceae) a phycobiont in the lichen family Lichinaceae. Phycologia 22:367–375.

Büdel, B., Wessels, D. C. J. 1991. Rock inhabiting blue-green algae/cyanobacteria from hot arid regions. Archiv für Hydrobiologie/Algological Studies 64:385–398.

Büdel, B., Lüttge, U., Stelzer, R., Huber, O., Medina, E. 1994. Cyanobacteria of rocks and soils of the Orinoco lowlands and the Guayana uplands, Venezuela. Botanic Acta 107:422–431.

Buell, H. F. 1938. A community of blue green algae in a Minnesota pond. Bulletin of the Torrey Botanical Club 65:377–396.

Burns, C. W., Stockner, J. G. 1991. Picoplankton in six New Zealand lakes: abundance in relation to season and trophic state. Internationale Revue der Gesamten Hydrobiologie 76:523–536.

Callieri, C., Bertoni, R., Amicucci, E., Pinolini, M. A. 1995. Picoplankton composition, size frequence distribution and carbon content in Lago Maggiore, Italy. Memorie dela Istituto Italiano di Idrobiologia 53:177–189.

Cameron, R. E. 1963. Algae of southern Arizona. Part I. Introduction—blue–green algae. Review of Algology 6:282–318.

Cameron, R. E., Morelli, F. A., Blank, G. B. 1965. Soil algae occurring in the valley of the 1000 smokes desert, Alaska. Transactions of the American Microscopical Society 84:151.

Carmichael, W. W. 1992. A review: Cyanobacterial secondary metabolites—the cyanotoxins. Applied Bacteriology 72:445–459.

Carmichael, W. W. 1997. The Cyanotoxins. Advances in Botanical Research 27:211–226.

Carr, N. G., Whitton, B. A., Eds. 1973. The biology of blue–green algae. Botanical Monographs 9. Blackwell, Oxford, 676 p.

Carr, N. G., Whitton, B. A., Eds. 1982. The biology of cyanobacteria. Botanical Monographs 19. Blackwell, Oxford, 688 p.

Castenholz, R. W. 1969a. Thermophilic blue–green algae and the thermal environment. Bacteriology Review 33:476–504.

Castenholz, R. W. 1969b. The thermophilic cyanophytes of Iceland and the upper temperature limit. Journal of Phycology 5:360–368.

Castenholz, R. W. 1970. Laboratory culture of thermophilic Cyanophytes. Schweizerische Zeitschrift für Hydrobiologie 32:538–551.

Castenholz, R. W. 1976. The effect of sulfide on the blue-green algae of hot springs. I. New Zealand and Iceland. Journal of Phycology 12:54–68.

Castenholz, R. W. 1977. The effect of sulfide on the blue–green of hot springs. II. Yellowstone National Park. Microbial Ecology 3:79–105.

Castenholz, R. W. 1992. Species usage, concept and evolution in the cyanobacteria (blue–green algae). Journal of Phycology 28:737–745.

Castenholz, R. W., 2001. Phylum BX. Cyanobacteria. Oxygenic Photosynthetic Bacteria. in: Boone D. R. & Castenholz R. W. eds., Bergey's Manual of Systematic Bacteriology, 2nd ed., Springer, New York, pp. 473–599.

Castenholz, R. W., Waterbury, J. B. 1989. Oxygenic photosynthetic bacteria (sect. 19), Group I. Cyanobacteria, in: Staley J. T., Ed., Bergey's manual of systematic bacteriology, Vol. 3. Williams & Wilkins, Baltimore, pp. 1710–1799.

Castenholz, R. W., Wickstrom, C. E. 1975. Thermal streams, in: Whitton, B.A., Ed., River ecology. Blackwell, Oxford, U.K., pp. 264–285.

Cepák, V. 1993. Morphology of DNA containing structures (nucleoids) as a prospective character in cyanophyte taxonomy. Journal of Phycology 29:844–852.

Chorus, I., Bartram, J., Eds. 1999. Toxic cyanobacteria in water. Spon, London, 416 p.

Clements, F. E., Shantz, H. L. 1909. A new genus of blue–green algae. Minnesota Botany Studies 4:133–135.

Codd, G. A. 1995. Cyanobacterial toxins: Occurrence, properties and biological significance. Water Science and Technology 32:149–156.

Cohen-Bazire, G., Bryant, D. A.1982. Phycobilisomes: Composition and structure, in: Carr, N. G., Whitton, B. A., Eds. 1982. The biology of cyanobacteria. Blackwell, Oxford, pp. 141–190.

Copeland, J. J. 1936. Yellowstone thermal Myxophyceae. Annals of the New York Academy of Science 36:1–232.

Corpe, W. A., Jensen, T. E. 1992. An electron microscopic study of picoplanktonic organisms from a small lake. Microbiological Ecology 24:187–197.

Croasdale, H. 1973. Freshwater algae of Ellesmere Island, N.W.T. National Museum of Natural Sciences (Ottawa). Publications in Botany 3:1–131.

Cronberg, G., Komárek, J. 1994. Planktic cyanoprokaryotes found in south Swedish lakes during the 12th International Symposium of Cyanophyte Research, 1992. Archiv für Hydrobiologie/Algological Studies 75:323–352.

Daily, W. A. 1942. The Chroococcaceae of Ohio, Kentucky, and Indiana. American Midland Naturalist 27:636–661.

Daily, W. A. 1946. Notes on the algae I, II. Butler University Botanical Studies 8:118–120.

Desikachary, T. V. 1959. Cyanophyta. Indian Council of Agricultural Research, New Delhi, 686 p.

De Toni, G. 1936. Noterelle di nomenclatura algologica. VIII. Terzo elenco di Missoficee omonime. Brescia 5.

DiCastri, F., Younèz, T. 1994. Diversities: Yesterday, today and a path towards the future. Biology International 29:3–23.

Doers, M. P., Parker, D. L. 1988. Properties of Microcystis aeruginosa and M. flos-aquae (Cyanophyta) in culture: Taxonomic implications. Journal of Phycology 24:502–508.

Drews, G., Weckesser, J. 1982. Function, structure and composition of cell walls and external layers, in: Carr, N. G., Whitton, B. A., Eds. The biology of cyanobacteria. Blackwell, Oxford, pp. 333–357.

Drews, G., Prauser, H., Uhlmann, O. 1961. Massenvorkommen von Synechococcus plancticus nov. spec., einer solitären planktischen Cyanophycee, in einem Abwasserteich. Archiv für Mikrobiologie 39:101–115.

Drouet, F. 1936a. Notes on the flora of Columbia, Missouri, III. Rhodora 38:191–195.

Drouet, F. 1936b. Myxophyceae of the G. Allan Hancock Expedition of 1934, collected by Wm. R. Taylor. Hancock Pacific Expedition 3:15–30.

Drouet, F. 1938. Myxophyceae of the Yale North India Expedition collected by G. E. Hutchinson. Transactions of the American Microscopical Society 57.

Drouet, F. 1942. Studies in Myxophyceae I. Field Museum of Natural History Publications, Botany Series 20:125–141.

Drouet, F., Daily, W. A. 1948. Nomenclatural transfers among coccoid algae. Lloydia 11:77–79.

Drouet, F., Daily, W. A. 1952. A synopsis of the coccoid Myxophyceae. Butler University Botany Studies 10:220–223.

Drouet, F., Daily, W. A. 1956. Revision of the coccoid Myxophyceae. Butler University Botany Studies 12:1–218.

Duthie, H. C., Socha, R. 1976. A checklist of the freshwater algae of Ontario, exclusive of the Great Lakes. Naturaliste Canadien (Quebec) 103:83–109.

Eguchi, M., Oketa, T., Miyamoto, N., Maeda, H., Kawai, A. 1996. Occurrence of viable photoautotrophic picoplankton in the aphotic zone of Lake Biwa, Japan. Journal of Plankton Research 18:539–550.

Elenkin, A. A. 1936, 1938, 1949. Monographia algarum cyanophycearum aquidulcium et terrestrium in finibus URSS inventarum. [Sinezelenye vodorosli SSSR.] Izd. AN SSSR, Moskva-Leningrad, 1,2(1–2): 1–1908. [Russian, Latin diagnoses and descriptions.]

Eskew, D. L., Ting, I. P. 1978. Nitrogen fixation by legumes and blue–green algal–lichen crusts in a Colorado desert environment. American Journal of Botany 65:850–856.

Fahnenstiel, G. L., Sicko-Goad, L., Scavia, D., Stoermer, E. F. 1986. Importance of picoplankton in Lake Superior. Canadian Journal of Fisheries and Aquatic Sciences 43:235–240.

Fairchild, E. C., Wilson, D. L. 1967. The algal flora of two Washington soils. Ecology 48:1053–1055.

Fallon, R. D., Brock, T. D. 1981. Overwintering of Microcystis in Lake Mendota. Freshwater Biology 11:217–226.

Fay, P. 1983. The blue–greens (Cyanophyta – Cyanobacteria). Studies in Biology 160. E. Arnold, London, pp. 1–88.

Fay P., Van Baalen, C., Eds. 1987 The cyanobacteria. Elsevier, Amsterdam, 543 p.

Fogg, G. E. 1986. Picoplankton. Proceedings of the Royal Society of London 228:1–30.

Fogg, G. E., Stewart, W. D. P., Fay, P., Walsby, A. E. 1973. The blue–green algae. Academic Press, London, 459 pp.

Forest, H. S., Weston, C. R. 1966. Blue-green algae from the Atacama desert of northern Chile. Journal of Phycology 2:163–164.

Feldmann, J. 1958. Les Cyanophycées marines de la Guadeloupe. Review Algologique 4:25–40.

Frémy, P. 1930a. Contribution a là Flore algologique du Congo Belge. Bulletin de Jardin Botanique 9:109–138.

Frémy, P. 1930b. Les Myxophycées de l'Afrique équatoriale française. Archives ole Botanique 3/2:1–508.

Frémy, P. 1933. Les Cyanophycées des Côtes d'Europe. Mémoire de Société Nationale de Sciences Naturelles et Mathematique Cherbourg 41:1–236.

Frémy, P. 1949. Cyanophyceae. Exploration du Parc National Albert, Mission H. Dumas, Fasc. 19:17–51.

Friedmann, E. I. 1971. Light and scanning electron microscopy of the endolithic desert algal habitat. Phycologia 10:411–428.

Friedmann, E. I. 1980. Endolithic microbial life in hot and cold deserts. Origins of Life 10:223–235.

Friedmann, E. I., Ocampo, R. 1985. Blue–green algae in arid cryptoendolithic habitats. Archiv für Hydrobiologie/Algological Studies 38/39:349–350.

Friedmann, E. I., Ocampo-Friedmann, R. 1984. Endolithic microorganisms in extreme dry environments: Analysis of a lithobiontic microbial habitat, in: Klug M. J., Reddy C. A., Eds., Current perspectives in microbial ecology. American Society of Microbiology, Washington, DC, pp.177–185.

Fulton, R. S., Paerl, H. W. 1987. Toxic and inhibitory effects of the blue–green alga *Microcystis aeruginosa* on herbivorous zooplankton. Journal of Plankton Research 9:837–855.

Garcia-Pichel, F., Belnap, J. 1996. Microenvironments and microscale productivity of cyanobacterial desert crusts. Journal of Phycology 32:774–782.

Garcia-Pichel, F., Castenholz, R. W. 1991. Characterization and biological implications of scytonemin, a cyanobacterial sheath pigment. Journal of Phycology 27:395–409.

Garcia Reina, G. 1997. Algae directory. Algologists, companies, culture collections and herbaria in European countries. Cooperation in Science and Technology – COST-49, Luxembourg, 235 pp.

Gardner, N. L. 1906. Cytological studies in Cyanophyceae. University of California, Publications in Botany 2:237–296.

Gardner, N. L. 1918. New Pacific coast marine algae III. University of California, Publications in Botany 6:455–486.

Gardner, N. L. 1927. New Myxophyceae from Porto Rico. Memoirs of the New York Botanical Gardens 7:1–144.

Garric, R. K. 1965. The cryoflora of the Pacific Northwest. American Journal of Botany 52:1–8.

Geitler, L. 1925. *Cyanophyceae*, in: Pascher's Süsswasser-Flora, Vol. 12. Fischer-Verlag, Jena, 455 p.

Geitler, L. 1932. *Cyanophyceae*, in: Rabenhorst's Kryptogamen-Flora, Vol. 14. Akademisches Verlagsgesellschaft, Leipzig, 1196 pp.

Geitler, L. 1942. *Schizophyta* (Klasse Schizophyceae), in: Engler, A., Prantl, K., Eds., Natürliche Pflanzenfamilien, Vol. 1b. Berlin, 232 p.

Geitler, L. 1960. *Schizophyceen*, in: Handbuch der Pflanzenanatomie, Vol. 6, Chap. 1. Nikolassee, Berlin, 131 p.

Geitler, L., Ruttner, F. 1935. Die Cyanophyceen der Deutschen limnologischen Sunda-Expedition, ihre Morphologie, Systematik und Ökologie. Archiv für Hydrobiologie, Supplement 14, 6:308–369, 371–483.

Gektidis, M., Golubić, S. 1996. A new endolithic cyanophyte/cyanobacterium: *Hyella vacans* sp. nov. from Lee Stocking Island, Bahamas. Nova Hedwigia 112:91–98.

Gibson, C. E., Smith, R. V. 1982. Freshwater plankton, in: Carr, N. G., Whitton, B. A., Eds. The biology of cyanobacteria. Blackwell, Oxford, pp. 463–489.

Gold-Morgan, M., Montejano, G., Komárek, J. 1994. Freshwater epiphytic cyanoprokaryotes from central Mexico. 2. Heterogeneity of the genus *Xenococcus*. Archiv für Protistenkunde 144:383–405.

Gold-Morgan, M., Montejano, G., Komárek, J. 1996. Freshwater epiphytic Chamaesiphonaceae from Central Mexico. Archiv für Hydrobiologie/Algological Studies 83:257–271.

Goldsborough, L. G., Robinson, G. G. C. 1996. Pattern in wetlands, in: Stevenson, R. J., Bothwell, M. L., Lowe, R. L., Eds., Algal ecology: Freshwater benthic ecosystems. Academic Press, San Diego, pp. 77–117.

Golubić, S. 1965. Zur Revision der Gattung *Gloeocapsa* Kützing (Cyanophyta). Schweizerische Zeitschrift für Hydrologie 27:218–232.

Golubić, S. 1967a. Algenvegetation der Felsen. Die Binnengewässer, Vol. 23, Schweizerbart'sche Verlagsbuchhandlung, Stuttgart, 183 pp.

Golubić, S. 1967b. Zwei wichtige Merkmale zur Abgrenzung der Blaualgengattungen. Schweizerische Zeitschrift für Hydrologie 29:176–184.

Golubić, S. 1980. Halophily and halotolerance in cyanophytes. Origins of Life 10:169–183.

Golubić, S., Perkins, R. D., Lukas, K. J. 1975. Boring microorganisms and microborings in carbonate substrates, in: Frey R. W., Ed., The study of trace fossils. Springer-Verlag, New York, pp. 229–269.

Golubić, S., Friedmann, E. I., Schneider, J. 1981. The lithobiontic ecological niche, with special reference to microorganisms. Sedimentary Petrology 51:475–478.

Golubić, S., Yun, Z., Campbell, S. E. 1985. Early evolution of morphological complexity in Procaryote (Cyanophyta, Cyanobacteria), in: Mlíkovský J., Novák V. J. P., Eds., Evolution and morphogenesis, Academia, Praha, pp. 355–368.

Gomont, M. 1892. Monographie des Oscillariées (Nostocacées homocystées). Annales de Sciences Naturales Botaniques, Series 7, 15:263–368; 16:91–264.

Gorham, P. R., Carmichael, W. W. 1988. Hazards of freshwater blue-greens (cyanobacteria), in: Lembi, C. A., Waaland, J. R., Eds., Algae and human affairs. Cambridge University Press, pp. 403–431.

Gorham, P. R., McLachlan, J., Hammer, U. T., Kim, W. K. 1964. Isolation and culture of toxic strains of *Anabaena flos-aquae* (Lyngb.) de Bréb. Internationale Vereinigung für Theoretische und Angewandte Limnologie 15:796–804.

Hällfors, G., Munsterhjelm, R. 1982. Some epiphytic Chamaesiphonales from fresh and brackish waters in southern Finland. Acta Botanica Fennica 19:147–176.

Hauschild, C. A., McMurter, H. J. G., Pick, F. R. 1991. Effect of spectral quality on growth and pigmentation of picocyanobacteria. Journal of Phycology 27:698–702.

Healy, F. P. 1982. Phosphate, in: Carr N. G., Whitton B. A., Eds., The biology of cyanobacteria. Botany Monographs 19. Blackwell, Oxford, pp. 105–124.

Hickel, B. 1981. *Cyanodictyon reticulatum* (Lemm.) Geitler (Cyanophyta) a rare planktonic blue–green alga refound in eutrophic lakes. Archiv für Hydrobiologie/Algological Studies 27:111–118.

Hietala, J., Lauren-Maatta, C., Walls, M. 1997. Sensitivity of *Daphnia* to toxic cyanobacteria: effects of genotype and temperature. Freshwater Biology 37:299–306.

Hindák, F. 1984. On the taxonomy of the cyanophycean genus *Rhabdogloea* Schröder (= *Dactylococcopsis* Hansg. sensu auct. post.). Archiv für Hydrobiologie/Algological Studies 35:121–133.

Hindák, F. 1996. Cyanophytes colonizing mucilage of chroococcal water blooms. Nova Hedwigia 112:69–82.

Hindák, F., Moustaka, M. P. 1988. Planktonic cyanophytes of Lake Volvi, Greece. Archiv für Hydrobiologie/Algological Studies 50–53:497–528.

Hirose, H., Hirano, M. 1981. Class Cyanophyceae, in: Hirose, H., Yamagishi, T., Eds., Illustrations of the Japanese fresh-water algae, 2nd ed., Uchidarokakuho, Tokyo, p. 1–151.

Hollenberg, G. J. 1939. Some new Myxophyceae from Southern California. Bulletin of the Torrey Botanical Club 66:489–494.

Hötzel, G., Croome, R. 1994. Long-term phytoplankton monitoring of the Darling River at Burtundy, New South Wales—incidence and significance of cyanobacterial blooms. Australian Journal of Marine and Freshwater Research 45:747–759.

Howe, M. A. 1924. Notes on algae of Bermuda and the Bahamas. Bulletin of the Torrey Botanical Club 51:351–359.

Howe, M. A. 1932. The geologic importance of the lime-secreting algae, with a description of a new travertine-forming organism. U.S. Geological Surveys Professional Papers 170:57–64, plates 19–23.

Hua, M. S., Friedmann, E. I., Ocampo-Friedmann, R., Campbell, S. B. 1989. Heteropolarity in unicellular cyanobacteria: Structure and development of *Cyanocystis violacea*. Plant Systematics and Evolution 164:17–26.

Huber-Pestalozzi, G. 1938. Das Phytoplankton des Süsswassers. Systematik und Biologie. *1*. Die Binnengewässer. Vol. 16. Schweizerbart'sche Verlagsbuchhandlung, Stuttgart, 342 p.

Humm, H. J., Wicks, S. R. 1980. Introduction and guide to the marine blue-green algae. Wiley, New York, 194 p.

Hunt, M. G., Floyd, G. L., Stout, B. B. 1979. Soil algae in field and forest environments. Ecology 60:362–375.

Jaag, O. 1941. Die Zellgrösse als Artenmerkmal bei den Blaualgen. Schweizerische Zeitschrift für Hydrologie 9:16–33.

Jaag, O. 1945. Untersuchungen über die Vegetation und Biologie der Algen des nackten Gesteins in den Alpen, im Jura und schweizerischen Mittelland. Beihefte zur Kryptogamenflora der Schweiz 9, 3, 560 p.

Jensen, T. E. 1984. Cyanobacterial cell inclusions of irregular occurrence: Systematic and evolutionary implications. Cytobios 39:35–62.

Jensen, T. E. 1985. Cell inclusions in the Cyanobacteria. Archiv für Hydrobiologie/Algological Studies 38/39:33–73.

Johansen, J. R. 1993. Cryptogamic crusts of semiarid and arid lands of North America. Journal of Phycology 29:140–147.

Johansen, J. R., Ashley, J., Rayburn, W. R. 1993. Effects of rangefire on soil algal crusts in semiarid shrub-steppe of the lower Columbia Basin and their subsequent recovery. Great Basin Naturalist 53:73–88.

Johansen, J. R., Rushforth, S. R. 1985. Cryptogamic soil crusts: seasonal variation in algal populations in the Tintic Mountains, Juab County, Utah, U.S.A. Great Basin Naturalist 45:14–21.

Jürgensen, M. F. 1973. Relationship between nonsymbiotic nitrogen fixation and soil nutrient status—a review. Journal of Soil Science 24:512–522.

Jüttner, F. 1987. Volatile organic substances, *in*: Fay P., Van Baalen C., Eds., The cyanobacteria. Elsevier, Amsterdam, pp. 453–469.

Kann, E. 1972. Zur Systematik und Ökologie der Gattung *Chamaesiphon* (Cyanophyceae) 1. Systematik. Archiv für Hydrobiologie/Algological Studies 8:117–171.

Kann, E. 1973. Zur Systematik und Ökologie der Gattung *Chamaesiphon* (Cyanophyceae) 2. Ökologie. Archiv für Hydrobiologie/Algology Studies 8:243–282.

Kato, T., Watanabe, M. 1993. Allozyme divergence of *Microcystis* strains from lake Kasumigaura. Proceedings of the International Phycology Forum, Tsukuba, pp. 1–73.

King, J. M., Ward, C. H. 1977. Distribution of edaphic algae as related to land usage. Phycologia 16:23–30.

Kiselev, I. A. 1947. K morfologii, ekologii, sistematike i geografičeskomu rasprostraneniju sinezelenoj vodorosli *Cyanothrix gardneri* (Frémy). Botaničeský Žurnal SSSR 32:111–118.

Komárek, J. 1958. Die taxonomische Revision der planktischen Blaualgen der Tschechoslowakei, *in*: Algologische Studien. Academia, Praha, pp. 10–206.

Komárek, J. 1969. Proposal for a uniform designation of cultivated algal strains. Regnum Vegetabile 60:64–73.

Komárek, J. 1970. Generic identity of the *"Anacystis nidulans"* strain Kratz-Allen/Bloom. 625 with *Synechococcus* Näg. 1849. Archiv für Protistenkunde 112:343–364.

Komárek, J. 1975. *Geitleribactron*—eine neue, *Chamaesiphon*-ähnliche Blaualgengattung. Plant Systematics and Evolution 123:263–281.

Komárek, J. 1976. Taxonomic review of the genera *Synechocystis* Sauv. 1892, *Synechococcus* Näg. 1849, and *Cyanothece* gen. nov. (Cyanophyceae). Archiv für Protistenkunde 118:119–179.

Komárek, J. 1984. Sobre las cianofíceas de Cuba: (3) Especies planctónicas que forman florecimientos de las aguas. Acta Botanica Cubana 19:1–33.

Komárek, J. 1985. Do all cyanophytes have a cosmopolitan distribution? Survey of the freshwater cyanophyte flora of Cuba. Archiv für Hydrobiologie/Algological Studies 38/39:359–386.

Komárek, J. 1989. Studies on the Cyanophytes of Cuba 7–9. Folia Geobotanica *et* Phytotaxonomica 24:131–206.

Komárek, J. 1993. Validation of the genera *Gloeocapsopsis* and *Asterocapsa (Cyanoprokaryota)* with regard to species from Japan, Mexico and Himalayas. Bulletin of the National Science Museum of Tokyo Series B 19:19–37.

Komárek, J. 1994. Current trends and species delimitation in the cyanoprokaryote taxonomy. Archiv für Hydrobiologie/Algological Studies 75:11–29.

Komárek, J. 1995. Studies on the Cyanophytes (Cyanoprokaryotes) of Cuba 10. New and little known Chroococcalean species. Folia Geobotanica *et* Phytotaxonomica 30:81–90.

Komárek, J. 1999. Intergeneric characters in unicellular cyanobacteria, living in solitary cells. Archiv für Hydrobiologie/Algological Studies 94:195–205.

Komárek, J., Anagnostidis, K. 1986. Modern approach to the classification system of cyanophytes, 2-Chroococcales. Archiv für Hydrobiologie/Algological Studies 43:157–226.

Komárek, J., Anagnostidis, K. 1989. Modern approach to the classification system of cyanophytes. 4. Nostocales. Archiüv für Hydrobiolgie/Algological Studies 56:247–345.

Komárek, J., Anagnostidis, K. 1995. Nomenclatural novelties in chroococcalean cyanoprokaryotes. Preslia (Prague) 67:15–23.

Komárek, J., Anagnostidis, K. 1998. Cyanoprokaryota 1. Teil: Chroococcales, *in*: Süsswasserflora von Mitteleuropa 19/1. Fischer Verlag, Stuttgart, 548 pp.

Komárek, J., Cepák, V. 1998. Cytomorphological characters supporting the taxonomic validity of *Cyanothece (Cyanoprokaryota)*. Plant Systematics and Evolution 210:25–39.

Komárek, J., Hindák, F. 1975. Taxonomy of the new isolated strains of *Chroococcidiopsis* (Cyanophyceae). Archiv für Hydrobiologie/Algological Studies 13:311–329.

Komárek, J., Hindák, F. 1988. Taxonomic review of natural populations of the cyanophytes from the *Gomphosphaeria* complex. Archiv für Hydrobiologie/Algological Studies 50–53:203–225.

Komárek, J., Hindák, F. 1989. The genus *Eucapsis (Cyanophyta/Cyanobacteria)* in Czechoslovakia. Acta Hydrobiologica 31:25–34.

Komárek, J., Komárková-Legnerová, J. 1992. Variability of some planktonic gomphosphaerioid cyanoprokaryotes in northern lakes. Nordic Journal of Botany 12:513–524.

Komárek, J., Komárková-Legnerová, J. 1993. *Radiocystis fernandoi*, a new planktonic cyanoprokaryotic species from tropical freshwater reservoirs. Preslia 65:355–357.

Komárek, J., Komárková-Legnerová, J., 2002. Contribution to the knowledge of planktic cyanoprokaryotes from central Mexico. Perslia (Prague) 74, in press.

Komárek, J., Lukavský, J. 1988. *Arthronema*, a new cyanophyte genus from Afro-Asian deserts. Archiv für Hydrobiologie/Algological Studies 50–53:249–267.

Komárek, J., Montejano, G. 1994. Taxonomic evaluation of several *Chlorogloea* species (Cyanoprocaryota) from inland biotopes. Archiv für Hydrobiologie/Algological Studies 74:1–26.

Komárek, J., Novelo, E. 1994. Little known tropical *Chroococcus* species (Cyanoprokaryotes). Preslia (Prague) 66:1–21.

Komárek, J., Ludvík, J., Pokorný, V. 1975. Cell structure and endospore formation in the blue–green alga *Chroococcidiopsis*. Archiv für Hydrobiologie/Algological Studies 12:205–223.

Komárek, J., Kopecký, J., Cepák, V. 1999. Generic characters of the simplest cyanoprokaryotes *Cyanobium*, *Cyanobacterium* and *Synechococcus*. Cryptogamie/Algologie 20:209–222.

Komárek, J., Komárková-Legnerová, J., Sant'Anna, C. L., Azevedo, M. T. P. Senna, P. A. C. 2002. Two common *Microcystis* species from tropical America. Cryptogamie/Algologie, 23(2):159–177.

Komárková, J. 2002. Cyanobacterial picoplankton and its colonial formations in two eutrophic canyon reservoirs (Czech Republic). Archiv für Hydrobiologie, 23:159–177.

Komárková, J., Šimek, K. 2003. Unicellular and colonial formations of picoplanktonic cyanobacteria under variable environmental conditions and predation pressure. Archiv für Hydrobiology/Algological Studies (Papers of Cyanobacterial Research 4) in press.

Komárková-Legnerová, J. 1991. *Epigloeosphaera*, a new cyanophyte genus from Nordic lakes. Archiv für Hydrobiologie/Algological Studies 62:7–12.

Komárková-Legnerová, J., Cronberg, G. 1994. Planktonic blue–green algae from lakes in South Scania, Sweden. Part I. Chroococcales. Archiv für Hydrobiologie/Algological Studies 72:13–51.

Komárková-Legnerová, J., Tavera, R. 1996. Cyanoprokaryota (cyanobacteria) in the phytoplankton of Lake Catemaco (Veracruz, Mexico). Archiv für Hydrobiologie/Algological Studies 83:403–422.

Kondrateva, N. V. 1968. Voprosy morfologii i sistematiki *Microcystis aeruginosa* Kuetz. emend. Elenk. i blizkich k nemu vidov [Problem of morphology and systematics of *Microcystis aeruginosa* Kuetz. emend. Elenk. and related species], *in*: Cvetenie vody. Akademia Nauk, URSR, Kiev, pp. 13–42.

Kondrateva, N. V., Kovalenko, O. V., Prichod'kova, L. P. 1984. Sin'o-zeleni vodorosti—Cyanophyta [Blue–green algae—Cyanophyta], *in*: Viznačnik Prisnovodnich Vodorostej URSR 1/1, Akademia Nauk, URSR, Kiev, 388 p.

Kosinskaja, E. K. 1948. Opredelitel' morskich sinezelenych vodoroslej [Identification key for marine blue–green algae] Izda Akademia Nauk SSSR, Moskva-Leningrad, 278 p.

Kováčik, L. 1988. Cell division in simple coccal cyanophytes. Archiv für Hydrobiologie/Algological Studies 50–53:149–190.

Krogmann, D. W., Buttala, R., Sprinkle, J. 1986. Blooms of cyanobacteria in the Potomac River. Plant Physiology 80:667–671.

Kubečková, K., Johansen, J. R., Warren, S. D. 2003. Development of immobilized cyanobacterial amendments for reclamation of microbiotic soil crusts. Archiv für Hydrobiologie/Algological Studies (Papers of Cyanobacterial Research 4) in press.

Kützing, T. F. 1849. Species Algarum. Brockhaus, Leipzig, 922 p.

Lawry, N. H., Simon, R. D. 1982. The normal and induced occurrence of cyanophycin inclusion bodies in several blue–green algae. Journal of Phycology 18:391–399.

LeCampion-Alsumard, T. and Golubić, S. 1985. *Hyella caespitosa* Bornet et Flahault and *Hyella balani* Lehman (Pleurocapsales, Cyanophyta). Archiv für Hydrobiologie/Algological Studies 38/39:115–118.

Lemmermann, E. 1899. Planktonalgen. Ergebnisse einer Reise nach dem Pacific. Abhandlung Naturwissenschaft Vereinigung in Bremen 16:313–398.

Lemmerman, E. 1905. Die Algenflora der Sandwich-Inseln. Botanisches fahrbuch für Systematik, Pflanzengeschichte und Pflanzenographie 34:607–663.

Lewin, R., Robinson, X. 1979. Greening of polar bears. Nature (London) 1234:278–280.

Lhotsky, O., Komárek, J. 1981. Unificated designation of cultivated algal strains (Abstract), in Proceedings of the IV International congress on Culture Collections, Brno, p. 13.

Lukas, K. J., Golubić, S. 1981. New endolithic cyanophytes from the North Atlantic ocean: I. *Cyanosaccus piriformis* gen. et sp. nov. Journal of Phycology 17:224–229.

Lukas, K. J., Golubić, S. 1983. New endolithic cyanophytes from the North Atlantic ocean: II. *Hyella gigas* Lukas et Golubić sp. nov. from the Florida continental margin. Journal of Phycology 19:129–136.

Lukas, K. J., Hoffman, E. J. 1984. New endolithic cyanophytes from the North Atlantic Ocean. III. *Hyella pyxis* Lukas & Hoffman sp. nov. Journal of Phycology 20:515–520.

Lukešová, A. 1993. Soil algae in four secondary successional stages on abandoned fields. Archiv für Hydrobiologie/Algological Studies 71:81–102.

MacEntree, F. J., Schreckenberg, S. G., Bold, H. C. 1972. Some observations on the distribution of edaphic algae. Soil Science 114:171–179.

Mann, N. H., Carr, N. G. 1992. Photosynthetic prokaryotes. Plenum Press, New York.

Maxwell, C. D. 1991. Floristic changes in soil algae and cyanobacteria in reclaimed metal-contaminated land at Sudbury, Canada. Water, Air, and Soil Pollution 60:381.

McCormick, P. V., Stevenson, R. J. 1998. Periphyton as a tool for ecological assessment and management in the Florida Everglades. Journal of Phycology 34:726–733.

Miller, S. R., Wingard, C. E., Castenholz, R. W. 1998. Effects of visible light and UV radiation on photosynthesis in a population of a hot spring cyanobacterium, a *Synechococcus* sp., subjected to high-temperature stress. Applied and Environmental Microbiology 64:3893–3899.

Montejano, G., Gold, M., Komárek, J. 1993. Freshwater epiphytic cyanoprokaryotes from central Mexico I. *Cyanocystis* and *Xenococcus*. Archiv für Protistenkunde 143:237–247.

Montejano, G., Gold, M., Komárek, J. 1997. Freshwater epiphytic cyanoprokaryotes from central Mexico III. The genus *Stichosiphon*. Archiv für Protistenkunde 148:3–16.

Montoya, H. T., Golubić, S. 1991. Morphological variability in natural populations of mat-forming cyanobacteria in the salinas of Huacho, Lima, Peru. Archiv für Hydrobiologie/Algological Studies 64:423–441.

Moore, G. T., Carter, N. 1923. Algae from lakes in the north-eastern part of North Dakota. Annals of Missouri Botanical Garden 10:393–422.

Nägeli, C. 1849. Gattungen einzelliger Algen. Neue Denkschrift Allgemeine Schweizerische Naturforschende Gesellschaft Zürich, 10:139 p.

Neilan, B. A., Jacobs, D., Del Lot, T., Blackall, L. L., Hawkins, P. R., Cox, P. T., Goodman, E. 1997. rRNA sequences and evolutionary relationships among toxic and nontoxic cyanobacteria of the genus *Microcystis*. International Journal of Systematic Bacteriology 47:693–697.

Norris, R. E. 1967. Algal consortisms in marine plankton, in Proceedings of the Seminar on Sea, Salt and Plants, pp. 178–189.

Paerl, H. W. 1996. A comparison of cyanobacterial bloom dynamics in freshwater estuarine and marine environments. Phycologia 35:25–35.

Paerl, H. W., Bowles, N. D. 1987. Dilution bioassays: Their application to assessments of nutrient limitation in hypereutrophic waters. Hydrobiologia 146:265–273.

Paerl, H. W., Millie, D. F. 1996. Physiological ecology of toxic aquatic cyanobacteria. Phycologia 35:160–167.

Palinska, K. A., Liesack, W., Rhiel, E., Krumbein, W. E. 1996. Phenotype variability of identical genotypes: The need for a combined approach in cyanobacterial taxonomy demonstrated on *Merismopedia*-like isolates. Archives of Microbiology 166:224–233.

Parker, D. L. 1982. Improved procedures for the cloning and purification of *Microcystis* cultures (Cyanophyta). Journal of Phycology 18:471–477.

Pick, F. R., Lean, D. R. S. 1987. The role of macronutrients (C, N, P) in controlling cyanobacterial dominance in temperate lakes. New Zealand Journal of Marine and Freshwater Research 21:425–434.

Platt, T., Li, W. K. W., Eds. 1986. Photosynthetic picoplankton. Canadian Bulletin of Fisheries and Aquatic Science 214:583 pp.

Prescott, G. W. 1962. Algae of the western Great Lakes area, 2nd ed. Brown, Dubuque, IA, 977 p.

Prescott, G. W., Croasdale, H. T. 1937. New or noteworthy freshwater algae of Massachusetts. Transactions of the American Microscopical Society 56:269–282.

Prescott, G. W., Vinyard, W. C. 1965. Ecology of Alaskan freshwater algae. V. Limnology and flora of Malikpuk Lake. Transactions of the American Microscopical Society 84:427–478.

Pringsheim, E. G. 1946. Pure cultures of algae, their preparation and maintenance. Cambridge University Press, 119 p.

Reed R. H., Warr, S. R. C., Kerby, N. W., Stewart, W. D. P. 1986. Osmotic shock-induced release of low molecular weight metabolites from free-living and immobilized cyanobacteria. Enzyme and Microbial Technology 8:101–104.

Reinhard, E. G. 1931. The plankton ecology of the upper Mississippi, Minneapolis to Winona. Ecological Monographs 1:395–464.

Rejmánková, E., Roberts, D. R., Manguin, S., Pope, K. O., Komárek, J., Post, R. A. 1996. *Anopheles albimanus* (Diptera: Culicidae) and cyanobacteria: An example of larval habitat selection. Population Ecology 25:1058–1067.

Reynolds, C. S., Descy, J.-P. 1996. The production, biomass and structure of phytoplankton in large rivers. Archiv für Hydrobiologie, Supplement 113:161–187.

Reynolds, C. S., Walsby, A. E. 1975. Water blooms. Biological Review 50:437–481.

Reynolds, C. S., Jaworski, G. H. M., Cmiech, H. A., Leedale, G. F. 1981. On the annual cycle of the blue–green alga *Microcystis aeruginosa* Kütz. Emend. Elenkin. Philosophical Transactions of the Royal Society of London, Series B. 293:419–477.

Rippka, R., Deruelles, J. B., Waterbury, J. B., Herdman, M., Stanier, R. Y. 1979. Generic assignments, strain histories and properties of pure cultures of cyanobacteria. Journal of General Microbiology 111:1–61.

Roussomoustakaki, M., Anagnostidis, K. 1991. *Cyanothece halobia*, a new planktonic chroococcalean cyanophyte from Hellenic heliothermal saltworks. Archiv für Hydrobiologie/Algological Studies 64:71–95.

Rudi, K., Skulberg, O. M., Jakobsen, K. S. 1998. Evolution of cyanobacteria by exchange of genetic material among phyletically related strains. Journal of Bacteriology 180:3453–3461.

Schiller, J. 1956. Die Mikroflora der roten Tümpel auf den Koralleninseln "Los Aves" in Karibischen Meer. Ergebnisse Deutscher Limnologischen Venezuela-Expedition 1952, 1:197–216.

Schulz-Baldes, M., Lewin, R. A. 1976. Fine structure of *Synechocystis didemni* (Cyanophyta/Chroococcales). Phycologia 15:1–6.

Senna, P. A. C., Peres, A. C., Komárek, J. 1998. *Coelomoron tropicalis*, a new cyanoprokaryotic species from São Paulo State, Brazil. Nova Hedwigia 67:93–100.

Setchell, W. A. 1924. American Samoa: Part I. Vegetation of Tutuila Island; Part II. Ethnobotany of the Samoas; Part III. Vegetation of Rose Atoll, 258–259 p.

Setchell, W. A., Gardner, N. L. 1905. Algae of northwestern America. University of California Publications in Botany 1:165–418.

Setchell, W. A., Gardner, N. L. 1919. The marine algae of the Pacific coast of North America, Part I. Myxophyceae. University of California Publications in Botany 8:1–138.

Setchell, W. A., Gardner, N. L. 1924. New marine algae from the Gulf of California. Proceedings of the California Academy of Science, Series 4, 12:695–949.

Setchell, W. A., Gardner, N. L. 1930. Proceedings of the California Academy of Science, Series 4, 19:118.

Sheath, R. G., Steinman. 1982. A checklist of freshwater algae of the Northwest Territories, Canada. Canadian Journal of Botany 60:1964–1997.

Shimmel, S. M., Darley, W. M. 1985. Productivity and density of soil algae in an agricultural system. Ecology 66:1439–1447.

Sinclair, C., Whitton, B. A. 1977. Influence of nutrient deficiency on hair formation in the Rivulariaceae. British Phycological Journal 12:297–313.

Skuja, H. 1939. Beitrag zur Algenflora Lettlands II. Acta Horti Botanici Universitatis Latviense 11/12:41–169.

Skuja, H. 1948. Taxonomie des Phytoplanktons einiger Seen in Uppland, Sweden. Symbolae Botanicae Upsalienses 9:1–399.

Skuja, H. 1964. Grundzüge der Algenflora und Algenvegetation der Fjeldgegenden um Abisko in Schwedisch-Lappland. Nova Acta Regiae Societatis Scientificae Upsaliensae Series 4, 18:1–465.

Šmajs, D., Šmarda, J. 1999. Cytomorphology of the smallest picoplanktonic Cyanobacteria. Archiv für Hydrobiologie/Algological Studies 94:333–351.

Šmarda, J. 1991. S-layer of chroococcal cell walls. Archiv für Hydrobiologie/Algological Studies 64:41–51.

Šmarda, J., Šmajs, D., Komrska, J., Krzyzanek, V. 2002. S-layers on cell walls of cyanobacteria. MICRON 33:257–277.

Smith, A. J. 1982. Modes of cyanobacterial carbon metabolism, in: Carr, N. G., Whitton, B. A., Eds., The biology of cyanobacteria. Blackwell, Oxford, pp. 47–85.

Smith, G. M. 1920. Phytoplankton of the Inland Lakes of Wisconsin. I. Wisconsin Geology and Natural History Surveys 57, Series on Science, 12:1–243.

Smith, G. M. 1925. The plankton algae of the Okoboji region. Transactions of the American Microscopical Society 45:156–223.

Smith, G. M. 1933. The freshwater algae of the United States. McGraw-Hill Book Co., New York, 716 p.

Smith, G. M. 1950. Fresh water algae of the United States of America, 2nd ed. McGraw-Hill, New York, 719 p.

Stanier, R. Y., Cohen-Bazire, G. 1977. Phototrophic prokaryotes: The cyanobacteria. Annual Review of Microbiology 31:225–274.

Starks, T. L., Shubert, L. E.,. R. Trainor, F. R. 1981. Ecology of soil algae: A review. Phycologia 20:65–80.

Starmach, K. 1966. Cyanophyta—sinice [Cyanophyta—blue–green algae]. Flora Słodkowodna Polski PWN, Warsaw, 753 p.

Starmach, K. 1973. Glony osiadłe w Wielkim Stawie w Dolinie Pięciu Stawów Polskich w Tatrach [Benthic algae of the Great Lake in the Valley of the Five Polish Lakes in the Tatra Mountains]. Fragmenta Floristica et Geobotanica 19(4):481–511.

Stein, J. R., Borden, C. A. 1979. Checklist of freshwater algae of British Columbia. Syesis 12:3–39.

Stockner, J. G. 1988. Phototrophic picoplankton: An overview from marine and freshwater ecosystems. Limnology and Oceanography 33:765–775.

Stockner, J. G., Shortreed, K. S. 1991. Autotrophic picoplankton: Community composition, abundance and distribution across a gradient of oligotrophic British Columbia and Yukon Territory lakes. Internationale Revue der Gesamten Hydrobiologie 76:581–601.

Sugawara, H., Miyazaki, S., Eds. 1999. World directory of colletions of cultures of microorganisms, 5th ed. WFCC-MIRCEN World Data Centre of Microorganisms, Mishima, Shizuoka, 140 p.

Suttle, C. A., Harrison, P. J. 1988. Ammonium and phosphate uptake rates, N:P supply ratios, and evidence for N and P limitation in some oligotrophic lakes. Limnology and Oceanography 33:186–202.

Tavera, R., Komárek, J. 1996. Cyanoprokaryotes in the volcanic lake of Alchichica, Puebla State, Mexico. Archiv für Hydrobiologie/Algological Studies 83:511–538.

Taylor, W. R. 1928. The marine algae of Florida with special reference to the Dry Tortugas. Papers of the Tortugas Laboratory of Carnegie Institution, Washington 25:1–219.

Teiling, E. 1941. Aeruginosa oder flos-aquae. Eine kleine *Microcystis*-Studie. Svensk Botanisk Tidskrift 35:337–349.

Thompson, R. 1938. A preliminary survey of the fresh-water algae of Eastern Kansas. Bulletin of the University of Kansas 39(11):5–83.

Tiffany, L. H. 1934. The plankton algae of the west end of Lake Erie. Contributions to Ohio State University 6:112 p.

Tilden, J. 1910. Minnesota algae. The Myxophyceae of North America and adjacent regions including Central America, Greenland, Bermuda, The West Indies and Hawaii. Botany Series, Minneapolis 8:1–328.

Utermöhl, H. 1958. Zur Vervollkommnung der quantitativen Phytoplankton-Methodik. Internationale Vereinigung für Limnologie, Mitteilungen 9:1–38.

Van Baalen, C. 1965. Mutation of the blue–green alga *Anacystis nidulans*. Science 149:70.

Van Baalen, C., O'Donnell, R. 1972. Action spectra for ultraviolet killing and photoreaction in the blue–green alga *Agmenellum quadruplicatum*. Photochemistry and Photobiology 15:269–274.

Venkataraman, G. S. 1969. The cultivation of algae. ICAR, New Delhi, 318 p.

Walsby, A. E. 1972. Structure and function of gas vacuoles. Bacteriological Reviews 36:1–32.

Walsby, A. E. 1978. The gas vesicles of aquatic prokaryotes. Symposium of the Society of General Microbiology 28:327–358.

Walsby, A. E. 1981. Cyanobacteria: Planktonic gas vacuolate forms, *in*: Starr M. P., Stolp, H., Trüper, H. G., Balows, A., Schlegel, H. G., Eds., The prokaryotes, Vol. 1. Springer-Verlag, Berlin, pp. 224–235.

Watanabe, M. M. 1999. Network approach to make biodiversity information accessible worldwide – as an example of microorganisms, *in*: Species 2000, Abstracts of the international joint workshop for studies on biodiversity, Tsukuba, p. 13.

Waterbury, J. B. 1989. Order Pleurocapsales Geitler 1925, emend. Waterbury and Stanier 1978, *in*: Staley J. T., Bryant, M. P., Pfennig, N., Holt, J. G., Eds., Bergey's Manual of Systematic Bacteriology, Vol. 3. Williams & Wilkins, Baltimore, pp. 1746–1770.

Waterbury, J. B., 1979. Wide spread occurrence of a unicellular, marine, planktonic cyanobacterium. Nature 277:293–294.

Waterbury, J. B., Rippka, R. 1989. Subsection I. Order Chroococcales Wettstein 1924, emend. Rippka *et al.*, 1979, *in*: Staley, J. T., Bryant, M. P., Pfennig, N., Holt, J. G., Eds., Bergey's Manual of Systematic Bacteriology, Vol. 3. Williams & Wilkins, Baltimore, pp. 1728–1746.

Waterbury, J. B., Stanier, R. Y. 1977. Two unicellular cyanobacteria which reproduce by budding. Archives of Microbiological 115:249–257.

Waterbury, J. B., Stanier, R. Y. 1978. Patterns of growth and development in pleurocapsalean cyanobacteria. Microbiology Reviews 42:2–44.

Weber van Bosse, A. A. 1925. Algues de l'expédition danoise aux îles Kei. Papers from Mortensen's Pacific Expedition 1914–1916.

Wehr, J. D. 1989. Experimental tests of nutrient limitation in freshwater picoplankton. Applied and Environmental Microbiology 55:1605–1611.

Wehr, J. D. 1990. Predominance of picoplankton and nanoplankton in eutrophic Calder lake. Hydrobiologia 203:35–44.

Wehr, J. D. 1992. Effects of experimental manipulations of light and phosphorus supply on competition among picoplankton and nanoplankton in an oligotrophic lake. Canadian Journal of Fisheries and Aquatic Sciences 50:936–945.

Wehr, J. D., Descy, J.-P. 1998. Use of phytoplankton in large river management. Journal of Phycology 34:741–749.

Wehr, J. D., Thorp, J. H. 1997. Impacts of navigation dams, tributaries and littoral zones on phytoplankton communities in the Ohio River. Canadian Journal of Fisheries and Aquatic Sciences 54:378–395.

Weisse, T. 1993. Dynamics of autotrophic picoplankton in marine and freshwater ecosystems. Microbial Ecology 13:327–370.

Wettstein, R. 1924. Handbuch der systematischen Botanik. Leipzig-Wien.

Whelden, R. M. 1941. Some observations on freshwater algae of Florida. Journal of the Elisha Mitchell Science Society 57:261–272.

Whelden, R. M. 1947. Algae, *in*: Polunin, N., Ed., Botany of Canadian eastern Arctic, II. Thallophyta and Bryophyta. Bulletin of the National Museum of Canada 97:13–127.

Whitford, L. A., Schumacher, G. J. 1969. A manual of the freshwater algae in North Carolina. North Carolina Agricultural Experimental Station Technical Bulletin 188:1–313.

Whitton, B. A. 1987. The biology of Rivulariaceae, *in*: Fay, P., Van Baalen, C., Eds., The cyanobacteria, Elsevier, Amsterdam, pp. 513–534.

Whitton, B. A. 1992. Diversity, ecology, and taxonomy of the cyanobacteria, *in*: Mann, N. H., Carr, N. G., Eds., Photosynthetic prokaryotes. Plenum Press, New York, pp. 1–51.

Whitton, B. A. 2002. Phylum Cyanophyta (Cyanobacteria). in: John, D. M., Whitton, B. A., Brock, A. J., Eds., The freshwater algal flora of the British Isles. An identification guide to freshwater and terrestrial algae. Cambridge University Press, pp. 25–122.

Whitton, B. A., Carr, N. G. 1982. Cyanobacteria: Current perspectives, *in*: Carr, N. G., Whitton, B. A., Eds., The biology of cyanobacteria. Blackwell, Oxford, pp. 1–8.

Whitton, B. A., Potts, M. 1982. Marine littoral, *in*: Carr, N. G., Whitton, B. A., Eds. 1982. The biology of cyanobacteria. Blackwell, Oxford, 515–542.

Whitton, B. A., Potts, M., Eds. 2000. The ecology of cyanobacteria: Their diversity in time and space, Kluwer, Boston.

Wilmotte, A., Stam, W. T. 1984. Genetic relationships among cyanobacterial strains originally designated as "*Anacystis nidulans*" and some other *Synechococcus* strains. Journal of General Microbiology 130:2737–2740.

Wojciechowski, I. 1971. Die Plankton-Flora der Seen in der Umgebung von Sosnowica (Ostpolen). Annals of the University M. Curie-Skłodowska, Lublin, 26:233–263.

Wood, H. C. 1869. Prodromus of a study of the fresh water algae of eastern North America. Proceedings of the American Philosophical Society 11:119–145.

Wood, H. C. 1872. A contribution to the history of the fresh-water algae of North America. Smithsonian Contributions to Knowledge 19(241):1–262.

Zalessky, M. M. 1926. Sur les nouvelles algues découvertes dans le sapropélogéne du lac Beloe. Revue Générale de Botanique 38:31–42.

Zohary, T., Breen C. M. 1989. Environmental factors favoring the formation of *Microcystis aeruginosa* hyperscums in a hypertrophic lake. Hydrobiologia 178:179–192.

Zohary, T., Robarts, R. D. 1990. Hyperscums and the population dynamics of *Microcystis aeruginosa*. Journal of Plankton Research 12:423–432.

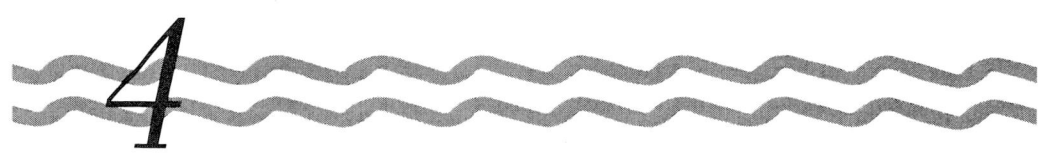

FILAMENTOUS CYANOBACTERIA

Jiří Komárek

Institute of Botany
Academy of Sciences of the
  Czech Republic
Faculty of Biological Sciences
University of South Bohemia
CZ-37982 Třeboň, Czech Republic

Hedy Kling

Freshwater Institute
Winnipeg, Manitoba
Canada R3T 2N6

Jaroslava Komárková

Hydrobiological Institute
Academy of Sciences of the
  Czech Republic
Faculty of Biological Sciences
University of South Bohemia
CZ-37005 České Budějovice
Czech Republic

I. Introduction
II. Morphology
    A. Cytology and Morphology
    B. Specialized Cells
    C. Reproduction
III. Ecology
IV. Methods

V. Key and Descriptions of Genera
    A. Key
    B. Descriptions of Genera
    Note Added in Proof
VI. Guide to Literature for Species
    Identification
    Literature Cited

I. INTRODUCTION

Filamentous cyanobacteria (blue–green algae, cyanoprokaryotes) include some of the most widely recognized and important freshwater algae in the world, many of which produce surface blooms, fix atmospheric nitrogen, and are important components of global carbon fixation (Fogg et al., 1973; Fay and Van Baalen, 1987; Whitton and Potts, 2000). Basic information on the biology and ecology of cyanobacteria is presented in Chapter 3 on coccoid forms; this chapter contains the information concerning the unique morphology and biology of filamentous blue–greens. The system that has been used for several decades for distinguishing genera of filamentous cyanobacteria is based on phenotypic characters that recently have been supported by ultrastructural and molecular data in many instances (Rippka et al., 1979; Anagnostidis and Komárek, 1988; Komárek and Anagnostidis, 1989). Current molecular analyses support the separation of non-heterocystous and heterocystous genera[1]; however, we can expect further changes on both the generic and the species levels for many of these organisms (Castenholz, 1992; Li and Watanabe, 1998; Rudi et al., 1998, 2000). The present chapter recognizes three orders of filamentous cyanobacteria—Oscillatoriales, Nostocales and Stigonematales — which can be clearly defined by several diagnostic features (see Chap. 3, Sect. V.A).

Several comprehensive reviews on filamentous cyanobacteria of North America have been published in the past. An early but thorough study on the distribution of blue–greens in North America is the monograph by Tilden (1910). Although largely using figures from other authors, with the list of exsiccates and very old records, it is a valuable view of the knowledge

---

[1] The term "heterocyte" is used in this chapter to refer to cells commonly known as "heterocysts"; they are not truly cysts.

of the North American cyanobacterial (= cyanoprokaryotic) flora from the beginning of this century. Several later studies present information on specific regions, including the upper Midwest (Wisconsin; Smith, 1920, 1950), Puerto Rico (Gardner, 1927), Yellowstone National Park (Copeland, 1936), the western Great Lakes area (Prescott, 1962), North Carolina (Whitford and Schumacher, 1969), and Illinois (Tiffany and Britton, 1952). Many of these past studies are still important sources of information. For example, in an account of Poulin's collections from Arctic Canada, Whelden (1947) recorded 108 cyanobacterial taxa (following Geitler, 1932), comprising 15 genera, with 65 species that were filamentous. A number of more recent studies in many parts of North America document a particularly rich cyanobacterial flora (e.g., Duthie and Socha, 1976; Stein and Borden, 1979; Sheath and Steinman, 1982). The taxonomic status of many species and consequently their nomenclature have changed over the years, which may affect estimates of cyanobacterial diversity in future studies (Rippka *et al.*, 1979; Anagnostidis and Komárek, 1988; Komárek and Anagnostidis, 1989; Castenholz, 1992). Nonetheless, previous published reports generally represent a useful source of information for current studies of the North American cyanobacterial microflora. In contrast, studies by Drouet and co-workers (Drouet and Daily, 1956; Drouet, 1968, 1973, 1981a, b) greatly oversimplified the taxonomy and classification of cyanobacteria and thus do not correspond to the true diversity in cyanobacterial assemblages in nature.

II. MORPHOLOGY

A. Cytology and Morphology

The internal cellular structure of filamentous cyanobacteria does not differ substantially from coccoid forms (Chap. 3, Sect. II.A and II.B). One important distinction is the arrangement of thylakoids inside cells, which differ from coccoid forms. Although there is wide variation in thylakoid patterns within the Chroococcales suggesting heterogeneity within that group, only three main types (with consistent variations) occur in the simplest filamentous order, the Oscillatoriales. These arrangements are parietal (family Pseudanabaenaceae; Figs. 1–3), radial (family Phormidiaceae; Figs. 12 and 15), and irregular (family Oscillatoriaceae; Fig. 9) (Anagnostidis and Komárek, 1988; Komárek and Cáslavská, 1991). Further, only one thylakoid type, irregular (similar to members of the family Oscillatoriaceae), exists in the Nostocales and Stigonematales.

In all representatives of the filamentous orders, cells are arranged in so-called trichomes, forming one physiological entity. Neighboring cells are connected through pores in their cross walls (microplasmodesms), the number and position of which are characteristic of the genera, and similar pores can sometimes be found on the outer cell walls, especially near the cross walls (Guglielmi and Cohen-Bazire, 1982a). Trichomes are capable of fragmentation (forming hormogonia) or complete disintegration into separate cells (Geitler, 1942, 1960; Watanabe and Komárek, 1989). Fragmentation proceeds after the formation of fine mucilaginous lamella between neighboring cells or via so-called necridic (sacrificial) cells (or necrids). These cells die off, after which the fragments separate (Geitler, 1960, 1982; Kondrateva, 1961; Anagnostidis and Komárek, 1988).

Trichomes may be uniseriate, with cell division exclusively perpendicular to the trichome axis, or multiseriate in more differentiated types (Stigonemataceae), where cells also have the ability to divide longitudinally or in more directions. Cells of many genera produce slimy, colloidal substances external to the cell wall, composed of hydrated polysaccharides (Drews, 1973; Martin and Wyatt, 1974; Jürgens and Weckesser, 1985; deVecchi and Grilli-Caiola, 1986). They are recognizable in the form of mucilaginous envelopes or homogeneous or layered sheaths of various kinds (e.g., fine, firm, diffuse, homogeneous or lengthwise, perpendicular or funnel-like lamellated, colorless or variously colored) surrounding the trichomes. Trichomes with sheaths are traditionally termed filaments. The presence or absence of sheaths, or the ability to produce them under special conditions, is probably a genetically determined characteristic.

The trichomes of various species may be morphologically uniform or exhibit a diversity of cell types. Species of some genera have the ability to form differentiated cells of various types at the ends of the trichome, which may have ecophysiological importance and may be used as a characteristic feature for identification of genera. Apical cells (one or more) may be narrower or wider (without thylakoids) or may be elongated. Apical cells in the family Rivulariaceae and in several genera of Nostocaceae (Nostocales) may become attenuated, elongated, and hyaline. These hairlike formations have been shown to be produced (or accentuated) under periods of inorganic phosphorus deficiency, and they may also be sites of alkaline phosphatase production (Livingstone *et al.*, 1983; Whitton, 1987). The morphological diversity of trichomes is particularly noticeable in polarized (e.g., *Calothrix* and *Rivularia*) and branched types (e.g., *Scytonematopsis* and *Fischerella*; Figs. 20B, 25, and 27), where the basal parts of the filament are morphologically distinct from the apical parts or branches. In contrast to hairs, other taxa have wider, shorter or spherical cells at the end of the filament, probably with a similar function (e.g., *Trichodesmium* and *Scytonema*). Facultative trichome motility is common in more simple forms or in segments of trichomes (motile hormogonia) that are liberated from the sheaths. Modifications of phototactic and other movements have been described in many species (Drews, 1959; Häder, 1974; Nultsch, 1974; Halfen, 1979; Whale and Walsby, 1984). For example, the genera *Geitlerinema* (motile) and *Jaaginema* (immotile) differ only in their motility (Anagnostidis and Komárek, 1988).

Two basic types of branching occur, depending on the type of cell division and filament (+ sheath) morphology. In false branching, trichomes divide (often with necridic cells or at heterocytes) within the sheath and one or both ends of the divided trichome diverge, breaking out the sheath. The resulting filament (e.g., *Tolypothrix* and *Scytonema*; Figs. 20–23) consists of trichomes, diverging singly or in pairs from a common sheath, giving the appearance of branches. Modifications of this branching type are characteristic of certain genera (e.g., Oscillatoriales: *Pseudophormidium*, *Blennothrix*, and *Plectonema*; Nostocales: *Tolypothrix*, *Hassallia*, *Scytonema*, and *Coleodesmium*). True branching occurs in more complex forms (Stigonematales), in which the cells are also capable of longitudinal division (Figs. 35–40). The laterally dividing cell alters its growth polarity (direction of cell division), and grows more or less perpendicular to the original trichome, forming a lateral branch that is physiologically connected to the original trichome. Branching type appears to be a stable character in several genera (e.g., *Brachytrichia*, *Hapalosiphon*, and *Stigonema*; Desikachary, 1959; Bourrelly, 1970; Martin and Wyatt, 1974; Golubić et al., 1996); this morphology may also be related to filament polarity. Distinctly polarized filaments attached by one end to the substrate and with clearly polarized apical growth occur within several groups of filamentous cyanobacteria (e.g., *Schizothrix* and *Rivularia*).

Filaments of some genera are often organized into different arrangements within colonies, and mucilage plays an important role in colony formation. Filaments of several planktonic and free-living metaphytic species can be arranged in clusters or fascicles, which are large enough to be recognizable in the field. Colonies with fasciculated (clustered) filaments are typical of several planktonic species in the genera *Trichodesmium* and *Aphanizomenon* and are radially arranged in spherical colonies in the genus *Gloeotrichia*. Periphytic species are capable of forming a large variety of mats and tufts on submerged as well as subaerial surfaces. Several species from the orders Nostocales and Stigonematales occur as slimy, irregular, or almost spherical macroscopic colonies, including well-known taxa in the genus *Nostoc*. Other types of macroscopic brushlike fascicles occur in water or subaerial habitats, such as members of the genus *Symploca* or several species of *Schizothrix* and *Symplocastrum*. Characteristic stromatolite and travertine formations are produced by species precipitating calcium carbonate in their sheaths and within colonial mucilage. Several species from the genera *Leptolyngbya*, *Schizothrix* (subgenus *Inactis*), *Calothrix*, and *Rivularia* belong to the most important travertine-forming filamentous cyanobacteria (Pentecost and Whitton, 2000; Stal, 2000).

B. Specialized Cells

Several members of orders within the filamentous cyanobacteria exhibit the greatest level of morphological and cellular differentiation. In the two most diversified orders, Nostocales and Stigonematales, two important types of specialized cells develop: heterocytes (heterocysts in old literature; see Komárek and Anagnostidis, 1989) and akinetes. Heterocytes develop from vegetative cells and may be solitary, in pairs, or several in a row (Fogg, 1949; Bahal and Talpasayi, 1972; Stewart, 1972; Wolk, 1982; Komárek and Anagnostidis, 1989). They produce cell walls, which are thick, multilayered, and apparently gas tight (Lang and Fay, 1971; Golecki and Drews, 1974; Stewart, 1980). During development, the thylakoid apparatus degrades and specific DNA rearrangements (e.g., *nif* genes) occur (Golden et al., 1985; Haselkorn, 1986). Heterocytes synthesize the enzyme nitrogenase, which enables fixation of gaseous nitrogen (N_2) from the atmosphere (or dissolved in water) under anaerobic conditions, which are maintained within the heterocyte (Winkenbach and Wolk, 1973; Wolk, 1973, 1982). The morphology (shape) of the heterocytes and their position in the trichomes are apparently genetically predetermined, but their frequency in the trichomes in the populations depends on the nitrogen supply in the environment, with frequencies of these cells along the trichome declining with greater levels of NH_4^+ or NO_3^- (Bahal and Talpasayi, 1972; Stewart, 1972; Kohl et al., 1987). In some genera, heterocytes develop exclusively from the apical (terminal) vegetative cells (e.g., *Cylindrospermum*, Fig. 30) or intercalary (e.g., *Anabaena*, Fig. 28) at roughly regular distances from one another (metameric), or occasionally in pairs (e.g., *Anabaenopsis*, Fig. 29A) or in short rows. In polarized filaments, they have an obligatory position forming basal cells (e.g., *Calothrix* and *Gloeotrichia*, Figs. 25 and 26).

Akinetes are resting cells that develop from solitary cells or after the fusion of two or more neighboring cells in several members of the Nostocales and Stigonematales (Geitler, 1932; Komárek and Anagnostidis, 1989; Anagnostidis and Komárek, 1990; Hindák, 2001). They often arise close to the end of the vegetative growth period, although their production is not obligatory (Rother and Fay, 1977). Akinetes possess thick cell walls, and an accumulation of photosynthetic assimilates and DNA increases during their development (Sutherland *et al.*, 1985a, b). These cells function as resting stages to survive harsh conditions, such as winter temperatures, drought, or frost, and can germinate to form new filaments when conditions improve (Huber, 1984; Cmiech *et al.*, 1986). The morphology of akinetes, their shape (spherical, ellipsoidal, cylindrical), size, position within the trichome, cell wall (epispore) characteristics (color, sculpture), and mode of germination apparently are genetically determined features (Komárek and Anagnostidis, 1989) and have been used as critical characters for distinguishing genera and species (e.g., Geitler, 1932; Desikachary, 1959). In some species in the genera *Anabaena*, *Aphanizomenon*, and *Anabaenopsis* (subfamily Anabaenoideae), akinetes develop in a defined configuration (up to five in a row) and in a regular position in relation to the heterocytes. Such akinetes can occur directly adjacent to the heterocytes or distant, within one to several vegetative cells from them. When mature, they are often much larger than the vegetative cells. In most species of *Nostoc*, *Trichormus*, and *Nodularia* (subfamily Nostocoideae), each vegetative cell may become an akinete, which develop in rows between the heterocytes and do not reach a size any larger than the vegetative cells (Komárek and Anagnostidis, 1989). Specialized cells with the same function as akinetes appear in the order Stigonematales; however, their morphology and position in the filaments are not as distinct as in true akinetes of Nostocales (Martin and Wyatt, 1974; Rippka *et al.*, 1979; Anagnostidis and Komárek, 1990).

C. Reproduction

In principle, cell division in filamentous cyanobacteria proceeds in the same way as it does in coccoid forms, usually perpendicular to the trichome axis. Among members of the Stigonematales, division can proceed in several planes. Reproduction can occur through solitary cells, which are liberated by several types of filamentous species, or by fragmentation of a filament and of a thallus. However, the most frequent mode of reproduction is fragmentation, forming distinct segments of trichomes, termed hormogonia. Fragmentation into hormogonia is simple in more primitive genera in the Pseudanabaenaceae, where trichomes disintegrate between neighboring cells via formation of thin lamella without necridic cells. In more complex filamentous taxa, hormogonia separate at the site of necridic cell production (Phormidiaceae and Oscillatoriaceae). Hormogonia can be of varying length, ranging from short filaments of several cells to fragments of single cells. Different types of hormogonia formation and their reproductive function are described in Anagnostidis and Komárek (1988). Longer hormogonia are sometimes motile and may contain gas vesicles (e.g., *Fischerella*). Fragments of ensheathed filaments are called hormocytes. Akinetes may also serve for reproduction (see Sect. II.B).

III. Ecology

The ecological requirements and importance of filamentous blue–greens and their geographical distribution were discussed in Chapter 3 (Sect. III.A–III.E) and are thoroughly reviewed in Whitton and Potts (2000). Their range of environments is as wide as any aquatic or terrestrial organisms on Earth. Many of the filamentous taxa are important members of lake and river phytoplankton, whereas others form thick, dense mats in a wide variety of benthic habitats. Still others form conspicuous floating masses on the water surface of various water bodies, or grow attached to stones, as epiphytes on aquatic angiosperms and other algae, epipelic and epipsammic on sediments or sand, respectively, and frequently colonize wet or moist soils. Some planktonic species are capable of forming water blooms in mesotrophic and eutrophic water bodies throughout the world. Important reviews that discuss ecological features of filamentous cyanobacteria are summarized in several important manuals (Jaag, 1945; Anagnostidis, 1961; Fogg *et al.*, 1973; Carr and Whitton, 1973; Golubić, 1980; Whitton and Potts, 2000). Aside from the general discussion of cyanobacterial ecology in Chapter 3, the most important ecological aspects of filamentous cyanobacteria in nature are as follows:

- Many are major components of autotrophic biomass production, forming a basis for aquatic food webs, with picoplanktonic, nanoplanktonic and bloom-forming species, particularly from the genera *Cyanobium*, *Planktothrix*, *Anabaena*, *Aphanizomenon*, *Nodularia*, and *Cylindrospermopsis* (Komárek, 1958, 1999; Kohl *et al.*, 1985; Watanabe, 1971; Kondrateva, 1972; Komárek *et al.*, 1993; Oliver and Ganf, 2000; Stockner *et al.*, 2000; Komárková, 2001).
- Filamentous cyanobacteria are often major components of attached communities in

submerged (littoral and benthic) habitats, in both standing and flowing waters, as well as on moist subaerial surfaces and on tree bark in tropical forests (Desikachary, 1959; Golubić, 1967a; Whitton, 1984; Rott and Pipp, 1999).
- Cyanobacteria are frequently the sole or dominant autotrophic organisms in extreme environments, including thermal, desert, and polar environments (Copeland, 1936; Anagnostidis, 1961; Golubić, 1980; Broady, 1984; Vincent, 2000; Ward and Castenholz, 2000; Wynn-Williams, 2000).
- Certain species produce toxins and allergic compounds, particularly in dense planktonic populations (Jackim and Gentile, 1968; Mahmood and Carmichael, 1986; Prinsep *et al.*, 1992; Codd, 1995; Carmichael, 1997; Falconer, 1998; Chorus and Bartram, 1999; Dow and Swoboda, 2000).
- A number of species is responsible for a large portion of global nitrogen fixation in aquatic and terrestrial ecosystems (e.g., tundra), and several species are used as "natural fertilizers" in rice fields (Fogg *et al.*, 1973; Wolk, 1973; Stewart, 1980; Fay, 1983; Van Baalen, 1987; Whitton, 2000).
- Particular species are responsible (at least in part) for lithogenic processes (travertine, stromatolites) and, in contrast, boring into limestone substrata (Prát, 1929; Carr and Whitton, 1973; Golubić *et al.*, 1975, 1981; Pentecost and Whitton, 2000; Stal, 2000).
- Cyanobacteria are key organisms in the colonization of soils and rocks during primary succession in terrestrial communities (Golubić, 1967a).
- Many cyanobacteria are used in biotechnology and in mass cultures (Seshadri and Jeeji-Bai, 1992; Belay *et al.*, 1994; Vonshak, 1997).
- Many cyanobacterial strains are important model laboratory organisms in the study of photosynthetic processes and physiological processes (Carr and Whitton, 1982; Fay and Van Baalen, 1987; Komagata, 1987; Sugawara and Miyazaki, 1999).

IV. METHODS

Methods of collecting, preparing and culturing filamentous cyanobacteria do not differ substantially from those described for coccoid and colonial forms (see Chap. 3).

V. KEY AND DESCRIPTIONS OF GENERA (FIGS. 1–40)

A. Key

Note: Many genera and species exhibit variable morphology without distinct morphological limits. Additional features that distinguish some genera exist only for ultrastructural and molecular characters, which cannot be used in the key.

1a.	Heterocytes[2] and/or akinetes never occur in trichomes: order Oscillatoriales	2
1b.	Heterocytes and/or akinetes develop commonly or occasionally in trichomes (if heterocytes are lacking, trichomes are morphologically complex with true branching or with akinetes)	35
2a.	Trichomes (without sheaths) very narrow and cylindrical; ≤3 μm wide (rarely up to 6 μm); cells sometimes with separated centroplasma and chromatoplasma (parietal arrangement of thylakoids)	3
2b.	Trichomes broader; width >3 μm, usually 4–16 μm (rarely to 60 μm); trichomes cylindrical to moniliform; cell content homogeneous or variably structured (thylakoids arranged radially or irregular)	16
3a.	Trichomes without sheaths or within simple, thin sheaths (when present, always one trichome per sheath), solitary or in mats; trichomes (or filaments) isopolar (both poles with same morphology) with exception (see Figs. 11b and 14): family, Pseudanabaenaceae	4
3b.	Sheaths wide, containing one or two or more trichomes, at least in a part of a filament; filaments mainly heteropolar: family Schizotrichaceae	15
4a.	Trichomes without individual sheaths, but may possess wide or diffuse, mucilaginous envelopes	5
4b.	Trichomes with distinct, thin, fine or firm sheaths	11
5a.	Trichomes straight, wavy, or irregularly coiled	6
5b.	Trichomes in regular, screwlike coils	10

[2] Also termed "heterocyst" (see Komárek and Anagnostidis, 1989).

6a.	Trichomes mainly short, curved or irregularly coiled, usually only few celled, disintegrating, sometimes enveloped by an indistinct, wide mucilaginous envelope; neighboring cells occasionally disorganized (Fig. 1A)	*Romeria*
6b.	Trichomes more or less cylindrical, multicelled, usually without wide mucilaginous envelopes	7
7a.	Trichomes solitary or in fine colonies, sometimes constricted at cross walls; cells sometimes with polar aerotopes (groups of gas vesicles)	8
7b.	Trichomes in larger clusters or in mats, cylindrical, usually not constricted at cross walls; cells always without aerotopes (occasionally with scattered or polar granules)	9
8a.	Trichomes constricted at cross walls (sometimes distinctly), solitary or in small clusters; cells usually without polar aerotopes (Fig. 1B)	*Pseudanabaena*
8b.	Trichomes unconstricted at cross walls, solitary, cylindrical; cells with polar and/or central aerotopes (Fig. 2A)	*Limnothrix*
9a.	Trichomes nonmotile, wavy, in clusters or mats (Fig. 2B)	*Jaaginema*
9b.	Trichomes motile, straight or slightly coiled, in thin mats (Fig. 3)	*Geitlerinema*
10a.	Coils free, long, and wide; trichomes always solitary (Fig. 4A)	*Glaucospira*
10b.	Coils usually joined one to another; trichomes in mats or among other algae (Fig. 4B)	*Spirulina*
11a.	Trichomes (filaments) isopolar (both poles with same morphology), free living (planktonic, metaphytic, or creeping on substratum), in clusters, or forming mats	12
11b.	Trichomes (filaments) heteropolar, with one end attached to the substratum	14
12a.	Filaments solitary in plankton and metaphyton, more or less short, straight, or irregularly or screwlike coiled (Fig. 5A)	*Planktolyngbya*
12b.	Filaments creeping on the substratum, in clusters or forming mats	13
13a.	Filaments solitary or in small groups, epiphytic or creeping on substratum (Fig. 5B)	*Leibleinia*
13b.	Filaments in clusters or in mats, irregularly coiled (Fig. 6A)	*Leptolyngbya*
14a.	Well-developed trichomes narrowed toward ends (terminal hairs may be present, sometimes separating) (Fig. 7A)	*Homoeothrix*
14b.	Filaments and trichomes cylindrical, never tapering toward ends (Fig. 6B)	*Heteroleibleinia*
15a.	Wide sheaths conically narrowed and closed at ends, individual trichomes occasionally enveloped by fine sheaths (Fig. 7B)	*Schizothrix*
15b.	Wide sheaths open at the ends, trichomes without own sheaths, usually tightly joined into fascicles or clusters (Fig. 8A)	*Trichocoleus*
16a.	Trichomes distinctly moniliform (deeply constricted at cross walls), more or less short, without sheaths, isopolar; cells barrel shaped or subspherical: family Borziaceae	17
16b.	Trichomes more or less cylindrical, long, sometimes constricted at the cross walls, but cells not barrel shaped or subspherical	18
17a.	Trichomes very short; with up to 8 (very rarely up to 16) cells (Fig. 8B)	*Borzia*
17b.	Trichomes long (up to over 600 µm); consisting commonly of more than eight cells (Fig. 9A)	*Komvophoron*
18a.	Cell length not less than one half width to ± isodiametric, or slightly longer than wide; protoplasts homogeneous or very finely striated or reticulate (radial thylakoid arrangement); trichomes 4–14 (18) µm wide: family Phormidiaceae	19
18b.	Cells very short, always shorter than one half cell width; protoplast slightly granulated (irregular arrangement of short, coiled thylakoids); trichomes 6–35 (60) µm wide: family Oscillatoriaceae	32
19a.	Trichomes typically in screwlike, free coils, always without sheaths; few species with aerotopes (groups of gas vesicles) in cells (Fig. 9B)	*Arthrospira*
19b.	Trichomes typically not in screwlike coils, with or without sheaths; individual genera always with or always without aerotopes	20
20a.	Trichomes solitary, planktonic, or forming fine mats on submersed substrata, usually pale reddish with finely but distinctly reticulate (keritomized) protoplasts, with typical widened thylakoids, lacking aerotopes (Fig. 11A)	*Tychonema*
20b.	Trichomes solitary, in clusters or in mats; if planktonic, then with aerotopes; protoplasts sometimes irregularly striated, but not distinctly reticulate	21
21a.	Filaments (trichomes) always without sheaths, cells with aerotopes, only planktonic species; trichomes solitary or in fascicles	22

21b.	Filaments usually with sheaths, cells without aerotopes and not planktonic (exception: few lyngbya species), in various habitats (metaphytic, benthic, subaerial); trichomes (filaments) solitary, clusters or mats	23
22a.	Filaments (trichomes) solitary (Fig. 10A)	*Planktothrix*
22b.	Filaments (trichomes) arranged in more or less parallel, microscopic fascicles (Fig. 10B)	*Trichodesmium*
23a.	Always only one trichome in a sheath	24
23b.	Several to many trichomes (rarely one in early stages) per sheath, sheaths sometimes slightly widened	28
24a.	Filaments with obligatory (usually common) false branching	25
24b.	Filaments without, or only exceptionally with very rare false branching	26
25a.	All trichomes cylindrical up to apex, without attenuated ends (Fig. 11B)	*Pseudophormidium*
25b.	Trichomes (branches) distinctly attenuated toward the ends (Fig. 18A)	*Ammatoidea*
26a.	Sheaths thick, lamellated, usually colored with sheath pigments (Fig. 13A)	*Porphyrosiphon*
26b.	Sheaths firm or thin, not distinctly lamellated or colored	27
27a.	Filaments in older mats joined (in fascicles) forming erect tufts (Fig. 12B)	*Symploca*
27b.	Filaments form flat (usually slimy) mats, not in fascicles arranged in tufts (Fig. 12A)	*Phormidium*
28a.	Sheaths sometimes anastomozing (joining), thin and firm; when joined, two trichomes per sheath in parallel arrangement (Fig. 13B)	*Lyngbyopsis*
28b.	Sheaths not anastomozing, sometimes widened, enclosing occasionally one, usually two to many trichomes, sometimes arranged in dense fascicles; trichomes within sheaths sometimes distant one from another	29
29a.	Widened sheaths contain (1) 2–several trichomes, which are usually slightly distant one from another (Fig. 14B)	*Dasygloea*
29b.	Trichomes within sheaths are usually tightly arranged in fascicles	30
30a.	Trichomes within a common sheath and with own individual sheaths; filaments in fascicles, often forming tufts in older mats (Fig. 14A)	*Symplocastrum*
30b.	Trichomes within common, widened sheath, but without individual sheaths	31
31a.	Sheaths hyaline, not stratified (Fig. 15A)	*Microcoleus*
31b.	Sheaths finely lengthwise striated (Fig. 15B)	*Hydrocoleum*
32a.	Trichomes in vegetative state always without sheaths (if formed, only under stress; desiccation, hypersaline conditions) (Fig. 16)	*Oscillatoria*
32b.	Trichomes in vegetative state always within distinct sheaths (only hormogonia and reproductive trichome segments can be without sheaths)	33
33a.	Filaments with typical "scytonematoid" false branching (common paired branches) (Fig. 18B)	*Plectonema*
33b.	Filaments without paired false branching	34
34a.	Filaments contain one trichome per sheath; forming mats (rarely solitary filaments in plankton with cells with aerotopes) (Fig. 17)	*Lyngbya*
34b.	Filaments typically arranged in parallel groups forming fascicles in common sheath; in fascicle-forming colonies or mats (Fig. 19)	*Blennothrix*
35a.	Trichomes never branched or only with false branching: order Nostocales	36
35b.	Trichomes (at least a portion) with true branching: order Stigonematales	60
36a.	Filaments with lateral false branches	37
36b.	Filaments always without any branching: family Nostocaceae	49
37a.	Filaments isopolar, solitary, entangled in clusters, or forming wooly mats with common lateral branches in pairs (V shape), or rarely with single (Y shape) branch (Fig. 20): family Scytonemataceae	38
37b.	Filaments typically heteropolar, solitary, forming fascicles or arranged parallel into flat or spherical colonies	39
38a.	Trichomes cylindrical up to the ends, apical ends of the trichomes (branches) not attenuated, end cells more or less rounded (Fig. 20A)	*Scytonema*
38b.	Trichomes (branches) attenuated toward the ends, sometimes tapering into long, multicellular hairs (Fig. 20B)	*Scytonematopsis*
39a.	Trichomes cylindrical, not (or slightly) attenuated toward ends, apical cells rounded; heterocytes basal and intercalary: family Microchaetaceae	40

39b.	Trichomes distinctly attenuated toward ends, may be elongated into long, thin, multicellular hairs; heterocytes mostly basal (rarely intercalary before false branching): family Rivulariaceae	45
40a.	Filaments long, lateral branches common and typically solitary, rarely (in older filaments) in twos ("scytonematoid"); heteropolar growth sometimes indistinct in old trichomes with numerous intercalary heterocytes	41
40b.	Filaments short, with or without solitary lateral branches; heterocytes basal or occasionally intercalary	44
41a.	Branches diverge distinctly from the main filament	42
41b.	Branches remain partly inside main sheath, forming fascicles with several trichomes arranged in parallel; branches diverge only at filament ends (Fig. 23)	*Coleodesmium*
42a.	Cells more or less uniform along length of trichome; sheaths thin or slightly widened and lamellated, branching common	43
42b.	Cells of variable length (shorter, longer than wide, or isodiametric), sometimes cylindrical, but barrel shaped near ends; sheaths very wide, funnel-like lamellated at ends; branching rare (Fig. 21B)	*Petalonema*
43a.	Cells mostly shorter than wide, false branches mainly at one side of filament; trichomes usually distinctly constricted at cross walls (Fig. 21A)	*Hassallia*
43b.	Cells isodiametric to longer than wide, false branching in several directions; trichomes cylindrical, constricted or unconstricted at cross walls (Fig. 22)	*Tolypothrix*
44a.	Simple trichomes cylindrical or slightly attenuated toward ends (Fig. 24B)	*Microchaete*
44b.	Simple trichomes widened distinctly toward the ends (Fig. 24A)	*Fortiea*
45a.	Heteropolar filaments simple; occur singly or in groups, but always distinctly separated one from another (Fig. 25A)	*Calothrix*
45b.	Heteropolar filaments distinctly branched (false branches); usually unified into mucilaginous colonies with parallel or radial arrangement	46
46a.	Filaments distinctly branched, in fascicles or clusters	47
46b.	Filaments unified more or less in parallel, forming flat or spherical mucilaginous colonies	48
47a.	Sheaths distinctly broader than trichome width, often lamellated and usually closed at the ends (Fig. 26B)	*Sacconema*
47b.	Sheaths not distinctly wider than the trichome, firm or gelatinous, lamellated and funnel-like, frayed at the ends (Fig. 25B)	*Dichothrix*
48a.	Colonies flattened or (young) hemispherical, sometimes large (millimeters to several square centimeters) with firm, individual sheaths; filaments arranged in parallel; akinetes absent (Fig. 27)	*Rivularia*
48b.	Colonies spherical with radially arranged filaments; sheaths confluent, gelatinizing; basal akinetes often form near end of growth period (Fig. 26A)	*Gloeotrichia*
49a.	Heterocytes absent; akinetes may develop at the end of growth season	50
49b.	Heterocytes present (absent rarely in some species with high nitrogen supply)	51
50a.	Akinetes (when present) solitary, distinctly larger than vegetative cells; cells more or less cylindrical; planktonic (Fig. 31A)	*Raphidiopsis*
50b.	Akinetes develop serially in rows, only slightly larger than the vegetative cells; cells mainly barrel shaped; planktonic, metaphytic, or in mats (Fig. 32B)	*Isocystis*
51a.	Heterocytes terminal, single pored; if intercalary, occur in pairs with trichomes fragmenting between them	52
51b.	Heterocytes mainly intercalary (terminal are exceptional), typically with pores on either side	54
52a.	Heterocytes develop primarily intercalary in pairs; trichomes soon disintegrate between them, heterocytes then appear terminal; mainly planktonic species (Fig. 29A)	*Anabaenopsis*
52b.	Heterocytes develop primarily from terminal cells; planktonic and/or forming mats	53
53a.	Trichomes forming benthic colonies, mats, epiphyton or metaphyton; akinetes always adjacent to terminal heterocytes; trichomes typically cylindrical (Fig. 30B)	*Cylindrospermum*
53b.	Trichomes planktonic and solitary; akinetes adjacent or slightly distant from heterocytes; trichomes slightly attenuated toward both ends or cylindrical (Fig. 30A)	*Cylindrospermopsis*
54a.	Trichomes somewhat asymmetrical in structure, with elongated and sometimes narrowed one to few terminal cells; akinetes elongated, cylindrical to long oval, rarely spherical; three species (from about 20 described) may form fascicles or clumps on the water surface (Fig. 29B)	*Aphanizomenon*

54b.	Trichomes with more or less regularly spaced heterocytes along trichome (metameric); apical cells undifferentiated from other vegetative cells; akinetes of various shapes	55
55a.	Akinetes adjacent to heterocytes or distant, solitary or up to five in a row, often formed after fusion of several vegetative cells; mature akinetes several times larger than vegetative cells	56
55b.	Akinetes positioned more or less between heterocytes, in long rows; mature akinete size only slightly larger than vegetative cells	57
56a.	Solitary planktonic trichomes, often with aerotopes (groups of gas vesicles), coiled or straight, or growing metaphytic and periphytic forming mats on substrata (Fig. 28)	*Anabaena*
56b.	Trichomes joined in macroscopic, benthic colonies; more or less spherical, saccate thallus with defined, firm surface (Fig. 31B)	*Wollea*
57a.	Cells (and akinetes) always shorter than wide; planktonic species in solitary trichomes (with aerotopes) or benthic species forming mats (Fig. 33)	*Nodularia*
57b.	Cells and akinetes always longer than wide	58
58a.	Trichomes in thin, firm sheaths (Fig. 31C)	*Aulosira*
58b.	Trichomes in colonies without firm sheaths, but sometimes within wide, mucilaginous envelopes in marginal parts of colonies	59
59a.	Trichomes more or less cylindrical, with constrictions at the cell walls, unified in fine, gelatinous, amorphous mats with diffuse surface (Fig. 32A)	*Trichormus*
59b.	Trichomes mainly moniliform, unified into firm, slimy colonies (often macroscopic) with a distinct, defined margin (Fig. 34)	*Nostoc*
60a.	Portions of filament or thallus multiseriate; composed of cell agglomerations or filaments dividing lengthwise and laterally; uniseriate portions moniliform	61
60b.	Filaments only uniseriate; ± cylindrical, with distinct true branching	64
61a.	Thallus formed from basal filaments (or cell agglomerations) and erect branches that are densely arranged and parallel (Fig. 35B)	*Stauromatonema*
61b.	Thallus formed from filaments with distinct lateral branches (sometimes only short) not arranged in dense parallel series; basal or main trichomes multiseriate or uniseriate; at least some cells dividing laterally	62
62a.	Basal filaments usually multiseriate, with rare true branching; long, uniseriate branches with false branching (like *Tolypothrix*): family Borzinemataceae (Fig. 37B)	*Schmidleinema*
62b.	Filaments only with (usually frequent) true branching	63
63a.	Basal filaments mainly multiseriate, true branches (mainly uniseriate) differ morphologically (Fig. 37A)	*Fischerella*
63b.	All trichomes (basal and erect) more or less of the same usually multiseriate character, often regularly narrowing toward ends; basal system multiseriate; some forming macroscopic masses (Fig. 36)	*Stigonema*
64a.	Thallus shrublike, with distinct heteropolar growth from base to apex; filaments successively branched into typically pseudodichotomous branches	65
64b.	Thallus composed of creeping, uniseriate filaments; filaments branched laterally: family Mastigocladaceae	68
65a.	Heterocytes develop very occasionally intercalary, or (in several genera) absent: family Loriellaceae	66
65b.	Heterocytes well developed, mainly lateral, at ends of short lateral branches or intercalary: family Nostochopsaceae (Fig. 39A)	*Nostochopsis*
66a.	Sheaths gelatinous, intensely encrusted with aragonite crystals, usually open at ends; inhabits caves (Fig. 38A)	*Geitleria*
66b.	Sheaths firm and smooth on the surface, without encrustations, closed at the ends; inhabits thermal waters	67
67a.	Sheaths wide, with distinct, funnel-like lamellation (Fig. 38C)	*Colteronema*
67b.	Sheaths thin, simple, fine, distinct, but not lamellated (Fig. 38B)	*Albrightia*
68a.	Sheaths, particularly in older filaments, thick and confluent, with trichomes in parallel arrangement (Fig. 40B)	*Thalpophila*
68b.	Sheaths simple, thin, fine or firm but never thick; forming confluent mass; trichomes not in parallel arrangement	69
69a.	Cells in main trichomes and branches isomorphic, more or less cylindrical or barrel shaped (Fig. 39B)	*Hapalosiphon*
69b.	Cells of different shapes in same thallus (e.g., short or elongated barrel shaped or cylindrical) (Fig. 40A)	*Mastigocladus*

B. Descriptions of Genera[3]

Oscillatoriales: Pseudanabaenaceae: Pseudanabaenoideae

Geitlerinema (Anagnostidis *et* Komárek) Anagnostidis (Fig. 3)

The thallus is thin, delicate, mostly bright blue–green, rarely violet or brown, diffuse, sometimes fascicle-like, usually forming thin macroscopic mats; occasionally isolated trichomes occur. Trichomes are (0.6) 1–4 (6.5) μm wide, usually in parallel arrangements, cylindrical, straight, sometimes slightly flexuous or (rarely) irregularly screwlike coiled (especially at the ends), without sheaths, mostly not constricted at the cross walls, rarely (more or less slightly) constricted, more or less gradually attenuated and bent or coiled at ends, rarely not attenuated and straight. Trichomes are motile, with intense gliding in the direction of the longitudinal axis, may be accompanied by distinctive clockwise or counterclockwise rotation, waving (oscillation) and circling. Cells are usually longer than wide, before division occasionally several times longer than wide; the cell content usually contains dispersed large cyanophycin granules or localized (apical) carotenoid bodies, without aerotopes; thylakoids are concentrically arranged, peripheral and parallel to longitudinal cell walls. Apical cells are conical, hooked, or bent, but mostly acuminate or rounded, occasionally spherical-capitate, or straight, cylindrical, and rounded. Reproduction occurs through disintegration of trichomes into motile hormogonia, without necridic cells.

Species of *Geitlerinema* (over 30 have been described) grow mostly in mats, on soils, or in various aquatic habitats on macrophytes or other substrata (mainly in unpolluted waters); they rarely occur as solitary filaments. *Geitlerinema splendidum* is the most common species in the periphyton of oligotrophic to mesotrophic waters in the north-temperate region; it is common in soft-water (granitic) lakes in eastern and western Canada (Bourrelly, 1966; Stein and Borden, 1979, as *Oscillatoria splendida*). *G. amphibium* (= *Oscillatoria amphibia*) is another widely distributed species (Prescott, 1951; Whitford and Schumacher, 1969; Stein and Borden, 1979). Other species are known from mineral waters and thermal springs: *G. jasorvense* and *G. acus* occur in Yellowstone National Park (Copeland, 1936); *G. claricentrosum* and *G. earlei* are reported from Puerto Rico (Gardner, 1927). The metaphytic species are likely to occur elsewhere in North America.

Jaaginema Anagnostidis *et* Komárek (Fig. 2B)

Trichomes are usually flexuous, solitary or entangled in clusters, or forming thin, membranaceous thalli without sheaths (exceptionally with very fine mucilaginous layers around the trichomes); they are always nonmotile. Trichomes are up to 3 μm wide; they are usually not constricted at the cross walls, sometimes slightly attenuated at the ends, and not capitate. Cells are cylindrical, mostly longer (up to 10×) than wide, rarely almost isodiametric; the cell content is homogeneous, without aerotopes; thylakoids are probably parietal. Apical cells are mostly rounded, occasionally conical, without calyptra, rarely with a thickened outer cell wall. Reproduction occurs by fragmentation of trichomes, without necridic cells.

Jaaginema is a little-known genus with more than 20 species, mainly benthic, in shallow water on sediments or aquatic plants in various types of water or in metaphyton. Several species are known from thermal or mineral springs in North America. *Jaaginema filiforme* (= *Oscillatoria filiforme*) was described from rivers in Yellowstone National Park by Copeland (1936).

Limnothrix Meffert (Fig. 2A)

Trichomes are solitary or in small, irregular fascicles or clusters, isopolar (both poles with the same morphology), usually free living, straight, slightly bent or flexuous, consisting of numerous cylindrical, mainly elongated cells. Trichomes are not constricted or very slightly constricted at indistinct, thin cross walls, 1–6 μm wide, cylindrical, not attenuated at ends, without or facultatively with fine, hyaline sheaths, without false branching, nonmotile or with reduced motility (slight trembling or gliding), often disintegrating. Cells are isomorphic, cylindrical, sometimes long, rarely slightly inflated, more or less isodiametric or frequently longer than wide, pale blue–green, blue–gray, yellowish, reddish, or pink; all cells are capable of dividing, with localized apical or central aerotopes, which may be lacking (depending on environmental conditions); thylakoids are mostly parietal; variable ratios of phycocyanin and phycoerythrin (recognizable mainly in cultures; Kohl and Nicklisch, 1981; Komárek, 1994). Apical cells are cylindrical, rounded or roundly flattened at ends, rarely conically attenuated and/or with pointed conical plasmatic protrusions, usually with one or few terminal aerotopes, without calyptra, not capitate. Cell division is perpendicular to the longitudinal axis, sometimes (rarely) asymmetrical. Daughter cells reach the original size before the next division. Repro-

---

[3] Drawings in figures are based on North American literature; however, documentation of several species was selected from non–North American papers. Localities are cited only in figures from North America. Figures of trichome structures and reproductive processes were selected according to various authors.

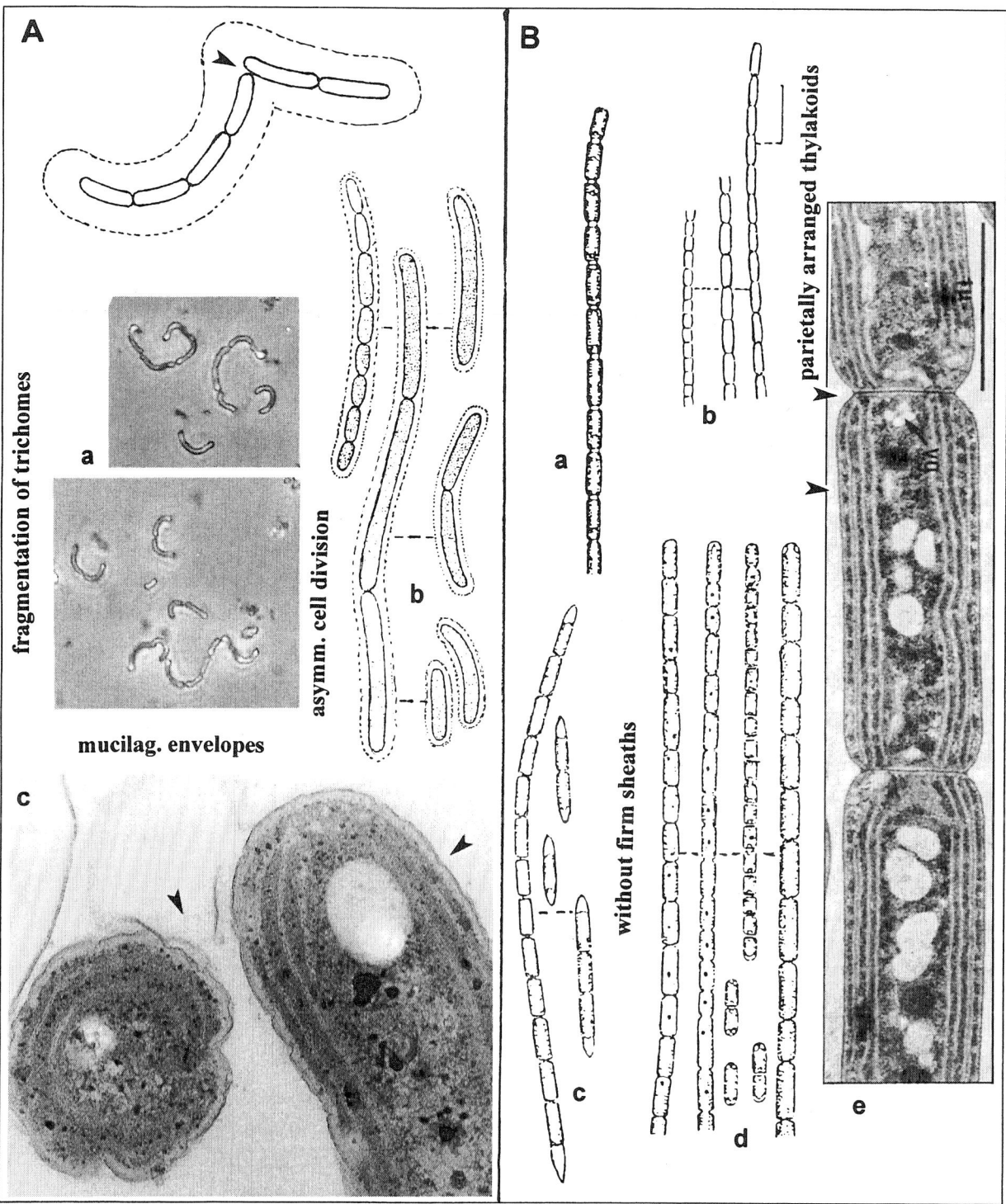

FIGURE 1 (A) *Romeria*: (a) *R. leopoliensis* (original, population from Europe); (b) *R. nivicola* (after Kol from Smith, 1950, from snow in Yellowstone National Park); (c) *R. leopoliensis*, crosswise and lengthwise sections of cells with characteristic position of thylakoids. (B) *Pseudanabaena*: (a) *P. catenata* (after Lauterborn from Geitler, 1932); (b) *P. limnetica* (bar = 10 μm; after Komárek, 1958); (c) *P. lonchoides* (after Anagnostidis, 1961, possible occurrence in thermal springs); (d) *P. galeata* (after Anagnostidis, 1961); (e) *Pseudanabaena* sp., lengthwise section (bar = 1 μm; after Guglielmi and Cohen-Bazire, 1984).

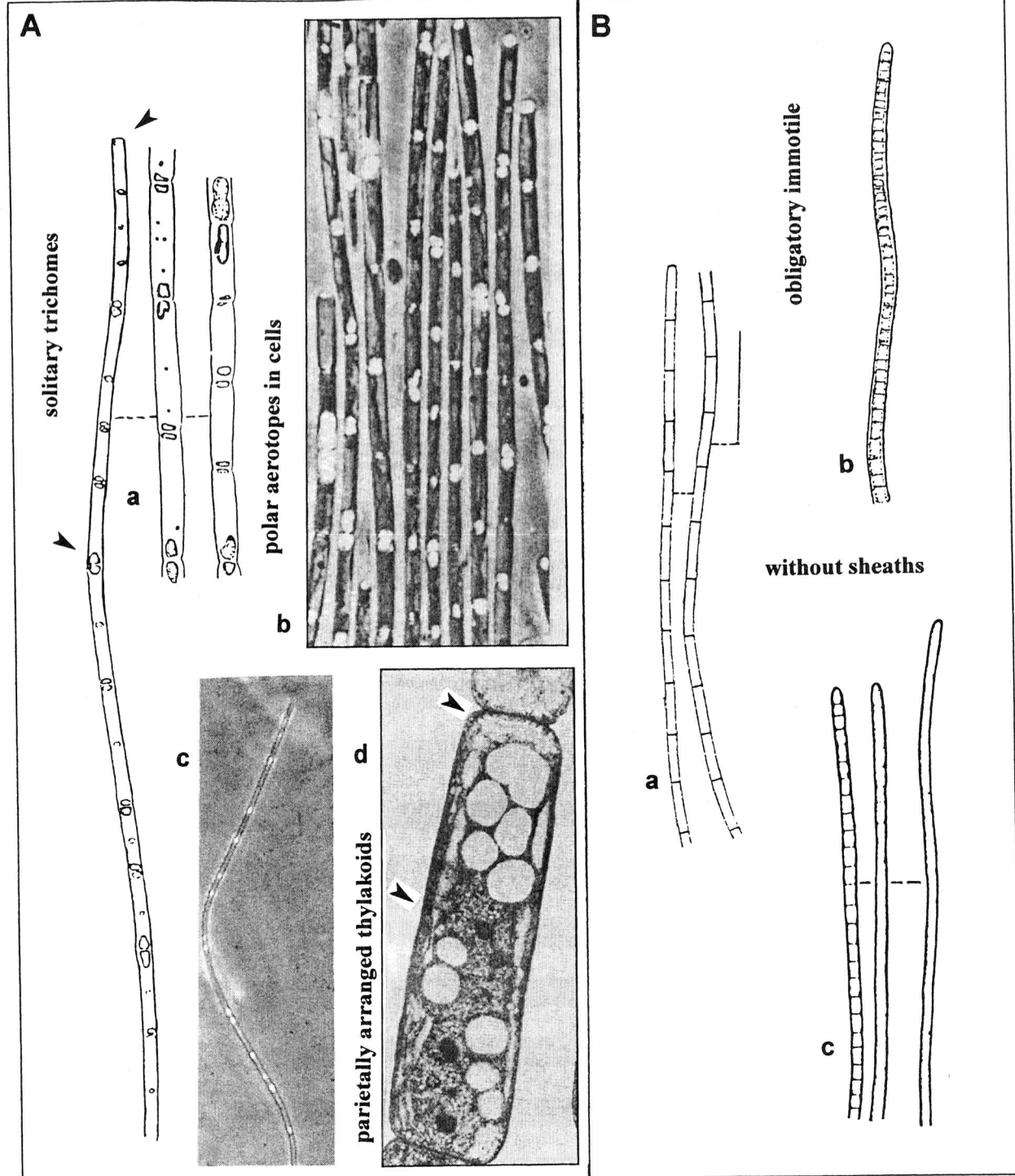

FIGURE 2 (A) *Limnothrix*: (a) *L. redekei* (after Van Goor and Skuja from Anagnostidis and Komárek, 1988); (b) *L. redekei* (after Meffert, 1988); (c) *Limnothrix* sp. (original by Komárková, from Florida); (d) *L. redekei*, lengthwise section (after Kalina from Anagnostidis and Komárek, 1988). (B) *Jaaginema*: (a) *J. neglecta* (bar = 10 μm; after Komárek, 1975); (b) *J. subtilissima* (after Böcher from Starmach, 1966); (c) *Jaaginema* sp. (after Copeland, 1936, from Yellowstone National Park, sub *Phormidium geysericola*).

duction occurs by disintegration of the trichomes into nonmotile hormogonia, without necridic cells.

Species of this genus occur in the plankton of mesotrophic to eutrophic lakes and reservoirs, sometimes developing metalimnetic maxima, mostly in temperate and northern areas (Whitton and Peat, 1969; Gibson, 1975; Meffert and Krambeck, 1977; Kohl and Nicklisch, 1981; Meffert, 1987, 1988; Kling, personal observation). There are about 20 revised species, but all *Pseudanabaena* species with polar aerotopes possibly belong to the genus *Limnothrix*. The most common species (*L. redekei*) is distributed throughout the temperate zone, where it occurs mainly in colder seasons. *L. redekei* was also one of the first species to respond to the artificial eutrophication of Lake 227 in the Experimental Lake Area in northwestern Ontario, Canada (recorded as *Oscillatoria redekei* by Kling and Holmgren, 1972). In large mesotrophic lakes, *L. redekei* may occur during spring and fall during turnover periods (Kling, personal observation). *L. vacuolifera* occurs in northern lakes in Scandinavia and Canada; the benthic *L. guttulata* was recorded from the United States (authors' records).

Pseudanabaena Lauterborn (Fig. 1B)

Trichomes are solitary or in fine mats, straight or curved, less frequently wavy, cylindrical; they are usually short, consisting of a very few to several cells, or long with many cells, usually with conspicuous constrictions at the cross walls, 1–3.5 µm wide, rarely unconstricted. Trichomes lack firm sheaths, but sometimes have wide, fine, diffuse mucilage. Apical cells are not differentiated, without calyptra or thickened outer cell wall. Motility is lacking or facultative, usually slow gliding, occurring also in separated unicells, probably without rotation. Cells are usually cylindrical with rounded ends, but sometimes almost barrel shaped, always longer than wide, rarely almost isodiametric (after division), with or without polar aerotopes (groups of gas vesicles—types with gas vesicles probably belong to the genus *Limnothrix*); thylakoids are concentric and parietal, parallel to the long axis; with one central perforation in the cross walls and/or multiple pores (300–500) near the cell poles. Cell division is exclusively by binary fission in one plane, perpendicularly to the long axis, sometimes asymmetrical. Reproduction occurs through production of one-celled to multicelled hormogonia or trichome fragmentation, without necridic cells.

Over 30 species have been described, several occurring in planktonic, metaphytic, periphytic, or benthic habitats in waters of different trophic status; a few occur on soil or within the mucilage of other algae or colonial rotifers (endogloeic). A few species are known from mineral waters and hot springs. Most reports of planktonic *Pseudanabaena* species are recorded as *Oscillatoria*, from which several species are also known from oligotrophic to eutrophic waters in North America (*O. limnetica* = *P. limnetica*, *P. mucicola*, *P. catenata*; authors' results). *P. thermalis* occurs in alkaline, thermal waters (*Oscillatoria amphigranulata* sensu Castenholz, 1976).

Romeria Koczwara in Geitler (Fig. 1A)

The thallus is microscopic, very fine, and filamentous or pseudofilamentous. Trichomes are solitary, usually short, fine, irregular and fragile, up to 3.5 µm wide, 1–8 (18–32) celled, rarely with more cells, usually semicircular curved, flexuous or irregularly screwlike coiled, with one or two or more (up to eight) helices, constricted at the cross walls, without a distinctive sheath, but with fine, more or less thick mucilaginous envelopes. The ends of neighboring cells are sometimes slightly shifted one from another. Usually one, rarely few irregularly localized nonmotile trichomes are present in a fine mucilaginous envelope. The mucilage is colorless, diffuse, and indistinct. All cells are of the same morphology, elongated, long cylindrical or barrel shaped; terminal cells are rounded at the apex. Thylakoids are parietally arranged. Cell division is transverse, symmetric, or slightly asymmetric. Reproduction occurs by trichome fragmentation into hormogonia or solitary cells.

Romeria species are mainly planktonic, living in clear oligotrophic to mesotrophic lakes and ponds, rarely in hypertrophic systems. Nineteen species have been described, most from the northern temperate zone; one species is marine (described from the Gulf of Mexico). From North America, Smith (1950) mentioned only the cryosestic *R. elegans* var. *nivicola* (= *R. nivicola*) from Yellowstone National Park, but several other U.S. species are known from the temperate zone (*R. alascense* from Alaska, *R. elegans*, *R. leopoliensis*, and marine *R. mexicana*) and from tropical regions (*R. heterocellularis* and *R. hieroglyphica*; Komárek, 2001). Several species are probably more widely distributed.

Oscillatoriales: Pseudanabaenaceae: Spirulinoideae

Glaucospira Lagerheim (Fig. 4A)

Filaments (trichomes) are solitary, short or slightly elongated, thin (up to 3 µm wide), without sheaths, regularly loosely screwlike coiled with wide and more or less long spirals, usually intensely motile (rotation), sometimes slightly flexible, cylindrical, not attenuated at ends, and not constricted at slightly visible cross walls. Cells are pale blue–green or yellowish, with homogeneous content, sometimes with few fine granules, probably always longer than wide. Cell division is per-

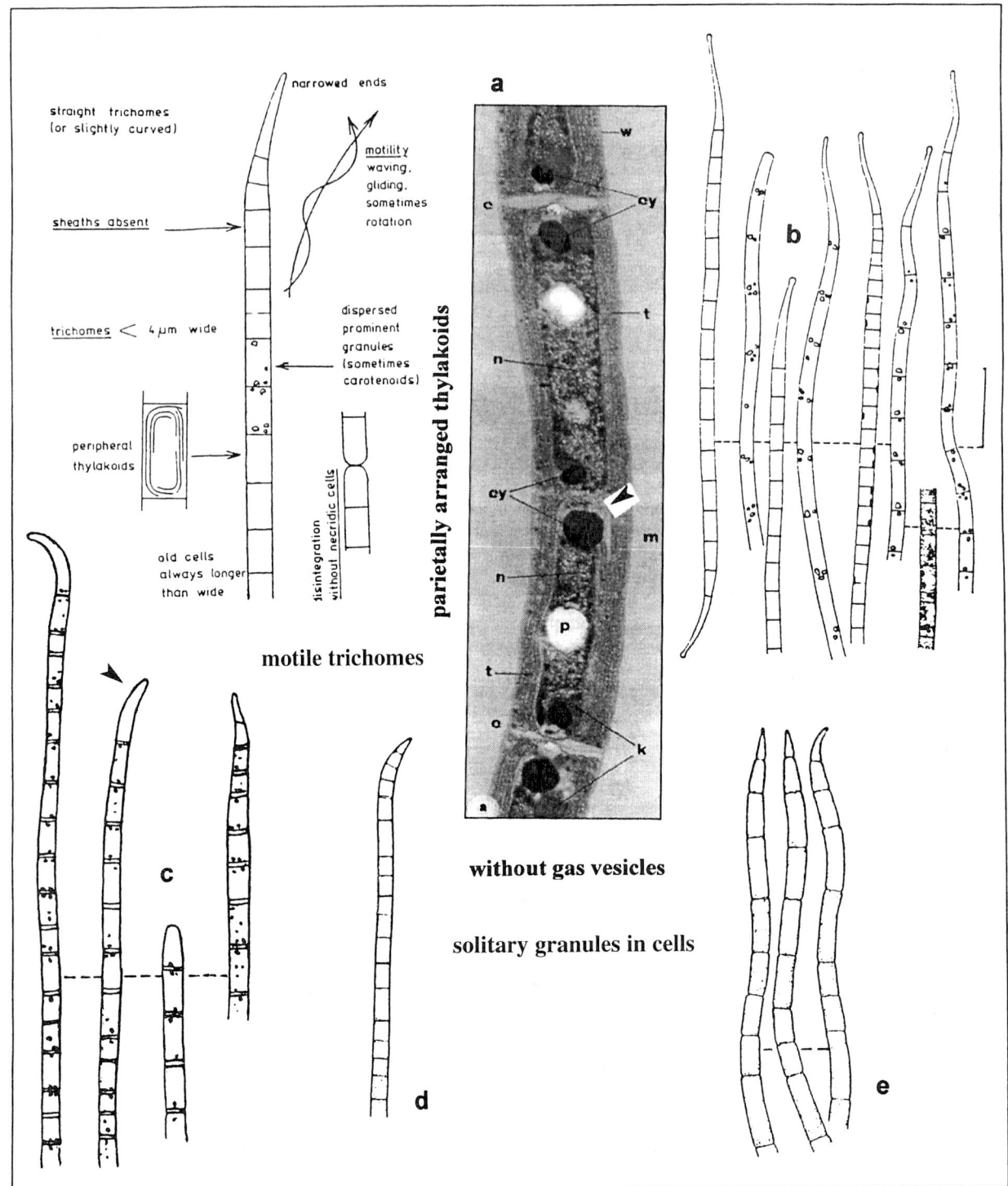

FIGURE 3 Geitlerinema: (a) G. unigranulatum, lengthwise section with polar cyanophycin granules and parietally arranged thylakoids (after Komárek and Azevedo, 2000); (b) G. splendidum (bar = 10 μm; after Komárek, 1975); (c) G. lemmermannii (after Tavera et al., 1994, from Mexico); (d) G. earlei (after Gardner, 1927, from Puerto Rico; sub Oscillatoria earlei); (e) G. claricentrosa (after Gardner, 1927, from Puerto Rico, sub Oscillatoria claricentrosa).

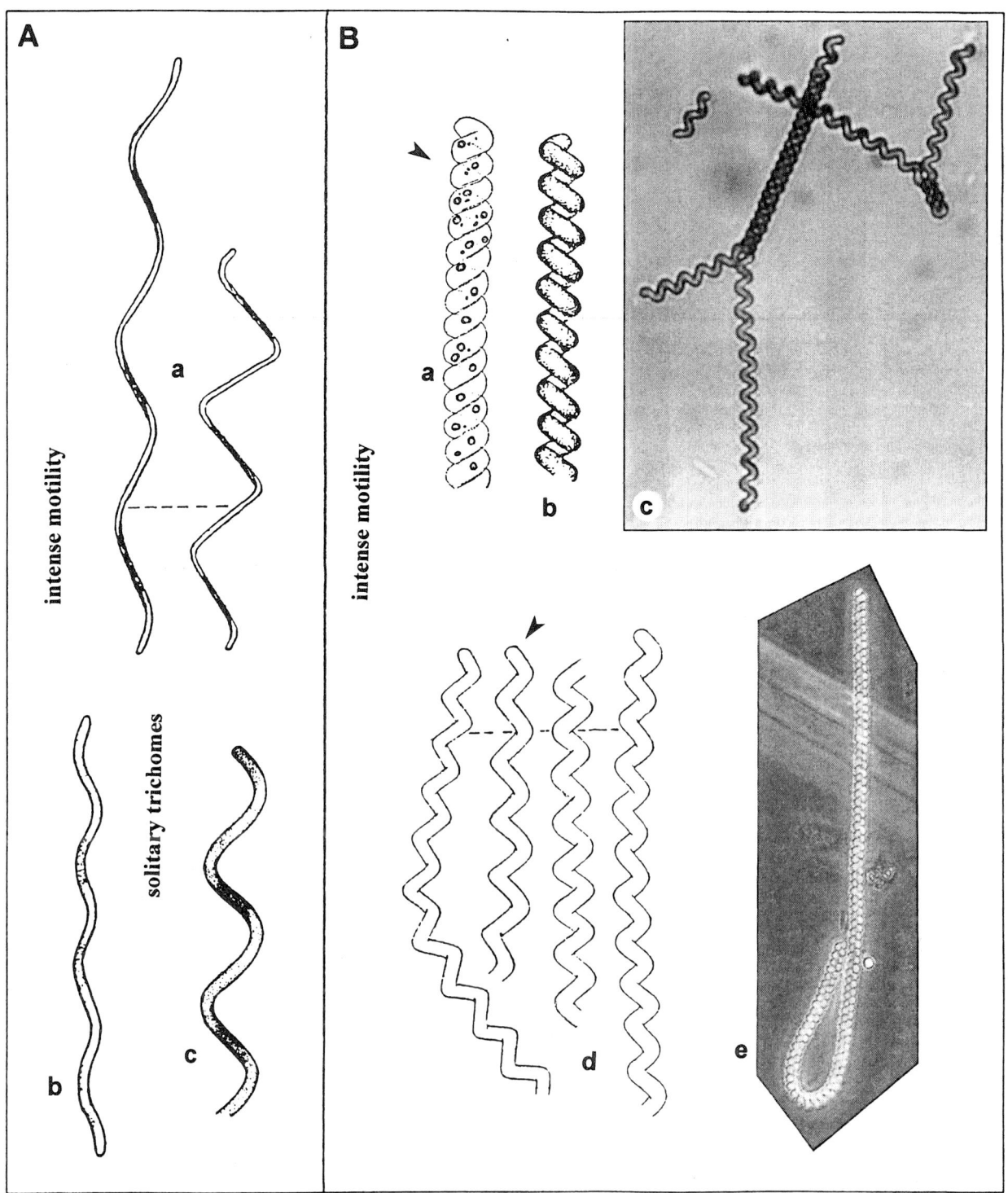

FIGURE 4 (A) *Glaucospira*: (a) *G. laxissima* (after G. S. West, 1907); (b) *Glaucospira* sp. (after Copeland, 1936, from Yellowstone National Park; sub *Spirulina caldaria* var. *magnifica*); (c) *Glaucospira* sp. (after Drouet, 1937, from Massachusetts; sub *Spirulina stagnicola*). (B) *Spirulina*: (a) *S. weissii* (after Drouet, 1942, from the United States); (b) *S. major* (after Geitler, 1932); (c) *Spirulina* sp. (from culture collection, the United States, original photo by J. D. Wehr, with permission); (d) *S. meneghiniana* (after Komárek, 1989, from Cuba); (e) *S. subsalsa* (original by Komárková, from the Everglades, Florida).

pendicular to the trichome axis. Reproduction probably occurs through trichome disintegration.

Glaucospira is a very poorly known and problematic genus; the cyanobacterial character of some of the described (more than five) species must be confirmed. The known species are mainly planktonic or metaphytic. Copeland (1936) described *G. yellowstonensis*, a slightly thermophilic species, from the surface of gelatinous bottom sediments in Yellowstone Lake. *Spirulina laxa* (which corresponds to the genus *Glaucospira*) was described from ponds in the Piedmont and central plains regions of North Carolina (Whitford and Schumacher, 1969) and from lakes in Ontario and British Columbia (Duthie and Socha, 1976; Stein and Borden, 1979). *Spirulina stagnicola*, described by Drouet (1937) from brackish waters in Massachusetts, also belongs to this genus.

Spirulina Turpin ex Gomont (Fig. 4B)

Trichomes are cylindrical, screwlike coiled, 0.3–7.5 μm wide, of variable length, solitary or forming fine, mucilaginous mats. Their color ranges from blue–green, olive green, gray–green, brownish to reddish or violet. Regular coiling occurs in all species, but ranges from loosely to tightly screwlike or helical; coils may be compressed or may have spaces between them. Rarely irregularities in coiling, variable curves or occasionally circle-like coils and straight portions, can occur. *Spirulina* is intensely motile; trichomes glide with rapid clockwise or counterclockwise rotation. Sheaths or mucilaginous envelopes are usually lacking; a fine slime is produced in some mats. Cells are not constricted at scarcely visible cross walls (not seen without staining in light microscope); end cells are usually not attenuated. Cells are typically isodiametric or longer than wide; the cell content is homogeneous, without aerotopes; a special pore and a perforation pattern is apparent in the cell walls (i.e., several rows on the concave side); thylakoids are arranged parallel to the longitudinal cell axis. Apical cells are rounded, hemispherical, without calyptra or thickened outer cell wall. Reproduction occurs via fragmentation into motile hormogonia, without necridic cells.

About 50 species have been described (about 20 revised; Anagnostidis and Golubić, 1966; Anagnostidis and Komárek, 2003). A widely conceived genus (Geitler, 1932) has been divided according to the morphology of filaments, ultrastructure and molecular sequencing into *Spirulina* and *Arthrospira* (Phormidiaceae); the existence of the separated genus *Arthrospira* has been confirmed by both morphological and genetic analyses (Tomaselli *et al.*, 1996; Mühling *et al.*, 1997). Several typical *Spirulina* species *sensu stricto* are benthic and occur in metaphyton, sometimes in heavily polluted habitats or among detritus. Other species grow in thermal and mineral springs or in saline lakes and ponds. Tilden (1910) recorded nine species of *Spirulina* from the United States (with *Arthrospira*). Smith (1920, 1950) noted three species, *S. maior*, *S. subtilissima* and *S. labyrinthiformis*; Whitford and Schumacher (1969) reported these plus *S. weissii*. North American records also include *S. gigantea*, *S. nordstedtii*, *S. major*, and *S. princeps* from northern and temperate lakes (Duthie and Socha, 1976; Stein and Borden, 1979; Sheath and Steinman, 1982).

Oscillatoriales: Pseudanabaenaceae: Leptolyngbyoideae

Leibleinia (Gomont) L. Hoffmann (Fig. 5B)

Filaments are solitary, waved or curved, 1.5–11 μm wide, with unique epiphytic habitat being attached to substrata along their length, or by a portion of it, later having both free ends, usually epiphytic, with a sheath, and very rarely with false branching. Sheaths are firm, thin, and colorless; trichomes are nonmotile. Cells are cylindrical, mainly pale blue–green or gray–blue, without aerotopes; they are usually longer than they are wide, with peripherally arranged thylakoids (?). Cells divide perpendicularly to the longitudinal axis; daughter cells approach the original size before subsequent division. Reproduction occurs via nonmotile hormogonia and hormocytes, separating without necridic cells, adhering to substrata lengthwise along their horizontal axis and growing upward at both poles.

The majority of the 14 species described grow epiphytically on filamentous algae and aquatic macrophytes. Both freshwater and marine species are known. The filament morphology is very simple and may often be mistaken for *Heteroleibleinia* or *Leptolyngbya*. Many more species of *Leibleinia* probably exist, but the genus is poorly known from North American habitats with many species likely identified as species of *Lyngbya*, *Phormidium*, or *Oscillatoria*, such as *L. calotrichicola* from Yellowstone National Park (Copeland, 1936) and marine *L. aeruginea* from Puerto Rico (Gardner, 1927).

Leptolyngbya Anagnostidis *et* Komárek (Fig. 6A)

Filaments are rarely solitary, usually loosely arranged in flakelike clusters or mats, free floating or attached to substrata, seldom in fascicles or forming compact colonies (thallus); they are more or less flexuous, finely undulating, occasionally nearly straight and long. Ends are usually neither attenuated nor capitate. Facultative, firm, thin, hyaline sheaths are present; branching is rare but may occur with occasional pseudobranches. Sheath frequency is species specific, possibly dependent on environmental factors.

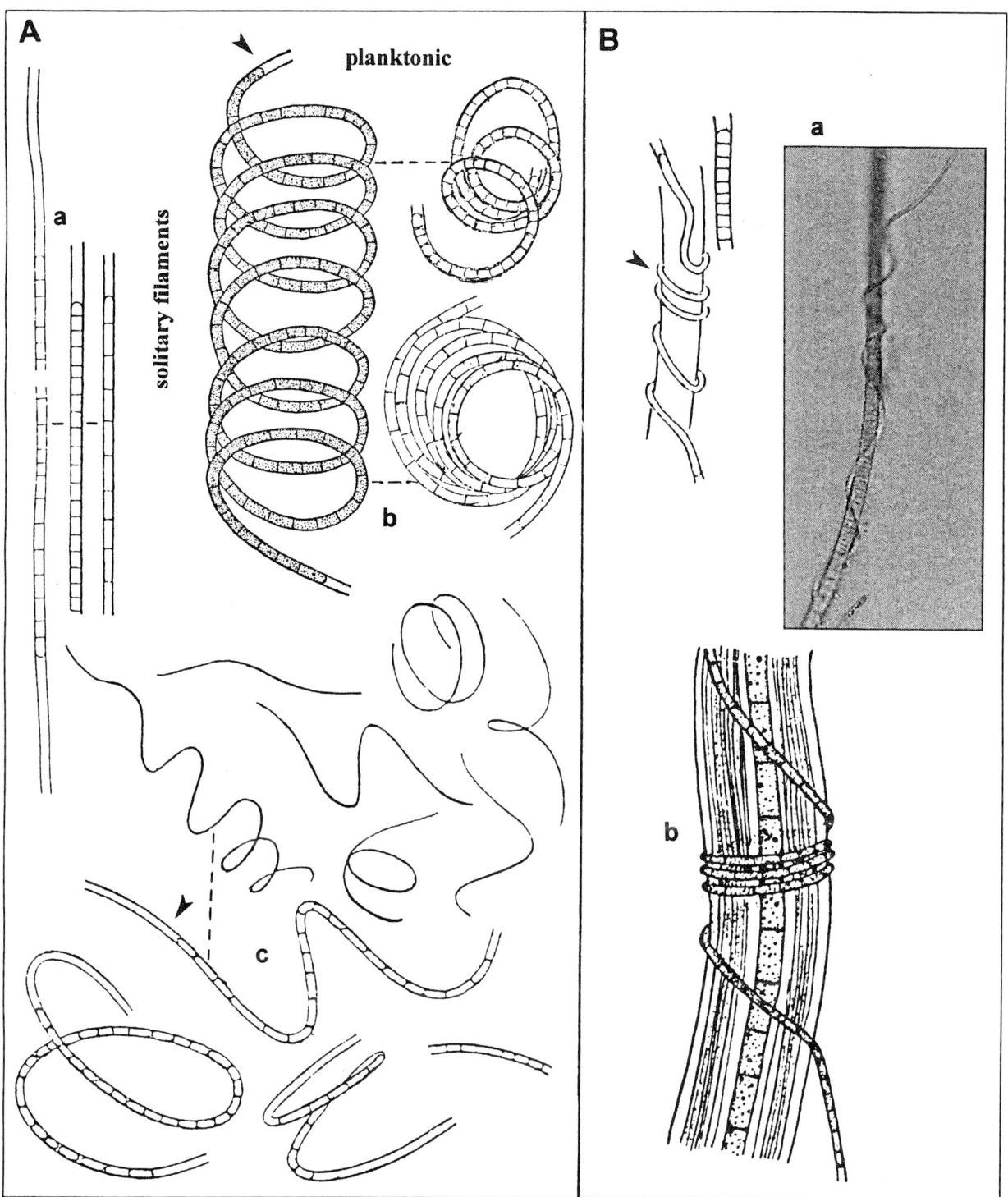

FIGURE 5 (A) *Planktolyngbya*: (a) *P. limnetica* (after Kondrateva, 1968; Hindák and Moustaka, 1988); (b) *P. contorta* (after Smith, 1950; Kondrateva, 1968); (c) *P. tallingii* (after Komárek and Kling, 1991). (B) *Leibleinia*: (a) *Leibleinia* sp. (original photo by Komárková, from the Everglades, Florida); (b) *L. calotrichicola* (after Copeland, 1936, from Yellowstone National Park).

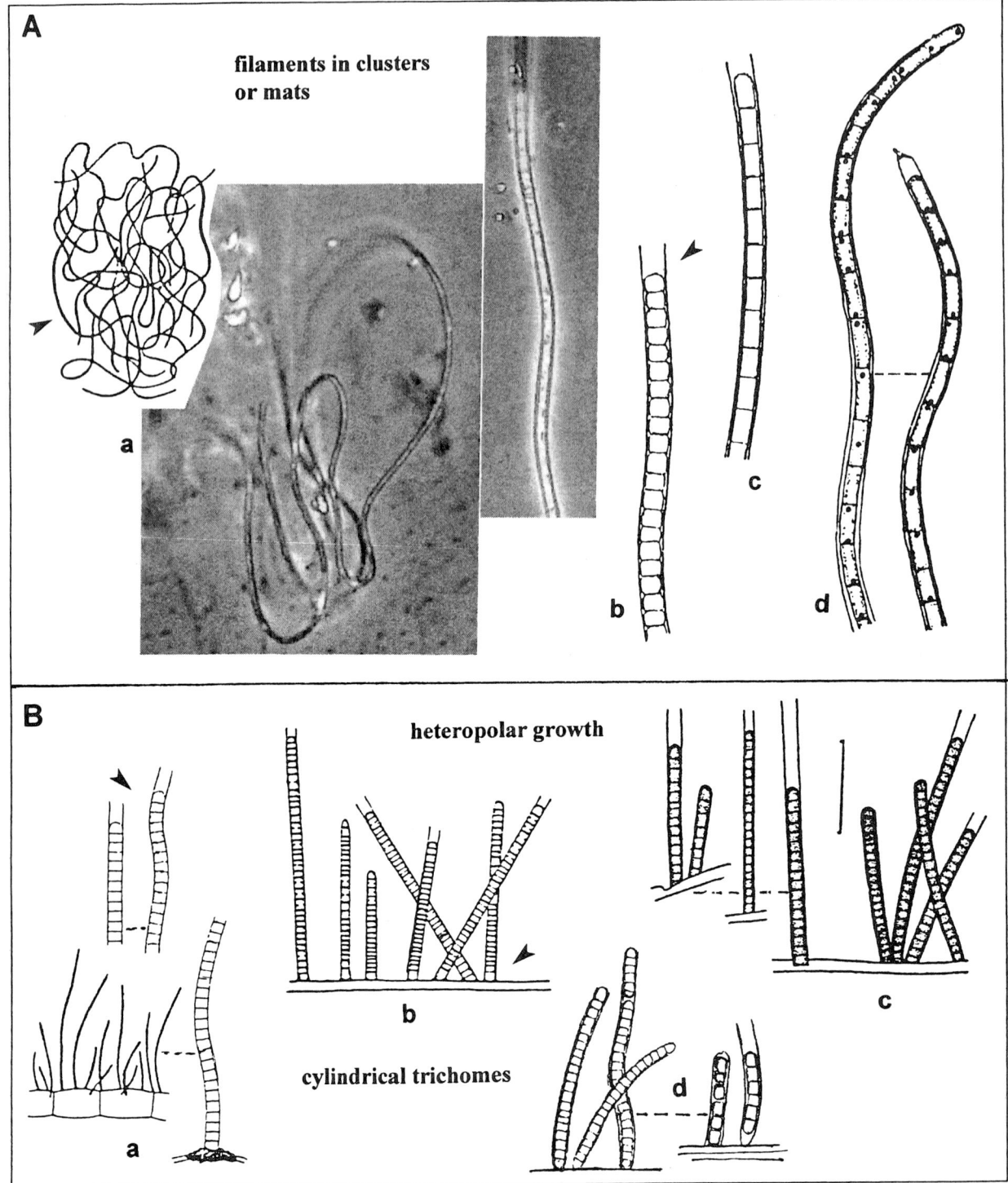

FIGURE 6 (A) *Leptolyngbya*: (a) *Leptolyngbya* sp. (typically coiled filaments in cluster; photos original by Komárková, from the Everglades, Florida); (b) *L. foveolarum* (after Komárek, 1988); (c) *L. nostocorum* (after Komárek, 1988); (d) *L. cartilaginea* (after Copeland, 1936, from Yellowstone National Park). (B) *Heteroleibleinia*: (a) *H. kuetzingii* (after Fott and Komárek, 1960); (b) *H. minor* (after Gardner, 1927, from Puerto Rico); (c) *H. pusilla* (bar = 10 μm; after Whelden, 1947, from Canadian Arctic); (d) *H. profunda* (reddish trichomes; after Tavera and Komárek, 1996, from volcanic lakes in Mexico).

Trichomes are 0.5–3.5 μm wide, motile (producing hormogonia) or nonmotile, or with slightly noticeable trembling. Cells are cylindrical, usually longer than wide, less frequently isodiametric or shorter than wide, usually with homogeneous contents, but often with recognizable chromatoplasma and centroplasma (= position of peripherally arranged thylakoids), without gas vesicles. Reproduction occurs via trichome fragmentation, with or without sacrificial cells (two subgenera); hormogonia range from nonmotile to barely motile, with trichome disintegration occurring from the apical portions of the trichomes.

The genus *Leptolyngbya* is one of the most common (and taxonomically most difficult) cyanobacterial genera, containing numerous morphotypes and ecotypes (species), which are very common in soils and in periphyton and metaphyton in a variety of freshwater and saline (marine) environments (Anagnostidis and Komárek, 1988; Albertano and Kováčik, 1994). Several species are known from thermal and mineral waters or grow subaerially on wet rocks. Identification of species (more than 140 have been described) is difficult due to indistinct morphological differences. North American species have not been well documented (usually under the names *Lyngbya* or *Phormidium*). Whitford and Schumacher (1969) reported *L. lagerheimii*, *L. subtilis*, *L. angustissima*, and *L. tenuis* under the generic name of *Phormidium*. *L. bijahensis*, *L. cartilaginea*, *L. geysericola*, *L. rubra*, *L. subterranea*, *L. vesiculosa* and *L. yellowstonensis* were recorded from mineral and thermal waters in Yellowstone National Park; *Schizothrix thermophila* also probably belongs in this genus (Copeland, 1936). A survey and revision of North American species in this genus is necessary.

Planktolyngbya Anagnostidis *et* Komárek (Fig. 5A)

Filaments are free living, solitary, free floating, straight, flexuous, wavy, or more or less spirally screwlike or irregularly coiled, with firm, thin, colorless sheaths, and very rarely false branched. Trichomes are nonmotile, cylindrical, isopolar (both poles with the same morphology), uniseriate, unconstricted or slightly constricted at the cross walls, not attenuated toward ends, with rounded apical cells (not capitate). Cells are cylindrical, up to 3 (5) μm wide, usually longer than wide, rarely more or less isodiametric, with peripherally arranged thylakoids and without aerotopes or with facultative polar solitary aerotopes and rounded end cells. All cells are capable of division. Reproduction occurs by fragmentation, without necridic cells (short hormogonia).

More than 15 species have been described from the plankton of large mesotrophic reservoirs and lakes, mainly from northern temperate zones (e.g., *P. limnetica* and *P. contorta*; Hindák, 1985; Cronberg and Komárek, 1994), but several species are specific to tropical and subtropical lakes (*P. tallingii* and *P. regularis*; Komárek and Kling, 1991; Komárková-Legnerová and Tavera, 1996; Komárek and Cronberg, 2001). Some *Planktolyngbya* species occur in North American reservoirs, but were previously recorded as the genus *Lyngbya*. Smith (1950) reported "*Lyngbya contorta*" from lakes in the United States. *Planktolyngbya capillaris*, *P. contorta*, *P. bipunctata*, and *P. limnetica* were reported in North America (authors' unpublished data). A review of North American species in this genus is necessary.

Oscillatoriales: Pseudanabaenaceae: Heteroleibleinioideae

Heteroleibleinia (Geitler) L. Hoffmann (Fig. 6B)

Filaments are solitary or in groups, heteropolar, individually attached by one end to substratum, rarely forming membranaceous, tuftlike layers with parallel sessile filaments; filaments are typically short, usually up to 100 μm long, rarely longer. Sheaths are thin, firm, and colorless. Trichomes may be constricted or not constricted at the cross-walls. Cells are usually isodiametric or slightly shorter or longer than wide. Apical cells are rounded, without calyptra or thickened outer cell wall. Reproduction occurs by disintegration of trichomes into motile hormogonia and nonmotile hormocytes, separating particularly from the apical parts of the trichomes; necridia present (?).

Heteroleibleinia does not differ substantially from thin forms of *Leibleinia* species except that the filaments are heteropolar, attached by one end to the substratum (main intergeneric feature). Almost all of the 30 species described are known from aquatic habitats worldwide, growing attached to different substrata in both marine and freshwater environments. Identification of species is difficult due to the small number of morphological characters. The genus is present in North American waters, but mainly mentioned under the previous generic names (usually *Lyngbya*); all species need to be accurately documented and their revisions are necessary, based on the current taxonomic criteria (Anagnostidis and Komárek, 1988). *H. pusilla* and *H. versicolor* are recorded (under *Lyngbya*) from several North American localities (Tilden, 1910; Duthie and Socha, 1976; Stein and Borden, 1979; Sheath and Steinman, 1982).

Homoeothrix (Thuret) Kirchner (Fig. 7A)

Filaments are solitary or forming mats, simple, not branched, or very rarely laterally branched (false branching), erect, solitary or in small, loose fascicles, attached to substrata basally, sometimes radially oriented

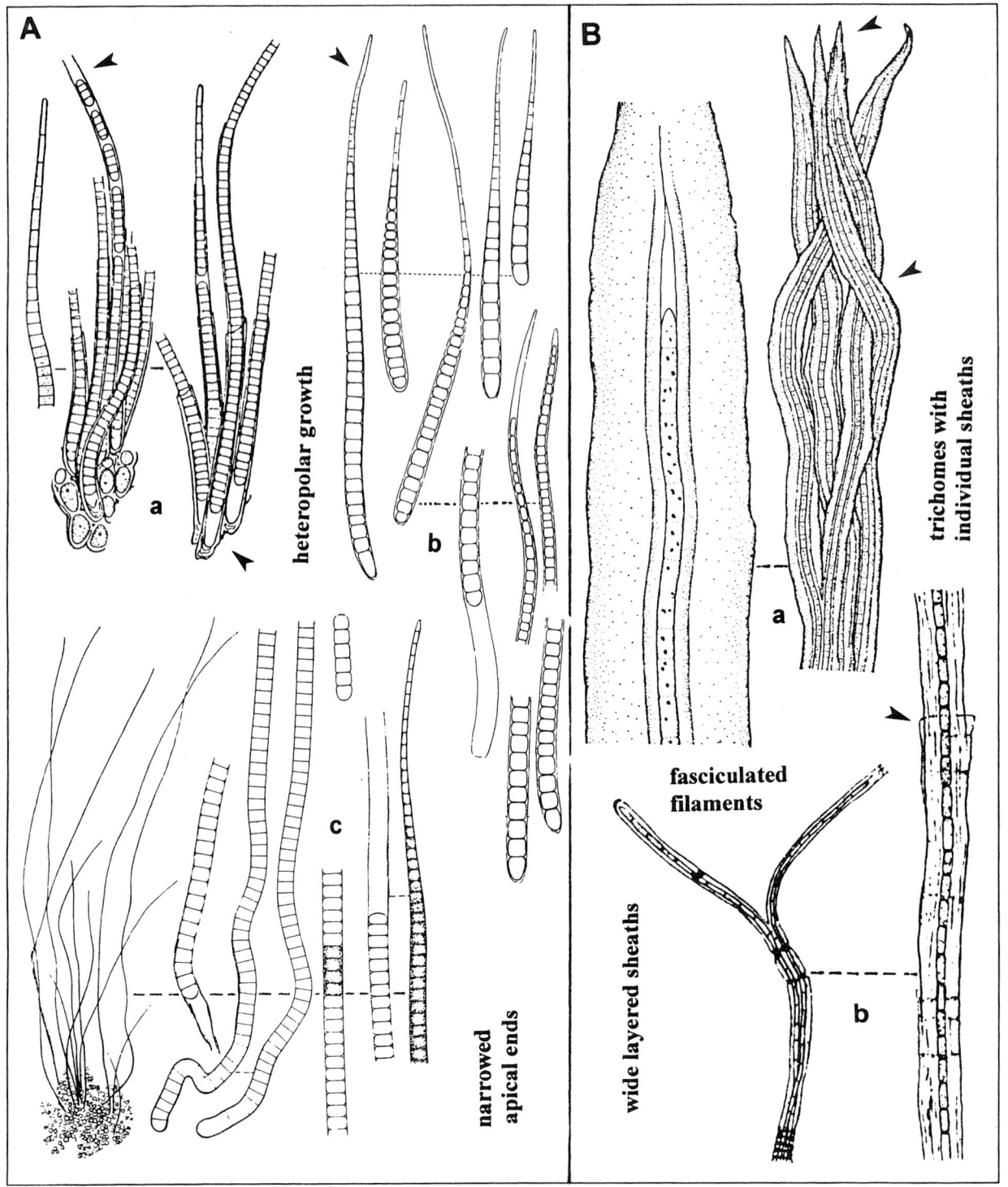

FIGURE 7 (A) *Homoeothrix*: (a) *H. janthina* (after Starmach, 1966); (b) *H. crustacea* (after Komárek and Kann, 1973); (c) *H. varians* (after Komárek and Kalina, 1965). (B) *Schizothrix*: (a) *S. violacea* (after Drouet, 1937, from the United States); (b) *S. constricta* (after Copeland, 1936, from Yellowstone National Park).

with bases in the center of the colony. Sheaths are thin, firm, hyaline or rarely slightly widened and lamellated, and yellowish in color. Trichomes are thin, straight or coiled, 3 (5, exceptionally to 7) μm wide, cylindrical, constricted or not constricted at the cross walls, tapering at ends, and sometimes elongated into thin, hyaline hairs (well-developed trichomes). Vegetative cells are short and cylindrical, bases of filaments of some species are enlarged (with wider and shorter cells); pale blue–green, olive green, or grayish. Reproduction occurs via hormogonia liberated from the upper part of trichomes after separation of the terminal hair.

All species (about 25 are valid) are known from submersed aquatic habitats, especially from stones or macrophytes in streams, rivers, and lakes; a few species inhabit the mucilage of other algae (endogloeic). Several taxa are reported from North America, for example, *H. varians* and *H. janthina* from eastern and western mountainous locations (Prescott, 1951; Stein and Borden, 1979; authors' results). Whitford and Schumacher (1969) recorded *H. janthina* (as *Amphithrix*) in rapids in the Piedmont region of North Carolina. However, *H. janthina* never forms encrustations, as it is sometimes reported, and also the marine habitats of this species (Tilden, 1910; Smith, 1950) are unlikely and may represent other species, e.g., (*H. crustacea* (?) and *H. violacea* (?). Whitford and Schumacher (1969) reported two other species (*H. stagnalis* and *H. crustacea*) from the pebbles of North Carolina streams (as *Leptochaete*).

Oscillatoriales: Schizotrichaceae

Schizothrix Kützing ex Gomont (Fig. 7B)

Filaments are solitary, free living in fascicles or attached and densely entangled forming a thallus; they are initially microscopic, but later form erect and polarized fascicles, with usually rich pseudobranched filaments attached by one end to stony substrata, or large, soft, thin or thick, fine or membranaceous layers. Two subgenera are recognized: In subgenus *Schizothrix*, fascicles are not encrusted and may be free floating or in mucilage of other algae; in subgenus *Inactis*, the thallus becomes firm, hard, sometimes spongy, later encrusted with calcium carbonate or calcified in basal parts. Encrusted colonies are usually lamellated, forming crustlike, hemispherical cushions or flat layers on stony substrata, 5 mm or more thick; the outside is warty, variously colored, gray, gray–brown, olive green, black–green, blue–green, or rusty reddish. Filaments of both subgenera are usually long, nearly straight or curved, sometimes densely aggregated, usually erect, with filaments parallel or occasionally radially arranged, sometimes forming ropelike tangles; filaments are rarely unbranched or sparsely, in tufts sometimes nearly pseudodichotomously pseudobranched, particularly at the ends. Sheaths are usually thickened and wide, rarely thin, soft to firm, often lamellated, delimited at the margin, rarely slightly diffused; they range from colorless to yellowish, gold–yellow, brownish, olive green or brownish, red or (rarely) violet or blue. The sheath surface ranges from smooth to uneven, rarely fibrous, ends attenuated or pointed, at free ends usually conically closed, rarely attenuated or funnel-like (facultative). Sheaths contain rarely one, usually several to numerous thin trichomes (facultatively changing in one species), often with more trichomes at the base, but with a single trichome in the upper part. Trichomes are nonmotile, distinctly, slightly or not constricted at the cross walls, and usually each with its own sheath. Cells are usually longer than they are wide, rarely almost isodiametric, with thylakoids probably only parietal. Apical cells are rounded, conically rounded, obtuse, or acutely conical, without a thickened outer cell wall or calyptra. Reproduction occurs via fragmentation into motile (?) hormogonia, without necridic cells.

More than 70 species are currently recognized. Many grow in the littoral zone of lakes or in streams, attached to substrata in wave-swept areas or in metaphyton among plants in wetlands and swamps. Several species form encrusted hemispherical colonies or layers on stones or rocks; other species colonize saline environments. There are a few subaerial species, some of which create characteristic brownish fascicles on wood or soil. Several species are recorded from Canada, the United States, the West Indies, Mexico, and Alaska, but few were recently taxonomically reclassified into other genera (e.g., *Symplocastrum*). Smith (1950) noted 30 species in the United States; several were also reported from British Columbia (Stein and Borden, 1979). Whitford and Schumacher (1969) reported 13 species from North Carolina, several of them growing on moist soil or rocks; *Schizothrix aikenensis* was collected from the epipelon of a small pool, whereas several others were found on rocks or bottom substrata in streams. Whelden (1947) and Sheath and Steinman (1982) reported several species from arctic and subarctic freshwater habitats. Tilden (1910) listed seven species in the United States, the West Indies, and Mexico. Copeland (1936) recorded several encrusting species from Yellowstone National Park and described a new species: *S. constricta*.

Trichocoleus Anagnostidis (Fig. 8A)

Filaments are mainly solitary and grow among other algae or cyanobacteria; they are rarely densely aggregated, forming prostrate thalli (mats). Filaments are either not or very rarely divaricated (spread apart like branches), containing a few to numerous cylindri-

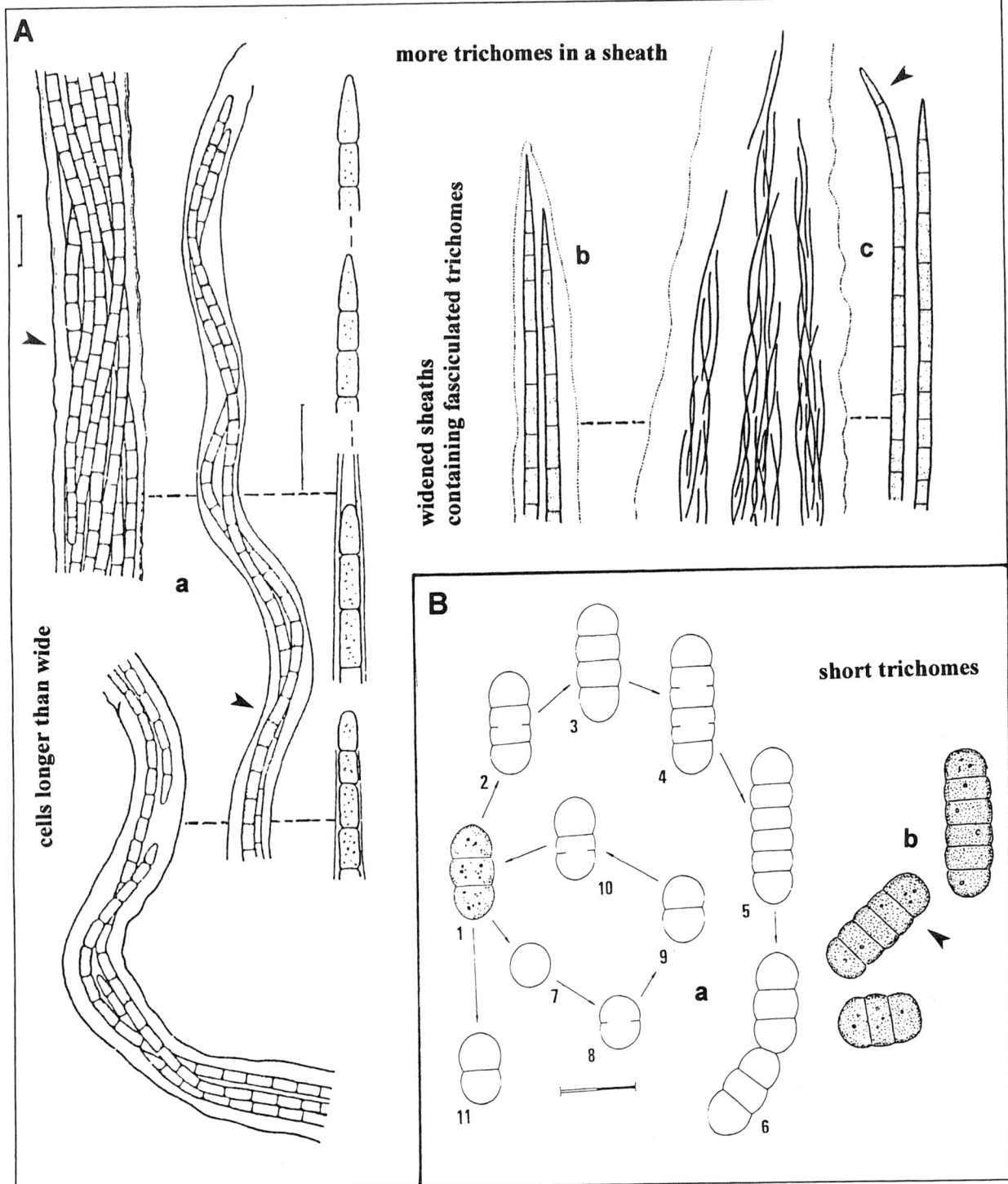

FIGURE 8 (A) *Trichocoleus*: (a) *T. erectiusculus* (bar = 10 μm; after Kondrateva, 1968); (b) *T. acutissimus* (after Gardner, 1927, from Puerto Rico); (c) *T. purpureus* (after Gardner, 1927, from Puerto Rico). (B) *Borzia*: (a) *B. trilocularis* (life cycle according to Bicudo, 1985, from Brazil); (b) *B. trilocularis* (after Gomont from Smith, 1950).

cal trichomes, arranged usually in parallel and forming dense fascicles. Sheaths are more or less cylindrical or (rarely) narrowed toward the ends, not or rarely lamellated, firm or mucilaginous, often diffuse, colorless, usually slightly or clearly distant from the trichome fascicles. The trichome width ranges from 0.5 to 2.5 (3) μm. Cells are always longer than they are wide. Apical cells are acutely conical, obtuse, or rounded, without calyptra or thickened outer cell wall. Reproduction occurs via hormogonia, which separate after disintegration from the ends of the trichomes. The genus was separated from *Microcoleus* based on cell morphology and cell structure (Anagnostidis, 2001).

Trichocoleus species occur in the periphyton or metaphyton in fresh and saline waters, although some species prefer alkaline substrata. About 16 species belong to this genus, almost all originally described as *Microcoleus* species. In North America, *T. sociatus* occurs mainly in freshwater environments (Whitford and Schumacher, 1969); *T. acutissimus*, *T. minor*, and "*Microcoleus purpureus*" were described from Puerto Rico (Gardner, 1927); *T. acutissimus* was recorded from wet rocks in Arizona, Florida, Jamaica, and Puerto Rico (Gardner, 1927). One species is benthic in shallow brackish pools in Arctic region.

Oscillatoriales: Borziaceae

Borzia Cohn ex Gomont (Fig. 8B)

Trichomes are solitary or aggregated in small groups, simple, very short, few celled, usually nonmotile, rarely motile or trembling, and constricted or unconstricted at the cross walls. Sheaths are usually lacking or sometimes present as a fine mucilage. Cells are cylindrical to barrel shaped, more or less isodiametric or slightly shorter or longer than wide. Thylakoids are coiled and probably spread over the cell volume. Apical cells are rounded. Reproduction occurs via fragmentation into nonmotile, few-celled hormogonia, without separation discs or necridia.

Seven species were recorded from the periphyton and metaphyton of clear, mostly small reservoirs. Various species grow among macrophytes, are benthic in lakes, are subaerial on calcareous substrata, or occur within the mucilage (endogloeic) of other cyanobacteria and algae (Anagnostidis and Komárek, 1988, 2001). *Borzia trilocularis* is fairly cosmopolitan, but distribution data are sparse; this species is reported from several localities in the United States (Daily, 1943; Smith, 1950; Taft and Taft, 1970). The distribution of other species has not been verified in North America.

Komvophoron Anagnostidis *et* Komárek (Fig. 9A)

Trichomes are solitary or agglomerated in small colonies enveloped by a fine mucilage, motile or nonmotile, straight or slightly flexuous, moniliform, simple, usually short or slightly elongated, up to 650 μm long, without firm sheaths. Cells are spherical or barrel shaped, up to 10 μm wide; thylakoids (known only in one species) are typically arranged in parallel arrays perpendicular to the cell walls. Aerotopes, heterocytes, and akinetes are absent. Reproduction occurs via fragmentation, without the formation of necridic cells, into nonmotile, rarely (occasionally) motile hormogonia.

About 15 species have been described, most of which are benthic and living in solitary trichomes or forming thin mats on the bottom (sandy, stony, epipelic) of unpolluted lakes, pools, reservoirs, and streams. Two species are known from thermal and mineral springs. *Komvophoron* is relatively new genus (Anagnostidis and Komárek, 1988) not reported by early authors or identified as *Pseudanabaena*. *Komvophoron* specimens were recently collected in North America from benthic habitats in several northern temperate lakes (Manitoba and Ontario, Canada, the authors, personal observation). *K. jovis* was described from thermal springs in Yellowstone National Park (Copeland, 1936), *K. groenlandicum* from an oligotrophic lake in Greenland near Narssaq (Nygaard, 1984, under *Isocystis* sp.). More detailed work is required.

Oscillatoriales: Phormidiaceae: Phormidioideae

Arthrospira Stizenberger ex Gomont (Fig. 9B)

Trichomes are solitary and free floating in the plankton or united into a fine, mostly slimy (diffuse margins), blue–green, olive-green, or reddish-brown thallus in the benthos. Trichomes are cylindrical, isopolar, regularly or rarely somewhat irregularly loosely spirally (screwlike) coiled. Trichomes are long or short, usually with relatively large spirals (width and height), sometimes attenuated at the ends, with variable coil width, not constricted or slightly constricted at the cross walls, usually nonmotile or rarely motile (gliding with clockwise or counterclockwise rotation). Sheaths are absent or facultatively present, fine and colorless. Cells are more or less isodiametric or shorter than wide, with visible cross walls in the trichomes, sometimes with aerotopes (groups of gas vacuoles, in planktonic species), with special pore and perforation patterns in the cell walls (one row of pores around cell) and whirl-like or radially arranged thylakoids. Apical cells are rounded or conical, occasionally with calyptras or thickened outer cell walls. No toxic strains are known. Reproduction occurs via trichome fragmentation into hormogonia or hormocytes with necridic cells. Cells divide perpendicularly to the horizontal axis.

About 16 species are known, several of which are freshwater benthic (*A. jenneri* and *A. platensis*); others are planktonic in subtropical and tropical saline lakes

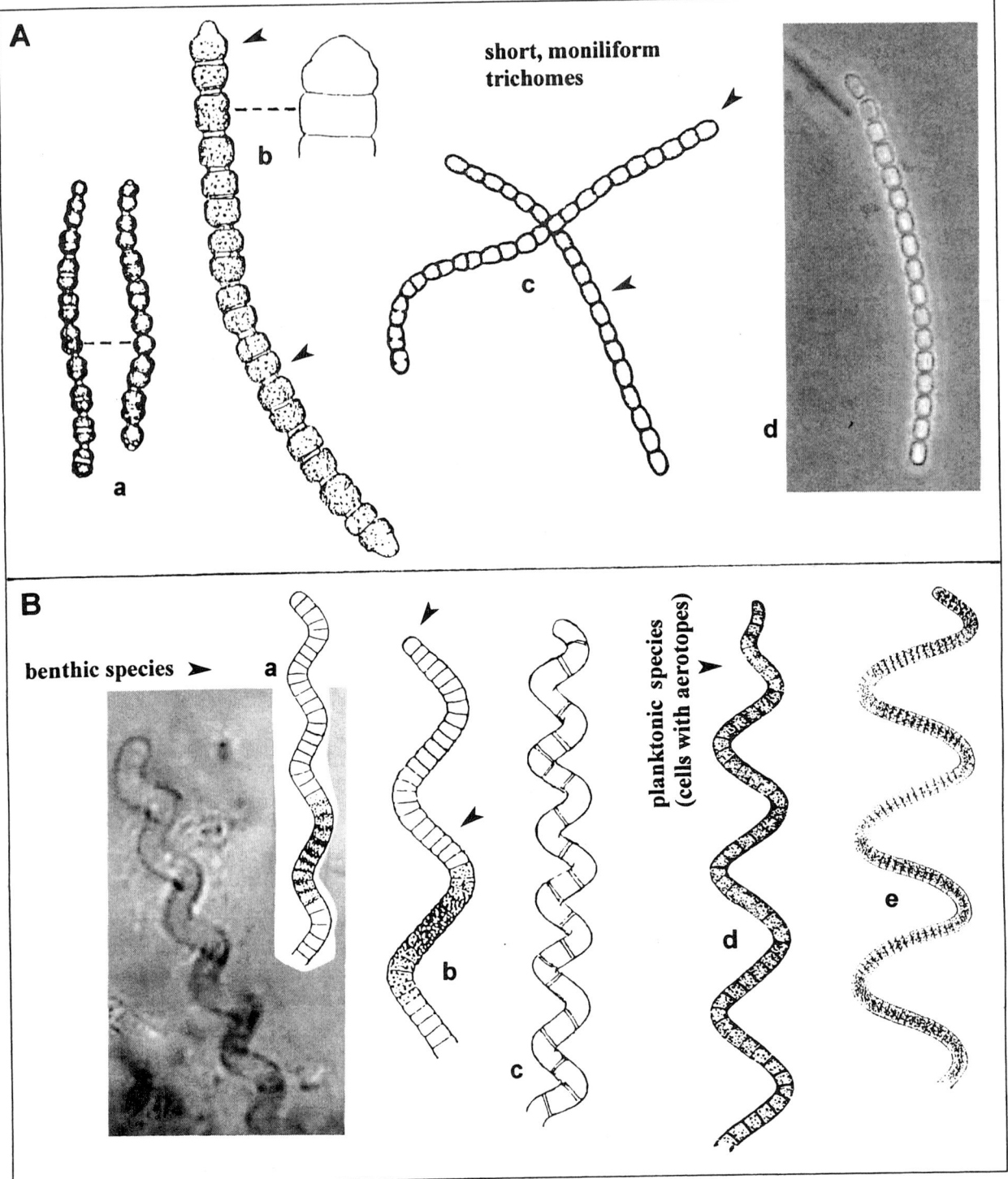

FIGURE 9 (A) *Komvophoron*: (a) *K. minutum* (after Skuja, 1948); (b) *K. schmidlei* (after Jaag, 1938); (c) *K. jovis* (after Copeland, 1936, from Yellowstone National Park); (d) *Komvophoron* sp. (original by Komárková, from the Everglades, Florida). (B) *Arthrospira*: (a) *A. jenneri* (photo original by Komárková from the Everglades, Florida); (b) *A. platensis* (after Komárek and Lund, 1990; redrawn from type material from La Plata, Argentina); (c) *A. skujae* (after Magrin *et al.*, 1997, from Brazil); (d) *A. maxima* (after Setchell and Gardner, 1917, from California); (e) *A. khannae* (after Drouet, 1942, from the United States).

(*A. maxima*) in southern California and Mexico. Two species, *A. maxima* and *A. fusiformis*, are used in mass cultivation, but usually designated as "*Spirulina platensis*." Tilden (1910) reported two species from the United States, whereas Smith (1950), Prescott (1962), Whitford and Schumacher (1969), and Stein and Borden (1979) reported the benthic species *A. jenneri* and/or *A. gomontiana* from several locations.

Phormidium Kützing ex Gomont (Fig. 12A)

The thallus is usually expanded, more or less fine, thin or cohesive, gelatinous, mucilaginous, cartilaginous, membranaceous, feltlike to leathery, attached to substrata or (secondary) free floating in masses, sometimes forming clusters, penicillate tufts, or (rarely) living in solitary filaments. Filaments vary in curvature; they are not pseudobranched, usually entangled, slightly to strongly waved or loosely and irregularly screwlike coiled. Sheaths occur facultatively (under unfavorable conditions) or almost obligatorily (frequency depending on environmental conditions); they are firm or thin, colorless, adherent to the trichome, not lamellated, sometimes slightly to intensely diffuse (rarely thickened and lamellated when old; generic revision may be needed). Trichomes are cylindrical, mostly long, (1.8) 2.5–11(15) µm wide, unconstricted or slightly to distinctly constricted at the cross walls, clearly motile inside the sheaths and outside the sheath (gliding, creeping, waving, trembling, with or without oscillation and rotation). Cells are typically isodiametric or shorter or longer than wide, without aerotopes (clusters of gas vesicles). Apical cells are pointed, narrowed, or rounded, with or without calyptra. Thylakoids are typically radially oriented within cells (cell content may appear netlike or striated). Cells divide by transverse fission, and each cell reaches its original size before the next division. Reproduction occurs via trichome disintegration into short or long, motile hormogonia, sometimes with biconcave necridic cells.

Phormidium is a very common genus, distributed worldwide, with nearly 200 described species. It forms mats on wet soil, mud, wetted rocks and macrophytes, and in standing and running waters. Several species are known from extreme environments (thermal springs and desert soils) a few species form travertine in springs, marl lakes, and streams (Prát, 1929). Tilden (1910) listed 27 species throughout North America from Alaska to Mexico, from Newfoundland to Florida. Smith (1950) mentioned 25 species in the United States. Of the most common species, *Phormidium inundatum* occurs on mud or rocks in lakes and ponds. *P. retzii* was the most commonly recorded macroscopic alga in North American stream sites (Sheath and Cole, 1992). Prescott (1962) reported 14 species in the western Great Lakes area; eight species were recorded from British Columbia (Stein and Borden, 1979). Other frequently reported species are *P. autumnale* and *P. uncinatum*, which colonize stones in streams and rivers, and *P. fonticolum* from cold stenotherm alpine and subarctic streams and seepages (authors' results); *P. minnesotense* and *P. favosum* are known from the periphyton and metaphyton among the aquatic plants in rivers and standing waters (Whitford and Schumacher, 1969).

Planktothrix Anagnostidis et Komárek (Fig. 10A)

Trichomes are solitary, free floating, more or less straight or slightly irregularly waved, isopolar (both poles with the same morphology), cylindrical, not constricted or slightly constricted at the cross walls, usually planktonic (rarely metaphytic), in massive blooms arranged into small, disintegrating irregular clusters, or diffuse in tight clumps, more or less long (up to 4 mm), (2.3)3–12(15) µm wide, nonmotile, but occasionally with inconspicuous trembling or gliding, rarely oscillation. Trichomes are slightly attenuated or not attenuated at the ends, sometimes with terminal calyptra. Mucilaginous envelopes or sheaths are usually lacking, occasionally (e.g., in culture) with fine, visible sheaths (one species has obligatory sheaths in nature as well); false branching is absent. Cells are cylindrical, rarely slightly barrel shaped, usually slightly shorter than wide or up to more or less isodiametric, rarely longer than wide. Thylakoids are typically radially arranged. Aerotopes (groups of gas vesicles) are distributed throughout the cells. Apical cells (when fully developed) are widely rounded or narrowed conical, sometimes with calyptra or with thickened outer cell wall. Segments of trichomes (with several cells) without aerotopes are a feature of the genus; these appear less pigmented than other parts of the trichome (probably diazocytes according to Bergman, 2002). PC:PE ratio was found stable in various species (without photoacclimation). Characteristic carotenoids include myxoxanthophyll and oscillaxanthin. Geosmin and toxins are present in several strains (Skulberg and Skulberg, 1985). Cells divide perpendicularly to the trichome. Reproduction occurs via disintegration into nonmotile hormocytes (with necridia). Hormogonia without gas vesicles probably overwinter in sediments.

Most species (about 15 have been described) are planktonic; a few form surface blooms and may be toxic (Skulberg and Skulberg, 1985; Skulberg *et al.*, 1993). Only *Planktothrix cryptovaginata* is known from the metaphyton of unpolluted pools (Skácelová and Komárek, 1989). *P. agardhii* probably has a cosmopolitan distribution (Prescott, 1951; Komárek, 1958; Duthie and Socha, 1976; Stein and Borden, 1979; Skulberg and Skulberg, 1985; Niiyama *et al.*, 1993;

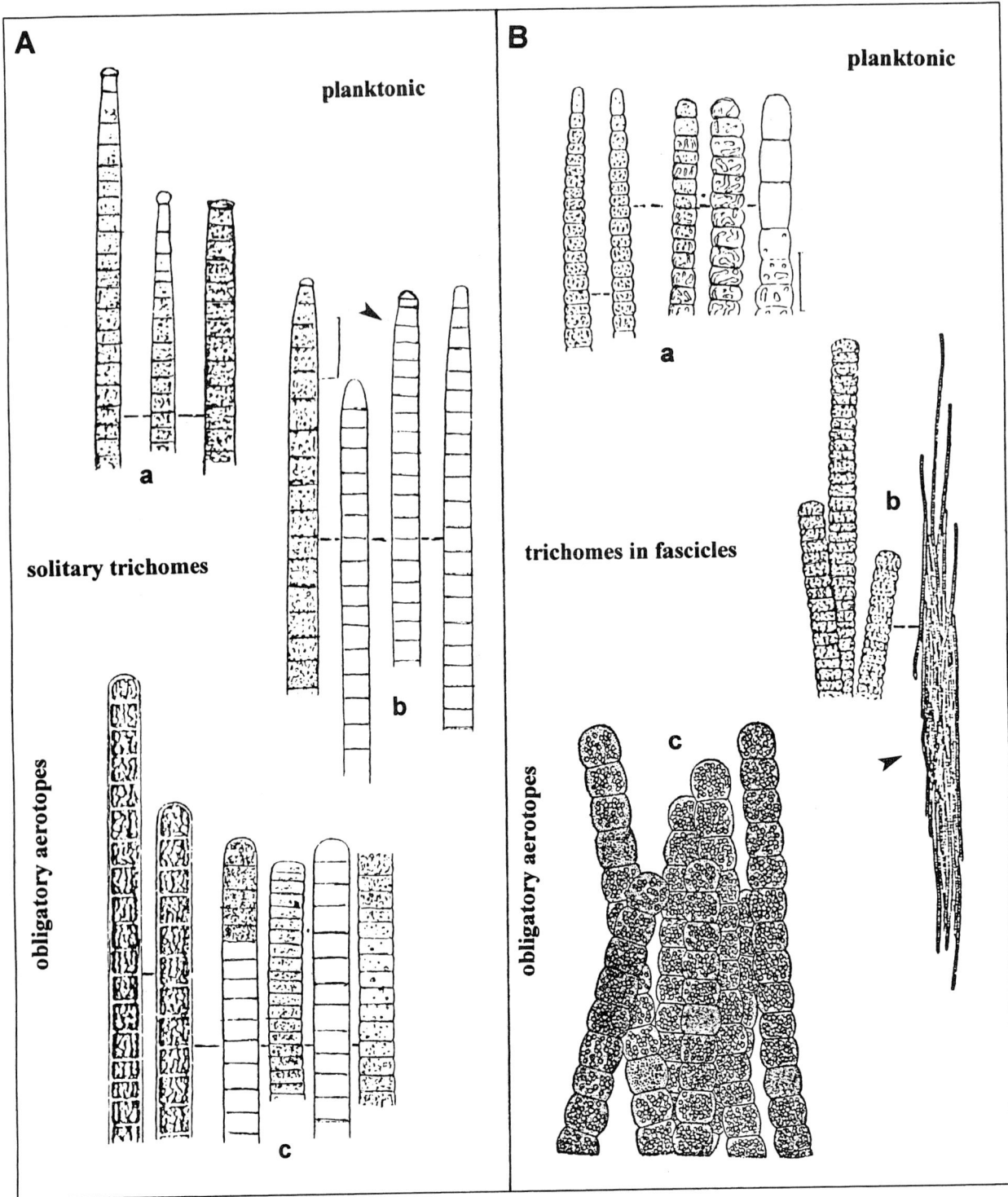

FIGURE 10 (A) *Planktothrix*: (a) *P. rubescens* (after Gomont, 1892); (b) *P. agardhii* (after Komárek, 1958); (c) *P. mougeotii* (after Skuja, 1948; Komárek, 1984). (B) *Trichodesmium*: (a) *T. lacustre* (after Nygaard, 1977); (b) *T. iwanoffianum* (after Nygaard, 1977); (c) *Trichodesmium* sp. (after Smith, 1920, from Wisconsin; sub *Trichodesmium lacustre*).

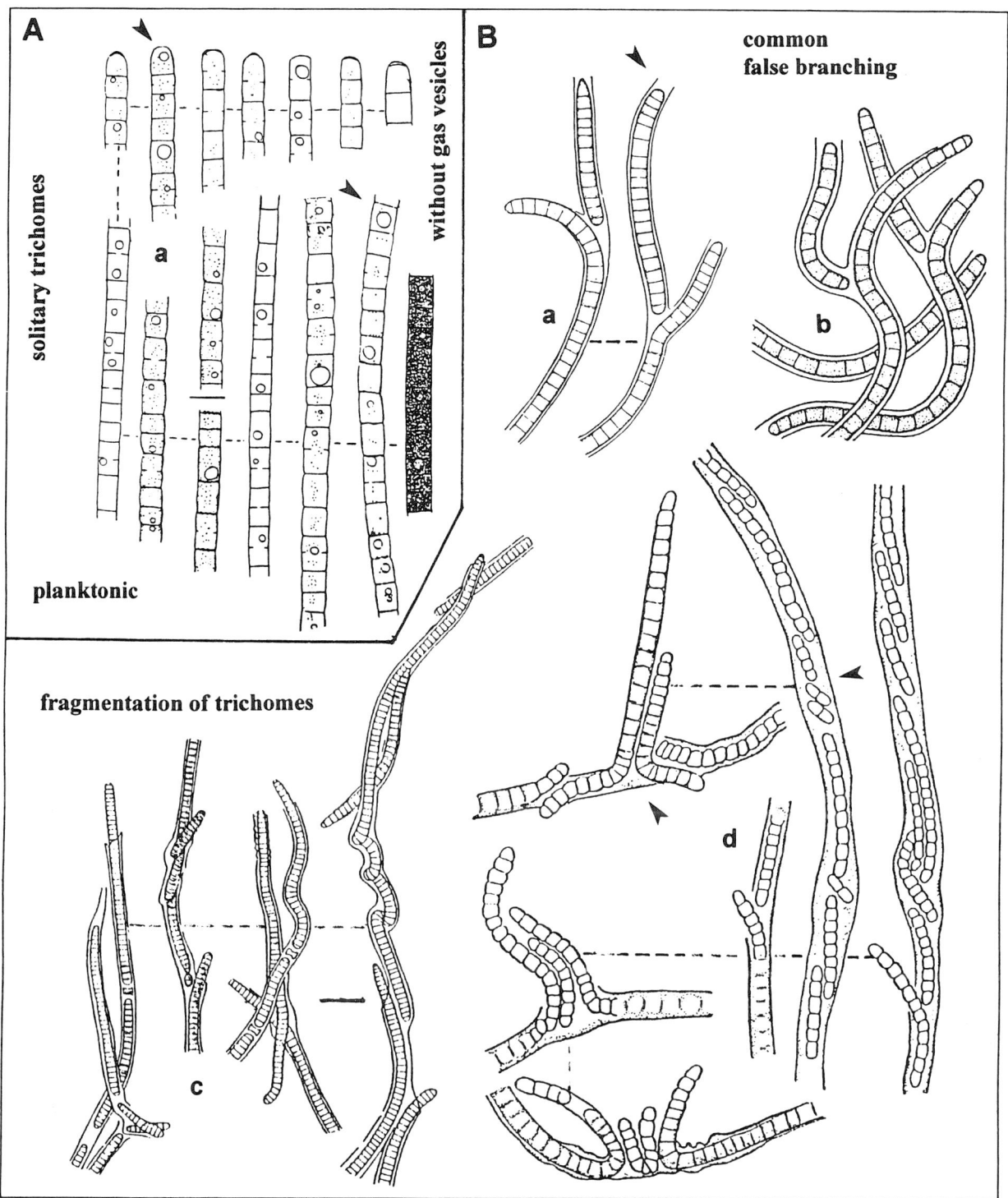

FIGURE 11 (A) *Tychonema*: (a) *T. bourrellyi* (after Lund, 1955). (B) *Pseudophormidium*: (a) *P. flexuosum* (after Gardner, 1927, from Puerto Rico; sub *Plectonema flexuosum*); (b) *P. murale* (after Gardner, 1927, from Puerto Rico; sub *Plectonema murale*); (c) *P. batrachospermi* (bar = 20 µm; after Starmach, 1957; sub *Plectonema batrachospermii*); (d) *P. edaphicum* (after Kondrateva, 1968, sub *Plectonema edaphicum*).

Komárek and Cronberg, 2001). In North America, several *Planktothrix* species are reported as species of *Oscillatoria*. Whitford and Schumacher (1969) reported *P. agardhii*, and *P. prolifica*. Prescott (1951) recorded *P. agardhii*, *P. prolifica* and *P. rubescens*. *P. agardhii* and *P. mougeotii* are also known from Mexico and Brazil (authors' unpublished records).

Porphyrosiphon Kützing ex Gomont (Fig. 13A)

Filaments are solitary among other algae or developed into an expanded and sometimes stratified blue–green, brown, or red thallus on various substrata. Filaments are contorted or undulating, rarely with single pseudobranches, containing one or, rarely, two trichomes in the sheaths. Sheaths are thick, firm, lamellated, and usually colored, red, reddish brown, purple, yellow, yellow–brown; they are rarely colorless. Young filaments have attenuated ends with closed sheaths; old sheaths have open ends, often characteristically widened or coiled and modified after fragmentation and hormogonia release. Trichomes are nonmotile, 6–10 μm wide, and constricted or not constricted at the cross walls. Cells are isodiametric or longer than wide. Apical cells are usually rounded or conical, without calyptra. Reproduction occurs by hormogonia released from the sheaths.

Most of the 20 species described are terrestrial or subaerial on wet rocks, often in mountains; others occur on tree bark. Some species colonize mud or form periphyton in clear freshwaters. *Porphyrosiphon notarisii* was recorded from North Carolina (Whitford and Schumacher, 1969; Tilden, 1910) and Ellesmere Island, Northwest Territories (Sheath and Steinman, 1982); it is probably widely distributed throughout North America (Smith, 1950), but the identity of different populations should be revised. Tilden (1910) listed *P. notarisii* from the West Indies; two other species probably occur in North America (*P. versicolor* and *P. fuscus*).

Pseudophormidium (Forti) Anagnostidis *et* Komárek (Fig. 11B)

Filaments are long or short and are usually aggregated into expanded mats, tufts, or clusters, often densely entangled, sometimes radially arranged, rarely solitary among other algae and cyanobacteria. Filaments vary in curvature and are heavily (obligately) pseudobranched; they commonly disintegrate into numerous trichome segments with single branches overlapping in a common sheath. Sheaths are always present; they are firm, rarely mucilaginous and diffuse, colorless or colored, and lamellated or not. Trichomes are up to 10 μm in width, usually distinctly constricted at the cross walls. Cells are more or less isodiametric or shorter or longer than wide; cells lack aerotopes. Apical cells are rounded to obtuse conical, without calyptra or thickened outer cell wall. Reproduction occurs via trichome fragmentation into nonmotile hormocytes or also possibly motile hormogonia (?).

Pseudophormidium contains mostly periphytic species, commonly growing on soil or creeping on stone surfaces or other substrata in unpolluted streams (Anagnostidis and Komárek, 1988, 2001). One species (*P. batrachospermi*) is endogloeic in the mucilage of algae (Starmach, 1957, as *Plectonema batrachospermi*). Although not yet recorded, some of the 14 described species likely occur in American freshwaters and soils (a species similar to *P. tenue* was recorded from Massachusetts, North Carolina, and Wisconsin; Tilden, 1910).

Symploca Kützing ex Gomont (Fig. 12B)

The thallus is composed of entangled or parallel filaments, forming compact, more or less wooly, often terrestrial or subaerial masses. Filaments are at first prostrate, irregularly curved, later mainly united into numerous, partly pseudobranched, erect (rarely prostrate) fascicles that arise from the thallus as erect, conical, often confluent tufts. Sheaths are thin or thick, firm, distinct, in the fascicles often mucilaginous and laterally slightly confluent or somewhat gelatinized, at the ends straight and slightly attenuated, containing one trichome, always open at the end. Trichomes are straight, often weakly attenuated, thin or up to 8 (14) μm wide. Cells are isodiametric or either shorter or longer than wide, probably with a radial arrangement of thylakoids. Apical cells are never capitate, but often with a thickened outer cell wall. Reproduction occurs by motile hormogonia.

Of the 70 species described, about 30 have been revised (Anagnostidis and Roussomoustakaki, 1985). Most species are terrestrial on wet soil or subaerophytic on rocks and mosses; some are marine. Smith (1950) listed *S. muscorum* on damp soils and moist cliffs, along with 11 other species from various locations in the United States. *S. thermalis*, *S. nemecii*, and *S. ciliata* were recorded from the edge of hot springs in Yellowstone National Park (Prát, 1929; Copeland, 1936). Whitford and Schumacher (1969) recorded *S. borealis*, *S. dubia*, and *S. muralis* on rocks and wet soil and at the bottom of drying pools in North Carolina. Other species reported from North America include *S. cartilaginea* (Gomont, 1892; Geitler, 1932), *S. cavernarum* (Copeland, 1936), and *S. kieneri* (Drouet, 1943a). Tilden (1910) listed nine species in the United States, Mexico, Greenland, and the West Indies.

Trichodesmium Ehrenberg ex Gomont (Fig. 10B)

Trichomes are planktonic, free floating, more or less straight, rarely solitary, usually in parallel or radial

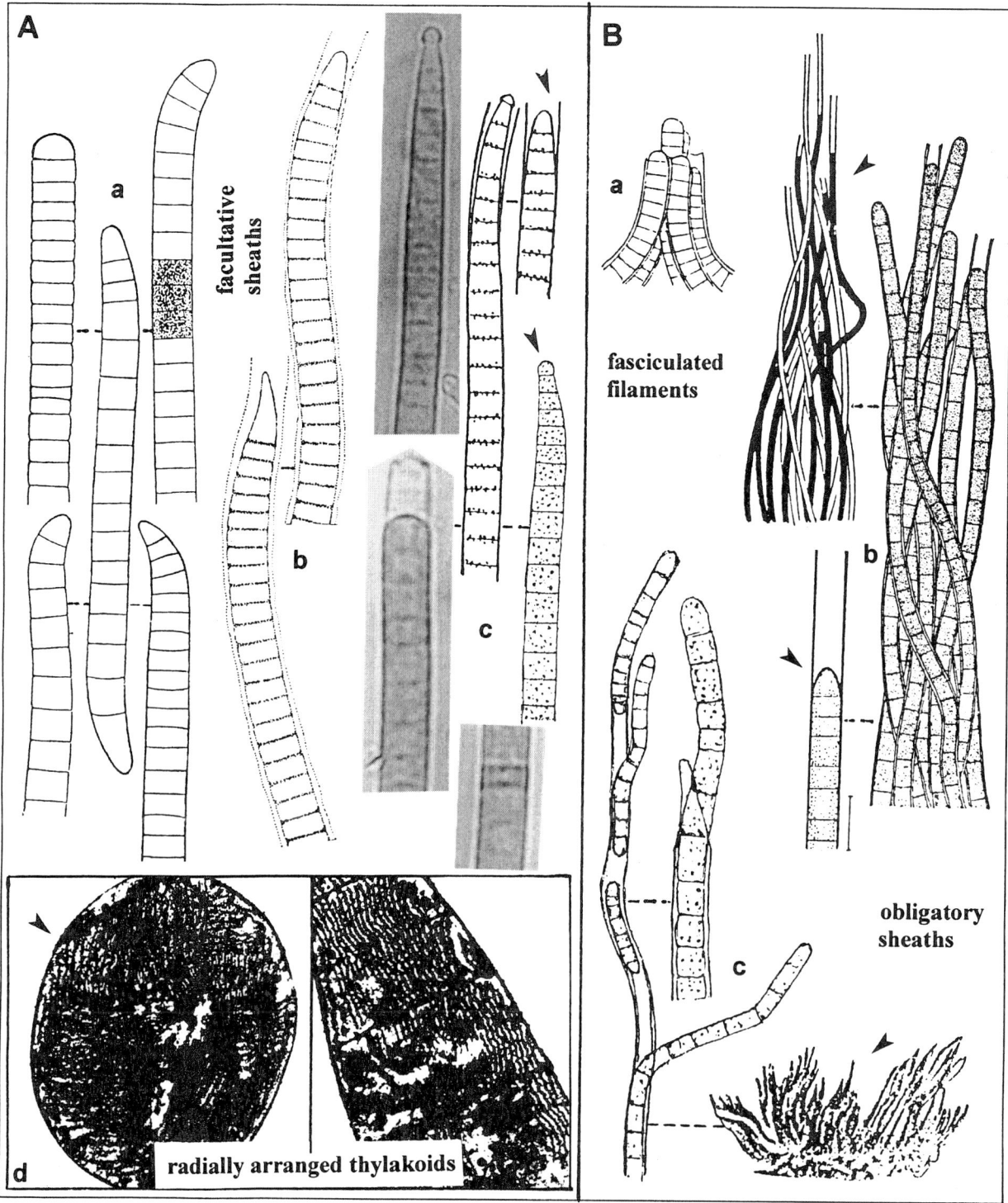

FIGURE 12 (A) *Phormidium*: (a) *P. formosum* (after Komárek, 1989, from Cuba); (b) *P. richardsii* (after Drouet, 1942, from the United States); (c) *P. autumnale* (after Komárek, 1988; Smith, 1950; microphoto originally by R. G. Sheath, with permission); (d) thylakoid arrangement in *Phormidium* cells (after Wolf from Komárek, 2001). (B) *Symploca*: (a) fascicles; (b) *S. muscorum* (bar = 10 μm; after Smith, 1950; Anagnostidis and Roussomoustakaki, 1985); (c) *S. hydnoides* (after Frémy, 1930).

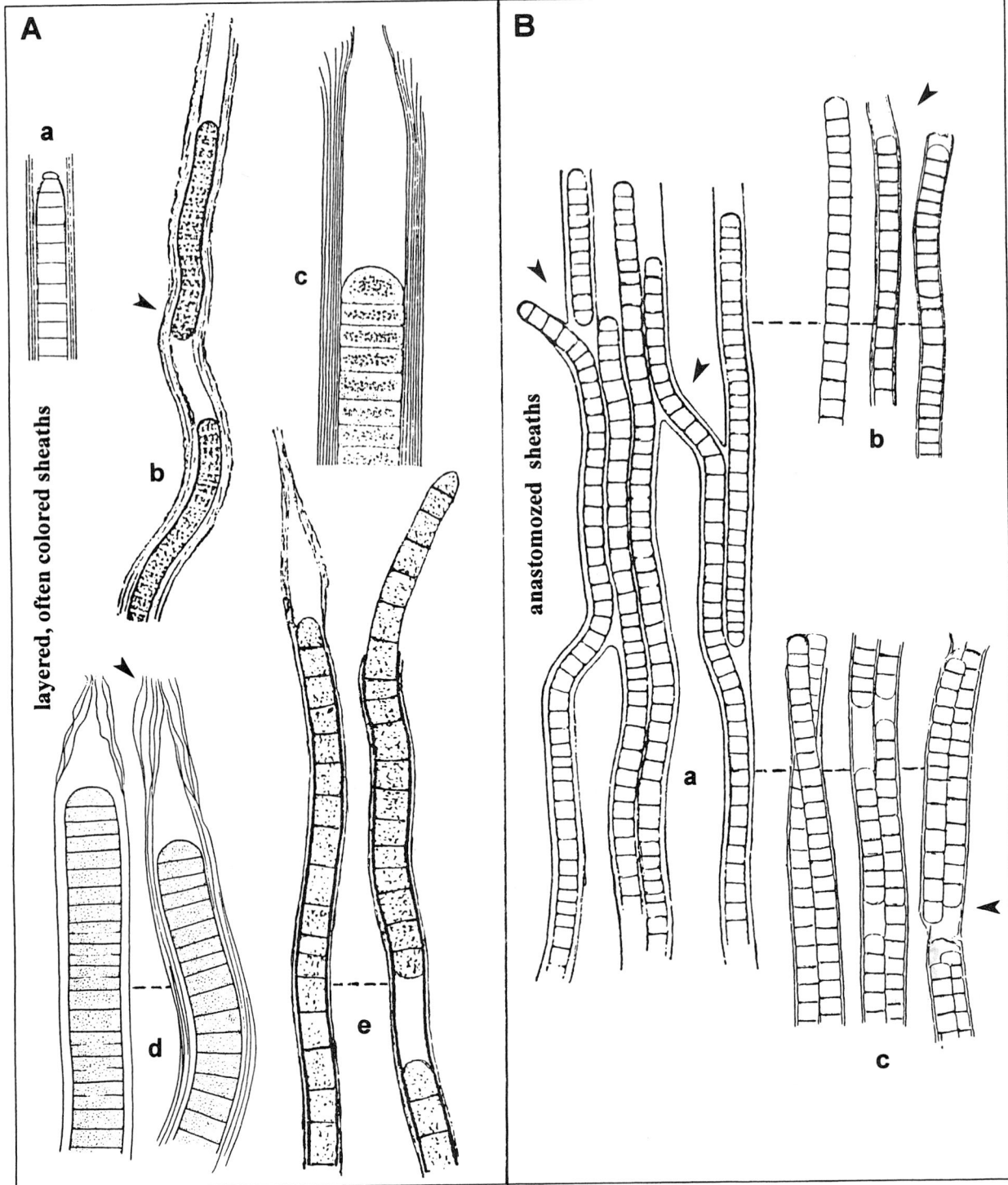

FIGURE 13 (A) *Porphyrosiphon*: (a) end of filament; (b) *P. versicolor* (after Frémy ex Geitler, 1932); (c) *P. notarisii* (Smith, 1950); (d) *P. robustus* (after Gardner, 1927, from Puerto Rico); (e) *P. fuscus* (after Frémy ex Geitler, 1932). (B) *Lyngbyopsis*: (a) *L. willei* (after Gardner, 1927 from Puerto Rico); (b–c) *L. willei* (after Komárek, 1989, from Cuba).

arrangements in colonies that form fascicles or flocculent masses, joined by diffuse mucilage. Trichomes occur without individual sheaths; they are straight or curved, rarely irregularly spirally twisted, slightly motile (inconspicuous gliding), 6–22 μm wide, cylindrical or with slightly tapering ends. Cells are typically isodiametric or slightly longer or shorter than wide, with homogeneous or finely granular content and with obligate aerotopes arranged irregularly throughout the blue–green or reddish protoplast. Apical cells are rounded or slightly capitate. Another distinguishing feature (from *Planktothrix*) is in the composition of fatty acids (Umezaki, 1974). Cell division occurs mainly in the meristematic zones in the middle of the trichomes; reproduction occurs via trichome fragmentation.

Trichodesmium is primarily a pelagic species in oceans, capable of forming extensive water blooms, but several species occur in freshwaters (sometimes considered more related to *Planktothrix*); 11 exclusively planktonic species have been described, but freshwater populations need taxonomic revision (Niiyama *et al.*, 1993). The freshwater *T. lacustre* (as *Oscillatoria lacustris*) is widespread but not abundant in North Carolina (Whitford and Schumacher, 1969) and in scattered locations in British Columbia (Stein and Borden, 1979). Smith (1920, 1950) mentioned the genus as being widely distributed throughout the United States. Tilden (1910) listed three marine species, two in the region of the West Indies and Central America.

Tychonema Anagnostidis et Komárek (Fig. 11A)

Trichomes are solitary or organized into fine mats, which may be benthic, tychoplanktonic, or planktonic. Trichomes are cylindrical, pale grayish–pinkish, purplish, reddish, or olive green, up to 5 mm long, 2–16 μm wide, without sheaths or with fine facultative mucilaginous sheaths, nonmotile (?) or with reduced motility (slightly trembling, gliding, or rotating), without false branching, usually not constricted at the cross walls. Filaments are straight or, occasionally, slightly irregularly coiled or more or less curved, not attenuated at ends. Cells are identical in morphology, cylindrical, more or less isodiametric, or shorter or longer (up to twice) than wide. Gas vesicles (and aerotopes) are always absent, but cells often contain prominent granules; cell content is pale and "alveolar" with keritomized chromatoplasma (seemingly vacuolated, but with radially arranged, widened thylakoids, and almost colorless; Komárek and Albertano, 1994). Apical cells are rounded, sometimes with thickened cell walls or narrow calyptras. Phycobilin content is variable (chromatic acclimation), mainly containing phycoerythrin (olive green to reddish brown); several characteristic carotenoids are present. Toxic strains are not known (Skulberg and Skulberg, 1985). Reproduction occurs via disintegration of trichomes into nonmotile hormocytes (motile hormogonia?).

Tychonema species are planktonic or benthic (altogether eight species have been described, but only *T. bourrellyi* and *T. bornetii* are well known and their generic characters proved). *Tychonema* is possibly a cold-stenotherm genus of northern temperate areas. *T. bourrellyi* is known to form metalimnetic maxima in plankton, mainly recorded in northern Europe (Lund, 1955; Skulberg and Skulberg, 1985), but also in Canada (Kling and Holmgren, 1972; Findlay and Kling, 1979).

Oscillatoriales: Phormidiaceae: Microcoleoideae

Dasygloea Thwaites ex Gomont (Fig. 14B)

Dasygloea occurs as solitary, creeping filaments growing among other algae or in mucilaginous mats on substrata. Sheaths are distinctly irregularly widened, colorless or colored, distinctly outlined, with a smooth surface, usually clearly or indistinctly lamellated (stratified), sometimes funnel-like widened, irregularly branched or unbranched, usually closed at the ends. Sheaths contain one or two (rarely several) parallel, distant trichomes. Trichomes are cylindrical, usually wavy or almost straight, constricted or unconstricted at the cross walls, with noncapitate ends, (3) 4–10 (12) μm wide, with isodiametric cells or slightly shorter or longer than wide. Reproduction occurs by hormogonia, separated by necridic cells.

The nine species described live on muddy sediments or are metaphytic in swamps and pools (Senna and Komárek, 1998). *D. lamyi* and *D. amorpha* are known from Jamaica and the United States (Pennsylvania) (Geitler, 1932; Starmach, 1966). *D. calcicola* and *D. yellowstonensis* were recorded from Yellowstone National Park from the periphyton on calcified encrustations near thermal springs (Copeland, 1936). Tilden (1910) listed other species from Pennsylvania.

Hydrocoleum Kützing (Fig. 15B)

Hydrocoleum is composed of solitary filaments or thallus usually composed of filaments joined in small, smooth, caespitose, spherical or cushion shaped, microscopic to macroscopic thalli, or forming extended amorphous, membranaceous, flat or cushion-like mats, rarely compact, often encrusted with $CaCO_3$. Filaments are usually straight, sometimes of variable curvature, rarely sparsely pseudobranched, more or less parallel (rarely radial), or joined to form tufts or erect ropelike fascicles. Sheaths are thick, mucilaginous, and smooth, colorless or yellowish, initially firm, usually limited, when older diffuse, with longitudinal striation (strati-

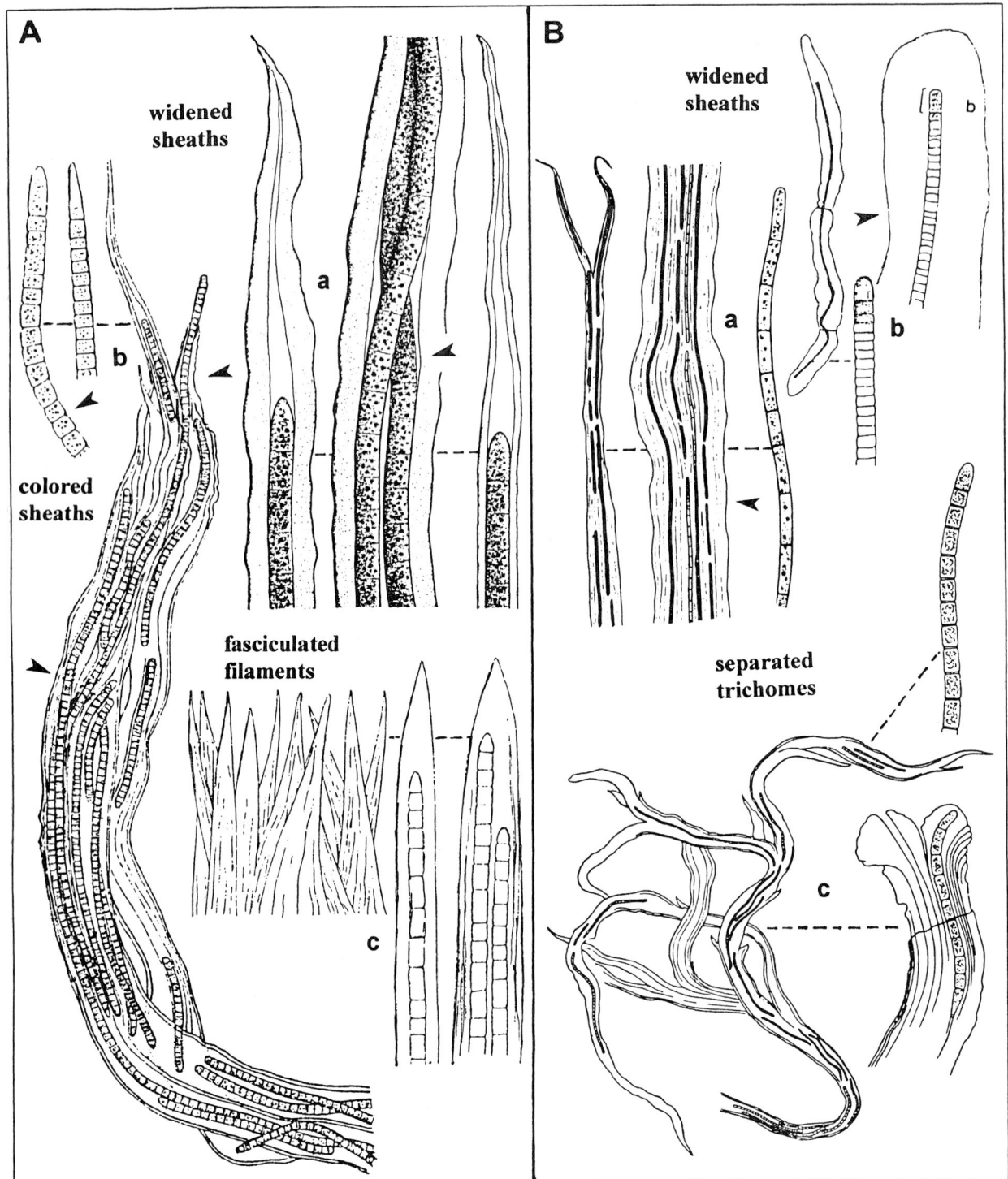

FIGURE 14 (A) *Symplocastrum*: (a) *Symplocastrum* sp. (after Drouet, 1939, from the United States; sub *Schizothrix wollei*); (b) *S. purpurascens* (after Gomont, 1892); (c) *S. parciramosa* (after Gardner, 1927, from Puerto Rico; sub *Hypheothrix parciramosa*). (B) *Dasygloea*: (a) *D. yellowstonensis* (after Copeland, 1936, from Yellowstone National Park); (b) *D. brasiliense* (after Senna and Komárek, 1998); (c) *D. lamyi* (after Gomont from Starmach, 1966).

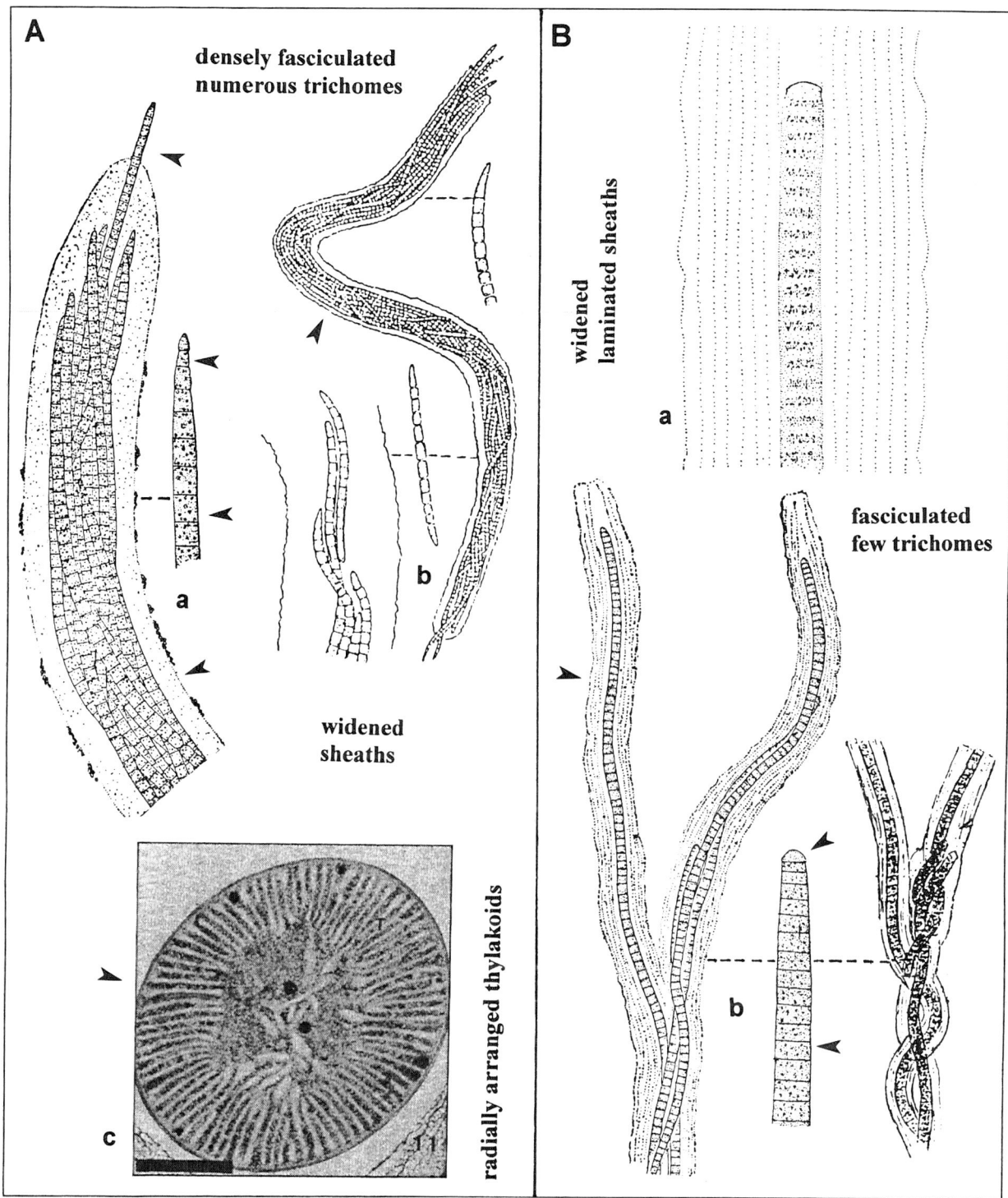

FIGURE 15 (A) *Microcoleus*: (a) *M. vaginatus* (after Smith, 1950, from the United States); (b) *M. chthonoplastes* (after Gomont from Kondrateva, 1968); (c) thylakoid position in cells (cross section in *Microcoleus chthonoplastes;* after Hernandéz-Mariné, 1996). (B) *Hydrocoleum*: (a) *H. groesbeckianum* (after Drouet, 1943, from eastern California); (b) *H. homeotrichum* (after Gomont, 1892; Smith, 1950).

fied), containing several (2–10) to many parallel and tightly fasciculated trichomes. Trichomes are straight, usually densely (rarely loosely) aggregated, more or less gradually attenuated at the ends. Cells are isodiametric or slightly shorter or longer than wide. Apical cells are capitate, often with a thickened outer cell wall or calyptra. Reproduction occurs via hormogonia.

Of the several species described, about 12 are periphytic in clear streams and rivers or clear cold lakes. Some are present in marine littoral areas or in the littoral areas of marl lakes (with precipitations in the sheaths). Whitford and Schumacher (1969) observed *H. homoeotrichum* on wet soil in cultivated fields, whereas Prescott (1962) noted one species from the western Great Lakes region. Smith (1950) mentioned that *H. homoeotrichum* appears in the United States in both aerial (?) and aquatic habitats. Tilden (1910) listed eight species from the Caribbean, the east coast of the United States, Pennsylvania, and Texas, but a few of them were later reclassified into the genus *Blennothrix* (Oscillatoriaceae).

Lyngbyopsis Gardner (Fig. 13B)

The thallus occurs in the form of macroscopic, semiglobose or flattened mats (in strata) on submersed stony substrata or with filaments grouped into prostrate layers. The sheaths are thin or slightly thickened, colorless, firm, and membranaceous, containing one or few parallel trichomes, each developing an individual sheath. Filaments are sometimes anastomozed by their sheaths and linked together. Pseudobranching occurs in one or both directions toward the ends of the longitudinal axis. Trichomes are isopolar, with straight, not capitate, ends. Reproduction occurs via hormogonia, after trichome disintegration through the formation of necridic cells.

Lyngbyopsis is a monotypic genus with the single species *Lyngbyopsis willei*. Gardner (1927) described *Lyngbyopsis* from Puerto Rico. It is also known from warm-water creeks in Cuba (Komárek, 1989).

Microcoleus Desmazières ex Gomont (Fig. 15A)

Filaments are solitary or in flat mats, usually prostrate on various substrata, and unbranched or sparsely pseudobranched. Sheaths are homogeneous, broad, occasionally indistinctly and irregularly longitudinally striated, mostly colorless, regularly cylindrical, firm, delimited or sometimes (when old) diffuse, occasionally transversely wrinkled, tapering and usually open at the ends, and containing numerous to many trichomes. The trichomes within sheaths are densely aggregated or ropelike contorted, nearly parallel and in tight fascicles, often extending beyond the sheath ends, without constrictions or slightly constricted at the cross walls, straight, and mostly attenuated at the ends without individual sheaths. Cells are cylindrical, more or less isodiametric, with radially arranged thylakoids and granular contents. Apical cells are usually subconical to acutely conical, less frequently capitate, sometimes with calyptra. Reproduction occurs via trichome disintegration and hormogonia production.

Several common species have been recorded from the littoral of clear lakes or slightly polluted pools; some species are benthic (in sediments or sand); other species are known from wet soils of different kinds, deserts, but also rivers, estuaries, and marine (haloplilic) environments, on wet rocks or mineral springs. More than 30 species were revised, after the genus *Trichocoleus* (Schizotrichaceae) was separated from *Microcoleus* (Anagnostidis and Komárek, 1988; Anagnostidis, 2001). Whitford and Schumacher (1969) mentioned seven species of which only *M. lacustris* grows in shallow freshwater; the others are known from wet soils and drying mud, very often in saline habitats. Smith (1950) listed a dozen freshwater species in the United States, and Tilden (1910) noted seven species from various areas of North America, including Alaska and the West Indies. Several other freshwater and marine species were described from locations in the United States (Setchell and Gardner, 1918, 1924; Drouet, 1943a) and Puerto Rico (Gardner, 1927).

Symplocastrum (Gomont) Kirchner ex Engler *et* Prantl (Fig. 14A)

The thallus is feltlike, in tufts, expanding to form velvet-like layers; it is composed of fascicles of parallel and tightly linked filaments. Trichomes have individual sheaths. Fascicles are mostly erect (rarely creeping), long (up to 3 cm), acutely pointed, and usually in closed fascicles enveloped by a common sheath. Enveloping sheaths are wide, colorless or colored, usually firm, sometimes lamellated, occasionally diffuse, variably branched, containing initially a few, later numerous trichomes. Trichomes are usually constricted at the cross walls. Cells are usually isodiametric or longer than wide. Reproduction occurs via hormogonia and thallus disintegration.

Symplocastrum contains species that resemble the members of the genus *Schizothrix* by their filament and thallus morphology; however, the trichomes are wider and their structure corresponds to the Phormidiaceae. Eleven revised species (usually referred to under the name *Schizothrix*) were recorded from North America, including the terrestrial *S. friesii* and *S. purpurascens*, *S. sauterianum* epiphytic on *Cladophora*, and *S. muelleri* in standing waters (Prescott, 1962; Sheath and Steinman, 1982). Tilden (1910) listed four species in the United States, Canada, and the West Indies. Whitford and

Schumacher (1969) recorded *S. penicillatum* growing on submerged concrete in waters in the coastal plains of North Carolina. Within *Symplocastrum* probably also belong taxa referred to as *Schizothrix acuminata*, *S. giuseppei*, *S. parciramosa*, and *S. telephoroides* from Puerto Rico (Gardner, 1927), and *S. acutissima*, *S. californica*, *S. chalybea*, *S. constricta*, *S. hancockii*, *S. mexicana*, *S. richardsii*, *S. rivularis*, *S. stricklandii*, and *S. taylori*, recorded from various locations in the United States and Mexico (Tilden, 1910).

Oscillatoriales: Phormidiaceae: Ammatoideoideae

Ammatoidea W. West *et* G. S. West (Fig. 18A)

Filaments occur in clusters, in tufts, or in indefinite strata, or solitary among other algae, often growing prostrate with ascending branches and tips, or curved in the middle with ascending ends, more or less irregularly coiled, isopolar, narrowed toward both ends, sometimes falsely branched. Sheaths are firm, thin or thick, usually lamellate (at least in parts), colorless or yellowish to brown, open at both ends. Trichomes are isopolar, rarely (young trichomes) heteropolar, usually distinctly gradually narrowed toward both ends and with elongated cells, usually forming colorless cylindrical hairs, without constrictions or slightly constricted at the cross walls in old parts and in hormogonia. Meristematic zones with dividing cells are present. Heteropolarity and akinetes are absent. Reproduction occurs via fragmentation of the trichomes with necridia, forming motile hormogonia. Hormogonia separate from both ends of the trichomes (rarely central) after separation of terminal hairs.

Filaments of *A. normannii* and *A. yellowstonensis* were observed by Copeland (1936) attached to the surface of gelatinous envelopes of other algae in Yellowstone National Park (see also Smith, 1950; Stein and Borden, 1979, under *Hammatoidea*).

Oscillatoriales: Oscillatoriaceae

Blennothrix Kützing ex Anagnostidis *et* Komárek (Fig. 19)

The thallus is mucilaginous, expanded or fasciculate and lengthened, cylindrical, filamentous or tufted, flaky, rarely hemispherical and cushion-like, up to 2 cm high, occasionally free floating, forming olive-green to black–green, bright to dull blue–green, blackish, rarely red–brown or black–violet masses. Filaments are straight or slightly undulating and entangled, in divaricated (spread apart like branches) or sparsely falsely branched fascicles or tufts, often with a special type of branching ("coleodesmoid," not "plectonematoid"; Watanabe and Komárek, 1989; compare Figs. 18B and 19). Sheaths are always present, thin or thick, mucilaginous, firm or diffuse, sometimes lamellated, colorless, frequently with transverse lamellation and constrictions, open at apex, containing one to several trichomes, tightly aggregated (rarely loose). Trichomes are 8–40 µm wide, usually cylindrical or attenuated and straight up to capitate ends, and usually unconstricted to slightly constricted at the cross walls. Cells are very short and discoid and divide by perpendicular cleavage in rapid succession. Apical cells are present, often with calyptra or thickened outer cell wall. Reproduction occurs via trichome disintegration and formation of hormogonia with necridic cells.

About 20 species have been described (usually under the name *Hydrocoleum*), which belong to two ecological groups: One colonizes clear mountain springs and streams; the other occurs in brackish and saline environments. Of North American species, *B. ganeshii* is common in mountain streams in central Mexico (Gold-Morgan *et al*., 1994; Gold-Morgan and Montejano, personal communication); *B. ravenelii*, *B. heterotricha*, and calcified *B. groesbeckiana* are known from clear standing and flowing waters (Tilden, 1910; Drouet, 1943a; Smith, 1950); *B. coerulea* was collected from mountain streams in Puerto Rico (Gardner, 1927). *B. glutinosa*, *B. majus*, and *B. mirifica* are known from marine habitats. Several other species occur in marine littoral and saline swamps in central America (e.g., *B. cantharidosma* and *B. comoides* in Florida, the Antilles, Bermuda, and Mexico; Geitler, 1932; Humm and Wicks, 1980, under *Hydrocoleus*).

Lyngbya C. Agardh ex Gomont (Fig. 17)

Filaments are straight or slightly undulating (several species are finely screwlike or coiled), rarely (a few free-floating species) solitary, mainly arranged in thin or thick, flat, compact, large, layered, leathery prostrate mats on the substrate, and very rarely false branched; they are usually wider than 6 µm. Sheaths are always present; only hormogonia and trichomes under extreme conditions leave the sheaths. Sheaths are attached to the trichome or slightly distant, firm, thin or thick, colorless or slightly yellow–brown or reddish (very rarely bluish), sometimes slightly lamellated, and containing one motile trichome. Trichomes are cylindrical and may or may not be constricted at the cross walls. Cells are short and discoid, always shorter than wide, and most without aerotopes (a few planktonic species have aerotopes). Apical cells usually have a thickened outer wall or calyptra. Reproduction occurs via trichome disintegration into usually short motile hormogonia, which often separate necridia formation.

Over 60 species have been revised and confirmed, several of which are cosmopolitan. Most form periphytic and benthic mats on different submersed

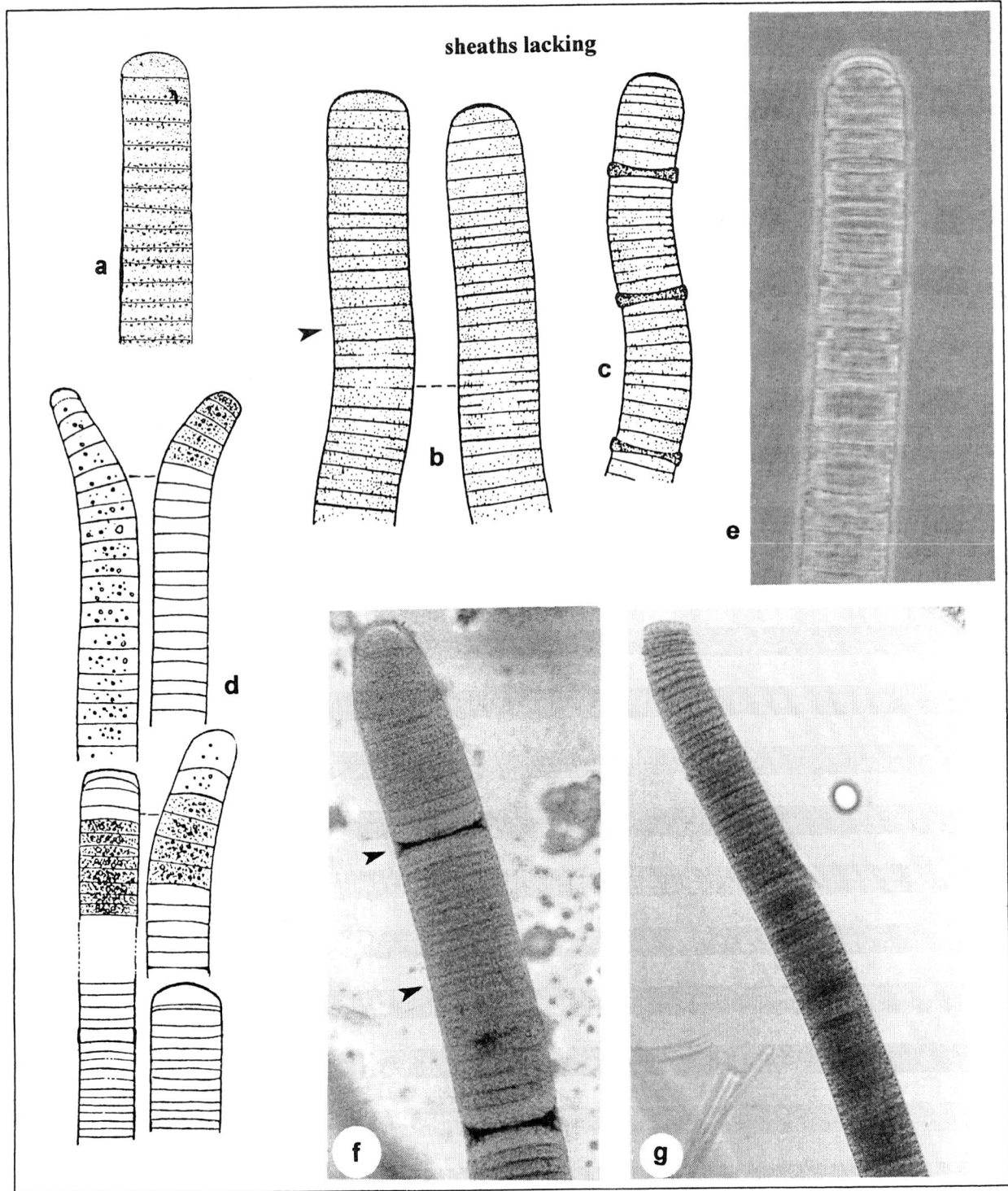

FIGURE 16 *Oscillatoria*: (a) *O. limosa* (after Smith, 1950, from the United States); (b) *O. obtusa* (after Gardner 1927 from Puerto Rico); (c) *O. refringens* (after Gardner, 1927, from Puerto Rico); (d) *O. jenensis* (after Komárek, 1989, from Cuba); (e) *O. sancta* (original photo by Kling, from central Canada); (f) *Oscillatoria* sp. (original photo by Komárková, from the Everglades, Florida); (g) *O. princeps* (original photo by R. G. Sheath, with permission).

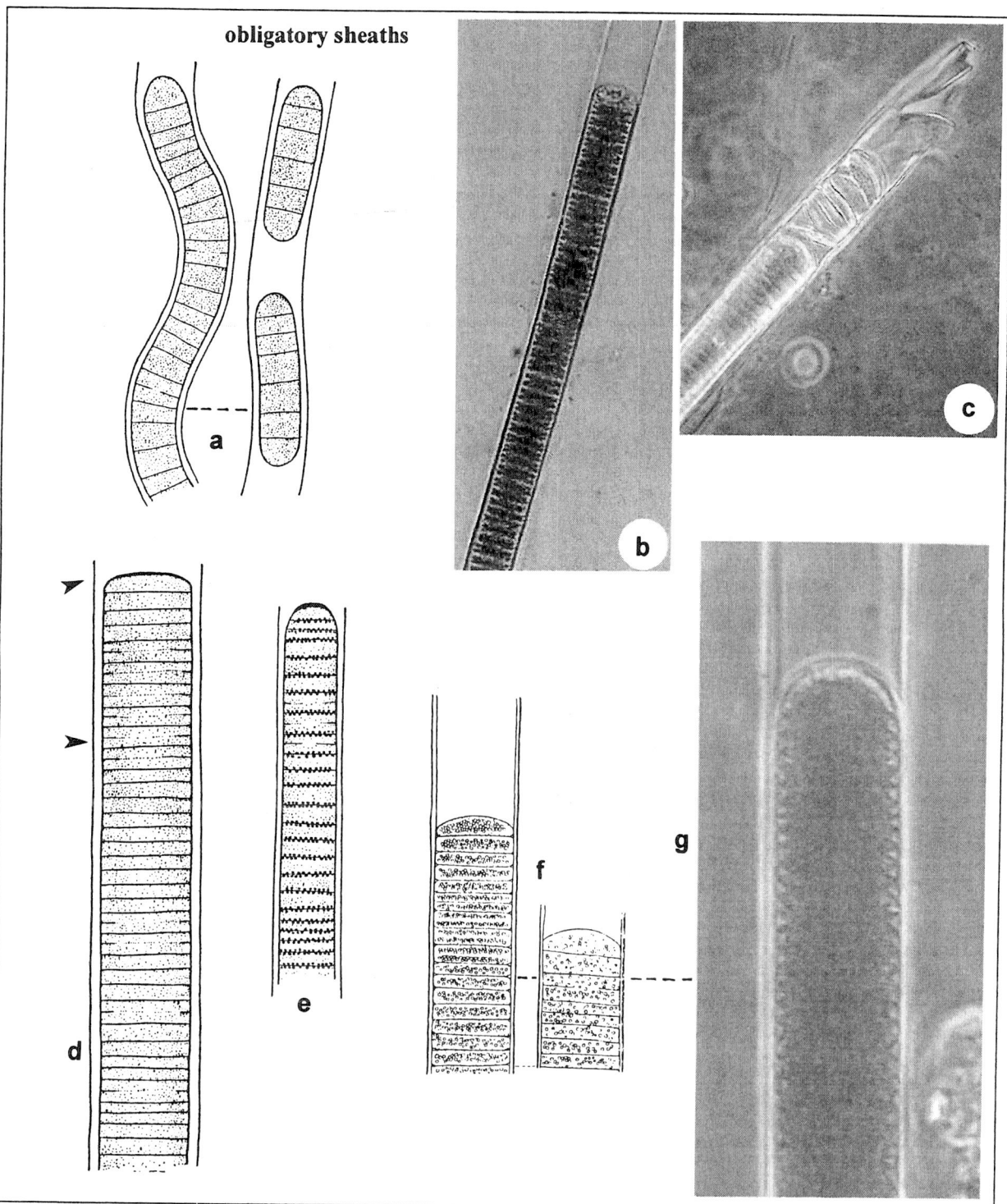

FIGURE 17 *Lyngbya*: (a) *L. splendens* (after Gardner, 1927, from Puerto Rico); (b–c) *Lyngbya* sp. (original photo by Komárková, from the Everglades, Florida); (d) *L. magnifica* (after Gardner, 1927, from Puerto Rico); (e) *L. intermedia* (after Gardner, 1927, from Puerto Rico); (f–g) *L. birgei* (after Smith, 1920; original photo by Kling, from Canadian lakes).

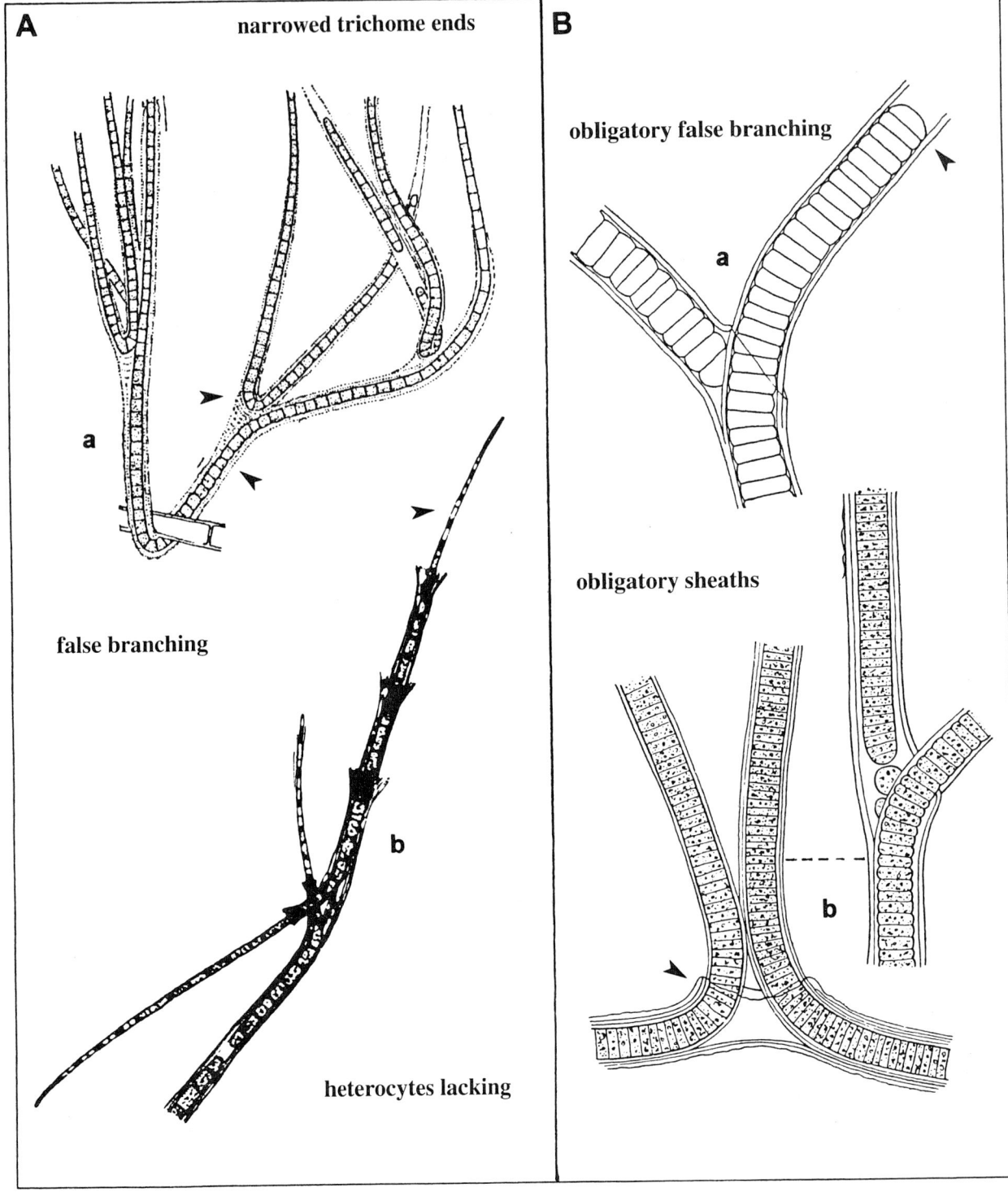

FIGURE 18 (A) *Ammatoidea*: (a) *A. normanii* (after W. West and G. S. West from Smith, 1950); (b) *A. yellowstonensis* (after Copeland, 1936, from Yellowstone National Park). (B) *Plectonema*: (a) *P. tomasinianum* (after Drouet, 1934, from Missouri; sub *P. tomasinianum* var. *gracile*); (b) *P. tomasinianum* (after Gomont from Kondrateva, 1968).

substrata in freshwater and brackish and marine waters. Some species are terrestrial or subaerial occurring on wet rocks; several of them are cosmopolitan. A few species (e.g., *L. birgei*, described from the United States) occur as solitary filaments in metaphyton and plankton. *Lyngbya maior*, *L. martensiana*, and *L. spirulinoides* are widely collected from periphyton in pools and streams (Tilden, 1910; Prescott, 1951; Whitford and Schumacher, 1969; Stein and Borden, 1979); *L. aestuarii*, *L. meneghiniana*, *L. salina*, and *L. confervoides* prefer saline waters (Geitler, 1932; Kosinskaja, 1948). Tilden (1910) listed 35 species, 29 of which are distributed over the North American continent; Smith (1950) recorded 35 species from the United States. However, many of these species now belong to other genera (e.g., *L. contorta* = *Planktolyngbya contorta*). Drouet (1934, 1942) described *L. hahatonkensis* from an artesian spring in the United States, along with periphytic *L. giuseppei* and *L. patrickiana*. The North American species require further research and revision.

Oscillatoria Vaucher ex Gomont (Fig. 16)

The thallus is usually flat, macroscopic, smooth, layered, rarely leathery, arranged in mats, less frequently in solitary trichomes. Trichomes are straight or slightly irregularly undulating, cylindrical, sometimes screwlike coiled at the ends, motile, gliding or oscillating in left-handed or right-handed rotation, usually wider than 6.8 µm (up to 70 µm), not constricted or constricted at the cross walls. Sheaths are usually absent, although they may occur under suboptimal conditions. Cells are short and discoid, always with lengths less than one half to one eleventh that of their widths; cell contents without aerotopes are homogeneous or sometimes contain large prominent granules. Cell division occurs in a rapid sequence transversely to the trichome axis. Reproduction occurs via trichome disintegration (sometimes completely) into short motile hormogonia, with the aid of necridia.

Almost 70 widely distributed, mainly benthic species have been described (revised; traditionally planktonic species with thinner trichomes and different ultrastructure now classified under the genera *Planktothrix*, *Pseudanabaena*, or *Limnothrix*). Oscillatoria occurs in mats on different substrata (mud, plants, stones, sand) in shallow water bodies or marshes and swamps. With greater biomass, parts of the mats dislodge from the substratum and form floating clusters on the water surface in shallow water bodies. Of the numerous *Oscillatoria* species, the most frequently mentioned are the cosmopolitan *O. limosa*, *O. sancta*, *O. princeps*, and *O. proboscidea*; these species grow epipsammic or among other algae in lakes, ponds, and pools and in slowly flowing rivers (Prescott, 1962; Whitford and Schumacher, 1969; Duthie and Socha, 1976; Stein and Borden, 1979; Sheath and Steinman, 1982). *O. ornata*, *O. curviceps*, and *O. anguina* have also been reported (Tilden, 1910; Prescott, 1951; Whitford and Schumacher, 1969). *O. rhamphoidea* was collected from drying pools with water plants in Winnipeg, Canada (authors' results). *O. limosa* and *O. tenuis* were recorded on rock and mud habitats in arctic freshwater lakes and streams (Whelden, 1947). Several species grow in marine coastal waters (*O. funiformis* and *O. margaritifera*). Copeland (1936) described *O. depauperata* from Yellowstone National Park. Smith (1950) mentioned 40 species definitely known from the United States and Tilden (1910) listed 43 species widely distributed widely throughout the North American continent; however, many of these species cited by previous authors have been classified into other genera (Anagnostidis and Komárek, 1988).

Plectonema Thuret ex Gomont (Fig. 18B)

The thallus forms expanded and usually compact tufts up to 2 cm high, attached to various substrata, or as free clusters with coiled filaments. Free-floating masses also occur, composed of variously coiled, usually densely entangled, or rarely almost straight and parallel filaments. Filaments are obligatorily falsely branched (sparsely to frequently), up to 120 µm wide; false branches occur singly or in clusters, sometimes *Scytonema*-like. Sheaths are firm, thin to thick, up to 4 µm wide, initially colorless, later sometimes yellow-brown, homogeneous or distinctly lamellated. Trichomes are 8–25 (up to 72?) µm wide and are nonmotile. Cells are short and discoid; apical cells are rounded, without (rarely with) calyptra. Heterocytes and akinetes are absent. Reproduction occurs mainly by hormogonia, separating from apical portions of the trichome by the help of necridia, rarely by whole trichome fragmentation. Hormogonia separate from both trichome ends (isopolar).

Of the many species described, only five have been confirmed (several have been transferred to other genera), occurring in the periphyton and metaphyton of clear, oligotrophic, and well-oxygenated ponds and pools; they are found less frequently in small streams or springs. Tilden (1910) and Smith (1950) recorded several species from the United States, most of which were attached to various submerged substrata and located on damp rocks or among mosses and liverworts; however, some were later reclassified into other genera. Whitford and Schumacher (1969) mentioned four species, with two freshwater species (*P. wollei* and *P. tomasinianum*). Recently, *P. wollei* from some North American lakes was found to contain toxins (Carmichael *et al.*, 1997).

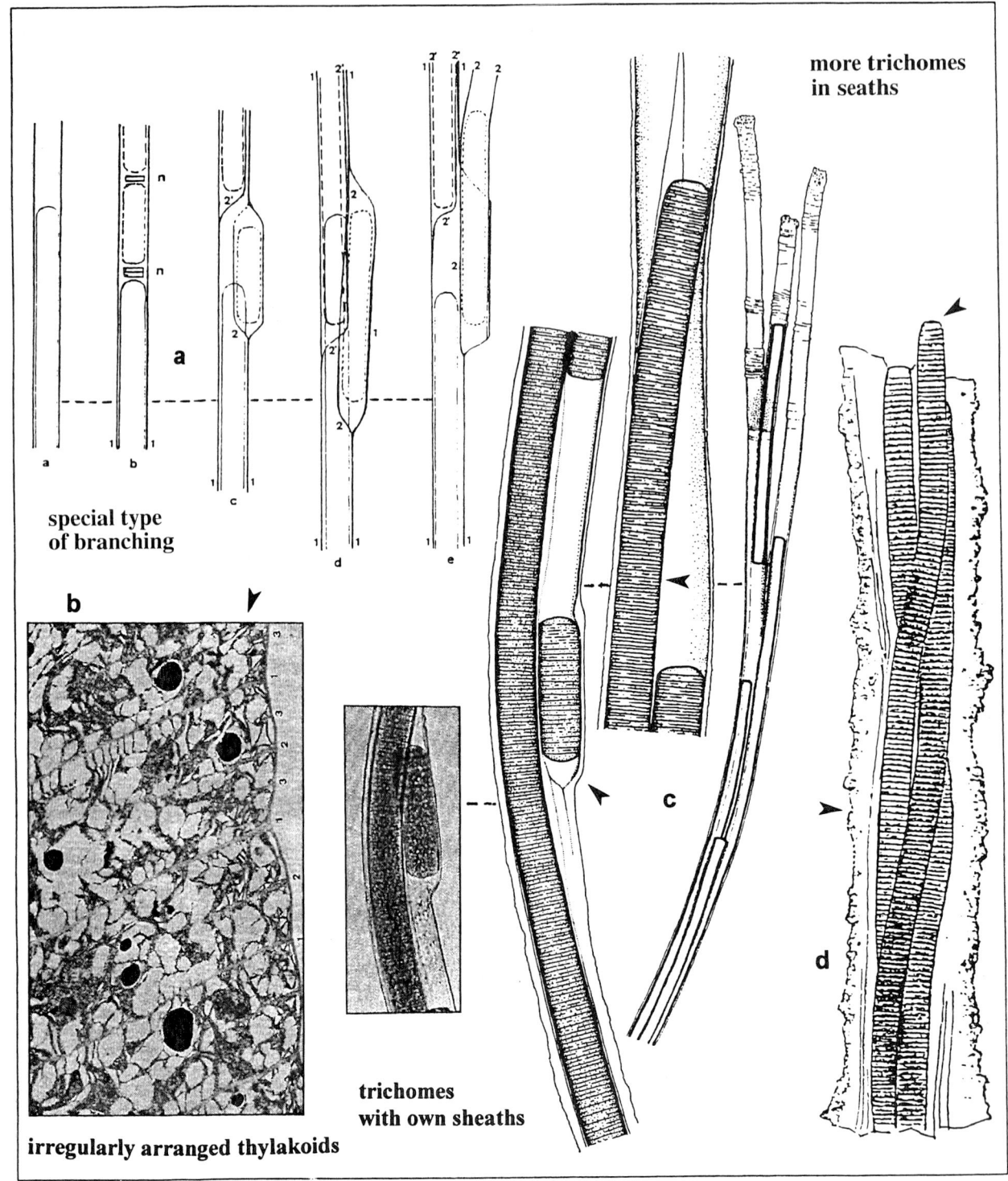

FIGURE 19 *Blennothrix*: (a) false branching; (b) position of thylakoids in cells; (c) *B. ganeshii* (after Watanabe and Komárek, 1989, from Himalayas; occurring also in Mexico mountains); (d) *B. cantharidosma* (after Gomont ex Geitler, 1932).

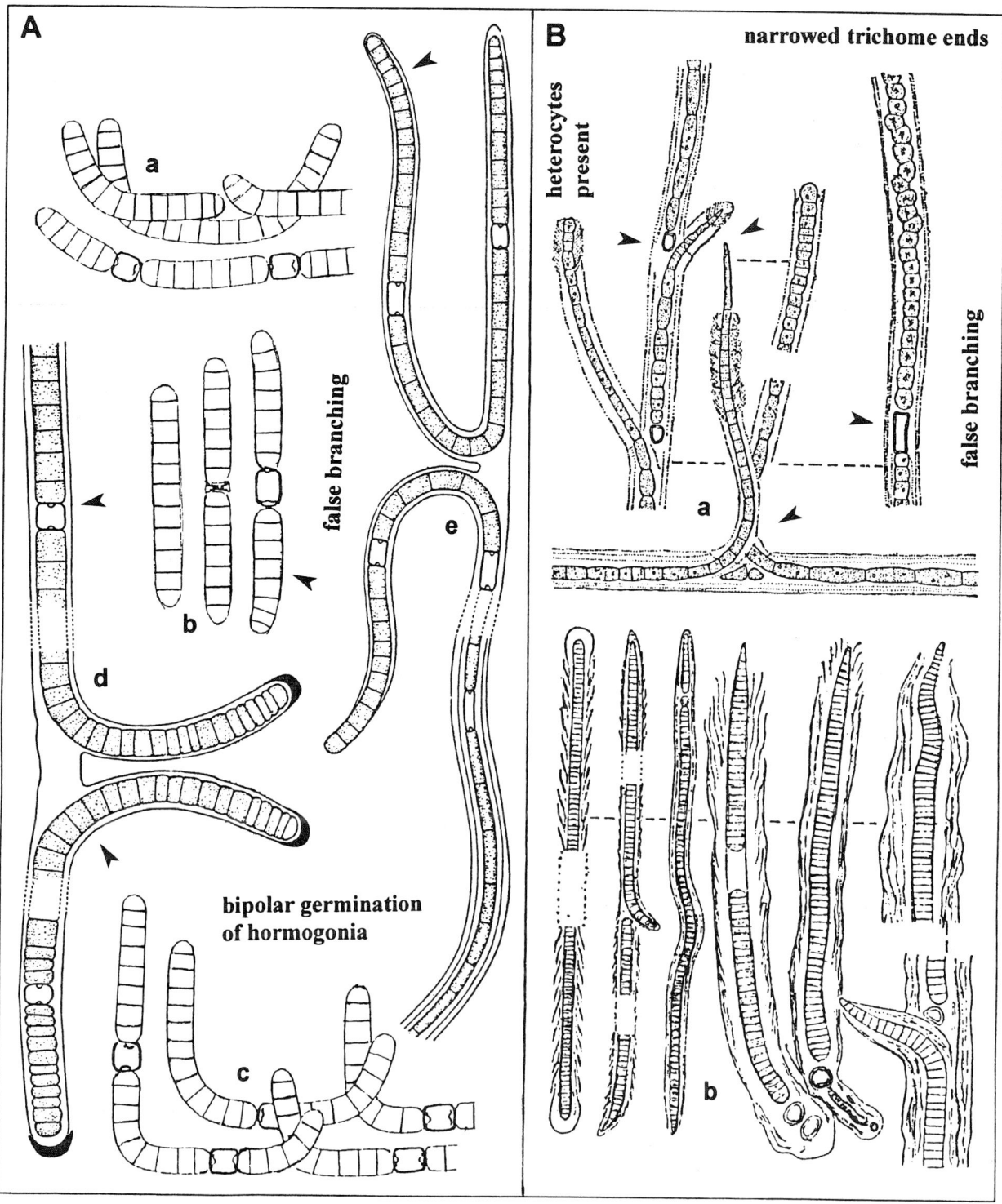

FIGURE 20 (A) *Scytonema*: (a–c) bipolar germination of hormogonia; (d) *S. capitatum* (after Gardner, 1927, from Puerto Rico); (e) *S. longiarticulatum* (after Gardner, 1927, from Puerto Rico). (B) *Scytonematopsis*: (a) *S. hydnoides* (after Copeland, 1936, from Yellowstone National Park); (b) *S. fuliginosa* (after Kosinskaja, 1926).

Nostocales: Scytonemataceae

Scytonema Agardh ex Bornet et Flahault (Fig. 20A)

The thalli form wooly mats or irregular clusters, yellow to brownish in color. Filaments are free floating or in fascicles, curved, sometimes densely coiled, and creeping on substratum. False branching is simple, usually into two branches (rarely in one branch). Branching often occurs after the formation of necridic cells, less frequently at heterocytes following disintegration of the trichome; the free ends of the trichomes squeeze through a rupture in the sheath and form new branches. Trichomes are isopolar (both poles with the same morphology), cylindrical, at their terminal parts cylindrical or sometimes widened, constricted or unconstricted at the cross walls, pale blue–green, olive green, or brownish, rarely violet. Sheaths are firm, lamellated or without visible structure, sometimes wide, often yellow–brown in color (scytonemin; Jaag, 1945; Geitler, 1960). Cells are cylindrical to barrel shaped, sometimes elongated in the central part of the trichome; apical cells are rounded. Heterocytes are solitary and intercalary. Akinetes have occasionally been reported, but not confirmed. Reproduction occurs via isopolar hormogonia released at both ends of the trichomes. Germination of hormogonia at both ends.

Scytonema species are mainly aerial or subaerial on alkaline substrata, wet rocks, wood, and soil, sometimes encrusted with calcium carbonate; some species grow in the periphyton in lakes (mainly calcareous) and at sea coasts. *Scytonema* is a common, variable genus, with over 100 described species, many of which are known only in tropical regions. Tilden (1910) reported 36 species from North America. Whelden (1947) noted *Scytonema crustaceum* embedded in calcareous precipitates inside the hollow nodes of a *Nostoc* colony, *S. myochrous* in a freshwater tundra pool (typically *S. myochrous* grows in travertine habitats), and *S. ocellatum* from the edge of an Arctic lake. Twenty species of *Scytonema* were reported by Smith (1950) primarily from subaerial habitats, but also from damp walls and rocky cliffs; one species has been found in algal mats in marshes in the Florida Everglades. *Diplocolon heppii*, which is considered synonymous with the genus *Scytonema* (Geitler, 1932, 1942), was recorded from Niagara Falls (Smith, 1950). Whitford and Schumacher (1969) reported 19 species in North Carolina, three of which (*S. arcangelii*, *S. crispum*, and *S. dubium*) were periphytic in freshwaters, whereas the rest were subaerial on wet soils and rocks. Several taxa were also reported from several locations across Canada (Duthie and Socha, 1976; Stein and Borden, 1979; Sheath and Steinman, 1982); *S. cincinnatum*, *S. ocellatum*, and *S. tolypothrichoides* were reported from the central United States (Tiffany and Britton, 1952).

Scytonematopsis E. Kiseleva (Fig. 20B)

The thalli are brownish, low, prostrate, and closely attached to the substrata or in clusters. Filaments are falsely branched, heteropolar when young, later usually isopolar, forming clusters with parallel or a bristle-like arrangement; old trichomes are isopolar. Branching (usually two) starts at necridic cells. Trichomes are usually constricted at the cell walls. Sheaths are firm, sometimes parallel lamellated and telescopic, hyaline or yellowish-brown in old filaments. Cells vary from short to long cylindrical or barrel shaped; several terminal cells are distinctly elongated, often forming a thick tapering hair, with the terminal cell bluntly or sharply pointed. Heterocytes are mostly intercalary and solitary. Akinetes have been described, but are rarely found. Reproduction occurs via hormogonia separated from filaments, which germinate from both ends.

This genus contains both periphytic and metaphytic species occurring in temperate mountain streams; one species (*S. hydnoides*) was described from thermal springs in Yellowstone National Park (Copeland, 1936); the type species *S. fuliginosa* was described from the Hawaiian islands (Tilden, 1910; Copeland, 1936).

Nostocales: Microchaetaceae: Tolypotrichoideae

Coleodesmium Borzì ex Geitler (Fig. 23)

Filaments are united into highly branched, mucilaginous, polarized blue–green or brownish fascicles, up to 1 cm high. Sheaths are firm, thin or thick, and often lamellated and opened at the apex, colorless or yellow–brown, containing one to several trichomes (with own sheaths). Trichomes are heteropolar with basal, elliptical heterocytes, parallel orientation, false branching, cylindrical, uniseriate, constricted at the cross walls, and sometimes slightly attenuated or widened toward the ends. False branching is often initiated at intercalary heterocytes. Cells are short, barrel shaped to cylindrical, with widely rounded end cells. Aerotopes are absent. Akinetes are rare. Cell division is perpendicular to the longitudinal axis of the trichome. Reproduction takes place by the production of hormogonia (separated by necridic cells), liberated from the sheaths.

Coleodesmium species are epiphytic on macrophytes or epilithic on stones in clear, unpolluted streams, especially in mountainous and northern areas. *Coleodesmium floccosum* (originally identified as *C. wrangelii*) has been recorded from Connecticut (Geitler, 1932; Komárek and Watanabe, 1990) and British Columbia (Stein and Borden, 1979); typical *C. wrangelii* occurs in subarctic to arctic waters (Elster *et al.*, 1997) and (as *Desmonema wrangelii*; Smith, 1950) in several other localities of North America. Whitford and Schumacher (1969) collected *C. wrangelii* (also as *Desmonema*) on concrete in a swift stream in North Carolina.

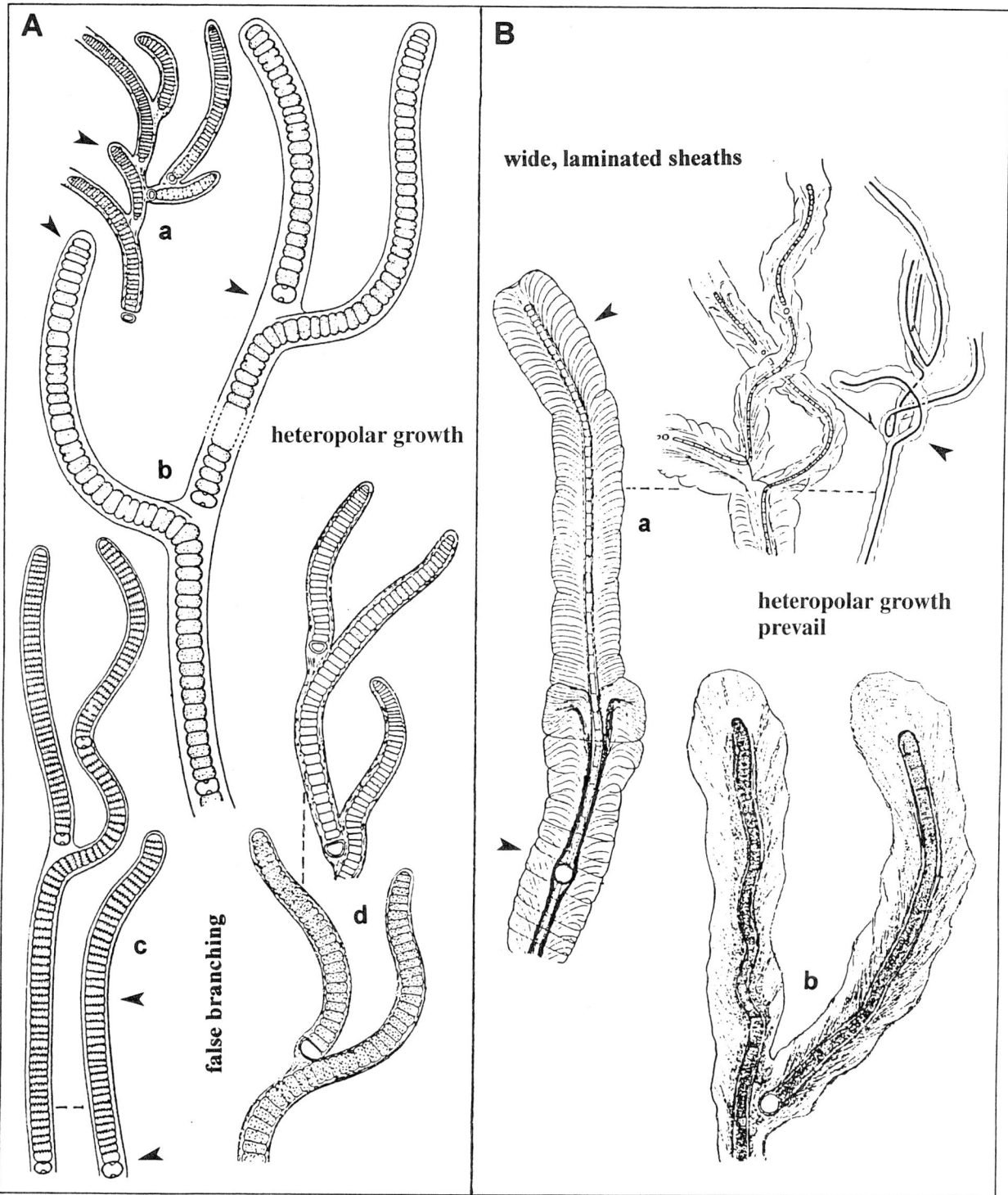

FIGURE 21 (A) *Hassallia*: (a) *Hassallia* false branching; (b) *H. discoidea* (after Gardner, 1927, from Puerto Rico); (c) *H. granulata* (after Gardner, 1927, from Puerto Rico); (d) *H. byssoidea* (after Frémy ex Kondrateva, 1968). (B) *Petalonema*; (a) *P. alatum* (after Komárek ex Fott, 1956; Kondrateva, 1968); (b) *P. involvens* (after Frémy ex Geitler, 1932).

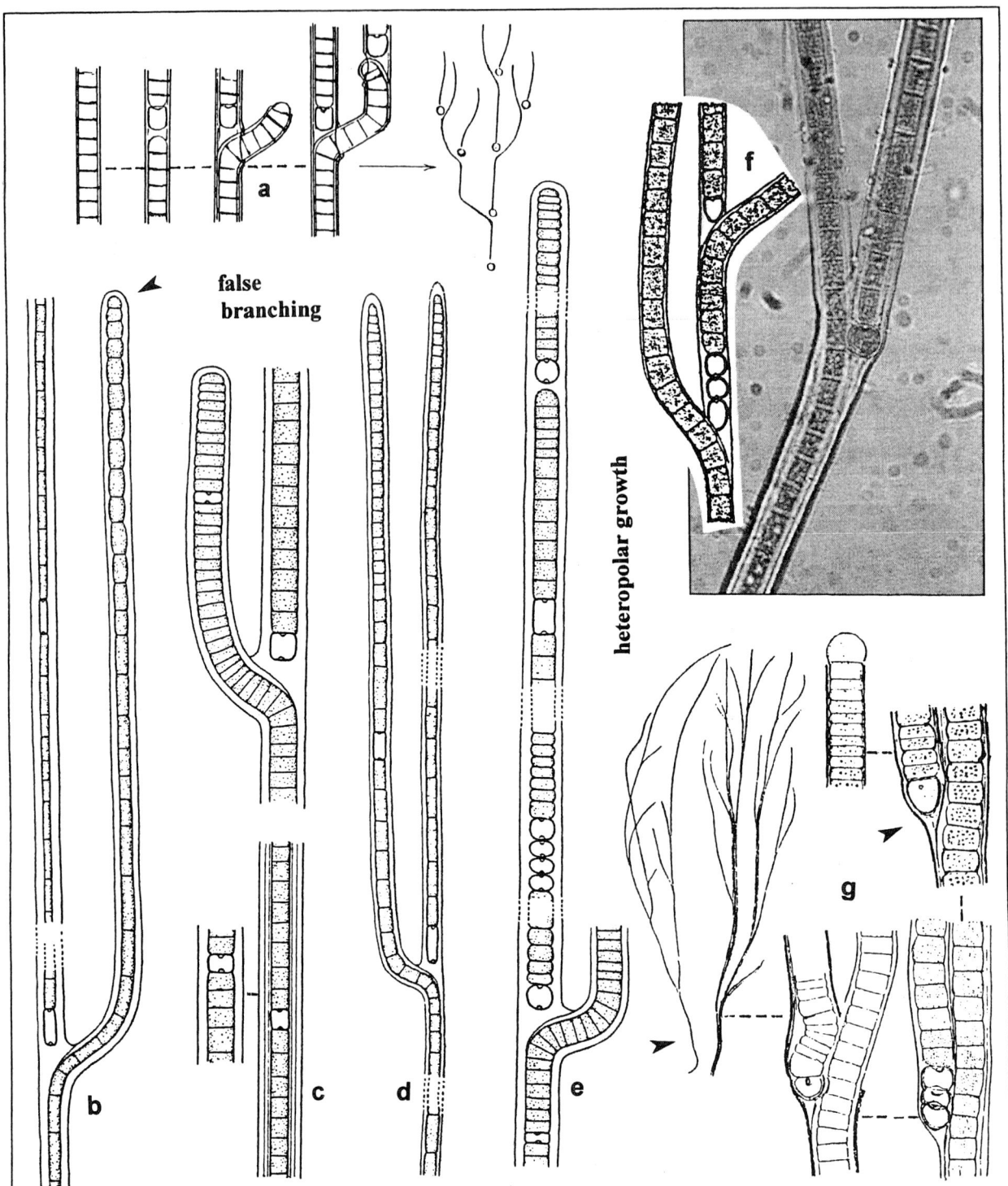

FIGURE 22 Tolypothrix: (a) branching; (b) *T. willei* (after Gardner, 1927, from Puerto Rico); (c) *T. robusta* (after Gardner, 1927, from Puerto Rico); (d) *T. papyracea* (after Gardner, 1927, from Puerto Rico); (e) *T. amoena* (after Gardner, 1927, from Puerto Rico); (f) *T. tenuis* (after Smith, 1950; original microphoto Komárková from the Everglades, Florida); (g) *T. penicillata* (after Geitler, 1932).

Hassallia Berkeley ex Bornet *et* Flahault (Fig. 21A)

The thalli are filamentous, forming low, bristle-like bundles or thin or wooly mats, or they are solitary, creeping along substrata. Rich false branching is due to many simple, solitary, short lateral branches, each starting at a heterocyte. Filaments are heteropolar, cylindrical, and curved. Sheaths are firm, thick and lamellated, sometimes yellowish brown in color. Trichomes are usually constricted at the cross walls, with end cells sometimes slightly attenuated or rounded, with granular content. Cells are barrel shaped to cylindrical when old, mostly shorter than wide to discoid. Heterocytes are mostly hemispherical or discoid, sometimes spherical, mostly single pored, situated at the trichome base or intercalary. Akinetes are unknown. Cells divide in meristematic zones at the ends of the trichomes, perpendicular to the main axis. Reproduction occurs via separation of entire branches and hormogonia.

Species occur in aerial or subaerial habitats on stones or tree bark or submersed. About 10 species are known; several occur in the United States and Canada, but are recorded mainly under *Tolypothrix* (e.g., subaerial *Hassallia byssoidea*; Tilden, 1910; Smith, 1950; Stein and Borden, 1979).

Petalonema Berkeley ex Kirchner (Fig. 21B)

Filaments are solitary or in small groups, heteropolar, often with a basal heterocyte, forming a creeping thallus on substrata, slightly curved or bent; false branching occurs next to heterocytes or after filament disintegration. Trichomes are cylindrical and narrow in comparison with the thick sheaths. Sheaths are very wide, firm, distinct, limited, funnel-like, lamellated, colorless to yellow–brown. Cells are elongated cylinders in the central parts of the trichomes, short and barrel shaped at the ends, with rounded apical cells. Heterocytes are spherical, solitary, situated at the trichome base, intercalary or at the basal point of the branch. Akinetes are absent. Cells divide in meristematic zones. Reproduction occurs via hormogonia from the end of the trichomes.

A common representative, *Petalonema alatum*, is known from subaerial or submersed habitats on calcareous substrata, particularly in temperate zones. About eight species were recorded from North America (under *Scytonema*; e.g., Tilden, 1910; Stein and Borden, 1979), but some require revision.

Tolypothrix Kützing ex Bornet *et* Flahault (Fig. 22)

Thalli form wooly mats, tufts, or caespitose colonies; they are grayish blue–green to yellowish or dark brownish in color. Young filaments are heteropolar with basal heterocytes and free apical ends, often very long. False branching begins next to the heterocyte, diverging singly from the main filament or forming two morphologically identical branches. Trichomes are uniseriate, with one or more heterocytes at the base, constricted or unconstricted at the cell walls, sometimes slightly widened toward the apex. Cells range from long cylindrical to short barrel shaped, sometimes with a few granules, blue–green, olive green, yellow to reddish. Cells in the subapical meristematic zones divide transversely to the trichome axis. Heterocytes are spherical, cylindrical, or discoid with one or two pores, situated intercalary, originally singly or in pairs, often at the base of branches. Akinetes are rarely seen. Reproduction occurs via hormogonia that separate from the filament ends or from a disintegrating part of a filament. Hormogonia germinate from one or both ends; growth later becomes heteropolar.

Species occur in unpolluted freshwaters attached to stones, macrophytes, and other algae, or forming mats in mineral springs, streams, and alkaline swamps or subaerial habitats. Many species are known from tropical and subtropical habitats (Gardner, 1927; Geitler, 1932; Desikachary, 1959). Forty species have been described, but there are difficulties with their identification (Hoffmann and Demoulin, 1985). *Tolypothrix* is more commonly encountered in aquatic habitats than the mainly terrestrial and subaerial *Scytonema*. Colonies originally formed in benthic habitats may later separate and float to the surface. Tilden (1910) listed 10 species distributed throughout the Arctic, the United States, and the West Indies. Whelden (1947) reported *T. bouteillei*, *T. distorta*, *T. penicillata*, *T. tenuis*, and *T. tenuis* f. *minor* (noted as a new taxon found on the bottom of a stream) from the Arctic. Smith (1950) reported *T. distorta*, *T. setchellii*, *T. lanata*, and *T. penicillata* from various areas in the United States, as have many other studies (e.g., Tiffany and Britton, 1952; Whitford and Schumacher, 1969; Duthie and Socha, 1976; Stein and Borden, 1979; Sheath and Steinman, 1982). Eleven species are recorded from North Carolina, mostly subaerial or in submersed periphyton (*T. limbata* and *T. rupestris*). *Tolypothrix* is mainly distinguished from *Hassallia* by its diverse thallus structure (Komárek and Anagnostidis, 1989).

Nostocales: Microchaetaceae: Microchaetoideae

Fortiea De Toni (Fig. 24A)

The thallus is filamentous and epiphytic or periphytic on macrophytes or stones. Filaments are solitary or in groups, heteropolar, characterized by basal heterocytes and free, widened apical ends, cylindrical, simple, solitary or in small groups. Filaments always contain only one trichome. Sheaths are firm, colorless, occasionally thick and lamellated length- and crosswise. Trichomes are cylindrical, constricted or unconstricted at the cross walls, widened toward ends, with the last

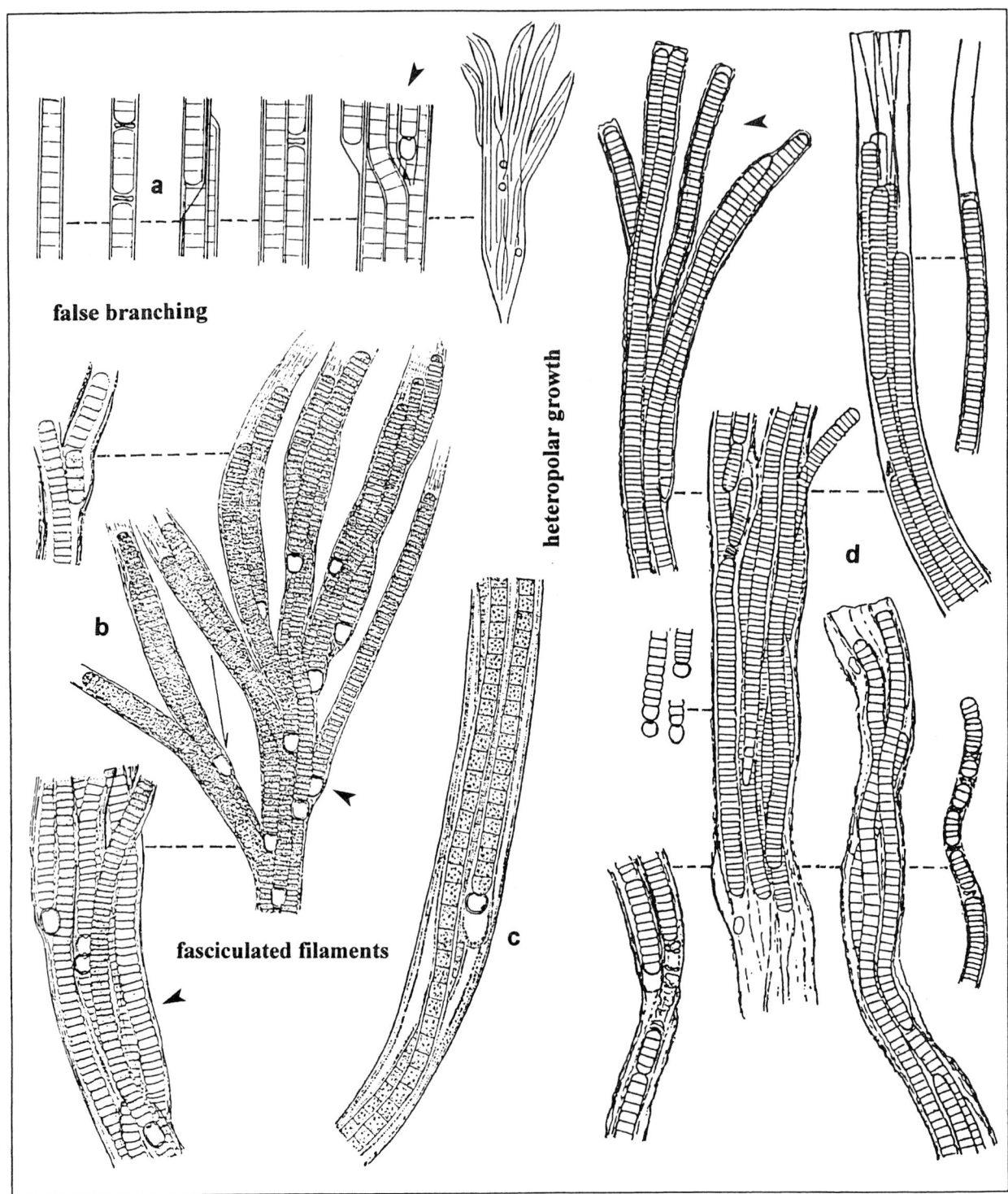

FIGURE 23 Coleodesmium: (a) false branching; (b) *C. wrangelii* (after Frémy, 1927; Bourrelly, 1970); (c) *Coleodesmium* sp. (after Smith, 1950, from the United States; sub *Desmonema wrangelii*); (d) *C. floccosum* (after Komárek and Watanabe, 1990, from Connecticut).

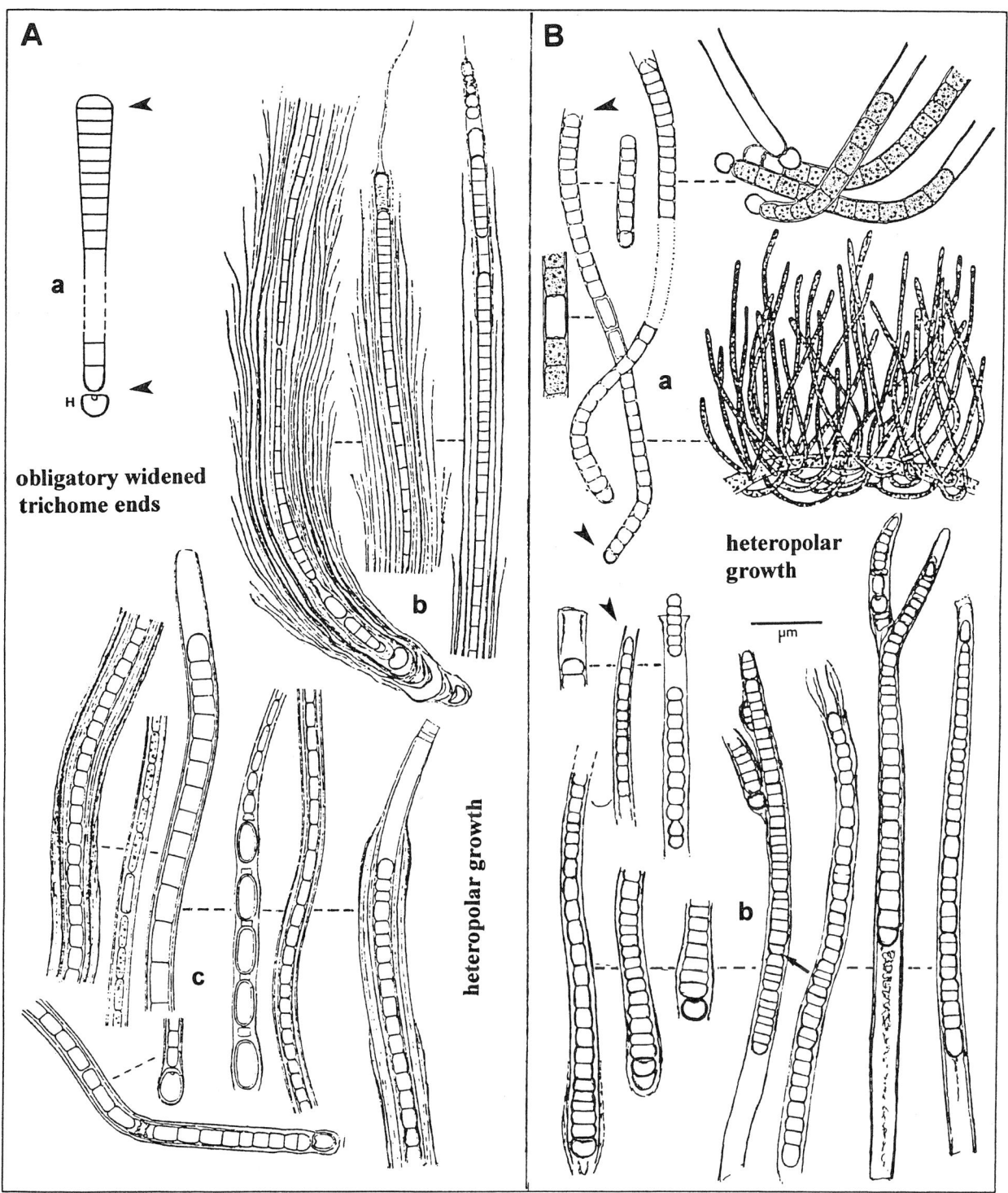

FIGURE 24 (A) *Fortiea*: (a) trichome; (b) *F. salinicola* (after Komárek, 1984, from Cuba); (c) *F. monilispora* (after Komárek, 1984, from Cuba). (B) *Microchaete*: (a) *M. tenera* (after Smith, 1950; Kondrateva, 1968); (b) *M. robinsonii* (bar = 20 µm; after Komárek, 1994, from southern Manitoba).

cell of the trichome widely rounded or spherical. Cells are cylindrical or barrel shaped and more elongated in the center of the trichome. Heterocytes are spherical, hemispherical, or cylindrical, obligatory at the base of the trichome and facultatively also intercalary, with one or two pores. Trichomes disintegrate near the heterocytes. Akinetes, if any, develop basally in a row. Reproduction occurs via hormogonia, which separate from the trichome with necridic cells, and by akinetes.

Most species are epiphytic on macrophytes or other algae in tropical alkaline pools and swamps and in moors in temperate zones; some are epilithic in streams and in waterfalls (in limestone regions). Several species (*F. bossei*, *F. monilispora*, and *F. salinicola*) occur in Central America (Komárek, 1984). Species within *Fortiea* resemble those in *Microchaete*, but they have widened apical ends. Data from North America need revision.

Microchaete Thuret ex Bornet *et* Flahault (Fig. 24B)

Thalli are filamentous, with filaments solitary or in small groups, attached to substrata or forming a thin mat on stones, heteropolar, cylindrical throughout the filament length. Trichomes, with a basal heterocyte, are constricted or unconstricted at the cross walls, cylindrical or slightly tapering toward ends, with meristematic zones at the ends. False branching is infrequent at intercalary heterocytes. Sheaths are distinct, firm, colorless, sometimes lamellated, and form open tubes. Cells are cylindrical or slightly barrel shaped, varying in length/width ratio; end cells are hemispherical or rounded. Heterocytes are ovoid, spherical, or cylindrical, in basal or intercalary position, with one or two pores. Akinetes develop facultatively. Reproduction occurs via hormogonia separating from the apical ends; germination of hormogonia is heteropolar and isopolar.

Most of about 20 species are epiphytic on algae, mosses, and plants in swamps and ponds, several are epilithic in streams, and a few occur only in tropical or marine environments (Geitler, 1932; Desikachary, 1959). Tilden (1910) listed four species from Canada, the United States, and the West Indies. Smith (1950) reported three species of *Fremyella* (synonymous with *Microchaete*; *F. diplosiphon*, *F. robusta*, and *F. tenera*) in standing waters in the United States; others also reported from various regions across North America (e.g., Whitford and Schumacher, 1969; Stein and Borden, 1979). Komárek (1994) described *Microchaete robinsonii* from a river in central Canada.

Nostocales: Rivulariaceae

Calothrix Agardh ex Bornet *et* Flahault (Fig. 25A)

The thallus is filamentous, attached to substratum basally forming bristle-like groups or thin mats. Filaments are heteropolar, with a wider basal part (with heterocytes and occasionally an associated akinete and/or with enlarged basal vegetative cells) and an apical portion forming an elongated, tapering, hairlike form. Heterocytes develop basally; false branching occurs occasionally with the formation of a separated trichome inside of its own sheath. Trichomes are constricted or unconstricted at the cross walls and always taper terminally. Sheaths are always present, firm, in some species lamellated or enlarged at the end, forming funnel-formed collars yellow to brownish in color (e.g., in *C. fusca*) or colorless. Depending on their position in the trichome, cells may be barrel shaped, cylindrical, or narrowly elongated toward the ends (= hairs), especially with nutrient limitation (Whitton, 1987). Aerotopes are absent from vegetative cells but may be present in hormogonia. Meristematic zones are known in several species. Heterocytes are ellipsoidal, spherical to hemispherical, mainly basal, sometimes intercalary (especially near false branches). Akinetes are ellipsoidal to cylindrical, appearing above basal heterocytes and developing from a vegetative portion of the trichome. Reproduction occurs via hormogonia (sometimes with aerotopes), which are released from the end of the filament after the end hair has separated.

Calothrix consists primarily of aquatic epilithic and epiphytic species (approximately 60); some species occur in marine littoral zones. *Calothrix* is taxonomically difficult genus. Tilden (1910) listed 39 species, 36 of which were found in the United States, Greenland, and the West Indies. Three species (*C. kawraiskii*, *C. stagnalis*, and *C. parietina*) were reported from freshwater habitats in Illinois (Tiffany and Britton, 1952). Smith (1950) referred to more than a dozen species growing attached to rocks or woods in flowing and standing waters in the United States. Whitford and Schumacher (1969) recorded 11 species from North Carolina. *Calothrix fusca* and *C. parietina*-complex are apparently widespread, whereas *C. braunii*, *C. donnellii*, *C. ascendens*, *C. elenkinii*, *C. epiphytica*, *C. juliana*, *C. scytonemicola*, *C. rivularis*, and *C. stellaris* were recorded less frequently. Whelden (1947) reported several species from the Canadian Arctic: *C. borealis* (from shallow streams), *C. braunii* (on rotting *Carex* leaves), *C. contarenii* (on rocks in a tidal pool), *C. parietina* (on stream boulders), and *C. pulvinata* (marine near the high-tide mark). Many other locations are known in Canada (Duthie and Socha, 1976; Stein and Borden, 1979; Sheath and Steinman, 1982).

Dichothrix Zanardini ex Bornet *et* Flahault (Fig. 25B)

The thallus is composed of filaments forming bristle-like groups attached to various substrata.

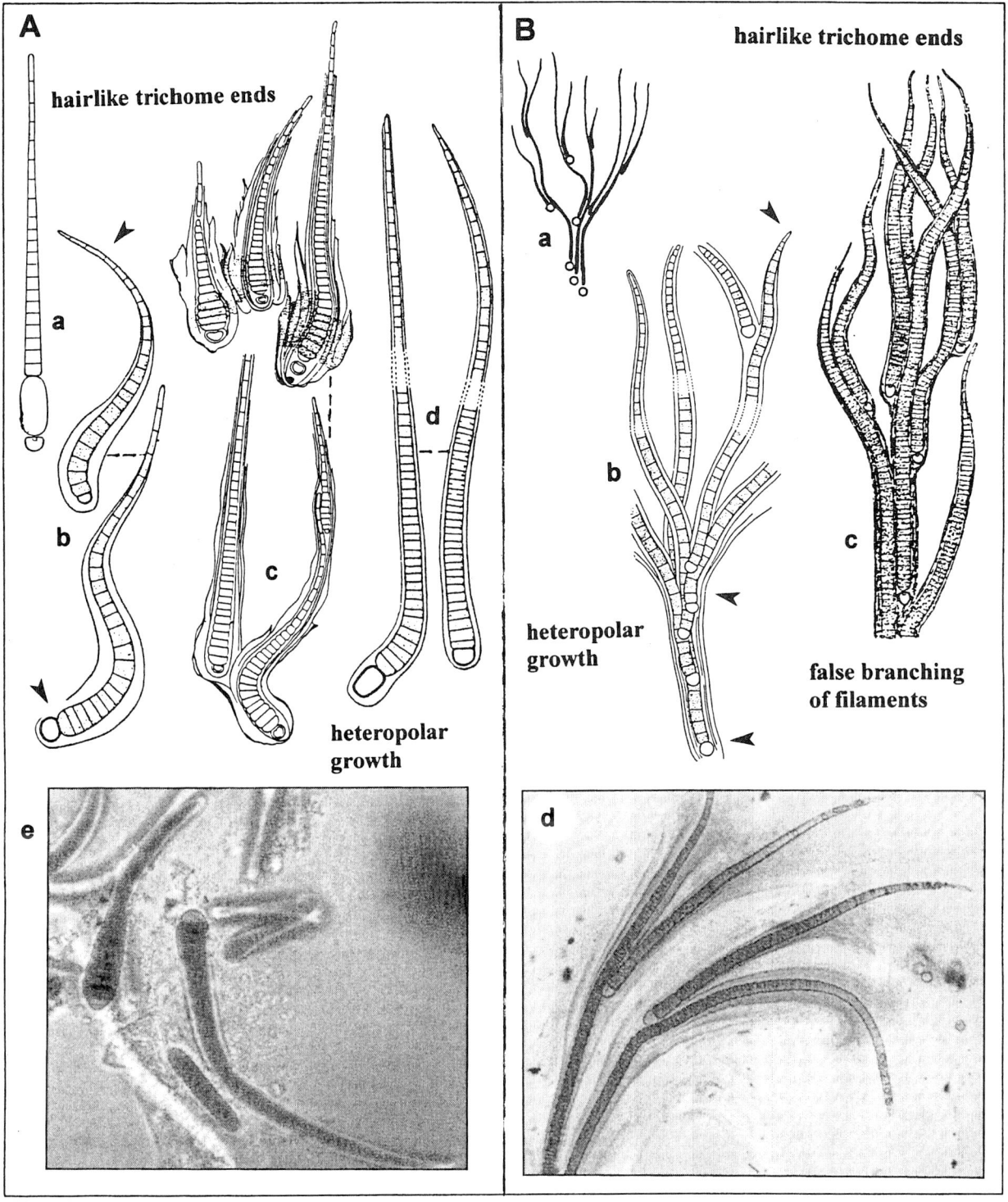

FIGURE 25 (A) *Calothrix*: (a) trichome; (b) *C. tenella* (after Gardner, 1927, from Puerto Rico); (c) *C. fusca* (after Starmach from Kondrateva, 1968); (d) *C. simplex* (after Gardner, 1927, from Puerto Rico); (e) *Calothrix* sp. (original photo by J. D. Wehr from the United States). (B) *Dichothrix*: (a) false-branched thallus; (b) *D. willei* (after Gardner, 1927, from Puerto Rico); (c) *D. orsiniana* (after Frémy, 1930); (d) *Dichothrix* sp. (original microphoto by J. D. Wehr, with permission).

Filaments are heteropolar with differentiated basal and apical portions and with strong lateral false branching beside the heterocytes. Branches form individual sheaths and diverge from the basal filament in parallel (at least in part) with the old sheath. Trichomes with basal heterocytes are constricted or unconstricted at the cross walls and taper toward hairlike ends. Sheaths are always present, firm, colorless and unlamellated or lamellated and yellow to brownish. Cells are short barrel shaped or short cylindrical; the apical region is composed of long, narrow, hyaline cells. Basal heterocytes are almost spherical to hemispherical, with single pores, occasional intercalary heterocytes with two pores; akinetes are unknown. Meristematic zones occur in several species. Reproduction occurs via hormogonia that separate from the filament by the help of necridic cells, following the separation of a terminal hair.

About 30 species have been described from lake periphyton on stones or macrophytes and in rapidly flowing streams (Geitler, 1932; Golubić, 1967a). Several species are known from the marine littoral zone. Tilden (1910) listed 13 freshwater species widely distributed throughout North America: *Dichothrix calcarea*, *D. hosfordii*, and *D. inyoensis* were described from the United States; other species (*D. baueriana*, *D. compacta*, *D. gypsophila*, and *D. orsiniana*) were found several times. Some of these species likely have a cosmopolitan distribution (Smith, 1950), but several are reported in various taxonomic concepts. Whelden (1947) noted six species of *Dichothrix* in a survey of marine and freshwater habitats in the Canadian Arctic: *D. baueriana* and *D. compacta* occurred in fast-flowing streams, *D. gypsophila* in lime-encrusted masses on a rock in a stream and in a freshwater pool, *D. hosfordii* in freshwater, *D. orsiniana* in marshy soil, and *D. rupicola* on rocks in a tidal pool. Villeneuve *et al.* (2001) reported an interesting species from benthic cyanobacterial mats in an Arctic lake (Ward Hunt Lake, Ellesmere Island); other sites are also known (Stein and Borden, 1979; Sheath and Steinman, 1982). Whitford and Schumacher (1969) reported five periphytic or epiphytic species (*D. compacta*, *D. gypsophila*, *D. meneghiniana*, *D. orsiniana*, and *D. spiralis*) and one subaerial species (*D. baueriana*) from North Carolina.

Gloeotrichia Agardh ex Bornet *et* Flahault (Fig. 26A)

Thalli typically occur as mucilaginous, ball-like or hemispherical colonies, enveloped by more or less firm, distinctly limited mucilage. Filaments are heteropolar with basal heterocytes and apical hairs, radially arranged within the colony inside a mucilaginous envelope. Colonies range from microscopic to several centimeters in diameter; they are free floating or attached to substrata, mainly aquatic plants. Sheaths are always present and distinct, but in some species they are gelatinized and confluent. Trichomes are constricted or unconstricted, tapering toward their ends to a narrow hair, straight, curved, or bent. False branching is rare. Branches grow parallel to the original trichome, reaching its full length, and are radially oriented in rows in colonies with the basal part of the filaments in the center. Cells are barrel shaped or cylindrical to hairlike, blue–green or reddish to dark brownish in color, with or without aerotopes. Heterocytes are spherical to hemispherical or ellipsoid, positioned basally (center of colony) or sometimes intercalary. Akinetes are cylindrical, elongated with rounded ends, adjacent to basal heterocytes, but inside the sheath. Cell division occurs mainly in meristematic zones in the vegetative portion of the trichome. Reproduction occurs via disintegration of trichomes, with hormogonia, which are released after the separation of apical hairs.

Of the 16 species described, two are planktonic with aerotopes (clusters of gas vesicles) in the cells, several are epiphytic on submerged water plants in pools, ponds, and littoral zones of lakes; they occur less frequently on stones, especially in mesotrophic and eutrophic waters (Kondrateva, 1968; Barbiero and Welch, 1992). All are exclusively freshwater species except planktonic *G. echinulata*, which can also occur in slightly brackish waters (eastern Baltic; Pankow, 1976). *Gloeotrichia echinulata* is a common species in many (mainly eutrophic) lakes and ponds in North America (e.g., Smith, 1950; Whitford and Schumacher, 1969; Duthie and Socha, 1976; Stein and Borden, 1979). Tiffany and Britton (1952) reported only two epiphytic species (*G. pisum* and *G. natans*). Smith (1950) mentioned several species, including *G. echinulata* (= *Rivularia planctonica*), *G. natans*, *G. pilgeri* (the latter from alkaline subtropical and tropical swamps), and *G. pisum* from localities in the United States.

Rivularia (Roth) Agardh ex Bornet *et* Flahault (Fig. 27)

The thallus is composed of filaments arranged in parallel, attached basally to substrata, forming small or large hemispherical colonies (sometimes hollow) when young, later becoming macroscopic (up to several centimeters) layered strata (hemispherical, flat, or irregular) and several millimeters thick; it is gelatinous or encrusted with $CaCO_3$. Filaments are heteropolar, with basally widened ends and apical hairlike sections. Sheaths are firm, mucilaginous, remaining inside the parental sheath after branching, often not reaching the end of the filaments. Trichomes are usually less cylindrical, constricted or unconstricted at the cross walls, with basal heterocytes, separating and branching at intercalary heterocytes, and hairlike narrowed toward the ends. Cells are barrel shaped or cylindrical,

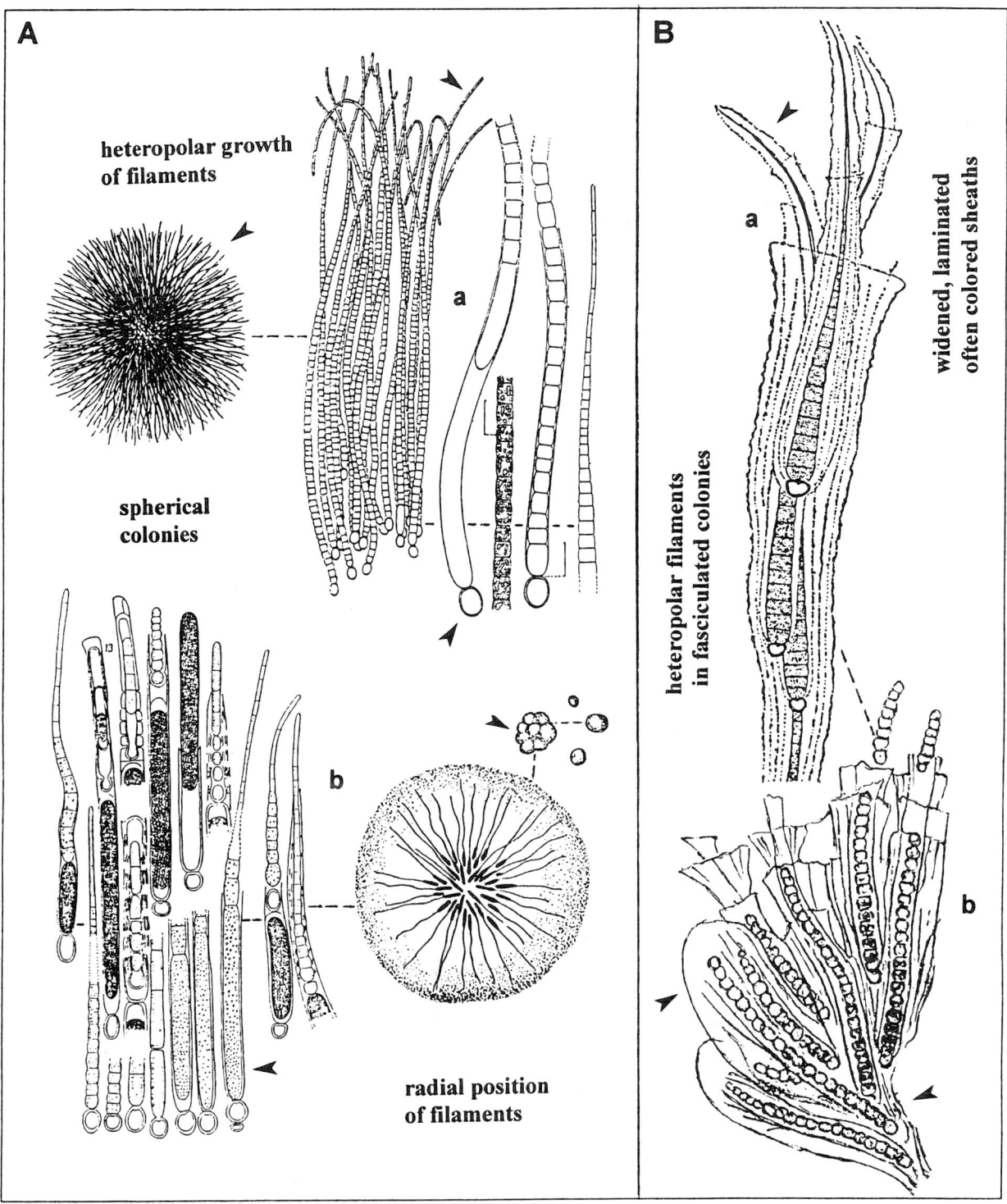

FIGURE 26 (A) *Gloeotrichia*: (a) planktonic *G. echinulata* (after Komárek, 1958); (b) *G. pisum* (after Kondrateva, 1968). (B) *Sacconema*: (a) *S. rupestre* (after Smith, 1950, from the United States); (b) *S. rupestre* (after Borzì, 1882).

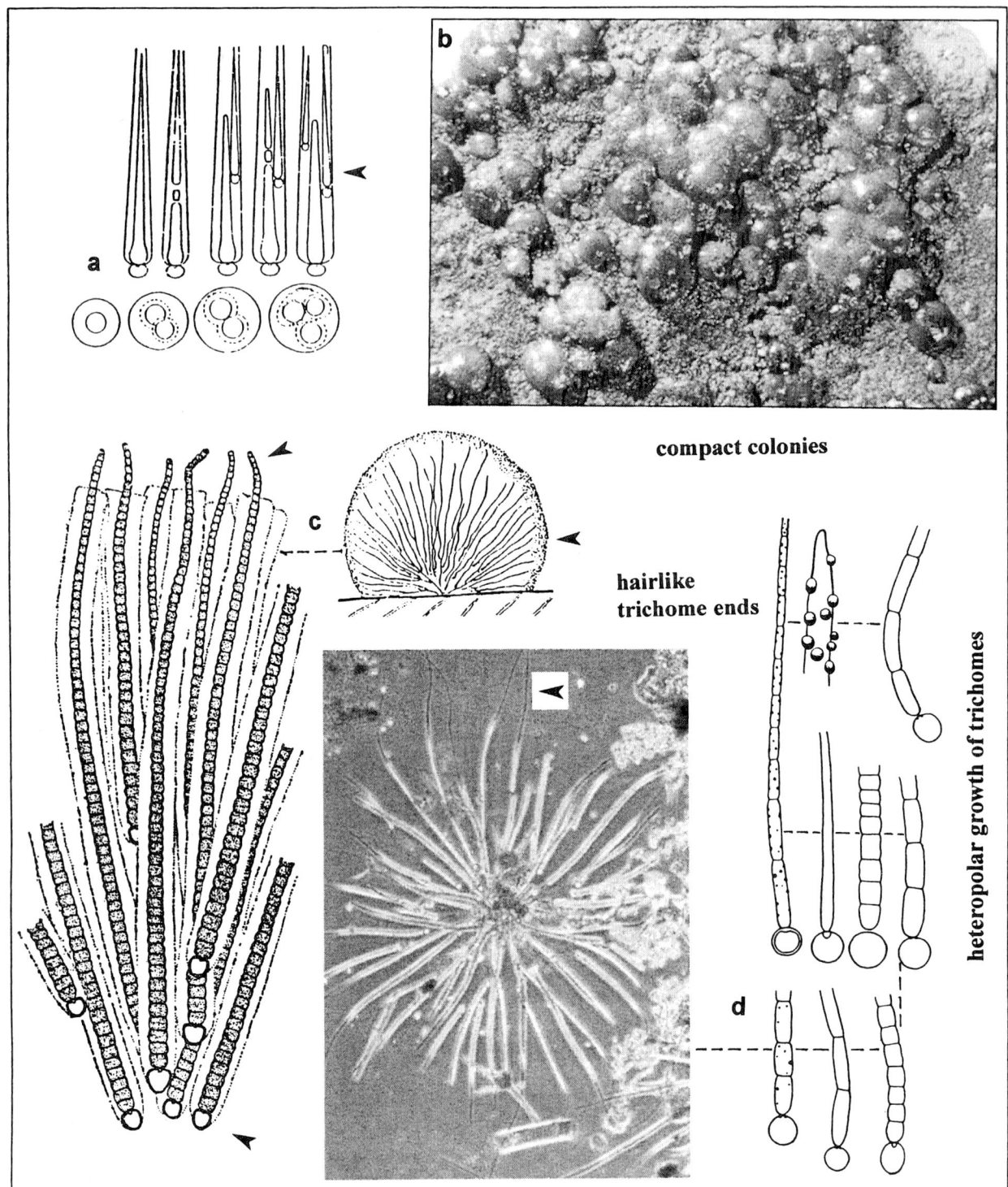

FIGURE 27 *Rivularia*: (a) trichome disintegration and false branching within colony; (b) *Rivularia* sp., colonies on stony substrata (original photo by J. D. Wehr, with permission); (c) *R. dura* (after Smith, 1950; Kondrateva, 1968); (d) *R. aquatica* (after Kondrateva, 1968, original microphoto by J. D. Wehr, with permission; probably young colony of *R. aquatica*).

without aerotopes, elongated and narrow at the apex. Heterocytes range from spherical to hemispherical, positioned basally or occasionally intercalary; akinetes are unknown. Meristematic zones are known in several species. Reproduction occurs via hormogonia, which often remain in the upper parts of the colonies, creating upper layers in the strata.

Of the 20 species described, most are epilithic or epiphytic, especially on calcareous substrata, occasionally forming travertine. Many species colonize in clear unpolluted standing and running waters, several in marine rocky littoral; some are $CaCO_3$ encrusted. *Rivularia* is widely reported genus in many floras and ecological studies. Smith (1950) reported several species (*R. biasolettiana, R. compacta, R. dura, R. globiceps, R. haematites,* and *R. minutula*) from the United States. Whelden (1947) reported four epiphytic and periphytic species (*R. biasolettiana, R. compacta, R. dura,* and *R. minutula* with a possible new variety) from High Arctic freshwater habitats, whereas *R. biasolettiana* and *R. globiceps* were also recorded on submersed freshwater substrata in North Carolina (Whitford and Schumacher, 1969). Tilden (1910) reported 19 species (five uncertain) distributed throughout the United States and in the West Indies; several species were reported from Ontario, British Columbia, and the Northwest Territories (Duthie and Socha, 1976; Stein and Borden, 1979; Sheath and Steinman, 1982). Several species are known from marine habitats. The calcified *R. dura* is mainly distributed in tropical and subtropical limestone regions, but Weldon (1947) also reported this species from the eastern Arctic (revision is needed).

Sacconema Borzì (Fig. 26B)

The thallus is composed of radially arranged or creeping filaments. Filaments are heteropolar, irregularly falsely branched, attached at the base to substrata or creeping among detritus and other algae, forming loose shrublike colonies. Trichomes are solitary or in pairs within a single sheath, uniseriate, heteropolar, constricted at the cross walls, with basal heterocytes. In the apical portion, cells are attenuated into elongated hair cells, which are cylindrical and hyaline. Sheaths are very thick, swollen, and widened at the apex (up to 10× broader than the trichome width). The sheath is initially closed, but later opened at the apex, intensely lamellated, funnel-like, and yellowish brown to brown in color. Heterocytes are basal, mainly spherical, but not well known. Cell division proceeds perpendicularly to the longitudinal axis of the trichome; meristematic zones are not well known. Reproduction most likely occurs via hormogonia.

Sacconema is most commonly subaerial on rock wall seepages and in rocky littoral communities. Tilden (1910) reported one species from Massachusetts; *S. rupestris* was reported from Kootenay Lake, British Columbia (Stein and Borden, 1979). The genus is rarely reported in the literature but has recently been observed in rock scrapings from the splash zones in Lake of the Woods, Ontario (Komárek, Kling, and Komárková, unpublished observations). *Sacconema* is likely present in more locations in North America; reports will probably increase with increasing studies of the algal flora of the littoral zones of northern lakes.

Nostocales: Nostocaceae

Anabaena Bory ex Bornet *et* Flahault (Fig. 28)

The thallus, or with solitary filaments, or is arranged in free clusters, in macroscopic mats, or in the tissues of aquatic plants. Trichomes are straight, curved, or regularly coiled, rarely in bundles with parallel-oriented filaments. Sheaths are never firm, in form of mucilaginous, hyaline and colorless, diffuse envelopes, occurring only in some species. Trichomes are constricted or unconstricted at the cross walls, uniseriate, isopolar (both poles with the same morphology). Cells are spherical, ellipsoidal, short or long cylindrical, sometimes bent (reniform), pale to bright blue–green or yellow–green; the planktonic species have aerotopes (subgenus *Dolichospermum*), occasionally with granular contents. Heterocytes are intercalary, solitary, at fairly regular intervals along filament (metameric), up to nine (rarely more) per filament. Akinetes are spherical, ellipsoidal, cylindrical, curved, intercalary, solitary or in groups of two to five, sometimes with a yellowish-colored epispore, in some species adjacent to heterocytes. Cells divide perpendicularly to trichome axis; meristematic zones are not noted. Reproduction occurs via trichome fragmentation and akinete production.

Anabaena is a common genus worldwide; about 110 species have been described, a large part of them planktonic (subgenus *Dolichospermum*). Some form dense surface blooms and some are recorded as toxic (Carmichael, 1992; Skulberg *et al.*, 1993; Codd, 1995; Chorus and Bartram, 1999). Species without aerotopes (clusters of gas vesicles) form mats in the littoral zone or cover sediments or aquatic plants in freshwater pools and ponds and in saline lakes. North American records include both benthic and planktonic species. Duthie and Socha (1976), Stein and Borden (1979), and Sheath and Steinman (1982) recorded about 18 or 19 taxa in the Canadian flora from Ontario, British Columbia, and the Northwest Territories, respectively. Tiffany and Britton (1952) reported nine species of *Anabaena* from Illinois. Smith (1950) mentioned 36 species in the United States, some of which produce water blooms. Whitford and Schumacher (1969)

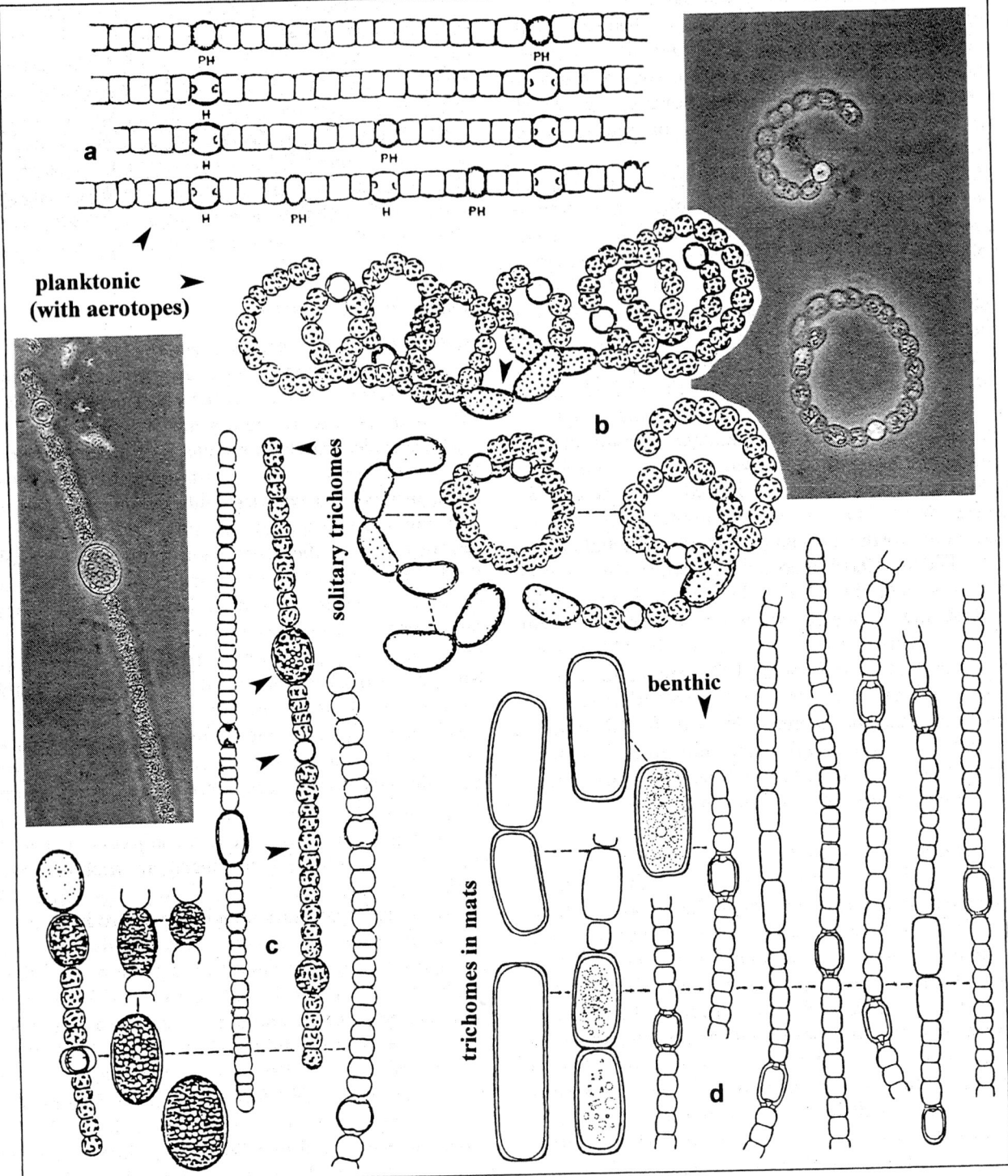

FIGURE 28 *Anabaena*: (a) origin of metameric trichomes in *Anabaena* (development of proheterocytes and heterocytes in regular distances); (b) planktonic *A. perturbata* (after Nygaard, 1949; original microphoto by Komárková); (c) planktonic *A. viguieri* (after Nygaard, 1949; original microphoto by Komárková); (d) benthic *A. oblonga* (after Komárek, 1989).

referred to 14 species both with and without aerotopes in North Carolina. Hill (1976a–c) described three new planktonic species from Minnesota lakes. Tilden (1910) listed 15 species (four "not well understood") as widely distributed throughout North America. Several interesting planktonic species from the United States (e.g., *A. mendotae*, and *A. perturbata*, which are common in the temperate zone) were described by Trelease (1889) and Hill (1976a, b, c). Several nonplanktonic *Anabaena* species were recently transferred into the genus *Trichormus* (see Komárek and Anagnostidis, 1989, and the *Trichormus* description in this chapter), which differs from *Anabaena* substantially by mode of akinete formation.

Anabaenopsis (Wołoszyńska) Miller (Fig. 29A)

The thallus is filamentous. Filaments are solitary or in free clusters, free floating, with or without constrictions at the cross walls; they are frequently bent, spirally coiled, or sigmoid, rarely straight. They lack sheaths, but have a mucilaginous, diffuse envelope. Trichomes have a characteristic pattern of metameric (regularly repeating) origin of pairs of heterocytes in the trichome: Trichomes break between mature heterocytes leaving filaments with "apical" heterocytes at both ends. Several pairs of young heterocytes can be situated at regular intervals over a filament. Cells are almost spherical, barrel shaped to cylindrical, bent or straight, pale blue–green to brownish, with facultative aerotopes. Heterocytes are spherical to hemispherical, intercalarly in pairs after asymmetrical division of two neighboring vegetative cells; they may be smaller or larger than the vegetative cells. Akinetes are spherical or ellipsoidal, bent or straight, solitary or several together, intercalary, often distant from heterocytes. All cells in the trichome are capable of division; division is asymmetric before heterocyte formation. Reproduction occurs via akinete production and trichome fragmentation.

Most of the 16 species described are planktonic, forming water blooms. The spirally coiled filaments occur in mesotrophic to eutrophic lakes or in alkaline and saline waters, sometimes with epizooic protozoans (e.g., *Vorticella* sp.; Canter-Lund and Lund, 1995). Few species live in the metaphyton of small temperate water bodies in tropical and subtropical regions. Several species were reported from Midwestern or higher conductivity U.S. reservoirs (Tilden, 1910; Smith, 1950; Whitford and Schumacher, 1969). This genus is rare in low-conductivity, low-nutrient boreal waters and apparently absent from arctic and subarctic waters (Duthie and Socha, 1976; Sheath and Steinman, 1982). However, species of the genus were reported in prairie lakes in central Canada (Kling, 1975).

Aphanizomenon Morren ex Bornet *et* Flahault
(Fig. 29B)

The thallus is filamentous. Filaments are free floating, solitary, or gathered in small or large fasciculated colonies with filaments in characteristically parallel arrangements (may appear as macroscopic flakes or like grass clippings on lake surface). Trichomes are isopolar, subsymmetrical, straight, sometimes with slightly curved or bent ends, cylindrical or narrowing toward the ends, with or without constrictions at the cross walls. Firm sheaths are absent, but the trichomes of several species are covered with fine, diffuse, mucilaginous, colorless envelopes. Cells are cylindrical or long barrel shaped, with quite variable width/length ratio, pale blue–green, with aerotopes (sometimes facultative). The end cells are often much longer than the central cells, and hyaline. In some species, they are cylindrical and rounded at the end without visible cell contents; in others, narrowed and pointed. Heterocytes are almost spherical, ellipsoidal, or cylindrical with two pores, always intercalary and solitary, usually only few (1–3) per trichome (rarely more). Akinetes are long or short cylindrical with rounded ends, or elliptical, rarely almost spherical, solitary or groups of two or three, adjacent to heterocytes or distant. Cell division occurs along the whole trichome with the exception of the end cells. Reproduction occurs via trichome disintegration and akinete germination.

About 20 species have been described, but some are not clearly defined; the position of species with narrowed and pointed apical cells is unresolved. Almost all are planktonic, often forming dense water blooms in temperate zones (*Aphanizomenon flos-aquae* in eutrophic lakes, reservoirs, regulated fish ponds, and cattle ponds). One species prefers brackish conditions or seawater (Baltic Sea). Few species are known only from tropical regions (Cronberg and Komárek, in press). The most noted species in North America are *A. flos-aquae* and *A. gracile*, although several other species have been reported (*A. issatschenkoi*, *A. skujae*, and *A. aphanizomenoides*). One species, *A. schindleri*, has only been identified from mesotrophic to eutrophic Canadian Shield lakes in northwestern Ontario (Kling *et al.*, 1994). Early authors (e.g., Tilden, 1910; Smith, 1950; Prescott, 1962), as well as Whitford and Schumacher (1969) and others (e.g., Duthie and Socha, 1976; Stein and Borden, 1979; Sheath and Steinman, 1982) mentioned only one species, *A. flos-aquae*, which is widely distributed throughout North America.

Aulosira Kirchner ex Bornet *et* Flahault (Fig. 31C)

The thallus is filamentous. Filaments are solitary or grouped in clusters, rarely in mats. Filaments are slightly

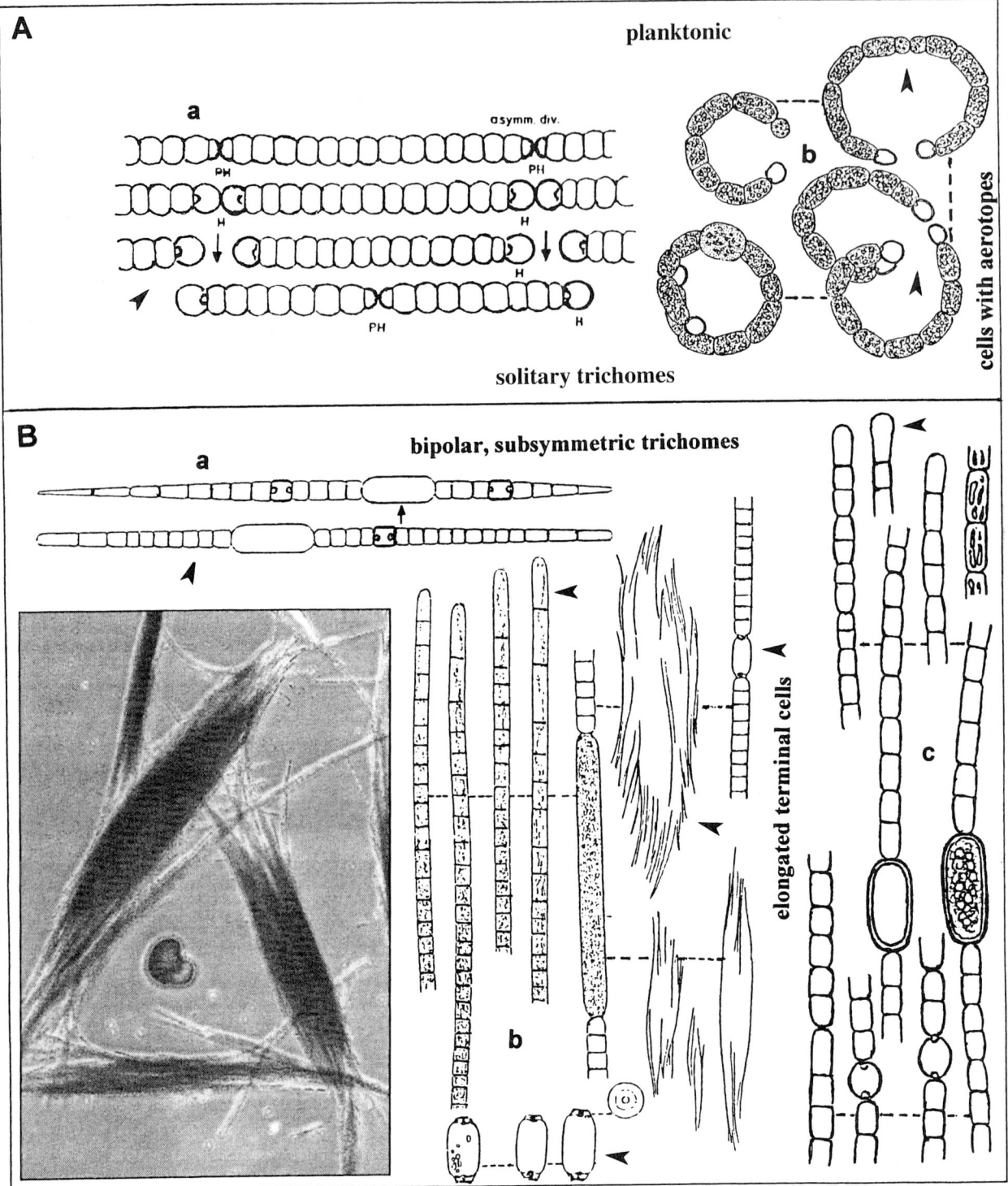

FIGURE 29 (A) *Anabaenopsis*: (a) origin of metameric trichomes in *Anabaenopsis* (development of two neighboring heterocytes in regular distances); (b) *A. elenkinii* (after Smith, 1950). (B) *Aphanizomenon*: (a) subsymmetrical trichomes; (b) *A. flos-aquae* (after Komárek, 1958; original microphoto by J. D. Wehr, with permission); (c) *A. schindleri* (after Kling et al., 1994, from central Canadian lakes).

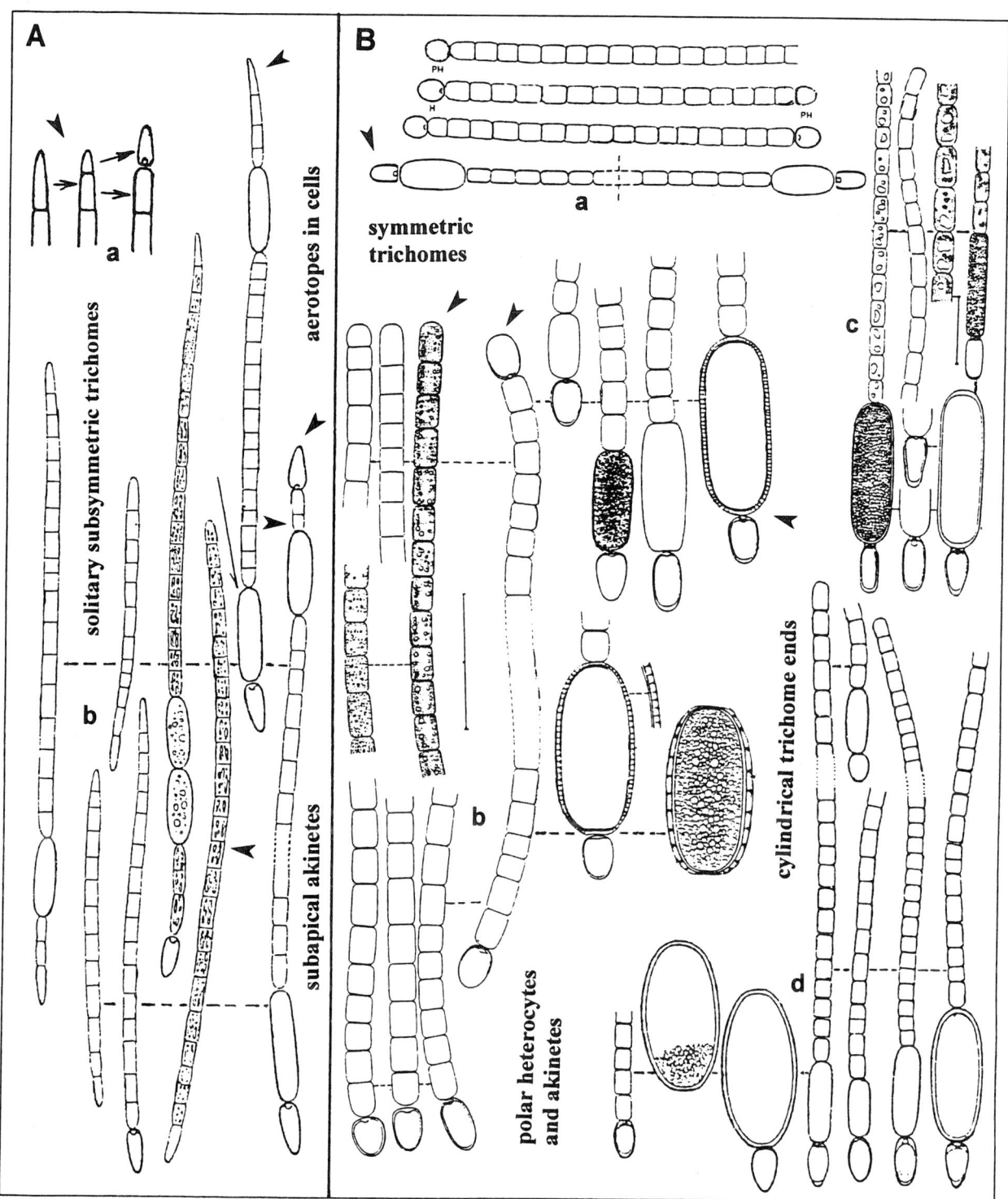

FIGURE 30 (A) *Cylindrospermopsis*: (a) development of terminal heterocytes; (b) *C. raciborskii* (after Horecká and Komárek, 1979). (B) *Cylindrospermum*: (a) development of symmetrical trichomes; (b) *C. stagnale* (bar = 20 μm; after Komárek, 1975); (c) *C. longisporum* (bar = 10 μm; after Komárek, 1975); (d) *C. minutissimum* (after Komárek, 1989, from Cuba).

irregularly coiled, sometimes oriented in parallel rows, with firm, distinct colorless sheaths enveloping one or occasionally two trichomes (young trichomes have no sheath), open at the end. Trichomes are of uniform width throughout their length, not attenuated at ends, with constrictions at the cross walls, uniseriate, isopolar, metameric (regularly repeating pattern), with several solitary heterocytes regularly arranged along trichome. Cells are cylindrical or barrel shaped, typically isodiametric, but becoming longer than wide with age, blue–green, pale grayish–blue, olive green, or reddish, occasionally (rarely) with aerotopes, often with prominent granules. Terminal cells are rounded. Heterocytes are spherical, oval or cylindrical, generally same width or slightly wider than vegetative cells. Akinetes are usually elongated, oval to cylindrical (rarely spherical), apoheterocytic (arise in rows between two heterocytes and develop successively toward a heterocyte), or sometimes irregularly situated. Cell division occurs perpendicularly to the trichome axis, with cells growing to the original size and shape before subsequent divisions; all cells are capable of division, with no meristematic zones. Reproduction is by hormogonia emerging from sheaths.

The species of the genus appear benthic or in the metaphyton of unpolluted lakes, ponds, and reservoirs, generally among or on submersed water plants or on sticks. Most species are geographically delimited, many to tropical regions (Frémy, 1930b; Geitler, 1932; Desikachary, 1959). Several species are important members of the microflora of rice paddies (e.g., *A. fertilissima*). Twenty species have been described but only six are well known. Smith (1920) noted the genus in widely separate localities, also reported from Wisconsin and British Columbia (Prescott, 1962; Stein and Borden, 1979). It is most likely present in tropical and subtropical regions as well (marshes in central America; authors' data).

Cylindrospermopsis Seenaya *et* Subba Raju (Fig. 30A)

The thallus is filamentous. Filaments are solitary, straight, bent, or coiled, free floating, narrowed toward the ends (in some species). Trichomes are isopolar or secondary heteropolar, subsymmetrical, with or without constrictions at the cross walls. Cells are cylindrical or barrel shaped, pale blue–green or yellowish, facultatively with aerotopes. End cells are often conical or bluntly or sharply pointed. Heterocytes are ovoid or conical, sometimes slightly curved in a droplike form, single pored, and always terminal; they develop after asymmetrical division of the end cells in one or in both ends of the filament. Akinetes are ellipsoidal or cylindrical, developing among vegetative cells, distant, or occasionally adjacent to apical heterocytes. Reproduction occurs via akinetes and trichome fragmentation.

The nine species described are planktonic, bloom forming, and common in tropical eutrophic freshwaters (pantropical) (Komárková, 1998; Komárek, 2001). The common *C. raciborskii* has been shown to be invasive into the warm regions of the temperate zone in Europe (Horecká and Komárek, 1979; Padisák, 1990, 1997; Dokulil and Mayer, 1996). *C. raciborskii* is a well-known species forming water blooms in tropical and subtropical water bodies (Komárková *et al.*, 1999) and producing toxins (Skulberg *et al.*, 1993). Occurrences of members of this genus apparently have increased in south-temperate and subtropical lakes (e.g., Florida) in the last few years (authors' data). However, Hill (1970a) lists an *Anabaenopsis* (= *Cylindrospermopsis*) *raciborskii* from southern Minnesota. Also his *Raphidiopsis mediterranea* (Hill, 1970b) could be a part of the developmental stage of *C. raciborskii* with only akinetes. A new species (*C. catemaco*) has been described from Mexico (Komárková-Legnerová and Tavera, 1996).

Cylindrospermum Kützing ex Bornet *et* Flahault (Fig. 30B)

The thallus is composed of fine or compact mucilaginous mats. Filaments are slightly curved or irregularly coiled, typically cylindrical throughout their length. Trichomes are without a firm sheath, but with a fine, colorless, homogeneous, and diffuse mucilaginous envelope. Trichomes are symmetrical, usually slightly constricted at the cell walls. Cells are cylindrical, more or less isodiametric, pale or bright blue–green or grayish, without aerotopes, occasionally granulated. Heterocytes are always terminal, developing from the end cells, ellipsoidal, ovoid, or conical, single pored, situated at one or both ends of the filaments. Akinetes are ellipsoidal, rarely spherical, developing near both ends of the filament at heterocytes, solitary or in rows up to seven, often with sculptured epispore. Cell division occurs throughout entire trichome; there areno meristematic zones. Reproduction occurs via akinete germination and fragmentation of trichomes into hormogonia.

Most of the 50 species described are benthic, epiphytic, or epilithic in unpolluted or slightly eutrophic waters, or from soils. Tilden (1910) noted eight common species, whereas Smith (1950) reported 11 species from terrestrial and aquatic habitats in the United States. Whitford and Schumacher (1969) distinguished 11 species based on differences in akinete shape. One aquatic species, *C. catenatum*, occurred in waters across North Carolina; others were subaerial. Tiffany and Britton (1952) reported three species of this genus from Illinois. A few species have been reported in scattered locations across Canada (Duthie and Socha,

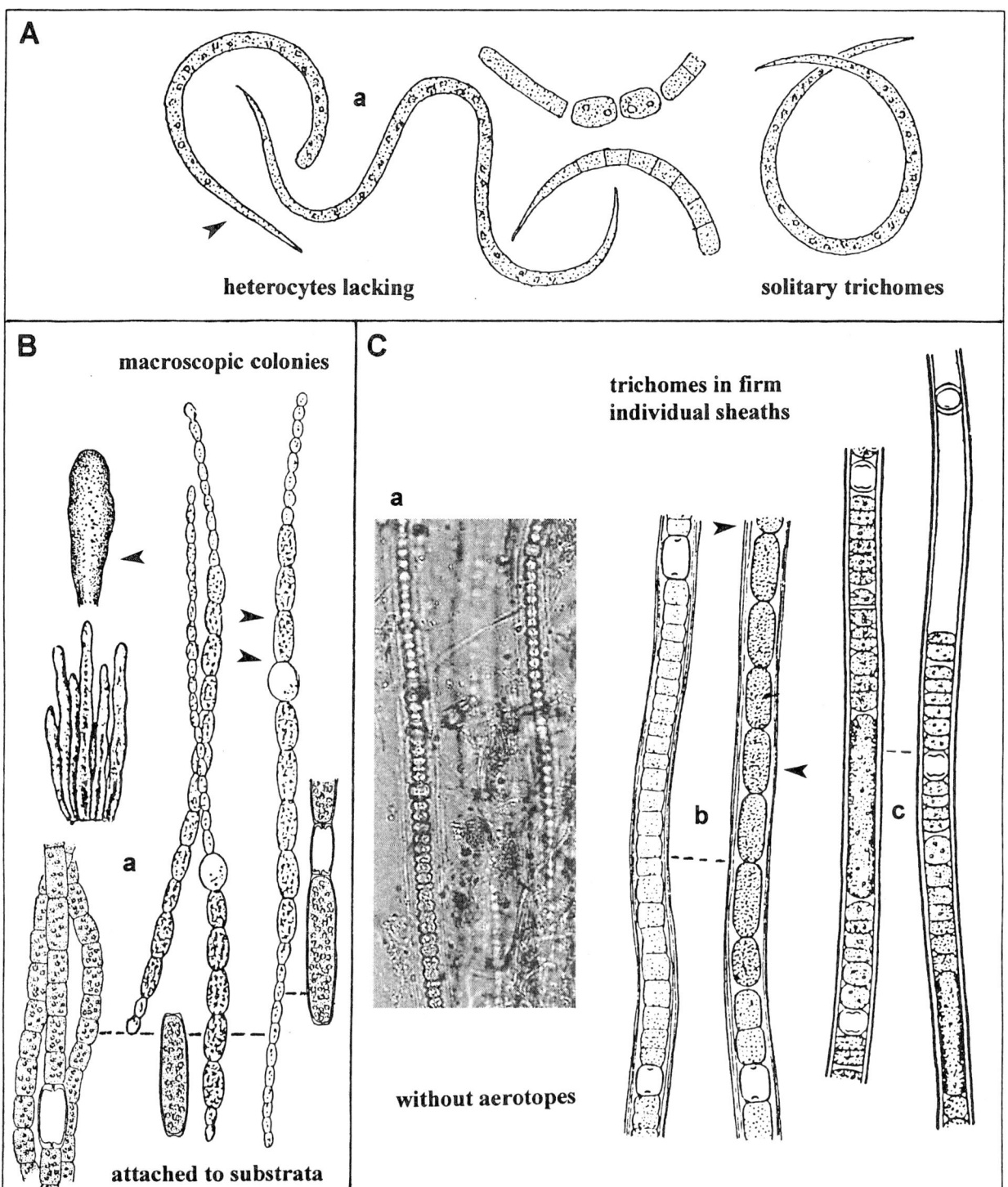

FIGURE 31 (A) *Raphidiopsis*: (a) *R. curvata* (after Fritsch and Rich from Geitler, 1932). (B) *Wollea*: (a) *W. saccata*, macroscopic colonies and solitary trichomes (after Wolle from Geitler, 1932, and Smith, 1950). (C) *Aulosira*: (a) *Aulosira* sp. (original photo by Komárková, from the Everglades, Florida); (b) *A. implexa* (after Geitler, 1932); (c) *A. laxa* (after Kondrateva, 1968).

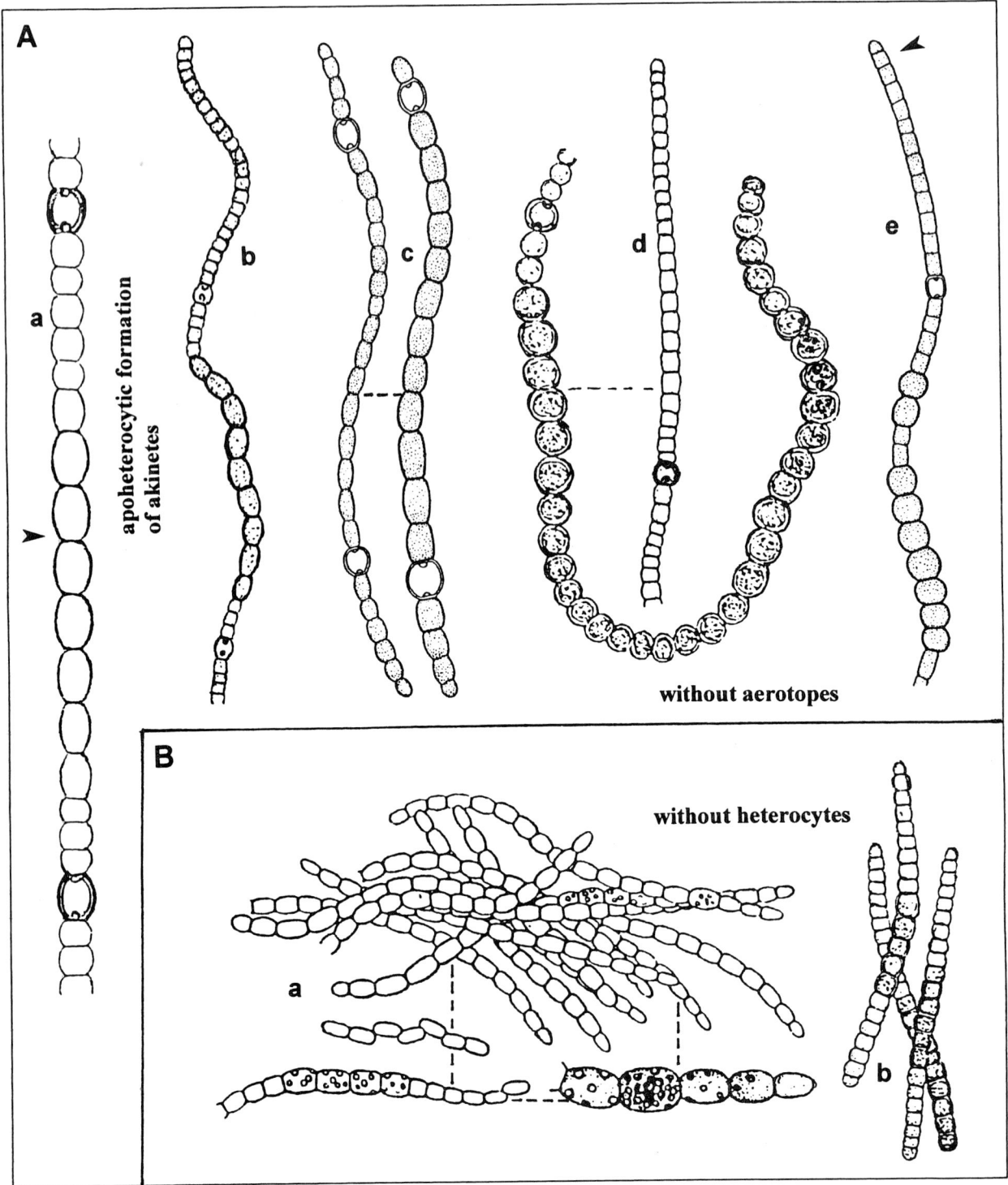

FIGURE 32 (A) *Trichormus*: (a) apoheterocytic akinete development between two heterocytes; (b) *T. variabilis* (after Frémy ex Geitler, 1932; sub *Anabaena variabilis*); (c) *T. luteus* (after Gardner, 1927, from Puerto Rico; sub *Anabaena lutea*); (d) *T. fertilissimus* (after Desikachary, 1959; sub *Anabaena fertilissima*); (e) *T. subtropicus* (after Gardner, 1927, from Puerto Rico; sub *Anabaena subtropica*). (B) *Isocystis*: (a) *I. planctonica* (after Starmach, 1966); (b) *I. infusionum* (after Hollerbach et al., 1953).

1976; Stein and Borden, 1979; Sheath and Steinman, 1982). Several interesting species are known from swampy habitats in the Caribbean (Komárek, 1989).

Isocystis Borzì ex Bornet *et* Flahault (Fig. 32B)

Isocystis is filamentous, growing in solitary filaments, in small macroscopic fascicles, or in indistinct microscopic gelatinous colonies attached to substrata. Filaments are enveloped by a fine, diffuse, unstructured, colorless, barely visible gelatinous envelope. Trichomes are uniseriate, simple curved, or coiled, isopolar, generally slightly attenuated toward both ends, and clearly constricted at the cross walls. The gelatinous sheaths are fine, diffuse, unstructured, colorless, and barely visible. Cells are barrel shaped, isodiametric, longer or shorter than wide, sometimes with solitary granules. Heterocytes are absent. Akinetes have thick cell walls, are slightly larger than vegetative cells, and are situated in apoheterocytic fashion (in rows in the middle of trichomes). All cells are capable of cell division, perpendicular to the trichome axis. Reproduction occurs via short hormogonia, which sometimes disintegrate into solitary cells, and akinete germination.

Species of *Isocystis* appear in the metaphyton or are free floating. Several species are periphytic in different freshwater habitats (springs and littoral areas of standing waters), one species is subaerial, colonizing wet walls, one is planktonic (Starmach, 1966), and one was described from soil (Geitler, 1932). Eight species have been described, of which five have been accepted as valid (Starmach, 1966; Komárek and Anagnostidis, 1989). The genus is not well known and probably rare (but present) in North America (not yet published).

Nodularia Martens ex Bornet *et* Flahault (Fig. 33)

This is a filamentous genus occurring as solitary filaments or in groups or clusters, occasionally (several benthic species) in mats. Filaments are isopolar (both poles with the same morphology), unbranched, more or less straight, curved, coiled, or irregularly spirally coiled with fine, diffuse, two-layered sheaths, open at both ends. Trichomes are uniseriate, cylindrical, occasionally (rarely) with slightly attenuated trichome ends, constricted at the cross walls, metameric (several heterocytes situated at regular intervals). Cells are short, barrel shaped, with length never exceeding width. Two groups of species (subgenera) differ cytologically and ecologically from each other; facultative gas vesicles occur only in planktonic species. Cell contents are yellowish, pale olive green, blue–green, or pinkish; thylakoids are irregularly coiled and distributed throughout the cell, sometimes peripheral. Heterocytes are typically identical in shape to vegetative cells. Akinetes are short, barrel shaped (shorter than wide) or spherical, but often with irregularities. All cells are capable of division, transverse to the trichome axis; trichomes lack meristematic zones. Reproduction occurs via hormogonia, disintegration of trichomes, and akinete germination.

After revision of the genus, only 9 of the 24 species (27 taxa) described were well defined (Nordin and Stein, 1980; Komárek *et al.*, 1993; Hayes *et al.*, 1997; Barker *et al.*, 2000). Four species with aerotopes (groups of gas vesicles) are planktonic (including type species *N. spumigena*), but other planktonic taxa occur in localities worldwide, including North and Central America. Species in this group sometimes form dense water blooms in inland lakes or reservoirs with higher salinity, in brackish coastal lakes and lagoons, or in the sea or estuaries of rivers (Huber, 1984; Blackburn and Jones, 1995; Tavera and Komárek, 1996; Perez *et al.*, 1999). Five species always lack aerotopes and are benthic; the most common, *N. harveyana* (cosmopolitan?), occurs in coastal or inland saline and mineral pools, as well as in marshes and lakes. *N. sphaerocarpa* (temperate) occurs in alkaline streams and in the littoral areas of lakes and pools, and *N. willei* (pantropical) occurs in rice fields, supplying an important source of nitrogen; it is highly probable that several species also colonize soils. Blooms of *Nodularia* have been noted in lakes in western North America, as well as in saline lakes from North American prairies (Nordin and Stein, 1980) and in Nevada (Galat *et al.*, 1990). Tilden (1910) recorded six species of the genus *Nodularia* in North America, Tiffany and Britton (1952) reported two species, and Smith (1950) mentioned all three species recognized prior to 1950 in the United States. Prescott (1962) reported two species from the western Great Lakes area, and several species have been reported from across Canada (Duthie and Socha, 1976; Stein and Borden, 1979; Sheath and Steinman, 1982).

Nostoc Vaucher ex Bornet *et* Flahault (Fig. 34)

The thallus is microscopic or macroscopic, gelatinous, spherical, or irregular gelatinous mats or flat colonies, smooth or warty at the surface, often with superficial mucilaginous exterior. Filaments are typically coiled, forming irregular, loose, or dense clusters, concentrated near the colony surface. The sheath is mucilaginous, firm, wide, and sometimes yellow to brownish, but visible only in young colonies and confluent with common mucilage. The mucilage of the colony is sometimes colored, yellowish green or brownish. Trichomes are isopolar, sometimes very long, curled inside the colony, often moniliform. Cells are barrel shaped or spherical, a uniform shape and size along trichome, pale to bright blue–green or olive green. Heterocytes are barrel shaped or spherical, soli-

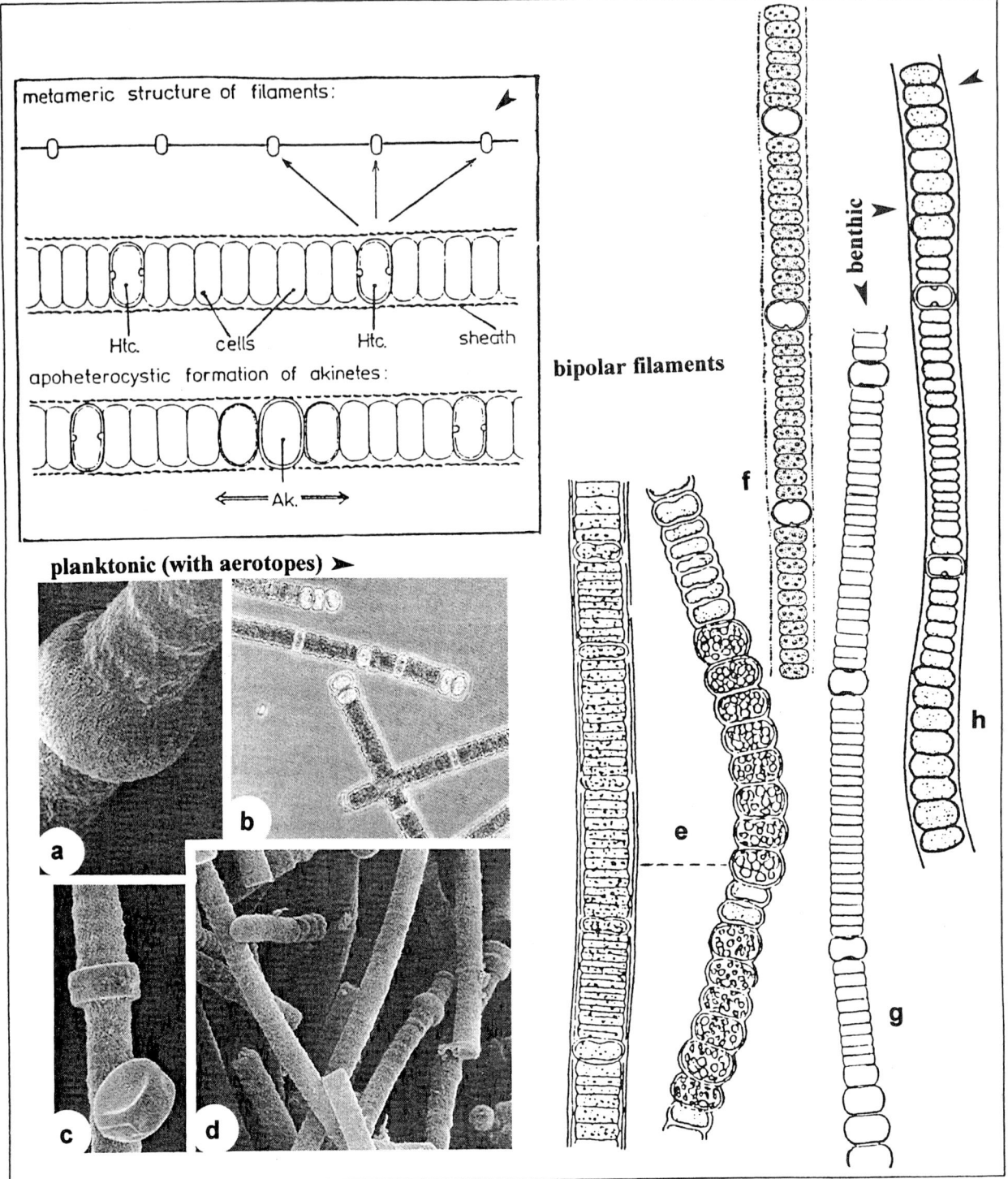

FIGURE 33 Nodularia: (a–d) N. cf. spumigena (after Pérez et al., 1999, from coastal lakes in Uruguay); (e) N. litorea (after Bornet and Thuret from Geitler, 1932); (f) N. baltica (after Smith, 1950, from the United States, sub N. spumigena var. minor); (g) N. harveyana (after Geitler, 1932); (h) N. willei (after Gardner, 1927, from Puerto Rico).

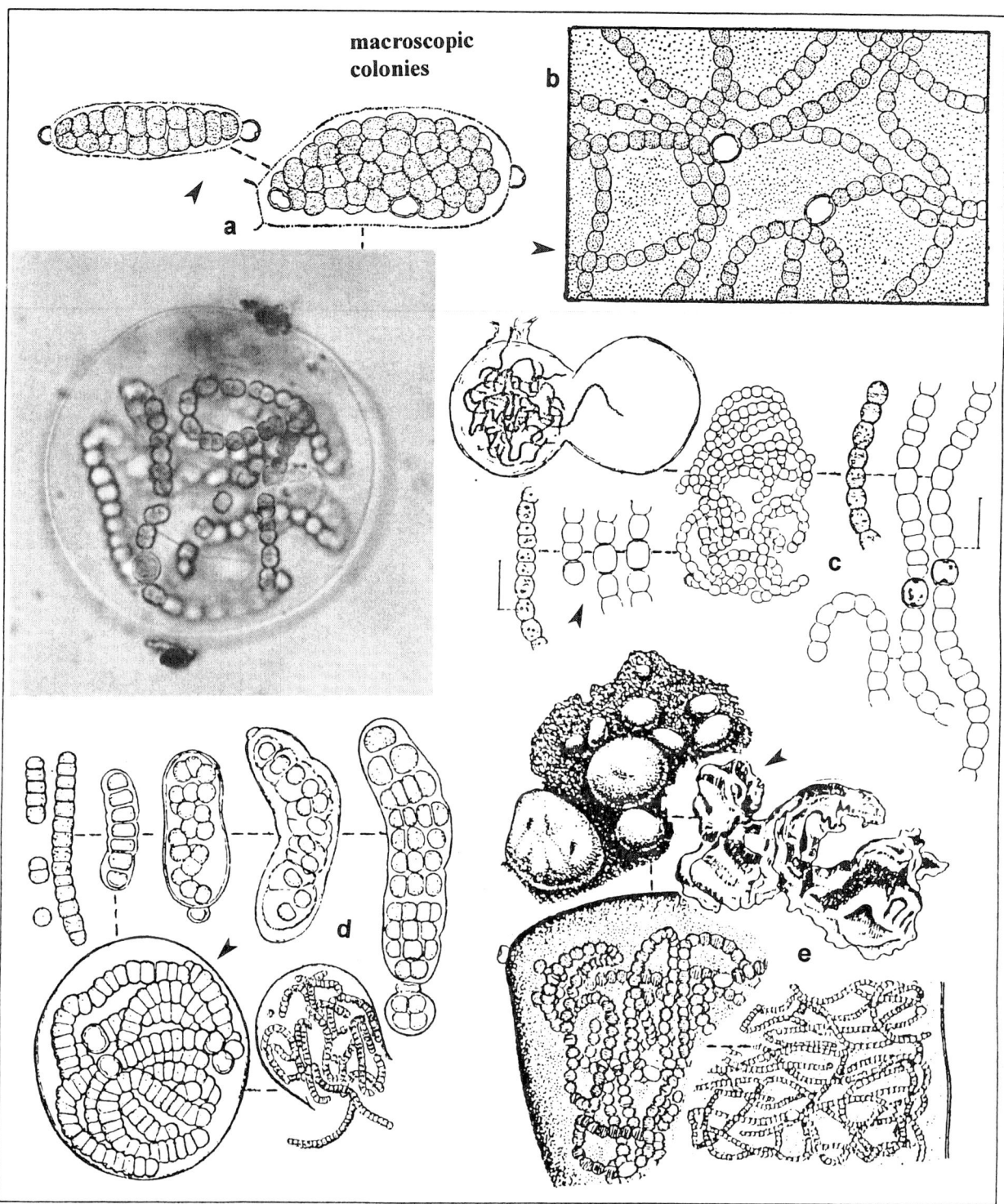

FIGURE 34 Nostoc: (a) Nostoc sp., typical initial stages (after Smith, 1950; original photo by R. G. Sheath, with permission); (b) typical position of trichomes in old colony (after Smith, 1950); (c) N. paludosum (after Komárek, 1975); (d) N. edaphicum (after Kondrateva, 1968); (e) N. commune (after Kondrateva, 1968).

tary, developing at ends of trichomes or intercalary. Akinetes are ellipsoidal, only slightly larger than vegetative cells, arising in rows between heterocytes. Cells divide perpendicularly to the trichome axis; meristematic zones are unknown. Colony morphology changes during development, and reproduction is specific for each subgenus; motile hormogonia develop between heterocytes, by akinete germination or filament disintegration.

Nostoc is common and widespread genus with more than 200 taxa described; a few species have been precisely revised (Mollenhauer, 1970). Species of *Nostoc* are mainly benthic, occurring in many epiphytic, epipelic, and epilithic habitats, in unpolluted lakes, ponds and pools, streams, and rivers, and on soils (including desert soils), some reaching 30 cm in diameter (Dodds *et al.*, 1995). *N. parmelioides*, with spherical colonies, grows on stones in streams and rivers and produces ear-shaped colonies after being occupied by aquatic midge larvae (Ward *et al.*, 1985; Dodds, 1989). A few *Nostoc* species are endophytic in fungi (*Geosiphon*), mosses, and ferns (Mollenhauer *et al.*, 1996; Mollenhauer and Mollenhauer, 1996), and phycobionts in lichens. At the time of Tilden (1910), the genus *Nostoc* comprised 31 species, 29 of which were found in North America. Whelden (1947) recorded several taxonomically uncertain species from the Arctic: *Nostoc aureum* (freshwater), *N. commune* (soils or tundra seepages), *N. kihlmanii* (floating in freshwater), *N. linckia* (in a drying pool), *N. minutum* (lake edges), *N. paludosum* (in marshes), *N. pruniforme* (floating in lakes), *N. sphaericum* (small pools and on rotting leaves in freshwater), and *N. sphaeroides* (among mosses). Smith (1950) recorded some species as mostly terrestrial, others as strictly aquatic, and still others as free floating in pools or attached to submersed substrata. Prescott (1962) recorded 14 species of the genus *Nostoc* from the western Great Lakes area; several species are widely reported across Canada (Duthie and Socha, 1976; Stein and Borden, 1979; Sheath and Steinman, 1982).

Raphidiopsis Fritsch *et* Rich (Fig. 31A)

The thallus is filamentous. Trichomes are solitary, free floating, without sheaths and mucilaginous envelopes, straight or screwlike coiled, unconstricted or slightly constricted at the cell walls. Cells are usually cylindrical, facultatively with aerotopes and granulations. Apical cells are conical, pointed, or bluntly pointed at the ends of the trichomes. Heterocytes are unknown. Akinetes are cylindrical to elongated ellipsoidal, intercalary, usually slightly off center in the trichomes. Reproduction occurs by trichome fragmentation and akinete germination.

The species are mainly planktonic. Hill (1970b) reported *R. mediterranea* from Minnesota, and the genus has been collected several times in Mexico (Komárková-Legnerová and Tavera, 1996). Smith (1950) and Taft and Taft (1970) mentioned only one species, *R. curvata* from Lake Erie. Whitford and Schumacher (1969) reported the same species in North Carolina, which has also been collected from standing and flowing water (irrigation ditches) in British Columbia (Stein, 1975; Stein and Borden, 1979).

Trichormus (Ralfs ex Bornet *et* Flahault) Komárek *et* Anagnostidis (Fig. 32A)

Thalli occur as macroscopic mats, in clusters and in strata; filaments are rarely solitary. Trichomes are uniseriate, straight, bent, or coiled, and constricted at the cross walls. Firm sheaths do not develop, but slimy envelopes are sometimes present. Cells are barrel shaped to cylindrical, pale blue–green, without aerotopes. Heterocytes are spherical to ellipsoidal, intercalary, in metameric (regularly repeating) configuration. Akinetes are oval to barrel shaped, rarely almost cylindrical, and develop in rows apoheterocytically between two heterocytes. One vegetative cell becomes one akinete. Reproduction occurs via hormogonia that separate from trichomes and via akinete germination.

Thirty-two species have been described (mainly formerly as *Anabaena*, which differ substantially by the strategy of akinete formation; Komárek and Anagnostidis, 1989). The majority of species colonize soils (often in rice fields) or mud; some are periphytic in the littoral of different waterbodies or live endophytic (known as "*Anabaena azollae*," "*Anabaena cycadearum*," and "*Nostoc gunnerae*"). Several species are tropical (e.g., *T. anomalus*, *T. doliolum*, and *T. fertilissimus*). *Trichormus* has been revised recently (Komárek and Anagnostidis, 1989) and thus may not be noted as such by earlier authors of North American studies. However, several species, listed as species of *Anabaena*, and recently recombined into *Trichormus*, are known from North America (e.g., *Trichormus* [*Anabaena*] *azollae* from Ontario; Duthie and Socha, 1976), and including the species *Anabaena* = *Trichormus variabilis* from soil habitats, reported from the United States by Whelden (1947), but from atypical localities. Several species living in intercellular spaces of vascular plants are also known from North and Central America: *T.* (*Anabaena*) *azollae* is a common symbiont of the aquatic fern *Azolla* (our results) and another species colonizes the roots of cycads (Mercado, 1977).

Wollea Bornet *et* Flahault (Fig. 31B)

Thalli are gelatinous, macroscopic colonies, with smooth surface, forming 5- to 10-cm tubelike (saccate) colonies or irregular balls, closed at apex, initially attached to a substrate and later free floating. Filaments are typically cylindrical. Trichomes are uniform along their length, more or less straight or slightly curved, uniseriate, unbranched, not attenuated or widened at the ends, highly constricted at the cross walls, with rounded end cells. Trichomes are irregular or almost parallel within a common diffuse mucilage. A fine pelicula (slimy but firm mucilage) is apparent on the surface of colonies, but sheaths around the trichomes are absent. Heterocytes are intercalary and solitary, and their position results in a metameric structure (regularly repeating pattern) along the trichomes. Spherical or oval akinetes arise in a short series, at both sides of the heterocytes. Cell division occurs by perpendicular binary fission. Reproduction is by hormogonia.

Wollea colonizes benthic habitats in freshwaters. The type species, *W. saccata*, is infrequently reported but known from several localities (mainly standing waters) in North America, including the Great Lakes region, Massachusetts, Minnesota, New Jersey, North Dakota, Louisiana, North Carolina, British Columbia, and Panama (Tilden, 1910; Prescott, 1962; Whitford and Schumacher, 1969; Stein, 1975). A second species (*W. bharadwajae*) was described from benthic habitats of ponds and paddy rice fields in India. Only these two species are known.

Stigonematales: Capsosiraceae

Besides the genus *Stauromatonema*, Smith (1950), Tilden (1910), and Whitford and Schumacher (1969) also mentioned a subaerophytic genus *Capsosira* (*C. brebissonii*) from New England and North Carolina. However, the original drawings do not correspond to the characteristics described by Frémy (Geitler, 1932), and North American records must be revised (see Fig. 35A).

Stauromatonema Frémy (Fig. 35B)

The thallus is firm and flat; the encrusting layer is composed primarily of basal creeping and coiled filaments and erect branches. Basal trichomes become multiseriate, sometimes disintegrating into coccoid-like masses from which grow secondary trichomes. Branches (true branching) are densely arranged and parallel, situated perpendicularly to substrata, uniseriate, and producing repeated V branches in successive layers. Sheaths are thin, firm, colorless, and closed at the apex of young filaments; they sometimes gelatinize and become slightly lamellated. Trichomes, constricted at the cell walls, are more or less cylindrical or slightly widened toward the ends, with rounded terminal cells. Cells are irregularly disclike or bluntly barrel shaped. Heterocytes are intercalary and barrel shaped. Akinetes are unknown. Cells divide in different planes in basal trichomes and in pseudoparenchymatous masses, but divide perpendicularly to the long axis in erect branches. Reproduction occurs via the liberation of monocytes (solitary reproduction cells) from the ends of branches after the sheaths are opened (rarely happens in intercalary cells). Hormogonia and hormocysts are unknown.

Thalli colonize stones in springs, creeks, and waterfalls, and rarely, littoral in lakes. Three species are tropical. *S. viride* has been found on the stones in an outlet from Lake Catemaco, Veracruz, Mexico (Carmona-Jiménez and Gold-Morgan, 1994).

Stigonematales: Stigonemataceae

Stigonema Agardh ex Bornet *et* Flahault (Fig. 36)

The thallus is filamentous, wooly or crusty, composed of free, coiled, true branched filaments, which are not diversified into morphologically different basal filaments and branches. Thalli are usually attached to substrata. Trichomes may be uniseriate or multiseriate (uniseriate mainly in young trichomes or ends of branches). Branches are thick, irregularly lateral, narrowed toward the ends, but with the apical cells often larger than others. Sheaths are thin or thick with a clear outer margin becoming wider with age, lamellated, and usually yellowish brown. Individual envelopes may develop around cells in the older parts of filaments. Trichomes disintegrate often in separated cells within the filaments. Cells are barrel shaped or irregularly rounded, usually connected by one pore ("pit connections") with one another (pores disappear in segments of some trichomes). The cell content is blue–green or olive green, usually with prominent solitary granules. Heterocytes are intercalary, solitary, occasionally lateral, and similar in form to adjacent vegetative cells. Akinetes are unknown. Clusters of coccoid cells occur occasionally. Cell division occurs in all planes; horizontal (crosswise) fission is most common, and meristematic zones are only present in sections where hormogonia arise. Reproduction occurs via hormogonia liberated from the ends of trichomes and branches. Hormogonia are uniseriate, morphologically different from trichomes, usually more cylindrical, and consist of two to many cells, sometimes with aerotopes.

Species of *Stigonema* attach to the substrata in standing and flowing waters or are subaerial or occur on soils (not common); they are distributed worldwide but apparently are more common in tropical regions (Gardner, 1927; Frémy, 1930b; Desikachary, 1959). Subaerial species colonize tree bark and rocks from lowlands to alpine regions (the common epilithic *S. minutum*, and *S. informe* from alpine wetted rocks are most known). Other species are known from the

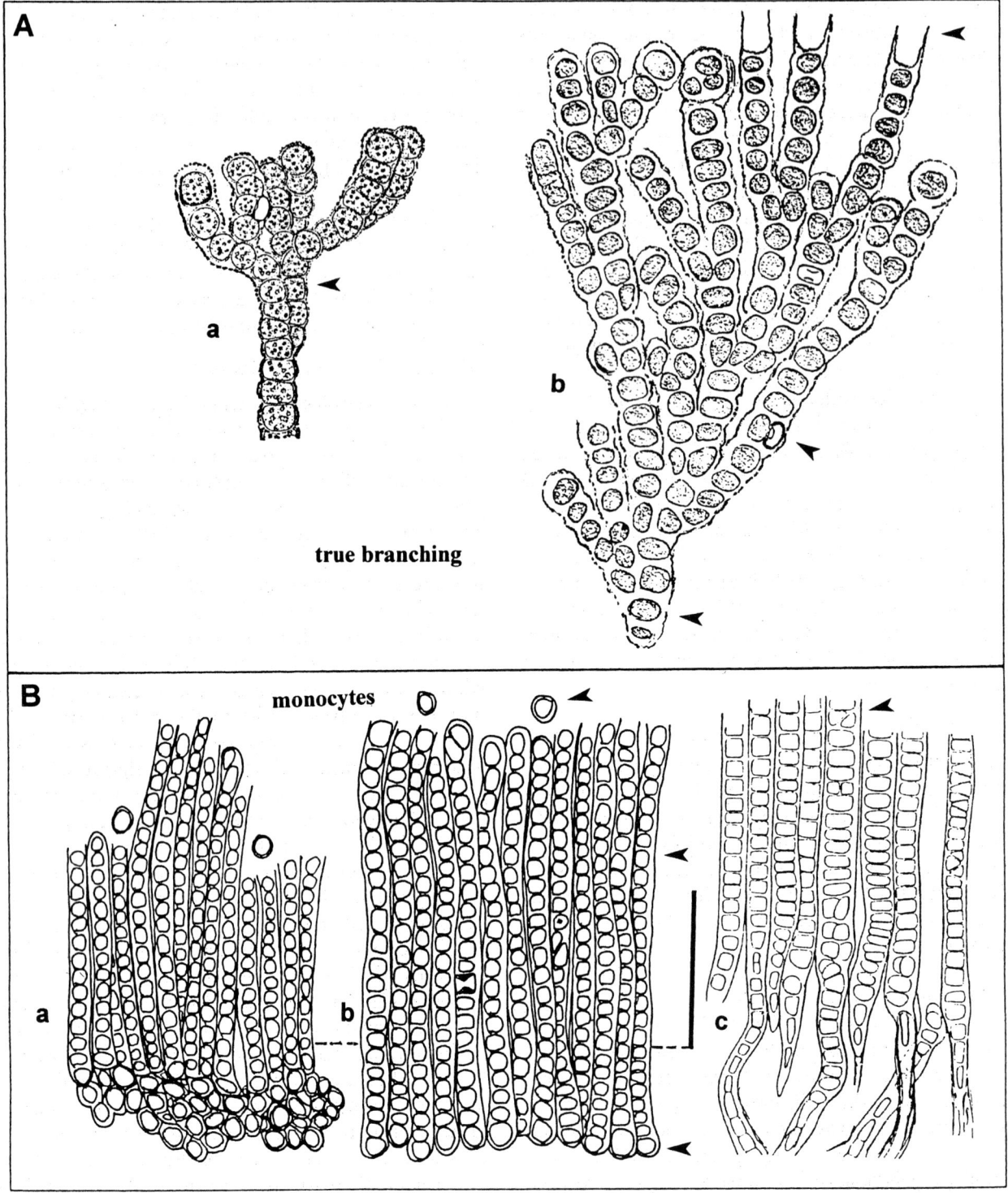

FIGURE 35 (A) *Capsosira*: (a) *C. brebissonii* (after Smith, 1950, from the United States); (b) *C. brebissonii* (after Frémy, 1930). (B) *Stauromatonema*: (a–b) *S. viride* (bar = 50 μm; after Carmona-Jimenez and Gold-Morgan, 1994, from Mexico); (c) *S. viride* (after Geitler and Ruttner, 1935).

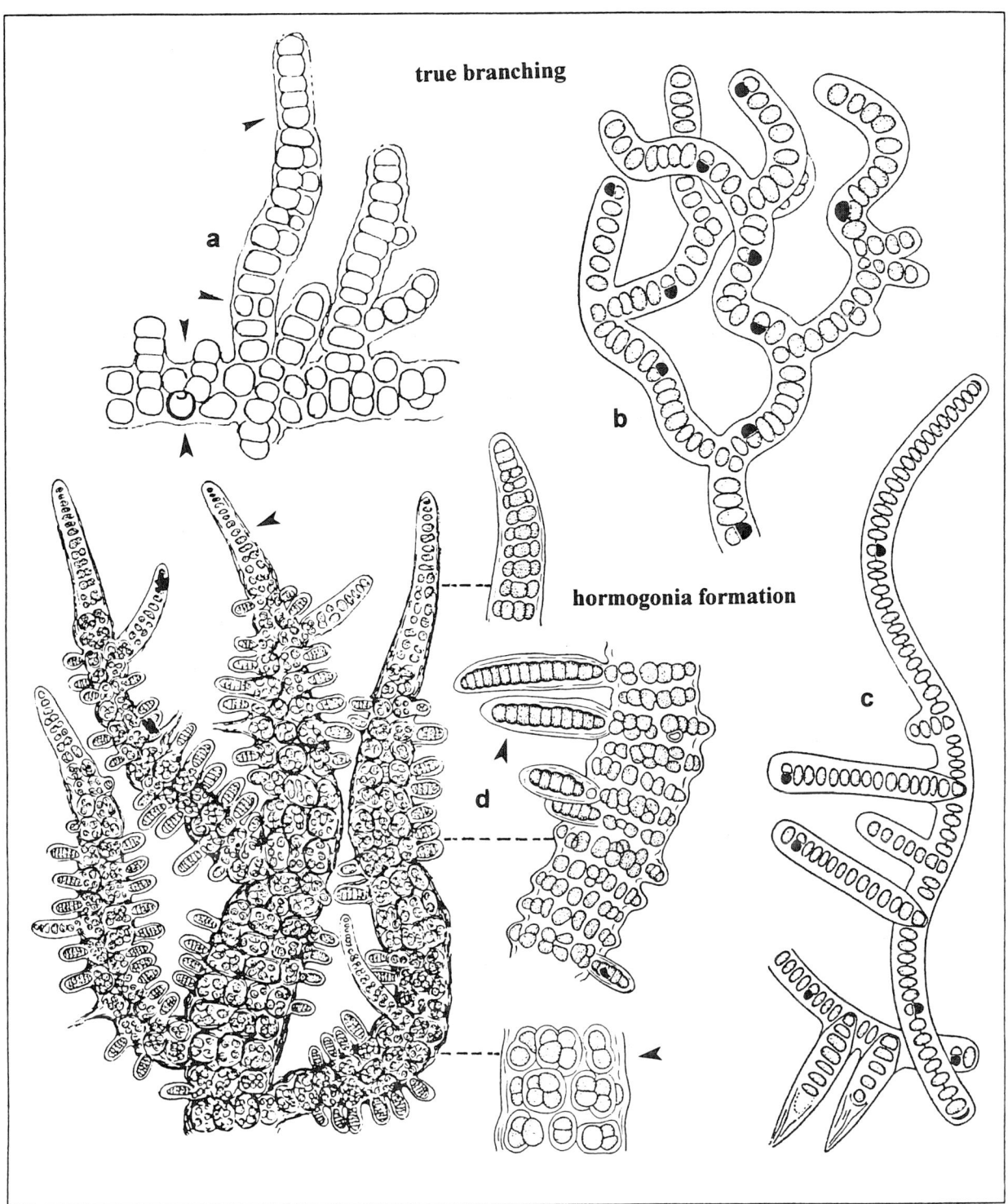

FIGURE 36 *Stigonema*: (a) thallus; (b) *S. congestum* (after Gardner, 1927, from Puerto Rico); (c) *S. elegans* (after Gardner, 1927, from Puerto Rico); (d) *S. mamillosum* (after Frémy and Kosinskaja from Kondrateva, 1968).

metaphyton or attached to wood and stones in pools, swamps, and wetlands. In tropical habitats, several species are important soil organisms (Desikachary, 1959). In North America and northern Europe, species can be found attached to rocks in oligotrophic lakes and mountain streams. There are many reports of this genus throughout North America, including Tilden (1910), who recorded nine *Stigonema* species of the ten described at that time in North America. Several species have been noted from the Arctic (Whelden, 1947): *Stigonema informe* (small freshwater pools and subaerial on wet granitic rocks), *S. mamillosum* (boulders in swift flowing streams), *S. minutum* (seepage rocks), *S. ocellatum* (stream), and *S. turfaceum* (peaty brook, tundra pool). In addition, Smith (1950) recorded *S. hormoides*, *S. panniforme*, and *S. thermale*. Prescott (1962) reported five species from the western Great Lakes area, whereas Whitford and Schumacher (1969) recorded 11 species in North Carolina; *S. hormoides* was the most common species in aquatic habitats with *S. ocellatum*. *S. mesentericum*, *S. minutum* var. *saxicola*, and *S. mirabile* were found on dripping rocks. At least nine taxa have been recorded from across Canada (Duthie and Socha, 1976; Stein and Borden, 1979; Sheath and Steinman, 1982).

Stigonematales: Fischerellaceae

Fischerella (Bornet *et* Flahault) Gomont (Fig. 37A)

The thallus is composed of morphologically diverse, uniseriate or multiseriate, usually creeping filaments and more or less erect uniseriate branches. Most species produce feltlike, rarely compact mats. Creeping trichomes with barrel-shaped cells sometimes occur within a gelatinous matrix. Trichomes are generally moniliform and enveloped by thick, sometimes from outside wavy or slightly lamellated, colored sheaths. Erect true branches (with T-type branching), usually unilateral, arise after longitudinal cell division in basal trichomes. Branches, primarily cylindrical, are generally composed of cylindrical elongated cells in colorless sheaths. Cell contents are slightly granular, with thylakoids distributed irregularly. Heterocytes are intercalary, subspherical in basal trichomes, and cylindrical in branches. Akinetes (known only in a few species) occur occasionally and irregularly in basal trichomes. Cell division occurs mainly via horizontal (crosswise), perpendicular fission. Reproduction occurs via uniseriate hormogonia separating from the ends of branches. Hormogonia liberated under wet or humid conditions, and are usually morphologically distinct from other cells in branches and contain aerotopes (gas vacuoles).

Most species are subaerial, growing on wet (often acidic, peaty) soils and rocks. Several species occur only in the tropical forests on mosses and the bark of trees. Two species, *F. ambigua* and *F. thermalis*, are reported in Tilden (1910). *Fischerella ambigua* is known from the United States, Canada, Mexico, the West Indies, and Hawaii as intermingled with large algae on moist rocks, wet earth, or the trees. *F. thermalis* is known from New Hampshire on damp stones and granite rocks and from thermal water in Hawaii and British Columbia (e.g., Stein and Borden, 1979). Prescott (1962) reported two species from the western Great Lakes region, and three were recorded from British Columbia (Stein and Borden, 1979). Whitford and Schumacher (1969) observed six species among the material collected from North Carolina, and three of these (*F. ambigua*, *F. letestui*, and *F. major*) were found on submerged substrata whereas the others were from subaerial habitats.

Stigonematales: Borzinemataceae

Schmidleinema De Toni (Fig. 37B)

Thalli consist of mats on various substrata or mixed with other algae, composed of basal prostrate creeping filaments and more or less erect branches. Basal creeping trichomes are torulous (thickened slightly), uniseriate or multiseriate, sometimes with true branching. Branches are cylindrical, uniseriate, erect, and usually repeatedly falsely branched (like *Tolypothrix*). Sheaths are usually (in parts) yellow–brown, distinctly outlined, thick and lamellated, closed or funnel-like or sometimes with a telescopic opening at the apex. Cells are barrel shaped or nearly cylindrical, pale blue–green to olive green, slightly or generally granular, sometimes with solitary distinct granules. Intercalary heterocytes are solitary or in pairs, usually single pored (prior to false branching), cylindrical, hemispherical or almost spherical. Akinetes are absent. Cell division occurs in several planes in older creeping filaments, perpendicular to the longitudinal axis in young trichomes and in branches; meristematic zones occur in the apical parts of branches. Reproduction occurs via motile hormogonia formed from the ends of branches and ensheathed hormocysts, differentiated in the basal or apical parts of the thallus. Hormogonia and hormocysts germinate on both sides of the heterocytes that develop in their center, becoming basal during later development.

Schmidleinema occurs in freshwater habitats, mainly tropical to subtropical, or subaerial. The type species, *Schmidleinema indicum*, is known from humid subaerial habitats such as on liverworts, tree trunks, and humid walls in India. Two other taxa (*S. cubanum* and *S. roberti-lamyi*) are known from alkaline swamps in the Antilles (Cuba and Guadeloupe; Komárek, 1989); *S. cubanum* is also known from the Everglades, Florida, and Belize.

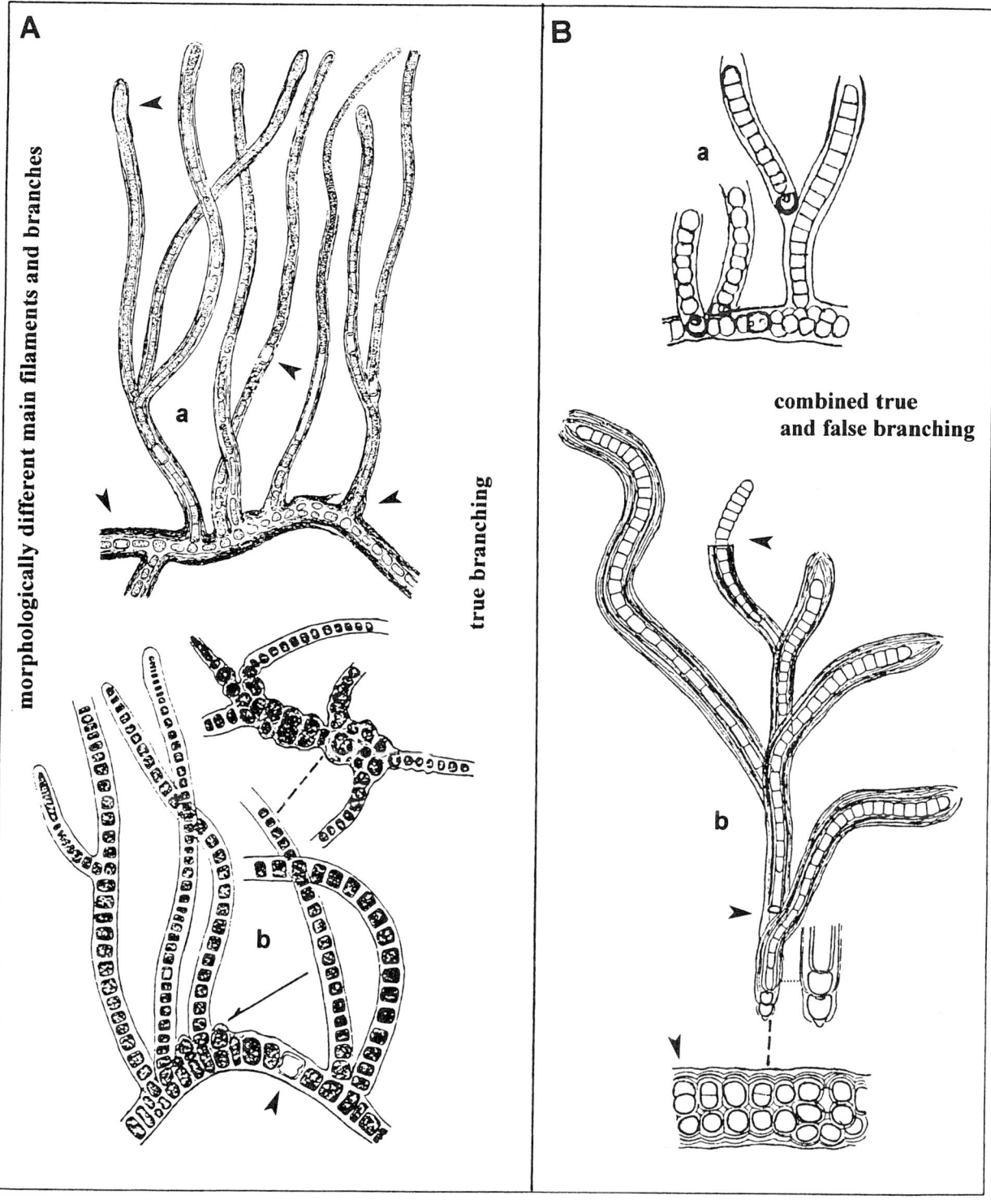

FIGURE 37 (A) *Fischerella*: (a) *F. thermalis* (after Frémy from Geitler, 1932); (b) *F. ambigua* (after Frémy from Geitler, 1932). (B) *Schmidleinema*; (a) thallus; (b) *S. cubanum* (after Komárek, 1989, from Cuba).

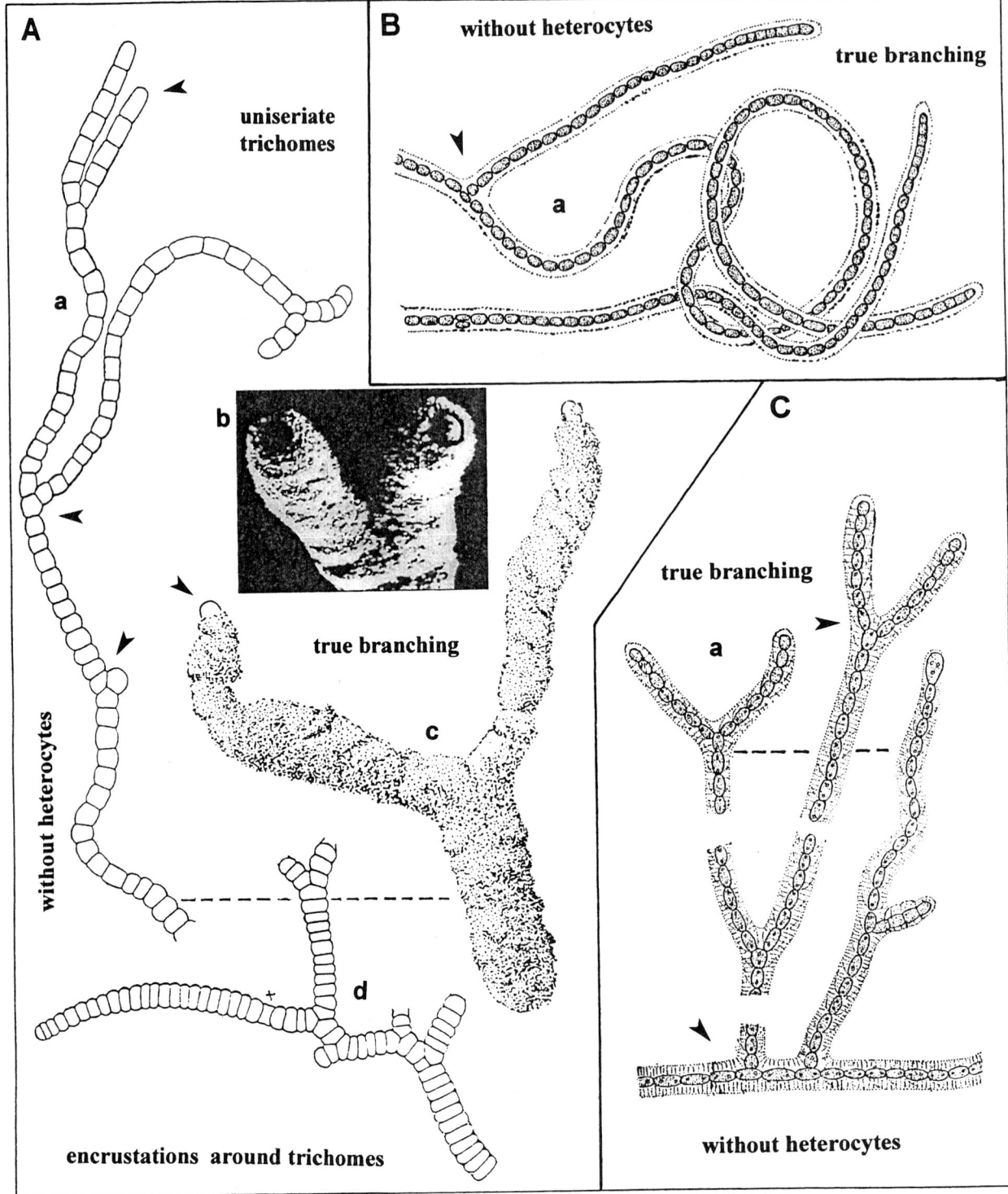

FIGURE 38 (A) *Geitleria*: (a–d) *G. calcarea* (after Friedmann, 1955, 1979; Couté, 1985); (B) *Albrightia*: (a) *A. tortuosa* (after Copeland, 1936 from Yellowstone National Park). (C) *Colteronema*: (a) *C. funebre* (after Copeland, 1936, from Yellowstone National Park).

Stigonematales: Loriellaceae

Albrightia Copeland (Fig. 38B)

The thallus consists of small clusters of free filaments. Filaments and branches are of the same morphology, cylindrical, single or true branched, flexuous, and irregularly and loosely tangled. Sheaths are firm, thick, homogeneous or layered. Trichomes are uniseriate and cylindrical and composed of moniliform rows of cells with deep constrictions at the cross walls. Growth is terminal, with branches of the same width and morphology as the main filaments and trichomes. True branching is either T type or pseudodichotomous V type. Cells are barrel shaped to cylindrical, narrowed at both ends, usually longer than wide, blue–green in color, with a granular content and sometimes prominent solitary granules. Terminal cells are more cylindrical and slightly elongated; end cells are rounded. Coccoid cells (two-celled groups) sometimes appear. Heterocytes and akinetes are unknown. Cell division is by horizontal (crosswise) or perpendicular fission. Reproduction occurs via few-celled hormogonia differentiated terminally at the ends of branches. Filaments disintegrate into persisting chroococcoid groups.

The type species *Albrightia tortuosa* is the only member of this genus. It grows in slightly thermal, alkaline springs in Yellowstone National Park among other cyanobacteria or on the surface of their thalli (Copeland, 1936; Smith, 1950). It may be present in other thermal streams in North America.

Colteronema Copeland (Fig. 38C)

The thallus is leathery or fibrous, membranous, and macroscopic, up to 1 mm in height, composed of true branched filaments. Filaments are usually crooked and roughly cylindrical. Trichomes are of identical morphology, more or less free, growing horizontally, densely crowded, slightly flexible, more or less cylindrical, torulous, uniseriate, distinctly constricted at the cross walls, and enclosed in colorless to yellow–brown, thick, firm, lamellated sheaths. Trichome branching is true (T and V types) with pseudodichotomous branches, arising after longitudinal division of the apical cells; lateral branching from the intercalary cells sometimes also occurs. Growth is apical. Cells are barrel shaped to ellipsoidal, subcylindrical, and longer than wide. End cells are rounded, sometimes slightly to noticeably enlarged. Heterocytes and akinetes have not been reported. Cell division occurs by horizontal (crosswise) perpendicular fission. Reproduction occurs via few-celled hormogonia separating as the ends of branches.

The single species (*Colteronema funebre*) is still known only from terrestrial, atmophytic (influenced by hot steam) habitats near thermal springs in Yellowstone National Park (Copeland, 1936).

Geitleria Friedmann (Fig. 38A)

The thallus is filamentous, loosely tufted, up to 2 mm thick, consisting of coiled creeping or erect filaments with irregular lateral and pseudodichotomous true branches (V and T types), with no prostrate basal system and no morphological differentiation into the main and lateral branches. Sheaths are firm, colorless, intensely lime encrusted, containing single trichomes. Trichomes are uniseriate, moniliform, and indistinctly to strongly constricted at the cross walls. Cells are generally barrel shaped or cylindrical, occasionally with narrow individual envelopes. Cell contents are granular, pale grayish–green in color with distinct chromoplasma and solitary granules. End cells are sometimes slightly widened and rounded. Thylakoids are distributed throughout the cell. Heterocytes and akinetes are absent. Cell division is horizontal (crosswise) or longitudinal (before true branching). Reproduction occurs via hormogonia (fragmented filament segments).

The habitat is aerial, subaerial, and epilithic, especially on calcareous rocks in caves. The type species, *Geitleria calcarea*, is probably distributed worldwide (e.g. Israel, France, Romania, Spain, ancient Yugoslavia, the United States, and the Cook Islands), but its occurrence is patchy; thus far, it has only been collected under specific ecological conditions, such as in aragonite caves (Friedmann, 1955, 1979). A second species, *G. floridana*, growing in similar habitats, is known only from Florida (Friedmann, 1979).

Stigonematales: Nostochopsaceae

Nostochopsis Wood ex Bornet *et* Flahault (Fig. 39A)

Thalli are gelatinous and attached to substrata, irregularly spherical or lobed with generally smooth mucilaginous surface, hollow center, up to 3.5 cm in diameter, bluish, olive green, or yellow–green. Common mucilage is usually homogeneous, colorless to yellowish brown. Relatively radially oriented filaments are present within the thallus; they are commonly true branched, slightly coiled, with fine, gelatinous, and diffuse (at margins), the colorless to brownish-yellow sheaths. Trichomes are always uniseriate, with typically isodiametric, barrel shaped or elongated (up to twice as long as wide), ellipsoid, blue–green cells. True lateral (T type or V type) branching is present, with long, cylindrical, multicelled branches terminating in slightly narrowed, rounded apical cells, or very short branches (with one to several cells) terminated by apical heterocytes. Heterocytes are always present, intercalary (bipored), lateral (single pored), or terminal (single pored). Akinetes are unknown. Cell division is horizontal (crosswise), sometimes in meristematic region, but primarily in apical trichome parts. Reproduction occurs via bipolar germinating hormogonia with short barrel-shaped cells.

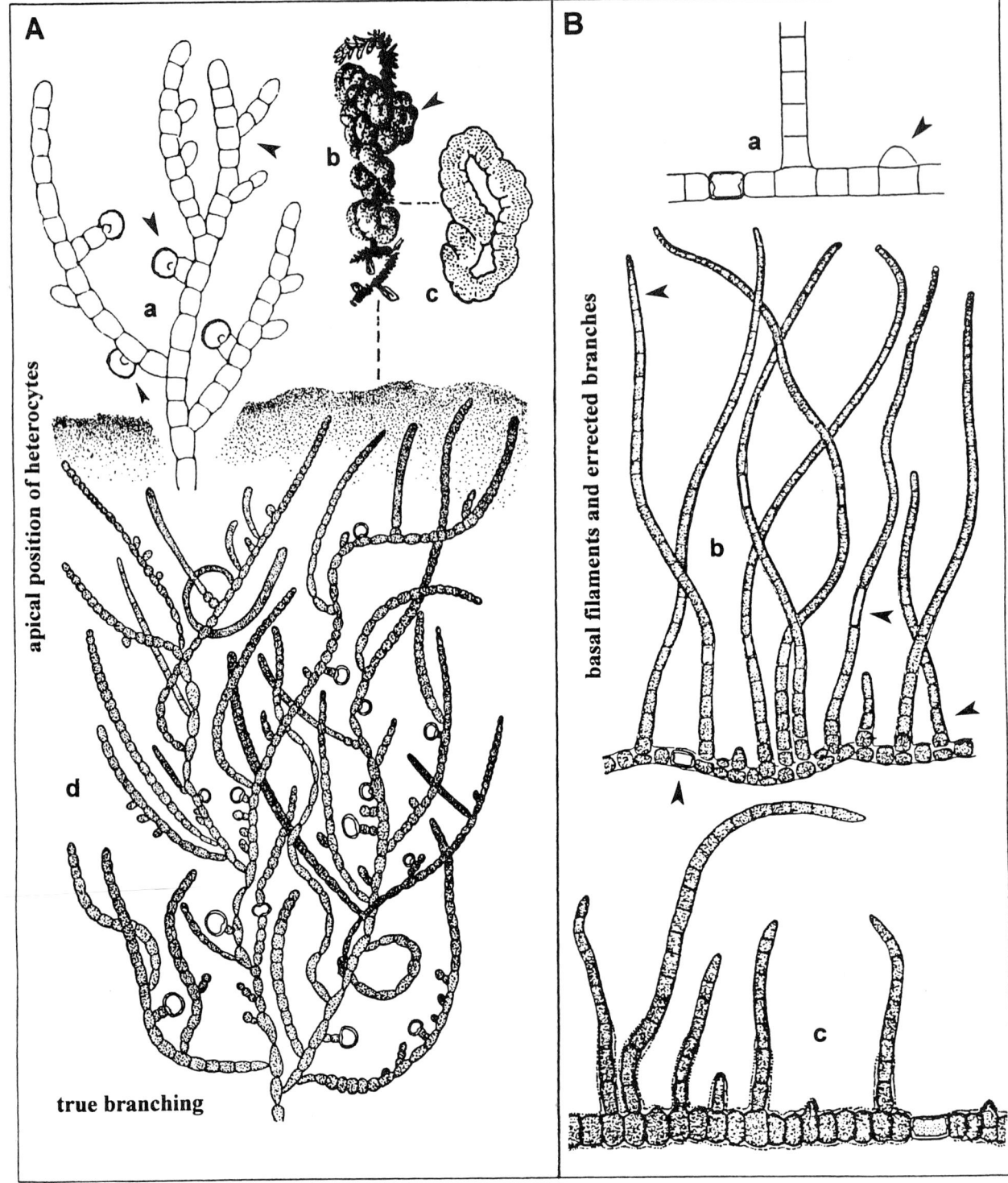

FIGURE 39 (A) *Nostochopsis*: (a) structure of thallus; (b–d) *N. lobata* (after Bornet and Frémy from Geitler, 1932). (B) *Hapalosiphon*: (a) true branching; (b) *H. hibernicus* (after Frémy from Geitler, 1932); (c) *H. welwitschii* (after Hirose and Hirano, 1981).

Only one freshwater species of *Nostochopsis* is known. It grows attached to mosses, stones, and submersed wood (rarely free floating when old) in unpolluted creeks and streams in temperate, subtropical, and tropical regions. One variable species, *N. lobata*, is referred by Smith (1950) to several findings in the eastern United States; Tilden (1910) listed it from Vermont. It has also been collected from several localities in the Greater Antilles (Cuba; Augusto Comas, personal communication).

Stigonematales: Mastigocladaceae

Hapalosiphon Nägeli ex Bornet *et* Flahault (Fig. 39B)

The thallus is filamentous, composed of free, coiled filamentous clusters, initially attached to the substratum, later floating free in the metaphyton. Filaments are irregularly curved with uniseriate trichomes (basal trichomes and branches), only rarely with some longitudinally dividing cells. True branching is usually lateral (T type), with no morphological differentiation in the main and lateral trichomes (branches rarely slightly narrower). Sheaths are colorless, thin, firm, and rarely indistinctly lamellated. Heterocytes are intercalary. Cells divide horizontally (crosswise) without meristematic zones. Reproduction occurs via hormogonia separated by necridic cells, usually at the ends of trichome branches.

Most species (about 15 have been described) occur in the littoral metaphyton of lakes and in swamps in tropical and temperate zones, usually with macrophytes. Several species prefer moors and peat bogs (*H. fontinalis* and *H. tenuis*). One species from the United States occurs in thermal waters (similar to *H. fontinalis*; Copeland, 1936), and two are subaerial. Smith (1950) recorded *H. aureus*, *H. flexuosus*, *H. fontinalis*, *H. hibernicus*, *H. pumilus*, and *H. welwitschii* in the United States. Prescott (1962) added *H. brasiliensis* and *H. confervaceus* to this list; most of these species (seven of eight) were also recorded from British Columbia, along with *H. delicatulus* and *H. intricatus* (Stein and Borden, 1979). Whitford and Schumacher (1969) recorded five species from freshwaters on submerged plants and wood and two from subaerial habitats. Tilden (1910) listed seven species from the United States and the West Indies.

Mastigocladus Cohn ex Kirchner (Fig. 40A)

The thallus is composed of usually densely tangled filaments forming soft, spongy mats, sometimes containing small carbonate crystals, varying from smooth or gelatinous at the surface to compact, occasionally layered, dirty blue–green to olive-green mass. Solitary filaments occur commonly among other cyanobacteria. Trichomes are uniseriate, irregularly coiled with thin, distinct, colorless sheaths becoming diffuse with age. Branching is true T type or less frequently reverse Y type, often unilateral (in some stages branching is rare or almost absent). Branches continually taper toward the ends. False branches are facultatively present but uncommon. Heterocytes are intercalary, solitary or in pairs (rare). Akinetes are rare, solitary in older parts of the trichomes. Polymorphic species sometimes occur in unbranched, moniliform filaments. Cell division is horizontal (crosswise) or longitudinal before branching. Reproduction occurs via hormogonia or disintegration of trichomes.

The type species, *Mastigocladus laminosus*, is a thermophilic species from hot springs throughout the world, including Yellowstone National Park (Castenholz and Wickstrom, 1975), British Columbia (Stein and Borden, 1979), Oregon (Jackson and Castenholz, 1975), and Antarctica (Broady, 1984, Broady *et al.*, 1987). It is most likely dependent on the following environmental conditions: 45–60°C, pH >7.5, low oxygen content, and low salinity. *M. laminosus* shows an extremely large degree of morphological variability, perhaps suggesting several morphotypes and genotypes.

Thalpophila Borzì (Fig. 40B)

The thallus is filamentous, forming mats on substrata, composed typically of parallel oriented or slightly coiling and creeping true branched filaments. Trichomes are slightly diversified, uniseriate, initially cylindrical, thin and constricted at the cross walls, not attenuated toward the ends, with rounded terminal cells, becoming torulous with age. Lateral T-type branching occurs with branches bending immediately in the direction of the original trichome. Sheaths are thick, lamellated, gelatinizing, from outside becoming confluent or merging with adjacent sheaths. Cells are barrel shaped or cylindrical with a fine granular texture. Heterocytes are uncommon, intercalary, barrel shaped, and solitary. Akinetes are small, forming rows in old trichomes, and embedded with a thick wall. Cell division is horizontal (crosswise). Reproduction occurs via solitary cells (monocytes), liberated from open sheaths. Hormogonia or hormocytes are unknown.

The type species, *Thalpophila cossyrensis*, was described from volcanic rocks near thermal springs (atmophytic) in Italy. Another species, *T. imperialis*, was described by Copeland (1936) from the calcareous substrata of thermal springs in Yellowstone National Park.

NOTE ADDED IN PROOF

A new genus of heterocytous filamentous cyanobateria (Nostocales, Scytonemataceae) was described from United States in 2002.

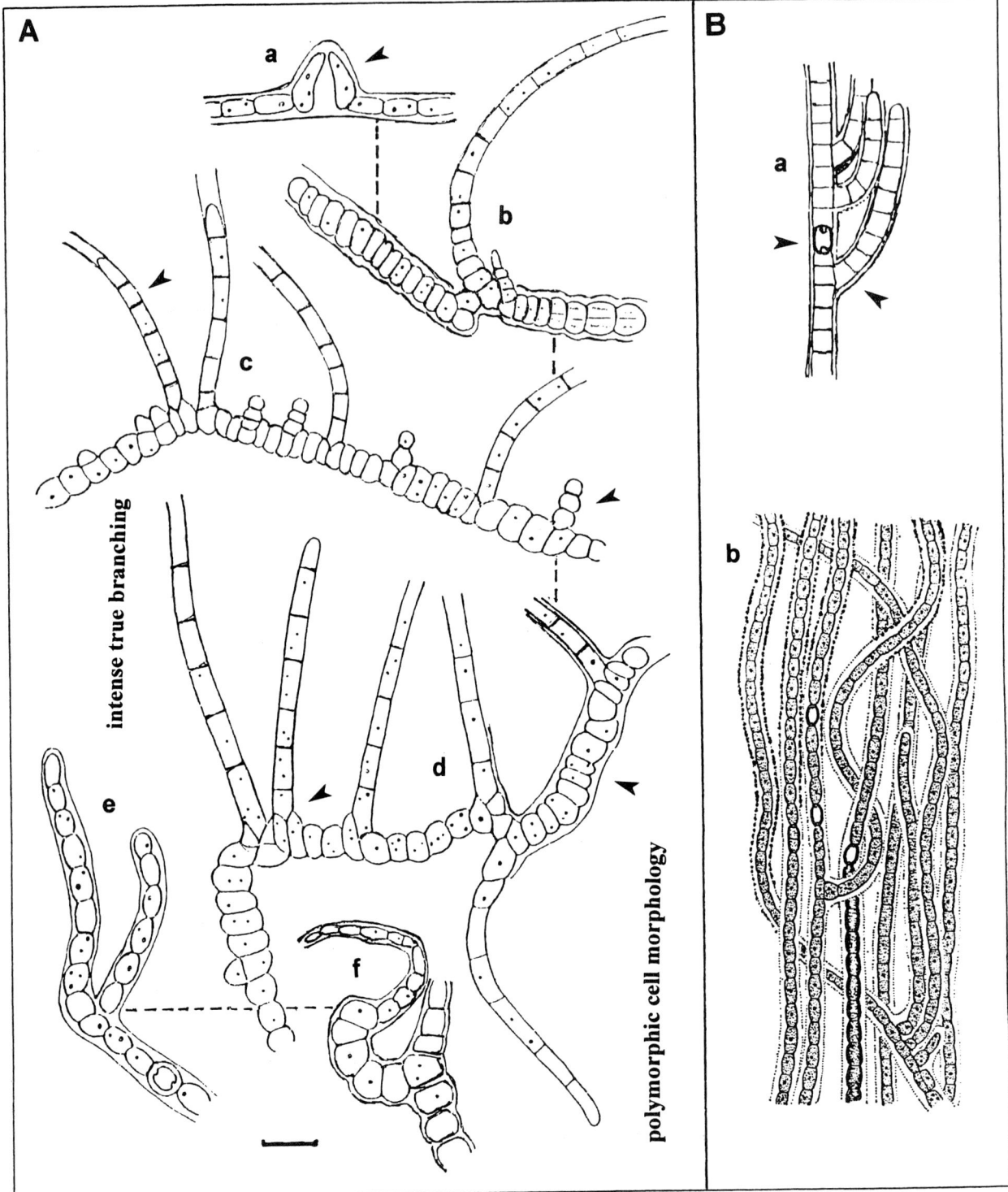

FIGURE 40 (A) *Mastigocladus*: (a–f) *M. laminosus* (bar = 10 μm; after Anagnostidis 1961). (B) *Thalpophila*: (a) true branching; (b) *T. imperialis* (after Copeland, 1936, from Yellowstone National Park).

Spirirestis Flechtner *et* Johansen (Figure 41)

Filaments heteropolar, free, forming tight, regular coils, more or less creeping, densely arranged in colonies, without external mucilage production, 6–20 μm in diameter. Sheats firm, thin or thick, sometimes lamellated. False branches double or single. Trichomes cylindrical, solitary within sheaths, without hairs, rounded at the ends. Cells shorter than broad. Heterocytes basal and intercalar, spherical to oval or appressed. Akinetes not observed. Genus supported by molecular analyses.

The single species, *S. rafaelensis*, was described from soils of a semi-arid juniper community in Utah, USA (Flechtner *et al.* 2002).

VI. GUIDE TO LITERATURE FOR SPECIES IDENTIFICATION

See Chapter 3, page 110 (this volume) for guide.

LITERATURE CITED

Albertano, P., Kováčik, L. 1994. Is the genus *Leptolyngbya* (Cyanophyte) a homogeneous taxon? Archiv für Hydrobiologie/Algological Studies 75:37–51.

Anagnostidis, K. 1961. Untersuchungen über die Cyanophyceen einiger Thermen in Griechenland. Institut für Systematische Botanik und Pflanzengeographie Thessaloniki, 322 p.

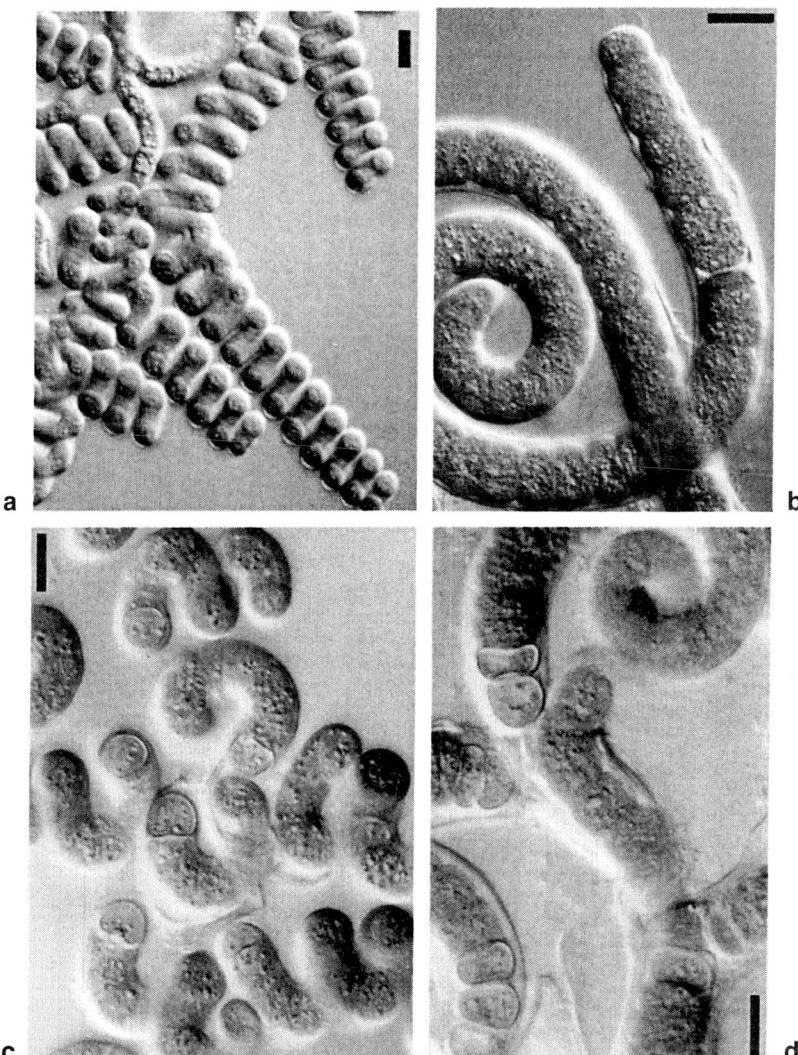

FIGURE 41 *Spirirestis rafaelensis* (a) Appearance of spiraling filaments; (b) filaments ends in relaxed coils, showing sheath, constrictions at crosswalls, and hormogonia; (c) heterocytes forming; (d) variability in heterocyte shape and false branching in relaxed spirals. Reprinted with permission from Flechtner *et al.*, 2002. © 2002 E. Schweizebart'sche Verlagsbuchlandlung. (mail@schweizbartide, www.schweizerbartide). All scales = 10 μm.

Anagnostidis, K. 2001. Nomenclatoric changes in cyanoprokaryotic order Oscillatoriales. Preslia (Prague) 3:359–373.

Anagnostidis, K., Golubić, S. 1966. Über die Ökologie einiger *Spirulina*-Arten. Nova Hedwigia 11:309–335.

Anagnostidis, K., Komárek, J. 1988. Modern approach to the classification system of cyanophytes. 3. Oscillatoriales. Archiv für Hydrobiologie/Algological Studies 50/53:327–472.

Anagnostidis, K., Komárek, J. 1990. Modern approach to the classification system of cyanophytes. 1. Introduction. Archiv für Hydrobiologie/Algological Studies 38/39:291–302.

Anagnostidis, K., Komárek, J. 2003. Cyanoprokaryota. 2. Oscillatoriales, in: Gerloff, J., Heynig, H., Mollenhauer. D., Eds., Süsswasserflora von Mitteleuropa, Vol. 19. Spektrum Verlag, Stuttgart, in press.

Anagnostidis, K., Roussomoustakaki, M. 1985. On the validity of the genus *Symploca* Kütz. ex Gom. Archiv für Hydrobiologie/Algological Studies 38/39:221–234.

Bahal, M., Talpasayi, E. R. S. 1972. Control of heterocyte development in *Anabaena ambigua* Rao, in: Desikachary, T. V., Ed., Taxonomy and biology of blue–green algae. University of Madras, Madras, India, pp. 197–202.

Barbiero, R. P., Welch, H. 1992. Contribution of benthic blue–green algal recruitment to lake populations and phosphorous translocation. Freshwater Biology 27:249–260.

Barker, G. L. A., Handley, B. A., Vacharapiyasophon, P., Stevens, J. R., Hayes, P. K. 2000. Allele-specific PCR shows that genetic exchange occurs among genetically diverse *Nodularia* (cyanobacteria) filaments in the Baltic Sea. Microbiology 146:2865–2875.

Belay, A., Kato, T., Ota, Y. 1996. *Spirulina* (*Arthrospira*): Potential application as an animal feed supplement. Journal of Applied Phycology 8:(4–5)303–311.

Bergman, B. 2002. Nitrogen-fixing Cyanobacteria in the open sea: How primitive organisms can survive a hostile environment, in: Solheim, B., Ventura, S., Wilmotte, A., Eds., Proceedings European Science Foundation CYANOFIX, Longyearbyen, Svalbard, 18 p.

Blackburn, S. I., Jones, G. J. 1995. Toxic *Nodularia spumigena* Mertens blooms in Australian waters: A case study from Orielton Lagoon, Tasmania, in: Lassus, P., Arzul, G., Erard-LeDenn, E., Gentien, P., Marcaillou-LeBaut, C., Eds., Harmful marine algal blooms, Paris, pp. 121–126.

Bornet, E., Flahault, C. 1886–1888. Révision des Nostocacées heterocystées contenues dans les principaux herbiers de France. Annales des Sciences Naturelles; Botanique 3:323–381, 4:343–373, 5:51–129, 7:177–262.

Bothe, H. 1982. Nitrogen fixation, in: Carr, N. G., Whitton, B. A., Eds., The biology of cyanobacteria. Blackwell Science, Oxford, pp. 87–104.

Bourrelly, P. 1966. Quelques algues d'eau douce du Canada. Internationale Revue der Gesamten Hydrobiologie 51:45–126.

Bourrelly, P. 1970. Les algues d'eau douce. III. Les algues bleues et rouges. Les Eugléniens, Peridiniens et Cryptomonadines. Boubée, Cie., Paris, 512 p.

Broady, P. 1984. Taxonomic and ecological investigations of algae on steam-warmed soil on Mt. Erebus, Ross Island, Antarctica. Phycologia 23:257–271.

Broady, P., Given, D., Greenfield, L., Thompson, K. 1987. The biota and environment of fumaroles on Mt. Melbourne, Northern Victoria Land. Polar Biology 7:97–113.

Canter-Lund, H., Lund, J. W. G. 1995. Freshwater algae: Their microscopic world explored. Biopress, Bristol, UK.

Carmichael, W. W. 1992. A review: Cyanobacterial secondary metabolites—the cyanotoxins. Applied Bacteriology 72:445–459.

Carmichael, W. W. 1997. The cyanotoxins. Advances in Botanical Research 27:211–226.

Carmichael, W. W., Ewans, W. R., Yin, Q. Q., Bell, P., Mocauklowski, E. 1997. Evidence for paralytic shellfish poisons in the freshwater cyanobacterium *Lyngbya wollei* (Farlow ex Gomont) comb. nov. Applied and Environmental Microbiology 63:3104–3110.

Carmona-Jimenéz, J., Gold-Morgan, M. 1994. New report for Mexico of *Stauromatonema viride* Frémy, 1930 (Capsosiraceae, Stigonematales). Cryptogamie/Algologie 15:287–296.

Carr, N. G., Whitton, B. A., Eds. 1973. The biology of blue–green algae. Blackwell Sci., Oxford, 676 p.

Carr, N. G., Whitton, B. A., Eds. 1982. The biology of cyanobacteria. Blackwell Science, Oxford, 688 pp.

Castenholz, R. W. 1969a. Thermophilic blue–green algae and the thermal environment. Bacteriological Review 33:476–504.

Castenholz, R. W. 1969b. The thermophilic cyanophytes of Iceland and the upper temperature limit. Journal of Phycology 5:360–368.

Castenholz, R. W. 1976. The effect of sulfide on the blue–green algae of hot springs. I. New Zealand and Iceland. Journal of Phycology 12:54–68

Castenholz, R. W. 1992. Species usage, concept and evolution in the cyanobacteria (blue–green algae). Journal of Phycology 28:737–745.

Castenholz, R. W. 2001. Phylum BX. Cyanobacteria. Oxygenic Photosynthetic Bacteria. in: Boone, D. R. & Castenholz, R. W. eds., Bergey's Manual of Systematic Bacteriology, 2nd ed., Springer, New York, pp. 473–599.

Castenholz, R. W., Wickstrom, C. E. 1975. In: Whitton, B. A., Ed., River ecology. Blackwell Science, Oxford, pp. 264–285.

Chorus, I., Bartram, J., Eds. 1999. Toxic cyanobacteria in water. Spon, London, 416 p.

Cmiech, H. A., Leedale, G. F., Reynolds, C. S. 1986. Morphological and ultrastructural variability of planktonic Cyanophyceae in relation to seasonal periodicity. II. *Anabaena solitaria*: Vegetative cells, heterocytes, akinetes. British Phycological Journal 21:81–92.

Codd, G. A. 1995. Cyanobacterial toxins: Occurrence, properties and biological significance. Water Science and Technology 32:149–156.

Copeland, J. J. 1936. Yellowstone thermal Myxophyceae. Annals of the New York Academy of Sciences 36:1–222.

Cronberg, G., Komárek, J. 1994. Planktic cyanoprokaryotes found in South Swedish lakes during the 12th International Symposium of Cyanophyte Research, 1992. Archiv für Hydrobiologie/Algological Studies 75:323–352.

Cronberg, G., Komárek, J. 2002. Some nostocalean cyanoprokaryotes from lentic habitats of central and southern Africa. Nova Hedwigia, in press.

Daily, W. A. 1943. First reports for the algae *Borzia*, *Aulosira*, and *Asterocystis* in Indiana. Butler University Botanical Studies 6:84–86.

Desikachary, T. V. 1959. Cyanophyta, in: ICAR Monographs on Algae, New Delhi, 686 p.

deVecchi, L., Grilli-Caiola, M. 1986. An ultrastructural and cytochemical study of *Anabaena* sp. (Cyanophyceae) envelopes. Phycologia 25:415–422.

Dodds, W. K. 1989. Photosynthesis of two morphologies of *Nostoc parmelioides* (cyanobacteria) as related to current velocities and diffusion patterns. Journal of Phycology 25:258–262.

Dodds, W. K., Gudder, D. A., Mollenhauer, D. 1995. The ecology of *Nostoc*. Journal of Phycology 31:2–18.

Dokulil, M., Mayer, J. 1996. Population dynamics and photosynthetic rates of *Cylindrospermopsis – Limnothrix* association in a highly eutrophic urban lake, Alte Donau, Vienna, Austria. Archiv für Hydrobiologie/Algological Studies 38:179–195.

Dow, C. S., Swoboda, U. K. 2000. Cyanotoxins, in: Whitton, B. A., Potts, M., Eds., The ecology of cyanobacteria: Their diversity in time and space, Kluwer Academic, Dordrecht, pp. 563–589.

Drews, G. 1959. Beiträge zur Kenntnis der phototaktischen Reaktionen der Cyanophyceen. Archiv für Protistenkunde 104:389–430.

Drews, G. 1973. Fine structure and chemical composition of the cell envelopes, *in*: Carr, N. G., Whitton, B. A., Eds., The biology of blue–green algae, Blackwell Science, Oxford. pp. 99–116.

Drouet, F. 1934. New or interesting Myxophyceae from Missouri. Botanical Gazette 95:695–701.

Drouet, F. 1937. Three American Oscillatoriaceae. Rhodora 35:277–280.

Drouet, F. 1938. The Oscillatoriaceae of southern Massachusetts. Rhodora 40:221–273.

Drouet, F. 1942. Studies in Myxophyceae. I. Botanical Series of the Field Museum of Natural History 20:125–141.

Drouet, F. 1943a. Myxophyceae of eastern California and western Nevada. Botanical Series of the Field Museum of Natural History 20:145–176.

Drouet, F. 1943b. New species and transfers in Myxophyceae. American Midland Naturalist 30:671–674.

Drouet, F. 1968. Revision of the classification of the Oscillatoriaceae. Monographs of the Academy of Natural Science, Philadelphia, 15:370 p.

Drouet, F. 1973. Revision of the Nostocaceae with cylindrical trichomes. Hafner, New York, 292 p.

Drouet, F. 1981a. Revision of the Stigonemataceae with a summary of the classification of blue–green algae. Nova Hedwigia Beihefte 66:1–221.

Drouet, F. 1981b. Summary of the classification of blue–green algae. Nova Hedwigia Beihefte 66:133–209.

Drouet, F., Daily, W. A. 1956. Revision of the coccoid Myxophyceae. Butler University Botanical Studies 12:1–218

Duthie, H. C., Socha, R. 1976. A checklist of the freshwater algae of Ontario, exclusive of the great lakes. Naturaliste Canadien 103:83–109.

Elster, J., Svoboda, J., Komárek, J., Marvan, P. 1997. Algal and cyanoprokaryote communities in a glacial stream, Sverdrup Pass, 79°N, central Ellesmere Island, Canada. Archiv für Hydrobiologie/ Algological Studies 85:57–93.

Falconer, I. R. 1998. Algal Toxins and Human Health, *in*: Hrubec, J., Ed., The handbook of environmental chemistry. Springer-Verlag, Berlin, pp. 53–82.

Fay, P. 1983. The blue–greens (Cyanophyta – Cyanobacteria). Arnold, London.

Fay, P., Van Baalen, C., Eds. 1987. The cyanobacteria. Elsevier, Amsterdam/New York/Oxford, 543 p.

Findlay, D. L., Kling, H. J. 1979. A species list and pictorial reference to the phytoplankton of central and northern Canada. Canadian Fisheries and Marine Services Management Report 1503: 619 p.

Flechtner, V. R., Boyer, S. L., Johansen, J. R., DeNoble, M. L. 2002. *Spirirestris rafaelensis* gen. et sp. nov. (Cyanophyceae), a new cyanobacterial genus from arid soils. Nova Hedwigia 74:1–24.

Fogg, G. E. 1949. Growth and heterocytes production in *Anabaena cylindrica* Lemm. II. In relation to carbon and nitrogen metabolism. Annals of Botany (London) 13:241–259.

Fogg, G. E., Stewart, W.D.P., Fay, P., Walsby, A.E. 1973. The blue–green algae. Academic Press, London/New York, 459 p.

Frémy, P. 1930a. Contribution a la flore algologique du Congo Belge. Bulletin du Jardin Botanique 9:109–138.

Frémy, P. 1930b. Les Myxophycées de l'Afrique équatoriale française. Archives de Botanique 2:1–508.

Friedmann, E. I. 1955. *Geitleria calcarea* n. gen. et n. sp. A new atmophytic lime incrusting blue–green alga. Botaniska Notiser 108:439–445.

Friedmann, E. I. 1979. The genus *Geitleria* (Cyanophyceae or Cyanobacteria): Distribution of *G. calcarea* and *G. floridana* n. sp. Plant Systematics and Evolution 131:169–178.

Galat, D. L., Verdin, J. P., Sims, L. L. 1990. Large-scale patterns of *Nodularia spumigena* blooms in Pyramid Lake, Nevada, determined from Landsat imagery: 1972–1986. Hydrobiologia 197:147–164.

Gardner, N. L. 1927. New Myxophyceae from Porto Rico. Memoirs of the New York Botanical Garden 7:1–144.

Geitler, L. 1932. Cyanophyceae, *in*: Rabenhorst's Kryptogamenflora von Deutschland, Österreich und Schweiz, Vol. 14. Akademisches Verlag, Leipzig, pp. 1–1196.

Geitler, L. 1942. Schizophyta (Klasse Schizophyceae), *in*: Engler, A., Prantl, K., Eds., Natürliche Pflanzenfamilien 1b, Berlin, 232 p.

Geitler, L. 1960. Schizophyceen, *in*: Handbuch der Pflanzenanatomie 6, 1. Spez. Teil, Berlin/Nikolassee, 131 p.

Geitler, L. 1982. Eine bemerkenswerte Oscillatoriacee, *Katagnymene accurata* n. sp. (Cyanophyceae). Plant Systematics and Evolution 140:293–306.

Gibson, C. E. 1975. Cyclomorphosis in natural populations of *Oscillatoria redekei* Van Goor. Freshwater Biology 5:279–286.

Golden, J. W., Robinson, S. J., Haselkorn, R. 1985. Rearrangement of nitrogen fixation genes during heterocyte differentiation in the cyanobacterium *Anabaena*. Nature 314:419–423.

Gold-Morgan, M., Montejano, G., Komárek, J. 1994. Freshwater epiphytic cyanoprokaryotes from central Mexico. 2. Heterogeneity of the genus *Xenococcus*. Archiv für Protistenkunde 144:398–411.

Golecki, J. R., Drews, G. 1974. Zur Structur der Blaualgenzellwand. Gefrierätzungersuchungen der normalen und extrahierten Zellwände von *Anabaena variabilis*. Cytobiology 8:213–227.

Golubić, S. 1967a. Algenvegetation der Felsen. Die Binnengewässer, Vol. 23. Stuttgart, 183 p.

Golubić, S. 1967b. Zwei wichtige Merkmale zur Abgrenzung der Blaualgengattungen. Schweizerische Zeitschrift für Hydrologie 29:176–184.

Golubić, S. 1980. Halophily and halotolerance in cyanophytes. Origins of Life 10:169–183.

Golubić, S., Focke, J. W. 1978. *Phormidium hendersonii* Howe: Identity and significance of a modern stromatolite building microorganism. Journal of Sedimentary Petrology 48:751–764.

Golubić, S., Friedmann, E. I., Schneider, J. 1981. The lithobiontic ecological niche, with special reference to microorganisms. Journal of Sedimentary Petrology 51:475–478.

Golubić, S., Hernandez-Mariné, M., Hoffmann, L. 1996. Developmental aspects of branching in filamentous cyanophyta/ cyanobacteria. Archiv für Hydrobiologie/Algological Studies 83:303–329.

Golubić, S., Perkins, R. D., Lukas, K. J. 1975. Boring microorganisms and microborings in carbonate substrates, *in*: Frey, R. W., Ed., The study of trace fossils. Springer-Verlag, New York, pp. 229–269.

Gomont, M. 1892. Monographie des Oscillariées (Nostocacées homocystées). Annales des Sciences Naturelles 7 – Botanique, Paris, 15:263–368, 16:91–264.

Guglielmi, G., Cohen-Bazire, G. 1982a. Structure et distribution des pores et des perforations de l'enveloppe de peptidoglycane chez quelques cyanobactéries. Protistologica 18:151–165.

Guglielmi, G., Cohen-Bazire, G. 1982b. Étude comparée de la structure et de la distribution des filaments extracellulaires ou fimbriae chez quelques cyanobactéries. Protistologica 18:167–177.

Häder, D.-P. 1974. Participation of two photosystems in the photophobotaxis of *Phormidium uncinatum*. Archives of Microbiology 96:255–266.

Halfen, L. N. 1979. Gliding movements, *in*: Haupt, W., Feinleib, M. E., Eds., Encyclopedia of plant physiology, N.S., Vol. 7, Physiology of movements, Springer-Verlag, Berlin, pp. 250–267.

Haselkorn, R. 1986. Organization of the genes for nitrogen fixation in photosynthetic bacteria and cyanobacteria. Annual Review of Microbiology 40:525–547.

Hayes, P. K., Barker, G. L. A., Walsby, A. E. 1997. The genetic structure of *Nodularia* populations. Abstracts of the Ninth International Symposium on Phototrophic Prokaryotes, Vienna, p. 187.

Hill, D. J. 1969. *Aphanizomenon elenkinii* Kissel. in Minnesota lakes. Journal of the Minnesota Academy of Science 36:55–57.

Hill, D. J. 1970a. *Anabaenopsis raciborskii* Woloszynska in Minnesota lakes. Journal of the Minnesota Academy of Science 36:80–82.

Hill, D. J. 1970b. A new form of *Raphidiopsis mediterranea* Skuja found in Minnesota lakes. Phycologia 9:73–77.

Hill, D. J. 1972. A new *Raphidiopsis* species (Cyanophyta, Rivulariaceae) from Minnesota lakes. Phycologia 11:213–215.

Hill, D. J. 1976a. A new species of *Anabaena* (Cyanophyta, Nostocaceae) from Minnesota lake I. Phycologia 15:61–64.

Hill, D. J. 1976b. A new species of *Anabaena* (Cyanophyta, Nostocaceae) from Minnesota lake II. Phycologia 15:65–68.

Hill, D. J. 1976c. A new species of *Anabaena* (Cyanophyta, Nostocaceae) from Minnesota lake III. Phycologia 15:69–71.

Hindák, F. 1985. The cyanophycean genus *Lemmermanniella* Geitler 1942. Algological Studies 40:393–401.

Hindák, F. 2001. Fotografický atlas mikroskopických siníc. [Microscopic cyanobacteria – microphotos]. Veda, Bratislava, 127 p.

Hoffmann, L., Demoulin, V. 1985. Morphological variability of some species of Scytonemataceae (Cyanophyceae) under different culture conditions. Bulletin de la Société Royale de Botanique de Belgique 118:189–197.

Horecká, M., Komárek, J. 1979. Taxonomic position of three planktonic blue-green algae from the genera *Aphanizomenon* and *Cylindrospermopsis*. Preslia (Prague) 51:289–312.

Huber, A. L. 1984. *Nodularia* (Cyanobacteriaceae) akinetes in the sediments of the Peel–Harvey estuaries, Western Australia: Potential inoculum source for *Nodularia* blooms. Applied and Environmental Microbiology 47:234–238.

Humm, H. J., Wicks, S. R. 1980. Introduction and guide to the marine bluegreen algae. Wiley, New York, 194 p.

Jaag, O. 1945. Untersuchungen über die Vegetation und Biologie der Algen des nackten Gesteins in den Alpen, im Jura und schweizerischen Mittelland. Beihefte zur Kryptogamenflora der Schweiz, Vol. 9, 560 p.

Jackim, E., Gentile, J. 1968. Toxins of a blue-green alga, similarity to saxitoxin. Science 162:915–916.

Jackson, J.E., Castenholz, R.W. 1975. Fidelity of thermophilic blue-green algae to hot springs habitats. Limnology and Oceanography 20:305–322.

Jürgens, U. J., Weckesser, J. 1985. The fine structure and chemical composition of the cell wall and sheath layers of cyanobacteria. Annales de l'Institut Pasteur: Microbiologie 136A:41–44.

Kling, H. 1975. Phytoplankton successions and species distribution in prairie pond of the Erickson-Elphinstone district and southwestern Manitoba. Canadian Fisheries and Marine Services Technical Report 512:1–31.

Kling, H., Holmgren, S. K. 1972. Species composition and seasonal distribution of phytoplankton in the Experimental Lakes Area, northwestern Ontario. Canadian Fisheries Research Board Report 337:1–56.

Kling, H. J., Findlay, D. L., Komárek, J. 1994. *Aphanizomenon schinderi* sp. nov.: A new nostocacean cyanoprocaryote from the Experimental Lakes Area, northwestern Ontario. Canadian Journal of Fisheries and Aquatic Sciences 51:2267–2273.

Kohl, J.-G., Nicklisch, A. 1981. Chromatic adaptation of the planktonic blue-green algae *Oscillatoria redekei* Van Goor and its ecological significance. Internationale Revue der Gesamten Hydrobiologie 66:83–94.

Kohl, J.-G., Dudel, G., Schlangstedt, M., Kuhl, H. 1985. Zur morphologischen und ökologischen Abgrenzung von *Aphanizomenon flos-aquae* und *A. gracile*. Archiv für Hydrobiologie/Algological Studies 38/39:271–272.

Kohl, J.-G., Schlangstedt, M., Dudel, G. 1987. Stabilization of growth during combined nitrogen starvation of the planktonic blue-green alga *Anabaena solitaria* by dinitrogen fixation, *in*: Proceedings from the Symposium Reservoir Limnology, České Budějovice, p. 6.

Komagata, K. 1987. World catalogue of algae. World Data Center, Riken, 146 p.

Komárek, J. 1958. Die taxonomische Revision der planktischen Blaualgen der Tschechoslowakei, *in*: Algologische Studien. Academia, Prague, pp. 10–206.

Komárek, J. 1984. Sobre las cianofíceas de Cuba. 3. Especies planktónicas que forman florecimientos de las aguas. Acta Botanica Cubana 19:1–33.

Komárek, J. 1989. Studies on the cyanophytes of Cuba. 7–9. Folia Geobotanica et Phytotaxonomica 24:131–206.

Komárek, J. 1994. Current trends and species delimitation in the cyanoprokaryote taxonomy. Archiv für Hydrobiologie/Algological Studies 75:11–29.

Komárek, J. 1999. Intergeneric characters in unicellular cyanobacteria, living in solitary cells. Archiv für Hydrobiologie/Algological Studies 94:195–205.

Komárek J. 2001. Review of the cyanoprokaryotic genus *Romeria*. Czech Phycology 1:5–19.

Komárek, J., Albertano, P. 1994. Cell structure of a planktonic cyanoprokaryote, *Tychonema bourrellyi*. Archiv für Hydrobiologie/Algological Studies 75:157–166.

Komárek, J., Anagnostidis, K. 1989. Modern approach to the classification system of cyanophytes. 4. Nostocales. Archiv für Hydrobiologie/Algological Studies 56:247–345.

Komárek, J., Čáslavská, J. 1991. Thylakoidal patterns in oscillatorialean genera. Archiv für Hydrobiologie/Algological Studies 64:267–270.

Komárek, J., Cronberg, G. 2001. Some chroococcalean and oscillatorialean Cyanoprokaryotes from southern African lakes, ponds and pools. Nova Hedwigia 73:129–60.

Komárek, J., Kling, H. 1991. Variation in six planktonic cyanophyte genera in Lake Victoria (East Africa). Archiv für Hydrobiologie/Algological Studies 61:24–45.

Komárek, J., Watanabe, M. 1990. Morphology and taxonomy of the genus *Coleodesmium* (Cyanophyceae/Cyanobacteria), *in*: Watanabe, M., Malla, S. R., Eds., Cryptogams of the Himalayas, Vol. 2. Central and Eastern Nepal. Tsukuba, pp. 1–22.

Komárek J., Hübel M., Hübel, H., Šmarda, J. 1993. The *Nodularia* studies 2. Taxonomy. Archiv für Hydrobiologie/Algological Studies 68:1–25.

Komárková, J. 1998. The tropical planktonic genus *Cylindrospermopsis* (Cyanophytes, Cyanobacteria). Anais 4. Congreso Latino-Americano, 2. Reunion Ibera-Americano, 7. Reunion Brasileiro de Ficologia 1:324–340.

Komárková, J. 2002. Cyanobacterial picoplankton and its colonial formations in two eutrophic canyon reservoirs (Czech Republic). Archiv für Hydrobiologie, in press.

Komárková, J., Ladares-Silva, R., Senna, P. A. C. 1999. Extreme morphology of *Cylindrospermopsis raciborskii* (Nostocales, Cyanobacteria) in the Lagoa do Peri, a freshwater coastal lagoon, Santa Catarina, Brazil. Archiv für Hydrobiologie/Algological Studies 94:207–222.

Komárková-Legnerová, J., Tavera, R. 1996. Cyanoprokaryota (Cyanobacteria) in the phytoplankton of Lake Catemaco (Veracruz, Mexico). Archiv für Hydrobiologie/Algological Studies 83:403–22.

Kondrateva, N. V. 1961. *Lyngbya aestuarii* (Mert.) Liebm. e salsis superficialibus Tauriae. Botaničeskie Materialy Otdela Sporovych Rastenij 14:75–82.

Kondrateva, N. V. 1968. Voprosy morfologii i sistematiki *Microcystis aeruginosa* Kuetz. emend. Elenk. i blizkich k nemu vidov [Problem of morphology and systematics of *Microcystis aeruginosa* Kuetz. emend. Elenk. and related species., *in*: Cvetenie vody, Kiev, pp. 13–42.

Kondrateva, N. V. 1972. Morfologija i sistematika gormogonievych vodoroslej, vyzyvajuščich "cvetenie" vody v Dnepre i dneprovskich vodochraniliščach [Morphology and systematics of hormogonal algae, causing the "water blooms" in the river Dnepr and Dnepr-reservoirs.] Izd. "Naukova dumka", Kiev, 150 p.

Kosinskaja, E. K. 1948. Opredelitel' morskich sinezelenych vodoroslej. [Identification key of marine cyanophytes.] Izd. AN SSSR, Moskva-Leningrad, 278 p.

Lang, N. J., Fay, P. 1971. The heterocytes of blue–green algae. II. Details of ultrastructure. Proceedings of the Royal Society of London Series B 178:193–203.

Li, R., Watanabe, M. M. 1998. The taxonomic studies of water-bloom forming species of Anabaena (Cyanobacteria) based on morphological, physiological, biochemical and genetic characteristics. Complete Abstracts of the Fourth International Conference on Toxic Cyanobacteria, Beaufort, NC, 67 p.

Li, R., Watanabe, M., Watanabe, M. M. 2000. Taxonomic studies of planktic species of Anabaena based on morphological characteristics in cultured strains. Hydrobiologia 438:117–138.

Livingstone, D., Khoja, T. M., Whitton, B. A. 1983. Influence of phosphorus on physiology of a hair-forming blue-green alga Calothrix parietina. British Phycological Journal 18:29–38.

Lund, J. W. G. 1955. Contribution to our knowledge of British algae. XIV. Three new species from the English Lake District. Hydrobiologia 7:219–229.

Mahmood, N. A., Carmichael, W. W. 1986. Paralytic shellfish poisons produced by the freshwater cyanobacterium Aphanizomenon flos-aquae NH-5. Toxicon 24:175–186.

Martin, T. C., Wyatt, J. T. 1974. Extracellular investments in blue–green algae with particular emphasis on the genus Nostoc. Journal of Phycology 10:204–210.

Meffert, M. E. 1987. Planktic unsheathed filaments (Cyanophyceae) with polar and central gas vacuoles. I. Their morphology and taxonomy. Archiv für Hydrobiologie, Suppl. 76:315–346.

Meffert, M. E. 1988. Limnothrix Meffert nov. gen. The unsheathed planktic cyanophycean filaments with polar and central gas vacuoles. Archiv für Hydrobiologie/Algological Studies 50–53:269–276.

Meffert, M. E., Krambeck, H. J. 1977. Planktonic blue–green alga of the Oscillatoria redekei group. Archiv für Hydrobiologie 82:231–239.

Mercado, A. 1977. Raíces coraloides de Microcycas calocoma: Estructura, desarrollo y endófitas presentes. Ciencias Biologicas, Academia de Ciencias, Cuba 1:3–40.

Mollenhauer, D. 1970. Beiträge zur Kenntnis der Gattung Nostoc Abhandlungen der Senckenbergischen Naturforschenden Gesellschaft 524:1–80.

Mollenhauer, D., Mollenhauer, R. 1996. Nostoc in symbiosis—taxonomic implications. Archiv für Hydrobiologie/Algological Studies 83:435–446.

Mollenhauer, D., Mollenhauer, R., Kluge, M. 1996. Studies on initiation and development of the partner association in Geosiphon pyriforme (Kütz.) v. Wettstein, a unique endocytobiotic system of a fungus (Glomales) and the cyanobacterium Nostoc punctiforme (Kütz.) Hariot. Protoplasma 193:3–9.

Mühling, M., Scott, M., Harris, N., Whitton, B. A. 1997. Characterization of Arthrospira and Spirulina strains: Phenotypic features. Abstracts of the Ninth International Symposium on Phototrophic Prokaryotes, Vienna, 188 p.

Niiyama, Y., Watanabe, M., Umezaki, I. 1993. Introduction of "Modern approach to the classification system of cyanophytes by K. Anagnostidis and J. Komárek". Japanese Journal of Phycology 41:55–67.

Nordin, R. N., Stein, J. R. 1980. Taxonomic revision of Nodularia (Cyanophyceae/Cyanobacteria). Canadian Journal of Botany 58:1211–1224.

Nultsch, W. 1974. Movements, in: Stewart, W. D. P., Ed., Algal physiology and biochemistry, Blackwell Science, Oxford, pp. 864–893.

Nygaard, G. 1984. Freshwater phytoplankton from the Nerssaq area, South Greenland. Botanisk Tidsskrift 73:191–238.

Oliver, R. L., Ganf, G. G. 2000. Freshwater blooms, in: Whitton, B. A., Potts, M. Eds., The ecology of cyanobacteria: Their diversity in time and space. Kluwer Academic, Dordrecht, pp. 149–194.

Padisák, J. 1990. Occurrence of Anabaenopsis raciborskii Wołosz. in the pond Tómalom near Sopron, Hungary. Acta Botanica Hungarica 36:163–165.

Padisák, J. 1997. Cylindrospermopsis raciborskii (Wołoszyńska) Seenayya et Subba Raju, an expanding, highly adaptive cyanobacterium: Worldwide distribution and review of its ecology. Archiv für Hydrobiologie 107:563–593.

Pankow, H. 1976. Algenflora der Ostsee. Vol. 2. Plankton. Fischer-Verlag, Stuttgart, 493 p.

Pentecost, A., Whitton, B. A. 2000. Limestones, in: Whitton, B. A., Potts, M., Eds., The ecology of cyanobacteria: Their diversity in time and space, Kluwer Academic, Dordrecht, pp. 257–279.

Perez, M.C., Bonilla, S., deLeón, L., Šmarda, J., Komárek, J. 1999. A bloom of Nodularia baltica-spumigena group (Cyanobacteria) in a shallow coastal lagoon of Uruguay, South America. Archiv für Hydrobiologie/Algological Studies 93:91–101.

Prát, S. 1929. Studie o biolithogenesi [Study about biolithogenesis] Nákladem České Akademie Věd a Umění. Prague, 187 p.

Prescott, G.W. 1951. Algae of the Western Great Lakes Area. Cranbrook Inst. Sci., Bull. 31, Bloomfield Hills, MI, 946 p.

Prescott, G. W. 1962. Algae of the Western Great Lakes Area, 2nd ed. W. C. Brown, Dubuque, IA, 977 p.

Prinsep, M. R., Caplan, F. R., Moore, R. E., Patterson, G. M. L., Honkanen, R. E., Boynton, A. L. 1992. Microcystin-LR from a blue–green alga belonging to the Stignonematales. Phytochemistry 31:1247–1248.

Rippka, R., Deruelles, J. B., Waterbury, J. B., Herdman, M., Stanier, R. Y. 1979. Generic assignments, strain histories and properties of pure cultures of cyanobacteria. Journal of General Microbiology 111:1–61.

Rother, J. A., Fay, P. 1977. Sporulation and the development of planktonic blue–green algae in two Salopian meres. Proceedings of the Royal Society of London Series B 196:317–332.

Rott, E., Pipp, E. 1999. Progress in the use of benthic algae for monitoring rivers in Austria, in: Prygiel, J., Whitton, B. A., Bukowska, J., Eds., Use of algae for monitoring rivers, Vol. 3. Agence de l'Eau Artoise-Picardie, Douai, France, pp. 110–112.

Rudi, K., Skulberg, O. M., Jakobsen, K. S. 1998. Evolution of cyanobacteria by exchange of genetic material among phyletically related strains. Journal of Bacteriology 180:3453–3461.

Rudi, K., Skulberg, O. M., Skulberg, R., Jakobsen, K. S. 2000. Application of sequence-specific labelled 16S rRNA gene oligonucleotide probes for genetic profiling of cyanobacterial abundance and diversity by array hybridization. Applied and Environmental Microbiology 66:4004–4011.

Senna, P. A. C., Komárek, J.1998. Dasygloea brasiliensis n. comb. (syn.: Lyngbya brasiliensis), Cyanoprokaryotes, from the central part ("cerrados") of Brazil. Archiv für Hydrobiologie/Algological Studies 88:1–16.

Seshadri, C. V., Jeeji-Bai, N., Eds. 1992. Spirulina. ETTA National Symposium, Shri AMM Murugappa Chettiar Research Centre (MCRC), Madras, 177 p.

Setchell, W. A., Gardner, N. L. 1918. New Pacific Coast algae. University of California Publications of Botany. Berkeley, CA.

Setchell, W. A., Gardner, N. L. 1924. New marine algae from the Gulf of California. California Academy of Science Series 4, 12:695–949.

Sheath, R. G., Cole, K. M. 1992. Biogeography of stream macroalgae in North America. Journal of Phycology 28:448–460.

Sheath, R. G., Steinman, A. D. 1982. A checklist of freshwater algae of

the Northwest Territories, Canada. Canadian Journal of Botany 60:1964–1997.
Skácelová, O., Komárek, J. 1989. Some interesting cyanophyte species from the Kutnar reserve (South Moravia, Czechoslovakia). Acta Musei Moraviae, Scientiae Naturales 74:101–116.
Skulberg, O. M., Skulberg, R. 1985. Planktonic species of *Oscillatoria* (Cyanophyceae) from Norway—characterization and classification. Archiv für Hydrobiologie/Algological Studies 38–39:157–174.
Skulberg, O. M., Carmichael, W. W. Codd, G. A., Skulberg, R. 1993. Taxonomy of toxic Cyanophyceae (cyanobacteria), *in*: Falconer, I. R., Ed., Algal toxins in seafood and drinking water. Academic Press, London, pp. 145–164.
Smith, G. M. 1920. Phytoplankton of the Inland Lakes of Wisconsin. I. Wisconsin Geological Natural History Survey 57, 12:1–243.
Smith, G. M. 1950. Freshwater algae of the United States, 2nd ed. McGraw-Hill, New York, 719 p.
Stal, L. J. 2000. Cyanobacterial mats and stromatolites, *in*: Whitton, B. A., Potts, M., Eds., The ecology of cyanobacteria, Kluwer Academic, Dordrecht, pp. 61–120.
Starmach, K. 1957. *Plectonema batrachospermi* sp. n. Acta Societatis Botanicorum Poloniae 26:565–568.
Starmach, K. 1966. *Cyanophyta—sinice* [Cyanophyta—blue–green algae.] Flora słodkow. Polski 2. Polskie Wydawnictwo Naukowe, Warsaw, 753 p.
Stein, J. R. 1975. Freshwater algae of British Columbia: The Lower Fraser Valley. Syesis 8:119–184.
Stein, J. R., Borden, C. A. 1979. Checklist of freshwater algae of British Columbia. Syesis 12:3–39.
Stewart, W. D. P. 1972. Heterocytes of blue–green algae, *in*: Desikachary, T. V., Ed., Taxonomy and biology of blue–green Algae. Univ. of Madras, Madras, India, pp. 227–235.
Stewart, W. D. P. 1980. Some aspects of structure and functions in N_2-fixing cyanobacteria. Annual Review of Microbiology 34:497–536.
Stockner, J. G., Callieri, C., Cronberg, G. 2000. Picoplankton and other non-bloom forming cyanobacteria in lakes, *in*: Whitton, B. A., Potts, M. 2000. Eds., The ecology of cyanobacteria: Their diversity in time and space, Kluwer Academic, Dordrecht, pp. 195–231.
Sugawara, H., Miyazaki, S., Eds. 1999. World directory of collections of cultures of microorganisms, 5th ed. WFCC-MIRCEN World Data Centre of Microorganisms, 140 p.
Sutherland, J. M., Reaston, J., Stewart, W. D. P., Herdman, M. 1985a. Akinetes of the cyanobacterium *Nostoc* PCC 7524: Macromolecular composition, structure and control of differentiation. Journal of General Microbiology 131:2855–2863.
Sutherland, J. M., Stewart, W. D. P., Herdman, M. 1985b. Akinetes of the cyanobacterium *Nostoc* PCC 7524: Morphological changes during synchronous germination. Archives of Microbiology 142:269–274.
Taft, C. E., Taft, C. W. 1970. The algae of western Lake Erie. Bulletin of the Ohio Biological Survey, N.S., Columbus, 185 pp.
Tavera, R., Komárek, J. 1996. Cyanoprokaryotes in the volcanic lake of Alchichica, Puebla State, Mexico. Archiv für Hydrobiologie/Algological Studies 83:511–538.
Tiffany, L. H., Britton, M. E. 1952. The algae of Illinois. Univ. of Chicago Press, Chicago.
Tilden, J. 1910. Minnesota algae. I. The Myxophyceae of North America and adjacent regions. Minneapolis, 328 p.
Tomaselli, L., Palandri, M. R., Tredici, M. R. 1996. On the correct use of the *Spirulina* designation. Archiv für Hydrobiologie/Algological Studies 83:539–548.

Umezaki, I. 1974. On the taxonomy of the genus *Trichodesmium*. Bulletin of the Plankton Society of Japan 20:93–100.
Van Baalen, C. 1987. Nitrogen fixation, *in*: Fay, P., Van Baalen, C., Eds., The cyanobacteria. Elsevier, Amsterdam, pp. 187–198.
Villeneuve V., Vincent, W. F., Komárek, J. 2001. Structure and microhabitat characteristics of cyanobacterial mats in an extreme high Arctic environment: Ward Hunt Lake (lat. 82.8 N). Nova Hedwigia Beihefte, 123:199–223.
Vincent, W. F. 2000. Cyanobacterial dominance in the polar regions, *in*: Whitton, B. A., Potts, M., Eds., The ecology of cyanobacteria: Their diversity in time and space, Kluwer Academic, Dordrecht, pp. 321–340.
Vonshak, A., Ed. 1997. *Spirulina platensis* (Arthrospira). Physiology, cell-biology and biotechnology. Taylor, Francis, London, 233 pp.
Ward, A. K., Dahm, C. N., Cummins, K. W. 1985. *Nostoc* (Cyanophyta) productivity in Oregon stream ecosystems: Invertebrate influences and differences between morphological types. Journal of Phycology 21:223–227.
Ward, D. M., Castenholz, R. W. 2000. Cyanobacteria in geothermal habitats, *in*: Whitton, B. A., Potts, M., Eds., The ecology of cyanobacteria: Their diversity in time and space. Kluwer Academic, Dordrecht, pp. 37–59.
Watanabe, M. 1971. The species of *Anabaena* from Hokkaido. Journal of Japanese Botany 46:263–276.
Watanabe, M., Komárek, J. 1989. New *Blennothrix*-species (Cyanophyceae/Cyanobacteria) from Nepal. Bulletin of the National Science Museum Series (Botany), 15:67–79.
Whale, G. F., Walsby, A. E. 1984. Motility of the cyanobacterium *Microcoleus chthonoplastes* in mud. British Phycological Journal 19:117–123.
Whelden, R.M. 1947. Algae. *In* Polunin, N. [Ed.] Botany of Canadian eastern Arctic, II. Thallophyta and Bryophyta. Bulletin of the National Museum of Canada 97:13–127.
Whitford, L. A., Schumacher, G. J. 1963. Communities of algae in North Carolina streams and their seasonal relations. Hydrobiologia 22:133–196.
Whitford, L. A., Schumacher, G. J. 1969. A manual of the freshwater algae in North Carolina. North Carolina Agricultural Experiment Station Technical Bulletin 188:1–313.
Whitton, B. A., Ed. 1984. Ecology of European rivers. Blackwell Sci., Oxford.
Whitton, B. A. 1987. The biology of Rivulariaceae, *in*: Fay, P., Van Baalen, C., Eds., The cyanobacteria. Elsevier, Amsterdam/New York/Oxford, pp. 513–534.
Whitton, B. A. 2000. Soils and rice-fields, *in*: Whitton, B. A., Potts, M., Eds., The ecology of cyanobacteria: Their diversity in time and space. Kluwer Academic, Dordrecht, pp. 233–255.
Whitton, B. A., Peat, A. 1969. On *Oscillatoria redekei* Van Goor. Archives of Microbiology 68:362–376.
Whitton, B. A., Potts, M., Eds., The ecology of cyanobacteria: Their diversity in time and space. Kluwer Academic, Dordrecht.
Winkenbach, F., Wolk, C. P. 1973. Activities of enzymes of the oxidative and the reductive pentose phosphate pathways in heterocytes of blue-green alga. Plant Physiology 52:480–483.
Wolk, C. P. 1973. Physiology and cytological chemistry of blue–green algae. Bacteriological Review 37:32–101.
Wolk, C. P. 1982. Heterocytes, *in*: Carr, N. G., Whitton, B. A., Eds., The biology of blue–green algae. Blackwell Sci., Oxford, pp. 359–386.
Wynn-Williams, D. D. 2000. Cyanobacteria in deserts—life at the limit? *in*: Whitton, B. A., Potts, M., Eds., The ecology of cyanobacteria: Their diversity in time and space. Kluwer Academic, Dordrecht, pp. 341–366.

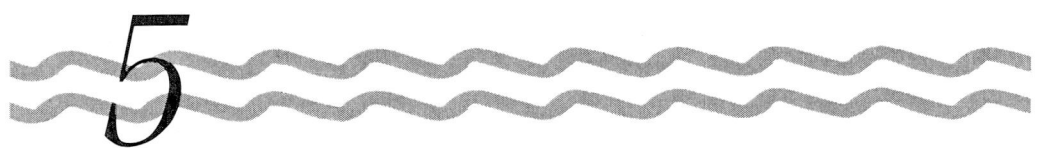

RED ALGAE

Robert G. Sheath

Office of Provost and Vice President for Academic Affairs
California State University, San Marcos
San Marcos, California 92096

I. Introduction
II. Diversity and Morphology
   A. Diversity
   B. Vegetative Morphology
   C. Reproduction
III. Ecology and Distribution
   A. Streams and Rivers
   B. Other Inland Habitats
IV. Collection and Preparation for Identification
V. Key and Descriptions of Genera
   A. Key
   B. Descriptions of Genera
VI. Guide to Literature for Species Identification
Literature Cited

The freshwater red algae in form, in physiology and in habitat may be considered as comprising an elite group of plants, long neglected by American botanists and thus all the more enchanting as representing a research frontier rich with the promise of happy days in the field and laboratory (Flint, 1970, p. 18).

I. INTRODUCTION

The Rhodophyta, the red algae, constitute a division of organisms that share the following combination of attributes: eukaryotic cells, lack of flagella, floridean starch, phycobiliprotein pigments (red and blue), unstacked thylakoids, and chloroplasts lacking an external endoplasmic reticulum (Woelkerling, 1990). They are primarily marine in distribution, with only approximately 3% of the over 5000 species occurring in truly freshwater habitats (Sheath, 1984). Most of the inland species of Rhodophyta are restricted to streams and rivers (lotic forms), although a few species are found in lakes and ponds (lentic forms) (Sheath and Hambrook, 1990). A small number of inland rhodophytes occurs in habitats other than typical freshwaters, such as hot springs (Doemel and Brock, 1971), soils (Geitler, 1932), caves (Nagy, 1965; Hoffmann, 1989), and even sloth hair (Wujek and Timpano, 1986).

Freshwater red algae in North America represent a widespread division, with members that range from the high Arctic to the tropical rainforest (Sheath and Hambrook, 1990). The number of species increases from the tundra to the tropics (6 to 25), similar to the trend seen for marine species of Rhodophyta.

II. DIVERSITY AND MORPHOLOGY

A. Diversity

Freshwater rhodophytes have a relatively low diversity compared to other major groups of algae. Analysis of the literature reveals a flora with 65 infrageneric taxa and 25 genera from North America (Table I). Among these taxa, 14 infrageneric taxa are known only from North America. The subclass Florideophycidae accounts for 52 infrageneric taxa, whereas the Bangiophycidae consists of 13 species. *Batrachospermum* is the most diverse genus, with 26, or 40%, of the total infrageneric taxa.

I am proposing in this chapter that there are three potential origins of the predominantly lotic forms: (1)

TABLE I Taxa, Habitats, Forms, Chloroplast Types and Reproduction of Inland Rhodophyta in North America

Taxon	Habitat	Form[c]	Chloroplast type[d]	Reproductive dissemination[e]
Bangiophycidae				
Porphyridiales				
Chroodactylon ornatum[a]	Lake, stream	pf	cs	ms
Chroothece mobilis	Cool spring	pf	cs	cd
Cyanidium caldarium	Hot spring	u	pd	cd
Flintiella sanguinaria	Cool spring	u	pl	cd
Kyliniella latvica	Stream	pf	pmd	cd
Porphyridium purpureum	Soil	u	cs	cd
P. sordidum	Soil	u	cs	cd
Rufusia pilicola[b]	Sloth hair	pf	pd	cd, es
Bangiales				
Bangia atropurpurea[a]	Lake, stream	ff	cs	ms
Compsopogonales				
Boldia erythrosiphon[b]	Stream	tu	pmd	ms
Compsopogon coeruleus	Stream, cool spring	ff	pmd	ms
C. prolificus	Stream	ff	pmd	ms
Compsopogonopsis leptocladus	Stream, cool spring	ff	pmd	ms
Florideophycidae				
Acrochaetiales				
Audouinella eugenea	Stream	tf	pmd	ms
A. hermannii	Stream	tf	pmd	ms, cs, ts
A. macrospora	Stream	tf	pmd	ms
A. pygmaea	Stream	tf	pmd	ms
A. tenella[b]	Stream	tf	pmd	ts
Balbianiales				
Rhododraparnaldia oregonica[b]	Stream	m	pmd	cs
Batrachospermales				
Batrachospermaceae				
Batrachospermum				
Section Contorta				
B. ambiguum	Stream	gf	pmd	cs
B. globosporum	Stream	gf	pmd	cs
B. intortum	Stream	gf	pmd	cs, ms
B. louisianae[b]	Stream	gf	pmd	cs
B. procarpum	Stream	gf	pmd	cs
Section Setacea				
B. androinvolucrum[b]	Stream	gf	pmd	cs
B. atrum	Stream	gf	pmd	cs
Section Virescentia				
B. elegans	Stream	gf	pmd	cs
B. helminthosum	Stream	gf	pmd	cs
Section Aristata				
B. macrosporum	Stream	gf	pmd	cs
Section Turfosum				
B. turfosum	Stream, pond, bog	gf	pmd	cs, ms
Section Hybrida				
B. virgato-decaisneanum	Stream	gf	pmd	cs
Section Batrachospermum				
B. anatinum	Stream, cool spring	gf	pmd	cs
B. arcuatum	Stream	gf	pmd	cs
B. boryanum	Stream, cool spring	gf	pmd	cs
B. carpocontortum[b]	Stream	gf	pmd	cs
B. carpoinvolucrum[b]	Stream, cool spring	gf	pmd	cs
B. confusum	Stream, cool spring	gf	pmd	cs
B. gelatinosum	Stream, cool spring	gf	pmd	cs
B. gelatinosum forma spermatoinvolucrum	Stream	gf	pmd	cs
B. heterocorticum[b]	Stream	gf	pmd	cs
B. involutum[b]	Stream, cool spring	gf	pmd	cs
B. pulchrum[b]	Stream	gf	pmd	cs

(Continues)

TABLE I (Continued)

Taxon	Habitat	Form[c]	Chloroplast type[d]	Reproductive dissemination[e]
B. skujae	Stream	gf	pmd	cs, ms
B. trichocontortum[b]	Stream	gf	pmd	cs
B. trichofurcatum[b]	Stream	gf	pmd	cs
Other genera				
Sirodotia huillensis	Stream, cool spring	gf	pmd	cs
S. suecica	Stream	gf	pmd	cs
Tuomeya americana[b]	Stream	pp	pmd	cs
Lemaneaceae				
Lemanea borealis	Stream	pp	pmd	cs
L. fluviatilis	Stream	pp	pmd	cs
L. fucina var. parva[b]	Stream	pp	pmd	cs
Paralemanea annulata	Stream	pp	pmd	cs
P. catenata	Stream	pp	pmd	cs
P. mexicana[b]	Stream	pp	pmd	cs
Thoreales				
Nemalionopsis tortuosa	Stream	ff	pmd	ms
Thorea hispida	Stream, cool spring	ff	pmd	ms, cs?
T. violacea	Stream, cool spring	ff	pmd	ms
Hildenbrandiales				
Hildenbrandia angolensis	Stream, cool spring	cr	pmd	g
Ceramiales				
Ballia prieurii[a]	Stream	ff	pmd	f?
Bostrychia moritziana[a]	stream	ff	pmd	st
B. radicans[a]	Stream	ff	pmd	st
B. tenella[a]	Stream	ff	pmd	st
Caloglossa leprieurii[a]	Stream	pp	pmd	f?
C. ogasawaerensis[a]	Stream	pp	pmd	f?
Polysiphonia subtilissima[a]	Stream, cool spring	ff	pmd	f?

[a] Potential invader from brackish/marine habitats.
[b] Unique to North America.
[c] pf, pseudofilament; u, unicell; ff, free filament; tu, tube; tf, tuft; m, mat; gf, gelatinous filament; pp, pseudoparenchymatous form; cr, crust
[d] cs, central stellate with pyrenoid; pd, single peripheral disc; pl, peripheral, intricate lamellate; pmd, peripheral, multiple disc.
[e] ms, monosporangia; cd, cell division; es, endospores; cs, carpospores; fs, tetraspores; g, gemmae; f, fragmentation; st, stichidia.

specialists that evolved early within the stream environment and are absent in other habitats, (2) generalists that occur in a wide range of other freshwater bodies, such as lakes and ponds, and (3) upstream migrants from estuaries. This proposal is an expanded version of that originally given by Skuja (1938). Of the 60 stream-inhabiting taxa, 50 appear to be specialists, whereas three are generalists and nine are potentially brackish/marine invaders (Table I). Note that there is overlap in two of the species, *Chroodactylon ornatum* and *Bangia atropurpurea*, in the last two categories. The other generalist species is *Batrachospermum turfosum* (Sheath et al., 1994c; Müller et al., 1998).

B. Vegetative Morphology

The red algae occurring in typical freshwater habitats tend to be macroscopic and benthic (as defined in Chapter 2) (Sheath and Hambrook, 1990). Nonetheless, these algae exhibit a smaller size range than do marine species with the majority (80%) of freshwater rhodophytes having a length range of 1–10 cm. Among the forms occurring in North America, there are 28 gelatinous filaments, 12 free filaments (individual filament without a gelatinous matrix) (e.g., Figs. 2B, 6H, I, and 7B, F, G), nine pseudoparenchymatous forms (tissue-like but composed of compacted filaments) (Figs. 4H, 5A–F, and 7C–E), five tufts (short radiating filaments without a common matrix) (Fig. 2F), two pseudofilaments (loose chains of cells held together with a common gelatinous matrix) (Fig. 1C–E), and one each of unicells (Fig. 1A), tubes (Fig. 1H), mats (flat plant body composed of tightly interwoven filaments), and crusts (flat thallus composed of compacted tiers of cells) (Fig. 6E, Table I). Species distributed in hot springs or soils are unicellular and *Rufusia* on sloth hair is pseudofilamentous. Among the 58 forms that have a filamentous construction, only three have multiaxial growth, the members of the Thoreales (Figs. 5G, H and 6A, B) (Sheath et al., 1993b); the rest are uniaxial, except for *Bangia*, which has a uniaxial base and multiaxial apex at maturity (Fig. 1F). Uniaxial filaments may be corticated with one or more layers of smaller cells covering those of the main axis (Figs. 6I and 7A, B, F, G).

The various morphological forms encounter the stress caused by flow in riverine habitats in various ways according to Sheath and Hambrook (1988, 1990). Crusts and short tufts occur within the boundary layer or at least in a region of reduced current velocity and hence avoid much of the flow-related stress. The remaining macroscopic species can be regarded as semierect, experiencing bending, tensile, and compressive forces and perhaps torsional stresses in flowing waters (Vogel, 1984). This group includes mucilaginous and nonmucilaginous filaments and pseudoparenchymatous forms and tubes. It would be expected that the semierect forms possess adaptive mechanisms to tolerate flow, such as branch reconfiguration and extension of thalli in high water motion (Sheath and Hambrook, 1988,1990).

Of the 65 infrageneric taxa of Rhodophyta in inland habitats in North America, 20 are reddish, while 45 are largely blue to olive in color. This trend contrasts with that of marine red algae, in which the great majority of species appear red in color. Nonetheless, there are a number of chloroplast morphologies among freshwater taxa: all members of the Florideophycidae have multiple discoidal or ribbon-like chloroplasts without pyrenoids (e.g., Fig. 2G); the bangiophycidean taxa have this type as well as central stellate chloroplasts with a pyrenoid (Fig. 1A), a peripheral lamellate structure, and a single peripheral disc (Table I). However, some of the species, which appear to have multiple discoidal or ribbon-like chloroplasts, may actually contain a complex, interconnected single chloroplast, such as *Compsopogon coeruleus* (Gantt et al., 1986). Like other red algae, the chloroplasts of freshwater species contain single thylakoids with phycobilisomes (granules consisting of the accessory pigments) on both sides (Sheath, 1984). The phycobilisomes of blue-colored species, such as *Porphyridium aerugineum* and *Compsopogon coeruleus*, tend to be hemidiscoidal in shape and predominated by the blue pigment phycocyanin (Gantt et al., 1986). In contrast, the phycobilisomes of the red-colored *Porphyridium purpureum* are larger and hemispherical and composed mostly of the red pigment phycoerythrin. Phycoerythrin and photosystem (PS) II activity appear to be absent from the pyrenoid of *P. purpureum* whereas PS I and ribulose-1,5-bisphosphate carboxylase/oxygenase (RuBisCO) activities can be detected in this structure (McKay and Gibbs, 1990). Chloroplasts of *Batrachospermum gelatinosum* develop from proplastids, which have a double-membraned envelope and a parallel outer photosynthetic thylakoid (Brown and Weier, 1968). This outer thylakoid functions in the production of additional ones to the interior. Thylakoids have been observed to be coiled in serial sections of *Cyanidium*, *Compsopogon*, and *Batrachospermum* (Pueschel, 1990). They can fragment and form dilated tubular units in some freshwater species, such as *Batrachospermum gelatinosum* and *Sirodotia suecica*, when subjected to reduced illumination (Sheath et al., 1979).

Another characteristic that is useful in analyzing the morphology of freshwater Rhodophyta is the external covering. Unicells and pseudofilaments typically have a gelatinous matrix surrounding the cells (e.g., Fig. 1C, D, G) which varies in thickness, depending on the age and physiological state of the organisms (Sheath, 1984). The gelatinous filamentous members of the Batrachospermales, such as *Batrachospermum* and *Sirodotia*, have distinct cell walls as well as an overall matrix surrounding the filament. The free filaments and pseudoparenchymatous forms have only cell walls. The gelatinous matrices of *Porphyridium* and *Batrachospermum* are complex mixtures of a variety of monomeric sugars, including galactose, glucose, and xylose (Craigie, 1990). The cell walls of a freshwater isolate of *Bangia atropurpurea* are similar to those of marine collections in having repeating water-soluble dissacharide units of agarose and porphyran and insoluble residues of galactose and mannose (Youngs et al., 1998). The cell walls of *Paralemanea annulata* have xylan as the major polysaccharide as well as cellulose in small quantities as the fibrillar components (Gretz et al., 1991). The amorphous component consists of a glucuronogalactan. Water-soluble cell wall polymers of freshwater *Bostrychia moritziana* are composed of a complex mixture, including methyl agarose and methyl porphyran (Youngs et al., 1998). The insoluble residues contain a mixture of galactose and glucose. Many freshwater red algal species exhibit differential staining of external coverings with Alcian Blue, particularly of mucilaginous layers, rhizoids, sporangia and spermatangia (Sheath and Cole, 1990).

C. Reproduction

Freshwater red algal species exhibit a diversity of reproductive types, particularly in terms of dissemination (Table I). Cell division is the major mode of population increase among the unicellular forms. During mitosis of *Porphyridium purpureum* and *Flintiella sanguinaria*, the nuclear envelope remains intact with polar openings, the spindle apparatus is composed of interdigitating half spindles, and the nuclear-associated organelle (NAO) is an electron-dense bipartite structure (Broadwater and Scott, 1994). Cell division in the other forms is generally the mechanism by which the thallus is expanded. Mitosis in *Batrachospermum anatinum* is similar to that of the unicells but also includes perinuclear endoplasmic reticula and a bipartite NAO that is composed of a small ring within a large one (Scott, 1983).

Monosporangia formation is the major form of asexual reproduction among the pseudofilamentous and filamentous taxa (Table I) and typically involves the formation of single spores that germinate back into the life history phase that produced them. In *Chroodactylon ornatum* and *Bangia atropurpurea*, spores are released by localized digestion of the common filamentous matrix (Fig. 1G) (Sheath, 1984). The order Compsopogonales is characterized by its method of monospore production, which involves the cleavage of a relatively small cell from a larger vegetative cell by oblique cell division (Fig. 2A, C, D) (Garbary et al., 1980). Monosporangia of the Acrochaetiales and Batrachospermales are specialized, enlarged, and obovoid cells typically produced at the apices of vegetative branches (Figs. 2G and 3I) (Sheath, 1984). Monosporangia can be regenerated after spore release by protrusion and cleavage of cytoplasm from the subtending cell in *Audouinella hermannii* (Hymes and Cole, 1983) and *Batrachospermum intortum* (Sheath et al., 1992). Monospores are also a mechanism by which certain life history phases perpetuate themselves in complex life history alternation, such as the "chantransia" phase (see below) of the Batrachospermales (Sheath, 1984). Another form of asexual spore reported from inland rhodophytes is the endospore in *Cyanidium caldarium* and *Rufusia pilicola* (Wujek and Timpano, 1986; Seckbach, 1991).

Sexual reproduction and life history alternation are known for many species of freshwater red algae, although these phenomena have not been conclusively demonstrated for freshwater members of the Bangiophycidae or the Hildenbrandiales of the Florideophycidae (Table I). In the freshwater *Audouinella* species for which the life history has been analyzed, the free-living gametophyte and tetrasporophyte are isomorphic, both having the same tuftlike morphology (Fig. 2F) (Drew, 1935; Necchi et al., 1993a). The haploid gametophyte produces the gametangia. The female gametangium, the carpogonium, is a colorless cell with an inflated base and narrow tip, the trichogyne (Fig. 2I). The male gametangium, the spermatangium, is also colorless, obovoid in shape, and releases one spermatium at a time (Fig. 2H). Spermatia attach to the trichogyne and one eventually fertilizes the carpogonium. The zygote divides into a microscopic diploid phase, the carposporophyte, which remains attached to the gametophytic stage until its deterioration (Fig. 2J). Carpospores germinate into the diploid tetrasporophyte, which, at maturity, forms tetrasporangia at the branch tips (Fig. 3A). Haploid tetraspores are formed by meiosis and germinate into the gametophytic stage, thereby completing the life history. In a small stream in Rhode Island, both the tetrasporangia and the carpogonia of *A. hermannii* are formed in a brief period of time, February and May, respectively (Korch and Sheath, 1989). It would appear that production of these structures in a short period of time is common in North America because gametangia and carposporophytes were observed in only 7 out of 75 collections from throughout the continent (Necchi et al., 1993a). *Audouinella tenella*, which has only been collected in California, has only been found to contain tetrasporangia (Necchi et al., 1993a). The blue-colored *Audouinella* species have not been observed to contain gametangia, carposporophytes, or tetrasporangia in 34 collections in North America (Necchi et al., 1993b). This finding may indicate that they are not, in fact, true *Audouinella* species, but rather one of the life history stages of the Batrachospermales, the "chantransia" (see below). Further substantiating this possibility, Pueschel et al. (2000) have demonstrated that isolates formerly classified as the blue-colored *Audouinella macrospora* were positioned in an 18S rRNA gene tree with samples of *Batrachospermum* and not with the freshwater red-colored *Audouinella hermannii*. In addition, the pit plugs of *A. macrospora* were also like those of the Batrachospermales.

Rhododraparnaldia oregonica has characteristics that are intermediate between the Acrochaetiales and the Batrachospermales, including reproductive structures (Sheath et al., 1994d). The carpogonia and spermatangia are similar in morphology to those of *Audouinella* (Fig. 3C) and fertilized carpogonia form microscopic carposporophytes (Fig. 3D). However, unlike the Acrochaetiales, the spermatangia are formed on specialized colorless stalks, rather than at the apices of vegetative branches (Fig. 3C). What is similar between *Rhododraparnaldia* and the Batrachospermales is the fact that the free-living life history stages are heteromorphic, being quite different in morphology. The semierect gametophytes are composed of a main axis with barrel-shaped axial cells that are distinctly larger in diameter than the more elongate lateral branch cells (Fig. 3B). The microscopic to small macroscopic "chantransia" stage (so named because it was originally thought to be a separate genus) contains simple branched filaments with no difference in diameter and no obvious main axis (Fig. 3E) (Sheath et al., 1994d). The latter stage is diploid and formed from the germination of carpospores. Where it differs from the tetrasporophyte of the Acrochaetiales described above is in the process of meiosis. In the "chantransia" stage, tetraspores are not produced but rather the haploid gametophyte is formed directly attached to this stage (Fig. 3E) (Sheath, 1984; Sheath et al., 1994d). In the freshwater red algal species for which this process has been studied, it appears that meiosis takes place in an apical cell of the "chantransia" filament; in each division, a residual nucleus is extruded into a lateral protrusion, which is then separated by wall formation (Sheath, 1984). The one remaining haploid nucleus forms the gametophyte. This life history alternation is typical of the Batrachospermales.

Fifty-nine percent of the freshwater Rhodophyta in North America belong to the order Batrachospermales or Thoreales (Table I). Like *Rhododraparnaldia*, the gametophyte is macroscopic and semierect, while the "chantransia" stage is often microscopic or composed of tiny tufts (Fig. 4C). The latter forms are morphologically similar to the bluish forms of *Audouinella* noted above. Sheath (1984) proposed that the "chantransia" stage of the Batrachospermales may be an evolutionary adaptation for population maintenance in the upper portions of drainage basins. This stage is typically perennial, seasonally producing the attached gametophyte (Yoshida, 1959; Sheath, 1984; Necchi, 1993). Therefore, the population can continue to proliferate upstream while colonizing downstream with the release of carpospores (Sheath, 1984). If these algae possessed the typical red algal life history, as exhibited by the Acrochaetiales, release of both tetraspores and carpospores would result in a gradual shift of populations downstream until they were solely in the larger trunk river, which is too deep, turbid, and sedimented to support the growth of most autotrophs. Raven (1993) demonstrated that the photosynthetic rates *in situ* of the "chantransia" stage of *Lemanea mamillosa* are one twentieth of those of the gametophyte and the former phase has a negligible role in provisioning the growing gametophyte. He also concluded that the key role of the "chantransia" stage is to occupy space throughout the year, including possible exposure during summer drawdown. This stage may also act in population dispersal through the production of monospores (Sheath, 1984; Raven, 1993). In North American temperate streams, gametophytes are typically present from late fall to late spring (Sheath and Hambrook, 1990).

Another key feature pertaining to the reproduction of the Batrachospermaceae of the Batrachospermales is the formation of relatively enlarged and persistent trichogynes of the carpogonia, compared to those of red algae from other orders (Figs. 3K–M and 4E, I) (Sheath, 1984). The larger surface area and longevity would enhance the probability of spermatia contact. Hambrook and Sheath (1991) demonstrated that the mean percentage fertilization rate for various species of *Batrachospermum* was 45–72%, including dioecious taxa. This rate may be obtained because spermatia are released into turbulent eddies downstream of rocks, where they are carried through the female plants numerous times as the water is moving back and forth (Sheath and Hambrook, 1990). Carpogonia are borne on carpogonial branches that may be little to highly differentiated from adjacent vegetative branches in the Batrachospermales (Fig. 3K, M). In the undifferentiated carpogonial branch of *Batrachospermum involutum*, the cells are uninucleate with abundant starch granules and several well-developed peripheral chloroplasts (Sheath and Müller, 1997). In contrast, the short carpogonial branch cells of *B. helminthosum* have no visible starch, chloroplasts are highly reduced, and cross walls break down among cells.

Freshwater populations of *Hildenbrandia* of the order Hildenbrandiales are typically vegetative and reproduce by gemmae, dense aggregations of cells formed in cavities in the thallus (Fig. 6F, G) (Starmach, 1969; Seto, 1977; Sherwood and Sheath, 2000b). The gemmae are eventually released from the thalli and germinate into new crusts, presumably of the same ploidy level.

Most of the freshwater members of the Ceramiales in North America have only been observed in their vegetative stages, including *Ballia*, *Caloglossa*, and *Polysiphonia* (Sheath et al., 1993c). They probably proliferate through fragmentation of the thallus and subsequent growth of the fragments. In contrast to *Ballia* in North American streams, inland populations from South America and Malaysia have been observed to contain monosporangia (Kumano, 1978; Couté and Sarthou, 1990; Necchi, 1995). In addition, brackish water populations of *Caloglossa ogasawaraensis* have been observed to contain gametangia and carposporophytes or tetrasporangia (Tanaka and Kamiya, 1993). A few freshwater populations of the genus *Bostrychia* form stichidia which are inflated, multichambered structures at the tips of vegetative branches (Fig. 6I). These chambers may form the tetrasporangia, but this has only been observed conclusively in collections of *B. moritziana* from Venezuela and Brazil (D'Lacoste and Ganesan, 1987; Kumano and Necchi, 1987; Sheath et al., 1993c).

III. ECOLOGY AND DISTRIBUTION

A. Streams and Rivers

Because 94% of the inland rhodophytes of North America occur in streams or rivers (Table I), this chapter will concentrate on this habitat. Much of the ecology of riverine red algae has been summarized by Sheath and Hambrook (1990) and I will synthesize the trends and give updated information here.

1. Patterns of Distribution

In North America, 51% of 1000 first- to fourth-order stream reaches surveyed contain red algae and 24% have two or more species (Sheath and Hambrook, 1990; Sheath and Cole, 1992). The maximum number of red algal species found per reach is six. The most widespread species is *Batrachospermum gelatinosum*, which occurs in about 13% of the streams examined. This species occurs from the polar desert on Ellesmere Island, Northwest Territories (80°N) to the southeastern coastal plain in Louisiana (37°N) (Vis *et al.*, 1996a; Vis and Sheath, 1997). Other widespread species include *Audouinella hermannii* (North Slope of Alaska to Georgia) (Necchi *et al.*, 1993a), *Batrachospermum helminthosum* (Washington and Maine to central Mexico) (Sheath *et al.*, 1994a), *B. turfosum* (North Slope of Alaska to central Mexico) (Sheath *et al.*, 1994d; Müller *et al.*, 1997), and *Lemanea fluviatilis* (central Alaska to Arkansas) (Vis and Sheath, 1992).

Among the taxa of lotic Rhodophyta in North America with more restricted patterns of distribution, there are some interesting trends. Members of the Compsopogonaceae, Thoreales, Hildenbrandiales, Ceramiales, and *Batrachospermum* section *Contorta* are largely restricted to warmer waters from south temperate to tropical streams (Vis *et al.*, 1992; Sheath *et al.*, 1992, 1993a, b). There are also taxa that are mostly in north temperate to tundra habitats, such as *Batrachospermum gelatinosum* forma *spermatoinvolucrum* and *Lemanea borealis* (Vis and Sheath, 1992, 1996, 1998). *Boldia* and *Tuomeya* have been collected only in eastern North America, ranging from Quebec to the southeastern United States (Howard and Parker, 1980; Sheath and Hymes, 1980; Kaczmarczyk *et al.*, 1992). There are a number of species that appear to be localized only in southwestern spring-fed streams, such as *Flintiella sanguinaria* (Ott 1976), *Chroothece mobilis* (Blinn and Prescott, 1976), and three members of *Batrachospermum* section *Batrachospermum* (*B. carpocontortum*, *B. carpoinvolucrum* and *B. involutum*) (Vis and Sheath, 1996; Sherwood and Sheath, 1999a). Some of the warm-water groups noted above are also distributed in spring-fed streams of southwestern North America.

2. Physical Factors

Riverine red algae exhibit a wide range of occurrence with respect to current velocity (Sheath and Hambrook, 1990). Nonetheless, most species are found in moderate flow regimes (mean 29–57 cm s^{-1}). Moderate flow enhances various aspects of metabolism, including productivity and pigment content (Thirb and Benson-Evans, 1982), growth (Whitford, 1960), respiration rate (Schumacher and Whitford, 1965), and phosphorus uptake level (Schumacher and Whitford, 1965). In addition, it has a positive influence on the ecology of these organisms, such as washout of loosely attached competitors (Whitton, 1975), constant replenishment of gases and nutrients (Hynes, 1970), and reduction of the boundary layers of depletion around the algal thallus (MacFarlane and Raven, 1985). Few taxa are typically localized at high current velocities (>1 m s^{-1}), the exceptions being *Lemanea* and *Paralemanea* of the Lemaneaceae (e.g. Everitt and Burkholder, 1991; Vis *et al.*, 1991). The morphology of some species, such as *Sirodotia delicatula*, can be altered under different flow regimes (Necchi, 1997). At high current velocities (132 cm s^{-1}), plants are denser, having shorter internodal lengths. Sheath and Hambrook (1988) calculated mean potential velocities (in cm s^{-1}) at which various morphological forms of red algae would break: tufts, 80; mucilaginous filaments, 160; and cartilaginous and pseudoparenchymatous thalli, 580.

The light regime, which includes changes in intensity, quality, and photoperiod, is one of the key factors affecting the distribution and seasonality of riverine Rhodophyta (Sheath and Hambrook, 1990). Illumination affects algal growth via photosynthesis, by processes indirectly related to photosynthesis, and by those processes unrelated to photosynthesis. In the case of freshwater red algae, distribution within a drainage basin and seasonality are determined by the photoregime established by the surrounding tree canopy. In a headwater Rhode Island stream containing *Batrachospermum boryanum*, the total illumination reaching the water surface is reduced by 90–99% on both sunny and cloudy days in a shaded reach compared with a nearby open segment (Kaczmarczyk and Sheath, 1991). There is a slight but significant increase in green light under the canopy and a corresponding increase in the red pigment phycoerythrin compared to the blue pigment phycocyanin. The action spectrum of collections from the canopied and open sites are similar and quite broad. Nonetheless, the populations of *Batrachospermum* mostly disappear during periods of peak canopy shading (Hambrook and Sheath, 1991). Likewise, many species of stream-inhabiting Rhodophyta exhibit a positive correlation to light and a negative one to temperature (Kremer, 1983; Sheath, 1984; Leukart and Hanelt, 1995). In addition, they tend to exhibit low saturating levels of illumination for photosynthesis (35–400 µmol photons m^{-2} s^{-1}).

Temperature regime influences the latitude, elevation, drainage basin distribution, as well as the seasonality of freshwater red algae (Sheath and Hambrook, 1990). Latitudinal patterns have been discussed above. Kremer (1983) concluded that some of the geographic patterns of riverine Rhodophyta are based on photosynthetic response to temperature. For example, the concentration of members of the Compsopogonaceae in warm waters can be explained by a maximum photosynthesis rate at 30–35°C. In large drainage basins, elevation and basin distribution patterns are interrelated; mean temperatures tend to increase from the source to the mouth, although the amplitude of diurnal fluctuations in temperature become less (Whitton, 1975). Israelson (1942) reported that most rhodophytes in Sweden were restricted to elevations less than 900 m above sea level. From our surveys of North America, we have observed a similar trend. Exceptions include the Lemaneaceae and some members of the Acrochaetiales, such as *Audouinella hermannii* and *A. tenella*, which can be abundant in montane streams (e.g., Necchi *et al.*, 1993a; Vis and Sheath, 1992). In temperate regions, most freshwater red algae exhibit maximum biomass, growth, and reproduction between late fall and early summer (Sheath and Hambrook, 1990), but in many cases this seasonality is more related to light penetration to the stream surface than to temperature (e.g., Hambrook and Sheath, 1991). Necchi (1993) noted a similar seasonality for batrachospermalean species in a tropical drainage basin in southeastern Brazil where a combination of lower temperature and reduced turbidity during the dry winter months promoted growth of macroscopic gametophytes. In contrast, *Compsopogon coeruleus* was present throughout the year and distribution was not related to temperature, but to current velocity in these Brazilian streams.

3. Chemical Factors

The interaction between pH and the form of inorganic carbon can greatly influence the productivity and distribution of freshwater Rhodophyta (Sheath and Hambrook, 1990). Although widespread species are found in a wide range of pH values, the majority occur in mildly acidic waters between pH 6 and 7. However, there are exceptions to this pattern, including *Bangia, Chroodactylon, Hildenbrandia,* and members of the Compsopogonaceae, Thoreales, and Ceramiales, which may be considered to be alkalophiles (Sheath, 1987; Sheath *et al.*, 1993a–c; Vis *et al.*, 1992; Vis and Sheath, 1993). The effect of pH can be attributed to the form of inorganic carbon available; some taxa, such as *Lemanea mamillosa,* have been shown to use only free CO_2 as a carbon source for photosynthesis, which is the predominant form at mildly acidic pH values (e.g., Raven *et al.*, 1994). Above pH 8, the proportion of free CO_2 drops below 2–5% and species occurring in these waters would require flow replenishment or use of alternative sources of inorganic carbon (Sheath and Hambrook, 1990). One species commonly distributed in high-pH waters is the crustose *Hildenbrandia rivularis*, which also utilizes CO_2 as a carbon source but may also use HCO_3^-, although this possibility has not been confirmed (Raven *et al.*, 1994).

Specific conductance and pH are related in that alkaline waters are high in ions, such as carbonates, and are buffered strongly above pH 8 (Sheath and Hambrook, 1990). In contrast, waters draining igneous rock catchment areas are less well buffered and usually more acidic. Four common freshwater red algal species in North America, *Audouinella hermannii, Batrachospermum gelatinosum, Lemanea fluviatilis,* and *Tuomeya americana*, exhibit a negative frequency distribution in relation to specific conductance (total ions); the greatest frequencies occur below < 100 µS cm^{-1} (Sheath and Hambrook, 1990). This pattern is due in part to the form of inorganic carbon. Those taxa that typically occur at higher pH values, as noted above, are also distributed at high conductance ranges. Likewise, species typically localized in hardwater streams constitute the same list given above for high pH values.

Freshwater red algae are found in a wide range of oxygen concentrations (0.2–21 mg L^{-1}), but there tends to be an increase in frequency of occurrence with higher concentrations (Sheath and Hambrook, 1990). To some extent, this relationship results from the occurrence of many species in the cooler months from late fall to late spring when oxygen solubility is highest. Nevertheless, freshwater Rhodophyta are not commonly associated with stagnant, organic-rich waters with very low oxygen contents.

Freshwater rhodophytes occur over a broad range of nutrient values, but they are more typically found in low to moderate nutrient regimes (e.g., PO_4^{3-} below detection to 100 µg L^{-1}) (Sheath and Hambrook, 1990). The common occurrence of red algae at low nutrient levels is partially due to flow replenishment and reduction of the boundary layer of depletion in riverine systems. In addition, many species form colorless hair cells that may be produced in response to nutrient deficiency, as is the case for some green algal filaments (Gibson and Whitton, 1987). Some researchers have employed rhodophytes for classification of streams; for example, in Austria *Hildenbrandia* is typical of lowland rivers with relatively high nutrients, whereas *Lemanea* is regarded as indicative of high-altitude streams with low nutrients (Pipp and Rott, 1994).

In general, freshwater red algae are localized in reasonably unpolluted waters and are infrequent to absent in streams and rivers that are organically enriched, greatly silted, or very high in inorganic nutrients (Sheath and Hambrook, 1990). However, *Lemanea fluviatilis* and *Bangia atropurpurea* appear to be tolerant of some heavy-metal pollution (Lin and Blum, 1977; Harding and Whitton, 1981). For example, *L. fluviatilis* can occur at aqueous concentrations of zinc up to 1.16 mg L^{-1}.

4. Biotic Factors

Thirty-eight riverine animals to date have been observed to ingest freshwater red algae, based on gut content or feeding studies (Sheath and Hambrook, 1990; Sheath *et al.*, 1995). These grazers include two amphipods and the larvae of six mayfly, thirteen caddisfly, six stonefly, six chironomids, one beetle, as well as two snails and two cyprinoid fish. Most animals remove small pieces of 5–20 cells and digest the cytoplasm, leaving the empty walls intact. The majority of these animals are polyphagous, consuming a wide variety of food matter, including detritus, leaf fragments, and other algal taxa. In grazing experiments done by Hambrook and Sheath (1987), it was observed that the preference of consumption was *Audouinella hermannii*, *Batrachospermum helminthosum*, followed by *Tuomeya americana*; this trend was likely based on increasing toughness of the thallus and reduction of protein content. Rosemond (1993) noted that irradiance, nutrients, and herbivore grazing simultaneously limited algal biomass, including *Audouinella* sp., in a small forested stream.

Cases of larvae and pupae from six caddisfly species have been observed to contain pieces of freshwater rhodophyte thalli in North America (Sheath *et al.*, 1995). Seven genera of Rhodophyta (*Batrachospermum*, *Bostrychia*, *Compsopogon*, *Compsopogonopsis*, *Lemanea*, *Paralemanea*, and *Tuomeya*), representing 13 species and 35 populations, have been observed in this association. Strips of the algae are fit together in a transverse, concentric, or spiralled fashion. In some of the associations, pieces of the rhodophyte are also found in the gut of the caddisfly. In addition, some of the algal strips remain viable in these cases. Three genera of chironomid larvae have also been observed to incorporate pieces of red algae in their cases in North America (Sheath *et al.*, 1996a). Five genera of Rhodophyta are used in this process, *Audouinella*, *Batrachospermum*, *Lemanea*, *Paralemanea*, and *Sirodotia*. The cases are tubular in shape with longitudinally oriented strips of algae held together with silken threads.

Competition for suitable substrata can occur among species of freshwater red algae or with other benthic algae at various levels (Sheath and Hambrook, 1990). Unicellular forms, such as *Flintiella*, microscopic stages, and low-growing forms, including the "chantransia" stage of the Batrachospermales, *Audouinella* and *Hildenbrandia*, are common components of the stream epilithic community. As such, they compete with a complex association of microalgae, usually dominated by diatoms during early colonization stages (e.g., Steinman and McIntire, 1986). Fritsch (1929) noted that *Hildenbrandia* thalli are often overgrown by diatoms and cyanobacteria in British streams. In later stages of succession, filamentous and stalked species can form a canopy, providing a competitive advantage for light and nutrient replenishment. The semierect forms fit into this category and are subjected to competition with macrophytes and other macroalgae. Bryophytes are frequent dominants in the upper reaches where red algae occur (e.g., Sheath *et al.*, 1986). The macrophytes are subjected to removal by flooding events, allowing new periphyton colonization. Therefore, lotic communities containing red algae are generally in a nonequilibrium state, consisting of most successional stages (Sheath and Hambrook, 1990).

On a geographic scale, certain freshwater rhodophyte species have distributional patterns that are correlated to that of other species (Sheath and Hambrook, 1990). For example, in tropical streams in North America, *Bostrychia*, *Compsopogon*, and *Hildenbrandia* frequently cohabit the same stream reaches. In temperate and boreal regions, *Audouinella hermannii* and members of the Lemaneaceae are frequently found together; the former taxon can be both epiphytic on Lemaneaceae and epilithic in these situations.

B. Other Inland Habitats

Soft-water ponds and bogs represent the second most common habitat to encounter inland Rhodophyta (Sheath, 1984). In particular, *Batrachospermum turfosum* is common in these habitats, and Yung *et al.* (1986) noted its broad occurrence in the northeastern regions of Canada and the United States. Müller *et al.* (1997) studied the phenology of a population of *B. turfosum* in a boreal pond in Newfoundland. The gametophyte is perennial with a peak cover in summer, which is correlated to water temperature and day length. Carpogonia and spermatangia are present throughout the year except for October and November when monospore production is predominant.

Two species, which can be classified as brackish/marine invaders of freshwaters, *Chroodactylon ornatum* and *Bangia atropurpurea*, are common species in hard-water sections of the lower Great Lakes (Sheath and Morison, 1982; Sheath, 1987; Müller et al., 1998). These species are absent from the low-ion waters of Georgian Bay, the North Channel, and Lake Superior (specific conductance < 200 µS cm^{-1}) (Sheath et al., 1988). Collections of Bangia from Lakes Ontario, Erie, Huron, Michigan, and Simcoe have identical DNA sequences for the nuclear gene coding for the small subunit of ribosomal RNA (18S rDNA), the chloroplast gene coding for the large subunit of RuBisCO (rbcL) and the spacer unit between rbcL, and the gene coding for the small subunit of RuBisCO (rbcS) (Müller et al., 1998). These sequences are also nearly identical to those from freshwater collections of Bangia from Europe, the Thames and Shannon Rivers, United Kingdom, and Lake Garda, Italy. Hence, it would appear that the North American freshwater Bangia arose by a single invasion from a European freshwater population, possibly by vector-assisted transport (e.g., ballast water of ships), rather than by migration from the Atlantic Ocean.

Cyanidium caldarium is an unusual species restricted to several acid hot springs in North America, including those in the United States, Mexico, and El Salvador (DeLuca et al., 1979). These springs exhibit the following range of conditions: pH 1–2.6 and temperature 25–44°C. Doemel and Brock (1971) examined growth properties of Cyanidium isolated from Yellowstone National Park and observed that the optimum pH was 2–3 (range 0.5–5.0) and the optimum temperature was 45°C (range 25–56°C). This species also forms surface and subsurface crusts in hot, acidic soils and river banks near thermal springs (Smith and Brock, 1973).

Species of Porphyridium can form gelatinous crusts on moist soils and decaying wood (Geitler, 1932; Ott, 1972). In these habitats, these species are reasonably desiccation resistant and shade tolerant (Hoffmann, 1989).

IV. COLLECTION AND PREPARATION FOR IDENTIFICATION

Because 94% of inland red algal taxa occur in streams or rivers, I will concentrate on collecting in this environment. While macroalgae can be distributed throughout a major drainage basin, they tend to be more common in the smaller channels of first- to fourth-order reaches (Sheath and Cole, 1992). Because only approximately 5% of stream channels examined in North America have 50% or more of the stream bottom covered by one or more species of red algae (Sheath and Hambrook, 1990), it is often necessary to actively search for these taxa. A view box, composed of a glass bottom and Plexiglas sides, is of great help in being able to observe the stream bottom during this search process. To attain a representative sampling of species, at least a 20-m length should be carefully examined using the view box, including a variety of flow regimes and substrata. For example, in a large, fast-flowing stream, members of the Acrochaetiales and Lemaneaceae may be found attached to rocks in more rapidly flowing portions in the midchannel, whereas members of the Batrachospermaceae may be localized in quiet side channels and pools, attached to a variety of substrata, such as logs and tree roots. Long forceps are quite useful for grabbing specimens in deep waters, particularly gelatinous filaments of the Batrachospermaceae. Razor blades are necessary to remove crustose forms. Other collecting equipment for consideration includes hip waders, diver's gloves for winter collecting, and various portable meters, such as pH, specific conductance, temperature, turbidity, and current.

Red algal specimens are best viewed shortly after collection in a live state using a combination of dissecting and compound microscopes. If they cannot be examined quickly, then they should be fixed in 2.5% histological-grade glutaraldehyde, buffered with a pinch of $CaCO_3$, and stored in a dark and cool environment. Under these conditions, they will maintain their morphology and pigmentation for several years. Other fixatives, such as formalin or Lugol's, cause more distortion, and drying of herbarium specimens results in considerable morphological damage. Generally, samples are sorted and initially viewed at low power with a dissecting microscope. Then a reproductive piece is removed, mounted on a microscope slide with cover slip, finely chopped with a sharp razor blade and squashed to obtain flat images, and then viewed at 200× or 400× in a compound microscope. It is necessary to find all of the key vegetative and reproductive features noted in the section below to achieve a proper identification.

V. KEY AND DESCRIPTIONS OF GENERA

A. Key

1a.	Thallus unicellular (Fig. 1A)	2
1b.	Thallus multicellular	4
2a.	Reproduction by endospores, hot springs	*Cyanidium*
2b.	Reproduction by cell division, other inland habitats	3
3a.	Parietal chloroplasts without pyrenoids	*Flintiella*
3b.	Axial chloroplast with pyrenoid	*Porphyridium*
4a.	Thallus pseudofilamentous (Fig. 1C–E)	5
4b.	Thallus crustose, filamentous, or pseudoparenchymatous	8
5a.	Axial chloroplast with pyrenoid	6
5b.	Parietal chloroplasts without pyrenoids	7
6a.	Few-celled colony, no branching (Fig. 1B)	*Chroothece*
6b.	Multiple celled with false branching (Fig. 1C)	*Chroodactylon*
7a.	Discoidal base (Fig. 1D), occurs in streams and rivers	*Kyliniella*
7b.	No discoidal base, in sloth hair	*Rufusia*
8a.	Thallus crustose (Fig. 6E)	*Hildenbrandia*
8b.	Thallus a monostromatic, hollow sac (Fig. 1H)	*Boldia*
8c.	Thallus filamentous or pseudoparenchymatous	9
9a.	Thallus filamentous	10
9b.	Thallus pseudoparenchymatous (tissue-like but with compacted filamentous construction)	21
10a.	Multiaxial filaments (Figs. 5G, H and 6A, B)	11
10b.	Uniaxial filaments (Figs. 2G and 3B)	12
11a.	Monosporangia on short branches at base of assimilatory filaments (Fig. 6C)	*Thorea*
11b.	Monosporangia on long branches at periphery of assimilatory filaments (Fig. 5I)	*Nemalionopsis*
12a.	Filament uniaxial at base and multiaxial at reproductively mature apices (Fig. 1F)	*Bangia*
12b.	Filament uniaxial throughout	13
13a.	Filaments non-corticated (Figs. 2G, I and 3B)	14
13b.	Filaments corticated (Figs. 2B, C and 7B)	16
14a.	No obvious main axis or reduction in diameter in lateral branches	*Audouinella*
14b.	Distinct main axis and lateral branches reduced in diameter	15
15a.	Lateral branches not compacted, spermatangial stalks (Fig. 3C)	*Rhododraparnaldia*
15b.	Lateral branches compacted and pinnate, no spermatangial stalks (Fig. 6H)	*Ballia*
16a.	Reproduction by monosporangia, which are cleaved obliquely from cortical cells (Fig. 2C)	17
16b.	Reproduction by transversely cleaved monosporangia and/or sexual only	18
17a.	Rhizoidal cells confined to filament base, cortical cells cuboidal (Fig. 2C)	*Compsopogon*
17b.	Rhizoidal cortication scattered throughout filament (Fig. 2E)	*Compsopogonopsis*
18a.	Thallus with whorls of determinate lateral branches (Figs. 3F–H and 4D)	19
18b.	Thallus with alternate or opposite lateral branches (Fig. 2G)	20
19a.	Carposporophytes a distinct mass of determinate gonimoblast filaments (Fig. 4A, B)	*Batrachospermum*

19b.	Carposporophytes indistinct, indeterminate gonimoblast filaments (Fig. 4F)	*Sirodotia*
20a.	Trichoblasts (hair cells) from cortical cells, no stichidia (Fig. 7F)	*Polysiphonia*
20b.	No trichoblasts, stichidia at branch tips (Fig. 6I)	*Bostrychia*
21a.	Thallus a branched flat blade with a distinct midrib (Fig. 7C, D)	*Caloglossa*
21b.	Thallus a cartilaginous tube with no midrib	22
22a.	Whorled lateral branches appearing like beads in a row at low magnification (Fig. 4H)	*Tuomeya*
22b.	Thallus tubular, sometimes with distinct nodal swellings (Fig. 5A)	23
23a.	Main axis corticated, spermatangia in rings around the diameter of the thallus (Fig. 5D)	*Paralemanea*
23b.	Main axis uncorticated, spermatangia in patches (Fig. 5A)	*Lemanea*

B. Descriptions of Genera (for a list of species in each genus, see Table I)

Porphyridiales: Unicellular Forms

Cyanidium L. Geitler

Cyandidium is composed of spherical unicells, rarely united in a common mucilaginous matrix, 1.5–6 μm in diameter. Each cell contains one blue–green, parietal, spherical to cuplike chloroplast without a pyrenoid, a mitochondrion, and a nucleus, but no vacuole. The chloroplast contains single, parallel, concentric thylakoids with phycobilisomes and predominant C-phycocyanin. Reproduction occurs by four endospores (2–3 μm). Sexual reproduction has not been reported.

This genus is widespread in acidic thermal areas. The temperature range *in situ* is 25–44°C; the pH range *in situ* 0.05–5.0 (Doemel and Brock, 1971; DeLuca *et al*., 1979).

Flintiella F. D. Ott in Bourelly

Flintiella is composed of spherical cells united in a gelatinous matrix, each with a massive, parietal, reddish chloroplast without a pyrenoid. The cell diameter ranges from 9 to 20 (45) μm. Reproduction occurs by cell division. Sexual reproduction is unknown.

This genus is found in Barton Springs, Austin, Texas, in autumn (Ott, 1976). Total dissolved solids, 513 mg L^{-1}; NO$_3^{-1}$, 301 mg L^{-1}; pH, 6.9.

Porphyridium Nägeli (Fig. 1A)

Porphyridium is composed of spherical to ovoid unicells with a stellate chloroplast and prominent central pyrenoid. The cell diameter is 5–10 μm in the exponential phase, 7–16 μm in the stationary phase. Cells are solitary, but often grouped into irregular colonies with an ill-defined mucilaginous matrix. Species are distinguished by chloroplast color. Reproduction occurs by cell division.

Porphyridium forms gelatinous coatings on various surfaces; it is widespread in freshwaters, brackish environments, and moist soils.

Porphyridiales: Multicelluar Forms

Chroodactylon Hansgirg (Fig. 1C)

Chroodactylon is composed of false-branched pseudofilaments of globose or elliptical cells enclosed in a broad, gelatinous sheath and arranged in an irregular uniseriate manner. Each cell contains a blue–green, stellate, axial plastid and a prominent pyrenoid. The cell diameter is 3–17 μm; the length is 6–28 μm. There is a linear correlation between the false branch number and the filament length (approximately 1 branch per 200 μm length). The maximum filament length is 1240 μm. Reproduction occurs by monospores and fragmentation. No sexual reproduction has been observed. *Stylonema* sp. has been reported from streams in central Mexico (Jimenez, 1999) based on multiaxial pseudofilaments, but most of the figures in this presentation appear to be *Chroodactylon* (Jimenez's figures 1b–d, g, h).

This genus is epiphytic on *Cladophora* and *Rhizoclonium* in the Laurentian Great Lakes and scattered streams from Ontario to Arizona (Vis and Sheath, 1993). *Chroodactylon* is epilithic on limestone cave walls in Kentucky (Nagy 1965). It occurs in freshwaters, largely hardwater, with a specific conductance of 170–540 μS cm^{-1} and a pH of 7.8–8.5.

Chroothece Hansgirg in Wittrock *et* Nordstedt (Fig. 1B)

Chroothece is composed of ellipsoidal to cylindrical cells, each with a broad, firm gelatinous envelope. The cell diameter is 20–30 μm; the length is 30–45 (50) μm. Cells contain an axial, stellate, blue–green to yellow–brown or orange chloroplast with a prominent pyrenoid. Cells are solitary or joined pole to pole into a few-celled colony. The basal pole, with a gelatinous sheath, extends into a lamellated stalk. Reproduction

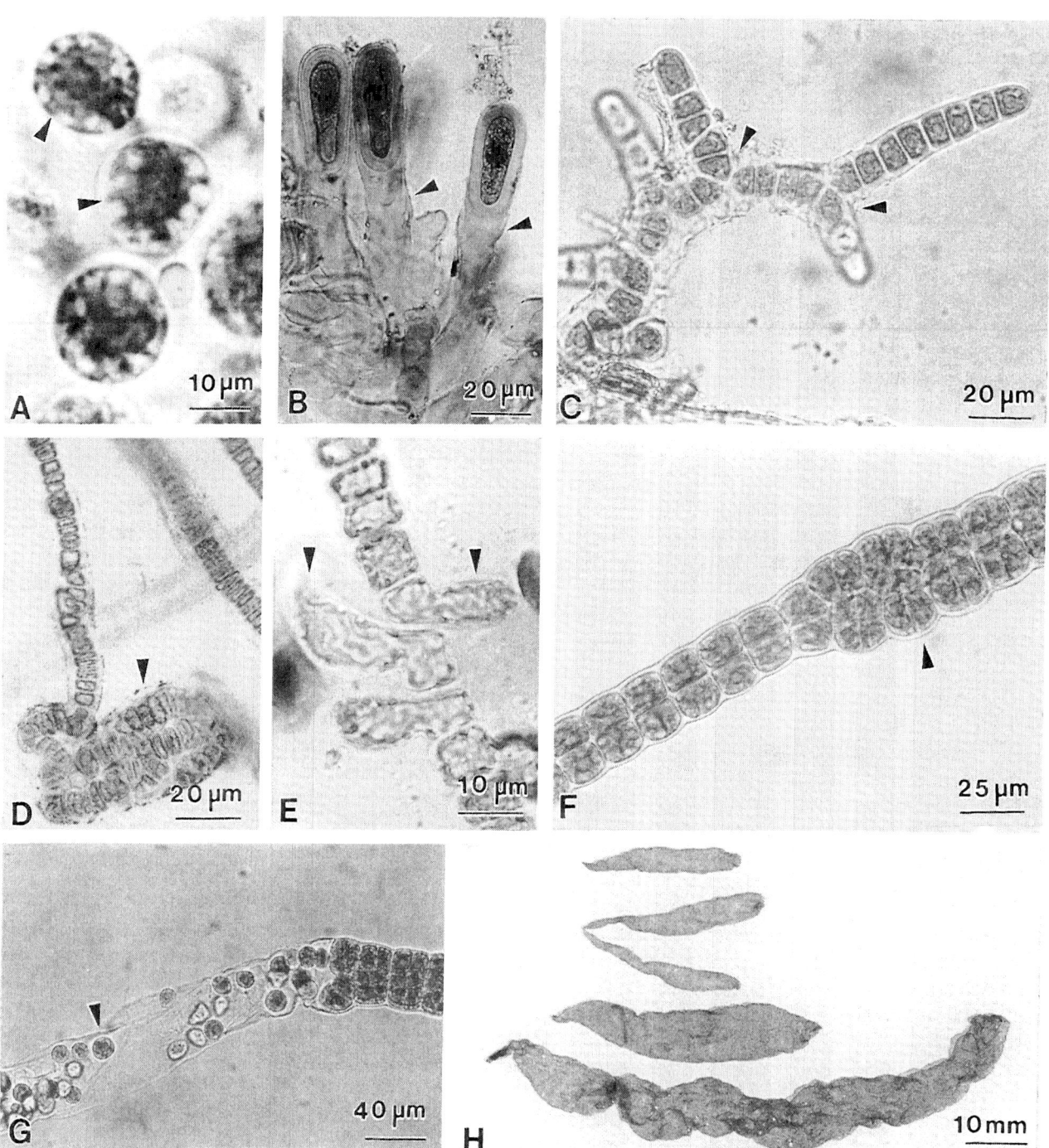

FIGURE 1 Freshwater bangiophycidean algae. (A) *Porphyridium purpureum*, unicellar form with axial stellate chloroplast with peripheral lobes (arrowheads). (B) *Chroothece* sp., elongated cells enclosed by a layered gelatinous stalk (arrowheads). (C) *Chroodactylon ornatum*, pseudofilamentous form with false branches (arrowheads). (D, E) *Kyliniella latvica*: (D) Pseudofilament extending from a discoid base (arrowhead). (E) Rhizoidal outgrowths (arrowheads) from upright cells. (F, G) *Bangia atropurpurea*: (F) Mature filament, which is biaxial toward the base and multiaxial at the apex (arrowhead). (G) Release of monospores (arrowhead) by localized digestion of matrix at apex. (H) *Boldia erythrosiphon*, macroscopic view showing saccate thallus. C–H reprinted from Sheath (1984) with permission.

occurs by cell division; the pyrenoid divides and then the cell undergoes transverse invagination. This species also forms resting akinetes. No sexual reproduction has been reported.

Chroothece is a rare component of freshwater streams, moist soils, and peat bogs. Populations occur in an Arizona spring at a temperature of 13°C, a pH of 7.2–7.8, and a specific conductance of 570 µS cm^{-1} (Blinn and Prescott, 1976).

Kyliniella Skuja (Fig. 1D, E)

Kyliniella is composed of unbranched pseudofilaments growing in gray–green clusters from a discoid, pseudoparenchymatous base. The maximum filament length is 2–3 cm. The cells are 18–19 µm in diameter and 5–17 µm long, surrounded by a mucilaginous envelope up to 16 µm wide. Cells are contiguous or separate, containing several parietal, blue–green discoid chloroplasts. Rhizoidal outgrowths [17–25 (150) µm long] occur at points of contact. Vegetative reproduction occurs by release of small fragments (hormogonia). Asexual reproduction occurs by monospores shed by expulsion through the sheath. Presumptive sexual reproduction has been reported in a New Hampshire population with small colorless spermatia and large, pigmented carpogonia with tubular projections. The fate of the zygote is unknown.

This genus is epiphytic on macrophytes in softwater streams in the northeastern United States (Rhode Island and New Hampshire), mostly in the summer and fall (Vis and Sheath, 1993).

Rufusia Wujek *et* Timpano

Rufusia is composed of branched pseudofilaments, composed of spherical or elliptical cells, each with several reddish–violet, parietal, discoid to band-shaped chloroplasts without pyrenoids. Cells are 5.5–15 × 3.5–10 µm. Cell division is apical and intercalary. Asexual reproduction occurs by endospores; vegetative propagation occurs by fragmentation. Sexual reproduction has not been observed.

Rufusia grows within the hair tissues and furrows of the two-toed sloth (*Choloepus*) and three-toed sloth (*Bradypus*) in Panama and Costa Rica (Wujek and Timpano, 1986). It is not found on adjacent vegetation.

Bangiales

Bangia Lyngbye (Figs. 1F, G)

Bangia is composed of filiform, unbranched cylinders of cells embedded in a firm gelatinous matrix. It is attached by down-growing rhizoids, usually in dense purple–black to rust-colored clumps. The initial uniaxial filament (diameter 10–30 µm) becomes largely multiaxial at maturity (the diameter is 60–6180 µm for freshwater filaments). The cell number and filament length are highly correlated in uniaxial filaments; filament lengths range from 0.2 to 35 cm. Vegetative cells contain a large, axial, stellate chloroplast with prominent pyrenoid. The apical region differentiates into packets of cells, the monosporangia.

This genus occurs in hard-water lakes in North America, particularly Lakes Ontario, Erie, Huron, Michigan, and Simcoe, as well as the upper St. Lawrence River (Müller *et al.*, 1998).

Compsopogonales

Boldia Herndon (Figs. 1H and 2A)

This genus occurs as a mauve–pink to reddish-brown (rarely olive green), hollow, monostromatic sac or tube, 1–20 (40–75) cm long and 0.1–2.0 cm in diameter. Vegetative cells are rectangular, 5–20 (45) µm in diameter, containing several peripheral, ribbon-like chloroplasts and a large central vacuole. Secondary filaments arise as outgrowths from vegetative cells, elongating between and above vegetative cells and eventually dividing to form monospores, 5–9 µm in diameter. Monospores germinate into a prostrate, monostromatic disc or an aggregation of creeping filaments. The disc or aggregation produces a cushion-like mound of cells that functions as a perennial holdfast, producing seasonally macroscopic thalli. Sexual reproduction has not been observed.

Boldia is localized in scattered streams in eastern North America, extending from central Alabama to Ontario and Québec. The range of ecological factors is as follows: current velocity 3–71 cm s^{-1}; pH, 6.1–8.5; specific conductance, 18–290 µS cm^{-1}, dissolved oxygen, 4.5 mg L^{-1} (saturation); temperature, 12–25°C. In the southern range, it is often associated with snails of the family Pleuroceridae with high manganese content in the shells, appearing in late winter and usually disappearing by early summer. In northern streams, it is largely epilithic, occurring throughout the summer (Howard and Parker, 1980; Sheath and Hymes, 1980).

Compsopogon Montagne in Bory and St. Vincent *et* Durieaux (Fig. 2B–D)

Compsopogon is composed of a branched, bluish to violet–green, uniaxial filament with older portions corticated. The small-celled cortex is produced by vertical division of axial cells into one to five layers. Plants are up to 20–50 cm long and 250–2000 µm in diameter. Axial cells are enlarged and are evident by slight constrictions in the older portions. Axial cells may break down, leaving hollow cylinders. This genus may be free floating or benthic. If attached, rhizoids form by outgrowths of lower cortical cells or a basal disc. Cortical cells 7–22 × 10–48 µm, contain several

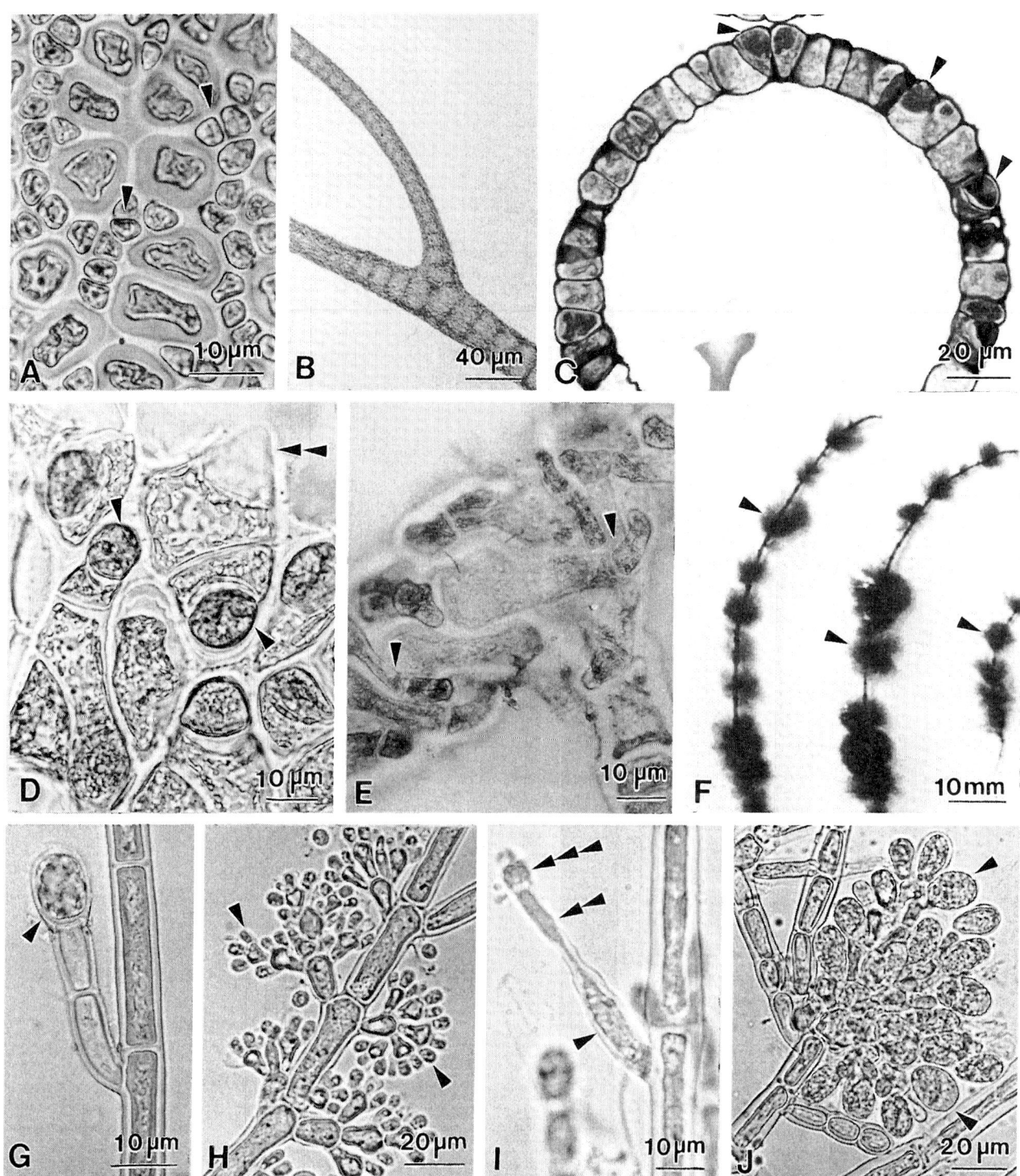

FIGURE 2 Freshwater bangiophycidean and acrochaetalian algae. (A) *Boldia erythrosiphon*, thallus surface with large vegetative cells and smaller monosporangia (arrowheads) between and above them. (B–D) *Compsopogon coeruleus*: (B) Branched uniaxial filament covered by small cortical cells. (C) Transverse section of mature thallus with hollow center resulting from deterioration of axial filament stained with Toluidine Blue O. Monosporangia are cleaved from the cortical cells and are evident as small, external cells (arrowheads). (D) Surface view with smaller, more densely pigmented monosporangia (arrowheads). An empty monosporangium (double arrowhead) that has released its contents is also included. (E) *Compsopogonopsis leptocladus*, portion of branched filament showing rhizoidal cortication (arrowheads). (F–J) *Audouinella*: (F) *A. hermannii*, macroscopic view of tufts (arrowheads) that are epiphytic on *Lemanea*. (G) *A. eugenea* with monosporangium (arrowhead) on short lateral branch. (H–J) *A. hermannii*: (H) Spermatangial clusters (arrowheads) at the tips of short lateral branches. (I) Carpogonium with cylindrical base (arrowhead), narrow trichogyne (double arrowhead), and attached spermatium (triple arrowhead). (J) Carposporophyte, a dense mass of gonimoblast filaments with obovoid carposporangia of branch tips (arrowheads). A, B, and F reprinted from Sheath (1984) with permission. C courtesy of T. Rintoul. G, I, and J reprinted from Necchi *et al.* (1993a) with permission.

peripheral, discoid chloroplasts. Reproduction occurs by fragmentation or monosporangia (9–28 μm long), which are cleaved from cortical cells by oblique, unequal cell division. The monospore divides into creeping, branched filament; a central cell eventually elongates vertically and divides to form the erect stage. Microaplanospores reported may represent spermatia but this is not confirmed.

Compsopogon is largely distributed in tropical to warm temperate streams; in North America, well-documented collections range from Virginia and Texas to Belize and the Caribbean islands (Vis *et al.*, 1992). Streams tend to be warm (13–27°C) and alkaline (pH, 7.3–8.6; specific conductance, 46–1880 μS cm^{-1}). Height, diameter, monosporangium number, branching, and basal disc presence are affected by current velocity. It is usually epilithic but can be epiphytic. *C. coeruleus* is epizoic on the parasitic copepod *Lernaea*, where it is attached to cyprinid fish in Mud River, Kentucky (Camburn and Warren, 1983). *Compsopogon* was recently found at a depth of 21 m in central Lake Huron of the Laurentian Great Lakes (Manny *et al.*, 1991).

Compsopogonopsis Krishnamurthy (Fig. 2E)

Compsopogonopsis is composed of a branched, bluish to olive-colored, uniseriate filament with older parts corticated. The cortex is initiated by rhizoid-like outgrowths, generally one layer but occasionally two or more. Thalli are 20–40 cm long and 0.2–1.0 mm in diameter. Cortical cells, 22–51 × 19–55 μm, contain several peripheral, discoid chloroplasts. Monosporangia are produced by cortical and uncorticated axial cells. Monospores (16–23 × 15–23 μm) germinate into a few-celled mass, one cell of which divides into the erect filament and the others contribute to the holdfast. Sexual reproduction has not been observed. The genus is considered by some researchers to be synonymous with *Compsopogon* (e.g., Rintoul *et al.*, 1999).

Compsopogonopsis occurs in scattered freshwater streams, which are typically warm (18–24°C) and alkaline (pH, 7.6–7.7; specific conductance, 1760–1880 μS cm s^{-1}). North American collections are restricted to New Mexico and Puerto Rico (Vis *et al.*, 1992).

Acrochaetiales

Audouinella Bory de St. Vincent (Figs. 2F–J and 3A)

Audouinella is composed of short, branched, uniaxial filaments, which typically grow in dense tufts, usually less than 1 cm in diameter but up to 2–3 cm. The filaments may be composed of erect and prostrate axes. Apices of erect axes often terminate with colorless hair cells. Cells contain either reddish or bluish, parietal, ribbon-like chloroplasts. The cell diameter is 6–26 μm. Filaments occur most commonly with monosporangia (5–38 μm in diameter) at the branch tips. Only reddish species have been observed to be sexual or tetrasporic. Colorless spermatangia (4–5 μm in diameter) occur in clusters at the branch tips; carpogonia have a cylindrical base and thin trichogyne (30 μm in length). Carposporophytes are spherical, compact mass of short gonimoblast filaments; carposporangia are obovoid (10 × 13 μm). Tetrasporangia are also formed at the branch tips (9 μm in diameter).

Audouinella is a widespread genus in streams, ranging from the North Slope of Alaska to Costa Rica (Necchi *et al.*, 1993a, b). *A. hermannii*, the most common species in North America, tends to occur in cool waters (11°C), with a low ion content (104 μS cm^{-1}) and mildly alkaline pH (7.5). *A. eugenea* and *A. pygmaea* are typical of warm streams of high ion content.

Balbianiales

Rhododraparnaldia Sheath, Whittick, *et* Cole
(Fig. 3B–E)

This genus is composed of crimson-colored filaments up to 15 cm long with barrel-shaped axial cells that have a distinctly larger diameter (17.3–30.1 μm) than that of the lateral branches (4.3–8.5 μm). Unique spermatangial stalks produce two types of spermatangia at their tips. The carpogonium is borne on an undifferentiated branch and has a swollen, cylindrical base and thin trichogyne. The carposporophyte consists of a spherical, compact mass of short gonimoblast filaments. The carposporangia are spherical up to 8 μm in diameter. Carpospores germinate into a "chantransia" phase with cells 5–7 × 16–38 μm; this phase produces gametophytes. DAPI-relative fluorescence values are twice as high for gonimoblast cells, carposporangia, and "chantransia" cells as for the gametophyte vegetative cells and gametangia.

The single species *R. oregonica* combines characteristics of both the Acrochaetiales and the Batrachospermales.

Rhododraparnaldia is found in two mountain streams in Oregon; the type locality has moderate current velocity (35–61 cm s^{-1}), temperatures of 8–11°C, a pH of 8.3, and a specific conductance of 30 μS cm^{-1} (Sheath *et al.*, 1994d).

Batrachospermales: Batrachospermaceae

Batrachospermum A.W. Roth (Figs. 3F–M and 4A–C)

Batrachospermum is composed of gelatinous gametophyte filaments, up to 40 cm long, with beaded appearance, varying from blue–green, olive, violet, gray to brownish. It contains a uniaxial central axis with large cylindrical cells; four to six pericentral cells produce repeatedly branched fascicles of limited growth. Rhizoid-like cortical filaments typically develop from

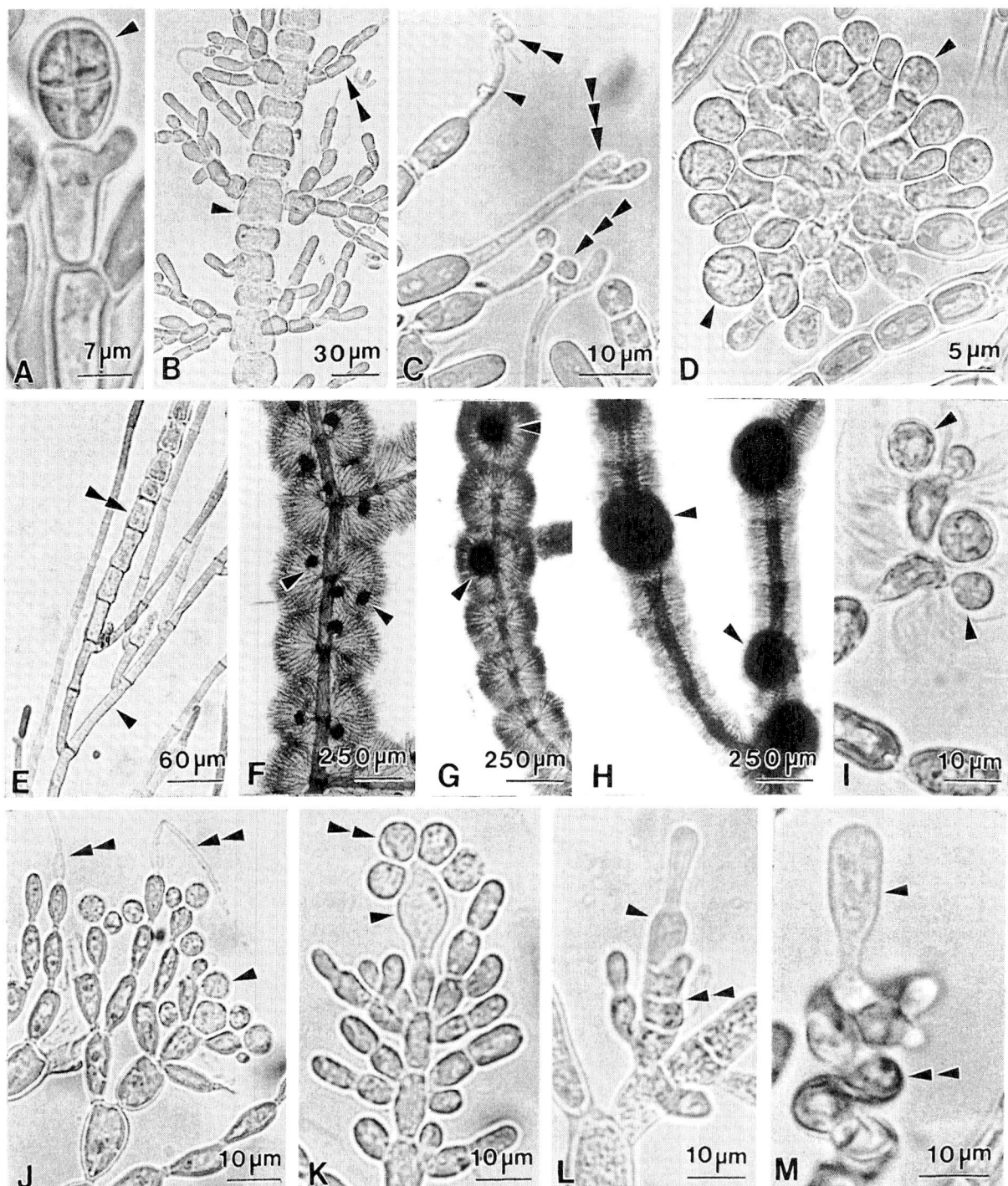

FIGURE 3 Freshwater members of the Acrochaetiales, Balbianiales, and Batrachospermaceae. (A) *Audouinella hermannii*, tetrasporangium (arrowhead) at the tip of a vegetative branch. (B–E) *Rhododraparnaldia oregonica*: (B) Branched filament with lateral branches, both opposite and alternate, arising from large barrel-shaped axial cells. (C) Carpogonia with cylindrical bases, thin trichogynes (arrowhead), and attached spermatia (double arrowhead). Spermatangia (triple arrowheads) are formed at the apex of colorless stalk cells. (D) Carposporophyte, consisting of compact gonimoblast filaments and carposporangia (arrowheads) at branch tips. (E) "Chantransia" phase (arrowhead) producing an attached gametophyte (double arrowhead). (F–M) *Batrachospermum*: (F) *B. gelatinosum*, branched filament with barrel-shaped whorls containing several carposporophytes (arrowheads). (G) *B. ambiguum*, branched filament with obovoid whorls and single axial carposporophytes (arrowheads). (H) *B. louisianae*, branched filament with confluent whorls and large carposporophytes (arrowheads) extending beyond whorls. (I) *B. intortum*, with monosporangia (arrowheads) at tips of vegetative branches. (J) *B. gelatinosum* forma *spermatoinvolucrum*, spermatangia (arrowhead) and hair cells (double arrowheads) at the tips of vegetative branches. (K) *B. anatinum*, with carpogonium with inflated trichogyne (arrowhead) with four attached spermatia (double arrowhead). The carpogonial branch is little differentiated from typical vegetative branches. (L) *B. helminthosum* with immature carpogonium (arrowhead). The carpogonial branch (double arrowhead) is composed of cells that are substantially smaller than those of adjacent vegetative branches. (M) *B. intortum* with immature carpogonium (arrowhead). The carpogonial branch (double arrowhead) is differentiated and twisted. A reprinted from Necchi *et al.* (1993a) with permission. B–E reprinted from Sheath *et al.* (1994d) with permisson of the International Phycological Society. F reprinted from Vis *et al.* (1996a) with permission. G–I and M reprinted from Sheath *et al.* (1992) with permission of the *Journal of Phycology*. J reprinted from Vis and Sheath (1996) with permission of the International Phycological Society. K reprinted from Vis *et al.* (1996b) with permission. L reprinted from Sheath *et al.* (1994a) with permission of the *Journal of Phycology*.

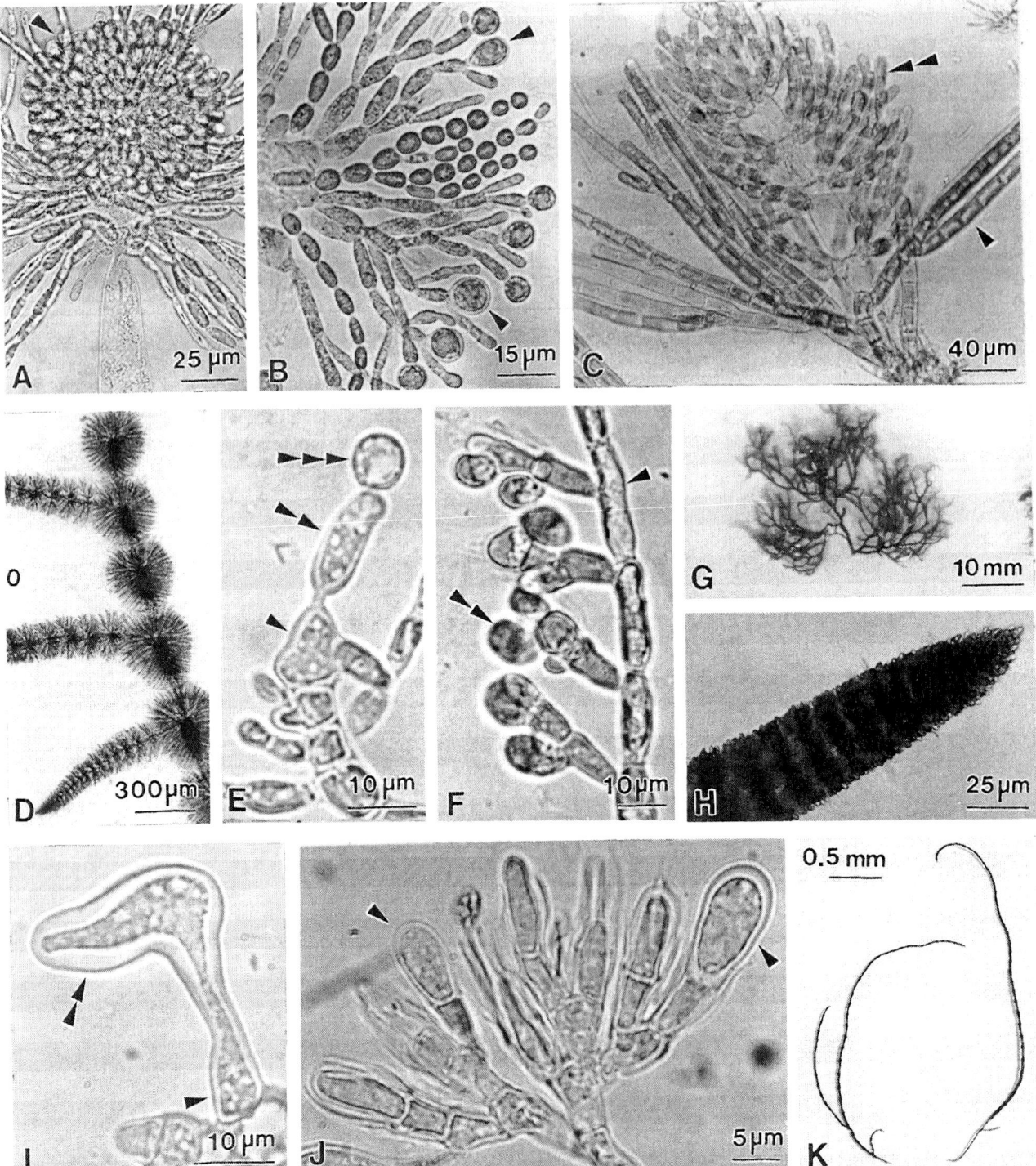

FIGURE 4 Members of the Batrachospermales. (A–C) *Batrachospermum*: (A) *B. gelatinosum* with a dense, spherical carposporophyte (arrowhead). (B) *B. globosporum* with loose carposporophyte with carposporangia (arrowheads) at tips of gonimoblast filaments. (C) *B. gelatinosum* "chantransia" stage (arrowhead) producing attached gametophyte (double arrowhead). (D–F) *Sirodotia suecica*: (D) Branched filament with obovoid whorls. (E) Carpogonium with base (arrowhead) with a protuberance, cylindrical trichogyne (double arrowhead), and attached spermatium (triple arrowhead). (F) Carposporophyte with indeterminate gonimoblast filament (arrowhead), and carposporangia (double arrowhead) at the tips of short lateral branches. (G–J) *Tuomeya americana*: (G) Macroscopic view of a mature plant, which is a well-branched, cartilaginous tube. (H) Branch apex showing confluent whorls. (I) Carpogonium with small base (arrowhead) and large trichogyne (double arrowhead) perpendicularly attached to a stalk. (J) Carposporophyte with prominent carposporangia (arrowheads) at the tips of short gonimoblast filaments. (K) *Lemanea fluviatilis*, macroscopic view showing thallus, which is tubular, with inflated nodes, branches, and a stalk. A reprinted from Vis *et al.* (1996a) with permission. B reprinted from Sheath *et al.* (1992) with permission of the *Journal of Phycology*. D–F reprinted from Necchi *et al.* (1993c) with permission of the *Journal of Phycology*, G, H, and K reprinted from Sheath (1984) with permission. I and J reprinted from Kaczmarczyk *et al.* (1992) with permission of the *Journal of Phycology*.

the lower side of the pericentral cells. Cortical filaments grow downward and ensheath axial cells, often producing secondary fascicle branches. Each fascicle cell contains several, ribbon-like, parietal chloroplasts with no pyrenoid. Few species form monosporangia in the gametophyte stage. Spermatangia bud off from terminal primary and secondary fascicle cells or in some species from involucral filaments of the carpogonial branch; they are spherical, colorless, and 4–8 µm in diameter. Carpogonial branches range from little modified to well differentiated with a twist in the section *Contorta*. Carpogonia with broad trichogynes are sometimes stalked on a small base containing the nucleus. Carposporophytes are generally a spherical or semispherical, compact or loose mass of gonimoblast filaments; carposporangia form at the apices. Carpospores germinate into the "chantrasia" stage, a crustose growth consisting of large basal cells and erect, the sparsely branched filaments. Filaments can form monosporangia or divide meiotically, producing an attached gametophyte and two residual cells.

Batrachospermum is a cosmopolitan genus occurring in moderately flowing, reasonably unpolluted streams (Sheath, 1984). *B. turfosum* is also common in bogs. Both euryphotic (e.g., *B. gelatinosum*) and shade or brown-water (e.g., *B. turfosum*) species exist. Filament fragments are common in the guts of grazing amphipods, insect larvae, and snails. *Batrachospermum* section *Contorta* is mostly concentrated in tropical to subtropical regions. *B. gelatinosum*, the only widespread species in tundra, tolerates a large range of conditions: temperature, 0–24°C; current velocity, 7–181 cm s^{-1}; pH, 4.1–8.2; specific conductance, 10–360 µS cm^{-1}; PO_4^{3-} < 1–4900 µg L^{-1} (Sheath and Hambrook, 1990).

Key to the North American Sections of Batrachospermum

1a.	Carpogonial branch with little or no differentiation from fascicles (Fig. 3K)	*Batrachospermum*
1b.	Carpogonial branch well differentiated	2
2a.	Carpogonial branch almost as elongate as fascicles	*Aristata*
2b.	Carpogonial branch much shorter than fascicles (Fig. 3L, M)	3
3a.	Carpogonial branch with one or more well-developed twists (Fig. 3M)	*Contorta*
3b.	Carpogonial branch straight or with slight curves	4
4a.	Carpogonial branches a combination of straight and curved in same filament	*Hybrida*[1]
4b.	Carpogonial branch always straight (Fig. 3L)	5
5a.	Mean mature whorl (with carposporophytes) diameter < 170 µm, mean fascicle cell number, <6	*Setacea*
5b.	Mean whorl diameter, > 300 µm; mean fascicle cell number, ≥ 7	6
6a.	Stalked trichogyne	*Virescentia*
6b.	Sessile trichogyne	*Turfosa*

[1]Vis and Entwisle (2000) proposed sinking this section into *Contorta* based on the positioning of *Batrachospermum virgato-decaisneanum* in *rbc*L gene trees.

Sirodotia Kylin (Fig. 4D–F)

Sirodotia is composed of attached, gelatinous gametophytic filaments, up to 17 cm long, with a beaded appearance varying from blue–green to yellow–green. It contains a uniaxial central filament with large, cylindrical cells; four to six pericentral cells produce repeatedly branched fascicles of limited growth. In most species, rhizoid-like cortical filaments develop from the lower side of the pericentral cells. Each fascicle cell contains several, ribbon-like, parietal chloroplasts with no pyrenoid. Spermatangia bud off from terminal fascicle cells; they are spherical, colorless, and 4–7 µm in diameter. Carpogonial branches are somewhat differentiated with small cells. Carpogonia with broad trichogyne are attached off-center to the base, the latter structure having a definite protrusion. The carposporophyte occurs as a branched indeterminant filament creeping along the main axis; carposporangia form at branch apices. Carpospores germinate into the "chantransia" stage, composed of branched, uniaxial filaments. Meiosis and monosporangia have not been observed. *Sirodotia* was considered to be a section of *Batrachospermum* by Necchi and Entwisle (1990), but this was questioned by Necchi *et al.* (1993c).

Sirodotia occurs in scattered, small, typically softwater (pH, 5.7–7.6; specific conductance, 10–140 µS cm^{-1}) streams in boreal to tropical environments; in North America, species range from northern Quebec and Newfoundland to central Mexico (Necchi *et al.*, 1993c). The most widespread species, *S. suecica*, has little increase in drag with increasing current velocity (20–80 cm s^{-1}) due to branch reconfiguration. The second species, *S. huillensis*, is apparently restricted to arid, southwestern habitats (Vis and Sheath, 1999).

Tuomeya W. H. Harvey (Fig. 4G–J)

Tuomeya has a densely branched, cartilaginous, and cylindrical gametophytic thallus, 1–5 (6.5) cm long. It ranges in color from blue–green to olive to black. The uniaxial filament is covered by two or three layers of cortical filaments; dense laterals arising from approximately six pericentral cells are of limited growth with outer cells fused. Axial cells are evident in mature branches by constrictions. Spermatangia form at the tips of the laterals. Carpogonia are asymmetrical with an irregularly broadened trichogyne attached obliquely or perpendicularly to a stalk and borne on a curved carpogonial branch derived from a pericentral cell. The carposporophyte occurs as a globular mass of filaments. Carpospores germinate into the branched, uniseriate "chantransia" stage in culture, but *in situ* gametophytes develop from an undifferentiated mass of cells. *Tuomeya* was considered to be a section of *Batrachospermum* by Necchi and Entwisle (1990), but this was questioned by Kaczmarczyk *et al.* (1992).

Tuomeya occurs in scattered freshwater streams in eastern North America, from Florida to Newfoundland, from fall to early summer. The range of conditions is as follows: temperature, 5–26°C; current velocity, 16–125 cm s^{-1}; pH, 4.7–7.6; specific conductance, 10–124 µS cm^{-1} (Kaczmarczyk *et al.*, 1992). It tolerates considerable stress before breaking (1780 ± 850 kN m^{-2}), stretching 22% in the process. *Tuomeya* is common in the guts of grazing amphipods and insect larvae; its cartilaginous structure and low protein content make it a little preferred food source.

Batrachospermales: Lemaneaceae

Lemanea Bory de St. Vincent (Figs. 4K and 5A, B)

Lemanea is composed of tufts of cartilaginous, tubular, pseudoparenchymatous gametophytic thalli, lacking cortical filaments, around a central uniseriate axis. T- or L-shaped ray cells are closely applied to the outer cortex. It is 1–40 cm long and 0.2–2.0 mm in diameter. It is blue–green to olive when young, becoming rusty-brown to black at maturity. Species are characterized by the presence of branching, by their diameter, and by the degree of basal constriction. Several parietal discoid chloroplasts occur in the outer cells only. Spermatangia develop as yellowish circular patches. Nearby small carpogonial branches are entirely internal except for a thin trichogyne that protrudes beyond the outer cell layer. Carposporophytes are microscopic, spherical masses of filaments forming large, ellipsoidal carpospores within the central cavity; they are released by thallus deterioration and germinate into branched, uniseriate "chantransia" filaments. The "chantransia" stage produces attached gametophytes seasonally after meiosis.

This genus is widespread in temperate and boreal streams and rivers with typically high current velocities (up to 2 m s^{-1}) (Sheath and Hambrook, 1990). Species have adapted to flow by developing dense turfs closely adherent to rocks and high breaking stress (910 ± 430 kN m^{-2} for *L. fluviatilis*). *Lemanea* is common at high elevations up to 1200 m. *L. fluviatilis* tolerates a temperature of 4–25°C, a pH of 4.1–8.2, and a specific conductance of 10–300 µS cm^{-1}. It occurs at low nutrient concentrations. In a Rhode Island river, growth and reproduction of *L. fluviatilis* gametophytes are confined to April–August, after which the thallus deteriorates and carpospores are released; between September and March, remnants persist.

Paralemanea Vis et Sheath (Fig. 5C–F)

Paralemanea is composed of tufts of cartilaginous, tubular, pseudoparenchymatous gametophytic thalli, with cortical filaments, around a central, axial filament and simple ray cells not abutting outer cortical cells. The mean length is 4.3–9.5 cm and the diameter is 0.5–0.7 mm. Species are characterized by the presence of branching, by their length, and by their diameter. Several parietal discoid chloroplasts occur in the outer cells only. Spermatangia develop as yellowish to brownish nodal rings. Small carpogonial branches are entirely internal except for a thin trichogyne that protrudes beyond the outer layer. Carpsoporophytes are small, spherical masses of filaments forming large, ellipsoidal carpospores in the central cavity; they are released by thallus deterioration and germinate into branched, uniaxial "chantransia" filaments. The "chantransia" stage produces attached gametophytes seasonally after meiosis.

Most populations of *Paralemanea* in North America are from the southeastern United States and northern California, but extend from central Mexico to New York (Vis and Sheath, 1992). This genus occurs under a wide range of conditions: mean current velocity, 18–110 cm s^{-1}; temperature, 4–17°C; pH, 5.5–8.6; specific conductance, 42–500 µS cm^{-1}. It is predominant in cool seasons in a North Carolina stream (Everitt and Burkholder, 1991).

Thoreales

Recent molecular sequence analyses have demonstrated that this family is not closely aligned with the Batrachospermaceae and Lemaneaceae and should be removed from the Batrachospermales (e.g., Vis *et al.*,

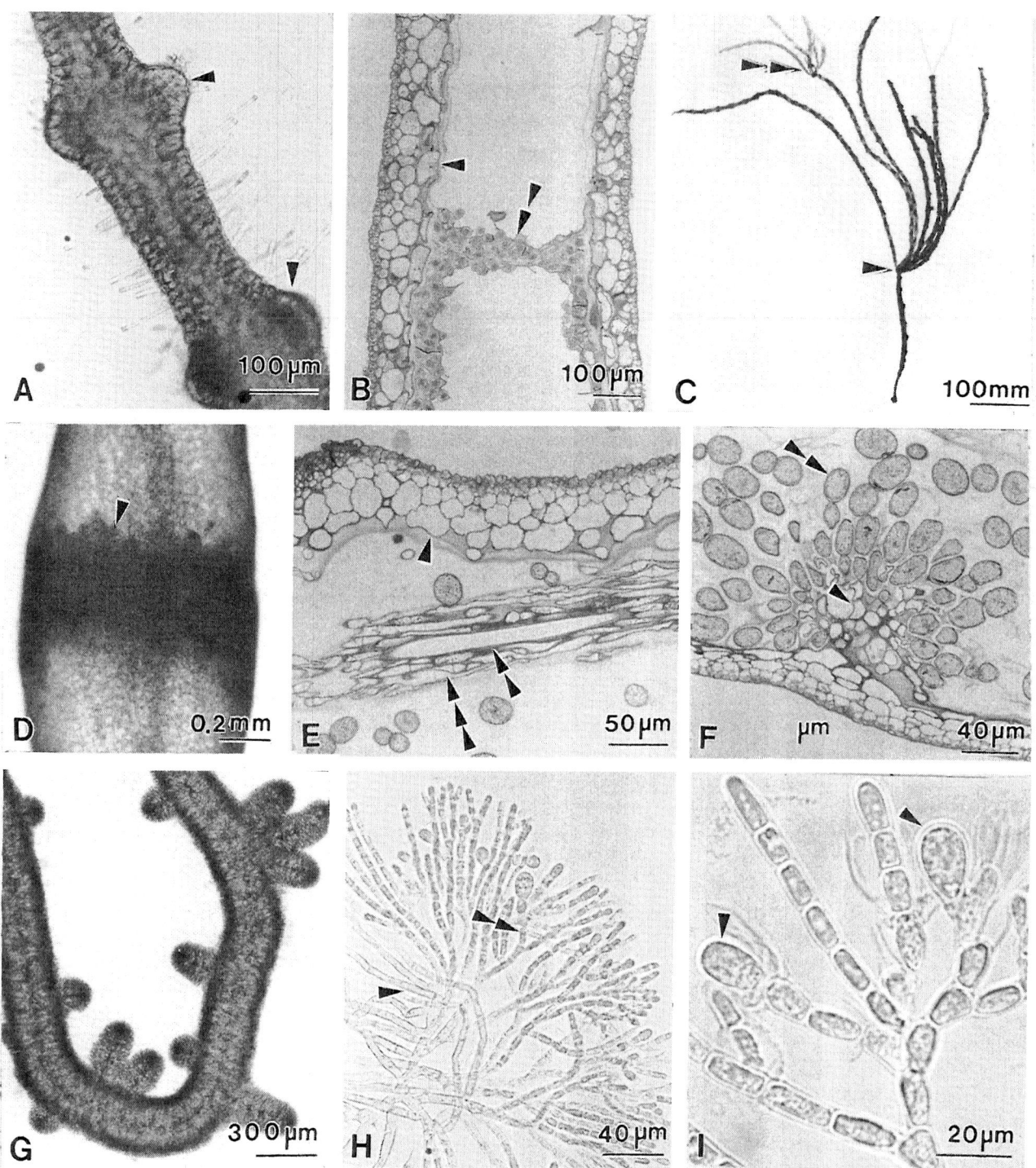

FIGURE 5 Members of the Lemaneaceae and Thoreales. (A, B) *Lemanea*: (A) *L. fluviatilis* with nodes obvious as a series of swellings will have patches of spermatangia (arrowheads). (B) Longitudinal Toluidine Blue O–stained longitudinal section showing outer cortical layer (arrowhead) and a central strand of carpospores (double arrowhead). (C–F) *Paralemanea*: (C) *P. mexicana*, macroscopic view with whorled branching (arrowhead), rebranching (double arrowhead), and tubular construction. (D) *Paralemanea* sp. with an obvious spermatangial ring (arrowhead). (E) *P. mexicana*, Toluidine Blue O–stained longitudinal section showing outer cortical layer (arrowhead), axial filament (double arrowhead), which is surrounded by inner cortical filaments (triple arrowhead). (F) *P. mexicana*, Toluidine Blue O–stained section showing carposporophyte (arrowhead) and large carpospores (double arrowhead). (G–I) *Nemalionopsis tortuosa*: (G) Branched, multiaxial filament. (H) Section of thallus showing colorless, central medulla (arrowhead) and photosynthetic, assimilatory branches (double arrowhead). (I) Monosporangia (arrowheads), which are formed at the tips of assimilatory filaments. A reprinted from Sheath (1984) with permission. B, E, and F reprinted from Sheath *et al.* (1996b) with permission. C and D reprinted from Vis and Sheath (1992) with permission of the International Phycological Society. I reprinted from Sheath *et al.* (1993b) with permission.

1998). We have proposed that the Thoreaceae be placed in a new order, the Thoreales (Sheath et al., 2000), Müller et al., 2002.

Nemalionopsis Skuja (Fig. 5G–I)

This genus contains a sparsely branched, cordlike thallus, which is burgundy to yellow–brown in color; it is composed of a central medullary region of interwoven, colorless filaments and an outer pigmented cortex of branched laterals of limited growth. The length is 5–30 (50) cm, the diameter is 700–1000 μm. It may be flattened or coiled in some portions. Monosporangia form at the outer tips of lateral filaments, 7–12 × 8–14 μm. The spore-bearing branch-to-vegetative branch length ratio is greater than 0.64. Monospores germinate into prostrate filaments forming monostromatic discs from which erect filaments arise. Sexual reproduction has not been observed.

Nemalionopsis is known from four freshwater streams in North America in Florida, Louisiana, and North Carolina with temperatures ranging from 13 to 22°C, current velocity of 29 cm s^{-1}, pH of 7.1–8.3, and specific conductance of 220 μS cm^{-1} (Sheath et al., 1993b).

Thorea Bory de St. Vincent (Fig. 6A–D)

Thorea is composed of branched gametophytic filaments, up to 20–200 cm long and 0.5–3 mm in diameter. It is composed of interwoven, colorless medullary filaments and dense, photosynthetic laterals of limited growth. It is olive green to reddish to black. Chloroplasts in assimilatory filaments are parietal and ribbon-like. Monosporangia are solitary or in clusters, formed at the base of the assimilatory filaments (spore-bearing branch-to-vegetative lateral length ratio < 0.3). Sexual reproduction is known for a few species. In *T. violacea*, spermatangia are borne on specialized branches near the base of assimilatory filaments, colorless, elliptical or obovoid, 8–10 × 4–7 μm. Carpogonia are conical with an elongated trichogyne 5–7 μm wide; the carpogonial branch is short and located at the base of assimilatory filaments. Carposporophytes are sparsely branched and compact. Carposporangia are terminal, 9–13 × 17–25 μm; carpospores germinate into branched, uniseriate "chantransia" filaments.

This genus is widespread in tropical to warm temperate, freshwater streams; in North America, it ranges from New York to Grenada (Sheath et al., 1993b). It typically occurs in alkaline waters (e.g., in North America, pH, 7.5–8.2; specific conductance, 180–500 μS cm^{-1}). An exception to this distribution pattern is a population in the Hudson River, New York, which occurs in temperatures ranging from 15 to 19°C, specific conductance of 44–136 μS cm^{-1} and pH of 6.6–7.2 (Pueschel et al., 1995).

Hildenbrandiales

Hildenbrandia Nardo (Fig. 6E–G)

This genus contains a bright-red uncalcified crustose thallus, which is composed of a single basal layer that gives rise to vertical files of cells. The thallus height varies from 23 to 182 μm. The cell dimensions, 2–8 × 4–10 μm, indicate that a single species, *H. angolensis*, exists in freshwater habitats of North America [*H. rivularis* appears to be largely restricted to Europe (Sherwood and Sheath, 2000a)]. Reproduction is largely by gemmae, dense aggregation of cells formed in the thallus, which are eventually released and germinate into new crusts. Gemma production continues year round in two spring-fed streams in Texas (Sherwood and Sheath, 2000b).

Although *H. angolensis* has been reported from Pennsylvania, well-documented collections occur in streams and springs from Texas in the north to Costa Rica in the south and throughout the Caribbean islands (Sheath et al., 1993a). These streams are mostly warm (14–27°C), alkaline in pH (7.0–8.6), and variable in current velocity (5–67 cm s^{-1}) and specific conductance (70–1558 μS cm^{-1}). The freshwater populations of *Hildenbrandia* in North America form a monophyletic clade in 18S rRNA gene trees distinct from marine populations of the genus (Sherwood and Sheath, 1999b).

Ceramiales

Ballia W. H. Harvey (Fig. 6H)

Ballia is composed of reddish filaments with a distinct main axis with hexagonal-shaped cells and smaller, pinnate, determinate lateral branches that may rebranch. The apical cell is typically quite long in *B. prieurii* (43–89 μm in length). Plants typically are small (3–15 mm). They can reproduce by the production of monosporangia. Choi et al. (2000) have proposed that marine species of this genus from Australia are not members of the Ceramiales, based on the presence of two pit plug cap layers and positioning in 18S rRNA gene trees. However, they were unable to resolve the status of the freshwater species, *B. prieurii*.

Three freshwater collections have been made in North America from Belize and Costa Rica, with slow to moderate current velocities (1–65 cm s^{-1}), pH of 7.6–7.8, specific conductance of 50–100 μS cm^{-1}, and temperatures of 19–22°C (Sheath et al., 1993c).

Bostrychia Montagne (Figs. 6I and 7A, B)

Bostrychia is composed of dark-reddish filaments with tiers of pericentral cells around the axial cells; the

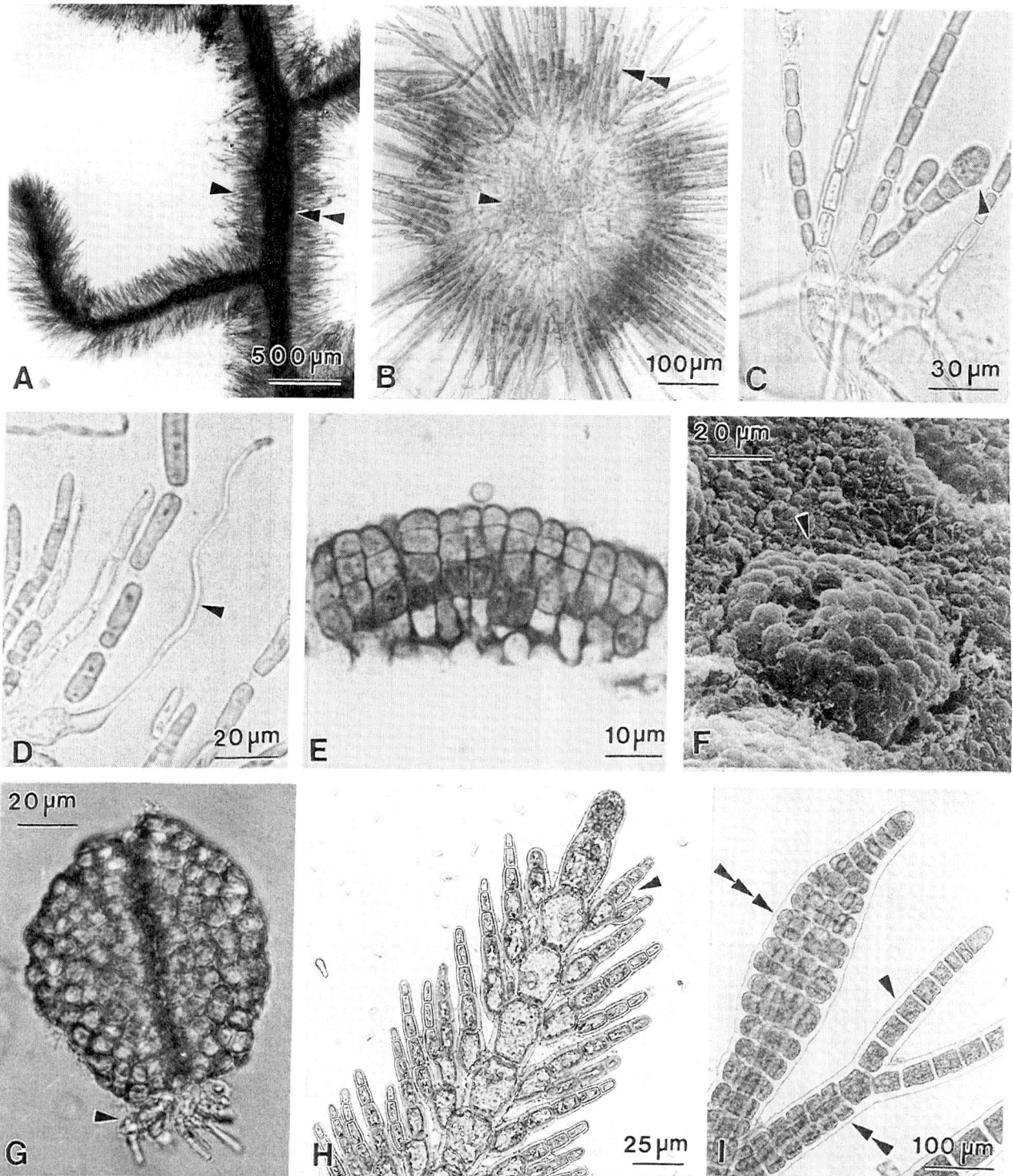

FIGURE 6 Freshwater members of the Thoreales, Hildenbrandiales, and Ceramiales. (A–D) *Thorea*: (A) *T. hispida*, branched, multiaxial filament, with an obvious outer, assimilatory layer (arrowhead) and inner medullary layer (double arrowhead). (B) *T. violacea*, transverse section showing colorless, central medulla (arrowhead) and outer, assimilatory filaments (double arrowhead). (C) *T. violacea* with monosporangium (arrowhead) on short branch at the base of assimilatory filament. (D) *T. violacea* carpogonium with thin, elongate trichogyne (arrowhead). (E–G) *Hildenbrandia angolensis*: (E) Toluidine Blue O–stained transverse section showing rows of erect filaments. (F) Scanning electron micrograph of thallus surface with gemmae (arrowhead) formed in cavities. (G) Released gemma with basal rhizoids for attachment (arrowhead). (H) *Ballia prieurii*, apex showing hexagonal axial cells and pinnate lateral branches that rebranch (arrowhead). (I) *Bostrychia moritziana*, apex with uniaxial branch tip, mature axis covered with pericentral cells (double arrowhead) and stichidium (triple arrowhead). A and D reprinted from Sheath *et al.* (1993b) with permission. B and C reprinted from Sheath (1984) with permission. E and F courtesy of A. Sherwood. H and I reprinted from Sheath *et al.* (1993c) with permission of the *Journal of Phycology*.

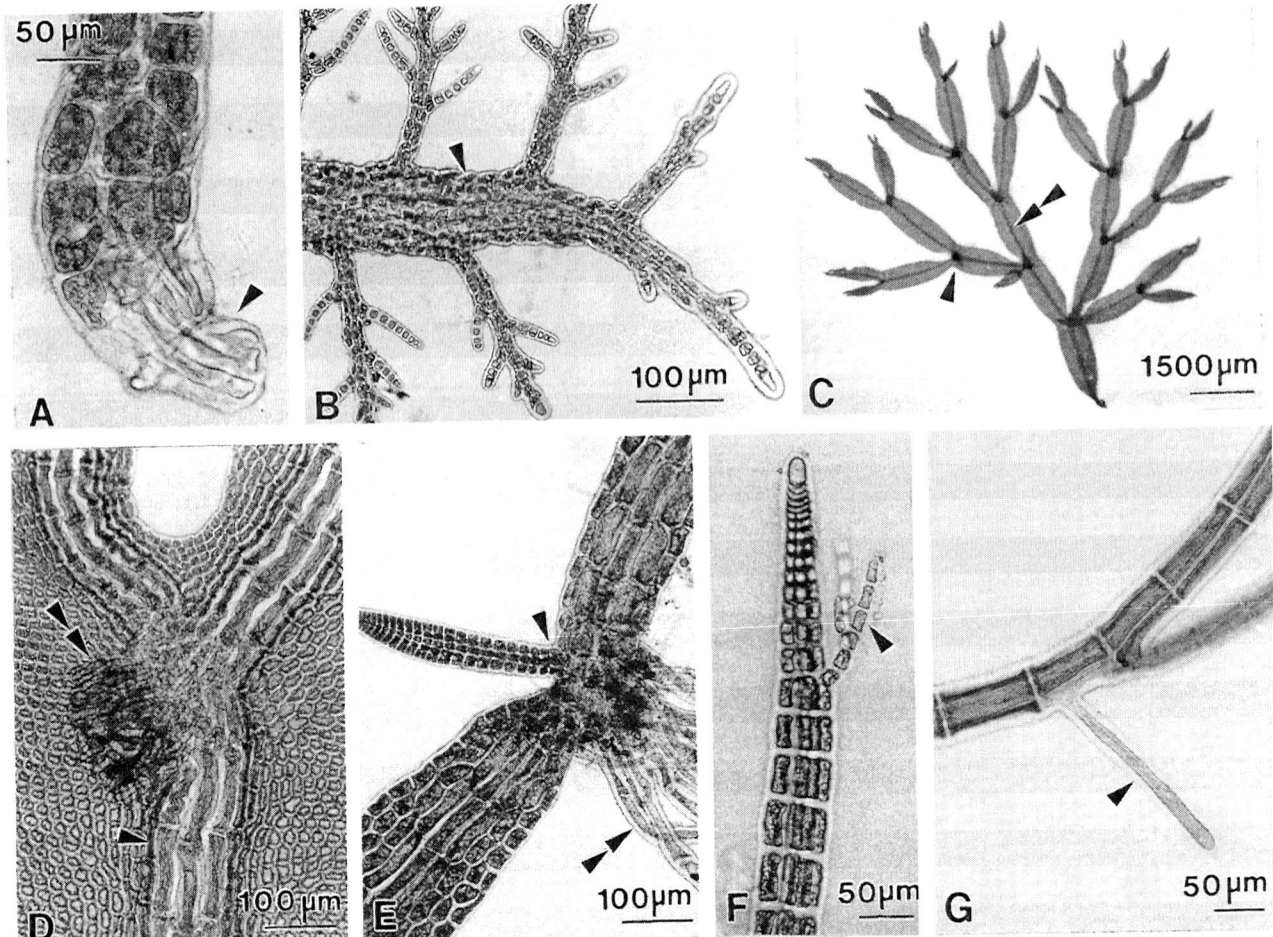

FIGURE 7 Freshwater members of the Ceramiales. (A, B) *Bostrychia*: (A) *B. moritziana* hapteron, specialized branch with colorless rhizoids (arrowhead) at tip. (B) *B. tenella* with an additional layer of cortication (arrowhead) outside of pericentral cells. (C–E) *Caloglossa*: (C) *C. leprieurii*, macroscopic view of flat blades with subdichotomous branching, constrictions at nodes (arrowhead), and obvious midrib (double arrowhead). (D) *C. leprieurii* showing cortical filaments of midrib area (arrowhead) and rhizoids (double arrowhead). (E) *C. ogasawaerensis* with new branch (arrowhead) and rhizoids arising at node. (F, G) *Polysiphonia subtilissima*: (F) Apex with axis surrounded by pericentral cells and trichoblast (arrowhead). (G) Rhizoid (arrowhead) arising from pericentral cell. A–G reprinted from Sheath *et al.* (1993c) with permission of the *Journal of Phycology*.

thallus may become additionally corticated to the outside of the pericentral cells (e.g., *B. tenella*). Specialized rhizoidal branches, the haptera, attach filaments to the substrata. Vegetative branching is bilateral; branches near the apex tend to incurve and there may be both long and short shoots. Only tetrasporangia have been observed in freshwater collections, which are formed in inflated, multichambered structures termed stichidia.

Freshwater populations in North America appear to be restricted to streams in the Caribbean islands which are warm (21–26°C), alkaline (pH 7.0–8.4), and range in specific conductance (56–440 µS cm^{-1}) (Sheath *et al.*, 1993c).

Caloglossa J. Agardh (Fig. 7C–E)

Caloglossa is composed of flat, dichotomously branched reddish blades with constrictions. A prominent midrib is evident and is composed of a broad axial row of cells surrounded by a cortex of elongated cells. The outer portions of the blade are monostromatic with an oblique series of hexagonal cells. Rhizoids arise at constrictions, either from the midrib area or from the peripheral layer of cells. Population spread in freshwaters is vegetative, although gametophytic and tetrasporic plants have been collected in brackish waters.

Two species, *C. lepreurii* and *C. ogasawarensis*, have been collected from streams in Puerto Rico and

Costa Rica, respectively. Current velocities are moderate (33–43 cm s^{-1}), temperature warm (23–24°C), pH alkaline (7.6–8.4), with a specific conductance of 100–200 µS cm^{-1} (Sheath *et al.*, 1993c).

Polysiphonia Greville (Fig. 7F, G)

Polysiphonia is composed of dark-reddish filaments with a single tier of pericentral cells around the axial cell. No freshwater collections have an additional layer of cortication but a few marine species do. Delicately branched hairs (trichoblasts) are formed in upper portions of the plant. Rhizoidal branches arise from pericentral cells. Freshwater samples have not been observed as being either sexual or tetrasporic.

This genus is common in marine and brackish habitats; only two populations of *P. subtilissima* have been collected in North American freshwaters in Florida and Jamaica (Sheath *et al.*, 1993c). These streams have moderate flow (25 cm s^{-1}), warm temperature (22–26°C), alkaline pH (7.7–7.8), and high specific conductance (1150–1840 µS cm^{-1}).

VI. GUIDE TO LITERATURE FOR SPECIES IDENTIFICATION

The following is a list of key North American references, each of which contains citations to older literature and those from other continents:

1. *Audouinella*—Necchi *et al.* (1993a, b)
2. *Ballia*—Sheath *et al.* (1993c)
3. *Bangia*—Sheath and Cole (1984), Müller *et al.* (1998)
4. *Batrachospermum*—Sheath *et al.* (1992, 1993d, 1994a–c), Sheath and Vis (1995), Vis *et al.* (1996a, b), Vis and Sheath (1996, 1997, 1998), Müller *et al.* (1997)
5. *Boldia*—Howard and Parker (1980), Sheath and Hymes (1980), Rintoul *et al.* (1999)
6. *Bostrychia*—Sheath *et al.* (1993c)
7. *Caloglossa*—Sheath *et al.* (1993c)
8. *Chroodactylon*—Vis and Sheath (1993)
9. *Chroothece*—Blinn and Prescott (1976)
10. *Compsopogon* - Vis *et al.* (1992), Rintoul *et al.* (1999)
11. *Compsopogonopsis*—Vis *et al.* (1992), Rintoul *et al.* (1999)
12. *Cyanidium*—Seckbach (1991)
13. *Flintiella*—Ott (1976)
14. *Hildenbrandia*—Sheath *et al.* (1993a), Sherwood and Sheath (1999b, 2000b)
15. *Kyliniella*—Vis and Sheath (1993)
16. *Lemanea*—Vis and Sheath (1992), Sheath *et al.* (1996b)
17. *Nemalionopsis*—Sheath *et al.* (1993b)
18. *Paralemanea*—Vis and Sheath (1992), Sheath *et al.* (1996b)
19. *Polysiphonia*—Sheath *et al.* (1993c)
20. *Porphyridium*—Ott (1972)
21. *Rhododraparnaldia*—Sheath *et al.* (1994d)
22. *Rufusia*—Wujek and Timpano (1986)
23. *Sirodotia*—Necchi *et al.* (1993c), Vis and Sheath (1999)
24. *Thorea*—Sheath *et al.* (1993b), Vis *et al.* (1998), Müller *et al.* (2002)
25. *Tuomeya*—Kaczmarczyk *et al.* (1992)

ACKNOWLEDGMENTS

I thank everyone who has worked on freshwater red algae in my laboratory and contributed to much of the knowledge summarized in this chapter, including JoAnn Burkholder, Lesley Campbell, Tracey Carlson, Wayne Chiasson, Dana Couture, Julie Hambrook, Bev Hymes, Don Kaczmarczyk, Judith Korch, Mollie Morison, Kirsten Müller, Orlando Necchi, Jr., Tara Rintoul, Troina Shea, Alison Sherwood, Al Steinman, Stacey Thompson, Kathy Van Alstyne, and Morgan Vis. In addition, collaborators from other laboratories and institutions have been very helpful in the various studies reported, such as Murray Colbo, Dave Larson, Tim Entwisle, Gary Saunders, and Alan Whittick. A special thank you to Kay Cole whose long-time collaboration has been an inspiration. Many collecting trips were made more enjoyable with the support and help of Mary Koske. Assistance in manuscript preparation from Toni Pellizzari and helpful reviews by Orlando Necchi, Jr., Morgan Vis, and John Wehr are also appreciated. Research support from NSERC, NSF, Universities of Guelph and Rhode Island, PCSP, and the Northern Training Program is gratefully acknowledged.

LITERATURE CITED

Blinn, D. W., Prescott, G. W. 1976. A North American distribution record for the rare Rhodophyceae, *Chroothece mobilis* Pascher and Petrova. American Midland Naturalist 96:207–210.

Broadwater, S. T., Scott, J. L. 1994. Ultrastructure of unicellular red algae, *in*: Seckbach, J., Ed., Evolutionary pathways and enigmatic algae: *Cyanidium caldarium*. Kluwer Academic, Dordrecht, pp. 215–230.

Brown, D. L., Weier, T. E. 1968. Chloroplast development and ultrastructure in the freshwater red alga *Batrachospermum*. Journal of Phycology 4:199–206.

Camburn, K. E., Warren, M. L., Jr. 1983. Epizoic occurrence of *Compsopogon coeruleus* (Rhodophyta) on *Lernaea* (Copepoda) from Kentucky fishes. Canadian Journal of Botany 61:3545–3548.

Choi, H.-G., Kraft, G. T., Saunders, G. W. 2000. Nuclear small-subunit rDNA sequences from *Ballia* spp. (Rhodophyta): Proposal of the Balliales ord. nov., Balliaceae fam. nov., *Ballia nana* sp. nov. and *Inkyuleea* gen. nov. (Ceramiales). Phycologia 39:272–287.

Couté, A., Sarthou, C. 1990. Révision des espèces d'eau douce du genre *Ballia* (Rhodophytes, Céramiales). Cryptogamie Algologie 11:265–279.

Craigie, J. S. 1990. Cell walls, *in*: Cole, K. M., Sheath, R. G., Eds., Biology of the red algae. Cambridge Univ. Press, Cambridge, UK, pp. 221–257.

D'Lacoste, V., Ganesan, E. K. 1987. Notes on Venezuelan freshwater algae I. Nova Hedwigia 45:263–281.

DeLuca, P., Gambardella, R., Merola, A. 1979. Thermoacidophilic algae in North and Central America. Botanical Gazette 140:418–27.

Doemel, W. N., Brock, T. D. 1971. The physiological ecology of *Cyanidium caldarium*. Journal of General Microbiology 67:17–32.

Drew, K. M. 1935. The life history of *Rhodochorton violaceum* (Kütz.) comb. nov. (*Chantrasia violacea* Kütz.). Annals of Botany 49:439–450.

Everitt, D. T., Burkholder, J. M. 1991. Seasonal dynamics of macrophyte communities from a stream flowing over granite flatrock in North Carolina, USA. Hydrobiologia 222:159–172.

Flint, L. H. 1970. Freshwater red algae of North America. Vantage Press, New York, 110 p.

Fritsch, F. E. 1929. The encrusting algal communities of certain fast-flowing streams. New Phytologist 28:165–196.

Gantt, E., Scott, J., Lipschultz, C. 1986. Phycobiliprotein composition and chloroplast structure in the freshwater red alga *Compsopogon coeruleus* (Rhodophyta). Journal of Phycology 22:480–484.

Garbary, D. J., Hansen, G. I., Scagel, R. F. 1980. A revised classification of the Bangiophyceae (Rhodophyta). Nova Hedwigia 33:145–166.

Geitler, L. 1932. *Porphyridium sordidum* n. sp., eine neue Süßwasserbangiale. Archiv für Protistenkunde 76:595–604.

Gibson, M. T., Whitton, B. A. 1987. Hairs, phosphatase activity and environmental chemistry in *Stigeoclonium, Chaetophora* and *Draparnaldia* (Chaetophorales). British Phycological Journal 22:11–22.

Gretz, M. R., Sommerfeld, M. R., Athey, P. V., Aronson, J. M. 1991. Chemical composition of the cell walls of the freshwater red alga *Lemanea annulata* (Batrachospermales). Journal of Phycology 27:232–240.

Hambrook, J. A., Sheath, R. G. 1987. Grazing of freshwater Rhodophyta. Journal of Phycology 23:656–662.

Hambrook, J. A., Sheath, R. G. 1991. Reproductive ecology of the freshwater red alga *Batrachospermum boryanum* Sirodot in a temperate headwater stream. Hydrobiologia 218:233–246.

Harding, J. P. C., Whitton, B. A. 1981. Accumulation of zinc, cadmium and lead by field populations of *Lemanea*. Water Research 15:301–319.

Hoffmann, L. 1989. Algae of terrestrial habitats. Botanical Review 55:77–105.

Howard, R. V., Parker, B. C. 1980. Revision of *Boldia erythrosiphon* Herndon (Rhodophyta, Bangiales). American Journal of Botany 67:413–422.

Hymes, B. J., Cole, K. M. 1983. The cytology of *Audouinella hermannii* (Rhodophyta, Florideophyceae). II. Monosporogenesis. Canadian Journal of Botany 61:3377–3385.

Hynes, H. B. N. 1970. The ecology of running waters. Liverpool Univ. Press, Liverpool.

Israelson, G. 1942. The freshwater Florideae of Sweden. Symbolae Botanicae Upsalienses 8(1):1–134.

Jimenez, J. C. 1999. Estudio floristico (taxonomico-ecologico-biogeografico) de las rodofitas de agua dulce en la region central de Mexico. Ph.D. thesis, Universidad Nacional Autonoma de Mexico.

Kaczmarczyk, D., Sheath, R. G. 1991. The effect of light regime on the photosynthetic apparatus of the freshwater red alga *Batrachospermum boryanum*. Cryptogamie Algologie 12:249–263.

Kaczmarczyk, D., Sheath, R. G., Cole, K. M. 1992. Distribution and systematics of the freshwater genus *Tuomeya* (Rhodophyta, Batrachospermaceae). Journal of Phycology 28:850–855.

Korch, J. E., Sheath, R. G. 1989. The phenology of *Audouinella violacea* (Acrochaetiaceae, Rhodophyta) in a Rhode Island stream (USA). Phycologia 28:228–236.

Kremer, B. 1983. Untersuchungen zur Ökophysiologie einiger Süsswasserrotalgen. Decheniana (Bonn) 136:31–42.

Kumano, S. 1978. Notes on freshwater red algae from West Malaysia. Botanical Magazine (Tokyo) 91:97–107.

Kumano, S., Necchi, O., Jr. 1987. Studies on the freshwater Rhodophyta of Brazil-5: Record of *Bostrychia radicans* (Montagne) Montagne f. *moniliforme* Post in freshwaters. Revista Brasileira de Biologia 47:437–440.

Leukart, P., Hanelt, D. 1995. Light requirements for photosynthesis and growth in several macroalgae from a small soft-water stream in the Spessart Mountains, Germany. Phycologia 34:528–532.

Lin, C. K., Blum, J. L. 1977. Recent invasion of a red alga *Bangia* in Lake Michigan. Journal of the Fisheries Research Board of Canada 34:2413–2416.

MacFarlane, J. J., Raven, J. A. 1985. External and internal CO_2 transport in *Lemanea*: Interactions with the kinetics of ribulose bisphosphate carboxylase. Journal of Experimental Botany 36:610–622.

Manny, B. A., Edsall, T. A., Wujek, D. F. 1991. *Compsopogon* cf. *coeruleus*, a new benthic red alga (Rhodophyta) in the Laurentian Great Lakes. Canadian Journal of Botany 69:1237–1240.

McKay, R. M. L., Gibbs, S. P. 1990. Phycoerythrin is absent from the pyrenoid of *Porphyridium cruentum*: Photosynthetic implications. Planta 180:249–256.

Müller, K. M., Vis, M. L., Chiasson, W. B., Whittick, A., Sheath, R. G. 1997. Phenology of a *Batrachospermum* population in a boreal pond and its implications for the systematics of section *Turfosa* (Batrachospermales, Rhodophyta). Phycologia 36:68–75.

Müller, K. M., Sheath, R. G., Vis, M. L., Crease, T. J., Cole, K. M. 1998. Biogeography and systematics of *Bangia* (Bangiales, Rhodophyta) based on the Rubisco spacer, *rbc*L gene, 18S rRNA gene sequences and morphometric analyses. I. North America. Phycologia 37:195–207.

Müller, K. M., Sherwood, A. R., Pueschel, C. M., Gutell, R. R., Sheath, R. G. 2002. A proposal for a new red algal order, the *Thoreales*. Journal of Phycology 38:807–820.

Nagy, J. P. 1965. Preliminary note on the algae of Crystal Cave, Kentucky. International Journal of Speleology 1:479–490.

Necchi, O., Jr. 1993. Distribution and seasonal dynamics of Rhodophyta in the Preto River Basin, southeastern Brazil. Hydrobiologia 250:81–90.

Necchi, O., Jr. 1995. Occurrence of the genus *Ballia* (Ceramiaceae, Rhodophyta) in freshwater in Brazil. Hoehnea 22:229–235.

Necchi, O., Jr. 1997. Microhabitat and plant structure of *Batrachospermum* (Batrachospermales, Rhodophyta) populations in four streams of São Paulo State, southeastern Brazil. Phycological Research 45:39–45.

Necchi, O., Jr., Entwisle, T. J. 1990. A reappraisal of generic and subgeneric classification in the Batrachospermaceae (Rhodophyta). Phycologia 29:478–488.

Necchi, O., Jr., Sheath, R. G., Cole, K. M. 1993a. Systematics of freshwater *Audouinella* (Acrochaetiaceae, Rhodophyta) in North America. 1. The reddish species. Algological Studies 70:11–28.

Necchi, O., Jr., Sheath, R. G., Cole, K. M. 1993b. Systematics of freshwater *Audouinella* (Acrochaetiaceae, Rhodophyta) in North America. 2. The bluish species. Algological Studies 71:13–21.

Necchi, O., Jr., Sheath, R. G., Cole, K. M. 1993c. Distribution and systematics of the freshwater genus *Sirodotia* (Batrachospermales, Rhodophyta) in North America. Journal of Phycology 29:236–243.

Ott, F. D. 1972. A review of the synonyms and taxonomic position of the algal genus *Porphyridium* Nägeli 1849. Nova Hedwigia 23:237–289.

Ott, F. D. 1976. Further observations on the freshwater alga *Flintiella sanguinaria* Ott in Bourrelly 1970 (Rhodophycophyta, Porphyridiales). Archiv für Protistenkunde 118:34–52.

Pipp, E., Rott, E. 1994. Classification of running-water sites in Austria based on benthic algal community structure. Verhandlungen Internationale Vereinigung für Theoretische und Angewandte Limnologie 25:1610–1613.

Pueschel, C. M. 1990. Cell structure, in: Cole, K. M., Sheath, R. G., Eds., Biology of the red algae. Cambridge Univ. Press, Cambridge, UK, pp. 7–41.

Pueschel, C. M., Sullivan, P. G., Titus, J. E. 1995. Occurrence of the red alga *Thorea violacea* (Batrachospermales: Thoreales) in the Hudson River, New York State. Rhodora 97:328–338.

Pueschel, C. M., Saunders, G. W., West, J. A. 2000. Affinities of the freshwater red alga *Audouinella macrospora* (Florideophyceae, Rhodophyta) and related forms based on SSU rRNA gene sequence analysis and pit plug ultrastructure. Journal of Phycology 36:433–439.

Raven, J. A. 1993. The roles of the *Chantransia* phase of *Lemanea* (Lemaneaceae, Batrachospermales, Rhodophyta) and of the "mushroom" phase of *Himanthalia* (Himanthaceae, Fucales, Phaeophyta). Botanical Journal of Scotland 46:477–485.

Raven, J. A., Johnston, A. M., Newman, J. R., Scrimgeour, C. M. 1994. Inorganic carbon acquisition by aquatic photolithotrophs of the Dighty Burn, U.K.: Uses and natural limitations of natural abundance measurements of carbon isotopes. New Phytologist 127:271–286.

Rintoul, T. C., Sheath, R. G., Vis, M. L. 1999. Systematics and biogeography of the Compsopogonales with emphasis on the freshwater families in North America. Phycologia 38:517–527.

Rosemond, A. D. 1993. Interactions among irradiance, nutrients, and herbivores constrain a stream algal community. Oecologia 94:585–594.

Schumacher, G. J., Whitford, L. A. 1965. Respiration and ^{32}P uptake in various species of freshwater algae as affected by a current. Journal of Phycology 1:78–80.

Scott, J. 1983. Mitosis in the freshwater red alga *Batrachospermum ectocarpum*. Protoplasma 118:56–70.

Seckbach, J. 1991. Systematic problems with *Cyanidium caldarium* and *Galdaria sulphuraria* and their implications for molecular biology studies. Journal of Phycology 27:794–796.

Seto, R. 1977. On the vegetative propagation of a freshwater red alga, *Hildenbrandia rivularis* (Liebm.) J. Ag. Bulletin of the Japanese Society of Phycology 25:129–136 [in Japanese].

Sheath, R. G. 1984. The biology of freshwater red algae. Progress in Phycological Research 3:89–157.

Sheath, R. G. 1987. Invasions into the Laurentian Great Lakes. Archiv für Hydrobiologie (Supplement) 25:165–187.

Sheath, R. G., Cole, K. M. 1984. Systematics of *Bangia* (Rhodophyta) in North America. I. Biogeographic trends in morphology. Phycologia 23:383–396.

Sheath, R. G., Cole, K. M. 1990. Differential Alcian Blue staining in freshwater Rhodophyta. British Phycological Journal 25:281–285.

Sheath, R. G., Cole, K. M. 1992. Biogeography of stream macroalgae in North America. Journal of Phycology 28:448–460.

Sheath, R. G., Hambrook, J. A. 1988. Mechanical adaptations to flow in freshwater red algae. Journal of Phycology 24:107–111.

Sheath, R. G., Hambrook, J. A. 1990. Freshwater ecology, in: Cole, K. M., Sheath, R. G., Eds., Biology of the red algae. Cambridge Univ. Press, Cambridge, UK, pp. 423–453.

Sheath, R. G., Hymes, B. J. 1980. A preliminary investigation of the freshwater red algae in streams of southern Ontario, Canada. Canadian Journal of Botany 58:1295–1318.

Sheath, R. G., Morison, M. O. 1982. Epiphytes on *Cladophora glomerata* in the Great Lakes and St. Lawrence Seaway with particular reference to the red alga *Chroodactylon ramosum* (=*Asterocytis smargdina*). Journal of Phycology 18:385–391.

Sheath, R. G., Müller, K. M. 1997. Ultrastructure of carpogonia and carpogonial branches of *Batrachospermum helminthosum* and *Batrachospermum involutum* (Batrachospermales, Rhodophyta. Phycological Research 45:177–181.

Sheath, R. G., Vis, M. L. 1995. Distribution and systematics of *Batrachospermum* (Batrachospermales, Rhodophyta) in North America. 7. Section *Hybrida*. Phycologia 35:431–438.

Sheath, R. G., Hellebust, J. A., Sawa, T. 1979. Effects of low light and darkness on structural transformations in plastids of the Rhodophyta. Phycologia 18:1–12.

Sheath, R. G., Burkholder, J. M., Hambrook, J. A., Hogeland, A. M., Hoy, E., Kane, M. E., Morison, M. O., Steinman, A. D., Van Alstyne, K. L. 1986. Characteristics of softwater streams in Rhode Island. III. Distribution of macrophytic vegetation in a small drainage basin. Hydrobiologia 140:183–191.

Sheath, R. G., Hambrook, J. A., Nerone, C. A. 1988. The benthic macro-algae of Georgian Bay, the North Channel and their drainage basin. Hydrobiologia 163:141–148.

Sheath, R. G., Vis, M. L., Cole, K. M. 1992. Distribution and systematics of *Batrachospermum* (Batrachospermales, Rhodophyta) in North America. 1. Section *Contorta*. Journal of Phycology 28:237–246.

Sheath, R. G., Kaczmarczyk, D., Cole, K. M. 1993a. Distribution and systematics of freshwater *Hildenbrandia* (Rhodophyta, Hildenbrandiales) in North America. European Journal of Phycology 28:115–121.

Sheath, R. G., Vis, M. L., Cole, K. M. 1993b. Distribution and systematics of the freshwater red algal family Thoreales in North America. European Journal of Phycology 28:231–241.

Sheath, R. G., Vis, M. L., Cole, K. M. 1993c. Distribution and systematics of freshwater Ceramiales (Rhodophyta) in North America. Journal of Phycology 29:108–117.

Sheath, R. G., Vis, M. L., Cole, K. M. 1993d. Distribution and systematics of *Batrachospermum* (Batrachospermales, Rhodophyta) in North America. 3. Section *Setacea*. Journal of Phycology 29:719–725.

Sheath, R. G., Vis, M. L., Cole, K. M. 1994a. Distribution and systematics of *Batrachospermum* (Batrachospermales, Rhodophyta) in North America. 4. Section *Virescentia*. Journal of Phycology 30:108–117.

Sheath, R. G., Vis, M. L., Cole, K. M. 1994b. Distribution and systematics of *Batrachospermum* (Batrachospermales, Rhodophyta) in North America. 5. Section *Aristata*. Phycologia 33:404–414.

Sheath, R. G., Vis, M. L., Cole, K. M. 1994c. Distribution and systematics of *Batrachospermum* (Batrachospermales, Rhodophyta) in North America. 6. Section *Turfosa*. Journal of Phycology 30:872–884.

Sheath, R. G., Whittick, A., Cole, K. M. 1994d. *Rhododraparnaldia oregonica*, a new freshwater red algal genus and species intermediate between the Acrochaetiales and the Batrachospermales. Phycologia 33:1–7.

Sheath, R. G., Müller, K. M., Larson, D. J., Cole, K. M. 1995. Incorporation of freshwater Rhodophyta into the cases of caddisflies (Trichoptera) from North America. Journal of Phycology 31:889–896.

Sheath, R. G., Müller, K. M., Colbo, M. H., Cole, K. M. 1996a. Incorporation of freshwater Rhodophyta into the cases of chironomid larvae (Chironomidae, Diptera) from North America. Journal of Phycology 32:949–952.

Sheath, R. G., Müller, K. M., Vis, M. L., Entwisle, T. J. 1996b. A re-examination of the morphology, ultrastructure and classification of genera in the Lemaneaceae (Batrachospermales, Rhodophyta). Phycological Research 44:233–246.

Sheath, R. G., Müller, K. M., Sherwood, A. R. 2000. A proposal for a new red algal order, the Thoreales. Journal of Phycology (Supplement) 36:62.

Sherwood, A. R., Sheath, R. G. 1999a. Seasonality of macroalgae and epilithic diatoms in spring-fed streams in Texas, USA. Hydrobiologia 390:73–82.

Sherwood, A. R., Sheath, R. G. 1999b. Biogeography and systematics of *Hildenbrandia* (Rhodophyta, Hildenbrandiales) in North America: Inferences from morphometrics, *rbc*L and 18S rRNA gene sequence analyses. European Journal of Phycology 34:523–532.

Sherwood, A. R., Sheath, R. G. 2000a. Biogeography and systematics of *Hildenbrandia* (Rhodophyta, Hildenbrandiales) in Europe: Inferences from morpometrics and *rbc*L and 18S rRNA gene sequence analyses. European Journal of Phycology 35:143–152.

Sherwood, A. R., Sheath, R. G. 2000b. Microscopic analysis and seasonality of gemma production in the freshwater red alga *Hildenbrandia angolensis*. Phycological Research 48:241–249.

Skuja, H. 1938. Comments on fresh-water Rhodophyceae. Botanical Review 4:665–676.

Smith, D. W., Brock, T. D. 1973. The water relations of the alga *Cyanidium caldarium* in soil. Journal of General Microbiology 79:219–231.

Starmach, K. 1969. Growth of thalli and reproduction of the red alga *Hildenbrandia rivularis* (Liebm.) J. Ag. Acta Societatis Botanicorum Poloniae 38:523–533.

Steinman, A. D., McIntire, C. D. 1986. Effects of current velocity and light energy on the structure of periphyton assemblages in laboratory streams. Journal of Phycology 22:352–361.

Tanaka, J., Kamiya, M. 1993. Reproductive structure of *Caloglossa ogasawaerensis* Okamura (Ceramiales, Rhodophyceae) in nature and culture. Japanese Journal of Phycology 41:113–121.

Thirb, H. H., Benson-Evans, K. 1982. The effect of different current velocities on the red alga *Lemanea* in a laboratory stream. Archiv für Hydrobiologie 96:65–72.

Vis, M. L., Entwisle, T. J. 2000. Insights into the phylogeny of the Batrachospermales (Rhodophyta) from *rbc*L sequence data of Australian taxa. Journal of Phycology 36:1175–1182.

Vis, M. L., Sheath, R. G. 1992. Systematics of the freshwater red algal family Lemaneaceae in North America. Phycologia 31:164–179.

Vis, M. L., Sheath, R. G. 1993. Distribution and systematics of *Chroodactylon* and *Kyliniella* (Porphyridiales, Rhodophyta) from North American streams. Japanese Journal of Phycology 41:237–241.

Vis, M. L., Sheath, R. G. 1996. Distribution and systematics *Batrachospermum* (Batrachospermales, Rhodophyta) in North America. 9. Section *Batrachospermum*: Description of five new species. Phycologia 35:124–134.

Vis, M. L., Sheath, R. G. 1997. Biogeography of *Batrachospermum gelatinosum* (Batrachospermales, Rhodophyta) in North America based on molecular and morphological data. Journal of Phycology 33:520–526.

Vis, M. L., Sheath, R. G. 1998. A molecular and morphological investigation of the relationship between *Batrachospermum spermatoinvolucrum* and *B. gelatinosum* (Batrachospermales, Rhodophyta). European Journal of Phycology 33:231–239.

Vis, M. L., Sheath, R. G. 1999. A molecular investigation of the systematic relationship among *Sirodotia* species (Batrachospermales, Rhodophyta) in North America. Phycologia 38:261–266.

Vis, M. L., Carlson, T. A., Sheath, R. G. 1991. Phenology of *Lemanea fucina* (Rhodophyta) in a Rhode Island river USA. Hydrobiologia 222:141–146.

Vis, M. L., Sheath, R. G., Cole, K. M. 1992. Systematics of the freshwater red algal family Compsopogonaceae in North America. Phycologia 31:564–575.

Vis, M. L., Sheath, R. G., Cole, K. M. 1996a. Distribution and systematics of *Batrachospermum* (Batrachospermales, Rhodophyta) in North America. 8a. Section *Batrachospermum*: *Batrachospermum gelatinosum*. European Journal of Phycology 31:31–40.

Vis, M. L., Sheath, R. G., Cole, K. M. 1996b. Distribution and systematics of *Batrachospermum* (Batrachospermales, Rhodophyta) in North America. 8b. Section *Batrachospermum*: Previously described species excluding *Batrachospermum gelatinosum*. European Journal of Phycology 31:189–199.

Vis, M. L., Saunders, G. W., Sheath, R. G., Dunse, K., Entwisle, T. J. 1998. Phylogeny of the Batrachospermales (Rhodophyta) inferred from *rbc*L and 18 ribosomal DNA gene sequences. Journal of Phycology 34:341–350.

Vogel, S. 1984. Drag and flexibility in sessile organisms. American Zoologist 24:37–44.

Whitford, L. A. 1960. The current effect and growth of fresh-water algae. Transactions of the American Microscopical Society 74:302–309.

Whitton, B. A. 1975. Algae, *in*: Whitton, B. A., Ed., River ecology. Univ. of California Press, Berkeley, pp. 81–105.

Woelkerling, W. J. 1990. An introduction, *in*: Cole, K. M., Sheath, R. G., Eds., Biology of the red algae. Cambridge Univ. Press, Cambridge, UK, pp. 1–6.

Wujek, D. E., Timpano, P. 1986. *Rufusia* (Porphyridiales, Phragmonemataceae), a new red alga from sloth hair. Brenesia 25/26:163–168.

Yoshida, T. 1959. Life-cycle of a species of *Batrachospermum* found in northern Kyushu, Japan. Japanese Journal of Botany 17:29–42.

Youngs, H. L., Gretz, M. R., West, J. A., Sommerfeld, M. R. 1998. The cell wall chemistry of *Bangia atropurpurea* (Bangiales, Rhodophyta) and *Bostrychia moritziana* (Ceramiales, Rhodophyta) from marine and freshwater environments. Phycological Research 46:63–73.

Yung, Y.-K., Stokes, P., Gorham, E. 1986. Algae of selected continental and maritime bogs in North America. Canadian Journal of Botany 70:1154–1156.

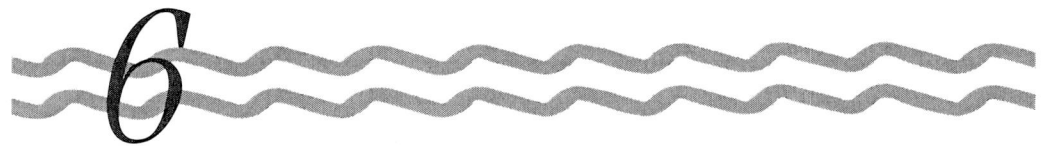

FLAGELLATED GREEN ALGAE

Hisayoshi Nozaki

Department of Biological Sciences
Graduate School of Science
University of Tokyo
Hongo, Bunkyo-ku, Tokyo 113-0033
Japan

I. Introduction
II. Diversity and Morphology
III. Ecology and Distribution
IV. Collection and Preparation for Identification
V. Key and Descriptions of Genera
    A. Key to Families of the Freshwater Green Flagellates
    B. Polyblepharidaceae
    C. Haematococcaceae
    D. Chlamydiomonadaceae
    E. Phacotaceae
    F. Volvocaceae
    G. Goniaceae
    H. Tetrabaenaceae
    I. Spondylomoraceae
VI. Guide to Literature for Species Identification
Literature Cited

I. INTRODUCTION

Green algae are organisms which are characterized by having chlorophylls *a* and *b* as the major photosynthetic pigments, starch located within the chloroplast as the major storage product and flagella of the whiplash (smooth) type (e.g., Bold and Wynne, 1985). They can be also distinguished from other eukaryotic algae in having two chloroplast membranes and stellate structure in the flagellar transition region (e.g., van den Hoek *et al.*, 1995). Members of the flagellate forms of the green algae were traditionally assigned to the order Volvocales or Chlamydomonadales of the Chlorophyceae (e.g., Smith, 1950). However, recent ultrastructural studies and molecular phylogenetic analyses of the green plants (green algae and land plants) suggest that some of the flagellated green algae should be classified in the Prasinophyceae or Pedinophyceae (see Norris, 1980; Moestrup, 1991; Friedl, 1997). Therefore, members of the flagellated green algae described in this chapter belong to the Chlorophyceae or Prasinophyceae, based on the current concepts of taxonomy/phylogeny of green algae (e.g., van den Hoek *et al.*, 1995; Friedl, 1997).

Although the flagellated chlorophycean algae can be divided into unicellular and colonial types of organization (Chlamydomonadales vs. Volvocales; see Mattox and Stewart, 1984), recent molecular phylogenetic analyses indicate that the colonial organisms represent polyphyletic status within the Volvocales (Buchheim *et al.*, 1994). *Volvox* is the organism which was first observed with a light microscope by van Leeuwenhoek (1700) within the Volvocales, and the generic name *Volvox* was established by Linnaeus (1758).

II. DIVERSITY AND MORPHOLOGY

About 100 genera with more than 1000 species are included within the freshwater green flagellates (Ettl, 1983). Among these taxa, 44 genera are known from North America. *Chlamydomonas* is the most diverse genus, with more than 400 described species worldwide (Ettl, 1983) and more than 20 species in North America.

The Volvocales usually have cell walls, loricae, or gelatinous matrices. The main component of the cell walls is glycoprotein, rather than cellulose (Harris, 1989). Ultrastructure of the cell walls reveals their median, tripartite structure. Such a structure is fundamentally recognized in the gelatinous matrices of the green flagellates, such as the colonial alga *Pandorina* and *Volvox* (Fulton, 1978; Kirk *et al.*, 1986). The vegetative cells have two or sometimes four or eight equal flagella of the whiplash type at the anterior end, a single nucleus in the center, and two contractile vacuoles generally located near the base of the flagella. The chloroplasts of most species are single and cup-shaped or they may appear H-shaped, asteroid, lamellate, or be divided into discoidal units (Iyengar and Desikachary, 1981; Ettl, 1983). The chloroplast envelope consists of two layers without a chloroplast endoplasmic reticulum. The pyrenoid and stigma (eyespot) are generally contained within the chloroplast. The major photosynthetic pigments include chlorophylls *a* and *b*. However, colorless, heterotrophic genera also exist. The major storage product of photosynthetic species is starch (α-1, 4 and α-1, 6 polymer of glucose). which is localized within the chloroplast and stains positively with iodine.

Asexual reproduction is accomplished by zoospore formation within the parental cell wall in the unicellular forms. In the colonial forms, all cells in a colony or some specialized (large) reproductive cells divide twice, three times or more (n) simultaneously to form four, eight or more (2^n) daughter cells which soon become a daughter colony with the cell number and shape both characteristic for a given species (autocolony formation). After the daughter colonies are released from the parental colony, the new colonies increase their size with their cell number unchanged. Naked unicellular species usually undergo bipartition or cell division during swimming. Sexual reproduction is either isogamous, anisogamous, or oogamous. In isogamy, two conjugating gametes are essentially identical in size, whereas in anisogamy the cell size of male gametes is smaller than that of female gametes. When the female gametes lack flagella, such cells and the male gametes are called sperm and eggs, respectively, representing oogamy. Although evolution from isogamy to oogamy through anisogamy can be recognized in relation to increase in colonial complexity within the colonial Volvocales (Nozaki and Ito, 1994; Kirk, 1998; Nozaki *et al.*, 2002), unicellular genera such as *Chlamydomonas* or *Chlorogonium* exhibit these three types of sexual reproduction (Ettl, 1983). After the conjugation, the zygote secretes a heavy cell wall to become a hypnozygote (thick-walled dormant zygote). After a period of dormancy, the hypnozygote undergoes meiosis and germinates to give rise to four haploid nuclei (Stein, 1958b; Coleman, 1959). In most members of the green flagellates, therefore, four gone cells (reproductive cells from germinating zygote) of identical size are formed on zygote germination. However, only one of the four haploid nuclei survives to form a single gone cell in the colonial Volvocales (Volvocaceae) (Coleman, 1959; Goldstein, 1964; Nozaki *et al.*, 1989).

Buchheim *et al.* (1996) and Nakayama *et al.* (1996b), on the basis of ribosomal (r) RNA gene sequence data, resolved that the volvocalean algae constitute an ancestral and nonmonophyletic (paraphyletic) group with closely related derived groups including the chlorococcalean and tetrasporalean species. In addition, other DNA sequence data, such as the *rbc*L gene (large subunit of ribulose-1, 5-bisphosphate carboxylase/oxygenase) gene suggest that several genera (including *Volvox*) of the colonial forms are nonmonophyletic (Nozaki *et al.*, 1995a, 1997a). Thus, further studies using both molecular and morphological data are needed to establish a natural taxonomic system for the flagellated green algae. Species of the volvocalean algae are generally delineated on their morphological differences. On the basis of intercrossings of heterothallic strains of the colonial species such as *Pandorina morum* (Coleman, 1959, 1977), multiple syngens (biological species) were demonstrated within the single morphological species. In addition, morphological species in the volvocalean algae may be nonmonophyletic or monophyletic depending upon the alga and the degree of morphological characterization of species (Nozaki *et al.*, 1997b, c).

III. ECOLOGY AND DISTRIBUTION

The green flagellates are found in a wide variety of freshwater habitats including lakes, ponds, rivers, rice paddies, and rainwater pools (e.g., Entwisle *et al.*, 1998) as well as on ice or snow (Hoham, 1980; Ling, 1996). The majority are free living, but some strains are endosymbiotic (Lembi, 1980). Water bodies with elevated levels of nutrients are often especially rich in Volvocales (Lembi, 1980; Kirk, 1998), although *Chlamydomonas* has been reported in the nutrient-poor lakes (Lembi, 1980). Heterotrophic and photoheterotrophic algae (e.g., *Polytoma*, *Astrephomene*, and *Pyrobotrys*) grow in highly eutrophic waters, such as sewage oxidation ponds (Silva and Papenfuss, 1953), cow pasture ponds (Stein, 1958a), and rice paddies (Nozaki, 1983). Such photoheterotrophic algae generally require acetate for their growth (Pringsheim and Weissner, 1960; Brooks, 1972) and often exhibit plasticity of pyrenoids depending upon growth conditions (e.g., Nozaki et al., 1994b, 1995, 1998b). Three species of *Pyrobotrys* require anaerobic conditions for their growth (Nozaki, 1986). Some species, such as *Tetrabaena socialis* (= *Gonium sociale*), are distributed from the temperate zone to Antarctica and one Antarctic isolate does not grow at 25°C (Nozaki and Ohtani, 1992). Snow or ice algae are known in the unicellular genera *Chloromonas*, *Chlamydomonas*, *Carteria*, *Chlainomonas*, and *Smithsonimonas* (Hoham, 1980; Ling, 1996). Some species of *Chlamydomonas* grow in the extremely acidic water (Ettl, 1983). Although the volvocalean algae do not usually exhibit predominant species, *Volvox* and *Haematococcus* sometimes form water blooms in lakes or reservoirs.

Many species are essentially worldwide in distribution (Iyengar and Desikachary, 1981; Ettl, 1983), although records from the tropics are sparse (Coleman, 1996). Extensive genetic and/or molecular analyses have been carried out on the colonial volvocalean species with worldwide distribution. *Pandorina morum* contains more than 20 sexually isolated groups (syngens) distributed throughout the world (Coleman, 1959, 1977) and their genetic differences are large (Coleman et al., 1994). Molecular phylogenetic analyses resolved that a large number of *Gonium pectorale* isolates from five continents are divided into six subclades, which are consistent with the interfertility between the isolates (Fabry et al., 1999). Although almost all of the well-studied genera of green flagellates are global in their distribution (Ettl, 1983), endemism can be considered in the colonial genus *Platydorina*, which has been reported only in the United States and Mexico (Kofoid, 1899; Harris and Starr, 1969; Ortega, 1984). Endemic distribution of species or infraspecific level is often recognized, particularly in *Volvox* (Smith, 1944; Nozaki, 1988).

IV. COLLECTION AND PREPARATION FOR IDENTIFICATION

General methods for collection and preservation of freshwater algae have been described by Smith (1950) and Entwisle et al. (1998). Collection of the vegetative cells or colonies of the green flagellates can be carried out using plankton nets. These cells should soon be identified while they are alive, since some morphological features diagnostic for genus or species level (e.g., contractile vacuoles, flagella) are unstable after collection. When fixation is needed, the samples should be fixed with 1–2% glutaradehyde (rather than formaldehyde) soon after collection, and then stored at 4°C before identification. For snow algae, such as *Chloromonas*, cells fixed with 1–2% OsO_4 will result in better cell morphology than those fixed with glutaradehyde.

V. KEY AND DESCRIPTIONS OF GENERA

A. Key to Families of the Freshwater Green Flagellates

1a.	Cell with a depression at the flagellar base and/or lacking cell walls, gelatinous matrix, or lorica	Polyblepharidaceae
1b.	Cell without a depression at the flagellar base and enclosed by cell walls, gelatinous matrix, or lorica	2
2a.	Cell with numerous protoplasmic strands radiating within the wall or gelatinous matrix	Haematococcaceae
2b.	Cell without numerous protoplasmic strands radiating within the wall or gelatinous matrix	3
3a.	Unicellular	4
3b.	Colonial	5
4a.	Cell with a wall	Chlamydomonadaceae
4b.	Cell with a lorica	Phacotaceae
5a.	Colony surrounded by a gelatinous matrix	6
5b.	Colony without a gelatinous matrix	Spondylomoraceae
6a.	Colony generally four-celled	Tetrabaenaceae
6b.	Colony cell number eight or more	7
7a.	Tripartite boundary of the gelatinous matrix surrounding each cell of the colony (cellular boundary); inversion absent during colony formation	Goniaceae
7b.	Tripartite boundary of the gelatinous matrix surrounding the whole colony (colonial boundary); inversion occurs during colony formation	Volvocaceae

B. Polyblepharidaceae

The flagellates of this family are unicellular and different from the typical green flagellates (e.g., *Chlamydomonas*) in that the cells lack cell walls (or lorica) and/or the protoplast exhibits depression at the flagellar base (Ettl, 1983). Asexual reproduction is generally affected by the bipartition of cells. Sexual reproduction is unknown in culture except for *Nephroselmis* (Suda et al., 1989). Based on the ultrastructure of cells, some genera traditionally assigned to the Polyblepharidaceae have been classified in the separate class of green algae, namely, the Prasinophyceae (Micromonadophyceae) or Pedinophyceae (Norris, 1980; Moestrup, 1991). Recent molecular phylogenetic analyses essentially support this separation (e.g., Friedl, 1997). However, this chapter follows the traditional taxonomic style here because of the easy identification/recognition of the green flagellates using the light microscope. Eight genera are known:

1a.	Cell with a single flagellum (Fig. 1)	*Pedinomonas*
1a.	Cell with two or more flagella	2
2a.	Cell with eight flagella (Fig. 2)	*Polyblepharides*
2b.	Cell two or four flagella	3
3a.	Cell with two flagella	4
3b.	Cell with four flagella	6
4a.	Flagella equal, inserted in the dorsal side of the platelike cell (Fig. 3)	*Mesostigma*
4b.	Flagella unequal, inserted in the edge of the platelike cell	5
5a.	Chloroplast lacking pyrenoid and stigma (Fig. 4)	*Scourfieldia*
5b.	Chloroplast containing pyrenoid and stigma (Fig. 5)	*Nephroselmis*
6a.	Cell spindle-shaped and curved (Fig. 6)	*Spermatozopsis*
6b.	Cell ovoid or ellipsoidal	7
7a.	Cells containing photosynthetic pigments; cell wall or theca present	8
7b.	Cells lacking photosynthetic pigments; cell wall or theca absent (Fig. 7)	*Polytomella*
8a.	Lateral margins of cell not flattened or winged (Fig. 8)	*Tetraselmis*
8b.	Lateral margins of cell flattened and winged (Fig. 9)	*Scherffelia*

Mesostigma Lauterborn (Fig. 3)

Disk-shaped cells are markedly compressed in the transapical axis, with two equal flagella inserted in the dorsal side of the cell. Cells without a wall are surrounded by gelatinous envelopes or numerous scales (Iyengar and Desikachary, 1981; Ettl, 1983). Two or more contractile vacuoles are present near the base of the flagella. Chloroplast is single and cup-shaped, with or without a pyrenoid and stigma. Asexual reproduction is by bipartition. Sexual reproduction has not been observed with certainty. Two species have been described. This genus is assigned to the Prasinophyceae (Norris, 1980), although recent phylogenetic analyses of actin genes show that *Mesostigma* is the earliest divergence within the Streptophyta (Charophyceae and land plants) (Bhattacharya et al., 1998; Karol et al., 2001). Based on the multiple chloroplast gene sequences, *Mesostigma* was resolved as the most basal organism within the green plants (Streptophyta plus Chlorophyta) (Lemieux et al., 2000), but this phylogenetic position is questioned (Karol et al., 2001).

M. viride and *M. grande* (Fig. 3) have been observed from the United States (e.g., FL, GA, IN, KY, OH) (Smith, 1950; Dillard, 1989).

Nephroselmis Stein (Fig. 5)

Cells are equatorially compressed, lacking a cell wall, with two unequal flagella inserted in the edge of the plate-like cell. A single contractile vacuole is located near the base of the flagella (Iyengar and Desikachary, 1981; Ettl, 1983). The chloroplast is single and cup-shaped, with a single basal pyrenoid and a stigma. The flagella and cell of this genus have a scaly covering (Suda et al., 1989). Asexual reproduction takes place by bipartition. Isogamous sexual reproduction has been confirmed in culture

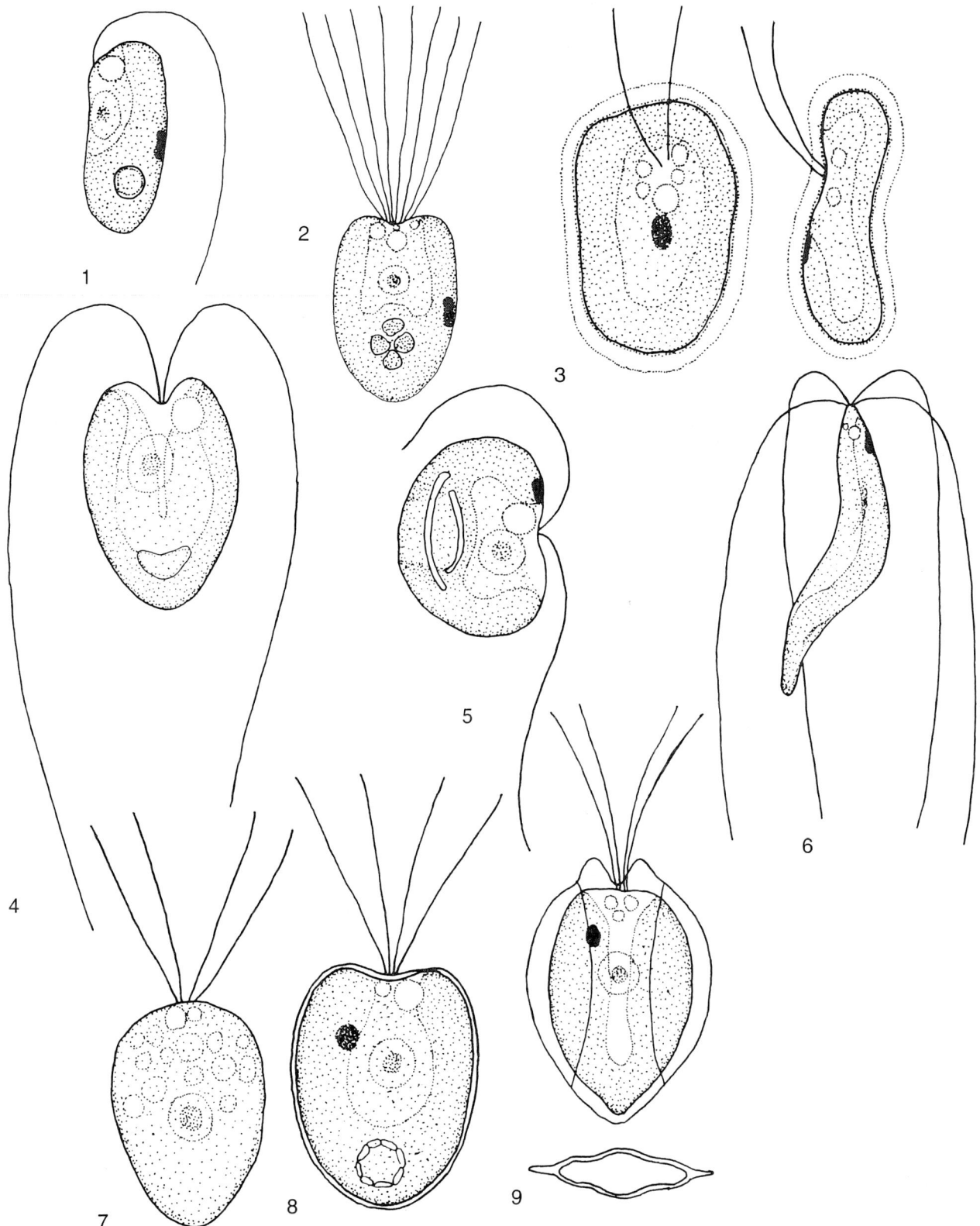

FIGURE 1 Uniflagellate vegetative cell of *Pedinomonas minor*. (× 5500.) FIGURE 2 *Polyblepharides singularis*, showing anterior depression and eight flagella. (× 1400.) FIGURE 3 Two views of vegetative cell of *Mesostigma grande*. (× 1600.) FIGURE 4 Vegetative cell of *Scourfieldia cordiformis*, showing unequal flagella. (× 4500.) FIGURE 5 Vegetative cell of *Nephroselmis olivacea*, showing anterior depression. (× 3000.) FIGURE 6 Quadriflagellate vegetative cell of *Spermatozopsis exsultans*, showing twisted cell body. (× 7300.) FIGURE 7 Naked vegetative cell of *Polytomella citrii* showing four flagella. (× 2400.) FIGURE 8 Vegetative walled cell of *Tetraselmis cordiformis*, showing anterior depression. (× 2000.) FIGURE 9 Two views of vegetative cell of *Scherffelia phacus*. (× 2500.)

(Suda et al., 1989). Two freshwater species have been described. *Nephroselmis* was recently assigned to the Prasinophyceae (Norris, 1980).

This genus was designated as *Heteromastix* Korshikov by Smith (1950), who recorded North American *H. angulata*, a taxonomic synonym of *Nephroselmis olivacea* (Fig. 5).

Pedinomonas Korshikov (Fig. 1)

Cells are compressed, ellipsoidal to subcircular in front view, without a cell wall. A single flagellum is posteriorly directed and inserted at the anterior pole of the cell (Iyengar and Desikachary, 1981; Ettl, 1983). A single contractile vacuole is present near the base of the flagellum. Chloroplast is single and parietal, with a single pyrenoid and a stigma. Asexual reproduction is by bipartition (Ettl, 1983). Sexual reproduction has not been observed with certainty. Approximately ten species are recognized in freshwater habitats (Ettl, 1983). This genus was recently assigned to the Pedinophyceae on the basis of ultrastructure of the flagellar apparatus (Moestrup, 1991).

P. maior, *P. rotunda*, and *P. minor* (Fig. 1) have been found in the United States (e.g., AL, FL) (Ettl, 1983; Dillard, 1989). This genus has also been collected in the Laurentian Great Lakes (Munawar and Munawar, 1981).

Polyblepharides Dangeard (Fig. 2)

Naked cells are ovoid to ellipsoidal, with eight equal flagella and two or four contractile vacuoles at the anterior pole. Chloroplast is single and massive, with a single pyrenoid surrounded by starch granules (Smith, 1950; Ettl, 1983). Asexual reproduction is by bipartition (Smith, 1950; Iyengar and Desikachary, 1981; Ettl, 1983). Although this genus is reported from various localities of the world, the existence of *Polyblepharides* has not been confirmed in culture (Bold and Wynne, 1985).

P. singularis (Fig. 2) and *P. fragariiformis* are known from the United States (e.g., FL, NC) (Smith, 1950; Ettl, 1983).

Polytomella Aragao (Fig. 7)

Naked cells are ovoid, ellipsoidal, or spherical, with four equal flagella inserted in the anterior papilla of the cell (Iyengar and Desikachary, 1981; Ettl, 1983). Two or four contractile vacuoles are present near the base of the flagella. Chloroplasts degenerate to become colorless leucoplasts. A single stigma may be present depending upon the species (Ettl, 1983). Pyrenoids are lacking. Asexual reproduction is by bipartition. Isogamous sexual reproduction is reported. More than ten species have been described (Ettl, 1983). On the basis of ultrastructure of flagellar apparatus and molecular phylogeny of 18S rRNA gene sequences, *Polytomella* was assigned to the Chlorophyceae (Nakayama et al., 1996a).

P. citrii (Fig. 7) and *P. agilis* have been reported from the United States (Smith, 1950).

Scherffelia Pascher (Fig. 9)

Walled cells are ovoid or ellipsoidal, compressed equatorially, with two large chloroplasts and four equal flagella inserted in the anterior depression of the cell. Lateral margins of the cell are flattened and winged (Ettl, 1983). The cell wall, or theca, is formed by the fusion of cell body scales characteristic of the Prasinophyceae. Two contractile vacuoles are present near the base of the flagella. The chloroplast lacks pyrenoids. Asexual reproduction is via formation of four zoospores within the theca. Sexual reproduction is unknown. Eight species were recognized (Ettl, 1983). Although Smith (1950) and Ettl (1983) classified this genus in the Chlamydomonadaceae, 18S rRNA gene sequence data suggest that *Scherffelia* is closely related to *Tetraselmis* (Steinkötter et al., 1994).

S. phacus Pascher (Fig. 9) is reported from the United States (e.g., OH, TN) (Smith, 1950).

Scourfieldia G. S. West (Fig. 4)

Naked cells are equatorially compressed, with two long unequal flagella inserted in the edge of the plate-like cell (Iyengar and Desikachary, 1981; Ettl, 1983). Two contractile vacuoles are present near the base of the flagella. Chloroplast is single and cup-shaped, lacking pyrenoid and stigma. Asexual reproduction is by bipartition. Sexual reproduction has not been observed with certainty. Three freshwater species have been described (Ettl, 1983). Although flagella and cell body of this genus lack scaly covering, *Scourfieldia* was recently assigned to the Prasinophyceae on the basis of ultrastructure of flagellar apparatus (see Moestrup, 1991).

S. cordiformis (Fig. 4) has been found in the United States (Dillard, 1989) and the Canadian province of Ontario (Duthie and Socha, 1976). The genus has also been reported from the arctic regions of Nunavut, Canada (Sheath and Steinman, 1982).

Spermatozopsis Korshikov (Fig. 6)

Naked cells are spindle-shaped and curved or twisted, with four equal flagella at the anterior end (Iyengar and Desikachary, 1981; Ettl, 1983). Two contractile vacuoles are located near the base of the flagella. A single chloroplast is linear, lying along the convex side of the cell, with a single stigma. Pyrenoid is lacking. Asexual reproduction is by bipartition. Sexual reproduction has not been observed in culture. Two species have been described (Ettl, 1983; Preisig and Melkonian, 1984) On the basis of ultrastructure of flagellar apparatus and molecular phylogeny of 18S rRNA gene sequences, *Spermatozopsis* was assigned to the Chlorophyceae (Friedl, 1997).

S. exsultans (Fig. 6) has been found in several location in of the United States (e.g., FL, KY, TN) (Smith, 1950; Dillard, 1989).

Tetraselmis Stein (Fig. 8)

Walled cells are ovoid or ellipsoidal, somewhat compressed equatorially, with four equal flagella inserted in the anterior depression of the cell. The cell wall or theca is formed by the fusion of cell body scales characteristic of the Prasinophyceae. Two contractile vacuoles are present near the base of the flagella in freshwater species (Ettl, 1983). The chloroplast is single and cup-shaped, with a single basal pyrenoid and a stigma. The pyrenoid may be absent depending upon the species (Ettl, 1983). Asexual reproduction is by bipartition within the theca. Sexual reproduction is unknown. Although *Tetraselmis* is generally found in marine water, seven freshwater species were recognized (Ettl, 1983).

T. cordiformis (Fig. 8) and *T. subcordiformis* is reported from the United States (e.g., FL, TN) (Smith, 1950; Ettl, 1983). Smith (1950) placed the latter species in the genus *Platymonas* (as *P. subcordiformis*) and he recorded North American *P. elliptica*, which is a taxonomic synonym of *T. cordiformis* (Ettl, 1983).

C. Haematococcaceae

Cells of members of this family have numerous protoplasmic strands radiating within the cell wall or gelatinous matrix (Ettl, 1983). The family may be unicellular (*Haematococcus*) or colonial (*Stephanosphaera*). In *Haematococcus*, asexual reproduction is accomplished by zoospore formation, whereas daughter colony formation take place in *Stephanosphaera*. Sexual reproduction is effected by spindle-shaped small isogametes in both genera (Pocock, 1960; Ettl, 1983).

1a. Unicellular (Fig. 10)..*Haematococcus*
1b. Colonial (Fig. 11)...*Stephanosphaera*

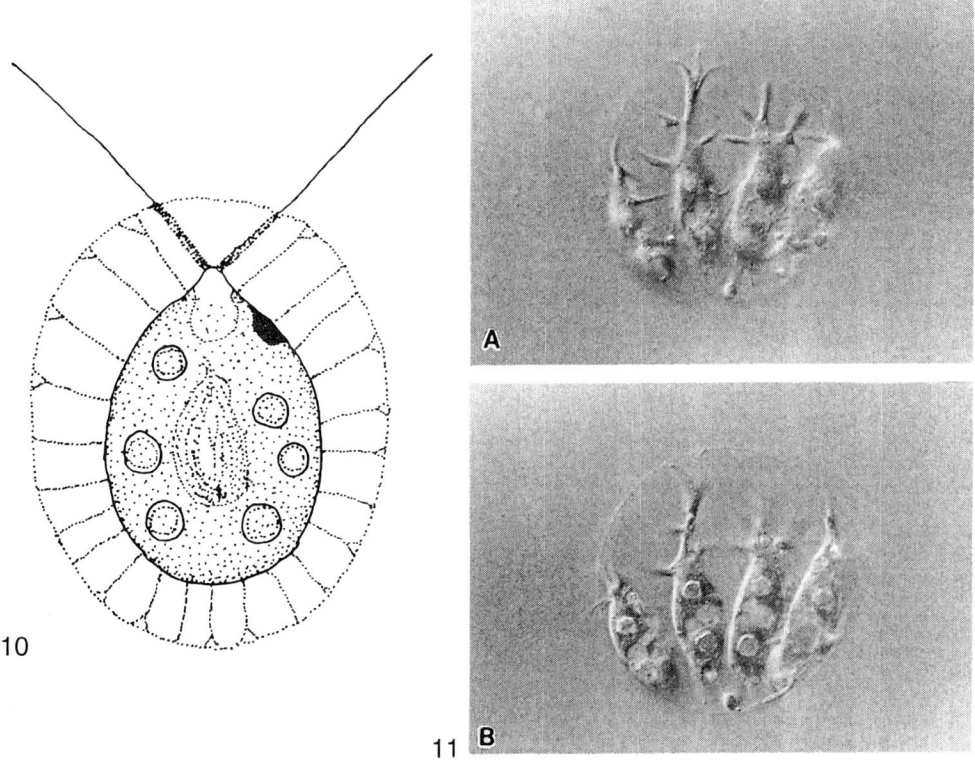

FIGURE 10 Vegetative cell of *Haematococcus pluvialis*. (× 900.) *FIGURE 11* Two views of vegetative colony of *Stephanosphaera pluvialis*. (× 590.)

Haematococcus C. A. Agardh (Fig. 10)

Cells are spherical, ellipsoidal, or pear-shaped, with two flagella and a cup-shaped chloroplast with a single or multiple pyrenoids (Pocock, 1960; Ettl, 1983; Thompson and Wujek, 1989). Protoplast is enclosed by a swollen, gelatinous cell wall and produces protoplasmic strands that extend through the wall. Accumulation of haematochrome often causes reddish color of the protoplasts. Sexual reproduction is by small, spindle-shaped isogametes. Six species are recognized (Ettl, 1983; Thompson and Wujek, 1989).

H. pluvialis (= *H. lacustris*) (Fig. 10) and *H. carocellus* are reported widely from the United States (Smith, 1950; Thompson and Wujek, 1989). The latter species was collected at least twice from Minnesota, but has not been studied in culture (Thompson and Wujek, 1989). The genus is also widespread in Canada, including British Columbia (Stein and Borden, 1979), Nunavut (Sheath and Steinman, 1982) and Ontario (Duthie and Socha, 1976). *H. pluvialis* is also reported from Mexico (Ortega, 1984).

Stephanosphaera Cohn (Fig. 11)

Colonies are spherical to ellipsoidal, containing eight elongate cells arranged in a ring within a gelatinous matrix. Cells have two equal flagella, a stigma, several contractile vacuoles, and a cup-shaped chloroplast (filling much of the cell) with several pyrenoids (Ettl, 1983). Protoplast exhibits numerous processes on the surface. Asexual reproduction occurs by autocolony formation by all the colonial cells. Sexual reproduction is isogamous (Ettl, 1983). Although this genus is sometimes classified in the family Volvocaceae (e.g., Bold and Wynne, 1985), a recent molecular phylogenetic study indicated that *Stephanosphaera* is closely related to *Haematococcus*, and separated from the volvocacean algae (Buchheim et al., 1994).

Stephanosphaera is monotypic, with *S. pluvialis* (Fig. 11), which has been observed in the United States (Smith, 1950).

D. Chlamydomonadaceae

The unicellular flagellates of this family have cell walls, chloroplasts, stigmata, and contractile vacuoles (Iyengar and Desikachary, 1981; Ettl, 1983). Asexual reproduction is accomplished by zoospore formation. Sexual reproduction is isogamous, anisogamous or oogamous (Ettl, 1983). Eight genera are known in North America (Figs. 12–19) and are distinguished by their flagellar number and cell form:

1a.	Cell with two flagella	2
1b.	Cell with four flagella	8
2a.	Cells lacking photosynthetic pigments (Fig. 15)	*Polytoma*
2b.	Cells containing photosynthetic pigments	3
3a.	Contractile vacuoles distributed throughout the cell surface (Fig. 16)	*Chlorogonium*
3b.	Contractile vacuoles positioned only near the base of the flagella	4
4a.	Two flagella remote from each other at the surface of the protoplast (Fig. 17)	*Gloeomonas*
4b.	Two flagella in close proximity to each other at the surface of the protoplast	5
5a.	Cell wall with protuberances or swollen	6
5b.	Cell wall without protuberances and not swollen	7
6a.	Cell wall with protuberances (Fig. 18)	*Lobomonas*
6b.	Cell wall without protuberances but swollen (Fig. 19)	*Vitreochlamys*
7a.	Chloroplasts containing pyrenoids (Fig. 12)	*Chlamydomonas*
7b.	Chloroplasts lacking pyrenoids (Fig. 13)	*Chloromonas*
8a.	Cell walls not swollen, tightly surrounding the protoplast (Fig. 14)	*Carteria*
8b.	Cell walls swollen, separated from the protoplast surface	*Chlainomonas*

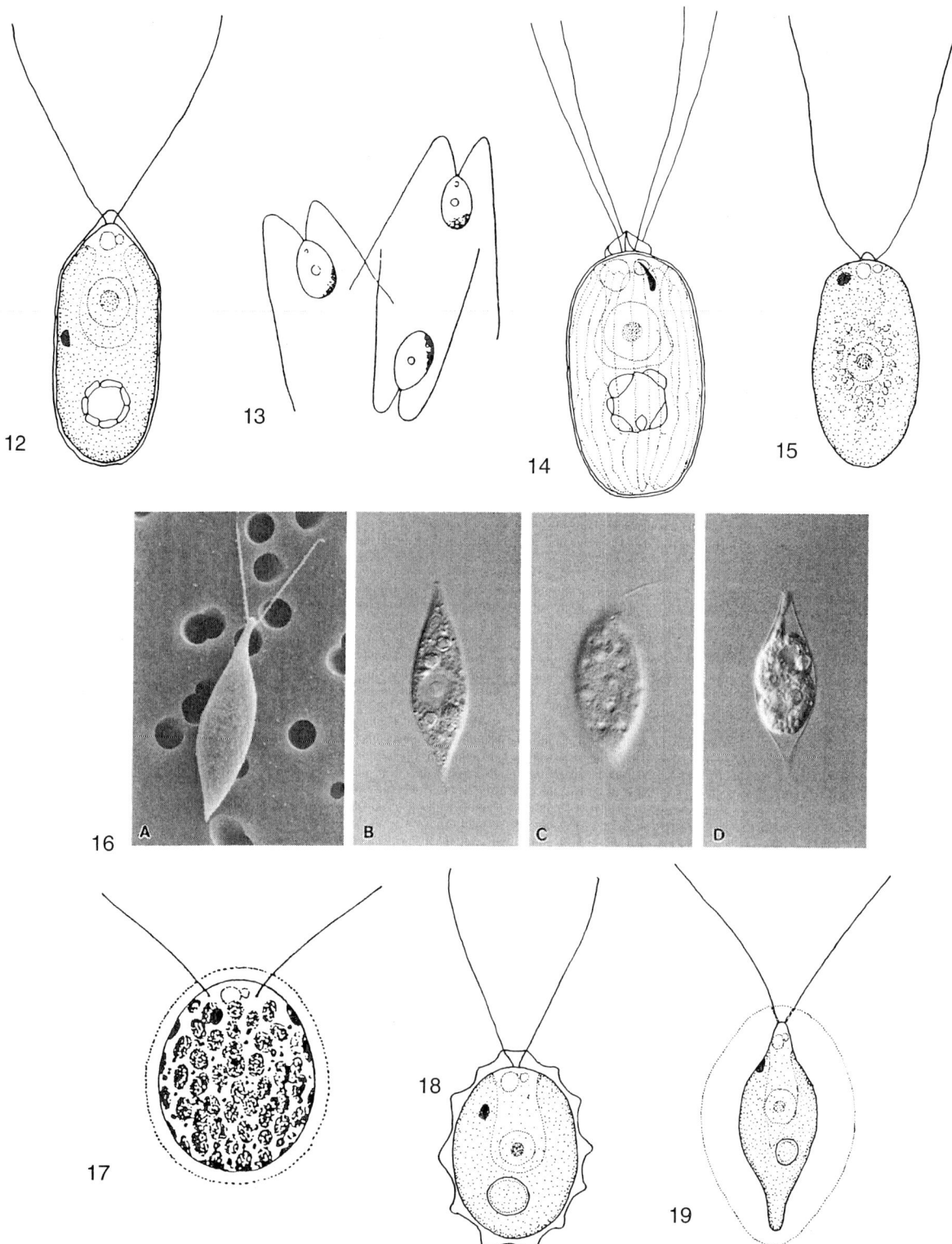

FIGURE 12 Vegetative cell of *Chlamydomonas sonowiae*, showing cell wall and two equal flagella. (× 2300.)
FIGURE 13 Vegetative cells of *Chloromonas minima*. (From Pascher, 1927.) (× 1900.) *FIGURE 14* Quadriflagellate vegetative cell of *Carteria eugametos*. (From Nozaki *et al.*, 1994a, reproduced by permission of International Phycological Society.) (× 1500.) *FIGURE 15* Vegetative cell of *Polytoma uvella*, lacking chloroplast. (× 1250.)
FIGURE 16 A–C: Vegetative cells of *Chlorogonium*. A: Scanning electron microscopy of *C. elongatum*. (× 1600.) B: Optical section of *C. euchlorum* cell. (× 690.) C: Surface view of *C. capillatum* cell, showing many contractile vacuoles. (× 690.) D: Two-celled stage of asexual reproduction in *C. euchlorum*. (From Nozaki *et al.*, 1998b, reproduced by permission of the *Journal of Phycology*.) (× 690.) *FIGURE 17* Vegetative cell of *Gloeomonas ovalis* (From Pascher, 1927). *FIGURE 18* Vegetative cell of *Lobomonas rostrata* (× 2400.)
FIGURE 19 Vegetative cell of *Vitreochlamys fluviatilis*, showing swollen cell wall (× 1150.)

Carteria Diesing (Fig. 14)

General features of the cells in this genus are essentially the same as in *Chlamydomonas* except for the typical four flagella (Ettl, 1983). Four contractile vacuoles and/or two stigmata are observed in some species. Asexual reproduction is by zoospore formation within the mother cell wall. Sexual reproduction is isogamous (Ettl, 1983). Aplanogamous sexual reproduction was recently reported in *C. eugametos* (Nozaki, 1994) (Fig. 14). More than 60 species have been described (Ettl, 1983). Lembi (1975), on the basis of the ultrastructure of flagellar apparatuses, recognized two distinct groups within the genus *Carteria*. Recent molecular phylogenetic analysis resolved such two clades of *Carteria* (Buchheim and Chapman, 1992; Buchheim *et al.*, 1996, 2002). Nozaki *et al.* (1994a), on the basis of SEM observations, characterized the swastika-shaped anterior papillae of the cell wall in one of these two groups of *Carteria*.

Twelve species of *Carteria* were observed in the United States (Smith, 1950; Dillard, 1989; Starr and Zeikus, 1993; Nozaki *et al.*, 1994a) as well as British Columbia, Canada (Stein and Borden, 1979), the Laurentian Great Lakes (Munawar and Munawar 1981), and Mexico (Ortega, 1984).

Chlainomonas Christen (not pictured)

Cells of this genus have a swollen or gelatinous cell wall separated from the protoplast surface but otherwise similar to those of *Carteria* cells. Three species have been described (Ettl, 1983). Asexual reproduction is accomplished by zoospore formation (Ettl, 1983).

Two species of this genus have been collected from snow in Canada and the United States; *C. rubra* in British Columbia, Canada, and Washington (Hoham, 1974a), and *C. kollii* in Oregon and Washington (Hoham, 1974b).

Chlamydomonas Ehrenberg (Fig. 12)

Cells are spherical, ovoid or ellipsoidal, with a cell wall, two equal flagella at the anterior pole and two contractile vacuoles at the base of the flagella (Ettl, 1983). The chloroplast is single and fills much of the cell, with a single or several pyrenoids. Stigma is generally single. Asexual reproduction is by zoospore formation within the mother cell wall. Sexual reproduction is isogamous, anisogamous, or oogamous. More than 400 species have been described (Ettl, 1983). It is very difficult to distinguish these morphological species, based on only field-collected materials. Analysis of rRNA gene sequences showed that *Chlamydomonas* was resolved as a basal, nonmonophyletic group within the Volvocales (Buchheim *et al.*, 1996).

Dillard (1989) listed 24 species of *Chlamydomonas* in the southeastern United States, such as *C. sonowiae* (Fig. 12) [excluding four *Chloromonas* species and including *Chlamydomonas tetragama* (as *Chlorogonium tetragamum*)]. The genus is also reported from across Canada (e.g., Duthie and Socha, 1976; Stein and Borden, 1979; Sheath and Steinman, 1982), as well as Mexico (Ortega, 1984), Guadelope (Bourrelly and Marguin, 1952), and Costa Rica (Haberyan *et al.*, 1995).

Chlorogonium Ehrenberg (Fig. 16)

Cells are spindle-shaped, elongate-ovoid, ovoid, or ellipsoidal, with a cell wall and two equal flagella at the anterior end (Nozaki *et al.*, 1998b). Contractile vacuoles are two or more, generally positioned in both of the anterior and posterior halves of the cell, but sometimes distributed in only the anterior portion of the cell. The chloroplast is parietal, lacking pyrenoids, or with one or more pyrenoids (Ettl, 1983). Asexual reproduction is by zoospore formation within the mother cell wall (Nozaki *et al.*, 1996a). During asexual reproduction, the first division is transverse without cross rotation of the parental protoplast. This type of initial cell division distinguishes *Chlorogonium* from *Chlamydomonas* (Ettl, 1980; Nozaki *et al.*, 1996a). Sexual reproduction may be isogamous, anisogamous, or oogamous. Recently, paedogamous sexual reproduction (conjugation of gametes within the gametangium) was observed in a culture of *C. capillatum* (Nozaki *et al.*, 1995). Ettl (1983) used the presence or absence of pyrenoids in vegetative cells for distinguishing species of *Chlorogonium* and he recognized 12 pyrenoid-lacking species and six pyrenoid-containing species. However, *Chlorogonium* strains completely lacking pyrenoids in the chloroplast of the vegetative cells have not previously been studied in culture (Nozaki *et al.*, 1998b). On the basis of ultrastructure and *rbc*L gene sequences, the genus *Chlorogonium* was resolved for at least two phylogenetically separated groups (Nozaki *et al.*, 1998b).

Five species assignable to *Chlorogonium* have been found in the United States (e.g., FL, LA, TN, VA, WV) (Smith, 1950; Dillard, 1989; Nozaki *et al.*, 1998b). This genus has been collected widely in Canada (Duthie and Socha, 1976; Stein and Borden, 1979; Sheath and Steinman, 1982) and Mexico (Ortega, 1984).

Chloromonas Gobi (Fig. 13)

Chloromonas is distinguished from *Chlamydomonas* only by the lack of pyrenoids in the chloroplasts. More than 100 species have been described (Ettl, 1983). Complete absence of pyrenoids in the chloroplasts was demonstrated by light and electron microscopy in four species of *Chloromonas* (Morita *et al.*, 1998). Buchheim *et al.* (1997), on the basis of 18S rRNA gene sequence data, showed that the genus

Chloromonas is nonmonophyletic within the Volvocales. Morita *et al.* (1999) and Nozaki *et al.* (2002a) resolved that *Chloromonas* and several species of *Chlamydomonas* constituted a closely related lineage. Pröschold *et al.* (2001) included several pyrenoid-containing species in the genus *Chloromonas*, mainly based on the 18S *r*RNA gene phylogeny.

Smith (1950) reported North American *Platychloris minima*, which should be assigned to *Chloromonas*, as *C. minima* (Fig. 13). Four other species assignable to this genus (*C. clathrata, C. platystigma, C. depauperata,* and *C. anglica*) were observed from the United States (e.g., GA, NC, SC, TN) (Dillard, 1989). The genus is also reported from Canada (e.g., Duthie and Socha, 1976; Stein and Borden, 1979; Sheath and Steinman, 1982) as well as Mexico (Ortega, 1984) (Haberyan *et al.*, 1995). In addition, *C. pinchiae, C. nivalis, C. brevispina, C. polypyera* and *C. granulosa* have been collected in snow in the United States and Canada (e.g., Hoham *et al.*, 1979, 1983).

Gloeomonas Klebs (Fig. 17)

Cells are spherical, ovoid, or ellipsoidal, with a thick cell wall or gelatinous envelope, two equal flagella at the anterior pole, and two contractile vacuoles near the base of the flagella (Iyengar and Desikachary, 1981; Ettl, 1983). The two flagella are remote from each other at the surface of the protoplast. Chloroplast is single and fills much of the cell or numerous discoid, lacking pyrenoids. A stigma may be present or absent depending on the species. Twelve species are recognized (Ettl, 1983). Asexual reproduction is by zoospore formation within the mother cell wall. Sexual reproduction is unknown. Vegetative cells of *G. kupfferi* have been examined by transmission electron microscopy (Domozych, 1989; Domozych and Nimmons, 1992).

G. ovalis (Fig. 17) has been found in the United States (e.g., MA, ME) (Smith, 1950) and British Columbia, Canada (Stein and Borden, 1979).

Lobomonas Dangeard (Fig. 18)

Cells are spherical, ovoid, or ellipsoidal, with a cell wall, two equal flagella at the anterior pole, and two contractile vacuoles at the base of the flagella. The cell wall has many protuberances on the surface and may be swollen (Iyengar and Desikachary, 1981; Ettl, 1983). The chloroplast is single and massive, with a single pyrenoid and a stigma. Pyrenoid may be absent, depending on the species. Thirteen species are recognized (Ettl, 1983). Asexual reproduction is by zoospore formation within the mother cell wall. Sexual reproduction is unknown.

L. rostrata (Fig. 18) has been found in the United States (e.g., NJ, VJ) (Smith, 1950; Dillard, 1989). This genus has also been reported from British Columbia, Canada (Stein and Borden, 1979).

Polytoma Ehrenberg (Fig. 15)

Cells of this genus are colorless, but otherwise similar to those of *Chlamydomonas* cells. More than 30 species have been described (Iyengar and Desikachary, 1981; Ettl, 1983). Asexual reproduction is via zoospore formation within the mother cell wall. Sexual reproduction is isogamous, anisogamous, or oogamous. On the basis of 18S rRNA gene sequence data, 13 *Polytoma* strains were resolved as a nonmonophyletic group (Rumpf *et al.*, 1996).

Members of this genus occur in water rich in organic matter (Ettl, 1983). *P. uvella* (Fig. 15) and *P. granuliferum* have been found in the United States (Moewus, 1959; Ettl, 1983).

Vitreochlamys Batko (Fig. 19)

Cells are spherical, ovoid, or ellipsoidal, with a cell wall, two equal flagella at the anterior pole, and two or three contractile vacuoles at the base of the flagella (Ettl, 1983). The cell wall is swollen or gelatinous, which is characteristic of the genus (Ettl, 1983; Nakazawa *et al.*, 2001). The chloroplast is single and fills much of the cell, with a single or multiple pyrenoids and a stigma. Pyrenoid may be absent, depending upon the species. Asexual reproduction is by zoospore formation within the mother cell wall. Sexual reproduction is unknown. Twenty-three species are recognized (Ettl, 1983). Although the name *Sphaerellopsis* Korshikov (1925) is generally used (e.g., Ettl, 1983), this name is a homonym of the fungus genus *Sphaerellopsis* M. C. Cooke (1883). Thus, Batko (1970) proposed *Vitreochlamys* as a *nomen novum* for *Sphaerellopsis* Korshikov (1925). The genus *Vitreochlamys* was resolved as three separate lineages basal to the Tetrasporales or the colonial Volvocales based on the *rbc*L gene phylogeny, representing the ancestral situation of these two orders (Nakazawa *et al.*, 2001).

V. fluviatilis (Fig. 19) and *Sphaerellopsis* (*Vitreochlamys*) *gelatinosa* have been collected in the United States (e.g., CA, KY, NC, TN, VA) (Smith, 1950; Dillard, 1989). The genus has been reported from Guadeloupe (Bourrelly and Manguin, 1952) and British Columbia, Canada (Stein and Borden, 1979), as *Sphaerellopsis*.

E. Phacotaceae

The unicellular, biflagellate algae in this family are characterized by having nonliving investments surrounding the protoplast or loricae (Bold and Wynne, 1985). The loricae are sometimes impregnated with iron and/or manganese salts rendering them brown in color. In some genera, the lorica is composed of two parts that separate at reproduction (Hepperle and Krienitz, 1997). The vegetative cell has a stigma, two contractile vacuoles at the base of the flagella and a cup-shaped chloroplast with a single or more pyrenoids. Asexual reproduction is by zoospore formation (Hepperle and Krienitz, 1997). Isogamous or anisogamous sexual reproduction is known in some species. Recent molecular phylogenetic analysis of the phatocacean algae indicates that *Dysmorphococcus* is separated from the clade composed of *Phacotus* and *Pteromonas* (Hepperle *et al.*, 1998). Nine genera are distributed in North America.

1a.	Cell with four flagella (Fig. 20)	*Pedinopera*
1b.	Cell with two flagella	2
2a.	Lorica composed of two overlapping halves	3
2b.	Lorica not composed of two overlapping halves	4
3a.	Lorica smooth and hyaline (Fig. 21)	*Pteromonas*
3b.	Lorica finely granulate and not hyaline (Fig. 22)	*Phacotus*
4a.	Posterior portion of lorica narrowed or protruded	5
4b.	Posterior portion of lorica broad or rounded	6
5a.	Compressed face of lorica with projections (Fig. 23)	*Wislouchiella*
5b.	Compressed face of lorica without projections (Fig. 24)	*Cephalomonas*
6a.	Cells more or less compressed; lorica not broad (Fig. 25)	*Thoracomonas*
6b.	Cells not compressed; lorica broad	7
7a.	Lorica very thin and smooth except for numerous spines or granules distributed on the surface (Fig. 26)	*Granuochloris*
7b.	Lorica thick and rough	8
8a.	Two flagella projected through a common opening of lorica (Fig. 27)	*Coccomonas*
8b.	Each flagellum projected through an individual opening of lorica (Fig. 28)	*Dysmorphococcus*

Cephalomonas Higinbotham (Fig. 24)

Loricae are compressed, rigid, and brittle, forming a broad anterior half and a narrowed posterior half. Cells have two equal flagella, two contractile vacuoles at the base of the flagella, a stigma, and a cup-shaped chloroplast with a single pyrenoid (Ettl, 1983). Asexual reproduction is by zoospore formation. Isogamous sexual reproduction is known. *Cephalomonas* is a monotypic genus with *C. granulata* (Ettl, 1983).

C. granulata (Fig. 24) has been reported from the United States (e.g., FL, KY, MD) (Smith, 1950; Dillard, 1989).

Coccomonas Stein (Fig. 27)

Loricae are thick and not compressed, impregnated with lime ($CaCO_3$) and iron compounds. Cells are ovoid to spherical in shape, with two equal flagella, two contractile vacuoles at the base of the flagella, a stigma, and a cup-shaped chloroplast with a single basal pyrenoid (Ettl, 1983). The two flagella are projected through a common opening of the lorica. Asexual reproduction is by zoospore formation. Sexual reproduction is unknown. Seven species are known in this genus (Ettl, 1983).

C. orbicularis (Fig. 27) has been recorded from the United States (e.g., FL, GA, KY, NC, SC, TN, VA) (Smith, 1950; Dillard, 1989), and from British Columbia, Canada (Stein and Borden, 1979).

Dysmorphococcus Takeda (Fig. 28)

Loricae are thick and with pores, sometimes becoming brown in color. Cells spherical or ovoid, with two equal flagella, two contractile vacuoles at the base of the flagella, a stigma, and a cup-shaped chloroplast with a single or multiple pyrenoids (Ettl, 1983). Each flagellum is projected through an individual opening of the lorica. Asexual reproduction is by zoospore formation. Sexual reproduction is unknown. Eight species are known in this genus (Ettl, 1983).

D. variabilis (Fig. 28) and *D. globosus* have been observed from several locations in the United States (e.g., AL, KY, OH, TN, VA, WV) (Bold and Starr, 1953; Dillard, 1989).

Granuochloris Pascher *et* Jahoda (Fig. 26)

Loricae are thin and not compressed, with numerous granules or spines distributed throughout the surface. Cells have two equal flagella, two contractile

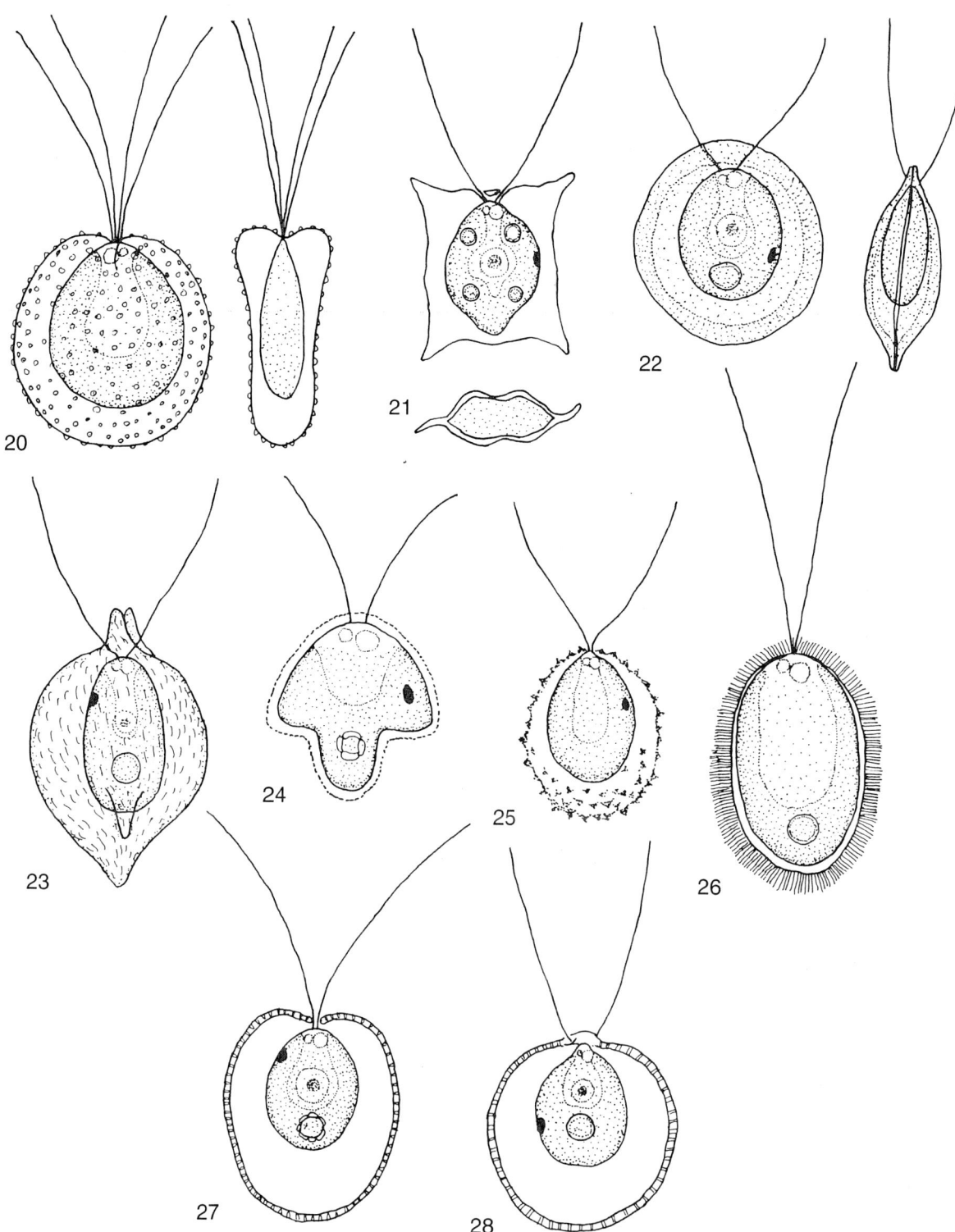

FIGURE 20 Two views of *Pedinopera granulosa* vegetative cell. (× 1000.) FIGURE 21 Two views of *Pteromonas aculeata* vegetative cell. (× 790.) FIGURE 22 Two views of vegetative cell of *Phacotus lenticularis*. (× 770.) FIGURE 23 Vegetative cell of *Wislouchiella planctonica*. (× 1370.) FIGURE 24 Vegetative cell of *Cephalomonas granulata*. (× 1800.) FIGURE 25 Vegetative cell of *Thoracomonas phacotoides*. (× 1900.) FIGURE 26 Vegetative cell of *Granulochloris spinifera*. (× 1700.) FIGURE 27 Vegetative cell of *Coccomonas orbicularis*, showing two flagella projecting through one anterior pore of the lorica. (× 1300.) FIGURE 28 Vegetative cell of *Dysmorphococcus variabilis*. (× 1500.)

vacuoles at the base of the flagella, a stigma, and a cup- or H-shaped chloroplast with a single pyrenoid (Fott, 1963). Reproduction is not known in detail. Three species are recognized in this genus (Ettl, 1983).

G. spinifera (Fig. 26) has been collected in Massachusetts (Fott, 1963).

Pedinopera Pascher (Fig. 20)

Loricae are compressed, with granulate surface, with or without longitudinal ridges. Cells are flattened and ovoid to pear-shaped, with four equal flagella and a cup-shaped chloroplast (Ettl, 1983). Pyrenoids/stigma may be present or absent, depending upon a given species. Although five species have been described, reproduction in this genus is unknown (Ettl, 1983).

P. granulosa (Fig. 20) and *P. rugulosa* have been collected in the United States (e.g., FL, NC, TN) (Dillard, 1989).

Phacotus Perty (Fig. 22)

Loricae are highly compressed and lens-shaped, composed of two parts (sometimes called shells), without pores. The loricae are usually dark brown colored and impregnated with lime, exhibiting a mineralized surface composed of many calcite crystals (Hepperle and Krienitz, 1997). Cells are flattened, with two equal flagella, two contractile vacuoles at the base of the flagella, a stigma, and a cup-shaped chloroplast. Pyrenoids may be single, multiple, or lacking depending upon a given species (Ettl, 1983). Asexual reproduction is by formation of 2–16 zoospores within a gelatinous sporangial sheath after separation of the two shells of the parental lorica (Hepperle and Krienitz, 1996). Sexual reproduction is rare and isogamous (Ettl, 1983). Calcification of *P. lenticularis* requires external supersaturation of calcium (Hepperle and Krienitz, 1997).

P. lenticularis (Fig. 22), *P. angustus*, *P. glaber*, and *P. subglobosus* have been collected from across in the United States (Smith, 1950; Dillard, 1989). The genus is also reported from Ontario and British Columbia, Canada (Duthie and Socha, 1976; Stein and Borden, 1979) and the Laurention Great Lakes (Munawar and Munawar, 1981).

Pteromonas Seligo (Fig. 21)

Loricae are compressed, forming projecting wings around the cell, composed of two hyaline shell-like portions joined at the wings (Iyengar and Desikachary, 1981; Ettl, 1983). Cells are flattened and pear-shaped, with two equal flagella, two contractile vacuoles at the base of the flagella, a stigma and a cup-shaped chloroplast. Pyrenoids may be single or multiple, which is species specific (Ettl, 1983). Asexual reproduction is by formation of two or four zoospores within the parental lorica. Sexual reproduction is isogamous. Approximately 20 species have been described (Ettl, 1983).

Five species of *Pteromonas*, *P. aculeata* (Fig. 21), *P. cordiformis*, *P. angulosa*, *P. sinuosa*, and *P. cruciata* have been collected in the United States (e.g., KY, NC, TN, WV) (Smith, 1950; Dillard, 1989).

Thoracomonas Korshikov (Fig. 25)

Loricae are somewhat compressed, verrucose, and thin. Cells have two equal flagella, two contractile vacuoles at the base of the flagella, a stigma, and a cup-shaped chloroplast with a single or multiple pyrenoids (Ettl, 1983). Asexual reproduction is by zoospore formation. Sexual reproduction is not known in detail. Four species are recognized in this genus (Ettl, 1983).

T. feldmanii and *T. phacotoides* (Fig. 25) have been observed in the United States (e.g., LA, TN) (Smith, 1950; Dillard, 1989).

Wislouchiella Skvortzow (Fig. 23)

Loricae are strongly compressed and finely verrucose, forming a broad wing-like expansion with two cylindrical projections in each compressed face (Ettl, 1983). Cells are ovoid in side view and rhomboidal in vertical view, with two equal flagella, two contractile vacuoles at the base of the flagella, a stigma, and a cup-shaped chloroplast with a single pyrenoid. Reproduction is unknown. *Wislouchiella* is a monotypic genus with *W. planctonica* (Ettl, 1983).

W. planctonica (Fig. 23) has been reported from the United States (Smith, 1950; Dillard, 1989) and British Columbia, Canada (Stein and Borden, 1979).

F. Volvocaceae

The coenobic colonial organisms (fixed number of cells) in which biflagellate *Chlamydomonas*-like cells are surrounded by a gelatinous matrix were traditionally assigned to the Volovocaceae and Astrephomenaceae (Bold and Wynne, 1985). However, the classification of these two families has recently become controversial. Nozaki and Kuroiwa (1992) removed the genus *Gonium* from the Volvocaceae and classified *Astrephomene* and *Gonium* in a single family, the Goniaceae, based on the vegetative ultrastructure of the extracellular (gelatinous) matrix. Subsequently, Nozaki and Ito (1994), on the basis of cladistic analysis of morphological data, resolved the four-celled species *Tetrabaena socialis* (*Gonium sociale*) as a sister group to the monophyletic group composed of other species in the Volvocaceae and the Goniaceae, and established a new family, the Tetrabaenaceae, for encompassing *T. socialis*. Furthermore, Nozaki *et al.* (1996b) assigned another four-celled colonial alga *Basichlamys sacculifera* to the Tetrabaenaceae, on the basis of the further cladistic analysis based on morphological data. Although the author follows the taxonomic concept of Nozaki *et al.* (1996b), the phylogenetic relationships and status of these three colonial families are uncertain in the phylogenetic analyses based on the *rbc*L and/or *atp*B (ATP synthase beta-subunit) gene sequence data except for the robust monophyly of the Tetrabaenaceae (Nozaki *et al.*, 1995a, 1997a, 1999). However, the recent molecular phylogenetic analyses using five chloroplast genes resolved the monophyly of the Volvocaceae as well as the Goniaceae (Nozaki *et al.*, 2000).

The volvocacean algae now can be characterized by having a tripartite colonial boundary of the extracellular matrix in vegetative colonies (Nozaki and Kuroiwa, 1992). In asexual reproduction, all of cells in the colony or only large reproductive cells (gonidia) divide successively to form a miniature of the parental colony (autocolony). The volvocacean genera exhibit a unique phenomenon called inversion or eversion just after the successive divisions during colony formation (see Smith, 1955). Sexual reproduction is either isogamous, anisogamous, or oogamous. In anisogamous and oogamous genera, spindle-shaped male gametes are produced as they are grouped, forming a hemispherical or flattened coenobic colony called a "sperm packet." This family includes seven genera (Figs. 29–35) which are distinguished by shape of the colony and cellular differentiation.

1a.	Colony flattened, with cells arranged in a single layer (Fig. 33)	*Platydorina*
1b.	Colony spheroidal, with cells arranged radially	2
2a.	Colony generally with up to 32 cells; differentiation of small somatic cells absent or incomplete	3
2b.	Colony generally with 64 or more cells; differentiation of small somatic cells complete	6
3a.	Maximum colony cell number 16; cellular envelopes absent	4
3b.	Maximum colony cell number 32; cellular envelopes present	5
4a.	Cells keystone-shaped or pear-shaped; colony contiguous in the center (Fig. 29)	*Pandorina*
4b.	Cells hemispherical or lenticular; colony hollow (Fig. 30)	*Volvulina*
5a.	Sexual reproduction isogamous (Fig. 31)	*Yamagishiella*
5b.	Sexual reproduction anisogamous with sperm packets (Fig. 32)	*Eudorina*
6a.	Colony generally 32- 64- or 128-celled, with 20–50% somatic cells (Fig. 34)	*Pleodorina*
6b.	Colony containing more than 500 cells, composed mostly of somatic cells (Fig. 35)	*Volvox*

Eudorina Ehrenberg (Fig. 32)

Colonies are ovoid, ellipsoidal, or cylindrical, containing 16 or 32 cells arranged radially in the periphery of a gelatinous matrix, forming a hollow sphere (Goldstein, 1964). Each cell of the colony is enclosed tightly by the fibrillar layer (cellular envelope) of the extracellular matrix of the colony (Nozaki and Kuroiwa, 1992). The matrix does or does not form individual sheaths, depending upon the species (Goldstein, 1964). Cells are ovoid or spherical, each with two equal flagella, a stigma, two contractile vacuoles at the base of the flagella, and a massive cup-shaped chloroplast with one (basal) or multiple pyrenoids (Goldstein, 1964). Stigmata of anterior cells are larger than in posterior cells. No differentiation between somatic and reproductive cells occurs except for *E. illinoisensis*, in which anterior four cells are small and facultatively somatic. Sexual reproduction is anisogamous with sperm packets, producing walled hypnozygotes (Goldstein, 1964). Upon germination, the zygote gives rise to a single or two biflagellate gone cell. *Eudorina* is cosmopolitan and contains about seven species (Goldstein, 1964). Although *Eudorina* was often confused with *Pleodorina*, it is now distinguished based of the absence of obligately somatic cells (Nozaki *et al.*, 1989). Molecular phylogenetic analyses resolved that the genus *Eudorina* is paraphyletic, exhibiting the ancestral situation of *Pleodorina* and *Volvox* (excluding section *Volvox*) (Nozaki *et al.*, 1995a, 1997a, b, 1999, 2000).

E. elegans is among the most frequently encountered species of green algae. Goldstein (1964), on the basis of cultured material originating from the United States and Canada, recognized five species assignable to *Eudorina*, *E. elegans* (Fig. 32), *E. unicocca*, *E. illinoisensis*, *E. cylindrica*, and *E. conradii*. Although Prescott (1955) described *E. interconnexa* from the Panama Canal Zone, Ettl (1983) questioned its existence. Several species have been recorded widely in the United States and Canada (Smith, 1950; Duthie and Socha, 1976; Stein and Borden, 1979; Sheath and Steinman, 1982; Whitford and Schumacher, 1984).

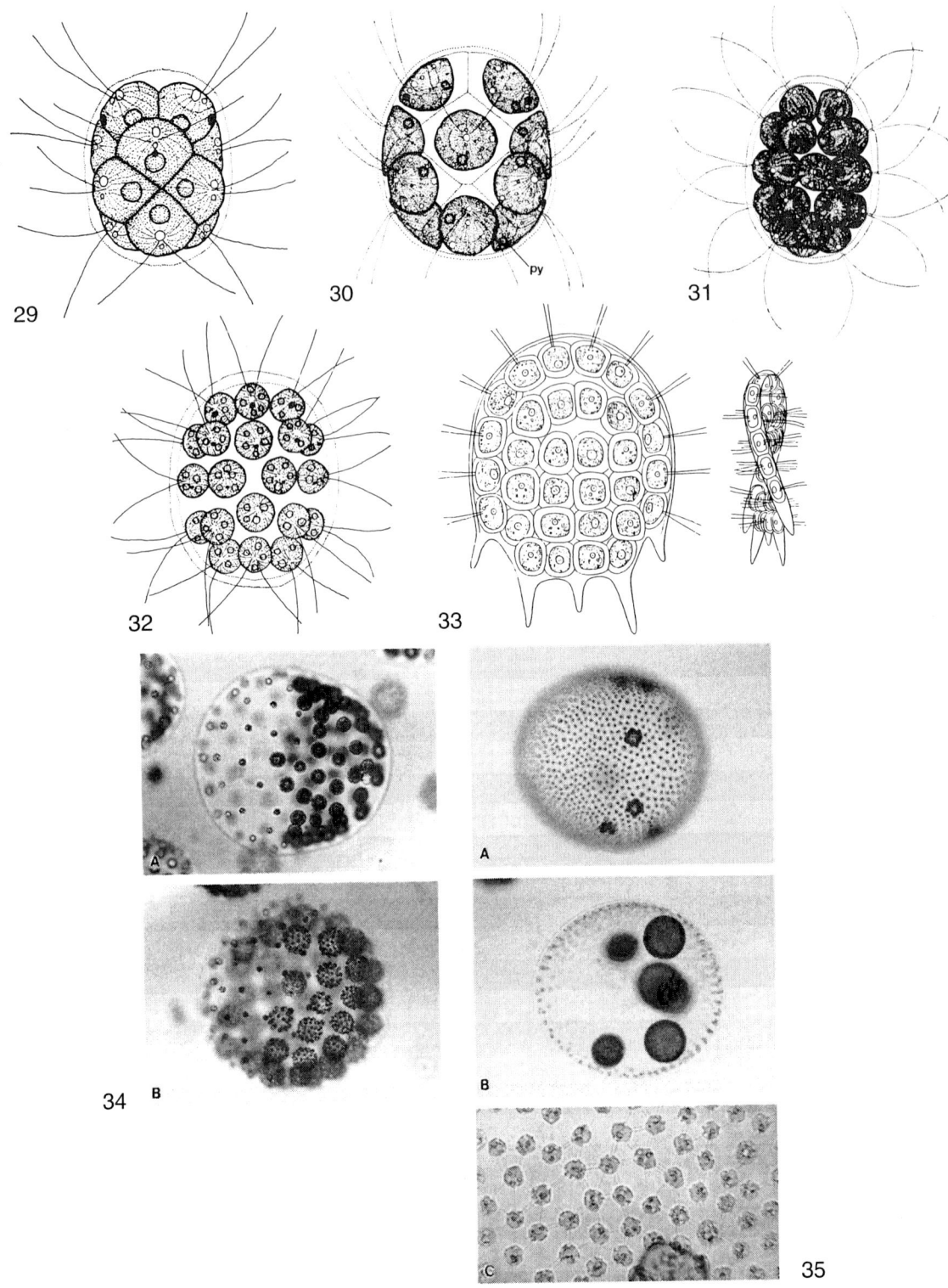

FIGURE 29 Vegetative colony of *Pandorina morum*, showing 16 cells compactly arranged. (From Nozaki, 1995, reproduced by permission of Koudan-Sha Scientific.) (× 520.) *FIGURE 30* Sixteen-celled vegetative colony of *Volvulina steinii*, showing pyrenoid (py) developing in the brim of the cup-shaped chloroplast. (From Nozaki, 1982, reproduced by permission of *Journal of Japanese Botany*.) (× 470.) *FIGURE 31* *Yamagishiella unicocca* vegetative colony, showing 32 cells loosely arranged. (From Nozaki, 1981, reproduced by permission of *Journal of Japanese Botany*.) (× 270.) *FIGURE 32* Vegetative colony of *Eudorina elegans*. (From Nozaki, 1995, reproduced by permission of Koudan-Sha Scientific.) (× 300.) *FIGURE 33* Two views of vegetative colony of *Platydorina caudata*. (From Kofoid, 1899.) (× 200, × 130.) *FIGURE 34* Light micrographs of *Pleodorina californica*. (× 85.) A: Vegetative colony. B: Daughter colony formation. *FIGURE 35* Light micrographs of *Volvox aureus*. A: Asexual colony with divided gonidia. (× 290.) B: Daughter colonies within the parental colony. (× 110.) C: Surface view of asexual colony. Note delicate cytoplasmic bridges between cells. (× 340.)

Pandorina Bory de St.-Vincent (Fig. 29)

Colonies are ovoid or ellipsoidal, containing eight or 16 cells compactly arranged radially in a gelatinous matrix (Ettl, 1983). Cells are keystone-shaped or ovoid, each with two equal flagella, a stigma, two contractile vacuoles at the base of the flagella, and a massive cup-shaped chloroplast with one basal or multiple pyrenoids (species dependent) (Nozaki and Kuroiwa, 1991). Stigmata in anterior cells are larger than in posterior cells. Sexual reproduction is isogamous, forming walled hypnozygotes (Coleman, 1959; Nozaki and Kuroiwa, 1991). Upon germination, zygotes give rise to single biflagellate gone cells. *Pandorina* is cosmopolitan in fresh waters (Coleman, 1959, 1977). Although this genus was frequently confused with *Eudorina*, it is now distinguished by the difference in structure of extracellualr matrix of the vegetative colony and in sexual reproduction (Nozaki, 1981; Nozaki and Kuroiwa, 1992). Recently, *Pandorina unicocca* was removed to the genus *Yamagishiella*, on the basis of its cellular envelopes within the colony (Nozaki and Kuroiwa, 1992). Although seven species were described, only *P. morum* and *P. colemaniae* seem reliable based on cultural studies (Coleman, 1959; Nozaki and Kuroiwa, 1991).

P. morum (Fig. 29) is broadly distributed throughout the United States (Coleman, 1959). *P. smithii* has been found in Wisconsin (Ettl, 1983). *P. morum* has also been reported widely from Canada (Duthie and Socha, 1976; Stein and Borden, 1979; Sheath and Steinman, 1982) as well as Mexico (Ortega, 1984) and Guadeloupe (Bourrelly and Marguin, 1952). *P. charkowiensis*, reported by Smith (1950), Thompson (1954) and Dillard (1989) should be assigned to *Eudorina* or *Yamagishiella*.

Platydorina Kofoid (Fig. 33)

Colonies are flattened, slightly twisted, and horseshoe-shaped with three to five posterior projections of a gelatinous matrix (Kofoid, 1899). Each colony contains 16 or 32 cells arranged in one layer and oriented in different directions in a gelatinous matrix. Cells are spherical or pear-shaped, each with two equal flagella, a stigma, two contractile vacuoles at the base of the flagella, and a massive cup-shaped chloroplast with a single basal pyrenoid (Kofoid, 1899). Colony formation involves inversion and subsequent intercalation (Harris and Starr, 1969). Sexual reproduction is anisogamous with sperm packets, forming walled hypnozygotes (Harris and Starr, 1969).

Platydorina is a monotypic genus with *P. caudata* (Fig. 33) and collected from sites in the United States (e.g., AL, FL, GA, KY, SC, TN) (Kofoid, 1899; Harris and Starr, 1969). The genus has been reported from Mexico, but no species was given (Ortega, 1984).

Pleodorina Shaw (Fig. 34)

Colonies are spherical, ovoid, or ellipsoidal, containing 32, 64, or 128 cells arranged radially at the periphery of a gelatinous matrix (Nozaki et al., 1989). Colonies have small, obligately somatic cells at the anterior pole, and large reproductive cells (gonidia) in the remaining portion. The matrix may or may not exhibit individual sheaths, depending upon a given species (Nozaki et al., 1989). Cells are spherical or ovoid, each with two equal flagella, a stigma, many contractile vacuoles on the cell surface, and a massive cup-shaped chloroplast. Chloroplasts of somatic cells have a single basal pyrenoid, whereas reproductive cells have multiple ones. Stigmata of anterior cells are larger than in posterior cells. Sexual reproduction is anisogamous with sperm packets, forming walled hypnozygotes (Nozaki et al., 1989). A single biflagellate gone cell is released from the germinating zygote. *Pleodorina* is cosmopolitan in freshwater and includes four species (Nozaki et al., 1989). *Pleodorina* and *Eudorina* are distinguished based on the presence or absence of obligately somatic cells respectively (Nozaki et al., 1989). Recent molecular phylogenetic analysis suggests that *P. indica* is separated from *P. californica* and *P. japonica* (Nozaki et al., 1997a).

P. californica (Fig. 34) has been recorded from the United States (e.g., FL, GA, TN, VA, WV) and the Panama Canal Zone (Goldstein, 1964) as well as from Ontario, Canada (Duthie and Socha, 1976). *P. indica*, originating from Mexico, has been studied by light and electron microscopy (Nozaki et al., 1989; Nozaki and Kuroiwa, 1992).

Volvox Linnaeus (Fig. 35)

Colonies are spherical, subspherical, ellipsoidal, or ovoid, containing 500–50,000 cells arranged radially at the periphery of a gelatinous matrix, forming a hollow sphere. Several to approximately 50 large reproductive cells (gonidia) are situated in posterior 1/2 to 2/3 of colony (Smith, 1944). Each cell is enclosed by a gelatinous sheath which is distinct or confluent, depending upon the species (Smith, 1944). Somatic cells are spherical, ovoid, or star-shaped, each with two equal flagella, two contractile vacuoles at the base of the flagella, and a cup-shaped chloroplast with a single pyrenoid. Cytoplasmic strands between cells are thick, thin, or absent, and this trait is species dependent (Smith, 1944). Stigmata in the anterior cells are larger than in posterior cells. Sexual reproduction is oogamous (Nozaki, 1988); in monoecious species, the sexual colony has both sperm packets and eggs. In dioecious species, the male colony contains androgonidia which divide successively into sperm packets; such males may be markedly reduced in size (dwarf male) or nearly as large as asexual colonies. The female colony has eggs, whose number is nearly the same as that of gonidia in asexual colonies (facultative female) or much larger (special female). After fertilization zygotes develop a heavy cell wall that may be ornamented with reticulation or spines (Smith, 1944). Upon germination, the zygote gives rise to a single biflagellate gone cell (Nozaki, 1988). This genus contains approximately 20 species, which are classified into four sections based on the differences in gelatinous matrix and cytoplasmic strands (Smith, 1944). Ultrastructure of the flagellar apparatus of two species of *Volvox* (Hoops, 1984) and molecular phylogenetic analyses based rRNA sequences (Larson *et al.*, 1992; Kirk, 1998) and internal transcribed spacer sequences (Coleman, 1999) indicate that *Volvox* is polyphyletic. Molecular phylogenetic analyses based on *rbcL* gene sequence data strongly suggest that section *Volvox* (=*Euvolvox*) is separated from the other three sections (Nozaki *et al.*, 1995a, 1997a, 1999). Results of chloroplast multigene phylogeny indicate that the genus *Volvox* represents four separate lineages (Nozaki *et al.*, 2002b).

Volvox is cosmopolitan in fresh waters and *V. aureus* (Fig. 35) is a widely reported species. Eleven species of *Volvox* have been found in the United States, namely, *V. aureus*, *V. globator*, *V. africanus*, *V. carteri*, [including *V. weismannia* (= *V. carteri* f. *weismannia*)], *V. perglobator*, *V. powersii*, *V. spermatosphaera*, *V. tertius*, *V. dissipatrix*, *V. prolificus*, and *V. rousseletii* (Smith, 1950; Dillard, 1989). The genus has also been widely reported from Canada (Duthie and Socha, 1976; Stein and Borden, 1979; Sheath and Steinman, 1982), Mexico (Ortega, 1984), and Guadeloupe (Bourrelly and Manguin, 1952). Starr (1970) described *V. pocockiae* based on cultured material originating from Mexico.

Volvulina Playfair (Fig. 30)

Colonies are ovoid or spherical, containing eight or 16 cells embedded in the periphery of a gelatinous matrix, forming a hollow structure (Pocock, 1954; Thompson, 1954; Stein, 1958a; Starr, 1962). Cells are lenticular or hemispherical, each with two equal flagella, a stigma, two contractile vacuoles at the base of the flagella or many contractile vacuoles scattered on the cell surface, a massive cup-shaped chloroplast without pyrenoids, or with one in the bottom or brim (Starr, 1962; Nozaki, 1982; Nozaki and Kuroiwa, 1990). Stigmata in anterior cells are larger than in posterior cells. Sexual reproduction is isogamous and walled hypnozygotes are formed (Stein, 1958a; Starr, 1962; Nozaki, 1982; Nozaki and Kuroiwa, 1990). Germinating zygotes give rise to a single biflagellate gone cell. *Volvulina* is cosmopolitan but rare. The current concept of *Volvulina* is based on the presence of hollow colonies with lenticular cells (Nozaki and Kuroiwa, 1990).

V. steinii is often collected from water rich in organic matter. Three reliable species were described (Nozaki and Kuroiwa, 1990). *V. steinii* (Fig. 30) has been observed in various localities across the United States (Thompson, 1954; Stein, 1958a; Carefoot, 1966). Starr (1962) described *V. pringsheimii* Starr based on cultured material originating from Texas. *V. steinii* has also been collected in British Columbia, Canada (Stein and Borden, 1979).

Yamagishiella Nozaki (Fig. 31)

Colonies are ovoid, ellipsoidal, or cylindrical, containing 16 or 32 cells arranged radially in the periphery of a gelatinous matrix, forming a hollow sphere (Rayburn and Starr, 1974). Each cell of the colony is enclosed tightly by the fibrillar layer (cellular envelope) of the extracellular matrix of the colony (Nozaki and Kuroiwa, 1992). Cells are ovoid or spherical, each with two equal flagella, a stigma, two contractile vacuoles at the base of the flagella, and a massive cup-shaped chloroplast with a single basal pyrenoid (Rayburn and Starr, 1974). Stigmata of anterior cells are larger than in posterior cells. No differentiation between somatic and reproductive cells occurs. Sexual reproduction is isogamous, producing walled hypnozygotes (Rayburn and Starr, 1974). Upon germination, the zygote gives rise to a single biflagellate gone cell. *Yamagishiella* is a monotypic genus with *Y. unicocca* (Fig. 31). This genus is distinguished from *Pandorina* by its cellular envelopes and 32-celled colonies (Nozaki and Kuroiwa, 1992). Although *Yamagishiella* differs from *Eudorina* by its isogamous sexual reproduction, the vegetative morphology and asexual reproduction characteristics of these two genera (especially *Y. unicocca* and *E. unicocca*) are indistinguishable. However, such a taxonomic problem may be resolved based on the analysis of *rbcL* gene sequence data (Nozaki *et al.*, 1998a).

Y. unicocca has been recorded from Indiana, Oregon, and Massachusetts (as *Pandorina unicocca*; Rayburn and Starr, 1974).

G. Goniaceae

This family was revived by Nozaki and Kuroiwa (1992) to comprise the two genera *Gonium* and *Astrephomene*. These two genera (Figs. 36 and 37) exhibit essentially the same colony structure in which each vegetative cell is enclosed by a tripartite boundary of the extracellular matrix (cellular boundary). This situation is essentially different from that of the Volvocaceae (colonial boundary). Asexual reproduction is by daughter colony formation without inver-

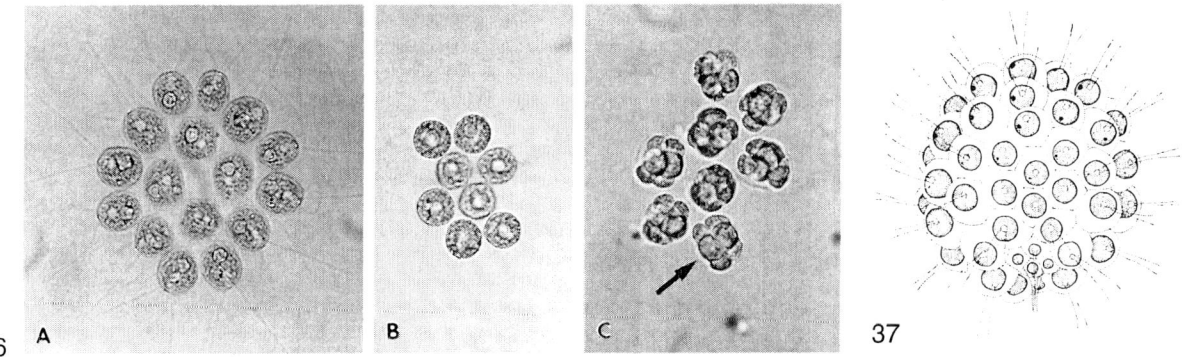

FIGURE 36 Light micrographs of *Gonium pectorale*. (× 340.) A: 16-celled colony. B: Eight-celled colony. C: Daughter colony formation. FIGURE 37 Vegetative colony of *Astrephomene gubernaculifera*. (From Nozaki, 1983, reproduced by permission of *Journal of Japanese Botany*.) (× 230.)

sion. All the vegetative cells in *Gonium* colonies undergo reproduction, whereas small somatic cells are present in the posterior pole of *Astrephomene* colonies.

Sexual reproduction is isogamous and three types of zygote germination are recognized in the Goniaceae (see Nozaki and Ito, 1994).

1a. Colony flattened (Fig. 36)..*Gonium*
1b. Colony spherical (Fig. 37)...*Astrephomene*

Astrephomene Pocock (Fig. 37)

Colonies are ovoid or subspherical, containing 32, 64, or 128 cells arranged radially at the periphery of a gelatinous matrix (Pocock, 1954). Colonies have two to several small somatic cells (rudder cells) at the posterior pole. Cells are nearly spherical or lenticular, each with two equal flagella, a stigma, many contractile vacuoles on the cell surface, and a massive cup-shaped chloroplast that lacks pyrenoids or develops one in the brim (Nozaki, 1983). Stigmata in the anterior cells are larger than in posterior cells. Each protoplast is enclosed by a gelatinous sheath (cellular boundary) and constitutive cells attach or connect by the fusion or attachment of these sheaths, forming a hollow colony (Nozaki, 1983). Sexual reproduction is isogamous and walled hypnozygotes are formed. On germination, the zygote gives rise to a single biflagellate gone cell (Brooks, 1966; Nozaki, 1983). *Astrephomene* is rare and grows in freshwater rich in organic matter. This genus contain two species: *A. gubernaculifera* (Fig. 37) and *A. perforata* (Nozaki, 1983). The two species differ in character of the gelatinous cellular sheaths, pyrenoids, and number of somatic cells (Nozaki, 1983). Some authors recognize the family Astrephomenaceae for only a single genus *Astrephomene*, based on having spheroidal colonies but lacking inversion during colony formation (Pocock, 1954).

A. gubernaculifera has been collected in the United States and Mexico (Stein, 1958a; Brooks, 1966).

Gonium O. F. Müller (Fig. 36)

Colonies are flattened, containing eight, 16, or 32 cells arranged in one layer and oriented in the same direction (Stein, 1958b; Ettl, 1983). The eight-celled colony exhibits a characteristic cell arrangement for a given species (Nozaki, 1989a). Cells are ovoid to angular, each with two equal flagella, a stigma, two contractile vacuoles at the base of the flagella, and a massive cup-shaped chloroplast with one or multiple pyrenoids (Pocock, 1955; Stein, 1958b; Nozaki, 1989a). Each protoplast is enclosed by a gelatinous sheath (cellular boundary), and the cells attaching or connecting to one another by the union or attachment of the sheaths form a colony. Sexual reproduction is isogamous, forming hypnozygotes with smooth walls (Stein, 1958b; Nozaki, 1989a). Germinating zygote produces four biflagellate gone cells which are joined in a colony (germ colony) (Stein, 1958b; Nozaki, 1989a), except for *G. multicoccum* (Nozaki and Ito, 1994). *Gonium* is cosmopolitan in fresh waters (Fabry *et al.*, 1999). *G. pectorale* (Fig. 36) is one of the most commonly encountered species of all chlorophytes. Five species have been studied in culture (Nozaki *et al.*, 1997a). Although cladistic analysis based on morphological data indicates that *Gonium* is paraphyletic (Nozaki and Ito, 1994), recent molecular phylogenetic analyses using combined data set from multiple chloroplast protein-coding gene sequences resolved that the genus is monophyletic (Nozaki *et al.*, 1999, 2000).

Stein (1965) demonstrated the existence of 33 sexual populations of *G. pectorale* from the United States and Canada. Pocock (1955) described *G. multicoccum* and *G. octonarium* based on material collected in the United States. Prescott (1942) described *G. discoideum* from Louisiana. Although *G. formosum* has been frequently collected in the United States (Smith, 1950; Dillard, 1989) and British Columbia, Canada (Stein and Borden, 1979), this species has not been previously studied in culture.

H. Tetrabaenaceae

The Tetrabaenaceae includes *Basichlamys* and *Tetrabaena*, both of which were sometimes assigned to the genus *Gonium* (e.g., Stein, 1959). However, this family differs from the Goniaceae in having vegetative colony composed of only four cells and reticulate zygote or hypnospore walls (Nozaki et al., 1996b). All the cells of the colony divide into a daughter colony in asexual reproduction. Sexual reproduction is isogamous. Based on the combined data set from *rbc*L and *atp*B gene sequences, *Basichlamys* and *Tetrabaena* are resolved as a close clade separated from the monophyletic group composed of the genus *Gonium* (Nozaki et al., 1999). Such separation between the Tetrabaenaceae and the genus *Gonium* was supported by the occurrence of the xanthophyll loroxanthin within the colonial Volvocales (Schagerl and Angeler 1998). Recently, the Tetrabaenaceae was resolved as the most basal lineage within the colonial Volvocales on the basis of the multigene phylogeny (Nozaki et al., 2000).

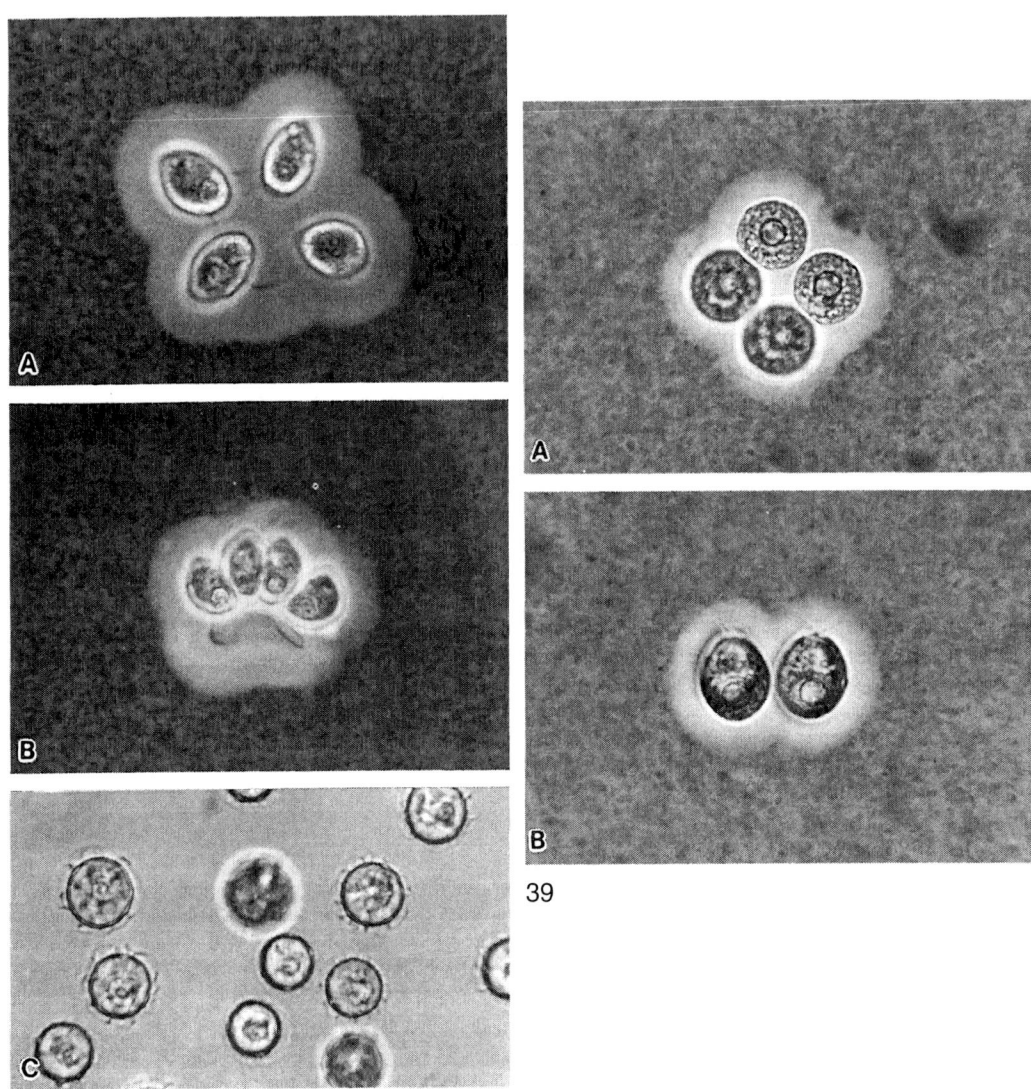

FIGURE 38 Light micrographs of *Basichlamys sacculifera*. (× 650.) A: Upper view of vegetative colony. India ink preparation. B: Side view of vegetative colony. India ink preparation. C: Aplanospores with reticulate walls. *FIGURE 39* Light micrographs of *Tetrabaena socialis* vegetative colonies. India ink preparation. (× 640.) A: Upper view of vegetative colony. B: Side view of vegetative colony.

1a.	Colonial cells separated from one another, but attached to their parental cellular sheath (sac) (Fig. 38)	.................*Basichlamys*
1b.	Colonial cells connecting to one another by the union or attachment of the cellular sheaths (Fig. 39)	............................*Tetrabaena*

Basichlamys Skuja (Fig. 38)

Colonies contain four cells attached to their parental cellular sheath (sac), forming a square (Stein, 1959). Cells are ovoid and somewhat asymmetrical, each with two equal flagella, a stigma, two contractile vacuoles at the base of the flagella, and a massive cup-shaped chloroplast with a single basal pyrenoid. Akinetes or hypnospores (thick-walled dormant cells) are sometimes formed (Stein, 1959). Sexual reproduction is isogamous. *Basichlamys* contains only a single species *B. sacculifera* (Fig. 38) and it is distinguished from *Tetrabaena* and *Gonium* in lacking connections of the cellular sheaths of the constitutive cells, which attach only to the sac. Some authors do not recognize *Basichlamys* and use *Gonium sacculiferum* (Stein, 1959).

This genus is rarely abundant, but it is apparently cosmopolitan in freshwaters (Stein, 1959). Stein (1959) studied morphology and reproduction of *B. sacculifera* (as *Gonium sacculiferum*) based on the cultured materials originating from Indiana, Minnesota, and California; it has also been reported from British Columbia (Stein and Borden, 1979).

Tetrabaena Fromentel (Fig. 39)

Colonies contain four cells attached to each other by the protuberances of their cellular sheath, forming a square (Stein, 1959; Nozaki and Ohtani, 1992). Cells are ovoid and somewhat asymmetrical, each with two equal flagella, a stigma, two contractile vacuoles at the base of the flagella, and a massive cup-shaped chloroplast with a single basal pyrenoid (Stein, 1959; Nozaki and Ohtani, 1992). Sexual reproduction is isogamous (Stein, 1959). This genus is rare but cosmopolitan in fresh water, occurring from temperate zones to the Antarctic (Nozaki and Ohtani, 1992). *Tetrabaena* contains only a single species, *T. socialis* (Fig. 39) and it is distinguished from *Gonium* in lacking colonies with more than four cells. Some authors do not recognize this genus and use *Gonium sociale* (Stein, 1959).

T. socialis has been found in various localities of the United States (e.g., FL, NC, SC, TN, VA) (Smith, 1950; Stein, 1959; Dillard, 1989), and Canada (Duthie and Socha, 1976; Stein and Borden, 1979).

I. Spondylomoraceae

Coenobic colonies of the Spondylomoraceae are composed of *Chlamydomonas*-like cells as in the Volvocaceae, the Goniaceae and the Tetrabaenaceae, but lack a gelatinous matrix surrounding the colony (Ettl, 1983; Bold and Wynne, 1985). Cells may be bi- or quadriflagellate (Ettl, 1983). Asexual reproduction takes place by autocolony formation by all the cells of the colony. Isogamous sexual reproduction is known for *Pyrobotrys* and *Pascherina* (Korshikov, 1928; Nozaki, 1986). Four genera in this family can be distinguished as follows:

1a.	Cells quadriflagellate (Fig. 40)	.................*Spondylomorum*
1b.	Cells biflagellate	2
2a.	Colonial cells interconnected to each other by elongate protuberances of the cell walls (Fig. 41)	...........*Chlorcorona*
2b.	Colonial cells connected to each other by the direct attachment of cell walls	3
3a.	Chloroplasts containing pyrenoids (Fig. 42)	.................*Pascherina*
3b.	Chloroplasts lacking pyrenoids (Fig. 43)	.................*Pyrobotrys*

Chlorcorona Fott (Fig. 41)

The colonies contain eight cells arranged in two parallel, alternating rhomboid tiers of four cells each, without encompassing gelatinous matrix (Ettl, 1983; Hoops and Floyd, 1982). The cells have a cell wall and are interconnected to each other by elongate protuberances of the walls. Each cell is ovoid and has two equal flagella, a stigma, two contractile vacuoles at the base of the flagella, and a cup-shaped chloroplast without pyrenoids (Ettl, 1983; Hoops and Floyd, 1982). Sexual reproduction is unknown. *Chlorcorona* is a monotypic genus with *C. bohemica* (Fig. 41), and rarely found in freshwater habitats of Europe and the United States. This genus was originally described by Fott (1949) as *Corone*. However, this generic name is a homonym of *Corone* (Hoffmannseg ex H. G. L. Reichenbach) Fourreau (1868) and *Corona* Lefébure et Chenevière (1938). Fott (1967) proposed *Chlorcorona* as a nomen novum for *Corone* Fott (1949).

Recently, *C. bohemica* was collected in Ohio and its flagellar apparatus has been observed by electron microscopy (Hoops and Floyd, 1982).

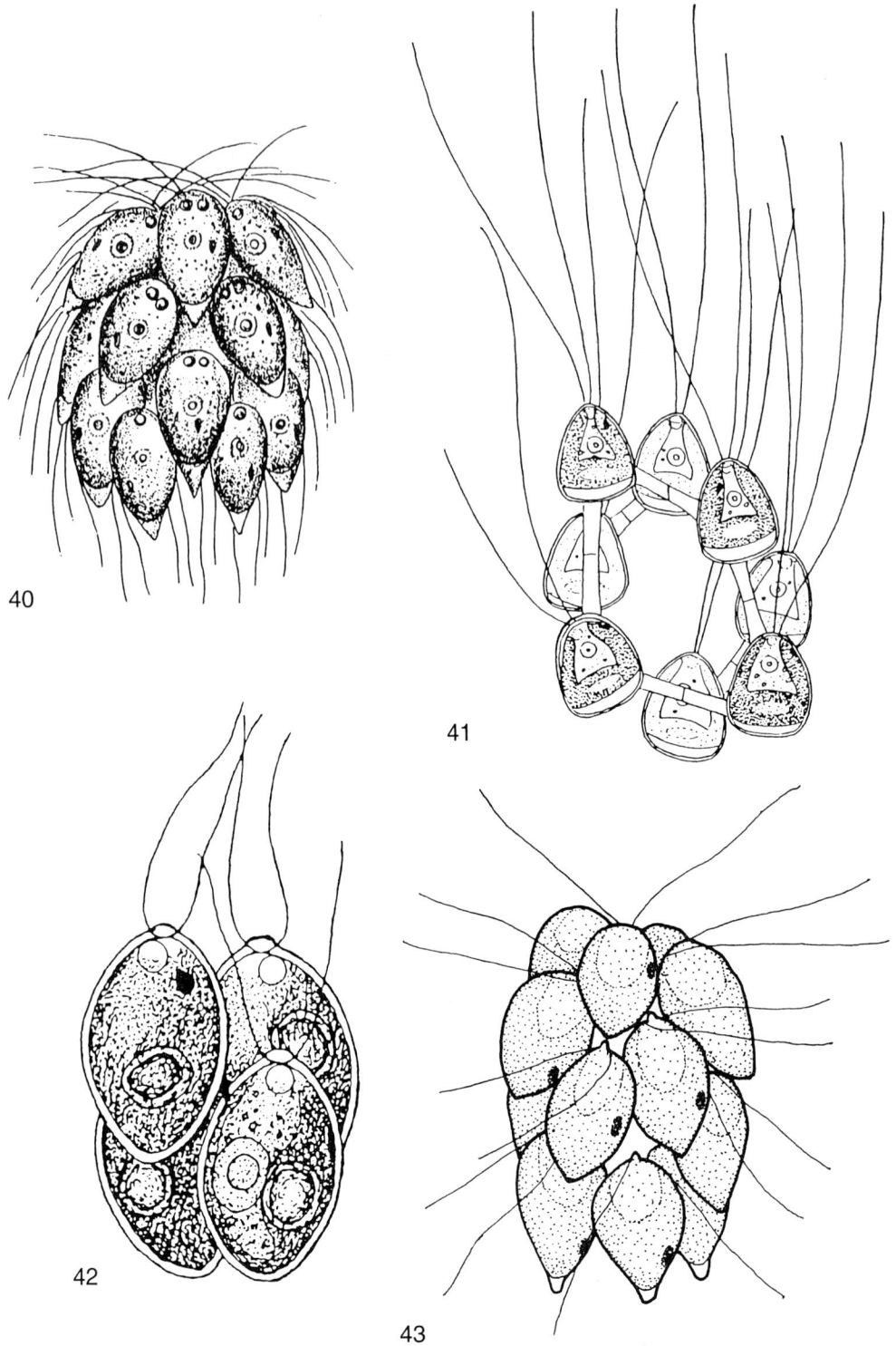

FIGURE 40 Vegetative colony of *Spondylomorum quaternarium*. (From Pascher, 1927.) (× 800.) FIGURE 41 Vegetative colony of *Chlorcorona bohemica*. (From Fott, 1949.) (× 1100.) FIGURE 42 Vegetative colony of *Pascherina tetras*. (From Korshikov, 1928.) (× 2100.) FIGURE 43 Vegetative colony of *Pyrobotrys casinoensis*. (From Nozaki, 1995, reproduced by permission of Koudan-Sha Scientific.) (× 950.)

Pascherina Silva (Fig. 42)

Colonies are mulberry-shaped, containing four cells arranged in two alternating tiers of two cells each, without encompassing gelatinous matrix (Korshikov, 1928; Smith, 1950). Cells are walled and ellipsoidal to ovoid, each with two equal flagella, a stigma, two contractile vacuoles at the base of the flagella, and a massive cup-shaped chloroplast with a single basal pyrenoid. Sexual reproduction is isogamous (Korshikov, 1928).

Pascherina is infrequently collected from cool, eutrophic habitats and only *P. tetras* (Fig. 42) has been described. Silva (1959) published *Pascherina* as a nomen novum for *Pascheriella* Korshikov (1928), an illegimate homonym of *Pascherella* (Conrad, 1926). *P. tetras* has been observed in the United States (Smith, 1950; Dillard, 1989).

Pyrobotrys Arnoldi (Fig. 43)

Colonies are star- or mulberry-shaped, containing four, eight or 16 cells arranged in two or four tiers, without encompassing gelatinous matrix (Nozaki, 1986). Cells are spherical, subspherical, ovoid, ellipsoidal, pear-shaped or irregularly pear-shaped, each with two equal flagella, two contractile vacuoles at the base of the flagella, and a massive cup-shaped chloroplast without pyrenoids. A stigma may be present in each cell of the colony. Cell walls are delicate with a papilla at the base of the flagella (Nozaki, 1986). In sexual reproduction, all the cells divide into four, eight, or 16 small, biflagellate isogametes that fuse to form planozygotes (Nozaki, 1986). Mature planozygotes are quadriflagellate with a large stigma and a species-dependent form (Nozaki, 1986). Mature aplanozygotes are spherical with a heavy cell wall. On germination, the zygote gives rise to four biflagellate gone cells released separately (Nozaki, 1989b). *Pyrobotrys* is cosmopolitan and found in fresh water rich in organic matter. This genus contains 12 described species. Some species require anaerobic conditions for growth in culture (Nozaki, 1986). *Pyrobotrys* also appears in the literature under the names *Uva* Playfair (1914) or *Chlamydobotrys* Korshikov, (1924). Silva (1972), however, resolved this nomenclatural confusion and used the name *Pyrobotrys* Arnoldi (1916).

P. casinoensis (Fig. 43) and *P. stellata* have been found in the United States (e.g., KY, VA) (Smith, 1950; Dillard, 1989). Planozygotes of *P. casinoensis* has been described as *Chlorobrachis gracilima* (e.g., Smith, 1950). Morphological description of *Chlorobrachis gracilima* by Smith (1950) seems to be incorrect; Figure 31 B is not "side view" of the alga (see Smith, 1950), but it is the immature stage of the alga (see Korshikov, 1925; Nozaki, 1986).

Spondylomorum Ehrenberg (Fig. 40)

Colonies are star- or mulberry-shaped, with eight or 16 cells arranged in four-celled tiers, without encompassing gelatinous matrix. Cells are walled and ovoid or pear shaped with a long posterior tail, each with four equal flagella, two contractile vacuoles at the base of the flagella, and a massive cup-shaped chloroplast without pyrenoids. Sexual reproduction is unknown. *Spondylomorum* is distinguished from other spondylomoraceaen genera by its quadriflagellate vegetative cells. This genus contains two described species and *S. quaternarium* (Fig. 40) has been recorded from various localities of the world (Huber-Pestalozzi, 1961). However, no culture studies have been carried out, and its existence has been questioned (Pringsheim, 1960).

S. quaternarium has been found in the United States (Smith, 1950; Dillard, 1989) and Ontario, Canada (Duthie and Socha, 1976).

VI. GUIDE TO LITERATURE FOR SPECIES IDENTIFICATION

Species identification of fixed materials is possible based on the taxonomic concepts of Smith (1950), Huber-Pestalozzi (1961), Ettl (1983), and Dillard (1989). Correct identification of species within the Volvocales often needs clonal cultured materials. Morphological characteristics of cultured materials observed under the controlled laboratory conditions provide stable and objective species diagnoses such as in *Eudorina* (Goldstein, 1964), *Pyrobotrys* (Nozaki, 1986), *Carteria* (Nozaki et al., 1994a), and *Chlorogonium* (Nozaki et al., 1998b). DNA sequence data, such as the *rbc*L gene, obtained from clonal cultures, may help confirm identification of species/genus within the volvocalean algae (see Nozaki et al., 1997a, 1998a, b). For general methods for clonal cultures of microalgae, see Stein (1975) and Starr and Zeikus (1993).

The following is a list of key recent references, each of which contains citations to older literature and those from other continents:

1. *Pedinomonas*—Iyengar and Desikachary (1981), Ettl (1983)
2. *Polyblepharides*—Iyengar and Desikachary (1981), Ettl (1983)
3. *Mesostigma*—Iyengar and Desikachary (1981), Ettl (1983), Dillard (1989)
4. *Scourfieldia*—Iyengar and Desikachary (1981), Ettl (1983), Dillard (1989)
5. *Nephroselmis*-Iyengar and Desikachary (1981), Ettl (1983)
6. *Spermatozopsis*—Iyengar and Desikachary (1981), Ettl (1983), Preisig and Melkonian (1984)

7. *Polytomella*—Iyengar and Desikachary (1981), Ettl (1983)
8. *Tetraselmis*—Ettl (1983)
9. *Scherffelia*—Iyengar and Desikachary (1981), Ettl (1983)
10. *Haematococcus*—Iyengar and Desikachary (1981), Ettl (1983), Thompson and Wujek (1989)
11. *Stephanosphaera*—Iyengar and Desikachary (1981), Ettl (1983)
12. *Chlamydomonas*—Iyengar and Desikachary (1981), Ettl (1983), Dillard (1989)
13. *Chloromonas*—Iyengar and Desikachary (1981), Ettl (1983), Pröschold *et al.* (2001)
14. *Carteria*—Iyengar and Desikachary (1981), Ettl (1983), Dillard (1989), Nozaki *et al.* (1994a)
15. *Polytoma*—Iyengar and Desikachary (1981), Ettl (1983)
16. *Chlorogonium*—Ettl (1983), Dillard (1989), Nozaki *et al.* (1998b)
17. *Gloeomonas*—Iyengar and Desikachary (1981), Ettl (1983)
18. *Lobomonas*—Iyengar and Desikachary (1981), Ettl (1983)
19. *Vitreochlamys*—Iyengar and Desikachary (1981), Ettl (1983), Dillard (1989), Nakazawa *et al.* (2001)
20. *Pedinopera*—Iyengar and Desikachary (1981), Ettl (1983)
21. *Pteromonas*—Iyengar and Desikachary (1981), Ettl (1983), Dillard (1989)
22. *Phacotus*—Iyengar and Desikachary (1981), Ettl (1983), Dillard (1989)
23. *Wislouchiella*—Iyengar and Desikachary (1981), Ettl (1983), Dillard (1989)
24. *Cephalomonas*—Iyengar and Desikachary (1981), Ettl (1983), Dillard (1989)
25. *Thoracomonas*—Iyengar and Desikachary (1981), Ettl (1983), Dillard (1989)
26. *Granulochloris*—Iyengar and Desikachary (1981), Fott (1963), Ettl (1983)
27. *Coccomonas*—Iyengar and Desikachary (1981), Ettl (1983), Dillard (1989)
28. *Dysmorphococcus*—Iyengar and Desikachary (1981), Ettl (1983), Dillard (1989)
29. *Pandorina*—Iyengar and Desikachary (1981), Ettl (1983), Dillard (1989), Nozaki and Kuroiwa (1991)
30. *Volvulina*—Iyengar and Desikachary (1981), Ettl (1983), Nozaki and Kuroiwa (1990)
31. *Yamagishiella*—Rayburn and Starr (1974), Nozaki and Kuoiwa (1992)
32. *Eudorina*—Iyengar and Desikachary (1981), Goldstein (1964), Ettl (1983)
33. *Platydorina*—Harris and Starr (1964)
34. *Pleodorina*—Iyengar and Desikachary (1981), Nozaki *et al.* (1989)
35. *Volvox*—Iyengar and Desikachary (1981), Smith (1944), Ettl (1983), Starr (1970), Dillard (1989)
36. *Gonium*—Iyengar and Desikachary (1981), Ettl (1983), Dillard (1989), Nozaki (1989)
37. *Astrephomene*— Nozaki (1983)
38. *Basichlamys*—Stein (1959), Iyengar and Desikachary (1981)
39. *Tetrabaena*—Stein (1959), Nozaki and Ohtani (1982)
40. *Spondylomorum*—Iyengar and Desikachary (1981), Ettl (1983)
41. *Chlorcorona*—Iyengar and Desikachary (1981), Ettl (1983)
42. *Pascherina*—Iyengar and Desikachary (1981), Ettl (1983), Dillard (1989)
43. *Pyrobotrys*—Iyengar and Desikachary (1981), Ettl (1983), Nozaki (1986)

ACKNOWLEDGMENTS

I would like to thank Dr. Robert Sheath who kindly added many non-U.S. references to the distributions of the flagellated green algal genera and gave me valuable comments on the manuscript.

LITERATURE CITED

Arnoldi, V. M. 1916. Ein neuer Organismus aus der Volvokazeenordnung: *Pyrobotris incurva. Rec. Act. Sc. Prof. Clement Timiryazeva* 51–58, 1 pl., Moscow (in Russian with German summary).

Batko, A. 1970. A new *Dangeardia* which invades motile Chlamydomonadaceous monads. *Acta Mycologica* 6:407–435.

Bhattacharya, D., Weber, K., An, S. S., Berning-Koch, W. 1998. Actin phylogeny identifies *Mesostigma viride* as a flagellate ancestor of the land plants. *Journal of Molecular Evolution* 47:544–550.

Bold, H. C., Starr, R. C. 1953. A new member of the Phacotaceae. *Bulletin Torrey Botany Club* 80:178–186.

Bold, H. C., Wynne, M. J. 1985. *Introduction to the Algae.* 2nd ed. Prentice-Hall. Inc., Englewood Cliffs, New Jersey, xiv + 720 pp.

Bourrelly, P., Manguin, E. 1952. *Algues d'Eau Douce de la Guadeloupe et Dépendances.* Sociéte d'Édition d'Enseignment Supérieur, Paris, 282 p.

Brooks, A. E. 1966. The sexual cycle and intercrossing in the genus *Astrephomene. Journal of Protozoology* 13:367–375.

Brooks, A. E. 1972. The physiology of *Astrephomene gubernaculifera. Journal of Protozoology* 19:195–199.

Buchheim, M. A., Chapman, R. L. 1992. Phylogeny of *Carteria* (Chlorophyceae) inferred from molecular and organismal data. *Journal of Phycology* 28:362–374.

Buchheim, M. A., McAuley, M. A., Zimmer, E. A., Theriot, E. C., Chapman, R. L. 1994. Multiple origins of colonial green flagellates from unicells: Evidence from molecular and organismal characters. Molecular Phylogenetic Evolution 3:322–343.

Buchheim, M. A., Lemieux, C., Otis, C., Gutell, R. R., Chapman, R. L., Turmel, M. 1996. Phylogeny of the Chlamydomonadales (Chlorophyceae): A comparison of ribosomal RNA gene sequences from the nucleus and the chloroplast. Molecular Phylogenetic Evolution 5:391–402.

Buchheim, M. A., Buchheim, J. A., Chapman, R. L. 1997. Phylogeny of *Chloromonas*: A study of 18S ribosomal RNA gene sequences. Journal of Phycology 33:286–293.

Buchheim, M. A., Buchheim, J. A., Carlson, T., Kugrens, P. 2002. Phylogeny of *Lobocharacium* (Chlorophyceae) and allies: A comparative study of 18S and 26S rDNA data. Journal of Phycology 38:276–383.

Carefoot, J. R. 1966. Sexual reproduction and intercrosing in *Volvulina steinii*. Journal of Phycology 2:150–156.

Coleman, A. W. 1959. Sexual isolation in *Pandorina morum*. Journal of Protozoology 6:249–264.

Coleman, A. W. 1977. Sexual and genetic isolation in the cosmopolitan algal species *Pandorina morum*. American Journal of Botany 64:361–368.

Coleman, A. W. 1996. Are the impacts of events in the Earth's history discernable in the current distributions of freshwater algae? *in*: Kristiansen, J., Ed., Biogeography of freshwater algae. Developments in Hydrobiology Vol. 118. Kluwer Academic Publishers, Dordrecht, pp. 137–142.

Coleman, A. W. 1999. Phylogenetic analysis of "Volvocaceae" for comparative genetic studies. Proceedings of the National Academy of Sciences of the United States of America 96:13892–13897.

Coleman, A. W., Suarez, A., Goff, L. 1994. Molecular delineation of species and syngens in the volvocalean green algae (Chlorophyta). Journal of Phycology 30:80–90.

Conrad, W. 1926. Recherches sur les flagellates de nos eaux saumâtres. II Chrysomonadines. Archiv für Protistenkunde 56:167–231.

Cooke, M. C. 1883. New American fungi. Grevillea 12:22–23.

Dillard, G. E. 1989. Freshwater Algae of the Southeastern United States. Part. 1. Chlorophyceae: Volvocales, Tetrasporales and Chlorococcales, *in*: Kies, L., Giessen, R. S., Eds., Bibliotheca Phycologia, Bd. 81. J. Cramer, Berlin, pp. 202, 37 pls.

Domozych, D. S. 1989. The endomembrane system and mechanism of membrane flow in the green algal flagellate *Gloeomonas*. I.) An ultrastructural analysis. Protoplasma 149:95–107.

Domozych, D. S., Nimmons, T. T. 1992. The contractile vacuole as an endocytic organelle of the chlamydomonad flagellate *Gloeomonas kupfferi* (Volvocales, Chlorophyta). Journal of Phycology 28:809–816.

Duthie, H. C., Socha, R. 1976. A checklist of the freshwater algae of Ontario, exclusive of the Great Lakes. Naturaliste Canadien (Que.) 103:83–109.

Entwisle, T. J., Sonneman, J. A., Lewis, S. H. 1998. *Freshwater Algae in Australia*. Sainty and Associates, Potts Point, vi–242 pp.

Ettl, H. 1980. Die taxonomische Abgrenzung der Gattung *Chlorogonium* Ehrenberg (Chlamydomonadales, Chlorophyta). Nova Hedwigia 33:709–722.

Ettl, H. 1983. Chlorophyta I. Phytomonadia, *in*: Ettl, H., Gerloff, J., Mollenhauer, D., Eds., *Süßwasserflora von Mitteleuropa*, Bd. 9. Gustav Fischer Verlag, Stuttgart, pp. xiv–807.

Fabry, S., Köhler, A., Coleman, A. W. 1999. Intraspecies analysis: comparison of ITS sequence data and gene intron sequence data with breeding data for a worldwide collection of *Gonium pectorale*. Journal of Molecular Biology 48:94–101.

Fott, B. 1949. Corone, a new genus of colonial Volvocales. Věstn. Král. České Společn. Nauk, Tr. Mat.-Přír. 2:1–9.

Fott, B. 1963. *Granulochloris spinifera* sp. nova. Phycologia 3:101–103.

Fott, B. 1967. Taxonomische Übertragungen und Namensänderungen unter unter den Algen. II. Chlorophyceae, Chrysophyceae und Xanthophyceae. Preslia (Prague) 39:352–364.

Fourreau, J. 1868. Catalogue des plantes qui croissent spontanément le long du cours du Rhone. Annals Society of Linnenian Lyon (Series 2) 16:301–404.

Friedl, T. 1997. The evolution of the green algae, *in*: Bhattacharya, D., Ed., Origin of Algae and their Plastids. Springer-Verlag, Wien, pp. 87–101.

Fulton, A. B. 1978. Colonial development in *Pandorina morum*. I. Structure and composition of the extracellular matrix. Dev. Biol. 64:224–235.

Goldstein, M. 1964. Speciation and mating behavior in *Eudorina*. Journal of Protozoology 1:317–334.

Haberyan, K. A., Umaña, G., Collado, V. C., Horn, S. P. 1995. Observations on the plankton of some Costa Rican lakes. Hydrobiologia 312:75–85.

Harris, D. O., Starr, R. C. 1969. Life history and physiology of reproduction of *Platydorina caudata* Kofoid. Archiv für Protistenkunde 111:138–155.

Harris, E. H. 1989. The Chlamydomonas Sourcebook. Academic Press, San Diego, xiv+780 p.

Hepperle, D. 1997. *Phacotus lenticularis* (Chlamydomonadales, Phacotaceae) zoospores require external supersaturation of calcium carbonate for calcification in culture. Journal of Phycology 33: 415–424.

Hepperle, D., Krientiz, L. 1996. The extracellular calcification of zoospores of *Phacotus lenticularis* (Chlorophyta, Chlamydomonadales). European Journal of Phycology 31:11–21.

Hepperle, D., Nozaki, H., Hohenberger, S., Huss, V. A. R., Morita, E., Krienitz, L. 1998. Phylogenetic position of the Phacotaceae within the Chlamydophyceae as revealed by analysis of 18S rDNA- and *rbc*L-sequences. Journal of Molecular Evolution 47:420–430.

Hoham, R. W. 1974a. New findings in the life history of the snow alga, *Chlainomonas rubra* (Stein et Brooke) comb. nov. (Chlorophyta, Volvocales). Syesis 7:239–247.

Hoham, R. W. 1974b. *Chlainomonas kolii* (Hardy et Curl) comb. nov. (Chlorophyta, Volvocales), revision of the snow alga, *Trachelomonas kolii* Hardy et Curl (Euglenophyta, Euglenales). Journal of Phycology 10:392–396.

Hoham, R. W. 1980. Unicellular chlorophytes (snow algae), *in*: Cox, E. R., Ed., Phytoflagellates. Developments in Marine Biology Vol. 2. Elsevier/North-Holland, New York, pp. 61–84.

Hoham, R. W., Roemer, S. C., Mullet, J. E. 1979. The life history and ecology of the snow alga *Chloromonas brevispina* comb. nov. (Chlorophyta, Volvocales). Phycologia 18:55–70.

Hoham, R. W., Mullet, J. E., Roemer, S. C. 1983. The life history and ecology of the snow alga *Chloromonas polyptera* comb. nov. (Chlorophyta, Volvocales). Canadian Journal of Botany 61:2416–2429.

Hoops, H. J. 1984. Somatic cell flagellar apparatuses in two species of *Volvox* (Chlorophyceae). Journal of Phycology 20:20–27.

Hoops, H. J., Floyd, G. L. 1982. Ultrastructure and taxonomic position of the rare volvocalean alga, *Chlorcorona bohemica*. Journal of Phycology 18:462–466.

Huber-Pestalozzi, G. 1961. Das Phytoplankton des Süsswassers. Teil 5. Chlorophyceae (Grünalgen) Ordnung: Volvocales, *in*: Thienemann, A., Ed., *Die Binnengewässer*, Vol. 16. E. Schweizerbart'sche Verlagsbuchhandlung (Nägele u. Obermiller), Stuttgart, 744 p.

Iyengar, M. O. P., Desikachary, T. V. 1981. Volvocales. Indian Council of Agricultural Research, New Delhi, 524 p.

Karol, K. G., McCourt, R. M., Cimino, M. T., Delwiche, C. F. 2001. The closest living relatives of land plants. Science 294:2351–2353.

Kirk, D. L. 1998. *Volvox*: Molecular Genetic Origins of Multicellularity and Cellular Differentiation. Cambridge Univ. Press, Cambridge, 381 p.

Kirk, D. L., Birchem, R., King, N. 1986. The extracellular matrix of *Volvox*: a comparative study and proposed system of nomenclature. Journal of Cell Science 80:207–231.

Kofoid, C. A. 1899. On *Platydorina*, a new genus of the family Volvocidae, from the plankton of the Illinois River. Bulletin of the Illinois State Laboratory of Natural History 5:419–440.

Korshikov, A. A. 1924. Zur Morphologie und Systematik der Volvocales. Arch. Russ. Protistol. 3: 45–56.

Korshikov, A. A. 1925. Beiträge zur Morphologie und Systematik der Volvocales. I. *Russ*. Archiv für Protistenkunde 4:153–197.

Korshikov, A. A. 1928. On two new Spondylomoraceae: *Pascheriella tetras* n. gen. et sp., and *Chlamydobotrys squarrosa* n. sp. Archiv für Protistenkunde 61:223–238.

Larson, A., Kirk, M. M., Kirk, D. L. 1992. Molecular phylogeny of the volvocine flagellates. Molecular Biology Evolution 9:85–105.

Lefébure, P., Cheneviére, E. 1938. Description et iconographie des Diatomees ou nouvelles. Bulletin de la Société Française de Microbiologie 7:8–12.

Lembi, C. A. 1975. The fine structure of the flagellar apparatus of *Carteria*. Journal of Phycology 11:1–9.

Lembi, C. A. 1980. Unicellular chlorophytes, *in*: Cox, E. R., Ed., Phytoflagellates. Developments in Marine Biology. Vol. 2. Elsevier/North-Holland, New York, pp. 5–59.

Lemieux, C., Otis, C., Turmel, M. 2000. Ancestral chloroplast genome in *Mesostigma viride* reveals an early branch of green plant evolution. Nature 403: 649–652.

Ling, H. U. 1996. Snow algae in the Windmill Island region, Antarctica, *in*: Kristiansen, J., Ed., Biogeography of freshwater algae. Developments in Hydrobiology Vol. 118. Kluwer Academic Publishers, Dordrecht, pp. 99–106.

Linnaeus, C. 1758. *Systema Naturae*. Regnum animale, ed. 10, Stockholm, 824 pp.

Mattox, K. R., Stewart, K. D. 1984. Classification of the green algae: a concept based on comparative cytology, *in*: Irvine, D. E. G., John, D. M., Eds., Systematics of the green algae. The Systemtics Association Special Vol. 27. Academic Press, London, pp. 29–72.

Moestrup, Ø. 1991. Further studies of presumedly primitive green algae, including description of Pedinophyceae class nov. and *Resultor* gen. nov. Journal of Phycology 27:119–133.

Moewus, F. 1959. Stimulation of mitotic activity by benzidine and kinetin in *Polytoma uvella*. Transactions American of Microscopical Society 78:295–304.

Morita, E., Abe, T., Tsuzuki, M., Fujiwara, S., Sato, N. Hirata, A., Sonoike, K., Nozaki, H. 1998. Presence of the CO_2-concentrating mechanism in some species of the pyrenoid-less algal genus *Chloromonas* (Volvocales, Chlorophyta). Planta 204:269–276.

Morita, E., Abe, T., Tsuzuki, M., Fujiwara, S., Sato, N., Hirata, A., Sonoike, K., Nozaki, H. 1999. Role of pyrenoids in the CO_2-concentrating mechanism: Comparative morphology, physiology and molecular phylogenetic analysis of closely related strains of *Chlamydomonas* and *Chloromonas* (Volvocales). Planta 208:365–372.

Munawar, M., Munawar, I. F. 1981. A general comparison of the taxonomic composition and size analyses of the phytoplankton of the North American Great Lakes. Internationale Vereinigung für Theoretische und Angewandte Limnologie Verhandlungen 21:1695–1716.

Nakayama, T., Watanabe, S., Inouye, I. 1996a. Phylogeny of wall-less green flagellates inferred from 18SrDNA sequence data. Phycology Research 44:151–161.

Nakayama, T., Watanabe, S., Mitsui, K., Uchida, H., Inouye, I. 1996b. The phylogenetic relationship between the Chlamydomonadales and Chlorococcales inferred from 18SrDNA sequence data. Phycology Research 44:47–55.

Nakazawa, A., Krienitz, L., Nozaki, H. 2001. Taxonomy of the unicellular green algal genus *Vitreochlamys* (Volvocales), based on comparative morphology of cultured material. European Journal of Phycology 36:113–128.

Norris, R. E. 1980. Prasinophytes, *in*: Cox, E. R., Ed., Phytoflagellates. Developments in Marine Biology Vol. 2. Elsevier/North-Holland, New York, pp. 85–145.

Nozaki, H. 1981. The life history of Japanese *Pandorina unicocca* (Chlorophyta, Volvocales). Japanese Journal of Botany 56: 65–72.

Nozaki, H. 1982. Morphology and reproduction of Japanese *Volvulina steinii* (Chlorophyta, Volvocales). Japanese Journal of Botany 57:105–113.

Nozaki, H. 1983. Morphology and taxonomy of two species of *Astrephomene* (Chlorophyta, Volvocales) in Japan. Japanese Journal of Botany 58:345–352.

Nozaki, H. 1986. A taxonomic study of *Pyrobotrys* (Volvocales, Chlorophyta) in anaerobic pure culture. Phycologia 25: 455–468.

Nozaki, H. 1988. Morphology, sexual reproduction and taxonomy of *Volvox carteri* f. *kawasakiensis* f. nov. (Chlorophyta) from Japan. Phycologia 27:209–220.

Nozaki, H. 1989a. Morphological variation and reproduction in *Gonium viridistellatum* (Volvocales, Chlorophyta). Phycologia 28:77–88.

Nozaki, H. 1989b. Morphology and zygote germination in heterothallic strains of *Pyrobotrys casinoensis* (Volvocales, Chlorophyta) from Japan. Phycologia 28:89–95.

Nozaki, H. 1994. Aplanogamous sexual reproduction in *Carteria eugametos* (Volvocales, Chlorophyta). Euroepan Journal of Phycology 29:135–139.

Nozaki, H. 1995. *Eudorina elegans*. *Pandorina morum*. *Pyrobotrys casinoensis*, *in*: Kojima, S., Sudo, R., Chihara M., Eds., Kankyou Biseibutsu Zukan (Illustrated Atlas of Environmental Microorganisms). Koudan-Sha Scientific, Tokyo, pp. 413–414, 418–420 (in Japanese).

Nozaki, H., Ito, M. 1994. Phylogenetic relationships within the colonial Volvocales (Chlorophyta) inferred from cladistic analysis based on morphological data. Journal of Phycology 30:353–365.

Nozaki, H., Krienitz, L. 2001. Morphology and phylogeny of *Eudorina minodii* (Chodat) Nozaki et Krienitz, comb. nov. (Volvocales, Chlorophyta) from Germany. European Journal of Phycology 36:23–28.

Nozaki, H., Kuroiwa, T. 1990. *Volvulina compacta* sp. nov. (Volvocaceae, Chlorophyta) from Nepal. Phycologia 29:410–417.

Nozaki, H., Kuroiwa, T. 1991. *Pandorina colemaniae* sp. nov. (Volvocaceae, Chlorophyta) from Japan. Phycologia 30: 449–457.

Nozaki, H., Kuroiwa, T. 1992. Ultrastructure of the extracellular matrix and taxonomy of *Eudorina*, *Pleodorina* and *Yamagishiella* gen. nov. (Volvocaceae, Chlorophyta). Phycologia 31:529–541.

Nozaki, H., Ohtani, S. 1992. *Gonium sociale* (Volvocales, Chlorophyta) from Antarctica. Japanese Journal of Phycology 40:267–271.

Nozaki, H., Kuroiwa, H., Mita, T., Kuroiwa, T. 1989. *Pleodorina japonica* sp. nov. (Volvocales, Chlorophyta) with bacteria-like endosymbionts. Phycologia 28:252–267.

Nozaki, H., Aizawa, K., Watanabe, M. M. 1994a. A taxonomic

study of four species of *Carteria* (Volvocales, Chlorophyta) with cruciate anterior papillae, based on cultured material. Phycologia 33:239–247.

Nozaki, H., Kuroiwa, H., Kuroiwa, T. 1994b. Light and electron microscopic characterization of two types of pyrenoids in *Gonium* (Goniaceae, Chlorophyta). Journal of Phycology 30:279–290.

Nozaki, H., Ito, M., Sano, R., Uchida, H., Watanabe, M. M., Kuroiwa, T. 1995a. Phylogenetic relationships within the colonial Volvocales (Chlorophyta) inferred from *rbc*L gene sequence data. Journal of Phycology 31:970–979.

Nozaki, H., Watanabe, M. M., Aizawa, K. 1995b. Morphology and paedogamous sexual reproduction in *Chlorogonium capillatum* (Volvocales, Chlorophyta). Journal of Phycology 31:653–661.

Nozaki, H., Aizawa, K., Watanabe, M. M. 1996a. Re-examination of two NIES strains labeled *Chlorogonium metamorphum* (Volvocales, Chlorophyta) from Japan. Nova Hedwigia, Beiheft 112:483–490.

Nozaki, H., Ito, M. Watanabe, M. M., Kuroiwa, T. 1996b. Ultrastructure of the vegetative colonies and systematic position of *Basichlamys* (Volvocales, Chlorophyta). European Journal of Phycology 31:67–72.

Nozaki, H., Ito, M., Sano, R., Uchida, H., Watanabe, M. M., Takahashi, H., Kuroiwa, T. 1997a. Phylogenetic analysis of *Yamagishiella* and *Platydorina* (Volvocaceae, Chlorophyta) based on *rbc*L gene sequences. Journal of Phycology 33:272–278.

Nozaki, H., Ito, M., Uchida, H., Watanabe, M. M., Kuroiwa, T. 1997b. Phylogenetic analysis of *Eudorina* species (Volvocaceae, Chlorophyta) based on *rbc*L gene sequences. Journal of Phycology 33:859–863.

Nozaki, H., Ito, M., Watanabe, M. M., Takano, H., Kuroiwa, T. 1997c. Phylogenetic analysis of morphological species of *Carteria* (Volvocales, Chlorophyta) based on *rbc*L gene sequences. Journal of Phycology 33:864–867.

Nozaki, H., Song, L.-R., Liu, Y.-D., Hiroki, M., Watanabe, M. M. 1998a. Taxonomic re-examination of a Chinese strain labeled "*Eudorina* sp." (Volvocaceae, Chlorophyta) in the Culture collection of freshwater algae at the Institute of Hydrobiology, Chinese Academy of Science, based on morphological and DNA sequence data. Phycological Research 46 (Suppl.):63–70.

Nozaki, H., Ohta, N., Morita, E., Watanabe, M. M. 1998b. Toward a natural system of species in *Chlorogonium* (Volvocales, Chlorophyta): A combined analysis of morphological and *rbc*L gene sequence data. Journal of Phycology 34:1024–1037.

Nozaki, H., Ohta, N., Takano, H., Watanabe, M. M. 1999. Reexamination of phylogenetic relationships within the colonial Volvocales (Chlorophyta): An analysis of *atp*B and *rbc*L gene sequences. Journal of Phycology 35:104–112.

Nozaki, H., Misawa, K., Kajita, T., Kato, M., Nohara, S., Watanabe, M. M. 2000. Origin and evolution of the colonial Volvocales (Chlorophyceae) as inferred from multiple, chloroplast gene sequences. Molecular Phylogenetics and Evolution 17:256–258.

Nozaki, H., Onishi, K., Morita, E. 2002a. Differences in pyrenoid morphology are correlated with differences in the *rbc*L genes of members of the *Chloromonas* lineage (Volvocales, Chlorophyceae). Journal of Molecular Evolution in press.

Nozaki, H., Takahara, M., Nakazawa, A., Kita, Y., Yamada, T., Takano, H., Kawano, S., Kato, M. 2002b. Evolution of *rbc*L group IA introns and intron open reading frames within the colonial Volvocales (Chlorophyceae). Molecular Phylogenetics and Evolution in press.

Ortega, M. M. 1984. *Catálogo de Algas Continentales Recientes de México*. Universidad Nacional Autónoma de México, México, 566 p.

Pascher, A. 1927. *Die Süßwasser-Flora Deutschlands, Österreichs und der Schweiz*. Heft 4: Volvocales=Phytomonadinae. Flagellatae IV=Chlorophyceae I. Gustav Fischer, Jena, 506 pp.

Playfair, G. I. 1914. Contributions to a knowledge of the biology of the Richmond River. Proceedings of the Linneian Society New South Wales 39:93–151.

Pocock, M. A. 1954. Two multicellular motile green algae, *Volvulina* Playfair and *Astrephomene*, a new genus. Transactions of the Royal Society of South Africa 34:103–127.

Pocock, M. A. 1955. Studies in the North American Volvocales. I. The genus *Gonium*. Madroño 13:49–80.

Pocock, M. A. 1960. *Haematococcus* in South Africa. Transactions of the Royal Society of South Africa 35:5–55.

Prescott, G. W. 1942. The fresh-water algae of Southern United States. II. The Algae of Louisiana, with descriptions of some new forms and notes on distribution. Transaction of American Microscopical Society 61:109–119.

Prescott, G. W. 1955. Algae of the Panama Canal and its tributaries. I. Flagellated organisms. Ohio Journal of Science 55:99–121.

Preisig, H. R., Melkonian, M. 1984. A light and electron microscopical study of the green flagellate *Spermatozopsis similis* sp. nov. Plant Systems Evolution 146:57–74.

Pringsheim, E. G. 1960. Zur Systematik und Physiologie der Spondylomoraceen. Österreichische Botanische Zeitschrift 107:425–438.

Pringsheim, E. G., Wiessner, W. 1960. Photoassimilation of acetate by green organisms. Nature 188:919.

Pröschold, T., Marin, B., Schlösser, U. G., Melkonian, M. 2001. Molecular phylogeny and taxonomic revision of *Chlamydomonas* (Chlorophyta). I. Emendation of *Chlamydomonas* Ehrenberg and *Chloromonas* Gobi, and description of *Oogamochlamys* gen. nov. and *Lobochlamys* gen. nov. Protist 152: 265–300.

Rayburn, W. R., Starr, R. C. 1974. Morphology and nutrition of *Pandorina unicocca* sp. nov. Journal of Phycology 10:42–49.

Rumpf, R. R., Vernon, D., Schreiber, D., Birky C. W. Jr. 1996. Evolutionary consequences of the loss of photosynthesis in Chlamydomonadaceae:phylogenetic analysis of *Rrn* 18 (18S rDNA) in 13 *Polytoma* strains (Chlorophyta). Journal of Phycology 32:119–126.

Schagerl, M., Angeler, D. 1998. The distribution of the xanthophyll loroxanthin and its systematic significance in the colonial Volvocales (Chlorophyta). Phycologia 37:79–83.

Sheath, R. G., Steinman, A. D. 1982. A checklist of freshwater algae of the Northwest Territories, Canada. Canadian Journal of Botany 60:1964–1997.

Silva, P. C. 1959. Remarks on algal nomenclature II. *Taxon* 8:62–63.

Silva, P. C. 1972. Remarks on algal nomenclature V. *Taxon* 21: 199–212.

Silva, P. C., Papenfuss, G. F. 1953. A Systematic Study of the Algae of Sewage Oxidation Ponds. State Water Pollution Control Board, Sacramento, Publ. 7, 35 p.

Smith, G. M. 1944. A comparative study of the species of *Volvox*. Transactions of the American Microscopical Society 63:265–310.

Smith, G. M. 1950. The Fresh-water Algae of the United States. 2nd ed. McGraw-Hill Book Company, New York. vii + 719 p.

Smith, G. M. 1955. Cryptogamic Botany. Vol. 1. McGraw-Hill, New York, ix–546 pp.

Starr, R. C. 1962. A new species of *Volvulina* Playfair. Archiv für Mikrobiologie 42:130–137.

Starr, R. C. 1970. *Volvox pocockiae*, a new species with dwarf males. Journal of Phycology 6:234–239.

Starr, R. C., Zeikus, J. A. 1993. UTEX-The Culture Collection of Algae at the University of Texas at Austin. Journal of Phycology (Suppl.) 29 (2):1–106.

Stein, J. R. 1958a. A morphological study of *Astrephomene guberna-

culifera and *Volvulina steinii*. American Journal of Botany 45:388–397.

Stein, J. R. 1958b. A morphologic and genetic study of *Gonium pectorale*. American Journal of Botany 45:664–672.

Stein, J. R. 1959. The four-celled species of *Gonium*. American Journal of Botany 46:366–371.

Stein, J. R. 1965. Sexual populations of *Gonium pectorale* (Volvocales). American Journal of Botany 52:379–388.

Stein, J. R. 1975. [Ed.] *Handbook of Phycological Methods*. Culture Methods and Growth Measurements. Cambridge Univ. Press, Cambridge, xii–448 pp.

Stein, J. R., Borden, C. A. 1979. Checklist of freshwater algae of British Columbia. *Syesis* 12:3–39

Steinkötter, J., Bhattacharya, D., Semmerlroth, I., Bibeau, C., Melkonian, M. 1994. Prasinophytes form independent lineages within the Chlorophyta: evidence from ribosomal RNA sequence comparisons. Journal of Phycology 30:340–345.

Suda, S., Watanabe, M. M., Inouye, I. 1989. Evidence for sexual reproduction in the primitive green alga *Nephroselmis olivacea* (Prasinophyceae). Journal of Phycology 25:596–600.

Thompson, R. H. 1954. Studies in the Volvocales. I. Sexual reproduction of *Pandorina charkowiensis* and observations on *Volvulina steinii*. American Journal of Botany 41:142–145.

Thompson, R. H., Wujek, D. E. 1989. *Haematococcus carocellus* sp. nov. (Haematococcaceae, Chlorophyta) from the United States. Phycologia 28:268–270.

van den Hoek, C., Mann, D. G., Jahns, H. M. 1995. Algae: An Introduction to Phycology. Cambridge Univ. Press, Cambridge, 627 pp.

van Leeuwenhoek, A. 1700. Concerning the worms in sheeps, livers, gnats and animalicula in the excrement of frogs. Philosophy Transaction of the Royal Society (London) 22:509–518.

Whitford, L. A., Schumacher, G. J. 1984. A Manual of Fresh-water Algae. Rev. Ed. Sparks Press, Raleigh, N.C.

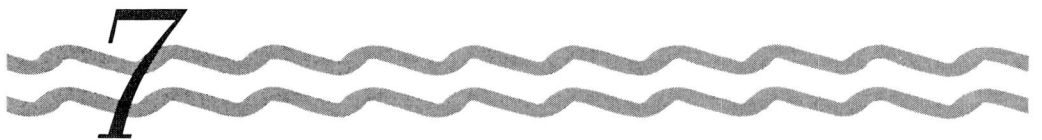

NONMOTILE COCCOID AND COLONIAL GREEN ALGAE

L. Elliot Shubert

Department of Botany
The Natural History Museum
London SW7 5BD
United Kingdom

I. Introduction
II. Diversity and Morphology
III. Ecology and Distribution
IV. Collection and Preparation for Identification
V. Key and Descriptions of Genera
  A. Key
  B. Descriptions of Genera
  C. Addendum
VI. Guide to Literature for Species Identification
Literature Cited

The systematic investigation of the green algae has so far been largely a descriptive process; this applies not only to the nineteenth century when only the light microscope was in use but also to the twentieth century with the electron microscope. (Round, 1984, p. 9)

I. INTRODUCTION

The nonmotile coccoid and colonial green algae belong to the division Chlorophyta. The Chlorophyta include a diversity of taxa (morphologically and ecologically), ranging from unicellular and freshwater taxa (e.g., *Chlamydomonas*; Chap. 6) to multicellular and marine taxa (e.g., *Ulva*). The nonmotile coccoid and colonial microscopic algae (hereafter referred to as "nonmotile greens") represent a large subgroup currently distributed over five classes encompassing nine different orders. They are commonly called little green balls, because they often appear as aggregations or clumps of green cells. At first glance a novice might think that these organisms all look the same, but using careful and detailed observation reveals that there are stable and discrete morphological characters that separate many of the genera from each other (Prescott, 1978; Komárek and Fott, 1983; Dillard, 1999). Nevertheless, numerous genera have been reassigned after detailed analysis with transmission electron microscopy (TEM) and/or molecular analysis (Watanabe and Floyd, 1989; Huss et al., 1999). Species identification is even more difficult because many species exhibit morphological variability or phenotypic plasticity, as well as motile stages in part of their life cycle. Although the taxa are nonmotile in the vegetative state (actively growing), some species produce flagellated stages or resting stages during part of their life cycle (Sect. II). Thus, for a definitive determination of species it may be necessary to culture the organism under controlled environmental conditions, to examine it at the ultrastructural level with TEM and/or scanning electron microscopy (SEM), or to use molecular methods.

The nonmotile greens include many ubiquitous species, which are found worldwide in a variety of habitats (see Sect. III). Whether distributed by air currents, birds, or the feet of animals, they are effective colonizers of denuded soils (e.g., disturbed by human activities), new soils (e.g., volcanic lava), and newly formed water bodies (e.g., refilled potholes and depressions formed by mechanical disturbance). Thus, they play an important role in primary and secondary successional processes.

There are approximately 200 recognized genera and thousands of described species in this group (Komárek and Fott, 1983; Ettl and Gärtner, 1995). Unfortunately, the proliferation of species names, such as in *Scenedesmus* (Hegewald and Silva, 1988), has created more confusion rather than a better understanding of the taxa. The problem arises because the classical taxonomic approach relies on differences in morphological characters of organisms collected from the field and the assumption is that these characters are stable. However, when these same organisms are isolated into axenic culture and grown under different environmental conditions (e.g., light, temperature, and nutrients) a range of morphological types can be expressed (Trainor, 1998). Phenotypic plasticity has been well documented for some taxa. The elucidation of the range of morphological types in taxa must be considered if we are to gain a better understanding of their systematics, phylogenetic relationships and eco-physiological roles.

Some of the algae are considered to be "green weeds" (e.g., *Chlorella*, *Chlorococcum Scenedesmus*, and *Desmodesmus*), because they can be easily cultured in the laboratory or can produce bloom conditions in the field. Indeed, the first report of an alga cultured from field material was that of *Chlorella* (Beijerinck, 1890).

II. DIVERSITY AND MORPHOLOGY

The vegetative cells of nonmotile greens are eukaryotic and either uni- or multinucleate (coenocytic). All are nonmotile in the vegetative state. When cell division occurs in coccoid and colonial algae, the progeny cells form their own walls. When the new cell or colony is released, the empty remains of the parent cell wall ("ghost cell") can be observed (e.g., *Desmodesmus* and *Pediastrum*). The dead remains of *Desmodesmus* and *Pediastrum* may be persistent for a considerable time in sediments, due to decay resistant compounds, such as algaenans and sporopollenin, and are still readily identifiable (Pickett-Heaps and Staehelin, 1975; Graham and Wilcox, 2000). In some cases these algae can withstand degradation for millions of years, and are associated with mudstones and oil deposits (Fleming, 1989; Gelin et al., 1997). In some taxa the remains of the parent wall become part of the progeny cells (e.g., *Dictyosphaerium*) and are distinctive enough to be used as a taxonomic character.

The vegetative cells of some taxa (uninucleate) are embedded in mucilage (homogenous or lamellated), and pseudoflagella (giving the appearance of flagella, but not functional) are often present and visible with the light microscope (LM). Other vegetative cells have a range of morphology, that is, unicells to colonies (uni- or multinucleated), and mucilage may be present, but there are no pseudoflagella. Some colonial forms have a well-developed mucilaginous sheath (Dillard, 1989).

Motile cells may be present in some genera (e.g., *Bracetococcus*, *Characium*, *Chlorococcum*, *Myrmecia*, and *Tetracystis*) and most are chlamydomonad in appearance (two whiplash flagella of nearly equal length, contractile vacuoles, basal bodies, persistent eye spot, and a cup-shaped chloroplast). The motile cells may function as zoospores (asexual) or gametes (sexual). If sexual reproduction occurs, it can be either isogamous (similar morphology) or anisogamous (dissimilar morphology). The nonmotile greens have a zygotic meiosis life cycle.

The nonmotile greens share characteristics common to all eukaryotic green algae, such as wall structure, pigments, storage products, and motility. They contain at least one plastid and are autotrophic. Some taxa are facultative heterotrophs. The dominant pigments are chlorophyll-*a* and -*b*, which give the cell a grass-green appearance. The light harvesting pigment complex is associated with proteins, which bind chlorophyll-*a* and -*b*, and accessory pigments (e.g., carotenoids). The accessory pigments may function to harvest light, but more typically provide protection from photo-oxidation and dispersal of excess energy. The major accessory pigment is lutein or a derivative, and β-carotene is always present. When a cell ages in the absence of essential nutrients, it may appear yellow to orange in color. *Chlorosarcinopsis* appears bright orange (presumably carotenoid production) when the culture is deficient in nitrogen (Starks and Shubert, 1979). There may be one to several chloroplasts per cell. The shape of the chloroplast is often used to distinguish among genera that have similar cell shapes. For example, the spherical unicells *Chlorococcum* and *Trebouxia* are distinguishable from one another because the former has a cup-shaped chloroplast and the latter has a star-shaped chloroplast (Dillard, 1989).

The storage product for members of this group is true starch, amylose, and amylopectin (α-1,4-linked polyglucans), and is found inside the chloroplasts. The starch can be detected with a drop of I_2KI, which turns dark blue–black. Staining for starch is a diagnostic tool for distinguishing green algae from morphologically similar algae that belong to other groups, but do not store starch. The starch (seen as whitish granules with the TEM) can often be observed surrounding the pyrenoid, a distinct spherical structure embedded in the chloroplast. There may be more than one pyrenoid or the pyrenoid is not always present (e.g., *Ankistrodesmus* and *Tetraedron*) or the pyrenoid is lacking (e.g., *Bracteococcus* and *Cerasterias*). It may be used as a taxonomic character to distinguish between some genera (e.g., *Palmodictyon* and *Tetraspora*; Dillard, 1989). Cells collected from nature may be full of food reserve, obscuring the plastid. This may make it difficult to determine plastid type, which is a problem if this character is necessary for identification to genus or species. Archibald and Bold (1970) solved this problem by culturing isolates in high levels of nitrogen Bold's Basal Medium (BBM). However, Blackwell *et al.* (1991) pointed out that "there are inherent dangers in assuming, *a priori*, that all members of a taxon will respond to particular growth conditions..." after they tried to grow *C. submarinum* in BBM and were unsuccessful. In the end, for an individual taxon, we have no data on possible modifications of plastid structure at low, intermediate, and high nitrogen levels. It is advisable to consider growth under standard conditions as a recommendation rather than an absolute requirement for determining a taxon. Nevertheless, culturing an organism may provide valuable information on its range of phenotypic plasticity (including the production of flagellated stages) and physiological tolerances under different environmental conditions.

Chlorococcum is considered a type genus for the chlorococcalean algal group (Bold and Wynne, 1978). Although this group of algae is nonmotile in the vegetative state, some taxa produce motile cells (planospores). Planospores may be asexual zoospores or sexual gametes. To form motile cells, the vegetative cell undergoes multiple cleavages of the protoplasm and the nucleus. The resulting division together with associated organelles becomes surrounded by a cytoplasmic membrane and a cell wall (sometimes they are "naked" because the wall is absent). Trainor (1978) described a number of possible outcomes when planospore production occurs, depending on the way in which they are released (e.g., singly through a pore or breakdown of a wall, or released as a group within a flexible membrane, the vesicle). A single planospore would be a unicell such as *Chlorococcum*, whereas a group of planospores would be unicells or a colony, such as *Hormotilopsis* and *Pediastrum*, respectively. If the planospores are retained within the parent cell wall and attach to each other, the organism would be colonial such as *Hydrodictyon*. These distinctions can be very helpful when observing live algae collected from the field or mixed cultures in the laboratory with respect to identification and understanding the reproductive process.

Alternatively, the method of formation of aplanospores (nonmotile cells) determines the morphological outcome (Trainor, 1978). Cells released singly result in a unicellular organism (e.g., *Ankistrodesmus*, *Chlorella*, and *Chlorococcum*), but if the progeny cells are retained and remain attached to each other, then a colony is produced (e.g., *Coelastrum* and *Scenedesmus*).

Chlorococcum is found in both aquatic and terrestrial habitats, and can produce both planospores and aplanospores in culture. Trainor (1978) suggested that sufficient nutrients and water might "trigger" the formation of planospores in *Chlorococcum*, whereas in older cultures or in a dry soil, aplanospores would develop. "Aplanospores have the ontogenetic potentiality for being flagellate and motile" (Bold and Wynne, 1978). Isogamous sexual reproduction occurs in *Chlorococcum* and the gametes are morphologically similar to zoospores, but function as sex cells.

In most representative taxa, the cells are surrounded by a cellulose cell wall. Some taxa may also have chitin or sporopollenin deposited on the wall. This gives added strength and is thought to help prevent desiccation (Graham and Wilcox, 2000). Some taxa have wall ornamentation, such as scales, a rough texture, thick walls with distinct layers, warts, ridges, and spines (Trainor, 1978). Some of these structures can be seen with the LM (e.g., ridges and spines), whereas others are visible only with TEM and SEM (e.g., scales and warts). Wall ornamentation is often used as a taxonomic character (e.g., number and position of spines); however, with some taxa this may not be a stable character, particularly at the species level. Trainor and Egan (1990a, b) demonstrated cyclomorphosis in an axenic clone of *Desmodesmus armatus* (as *Scenedesmus armatus*), which exhibited a variety of spine patterns that ranged from two spines on each of the terminal cells, to one spine on each of the terminal cells, to no spines on any cells (spineless). Some of these morphs could be identified as different species if spine number and pattern were assumed to be stable characters. Thus, caution is advised when making species identifications from field material, because some taxa (e.g., *Chlorococcum*, *Desmodesmus*, and *Scenedesmus*) have a variable morphology. To complicate matters, unicellular morphs of *Desmodesmus* have been misidentified as other genera: *Lagerheimia* (Hegewald and Schmidt, 1987, 1991; Trainor and Egan, 1990c; Trainor, 1991), *Franceia*, or *Oocystis*. This is unfortunate and is the consequence of relying solely on field material for identification (Smith, 1950) rather than using axenic cultures or considering cyclomorphosis and phenotypic plasticity (Hegewald and Silva, 1988). Subsequent investigations, beginning in the 1960s, by Trainor and co-workers (summarized in Trainor, 1998) unequivocally demonstrated that the colony form of *Desmodesmus* could produce unicells. The unicells are formed within the parent wall and are released as new progeny. Alternatively, unicells can produce more unicells or colonies, depending on the environmental conditions

(Trainor, 1998). Shubert (1975) described a new species, *Scenedesmus trainorii*, based on a strain (UTEX 1588) that showed phenotypic plasticity. The alternation between colonies and unicells was mediated by modifying the concentration of phosphorus (Shubert and Trainor, 1974). Forma designation was used to differentiate between the unicellular stage, f. *trainorii*, and the colonial stage, f. *quadricauda*. Komárek and Fott (1983) dropped the forma designation because they considered that the ecomorphs had no taxonomic significance. In rejecting the forma designation, they are ignoring morphological plasticity in response-known environmental variables. Hegewald and Silva (1988) considered f. *quadricauda* to be an invalid name because it shared the type strain with f. *trainorii* and was a typical form (e.g., four-celled stage), thus required no taxonomic recognition. The decisions by Komárek and Fott (1983) and Hegewald and Silva (1988) imply that forma is a redundant rank, but the International Botanical Code permits the use of forma to discriminate between contrasting morphs within a species. My opinion is that the genera *Franceia* and *Lagerheimia* require revalidation as distinct taxa using modern taxonomical approaches, such as combining axenic culture work with ultrastructural and molecular analysis.

A variety of morphological characters have been used to define genera, including shape of the vegetative cell, structure and position of the chloroplast, number of nuclei per cell, presence or absence of pyrenoids, and presence or absence of flagellated cells (Starr, 1955). In theory this makes sense, because a certain combination of characters should describe a genus. For example, a walled planospore with flagella of equal length, which is the reproductive cell of an organism with a cup-shaped chloroplast, was called *Chlorococcum*. Alternatively, a naked planospore from a cell with a netlike plastid was called *Spongiochloris*. However, despite being insightful, this does not always work in practice. As Trainor (1978) aptly pointed out, it could be difficult to identify plastid type if food reserves obscure it and/or flagellated stages might not always be observed, especially from live field material. So, do we make identification on a few live cells or fixed material? Before this question is answered, let us consider a few examples.

Starr (1955) described a new monospecific genus, *Neochloris*, using a variety of morphological characters. Subsequently, 12 new species were added. Much later, Watanabe and Floyd (1989) studied nine species of *Neochloris* and divided them into three groups based on morphological characters. They investigated the ultrastructure of the motile cells by focusing on the flagellar apparatus (FA) and discovered that the basal bodies could be arranged three different ways: clockwise, (CW), counterclockwise (CCW), and directly opposite (DO). Using this stable FA character, the species did not group in the same way. Some species were retained with *Neochloris* (DO) and some were assigned to other genera [*Chlorococcopsis* (CW) and *Parietochloris* (CCW); Watanabe and Floyd, 1989]. Subsequently, Lewis et al. (1992) analyzed the same taxa for 18S rRNA and the results were concordant with the designation by Watanabe and Floyd (1989). However, examination of other characters in *Neochloris* and related taxa has created uncertainty (Deason et al., 1991; Kouwets, 1995). The classification of former species of *Neochloris* remains unresolved (Trainor and Morales, 1999).

Lewis (1997) applied molecular techniques (18S rRNA) to *Bracteococcus* to determine its placement in the "larger picture of chlorophycean algae" because the motile cells have a FA of unusual orientation (CW). Lewis' (1997) 18S sequence data confirmed that all nine species analyzed supported a clade with bootstrap support of 95% and were monophyletic, which demonstrated that the morphological characters used to distinguish the genus from other chlorococcalean algae were "reliable." However, *Bracteococcus* has been interpreted to have CW basal body orientation and the 18S data pointed toward taxa with DO basal body orientation. Also, two *Chlorococcum* species did not form a clade, and the multinucleated zoosporic CW taxa *Spongiochloris* and *Protosiphon* "nested" with the uninucleated zoosporic CW taxa (Lewis, 1997). These data suggest that molecular characterization (e.g., 18S rRNA gene sequence alone) may not support the present classification based on morphological characters. A suite of stable characters (morphological, ultrastructural, biochemical, and molecular) may be required for the determination of some specific taxa.

Trainor et al. (1976) proposed that *Scenedesmus* be divided into two groups—the *obliquus* or nonspiny type and the spiny type—based on morphological (LM and TEM/SEM) and reproductive characters. However, they took the conservative approach to taxonomy and postponed formal establishment of generic level taxa until more data were available. Kessler et al. (1997) reported that biochemical and physiological properties were not suitable for species differentiation of *Scenedesmus*, but they did identify two subgenera of *Scenedesmus* (*Scenedesmus*, nonspiny type and *Desmodesmus*, spiny type) using 16S rRNA gene sequence, DNA base composition, and DNA/DNA hybridization analysis. An et al. (1999) analyzed a more conserved region of DNA (ITS-2 rDNA sequence) in selected species of *Scenedesmus* and resolved a distinct separation between the subgenera, *Scenedesmus* and *Desmodesmus*, and proposed the recognition of both groups as distinct genera. The ITS-2 gene sequence analysis was congruent with cell wall ultrastructure, but not with other morphological features (An *et al.*, 1999). Thus, molecular data confirmed what had been originally proposed using morphological and reproductive characters.

Taxa of the unicellular green alga *Chlorella* have been the organisms of choice for a variety of physiological and biochemical studies, including photosynthesis and nitrate reduction. However, the lack of obvious morphological characters combined with asexual reproduction only has created problems in delineating species. Numerous methods have been used, such as nutritional requirements, numerical classification strategies, combination of morphological and structural features, serological cross reactions, ultrastructure and chemical composition of the cell wall, pyrenoid ultrastructure, and a combination of biochemical and physiological characters (reviewed by Huss *et al.*, 1999). The most convincing taxonomical scheme to date is based on a multimethod approach that compares 18S rRNA gene sequences, DNA base composition, and DNA hybridization values, combined with biochemical, physiological, and ultrastructure characters (Huss and Sogin, 1990; Huss *et al.*, 1999). The results of Huss and co-workers showed that 19 *Chlorella* taxa were polyphyletic and dispersed over two classes of green algae, and they proposed that only four species be kept in the genus (glucosamine as a dominant cell wall component and presence of a double thylakoid bisecting the pyrenoid matrix), the other species being assigned to other genera (production of secondary carotenoids under nitrogen-deficient conditions (Huss *et al.*, 1999).

Molecular and TEM investigations of nonmotile greens have shown that many genera are polyphyletic. It is beyond the scope of this chapter to explore the systematics and evolution of green algae and phylogenetic relationships in detail, and the reader is advised to consult Mattox and Stewart (1984), Kantz *et al.* (1990), Wilcox *et al.* (1992), Floyd *et al.* (1993), Nozaki (1993), Friedl and Zeltner (1994), O'Kelly *et al.* (1994), Friedl (1995, 1997), Melkonian and Surek (1995), Nakayama *et al.* (1996), Booton *et al.* (1998), Hanagata (1998), Krienitz *et al.* (1999), Preisig (1999), and Sluiman and Guihal (1999).

Thus, an extension of the classical approach to the taxonomy of nonmotile greens might use the following holistic protocol: Initial identification of field material (ideally live) is determined with the LM; cells are isolated into axenic culture, grown under different environmental conditions, and phenotypic plasticity, and biochemical and physiological changes are recorded; ultrastructural features are documented with confocal laser scanning microscope (CLFM), TEM, and/or SEM; and molecular analysis is conducted for the final determination of species (Trainor and Morales, 1999). However, in the not too distant future this entire procedure might be abbreviated by simply placing a drop of water from a pond, lake, or ocean on a surface containing fluorescent nucleic acid probes for identification of the major species present (Miller and Scholin, 1998).

Nonmotile colonial or tetrasporalean algae often resemble the motile colonial algae. Bold and Wynne (1978) identified a problem with *Gloeococcus* "in which the flagellate cells move slowly within nonmotile colonies; this occurs also in the palmelloid phases of *Chlamydomonas*." Further, they stated that "the decision where to classify a given organism may be difficult and is often subjective insofar as some tetrasporalean algae are concerned." The holistic approach, as previously described, may give a definitive answer

Despite the variable morphology, the inconsistent appearance of reproductive cells, and the similarity of morphological forms, it is possible to identify most genera of nonmotile coccoid and colonial green algae (live or fixed) with a classical taxonomic key and a LM. When making microscopic observations (preferably using phase contrast or differential interference contrast with a 40× or 60× dry objective or 100× oil objective), a series of questions should be asked to determine the genus, such as what is the cell shape, what is the chloroplast structure, are pyrenoids present or absent, are flagellated cells present or absent, are there any unusual features of the cell, are there distinguishing external wall structures, is a mucilaginous sheath present, is it free-living or attached, what is its habitat? With experience, researchers will gain the confidence to successfully identify many of the nonmotile green algae found in North America. TEM and SEM have been used for the identification of species, but it is beyond the scope of this chapter to discuss this approach. The reader is encouraged to consult Pickett-Heaps (1975), who has produced an extensive treatise on the cell biology and ultrastructure of green algae.

III. ECOLOGY AND DISTRIBUTION

The nonmotile greens are ubiquitous and widely distributed in aquatic habitats throughout the North American continent. Aquatic habitats where these algae are found include temporary pools of water, waterfalls, ponds and lakes (e.g., *Ankistrodesmus*, *Coelastrum*, and *Scenedesmus*), rivers (e.g., *Actinastrum* and *Dictyosphaerium*), slow flowing streams (e.g., *Hydrodictyon*), marshes, and estuaries. Nonmotile greens are primarily a component of the plankton community (e.g., *Desmodesmus*, *Golenkinia*, *Pediastrum*, and

Scenedesmus). They also may be found growing attached to rocks in lakes (e.g., *Apatococcus* and *Desmococcus*), on leaf surfaces or on other algae (e.g., *Apiocystis* and *Characium*), and/or on the sediment surface. In the temperate zone, they are most abundant in freshwater ecosystems during the summer, when light and temperature are near their seasonal maximum and nutrients become limiting (e.g., N and P). Reynolds (1984) characterized freshwater plankton from temperate latitudes using trophic status and showed that green algae may be important in phytoplankton dynamics from spring through to autumn. Happey-Wood (1988) reviewed the ecology of freshwater planktonic green algae. In high altitude lakes, plankton may be detectable in the spring under considerable cover of ice and snow, and small coccoid greens (e.g., *Chlorella* and *Coccomyxa*) are often the most common components. During the summer, in oligotrophic lakes, the smaller sized fractions of unicellular green algae appear as the dominant organisms (e.g., *Sphaerocystis* and *Gloeocystis*). In contrast, in well-mixed shallow eutrophic lakes, larger green algae (e.g., *Coelastrum*, *Dictyosphaerium*, and *Pediastrum*) may dominate during the summer. With respect to the seasonal successional process in oligotrophic to mesotrophic lakes, nonmotile greens appear to be restricted to a relatively short growth period defined by a narrow range of environmental conditions within which to successfully compete with a mixed assemblage of phytoplankton (Happey-Wood, 1988). This restricted period occurs during the early stratification of the lake when light is not limiting, and turbulence and turbidity (e.g., suspended solids or shading by other algae) are minimal. Once stratification is stabilized, the nonmotile greens begin to sink and decline in the water column, and sedimentation increases. Thus, their survival in the euphotic zone is dependent on the residence time together with the available nutrients in the epilimnion (Happey-Wood, 1988). Nonmotile greens have numerous adaptations to reduce sinking rates, including a mucilaginous sheath, wall ornamentation (e.g., bristles and spines), number of cells in a colony, and a shape other than spherical (Reynolds, 1984). Some nonmotile green algae (e.g., *Chlorococcum submarinum* and *Chlorella*) are capable of growing in saline conditions (e.g., inland saline lakes, estuaries, and marine coastal habitats; Shubert, 1998).

Subaerial habitats also support many nonmotile green taxa, which may be easily observed growing on tree bark, wood fences, stone walls, gravestones, plaster walls, and so forth; in fact, any surface that holds some moisture. In extreme cases, this may result in damage to the surface over time (e.g., frescoes). Also, nonmotile green algae grow symbiotically with fungi in lichens (e.g., *Trebouxia*, a phycobiont of many lichens, e.g., *Cladonia*), and with plants and animals (e.g., *Chlorella* in *Hydra viridis*, *Paramecium bursaria*, and *Spongilla lacustris*).

Soil is a common habitat for nonmotile green algae (e.g., *Apatococcus*, *Chlorococcum*, *Chlorosarcinopsis*, and *Tetracystis*), which are widely distributed in a variety of soil types and microclimates (Starks *et al.*, 1981). *Chlorococcum* is a common taxon found in soils. Archibald and Bold (1970) differentiated species of *Chlorococcum* by isolating clones into liquid culture medium (BBM) and scoring their response to various physiological/biochemical attributes (e.g., sensitivity to antibiotics, production of extracellular proteases, and amylase). Although this approach is scientifically sound, it can be problematical and appears to be little used by contemporary phycologists. Some algae grow on the surface of moist, bare soil (e.g., *Protosiphon*). As the soil dries, *Protosiphon* becomes orange–red (Bold and Wynne, 1978).

IV. COLLECTION AND PREPARATION FOR IDENTIFICATION

Collecting planktonic nonmotile coccoid and colonial green algae is rather easy and straightforward. Whereas they are suspended in the water column, a plankton net with a mesh size of 20 µm is adequate for most taxa (nannoplankton and net plankton). A bottle sample can be taken, which is less concentrated and may require time to settle, but does allow quantification. Mosses, and stems and leaves of aquatic plants can be collected or hand squeezed for analysis of epiphytes. Algae on damp rock surfaces, walls, or bark can be removed with a penknife or other sharp instrument. Water from the habitat or distilled water should be added to the specimen vial. Soil samples can be collected with a trowel and stored in sterile plastic bags.

Algae generally survive well after collection if the container is clear glass, illuminated, and kept cool (15–20°C) in an environmental growth chamber or on a window ledge with a northern exposure. Alternatively, culturing field material may yield a range of morphologies and/or reproductive structures for some genera (e.g., *Chlorococcum*, *Desmodesmus*, and *Pediastrum*) that may aid in identification of species. Stein (1973) described isolation and culture methods and media for growing algae. In addition, Prescott (1978) and Dillard (1999) may be consulted for additional information on the collection and preservation of algae.

Wet mounts of fresh samples for microscopic analysis give the best results when a number 1 coverslip and at least 400× magnification are used. Resolution can be enhanced with phase contrast or differential interference contrast. A 60× dry objective is preferable, because superior detail of the cells can be observed without the use of oil. If greater resolution is required (which is often the case with *Chlorella* and other extremely small green unicells), a 100× oil immersion objective may be used. For enhanced resolution, it is recommended that an additional drop of oil be placed on the condenser lens and the condenser carefully raised until the oil makes contact with the bottom of the slide. For health and safety reasons, the microscope oil should be labeled as PCB-free.

If live material is not observed within 2–3 days after collection, it is recommended that the sample be fixed. The most common fixative is Lugol's solution [2 g KI + 1 g resublimed I +300 mL distilled water (DW); it can be acidified with 10 mL glacial acetic acid or made neutral or slightly basic with 1 g of sodium acetate; it should be stored in a dark bottle at room temperature]. Lugol's is very good for preserving small algae and flagellates. Lugol's will stain starch dark blue–black, which is a simple test to separate the Chlorophyta from the Chrysophyta (which do not have true starch), because some look morphologically similar to the green algae. Add just enough drops of Lugol's to give a pale straw color to the sample and store samples in a cool dark place. M^3 fixative is superior to Lugol's and is excellent for long preservation. M^3 fixative is made by adding 1 g I_2 + 0.5 g KI + 5 mL glacial acetic acid +25 mL formalin +100 mL DW (R. Meyer, personal communication). Another simple fixative is 2–4% neutralized formalin. If samples are to be observed with epifluorescence, TEM, and/or SEM, EM grade glutaraldehyde (2.5–5% per volume) is the best choice. A cautionary note, glutaraldehyde is a strong fixative and human exposure should be minimized (e.g., wet mounts and light microscopy), because it can fix eye tissue. Another good fixative is FAA (a ratio of 10:7:2:1 parts of 95% ethanol: DW: formalin: acetic acid).

V. KEY AND DESCRIPTIONS OF GENERA

A. Key

This dichotomous taxonomic key has been constructed to be "user friendly" and is based on morphological characters seen with the LM. For some taxa, especially the spherical unicells, identification from field material may be difficult. Isolation and culturing of the organism may be required to observe motile cells.

1a.	Cells forming macroscopic growths	2
1b.	Cells solitary and/or not forming macroscopic growths	8
2a.	Free-floating	3
2b.	Usually attached	4
3a.	Cells elongate, forming a netlike reticulum (Fig. 15B)	*Hydrodictyon*
3b.	Cells spherical, often with several long, fine pseudoflagella, arranged in groups of two or four; fragments of old parental walls may be visible (Fig. 24C)	*Schizochlamydella*
4a.	Colony leaflike or a gelatinous sac or matrix	5
4b.	Colony not as above, but with cells embedded in stratified mucilage	7
5a.	Colony leaflike, cells in packets of two at the surface, lacking pseudoflagella (Fig. 21E)	*Phyllogloea*
5b.	Colony a gelatinous sac or matrix	6
6a.	Cells arranged in packets of four (occasionally pairs) arranged around the edge of the colony, each cell with two pseudoflagella (Fig. 26F)	*Tetraspora*
6b.	Cells not in packets of four, cells feebly motile within gelatinous matrix (Fig. 13C)	*Gloeococcus*
7a.	Cells spherical, embedded in short, dichotomously branching, stratified gelatinous stalks, chloroplasts stellate (Fig. 15A)	*Hormotila*
7b.	Cells ellipsoidal, embedded in extensive, anastomosing, stratified gelatinous strands, chloroplasts cup-shaped (Fig. 14A)	*Gloeodendron*
8a.	Cells growing within plants (endophytic) or animals (endozooic; including egg masses)	9
8b.	Cells not growing endophytically or endozooically	13
9a.	Cells growing in animals or egg masses	10
9b.	Cells growing among surficial cells of plants	11

10a.	Small spherical cells with a cup- or platelike chloroplast in protozoa or invertebrates (Fig. 5B)	*Chlorella*
10b.	Cells spherical to ellipsoidal with an axial, irregularly lobed chloroplast, in amphibian egg masses (often giving these a green color; Fig. 19D)	*Oophilia*
11a.	Largely colorless, threadlike growth on and within *Sphagnum*, chloroplasts concentrated in expanded ends of cells (Fig. 21D)	*Phyllobium*
11b.	Growth not threadlike, cells rounded	12
12a.	Cells irregularly oval; chloroplast cuplike in young cells, massive and diffuse in older cells, with thick lamellate walls, growing between host cells of *Lemna* and mosses (Fig. 5D)	*Chlorochytrium*
12b.	Cells globular or flask-shaped, with rhizoidal lobes and extensions; parasitic within Ambrosia; protoplast reddish orange (Fig. 24A)	*Rhodochytrium*
13a.	Cells or microscopic colonies attached to a substratum	14
13b.	Cells or microscopic colonies not attached to a substratum	30
14a.	Colonies or aggregations of cells	15
14b.	Cells attached as individuals, without a communal point of attachment	20
15a.	Cells attached at end of branching mucilaginous stalks, forming treelike colonies; sometimes attached to small crustacea or insect larvae (Fig. 5A)	*Chlorangiella*
15b.	Cells not on branching stalks, occurring in an amorphous matrix, mucilaginous vesicle, or forming disklike thallus on the substratum	16
16a.	Colony mucilaginous, often amorphous or bulbous	17
16b.	Colony not mucilaginous, closely appressed to substratum	19
17a.	Colony pear-shaped or bulbous vesicle containing cells in pairs or fours with pseudoflagella projecting well beyond the colony envelope (Fig. 2C)	*Apiocystis*
17b.	Colony amorphous, cells lacking pseudoflagella	18
18a.	Cells embedded in concentrically stratified mucilage; may contain hematochrome (red pigment); often terrestrial (Fig. 20A)	*Palmella*
18b.	Cells not embedded in stratified mucilage; aquatic (Fig. 20C)	*Palmellopsis*
19a.	Cells forming a pseudoparenchymatous small circular disk, bearing long erect pseudoflagella; chloroplast cup-shaped (Fig. 4C)	*Chaetopeltis*
19b.	Cells growing as a single layer of cells or pseudoparenchymatous disk of filaments; pseudoflagella absent; chloroplast parietal (Fig. 22D)	*Protoderma*
20a.	Cells saccate, one end forming a rhizoid-like extension, the other end bulbous; chloroplast reticulate (Fig. 22E)	*Protosiphon*
20b.	Cells not as above, attached by well-defined stalk or over large part of cell surface	21
21a.	Cells with a well-defined stalklike extension for attachment	22
21b.	Cells flattened onto substratum or attached by a broad face or pad	25
22a.	Cells elongate–oval to fusiform, attached by a broad disk formed external to the cell wall (Fig. 4E)	*Characiochloris*
22b.	Attachment by stalk or pad that is part of the cell wall	23
23a.	Cells small, with a very fine tapering attachment; epiplanktonic; cup-shaped chloroplast with a single pyrenoid (Fig. 26A)	*Stylosphaeridium*
23b.	Cells with a flattened pad where they attach	24
24a.	Cells spherical, elliptical, or elongate with one to many disk-shaped chloroplasts lacking pyrenoids (Fig. 4F)	*Characiopsis*
24b.	Cells straight, slightly crescent-shaped, or sigmoid, with one to several parietal chloroplasts with pyrenoids; may be epiplanktonic (Fig. 4G)	*Characium*
25a.	Cells more or less spherical attached by a broad base or broad mucilage pad	26
25b.	Cells ellipsoidal to flattened, or in a flattened casing	27
26a.	Cells with anterior end down, attached by wide mucilage pad and enclosed in brown sheath; chloroplast in upper part of cell (Fig. 17A)	*Malleochloris*

26b.	Cells on broad attaching base; stalk almost lacking, thick brown sheath around, and much larger than, protoplast (Fig. 6A) ....*Chlorophysema*
27a.	Cells in angular casing or with angular outline...28
27b.	Cells rounded (not angled) in face view..29
28a.	Cell with a cup-shaped chloroplast enclosed in flattened quadrangular casing with pores at corners through which pseudoflagella emerge (Fig. 22C)...*Porochloris*
28b.	Cells with octagonal outline and bipartite wall (Fig. 19A)..*Octagoniella*
29a.	Cells lenticular, attached to submerged leaves of higher plants; cell wall thin against host, but thicker on free face (Fig. 10B) ....*Ectogeron*
29b.	Cells ellipsoidal, occurring in smooth, rounded casings with brownish encrustation around edge, on filamentous algae (Fig. 5C) ....*Chloremys*
30a.	Cells with spines or protuberances...31
30b.	Cells lacking spines or protuberances..52
31a.	Cells solitary..40
31b.	Cells aggregated or forming colonies or coenobia...32
32a.	Cells regularly arranged, in pairs, fours, or multiples of four...33
32b.	Cells irregularly arranged colonies, sometimes in mucilage or cells attached to each other by strands......................................37
33a.	Cells arranged in pairs..34
33b.	Cells usually fours or multiples of four, or if pairs, with few (<2) spines per cell..35
34a.	Cells ovoid, aligned side by side, with many needle-like spines projecting from outer margins (Fig. 8B).............................*Dicellula*
34b.	Cells pear-shaped, narrow end extended into perpendicular outgrowths that link cells; long fine spines projecting more or less parallel to cell axis (Fig. 20E)..*Paradoxia*
35a.	Cells ellipsoidal or ovoid, arranged in coenobia (2, 4, 8, or 16 cells) with long axes parallel; spines on terminal and/or medial cells (occasionally lacking; Fig. 8A)..*Desmodesmus*
35b.	Cells crescent-shaped to triangular, or triangular to heart-shaped or ovoid, arranged about an axial point...................................36
36a.	Cells forming a flattened plate, with a hole at the center, each cell with one to four slender outwardly projecting spines (Fig. 27A) ....*Tetrastrum*
36b.	Cells not forming a flattened plate, radiating from an axial point; cell poles extended into horns (Fig. 16C)..............*Lauterborniella*
37a.	Cells kidney-shaped, pear-shaped, wedgelike, or pyramidal, attached by strands and forming globular colonies; each cell has two to four outwardly directed spines (Fig. 25C)...*Sorastrum*
37b.	Cells spherical to ovoid, not attached by strands..38
38a.	Cells closely aggregated, each with a single cup-shaped plastid and one to several delicate spines (Fig. 17B)..................*Micractinium*
38b.	Cells more loosely associated, often with more than one chloroplast..39
39a.	Cells spherical, each with two to eight long tapering spines, arranged within mucilage envelope including fragments of parental walls (Fig. 20B)..*Palmellochaete*
39b.	Cells ovoid, with many short spines, especially at poles, loosely aggregated in mucilage (Fig. 22F).........................*Pseudobohlinia*
40a.	Cells solitary, with two or four strong projections, much longer than cell...41
40b.	Cells solitary, with more numerous finer spines or if with four points, not longer than cell diameter.......................................44
41a.	Cells with two projections..42
41b.	Cells with four projections...43
42a.	Cells small ellipsoidal with two very long, hollow spines, one at each pole, oriented along long axis of cell (Fig. 8C).......*Diacanthos*
42b.	Cells spindle-like, straight or curved, tapering to long spines, one of which may be divided at the end (Fig. 24D)...........*Schroederia*
43a.	Cells pyramidal, with thick spine from each corner, one or more parietal chloroplasts (Fig. 27D)......................................*Treubaria*
43b.	Cells spherical with a cup-shaped chloroplast, spines often brown and forked at apices (Fig. 19E)............................*Pachycladella*

44a.	Cells angular	45
44b.	Cells not angular	46
45a.	Cells rectangular or five-sided, with squared-off angles bearing tufts of long, tapering spines (Fig. 22B)	*Polyedriopsis*
45b.	Cells bilaterally symmetrical, each half triangular or pyramidal, angles extended into stout spines (Fig. 26D)	*Tetraedron*
46a.	Cells ellipsoidal with somewhat curved spines	47
46b.	Cells spherical with straight, tapering spines	48
47a.	Cells with spines emerging from the poles (Fig. 16A)	*Lagerheimia*
47b.	Cells with spines emerging over entire surface of the cell (Fig. 12D)	*Franceia*
48a.	Cells with short spines or irregularly sculptured over surface of cell (Fig. 27E)	*Trochiscia*
48b.	Cells with longer spines, as great as or greater than cell diameter	49
49a.	Spines with a distinct thickened basal section narrowing abruptly to a fine bristle (Fig. 1A)	*Acanthosphaera*
49b.	Spines otherwise, either tapering gradually to tip or very fine throughout	50
50a.	Cells with long, stout, clear spines, broad at base, tapering to sharp point (Fig. 10A)	*Echinosphaerella*
50b.	Cells with finer spines	51
51a.	Cells spherical to slightly ovoid, spines thickened at base, tapering to apex (Fig. 14D)	*Golenkiniopsis*
51b.	Cells spherical, spines long, slender (sometimes forming a false colony via interlocking setae; Fig. 14C)	*Golenkinia*
52a.	Cells triangular or pyramidal; angles extending into long tapering, rather irregular processes (Fig. 4C)	*Cerasterias*
52b.	Cells not as above	53
53a.	Cells markedly longer than wide	54
53b.	Cells spherical, ovoid, or ellipsoidal, usually less than two times longer than broad	68
54a.	Cells curved, crescent-, or kidney-shaped with similar rounded apices	55
54b.	Cells usually more or less straight, but if curved, with pointed apices	58
55a.	Cells in two- or four-celled flat colonies; convex surfaces apposed; nonmucilaginous (Fig. 9D)	*Didymogenes*
55b.	Cells aggregated in larger numbers, embedded in mucilage	56
56a.	Colony comprising 4–8(16) curved cells within the mucilaginous, expanded parental wall (Fig. 18D)	*Nephrocytium*
56b.	Colony comprising markedly crescent-shaped cell	57
57a.	Cells grouped in fours or eights, one pair linked at both their apices, the other two linked at one end and vertical to first pair (Fig. 26E)	*Tetrallantos*
57b.	Cells arranged irregularly within gelatinous sheath (Fig. 16A)	*Kirchneriella*
58a.	Cells usually in groups of four or multiples of four	59
58b.	Cells not usually in groups of four	61
59a.	Cells forming flattened coenobia, arranged in linear or alternating series; cells may be tapered at ends, but not ending in spines (Fig. 24B)	*Scenedesmus*
59b.	Cells aligned with long axis parallel, but not forming flat coenobia	60
60a.	Cells enclosed in mucilaginous envelope (Fig. 23B)	*Quadrigula*
60b.	Cells not enclosed in mucilage (Fig. 26.C)	*Tetradesmus*
61a.	Cells strongly curved or crescent-shaped	62
61b.	Cells not strongly curved or crescent-shaped	63
62a.	Cells much longer than broad, curved to weak crescent-shaped; aggregated in irregular bundles, sometimes entwined (Fig. 2A)	*Ankistrodesmus*
62b.	Cells longer than broad, strongly crescent-shaped; cells not entwined (Fig. 25A)	*Selenastrum*
63a.	Cells spindle-shaped, forming stellate colonies (usually eight cells), with narrower outwardly pointing apices (Fig. 1C)	*Actinastrum*

63b.	Cells not forming star-shaped colonies...64	
64a.	Cells arranged in well-defined mucilage colonies...65	
64b.	Cells not enclosed in well-defined mucilage..66	
65a.	Cells straight to slightly bent, with rounded ends, irregularly arranged in colonies within clearly defined mucilage (Fig. 12B) ....*Fottea*	
65b.	Cells arranged in pairs or fours, tapered at poles (Fig. 10C)...*Elakatothrix*	
66a.	Cells more or less spindle-shaped, narrowed at one or both ends in an extension of the cell wall (Fig. 15D)................*Keratococcus*	
66b.	Cells very long and narrow, or spindle-shaped but without apical extension..67	
67a.	Cells very long and narrow, needle-like, single chloroplast with an axial row of pyrenoids (Fig. 6C)..........................*Closteriopsis*	
67b.	Cells spindle-shaped, up to six times as long as broad; pyrenoids lacking (Fig. 17C)..*Monoraphidium*	
68a.	Cells in pairs or forming flattened coenobia, at least the outer cells of which usually lobed...69	
68b.	Cells not lobed, solitary or forming other types of colony..70	
69a.	Cells in flattened pairs, joined along a straight margin; outer margins lobed (Fig. 11B)...*Eustropsis*	
69b.	Cells forming flattened more or less circular coenobia containing at least four cells; outer cells usually lobed or with processes (Fig. 21C)..*Pediastrum*	
70a.	Cells of various shapes (usually) in groups of four or multiples of four..71	
70b.	Cells not in groups of four..88	
71a.	Cells more or less spherical with pseudoflagella...72	
71b.	Cells without pseudoflagella..74	
72a.	Cells with one or two pseudoflagella, each cell containing an endosymbiotic cyanelle (Fig. 3B)..................................*Gloeochaete*	
72b.	Cells without cyanelles, each with two long pseudoflagella..73	
73a.	Cells with extremely long pseudoflagella, 20 times longer than cell, emerging from mucilaginous colony (Fig. 12C)............*Fottiella*	
73b.	Cells with shorter pseudoflagella, about four or five times longer than cell; products of cell division tend to stay together in tetrads and form complexes of colonies (Fig. 21A)..*Paulschulzia*	
74a.	Cells retained within the persistent parent wall...75	
74b.	Cells not retained within the parent wall, although wall fragments may persist..76	
75a.	Cells spherical to ellipsoidal, arranged cruciately within the parent wall and separated by dark colored deposits within the colony (Fig. 14B)..*Gloeotaenium*	
75b.	Cells ellipsoidal, arranged with long axes parallel in the colony and at the two ends of the parent cell; colonies linked by mucilaginous remnants (Fig. 23D) ....*Rayssiella*	
76a.	Remains of parent cell walls visible linking cells within colony..77	
76b.	Cells not linked by remains of parent cell walls (but may be entirely enclosed within gelatinous remnants of wall).....................83	
77a.	Colony lacking mucilaginous sheath..78	
77b.	Colony enclosed by mucilage..79	
78a.	Cells ovoid, ellipsoidal, or lunate, linked in fours by parental cell wall strands; cells with a winglike projection on outer face (Fig. 6F) ....*Coronastrum*	
78b.	Cells ovoid, or kidney- to heart-shaped, arranged in fours on branched parental wall fragments (Fig. 9E)............*Dimorphococcus*	
79a.	Colonies of four broadly ellipsoidal cells linked by flattened shelllike parental wall (Fig. 23A)............................*Quadricoccus*	
79b.	Colonies larger, containing many more cells..80	
80a.	Cells grouped in twos or fours into irregular colonies, with wall fragments around groups; cells with fine pseudoflagella (Fig. 24C)..*Schizochlamydella*	
80b.	Cells linked by wall fragments, lacking pseudoflagella..81	
81a.	Cells spherical to broadly oval, attached by wall fragments from a common center; chloroplast cup-shaped (Fig. 9C) ....*Dictyosphaerium*	

81b.	Cells not linked by wall fragments from a common center; chloroplast parietal	82
82a.	Cells spherical, arranged in fours or eights, linked by looplike wall fragments (Fig. 27F)	*Westella*
82b.	Cells spherical to ovoid, 4–16 in colony; loosely associated (not contiguous); connected by threads of parental wall (Fig. 8E)	*Dictyochlorella*
83a.	Cells arranged in flattened colonies	84
83b.	Cells not in flattened colonies	85
84a.	Cells ellipsoidal to oval, irregularly arranged in flattened colonies (Fig. 9F)	*Dispora*
84b.	Cells ellipsoidal, ovoid to triangular, joined in groups of fours within flattened colonies (Fig. 6G)	*Crucigenia*
85a.	Cells ellipsoidal with two plastids, held within a thin-walled mucilaginous vesicle (Fig. 19B)	*Oocystidium*
85b.	Cells with a single chloroplast, forming larger colonies	86
86a.	Cells spherical, arranged at corners of eight-celled hollow cubes linked by a framework of gelatinous strands to form extensive colony (Fig. 21B)	*Pectodictyon*
86b.	Cells not arranged in cubes or linked by gelatinous strands	87
87a.	Cells elongate, ovoid, or wedge-shaped, arranged in radiating clusters at the edge of the colony (Fig. 13A)	*Gloeoactinium*
87b.	Cells spherical to broadly ellipsoidal, arranged in cruciform groups of four in a mucilage envelope containing radiating fibrils (Fig. 23C)	*Radiococcus*
88a.	Cells retained within the parent wall in pairs or multiples of two	89
88b.	Cells not retained within the parent wall as above	91
89a.	Cells ellipsoidal with pointed apices, retained in enlarged parent wall, cells with one to several chloroplasts (Fig. 19C)	*Oocystis*
89b.	Cells ellipsoidal with rounded ends, forming more irregular colonies enclosed within mucilage	90
90a.	Cells broadly ellipsoidal arranged in pairs within parental cell wall material at end of dichotomously branching strands (Fig. 16D)	*Lobocystis*
90b.	Cells sausage-shaped, connected by fine threads of parental material (Fig. 27B)	*Tomaculum*
91a.	Cells in cuboidal packets or forming pseudoparenchymatous clumps, occasionally developing short filaments	92
91b.	Cells not in packets or pseudoparenchymatous	95
92a.	Cells forming cuboidal groups held together by gelatinized parent wall within mucilaginous masses (Fig. 5F)	*Chlorokybos*
92b.	Cells not held together by gelatinized parent wall	93
93a.	Cells ellipsoidal spherical to spherical with cup-shaped chloroplast containing a pyrenoid (like *Chlorococcum*), arranged in groups (pairs, tetrads); sometimes cells within cells (Fig. 26B)	*Tetracystis*
93b.	Cells not as above, forming more irregular packets or short filaments	94
94a.	Cells rounded to spherical or angularly flattened, forming irregular cell packets by consecutively perpendicular divisions; chloroplast parietal becoming partly removed from cell surface; folded to bilobed; pyrenoid lacking (Fig. 2B)	*Apatococcus*
94b.	Cells forming cuboidal packets or short branched uniseriate filaments; chloroplast relatively small, parietal, with a small naked pyrenoid (difficult to see; Fig. 7C)	*Desmococcus*
95a.	Cells spherical to ovoid densely arranged within semi-opaque, golden brown, mucilage; cell contents of older cells often obscured by accumulated oil (Fig. 3C)	*Botryococcus*
95b.	Cells not in semi-opaque mucilage; cell contents visible	96
96a.	Cells spherical, irregularly arranged in linear series in netlike or branched but nonlamellate mucilage tubes (Fig. 20D)	*Palmodictyon*
96b.	Cells not living in tubelike colonies	97
97a.	Cells linked to form a hollow sphere	98
97b.	Cells not linked to form a hollow sphere	99
98a.	Cells spherical to polygonal, united to form a hollow sphere (usually containing 4–8–16–32 cells) by interconnecting protuberances (Fig. 6E)	*Coelastrum*

98b.	Cells spherical, united within a wide colorless mucilage envelope to form a spherical colony of abutting cells (Fig. 22A) ....*Planktosphaeria*
99a.	Colonies formed of numerous ovoid, spindle-shaped or pear-shaped cells radiately arranged in mucilage..............................100
99b.	Cells not forming such colonies..101
100a.	Cells pear-shaped, with broad anterior end toward colony periphery, narrow end toward center; chloroplast parietal in the anterior end with a pyrenoid (Fig. 3A) ....*Askenasyella*
100b.	Cells ovoid to spindle-shaped in stellate clusters at the end of radiating gelatinous threads; chloroplast parietal; pyrenoid lacking (Fig. 1B)..*Actidesmium*
101a.	Cells ellipsoidal to cylindrical, often small and enclosed in mucilage, which may be colored...102
101b.	Cells not tending to be cylindrical...105
102a.	Cells very small, <4.5 µm diameter, solitary or in pairs enclosed within lamellate sheaths in a mucilaginous mass (Fig. 17E) ....*Nannochloris*
102b.	Cells larger, if solitary in colored sheath or colored cell wall..103
103a.	Cells ellipsoidal to cylindrical, tending to lie more or less parallel to each other within colonial mucilage envelope (Fig. 6D) ....*Coccomyxa*
103b.	Cells solitary; sheath or cell wall colored..104
104a.	Cells ellipsoidal to cylindrical, with parietal chloroplast(s); cell wall yellowish, often impregnated with iron salts or irregular brown warts (Fig. 25B)..*Siderocelis*
104b.	Cells ellipsoidal with cup-shaped chloroplast in thick, yellow or brown gelatinous sheath, encrusted with iron salts (Fig. 25D) ....*Sphaerellocystis*
105a.	Cells usually ovoid or ellipsoidal, in colonies of two- to four to eight to numerous cells, with a broad firm envelope, some with concentric layers of mucilage (Fig. 13D)...*Gloeocystis*
105b.	Cells not as above..106
106a.	Cells large (50–300 µm diameter) with many chloroplasts...107
106b.	Cells smaller, with one or few chloroplasts...108
107a.	Cells spherical, ellipsoidal to pear-shaped; cell wall irregularly thickened, sometimes lamellate; chloroplasts parietal, each with one to many pyrenoids (Fig. 11C)..*Excentrosphaera*
107b.	Cells spherical, with or without mucilaginous envelope; chloroplasts arranged in radiating strands from center of cell, each with a pyrenoid (Fig. 11A)...*Eremosphaera*
108a.	Cells usually solitary, spherical...109
108b.	Cells not spherical, usually ellipsoidal or with distinctive wall ornamentation...125
109a.	Floating cells forming clusters; cells with thick wall with alveolae that refract the light; chloroplast parietal, platelike (Fig. 15E) ....*Keriochlamys*
109b.	Cells without thick alveolate wall..110
110a.	Cells with cup-shaped chloroplast..111
110b.	Cells with other type of chloroplast; stellate, lobed, reticulate, or fragmented...116
111a.	Cells lacking pyrenoid, sometimes found in packets; cell wall thin, slightly thickened in older cells (Fig. 6B)..............*Chlorosarcina*
111b.	Cells with a pyrenoid...112
112a.	Cells tending to be grouped within mucilage...113
112b.	Cells more usually solitary, not in obvious mucilage layers...114
113a.	Cells usually in 4–32 celled groups, always embedded in common mucilage, often with additional slime sheaths; pyrenoid with one or two thick starch layers (Fig. 12A) ....*Fasciculochloris*
113b.	Cells in colonies of 32–64 cells scattered throughout colonial mucilage, with intermingled progeny colonies of smaller cells (Fig. 25E)...*Sphaerocystis*
114a.	Cells very small (5 µm diameter), released in fours from parent cell (Fig. 5B)...*Chlorella*
114b.	Cells larger (at least 10 µm diameter), forming numerous ellipsoidal aplanospores or flagellated zoospores............................115

115a.	Cells sometimes gregarious; colonial mucilage thin and not very obvious; zoospores become spherical over several days (Fig. 5E)	....*Chlorococcum*
115b.	Cells solitary; zoospores become spherical immediately on quiescence (Fig. 18B)	....*Neochloris*
116a.	Cells with numerous parietal chloroplasts	117
116b.	Cells with a single chloroplast (but may have holes or be reticulate)	118
117a.	Cells containing several to many parietal, lens-shaped chloroplasts without pyrenoids (resembles *Planktosphaeria*; Fig. 4A)	....*Bracteococcus*
117b.	Cells coenocytic, containing segments each with a divided chloroplast with polygonal pieces lacking pyrenoids (Fig. 9B)	....*Dictyococcus*
118a.	Cells with star-shaped or lobed chloroplast	119
118b.	Cells with a parietal chloroplast	121
119a.	Cells united into colonies of 2–16; cells enclosed in broad lamellate envelope; chloroplast star-shaped with radiating lobes flattened against the wall; one central pyrenoid (Fig. 3B)	....*Asterococcus*
119b.	Cells not enclosed in broad lamellate envelope	120
120a.	Cells coenocytic, with smooth cell wall, thick in mature cells; chloroplast star-shaped, lobed, radiating from a thick central piece containing a very large pyrenoid (Fig. 1D)	....*Actinochloris*
120b.	Cells uninucleate; chloroplast stellate, irregularly lobed with one to several pyrenoids; lichen phycobiont (Fig. 27C)	.........*Trebouxia*
121a.	Cells without any pyrenoids	122
121b.	Cells with one to several pyrenoids	123
122a.	Cells with thin smooth cell wall that does not become thickened in mature cells; hollow reticulate chloroplast (Fig. 8D)	....*Dictyochloris*
122b.	Cells wall thin and compact when young but with some thickenings, somewhat thicker in mature cells, hollow reticulate chloroplast (Fig. 9A)	....*Dictyochloropsis*
123a.	Cells with one to several pyrenoids; chloroplast a parietal reticulum with inwardly directed portions (Fig. 25F)	....*Spongiochloris*
123b.	Cells with a single pyrenoid	124
124a.	Cells with irregularly lobed parietal chloroplast when young, becoming a hollow sphere with holes and incisions in mature cells; off-center pyrenoid (Fig. 18C)	....*Neospongiococcum*
124b.	Cells with parietal more or less open cup-shaped chloroplast with off-center pyrenoid when young, becoming central when mature (Fig. 7A)	....*Deasonia*
125a.	Cells in a casing	126
125b.	Cells not in a casing	127
126a.	Cells spherical to ellipsoidal; solitary; enclosed in a longitudinally ribbed casing, tapering to each end, resembling a seed pod (Fig. 7B)	....*Desmatractum*
126b.	Cells hemispherical with a depressed apex within an ovoid to spherical, gelatinous membrane-like casing; cells with cyanelles (Fig. 4D)	....*Chalarodora*
127a.	Cells with axial chloroplast	128
127b.	Cells with parietal, lobed, or divided chloroplast	130
128a.	Cells ovoid with a flotation cap at the narrow end (Fig. 18A)	....*Nautococcus*
128b.	Cells with some sort of irregularly thickened walls	129
129a.	Cells spherical to irregular pear-shaped; wall with a lateral thickening on one side projecting into the cell (Fig. 17D)	........*Myrmecia*
129b.	Cells irregularly spherical or ellipsoidal; walls with irregular external thickenings; chloroplast with peripheral processes that may appear like several irregular chloroplasts (Fig. 15C)	....*Kentrosphaera*
130a.	Cells ovoid, ellipsoidal, broadly oval with 6–12 fine longitudinal ribs on the outer surface (may be difficult to see), converging to form small papilla-like thickenings (Fig. 24F)	....*Scotiellopsis*
130b.	Cells ellipsoidal to oval; cell wall relatively thick with thick, longitudinal or slightly spiral ribs on the outer surface (Fig. 24E)	....*Scotiella*

B. Descriptions of Genera[1]

Whereas over 200 genera are recognized as belonging to the nonmotile coccoid and colonial green algae, representative taxa that illustrate a range of morphologies and ecological habitats are described. More recent taxonomic references are used for the most current accepted taxonomic name, and synonyms and basionyms are given to indicate previous taxonomic designations. Please keep in mind that a taxonomic designation does not mean that it is accepted by all investigators and that the name could change as new data become available (e.g., reproductive, ultrastructural, and molecular).

Acanthosphaera Lemmermann (Fig. 1A)

Cells are solitary, spherical, and have long and slender spines that have a distinct thickened basal section that abruptly narrows to a fine bristle. The chloroplast is parietal and platelike, with one pyrenoid. Found in the phytoplankton in ponds and lakes. Widely distributed; noted in scattered reports (Ontario, Great Lakes, central Canada, British Columbia, western, midwest, southeastern United States, the Caribbean and Central America)

Actidesmium Reinsch (Fig. 1B)

Colony comprises numerous cells. Cells are ovoid to spindle-shaped, in star-shaped clusters of 4–8–16 cells. Cell clusters occur at the end of radiating gelatinous strands (old parent walls), have one parietal chloroplast, and no pyrenoid. Found in metaphyton of ditches, bogs, ponds, and lakes. Not widely reported (central Canada, British Columbia, and western and southeastern United States).

Actinastrum Lagerheim (Fig. 1C)

Colony is (4)–8–(16)-celled. The cells are much longer than broad, cylindrical, cigar-shaped, or elongated, radiating from a common center; compound coenobia often are formed. The colony is not enclosed in a gelatinous sheath, and the chloroplast is elongate, parietal, and has one pyrenoid. Found in phytoplankton or metaphyton of ditches, bogs, ponds, and lakes. Very common genus and widely reported from most sections of North America.

Actinochloris Korschikoff (=*Radiosphaera* Snow *sensu* Starr) (Fig. 1D)

Solitary cell (coenocyte), broadly elliptical or broadly ovoid when young, always spherical when mature. The cell wall is smooth, hyaline, and firm, and thick in mature cells. The chloroplast is stellate-lobed, with a large, thick central piece, from which the numerous radial, often rather irregularly arranged lobes extend to the periphery of the cell. The individual lobes are often additionally more or less branched, sometimes bushlike, ending just under the cell wall and somewhat wider. One very large, simple or composite pyrenoid lies in the central portion of the chloroplast. Cells can form biflagellated zoospores. Habitats are subaerial to terrestrial. Not widely reported (Caribbean, Central America, and southeastern United States).

Ankistrodesmus Corda [=*Raphidium* Kützing] (Fig. 2A)

Cells are solitary or loosely clustered in bundles (some may be spirally twisted about one another) or in tufts or intermingled with other algae, they have no mucilage envelope, and are needle-like, crescent-shaped, or narrowly tapering toward each end, sometimes straight, usually curved. The chloroplast is parietal, and has or does not have pyrenoid. Found in phytoplankton of ponds and lakes, and metaphyton of ditches, ponds, and lakes. One of the most common genera of coccoid green algae; recorded in all regions throughout North America.

Apatococcus Brand em. Geitler [=*Pleurococcus* p.p. *sensu* auct. *Protococcus* p.p. *sensu* auct; *Pleurastrum* p.p. *sensu* Printz] (Fig. 2B)

Cells are rounded to spherical or somewhat angularly flattened, in more or less irregular cell packets formed through divisions in two to three consecutively perpendicular directions or multilayered pseudoparenchymatous cell aggregates. Occasionally the beginning of a short thread formation can be observed. The cell wall has a smooth to rough surface, that is somewhat thickened in older cells. A parietal chloroplast is found in young cells; in older cells, it is partly removed from the cell wall, folded on the surface and narrowed at the center to bilobed or bipartite, and lacking a pyrenoid. Infrequently reported, but possibly very common, it is a subaerial alga that forms part of the *Protococcus–Pleurococcus* community on tree bark (classified by some in the Chaetophorales; listed also in Chap. 8); distribution unknown.

Apiocystis Nägeli (Fig. 2C)

Colony is a microscopic, pear-shaped or irregularly bulbous mucilaginous vesicle, with spherical cells in pairs or in fours and two pseudoflagella extending well beyond the colony envelope. The chloroplast is cup-shaped and has one pyrenoid. The colony is attached at the narrow end to filamentous algae or aquatic plants; infrequently reported but fairly widespread (central and western Canada, most regions of the United States, the Caribbean, and Central America).

---

[1] All scale bars in the figures equal 10 μm unless otherwise indicated.

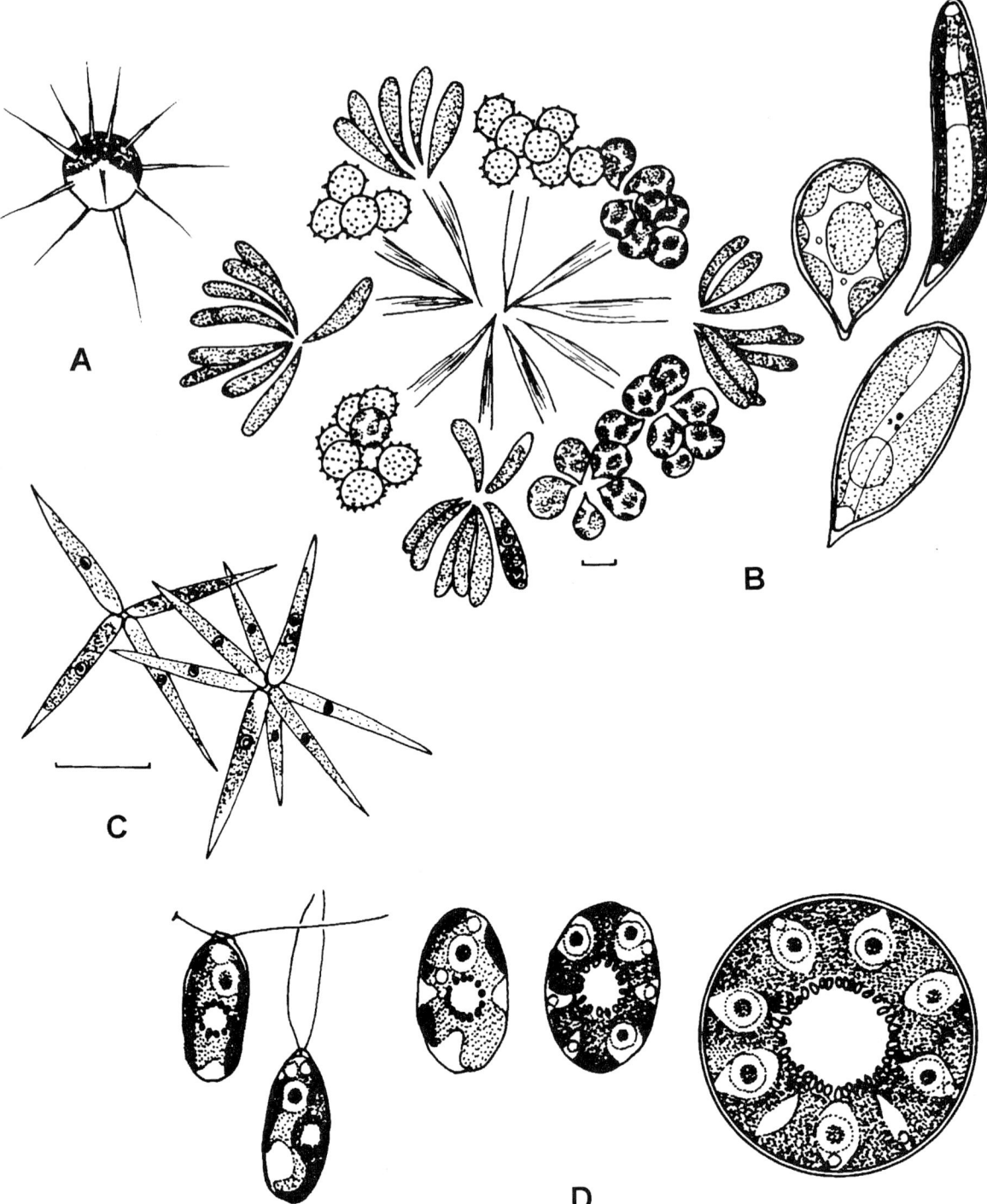

FIGURE 1 A. *Acanthosphaera zachariasii*. B. *Actidesmium hookeria*. C. *Actinastrum hantzchii*. D. *Actinochloris sphaerica*. A from Prescott, G. (1978) How to know the freshwater algae, 3rd ed. Brown, Dubuque, IA. This material is reproduced with permission of The McGraw-Hill Companies; B and C from Bourrelly (1966) with permission; D from Ettl and Gärtner (1995) with permission.

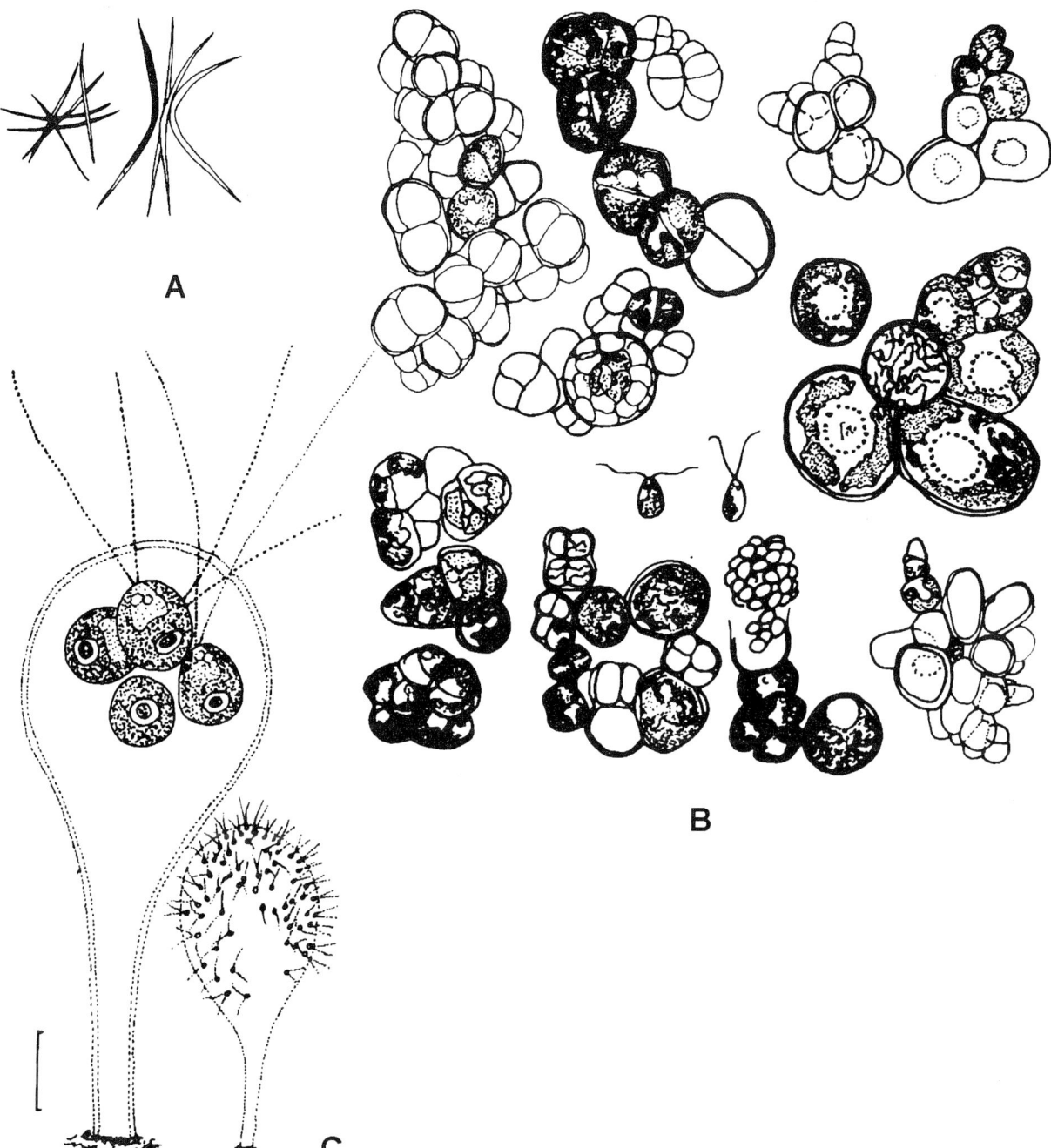

FIGURE 2 A. *Ankistrodesmus falcatus*. B. *Apatococcus lobatus*. C. *Apiocystis brauniana*. A from Prescott (1978) with permission; B, from Ettl and Gärtner (1995) with permission; C from Bourrelly (1966) with permission.

Askenasyella Schmidle (Fig. 3A)

Colony of numerous cells within a globose to irregularly shaped mucilage envelope. Cells are pear-shaped, where the broad anterior end points toward the envelope surface and the narrow posterior end points toward center. Cells are radially arranged, each with a fine protoplasmic thread running from the apex to the chloroplast. There is a parietal chloroplast in the anterior end and one pyrenoid. Rarely recorded; found in some locations in the southeastern United States.

Asterococcus Scherffel (Fig. 3B)

Cells are spherical to broadly ellipsoidal, united into colonies of 2–4–8–16 cells, rarely solitary. The colony envelope is broad and sometimes lamellated. The chloroplast is star-shaped with radiating lobes flattened against the wall and with one central pyrenoid. Found in phytoplankton in ponds and lakes; also occurs in soils. Common and widely reported, but rarely abundant in most regions of North America (Great Lakes region, central Canada, British Columbia, Ontario, Northwest Territories, all regions of the United States, the Caribbean, and Central America).

Botryococcus Kützing (Fig. 3C)

Colony consists of numerous spherical to ovoid cells densely arranged within copious semi-opaque (golden brown) mucilaginous lumps. Parental cell wall fragments, if present, usually are mucilaginized and indistinct. Older cells usually have large amounts of reserve food material (oil) such that the cell contents are obscured. Colonies frequently are compounded by interconnecting strands of tough mucilage between clusters of cells. Cells have a cup-shaped plastid with a naked pyrenoid-like body. Found in phytoplankton and metaphyton of ponds and lakes. Uncommon and geographically widespread, but rarely abundant; from locations in central and eastern United States, British Columbia, Ontario, Nunavut, and Great Lakes region.

Bracteococcus Tereg (Fig. 4A)

Spherical cells occur singly, resembling *Planktosphaeria*, contain several to many parietal, lens-shaped chloroplasts that lack pyrenoids; can form biflagellated zoospores. Occurs in soils. Not widely reported (southeastern United States, the Caribbean, and Central America).

Cerasterias Reinsch (Fig. 4B)

Unicellular, triangular or pyramidal cells, the angles extending into long tapering processes that are all in one plane or more than one plane. The chloroplast is parietal and a pyrenoid is absent. Occurs in plankton in ponds and lakes. Not widely reported (Great Lakes, Ontario, central and eastern Canada, and western, Midwestern, and northeastern United States).

Chaetopeltis Berthold (Fig. 4C)

Colony is a pseudoparenchymatous mass of rounded or closely appressed angular cells forming relatively small, circular disks. Cells have a variable number of pseudoflagella that are long and erect. The chloroplast is cup-shaped and has one pyrenoid. Occurs as an epiphyte on filamentous algae and aquatic plants. Not widely reported; observed in Ontario, and central and southeastern United States.

Chalarodora Pascher (Fig. 4D)

Cells are hemispherical with a depressed apex contained within a gelatinous membrane-like casing that is ovoid to spherical. A few cyanelles form a parietal ribbon around the edge of the cell, and numerous starch grains are dispersed within the cytoplasm. Rarely reported.

Characiochloris Pascher (Fig. 4E)

Cells are elongate–oval to fusiform, pointed anteriorly and narrowed posteriorly to a broad attachment disk that is formed external to the cell wall. Usually several to many narrow chloroplasts. Has parietal plates and may form biflagellated zoospores; epiphytic. Possibly rare; collected from small ponds in Texas (Mosto-Cascallar, 1987) and mentioned as "expected" in southeastern United States (Dillard, 1989).

Characiopsis Borzi (Fig. 4F)

Solitary spherical, elliptical or elongate shaped cells attached to the substratum by a small pad or stalk; distal end may extend into a narrow tip. One to many disk shaped chloroplasts; pyrenoids and starch are absent. Grows on filamentous algae or microfauna. Infrequently recorded (central Canada, western, northeastern, and southeastern United States, the Caribbean, and Central America); now classified in the yellow–green algae (Tribophyceae; Mischococcales; Pizarro, 1995; see Chap. 11).

Characium A. Braun in Kützing [=*Korschikoviella* Silva, p.p. = *Pseudocharacium* Korschikoff, p.p.] (Fig. 4G)

Cells are solitary (sometimes clustered), slightly crescent-shaped, usually straight or nearly so and sometimes S-shaped or sigmoid with poles drawn out into fine points. Usually attached to substratum by a slender stalk with a basal disk. The chloroplast has one to several parietal plates, often diffuse in older cells,

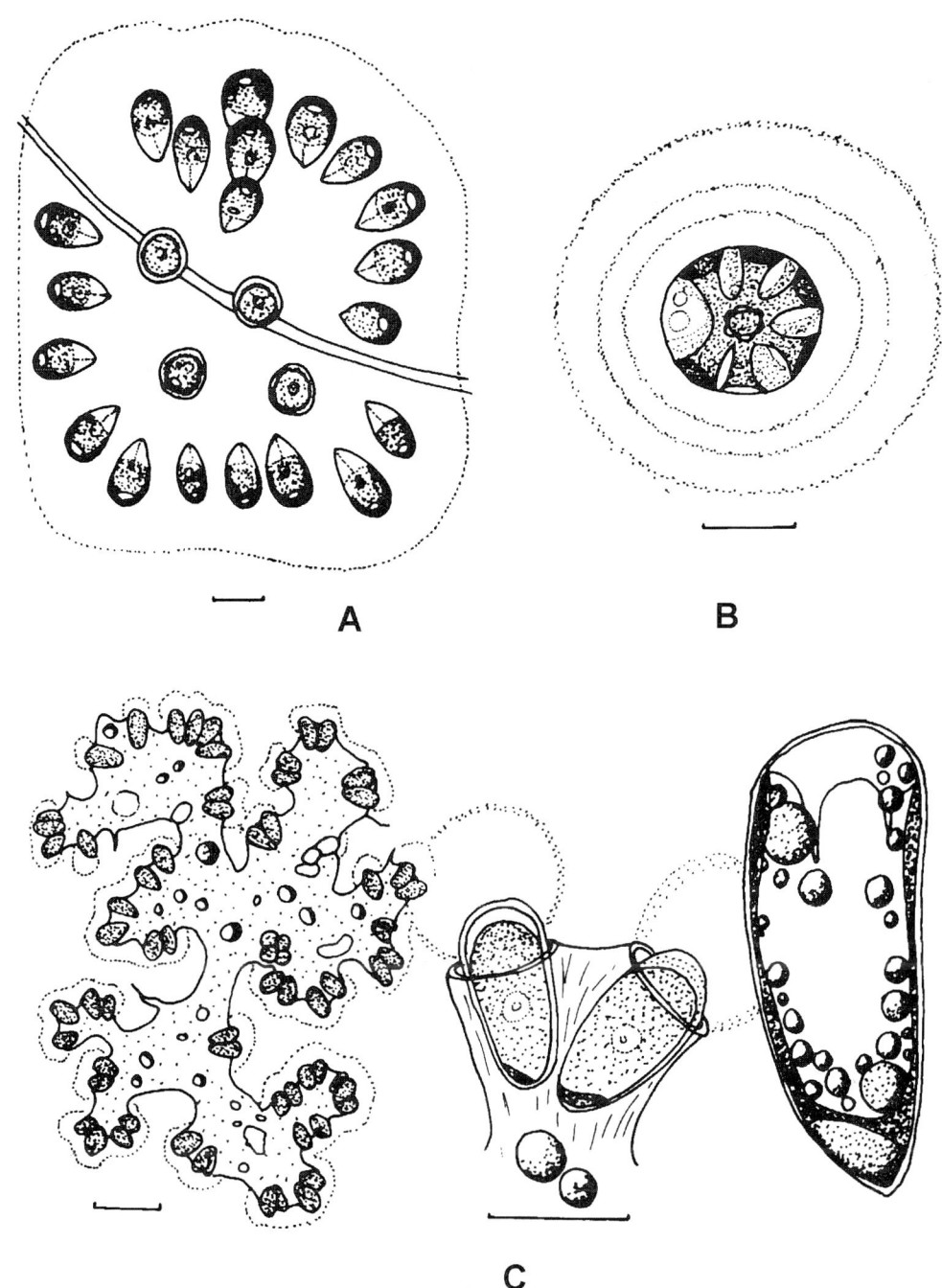

FIGURE 3 A. *Askenasyella chlamydopus*. B. *Asterococcus superbus*. C. *Botryococcus braunii*. A from Prescott (1978) with permission; B, from Ettl and Gärtner (1995) with permission; C from Bourrelly (1966) with permission.

and one to several pyrenoid(s). Often epiphytoplanktonic or epizooplanktonic. Very common and widely reported from all regions of North America.

Chlorangiella De Toni [=*Chlorangium* Stein non *Chlorangium* Link] (Fig. 5A)

Ellipsoidal cells borne at the apices of branching mucilaginous stalks, united into treelike colonies. Chloroplasts are one or two longitudinal bands or cup-shaped, with or without a pyrenoid. Often attached to small crustacea or insect larvae. Uncommon; reported from southeastern United States.

Chlorella Beijerinck [=*Palmellococcus* Chodat (without a pyrenoid) =*Zoochlorella* Brandt (endosymbionts in various animals)] (Fig. 5B)

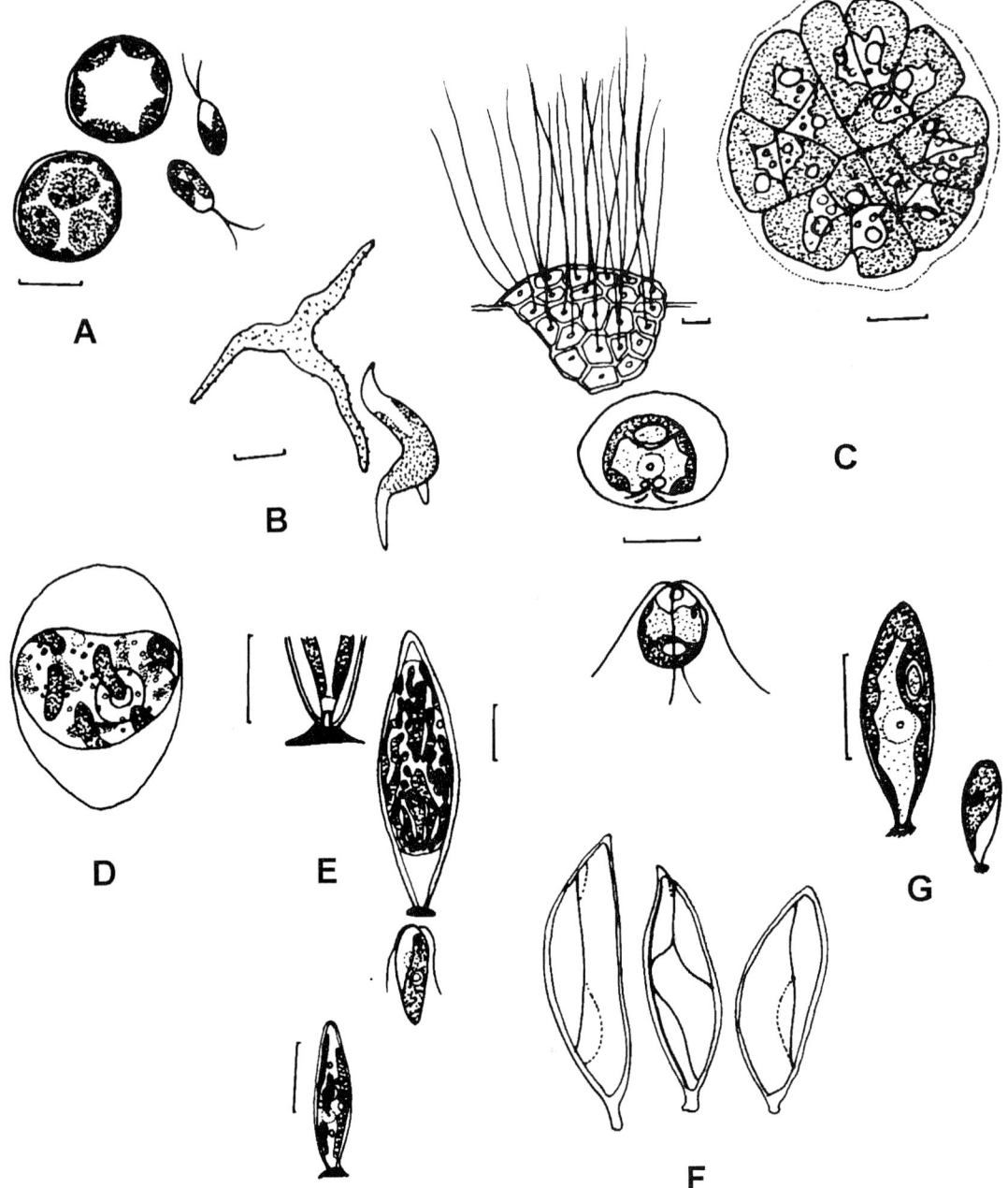

FIGURE 4 A. *Bracteococcus minor*. B. *Cerasterias irregularis*. C. *Chaetopeltis orbicularis*. D. *Chalarodora azurea*. E. *Characiochloris characiodes*. F. *Characiopsis minuta*. G. *Characium sieboldii*. A–E and G from Bourrelly (1966) with permission; F from Ettl and Gärtner (1995) with permission.

Cells are spherical, ovoid, or ellipsoid, solitary or clustered. The chloroplast is cup-shaped or platelike, with or without a pyrenoid. Found in phytoplankton of ponds and lakes, and also on soil or moist subaerial substrata or inhabiting ciliated protozoa, *Hydra*, *Spongilla* or other microfauna. Common and widely reported from all regions of North America.

Chloremys Pascher (Fig. 5C)

Ellipsoidal cell in a casing (smooth, rounded like a skullcap, and with a brownish encrustation around the periphery). Single chloroplast has a pyrenoid and two contractile vacuoles. Found attached to filamentous algae. Possibly rare; mentioned as "expected" in southeastern United States (Dillard, 1989).

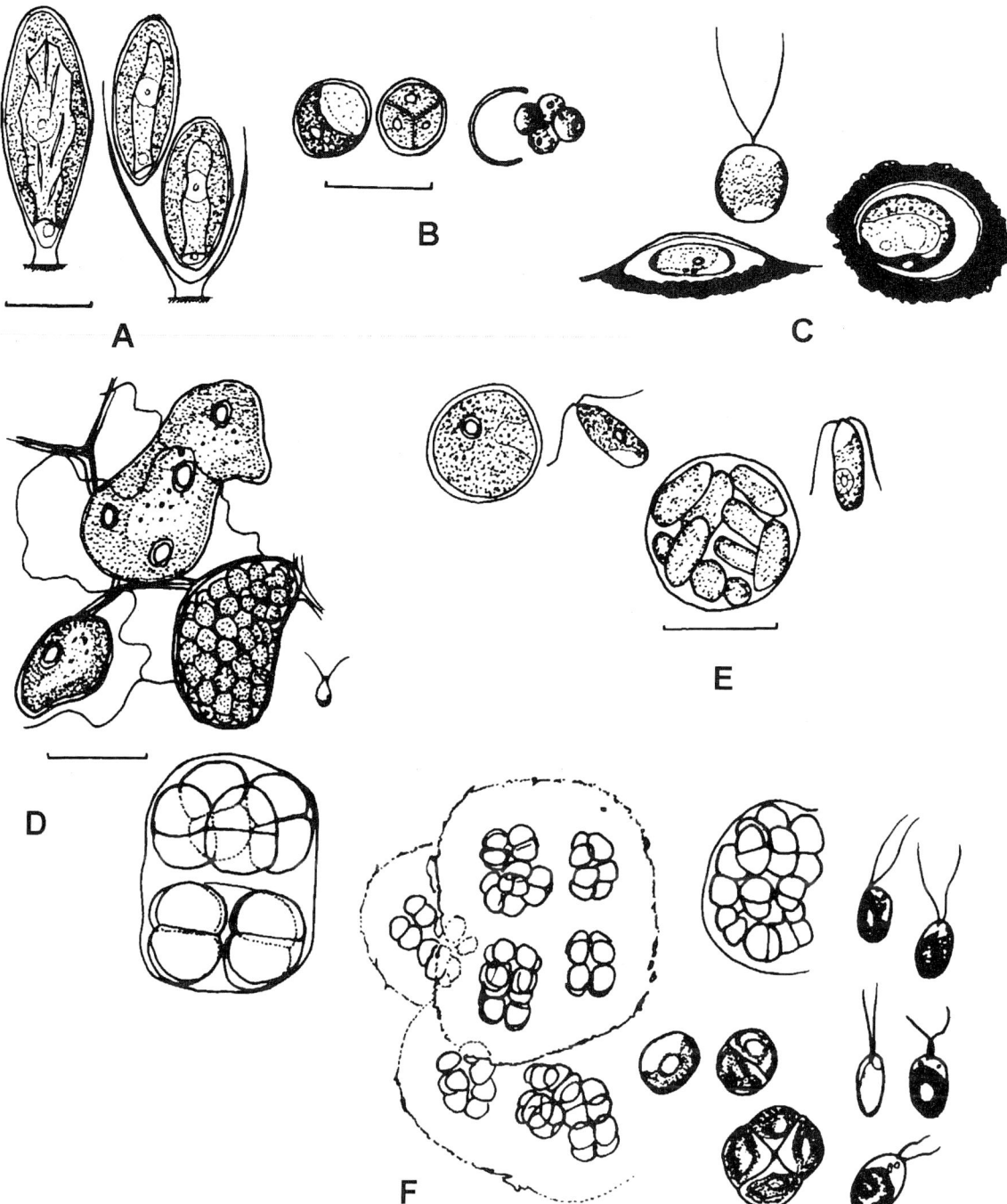

FIGURE 5 A. *Chlorangiella pygmaea*. B. *Chlorella vulgaris*. C. *Chloremys sessilis*. D. *Chlorochytrium lemnae*. E. *Chlorococcum wimmeri*. F. *Chlorokybus atmophyticus*. A–E from Bourrelly (1966) with permission; F from Ettl and Gärtner (1995) with permission.

Chlorochytrium Cohn [=*Scotinosphaera* Klebs] (Fig. 5D)

Unicellular, irregularly oval; cells of mature cells typically are thick and lamellate. The chloroplasts of young cells are cup-shaped, becoming massive and diffuse (netlike) in older cells, and have one to several pyrenoids. Endophytic in tissues of *Lemna* (duckweed) and mosses. Uncommon (central and western Canada and the United States, and south- and northeastern United States).

Chlorococcum Meneghini (Fig. 5E)

Cells are spherical, solitary or sometimes gregarious; colonial mucilage is thin and sometimes not evident. The chloroplast is cup-shaped, parietal, and has one pyrenoid. Can form biflagellated zoospores. Most often collected from soils or subaerial habitats. Fairly common and widely distributed in most regions of North America, including the Caribbean and Central America.

Chlorokybus Geitler (Fig. 5F)

Cells are elliptical, hemispherical, or rounded–polygonal to almost spherical, with distinct solid cell walls. The thallus is formed of microscopic, almost amorphous, homogeneous, soft mucilage, with up to 100 or more cells that typically lie in cuboidal groups of two to four to eight, butted up against each other. Cell groups are held together by the persistent, gelatinized parent cell wall. The chloroplast is bowllike, with a thickened basal part, often slightly lobed at the edge; two types of pyrenoid are present: the larger more central pyrenoid has starch grains, whereas the smaller naked pyrenoid (pseudopyrenoid) always lies below on the edge of the chloroplast; starch grains lie more or less parallel to each other. Can form biflagellated zoospores. Subaerial. Rarely reported.

Chlorophysema Pascher (Fig. 6A)

Cells are irregularly globular, sedentary on a broad attaching base, stalk practically wanting, with a thick, brown sheath enclosing a protoplast that does not fill it. The protoplast contains a cup-shaped chloroplast, an eye spot, and a contractile vacuole (giving the general appearance of a shelled zoospore). Epiphytic and epizoic. Uncommon (Caribbean and Central America).

Chlorosarcina Gerneck em. Vischer
[=*Pleurastromsarcina* Sluiman & Bloomers] (Fig. 6B)

Cells are solitary or more often in cuboidal, ball-like, cell packets. The cells are flattened against each other, the cell wall is thin, but slightly thickened in older cells. The chloroplast is parietal, bowllike, and lacks a pyrenoid. Spiny resting stages. Found among epidermal cells of aquatic plants, but sometimes found free-living. *Chlorosarcinopsis* is similar but has pyrenoids and colonizes soils. A few scattered reports from western, midwestern, and southeastern United States locations.

Closteriopsis Lemmermann (Fig. 6C)

Cells are solitary, needle-like, acute apices, with no mucilaginous envelope. The chloroplast is single and not interrupted at the midregion, with an axial row of 12–(16) pyrenoids. Found in phytoplankton of ponds and lakes. Fairly common, with collections from most locations across North America, but possibly not in far northern localities.

Coccomyxa Schmidle (Fig. 6D)

Colony of numerous cells that are ellipsoidal to cylindrical and irregularly distributed within a colonial mucilage envelope with some tendency for their long axes to be parallel. Cells are united by the confluence of individual mucilage envelopes. The chloroplast is parietal and has one pyrenoid. May grow as a free-living aerial alga, in plankton (Boucher *et al.*, 1984), attached to wood or on soil, and as an epiphyte on or endophyte in lichens. Scattered reports from a variety of habitats and regions, including eastern, central, and western United States and Canada, southeastern United States, and the Nunavut.

Coelastrum Nägeli (Fig. 6E)

Colony is 4–8–16–32–(64)-celled, spherical and hollow. Cells are spherical to polygonal, adjoined by interconnecting protuberances of the mucilaginous cell sheaths, with or without ornamentation, but lacking sharp spines or processes. The chloroplast is parietal and has one pyrenoid. Common member of the phytoplankton in mesotrophic or eutrophic ponds and lakes. A mostly planktonic genus, reported from all regions of North America, including locations in the Caribbean and Central America.

Coronastrum Thompson (Fig. 6F)

Colony is four-celled or multiples of four, quadrate, flattened, and separated by parental cell wall strands. Colonies are solitary or interconnected by gelatinous strands. Cell walls form 8–16-celled multiple coenobia. Cells are incompletely globose, ovoid, ellipsoid, or lunate, bearing a distinctive winglike scale projection on the outer face (parental cell wall remnant). The chloroplast is parietal and has one pyrenoid. Found in phytoplankton of ponds and lakes. An uncommon or underrecorded genus reported from northeastern and southeastern United States and the Caribbean region.

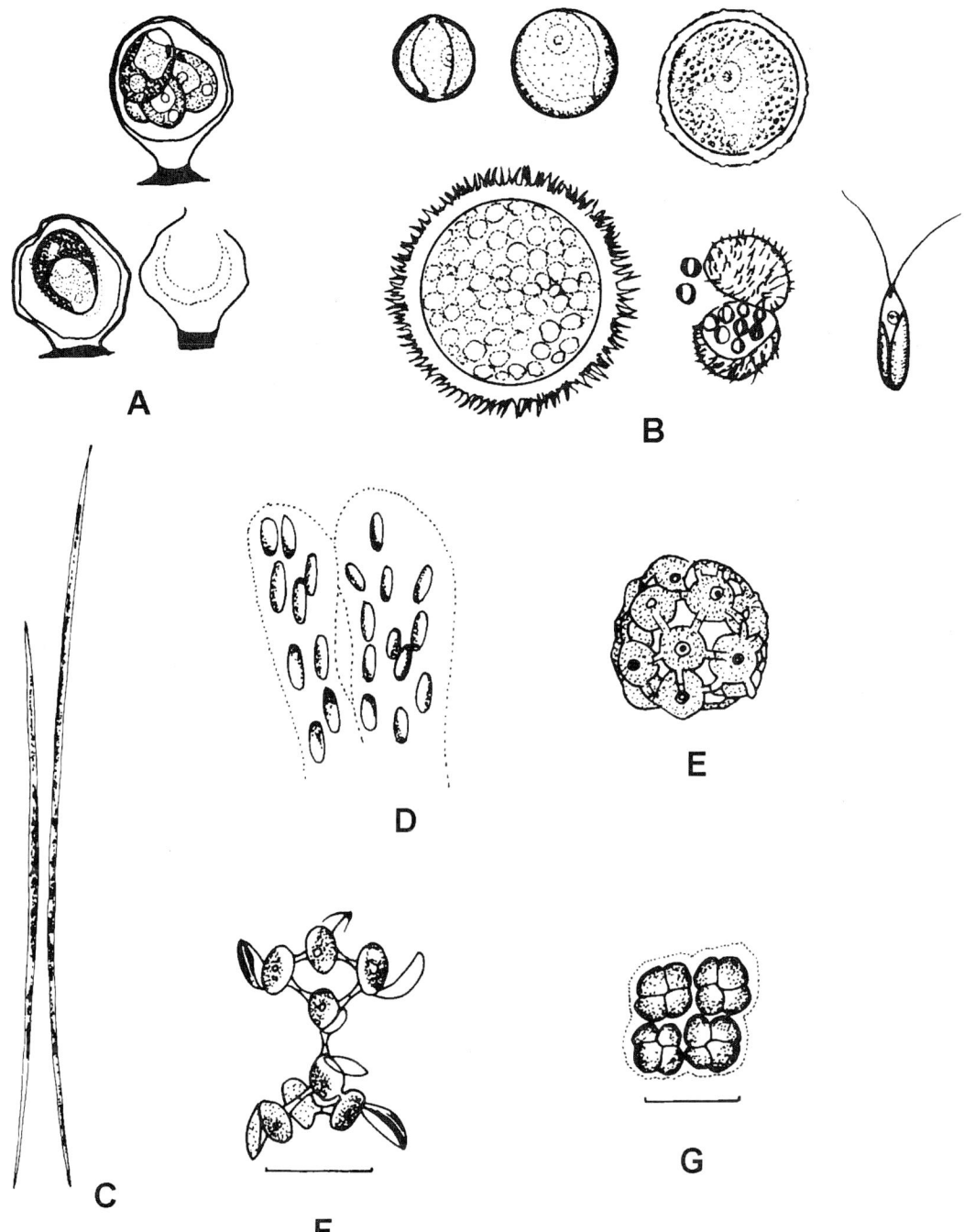

FIGURE 6 A. *Chlorophysema contractum*. B. *Chlorosarcina brevispinosa*. C. *Closteriopsis longissima*. D. *Coccomyxa dispar*. E. *Coelastrum reticulum*. F. *Coronastrum aestivale*. G. *Crucigenia quadrata*. A and E–G from Bourrelly (1966) with permission; B, from Ettl and Gärtner (1995) with permission; C and D from Prescott (1978) with permission.

Crucigenia Morren (Fig. 6G)

Colony is four-celled, quadrate, and flattened. Colonies are solitary or adjoined to form 8–16-celled multiple coenobia. Cells in the surface view are ellipsoid to ovoid, triangular, or rectangular. The chloroplast is parietal, with or without a pyrenoid. Found in phytoplankton of ponds and lakes. Often recorded but rarely abundant genus from phytoplankton or metaphyton; reported from arctic to tropical regions of North America

Deasonia Ettl *et* Komárek (Fig. 7A)

Single cell (coenocyte) that has elliptical young cells or ovoid mature cells, always spherical with a smooth and mostly thickened cell wall, and with bifurcations and indentations in mature cells. Ultimately the cells becoming reticulate, sometimes clearly striped or ridged on the outer surface. The chloroplast is parietal, a more or less an open cup in young cells and has holes. The pyrenoid in young cells is positioned off-center in compact chloroplasts and moves to the center in mature cells. There is a starch sheath around the pyrenoid that is continuous cup-shaped or articulated. Vegetative stages are multinucleate. "Storage" vacuoles occasionally are present. Can form biflagellated zoospores. Occurs in soil. Rarely recorded.

Desmatractum West *et* West (Fig. 7B)

Cells are solitary, spherical to ellipsoidal, and enclosed in a longitudinally ribbed sheath that tapers to each end (often appearing like a minute seed pod with a single, globular seed). The chloroplast is broad and parietal, with one pyrenoid. Found in the phytoplankton and metaphyton of ditches, bogs, ponds, and lakes. Recorded from several disjunct regions of North America, including northeast and southeastern United States, British Columbia, and Caribbean freshwater locations.

Desmococcus Brand em. Vischer [=*Pleurococcus* p.p. *sensu auct*, *Protococcus* p.p. *sensu auct.* Brand]
(Fig. 7C)

Cells united in cuboidal packets, but also forming short branched, uniseriate pseudofilamentous threads. The chloroplast is relatively small, parietal, and trough-like, with one, small, often obscure, naked pyrenoid. Subaerial. Frequently reported (usually as *Protococcus* species) from suitable habitats in several regions, including the Great Lakes, central and western Canada, south- and northeastern United States, and in the Caribbean.

Desmodesmus An, Friedl, *et* Hegewald (Fig. 8A)

Unicells or coenobia 2-4-8-16-celled, with long axes of cells parallel, laterally adjoined, and arranged in a single linear or alternating series. Cells are ellipsoidal to ovoid, and spines usually are present on the terminal cells and/or medial cells, but may be entirely absent. The cell wall may have ridges, warts, or nets. The chloroplast is parietal, usually with one pyrenoid. Found in the phytoplankton of ponds and lakes. An extremely common genus (and occasionally abundant), but previously recorded as one of the spine-producing species of *Scenedesmus* (e.g., *S. armatus*), reported from all regions of North America.

Diacanthos Korschikoff (Fig. 8C)

Cells are solitary, free, ellipsoidal, 10 μm long, and have a very long hollow spine at each pole, oriented along the long axis of the cell. The wall is thin, colorless, and without a mucilaginous sheath. A single, parietal plastid with a pyrenoid is present. Planktonic; North America.

Dicellula Swirenko (Fig. 8B)

Cells are oval and the walls are beset with needle-like spines with thickened bases. Cells occur in pairs, side by side, and have parietal chloroplasts with pyrenoids. Found in the phytoplankton of ponds and lakes. Uncommon; reported from a few scattered localities, including the Caribbean.

Dictyochlorella Silva [=*Dictyochloris* Korschikoff *non Dictyochloris* Vischer] (Fig. 8E)

Colony of 4–16 cells embedded in an amorphous mucilage envelope. Cells are spherical to ovoid, not contiguous within the colony, and connected by fine threads of parental cell wall material. Cells have a parietal chloroplast, without a pyrenoid and may form zoospores. Found in plankton. Uncommon. Reported from a few localities, including sites in the southeastern United States.

Dictyochloris Vischer [*non Dictychloris* Korschikoff] (Fig. 8D)

Single cells (coenocyte) are spherical, with thin, smooth cell walls that do not become obviously thicker when mature. The chloroplast is netlike and perforated, lacks pyrenoids, and is parietal in young cells, later in mature cells forming an irregular three-dimensional network. Subaerial and terrestrial. Uncommon; reported (as *Dictyochlorella*) in several locations in southeastern United States.

Dictyochloropsis Geitler em. Tschermak-Woess
(= *Myrmecia* p.p. *sensu* Tschermak-Woess) (Fig. 9A)

Cells are solitary, ellipsoidal, kidney-shaped, but mostly spherical. The cell wall is thin or compact,

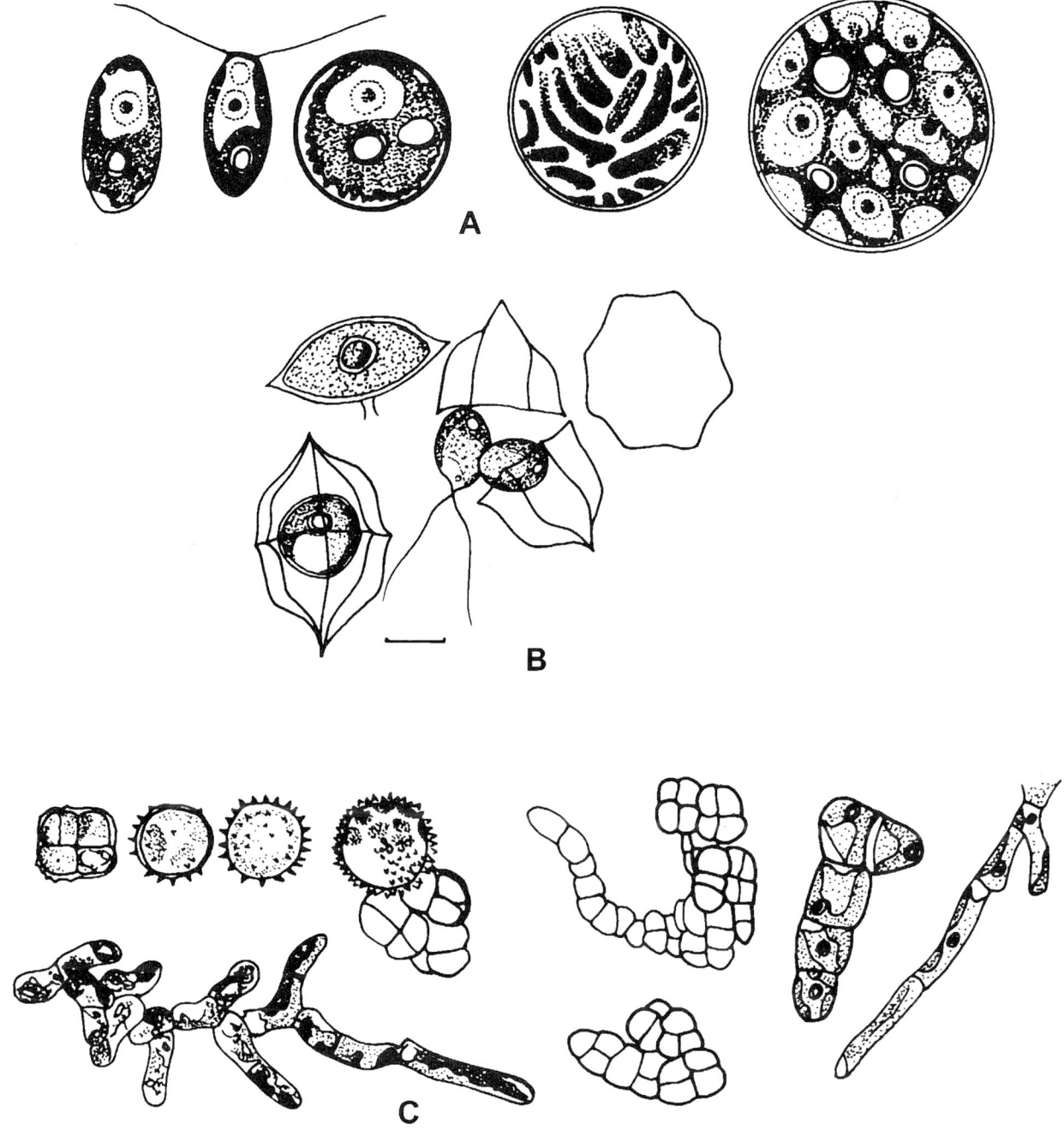

FIGURE 7 A. *Deasonia granata.* B. *Desmatractum bipyramidatum.* C. *Desmococcus olivaceus.* A and C from Ettl and Gärtner (1995) with permission; B from Bourrelly (1966) with permission.

smooth, and sometimes has slight local thickenings; in a few cases have mucilage sheaths. The chloroplasts in young cells are simply formed, potlike, and also irregular. The parietal part is bound to the central portion in different ways: in mature cells it is a thick, hollow sphere to netlike; toward the periphery it divides into ever more markedly articulate ridges, cords, and lobes, to three-dimensional netlike, and has no pyrenoid. Can form biflagellated zoospores. Generally aerophytic or terrestrial. Rarely recorded.

Dictyococcus Gerneck em. Vischer (Fig. 9B)

Single, spherical cell (coenocyte) with smooth, thin cell wall that becomes thickened in older cells. The protoplast comprises several abutting segments, separated only by a plasma membrane. The number of

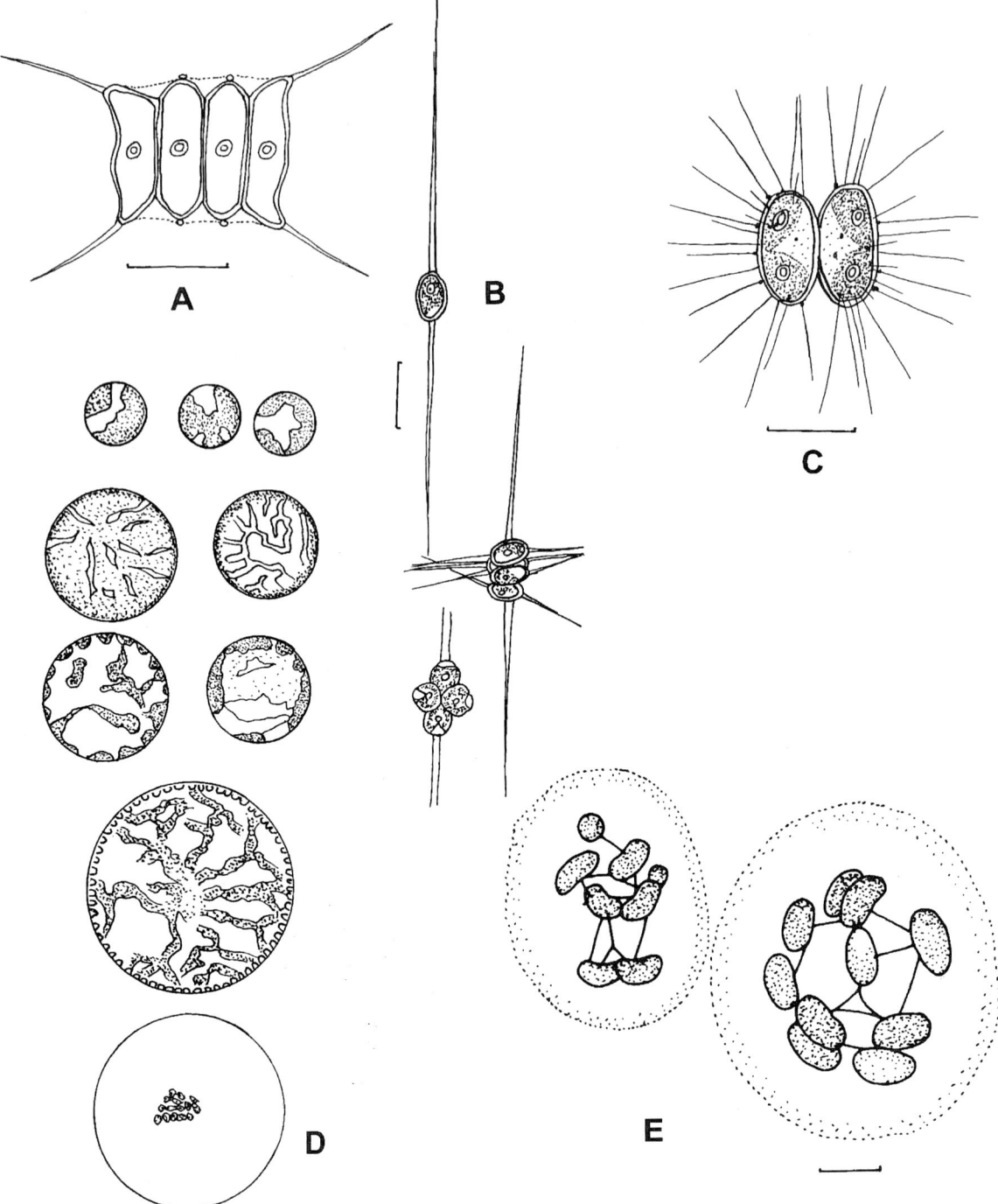

FIGURE 8 A. *Desmodesmus protuberans.* B. *Diacanthos belenophorus.* C. *Dicellula planctonica.* D. *Dictyochloris fragrans.* E. *Dictyochlorella reniformis.* A–C and E from Bourrelly (1966) with permission; D from Ettl and Gärtner (1995) with permission.

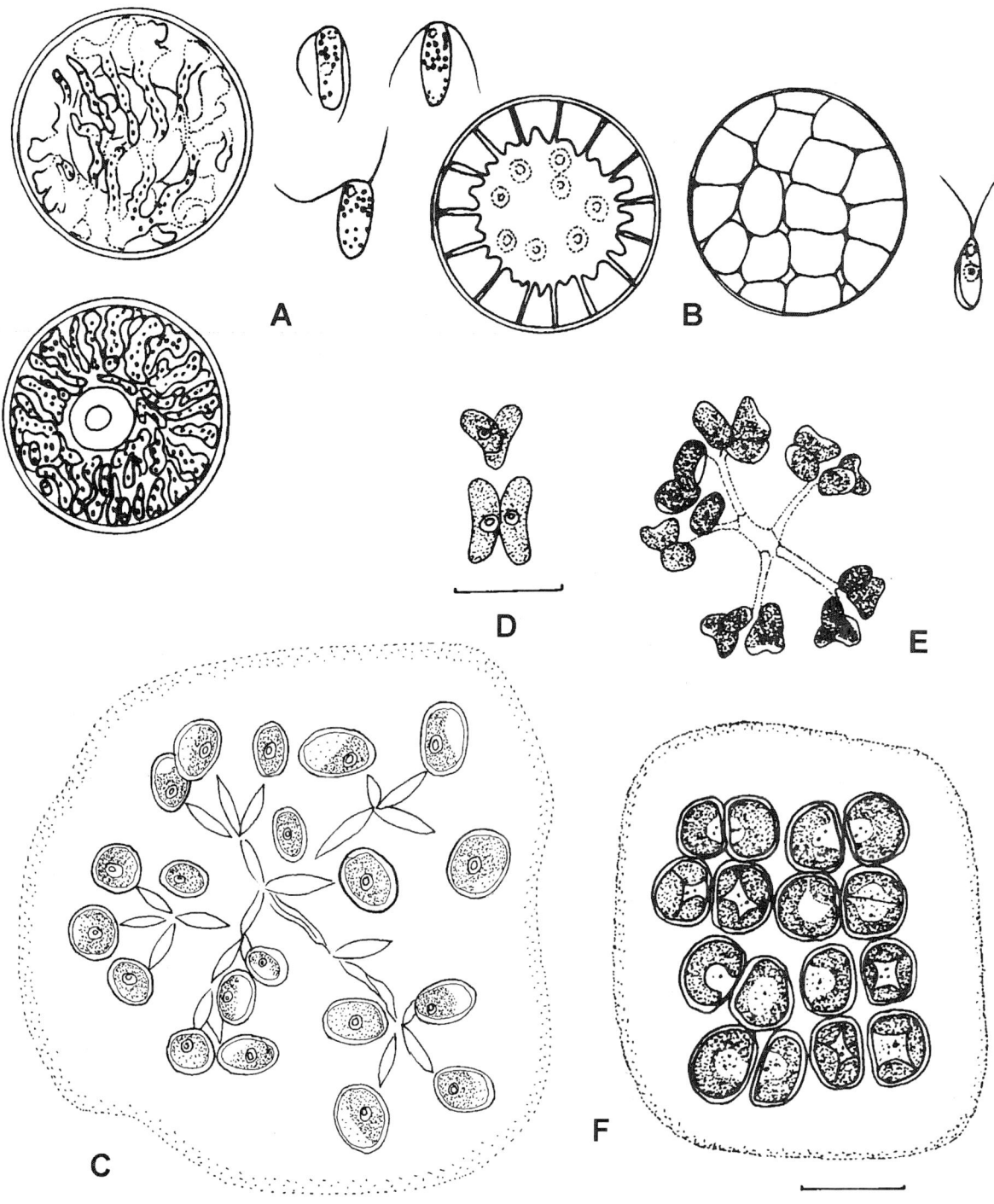

FIGURE 9 A. *Dictyochloropsis splendida*. B. *Dictyococcus varians*. C. *Dictyosphaerium pulchellum*. D. *Didymogenes palatina*. E. *Dimorphococcus lunatus*. F. *Dispora crucigenoides*. A and B from Ettl and Gärtner (1995) with permission; C, D, and F from Bourrelly (1966) with permission; E from Prescott (1978) with permission.

segments increases during the growth of the organism; each segment has a parietal chloroplast that is divided into several pieces, polygonal in outline, thick, bent toward the inside along the margins like a seam, but lacking pyrenoids. Can form biflagellated zoospores. Occurs in soils. Rarely recorded.

Dictyosphaerium Nägeli (Fig. 9C)

Cells are spherical to broadly oval, in clusters of four, attached by parental cell wall fragments radiating from a common center. Sometimes an indefinite clear mucilaginous envelope can be discerned. The chloroplast is cup-shaped, parietal, and covers most of the wall; there is a single pyrenoid. Found in the phytoplankton of ponds, lakes, and rivers, and the metaphyton of ditches and ponds. Very common in lakes and ponds, and frequently reported (but rarely abundant) throughout North America, from arctic to tropical localities.

Didymogenes Schmidle (Fig. 9D)

Colony is two to four-celled and flat. Cells are curved and apices are rounded, with convex surfaces apposed; may or may not have spines or granules. The chloroplast is parietal and there is one pyrenoid. Planktonic. A few reports, mainly in the southeastern United States.

Dimorphococcus A. Braun (Fig. 9E)

Cells are arranged in groups of four on the branched parental cell wall fragments. Cell groups consists of two cell shapes: two ovoid and two kidney- to heart-shaped. The colony has no gelatinous sheath. Young cells have a single parietal chloroplast with one pyrenoid; old cells have chloroplasts that completely fill the cell and the pyrenoid often is obscured by starch grains. Found in the phytoplankton and metaphyton of bogs, ponds, and lakes and intermingled with non-planktonic algae. Fairly common and reported from most regions in North America, including eastern, central and western United States and Canada, and the Caribbean.

Dispora Printz (Fig. 9F)

Platelike (2)–4–32–(128)-celled, colony flattened and in one layer, with a mucilaginous envelope. Cells are broadly ellipsoid to oval, somewhat irregularly arranged in groups of four within the envelope. The chloroplast is parietal and there is no pyrenoid. Found in the metaphyton of ditches, ponds, and lakes, especially in low pH conditions. Uncommon; reported from western United States and Canada, southeastern United States, and Caribbean/Central American localities.

Echinosphaerella G. M. Smith (Fig. 10A)

Cells are spherical and solitary, bearing numerous long, stout, clear spines that are broad at the base and tape to a sharp point, and are regularly distributed over the cell wall. The chloroplast is cup-shaped and has one pyrenoid. Found in plankton. Uncommon or possibly underreported; from localities in eastern and western Canada, and southeastern United States.

Ectogeron Dangeard [=*Eremotyl* Geitler] (Fig. 10B)

Cells are solitary, more or less rounded in outline, but flattened (lenticular). The wall is thin against the substratum but thickened on its free face. A parietal plastid surrounds the entire cell and there are numerous pyrenoids. Occurs attached to submerged leaves of higher aquatic plants. Rarely recorded or possibly overlooked; mentioned as "expected within the southeastern United States." by Dillard (1989).

Elakatothrix Wille [=*Raphidium* Kützing p.p. *sensu auct.*, *Spirotaenia* Brébisson] (Fig. 10C)

Colony is two- to four- to many-celled, with one or both cell poles tapering to a slender point. Cells are arranged in a colony with their long axes more or less parallel, and the chloroplasts are parietal and have one or two pyrenoids. Found in phytoplankton and metaphyton of ponds and lakes. A common genus in nutrient-rich water bodies with other members of Chlorococcales. Widely reported from most localities across North America; not know from arctic regions.

Eremosphaera De Bary (Fig. 11A)

Cells are spherical (sometimes angular in face view), large (up to 300 µm in diameter), and solitary (two to four cells sometimes retained within parental cell wall), with or without a mucilaginous envelope. Chloroplasts are numerous and parietal, disklike to irregular in shape, lumpy with starch grains, and arranged in radiating strands from the center of the cell; each has a pyrenoid. Found in the metaphyton of ditches, bogs, ponds, and lakes, particularly in conditions of low pH, associated with desmids. Common and frequently reported from most parts of North America, including eastern, central, and western United States and Canada, and the Caribbean.

Eustropsis Lagerheim (Fig. 11B)

Two-celled colonies. The apposed bases of the two cells are flattened and the distal ends are deeply lobed. Has diffuse chloroplasts, each cell containing a pyrenoid. Found in the phytoplankton and metaphyton of bogs, ponds, and lakes. Uncommon, but reported in widely scattered geographic locations; in Canada from British Columbia to Ontario, and in the United States, in central to western locations.

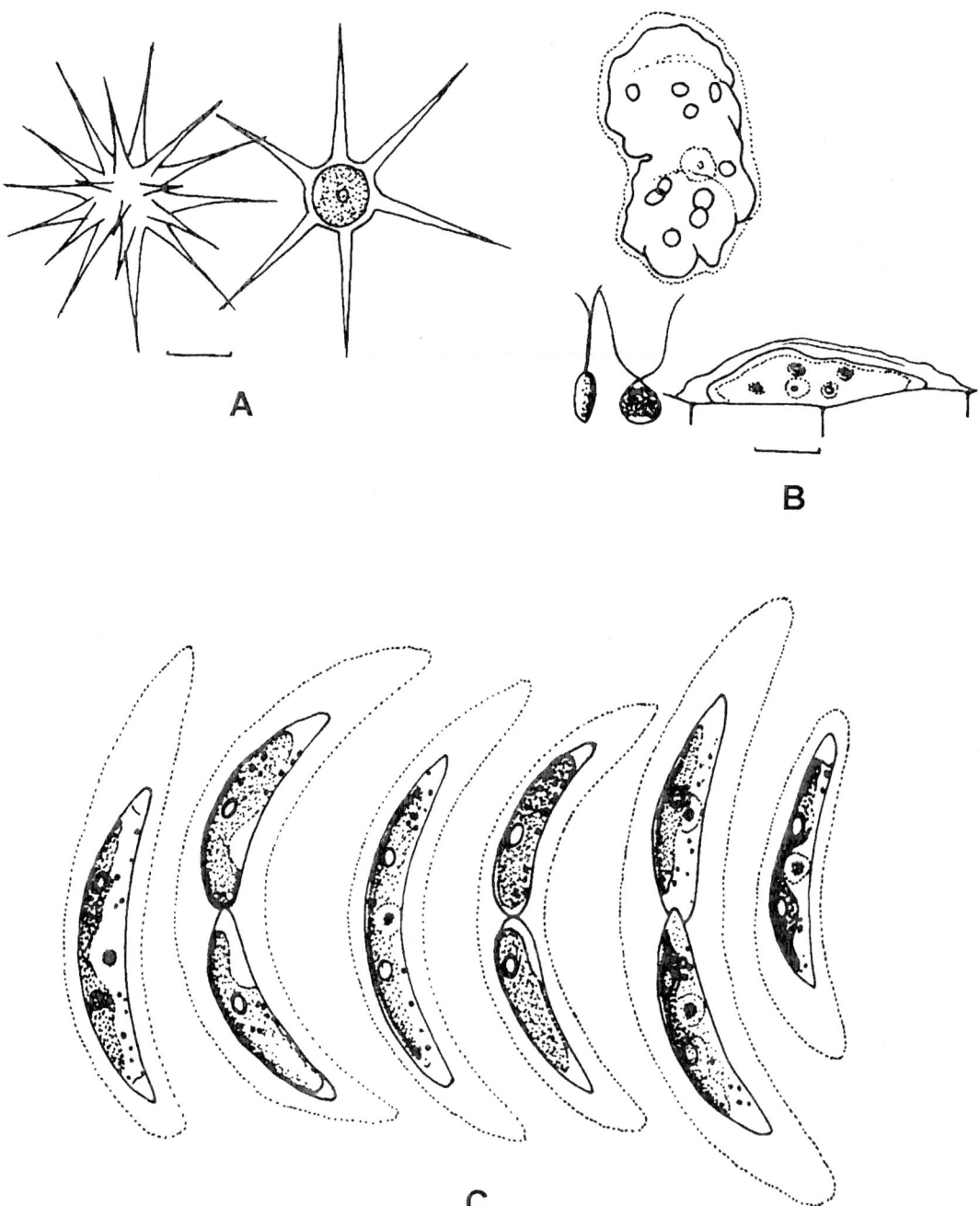

FIGURE 10 A. *Echinosphaerella limnetica*. B. *Ectogeron elodeae*. C. *Elakatothrix inflexa*. A and B from Bourrelly (1966) with permission; C. from Ettl and Gärtner (1995) with permission.

Excentrosphaera G. Moore (Fig. 11C)

Cells are spherical, ellipsoidal to pearlike, and solitary. The cell walls are irregularly thickened and sometimes lamellate. Chloroplasts are numerous, cone-shaped, closely appressed, and directed inwardly from the parietal position against the wall; each has one to several pyrenoids. Found in the metaphyton of bogs, ponds, and lakes, particularly in conditions of low pH.

Infrequently reported from sites in western, central, northeastern, and southeastern United States.

Fasciculochloris McLean *et* Trainor (Fig. 12A)

Cells are single or in 4–32-celled groups, often also in cuboidal to irregularly packet-like colonies, formed by earlier divisions in two or three directions. Always embedded in common mucilage. Young cells are ovoid

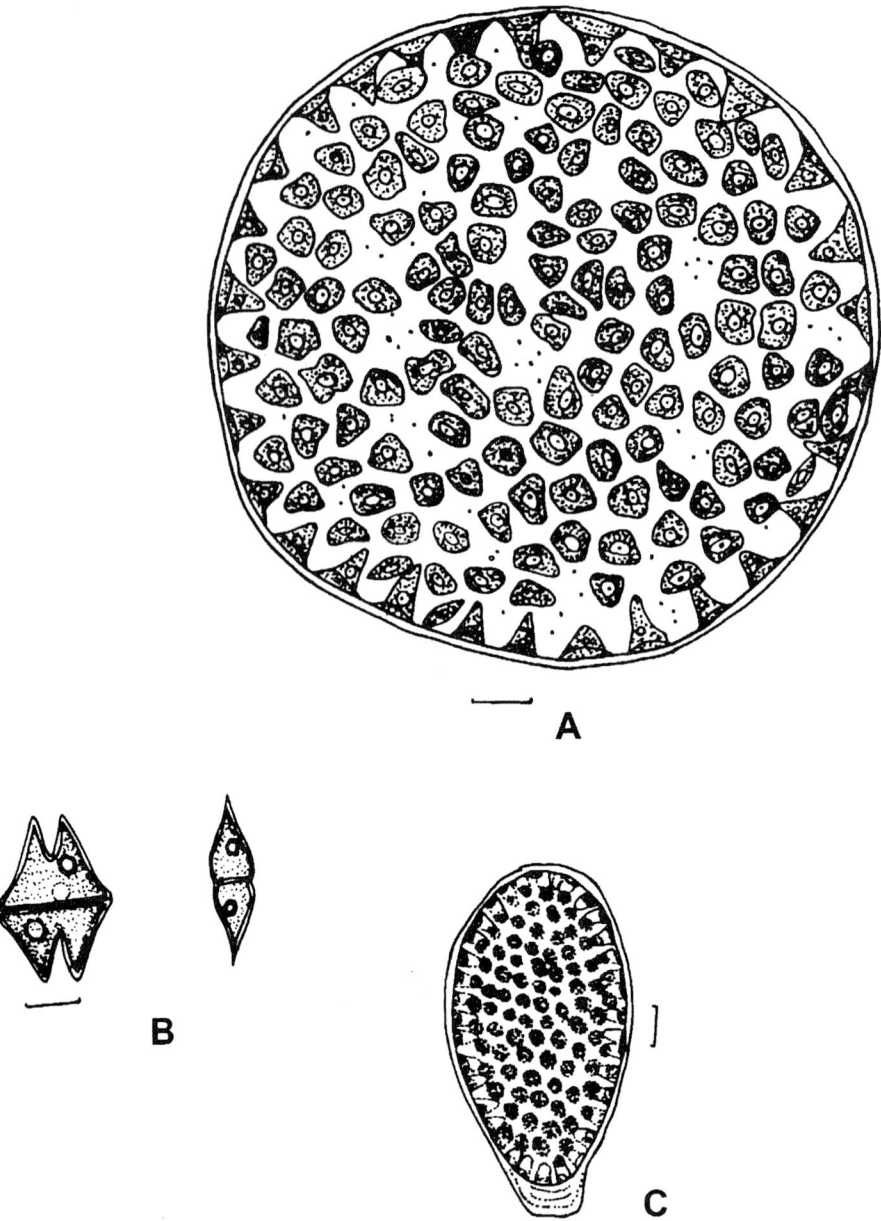

FIGURE 11 A. *Eremosphaera viridis*. B. *Eustropsis richteri*. C. *Excentrosphaera viridis*. A–C from Bourrelly (1966) with permission.

elliptical; mature cells are broadly elliptical to spherical, slightly flattened on abutting sides, with smooth, thin or slightly thickened cell walls, often surrounded by additional, more or less thick, homogeneous slime sheaths. Chloroplast in young cells is channeled to cuplike; later clearly potlike, with deep incisions, strongly and irregularly lobed in adult cells. Cells have one pyrenoid with a starch covering of two or less large starch layers; sometimes pyrenoid doubling occurs. In older cells the chloroplast commonly is obscured due to overlying starch and oil accumulation. Can form biflagellated zoospores. Found in soils. Rarely recorded and distribution unknown.

Fottea Hindák (Fig. 12B)

Cells usually single, but can be in colonies of 4–16 or more, irregularly arranged in homogeneous, but clearly defined mucilage. The form of the colony is variable, usually irregular, sometimes oval to round. Cells in the interior of the colony often form short threads or chains. Individual cells are slightly or many times longer than broad, straight to slightly bent, and

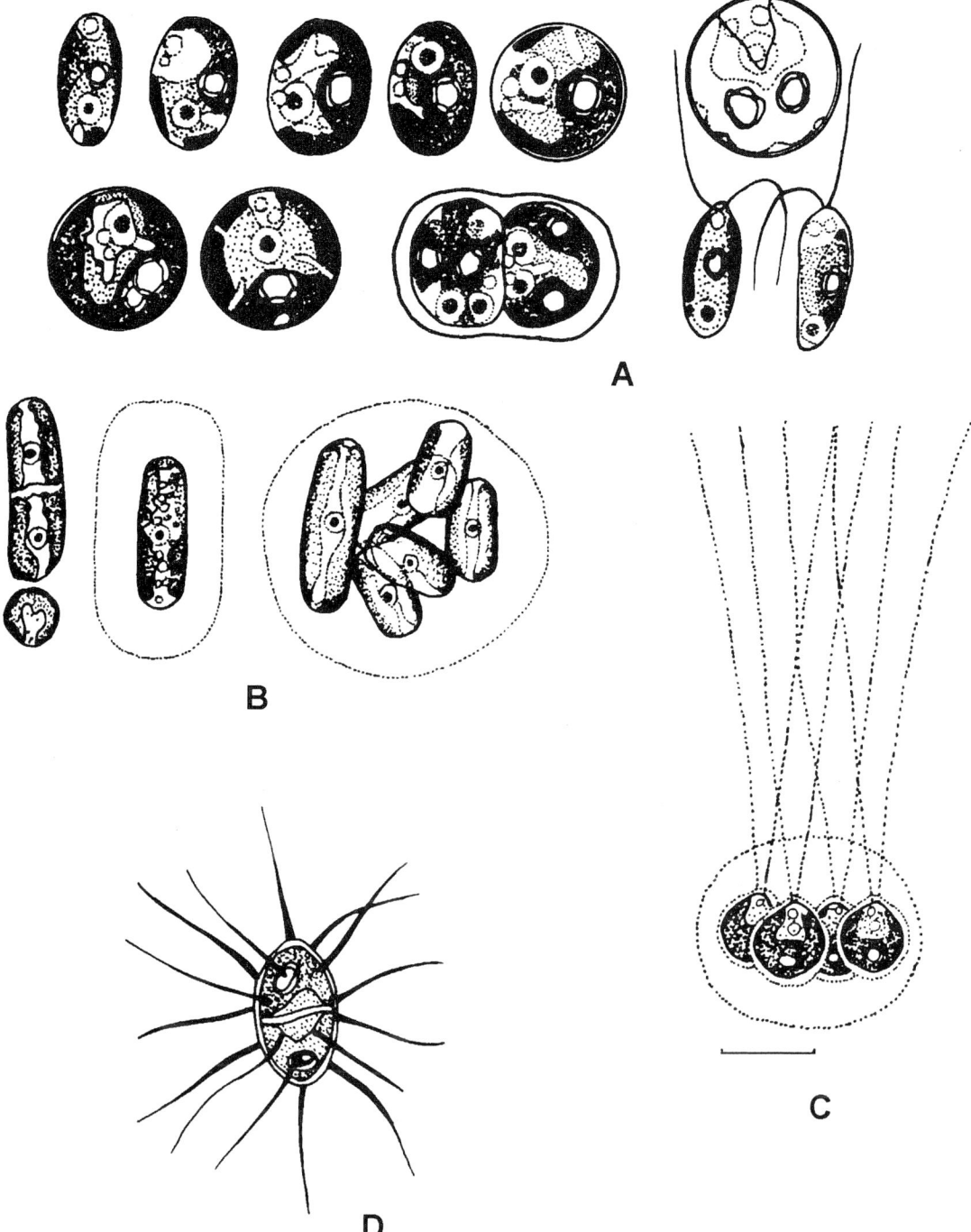

FIGURE 12 A. *Fasciculochloris boldii*. B. *Fottea cylindrica*. C. *Fottiella quadrangularis*. D. *Franceia droescheri*. A and B from Ettl and Gärtner (1995) with permission; C and D from Bourrelly (1966) with permission.

have broadly rounded ends. Chloroplasts are curved or channel-like, and may or may not have pyrenoids. Terrestrial or aerophytic. Rarely recorded.

Fottiella Ettl (Fig. 12C)

Very similar to *Tetraspora*, but with small spherical or irregularly lobed colonies that contain fewer cells with extremely long pseudoflagella, which are 20–30 times longer than the cell and emerge from the mucilaginous coenobium. Found in ditches, ponds, and lakes. Rarely recorded.

Franceia Lemmermann [=*Bohlinia* Lemmermann] (Fig. 12D)

Cells are ellipsoidal and solitary, bearing numerous delicate spines uniformly distributed over the surface; sometimes the cells are arranged side by side because of interlocking spines. The one to four chloroplasts are parietal and pyrenoids may be present. Rare in phytoplankton of ponds and lakes. Appears exactly like the unicell stage of the coenobial morph of *Desmodesmus*, which has lateral spines on the terminal cells; hence, it is likely a unicellular stage of *Desmodesmus*. Possibly uncommon or underreported, but known from several locations in the Great Lakes region, British Columbia and eastern (temperate) Canada, the southeastern United States, and the Caribbean.

Gloeoactinium G. M. Smith (Fig. 13A)

Gelatinous globular colony. Cells are elongate-ovoid to wedge-shaped, in radiating clusters, and arranged mostly in groups of four, apposed at their bases. Cell groups occur at the periphery of a wide, homogenous mucilage envelope. Cells have one parietal chloroplast and no pyrenoid. Found in plankton of ponds and lakes. Few reports, but in several different regions, including (temperate) western and eastern Canada, and southeastern United States.

Gloeochaete Lagerheim (Fig. 13B)

Colony of two to four to as many as eight spherical cells each with an endosymbiotic cyanelle. Cells are spherical with one to two long pseudoflagella. Occurs as an epiphyte. An infrequently recorded alga; reported from southeastern United States and British Columbia.

Gloeococcus A. Braun (Fig. 13C)

Noncoenobic, spherical, nonmotile colonies that contain feebly motile, biflagellate, *Chlamydomonas*-like cells embedded in a gelatinous matrix. Some species are aquatic and produce apple-sized colonies; others are found in soils. A few reports from eastern United States and Canada, and western United States.

Gloeocystis Nägeli (Fig. 13D)

Cells are spherical, ovoid or ellipsoid and united into colonies of two to four to eight to numerous cells; rarely solitary. The colony envelope is broad and firm; some have concentric layers of mucilage. The chloroplast is generally cup-shaped and has one pyrenoid. Found in the metaphyton of ditches, bogs, ponds, and lakes. A fairly common, widespread genus reported from at least a few localities in nearly all parts of North America, with the possible exception of arctic habitats.

Gloeodendron Korschikoff (=*Schizodictyon* Thompson) (Fig. 14A)

Thallus micro- to macroscopic, consisting of branched, gelatinous strands that are freely anastomosing and lamellate transversely, enclosing spherical, oval, or globular cells that are sparsely distributed throughout the length of the tube in linear pairs (the thallus is faintly colored). Can form biflagellated zoospores. Possibly rare; regarded as "expected" in the southeastern United States (Dillard, 1989).

Gloeotaenium Hansgirg (Fig. 14B)

Colony is spherical in broad view, consisting of two to four to as many as eight spherical to ellipsoidal, contiguous crosslike cells arranged within the persistent parental cell wall. Cells are separated within the colony by dark colored deposits that are I-shaped in two-celled colonies and X-shaped in four-celled colonies. The chloroplast is massive and parietal. Found in the metaphyton of ditches, ponds, and lakes, particularly acidic water. Fairly common; reported from central, southeastern, and northeastern United States, eastern Canada, British Columbia, and the Caribbean.

Golenkinia Chodat (Fig. 14C)

Cells are spherical, solitary, and free-floating, long, and have slender, tapering spinelike setae covering the outer wall. Sometimes cells form a "false colony" by interlocking the setae. Member of the phytoplankton in rivers, ponds, and lakes. Common and occasionally abundant in plankton. Reported from most regions of North America, with the possible exception of arctic areas.

Golenkiniopsis Korschikoff (Fig. 14D)

Cells are solitary, spherical to slightly ovoid, and have spines that are thickened at the base or that taper from the base to the apex. The chloroplast is parietal and cup-shaped, and has one spherical to ellipsoidal pyrenoid. Found in plankton. Infrequently reported from localities in southeastern United States and western Canada.

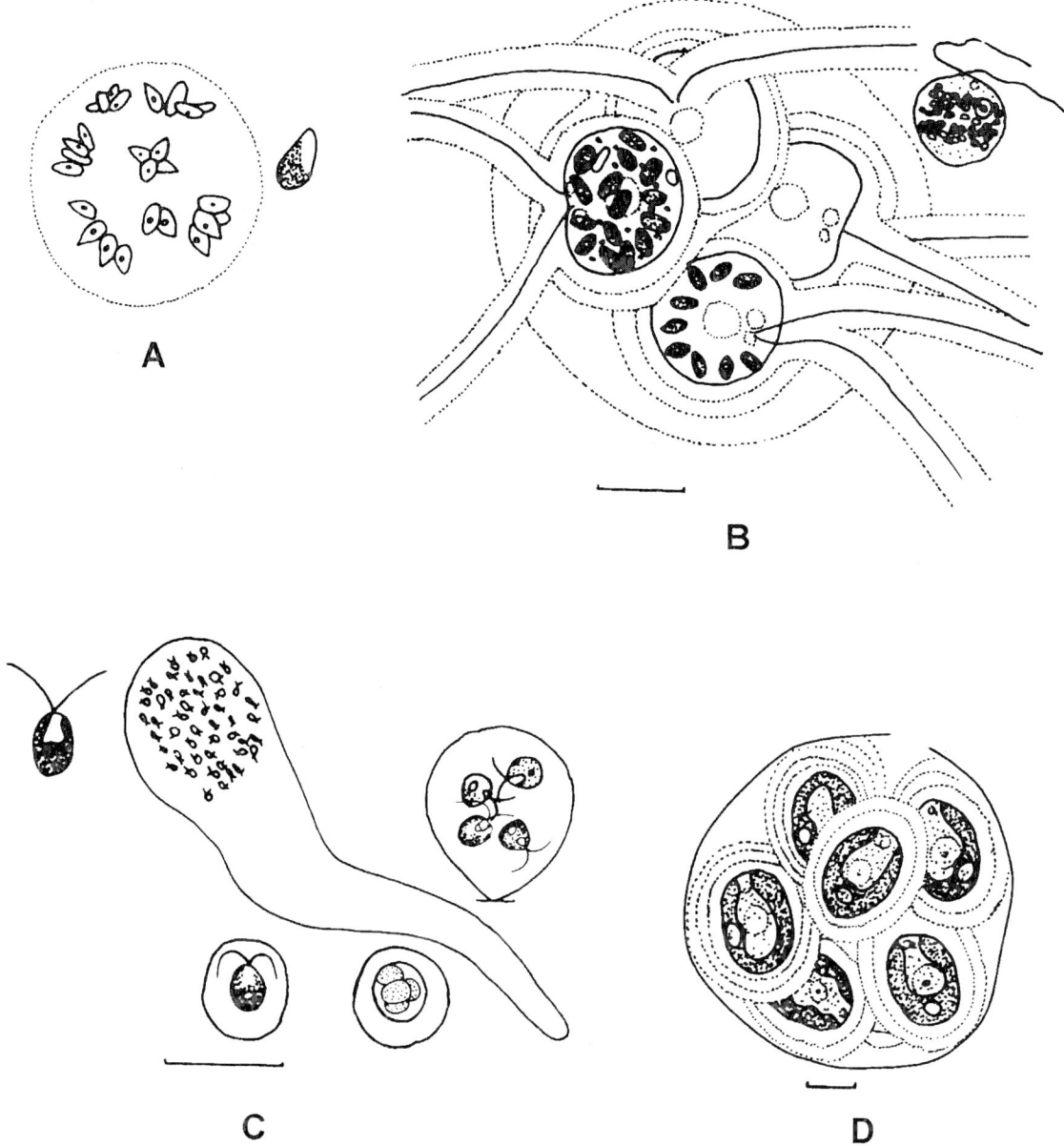

FIGURE 13 A. *Gloeoactinium limneticum*. B. *Gloeochaete wittrockiana*. C. *Gloeococcus pyriformis*. D. *Gloeocystis bacillus*. A from Prescott (1978) with permission; B and C from Bourrelly (1966) with permission.

Hormotila Borzi (Fig. 15A)

Cells are embedded in dichotomously branching, stratified gelatinous stalks. Forms biflagellate zoospores. Note: *Hormotilopsis* Trainor and Bold has an identical morphology to *Hormotila*, but the zoospores of the former are quadriflagellate. Terrestrial. Uncommon. Reported from central and eastern United States, eastern Canada, British Columbia, and the Caribbean.

Hydrodictyon Roth (Fig. 15B)

Colony is a macroscopic, saclike network consisting of numerous cylindrical cells that are multinucleate; each cell is attached to two others at end walls to form five or six sided meshes, organized into a network. The chloroplast in young cells is a parietal plate that has one pyrenoid, later becoming reticulate with numerous pyrenoids. Forms biflagellated zoospores. Floating mass

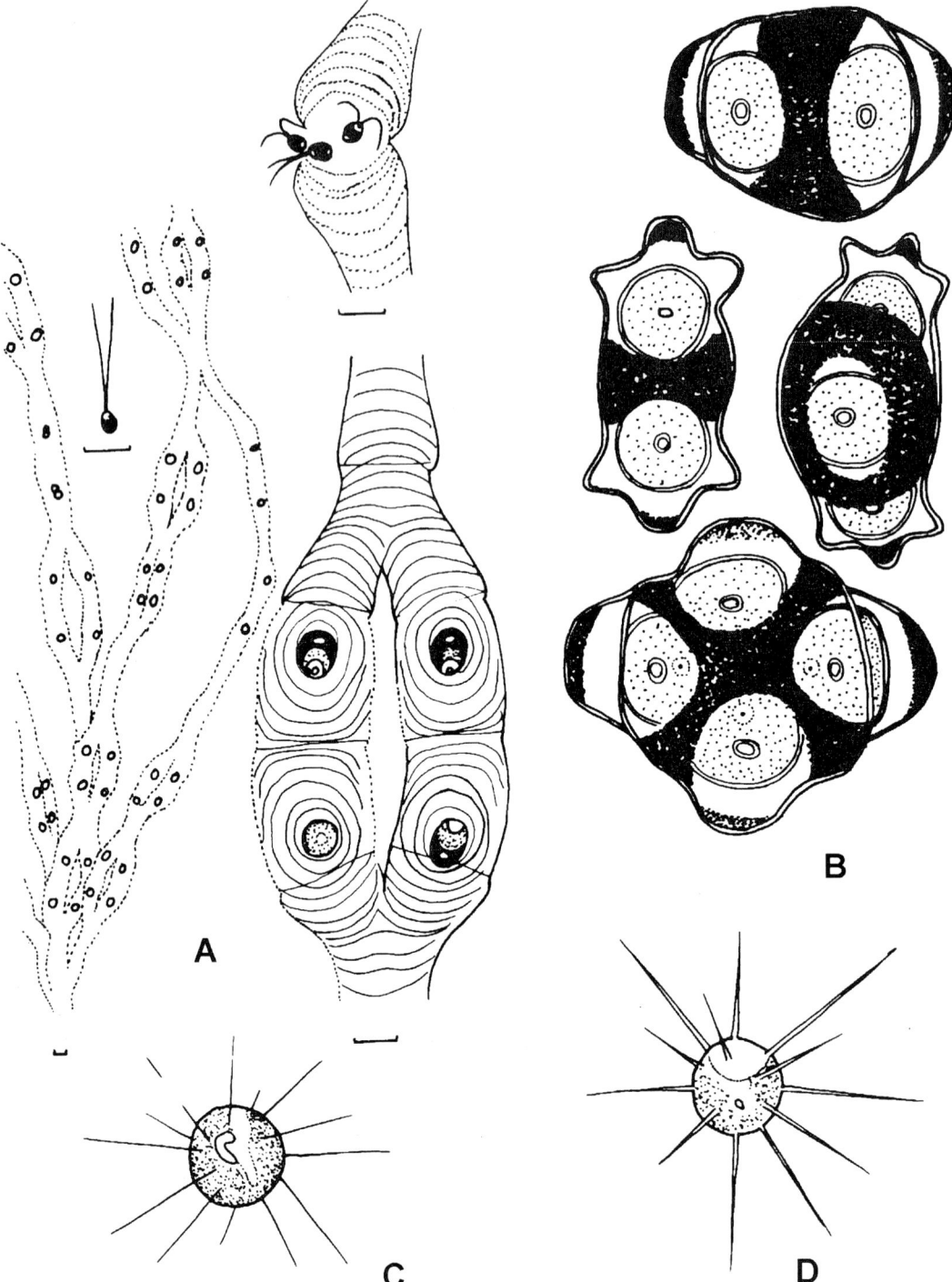

FIGURE 14 A. *Gloeodendron catenatum*. B. *Gloeotaenium loitlebergerianum*. C. *Golenkinia radiata*. D. *Golenkiniopsis solitaria*. A and B from Bourrelly (1966) with permission; C and D from Prescott (1978) with permission.

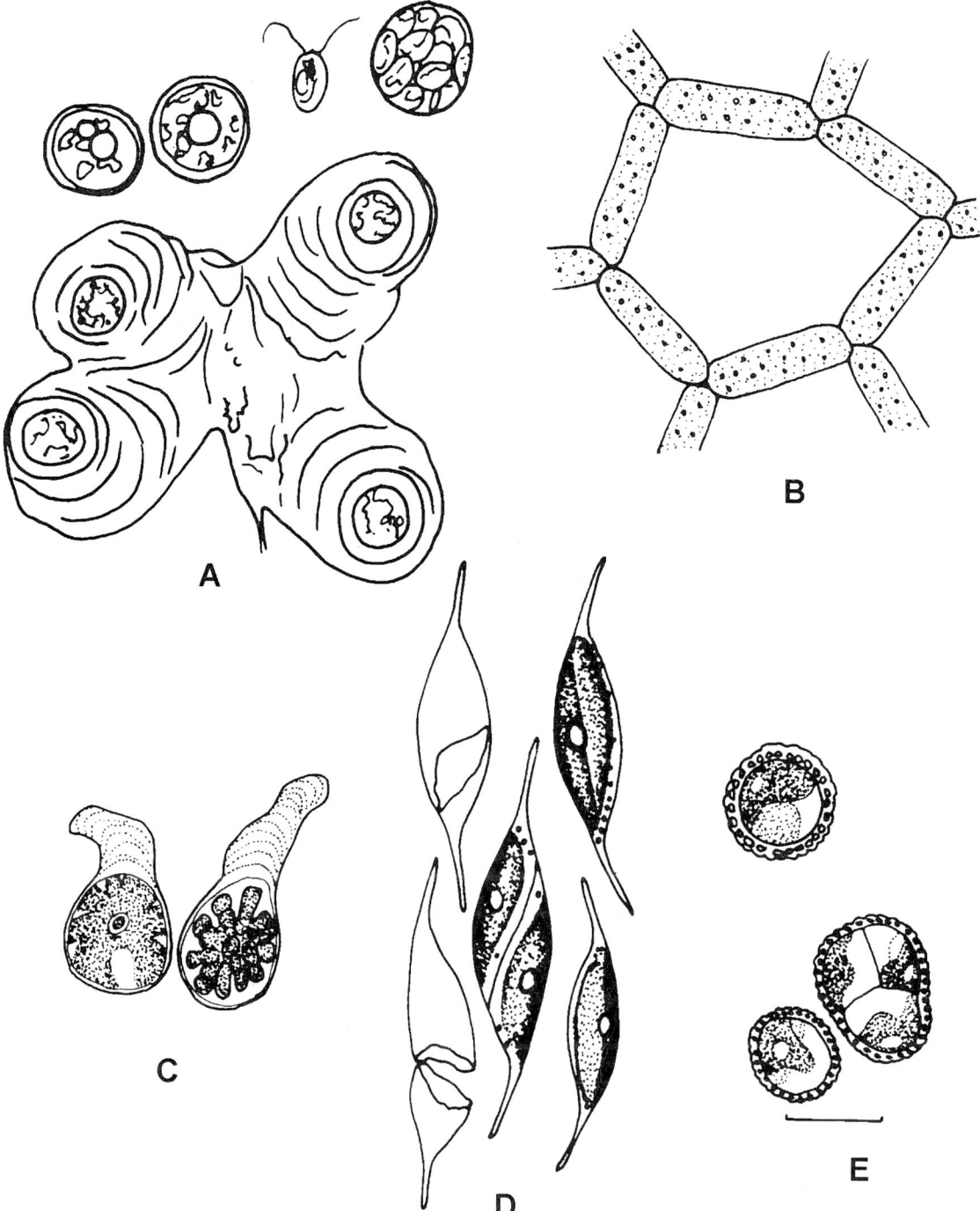

FIGURE 15 A. *Hormotila blennista*. B. *Hydrodictyon reticulatum*. C. *Kentrosphaera facciolae*. D. *Keratococcus bicaudatus*. E. *Keriochlamys styriaca*. A and D from Ettl and Gärtner (1995) with permission; B, C, and E from Bourrelly (1966) with permission.

in ponds and lakes. Common and occasionally very abundant or forming nuisance growths (see Chap. 24). Reported from habitats in most temperate to tropical regions; possibly absent from colder biomes.

Kentrosphaera Borzi (Fig. 15C)

Cells are irregularly globose or ellipsoidal and the walls often have irregular thickenings in certain portions. The chloroplast is axial and has numerous palisade-like processes at the periphery (may appear to have several irregularly shaped chloroplasts due to the flattened processes), and produces a large number of zoospores and also spherical aplanospores. Usually occurs on damp soil, but may also be aquatic and intermingled with or in the gelatinous envelopes of other algae. Reported from a few localities in northeastern, central, and western United States, and (temperate) eastern and western Canada.

Keratococcus Pascher [=*Ourococcus* Grobety *non Urococcus* Kützing; =*Dactylococcus* Nägeli *sensu* Hansgirg] (Fig. 15D)

Cells are single or in groups, more or less spindle-shaped, narrowed at one or both ends in a colorless extension of the cell wall; rarely are they rounded or pointed at one end. Cells are straight or curved to bent, and the cell wall is smooth and has no mucilage covering. The chloroplast is delicate, parietal, troughlike to convex, and has a single pyrenoid. Asexual reproduction via two (more rarely four) autospores. Some taxa are aerophytic or terrestrial. Scattered reports (mostly as *Ourococcus*) from northeastern, southeastern, central, and western United States, and central–western Canada.

Keriochlamys Pascher (Fig. 15E)

Free-floating globular or semiglobular clustered (rarely solitary) cells are readily identified by the thick wall, which has alveolae that refract light. Each cell has a parietal, platelike chloroplast and one pyrenoid. Produces two or four autospores with homogenous walls. Found in phytoplankton. Possibly rare; regarded as "expected" in the southeastern United States (Dillard, 1989).

Kirchneriella Schmidle (Fig. 16A)

Colony (cells are seldom solitary) is spherical to ovoid, consisting of 4 to 16 to many crescent-shaped, curved, spindle-like or elongate cylindrical cells; the apices sometimes nearly touch. The cells are irregularly arranged within a gelatinous sheath (expanded parental cell wall). The chloroplast is parietal, contiguous at the convex side of the cell, and may have a pyrenoid. Occurs in the phytoplankton or metaphyton of ponds and lakes. Widespread; reported across most temperate localities in North America (lacking from arctic habitats) south to Central America.

Lagerheimia Chodat [=*Chodatella* Lemmermann] (Fig. 16B)

Cells are spherical, ovoid, ellipsoid, or lemon-shaped, and solitary, bearing long needle-like spines at the poles. One to four chloroplasts parietal and each has one pyrenoid. Found in the phytoplankton of rivers, ponds, and lakes. This species looks like the unicellular stage of the coenobial morph of *Desmodesmus*, which does not have lateral spines on the terminal cells. This may be considered a unicellular stage of the *Desmodesmus* life cycle. Many reports and widespread in all regions of North America (sometimes recorded as *Chodatella* or stages of *Desmodesmus/Scenedesmus*).

Lauterborniella Schmidle (Fig. 16C)

Colony is four-celled. The cells are crescent-shaped to somewhat triangular, arranged in fours about an axial point, and have concave surfaces directed outward and poles extended into horns. The chloroplast is parietal and there is one pyrenoid. Planktonic. Rarely recorded; known from southeastern United States.

Lobocystis Thompson (Fig. 16D)

Colonial; cells are broadly elliptical to broadly ovoid, arranged in pairs enclosed by a sheath (parental cell wall material) and occurring at the ends of dichotomous or V-shaped strands. The entire colony (multiples of four cells) usually is enclosed by a gelatinous sheath. One or two parietal chloroplast(s) are present, each with one pyrenoid. Planktonic. Apparently rare; a few locations in Mississippi and North Carolina (Dillard, 1989).

Malleochloris Pascher (Fig. 17A)

Cells solitary and spherical, with the anterior end down and a broad mucilage pad as wide as or wider than the cell and enclosed in a brown sheath. The chloroplast is cup-shaped or urnlike and located in the upper (posterior) portion of the cell, which has one pyrenoid. May form quadriflagellated zoospores. Attached to substrata (often filamentous green algae such as *Cladophora*). Uncommon; a few reports in general floras and from the southeastern United States.

Micractinium Fresenius [=*Richteriella* Lemmermann; *Errerella* Conrad] (Fig. 17B)

Colony is triangular to pyramidal, consisting of a cluster of four spherical to broadly ellipsoid cells, each bearing one to several long delicate tapering spines; coenobia often compound, consisting of up to

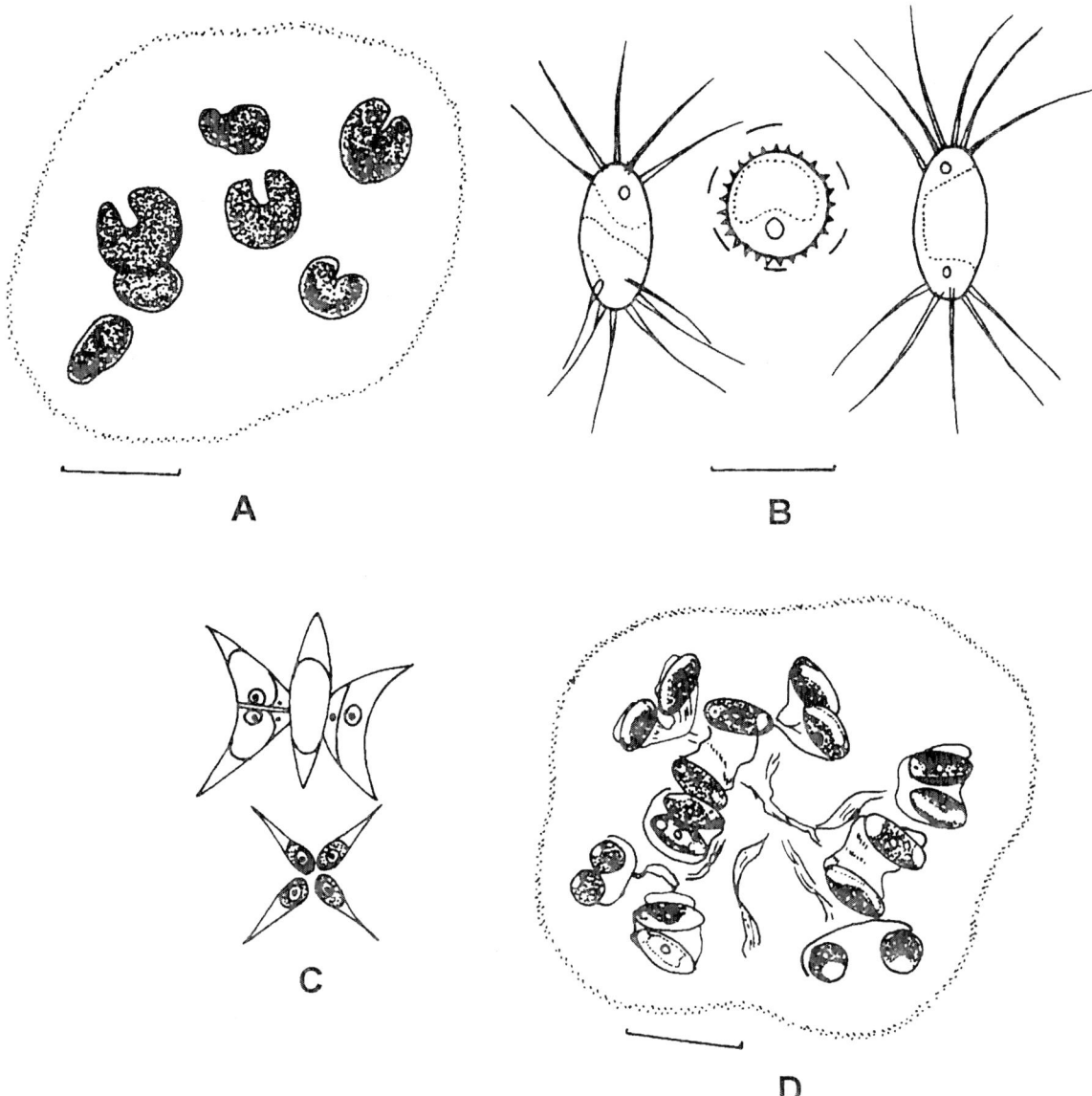

FIGURE 16 A. *Kirchneriella obesa*. B. *Lagerheimia* sp. C. *Lauterborniella elegantissima*. D. *Lobocystis dichotoma* var. *mucosa*. A–D from Bourrelly (1966) with permission.

128–(256) cells. The chloroplast is cup-shaped and has one pyrenoid. Occurs in the phytoplankton of rivers, ponds, and lakes. Uncommon, but possibly underreported; reported from most regions of North America, including subarctic, temperate, and tropical regions.

Monoraphidium Komárková-Legnerová
[=*Ankistrodesmus* Corda p.p. *sensu auct*. Komárek *et* Fott] (Fig. 17C)

Cells are single, rarely attached at one end, and more or less spindle-shaped, straight, curved, sigmoidal, or spirally twisted. Both cell ends are equally pointed or curved. The cell wall is thin and smooth with no mucilage sheath. The chloroplast is single and parietal, and has no pyrenoid. Reproduction is via four to eight autospores. Some terrestrial taxa. Common (possibly confused in some studies with *Ankistrodesmus*); reported in several general floras, as well as from habitats in northeastern, central, and western United States, western Canada, and the Caribbean.

Myrmecia Printz [=*Lobococcus* Reisgel; *Pulchrasphaera* Deason] (Fig. 17D)

Cells are solitary and spherical to irregularly pear-shaped. The cell wall is with a lateral lobe (protuberance on one side) and asymmetrical. The chloroplast is massive, axial, irregularly lobed, and has no pyrenoid.

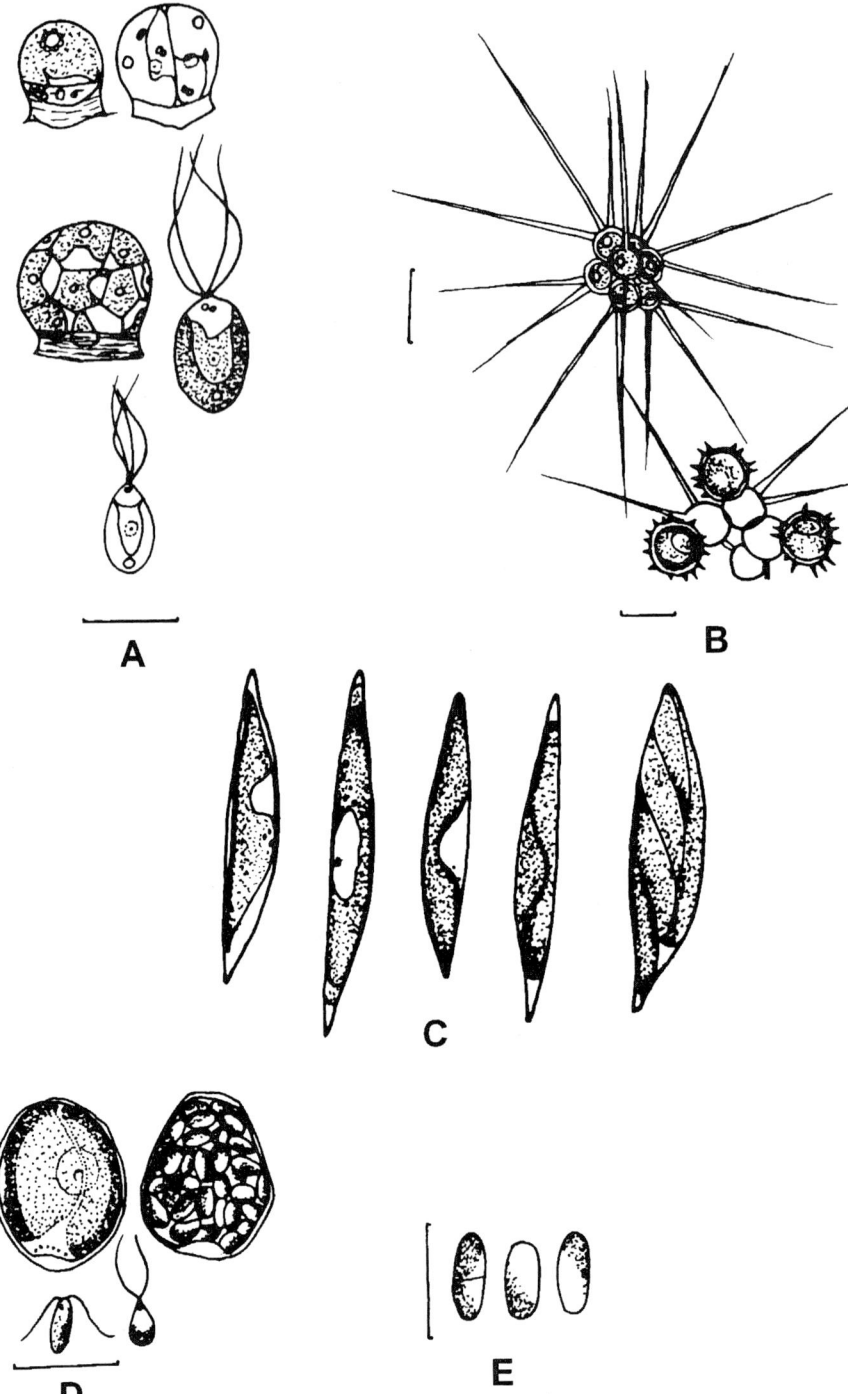

FIGURE 17 A. *Malleochloris sessilis*. B. *Micractinium pusillum*. C. *Monoraphidium pusillum*. D. *Myrmecia pyriformis*. E. *Nannochloris bacillaris*. A, B, D, and E from Bourrelly (1966) with permission; C from Ettl and Gärtner (1995) with permission.

Can form biflagellated zoospores. Subaerial and aquatic habitats. Uncommon; recorded in a few general floras and from southeastern United States.

Nannochloris Naumann [=*Diogenes* Pennington] (Fig. 17E)

Cells are solitary or in pairs, subspherical to subcylindrical, and small (less than 4.5 µm in diameter). The cells are enclosed by lamellate sheaths and usually occur in large numbers within a gelatinous mass. The chloroplast is simple, parietal, platelike at one or both ends, and has one naked pyrenoid. Planktonic. Uncommon, but possibly overlooked, especially very small species (mistaken for small coccoid cyanobacteria). Recorded from scattered regions in the United States and Canada.

Nautococcus Korschikoff (Fig. 18A)

Cells are egg-shaped and have a flotation cap at the narrow end. The chloroplast is axial and has a central pyrenoid. Can form biflagellated zoospores. Neustonic, floating beneath the surface film of water. Rarely recorded.

Neochloris Starr (Fig. 18B)

Unicellular cells, that have a parietal chloroplast and one to several pyrenoids. Aplanospore formation; biflagellated zoospores become spherical immediately upon quiescence. Infrequently reported from both aquatic and soil habitats; most records and new species were based on culture isolations from soils and ponds in Texas and Cuba (Archibald, 1973).

Neospongiococcum Deason [=*Spongiococcum* Deason p.p. *sensu* Deason] (Fig. 18C)

Cells are single or in groups, young cells are ellipsoidal, ovoid, or slightly irregularly lemon-shaped; older cells are spherical, with a smooth, colorless, mostly thickened cell wall. The chloroplast is adjacent to the wall; in young cells it is a more or less open cup with irregularly lobed and dissected opening; in older cells the hollow sphere of the chloroplast has holes, bifurcations, and incisions (appears disintegrated or perforated into connected ribs, ridges, and transverse bars; finally spongelike). The greater part is a compact chloroplast in which an off-center pyrenoid lies. The starch sheath is a hollow sphere or articulated. Can form biflagellated zoospores. Found in soils. Rarely reported; a few locations in the United States and Caribbean region.

Nephrocytium Nägeli [=*Gloeocystopsis* G. M. Smith] (Fig. 18D)

Colony is spherical to ovoid, consisting of 4–8–16 asymmetrical, kidney-shaped to oblong–ellipsoid cells retained within the mucilaginous, expanded parental wall. The chloroplast is laminate and parietal, sometimes massive and diffuse in old cells, and has one pyrenoid. Found in the phytoplankton and metaphyton of ponds and lakes. Fairly common; reported from suitable habitats in northeastern, southeastern, central, and western United States, eastern, central, and western (temperate) Canada, and Central America.

Octogoniella Pascher (Fig. 19A)

Cells are solitary with an octagonal outline, flattened on the substratum with a large, bipartite wall. The cytoplasm is globose with a cuplike plastid that contains one or two pyrenoids. Can form biflagellated zoospores. Epiphytic on *Sphagnum*. Rarely recorded; mentioned as "expected" in the southeastern United States by Dillard (1989).

Oocystidium Korschikoff (Fig. 19B)

Cells are solitary, ellipsoidal, and enclosed within a thin-walled vesicle of aqueous mucilage. Cells have two symmetrically arranged, cuplike plastids and a pyrenoid; free-living. Possibly rare; mentioned as "expected" in southeastern United States (Dillard, 1989).

Oocystis A. Braun (Fig. 19C)

Cells are ovoid, ellipsoid, or more commonly lemon-shaped, with rounded, sometimes thickened poles, usually colonial (2–16 cells) within persistent, enlarged parental cell wall (two to three generations of parental cell walls may be enclosed in the original parental cell wall, which enlarges so that it often appears as a gelatinous sheath); occasionally solitary. The cells have one to several chloroplasts, usually parietal, variably shape, and with or without a pyrenoid. Found in the phytoplankton and metaphyton of ditches, bogs, ponds, and lakes. A very common genus with several common species, collected from a variety of water bodies in all regions of North America.

Oophilia Lambert in Printz (Fig. 19D)

Cells are spherical to ellipsoidal. The chloroplasts in young cells are axial and irregularly lobed, in older cells becoming fragmented and parietal, and have a pyrenoid. Typically occur within the gelatinous envelope of salamander (*Amblystoma*) or frog (*Rana*) egg masses (giving the egg masses a distinct green color). Regarded by some as a nonmotile stage of a species of *Chlamydomonas* (Goff and Stein, 1978). Uncommon from scattered locations; most likely overlooked due to its unusual habitat; perhaps widespread where amphibian egg masses occur. Recorded from northeast-

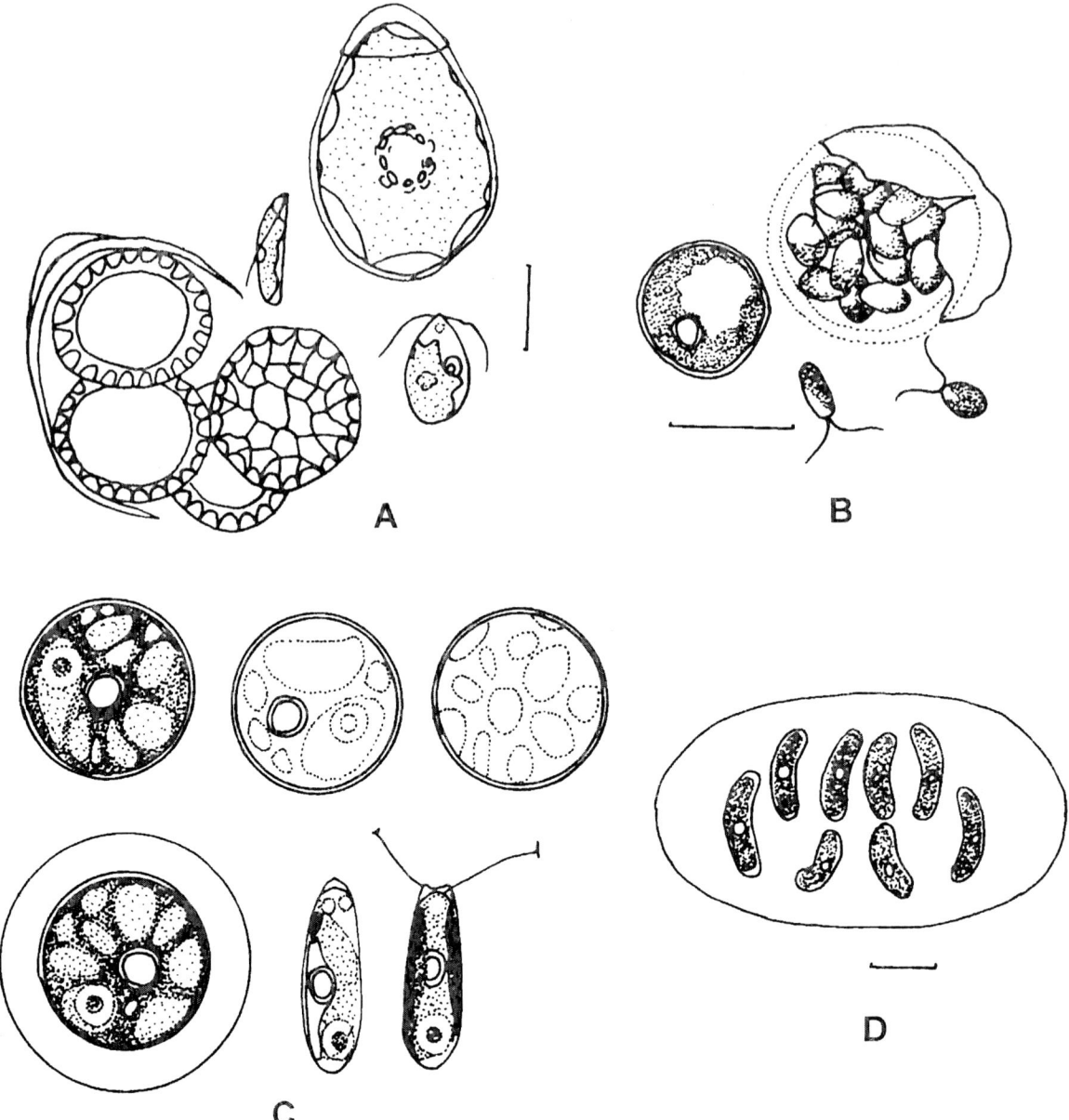

FIGURE 18 A. *Nautococcus piriformis*. B. *Neochloris aquatica*. C. *Neospongiococcum gelatinosum*. D. *Nephrocytium agardhianum*. A, B, and D from Bourrelly (1966) with permission; C from Ettl and Gärtner (1995) with permission.

ern, southeastern, central, and western United States as well as western Canada.

Pachycladella P. Silva [=*Pachycladon* G.M. Smith *non Pachycladon* Hooker] (Fig. 19E)

Cells are solitary and spherical, bearing four stout radiating appendages (brown colored), tapering to blunt or two-forked apices. The chloroplast is cup-shaped and has one pyrenoid. Found in phytoplankton. Uncommon. Recorded in a few general floras, and from the Great Lakes and southeastern United States regions.

Palmella Lyngbye (Fig. 20A)

Cells are embedded in a common matrix (concentrically stratified individual matrices). The colony is irregularly shaped, consisting of numerous, spherical to broadly ellipsoidal cells. The cells, solitary or in pairs, are *Chlamydomonas*-like in organization, but the vegetative cells lack flagella and contractile vacuoles.

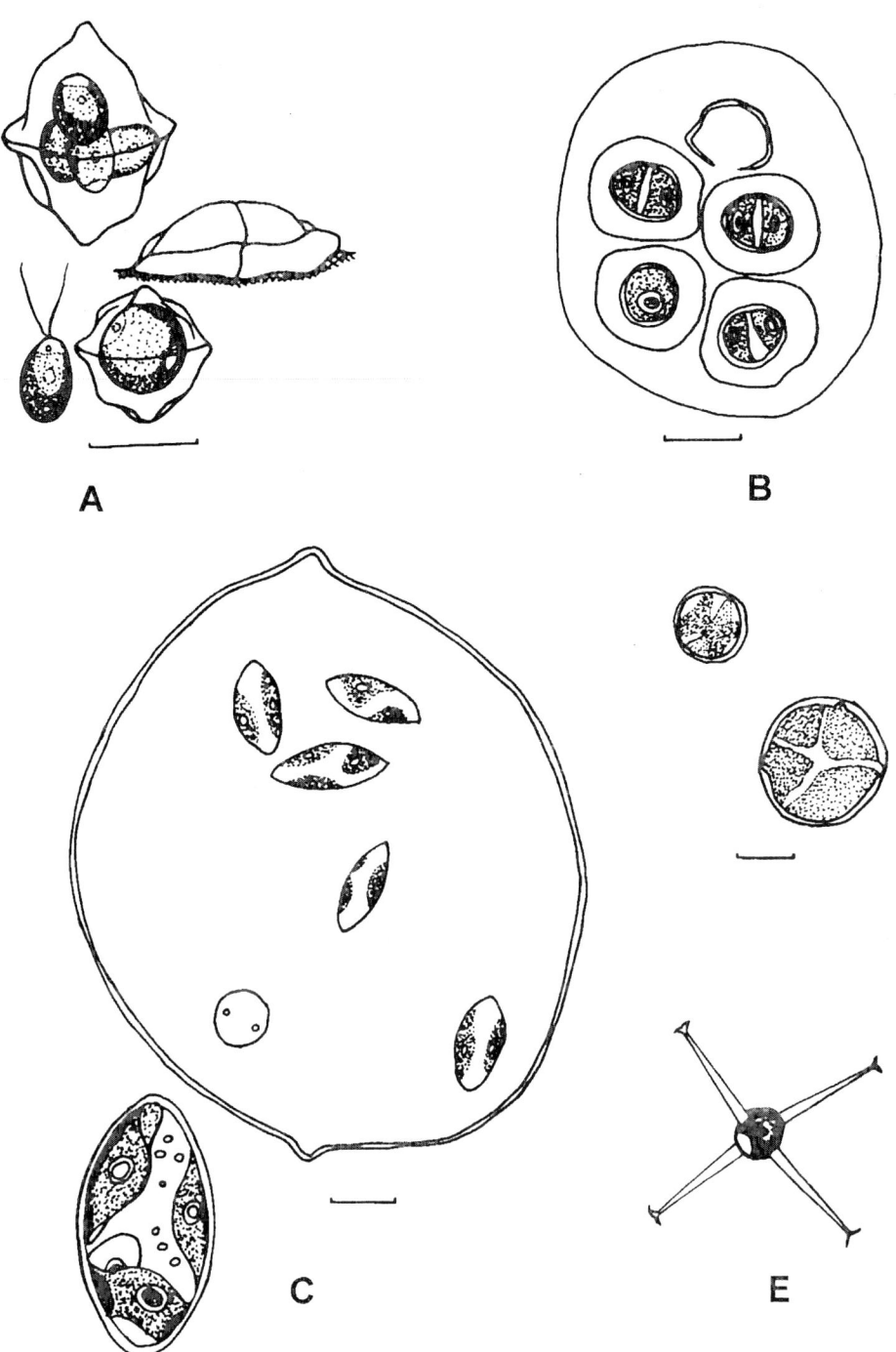

FIGURE 19 A. *Octogoniella sphagnicola*. B. *Oocystidium ovale*. C. *Oocystis lacustris*. D. *Oophilia amblystomalis*. E. *Pachycladella umbrinus*. A–D from Bourrelly (1966) with permission; E from Prescott (1978) with permission.

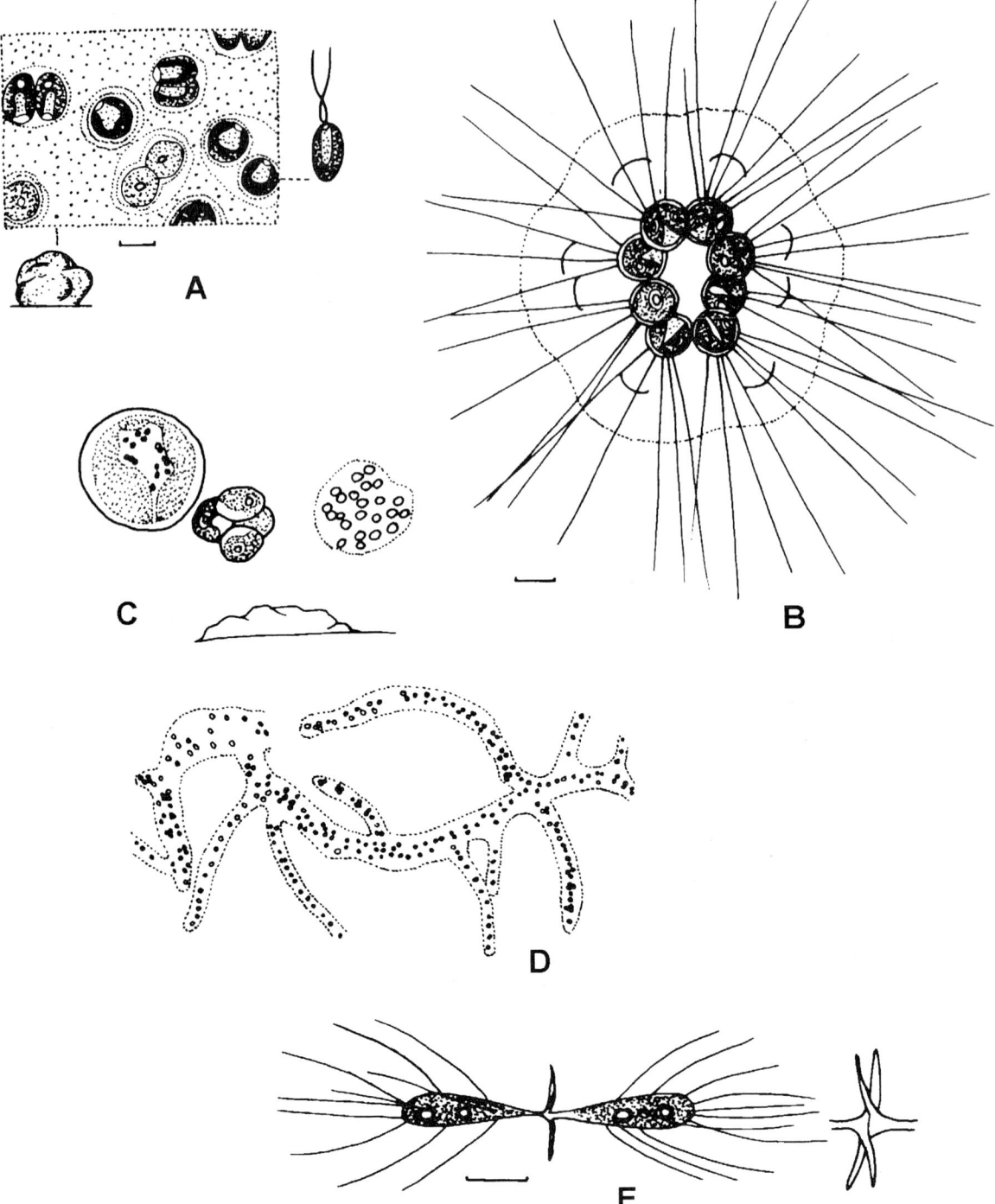

FIGURE 20 A. *Palmella miniata* var. *aequalis*. B. *Palmellochaete tenerrima*. C. *Palmellopsis gelatinosa*. D. *Palmodictyon varium*. E. *Paradoxia multiseta*. A, B, D, and E from Bourrelly (1966) with permission; C from Prescott (1978) with permission.

Chlorophyll is often masked by red pigment (hematochrome). One pyrenoid. Found as a gelatinous amorphous mass on damp soil or dripping rocks. Common; several records from the most temperate regions of North America (possibly absent from arctic locations) and the Caribbean.

Palmellochaete Korschikoff (Fig. 20B)

Colony consisting of four to eight spherical cells within a mucilage envelope containing fragments of parental cell walls. The cells bear two to eight long tapering spines, largely in the same plane. Each cell has two to four chloroplasts, each with one pyrenoid. Planktonic. Apparently rare; recorded from lakes in North Carolina (Whitford and Schumacher, 1984).

Palmellopsis Korschikoff [=*Palmella* Lyngbye p.p. sensu auct. Groover et Bold, Ettl et Gärtner] (Fig. 20C)

Amorphous gelatinous colonies are either planktonic or adherent to aquatic substrates. The cells have cup-shaped chloroplasts, a single pyrenoid, and two contractile vacuoles. Mucilage is not lamellate as in the similar genus *Palmella*. Aquatic and free-floating; sometimes attached. Uncommon; recorded from a few locations in eastern, central, and western United States and central–western (temperate) Canada.

Palmodictyon Kützing [=*Palmodactylon* Kützing in Nägeli] (Fig. 20D)

Cells are spherical, multiseriately arranged in a linear series, within unbranched, netlike or branched, nonlamellate mucilaginous tubes that may be unbranched or branched; sometimes treelike. Found in the metaphyton of ditches, bogs, ponds, and lakes. Fairly common, but rarely abundant; recorded from many locations in northeastern, southeastern, central, and western United States, and eastern, central, western, and northwestern Canada.

Paradoxia Swirenko (Fig. 20E)

Cells are pear-shaped, tapering into a beak at the apex, and have long fine spines more or less parallel to the axis of the cell. The apex is extended into two sigmoid perpendicular outgrowths that link cells in pairs. The chloroplast has two pyrenoids. Planktonic. Rarely recorded; reported from the Great Lakes area (Bay of Quinte).

Paulschulzia Skuja (Fig. 21A)

Colonies are spherical or subspherical, composed of cells each of which has two contractile vacuoles and two very long pseudoflagella (sometimes more apparent after staining with dilute methylene blue and colonial matrices in India ink). Tendency for the four products of cell division to persist as tetrads and form complexes of colonies. Planktonic. Uncommon; a few scattered reports from the western United States, and eastern and western (temperate) Canada.

Pectodictyon Taft (Fig. 21B)

Colony is an eight-celled hollow cube. The cells are spherical and located at the corners of the mucilage framework, with cells interconnected by stout gelatinous strands, often forming compound coenobia. The chloroplast is parietal and it has one pyrenoid. Occurs in the phytoplankton of ponds and lakes. Uncommon; recorded from western and southeastern United States.

Pediastrum Meyen (Fig. 21C)

Colony is flat–circular (sometimes irregularly subcircular), consisting of a single platelike layer of cells (minimum of four, multiples of two). Peripheral cells have one to two lobes or processes or no processes, and usually differ in shape from those within. Cells of the colony are contiguous or the colony is perforate. The cell walls are smooth or variously ornamented. The chloroplast is parietal and has one pyrenoid. Can form biflagellated zoospores. Found in the phytoplankton of rivers, ponds, and lakes. Extremely common with other members of Chlorococcales, especially (but not exclusively) in nutrient-rich waters; recorded from all regions of North America.

Phyllobium Klebs (Fig. 21D)

Almost colorless branched tubes with globose swellings at the tips. The contents of the tube accumulate at the tips of branches to form thick-walled akinetes. Cells have numerous ellipsoidal chloroplasts, radially arranged in the apical swellings. Occurs as an endophyte growing on and among the leaf cells of *Sphagnum*. Uncommon; recorded in a few general floras. Distribution unknown.

Phyllogloea Silva (Fig. 21E)

Colonies are large and leaflike. The cells are spherical and grouped in pairs or groups of four in a stratified mucilage formed by the gelatinization of maternal envelopes. The cup-shaped plastid has a single pyrenoid. Can form quadriflagellated zoospores. Attached to *Typha*. Rarely recorded.

Planktosphaeria G. M. Smith (Fig. 22A)

Colony 8–16 cells. The cells are spherical and united within a wide, colorless mucilage envelope that is often very thin (cells are rarely solitary). Cells within a colony usually are contiguous. The chloroplasts (in mature cells) are several parietal disks (angular, five-

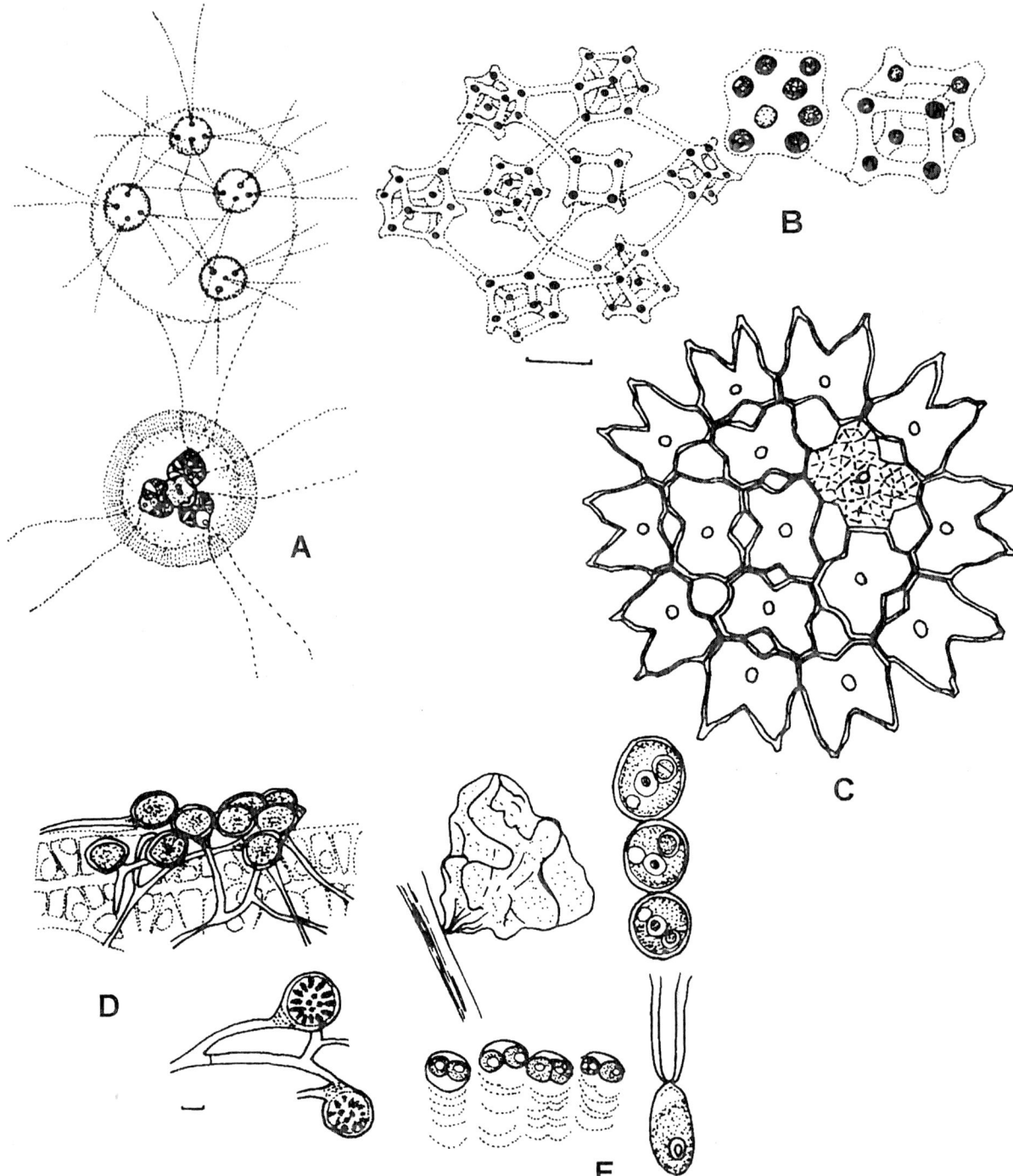

FIGURE 21 A. *Paulschulzia pseudovolvox*. B. *Pectodictyon cubicum*. C. *Pediastrum duplex* var. *typicum*. D. *Phyllobium sphagnicola*. E. *Phyllogloea fimbriatum*. A– E from Bourrelly (1966) with permission.

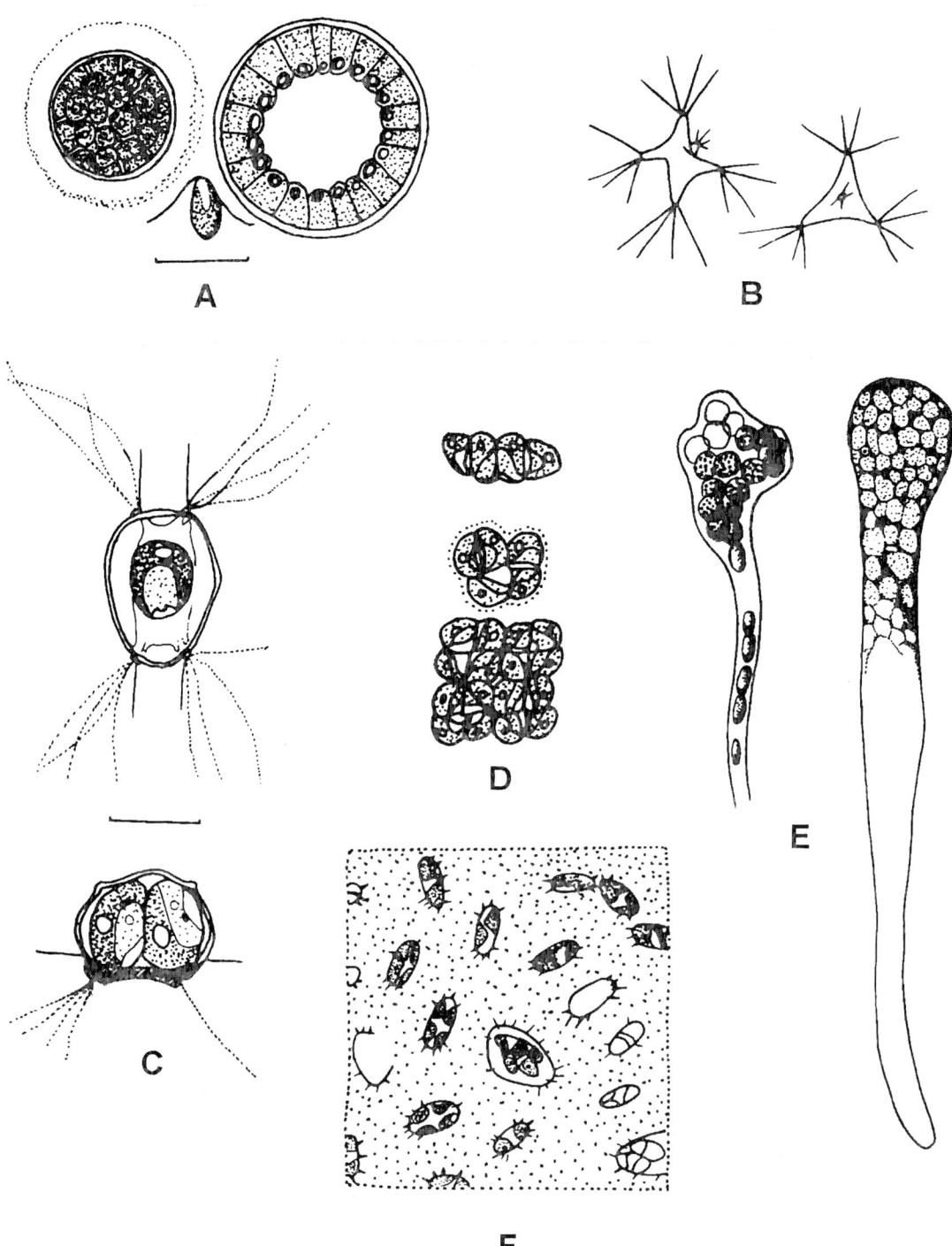

FIGURE 22 A. *Planktosphaeria gelatinosa*. B. *Polyedriopsis spinulosa*. C. *Porochloris filamentarum*. D. *Protoderma sarcinodeum*. E. *Protosiphon botryoides*. F. *Pseudobohlinia americana*. A, C, and F from Bourrelly (1966) with permission; B from Prescott (1978) with permission; D and E from Ettl and Gärtner (1995) with permission.

sided), each with one pyrenoid. Can form biflagellated zoospores. Found in the metaphyton of ditches, bogs, ponds, and lakes. Common and reported from most regions, including northeastern, southeastern, central, and western United States, eastern, central, and western (temperate) Canada, and the Caribbean.

Polyedriopsis Schmidle (Fig. 22B)

Cells are solitary (sometimes clustered), rectangular or five-sided, with squared off, blunt angles, bearing a tuft of 3–10 long tapering spines at the angles of the cell. The chloroplast is a large parietal plate with one pyrenoid. Can form biflagellated zoospores. Planktonic. Uncommon; a few scattered localities from northeastern, southeastern, and central United States, the Great Lakes region, and Central America.

Porochloris Pascher (Fig. 22C)

The cell surrounds itself with a colorless or brown, flattened casing that is quadrangular in apical view. Four pseudoflagella emerge from each of four pores at the angles of the casing. Cells contain a cup-shaped plastid that has a basal pyrenoid. An epiphyte on mosses, *Sphagnum*, and filamentous algae. Possibly rare; regarded as "expected" in southeastern United States (Dillard, 1989).

Protoderma Kützing (Fig. 22D)

Thallus as an attached, single layer of cells or pseudoparenchymatous disk of horizontally growing filaments, closely arranged and semiradiate. The filaments are irregularly branched and often indeterminate. The chloroplast is parietal and has one pyrenoid. An epiphyte on submerged macrophytes and filamentous algae. Common, but possibly overlooked growing amongst other epiphytic algae; northeastern, southeastern, central, and western United States, eastern and western temperate Canada, and Central America.

Protosiphon Klebs (Fig. 22E)

Green, bladder-like to tubular aerial portion, and colorless rhizoidal below-ground portion: Coenocytic and has a single parietal chloroplast with several irregularly shaped perforations. Mature cells may have pyrenoids. Found on muddy banks of streams and ponds, and the surface of damp soils. Uncommon; a few scattered localities in the Great Lakes region, southeastern and northeastern United States, eastern Canada, and Central America.

Pseudobohlinia Bourrelly (Fig. 22F)

Colony is irregularly shaped, consisting of numerous ellipsoidal to ovoid cells that bear many short spines, particularly at the poles. Cells have one to four to as many as eight chloroplasts and each has one pyrenoid. Rarely recorded; reported from the southeastern United States.

Quadricoccus Fott [=*Tetratomococcus* Korschikoff] (Fig. 23A)

Colony is (2)–4–(64)-celled. The cells are broadly ellipsoidal to cylindrical with rounded poles, arranged in groups of four, and attached to flattened, shell-like parental cell wall remnant. Cells have one chloroplast and one pyrenoid. Planktonic. Uncommon; reported from the southeastern United States and in the Caribbean.

Quadrigula Printz (Fig. 23B)

Colony of two-to four to eight elongate–cylindrical or cigar-shaped cells. The long axes of the cells parallel (bundle with) the mucilaginous envelope within which they are enclosed (parental wall); the poles are moderately acute or sharply rounded. Single parietal chloroplast. Planktonic. Common and widespread; reported from at least a few localities in all regions of North America.

Radiococcus Schmidle (Fig. 23C)

Colony of 4–8–16–(32) cells within a mucilage envelope includes fine, radiating fibrils. The cells are spherical to broadly ellipsoidal, arranged in cruciform clusters of four. Found in the phytoplankton of ponds and lakes. Uncommon; a few scattered localities (eastern–central and southeastern United States and British Columbia).

Rayssiella Edelstein *et* Prescott (Fig. 23D)

Colony of 4–16 ellipsoidal, oblong to kidney-shaped cells enclosed by the parental cell wall; cells often arranged in groups of four, with long axes parallel and at the two poles of the parental cell. Several colonies are connected by mucilaginous remnants of parental cell walls. Autospore formation occurs in some. Planktonic. Uncommon or possibly rare; reported from the southeastern United States.

Rhodochytrium Lagerheim (Fig. 24A)

Unicellular. The cell is wall thick, globular or flask-shaped with rhizoidal lobes and extensions. The chloroplast is massive and indefinite, densely packed with starch. The protoplast is reddish orange. Can form biflagellated zoospores. Endophytic (parasitic within *Ambrosia*, ragweed). Uncommon; reported in a few genera floras and from the southeastern United States.

Scenedesmus Meyen (Fig. 24B)

Colony is 2–4–8–16–(32)-celled, flattened, with long axes of cells parallel, laterally adjoined and

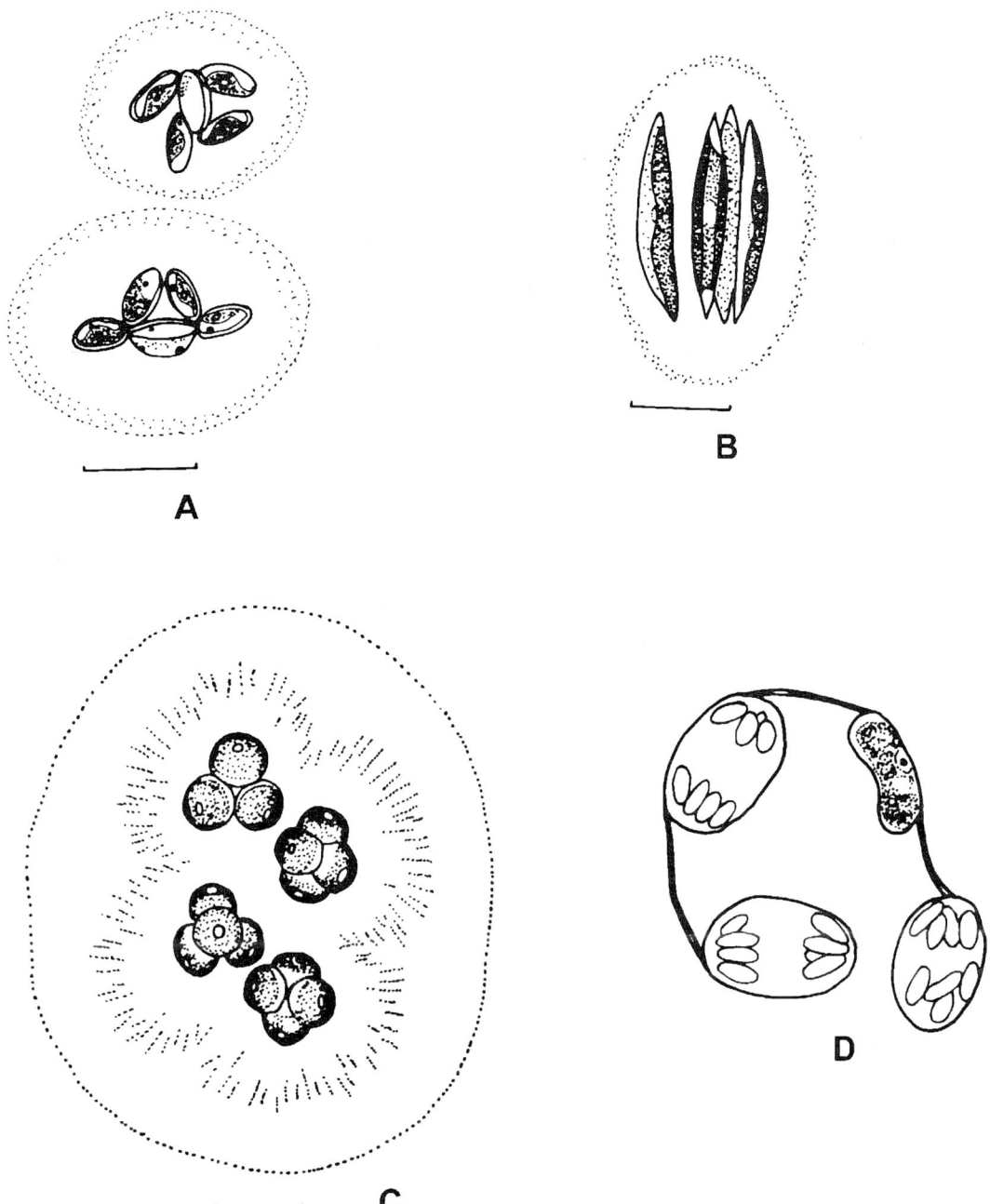

FIGURE 23 A. *Quadricoccus verrucosus*. B. *Quadrigula closteroides*. C. *Radiococcus nimbatus*. D. *Rayssiella hemisphaerica*. A–C from Bourrelly (1966) with permission; D from Prescott (1978) with permission.

arranged in single linear or alternating series. Cells are ellipsoidal, ovoid, or crescent-shaped or tapering toward each end. The cell wall is smooth and spines are absent. The chloroplast is parietal and usually has one pyrenoid. Found in the phytoplankton of rivers, ponds, and lakes. Probably the most commonly reported genus of coccoid green algae worldwide, and frequently abundant in nutrient-rich (especially high inorganic N) waters; commonly co-occurs with other genera from the Chlorococcales (spine-producing species, such as *S. armatus*, now classified in the genus *Desmodesmus*). Species in this genus have been reported from all regions of North America from arctic to tropical biomes.

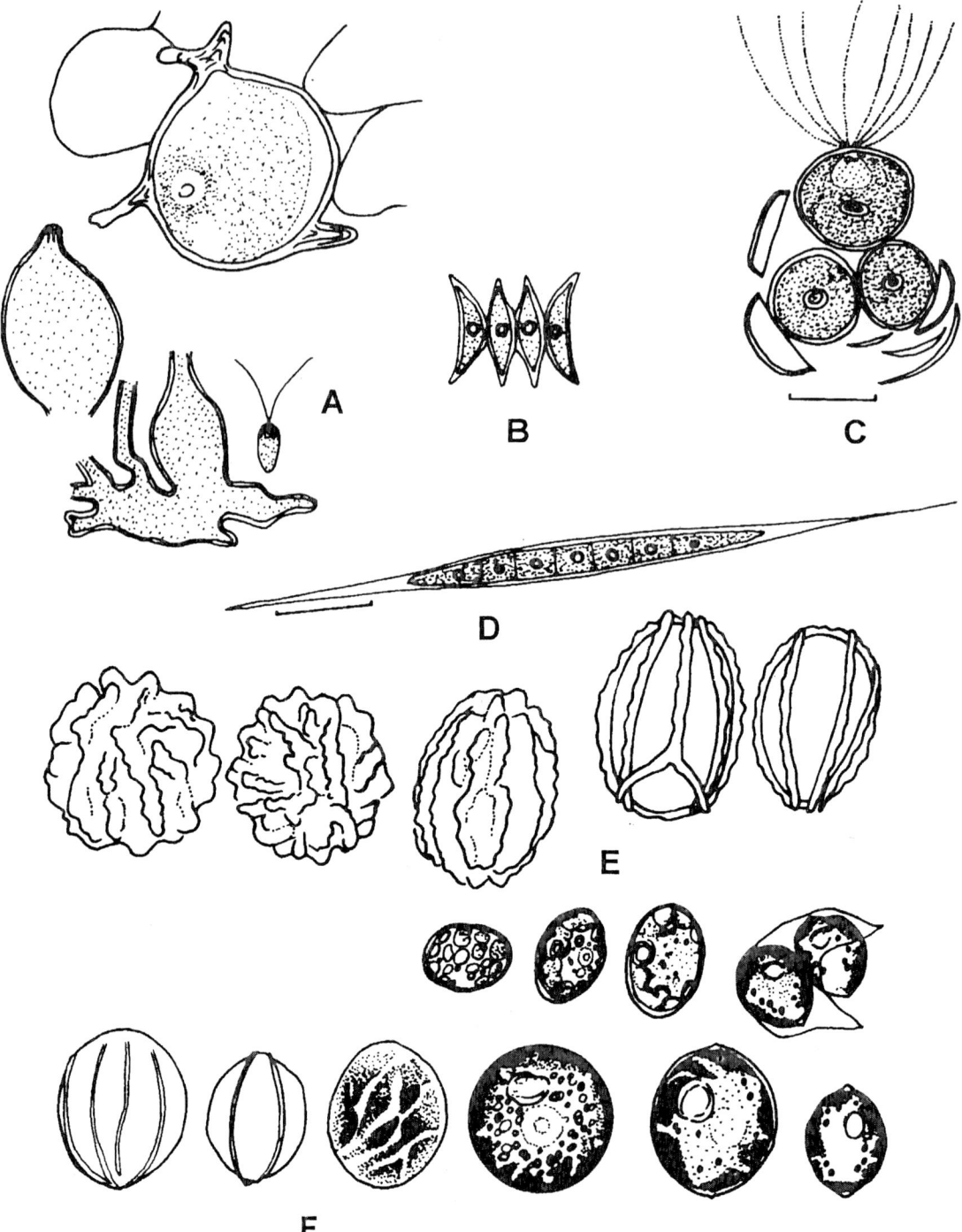

FIGURE 24 A. *Rhodochytrium spilanthidis*. B. *Scenedesmus acutus*. C. *Schizochlamydella gelatinosa*. D. *Schroederia setigera*. E. *Scotiella tuberculata*. F. *Scotiellopsis rubescens*. A from Prescott (1978) with permission; B, E, and F from Ettl and Gärtner (1995) with permission; C and D from Bourrelly (1966) with permission.

Schizochlamydella Korschikoff [=*Schizochlamys* Braun in Kützing *sensu* Komárek *et* Fott] (Fig. 24C)

Colony is an irregularly shaped floating mass, usually microscopic, but sometimes macroscopic, consisting of numerous spherical cells arranged in groups of two to four. Semicircular fragments of old parental cell walls are evident within the colony envelope. The cells often have several (up to 16) long, fine pseudoflagella, one to two platelike chloroplasts that are parietal, but appear massive and indistinct, oil rather than starch, and one pyrenoid. Planktonic. Common (although rarely abundant) and frequently reported (mostly as *Schizochlamys*) from most regions of North America.

Schroederia Lemmermann (Fig. 24D)

Cells are solitary, needle-like to tapering toward each end, straight or curved, and the poles extend into long spines, one of which may divide terminally into two parts. The chloroplast is a parietal plate and there are one to several pyrenoid(s). Found in the phytoplankton of ponds and lakes. Widely reported and common in the plankton of lakes and ponds in most regions of North America.

Scotiella Fritsch (Fig. 24E)

Cells are spindle-shaped, ellipsoidal to broadly oval, but polygonal in optical section, living individually. The cell wall is relatively thick, sometimes has a small papilla-like thickening at the cell pole; thick, ledgelike ribs run or slightly spiral the length of the cell from this thickening. The chloroplast is more or less parietal, mostly covering the entire cell or lobed at the margin, and is divided in some taxa and forms numerous closely crowded sections; has one or more pyrenoids. Subaerial. Uncommon, but reported from widely separated regions, including southeastern and central United States and British Columbia.

Scotiellopsis Vinatzer (Fig. 24F)

Cells are solitary or, rarely, in small groups. Cells are spindle-, lemon-, or egg-shaped, or ellipsoidal to broadly oval and lack mucilage. The cell wall has 6–12–(16) fine ribs (difficult to see in LM); at the poles the ribs form small papilla-like thickenings where they converge. The chloroplast is single and parietal, and has a pyrenoid. In young cells the plastid is entire, becoming lobed to divided, with many uneven and crowded disks in older cells. Subaerial. Rarely reported.

Selenastrum Reinsch (Fig. 25A)

Nonmucilaginous colony of 4–16–32 strongly crescent-shaped cells with acute apices, arranged such that dorsal or convex walls are adjacent. The cells are clustered but not entangled. The chloroplast is laminate and parietal, and may have pyrenoids. Found in the phytoplankton of ponds and lakes. Widely reported in general floras and from most regions, including the Great Lakes, northeastern, southeastern, central, and western United States, eastern, central, and western Canada, and Central America. Some species co-occur with other members of the Chlorococcales in nutrient-rich waters (e.g., *Pediastrum*, *Scenedesmus*, and *Oocystis*), although they are possibly less common than other genera in this order.

Siderocelis Fott (Fig. 25B)

Cells are solitary, free, small, ellipsoidal, spherical, or cylindrical with rounded apices, enclosed within a mucilaginous sheath. The cell wall is yellowish, impregnated with iron salts or ornamented with projecting, often irregular, brown warts. The cells have a small number (one to four) of parietal plastids that lack pyrenoids. Rarely reported, but known from Central America.

Sorastrum Kützing (Fig. 25C)

Colony is globular, consisting of kidney-shaped, pear-shaped, wedgelike, or pyramid-like cells, loosely or compactly arranged. The cells are attached by radiating, stout gelatinous strands to a common central mucilage body. The outward and free surface of the cells (has two to four) stout outwardly directed spines (horns). The chloroplast is parietal and has one pyrenoid. Occurs in the phytoplankton of ponds and lakes, and the metaphyton of ditches, bogs, and ponds. Fairly common and occasionally abundant; reported from most regions of North America, from Central America through the temperate biomes, to northern Canada (Great Slave Lake).

Sphaerellocystis Ettl (Fig. 25D)

Cells are solitary, not attached, ellipsoidal or spherical, enclosed in a large, yellow or brown, gelatinous sheath, and encrusted with iron salts. The cytoplasmic structure resembles *Chlamydomonas* with an entire or lobed, cup-shaped chloroplast, one pyrenoid, and two contractile vacuoles, but no stigma. Can form biflagellated zoospores. Planktonic. Rare; reported from North American prairie soils (*S. aplanospora*; Nichols *et al.*, 1991), although other described species from outside the continent are from aquatic habitats (pools and bogs; Hindak and Hindakova, 1992).

Sphaerocystis Chodat [=*Palmellocystis* Korschikoff] (Fig. 25E)

Colony is spherical, consisting of 32–64 spherical cells scattered throughout the colonial mucilage; several progeny colonies of smaller cells are inter-

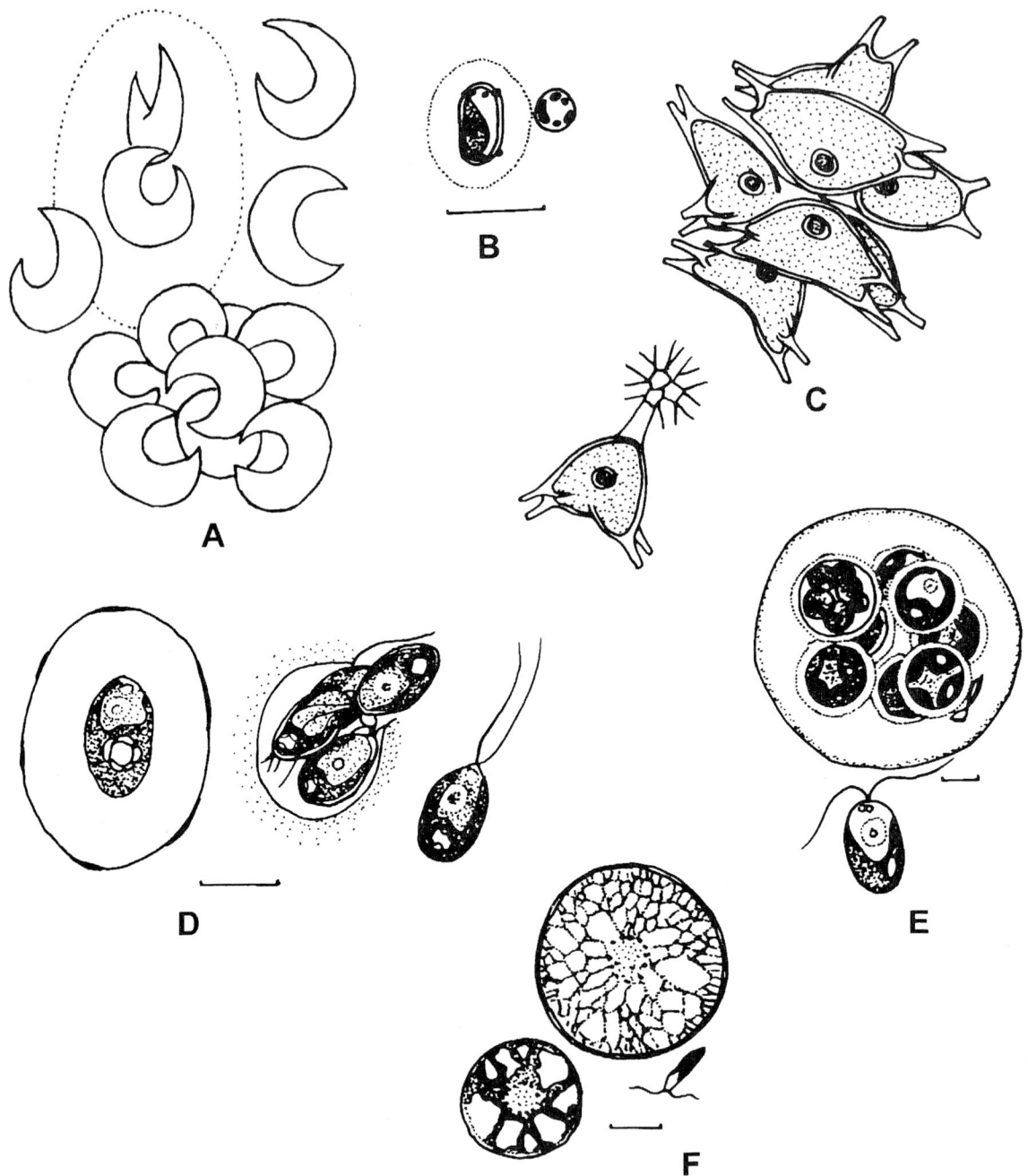

FIGURE 25 A. *Selenastrum capricornatum*. B. *Siderocelis minutissimus*. C. *Sorastrum spinulosum*. D. *Sphaerellocystis ellipsoidea*. E. *Sphaerocystis schroeteri*. F. *Spongiochloris spongiosa*. G. *Spongiococcum alabamense*. A from Komárek, and Fott (1983) with permission; B–G from Bourrelly (1966) with permission.

mingled. The chloroplast is cup-shaped, becoming massive in older cells, and there is one pyrenoid. Can form biflagellated zoospores. Found in the phytoplankton and metaphyton of ditches, ponds, and lakes. Fairly common and reported in several general floras, as well as in regional reports from the Great Lakes region, northeastern, southeastern, central, and western United States, eastern and western Canada, including the Northwest Territories, and the Caribbean.

Spongiochloris Starr (Fig. 25F)

Cells are spherical or subspherical. The chloroplast is a parietal reticulum with inward projecting, spongy portions and one to several pyrenoids. Can form biflagellated zoospores. Found in soil. Rare; reported from an air sample apparently in North Carolina (Whitford and Schumacher, 1984).

Spongiococcum Deason (Fig. 25G)

Cells are solitary or stay together in pairs, tetrads, or small aggregates, young cells are ellipsoidal and mature cells are spherical to broadly ovoid or broadly elliptical, and have a smooth distinct wall with no obvious thickening. The chloroplast essentially is bowllike, with obvious incisions and channeled to cuplike, and has one pyrenoid. Can form biflagellated zoospores. Found in soils. Rarely reported; from soils (in Alabama as *Neospongiococcum*; Deason, 1959), possibly also in the Caribbean.

Stylosphaeridium Geitler et Gimesi in Geitler (Fig. 26A)

Cells are solitary, often clustered, and spherical to ovoid. The chloroplast is massive and cup-shaped, lying along the upper (outward) wall, and has one pyrenoid. The cells are attached to substratum by a slender, tapering stalk. Usually epiphytoplanktonic (*Anabaena*, *Coelosphaerium*) or epizooplanktonic. A few scattered reports from northeastern, southeastern, central, and western United States, and western Canada.

Tetracystis Brown et Bold (Fig. 26B)

The cell structure is *Chlorococcum*-like, but not solitary; rather usually in cell groups (pairs, tetrads). Single young cells are ellipsoidal or ovoid; in a mature condition they are ellipsoidal–spherical to spherical. The cell wall is thin or thick, sometimes with uni- or bipolar thickening, smooth, and hyaline. The chloroplast is essentially potlike, often massive, but perforated by slits or dissected lobes, or divided by coarse or finely stellate rays, with a thickened basal part; a single or multiple pyrenoid is located in the basal region, surrounded by two hemispherical or numerous disklike starch grains. Can form biflagellated zoospores. Typically found in soils. Rarely recorded; some reports from soils in North Carolina (e.g., Whitford and Schumacher, 1984) and in the Caribbean.

Tetradesmus G. M. Smith (Fig. 26D)

Colony is 2–(4)-celled. The cells taper toward each end or are cylindrical or spherical in the vertical view and quadrately arranged; in the face view the cells are in two planes with their longitudinal axes parallel, the adjoined walls straight and in contact most of their length, the outer free wall straight or concave, and the poles directed away from the colony center. The chloroplast is parietal and there is one pyrenoid. Found in the phytoplankton of ponds and lakes. Uncommon; reported from a few locations in northeastern, southeastern, central, and western United States and western Canada.

Tetraedron Kützing (Fig. 26C)

Cells are solitary, symmetrically triangular or pyramidal, and three to four to five angled with angles simple or slightly extended, with a stout spine at each angle; in some, the spines have an inflated base and are longer than the diameter of the cell. The cells have one parietal chloroplast and one pyrenoid. Found in phytoplankton and metaphyton. Common and widely distributed genus in a variety of temperate lakes and ponds, and from tropical habitats in Central America north to arctic sites on Ellesmere Island (Canada).

Tetrallantos Teiling (Fig. 26E)

Colony is four- to eight-celled. Cells are strongly curved, sausage-shaped with rounded ends, and arranged in groups of four, where two cells lie in one plane linked at both their apices and the other two are linked one at each end and curved vertically to the first pair. Fragments of the parental wall persist as interconnecting or radiating threads within the colonial mucilage. The chloroplast is parietal and has one pyrenoid. Found in the phytoplankton and metaphyton of ditches, bogs, ponds and lakes. Possibly rare, but encountered in several ponds, and in river plankton from several sites in North Carolina (Whitford and Schumacher, 1984); also from Cuba (Comas, 1996).

Tetraspora Link (Fig. 26F)

Colony is a macroscopic, spherical, saclike, balloon-like, or irregularly expanded mass (intestine-like) of spherical cells, usually arranged in groups of two to four at the periphery. Cells have two pseudoflagella extending slightly, if at all, beyond the colony envelope. The chloroplast is cup-shaped and has with one pyrenoid. Attached to submerged substrata or floating as colonial clumps, they form massive sheetlike skeins in ditches, bogs, ponds, and lakes. Extremely common and widespread genus reported from all parts of North America; several species more common in springs or during cold temperatures.

Tetrastrum Chodat (Fig. 27A)

Colony is four-celled and flat, with a small central opening. Cells are ovoid. triangular, or heart-shaped, usually bearing one to four slender spines or warts on the free face; sometimes smooth. One to four chloroplasts are parietal and discoid; with or without a pyrenoid. Found in the phytoplankton or metaphyton of ponds and lakes. Common; reported from most

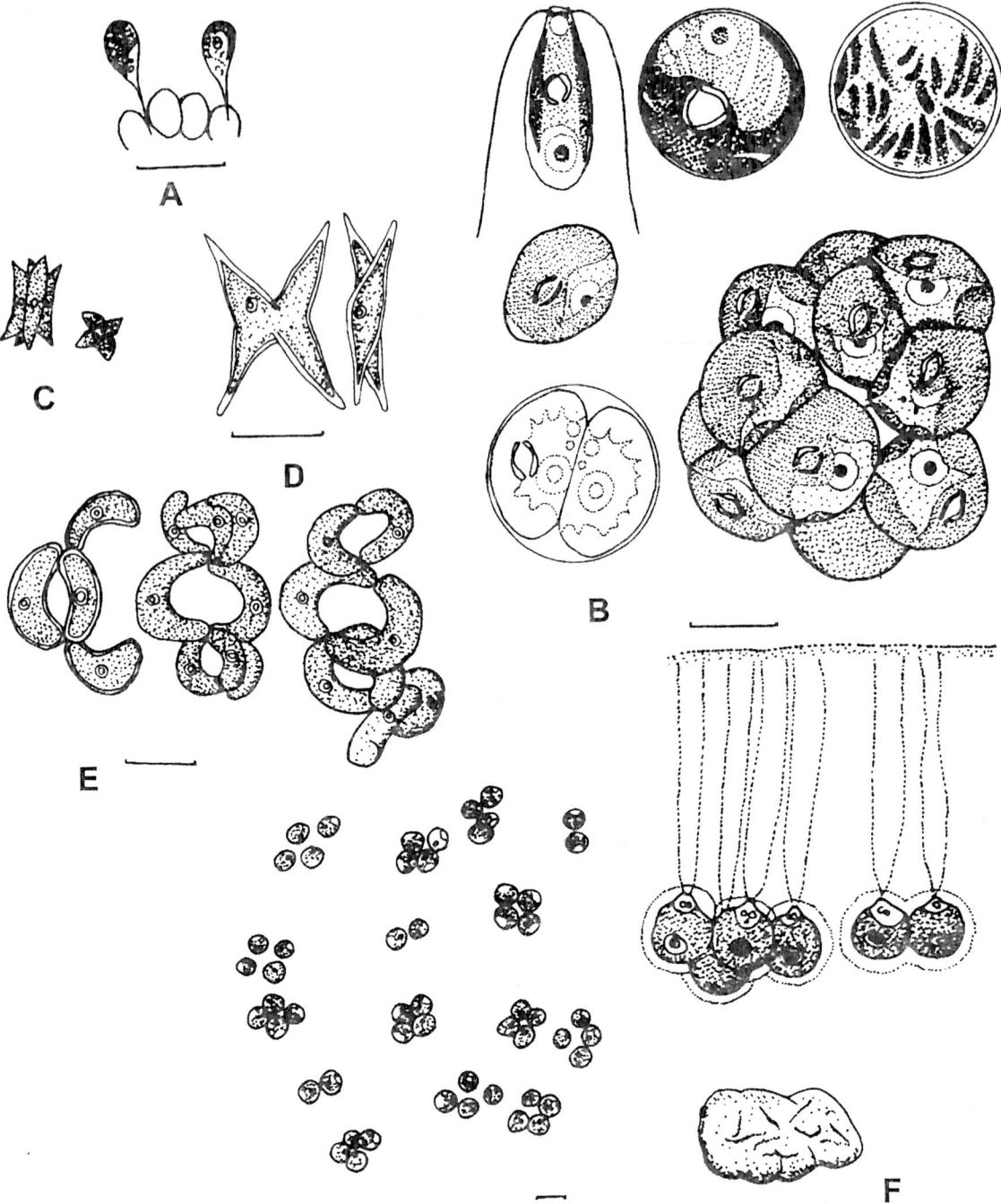

FIGURE 26 A. *Stylosphaeridium stipitatum*. B. *Tetracystis texensis*. C. *Tetradesmus wisconsinensis*. D. *Tetraedron victoriae*. E. *Tetrallantos lagerheimii*. F. *Tetraspora gelatinosa*. A and C–F from Bourrelly (1966) with permission; B from Ettl and Gärtner (1995) with permission.

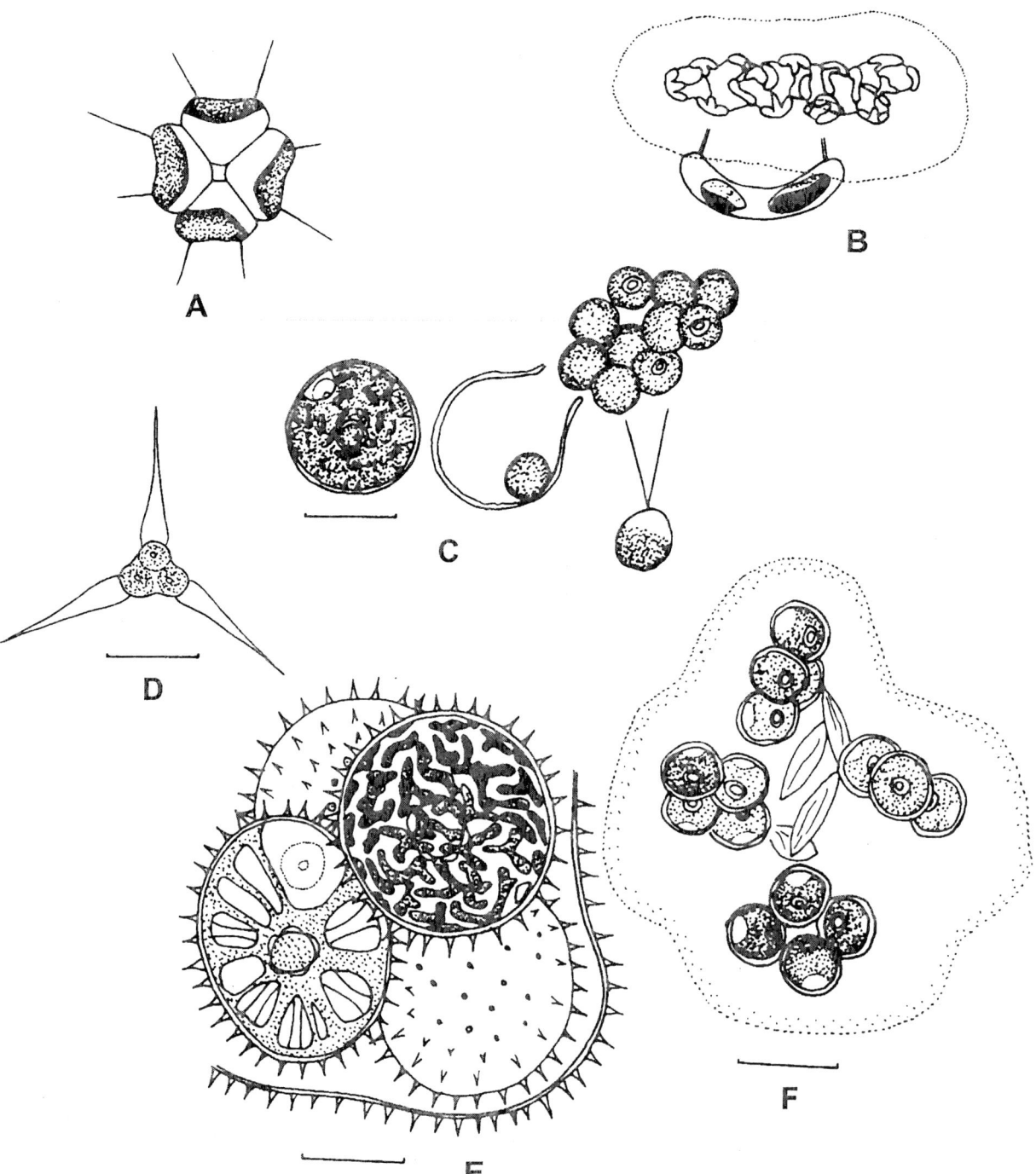

FIGURE 27 A. *Tetrastrum heterocanthum*. B. *Tomaculum catenatum*. C. *Trebouxia parmeliae*. D. *Treubaria triappendiculata*. E. *Trochiscia hystrix*. F. *Westella botryoides*. A and B from Prescott (1978) with permission; C–F from Bourrelly (1966) with permission.

locations including the Great Lakes region, northeastern, central, southeastern, and western United States, western Canada, and the Caribbean.

Tomaculum Whitford (Fig. 27B)

Colony of up to 20 cells embedded in a broad, spherical to ovoid clear mucilage envelope. The cells are sausage-shaped with rounded ends, connected to each other by fine threads of parental cell wall material. The cells have one to two parietal chloroplasts, each with a pyrenoid. Planktonic. Uncommon; reported from North Carolina (Whitford and Schumacher, 1984) and the Caribbean.

Trebouxia Puymaly (Fig. 27C)

Cells are spherical to ovoid, sometimes pear-like. The chloroplast is star-shaped and irregularly lobed, and has one to several pyrenoid(s). Can form biflagellated zoospores. Typically occurs as a lichen phycobiont (e.g., *Cladonia*). Common symbiont in lichens and underreported in aquatic treatises; from several various habitats in the southeastern United States, and eastern and western (temperate) Canada.

Treubaria Bernard (Fig. 27D)

Cells are pyramidal, solitary, and three to five angled, each angle bearing a long, stout spine. One to several chloroplasts in mature cells parietal and one to four pyrenoid(s). Planktonic. Apparently common; in many locations within the Great Lakes region, northeastern, southeastern, central, and western United States, eastern (temperate) Canada, and the Caribbean.

Trochiscia Kützing (Fig. 27E)

Cells are spherical to subspherical and solitary (sometimes gregarious). The cell wall is thick and bears spines or is irregularly and variously sculptured (e.g., ridges and sharp spines). The chloroplast is parietal, usually lobed, and it has one pyrenoid (spine-bearing species) or one to several parietal disks and has one pyrenoid (sculptured species). Occurs in the metaphyton and phytoplankton of ditches, bogs, ponds, and lakes, particularly acid waters. Some species occur on snow; also terrestrial. Fairly common, with different species colonizing aquatic, terrestrial, and snow habitats; reported from most regions, including arctic, temperate, and tropical biomes.

Westella de Wildeman (Fig. 27F)

Colony of 30–80–(100) spherical cells with no mucilage envelope. Cells are arranged in groups of four or eight, and the groups are held together by looplike fragments of the parental cell wall. The chloroplast is parietal. Found in the phytoplankton of ponds and lakes. Common; recorded from most regions of North America in largely aquatic (planktonic) habitats.

C. Addendum

The following list catalogs of other nonmotile unicellular or colonial genera (mostly subaerial or terrestrial) that have been reported or are suspected in the region, but are not commonly encountered in aquatic samples. For more information on these genera, consult Bourrelly (1966), Komárek and Fott (1983), and Ettl and Gartner (1995), and also the literature for species identification given in Section VI.

Ankyra Fott
Apodochloris Komárek
Ascochloris Bold and Mac Entee
Auxenchlorella Shihara and Krauss
Axilococcus Deason and Herndon
Axilosphaera Cox and Deason
Borodinella Miller
Borodinellopsis Dykstra
Botryokoryne Reisigl
Characiopodium Floyd and S. Watanabe
Charaiosiphon Iyengar
Chlamydocapsa Fott
Chlamydopodium Ettl and Komárek
Chondrosphaera Skuja
Chlorococcopsis S. Watanbe and Floyd
Chlorolobion Korschikoff
Chloroplana Gollerbach
Chlorosarcinopsis Herndon
Chlorosphaeropsis (Vischer) Herndon
Chloroteraedron Mac Entee *et al.*
Choricystis (Skuja) Fott
Coelochlamys Korschikoff
Coenochloris Korschikoff
Coenocystis Korschikoff
Cystomonas Ettl and Gärtner
Dactylothece Lagerheim
Desmotetra Deason and Floyd
Diplosphaera Bialosuknia em. Vischer
Ducellieria Teiling
Ecballocystis Bohlin
Elliptochloris Tschermak-Woess
Enallax Pascher
Ettila Komárek
Fernandinella Chodat
Follicularia Miller
Friedmannia Chantanchat and Bold
Graesiella Kalina and Puňcochářova
Györffiana Kol and F. Chodat
Halochlorellla Dangeard
Hemichloris Tschermak-Woess and Friedmann

Heterotetracystis Cox and Deason
Ignatius Bold and Mac Entee
Inoderma Kützing
Interfilium Chodat and Topali
Iwanoffia Pascher
Lautosphaeria Deason and Herndon
Lobosphaera Reisigl
Lobospheropsis Reisigl
Machrochloris Korschikoff
Muriella J. B. Petersen
Muriellopsis Reisigl
Parietochloris Reisigl
Phaseolaria Printz
Pilidiocystis Bohlin
Planktospherella Reisigl
Planophila Gerneck
Podohedra Düringer
Poloidion Pascher
Prasiococcus Vischer
Prasiolopsis Vischer
Pseudochlorella Lund
Pseudochlorococcum Archibald
Pseudococcomyxa Korschikoff
Pseudodictyochloris Vinatzer
Pseudodictyosphaerium Hindák
Pseudoplanophila Ettl and Gärtner
Pseudospongiococcum Gromov and Mamkaeva
Pseudotetracystis Arneson
Pseudotrochiscia Vinatzer
Pulchrasphaera Deason
Raphidonemopsis Deason
Rhopalocystis Schussing
Saturnella Mattauch and Pascher
Thelesphaera Pascher
Trochisciopsis Vinatzer

VI. GUIDE TO LITERATURE FOR SPECIES IDENTIFICATION

Taxonomic keys to the genera of the nonmotile greens that cover North American taxa include Smith (1950), Prescott (1962, 1978), and Dillard (1999). References that are geographically defined, include Prescott (1962) for the Western Great Lakes area, Dillard (1989) for the southeastern United States, and Comas (1996) for Cuba. For the physiological and biochemical characterization of species of *Chlorella*, consult Kessler and Huss (1992). As a general caution, be advised to consult several references for species determination to ensure good agreement between authors. There are valuable taxonomic treatments that apply primarily to the European flora that include keys to the species (Bourrelly, 1966, 1988; Komárek and Fott, 1983; Ettl and Gärtner, 1995; Andreyeva, 1998). Whereas many taxa are cosmopolitan, all the preceding references are valuable resources for the determination of North American taxa.

ACKNOWLEDGMENTS

I thank the numerous undergraduate and graduate students who have studied under my direction and worked in my laboratory during my tenure at the University of North Dakota, most notably, David Bolt, Paul Johnson, Steve Mercil, Wayne Moe, Bruce Pankratz, Charles Pederson, Thomas Starks, and Paul Tomasek. Numerous work–study students at the Natural History Museum provided assistance with literature searches and photocopying, including Melven Comber, Richard Spendlove, Susan Trendell, and Rebecca Yates. A special thank you is extended to Francis R. Trainor for his inspiration and for introducing me to the nonmotile green algae. Technical assistance at the Natural History Museum was provided by Anne Hume, Harold Taylor, and Kevin Webb. The artwork was completed by Phil Rye. I particularly appreciate the encouragement, support, and assistance with manuscript preparation, including the key and German and French translations, that I received from Eileen J. Cox; and critical reviews by Alan Harrington, F. R. Trainor, and an unnamed person. The artwork was supported by the Department of Botany, The Natural History Museum, London.

LITERATURE CITED

An, S. S., Friedl, T., Hegewald, E. 1999. Phylogenetic relationships of *Scenedesmus* and *Scenedesmus*-like coccoid green algae as inferred from ITS-2 rDNA sequence comparisons. Plant Biology 1:418–428.

Andreyeva, V. M. 1998. Terrestrial and aerophilic algae (Chlorophyta: Tetrasporales, Chlorococcales, Chlorosarcinales). Nauka, St. Petersburg, 351 p.

Archibald, P. 1973. The genus *Neochloris* (Chlorophyceae, Chlorococcales). Phycologia 12:187–193.

Archibald, P. A., Bold, H. C. 1970. Phycological studies XI. The genus *Chlorococcum* Meneghini. Publication. 7015, The University of Texas, 115 p.

Beijerinck, M. W. 1890. Cultureversuche mit Zoochlorellen Lichen gonidien und anderen niederen Algen. Botanie Zeitschrift Leipzig 48:757–768.

Blackwell, J. R., Cox, E. J., Gilmour, D. J. 1991. The morphology and taxonomy of *Chlorococcum submarinum* (Chlorococcales) isolated from a tidal rock pool. British Phycological Journal. 26:133–139.

Bold, H. C., Wynne, M. J. 1978. Introduction to the algae. Structure and reproduction. Prentice–Hall, Englewood Cliffs, NJ, 706 pp.

Booton, G. C., Floyd, G. L., Fuerst, P. A. 1998. Polyphyly of tetrasporalean green algae inferred from nuclear small-subunit ribosomal DNA. Journal of Phycology 34:306–311.

Boucher, P., Blinn, D. W., Johnson, D. B. 1984. Phytoplankton ecology in an unusually stable environment (Montezuma Well, Arizona, U.S.A.) Hydrobiologia 119:149–160.

Bourrelly, P. 1966. Les algues d'eau douce. Initiation à la systématique. Tome I: Les algues vertes. Éditions N. Boubée & Cie, Paris, 511 p.

Bourrelly, P. 1988. Compléments les algues d'eau douce. Initiation à la systématique. Tome I: Les algues vertes. Société Nouvelle des Éditions Boubée, Paris, 182 p.

Comas, A. 1996. Las Chlorococcales Duliacuícolas de Cuba. Bibliotheca Phycologia, 99. Cramer, Stuttgart, 192 p., 65 plates.

Deason, T. 1959. Three Chlorophyceae from Alabama soil. American Journal of Botany 46:572–578.

Deason, T. R., Silva, P. C., Watanabe, S., Floyd, G. L. 1991. Taxonomic status of the green alga genus *Neochloris*. Plant Systems Evolution 177:213–219.

Dillard, G. E. 1989. Freshwater algae of the southeastern United States. Part 1. Chlorophyceae: Volvocales, Tetrasporales and Chlorococcales. Bibliotheca Phycologica 81. Cramer, Stuttgart, 202 p., 37 plates.

Dillard, G. E. 1999. Common freshwater algae of the United States. Cramer, Stuttgart, 173 p.

Ettl, H., Gärtner, G. 1995. Syllabus der Boden-, Luft- und Flechtenalgen. Fischer Verlag, Stuttgart, 721 p.

Fleming, R. F. 1989. Fossil *Scenedesmus* (Chlorococcales) from the Raton formation, Colorado and New Mexico, USA. Review of Paleobotany and Palynology 59:1–6.

Floyd, G. K., Watanabe, S., Deacon, T. R. 1993. Comparative ultrastructure of the zoospores of eight species of *Characium* (Chlorophyta). Archiv für Protistenkunde 143:63–73.

Friedl, T. 1995. Inferring taxonomic positions and testing genus level assignments in coccoid green lichen algae: a phylogenetic analysis of 18S ribosomal RNA sequences from *Dictochloropsis reticulata* and from members of the genus *Myrmecia* (Chlorophyta, Trebouxiophyceae cl. nov.). Journal of Phycology 31:632–639.

Friedl, T. 1997. The evolution of the green algae. Plant Systems Evolution [Supplement] 11:87–101.

Friedl, T., Zeltner, C. 1994. Assessing the relationships of some coccoid green lichen algae and the Microthamniales (Chlorophyta) with 18S ribosomal RNA gene sequence comparisons. Journal of Phycology 30:500–506.

Gelin, F., Boogers, I., Noordelos, A. A. M., Sinninghe Damsté, J. S., Riegman, R., de Leeuw, J. W. 1997. Resistant biomacromolecules in marine microalgae of the classes Eustigmatophyceae and Chlorophyceae: Geochemical implications. Organic Geochemistry 26:659–675.

Goff, L. J., Stein, J. R. 1978. Ammonia: Basis for algal symbiosis in salamander egg masses. Life Science 22:1463–1468.

Graham, L. E., Wilcox, L. W. 2000. Algae. Prentice–Hall, Upper Saddle River, NJ, 640 p.

Hanagata, N. 1998. Phylogeny of the subfamily Scotiellocystoidae (Chlorophyceae, Chlorophyta) and related taxa inferred from 18S ribosomal RNA gene sequence data. Journal of Phycology 34:1049–1054.

Happey-Wood, C. M. 1988. Ecology of freshwater planktonic green algae, in: Sandgren, C. D., Ed., Growth and reproductive strategies of freshwater phytoplankton. Cambridge University Press, Cambridge, UK, pp. 175–226.

Hegewald, E., Silva, P. C. 1988. Annotated catalogue of *Scenedesmus* and nomenclaturally related genera, including original descriptions and figures. Bibliotheca Phycologica. 80. Cramer, Stuttgart, 587 p., 900 plates.

Hegewald, E., Schmidt, A. 1987. Untersuchengen an Isolaten und Freilandmaterial der Gattung *Lagerheimia*, Chlorophyta. Archiv für Hydrobiologie Supplement 73:523–558.

Hegewald, E., Schmidt, A. 1991. *Lagerheimia hindakii* is not the unicellular stage of a *Scenedesmus*. Journal of Phycology 27:555.

Hindak, F., Hindakova, A. 1992. New observations on *Sphaerellocystis pallens* Ettl (Chlorophyceae, Tetrasporales). Biologie Bratislava 47:529–537.

Huss, V. A. R., Sogin, M. L. 1990. Phylogenetic position of some *Chlorella* species within the Chlorococcales based upon complete small-subunit ribosomal RNA sequences. Journal of Molecular Evolution 31:432–442.

Huss, V. A. R., Frank, C., Hartmann, E. C., Hirmer, M., Kloboucek, A., Seidel, B. M., Wenzler, P., Kessler, E. 1999. Biochemical, taxonomy and molecular phylogeny of the genus *Chlorella* sensu lato (Chlorophyta). Journal of Phycology 35:587–598.

Kantz, T. S., Theriot, E. C., Zimmer, E. A., Chapman, R. L. 1990. The Pleurastrophyceae and Micromonadophyceae: A cladistic analysis of nuclear rRNA sequence data. Journal of Phycology 26:711–721.

Kessler, E., Huss, V. A. R. 1992. Comparative physiology and biochemistry and taxonomic assignment of the *Chlorella* (Chlorophyceae) strains of the culture collection of the University of Texas at Austin. Journal of Phycology 28:550–553.

Kessler, E., Schäfer, M., Hümmer, C., Kloboucek, A., Huss, V. A. R. 1997. Physiological, biochemical, and molecular characters for the taxonomy of the subgenera of *Scenedesmus* (Chlorococcales, Chlorophyta). Botanic Acta 110:244–250.

Komárek, J., Fott, B. 1983. In: Huber–Pestalozzi, G., Ed., Das Phytoplankon des Süßwassers Systematik und Biologie 7, Teil 1. Hälfte, Die Binnengewasser, Band XVI. Schweizerbart'sche Verlagsbuchhandlung, Stuttgart, 1043 p.

Kouwets, F. A. C. 1995. Comparative ultrastructure of sporulation in six species of *Neochloris* (Chlorophyta). Phycologia 34:486–500.

Krienitz, L., Takeda, H., Hepperle, D. 1999. Ultrastructure, cell wall composition, and phylogenetic position of *Pseudodictyosphaerium jurisii* (Chlorococcales, Chlorophyta) including comparison with other picoplanktonic green algae. Phycologia 38:100–107.

Lewis, L. A. 1997. Diversity and phylogenetic placement of *Bracteococcus* Tereg (Chlorophyceae, Chlorophyta) based on 18S ribosomal RNA gene sequence data. Journal of Phycology 33:279–285.

Lewis, L. A., Wilcox, L. W., Fuerst, P. A., Floyd, G. L. 1992. Concordance of molecular and ultrastructural data in the study of zoosporic Chlorococclean green algae. Journal of Phycology 28:375–380.

Mattox, K. R., Stewart, K. D. 1984. Classification of the green algae: A concept based on comparative cytology, in: Irvine, D. E. G., John, D. M., Eds, Systematics of the green algae. Academic Press, London, 29–72 p.

Melkonian, M., Surek, B. 1995. Phylogeny of the Chlorophyta: Congruence between ultrastructural and molecular evidence. Bulletin de Société Zoologie France 120:191–208.

Miller, P. E., Scholin, C. A. 1998. Identification and enumeration of cultured and wild *Pseudo-Nitzschia* (Bacillariophyceae) using species-specific LSU rRNA-targeted fluorescent probes and filter-based whole cell hybridization. Journal of Phycology 34:371–382.

Mosto-Cascallar, P. 1987. The genus *Characiochloris* Pascher (Chlorophyceae, Tetrasporales). Physis Seccion B Las Aguas Continentales y su Organismos 45:41–52

Nakayama, T., Watanabe, S., Mitsui, K., Uchida, H., Inouye, I. 1996. The phylogenetic relationships between the Chlamydomonadales and Chlorococcales inferred from 18S rDNA sequence data. Phycological Research 44:47–86.

Nichols, H. W., Nichols, M. S., Thomas, C. M., Deacon, J. S., Veith, M. 1991. A new *Sphaerellocystis* (Palmellopsidaceae, Tetrasporales) from prairie soils. Archiv für Protistenkunde 140:157–170.

Nozaki, H. 1993. Morphology, reproduction and taxonomy of *Characiochloris sasae* sp. nov. (Chlorophyta) from Japan. Phycologia 32:129–135.

O'Kelly, C. J., Watanabe, S., Floyd, G. L. 1994. Ultrastructure and phylogenetic relationships of Chaetopeltidales ord. Nov. (Chlorophyta, Chlorophyceae). Journal of Phycology 30:118–128.

Pickett-Heaps, J. D. 1975. Green algae: Structure, reproduction and evolution in selected genera. Sinauer, Sunderland, MA, 606 pp.

Pickett-Heaps, J. D., Staehelin, L. 1975. The ultrastructure of *Scenedesmus* (Chlorophyceae). II. Cell division and colony formation. Journal of Phycology 11:186–202.

Pizzaro, H. 1995. The genus *Characiopsis* Borzi (Mischococcales, Tribophyceae). Taxonomy, biogeography and ecology. Bibliotheca Phycologica 98:1–146.

Preisig, H. R. 1999. Systematics and evolution of the algae: Phylogenetic relationships of taxa within the different groups of algae, *in*: Esser, K., Kadereit, J. W., Lüttge, U., Runge, M., Ed., Systematics and comparative morphology. Progress in Botany, Volume 60, Springer-Verlag, Berlin, pp. 369–412.

Prescott, G. W. 1962. Algae of the western Great Lakes area, rev. ed. Brown, Dubuque, IA, 977 p.

Prescott, G. W. 1978. How to know the freshwater algae, 3rd ed. Brown, Dubuque, IA, 293 p.

Reynolds, C. S. 1984. The ecology of freshwater plankton. Cambridge University Press, Cambridge, UK.

Round, F. E. 1984. The systematics of the Chlorophyta: An historical review leading to some modern concepts [taxonomy of the Chlorophyta III], *in*: Irvine, D. E. G., John, D. M., Eds., Systematics of the green algae. Academic Press, London, pp. 1–27.

Shubert, L. E. 1975. *Scenedesmus trainorii* sp. nov. (Chlorophyta, Chlorococcales): A polymorphic species. Phycologia 14:177–182.

Shubert, L. E. 1998. The effect of the Sea Empress oil spill on the distribution and density of intertidal microscopic algae on the Welsh coast, *in*: Edwards, R., Sime, H., Eds., The Sea Empress oil spill conference proceedings. Chartered Institution of Water and Environmental Management, UK, pp. 411–422.

Shubert, L. E., Trainor, F. R. 1974. *Scenedesmus* morphogenesis: Control of the unicell stage with phosphorus. British Phycological Journal 9:1–7.

Sluiman, H. J., Guihal, C. 1999. Taxonomic studies on the genus *Pleurastrum* (Pleurastrales, Chlorophyta). I. The type species, *P. insigne*, rediscovered and isolated from soil. Phycologia 29:133–138.

Smith, G. M. 1950. The freshwater algae of the United States, 2nd ed. McGraw–Hill, New York, 719 p.

Starks, T. L., Shubert, L. E. 1979. Algal colonization on a reclaimed surface mined area in western North Dakota, *in*: Wali, M. K., Ed., Ecology and coal resource development. Pergamon, Elmsford, NY, pp. 652–660.

Starks, T. L., Shubert, L. E., Trainor, F. R. 1981. Soil algae: A review. Phycologia 20:65–80.

Starr, R. 1955. Comparative study of *Chlorococcum meneghini* and other spherical, zoospore-producing genera of the Chlorococcales. Indiana University Press, Bloomington, 111 p.

Stein, J. R., Ed. 1973. Handbook of phycological methods. Culture methods and growth measurements. Cambridge University Press, Cambridge, UK, 448 p.

Trainor, F. R. 1978. Introductory phycology. Wiley, New York, pp. 69–90.

Trainor, F. R. 1991. *Scenedesmus* plasticity: Facts and hypotheses. Journal of Phycology 27:555–556.

Trainor, F. R. 1998. Biological aspects of *Scenedesmus* (Chlorophyceae)—Phenotypic plasticity. Nova Hedwigia 117. Cramer, Stuttgart, 367 p.

Trainor, F. R., Egan, P. 1990a. The implications of polymorphism for the systematics of *Scenedesmus*. British Phycology Journal 25:275–279.

Trainor, F. R., Egan, P. 1990b. Phenotypic plasticity in *Scenedesmus* (Chlorophyta) with special reference to *Scenedesmus armatus* unicells. British Phycological Journal 29:461–469.

Trainor, F. R., Egan, P. 1990c. *Lagerheimia hindakii* is the unicellular stage of *Scenedesmus*. Journal of Phycology 26:535–539.

Trainor, F. R., Morales, E. A. 1999. Recent trends in the taxonomy of selected chlorococcalean algae, *in*: Vidyavati, Mahato, A. K., Eds., Recent trends in algal taxonomy, Vol. 1. Taxonomical issues. APC Publications, New Delhi, pp. 67–100.

Trainor, F. R., Cain, J. R., Shubert, L. E. 1976. Morphology and nutrition of the colonial green alga *Scenedesmus*:80 years later. Botanical Review 42:5–25.

Watanabe, S., Floyd, G. 1989. Comparative ultrastructure of the zoospores of nine species of *Neochloris* (Chlorophyta). Plant Systems Evolution 168:195–219.

Whitford, L. A., Schumacher, G. J. 1984. Manual of freshwater algae in North Carolina. Sparks Press, Raleigh, North Carolina.

Wilcox, L. W., Lewis, L. A., Fuerst, P. A., Floyd, G. L. 1992. Assessing the relationships of autosporic and zoosporic chlorococcalean algae with 18S rDNA sequence data. Journal of Phycology 28:381–386.

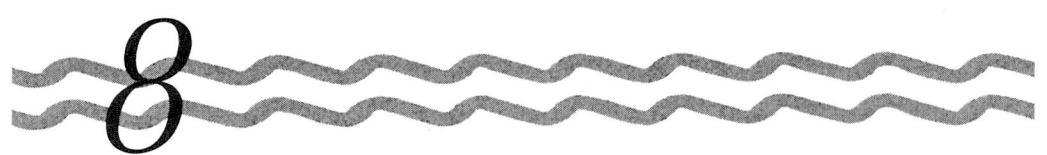

FILAMENTOUS AND PLANTLIKE GREEN ALGAE

David M. John

Department of Botany
The Natural History Museum
London SW7 5BD,
United Kingdom

I. Introduction
II. Diversity and Morphology
    A. Morphology and Cell Structure
    B. Reproduction
III. Classification of Green Algae
IV. Ecology and Distribution
    A. Aquatic Habitats
    B. Soil and Subaerial Habitats
V. Collection and Preparation of Samples
VI. Key and Descriptions of Genera
    A. Key to Groups
    B. Key to Genera by Group
    C. Descriptions of Genera
VII. Guide to Literature for Species Identification
Literature Cited

I. INTRODUCTION

The green algae (Division Chlorophyta) possess chlorophyll *a* and *b* within a double membrane-bound chloroplast. True starch is stored in the chloroplast and forms the principal photosynthetic product. The cell wall is normally composed of cellulose, and flagellated stages, such as zoospores or gametes, are usually a part of the life history. Chlorophytes demonstrate a considerable amount of morphological variation, ranging from microscopic flagellated unicells to complex macroscopic thalli showing varying degrees of morphological differentiation. Traditionally the classification is based upon thallus organization with the green algae including a wide range of structural forms: single-celled and motile (flagellated), nonmotile (coccoid), colonial, filamentous, parenchymatous, and coenocytic (multinucleate). Considered in this chapter are those green algal orders in which the majority of the genera are filamentous (includes coenocytic forms), parenchymatous (tissue-like and often tubular or membrane-like in form) or possess a macroscopic plantlike thallus. Of the groups covered here, the Charales (charophytes) are the most plantlike and the order is sometimes considered in accounts of higher aquatic plants. The charophytes are commonly referred to as 'stoneworts' or 'brittleworts' since many are lime-encrusted ($CaCO_3$, especially *Chara*). Of the major orders of largely filamentous green algae the Zygnematales (reproduce by conjugation) is the only one excluded (see Chapter 9).

II. DIVERSITY AND MORPHOLOGY

The green algae contain over 500 genera and approximately 15,000 species (John, 1994), thus making it the largest algal division. The large majority (over 80%) live in freshwater with the remainder occupying habitats that include seawater/ brackish-water and soil or soil-free surfaces. Most are free living, but there are a few that are parasitic or symbiotic (Gartner, 1992; Chapman and Waters, 1992). In the orders Charales and Oedogoniales, almost all are confined to fresh water, unlike most other orders (e.g., orders Ulvales, Cladophorales) where the majority are marine.

A. MORPHOLOGY AND CELL STRUCTURE

Chloroplasts in the chlorophyte orders considered here vary greatly in size and shape. Commonly there is a single chloroplast lying against the inner cell wall and these vary from disk- or platelike, ringlike, or netlike (Hoek et al., 1995). There are many exceptions that include the Prasiolales, where the centrally positioned chloroplast is marginally lobed. Often the chlorophyll in terrestrial algae is masked by red carotenoid pigments (e.g., astaxanthin), which are sometimes dissolved in lipid droplets within the cell. Each chloroplast normally contains one to several pyrenoids (proteinaceous bodies) surrounded by a sheath of starch; pyrenoid number is a character of taxonomic importance. Its absence has to be viewed with caution, since they are known to arise *de novo* and are not always visible using the light microscope (LM). Taxonomically important characters are associated with ultrastructural features of the chloroplast and pyrenoid. Of the orders considered, only the Cladophorales, Siphonales, and Sphaeropleales are multinucleate. In the Siphonales, cross walls develop only during reproduction and thus the vegetative thallus is nonseptate (acellular) and multinucleate.

The cellulose cell wall of green algae has a structural fibrillar component and an amorphous matrix component; the latter is predominantly on the outer wall and often forms a slimy or mucilaginous outer layer. In a few orders the outermost layer is very hard and provides a suitable attachment surface for many algal epiphytes. Of the algae considered here, only the Trentepohliales and the Charales possess sporopollenin, a very hard and phenol-containing substance present in higher plant spore walls (Good and Chapman, 1978). Other green algae have decay-resistant wall compounds which differs chemically from sporopollenin in consisting of polymers of unbranched hydrocarbons and lacking phenolic groups (Gelin et al., 1997). Some green algae are characterized by having a diffluent or firm mucilaginous sheath or envelope.

B. REPRODUCTION

Vegetative reproduction normally takes place through fragmentation and simple cell division (Hoek et al., 1995). Asexual reproduction frequently involves biflagellate or quadriflagellate zoospores, except in the Oedogoniales where there is a ring of small cilia at the anterior pole. The cell functioning as a zoosporangium is similar in size to a vegetative cell or swollen and of a different form. Within a zoosporangium, one or more zoospores develop and are released through a pore(s) or by gelatinization of the wall. In some species, thick-walled, nonmotile spores, known as aplanospores, are produced. These have the ontogenetic possibility of developing into zoospores and thus are usually regarded as zoospores whose development has become arrested. Commonly formed are thick-walled resting vegetative cells known as akinetes; sometimes these are in series.

Sexual reproduction ranges from isogamy or anisogamy to oogamy (Hoek et al., 1995). In isogamy, the gametes are morphologically similar, and in anisogamy, the gametes differ in size; often gametes only fuse with those produced by a different individual. Oogamy is the most advanced form of reproduction, with the egg cell or oogonium retained and fertilized on the parent plant (e.g., Charales, Oedogoniales, Coleochaetales). Oogamy and other features (more than four meiospores per zygote) have led to the Coleochaetales being regarded as closely related to the ancestors of land plants (Graham and Wilcox, 2000). In some orders, the zygote resulting from fertilization develops a wall of sporopollenin whose ornamentation can provide taxonomically important characters (e.g., Charales, Leitch et al., 1990). For further information on reproduction in the chlorophytes, consult modern phycological textbooks such as Graham and Wilcox (2000) and Hoek et al. (1995).

Our knowledge of the life history of many green algae is still incomplete since ploidy levels of different stages are often unknown. In most of the green algal orders considered here, the only diploid stage is believed to be the resting zygote, a haplobiontic type of life history (John, 1994). Often these thick-walled spores require a period of obligate dormancy and this is considered an adaptation to freshwater life since many green algae have to survive seasonally adverse conditions. Only a few of the orders (e.g., Cladophorales, Trentepohliales) are known to have a regular alternation of diploid and haploid generations (diplohaplontic life cycles).

III. CLASSIFICATION OF GREEN ALGAE

The four genera of green algae (*Ulva*, *Conferva*, *Chara*, *Volvox*) mentioned in *Species Plantarum* (Linnaeus, 1753) were placed in the class Cryptogamia. It was not until Harvey (1836) created the Chlorospermeae that the green algae were recognized as a natural and separate assemblage. Pascher (1914, 1931) is credited with establishing thallus organization level as one of the most important characters for recognizing higher green algal groups. Since the 1950's three different approaches have contributed most to our understanding of green algal systematics and have led to a re-examination of phylogenetic relations. The most important resulted from ultrastructural research on cell division patterns (mitosis and cytokinesis) and flagellar organization. Cell division pattern and flagellar organization now provide the principal characters used for defining the Division (or Phylum) Chlorophyta, along with its classes and orders.

The first modern synthesis of new ultrastructural data and biochemical features led to the separation of two classes: Chlorophyceae *sensu* Stewart & Mattox and Charophyceae *sensu* Stewart and Mattox (1975; Pickett-Heaps, 1975). Later, two further classes were proposed, the Ulvophyceae (Stewart and Mattox, 1978) and Class Prasinophyceae (Mattox and Stewart, 1985). A fifth class has been recognized, the Pleurastrophyceae, but with the removal of the genus *Pleurastrum* it has been renamed the Trebouxiophyceae (Friedl, 1998). The five-class scheme has been tested several times and refined based primarily on further information on the ultrastructure of motile cells and analysis of molecular data (mostly small unit rDNA) (Friedl, 1998; McCourt, 1995; Booton *et al.*, 1998). Two major lineages in the chlorophytes are accepted at present, one consisting mostly of microscopic freshwater forms and larger green marine algae (Chlorophyceae, Ulvophyceae, Trebouxiophyceae, Prasinophyceae), and the other lineage (class Charophyceae) of species more closely related to higher green plants than to other chlorophytes (Graham and Wilcox, 2000). The Charophyceae includes the Charales (charophytes) and the Coleochaetales; the latter order contains filamentous forms with an advanced form of oogamous reproduction and yet placed traditionally in the Chaetophorales.

The classification scheme adopted by Hoek *et al.* (1995) is based upon new molecular and ultrastructural data in addition to characters associated with reproduction, vegetative morphology, and biochemistry. Twelve new classes (including the Chlamydophyceae) were recognized of which the following contain filamentous (including siphonaceous), parenchymatous or "plantlike" green algae:

- Chlorophyceae (Oedogoniales, Chaetophorales)
- Ulvophyceae (Ulotrichales, as Codiolales; Ulvales)
- Cladophorophyceae (Cladophorales)
- Trentepohliophyceae (Trentepohliales)
- Zygnematophyceae (Zygnematales, Desmidiales)
- Klebsormidiophyceae (Klebsormidiales, Coleochaetales)
- Charophyceae (Charales)

The more soundly based green algal orders (e.g., Charales, Oedogoniales, Trentepohliales, Cladophorales, and Zygnematales) have been supported by molecular and ultrastructural data. Other orders defined largely on thallus organization have proved to be very heterogeneous so that very similar morphological forms may fall within different evolutionary lineages, for example, the Ulotrichales (unbranched filaments), Chaetophorales (branched, often heterotrichous), and Ulvales (parenchymatous), suggesting that similarity in thallus organization may be the result of convergent evolution. For example, genera traditionally assigned to the order Ulotrichales [e.g., *Ulothrix*, *Uronema*, *Klebsormidium* (= *Hormidium*)] belong to different orders and classes when ultrastructural characters are taken into account.

The various newly proposed schemes for the classification of the Division Chlorophyta (e.g., Hoek *et al.*, 1995; Friedl, 1998; Graham and Wilcox, 2000) are not adopted here because the classification is still in a state of flux with no consensus likely to emerge until ultrastructural and molecular studies are undertaken on a considerably greater number and range of taxa. Many taxa are putatively placed and some will undoubtedly be transferred to other orders and classes when investigated using modern methods. Therefore, the position of genera for which there exists no information on cell division and/or flagellar organization must be considered uncertain (*incertae sedis*). For the purposes of this chapter, all genera are placed within the framework of a traditional classificatory scheme. It is tempting to adopt the solution suggested by Hoek *et al.* (1995), namely to assign a genus to the order or class to which they might appear to belong on the basis of light microscope observations. The problem of adopting such a solution has been addressed already, namely that organization level might give a misleading impression of relationships (see comments concerning genera traditionally placed within the Ulotrichales).

The traditional classification of the green algae adopted by Dillard (1989) and modified from that used by Prescott (1962), Bourrelly (1988), and others is followed. It has been decided not to recognize families since many of the same problems associated with orders apply equally to them. Genera are arranged alphabetically under each of the orders.

IV. ECOLOGY AND DISTRIBUTION

A. Aquatic Habitats

The majority of the filamentous and 'plantlike' green algae are aquatic and live attached to hard surfaces, such as rocks, man-made objects (plastic, glass, concrete channels, etc.), aquatic plants (flowering plants, mosses, macroalgae) and various animals (mollusks, turtles, fish). Frequently, larger filamentous forms are not only attached to submerged vegetation but entangled with it or else are free-floating on or just beneath the water surface. Water plants that frequently provide attachment surfaces for filamentous algae include

Najas, Myriophyllum, Ceratophyllum, rushes, and sedges. Many filamentous algae growing on water plants are also capable of growing on other surfaces, as has been shown to be the case when artificial surfaces have been placed in aquatic habitats and examined after being retrieved at different time intervals (John and Moore, 1985; Sabater et al., 1998). Associated with surfaces are diverse assemblages of attached or unattached microscopic forms that only become noticeable to the naked eye when present in vast numbers. Generally the assemblage is dominated by green algae or diatoms along with sessile and moving animals. The term "aufwuchs," or periphyton, is used for this assemblage of microscopic algae whether on submerged plants, rocks, or other surfaces (Chapter 2, Section II–F). The filamentous algae forming the periphyton include crustose forms (e.g., *Aphanochaete, Protoderma*) and others having erect and/or prostrate systems of branches (e.g., *Stigeoclonium, Chaetophora, Draparnaldia*). Some species perforate soft limestone and are partly (*Gomontia perforans*) or completely (*Gongrosira debaryana*) endolithic.

The majority of aquatic algae require hard attachment surfaces. Unconsolidated surfaces, such as mud, sand, or gravel, are normally too unstable for such forms. Some form mats overlying muddy bottoms and usually develop in deep or otherwise quiet water. For example, a band of *Dichotomosiphon tuberosa* is known from a depth of about 11–17 m in the Straits of Mackinac area of Lake Michigan and Lake Huron (Henson, 1984). The only green algal group well adapted to growing on beds of sand or silt are the charophytes or stoneworts. In marl-rich lakes or ponds recently cleared of higher plant vegetation, the charophytes may form extensive underwater "meadows". In such charophyte-dominated water bodies, plankton numbers tend to be low and the water very clear. If left unmanaged, small ponds containing charophytes become, in time, dominated by competitively superior aquatic macrophytes (Hutchinson, 1975).

One of the most common and important filamentous green alga in freshwater is *Cladophora*, commonly known as "cotton-mat" or "blanket weed" because dead and dying stranded masses turn gray or brown and resemble pieces of a woolen blanket (Canter-Lund and Lund, 1995; Chapter 24). These common names are usually taken as referring to mats of living *Cladophora*, but these are likely to contain and sometimes to be dominated by other filamentous green algae (particularly *Oedogonium* and *Spirogyra*; Chapter 9). Many secondarily detached green algae (including broken filaments) often become a nuisance when forming large free-floating mats in drainage ditches, ornamental ponds, and other small water bodies. *Cladophora* grows in greatest abundance along the shallow margins of nutrient-enriched lakes and on rocks in streams and faster-flowing rivers. *Cladophora glomerata* and the unbranched filaments of *Ulothrix zonata* are the dominant algae in the marginal shallows of nutrient-enriched lakes such as the Laurentian Great Lakes (Garwood, 1982) where *Cladophora* has become a serious problem alga. Dead and dying plants are known to accumulate in vast quantity along downwind shores and in sheltered bays thus reducing the amenity value of an area and profoundly affecting the local ecology. The blanket weed can also cause similar problems when it accumulates in the quiet backwater of lowland rivers. For a more comprehensive account of problems caused by filamentous and other algae, see Chapter 23.

Some filamentous algae, such as *Cladophora glomerata*, are reported to disappear from quiet backwaters during the warmer summer months and only continue to survive in fast-flowing reaches during this time. Its morphology depends on habitat conditions; plants growing in still water conditions are bushy since the branches are short and arise at obtuse angles, whereas in swiftly flowing water the filaments are long and branch only infrequently at acute angles. Filamentous algae, such as *Cladophora* and *Oedogonium*, often attain their greatest size in streams and fast-flowing rivers, possibly a reflection that in flowing water there is an accentuated gradient across the laminar boundary layer of water allowing for the rapid exchange of gases and nutrients (Whitford, 1960; Raven, 1992). An unusual growth form is reported from some lakes resulting from filaments growing out radially to produce a mossy ball (often 10 cm or more in diameter) commonly referred to as "*Cladophora*-balls" (Kindle, 1934; Daily, 1952).

Only a few filamentous green algae are endophytic and invade the tissues of water plants. One of the most common of these endophytes is *Chlorochytrium lemnae*, normally associated with the duckweed *Lemna*, but is also known in the tissues of *Ceratophyllum demersum, Elodea* spp., and certain liverworts (Hutchinson, 1975). Filamentous green algae occur on animals, notably mollusks, turtles, and fish. One of the most well known is *Basicladia* whose filaments form a feltlike covering on the backs of snapping turtles when present in abundance. Other green algae reported on turtles include *Rhizoclonium hieroglyphicum, Dermatophyton*, and *Cladophora* (Edgren et al., 1953); the latter genus and an unidentified chaetophoralean alga were discovered by Vinyard (1955) growing on salmon in Lake Texoma, Oklahoma. Some of those growing on turtles also occur on alligators which according to Prescott (1983) are often "...veritable algal gardens and offer a variety of species for the less timid collector".

Sometimes microscopic filamentous algae are swept into the plankton, otherwise only a few are to be regarded as truly planktonic. The only planktonic filamentous genus is *Helicodictyon*, which sometimes forms sulfur-yellow colored blooms during the autumn in the acid and slightly brown water of ponds and lakes on the coastal plain along the eastern seaboard of the United States (Whitford and Schumacher, 1969).

B. Soil and Subaerial Habitats

In terrestrial environments, algae grow on or within soil, rocks, stones, snow, animals, and plants. Filamentous algae sometimes develop on the surface of soil where they may form macroscopic growths when in quantity (e.g., *Fritschiella tuberosa*, *Klebsormidium flaccidum*, *Protosiphon botryoides*). Other soil-living algae have been discovered only when moist soil or soils mixed with nutrient media are cultured in the laboratory. Such algae are often defined upon characteristics observed in laboratory culture and frequently cannot be recognized by an examination of field samples [e.g., *Pseudoschizomeris*, *Filoprotococcus* (=*Trichosarcina*), *Hazenia*].

Damp stones, wooden fencing, greenhouse glass, roofing tiles, and tree bark are frequently discolored by growths of green algae forming the so-called subaerial or aero-terrestrial algal community. The almost ubiquitous powdery green layer, often referred to as the *Protococcus-Pleurococcus* community, is characterized by *Desmococcus* and *Apatococcus*. These two genera grow as solitary cells or small cell packets, although *Desmococcus* is capable of producing short filaments under very moist conditions; traditionally the genera are placed in the Order Chaetophorales. Another very characteristic genus on moist rock surfaces (often on steep cliffs), wood, tree trunks, leaves, and other subaerial surfaces is *Trentepohlia*. Its distinctive orange or reddish color and often feltlike growth readily distinguishes it from the *Protococcus–Pleurococcus* community. Subaerial algae sometimes grow in close association with fungi to form what is sometimes referred to as a "protolichen". *Trentepohlia* is a common algal component (phycobiont) of lichens, which consists of an intimate symbiotic association between a fungus and one or more algae or cyanobacteria. Lichen-algae are outside the scope of this chapter, although species belonging to some of the genera considered here (e.g., *Dilabifilum*, *Leptosira*, *Cephaleuros*) are known to be the algal partner of some lichens (Gartner, 1992). An unusual subaerial alga is *Trichophilus welcheri* whose filaments commonly grow amongst the hair scales of two- and three-toed sloths (*Bradypus*) in the rainforests of Central America (Thompson, 1972a). Often its hairs are pinkish in color due to the presence of another alga, *Cyanoderma bradypodis*.

In the humid tropics various filamentous algae are common on tree trunks and associated with leaves as epiphytes or endophytes. Some produce a red carotenoid pigment that masks the green color of the chlorophyll pigments when present in quantity. Most of these algae are epiphytic (e.g., *Trentepohlia*, *Phycopeltis*, *Stromatochroon*) or endolithic with a few presumed to be parasites or semi-parasites (Thompson and Wujek, 1997). One such parasitic species is *Cephaleuros virescens*, whose largely endophytic branches cause discoloration and death of leaf tissues in a wide range of tropical and subtropical plants (Chapman and Waters, 1992). It is a serious disease of certain commercial plants and causes the red rust of tea plants. *Phyllosiphon* is responsible for the yellow or red spots in the tissues of *Arisaema*, a vascular plant commonly known as Jack-in-the-Pulpit (Smith, 1950).

An extreme environment for algae is the surface of more or less permanent snowfields. The only filamentous green alga among other forms associated with this inhospitable environment is *Raphidonema* (Hoham, 1973). Often snow banks containing large quantities of algae are streaked green, yellow, or reddish depending on the dominant alga and the extent to which the green chlorophylls are masked by the red carotenoid pigments (see also Chapter 2; Section VI–B).

V. COLLECTION AND PREPARATION OF SAMPLES

Macroscopic growths of filamentous and parenchymatous algae, as well as charophytes, can be collected by hand from the littoral zone of lakes or from the surface if free-floating. In deeper water, snorkel or SCUBA-living is necessary unless a remote sampler is used, such as a grapnel or dredge. A weighted three- or four-pronged grapnel hook is ideal for sampling deeper-water charophytes. Small individual colonies or mats of softer filamentous algae need to be collected with a fine forceps and placed separately in vials or collecting jars. Often many interesting filamentous forms grow as a feltlike or fuzzy covering over stones, grasses, tree roots, and aquatic plants (e.g., *Myriophyllum*, *Najas*, *Potamogeton*, water lillies), including the submerged portions of emergent plants such as sedges or rushes. A small portion can be removed by means of a knife or scalpel and put in a suitable collecting vessel. Stones or other hard surfaces are best scraped to remove the surface film of algae. A simple and effective method of removing the algal film from rushes or sedges is to place the forefinger around the stem and to move it upward while at the same time pressing the thumbnail against the surface. Rotting pieces of wood or soft limestone rocks should be examined and sampled since they might contain algae that penetrate the surface.

Many methods that involve scraping a surface often fail to adequately sample those green algae forming minute crusts or having both prostrate and erect filaments. Often it is not easy to detect these algae while examining portions of water plants unless the underlying cells have died and become almost opaque. Endophytic algae, such as *Chlorochytrium*, are most readily detected in *Lemna* leaves that are dead or dying since its spots become more evident when the leaves are no longer green. *Cladophora* and *Oedogonium* have very hard walls and as a result are frequently heavily epiphytized by many minute filamentous forms. Glass bottles, pieces of glass, plastic bags, or similar transparent surfaces should be examined. On these artificial surfaces it is possible to detect very minute forms that are likely to be common on natural surfaces nearby. Of course, it is possible to provide substrata for colonization by these minute algae. Glass slides, plastic Petri dishes, or plastic bags are suitable surfaces and become colonized by such algae if left in the water for at least two weeks. It must be borne in mind that such artificial surfaces are selective and so not all small benthic forms might be sampled in this way.

Various algae growing on subaerial surfaces appear reddish or orange in color and are usually readily collected by scraping rock surfaces or by collecting discolored snow, especially from permanent snow banks. Subaerial algae are especially common in the humid tropics and grow as crusts, tufts, or mats on twigs, trunks, and leaves of trees and other plants. These algae should be collected along with the underlying surface to which they are attached. Various animals have filamentous or crustose algae growing on their shells and carapace (e.g., snapping turtles) and these usually have to be removed by a forceps or scraping. Sometimes soil algae form conspicuous surface mats if present in quantity. Otherwise, many soil algae can only detect and examined when soil is mixed with nutrients and cultured under laboratory conditions ("enrichment cultures"). Many of the filamentous soil algae mentioned are poorly defined and their taxonomy is based on features only observed in culture. Relatively few species of filamentous green algae are recorded from the plankton. Many of those recorded have been swept into the plankton by water turbulence. Such algae are collected along with truly planktonic forms in net tows.

Whenever possible, algae should be examined in the living state. Filamentous algae can often survive in reasonable condition for several days provided they are not kept in direct sunlight and in a reasonably large volume of water. Sterile specimens belonging to the Oedogoniales or other groups sometimes become fertile under such conditions. To investigate morphological variation and life history characters, it is desirable to grow them in laboratory culture. For long-term preservation, 4% formalin is ideal, although FAA (formalin: acetic acid: ethyl alcohol) is considered more suitable for flagellated stages since flagella are less likely to be lost than when formalin is used alone. If freshly collected or preserved macroscopic forms are to be sent through the mail, all surplus liquid should be drained and the specimen placed in a securely sealed polyethylene bag. Such forms can be preserved by drying on herbarium sheets, with muslin or cloth used to prevent them adhering to the drying papers. Charophytes are relatively coarse and do not generally adhere to drying papers and thus have to be tied by thread or held by cloth tape to the herbarium sheets. Often care is needed when drying since heavily $CaCO_3$-incrusted charophytes readily disintegrated if too much pressure is applied to a herbarium press.

VI. KEY AND DESCRIPTIONS OF GENERA

These dichotomous generic-level identification keys to filamentous and "plantlike" green algae are artificial and based upon morphological characters observed with the light microscope; only when absolutely necessary are reproductive, ecological, or more obscure characters used. Some characters are not absolute (e.g., free-living or attached, solitary, or colonial), making it necessary on occasion to "back track" and follow an alternative proposition. There are genera for which a single character applies only to some of its species and for this reason a genus might appear two or three times in the key (indicated as "in part"). The use of special descriptive terms is kept to a minimum. All generic determinations arrived at using this key need to be confirmed by reference to detailed descriptions and illustrations (see "Guide to the Literature for Species Identification").

A. Key to Groups*

1a.	Macroscopic and consisting of whorls of branches bearing divided or undivided whorls of branchlets	GROUP I
1b.	Microscopic, or if macroscopic then of another form	2
2a.	Parenchymatous, forming tubular branched thalli or flat plates of cells, often membrane- or bladelike	GROUP II
2b.	Unicellular or filamentous	3
3a.	Unicellular (sometimes form 2- or 4-celled packets), associated with surfaces or within tissues of plants or animals, sometimes growing together but each cell with its own attachment	GROUP V**
3b.	Filaments simple or branched	4
4a.	Filaments unbranched or possess rhizoidal lateral branches, uniseriate or rarely multiseriate above	GROUP III***
4b.	Filaments branched, often a prostrate and an erect system present, always uniseriate.	GROUP IV

* *Elakatothrix* does not fall into any of the broad groups; rows of separate or loosely contiguous cells form "pseudofilaments" lying within a broad mucilaginous envelope (often planktonic).

** Unicellular algae included because traditionally placed in orders where most other genera are filamentous.

*** Some genera of zygnematalean algae are unbranched and filamentous although with very distinctive chloroplasts and reproduce by conjugation (see Chapter 9).

B. Key to Genera by Group

1. Group I (Order Charales)

1a.	Branches corticated, very rarely ecorticate throughout or partially corticate, with spine cells usually present; branchlets undivided and with bract cells at nodes (Fig. 9A)	*Chara*
1b.	Branches usually ecorticate, without spines	2
2a.	Branchlets undivided; single-celled outgrowths from outermost cells of the branch nodes (stipulodes), form a single ring, downward-pointing	*Lamprothamnium*
2b.	Branchlet divided (sometimes only minutely at apex); outgrowths from outermost cells on the branch nodes absent	3
3a.	Branchlets forked one or more times, each unicellular ray similar (Fig. 9B)	*Nitella*
3b.	Branchlets divided into unequal, multicellular rays (Fig. 9C)	*Tolypella*

2. Group II (Orders Ulvales, Prasiolales, in part)

1a.	Thallus tubular and hollow (Fig. 7A)	*Enteromorpha*
1b.	Thallus flat and membrane-like	2
2a.	Single-layered initially, multilayered at onset of reproduction, with cells in pairs or in larger groups and separated by mucilage; cells each with a stellate chloroplast; terrestrial (Fig. 7E)	*Prasiola*, in part
2b.	Single-layered and parenchymatous; cells each with a cup-shaped chloroplast; aquatic (Fig. 7D)	*Monostroma*

3. Group III (Orders Oedogoniales, Sphaeropleales, Microsporales, Ulotrichales in part, Cylindrocapsales, Cladophorales in part).

1a.	Filaments uniseriate initially and later becoming multiseriate	2
1b.	Filaments uniseriate throughout	5
2a.	Cells quadrate to shorter than broad where filaments uniseriate, each with a stellate chloroplast	*Prasiola* (*Rosenvingrella* stage)
2b.	Cells cylindrical where filaments uniseriate, each with a broad parietal and band-shaped chloroplast which sometimes almost encircles cell	3
3a.	Cells quadrate and bricklike where filament multiseriate, thick-walled, each containing several pyrenoids; aquatic (Fig. 7C)	*Schizomeris*
3b.	Cells spherical or in sarcinoid groups where multiseriate, thin-walled, each with a single pyrenoid; soil environments	4
4a.	Filaments mostly uniseriate, only final few cells in more than one series and spherical in shape (Fig. 4P)	*Pseudoschizomeris*

4b.	Filaments more extensively multiseriate, often cells angular and grouped in packets (sarcinoid) (Fig. 7B)	*Filoprotococcus*
5a.	Cells often having distally one or several ringlike growths or ridges (cap cells) (Fig. 8C)	*Oedogonium*
5b.	Cells never possess ringlike growths or ridges	6
6a.	Thallus consisting of very long tubular cells containing cytoplasmic units separated by vacuoles and bandlike chloroplasts (Fig. 6E)	*Sphaeroplea*
6b.	Thallus distinctly filamentous	7
7a.	Cell walls composed of two overlapping H-shaped sections, often apparent at end of a filament; pyrenoids absent (Fig. 4E)	*Microspora**
7b.	Cell walls never so constructed; pyrenoids present or absent	8
8a.	Cells containing one or two protoplasts, with mucilage between protoplast and cross wall often with transverse lamellations (Fig. 4B)	*Binuclearia*
8b.	Cells without separate protoplasts	9
9a.	Chloroplast dense and netlike formed of interconnecting discs each with a pyrenoid	10
9b.	Chloroplasts usually parietal and platelike, often with a single pyrenoid	12
10a.	Filaments less than 40 μm in diameter; occasionally with 1- to 3-celled rhizoidal branches arising at right angles to main axis (Fig. 5A)	*Rhizoclonium*
10b.	Cells usually more than 40 μm in diameter, without rhizoidal branches	11
11a.	Cells less than twice as long as broad, cylindrical to barrel-shaped, slightly to markedly constricted at cross walls	*Chaetomorpha*
11b.	Cells more than twice as long as broad, cylindrical, without constrictions at cross walls	*Cladophora* (unbranched form)
12a.	Filaments not enclosed within a mucilaginous sheath	13
12b.	Filaments enclosed within a mucilaginous sheath	21
13a.	Filaments readily fragmenting into short lengths (< 10 cells in number) or solitary cells	14
13b.	Filaments not readily fragmenting	18
14a.	Filaments forming zigzag growths on surfaces; cells spherical, oval or cylindrical with short, blunt apices (Fig. 4D)	*Stichococcus*, in part
14b.	Filaments not forming such a pattern; cells spindle-shaped, elongate cylindrical, or cylindrical with rounded apices	15
15a.	Filaments readily dissociate into solitary, cylindrical cells with bluntly rounded apices (Fig. 4O)	*Gloeotilopsis*
15b.	Filaments attached or dissociate into short lengths of filament with cells of a different shape	16
16a.	Cells or short filaments attached by a small disc; on soils (Fig. 4N)	*Raphidonemopsis*
16b.	Cells not attached, free-living; rarely associated with soil	17
17a.	Cells spindle-shaped very elongated; aquatic and terrestrial (Fig. 4J)	*Koliella*
17b.	Cells elongate-cylindrical and apices acuminate or gradually tapering to a point; terrestrial, only associated with snow (Fig. 4K)	*Raphidonema*
18a.	Chloroplast filling cells and structure indecipherable due to density of food storage (principally starch) material; cells ovoid and thick-walled; reproduction oogamous, with reproductive cells swollen and red or orange (Fig. 6I)	*Cylindrocapsa*
18b.	Chloroplast and cells of another form; reproduction not oogamous and without swollen or colored reproductive organs	19
19a.	Filaments terminating in an asymmetrically pointed apical cell (Fig. 4A)	*Uronema*
19b.	Filaments without a pointed apical cell	20
20a.	Chloroplast parietal, platelike, usually lobed, partly (often > 80%) or fully encircling cell (Fig. 4F)	*Ulothrix*
20b.	Chloroplast parietal, elliptical or platelike, not lobed marginally, normally encircling 80% or less of cell (Fig. 4C)	*Klebsormidium*
21a.	Cells shorter than broad	22
21b.	Cells longer than broad	23
22a.	Cells ellipsoidal to subquadrate, always contiguous within an uninterrupted mucilaginous sheath; some species have two overlapping helmet-like pieces resulting in a rim around mid-region (Fig. 4G)	*Radiofilum*

22b. Cells oval or oblong, never with a rim, often an interrupted row of 2 or 4 cells within a series of loosely attached mucilaginous sheaths (Fig. 4M)..................*Hormidiopsis*

23a. Cells in a contiguous row; pyrenoids absent (Fig. 4L)..................*Gloeotila*

23b. Cells separated, often equidistant and in pairs; pyrenoids present (Fig. 4I)..................*Geminella*

*Similar wall structure to the yellow-green genus *Tribonema* (see Chapter 11), but differs in several characters including having netlike rather than platelike chloroplasts

Group IV (Chaetophorales in part, Coleochaetales, Oedogoniales in part, Cladophorales in part, Siphonales in part)

1a. Filaments perforating limestone rock, wood, or mollusk shells (Fig. 1H)..................*Gomontia*

1b. Filaments not perforating surfaces..................2

2a. Filaments lime-encrusted..................3

2b. Filaments not lime-encrusted..................4

3a. Thalli of loose and often unilaterally branched filaments, with pairs of short, green cells alternating with longer almost colorless cells (Fig. 3F)..................*Chlorotylium*

3b. Thalli of compact filaments and irregularly divided filaments, with all cells green and no such cellular differentiation (Fig. 3J)..................*Gongrosira*, in part

4a. Thalli tubular or threadlike and contain many nuclei (coenocytic), cross walls only associated with reproduction..................5

4b. Thalli multicellular, with each cell uninucleate or multinucleate..................6

5a. Terrestrial, parasitic in higher plants (especially members of family Araceae) (Fig. 6H)..................*Phyllosiphon*

5b. Aquatic, free-living, repeatedly dichotomously divided with constrictions at base of each fork (Fig. 6F)..................*Dichotomosiphon*

6a. Associated with soil, growing on or partly subterranean; terminal cells often conical..................7

6b. Aquatic or subaerial; terminal cells rarely conical..................8

7a. Filaments green above soil surface and colorless below; cells each having a single chloroplasts and terminal cells without a cap (Fig. 3H)..................*Fritschiella*

7b. Filaments all usually on surface and only rhizoids sometimes colorless; cells having a netlike chloroplast and terminal cells sometimes having a distinctive cap (Fig. 8B)..................*Oedocladium*

8a. Thalli as disklike expansions on or within the carapace of turtles..................9

8b. Thalli disklike or otherwise, never associated with turtles..................10

9a. Cells each possess a single nucleus and pyrenoid, rare on turtles (only *Protoderma involvens*) (Fig. 1G)..................*Protoderma involvens*

9b. Cells each possess several nuclei and one to several pyrenoids, rare on surfaces other than turtles (Fig. 5D)..................*Dermatophyton*

10a. Epizoic among the hair scales of the sloth (Fig. 1B)..................*Trichophilus*

10b. Aquatic or terrestrial on vascular plants and soil-free surfaces..................11

11a. Endophytic, parasitic, or semi-parasitic, within the intercellular spaces of vascular plants and giving rise to erect filaments bearing sporangia (Fig. 6A)..................*Cephaleuros*

11b. Epiphytic, on other surfaces and of a different form than above..................12

12a. Subaerial on trees and other terrestrial surfaces..................13

12b. Aquatic..................16

13a. Thalli grow within the substomatal cavity of vascular plants, consisting of a bulbose cell giving rise to a few-celled stalk (Fig. 6D)..................*Stomatochroon*

13b. Thalli not within the substomatal cavity and of another form..................14

14a. Closely adjoined filaments forming epiphytic disks on the leaves of higher plants (Fig. 6B)..................*Phycopeltis*

14b. Felty mats or tufts of filaments with branches arising from middle of cell and at right angles to main axis, growing on vascular plants and other surfaces (Fig. 6C)..................15

15a. Sporangia always solitary, globular to kidney-shaped, and papilla pore basal; gametangia sometimes lateral on the sporangium rather than apical; gametangia lateral or terminal..................*Printzina*

15b.	Sporangia often grouped, ovoid and papilla pore apical; gametangia always terminal (Fig. 6C)	*Trentepohlia*
16a.	Thalli macroscopic, spherical, cushion-like, or of irregularly divided, nodulose, somewhat flattened branches formed of filaments embedded in a firm or soft mucilage (Fig. 3I)	*Chaetophora*
16b.	Thalli not forming macroscopic growths with a definite outline	17
17a.	Filaments closely aggregated and pseudoparenchymatous, at least in part	18
17b.	Filaments free from one another throughout and openly branched	27
18a.	Erect filaments absent and prostrate system forming a single- or multi-layered crust	19
18b.	Erect and prostrate filaments, or only with erect filaments	23
19a.	Cells each containing several chloroplasts; crusts multi-layered in center (Fig. 1A)	*Pseudulvella*
19b.	Cells each containing a single parietal chloroplast; crusts single-layered throughout	20
20a.	Bristles or fine hairs absent	21
20b.	Bristles or hairs present	22
21a.	Zoospores biflagellate (Fig. 1G)	*Protoderma*
21b.	Zoospores quadriflagellate (Fig. 1E)	*Chamaetrichon*
22a.	Bristles with a basal sheath, often sheath persists after loss of bristle (Figs. 2I, J)	*Coleochaete*, in part
22b.	Bristles without a basal sheath (Fig. 2K)	*Aphanochaete*, in part
23a.	Prostrate system concentric in appearance due to alignment of cross wall and chloroplasts of adjacent cells; erect system very reduced	*Stigeoclonium*, in part
23b.	Prostrate system and erect system of another form	24
24a.	Prostrate system multi-layered (cushion-like) and bears sparingly divided erect filaments; zoospores biflagellate	25
24b.	Prostrate system single-layered and bears a poorly to profusely developed erect system; zoospores quadriflagellate where known	26
25a.	Pyrenoids evident; sometimes lime-encrusted (Fig. 3J)	*Gongrosira*, in part
25b.	Pyrenoids absent (if present not visible with LM); never lime-encrusted (Fig. 1J)	*Leptosira*, in part
26a.	Reproduction only by akinetes and aplanospores (Fig. 3K)	*Dilabifilum*
26b.	Reproduction by zoospores	*Pseudendoclonium*
27a.	Branches bearing bristles or hairs	28
27b.	Branches without such projections	31
28a.	Prostrate system only, consists of simple or irregularly divided filaments (Fig. 2K)	*Aphanochaete*, in part
28b.	Erect system only	29
29a.	Bristle formed as extension of lateral wall at distal end of cell, with no cross wall at base (Fig. 3E)	*Fridaea*
29b.	Bristles not as above	30
30a.	Less than six-celled and branches rudimentary or absent; minute epiphyte within mucilage of *Gloeotrichia* (cyanobacterium) (Fig. 2B)	*Thamniochaete*, in part
30b.	More than six-celled and branching in one plane; never embedded in mucilage (Fig. 8A)	*Bulbochaete*
31a.	Thalli consist of short filaments often bearing 1- or 2-celled short branchlets terminating in truncate cells; soil alga (Fig. 1I)	*Hazenia*
31b.	Thalli of another form; aquatic algae	32
32a.	Filaments peculiarly twisted, indistinctly branched, enclosed within a tough oval or spherical mucilaginous envelope (Fig. 1M)	*Helicodictyon*
32b.	Filaments distinctly branched, without a mucilaginous envelope	33
33a.	Freely branching erect system only and some genera attached by a disk-shaped holdfast or rhizoids	34
33b.	Freely branching erect and prostrate system, or only a prostrate system	42

34a.	Microscopic (<100 μm in height), often freely branched with cross wall distal to point of branching; pyrenoids absent (Fig. 3L) ....*Microthamnion*	
34b.	Macroscopic, coarse or delicate and gelatinous...35	
35a.	Thalli enclosed within a delicate mucilage envelope, with distinct main branches bearing oppositely, alternately, or whorls of narrower-celled branchlets...36	
35b.	Thalli without a mucilaginous envelope and branching of another form...37	
36a.	Cells of main branches similar throughout (Figs. 3C, D)..*Draparnaldia*	
36b.	Cells of main branches of two types: cylindrical or barrel-shaped and very short from which arise lateral branchlets (Fig. 3B) ....*Draparnaldiopsis*	
37a.	Thalli usually soft and main axes <50 μm in diameter; cells each having a parietal platelike chloroplast, pyrenoid and nucleus (Fig. 3A)..*Stigeoclonium*, in part	
37b.	Thalli usually coarse and main branches usually over 50 μm in diameter; cells each having a netlike chloroplast and many pyrenoids and nuclei..38	
38a.	Branching confined to base of main filaments; commonly on the backs of snapping turtles (Fig. 5F)...................*Basicladia*	
38b.	Branching of another form; not growing on turtles...39	
39a.	Branches arising immediately below cross wall; akinetes rarely produced...40	
39b.	Branches sometimes arising a short distance below cross wall; akinetes common and large, thick-walled, intercalary, or terminal 41	
40a.	Cells of main axes considerable larger than those of short branchlets, walls not lamellate, each having a parietal band-shaped chloroplast and one or a few pyrenoids (Fig. 3G)...*Cloniophora*	
40b.	Cells of main axes and branchlets of similar size, walls thick and lamellated, each having a netlike chloroplast and large numbers of pyrenoids (Fig. 5C)..*Cladophora*	
41a.	Akinetes spherical and in series; inland saline environments (Fig. 5B)...*Ctenocladus*	
41b.	Akinetes oval or barrel-shaped, solitary; freshwater environments (Fig. 5E)...*Pithophora*	
42a.	Prostrate system only; minute endo- or epiphytes...43	
42b.	Prostrate and an erect system; never endophytic...44	
43a.	Filaments sometimes highly branched, with cell thin-walled and not bearing bristles; epiphytic on *Cladophora* and *Rhizoclonium* (Fig. 1K)...*Entocladia*	
43b.	Subfilamentous mass of 2–6 cells, with cells having thick, lamellated walls and each bearing 8–12 long, flexuous bristles; epiphytic on vascular aquatic plants (Fig. 2D)...*Polychaetophora*, in part	
44a.	Plants growing within mucilage of other algae; filaments bearing basally swollen hairs, or these arising laterally on short projections (Fig. 2F)..*Chaetonema*	
44b.	Plants not growing within mucilage of other alga; hairs developing terminally..45	
45a.	Prostrate system bearing 5- to 8-celled, simple erect branches tapering to fine points and slender, multicellular hairs or bristles (Fig. 2H)..*Pseudochaete*	
45b.	Prostrate and erect systems usually well-developed, sometimes branches giving rise terminally to long hairs (Fig. 3A) ....*Stigeoclonium*, in part	

5. Group V (Chaetophorales, Siphonales in part)

1a.	Endophytic, consisting of irregularly oval cells within the duckweed *Lemna* (Fig. 1N)..............................*Chlorochytrium*	
1b.	Free-living, aquatic or terrestrial..2	
2a.	Cells bearing simple or branched bristles or hairs; aquatic algae..3	
2b.	Cells without bristles or hairs; usually soil or subaerial algae...7	
3a.	Cells loaf-shaped, bearing 1–2 very fine, forked bristles (Fig. 2C)..*Dicranochaete*	
3b.	Cells of another shape, each bearing one or more unbranched bristles..4	
4a.	Cells bearing sheathed hairs...5	

4b.	Cells bearing unsheathed hairs	6
5a.	Cells subglobose, depressed, embedded in mucilage and not connected by tubes, bearing a number of radiating bristles	*Conochaete*
5b.	Cells spherical, sometimes connected by mucilage tubes, each bearing a single, sheathed bristle (Fig. 2A)	*Chaetosphaeridium*
6a.	Cell walls thick and lamellate, with 8–12 bristles arising from each cell (Fig. 2D)	*Polychaetophora*
6b.	Cell walls thin, with 2–4 long, flexuose bristles arising from each cell (Fig. 2E)	*Oligochaetophora*
7a.	Swollen, multinucleate vesicle growing on surface of moist soil, with downward-growing colorless rhizoids; soil habitats (Fig. 6G)	*Protosiphon*
7b.	Spherical and uninucleate cells, often as 2- or 4-celled cuboidal packets; subaerial or isolated from soil or stagnant water	8
8a.	Isolated from surface of damp soil or stagnant water; cells with one or more pyrenoids (Fig. 1F)	*Pleurastrum*
8b.	Growing as a green powdery layer on various terrestrial surfaces; cells with or without a pyrenoid	9
9a.	Cells each with a small pyrenoid (Fig. 1C)	*Desmococcus*
9b.	Cells without a pyrenoid (Fig. 1D)	*Apatococcus*

C. Descriptions of Genera (Figs. 1–9[1])

Order Chaetophorales

Filaments are uniseriate, branched, consist of erect and prostrate systems of filaments, sometimes erect branches absent or very reduced and prostrate system pseudoparenchymatous and well developed, with or without hairs or bristles. Cells are uninucleate, each with a parietal chloroplast and one to several pyrenoids.

Apatococcus V. F. Brand (Fig. 1D)

Unicellular, cells often divided in two or three planes to form irregular, cuboidal packets, sometimes forming short uniseriate filaments. Cells each with a parietal and often lobed chloroplast, pyrenoids absent. Asexual reproduction by biflagellate and slightly flattened zoospores, 8–32(–64) produced per cell. Autospores spherical. Sexual reproduction is unknown.

Very common and widespread subaerial alga, usually on the bark of trees where it forms part of the ubiquitous *Protococcus-Pleurococcus* community and often confused with other members of it.

The current consensus is that there exists only a single very polymorphic species, whereas formerly several were recognized (Ettl and Gartner, 1995). Retained in the Chaetophorales where traditionally placed but on ultrastructural grounds it should be assigned to the Chlorococcales (Hoek et al., 1995).

Aphanochaete A. Braun (Fig. 2K)

Filaments are creeping and uniseriate; unbranched or irregularly branched, with cells sometimes bearing on upper surface one to several, long, unicellular, basally-swollen bristles. Cells are cylindrical or inflated, each possessing a parietal, disc-shaped chloroplast, with one to several pyrenoids. Asexual reproduction is by quadriflagellate zoospores or aplanospores. Sexual reproduction is oogamous: oosphere is large with a stalk, male gametes are quadriflagellate. The zygote is thick-walled and contains reddish to yellowish oil droplets.

Aphanochaete is very widespread and a common epiphyte on submerged aquatic plants (e.g., *Myriophyllum*, *Sagittaria*, *Najas*, *Chara*) and filamentous algae (particularly *Oedogonium*, *Cladophora*, *Rhizoclonium*, *Vaucheria*), growing commonly in hard and often eutrophic water in a wide range of still and flowing-water habitats.

There are six species of which four are considered valid and the other two are doubtful (Tupa, 1974). The genus is in need of further revision.

Chaetonema Nowakowski (Fig. 2F)

Filaments are uniseriate and irregularly divided, with some short side branches perpendicular to main axis and arising from the middle of the cell, often terminating in a hair, sometimes basally swollen hairs borne on short branches. Cells are cylindrical, each possessing a parietal, platelike chloroplast encircling more than half of the cell, with one or two pyrenoids. Asexual reproduction is by two quadriflagellate zoospores produced in each cell, flagella of equal length. Sexual reproduction is oogamous: eight biflagellate male gametes are in each swollen gametangium and the female gamete is in an enlarged vegetative cell.

Chaetonema grows in association with the mucilaginous sheath of various freshwater algae, including *Draparnaldia*, *Chaetophora*, *Tetraspora*, and *Batrachospermum*. There are only two species, both of which occur in North America.

It has been suggested that *Chaetonema ornatum*

---

[1] All scale bars in the figures equal 10 μm unless otherwise indicated.

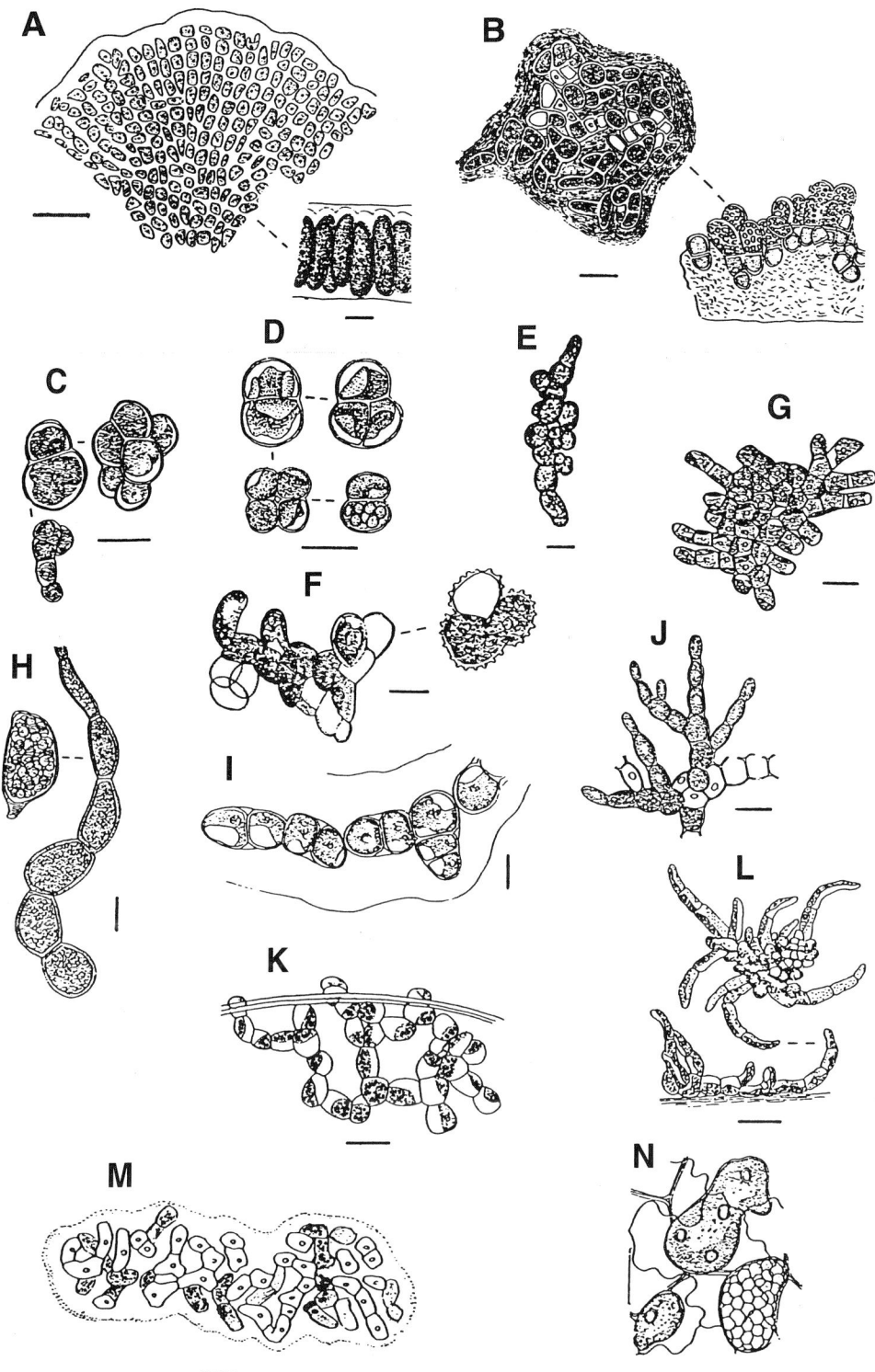

FIGURE 1 (A) *Pseudulvella americana*, surface view and section through thallus (after Snow, 1899). (B) *Trichophilus welcheri*, surface view and side view showing terminal sporangia (after Weber-van Bosse, 1887). (C) *Desmococcus olivaceus*, packets of cells and short filament (after Prescott, 1951). (D) *Apatococcus lobatus* (after Printz, 1964). (E) *Chamaetrichon capsulatum* (redrawn after Tupa, 1974). (F) *Pleurastrum insigne* (after Chodat, 1894). (G) *Protoderma viridae* (after Smith, 1933). (H) *Gomontia holdenii* (after Smith, 1933). (I) *Hazenia mirabilis* (after Bold, 1955). (J) *Leptosira mediciana* (after Borzi, 1883). (K) *Entocladia polymorpha* (after Prescott, 1951). (L) *Pseudendoclonium basiliense* (after Vischer, 1926). (M) *Helicodictyon planktonicum* (after Whitford, 1956). (N) *Chlorochytrium lemnae* (after Klebs, 1881).

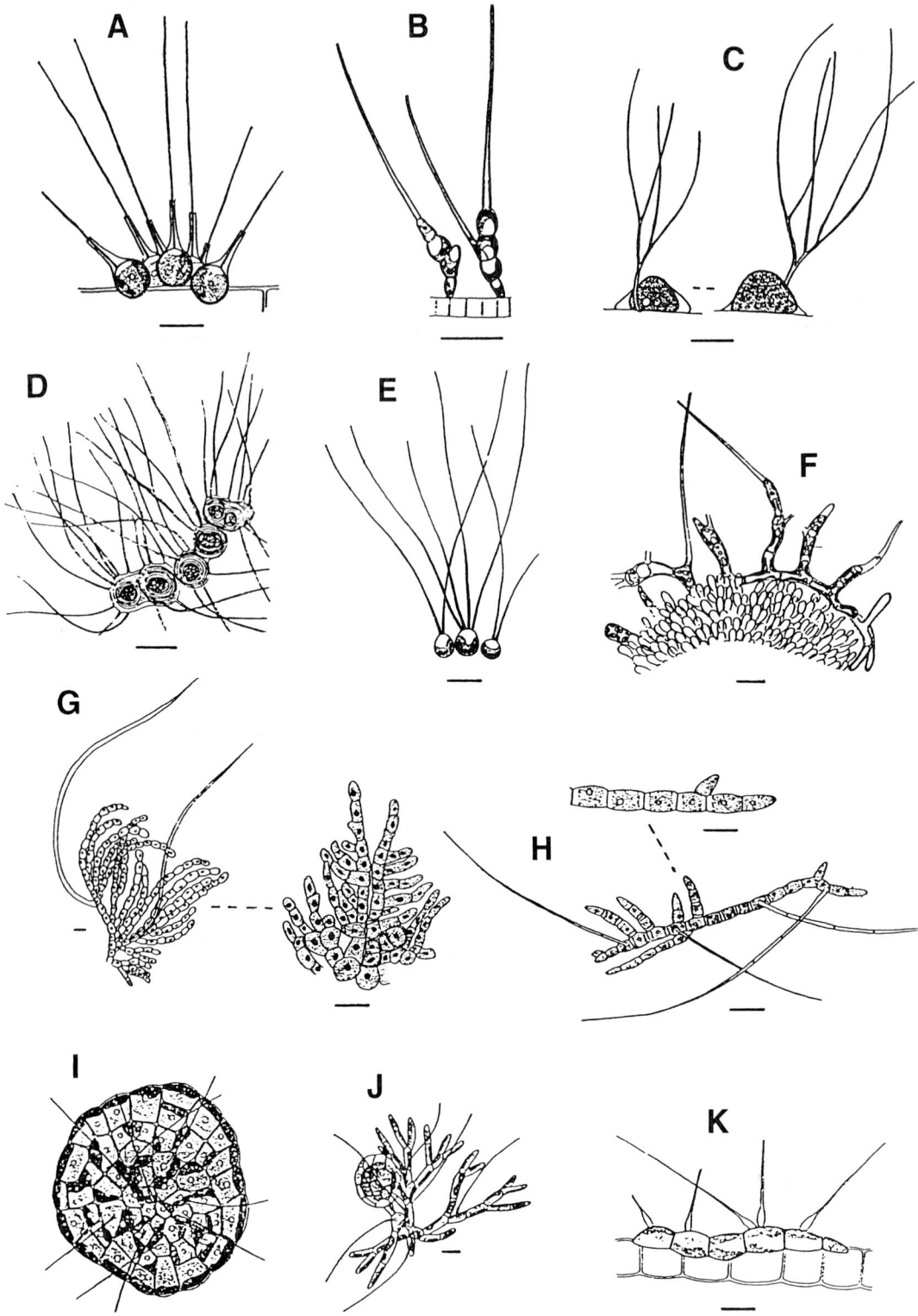

FIGURE 2 (A) *Chaetosphaeridium globosum* (after Smith, 1933). (B) *Thamniochaete huberi* (after Gay, 1891). (C) *Dicranochaete reniformis* (after Smith, 1933). (D) *Polychaetophora lamellosa* (after W. and G. S. West, 1903). (E) *Oligochaetophora simplex* (after West, 1911). (F) *Chaetonema irregulare* (after Huber, 1892). (G) *Trichodiscus elegans*, erect filaments with hairs and prostrate system without hairs (after Welsford, 1912). (H) *Pseudochaete gracilis* (after Smith, 1933). (I) *Coleochaete scutata* (after Smith, 1933). (J) *Coleochaete pulvinata* (after Smith, 1933). (K) *Aphanochaete repens* (after Smith, 1933).

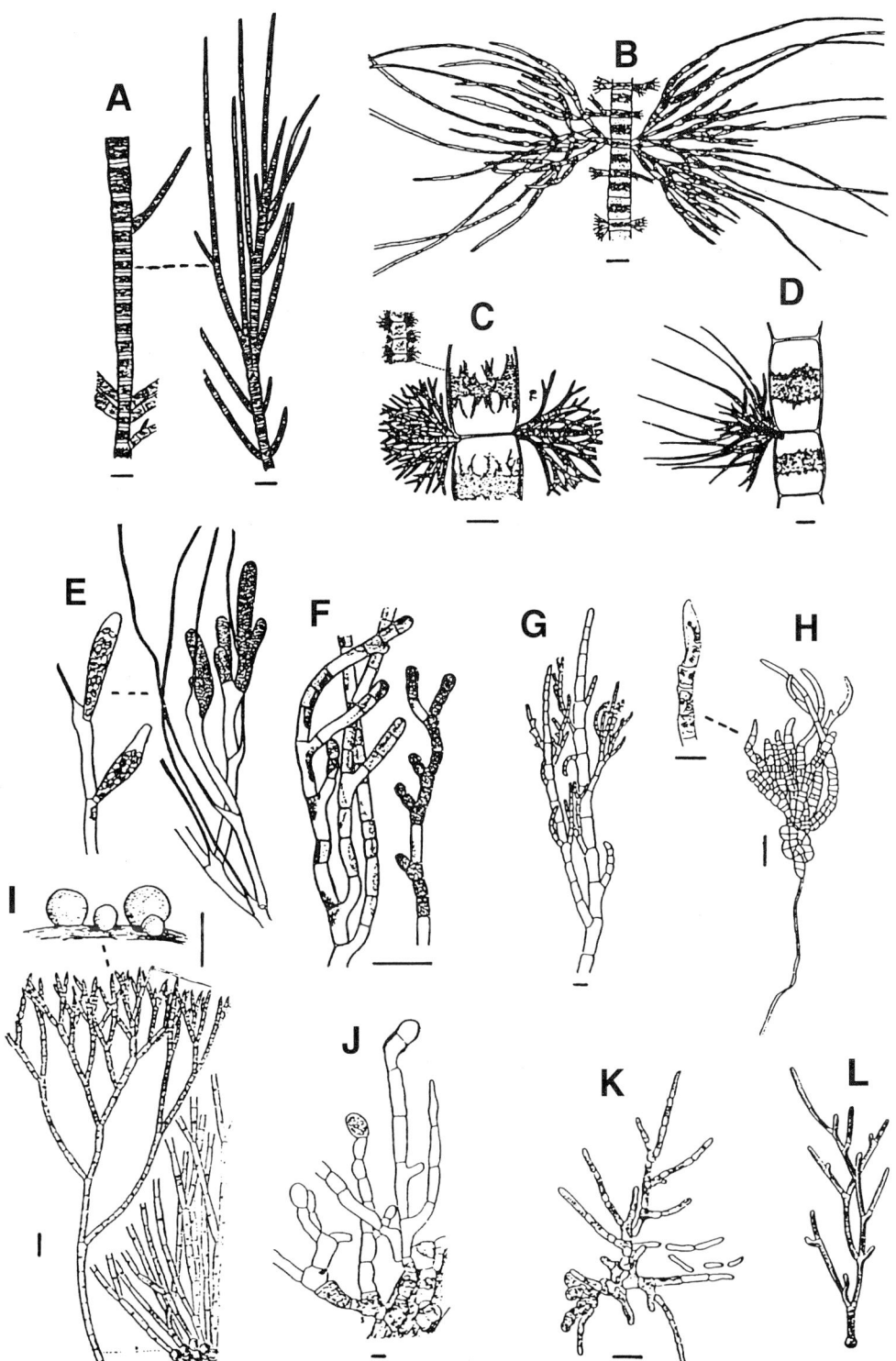

FIGURE 3 (A) *Stigeoclonium lubricum* (after Hazen, 1902). (B) *Draparnaldiopsis alpinis* (after Smith and Klyver, 1929). (C) *Draparnaldia ravenelii* (after Tiffany and Britton, 1952). (D) *Draparnaldia glomerata* (after Hazen, 1902). (E) *Fridaea torrenticola*, vegetative filament and one bearing bottle-shaped sporangia (modified after Smith, 1933). (F) *Chlorotylium cataractum*, vegetative filament and filament on right containing akinetes (after Smith, 1933). (G) *Cloniophora spicata* (after Prescott, 1983). (H) *Fritschiella tuberosa* (after Iyengar, 1932). (I) *Chaetophora elegans*, habit (after Prescott, 1951) and terminal and basal portions of colony (after Hazen, 1902). (J) *Gongrosira debaryana* (after Prescott, 1951). (K) *Dilabifilum printzi* (after Vischer, 1933). (L) *Microthamnion kuetzingianum* (after Tiffany, 1937).

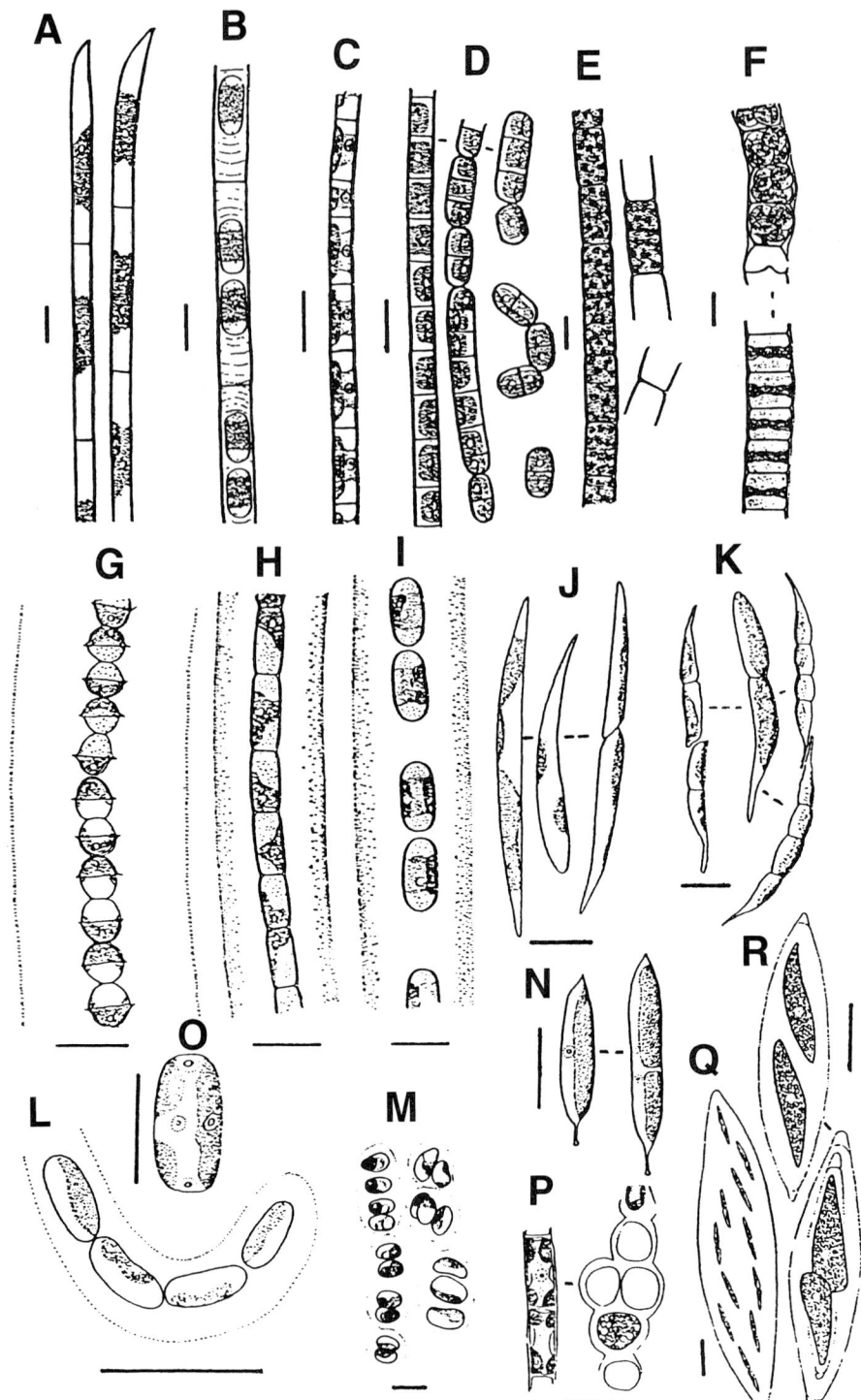

FIGURE 4 (A) *Uronema elongatum* (after Smith, 1933). (B) *Binuclearia tatrana* (after Smith, 1933). (C) *Klebsormidium klebsii* (after Smith, 1933). (D) *Stichococcus subtilis* (after Smith, 1933). (E) *Microspora willeana* (modified after Smith, 1933). (F) *Ulothrix zonata*, vegetative filament and one with the cells containing zoospores (after Smith, 1933). (G) *Radiofilum conjunctivum* (modified after Smith, 1933). (H) *Geminella minor.* (after Smith, 1933). (I) *Geminella interrupta* (modified after Smith, 1933). (J) *Koliella* sp. (after Prescott, 1983). (K) *Raphidonema nivale* var. *taylorii* (after Kol, 1938). (L) *Gloeotila contorta* (after Prescott, 1983). (M) *Hormidiopsis ellipsoideum* (after Prescott, 1983). (N) *Raphidonemopsis sessilis* (after Deason, 1969). (O) *Gloeotilopsis sterile* (after Deason, 1969). (P) *Pseudoschizomeris caudata*, uniseriate lower portion and multiseriate uppermost portion of filament (after Deason and Bold, 1960). (Q) *Elakatothrix gelatinosa* (after Smith, 1933). (R) *Elakatothrix viridis* Snow (after Smith, 1933).

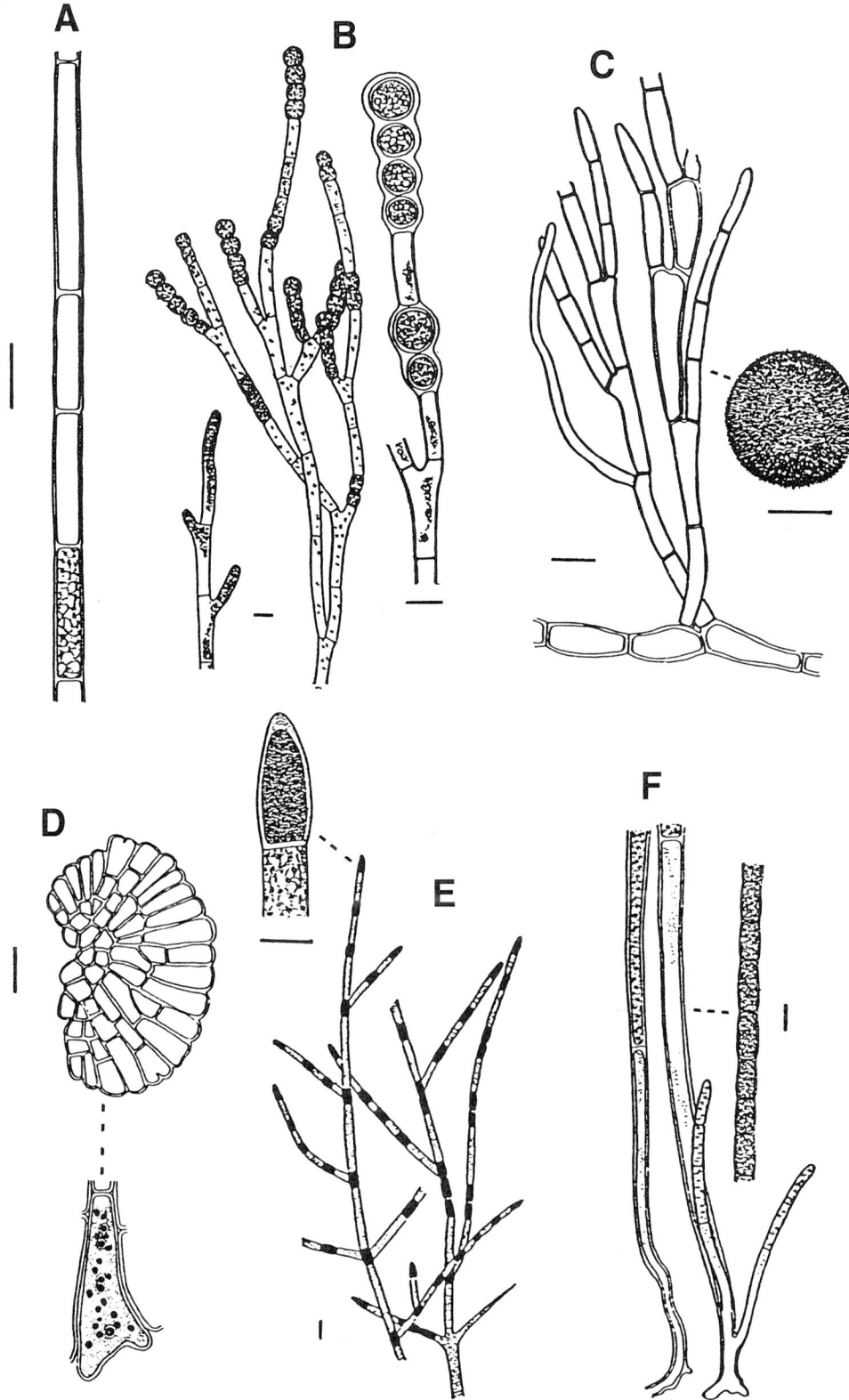

FIGURE 5 (A) *Rhizoclonium hieroglyphicum* (after Prescott, 1951). (B) *Ctenocladus circinnatus*, some branches with series of akinetes (after Smith, 1933). (C) *Cladophora glomerata*, portion of branch and a "*Cladophora*" ball (after Smith, 1933). (D) *Dermatophyton radians*, disklike thallus and close up of one of the multinucleate cells (after Feldmann, 1939). (E) *Pithophora oedogonia*, ortion of filament with akinete and close up of a terminal akinetes (after Smith, 1933). (F) *Basicladia chelonum* (after Smith, 1933).

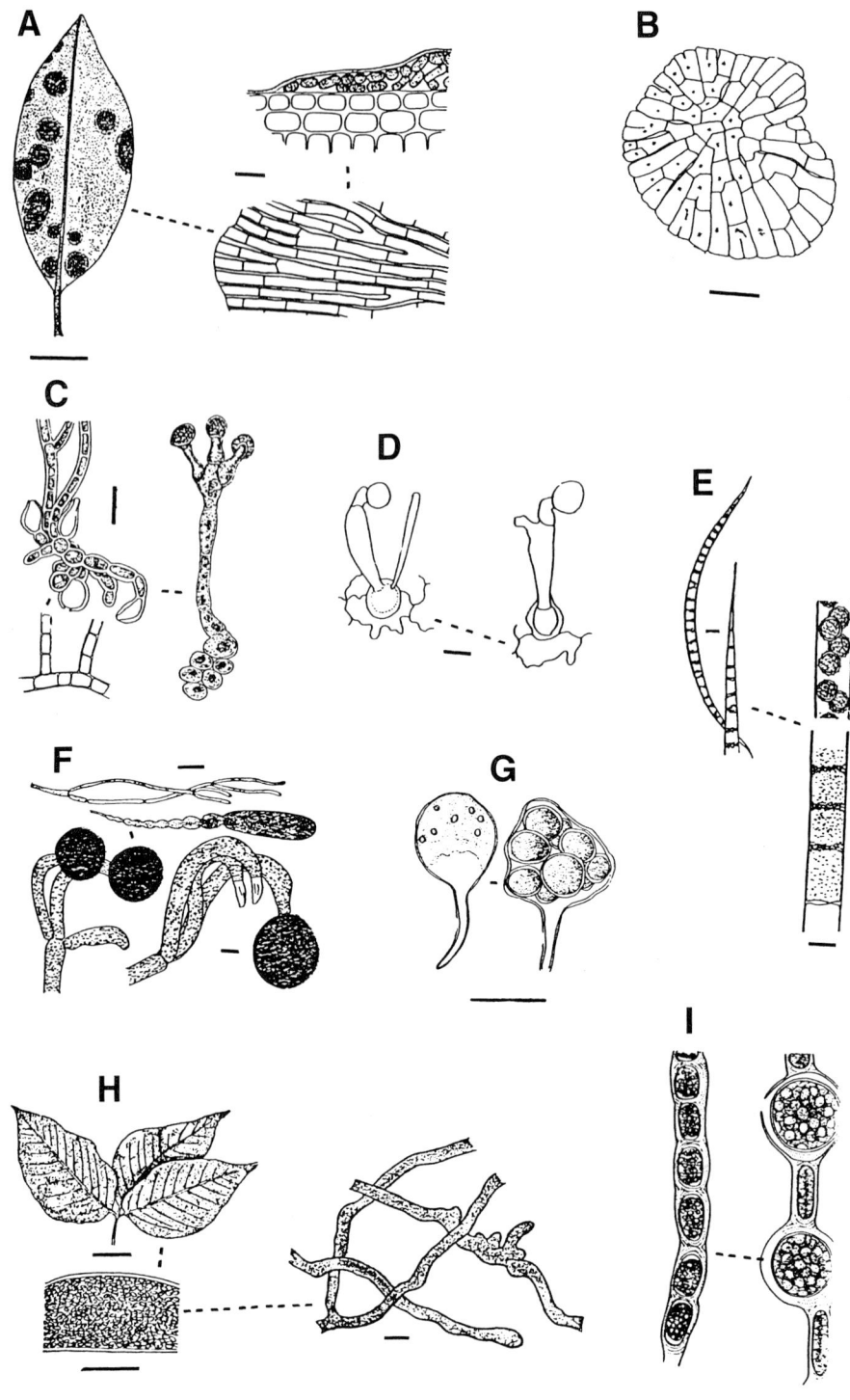

FIGURE 6 (A) *Cephaleuros virescens*, crusts of *Cephaleuros* on a *Magnolia* leaf and closer view of a crust and a section through a crust (after Smith, 1933). (B) *Phycopeltis arundinacea* (after Bourrelly, 1968). (C) *Trentepohlia aurea* (after Gobi, 1871). (D) *Stomatochroon lagerheimii* (after Palm, 1934). (E) *Sphaeroplea annulina*, portions of vegetative thallus and of one containing spores whose walls are thick and ornamented (after Prescott, 1951). (F) *Dichotomosiphon tubersosus*, tubular thallus and portions possessing sex organs (below left and right) as well an elongate akinetes (after Smith, 1933). (G) *Protosiphon botryoides*, young thallus and one containing thick-walled resting spores (after Bourrelly, 1967). (H) *Phyllosiphon arisari*, leaf of *Arisaema triphyllum* infected by *Phyllosiphon* and portions of thallus (after Smith, 1933). (I) *Cylindrocapsa geminella*, vegetative filaments and another containing oosporangia (modified after Prescott, 1951).

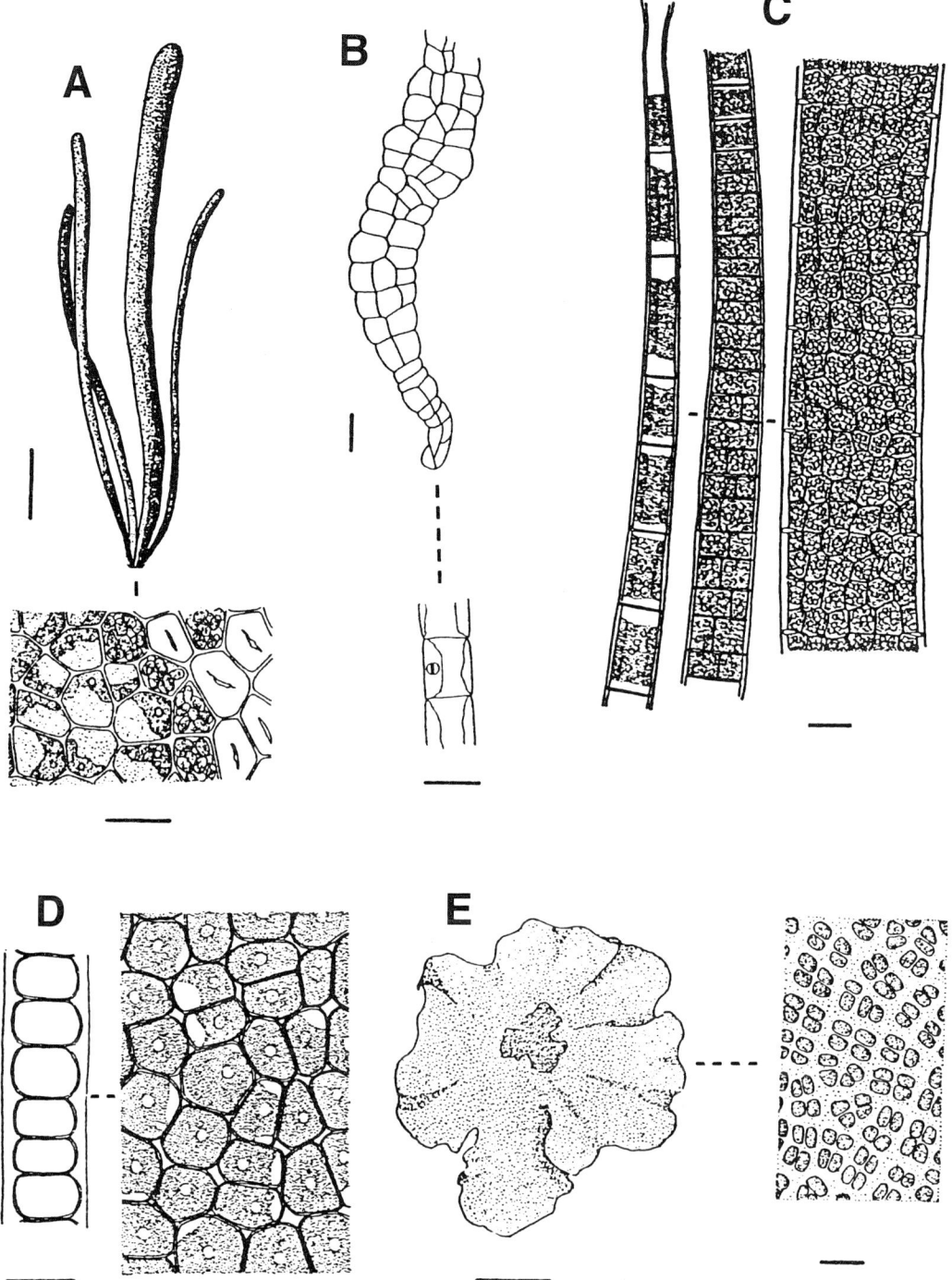

FIGURE 7 (A) *Enteromorpha intestinalis*, habit and section through tubular thallus (after Kützing, 1853–1856). (B) *Filoprotococcus polymorphum* (after Prescott, 1983). (C) *Schizomeris leibleinii* (after Smith, 1933). (D) *Monostroma latissimum* (after Smith, 1933). (E) *Prasiola mexicana*, habit and a portion of thallus (modified after Smith, 1933).

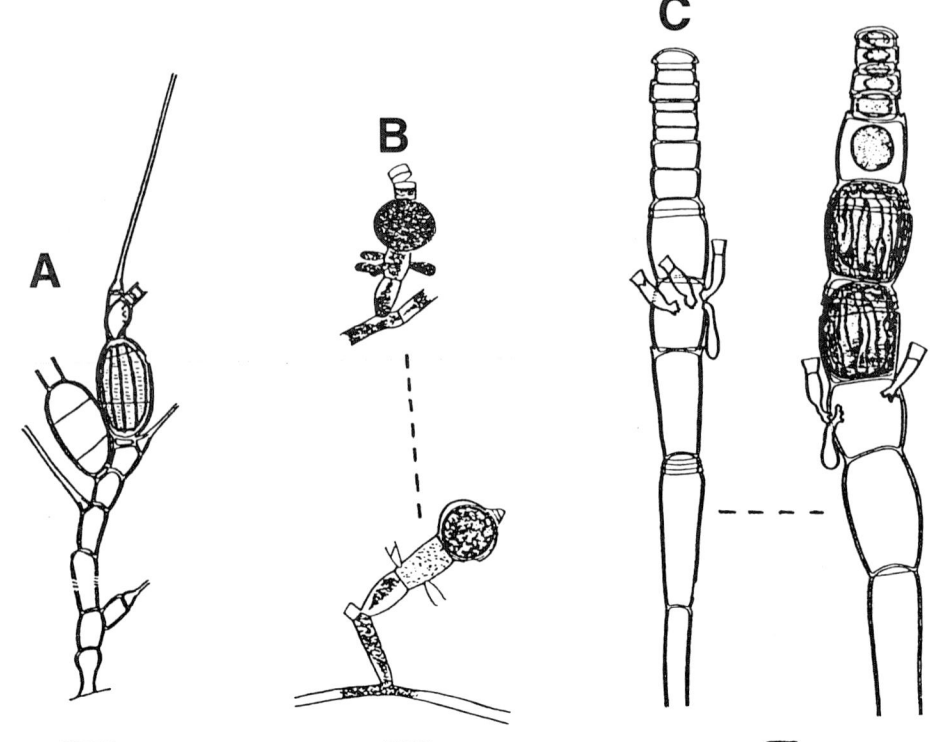

FIGURE 8 (A) *Bulbochaete minor*, portion of filament with swollen and ornamented zygospore (after Tiffany and Britton, 1952). (B) *Oedocladium hazenii*, portion of filaments with dwarf males and unfertilized oogonium (above) and of an oogonium containing a thick-walled zygospore (below) (after Smith, 1933). (C) *Oedogonium croasdaleae*, filament with dwarf males (left) and other filaments having dwarf males and two ornamented zygospores (right) (after Jao, 1934).

should be transferred to Thompson's doubtful genus *Chaetonemopsis* (Thompson, 1972b).

Chaetophora F. Schrank (Fig. 3I)

Chaetophora forms macroscopic growths, spherical, tubercular, arbuscular, or hemispherical enveloped in soft or firm mucilage, consisting of uniseriate filaments; a prostrate system of filaments is little-developed but erect system highly branched, intertwined filaments with each tapering to a blunt point or terminating in a long, multicellular hair. Cells contain a parietal chloroplast, sometimes bandlike, one to several pyrenoids. Zoospores are quadriflagellate; gametes are biflagellate and isogamous. Akinetes are generally produced in upper cells of branches and are brown in color.

Chaetophora grows on submerged aquatic plants (including tree roots) and a wide variety of other submerged surfaces in springs, pools, ponds, and lakes, as well as in streams and other relatively fast-flowing water courses. Many species are most abundant in winter and spring, evidently preferring cooler water conditions. Of the 12 species, at least four are known from North America.

Re-investigation is required as the species is distinguished principally on such unreliable criteria as the macroscopic form of the thallus and the nature of branching of the filaments forming it.

Chaetosphaeridium Klebahn (Fig. 2A)

The *Chaetosphaeridium* consists of spherical to flask-shaped cells, solitary or in dense groups, sometimes within a common mucilage envelope or connected in series by a tubular mucilaginous elongation, each cell bearing a long, simple, and prominently basally sheathed bristle. Cells contain one or two parietal, plate-like chloroplasts, and a single pyrenoid. Zoospores biflagellate, two or four per cell. Sexual reproduction is oogamous, oogonia are very swollen and male gametes biflagellate.

Chaetosphaeridium often grows on mosses and other submerged surfaces, rarely endophytic, frequently becomes dislodged and free-floating along with the plankton in ponds and creeks.

In most recent accounts, the genus is placed in the order Coleochaetales based on its ultrastructural features.

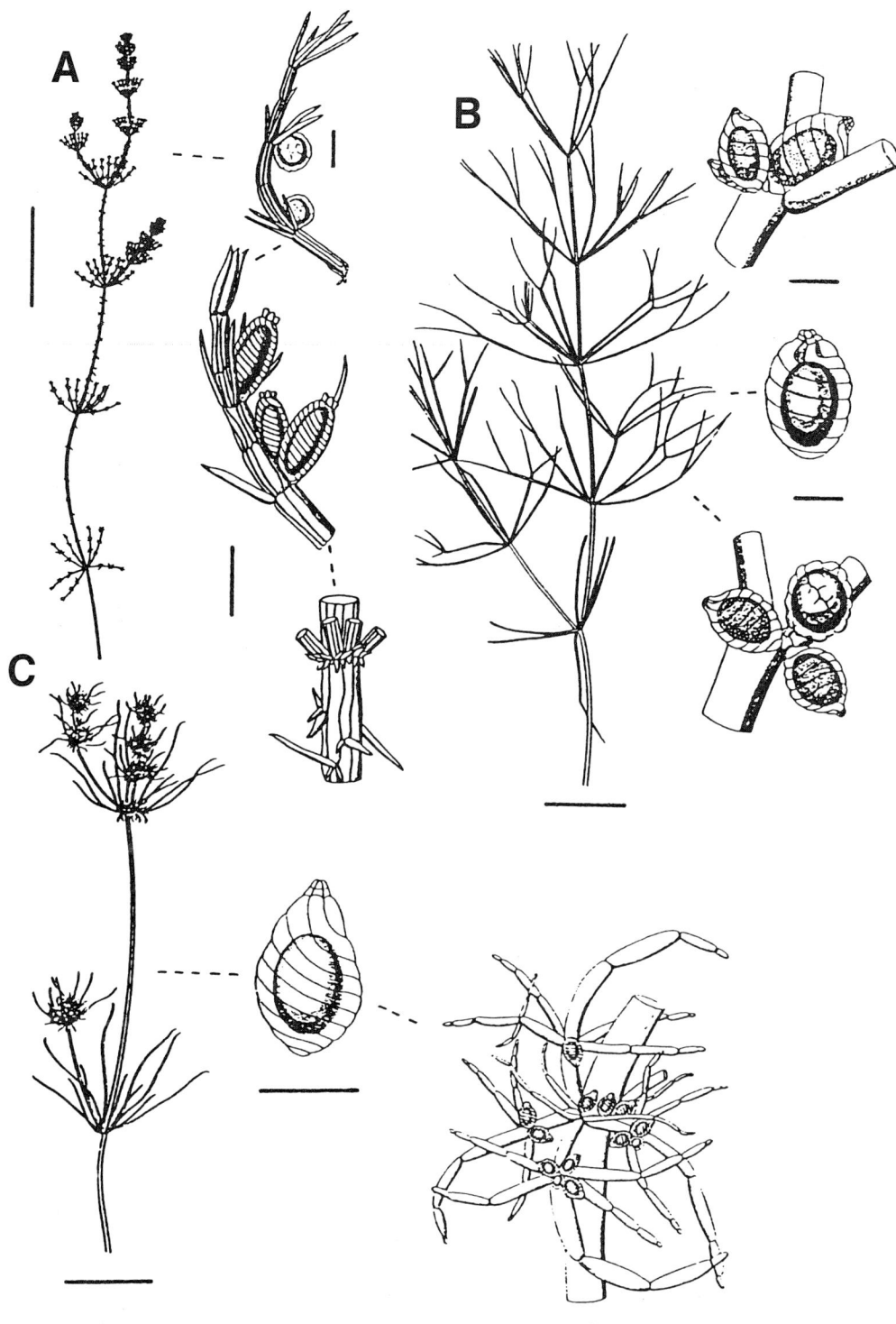

FIGURE 9 (A) *Chara canescens*, habit of alga, branchlet of a male and a female individual, and a node showing the corticated main axis with its bract cells and stipulodes arranged in 2 rows (after Wood and Imahori, 1964). (B) *Nitella flexilis*, habit of alga, branchlets with clustered oogonia or conjoined oogonia and antheridia, and an oogonium with a 2-tiered corona (after Wood and Imahori, 1964). (C) *Tolypella nidifica*, habit, portion of node with fertile branchlets, and an oogonium with a 2-tiered corona (Wood and Imahori, 1964).

Chamaetrichon Tupa (Fig. 1E)

Filaments are uniseriate, creeping, and simple or with a few side branches. Cells each contain a parietal chloroplast and a single pyrenoid. Asexual reproduction is by naked, symmetrical, quadriflagellate zoospores, each with a parietal chloroplast, pyrenoid, stigma, and two contractile vacuoles. Aplanospores are known.

This is a little-known genus, with the only traced record accompanying Tupa's description of the type (*C. capsulatum*) (Tupa, 1974). She isolated it from the surface of a submerged liverwort and the leaves of *Sagittaria* growing in a shallow drainage area of Double Lake, Sam Houston National Forest, near Coldspring, San Jacinto County, Texas.

It is structurally similar to the genus *Protoderma* but differs from it in producing quadriflagellate rather than biflagellate zoospores.

Chlorochytrium Cohn (Fig. 1N)

Chlorochytrium is endophytic and consists of solitary and spherical cells, irregularly curved or lobed, often with walls thick and lamellate. Cells each possess a chloroplast whose arms radiate from center towards the periphery where they expand into broad parietal lobes or a complete parietal layer, with one or several scattered pyrenoids. Reproduction is by biflagellate zoospores and gametes; aplanospores and akinetes known.

Chlorochytrium commonly develops in the intercellular spaces of various aquatic plants (especially the duckweed *Lemna*), or is associated with soil or other algae (including cyanobacteria); it is mostly within animal and plant tissues, only more rarely reported free-living. Five species have been reported from North America.

The taxonomic status of *Chlorochytrium* needs reevaluating since some species have been demonstrated to be life history stages of other filamentous algae (Chapman and Waters, 1992). Its exact taxonomic placement remains uncertain; often it has been regarded as belonging to the Chlorococcales. The genus is in need of revision.

Chlorotylium Kützing (Fig. 3F)

Chlorotylium forms hemispherical to irregularly cushion-shaped clumps, lime-encrusted, and consists of uniseriate and almost wholly unilaterally branched filaments. Cells are of two types: short cells often in pairs, each possessing a green-colored or often reddish parietal chloroplast and a pyrenoid, alternating with longer and sometimes colorless cells, with each containing a pyrenoid.

It is most commonly found in flowing water especially streams and areas of rapids. Of the five known species, at least two occur in North America.

Cloniophora Tiffany (Fig. 3G)

The filaments are uniseriate, erect and with downwardly growing rhizoidal branches, the secondary branches considerably smaller than the primary branches and irregularly arranged, with the terminal cell conical or its apex blunt. Cells are cylindrical, inflated, or capitate, each possessing a large parietal chloroplast and one to several pyrenoids. Zoospores are reported in one species, biflagellate, arising in cells of lateral branches.

Cloniophora is often attached to rock surfaces in usually fast-flowing water; it is also reported coastally from freshwater seepages well above the high tide. Islam (1961) suggests that it probably grows in habitats where the salt content is higher than usual fresh water. It is generally considered to be a subtropical or tropical genus, and is widely distributed in Central America and the Caribbean and most common along the eastern coast of the United States (Florida to New Brunswick).

Coleochaete Brébisson (Fig. 2I, J)

Filaments are uniseriate, dichotomously, or irregularly divided, consisting of only a prostrate system or a prostrate and an erect system, often the latter system is sometimes initially unbranched; prostrate forms have loose and spreading or laterally coalescing filaments forming a pseudoparenchymatous disklike expansion. Cells each contain a single, parietal, platelike chloroplast, one or two large pyrenoids, often bearing long, basally sheathed bristles. Zoospores biflagellate, one per cell. Sexual reproduction is oogamous: male gametes biflagellate, one per cell; oogonia are flask-shaped and bear a long trichogyne.

Coleochaete is frequently epiphytic (rarely endophytic) on submerged aquatic macrophytes, such as the submerged stems of cattail (*Typha*), and macroalgae, including the stonewort *Nitella*. It is also conspicuous on artificial surfaces including glass and crockery and frequently forms disklike colonies on the glass sides of aquaria (Prescott, 1983).

For a review of the genus, see Szymanska and Spalik (1993). Placed in the order Coleochaetales based on ultrastructural details.

Conochaete Klebahn (not pictured)

Conochaete grows as gelatinous cushions consisting of loose clusters of subglobose cells, frequently depressed, each cell bears a few long and delicate bristles arising apically on a mamillate wall protuberance or with an elongated gelatinous sheath. Cells each possess one or two parietal chloroplasts and a single pyrenoid, an oil droplet is often evident. Zoospores are usually four or eight per cell, the flagella number is unknown. Sexual reproduction is also unknown.

Only rarely reported, *Conochaete* is normally epiphytic on aquatic plants in upland areas. It is placed in the order Coleochaetales based on ultrastructural details.

Ctenocladus A. Borzi (Fig. 5B)

Ctenocladus is gelatinous and globular, consisting of erect uniseriate filaments, more or less radiating, with branches arising unilaterally. Cells are cylindrical and elongated, each containing a parietal chloroplast and one to several pyrenoids. Asexual reproduction is by biflagellate, micro- and macrozoospores arising in terminal cell or a series of cells. Quadriflagellate zoospores are reported in two undetermined species. Sexual reproduction is isogamous and gametes are biflagellate. Akinetes are spherical to subspherical, thick-walled, terminal, or in series.

Ctenocladus grows on rocks and higher plants (usually *Salicornia*) in inland saline environments, especially ponds and the marginal shallows of lakes. It is known mainly from more arid parts of western North America. According to Blinn and Stein (1970), its distribution pattern in North America suggests that migratory waterfowl play an important role in its spread. It often occurs where the Solonchak and Solonetz soil groups are present; these soils contain a high percentage of high exchangeable sodium. It is one of the more important food sources for brine flies in saline lakes (Herbst and Castenholz, 1994).

All the available evidence suggests that *Ctenocladus* consists of a single very polymorphic species, *C. circinnatus*.

Desmococcus F. Brand (Fig. 1C)

Desmococcus forms solitary or sarcinoid packets of cells (see also Chap. 7), occasionally as short uniseriate filaments. Cells are normally spherical, each possessing a parietal chloroplast and a single pyrenoid. Aplanosporangia are large, spherical, and the surface of each spore is warty or punctate.

According to Smith (1933), it is "the commonest green alga in the world," and often forms a green coating along with *Apatococcus* on soil-free and damp surfaces including the trunks of trees, stone or brick walls, and wooden fencing; it is often very abundant in both shaded and polluted habitats where lichens are less uncommon.

In the past, considerable taxonomic confusion has surrounded it and closely related genera (e.g., *Apatococcus*, *Chlorococcum*, *Pleurococcus*). It is retained in the Order Chaetophorales, but has been placed in the Ctenocladales.

Dicranochaete Hieronymus (Fig. 2C)

Dicranochaete is solitary, or more rarely, cells united in short rows consisting of spherical or kidney-shaped cells with the apical portion of each usually ornamented and forming an operculum, laterally bearing one or more repeatedly dichotomously divided bristles. Cells each containing a single chloroplast in the form of an inverted parietal cup, with or without a few pyrenoids. Zoospores biflagellate, four to 32 per sporangium. Gametes biflagellate and fuse to form a quadriflagellate zygote; the resting spore thick-walled.

Rare, reported as an epiphyte on filamentous algae and submerged aquatic plants in lakes.

Placed in the order Coleochaetales based on details of its ultrastructure.

Dilabifilum Tschermak-Woess (Fig. 3K)

The filaments are uniseriate, consisting of a creeping primary system bearing secondary or tertiary branches and erect ones. Cells are cylindrical in younger filaments, usually swelling to form akinetes in aging cultures; on agar it develops into dark green mass of coccoid cells. Cells each contain a parietal chloroplast and a pyrenoid. Zoospores quadriflagellate, produced in swollen intercalary cells. Aplanospores 16 or 32 per cell.

Isolated from various surfaces (e.g., lichens, mollusk shells, calcareous crusts, stones), it is collected in freshwater and marine (intertidal) habitats (more frequently recorded from the former). It is only known from Europe and North America.

Dilabifilum is a rather ill-defined genus in which reports of sporangia and quadriflagellate zoospores require further confirmation; asexual reproduction has yet to be discovered in two species.

Draparnaldia Bory de St.-Vincent (Figs. 3C, D)

A soft mucilaginous envelope encloses the erect uniseriate filaments of *Draparnaldia* that are attached basally by rhizoidal branches; the principal erect branches bear oppositely, alternately, or in whorls, lateral clusters or tufts of smaller-celled and much divided secondary-branches, each terminating in a blunt cell or multicellular hair. Cells of main branches are barrel-shaped or cylindrical, each containing a large parietal, band-shaped, entire or reticulate chloroplast with a smooth or laciniate margin, and several pyrenoids. Cells of secondary branches each possess a single, laminate chloroplast covering almost the entire wall, with one to three pyrenoids. Asexual reproduction is by biflagellate zoospores and thick-walled aplanospores. Sexual reproduction is isogamous by quadriflagellate gametes.

Draparnaldia is most frequently encountered in pristine upland streams and spring areas. It is more common and abundant in cooler waters, hence it is more likely to be encountered in the spring and autumn months in temperate areas. About 20 species are known, of which at least five are reported from North America.

The genus exhibits considerable morphological plasticity in response to differences in the environment, leading to considerable doubt concerning the validity of characters traditionally used for taxonomic discrimination. In nutrient-enriched water, species of *Stigeoclonium* are capable of displaying "draparnaldioid" features (Simons *et al*., 1986), but are unlikely to be mistaken for a *Draparnaldia* due to the absence in *Stigeoclonium* of distinctive clusters of secondary branches and the smaller diameter of main axes (< 30 μm).

Draparnaldiopsis G. M. Smith *et* Klyver (Fig. 3B)

Similar to *Draparnaldia*, but the principal branches of *Draparnaldiopsis* are rarely divided, consisting of alternating short and long cells; clusters of secondary branches arise opposite one another or in whorls and only from short cells, basal cells of secondary branches are usually di- or trichotomously branched, each branch generally terminates in narrowly obtuse cells or a long hair. Cells of secondary branches are cylindrical, inflated, or fusiform, each with a bandlike or reticulate chloroplast as in *Draparnaldia*. Zoospores are often produced in cells of main axes and gametes in cells of secondary branches. Zoospores are of different sizes: larger zoospores quadriflagellate (two flagella of different length), smaller ones biflagellate. Gametes are isogamous and biflagellate.

Only two species are known from North America, both from lakes in the western United States.

Entocladia Reinke (Fig. 1K)

The *Entocladia* filaments are uniseriate, creeping, and irregularly divided, sometimes coalescing to form a compact pseudoparenchymatous and one-celled layer prostrate expansion; branches are occasionally short and grow mostly by division of bluntly pointed terminal cells, with some species bearing from the upper cell wall a long, tapering bristle (sometimes basally swollen). Cells each contain a parietal, platelike chloroplast, with one or more pyrenoids. Zoospores quadriflagellate. Sexual reproduction is by biflagellate isogametes.

It is endo- or epiphytic on or within the wall layer of filamentous green algae such as *Cladophora*, *Rhizoclonium*, and *Pithophora*, or growing among the cells of the red alga *Lemanea*. Some species have been regarded as mere juvenile stages in the development of *Stigeoclonium*. Most species are marine rather than freshwater.

The exact taxonomic placement of freshwater representatives remains uncertain, as does the relationship of these to the marine species. The genus is in need of further investigation.

Fridaea W. Schmidle (Fig. 3E)

Fridaea forms hemispherical, compact, heavily lime-encrusted tufts, usually yellow-green in color, consisting of a prostrate and erect system of branched uniseriate filaments bearing tapering bristles as an extension of the distal cross wall of a cell, occasionally bristles separated from the cell by a wall. Cells are cylindrical and elongated, each possessing a very pale green, parietal chloroplast confined to the distal end, with one to four pyrenoids present. Cells interpreted as zoosporangia arise laterally, elongate and bottle-shaped, zoospores or swarmers unknown. Akinetes is uncertain.

Only known from moving water, it very rarely is recorded, but when present it is often abundant. The genus is monospecific.

Fritschiella Iyengar (Fig. 3H)

The filaments of *Fritschiella* are uniseriate, consisting of colorless downward-growing rhizoidal filaments and prostrate subterranean filaments, the latter often crowded and irregularly divided. Above ground the tufted, irregularly branched erect filaments have cells becoming longer and more elongated toward the apex, the terminal cell is conical. Cells above soil each contain a parietal chloroplast and several pyrenoids. Reproduction is confined to subterranean filaments. Zoospores are of three types: quadriflagellate macro- and microzoospores, and biflagellate microzoospores. Gametes are biflagellate and isogamous.

Fritschiella is terrestrial, growing more or less gregariously on moist soil or silt including drying rainwater puddles. It is known from the southwest United States.

Gomontia Bornet *et* Flahault (Fig. 1H)

The filaments of *Gomontia* are uniseriate, consisting of irregularly branched masses lying immediately below the surface of mollusk shells, wood, or limestone, sometimes densely crowded and pseudoparenchymatous, with downward-growing rhizoidal filaments. Cells are cylindrical, ovoid or polygonal, walls are thick and lamellate, each containing a dense laminate and sometimes netlike or lobed chloroplast. Several pyrenoids and nuclei are present. Zoosporangia and aplanosporangia are formed by enlargement of cells of short erect branches; the number of flagella is uncertain (variously reported to be two or four). Akinetes are produced.

Gomontia perforates the shells of mollusks, the carapace of turtles, limestone, submerged wood, or

"*Cladophora* balls." Most species are marine, with only two reported in North America from non-marine environments.

Gongrosira Kützing (Fig. 3J)

Commonly cushion-like, sometimes lime-encrusted, *Gongrosira* consists of a prostrate and erect system of uniseriate filaments, the prostrate system single to multilayered, loose or pseudoparenchymatous, giving rise to an erect system usually of short branches terminating in blunt tips. Cells are cylindrical or somewhat inflated, sometimes with thick and lamellate walls, each with a prominent parietal chloroplast and one to several pyrenoids. Zoosporangia are terminal or intercalary; zoospores are biflagellate. Aplanospores or akinetes are produced, the latter reddish in color.

It often forms a greenish layer or cushion-like mass on stones, gastropods, aquatic macrophytes (e.g., *Potamogeton*, *Myriophyllum*, *Najas*, *Chara*) and other hard surfaces, and is frequently most evident along the shallow margins of ponds, lakes, and rivers; it is only rarely recorded from soil.

Many species are of doubtful validity, with considerable overlap of morphological characters considered of taxonomic importance. Few species have been examined in laboratory culture (e.g., *G. debaryana*, *G. scourfieldia*), with one species (*G. pseudoprostrata*) defined only on culture-based characters. Some species are poorly known or doubtful, and others have been transferred to other chaetophoralean genera. It is retained here in the Chaetophorales, but sometimes is placed in the order Ctenocladales.

Hazenia H. C. Bold (Fig. 1I)

Hazenia has a tubelike sheath of soft mucilage containing irregularly branched uniseriate or partly multiseriate filaments; short lateral branches terminate in a conical cell. Cells are subquadrate to globose, not always in close contact, each contains a parietal chloroplast and a pyrenoid. Gametes are biflagellate and isogamous. Zoospores are unknown.

Only known from soil. *Hazenia* is monospecific.

Helicodictyon L. A. Whitford *et* G. J. Schumacher (Fig. 1M)

A spherical or irregularly shaped mucilaginous envelope surrounds the peculiarly curved or twisted and indistinctly branched filaments of *Helicodictyon*. Cells are short, curved, or irregularly lobed, each possess one to three short, parietal, platelike chloroplasts and a single pyrenoid. Vegetative cells become transformed into single biflagellate zoospores.

Helicodictyon is sometimes present in sufficient quantity in ponds (including fish hatcheries) and lakes to form a sulfur-yellow bloom, especially in autumn. It is most commonly reported in acid and slightly brown water along the coastal plain of the Atlantic seaboard of the United States, and is monospecific.

Older colonies are reported (Biebel, 1968; Whitford and Schumacher, 1969) to often have a large gas vacuole trapped within the center. It has been suggested that it may be more closely allied to the Order Chlorococcales than the Chaetophorales.

Leptosira A. Borzi (Fig. 1J)

Leptosira has cushion-like tufts consisting of erect and prostrate systems of uniseriate filaments, irregularly to subdichotomously branched, with side branches short and terminating in variously shaped cells. Cells are spherical, barrel-shaped, elliptical, or irregularly shaped, each possess a parietal chloroplast filling the lumen; starch is present, with pyrenoids not usually evident (sometimes masked by starch). Reproduction is by biflagellate zoospores and biflagellate isogametes. Aplanospores are spherical, four to several in each sporangium.

It grows epiphytically on plants such as mosses in springs and ponds, rarely abundant but widely distributed in North America. Only one species is commonly reported in North America of the five to six currently recognized.

Some uncertainty surrounds the validity of this genus. It very closely resembles *Pleurastrum* but differs from it by its indistinct pyrenoid (see the note under *Pleurastrum*). Wujek (1971) considered *Leptosira* to be identical to the monospecific genus *Leptosiropsis*, the latter originally separated from *Leptosira* by having an evident pyrenoid. A pyrenoid is normally present but is often only observed when examined with the electron microscope, or is sometimes obscured by a mass of starch (Lokhorst and Rongen, 1994).

Microthamnion Nägeli (Fig. 3L)

Microthamnion is microscopic, attached by a mucilage pad and consists of an erect system of uniseriate filaments, each side branch arise immediately below the cross wall, and its first cross wall is often remote from its point of origin, terminates in an obtuse cell. Cells are cylindrical, each contains a parietal chloroplast and no pyrenoid. Zoospores are biflagellate, anteriorly projecting outward to form a neck. There is doubt regarding sexual reproduction and formation of akinetes and aplanospores.

Microthamnion occurs on aquatic macrophytes, rotting wood, stones and other surfaces in the shallows of pools, slow-flowing streams, and ditches. It is also common in springs and grows on peat in bog pools. Sometimes it is entrapped in water film and is associated with soil or wet carpets of moss. Of those species present in organically polluted habitats (e.g., sewage treatment plant filters), it develops in such abundance as to form macroscopic growth. Only three species are currently recognized (see below), one known from North America.

The considerable morphological variation exhibited by the genus has given rise to taxonomic problems because of the unreliability of the trivial vegetative features traditionally used to define subgeneric taxa. Only three species are recognized by John and Johnson (1987), and of these, two are regarded as doubtful with *Microthamnion kuetzingianum* and *M. strictissimum* considered to be conspecific. Curved branches commonly develop on plants entrapped in the surface water film. It is traditionally placed in the Chaetophorales, but its correct position remains uncertain, although it is placed by Melkonian (1990) in its own order, the Microthamniales.

Oligochaetophora G. S. West (Fig. 2E)

Oligochaetophora has a mucilaginous envelope surrounding the loose clusters of two to six subglobose or ovoid cells. The walls are very thin and homogeneous, each cell bears two to four unbranched, spinelike bristles. Each cell has a single parietal chloroplast and two or three starch grains. Reproduction is unknown.

It is rare and epiphytic on submerged aquatic plants, monospecific, and is placed in the order Coleochaetales based on ultrastructure details.

Pleurastrum R. Chodat (Fig. 1F)

The cells are solitary or form sarcinoid packets each of a tetrad of cells, or as short-branched filaments. Cells are also spherical, each containing a parietal chloroplast (often lobed) and one or more pyrenoids. Reproduction is by zoospores, aplanospores, and akinetes. Zoospores are naked, biflagellate, variable in shape, and often asymmetrically flattened.

Pleurastrum has been collected and isolated into culture from standing water (often stagnant) and the surface of damp soil.

A chemotaxonomic investigation has concluded it to be impossible to separate certain taxa of *Pleurastrum* and *Leptosira*. These two genera are morphologically very similar and separation is based solely on the absence of a pyrenoid in *Leptosira*, a distinction that is no longer valid (see *Leptosira*). Ultrastructural investigation of *Pleurastrum* has revealed the presence of two types of pyrenoid structure: one type traversed by several membranes and the other by a single membrane. Seven species are currently recognized and separation is only considered possible if they are compared in culture.

Polychaetophora W. West *et* G. S. West (Fig. 2D)

Polychaetophora is solitary, or more usually, two to six subglobose, ellipsoid, or ovoid cells clustered together to form a pseudofilamentous mass, each cell with a very thick lamellate wall, sometimes with lamellate outgrowths, bearing eight to 12 delicate, flexuose, and unbranched bristles. Cells each have a single parietal chloroplast, often indistinct at the margins, with oil droplets and no pyrenoid(s).

Occurrence is rare, usually growing on aquatic macroalgae, and is monospecific. See remarks under *Oligochaetophora*. It is placed in the order Coleochaetales, based on ultrastructure details.

Protoderma Kützing (=*Ulvella* Crouan) (Fig. 1G)

The filaments of *Protoderma* are uniseriate, closely coalesce to form a pseudoparenchymatous, single-layered expansion with filaments free from one another towards the margin. Cells towards the center are usually polygonal and those towards the margin are more cylindrical, each possess a parietal platelike or flattened chloroplast, with or without a pyrenoid. Zoospores are biflagellate and generally form in central cells. Aplanospores are elliptical or spherical. The palmelloid stage is formed by repeated division of the central cells.

Protoderma is common in streams growing on rocks and the stems of submerged aquatic plants, including filamentous macroalgae, such as *Cladophora*.

It is almost impossible to separate *Protoderma* from other related genera, especially species in other genera in which the erect system is very reduced (e.g., *Pseudendoclonium prostratum*). Often separation is only possible if the flagellar number of the zoospores is known. Some authorities consider the genus to be monotypic with *Protoderma beesleyi* regarded as belonging to *Ulvella*, a largely marine genus to which it was originally attributed. The exact placement of the genus is problematic, with ultrastructural findings indicating that those species examined should be transferred to the order Ulvales. Until further species are examined, it is retained in the Chaetophorales, the order to which it has been traditionally assigned.

Pseudendoclonium Wille (Fig. 1L)

Filaments are uniseriate. An irregularly branched prostrate and erect system are present, with prostrate filaments loose and spreading or pseudoparenchymatous. Cells are cylindrical or inflated, each possessing a parietal chloroplast and a single pyrenoid. Zoospores are quadriflagellate. Aplanospores and zoospores often are produced in two- or four-celled sarcinoid packets of swollen cells, which are pleurococcoid in appearance. Akinetes are known.

Pseudendoclonium is isolated from stones and the surface of various submerged plants (e.g., *Ceratophyllum, Lemna, Vallisneria, Myriophyllum, Najas, Potamogeton*) and grows in a wide variety of water bodies including ponds, springs and slow-flowing rivers; it is rarely isolated from soil.

It is essential to know the flagellar number of the zoospores, otherwise it is impossible to separate it from species of *Gongrosira*. All the freshwater species have been described from isolates studied in laboratory culture. Square or diamond-shaped surface scales have been discovered on the zoospores of *P. basiliense* var. *brandii* and *P. akinetum*, but not on those of the marine species *P. submarinum*. Inconsistencies in ultrastructural features of the few species so far investigated inevitably cast doubt on whether the genus represents a natural grouping. Fine structural details of the flagellated stage have led to the opinion that it should be placed in the order Ulotrichales.

Pseudochaete W. West et G. S. West (Fig. 2I)

The filaments of *Pseudochaete* are uniseriate, prostrate, short, bearing from the sidewalls of some cells unbranched tapering filaments about five to eight cells in length. Cells are usually cylindrical, each possess a parietal chloroplast and a single pyrenoid only in cells of prostrate system. Zoosporogenesis and sexuality are unknown.

Only two species are known, both rare and often on submerged plants, logs, and on cattail stems.

Genus is incompletely known and doubtful. It has been suggested that the type species (*P. gracilis*) might be a developmental stage of *Stigeoclonium*.

Pseudulvella Wille (Fig. 1A)

Filaments are uniseriate, pseudoparenchymatous with radially arranged cells and several cells in thickness in the center of a disklike or somewhat irregular expansion, enveloped in mucilage and bearing a few hyaline bristles. Cells each possess a parietal chloroplast, netlike or divided into several oval bodies and a single pyrenoid. Zoospores are quadriflagellate, generally formed toward the center.

Only one freshwater species, *P. americana*, is reported on rare occasions from the United States (e.g., Michigan, Iowa).

Schizomeris Kützing (Fig. 7C)

Schizomeris has macroscopic filaments, unbranched, basally uniseriate, and becoming multiseriate above to form a solid parenchymatous cylinder with ringlike constrictions at intervals along its length. Cells are long and cylindrical toward the base, quadrate to angular above where multiseriate, thick-walled, each containing a parietal chloroplast encircling two-thirds of the cell, and toward the base cells possessing several pyrenoids. Zoospores are quadriflagellate, formed in multiseriate portions.

It grows as dark green, coarse clumps and is widespread but not common. Known from quiet water and in or near waterfalls. *Schizomeris* is often most common in fairly eutrophic water, including polluted environments such sewage sludge or shallow lakes close to discharge from sewage treatment works. Only two species are known.

The view that the genus represents a growth form of *Stigeoclonium* has not received general acceptance. On ultrastructural grounds, its affinities lie with the order Chaetophorales *sensu stricto*, although it is morphologically similar to members of the Ulvales.

Stigeoclonium Kützing (Fig. 3A)

Filaments are uniseriate with variously developed prostrate and erect systems: erect filaments alternately, oppositely or dichotomously branched, occasionally whorled or irregularly arranged, apices acute, narrowly obtuse, or each bearing a multicellular hyaline hair; prostrate filaments are creeping or rhizoidal, occasionally forming a pseudoparenchymatous disklike expansion. Cells are cylindrical or inflated, thick or thin-walled, each contain a single, parietal chloroplast, and one to several pyrenoids. Zoospores are quadriflagellate and of two sizes: micro- and macrozoospores. Gametes are isogametes and biflagellate or quadriflagellate.

Stigeoclonium often occurs as delicate tufts or mats on rocky surfaces and on submerged aquatic plants. It grows over a very wide habitat range, although is often especially common in swiftly flowing streams and rivers. It is sometimes very abundant in polluted rivers (e.g., immediately below sewage treatment works) including waters with relatively high levels of heavy metals where *Cladophora* may be absent.

This is an extremely polymorphic genus in which considerable doubt attaches to the validity of many described species. Species are best delineated only after growth in carefully defined culture conditions (Cox and Bold, 1973). Some modern culture-based studies (Simons *et al.*, 1993) have recognized just three species and consider the majority of the 30 or more recognized species to be environmentally induced forms.

Thamniochaete F. Gay (Fig. 2B)

Filaments of *Thamniochaete* are uniseriate, creeping or erect branches are few and short, terminal and sometimes other cells also bear long and often basally inflated bristles or short spinelike projections. Cells are irregularly cylindrical, barrel-shaped, or inflated, each containing a parietal chloroplast and a single pyrenoid. One species produces ovoid or globose, biflagellate zoospores. Akinetes are known.

Thamniochaete grows as an epiphyte on larger algae or hard surfaces, usually in moving water. Only one of the three known species so far reported is from North America.

One species of *Endoclonium*, *E. rivulare*, is often placed in *Thamniochaete*.

Trichodiscus Welsford (Fig. 2G)

Trichodiscus is a disklike thallus reaching about 1 mm in diameter, consisting of branched uniseriate filaments; prostrate filaments radiate and centrally form a pseudoparenchymatous disklike expansion and are free at the margin, bear short erect branches and colorless (above the base) multicellular hairs arise from the center. Cells each have a lobed, parietal chloroplast and a single pyrenoid. Zoospores are biflagellate, produced in central cells of pseudoparenchymatous disklike expansion. Akinetes are produced by rounding up of cells.

This is a dubious genus, with its only species (*T. elegans*) discovered in the U.K. on the sides of a glass jar in which was grown an *Azolla* plant imported from North Carolina. It closely resembles some species of *Stigeoclonium* species (e.g., *S. farctum*) from which it is doubtfully distinguished by having somewhat inflated sporangia and gametangia. Due to the doubt attached to the genus and its putative North America record, it is not keyed out here.

Trichophilus Weber-van Bosse (Fig. 1B)

The filaments of *Trichophilus* are uniseriate and multiseriate, prostrate, irregularly divide to give short branches, partly pseudoparenchymatous. Cells are thick-walled, each contain a parietal chloroplast and a difficult-to-discern pyrenoid. Zoospores quadriflagellate and of two sizes.

It grows on or within the hair scales of sloths and on the shells of freshwater mollusks. Only three species are known.

Trichophilus closely resembles the genus *Entocladia* Reinke.

Order Charales

Order Charales is macroscopic, consisting of algae with creeping rhizoidal branches from which arise erect branches of limited growth, each bearing whorls of secondary branches (branchlets) of limited growth. It has complex and unique type of advanced oogamous reproduction. Only three of the seven genera are known from North America. The genus *Nitellopsis* is so far only reported from South America.

Chara L. (Fig. 9A)

Chara is macroscopic, usually lime-encrusted, with similar organization to other members of the order. Branches of a unlimited growth composed of elongated, single-celled internodes and multicellular nodes, with branchlets (rarely branches of unlimited growth) arising from the short cells of nodes; the cortex is single-layered over internodal cells, rarely partially or ecorticate, showing varying degrees of development of primary, secondary, and tertiary cell rows, with the primary row always distinguished by the presence of spine cells. Branchlets are undivided, subtended by a single or double ring of unicellular outgrowths (stipulodes); simplified cortication is over lower internodal cells; rings of unicellular bract-cells develop at nodes; the terminal segment is single-celled or in chains not separated by nodal cells. Reproduction is by an advanced form of oogamy, the oogonium always above the antheridium when together on a branchlet (monoecious species). Oosporangium each consists of an oogonium surrounded by eight spirally twisted sterile cells, bearing a crown of five cells. Antheridia are usually spherical consisting of eight shield-shaped cells borne on the end of a short stalk.

Thalli are often large (up to 1 m) and coarse, especially when heavily lime-encrusted. It is most abundant in hard water or alkaline ponds and lakes (also in the Great Lakes), where it can form extensive underwater meadows. It is found occasionally in the shallows of slow-flowing rivers and in spring seepage areas and known to grow to depths as great as 12 m (see also Chapter 2, Section II.F-2).

Often *Chara* species have a strong odor, hence its common name in North America of skunkweed or muskweed.

Lamprothamnium J.Groves (not pictured)

Lamprothamnium is macroscopic, with or without lime-encrustation and similar organization to other members of the order. Branches of limited growth are ecorticated. Branchlets are undivided, subtended by a single whorl of unicellular outgrowths (stipulodes), each acuminate and downwardly pointing, with rings of more or less equal unicellular bract-cells developing at the nodes. Oogonia arise below the antheridia at the branchlet nodes, each bears a crown of five cells in the form of a single ring, not compressed.

It is often large (reaching to c. 50 cm) and most common in coastal lakes and lagoons where the water is brackish. Considerable doubt attaches to *L. longifolium* and its variety *buckellii*, the only *Lamprothamnion* known from North America. It differs from all other members of the genus in being dioecious and having rudimentary cortical and spine cells. It is reported from a few widely scattered locations including British Columbia in Canada and Kansas in the United States.

Nitella C. Agardh (Fig. 9B)

Nitella is macroscopic, only lightly lime-encrusted or not at all; it is similar in organization to other members of the order but not as erect as *Chara*. Branches of limited growth ecorticate, hence spines are absent. Branchlets are not subtended by stipulodes and bract-cells are absent, forked one or more times into similar single-celled rays and/or one- to three-celled ultimate rays (sometimes divided two, three, or four times). Oogonia are produced at the forking of branchlets and occasionally at the base of a whorl of branchlets, with oogonia arising laterally and two to three together and antheridia solitary and terminal. Oogonia are with a crown of 10 cells in two tiers and laterally compressed

It is widely distributed and usually more common than *Chara* in softer water areas and acid lakes, including bog lakes where the water is stained brown. *N. flexilis* is common in streams throughout temperate regions of North America (Sheath and Cole, 1992).

Tolypella (A. Braun) A. Braun (Fig. 9C)

Macroscopic and not lime-encrusted, *Tolypella* is similar in structure to other charophytes, although less symmetrical, with dense heads and whorls of short branches and longer more disorganized branches. Branches ecorticate and hence are without spines. Branchlets have stipulodes or bract cells and are divided into unequal and multicellular rays. Oogonia and antheridia are in groups, often on the lower nodes of the branchlets or at their base, the oogonia having a corona of 10 cells in two tiers and not compressed.

Often somewhat scraggy looking plants. *Tolypella* often grows solitary in pools, ponds, ditches, shallows of hardwater lakes (including the Great Lakes), and slow-flowing streams. It is widely distributed in North America and occurs as far north as Newfoundland.

Order Cladophorales

Filaments are uniseriate, simple or branched, rarely forming a prostrate disklike expansion, more commonly erect and attached by rhizoids. Cells are quadrate to cylindrical, sometimes swollen, multinucleate, each containing a parietal and netlike chloroplast (more rarely numerous disklike chloroplasts), with numerous pyrenoids.

Basicladia Hoffman *et* Tilden (Fig. 5F)

Basicladia has a prostrate rhizoidal filament giving rise to a coarse erect filament only sparingly branched in the basal region. Cells are cylindrical toward the base and several times longer than broad, becoming gradually broader, shorter, and often barrel shaped toward the apex, walls are thick and lamellated, each possessing a parietal netlike chloroplast and numerous pyrenoids. Zoospores are biflagellate and produced in upper cells.

It is commonly epizoic on snails and on the upper carapace of several turtle species (occasionally on the head and tail), especially snapping turtles with the expression 'mossback' used for those animals with a thick algal growth. It is capable of growth on other surfaces provided they are hard and rough. Only four species comprise the genus with the two recorded from North America considered less common in northern parts of the United States than in the southern. For a review of its ecology, see Hutchinson (1975: 562-564) and Colt *et al.* (1995).

Chaetomorpha Kützing

Chaetomorpha is macroscopic, consisting of unbranched filaments, growth is intercalary, free living or attached basally by rhizoids. Cells are usually cylindrical to barrel shaped, walls lamellate, each with a parietal, netlike chloroplast and numerous pyrenoids and nuclei. Zoospores are bi- or quadriflagellate.

This is typically a marine genus with considerable doubt attaching to the report of it from freshwater in North America. The only report is that of *Chaetomorpha heningsii* growing in a roadside ditch in Washburn County, Wisconsin (Faridi, 1962).

Cladophora Kützing (Fig. 5C)

Macroscopic, free-living or attached by a disklike holdfast and/or downward growing rhizoids, consisting of erect or prostrate systems of filaments, sparsely to profusely branched, *Cladophora* is typically dichotomously or sub-dichotomously branched, with branches originating immediately below cross wall; growth is apical and/or intercalary. Cells are cylindrical to barrel shaped, each with a parietal and netlike chloroplasts, many nuclei and numerous, often bilenticular pyrenoids. Zoospores are bi- or quadriflagellate. Gametes are biflagellate and isogamous. Akinetes are rarely produced.

It is principally a marine genus, yet *Cladophora glomerata* is one of the most widely distributed and abundant macroalgae in fresh waters world-wide. It exhibits considerable morphological variation depending on environmental conditions, thus making identification difficult with many species records of doubtful validity. It is commonly attached but on occasion free-floating or lying as loose masses over sediments, frequently as long streamers attached to rocks in slow-flowing streams, and often forming entangled growths along lake shores; it is especially abundant in eutrophic habitats (e.g., parts of the Great Lakes) providing there is relatively low heavy metal contamination. One species is known to occur to a depth of 50 m in Lake Ontario. In a few North American lakes (e.g., in Massachusetts, New York, Indiana), it forms spherical to subspherical growths (ca. 2–10 cm in diameter) known as 'Cladophora balls' (Daily, 1952; Kindle, 1934). It is also a nuisance alga commonly known as 'blanket weed' or 'cotton-mat' alga (see Chapter 24); the term might equally apply to other mat-forming green algae (e.g., *Oedogonium, Spirogyra*).

Considerable uncertainty surrounds the separation of many species whose identification often depends on combinations of overlapping character states. Culture studies have demonstrated that many earlier recognized species are merely growth forms or environmental variants (Hoek, 1982). Due to its world-wide importance, it has been the subject of several reviews (e.g., Whitton, 1970; Dodds and Gudder, 1992).

Dermatophyton A. Peter (Fig. 5D)

Epizoic, disklike, consisting of radiating rows of cells, *Dermatophyton* is polystromatic in its center and single-layered and dichotomously divided toward the margin. Cells are thick-walled, quadrate or subquadrate in the center, and more elongated close to the margin, each contain several nuclei, a parietal chloroplast and one to several pyrenoids. Zoospores are biflagellate, produced from flask-shaped cells in the center.

Thalli form deep green, irregular crusts reaching 5 mm across, growing on and penetrating cracks into the carapace of freshwater turtles where it develops between the layers of the lamellae. It grows on various turtles and sometimes accompanies the green alga *Basicladia*. It is monospecific.

It was transferred from Chaetophorales, where it was often placed due to presence of several nuclei in each cell.

Pithophora Wittrock (Fig. 5E)

Pithophora is macroscopic, consisting of freely branched filaments, lateral or occasionally oppositely branched, often side branches originate a short distance below the cross wall; occasionally unicellular or multicellular rhizoidal branches develop from some branch tips or base of filaments. Cells are cylindrical, often thick-walled, each with a parietal and netlike chloroplast as well as several pyrenoids and nuclei. Akinetes are thick-walled, barrel-shaped, or oval, terminal or intercalary in position.

It is known from rapids in rivers but is more common in ponds and lakes where it forms dense free-floating mats in shallow littoral areas; it is similar to *Cladophora* in that its prolific growth leads to lake management problems. Often somewhat coarser in texture than *Cladophora*, for this reason *Pithophora* is commonly known as the 'horsehair' alga. It is considered to be a tropical or subtropical genus but is widely distributed in temperate regions, especially in the eastern United States. Eight species are known from the United States alone.

Rhizoclonium Kützing (Fig. 5A)

Rhizoclonium consist of uniseriate unbranched filaments, occasionally short, unicellular, or few-celled rhizoidal branches arise at right angles to axis. Cells are usually more than twice as long as broad, sometimes with walls thick and lamellate, each with a parietal, reticulate chloroplast and several pyrenoids and nuclei. Reproduction is by fragmentation or akinetes. Zoospores biflagellate and gametes quadriflagellate.

It intermingles with other filamentous green algae usually as tangled mats in quiet stagnant water or as ropelike strands in flowing water. *Rhizoclonium hieroglyphicum* is the most common of the five freshwater species recorded from North America.

Order Cylindrocapsales

Filaments are uniseriate and unbranched, often becoming biseriate or multiseriate with age. See further details under *Cylindrocapsa*.

Cylindrocapsa Reinsch (Fig. 6I)

Filaments are unbranched, initially uniseriate, often becoming multiseriate or forming irregular, pseudoparenchymatous masses of cells. Cells are cylindrical, elliptical, rounded, oval, or subspherical, sometimes arranged in pairs, surrounded by a thick, rough, lamellated mucilaginous sheath, each containing an axial or stellate chloroplast in young filaments, sometimes chloroplast parietal, filling cell and often obscured by starch, with a single and often indistinct pyrenoid. Asexual reproduction is by biflagellate zoospores and aplanospores. Sexual reproduction is oogamous and monoecious: oogonia are greatly swollen and spherical or oval, dark green, with a thick lamellated wall; antheridia are in a series, two to four in each enlarged vegetative cell, producing two spindle-shaped biflagellate gametes. The zygote is spherical, thick-walled, and bright reddish when mature.

It is often attached but commonly become free floating and frequently intermingled with other filamentous forms growing in soft water lakes and bogs.

It is often assigned to its own order (occasionally to the Ulotrichales) although regarded by Hoek *et al*.

(1995) to be a filamentous representative of the Chlorococcales based on its type of mitosis and cellular division.

Order Microsporales

Filaments are uniseriate and unbranched, with separation of cells taking place at the point of weakness in the mid region of the wall, often resulting in the formation of 'H-shaped' portions on dissociation of filaments. If the filaments are correctly interpreted as being a row of autospores, then it is closely related to the Chlorococcales. According to Lokhorst (1999), ultrastructural details of the flagella apparatus provide yet further evidence that *Microspora* belongs to this order.

Microspora Thuret (Fig. 4E)

Free-living or basally attached at least initially, *Microspora* has a bipartite wall structure usually evident. Cells are quadrate, cylindrical, or slightly swollen, sometimes constricted at the cross wall, walls are often lamellate, each with a parietal chloroplast and perforated to produce a fine or coarse net, without pyrenoids. Reproduction is by biflagellate zoospores and biflagellate isogametes. Aplanospores and akinetes produced.

It is widely distributed in aquatic habitats, particularly among other filamentous algae in pools, streams, and ditches, especially during cooler times of the year. Some species are frequent in low pH environments, including bogs, marshes, and acid-mine drainage water.

Many characters traditionally considered of taxonomic importance are unreliable and are known to exhibit considerable variability within the same species depending on environmental conditions. Lokhorst (1999) has attempted to develop sounder species concepts by testing in laboratory culture the reliability of those morphological characters traditionally used to define species. It is possible to confuse *Microspora* with the Tribophyte (xanthophyte) *Tribonema* (see chapter 11), which is readily distinguished by having two or more, pale green or golden color, disklike chloroplasts and no starch.

Order Oedogoniales

Filaments are uniseriate, branched or unbranched, with or without bristles, often characterized by having one or more ringlike thickenings ('cap cells') at the distal end of some cells. It has a complex and unique form of oogamous reproduction.

The three genera represent some of the most comprehensively researched in North America thanks to the classic work of Tiffany (1930) on the Oedogoniales of the United States.

Bulbochaete C. Agardh (Fig. 8A)

The filaments of *Bulbochaete* are unilaterally branched, usually attached, consisting of cells broader at the upper ends and the majority bearing a long, basally swollen, colorless bristle. Cells are broader at the upper end, each containing a parietal, netlike chloroplast, with several pyrenoids. Zoospores have an apical ring of flagella, one produced per cell. Sexual reproduction is oogamous: large swollen female oogonia, boxlike antheridia, or dwarf male plant (short filaments) grow on or near the oogonia; the zygote is thick-walled and sometimes distinctively ornamented.

It is commonly epiphytic on submerged plants, submerged portions of overhanging and larger filamentous algae in slow-flowing water and pools and ponds. Of almost 60 species known, over half occur in the United States.

Oedocladium Stahl (Fig. 8B)

The filaments of *Oedocladium* are profusely branched, consisting of prostrate and erect portions; rhizoidal branches are produced in terrestrial and aquatic environments with these colorless in former, often with the apical cell conically pointed and having a distinct cap. Cells are cylindrical, each with a parietal and netlike chloroplast, with several pyrenoids. Reproduction is similar to that of *Oedogonium*. Akinetes usually developing on rhizoidal branches, solitary or in short chains and reddish in color.

It is most common along moist stream banks and on sandy loam, often growing together with moss protonema, thalloid liverworts, *Vaucheria*, and other soil algae.

Oedogonium Link (Fig. 8C)

Oedogonium filaments are unbranched, usually attached and without bristles. Cells are cylindrical, sometimes slightly broader at the anterior end, characterized by one or more ringlike caps immediately below the cross wall, each containing a parietal, netlike chloroplast and several pyrenoids. Zoospores have a distinctive apical ring of many flagella, one produced per cell. Sexual reproduction is oogamous: large swollen female oogonia, boxlike antheridia or some species with dwarf male plants (short filaments) grow on or close to the oogonia; zygote is thick-walled and sometimes distinctively ornamented.

It is free-living and sometimes epiphytic on submerged plants growing in similar aquatic habitats to *Bulbochaete*. Over half of the more than 250 known species are recorded from the United States.

Species-level identification is very difficult since they are based mainly on reproductive characters. Only rarely is the alga discovered in the reproductive condition.

Order Prasiolales (=Schizogoniales)

Filaments are uniseriate or multiseriate and form solid cylinders or expanded sheets of cells. See the description of *Prasiola* for further details.

Prasiola C. Agardh (Fig. 7E)

Prasiola is terrestrial or aquatic, growing as two morphological forms: single-layered (rarely two-layered), lobed, or ruffled and membrane-like, or as unbranched filaments (*Rosenvingiella*-stage). Flat expansions consist of quadrate or polygonal cells arranged in square to rectangular blocks, sometimes in smaller groups of four, often shortly stalked and attached by hairlike rhizoids. The filamentous stage is initially uniseriate, often becoming multiseriate, attached by rhizoids. Cells each contain a large, stellate chloroplast (both forms) and a single central pyrenoid. Reproduction is by nonmotile aplanospores produced in areas of thickening in membrane-like forms. Sexual reproduction is by oogamy: reproductive tissues (haploid) develops in the upper part of membrane-like stage, some areas producing biflagellate male gametes and others nonmotile egg cells.

Prasiola occurs in wide range of habitats that include terrestrial, marine, and freshwater. Some species grow on moist soil, rocks, stones, old walls, and trunks of trees, and others in cold, swift-flowing mountain streams. This genus is abundant in terrestrial habitats rich in nitrogenous compounds, and hence its frequent association with the guano in bird roosts, as well as the base of urban walls and trees where animals frequently urinate. Other species (e.g., *P. mexicana*) colonize rocks in clear, swiftly flowing streams (Wehr and Stein, 1985; Sheath and Cole, 1992). The genus occurs over a wide geographical range from Arctic Canada in the north of North America to the tropics.

Considerable uncertainty surrounds the relationship between *Prasiola* and *Rosenvingiella* with some authors (e.g., Edwards, 1975) considering the latter to be a growth form of *Prasiola*. Rindi et al. (1999) observed a *Rosenvingiella* stage of *Prasiola* in Galway, Ireland and also recognize as distinct the genus *Rosenvingiella*; molecular studies are needed to resolve whether the latter is a separate genus. In the current treatment, it has been decided to regard *Prasiola* and *Rosenvingiella* as separate genera, but to recognize that the former has a *Rosenvingiella* stage. Uncertainty surrounds the exact placement of the genus in the classification of the green algae, although *Prasiola* and *Rosenvingiella* form a well-defined clade in molecular trees (Sherwood et al., 2000).

Rosenvingiella P. C. Silva

This is a terrestrial alga growing as unbranched, uniseriate, or multiseriate filaments attached at intervals by rhizoids, or parenchymatous, cylindrical and sometimes constricted, with cells sometimes in four-celled groups; chloroplasts are stellate and possess a single central pyrenoid. Reproduction is by nonmotile asexual spores.

It grows in many of the same habitats as *Prasiola*.

Considerable uncertainty surrounds the relationship between *Prasiola* and *Rosenvingiella* (see comments under *Prasiola*).

Order Sphaeropleales

Filaments are uniseriate, unbranched, consisting of exceedingly long cells whose cytoplasm is divided by numerous large vacuoles into separate cytoplasmic units. Cytoplasmic units each are multinucleate and contain several a band-shaped or disklike chloroplasts.

Sphaeroplea C. Agardh (Fig. 6E)

Sphaeroplea cells are 2 to 60(-90) times longer than broad, divided into several large central cavities by a succession of transverse units of cytoplasm, each multinucleate unit containing several narrow, bandlike chloroplasts (occasionally perforate with age) and many pyrenoids. Reproduction is by fragmentation and production of biflagellate zoospores. Sexual reproduction is oogamous and vegetative cells do not change shape on becoming reproductive cells, monoecious, or dioecious. Oogonia are spherical, often lying in a single or double series, each having a conspicuous receptive spot; antheridia produces up to 300, spindle-shaped, reddish, biflagellate antherozoids. Zygotes have thick and ornamented walls, bright red when mature.

Free-floating and of sporadic occurrence, it is specially common in seasonally flooded grasslands and areas of gravel. If *Sphaeroplea* is present in considerable abundance, it causes the discoloration of the water on producing its red-colored zygotes. Only two species are reported from the United States.

Species are distinguished mainly upon features associated with zygote shape and ornamentation of its thickened wall (Ramanathan, 1964).

Order Trentepohliales

The filaments are uniseriate, branched, sometimes coalescing to form a pseudoparenchymatous and discoid expansion. Cells are each uninucleate, often red or reddish-orange due to an accumulation of haematochrome or astaxanthin, with a reticulate chloroplast or several discoid chloroplasts and without pyrenoids.

Cephaleuros Kunze (Fig. 6A)

Cephaleuros is epiphytic and endophytic on leaves, stems, and fruits of terrestrial plants. It consists of regularly branched coalescing filaments forming a prostrate and a one to several-layered pseudoparenchymatous mass from which downward-growing, irregularly branched filaments penetrate the deeper tissues and upward-growing, unbranched filaments terminate in a hair or group of sporangia. Cells are cylindrical, regularly or irregularly shaped, each possess a parietal and a netlike chloroplast or several discoid chloroplasts, without pyrenoids. Asexual reproduction is by zoosporangia produced in clusters terminally on erect hairs, zoospores biflagellate. Sexual reproduction is by enlargement of cells in the pseudoparenchymatous mass and produce biflagellate gametes.

It is considered a parasite or semi-parasite of terrestrial plants (*Magnolia*, *Rhododendron*, banana, tea, etc.) in the tropics and subtropics where it occurs on the surface, beneath the cuticle, and/or within the intercellular spaces and leads to the discoloration and degeneration of the tissues of the host. Often the parasitized areas are gray-green or reddish-orange in color and on the surface it forms cushion-like growths and has a velvety appearance. It is known to occur on many host species with Holcomb (1986) recording it from over 200 plant taxa in and near Louisiana. A serious parasite on some plants, it results in economic losses (especially to tea growers), but infections are usually easily controlled. For further information, see Chapman and Waters (1992) and Thompson and Wujek (1997).

Phycopeltis Millardet (Fig. 6B)

Phycopeltis is epiphytic on leaves of terrestrial plants. It consists of regularly or irregularly branched, coalescing filaments forming a prostrate, single-layered, pseudoparenchymatous discoid, lobed or irregularly shaped crust. Cells are cylindrical, each possess a parietal, netlike chloroplast or many irregularly shaped and platelike chloroplasts, often orange or yellowish due to haematochrome pigment, with a single pyrenoid. Asexual reproduction is by quadriflagellate zoospores, produced in sporangia arising on curved, one to multicellular stalks. Sexual reproduction is by biflagellate isogametes formed in terminal or intercalary gametangia.

It forms yellowish-green to deep orange colored expansions on leaves, most common in the humid tropical and subtropical regions where it causes commercial losses to cultivated plants.

Printzina Thompson *et* Wujek (not pictured)

Printzina is epiphytic. It consists of a widely spreading network of prostrate, branched filaments bearing few to many erect branches if present. Cells are cylindrical, barrel-shaped, globular, or elliptical, with numerous discoid chloroplasts, green color usually obscured by haematochrome, with pyrenoids absent. Reproduction is similar to that in *Trentepohlia* but differs from it by having globular to kidney-shaped rather than ovoid sporangia, the papilla pore basal on the sporangium rather than apical, sporangia lateral and always solitary and never grouped, and gametangia lateral or terminal rather than terminal.

It grows on the leaves and bark of trees growing in deep shade and high humidity and is often green in color.

The type species of the genus (*P. ampla*) is known from Columbia and Costa Rica where it grows on trees belonging to the families Melastomaceae, Mimosaceae, and Monimiaceae. In creating the genus, Thompson and Wujek (1992) transferred to it nine *Trentepohlia* species.

Stomatochroon Palm (Fig. 6D)

Subaerial and epiphytic, *Stomatochroon* consists of a lobed basal cell within a stomatal chamber, from the basal cell arises a cylindrical and erect, few-celled branch or hair bearing one or more sporangia. The basal cell contains chloroplasts in the lower part and the erect cell is brick red in color due to presence of carotenoid pigments.

It usually grows on the leaves of broad-leaved evergreen plants in humid tropical regions where it causes wounding responses and local damage when its growth is prolific (Chapman and Waters, 1992; Thompson and Wujek, 1997).

Trentepohlia Martius (Fig. 6C)

Trentepohlia is subaerial. It consists of variously developed prostrate and erect systems of sparingly to profusely branched filaments, irregular, alternate, or rarely oppositely divided, sometimes erect system very reduced, often terminal branches gradually attenuated and apical cells sometimes bearing a cellulose cap. Unicellular and cylindrical hairs are sometimes present. Cells are cylindrical, ellipsoidal, or beadlike, rarely more than twice as long as broad, walls are thick and lamellate, each with numerous, discoid or band-shaped chloroplast; chlorophyll usually is obscured by red or orange carotenoid pigments; pyrenoids are absent. Asexual reproduction is by quadriflagellate zoospores produced in terminal sporangia, or of two types of stalked lateral sporangia: 'funnel sporangia' consists of an apical portion of a terminal cell and a septum lying between two superimposed ringlike thickenings, and a sporangium with a bent stalk. Sexual reproduction is by biflagellate gametes produced in lateral or terminal, spherical gametangia. Akinetes are thick-walled and generally in series.

It forms greenish, yellowish-orange, or orange feltlike patches on leaves, twigs, stone walls, wooden fencing, and poles, and is often very extensive on rocky cliffs and moist soil in ravines and the damper sides of tree trunks. It is the second most common algal component of lichens where almost impossible to identify to species since it is frequently only present as single cells or short filaments without reproductive organs.

The genus *Physolinum* is now usually considered a synonym of *Trentepohlia*.

Order Siphonales

Filaments are tubular, with a single multinucleate cell, rarely divided by transverse walls, containing numerous discoid chloroplasts; pyrenoids are present or absent.

Dichotomosiphon Ernst (Fig. 6F)

Filaments of *Dichotomosiphon* are coenocytic and tubular, repeatedly dichotomously divided and transverse constrictions at the base of each dichotomy and additional ones often between successive branch divisions. Branches possess numerous discoid or ellipsoidal chloroplasts, often absent from rhizoidal branches at plant base; pyrenoids are absent. Asexual reproduction is by large akinetes produced terminally on lateral rhizoidal branches. It is oogamous, with gametangia borne on di-, tri- or tetrachotomously divided branches: oogonia are very large, yellowish, spherical, and arise on a curved supporting branch, antheridia are the same diameter as the supporting branch, curved and borne immediately below oogonia. Zygotes possess a smooth thick wall.

It grows as dense, tangled tufts or mats on streams and lake beds where it may occur in depths as great as 20 m; it often becomes buried in soft sediment (always sterile in deep water).

Phyllosiphon Kühn (Fig. 6H)

Phyllosiphon is endophytic on various members of the higher plant Family Araceae. It consists of coenocytic tubular branches (multinucleate and noncellular), profusely dichotomously or irregularly divided and interwoven with one another. Branches contain numerous discoid or elliptical chloroplasts except at apices; pyrenoids are absent. Reproduction is by many small ellipsoidal aplanospores.

It forms green patches within leaves and stems and often induces discoloration of the area of host tissue infected. Known from northern and eastern parts of the United States growing within the tissues of *Arisaema triphyllum*, it is commonly known as Jack-in-the-Pulpit (Smith, 1933, 1950). For further information, see Chapman and Waters (1992).

Protosiphon Klebs (Fig. 6G)

Protosiphon is a soil alga having above ground a swollen sacklike vesicle and an elongate, colorless, rhizoid-like portion extending downward into the soil. Cells are multinucleate, possessing a netlike chloroplast and several pyrenoids. Reproduction is by many biflagellate zoospores and isogametes produced within the vesicle. Cytoplasm sometimes divides in multinucleate portions and these develop into a thick-walled resting stage (known as coenocysts).

It is considered to be fairly common and widespread on moist soil where it often forms conspicuous green or orange patches.

Order Ulotrichales

Filaments are uniseriate, rarely multiseriate, unbranched, sometimes exhibiting basal-apical polarity when attached. Cells usually are quadrate or cylindrical, uninucleate, each possess a single, parietal, laminate, or broadly discoid chloroplast, bandlike or more usually only partially encircling cell, with or without pyrenoids.

This is a very heterogeneous order, with several genera now placed in other orders based on characters associated with mitosis and cellular division as well molecular data (e.g., *Uronema*, *Klebsormidium*; see comments below). The affinities or many genera remain unclear and, therefore, most are retained in this order.

Binuclearia Wittrock (Fig. 4B)

The filaments consist of long cells usually containing pairs of oval or subspherical protoplasts. There are spaces between protoplasts and the cross wall have transverse lamellations of a gelatinous material, each protoplast with a single laminate chloroplast, incompletely or completely encircling cell, usually with pyrenoid (rarely absent).

It often occurs intermingled with other filamentous algae and submerged aquatic macrophytes, especially in softwater ponds and bogs.

Elakatothrix Wille (Fig. 4Q,R)

Elakatothrix is solitary or colonial, with cells contiguous and arranged in pseudofilaments embedded in a spindle-shaped and unstructured mucilaginous sheath, with the sheath margin definite or somewhat irregular in outline. Cells are spindle-shaped, cylindrical, oval, or elliptical, apices blunt or pointed, walls thin and hyaline, each contain one or three parietal, laminate, cup- or girdle-shaped, straight or spiral chloroplasts, and one or two pyrenoids. Asexual reproduction is by aplanospores. Sexual reproduction is unknown.

It is a widely distributed planktonic alga, although its juveniles are sometimes epiphytic (Lokhorst, 1991). Of the five known species, only one is known only from the United States, namely *Elakatothrix americana*.

It is often now placed in the Order Klebsormidiales based on ultrastructural features.

Filoprotococcus (H. W. Nichols *et* H. C. Bold) Thompson *et* Wujek (Fig. 7B)

Filaments are initially uniseriate, soon becoming multiseriate and transformed into a series of sarcinoid packets of cells which eventually dissociate. Cells each contain a parietal chloroplast and a single pyrenoid. Zoospores are quadriflagellate, produced singly or two to four in each cell of multiseriate filaments and sarcinoid packets. Sexual reproduction is unknown.

It has been isolated from cultures of soil from Texas and *Tolypella* collected from a marsh near Lawrence, Kansas (Thompson and Wujek, 1996). It is probably more widely distributed than records indicate.

Thompson and Wujek (1996) transferred *Trichosarcina polymorphum* to this genus although recognizing minor differences between the two (e.g., two to four zoospores per cell in the plants examined by them from Kansas and only one in those described from soil cultures).

Geminella Turpin (Figs. 4H, I)

Geminella filaments consist of cells separated but in a loose linear arrangement and equidistant, lying in pairs or contiguous, surrounded by a thick, cylindrical mucilaginous sheath. Cells are generally longer than broad, cylindrical with rounded apices, inflated and ellipsoidal, oval or barrel-shaped, each contain a parietal, girdle-like or laminate chloroplast only partially encircling the cell, usually with a single pyrenoid. Fragmentation common, as are thick, brown-walled akinetes.

It grows attached or free-floating in often soft, more acid water areas where desmids tend also to be abundant.

Most authors have chosen to maintain it as distinct from *Gloeotila*, a genus accommodating species having a strict contiguous arrangement of cells and lacking pyrenoids (Lokhorst, 1991). Several *Geminella* and/or *Gloeotila* species lacking pyrenoids and a mucilaginous sheath have been transferred by Hindák (1996) to the genus *Stichococcus*. It is traditionally placed in the order Ulotrichales, but is more closely related to the Chlorococcales based upon features associated with mitosis and cellular division (Hoek *et al.*, 1995).

Gloeotila Kützing (Fig. 4L)

Filaments of *Gloeotila* consist of loosely adherent cells with a tendency to break into short lengths, surrounded by a homogeneous mucilaginous sheath. Cells are generally longer than broad, oblong or ellipsoidal, and blunt at apices, each with a parietal, laminate or girdle-like chloroplast encircling most of the cell, pyrenoids are absent. Zoospores are biflagellate.

It grows attached, free-floating and planktonic, or on soil.

Sometimes it is considered congeneric with *Geminella*, but is usually separated on the absence of pyrenoids and its propensity to fragment. See comments under *Geminella*.

Gloeotilopsis Ramanathan (Fig. 4O)

Cells of *Gloeotilopsis* are solitary, rarely loosely attached, without a mucilaginous sheath. Cells are cylindrical and bluntly rounded at apices, each with a parietal chloroplast encircling most of the cell and a single pyrenoid.

The North American species is only known from a culture of soil from Dauphin Island, Alabama.

It is a little known and doubtful genus.

Hormidiopsis Heering (Fig. 4M)

Hormidiopsis filaments consist of a series of loosely attached mucilaginous sheaths; within each sheath lies an interrupted row of two or four cells. Cells are oval or oblong, shorter than broad, each possess a parietal chloroplast encircling part of the cell, with or without a single pyrenoid.

It is associated with soil and intermingled with algae in small boggy pools. Considerable doubt surrounds the two known species, with the type (*H. crenulatum*) a subaerial alga not reported to have a pyrenoid. The genus *Hormidiopsis* is possibly congeneric with *Ulothrix* or *Klebsormidium* (see remarks in Prescott, 1983; Bourrelly, 1988).

Klebsormidium Silva, Mattox *et* Blackwell (Fig. 4C)

Klebsormidium filaments are without differentiation into a basal or apical cell. Cells are cylindrical or beadlike (doliform), walls thin or thickened and often lamellated, if thickened then rough (crenulate and/or verrucose), sometimes cross walls are surrounded by H-shaped pieces, each with a parietal, elliptical, discoid or girdle-shaped chloroplast normally encircling less than 80% of the cell, typically with a single pyrenoid. Zoospores are biflagellate and produced in unspecialized cells and released through a pore. Aplanospores and thick-walled akinetes are produced.

Often one of the more abundant green filamentous algae in temperate streams, it is frequently regarded as a seasonal annual, sometimes growing intermingled with other filamentous forms. It is commonly subaerially on wet soil, dripping rocks, or seepage areas and when dry often developing thick H-shaped walls. Some species are especially abundant in acid environments where the water is contaminated by high concentrations of heavy metals.

Characters traditionally used to distinguish this species are known to be influenced by environmental conditions. According to Lokhorst (1996), for accurate and reliable identification it is desirable to examine material after growing under controlled laboratory conditions. About 15 species are known.

It is usually placed in the Order Klebsormidiales based on ultrastructural details (in older literature as *Hormidium*, but now invalid).

Koliella Hindák (Fig. 4J)

The cells of *Koliella* are solitary, usually detaching immediately after dividing and, only more rarely, daughter cells remain in pairs for a time to form short pseudofilaments before fragmenting. Cells are spindle-shaped, filiform, or very elongated, straight or curved, apices obtuse, rounded, or gradually tapering to form very long points, each contain a parietal, laminate or girdle-shaped, straight, or spiral chloroplast, with or without pyrenoids and oil droplets present. Zoospores are biflagellate. Sexual reproduction is oogamous, with details uncertain. Akinetes are produced.

Usually considered to be widely distributed in freshwater plankton, it is occasionally periphytic, epilithic or associated with ice and snow banks. The only North American record traced so far is a mention of "*Koliella* sp." by Prescott (1983) based upon material collected from snow banks in the Olympic Mountains, Washington.

It is very similar to *Raphidonema*, the two possibly being congeneric.

Koliella is placed in the Order Klebsormidiales based upon details of its ultrastructure.

Pseudoschizomeris Deason et H. C. Bold (Fig. 4P)

Pseudoschizomeris filaments are unbranched in young cultures; with age it develops distally more than one series of spherical cells, eventually producing much mucilage and becoming palmelloid. Cells each possess a parietal chloroplast encircling most of the cell and a single pyrenoid.

It is only known from a culture of Texas soil and is monospecific.

Radiofilum Schmidle (Fig. 4G)

Radiofilum filaments are usually unbranched but occasionally falsely branched, consisting of a row of cells whose wall is of two helmet-shaped halves with a ringlike transverse rim formed where they join, and surrounded by a thick mucilaginous sheath. Cells are spherical, subspherical, or ellipsoidal, each possess a parietal, cup-shaped chloroplast lying adjacent to the cross wall, with a single pyrenoid. Reproduction is by fragmentation.

It occurs in various aquatic habitats including pools and ponds. Three species are known to be widely distributed in the United States.

It is traditionally assigned to the order Ulotrichales, but placed in the Chlorococcales by Hoek *et al.* (1995).

Raphidonema Lagerheim (Fig. 4K)

The filaments of *Raphidonema* are two to 32 cells in length. Cells are cylindrical, only apical cells acuminate or gradually tapering to a point, walls thin, each containing a parietal, laminate or girdle-shaped chloroplast, with or without pyrenoids. Zoospores are biflagellate. Sexual reproduction is oogamous: male gametes biflagellate (one per cell) and vegetative cells functions as a female gamete.

Some species are found in aquatic habitats and others are associated with snow or ice, the latter of which become green when the alga is present in quantity. About 20 species are recorded.

It is placed in the Order Klebsormidiales based upon ultrastructural details.

Raphidonemopsis Deason (Fig. 4N)

The cells of *Raphidonemopsis* are solitary or form two-celled filaments, with a small attachment disk and are terminally bluntly pointed. Cells gradually narrow to the base and apex, walls thin, each containing a parietal, laminate chloroplast and a single pyrenoid.

It is known only from soil enrichment cultures.

This is a doubtful genus which is principally distinguished from *Raphidonema* by its attachment to a surface.

Stichococcus Nägeli (Fig. 4D)

Cells are solitary or form a readily dissociating filament, without a mucilaginous sheath. Cells cylindrical and often elongated with rounded apices, sometimes short, spherical to slightly oval, thin-walled, each containing a single parietal, laminate chloroplast often covering only a small part of the cell, normally without a pyrenoid. Reproduction is by fragmentation and cell division.

It has a very wide ecological amplitude, ranging from soil-free surfaces to freshwater and marine habitats, although it is most common on tree trunks, wooden fencing, or damp soil, and sometimes associated with other subaerial green algae (e.g., *Desmococcus* and *Apatococcus*).

It is placed in the Order Klebsormidiales based on ultrastructural details.

Ulothrix Kützing (Fig. 4F)

The filaments of *Ulothrix* are attached by a single basal cell or rhizoidal basal cell, or by rhizoids arising from other cells. Cells are cylindrical, sometimes beadlike (doliiform), often longer than broad, walls thin or thick and occasionally lamellated or roughened, with "H-shaped" pieces, each has a parietal, girdle-shaped, and more usually a lobed chloroplast, encircling all or normally over three quarters of the cell, with one or more pyrenoids. Zoospores are quadriflagellate. Gametes are biflagellate and isogamous. Aplanospores and thick-walled akinetes are produced. Unicellular stage in life history, the *Codiolum*-stage.

It grows attached or free-living on damp soil, cliffs wetted by spray, as well as in stagnant water; it is less common in rapidly flowing water or on stones and wood along the margins of wave-splashed lakeside shores, with the possible exception of the Great Lakes.

Uronema Lagerheim (Fig. 4A)

Uronema filaments are attached by a basal cell, often terminating in a slightly curved and asymmetrically pointed apex. Cells are usually cylindrical, each with a parietal, laminate chloroplast partially encircling the cell, one to four pyrenoids. Zoospores are quadriflagellate, one or two per sporangium.

Most grow epiphytically on aquatic plants in various freshwater bodies. About 10 species have been described.

Taxonomic position of this little-known genus remains uncertain. It is retained in the order to which it has been traditionally assigned, but ultrastructural features of cell division and the basal body of flagellated stages suggest that it should be assigned to the Chaetophorales *sensu stricto* (Hoek *et al.*, 1995).

Order Ulvales

Order Ulvales is membrane-like or tubular and normally branched, single-layered, and parenchymatous. Cells each possess a single nucleus, a parietal laminate chloroplast and one to several pyrenoids.

Enteromorpha Link (Fig. 7A)

Enteromorpha is macroscopic, consisting of attached elongated, tubular sacks, usually branched and sometimes with numerous proliferations, initially attached by a basal rhizoidal cell or proliferations developing from its base. Cells are quadrate or cylindrical, often angular by mutual compression, each possess a parietal, laminate chloroplast, and one or several pyrenoids. Zoospores are quadriflagellate. Sexual reproduction is by biflagellate gametes.

The genus grows often initially attached when growing on submerged objects in flowing-water or occasionally in ponds, it frequently becomes free-floating, inflated, and intestiniform (intestine-like). Only one species (*Enteromorpha flexuosa*) is considered to be freshwater, with many others marine and brackish-water. Many North American inland records (e.g., Silver Springs, Wyoming County, western New York State; see Catling and McKay, 1980; salt lakes in Kansas; Ranch, 1981) are of species usually reported from marine habitats (*Enteromorpha intestinalis* and *E. prolifera*). These plants were collected growing on rocks in salt springs or in streams and lagoons receiving effluent from salt mines.

Monostroma Thuret (Fig. 7D)

Monostroma is macroscopic, initially sacklike, and later splitting to form a single-layered membrane, parenchymatous or cells rounded and grouped in fours or separated by mucilage, commonly attached by rhizoidal protuberances. Cells are angular by compression or rounded, each with a single, parietal chloroplast encircling most of the cell and a single pyrenoid. Reproduction is by quadriflagellate zoospores and biflagellate anisogametes. *Codiolum*-stage in life history. Various types of life history are known in genus.

Only one or two species are known exclusively from freshwater habitats, otherwise it is a marine and brackish-water genus. It often grows on rocks or lodged driftwood in swift-flowing streams and rivers (Taft, 1964); it is also reported from standing water in Arctic Canada.

It is retained in the Ulvales where it is traditionally placed, but Hoek *et al.* (1995) consider it to be a multicellular, thalloid member of the Order Codiolales (= Ulotrichales) based on ultrastructural features associated with two marine species so far examined.

VII. GUIDE TO LITERATURE FOR SPECIES IDENTIFICATION

The majority of freshwater algal genera have a world-wide distribution. For this reason, floristic works covering other regions of the world can be used in identifying North American algae. Many of the more useful and comprehensive works dealing with green algae are frequently in languages other than English. For example, Starmach's series of volumes on the freshwater algae of Poland are very useful since coverage is world-wide. These are written in Polish, although the volume dealing with filamentous green algae (Starmach, 1972) has a key in English. Very few libraries or institutions hold more than a fraction of these works, some of which are difficult to use even by specialists possessing a working knowledge of the language. Most of the most useful monographs, flora's and identification guides on freshwater algae are listed in "Key Works to the Fauna and Flora of the British Isles and North-western Europe" (Sims *et al.*, 1988).

The majority of the green algal genera considered in this chapter are mentioned in Smith's classical work *Fresh-water Algae of the United States* (Smith, 1933, 1950) and various regional floristic works (e.g., Prescott, 1951; Tiffany and Britton, 1952; Whitford and Schumacher, 1963). The regional floristic works are useful for species level identification since they contain descriptions, keys, and are profusely illustrated. Sometimes the charophytes are not included in these treatments, one of the few major algal groups for which there exists a complete world-wide review. This review by Wood and Imahori (1964, 1965) still remains the most comprehensive treatment of the group and includes much data on North American charophytes. The first author also published a useful guide to the charophytes of North America, Central America, and the West Indies (Wood, 1967) and 20 years earlier had reviewed the genus *Nitella* in North America (Wood, 1948). One of the first comprehensive treatments of North American green filamentous algae was Hazen's monograph *The Ulothricaceae and Chaetophoraceae of the United States* (Hazen, 1902) with many considered in *The Green Algae of North America* by Collins (1904). Another monographic treatment was Tiffany's monumental work on North American Oedogoniales (Tiffany, 1937) following an earlier publication dealing with the order world-wide (Tiffany, 1930). Unfortunately, all these useful identification works are unobtainable except through second-hand booksellers. The most useful modern series of identification guides for North American algae has been written by Dillard and covers the southeastern part of the United States. The volume dealing with the filamentous green algae was published in 1989 and like others in the series mentions taxa known elsewhere in the United States. The most useful and comprehensive volume on filamentous and parenchymatous freshwater green algae is undoubtedly *Die Chaetophoralean der Binnengewässer* (Printz, 1964). Like the major of works mentioned above, it has been long out of print and only obtainable through the secondhand book trade.

The following is a list of works dealing with North American genera; often these cite references to earlier literature. Sometimes no modern treatment exists, and the most comprehensive account is to be found in earlier, sometimes obscure, and often out of print works (e.g., Printz, 1964; Smith, 1950; Starmach, 1972). Otherwise, identification has to be based on keys, descriptions, and illustrations in more general floristic works such as John *et al.*, (2002).

1. *Apatococcus*—Ettl and Gartner (1995), Gartner and Ingolic (1989), Prescott (1983), Printz (1964).
2. *Aphanochaete*—Dillard (1989), Prescott (1962), Printz (1984), Starmach (1972), Tupa (1974).
3. *Basicladia*—Dillard (1989), Prescottt (1983).
4. *Binuclearia*—Dillard (1989), Prescott (1983).
5. *Bulbochaete*—Dillard (1989), Mrozinska (1985), Tiffany (1930, 1937, 1944).
6. *Cephaleuros*—Dillard (1989), Printz (1964), Thompson and Wujek (1997).
7. *Chaetomorpha*—Faridi (1962).
8. *Chaetonema*—Dillard (1989), Prescott (1983), Printz (1964), Tupa (1974).
9. *Chaetophora*—Dillard (1989), Prescott (1983), Printz (1964), Stamarch (1972).
10. *Chaetosphaeridium*—Dillard (1989), Prescott (1983), Printz (1964).
11. *Chamaetrichon*—Tupa (1974).
12. *Chara*—Wood (1967), Wood and Imahori (1964, 1965).
13. *Chlorochytrium*—Chapman and Waters (1992), Prescott (1983).
14. *Chlorotylium*—Dillard (1989), Prescott (1983), Printz (1964).
15. *Cladophora*—Hoek (1963, 1982).
16. *Cloniophora*—Dillard (1989), Islam (1961), Prescott (1983), Starmach (1972).
17. *Coleochaete*—Dillard (1989), Prescott (1983), Printz (1964), Szymanska and Spalik (1993).
18. *Conochaete*—Prescott (1983), Printz (1964), Smith (1933, 1950).
19. *Ctenocladus*—Blinn and Stein (1970), Prescott (1983).
20. *Cylindrocapsa*—Dillard (1989).
21. *Dermatophyton*—Prescott (1983), Printz (1964), Smith (1933, 1950), Starmach (1972).
22. *Desmococcus*—Ettl and Gartner (1995), Gartner and Ingolic (1989), Printz (1964).
23. *Dicranochaete*—Dillard (1989), Matula (1992), Prescott (1983), Printz (1964).
24. *Dichotomosiphon*—Dillard (1989), Prescott (1983).
25. *Dilabifilum*—Johnson and John (1990).
26. *Draparnaldia*—Dillard (1989), Forest (1976), Prescott (1983), Printz (1964), Starmach (1972).
27. *Draparnaldiopsis*—Forest (1976), Printz (1964), Prescott (1983), Starmach (1972).

28. *Entocladia*—Dillard (1989), Prescott (1983), Printz (1964), Smith (1933, 1959).
29. *Elakatothrix*—Prescott (1983), Printz (1964).
30. *Entermorpha*—Prescott (1983), Bliding (1963).
31. *Filoprotococcus* (=*Trichosarcina*)—Nichols and Bold (1965), Prescott (1983), Thompson and Wujek (1996).
32. *Fridaea*—Prescott (1983), Printz (1964).
33. *Fritschiella*—Prescott (1983), Printz (1964).
34. *Geminella*—Dillard (1989), Prescott (1983).
35. *Gloeotila*—Printz (1964), Ramanthan (1964), Smith (1933, 1950).
36. *Gloeotilopsis*—Ramanthan (1964).
37. *Gomontia*—Dillard (1989), Prescott (1983).
38. *Gongrosira*—Dillard (1989), Printz (1964), Starmach (1972), Tupa (1974).
39. *Hazenia*—Bold (1958), Prescott (1983).
40. *Helicodictyon*—Dillard (1989), Biebl (1968), Prescott (1983), Whitford and Schumacher (1969).
41. *Hormidiopsis*—Prescott (1983).
42. *Klebsormidium*—Dillard (1989), Ettl and Gartner (1995), Lokhorst (1996), Ramanthan (1967).
43. *Koliella*—Hindak (1983), Prescott (1983).
44. *Leptosira*—Dillard (1989), Ettl and Gartner (1995), Starmach (1972), Steil (1944), Tupa (1974), Wujek (1971).
45. *Microspora*—Dillard (1989), Prescott (1983), Lokhorst (1999), Ramanthan (1964).
46. *Microthamnion*—Dillard (1989), John and Johnson (1987), Printz (1964).
47. *Monostroma*—Bliding (1968), Prescott (1983), Taft (1964).
48. *Nitella*—Wood (1948, 1967), Wood and Imahori (1964, 1965).
49. *Oedocladium*—Dillard (1989), Ettl and Gartner (1995), Mrozinska (1985), Tiffany (1930, 1937).
50. *Oedogonium*—Dillard (1989), Hoffman (1967), Tiffany (1930, 1937), Yung *et al.* (1986).
51. *Oligochaetophora*—Prescott (1983), Printz (1964).
52. *Phycopeltis*—Printz (1994), Prescott (1983).
53. *Phyllosiphon*—Dillard (1989), Prescott (1983).
54. *Pithophora*—Dillard (1989), Prescott (1983).
55. *Pleurastrum*—Groover and Bold (1969), Starmach (1972), Tupa (1974).
56. *Polychaetophora*—Prescott (1983), Smith (1933, 1950).
57. *Prasiola*—Ettl and Gartner (1995), Rindi *et al.* (1999), Smith (1933, 1950).
58. *Printzia*—Thompson and Wujek (1992).
59. *Protoderma*—Dillard (1989), Printz (1964), Starmach (1972), Tupa (1974).
60. *Protosiphon*—Dillard (1989), Whitford and Schumacher (1967).
61. *Pseudendoclonium*—Dillard (1989), John and Johnson (1989), Printz (1964), Tupa (1974).
62. *Pseudochaete*—Prescott (1983), Smith (1933, 1950).
63. *Pseudoschizomeris*—Deason and Bold (1960), Ettl and Gartner (1995), Prescott (1983).
64. *Pseudulvella*—Dillard (1989), Prescott (1983), Printz (1964).
65. *Radiofilum*—Dillard (1989), Prescott (1983), Printz (1964).
66. *Raphidonema*—Dillard (1989), Hindák (1963), Hoham (1973), Prescott (1983), Printz (1964).
67. *Raphidonemopsis*—Deason (1969), Ettl and Gartner (1995).
68. *Rosenvingiella*—Edwards (1975).
69. *Rhizoclonium*—Dillard (1989), Prescott (1983).
70. *Schizomeris*—Campbell and Sarafis (1972), Dillard (1989), Prescott (1983), Printz (1964).
71. *Sphaeroplea*—Dillard (1989), Prescott (1983).
72. *Stichococcus* – Dillard (1989), Ettl and Gartner (1995), Hindák (1996), Printz (1964), Ramanthan (1964).
73. *Stigeoclonium*—Dillard (1989), Cox and Bold (1966), Islam (1963), Printz (1964), Starmach (1974).
74. *Stomatochroon*—Thompson and Wujek (1997).
75. *Thamniochaete*—Dillard (1989*)*, Prescott (1983), Printz (1964), Starmach (1974).
76. *Tolypella*—Wood (1946, 1967), Wood and Imahori (1964, 1965).
77. *Trentepohlia*—Dillard (1989), Ettl and Gartner (1995), Printz (1964), Starmach (1972).
78. *Trichodiscus*—Printz (1964), Starmach (1974).
79. *Trichophilus*—Bourrelly (1988), Printz (1964), Starmach (1974).
80. *Ulothrix*—Dillard (1989), Lokhorst (1979), Printz (1964), Starmach (1972).
81. *Uronema*—Dillard (1989), Ettl and Gartner (1995), Prescott (1983), Printz (1964), Starmach (1972).

LITERATURE CITED

Biebel, P. 1968. Reproduction in *Helicodictyon planctonicum* (Whitford) Whitford and Schumacher. Journal of Phycology 4:55–58.

Bliding, C. 1963. A critical survey of European taxa in Ulvales. Part I. *Capsosiphon, Percursaria, Blidingia, Enteromorpha.* Op. Botany Society of Botany Lund 9:1–160.

Bliding, C. 1968. A critical survey of European taxa in Ulvales. Part II. *Ulva, Ulvaria, Monostroma, Kornmannia.* Botaniska Notiser 121:535–629.

Blinn, D. W., Stein, J. R. 1970. Distribution and taxonomic reappraisal of *Ctenocladus* (Chlorophyceae, Chaetophorales). Journal of Phycology 6:101–105.

Bold, H. C. 1958. Three new chlorophycean algae. American Journal of Botany 45:737–743.

Booton, G. C., Floyd, G. L., Fuerst, P. A. 1998. Origins and affinities of the filamentous green algal orders Chaetophorales and Oedogoniales based on 18s rRNA gene sequences. Journal of Phycology 34:312–318.

Bourrelly, P. 1988. *Compléments les algues d'eau douce Initiation à la Systématique* Tome I: *Les Algues Vertes*. 1. Boubée et Cie, Paris, 182 pp.

Campbell, E. O., Sarafis, V. 1972. *Schizomeris*—a growth form of *Stigeoclonium tenue* (Chlorophyta: Chaetophoraceae). Journal of Phycology 8:276–282.

Canter-Lund, H., Lund J. W. G. 1995. Freshwater algae: Their microscopic world explored. BioPress, Bristol, 360 p.

Catling, P. M., McKay, S. M. 1980. Halophytic plants in southern Ontario. Canadian Field Naturalist 94:248–258.

Chapman, R. L., Waters, D. A. 1992. Epi- and endobiotic chlorophytes, in: Reiser, W., Ed., Algae and symbioses: Plants, animals, fungi, viruses, interactions explored. Biopress, Bristol, pp. 620–639.

Colt, L. C., Jr., Saumure, R. A., Jr., Baskinger, S. 1995. First record of the algal genus *Basicladia* (Chlorophyta, Cladophorales) in Canada. Canadian Field Naturalist 109:454–455.

Cox, E. R., Bold, H. C. 1966. Phycological studies. VII. Taxonomic investigations of *Stigeoclonium*. University of Texas Publications 6618:1–167.

Daily, F. K. 1952. '*Cladophora* balls' collected in Steuben County, Indiana. Butler University Botanical Studies 10:141–143.

Deason, T. R. 1969. Filamentous and colonial soil algae from Dauphin Island, Alabama. Transactions of American Microscopical Society 88:109–112.

Deason, T. R., Bold, H. 1960. Phycological Studies. I. Exploratory studies of Texas soil algae. University of Texas Publications. No. 6022. Austin, Texas. 71 p.

Dillard, G. E. 1989. Freshwater algae of the Southeastern United States. Part 2, Chlorophyceae: Ulothrichales, Microsporales, Cylindrocapsales, Sphaeropleales, Chaetophorales, Cladophorales, Schizogoniales, Siphonales and Oedogoniales. Bibliotheca Phycologica 83:1–163.

Dodds, W. K., Gudder, D. A. 1992. The ecology of *Cladophora*. Journal of Phycology 28:415–427.

Blinn, D. W, Stein, J. R. 1970. Distribution and taxonomic reappraisal of *Ctenocladus* (Chlorophyceae, Chaetophorales). Journal of Phycology 6:101–105.

Edgren, R. A., Egren, M. K., Tiffany, L. H. 1953. Some North American turtles and their zooepiphytic algae. Ecology 34:733–740.

Edwards, P. 1975. Evidence for a relationship between the genera *Rosenvingiella* and *Prasiola* (Chlorophyta). British Phycological Journal 10:291–297.

Ettl, H., Gartner, G. 1995. *Syllabus der Boden-, Luft- und Flechtenalgen*. G. Fischer, Stuttgart, 721 p.

Faridi, M. A. F. 1962. *Chaetomorpha heningsii* P. Richter in the U.S. Transactions of Kansas Academy Sciences 65:420–421.

Forest, H.S. 1956. A study of the genera *Draparnaldia* Bory and *Draparnaldiopsis* Smith and Klyver. Castanea 21:1–29.

Friedl, T. 1998. The evolution of the green algae. Plants and Systems Evolution Supplement 11:87–101.

Gartner, G. 1992. Taxonomy of symbiotic eukaryotc algae, in: Reiser, W., Ed., Algae and symbioses: Plants, animals, fungi, viruses, interactions explored. Biopress Ltd., Bristol, pp. 325–338.

Gartner, G., Ingolic, E. 1989. A contribution to the knowledge of *Apatococcus lobatus* (Chlorophyta, Chaetophorales, Leptosiroideae). Plant Systematics and Evolution 164:133–143.

Garwood, P. E. 1982. Ecological interactions among *Bangia*, *Cladophora* and *Ulothrix* along the Lake Erie shoreline. Journal of Great Lakes Research 8:54–60.

Gelin, F., Boogers, I., Noordelos, A. A. M., Sinnighe Damsté, J. S., Riegman, R., Leeuw, J. W. 1997. Resistant macromolecules in marine macroalgae of the classes Eustigmatophyceae and Chlorophyceae: geochemical implications. Organic Geochemistry 26:659–675.

Good, B. H., Chapman, R. L. 1978. The ultrastructure of *Phycopeltis* (Chroolepidaceae: Chlorophyta). I. Sporopollenin in the cell. American Journal of Botany 65:27–33.

Graham, L. E., Wilcox, L. W. 2000. Algae. Prentice Hall, Upper Saddle River, NJ, 640 p.

Harvey, W. H. 1836. Algae, in: Mackay, J. T., Ed., *Flora Hibernica*. 2(3), William Curry, Dublin, Ireland, pp. 157–254.

Hazen, T. E. 1902. The Ulothrichaceae and Chaetophoraceae of the United States. Memoirs of the Torrey Botany Club 11:135–250.

Henson, E. B. 1984. Notes on the benthic alga *Dichotomosiphon* from the Straits of Mackinac area of Lakes Michigan and Huron. Journal of Great Lakes Research 10:85–89.

Herbst, D. B., Castenholz, R. W. 1994. Growth of the filamentous green alga *Ctenocladus circinnatus* (Chaetophorales, Chlorophyceae) in relation to environmental salinity. Journal of Phycology 30:588–593.

Hindák, F. 1963. Systematik der Gattungen *Koliella* gen. nov. und *Raphidonema* Lagerh. Nova Hedwigia 6:95–125.

Hindák, F. 1996. New taxa and nomenclatural changes in the Ulotrichineae (Ulotrichales, Chlorophyta). Biologia 51(4):357–364.

Hoek, C. van den. 1982. A taxonomic revision of the American species of *Cladophora* (Chlorophyceae) in the North Atlantic Ocean and their geographic distribution. Verhandelingen der Koninklijke Nederlandse Akademie van Wetenschappen, Afdeling Naturankunde Tweede Sectie Holcomb 78:1–236.

Hoek, C. van den, Mann, D. G., Jahns, H. M. 1995. Algae, An Introduction to Phycology. George Thieme, Stuttgart. 623 pp.

Hoffman, L. R. 1967. Four new species of *Oedogonium*. Canadian Journal of Botany 45:405–412.

Hoham, R. W. 1973. Pleimorphism in the snow alga, *Raphidonema nivale* Lagerh. (Chlorophyta), and a revision of the genus *Raphidonema* Lagerh. Syesis 6:255–263.

Holcomb, G. E. 1986. Hosts of the parasitic alga *Cephaleuros virescens* in Lousiana and new host records for the continental United States. Plant Disease 70:1080–1083.

Hutchinson, G. E. 1975. A Treatise on Limnology. Vol. III Limnological Botany. J. Wiley, Sons, New York. 660 pp.

Islam, A. K. M. N. 1961. The genus *Cloniophora* Tiffany. Revue Algologie 6:7–32.

Islam, A. K. M. N. 1963. A revision of the genus *Stigeoclonium*. Nova Hedwigia, Beiheft 10:1–164.

John, D. M. 1994. Alternation of generations in algae: its complexity, maintenance and evolution. Biology Review 69:275–291.

John, D. M., Johnson, L. R. 1987. Observations on the developmental morphology, growth rate, and reproduction of *Microthamnion kuetzingianum* Naegeli (Pleurastraceae, Pleurastrales) in culture and a taxonomic assessment of the genus. Nova Hedwigia 44:25–53.

John, D. M., Johnson, L. R. 1989. A cultural assessment of the freshwater species of *Pseudendoclonium* Wille (Ulotrichales, Ulvophyceae, Chlorophyta). Algological Studies 82:79–112.

John, D. M., Moore, J. A. 1985. Observations on the phytobenthos of the freshwater Thames II. The floristic composition and distribution of the smaller algae sampled using artificial surfaces. Archiv für Hydrobiologie 103:83–97.

John, D. M., Whitton, B. A., Brook, A. J. 2002. The Freshwater Algal Flora of the British Isles. Cambridge University Press, Cambridge, 702 p.

Johnson, L. R., John, D. M. 1990. Observation on *Dilabifilum* (Class Chlorophyta, Order Chaetophorales *sensu stricto*) and allied genera. British Phycological Journal 25:53–61.

Kindle, E. M. 1934. Concerning 'Lake Balls', '*Cladophora* Balls and 'Coal Balls'. American Midland Naturalist 15:752–760.

Leitch, A. R., John, D. M., Moore, J. A. 1990. The oosporangium of the Characeae (Chlorophyta, Charales). Prog. Phycology Research 7:213–268.

Linnaeus, C. 1753. *Species Plantarum*. Vols. 1 and 2. Laurenti Salvii, Stockholm.

Lokhorst, G. M. 1978. Taxonomic studies on the marine and brackish-water species of *Ulothrix* (Ulothricales, Chlorophyceae) in Western Europe. Blumea 24:191–299.

Lokhorst, G. M. 1991. Synopsis of genera of Klebsormidiales and Ulotrichales. Cryptogamie Botany 2/3:274–288.

Lokhorst, G. M. 1996. Comparative taxonomic studies on the genus *Klebsormidium* (Charophyceae) in Europe. Cryptogamic Studies 5:1–132.

Lokhorst, G. M. 1999. Taxonomic study of the genus *Microspora* Thuret (Chlorophyceae), an integrated field, culture and herbarium analysis. Algological Studies 128:1–38.

Lokhorst, G. M., Rongen, G. P. J. 1994. Comparative ultrastructural studies of division processes in the terrestrial green alga *Leptosira erumpens* (Deason, Bold) Lukešova confirm the ordinal status of the Pleurastrales. Cryptogamie Botany 4:394–409.

McCourt, R. M. 1995. Green algal phylogeny. Trends Ecology Evolution 10:159–163.

Mattox, K. R., Stewart, K. D. 1985. Classification of the green algae: a concept based on comparative cytology, *in:* Irvine, D. E. G., John, D. M., Eds., Systematics of green algae. Academic Press, London, Orlando, pp. 29–72.

Matula, J. 1992. *Dicranochaete* species (Chlorophyceae, Gloeodendrales) in peat bogs of lower Silesia (South-west Poland). Algology Studies 93:63–72.

Melkonian, M. 1990. Chlorophyte orders of uncertain affinities: Order Microthamniales, *in:* Margulis, L., Corliss, J. O., Melkonian, M., Chapman, D. J., Eds., Handbook of Protoctista. John, Barlett, Boston, pp. 652–654.

Mrozinska, T. 1995. *Süsswasserflora von Mitteleuropa, Chlorophyta VI, Oedogoniophyceae: Oedogoniales.* 14. G. Fisher, Stuttgart, 624 p.

Nichols, H. W., Bold, H. C. 1965. *Trichosarcina polymorpha* gen. et sp. nov. Journal of Phycology 1:34–38.

Pascher, A. 1914. Über Flagellaten und Algen. Berichte der Deutschen Botanischen Gesellschaft 32:136–160.

Pascher, A. 1931. Systematische Übersicht über die mit Flagellaten in Zusammenhang stehenden Algenreichen und Versuch einer Einreihung dieser Algenstämme in die Stämme des Pflanzenreichs. Beihefte zur Botanischen Zeitschrift 48:317–332.

Pickett-Heaps, J. D. 1975. *Green Algae: Structure, Reproduction and Evolution in Selected Genera.* Sinauer Associates, Inc., Sunderland, Massachusetts. 606 pp.

Potter, M. C. 1888. Note on an alga (*Dermatophyton radicans*, Peter) growing on the European tortoise. Journal of Linnean Society of Botany 24:251–254.

Prescott, G. W. 1951. Algae of the Western Great Lakes area exclusive of desmids and diatoms. Cranbrook Inst. Sci., Bloomfield Hills, Michigan, 946 pp.

Prescott, G. W. 1962. Algae of the Western Great Lakes area with an illustrated key to the genera of desmids and freshwater diatoms. Revised ed. W.C. Brown, Dubuque, Iowa, 977 p.

Prescott, G. W. 1983. How to Know the Freshwater Algae. 3rd Edn. Brown, Iowa. 293 p.

Printz, H. 1964. Die Chaetophoralen der Binnengewasser (eine systematische Ubersicht). Hydrobiologia 24:1–376.

Ranch, D. C. 1981. *Enteromorpha*, a marine alga in Kansas. Transactions of Kansas Academy Sciences 84:228–230.

Ramanathan, K. R. 1964. *Ulotrichales*. ICAR, New Dehli, 188 p.

Raven, J. 1992. How benthic macroalgae cope with fresh-water-resource acquisition and retention. Journal of Phycology 28:133–146.

Rindi, F., Guiry, M. D., Barbiero, R. 1999. The marine and terrestrial Prasiolales (Chlorophyta) of Galway City, Ireland: a morphological and ecological study. Journal of Phycology 35:469–482.

Round, F. E. 1984. The systematics of the Chlorophyta: An historical review leading to some modern concepts [Taxonomy of the Chlorophyta III]. *in:* Irvine, D. E. G., John, D. M., Eds., Systematics of the green algae. Academic Press, Orlando, London, pp. 1–27.

Sabater, S., Gregory, S. V., Sedell, J. R. 1998. Community dynamics and metabolism of benthic algae colonizing wood and rock substrata in a forest stream. Journal of Phycology 34:561–567.

Sarma, P. 1986. The freshwater Chaetophorales of New Zealand. Nova Hedwigia, Beiheft 58:1–169.

Sheath, R. G., Cole, K. A. 1992. Biogeography of stream macroalgae in North America. Journal of Phycology 28:448–460.

Sherwood, A. R., Garbary, D. J., Sheath, R. G. 2000. Assessing the phylogenetic position of the Prasiolales (Chlorophyta) using rbcL and 18S rRNA gene sequence data. Phycologia 39:139–146.

Simons, J., Beem, A. P. van, Vries, de P. J. R. 1986. Morphology of the prostrate thallus of *Stigeoclonium* (Chlorophyceae, Chaetophorales) and its taxonomic implications. Phycologia 25:210–220.

Sims, R. W., Freeman, P., Hawksworth, D. L. Eds. 1988. Key Works to the Fauna and Flora of the British Isles and North-western Europe, 5th ed., Association Special Volume No. 33, Clarendon Press, Oxford, U.K., 312 p.

Smith, G. M. 1933. The Freshwater Algae of the United States, McGraw-Hill, New York, 716 pp.

Smith, G. M. 1950. The Freshwater Algae of the United States, 2nd Edn., McGraw-Hill, New York, 719 pp.

Starmach, K. 1972. Chlorophyta III. Ulotrichales, Ulvales, Prasiolales, Sphaeropleales, Cladophorales, Chaetophorales, Trentepohliales, Siphonales, Dichotomosiphonales, *in:* Starmach, K., Sieminska, J., Eds., Flora Slodkowodna Polski [Freshwater Flora of Poland] 10. Polska Akademia Nauck, Warsaw, pp. 1–714.

Steil, W.N. 1944. *Leptosira mediciana* Borzi. Bulletin of the Torrey Botany Club 71:507–511.

Stewart, K. D., Mattox, K. R. 1975. Comparative cytology, evolution and classification of the green algae with some consideration of the origin of other organisms with chlorophylls *a* and *b*. Botany Review 41:105–135.

Stewart, K. D., Mattox, K. R. 1978. Structural evolution in the flagellated cells of green algae and land plants. Biosystems 10:145–152.

Surek, B., Beemelmanns, U., Melkoniam, M., Bhattacharya, D. 1994. Ribosomal RNA sequence comparisons demonstrate an evolutionary relationship between Zygnematales and charophytes. Plant Systematics and Evolution 191:171–181.

Szymanska, H., Spalik, K. 1993. Typification of names enumerated in Pringsheim monograph of *Coleochaete* (Charophyceae). Algological Studies 70:29–37.

Taft, C. E. 1964. The occurrence of *Monostroma* and *Enteromorpha* in Ohio. Ohio Journal of Science 64:272–274.

Thompson, R. H. 1972a. Algae from the hair of the sloth *Bradypus*. Journal of Phycology Supplement 8:2.

Thompson, R. H. 1972b. On the genus *Chaetonemopsis* Gauthier-Lievre. Journal of Phycology Supplement 8:9.

Thompson, R. H., Wujek, D. E. 1996. *Printzia* gen. nov. (Trentepohliaceae), including a description of a new species. Journal of Phycology 28:232–237.

Thompson, R. H., Wujek, D. E. 1997. Trentepohliales: *Cephaleuros*, *Phycopeltis*, and *Stromatochroon*: Morphology, taxonomy, and ecology. Science Publishers, Enfield, New Hampshire, 149 pp.

Tiffany, L. H. 1930. The Oedogoniaceae: A Monograph—Including All the Known Species of the Genera Bulbochaete, Oedocladium and Oedogonium. Ohio State University, Ohio, 253 pp.

Tiffany, L. H. 1937. Oedogoniales: Oedogoniaceae. North America Flora 11:1–85.

Tiffany, L. H. 1944. The Oedogoniales of Florida. American of Midland Naturalist 32:98–136.

Tiffany, L. H., Britton, M. E. 1952. The Algae of Illinois. University of Chicago Press, Chicago,

Tupa, D. D. 1974. An investigation of certain chaetophoralean algae. Beih. Nova Hedwigia 46:1–155.

Vinyard, W. C. 1955. Epizoophytic algae from mollusks, turtles, and fish in Oklahoma. Proceedings of Oklahoma Academy Sciences 34:63–65.

Wehr, J. D., Stein, J. R. 1985. Studies on the autecology and biogeography of the freshwater brown alga *Heribaudiella fluviatilis* (Aresch.) Sved. Journal of Phycology 21:81–93.

Whitford, L. A. 1969. The current effect and growth of freshwater algae. Transaction of American Microscopical Society 79:302–309.

Whitford, L. A., Schumacher, G. J. 1969. A Manual of the Freshwater Algae in North Carolina. Technical Bulletin North Carolina Agricultural Research Station 188:1–313.

Whitton, B. A. 1970. Biology of *Cladophora* in fresh waters. Water Research 4:457–476.

Wood, R. D. 1948. A review of the genus *Nitella* (Characeae) of North America. Farlowia 3:331–398.

Wood, R. D. 1967. Charophytes of North America: A guide to the species of Charophyta of North America, Central America and the West Indies. University of Rhode Island Bookstore (Stella's Printing), Kingston, Rhode Island. 72 p.

Wood, R. D., Imahori, K. 1964. A Revision of the Characeae: Volume II, Iconograph of the Characeae. Verlag Von J. Cramer, Weinheim, Germany. 395 icones with index.

Wood, R. D., Imahori, K. 1965. A Revision of the Characeae: Volume I, Monograph of the Characeae. Verlag Von J. Cramer, Weinheim, Germany. 904 p.

Wujek, D. 1971. Light and electron microscope observations on the pyrenoid of the green alga *Leptosira*. Michigan Academy of Sciences 3:59–62.

Yung, Y. K., Sawa, T., Stokes, P. M. 1986. Oedogoniaceae (Chlorophyta) of southern Ontario, Canada. I. *Oedogonium* of Chub Lake. Nova Hedwigia, Beiheft 43:357–366.

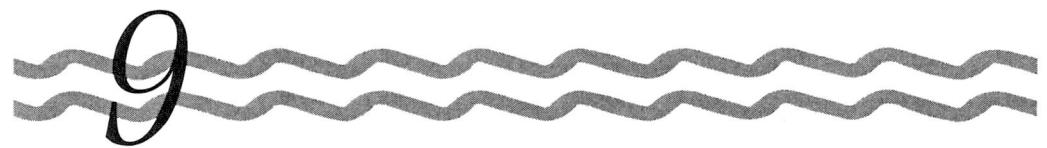

CONJUGATING GREEN ALGAE AND DESMIDS

Joseph F. Gerrath

Department of Botany
University of Guelph
Guelph, Ontario, Canada N1G 2W1

I. Introduction
II. Diversity and Morphology
   A. Diversity
   B. Morphology
   C. Reproduction
III. Ecology and Distribution
IV. Collection and Preparation for Identification
V. Key and Descriptions of Genera
   A. Key
   B. Descriptions of Genera of Zygnematales
   C. Descriptions of Genera of Desmidiales
VI. Guide to Literature for Species Identification
Literature Cited

I. INTRODUCTION

The conjugating green algae are a group of about 4000 species that are characterized by the combination of two features: (1) the absence of any flagellated cells in the life history and (2) the sexual process, called conjugation, which involves the fusion of amoeboid gametes. (Hoshaw *et al.*, 1990). Recent revisions of the green algae based largely on the structure of flagellated cells and details of nuclear and cell division, have created problems in deciding the proper taxonomic position of the conjugating algae at the division or class level. Some investigators (Round, 1963, 1971; van den Hoek *et al.*, 1995) consider the absence of flagellated cells to be so fundamental that the group should be considered a separate division (Zygnemaphyta, Conjugatophyta, or Gamophyta), or at least a separate class within the Division Chlorophyta (as Zygnemaphyceae, Zygnematophyceae, or Conjugatophyceae). Other researchers point to the similarity between charophytes (e.g., charophytes and *Coleochaete*) and conjugating green algae with regard to the processes of nuclear and cell divison, and they combine these two groups (and several other minor groups) within one class, the Charophyceae (e.g, Lee, 1999).

There is also no agreement on how to subdivide the conjugating green algae into orders and families. The group that comprises common and widespread green algae, such as *Spirogyra* and the desmids, was long treated as a single order Zygnematales of the green algae (e.g., Smith, 1950). Some authorities (e.g., Lee, 1999) maintain Smith's treatment, that is, one order—Zygnematales (or sometimes Zygnemales). In the last three decades, however, there has been a trend to classify the group into two orders—Zygnematales and Desmidiales (Růžička, 1977; Brook, 1981; van den Hoek *et al.*, 1995). In the latter scheme, which is used in this chapter, the order Zygnematales contains about 1000 species in which the cells have a one-piece cell wall, whereas the approximately 3000 species in the order Desmidiales have cell walls made up of two or more pieces. In the past, and even in a more recent phycological text (Lee, 1999), unicellular members of the Zygnematales were called saccoderm desmids and treated as a separate family, the Mesotaeniaceae (Prescott *et al.*, 1972), whereas the filamentous members were put into the Zygnemataceae. However, molecular studies have shown that unicellular and filamentous genera with similar chloroplast morphology are closely related (McCourt *et al.*, 1995; Park *et al.*,

1996). Thus it is more logical to combine these families into one—the Zygnemataceae, which contains both unicellular and filamentous algae.

II. DIVERSITY AND MORPHOLOGY

A. Diversity

The conjugating green algae have a relatively high diversity compared with other groups of green algae that occur in North America. The North American flora of the order Zygnematales comprises 15 genera and 296 species. The most diverse genus is *Spirogyra*, which has 145 of its 386 species occurring in North America. Other diverse genera are *Mougeotia*, with 117 species, 53 of which occur in North America, and *Zygnema*, with 120 species, 38 of which are reported in North America.

The North American flora of the order Desmidiales comprises 31 genera and approximately 950 species. The most diverse genera are *Cosmarium*, with 420 of its more than 1000 species occurring in North America, and *Staurastrum*, with 320 of its approximately 800 species recognized from North America.

B. Morphology (Figs. 1–93)

The conjugating green algae typically either are unicellular or have cells united in uniseriate (unbranched) filaments. In some unicellular genera (e.g., *Mesotaenium*), cells become aggregated within a common mucilage, thus forming irregular colonies (Boney, 1980, 1982). A few desmid genera (*Oocardium*, *Cosmocladium*, and *Heimansia*) have unusual morphology. *Oocardium* (Figs. 28 and 29) has cells positioned at the ends of branched gelatinous stalks within calcareous tubes attached to substrata (Smith, 1950). *Cosmocladium* (Fig. 30 and *Heimansia* (Fig. 91) are sometimes referred to as "colonial" genera, but actually form branched filaments in which the cells are widely separated and held together either by fibrillar strands secreted by adjacent cells (*Cosmocladium*; Gerrath, 1970) or by primary cell wall remnants (*Heimansia*; Coesel, 1993).

The order Zygnematales is distinguished from the order Desmidiales, because it lacks segmentation of the cell wall into two or more pieces and the specialized cell wall pores that characterize the latter order. Filamentous Zygnematales have cylindrical cells. The appressed end walls of adjacent cells may be flat (Fig. 1) or may be folded (replicate, Fig. 14). Cells of the unicellular Zygnematales are also circular in cross section, but often taper toward narrowed apices. In all members of the Zygnematales the cells have a single nucleus that may be in a central or parietal position, and usually is equidistant between the ends of the cell. If it is central, the nucleus may occur within a broad cytoplasmic bridge or be kept in position by narrow cytoplasmic strands that extend inward from the peripheral cytoplasm. Depending on the genus and species, there may be one to several chloroplasts in each cell. With respect to morphology and position within the cell, there are three main types of chloroplast found in the Zygnematales. One or more parietal, ribbon-like chloroplasts occur in several genera (e.g., *Spirogyra*, *Spirotaenia*, and *Sirogonium*; Smith, 1950) and often are disposed in a helix (Figs. 2, 13, and 52). Axial platelike or ribbon-like chloroplasts occur in *Mougeotia* and other genera (Smith, 1950). Platelike chloroplasts have been shown to orient themselves toward or away from light sources (Haupt and Schönbohm, 1970; Haupt, 1972; Schöbohm, 1972). Pyrenoids are usually present in both types of ribbon-like chloroplasts. The third kind of chloroplast is axial and appears stellate in cross section (e.g., *Zygnema* and *Netrium*; Smith, 1950). Usually two chloroplasts occur in each cell and the nucleus is situated in a cytoplasmic bridge between the chloroplasts. One or more pyrenoids are present in the axial core of the chloroplast. Cells and filaments of the Zygnematales often secrete narrow or broad mucilaginous envelopes, which make field collections feel slimy to the touch.

Most of the Desmidiales have a median constriction that defines the narrow isthmus that joins the two halves (semicells) of the cell (Fig. 22). The name "desmid" (based on the Greek *desmos*, meaning bond) comes from the fact that early researchers thought that they were made up of two cells joined together (Brook, 1981). The semicells are each covered by a separate piece of cell wall pierced by many cylindrical pores through which a narrow or broad sheath may be secreted. The semicell walls overlap at the narrowed isthmus region. The two semicells that compose the desmid cell usually have almost identical morphology (often a "mirror-image" relationship), even in those few genera that lack a median constriction. Because of the peculiar way in which daughter cells are formed after cell division, the two cell wall segments and the semicell morphologies that they define are of different ages. Each daughter cell receives from the parent cell one intact semicell with its wall segment and must regenerate a new semicell of similar morphology that has a completely new wall segment on it (Brook, 1981). This process, termed semicell morphogenesis, involves the gradual expansion of the new semicell to a size and form more or less identical to that of the parental semicell (Figs. 22–26). During this expansion

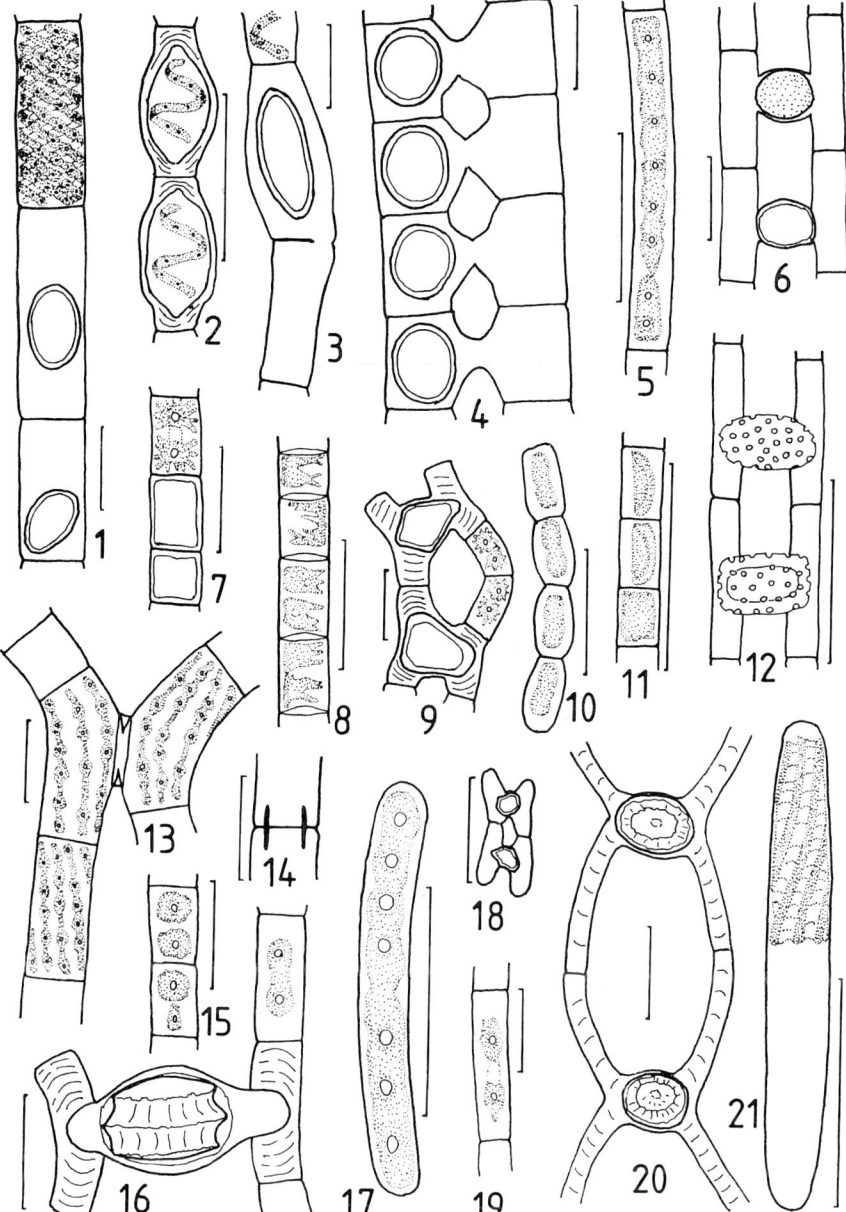

FIGURE 1 *Spirogyra wrightiana*. Vegetative cell and aplanospores. Redrawn from Transeau (1951). FIGURE 2 *Spirogyra juergensii*. Akinetes. Redrawn from Transeau (1951). FIGURE 3 *Spirogyra* sp. Lateral conjugation with zygospore in one gametangium. FIGURE 4 *Spirogryra* sp. Scalariform conjugation with zygospores within gametangia. FIGURE 5 *Mougeotia*. Vegetative cell of filament. FIGURE 6 *Zygnema conspicuum*. Scalariform conjugation with zygospores in conjugation tube. Redrawn from Transeau (1951). FIGURE 7 *Zygnema frigidum*. Vegetative cell and akinetes. Redrawn from Transeau (1951). FIGURE 8 *Entransia fimbriata*. Vegetative cells. Redrawn from Transeau (1951). FIGURE 9 *Zygnemopsis decussata*. Vegetative cells and conjugation. Redrawn from Transeau (1951). FIGURE 10 *Ancylonema nordenskioeldii*. Vegetative cells. Redrawn from Kol (1942). FIGURE 11 *Mougeotiopsis calospora*. Vegetative cells. Redrawn from Transeau (1951). FIGURE 12 *Mougeotiopsis calospora*. Zygospores. Redrawn from Transeau (1951). FIGURE 13 *Sirogonium illinoiense*. Early stage of conjugation. Redrawn from Transeau (1951). FIGURE 14 Replicate end walls of cells of *Spirogyra*. Original. FIGURE 15 *Pleurodiscus borinquinae*. Vegetative cells. Redrawn from Transeau (1951). FIGURE 16 *Debarya smithii*. Vegetative cell and zygospore. Redrawn from Transeau (1951). FIGURE 17 *Roya obtusa*. FIGURE 18 *Ancylonema nordenskioeldii*. Conjugation. Redrawn from Kol (1942). FIGURE 19 *Zygogonium ericetorum*. Vegetative cells. Redrawn from Transeau (1951). FIGURE 20 *Debarya glyptosperma*. Conjugation. Redrawn from Transeau (1951). FIGURE 21 *Polytaenia trabeculata*, showing detail of one chloroplast. Scales in Figs. 1, 2, 5, and 11–13, 100 µm; in Figs. 3, 4, 6–10, and 14–21, 50 µm.

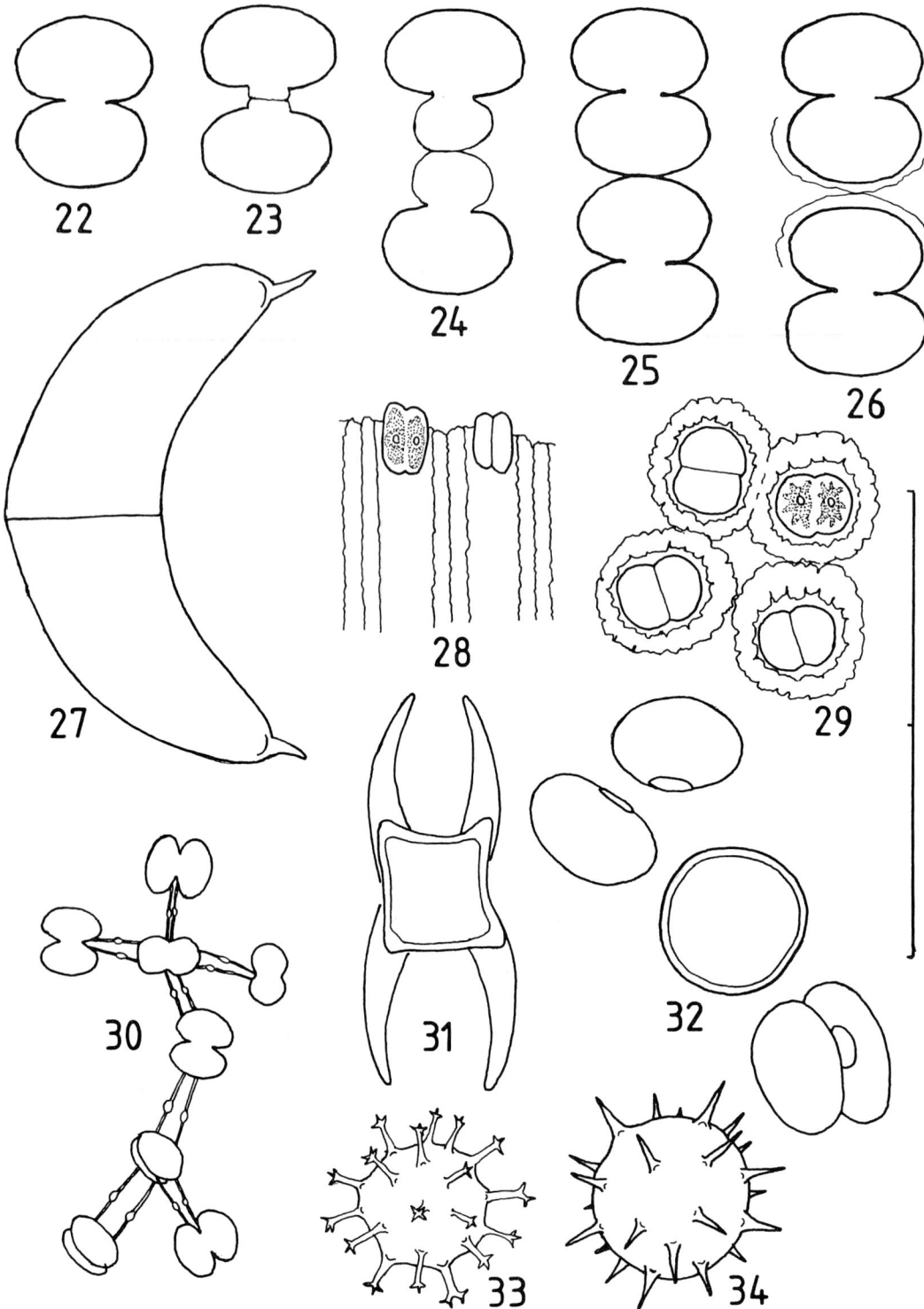

FIGURES 22–26 Series of drawings that illustrate semicell morphogenesis in a desmid (*Cosmarium* sp.). Fig. 22, vegetative cell before cell division; Fig. 23, cell divided; growth of new semicells of daughter cells is beginning; Fig. 24, new semicells about half of mature size; Fig. 25, new semicells fully formed; Fig. 26, casting off of primary wall from new semicells. *FIGURE 27 Spinoclosterium cuspidatum. FIGURES 28 AND 29 Oocardium stratum.* Longitudinal and cross sections of calcareous tubes with vegetative cells. Redrawn from Fritsch (1935). *FIGURE 30 Cosmocladium saxonicum.* Small colony. *FIGURE 31 Closterium acutum.* Quadrate zygospore with attached empty gametangial semicells. *FIGURE 32 Cosmarium* sp. Spherical zygospore and empty gametangial semicells. *FIGURE 33* Desmid zygospore with furcate spines. *FIGURE 34* Desmid zygospore with simple spines. Scale: 100 µm.

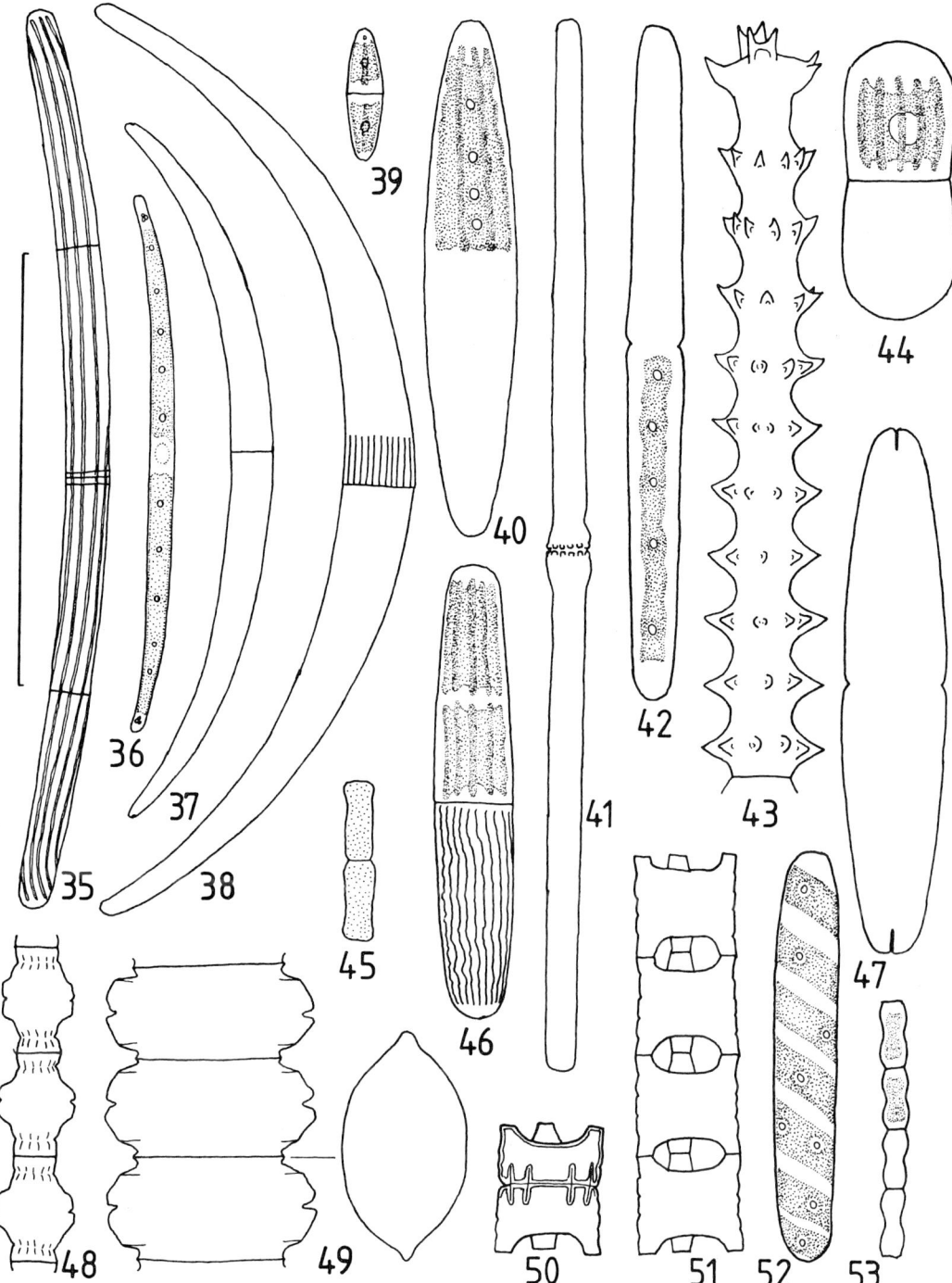

FIGURE 35 *Closterium angustatum* with girdle and connecting bands and longitudinal costae on cell wall. FIGURE 36 *Closterium gracile* with axial chloroplasts and terminal gypsum granules. FIGURE 37 *Closterium dianae*. Markedly curved, smooth-walled species. FIGURE 38 *Closterium archerianum*. Markedly curved species with striate cell wall (striae shown only near median suture). FIGURE 39 *Closterium navicula*. Tiny straight, smooth-walled species. FIGURE 40 *Closterium closterioides*. Large straight, smooth-walled species. FIGURE 41 *Docidium baculum*. FIGURE 42 *Haplotaenium minutum*. FIGURE 43 *Triploceras gracile*. Semicell. FIGURE 44 *Actinotaenium rufescens*. FIGURE 45 *Penium exiguum* with granulate cell wall. FIGURE 46 *Penium spirostriolatum* with striate cell wall. FIGURE 47 *Tetmemorus laevis*. FIGURE 48 *Bambusina brebissonii*. FIGURE 49 *Desmidium grevillii*. Lateral and apical views. FIGURE 50 *Desmidium baileyi*. Divided cell showing replicate walls of developing semicells. FIGURE 51 *Desmidium baileyi*. Filament. FIGURE 52 *Spirotaenia condensata*. FIGURE 53 *Groenbladia undulata*. Scale: 100 μm.

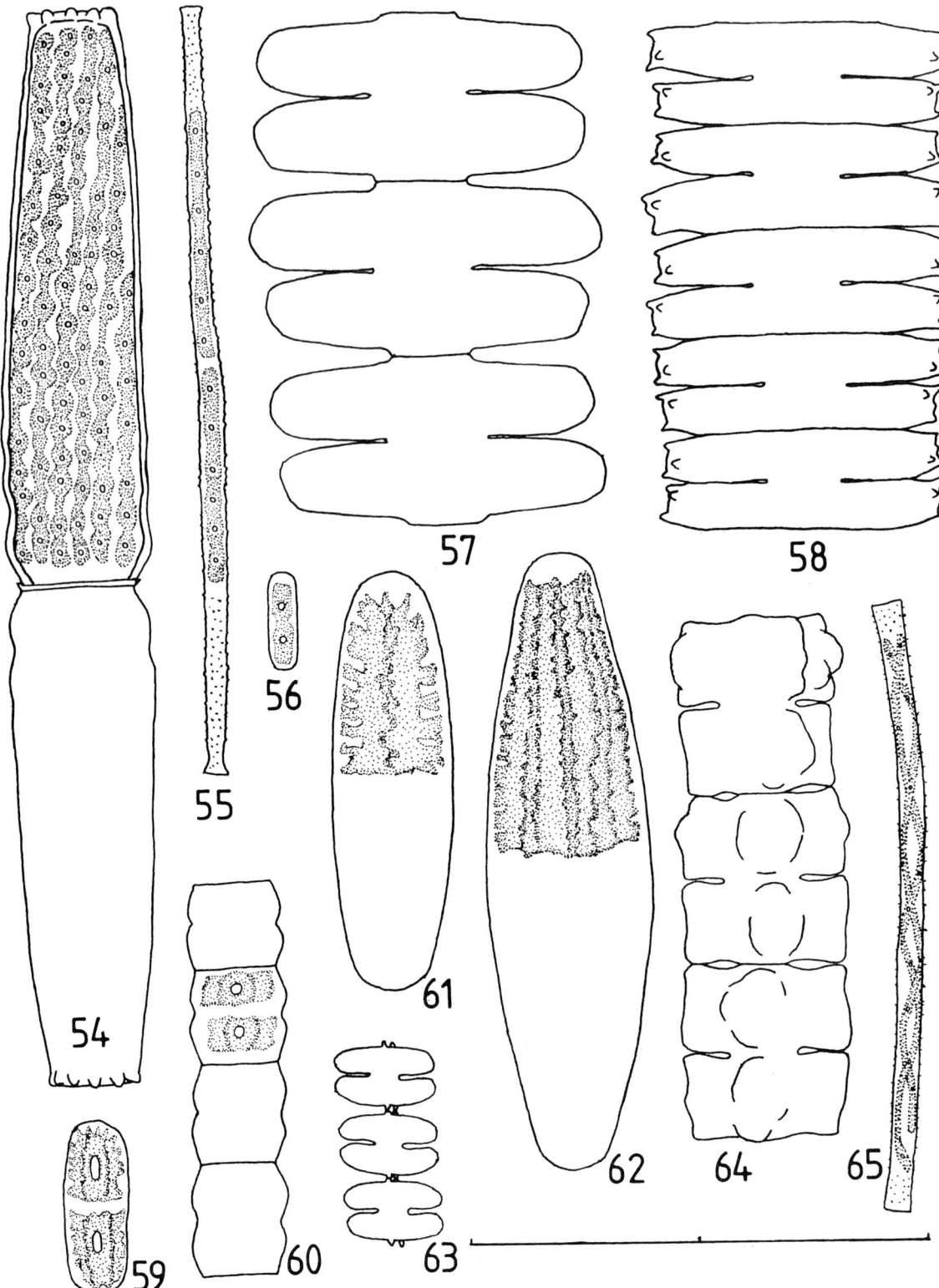

FIGURE 54 *Pleurotaenium coronatum.* FIGURE 55 *Gonatozygon brebissonii.* FIGURE 56 *Mesotaeniium endlicherianum.* FIGURE 57 *Spondylosium pulchrum.* FIGURE 58 *Spondylosium rectangulare.* FIGURE 59 *Cylindrocystis brebissonii.* FIGURE 60 *Hyalotheca dissiliens.* FIGURE 61 *Netrium oblongum.* FIGURE 62 *Netrium digitus.* FIGURE 63 *Sphaerozosma vertebratum* var. *latius.* FIGURE 64 *Phymatodocis nordstedtiana.* FIGURE 65 *Genicularia elegans.* Scale: 100 µm.

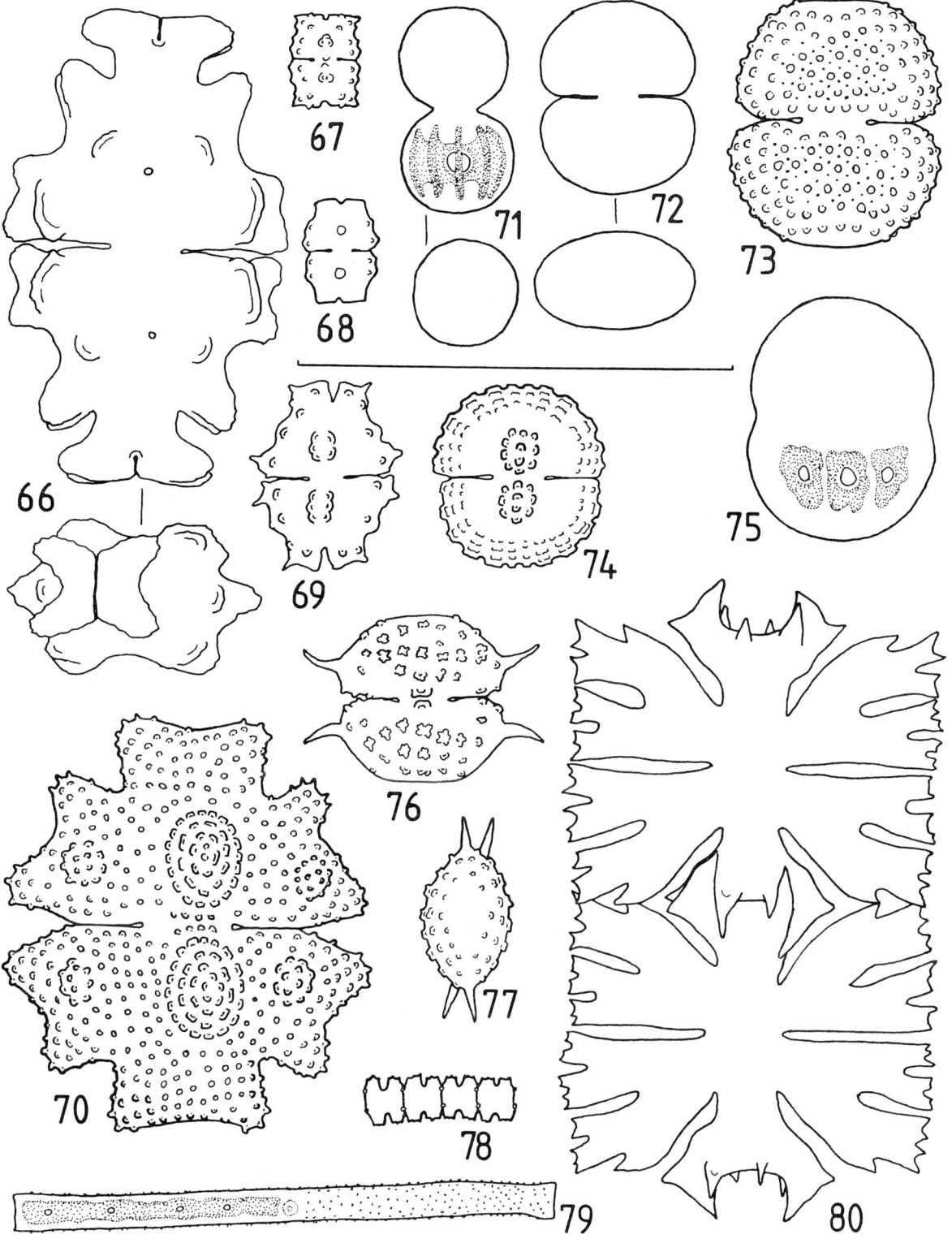

FIGURE 66 Euastrum humerosum. *FIGURE 67* Euastrum boldtii. *FIGURE 68* Euastrum pseudoboldtii. *FIGURE 69* Euastrum divaricatum. *FIGURE 70* Euastrum verrucosum. *FIGURE 71* Cosmarium contractum. Frontal and apical views. *FIGURE 72* Cosmarium montrealense. Frontal and apical views. *FIGURE 73* Cosmarium margaritatum. *FIGURE 74* Cosmarium quadrifarium f. hexastichum. *FIGURE 75* Cosmarium pseudoconnatum. *FIGURE 76* Spinocosmarium quadridens. Frontal view. *FIGURE 77* Spinocosmarium quadridens. Apical view. *FIGURE 78* Teilingia granulata. Filament. *FIGURE 79* Gonatozygon monotaenium. *FIGURE 80* Micrasterias foliacea. Two cells of filament. Scale: 100 µm.

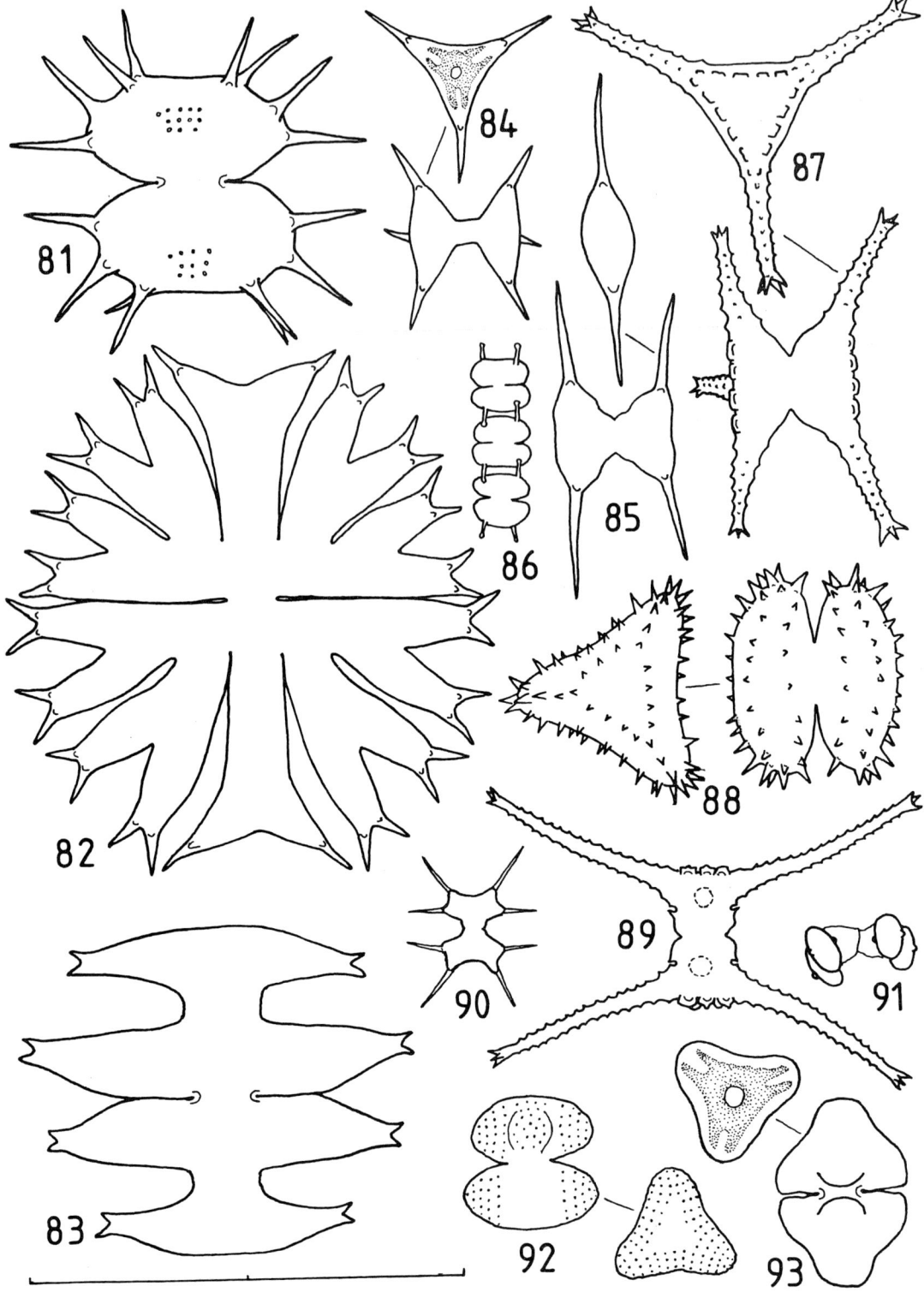

FIGURE 81 Xanthidium hastiferum. *FIGURE 82* Micrasterias johnsonii var. ranoides. *FIGURE 83* Micrasterias pinnatifida. *FIGURE 84* Staurodesmus cuspidatus. Frontal and apical views. Triradiate species. *FIGURE 85* Staurodesmus subtriangularis. Frontal and apical views. Biradiate species. *FIGURE 86* Onychonema filiforme. Filament. *FIGURE 87* Staurastrum anatinum f. curtum. Frontal and apical views. *FIGURE 88* Staurastrum claviferum. Frontal and apical views. *FIGURE 89* Staurastrum bioculatum; biradiate species. *FIGURE 90* Octacanthium octocorne. *FIGURE 91* Heimansia pusilla. Two-celled colony. *FIGURE 92* Staurastrum turgescens. Frontal and apical views. *FIGURE 93* Staurastrum trihedrale. Frontal and apical views. Scale: 100 μm.

the new semicell is covered by a thin, extensible primary wall (Figs. 23 and 24). When new semicell expansion is complete, the primary wall serves as a template for the deposition of a secondary wall, which is much thicker and contains plugs of an unknown material that traverse the wall. These plugs disintegrate to open up cylindrical pores through the wall. Except at the isthmus, where the primary and secondary wall layers are fused, these wall layers are separated by a narrow space. When semicell morphogenesis is completed, the primary wall is cast off, either as a single structure or as many small pieces (Fig. 26). This process is often aided by the secretion of gelatinous material through the newly functional pores in the secondary wall. Because the two semicells of a desmid cell are formed at different times, it is not surprising that there is often some difference in the morphology of the two. Cells with differing semicell morphologies are often useful in determining the range of morphological variability that can occur in a single taxon. A marked change in morphology of the new semicell compared with that of the parental semicell sometimes occurs, resulting in a cell in which the two semicells may, in extreme cases, have characteristics of different genera. There are a few desmid genera that always have semicells of markedly different morphology (Brook, 1981), but these do not occur in North America.

Some genera (e.g., *Penium* and *Closterium*) do not have a median constriction and several of the species in these genera have three or more segments in the cell wall; the middle pieces are being called girdle bands, connecting bands, or pseudo-girdle bands (Krieger, 1933; Fritsch, 1935; Růžička, 1977). Girdle bands are elongate wall segments interpolated between the terminal wall pieces (Fig. 35). There is a maximum of two girdle bands on a cell and their presence is an important character for keying some species (Růžička, 1977). Connecting bands are very short wall segments and there may be several present on a cell (Fig. 35). These bands arise as a result of the slight change in the position of the division wall at each cell division from the median suture line between old and young semicells to a position a short distance inside the younger semicell. Pseudo-girdle bands, formed by some *Closterium* and *Penium* species, are thought to be merely elongated transverse bands (Růžička, 1977). They are often of different lengths on a single cell (this never occurs with true girdle bands on fully mature cells) and there are often more than two on a cell. Sometimes it is not easy to distinguish girdle bands from pseudo-girdle bands (Růžička, 1977), leading to confusion in attempts to identify species.

The taxonomy of the Desmidiales is based almost entirely on the external morphology of the vegetative cells. The morphological diversity exhibited within the group is quite remarkable. Because of the peculiar structure and symmetry of the desmid cell, its form appears different when viewed from different directions. In what are termed biradiate desmid cells (Teiling, 1950; Figs. 72 and 85), there are three axes (or planes) of symmetry, all of which pass through the center of the isthmus. The first axis goes from side to side through the isthmus and the sinuses, which define this constricted area on both sides of the cell. Viewing the cell along this axis is gives the lateral (or side) view of the cell. The second (apical or vertical) axis goes from the middle of one semicell apex through the middle of the isthmus to the middle of the other semicell apex. Viewing the cell along this second axis gives the apical (or end) view of the cell (e.g., *Spinocosmarium*; Fig. 77). The third axis is perpendicular to the first two axes, running from the front to the back of the cell through the middle of the isthmus. Viewing the cell along this third axis gives the frontal (or face) view of the cell (e.g., *Spinocosmarium*; Fig. 76). A biradiate cell is compressed or flattened in the plane defined by the first two axes, and mirror-image symmetry usually exists with respect to the three planes defined by the three axes described. In frontal view (along the third axis), the cell appears broadest and the front of both semicells is fully visible (e.g., *Euastrum*; Fig. 70; *Micrasterias*; Fig. 82). In lateral view (along the first axis) it appears narrow and, again, both semicells are visible. In apical view (along the second axis), only the end of one semicell is visible. In a biradiate cell, the shape of the cell in apical view is usually elliptical (e.g., *Cosmarium*; Fig. 72) to elongate fusiform (e.g., *Micrasterias*), reflecting the degree of flattening of the cell (Teiling, 1950).

Symmetry patterns other than biradiate occur in several desmid genera (Teiling, 1950) and often are used to define these genera (e.g., *Staurodesmus*; Fig. 84; *Staurastrum*; Fig. 87). These patterns are clearest when cells are seen in apical view. If the apical view is triangular or exhibits three long, radiating processes 120° apart, the cell is said to be triradiate (Figs. 87, 88, 92, and 93). Similarly, if the apical view is four-angled, the cell is quadriradiate (usually written four-radiate). Radiation patterns up to 12-radiate have been described. Finally, there are some desmid genera, such as *Penium* and *Pleurotaenium* (Figs. 46 and 54), that have cylindrical cells (a circular outline in apical view): these cells are usually referred to as omniradiate. A more complete discussion of desmid radiation patterns can be found in Teiling (1950).

Desmids usually have at least two chloroplasts, one in each semicell, but there are many species that have several chloroplasts in a semicell. Chloroplasts similar to those that occur in the order Zygnematales are also found in the Desmidiales (Teiling, 1952): parietal ribbons (e.g., *Pleurotaenium*; Fig. 54), axile ribbons (e.g., *Haplotaenium*; Fig. 42), and axial chloroplasts with stellate cross section (e.g., *Penium*, *Closterium*, and *Triploceras*; Fig. 46). However, most desmids have chloroplast forms that are difficult to categorize (Teiling, 1952). Often the chloroplast conforms to the shape of the semicell; it has narrow lobes that fit into hollow processes on the semicell or angular form if the semicell itself is angular in apical view (Fig. 84). One or more pyrenoids usually are present in the chloroplasts of desmids. Further information about desmid chloroplasts, including a discussion of possible evolutionary patterns, is presented in Teiling (1952). There is a single nucleus in each cell and it is located in the isthmus of constricted cells or centrally positioned in unconstricted cells.

C. Reproduction

Sexual reproductive morphology is used extensively in identification of genera and species in the Zygnematales, but rarely is used for the Desmidiales. The sexual process is called conjugation and there are several variations of the process that contribute to the diversity of this group of algae (Hoshaw et al., 1990). Conjugation is perhaps most familiar in *Spirogyra*, which has been used for many decades in teaching laboratories to demonstrate this kind of sexual reproduction. In *Spirogyra* conjugation occurs in some species between gametangia in separate filaments (scalariform conjugation; Fig. 4) and in other species between gametangia that are adjacent cells of the same filament (lateral conjugation; Fig. 3). There are a few species that use both methods (Kadlubowska, 1972). In scalariform and lateral conjugation, both gametangia produce outgrowths that fuse to provide a continuous pathway for gamete transfer, the conjugation tube (Fritsch, 1935; Hoshaw et al., 1990). In some species of *Spirogyra*, the gamete in one gametangium becomes amoeboid and moves through the conjugation tube to fuse with the other gamete, so that the zygote is completely within the latter gametangium (Figs. 3 and 4). The zygote subsequently produces a thick wall made up of at least three layers, and becomes a zygospore that can remain dormant and resistant to adverse environmental conditions for a considerable time. In other species and genera, both gametes become amoeboid and fusion (and zygospore formation) occurs in the conjugation tube (Figs. 6, 12, and 16).

There are many species of conjugating green algae in which conjugation occurs without the formation of a conjugation tube (Fritsch, 1935). In some filamentous Zygnematales, such as the genus *Sirogonium*, conjugating filaments make direct contact and fusion of the gametangia occurs at the point of contact (Fig. 13). Most desmids also lack a conjugation tube (Fritsch, 1935). The gametangial cells come close together and secrete a common mucilage around themselves. The gametangia then split open at the isthmus and gametes from both gametangia move together to fuse at a position midway between the empty gametangial walls (Fig. 32). The empty gametangial semicells often remain in close proximity to the mature zygospore (sometimes even attached to it; Fig. 31).

Zygospore shape and wall structure are important features used in identifying species in the Zygnematales, but rarely are used in the Desmidiales (Fritsch, 1935). Zygospores of Zygnematales may be spherical (Fig. 6), ellipsoid (Fig. 3), rectangular (Fig. 9), or lenticular (Fig. 16), and may be colored yellow, brown, purple, blue, or black. Zygospores of Desmidiales have a similar range of forms, but some genera produce zygospores that are covered by radiating spines (Figs. 33 and 34) or by mamillate protuberances. The zygospore wall of conjugating green algae has three major layers (exospore, mesospore, and endospore) and it is the structure of the mesospore surface that must be examined to make species determinations in several genera of the Zygnematales. (Transeau, 1926; Fritsch, 1935; Hoshaw et al., 1990). The mesospore surface may be smooth (Fig. 3) or may be ornamented or sculptured in various ways (pits, ridges, etc.; Fig. 4 and 16). The infrequent occurrence of zygospores in field collections is a major problem for researchers attempting to identify species, especially those of the filamentous Zygnematales.

The formation of asexual spores is known to occur in the conjugating algae, but it is more commonly observed in the filamentous Zygnematales than in other taxa. There are even some species in which this is the preferred method of reproduction (Transeau, 1951). Asexual spores may be placed in one of three categories, akinetes, aplanospores, and parthenospores, depending on the details of their development. Akinetes develop from vegetative cells by the deposition of thick layers of cell wall material that is attached to the inner surface of the vegetative cell wall (Figs. 2 and 7). During the development of aplanospores, the protoplast shrinks and a complete new wall, separated from the vegetative cell wall, is formed around it (Fig. 1; Transeau, 1951). The morphology of akinetes and aplanospores usually differs from that of zygospores of the same taxon. Parthenospores form in association with sexual reproduction and usually are produced in cells that have failed to pair up with compatible sexual partners. The process of formation of parthenospores is identical to that of aplanospores, resulting in a spore that often resembles a zygospore but is smaller (Transeau, 1951).

III. ECOLOGY AND DISTRIBUTION

A broad range of habitat preferences is exhibited by various members of the conjugating green algae. In North America, all genera are found in freshwater or subaerial habitats (Hoshaw & McCourt, 1988). Most species of the filamentous Zygnematales belong to the three largest genera, *Spirogyra*, *Zygnema*, and *Mougeotia*. In a series of collecting trips from 1982 to 1984 along U.S. highways in the eastern, midwest, mountain, and south-coast states (McCourt *et al.*, 1986), over 95% of the strains collected belonged to these three genera (only two other genera, *Sirogonium* and *Zygogonium*, were collected). These cosmopolitan genera occur most often in stagnant water of roadside ditches, ponds, and lakes, where they may grow alone or mingled with other algae in benthic or floating masses. Species that form floating mats show rapid growth during early spring and summer, and the mats disappear during late summer. Some species in these genera occur in flowing water, where the filaments may attach to a substratum by means of rhizoid-like outgrowths from cells near the base of the filament (Fritsch, 1935; Israelson, 1949).

Other filamentous genera of the Zygnematales, except *Zygogonium*, are rarely encountered, but occur in the same types of habitats that support the growth of *Spirogyra*, *Mougeotia*, and *Zygnema* (Transeau, 1951). The infrequent occurrence of the distinctive zygospores and associated sexual morphology of several of these genera contributes to their rarity, because they have vegetative morphology similar to that of the common genera and most often are recorded as belonging to those genera. *Pleurodiscus* is known only from Puerto Rico, *Entransia* only from Nova Scotia, and *Mougeotiopsis* thus far only from Michigan, Wisconsin (Transeau, 1951), North Carolina (Whitford and Schumacher, 1984), and British Columbia (Stein, 1975). *Zygogonium* differs from the other genera in showing a distinct preference for acidic, aquatic or subaerial habitats (pH 2.4–4.5; Smith, 1950; Hargreaves *et al.*, McCourt *et al.*, 1986; Hoshaw and McCourt, 1988). It is usually collected from moist soil, rocks, or peat (Hosiaisluoma, 1975; Yung *et al.*, 1986), but is known to form floating mats in ponds near thermal springs in Yellowstone Park and to form benthic masses in dilute lakes impacted by acid precipitation. *Zygogonium* also occurs in about 20% of samples taken from English streams influenced by acid mine drainage, causing extreme acidity (pH less than 3.0; Hargreaves *et al.*, 1975).

With the exception of *Ancylonema* and one species of *Mesotaenium*, which are algae that grow on glacial ice (Kol, 1942, 1944, 1964), unicellular Zygnematales generally are found in aquatic and subaerial habitats that are at least moderately acidic (pH 4.1–6.5; Brook, 1981). *Cylindrocystis*, *Mesotaenium*, and *Roya* often are collected from moist soil or rock, although they can also occur in ponds (Smith, 1950; Brook, 1981). In subaerial situations these genera often form small or large masses of cells embedded in a copious matrix of pectic material. This matrix may aid water retention during dry periods, permitting vegetative cells to survive desiccation (Boney 1980, 1981, 1982). *Mesotaenium* also may be found among the leaves of mosses in moist and aquatic environments, again as a mass of cells within a gelatinous matrix (Ohtani, 1986). *Netrium* and *Spirotaenia* are common in *Sphagnum* bogs or in ponds surrounded by marginal mats of *Sphagnum*. Both genera occur as unicells, but *Spirotaenia* also occurs as clusters of cells within a common mucilage (Prescott *et al.*, 1972).

In general, the Desmidiales are found in oligotrophic to mesotrophic freshwater lakes, ponds, and streams (low conductance and low calcium content). Pools in bogs and fens are also rich in desmids (Howell and South, 1981; Yung *et al.*, 1986). The greatest known diversity of desmids in North America occurs in the southeastern coastal plain, the Canadian Shield, and the Pacific coast. Within individual lakes or ponds, the greatest diversity and numbers of desmids occur along the sediment surface in shallow areas or among aquatic vascular plants such as *Utricularia* (Hoshaw *et al.*, 1990; Gerrath, 1993). The pH of such areas is usually in the range of 4.0–7.0. More acidophilic desmids inhabit *Sphagnum* bogs or lakes that have a marginal area of bog. In North America, desmids are more common in bogs closer to the ocean than those further inland, and are more abundant and diverse in bogs that have an area of open water (Hosiaisluoma, 1975; Yung *et al.*, 1986). Collections in suitable aquatic habitats often have exceedingly diverse desmid assemblages. It is not unusual to identify from 150 to more than 200 different taxa from a single collection (Gerrath, unpublished data).

A relatively low number of desmid species are truly planktonic. These desmids usually are characterized by the presence of long processes (e.g., *Staurastrum*; Fig. 87), long spines (e.g., *Staurodesmus* and *Xanthidium*; Figs. 81 and 85), or very long cells (e.g., *Triploceras*; Fig. 43). Because of their relatively slow growth rates (cell divisions every 2–5 days), planktonic desmids usually reach maximum density in late summer or autumn (Happey-Wood, 1988). The presence of certain desmids, even in low numbers, is considered to be a good indicator of mildly acidic, oligotrophic conditions (Coesel, 1983). Examples of such indicators include *Staurodesmus crassus*, *S. cuspidatus*, *S. sellatus*, *S. extensus* var. *joshuae*, *S. triangularis* var. *limneticus*, and *Staurastrum longipes* (Rosén, 1981; Willén, 1992). Fewer desmids occur as phytoplankters in eutrophic lakes, usually belonging to the genera *Closterium*, *Cosmarium*, and *Staurastrum*, and these may be useful as indicator species for eutrophic conditions (Coesel, 1983). An example is *Closterium aciculare*, which usually occurs sparsely among cyanobacterial blooms in eutrophic lakes with pH 6.7–8.5 (Růžička, 1977). This desmid is unusual in because it lacks nitrate reductase activity (Coesel, 1991) and requires the presence of ammonia in eutrophic waters for continued growth. The relative importance of planktonic desmids compared to other planktonic algal groups has been used to define quotients that indicate the trophic status of lakes (Brook, 1981). Nygaard's compound phytoplankton quotient (the ratio of the sum of the number of species of Cyanobacteria, Chlorococcales, centric diatoms, and euglenoids to the number of desmid species) is one example of such a quotient (Brook, 1981). Obviously a diverse assemblage of planktonic desmids will produce a low quotient value of 2 or less, an indication of oligotrophic water.

The desmid genus *Oocardium* (Figs. 28 and 29) is unusual because it forms tiny calcareous incrustations attached to twigs or stones in swiftly flowing streams or near waterfalls that have water of high calcium content (Fritsch, 1935; Brook, 1981; Pentecost, 1991). The cells sit at the ends of branched gelatinous strands within calcareous tubes. This genus rarely is collected and possibly is overlooked in suitable sites, but is known to occur in California and in the northeastern United States.

Seasonality in abundance and reproduction is a feature of the ecology of many conjugating green algae (Hoshaw, 1968; Gerrath, 1993). Although there are several species that may be considered to be perennials, many conjugating algae are collected only during certain periods of the year or are most abundant during a certain season. Transeau (1916) concluded that most of the filamentous Zygnematales he studied were either spring or summer annuals. These algae also demonstrated peaks of sexual reproduction during or toward the end of these seasons. Desmids also show peaks in abundance in lakes and streams. Depending on the species, these peaks may occur during the spring, summer, autumn, and, in rare cases, winter (Gerrath, 1993). Some planktonic desmids exhibit two peaks in abundance during the year (usually spring and autumn). Seasonal peaks in abundance of desmids also were found in a Rhode Island stream (Burkholder and Sheath, 1984). The genus *Hyalotheca* was dominant during the spring and other desmids were more abundant during the summer and autumn.

Sexual reproduction rarely is observed in natural populations of conjugating algae, both for Zygnematales and for Desmidiales. It is more commonly seen in temporary water bodies than in permanent ones (Hoshaw and McCourt, 1988; Gerrath, 1993). Chemical substances are known to be involved in the induction of conjugation tubes in *Spirogyra* (Grote, 1977), and in the processes of cell aggregation and pairing, gametangial formation, and gamete release in *Closterium* (reviewed in Gerrath, 1993). Grote (1977) found that a pH range of 7.0–8.0 was crucial for induction of zygote formation in *Spirogyra majuscula*, but pH requirement has not been studied in other genera or species. Tiftickjian and Rayburn (1986) noted that in *Mesotaenium kramstai* both calcium and magnesium (>1.0 mM) were essential for the pairing and conjugation of gametangia, and they suggested that if this were a widespread requirement in conjugating algae, it could explain the rarity of desmid zygotes in soft-water habitats.

The Zygnematales and Desmidiales serve as hosts for many parasitic fungi, especially chytrids. There is some indication that host specificity exists, although there are few experimental studies to support this conclusion. *Rhizophydium sphaerocarpum* preferentially parasitizes certain strains of *Spirogyra* (Barr and Hickman, 1967). It is interesting to note that this chytrid also infects the snow alga *Ancylonema nordenskioeldii*, another member of the Zygnematales (Kol, 1942). Host range studies on the desmid genera *Closterium* and *Micrasterias* indicate that host specificity of certain chytrids could be a useful taxonomic criterion to delimit desmid species (Brook, 1981). Chytrid parasitism also may have a role in the control of desmid populations. Large populations of planktonic desmids occasionally are reported and such populations often are heavily parasitized by chytrids, resulting in the death of a significant proportion of the population (Brook, 1981).

The biogeography of desmids was reviewed by Coesel (1996). With respect to the North American desmid flora, he noted that it is made up of several biogeographical elements. Many North American species are found also in temperate Eurasia and a large number can be considered to be arctic–alpine in distribution, although many of the latter group can be found in temperate regions. Many species in both of these groups seem to be restricted to the northern hemisphere. Several species or genera (e.g., *Phymatodocis*, *Spino-*

closterium, and *Triploceras*) that are considered to be tropical desmids have spread as far north as eastern Canada and Alaska. This northward migration is thought to be related to the migration routes of waterfowl. Desmid species endemic to North America, according to our present state of knowledge, seem to be numerous. Examples of endemics include *Cosmarium eloiseanum*, *Micrasterias muricata*, and *Spinocosmarium quadridens*.

IV. COLLECTION AND PREPARATION FOR IDENTIFICATION

Any collection method suitable for sampling freshwater algae can be used for the conjugating green algae. Planktonic forms can be collected with a net of suitable mesh size (10–25 μm). Floating mats of filamentous algae can be collected by hand, and algae among aquatic vascular plants often can be obtained by squeezing a plant mass over a container to collect the material (including algae) that is exuded. Flocculant or mucilagenous material on plant stems and at the surface of sediments in lakes and ponds can be collected using a small laboratory pipette or a larger, plastic or glass, poultry baster. Samples should be kept cool if they are to be examined alive after collection and on return to the laboratory.

Samples of filamentous Zygnematales can be kept alive for several weeks in the hope that sexual stages necessary for species identification will develop. Sometimes placing collected material in a low-nitrogen medium helps to induce sexual reproduction. Specimens of Desmidiales can be examined live or can be fixed for later identification using formalin (3%), FAA (formalin/acetic acid/alcohol; Berlyn and Miksche, 1976), or any fixative used for electron microscopy.

For a description of suitable culture media for growing filamentous Zygnematales, consult Hoshaw (1968). Culture media for growing desmids are listed in Gerrath (1993). Because mature zygotes are required for identification of Zygnematales, induction of sexual reproduction can be attempted. In most cases, high light levels and long day length have been used for experimental induction of sexuality in culture. Another strategy that is often successful is to transfer cultured algae into a low-nitrogen medium and to combine this treatment with an increased concentration of carbon dioxide (Hoshaw and McCourt, 1988; Gerrath, 1993)

Whereas the taxonomy of desmids relies almost exclusively on the morphology of vegetative cells, the identification of desmids often requires cells to be examined from both frontal and apical views. Making temporary slides with cells mounted in glycerol or glycerine jelly is a means to facilitate shifting cells from one view to the other. Cells can be oriented for proper viewing by carefully moving the cover glass (glycerin jelly mounts must first be heated to liquefy the medium). Drawings of desmid cells that show details of semicell shape and ornamentation should be made to document the record and to make keying at a later date easier. The use of a camera lucida or prismatic drawing attachment is the best way to achieve accurate drawings. Light microscope photographs of desmids are generally less useful for identification, because the shallow focal plane does not permit adequate viewing of the complete ornamentation on many desmids. The use of scanning electron microscopy, although more complex and time-consuming, results in excellent photographs of external cell morphology (Couté and Tell, 1981).

V. KEY AND DESCRIPTIONS OF GENERA

A. Key

1a.	Cells united into long uniseriate filaments (rarely branching; if so, then branches usually only one to two cells)................................2
1b.	Cells solitary, aggregated within a gelatinous matrix or calcareous crust (rarely forming very short filaments; Desmidiales in part)................................23
2a.	Cell wall consists of one continuous piece, usually without any ornamentation; cell without median constriction (Zygnematales)................................3
2b.	Cell wall consists of two overlapping pieces of equal size; wall usually with pores, tubercles, spines, or striae; cell usually with median constriction (Desmidiales in part)................................13
3a.	Chloroplast a parietal, spiralling (or nearly straight) ribbon (1–16 per cell)................................4
3b.	Chloroplast of another form or if ribbon-like, not spiralling................................5
4a.	During conjugation, isogamous gametangia are united by a distinct conjugation tube; chloroplast spiralling more than one-half a complete turn (Figs. 1, 3, and 4)................................ *Spirogyra*

4b.	During conjugation, isogamous gametangia (sometimes anisogamous) contact one another (no conjugation tube formed); chloroplast spiral or nearly straight, not more than one-half a turn (Fig. 13)	*Sirogonium*
5a.	Chloroplast discoid or platelike	6
5b.	Chloroplast stellate or a nonspiralling ribbon	7
6a.	Two chloroplasts per cell, each with a pyrenoid (Fig. 15)	*Pleurodiscus*
6b.	One chloroplast per cell, without a pyrenoid (Figs. 11 and 12)	*Mougeotiopsis*
7a.	Chloroplast axial, stellate, or globular, with a central pyrenoid	8
7b.	Chloroplast axial or parietal, ribbon-like	10
8a.	Chloroplast stellate	9
8b.	Chloroplast a globular mass (Fig. 19)	*Zygogonium*
9a.	Gametangium filled with stratified gel as gametes contract and fuse (Fig. 9)	*Zygnemopsis*
9b.	Gametangium not filled with stratified gel (Fig. 8)	*Zygnema*
10a.	Chloroplast parietal, one to two per cell (Fig. 8)	*Entransia*
10b.	Chloroplast axial	11
11a.	Cell with two chloroplasts (Figs. 55 and 79)	*Gonatozygon*
11b.	Cell with a single chloroplast	12
12a.	Gametangium becomes filled with stratified gel as gametes contract and fuse (Figs. 16 and 20)	*Debarya*
12b.	Gametangium not filled with stratified gel (Fig. 5)	*Mougeotia*
13a.	Cells with two or four small apical tubercles or elongate apical processes	14
13b.	Cells without such structures	16
14a.	Each cell apex has four small tubercles (Fig. 78)	*Teilingia*
14b.	Each cell apex has two tubercles or processes	15
15a.	Tubercles/processes close together at middle of apex (Fig. 63)	*Sphaerozosma*
15b.	Tubercles/processes far apart (Fig. 86)	*Onychonema*
16a.	Cells elongate (length/width greater than 4), cylindrical (Fig. 54)	*Pleurotaenium*
16b.	Cells relatively short, cylindrical or flattened	17
17a.	Cells very flat, with deeply divided lateral lobes (Fig. 80)	*Micrasterias foliacea*
17b.	Cells not as above	18
18a.	Cells form a replicate septum during cell division and semicell morphogenesis (Fig. 50)	19
18b.	Cells never form a replicate septum	20
19a.	Cell wall with longitudinal striae near apex; apical view circular (Fig. 48)	*Bambusina*
19b.	Cell wall without longitudinal striae; apical view broadly elliptical, elongate, or three- to four-angled (Figs. 49 and 51)	*Desmidium*
20a.	Cells have a deep median constriction	21
20b.	Cells have a very shallow median constriction	22
21a.	Cells flattened (biradiate) or triradiate (apical view is three-angled; Figs. 57 and 58)	*Spondylosium*
21b.	Cells four- or five-angled in apical view (Fig. 64)	*Phymatodocis*
22a.	Chloroplast stellate in apical view (Fig. 60)	*Hyalotheca*
22b.	Chloroplast platelike or ribbon-like (Fig. 53)	*Groenbladia*
23a.	Cells at tips of branched gelatinous stalks within a calcified crust in calcium-rich streams (Figs. 28 and 29)	*Oocardium*
23b.	Cells solitary or aggregated in a different way	24

24a.	Cell wall made up of one continuous piece, without ornamentation or bearing granules or spines; cells without median constriction	25
24b.	Cell wall made up of two or more overlapping pieces; wall usually with pores and often with tubercles, spines, or striae; cells constricted or unconstricted	34
25a.	Chloroplast ribbonlike or platelike	26
25b.	Chloroplast with central axis and radiating lateral ridges	31
26a.	Chloroplast a parietal spiralling ribbon	27
26b.	Chloroplast an axial ribbon or plate	28
27a.	Cell wall with granules (Fig. 65)	*Genicularia*
27b.	Cell wall smooth (Fig. 52)	*Spirotaenia*
28a.	Cell wall with granules or spines (Fig. 55)	*Gonatozygon*
28b.	Cell wall smooth	29
29a.	Cells in short fragile filaments (up to about 16 cells; easiliy fragmenting; growing on snow or ice of glaciers (Fig. 10 and 18)	*Ancylonema*
29b.	Cells solitary or in gelatinous masses; not on snow or ice	30
30a.	Cells usually short and straight, chloroplast with one to two pyrenoids (Fig. 56)	*Mesotaenium*
30b.	Cells usually elongate and curved, chloroplast with several pyrenoids (Fig. 17)	*Roya*
31a.	Chloroplast with spiralling lateral ridges (Fig. 21)	*Tortitaenia*
31b.	Chloroplast ridges not spiralling	32
32a.	Cell slightly curved; chloroplast in cross section with two to four lobes (Fig. 17)	*Roya*
32b.	Cell straight; chloroplast with many lobes in cross-section	33
33a.	Outer margin of chloroplast lateral ridges with complex lobing (Figs. 61 and 62)	*Netrium*
33b.	Chloroplast ridges without complex lobing (Fig. 59)	*Cylindrocystis*
34a.	Cells usually more than five times longer than broad	35
34b.	Cells usually less than five times longer than broad	46
35a.	Cells with shallow to marked median constriction	36
35b.	Cells without median constriction	41
36a.	Semicell apex truncate or broadly rounded	37
36b.	Semicell apex incised, notched, or two- to four-lobed	40
37a.	Cell wall with small spines, granules, or striae (Figs. 45 and 46)	*Penium* in part
37b.	Wall without granules or striae; spines, if present, are in whorls along cell or at apex of cell	38
38a.	Semicell with a whorl of small tubercles at base (Fig. 41)	*Docidium*
38b.	Semicell without basal tubercles	39
39a.	Chloroplasts parietal, ribbon-like, several per semicell (Fig. 54)	*Pleurotaenium*
39b.	Chloroplasts axial, ribbon-like, one per semicell (Fig. 42)	*Haplotaenium*
40a.	Semicell apex two- to four-lobed, each lobe tipped with one to two small spines (Fig. 43)	*Triploceras*
40b.	Semicell apex notched or incised (Fig. 47)	*Tetmemorus*
41a.	Cells curved, at least near semicell apices; cell width narrowing toward apices; cell wall smooth or with longitudinal striae or costae	42
41b.	Cells straight, cylindric or narrowed toward apices; cell wall smooth, granulate, or with tiny spinules	43
42a.	Cell with thick spine at each apex (Fig. 27)	*Spinoclosterium*
42b.	Cell apices without spines (Figs. 35–38)	*Closterium*

43a.	Cells usually more than six times longer than broad; apices often broadened..	44
43b.	Cells more than three times longer than broad; apices rarely broadened..	45
44a.	Chloroplast axial, not spiralling (Fig. 79)..	*Gonatozygon*
44b.	Chloroplast parietal, spiralling (Fig. 65)..	*Genicularia*
45a.	Cell wall smooth (Figs. 39 and 40)..	*Closterium* in part
45b.	Cell wall with striae or granules (Figs. 45 and 46)..	*Penium* in part
46a.	Cells in colonies (short filaments) of 4–16 cells; cells separated from one another, but united by narrow strands..................	47
46b.	Cells solitary, except just after cell division..	48
47a.	Uniting strand produced by pores at one semicell base, overlapping midway between cells with strand from sister cell of previous cell division (Fig. 30)..	*Cosmocladium*
47b.	Uniting strand composed of primary walls of a pair of sister cells from previous division (Fig. 91)..........................	*Heimansia*
48a.	Cell outline in apical view circular (omniradiate)..	49
48b.	Cell outline in apical view oval, elliptical, fusiform, or angular..	50
49a.	Single chloroplasts in each semicell, axial, stellate in apical view, or several parietal ribbons (Fig. 44).........................	*Actinotaenium*
49b.	Two to four chloroplasts in each semicell, not stellate, occupying central part of the semicell (Figs. 71–75)............	*Cosmarium* in part
50a.	Cells biradiate (apical view oval, elliptical, or fusiform)..	51
50b.	Cells 3–12-radiate (apical view 3–12-angled)..	60
51a.	Semicell with long hollow processes (Fig. 89)..	*Staurastrum* in part
51b.	Semicell without long hollow processes	52
52a.	Semicell apex with median incision or notch..	53
52b.	Semicell apex without incision or notch..	55
53a.	Cell in front view with three distinct lobes (apical and two lateral) or with angular outline..	54
53b.	Cell in front view not as above (Fig. 47)..	*Tetmemorus*
54a.	Cell very flattened; semicell often with several deep narrow incisions on lateral lobes (Figs. 82 and 83).........................	*Micrasterias*
54b.	Cell not very flattened; semicell with angular outline or, if three-lobed, lateral lobes with shallow, open incisions (Figs. 66–70)..	*Euastrum*
55a.	Semicell with 2–16 spines (often in pairs) around margin in front view..	56
55b.	Semicell lacking spines (or with many spines not arranged in pairs; Figs. 71 and 72)...	*Cosmarium* in part
56a.	Semicell with 4–16 spines around margin..	57
56b.	Semicell with two spines around margin..	59
57a.	Semicell with four spines; cell wall without ornamenting verrucae; center of semicell face not incrassate (no thickened area of cell wall; Fig. 90)..	*Tetracanthium*
57b.	Semicell with 4–16 spines; cell wall with rows of verrucae or incrassate (thickened) central area.................................	58
58a.	Cell wall with several rows of verrucae or, if lacking verrucae, with one large round granule on each semicell near isthmus (Figs. 76 and 77)..	*Spinocosmarium* in part
58b.	Cell wall lacking verrucae or with verrucae only near center of semicell; large granule at isthmus seldom present (Fig. 81)	*Xanthidium*
59a.	Cell wall with several rows of verrucae or, if lacking verrucae, with one large round granule on each semicell near isthmus (Figs. 76 and 77)..	*Spinocosmarium* in part
59b.	Cell wall without verrucae (Fig. 85)..	*Staurodesmus* in part
60a.	Semicell with three or more long hollow processes (Fig. 87)..	*Staurastrum*
60b.	Semicell without hollow processes..	61
61a.	Semicell with many spines or verrucae, regularly or irregularly disposed (Fig. 88)..	*Staurastrum* in part

61b.	Semicell with smooth wall or with spines only at angles of semicells	62
62a.	Semicell angles each bearing only one spine (Fig. 84)	*Staurodesmus* in part
62b.	Semicell angles with more than one spine or lacking spines (Figs. 88, 92, and 93)	*Staurastrum* in part

B. Descriptions of Genera of Zygnematales

Ancylonema Berggren (Figs. 10 and 18)

Cells of this monotypic genus are solitary or loosely united into short filaments (up to 16 cells). Cells are cylindric, two to three times longer than wide, with rounded or truncate (in filament) ends, a smooth cell wall, and brownish purple or reddish violet cell contents. The chloroplast is a narrow, parietal band, containing one or two pyrenoids. A conjugation tube develops during sexual reproduction and the spherical zygospore develops within the tube (Fig. 18).

There is one species, *A. nordenskioeldii*, that is known only from permanent patches of snow and ice, Recorded from Greenland and other arctic islands, western North America, Switzerland, and Antarctica (Prescott *et al.*, 1972); common on glacial ice in the mountains of western North America (Kol, 1942, 1944, 1964).

Cylindrocystis Meneghini *ex* De Bary (Fig. 59)

Cells are usually solitary, but members of this genus are known to form short filaments or aggregations of cells within a common gelatinous matrix. Cells are elliptic to elongate–cylindric, straight or curved, with broadly rounded ends. There are two chloroplasts per cell, each of which has an axial core with a large, often elongate pyrenoid and several lateral ridges or lobes (stellate form in cross section). The nucleus is central between the chloroplasts. A broad conjugation tube is formed during sexual reproduction and the zygospore is formed inside the tube. Mature zygospores are required for identification of most species.

The genus contains about a dozen species that usually are collected from moist acidic soils or from wet cliffs. The genus also occurs in acidic aquatic or bog habitats, and occasionally is collected from alkaline (hard) waters. *C. brebissonii* and *C. crassa* are the most widely distributed of the six species in North America (Prescott *et al.*, 1972).

Debarya Wittrock (Figs. 16 and 20)

The vegetative cell structure of this genus is identical with that of *Mougeotia* (Fig. 5), and the two genera can be distinguished only on the basis of sexual structures (gametangia and zygospores). Both genera have elongate cylindrical cells, usually more than three times longer than broad, united into a uniseriate filament. The chloroplast is an axial ribbon that contains several pyrenoids. The nucleus is in the middle of the cell, usually on one side of the chloroplast. During sexual reproduction *Debarya* is distinguished from *Mougeotia* by having the entire gametangial protoplast incorporated into the gamete, by lacking the special walls that separate the young zygote from the gametangia in *Mougeotia*, and by filling the gametangia with stratified pectic material (Figs. 16 and 20).

Both genera occur in similar habitats (generally in small ponds or ditches), but *Debarya* is much rarer. Three of the ten known species are recorded from North America (Transeau, 1925, 1951; Prescott, 1962; Kadlubowska, 1972).

Entransia Hughes (Fig. 8)

Since sexual reproduction has never been observed in this monotypic genus, its true taxonomic position remains in doubt. It differs from other filamentous Zygnematales because it has one or two laminate, fimbriate chloroplasts in each cell of the filament.

The genus is known only from the type locality, an artificial lake near Charleston, Nova Scotia (Transeau, 1951).

Mesotaenium Nägeli (Fig. 56)

Cells in this genus may be solitary, but more often are aggregated within a common gelatinous matrix that may have either a homogeneous or a layered structure. Cells are straight or slightly curved, range from long to short cylindric, and have broadly rounded ends. The chloroplast (usually one, but sometimes two per cell) is an axial plate or band (rarely in a parietal position) and contains one or two pyrenoids. During sexual reproduction a broad conjugation tube is formed, in which the zygospore develops.

This genus occurs mostly in subaerial habitats, often as small gelatinous masses among mosses or leafy liverworts, or on moist soil or rocks. It is occasionally collected from oligotrophic aquatic habitats or bogs. Only *M. degreyi* is reported from hard-water habitats. *M. berggrenii* occurs on snow and ice in arctic–alpine regions (Kol, 1942, 1964). The genus contains 12 species, 10 of which occur in North America (Prescott *et al.*, 1972).

Mougeotia C. A. Agardh (Fig. 5)

The vegetative cell structure of this genus is identical with that of *Debarya*, and the two genera can be distinguished only on the basis of sexual structures (gametangia and zygospores). Members of this genus form uniseriate filaments that sometimes attach to a solid substratum by means of rhizoid-like outgrowths from some of the basal cells. The cylindrical cells are usually more than four times longer than broad and contain a single (rarely two) chloroplast of the axial ribbon type. Pyrenoids are numerous and may be scattered or arranged in a single row along the chloroplast. The nucleus is usually in the middle of the cell beside the chloroplast. This genus is distinguished from *Debarya* on the basis of the following features of reproductive morphology: The formation of a gamete usually leaves behind remnants of the gametangial protoplast (entire protoplast is incorporated into the gamete in *Debarya*). The young zygospore is separated from the gametangia by newly formed, special walls (not formed in *Debarya*). Finally, the gametangia of *Mougeotia* do not become filled with a stratified pectic deposit, as occurs in *Debarya*. Mature zygospores must be present to identify species of this genus. The formation of aplanospores occurs frequently in some species.

A widely reported genus and very common in lakes, ponds, and rivers. Occasionally forms blooms in the littoral zone of lakes undergoing acidification (Turner et al., 1995). Kadlubowska (1972) included 117 species in *Mougeotia*, 53 of which are recorded from North America.

Mougeotiopsis Palla (Fig. 11)

This rare monotypic genus has relatively short cylindrical cells (length less than twice the width) united in a uniseriate filament. The chloroplast is platelike and axial, but sometimes is folded over, resembling the chloroplast of *Ulothrix*. The main character that distinguishes this genus from *Mougeotia* and *Debarya* is the absence of pyrenoids in the chloroplast. During sexual reproduction this genus resembles *Debarya* because it has the entire gametangial protoplast incorporated into the gamete and lacks special walls that separate the young zygospore from the gametangia. It resembles *Mougeotia* that it has empty gametangia (not filled with stratified pectic substances).

M. calospora is known from few locations, including Europe, British Columbia (Stein, 1975), North Carolina (Whitford and Schumacher, 1984), and the central United States (Michigan and Wisconsin; Transeau, 1951).

Netrium (Nägeli) Itzigssohn *et* Rothe
(Figs. 61 and 62)

Cells are solitary and range in form from elongate–cylindric to elliptic or fusiform, with broadly rounded or truncate ends. The chloroplasts (two or four per cell) are stellate in cross section, with 6–12 longitudinal ridges extending laterally from an axial core that contains several pyrenoids (lateral ridges may also have scattered pyrenoids). The margin of each lateral ridge is cut up into many small lobes, producing a complex, frilled appearance. The nucleus is in the center of the cell between the chloroplasts.

The genus contains ten species and four of which are known from North America (Prescott *et al.*, 1972). It is often abundant in acidic, oligotrophic aquatic habitats or in *Sphagnum* bogs. Only *N. minus* has been recorded from an alkaline habitat (alpine pool, pH 8.0; Prescott *et al.*, 1972).

Pleurodiscus Lagerheim (Fig. 15)

This genus is distinguished from all other filamentous Zygnematales by the presence of two axial plate-like chloroplasts, each with a single pyrenoid, in the relatively short cells of the filament (cell length less than twice width). The nucleus lies in a cytoplasmic bridge between the chloroplasts. Scalariform conjugation and gamete formation are similar to that of *Mougeotia*. The zygospore is formed in the conjugation tube.

Only one species has been reported in the Western Hemisphere, *P. borinquinae* from Puerto Rico (Tiffany, 1936; Smith, 1950; Transeau, 1951).

Roya West *et* West (Fig. 17)

Cells are solitary, elongate–cylindric, usually curved, with broadly rounded or truncate ends. The chloroplast is either an axial ribbon or an axial rod that is triangular or quadrangular in cross section; there is an axial row of 4–12 pyrenoids. Usually there is one chloroplast per cell, but cells with two chloroplasts are known to occur. The chloroplast is often notched in the middle to accommodate the nucleus. Sexual reproduction begins with the formation of a conjugation papilla on each gametangium. A circular pore develops on the papilla, through which the gamete escapes. The gametes fuse and the zygospore develops midway between the empty gametangial walls.

Usually *Roya* is found in low numbers in oligotrophic, aquatic habitats or in *Sphagnum* bogs. There are occasional records from subaerial habitats (on dripping wet rocks or among mosses) in alpine environments (Prescott *et al.*, 1972). All four known species occur in North America.

Sirogonium Kützing (Fig. 13)

The uniseriate filaments of this genus have cells that range from one to more than 5 times longer than broad. Cells are similar to those of *Spirogyra* in containing 2–10 parietal, ribbon-like chloroplasts. However, the chloroplasts are either parallel to the cell axis or, if spiralling, rarely exhibit more than half a turn of the spiral from one end of the cell to the other. During sexual reproduction, cells lack the formation of a definite conjugation tube. Instead, the conjugating filaments become geniculate so that the gametangia can come into direct contact (Fig. 13). A perforation develops to allow passage of one gamete into the other gametangium. Mature zygospores are necessary for species identification.

The genus contains 17 species (Kadlubowska, 1972), 6 of which are found in North America. *S. floridana* has been collected only in Florida (Transeau, 1951); the other species are widespread, but rarely collected (e.g., Stein, 1975).

Spirogyra Link (Figs. 1–4)

Cells of the uniseriate filaments of this genus may be one to more than six times longer than broad. Depending on the species, each cell contains one or more (up to 16) spiral, parietal, ribbon-like chloroplasts, each containing several pyrenoids. Usually the chloroplast makes several turns of the spiral from one end of the cell to the other. In species with several chloroplasts per cell or with very short cells there may be less than one complete turn in the chloroplast spiral. The nucleus commonly is positioned in the middle of the cell, suspended by several narrow cytoplasmic strands that attach to the peripheral layer of cytoplasm. Sometimes filaments are attached to a substratum by rhizoid-like outgrowths from basal cells. In most species, adjacent cells of the filament abut along flat transverse end walls. In a few species, however, the end walls become folded (replicate). Sexual reproduction in many species occurs by scalariform conjugation (involving cells from two filaments). In other species, conjugation is lateral, involving adjacent cells of a single filament. During both processes a definite conjugation tube is formed between the gametangia. The zygospore always forms inside one of the gametangia. Mature zygospores are necessary for species identification. Parthenospores (from unmated gametes) and aplanospores are known to be formed by some species, but they are rare. Recent research has indicated that polyploid species complexes exist in this genus (Hoshaw *et al.*, 1985, 1987; Wang *et al.*, 1986). These studies imply that species with broad filaments may be polyploids of species with narrow filaments.

An extremely common and occasionally abundant genus in standing flowing waters. Most species are collected from permanent or temporary, stagnant water habitats with neutral or slightly acidic pH values, where they are mixed with other filamentous algae in benthic or floating masses. Kadlubowska (1972) listed 386 species in the genus, 145 of which have been recorded from North America.

Spirotaenia Brébisson ex Ralfs (Fig. 52)

Cells may be solitary or grouped within a common gelatinous matrix (often paired following cell division). Cells are elliptic, cylindric, or fusiform, with broadly rounded to acutely pointed ends. The chloroplast is a parietal, spiralling band that contains several pyrenoids. The nucleus usually lies along the inner surface of the chloroplast midway between the ends of the cell. Sexual reproduction by conjugation is known for only 2 species. Conjugating cells come together within a broad gelatinous sheath, and conjugation occurs without the formation of a conjugation tube. In one species, each gametangial cell divides before conjugation, resulting in the formation of a pair of zygospores.

Most of the 20 species (14 in North America; Prescott *et al.*, 1972) are rarely collected, but *S. condensata* is a cosmopolitan species that occurs in acidic aquatic habitats or *Sphagnum* bogs. Members of the genus are collected occasionally from moist subaerial habitats.

Tortitaenia (Brook) Brook (Fig. 21)

Cells are solitary and cylindrical to spindle-shaped, with rounded to truncate ends. There is usually a single chloroplast per cell. It has a solid axis that contains two pyrenoids and up to eight lateral ridges, which spiral around the axis. In some species, the apical end of the chloroplast is red. The nucleus is localized near the middle of the cell but positioned to one side of the chloroplast. Since zygospores are reported only for one species, *T. obscura*, little is known about sexual reproduction in this genus.

Tortitaenia contains five species that previously were placed in the genus *Spirotaenia*, and later in the nomenclaturally illegitimate genus *Polytaenia* (Brook, 1997, 1998). Four species are known from North America.

Zygnema C.A. Agardh (Figs. 6 and 7)

The uniseriate filaments of this genus comprise cylindrical cells that are one to five times longer than broad. Rhizoidal extensions of basal cells for attachment to a substratum are formed rarely. There are two stellate chloroplasts per cell, each with a central pyrenoid. The nucleus is centrally positioned in a cytoplasmic bridge between the chloroplasts. Sexual reproduction in most species is by scalariform conjugation (Fig. 6), although lateral conjugation occurs in some species. In both processes, a definite conjugation tube is formed. In some species, zygospores are formed inside one of the gametangia, whereas in other species, the zygospore develops in the conjugation tube. This genus is distinguished from *Zygnemopsis* by the absence of layered pectic material in the gametangia. Mature zygospores are necessary for species identification. Formation of aplanospores and akinetes (Fig. 7) is known to occur, but is rare.

Zygnema is a widespread and occasionally abundant alga that usually occurs in neutral to slightly acidic, lentic habitats (ditches, ponds, and lakes), but is also collected frequently from streams. Although it sometimes forms essentially unialgal masses, it is more often found mixed with other filamentous green algae, such as *Spirogyra* or *Mougeotia*. Kadlubowska (1972) listed 120 species in the genus, 38 of which are known from North America.

Zygnemopsis (Skuja) Transeau (Fig. 9)

The structure of the filaments in this genus is essentially identical to that in the genus *Zygnema*. *Zygnemopsis* is distinguished from *Zygnema* by the formation of a layered pectic deposit that fills the gametangia during conjugation. Scalariform conjugation appears to be the rule in this genus, and the zygospore is formed in the broad conjugation tube. Mature zygospores are necessary for identification of species. The formation of parthenospores from unmated gametes is known to occur.

This genus is collected rarely, possibly due to the absence of sexual stages that distinguish it from *Zygnema*. It occurs in the shallow water of lake margins and swamps. Kadlubowska (1972) listed 38 species in the genus, 7 of which occur in North America.

Zygogonium Kützing (Fig. 19)

The uniseriate filaments of this genus contain cylindrical cells that range in length from one-half to several times the width. The cell wall is sometimes greatly thickened and brownish, and the protoplast occasionally is colored purple. There are two chloroplasts per cell, and the nucleus is situated between them. The chloroplasts usually are described as "cushion-shaped" with a central pyrenoid, but they sometimes have irregular lobing and resemble the stellate chloroplasts of *Zygnema*. Formation of aplanospores, akinetes, or parthenospores is observed frequently in certain species (Transeau, 1933, 1951). Conjugation is usually scalariform, but lateral conjugation is also known in this genus. In both processes, a special wall is formed that separates the developing zygospore from the rest of the gametangia. There is usually a residue of the gametangial protoplast that was not incorporated into the gamete. Zygospores invariably are formed in the conjugation tube and are essential for the identification of the species.

Zygogonium usually occurs in subaerial habitats, such as moist acidic soils and rock, on the surface of peat in bogs, or in the littoral zone of soft-water lakes undergoing acidification (Turner *et al.*, 1995) Kadlubowska (1972) listed 25 species in the genus, four of which are known from North America.

C. Descriptions of Genera of the Desmidiales

1. Family Closteriaceae

Closterium Nitzsch *ex* Ralfs (Figs. 31 and 35–40)

Cells of this unicellular genus are usually elongate–cylindric and curved (at least at the ends). A few species have elongate–fusiform, straight cells (Figs. 39 and 40). Cells gradually narrow toward both apices. Apices may be acutely pointed, rounded, or truncate. The cell wall may be smooth (Fig. 37), have fine longitudinal striae (Fig. 38), or have coarse costae (Fig. 35). The wall may be colorless or yellow to brown (sometimes only at the ends of the cell). Girdle bands (extra wall sections) are present in some species (Fig. 35). There are two (rarely four) chloroplasts per cell that are axial, elongate, and stellate in cross section, with one to many axial or scattered pyrenoids. The nucleus is central and between the chloroplasts. At each end of the cell there is a vacuole that contains one or more granules of calcium sulfate (which may appear to be in motion). Conjugation may occur between cells with mature morphology or between cells from a recent division that have an immature morphology resulting from the incomplete expansion of their newly formed semicells. There is usually no conjugation tube. Both gametangial cells split open at the median suture and the gametes partially or completely move out to fuse between the empty gametangial walls (Fig. 31). Some species are known to form aplanospores or parthenospores.

Closterium is collected most often from the benthos or periphyton of acidic, oligotrophic lakes and ponds; it occurs more rarely in alkaline, eutrophic environments. *C. aciculare* and *C. acutum* are planktonic in eutrophic waters, often among cyanobacterial blooms. There are about 140 species in this genus, 88 of which are found in North America (Prescott *et al.*, 1975).

Spinoclosterium Bernard (Fig. 27)

Cells of this monotypic genus are solitary, elongate–fusiform, and strongly curved (lunate), resembling some species of *Closterium*. This genus is distinguished from *Closterium* by the presence of a solid, stout spine on each end of the cell. The cell wall is smooth and colorless, comprising two pieces that meet at a median suture line. There are two axial chloroplasts per cell that possess several longitudinal ridges and many scattered pyrenoids. The nucleus is at the middle of the cell in a cytoplasmic bridge between the chloroplasts.

S. cuspidatum has been collected several times in North America (Prescott *et al.*, 1975; Stein, 1975). It is usually rare in samples taken from acidic, oligotrophic ponds and lakes.

2. Family Peniaceae

Genicularia De Bary (Fig. 65)

Cells of this genus are elongate–cylindric with truncate ends and may be solitary or united in short filaments. The cell wall has scattered granules. There are two or three narrow, parietal, spiralling, ribbon-shaped chloroplasts per cell, each with numerous pyrenoids. The nucleus is positioned near the middle of the cell. Sexual reproduction begins with the gametangia becoming geniculate before a narrow conjugation tube forms. Spherical zygospores are formed within the conjugation tube.

Both of the known species occur in North American acidic, soft-water habitats, but are extremely rare (Stein and Gerrath, 1969; Prescott *et al.*, 1972).

Gonatozygon De Bary (Figs. 55 and 79)

Cells may be solitary or may be loosely united into short uniseriate filaments. Cells are usually narrow, elongate–cylindric to elongate–fusiform, with truncate ends. The cell wall may be smooth or have scattered granules, setae, or tiny spines. There is one (sometimes two) axial ribbon-like chloroplast per cent containing several pyrenoids. The nucleus is localized at the middle of the cell beside the chloroplast or between the chloroplasts if two are present. Sexual reproduction by conjugation is known for three of the nine species. Gametangia become geniculate before a narrow conjugation tube forms between them. Zygospores are spherical and are formed within the conjugation tube. *Gonatozygon* is much more common than *Genicularia*.

Five species occur in North America (Prescott *et al.*, 1972), usually in acidic, oligotrophic lakes and ponds or in *Sphagnum* bogs.

Penium Brébisson *ex* Ralfs (Figs. 45 and 46)

Cells are solitary, short–cylindric to elongate–cylindric, with broadly rounded or truncate ends. The cell wall may be made up of two pieces that meet at the median suture or may have one or more extra pieces (girdle bands) interpolated between the two. The wall may have striae, pores, granules, or spines, depending on the species. There are two (rarely four) axial chloroplasts per cell, each of which is stellate in cross section and contains one or two axially positioned pyrenoids. Terminal vacuoles that contain small crystals are present in some species. The nucleus is at the middle of the cell between the chloroplasts. During sexual reproduction gametangia come together within a mucilage envelope. The gametangial cells split open to release the gametes, which fuse between gametangia. Zygospores are spherical to ellipsoid (occasionally angular), and usually smooth walled.

A few species are cosmopolitan, but most are rare. *Penium* usually occurs in acidic, oligotrophic ponds or lakes, or in bogs. The genus contains about 20 species. Prescott *et al.* (1975) listed 13 species in North America. However, four of these species (*P. didymocarpum*, *P. phymatosporum*, *P. silvae-nigrae*, and *P. spinospermum*) have since been transferred to *Actinotaenium* (Coesel and Delfos, 1986; Kouwets and Coesel, 1984), leaving only nine species in North America.

3. Family Desmidiaceae

Actinotaenium (Nägeli) Teiling (Fig. 44)

Cells are solitary, short to elongate, fusiform to cylindric, with broadly rounded or truncate ends and a very shallow median constriction. In apical view the cell outline is circular. The cell wall may be smooth or may have scattered pores or scrobiculations. There is usually one chloroplast per semicell, stellate in end view, and a central pyrenoid (rarely two or three in an axial row). The nucleus is centrally located between the chloroplasts. Sexual reproduction by conjugation is known for several species. Zygospores are formed between the gametangial cells. Mature zygospores may be spherical or quadrate with smooth walls, or spherical to subspherical with broad mamillate protuberances or sharp spines. The empty gametangial semicell walls often remain attached to corners of quadrate zygospores. Paired zygospores are known to be formed in some species, such as *A. diplosporum*.

There are about 50 species in the genus, many of which have a worldwide distribution. Prescott *et al.* (1981) listed 30 species from North America and, as noted for *Penium*, four additional species that occur in North America have been transferred into the genus. *Actinotaenium* usually is collected from acidic, oligotrophic lakes and ponds or *Sphagnum* bogs. *A. curtum* and *A. perminutum* both commonly subaerial, also occur in North America.

Bambusina Kützing (Fig. 48)

The uniseriate filaments of *Bambusina* contain small, barrel-shaped cells that have a shallow median constriction (isthmus). Cells in apical view are circular, except for two or three low protuberances that occur near the isthmus. The cell wall has transverse rows of inconspicuous pores and usually faint longitudinal striations near the cell apices. There is a single chloroplast per semicell, stellate in end view, with a central pyrenoid. The nucleus is in the middle of the cell between the chloroplasts. During the early stages of semicell morphogenesis following cytokinesis, a replicate (folded) cell wall is formed. The primary wall layer, which separates the two daughter cells, develops a cylinder that extends into each daughter cell. Secondary wall layers for both daughter semicells are deposited on each side of this primary wall template, producing a folded (replicate) region around the projecting primary wall cylinder. Semicell morphogenesis in each daughter cell is completed by the expansion of the new semicells, which tears apart the primary wall cylinder as the plicate secondary wall unfolds. On a fully formed cell the position of the secondary wall fold usually can be recognized as a shallow groove a short distance from the apex. During sexual reproduction, one of the conjugating filaments usually breaks apart into individual cells, which then pair with cells of the other filament. The gametes are released to form zygospores between each pair of empty gametangia. Mature zygospores are spherical to ellipsoidal with smooth or mamillate walls.

Five species are known to occur in North American (Croasdale et al., 1983). The genus usually is found mixed with other filamentous algae in acidic, oligotrophic ponds and lakes.

Cosmarium Corda ex Ralfs (Figs. 22–26, 32, and 71–75)

This is the largest desmid genus, containing more than 1000 species. It is difficult to make generalizations about the morphological diversity that is exhibited wihin the genus. Cells may be minute (< 10 μm) or relatively large (< 200 μm) and there may be a shallow (Fig. 75) or deep (Fig. 73) median constriction (isthmus). Semicells in frontal view may be rounded, semicircular, reniform, pyramidate, quadrate, or some other shape. The semicell margin may be entire or have undulations. Most species have compressed (biradiate) cells, so that the apical view is elongate–elliptical (Fig. 72). However, there are species in which the apical view of the cell is almost circular (Fig. 71). Triradiate forms are known to develop in collections and cultures. The cell wall is many species is smooth with scattered pores, but in other species the wall is ornamented with small or large granules, emarginate verrucae, round or triangular pits, or short spinules (Figs. 73 and 74). The ornamentation of the central part of the semicell face often differs from that of the marginal area (Fig. 74). Some species have one chloroplast in each semicell, but others have several. One or more pyrenoids are present in each chloroplast. The nucleus is invariably localized in the isthmus. Sexual reproduction starts when the conjugating cells come together within a mucilaginous envelope. The semicells split at the isthmus, releasing the amoeboid gametes, which fuse between the empty gametangia (Fig. 32). Mature zygospore are usually spherical with short, acute or truncate spines (sometimes with furcate tips on the spines).

Cosmarium is widespread in North America and most common in acidic, oligotrophic, aquatic environments. Some species occur in subaerial habitats or in alkaline, eutrophic ponds and lakes. There are many cosmopolitan species. Prescott et al. (1981) listed 420 species that occur in North America.

Cosmocladium Brébisson (Fig. 30)

This is one of the few desmid genera that is considered to be colonial. Cells are essentially identical in form to some smooth-walled *Cosmarium*, that is, they are biradiate with a narrow or broad isthmus. There is one chloroplast per semicell and it contains a central pyrenoid. The nucleus is positioned in the isthmus between the chloroplasts. The cells are united into colonies by connecting strands secreted by special groups located near the base of the semicell. Each connecting strand is made up of two overlapping parts, one that originates from the pore group on one cell and the other from a similar pore group on the adjacent cell of the colony (Gerrath, 1970). Colonies often have a broad mucilaginous envelope secreted through scattered cell wall pores on all cells of the colony. Sexual reproduction by conjugation is known in two of the three or four species. Mature zygospores are spherical and bear several short, stout, acute spines.

The genus is widespread, but rarely collected, in acidic, oligotrophic, aquatic habitats in Europe and North America; isolated records exist from Asia and New Zealand. Recently, species that have cells united by strands made up of persistent primary cell walls have been removed to the new genus *Heimansia* Coesel (1993). *C. constrictum*, *C. saxonicum*, and the doubtful species, *C. pulchellum*, are recorded from North America (Prescott, et al., 1981).

Desmidium C. Agardh ex Ralfs (Figs. 49–51)

Cells are united into long uniseriate filaments, either by the apposition of the entire flat apical surface of adjacent cells or by the apposition of the flat surfaces of two to five narrow apical processes (the number is dependent on the morphology of the cells). Cells may be longer or shorter than broad, with a distinct or indistinct median constriction. Cells are three- to five-angled (Fig. 51) or elliptical (Fig. 49; biradiate) in apical view. The cell angles are usually slightly offset on each semicell, producing a helical pattern of angles from one end of the filament to the other. The cell wall is smooth and has many pores in transverse rows or scattered. There is one chloroplast per semicell, which is stellate in end view. Depending on the species, it may contain a central pyrenoid or have a pyrenoid in each lobe. The nucleus is localized in the isthmus, between the chloroplasts. *Desmidium* is similar to *Bambusina* in the formation of a replicate secondary wall during the processes of cell division and new semicell morphogenesis (Fig. 50). During conjugation, there is a definite conjugation tube formed, in which the zygospore is usually located. Mature zygospores are spherical to ellipsoid and are smooth-walled or bear rounded warts.

The genus contains 20 species, 16 of which are known from North America (Croasdale *et al.*, (1983), usually occurring in oligotrophic, acidic lakes, ponds, and bogs.

Docidium Brébisson *ex* Ralfs (Fig. 41)

Cells are solitary, elongate–cylindrical (length 7.5–30 times the width) with straight or undulate sides and a shallow median constriction. Plications or small granules on each side of isthmus are diagnostic for the genus. Each semicell usually has a basal swelling and the truncate apex is often also broadened. There is a single axial chloroplast per semicell, stellate in cross section and containing several pyrenoids. The nucleus is situated in the isthmus. As far as can be determined from the scattered desmid literature, sexual reproduction is unknown in this genus. Cell division, the only known method of reproduction, is typical of that found in other desmids.

Several of the eight known species are cosmopolitan, but *Docidium* is usually rare in collections. It is regarded as acidophilic and is often associated with *Sphagnum* in lakes and ponds with pH as low as 3.5. Five species are recorded from North America (Prescott *et al.*, 1975).

Euastrum Ehrenberg *ex* Ralfs (Figs. 66–70)

Cells are solitary, usually longer than broad, with a deep median constriction. Cells are moderately compressed (biradiate). Each semicell usually has distinct apical and lateral lobes. The apical lobe has an emarginate outline (Fig. 70) or an apical incision (Fig. 66). The face of a semicell usually has one or more rounded protuberances, which are detected more easily in lateral or apical view. The cell wall may be smooth with scattered pores or variously ornamented with granules, verrucae, or short spines (Figs. 69 and 70). There is usually a single chloroplast per semicell that contains one or more pyrenoids. The nucleus is positioned in the isthmus. Sexual reproduction by conjugation is known to occur in several species. Conjugating cells pair within a gelatinous envelope and gametes are released to fuse between gametangia. Mature zygospores are globose to ellipsoidal with many short, acute spines or mamillate protuberances.

Several species of *Euastrum* are cosmopolitan and common in any single collection. The genus usually occurs in acidic oligotrophic aquatic habitats or bogs. Some small-celled *Euastrum* with emarginate apices cannot be distinguished with absolute confidence from *Cosmarium*. There also is no clear distinction between *Micrasterias* and *Euastrum*, so the placement of certain taxa may be arbitrary. The genus contains approximately 256 species, 116 of which have been recorded from North America (Prescott *et al.*, 1977).

Groenbladia Teiling (Fig. 53)

Cells are united into short or long uniseriate filaments. Cells are elongate (two to nine times longer than wide), cylindrical or narrowing toward the ends. There is usually a slightly wider basal region on each semicell on either side of the shallow median constriction. Each cell has a single axial, band-shaped chloroplast containing one to eight pyrenoids. The nucleus is near the middle of the cell and beside the chloroplast. Sexual reproduction by conjugation known in two of the five species. Gametangia become geniculate and produced a broad conjugation tube in which gametes fuse. Mature zygospores are quadrangular with round to acutely pointed corners. The formation of yellowish, elliptical aplanospores has been recorded for *G. neglecta*.

Groenbladia usually occurs sparsely among other desmids on sediments or among periphyton of acidic, oligotrophic lakes and ponds. Four species are known from North America (Croasdale *et al.*, 1983).

Haplotaenium Bando (Fig. 42)

Cells are solitary, elongate–cylindrical with a shallow median constriction. Semicells have a slight basal swelling and a rounded to truncated apex. The semicell apex lacks the terminal vacuole present in *Pleurotaenium*. The cell wall appears smooth or finely porose. This genus was segregated from *Pleurotaenium* on the basis of chloroplast number and shape (Bando, 1988). *Pleurotaenium* has a single chloroplast in each semicell and it is a flat axial band or is axial with irregular lateral laminae, with a central row of 2–15 pyrenoids.

Two of the three species are widespread, usually in acidic, oligotrophic lacustrine environments or in swamps. Two species, *H. minutum* and *H. sceptrum*, occur in North America.

Heimansia Coesel (Fig. 91)

Cells of the colony remain attached following cell division by the persistent primary walls produced on the newly developing semicells. The small biradiate cells of this genus are similar to those of some species of *Cosmarium*. Cells are deeply constricted and each semicell has a single chloroplast that contains a central pyrenoid. Sexual reproduction is unknown in the genus. All species in this genus were formerly placed in the genus *Cosmocladium*, but were segregated as a new genus by Coesel (1993) because of the nature of the strands that hold the cells together in a colony.

Prescott *et al.* (1981) listed *Cosmocladium pusillum*, *C. tuberculatum*, and *C. tumidum*, which are all now considered to belong to *Heimansia*, from North America. The genus is rare but widespread, occurring in both soft- and hard-water ponds and lakes.

Hyalotheca Ehrenberg *ex* Ralfs (Fig. 60)

Cells in this genus are more or less cylindrical and united at broad truncate ends into long uniseriate filaments. Depending on the species, there may or may not be a slight median constriction. Some species have small protuberances near the base of each semicell (two on opposite sides or three equally spaced around cell). The cell wall is smooth and has pores arranged in several transverse rows. A narrow or broad gelatinous sheath is usually present around the filament. There are two chloroplasts per cell, stellate in end view, each containing a central pyrenoid. The nucleus is at the middle of the cell between the chloroplasts. Sexual reproduction is known in three of the six species. Compatible filaments come together and their cells dissociate before gametangial pairing occurs. The spherical zygospore is formed within the broad conjugation tube. Aplanospore formation is known in two species.

H. dissiliens and *H. mucosa* are cosmopolitan and often abundant in collections. The genus usually occurs in acidic, oligotrophic lakes, ponds, swamps, and streams. Croasdale *et al.* (1983) listed three species from North America.

Micrasterias Ralfs (Figs. 80, 82, and 83)

Cells are usually large and solitary (*M. foliacea* is filamentous; Fig. 80), much compressed, with a very deep median constriction. Other shallow or deep incisions divide each semicell into an apical (polar) lobe and two lateral lobes. Each of the lateral lobes is usually further subdivided by deep incisions (Fig. 92). The polar lobe often bears a pair of divergent processes (Fig. 83). Some species have small or large protuberances or hollow facial processes, usually near the base or middle of each semicell. The cell wall may be smooth with scattered pores, or have numerous granules or spinules, either covering most of the cell surface, restricted to certain lobes, or restricted to the margins of incisions. There is usually one chloroplast per semicell, containing few to numerous, scattered pyrenoids. The nucleus is localized in the narrow isthmus.

Several species are cosmopolitan and may be abundant in collections. *Micrasterias* may be in the plankton or periphyton of periphyton of acidic, oligotrophic or dystrophic lakes and ponds, or in swamps or mires. There is no clear distinction between *Micrasterias* and *Euastrum*, so the placement of certain taxa may be arbitrary. The genus contains about 75 species, 40 of which are known from North America (Prescott *et al.*, 1977).

Octacanthium (Hansgirg) Compère (Fig. 90)

This genus was erected by Compère (1996) to include those biradiate species of the illegitimate genus *Arthrodesmus* Ralfs and those species of the genus *Xanthidium* that have four or six spines arranged in one plane on each semicell. It is distinguished from biradiate *Staurodesmus* species (Fig. 85), which have two spines per semicell, and from *Xanthidium* species (Fig. 81), which usually have paired spines (not in one plane) at two or more locations on a semicell. There is usually a single chloroplast, containing a central pyrenoid, in each semicell.

O. octocorne is cosmopolitan and often common in collections. This genus is usually collected from acidic, oligotrophic lakes and ponds. Compère (1996) included nine species, six of which occur in North America.

Onychonema Wallich (Fig. 86)

Cells are small, compressed (biradiate) with a deep median constriction, and are united in a uniseriate filament. The semicell apex bears two, widely separated, long hollow processes that are asymmetrically disposed: one extends over the front face of the adjacent cell and the other over the rear face. A single chloroplast with a central pyrenoid occupies each semicell. The nucleus is positioned in the narrow isthmus. Some taxonomists follow Teiling (1957) and reduce this genus to synonymy with *Sphaerozosma*, which is distinguished by having the apical processes close together on the semicell apex.

The three reported North American species occur on acidic, oligotrophic ponds and lakes.

Oocardium Nägeli (Figs. 28 and 29)

Cells of this monotypic genus are usually laterally asymmetric, slightly compressed, with a shallow median constriction. Cells are at the ends of branched gelatinous stalks within encrusting calcareous tubes, forming macroscopic lumps attached to solid substrates in calcium-rich streams or waterfalls. There is usually one axial chloroplast that contains a central pyrenoid in each semicell (sometimes pressed to one side of semicell).

Oocardium is rare but widespread in North America (Prescott et al., 1981), occurring in California and in the northeastern United States.

Phymatodocis (Fig. 64)

Cells of this filamentous genus are quadrangular in lateral view, with a moderately deep median constriction. In apical view, each semicell is four-lobed (rarely three- or five-lobed). Lobing is symmetric, except in *P. irregulare* (two long and two short lobes). There is one axial chloroplast per semicell containing one or two pyrenoids. Two broad, flattened lobes extend into each lateral lobe of the semicell from the central mass of the chloroplast. The nucleus lies in the isthmus. Sexual reproduction results in quadrangular zygospores.

All of the three or four species are rare, usually occurring in acidic, oligotrophic lakes in tropical or subtropical regions. *P. alternans* Nordstedt has been collected in Florida, whereas *P. nordstedtiana* is distribued along the eastern region of North America from Mississippi to Quebec and Ontario (Croasdale et al., 1983).

Pleurotaenium Nägeli (Fig. 54)

Cells are usually solitary (some taxa are filamentous), mostly elongate–cylindrical with a shallow median constriction. Semicells have a basal swelling and a truncated apex. The semicell apex may be smooth or may have a ring of round or conical warts, or short spines. Near the semicell apex there is usually a terminal vacuole that contains granules of unknown composition (possibly calcium sulfate crystals, similar to those in *Closterium*). Some species have several whorls of mamillate protuberances, smooth or bearing spines, on each semicell. Several parietal, ribbon-like chloroplasts are present in each semicell and each chloroplast contains many pyrenoids. Former species with a single, axial band-shaped chloroplast have been removed to the recently described genus *Haplotaenium* (Bando, 1988). The nucleus is located in the isthmus.

Several species of *Pleurotaenium* are cosmopolitan, usually occurring in acidic, oligotrophic lacustrine environments or in swamps. There are about 50 species in the genus, 29 of which are known from North America (Prescott et al., 1975).

Sphaerozosma Ralfs (Fig. 63)

Cells of this filamentous desmid are small, compressed (biradiate), and have deep median constriction. Near the middle of each semicell apex is a pair of obliquely disposed, rodlike processes that overlap similar processes on the adjacent cell of the filament.

One chloroplast, with a central pyrenoid, is present in each semicell. Zygospores are spherical, and may be smooth or have many spines.

Two or three species occur in North America, usually in acidic, oligotrophic lacustrine environments (Croasdale et al., 1983).

Spinocosmarium Prescott *et* Scott (Figs. 76 and 77)

Cells are solitary and compressed (biradiate), with a deep median constriction. Each semicell possesses a pair of laterally directed, simple or furcate spines. The type species also may have one or two spines at each angle of the flattened semicell apex. There is a large granule just above the isthmus and the semicell face also has many simple to emarginate warts disposed in irregular vertical and horizontal lines. Each semicell contains a single chloroplast with a central pyrenoid. The nucleus is located in the isthmus.

Spinocosmarium is endemic to North America, usually occurring in acidic, oligotrophic lakes and ponds. *S. quadridens* is widespread in eastern North America and also has been collected in British Columbia and Alaska; *S. laconiense* is very rare in a few eastern locations (Stein, 1975; Prescott et al., 1982).

Spondylosium Brébisson *ex* Kützing (Figs. 57 and 58)

Cells of the uniseriate filament may be small or large, and are compressed (biradiate) or triradiate in end view, with a deep median constriction. Filaments are commonly unattached but *S. pulchellum* is sometimes attached by a gelatinous stalk to aquatic vascular plants. Adjacent cells of filaments attach along part or all of the semicell apex. The genus is distinguished from other filamentous genera by lack of apical processes and replicate walls following cell division. There is one chloroplast per semicell, with a central mass and two flattened lateral lobes (visible in apical view only) on each side in biradiate cells, or into each angle of triradiate cells. Pyrenoids may be single and in the central mass of the chloroplast, or several located in the lateral lobes. The nucleus is located in the isthmus.

There are 34 species, 14 of which are known from North America (Croasdale et al., 1983). Several species are cosmopolitan and common in collections, usually occurring in acidic, oligotrophic lakes and ponds.

Staurastrum Meyen *ex* Ralfs (Figs. 33, 34, 87–89, 92, and 93)

Cells are small to large, 2- to 12-radiate in end view, with a shallow or deep median constriction. Many species have long hollow processes on each semicell (Figs. 87 and 89; the number of processes is related to the degree of radiation pattern). The processes usually have two or more small terminal spines and often one or more series of denticulations, spines, or verrucae along the processes and on the apex and body of the central axis of the semicell. In species that lack long hollow processes, the semicell angles may be rounded, truncate, or have short processes. The cell wall in these species may be smooth (Fig. 93) or may exhibit rows of small granules or spinules (Figs. 88 and 92). There is usually one multilobed chloroplast per semicell, with a central pyrenoid, but several other chloroplast forms and pyrenoid arrangements occur in this genus. The nucleus is in the isthmus. Zygospores are usually spherical and bear many long narrow spines that are often multifurcate at the tip (Figs. 33 and 34).

Many of the approximately 800 species are cosmopolitan; and often common in collections. Species with long processes are often planktonic and may exhibit considerable morphological variability in radiation pattern. Species mostly occur on sediments or among periphyton in acidic, oligotrophic lakes, ponds, and swamps. Prescott *et al.* (1982) included 320 North American species, a number that excludes 36 species that are currently placed in the genus *Staurodesmus*.

Proposals to segregate groups of species into distinct genera generally have been ignored because of the difficulty in establishing clear generic limits. Former smooth-walled *Staurastrum* taxa with a single spine (or thickened cell wall) at each angle are now included in the genus *Staurodesmus*. Palamar-Mordvintseva (1976) proposed to segregate *Staurastrum* species without long, hollow processes into several genera, but her scheme has not been accepted outside Russia and the Ukraine. Her new genus *Cylindriastrum* would include species with elongate angular–cylindrical cells. *Cosmoastrum* would include species with a smooth wall, or with rows of granules or spinules over the entire semicell wall or only around semicell angles. *Raphidiastrum* would include species with two or more relatively large spines at each angle and those with a single angular spine that also have rows of granules or spinules on the cell wall. Species with long hollow processes would remain in a much reduced genus *Staurastrum*.

Staurodesmus Teiling (Figs. 84 and 85)

Cells are solitary, small to medium sized, with a shallow or deep median constriction. The apical (end) view of a cell may be elliptical (biradiate), triangular or multiangular. Semicell angles each bear a single stout or tiny spine, a small granule, or merely a thickening of the wall. The wall elsewhere is unornamented, but contains scattered pores through which a broad gelatinous sheath is often produced. Each semicell usually has one axial chloroplast, which has lobes that extend toward each angle, and contains one or two pyrenoids. The nucleus is positioned in the isthmus. Zygospores, where known, are spherical and bear several narrow spines.

The genus comprises about 100 species, several of which are widespread and may be common in collections. *Staurodesmus* usually occurs in the plankton or periphyton of acidic, oligotrophic aquatic environments. This genus combines most members of the illegitimate genus *Arthrodesmus* with species of *Staurastrum* that have monospinous angles. It cannot be clearly distinguished from *Staurastrum* or *Xanthidium* because intergrading forms exist.

Teilingia Bourrelly (Fig. 78)

In this filamentous genus, cells are usually small, compressed (biradiate), and have a deep median constriction. Cells are united into short or long filaments and the joining of adjacent cells is thought to involve the apposition of four small, rectangularly disposed granules that occur on each end of the cell. There is one chloroplast per semicell that contains a central pyrenoid. The nucleus is localized in the isthmus.

Three of the seven known species have a cosmopolitan distribution, usually occurring in acidic, oligotrophic lakes and ponds. Five species are known from North America (Croasdale *et al.*, 1983).

Tetmemorus Ralfs (Fig. 47)

Cells are solitary, four to nine times longer than wide, with a shallow median constriction. Semicells gradually narrow toward the ends of the cell and there is a deep, closed incision in each semicell apex. Cells are circular to elliptic in end view. The cell wall is smooth, either with scattered pores or with longitudinal lines of elongate scrobiculations. There is one chloroplast per semicell, stellate in end view, containing one or more pyrenoids in an axial row. Zygospores are spherical to ovoid with a smooth or variously sculptured wall.

Three of the six species are known from North America (Prescott *et al.*, 1975). The genus is considered to be acidophilic, most commonly occurring among *Sphagnum* in bogs or in small ponds with marginal mats of *Sphagnum*. It is occasionally collected from subaerial habitats (among mosses or on moist substrata).

Triploceras (Bailey *ex* Ralfs) Bailey (Fig. 43)

Cells are solitary, elongate–cylindrical with a relatively shallow median constriction. This genus differs from other elongate desmids in that it possesses several transverse whorls of protuberances, each ending in either a short spine or a broad truncate or emarginate verruca. Cell ends are two- to four-lobed and each apical lobe usually bears two spines. There is one axial chloroplast per semicell, stellate in cross section, with an axial row of pyrenoids. The nucleus is located in the isthmus. Zygospores are spherical and bear many bi- or trifurcate spines.

The genus is considered to have a pantropical distribution. Two of the three species are widespread in North America, usually found in acidic, oligotrophic lakes and ponds as far north as Quebec in the east and Alaska in the west (Prescott et al., 1975).

Xanthidium Ehrenberg *ex* Ralfs (Fig. 81)

Cells are solitary, slightly compressed (biradiate), with a deep median constriction. Each semicell usually bears four or more, simple or furcate, short or long, marginal spines. The form and disposition of spines is a major morphological feature used for species characterization. The middle of each semicell face may be smooth or exhibit one of the following features: a ring or line(s) of conspicuous pores, a central incrassate area (often brown colored), a line of small verrucae, or a protuberance bearing short or long spines. Each semicell has two or four chloroplasts, each containing a single pyrenoid. The nucleus is localized in the isthmus. Zygospores are spherical and may or may not bear many simple or furcate spines.

Several of the approximately 115 species are cosmopolitan, usually collected from acidic, oligotrophic ponds and lakes. There are 29 species known from North America (Prescott et al., 1982). Note that *X. controversum* has been transferred to the genus *Octacanthium* (Compère, 1996).

VI. GUIDE TO LITERATURE FOR SPECIES IDENTIFICATION

The most recent monograph for the identification of species of filamentous Zygnemales is that of Kadlubowska (1972), which is in Polish. The older monograph by Transeau (1951) is in English, and contains keys and descriptions of species likely to be found in North America. Species identified using Transeau (1951) should be cross-checked with those in Kadlubowska to make sure that the nomenclature is consistent. For the unicellular Zygnematales (saccoderm desmids), the monograph by Prescott et al. (1972) in the North American Flora series provides keys and descriptions of all taxa, except for the new genus, *Tortitaenia* (Brook, 1998). The recent revision of the genus *Netrium* by Ohtani (1990) also should be consulted.

A five-volume synopsis of North American Desmidiales will enable researchers to identify species of this group of conjugating algae (Croasdale et al., 1983; Prescott et al., 1975, 1977, 1981, 1982). One major problem with this synopsis results from refusal of the authors to recognize the genus *Staurodesmus*. Instead, they continued to maintain the genus *Arthrodesmus*, which now is considered illegitimate (Bicudo, 1984; Christensen, 1987), and to place three- to six-radiate species of *Staurodesmus* into the genus *Staurastrum*. In most cases, however, the nomenclaturally correct *Staurodesmus* taxon is found in the synonymy. Teiling (1948, 1967) also can be consulted for descriptions of *Staurodesmus* taxa, but this paper does not contain a key. The outdated monograph of British desmids (West and West, 1904, 1905, 1908, 1912; West et al., 1923) contains descriptions and illustrations of many cosmopolitan desmids. For descriptions of three genera described since the publication of the North American synopsis, consult Bando (1988; *Haplotaenium*), Coesel (1993; *Heimansia*), and Compère (1996; *Octacanthium*).

ACKNOWLEDGMENTS

A special thank you is extended to Janet R. Stein, who first inspired me to undertake research among the conjugating green algae. Many thanks to Jean Gerrath for her patience and assistance during collection trips. The helpful suggestions from the reviewers, Alan J. Brook and Rick McCourt, and the editor, Bob Sheath, are also acknowledged.

LITERATURE CITED

Bando, T. 1988. *Haplotaenium*, a new genus separated from *Pleurotaenium* (Desmidiaceae). Journal of Japanese Botany 63:169–178.

Barr, D. J. S., Hickman, C. J. 1967. Chytrids and algae. I. Host–substrate range, and morphological variation of species of *Rhizophidium*. Canadian Journal of Botany 45:423–430.

Berlyn, G. P., Micksche, J. P. 1976. Botanical microtechnique and cytochemistry. Iowa State University Press, Ames, IA, 226 p.

Bicudo, C. E. M. 1984. Proposal for the conservation of the generic name Arthrodesmus Archer. Taxon 33:107–108.

Boney, A. D. 1980. Water retention and radiation transmission by gelatinous strata of the saccoderm desmid *Mesotaenium chlamydosporum* De Bary. Nova Hedwigia 33:949–970.

Boney, A. D. 1981. Mucilage: The ubiquitous algal attribute. British Phycological Journal 16:115–132.

Boney, A. D. 1982. Living in mucilage: The saccoderm desmid *Mesotaenium*. Glasgow Naturalist 20:237–243.

Brook, A. J. 1981. The biology of desmids. Botanical Monographs, Vol. 16. Blackwell, Oxford, UK, 276 p.

Brook, A. J. 1997. The proposed establishment of a new desmid genus *Polytaenia*, previously the sub-genus *Polytaenia* of the genus *Spirotaenia*, and a description of a new species, *P. luetkemuelleri*. Quekett Journal of Microscopy 38:7–14.

Brook, A. J. 1998. *Tortitaenia* nom. nov. pro *Polytaenia* Brook, a name of a genus of saccoderm desmids. Quekett Journal of Microscopy 38:146.

Burkholder, J. M., Sheath, R. G. 1984. The seasonal distribution, abundance and diversity of desmids (Chlorophyta) in a softwater, north temperate stream. Journal of Phycology 20:159–172.

Christensen, T. 1987. Nomenclature. Report of the committee for algae. Taxon 36:66–69.

Coesel, P. F. M. 1983. The significance of desmids as indicators of the trophic status of freshwaters. Schweizerische Zeitschrift für Hydrologie 45:388–394.

Coesel, P. F. M. 1991. Ammonium dependency in *Closterium acicularare* T. West, a planktonic desmid from alkaline, eutrophic waters. Journal of Plankton Research 13:913–922.

Coesel, P. F. M. 1993. Taxonomic notes on Dutch desmids. II. Cryptogramie Algologie 14:105–114.

Coesel, P. F. M. 1996. Biogeography of desmids. Hydrobiologia 336:41–53.

Coesel, P. F. M., Delfos, A. 1986. New and interesting cases of conjugating desmids from Lapland. Nordic Journal of Botany 6:363–371.

Compère, P. 1996. *Octacanthium* (Hansgirg) Compère, a new generic name in the Desmidiaceae. Nova Hedwigia, Beiheft 112:501–507.

Couté, A., Tell, G. 1981. Ultrastucture de la paroi cellulaire des Desmidiacées au microscope électronique à balayage. Nova Hedwigia, Beiheft 68:3–228.

Croasdale, H. T., Bicudo, C., Prescott, G. W. 1983. A synopsis of North American desmids. Part II. Desmidiaceae: Placodermae. Section 5. University of Nebraska Press, Lincoln, 117 p.

Fritsch, F. E. 1935. The structure and reproduction of the algae, Vol. 1. Cambridge University Press, Cambridge, UK, 791 p.

Gerrath, J. F. 1970. Ultrastructure of the connecting strands in *Cosmocladium saxonicum* de Bary (Desmidiaceae) and a discussion of the taxonomy of the genus. Phycologia 9:209–215.

Gerrath, J. F. 1993. The biology of desmids: A decade of progress. Progress in Phycological Research 9:79–192.

Grote, M. 1977. Untersuchungen zum Kopulationsverlauf bei der Grünalge *Spirogyra majuscula*. Protoplasma 91:71–82.

Happey-Wood, C. M. 1988. Ecology of freshwater planktonic green algae, *in*: Sandgren, C. D., Ed., Growth and reproductive strategies of freshwater phytoplankton. Cambridge University Press, Cambridge, UK, pp. 175–226.

Hargreaves, J. W., Lloyd, E. J. H., Whitton, B. A. 1975. Chemistry and vegetation of highly acidic streams. Freshwater Biology 5:563–576.

Haupt, W. 1972. Perception of light direction in oriented displacement of cell organelles. Acta Protozoologica 11:179–188.

Haupt, W., Schönbohm, W. 1970. Light-oriented chloroplast movements, *in*: Halldal, P., Ed., Photobiology of microorganisms, Wiley, New York, pp. 283–307.

Hoshaw, R. W. 1968. Biology of filamentous conjugating algae, *in*: Jackson, D. F., Ed., Algae, man and the environment. Syracuse University Press, Syracuse, NY, pp. 135–184.

Hoshaw, R. W., McCourt, R. M. 1988. The Zygnemataceae (Chlorophyta): A twenty-year update of research. Phycologia 27:511–548.

Hoshaw, R. W., Wang, J. C., McCourt, R. M., Hull, H. M. 1985. Ploidal changes in clonal cultures of *Spirogyra communis* and implications for species definition. American Journal of Botany 72:1005–1011.

Hoshaw, R. W., Wells, C. V., McCourt, R. M. 1987. A polyploid species complex in *Spirogyra maxima* (Chlorophyta, Zygnemataceae), a species with large chromosomes. Journal of Phycology 23:267–273.

Hoshaw, R. W., McCourt, R. M., Wang, J. C. 1990 Phylum Conjugaphyta, *in*: Margulis, L., Carliss, J. O., Melkonian, M., Eds., Handbook of Protoctista. Jones and Bartlett, Boston, pp. 119–131.

Hosiaisluoma, V. 1975. Muddy peat algae of Finnish raised bogs. Annales Botanici Fennici 12:63–72.

Howell, E. T., South, G. R. 1981. Population dynamics of *Tetmemorus* (Chlorophyta, Desmidiaceae) in relation to a minerotrophic gradient on a Newfoundland fen. British Phycological Journal 16:297–312.

Israelson, G. 1949. On some attached Zygnemales and their significance in classifying streams. Botaniska Notiser 21:313–358.

Kadlubowska, J. Z. 1972. Flora Słodkowodna Polski, Vol. 12A. Chlorophyta. V. Conjugales. Zygnemaceae, Zrostnicowate. Polska Akademia Nauk, Krakow, 431 p.

Kol, E. 1942. The snow and ice algae of Alaska. Smithsonian Miscellaneous Collections 101(16):1–36.

Kol, E. 1944. Vergleich der Kryovegetation der nördlichen und südlichen Hemisphäre. Archiv für Hydrobiologie 40:835–846.

Kol, E. 1964. Cryobiological research in the Rocky Mountains. Archiv für Hydrobiologie 60:278–285.

Kouwets, F., Coesel, P. 1984. Taxonomic revision of the Conjugatophycean family Peniaceae on the basis of cell wall ultrastructure. Journal of Phycology 20:555–562.

Krieger, W. 1933. Rabenhorst's Kryptogamenflora von Deutschland, Österreich und der Schweiz. Bd. 13. Conjugatae. Abteil. 1. Die Desmidiaceen. Teil 1. Akademie Verlagsgesellschaft, Leipzig, 223 p.

Lee, R. E. 1999. Phycology, 3rd ed. Cambridge University Press, Cambridge, UK, 614 p.

McCourt, R. M., Hoshaw, R. W., Wang, J. C. 1986. Distribution, morphological diversity, and evidence for polyploidy in North American Zygnemataceae. Journal of Phycology 22:307–313.

McCourt, R. M., Karol, K. G., Kaplan, S., Hoshaw, R. W. 1995. Using *rbc*L sequences to test hypotheses of chloroplast and thallus evolution in conjugating green algae (Zygnematales, Charophyceae). Journal of Phycology 31:989–995.

Ohtani, S. 1986. Epiphyhtic algae on mosses in the vicinity of Syowa Station, Antarctica. Memoirs of the National Institute of Polar Research, Special Issue 44:209–219.

Ohtani, S. 1990. A taxonomic revision of the genus *Netrium* (Zygnematales, Chlorophyceae). Journal of Science, Hiroshima University, Series B, 23:1–51.

Palamar-Mordvintseva, G. 1976. Novi rodi Desmidiales. Ukrainski Botanichni Zhurnal 33(4):396–398.

Park, N. E., Karol, K. G., Hoshaw, R. W., McCourt, R. M. 1996. Phylogeny of *Gonatozygon* and *Genicularia* (Gonatozygaceae, Desmidiales) based on rbcL sequences. European Journal of Phycology 31:309–313.

Pentecost, A. 1991. A new and interesting site for the calcite-encrusted desmid *Oocardium stratum* Naegeli in the British Isles. British Phycological Journal 26:297–301.

Prescott, G. W. 1962. Algae of the western Great Lakes area, rev. ed. Brown, Dubuque, IA, 977 p.

Prescott, G. W., Croasdale, H. T., Vinyard, W. C. 1972. Desmidiales, Part 1. Saccodermae, Mesotaeniaceae. North American flora, Vol. II, Chap. 6, University of Nebraska Press, Lincoln, 84 p.

Prescott, G. W., Croasdale, H. T., Vinyard, W. C. 1975. A synopsis of North American desmids. Part II. Desmidiaceae: Placodermae. Section 1. University of Nebraska Press, Lincoln, 275 p.

Prescott, G. W., Croasdale, H. T., Vinyard, W. C. 1977. A synopsis of

North American desmids. Part II. Desmidiaceae: Placodermae. Section 2. University of Nebraska Press, Lincoln, 413 p.

Prescott, G. W., Croasdale, H. T., Vinyard, W. C., Bicudo, C. 1981. A synopsis of North American desmids. Part II. Desmidiaceae: Placodermae. Section 3. University of Nebraska Press, Lincoln, 720 p.

Prescott, G. W., Bicudo, C., Vinyard, W. C. 1982. A synopsis of North American desmids. Part II. Desmidiaceae: Placodermae. Section 4. University of Nebraska Press, Lincoln, 700 p.

Rosén, G. 1981. Phytoplankton indicators and their relations to certain chemical and physical factors. Limnologica 13:263–290.

Round, F. E. 1963. The taxonomy of the Chlorophyta. British Phycological Bulletin 2:224–235.

Round, F. E. 1971. The taxonomy of the Chlorophyta. II. British Phycological Journal 6:235–264.

Růžička, J. 1977. Die Desmidiaceen Mitteleuropas, Band. 1, Lief. 1. Schweizerbart'sche Verlagsbuchandlung Stuttgart, 291 p.

Schönbohn, W. 1972 Experiments on the mechanism of chloroplast movements in light-oriented chloroplast movements. Acta Protozoologie 11:211–223.

Smith, G. M. 1950. The fresh-water algae of the United States. 2nd ed. McGraw–Hill, New York, 719 p.

Stein, J. R. 1975. Freshwater algae of British Columbia: The lower Fraser Valley. Syesis 8:119–184.

Stein, J. R., Gerrath, J. F. 1969. Freshwater algae of British Columbia: The Queen Charlotte Islands. Syesis 2:213–226.

Teiling, E. 1948. *Staurodesmus*, genus novum. Containing monospinous desmids. Botany Notices 101:49–83.

Teiling, E. 1950. Radiation in desmids, its origins and consequences as regards taxonomy and nomenclature. Botaniska Notiser 103:299–327.

Teiling, E. 1952. Evolutionary studies on the shape of the cell, and the chloroplasts in desmids. Botaniska Notiser 105:264–306.

Teiling, E. 1957. Morphological investigations of asymmetry in desmids. Botaniska Notiser 111:49–82.

Teiling, E. 1967. The desmid genus *Staurodesmus*. A taxonomic study. Arkiv für Botanik 6:467–629.

Tiffany, L. H. 1936. Wille's collection of Puerto Rican fresh-water algae. Brittonia 2:165–176.

Tiftickjian, J. D., Rayburn, W. R. 1986. Nutritional requirements for sexual reproduction in *Mesotaenium kramstai* (Chlorophyta). Journal of Phycology 22:1–8.

Transeau, E. N. 1916. The periodicity of fresh-water algae. American Journal of Botany 3:121–133.

Transeau, E. N. 1925. The genus *Debarya*. Ohio Journal of Science 25:193–201.

Transeau, E. N. 1926. The genus *Mougeotia*. Ohio Journal of Science 26:311–338.

Transeau, E. N. 1933. The genus *Zygogonium*. Ohio Journal of Science 33:156–162.

Transeau, E. N. 1951 The Zygnemataceae. Ohio State University Press, Columbus, 327 p.

Turner, M. A., Howell, E. T., Robinson, G. G. C., Brewster, J. F., Sigurdson, L. J., Findlay, D. L. 1995. Growth characteristics of bloom-forming filamentous green algae in the littoral zone of an experimentally acidified lake. Canadian Journal of Fisheries and Aquatic Science 52:2251–2263.

van den Hoek, C., Mann, D. G., Jahns, H. M. 1995. Algae. An introduction to phycology. Cambridge University Press, Cambridge, UK, 623 p.

Wang, J. C., Hoshaw, R. W., McCourt, R. M. 1986. A polyploid species complex of *Spirogyra communis* (Chlorophyta) occurring in nature. Journal of Phycology 22:102–107.

West, W., West, G. S. 1904. A monograph of the British Desmideaceae, Vol. 1. Ray Society, London, 224 p.

West, W., West, G. S. 1905. A monograph of the British Desmideaceae, Vol. 2. Ray Society, London, 206 p.

West, W., West, G. S. 1908. A monograph of the British Desmideaceae, Vol. 3. Ray Society, London, 274 p.

West, W., West, G. S. 1912. A monograph of the British Desmideaceae, Vol. 4. Ray Society, London, 191 p.

West, W., West, G. S., Carter 1923. A monograph of the British Desmideaceae, Vol. 5. Ray Society, London, 300 p.

Whitford, L. A., Schumacher, G. J. 1984. A manual of freshwater algae, rev. ed. Sparks Press, Raleigh, NC.

Willén, E. 1992. Planktonic green algae in an acidification gradient of nutrient-poor lakes. Archiv für Protistenkunde 141:47–65.

Yung, Y.-K., Stokes, P., Gorham, E. 1986. Algae of selected continental and maritime bogs in North America. Canadian Journal of Botany 64:1825–1833.

10

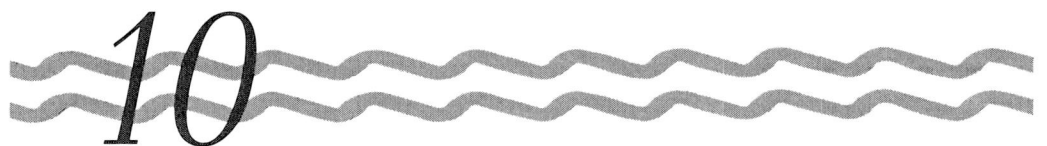

PHOTOSYNTHETIC EUGLENOIDS

James R. Rosowski
School of Biological Sciences
College of Arts and Sciences
University of Nebraska–Lincoln
Lincoln, Nebraska 68588-0118

I. Introduction
II. Diversity and Morphology
   A. Diversity
   B. Morphology and Cellular Structure
   C. Reproduction
III. Ecology and Distribution
IV. Collection, Culturing, and Preparation for Identification
V. Key to and Descriptions of North American Genera
   A. Key
   B. Descriptions of Genera
VI. Guide to Literature for Species Identification
Literature Cited

Our attention then shall first be given to some elegant creatures of a brilliant translucent green hue which are gracefully gliding about. They are of the genus *Euglena*, so called because each is furnished with a very conspicuous spot of a clear red hue, situated near the head, which Ehrenberg, on account of its resemblance to the lowest forms of eyes in the *Rotifera*, that are somewhat similar in colour and appearance, pronounced to be an organ of vision. More recent physiologists, however, doubt the correctness of the conclusion. (Gosse, 1859, p. 444.)

I. INTRODUCTION

The algae of this chapter are unicellular photosynthetic flagellates that are well recognized as euglenoids because of their conspicuous orange-red stigma and striated cell surface, the pellicle. Thus, they are obvious components of algal samples from plankton tows and handgrab samples. Nonetheless, those most representative of the group are not so obvious, because euglenoids as a whole are predominantly colorless, nonphotosynthetic flagellated protozoans. They are classified in the kingdom Protozoa, phylum Euglenozoa (Cavalier-Smith, 1981, 1993) along with 12 other phyla (Cavalier-Smith, 1998), or among 45 phyla when in the division Euglenophyta, kingdom Protista (Corliss, 1984). They also have been classified in the phylum Euglenida, class Euglenophyceae (Margulis *et al.*, 1990), and in the phylum Sarcomastigophora, order Euglenida (Lee and Capriulo, 1990). Those taxa that are nonphotosynthetic and have a feeding apparatus, a cytostome with cytopharynx, are phagotrophs and derive their heterotrophic nutrition from insoluble particles (e.g., bacteria, detritus). Those species that are nonphotosynthetic and lack this feeding apparatus are osmotrophs and require soluble (dissolved) nutrients. Green species are photoauxotrophs; that is, they require light, inorganic nutrients, and one or more vitamins to function. Some species combine two of these modes of nutrition. A vestigial feeding apparatus has now been documented in members of 5 genera of photosynthetic euglenoids (see Sect. II.B.5) and a functional feeding apparatus may occur in two photosynthetic species of *Phacus*.

Among the protozoa with photosynthetic species that have been considered algae are the chlorarachnids, the dinoflagellates, and the euglenoids, whereas the rest of the algae are distributed in the kingdoms Bacteria, Plantae, or Chromista (*sensu* Cavalier-Smith, 1998, in his revised, six-kingdom system). In addition, certain heterotrophic parasites of the protistan phylum Apicomplexa, including one that causes malaria, appear to have had ancestors with functioning plastids (Monastersky, 1998; Graham and Wilcox, 2000) and thus the legacy of the photosynthetic protists may be broader than is now appreciated. In any case, our present knowledge has gained much from the book-length consideration of their ultrastructure and general biology (Leedale, 1967b; Buetow, 1968) that brought them once again to the attention of phycologists and protozoologists and ushered in their serious study by cellular and molecular biologists.

The present treatment of euglenoids (= euglenids, euglenins) focuses mostly on those genera that have photosynthetic members. The green forms are conspicuous in having an orange to red stigma (= eyespot), a striated surface called the pellicle, and one or two (rarely three or four) flagella that are emergent from the flagellar canal (part of the gullet of the older literature). These features, along with their rotatory swimming, the unusual crawling behavior of some, and the ability of many to rapidly change shape, have resulted in the euglenoids being recognized since the earliest days of microscopy. For example, according to D. S. Kellicott [in J. S. Kingsley's *The Riverside Natural History* (Kingsley, 1888, p. 31)], "In 1696, Mr. John Harris described what is undoubtedly *Euglena viridis*." Even earlier, the founding father of protozoology and inventor of the first practical microscope, Anthony van Leeuwenhoek, in 1674, likely observed the same species (Corliss, 1975). Today, this species has provided the first evidence of virus-like particles in a euglenoid (Shin and Boo, 1999).

There are at least seven divisions or classes of algae that, like the euglenoids, appear grass-green when examined in white, transmitted light and have the same green pigments, chlorophylls *a* and *b*, as do all photosynthetic embryophytes (the embryo-producing, green land plants). In only a few species is the chlorophyll often masked by a red pigment (Prescott, 1978; Philipose, 1982; Khan, 1993). The most primitive green algae, the prochlorophytes, are more common than originally thought and now are known to include members not closely related (Graham and Wilcox, 2000), including some with phycobilins (Penno *et al.*, 2000). Such organisms are now placed in the blue-green algae or cyanobacteria (Lee, 1999; Graham and Wilcox, 2000), members of which are without nuclei, chloroplasts, or flagella and belong to the kingdom Eubacteria or Monera (and thus are prokaryotes). The other six groups all have nucleated members (and thus are eukaryotes).

The algae that are green may be broadly grouped on the basis of the chemistry and position of their particulate food reserves, which reflect their evolutionary origins. Unlike most other "green algae," euglenoids and chlorarachnids do not react positively with Lugol's I_2KI (i.e., food storage particles do not turn black or brown-black). Their particulate stored food reserve, the noncarbohydrate polysaccharide paramylon, is always in the cytoplasm rather than within the chloroplast (Leedale, 1967b; Hibberd, 1990; van den Hoek *et al.*, 1995). In some euglenoids the paramylon specifically surrounds pyrenoids or aggregates in the cytoplasm near pyrenoids (which are always inside the chloroplasts). The Prasinophyceae, Chlorophyceae, Charophyceae, and Ulvophyceae (grouped in the Chlorophyta in older literature, or lumped and split in other ways today) show a positive I_2KI starch test for particles of starch, all of which are found only inside their chloroplasts.

The name *Euglena* comes from the Greek and means "eyeball organism," and early on these organisms were known among the Infusoria as "the blood-red eye" animalcules (Brocklesby, 1851). Indeed, the "red eye" is still a key feature to initial recognition of the entire group of green euglenoids and is found in some that would otherwise be colorless (e.g., *Khawkinea*). That is, their single, isolated, bright-orange to red stigma is typically conspicuous in its relatively large size and in its lack of association with any chloroplast, features which, in tandem, separate the green euglenoids from all other green flagellated algae and their reproductive cells (gametes and spores).

Although discovered in the 17th century, euglenoids did not receive serious study for more than another 100 years, at about the time when the cell theory was being proposed. Beginning in the 1830s, Christian Gottfried Ehrenberg, "one of the most prolific and influential protozoologists of early times" (Corliss, 1989), named and formally described several genera and species of green euglenoids: *Euglena* (Ehrenberg, 1830, 1838), *Colacium* (Ehrenberg, 1833, 1838), *Trachelomonas* (Ehrenberg, 1833), and *Cryptoglena* (Ehrenberg, 1831). The last genus turned out to be so rarely recognized or found, and unclear in its euglenoid features, that it took 147 years for it to be vindicated as a euglenoid (cf. Leedale, 1967b, p. 69; Rosowski and Lee, 1978). Ehrenberg's 547-page folio *magnum opus* (Ehrenberg, 1838) has 64 hand-colored copper-engraved plates with such remarkable precision in the recognition and drawing of detail that it considerably advanced the knowledge of the protozoa and microscopic algae, then collectively referred to as the Infusoria. Soon after this color-illustrated folio appeared, and as compound light microscopes became more available, the parlor study of life in pond water and other microscopic wonders became the hobby among many well-to-do Victorians (Barber, 1980). In fact, "evenings at the microscope" developed into a passion for some that lasted for decades (Gosse, 1859, 1896). This focused interest of lay persons, clergy, and medical doctors alike lead to the wide dissemination of the knowledge of cellular features of microorganisms. *Euglena* appears in one of the earliest encyclopedic treatments of natural history in North America, which begins with the mammals and descends to the protozoa (Goodrich, 1859; euglenoids were classified in the division Protozoa, along with other "mouthless Infusoria"). In the early 20th century the euglenoids were removed from the Infusoria (which then included only the ciliated protozoa) and were placed in the Mastigophora with other protozoan flagellates (Conn and Edmondson, 1918; Calkins, 1926), and included 8 of the 10 photosynthetic genera that we consider here. For those who grew up thinking of the protozoa as "one-celled animals," it might come as a surprise to discover that *The Biology of the Protozoa* (Calkins, 1926) has a section "Plastids of the Protozoa" (p. 26). By the mid-20th century euglenoids were claimed by phycologists because the close relationship between the green and colorless forms was appreciated thus placing them in the realm of algae (Fritsch, 1945; Smith, 1950).

A more recent taxonomic treatment of the protozoa (Taylor and Sanders, 1991) excludes details of the euglenoids below the level of order (Euglenida), as likewise the present treatment excludes taxonomic treatment of the colorless euglenoids. Nonetheless, the consideration of both heterotrophic and photoauxotrophic euglenoids as protozoa continues by some authors (Pennak, 1989; Patterson and Hedley, 1992), and currently the euglenoids are entrenched in the zoological literature within the Euglenozoa along with the Kinetoplastidea (Fig. 1F, G) or Kinetoplastida, (Kivic and Walne, 1984; Vickerman, 1990; Vickerman *et al.*, 1991) in the kingdom Protista or Protozoa (Corliss, 1984, 1991b; Cavalier-Smith, 1993, 1998, 1999). [Note, however, that Cavalier-Smith (1993, p. 956) suggests dropping the name Protoctista and its equivalent taxon Protista, *sensu* Margulis (1974), on the basis of an erroneous attribution of the former and excessive diversity within the latter.] No longer is there much debate on whether euglenoids are algae or protozoa, for they still may be considered both (and are taxonomically ambiregnal; Corliss, 1995). The green euglenoid's cytoplasm appears to have originated from Kinetoplastida protozoa whereas its chloroplasts were acquired either by uptake of chloroplasts or more likely by whole "green algae" through secondary endosymbiosis (Gibbs, 1978, 1981; Cavalier-Smith, 1993, 1998, 1999; McFadden, 2001) by the phagotrophic feeding apparatus of colorless euglenoids or their ancestors (Fig. 1F, G). After acquiring photosynthetic organelles, this phagotrophic apparatus (cytostome–cytopharynx, Fig. 1E) became greatly reduced in size and vestigial as in *Tetreutreptia* (Triemer and Lewandowski, 1994), *Eutreptia* (Solomon *et al.*, 1991), *Euglena* (Surek and Melkonian, 1986), *Colacium* (Willey and Wibel, 1985a, b), and *Cryptoglena* (Owens *et al.*, 1988), although variability in these apparati and associated structures make it difficult to determine homologies (Triemer and Farmer, 1991a). In the case of the extant, heterotrophic euglenoids, some might be derived from photosynthetic euglenoids that have lost their photosynthetic ability while retaining the pigmented stigma and parabasal body photoreceptor (e.g., *Khawkinea*), but others may never have had a photosynthetic ancestor [thus are without a stigma or paraflagellar body, e.g., *Distigma* (Dawson and Walne, 1994)]. The origins of the osmotrophic *Astasia* (no stigma) and *Khawkinea* (stigma; Angeler, 2000) with respect to photoauxotroph and chloroplast-containing *Euglena* species continues to be a source of intense study and debate (Bodyl, 1996; Linton *et al.*, 1999; Linton and Triemer, 1999b).

Research on the manner in which nuclear-encoded chloroplast proteins move toward and are incorporated into chloroplasts is providing new evidence on their evolutionary origin. Such information may one day elucidate the manner in which the chloroplast envelope membranes were derived (Haüber *et al.*, 1994; Nakamura, 1994), which is key to understanding the origins of the chloroplasts themselves. For example, although the importation of protein is directly from the cytoplasm into the simple double membrane-bound cyanelles of the glaucophytes, and the chloroplasts of the rhodophytes, green algae, and higher plants, in euglenoids, which have three separate chloroplast membranes in some thin sections (Gibbs, 1978, 1981), chloroplast proteins move in transport vesicles from the endoplasmic reticulum (ER) to the Golgi apparatus (GA), and then transport vesicles derived from the GA fuse with the outermost envelope membrane of the chloroplast (Schwartzbach *et al.*, 1998). Import across the remaining two envelope membranes of the euglenoid chloroplast appears to use the same presequence targeting signals as proposed for import across the two envelope membranes of the glaucophytes, rhodophytes, green algae, and higher plants (Sulli *et al.*, 1999). Such evidence supports the hypothesis that simple chloroplasts (two membrane envelopes) arose from a single, primary, endosymbiotic event whereas complex chloroplasts (three or four membrane envelopes) arose from secondary or tertiary endosymbiotic events. Cavalier-Smith (1999) has proposed that the source of the

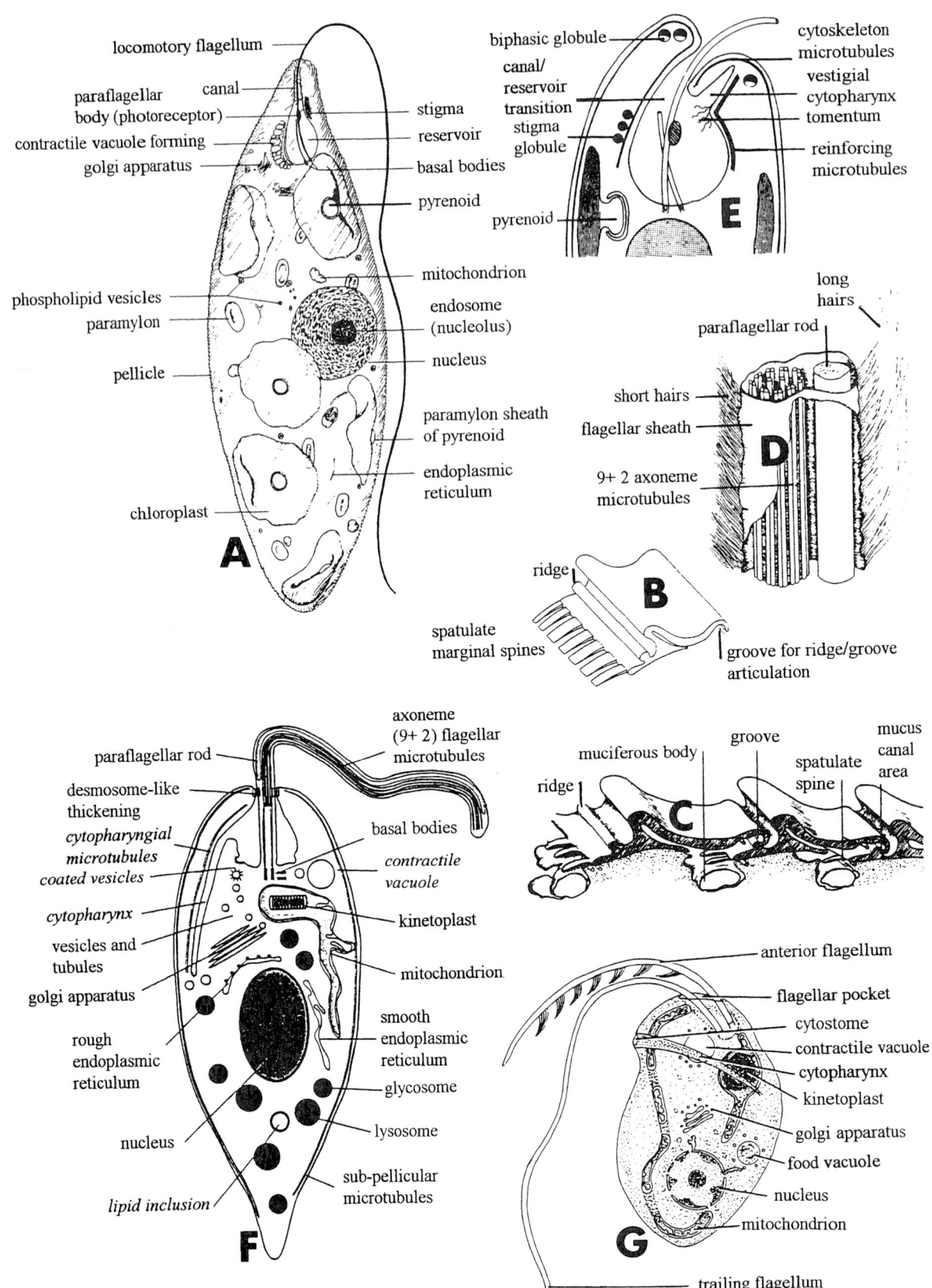

FIGURE 1 Euglenoids and Kinetoplastida protozoans, not to scale. (A) *Euglena* morphology, schematic diagram. Adapted from Carolina Biological Supply Co., Bioreview Sheet 8254, with permission. (B) *Euglena*. Deduced shape of a short piece of one pellicular strip. From Leedale (1964), with permission. (C) Ridge and groove articulation of pellicular strips along their lateral edges. Courtesy of Carolina Biological Supply Co., Bioreview Sheet 8254, with permission. (D) *Euglena* emergent flagellum, ultrastructure representation. Courtesy of Carolina Biological Supply Co., Bioreview Sheet 8254, with permission. (E) Generalized representation of reservoir and vestigial cytopharynx of *Colacium* spp. Willey and Wibel (1985b), with permission. (F) Generalized Kinetoplastida protozoan; italicized structures are not found in all species. From Clayton *et al*. (1995), with permission. (G) The Kinetoplastida protozoan *Bodo sultans*. From Farmer (1980), with permission.

chloroplast was the same for the euglenoids as for the chlorarachnids, originally a photophagotrophic cell, with its former phagosome membrane (vacuolar) becoming the outer chloroplast membrane of its chloroplast endosymbiont. The food reserve of the chlorarachnids is in a host vesicle and is likely paramylon, as in the euglenoids (Hibberd, 1990; van den Hoek et al., 1995). With regard to this host cell and its subsequent evolutionary progeny, Linton et al. (1999) determined that euglenoids form a monophyletic clade, with phagotrophic species diverging before osmotrophic and phototrophic species, and biflagellates diverging prior to uniflagellates in the latter group.

Both the euglenoids and members of the Kinetoplastida (Fig 1F; Farmer, 1980; Cavalier-Smith, 1993) have been proposed to be derived from the bodonids (Fig. 1G; Willey et al., 1988), or all three from a common heterotrophic ancestor (Triemer and Farmer 1991a), based on similarities of their cytoskeletons and on their flagellar, feeding, and mitotic apparati (Triemer and Farmer, 1991b). Among likely ancestral forms having the Type I feeding apparatus would be *Bodo* (Fig. 1D) in the Kinetoplastida and *Petalomonas* in the euglenoids (Triemer and Farmer, 1991b). Studies of ribosomal RNAs among genera of euglenoids suggest that heterotrophs with parallel pellicular striations arose before those with helical striations, and support an interpretation of the morphological data that euglenoids and the Kinetoplastida arose from a common ancestor, a phagotroph with two flagella (Montegut-Felkner and Triemer, 1997; Linton et al., 1999). Finally, rRNA gene sequences of eukaryotes in general suggest that the colorless euglenoids are the oldest of all eukaryotic algae, displacing the red algae for this distinction (Hori and Osawa, 1987; Stiller and Hall, 1997). According to Cavalier-Smith (1991), euglenoids likely developed from a quadriflagellate phagotrophic member of the Parabasalia (which lack mitochondria, but evolved from the Percolozoa that have them) through loss of two flagella. Significantly, euglenoids have discoidal mitochondrial cristae in face view (but see *Diplonema*, Triemer and Farmer, 1991b) as do the Percolozoa (the latter lack Golgi but acquired mitochondria endosymbiotically). Discoidal mitochondrial cristae are not present in other major algal groups (Graham and Wilcox, 2000, p. 134). Because the euglenoids have Golgi, Cavalier-Smith (1991) suggests that the Euglenozoa diverged from the Parabasalia after the evolution of the dictyosome.

As previously stated, euglenoids are largely colorless and appear to have acquired chloroplasts late in their evolution (Whatley, 1993; Melkonian, 1996). In this chapter are considered only those genera that became photosynthetic. Nonetheless it is important to point out key references on heterotrophic euglenoids because they constitute perhaps two-thirds of the species and are closely related to the photoauxotrophs.

General taxonomic information on heterotrophic euglenoids is available in Leedale (1967b), Pringsheim (1963), Bourrelly (1970), Starmach (1983), Margulis et al. (1990), and Shi (1999). Larsen and Patterson (1991) describe 24 heterotrophic euglenoid genera currently recognized worldwide; 10 new taxa of colorless euglenoids from China are described by Shi (1998). Fortunately, for those interested in identifying the nonphotosynthetic euglenoid genera for North America, there are descriptions, photographs, and line drawings of 12 genera and 22 taxa from the southeastern United States by Wołowski and Walne (1997); earlier, Smith (1950) described 12 other genera of colorless members for the United States. Linton and Triemer (1999a) consider the importance of the structural aspects of the feeding apparatus in elucidating heterotrophic euglenoid relationships, and also stress the need for more information on the flagellar apparatus of additional species (graphically compared by Walne and Dawson, 1993). For an introduction to the phagotrophy of protozoa on phototrophic algae, and for an appreciation of the intricate beauty of the photosynthetic euglenoids themselves, see the luminous color photographs in Canter-Lund and Lund (1995).

Finally, Walne and Kivic (1990) consider fossil euglenoids and provide a useful discussion on the commercial uses of euglenoids in agriculture, biology, and medicine. *Euglena gracilis* strain Z is the first euglenoid to have its chloroplast DNA genome sequenced (Hallick et al., 1993). *Euglena gracilis* strain Z is used in medical science as a bioassay for vitamin B_{12} in blood plasma, and it may one day become useful in aquaculture because of the significant enhancement of its long chain, unsaturated fatty acid profile (particularly with DHA), making it a nutritious food item for the mass production of cultured marine fish larvae (Hayashi et al., 1993).

II. DIVERSITY AND MORPHOLOGY

A. Diversity

There are 13 generally accepted genera of photosynthetic euglenoids worldwide. More than 800 species of heterotrophs and autotrophs have been named (Huber-Pestalozzi, 1955), with the total species estimated at 2000 (Norton et al., 1996). Ten pigmented genera have been reported in fresh waters of North America (Smith, 1950; Prescott, 1978; Dillard, 2000), but only five are common enough to have been treated in a general text on aquatic invertebrates (Pennak,

1989). Genera and species of euglenoids have been established on the basis of light and electron microscopy of vegetative cells and the chemistry of pigments and food reserves, for there is no sexuality in euglenoids yet verified (Leedale, 1967b). Given the great cell-size range and morphological variation now recognized among euglenoid populations throughout the world [see species cell-size ranges in Zakryś (1986) and Zakryś and Walne (1994)], it is likely that there are far fewer species than have been described. For example, the examination of populations of closely related species in nature and in culture invariably leads to the lumping of some species based on morphology (Pringsheim, 1956). As molecular and standard cladistic analyses yield further refinements of the species concept (Shi, 1996a; Zakryś, 1997b), even fewer taxa could result. For example, 14 intraspecific taxa and 4 previously separate species appear to be synonyms of *Euglena agilis* (Zakryś, 1997a). On the other hand, this reduction in species number by lumping is somewhat offset by discovery and description of new species, which occurs regularly with serious study of this cosmopolitan group (Zakryś, 1994; Zakryś and Walne, 1994; Shi, 1995; Bicudo and Wołowski, 1998; Thérézien, 1999). With regard to nomenclature, there are various viewpoints for describing and naming these protists (Corliss, 1990, 1991a, 1995), which are considered by Margulis (Margulis *et al.*, 1990) "no more one-celled animals and one-celled plants than people are shell-less multicellular amebas."

Of the 10 genera considered here, 7 are naked (no lorica): *Euglena* (1 emergent flagellum; generally with a flexible body); *Cryptoglena* (1 emergent flagellum; cell with a deep longitudinal ventral sulcus); *Phacus* (1 subapically emergent flagellum; cells rigid, flattened, and leaflike); *Lepocinclis* (1 apically emergent flagellum; cells rigid and circular in median transverse view); *Eutreptia* (2 equal-length, emergent flagella); *Euglenamorpha* (3 emergent flagella); and the epizoic genus *Colacium* (cells attach by their anterior, mostly to aquatic arthropods, both larvae and adults). Also included here are the 3 loricate genera: *Ascoglena* (attached posteriorly to nonmotile substrata); *Trachelomonas* (unattached; motile cells usually with a rigid, ovoid, reddish-brown lorica); and *Strombomonas* (unattached; motile cells often urn-shaped; loricas flexible, slightly brown, yellow, or clear). There is one marine loricate euglenoid genus *Klebsiella* Pascher 1931 (posteriorly sessile; lorica not surrounding the anterior cell apex). However, this latter taxon is apparently unreported since Pascher's original description (Leedale, 1967b). All species of *Eutreptiella* (2 emergent flagella, unequal in length) thus far are marine (Leedale, 1967b), as is the quadriflagellate *Tetreutreptia* (McLachlan *et al.*, 1994). The endosymbiont of the dinoflagellate *Noctiluca*, named *Protoeuglena* Subramanyan (cited and figured in Prescott, 1984), was considered by Leedale (1967b, p. 69) of doubtful affinity to euglenoids (with no paramylon and with the stigma in a chloroplast) and subsequently was ultrastructurally shown to be a prasinomonad, *Pedinomonas noctilucae* (Taylor, 1990).

B. Morphology and Cellular Structure

The characteristic morphological features of green euglenoids at the level of bright-field microscopy are described in this section, as well as certain ultrastructural features basic to interpreting cellular diversity and developing concepts in phylogeny. (See Figs. 1–24.) Unique generic features are presented in Section V, along with a key to the 10 green, freshwater, North American genera, both common and rare.

1. Cell Morphology

The euglenoids are ovoid, spindle-shaped, or flattened single cells (unicells), of various transverse shapes depending on the genus or species. In the spindle-shaped forms (*Euglena*) a cell may be straight (Fig. 5J) or twisted (Fig. 6E, F), even triradiate (Fig. 5R, S), to circular or laterally compressed in median, transverse-sectional view (*Phacus*, Fig. 13B, C). Most *Euglena*, *Colacium*, and *Trachelomonas* species can change cell shape (metaboly). In leaflike forms (*Phacus*), cells can be highly laterally compressed and do not change shape. No euglenoids have true cell walls, but some genera (*Phacus* and *Lepocinclis*) nevertheless have a rigid cell surface, the pellicle, and are naked as well (no lorica). The common enclosed genera (*Trachelomonas*, *Strombomonas*) have a mucilaginous covering outside the plasma membrane called the lorica, and in the former genus it may be dark brown, highly mineralized, and often spiny (Figs. 16–23). Species of *Colacium* secrete mucilaginous stalks at their canal end that, after repeated cell divisions, result in colonies like those of dichotomously branched, stalked diatoms (Figs. 9 and 11D, E). However, these stalks are not a regular or even consistent feature of all species, and no *Colacium* species, or other euglenoids, are considered colonial although they may form palmelloid masses in a nonmotile phase (*E. myxocylindracea*, Bold and MacEntee, 1973; *Colacium libellae*, Rosowski and Willey, 1975).

The figures in Huber-Pestalozzi (1955), Bourrelly (1970), Starmach (1983), Tell and Conforti (1986), and Shi (1999) provide a good morphological overview and starting point for the study of diversity of green euglenoid species, and Ettl and Popovský (1986)

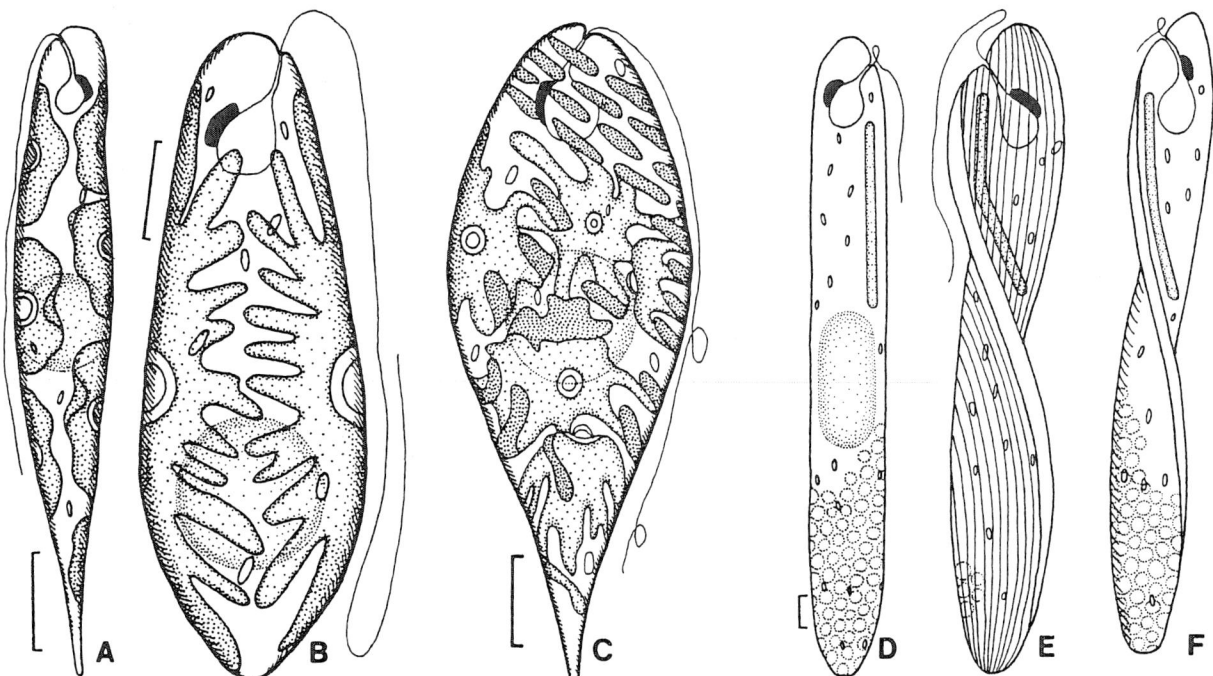

FIGURE 6 Euglena morphology and diversity, continued: (A) *E. caudata*, swimming cell; (B) *E. jirovecii*; (C) *E. sociabilis*, showing chloroplast lobes extending to pellicle (see also Fig. 2D); (D)–(F) *E. truncata* var. *baculifera* [(D) cellular detail; (E) pellicle detail; (F) swimming cell]. From Zakryś and Walne (1994), with permission. Scale bars = 10 μm.

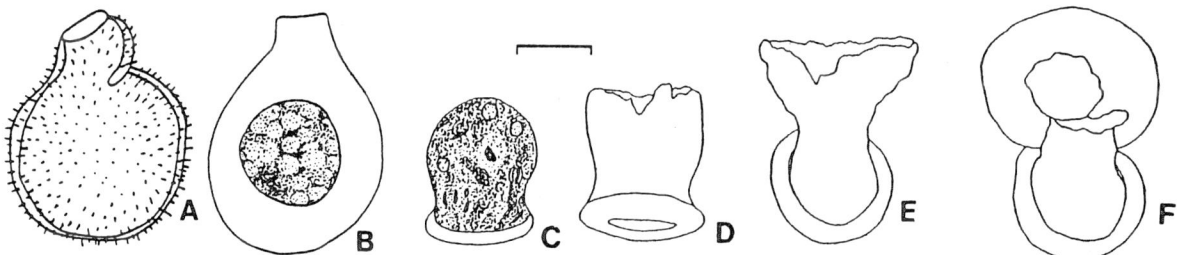

FIGURE 7 Euglena cysts: (A) and (B) Cysts of *Euglena* sp.; (C)–(F) cysts of *E. tuba*, scale bar = 2 μm [(C) tuba-like cyst; (D)–(F) deserted cyst walls]. A and B from Prescott (1955), with permission. C–F from Johnson (1944), with permission.

provide a useful discussion of the taxonomic issues of algae in general. The works of Pringsheim (1948, 1953a, b, 1956) remain required reading for euglenoid biologists. For discussions of euglenoid evolutionary stock and useful criteria for assessing evolutionary trends, see Willey and Wibel (1985b), Willey *et al.* (1988), Triemer and Farmer (1991b), Kuźnicki and Walne (1993), Dawson and Walne (1994), Montegut-Felkner and Triemer (1997), Schwartzbach *et al.* (1998), Linton *et al.* (1999), and Linton and Triemer (1999a, b). For a review of an extensive historical literature considering the protozoa as acellular, rather than as cellular, organisms, see Corliss (1989).

2. Stigma

The most obvious feature of living, pigmented, light-grown euglenoids is their single, bright, orange-to-red anterior stigma (eyespot), which is on average 10 times larger than the stigma of chlorophycean green algae (Walne, 1971). It is always outside of chloroplasts (Fig. 1A), lateral to the reservoir of the flagellar pocket (Fig. 3, and Walne, 1971), takes on the curved

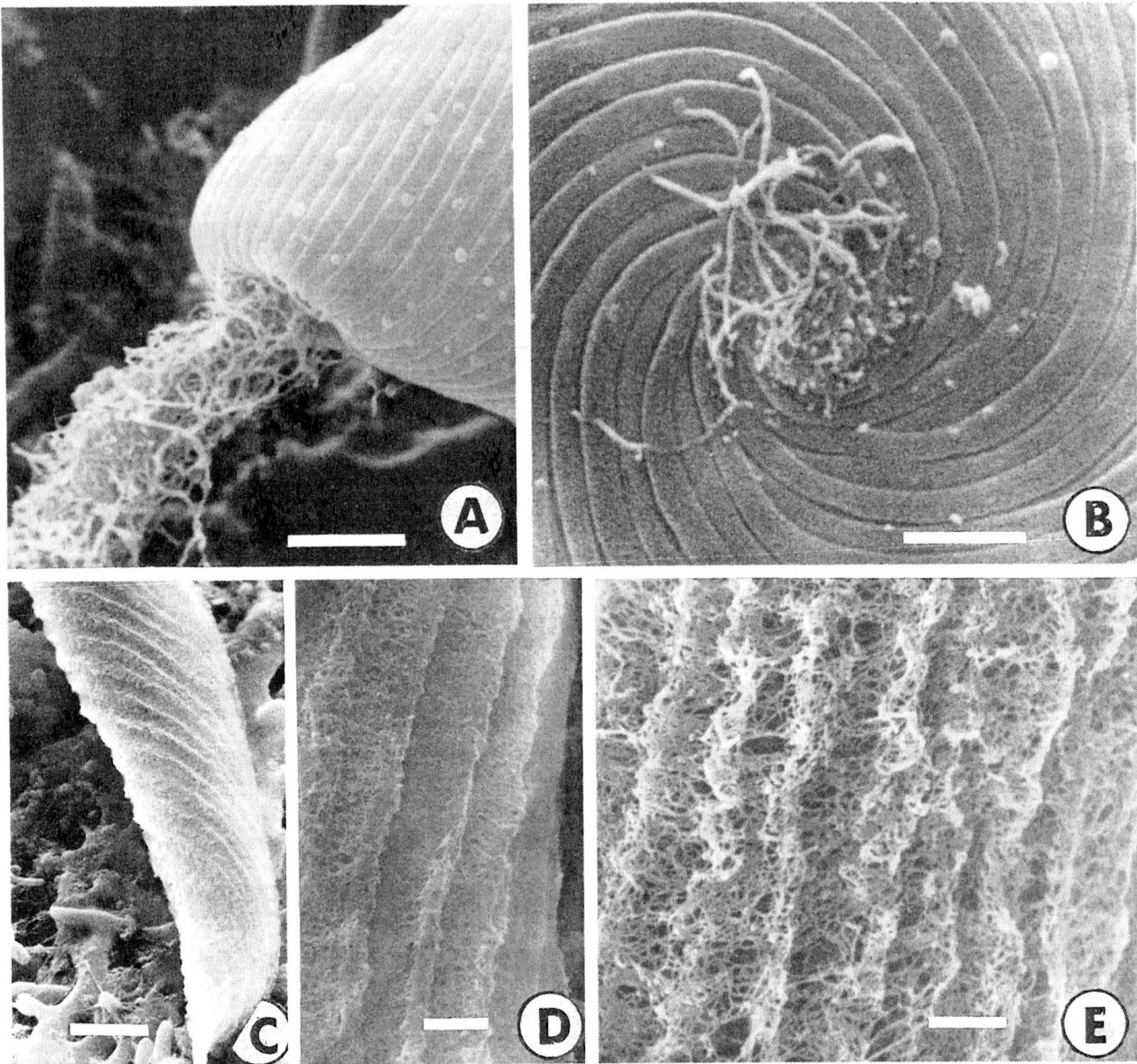

FIGURE 8 (A)–(E) *Euglena gracilis* Z strain; scanning electron micrographs: (A) Motile cell apex showing subapical emergent flagellum with hairs, and pellicle without mucilage strands; scale bar = 1 μm; (B) mucilage strands on posterior tip of a contracted, nonmotile cell strands; scale bar = 1 μm; (C)–(E) *E. tripteris*, creeping cells [(C) Mucilage obscures the pellicular strips and their articulation; scale bar = 5 μm; (D) densest mucilage above the edge of each pellicular strip pair; scale bar = 1 μm; (E) mucilage appears as waves above the pellicular strips, which are invisible; scale bar = 1 μm]. From Rosowski (1977), with permission.

shape of the pocket (reservoir) to which it is appressed (Figs. 3, 5, and 6), and is on the same side as, and anterior to, the contractile vacuole (not as drawn in Fig. 1A). The stigma is composed of tiny, and usually individual, tightly membrane-bound orange-to-red carotenoid globules, chemically composed of more than two dozen kinds (including astaxanthin and/or echineone). These globules are not in an orderly or definite arrangement within their clustered unit, although small filaments connect some of them to microtubules associated with the reservoir cytoskeleton (Kuźnicki *et al.*, 1990). The stigma of motile green euglenoids is opposite the paraflagellar body (a swelling) of the emergent flagellum in the common genera (Fig. 1A). This positioning has suggested to some observers (Walne *et al.*, 1998) that together they

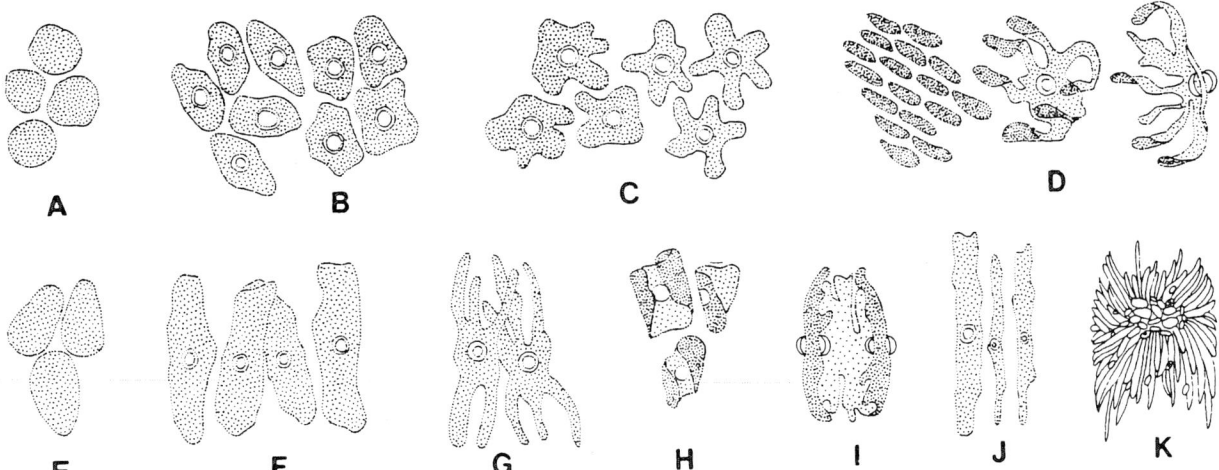

FIGURE 2 *Euglena* chloroplast morphology, (A–K): (A) Disc-shaped, smooth margin, no pyrenoids; (B) disc-shaped, slightly sinuous margin, central pyrenoid; (C) disc-shaped, sinuous margin, central pyrenoid; (D) left figure is surface view of tip of fringed margin, center figure is front view, right figure is side view showing double-sheathed pyrenoid; (E) ovate, no pyrenoid; (F) lobate, slightly sinuous; (G) elongate, fringed, central pyrenoid; (H) U-shaped, slightly sinuous margin, central pyrenoid; (I) fringed margin, two double-sheathed pyrenoids; (J) spatulate, central pyrenoid; (K) many ribbon-shaped, short to long, likely each with terminal, central pyrenoid (one per chloroplast), paramylon-covered = paramylon center. From Batko and Zakryś (1995), with permission.

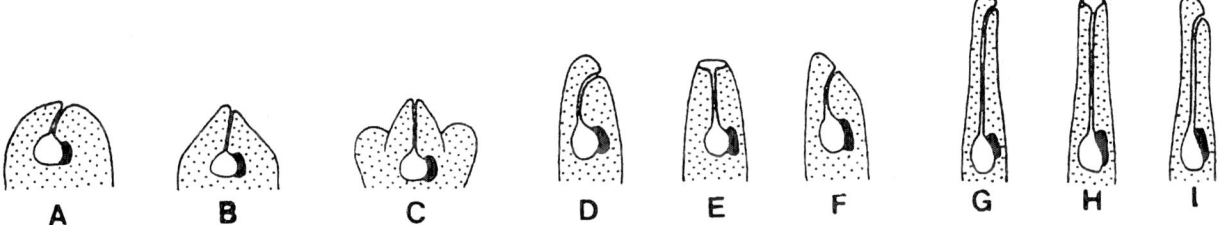

FIGURE 3 *Euglena* anterior apex morphology, (A–I): (A) Wide, rounded; (B) wide, acute; (C) wide, crown-shaped; (D) narrow, rounded; (E) narrow, truncated; (F) narrow, oblique truncated; (G) elongated, rounded; (H) elongated, truncated; (I) elongated, oblique truncated. From Batko and Zakryś (1995), with permission.

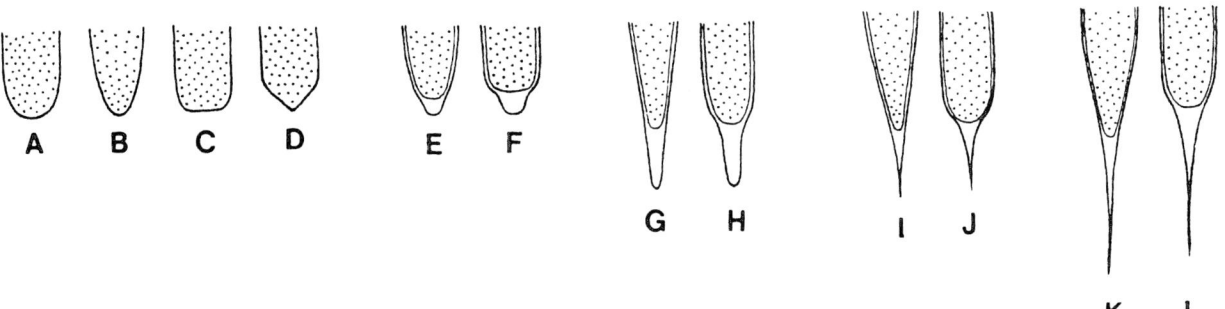

FIGURE 4 *Euglena* posterior cell apex morphology, (A–L): (A) Wide, rounded; (B) tapered, rounded; (C) wide, truncated; (D) wide, acute; (E) tapered, peg-tipped; (F) wide, peg-tipped; (G) narrow and tapered, snout-shaped and blunt; (H) wide, snout-shaped and blunt; (I) evenly tapering to a short point; (J) wide, tapering to a short point; (K) evenly tapering to a long point; (L) wide, tapering to a long point. From Batko and Zakryś (1995), with permission.

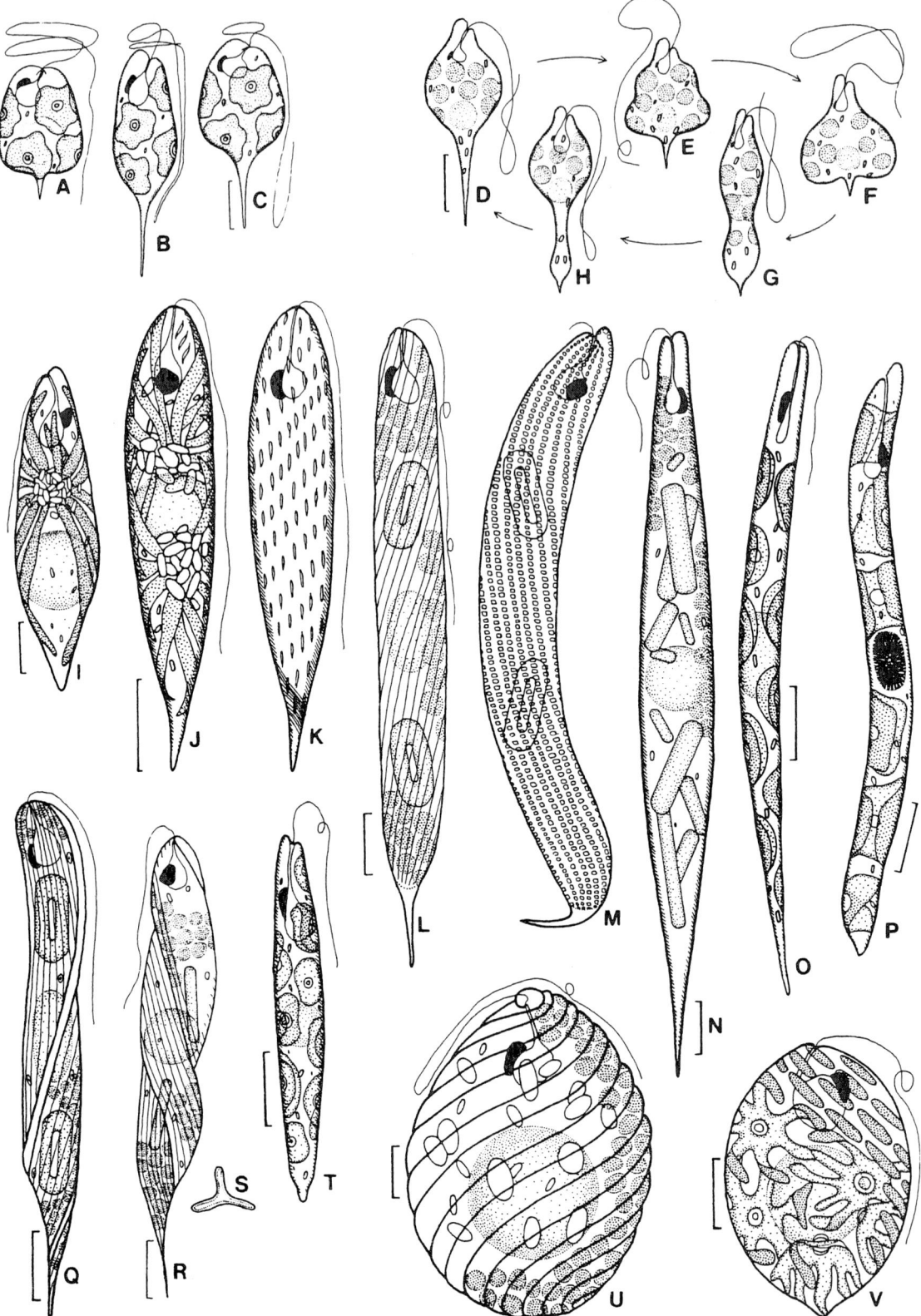

FIGURE 5 *Euglena* cell morphology and diversity, (A–H), metaboly in *Euglena*: (A–C) Metaboly in *E. clavata* [(A) swimming cell; (B) and (C) during metaboly]; (D–H) sequence of metaboly in *E. repulsans*. *Euglena* diversity, (I–V): (I) *E. viridis*, numerous chloroplasts radiate from a paramylon center; (J and K) *E. chadefaudii* [(J) chloroplasts aggregate in two paramylon centers; (K) surface view showing regular arrangement of trichocysts]; (L) and (M) *E. spirogyra* [(L) numerous discoid chloroplasts without pyrenoids; (M) pellicle with regular rows of exterior, mineralized ornamentation]; (N) *E. acus*; (O) *E. adhaerens*; (P) *E. mutabilis*; (Q) *E. oxyuris*; (R) and (S) *E. tripteris* [(S) transverse section]; (T) *E. gracilis*; (U) *E. texta*, not metabolic; (V) *E. oblonga*, tips of ribbon-shaped chloroplast lobes extend to the pellicle (see also Fig. 2D). From Zakryś and Walne (1994), with permission. Scale bars = 10 μm.

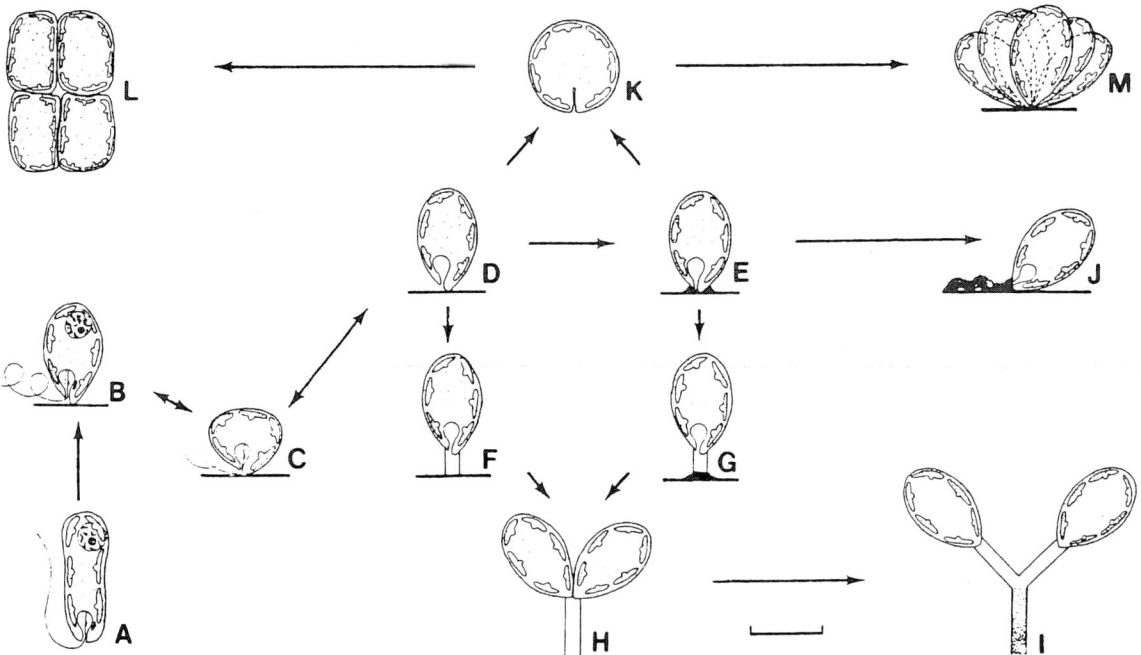

FIGURE 9 Life cycle of *Colacium vesiculosum* Ehr, the most common green epibiont on microcrustaceans; from studies of clonal cultures in soil–water pea medium; all are side views except L, which is top view; scale bar = 10 µm: (A) Motile cell, not common in nature but common in culture; (B) motile cell attaches, flagella becomes motionless; (C) cell undergoes metaboly before flagellum disappears; (D) 30 seconds after C; (E) attached cell produces brownish holdfast of various shapes; (F) the cell at D produces a clear stalk without holdfast; (G) a clear stalk is produced following holdfast formation; (H) cell at F or G divides and remains attached to the clear stalk; (I) each sibling cell produces its own stalk attached to the parent stalk. The original stalk may then turn brown below the branch; (J) the cell at E produces ropelike incrustations; (K) attached cells may not produce stalks but instead divide, beginning at the attached end; (L) attached cells enter the palmella stage, divide, and spread out in a one-layer sheet; (M) attached cells divide but sometimes fail to move over the substratum. From Rosowski and Kugrens (1973), with permission.

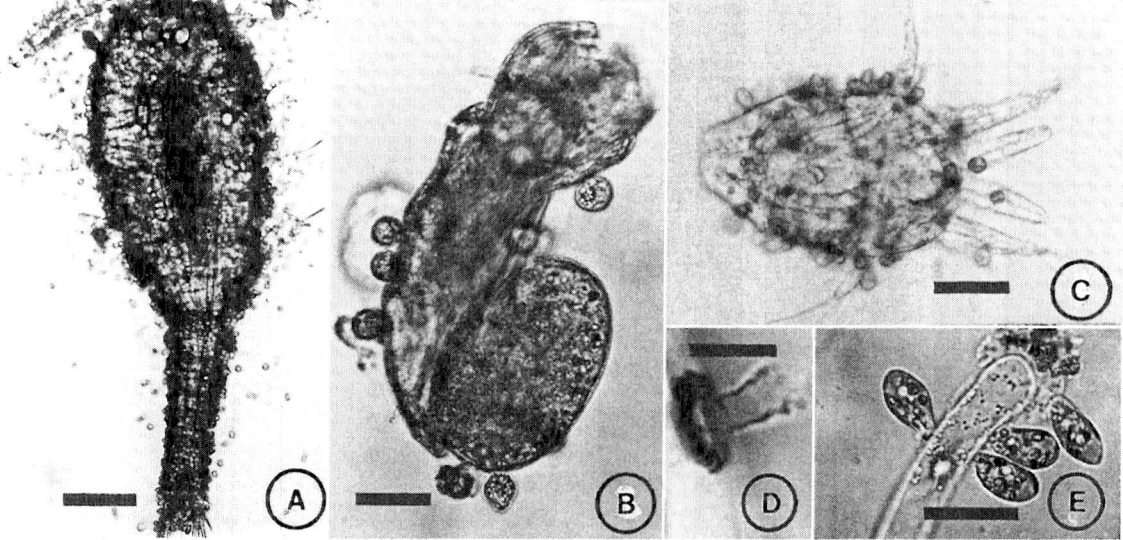

FIGURE 10 *Colacium vesiculosum* from nature or culture, (A)–(E): (A) *Colacium* on *Cyclops* sp., natural collection, 2°C; scale bar = 100 µm; (B) *Colacium* contracted on cuticle of *Keratella* sp., scale bar = 25 µm; (C) *Colacium* on nauplius, 5°C; scale bar = 50 µm; (D) Brown holdfast with clear stalk, natural collection, 4°C; scale bar = 5 µm; (E) *Colacium* sp. attached to *Mougeotia* sp., from 2-week-old biclonal culture; scale bar = 10 µm. From Rosowski and Kugrens (1973), with permission.

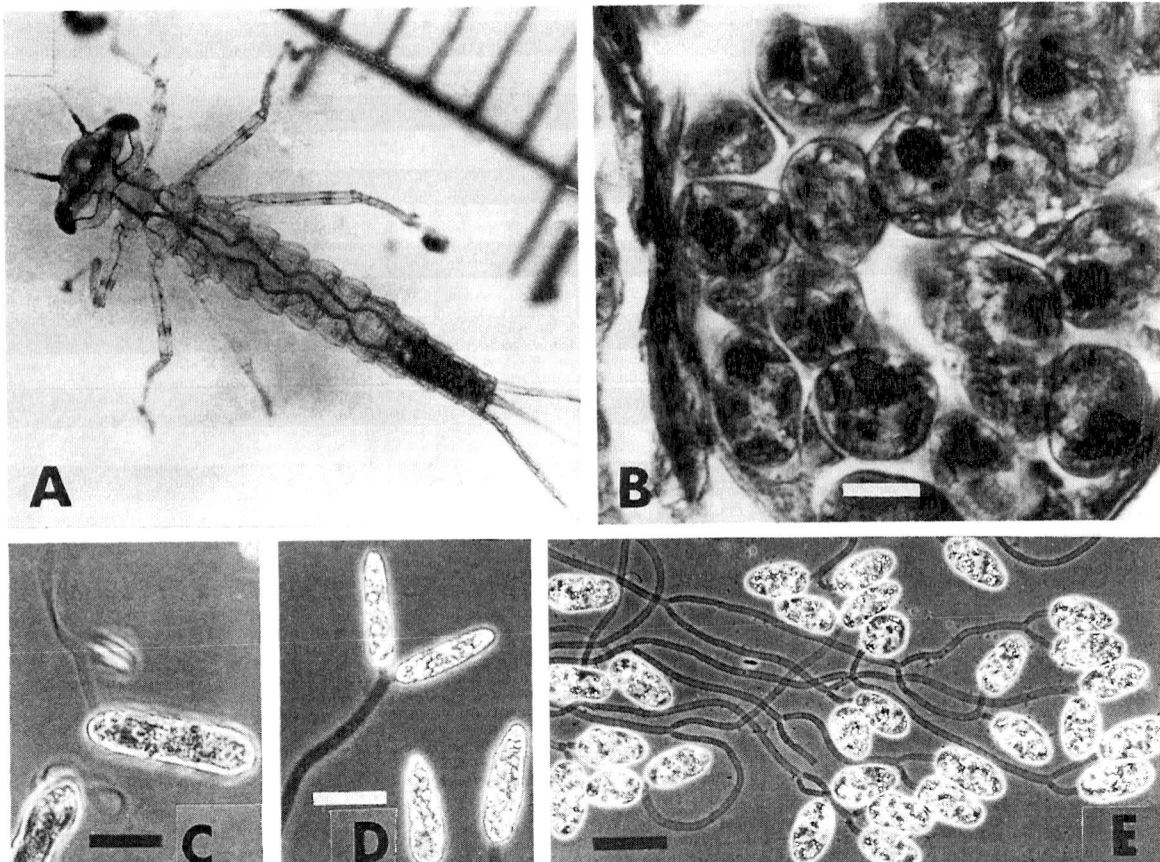

FIGURE 11 *Colacium libellae*, (A)–(E): (A) and (B) *Colacium libellae* in rectum of the damselfly *Ischnura verticalis*, bright-field microscopy [(A) rectum has a 2-mm-long dark area (green plug) of *C. libellae*; (B) *C. libellae* palmellae from the rectum, Feulgen fast-green stained; scale bar = 10 μm]; (C)–(E) cultured cells of *C. libellae*, phase contrast microscopy [(C) motile cells; scale bar = 15 μm; (D) stalked, elongated cells on 4-μm-wide stalk, log-phase culture; scale bar = 20 μm; (E) ovoid cells on long stalks from a 3-month-old culture, attached to glass; scale bar = 20 μm]. From Rosowski and Willey (1975), with permission.

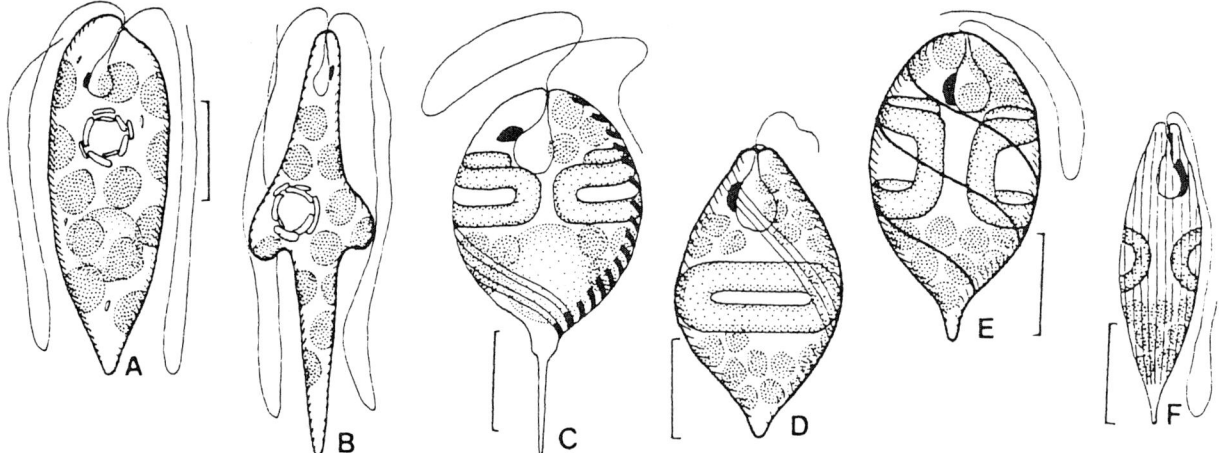

FIGURE 12 Euglenoid morphology, continued: (A) and (B) *Eutreptia globulifera* [(A) swimming cell; (B) cell in metaboly]; (C)–(F) *Lepocinclis* spp., chloroplasts discoid and without pyrenoids [(C) *L. capito*, pellicular strips shown in their entirety for bottom two rows and as dark abbreviated black bands along the right side above those rows; (D) *L. fusiformis*; (E) *L. ovum*; (F) *L. marssonii*]. From Zakryś and Walne (1994), with permission. Scale bars = 10 μm.

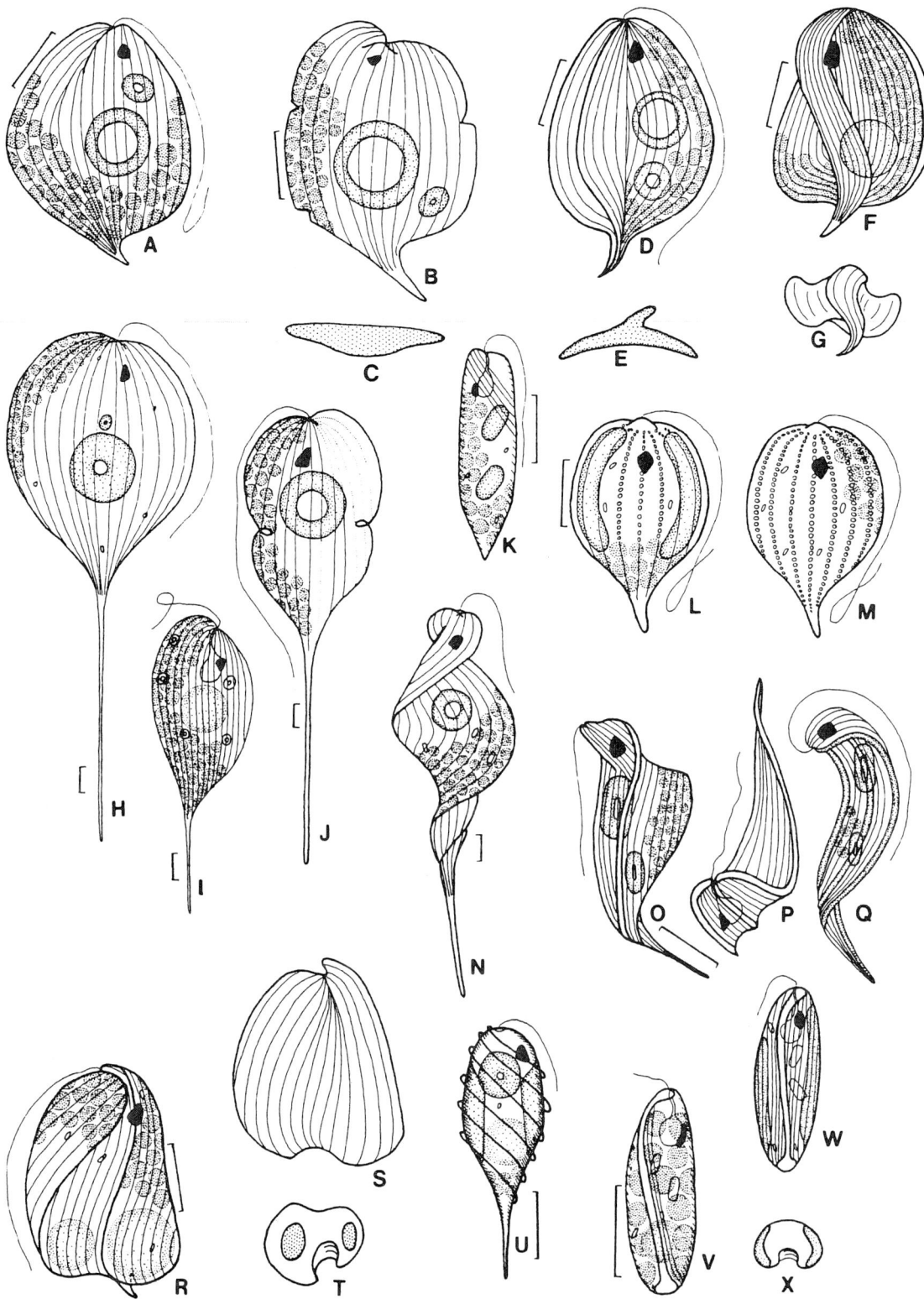

FIGURE 13 Morphology of *Phacus* spp., (A)–(X) Chloroplasts mostly discoid, numerous, and without pyrenoids: (A) *Ph. pleuronectes*; (B) and (C) *Ph. orbicularis* var. *undulatus* [(C) Transverse section]; (D) and (E) *Ph. triqueter* [(E) Transverse section]; (F) and (G) *Ph. mariana* [(G) posterior view]; (H) *Ph. longicauda*; (I) *Ph. elegans*; (J) *Ph. longicauda* var. *insecta*; (K) *Ph. polytrophos*; (L) and (M) *Ph. monilata* [(L) Internal detail; (M) pellicle ornamentation]; (N) *Ph. longicauda* var. *tortus;* (O)–(Q) *Ph. trimarginatus* [(O) internal detail; (P) pellicle features; (Q) swimming cell]; (R)–(T) *Ph. curvicauda* fa. *anomalus* (R) Internal features showing numerous discoid chloroplasts; (S) Pellicular strips of the pellicle; (T) Transverse section]; (U) *Ph. pyrum*; (V)–(X) *Ph. agilis* (V) Internal features showing numerous discoid chloroplasts; (W) Pellicle view showing two lateral, massive paramylon grains plus much smaller paramylon grains; (X) Transverse section showing ventral sulcus and lateral paramylon grains]. From Zakryś and Walne (1994), with permission. Scale bars = 10 μm.

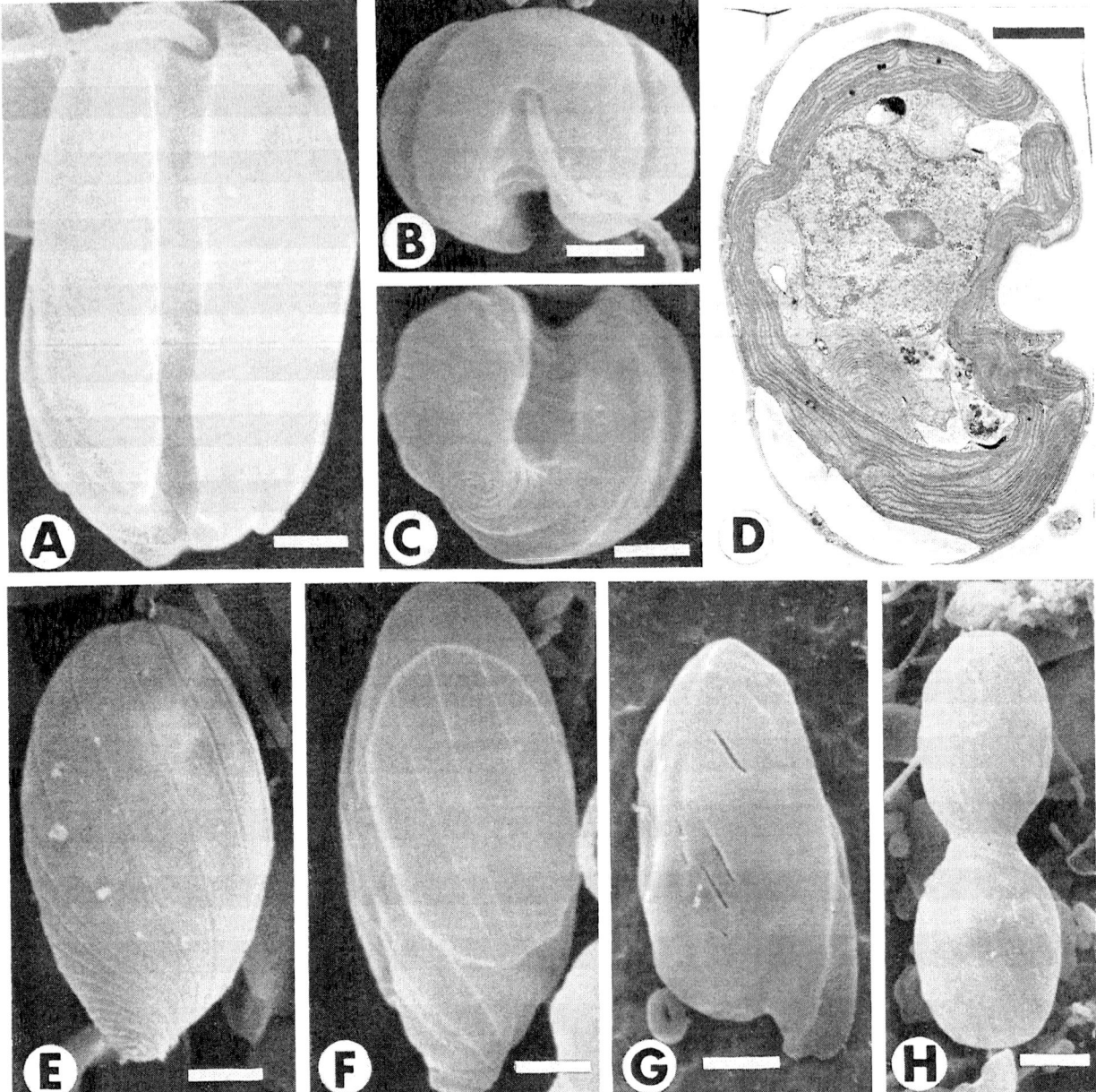

FIGURE 14 Cryptoglena pigra, (A)–(H), scanning and transmission electron micrographs: (A) Ventral cell surface showing emergent flagellum at top and longitudinal sulcus; scale bar = 2 μm; (B) anterior view showing sulcus and venturi-like flanges caused by lateral, shieldlike paramylon grains; scale bar = 2 μm; (C) posterior apical view showing sulcus and reduction in size of the pellicular strips as they twist towardsthe apex; scale bar = 2 μm; (D) transverse view of cell showing single chloroplast capped laterally by paramylon grains; scale bar = 1 μm; (E) dorsal view or pellicle, cell anterior toward the top; scale bar = 2 μm; (F) lateral view of the pellicle, cell anterior toward the top; shieldlike paramylon grain is outlined by the pellicle; scale bar = 2 μm; (G) shieldlike paramylon grain isolated from whole cells; smaller paramylon grain above the letter G; scale bar = 2 μm; (H) sibling cells about to separate following mitosis; scale bar = 3 μm. B–D, Rosowski and Lee (1978), with permission; others original.

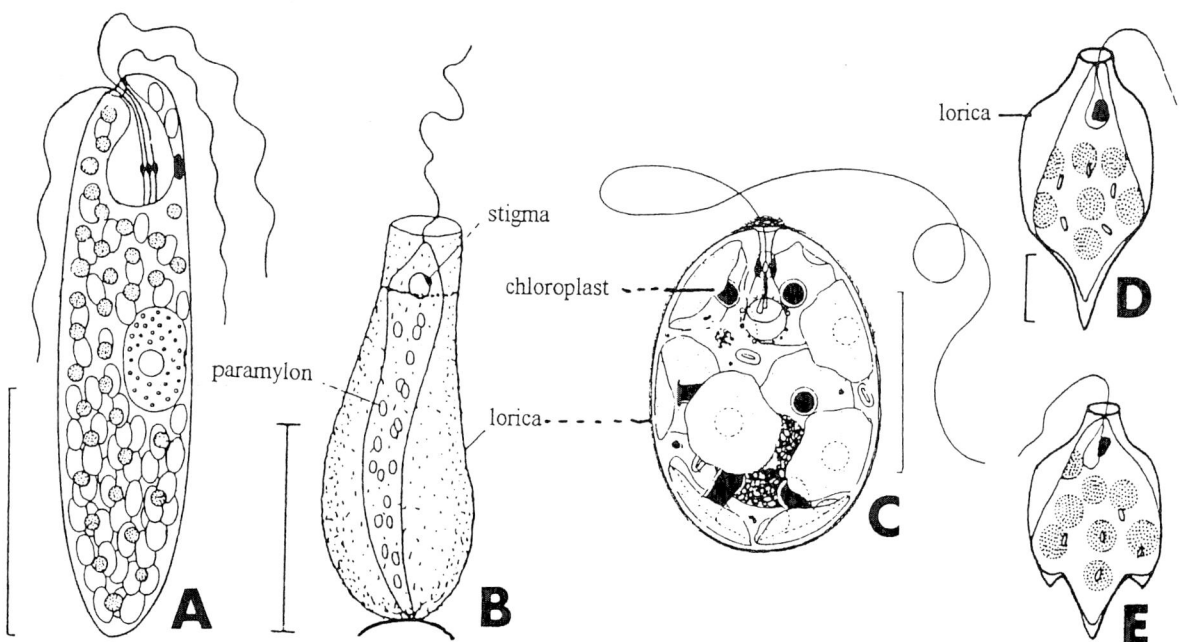

FIGURE 15 Euglenoid morphology, continued, (A)–(E): (A) *Euglenamorpha hegneri*, each of the three flagella has a paraflagellar body opposite the single stigma; scale bar = 25 µm; chloroplasts are discoid and numerous; paramylon grains are oval and numerous; (B) *Ascoglena vaginicola*; scale bar = 25 µm; (C) *Trachelomonas grandis*; inner projecting pyrenoids (black in figure); scale bar = 25 µm; (D) and (E) *Strombomonas urceolata* [(D) swimming cell; (E) in metaboly, the lorica flexes; scale bar = 10 µm]. A–C from Leedale (1967b), with permission. D–E from Zakryś and Walne (1994), with permission.

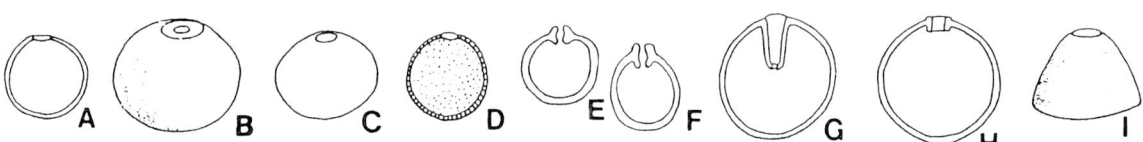

FIGURE 16 Loricas of species of *Trachelomonas*., (A)–(I): (A) *T. volvocina*, × 797; (B) and (C) *T. volvocina* var. *compressa* [(B) × 1275; (C) × 956]; (D) *T. volvocina* var. *punctata* × 1275; (E) and (F) *T. volvocina*, × 638; (G) and (H) *T. varians*, × 752; (I) *T. triangularis*, × 1275. From Prescott (1962), with permission.

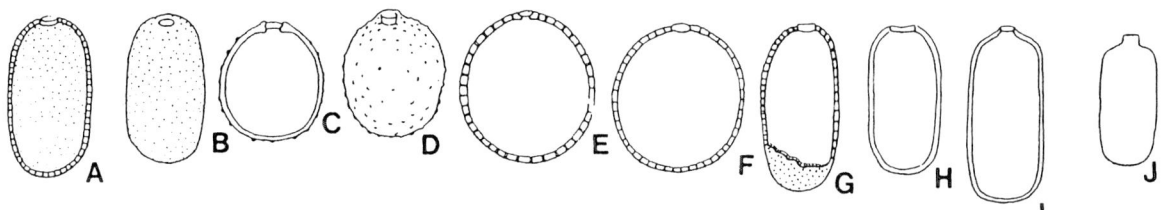

FIGURE 17 Loricas of species of *Trachelomonas*, (A)–(J): (A) and (B) *T. lacustris*, × 797; (C) and (D) *T. kelloggii*, × 797; (E) *T. rotunda*, × 797; (F) *T. intermedia*, × 797; (G) *T. erecta*, × 765; (H) *T. lacustris*, × 797; (I) *T. cylindrica*, × 797; (J) *T. dubia*, × 797. From Prescott (1962), with permission.

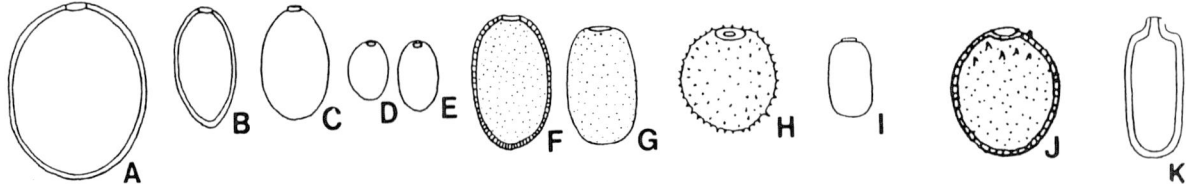

FIGURE 18 Loricas of species of *Trachelomonas*, (A)–(K): (A) *T. dybowskii*, × 1,275; (B) and (C) *T. pulcherrima*, × 797; (D) and (E) *T. pulcherrima* var. *minor*, × 892; (F) and (G) *T. abrupta*, × 797; (H) *T. robusta*, × 956; (I) *T. cylindrica*, × 797; (J) *T. acanthostoma*, × 797; (K) *T. dubia*, × 797. From Prescott (1962), with permission.

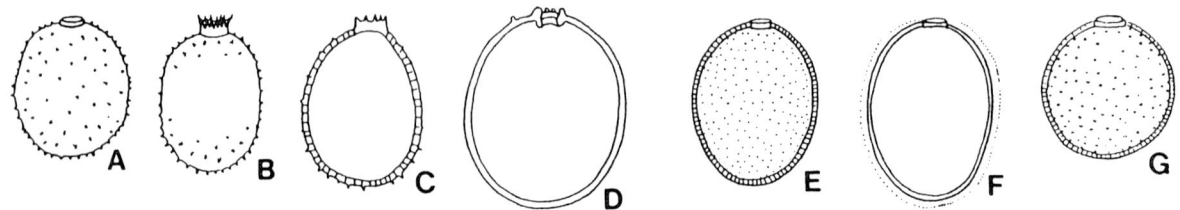

FIGURE 19 Loricas of species of *Trachelomonas*, (A)–(G): (A) *T. hispida*, × 752; (B) *T. hispida* var. *coronata*, × 752; (C) *T. hispida* var. *crenulatocollis* fa. *recta*, × 765; (D) *T. hispida* var. *papillata*, × 829; (E) and (F) *T. hispida* var. *punctata*, × 797; (G) *T. acanthostoma*, × 956. From Prescott (1962), with permission.

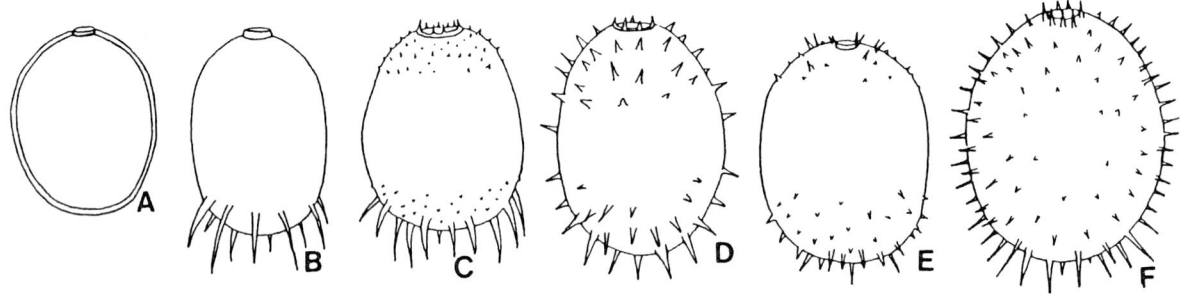

FIGURE 20 Loricas of species of *Trachelomonas*, (A)–(F): (A) *T. armata* fa. *inevoluta*, × 765; (B) *T. armata*, × 956; (C) *T. armata* var. *longispina*, × 892; (D) *T. armata* var. *steinii*, × 956; (E) *T. superba* var. *swirenkiana*, × 892; (F) *T. superba* var. *spinosa*, × 956. From Prescott (1962), with permission.

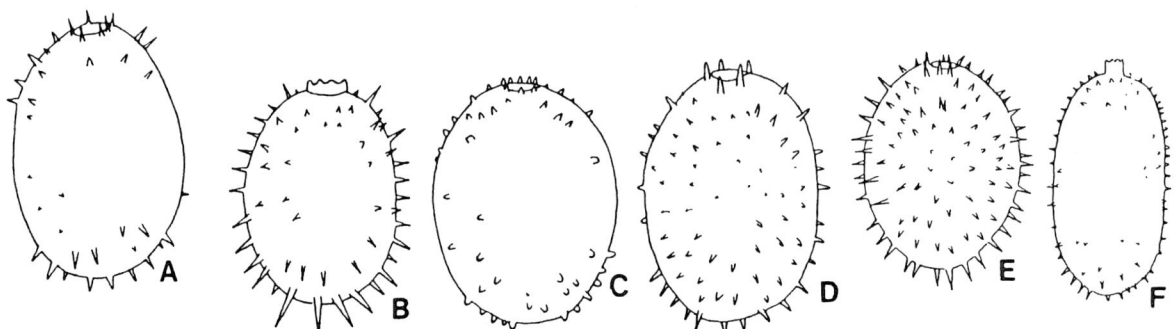

FIGURE 21 Loricas of species of *Trachelomonas*, (A)–(F): (A) and (B) *T. superba* var. *swirenkiana* [(A) × 797; (B) × 956]; (C) *T. superba*, × 956; (D) *T. superba* var. *duplex*, × 956; (E) *T. horrida*, × 956; (F) *T. sydneyensis*, × 765. From Prescott (1962), with permission.

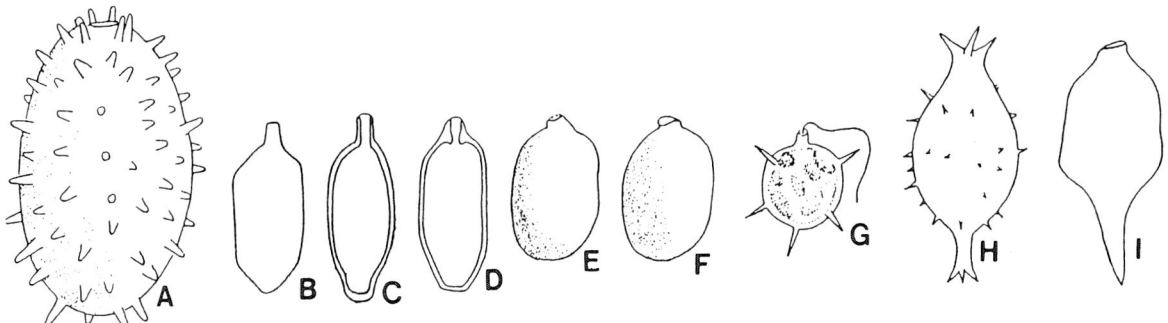

FIGURE 22 Loricas of species of *Trachelomonas*, (A)–(I): (A) *T. spectabilis*, × 765; (B) *T. hexangulata*, × 765; (C) *T. hexangulata*; (E) and (F) *T. playfairii*, × 765; (G) *T. aculeata* fa. *brevispinosa*, × 650; (H) *T. speciosa*, × 765; (I) *T. girardiana*, × 765. From Prescott (1962), with permission.

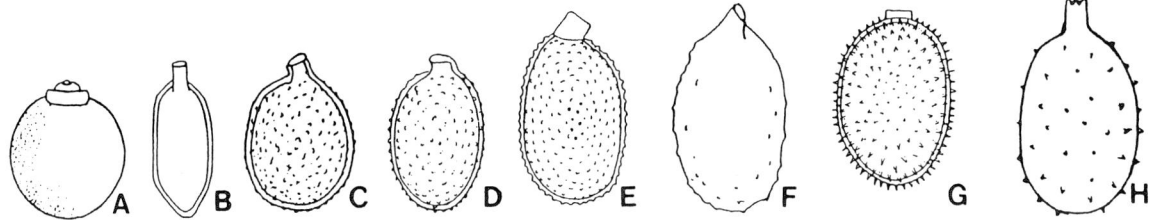

FIGURE 23 Loricas of species of *Trachelomonas*, (A)–(H): (A) *T. mammillosa*, × 765; (B) *T. hexangulata*, × 765; (C) *T. scabra* var. *longicollis*, × 956; (D)–(F) *T. similis* Stokes [(D) × 1020; (E) × 1275; (F) × 1275]. (G) *T. charkowiensis*, × 956; (H) *T. bulla*, × 756. From Prescott (1962), with permission.

are part of a light-sensing, motile-cell-directing apparatus. When *E. gracilis* is grown for long periods in the dark, the stigma color fades to yellow (Kivic and Vesk, 1972). The stigma apparently does not arise *de novo* (Jahn, 1946), and sibling cells each contain a stigma before their complete separation occurs (*Euglena deses*, Leedale, 1967b, Fig. 93). A stigma is some-times missing in species of *Phacus* (Smith, 1950; Prescott, 1962).

3. Pellicle

The next most obvious and useful feature for recognizing the larger, naked-cell euglenoid species (those without a lorica) is the striated cell surface, the pellicle (Leedale, 1964), the term recommended by Preisig *et al.* (1994) and used in texts by van den Hoek *et al.* (1995), Lee (1999), and Graham and Wilcox (2000) and in a comparative study of the patterns it forms (Leander and Farmer, 2000). The term periplast, however, also has long been used to include the striated pellicular strips and the plasma membrane of the cell surface, by Pringsheim (1948) and others, for example, Smith (1950), Prescott (1962), Rosowski and Lee (1978), Starmach (1983), Zakryś (1986), and Zakryś and Walne (1994). As defined here, the pellicle is the characteristic cell surface unit (of the cortical or peripheral protoplasmic layer) found in all euglenoid species. It is composed of a covering of the plasma membrane over narrow-to-broad, elastic, much elongated and continuous, ribbon-like, proteinaceous pellicular strips (the epiplasm or membrane skeleton) and includes associated microtubules and endoplasmic reticulum (Sommer, 1965; Leander and Farmer, 2000, 2001). These strips are held together partly from over- and under-lapping along their lateral edges [shown diagrammatically in Fig. 1B, C, but see Bouck and Ngô (1996) for more ultrastructural details]. Longitudinally aligned microtubules (at least one per articulation) are adjacent to the overlap area, with fibrous connections to the pellicular strips (Bouck and Ngô, 1996). Lateral spatulate spines or projections (Fig. 1B) occur in some cases and are associated with ER cisternae, and together create the surface complex (Bouck and Ngô, 1996) or pellicular apparatus. Species with pellicular strips obvious in light microscopy are termed striated. They may be of a constant number for a given species (e.g., 40 in *Euglena gracilis*, Bouck and Ngô, 1996; Leander and Farmer, 2000), and they may vary greatly in width between species (cf. Fig. 5R, U; e.g., 1.05 µm in *E. texta* to 0.24 µm in *E. gracilis*; Dragos *et al.*, 1997) and in number in a given cell location (Leander and Farmer, 2000). In species that are small or with thin strips,

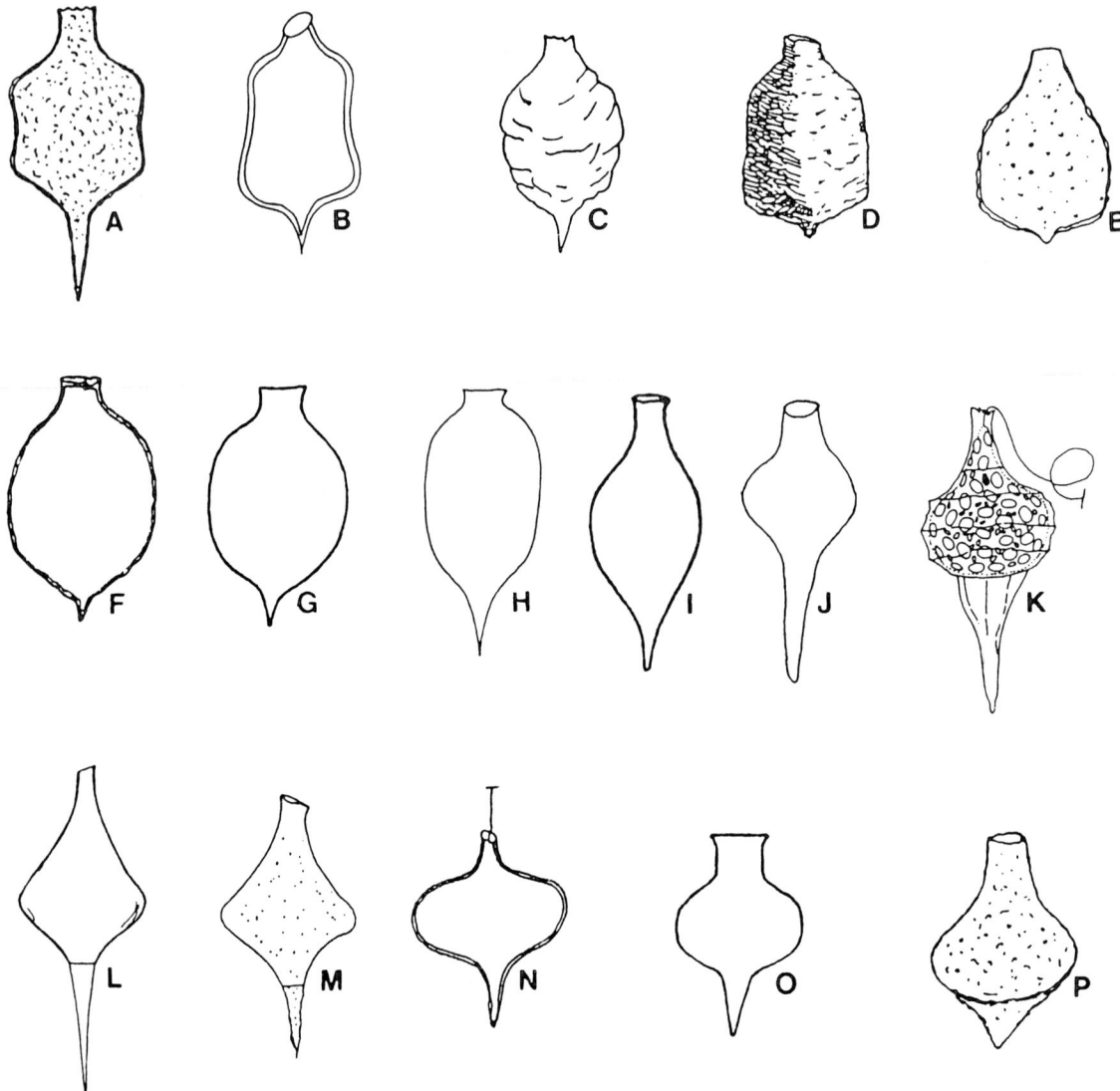

FIGURE 24 Loricas of species of *Strombomonas*, (A)–(P); all cell sizes for figures can be found in Dillard (2000, with permission), and figures are after other authors: (A) *S. giardiana*; (B) *S. acuminata*; (C) *S. tambowika*; (D) *S. verricosa*; (E) *S. verricosa* var. *zmiewika*; (F) *S. deflandrei*; (G) *S. ovalis*; (H) *S. urceolata*; (I) *S. fluviatilis*; (J) *S. longicauda*; (K) *S. lackeyi*; (L) *S. ensifera*; (M) *S. gibberosa*; (N) *S. volgensis*; (O) *S. rotunda*; (P) *S. schauinslandii*.

individual strips may be unresolvable with light microscopy and thus are not figured in line drawings; such species may be referred to as having a smooth pellicle (e.g., *Euglena sima*, Zakryś, 1986). Nonetheless, striations invisible with light microscopy can be demonstrated with scanning electron microscopy (as in *Cryptoglena pigra*, Fig. 14E). To avoid confusion, the pellicular striations themselves are best not referred to as ornamentation, because that term has been used for the mineralized deposits that occur on the pellicular surface in several genera (e.g., *Euglena*, Fig. 5M, and *Phacus*, Fig. 13L, M).

The degree of euglenoid movement (elasticity or metaboly when cells are actively changing shape) has a basis in the ultrastructure of the pellicular strips. Species with thick, wide pellicular strips and flat, tapering, and spatulate lateral projections have a rigid pellicle and are incapable of or exhibit only slow metaboly, whereas those with thin, narrow strips and small or no lateral strip projections exhibit fast metaboly (Dragos et al., 1997). Strips may be slightly to greatly twisted along the cell's apical axis, creating a spiral (Figs. 5L and 6F, striations mostly omitted), or parallel and straight with respect to the long axis

(Figs. 12F and 13J), and in heterotrophic forms these show no metaboly (aplastic, Triemer and Farmer, 1991b). Pellicular strips bend into the canal opening of the flagellar pocket but do not line the reservoir although their subtending microtubules do extend into the reservoir (Triemer and Farmer, 1991b). When cells divide, progeny get half the number of strips from the parent and, produce new strips between the parent strips (Bouck and Ngô, 1996). The cell posterior is often pointed and may become rigid from the twisting of the pellicular strips (Rosowski, 1977), and reduced in width (Figs. 6E, 13D) and in number (Fig. 8B) to form a caudus (Fig. 4G–L) which may change length during metaboly (Fig. 5A–H). In one common green species, *E. spirogyra*, the pellicle is ornamented on its outer surface with regularly spaced, mineralized warts (Fig. 5M); similar warts on the pellicle are found in *Phacus* (Fig. 13L, M). In the phagotrophic genera the pellicle with parallel and longitudinal strips appears to have arisen evolutionarily before those with helically arranged strips (Montegut-Felkner and Triemer, 1997). Nineteen morphological characters of the pellicle have been found useful in phylogenetic analysis (Leander and Farmer, 2001).

4. Flagellar Pocket

At the anterior end of a cell is a deep and usually conspicuous, permanently inflated invagination, the flagellar pocket or pouch (= ampulla or vestibule of van den Hoek *et al.*, 1995), which has two (often one vestigial, Fig. 1A), three (15A), or four flagella (in the green marine species *Tetreutreptia pomquetensis*, McLachlan *et al.*, 1994, 1999) attached at its base. The pocket is pyriform and consists of a bulbous reservoir with a neck (the canal) that opens to the outer cell surface. The position of the canal opening with respect to the cell apex determines whether or not locomotory flagella are considered apically emergent (*Lepocinclis*, Fig. 12C–F) or subapically emergent (all other photosynthetic genera). A contractile vacuole is always in a fixed area near the stigma and empties into the reservoir. The old term "gullet" for the flagellar canal or reservoir is mostly avoided now because the pocket has not been found to be involved in the direct uptake of macroparticulate food, although pinocytotic (plasma membrane) uptake of large protein molecules within the reservoir has been demonstrated in *Euglena* (Kivic and Vesk, 1974) and in trypanosomes (Webster, 1989).

5. Phagotrophic Ingestion Apparatus

Of evolutionary significance is that, in green *Tetreutreptia* (Triemer and Lewandowski, 1994), *Eutreptia* (Solomon *et al.*, 1991), *Euglena* (Surek and Melkonian, 1986), *Colacium* (Willey and Wibel, 1985a, b), and *Cryptoglena* (Owens *et al.*, 1988), there is a vestigial phagotrophic ingestion apparatus in the form of an opening, the cytostome, to a small elongated apparatus of various and often complex morphologies, the cytopharynx or pocket, located near the canal base–reservoir interface (Fig. 1E). In the photosynthetic species *Phacus trypanon* (Shin and Boo, 2001) and *P. pleuronectes* (Shin *et al.*, 2001) ultrastructural evidence suggests an ingestion apparatus similar to those in colorless phagotrophic euglenoids.

Triemer and Farmer (1991a) describe four basic types of such feeding structures in the euglenoids as a whole, with some species of heterotrophs having two of these together (Triemer and Farmer, 1991b) suggesting the retention of the primitive form while evolving another more elaborate feeding apparatus. These feeding structures, vestigial in appearance in some photoauxotrophs and more elaborate and fully functional in heterotrophs, are differentiated in part based on the number and organization of their reinforcing microtubules (MTRs), from simple rods in the Type I feeding apparatus to various groups of configurations of microtubules into vanes in Types II–IV (Linton and Triemer, 1999a). Among the heterotrophs, those with the Type III feeding apparatus can engulf eukaryotic cells as did the original phagotroph that incorporated a eukaryotic green alga for its chloroplast source, whereas Types I, II, and IV feed on prokaryotic cells (Triemer and Farmer, 1991a). Triemer and Farmer (1991a) note that "in spite of the supporting data, the case for homology between reservoir pockets and cytostomes is not clear as it may first seem." The reader is referred to the original comparative studies and reviews (Triemer and Farmer, 1991a, b; Linton and Triemer, 1999a) for further details.

6. Contractile Vacuole

Freshwater euglenoids have a single vacuole, which may be conspicuous at the time it is near its maximum expansion and about to contract. Through periodic contraction, the contractile vacuole eliminates osmotically collected water and then is reformed by a progressive coalescence of smaller vacuoles (Huber-Pestalozzi, 1955, p. 8, Fig. 1; Fig. 1A). The vacuolar expansion–contraction cycle occurs once every 20–30 seconds depending mostly on the temperature. Parasitic and saltwater euglenoid species lack contractile vacuoles (Leedale, 1967b).

7. Flagella

Features of the flagella, including number, morphology, length, insertion position, and behavior have long been recognized as important taxonomic criteria (Klebs, 1883). The primitive number of flagella in green euglenoids has long been considered two (Leedale, 1967b) emergent from the canal as in *Eutreptia* (Fig. 9A, B), but the five most common freshwater genera of North America have only one emergent flagellum, and it is locomotory (Jahn and Bovee, 1968). This difference is because the second flagellum is reduced to a short, nonemergent stub that may adhere to the emergent flagellum near its base, where its tapering, distal tip terminates opposite the paraflagellar body (swelling) of the emergent flagellum (Leedale, 1967b). Evidence suggests that the emergent flagellum of *Euglena* may contain the photoreceptor pigment rhodopsin (Gualtieri *et al.*, 1992; Gualtieri, 1993; Walne and Gualtieri, 1994; Barsanti *et al.*, 1997), in the form of a three-dimensional protein crystal that creates a swelling (Fig. 1A), the paraflagellar body, opposite the stigma (Walne and Dawson, 1993; Walne *et al.*, 1998). When there is only one emergent flagellum, it is considered to be dorsal, and the nonemergent, very short second flagellum, ventral. Both arise opposite basal bodies during cell division. The emergent flagella all are locomotory and have two rods that traverse most of their length (Figs. 1A, D, E). One is the axoneme (typical 9+2 array of microtubules) and the other the paraxonemal rod (= paraflagellar rod or paraxial rod) (Andersen *et al.*, 1991; Walne and Dawson, 1993, respectively), the latter rod often making them wider than flagella of other algae that lack this structure. The length of an emergent (locomotory) flagellum is usually described in comparison to the length of the cell and may be of taxonomic importance (Batko and Zakryś, 1995, p. 12). In *Euglena*, this emergent flagellum has two helical arrangements of hairs on its surface, of two types: tufts of three to four fine hairs, 2–3 μm long, form one continuous row; and tufts of finer, more numerous and shorter hairs form another continuous row next to it (stylized in Fig. 1D). The total number of these hairs is estimated at 30,000 per flagellum [see Bouck *et al.* (1978) and Bouck (1982), for ultrastructural details]. This arrangement and number are obscured by typical techniques for scanning electron microscopy (Fig. 8A) which artifactually anastomose the hairs and make them thicker (Rosowski, 1977; Rosowski and Glider, 1977). The details of flagellar ultrastructure have been useful in solidifying the evolutionary relationship of euglenoids and the Kinetoplastida (Walne and Dawson, 1993).

8. Chloroplasts, Pyrenoids, and Paramylon

The shape, number, and position of the chloroplasts along with the presence or absence of pyrenoids traditionally have been key taxonomic features (Fig. 2A–K), along with cell size and apical shapes (Figs. 3 and 4) (Pringsheim, 1956; Batko and Zakryś, 1995). In addition, paramylon granules (β-1:3-linked glucan, Leedale, 1967a) often are distinctive in size and shape for certain genera and species. For example, they are donut-like in *Phacus* (Fig. 13), of two morphologies (Figs. 5U and 6D) in species of *Euglena* without pyrenoids (subgenus *Discoglena*, Zakryś, 1986; Zakryś and Walne, 1994), and of two vastly different sizes in *Cryptoglena* (Fig. 14G). The paramylon may or may not be associated with pyrenoids when the latter are present. In some taxonomic keys, the number and shape of paramylon granules are used to separate groups of species (Philipose, 1982). Evidence has been presented (Gibbs, 1978, 1981) that the chloroplast has three membranes; in areas with fewer than three membranes (most of the chloroplast surface in most studies), it has been assumed that two or more membranes have tightly fused. The outside membrane lacks ribosomes, and various origins for this membrane have been hypothesized (Haüber *et al.*, 1994; Cavalier-Smith, 1998) depending on how the original green photosynthetic unit was acquired endosymbiotically. However, Cavalier-Smith (1999) argues that the uptake of a whole green alga rather than just a chloroplast is the best supported hypothesis. Internally, the green photosynthetic lamellae, the thylakoids, are appressed in groups of three forming a band (Leedale, 1967b, 1982). The chloroplasts can be discoid, numerous, and without pyrenoids (Fig. 5L, R), or there can be as few as one chloroplast, axial and stellate (Fig. 5I), incised (Wołowski, 1993), parietal and U-shaped (Figs. 5P and 14D), or of other shapes (Fig. 2). Ultrastructural studies (Walne *et al.*, 1986; Zakryś and Walne, 1998a) have clarified our understanding of the paramylon center of some euglenoids, a region that is difficult to interpret structurally using light microscopy. Often it is unclear if just one or many chloroplasts contribute to the center of certain "stellate chloroplasts." This ambiguity is a result of paramylon granules near clusters of chloroplast ribbons (Fig. 5I, J). The pyrenoid(s) at the cluster may not be described (e.g., *Euglena chadefaudii* and *E. geniculata*, Zakryś and Walne, 1994) because they cannot be observed in that cluster (Fig. 5J). That is, pyrenoids themselves are colorless and thus transparent (Rosowski and Hoshaw, 1971), but, being largely protein, are readily stained (Rosowski and Hoshaw, 1970). Electron microscopy has shown them to be penetrated by single, double, or triple appressed thylakoid units (Zakryś and Walne, 1998a). Some stellate chloroplasts are ribbon-like with a single central pyrenoid (Zakryś and Walne, 1998a), but again, when many chloroplast ribbons radiate from a center obscured by a covering of clustered paramylon granules, the number of individual chloroplasts is difficult or impossible to discern with light microscopy (Fig. 5K). For example, the paramylon center in many *Eutreptiella* (Walne *et al.*, 1986) have from 6 to more than 10 straplike chloroplasts, and their single terminal pyrenoids cluster together and create centers around which paramylon occurs (chloroplasts may later disperse from their center clusters in older cultures, suggesting there are many rather than one with ribbon-like projections). Transmission electron microscopy will continue to be necessary, in some cases, to verify that axial, stellate chloroplasts are either single or formed from many separate chloroplasts clustered at their pyrenoids. Finally, although pyrenoid shape and position are often characteristic of a species (see variations within *Euglena*, Pringsheim, 1956), and their absence can be an important criterion as well, pyrenoids may or may not be present in different populations of the same species (e.g., *Euglena adhaerens*, Zakryś and Walne, 1994, p. 80).

9. Extrusive Organelles (Extrusomes)

In the peripheral cytoplasm of protists are membrane-bound organelles that can discharge their contents slowly (or themselves in some cases), or quickly and forcefully (Hausmann, 1978). Pringsheim (1956) described two forms, spherical and fusiform, but stated that "spirally arranged mucus bodies, at first taken as spherical, sometimes turned out to be spindles on more careful inspection, but occurrences not fitting into one of the two groups may exist." Zakryś (1986) noted in *Euglena tristella* that "trichocysts underlie in the spiral rows the periplast (without staining they are visible as points, but stained with neutral red—as elongate, spindle-shaped bodies)." In another study (Zakryś and Walne, 1998b), mucocysts in *E. oxyuris* previously reported as spherical with light microscopy turned out to be tubular and much elongated with electron microscopy. Although the shape of mucus bodies is used to distinguish *E. viridis* (spherical) from *E. stellata* (fusiform), both called trichocysts by Zakryś and Walne (1998b), they question these distinctions. Hausmann (1978) reviewed and characterized extrusive organelles among the protozoa and found a high degree of variation and diverse ultrastructure among them. In the euglenoids, the ultrastructure, function, and chemistry of these bodies have yet to be adequately characterized to relate form with function or to apply a specific name to an extrusome that would exclude the usage of that name for other extrusive structures.

The most commonly described extrusive organelle in euglenoids has been the mucus or muciferous body (MB; Fig. 1C). Muciferous bodies are single membrane-bound bags, spherical or pear-shaped, that lie in a row between adjacent pellicular strips (between the striae). These organelles open to the cell surface through a canal that is the point of discharge of amorphous contents (Leedale, 1964, 1982), and this material trails from the canal openings, creating contorted filaments (Rosowski, 1977; Rosowski and Willey, 1977). Muciferous bodies are aligned linearly and follow the path set by the position of articulation of adjacent pellicular strips (Leedale, 1964, 1982; Leander and Farmer, 2000). Thus MB may be arranged spirally or in rows parallel to the long axis of a cell and are sparse or abundant depending on the species. Muciferous bodies occur in many euglenoids and are responsible for the production of the slimy mucilaginous coating of certain species, particularly the crawling species, and may be abundant when cells are in the palmelloid condition (e.g., *E. myxocylindracea*, Bold and MacEntee, 1973; Rosowski, 1977). Neutral red, methylene blue, brilliant cresyl blue, and ruthenium red will reveal muciferous bodies in living cells, as will iodine, but fusiform mucus bodies (trichocysts) "do not seem to stain with iodine but become blue with brilliant cresyl blue" (Pringsheim, 1956, p. 13). Occurrence of ornamentation on the external pellicle surface in euglenoids (e.g., in *Euglena*, *Trachelomonas*, *Strombomonas*) is a result of external mineralization in an organic, nucleating matrix originating from MB (Barnes *et al.*, 1986; Dunlap *et al.*, 1986). Some mineralization occurs in the basal regions of stalks and mucilaginous attachment pads of *Colacium vesiculosum* (Rosowski and Kugrens, 1973; Figs. 9E, G, J, and 10D).

The term mucocyst was created to describe extrusive organelles (protrichocysts, mucoid trichocysts, mucigenic bodies, secretory ampules) that produce an amorphous material (Tokuyasu and Scherbaum, 1965), but it is unclear that its use in some cases implies a body different from the muciferous body. Like muciferous bodies, it releases mucilage to the pellicular surface from typically pyriform bodies (Zakryś and Walne, 1998b). The term has been used for the elongated extrusive organelles of *E. oxyuris* (Zakryś and Walne, 1998b) that are not ejectile organelles like trichocysts (Hausmann, 1978). However, it also has been used both for single membrane-bound vesicles with amorphous contents and for the rodlike, tubular structures in the phagotrophic euglenoid *Peranema trichophorum* that upon discharge reveal attached apical filaments (Hilenski and Walne, 1983).

The term trichocyst (spindle trichocysts, Hausmann, 1978) refers to much elongated, fast-release ejectile organelles that lie in the same position as muciferous bodies (Fig. 5K). When they occur, their patterns are important criteria in defining species. For example, they may be regularly distributed in rows (*Euglena tristella*, *E. splendens* Zakryś, 1986; *E. chadefaudii*, Zakryś and Walne, 1994), irregular in pattern (*E. polymorpha*, Zakryś and Walne, 1994), or in groups (*E. walnei*, Zakryś,

1994). Much variation in ultrastructure undoubtably occurs among organelles called trichocysts. In this regard, it would be useful to ultrastructurally compare the extrusomes of *E. velata* with *E. granulata*, which are described from light microscopy as rodlike in the former and fusiform in the latter (Johnson, 1968). Extrusive organelles of euglenoids have yet to receive detailed comparative study, at the levels of light and electron microscopy; thus, clear distinctions among these peripheral organelles cannot be made.

10. Palmella

Palmella (= palmelloid) stages are reported for *Euglena* and *Colacium* (Jahn, 1946). Nonmotile cells of *Euglena* that become covered with mucilage and undergo repeated division in this condition are said to be in a palmelloid stage (e.g., *E. viridis*, *E. stellata*, *E. schmitzii*, *E. pisciformis*, *E. oblonga*, *E. sociabilis*, *E. splendens*, *E. mutabilis*, *E. proxima*, Zakryś, 1986; *E. oblongata*, *E. sanguinea*, *E. rubra*, Zakryś and Walne, 1994). Palmelloid cells may move in a creeping manner before becoming flagellated again. In *E. myxocylindracea*, a soil species, cells rarely have an emergent flagellum and are mostly in a palmelloid stage throughout their life history (Bold and MacEntee, 1973) and these palmelloid cells are covered with mucilage (Rosowski, 1977). Cells of *Colacium vesiculosum* also remain in a palmelloid condition as they divide on the cuticle of aquatic arthropods, particularly microcrustaceans (Jahn, 1951; Rosowski and Kugrens, 1973). However, they may not become covered with mucilage (Rosowski and Kugrens, 1973; Willey *et al.*, 1973), and the same is true for palmellae of *E. pisciformis* in culture (Zakryś *et al.*, 1996). Nonetheless, *C. libellae* occurs as mucilaginous palmellae in the hindgut of damselfly nymphs (Rosowski and Willey, 1975). Interestingly, palmelloid stages are not reported for heterotrophic euglenoids (Jahn, 1946; Larsen and Patterson, 1991).

11. Cysts

Cysts of three types have been described in euglenoids: protective (thick walled); reproductive (thin walled; cell division occurs within them); and temporary, transitory, or resting (thick walled but not completely closed, Jahn, 1946). Unfortunately, these stages are incompletely described for euglenoid species from North America or elsewhere. Johnson (1944) found cysts in 28 of 41 species of *Euglena* from Iowa, but he presented figures for only one species (Fig. 5C–F). Prescott (1955) provided two figures of cysts he attributed to *Euglena* (Fig. 7A, B). We have few morphological details about cysts, and the fact that cysts are produced by euglenoids is generally unknown except by specialists. In addition, we know little of the conditions that induce cyst formation, but Jahn (1951) notes that cysts form in *E. deses* at 1–4°C. When spherical *Euglena* cysts have been collected from sediments in the winter (Capitol Beach Lake, Lincoln, NE) and identified by their large pigmented stigma, they soon showed metaboly within their thick, mucilaginous covering (Rosowski, unpublished). Cysts have been reported for *Euglena*, *Phacus*, *Trachelomonas*, and *Eutreptia* (Jahn, 1946), and although usually spherical or with eccentric walls (Bold and McIntee, 1973), they may be pyriform and tubular as well [*E. orientalis*; see also Philipose (1982) for numerous diagrams of cysts of *E. tuba*]. In *E. gracilis*, cysts are formed from Golgi secretions (Triemer, 1980), but in *E. myxocylindracea* and *E. tripteris* it appears that muciferous bodies between the pellicular strips directly contribute this material (Rosowski, 1977). Escape from the mucilaginous cyst covering may occur without prior cell division and involves metaboly followed by rupture of the covering (Johnson, 1944). In a microcosm study of *Eutreptiella gymnastica*, Olli (1996) found that cysts accumulated in detrital material and never were greater than 0.5% of the cells found in the water column above. The general role of cysts in free-living protozoa has been considered by Corliss and Esser (1974).

12. Metaboly

Bovee (1982) characterized the movements of *Euglena* as (1) swimming, (2) contraction, (3) crawling, and (4) gliding. Contraction, also known as euglenoid movement (Leedale, 1967b) or metaboly, results from cells quickly becoming spherical from an end-to-end contraction followed by a quick expansion, and/or from extending surface protuberances in an amoeboid fashion, or otherwise actively contorting themselves in a variety of shapes through various peristaltic-like motions (Fig. 5A–H). Cells attached posteriorly by mucilage (Fig. 8B) will also show coiling, twisting, or rotary movements (Bovee, 1982). Motile, palmelloid, and attached cells may soon begin metaboly when examined with light microscopy (e.g., *Colacium* on zooplankters, Rosowski and Kugrens, 1973); however, euglenoids with a rigid pellicle do not usually exhibit any metaboly (Fig. 5U).

13. Nucleus

The single nucleus of undividing cells is generally central or posterior in position in a particular species (Fig. 5I, J, L, N, R, U, V), has one or more conspicuous endosomes (nucleoli, Fig. 1A), and is usually visible in living cells without staining. Before division it moves forward toward the reservoir of the flagellar pocket (Mignot *et al.*, 1987). Mitosis precedes cytokinesis (for cell surface view see Fig. 11H). The nucleus has a double membrane, the outer of which is unassociated with the chloroplast. This double nuclear membrane remains intact throughout mitosis, whereas the nucleus itself becomes constricted (dumbbell shape) prior to cleavage, in a fission-like process. The chromosomes are condensed at interphase (unusual, but as in the dinoflagellates) and, with transmission electron microscopy, some appear attached to the inner membrane of the nucleus at this time. During mitosis, from metaphase to telophase, the nucleolus (endosome) generally persists and becomes broadly elongated and dumbbell-shaped like the nucleus itself. The mitotic spindle is perhaps unique among protists and consists not of individual microtubules but of groups of about 12 microtubules creating subspindles, some attached to chromosomes but others not (Triemer and Farmer, 1991b). The intranuclear spindle is aligned at a right angle to the long axis of the cell prior to cell division, and pairs of basal bodies, which have already replicated flagella prior to cytokinesis, become associated with the separate poles of the elongated nucleus (see van den Hoek et al., 1995, Fig. 5). Chromosomes attached by their kinetochores to microtubules form a loose metaphase plate; then, during anaphase, they move to the poles as the nucleus becomes elongated and not from a shortening of the chromosomal-connected microtubules. Finally, the chromosomes are partitioned into two sibling nuclei through constriction of the intact nuclear membrane. Longitudinal cytokinesis, beginning at the flagellar pocket, then cleaves the cell into sibling offspring (Mignot et al., 1987). Because of the conservative nature of cell division in the course of evolution, the details of mitosis can be quite important in understanding the taxonomic position of enigmatic species (e.g., *Diplonema*, Triemer, 1992).

C. Reproduction

Cell division is apparently the only mode of reproduction in euglenoids because earlier suggestions of sexuality are unconfirmed (Leedale, 1967b). Euglenoids are thus uniparental; that is, each individual arises only through mitosis and cytokinesis of a parent cell, and the offspring are thus clones of one another. Cell division, which occurs in motile, creeping, palmelloid, or cyst stages (or in most of these stages depending on the species), starts anteriorly at the reservoir base and proceeds posteriorly until sibling cells separate (Fig. 9H). Johnson (1944) reported cell division in the cysts of several species of *Euglena*, mostly in thin-walled cysts (e.g., *E. rubra, E. gracilis, E. splendens*) but in one case, also thick-walled cysts (*E. polymorpha*). In *Colacium*, after each cell division, some individuals remain nonmotile and secrete mucilaginous stalks at their canal end (Figs. 9F–I and 11D, E), which moves them away from their substratum (Rosowski and Kugrens, 1973; Willey et al., 1977). With repeated cell divisions, sibling cells, by continuous mucilage secretion, will separate, with each cell connected at the growing apex of its own stalk forming a dichotomously branched colony.

III. ECOLOGY AND DISTRIBUTION

Green euglenoids occur in most standing, illuminated waters, creating green puddles in cow hoof prints (Rosowski, personal observation, Orono, ME), also becoming major components of ponds and our largest lakes (Prescott, 1962, 1978), as well as rivers (Round and Palmer, 1966; Palmer and Round, 1965; Zakryś, 1986, e.g., *Euglena viridis, E. schmitzii, E. polymorpha*; Conforti, 1991; Conforti and Pérez, 2000). Euglenoids are particularly abundant in eutrophic lakes and reservoirs surrounded by agricultural land or in landscapes that were highly productive (e.g., grasslands) even before the major impact of agricultural practices (Edmondson, 1969; Hutchinson, 1969). In such localities they occupy waters fringed with vascular plants, especially where an accumulation of plant litter occurs in the water. In addition, euglenoids may reach bloom proportions on a regular basis in newly created impoundments over garbage landfills (Rosowski, personal observations) and in ponds and slow-flowing drainage ditches exposed to animal dung, although species diversity may be low in such nutrient-enriched situations (Lackey, 1968; Palmer, 1980).

Euglenoids thus thrive in waters enriched (polluted) with organic matter and are collected in open water, or on mud or sediments associated with those waters. Knowing their preference for such organically rich habitats, however, does not necessarily mean that one can predict their occurrence, as pointed out by Lackey (1968).

Although species of *Euglena* have received the most attention in studies on the ecology of euglenoids, much of our ecological information has come from taxonomic studies where ecological questions were secondary. Therefore, we have only a cursory knowledge of euglenoid ecology at this time. For most genera of photosynthetic euglenoids, there are few specific ecological studies except those on *Colacium*, which will be considered at the end of this review.

Lackey (1968) is the only lengthy review of the ecology of species of *Euglena*. There is a brief review of *Euglena* in India including a discussion of the red pigmented species, particularly *E. tuba* (Philipose, 1982). As for worldwide distribution, a summary of *Euglena* species in Poland includes distributional data for many of the 35 species considered (Zakryś, 1986), and similar data for other euglenoid genera are found in Zakryś and Walne (1994). A general treatment of euglenoids as indicators of water quality is covered by Palmer (1969, 1980) with some species being indicators of organic water pollution because of their occurrence in sewage oxidation–stabilization ponds (species in the genera *Euglena*, *Trachelomonas*, *Phacus*, and *Lepocinclis*). In a study of the phytoplankton of sewage effluent from a city to a marine harbor, Sarojini (1994) found *E. acus* to be the dominant among 119 species identified, in two out of four sampling stations. *Colacium vesiculosum* has been found from February through December in sewage oxidation ponds, and in July and August formed water blooms (Matviyenko, 1972). *Euglena gracilis* has shown some promise in purification of waste water (sewage, swine manure) when grown at pH 4, at 30–35°C (Waygood et al., 1980). It is odd that no euglenoids were reported from any collections of algae from 138 temple tanks in southern Kerala, India, described as highly eutrophic due to organic wastes (Maya et al., 2000). An excellent one-page summary treats general ecological aspects of all euglenoids (Walne and Kivic, 1990).

According to Pringsheim (1956), several species of *Euglena* (*E. obtusa*, *E. mutabilis*, *E. deses*) may occupy the same epipelic region and do not seem to move with the aid of flagella (and may have none emergent); others readily change from creeping to swimming (*E. geniculata* and *E. tristella*); still others are continuously swimming (*E. acus*, *E. spirogya*, *E. tripteris*). Then there are those living exclusively in permanent waters (*E. gracilis*, *E. acus*, *E. spirogyra*, *E. oxyuris*), or on mud that has lost most of its water *(E. sociabilis, E. pisciformis, E. viridis, E. proxima)* where they survive as cysts (Pringsheim, 1956). A few species with creeping, swimming, and palmelloid stages (*E. myxocylindracea*) have had their surfaces documented with scanning electron microscopy (Rosowski, 1977; Fig. 8A–E).

The well-known association and often periodic dominance of green euglenoids in waters containing high levels of dissolved organic material results in part from their being photoauxotrophs rather than photoautotrophs. Species studied (mostly *Euglena*) have both a vitamin B_{12} (cyanocobalamin) and a B_1 (thiamine) requirement (Provasoli and Pintner, 1953; Provasoli, 1958), whereas one species requires only thiamine and another only biotin (Provasoli, 1969). In nature, these vitamins become available from the bacteria that utilize organic compounds in nutrient-rich waters. Pringsheim (1956) reviewed the history of the culture of euglenoids. Some green *Euglena* and *Colacium* species are facultative heterotrophs and will grow bacteria-free in the dark on certain carbon sources, and these substrates may enhance their photosynthetic growth as well (Leedale, 1967a). Such species are also known as "acetate flagellates" from their ability to obtain carbon from sodium acetate, or from a variety of organic acids and alcohols (ethanol), but such ability is strain specific (Cramer and Meyers, 1952). *Euglena gracilis* strain Z is known to synthesize and excrete vitamin E in significant amounts (Tani and Tsumura, 1989), as well as biotin (Baker et al., 1981).

Nitrogen is supplied to green euglenoids in growth media as either ammonium sulfate or ammonium phosphate (Walne and Kivic, 1990) because of the apparent preference of some species for nitrogen in the form of ammonia (Cramer and Meyers, 1952; Munawar, 1972). For example, in sewage treatment ponds near Auckland City, New Zealand, growth of *E. acus* responded positively to high levels of NH_4^+ except when associated with small-celled Chlorococcales, which appeared to out-compete them (Haughey, 1970). However, other species seem to favor habitats that include NO_3^- (*E. oxyuris* var. *charkowiensis* and *E. pisciformis*, Philipose, 1982). In a study on inorganic nutrients in seven natural communities, Kim and Boo (1998) observed that growth of a small morphotype of *E. geniculata* was positively correlated with high NO_3^- (negatively with NH_4^+), whereas growth of the larger form was positively correlated with NH_4^+ (negatively with NO_3^-). Dunlap and Walne (1987) reported that an increase NH_4NO_3 concentration in cultures of *Trachelomonas lefevrei* caused an increase in Fe and Mn in the lorica. Graham and McCoy (1974) demonstrated that growth of xenic cultures of *T. hispida* responded positively to increasing levels of NO_3^-, and although NH_4^+ may have been produced by bacteria in those cultures, data for NH_4^+ indicated insignificant amounts over time.

Phosphate uptake in *Euglena gracilis* strain Z shows some interesting features in culture studies. In cells that are dividing once a day on a 14:10-hour light–dark cycle, P uptake occurs in the light in a nonlinear fashion within the first 2 hours, peaking in the middle of the light period under saturated P levels, but with no uptake in the dark phase thus suggesting a rhythmic uptake periodicity (Chisholm and Stross, 1976a). In a further study, this time in P-limited cultures, diel uptake of P occurred in a linear fashion in the first hour and peaked at the end of the light period (Chisholm and Stross, 1976b). Because P incorporation occurred in nondividing, stationary-phase cultures ($< \frac{1}{3}$ cells dividing per day), periodicity appears independent of cell division. With P uptake perhaps tied to photosynthesis in P-saturated cells more than in P-limited cells, we see a mechanism by which *E. gracilis* has some flexibility in P uptake that "could influence the outcome of competitive interactions among the phytoplankton in environments with oscillating nutrient supply rates" (Chisholm and Stross, 1976b).

Lackey (1968) noted the paucity of data on *Euglena* ecology and suggested three reasons: (a) questions concerning validity of certain species, (b) infrequent reports on occurrence of many species, and (c) lack of detailed ecological observations. This situation has changed little since his review. Indeed, the species concept (Zakryś et al., 1997) and population variation for species of *Euglena*, the most common and widely studied taxon, are only now being investigated by modern techniques of DNA analysis (Zakryś et al., 1996; Zakryś, 1997a, b). Distribution data, although available from floristic studies, rarely incorporate quantitative analysis (but see Shi, 1996a). Lackey (1968) noted that it is difficult to determine the ecological status of certain species of *Euglena* because their occurrence is documented "only a few times." Although Lackey (1968) found 12 *Euglena* species to occur on all continents, the reports were often from floristic studies with little ecological information. Some easily recognized species such as *E. pisciformis*, *E. gracilis*, and *E. mutabilis* are reported as widespread in Poland (Zakryś, 1986) and the southeastern United States (Zakryś and Walne, 1994). Of 36 species of *Euglena* found in the southeastern United States, 53% were specifically stated to be common, cosmopolitan, or ubiquitous on a worldwide basis, whereas 22% were rare; of 5 species of *Lepocinclis*, 80% were common and 20% rare; of 14 species of *Phacus*, 50% were cosmopolitan and 22% were rare (Zakryś and Walne, 1994).

With regards to acidity, Zakryś and Walne (1994) cite a pH range for *E. mutabilis* of 1–8.6, and Hargreaves and Whitton (1976) showed this species to be more abundant in acid streams at pH 2.6–2.9 than pH 3.0–3.3, with more rounded, nonmotile cells at pH 1.3–1.5. Lackey (1968) reported *E. mutabilis* "in alkaline waters, acid mine waters, soft-water ponds, and iron seeps," whereas Lund reported that it prefers waters below pH 5, being particularly abundant on mud when it is "strongly acid." This species has also shown a tolerance for high levels of zinc in acidic waters (Say and Whitton, 1980), and can scavenge zinc from the environment (Kempner and Miller, 1972). *Euglena gracilis* was shown to have a pH range of 3.65–9.38 and *Trachelomonas grandis* a pH range of 6.3–8.43 (Moss, 1973), whereas other species "cannot grow well at high or low pH" (Jahn et al., 1979).

Many species of *Euglena* are "interface dwellers" (Lackey, 1968) and, as such, are associated with scums on the air–water interface (surface microlayer), on the surface of sediments, or as green clouds above the sediments (Round, 1984). For example, *Trachelomonas volvocina* and *T. pulcherrima* were collected in the first 2 cm of sediment (epipelic) samples and were the dominants among the algae in August in a pond on Baffin Island, in the Canadian eastern arctic (Moore, 1979).

As a result of the ability to form cysts, certain genera of euglenoids appear in soil samples (Jahn, 1946). *Euglena* sp. has been isolated and cultured from atmospheric samples (Schlichting, 1964), raising the possibility that airborne cysts may aid species dispersal. Viable *Phacus* sp. and *Euglena* sp. have been cultured from the feet of waterfowl, suggesting (as in other algal species) that birds disperse them (Schlichting, 1960). A very brief note (Kiener, 1944) about green *Euglena* in snow in Nebraska gives no description of the cells nor a location and thus cannot be evaluated; however, what may be a colorless euglenoid, *Notosolenus*, has been illustrated and reported from snow in the U. S. Southwest (Hoham and Blinn, 1979) and in bogs (Leedale, 1967b).

Among the more interesting species of *Euglena* are the red bloom-formers. *Euglena sanguinea* is reported as permanently blood red (Prescott, 1978) from haematochrome granules that mask the chlorophyll (Zakryś and Walne, 1994). In a bloom situation cells may occupy the upper water surface and then migrate downward by late afternoon and evening (Lackey, 1968), where they can then form "red mud" (Xavier et al., 1991). *Euglna rubra* forms red surface scums on water as well (Zakryś and Walne, 1994). The conditions that bring on these blooms have not been thoroughly investigated, although the pigment is believed to be formed in response to warm water and maximum solar irradiation (Philipose, 1982; Xavier et al., 1991). In another study it has been shown that, in the dark, *Euglena gracilis* strain Z moves towards the water's surface by negative gravitaxis (Häder, 1987). At low-light flux rates it moves toward the light, and at high light flux rates away from the light and can protect itself from UVB radiation for short periods of time by assuming a spherical shape and becoming immobile (Gerber and Häder, 1993).

An unusual phenomenon is the vertical migration of *Euglena obtusa*, which, along with some other algae, cover the banks of the River Avon, at Bristol, England. *Euglena obtusa* occurs on the exposed mud surface at low tide at a density in excess of 10^5 cells/cm^3, turning the surface green or yellowish brown due to the accumulation of paramylon; diatoms are abundant as well (Palmer and Round, 1965; Round and Palmer, 1966). When the tide returns 7.5 h later *E. obtusa* migrates from the surface back into the mud. In a laboratory situation, this diel migration cycle continues for several weeks under continuous illumination, but does not persist in continuous darkness. Under natural conditions, when the tide comes in, the water is so turbid that it effectively transforms the diurnal cycle "into one of tidal frequency" (Palmer and Round, 1965). That is, the effect of the turbid waters of high tide is to darken the epipelon and to cause the organisms to migrate downward (Palmer and Round, 1965). In the laboratory, this migration cycle occurs unaltered between 5 and 15°C but is greatly reduced at 2°C. However, an increase in temperature from 2°C to 12°C will cause a backward shift; that is, it will rephase the migration cycle. Thus, both light and temperature appear to be effective in entraining the biological clock of *E. obtusa* and affecting cell migration within surface sediments (Round and Palmer, 1966).

The more recently discovered green *Tetreutreptia* (McLachlan *et al.*, 1994) is described from a marine habitat and deserves mention because of some unusual attributes. It grows best at 5°C and is distinctive in having four emergent flagella, two long and two short. It is highly unusual for a euglenoid in having 20% of its fatty acids as EPA and DHA, in a ratio of about 2:1 (McLachlan *et al.*, 1994, 1999). This high percentage of long-chain, polyunsaturated fatty acids may relate to the low temperatures at which it lives, perhaps providing membrane flexibility. *Tetreutreptia* has the basic body plan of *Euglena* and possesses a remnant feeding apparatus like that of certain phagotrophic euglenoids (Triemer and Lewandowski, 1994).

This discussion has thus far considered mainly species of *Euglena*, the only genus for which much specific ecological data exists. Still, researchers have determined for very few species of *Euglena* or other genera where, within a body of water, a particular species spends the phases of its life history (e.g., Lund, 1942; Palmer and Round, 1965; Olli, 1996). Indeed, Johnson's (1944) conclusion that for *Euglena* there is "a need for life history studies" is as true today as it was then, and applies to most all euglenoid genera.

Colacium is unique among euglenoids in that all but one species (*C. gojdicsae*) establishes itself on the cuticle (exoskeleton) of aquatic arthropods by an adhesive pad (Rosowski and Kugrens, 1973), while having a free-swimming phase as well. For example, in the fall, in an interdunal pond in northern Indiana, *C. libellae* (Rosowski and Willey, 1975) swims into the anus of the larvae of the damselflies *Enallagma civile* and *Ischnura verticalis* (Fig. 11A) and attaches itself anterior-end first to the lining of the hindgut; this condition occurs in 35–50% of the larvae in nature (Willey, 1972). At the first molt, in the spring, this green lining of cells is shed in a mass still attached to the inner surface of the exoskeleton, also called the castskin when discarded (Willey, 1972); presumably, the attached cells become motile again and seek out new habitat for reattachment. This *Colacium* species is thus a specialist as is *C. calvum*, which apparently prefers to attach to the cladoceran *Daphnia pulex* or other *Daphnia* species (Chiavelli *et al.*, 1993), whereas *C. vesiculosum* selects as suitable substrata a much broader array of invertebrate genera (Rosowski and Kugrens, 1973; Threlkeld *et al.*, 1993), and *Daphnia* is not its first choice (Chiavelli *et al.*, 1993).

There are two ways in which cells of *Colacium* increase their epibiont numbers on the surface of a zooplankter, either by attachment in the motile phase or by cell division after attachment. At each molt of the host the attached *Colacium* cells are discarded with the exuvia so that the new cuticle is initially free of epibionts (Al-Dhaheri and Willey, 1996). At the beginning of an intermolt period colonization by motile cells of *C. vesiculosum* on the new surface of *Daphnia pulex* is most important for increasing cell density, whereas cell division of attached cells is more important at the end of that period (Al-Dhaheri and Willey, 1996).

In an experimental study, Bartlett and Willey (1998) showed that the feeding of *Daphnia laevis* on free-living *C. vesiculosum* improves the fitness of the *Daphnia*, and that the cuticle of living specimens is a preferred substratum for attachment of *Colacium* compared to their exuviae. They suggested that the loss of the motile cells in the plankton by herbivory is offset by cell division that occurs when *C. vesiculosum* is attached to the exoskeleton of the host during its intermolt period. The presence of green *Colacium* on zooplankters increases the zooplankters' susceptibility to capture by both planktivorous fish (Willey and Cantrell, 1990) and salamanders (Threlkeld and Willey, 1993). This likely results from an increased visibility and apparent size (see Threlkeld *et al.*, 1993). Predation risk for the arthropod also may increase by greater frictional drag and hence an inability to escape predators (Threlkeld and Willey, 1993; Willey *et al.*, 1993).

IV. COLLECTION, CULTURING, AND PREPARATION FOR IDENTIFICATION

Methods for collecting euglenoids are not greatly different than those for other flagellated algae, and thus the general techniques as given in Prescott (1978) or Whitford and Schumacher (1984) can be consulted by the beginner. The most common method of collecting pond euglenoids is from the shoreline with a hand-thrown plankton net, typically of Nitex™ monofilament of 20–153 µm net mesh size, for tychoplanktonic and euplanktonic species; mesh size is sometimes a function of plankton particle size, how quickly the net becomes clogged, or simply availability. Samples of the microbenthos (on clay, silt, or sandy soils) are collected with poultry basters, with water-fill tubes sold for car batteries, or by the handgrab method for algae associated with floating or submerged vegetation and detritus. It is useful to stir the water above sediments to collect surface-dwelling euglenoids and presumably cysts as well. In winter, mucilage-encysted euglenoids can be found in plankton tows below ice and above the sediments when sediments presumably are stirred in the collection process (Rosowski, personal observation). In warmer weather, squeezes of aquatic plants and debris from the shoreline of ponds or lakes is a means of obtaining littoral species. Samples (water, sediment surface, plant litter) are transferred with a few milliliters of liquid to plastic sandwich bags, clear plastic silver-dollar coin storage containers, or baby food jars because of their small size and availability. The volume of liquid in the samples should be half or less of the volume of the container, because oxygen demand in concentrated samples is great, particularly on warm days or with polluted samples, where anoxia is to be avoided. Containers should be labeled with the critical information required such as location, microhabitat, date, time, sample number, temperature, and other parameters deemed significant. Samples in bags or other containers are loosely capped and placed in an ice chest and kept cool until examination. Concentration of samples (e.g., 8 mL) by centrifugation in the laboratory in a table-top centrifuge (run at maximum speed) saves considerable time in the discovery, observation, and identification stages, because cell density is greatly increased with minimal damage to euglenoid cells. In eutrophic ponds with surface or planktonic blooms, centrifugation is likely unnecessary. Fresh samples may be kept in a refrigerator (4°C) for short periods of time (2–3 days), in which case the lids of their containers should be loose for gas exchange. Walk-in coolers with continuous or programed illumination appear to keep samples considerably longer (weeks or months) than refrigerators that light up only when the door is opened. Perhaps for up to a week such latter samples are viable for making clonal isolations, but viability rapidly declines over time and is dependent on the temperature and cell densities. Also, morphology in stored field samples may be altered drastically with time so studies are best done soon after samples are collected unless paramylon granules are too dense, in which case storage of samples in the dark will eventually diminish their number.

Samples may be placed on dim (40 watt) fluorescent light racks and do best in cool locations; 15°C seems ideal, or at least a temperature several degrees less than the temperature at which the sample was collected (unnecessary in winter or early spring, Rosowski, personal observation). In a refrigerator, samples should be kept on top shelves, with lids ajar for gas exchange. At 25°C (room temperature) samples may be viable for only a couple of days if too concentrated. Once back in the laboratory, field samples in small containers can be transferred to pie plates and held at room temperature for a day or two because of the expanded surface area they provide for gas exchange (G.W. Prescott, pers. comm., 1972).

Typically, the culture of green euglenoids begins by using a soil–water medium but requires one step in media preparation not needed for other green algae, that of adding a single pea cotyledon or seed to the soil–water bottle prior to steam pasteurization. All photosynthetic euglenoids thus far investigated have a need for one or more vitamins (photoauxotrophs) and their special vitamin needs are fulfilled by the bacteria active in their environment; bacterial growth over the pea cotyledon surface produces those vitamins. General culture techniques for clonal isolation and axenic culture are found in Hoshaw and Rosowski (1973) and Starr and Zeikus (1993). Transferring several pipettes of a natural collection into a soil–water pea bottle and waiting a week or two often results in species shifts and the appearance of species not originally noted. Cells isolated from such "slop cultures" have already found their new environment suitable for growth and thus are preselected for growth in the same medium after their clonal isolation.

Observations of euglenoids for identification and description are usually performed with newly collected field samples, using bright field, dark field or Nomarski optics; inverted microscopes are sometimes used when enumeration of species is the goal. Recently immobilized cells yield the most accurate measurements. Fixing cells in 2% glutaraldehyde will retain their original size and shape and the position of internal structures unlike fixation with Lugol's iodine solution (Zakryś, 1986), which was more commonly used in the past (Prescott, 1978).

The best images of internal cellular details are usually obtained when the fresh slide preparation is allowed to dry out under a coverslip, causing air bubbles to close in near the specimens. This procedure causes the cells to flatten, unless sand grains are present in the preparation and elevate the coverslip. In determining the limits of extracellular mucilage, particularly of stalks of *Colacium* or mucilaginous sheaths around palmelloid cells of *Euglena*, a dilute India Ink solution is applied to one side of the coverslip and drawn over the material by removing water with a paper towel from the opposite side of the coverslip. This negative staining method outlines the limits of the most watery sheaths. Neutral red will highlight the nucleus as well as the muciferous bodies and trichocysts of the peripheral cytoplasm. Another value of this dye is that it is vital and thus cells remain alive and contractile vacuole activity near the reservoir also may be observed. Staining procedures have been developed for revealing the pyrenoids of algae including euglenoids (proprionocarmine, Rosowski and Hoshaw, 1970) and the stalks of *Colacium* (PAS and Alcian Blue, Willey *et al.*, 1977).

Additional suggestions about collecting, observing, and photographing flagellates are presented by Patterson and Hedley (1992). Permanent and semipermanent microscope slides can be made in a variety of ways (Prescott, 1978). Cultures of euglenoids may be obtained from numerous permanent culture collections (Starr and Zeikus, 1993; and see Andersen, 1996, or Norton *et al.*, 1996, for lists of algal culture collections).

V. KEY AND DESCRIPTIONS OF NORTH AMERICAN GENERA

A. Key: Freshwater Genera of Photosynthetic Euglenoids

1a.	Cells attached to aquatic arthropod cuticles (mostly microcrustacea), singly or in palmelloid colonies; sometimes on branched, mucilaginous stalks; become metabolic (Figs. 9–11)	*Colacium*
1b.	Cells motile and solitary, or if in a palmella stage not on arthropod cuticles	2
2a.	Cells with a lorica (case, envelope), often dark-brown colored	3
2b.	Motile cells without a lorica (naked) and never developing one	4
3a.	Lorica attached posteriorly, with anterior opening the width of the cell (Fig. 15 B)	*Ascoglena*
3b.	Loricate cells unattached and motile; anterior lorica opening $< \frac{1}{4}$ width of lorica	5
4a.	Motile cells with one emergent flagellum	6
4b.	Motile cells with two or more emergent flagella	7
5a.	Lorica light to dark reddish brown, rigid and brittle when highly pigmented; cell and lorica ovoid or sometimes spherical; collar, if present, abrupt (Figs. 16–23)	*Trachelomonas*
5b.	Lorica clear or yellowish, flexible, typically tapering to both apices; collar base tapered, not abrupt (Figs. 15D, E and 24)	*Strombomonas*
6a.	Motile cells with a deep ventral sulcus when viewed apically (Fig. 14)	*Cryptoglena*
6b.	Motile cells without a deep sulcus (apex-to-apex longitudinal groove)	8
7a.	Motile cells with two emergent flagella of equal length no stage in tadpole (Fig. 12A, B)	*Eutreptia*
7b.	Cells with three emergent flagella; palmelloid in tadpole rectum (Fig. 15A)	*Euglenamorpha*
8a.	Cells may change shape, canal opens subapically (Figs. 3, 5, and 6)	*Euglena*
8b.	Cells with a firm pellicle (surface usually striated)	9
9a.	Cells circular in apical view, or if otherwise, pellicle firm	10
9b.	Cells laterally flattened, leaf-shaped, and perhaps twisted as well (Fig. 13)	*Phacus*
10a.	Subapical canal opening (Fig. 3)	*Euglena*
10b.	Apical canal opening; cells transversely circular, not flattened (Fig. 12C–F)	*Lepocinclis*

B. Desciptions of Genera

The following 10 photosynthetic euglenoid genera are reported from fresh waters of North America (Smith, 1950; Prescott, 1978; Zakryś and Walne, 1994; Dillard, 2000). Generic descriptions include primarily those features useful in identifying or characterizing genera and species with light microscopy. General features, including cell ultrastructure, were previously discussed in Section II.B. Drawings are mostly based on collections from North America, for as noted by Komárek (1991), figures from elsewhere could yield morphotypes that do not exist in North America, perhaps creating confusion. For descriptions of 13 genera and 22 taxa of colorless euglenoids from the southeastern United States, see Wołowski and Walne (1997).

Ascoglena Stein 1878 (Fig. 15B)

Members of this genus have a *Euglena*-like cell, but are enclosed in a narrow, cup-shaped, collarless lorica attached to a stationary substratum at its posterior. The anterior apical end of the lorica is as broad as the cell width and extends beyond the cell (protoplast) apex (unlike *Klebsiella*, Leedale, 1967b). A newly divided cell that leaves the parent lorica is naked (stages are shown in Smith, 1950); after attachment it secretes a yellowish lorica that turns reddish-brown at maturity, except perhaps at the anterior. The lorica remains free of the pellicle except at the base, where it is firmly attached (Fig. 15B). The canal opening of the flagellar pocket is subapical, and there is a single emergent flagellum. Chloroplasts are numerous and discoid; pyrenoids may or may not be present. Cell division has been described in sessile cells. *Ascoglena* is rare.

An unidentified species was reported from British Columbia (Stein and Borden, 1979). *Ascoglena vaginicola* Stein 1878 is the only species reported from North America (from Montana, Michigan, and Ohio, Prescott, 1978), of two taxa worldwide.

Colacium Ehrenberg 1838 (Figs. 9–11)

Recognized by Fritsch (1945), Bourrelly (1970), and Smith (1950), species such as *C. vesiculosum* are commonly found on arthropods, particularly microcrustacea (Rosowski and Kugrens, 1973; Prescott, 1978; Threlkeld *et al.*, 1993). The cells also can be identified in their unattached motile stage once the morphology of attached cells, their inner projecting pyrenoids, and their colorless anterior end are recognized. The motile cells of *C. vesiculosum* are slightly constricted in the middle unlike most attached cells (Rosowski and Kugrens, 1973; Figs. 9A and 11A). They sometimes occur unattached in bloom proportions (Green, 1953; Matviyenko, 1972; Rosowski, unpublished, Capitol Beach Lake, Lincoln, NE). However, *C. vesiculosum* is most readily found and identified attached anteriorly by a small mucilaginous pad or stalk to the surface of larvae and adults of ostracods (*Moina*), rotifers (*Keratella, Brachionus*), cladocerans *(Daphnia)*, copepods *(Cyclops)*, and much less frequently on the faster amphipods. *Colacium libellae* is found in the rectum of tadpoles and mayfly nymphs (Rosowski and Willey, 1975), and *C. calvum* is a large cell attached on the post-abdomen around the anus of *Daphnia* (Willey, 1982). *Colacium vesiculosum* usually occurs epizoically singly or in palmelloid groups, and much less frequently as stalked colonies on cyclopoid copepods or cladocerans, although this latter morphology is more obvious. Their dichotomously branched stalks develop when single cells become attached (Ward and Willey, 1981), divide, and each of the offspring secretes a mucilaginous stalk. All species of *Colacium* examined thus far have many colorless globules, unstained by neutral red, in a ring around the canal opening. These globules can be observed with careful light microscopy but are most readily identified with transmission electron microscopy, where they stain with two densities within the same globule (= biphasic, Fig. 1E; Willey, 1980). In nature, cell division most often produces a palmelloid covering of sessile cells on their host, but in culture *C. vesiculosum* may also colonize other substrata (e.g., glass microscope slides or chitinous exoskeletons of insects placed there). The motile and sessile stages are capable of metaboly; a single flagellum becomes emergent from a subapical canal prior to detachment of a cell from its substratum. A stigma is present. The numerous discoidal chloroplasts have central, inner projecting pyrenoids. Paramylon granules cap the inner projecting pyrenoids but occur elsewhere away from pyrenoids. No cysts are reported.

This genus is widely distributed, but it is most often found on zooplankters, both their naupliar and adult stages, attached on the exoskeleton as single pads, palmellae or sometimes on dichotomously branched stalks. *Colacium* is reported for western Labrador (Duthie and Ostrofsky, 1975), British Columbia (Stein and Borden, 1979), and Ontario (Duthie and Socha, 1976). Four species have been found in the United States: *C. vesiculosum* (Rosowski and Kugrens, 1973; Willey *et al.*, 1993; Dillard, 2000), *C. calvum* (Willey, 1982), *C. libellae* (Willey, 1972; Rosowski and Willey, 1975), and *C. gojdicsae* (Rosowski and Kugrens, 1973). Two species have been reported from Mexico (Ortega, 1984).

Cryptoglena Ehrenberg 1831 (Fig. 14)

The small, motile, naked ovoid cells of this genus are always somewhat laterally compressed and have a single, subapical emergent flagellum as long as or longer than the cell. The pellicle is rigid (no metaboly) with broad pellicular strips (but not observable with light microscopy), with a deeply grooved sulcus extending in more or less a straight line along the length of the ventral cell surface (Fig. 14A–D; Rosowski and Lee, 1978). This groove is best recognized when the cell is viewed apically (Fig. 14B, C). In the only species studied with electron microscopy, there is a single, shieldlike chloroplast (Fig. 14D; Rosowski and Lee, 1978), originally and long interpreted and illustrated as being two (Ehrenberg, 1831; Fritsch, 1945; Smith, 1950; Prescott, 1978). The thick lateral lobes of the chloroplast are continuous with a thinner middle opposite the sulcus, making it appear as two chloroplasts when viewed with bright-field microscopy (Rosowski and Lee, 1978). The chloroplast may be lobed in the area near the reservoir and canal (Owens et al., 1988). There are two large, lateral paramylon grains that are shieldlike (Fig. 14G), and these create a flange on each side of the cell (Fig. 14B, F). There are also much smaller paramylon granules (Fig. 14G) that are elongated and typical of many euglenoids (Rosowski and Lee, 1978) such as *Euglena*. A small stigma is present.

Although *C. pigra* is reported for North America from several localities (Smith, 1950; Stein and Borden, 1979; Dillard, 2000) and is occasionally encountered (Prescott, 1931; Thompson, 1938; Meyer and Brook, 1969), it is considered rare.

Euglena Ehrenberg 1830 (Figs. 1A–D, 2–7)

The most common euglenoid genus worldwide, reference is often made to it when describing other genera. Motile cells are ovoid-cylindrical to narrowly fusiform (Fig. 5), frequently with a narrow, elongated, posterior caudus (Fig. 4G–L). Cells may exhibit metaboly while motile (Fig. 3A–H), and many species are capable of metaboly while in a creeping or palmelloid condition. The pellicle is firm in species belonging to the subgenus *Rigida* (Pringsheim, 1956), and these species show minimal metaboly under normal conditions. In *E. spirogyra*, the pellicle has helical, regularly arranged surface warts mostly containing iron (Dawson et al., 1988). Subtending the articulation of pellicular strips may be extrusive organelles called muciferous bodies, mucocysts or trichocysts, and their presence and patterns are of taxonomic significance. All green cells have an orange to red stigma unassociated with chloroplasts. There is a single emergent flagellum from a ventral canal opening, and a very short second flagellum retained in the reservoir (not often apparent with light microscopy), with its tip opposite the paraflagellar swelling of the emergent flagellum. In apical view most species are circular but others are laterally compressed, twisted, and twisted and triradiate (Figs. 3Q–S and 4E, F). Most species appear green, but the green color in some species is masked by a red pigment called haematochrome (*E. sanguinea*, *E. tuba*, and *E. rubra*). In *E. rubra*, when haematochrome granules are concentrated in the center along the longitudinal axis of cells, the cells and the scum are green, but when they are dispersed throughout the cells, the cells and the scum are red (Johnson and Jahn, 1942). Philipose (1982) observed that the red species *E. tuba* has haematochrome pigments dispersed throughout the cell while under bright light but "while in fading light they recede to the hind end," and their cysts as well have haematochrome in the rear.

Chloroplast number ranges from one to many, with or without pyrenoids. Chloroplast position and features distinguish three subgenera: one or more axial, stellate chloroplasts (subgenus *Euglena*); several to many parietal chloroplasts each with one central, bilateral pyrenoid (subgenus *Calliglena*); and small, numerous, disc-shaped parietal chloroplasts without pyrenoids (subgenus *Discoglena*, Batko and Zakryś, 1995; Zakryś, 1986; Zakryś and Walne, 1994). Pyrenoids can be free of cytoplasmic associated paramylon granules, protrude from both sides of the chloroplast (bilateral = two-sided), be sheathed with paramylon (double-sheathed, Fig. 2D, I), or protrude only interiorly (unilateral = one-sided) and be sheathed with paramylon (single-sheathed, Zakryś and Walne, 1994; Batko and Zakryś, 1995). Cell division has been described in motile and palmelloid cells, and in certain cyst types (Johnson, 1944; Jahn, 1951; Philipose, 1982). Typically cysts are spherical, surrounded by mucilage, perhaps opened on one end (Fig. 7; Gojdics, 1953; Prescott, 1955; Philipose, 1982; Kahn, 1993).

Euglena mostly occurs in still waters of puddles, ponds, and lakes, especially in waters with high levels of organic nutrients (from animal wastes or aquatic plants), but it may occur in and on sediments of river banks (Palmer and Round, 1965; Round and Palmer, 1966; Say and Whitton, 1980). *Euglena* inhabits shaded or sunny areas, in hard or soft waters, of low pH (0.9) to high pH (over 8.0). Species can be found in the psammon or on the surface of sediments, but most have an active motile phase in the plankton. Ten species are reported from Ontario (Duthie and Socha, 1976), 3 from the Northwest Territories (N.W.T./Nunavut, Sheath and Steinman, 1982), and 18 from British Columbia (Stein and Borden, 1979). See also reports from Costa Rica (Haberyan et al., 1995), Mexico (Ortega, 1984), and Guadeloupe (Bourrelly and Manguin, 1952).

Johnson (1944) reported 41 species for Iowa (the first of the 41 was actually *Colacium*); among others, 36 *Euglena* taxa have been reported from the southeastern United States by Zakryś and Walne (1994) and 67 taxa by Dillard (2000); in the past, 50 (Smith, 1950) to 60 species have been reported for the United States as a whole (Prescott, 1978).

Euglenamorpha Wenrich 1924 (Fig. 15A)

Cells have the body plan and metaboly of *Euglena*, but it is the only euglenoid with three emergent flagella, each with a paraflagellar swelling. The canal opening of the flagellar pocket is subapical. There is a single stigma at the level of the three flagellar swellings (Fig. 15A). *Euglenamorpha* has numerous, small, parietal discoidal chloroplasts without pyrenoids; paramylon grains are evident.

The genus has been reported by Hegner (1922, 1923) and Wenrich (1923, 1924) in the North Atlantic states. Found exclusively in the rectum of the tadpoles of *Rana pipiens* and *Rana clamitans*. A colorless form was also described (Wenrich, 1923). No further reports have appeared.

Eutreptia Perty 1852 (Fig. 12A, B)

Species in this genus are thought to be quite closely related to brackish and marine species of *Eutreptiella* (Dawson and Walne, 1991), and for that reason literature related to both genera is cited here. Morphology is similar to *Euglena*; *Eutreptia* is highly metabolic when swimming, but maintains a truncate apex in *Eutreptia viridis* (Prescott, 1978). It is distinguished from *Euglena* in having two nearly equal or clearly subequal flagella; only the dorsal flagellum has a flagellar swelling. A stigma is present opposite the flagellar swelling. Numerous discoidal chloroplasts lacking pyrenoids are present in the freshwater species *E. viridis* (Smith, 1950) and *E. globulifera* (Zakryś and Walne, 1994). The marine species *E. pertyi* (Dawson and Walne, 1991) possesses 25–30 chloroplast ribbons that radiate from a large central pyrenoid. *Eutreptiella eupharyngea* has two stellate clusters of chloroplasts and each chloroplast has a terminal pyrenoid creating a center cluster and giving a light microscope appearance of there being only one chloroplast, highly stellate (Walne et al., 1986). The finely striated pellicle may be quite mucilaginous in *Eutreptiella eupharyngea*. *Eutreptia pertyi* has an encysted stage, and the flagellar swelling is retained on the stub of the dorsal flagellum (Dawson and Walne, 1991); others report cysts as well (Jahn, 1946; Smith, 1950).

According to Smith (1950), *Eutreptia viridis* "has been recorded from several widely separated stations in this country." Whitford and Schumacher (1969) report *E. viridis* for North Carolina; *E. globulifera* and *E. viridis* have been reported for the southeastern United States (Dillard, 2000).

Lepocinclis Perty 1849 (Fig. 12C–F)

This genus is characterized by motile, naked cells, which are radially symmetrical (never compressed) in a transverse optical section, with a rigid pellicle (no metaboly) and conspicuous longitudinal or spirally arranged pellicular strips. It has a single, emergent flagellum, usually longer than the cell, from an apical canal opening that is diagnostic (other green euglenoid genera have a subapical canal opening). The cells are ovoid, broadly ellipsoidal, or fusiform in shape, with a short or long caudus (posterior pole). A stigma is appressed to the reservoir. The chloroplasts are usually numerous, small, parietal, discoid, and without pyrenoids. Most species have two (Philipose, 1984), rarely four (Prescott, 1962, *L. playfairiana*), large, circular, and distinctive (donut-like) parietal paramylon grains that are opposite one another and shieldlike, or folded or laterally curved on each side. Cysts and palmelloid stages have not been reported (Smith, 1950; Leedale, 1967b).

Lepocinclis is collected in plankton tows, but generally "not found in euplankton but occur among dense growth of algae in shallow bays, swamps, and in ponds" (Prescott, 1962). It often occurs with *Euglena* and *Phacus* (Prescott, 1978). Stein and Borden (1979) report *L. ovum* and *L. salina* in British Columbia, Ortega (1984) *L. ovum* and *L. texta* in Mexico, and Duthie and Socha (1976) *L. ovum* in Ontario. Smith (1950) reported 15 species for the United States. Prescott (1962) reported 7 species in the western Great Lakes region; Dillard (2000) reported 14 species for the southeastern United States.

Phacus Dujardin 1841 (Fig. 13)

The cells are solitary and the pellicle rigid (no metaboly), ovoid to fusiform, often twisted, and much compressed (being platelike or leaflike), with a straight or slightly bent caudus of variable length depending on the species. Observations from several views are useful in characterizing the shape of species (Kirjakov, 1998). The pellicular strips are longitudinal (pole-to-pole) or spiral, and the outside pellicle surface is ornamented in some species. Cells may be so compressed and transparent that the pellicle striations from one side are visible through the other, giving the appearance of cross-hatching. Often there is a pronounced lateral flange, called the keel (Fig. 13E), which projects ventrally (cells triradiate in transverse optical view). Cells are motile, naked, and have a single emergent flagellum from a canal that is slightly to clearly subapical. Some species have discoidal chloroplasts that are small, numerous, and without pyrenoids, whereas others have large discoidal chloroplasts with pyrenoids. However, *P. chloroplastes* and *P. chloroplastes* fa. *incisa* are unusual in having several much elongated (straplike) chloroplasts aligned with the long axis of the cell (Prescott, 1962). There are one or two large paramylon bodies, or several circular or elongated donut-shaped disks or rods. A stigma may or may not be present. Encysted cells have been reported (Lund, 1942; Smith, 1950).

According to Smith (1950), *Phacus* "does not have the same preference for stagnant waters as does *Euglena*." *Phacus* is found in swamps, ditches, and ponds, euplanktonic and tycoplanktonic (Prescott, 1962). Thirty-two species have been recorded for the United States (Smith, 1950). Dillard (2000) found 63 taxa in the southeastern United States. In Canada, 11 species have been identified from Ontario (Duthie and

Socha, 1976) and 17 from British Columbia (Stein and Borden, 1979). Ten species have been summarized from Mexico (Ortega, 1984).

Strombomonas Deflandre 1930 (Figs. 15D, E, 24A–P)

Cells are of the *Euglena* form (e.g., *S. taiwanensis* var. *bigeonii*, Thérézien, 1999), but the lorica of most *Strombomonas* species (and its internal cell), starting from the posterior, gradually tapers toward the anterior apical opening, which typically lacks a sharply defined collar. That is, the collar tapers to a somewhat narrower, straight or slightly flaring neck resulting in a lorica that is spindle-, pear-, or urn-shaped (Fig. 24A–P). This shape is distinguished from most *Trachelomonas* species that show no flat or concave tapering of the lorica toward the anterior apical opening. Also, unlike most *Trachelomonas*, *Strombomonas* is reported to have a clear, yellow or slightly brown lorica, which is soft rather than brittle (largely unmineralized and thus can change shape during metaboly, Fig. 15E) and is generally without ornamentation, but may be observed "with wrinkles or folds or with transverse furrows" (Dillard, 2000). Irregular, granular particles may collect on the external lorica surface (Fig. 24A, E, P), probably sand grains based on a high silica content (Conforti *et al.*, 1994). *Strombomonas* was not recognized by Fritsch (1945) or Smith (1950), but was by Pringsheim (1953a, b), who accepted Deflandre's (1930) creation of this genus, calling it "a decision fully justified by my own experience" (Pringsheim, 1953a, p.104). It is also recognized by Tell and Conforti (1986), Zakryś and Walne (1994), and Dillard (1999, 2000). Prescott (1978) stated that "the genus *Strombomonas* includes the *Trachelomonas* species which have a pale tan or nearly colorless lorica." Pringsheim (1953a) noted that *Strombomonas* "never forms" brown loricas, but does form brown "excrescences or warts." It is important to note that Dunlap *et al.* (1986) observed the loricas of *S. conspersa* to be hyaline or mineralized with Mn and similar in structure and elemental composition to those of *Trachelomonas*; they recommended that this species be returned to its original taxonomic position as *T. conspersa*. Metaboly in *Strombomonas* is pronounced (Fig. 15D, E), surpassing that of *Euglena* (Pringsheim, 1953b, p. 253). The chloroplasts lack pyrenoids or have inner projecting pyrenoids as in many *Trachelomonas* species (Pringsheim, 1953b). The shape of the paramylon granules may be useful in the separation of species with similar loricas (Thérézien, 1999). There is a single, slightly subapical emergent flagellum. A stigma is present. No cysts are reported by Dillard (2000). Further clarification of criteria for separation of *Strombomonas* from *Trachelomonas* is needed because some species can be classified in both genera depending on the criteria considered (e.g., Nudelman *et al.*, 1998). Therefore, the taxonomic key in this chapter separating those two genera may break down on occasion.

As in *Trachelomonas*, all known species are freshwater. *Strombomonas ovalis* is reported from the N.W.T./Nunavut (Sheath and Steinman, 1982), *S. urceolata* from the southeastern United States (Zakryś and Walne, 1994), and *S. costata* from Mexico (Ortega, 1984). Dillard (2000) reports 15 species from the southern United States and all are illustrated here (Fig. 24A–P).

Trachelomonas Ehrenberg 1833 (Figs. 15C, 16–23)

The cells of this genus are of the *Euglena* form but each mature cell is enclosed in a lorica (= envelope, shell, or test) and is free-swimming unlike *Ascoglena*, which attaches posteriorly. The shell-like lorica is spherical to ovoid, with a small, single, circular opening (the anterior apical opening), usually with a distinct and abrupt collar through which a long, locomotory flagellum emerges (Fig. 15C). The cell can exhibit metaboly within the rigid lorica and may not assume the shape of the lorica when motile. After cell division, the two products of division line up one behind the other along the longitudinal lorica axis, and presumably one or both of the sibling protoplasts emerge naked from the apical opening, swim off, and each secretes a new lorica (arguments for and against one of the offspring using the old lorica are made by Pringsheim, 1953a). The lorica requires iron to develop and is colorless to red-brown at maturity depending on the manganese levels (Pringsheim, 1953a). The lorica may be papillate or spiny and is typically so brittle from mineralization that it easily cracks under the pressure of a coverslip. The surface projections of the lorica are of various lengths and shapes and retain their general form when air dried for scanning electron microscopy (SEM; Rosowski *et al.*, 1975a) and when specimens are critically point dried for SEM (Rosowski *et al.*, 1975b). Scanning electron microscopy reveals lorica features in air-dried specimens not observed with light microscopy (Rosowski *et al.*, 1975a); critically point-dried specimens for SEM that are properly coated with heavy metals yield even more features than air-dried specimens (Rosowski *et al.*, 1975b, 1981). The lorica is somewhat transparent (hyaline) when first formed but may be so opaque at maturity when mineralized with manganese and iron oxides that often one cannot see the protoplast within (Rosowski, personal observation). Manganese is the predominant element of loricas that are golden to dark brown and have a microcrystalline architecture with SEM, whereas iron predominates in areas of the lorica that are hyaline with light microscopy and that appear granular with SEM (Dunlap et al., 1983; Barnes et al., 1986). This mineralized material is dissolved in acids that would not affect silica, which is one way to distinguish siliceous stomatocysts of the Chrysophyceae from the loricas of *Trachelomonas* [stomatocysts misidentified as *Trachelomonas* loricas by Palmer (1902) are discussed in Rosowski and Couté (1996)]. Rod-shaped bacteria may attach apically on the lorica of some species and when critically point dried they may resemble spines (Rosowski and Langenberg, 1994; Rosowski and Couté, 1996). There is great difficulty in delimiting some species of *Trachelomonas* because of the extreme polymorphism of the lorica within a single population, including size, shape, and the absence or presence and distribution of ornamentation such as spines (Bicudo and De-Lamonica-Freire, 1993). Conforti and Ruiz (2000) also noted that, in a population of *T. spirillifera*, four different lorica surfaces occurred, with iron but not manganese increasing in abundance as the surface became more highly ornamented (and mineralized).

The chloroplasts (1–15) may be small and discoid without pyrenoids, in which case the paramylon grains are scattered; or there may be small, discoid or larger shield-shaped chloroplasts with inner projecting pyrenoids (with or without caps of paramylon, Fig. 15C); or there may be discoidal chloroplasts with bilateral pyrenoids in which case paramylon grains cap the protruding portions (Pringsheim, 1953a, b), which are called double-sheathed (Leedale, 1967b). Although species traditionally have been described on the basis of lorica morphology (Prescott, 1962; Conforti, 1999; Figs. 16–23), Pringsheim (1956) makes a case for the use of chloroplast features as well. Cysts are reported (Jahn, 1946).

All species of *Trachelomonas* are freshwater, and blooms may turn the water brown (Fritsch, 1945). Sheath and Steinman (1982) report 5 species from N.W.T./Nunavut, Stein and Borden (1979) 23 species from British Columbia, Duthie and Socha (1976) 23 species from Ontario, and Ortega (1984) 12 species from Mexico. Species occur in Guadeloupe (Bourrelly and Manguin, 1952) and Costa Rica as well (Haberyan et al., 1995).

Smith (1950) reported about 70 species for the United States. Prescott (1962) reported 50 species for the Great Lakes and noted (Prescott, 1978) that *Trachelomonas* often occurs with *Euglena* and *Phacus*. Whitford and Schumacher (1969) identified 22 species from North Carolina but believe they represent only about half of the species for the state. Dillard (2000) recognizes 85 species for the southeastern United States.

VI. GUIDE TO LITERATURE FOR SPECIES IDENTIFICATION

The cosmopolitan and ubiquitous nature of euglenoids, yet the apparent rarity of a significant number of species (some reported only a few times and continents apart), requires that the journal literature published on all continents be consulted before describing new species. Even for species described for North America, this survey requires examination of journals published outside the continent, where such studies are often published (e.g., Zakryś and Walne, 1994, Wołowski and Walne, 1997). General studies that include a significant number of species or have notable references are Walton (1915), Playfair (1921), Jahn (1946), Smith (1950), Gojdics (1953), Suxena (1955), Pringsheim (1956), Prescott (1931, 1962), Bourrelly (1970), Taft and Taft (1971), Asaul (1975), Reinke (1979), Prescott and Dillard (1979), Stein and Borden (1979), Sheath and Steinman (1982), Starmach (1983), Bold and Wynne (1985), Tell and Conforti (1986), and Dillard (2000). The book on euglenoids of China by Shi (1999) includes photosynthetic and colorless species; although in Chinese, it considers 27 genera and has 86 excellent plates of original and detailed cell drawings, and a few scanning electron and light micrographs. The earlier work by Bourrelly (1970) deserves special mention because it is often overlooked by reviewers of euglenoids and is rarely cited with euglenoids in general algal textbooks published in English. The book cover title, and often the exact citation to the text, is *Les algues d'eau douce. Algues blues et rouges*, but the text title page itself also reveals the inclusion of euglenoids, dinoflagellates, and cryptomonads, for a total of five chapters.

The following publications are useful in identifying species of the photosynthetic genera described in this chapter:

1. *Ascoglena*—Stein (1878), Smith (1950), Huber-Pestalozzi (1955), Leedale (1967b).
2. *Colacium*—Bourrelly (1970), Rosowski and Kugrens (1973), Willey (1980, 1982), Tell and Conforti (1986).
3. *Cryptoglena*—Huber-Pestalozzi (1955), Rosowski and Lee (1978), Owens et al. (1988), Tell and Conforti (1986).
4. *Euglena*—Conrad and van Meel (1952), Gojdics (1953), Huber-Pestalozzi (1955), Pringsheim (1956), Tell and Conforti (1986), Zakryś (1986), Zakryś and Walne (1994), Batko and Zakryś (1995), Dillard (2000).
5. *Euglenamorpha*—Hegner (1922, 1923), Wenrich (1923, 1924), Leedale (1967b).

6. *Eutreptia*—Huber-Pestalozzi (1955), Leedale (1967b), Kato (1994), Zakryś and Walne (1994), Dillard (2000).
7. *Lepocinclis*—Conrad (1934), Huber-Pestalozzi (1955), Prescott (1962), Tell and Conforti (1986), Zakryś and Walne (1994), Dillard (2000).
8. *Phacus*—Pochmann (1942), Allegre and Jahn (1943), Huber-Pestalozzi (1955), Suxena, 1955 Prescott (1962), Tell and Conforti (1986), Wołowski (1992), Zakryś and Walne (1994), Shi (1995), Dillard (2000), Conforti and Ruiz (2002).
9. *Strombomonas*—Conrad and van Meel (1952), Pringsheim (1953a, b), Huber-Pestalozzi (1955), Kirjakov (1983), Tell and Conforti (1984, 1986), Philipose (1988), Conforti et al. (1994), Conforti and Joo (1994), Couté and Thérézien (1994), Zakryś and Walne (1994), Yamagishi and Couté (1995), Shi and Jao (1998), Thérézien (1999), Conforti and Pérez (2000), Dillard (2000).
10. *Trachelomonas*—Deflandre (1926), Conrad and van Meel (1952), Pringsheim (1953a, b), Huber-Pestalozzi (1955), Singh (1956), Tell and Conforti (1986), Philipose (1988), Conforti and Joo (1994), Couté and Thérézien (1994), Zakryś and Walne (1994), Shi and Jao (1998), Conforti (1999), Dillard (2000).

LITERATURE CITED

Al-Dhaheri, R. S., Willey, R. L. 1996. Colonization and reproduction of the epibiotic flagellate *Colacium vesiculosum* (Euglenophyceae) on *Daphnia pulex*. Journal of Phycology 32:770–774.

Allegre, C. F., Jahn, T. L. 1943. A survey of the genus Phacus Dujardin (Protozoa: Euglenoidina). Transactions of the American Microscopical Society 62:233–244.

Andersen, R. A. 1996. Algae, *in*: Hunter-Cevera, J. C., Belt, A., Eds., Maintaining cultures for biotechnology and industry. Academic Press, San Diego, pp. 29–64.

Andersen, R. A., Barr, D. J. S., Lynn, D. H., Melkonian, M., Moestrup, Ø., Sleigh, M. A. 1991. Terminology and nomenclature of the cytoskeletal elements associated with the flagellar/ciliary apparatus in protists. Protoplasma 164:1–8.

Angeler, D. G. 2000. A light microscopical and ultrastructural investigation and validation of *Khawkinea perti* comb. nova (Euglenophyta). Algological Studies 96:89–103.

Asaul, Z. I. 1975. Vizacnik evglenovich vodoroslej URSR [Key for determination of the euglenophytes of the URSR]. Naukova Dumka, Kiev, 407 p.

Baker, E. R., McLaughlin, J. J. A., Hutner, S., DeAngelis, B., Feingold, S., Frank, O., Baker, H. 1981. Water-soluble vitamins in cells and spent culture supernatants of *Poteriochromonas stipitata*, *Euglena gracilis*, and *Tetrahymena thermophila*. Archives of Microbiology 129:310–313.

Barber, L. 1980. The heyday of natural history. Cope, London, 320 p.

Barnes, L. S., Walne, P. L., Dunlap, J. R. 1986. Cytological and taxonomic studies of Euglenales. I. Ultrastructure and envelope elemental composition in *Trachelomonas*. British Phycological Journal 21:387–397.

Bartlett, R., Willey, R. 1998. Epibiosis of *Colacium* on *Daphnia*. Symbiosis 25:291–299.

Barsanti, L., Passarelli, V., Walne, P. L., Gualtieri, P. 1997. In vivo photocycle of the *Euglena gracilis* photoreceptor. Biophysical Journal 72:545–553.

Batko, A., Zakryś, B. 1995. Numerical proof of the subgeneric classification of *Euglena*. Algological Studies 79:1–18.

Bicudo, C. E. de M., De-Lamonica-Freire, E. M. 1993. *Trachelomonas armata* (Euglenophyceae): An evaluation of the diagnostic features in the species. Algological Studies 69:57–66.

Bicudo, C. E. de M., Wołowski, K. 1998. *Trachelomonas alabamensis*, a new species of Euglenophyte from the Talladega wetland pond, southeastern U.S.A. Algological Studies 88:23–28.

Bodyl, A. 1996. Is the origin of *Astasia longa* an example of the inheritance of acquired characteristics? Acta Protozoologica 35:87–94.

Bold, H. C., MacEntee, F. J. 1973. Phycological notes. II. *Euglena myxocylindracea* sp. nov. Journal of Phycology 9:152–156.

Bold, H. C., Wynne, M. J. 1985. Introduction to the algae, 2nd ed. Prentice Hall, Englewood Cliffs, NJ, 720 p.

Bouck, G. B., Rogalski, A., Valaitis, A. 1978. Surface organization and composition of *Euglena*. II. Flagellar mastigonemes. Journal of Cell Biology 77:805–26.

Bouck, G. B. 1982. Flagella and the cell surface, *in*: Buetow, D. E., Ed., The biology of *Euglena*, Vol. III. Academic Press, New York, pp. 29–51.

Bouck, G. B., Ngô, H. 1996. Cortical structure and function in euglenoids with reference to trypanosomes, ciliates, and dinoflagellates. International Review of Cytology 169:267–318.

Bourrelly, P. 1970. Les algues d'Eau douce, Vol. III. Les algues bleues et rouges. Les eugléniens, peridiniens et cryptomonadines. Boubée, Paris, pp. 117–184.

Bourrelly, P., Manguin, E. 1952. Algues d'eau douce de la Guadeloupe et depandances. Société d'Edition d'enseignement Superieur, Par 3, 277 p.

Bovee, E. C. 1982. Movement and locomotion of *Euglena*, *in*: Buetow, D. E., Ed., The biology of *Euglena*, Vol. III. Academic Press, New York, pp. 143–168.

Brocklesby, J. 1851. Views of the microscopic world. Pratt, Woodford, Co., New York, 146 p.

Buetow, D. E., Ed. 1968. The biology of *Euglena*, Vol. 1. General biology and ultrastructure. Academic Press, New York, 361 p.

Calkins, G. N. 1926. The biology of the Protozoa. Lea, Febiger, New York, 623 p.

Canter-Lund, H., Lund, J. W. G. 1995. Freshwater algae. Their microscopic world explored. Biopress, Bristol, UK, 360 p.

Cavalier-Smith, T. 1981. Eukaryotic kingdoms, seven or nine? Biosystems 14:461–481.

Cavalier-Smith, T. 1991. Cell diversification in heterotrophic flagellates, *in*: Patterson, D. J., Larsen, J., Eds., The biology of free-living heterotrophic flagellates. Clarendon, Oxford, Systematics Association Special Volume 45:113–131.

Cavalier-Smith, T. 1993. Kingdom Protozoa and its 18 phyla. Microbiological Reviews 57:953–94.

Cavalier-Smith, T. 1998. A revised six-kingdom system of life. Biological Reviews 73:203–266.

Cavalier-Smith, T. 1999. Principles of protein and lipid targeting in secondary symbiogenesis: Euglenoid, dinoflagellate, and sporozoan plastid origins and the eukaryote family tree. Journal of Eukaryotic Microbiology 46:347–366.

Chiavelli, D. A., Mills, E. L., Threlkeld, S. . 1993. Host preference, seasonality, and community interactions of zooplankton epibionts. Limnology and Oceanography 38:574–583.

Chisholm, S. W., Stross, R. G. 1976a. Phosphate uptake kinetics in *Euglena gracilis* (Z) (Euglenophyceae) grown on light/dark cycles. I. Synchronized batch cultures. Journal of Phycology 12:210–217.

Chisholm, S. W., Stross, R. G. 1976b. Phosphate uptake kinetic in *Euglena gracilis* (Z) (Euglenophyceae) grown in light/dark cycles. II. Phased PO_4-limited cultures. Journal of Phycology 12:217–222.

Clayton, C., Häusler, T., Blattner, J. 1995. Protein trafficking in kinetoplastid protozoa. Microbiological Reviews 59:325–344.

Conforti, V. 1991. Taxonomic study of the Euglenophyta of a highly polluted river of Argentina. Nova Hedwigia 53:73–98.

Conforti, V. 1999. A taxonomic and ultrastructural study of *Trachelomonas* Ehr. (Euglenophyta) from subtropical Argentina. Cryptogamie Algologie 20:167–207.

Conforti, V., Joo, G.-J. 1994. Taxonomic and ultrastructural study of *Trachelomonas* Ehr. and *Strombomonas* Defl. (Euglenophyta) from oxbow lakes in Alabama and Indiana (U.S.A.). Cryptogamie Algologie 15:267–286.

Conforti, V., Pérez, M. del C. 2000. Euglenophyceae of Negro River, Uruguay, South America. Algological Studies 97:59–78.

Conforti, V., Ruiz, L. 2000. Morphological study of the lorica of *Trachelomonas spirillifera* Schkorbatov. Algological Studies 98:109–118.

Conforti, V., Ruiz, L. 2002. Euglenophytes from Chunam reservoir (South Korea) I. *Euglena* Ehr., *Lepocinclis* Perty and *Phacus* Duj. Algological Studies 104:81–96.

Conforti, V., Walne, P. L., Dunlap, J. R. 1994. Comparative ultrastructure and elemental composition of envelopes of *Trachelomonas* and *Strombomonas* (Euglenophyta). Acta Protozoologica 33:71–78.

Conn, H. W., Edmondson, C. H. 1918. Flagellate and ciliate protozoa, in: Ward, H. B., Whipple, G. C., Eds., Fresh-water biology. Stanhope Press, Boston, pp. 238–270.

Conrad, W. 1934. Matériaux pour une monographie due genre *Lepocinclis* Perty. Archiv für Protistenkunde 82:203–249.

Conrad, W., van Meel, L. 1952. Matériaux pour une monographie de *Trachelomonas* Ehrenberg, C., 1934, *Strombomonas* Deflandre, G. 1930 et *Euglena* Ehrenberg, C. 1832, genres d'Euglénacées. Institut royal des sciences naturelles de Belgique, Mémoire No. 124, pp. 1–175, + 19 plates and captions.

Corliss, J. O. 1975. Three centuries of protozoology: A brief tribute to its founding father, A. van Leeuwenhoek of Delft. Journal of Protozoology 22:3–7.

Corliss, J. O. 1984. The kingdom Protista and its 45 phyla. Biosystems 17:87–126.

Corliss, J. O. 1989. The protozoon and the cell: A brief twentieth-century overview. Journal of the History of Biology 22:307–323.

Corliss, J. O. 1990. Towards a nomenclatural protist perspective, in: Margulis, L., Melkonian, M., Chapman, D. J., Eds., Handbook of protoctista. Jones & Barlett, Boston, pp. xxv–xxx.

Corliss, J. O. 1991a. Problems in cytoterminology and nomenclature for the protists, in: Huang, L. H., Ed., Advances in culture collections, Vol. 1. U.S. Federation for Culture Collections, pp. 23–37.

Corliss, J. O. 1991b. Introduction to the Protozoa, in: Harrison, F. W., Corliss, J. O., Eds., Microscopic anatomy of invertebrates, Vol. 1. Protozoa. Wiley, New York, pp. 1–12.

Corliss, J. O. 1995. The ambiregnal protists and the codes of nomenclature: A brief review of the problem and of proposed solutions. Bulletin of Zoological Nomenclature 52:11–7.

Corliss, J. O., Esser, S. C. 1974. Comments on the role of the cyst in the life cycle and survival of free-living protozoa. Transactions of the American Microscopical Society 93:578–93.

Couté, A., Thérézien, Y. 1994. Nouvelle contribution à l'étude des Euglénophytes (Algae) de l'Amazonie bolivienne. Nova Hedwigia 58:245–272.

Cramer, M., Myers, J. 1952. Growth and photosynthetic characteristics of *Euglena gracilis*. Archiv für Mikrobiologie 17:384–402.

Dawson, N. S., Walne, P. L. 1991. Structural characterization of *Eutreptia pertyi* (Euglenophyta). I. General description. Phycologia 30:287–302.

Dawson, N. S., Walne, P. L. 1994. Evolutionary trends in euglenoids. Archiv für Protistenkunde 144:221–25.

Dawson, N. S., Dunlap, J. R., Walne, P. L. 1988. Structure and elemental composition of pellicular warts of *Euglena spirogyra* (Euglenophyceae). British Phycological Journal 23:61–69.

Deflandre, G. 1926. Monographie du genre *Trachelomonas* Ehr. Imprimerie André Lesot, Nemours, 162 p. + 15 plates.

Deflandre, G. 1930. *Strombomonas*, nouveau genre d'Euglénacées (*Trachelomonas* Ehrbg. *pro parte*). Archiv für Protistenkunde 69:551–614.

Dillard, G. E. 1999. Common freshwater algae of the United States. An illustrated key to the genera (excluding the diatoms). Cramer, Stuttgart, 173 p.

Dillard, G. E. 2000. Freshwater algae of the southeastern United States. Part 7. Pigmented Euglenophyceae. Bibliotheca Phycologica, Vol. 106. Cramer, Stuttgart, 134 p. + 20 plates and captions.

Dragos, N., Péterfi, L. S., Popescu, C. 1997. Comparative fine structure of pellicular cytoskeleton in *Euglena* Ehrenberg. Archiv für Protistenkunde 148:277–285.

Dujardin, F. 1841. Histoire naturelle des zoophytes infusoires. Roret, Paris, 684 p.

Dunlap, J. R., Walne, P. L. 1985. Fine structure and biomineralization of the mucilage in envelopes of *Trachelomonas lefevrei* (Euglenophyceae). Journal of Protozoology 32:437–441.

Dunlap J. R., Walne, P. L. 1987. Variations in envelope morphology and mineralization in *Trachelomonas lefevrei* (Euglenophyceae). Journal of Phycology 23:556–564.

Dunlap, J. R., Walne, P. L., Bentley, J. 1983. Microarchitecture and elemental spatial segregation of envelopes of *Trachelomonas lefevrei* (Euglenophyceae). Protoplasma 117:97–106.

Dunlap, J. R., Walne, P. L., Kivic, P. A. 1986. Cytological and taxonomic studies of the Euglenales. II. Comparative microarchitecture and cytochemistry of envelopes of *Strombomonas* and *Trachelomonas*. British Phycological Journal 21:399–405.

Duthie, H. C., Ostrofsky, M. L. 1975. Freshwater algae from western Labrador. II. Chlorophyta and Euglenophyta. Nova Hedwigia 26:253–268.

Duthie, H. C., Socha, R. 1976. A checklist of the freshwater algae of Ontario, exclusive of the Great Lakes. Naturaliste Canadien 103:83–109.

Edmondson, W. T. 1969. Eutrophication in North America, in: (Anonymous) Eutrophication: Causes, consequences, correctives. National Acad. Sci., Washington, DC, pp. 124–149.

Ehrenberg, C. G. 1830. Neue Beobachtungen über blutartige Erscheinungen in Ägypten, Aravien und Sibirien, nebst einer Übersicht und Kritik der früher bekannten. Annalen der Physik 18:477–514.

Ehrenberg, C. G. 1831. Über die Entwickelung und Lebensdauer der Infusionsthiere; nebst ferneren Beiträgen zu einerVergleichung ihrer organischen Systeme. Physikalische Abhandlungen der königlicher Akademie der Wissenschaften Berlin (1832):1–154.

Ehrenberg, C. G. 1833. Dritter Beitrag zur Erkenntniss grosser Organisation in der Richtung des kleinsten Raumes. Physikalische Abhandlungen der Akademie der Wissenschaften Berlin (1835):145–336.

Ehrenberg, C. G. 1838. Die Infusionsthierchen als vollkommene Organismen. Verlag von Leopold Voss, Leipzig, 547 p. + 64 plates.
Ettl, H., Popovský, J. 1986. Current problems in the taxonomy of algae. Archiv für Hydrobiologie, Supplement 73:1–20.
Farmer, J. N. 1980. The protozoa. Mosby, St. Louis, 732 p.
Fritsch, F. W. 1945. The structure and reproduction of the algae, Vol 1. Cambridge Univ. Press, New York, 791 p.
Gerber, S., Häder, D.-P. 1993. Effects of solar irradiation on motility and pigmentation of three species of phytoplankton. Environmental and Experimental Botany 33:515–521.
Gibbs, S. P. 1978. The chloroplasts of *Euglena* may have evolved from symbiotic green algae. Canadian Journal of Botany 56:2883–2889.
Gibbs, S. P. 1981. The chloroplasts of some algal groups may have evolved from endosymbiotic eukaryotic algae, in: Fredrick, J., Ed., Origins and evolution of eukaryotic intracellular organelles. Annals of the New York Academy of Sciences 361:193–218.
Gojdics, M. 1953. The genus *Euglena*. Univ. of Wisconsin Press, Madison, 268 p.
Goodrich, S. G. 1859. Illustrated natural history of the animal kingdom, Vol II. Derby, Jackson, New York, 680 p.
Gosse, P. H. 1859. Evenings at the microscope; or, Researches among the minuter organs and forms of animal life. Appleton, New York, 480 p.
Gosse, P. H. 1896. Evenings at the Microscope; or, Researches Among the Minuter Organs and Forms of Animal Life. D. Appleton and Co., New York, 480 p.
Graham, L. E., Wilcox, L. W. 2000. Algae. Prentice Hall, Upper Saddle River, NJ, pp. 154–168.
Graham, T. P., McCoy, J. J. 1974. The growth effects of varying concentrations of nitrate and phosphate on the eugleneoid *Trachelomonas hispida*. Bios 45:74–79.
Green, J. 1953. A swarm of *Colacium*. Journal of the Quekett Microscopical Club 3:510–511.
Gualtieri, P. 1993. *Euglena gracilis*: Is the photoreception enigma solved? Journal of Photochemistry and Photobiology, B: Biology 19:3–14.
Gualtieri, P., Pelosi, P., Passarelli, V., Barsanti, L. 1992. Identification of a rhodopsin photoreceptor in *Euglena gracilis*. Biochimica et Biophysica Acta 1117:55–59.
Haberyan, K. A., Umaña, G. V., Collado, C., Horn, S. P. 1995. Observations on the plankton of some Costa Rican lakes. Hydrobiologia 312:75–85.
Häder, D.-P. 1987. Polarotaxis, gravitaxis and vertical phototaxis in the green flagellate, *Euglena gracilis*. Archives of Microbiology 147:179–183.
Hallick, R. B., Hong, L., Drager, R. G., Favreau, M. R., Monfort, A., Orsat, B., Spielmann, A., Stutz, E. 1993. Complete sequence of *Euglena gracilis* chloroplast DNA. Nucleic Acids Research 21:3537–3544.
Hargreaves, J. W., Whitton, B. A. 1976. Effect of pH on growth of acid stream algae. British Phycological Journal 11:215–223.
Haüber, M. M., Müller, S. B., Speth, V., Maier, U.-G. 1994. How to evolve a complex plastid? A hypothesis. Botanica Acta 107:383–386.
Haughey, A. 1970. Notes on *Euglena acus* Ehrenberg from sewage treatment ponds. British Phycological Journal 5:97–102.
Hausmann, K. 1978. Extrusive organelles in protists. International Review of Cytology 52:197–276.
Hayashi, M., Toda, K., Kitaoka, S. 1993. Enriching *Euglena* with unsaturated fatty acids. Bioscience, Biotechnology and Biochemistry 57:352–353.
Hedley, S., Patterson, D. J. 1992. Free-living freshwater Protozoa. A color guide. CRC Press, Boca Raton, FL, 223 p.
Hegner, R. W. 1922. Frog and toad tadpoles as sources of intestinal protozoa for teaching purposes. Science (Washington, DC) 56:439–441.
Hegner, R. W. 1923. Observations and experiments on Euglenoidina in the digestive tract of frog and toad tadpoles. Biological Bulletin (Woods Hole) 45:162–180.
Hibberd, D. J. 1990. Phylum Chlorarachnida, in: Margulis, L., Melkonian, M., Chapman, D. J., Eds., Handbook of protoctista. Jones & Bartlett, Boston, pp. 288–292.
Hilenski, L. L., Walne, P. L. 1983. Ultrastructure of mucocysts in *Peranema trichoporum* (Euglenophyceae). Journal of Protozoology 30:491–496.
Hoham, R. W., Blinn, D. W. 1979. Distribution of cryophilic algae in an arid region, the American Southwest. Phycologia 18:133–145.
Hori, H., Osawa, S. 1987. Origin and evolution of organisms as deduced from 5s ribosomal RNA sequences. Molecular Biology and Evolution 4:445–472.
Hoshaw, R. W., Rosowski, J. R. 1973. Methods for microscopic algae, in: Stein, J. R., Ed., Handbook of phycological methods. Culture methods and growth measurements. Cambridge Univ. Press, New York, pp. 53–68.
Huber-Pestalozzi, G. 1955. Das Phytoplankton des Süsswassers, in: Thienemann, A., Ed., Die Binnegewässer, Vol. XVI, Part 4. Euglenophyceen. Schweizerbart'sche Verlagsbuchhandlung, Stuttgart, 606 p. + 114 plates.
Hutchinson, G. E. 1969. Eutrophication, past and present, in: (Anonymous) Eutrophication: Causes, consequences, correctives. Nat. Acad. Sci., Washington, DC, pp. 17–28.
Inagaki, Y., Hayashi-Ishimaru, Y., Ehara, M., Igarashi, I., Ohama, T. 1997. Algae or protozoa: Phylogenetic position of euglenophytes and dinoflagellates as inferred from mitochondrial sequences. Journal of Molecular Evolution 45:295–300.
Jahn, T. L. 1946. The euglenoid flagellates. Quarterly Review of Biology 21:246–274.
Jahn, T. L. 1951. Euglenophyta, in: Smith, G. M., Ed., Manual of phycology. Chronic Botanica, Waltham, MA, pp. 69–81.
Jahn, T. L., Bovee, E. C. 1968. Locomotive and motile response in *Euglena*, in: Buetow, D. E., Ed., The biology of *Euglena*, Vol. 1. Academic Press, New York, pp. 45–108.
Jahn, T. L., Bovee, E. C., Jahn, F. F. 1979. How to know the protozoa. 2nd ed. Brown, Dubuque, IA, 279 p.
Johnson, L. P. 1944. Euglenae of Iowa. Transactions of the American Microscopical Society 63:97–135.
Johnson, L. P. 1968. The taxonomy phylogeny, and evolution of the genus *Euglena*, in: Buetow, D. E., Ed., The biology of *Euglena*, Vol. 1. Academic Press, New York, pp. 1–25.
Johnson, L. P., Jahn, T. L. 1942. Cause of the green–red color change in *Euglena rubra*. Physiological Zoology 15:89–94.
Kato, S. 1994. Three species of *Eutreptia* (Euglenophyceae) from Japan. Japanese Journal of Phycology 42:221–226.
Kempner, E. S., Miller, J. H. 1972. The molecular biology of *Euglena gracilis*. VII. Inorganic requirements for a minimal culture medium. Journal of Protozoology 19:343–346.
Khan, M. A. 1993. Occurrence of a rare euglenoid causing red-bloom in Dal Lake waters of the Kashmir Himalaya. Archiv für Hydrobiologie 127:101–103.
Kiener, W. 1944. Green snow in Nebraska. Proceedings of the Nebraska Academy of Science 54:12.
Kim, J. T., Boo, S. M. 1998. Morphology, population size and environmental factors of two morphotypes in *Euglena geniculata* (Euglenophyceae) in Korea. Algological Studies 91:27–36.
Kingsley, J. S., Ed. 1888. The riverside natural history, Vol. I. Lower invertebrates. Houghton Mifflin, New York, 425 p.

Kirjakov, I. K. 1983. Le genre *Strombomonas* Deflandre (Euglenophyta) en Bulgarie. Cryptogamie Algologie 4:127–139.

Kirjakov, I. K. von. 1998. Eine neue Form von *Phacus curvicauda* Swirenko 1915 (Euglenophyta) und kritische Bemerkungen über ihre Grösse und Struktur. Algological Studies 88:29–36.

Kivic, P. A., Vesk, M. 1972. Structure and function in the euglenoid eyespot apparatus: The fine structure, and response to environmental changes. Planta 105:1–14.

Kivic, P. A., Vesk, M. 1974. Pinocytotic uptake of protein from the reservoir in *Euglena*. Archives of Microbiology 96:155–159.

Kivic, P. A., Walne, P. L. 1984. An evaluation of a possible relationship between Euglenophyta and Kinetoplastida. Origins of Life 13:269–288.

Klebs, G. 1883. Über die organisation einiger Flagellatengruppen und ihre Beziehungen zu Algen und Infusorien. Untersuchungen Aus Dem Botanisches Institut zu. Tübingen 1:233–362.

Komárek, J. 1991. New books, review. Algological Studies 62:143.

Kuźnicki, L., Walne, P. L. 1993. Protistan evolution and phylogeny: Current controversies. Acta Protozoology 32:135–140.

Kuźnicki L., Mikołajczyk, E., Walne, P. L. 1990. Photobehavior of euglenoid flagellates: Theoretical and evolutionary perspectives. Critical Reviews on Plant Scienes 9:343–369.

Lackey, J. B. 1968. Ecology of *Euglena*, in: Buetow, D. E., Ed., The biology of *Euglena*, Vol 1. Academic Press, New York, pp. 27–44.

Larsen, J., Patterson, D. J. 1991. The diversity of heterotrophic euglenids, in: Patterson, D. J., Larsen, J., Eds., The biology of free-living heterotrophic flagellates. Clarendon, Oxford, pp. 205–217.

Leander, B. S., Farmer, M. A. 2000. Comparative morphology of the euglenid pellicle. I. Patterns of strips and pores. Journal of Eukaryotic Microbiology 47:469–479.

Leander, B. S., Farmer, M. A. 2001. Evolution of Phacus (Euglenophyceae) as inferred from pellicle morphology and SSU rDNA. Journal of Phycology 37:143–159.

Lee, J. J., Capriulo, G. M. 1990. The ecology of marine protozoa: Aan overview, in: Capriulo, G. M., Ed., Ecology of marine Protozoa. Oxford Univ. Press, New York, pp. 3–45.

Lee, R. E. 1999. Phycology, 3rd ed. Cambridge Univ. Press, Cambridge, UK, 614 p.

Leedale, G. F. 1964. Pellicle structure in *Euglena*. British Phycological Bulletin 2:291–306.

Leedale, G. F. 1967a. Euglenida/Euglenophyta. Annual Review of Microbiology 21:31–48.

Leedale, G. F. 1967b. Euglenoid flagellates. Prentice Hall, Englewood Cliffs, NJ, 242 pp.

Leedale, G. F. 1982. Ultrastructure, in: Buetow, D. E., Ed., The biology of *Euglena*, Vol. III. Academic Press, New York, pp. 1–27.

Linton, E. W., Triemer, R. E. 1999a. Reconstruction of the feeding apparatus in *Ploeotia costata* (Euglenophyta) and its relationship to other euglenoid feeding apparatuses. Journal of Phycology 35:313–324.

Linton, E. W., Triemer, R. E. 1999b. Analysis of the genus *Euglena* using SSU rDNA. Journal of Phycology Supplement 35:20.

Linton, E. W., Hittner, D., Lewandowski, C., Auld, T., Triemer, R. E. 1999. A molecular study of euglenoid phylogeny using small subunit rDNA. Journal of Eukaryotic Microbiology 46:217–223.

Lund, J. W. G. 1942. The marginal algae of certain ponds, with special reference to the bottom deposits. Journal of Ecology 30:245–283.

Margulis, L. 1974. Five-kingdom classification and the origin and evolution of cells. Evolutionary Biology 7:45–78.

Margulis, L., Corliss, J. O., Melkonian, M., Chapman, D. J., Eds., 1990. Handbook of protoctista. Jones & Barlett, Boston, 914 p.

Matviyenko, A. M. 1972. Epizoic algae in sewage. Hydrobiological Journal 8(2):41–45.

Maya, S., Prameela, S. K., Menon, V. S. 2000. A preliminary study on the algal flora of temple tanks of southern Kerala. Phykos 38:77–83.

McFadden, G. I. 2001. Primary and secondary endosymbiosis and the origin of plastids. Journal of Phycology 37:951–959.

McLachlan, J. L., Seguel, M. R., Fritz, L. 1994. *Tetreutreptia pomquetensis* gen. et sp. nov. (Euglenophyceae): A quadriflagellate, phototrophic marine euglenoid. Journal of Phycology 30:538–544.

McLachlan, J. L., Curtis, J. M., Boutilier, K., Keusgen, M., Seguel, M. R. 1999. *Tetreutreptia pomquetensis* (Euglenophyta), a psychrophilic species: Growth and fatty acid composition. Journal of Phycology 35:280–286.

Melkonian, M. 1996. II. Systematics and evolution of the algae: Endocytobiosis and evolution of the major algal lineages. Progress in Botany 57:281–311.

Meyer, R. L., Brook, A. J. 1969. Freshwater algae from Itasca state park, Minnesota. III. Pyrrhophyta and Euglenophyta. Nova Hedwigia 18:367–382.

Mignot, J. P., Brugerolle, G., Bricheux, G. 1987. Intercalary strip development and dividing cell morphogenesis in the euglenoid *Cyclidiopsis acus*. Protoplasma 139:51–65.

Monastersky, R. 1998. The rise of life on earth. National Geographic 193:54–81.

Montegut-Felkner, A. E., Triemer, R. E. 1997. Phylogenetic relationships of selected euglenoid genera based on morphological and molecular data. Journal of Phycology 33:512–519.

Moore, J. W. 1979. Benthic algae of southern Baffin Island. II. The epipelic communities in temporary ponds. Journal of Ecology 62:809–819.

Moraczewski, I. R., Zakrys, B. 1992. Non-Linnaean classification of the genus *Euglena*. Algological Studies 67:59–68.

Moss, B. 1973. The influence of environmental factors on the distribution of freshwater algae: An experimental study. II. The role of pH and the carbon dioxide–bicarbonate system. Journal of Ecology 61:157–177.

Munawar, M. 1972. Ecological studies of Euglenineae in certain polluted and unpolluted environments. Hydrobiologia 39:307–320.

Nakamura, H. 1994. Origin of eukaryota from cyanobacterium: membrane evolution theory, in: Seckbach J., Ed., Evolutionary pathways and enigmatic algae: *Cyanidium caldarium* (Rhodophyta) and related cells. Kluwer Academic Publishers, Dordrecht. pp. 3–18.

Norton, T. A., Melkonian, M., Andersen, R. A. 1996. Algal biodiversity. Phycologia 35:308–326.

Nudelman, M. A., Lombardo, R., Conforti, V. 1998. Comparative analysis of envelopes of *Trachelomonas argentinensis* (Euglenophyta) from different aquatic environments in South America. Algological Studies 89:97–105.

Olli, K. 1996. Resting cyst formation of *Eutreptiella gymnastica* (Euglenophyceae) in the northern coastal Baltic Sea. Journal of Phycology 32:535–542.

Ortega, M. M. 1984. Catálogo de algas continentales recientes de México. Universidad Nacional Autónoma de México, 561 p.

Owens, K. J., Farmer, M. A., Triemer, R. E. 1988. The flagellar apparatus and reservoir/canal cytoskeleton of *Cryptoglena pigra* (Euglenophyceae). Journal of Phycology 24:520–528.

Palmer, C. M. 1969. A composite rating of algae tolerating organic pollution. Journal of Phycology 5:78–82.

Palmer, C. M. 1980. Algae and water pollution. Castle House Publ., UK, 123 p.

Palmer, J. D., Round, F. E. 1965. Persistent, vertical-migration rhythms in benthic microflora. I. The effect of light and temperature on the rhythmic behaviour of *Euglena obtusa*. Journal of Marine Biological Association of the United Kingdom 45:567–582.

Palmer, T. C. 1902. Five new species of *Trachelomonas*. Proceedings of the Academy Natural Sciences of Philadelphia 54:791–795 + 1 plate.

Pascher, A. 1931. Über die Verfestigung des Protoplasten im Gehäuse einer neuen Euglenine (*Klebsiella*). Archiv für Protistenkunde 73:325–322.

Patterson, D. J., Hedley, S. 1992. Free-living freshwater Protozoa. A color guide. CRC Press, Boca Raton, FL, 223 p.

Pennak, R. W. 1989. Freshwater invertebrates of the United States. Protozoa to Mollusca, 3rd ed. Wiley, New York, 628 p.

Penno, S., Campbell, L, Hess, W. R. 2000. Presence of phycoerythrin in two strains of *Prochlorococcus* (Cyanobacteria) isolated from the subtropical north Pacific Ocean. Journal of Phycology 36:723–729.

Perty, M. 1849. Über verticale Verbreitung mikroskopischer Lebensformen. Mittheilungen der naturforschenden Gesellschaft in Bern 146–149:17–45.

Perty, M. 1852. Zur Kenntnis kleinster Lebensformen nach Bau, Funktionen, Systematik, mit Spezialverzeichnis der in der Schweiz beobachteten Arten. Verlag von Jent und Reinert, Bern.

Philipose, M. T. 1982. Contributions to our knowledge of Indian algae III. Euglenineae Part 1. The genus *Euglena* Ehrenberg. Proceedings, Plant Sciences (Indian Academy of Sciences) 91:551–599.

Philipose, M. T. 1984. Contributions to our knowledge of Indian algae. III. Euglenineae Part 2. Proceedings, Plant Sciences (Indian Academy of Sciences) 91:503–552.

Philipose, M. T. 1988. Contributions to our knowledge of Indian algae. III. Euglenineae Part 3. The genera *Trachelomonas* and *Strombomonas* Deflandre. Proceedings, Plant Sciences (Indian Academy of Sciences) 98:317–394.

Playfair, G. 1921. Australian freshwater flagellates. Proceedings of the Linnaean Society of the New South Wales 3:99–146 + plates 1–9.

Pochmann, A. 1942. Synopsis der Gattung Phacus. Archiv für Protistenkunde 95:81–252.

Preisig, H. R., Anderson, O. R., Corliss, J. O., Moestrup, Ø., Powell, M. J., Roberson, R. W., Wetherbee, R. 1994. Terminology and nomenclature of protist cell surface structures. Protoplasma 181:1–28.

Prescott, G. W. 1931. University of Iowa Studies in Natural History, Vol.13, Iowa algae. Univ. of Iowa Press, Iowa City, 235 p.

Prescott, G. W. 1955. Algae of the Panama Canal and its tributaries. I. Flagellated organisms. Ohio Journal of Science 15:99–121.

Prescott, G. W. 1962. Algae of the western Great Lakes area, revised ed. Brown, Dubuque, IA, 977 p.

Prescott, G. W. 1978. How to know the freshwater algae, 3rd ed. Brown, Dubuque, IA, 293 p.

Prescott, G. W. 1984. The algae: A review, reprint with corrections. Bishen Singh Mahendra Pal Singh, Dehra Dun, India, and Otto Koeltz Sci. Pub., Koenigstein, Germany, 436 p.

Pringsheim, E. G. 1948. Taxonomic problems in the Euglenineae. Biological Reviews 23:46–61.

Pringsheim, E. G. 1953a. Observations on some species of *Trachelomonas* grown in culture. New Phytologist 52:93–113.

Pringsheim, E. G. 1953b. Observations on some species of *Trachelomonas* grown in culture. New Phytologist 52:238–266.

Pringsheim, E. G. 1956. Contributions towards a monograph of the genus *Euglena*. Nova Acta Leopoldina 18:1–168.

Pringsheim, E. G. 1963. Farblose Algen. Ein Beitrag zur Evolutionsforschung. Fischer, Stuttgart, 471 p.

Provasoli, L. 1958. Nutrition and ecology of protozoa and algae. Annual Review of Microbiology 12:279–308.

Provasoli, L. 1961. Micronutrients and heterotrophy as possible factors in bloom production in natural waters, *in*: Transactions of the Seminar Algae and Metropolitan Wastes. R. A. Taft Sanit. Engr. Center Rep. W 61-3, pp. 48–56.

Provasoli, L. 1969. Algal nutrition and eutrophication, *in*: (Anonymous) Eutrophication: Causes, consequences, correctives. Pub. 1700, Natl. Acad. Sci., Washington, DC, pp. 574–593.

Provasoli, L., Pintner, I. J. 1953. Ecological implication of *in vitro* nutritional requirements of algal flagellates. Annals of the New York Academy of Sciences 56:839–851.

Reinke, D. C. 1979. A preliminary checklist of Kansas algae, *in*: Brooks, R. E., Ed., Reports of the State Biological Survey of Kansas, No. 23, 70 p.

Rosowski, J. R. 1977. Development of mucilaginous surfaces in euglenoids. II. Flagellated, creeping and palmelloid cells of *Euglena*. Journal of Phycology 13:323–328.

Rosowski, J. R., Couté, A. 1996. Bacteria on the lorica of *Trachelomonas* occur in nature, not just in culture. Journal of Phycology 32:697–698.

Rosowski, J. R., Glider, W. V. 1977. Comparative effects of metal coating by sputtering and by vacuum evaporation on delicate features of euglenoid flagellates. Scanning Electron Microscopy IITRI/SEM/I 471–480.

Rosowski, J. R., Hoshaw, R. M. 1970. Staining algal pyrenoids with carmine after fixation in an acidified hypochlorite solution. Stain Technology 45:293–298.

Rosowski, J. R., Hoshaw, R. M. 1971. Results of an attempt to obtain pyrenoids of *Zygnema* by bulk-isolation methods. Journal of Phycology 7:312–316.

Rosowski, J. R., Hoagland, K. D., Roemer, S. C., Lee, K. W. 1981. Improving the image of delicate and complex biological surfaces. Scanning 4:181–187.

Rosowski, J. R., Kugrens, P. 1973. Observations on the euglenoid *Colacium* with special reference to the formation and morphology of attachment material. Journal of Phycology 9:370–383.

Rosowski, J. R., Langenberg, W. G. 1994. The near-spineless *Trachelomonas grandis* (Euglenophyceae) superficially appears spiny by attracting bacteria to its surface. Journal of Phycology 30:1012–1022.

Rosowski, J. R., Lee, K. W. 1978. *Cryptoglena pigra*: A euglenoid with one chloroplast. Journal of Phycology 14:160–166.

Rosowski, J. R., Willey, R. L. 1975. *Colacium libellae* sp. nov. (Euglenophyceae), a photosynthetic inhabitant of the larval damselfly rectum. Journal of Phycology 11:310–315.

Rosowski, J. R., Willey, R. L. 1977. Development of mucilaginous surfaces in euglenoids. I. Stalk morphology of *Colacium mucronatum*. Journal of Phycology 13:16–21.

Rosowski, J. R., Vadas, R. L., Kugrens, P. 1975a. Surface configuration of the lorica of the euglenoid *Trachelomonas* as revealed with scanning electron microscopy. American Journal of Botany 62:48–57.

Rosowski, J. R., Walne, P. L., West, L. K. 1975b. Comparative effects of critical point and air-drying on the morphology of the rigid mucilaginous coating (lorica) of *Trachelomonas* (Euglenophyceae). Micron 5:321–339.

Round, F. E. 1984. The ecology of algae. Cambridge Univ. Press, New York, 653 p.

Round, F. E., Palmer, J. D. 1966. Persistent, vertical-migration rhythms in benthic microflora. II. Field and laboratory studies on diatoms from the banks of the river Avon. Journal of the Marine Biological Association U.K. 46:191–214.

Sarojini, Y. 1994. Composition, abundance and distribution of phytoplankton in sewage and receiving habour water at Visakhapatnam. Phykos 33:137–146.

Say, P. J., Whitton, B. A. 1980. Changes in flora down a stream showing a zinc gradient. Hydrobiologia 76:255–262.

Schlichting, H. E., Jr. 1960. The role of waterfowl in the dispersal of algae. Transactions of the American Microscopical Society 79:160–166.

Schlichting, H. E., Jr. 1964. Meteorological conditions affecting the dispersal of airborne algae and protozoa. Lloydia 27:64–78.

Schwartzbach, S. D, Osafune, T., Löffelhardt, W. 1998. Protein import into cyanelles and complex chloroplasts. Plant Molecular Biology 38:247–263.

Sheath, R. G., Steinman, A. D. 1982. A checklist of freshwater algae of the Northwest Territories, Canada. Canadian Journal of Botany 60:1964–97.

Shi, Z. 1995. New taxa of Euglenophyta from China. Chinese Journal of Oceanology and Limnology 13:348–353.

Shi, Z. 1996a. 6. Quantitative analysis on euglenoid distribution in seven regions of China, in: Kristiansen, Ed., Biogeography of freshwater algae. Hydrobiologia 336:55–65.

Shi, Z. 1996b. Cladistic analysis of euglenoids. Acta Phytotaxonomica Sinica 34:265–275.

Shi, Z. 1998. New taxa of colourless euglenoids from China. Oceanologica et Limnologica Sinica 29:261–268.

Shi, Z. 1999. Flora alarum sinicarum aquae dulcis, Vol. VI. Euglenophyta. Science Press, China, 414 p.

Shi, Z., Jao, C. 1998. New taxa of euglenoids with lorica from China. Acta Hydrobiologica Sinica 22:62–70.

Shin, W., Boo, S. M. 1999. Virus-like particles in both nucleus and cytoplasm of *Euglena viridis* (Euglenophyceae). Algological Studies 95:125–131.

Shin, W., Boo, S. M. 2001. Ultrastructure of *Phacus trypanon* (Euglenophyceae) with an emphasis on striated fiber and microtubule arrangement. Journal of Phycology 37:95–105.

Shin, W., Boo, S. M., Triemer, R. E. 2001. Ultrastructure of the basal body complex and putative vestigial feeding apparatus in *Phacus pleuronectes* (Euglenophyceae). Journal of Phycology 37:913–921.

Singh, K. P. 1956. Studies in the genus *Trachelomonas*. I. Description of six organisms in cultivation. American Journal of Botany 43:258–366.

Smith, G. M. 1950. The fresh-water algae of the United States, 2nd ed. McGraw-Hill, New York, 719 p.

Solomon, J. A., Walne, P. L., Dawson, N. S., Willey, R. L. 1991. Structural characterization of *Eutreptia* (Euglenophyta). II. The flagellar root system and putative vestigial cytopharynx. Phycologia 30:402–414.

Sommer, J. R. 1965. The ultrastructure of the pellicle complex of *Euglena gracilis*. Journal of Cell Biology 24:253–257.

Starmach, K. 1983. Euglenophyta—Eugleniny, in: Starmach, K, Siemińska, J., Eds., Flora Słodkowodna Polski [Freshwater flora of Poland], Vol. 3, Warszawa, 594 p.

Starr, R. C., Zeikus, J. A. 1993. UTEX. The culture collection of algae at the University of Texas at Austin. Journal of Phycology Suppl. 29:1–106.

Stein, F. R. 1878. Der Organismus der Infusionsthiere. Part 3. Der Organismus der Flagellaten. Hälfte, Leipzig, 154 p.

Stein, J. R., Borden, C. A. 1979. Checklist of freshwater algae of British Columbia. Syesis 12:3–39.

Stiller, J. W., Hall, B. D. 1997. The origin of red algae: Implications for plastid evolution. Proceedings of the National Academy of Sciences of the United States of America 94:4520–4525.

Sulli, C., Fang, Z. W., Muchhal, U., Schwartzbach, S. D. 1999. Topology of *Euglena* chloroplast protein precursors within endoplasmic reticulum to golgi to chloroplast transport vesicles. Journal of Biological Chemistry 274:457–463.

Surek, B., Melkonian, M. 1986. A cryptic cytostome is present in *Euglena*. Protoplasma 133:39–49.

Suxena, M. R. 1955. Fresh-water Euglenineae from Hyderabad, India, I. Journal of the Indian Botanical Society 34:429–450.

Taft, C. E., Taft, C. W. 1971. The algae of western Lake Erie. Bulletin/Ohio Biological Survey 1:189 p.

Tani, Y., Tsumura, H. 1989. Screening for tocopherol-producing microorganisms and α-tocopherol production by *Euglena gracilis* Z. Agricultural and Biological Chemistry 53:305–312.

Taylor, F. J. R. 1990. Symbionts in marine protozoa, in: Capriulo, G. M., Ed., Ecology of marine Protozoa. Oxford Univ. Press, New York, pp. 323–340.

Taylor, W. D., Sanders, R. W. 1991. Protozoa, in: Thorp, J. H., Covich, A. P., Eds., Ecology and classification of North American freshwater invertebrates. Academic Press, San Diego, pp. 37–93.

Tell, G., Conforti, V. 1984. Ultrastructura de la lóriga de cuatro especies de *Strombomonas* Defl. (Euglenophyta) en M.E.B. Nova Hedwigia 40:123–131.

Tell, G., Conforti, V. 1986. Euglenophyta pigmentadas de la Argentina. Bibliotheca Phycologica, Vol. 75. Cramer, Berlin, 301 p.

Thérézien, Y. 1999. Strombomonas taiwanensis var. bigeonii nova var. (Euglenophyta, Euglenophyceae). Algological Studies 92:11–18.

Thompson, R. H. 1938. A preliminary survey of the freshwater algae of eastern Kansas. University of Kansas Science Bulletin 25:5–83.

Threlkeld, S. T., Willey, R. L. 1993. Colonization, interaction, and organization of cladoceran epibiont communities. Limnology and Oceanography 38:584–591.

Threlkeld, S. T., Chiavelli, D. A., Willey, R. L. 1993. The organization of zooplankton epibiont communities. Trends in Ecology and Evolution 8:317–321.

Tokuyasu, K., Scherbaum, O. H. 1965. Ultrastructure of mucocysts and pellicle of *Tetrahymena pyriformis*. Journal of Cell Biology 27:67–81.

Triemer, R. 1980. Role of golgi apparatus in mucilage production and cyst formation in *Euglena gracilis* (Euglenophyceae). Journal of Phycology 16:46–52.

Triemer, R. E. 1992. Ultrastructure of mitosis in *Diplonema ambulator* Larsen and Patterson (Euglenozoa). European Journal of Protistology 28:398–404.

Triemer, R. E., Farmer, M. A. 1991a. An ultrastructural comparison of the mitotic apparatus, feeding apparatus, flagellar apparatus and cytoskeleton in euglenoids and kinetoplastids. Protoplasma 164:91–104.

Triemer, R. E., Farmer, M. A. 1991b. The ultrastructural organization of the heterotrophic euglenids and its evolutionary implications, in: Patterson, D. J., Larsen, J., Eds., The biology of free-living heterotrophic flagellates. Clarendon, Oxford, Systematics Association Special Volume 45:185–201.

Triemer, R. E., Lewandowski, C. L. 1994. Ultrastructure of the basal apparatus and putative vestigial feeding apparatuses in a quadriflagellate euglenoid (Euglenophyta). Journal of Phycology 34:28–38.

van den Hoek, C., Mann, D. G., Jahns, H. M. 1995. Algae. An introduction to phycology. Cambridge Univ. Press, Cambridge, UK, 623 pp.

Vickerman, K. 1990. Phylum Zoomastigina. Class Kinetoplastida, in: Margulis, L., Corliss, J. O., Melkonian, M. J., Chapman, D. J., Eds., Handbook of protoctista. Jones & Bartlett, Boston, pp. 215–238.

Vickerman, K., Brugerolle, G., Mignot, J.-P. 1991. Mastigophora, *in*: Harrison, F. W., Corliss, J. O., Eds., Microscopic anatomy of invertebrates, Wiley, New York, pp. 13–159.

Walne, P. L. 1971. Comparative ultrastructure of eyespots in selected euglenoid flagellates, *in*: Parker, B. C., Brown, R. M., Jr., Eds., Contributions in phycology. Allen Press, Lawrence, KS, pp. 107–120.

Walne, P. L., Dawson, N. S. 1993. A comparison of paraxial rods in the flagella of euglenoids and kinetoplastids. Archiv für Protistenkunde 143:177–194.

Walne, P. L., Gualtieri, P. 1994. Algal visual proteins: An evolutionary point of view. Critical Reviews in Plant Science13:185–197.

Walne, P. L., Kivic, P. A. 1990. Phylum Euglenida, *in*: Margulis, L., Corliss, J. O., Melkonian, M. J., Chapman, D. J., Eds., Handbook of protoctista. Jones & Bartlett, Boston, pp. 270–287.

Walne, P. L., Möestrup, O., Norris, R. E., Ettl, H. 1986. Light and electron microscopical studies of *Eutreptiella eupharyngea* sp. nov. (Euglenophyceae) from Danish and American waters. Phycologia 25:109–126.

Walne, P. L., Passarelli, V., Lenzi, P., Barsanti, L., Gualtieri, P. 1998. Rhodopsin: A photopigment for phototaxis in *Euglena gracilis*. Critical Reviews in Plant Science 17:559–574.

Walton, L. B. 1915. A review of the described species of the order Euglenoidina Bloch. Class Flagellata (Protozoa) with particular reference to those found in the city water supplies and in other localities of Ohio, *in*: Ohio Biological Survey, Vol.1 (Bulletin 4). Ohio State University, pp. 343–459.

Ward, K. A., Willey, R. L. 1981. The development of a cell-substrate attachment system in a euglenoid flagellate. Journal of Ultrastructure Research 74:165–174.

Waygood, F. R., Hussain, A., Godavari, H. R., Tai, Y. C., Badour, S. S. 1980. Purification and reclamation of farm and urban wastes by *Euglena gracilis*: photosynthetic capacity, effect of pH hydrogen-ion concentration, temperature acetate and whey. Ecological and Biological Series A Environmental Pollution. 23:179–215.

Webster, P. 1989. Endocytosis by African trypanosomes. I. Three-dimensional structure of the endocytic organelles in *Trypanosoma brucei* and *T. congolense*. European Journal of Cell Biology 49:295–302.

Wenrich, D. H. 1923. Variations in *Euglenomorpha hegneri* n.g., n.sp., from the intestine of tadpoles. Anatomical Record 24:370–371.

Wenrich, D. H. 1924. Studies on *Euglenamorpha hegneri* n.g., n.sp., a euglenoid flagellate found in tadpoles. Biological Bulletin (Woods Hole) 47:149–174.

Whatley, J. M. 1993. The endosymbiotic origin of chloroplasts. Review of Cytology 144:259–299.

Whitford, L. A., Schumacher, G. J. 1969. A manual of the freshwater algae in North Carolina. North Carolina Agricultural Experiment Station, Technical Bulletin 188, 313 p.

Whitford, L. A., Schumacher, G. J. 1984. A manual of freshwater algae, revised ed. Sparks Press, Raleigh, NC, 337 pp.

Willey, R. L. 1972. The damselfly (Odonata) hindgut as host organ for the euglenoid flagellate *Colacium*. Transaction of the American Microscopical Society 91:585–593.

Willey, R. L. 1980. Proposed new identification character for the genus *Colacium* (Euglenophyceae). Journal of Phycology 16:143–146.

Willey, R. L. 1982. The synonymy of *Colacium calvum* Stein and *Colacium physeter* Fott (Euglenophyceae). Phycologia 21:173–177.

Willey, R. L., Cantrell, P. A. 1990. Epibiotic euglenoid flagellates increase the susceptibility of some zooplankton to fish predation. Limnology and Oceanography 35:952–959.

Willey, R. L., Threlkeld, S. T. 1993. Organization of crustacean epizoan communities in a chain of subalpine ponds. Limnology and Oceanography 38:623–627.

Willey, R. L, Wibel, R. G. 1985a. The reservoir cytoskeleton and a possible cytostomal homologue in *Colacium* (Euglenophyceae). Journal of Phycology 21:570–577.

Willey, R. L., Wibel, R. G. 1985b. A cytostome/cytopharynx in green euglenoid flagellates (Euglenales) and its phylogenetic implications. BioSystems 18:369–376.

Willey, R. L., Durban, E. M., Bowen, W. R. 1973. Ultrastructural observations of a *Colacium* palmella: The reservoir, eyespot, and flagella. Journal of Phycology 9:211–215.

Willey, R. L., Ward, K., Russin, W., Wibel, R. G. 1977. Histochemical studies of the extracellular carbohydrate of *Colacium mucronatum* (Euglenophyceae). Journal of Phycology 13:349–353.

Willey, R. L., Walne, P. L., Kivic, P. 1988. Phagotrophy and the origins of the euglenoid flagellates. Critical Reviews in Plant Science 7:303–339.

Willey, R. L., Willey, R. B., Threlkeld, S. T. 1993. Planktivore effects on zooplankton epibiont communities: Epibiont pigmentation effects. Limnology and Oceanography 38:1818–1822.

Wołowski K. 1992. Occurrence of Euglenophyta in the Třeboň biosphere reserve (Czechoslovakia). Algological Studies 66:73–98.

Wołowski K. 1993. *Euglena ettlii* Wołowski sp. nova (Euglenophyceae). Archiv für Prostenkunde 143:173–176.

Wołowski, K., Walne, P. L. 1997. Euglenophytes from the Southeastern United States I. Colorless species. Algological Studies 86:109–135.

Xavier, M. B., Mainardes-Pinto, C. S. R., Takino, M. 1991. *Euglena sanguinea* Ehrenberg bloom in a fish-breeding tank (Pindamonhangaba, São Paulo, Brazil). Algological Studies 62:133–142.

Yamagishi, T., Couté, A. 1995. *Strombomonas taiwanensis* nov. sp. (Euglenophyta, Euglenophyceae). Cryptogamie Algologie 16:255–262.

Zakryś, B. 1986. Contribution to the monograph of Polish members of the genus *Euglena* Ehrenberg 1830. Nova Hedwigia 42:491–540.

Zakryś, B. 1994. *Euglena walnei* sp. nova (Euglenophyta)—a new species from Alabama (United StatesA). Algological Studies 72:9–11.

Zakryś, B. 1997a. On the identity and variation of *Euglena agilis* Carter (= *E. pisciformis* Klebs). Algological Studies 86:81–90.

Zakryś, B. 1997b. The taxonomic consequences of morphological and genetic variability in *Euglena agilis* Carter (Euglenophyta): Species or clones in *Euglena*? Acta Protozoologica 36:157–169.

Zakryś, B., Walne, P. L. 1994. Floristic, taxonomic and phytogeographic studies of green Euglenophyta from the southeastern United States, with emphasis on new and rare species. Algological Studies 72:71–114.

Zakryś, B., Walne, P. L. 1998a. Comparative ultrastructure of chloroplasts in the subgenus *Euglena* (Euglenophyta): Taxonomic significance. Cryptogamie Algologie19:3–18.

Zakryś, B., Walne, P. L. 1998b. Ultrastructure of mucocysts in *Euglena oxyuris* Schmarda (Euglenophyceae). Algological Studies 88:125–133.

Zakryś, B., Kucharski, R., Moraczewski, I. 1996. Genetic and morphological variability among clones of *Euglena pisciformis* based on RAPD and biometric analysis. Algological Studies 81:1–21.

Zakryś, B., Moraczewski, I., Kucharski, R. 1997. The species concept in *Euglena* in light of DNA polymorphism analyses. Algological Studies 86:51–79.

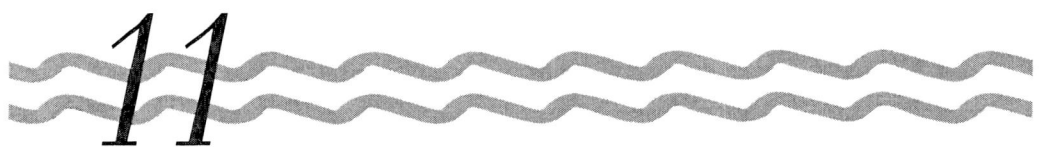

EUSTIGMATOPHYTE, RAPHIDOPHYTE, AND TRIBOPHYTE ALGAE

Donald W. Ott
Carla K. Oldham-Ott

Department of Biology
University of Akron
Akron, Ohio 44325

I. General Introduction
II. Eustigmatophytes
   A. Introduction
   B. Diversity and Morphology
   C. Ecology and Distribution
   D. Key and Descriptions of Genera
III. Raphidophytes
   A. Introduction
   B. Diversity and Morphology
   C. Ecology and Distribution
   D. Key and Descriptions of Genera
IV. Tribophytes
   A. Introduction
   B. Diversity and Morphology
   C. Ecology and Distribution
   D. Key and Descriptions of Genera
V. Collection and Preparation for Identification
VI. Guide to Literature for Species Identification
Literature Cited

I. GENERAL INTRODUCTION

Along with diatoms, chrysophytes, synurophytes, brown algae, and several groups of colorless protists, the three algal classes covered in this chapter belong to a group of diverse but poorly known organisms variously known as heterokonts (van den Hoek et al., 1995), stramenopiles (Patterson, 1989), or ochrophytes (Cavalier-Smith and Chao, 1996; Graham and Wilcox, 2000), when excluding non-photosynthetic organisms. As such, most members of the Eustigmatophyceae, Raphidophyceae, and Tribophyceae share the following characteristics (Bold and Wynne, 1985; van den Hoek et al., 1995; Graham and Wilcox, 2000): (1) Flagellated cells possess a long forward-directed tinsel (hairy) flagellum and a shorter backward-directed smooth flagellum. (2) The tinsel flagellum has two rows of tripartite tubular hairs known as mastigonemes, which are composed of glycoproteins. (3) A transitional helix is usually present between the shaft of the flagellum and the basal body (absent in raphidophytes and in *Vaucheria*). (4) Plastids are surrounded by four membranes. (5) Thylakoids are in stacks of three found traversing the entire length of the plastid; a girdle lamella is present beneath the plastid membrane (absent in the eustigmatophytes and some tribophytes). (6) Plastids contain chlorophyll-*a* and chlorophyll-*c* (*c* absent in eustigmatophytes); chlorophyll-*b* is not present. (7) The photosynthetic storage product is chrysolaminarin, commonly reported as oils or fat droplets; starch is not present. (8) Flagellar swelling and eye spot (stigma) are present in the plastids of most members; eyespot occurs outside the plastid in the eustigmatophytes and in one species of the tribophyte alga *Chloromeson*; eyespot is absent in raphidophytes, zoospores of some flagellated tribophyte species (R. L. Meyer, personal communication), and *Vaucheria*.

In contrast to other heterokont algae, the plastids of organisms included in the Eustigmatophyceae, the Raphidophyceae (freshwater representatives), and the Tribophyceae contain the accessory pigment vaucheriaxanthin, so far found in no other algal groups (van den Hoek *et al.*, 1995). They also lack fucoxanthin, causing them to appear yellow–green or greenish, often being mistaken for green algae. These statements do not imply that these three groups are closely related (Ariztia *et al.*, 1991; Cavalier-Smith and Chao, 1996; Daugbjerg and Andersen, 1997; Potter *et al.*, 1997). The phylogenetic relationships among these algae have yet to be satisfactorily resolved.

II. EUSTIGMATOPHYTES

A. Introduction

Long considered members of the Tribophyceae, the algae now known as eustigmatophytes were moved to their own class when Hibberd and Leedale (1970, 1971a, 1972) discovered that these organisms differed from tribophytes structurally and in their pigment composition. In those genera with motile cells, the motile cells have an eyespot outside of the plastid (hence the name) and have one or two flagella, although two basal bodies are present (Hibberd, 1990b). Eustigmatophytes lack chlorophyll-*c* (Guillard and Lorenzen, 1972) and have violaxanthin and vaucheriaxanthin as accessory pigments (van den Hoek *et al.*, 1995; Graham and Wilcox, 2000).

B. Diversity and Morphology

The Eustigmatophyceae is a small class, presently containing eight genera, although it is probable that many more tribophyte algae will be recognized as eustigmatophytes after further study (Andersen *et al.*, 1998). Hibberd (1981) recognized a single order in the Eustigmatophyceae, the Eustigmatales, containing four families. Members of the Eustigmataceae have one emergent flagellum (*Eustigmatos*, *Pseudostaurastrum*, *Vischeria*), whereas those in the Pseudocharaciopsidaceae (*Ellipsoidion*, *Pseudocharaciopsis*) have two emergent flagella. Species classified in the Chlorobotryaceae (*Chlorobotrys*) and Monodopsidaceae (*Monodopsis*, *Nannochloropsis*) do not produce zoospores.

The vegetative morphology of the Eustigmatophyceae is very simple: Cells are spherical, oval, elliptical, or irregularly shaped coccoid unicells (Figs. 1A–K); the coccoid cells are encountered singly, in pairs, or as colonies (Hibberd, 1990b). Vegetative cells have walls, but their chemical components are unclear at this time (Santos, 1996). The extraplastidal eyespot is not surrounded by a membrane (Bold and Wynne, 1985; Lee, 1989). Eustigmatophytes have usually one (Andersen *et al.*, 1998) or more parietal plastids, containing chlorophyll-*a* and lacking girdle lamellae. Reproduction in the Eustigmatophyceae is by means of walled autospores (nonmotile spores without the potential to produce flagella) or naked (nonwalled) zoospores (flagellated asexual reproductive cells) (Bold and Wynne, 1985; Hibberd, 1990b). Sexual reproduction has not been observed in eustigmatophyte algae. The plastids of vegetative cells of most eustigmatophytes contain a pyrenoid, but zoospores do not (Hibberd, 1981). The large eyespot is at the anterior end of the cell, is not associated with the plastid (Hibberd, 1981), and is structurally like those found in euglenoids, though it appears that the photoreceptive mechanisms of the eyespot differ from those of the euglenoids (Santos *et al.*, 1996).

The zoospores of some eustigmatophytes have a single anterior flagellum (Fig. 1F) bearing mastigonemes. The shorter flagellum, when present, is smooth and laterally or posteriorly directed. A single plastid is present in the zoospores (Hibberd, 1990b).

C. Ecology and Distribution

The typically coccoid cells of eustigmatophyte genera are frequently encountered in the euplankton (true plankton community of open water) and the tychoplankton (occasional plankton, moved into the area) (Tarapchak, 1972; Sheath and Hellebust, 1978). They are also often a component of the metaphyton (found between or among vascular plants and other algae) (Whitford and Schumacher, 1969; R. L. Meyer, personal communication). Less often, eustigmatophytes may be epiphytic (growing on other plants) (Schumacher *et al.*, 1966) or epilithic (growing on rocks) (Poulton, 1930). Terrestrial habitats also contain some eustigmatophytes, as they have been collected from the soil (Hibberd, 1990b; Graham and Wilcox, 2000).

Aquatic habitats in which eustigmatophytes have been collected include ponds, pools, bogs, streams, and lakes. Dystrophic (nutrient-poor acidic waters) and mesotrophic (waters with a moderate amount of nutrients present) conditions coupled with low temperatures (0–15°C) seem to be preferred by many eustigmatophytes (R. L. Meyer, personal communication), although Tarapchak (1972) has collected them in waters with temperatures from 17.5 to 30°C and Transeau (1913) reports finding *Chlorobotrys regularis* in a small Michigan pool in August.

Based on the current literature, eustigmatophytes seem to have a sparse and limited distribution. Perhaps because of their small size, reports of eustigmatophytes in algal checklists are rare. Eustigmatophyte genera have been collected in the Arctic (Sheath and Hellebust,

1978), Canada (Stein, 1975; Duthie and Socha, 1976; Stein and Borden, 1978), Minnesota (Tarapchak, 1972), Wisconsin (Smith, 1920), Connecticut and Massachusetts (Poulton, 1930; Andersen et al., 1998), North Carolina (Schumacher et al., 1966; Krienitz et al., 2000), South Carolina (Goldstein and Manzi, 1976), Texas (Andersen et al., 1998), and Alaska, Arkansas, and Montana (R. L. Meyer, personal communication).

D. Key and Descriptions of Genera

1. Key

1a.	Cells spherical	2
1b.	Cells oval, elliptical, ellipsoidal, or irregularly shaped	5
2a.	Cell wall smooth	3
2b.	Cell wall ornamented (Fig. 1 K)	*Vischeria*
3a.	Cells in pairs or colonies (Fig. 1A, B)	*Chlorobotrys*
3b.	Cells usually solitary, occasionally in pairs	4
4a.	Zoospores never produced, cells less than 5 mm in diameter (Fig. 1J)	*Nannochloropsis*
4b.	Reproduction by zoospores (Figs. 1D–F)	*Eustigmatos*
5a.	Cells ellipsoidal, with no stipe or disc (Fig. 1C)	*Ellipsoidion*
5b.	Cells oval, elliptical, or irregularly shaped	6
6a.	Cells asymmetrical (Fig. 1G)	*Monodopsis*
6b.	Cells symmetrical	7
7a.	Cells tetrahedral (Fig. 1I)	*Pseudostaurastrum*
7b.	Cells oval to elliptical, usually with small stipe or attachment disc (Fig. 1H)	*Pseudocharaciopsis*

2. Descriptions of Genera[1]

Chlorobotrys Bohlin 1901 (Fig. 1A, B)

Cells are spherical to globose; one to two cells are surrounded by a mucilaginous sheath in colonies of 2–32 cells. No zoospores have been reported. At this time, *C. stellata* and *C. regularis* are the only eustigmatophyte species in this genus; *C. regularis* is synonymous with *Chlorococcum regulare*. *Chlorobotrys* is euplanktonic and tychoplanktonic in acid bogs and fens (R. L. Meyer, personal communication), in ponds and lakes of Ontario (Duthie and Socha, 1976), British Columbia (Stein, 1975; Stein and Borden, 1978), Minnesota (Tarapchak, 1972), Wisconsin (Smith, 1920), and Michigan (Transeau, 1913), and attached to stones and greenhouse pots in Connecticut and Massachusetts (Poulton, 1930).

Ellipsoidion Pascher 1937 (Fig. 1C)

Cells are solitary, ellipsoidal, tapering to a point at each end; there are three to four plastids per cell. Reproduction occurs by autospores. Hibberd (1990b) has renamed *E. acuminatum* as *Pseudocharaciopsis ovalis*, although van den Hoek et al. (1995) refer to it as *Ellipsoidion acuminatum*. *Ellipsoidion* has been observed in mesotrophic and dystrophic ponds and pools; it is metaphytic among vascular plants and other algae in North America, especially in acid bogs and swamps (Prescott, 1978; R. L. Meyer, personal communication). It is epiphytic on leaves in greenhouses in North Carolina (Schumacher et al., 1966).

Eustigmatos Hibberd 1981 (Fig. 1D–F)

Cells are spherical, with a smooth cell wall and usually one parietal plastid. Reproduction occurs by zoospores with a single flagellum. It is usually collected from soil. *Eustigmatos vischeri* is synonymous with *Pleurochloris commutata* (Hibberd, 1990b). *Eustigmatos magnus* is synonymous with *Pleurochloris polyphem* (Hibberd, 1990b) and *P. magna* (van den Hoek et al., 1995). *Eustigmatos* has been collected in Minnesota, Alaska, and Arkansas (R. L. Meyer, personal communication) and, as *Pleurochloris commutata*, in a bog in Minnesota (Tarapchak, 1972).

Monodopsis Hibberd 1981 (Fig. 1G)

Cells are solitary, 5–10 µm long, and asymmetrical. Zoospores are not present. *Monodopsis subterranea* is synonymous with *Monodus subterraneus*. *Monodopsis* is metaphytic and epiphytic in dystrophic ponds and

---

[1] All scale bars in the figures equal 10 µm unless otherwise indicated.

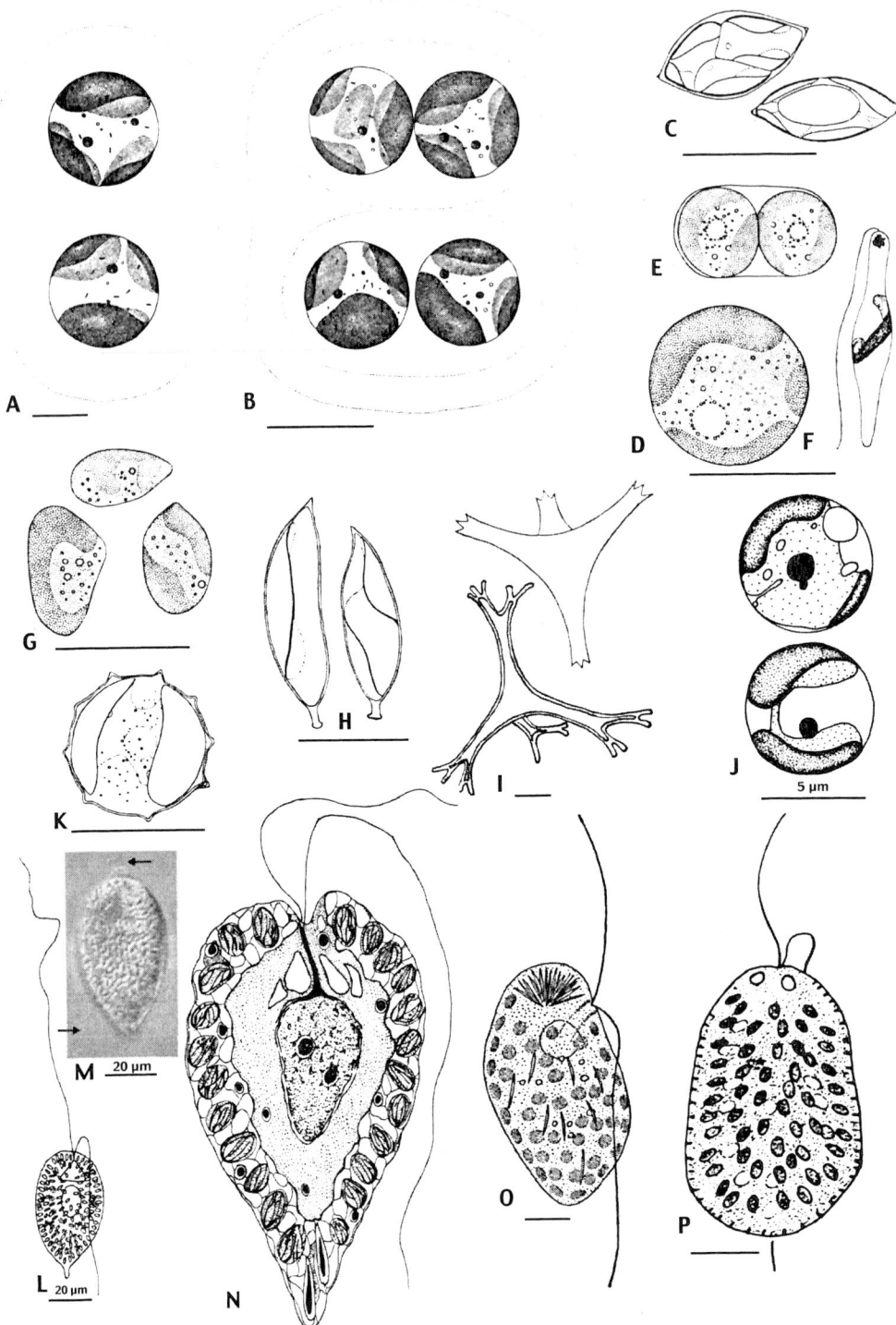

FIGURE 1 Freshwater members of the Eustigmatophyceae and Raphidophyceae. (A, B) *Chlorobotrys regularis*; groups of two to four cells surrounded by mucilage. (C) *Ellipsoidion acuminatum* vegetative cells. (D–F) *Eustigmatos magnus*: (D) Autospore. (F) Zoospore. (G) *Monodopsis subterranea* asymmetrical vegetative cells. (H) *Pseudocharaciopsis minuta* vegetative cells with small stipes. (I) *Pseudostaurastrum limneticum* vegetative cells. (J) *Nannochloropsis oculata* vegetative cells. (K) *Vischeria stellata* vegetative cell. (L–N) *Gonyostomum*: (L, N) Flagellated cells. (M) Light micrograph of *Gonyostomum* sp. collected from Singer Lake Bog, Ohio; arrows showing flagella. (O) *Merotrichia capitata* flagellated cell with trichocysts in a cluster at the anterior end of the cell. (P) *Vacuolaria virescens* flagellated cell. A–B reprinted from Tarapchak (1972), C–I, K from Ettl (1978), J from Andersen *et al.* (1998), L, O from Smith (1950), M by authors, N from Bold and Wynne (1985), and P from Prescott (1978), all with permission.

pools, often in association with *Sphagnum*, sedges, and/or *Typha* in Minnesota, Alaska, Arkansas, and Montana throughout the year at a pH of 4.5–6.8 and a temperature of 0–15°C. (R. L. Meyer, personal communication). *Monodopsis* has been collected from a rock in a stream in Connecticut (Andersen *et al.*, 1998).

Nannochloropsis Hibberd 1981 (Fig. 1J)

Cells are spherical to ovoid, 2–4 μm in diameter. Cells have a single plastid lacking a pyrenoid. Zoospores are not present. This genus is easily confused with the green alga *Chlorella*, but lacks a positive starch test (Sect. V). *Nannochloropsis occulata* is synonymous with *Nannochloris occulata*. Species delineations of *Nannochloropsis* in all probability cannot be made using light microscopy (Andersen *et al.*, 1998). *Nannochloropsis* has been collected from ponds in North Carolina (Krienitz *et al.*, 2000).

Pseudocharaciopsis Lee *et* Bold 1974 (Fig. 1H)

Cells are ovoid to elliptical, usually with a small stipe or attachment disc. Reproduction occurs by biflagellate zoospores. This genus is easily confused with *Characiopsis* (Tribophyceae) (Fig. 11A, B) and *Characium* (Chlorophyceae). It may be identified by the conspicuous stalked pyrenoid, visible using light microscopy (Graham and Wilcox, 2000). *Pseudoharaciopsis minuta* is synonymous with *Characium minutum*, *Characiopsis minuta*, and *Pseudocharaciopsis texensis*. *Pseudocharaciopsis ovalis* is synonymous with *Monodus ovalis*, *Characiopsis ovalis*, and *Ellipsoidion acuminatum* (see *Ellipsoidion*). *Pseudocharaciopsis* has been observed in a creek in Texas (Andersen *et al.*, 1998) and is tychoplanktonic in a tundra pond of the Northwest Territories, Canada, as *Characiopsis minuta* (Sheath and Hellebust, 1978).

Pseudostaurastrum Chodat 1921 (Fig. 1I)

Pseudostaurastrum limneticum, recently transferred to the Eustigmatophyceae (Schnepf *et al.*, 1996), has tetrahedral cells, deeply incised and having four projections at the end of each arm. The zoospores have one emergent flagellum and do not contain the typical extraplastidal eyespot. *Pesudostaurastrum* occurs in dystrophic ponds and pools and mesotrophic ponds and lakes. It is euplanktonic in open water and metaphytic among *Sphagnum* and/or sedges in Minnesota and Alaska at a pH of 4.5–6 and a temperature of 0–8°C. (R. L. Meyer, personal communication).

Vischeria Pascher 1938 (Fig. 1K)

Cells are spherical, polyhedral, or irregularly shaped with various ridges and projections, 7–20 μm in size. There is one to several plastids in each cell. Reproduction occurs by autospores and uniflagellate zoospores. *Vischeria stellata* is synonymous with *Chlorobotrys stellata*. *Vischeria punctata* is synonymous with *V. stellata*. *Vischeria helvetica* is synonymous with *Polyedriella helvetica* (Hibberd, 1990b); van den Hoek *et al.* (1995) refer to this species as *P. helvetica*. *Vischeria* is euplanktonic, tychoplanktonic, and metaphytic in bogs, dystrophic and mesotrophic ponds, streams, and the protected littoral zone in lakes. It is associated with *Sphagnum*, *Typha*, *Nuphar*, *Nymphaea*, and sedges in waters abundant in dissolved organic matter. It has been collected in Minnesota and Alaska, most abundant in spring and fall, at a water temperature of 5–12°C, as high as 15°C, and in North Carolina and South Carolina (Schumacher *et al.*, 1966; Tarapchak, 1972; Goldstein and Manzi, 1976; Prescott, 1978; R.L. Meyer, personal communication).

III. RAPHIDOPHYTES

A. Introduction

The Raphidophyceae are also heterokonts, with the flagella arising from a shallow invagination or gullet (Fig. 1M) at or near the cell apex. Whereas marine raphidophytes contain accessory pigments such as those found in the chrysophytes and brown algae, the freshwater genera have pigments common to those of the Tribophyceae—diadinoxanthin, heteroxanthin, and vaucheriaxanthin (van den Hoek *et al.*, 1995; Graham and Wilcox, 2000).

B. Diversity and Morphology

Freshwater raphidophytes are a small group, containing only three genera, in the single order Raphidmonadales. *Gonyostomum*, *Merotrichia*, and *Vacuolaria* have all been reported from North American collections. These organisms are naked (having no cell wall), 30- to 80-μm-long unicellular flagellates (Heywood, 1990); the anterior flagellum, responsible for the cell's movements, bears mastigonemes, whereas the long, trailing flagellum is smooth. The flagellar basal bodies are connected to the nucleus by a kinetid or rhizostyle (Heywood, 1980, 1989, 1990; Mignot, 1967, 1976). Palmelloid (nonmotile cells embedded in a common mucilage, potentially capable of forming flagella) stages have also been observed (Heywood, 1978a). Numerous green to greenish-yellow plastids lacking pyrenoids are present in the periphery of the cell. Eyespots are lacking; contractile vacuoles, trichocysts, and mucocysts are present (Mignot, 1967; Fott, 1971; Heywood, 1978b; Graham and Wilcox, 2000). A large Golgi body is easily observed at the anterior end of the large nucleus (Mignot, 1967, 1976; Heywood, 1980, 1990). Reproduction seems to be by means of longitudinal division; meiosis has not been reported. Cysts of *Gonyostomum* have been observed in old cultures (Drouet and Cohen, 1935; Cronberg *et al.*, 1988).

Five marine and three freshwater genera of raphidophytes are commonly recognized, although some authorities recognize fewer or more taxa and some include non-photosynthetic organisms (Heywood, 1990).

C. Ecology and Distribution

Lepisto *et al.* (1994) observed that dystrophic and eutrophic conditions seem to be favored by raphidophytes. They are frequently found in neutral or acidic waters (pH 3.2–7), as plankton, associated with aquatic macrophytes, or in the layer just above the sediments (Stein, 1975; Heywood, 1990). Studies of these algae in Japan show that at least some freshwater raphidophytes cannot survive at pH 8.0 or higher (Kato, 1991). *Gonyostomum* occurs in standing water, lakes, ponds, and rivers but most frequently colonizes *Sphagnum* bogs. It is reported to form nuisance blooms, especially during the summer months, often comprising over 90% of the phytoplankton biomass in small lakes in Ohio (Havens, 1989) and floodplain pools in Nordic countries (Pithart *et al.*, 1997).

Collections and descriptions of these algae are rare, but they seem to have a wide distribution. *Gonyostomum* is the most commonly collected of the three and has been reported from British Columbia and Ontario (Stein, 1975; Duthie and Socha, 1976; Stein and Borden, 1978) to the southeastern United States (Prescott, 1978). The less common *Vacuolaria* has been collected from British Columbia (Stein, 1975; Stein and Borden, 1978) to North Carolina (Whitford and Schumacher, 1969; Whitford, 1979). *Merotrichia* has been reported by Whitford and Schumacher (1969) and Prescott (1978).

D. Key and Descriptions of Genera

1. Key

1a.	Cells not flattened, rigid, trichocysts scattered but many in a cluster at anterior end of cell, flagella subapical to lateral (Fig. 1O)	*Merotrichia*
1b.	Cells flattened in side view, may change shape while swimming, numerous plastids and trichocysts present at cell periphery, flagella apical	2
2a.	Cells ovoid to spherical, may be narrowed posteriorly, gullet triangular (Fig. 1L–N)	*Gonyostomum*
2b.	Cells ovoid to pear shaped, may be narrowed anteriorly, usually two or more contractile vacuoles present (Fig. 1P)	*Vacuolaria*

2. Descriptions of Genera

Gonyostomum Diesing (Fig. 1L–N)

Cells are ovoid to spherical, flattened in side view, may be narrowed posteriorly, 36–100 μm in length. Cells may exhibit metaboly (change shape while swimming). Numerous plastids and trichocysts are present at cell periphery. Flagella arise from the triangular gullet; flagella are usually as long as the cell body, with one directed forward and one trailing. A typical habitat for *Gonyostomum* is described by Havens (1989) as having a summer high temperature of 29°C and a pH of 4.4–4.6. *Gonyostomum depressum*, *G. latum*, and *G. semen* have been collected from acid bogs, lakes, ponds, and rivers at a pH of 4.4–6.6 and a temperature of 11–29°C (Drouet and Cohen, 1935; Cowles and Brambel, 1936; Lackey, 1942; Stein, 1975; Prescott, 1978; Havens, 1989). *Gonyostomum* species have been reported from Ontario (Duthie and Socha, 1976), British Columbia (Stein, 1975; Stein and Borden, 1978), Ohio (Havens, 1989; authors' personal observations), Massachusetts (Drouet and Cohen, 1935), Tennessee (Lackey, 1942), North Carolina (Whitford and Schumacher, 1969), Minnesota, Montana, Alaska, and Arkansas (R. L. Meyer, personal communication, and the southeastern United States (Prescott, 1978).

Merotrichia Skuja (Fig. 1O)

Cells are ovoid, circular in cross section, and may be slightly narrowed posteriorly. Cells are not flattened in side view and do not exhibit metaboly. Trichocysts are scattered throughout the cell but also clustered at the anterior end. The gullet is circular; flagella are subapical to lateral in insertion. *Merotrichia capitata* has been collected from small pools, bogs, and swamps in North Carolina (Whitford and Schumacher, 1969) and dystrophic ponds and *Typha* marshes in Minnesota, Alaska, and Arkansas (R. L. Meyer, personal communication).

Vacuolaria Senn (Fig. 1P)

Cells are ovoid to pear shaped, narrowest at

the anterior end. Cells are flattened in side view and exhibit metaboly. Numerous plastids and trichocysts are present at periphery. Two or more apical contractile vacuoles are usually evident. Flagella are apical. *Vacuolaria* occurs in ditches, swamps, and ponds; it is sometimes associated with *Lemna* in British Columbia and North Carolina (Whitford and Schumacher, 1969; Stein, 1975; Stein and Borden, 1978; Prescott, 1978) and occurs in dystrophic ponds and *Larix* bogs in Minnesota (R. L. Meyer, personal communication).

IV. TRIBOPHYTES

A. Introduction

The yellow–green algae or Tribophyceae (Hibberd, 1981) are also known as the Xanthophyceae. These algae have both flagella apically or subapically inserted. Tribophytes have vaucheriaxanthin, diatoxanthin, diadinoxanthin, and heteroxanthin as their main accessory pigments (Hibberd, 1980; van den Hoek et al., 1995).

B. Diversity and Morphology

1. Diversity

Most authors classify the Tribophyceae in six (Bold and Wynne, 1985; Lee, 1989) or seven (van den Hoek et al., 1995) orders, based on vegetative characteristics. At present, it is estimated that there are over 100 genera, containing 600 species (van den Hoek et al., 1995). Relatively few of these species have been seen since their original description (Pascher, 1937–1939; Ettl, 1978; Hibberd, 1990a). Perhaps one reason for their relative obscurity is that approximately two-thirds of these species are coccoid unicells that often occur in small numbers (Hibberd, 1980). It has also been suggested that some taxa are delicate enough to be lost when preserved (Stein and Gerrath, 1969). Thus far, 101 tribophyte genera have been collected in the freshwater and soil habitats of North America. R. L. Meyer (1969, personal communication), Meyer et al. (1970), Stein (1975), Stein and Borden (1978), and Tarapchak (1972) have reported the occurrence of tribophytes extensively.

Most tribophyte algae are not found in great numbers in the field with the exceptions of *Tribonema* (Colt, 1974; Stein, 1975) (Fig. 13M–S) and *Vaucheria* (Fig. 14D–K), which are often seen in great quantities. *Botrydium* (Fig. 14A–C), although cosmopolitan in distribution (Poulton, 1930), is often overlooked as it is usually collected from soil or mud (Transeau, 1913; Stein, 1975; LaRivers, 1978). *Ophiocytium* (Fig. 11O–S) and *Characiopsis* (Fig. 10A, B) are also often collected in North America (Poulton, 1930; Prescott, 1931; Taylor, 1934; Duthie and Socha, 1976; Duthie et al., 1976). *Mischococcus* (Fig. 11E) is also mentioned in collections quite frequently (Smith, 1916; Meyer et al., 1970; Stein, 1975; Stein and Borden, 1978; Sheath and Hellebust, 1978; Sheath and Steinman, 1981).

2. Morphology

The uninucleate filamentous or coenocytic (multinucleate) filamentous forms are the most conspicuous of the tribophyte genera. *Vaucheria* (Fig. 14D–K) is often seen forming dense mats on damp soil or in water, whereas *Tribonema* (Fig. 13M–S) filaments often form entangled floating masses in ponds, ditches, rivers, and lakes (Tiffany, 1937; Prescott, 1962; Woodson and Holoman, 1964; Colt, 1974; Stein, 1975). The vast majority of tribophyte taxa collected are, however, greenish coccoid unicells. These nonflagellated walled forms include spherical genera such as *Chloridella* (Fig. 5L, M), and *Ophiocytium* (Fig. 11O–S), which has elongate, cylindrical, sometimes coiled cells. Rhizopodal (amoeboid) (Fig. 3A–T), monadoid (unicellular flagellates) (Fig. 2A–W), and palmelloid (solitary or colonial, embedded in a gelatinous matrix or attached with a gelatinous disc) (Fig. 4A–R) morphologies also occur in the Tribophyceae (Ettl, 1978; Hibberd, 1990a). Classification of the Tribophyceae has traditionally been based on morphology. The classification used in this chapter is based on that of Ettl (1978) and Rieth (1980). Seven orders are represented corresponding to monadoid (Chloramoebales), rhizopodal (Rhizochloridales), palmelloid (Heterogloeales), coccoid (Mischococcales), filamentous (Tribonematales), coenocytic vesiculate (Botrydiales), and coenocytic filamentous (Vaucheriales) forms (Table I). Recent analyses of these taxonomic groups using molecular techniques have led to questions as to whether general morphological terms should be used to classify tribophyte algae (Daugbjerg and Andersen, 1997; Bailey and Andersen, 1998). It is likely that future systematic studies will bring about revised taxonomic schemes for the Tribophyceae.

A parallelism in vegetative morphology is found in some genera of the Tribophyceae, Chrysophyceae, and Chlorophyceae (Smith, 1950; Rhodes and Stofan, 1967; Whitford and Schumacher, 1969; Bold and Wynne, 1985). Examples of unicellular flagellates in these groups include *Heterochloris* (Tribophyceae, Fig. 2M–P), *Ochromonas* (Chrysophyceae; Chapt. 12), and *Chlamydomonas* (Chlorophyceae; Chapt. 6). Coccoid, palmelloid, filamentous, and coenocytic morphologies also exist in all three classes. Rhizopodal morphologies are seen in the Tribophyceae and the Chrysophyceae, but are absent in the Chlorophyceae (Bold and Wynne, 1985).

TABLE I Taxonomy and Morphology of the Freshwater Tribophyceae in North America

Order/morphology	Family	Genus
Chloramoebales Monadoid. Vegetative cells solitary, motile by means of one or two heterokont flagella; cell wall absent.	Chloramoebaceae	*Ankylonoton* *Chloramoeba* *Chlorokardion* *Chloromeson* *Heterochloris* *Nephrochloris* *Polykyrtos*
Rhizochloridales Rhizopodal. Vegetative cells amoeboid with pseudopods, cell wall absent; lorica may be present.	Rhizochloridaceae Stipitococcaceae Myxochloridaceae	*Rhizochloris* *Rhizolekane* *Stipitococcus* *Stipitoporos* *Chlamydomyxa* *Myxochloris*
Heterogloeales Palmelloid. Vegetative cells nonmotile with contractile vacuoles and eyespot, cell wall absent or present; mucilage present in some.	Heterogloeaceae Malleodendraceae Pleurochloridellaceae Characidiopsidaceae	*Gloeochloris* *Helminthogloea* *Heterogloea* *Malleodendron* *Pleurochloridella* *Characidiopsis*
Mischococcales Coccoid. Vegetative cells nonmotile, lacking contractile vacuole and eyespot; cell wall distinct.	Pleurochloridaceae Botrydiopsidaceae Botryochloridaceae	*Akanthochloris* *Arachnochloris* *Aulakochloris* *Bracchiogonium* *Chlorallantus* *Chlorarkys* *Chloridella* *Chlorocloster* *Chlorogibba* *Diachros* *Ellipsoidion* *Endochloridion* *Goniochloris* *Isthmochloron* *Keriosphaera* *Meringosphaera* *Monallantus* *Monodus* *Nephrodiella* *Pleurochloris* *Pleurogaster* *Polyedriella* *Polygoniochloris* *Prismatella* *Pseudopolyedriopsis* *Pseudostaurastrum* *Pseudotetraedron* *Rhomboidella* *Sklerochlamys* *Tetraedriella* *Tetraplektron* *Trachychloron* *Trachycystis* *Trachydiscus* *Botrydiopsis* *Excentrochloris* *Perone* *Botryochloris* *Chlorellidiopsis* *Chlorellidium*

(*Continues*)

TABLE I (Continued)

Order/morphology	Family	Genus
		Dichotomococcus
		Ducelliera
		Heterodesmus
		Ilsteria
		Raphidiella
		Sphaerosorus
		Tetraktis
	Gloeobotrydaceae	Asterogloea
		Chlorobotrys
		Gloeobotrys
		Gloeoskene
		Gloeosphaeridium
		Merismogloea
	Gloeopodiaceae	Gloeopodium
	Mischococcaceae	Mischococcus
	Characiopsidaceae	Characiopsis
		Chlorokoryne
		Chlorothecium
		Chytridiochloris
		Dioxys
		Hemisphaerella
		Lutherella
		Peroniella
	Chloropediaceae	Chloropedia
	Centritractaceae	Bumilleriopsis
		Centritractus
	Ophiocytiaceae	Ophiocytium
Tribonematales Filamentous. Algal body a chain of usually uninucleate cells.	Neonemataceae	Chadefaudiothrix
		Neonema
	Tribonemataceae	Bumilleria
		Heterothrix
		Tribonema
	Heterodendraceae	Heterodendron
	Heteropediaceae	Aeronemum
		Capitulariella
		Chaetopedia
		Fremya
		Heterococcus
		Heteropedia
Botrydiales Coenocytic. Algal body coenocytic, vesiculate.	Botrydiaceae	Botrydium
Vaucheriales Coenocytic. Algal body coenocytic, filamentous.	Vaucheriaceae	Vaucheria

Based on Ettl (1978), Rieth (1980), and Hibberd (1990a).

In those tribophyte genera with walls, the walls of some genera may be "delicately sculptured" (Hibberd, 1990a) (e.g., *Chlorallantus*, Fig. 5H, I; *Polygoniochloris*, Fig. 6K). In many tribophytes, the wall may be composed of two overlapping pieces (e.g., *Chloromeson*, Fig. 2L; *Diachros*, Fig. 5Q; *Goniochloris*, Fig. 5U, V; *Pseudotetraedron*, Fig. 7A, B), although this characteristic is not always easily visible without phase or interference-contrast microscopy. The walls of some of the filamentous genera consist of interlocking H-shaped pieces, easily observed at the ends of broken filaments. *Tribonema* (Fig. 13M–S) and *Microspora* (Chlorophyceae; Chapt. 8) both exhibit this feature. The cell wall of *Vaucheria* is cellulosic (Parker *et al.*, 1963), and it is thought the cell walls of other tribophyte algae are also composed of cellulose, with some also having their walls impregnated with silica (van den Hoek et al., 1995; Graham and Wilcox, 2000).

Plastids are usually discoid and parietal in tribophyte algae. Members of the Tribophyceae usually have a greater number of plastids than do most green algae (Hibberd, 1980) and these plastids vary little in form, unlike those in green algae (Ettl, 1978). Girdle lamellae are present in the plastids of tribophyte algae so far investigated with electron microscopy, excluding the genera *Bumilleria* (Fig. 13A–D) and *Bumilleriopsis* (Fig. 11G, H) (Massalski and Leedale, 1969; Deason, 1971a; Hibberd and Leedale, 1971b). Pyrenoids are present in the plastids of some genera (e.g., *Ankylonoton*), absent in some (e.g., *Chloramoeba*), and, in some genera, may be present or absent depending on the species (e.g., *Chlorocloster* and *Vaucheria*) (Ettl, 1978).

3. Reproduction

Asexual reproduction in tribophyte algae can occur by means of simple fragmentation, bilateral division, akinetes (vegetative cells with thickened cell walls; allow survival under adverse environmental conditions), aplanospores (nonmotile spores with the potential to produce flagella), autospores (nonmotile spores without the potential to produce flagella), statospores (resting cysts, usually composed of two pieces of approximately equal size in tribophytes), and zoospores (flagellated asexual reproductive cells). Sexual reproduction has been observed only in *Botrydium*, *Tribonema*, and *Vaucheria*.

Fragmentation (*Tribonema*, Fig. 13P) occurs in filamentous forms. Bilateral division is an asexual reproductive strategy found in monadoid, rhizopodal, and palmelloid forms (Hibberd, 1980). Akinetes generally occur in filamentous genera (Lee, 1989), such as *Tribonema* (Smith, 1950), but have also been observed in *Botrydium*, possibly in *Ophiocytium* (Poulton, 1930), and we have observed them in *Vaucheria* under suboptimal culture conditions. Aplanospores have been observed in coccoid taxa such as *Botrydiopsis* (Luther, 1899) and *Ophiocytium* (Bold and Wynne, 1985), in filamentous taxa such as *Tribonema* (Lagerheim, 1889) and *Bumilleria* (Poulton, 1930; Smith, 1950), and in the coenocytic *Vaucheria*. Autospores seem to be common only in the coccoid forms and have been observed in many of these genera, including *Botrydiopsis* (Poulton, 1930), *Diachros* (Fig. 5Q), and *Monallantus* (Pascher, 1937–1939). Endogenous statospores occur in only a few rhizopodal and monadoid taxa (Pascher, 1937–1939) such as *Chloromeson* (Fig. 2L) (Ettl, 1978).

Zoospores (*Characidiopsis*, Fig. 4B; *Endochloridion*, Fig. 5T; *Tribonema*, Fig. 13S) have been observed in many tribophyte genera, including coccoid, palmelloid, filamentous, and coenocytic forms. One or more plastids (usually two) are present in the zoospores of tribophytes (Hibberd, 1990a). Most taxa have an eyespot at the anterior end of one plastid (Hibberd, 1980). Ultrastructural details of zoospores are described in Massalski and Leedale (1969), Deason (1971a, b), Hibberd and Leedale (1971b), and O'Kelly (1989). The zoospores of *Vaucheria* (Fig. 14F) differ from those of other tribophyte algae in that they are large compound zoospores (= synzoospores) with many pairs of unequal, smooth flagella (Koch, 1951; Greenwood, 1959; Ott and Brown, 1974). The numerous plastids do not have eyespots (Ott and Brown, 1974).

Sexual reproduction has been described in only three genera of tribophyte algae. *Tribonema* is reported to be isogamous (Smith, 1920; Ross, 1954; Bold and Wynne, 1985), whereas *Botrydium* may be isogamous or anisogamous (Iyengar, 1925; Moewus, 1940). *Vaucheria* is oogamous (Oltmanns, 1895; Davis, 1904; Mundie, 1929; Couch, 1932); antheridia and oogonia are formed either directly on the filament (Fig. 14D, H–K) or on gametophores (Fig. 14G). Most species are monoecious, but some brackish-water species are dioecious (Ott and Hommersand, 1974). Spermatozoids of *Vaucheria* are unlike the typical tribophycean motile cells in their flagellar root structure (Moestrup, 1982; Preisig, 1989), lack of a transitional helix (Moestrup, 1982), and lack of plastids (Ott and Brown, 1978).

C. Ecology and Distribution

Morphological forms in the Tribophyceae often correspond with the habitats in which they are found. The monadoid, or unicellular flagellated, genera (Chloramoebales) are often metaphytic among submersed vascular plants (Prescott, 1978; R. L. Meyer, personal communication). Many of the rhizopodal or amoeboid genera (Rhizochloridales) are also metaphytic, but may be epiphytic (attached to vascular plants or other algae) or endophytic (living within plant cells, most often *Sphagnum*) (Poulton, 1930; Prescott, 1931, 1944, 1978; Tarapchak, 1972; R. L. Meyer, personal communication). Palmelloid genera (Heterogloeales) may be euplanktonic, metaphytic, or epiphytic in association with aquatic vascular plants, bryophytes, and filamentous algae (Stein, 1975; R. L. Meyer, personal communication). Coccoid tribophytes (Mischococcales) are ubiquitous. They may be euplanktonic, tychoplanktonic, neustonic (on the water surface), metaphytic, or epiphytic on vascular plants and filamentous algae (Lowe, 1927; Poulton, 1930; Hirsch and Palmer, 1958; Prescott and Vinyard, 1965; Schumacher et al., 1966; Meyer et al., 1970; Tarapchak, 1972; Duthie and Socha,

1976; Colt, 1974; Dillard *et al.*, 1976; Goldstein and Manzi, 1976; Sheath and Hellebust, 1978). Less common are genera that are epilithic (growing on rocks), epipelic (growing on mud, silt, or soil), or epizooic (growing on animals) (Poulton, 1930; Schumacher *et al.*, 1963, 1966; Whitford and Schumacher, 1969). The filamentous genera (Tribonematales) may be euplanktonic, tychoplanktonic, metaphytic, neustonic, epiphytic, epilithic, epipelic, or, rarely endophytic (Prescott and Vinyard, 1965; Whitford and Schumacher, 1969; Croasdale, 1973; Stein, 1975; R. L. Meyer, personal communication). The coenocytic vesiculate *Botrydium* (Botrydiales) is virtually always epipelic (Thompson, 1938; Stein, 1975; LaRivers, 1978). The coenocytic filamentous *Vaucheria* (Vaucheriales) is most often epipelic or epilithic, either aquatic or terrestrial, but it may also be tychoplanktonic (Prescott, 1931; Tiffany, 1937; Thompson, 1938; Harris, 1964; Woodson, 1962; Colt, 1985).

Tribophyte algae occupy a wide variety of habitats. They have been collected from temporary pools, swamps, ponds, lakes, ditches, streams, rivers, and on mud or soil, but little is known of the physical and chemical factors influencing the occurrence of the many members of the group. Current velocity, light regime, specific conductance, oxygen, pH, and nutrient concentration ranges have not been established for most members of this group in the field. However, some data on light and nutrient requirements for culturing *Vaucheria* and a few other tribophyte algae has been published (League and Greulach, 1955; Schneider *et al.*, 1993, 1999). Many *Vaucheria* species grow well in culture at 10–15°C; collections in the field are often most successful between fall and late spring, but also from under ice (Ott and Oldham-Ott, unpublished data). Some species of *Vaucheria* are habitat specific: *Vaucheria fontinalis* (Fig. 14J) and *V. geminata* (Fig. 14G) are usually found in the shallow waters of lakes and ponds, whereas *V. dillwynii* (Fig. 14H) and *V. aversa* usually occur on mud in ditches and bogs. Other *Vaucheria* species are not as particular. *Vaucheria bursata* (Fig. 14D) has been collected from all of the habitats described above (Ott and Oldham-Ott, unpublished data).

In general, tribophytes occur most often under dystrophic (Round, 1981) and mesotrophic conditions. Meyer (personal communication) writes that the "greatest abundance and diversity of taxa occurs under conditions in which the water temperatures range from 0 to 20°C. Some taxa may be present above this temperature. Waters stained with organic acids and having pH values within the range from 4.5 to 7.2 typically have the most diverse flora. The greatest numbers of taxa have been reported from communities enriched with dissolved organic matter. These substances are usually associated with low nutrient concentrations. Light seems not to be a controlling factor since the organisms can be found in abundance during twenty-four hour daylight in the boreal summer and the abbreviated day length of the southern latitudes. In Minnesota, tribophytes may be found under the ice and throughout the open water season if the temperature remains moderate. In lower latitudes, such as Arkansas, the taxa are most abundant during the spring and fall as well as in the free water under the ice. In Alaska, the taxa may be present under the ice, but snow frequently limits the amount of light available. However, tribophytes are present throughout the period of open water. The distribution in Florida is disjunct. Where cool water springs are present, they can be collected year around. At other sites, elevated temperatures may be a limiting factor."

There are, however, reports of tribophytes collected from habitats with higher temperatures, pH values, and nutrient conditions than mentioned above. The few published instances of tribophytes occurring at higher temperatures include *Stipitococcus* at 21.5–23.5°C (Stein and Gerrath, 1969), *Ophiocytium* at 27°C (Woodson and Holoman, 1964), and *Tribonema* at 10.5–24.5°C (Woodson and Afazal, 1976). Tarapchak (1972) collected over 50 tribophytes in waters ranging from 17.5 to 30°C. Several instances of tribophytes in habitats with pH values higher than 7.2 have been reported: *Stipitococcus*, 8.2 (Woodson and Holoman, 1964); *Arachnochloris*, *Chlorallantus*, *Gloeobotrys*, *Mischococcus*, *Characiopsis*, *Lutherella*, *Ophiocytium*, and *Tribonema*, 6.3–8.09 (Prescott and Vinyard, 1965); *Gloeobotrys*, *Ophiocytium*, and *Tribonema*, 6.5–9.0 (Durrell and Norton, 1960); and *Tribonema*, 7.6–8.9 (Sheath *et al.*, 1996; Sheath and Müller, 1997). Nutrient conditions may exceed the typical low nutrient conditions; *Goniochloris*, *Mischococcus*, *Tetraplektron* (Meyer, 1969; Meyer *et al.*, 1970), *Polyedriella*, *Polygoniochloris* (R. L. Meyer, personal communication), and other tribophyte algae have been collected from eutrophic waters.

The distribution of the tribophyte genera most often collected in North America (*Vaucheria* and *Tribonema*) is cosmopolitan; they have been identified in collections from as far north as the Arctic (Prescott, 1953; Polunin, 1954; Ross, 1954; Prescott and Vinyard, 1965; Sheath and Hellebust, 1978; Sheath *et al.*, 1986; Sheath and Müller, 1997) to as far south as Puerto Rico (Tiffany and Britton, 1944). *Botrydium*, *Bumilleria*, *Characiopsis*, *Mischococcus*, and *Ophiocytium* have also been reported from widespread locations in North America. R. L. Meyer (personal communication) reports a wide distribution for tribophyte algae in

North America, ranging from northern Alaska through California, east to New England, and south to the Everglades of Florida; collections were also made in the Canadian Rockies, the prairies, the Experimental Lakes Area (Ontario), and Quebec.

Many of the less common tribophyte algae have most often been collected from northern habitats. *Akanthochloris*, *Arachnochloris*, *Chlamydomyxa*, *Chlorobotrys*, *Gloeobotrys*, *Heterogloea*, *Peroniella*, and *Stipitococcus* have been identified for the most part in collections from Canada and the northern United States (Smith, 1916, 1918; Moore and Carter, 1926; Lowe, 1927; Poulton, 1930; Prescott, 1931; Hughes, 1948; Vinyard, 1958; Durrell and Norton, 1960; Stein and Gerrath, 1969; Tarapchak, 1972; Stein, 1975; Duthie and Socha, 1976; Gaufin et al., 1976; Stein and Borden, 1978; LaRivers, 1978; Whitford, 1979; Sheath and Steinman, 1981) although some of these genera have also been found as far south as Arkansas (Meyer, 1969; Meyer et al., 1970), North Carolina (Schumacher et al., 1963; Whitford and Schumacher, 1969), South Carolina (Jacobs, 1968), Tennessee (Lackey, 1942), and Virginia (Woodson and Holoman, 1964). In contrast, *Centritractus* has been collected most often in warmer locations, including Jamaica (Hegewald, 1976), Puerto Rico (Tiffany and Britton, 1944), and many southern United States locations (Camburn, 1982). Unfortunately, many of the genera included in this chapter have been seldom collected (some only once) and few clear distribution patterns can be determined for the vast majority of the tribophyte algae.

D. Key and Descriptions of Genera

1. Key

The following is based in part on Bourrelly (1968, 1981), Ettl (1978), Prescott (1978), Hibberd (1990a), and R. L. Meyer (personal communcation).

1a.	Filamentous organization, cells usually uninucleate (Figs. 12 and 13) (Tribonematales)	7
1b.	Nonfilamentous organization or thalli coenocytic (multinucleate cells, may be filamentous or otherwise)	2
2a.	Coenocytic organization	3
2b.	Cells uninucleate	4
3a.	Algal body coenocytic vesicles (Fig. 14A–C) (Botrydiales)	*Botrydium*
3b.	Algal body a coenocytic filament (Fig. 14D–K) (Vaucheriales)	*Vaucheria*
4a.	Vegetative cells solitary, cell wall absent, motile by means of 1–2 heterokont flagella (Fig. 2) (Chloramoebales)	17
4b.	Vegetative cells not flagellated	5
5a.	Cells naked, amoeboid, with pseudopods, sometime loricate, solitary or colonial (Fig. 3) (Rhizochloridales)	23
5b.	Cells not amoeboid, no lorica evident	6
6a.	Vegetative cells palmelloid (nonmotile), solitary or colonial, sometimes surrounded by mucilage, with contractile vacuole and eyespot (Fig. 4) (Heterogloeales)	28
6b.	Cells nonmotile, having a distinct cell wall, without contractile vacuole and eyespot (Figs. 5–11) (Mischococcales)	33
7a.	Filaments or pseudofilaments uniseriate to multiseriate, usually surrounded by mucilage	8
7b.	Filaments not surrounded by mucilage	9
8a.	Filaments uniseriate, individual cells globular to cylindrical (Fig. 13G–L)	*Neonema*
8b.	Pseudofilaments uniseriate or multiseriate; cells ellipsoidal/elongate (Fig. 12C–E)	*Chadefaudiothrix*
9a.	Filaments unbranched, uniseriate	10
9b.	Filaments branched, uniseriate to multiseriate	11
10a.	Filaments short, cells short-cylindrical, markedly constricted at cross walk, no H-shaped pieces evident (Fig. 13E)	*Heterothrix*
10b.	Filaments short or long, cells short-cylindrical to quadrate, not inflated (slightly so during zoospore release), H-shaped pieces rarely evident (Fig. 13A, D)	*Bumilleria*
10c.	Filaments usually long, cells elongate-cylindrical, usually thick-walled, H-shaped pieces evident (Fig. 13M–R)	*Tribonema*
11a.	Thallus of uniseriate filaments with dendroid branching; attached by large basal cell (Fig. 12N)	*Heterodendron*
11b.	Thallus pseudoparenchymatous or irregularly branched with no large basal cell	12

12a.	Thallus pseudoparenchymatous	13
12b.	Thallus irregularly branched with no large basal cell	15
13a.	Terminal cells of filaments tapered at apex (Fig. 12I)	*Chaetopedia*
13b.	Terminal cells of filaments not tapered	14
14a.	Terminal cells of filaments nearly spherical (Fig. 12A, B)	*Aeronemum*
14b.	Terminal cells of filaments variously shaped (Fig. 12O)	*Heteropedia*
15a.	Filaments usually very short, with some cells solitary (Fig. 12J–L)	*Fremya*
15b.	Short, irregularly branched filaments, but never with solitary cells	16
16a.	Branched filaments, cells cylindrical, 3–7 plastids per cell (Fig. 12F–H)	*Capitulariella*
16b.	Branched filaments, terminal cells tapered, one plastid per cell (Fig. 12M)	*Heterococcus*
17a.	Cells uniflagellate (Fig. 2Q–T)	*Nephrochloris*
17b.	Cells biflagellate	18
18a.	Flagella subapically inserted (Fig. 2A, B)	*Ankylonoton*
18b.	Flagella apically inserted	19
19a.	Cells exhibiting metaboly	20
19b.	Cells rigid	21
20a.	Cells with 2–6 small, parietal plastids (Fig. 2C, D)	*Chloramoeba*
20b.	Cells with two large parietal plastids (Fig. 2M–P)	*Heterochloris*
21a.	Cells with more than one plastid (Fig. 2E–G)	*Chlorokardion*
21b.	Cells with one plastid	22
22a.	Cell outline regular (Fig. 2H–L)	*Chloromeson*
22b.	Cell outline irregular, lobate (Fig. 2U–W)	*Polykyrtos*
23a.	Cells loricate	24
23b.	Cells without lorica	26
24a.	Loricate cells at end of long stipe	25
24b.	Stipe absent or not visible (Fig. 3L–N)	*Rhizolekane*
25a.	Cells with 1–3 plastids (Fig. 3O–S)	*Stipitococcus*
25b.	Cells with 4–5 plastids (Fig. 3T)	*Stipitoporos*
26a.	Amoeboid cells, usually endophytic in or epiphytic on *Sphagnum*	27
26b.	Amoeboid cells, free living (Fig. 3D, E)	*Rhizochloris*
27a.	Cell walls usually hyaline, cytoplasm granular (Fig. 3A–C)	*Chlamydomyxa*
27b.	Cells otherwise (Fig. 3F–K)	*Myxochloris*
28a.	Cells solitary, with distinct cell wall	29
28b.	Cells massed or colonial, colonies large or just small clusters of cells	30
29a.	Cells unattached, not surrounded by mucilage (Fig. 4P–R)	*Pleurochloridella*
29b.	Cells attached with stipe and gelatinous disc (Fig. 4A–D)	*Characidiopsis*
30a.	Cells attached with thick mucilaginous stipe, forming treelike colonies (Fig. 4M–O)	*Malleodendron*
30b.	Cells without stipe	31
31a.	Colonial cells in tall, mucilaginous sheaths (Fig. 4G, H)	*Helminthogloea*
31b.	Mucilaginous sheath irregular, spherical, or ellipsoidal	32
32a.	Mucilaginous thallus spherical or ellipsoidal, cells regularly arranged around periphery of thallus (Fig. 4E, F)	*Gloeochloris*

32b.	Mucilaginous thallus irregular, surrounding small cluster of cells, cells dispersed randomly within thallus (Fig. 4I–L)	*Heterogloea*
33a.	Cells united on thin, mucilaginous stipes, dendritic (forming treelike colonies) (Fig. 11E) (Mischococcaceae)	*Mischococcus*
33b.	Unicellular, cells in small clusters, or colonial but not forming treelike colonies	34
34a.	Cells elongate, cylindrical, with two or more nuclei, solitary or colonial (Fig. 11O–S) (Ophiocytaceae)	*Ophiocytium*
34b.	Cells either elongate and uninucleate or of other shapes, solitary or colonial	35
35a.	Cells prominently elongate, uninucleate, solitary (Fig. 11G–K) (Centritractaceae)	36
35b.	Cells of other shapes, solitary or colonial	37
36a.	Cells cylindrical, straight, or slightly bent with long spine at each end (Fig. 11I–N)	*Centritractus*
36b.	Cells cylindrical, straight or curved, without spines (Fig. 11G, H)	*Bumilleriopsis*
37a.	Cells in flat colonies, attached without stipe or disc (Fig. 11F) (Chloropediaceae)	*Chloropedia*
37b.	Cells in other types of colonies or solitary	38
38a.	Cells in thick, stratified, mucilaginous stalks, sometimes branched (Fig. 11A–D) (Gloeopodiaceae)	*Gloeopodium*
38b.	Cells in other types of colonies or solitary	39
39a.	Older cells large, with many nuclei and plastids; spherical or irregularly shaped; cell wall unornamented (Fig. 7R–W) (Botrydiopsidaceae)	40
39b.	Cells solitary or colonial, mostly uninucleate	42
40a.	Cells epiphytic or amoeboid (Fig. 7V, W)	*Perone*
40b.	Cells not epiphytic; cells spherical or irregularly ellipsoidal	41
41a.	Cells spherical (Fig. 7R)	*Botrydiopsis*
41b.	Cells irregularly ellipsoidal (Fig. 7S–U)	*Excentrochloris*
42a.	Cells attached to substratum directly with mucilaginous disc or by stipelike extension of cell wall; cells solitary or in twos or fours (Fig. 10) (Characiopsidaceae)	43
42b.	Cells either attached to one another or embedded in a common mucilage; solitary and free living or colonial	50
43a.	Cells pyriform (Fig. 10C)	*Chlorokoryne*
43b.	Cells either hemispherical, stipulate, with thin threadlike stipe, with basal disc, or directly attached to substratum	44
44a.	Cells hemispherical (Fig. 10K, L)	*Hemisphaerella*
44b.	Cells either stipulate, with thin threadlike stipe, with basal disc, or directly attached to substratum	45
45a.	Cells stipulate	46
45b.	Cells attached by thin threadlike stipe, basal disc, or directly to substratum	48
46a.	Stipe short, cells irregularly fusiform, often flattened triangles (Fig. 10G–J)	*Dioxys*
46b.	Stipe long, cells spherical, ovoid, ellipsoidal, or fusiform	47
47a.	Cells spherical, stipe threadlike, attached to other algae (Fig. 10O, P)	*Peroniella*
47b.	Cells ovoid to ellipsoidal on stipe with small basal disc (Fig. 10A, B)	*Characiopsis*
48a.	Cells cylindrical to spindle shaped, attached with short stalk; cell wall usually in two pieces of unequal size (Fig. 10D, E)	*Chlorothecium*
48b.	Cells attached directly to substratum	49
49a.	Cells long, cylindrical to fusiform, usually on filamentous algae (Fig. 10F)	*Chytridiochloris*
49b.	Cells spherical to upright ovals, solitary or in groups of two to four (Fig. 10M, N)	*Lutherella*
50a.	Cells colonial	51
50b.	Cells solitary, free living	65
51a.	Cells attached to one another (Fig. 8) (Botryochloridaceae)	52
51b.	Cells embedded in a common mucilage (Fig. 9) (Gloeobotrydaceae)	59

52a.	Cells attached in chains of 2–8 cells (Fig. 8I)	Heterodesmus
52b.	Cells attached in other groups of two, tetrads, spherical colonies, or unorganized clusters	53
53a.	Cells elongate, fusiform, 18–30 μm in length in groups of twos, fours, or solitary (Fig. 8L)	Raphidiella
53b.	Cells spherical, hemispherical, cylindrical, recurved ovals, or pyramidal	54
54a.	Cells spherical in spherical colonies of 8–32 cells (Fig. 8M)	Sphaerosorus
54b.	Cells in groups of two, tetrads, or unorganized clusters	55
55a.	Cells cylindrical, in pyramidal tetrads (Fig. 8N)	Tetraktis
55b.	Cells spherical to globose or hemispherical	56
55c.	Cells recurved ovals or pyramidal	58
56a.	Cells hemispherical, in tetrads, singly or in unorganized cushion-like clusters, epiphytic on filamentous algae or submerged vegetation (Fig. 8B, C)	Chlorellidiopsis
56b.	Adult cells spherical to globose	57
57a.	Cells in clusters of 10–100 cells (Fig. 8A)	Botryochloris
57b.	Cells 8–25 μm, varying in shape due to compression, in tetrads; or as adults, spherical and sometimes solitary (Fig. 8D–F)	Chlorellidium
57c.	Cells 4–14 μm, spherical to rarely ellipsoidal, in tetrads (Fig. 8J, K)	Ilsteria
58a.	Cells elongate, recurved ovals in groups of four attached to other groups of four (Fig. 8G)	Dichotomococcus
58b.	Cells pyramidal with spine at distal end, attached to other cells by cylindrical processes in spherical to globular colonies of various size (Fig. 8H)	Ducelliera
59a.	Cells at periphery of gelatinous matrix (Fig. 9D–H)	Chlorosaccus
59b.	Cells otherwise embedded in common mucilage	60
60a.	Cells ornamented with spines (Fig. 9A, B)	Asterogloea
60b.	Cells smooth	61
61a.	One to two cells within mucilaginous sheath	62
61b.	More than two cells in mucilaginous sheath	63
62a.	Cells spherical, in concentric layers of mucilage (Fig. 9C)	Chlorobotrys
62b.	Cells ellipsoidal (Fig. 9M, N)	Merismogloea
63a.	Colony spherical (Fig. 9L)	Gloeosphaeridium
63b.	Colony spherical	64
64a.	Cells with 2–3 parietal plastids (Fig. 9I)	Gloeobotrys
64b.	Cells with 2–3 central plastids (Fig. 9J, K)	Gloeoskene
65a.	Cells spherical, not dorsiventrally flattened	66
65b.	Cells triradiate (three armed), renate (kidney bean shaped), lunate, fusiform, ovate, ellipsoidal, elliptical, cylindrical, oblong, pyramidal, polyhedral, polygonal, tetragonal, triangular, quadraradiate (four armed), asymmetrical, rhomboidal, rectangular, four to six sided, or with various shapes, not dorsiventrally flattened	74
65a.	Cells dorsiventrally flattened; may be spherical, elliptical, fusiform, rectangular, hexagonal, or polygonal (Fig. 6K, L)	Polygoniochloris
66a.	Cell wall smooth (but may be encrusted by mineral deposits)	67
66b.	Cell wall ornamented; may have spines, depressions, or striae	69
67a.	Cells 5–13 μm in diameter	68
67b.	Cells 10–12 μm up to 20 mm in diameter, mother-cell wall fragments persistent (Fig. 5Q)	Diachros
68a.	Plastids without pyrenoids (Fig. 5L, M)	Chloridella
68b.	Plastids with pyrenoids (Fig. 6F, G)	Pleurochloris

69a.	Cells 4–12 µm, with a small number of irregularly arranged spines up to 50 µm long depending on species; spines may be of different lengths.	*Meringosphaera*
69b.	Cells 6–14 µm, with numerous regularly arranged spinelike projections, all approximately the same length.	*Akanthochloris*
69c.	Cells striate, punctate, or with circular depressions.	70
70a.	Cells appearing striate or punctate, cells 10–22 µm in diameter (Fig. 7D).	*Sklerochlamys*
70b.	Cell wall with circular depressions.	71
71a.	Circular depressions of cell wall forming a polygonal network (Fig. 5J, K).	*Chlorarkys*
71b.	Cell wall without polygonal network.	72
72a.	Plastids central (Fig. 5S, T).	*Endochloridion*
72b.	Plastids parietal.	73
73a.	Cells with one large, lobed plastid (Fig. 5D).	*Arachnochloris*
73b.	Cell wall appearing striated (Fig. 7N).	*Trachycystis*
73c.	Circular depressions, shallow, regular, giving cells an uneven appearance (Fig. 6A).	*Keriosphaera*
74a.	Cells triradiate (three armed) with no central body (Fig. 5G).	*Bracchiogonium*
74b.	Cells renate, fusiform, lunate, ovate, ellipsoidal, elliptical, cylindrical, oblong, pyramidal, polyhedral, polygonal, tetragonal, triangular, asymmetrical, rhomboidal, rectangular, four to six sided, or with various shapes.	75
75a.	Cells renate (kidney bean shaped), lunate, or narrowly or broadly fusiform.	76
75b.	Cells ovate, ellipsoidal, elliptical, cylindrical, oblong, pyramidal, polyhedral, polygonal, tetragonal, triangular, asymmetrical, rhomboidal, rectangular, four to six sided, or with various shapes.	78
76a.	Cells elongated, fusiform, sickle shaped, or twisted without spines (Fig. 5N).	*Chlorocloster*
76b.	Cells broadly fusiform, may have spines (Fig. 6H).	*Pleurogaster*
76c.	Cells renate or lunate.	77
77a.	Cells broadly crescent shaped to hemispherical, with irregular margins (Fig. 5O, P).	*Chlorogibba*
77b.	Cells renate to luniform, with smooth margins (Fig. 6E).	*Nephrodiella*
78a.	Cells ovate, ellipsoidal, elliptical, cylindrical, or oblong.	79
78b.	Cells pyramidal, polyhedral, polygonal, tetragonal, triangular, asymmetrical, rhomboidal, rectangular, four to six sided, or with various shapes.	82
79a.	Cell wall smooth.	80
79b.	Cell wall ornamented (spines, striae, or depressions).	81
80a.	Cells ovate to ellipsoidal (Fig. 5R).	*Ellipsoidion*
80b.	Cells cylindrical to oblong (Fig. 6C).	*Monallantus*
81a.	Cell wall heavily ornamented, without striae; cell wall bipartite (Fig. 5H, I).	*Chlorallantus*
81b.	Cell wall ornamented so as to appear striated (Fig. 5E, F).	*Aulakochloris*
81c.	Cell wall neither bipartite nor striated (Fig. 7L, M).	*Trachychloron*
82a.	Cells pyramidal, polyhedral, polygonal, tetragonal, or triangular.	83
82b.	Cells asymmetrical, rhomboidal, rectangular, four to six sided, or with various shapes.	88
83a.	Cells polygonal, with (usually) long spines at each apex (Fig. 6N, O).	*Pseudopolyedriopsis*
83b.	Cells with short or no spines.	84
84a.	Cell wall regularly ornamented, forming a definite pattern.	85
84b.	Cell wall smooth or not regularly ornamented.	86
85a.	Cells triangular, sometimes twisted (Fig. 5U, V).	*Goniochloris*
85b.	Cells pyramidal or tetragonal (Fig. 7E–H).	*Tetraedriella*

86a.	Cells tetrahedral (Fig. 7I–K)	*Tetraplektron*
86b.	Cells polygonal (Fig. 6I, J)	*Polyedriella*
86c.	Cells triangular, polyhedral, pyramidal, or deeply lobed so as to appear quadraradiate	87
87a.	Cells quadraradiate (four armed) with all arms in one plane (Fig. 5W, X)	*Isthmochloron*
87b.	Cells triangular, polyhedral, or pyramidal (Fig. 6P–R)	*Pseudostaurastrum*
88a.	Cells asymmetrical (Fig. 6D)	*Monodus*
88b.	Cells symmetrical	89
89a.	Cells roundly rhomboidal (Fig. 7C)	*Rhomboidella*
89b.	Cells rectangular, with spines (Fig. 7A, B)	*Pseudotetraedron*
89c.	Cells four to six sided, corners acute (Fig. 6M)	*Prismatella*

Note: *Trachydiscus* (Fig. 7O–Q) is not included in this key. Generic distinctions among the Pleurochloridaceae are based primarily on cell shape: *Trachydiscus* may be spherical, elliptical, fusiform, rectangular, hexagonal, or polygonal.

2. Descriptions of Genera

Chloramoebales

Ankylonoton Pascher 1932 (Fig. 2A, B)

Cells are solitary with typical heterokont flagellation; cells are 13–15 µm, up to 22 µm in length. Flagella are inserted subapically. Contractile vacuoles occur near the base of the flagella. A pyrenoid has been observed in the single, parietal plastid. Bilateral division has been observed. Two species have been found in North America: *A. pyreniger* and *A. salinis* (formerly *Nephrochloris salina*; R. L. Meyer, personal communication). *Ankylonoton* is metaphytic in dystrophic and mesotrophic ponds and pools in association with vascular vegetation; it is rare, but has been collected fall through winter in Minnesota and into summer in Alaska (R. L. Meyer, personal communication).

Chloramoeba Bohlin 1887 (Fig. 2C, D)

This genus is expected in North America, but it has yet to be observed (Prescott, 1978). Meyer (personal communication) reports collecting a possible zoospore of *Chloramoeba*. Cells are solitary, with typical heterokont flagellation, exhibiting metaboly (changing shape while in motion). Several (2–6) discoid, small, parietal plastids without eyespots or pyrenoids are present. Pseudopodia are sometimes evident. *Chloramoeba* often associated with aquatic vascular plants.

Chlorokardion Pascher 1930 (Fig. 2E–G)

Cells are solitary with typical heterokont flagellation; cells are 12–18 µm in length. Three to four discoid plastids are present; contractile vacuoles have been observed. Reproduction occurs by bilateral division. *Chlorokardion* is metaphytic in dystrophic and mesotrophic ponds and pools in association with vascular vegetation. *Chlorokardion pleurochloron* was collected in Minnesota, Alaska, and, associated with *Typha*, in Arizona (R. L. Meyer, personal communication).

Chloromeson Pascher 1930 (Fig. 2H–L)

Cells are solitary with typical heterokont flagellation, apically inserted; cells are 8–13 µm in length. A single platelike plastid is present. Amoeboid stages and endogenous statospores (Fig. 2L) have been observed. *Chloromeson agile* lacks an eyespot. *Chloromeson parvum* is probably a eustigmatophyte, as it has a straplike stigma free in the cytoplasm (R. L. Meyer, personal communication). *Chloromeson* is metaphytic in dystrophic and mesotrophic ponds and pools in association with vascular vegetation. *Chloromeson agile* and *C. viridis* were collected in Minnesota and Alaska (R. L. Meyer, personal communication).

Heterochloris Pascher 1925 (Fig. 2M–P)

Cells are solitary with typical heterokont flagellation; cells are 6–18 µm long. Cells exhibit metaboly (changing shape while in motion). Two discoid plastids are present in each cell. Amoeboid stages with rhizopodal processes have been observed. An eyespot is lacking in *H. mutabilis*. *Heterochloris* is metaphytic in dystrophic and mesotrophic ponds and pools and often associated with *Sphagnum*, *Carex*, and *Scirpus* mats, as well as *Typha* stands. *Heterochloris mutabilis* and *H. viridis* were collected from fall to late spring in Minnesota (R. L. Meyer, personal communication).

Nephrochloris Geitler *et* Gimesi 1925 (Fig. 2Q–T)

Cells are solitary with single anteriorly directed flagellum (in *N. incerta*); cells are 4–6 µm long, exhibiting metaboly (changing shape while in motion). A contractile vacuole is present at the base of the flagellum. One large parietal plastid is present, although it may appear as if there are two plastids, depending on the view. Amoeboid stages have been observed. *Nephrochloris* is

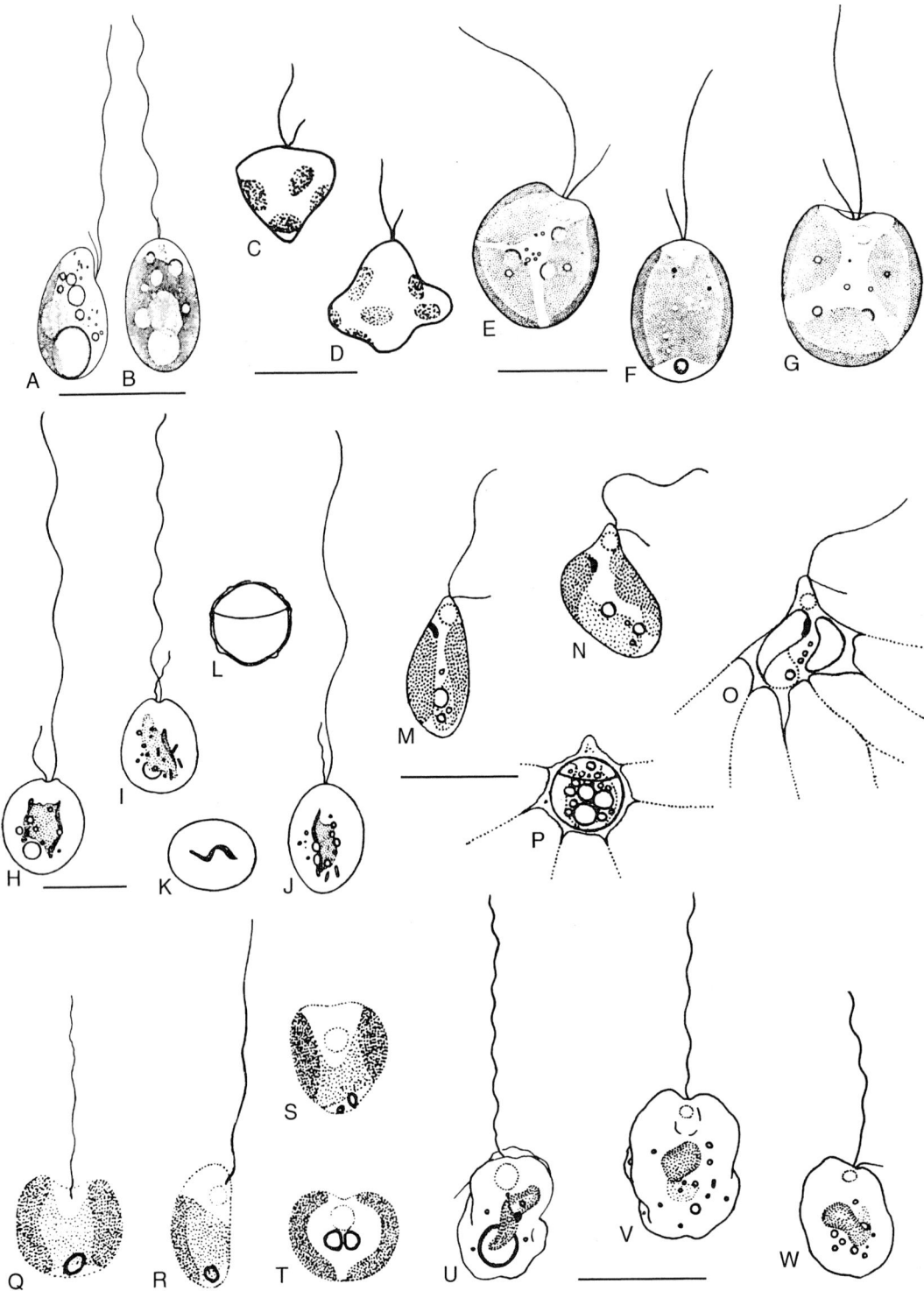

FIGURE 2 Freshwater members of the Chloramoebales. (A, B) *Ankylonoton pyreniger*: (A) Lateral view, showing subapical insertion of flagella. (B) Dorsal view. (C, D) *Chloramoeba heteromorpha*, with cell exhibiting metaboly. (E–G) *Chlorokardion pleurochloron*, dorsal, lateral, and ventral views. (H–L) *Chloromeson agile*: (K) Median view. (L) Statospore; unlike statospores in the Chrysophyceae, statospores of tribophyte algae are composed of two pieces of approximately equal size. (M–P) *Heterochloris viridis*: (O) Amoeboid stage. (P) Cyst formation. (Q–T) *Nephrochloris incerta*; the single flagellum may not always be visible. (U–W) *Polykyrtos vitreus*, showing the irregular shape of the cells. (V) The posteriorly directed flagellum is not always visible. A, B, E–W (and all following figures from Ettl, 1978) reprinted from: Ettl, H. Xanthophyceae. I, *in*: Ettl, H., Gerloff, J., and Heynig, H., Eds., Süsswasserflora von Mitteleuropa, Vol. 3, Gustav Fischer Verlag © Spektrum Akademischer Verlag, Heidelberg, 1978, pp. 58–508, all with permission. C, D from Prescott (1978) with permission.

metaphytic, epiphytic, and endophytic in the air cells of *Sphagnum* and in dystrophic and mesotrophic ponds and pools. *Nephrochloris incerta* was collected in early spring and late fall in Minnesota and Alaska (R. L. Meyer, personal communication).

Polykyrtos Pascher 1937 (Fig. 2U–W)

Cells are solitary and irregularly shaped; cells are approximately 10 mm long. Posteriorly directed flagellum are extremely short and often not visible. One or two contractile vacuoles are present at the base of the flagella. One small plastid is present. Amoeboid stages and longitudinal division have been observed. *Polykyrtos* is metaphytic in dystrophic and mesotrophic ponds and pools in association with vascular and bryophyte vegetation. Although common, it has been collected from late fall to early spring in Minnesota and Alaska (R. L. Meyer, personal communication).

Rhizochloridales

Chlamydomyxa Archer 1875 (Fig. 3A–C)

Cell walls are hyaline; the cytoplasm is granular. There are two principal forms: (1) an amoeboid, multinucleate stage with numerous plastids and with or without rhizopodal processes that has been found crawling over *Sphagnum* and (2) cysts seen in or on the cells of *Sphagnum*. Reproduction occurs by zoospores. This organism is reported to be free living, epilithic, or endophytic; collections from North America have been obtained from within the cells of aquatic plants, most often *Sphagnum* (Prescott, 1978). *Chlamydomyxa* has been ollected from Minnesota (Tarapchak, 1972).

Myxochloris Pascher 1931 (Fig. 3F–K)

This organism has a complex life cycle consisting of a multinucleate, amoeboid stage (Fig. 3I), uninucleate zoospores (Fig. 3H), and endogenous, exogenous, and secondary cysts. Amoeboid cells are naked with many plastids. *Myxochloris* is metaphytic in dystrophic and mesotrophic ponds and pools and endophytic in the air cells of *Sphagnum*. *Myxochloris sphagnicola* is widely distributed throughout Minnesota, Alaska, and Arkansas (R. L. Meyer, personal communication).

Rhizochloris Pascher 1918 (Fig. 3D, E)

Vegetative cells are amoeboid, 8–10 or 12–18 μm long (depending on species), with long pseudopods. This genus is free living. It is heterotrophic, ingesting bacteria or small algae. Two to ten plastids without pyrenoids are present; reproduction occurs by endogenous cysts. *Rhizochloris* is endophytic in *Sphagnum* air cells, but also metaphytic and neustonic in dystrophic and mesotrophic ponds and pools. *Rhizochloris mirabilis* and *R. stigmatica* are widely distributed in the boreal forest of Minnesota and Alaska and in the tundra of Alaska (R. L. Meyer, personal communication).

Rhizolekane Pascher 1932 (Fig. 3L–N)

Vegetative cells are amoeboid, with long pseudopods, resting in a lorica. The stipe is absent or not visible. No eyespot or pyrenoids have been observed in the plastid. *Rhizolekane* is epiphytic on filamentous algae associated with vascular vegetation in dystrophic and mesotrophic ponds and pools. *Rhizolekane sessilis* typically occurs in openings within or along the margins of floating bogs and fens in Minnesota and Alaska (R. L. Meyer, personal communication).

Stipitococcus West et West 1898 (Fig. 3O–S)

Vegetative cells are amoeboid with long pseudopods, in a lorica on a stipe; cells do not completely fill the lorica. A basal disc may attach the stipe to the substrate. There are one to three plastids per cell. Zoospores have been observed. *Stipitococcus* is usually epiphytic on filamentous algae, such as members of the Zygnemataceae (Poulton, 1930; Prescott, 1931), but has also been collected on *Oedogonium*, members of the Desmidiaceae such as *Hyalotheca*, and other filamentous desmids (Prescott, 1931, 1944). There are many reports of this alga in North America, ranging from Nova Scotia (Hughes, 1948) to South Carolina (Jacobs, 1968).

Stipitoporos Ettl 1965 (Fig. 3T)

Vegetative cells are amoeboid, with long rhizopodal processes extending through openings in the lorica. A stipe and basal disc are present. Four to five discoid plastids without pyrenoids or eyespot are present in each cell. Reproduction by division of the protoplasm has been reported. *Stipitoporos* is epiphytic on filamentous algae associated with vascular vegetation in dystrophic and mesotrophic ponds and pools. It is typically collected from the margins of floating bogs and fens in Minnesota and Alaska (R. L. Meyer, personal communication).

Heterogloeales

Characidiopsis Pascher 1938 (Fig. 4A–D)

Cells are stalked and ellipsoidal to spindle shaped with a distinct cell wall; the stalk is attached by a gelatinous disc. Cells are 12–16 μm without a stipe. Contractile vacuoles and eyespot are present. There are two to four plastids per cell. Reproduction occurs by zoospores. This genus is easily confused with *Characiopsis* (Tribophyceae), *Pseudocharaciopsis* (Eustigmatophyceae), and *Characium* and *Pseudocharacium* (Chlorophyta). *Characidiopsis* is epiphytic on macrophytes in dystrophic and mesotrophic ponds and pools

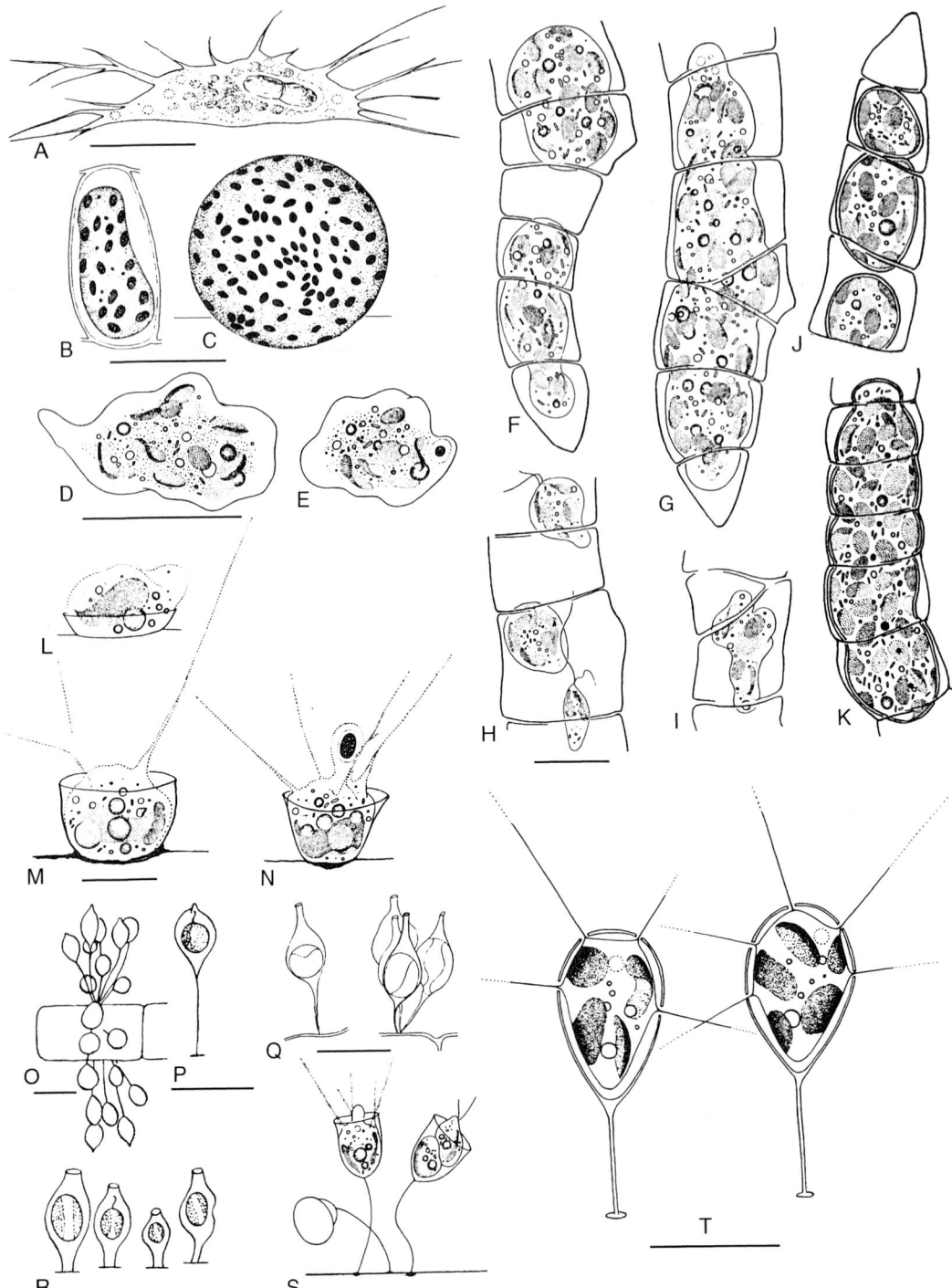

FIGURE 3 Freshwater members of the Rhizochloridales. (A–C) *Chlamydomyxa labyrinthuloides*: (A) Amoeboid stage. (B) Parasitic plasmodium. (C) Mature cyst. (D, E) *Rhizochloris mirabilis*; this free-living amoeboid organism has numerous plastids. (F–K) *Myxochloris sphagnicola*: (F, G) Plasmodia in water-holding cells of *Sphagnum*. (H) Zoospores. (I) Young plasmodium. (J, K) Akinete formation in the water-holding cells of *Sphagnum*. (L–N) *Rhizolekane sessilis*: (L) Young cell. (M) Cell with projecting rhizopods. (N) Cell with large pseudopodium containing food vesicle. (O.-S) *Stipitococcus*. Loricate cells epiphytic on filamentous algae: (O, P) *S. apiculata*. (Q) *S. vasiformis*. (R) *S. crassistipitatus*. (S) *S. vas*; cell on the right producing zoospores. (T) *Stipitoporos polychloris*, with long rhizopodal processes extending through openings in the lorica. A, D–N, Q–T reprinted from Ettl (1978), B, C from Tarapchak (1972), O, P from Prescott (1962), all with permission.

(R. L. Meyer, personal communication); it has been collected in British Columbia (Stein, 1975; Stein and Borden, 1978) and Minnesota (Tarapchak, 1972).

Gloeochloris Pascher 1932 (Fig. 4E, F)

Colonial cells are embedded in a small, spherical or ellipsoidal mucilaginous thallus. Cells are regularly arranged around the periphery of the thallus. Each cell contains two to four parietal plastids. Zoospores may have one or two flagella; endogenous cysts have been noted. *Gloeochloris* is euplanktonic and metaphytic in dystrophic and mesotrophic ponds, pools, fens, and bogs in association with vascular vegetation and aquatic bryophytes. *Gloeochloris smithiana* is frequently associated with *Typha* stands in Minnesota, Alaska, and Arkansas (R. L. Meyer, personal communication).

Helminthogloea Pascher 1932 (Fig. 4G, H)

Colonial cells are embedded in a tall, mucilaginous thallus, occasionally branched. There are usually two parietal plastids per cell. *Helminthogloea* is epiphytic on macrophytes in dystrophic and mesotrophic ponds and pools of Minnesota, Alaska, and Arkansas (R. L. Meyer, personal communication). It has been collected from a creek in North Carolina (Whitford, 1979).

Heterogloea Pascher 1930 (Fig. 4I–L)

Cells are naked and spherical to subspherical. Cells are solitary or in groups dispersed randomly within an extensive, irregular fluid mucilage. One to two parietal plastids and contractile vacuoles are visible in the cells. Zoospores and cysts have been found. *Heterogloea* is euplanktonic and epiphytic in dystrophic and mesotrophic ponds and pools (R. L. Meyer, personal communication) and lakes, bogs, and fens in British Columbia (Stein, 1975; Stein and Borden, 1978) and Minnesota (Tarapchak, 1972).

Malleodendron Pascher 1937 (Fig. 4M–O)

Malleodendron is composed of gelatinous, attached colonies, dichotomously branched, forming treelike colonies. One to two thick-walled (Prescott, 1978) cells are present at the end of each stalk (although Hibberd, 1990b, states that the cells are naked). Reproduction occurs by zoospores. Meyer (1995) has recommended that, based on pigmentation, statocysts, and other characters, *M. caespitosum* be transferred to the Chrysophyceae as a new genus, *Chrysomallus*. *Malleodendron gloeopus* is epiphytic on aquatic bryophytes and filamentous algae in dystrophic and mesotrophic ponds and pools of Minnesota, Alaska, and Arkansas (R. L. Meyer, personal communication). It has also been collected from a ditch, associated with *Lemna*, in British Columbia (Stein, 1975).

Pleurochloridella Pascher 1937 (Fig. 4P–R)

Vegetative cells are 10–14 or 15–20 µm in diameter, depending on the species, with a distinct cell wall. Cells are unattached and not surrounded by mucilage. Contractile vacuoles and (usually) two plastids are present, without eyespot or pyrenoids. Reproduction occurs by zoospores or autospores. Based on the finding of fucoxanthin and heteroxanthin, along with *rbc*L sequences and other observations in this genus, Bailey *et al.* (1998) have transferred this genus and several of the Chrysophyceae to a new class, the Phaeothamniophyceae. *Pleurochloridella* is metaphytic among vascular vegetation and filamentous algae in dystrophic and mesotrophic ponds, pools, and lakes. It is most abundant during the late spring through early fall in Minnesota and Alaska and in the fall, winter, and spring in Arkansas (R. L. Meyer, personal communication).

Mischococcales: Pleurochloridaceae

Akanthochloris Pascher 1930 (Fig. 5A–C)

Cells are solitary, spherical, and 6–14 µm in diameter. Cells are thick walled with spinelike projections, regularly arranged and with all spines being approximately of the same length. There are one to eleven parietal plastids. Reproduction occurs by autospores and zoospores. *Akanthochloris* has been collected from a seep in North Carolina (Schumacher *et al.*, 1963), in lakes, bogs, and fens in Minnesota (Tarapchak, 1972), and in Montana (as *Groenlandiella*) (Prescott and Dillard, 1979).

Arachnochloris Pascher 1930 (Fig. 5D)

Cells are solitary, spherical, with size depending on the species, generally 7–18 µm in diameter. The cell wall is thick, with circular depressions in the wall. There is usually one large, lobed, parietal plastid per cell. Reproduction occurs by autospores and zoospores. *Arachnochloris* is metaphytic among other algae (Whitford and Schumacher, 1969) in fens, bogs, and lakes in the Arctic (Prescott and Vinyard, 1965), Montana (Gaufin *et al.*, 1976), Minnesota (Tarapchak, 1972), and North Carolina (Whitford and Schumacher, 1969).

Aulakochloris Pascher 1930 (Fig. 5E, F)

Cells are solitary and elliptical; most species are 10–18 µm. The cell wall is ornamented, forming striations. Parietal plastids are present, without pyrenoids. Reproduction occurs by autospores. *Aulakochloris* is metaphytic in association with *Sphagnum* and *Typha* in dystrophic and mesotrophic ponds of Minnesota, Alaska, and Arkansas (R. L. Meyer, personal communication).

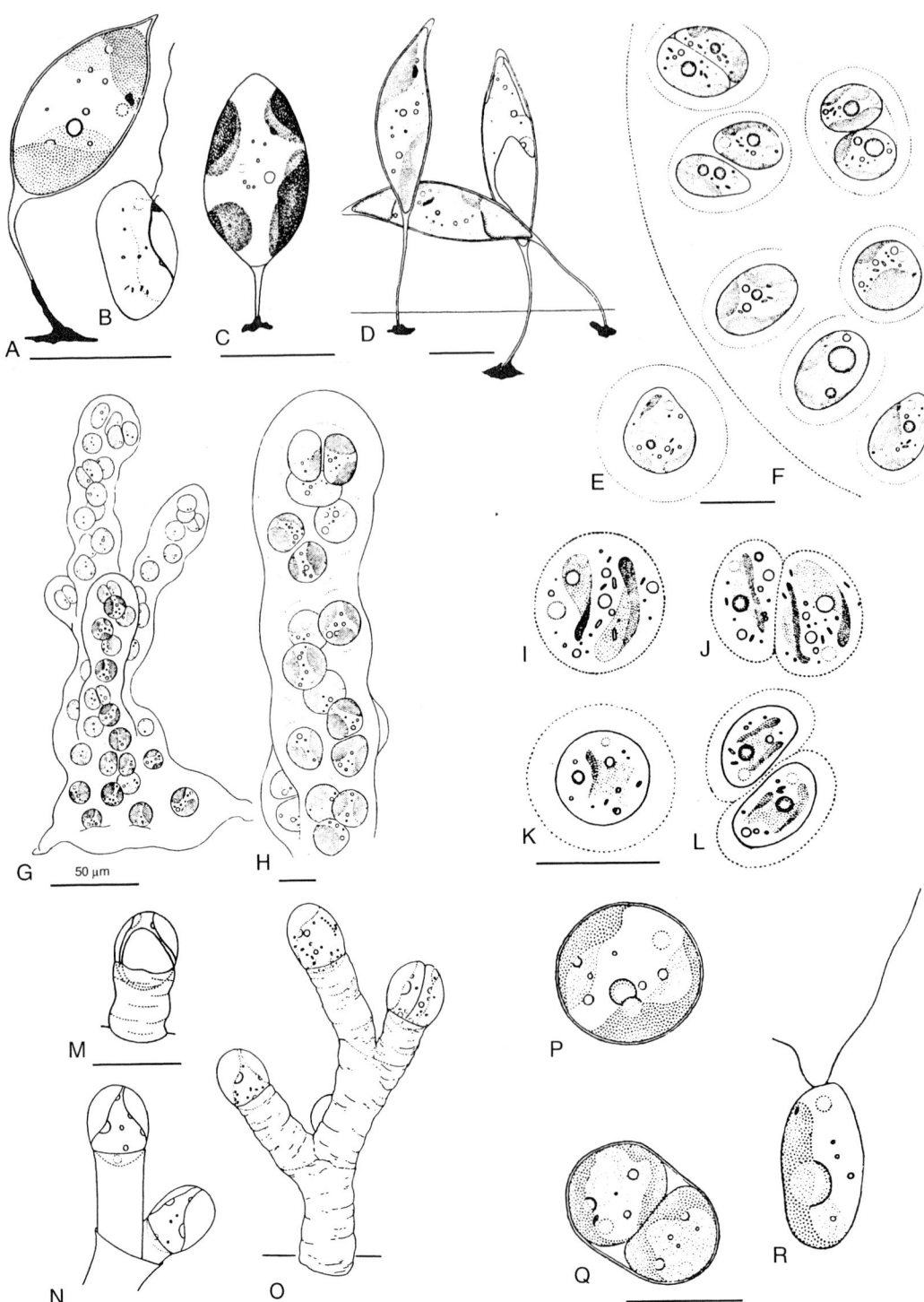

FIGURE 4 Freshwater members of the Heterogloeales. (A–D) *Characidiopsis*; stalked cells attached with gelatinous discs: (A, B) *C. acuta*: (B) Zoospore. (C) *C. ellipsoidea*. (D) *C. elongata*. (E, F) *Gloeochloris planctonica*: (E) Single cell. (F) Colonial cells, regularly arranged around the periphery of the mucilaginous thallus. (G, H) *Helminthogloea ramosa*; colonial cells embedded in a tall mucilaginous thallus: (I–L) *Heterogloea*; solitary cells and groups of cells are dispersed randomly in a thin mucilaginous thallus: (I, J) *H. endochloris*. (K, L) *H. minor*. (M–O) *Malleodendron gloeopus*; gelatinous, treelike colonies, with one or two cells at the end of each stalk. (P–R) *Pleurochloridella vacuolata*; unattached cells with no surrounding mucilage: (P) Vegetative cell. (Q) Zoospore. (R) Autospores. A, B, D–R reprinted from Ettl (1978), C from Tarapchak (1972), all with permission.

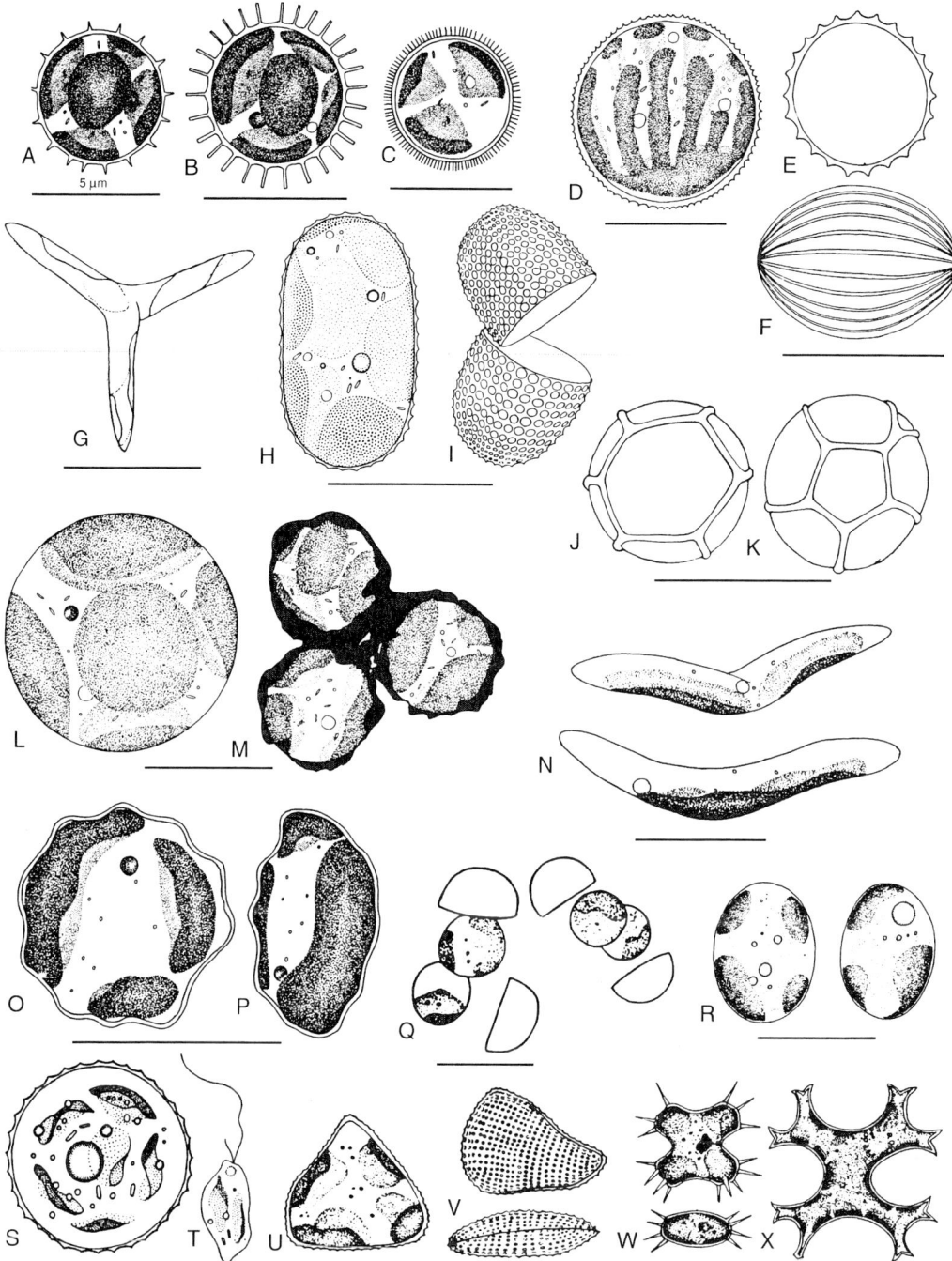

FIGURE 5 Freshwater members of the Mischococcales. Family Pleurochloridaceae. (A–C) *Akanthochloris*; the spines are regularly arranged and all of the same length: (A) *A. brevispinosa*. (B) *A. bacillifera*. (C) *A. scherffelii*. (D) *Arachnochloris major*. (E, F) *Aulakochloris striata*: (E) End view. (F) Lateral view, showing regular striations. (G) *Bracchiogonium ophiaster*, triradiate vegetative cell. (H, I) *Chlorallantus oblongus*: (I) Showing sculptured walls. (J, K) *Chlorarkys reticulata* showing sculptured walls forming polygonal network. (L, M) *Chloridella*: (L) *C. neglecta*, solitary cell. (M) *C. ferruginea*, cluster of cells encrusted with mineral deposits. (N) *Chlorocloster angulus* vegetative cells. (O, P) *Chlorogibba trochisciaeformis*, showing the irregular contours of the cells: (O) Dorsal view. (P) Lateral view. (Q) *Diachros simplex*; after autosporulation, small clusters of cells and mother cell walls are visible. (R) *Ellipsoidion stellatum* vegetative cells. (S, T) *Endochloridion polychloron*: (S) Vegetative cell. (T) Zoospore. (U, V). *Goniochloris sculpta*: (U) Triangular cell. (V) Sculptured depressions in walls in dorsal and lateral views. *Goniochloris* appears as a biconvex lens in lateral view. (W, X) *Isthmochloron*; deeply lobed vegetative cells with all arms in one plane: (W) *I. trispinatum*. (X) *I. lobulatum*. A–D, L–P reprinted from Tarapchak (1972), E–K, R–T, W–X from Ettl (1978), Q, U–V from Prescott (1978), all with permission.

Bracchiogonium (Pascher) Ettl 1965 (Fig. 5G)

Cells are solitary, triradiate, with one to three plastids. *Bracchiogonium* is metaphytic in dystrophic and mesotrophic ponds, pools, and seeps in Minnesota, Alaska, and Arkansas (R. L. Meyer, personal communication).

Chlorallantus Pascher 1930 (Fig. 5H, I)

Cells are solitary, oblong to broadly ellipsoidal, with sculptured bipartite walls. Walls have circular depressions and often spiny thickenings between depressions. Two to many plastids are present. Reproduction occurs by autospores and zoospores. This genus is sometimes referred to incorrectly as *Chlorallanthus*. *Chlorallantus* is planktonic in an Arctic lake (Prescott and Vinyard 1965), bogs and fens in the boreal forest of Minnesota, and a tundra pond in Alaska (R. L. Meyer, personal communication).

Chlorarkys Pascher 1939 (Fig. 5J, K)

Cells are solitary, spherical, and 6–9 µm in diameter. Walls contain depressions; thickenings between the depressions form a polygonal network. *Chlorarkys* is metaphytic in dystrophic ponds and pools in Minnesota, Alaska, and Arkansas (R. L. Meyer, personal communication).

Chloridella Pascher 1932 (Fig. 5L, M)

Cells are solitary or grouped in clusters, spherical, and 5–13 µm in diameter. Clusters may be encrusted by mineral deposits. There are usually four to seven plastids, generally without pyrenoids. *Chloridella* superficially resembles *Chlorella* (Chlorophyta). Reproduction has long been thought to occur only by autospores: This was the only characteristic separating this genus from *Pleurochloris*, which forms autospores and zoospores. However, Tarapchak (1972) observed zoospores in *Chloridella*, rendering this character inadequate for separating the two genera. *Chloridella* is metaphytic among filamentous algae. It has been reported from Minnesota lakes, bogs, and fens (Tarapchak, 1972) and eutrophic ponds (Meyer, 1969).

Chlorocloster Pascher 1925 (Fig. 5N)

Cells are solitary or in groups of two to four, elongated, sickle shaped, or twisted, tapering toward the poles, and 8–30 µm in length. There are usually one to four plastids, with or without pyrenoids, depending on species. The cell wall is thin and smooth. Reproduction occurs by autospores. *Chlorocloster* has been collected from lakes, rivers, fens, and bogs in British Columbia (Stein, 1975; Stein and Borden, 1978) and Minnesota (Tarapchak, 1972).

Chlorogibba Geitler 1928 (Fig. 5O, P)

Cells are solitary, broadly crescent shaped with irregular contours. Cells may appear spherical in top view (Fig. 5O). Cells are 7–16 µm in diameter with three to five parietal plastids. Reproduction occurs by autospores or zoospores. *Chlorogibba* is relatively rare. It has been collected from Minnesota lakes, bogs, and fens (Tarapchak, 1972).

Diachros Pascher 1937 (Fig. 5Q)

Cells are solitary (or in small clusters after autospore liberation), spherical, with a thin cell wall. During autosporogenesis, the original cell wall separates into two sections to release the autospores; the mother cell wall is persistent. Cells are 10–12 µm, up to 20 µm, in diameter, with one to three plastids. *Diachros* is metaphytic in dystrophic, mesotrophic, and eutrophic lakes, ponds, and pools in Minnesota, Alaska, and Arkansas (R. L. Meyer, personal communication) and British Columbia (Stein and Borden, 1978).

Ellipsoidion Pascher 1937 (Fig. 5R)

Cells are solitary, ellipsoidal to ovoid, and 3–20 µm in diameter, depending on the species. The cell wall is smooth and colorless. There are one to six plastids, with or without pyrenoids. *E. acuminatum* has been placed in the Eustigmatophyceae and Hibberd (1990a) has renamed this taxon as *Pseudocharaciopsis ovalis*. *Ellipsoidion* is epiphytic on leaves in greenhouses in North Carolina (Schumacher *et al.*, 1966) and has been collected from fens, bogs, and lakes in Minnesota, Alaska, and Arkansas (R. L. Meyer, personal communication, Tarapchak 1972).

Endochloridion Pascher 1930 (Fig. 5S, T)

Cells are solitary, spherical to globose, and 7–16 µm in diameter. The cell wall is thick, sometimes appearing serrated, with rounded depressions. One to five centrally located plastids are usually present. This genus is distinguished from *Arachnochloris* by the position of the plastids. *Endochloridion* has been collected from ponds, pools, lakes, bogs, and fens in Minnesota (Tarapchak, 1972).

Goniochloris Geitler 1928 (Fig. 5U, V)

Cells are solitary or, occasionally, in groups of two to four, triangular, sometimes twisted. Cells are 4–20 µm, appearing as a biconvex lens from the side. Walls are thick, with or without pits, points, or spines, but regularly ornamented. Two to five plastids are present per cell. Reproduction occurs by autospores. *Goniochloris* is metaphytic in dystrophic and mesotrophic ponds and pools in Arkansas (Meyer, 1969;

Meyer et al., 1970; R. L. Meyer, personal communication). It has also been collected in South Carolina (Jacobs, 1971) and Jamaica (Hegewald, 1976).

Isthmochloron Skuja 1948 (Fig. 5W, X)

Cells are solitary, deeply lobed so as to appear quadraradiate, with all four arms in one plane. Walls are variously ornamented, with or without spines. Plastids range from a few to many, without pyrenoids. Reproduction occurs by autospores. This genus resembles *Staurastrum* (Chlorophyta) and *Pseudotetraedron* and *Pseudostaurastrum* (Tribophyceae). Bourrelly (1968, 1981) does not recognize *Isthmochloron* as a separate genus, including it with *Pseudostaurastrum*. The two reports of this genus in North America are from Arkansas, where it is metaphytic in dystrophic ponds and pools (R. L. Meyer, personal communication), and Ontario (Duthie and Socha, 1976).

Keriosphaera Pascher 1939 (Fig. 6A)

Cells are solitary, spherical, with regular depressions in the cell wall. This genus is similar in appearance to *Arachnochloris*, *Chlorarkys*, and *Trachycystis*; some authors consider them to be members of the same genus, usually *Arachnochloris*. *Keriosphaera* is metaphytic in dystrophic and mesotrophic ponds and pools in Minnesota, Alaska, and Arkansas (R. L. Meyer, personal communication).

Meringosphaera Lohman em. Pascher 1932 (Fig. 6B)

Cells are solitary and spherical; cell walls are smooth or with spines, up to 50 µm in length. Spines may be of different lengths. One to many plastids are present, with or without pyrenoids. Reproduction occurs by autospores. *Meringosphaera* is euplanktonic and metaphytic in dystrophic, mesotrophic, and eutrophic lakes, ponds, and pools in Arkansas (R. L. Meyer, personal communication) and a Montana lake (Gaufin et al., 1976).

Monallantus Pascher 1939 (Fig. 6C)

Cells are solitary, cylindrical to oblong, and 5–11 µm in length. One to three parietal plastids are present. Cells are often found in small groups. Reproduction occurs by autospores and zoospores. *Monallantus* has been reported from British Columbia (Stein, 1975), Minnesota lakes, bogs, and fens (Tarapchak, 1972), and an organically rich pond in Arkansas (Meyer, 1969).

Monodus Chodat 1913 (Fig. 6D)

Cells are solitary, asymmetrical, and usually rounded at one pole and pointed at the other. Cells are 5–25 µm long. One or more parietal plastids are present, with or without pyrenoids. Reproduction occurs by autospores and zoospores; aplanospores and cysts have also been noted. Two species have been transferred to the Eustigmatophyceae: *M. ovalis* as *Pseudocharaciopsis ovalis* and *M. subterraneus* as *Monodopsis subterranea*. It is possible that other species may also be eustigmatophytes (Ehara et al., 1997). *Monodus* is euplanktonic and metaphytic in dystrophic, mesotrophic, and eutrophic ponds, pools, seeps, and lakes in Minnesota, Alaska, and Arkansas (R. L. Meyer, personal communication).

Nephrodiella Pascher 1937 (Fig. 6E)

Cells are solitary except during autosporulation, renate to luniform, and 5–21 µm long. The cell wall is thin, smooth, and colorless. One to two parietal plastids are present. Reproduction occurs by autospores and (rarely) zoospores. *Nephrodiella nana* (= *Nephrodiella brevis*) has been transferred to the Chlorophyceae as *Monoraphidium nanum* (Hindak, 1981). Three species have been collected in Minnesota (Tarapchak, 1972) and five in Arkansas (R. L. Meyer, personal communication).

Pleurochloris Pascher 1925 (Fig. 6F, G)

Cells are solitary or, occasionally, in groups, spherical, and 5–13 µm in diameter. One or more large plastids are present, usually with pyrenoids. Reproduction occurs by autospores (Tarapchak, 1972) and zoospores. Three species have been transferred to the Eustigmatophyceae: *P. commutata*, *P. magna*, and *P. polyphem* as species of *Eustigmatos*. *Pleurochloris* is euplanktonic and metaphytic in dystrophic, mesotrophic, and eutrophic lakes, ponds, pools, and seeps in Minnesota, Alaska, and Arkansas (R. L. Meyer, personal communication; Tarapchak, 1972).

Pleurogaster Pascher 1937 (Fig. 6H)

Cells are solitary, irregularly fusiform, narrowed to pointed at poles, and may be concave on one or both sides; cells are 12–16 µm in length. The cell wall is thick. One to four plastids are present without pyrenoids. Reproduction occurs by autospores and zoospores. *Pleurogaster* has been collected in ponds, bogs, and lakes in Alaska (R. L. Meyer, personal communication), Minnesota (Tarapchak, 1972), and North Carolina (Whitford, 1979).

Polyedriella Pascher 1930 (Fig. 6I, J)

Cells are solitary, polygonal, 7–20 µm in diameter, with or without spines. One or more parietal plastids are present. Reproduction occurs by autospores and zoospores. *P. helvetica* has been transferred to the Eustigmatophyceae as *Vischeria helvetica*. *Polyedriella* is euplanktonic and metaphytic in dystrophic, meso-

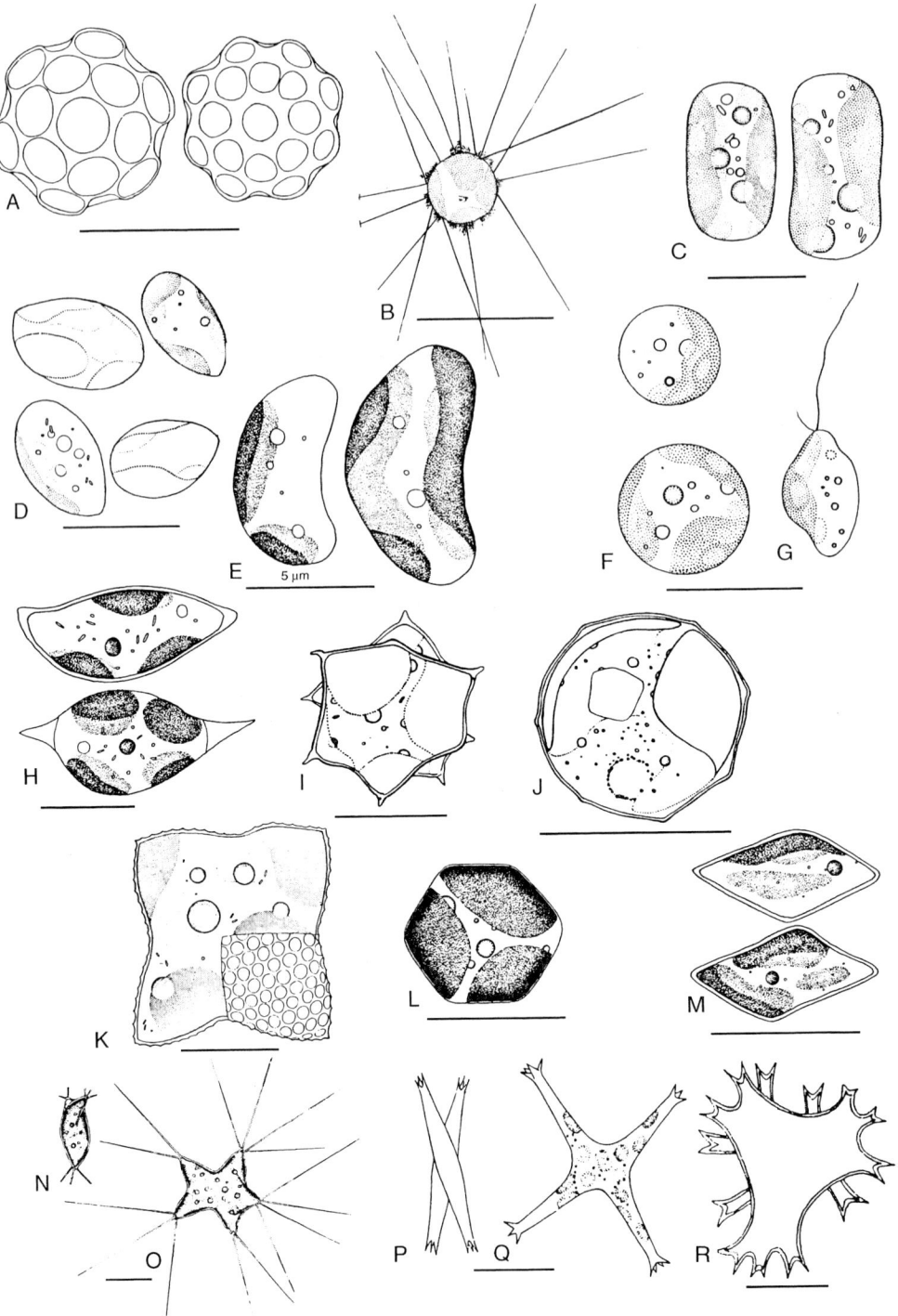

FIGURE 6 Freshwater members of the Mischococcales. Family Pleurochloridaceae. (A) *Keriosphaera gemma* showing sculptured walls. (B) *Meringosphaera tenerrima*, with spines of different lengths. (C) *Monallantus pyreniger* vegetative cells. (D) *Monodus chodatii* asymmetrical vegetative cells. (E) *Nephrodiella phaseolus* renate vegetative cells. (F, G) *Pleurochloris pyrenoidosa*: (F) Vegetative cell. (G) Zoospore. (H) *Pleurogaster lunaris*; top cell narrowed at poles, bottom cell with spines at poles. (I, J) *Polyedriella*: (I) *P. aculeata* vegetative cell with spines. (J) *P. helvetica* vegetative cell without spines. (K, L) *Polygoniochloris*: (K) *P. tetragona* with cutout portion showing sculpturing of the wall. (L) *P. regularis* vegetative cell. (M) *Prismatella hexagona* vegetative four-sided cell. (N, O) *Pseudopolyedriopsis skujae* polygonal cells with spines at apices. (P–R) *Pseudostaurastrum* vegetative cells: (P, Q) *P. hastatum*. (R) *P. enorme*. A–D, F–G, I–L, N–R reprinted from Ettl (1978), E, H, M from Tarapchak (1972), all with permission.

trophic, and eutrophic ponds, pools, lakes, and seeps of Minnesota, Alaska, and Arkansas (R. L. Meyer, personal communication).

Polygoniochloris Ettl 1965 (Fig. 6K, L)

Cells are solitary, spherical, cuboidal, or polygonal, dorsiventrally flattened, and 7–20 μm in diameter. Walls are smooth or ornamented. The cells contains many plastids. Reproduction occurs by autospores and zoospores. *Polygoniochloris* is euplanktonic and metaphytic in dystrophic, mesotrophic, and eutrophic ponds, pools, lakes, and seeps of Minnesota, Alaska, and Arkansas (R. L. Meyer, personal communication).

Prismatella Pascher 1937 (Fig. 6M)

Cells are solitary, four to six sided, and 14–20 μm. The cell wall is thick and unornamented. Two to three plastids are present, without pyrenoids. Reproduction occurs by autospores. *Prismatella hexagona* was collected in Minnesota (Tarapchak, 1972).

Pseudopolyedriopsis Gollerbach 1962 (Fig. 6N, O)

Cells are solitary and polygonal; apices are ornamented with two to six (usually) long spines. Numerous discoidal parietal plastids are present, without pyrenoids. *Pseudopolyedriopsis* is euplanktonic and metaphytic in dystrophic and mesotrophic lakes, ponds, and pools of Minnesota, Alaska, and Arkansas (R. L. Meyer, personal communication).

Pseudostaurastrum Chodat 1921 (Fig. 6P–R)

Cells are solitary, triangular, polyhedral, or pyramidal; they may be lobed with branched arms, with or without spines at apices. Walls are smooth or ornamented. Numerous plastids are present, without pyrenoids. Reproduction occurs by autospores and zoospores. *P. limneticum* has been transferred to the Eustigmatophyceae (Schnepf *et al.*, 1996). *Pseudostaurastrum* is metaphytic in dystrophic and mesotrophic ponds and pools. It has been collected in the Northwest Territories, Canada (Sheath and Steinman, 1981), Minnesota, Alaska, and Arkansas (R. L. Meyer, personal communication), and Jamaica (Hegewald, 1976).

Pseudotetraedron Pascher 1912 (Fig. 7A, B)

Cells are solitary and rectangular, with a spine at each corner. They are elliptical in side view (Fig. 7B). The wall is in two sections. Many plastids are present. Reproduction occurs by autospores and cysts. *Pseudotetraedron* is euplanktonic and metaphytic in dystrophic, mesotrophic, and eutrophic ponds, pools, seeps, rivers, and lakes in Minnesota, Alaska, and Arkansas (R. L. Meyer, personal communication), Iowa (Prescott, 1931), North Carolina (Whitford and Schumacher, 1969), and South Carolina (Goldstein and Manzi, 1976).

Rhomboidella Pascher 1937 (Fig. 7C)

Cells are solitary, irregularly rhomboidal, and 12–18 μm long. Cell walls are unornamented. Three to four discoidal plastids are present. Reproduction occurs by autospores. *Rhomboidella* is metaphytic in dystrophic, mesotrophic, and eutrophic ponds, pools, and seeps in Minnesota, Alaska, and Arkansas (R. L. Meyer, personal communication).

Sklerochlamys Pascher 1937 (Fig. 7D)

Cells are solitary, spherical, and 10–22 μm in diameter. The cell wall may appear striated or punctate; the wall is in two pieces after autosporulation. One or more plastids are present, without pyrenoids. Reproduction occurs by autospores and zoospores. *Sklerochlamys* is metaphytic in dystrophic and mesotrophic pools in Minnesota, Alaska, and Arkansas (R. L. Meyer, personal communication).

Tetraedriella Pascher 1930 (Fig. 7E–H)

Cells are solitary and pyramidal or tetragonal. Walls are ornamented by regularly arranged rows of depressions, in contrast to *Pseudostaurastrum*, which has similar cell shapes but does not have the regular ornamentation found in *Tetraedriella*. Several to many parietal plastids are present. Reproduction occurs by autospores and zoospores. *Tetraedriella* is metaphytic in dystrophic, mesotrophic, and eutrophic lakes, ponds, pools, and seeps in Minnesota, Alaska, and Arkansas (R. L. Meyer, personal communication), Montana (Gaufin *et al.*, 1976), and Kentucky (as *Tetragoniella gigas* = *Tetraedriella regularis*) (Dillard *et al.*, 1976).

Tetraplektron Fott 1957 (Fig. 7I–K)

Cells are solitary, regularly to irregularly tetrahedral, and often four rayed. The cell contains many plastids. Reproduction occurs by autospores and zoospores. *Tetraplektron* is metaphytic in dystrophic, mesotrophic, and eutrophic lakes, ponds, and pools (R. L. Meyer, personal communication). It has rarely been collected, but was reported from an organically rich pond in Arkansas (Meyer *et al.*, 1970).

Trachychloron Pascher 1939 (Fig. 7L, M)

Cells are solitary, ovate to irregularly ellipsoidal, and 8–18 μm long. One or more parietal plastids are present, with or without pyrenoids. Reproduction occurs by autospores. *Trachychloron* has been collected in British Columbia (Stein and Borden, 1978), Ontario (Duthie and Socha, 1976), and bogs and lakes in Minnesota (Tarapchak, 1972).

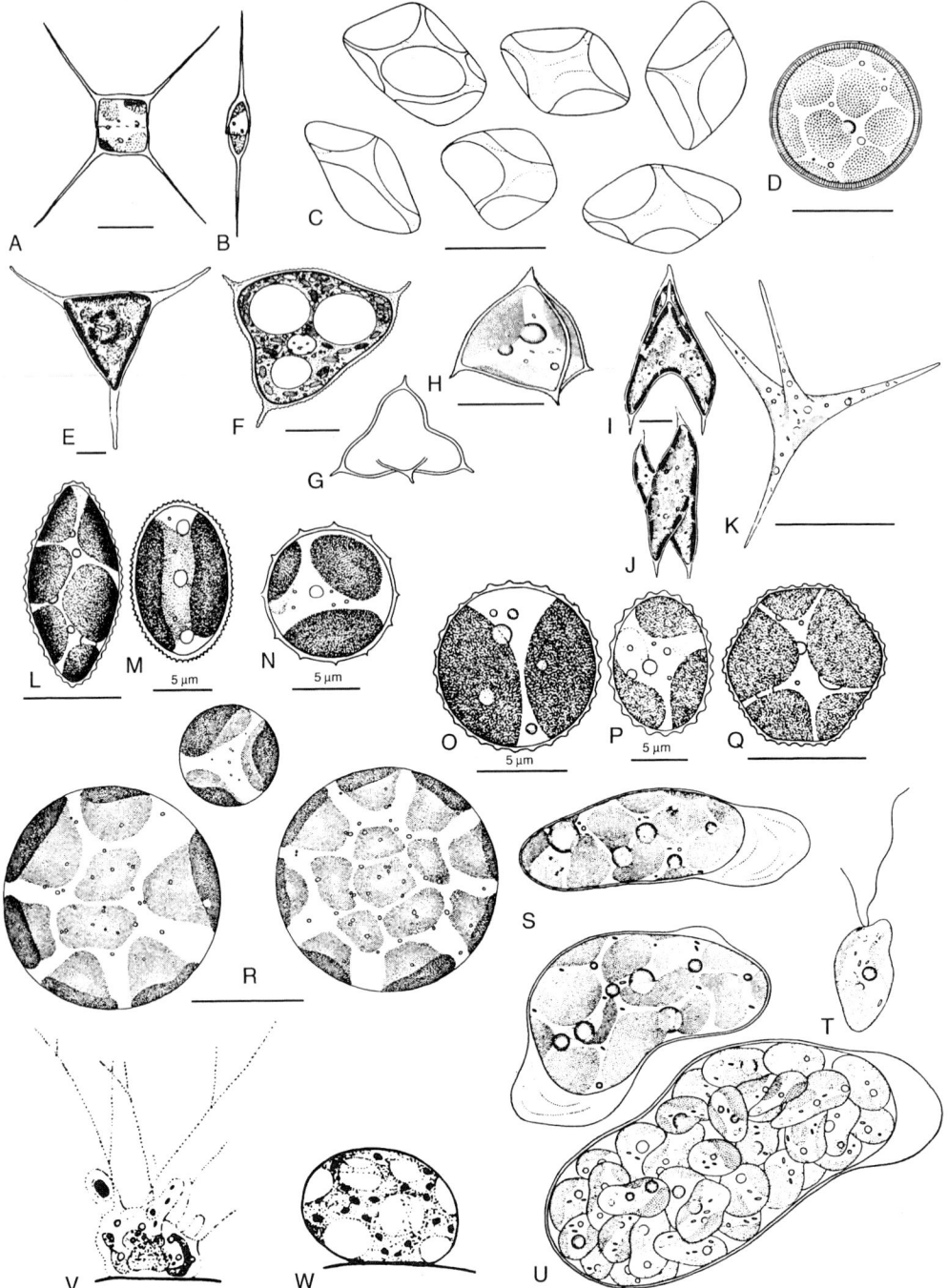

FIGURE 7 Freshwater members of the Mischococcales. (A–Q) Family Pleurochloridaceae. (R–W) Family Botrydiopsidaceae. (A, B) *Pseudotetraedron neglectum*: (A) Dorsal view of cell. (B) Ventral view of cell. (C) *Rhomboidella oblique* vegetative cells. (D) *Sklerochlamys pachyderma* vegetative cell. (E–H) *Tetraedriella*: (E) *T. spinigera*; note fourth spine in center of cell in dorsal view. (F, G) *T. regularis* vegetative cells. (H) *T. Limbata* vegetative cell. (I–K) *Tetraplektron* vegetative cells: (I, J) *T. torsum*. (K) *T. tribulus*. (L, M) *Trachychloron* vegetative cells. (L) *T. fusiforme*. (M) *T. depauperatum*. (N. *Trachycystis subsolitaria* vegetative cell: (O–Q) *Trachydiscus* showing varying shapes of vegetative cells: (O) *T. lenticularis*. (P) *T. ellipsoideus*. (Q) *T. sexangulatus*. (R) *Botrydiopsis arhiza* spherical vegetative cell. (S–U) *Excentrochloris gigas*: (S) Adult vegetative cell. (T) Zoospore. (U) Zoospore formation. (V, W). *Perone dimorpha*: (V) Free-living amoeboid stage with long pseudopods. (W) Globose epiphytic stage. A, B reprinted from Whitford, L. A., Schumacher, G. J. 1969. *A Manual of the Fresh-water Algae in North Carolina*. North Carolina Agricultural Research Service, North Carolina State University. C–L, O–Q, S–U reprinted from Ettl (1978), M–N, R from Tarapchak (1972), V, W from Prescott (1978), all with permission.

Trachycystis Pascher 1939 (Fig. 7N)

Cells are solitary or temporarily clustered after autosporulation. They are spherical or slightly flattened when abutting another cell. The cell wall is thick and ornamented with deep depressions. Plastids range from a few to many. Reproduction occurs by autospores. *Trachycystis subsolitaria* has been collected from an acid bog in Minnesota (Tarapchak, 1972).

Trachydiscus Ettl 1964 (Fig. 7O–Q)

Cells are solitary and spherical, elliptical, fusiform, rectangular, or hexagonal. Walls are heavily ornamented. Two or more parietal plastids are present, without pyrenoids. Reproduction occurs by autospores. *Trachydiscus* is euplanktonic and metaphytic in dystrophic, mesotrophic, and eutrophic lakes, ponds, pools, and seeps in Minnesota, Alaska, and Arkansas (R. L. Meyer, personal communication).

Mischococcales: Botrydiopsidaceae

Botrydiopsis Borzi 1889 (Fig. 7R)

Cells are solitary or in clusters and spherical. Young cells are 8–10 μm in diameter, up to 150 μm in older cells. The cell wall occurs in one piece and is smooth. Young cells have a single parietal plastid; older cells have many plastids and nuclei. Reproduction occurs by autospores and zoospores; aplanospores also have been observed (Moore and Carter, 1926). There have been many reports of this genus in North America; it has been collected from lakes and pools n British Columbia (Stein, 1975; Stein and Borden, 1978), Montana (Prescott and Dillard, 1979), Minnesota (Tarapchak, 1972), Massachusetts (Poulton, 1930), and North Carolina (Whitford, 1979). It has also been collected from soil in Missouri (Moore and Carter, 1926) and Nevada (LaRivers, 1978).

Excentrochloris Pascher 1937 (Fig. 7S–U)

Cells are solitary and irregularly ellipsoidal. The cell wall is in one piece and smooth. Young cells have one or two plastids; adult cells have many nuclei and parietal plastids. Reproduction occurs by autospores and zoospores. *Excentrochloris* is euplanktonic and metaphytic in dystrophic, oligotrophic, mesotrophic, and eutrophic ponds, pools, and seeps in Minnesota, Alaska, and Arkansas (R. L. Meyer, personal communication).

Perone Pascher 1932 (Fig. 7V, W)

This alga has a complex life cycle (Bourrelly, 1968, 1981) with a solitary, globose stage epiphytic on or in *Sphagnum* or other plants (Fig. 7W) and a free-living amoeboid stage in which the cells have long pseudopods (Fig. 7V). The cell wall is in one piece and smooth. Older cells are large, with many plastids and nuclei. Reproduction occurs by zoospores. Cells referred to as the genus *Leuvenia* are possibly one of the stages in the life cycle of *Perone*. *Perone* has been collected in British Columbia (Stein and Gerrath, 1969; Stein and Borden, 1978), Montana (as *Leuvenia*) (Prescott and Dillard, 1979), and Minnesota (as *Leuvenia natans*) (Tarapchak ,1972).

Mischococcales: Botryochloridaceae

Botryochloris Pascher 1930 (Fig. 8A)

Cells are spherical to globose and aggregated in clusters of 10–100 cells. Colonial mucilage is not evident. Individual cells are 3–30 μm in diameter, depending on the species, with one or more parietal plastids without pyrenoids. *Botryochloris* is epiphytic in a bog in Labrador (Duthie et al., 1976) and has been collected from *Sphagnum* bogs and *Carex* fens in Minnesota (R. L. Meyer, personal communication, Tarapchak, 1972).

Chlorellidiopsis Pascher 1939 (Fig. 8B, C)

Cells may occur singly or in unorganized cushion-like clusters. Individual cells are 8–14 μm, hemispherical, and in tetrads. Two or more parietal plastids are present without pyrenoids. Reproduction occurs by autospores and zoospores. Bourrelly (1968, 1981) and Prescott (1978) suggested uniting this genus with *Chlorellidium*. *Chlorellidiopsis* is epiphytic on filamentous algae or submerged vegetation (Prescott, 1978) in North Carolina (Whitford and Schumacher, 1969), Montana (Prescott and Dillard, 1979), and Ontario (Duthie and Socha, 1976).

Chlorellidium Vischer and Pascher 1937 (Fig. 8D–F)

Cells are 8–25 μm. Adult cells are solitary; young cells are grouped in clusters of tetrads, varying in shape due to compression. Cells have 5–10 plastids. Reproduction occurs by autospores and zoospores. Tarapchak (1972) distinguished this genus from *Chlorellidiopsis* by the number and shape of the plastids and by the shape of mature cells. *Chlorellidium* is epiphytic on filamentous algae (Prescott, 1978). Autospores have been collected on *Sphagnum* and submerged vegetation in Minnesota (Tarapchak, 1972).

Dichotomococcus Korchikoff 1928 (Fig. 8G)

Cells are elongate, recurved ovals, usually in groups of four, adherent to other groups of four; cells are enclosed in mucilage. Individual cells have one parietal plastid without pyrenoids.

Bourrelly (1968, 1981) reported that chlorophyll-*b* was found in one species of this genus (*D. lunatus*), indicating that this species should properly be moved

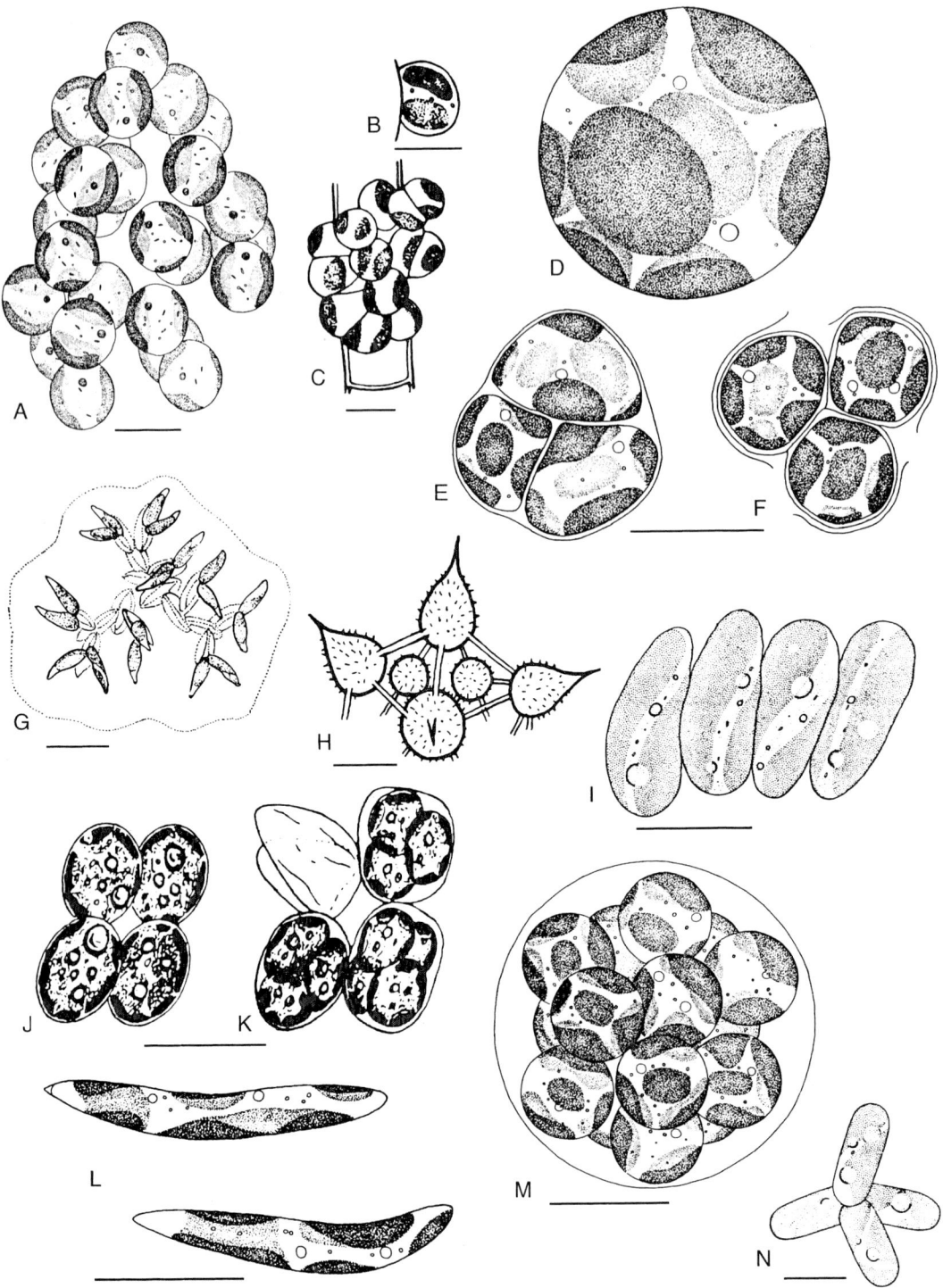

FIGURE 8 Freshwater members of the Mischococcales. Family Botryochloridaceae. (A) *Botryochloris cumulata* aggregation of vegetative cells. (B, C) *Chlorellidiopsis separabilis*: (B) Solitary vegetative cell. (C) Cluster of vegetative cells. (D–F) *Chlorellidium tetrabotrys*: (D) Vegetative cell. (E) Amorphous mass of tetrad autospores. (F) Autospore release. (G) *Dichotomococcus elongatus*; groups of four vegetative cells adherent to other groups of four. (H) *Ducelliera chodati* colony of vegetative cells joined by cylindrical processes. (I) *Heterodesmus bichloris*; a chain of four vegetative cells. (J, K) *Ilsteria quadrijuncta*: (J) Tetrad of vegetative cells. (K) Release of autospores. (L) *Raphidiella fascicularis* vegetative cells. (M) *Sphaerosorus coelastroides* spherical colony. (N) *Tetraktis aktinastroides* quartet of vegetative cells. A, D–F, L–M, reprinted from Tarapchak (1972), B–C, G from Prescott (1978), H from Whitford and Schumacher (1969), I–K, N from Ettl (1978), all with permission.

to the Chlorophyta. *Dichotomococcus* is euplanktonic and metaphytic in dystrophic, oligotrophic, mesotrophic, and eutrophic lakes, ponds, pools, and seeps in Minnesota, Alaska, and Arkansas (R. L. Meyer, personal communication).

Ducelliera Teiling 1957 (Fig. 8H)

Cells are composed of spherical to globular colonies, connected by cylindrical processes. Individual cells are pyramidal with a spine at the distal end. Numerous parietal discoidal plastids are present without pyrenoids. The method of reproduction is unknown. This genus resembles *Coelastrum* (Chlorophyta). *Ducelliera* is metaphytic among other algae in ponds (Whitford and Schumacher, 1969). It has been collected in British Columbia (Stein and Gerrath, 1969; Stein and Borden, 1978), Ontario (Duthie and Socha, 1976), and Montana (Wujek and Wee, 1984) and from dystrophic ponds and bogs in Minnesota (R. L. Meyer, personal communication).

Heterodesmus Ettl 1956 (Fig. 8I)

Cells are elliptical, ovoid, or rhomboidal in chains of two to eight cells. Individual cells are 5–15 μm long with one or more parietal plastids without pyrenoids. Reproduction occurs by autospores. This genus resembles *Scenedesmus* (Chlorophyta). *Heterodesmus* is euplanktonic and metaphytic in dystrophic, oligotrophic, mesotrophic, and eutrophic lakes, ponds, pools, and seeps in Minnesota, Alaska, and Arkansas (R. L. Meyer, personal communication).

Ilsteria Skuja et Pascher 1939 (Fig. 8J, K)

Cells are spherical, rarely ellipsoidal, and grouped in tetrads. Individual cells are 4–14 μm in diameter, with one or more plastids without pyrenoids. Reproduction occurs by autospores. *Ilsteria* is euplanktonic and metaphytic in dystrophic and mesotrophic ponds and pools (R. L. Meyer, personal communication).

Raphidiella Pascher 1938 (Fig. 8L)

Cells are elongate, fusiform, and solitary or in groups of two or four. Individual cells are 18–30 μm long, with two or more parietal plastids. The cell wall is usually thin. Reproduction occurs by autospores. *Raphidiella* has been collected from a bog in Minnesota (Tarapchak, 1972).

Sphaerosorus Pascher 1939 (Fig. 8M)

Cells are spherical to globose in spherical colonies of 8–32 cells. Individual cells have three to many parietal plastids. Reproduction occurs by autospores. Bourrelly (1968, 1981) unites this genus with *Botryochloris*, but R. L. Meyer (personal communication) notes that *Botryochloris* has random planes of division during cleavage, whereas *Sphaerosorus* does not. The cleavage pattern of *Sphaerosorus* results in a hollow sphere with no cell in the center. *Sphaerosorus* has been collected from *Sphagnum* bogs and *Carex* fens in Minnesota (R. L. Meyer, personal communication, Tarapchak, 1972).

Tetraktis Pascher 1938 (Fig. 8N)

Tetraktis is composed of quartets of cylindrical cells, 9–15 μm long. One or more parietal plastids are present, without pyrenoids. Reproduction occurs by autospores. *Tetraktis* is euplanktonic and metaphytic in dystrophic and mesotrophic ponds and pools in Minnesota, Alaska, and Arkansas (R. L. Meyer, personal communication).

Mischococcales: Gloeobotrydaceae

Asterogloea Pascher 1930 (Fig. 9A, B)

Cells are spherical, colonial, and in a gelatinous envelope. Colonies are irregular, 1–50 mm. Cells are ornamented with spines; two or more parietal, discoidal plastids are present, without pyrenoids. Reproduction occurs by autospores and zoospores. *Asterogloea* is euplanktonic, epiphytic, and metaphytic in dystrophic, oligotrophic, mesotrophic, and eutrophic lakes, ponds, pools, and seeps (R. L. Meyer, personal communication).

Chlorobotrys Bohlin 1901 (Fig. 9C)

Cells are spherical to ovoid; one to two cells are contained in a concentric mucilaginous sheath and arranged in colonies of 2–32 cells. Individual cells are 11–22 μm in diameter, with a variable number of parietal plastids without pyrenoids (Smith, 1920). The cell wall is thick and sometimes reddish in color. Reproduction occurs by autospores. *C. stellata* and *C. regularis* have been transferred to the Eustigmatophyceae. *Chlorobotrys* is euplanktonic and tychoplanktonic in ditches, streams, and lakes in the Arctic (Prescott and Vinyard, 1965), Montana (Wujek and Wee, 1984), Minnesota (Tarapchak, 1972), and Wisconsin (Smith, 1918). It is epipelic on damp clay soil in Arkansas (Meyer, 1969).

Chlorosaccus Luther 1899 (Fig. 9D–H)

Cells are solitary or in twos and fours at the periphery of a gelatinous matrix. Individual cells are ovate or pyriform with two to eight parietal plastids. Reproduction occurs by autospores, zoospores, and akinetes. *Chlorosaccus* is epiphytic or free floating. *Chlorosaccus fluidus* has been collected from streams and the wave zones of ponds in North Carolina (Schumacher et al., 1963).

Gloeobotrys Pascher 1930 (Fig. 9I)

Cells are spherical or ellipsoidal, grouped in twos or fours in a spherical, gelatinous mass. Individual cells are 4–12 µm, with a small number (usually two or three) parietal plastids. Reproduction occurs by autospores and zoospores. In appearance, this genus is much like *Chlorobotrys*, but differentiated by the concentric mucilaginous sheath around *Chlorobotrys*. *Gloeobotrys* is tychoplanktonic in lakes (Prescott and Vinyard, 1965), bogs, and creeks in the Arctic (Sheath and Steinman, 1981), British Columbia (Stein, 1975; Stein and Gerrath, 1969; Stein and Borden, 1978), Ontario (Duthie and Socha, 1976), Colorado (Durrell and Norton, 1960), and Lake Superior (Munawar and Munawar, 1978).

Gloeoskene Fott 1957 (Fig. 9J, K)

Cells are spherical, grouped in a mucilaginous thallus. Two to three centrally located plastids are present, without pyrenoids. Reproduction occurs by autospores. This genus looks much like *Gloeobotrys*, except for the central plastids, the lack of zoospores, and the persistence of the original cell wall after autosporulation. *Gloeoskene* is euplanktonic, epiphytic, and metaphytic in dystrophic, mesotrophic, and eutrophic ponds, pools, and seeps in Minnesota, Alaska, and Arkansas (R. L. Meyer, personal communication).

Gloeosphaeridium Pascher 1939 (Fig. 9L)

Cell are grouped in lobular colonies of 4–32 cells, enclosed in mucilage. Individual cells have three to six plastids without pyrenoids. Reproduction occurs by autospores and zoospores. *Gloeosphaeridium* is euplanktonic and metaphytic in dystrophic, oligotrophic, mesotrophic, and eutrophic lakes, lacustrine reservoirs, ponds, and pools in Minnesota, Alaska, and Arkansas (R. L. Meyer, personal communication).

Merismogloea Pascher 1938 (Fig. 9M, N)

Cells are ellipsoidal in colonies of 2–32 cells. Individual cells are 6–10 µm long, with a variable number of plastids without pyrenoids. The cell wall is thick and unsculptured. In appearance, this genus is much like *Chlorobotrys* except for the ellipsoidal shape of the cells. *Merismogloea* is euplanktonic and metaphytic in dystrophic, oligotrophic, and eutrophic lakes, ponds, and pools in Minnesota, Alaska, and Arkansas (R. L. Meyer, personal communication).

Mischococcales: Gloeopodiaceae

Gloeopodium Pascher 1938 (Fig. 11A–D)

Cells are ovoid to ellipsoidal, at the ends of stratified, gelatinous, sometimes branched stalks. Cells have a small number of parietal plastids. Reproduction occurs by autospores and zoospores. *Gloeopodium* is epilithic in streams and rivers in North Carolina and South Carolina (Schumacher *et al.*, 1963, 1966; Dillard, 1967).

Mischococcales: Mischohoccaceae

Mischococcus Nägeli 1849 (Fig. 11E)

Cells are spherical to ellipsoidal, in groups of two and four at the ends of mucilaginous stalks. The thallus is composed of branched, gelatinous tubes, forming treelike colonies. One or more parietal plastids are present, without pyrenoids. Reproduction occurs by autospores and zoospores. *Mischococcus* is tychoplanktonic and epiphytic on filamentous algae in ditches, ponds, and lakes in the Arctic (Ross, 1954; Prescott and Vinyard, 1965; Sheath and Hellebust, 1978; Sheath and Steinman, 1981), British Columbia (Stein 1975; Stein and Borden, 1978), South Carolina (Jacobs, 1968), and Arkansas (Meyer *et al.*, 1970).

Mischococcales: Characiopsidaceae

Characiopsis Borzi 1895 (Fig. 10A, B)

Cells are solitary, in groups, or massed, ovoid to ellipsoidal or fusiform, on a stipe with a small basal disc (sometimes not visible). Individual cells are 7–26 µm long without a stipe; one or more parietal plastids are present, without pyrenoids. Reproduction occurs by zoospores and, rarely, aplanospores and autospores. In appearance, this genus is much like *Characium* (Chlorophyceae); an iodine test for the presence of zoospores will differentiate between the two genera. *Characiopsis* is euplanktonic, tychoplanktonic, and epiphytic on filamentous algae, vascular plants, and the exoskeletons of microcrustaceans in ponds, pools, ditches, lakes, bogs, and sloughs in the Arctic (Ross, 1954; Prescott and Vinyard, 1965), Canada (Rawson, 1956; Sheath and Steinman, 1981), Labrador (Duthie *et al.*, 1976), Nova Scotia and New Brunswick (Hughes, 1948), Montana (Prescott and Dillard, 1979), Minnesota (R. L. Meyer, personal communication), Iowa (Prescott, 1931), Oklahoma (Vinyard, 1958), Utah (Benson and Rushforth, 1975), and Cuba (Margalef, 1947).

Chlorokoryne Pascher 1938 (Fig. 10C)

Cells are solitary or in groups, attached, pyriform, with the proximal end greatly elongated in some species. Cells have many parietal plastids; the cell wall is smooth or ornamented. The method of reproduction is unknown, but Ettl (1978) believes this genus probably reproduces by zoospores. *Chlorokoryne* is epiphytic on other algae, including the Chaetophorales, Oedogoniales, Cladophorales, and other tribophytes, in dystrophic, oligotrophic, and mesotrophic ponds and

FIGURE 9 Freshwater members of the Mischococcales. Family Gloeobotrydaceae. (A, B) *Asterogloea gelatinosa*: (A) Spiny vegetative cell. (B) Irregular colony of cells. (C) *Chlorobotrys simplex*, with concentric mucilaginous sheath. (D–H) *Chlorosaccus fluidus*: (D) Colony profile with cells at periphery of colony. (E) Zoospores. (F–H) Different cell arrangements in the colonies. (I) *Gloeobotrys limneticus* vegetative cells in mucilaginous sheath. (J, K) *Gloeoskene turfosa*: (J) Autospore release. (K) Vegetative cells with centrally located plastids. (L) *Gloeosphaeridium firmum* vegetative cells in globular colony. (M, N) *Merismogloea polychloris*: (N) Vegetative ellipsoidal cell. A–N reprinted from Ettl (1978) with permission.

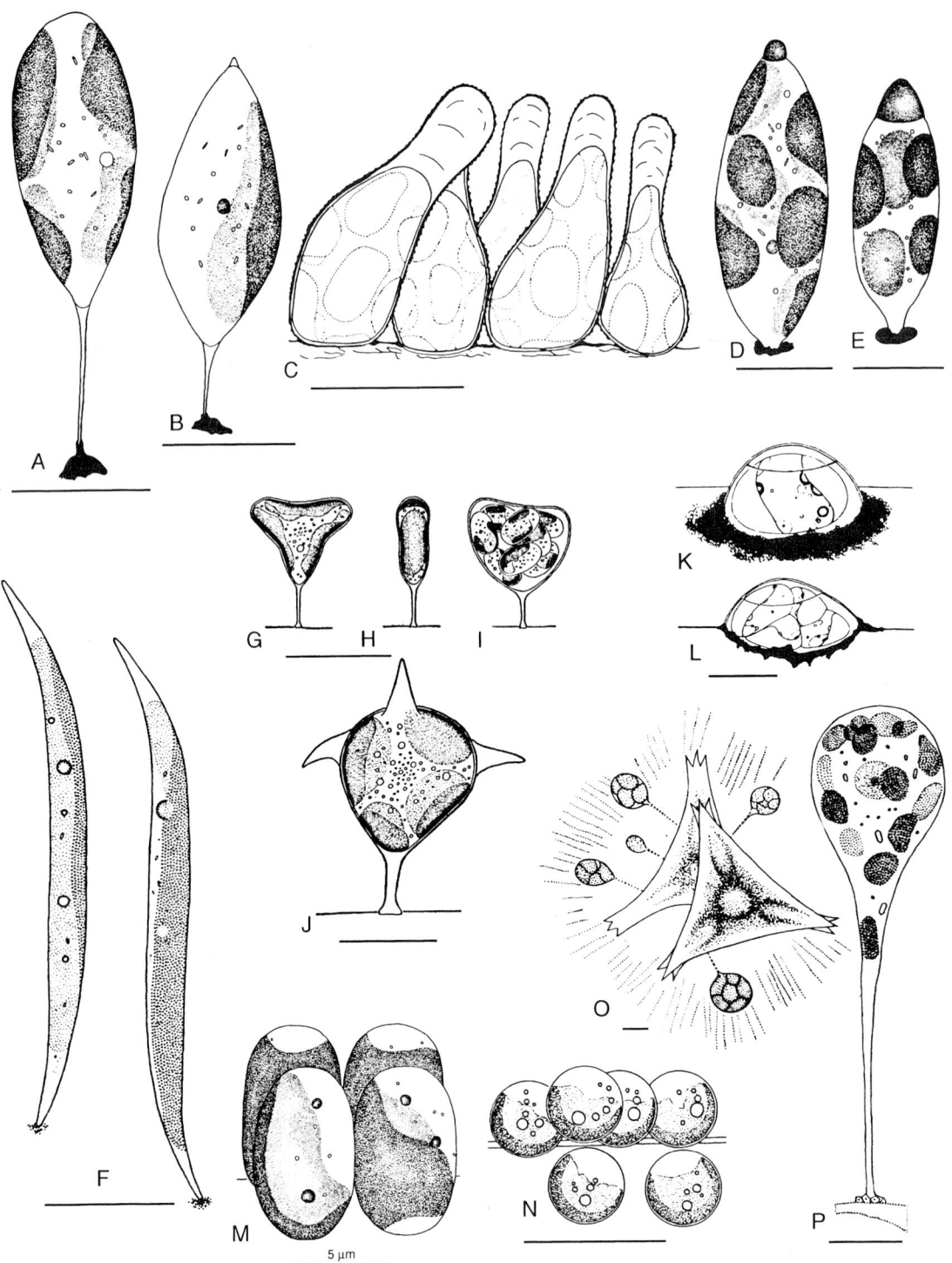

FIGURE 10 Freshwater members of the Mischococcales. Family Characiopsidaceae. (A, B) *Characiopsis*: (A) *C. pyriformis* stipulate vegetative cell. (B) *C. acuta* stipulate vegetative cell. (C) *Chlorokoryne petrovae* epiphytic vegetative cells. (D, E) *Chlorothecium*: (D) *C. capitatum* vegetative cell with very short stalk. (E) *C. crassiapex* vegetative cell. (F) *Chytridiochloris acus* vegetative cells. (G–J) *Dioxys*: (G–I) *D. inermis*: (H) Lateral view. (I) Aplanospore formation. (J) *D. tricornuta*. (K, L) *Hemisphaerella operculata*, with bases encrusted with mineral deposits; note bipartite cell wall, with the top half being smaller than the bottom. (M, N) *Lutherella*: (M) *L. adhaerens*, showing group of ovoid vegetative cells. (N) *L. globulosa*, showing group of spherical vegetative cells. (O, P) *Peroniella hyalothecae*, with slender, threadlike stipes: (O) Vegetative cells epiphytic on *Staurastrum*. A, B, D, E, M reprinted from Tarapchak (1972), C, F, G–L, N–P from Ettl (1978), all with permission.

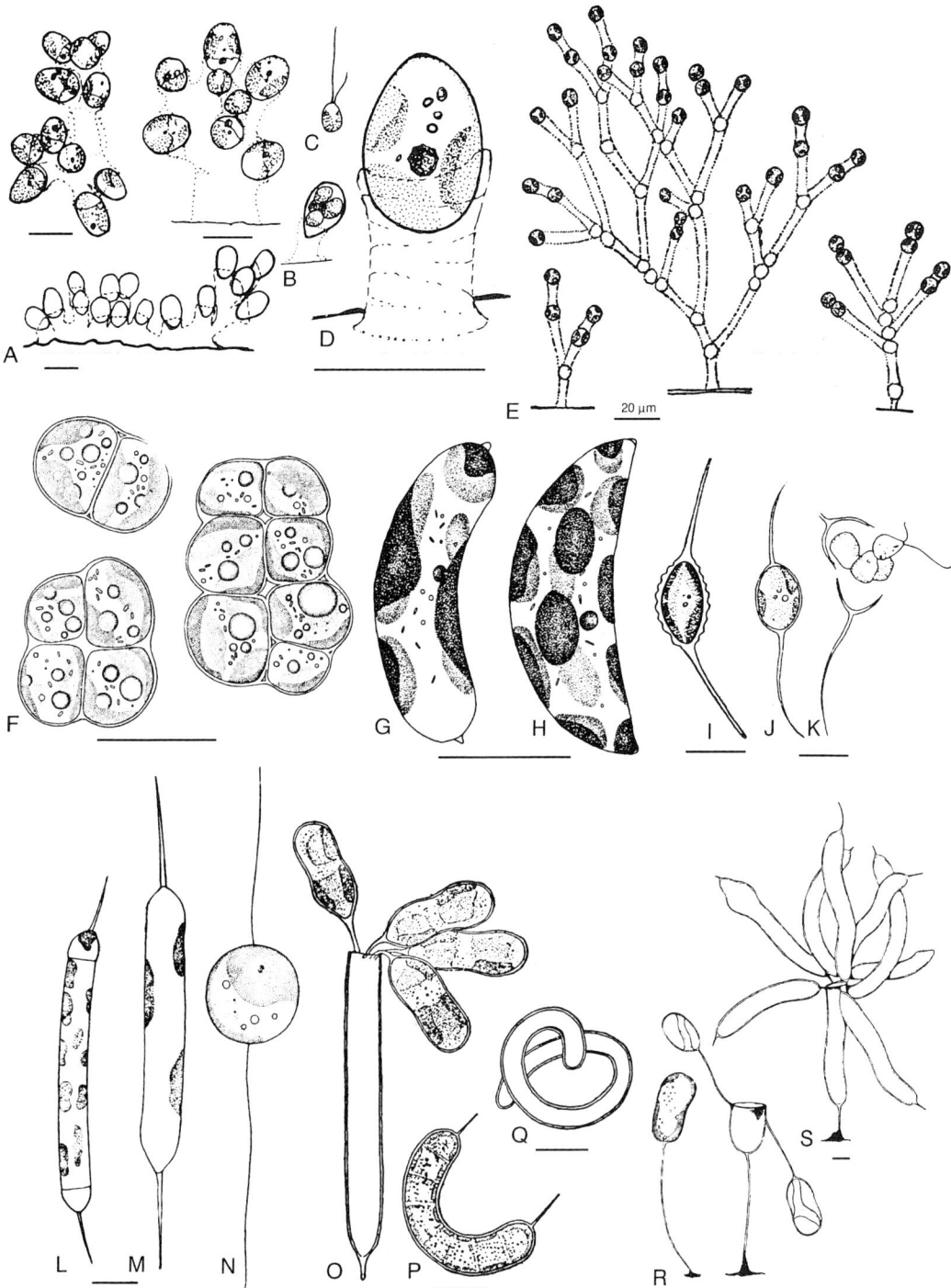

FIGURE 11 Freshwater members of the Mischococcales. (A–D) Family Gloeopodiaceae: *Gloeopodium rivulare*: (A) Different views of small epilithic colonies. (B) Autospore formation. (C) Zoospore. (D) Single cell at end of mucilaginous stalk. (E) Family Mischococcaceae: *Mischococcus confervicola*, treelike colonies with vegetative cells at ends of stalks. (F) Family Chloropediaceae: *Chloropedia plana*, flat epiphytic colonies. (G–N) Family Centritractaceae: (G, H) *Bumilleriopsis*: (G) *B. biverruca* vegetative cell. (H) *B. closterioides* vegetative cell. (I–N) *Centritractus*: (I) *C. brunneus*. (J, K) *C. ellipsoideus*: (K) Zoospore release. (L, M) *C. belenophorus*. (N) *C. globulosus*. Note varying shapes of vegetative cells, depending on species. (O–S) Family Ophiocytiaceae: *Ophiocytium* vegetative cells: (O) *O. arbusculum*. (P) *O. capitatum* Wolle. (Q) *O. parvulum*. (R) *O. longipes*. (S) *O. mucronatum*. A–D reprinted from Schumacher et al. (1963), E, O–Q from Smith, G. M. (1950). *The Fresh-water Algae of the United States*, 2nd ed. McGraw–Hill, New York, F, I–K, N, R, S from Ettl (1978), G, H from Tarapchak (1972), L, M from Prescott (1962), all with permission.

pools in Minnesota, Alaska, and Arkansas (R. L. Meyer, personal communication).

Chlorothecium Borzi 1885 (Fig. 10D, E)

Cells are solitary, cylindrical to spindle shaped, and attached by a short stalk. Cells have one or more parietal plastids without pyrenoids. The cell wall is relatively thick; during sporogenesis, the wall is in two pieces, with the upper half usually being smaller. Reproduction occurs by autospores, zoospores, and aplanospores. *Chlorothecium* has been collected from ponds and lakes in British Columbia (Stein, 1975; Stein and Borden, 1978) and Minnesota (Tarapchak, 1972).

Chytridiochloris Jane 1942 (Fig. 10F)

Cells are solitary, cylindrical to fusiform, and usually attached to filamentous algae. Cells are 30–45 µm long, with one or more parietal plastids. Reproduction occurs by zoospores. This genus is synonymous with *Harpochytrium* in some references. The species *C. viridis* has been moved to the Chlorophyta, based on its flagellar structure (Bourrelly, 1968, 1981). As *Harpochytrium*, it is epiphytic on *Oedogonium* in a swamp pool (Schumacher et al., 1966) and in flowing water in British Columbia (Stein, 1975).

Dioxys Pascher 1932 (Fig. 10G–J)

Cells are variable fusiform, often resembling a flattened triangle attached with a short stipe. There are a small number of parietal plastids in each cell. Reproduction occurs by zoospores. *Dioxys* is epiphytic in dystrophic, oligotrophic, mesotrophic, and eutrophic lakes, lacustrine reservoirs, ponds, and pools in Minnesota, Alaska, and Arkansas and is often observed with *Sphagnum*, *Typha*, *Scirpus*, and *Carex* (R. L. Meyer, personal communication). It has been collected from the lorica of *Difflugia* and may also be planktonic (Whitford and Schumacher, 1969).

Hemisphaerella Pascher 1939 (Fig. 10K, L)

Cells are solitary, hemispherical, with the base often covered with mineral deposits. Cells have a small number of parietal plastids. The cell wall is in two pieces, with the top half being smaller than the lower half. Reproduction occurs by zoospores. *Hemisphaerella* is epilithic and epiphytic on filamentous algae or vascular plants in North Carolina (Whitford and Schumacher, 1969). It is epiphytic in dystrophic and mesotrophic ponds and pools in Minnesota, Alaska, and Arkansas and is often associated with *Sphagnum*, *Typha*, *Scirpus*, and *Carex* (R. L. Meyer, personal communication).

Lutherella Pascher 1930 (Fig. 10M, N)

Cells are spherical to ovoid, solitary or in groups of two to four, and attached to the substrate. Cells have two (usually) parietal plastids. The cell wall is in two pieces during sporogenesis. Reproduction occurs by autospores and zoospores. *Lutherella* is epiphytic on filamentous algae or *Sphagnum* in lakes and bogs in the Arctic (Prescott and Vinyard, 1965), British Columbia (Stein and Borden, 1978), and Minnesota (Tarapchak, 1972).

Peroniella Gobi 1886–1887 (Fig. 10O, P)

Cells are spherical to ovoid, 6–9.5 µm long, with a long, slender threadlike stipe; the stipe is 8–10 µm long (Smith, 1920). Cells have a variable number of parietal plastids without pyrenoids. Reproduction occurs by zoospores. *Peroniella* is planktonic, tychoplanktonic, or epiphytic on other algae in pools and lakes in the Arctic (Prescott and Vinyard, 1966), British Columbia (Stein and Borden, 1978), southern Quebec (Lowe, 1927), Iowa (Prescott, 1931), North Carolina (Schumacher et al., 1966), and Oklahoma (Vinyard, 1965). *Peroniella hyalothecae* was thought to occur only on the filamentous desmid *Hyalotheca* (Prescott, 1931, 1978), but has been observed epiphytic on *Staurastrum* (Ettl, 1978). *Peroniella planctonica* has been found in the gelatinous envelope of *Sphaerozosma* filaments (Smith, 1920).

Mischococcales: Chloropediaceae

Chloropedia Pascher 1930 (Fig. 11F)

Cells are globular and attached to the substrate without a stipe or disc in twos and fours, forming yellowish, crustlike flat colonies. Individual cells have one or more parietal plastids. Reproduction occurs by autospores and zoospores. *Chloropedia* is epiphytic in dystrophic and mesotrophic ponds and pools in Minnesota, Alaska, and Arkansas and often found with *Sphagnum*, *Typha*, *Scirpus*, and *Carex* (R. L. Meyer, personal communication).

Mischococcales: Centritracttaceae

Bumilleriopsis Printz 1914 (Fig. 11G, H)

Cells are solitary or incidentally clustered; they are prominently elongate, cylindrical, straight, or curved. Cells are 10–120 µm in length, uninucleate, with several to many parietal plastids. Reproduction occurs by autospores and zoospores. *Bumilleriopsis* has been collected from lakes and bogs in Montana (Wujek and Wee, 1984) and Minnesota (Tarapchak, 1972).

Centritractus Lemmermann 1900 (Fig. 11I–N)

Cells are solitary; they are prominently elongate, cylindrical, straight, or slightly bent, with a long spine at each apex. Cells are uninucleate, with one or more parietal plastids; the cell wall is in two sections. Reproduction occurs by zoospores. *Centritractus* is planktonic and metaphytic (often with *Lemna*, *Potamogeton*, and *Ceratophyllum*) in dystrophic and mesotrophic ponds and small lakes in Minnesota, Alaska, and Arkansas (R. L. Meyer, personal communication) and rivers, ponds, and pools in Wisconsin, Michigan, West Virginia, Kentucky, Alabama, Florida (Camburn, 1982), Ohio (Hirsch and Palmer, 1958), Tennessee (Lackey, 1942), North Carolina (Schumacher et al., 1963), Jamaica (Hegewald, 1976), and Puerto Rico (Tiffany and Britton, 1944).

Mischococcales: Ophiocytiaceae

Ophiocytium Nägeli 1849 (Fig. 11O–S)

Cells are solitary or colonial, free living or epiphytic; they are elongate, cylindrical, straight, curved, or sometimes coiled. Cells have two or more nuclei. Cells may or may not have spines; several to many parietal plastids are present, without pyrenoids. Reproduction occurs by autospores, zoospores, and aplanospores. *Ophiocytium majus* may possibly be a eustigmatophyte, based on molecular studies of the mitochondrial *cox*I gene (Ehara et al., 1997). *Ophiocytium* is euplanktonic, tychoplanktonic, and metaphytic among vascular plants and epiphytic on filamentous algae and vascular plants in ponds, pools, lakes, rivers, creeks, and swamps in the Arctic (Prescott and Vinyard, 1965), throughout Canada (Hughes, 1948; Croasdale, 1973; Stein, 1975; Stein and Borden, 1978; Sheath and Hellebust, 1978; Sheath and Steinman, 1981), in the eastern half of the continental United States from Michigan (Transeau, 1917), Minnesota (Tarapchak, 1972), and Ohio (authors' personal observations) south to Florida (Whelden, 1941) and Bermuda (R. L. Meyer, personal communication), in the western United States in Colorado (Durrell and Norton, 1960), Nevada (LaRivers, 1978), Montana (Gaufin et al., 1976), Arkansas (Meyer, 1969; Meyer et al., 1970), Utah (Benson and Rushforth, 1975), Oklahoma (Vinyard, 1958, 1966), and in Puerto Rico (Tiffany and Britton, 1944), and Jamaica (Hegewald, 1976).

Tribonematales

Aeronemum Snow 1912 (Fig. 12A, B)

The thallus is composed of branched filaments; it is pseudoparenchymatous and microscopic. Cells are cylindrical, uninucleate, with several discoid plastids. The terminal cells of filaments are nearly spherical (Fig. 12A). Reproduction occurs by zoospores and aplanospores. This genus is synonymous with *Monocilia* and *Heterococcus*. Several researchers (Bourrelly, 1968, 1981; R. L. Meyer, personal communication) consider *Aeronemum*, *Capitulariella*, *Chaetopedia*, and *Heteropedia* to be questionable genera. *Aeronemum* is epiphytic on *Potamogeton*, *Nymphaea*, *Nuphar*, *Typha*, and *Scirpus* in dystrophic and mesotrophic ponds and pools in Minnesota, Alaska, and Arkansas (R. L. Meyer, personal communication).

Bumilleria Borzi 1888 (Fig. 13A–D)

Bumilleria is composed of unbranched filaments of cylindrical to quadrate cells with one or more parietal plastids. H-shaped pieces of walls are evident only at broken ends. Reproduction occurs by zoospores. Ettl (1978) includes *Pseudobumilleriopsis* in this genus. *Bumilleria* is tychoplanktonic and metaphytic among other filamentous algae in creeks, rivers, and ponds in British Columbia (Stein and Gerrath, 1969; Stein and Borden, 1978), Connecticut (Poulton, 1930), South Carolina (Jacobs, 1971), and California and Nevada (LaRivers, 1978).

Capitulariella Pascher 1944 (Fig. 12F–H)

The microscopic thallus is composed of branched filaments. Cells are cylindrical, with three to seven parietal plastids without pyrenoids. Reproduction occurs by zoospores. This genus is synonymous with *Heteropedia* p.p. *sensu* Bourrelly. These two genera differ only in that the filaments of *Heteropedia* are pseudoparenchymatous; in addition, the cells of *Heteropedia* differ from those of *Capitulariella* in the number of plastids. *Capitulariella* is epiphytic in dystrophic ponds and pools in Minnesota, Alaska, and Arkansas (R. L. Meyer, personal communication).

Chadefaudiothrix Bourrelly 1957 (Fig. 12C–E)

Pseudofilaments are uniseriate or multiseriate, surrounded by mucilage. Cells are ellipsoidal to elongated, usually separated from each other. Cells have one or more parietal plastids without pyrenoids. Reproduction occurs by cell division. *Chadefaudiothrix* has been found attached to solid surfaces in swamp pools in North Carolina (Schumacher et al., 1966; Whitford and Schumacher, 1969).

Chaetopedia Pascher 1939 (Fig. 12I)

Chaetopedia is composed of uniseriate, branched filaments forming a pseudoparenchyma. Cells are cylindrical with one or more parietal plastids. Terminal cells of filaments are tapered at the apex. Reproduction occurs by autospores and zoospores. This genus is synonymous with *Heteropedia* Pascher p.p. *sensu* Bourrelly. Epiphytic in ponds and pools. It is also epiphytic on vascular plants (*Potamogeton*, *Nymphea*, *Nuphar*, *Typha*, *Scirpus*, etc.), *Sphagnum*, and filamentous algae in dystrophic ponds and pools in Minnesota, Alaska, and Arkansas (R. L. Meyer, personal communication).

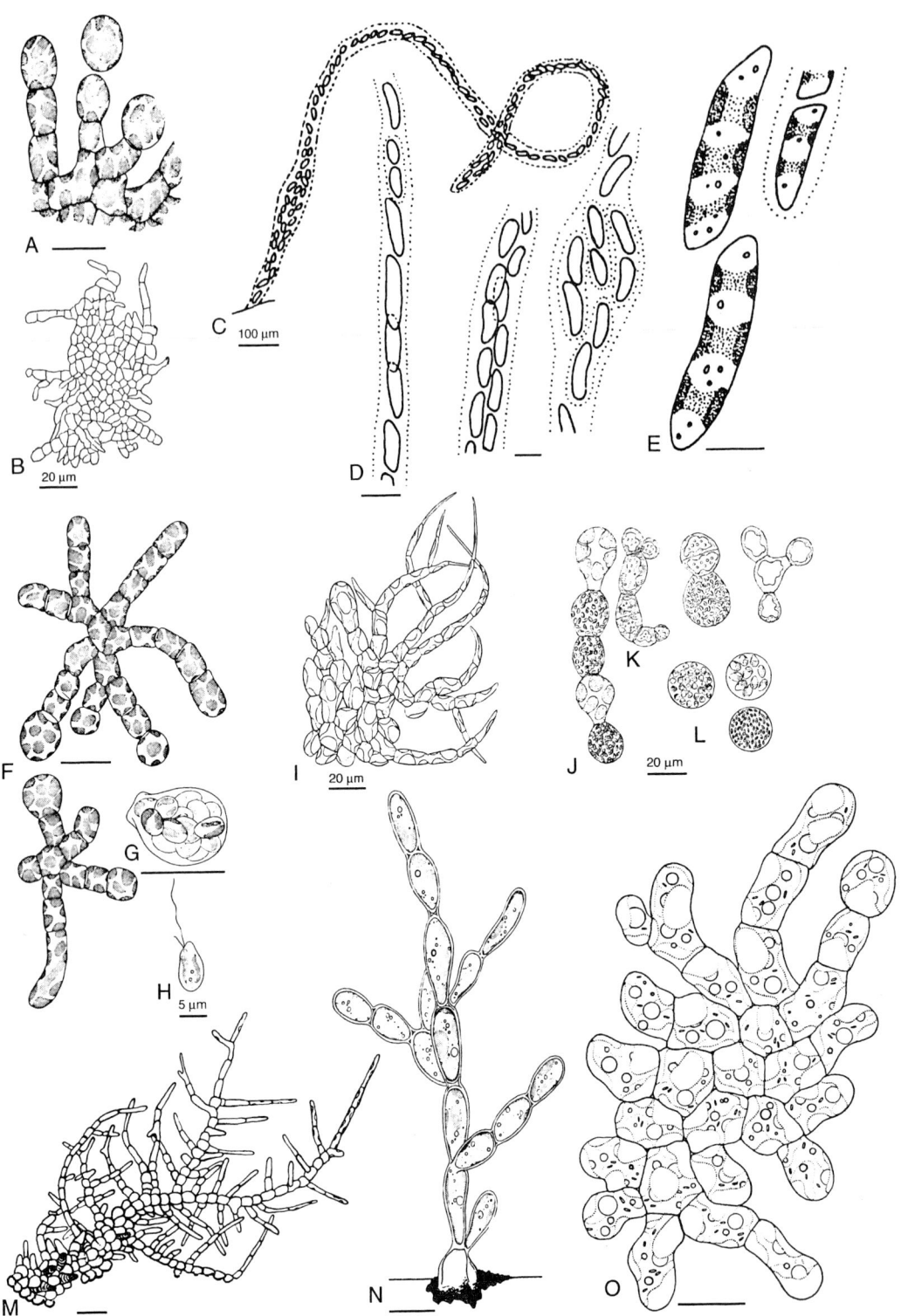

FIGURE 12 Freshwater members of the Tribonematales. (A, B) *Aeronemum polymorphum*, pseudoparenchymatous thallus: (A) Cells at tips of filaments are nearly spherical. (C–E) *Chadefaudiothrix gallica*: (C, D) Pseudofilaments surrounded by mucilage. (E) Individual cells. (F–H) *Capitulariella radians*: (G) Zoosporangium. (H) Zoospore. (I) *Chaetopedia stigeoclonioides*, with tapered cells terminating filaments. (J–L) *Fremya sphagni*: (J, K) Short, sometimes branched, filaments. (L) Solitary spherical cells, one of which is forming zoospores. (M) *Heterococcus ramosissimum* branched filaments. (N) *Heterodendron pascheri*, with dendroid branching and attachment with basal disc. (O) *Heteropedia polychloris* branched filaments. A–B, F–O reprinted from Ettl (1978), C–E from Schumacher et al. (1966), all with permission.

Fremya P. A. Dangeard 1934 (Fig. 12J–L)

Cells are spherical and solitary or occur in irregularly branched filaments. Cells contain many plastids without pyrenoids. Reproduction occurs by zoospores. Meyer (personal communication) states he does not assign species names to this genus in the field, as he has produced several morphotypes from cultures of axenic isolates. *Fremya* is epiphytic, epilithic, and endophytic, often in *Sphagnum*, in dystrophic, oligotrophic, mesotrophic, eutrophic, and hypereutrophic lakes, lacustrine reservoirs, ponds and pools in Minnesota, Alaska, and Arkansas (R. L. Meyer, personal communication).

Heterococcus Chodat 1908 (Fig. 12M)

Filaments are uniseriate or pleuriseriate and irregularly branched. Apical cells are often tapered. Cells are often dissociated; each cell has one parietal plastid without a pyrenoid. Reproduction occurs by zoospores and aplanospores. *Heterococcus* is epiphytic on vascular plants (*Potamogeton, Nymphea, Nuphar, Typha, Scirpus*, etc.), *Sphagnum*, and filamentous algae in dystrophic ponds, and pools in Minnesota, Alaska, and Arkansas (R. L. Meyer, personal communication), British Columbia (Stein, 1975; Stein and Borden, 1978), and attached to *Difflugia* in North Carolina (Whitford and Schumacher, 1969).

Heterodendron Steinecke 1932 (Fig. 12N)

The thallus is composed of uniseriate filaments with dendroid branching, attached by a large basal cell. Cells are 8–18 μm long, with one or more parietal plastids. Reproduction occurs by autospores and zoospores. *Heterodendron* is epiphytic on vascular plants (*Potamogeton, Nymphea, Nuphar, Typha, Scirpus*, etc.), *Sphagnum*, and filamentous algae in dystrophic ponds and pools in Minnesota, Alaska, and Arkansas (R. L. Meyer, personal communication).

Heteropedia Pascher 1939 (Fig. 12O)

The microscopic thallus is composed of branched, uniseriate filaments; it is sometimes pseudoparenchymatous. Cells are cylindrical, globular, or polygonal with one or more parietal plastids. Reproduction occurs by autospores and zoospores. *Heteropedia* is epiphytic on vascular plants (*Potamogeton, Nymphea, Nuphar, Typha, Scirpus*, etc.), *Sphagnum*, and filamentous algae in dystrophic and eutrophic ponds and pools in Minnesota, Alaska, and Arkansas (R. L. Meyer, personal communication).

Heterothrix Pascher 1932 (Fig. 13E, F)

Short, unbranched filaments are constricted at the cross walls or seen as dissociated cells. Cells are cylindrical, with one or two parietal plastids. The cell wall is in one section, except during zoosporogenesis, when the bipartite nature of the cell wall is visible. *Heterothrix* is euplanktonic and metaphytic in dystrophic, oligotrophic, mesotrophic, and eutrophic ponds, pools, and seeps in Minnesota, Alaska, and Arkansas (R. L. Meyer, personal communication).

Neonema Pascher 1925 (Fig. 13G–L)

Filaments are uniseriate, surrounded by mucilage. Cells are globular, cuboidal, to cylindrical, usually separated from each other. Cells contain a small number of parietal plastids. Reproduction occurs by cell division and zoospores. *Neonema* is metaphytic and epiphytic in dystrophic, oligotrophic, mesotrophic, and eutrophic ponds, pools, and seeps in Minnesota, Alaska, and Arkansas (R. L. Meyer, personal communication).

Tribonema Derbes and Solier 1956 (Fig. 13M–S)

Unbranched, uniseriate filaments are composed of long, cylindrical cells with one to many parietal plastids. The cell wall is in two sections, with H-shaped pieces visible upon close inspection. Reproduction occurs by zoospores, aplanospores, and cysts. Sexual reproduction occurs by isogametes. *Tribonema affine* has been collected in water with a pH as high as 9.0 (Durrell and Norton, 1960) and *T. viride* with a pH as high as 8.9 (Sheath *et al.*, 1996). A very common alga, *Tribonema* is tychoplanktonic and metaphytic among bryophytes and vascular plants in ditches, ponds, pools, lakes, rivers, seeps, streams, and creeks from the Arctic (Ross, 1954; Prescott and Vinyard, 1965; Sheath and Müller, 1997), throughout Canada (Lowe, 1927; Taylor, 1928, 1934; Hughes, 1948; Rawson, 1956; Stein, 1975; Stein and Gerrath, 1969; Stein and Borden, 1978; Croasdale,1973; Sheath and Hellebust, 1978; Sheath and Steinman, 1981; Sheath *et al.*, 1986, 1996), in the eastern United States from the Great Lakes (Tiffany, 1937; Damann, 1945, Munawar and Munawar, 1978) south through Massachusetts and Connecticut (Poulton, 1930; Colt, 1974), Ohio (Rhodes and Terzis, 1970), Kentucky (McInteer, 1930, 1939; Dillard *et al.*, 1976), Tennessee (Silva and Sharp 1944), Virginia (Lewis *et al.*, 1933; Woodson and Holoman, 1964; Woodson and Wilson, 1973), North Carolina (Schumacher *et al.*, 1966), and South Carolina (Jacobs, 1971); it has also been collected in Bermuda (R. L. Meyer, personal communication). In the western United States, *Tribonema* has been collected in Montana (Prescott and Dillard, 1979), Minnesota (Drouet, 1954), Wisconsin (Smith, 1916), Iowa (Prescott, 1931), Kansas (Thompson, 1938), Arkansas (Meyer, 1969; Meyer *et al.*, 1970), Utah (Benson and Rushforth, 1975), Oklahoma (Vinyard, 1958), and

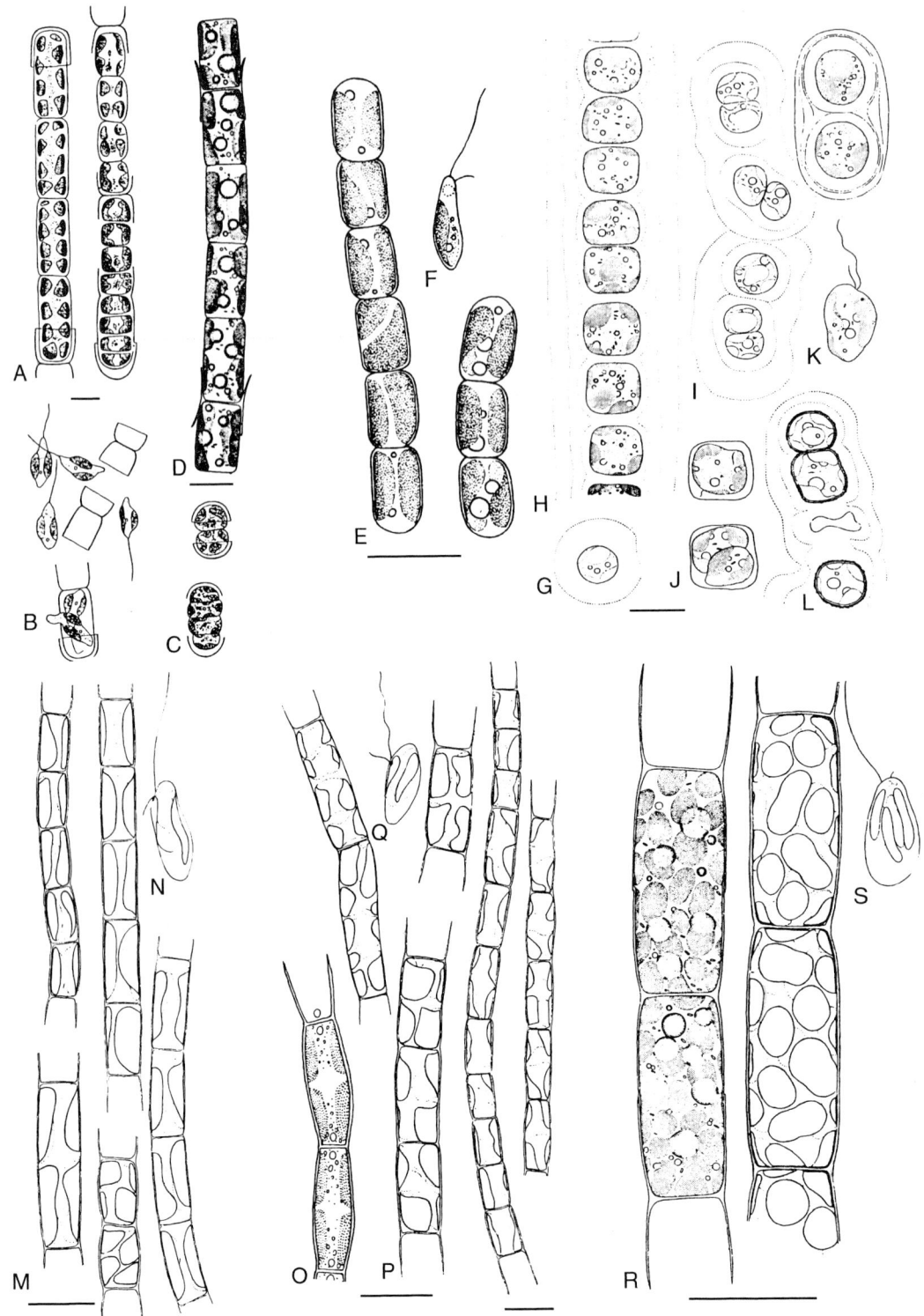

FIGURE 13 Freshwater members of the Tribonematales. (A–D) *Bumilleria*: (A–C) *B. sicula*: (B) Zoospore release. (C) Zoospore formation. (D) *B. klebsiana* vegetative filament. (E, F) *Heterothrix exilis*: (E) Vegetative filament. (F) Zoospore. (G–L) *Neonema quadratum*: (G) Single cell. (H) Vegetative filament, showing most cells separate from one another. (I) Filament fragmenting. (K) Zoospore. (L) Akinete formation. (M–S) *Tribonema*: (M, N) *T. aequale*: (N) Zoospore. (O–Q) *T. regulare*: (O) Enlarged portion of filament. (P) Vegetative filaments. (Q) Zoospore. (R, S) *T. viride*: (S) Zoospore. A–S reprinted from Ettl (1978) with permission.

California, Nevada, and Oregon (LaRivers, 1978). *Tribonema* has been collected as far south as Puerto Rico (Tiffany and Britton, 1944).

Botrydiales

Botrydium Wallroth 1815 (Fig. 14A–C)

The terrestrial thallus is composed of small, coenocytic vesicles. Many discoid plastids and underground rhizoidal extensions are present. Reproduction occurs by aplanospores, autospores, zoospores, and hypnospores. Sexual reproduction is by isogametes or anisogametes. *Botrydium* is epipelic on mud, damp soil, and the bottoms and banks of creeks and rivers in British Columbia (Stein, 1975; Stein and Borden, 1978), Connecticut and Massachusetts (Poulton, 1930), Ohio and Kentucky (McInteer, 1930, 1939), Virginia (Lewis *et al.*, 1933), South Carolina (Goldstein and Manzi, 1976), Illinois (Transeau, 1913), Missouri (Hayden, 1910), Iowa (Prescott, 1931), and California and Nevada (LaRivers, 1978).

Vaucheriales

Vaucheria de Candolle 1801 (Fig. 14D–K)

Vaucheria is commonly known as "water felt"; the sparsely branched or unbranched coenocytic filaments often form feltlike mats. The often macroscopic cells contain a large number of discoid plastids, with or without pyrenoids. Reproduction occurs by zoospores, aplanospores, and akinetes. Sexual reproduction is oogamous, with the filaments being monoecious or dioecious, depending on the species. Galls induced by the rotifer *Proales werneckii* have been observed on more than a dozen species of *Vaucheria* (Verb *et al.*, 1999). *Vaucheria* is epipelic and tychoplanktonic in ditches, streams, seeps, rivers, ponds, swamps, lakes, and brackish-water coastal habitats (Ott and Hommersand, 1974). Some species of *Vaucheria* are habitat-specific (see Sect. IV.C). *Vaucheria* has been collected in the Arctic (Prescott, 1953; Polunin, 1954; Sheath and Hellebust, 1978) and Canada (Stein, 1975; Stein and Gerrath, 1969; Stein and Borden, 1978; Duthie and Socha, 1976; Sheath and Steinman, 1981; Sheath *et al.*, 1986). In the continental United States, we have collected *Vaucheria* in Massachusetts, Michigan, Ohio, Pennsylvania, Tennessee, North Carolina, Georgia, Alabama, Louisiana, Florida, Iowa, and California. It has also been collected in Maine, Rhode Island, Alaska, New Jersey, and Michigan (Prescott, 1963; Colt, 1985), Connecticut (Collins, 1905; Colt, 1974; Schneider *et al.*, 1999; Schneider and Lane, 2000), New York (Blum, 1951), West Virginia (Fling, 1939), Virginia (Woodson and Holoman, 1964; Woodson and Wilson, 1973), Kentucky (McInteer, 1930, 1939; Harris, 1964), South Carolina (Dillard, 1967; Goldstein and Manzi, 1976), Oklahoma (Vinyard, 1958), Kansas (Thompson, 1938), Montana (Prescott and Dillard, 1979), Minnesota (Drouet, 1954), Illinois (Transeau, 1913), Missouri (Hayden, 1910) and Arkansas (Drouet, 1933, Meyer *et al.*, 1970), Nevada and Utah (Benson and Rushforth, 1975; LaRivers, 1978) and as far south as Puerto Rico (Tiffany and Britton, 1944) and Trinidad (West, 1904).

V. COLLECTION AND PREPARATION FOR IDENTIFICATION

Bogs and swamps are some of the best collecting sites for raphidophytes, although *Vacuolaria* may also be generally collected from ditches (Prescott, 1978) and from the layer of water above the mud (Heywood, 1990). The collection of aquatic plants may yield representatives of this algal class. In particular, squeezing of *Sphagnum* is a good source for *Gonyostomum* (Heywood, 1990). Collections of *Gonyostomum* may be more successful in the early morning hours (Drouet and Cohen, 1935; Cowles and Bramble, 1936).

Tribophytes may be collected from a wide variety of habitats, as described in Section IV.C, as may eustigmatophytes. Macroscopic mats of *Vaucheria* and colonies of *Botrydium* can be cut from the soil using a pocket knife. Soil samples may yield microscopic eustigmatophyte and tribophyte taxa such as *Botrydiopsis*, *Chlorobotrys*, and *Eustigmatos*. Filamentous forms can be gathered by hand near the shoreline or may be sampled farther out with the aid of a golf-ball retriever modified with wire or screen mesh over the open end. A plankton net is useful for gathering samples in open water. Many of the metaphytic and epiphytic forms may be collected by gathering vascular plants, algal mats, dead leaves, small sticks, stones, and so on from the shallow waters. Artificial substrata, such as glass sides, are also useful in collecting some algae (Dillard, 1999). A "slurper" (large pipette and bulb) can collect an interesting assemblage of nearshore bottom dwellers.

Collections may be placed in closable plastic bags or small vials for transport. Placing the bags or vials in a cooler with ice is advisable in warm weather or if the samples are not to be immediately taken to the laboratory. It is best to examine collections as soon as possible (Stein and Gerrath, 1969). If this is not possible, they can be kept in covered finger bowls for a short time; the temperature of the collection site should be maintained in a culture chamber. For longer storage, samples may be preserved in a solution of 1–2% buffered histological or EM-grade glutaraldehyde. Lugol's iodine is another option and has the

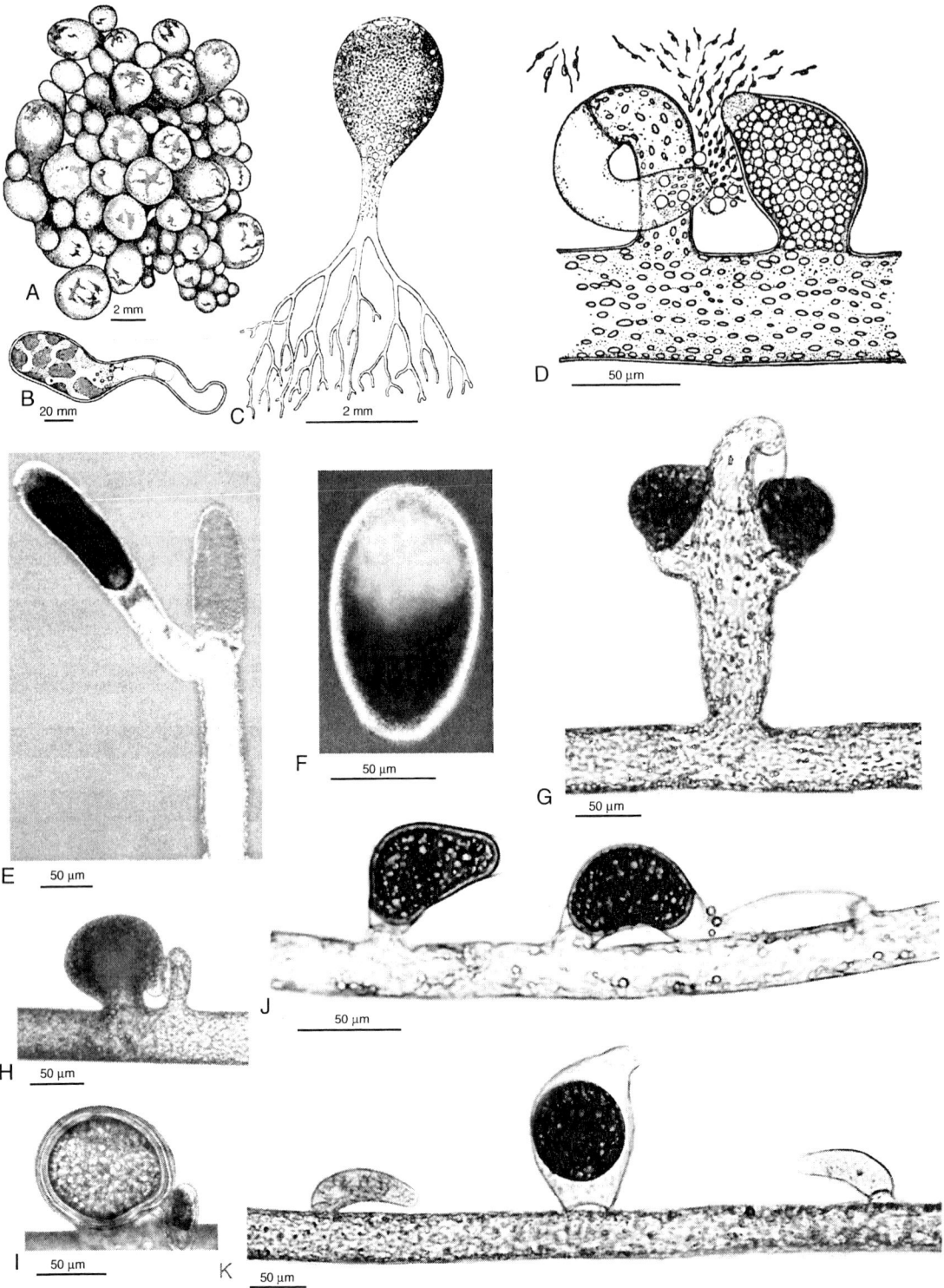

FIGURE 14 Freshwater members of the Botrydiales and Vaucheriales. (A–C) *Botrydium granulatum*: (A) Thallus of coenocytic vesicles. (B) Single germinating cell with small rhizoid. (C) Vesicle with rhizoidal extensions. (D–K) *Vaucheria*: (D–F) *V. bursata*: (D) Antheridium releasing spermatozoids adjacent to sessile oogonium. (E) DIC light micrograph of mature zoosporangium with empty zoosporangium to the right. (F) DIC light micrograph of zoospore. (G) *V. geminata*; light micrograph of two oospores between an empty antheridium on a gametophore. (H–I) *V. dillwynii*: (H) Light micrograph of mature oogonium. (I) Light micrograph of oogonium with an oospore with many wall layers. (J) *V. fontinalis*; light micrograph of two oogonia with oospores next to an empty antheridium. (K) *V. aversa*; light micrograph of single oogonium with an oospore between two empty antheridia. A–C reprinted from Ettl (1978), D from Smith (1950), E–K by authors, all with permission.

added benefit of differentiating green from tribophyte and eustigmatophyte taxa, although this may cause more distortion than glutaraldehyde. Lugol's solution has been successfully used to preserve raphidophytes (Islam and Khondker, 1994) for further study; glutaraldehyde may also be used. If samples are to be dried for herbarium use, it is imperative that they not be pressed, as this destroys many morphological features.

The identification of the Eustigmatophyceae is especially problematic. Transmission electron microscopy, gene sequencing, and/or pigment analysis may be necessary to definitively identify certain eustigmatophyte algae (Andersen et al., 1998; Graham and Wilcox, 2000).

Raphidophytes may be separated from some other unicellular algae for identification by phototaxis, as many migrate vertically depending on light and temperature conditions (Cowles and Brambel, 1936; Spencer, 1971; Heywood, 1990). *Gonyostomum*, for example, will rise to the surface as the sample stands (Heywood, 1973), although this is also true of dinoflagellates and haptophytes.

The identification of many tribophyte algae is difficult. A negative iodine test is a good indicator that the alga in question is tribophycean and not chlorophycean, but this characteristic may be difficult to see in many of the small coccoid forms or if few storage products are present. According to Whitford and Schumacher (1969), hot HCl will give a blue–green color to the plastids of tribophyte algae. A positive reaction with Schiff's reagent is a useful diagnostic tool, but all species of tribophyte algae may not react (Hibberd, 1980). R. L. Meyer (personal communication) suggests it is critical to examine plastid morphology carefully. The plastids of tribophyte algae will be disc shaped and parietal and will appear smooth and somewhat shiny, whereas the plastids of some greens are cup shaped or axial and will have a fine to coarse granular appearance.

For some species, it is necessary to examine reproductive features to ensure proper identification; this is especially true of *Vaucheria*, which is often sterile when collected. Species delineations of *Vaucheria* are based on antheridial and oogonial characters. Vegetative samples of *Vaucheria* may have to be placed in covered finger bowls or culture dishes, covered with water taken from the collection site or culture media (Provasoli 1968), and kept in a growth chamber at 10–15°C. A longer photoperiod (approximately 18 hours of light) appears to induce gametogenesis more quickly (League and Greulach, 1955; Schneider et al., 1993). For long-term study of any algal collection, maintenance and/or isolation of taxa in suitable liquid, agar, or "agar-slush" culture media is often advisable.

VI. GUIDE TO LITERATURE FOR SPECIES IDENTIFICATION

Excellent descriptions of many species may be found in older literature, but the classification of some taxa may not follow current schemes. Many of the Eustigmatophyceae, for example, are described in Bourrelly (1968, 1981), Ettl (1978), and Pascher (1937–1939) as tribophyte (xanthophyte or heterokont) algae. The Raphidophyceae may be called the Chloromonadophyceae in older literature. Several of the Tribophyceae were originally described as green algae. Smith (1920) included several tribophyte taxa in his descriptions of *Tetraedron*: *T. regulare* is actually *Tetraedriella regularis*, *T. enorme* = *Pseudostaurastrum enorme*, *T. limneticum* = *Pseudostaurastrum limneticum*, *T. hastatum* = *Pseudostaurastrum hastatum*, *T. trigonum* = *Goniochloris fallax*. *Vaucheria* was long considered a member of the Chlorophyta and older descriptions of *Vaucheria* species are included with siphonous green algae.

Eustigmatophyceae
  Hibberd, 1981, 1990b
  Santos, 1996
Nannochloropsis
  Andersen et al., 1998
Raphidophyceae
  Fott, 1968
Tribophyceae
  Bourrelly, 1968, 1981
  Ettl, 1978.
  Pascher, 1937–1939
  Rieth, 1980
Characiopsis
  Pizarro, 1995
Heterococcus
  Lokhorst, 1992
Vaucheria
  Blum, 1972
  Christensen, 1968, 1969, 1987a, b
  Entwisle, 1987, 1988a, b
  Rieth, 1980
  Schneider et al., 1999
  Venkataraman, 1961

ACKNOWLEDGMENTS

We thank Dr. Richard L. Meyer for his significant contributions to this chapter. His vast understanding of the morphology and ecology of the Tribophyceae has added immeasurably to the knowledge of this little-known group of algae. We also thank the staff of the University of Akron audiovisual services, especially Stephen Allen.

LITERATURE CITED

Andersen, R. A., Brett, R. W., Potter, D., Sexton, J. P. 1998. Phylogeny of the Eustigmatophyceae based upon 18S rDNA, with emphasis on *Nannochloropsis*. Protist 149:61–74.

Ariztia, E. V., Andersen, R. A., Sogin, M. L. 1991. A new phylogeny for chromophyte algae using 16S-like rRNA sequences from *Mallomonas papillosa* (Synurophyceae) and *Tribonema aequale* (Xanthophyceae). Journal of Phycology 29:701–715.

Bailey, J. C., Andersen, R. A. 1998. Phylogenetic relationships among nine species of the Xanthophyceae inferred from *rbc*L and 18S rRNA gene sequences. Phycologia 37:458–466.

Bailey, J. C., Bidigare, R. R., Christensen, S. J., Andersen, R. A. 1998. Phaeothamniophyceae *classis nova*: A new lineage of chromophytes based upon photosynthetic pigments, *rbc*L sequences analysis and ultrastructure. Protist 149:245–263.

Benson, C. E., Rushforth, S. R. 1975. The algal flora of Huntington Canyon, Utah, U.S.A. Bibliotheca Phycologica 18.

Blum, J. L. 1951. Notes on Vaucheriaceae, with particular reference to western New York. Bulletin of the Torrey Botanical Club 78:441–448.

Blum, J. L. 1972. Series II. North American flora. Part 8. Vaucheriaceae. New York Botanical Garden, New York.

Bold, H. C., Wynne, M. J. 1985. Introduction to the algae. Prentice Hall, Englewood Cliffs, New Jersey.

Bourrelly, P. 1968, 1981. Les algues d'eau douce. Initiation à la systématique. Tome II: Les algues jaunes et brunes. Chrysophycées, Pheophycées, Xanthophycées et Diatomées. Boubée et Cie, Paris.

Camburn, K. E. 1982. The occurrence of thirteen algal genera previously unreported from Kentucky. Transactions of the Kentucky Academy of Science 43:72–79.

Cavalier-Smith, T., Chao, E. E. 1996. 18S rRNA sequence of *Heterosigma carterae* (Raphidophyceae), and the phylogeny of heterokont algae (Ochrophyta). Phycologia 35:500–510.

Christensen, T. 1968. *Vaucheria* types in the Dillenian Herbaria. British Phycological Bulletin 3:463–69.

Christensen, T. 1969. *Vaucheria* collections from Vaucher's region. Royal Danish Academy of Sciences and Letters, Copenhagen.

Christensen, T. 1987a. Some collections of *Vaucheria* (Tribophyceae) from south-eastern Australia. Australian Journal of Botany 35:617–629.

Christensen, T. 1987b. Seaweeds of the British Isles, Vol. 4. Tribophyceae (Xanthophyceae). British Museum (Natural History), London.

Collins, F. S. 1905. Phycological notes of the late Isaac Holden. II. Rhodora 7:222–243.

Colt, L. C. 1974. Some algae of the Connecticut River, New England, U.S.A. Nova Hedwigia 25:195–209.

Colt L. C. 1985. *Vaucheria undulata* Jao again in New England. Rhodora 87:597–599.

Couch, J. N. 1932. Gametogenesis in *Vaucheria*. Botanical Gazette 94:272–296.

Cowles, R. P., Brambel, C. E. 1936. A study of the environmental conditions in a bog pond with special reference to the diurnal vertical distribution of *Gonyostomum semen*. Biological Bulletin (Woods Hole) 71:286–298.

Croasdale, H. 1973. Freshwater algae of Ellesmere Island, N.W.T. Publications in botany, No. 3, National Museum of Natural Sciences, Ottawa.

Cronberg, G., Lindmark, G., Bjork, S. 1988. Mass development of the flagellate *Gonyostomum semen* (Raphidophyta) in Swedish forest lakes—an effect of acidification? Hydrobiologia 161:217–236.

Damann, K. E. 1945. Plankton studies of Lake Michigan. 1. Seventeen years of plankton data collected at Chicago, Illinois. American Midland Naturalist 34:769–796.

Daugbjerg, N., Andersen, R. A. 1997. A molecular phylogeny of the heterokont algae based on analyses of chloroplast-encoded *rbc*L sequence data. Journal of Phycology 33:1031–1041.

Davis, B. M. 1904. Oogenesis in *Vaucheria*. Botanical Gazette 38:81–99.

Deason, T. R. 1971a. The origin of flagellar hairs in the xanthophycean alga *Pseudobumilleriopsis pyrenoidosa*. Transactions of the American Microscopical Society 90:441–448.

Deason, T. R. 1971b. The fine structure of sporogenesis in the xanthophycean alga *Pseudobumilleriopsis pyrenoidosa*. Journal of Phycology 7:101–107.

Dillard, G. E. 1967. The freshwater algae of South Carolina. I. Previous work and recent additions. Journal of the Elisha Mitchell Scientific Society 83:128–131.

Dillard, G. E. 1999. Common freshwater algae of the United States. An illustrated key to the genera (excluding the diatoms). Cramer, Berlin.

Dillard, G. E., Moore, S. P., Garret, L. S. 1976. Kentucky algae II. Transactions of the Kentucky Academy of Science 37:20–25.

Drouet, F. 1933. Algal vegetation of the large Ozark Springs. Transactions of the American Microscopical Society 52:83–100.

Drouet, F. 1954. A preliminary study of the algae of northwestern Minnesota. Journal of the Minnesota Academy of Science 22:116–38.

Drouet, F., Cohen, A. 1935. The morphology of *Gonyostomum semen* from Wood's Hole, Massachusetts. Biological Bulletin 68:422–439.

Durrell, L. W., Norton, C. 1960. Phytoplankton of lakes of Grand Mesa, Colorado. Transactions of the American Microscopical Society 79:91–97.

Duthie, H. C., Socha, R. 1976. A checklist of the freshwater algae of Ontario, exclusive of the Great Lakes. Naturaliste Canadien (Quebec) 103:83–109.

Duthie, H. C., Ostrofsky, M. L., Brown, D. J. 1976. Freshwater algae from western Labrador IV. Chrysophyta, Xanthophyta, Pyrrophyta, Cryptophyta. Nova Hedwigia 27:909–917.

Ehara, M., Hayashi-Ishimaru, Y., Inagaki, Y., Ohama, T. 1997. Use of a deviant mitochondrial genetic code in tribophyte algae as a landmark for segregating members within the phylum. Journal of Molecular Evolution 45:119–124.

Entwisle, T. J. 1987. An evaluation of taxonomic characters in the subsection Sessiles, section Corniculatae, of *Vaucheria* (Vaucheriaceae, Chrysophyta). Phycologia 26:297–321.

Entwisle, T. J. 1988a. An evaluation of taxonomic characters in the *Vaucheria prona* complex (Vaucheriaceae, Chrysophyta). Phycologia 27:183–200.

Entwisle, T. J. 1988b. A monograph of *Vaucheria* (Vaucheriaceae, Chrysophyta) in south-eastern mainland Australia. Australian Systematic Botany 1:1–77.

Ettl, H. 1978. Xanthophyceae I, *in*: Ettl, H., Gerloff, J., Heynig, H.,Eds., Süsswasserflora von Mitteleuropa, Vol. 3. Fischer, Stuttgart/New York.

Fling, E. M. 1939. One hundred algae of West Virginia. Castanea 4:11–25.

Fott, B. 1968. Klasse: Chloromonadophyceae, *in*: Huber-Pestalozzi, G., Ed., Das Phytoplankton des Süsswassers, 2nd ed., Vol. 3. Schweizerbart, Stuttgart.

Fott, B. 1971. Algenkunde. Fischer, Jena, Germany.

Gaufin, A. R., Prescott, G. W., Tibbs, J. F. 1976. Limnological studies of Flathead Lake Montana: A status report. EPA-600/3-76-039, U.S. Enviromental Protection Agency, Washington, DC.

Goldstein, A. K., Manzi, J. J. 1976. Additions to the freshwater algae

of South Carolina. Journal of the Elisha Mitchell Scientific Society 92:9–13.
Graham, L. E., Wilcox, L. W. 2000. Algae. Prentice Hall, Upper Saddle River.
Greenwood, A. D. 1959. Observations on the structure of the zoospore of *Vaucheria*. II. Journal of Experimental Botany 10:55–68.
Guillard, R. R. L., Lorenzen, C. J. 1972. Yellow-green algae with chlorophyllide *c*. Journal of Phycology 31:774–777.
Harris, D. O. 1964. Occurrence of a variety of *Vaucheria pseudohamata* Prescott. Transactions of the American Microscopical Society 83:406–409.
Havens, K. E., III. 1989. Seasonal succession in the plankton of a naturally acidic, highly humic lakes in Northeastern Ohio, USA. Journal of Plankton Research 11:1321–1327.
Hayden, A. 1910. The algal flora of the Missouri Botanical Garden. Annals of the Missouri Botanical Garden 22:25–48.
Hegewald, E. 1976. A contribution to the algal flora of Jamaica. Nova Hedwigia 28:45–69.
Heywood, P. 1973. Nutritional studies on the Chloromonodophyceae: *Vacuolaria virescens* and *Gonyostomum semen*. Journal of Phycology 9:156–159.
Heywood, P. 1978a. Ultrastructure of mitosis in the Chloromonadophycean alga *Vacuolaria virescens*. Journal of Cell Science 31:37–51.
Heywood, P. 1978b. Osmoregulation in the alga *Vacuolaria virescens*. Structure of the contractile vacuole and the nature of its association with the Golgi apparatus. Journal of Cell Science 31:213–224.
Heywood, P. 1980. Chloromonads, *in*: Cox, E. R., Ed., Phytoflagellates. Elsevier, New York, pp. 351–379.
Heywood, P. 1989. Some affinities of the Raphidophyceae with other chromophyte algae, *in*: Green, J. C., Leadbeater, B. S. C., Diver, W. L., Eds., The chromophyte algae: Problems and perspectives. Clarendon, Oxford, pp. 279–293.
Heywood, P. 1990. Phylum Raphidophyta, *in*: Margulis, L., Corliss, J. O., Melkonian, M., Chapman, D. J., Eds., Handbook of Protoctista. Jones & Bartlett, Boston, pp. 318–325.
Hibberd, D. J. 1980. Xanthophytes, *in*: Cox, E. R., Ed., Phytoflagellates. Elsevier, New York, pp. 243–271.
Hibberd, D. J. 1981. Notes on the taxonomy and nomenclature of the algal classes Eustigmatophyceae and Tribophyceae (synonym Xanthophyceae). Botanical Journal of the Linnean Society 82:93–119.
Hibberd, D. J. 1990a. Xanthophyta, *in*: Margulis, L., Corliss, J. O., Melkonian, M., Chapman, D. J., Eds., Handbook of Protoctista. Jones & Bartlett, Boston, pp. 686–697.
Hibberd, D. J. 1990b. Eustigmatophyta, *in*: Margulis, L., Corliss, J. O., Melkonian, M., Chapman, D. J., Eds., Handbook of Protoctista. Jones & Bartlett, Boston, pp. 326–333.
Hibberd, D. J., Leedale, G. F. 1970. Eustigmatophyceae—a new algal class with unique organization of the motile cell. Nature 225:758–760.
Hibberd, D. J., Leedale, G. F. 1971a. A new algal class—the Eustigmatophyceae. Taxon 20:523–525.
Hibberd, D. J., Leedale, G. F. 1971b. Cytology and ultrastructure of the Xanthophyceae. II. The zoospore and vegetative cell of coccoid forms, with special reference to *Ophiocytium majus* Naegeli. British Phycological Journal 6:1–23.
Hibberd, D. J., Leedale, G. F. 1972. Observations on the cytology and ultrastructure of the new algal class, Eustigmatophyceae. Annals of Botany 36:49–71.
Hindak, F. 1981. Chlorococcal algae. Chlorophyceae 2. Biologicke Prace. 26:7–194.
Hirsch, A., Palmer, C. M. 1958. Some algae from the Ohio River drainage basin. Ohio Journal of Science 58:375–384.
Hughes, E. O. 1948. New fresh-water Chlorophyceae from Nova Scotia. American Journal of Botany 35:424–427.
Islam, A. K. M. N., Khondker, M. 1994. New records of algae from Bangladesh. IV. *Heteromastix* and *Gonyostomum*. Bangladesh Journal of Botany 23:199–203.
Iyengar, M. O. P. 1925. Note on two new species of *Botrydium* from India. Journal of the Indian Botanical Society 4:193–201.
Jacobs, J. E. 1968. A preliminary check list of fresh-water algae in South Carolina. Journal of the Elisha Mitchell Scientific Society 84:454–457.
Jacobs, J. E. 1971. A preliminary taxonomic survey of the freshwater algae of the Belle W. Barunch plantation in Georgetown County, South Carolina. Journal of the Elisha Mitchell Scientific Society 87:26–30.
Kato, S. 1991. Geographic distribution of freshwater raphidophycean algae in Japan and the effect of pH on their growth. Japanese Journal of Phycology 39:179–184.
Koch, W. J. 1951. A study of the motile cells of *Vaucheria*. Journal of the Elisha Mitchell Scientific Society 67:123–131.
Krienitz, L., Hepperle, D., Stich, H., Weiler, W. 2000. *Nannochloropsis limnetica* (Eustigmatophyceae), a new species of picoplankton from freshwater. Phycologia 39:219–227.
Lackey, J. B. 1942. The plankton algae and protozoa of two Tennessee rivers. American Midland Naturalist 17:191–202.
Lagerheim, G. 1889. Studien über die Gattungen *Conferva* und *Microspora*. Flora 72:179–210.
LaRivers, I. 1978. Algae of the Western Great Basin. Publication 50008, Bioresources Center, Desert Research Institute, University of Nevada.
League, E. A., Greulach, V. A. 1955. Effects of daylength and temperature on the reproduction of *Vaucheria sessilis*. Botanical Gazette 117:45–51.
Lee, R. E. 1989. Phycology. Cambridge Univ. Press, Cambridge, UK.
Lepisto, L., Antikainen, S., Kivinen, J. 1994. The occurrence of *Gonyostomum semen* (Her.) Diesing in Finnish lakes. Hydrobiologia 273:1–8.
Lewis, I. F., Zirkle, C., Patrick, R. 1933. Algae of Charlottesville and vicinity. Journal of the Elisha Mitchell Scientific Society 48:207–222.
Lokhorst, G. M. 1992. Taxonomic studies in the Genus *Heterococcus*. Fischer, Stuttgart.
Lowe, C. W. 1927. Some freshwater algae of southern Quebec. Transactions of the Royal Society of Canada 5:291–316.
Luther, A. 1899. Über *Chlorosaccus* eine neue Gattung der Süsswasseralgen, nebst Bemerkungen zur Systematik verwandter Algen. Bihand till Kongliga Svenska Vetenskaps Academiens Handlingar 24. 13:1–22.
Margalef, R. 1947. Algas de agua dulce de la laguna de Ariguanabo (Isla de Cuba). Publicaciones de Instituto de Biologica Applicada 4:79–89.
Massalski, A., Leedale, G. F. 1969. Cytology and ultrastructure of the Xanthophyceae. I. Comparative morphology of the zoospores of *Bumilleria sicula* Borzi and *Tribonema vulgare* Pascher. British Phycological Journal 4:159–180.
McInteer, B. B. 1930. Preliminary report of the algae of Kentucky. Ohio Journal of Science 30:131–142.
McInteer, B. B. 1939. A check list of the algae of Kentucky. Castanea 4:27–37.
Meyer, R. L. 1969. The freshwater algae of Arkansas. I. Introduction and recent additions. Proceedings of the Arkansas Academy of Science 23:145–156.
Meyer, R. L. 1995. Transference of *Malleodendron caespitosum* Thompson to *Chrysomallus* gen. nov. Journal of Phycology (Supplement) 31:20.
Meyer, R. L., Wheeler, J. H., Brewer, J. R. 1970. The freshwater

algae of Arkansas. II. New additions. Proceedings of the Arkansas Academy of Science 24:32–35.

Mignot, J. P. 1967. Structure et ultrastructure de quelques Chloromonadines. Protistologica 3:5–23.

Mignot, J. P. 1976. Compléments a l'étude des Chloromonadines. Ultrastructure de *Chattonella subsalsa* Biecheler flagella d'eau saumâtre. Protistologica 12:279–293.

Moestrup, Ø. 1982. Phycological reviews 7. Flagellar structure in algae: a review, with new observations particularly on the Chrysophyceae, Phaeophyceae (Fucophyceae), Euglenophyceae, and *Reckertia*. Phycologia 21:427–528

Moewus, F. 1940. Über Sexualität von *Botrydium granulatum*. Biologisches Zentralblatt 60:484–498.

Moore, G. T., Carter, N. 1926. Further studies on the subterranean algal flora of the Missouri Botanical Garden. Annals of the Missouri Botanical Garden 13:101–140.

Munawar, M., Munawar, I. F. 1978. Phytoplankton of Lake Superior 1973. Journal of Great Lakes Research 4:415–442.

Mundie, J. R. 1929. Cytology and life history of *Vaucheria geminata*. Botanical Gazette 87:397–410.

O'Kelly, C. J. 1989. Reconstructions from serial sections of *Heterococcus tectiformis* (Tribophyceae = Xanthophyceae) zoospores, with emphasis on the flagellar apparatus. Cryptogamic Botany 1:50–69.

Oltmanns, F. 1895. Über die Entwicklung der Sexualorgane bei *Vaucheria*. Flora 80:388–420.

Ott, D. W., Brown, R. M. 1974. Developmental cytology of the genus *Vaucheria*. II. Sporogenesis in *V. fontinalis* (L.) Christensen. British Phycological Journal 9:333–351.

Ott, D. W., Brown, R. M. 1978. Developmental cytology of the genus *Vaucheria*. IV. Spermatogenesis. British Phycological Journal 13:69–85.

Ott, D. W., Hommersand, M. H. 1974. Vaucheriae of North Carolina. I. Marine and brackish water species. Journal of Phycology 10:373–385.

Parker, B. C., Preston, R. E., Fogg, G. E. 1963. Studies of the structure and chemical composition of the cell walls of Vaucheriaceae and Saprolegniaceae. Proceedings of the Royal Society of London Series B 158:435–445.

Pascher, A. 1937–1939. Heterokonten, in: Rabenhorst, L., Ed., Kryptogamen-Flora von Deutschland, Österreich und der Schweiz, Lieferung 2, Band XI. Akad. Verlagsgesellschaft, Leipzig.

Patterson, D. J. 1989. Stramenopila: Chromophytes from a Protistan Perspective, in: Green, J. C., Leadbeater, B. S. C., Diver, W. L., Eds., The Chromophyte algae: Problems and perspectives. Clarendon, Oxford, pp. 357–379.

Pithart, D., Pechar, L., Mattsson, G. 1997. Summer blooms of raphidophyte *Gonyostomum semen* and its diurnal vertical migration in a floodplain pool. Archiv für Hydrobiologie 119:119–133.

Pizarro, H. 1995. The genus *Characiopsis* Borzi (Mischococcales, Tribophyceae). Taxonomy, biogeography and ecology. Bibliotheca Phycologica, 98.

Potter, D., Saunders, G. W., Andersen, R. A. 1997. Phylogenetic relationships of the Raphidophyceae and Xanthophyceae as inferred from nucleotide sequences of the 18S ribosomal RNA gene. American Journal of Botany 84:966–972.

Poulton, E. M. 1930. Further studies on the Heterokontae: some Heterokontae of New England, U.S.A. New Phytologist 29:1–26.

Polunin, N. 1954. The cryptogamic flora of the Artic. Botanical Review 20:361–399.

Preisig, H.R. 1989. The flagellar base ultrastructure and phylogeny of chromphytes, in: Green, J. C., Leadbeater, B. S. C., Diver, W. L., Eds., The Chromophyte algae: Problems and perspectives. Clarendon, Oxford, pp. 167–188.

Prescott, G. W. 1931. Iowa algae. University of Iowa Studies 13:1–235.

Prescott, G. W. 1944. New species and varieties of Wisconsin algae. Farlowia 1:347–385.

Prescott, G. W. 1953. Preliminary notes on the ecology of freshwater algae in the arctic slope, Alaska, with description of some new species. American Midland Naturalist 50:453–470.

Prescott, G. W. 1962. Algae of the western Great Lakes area. Cranbrook Institute of Science, Bloomfield Hills, MI.

Prescott, G. W. 1963. Ecology of Alaskan freshwater algae II. Introduction: General considerations. Transactions of the American Microscopical Society 82:83–142.

Prescott, G. W. 1978. How to know the freshwater algae. Brown, Dubuque, IA.

Prescott, G. W., Dillard, G. E. 1979. A checklist of algal species reported from Montana 1891 to 1977; in: Proceedings of the Montana Academy of Science, Vol. 38, Monograph 1.

Prescott, G.W., Vinyard, W. C. 1965. Ecology of Alaskan freshwater algae. V. Limnology and flora of Malikpuk Lake. Transactions of the American Microscopical Society 84:427–478.

Provasoli, L. 1968. Media and prospects for cultivation of marine algae, in: Watanabe, A., Hattori, A., Eds., Cultures and collections of algae. Proceedings of the U.S.–Japan conference. Japanese Society of Plant Physiologists, Hakone, pp. 63–75.

Rawson, D. S. 1956. The net plankton of Great Slave Lake. Journal of the Fisheries Research Board of Canada 13:53–127.

Rieth, A. 1980. Xanthophyceae II (Vaucheriales), in: Ettl, H., Gerloff, J., Heynig, H., Eds., Süsswasserflora von Mitteleuropa, Vol. 4. Fischer, Stuttgart/New York.

Rhodes, R. G., Stofan, P. E. 1967. Tetraspora, Chlorosaccus, and Phaeosphaera, a unique example of parallel evolution in the algae. Journal of Phycology 3:87–89.

Rhodes, R. J., Terzis, A. J.1970. Some algae of the upper Cuyahoga River system in Ohio. Ohio Journal of Science 70:295–299.

Ross, R. 1954. III. Algae: Planktonic. Botany Reviews 20:400–416.

Round, F. E. 1981. The ecology of algae. Cambridge Univ. Press, Cambridge, UK.

Santos, L. M. A. 1996. The Eustigmatophyceae: Actual knowledge and research perspectives. Nova Hedwigia 112:391–405.

Santos, L. M. A., Melkonian, M., Kreimer, G. 1996. A combined reflection confocal laser scanning, electron and fluorescence microscopy analysis of the eyespot in zoospores of *Vischeria* spp. (Eustigmatales, Eustigmatophyceae). Phycologia 35:299–307.

Schneider, C. W., Lane, C. E. 2000. Two species of *Vaucheria* new for New England, *V. lii* and *V. racemosa*. Northeastern Naturalist 7:25–32.

Schneider, C. W., MacDonald, L. A., Cahill, J. F., Heminway, S. W. 1993. The marine and brackish water species of *Vaucheria* (Tribophyceae, Chrysophyta) from Connecticut. Rhodora 95:97–112.

Schneider, C. W., Lane, C. E., Norland, A. 1999. The freshwater species of *Vaucheria* (Tribophyceae, Chrysophyta) from Connecticut. Rhodora 101:234–263.

Schnepf, E., Niemann, A., Christian, W. 1996. *Pseudostaurastrum limneticum*, a eustigmatophycean alga with astigmatic zoospores: Morphogenesis, fine structure, pigment composition and taxonomy. Archiv für Protistenkunde 146:237–249.

Schumacher, G. J., Bellis, V. J., Whitford, L. A. 1963. Additions to the fresh-water algae in North Carolina. VI. Journal of the Elisha Mitchell Scientific Society 79:22–26.

Schumacher, G. J., Kim, Y. C., Whitford, L. A., Dillard, G. E. 1966. Additions to the fresh-water algae in North Carolina. VII. Journal of the Elisha Mitchell Scientific Society 82:131–138.

Sheath, R. G., Hellebust, J. A. 1978. Comparison of algae in the euplankton, tychoplankton, and periphyton of a tundra pond. Canadian Journal of Botany 56:1472–1483.

Sheath, R. G., Müller, K. M. 1997. Distribution of stream macroalgae in four high arctic drainage basins. Arctic 50:355–364.

Sheath, R. G., Steinman, A. D. 1981. A checklist of freshwater algae of the Northwest Territories, Canada. Canadian Journal of Botany 60:1964–1997.

Sheath, R. G., Morison, M. O., Korch, J. E., Kaczmarczyk, D., Cole, K. M. 1986. Distribution of stream macroalgae in south-central Alaska. Hydrobiologia 135:259–269.

Sheath, R. G., Vis, M. L., Hambrook, J. A., Cole, K. M. 1996. Tundra stream macroalage of North America: Composition, distribution, and physiological adaptations. Hydrobiologia 336:67–82.

Silva, H., Sharp, A. J. 1944. Some algae of the southern Appalachians. Journal of the Tennessee Academy of Science 19:337–345.

Smith, G. M. 1916. A preliminary list of algae found in Wisconsin lakes. Transactions of the Wisconsin Academy of Sciences, Arts and Letters 18:531–565.

Smith, G. M. 1918. A second list of algae found in Wisconsin lakes. Transactions of the Wisconsin Academy of Sciences, Arts and Letters 19:614–654.

Smith, G. M. 1920. Phytoplankton of the inland lakes of Wisconsin. Bulletin 57, Wisconsin Geological and Natural History Survey, Madison.

Smith, G. M. 1950 The Fresh-water algae of the United States, 2nd ed. McGraw-Hill, New York.

Spencer, L. B. 1971. A study of *Vacuolaria virescens* Cienkowski. Journal of Phycology 7:274–279.

Stein, J. R. 1975. Freshwater algae of British Columbia: The lower Fraser Valley. Syesis 8:119–184.

Stein, J. R., Borden, C. A. 1978. Checklist of freshwater algae of British Columbia. Syesis 12:3–39.

Stein, J. R., Gerrath, J. F. 1969. Freshwater algae of British Columbia: The Queen Charlotte Islands. Syesis 2:213–226.

Tarapchak, S. J. 1972. Studies on the Xanthophyceae of the Red Lake Wetlands, Minnesota. Nova Hedwigia 23:1–43.

Taylor, W. R. 1928. Alpine algae of the mountains of British Columbia. Ecology 9:343–348.

Taylor, W. R. 1934. The freshwater algae of Newfoundland. Michigan Academy of Science 20:185–229.

Thompson, R. H. 1938. A preliminary survey of the fresh-water algae of eastern Kansas. University of Kansas Scientific Bulletin 25:5–83.

Tiffany, L. H. 1937. The filamentous algae of the west end of Lake Erie. American Midland Naturalist 18:911–951.

Tiffany, L. H., Britton, M. E. 1944. Freshwater Chlorophyceae and Xanthophyceae from Puerto Rico. Ohio Journal of Science 44:39–50.

Transeau, E. N. 1913. The periodicity of algae in Illinois. Transactions of the American Microscopical Society 32:31–40.

Transeau, E. N. 1917. The algae of Michigan. Ohio Journal of Science 17:217–232.

van den Hoek, C., Mann, D. G., Jahns, H. M. 1995. Algae. An introduction to phycology. Cambridge Univ. Press, Cambridge, UK.

Venkataraman, G. S. 1961. *Vaucheriaceae*. Indian Council of Agricultural Research, New Delhi.

Verb, R. G., Vis, M. L., Ott, D. W., Wallace, R. L. 1999. New records of *Vaucheria* species (Xanthophyceae) with associated *Proales werneckii* (Rotifera) from North America. Cryptogamie, Algologie 20:67–73.

Vinyard, W. C. 1958. The algae of Oklahoma (exclusive of diatoms). Ph.D. dissertation, Michigan State University.

Vinyard, W. C. 1966. Additions to the algal flora of Oklahoma. Southwestern Naturalist 11:196–204.

West, G. S. 1904. West Indian freshwater algae. Journal of Botany 42:281–295.

Wheldon, R. M. 1941. Some observations of freshwater algae of Florida. Journal of the Elisha Mitchell Scientific Society 57:262–272.

Whitford, L. A. 1979. Additions to the freshwater algae in North Carolina. IX. Journal of the Elisha Mitchell Scientific Society 95:42–47.

Whitford, L. A., Schumacher, G. J. 1969. A manual of the freshwater algae in North Carolina. North Carolina Agricultural Experiment Station.

Woodson, B. R. 1962. Research in the algae of the James River Basin. Virginia State Collge Gazette 68:39–44.

Woodson, B. R., Afazal, M. 1976. The taxonomy and ecology of algae in the Appomattox River, Chesterfield County, Virginia. Virginia Journal of Science 27:5–9.

Woodson, B. R., Holoman, V. 1964. A systematic and ecological study of algae in Chesterfield County, Virginia. Virginia Journal of Science 15:52–70.

Woodson, B. R., Wilson, W. 1973. A systematic and ecological survey of algae in two streams of Isle of Wight County, Virginia. Castanea 38:1–18.

Wujek, D. E., Wee, J. L. 1984. New, rare, and unusual algae from Montana. Northwest Science 58:213–221.

12

CHRYSOPHYCEAN ALGAE

Kenneth H. Nicholls*
Aquatic Science Section
Ontario Ministry of the Environment

Daniel E. Wujek
Department of Biology
Central Michigan University
Mt. Pleasant, Michigan 48859

I. Introduction
  A. History
  B. Characteristics
  C. Molecular Systematics
II. Diversity and Morphology
  A. Diversity
  B. Vegetative Morphology
  C. Reproduction
III. Ecology
  A. Distribution
  B. Habitat Characteristics

IV. Collection and Preparation for Identification
  A. Collection
  B. Preparation
V. Key and Descriptions of Genera
  A. Key to Genera
  B. Descriptions of Genera
VI. Guide to Literature for Species Identification
Literature Cited

I. INTRODUCTION

A. History

Anthophysa vegetans is distinguished as the first chrysophyte ever described [as *Volvox vegetans* in Müller's (1786) *Animalcula Infusoria*]. Over the next one-half century, light microscopic investigations of chrysophytes moved rapidly forward (Kristiansen, 1995a) and included descriptions of species in the genera *Dinobryon* and *Synura* (Ehrenberg, 1838), many of which are still valid today.

The Chrysophyceae or chrysomonads were first recognized as a distinct group by Stein (1878), as the family Chrysomonadaceae, but Klebs (1893a, b) was among the first to realize the group contained more than flagellates. Pascher (1914) erected the class Chrysophyceae, based primarily on a morphological series paralleling those observed within the green algae, which no longer emphasized the taxonomic significance of flagellation. Hibberd (1976a) formally typified this class of golden-brown algae. Other groups, once considered as members of the class Chrysophyceae, have since been placed into more recently erected classes [Prymnesiophyceae (=Haptophyceae) (Christensen, 1962, 1964), Synurophyceae (Andersen, 1987), Pedinellophyceae (Kristiansen, 1990), and Pelagophyceae (Andersen et al., 1993)], separated as distinct classes because of their special cellular features. Andersen and coworkers (Andersen et al., 1998a, b, c; Bailey et al., 1998) using molecular data, pigments, and ultrastructure have separated additional genera from the Chrysophyceae and have proposed yet another class, the Phaeothamniophyceae.

Kristiansen (1995a) presents a history of the development of research on and early investigators of the group of organisms traditionally referred to as the Chrysophyceae. Preisig (1995) and Moestrup (1995) discuss historical and modern interpretations of chrysophyte classification. Preisig (1995) also reviews past schemes of classification based mainly on flagellation

*Present address: S-15 Concession 1, RR #1 Sunderland, Ontario, Canada, L0C 1H0

and life forms, and proposes a new model for chrysophycean classification. It is his taxonomic treatment of the Chrysophyceae that we follow in this chapter.

Although many individuals have contributed to our knowledge of North American golden-brown algae, several individuals have proved instrumental in building a framework for subsequent researchers. In the United States, these include G. W. Prescott, G. J. Schumacher, G. M. Smith, R. H. Thompson, and L. A. Whitford. Other pioneering works include Bachmann (1921) and Nygaard (1978) for Greenland, and Ortega's (1984) compilation of Mexican algae. Perhaps the earliest report of a chrysophyte in North America is Stokes' (1886) description of *Stylobryon abbotti*.

B. Characteristics

The chrysophytes, the motile representatives of which are sometimes called chrysomonads, have vegetative or reproductive cells that are heterokont (flagella of unequal length and different function) with the two flagella inserted somewhat perpendicular to each other near the apex of the cell (Fig. 1). One, a short smooth flagellum, emerges at about a 45° angle to the longer

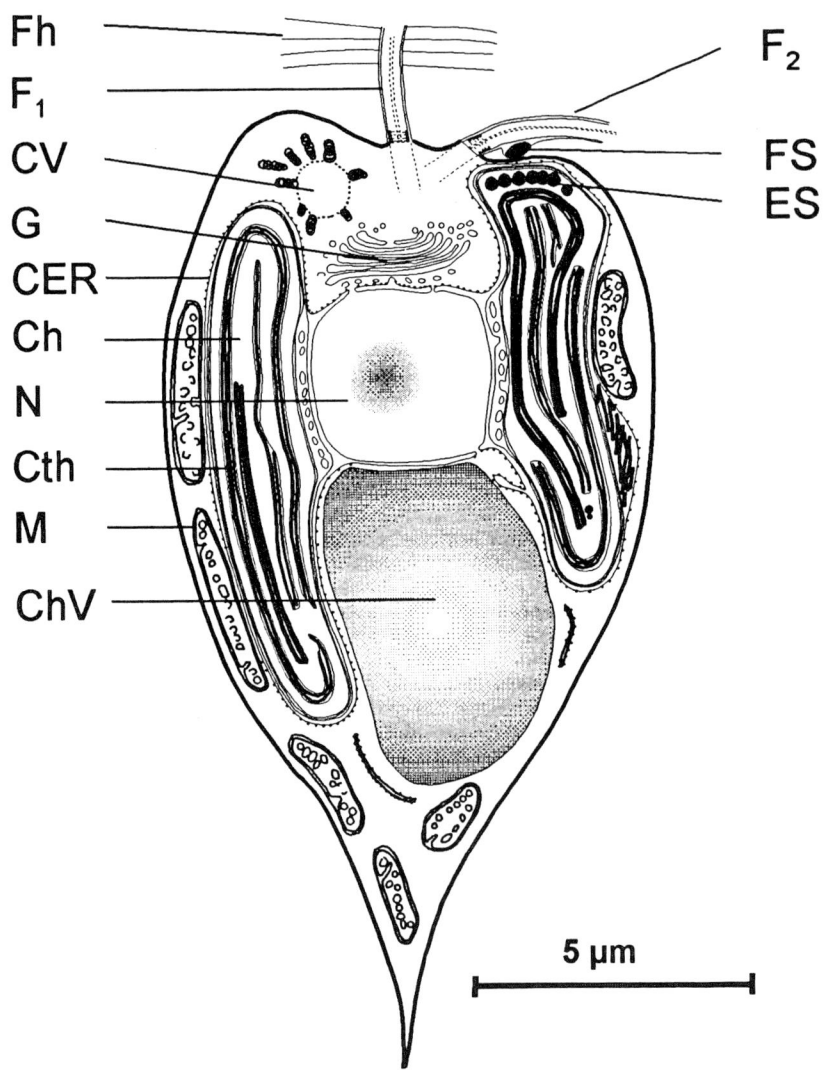

FIGURE 1 Basic organization of a typical chrysophyte cell (*Ochromonas*) based on an EM section, keyed as follows: Fh, flagellar hairs; F_1, anteriorly directed hairy flagellum; F_2, laterally directed smooth flagellum; FS, flagellar swelling; ES, eyespot; CV, contractile vacuole; G, Golgi body; CER, chloroplast endoplasmic reticulum; Ch, chloroplast; N, nucleus; Cth, chloroplast thylakoid; M, mitochondrion; ChV, chrysolaminarin vesicle. Redrawn from Gibbs (1981), *International Review of Cytology* 72:49–99, © Academic Press, Orlando, FL, with permission.

flagellum and is directed laterally or posteriorly. The other flagellum is directed anteriorly and is a long, pleuronematic flagellum (tinsel, hairy, or flimmer flagellum) bearing two rows of tripartite hairs attached through the flagellar membrane to specific outer doublets. In *Epipyxis*, Wetherbee et al. (1988) using image enhanced microscopy has shown that, during cell division, predivision cells produce two new long flagella while transforming the parental long flagellum into a new short flagellum, achieving what has been termed flagellar heterogeneity (Melkonian et al., 1987). In some genera (e.g., *Chromulina*) the short flagellum is greatly reduced and is not visible in the light microscope (Hibberd, 1976a).

Chrysophytes obtain their energy and nutrients by photosynthesis and/or heterotrophy. In the latter, they take up food either by engulfing particulate matter, such as bacteria or other protists (phagotrophy), or by absorbing complex organic molecules (osmotrophy). Up to 36% of the pigmented flagellates (predominately chrysophytes) in both oligotrophic and eutrophic waters have been shown to ingest bacteria-sized particles (Sanders et al., 1990).

The long flagellum has been shown to be used as a feeding apparatus for phagotrophic species (Bird and Kalff, 1986, 1987). Andersen and Wetherbee (1992) and Wetherbee and Andersen (1992) demonstrated the formation of a feeding cup in *Epipyxis pulchra* that engulfs food particles captured between the flagella. Prey gathered by rapid beats of the long flagellum is directed by a strong water current toward the cell while the short flagellum moves slowly. Both flagella then rotate the food item before selecting or rejecting it. The cup, consisting of microtubules, along with the enclosed food particle, is retracted back into the cell. In other species such as *Chrysosphaerella* and *Ochromonas*, particle feeding may not involve such complex morphogenesis (Andersen and Wetherbee, 1992).

C. Molecular Systematics

The technical advancements in isolating and characterizing nucleic acids have impacted all areas of biology (Wee, 1996; Wee et al., 1996). The early molecular applications to protistan biology were at the division or class level (Olsen, 1990; Coleman and Goff, 1991; Saunders et al., 1995). Presently, work "...at the lower taxonomic levels appears to be driven by economic (toxic or taste/odor-causing blooms) and environmental (biodiversity, global warming)" concerns (Wee et al., 1996) or to aid in resolving classification issues (Andersen et al., 1999).

Molecular data, in combination with ultrastructural and pigment data, prompted Andersen and coworkers to move such chrysophyte genera as *Chrysoclonium*, *Phaeoschizochlamys*, *Phaeothamnion*, *Sphaeridiothrix*, *Stichogloea*, *Tetrachrysis*, and *Tetrasporopsis* to their proposed new class, the Phaeothamniophyceae (Andersen et al., 1998b; Bailey et al., 1998). Also transferred to this new class is *Pleurochloridella*, an organism traditionally assigned to the Tribophyceae (Xanthophyceae, Chap. 11). In addition to specific ultrastructural characteristics (such as zoospore basal bodies forming an angle of ca. 145°, the presence of a multiple transitional helix, and numerous electron-dense vesicles at the periphery of the cell), the pigments fucoxanthin and heteroxanthin are found together in no other chromophyte algal class.

Wee (1996), Wee et al. (1996), Daugbjerg and Andersen (1997a, b), Medlin et al. (1997), Andersen et al. (1998a), and Caron et al. (1999) have explored the impacts of molecular analyses on the phylogeny of chrysophytes and other protistan groups, the applications of methodology, and emerging trends. One such trend is the identification of species from natural water samples. Development and application of oligonucleotide probes are currently underway for both visualizing and enumerating individual species within mixed natural assemblages (Lim et al., 1993, 1996, 1999); however, such techniques have not yet achieved a level of specificity and simplicity that permits their universal application.

II. DIVERSITY AND MORPHOLOGY

A. Diversity

When compared to the more widely investigated European flora, the North American chrysophyte flora is under-represented. Most known taxa are photosynthetic, but a few are colorless. There are various levels of cellular organization represented in the Chrysophyceae. (See Figs. 2–12.) Many genera are unicellular (e.g., *Spiniferomonas*; Figs. 3H, 8C) or colonial (e.g., *Uroglenopsis*; Figs. 4H, 8B), whereas some forms live within a protective case termed a lorica (*Dinobryon*; Figs. 3A, 5J). However, a relatively small number have a simple multicellular organization such as simple filaments, with or without mucilage (e.g., *Tetrachrysis*; Figs. 4F, 12A), or have a pseudoparenchymatous thallus (e.g., *Phaeoplaca*; Fig. 9I). Nonmotile forms can be planktonic or sessile. Unicellular forms may be flagellated, rhizopodal, palmelloid, or coccoid and in the same organism may assume more than one of these forms during its life cycle. A hollow stalk of chitinous microfibrils forms the base of a cuplike lorica for *Poterioochromonas* (Fig. 5K) and anchors this organism to the substratum. Still others possess an attenuated or filamentous cytoplasmic extension that attaches

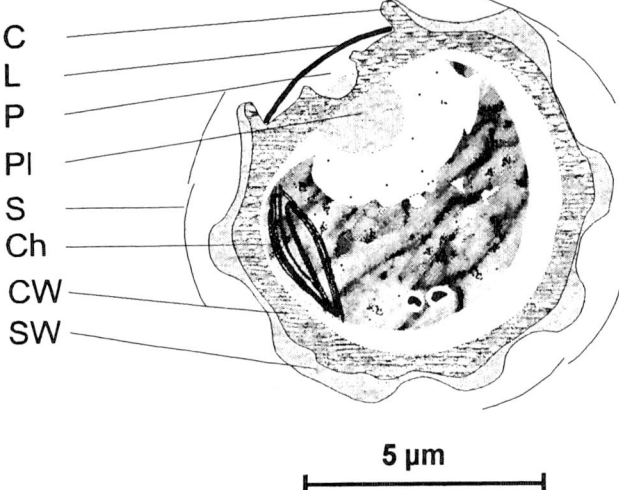

FIGURE 2 Stomatocyst of *Spiniferomonas bourrellyi* Takahashi, keyed as follows: C, collar; L, lid; P, pore; Pl, plug; S, scale; Ch, chloroplast; CW, inner cyst wall; SW, secondary layer. Redrawn from, Skogstad and Reymond (1989), Beihefte zur Nova Hedwigia 95:71–79, © Schweizerbart'sche Buchhandlung (http://www.schweizerbart.ue) with permission.

the posterior end of the cell to various substrata (e.g., *Paraphysomas vestita*; Fig. 5C) or within a lorica [*Epipyxis*; Fig. 5G(*ii*)]. A gelatinous matrix (e.g., *Chrysosphaerella*; Fig. 7F) or cytoplasmic threads (e.g., *Uroglena*; Fig. 8A) hold the cells of these colonial forms together (Wujek 1976; Preisig *et al.*, 1991).

B. Vegetative Morphology

General chrysophycean cellular organizations have been reviewed by Hibberd (1976a), Kristiansen (1986), Moestrup and Andersen (1995), Hoek *et al.* (1995), and Wujek (1996, 1999). The basic organization of a typical chrysophyte cell is depicted in Figure 1. The golden-brown color of most chrysophycean chloroplasts is due to the accessory xanthophyll pigment fucoxanthin, which masks chlorophylls *a* and *c* ($c_1 + c_2$) (Andersen and Mulkey, 1983; Jeffrey, 1989). The DNA within each chloroplast is generally arranged in a ring. Cells typically possess one or two chloroplasts that contain regularly arranged three-thylakoid lamellae, one of which, the girdle lamella, is continuous around the edge of the plastid (Hibberd, 1976a; Fig. 1). Other xanthophylls include diatoxanthin and diadinoxanthin in addition to carotenes. Food reserves include lipids and the main polysaccharide, a β-1-3-linked glucan, chrysolaminarin (=leucosin).

Within each chrysophycean cell, the plastids are surrounded by an endoplasmic reticulum (ER) that is usually continuous with the nuclear envelope (Fig. 1).

It may serve structurally to keep organelles properly positioned in the cell, particularly where the flagellar apparatus and plastid must be aligned for the photoreceptor. Lee and Kugrens (1998) have shown the adaptive advantage of an organism possessing a chloroplast ER, namely, the ability to sequester low dissolved CO_2 concentrations.

The silica-scaled genera (only four) are placed within the family Paraphysomonadaceae, one of which, *Paraphysomonas*, is colorless. The Synurophyceae (Chap. 14), all of which have silica scales, were once classified with the chrysophytes but are now considered a separate class (Andersen, 1987), although, more recently, this recognition as a distinct class has been questioned (Andersen *et al.*, 1999). Within the Chrysophyceae, silica scales (if present) are homopolar, are radially symmetrical, and often have a central spine. They differ markedly from the scales observed in the Synurophyceae—which are usually heteropolar, with the proximal and distal ends bearing different ornamentation.

Unusual scales with an interwoven fibrillar nature have been observed in the colonial chrysophyte *Lepidochrysis* (Ikävalko *et al.*, 1994), a genus not yet found in the Americas. Organic scales of taxonomic value are observed in other chrysophytes. Such scales of varying morphologies have been reported from both the surface of the cell and the flagella in *Chrysolepidomonas* (Peters and Andersen, 1993). The occurrence of these organic flagellar scales helped justify the erection of the family Chrysolepidomonadaceae (Peters and Andersen, 1993).

In the families Bicosoecaceae [not always placed within the Chrysophyceae; see Moestrup (1995)] and Dinobryaceae, and in some genera in the family Chromulinaceae, the protoplast is surrounded by an envelope or lorica. The lorica shape under light microscopy (LM) is a character of great importance and is a basis for distinction at both the genus and the species level. In most genera, the lorica has been shown to be an interwoven case of microfibrils (Karim and Round, 1967; Belcher, 1968; Kristiansen, 1969, 1972). However, in *Epipyxis*, the lorica is composed of distinct imbricate scales (Hilliard and Asmund, 1963).

Herth and Zugenmaier (1979), using shadow-cast and negatively stained preparations, have demonstrated the fibrillar structure of *Dinobryon* loricas and its arrangement in bands. They also confirmed and extended the early claim of Klebs (1893a, b) regarding the cellulosic–proteinaceous nature of the lorica and suggested that the composition of the *Dinobryon* lorica was comparable to the cellulose- and protein-containing scales of *Pleurochrysis* (Brown *et al.*, 1973; Herth *et al.*, 1975) and *Chrysochromulina* (Allen and Northcote, 1975; see also Chap. 13).

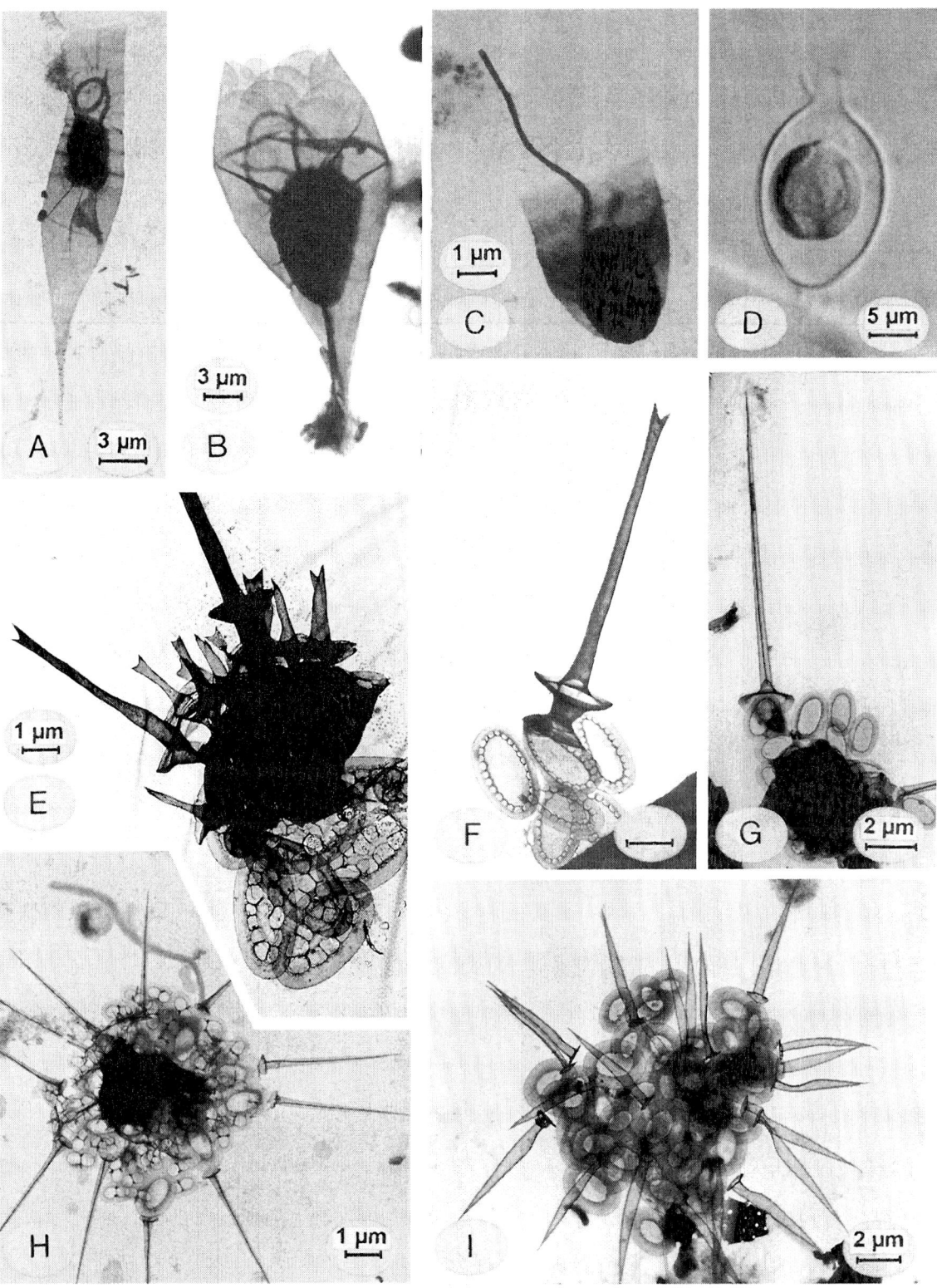

FIGURE 3 Some loricate and silica-scaled chrysophytes: (A) *Dinobryon suecicum*, TEM of dried specimen; (B) *Epipyxis tabellariae*, TEM of dried specimen; (C) *Pseudokephyrion auroreum*, TEM of dried specimen; (D) *Derepyxis ollula*, LM of living specimen; (E) single cell and associated spine- and plate-scales of *Chrysosphaerella longispina*. (dried specimen, TEM); (F) single spine-scale and group of plate-scales (dried specimen, TEM) of *Chrysosphaerella brevispina*; (G) *Spiniferomonas silverensis*, showing two types of scales (dried specimen, TEM); (H) single cell of *Spiniferomonas cornuta* with investiture of silica scales of three types (dried specimen, TEM); (I) the two types of silica scales in *Spiniferomonas abei*.

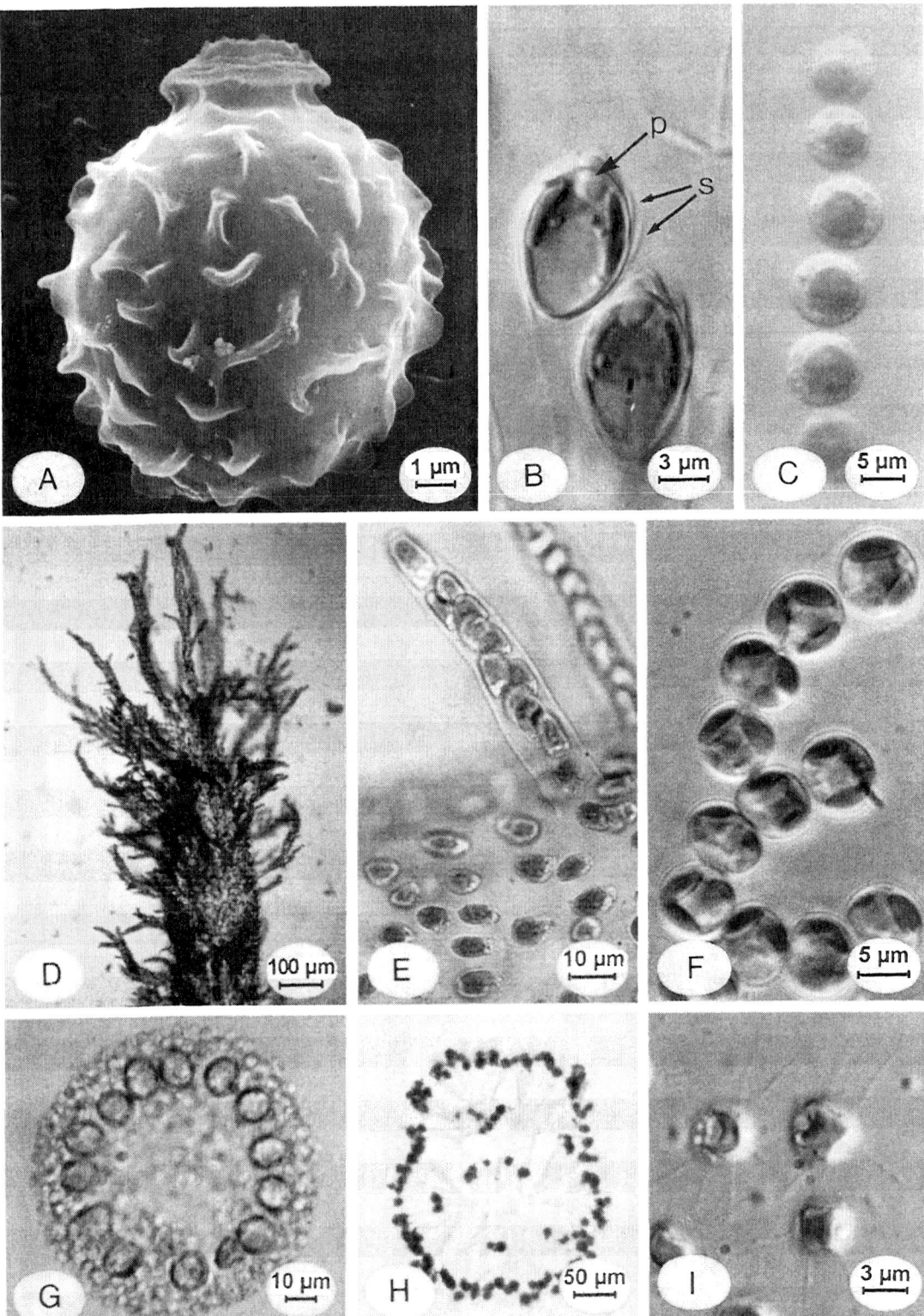

FIGURE 4 Stomatocysts and vegetative stages of several chrysophytes: (A) SEM of a stomatocyst of *Ochromonas sphaerocystis*; (B) stomatocysts of living specimens of *Epipyxis ramosa*, showing the plug (p) stopping the pore and two pectin scales (s) from the lorica; (C) *Chrysidiastrum catenatum*, colony of six cells; (D) apical portion of a colony of *Hydrurus foetidus*; (E) Organization of cells in the thallus of *Hydrurus foetidus*; (F) colony shape (cells in groups of four) in *Tetrachrysis dendroides*; (G) single colony of *Chrysostephanosphaera globulifera*; (H) Organization of the gelatinous connectives linking cells in a colony of *Uroglenopsis* (stained with methylene blue); (I) cells of *Chrysamoeba mikrokonta* showing finely beaded filopodia.

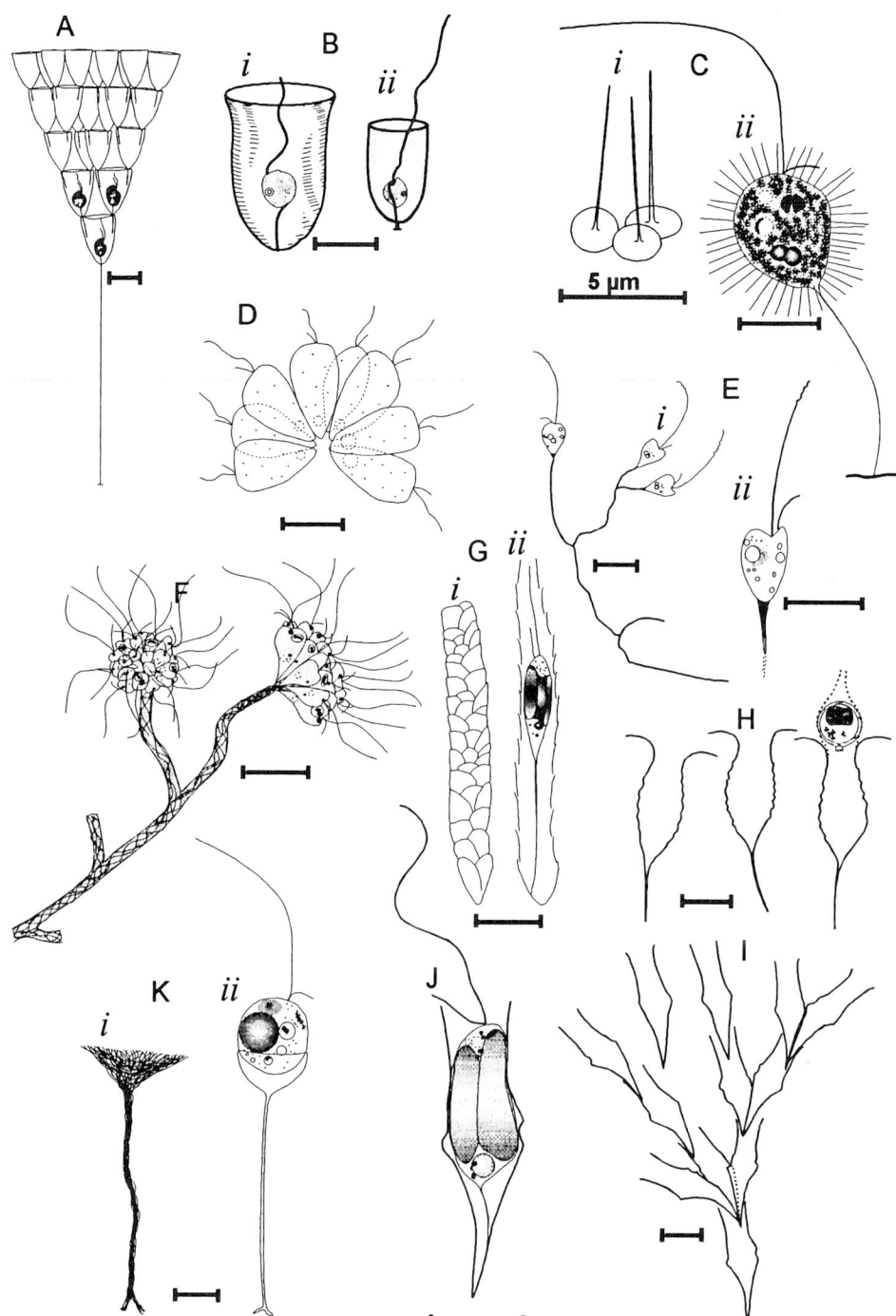

FIGURE 5 Some loricate and/or colonial chrysophytes: (A) *Stylobryon abbotti*; redrawn from Stokes (1885); (B) *Bicosoeca* spp. [(*i*) *B. kenaiensis* Hilliard; (*ii*) *B. borealis*]; (C) *Paraphysomonas vestita* [(*i*) scales; (*ii*) single cell]; (D) *Spumella sociabilis*; (E) *Dendromonas cryptostylis*; (F) *Anthophysa vegetans*; (G) *Epipyxis ramosa*; (H) *Dinobryon dillonii*; (I) and (J) *Dinobryon divergens* [(I) colony; (J) single cell]; (K) *Poterioochromonas malhamensis* [(*i*), fibrillar structure of lorica; (*ii*) single cell]; [Reproduced (redrawn) with permission from B, Hilliard (1971a), Archiv für Protistenkunde 113:98–122, D, Stein (1975), Syesis 8:119–184; E, Kouwets (1980), Cryptogamie Algologie 4:293–309, © Springer-Verlag; F, Whitford and Schumacher (1969), "A Manual of the Fresh-water Algae in North Carolina," Technical Bulletin 188, NC Agricultural Experimental Station; G, Hilliard and Asmund (1963), Hydrobiologia 22:331–397, with kind permission from Kluwer Academic Publishers; H, Nicholls (2000), Phycologia 39:134–138; I, J, Ahlstrom (1937), Transactions of the American Microscopical Society 56:139–159, Figs. 1–20, Pl. 1, and Figs. 1–18, Pl. 2; K, Peterfi (1969), Nova Hedwigia 16:94–103, © Borntraeger, Stuttgart.] All scale bars = 10 μm unless indicated otherwise.

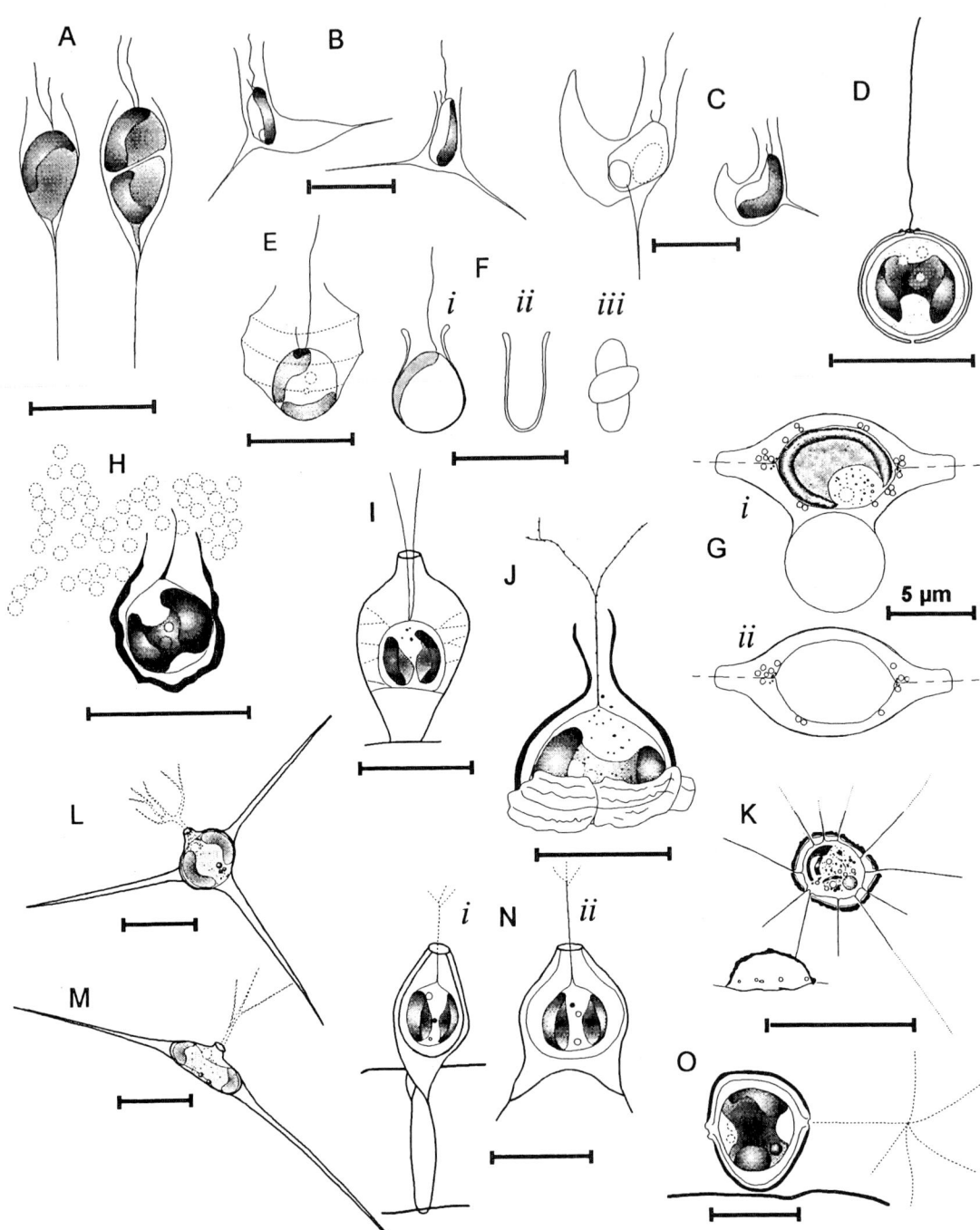

FIGURE 6 Several species of loricate rhizopodial and flagellated chrysophytes: (A) *Stylochrysalis aurea*; (B) and (C) *Chrysolykos* spp. [(B) *C. skujae*; (C) *C. planktonicus*]; (D) *Chrysococcus minutus*; (E) *Pseudokephyrion alaskanum*; (F) *Kephyrion obliquum*; [(*i*) plan view; (*ii*) lateral view; (*iii*) apical view]; (G) *Chrysoamphipyxis canadensis*; [(*i*) lateral view; (*ii*) dorsal view]; (H) *Lepochromulina bursa*; (I) *Derepyxis dispar*; (J) *Lagynion macrotrachelum* var. *oedotrachelum*; (K) *Stephanoporos sphagnicola*; (L) and (M) *Bitrichia* spp. [(L) *B. ollula*; (M) *B. chodati*; original]; (N) *Chrysopyxis stenostoma*; [(*i*) lateral view; (*ii*) plan view]; (O) *Chrysoamphitrema nygaardii*. [Reproduced (redrawn) with permission from A. Bachman (1908), Archiv für Hydrobiologie 3:1–90; B, C, Nauwerck (1979), Botaniska Notiser 132:161–183; D, Meyer (1971), Proceedings of the Arkansas Academy of Sciences 25:31–37, E, F, Hillard (1966), Hydrobiologia 28:553–576, with kind permission from Kluwer Academic Publishers; G, Nicholls (1987), Journal of Phycology 23:499–501; H, J, O, Ellis-Adam (1983), Acta Botanica Neerlandica 32:1–23; I, N, Whitford and Schumacher (1969), "A Manual of the Fresh-water Algae in North Carolina," Technical Bulletin 188, NC Agricultural Experimental Station; K, Pascher (1940a), Archiv für Prostistenkunde 93:331–349, L, Nicholls (1981b), Phycologia 20:131–137.] All scale bars = 10 µm unless indicated otherwise.

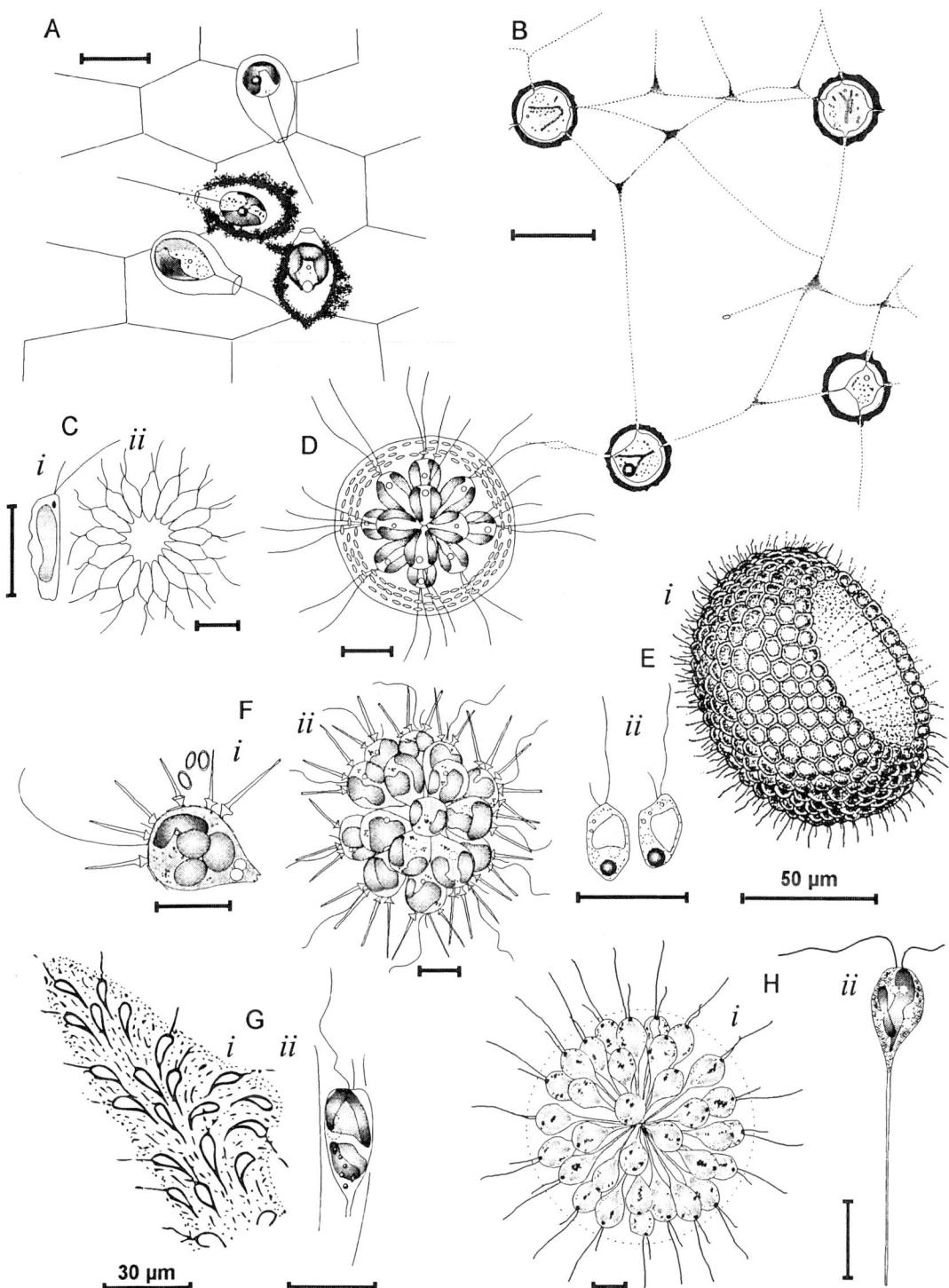

FIGURE 7 Two sessile and six motile chrysophyte species: (A) *Kybotion eremita*; (B) *Heliapsis mutabilis*; (C) *Cyclonexis annularis*; (D) *Syncrypta volvox*; (E) *Uroglenopsis turfosa* (*Eusphaerella turfosa*); [(*i*) colony; (*ii*) single cells]; (F) *Chrysosphaerella brevispina* [(*i*) single cell showing elliptical plate-scales and spine scales; (*ii*) colony]; (G) *Chrysoxys maior* [(*i*) single cell in sheath; (*ii*) colony]; (H) *Synuropsis gracilis* [(*i*) colony; (*ii*) single cell]. [Reproduced (redrawn) with permission from: A, Pascher (1940a), Archiv für Prostistenkunde 93:331–349; B, Pascher (1940b), Archiv für Prostistenkunde 94:295–310; C, D, Whitford and Schumacher (1969), "A Manual of the Fresh-water Algae in North Carolina," Technical Bulletin 188, NC Agricultural Experimental Station; E, Skuja (1948), Symbol. Bot. Upsal. 9:1–399; F, H, Korshikov (1942), Archiv für Prostistenkunde 95:22–44; G, Ellis-Adam (1983), Acta Botanica Neerlandica 32:1–23.] All scale bars = 10 μm unless indicated otherwise.

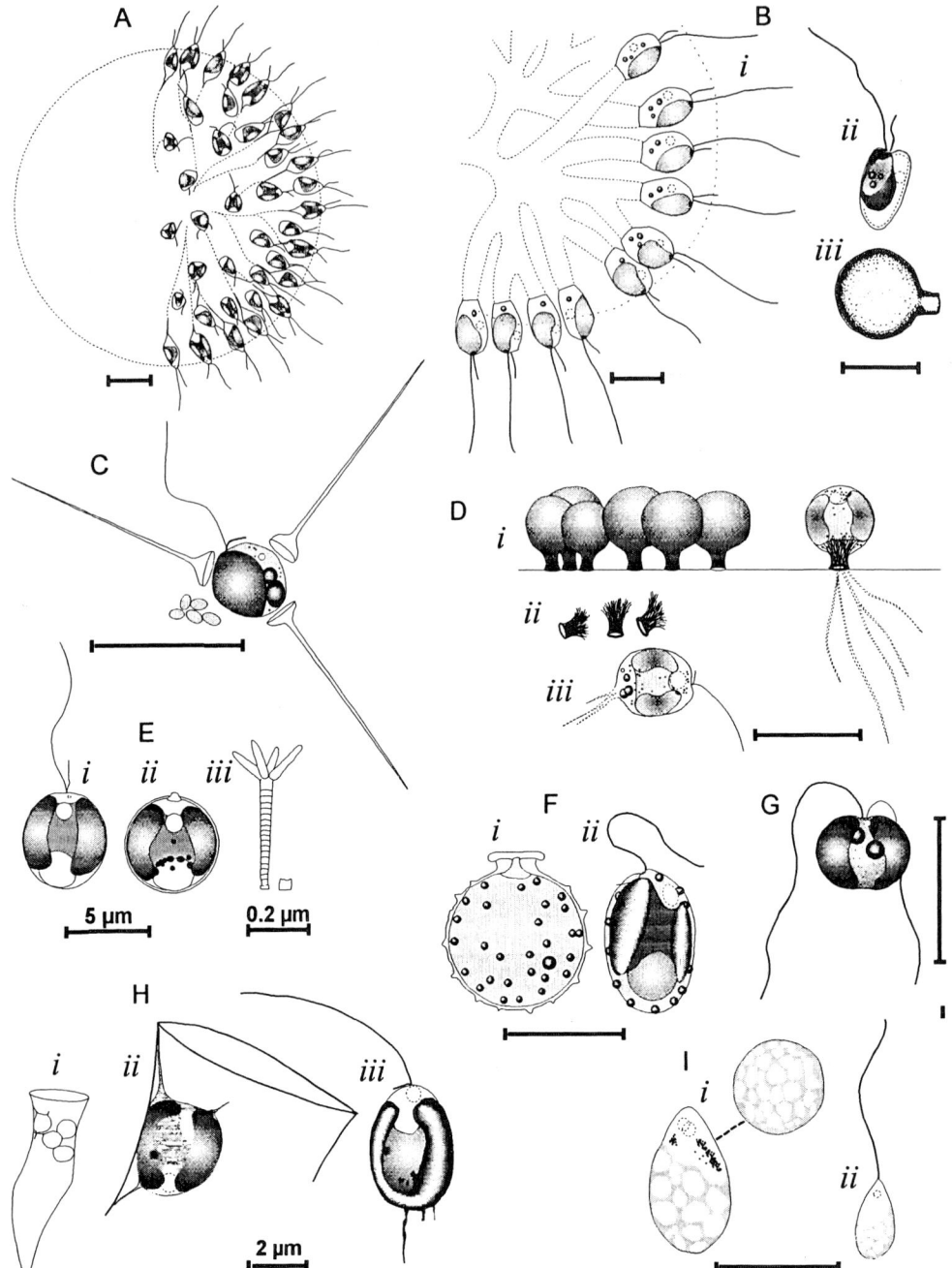

FIGURE 8 Colonial and solitary chrysomonads, with and without scales: (A) *Uroglena volvox*; (B) *Uroglenopsis botrys*. [(*i*) portion of colony showing mucilaginous intercellular connectives; (*ii*) single cell; (*iii*) stomatocyst]; (C) *Spiniferomonas bourrellyi*; (D) *Chromophyton rosanofii* [(*i*) cells at the air-water interface; (*ii*) three disassociated "pseudocyst" collars; (*iii*) flagellated amoeboid cell]; (E) *Chrysolepidomonas dendrolepidota* [(*i*) flagellated cell; (*ii*) stomatocyst; (*iii*) organic scales]; (F) *Ochromonas sphaerocystis* [(*i*) stomatocyst; (*ii*) single cell]; (G) *Erkenia subaequiciliata*; (H) *Rhizoochromonas endoloricata* [(*i*) amoeboid cells in *Dinobyron* lorica; (*ii*) flagellated cell]; (I) *Chrysapsis* spp. [(*i*) *C. fenestrata*; (*ii*) *C. agilis*.]; [Reproduced (redrawn) with permission from A, Whitford and Schumacher (1969), "A Manual of the Fresh-water Algae in North Carolina," Technical Bulletin 188, NC Agricultural Experimental Station; B, Matvienko (1954), Presn. Vodor. SSR, 3; C, Skogstad and Reymond (1989), Beihefte zur Nova Hedwigia 95:71–79, Fig. 6, © Schweizerbart'sche Buchhandlung (http://www.schweizerbart.de); D, Couté (1983), Protistologica 19:393–416; © Fischer Verlag, Jena; E, Peters and Andersen (1993), Journal of Phycology 29:469–475; F, Andersen (1982), Phycologia 21:390–398; G, Pavoni (1963), Schweizerische Zeitschrift für Hydrologie 25:219–341, © Birkhäuser, Basel; H, Nicholls (1990), Journal of Phycology 26:558–563; I, Pascher (1910), Monogr. Abhandl. Intern. Rev. Hydrobiol. 1:7–66.] All scale bars = 10 µm unless indicated otherwise.

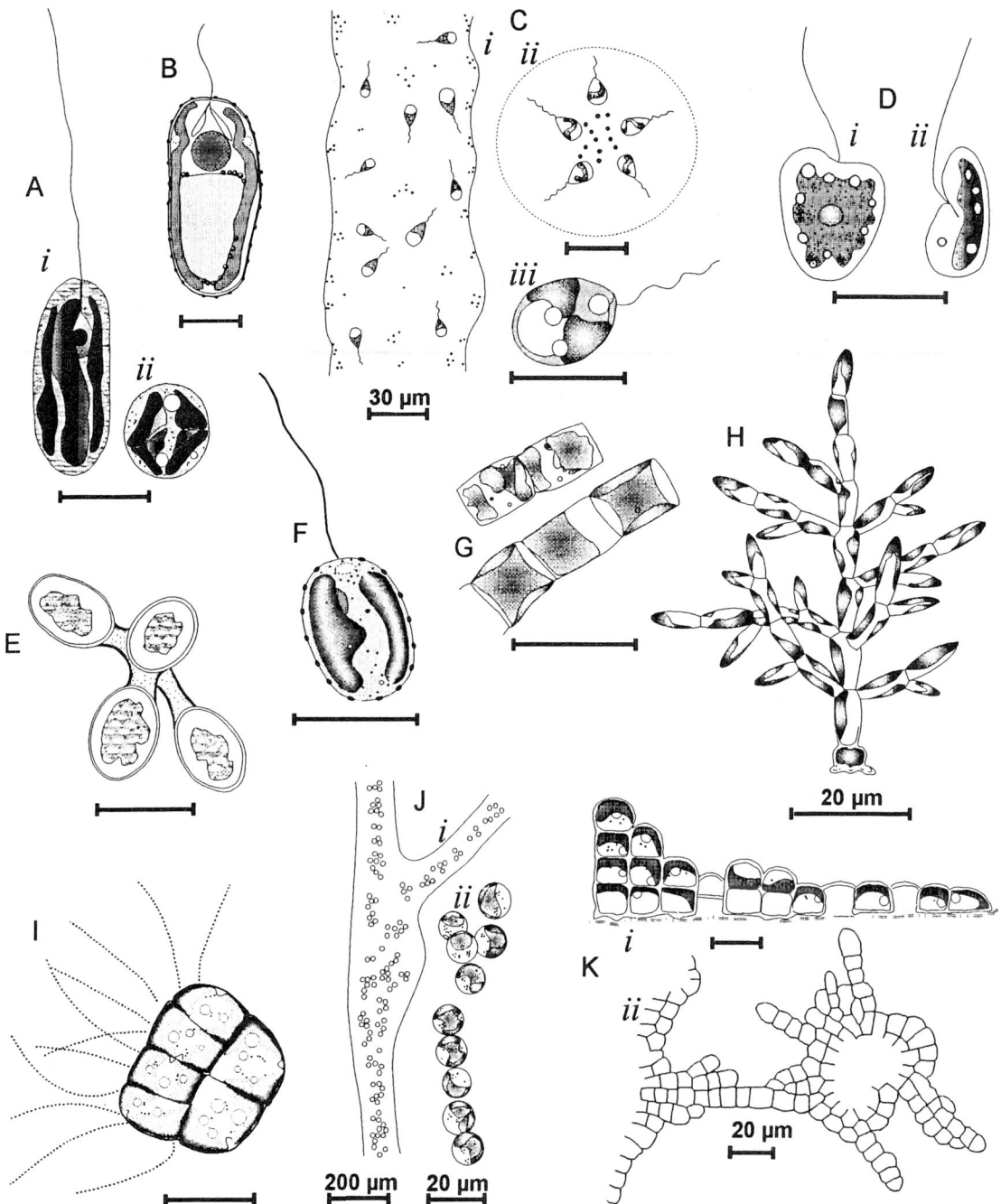

FIGURE 9 (A) *Amphichrysis compressa* [(*i*) lateral view of cell; (*ii*) polar view of cell]; (B) *Microglena butcheri*; (C) *Saccochrysis piriformis* [(*i*) cells in tubular mucilaginous colony; (*ii*) cross-section of colony; (*iii*) single cell]; (D) *Monochrysis vesiculifera* [(*i*) dorsal view; (*ii*) lateral view]; (E) *Stichogloea doederleinii*; (F) *Chromulina stellata*; (G) *Stichochrysis immobilis*; (H) *Phaeothamnion confervicola*; (I) *Phaeoplaca thallosa*; (J) *Phaeosphaera gelatinosa* [(*i*) colony; (*ii*) cells]; (K) *Phaeodermatium rivulare* [(*i*) cross-section of thallus; (*ii*) dorsal view of thallus]. [Reproduced (redrawn) with permission from: A, Korshikov (1929), Archiv für Prostistenkunde 67:254–290; B, Belcher (1966), Hydrobiologia 27:65–69, Fig. 1, with kind permission from Kluwer Academic Publishers; C, Andersen (1986), in J. Kristiansen and R.A. Andersen, Eds., *Chrysophytes: Aspects and Problems*, Chap. 8, p. 109, © Cambridge University Press; D, [as "*Pheaster*" (*Phaeaster*) *vesiculosa*]; F, Whitford and Schumacher (1969), "A Manual of the Fresh-water Algae in North Carolina," Technical Bulletin 188, NC Agricultural Experimental Station; E, [redrawn from Schmidle (1902), as *Oodesmus oederleinii*]; Hedwigia 42:150–163; G, Prinsheim (1955), Arhiv für Mikrobiologie 21:401–410; H, J, K, Pascher (1925), Archiv für Protistenkunde 52:489–464; I, Dop (1978).] All scale bars = 10 µm unless indicated otherwise.

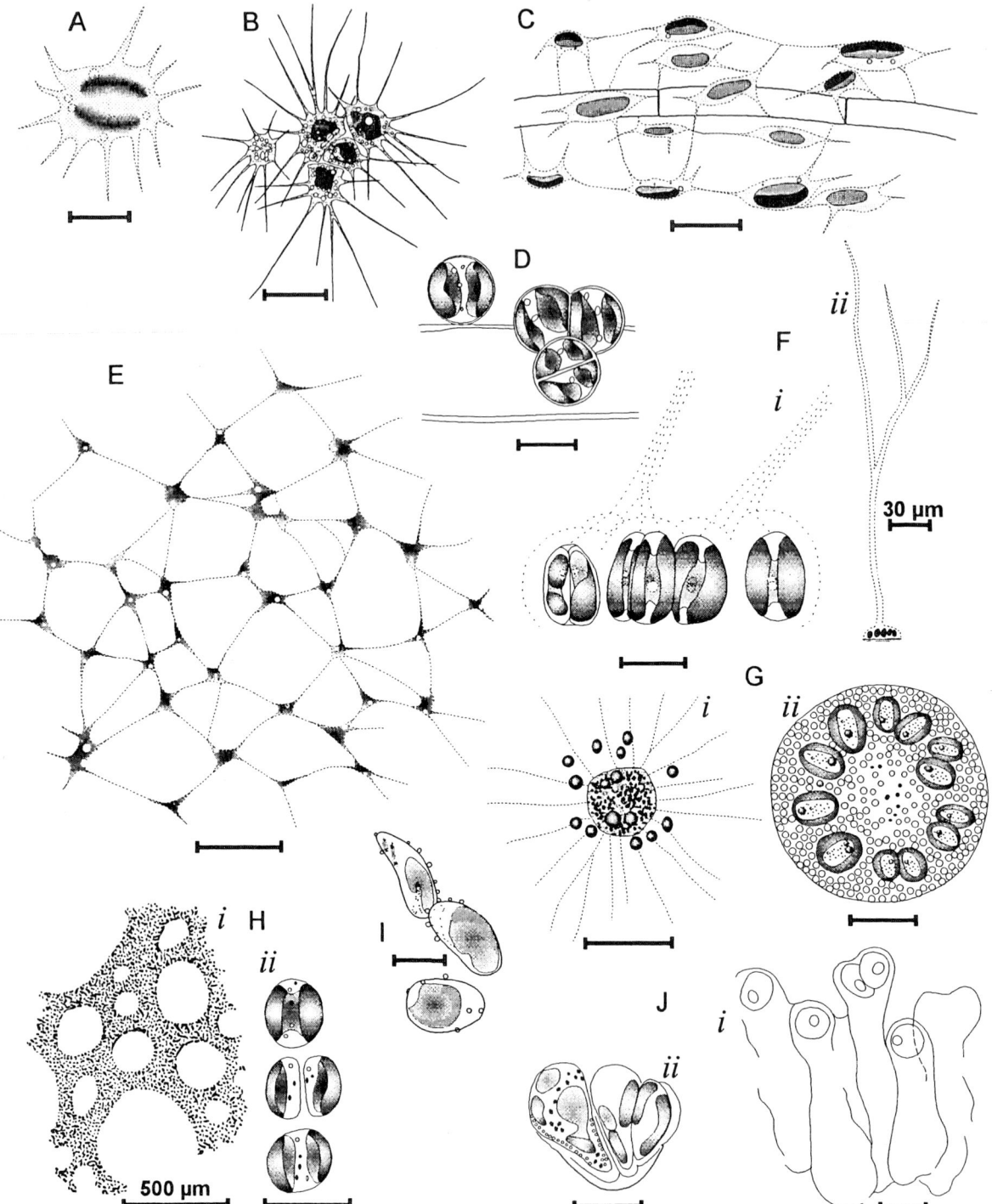

FIGURE 10 Some species of colonial and solitary chrysophytes: (A) *Chrysamoeba radians*; (B) *Rhizochrysis scherffelii*; (C) *Chrysidiastrum epiphyticum*; (D) *Chrysosphaera nitans*; (E) *Chrysarachnion insidians*; (F) *Chrysochaete britannica* [(i) colony of five cells in mucilaginous sheath; (ii) branching sheath]; (G) *Chrysostephanosphaera globulosa* [(i) single cell; (ii) older colony]; (H) *Dermatochrysis reticulata* [(i) perforated colonial sheet; (ii) single cells]; (I) *Selenophaea granulosa*; (J) *Tetrasporopsis fuscescens* [(i) peripheral location of cells in mucilaginous colony; (ii) cell division]. [Reproduced (redrawn) with permission from: A, D, H (as *Tetrasporopsis perforata*), Whitford and Schumacher (1969), "A Manual of the Fresh-water Algae in North Carolina," Technical Bulletin 188, NC Agricultural Experimental Station; B, E, G, Pascher (1917), Archiv für Protistenkunde 37:15–30; I. Chodat (1922), Bulletin de la Société Botanique de Genève 13:66–114; J, Tschermak-Woess (1980), Plant Systematics and Evolution 133:121–133, © Springer-Verlag, Vienna and Tschermak-Woess and Kusel-Fetzmann (1982), Archiv für Protistenkunde 142:157–165.] All scale bars = 10 μm unless indicated otherwise.

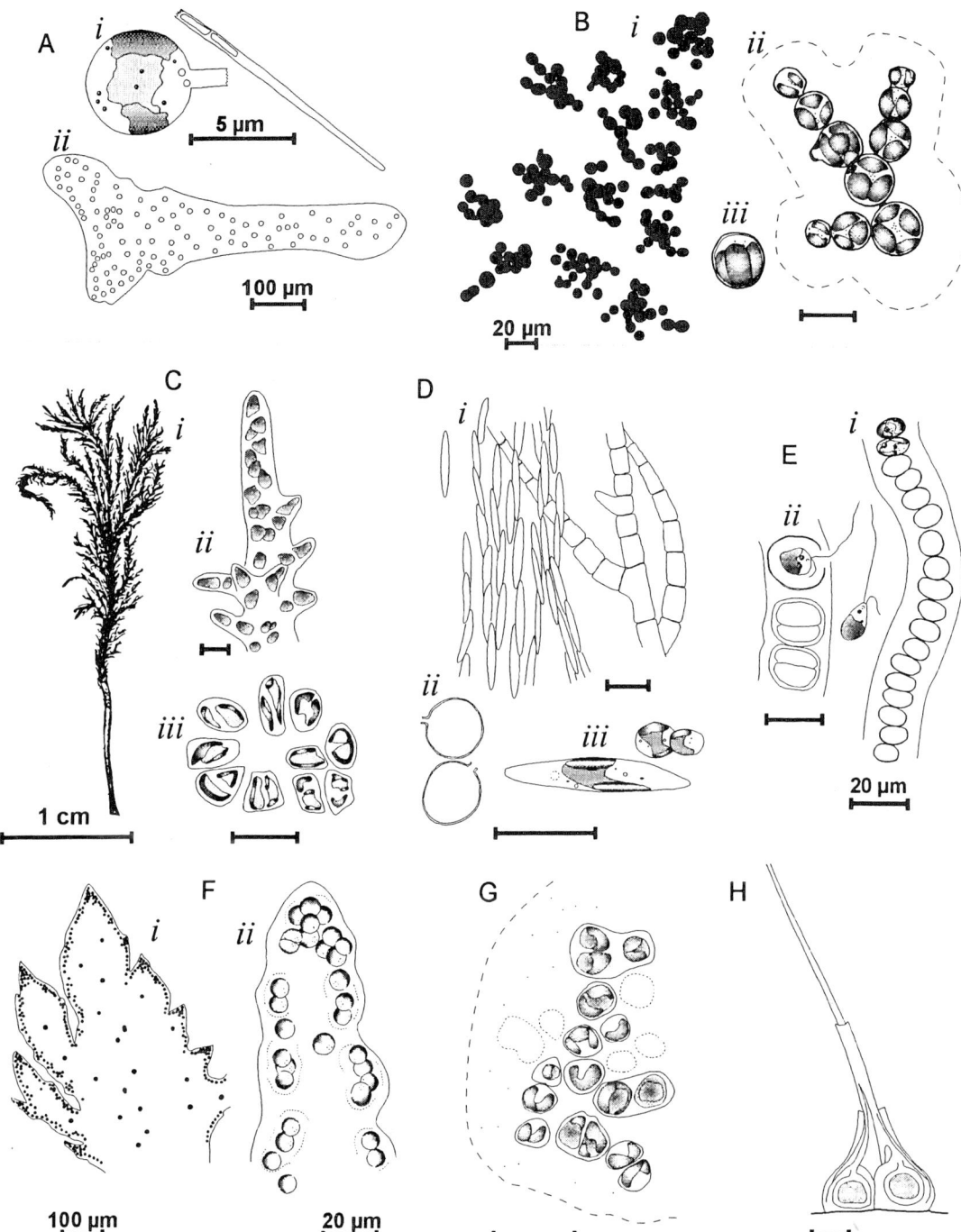

FIGURE 11 Eight species of colonial chrysophytes: (A) *Bourrellia skuja* [(*i*) single cell; (*ii*) colony]; (B) *Chrysocapsopsis rupicola* [(*i*) arrangement of cells in the thallus; (*ii*) and (*iii*) single cells, showing range in size and budding]; (C) *Hydrurus foetidus* [(*i*) macroscopic thallus; (*ii*) and (*iii*) arrangement of cells in mucilaginous thallus]; (D) *Eirmodesmus phaeotilus* [(*i*) arrangement of colonial cells epiphytic on *Draparnaldia*; (*ii*) stomatocysts; (*iii*) single cells]; (E) *Sphaeridiothrix compressa* [(*i*) arrangement of cells in thallus; (*ii*) zoospore release]; (F) *Celloniella palensis* [(*i*) and (*ii*) placement of cells around the periphery of lobed mucilaginous sheath]; (G) *Phaeogloea mucosa*; (H) *Naegeliella flagellifera*. [Reproduced (redrawn) with permission from: A, Dillard (1970), Journal of the Elisha Mitchell Science Society 86:128–130; B, Thompson and Wujek (1998a), Phycological Research 46:165–168; C, H, Pascher (1925), Archiv für Protistenkunde 52:489–464; D, Whitford (1970), Phycologia 9:195–197; E, Andrews (1970), Journal of Phycology 6:133–136; F, Pascher (1929), Archiv für Protistenkunde 68:637–688; G, Stein (1975), Syesis 8:119–184.] All scale bars = 10 µm unless indicated otherwise.

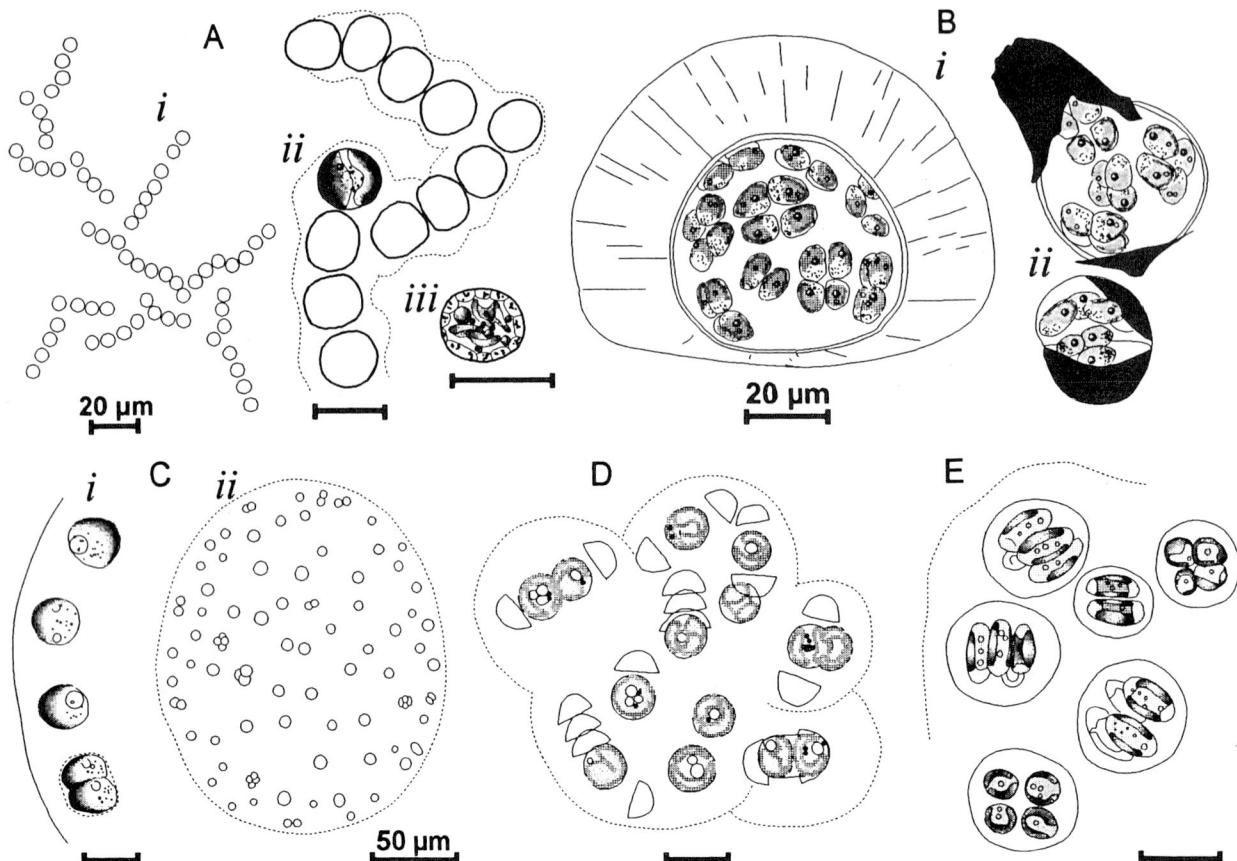

FIGURE 12 Five species of mucilaginous colonial chrysophytes: (A) *Tetrachrysis dendroides* [(*i*) and (*ii*) arrangement of cells in thallus; (*iii*) single cell]; (B) *Chalkopyxis tetrasporoides* [(*i*) palmelloid thallus; (*ii*) germination of "pseudocysts"]; (C) *Chrysocapsa planktonica* [(*i*) cells at edge of colony; (*ii*) colony]; (D) *Phaeoschizochlamys mucosa*; (E) *Chrysosaccus incompletus*. [Reproduced (redrawn) with permission from: A, Dop (1980), Acta Botanica Neerlandica 29:65–86; B, Pascher (1931), Archiv für Protistenkunde 73:73–103; C, E, Pascher (1925), Archiv für Protistenkunde 52:489–464; D, Nicholls (1984a), Phycologia 23:213–221.] All scale bars = 10 µm unless indicated otherwise.

In some genera, for example, *Chrysococcus* (Fig. 6D), the lorica may be colored with deposits of manganese or iron (Belcher, 1968, 1969). In *Chrysocrinus*, these deposits can be caused by oxidation resulting from photosynthesis (Schoonoord and Ellis-Adams, 1984). Based on analytical electron microscopy, Dunlap *et al.* (1987) reported the first conclusive evidence of both iron and manganese mineralization in loricae of chrysophycean organisms (see also Preisig, 1986).

Chromulina has only one flagellum visible in LM. This distinguishes it from *Ochromonas*, which has two. Some species of both genera have the ability to alter their shape (i.e., they are "metabolic" or "plastic"). Some species of *Ochromonas* produce clublike or threadlike extensions (pseuopodia or filopodia). In *Chrysamoeba* (Fig. 4I), cell shape may be highly variable owing to active production of filopodia. *Rhizoochromonas* (Fig. 18H) has an amoeboid stage after invasion of the lorica of a *Dinobryon* species.

C. Reproduction

A unifying characteristic of the Chrysophyceae is their ability to produce "resting stage" cells termed stomatocysts, statospores, or siliceous-walled cysts (Figs. 2, 4A, B), which are formed endogenously. For many taxa the morphology of their stomatocysts is species specific. The cyst wall of the bicosoecids is believed to be organic and hence provides one argument for possible removal of this group from the Chrysophyceae. Cienkowski (1870) was the first to describe golden-brown algae that formed cysts. Stomatocyst shapes are highly variable, but most are ovoid or spherical. The wall may be ornamented or smooth. A pore from which the germling alga (vegetative cell or zoospore) will emerge can be plugged with a nonsilicified stopper (Figs. 2, 4B). The pore may possess a collar (Figs. 2, 4A). In *Dinobryon cylindricum*, Sandgren (1983) demonstrated that the spines on the cyst walls, and/or the collar lengths, varied when clones of the same population were exposed to different temperatures.

There are many descriptions of chrysophyte stomatocysts based on observations from natural or cultured material. However, most stomatocysts described to date have not been associated with a known species; a majority have been classified only as to morphotype (Duff et al., 1995). Because of this, they have been of limited use in taxonomy unless based on or clearly referred to living material. Only in some of the nonsilicified genera such as *Uroglena* and *Ochromonas* is cyst morphology necessary for species identification (Bourrelly, 1963).

Hibberd (1977) was the first to study ultrastructurally the endogenous formation of stomatocysts, in *Ochromonas tuberculata*. This study was followed by others on the stomatocysts of *Uroglena* (Esser and Valkenburg, 1977), *Dinobryon* (Sandgren, 1980a, b), and *Spiniferomonas* (Skogstad and Reymond, 1989). These studies have shown that the cyst is produced within a cytoplasmic silica deposition vesicle. The extracystic cytoplasm contains plastid material, mitochondria, a contractile vacuole, and smaller vesicles. Stomatocysts may develop from vegetative cells or as a result of sexual reproduction (Fott, 1959; Sandgren and Flanagin, 1986).

Asexual reproduction is by cleavage, by division of a cell into several daughter cells that are liberated as zoospores, or by autospores. Fragmentation occurs in the colonial forms (e.g., *Uroglenopsis*). Where sexual reproduction has been observed, it takes place by fusion of isogametes that do not markedly differ from vegetative cells, especially in the loricate genera, but anisogamous sexual reproduction has been reported in *Dinobryon cylindricum* (Sandgren, 1981), which is also capable of autogamous sexual reproduction (Gayral et al., 1972). Meiosis is thought to be zygotic.

III. ECOLOGY

A. Distribution

Chrysophyceae have been collected on every continent except Antarctica. They are almost entirely freshwater inhabitants, although a few species are found in brackish or marine waters, and some of these apparently grow equally well in seawater and freshwater (e.g., *Paraphysomonas butcheri*, *P. imperforata*). Chrysophytes were long thought to be found chiefly in the north temperate zone, but studies on tropical lakes and ponds have reported a considerable chrysophycean flora (Cronberg, 1989, 1996; Saha and Wujek, 1990; Wujek and Saha, 1995, 1996; Hansen, 1996).

In general, chrysophytes are associated with softwater lakes and ponds that are low or moderate in productivity, are low in alkalinity and conductivity (specific conductance < 50 µS cm^{-1}), and have a pH in a range of about 6–7 (Sandgren, 1988). Chrysophytes are also common in dilute, brown (high humic acids) waters. Their ability to compete for phosphorus when it is limiting in these lakes allows them to achieve dominance over other algae under these conditions. Countless numbers of such lakes exist throughout the northeastern United States and central, eastern, and northern parts of Canada as well as in parts of Europe and Scandinavia. It is instructive to compare phytoplankton compositional data from North American lakes with data from elsewhere, especially northern Europe and Scandinavia [see Morris (1980) and Round (1981) for general reviews]. Typical of these lakes are those of the Experimental Lakes Area (ELA) in northwestern Ontario, which have ice cover from December to May and support a phytoplankton composition very similar to that of alpine lakes (Schindler et al., 1973). The ELA lakes are dilute, softwater lakes with very low nutrient levels and are dominated by Chrysophyceae all year round: species of *Chromulina*, *Chrysococcus*, *Kephyrion*, and *Pseudokephyrion* during the winter; species of *Chrysolykos*, *Dinobryon*, *Pseudokephyrion*, and *Kephyrion* during ice-free periods. In a group of 35 Ontario lakes with ice-free total phosphorus (TP) concentrations ranging from less than 15 µg L^{-1} to about 80 µg L^{-1}, Nicholls et al. (1977) showed that chrysophytes constituted greater than 30% of total phytoplankton cell volume at the low TP concentrations and less than 2% in the high TP lakes.

More recent surveys of chrysophytes from the northern and southern hemispheres have contributed greatly to our understanding of global distribution of certain chrysophytes, especially the silica-scaled Paraphysomonadaceae that is now known to be so well represented in Europe, Scandinavia, and at north temperate latitudes in North America by the genera *Paraphysomonas* and *Spiniferomonas*. Nygaard (1978) was the first to apply electron microscopy (EM) to phytoplankton samples from Greenland, but his investigation revealed only one *Paraphysomonas* species (*P. vestita*) and no *Spiniferomonas* species. Other poorly represented chrysophyte genera included *Chrysosphaerella* (neither of the common species *C. longispina* or *C. brevispina* was observed), and only one species each of *Kephyrion*, *Chrysolykos*, and *Epipyxis* was observed. Significantly, he did find several species of *Dinobryon*, including six previously undescribed species, two new forms, and one new variety, many of which still have not been found elsewhere in the world. Subsequent surveys of Greenland chrysophytes have revealed the presence of the most common silica-scaled forms [*C. longispina*, *C. brevispina*, *P. vestita*, *P. imperforata*, *S. trioralis*, *S. silverensis*, and *S. abei* (Jacobsen, 1985; Kristiansen, 1992; Wilken *et al.*, 1995)]. Other far northern surveys include those of McKenzie and Kling (1989) and Hällfors and Hällfors (1988). These papers demonstrate that studies undertaken with electron microscopy suddenly reveal the presence of many chrysophyte taxa hitherto not known for that region. For example, Sheath and Munawar (1974) found that the Chrysophyceae comprised only 5% of the total number of phytoplankton species in Hannah Lake at latitude 65°17'N, but their investigative methods precluded detection of the small silica-scaled paraphysomonadaceans now known to exist in this region (McKenzie and Kling, 1989). As more investigations are completed using these modern methods, it is becoming apparent that many chrysophyte taxa that once represented unique reports for particular geographical areas are in fact widely distributed in many parts of the world. More than half of the 73 silica-scaled chrysophytes found during the Finnish survey of 141 water bodies by Hällfors and Hällfors (1988) were first records for Finland; five of these were new for Europe. A decade later, most of these taxa were known from several European and North American locations. Similarly, Vørs *et al.* (1990) found 29 taxa of *Paraphysomonas* in Danish lakes and ponds, thus contributing to the growing realization that many of these species occur worldwide. These authors suggested that the apparent absence of a species from a particular geographic area implies more an inadequate sampling regime rather than any reasons relating to biogeography. Many, if not most, of the 49 known species of *Paraphysomonas* apparently have a cosmopolitan distribution and can be found wherever local environmental conditions are suitable. The ability to form stomatocysts that allow cells to survive drought and other adverse environmental conditions, their heterotrophic mode of nutrition, their small size and potential for rapid growth, and their passive transport in winds and in the plumage of waterfowl are all possible mechanisms leading to global distribution.

Still, there are some patterns emerging in global distribution. Kristiansen (1992) suggests that further studies at more northerly (sub-Arctic) locations are likely to reveal a reduction in the number of taxa with increasing latitude. Vigna and Kristiansen (1996) surveyed the chrysophytes of Tierra del Fuego (Argentina). Many taxa found at this southerly location (latitude 54°S) have global distribution; others fall into a temperate (both north and south) latitudinal zone that Kristiansen and Vigna (1996) suggest relates to the similar climates of both zones. They also conclude that a unique Antarctic chrysophyte flora does not exist.

There are also inconsistencies in the known distributions of several species. Why were no chrysophytes found in a survey of 13 ponds in Mexico (Kristiansen and Tong, 1995)? Why are some *Paraphysomonas* species such as *P. vestita* and *P. foraminifera* so abundant and distributed worldwide, whereas others such as *P. sediculosa*, *P. campanulata*, *P. sideriophora*, and *P. sigillifera* are still known only from their original localities despite many years of EM analysis of samples from all over the world? Are members of the latter group so recently evolved that they have not had time to disperse? Do they have very specific environmental requirements and tolerances that have slowed range expansion? Similar questions can be asked about the scaleless chrysophytes. Nygaard's (1978) Greenland *Dinobryon* species and Hilliard and Asmund's (1963) Alaskan *Epipyxis* species are two examples of large blocks of species not reported since their original discoveries. The intervening two to four decades should have been enough time to discover these taxa in other parts of the world if they do indeed exist outside their original type localities.

Some of the less-known loricate genera, such as *Pseudokephyrion* and *Kephyrion* spp., although sometimes abundant in plankton samples, but likely often overlooked because of their small size (nearly all are ≤10 μm long or broad), have received scant study and little is known of their biogeography outside of Alaska (Hilliard, 1966, 1967). An unusual habitat for another loricate genus, *Chrysococcus*, was reported by Lackey (1938), who reported nine species from the Scioto River in Ohio, one of which, *C. rufescens*, attained densities of more than 400 cells mL^{-1}. Rivers, as habitats for chrysophytes in North America, have generally been overlooked.

B. Habitat Characteristics

1. Trophic Status

Quantitative ecological data on environmental variables responsible for defining chrysophyte niches are only now beginning to be collected. The earlier notion that chrysophytes were exclusively found in oligotrophic lakes can no longer be supported (Kristiansen, 1988). Vørs *et al.* (1990) observed the greatest diversity of *Paraphysomonas* in eutrophic mill-ponds (one such pond contained 25 *Paraphysomonas* taxa!). Nicholls (1995) reviewed experiences with lake fertilization and summarized several instances where experimental enrichment with N and P led to increased abundance of certain chrysophytes, although the general trend was for proportionately lower representation by chrysophytes after such alterations and is in agreement with more recent research. Reynolds *et al.* (1998) increased the available phosphate-P from <1 µg L^{-1} to 20–28 µg L^{-1} (based on calculated dose) in oligotrophic acidic Seathwaite Tarn in the English Lake District. As a result, the chrysophytes dominating the system prior to the experiment (*Chrysolykos*, *Dinobryon*, *Kephyrion*, and *Monochrysis* spp.) made up a much smaller proportion of the total phytoplankton biomass. Similarly, Cottingham *et al.* (1998) fertilized three similar lakes in Wisconsin in 1993 and 1994. Their data also included two years of prefertilization phytoplankton monitoring and four years of data from a nearby unfertilized reference lake. The unfertilized lakes included chrysophytes such as *Stichogloea olivacea*, *Uroglena* spp., *Chrysosphaerella longispina*, *Dinobryon sertularia*, *D. divergens*, and *Kephyrion* sp. among the most common phytoplankton species. All of these declined in relative abundance after fertilization.

The role of nutrients as stimulatory factors in chrysophyte blooms is poorly understood. Chemostat studies such as those used for diatoms (Kilham and Kilham, 1978; Kilham and Tilman, 1979; Tilman and Kilham, 1976; Tilman *et al.*, 1982, 1986) are urgently needed for several common species of *Dinobryon*, *Urogena*, and *Uroglenopsis*.

2. Temperature

Temperature may influence the occurrence of chrysophytes in eutrophic lakes. Species able to grow quickly at temperatures below the temperature optima of other typical eutrophic algae (e.g., certain euglenoids and blue-green algae) may successfully compete for nutrients during the cooler seasons. Felip *et al.* (1995) observed that chrysophytes (*Chrysolykos* and *Ochromonas*) were an important component of the microbial community of slush layers of the winter cover lakes of the Pyrenees and the Alps. Wiedner and Nixdorf (1998) reported that the eutrophic lake Melangsee was dominated by blue-green algae during the ice-free period, but chrysophytes (*Uroglena volvox* and *Dinobryon* spp.) dominated under ice cover.

Ito and Takahashi (1982) measured the abundances of *Spiniferomonas* over a 24-month period in two ponds, one eutrophic and the other oligotrophic. Five *Spiniferomonas* taxa were common to both ponds—of which *S. trioralis*, *S. bourrellyi*, and *S. bilacunosa* were the most abundant. *Spiniferomonas* was present year round in the oligotrophic pond at a pH range of 5.8–6.7 but was present in the eutrophic pond only between October and April when pH was below 7.1 and water temperature was lower. Peak abundances in the oligotrophic pond were in the order of 400–800 cells mL^{-1} during spring (March–May). Suykerbuyk *et al.* (1995) studied a small eutrophic pond in The Netherlands (TP = 40 µg L^{-1}; specific conductance = 277 µS cm^{-1}). *Paraphysomonas vestita* and *S. trioralis* were present throughout the year but at low densities. *Chrysosphaerella* spp. (mainly *C. brevispina*) had highest concentrations in fall, winter, and spring, thus confirming the conclusions of Roijackers (1983) and Siver (1993), who also observed that *C. brevispina* was most abundant during the cooler part of the year.

Cooler water temperature may also help to exacerbate the offensive odors related to *Dinobryon* and *Uroglena* blooms in water supplies (Nicholls, 1995). Rashash *et al.* (1995) identified the "fishy" odor produced by *Dinobryon cylindricum* as 2t,4c,7c-decatrienal. Most of this compound (90%) was retained within the cells during growth in laboratory culture and the highest rate of odor production occurred during log phase growth. Environmental conditions, such as low temperature, that prolong the log phase of growth in natural systems would likely prolong the incidence of odor production and associated impairment of water use.

Dinobryon spp. are typical summer members of the plankton community in a wide variety of lake types (Heinonen, 1980). *Dinobryon* contributed 50% of the phytoplankton biomass in Canadian Shield lakes (Ostrofsky and Duthie, 1975). The genus has been a subject of many intensive studies, either autecological or as part of a phytoplankton community study, beginning in North America with Ahlstrom (1937).

Observations on the growth characteristics of *D. suecicum* and *D. borgei* were made by Hilliard (1968) throughout a one-year cycle from two different habitats in Alaska, a small pond and a large lake. Hilliard reported that in each habitat the length of the caudal spine for both species showed an inverse relationship with temperature; that is, spines were longer in cells collected during the early spring and late fall than those observed during the warmer summer months. He further noted that the lorica did not show the pronounced elongation observed in spines, although it was more linear during the colder periods and tended to be shorter and broader during the summer. His study showed that environmental factors (temperature in his study) should not be justification for recognizing the varietal forms *longispinum* and *elongata*, respectively, in the two species. This phenomenon of cyclomorphosis has not been thoroughly investigated for most

Chrysophyceae. However, Sandgren (1983) noted that the density and morphology of spines on the cyst were related to clonal history of *Dinobryon cylindricum*, whereas the length of spines and of the cyst collar were influenced by the temperature at the time of encystment. Some of these morphological differences within a taxon may have ecological ramifications relating to increased buoyancy and resistence to grazing by zooplankton.

In a 58-month study, Siver and Hamer (1992) developed an inference model for the seasonal periodicity of 16 species of scaled Chrysophyceae (*Chrysosphaerella*, *Paraphysomonas*, *Spiniferomonas*). Although specific conductance and pH were important variables in controlling the changes in species composition, water temperature was the most important environmental variable. They also concluded that *C. brevispina* was a cold-water organism, whereas *C. longispina* occurred most often in warmer summer waters. Data were inconclusive for the majority of the 16 species, but clearly, some species (*S. trioralis*, *S. serrata*, *S. bourrellyi*) did not exhibit any specific seasonality. Similar studies are needed for other North American Chrysophyceae of the genera *Dinobryon*, *Uroglena*, and *Uroglenopsis*.

One of the most dramatic examples of a cold-water stenotherm is the mountain-stream-dwelling chrysophyte *Hydrurus foetidus*. This macroscopic, brown, gelatinous, unpleasant-smelling alga is relatively abundant in both the eastern and western mountain streams of North America. The gelatinous envelope in which the cells are embedded is exceedingly tough and the plant frequently covers the entire surface of submerged rocks and has caused more than one hiker to loose his or her footing when crossing a stream. It normally begins to disappear when water temperatures rise much above 10°C. Squires *et al.* (1973) have shown that the species declined significantly in early June when the water temperatures increased, not reappearing until December when temperatures dropped to 1°C. Other reqirements for this species apparently include relatively low pH and bright sunlight (Parker *et al.*, 1973).

Another cold-water stream chrysophyte that forms a mucilaginous, filamentous-like thallus is *Phaeodermatium*. It forms thin, mainly one cell thick, disclike to circular crusts whose cells are closely appressed to the substrate obscuring their radiating filamentous nature. As is true of *Hydrurus*, any cell may develop a flagellum and swim away or be carried downstream to settle on a new substrate to grow into a new thallus.

3. Acid Status, pH

Acid neutralizing capacity and pH may influence the proportional abundance and numbers of chrysophytes in lakes. Generally, chrysophytes contribute less to total phytoplankton biomass as lakes acidify and regain a dominant or subdominant role as lakes are neutralized or otherwise return to higher alkalinity values. Findlay and Kasian (1996) manipulated the pH of an experimental lake in northwestern Ontario with H_2SO_4 over a 20-year period to mimic acidification (decrease of pH from 6.7 to 5.0) and then neutralization (rise in pH to 6.7). *Dinobryon bavaricum*, *D. sertularia*, and *Uroglena volvox* declined during the acidification phase but then increased when pH rose to preacidification levels. Nicholls *et al.* (1992) reported that several chrysophycean genera increased their representation in the total phytoplankton during the "recovery" of acidified lakes in the Sudbury region of Ontario.

Still, there are exceptions to these general patterns. Krienitz *et al.* (1997) reported that *Dinobryon pediforme* was a dominant element of the phytoplankton of the acidic bog lake Große Fuchskuhle in Germany (pH = 4.4–4.7). However, after division of the lake into four basins with plastic curtains prior to food-web manipulation experiments (and a wider range of pH values, 4.2–6.1), the lake produced a bloom (1000 colonies mL^{-1}) of *Uroglenopsis articulatus*. Nixdorf *et al.* (1998) found that an *Ochromonas* species dominated the phytoplankton of several highly acidic lakes in Lusatia, Germany, at pH values of only 2.5–3.5. Albertano *et al.* (1994) isolated an *Ochromonas* species from acid environments in Italy and determined its pH tolerance was 1.8–6.5 in laboratory culture. One of the most acidophilic algal species known is the chrysophyte *Ochromonas vulcania* which can grow at pH values as low as 1.0 (Gromov *et al.*, 1988, cited in Albertano *et al.*, 1994).

Other chrysophyte taxa appear to be indifferent to the acid–pH status of their habitats. Siver (1988a) reported that 13 *Spiniferomonas* species accounted for 37% of 113 phytoplankton samples from 33 lakes and ponds in Connecticut. The three most commonly reported taxa were *S. trioralis*, *S. bilacunosa*, and *S. bourrellyi*, which confirmed Nicholls' (1981a) findings from Ontario. In Connecticut, *S. trioralis* and *S. bilacunosa* were found in samples from water bodies having a pH range of 5.85–8.0. The total phosphorus range for *S. trioralis* occurrence was 1–27 $\mu g\,L^{-1}$ and for *S. bilacunosa*, 6.7–48 $\mu g\,L^{-1}$. Widest pH tolerances were for *S. coronacircumspina*, which was found in waters with pH ranging between 5.6 and 9.1 and total P from 1 to 27 $\mu g\,P\,L^{-1}$. These broad tolerances for key ecosystem variables help to explain the worldwide distribution of these taxa in a wide variety of habitat types. Environmental tolerances for other *Spiniferomonas* species are less conclusively established because of their infrequent occurrence. Other studies from elsewhere in the world show similar findings. Roijackers and Kessels (1986) surveyed 50 surface water sites in the southeastern part of The Netherlands. *Spiniferomonas trioralis* and *P. vestita* were very common taxa and were found over pH ranges of 4.5–8.1 and 3.9–8.9, respectively. *Polylepidomonas vacuolata*, *S. cornuta*, *S. abei*, and 11 other taxa of *Paraphysomonas* were too infrequently observed to establish clear pH preferences.

Although environmental tolerance data for *Paraphysomonas* taxa is nearly nonexistent, only a little data exists for the nonscaled chrysophytes. Eloranta (1989) examined the water quality characteristics (color, pH, total P, total N, lake size) of hundreds of Finnish lakes containing *Dinobryon* species. The most common species included *D. acuminatum*, *D. bavaricum*, *D. borgei*, *D. cylindricum*, and *D. divergens*. On the whole, species of *Dinobryon* were not especially useful indicators in Finnish lakes, perhaps owing to the narrow range of most environmental variables represented in this lake set. One exception was *D. pediforme*, which occurred only in softwater and acidic lakes. Lepistö and Rosenström (1998) examined the summer phytoplankton composition of 38 lakes located between 60°N and 69°N in Finland with specific reference to community types that might characterize four types of boreal lake: eutrophic, oligotrophic, dystrophic, and acidic. *Dinobryon divergens* was abundant only in the clear lakes; *Uroglena americana* and unidentified ochromonads dominated in the dystrophic lakes.

4. Light and Mixotrophy

Despite the crucial role of light in photosynthesis and growth, its effect on freshwater chrysophytes remains understudied, especially in comparison to the large number of publications on abiotic and nutrient effects. Although many diatoms, cyanobacteria, and rhodophytes may fare better than chrysophytes under low light, most taxa grow best under moderately high irradiance (e.g., 200–800 $\mu mol\ m^{-2}\ s^{-1}$) (Hill, 1996, and literature cited therein). Autecological light requirements of planktonic and benthic chrysophyte species are essentially unknown [but see Parker *et al.*'s (1973) field experiments with transplanted *Hydrurus*] because unialgal growth versus irradiance measurements have rarely been made. The interaction of other potential limiting factors (e.g., nutrients) with light may be as important as the effect of light alone, and our knowledge of benthic algal ecology would benefit from multifactor experiments that explore these interactions.

The ability of many chrysophytes to utilize particulate carbon may impart some competitive advantage over other algae that are strict autotrophs (light-dependent inorganic C assimilation). The phenomenon is potentially very significant physiologically and ecologically. Olrik (1998) lists 16 chrysophyte genera known to be mixotrophic (ability to obtain C by both phototrophy and phagotrophy). Species of *Ochromonas* and *Chromulina* are capable of consuming up to 200% of their cellular weight in bacterial C per day (Salonen and Jokinen, 1988). Mixotrophic chrysophytes may not be dependent on this mode of nutrition for C, but rather for P because the C:P ratio in algae is much higher than it is in bacteria (Olrik, 1998).

McKenzie *et al.* (1995) demonstrated in a NW Atlantic habitat (coastal Newfoundland) that *Dinobryon balticum* exhibited two subsurface peaks, at 5 m (temperature = 8°C) and at 40 m (temperature = 4°C). Cells at the greater depth (low light and low temperature) were ingesting particles (phagotrophy) at twice the rate of those cells at 5 m. Jones and Rees (1994) undertook particle ingestion experiments with cultures of both *D. sertularia* and *D. divergens* on the effects of irradiance, and with *D. sertularia* on effects of temperature over a range of 5–30°C. There was a strong temperature effect with a maximum uptake at 20°C. There was a weak irradiance effect with ingestion rates only slightly higher during the dark phase. They concluded that photosynthesis is the primary mode of C acquisition and that phagotrophy probably represents an adaptation for obtaining some other essential nutrient. One further example of this might be *Ochromonas monicis*, which is a particle feeder that both utilizes flagella for food capture and has endosymbiotic bacteria that may allow the flagellate to survive in vitamin-depleted water. Phagotrophism is initiated when major nutrients in the water are depleted (Doddema and van der Veer, 1983).

Although the costs and benefits of mixotrophy are still not clearly understood, conceptual models for six physiological types of mixotrophs were developed by Stoecker (1998) that predict the functional relationships of phototrophy and phagotrophy to availability of light, nutrients, and particulate food. Some chrysophytes (e.g., *Ochromonas*) seem to fit into two or more of these models, suggesting that phagotrophy may serve more than one purpose for at least some mixotrophs. More experimental work in this area is needed to understand more fully the ecological implications of mixotrophy.

In summary, the most frequently sampled chrysophycean habitats have been lakes, ponds, and bogs. Fewer collections have been made from streams, rivers, soil, and subaerial habitats. Broad ranges of environmental factors are apparently influencing the distribution, abundance, and community structure of chrysophytes. Some patterns have emerged from data collected more recently as regards chrysophyte indicators of acidity, trophic state, and temperature. However, on the whole, the occurrence of most chrysophyte species has generally not been explained satisfactorily by quantitative ecosystem data. Several species appear to be able to tolerate a wide range of environmental conditions and are very common throughout the world. The complexity of controlling variables is illustrated by the sometimes very low amount of variance in species composition that can be explained by traditional limnological descriptors. For example, in ponds in the high Canadian Arctic, Douglas and Smol (1995) observed only 26% of the total variance in benthic diatom species composition could be accounted for by the measured environmental and habitat variables such as major ions, pH, and substrate type. Algal communities develop as a result of a variable capacity of different species to colonize, grow, compete, tolerate multiple stresses, and resist loss processes. The net result is the production of different community structures in different habitats (Cox, 1990). Clearly, more quantitative and experimental data on natural systems are needed to answer many questions about chrysophyte species distributions and abundances. More attention paid to "non traditional limnological variables" such as hydrologic factors, meteorologic (e.g., wind and turbulence effects) factors, cell and colony motility and depth regulation, and the interaction of other protistan and microcrustacean grazing impacts might provide new hypotheses, predictions, and models of chrysophyte ecology.

IV. COLLECTION AND PREPARATION FOR IDENTIFICATION

A. Collection

Plankton can be collected using 5-, 10-, or 20-μm mesh nets, or a whole water sample can be concentrated for enumeration by settling (Lund *et al.*, 1958) or by filtration through membrane filters (McNabb, 1960). Whole water samples can be collected using a Van Dorn bottle or Kemmerer sampler. Because chrysophytes can sometimes determine their own position in the water column (Happey and Moss, 1967; Happey-Wood, 1976) and can form subsurface population peaks (Sandgren, 1988), tows may have to be made through portions of the water column. Messenger activated closing nets or vertical tubes with flow checkvalves (Nicholls, 1979) can serve to sample subsurface populations.

Attached forms are removed by scraping submerged aquatic plants, rocks, or woody debris. Techniques and samplers for quantitative sampling of attached forms have not been perfected. Squeezing or shaking vascular plants or *Sphagnum* into a container or through a plankton net works well for forms that become dislodged by the squeezing–shaking process.

Many species of neustonic chrysophytes can develop in the surface microlayer of quiet woodland ponds and small lakes (Nicholls, 1995). This community is best observed after letting a plankton sample come to "rest" in the laboratory for a few hours. Touching the surface of the water lightly with a coverslip and then mounting on a microscope slide will usually reveal neustonic species if they are present.

B. Preparation

Field collections containing live chrysophytes survive better the return trip to the laboratory if they are placed on ice. In the laboratory they should then be kept refrigerated (ca. 4°C). Depending on how the sample is to be processed, collections may be fixed in the field (or laboratory) using a variety of fixatives. No single fixative works well for all chrysophytes, although the best fixative will prevent colonial forms from breaking apart. Common fixatives include the following: (1) Lugol's or Acid-Lugol's (10 g iodine + 20 g potassium iodide + 20 g glacial acetic acid in 200 mL distilled water; add 1 mL to 1000 mL sample); (2) Transeau's solution (6 parts water, 3 parts 95% alcohol, and 1 part formalin); (3) 2% buffered glutaraldehyde; (4) FAA (formaldehyde–acetic acid–alcohol); and (5) 5% formalin. Wee (1983) lists several other fixatives.

Light microscopy can be used in most cases to identify chrysophycean genera. The importance of examining a live sample, particularly if it represents a duplicate of a companion sample that has been fixed for enumeration, cannot be overemphasized. For general LM observations, regardless of the form of collection, letting the live sample sit for a day in a north-facing window in a lightly covered container facilitates sampling motile or settled forms by pipette. Enhancing optics [e.g., Nomarski, differential interference contrast (DIC), or phase contrast] coupled with the proper choice of mount will optimize image quality. Staining [e.g., 1% methylene blue or Jensen's stain; Nicholls (1978)] can, for example, allow observation of the arrangement of lorica scales in *Epipyxis*, facilitate length measurements of flagella, permit visualization of intercellular connectives in *Uroglena* and *Uroglenopsis*, and even reveal the flimmer hairs on the long flagellum of some taxa.

By far the largest number of genera within the Chrysophyceae (*sensu stricto*) are without scales, but the genera covered with silica scales and organic structures have received more attention than others (Kristiansen, 1995b). Although written for the Synurophyceae, Wee's (1983) LM methods of collection, fixation, and staining also have application for the Chrysophyceae. Use of stains (above), DIC, or phase contrast will allow detection of scales in many species of *Chrysosphaerella*, *Paraphysomonas*, and *Spiniferomonas*. However, for unequivocal (species-level) identification of these taxa (with the possible exception of *C. longispina*, *C. brevispina*, *S. serrata*, *S. minuta*, *S. coronacircumspina*, *S. takahashii*, and *P. vestita*, for which critical LM might suffice), electron microscopy is necessary. Contrast in images produced by EM of dried whole mounts can be improved by shadow-casting in an evaporator at an angle of about 20° with platinum and/or paladium, a technique usually required for identification of most of the species in the Paraphysomonadaceae (Moestrup and Thomsen, 1980).

V. KEY AND DESCRIPTIONS OF GENERA

A. Key to Genera

The elements of this key stress the structure of the vegetative stages and are based as much as possible on features seen with a light microscope. The key is restricted to those chrysophyte genera reported thus far from North America. Although species of *Chrysosaccus*, *Chrysocapsa*, *Phaeogloea*, *Phaeosphaera*, *Stichogloea*, and *Selenophaea* have appeared in published checklists of algal species from various regions of North America (e.g., Meyer and Brook, 1969; Jacobs, 1971; Duthie and Socha, 1976; Duthie and Ostrofsky, 1978; Stein and Borden, 1979), few have been convincingly described with photographs, drawings, and critical text; consequently, some species of the above-mentioned taxa are of doubtful North American occurrence. Also included in the key are several genera that might be better classified in the proposed new class, Phaeothamniophyceae (Andersen et al., 1998b; Bailey et al., 1998); these have been marked with the Superscript "2" in Sect. V.B.

1a.	Cells colorless (lacking chloroplasts)	2
1b.	Cells with chloroplasts	7
2a.	Cells housed in a lorica	3
2b.	Cells not housed in a lorica	4
3a.	Cells with heterokont flagellation, attached to the lorica base by a cytoplasmic filament (Fig. 5A)	*Stylobryon*
3b.	Cells with two flagella, one of which is directed posteriorly as an attachment to the lorica base (Fig. 5B)	*Bicosoeca*
4a.	Cells solitary, not joined together forming colonies	5
4b.	Cells joined together forming colonies	6
5a.	Cells with a covering of scales (Fig. 5C)	*Paraphysomonas*
5b.	Cells not covered in scales (Fig. 5D)	*Spumella*
6a.	Cell attachments branching (dendroid) (Fig. 5E)	*Dendromonas*
6b.	Cell attachments radiate from a central point of attachment (Fig. 5F)	*Anthophysa*
7a.	Cells housed in a lorica	8
7b.	Cells not housed in a lorica	26
8a.	Cells with two flagella of unequal length (light microscopy)	9
8b.	Cells with one visible flagellum, two subequal flagella, or no flagella	10
9a.	Lorica comprising overlapping pectin plates (stain pink with methylene blue) (Figs. 3B, 4B, 5G)	*Epipyxis*
9b.	Lorica not comprising overlapping plates, but of a single continuous entity	11
10a.	Cells with one visible (LM) flagellum	15
10b.	Cells with rhizopodia or two subequal flagella	17
11a.	Loricate cells planktonic, either solitary or colonial	12
11b.	Loricate cells attached to a substratum	13

12a.	Cells in branching colonies (but some solitary species) (Figs. 3A, 5H–J)	*Dinobryon* (in part)
12b.	Cells solitary	14
13a.	Short flagellum much shorter than long flagellum (Fig. 5K)	*Poterioochromonas*
13b.	Short flagellum nearly as long as long flagellum (Fig. 6A)	*Stylochrysalis*
14a.	Lorica bilaterally symmetric (Fig. 6B, C)	*Chrysolykos*
14b.	Lorica radially symmetric (Figs. 3C, 6E)	*Pseudokephyrion*
15a.	Lorica with wide mouth (similar to cell width)	16
15b.	Lorica usually thick-walled and spherical; flagellum emerges through a pore in the lorica (Fig. 6D)	*Chrysococcus*
16a.	Cells planktonic (Fig. 6F)	*Kephyrion*
16b.	Cells epiphytic, usually with a cloud of symbiotic bacteria in the lorica mouth (Fig. 6H)	*Lepochromulina*
17a.	Cells with rhizopodia	18
17b.	Cells with two subequal flagella (Figs. 3D, 6I)	*Derepyxis*
18a.	Cells planktonic, lorica with elongated spines (Figs. 6L, M)	*Bitrichia*
18b.	Cells epiphytic, lorica not spined	19
19a.	Lorica with basal attachment structure (band or stipe)	20
19b.	Lorica without basal attachment structure	21
20a.	Lorica attached to substratum by a stipe (stalklike extension of the lorica base)	*Stylochrysalis* (see 13, above)
20b.	Lorica attached to substratum by a band	22
21a.	Lorica with a single opening or pore	23
21b.	Lorica with more than one opening or pore	24
22a.	Lorica with single apical opening (Fig. 6N)	*Chrysopyxis*
22b.	Lorica with two lateral openings (Fig. 6G)	*Chrysoamphipyxis*
23a.	Lorica axis vertical to substratum, opening apical (Fig. 6J)	*Lagynion*
23b.	Lorica axis horizontal to substratum, opening lateral (Fig. 7A)	*Kybotion*
24a.	Lorica openings two, lateral, opposite (Fig. 6O)	*Chrysoamphitrema*
24b.	Lorica openings more than two	25
25a.	Cells colonial, joined by rhizopodia (Fig. 7B)	*Heliapsis*
25b.	Cells not joined by rhizopodia (Fig. 6K)	*Stephanoporos*
26a.	Cells with two flagella of unequal length (LM)	27
26b.	Cells with only one or two long flagella visible with LM, or cells in filaments, coccoid, palmelloid, rhizopodial, or amoeboid	39
27a.	Cells attached together or grouped in colonies	28
27b.	Cells solitary	35
28a.	Cells with siliceous plate- and/or spine-scales (Figs. 3E, F, 7F)	*Chrysosphaerella*
28b.	Cells not with siliceous scales	29
29a.	Cells of the colony tightly pressed together	30
29b.	Cells of the colony not pressed together, but loosely arranged or occupying gelatinous tubes	33
30a.	Cells grouped in a ring, easily separated (Fig. 7C)	*Cyclonexis*
30b.	Cells forming a hollow sphere or radially arranged and joined by basal extensions	31
31a.	Cells forming a hollow spherical colony with an open end (Fig. 7E)	*Eusphaerella*
31b.	Cells radially arranged in mucilaginous colonies	32
32a.	Cells joined at their bases by short branching tubular connectives (Fig. 7H)	*Synuropsis*

32b.	Cells in colony lacking connectives (Fig. 7D)	*Syncrypta*
33a.	Cells occupying tubular cavities in a lobed gelatinous matrix (Fig. 7G)	*Chrysoxys*
33b.	Cells joined at the center of the colony by long tubular connectives or cytoplasmic filaments	34
34a.	Cells tapered posteriorly and united by branching cytoplasmic threads (Fig. 8A)	*Uroglena*
34b.	Cells truncate posteriorly and united by branching tubular connectives (Figs. 4H, 8B)	*Uroglenopsis*
35a.	Cells covered with scales	36
35b.	Cells not covered with scales	37
36a.	Scales organic (Fig. 8E)	*Chrysolepidomonas*
36b.	Scales siliceous (Figs. 3G–I, 8C)	*Spiniferomonas*
37a.	Cells forming an "oily" sheen (neustonic) on the surface of quiet ponds, tanks, etc. (Fig. 8D)	*Chromophyton*
37b.	Cells planktonic or occupying *Dinobryon* loricae	38
38a.	Cells planktonic (Fig. 8F)	*Ochromonas*
38b.	Cells occupying *Dinobryon* loricae (Fig. 8H)	*Rhizoochromonas*
39a.	Cells with one flagellum or two subequal flagella	40
39b.	Cells without flagella[1]	46
40a.	Cells with one flagellum visible in LM[1]	41
40b.	Cells with two flagella of equal or subequal length (Fig. 8G)	*Erkenia*
41a.	Chloroplasts reticulate (Fig. 8I)	*Chrysapsis*
41b.	Chloroplasts cup- or band-shaped	42
42a.	Chloroplasts four, symmetrically positioned (Fig. 9A)	*Amphichrysis*
42b.	Chloroplasts one or two	43
43a.	Cells colonial in mucilage (Fig. 9C)	*Saccochrysis*
43b.	Cells not colonial in mucilage	44
44a.	Cells strongly flattened, flagella inserted ventrally (Fig. 9D)	*Monochrysis*
44b.	Cells not strongly flattened, but spherical or ovoid with apical flagella	45
45a.	Cells covered with organic scales (Fig. 9B)	*Microglena*
45b.	Cells not covered with organic scales (Fig. 9F)	*Chromulina*
46a.	Cells in filaments (no mucilage)	47
46b.	Cells not in mucilage-free filaments	48
47a.	Filaments short, unbranched (Fig. 9G)	*Stichochrysis*
47b.	Filaments dendroid (short branching), attached to substratum with rounded basal cells (Fig. 9H)	*Phaeothamnion*
48a.	Cells in parenchymatous thallus	49
48b.	Cells not in parenchymatous thallus	50
49a.	Cell arrangement rectangular (Fig. 9I)	*Phaeoplaca*
49b.	Cell arrangement circular, disclike (Fig. 9K)	*Phaeodermatium*
50a.	Cells rhizopodial or in mucilaginous colonies	51
50b.	Cells not rhizopodial and not in mucilaginous colonies	73
51a.	Cells rhizopodial or amoeboid or with pseudocilia	52
51b.	Cells in mucilaginous colonies	57
52a.	Cells solitary	53
52b.	Cells colonial	54

53a.	Cells with short (≤ 1/2 cell diameter) undulating flagellum (Figs. 4I, 10A)	*Chrysamoeba*
53b.	Cells without flagella (Fig. 10B)	*Rhizochrysis*
54a.	Cells connected by rhizopodia	55
54b.	Cells not connected by rhizopodia	56
55a.	Cells forming linear chains (Fig. 10C)	*Chrysidiastrum*
55b.	Cells forming reticulate colonies (Fig. 10E)	*Chrysarachnion*
56a.	Cells producing long pseudocilia or rhizopodia encased in mucilage (Fig. 10F)	*Chrysochaete*
56b.	Cells producing pseudocilia or rhizopodia surrounded by tiny spherules (symbiotic bacteria?) (Figs. 4G, 10G)	*Chrysostephanosphaera*
57a.	Thallus monostromatic, often perforated, sheetlike or cuplike	58
57b.	Thallus not monostromatic	59
58a.	Cell walls and stigmata absent; contractile vacuoles present (Fig. 10H)	*Dermatochrysis*
58b.	Cell walls and stigmata present; contractile vacuoles absent (Fig. 10J)	*Tetrasporopsis*
59a.	Cells with a tubular spinelike extension (Fig. 11A)	*Bourrellia*
59b.	Cells lacking spinelike extensions	60
60a.	Thallus macroscopic (up to several centimeters), highly branched or lobed	61
60b.	Thallus microscopic (up to a few millimeters)	62
61a.	Cells restricted to the periphery of the mucilage (Fig. 11F)	*Celloniella*
61b.	Cells dispersed throughout the mucilage (Figs. 4D, E, 11C)	*Hydrurus*
62a.	Cells fusiform, grouped in parallel bundles (Fig. 11D)	*Eirmodesmus*
62b.	Cells not fusiform	63
63a.	Cells in the center of mucilage mass, in clumps or in series	64
63b.	Cells not restricted to the center of the mucilage	70
64a.	Cells in linear series	65
64b.	Cells not in linear series	67
65a.	Cells in short filaments emanating from a clump of cells, cells smaller at the ends of filaments (Fig. 11B)	*Chrysocapsopsis*
65b.	Cells not greatly different in size and not emanating from clumps of cells	66
66a.	Linear series of usually >8 cells, flattened pole-to-pole, obovate-subspherical in shape (Fig. 11E)	*Sphaeridiothrix*
66b.	Cells spherical, in series of usually 4 cells (Figs. 4F, 12A)	*Tetrachrysis*
67a.	Cells with mucilaginous hair- or tube-like extension, with no central cytoplasmic filament (Fig. 11H)	*Naegeliella*
67b.	Cells grouped in 2's or 4's lacking mucilaginous tubes	68
68a.	Mother cell wall remnants in two halves (Fig. 12D)	*Phaeoschizochlamys*
68b.	Mother cell wall remnants not evident	67
69a.	Mucilage in inner (clear) and outer (radially striated) zones; often with remnants of pseudocyst wall (dark brown) (Fig. 12B)	*Chalkopyxis*
69b.	Mucilage not divided into inner and outer zones, cells lacking walls (Fig. 12E)	*Chrysosaccus*
70a.	Mucilage spherical	71
70b.	Mucilage irregularly shaped or cylindrical	72
71a.	Cells restricted to periphery of mucilage (Fig. 12C)	*Chrysocapsa*
71b.	Cells distributed without order (Fig. 11G)	*Phaeogloea*
72a.	Mucilage cylindrical, cells distributed without order (Fig. 9J)	*Phaeosphaera*

72b.	Mucilage not cylindrical, cells ellipsoidal and interconnected with bands of mucilage (Fig. 9E)	*Stichogloea*
73a.	Cells reniform, cell wall papillate (Fig. 10I)	*Selenophaea*
73b.	Cells spherical, solitary or in small clumps (often epiphytic) (Fig. 10D)	*Chrysosphaera*

[1] Omitted from this branch of the key (at key dichotomies 39 and 40) is the genus *Chrysamoeba*, which has a very short flagellum usually visible in LM. This genus is keyed through its more obvious rhizopodial features (key dichotomy 26).

B. Descriptions of Genera

Amphichrysis Korshikov (Fig. 9A)

The only known species of this rare flagellate genus (*A. compressa* Korsh.) is distinctive for its large size (cells about 30 µm long), its metabolic shape, four ribbon-like chloroplasts and two to four stigmata. Its single visible flagellum is about as long as the cell.

Anthophysa B. de Saint-Vincent (Fig. 5F)

Colonies of these colorless *Ochromonas*-like pyriform cells form clusters attached at their posterior ends. Cell clusters are loosely attached to the ends of branching stalks colored reddish brown by impregnated iron salts. The stalk is markedly thicker in the basal region. "Empty" stalks are frequently found because cell colonies easily break free and swim away. Only one species is known, *A. vegetans*, and has the distinction of being the first chrysophyte named [but under the name *Volvox vegetans* (Müller, 1786)]. It commonly occurs among plant debris in shallow lakes and ponds, or in the littoral zones of larger lakes, especially in spring.

Bicosoeca J. Clark (Fig. 5B)

Protoplast is housed in a cylindrical or cone-shaped lorica, the shape of which allows determination of the species. The cell has two flagella, one of which is smooth (hairless) and is directed posteriorly in a groove in the cell to the bottom of the lorica where it is attached and thus anchors the cell inside its lorica. The other anterior flagellum has flimmer hairs, and in most freshwater species it extends well beyond the mouth of the lorica where it is utilized in the capture of food particles. Hilliard (1971a) has described several of the common North American species.

Bitrichia Woloszynska (=*Diceras* Reverdin) (Fig. 6L, M)

Cell is housed in a fusiform lorica with two or three long spinelike extensions. The lorica has a pore through which emerge branched rhizopodia. There are no flagellated stages. *Bitrichia chodatii* and *B. longispina* are common in the plankton of softwater lakes; *B. ollula* is less common (Nicholls, 1981b).

Bourrellia Dillard (Fig. 11A)

Cells are spherical with a long tubular extension and are embedded in a gelatinous matrix. There are one parietal chloroplast and two contractile vacuoles near the base of the spine. The only known species of this unusual genus, *B. skujae* Dillard, was described from a farm pond in South Carolina (Dillard, 1970).

Celloniella Pascher (Fig. 11F)

Cells are embedded in the periphery of a columnar or dendroid mucilaginous matrix which is attached to a substrate. Reproduction is by cell division and production of *Chromulina*-like zoospores. The only known species, *C. palensis* Pascher, has distinctive stomatocysts with curved bladelike wings criss-crossing the surface; rare.

Chalkopyxis Pascher (Fig. 12B)

Colonies may be either attached or free-floating and consist of groups of two or four cells embedded in the central zone of a bizoned mucilagenous matrix. The outer zone is devoid of cells but is radially striated. Cells produce pseudocysts with thickened brownish colored polar caps that remain behind after release of the daughter cells. The only species, *C. tetrasporoides* Pascher, is rare but has been reported from North America by Dillard and Crider (1970).

Chromophyton Woronin emend. Couté (Fig. 8D)

Cells are ellipsoidal to spherical with a single chloroplast, a pyrenoid, and one or two anterior contractile vacuoles but no stigma. Cells have typical ochromonad heterkont flagellation and the posterior region of the cell is often slightly amoeboid. The most distinctive characteristic of this genus is its ability to form resting stages called pseudocysts in the epineuston of small ponds and woodland pools. High concentrations of these pseudocysts impart a golden iridescence to the surface of the water. Cell division takes place within the pseudocysts, which on fracturing to release the daughter cells leave behind the detached short stalk of the pseudocyst. *Chromophyton* is relatively common and may be locally abundant in pools.

Chromulina Cienkowski (Fig. 9F)

The cell has one visible emergent flagellum and, depending on the species, one or two chloroplasts, and a pyrenoid; a stigma may or may not be present. This is a large genus; the identification of some species is facilitated by the presence of stomatocysts. It differs from *Ochromonas* in having only one flagellum visible by LM (a second very short or vestigial flagellum is likely present in all species). There are many common species.

Chrysamoeba Klebs (Figs. 4I, 10A)

Cells are amoeboid with fine pseudopodia or rhizopodia extended in the same plane. A single short flagellum is visible and is detected in living cells by its slow undulations. There are a few common species, especially in small ponds.

Chrysapsis Pascher (Fig. 8I)

Cells are 8–15 µm, spherical to ovoid with one visible flagellum and a reticulate chloroplast. The "stringy", perforated appearance of the chloroplast of this genus separates it from *Chromulina*. This is a rare genus.

Chrysarachnion Pascher (Fig. 10E)

Like *Chrysidiastrum*, cells are joined by their rhizopodia, but *Chrysarachnion* forms reticulate colonies rather than linear chains. This is a rare genus.

Chrysidiastrum Lauterborn (Fig. 10C)

Most commonly found in the neuston of small sheltered pools, *Chrysidiastrum* consists of *Rhizochrysis*-like cells joined in a linear chain by their rhizopodia. *Chrysidiastrum catenatum* is probably the most common North American species in this group of rhizopodial algae (e.g., Prescott, 1962; see also Whitford, 1969).

Chrysoamphipyxis Nicholls (Fig. 6G)

Like *Chrysopyxis*, this monospecific genus (only known species is *C. canadensis*) possesses a ringed base for attachment to the substratum, but is distinguished by two lateral openings in the lorica through which the rhizopodia emerge (*Chrysopyxis* has one apical opening). The cell has a single lobed chloroplast and one or two basal contractile vacuoles. Clusters of small spherules occur in the lorica near the bases of the rhizopodia (see also *Lepochromulina*). This is a rare epiphyte.

Chrysoamphitrema Scherffel (Fig. 6O)

The protoplast possesses one or two chloroplasts and contractile vacuoles and is housed within a lorica that is often yellowish-brown. The lorica has two opposite pores through which the rhizopodia from the protoplast extend (resembles a *Chrysoamphipyxis* without the basal ringlike band for attachment). This is a rare genus; look for it on filamentous chlorophytes.

Chrysocapsa Pascher (Fig. 12C)

Cells with a single parietal chloroplast and one or more contractile vacuoles are located in the peripheral zone of a spherical gelatinous matrix. Zoospores have one visible flagellum. This is a rare genus.

Chrysocapsopsis Thompson *et* Wujek (Fig. 11B)

Cells of the only known species (*C. rupicola*) are 4.2–9.1 µm in diameter and are arranged in short branching filaments formed by budding within a gelatinous matrix. Cells are smaller toward the distal ends of the branches. There are two to six chloroplasts in each cell. Aplanospore production is also known. *Chrysocapsopsis rupicola* has both epiphytic and planktonic forms in its life history; this, with its unique budding method of vegetative reproduction, distinguishes it from *Chrysocapsella* and *Tetrachrysis* (Thompson and Wujek, 1998a). This is a rare genus.

Chrysochaete Rosenberg (Fig. 10F)

Cells are grouped in tight packets usually epiphytic in habit but also sometimes free-floating. From each cell extends a long pseudocilium or thin rhizopod that is encased in mucilage and branched in the distal portion. Cells have one or two chloroplasts, a pyrenoid, and one or more contractile vacuoles. Dop (1978) has described the essential differences between *Chrysochaete* and *Phaeoplaca* based on wild and laboratory-cultured material. *Chrysochaete* is uncommonly attached to submerged stones and to leaves and stems of aquatic plants.

Chrysococcus Klebs (Fig. 6D)

The loricae of members of this genus are composed of pectic material including compounds of iron and/or calcium. There are both a long and a short flagellum, but the short flagellum is usually not detectable with light microscopy. The long flagellum emerges through a pore in the lorica (hence distinguishing members of this genus from *Kephyrion*, which has a much larger opening). Lorica shape and size and presence or absence of stigma and pyrenoids are the main features distinguishing species, all of which are planktonic. There are several common species.

Chrysolepidomonas Peters *et* Andersen (Fig. 8E)

The only known freshwater species of this genus is *C. dendrolepidota*. Cells are ovoid to ellipsoidal (5.5–6.5 µm by 7–8 µm and contain a single parietal chloroplast with a stigma. There are one or two anterior-positioned contractile vacuoles and typical heterokont flagellation. Distinctively, cells are covered with small, organic, canistrate (about 0.04 µm high), and dendritic (about 0.5 µm high) scales, which can only be resolved by electron microscopy. This is a rare genus.

Chrysolykos Mack (incl. *Chrysoikos* Willén) (Fig. 5B, C)

The *Dinobryon*-like cell is housed in a lorica of interwoven cellulosic microfibrils, but the loricae are bilaterally symmetric, often with projecting spinelike extensions. Members of this genus are easily overlooked owing to their small size and the highly transparent nature of the lorica. Use dried or stained (e.g., Jensen's stain) preparations for optimal detection. Three species are known from North America (Nicholls, 1981b).

Chrysosphaera Pascher emend. Bourrelly (Fig. 10D)

Cells are not encased in mucilage but are aggregated in small clumps and are usually epiphytic. Reproduction is by cell division, autospores, or *Chromulina*-like zoospores. *Epichrysis* Pascher (=*Phaeocapsa* Korsch.) and *Chrysobotrys* Pascher are included here in *Chrysosphaera* (Bourrelly, 1957). There are a few common species.

Chrysopyxis Stein (Fig. 6N)

This is a loricate sessile organism with one or two chloroplasts, one or more contractile vacuoles, and a branched rhizopod that extends through a constricted mouth opening at the apex of the lorica. The base of the lorica forms a thin band that encircles the substratum. The circular ringlike base of the lorica is difficult to see when attached to its substrate but is highly distinctive in unattached specimens. Kristiansen (1969, 1972) showed the microfibrillar nature of the lorica of *Chrysopyxis*. Species of this genus are invariably attached to filamentous algae (*Chaetophora*, *Oedogonium*, *Zygogonium*, etc.) by the lorica band, which encircles the host algal filament. At least three species are known from North America (King, 1984). *Chrysopyxis* may be locally abundant.

Chrysosaccus Pascher (Fig. 12E)

Cells are in groups of (usually) four, with one to four chloroplasts, and are embedded in a mucilaginous matrix of variable shape. Reproduction is by cell division; zoospores are not known. Cell shape (e.g., spherical or fusiform), presence of contractile vacuoles, and number of chloroplasts are important features for distinguishing among the three known species. This is a rare genus.

Chrysosphaerella Lauterb. emend. Korshikov (Figs. 3E, F, 7F)

Pyriform cells are joined at their posterior attenuated ends, forming spheroidal free-swimming clusters. Cells have heterokont flagellation and outward-directed siliceous spine-scales. Also covering the cell surface are plate-scales (circular or elliptical) with submarginal pits or markings, most easily detected with phase contrast microscopy in dried specimens. Some authors include in *Chrysosphaerella* those solitary (noncolonial) heterokont species of *Spiniferomonas* with "*Chrysosphaerella*-like" spine- and plate-scales, but there are many other examples in the Chrysophyceae and Synurophyceae where solitary versus colonial cell habit serves to separate morphologically similar taxa at the genus level (Nicholls, 1984c). To maintain consistency, solitary *Chrysosphaerella*-like species should be classified as *Spiniferomonas*. This is a common genus.

Chrysostephanosphaera Scherffel (Figs. 4G, 10G)

Cells form a ring in a gelatinous matrix surrounded by and interspersed with symbiotic bacteria. Cells are spherical and slightly plastic with one or more chloroplasts and basal (toward the center of the colony) contractile vacuoles. From the apical ends of cells extend long branched rhizopods. Zoospores are of the *Chromulina* type. Thompson and Wujek (1998b) give an account of *C. globulifera* Scherffel based on wild and cultured material from Maryland, Kansas, and Michigan and suggest that it might be the palmelloid stage of *Cyclonexis annularis*. *Chrysostephanosphaera* is a rare genus.

Chrysoxys Skuja (Fig. 7G)

This genus consists of *Dinobryon*-like cells, 20–30 µm long, occupying tubular cavities in a lobed gelatinous matrix reaching up to 500 µm in length. Like *Dinobryon* and *Epipyxis*, the cells are attached posteriorly by a thin contractile stipe. This is a rare genus.

Cyclonexis Stokes (Fig. 7C)

Originally described from a New Jersey bog by Stokes (1886), *C. annularis* has been found in several other North American locations (Thompson and Wujek, 1998b). This beautiful but fragile species (colonies and cell structure are disrupted easily by "normal" phycological methods) consists of ringlike colonies of cells that have two chloroplasts, two anterior contractile vacuoles, and typical heterokont flagellation. Cells abut along their lateral margins and, because of their truncate-ovoid shape, form wide ringlike colonies. The long axis of individual cells lies at an angle of about 30° to the plane of the colony, so the whole colony takes on the appearance of a short frustrum; however, colony shape can become quite irregular as a result of cell division of only a few of the cells in the colony (Thompson and Wujek, 1998b).

Dendromonas Stein (=*Monadodendron* Pascher)
(Fig. 5E)

This genus consists of *Spumella*-like cells (colorless ochromonad), but which form colonies that consist of cells at the end of dichotomously branched stalks, all in the same plane. Note the distinction between *Dendromonas* and *Pseudodendromonas* Bourrelly; the latter genus is not a chrysophyte but a colonial relative of the zooflagellate *Cyathobodo* (Hibberd, 1976b). *Dendromonas* is not common.

Derepyxis Stokes (Figs. 3D, 6I)

The cell is housed in a vase-shaped lorica with a restricted mouth opening and with a short stalklike base (in some species) for attachment to submerged vegetation. As in *Lepochromulina*, small clusters of loricae may be found, but these lack the clouds of spherules associated with the lorica mouth area of *Lepochromulina*. This rarely reported genus is in need of a thorough reinvestigation, especially as regards flagellar structure, because one species (*Derepyxis anomala* Korsh.) was described with rhizopodia, not the two subequal flagella usually ascribed to this genus. Two subequal flagella would suggest the presence of a haptonema and the reclassification of this taxon as a haptophyte. About a dozen species have been described on the basis of lorica morphology.

Dermatochrysis Entwisle *et* Andersen (Fig. 10J)
See *Tetrasporopsis*.

Dinobryon Ehrenberg (Figs. 3A, 5H–J)

Cells have heterokont flagellation, one or two chloroplasts, one or more contractile vacuoles, and a stigma. The protoplast is housed in a cylindrical to funnel-shaped lorica made of interwoven cellulose microfibrils. Species are determined on the basis of lorica and colony morphology. The truly solitary (noncolonial) species fall into the subgenus *Dinobryopsis*; the colonial species are assigned to the subgenus *Eudinobryon*. Some species of *Dinobryon* (e.g., *D. suecicum*, *D. borgei*, *D. tubaeforme*, *D. dilatatum*, *D. attenuatum*) are all planktonic solitary species distinguished from *Pseudokephyrion* by virtue of their attenuate lorica bases (*Pseudokephyrion* species have bluntly rounded lorica bases). Ahlstrom (1937), Hilliard (1968, 1971a), and Nicholls (2000) describe several species known from across North America. Several species are very common.

Eirmodesmus Whitford (Fig. 11D)

The single species of this genus (*E. phaeotilus*) has elongated ovoid or fusiform cells grouped in parallel bundles within a mucilage layer associated with other algae (e.g., *Tetraspora*, *Draparnaldia*). The one or two parietal chloroplasts lack a pyrenoid. Zoospores are of the *Chromulina* type. *Eirmodesmus phaeotilus* is known from acidic streams at water temperatures below 10°C. This is a rare genus.

Epipyxis Ehrenberg emend. Hilliard *et* Asmund (=*Hyalobryon* Lauterborn) (Figs. 3B, 4B, 5G).

Cells are very much like *Dinobryon*, but are housed within a tubular lorica comprising overlapping thin plates. Some species form branching or bushy colonies. The lorica plates or scales were earlier misinterpreted as bands or rings in the loricae of *Hyalobryon* species. The scales are of pectin and stain pink with methylene blue, thus affording ready detection of the genus in a mixed phytoplankton sample. Species of *Epipyxis* are determined by lorica shape and colony formation. See Hilliard and Asmund (1963) for descriptions of many North American species.

Erkenia Skuja (Fig. 8G)

Cells are flattened, ellipsoidal, 4–6 μm, with two chloroplasts, two contractile vacuoles, a stigma, and two subequal flagella. During swimming, one flagellum is directed anteriorly, whereas the other trails behind. *Erkenia* needs reinvestigation with electron microscopy. It is possibly a haptophyte, not a chrysophyte, because it is very reminiscent of *Chrysochromulina parva* without a haptonema. A haptonema may have been overlooked by the few investigators who have reported *Erkenia*.

Eusphaerella Skuja (Fig. 7E)

Colonies of tightly fitting cells with heterokont flagella directed outward, forming a hemispherical colony with a hollow center. The free-swimming colonies may attain diameters of 400 μm. *E. turfosa* has recently been placed in the genus *Uroglenopsis* (Wujek and Thompson, 2002).

Heliapsis Pascher (Fig. 7B)

Loricate cells of this genus form colonies by interconnecting rhizopodia extending from two or more pores in the lorica wall. Like several other loricate rhizopodial genera, *Heliapsis* is epiphytic on algal or detrital material. This is a rare genus.

Hydrurus C. Agardh (Figs. 4D, E, 11C)

Hydrurus forms macroscopic mucilaginous mosslike colonies on rocks and other firm substrata up to 0.3 m in length. Cells are spherical to ellipsoidal, 8–12 μm long, and are embedded in the central part of the matrix. Zoospores are distinctive in their metabolic but basic tetrahedral shape and have one long and one short flagellum. *Celloniella* and *Hydrurus* are macroscopically similar; microscopically, the different organization of cells in the gelatinous matrix, the shapes of the vegetative cells and zoospores (Vesk *et al.*, 1984; Hoffman *et al.*, 1986), and stomatocysts serve to distinguish them. In North America, the only known species of *Hydrurus* (*H. foetidus*) is better known than *C. palensis*; it produces offensive odors and is found in clear, cold flowing waters in full sunlight (Parker *et al.*, 1973; Lakshminarayana and Devi, 1975).

Kephyrion Pascher (Fig. 6F)

Species of this genus can best be described as *Chrysococcus*-like protoplasts (very reduced second flagellum) housed in a *Pseudokephyrion*-like lorica. With light microscopy, the second flagellum in *Pseudokephyrion* is short but visible (may require drying or staining to detect); in *Kephyrion*, the second flagellum can be detected only in sections examined by transmission electron microscopy. Like some *Pseudokephyrion* species, the lorica of some *Kephyrion* taxa are variously embellished with flanges, flares, and ridges that apparently are species-specific. The genus is common in the spring plankton of oligotrophic to mesotrophic lakes. Hilliard (1966, 1967) and Meyer (1971) provide descriptions of North American species. There are several common species, especially in lakes during spring.

Kybotion Pascher (Fig. 7A)

The loricate vegetative cells of this genus are attached to the substratum (aquatic vegetation) along the long axis of the lorica. This is a rarely reported genus with three or four species resembling a *Lagynion* on its side. Zoospores are uniflagellate.

Lagynion Pascher (Fig. 6J)

These are rhizopodial cells with one or two chloroplasts and one or two contractile vacuoles; they are housed in a lorica usually with a flattened, bulbous base attached to the substratum (often filamentous algae). There is no attachment between protoplast and interior of the lorica. Prescott (1962) reported five species from the western Great Lakes region. Note the difference between this genus and *Kybotion*, which is essentially a *Lagynion* attached to the substratum along its lateral edge.

Lepochromulina Scherffel (Fig. 6H)

The loricae of species in this genus are short and cylindrical with rounded bases and are attached to filamentous algae or detrital material, often in small clusters of three or more. Cells have one visible flagellum and are not attached to the inside of the lorica. The most distinctive feature is the cloud of spherules or symbiotic bacteria that surround the lorica mouth. Although widely believed to be symbiotic bacteria, Hibberd (1983) has shown that clusters of spherules of similar appearance in *Spongomonas* and *Phalansterium* (although not chrysophytes) are endogenously produced in cytoplasmic vesicles. Similar origins might be suggested for the spherules associated with *Lepochromulina*, *Chrysoamphipyxis*, and *Chrysostephanosphaera*. *Lepochromulina* may be locally abundant.

Microglena (O.F.M.) Ehrenberg (Fig. 9B)

Cells are solitary with one or two chloroplasts, one visible flagellum, a stigma, and several small contractile vacuoles emptying into a large one at the anterior end of the cell. The periplast may be encased in a layer of organic scales [per *M. butcheri* (Belcher, 1966)]. Also present in the periphery of the cell of *M. punctifera* (O.F.M.) Ehr., the only known North American species (Wujek, 1967), are many siliceous lenslike nodules arranged in an irregular fashion. This is a rare genus.

Monochrysis Skuja emend. Bourrelly (Fig. 9D)

Cells are ellipsoidal and strongly flattened with a single visible flagellum emerging from the ventral (concave) side. One or two chloroplasts and one or two contractile vacuoles are present; a stigma may be present depending on the species. Two species of *Monochrysis* have been reported from North America (Whitford and Schumacher, 1969) as species of *Phaeaster*. This is a rare genus.

Naegeliella Correns (Fig. 11H)

Ovoid cells with a single golden-brown chloroplast (no pyrenoid) are grouped in the center of a mucilaginous matrix from which is extended a series of gelatinous tubes that are fused near their origins above the cells but are separated at the distal ends. Unlike the similar structures found in *Chrysochaete* there are no cytoplasmic strands extending from the cell up through the gel tubes. Other differences between these two genera include the production of *Ochromonas*-like zoospores in *Naegeliella* (zoospores of *Chrysochaete* have only one visible flagellum). This is a rare genus.

Ochromonas Wyssotzki (Figs. 4A, 8F)

Vegetative cells are spherical to ovoid, 2–30 μm in diameter with one long flimmer flagellum and one short smooth flagellum. Typically there are one or two golden-brown chloroplasts and one or more contractile vacuoles, and a stigma may or may not be present.

Ochromonas is considered by many to be the "model" genus of the Chrysophyceae. Except for a few species with distinctive morphology, identification of the more than 80 known freshwater species is difficult but is sometimes aided by the presence of stomatocysts with distinctive morphology (Fig. 4A). There are several common species.

Paraphysomonas de Saedeler (=*Physomonas* Kent) (Fig. 5C)

Cells are solitary, lack chloroplasts, and are attached to a substratum by a thin flexible stalk or are free-swimming (one long flimmer flagellum and one short smooth flagellum). Cells are covered in siliceous scales, the presence of which in some species can be detected in dried cells with light microscopy at high magnification using staining (e.g., Jensen's stain) or phase contrast optics. Electron microscopy is required for resolution of scale structure necessary for species identification. Kling and Kristiansen (1983), Nicholls (1981b, 1984a, b, 1985, 1988), and Wujek (1983) include EM micrographs of all known North American species. There are several common species.

Phaeodermatium Hansgirg (Fig. 9K)

The only known species, *P. rivulare* Hansgirg, grows as a macroscopic thallus attached to stones in fast-flowing streams. The thallus is generally circular in outline and is several cells thick near the center but decreases in thickness toward the margin. Reproduction is by cell division and by production of highly plastic *Hydrurus*-like zoospores that have a single flagellum and strongly angular shape. This is a rare genus.

[2]*Phaeogloea* Chodat (Fig. 11G)

The only known species, *P. mucosa*, consists of a spherical mucilaginous thallus 50–60 µm in diameter within which the spherical cells (5–7 µm diameter, one parietal chloroplast) are randomly distributed. Reproduction is by production of two to four autospores or biflagellated zoospores lacking stigmata. Autospore release results in mother cell wall remnants in the mucilage. This is a rare genus.

Phaeoplaca Chodat (Fig. 9I)

Cells contain one or two brownish-yellow chloroplasts, basal contractile vacuoles, and a pyrenoid. Thick-walled cells are rounded-rectangular in outline with cell height greater than width and are regularly arranged in a parenchymatous platelike thallus with epiphytic or epilithic habit. Reproduction is by cell division and production of *Chromulina*-like zoospores. This is a rare genus.

[2]*Phaeoschizochlamys* Lemmermann (Fig. 12D)

Cells are spherical with one or two chloroplasts and are grouped in clusters of usually two, four, or eight in the center of a gelatinous matrix. Cell division results in the production of two cup-shaped mother cell wall remnants that remain in the colonial matrix (the main identifying feature for this genus). Zoospores with a single visible flagellum are also produced. This is a rare genus.

Phaeosphaera West et West (Fig. 9J)

The colony consists of spherical cells with a single parietal chloroplast that are irregularly disposed in a gelatinous matrix forming a subcylindrical thallus with weak branching. Reproduction is by cell division; zoospores are not known. This is a rare genus.

[2]*Phaeothamnion* Lagerheim (Fig. 9H)

Cells are ellipsoidal with one to several parietal chloroplasts lacking pyrenoids. The thallus consists of a basal cell attached to a substratum and a main upright axis with side branches forming small treelike colonies. Cells produce four or eight *Ochromonas*-like zoospores. Earlier work suggested the presence of a palmelloid stage, but Dop (1980) showed that this "palmelloid stage" is really the separate genus *Tetrachrysis*. There would appear to be little difference between *Phaeothamnion* and *Chrysoclonium*. *Chrysoclonium* (not reported from North America) is poorly known and needs reinvestigation, especially as regards its mode of reproduction, its basal cell morphology, and its free-floating (not attached) habit. This is a rare genus.

Poterioochromonas Sherffel (Fig. 5K)

The *Ochromonas*-like cell with a very short second flagellum is housed within a cup-shaped lorica at the end of a long hollow stalk (best seen in dried or stained specimens or in those living specimens observed with phase contrast microscopy). The lorica is constructed of chitinous microfibrils (Herth et al., 1977). The protoplast lacks a basal stalk for attachment inside the lorica (thus distinguishing this cell form from that of, e.g., *Dinobryon* and *Epipyxis*). The validity of the closely related genus *Arthropyxis* Pascher (identical to *Poterioochromonas* except that it is colorless) is questionable because *Poterioochromonas* is known to lose its chloroplasts under certain culture conditions (Wolken and Palade, 1952). Laboratory cultured species of this genus have been widely used in physiological and toxicological investigations (Boxhorn et al., 1998; Leeper and Porter, 1995). It is most common in small sheltered ponds and temporary pools. One or two species may be locally abundant in the neuston.

Pseudokephyrion Pascher emend. Schmid (incl. *Kephyriopsis* Pascher and Ruttner) (Figs. 3C, 6E)

The *Dinobryon*-like cell is housed in a lorica of interwoven cellulosic microfibrils that is radially symmetric (unlike the bilateral symmetry of *Chrysolykos*) and is often impregnated with iron salts, which render some forms a yellowish-brown color. *Pseudokephyrion* is distinguished from *Dinobryon* (*Dinobryopsis*) by a narrowing of the lorica mouth, lack of any attenuation in the basal region, and much smaller size (typically <10 μm long). Species are determined on the basis of lorica shape. *Pseudokephyrion* Pascher includes the genus *Kephyriopsis* Pascher and Ruttner; the features used previously to distinguish these two genera (flagellar length and insertion, chloroplast structure, and lorica morphology) are more appropriately applied to species of *Pseudokephyrion*. Hilliard (1967) and Meyer (1971) provide descriptions of North American species. There are several common species.

Rhizochrysis Pascher (Fig. 10B)

This is essentially a *Chrysamoeba* without a flagellum. Species are distinguished on the basis of cell size, rhizopod and chloroplast number and shape, and stomatocyst morphology. There exists some uncertainty about the validity of this genus. Starmach (1985) has included all *Rhizochrysis* species in *Chrysamoeba*. King (1983) reviewed the taxonomy of some North American species of these two genera.

Rhizoochromonas Nicholls (Fig. 8H)

The vegetative stage of this genus is sessile with threadlike rhizopodia emerging at several sites around the cell. The motile stage has two flagella of unequal length; there is no stigma. Nicholls (1990) has described the only known species, *R. endoloricata*, which grows inside *Dinobryon* loricae. This is a rare genus.

Saccochrysis Korshikov *emend.* Andersen (Fig. 9C)

Cells with one visible flagellum are randomly (rarely radially) disposed in a gelatinous envelope within which they are constantly mobile. Because the flagella do not extend beyond the matrix into the aqueous environment, the colony floats passively. The chloroplast is single and band-shaped. Reproduction is by vegetative cell division and release of spindle-shaped zoospores. Andersen (1986) provided the first report of *Saccochrysis piriformis* from several sites in North America with an amended description including a rather complicated life cycle.

[2]*Selenophaea* Chodat (Fig. 10I)

The only known species, *S. granulosa* Chodat, is inadequately known; nothing is known of stomatocyst morphology and reproduction. The vegetative stage consists of planktonic, solitary cells with a somewhat reniform shape and a cell envelope with wartlike ornamentations. This is a rare genus.

[2]*Sphaeridiothrix* Pascher *et* Vlk (Fig. 11E)

Cells are spherical to ellipsoidal and are slightly flattened at the poles near their abutting surfaces. Cells contain a single lobed chloroplast and are aligned within a stiff gelatinous matrix in an unbranched chain of cells. The thallus is attached to the substratum by a short extension of the basal part of the sheath. Zoospores are of the *Ochromonas* type. Variations in form of wild and cultured material from North America were described by Andrews (1970); see also Dop (1980). This is a rare genus.

Spiniferomonas Takahashi (Figs. G–I, C)

Cells are 3–10 μm in diameter, of typical ochromonad structure (one or two chloroplasts, one or two contractile vacuoles, stigma) and are covered with a single type of siliceous spine-scale and either one or two differently structured plate-scales. See also the comments under *Paraphysomonas* regarding LM and EM detection of scales and species identification. Siver (1988a) provides a North American accounting of most of the known species. There are several common species.

Spumella Cienkowsky (=*Heterochromonas* Pascher; =*Monas* O.F. Müller) (Fig. 5D)

This genus is essentially a colorless *Ochromonas*, at least one species of which has a leucoplast (Belcher and Swale, 1976). Other features include possible presence of a stigma and formation of stomatocysts. This is a rare genus.

Stephanoporos Conrad and Pascher emend. Bourrelly (Fig. 6K)

Protoplasts of this genus are housed within a globular or ellipsoidal lorica that has multiple pores arranged in a line around the equatorial region of the lorica and through which rhizopodia emerge. The lorica may be attached to a substratum or free-floating. Unlike *Heliapsis*, the rhizopodia are not interconnected with those of other cells. This is a rare genus.

Stichochrysis Pringsheim (Fig. 9G)

The only known species, *S. immobilis*, exists as short (2–8 cells) unbranched filaments. Cells are 7–10 μm long and about 5 μm wide with one or two parietal chloroplasts. Reproduction is by cell division; no zoospores or stomatocysts are known. This is a rare genus.

[2]*Stichogloea* Chodat (Fig. 9E)

Cells with one or two chloroplasts are ellipsoidal and grouped end-to-end in rough tetrahedral arrangements within a gelatinous free-floating matrix. Mucilaginous connectives linking cells are sometimes present. Reproduction is by autospores or *Ochromonas*-like zoospores. The genus is not uncommon.

Stylobryon Fromentel (Fig. 5A)

The protoplast lacks chloroplasts and is housed within a lorica, attached at its posterior end by a stalk. Loricae form branching colonies. Cells possess one long and one short flagellum. The genus *Stylobryon* Fromentel (1874) exists on the basis of a very imprecise description that included only one flagellum. Stokes (1885) described *S. abbotti* with two flagella of unequal length. For the present, Stokes' concept of the genus (i.e., a colorless *Dinobryon*) is perpetuated here. This is a rare genus.

Stylochrysalis Stein (Fig. 6A)

The protoplast with one or two chloroplasts and one or two contractile vacuoles is housed in a very thin-walled lorica with a restricted mouth opening. The lorica has a long delicate stipe, which is attached to the substratum (usually a filamentous alga). Mature cells may have rhizopodia extending through the lorica mouth or two subequal flagella. The flagellate form may be indicative of zoospore formation. Skuja (1948) described *Stylochrysalis parasitica* with rhizopodia that may have been misinterpreted as flagella in the original description (Stein, 1878). If flagella are truly absent in the loricate nonzooid stage, then this genus should perhaps be combined with *Stylococcus* Chodat (a "biflagellated *Stylochrysalis*"), over which it has nomenclatural priority. This is a rare genus.

Syncrypta Ehrenberg emend. Bourrelly (Fig. 7D)

Species of this genus consist of pyriform cells forming globular free-swimming colonies with a *Synura*-like appearance (but without siliceous scales). Cells lack the interconnecting mucilaginous strands that join *Uroglena* or *Uroglenopsis* cells within a colony. Also the intercellular distances in *Uroglena* or *Uroglenopsis* colonies are greater than in *Syncrypta*. The whole colony is engulfed in a mucilagenous matrix. The difference between *Syncrypta* and *Lepidochrysis* Ikävalko, Kristiansen, and Thomsen is that the latter genus is surrounded by an investiture of scales of organic fibrillar nature. See Ikävalko *et al.* (1994) and Wujek and Thompson (2001) for a review of the taxonomy of this and related genera. The few North American reports of *Syncrypta* do not negate the likelihood that this genus is really a scaleless growth form of *Synura*.

Synuropsis Schiller emend. Wujek *et* Thompson (=*Volvochrysis* Schiller; = *Synochromonas* Korshikov) (Fig. 7H)

Cells are spherical to pyriform-oval with two to eight or more yellow-brown chloroplasts with or without a stigma and are arranged in a *Synura*-like mucilaginous colony. Cells have heterkont flagellation and are joined in the center of the colony by a close-branching system of gelatinous "tubes" generally only visible if impregnated with iron salts or after staining with methylene blue (Wujek and Thompson, 2001). This is a rare genus.

[2]*Tetrachrysis* Dop (Figs. 4F, 12A)

Cells are spherical or ovoid with one or two chloroplasts and are enclosed in a tubular gelatinous matrix. They are arranged in linear series of two to four cells, forming a branching thallus. Reproduction is by cell division and by production of *Ochromonas*-like zoospores. One of the two known species of this genus, *T. dendroides*, was reported from North America by Smith (1950) as the palmelloid stage of *Phaeothamnion confervicolum* (Dop, 1980). This is a rare genus.

[2]*Tetrasporopsis* (De Toni) Lemmermann and *Dermatochrysis* Entwisle *et* Andersen (Fig. 10J)

These genera consist of a monostromatic thallus (single layer of cells in mucilage) growing attached to submerged substrata often in flowing water. Entwistle and Andersen (1990) reviewed European and Australian material and created the new genus *Dermatochrysis* for those *Tetrasporopsis*-like specimens that have contractile vacuoles, but lack a cell wall and a stigma; *Tetrasporopsis* has a cell wall and a stigma, but lacks contractile vacuoles. It is possible that all North American reports of *Tetrasporopsis* can be referred to *Dermatochrysis*, leaving some uncertainty as to whether or not *Tetrasporopsis* has been found in North America.

Uroglena Ehrenberg (Fig. 8A)

Cells are regularly but widely spaced in globular free-swimming colonies, propelled by flagella directed outward at the periphery of the colony. The short flagellum is about one-half the length of the longer flimmer flagellum. Cells are joined at their attenuated

---

[2]On the basis of pigmentation, nuclear-encoded SSU rDNA analysis, and ultrastructure of vegetative cells and motile stages, Andersen *et al.* (1998c) and Bailey *et al.* (1998) have suggested moving the genera *Chrysapion*, *Chrysoclonium*, *Chrysodictyon*, *Phaeobotrys*, *Phaeogloea*, *Phaeoschizochlamys*, *Phaeothamnion*, *Selenophaea*, *Sphaeridiothrix*, *Stichogloea*, *Tetrachrysis*, *Tetrapion*, and *Tetrasporopsis* to the new class Phaeothamniophyceae.

bases by a branching system of cytoplasmic threads. A stigma is usually very evident; there are one or two chloroplasts, of a light golden color. Species are determined on the basis of stomatocyst morphology. See Wujek and Thompson (2002) for a revision of this and related genera. A few species may form "blooms" in lakes, coloring the water a golden-brown and imparting "fishy" odors.

Uroglenopsis Lemmermann (Figs. 4H, 8B)

The genus is like *Uroglena* in colony shape; however, the cells are truncate at the posterior end and are joined by a branching system of gelatinous stalks. The short flagellum is at most one-fourth the length of the long flagellum (Wujek, 1976; Wujek and Thompson, 2002). There are a few common species.

VI. GUIDE TO LITERATURE FOR SPECIES IDENTIFICATION

Two general references on Chrysophyceae are by Bourrelly (1957, 1981) (in French), which are valuable for genus identifications. A series emanating from previous international chrysophyte symposia should also be consulted (Kristiansen and Andersen, 1983; Kristiansen *et al.*, 1989; Sandgren *et al.*, 1995; Kristiansen and Cronberg, 1996).

Useful references for identifying species include Huber-Pestalozzi (1941) (in German), Matvienko (1954) (in Russian), Pascher (1939) (in German), Prescott (1962), and Whitford and Schumacher (1973). The most contempory reference for the Chrysophyceae is Starmach (1985) (in German); understandably, it lacks information on the more recently described species. Another shortcoming of this volume is the omission of the entire genus *Paraphysomonas*. Because no inclusive study of North American Chrysophyceae has been published, we recommend Starmach (1985) as the basic manual for identification of species with augmentation by Thomsen *et al.* (1981) and Preisig and Hibberd (1982a, b, 1987) for *Paraphysomonas*.

References containing descriptions of North American chrysophycean species *not included* in Starmach (1985) are listed next. (see also references cited in Sect. V.B):

Chrysoamphipyxis—Nicholls (1987)
Chrysocapsopsis—Thompson and Wujek (1998a)
Chrysolepidomonas—Peters and Andersen (1993)
Dinobryon—Ahlstrom (1937), Nygaard (1978), Nicholls (2000)
Kephyrion—Nicholls (1981c)
Paraphysomonas—Kling and Kristiansen (1983), Nicholls (1984b), Wujek and Gardiner (1985)
Rhizoochromonas—Nicholls (1990)
Spiniferomonas—Nicholls (1984a, c, d, 1989), Siver (1987, 1988b), Wujek and Bland (1988)
Synuropsis—Wujek and Thompson (2001)
Uroglena and *Uroglenopsis*—Wujek and Thompson (2002)

ACKNOWLEDGMENTS

D.E.W. thanks Dr. James Gillingham, Director of the Central Michigan University Biological Station, for use of the station's facilities where a portion of this chapter was written.

LITERATURE CITED

Ahlstrom, E. H. 1937. Studies on variability in the genus *Dinobryon* (Mastigophora). Transactions of the American Microscopical Society 56:139–159.

Albertano, P., Pinto, G., Pollio, A. 1994. Ecophysiology and ultrastructure of an acidophilic species of *Ochromonas* (Chrysophyceae, Ochromonadales). Archiv für Protistenkunde 144:75–82.

Allen, D. M., Northcote, D. H. 1975. The scales of *Chrysochromulina chiton*. Protoplasma 83:389–412.

Andersen, R. A. 1982. A light and electron microscopical investigation of *Ochromonas sphaerocystis* Matvienko (Chrysophyceae): The statospore, vegetative cell and its peripheral vesicles. Phycologia 21:390–398.

Andersen, R. A. 1986. Some new observations on *Saccochrysis piriformis* Korsh. emend. Andersen (Chrysophyceae), *in*: Kristiansen, J., Andersen, R. A., Eds., Chrysophytes: Aspects and problems. Cambridge Univ. Press, Cambridge, UK, pp. 107–118.

Andersen, R. A. 1987. Synurophyceae classis nov., a new class of algae. American Journal of Botany 74:337–353.

Andersen, R. A., Mulkey, T. J. 1983. The occurrence of chlorophylls c_1 and c_2 in the Chrysophyceae. Journal of Phycology 19:289–294.

Andersen, R. A., Wetherbee, R. 1992. Microtubules of the flagellar apparatus are active during prey capture in the chrysophycean alga *Epipyxis pulchra*. Protoplasma 166:8–20.

Andersen, R. A., Saunders, G. W., Paskind, M. P., Sexton, J. P. 1993. The ultrastructure and 18s rRNA gene sequence for *Pelagomonas calceolus* gen. et sp. nov., and the description of a new algal class, the Pelagophyceae classis. nov. Journal of Phycology 29:701–715.

Andersen, R. A., Brett, R. W., Potter, D., Sexton, J. P. 1998a. Phylogeny of the Eustigmatophyceae based on 18s rDNA, with emphasis on *Nannochloropsis*. Protist 149:61–74.

Andersen, R. A., Potter, D., Bidigare, R. R., Latasa, M., Rowan, K., O'Kelly, C. J. 1998b. Characterization and phylogenetic position of the enigmatic golden alga *Phaeothamnion confervicola*: Ultrastructure, pigment composition and partial SSU RDNA sequence. Journal of Phycology 34:286–298.

Andersen, R. A., Potter, D., Daugberg, N., Bailey, J. C. 1998c. Phylogeny of Chrysophyceae using 18s and rbcL sequences, with comments on ultrastructure and classification. Journal of Phycology (Supplement) 34:3.

Andersen, R. A., van de Peer, Y., Potter, D., Sexton, J. P., Kawachi, M., LaJeunesse, T. 1999. Phylogenetic analysis of the SSU rRNA from members of the Chrysophyceae. Protist 150:71–84.

Andrews, H. T. 1970. Morphology and taxonomy of the euryhaline chrysophyte, *Sphaeridiothrix compressa*. Journal of Phycology 6:133–136.

Bachmann, H. 1908. Vergleichend Studien über das Phytoplankton von Seen Schottlands und der Schweiz. Archiv für Hydrobiologie 3:1–90.

Bachmann, H. 1921. Beiträge zur Algenflora des Süsswassers von Westgrönland. Mitteilungen der Naturforschlichen Gesellschaft Luzern 8:1–181.

Bailey, J. C., Bidigare, R. R., Christensen, S. J., Andersen, R. A. 1998. Phaeothamniophyceae classis nova: A new lineage of chromophytes based upon photosynthetic pigments, *rbc*L sequence analysis and ultrastructure. Protist 149:245–263.

Belcher, J. H. 1966. *Microglena butcheri* nov. sp., a flagellate from the English Lake District. Hydrobiologia 27:65–69.

Belcher, J. H. 1968. Lorica construction in *Pseudokephyrion pseudospirale* Bourrelly. British Phycological Bulletin 3:495–499.

Belcher, J. H. 1969. A morphological study of the phytoflagellate *Chrysococcus rufescens* Klebs in culture. British Journal of Phycology 4:105–117.

Belcher, J. H., Swale, E. M. F. 1976. *Spumella elongata* (Stokes) nov. comb., a colourless flagellate from soil. Archiv für Protistenkunde 118:215–220.

Bird, D. F., Kalff, J. 1986. Bacterial grazing by planktonic algae. Science 231:493–495.

Bird, D. F., Kalff, J. 1987. Algal phagotrophy regulating factors and importance relative to photosynthesis in *Dinobryon* (Chrysophyceae). Limnology and Oceanography 32:277–284.

Bourrelly, P. 1957. Recherches sur les Chrysophycées. Revue d'Algologie 1:1–412.

Bourrelly, P. 1963. Loricae and cysts in the Chrysophyceae. Annals of New York Academy Sciences 108: 421–429.

Bourrelly, P. 1981. Les algues d'eau douce, les algues jaunes et brunes, Vol 2. Boubée, Paris.

Boxhorn, J. E., Holden, D. A., Boraas, M. E. 1998. Toxicity of the chrysophyte flagellate *Poterioochromonas malhamensis* to the rotifer *Brachionus angularis*. Hydrobiologia 387:283–287.

Brown, R. M., Herth, W., Franke, W. W., Romanovicz, D. 1973. The role of the Golgi apparatus in the biosynthesis and secretion of a cellulosic glycoprotein in Pleurochrysis: A model system for the synthesis of structural polysaccharides, *in*: Loewus, F., Ed., Biogenesis of plant cell wall polysaccharides. Academic Press, New York, pp. 207–257.

Caron, D. A., Lim, E. L., Dennett, M. R., Gast, R. J., Kosman, C., DeLong, E. F. 1999. Molecular phylogenetic analysis of the heterotrophic chrysophyte genus *Paraphysomonas* (Chrysophyceae), and the design of rRNA-targeted oligonucleotide probes for two species. Journal of Phycology 35:824–837.

Chodat, R. 1922. Materiaux pour l'histoire des algues de la Suisse, I–IX. Bulletin de la Société Botanique de Genève 13:66–114.

Christensen, T. 1962. Systematisk botanik II, Vol. 2. Alger. Munksgaard, Copenhagen, 178 p.

Christensen, T. 1964. The gross classification of algae, *in*: Jackson, D. F., Ed., Algae and man. Plenum, New York, pp. 59–64.

Cienkowski, C. 1870. Über Palmellaceen und einige Flagellaten. Archiv für Mikroskopische Anatomie 6:421–438.

Coleman, A. W., Goff, L. J. 1991. DNA analysis of eukaryotic algal species. Journal of Phycology 27:463–473.

Cottingham, K. L., Carpenter, S. R., St. Amand, A. L. 1998. Responses of epilimnetic phytoplankton to experimental nutrient enrichment in three small seepage lakes. Journal of Plankton Research 20:1889–1914.

Couté, A. 1983. Ultrastructure de *Chromophyton rosanofii* Woronin emend. Couté et *Chr. vischeri* (Bourrel.) nov comb. (Chrysophyceae, Ochromonadales, Ochromonadaceae). Protistologica 19:393–416.

Cox, E. J. 1990. Studies on the algae of a small softwater stream. I. Occurrence and distribution with particular reference to the diatoms. Archiv für Hydrobiologie, Supplement 83:525–552.

Cronberg, G. 1989. Scaled chrysophytes from the tropics. Beihefte zur Nova Hedwigia 95:191–232.

Cronberg, G. 1996. Scaled chrysophytes from the Okavango Delta, Botswana, Africa. Beihefte zur Nova Hedwigia 114:91–108.

Daugbjerg, N., Andersen, R. A. 1997a. A molecular phylogeny of the heterokont algae based on analyses of chloroplast-encoded *rbc*L sequence data. Journal of Phycology 33:1031–1041.

Daugbjerg, N., Andersen, R. A. 1997b. Phylogenetic analyses of the *rbc*L sequences for Haptophyceae and heterokont algae suggest their chloroplasts are unrelated. Molecular Biology and Evolution 14:1242–1251.

Dillard, G. E. 1970. *Bourrellia*, a new genus (Chrysophyceae) from South Carolina. Journal of the Elisha Mitchell Science Society 86:128–130.

Dillard, G. E., Crider, S. B. 1970. Kentucky algae, I. Transactions of the Kentucky Academy of Sciences 31:66–72.

Doddema, H., van der Veer, J. 1983. *Ochromonas monicis* sp. nov., a particle feeder with bacterial endosymbionts. Cryptogamie Algologie 4:89–97.

Dop, A. J. 1978. Systematics and morphology of *Chrysochaete brittannica* (Godward) Rosenberg and *Phaeoplaca thallosa* Chodat (Chrysophyceae). Acta Botanica Neerlandica 27:35–60.

Dop, A. J. 1980. The genera *Phaeothamnion* Lagerheim, *Tetrachrysis* gen. nov. and *Sphaeridothrix* Pascher et Vlk (Chrysophyceae). Acta Botanica Neerlandica 29:65–86.

Douglas, M. S. V., Smol, J. P. 1995. Periphytic diatoms assemblages from high Arctic ponds. Journal of Phycology 31:60–69.

Duff, K. E., Zeeb, B. A., Smol, J. P. 1995. Atlas of chrysophycean cysts. Developments in Hydrobiology, Vol. 99. Kluwer Academic, Dordrecht.

Dunlap, J. R., Walne, P. L., Preisig, H. R. 1987. Manganese mineralization in chrysophycean loricas. Phycologia 26:394–396.

Duthie, H. C., Ostrofsky, M. L. 1978. Additions to the freshwater algae of Labrador. Nova Hedwigia 29:237–242.

Duthie, H. C., Socha, R. 1976. A checklist of the freshwater algae of Ontario, exclusive of the Great Lakes. Nature Canada (Quebec) 103:83–109.

Ehrenberg, C. G. 1838. *Die Infusionsthierchen als vollkommene Organismen*. Verlag von Leopold Voss, Leipzig, 547 p.

Ellis-Adam, A. C. 1983. Some new and interesting benthic Chrysophyceae from a Dutch moorland pool complex. Acta Botanica Neerlandica 32:1–23.

Eloranta, P. 1989. Ecological studies. On the ecology of the genus *Dinobryon* in Finnish lakes. Beihefte zur Nova Hedwigia 95:99–109.

Entwisle, T. J., Andersen, R. A. 1990. A re-examination of *Tetrasporopsis* (Chrysophyceae) and the description of *Dermatochrysis* gen. nov. (Chrysophyceae): A monostromatic alga lacking cell walls. Phycologia 29:263–274.

Esser, S. C., Valkenburg, S. D. 1977. The fine structure of vegetative and statospore forming cells of *Uroglena volvox* Ehrenberg. Journal of Phycology (Supplement) 13:20.

Felip, M., Sattler, B., Psenner, R., Catalan, J. 1995. Highly active microbial communities in the ice and snow cover of high mountain lakes. Applied and Environmental Microbiology 61:2394–2401.

Findlay, D. L., Kasian, S. E. M. 1996. The effect of incremental pH recovery on the Lake 223 phytoplankton community. Canadian Journal of Fisheries and Aquatic Sciences 53: 856–864.

Fott, B. 1959. Zur Frage der Sexualität bei den Chrysomonaden. Nova Hedwigia 1:115–130.

Fromentel, E. de. 1874. Études sur les Microzoaires ou Infusoires proprement dits. Paris.

Gayral, P., Haas, C., Lepailleur, H. 1972. Alternance morpholgique de générations et alternance de phases chez les Chrysophycées. Société Botanique Française 1972:215–230.

Gibbs, S. P. 1981. The chloroplast endoplasmic reticulum: Structure, function and evolution significance. International Review of Cytology 72:49–99.

Gromov, B. V., Mamkayeva, K. A., Bobina, V. D. 1988. Acidophilic Ochromonas (Chrysophyceae) from sulphuric spring on the Kunashir Island. Akademia Nauk USSR, Izvestia, Seriya Biologia 2:293–296.

Hällfors, G., Hällfors, S. 1988. Records of chrysophytes with siliceous scales (Mallomonadaceae and Paraphysomonadaceae) from Finnish inland waters. Hydrobiologia 161:1–29.

Hansen, P. 1996. Silica-scaled Chrysophyceae and Synurophyceae from Madagascar. Archiv für Protistenkunde 147:145–172.

Happey, C., Moss, B. 1967. Some aspects of the biology of Chrysococcus diaphanus in Abbot's Pond Somerset. British Phycological Bulletin 3:269–279.

Happey-Wood, C. M. 1976. Vertical migration pattern in phytoplankton of mixed species composition. British Phycological Journal 11:355–369.

Heinonen, P. 1980. Quantity and composition of phytoplankton in Finnish inland waters. Public Water Resources Institute, National Board of Waters, Finland 37:1–91.

Herth, W., Zugenmaier, P. 1979. The lorica of Dinobryon. Journal of Ultrastructure Research 69:262–272.

Herth, W., Kuppel, A., Brown, R. M. 1975. Chitinous fibrils in the lorica of the flagellate chrysophyte Poterioochromonas stipitata (syn. Ochromonas malhamensis). Cytobiologia 10:268–284.

Herth, W., Kuppel, A., Schnepf, E. 1977. Chitinous fibrils in the lorica of the flagellate chrysophyte Poteriochromonas stipitata (syn. Ochromonas malhamensis). Journal of Cell Biology 73:311–321.

Hibberd, D. J. 1976a. The ultrastructure and taxonomy of the Chrysophyceae and Prymnesiophyceae (Haptophyceae): A survey with some new observations on the ultrastructure of the Chrysophyceae. Botanical Journal of Linnaean Society 72:55–80.

Hibberd, D. J. 1976b. Observations on the ultrastructure of three new species of Cyathobodo Petersen et Hansen (C. salpinx, C. intricatus and C. simplex) and on the external morphology of Pseudodendromonas vlkii Bourrelly. Protistologica 12:249–261.

Hibberd, D. J. 1977. Ultrastructure of cysts formation in Ochromonas tuberculata (Chrysophyceae). Journal of Phycology 13309–320.

Hibberd, D. J. 1983. Ultrastructure of the colonial colourless zooflagellates Phalansterium digitatum Stein (Phalansteriida ord. nov.) and Spongomonas uvella Stein (Spongomonadida ord. nov.). Protistologica 19:523–535.

Hill, W. 1996. Effects of light, in: Stevenson, R. J., Bothwell, M. L., Lowe, R. L., Eds., Benthic algal ecology in freshwater ecosystems. Academic Press, San Diego, pp. 121–148.

Hilliard, D. K. 1966. Studies on Chrysophyceae from some ponds and lakes in Alaska. V. Notes on the taxonomy and occurrence of phytoplankton in an Alaskan pond. Hydrobiologia 28:553–576.

Hilliard, D. K. 1967. Studies on Chrysophyceae from some ponds and lakes in Alaska. VII. Notes on the genera Kephyrion, Kephyriopsis, and Pseudokephyrion. Nova Hedwigia 14:39–56.

Hilliard, D. K. 1968. Seasonal variation in some Dinobryon species (Chrysophyceae) from a pond and a lake in Alaska. Oikos 19:28–38.

Hilliard, D. K. 1971a. Notes on the occurrence and taxonomy of some planktonic chrysophytes in an Alaskan lake, with comments on the genus. Archiv für Protistenkunde 113:98–122.

Hilliard, D. K. 1971b. Observations on the lorica structure of some Dinobryon species (Chrysophyceae), with comments on related genera. Österreichische Botanische Zeitschrift 119:25–40.

Hilliard, D. K., Asmund, B. 1963. Studies on Chrysophyceae from some ponds and lakes in Alaska. II. Notes on the genera Dinobryon, Hyalobryon and Epipyxis with descriptions of new species. Hydrobiologia 22:331–397.

Hoek, C. van den, Mann, D. G., Jahns, H. M. 1995. Algae, an introduction to phycology. Cambridge Univ. Press, Cambridge, UK.

Hoffman, L. R., Vesk, M., Pickett-Heaps, J. D. 1986. The cytology and ultrastructure of zoospores of Hydrurus foetidus (Chrysophyceae). Nordic Journal of Botany 6:105–122.

Huber-Pestalozzi, G. 1941. Das Phytoplankton des Süsswassers. Die Binnengewässer. Vol. XVI, 2, Chrysophyceen, farblose Flagellaten, Heterokonten. Schweizerbart'sche Verlagsbuchhandlung, Stuttgart.

Ikävalko, J., Kristiansen, J., Thomsen, H. A. 1994. A revision of the taxonomic position of Syncrypta glomifera (Chrysophyceae), establishment of a new genus Lepidochrysis, and observations on the occurrence of L. glomifera comb. nov. in brackish waters. Nordic Journal of Botany 14:339–344.

Ito, H., Takahashi, E. 1982. Seasonal fluctuation of Spiniferomonas (Chrysophyceae, Synuraceae) in two ponds on Mt. Rokko, Japan. Japanese Journal of Phycology 30:272–278.

Jacobs, J. E. 1971. A preliminary taxonomic survey of the freshwater algae of the Bell W. Baruch Plantation in Georgetown County, South Carolina. Journal of the Elisha Mitchell Science Society 87:26–30.

Jacobsen, B. 1985. Scale-bearing Chrysophyceae (Mallomonadaceae and Paraphysomadaceae) from West Greenland. Nordic Journal of Botany 5:381–395.

Jeffrey, S. W. 1989. Chlorophyll c pigments and their distribution in chromphyte algae, in: Leadbeater, B. S. C., Diver, W. L., Ed., The chromophyte algae: Problems and perspectives. Clarendon, Oxford, pp. 13–36.

Jones, R. I., Rees, S. 1994. Influence of temperature and light on particle ingestion by the freshwater phytoflagellate Dinobryon. Archiv für Hydrobiologie 132:203–211.

Karim, A. G., Round, F. E. 1967. Microfibrils in the lorica of the freshwater alga Dinobryon. New Phytologist 66:409–412.

Kilham, P., Tilman, D. 1979. The importance of resource competition and nutrient gradients for phytoplankton ecology. Beihefte zur Archiv für Hydrobiologie 13:110–119.

Kilham, S. S., Kilham, P. 1978. Natural community bioassays: Predictions of results based on nutrient physiology and competition. International Vereinigung für Theoretische und Angewandte Limnologie Verhandlungen 20:68–74.

King, J. M. 1983. A report on the occurrence of Chrysamoeba radians Klebs (Chrysophyceae) in Kentucky. Transactions of the Kentucky Academy of Sciences 44:159–161.

King, J. M. 1984. The occurrence of Chrysopyxis urna (Chrysophyceae) in the United States. Transactions of the American Microscopical Society 103:317–319.

Klebs, G. 1893a. Flagellatenstudien. I. Zeitschrift für Wissenschaft, Zoologie 55:265–351.

Klebs, G. 1893b. Flagellatenstudien. II. Zeitschrift für Wissenschaft, Zoologie 55:353–445.

Kling, H. J., Kristiansen, J. 1983. Scale bearing Chrysophyceae (Mallomonadaceae) from central and northern Canada. Nordic Journal of Botany 3:269–290.

Korshikov, A. A. 1929. Studies on the Chrysomonads I. Archiv für Protistenkunde 67:253–290.

Korshikov, A. A. 1942. On some new or little known flagellates. Archiv für Protistenkunde 95:22–44.

Kouwets, F. A. C. 1980. Floristic and ecological notes on some little known unicellular and colony-forming algae from a Dutch moorland pool complex. Cryptogamie Algologie 4:293–309.

Krienitz, L., Hehmann, A., Casper, S. J. 1997. The unique phytoplankton community of a highly acidic bog lake in Germany. Nova Hedwigia 65:411–430

Kristiansen, J. 1969. Lorica structure in *Chrysolykos* (Chrysophyceae). Svensk Botanisk Tidskrift 64:162–168.

Kristiansen, J. 1972. Studies on the lorica structure in Chrysophyceae. Svensk Botanisk Tidskrift 66:184–190.

Kristiansen, J. 1986. The ultrastructural bases of chrysophyte systematics and phylogeny. Critical Reviews in Plant Science 4:149–211.

Kristiansen, J. 1988. Seasonal occurrence of silica-scaled chrysophytes under eutrophic conditions. Hydrobiologia 161:171–184.

Kristiansen, J. 1990. Phylum Chrysophyta, in: Margulis, L., Corliss, J. O., Melkonian, M., Chapman, D. J., Eds., Handbook of Protoctista. Jones & Bartlett, Boston.

Kristiansen, J. 1992. Silica-scaled chrysophytes from West Greenland: Disko Island and the Søndre Strømfjord region. Nordic Journal of Botany12:525–536.

Kristiansen, J. 1995a. History of chrysophyte research: Origin and development of concepts and ideas, in: Sandgren, C. D., Smol, J. P., Kristiansen, J., Eds., Chrysophyte algae: Ecology, phylogeny and development. Cambridge Univ. Press, Cambridge, UK, pp. 1–22.

Kristiansen, J. 1995b. Silica structures in the taxonomy and identification of scaled chrysophytes. Beihefte zur Nova Hedwigia 112:355–365.

Kristiansen, J., Andersen, R. A. 1983. Chrysophytes: Aspects and problems. Cambridge Univ. Press, Cambridge, UK.

Kristiansen, J., Cronberg, G. 1996. Chrysophytes, Progress and New Horizons. Beihefte zur Nova Hedwigia, Vol. 114 Cramer, Stuttgart.

Kristiansen, J., Tong, D. 1995. A contribution to the knowledge of the silica-scaled chrysophytes in Mexico. Algological Studies 77:1–6.

Kristiansen, J., Vigna, M. S. 1996. Bipolarity in the distribution of silica-scaled chrysophytes. Hydrobiologia 336:121–126.

Kristiansen J., Cronberg, G., Geissler, U. 1989. Chrysophytes, Development and Perspectives. Beihefte zur Nova Hedwigia, Vol. 95 Cramer, Stuttgart.

Lackey, J. B. 1938. Scioto River forms of *Chrysococcus*. American Midland Naturalist 20:619–623.

Lakshminarayana, J. S. S., Devi, J. S. 1975. Record of the alga *Hydrurus foetidus* (Vill.) Trèv. from Labrador. The Canadian Field Naturalist 89:186–188.

Lee, R. E., Kugrens, P. 1998. The possible adaptive advantage of chloroplast E.R.—the ability to out compete at low dissolved CO_2 concentrations. Journal of Phycology (Supplement) 34:31.

Leeper, D. A., Porter, K. G. 1995. Toxicity of the mixotrophic chrysophyte *Poterioochromonas malhamensis* to the cladoceran *Daphnia ambigua*. Archiv für Hydrobiologie 134:207–222.

Lepistö, L., Rosenström, U. 1998. The most typical phytoplankton taxa in four types of boreal lakes. Hydrobiologia 369/370:89–97.

Lim, E. E., Amaral, L., Caron, D. A., DeLong, E. F. 1993. Application of rRNA-based probes for observing marine nanoplanktonic protists. Applied and Environmental Microbiology 59:1647–1655.

Lim, E. E., Caron, D. A., DeLong, E. F. 1996. Development and field application of a quantitative method for examining natural assemblages of protists using oligonucleotide probes. Applied and Environmental Microbiology 62:1416–1423.

Lim, E. E., Caron, D. A., Dennett, M. R. 1999. The ecology of *Paraphysomonas imperforata* based on studies employing oligonucleotide probe identification in coastal water samples and enrichment culture. Limnology and Oceanography 44:37–51.

Lund, J. W. G., Kipling, C., Le Cren, E. D. 1958. The inverted microscope method of estimating algal numbers and the statistical basis of estimation by counting. Hydrobiologia 11:143–170.

Matvienko, A. M. 1954. Zolotistie *Vodorilso* (Chrysophyte) Opredelitel Presnovodynch Vodorosliej SSR, 3, Moscow 365 p.

McKenzie, C., Kling, H. 1989. Scale-bearing Chrysophyceae (Mallomonadaceae and Paraphysomonadaceae) from Mackenzie Delta area lakes, Northwest Territories, Canada. Nordic Journal of Botany 9:103–112.

McKenzie, C., Deibel, D., Paranjape, M., Thompson, R. J. 1995. The marine mixotroph *Dinobryon balticum* (Chrysophyceae): Phagotrophy and survival in a cold ocean. Journal of Phycology 31:19–24.

McNabb, C. D. 1960. Enumeration of freshwater phytoplankton concentrated on the membrane filter. Limnology and Oceanography 5:57–61.

Medlin, L. K., Kooistra, W. H. C. F., Potter, D., Saunders, G. W., Andersen, R. A. 1997. Phylogenetic relationships of the 'golden algae' (haptophytes, heterokont chromophytes) and their plastids. Plant Systematics and Evolution (Supplement) 11:187–219.

Melkonian, M., Robenek, H., Reize, I. B., Preisig, H. 1987. Maturation of a flagellum/basal body requires more than one cell cycle in algal flagellates: Studies on *Nephroselmis olivacea* (Prasinophyceae), in: Wiesser, W., Robinson, D. G., Starr, R. C., Eds., Algal development, molecular and mellular aspects. Springer-Verlag, Berlin.

Meyer, R. L. 1971. Notes on the algae of Arkansas. 1. *Chrysococcus*, *Kephyrion*, *Kephyriopsis*, *Pseudokephyrion* and *Stenokalyx*. Arkansas Academy of Sciences Proceedings 25:31–37.

Meyer, R. L., Brook, A. J. 1969. Freshwater algae from the Itasca State Park, Minnesota. II. Chrysophyceae and Xanthophyceae. Nova Hedwigia 17:105–112.

Moestrup, Ø. 1995. Current status of chrysophyte 'splinter groups': Synurophytes, pedinellids, silicoflagellates, in: Sandgren, C. D., Smol, J. P., Kristiansen, J., Eds., Chrysophyte algae: Ecology, phylogeny and development. Cambridge Univ. Press, Cambridge, UK, pp. 75–91.

Moestrup, Ø., Andersen, R. A. 1995. Organization of heterotrophic heterokonts, in: Patterson, D. J., Larsen, J., Eds., The biology of free-living heterotrophic flagellates. Clarendon, Oxford, pp. 333–360.

Moestrup, Ø., Thomsen, H. A. 1980. Preparation of shadow-cast whole mounts, in: Gantt, E., Ed., Handbook of phycological methods, development and cytological methods. Cambridge Univ. Press, Cambridge, UK, pp. 385–390.

Morris, I. 1980. The physiological ecology of phytoplankton. Univ. of California Press, Berkeley.

Müller, O. F. 1786. Animalcula Infusoria, fluviatilia et marina. Hafniae et Lipsiae, Copenhagen.

Nauwerck, A. 1979. Zur Gattung *Chrysolykos* Mack. Botaniska Notiser 132:161–183.

Nicholls, K. H. 1978. Jensen staining, a neglected tool in phycology. Transactions of the American Microscopical Society 97:129–132.

Nicholls, K. H. 1979. A simple tubular phytoplankton sampler for vertical profiling in lakes. Freshwater Biology 9:85–89.

Nicholls, K. H. 1981a. *Spiniferomonas* (Chrysophyceae) in Ontario lakes including a revision and descriptions of two new species. Canadian Journal of Botany 59:107–117.

Nicholls, K. H. 1981b. Six Chrysophyceae new to North America. Phycologia 20:131–137.

Nicholls, K. H. 1981c. *Kephyrion hilliardii* sp. nov. and *Pseudokephyrion millerensis* sp. nov., two new members of the Chrysophyceae. British Phycological Journal 16:241–245.

Nicholls, K. H. 1984a. Eight Chrysophyceae new to North America. Phycologia 23:213–221.

Nicholls, K. H. 1984b. *Paraphysomonas sediculosa* sp. nov. and *Paraphysomonas campanulata* sp. nov., new freshwater members of the Chrysophyceae. British Phycological Journal 19:230–244.

Nicholls, K. H. 1984c. *Spiniferomonas septispina* sp. nov. and *S. enigmata* sp. nov., two new algal species confusing the distinction between *Spiniferomonas* and *Chrysosphaerella* (Chrysophyceae). Plant Systematics and Evolution 148:113–117.

Nicholls, K. H. 1984d. Descriptions of *Spiniferomonas silverensis* sp. nov. and *S. minuta* sp. nov. and an assessment of form variation in their closest relative *S. trioralis* Tak. (Chrysophyceae). Canadian Journal of Botany 62:2329–2335.

Nicholls, K. H. 1985. Five *Paraphysomonas* species (Chrysophyceae) new to North America, with notes on three other rarely reported species. Canadian Journal of Botany 63:1208–1212.

Nicholls, K. H. 1987. *Chrysoamphipyxis* gen. nov.: A new genus in the Stylococcaceae (Chrysophyceae). Journal of Phycology 23:499–501.

Nicholls, K. H. 1988. *Paraphysomas caelifrica* new to North America and an amended description of *Paraphysomas subrotacea* (Chrysophyceae). Canadian Journal of Botany 67:2525–2527.

Nicholls, K. H. 1989. *Spiniferomonas genuiformis* and *Spiniferomonas alata* (Chrysophyceae): Taxonomic implications of form variation. Canadian Journal of Botany 67:1294–1297.

Nicholls, K. H. 1990. Life history and taxonomy of *Rhizochromonas endoloricata* gen. et sp. nov., a new freshwater chrysophyte inhabiting *Dinobryon* loricae. Journal of Phycology 26:558–563.

Nicholls, K. H. 1995. Chrysophyte blooms in the plankton and neuston of marine and freshwater systems, *in*: Sandgren, C. D., Smol, J. P., Kristiansen, J., Eds., Chrysophyte algae: Ecology, phylogeny and development. Cambridge Univ. Press, Cambridge, UK, pp. 181–213.

Nicholls, K. H. 2000. Three new freshwater species of *Dinobryon* (Chrysophyceae). Phycologia 39:134–138.

Nicholls, K. H., Carney, E. C., Robinson, G. W. 1977. Phytoplankton of an inshore area of Georgian Bay, Lake Huron, prior to reductions in phosphorus loading. Journal of Great Lakes Research 3:79–92.

Nicholls, K. H., Nakamoto, L., Keller, W. 1992. Phytoplankton of Sudbury area lakes (Ontario) and relationships with acidification status. Canadian Journal of Fisheries and Aquatic Sciences 49 (Supplement 1):40–51.

Nixdorf, B., Mischke, U., Leßmann, D. 1998. Chrysophytes and chlamydomonads: Pioneer colonists in extremely acidic mining lakes (pH <3) in Lusatia (Germany). Hydrobiologia 369/370: 315–327.

Nygaard, G. 1978. Freshwater phytoplankton from the Narssaq area, South Greenland. Botanisk Tidsskrift 73:191–238.

Olrik, K. 1998. Ecology of mixotrophic flagellates with special reference to Chrysophyceae in Danish lakes. Hydrobiologia 369/370: 329–338.

Olsen, J. L. 1990. Nucleic acids in algal systematics. Journal of Phycology 26:209–214.

Ortega, M. M. 1984. Catálogo de algas continentales recientes de México. Univ. Nacional Autónoma de México, Mexico City.

Ostrofsky, M. L., Duthie, H. 1975. Primary productivity and phytoplankton of lakes on the Eastern Canadian Shield. International Vereinigung für Theoretische und Agenwandte Limnologie Verhandlungen 19:732–738.

Parker, B. C., Samsel, G. E., Prescott, G. W. 1973. Comparison of microhabitats of macroscopic subalpine stream algae. American Midland Naturalist 90:143–153.

Pascher, A. 1910. Chrysomonaden aus dem Hirschberger Monographien und Abhandlungen zur Internationalen Revue der gesamten Hydrobiologie und Hydrographie 1:7–66.

Pascher, A. 1914. Über Flagellaten und Algen. berichte der Deutschen Botanischen Gesellschaft 32:136–160.

Pascher, A. 1917. Rhizopodialnetze als Fangvorrichtung bei einer plasmodialen Chrysomonade. Archiv für Protistenkunde 37:15–30.

Pascher, A. 1925. Die braune Algenreihe der Chrysophyceen. Archiv für Protistenkunde 52:489–464.

Pascher, A. 1929. Über die Beziehungen zwischen Lagerform und Standortsverhältnissen bei einer Gallertalge (Chrysocapsales). Archiv für Protistenkunde 68:637–668.

Pascher, A. 1931. Über eigenartige zweischalige Dauerstadien bei zwei tetrasporalen Chrysophyceen (Chrysocapsalen). Archiv für Protistenkunde 73:73–103.

Pascher, A. 1939. Heterokonten, *in*: Rabenhorst, L., Ed., Kryptogamenflora, 2nd ed. Acad. Verlagsges, Leipzig.

Pascher, A. 1940a. Rhizopodiale Chrysophyceen. Archiv für Protistenkunde 93:331–349.

Pascher, A. 1940b. Filarplasmodiale Ausbildungen bei Algen. Archiv für Protistenkunde 94:295–310.

Pavoni, M. 1963. Die Bedeutung des Nannoplanktons im Vergleich zum Netzplankton. Schweizerische Zeitschrift für Hydrologie 25:219–341.

Peterfi, L. 1969. The fine structure of *Poterioochromonas malhamensis* (Pringsheim) comb. nov. with special reference to the lorica. Nova Hedwigia 16:94–103.

Peters, M. C., Andersen, R. A. 1993. The fine structure and scale formation of *Chrysolepidomonas dendrolepidota* gen. et sp. nov. (Chrysolepidomonadaceae fam. nov., Chrysophyceae). Journal of Phycology 29:469–475.

Preisig, H. R. 1986. Biomineralization in the Chrysophyceae, *in*: Leadbeater, B. S. C., Riding, R., Eds., Biomineralization in lower plants and animals. Oxford Univ. Press, Oxford, pp. 327–344.

Preisig, H. R. 1995. A modern concept of chrysophyte classification, *in*: Sandgren, C. D., Smol, J. P., Kristiansen, J., Eds., Chrysophyte algae: Ecology, phylogeny and development. Cambridge Univ. Press, Cambridge, UK, pp. 46–74.

Preisig, H. R., Hibberd, D. J. 1982a. Ultrastructure and taxonomy of *Paraphysomonas* (Chrysophyceae) and related genera 1. Nordic Journal of Botany 2:397–420.

Preisig, H. R., Hibberd, D. J. 1982b. Ultrastructure and taxonomy of *Paraphysomonas* (Chrysophyceae) and related genera 2. Journal of Botany 2:601–638.

Preisig, H. R., Hibberd, D. J. 1987. Validation of *Paraphysomonas diademifera* and *Polylepidomonas vacuolata* (Chrysophyceae). Journal of Botany 7:497.

Preisig, H. R., Vørs, N., Hällfors, G. 1991. Diversity of heterokont flagellates, *in*: Patterson, D. J., Larsen, J., Eds., The biology of free-living flagellates. Clarendon, Oxford, pp. 361–399.

Prescott, G. W. 1962. Algae of the western Great Lakes area, 2nd ed., W.C. Brown, Dubuque, IA, 977 p.

Pringsheim, E. G. 1955. Kleine Mitteilungen über Flagellaten und Algen. Archiv für Mikrobiologie 21:401–410.

Rashash, D. M. C., Dietrich, A. M., Hoehn, R. C., Parker, B. C. 1995. The influence of growth conditions on odor-compound production by two chrysophytes and two cyanobacteria. Water Science Technology 31:165–172.

Reynolds, C. S., Jawarski, G. H. M., Roscoe, J. V., Hewitt, D. P., George, D.G. 1998. Responses of the phytoplankton to a deliberate attempt to raise the trophic status of an acidic, oligotrophic mountain lake. Hydrobiologia 369/370:127–131.

Roijackers, R. M. M. 1983. Development and succession of scale-bearing chrysophyceae in two shallow freshwater bodies near Nijmegen, The Netherlands, in: Kristiansen, J., Andersen, R. A., Eds., Chrysophytes: Aspects and problems. Cambridge Univ. Press, Cambridge, UK, pp. 241–258.

Roijackers, R. M. M., Kessels, H. 1986. Ecological characteristics of scale-bearing Chrysophyceae from The Netherlands. Nordic Journal of Botany 6:373–385.

Rosenberg, M. 1941. *Chrysochaete*, a new genus of the Chrysophyceae allied to *Naegeliella*. New Phytologist 40:304–15.

Round, F. E. 1981. The ecology of algae. Cambridge Univ. Press, Cambridge, UK.

Saha, L. C., Wujek, D. E. 1990. Scaled chrysophytes from Northeastern India. Nordic Journal of Botany 10:343–357.

Salonen, K., Jokinen, S. 1988. Flagellate grazing on bacteria in a small dystrophic lake. Hydrobiologia 169:203–209.

Sanders, R. W., Porter, K. G., Caron, D. A. 1990. Relationship between phototrophy and phagotrophy in the mixotrophic chrysophyte *Poterioochromonas malhamensis*. Microbial Ecology 19:97–109.

Sandgren, C. D. 1980a. An ultrastructural investigation of resting cysts formation in *Dinobryon cylindricum* (Chrysophyceae, Chrysophyta). Protistologica 16:259–276.

Sandgren, C. D. 1980b. Resting cyst formation in selected chrysophyte flagellates: An ultrastructural survey including a proposal for the phylogenetic significance of interspecific variations in the encystment process. Protistologica 16:289–303.

Sandgren, C. D. 1981. Characteristics of sexual and asexual resting cyst (statospore) formation in *Dinobryon cylindricum* Imhof (Chrysophyta). Journal of Phycology 17:199–210.

Sandgren, C. D. 1983. Morphological variability in populations of chrysophycean resting cysts. I. Genetic (interclonal) and encystment temperature effects on morphology. Journal of Phycology 19:64–70.

Sandgren, C. D. 1988. The ecology of chrysophyte flagellates: Their growth and perennation strategies as freshwater phytoplankton, in: Sandgren, C. D., Ed., Growth and reproductive strategies of freshwater phytoplankton. Cambridge Univ. Press, Cambridge, UK, pp. 9–104.

Sandgren, C. D., Flanagin, J. 1986. Heterothallic sexuality and density dependent encystment in the chrysophycean alga *Synura petersenii* Korsh. Journal of Phycology 22:206–212.

Sandgren, C. D., Smol, J. P., Kristiansen, J. 1995. Chrysophyte algae: Ecology, phylogeny and development. Cambridge Univ. Press, Cambridge, UK.

Saunders, G. W., Potter, D., Paskind, M. P., Andersen, R. A. 1995. Cladistic analyses of combined traditional and molecular data sets reveal an algal lineage. Proceedings of the National Academy of Sciences 92:244–248.

Schindler, D. W., Kling, H., Schmidt, R. V., Prokopowich, J., Frost, V. E., Reid, R. L., Capel, M. 1973. Eutrophication of Lake 227 by addition of phosphate and nitrate: the second, third and fourth years of enrichment 1970, 1971, 1972. Journal of the Fisheries Research Board of Canada 30: 1415–1440.

Schmidle, W. 1902. Notizen zu einigen Süsswasseralgen. Hedwigia 42:150–163.

Schoonoord, M. P., Ellis-Adams, A. C. 1984. *Chrysocrinus homorarius* spec. nov. (Stylococcaceae, Chrysophyceae): Some stages of its lorical development and its distribution on the substrate. Acta Botanica Neerlandica 33:399–418.

Sheath, R., Munawar, M. 1974. Phytoplankton composition of a small subarctic lake in the Northwest Territories, Canada. Phycologia 13:149–161.

Siver, P. A. 1987. *Spiniferomonas breakneckii* sp. nov., a new species of freshwater Chrysophyceae. British Phycological Journal 22:97–100.

Siver, P. A. 1988a. The distribution and ecology of *Spiniferomonas* (Chrysophyceae) in Connecticut (USA). Nordic Journal of Botany 8:205–212.

Siver, P. A. 1988b. *Spiniferomonas triangularis* sp. nov., a new silica-scaled freshwater flagellate (Chrysophyceae, Paraphysomonadaceae). British Phycological Journal 23:379–383.

Siver, P. A. 1993. Morphological and ecological characteristics of *Chrysosphaerella longispina* and *C. brevispina* (Chrysophyceae). Nordic Journal of Botany 13:343–351.

Siver, P. A., Hamer, J. S. 1992. Seasonal periodicity of Chrysophyceae and Synurophyceae in a small New England lake: Implications for paleolimnological research. Journal of Phycology 28:186–198.

Skogstad, A., Reymond, O. L. 1989. An ultrastructural study of vegetative cells, encystment, and mature statospores in *Spiniferomonas bourrellyi* (Chrysophyceae). Beihefte zur Nova Hedwigia 95:71–79.

Skuja, H. 1948. Taxonomie des Phytoplanktons einiger Seen in Uppland, Schweden. Symbolae Botanicae Upsaliensis 9:1–399.

Smith, G. M. 1950. The freshwater algae of the United States. McGraw-Hill, New York.

Squires, L. E., Rushforth, S. R., Endsley, C. J. 1973. An ecological survey of the algae of Huntington Canyon, Utah. Brigham Young University Science Bulletin 18:1–87.

Starmach, K. 1985. Chrysophyceae and Haptophyceae, in: Ettl, H., Gerloff, J., Heynig, H., Mollenhauer, D., Eds., Süsswasserflora von Mitteleuropa, Vol. 1. Fischer Verlag, Stuttgart.

Stein, F. von 1878. Der Organismus der Infusionsthiere, Part 3. Engelman, Leipzig.

Stein, J. R. 1975. Freshwater algae of British Columbia: The Lower Fraser Valley. Syesis 8:119–184.

Stein, J. R., Borden, C. A. 1979. Checklist of freshwater algae of British Columbia. Syesis 12:3–39.

Stoecker, D. K. 1998. Conceptual models of mixotrophy in planktonic protists and some ecological and evolutionary implications. European Journal of Protistology 34:281–290.

Stokes, A. C. 1885. Notes on some apparently undescribed forms of fresh-water infusoria. No. 2. American Journal of Science 29:313–328.

Stokes, A. C. 1886. Notices of new fresh-water infusoria. Proceedings of the American Philosophical Society 23:562–568.

Suykerbuyk, R. E. M., Roijackers, R. M. M., Houtman, S. S. J. 1995. Ecological characteristics of scale-bearing chrysophytes in a small eutrophic pond in The Netherlands. Nordic Journal of Botany 15:665–676.

Thompson, R. H., Wujek, D. E. 1998a. *Chrysocapsopsis rupicola*, a new genus and species in the Chrysophyceae. Phycological Research 46:165–168.

Thompson, R. H., Wujek, D. E. 1998b. The genera *Cyclonexis* Stokes and *Chrysostephanosphaera* Scherffel (Chrysophyceae). Nordic Journal of Botany 18:627–632.

Thomsen, H. A., Zimmermann, B., Moestrup, Ø., Kristiansen, J. 1981. Some new freshwater species of *Paraphysomonas* (Chrysophyceae). Nordic Journal of Botany 1:559–581.

Tilman, D., Kilham, S. S. 1976. Phosphate and silicate growth and uptake kinetics of the diatoms *Asterionella formosa* and *Cyclotella meneghiniana* in batch and semicontinuous cultures. Journal of Phycology 12:375–383.

Tilman, D., Kilham, S. S., Kilham, P. 1982. Phytoplankton community ecology: The role of limiting nutrients. Annual Review of Ecology and Systematics 13:349–372.

Tilman, D., Kiesling, R., Sterner, S. S., Johnson, F. A. 1986. Green, bluegreen and diatom algae: taxonomic differences in competitive ability for phosphorus, silicon and nitrogen. Archiv für Hydrobiologie 106:473–485.

Tschermak-Woess, W. 1980. Zur Kenntnis von *Tetrasporopsis fuscescens*. Plant Systematics and Evolution 133:121–133.

Tschermak-Woess, E., Kusel-Fetzman, E. 1992. A new find of *Tetrasporopsis fuscescens* (A. Braun ex Kützing) Lemmermann (Chrysophyta) in Austria, and some additional observations. Archiv für Protistenkunde 142:157–165.

Vesk, M., Hoffman, L. R., Pickett-Heaps, J. D. 1984. Mitosis and cell division in *Hydrurus foetidus* (Chrysophyceae). Journal of Phycology 20:461–470.

Vigna, M. S., Kristiansen, J. 1996. Biogeographic implications of new records of scale-bearing chrysophytes from Tierra del Fuego (Argentina). Archiv für Protistenkunde 147:137–144.

Vørs, N., Johansen, B., Havskum, H. 1990. Electron microscopical observations on some species of *Paraphysomonas* (Chrysophyceae) from Danish lakes and ponds. Nova Hedwigia 50:337–354.

Wee, J. L. 1983. Specimen collection and preparation for critical light microscope examination of Synuraceae (Chrysophyceae). Transactions of the American Microscopical Society 102:68–78.

Wee, J. L. 1996. Molecular investigations of heterokont comparative biology: Recent and emerging trends. Beihefte zur Nova Hedwigia 114:7–27.

Wee, J. L., Hinchey, J. M., Nguyen, K. X., Kores, P., Hurley, D. L. 1996. Investigating the comparative biology of the heterokonts with nucleic acids. Journal of Eukaryotic Microbiology 43:106–112.

Wetherbee, R., Andersen, R. A. 1992. Flagella of a chrysophycean alga play an active role in prey capture and selection. Direct observations on *Epipyxis pulchra* using image enhanced video microscopy. Protoplasma 166:1–7.

Wetherbee, R., Platt, S. J., Beech, P. L., Pickett-Heaps, J. D. 1988. Flagellar transformation in the heterokont *Epipyxis pulchra* (Chrysophyceae): Direct observations using image enhanced light microscopy. Protoplasma 145:47–54.

Whitford, L. A. 1969. *Chrysidiastrum epiphyticum* sp. nov. (Chrysophyceae). Phycologia 8:199–200.

Whitford, L. A. 1970. *Eirmodesmus phaeotilus* gen. et sp. nov. (Chrysophyceae). Phycologia 9:195–197.

Whitford, L. A., Schumacher, G. J. 1969. A manual of the freshwater algae in North Carolina. Technical Bulletin No. 188. The North Carolina Agricultural Experiment Station.

Whitford, L. A., Schumacher, G. J. 1973. A manual of fresh-water algae. Sparks Press, Raleigh.

Wiedner, C., Nixdorf, B. 1998. Success of chrysophytes, cryptophytes and dinoflagellates over blue-greens (cyanobacteria) during an extreme winter (1995/96) in eutrophic shallow lakes. Hydrobiologia 369/370:229–235.

Wilken, L. R., Kristiansen, J., Jürgensen, T. 1995. Silica-scaled chrysophytes from the peninsula of Nuussuaq/Nûgsssuaq, West Greenland. Nova Hedwigia 61: 355–66.

Wolken, J. J., Palade, G. E. 1952. Fine structure of chloroplasts in two flagellates. Nature 170:114–115.

Wujek, D. E. 1967. *Microglena punctifera* (O.F.M.) Ehrenberg in the United States. Transactions of the American Microscopical Society 86:340–341.

Wujek, D. E. 1976. Ultrastructure of flagellated chrysophytes. II. *Uroglena* and *Uroglenopsis*. Cytologia 41:665–670.

Wujek, D. E. 1983. A new fresh-water species of *Paraphysomonas* (Chrysophyceae: Mallomonadaceae). Transactions of the American Microscopical Society 102:165–168.

Wujek, D. E. 1996. Chrysophyte ultrastructure and taxonomy: A mini review. In Chaudhary, B. R., Agrawal, S. B., Eds., Cytology, genetics and molecular biology of algae. SPB Academic, Amsterdam, pp. 21–36.

Wujek, D. E. 1999. Recent and emerging trends in chrysophycean taxonomy, in: Vidyavati, Mahato, A. K., Eds., Recent trends in algal taxonomy, Vol 1. Taxonomical issues. APC Publication, New Delhi, pp. 225–236.

Wujek, D. E., Bland, R. G. 1988. *Spiniferomonas* and *Mallomonas*: Descriptions of two new taxa of Chrysophyceae. Transactions of the American Microscopical Society 107:301–304.

Wujek, D. E., Gardiner, W. E. 1985. Chrysophyceae (Mallomonadaceae) from Florida. II. New species of *Paraphysomonas* and the prymnesiophyte *Chrysochromulina*. Florida Scientist 47:161–170.

Wujek, D. E., Saha, L. C. 1995. The genus *Paraphysomonas* from Indian rivers, ponds and tanks, in: Sandgren, C. D., Smol, J. P., Kristiansen, J., Eds., Chrysophyte algae: Ecology, phylogeny and development. Cambridge Univ. Press, Cambridge, UK, pp. 373–384.

Wujek, D. E., Saha L. C. 1996. Scale-bearing chrysophytes (Chrysophyceae and Synurophyceae) from India. II. Beihefte zur Nova Hedwigia 112:367–377.

Wujek, D. E., Thompson, R. H. 2001. The chrysophycean genera *Synuropsis* Schiller, *Volvochrysis* Schiller, *Synochromonas* Korshikov, *Pseudosynura* Kisselew, *Pseudosyncrypta* Kisselew, *Chrysomoron* Skuja and *Syncrypta* Ehrenberg. Transactions of the Kansas Academy of Science 104:71–78.

Wujek, D. E., Thompson, R. H. 2002. The genera *Uroglena*, *Uroglenopsis* and *Eusphaerella* (Chrysophyceae). Phycologia 41: 293–305.

13

HAPTOPHYTE ALGAE

Kenneth H. Nicholls*

Aquatic Science Section
Ontario Ministry of the Environment

I. Introduction
II. Diversity and Morphology
III. Ecology and Distribution
IV. Collection and Preparation for Identification
V. Key and Descriptions of Genera
   A. Key
   B. Descriptions of Genera
VI. Guide to Literature for Species Identification
Literature Cited

I. INTRODUCTION

Christensen (1962) split off the Haptophyceae as a class separate from the Chrysophyceae primarily on the basis of the presence of a unique flagellum-like organ called a haptonema (Parke *et al.*, 1955), which has no homologue in any other protistan group. Later, in response to a recommendation in the International Code of Botanical Nomenclature (Article 16) that names above the level of genus should have a genus name as their root, Hibberd (1976a) replaced Haptophyceae with Prymnesiophyceae (root = *Prymnesium*). The system followed here is based on that proposed by Edvardsen *et al.* (2000), wherein the Prymnesiophyceae, one of two classes in the Haptophyta, is composed of four orders: Phaeocystales, Prymnesiales, Isochrysidales, and Coccolithales. The other haptophyte class is the Pavlovophyceae with the single order Pavlovales and for which no representatives have been reported from freshwater environments in the Western Hemisphere.

The best known representatives of the Prymnesiophyceae are the marine coccolithophorids, which, during the Late Cretaceous, were extremely diverse and abundant, contributing to massive chalk deposits resulting from sedimentation of their calcareous scales (coccoliths) that cover the cell exterior. Present-day blooms of coccolithophorids in the sea are of sufficient magnitude to influence significantly global carbon and sulfur cycles and world climate (Westbroek *et al.*, 1993; Malin *et al.*, 1994). The birefringent reflectance from coccoliths in such blooms is regularly captured by satellite imagery and has led to the development of remote-sensing methods for determining coccolith density (Balch *et al.*, 1991; Brown and Yoder, 1993). Freshwater coccolithophorids, virtually restricted to the single species *Hymenomonas roseola* Stein, provide no similarly spectacular phenomena.

Identification of coccolithophorids is based mainly on the ultrastructure of the crystallization of the two main types of coccoliths (heterococcoliths, coccoliths formed within a specialized Golgi body; and holococcoliths, coccoliths formed at the plasma membrane), which, with electron microscopy (EM), reveal their varied and beautifully complex structure (see, e.g., the scanning EM photomicrographs in Faber and Preisig, 1994). Several subtypes (e.g., areoliths, zygoliths, tremaliths, cricoliths, among others) of these two main types of coccoliths are now recognized (Jordan *et al.*,

---

*Present address: S-15 Concession 1, RR #1 Sunderland, Ontario, Canada, L0C 1H0

1995). Fresnel (1994) has recently demonstrated a heteromorphic life cycle for two species of marine *Hymenomonas* that have both coccolith-bearing and organic scale-bearing (coccoliths absent) stages in their life cycles. Only one freshwater coccolithophorid, *Hymenomonas roseola*, is known from North America (Lackey, 1939; Meyer and Brook, 1969; Stoermer and Sicko-Goad, 1977; Whitford, 1979). Although not colonizing freshwater, *Pleurochrysis carterae* var. *dentata* was discovered in a saline pond in New Mexico far removed from any coastal marine influence (Johansen et al., 1988); this taxon has at least three morphologically distinct stages in its life history. The putative coccoliths of the freshwater species, *Hymenomonas prenanti*, were shown by Nicholls (1979) to be siliceous, not calcareous; this taxon is a superfluous synonym for the noncoccolithophorid *Gyromitus disomatus*.

Other economically important haptophytes include about a dozen species of ichthyotoxin producers in the genera *Chrysochromulina*, *Phaeocystis*, and *Prymnesium* (Moestrup and Thomsen, 1995). One species (*Prymnesium parvum* Carter) is a confirmed toxin producer in rivers impacted by seawater incursions (Holdway et al., 1978), in brackish embayments (Holmquist and Willén (1993), and in high-electrolyte aquaculture ponds (Guo et al., 1996). Anecdotal observations of dead aquatic organisms associated with freshwater blooms of *Chrysochromulina breviturrita* (Nicholls et al., 1982) and *C. parva* (Hansen et al., 1994) beg for further laboratory investigation of the potential for toxin production by these two freshwater species. Simonsen and Moestrup (1997) found no evidence for toxicity of *C. parva* to the brine shrimp, *Artemia salina*. Like some other toxic algae, expression of toxicity may depend on environmental factors and growing conditions, especially phosphorus deficiency (Larsen et al., 1993). Old cultures of some otherwise apparently nontoxic species have demonstrated toxicity in laboratory settings (Moestrup, 1994).

This chapter will review the morphology, taxonomy, ecology, and important diagnostic features necessary for identification of the freshwater haptophytes with emphasis on North American species.

II. DIVERSITY AND MORPHOLOGY

In the recent past, the freshwater haptophyte algae were envisioned by some as a relatively diverse group of mainly isokont biflagellates contained in the subclass Isochrysophycideae (Bourrelly, 1981), which included several anomalous genera, many which have been reexamined with modern methods and placed in more appropriate phylogenies elsewhere in the Protista (e.g., Hibberd, 1976b, 1983, 1985). The contemporary concept of the Haptophyta encompasses over 300 marine species (Jordan and Green, 1994; Jordan and Chamberlain, 1997), but the group is represented in freshwaters by fewer than a dozen species within the genera *Hymenomonas*, *Chrysochromulina*, *Acanthoica*, *Anacanthoica*, *Pavlova*, *Diacronema*, and *Exanthemachrysis*. Additionally, a species of *Pleurochrysis* has been reported from a shallow saline pond in New Mexico (Johansen et al., 1988). The genus *Prymnesium* is well represented in some marine coastal and brackish-water lagoons, but at this time there are no confirmed reports of occurrence of any *Prymnesium* species in strictly freshwater habitats of North America. Moestrup (1994) has questioned the validity of the report by James and de la Cruz (1989) of *P. parvum* in the Pecos River, Texas, since electron microscopy was not used for identification.

Cell biology and ultrastructure (Fig. 1) of haptophyte algae have been thoroughly reviewed in Green and Leadbeater (1994) and Jordan et al. (1995). The most distinctive feature of most members of the Haptophyta is the haptonema which, in most genera, is inserted between two smooth (nontinsel) flagella of equal or subequal length (but two marine species of *Chrysochromulina* are quadriflagellate!). Haptonema lengths range from a vestigial nonemergent structure in the genus *Imantonia* (Green and Pienaar, 1977) to a remarkably flexible filament, coiling and uncoiling and extending in length to about 15 times the diameter of the cell in the freshwater *Chrysochromulina parva* Lackey (Fig. 3B) and in some marine species (e.g., *Chrysochromulina strobilus*). In others, such as *Pleurochrysis carterae* var. *dentata* Johansen et Doucette, it is found only in the noncoccolith motile stage as a short bulbous structure at the base of two distinctly unequal flagella (Fig. 5A). Its structure differs from a flagellum in its lack of an axial pair of microtubules. Also, the seven peripheral tubules are not paired as they are in flagella and cilia and may decrease in number distally; they are surrounded by a three-layered membrane system. The haptonema of some of the larger marine species has been implicated in phagotrophic prey capture (Kawachi and Inouye, 1995). Other essential features of typical haptophyte cells include a nuclear envelope with an outer membrane that is continuous with the chloroplast endoplasmic reticulum (ER) and a peripheral ER that extends up into the haptonema but is discontinuous at the flagellar bases. A Golgi body located near the flagellar basal bodies produces scales that cover the exterior of the cell in many species. Haptophyte chloroplasts (one or two in number) have chlorophyll-c_1, -c_2, and -c_3, (the Chrysophyceae lack c_3

and the Synurophyceae lack c_2 and c_3; Andersen, 1989). Also usually present is a layer of organic (e.g., *Chrysochromulina*) and/or calcified (e.g., *Hymenomonas*) scales covering the cell. Flagellar autofluorescence, a recently discovered phenomenon in chromophyte algae, was reported for the proximal part of one of the isokont flagella in seven species (four genera) of the Prymnesiales and Isochrysidales but was not detected in coccolithophorids or in related members of the Pavlovales (Pavlovophyceae) (Kawai and Inouye, 1989). No haptophytes have siliceous scales like those found in the Synurophyceae or Paraphysomonadaceae (Chrysophyceae).

III. ECOLOGY AND DISTRIBUTION

Without doubt, *Chrysochromulina parva* is the most common haptophyte in lakes, ponds, and rivers. Based on reports from England, Europe, Scandinavia, North and South America, Japan, and China, *C. parva* would appear to have a worldwide distribution in a wide variety of freshwater habitats (Parke *et al.*, 1962; Heynig, 1963; Thompson and Halicki, 1965; Kristiansen, 1971; Diaz and Lorenzo, 1990; Ito, 1989; Wujek and Saha, 1991; Wei, 1996). I have even found it "blooming" in an abandoned backyard swimming pool where Secchi disc visibility had been reduced to 40 cm!

The distribution of *C. parva* is not limited to small bodies of water; it has a wide distribution in the Laurentian Great Lakes (Munawar and Munawar, 1982). In Ontario waters of Lake Erie and southern Georgian Bay, at densities as high as 10 million cells L^{-1}, it has occasionally been a dominant element of the phytoplankton, comprising up to 60% of the total phytoplankton density (as Areal Standard Units; Ontario Ministry of the Environment unpublished data). Parke *et al.* (1962) reported maximal densities of about 20 million cells L^{-1} for some of the lakes in the English Lake District. Restriction of peak densities to spring is a common feature of the Great Lakes and English Lake District lakes. Similarly, Ito (1989) found densities as high as 59 million cells L^{-1} in a Japanese pond in spring; however, Kristiansen (1971) reported a density of 50 million cells L^{-1} from a Danish pond in July when the water temperature was 16°C and Pollinger (1986) found maximal development of *C. parva* at 14–18°C. The highest density thus far reported for *C. parva* was for a small Danish lake where more than 600 million cells L^{-1} were counted in June of 1991 when the water temperature was 14°C (Hansen *et al.*, 1994). Based on the lack of reports of *C. parva* from oligotrophic waters, Ito (1989) concluded that it is a eutrophic species, with a temperature optimum below 20°C.

C. parva is, however, a common element of the phytoplankton in small oligotrophic softwater lakes of Ontario's Precambrian Shield (Findlay and Kasian, 1987; Nicholls *et al.*, 1992), where specific conductance is about one order of magnitude lower and dissolved nitrogen and total phosphorus concentrations are generally less than half those found in the lower Great Lakes. In a softwater lake in central Norway, Reinertsen (1982) observed that *C. parva* became a dominant after whole lake fertilization. Similarly, Holmgren (1984) reported a dramatic increase in numbers of *C. parva* in a subarctic lake in northern Sweden after fertilization with nitrogen and phosphorus. This species nearly doubled its biomass every second day during a period of warm weather in July, reaching a peak biomass of 2.5 mg L^{-1} (equivalent to about 75 million cells L^{-1}).

Unlike *C. parva*, the distributions of *C. breviturrita* and *C. laurentiana* are restricted to neutral and acidic lakes in northeastern North America. Neither species has been found in the Great Lakes. At modest cell densities of 500–9000 cells mL^{-1} in moderately acidic lakes, *C. breviturrita* has produced "rotten cabbage," "dead animal," or "garbage dump" types of odors with severe implications for lake use by humans (Nicholls *et al.*, 1982). This prompted a series of exhaustive laboratory investigations, resulting in considerable knowledge of the physiological ecology of *C. breviturrita*. Wehr *et al.* (1985) determined in laboratory cultures that *C. breviturrita* had an optimum pH range for growth of 5.5–6.9; it was unable to survive above pH 7. This pattern, combined with its absolute requirement for selenium as selenate, selenite, dimethylselenide, and/or selenomethionine (Wehr and Brown, 1985), suggests it might be an early indicator of lake acidification because coal-fired electricity generating plants are a source of both acids and selenium. *C. breviturrita* can utilize only ammonium-N (not nitrate) as its inorganic nitrogen source; this phenomenon is perhaps a physiological adaptation that may help to ensure optimal pH of its habitat through H^+ generation associated with NH_4^+ assimilation (Wehr *et al.*, 1987).

Chrysochromulina breviturrita also produces subsurface peaks in its vertical distributions in lakes (Nicholls *et al.*, 1982; O'Grady and Brown, 1989), which may reflect its intolerance of high irradiances at lake surfaces in summer. Exposure to the high light intensities found at the lake surface can be lethal to cells deficient in nitrogen and may explain the absence of blooms of *C. breviturrita* in strongly acidic, clearwater lakes (O'Grady and Brown, 1989).

The other two freshwater species, *C. laurentiana* and *C. inornata*, have very limited known distributions in Ontario and Florida, respectively (Kling, 1981; Wujek and Gardiner, 1985); hence, little can be said about their environmental tolerances and preferences. A fifth freshwater species, still undescribed but illustrated by Whitford (1979), is clearly different from the other known freshwater species by virtue of its haptonema, which is about equal in length to the flagella. It awaits rediscovery and examination of its scales by electron microscopy.

Moestrup (1994) reviewed the world distribution, toxicity, and likely synonomy of *Prymnesium parvum* and *P. saltans*. Freshwater occurrences are apparently restricted to one site in Germany with high sulfate and chloride concentrations. Most marine species have clear tendencies to bloom in brackish-water embayments and estuaries. *Prymnesium parvum* has caused fish toxicity in widely separated parts of the world. Among the best documented cases are the experiences in the Norfolk Broads in eastern England, especially in habitats influenced by the River Thurne, which is brackish, even in its headwaters, owing to its proximity to the sea. At this location, nutrient enrichment (sources included sewage effluents and excrement from a large gull population) was probably an important factor contributing to toxic *Prymnesium* blooms (cell densities exceeding 10^4 cells mL^{-1}). Salinity appears to be the primary factor controlling distribution of this species; cultures died at salinities less than 1.12‰, but showed good growth in the salinity range 5.6–22.5‰ (3125–12,500 mg Cl L^{-1}; Holdway *et al.*, 1978). Johansen *et al.* (1988) isolated *Pleurochrysis carterae* var. *dentata* from a saline pond in New Mexico, which grew best in laboratory cultures at a specific conductance of 25 mS/cm with high concentrations of divalent cations.

The presumed main mode of nutrition of the freshwater haptophytes is autotrophism, but the possibility of mixotrophy in freshwater haptophytes has not been as thoroughly investigated as for marine species. Based on laboratory investigations of cultured material, *C. parva* is not phagotrophic (Parke *et al.*, 1962), in contrast to many larger, spinose marine species of this genus (Jones *et al.*, 1994).

IV. COLLECTION AND PREPARATION FOR IDENTIFICATION

Because all freshwater haptophytes are planktonic, field sampling methods for phytoplankton (Vollenweider, 1969; Wetzel and Likens, 1991) will be applicable to this group. Freshwater haptophytes, like most other nanoflagellates, present some difficulties for fixation and preservation of cell attributes necessary for identification. Even dilute buffered formalin may cause flagella and haptonemata to detach. Osmic acid (in the ratio of 1 : 1 or 0.5 : 1 with plankton suspension) is a good fixative, but specimens must be postfixed with gluteraldehyde or formalin and then washed thoroughly with water, because excess osmium blackens the cell contents over time; see Lund (1942) for details of the osmic acid method. Lugol's solution (200 mL distilled water, 10 g I, 20 g KI, 20 g glacial acetic acid) and 1% gluteraldehyde are also suitable fixatives. If phytoplankton samples contain coccoliths (e.g., *Hymenomonas roseola*), the acid should be omitted from the Lugol's solution because it will destroy these calcified structures over time. Material fixed with Lugol's solution will often deteriorate on standing over several weeks in daylight. This light-induced loss of iodine is often followed by infestation by aquatic fungi, which can essentially render the sample useless. Samples fixed with Lugol's solution should be kept in the dark or, if intended to be retained over many years, should be preserved with formalin (add 3–4 drops of 37% formaldehyde to 25 mL Lugol's fixed water suspension of algae). Store samples in glass vials with soft-plastic-lined screw caps (to curb evaporative losses). I have found flagella and haptonema intact on cells of *Chrysochromulina breviturrita* 25 years after having been fixed, preserved, and stored in this manner.

For especially dilute samples, concentrate fixed cells by sedimentation. For living material, centrifugation and filtration pressure differentials exceeding 3 kPa should not be used in order to avoid a tendency for cell rupture (Bloem and Bär Gilissen, 1988). Whenever possible, living specimens should be examined with a light microscope at high magnification (preferably with phase contrast and/or differential interference contrast) in order to observe the swimming behavior, orientation, and movement of flagella and haptonema and the presence of contractile vacuoles and other organelles prone to disruption or distortion by commonly used fixatives and preservatives.

Detection of haptonemata and flagella can be difficult with ordinary bright-field microscopy. Detection is greatly facilitated in specimens dried directly onto a coverglass (no mounting medium) and viewed with phase contrast and an oil immersion high-power objective (oiled to the clean side of the coverglass). A coverglass holder fashioned out of sheet plastic or metal with the dimensions of a standard glass slide will allow the coverglass to be manipulated on the microscope stage. Identification to species level depends on electron microscopy of organic scales, so fixation of living collections intended for this purpose should follow standard transmission electron microscopy (TEM) protocol for maximizing observations of scale structure (consult Moestrup and Thomsen, 1980).

V. KEY AND DESCRIPTIONS OF GENERA

A. Key

1a.	Cell covered with calcified scales (coccoliths)	2
1b.	Cell not covered with coccoliths	3
2a.	Coccoliths of a single type	4
2b.	Coccoliths of two types	*Anacanthoica*[1]
3a.	Flagella about equal in length, apically inserted	5
3b.	Flagella unequal in length, laterally inserted	6
4a.	Coccoliths flattened with a C-like thickening	*Acanthoica*[1]
4b.	Coccoliths elliptical or circular with a columnar rim and an open center	7
5a.	Haptonema length less than three quarters of cell diameter with a "sticky" apex	*Prymnesium*[2]
5b.	Haptonema length about equal to or much longer than cell diameter (Figs. 3 and 4)	*Chrysochromulina*[3]
6a.	Flagellar scales covering long flagellum	*Pavlova*[1]
6b.	Flagellar scales absent	8
7a.	Coccoliths of the tremalith type (Fig. 2B, C)	*Hymenomonas*
7b.	Coccoliths of the cricolith type (Fig. 5C, D)	*Pleurochrysis*[4]
8a.	Both flagella covered with fine nontubular hairs	*Diacronema*[1]
8b.	Flagella hairless	*Exanthemachrysis*[1]

[1] *Acanthoica schilleri* Conrad, *Anacanthoica ornata* (Conrad) Bourrelly, *Pavlova granifera* (Mack) Green, *Exanthemachrysis noctivaga* (Kalina) Gayral and Fresnel, and *Diacronema vlkianum* Prauser emend. Green and Hibberd are freshwater haptophytes yet to be reported from the Western Hemisphere.
[2] *P. parvum* is widespread in brackish water, but at this time there are no confirmed reports of its occurrence in freshwater in North America.
[3] This distinction between *Prymnesium* and *Chrysochromulina* is rather arbitrary; there is a need to review critically the validity of recognizing both genera as presently constituted. Further research on small subunit rDNA sequences may help to clarify generic classification of species in these groups (Simon *et al.*, 1997; Edvardsen *et al.*, 2000).
[4] One species (*P. carterae* var. *dentata*) reported from an inland saline pond in New Mexico (Johansen *et al.*, 1988) has a naked (coccoliths absent), haptonema-bearing motile stage, as well as a coccolith-bearing motile stage lacking a visible haptonema and a pseudofilamentous benthic stage in its life cycle.

B. Descriptions of Genera (North American, freshwater only)

Chrysochromulina Lackey

Cells are spherical or ovoid or flattened with a slight concavity (*C. parva*); they have two flagella of equal or subequal length and one haptonema three quarters of the cell diameter in length or longer. There are two or four brownish–yellow chloroplasts, a pyrenoid, and a prominent contractile vacuole. The cell is covered in minute organic scales (resolved only by transmission electron micropcopy) of one type (*C. parva*, *C. laurentiana*, and *C. inornata*) or of two types (*C. breviturrita*). There are four freshwater species as follows.

Key to Freshwater Chrysochromulina Species

1a.	Cell with haptonema length about 10–12 times the cell diameter	*C. parva*
1b.	Cell with haptonema length about equal to the cell diameter	2
2a.	Cell with two types of scales	*C. breviturrita*
2b.	Cell with a single scale type	3
3a.	Scales with no surface ornamentation	*C. inornata*
3b.	Scale surface ornamented with a series of parallel ridges	*C. laurentiana*

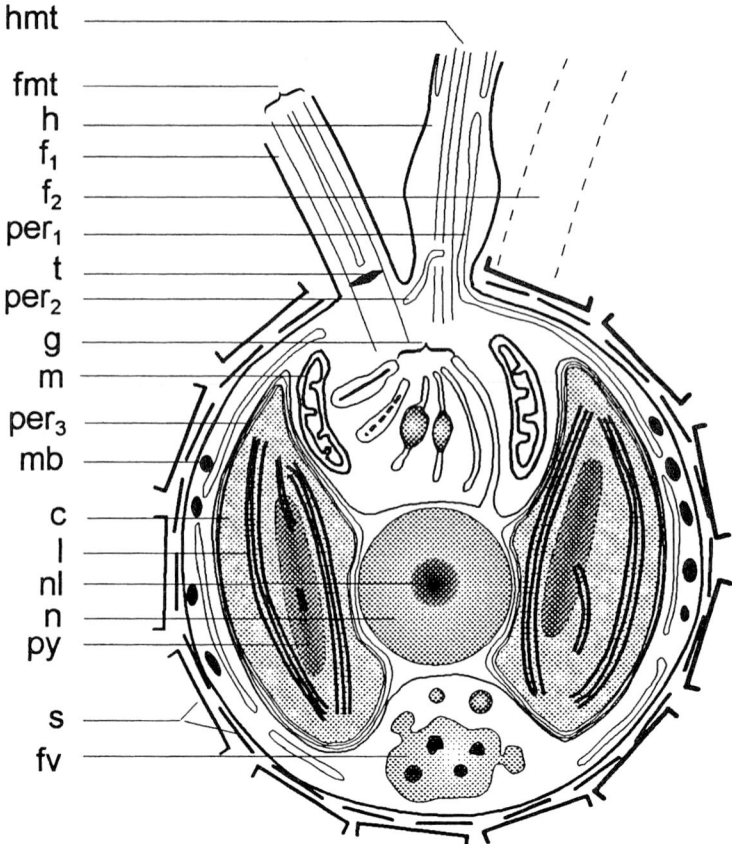

FIGURE 1 Section through a typical prymnesiophycean cell showing the principal organelles; keyed as follows: hmt, haptonema microtubules; fmt, flagellum microtubules; h, haptonema; f_1, first flagellum; f_2, second flagellum; per_1, peripheral endoplasmic reticulum; t, transitional plate; per_2, finger-like projection of endoplasmic reticulum; g, Golgi body with dilated cisternae and nascent scales; m, mitochondrion; per_3, periplastid endoplasmic reticulum; mb, muciferous body; c, chloroplast; l, chloroplast lamellae composed of three thylakoids; nl, nucleolus; n, nucleus; py, pyrenoid; s, body scales; fv, vacuole containing ingested material. Note: Schematic only, not to scale. Redrawn from Jordan et al. (1995).

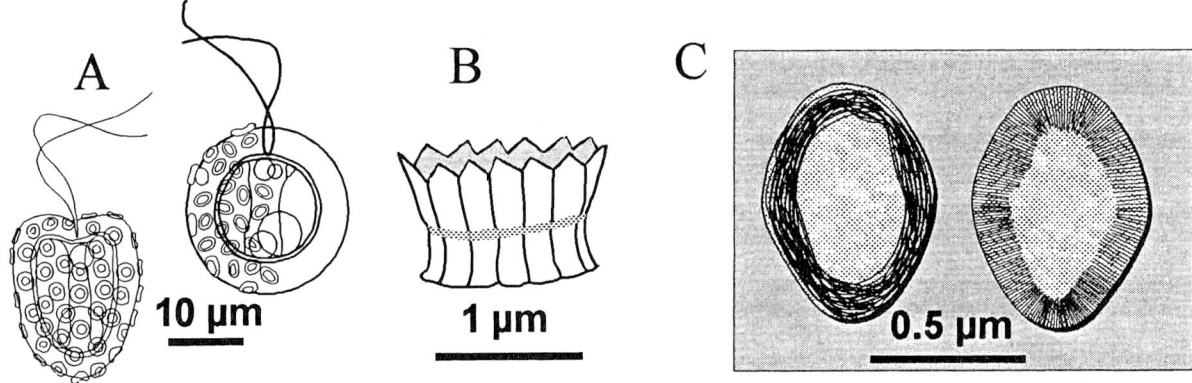

FIGURE 2 Hymenomonas roseola (= H. coccolithophora, H. danubiensis, H. scherffelli, Pontosphaera stagnicola; Bourrelly, 1981). (A) Two whole cells; redrawn from Conrad (1928). (B) Single coccolith; redrawn from Braarud (1954). (C) Organic scales; redrawn from Manton and Peterfi (1969).

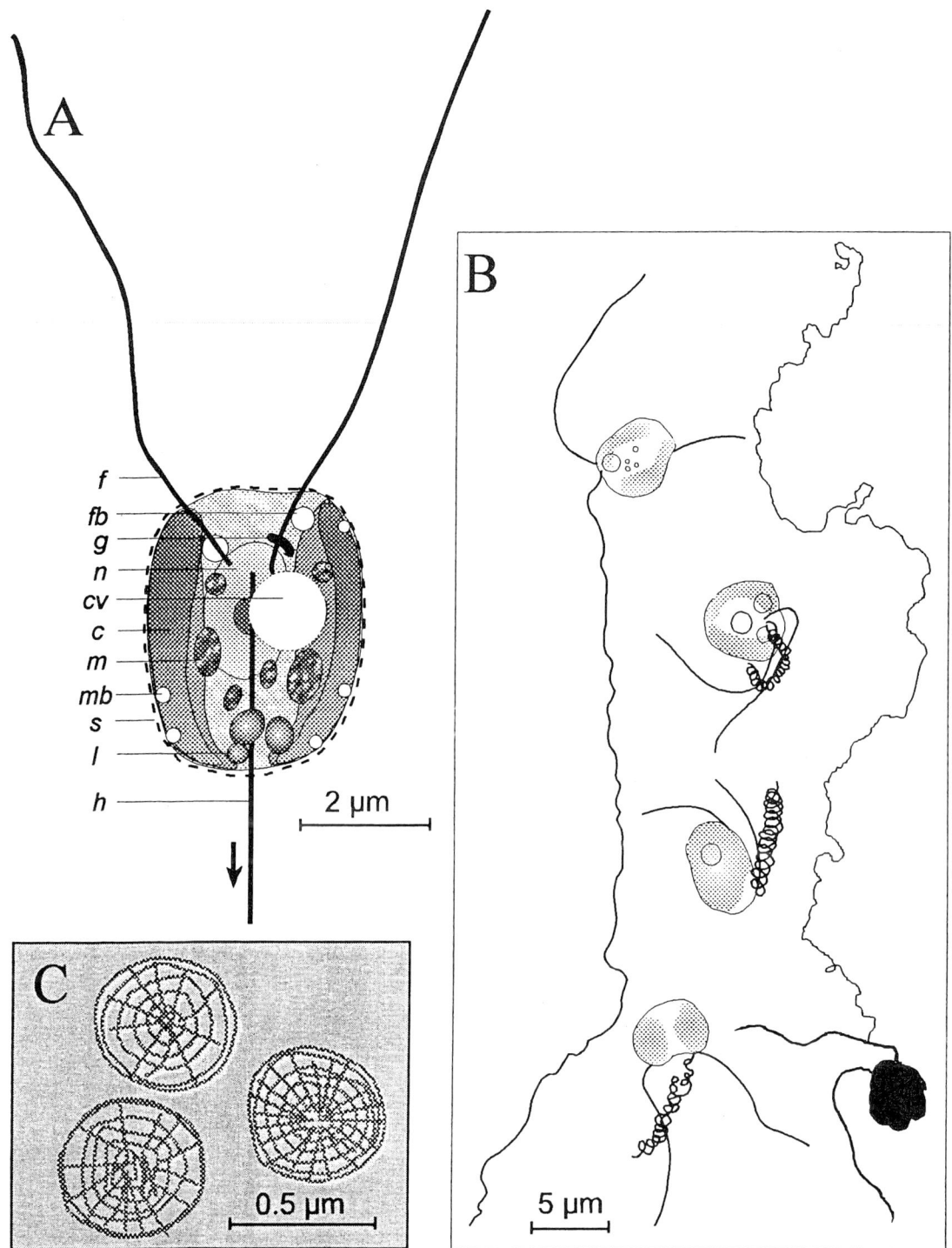

FIGURE 3 *Chrysochromulina parva* Lackey; redrawn from Parke *et al.* (1962). (A) Major cell organelles: f, flagellum; fb, lipoid globule; g, Golgi body; n, nucleus; cv, contractile vacuole; c, chloroplast; m, mitochondrion; mb, muciferous body; s, scale layer; l, leucosin vesicle; h, haptonema (proximal part). (B) Three cells with coiled haptonemata and two cells with extended haptonemata. (C) Three body scales.

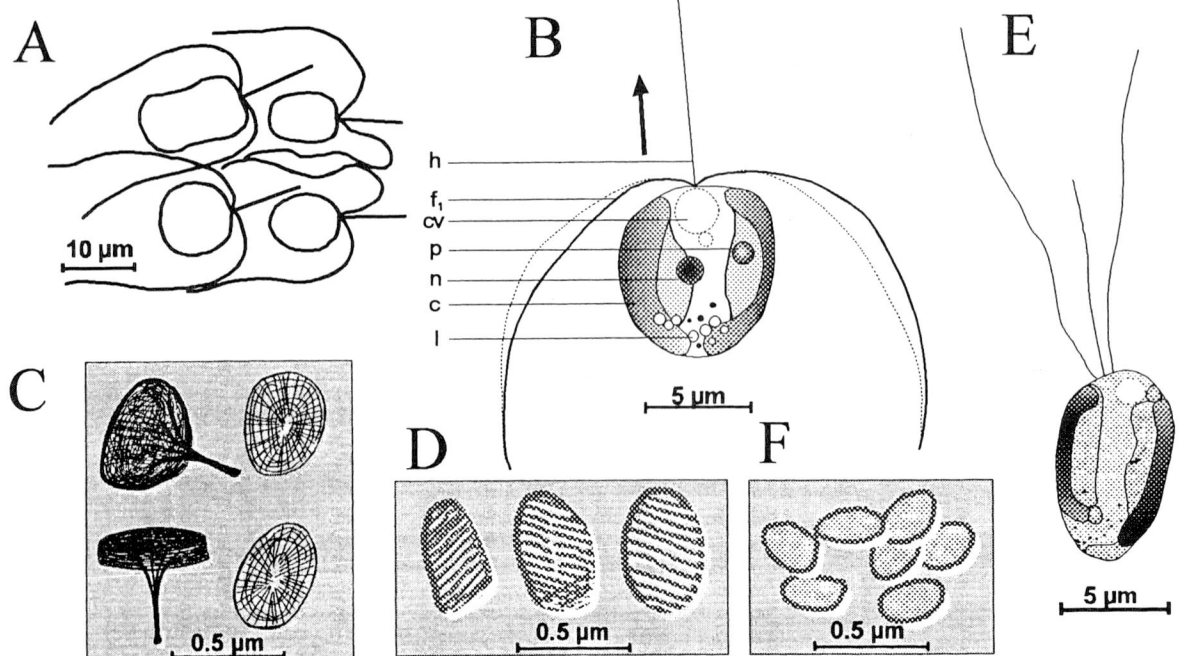

FIGURE 4 Three freshwater *Chrysochromulina* species. (A–C) *C. breviturrita*: (A) Four differently shaped cells, each with characteristically straight haptonema. (B) Cell in forward swimming mode (propulsion by vibrating flagella, indicated by dotted outline of flagella): h, haptonema; f_1, first flagellum; cv, contractile vacuole; p, pyrenoid; n, nucleus; c, chloroplast. (C) Organic body scales: two plate scales on right and two spine scales on left; redrawn from Nicholls (1978) and Brown et al. (1986). (D) Organic scales of *C. laurentiana*; redrawn from Nicholls (1978) and Kling (1981). (E, F) *Chrysochromulina inornata*: (E) Light microscopic view of whole cell; (F) Organic body scales; redrawn from Wujek and Gardiner (1985).

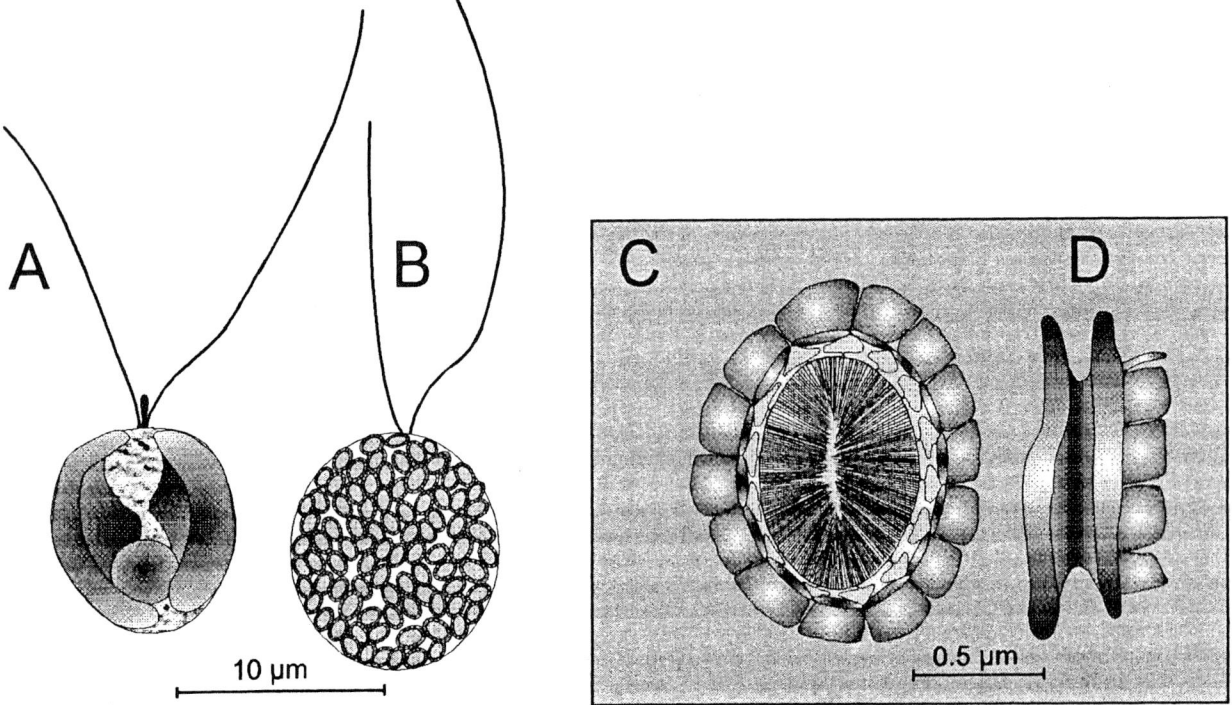

FIGURE 5 *Pleurochrysis carterae* var. *dentata*. (A) Noncoccolith-bearing motile stage with short bulbous haptonema; (B) Coccolith-bearing cell lacking visible haptonema, (C) Coccolith (cricolith) showing calcified peripheral "teeth" and organic baseplate ornamented with radial striae; (D) Lateral view of cricolith showing pulley-like structure.

Hymenomonas von Stein emend. Gayral *et* Fresnel

Cells are spherical or ovoid. They may or may not be motile; motile cells have two flagella of equal or subequal length and a very short bulbous haptonema. Cells are covered in coccoliths of one type (tremaliths) consisting of an organic baseplate and a short tubelike structure of fused calcite elements with pointed tops. An alternating life cycle of coccolith-bearing (diploid) and organic scale-bearing (no coccoliths; haploid) stages is known for marine species and possibly for *H. roseola*, the only confirmed freshwater species.

VI. GUIDE TO LITERATURE FOR SPECIES IDENTIFICATION

Chrysochromulina parva: Parke *et al*. (1962)
Chrysochromulina breviturrita: Nicholls (1978) and Brown *et al*. (1986)
Chrysochromulina laurentiana: Kling (1981)
Chrysochromulina inornata: Wujek and Gardiner (1985)
Hymenomonas roseola: Braarud (1954) and Manton and Peterfi (1969)

LITERATURE CITED

Andersen, R. A. 1989. The Synurophyceae and their relationship to other golden algae. Beihefte zur Nova Hedwigia 95:1–26.

Balch, W. M., Holligan, P. M., Ackleson, S. G., Voss, K. J. 1991. Biological and optical properties of mesoscale coccolithophore blooms in the Gulf of Maine. Limnology and Oceanography 36:629–643.

Bloem, J., Bär Gilissen, M.-J. B. 1988. Fixing nanoflagellates. Archiv für Hydrobiologie (Supplement) 31:275–280.

Bourrelly, P. 1981. Les algues d'eau douce, Tome II. Les algues jaunes et brunes, 2nd ed. Boubée et Cie, Paris.

Braarud, T. 1954. Coccolith morphology and taxonomic position of *Hymenomonas roseola* Stein and *Syracosphaera carterae* Braarud & Fagerland. Nytt Magasin for Botanikk 3:1–4 + 2 plates.

Brown, C. W., Yoder, J. A. 1993. Blooms of *Emiliania huxleyi* (Prymnesiophyceae) in surface waters of the Nova Scotian Shelf and the Grand Bank. Journal of Plankton Research 15:1429–1438.

Brown, L. M., Smith, R. J., Shivers, R. R., Day, A. W. 1986. A reexamination of the surface scales of *Chrysochromulina breviturrita* Nicholls (Prymnesiophyceae). Phycologia 25:572–575.

Christensen, T. 1962. Alger, *in*: Böcher, T. W., Lange, M., Sørensen, T., Eds., Botanik, Vol. 2. Systematisk botanik. Munksgaard, Copenhagen, 178 p.

Conrad, W. 1928. Sur les coccolithophoracées d'eau douce. Archiv für Protistenkunde 63:58–66.

Diaz, M. M., Lorenzo, L. E. 1990. *Chrysochromulina parva* Lackey (Prymnesiophyceae) new for South America. Algological Studies 60:19–24.

Edvardsen, B., Eikrem, W., Green, J. C., Andersen, R. A., Moon-Van der Staay, S. Y., Medlin, L. K. 2000. Phylogenetic reconstruction of the Haptophyta inferred from 18s ribosomal DNA sequences and available morphological data. Phycologia 39:19–35.

Faber Jr., W. W., Preisig, H. R. 1994. Calcified structures and calcification in protists. Protoplasma 181:78–105.

Findlay, D. L., Kasian, S. E. M. 1987. Phytoplankton community responses to nutrient addition in Lake 226, Experimental Lakes Area, northwestern Ontario. Canadian Journal of Fisheries and Aquatic Sciences 44:35–46.

Fresnel, J. 1994. A heteromorphic life cycle in two coastal coccolithophorids, *Hymenomonas lacuna* and *Hymenomonas coronata* (Prymnesiophyceae). Canadian Journal of Botany 72:1455–1462.

Green, J. C., Leadbeater, B. S. C., Eds. 1994. The haptophyte algae. Systematics Association Special Volume No. 51. Clarendon, Oxford.

Green, J. C., Pienaar, R. N. 1977. The taxonomy of the order Isochrysidales (Prymnesiophyceae) with special reference to the genera *Isochrysis* Parke, *Dicrateria* Parke and *Imantonia* Reynolds. Journal of the Marine Biological Association of the U.K. 57:7–17.

Guo, M., Harrison, P. J., Taylor, F. R. J. 1996. Fish kills related to *Prymnesium parvum* N. Carter (Haptophyta) in the People's Republic of China. Journal of Applied Phycology 8:111–117.

Hansen, L. R., Kristiansen, J., Rasmussen, J. V. 1994. Potential toxicity of freshwater *Chrysochromulina* species *C. parva* (Prymnesiophyceae). Hydrobiologia 287:157–159.

Heynig, H. 1963. *Chrysochromulina parva* Lackey im Plankton Mitteldeutschlands. Archiv für Protistenkunde 106:453–455.

Hibberd, D. J. 1976a. The ultrastructure and taxonomy of the Chrysophyceae and Prymnesiophyceae (Haptophyceae): A survey with some new observations on the ultrastructure of the Chrysophyceae. Botanical Journal of the Linnean Society 72:55–80.

Hibberd, D. J. 1976b. The fine structure of the colonial colourless flagellate *Rhipidodendron splendidum* Stein and *Spongomonas uvella* Stein with special reference to the flagellar apparatus. Journal of Protozoology 23:374–385.

Hibberd, D. J. 1983. Ultrastructure of the colonial colourless zooflagellates *Phalansterium digitatum* Stein (Phalansteriida ord. nov.) and *Spongomonas uvella* Stein (Spongomonadida ord. nov.). Protistologica 19:523–535.

Hibberd, D. J. 1985. Observations on the ultrastructure of new species of *Pseudodendromonas* Bourrelly (*P. operculifera* and *P. insignis*) and *Cyathobodo* Petersen and Hansen (*C. peltatus* and *C. gemmatus*), Pseudodendromonadida ord. nov. Archiv für Protistenkunde 129:3–11.

Holdway, P. A., Watson, R. A., Moss, B. 1978. Aspects of the ecology of *Prymnesium parvum* (Haptophyta) and water chemistry in the Norfolk Broads, England. Freshwater Biology 8:295–311.

Holmgren, S. K. 1984. Experimental lake fertilization in the Kuokkel area, northern Sweden. Phytoplankton biomass and algal composition in natural and fertilized subarctic lakes. Internationale Revue der Gesamten Hydrobiologie 69:781–817.

Holmquist, E., Willén, T. 1993. Fish mortality caused by *Prymnesium parvum*. Vatten 49:110–115.

Ito, H. 1989. Seasonal fluctuation of *Chrysochromulina parva* (Prymnesiophyceae) in four ponds and lakes in the Kinki district, Japan. Japanese Journal of Phycology 37:117–122.

James, T. L., de la Cruz, A. 1989. *Prymnesium parvum* Carter (Chrysophyceae) as a suspect of mass mortalities of fish and shellfish communities in Western Texas. Texas Journal of Science 41:429–430.

Johansen, J. R., Doucette, G. J., Barclay, W. R., Bull, J. D. 1988. The morphology and ecology of *Pleurochrysis carterae* var. *dentata*

var. nov. (Prymnesiophyceae), a new coccolithophorid from an inland saline pond in New Mexico, USA. Phycologia 27:78–88.
Jones, H. L. J., Leadbeater, B. S. C., Green, J. C. 1994. Mixotrophy in haptophytes, *in*: Green, J. C., Leadbeater, B. S. C., Eds., The haptophyte algae. Systematics Association Special Volume No. 51. Clarendon, Oxford, pp. 247–263.
Jordan, R. W., Chamberlain, A. H. L. 1997. Biodiversity among haptophyte algae. Biodiversity and Conservation 6:131–152.
Jordan, R. W., Green, J. C. 1994. A check-list of the extant Haptophyta of the world. Journal of the Marine Biological Association of the U.K. 74:149–174.
Jordan, R. W., Kleinjne, A., Heimdal, B. R., Green, J. C. 1995. A glossary of the extant Haptophyta of the world. Journal of the Marine Biological Association of the U.K. 75:769–814.
Kawachi, M., Inouye, I. 1995. Functional roles of the haptonema and the spine scales in the feeding process of *Chrysochromulina spinifera* (Fournier) Pienaar et Norris (Haptophyta = Prymnesiophyta). Phycologia 34:193–200.
Kawai, H., Inouye, I. 1989. Flagellar autofluorescence in forty-four chlorophyll *c*–containing algae. Phycologia 28:222–227.
Kling, H. J. 1981. *Chrysochromulina laurentiana*: An electron microscopic study of a new species of Prymnesiophyceae from Canadian Shield lakes. Nordic Journal of Botany 1:551–555.
Kristiansen, J. 1971. A Danish find of *Chrysochromulina parva* (Haptophyceae). Botanisk Tidsskrift 66:33–37.
Lackey, J. B. 1939. Notes on plankton flagellates from the Scioto River. Lloydia 2:128–143.
Larsen, A., Eikrem, W., Paasche, E. 1993. Growth and toxicity in *Prymnesium patelliferum* (Prymnesiophyceae) isolated from Norwegian waters. Canadian Journal of Botany 71:1357–1362.
Lund, J. W. G. 1942. Contributions to our knowledge of British Chrysophyceae. New Phytologist 41:274–292.
Malin, G., Liss, P. S., Turner, S. M. 1994. Dimethyl sulfide: Production and atmospheric consequences, *in*: Green, J. C., Leadbeater, B. S. C., Eds., The haptophyte algae. Systematics Association Special Volume No. 51. Clarendon, Oxford, pp. 303–320.
Manton, I., Peterfi, L. S. 1969. Observations on the fine structure of coccoliths, scales and the protoplast of a freshwater coccolithophorid, *Hymenomonas roseola* Stein, with supplementary observations on the protoplast of *Cricosphaera carterae*. Proceedings of the Royal Society of London Series B 172:1–15.
Meyer, R. L., Brook, A. J. 1969. Freshwater algae from the Itasca State Park, Minnesota II. Chrysophyceae and Xanthophyceae. Nova Hedwigia 17:105–112.
Moestrup, Ø. 1994. Economic aspects: "Blooms," nuisance species, and toxins, *in*: Green J. C., Leadbeater, B. S. C., Eds., The haptophyte algae. Systematics Association Special Volume No. 51. Clarendon, Oxford, pp. 265–285.
Moestrup, Ø., Thomsen, H. A. 1980. Preparation of shadow-cast whole mounts, *in*: Gantt, E., Ed., Handbook of phycological methods: Development and cytological methods. Cambridge Univ. Press, Cambridge, UK, pp. 385–390.
Moestrup, Ø., Thomsen, H. A. 1995. Taxonomy of toxic haptophytes (prymnesiophytes), *in*: Hallegraeff, G. M., Anderson, D. M., Cembella A. D., Eds., Manual on harmful marine microalgae. IOC Manuals and Guides No. 33, UNESCO, Paris, pp. 319–338.
Munawar, M., Munawar, I. F. 1982. Phycological studies in lakes Ontario, Erie, Huron, and Superior. Canadian Journal of Botany 60:1837–1858.
Nicholls, K. H. 1978. *Chrysochromulina breviturrita* sp. nov., a new freshwater member of the Prymnesiophyceae. Journal of Phycology 14:499–505.
Nicholls, K. H. 1979. Is *Hymenomonas prenanti* Lecal (Prymnesiophyceae) really the colourless flagellate *Gyromitus disomatus* Skuja? Phycologia 18:420–423.
Nicholls, K. H., Beaver, J. L., Estabrook. R. H. 1982. Lakewide odours in Ontario and New Hampshire caused by *Chrysochromulina breviturrita* Nich. (Prymnesiophyceae). Hydrobiologia 96:91–95.
Nicholls, K. H., Nakamoto, L., Keller, W. 1992. Phytoplankton of Sudbury area lakes (Ontario) and relationships with acidification status. Canadian Journal of Fisheries and Aquatic Sciences 49:40–51.
O'Grady, K., Brown, L. M. 1989. Growth inhibition by high light intensities in algae from lakes undergoing acidification. Hydrobiologia 184:201–208.
Parke, M., Manton, I., Clarke, B. 1955. Studies on marine flagellates II. Three new species of *Chrysochromulina*. Journal of the Marine Biological Association of the U.K. 34:579–609.
Parke, M., Lund, J. W. G., Manton, I. 1962. Observations on the biology and fine structure of the type species of *Chrysochromulina* (*C. parva* Lackey) in the English Lake District. Archiv für Mikrobiologie 42:333–352.
Pollinger, U. 1986. Phytoplankton periodicity in a subtropical lake (Lake Kinneret, Israel). Hydrobiologia 138:127–138.
Reinertsen, H. 1982. The effect of nutrient addition on the phytoplankton community of an oligotrophic lake. Holarctic Ecology 5:225–252.
Simon, N., Brenner, J., Edvardsen, B., Medlin, L. K. 1997. The identification of *Chrysochromulina* and *Prymnesium* species (Haptophyta, Prymnesiophyceae) using fluorescent of chemiluminescent oligonucleotide probes: A means for improving studies on toxic algae. European Journal of Phycology 32:393–401.
Simonsen, S., Moestrup, Ø. 1997. Toxicity tests in eight species of *Chrysochromulina* (Haptophyceae). Canadian Journal of Botany 75:129–136.
Stoermer, E. F., Sicko-Goad, L. 1977. A new distribution record for *Hymenomonas roseola* Stein (Prymnesiophyceae, Coccolithophoraceae) and *Spiniferomonas trioralis* Takahashi (Chrysophyceae, Synuraceae) in the Laurentian Great Lakes. Phycologia 16:355–358.
Thompson, R. H., Halicki, P. J. 1965. *Chrysochromulina parva* in eastern Kansas. Transactions of the American Microscopical Society 84:14–17.
Vollenweider, R. A. 1969. A manual on methods for measuring primary production in aquatic environments. IBP Handbook No. 12. Blackwell Sci., Oxford.
Wehr, J. D., Brown, L. M. 1985. Selenium requirement of a bloom-forming planktonic alga from softwater and acidified lakes. Canadian Journal of Fisheries and Aquatic Sciences 42:1783–1788.
Wehr, J. D., Brown, L. M., O'Grady, K. 1985. Physiological ecology of the bloom-forming alga *Chrysochromulina breviturrita* (Prymnsiophyceae) from lakes influenced by acid precipitation. Canadian Journal of Botany 63:2231–2239.
Wehr, J. D., Brown, L. M., O'Grady, K. 1987. Highly specialized nitrogen metabolism in a freshwater phytoplankter, *Chrysochromulina breviturrita*. Canadian Journal of Fisheries and Aquatic Sciences 44:736–42.
Wei, Y. 1996. *Chrysochromulina parva* Lackey (1939): new record in China and its seasonal fluctuation in Lake Donghu, Wuhan. Acta Hydrobiologica Sinica 20:317–321.
Westbroek, P., Brown, C. W., van Bleijswijk, J., Brownlee, C., Brummer, G., Conte, M., Egge, J., Fernández, E., Jordan, R., Knappertsbusch, M., Stefels, M., Veldhuis, M., van der Wal, P., Young, J. 1993. A model system approach to biological climate forcing: An example of *Emiliania huxleyi*. Global and Planetary Change 8:27–46.

Wetzel, R. G., Likens, G. E. 1991. Limnological analyses, 2nd ed., Springer-Verlag, New York, 391 p.

Whitford, L. A. 1979. Additions to the freshwater algae in North Carolina. IX. Journal of the Elisha Mitchell Scientific Society 95:42–47.

Wujek, D. E., Gardiner, W. E. 1985. Chrysophyceae (Mallomonadaceae) from Florida. II. New species of *Paraphysomonas* and the prymnesiophyte *Chrysochromulina*. Florida Scientist 48:59–63.

Wujek, D. E., Saha, L. C. 1991. *Chrysochromulina parva* Lackey (Prymnesiophyceae) a new record from India. Phykos 30:169–171.

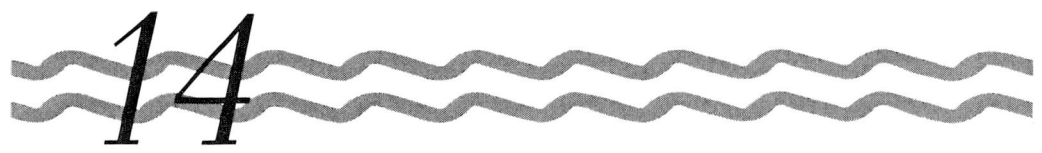

14 SYNUROPHYTE ALGAE

Peter A. Siver

Botany Department
Connecticut College
New London, Connecticut, 06320

I. Introduction
II. Diversity and Morphology
   A. General Characteristics
   B. Scale and Bristle Anatomy
   C. Arrangement of Scales on the Cell Surface
   D. Cysts Life Cycle, and Cell Division
   E. Taxonomic and Phylogenetic Considerations
III. Ecology and Distribution
   A. General Comments
   B. Primary Habitat Requirements
   C. Distribution along a pH Gradient
   D. Distribution along a Dissolved Salt Gradient
   E. Distribution along a Trophic Gradient
   F. Distribution along a Temperature Gradient
IV. Collection and Preparation for Identification
   A. Collection of Samples
   B. Preparing Samples for Observation with Light Microscopy
   C. Preparing Samples for Observation with Electron Microscopy
   D. Storage of Samples and Preparations of Scales and Bristles
V. Keys to Genera and Common Species from North America
   A. Introduction to Keys
   B. Key to Genera
   C. Descriptions of Genera
   D. Key to Common Species of *Mallomonas*
   E. Key to Common Species of *Synura*
VI. Guide to Literature for Species Identification
Literature Cited

This chapter is dedicated to my mother, Helen Jean Siver. Thanks, Mom

I. INTRODUCTION

The Synurophyceae is a relatively new class of algae based primarily on differences in biochemical and ultrastructural characteristics (Cavalier-Smith, 1986; Andersen, 1987). The organisms contained in the Synurophyceae were previously placed in the family Mallomonadaceae Diesing (*sensu* Hibberd, 1976; Bourrelly, 1981), also referred to as the Synuraceae (e.g., Wee, 1982; Starmach, 1985), in the class Chrysophyceae. Wee (1982) listed 10 genera within the family Synuraceae (or Mallomonadaceae): *Synura, Microglena, Mallomonas, Chrysosphaerella, Conradiella, Paraphysomonas, Mallomonopsis, Chlorodesmus* (also known as *Catenochrysis* and *Phillipsiella*), *Chrysodidymus*, and *Spiniferomonas*. All genera within the Mallomonadaceae possessed a cell covering composed of siliceous scales and commonly were referred to as scaled chrysophytes.

In a review of the class Chrysophyceae, Hibberd (1976) recognized that the ultrastructure of some genera in the Mallomonadaceae deviated substantially from that of *Ochromonas*, the genus considered as the type for the class. Preisig and Hibberd (1983, 1986) removed *Paraphysomonas, Spiniferomonas*, and related genera from the Mallomonadaceae based on differences in cell structure and placed them into a new family, the Paraphysomonadaceae. Preisig and Hibberd clearly demonstrated that organisms in the Paraphysomonadaceae had a cell structure that was similar to that of *Ochromonas* and *Chromulina*, but quite different from other members of the Mallomonadaceae, namely *Mallomonas* and *Synura*. Shortly thereafter, Andersen (1987) moved *Mallomonas* and *Synura* out of the class Chrysophyceae and placed them into a new class, the Synurophyceae.

There are currently four genera recognized within the Synurophyceae. Three of the genera, *Mallomonas*, *Synura*, and *Chrysodidymus*, are reported commonly from North America, and contain about 160 recognized species or subspecific taxa (Moestrup, 1995). *Tessellaria*, a fourth genus more recently assigned to the Synurophyceae, is known from Australian waters (Tyler *et al.*, 1989; Pipes *et al.*, 1991), but has not been reported from North America. A fifth genus, *Mallomonopsis*, was originally separated from *Mallomonas* on the basis of possessing two, rather than one, emergent flagella (Matvienko, 1941; Wujek and Timpano, 1984). Belcher's (1969) recommendation to combine these two genera was supported by Andersen's (1987) conclusion that the occurrence of two basal bodies and two flagella appeared to be universal among members of the genus *Mallomonas*. Such a proposal was followed in many subsequent taxonomic works, including Momeu and Péterfi (1979), Asmund and Kristiansen (1986), and Siver (1991a), and has been supported by phylogenetic analyses using scale covering characters and molecular data (Lavau *et al.*, 1997). *Chlorodesmus hispidus* was shown by Calado and Rino (1994) to be a morphological form of, and synonymous with, *Synura spinosa*. As noted by Wee (1982), Kristiansen (1988a), and Leadbeater and Barker (1995), the taxonomic positions of the rarely observed genera *Conradiella* and *Microglena* remain unclear. Kristiansen (1988a) has suggested that *Conradiella*, originally described by Pascher (1925) as a unicellular organism surrounded by transverse rings of silica, may actually represent a misidentified *Mallomonas*.

II. DIVERSITY AND MORPHOLOGY

A. General Characteristics

All synurophytes are motile flagellates that consist of either a unicellular or colonial habit (Figs. 1–3). Cells contain one bilobed or two golden-colored chloroplasts that generally are positioned such that their long axes are parallel to each other and to the long axis of the cell (Fig. 1). All species possess siliceous scales that either surround individual cells (*Chrysodidymus*, *Mallomonas*, and most species of *Synura*) or are situated in multiple layers around the colony (*Synura lapponica* and *Tessellaria*; Skuja, 1956; Wujek and Wee, 1983; Tyler *et al.*, 1989; Siver and Glew, 1990; Figs. 1–3). Each cell has two parallel aligned and apically inserted flagella that emerge either from a pore in the anterior end of the scale coat (Fig. 3E) or through the scale matrix on the surface of the colony (Fig. 3F). For all species of *Mallomonas*, except for taxa within the sections Mallomonopsis, Multisetigerae, and Papillosae, only a single emergent flagellum can be seen with light microscopy (LM; Asmund and Kristiansen, 1986; Siver, 1991a). The part of the cell with the flagella is referred to as the anterior end and the nonflagellated end is denoted the posterior end.

Although some species of *Mallomonas* are spherical unicells, most possess ovoid or ellipse-shaped cells (Figs. 1A and B, and 3A–D; Siver, 1991a). In addition to siliceous scales, species of *Mallomonas* also have elongated siliceous structures called bristles that are associated with all or selected scales on the cell covering (Figs. 1A–E, and 3A and D). Only one species of *Mallomonas*, *M. adamas*, is thought to lack bristles (Lavau *et al.*, 1997). The bristles are tucked under the distal ends of the scales in such a way that when the cells are actively swimming the bristles become streamlined and parallel with the long axis of the cell. Common morphologies for *Mallomonas* taxa include cells where the bristles are (a) restricted to the anteriormost scales that immediately surround the flagellar pore (e.g., *M. dickii*; Fig. 13C), (b) restricted to scales on the anterior end of the cell (e.g., *M. tonsurata*; Fig. 1E), (c) distributed over most of the cell (e.g., *M. transsylvanica*; Fig. 3A), and (d) where distinctly different types of bristles are found on different parts of the scale coat (e.g., *M. hamata*). In addition, many species of *Mallomonas* have scales with spines that are positioned on the posterior end of the cell (Fig. 3B). Spines can range from being quite small, less than 1 µm (e.g., *M. dickii*) to over 10 µm in length (e.g., *M. torquata*; Siver, 1991a). Most species of *Mallomonas* have cells that range in size from 10 to 50 µm long and 5 to 20 µm wide, although larger and smaller cells can be observed (Siver, 1991a). Size records and morphological details for many species can be found in Takahashi (1978), Wee (1982), Asmund and Kristiansen (1986), and Siver (1991a, b). *Mallomonas* is a cosmopolitan genus (see Asmund and Kristiansen, 1986) that is observed in collections throughout North America (Nicholls, 1988a; Siver, 1991a).

Species of *Chrysodidymus*, *Synura*, and *Tessellaria* are colonial flagellates where each cell of the colony has two clearly visible emergent flagella. Colonies of *Chrysodidymus* consist of two somewhat elongated cells that are attached at their posterior ends and face

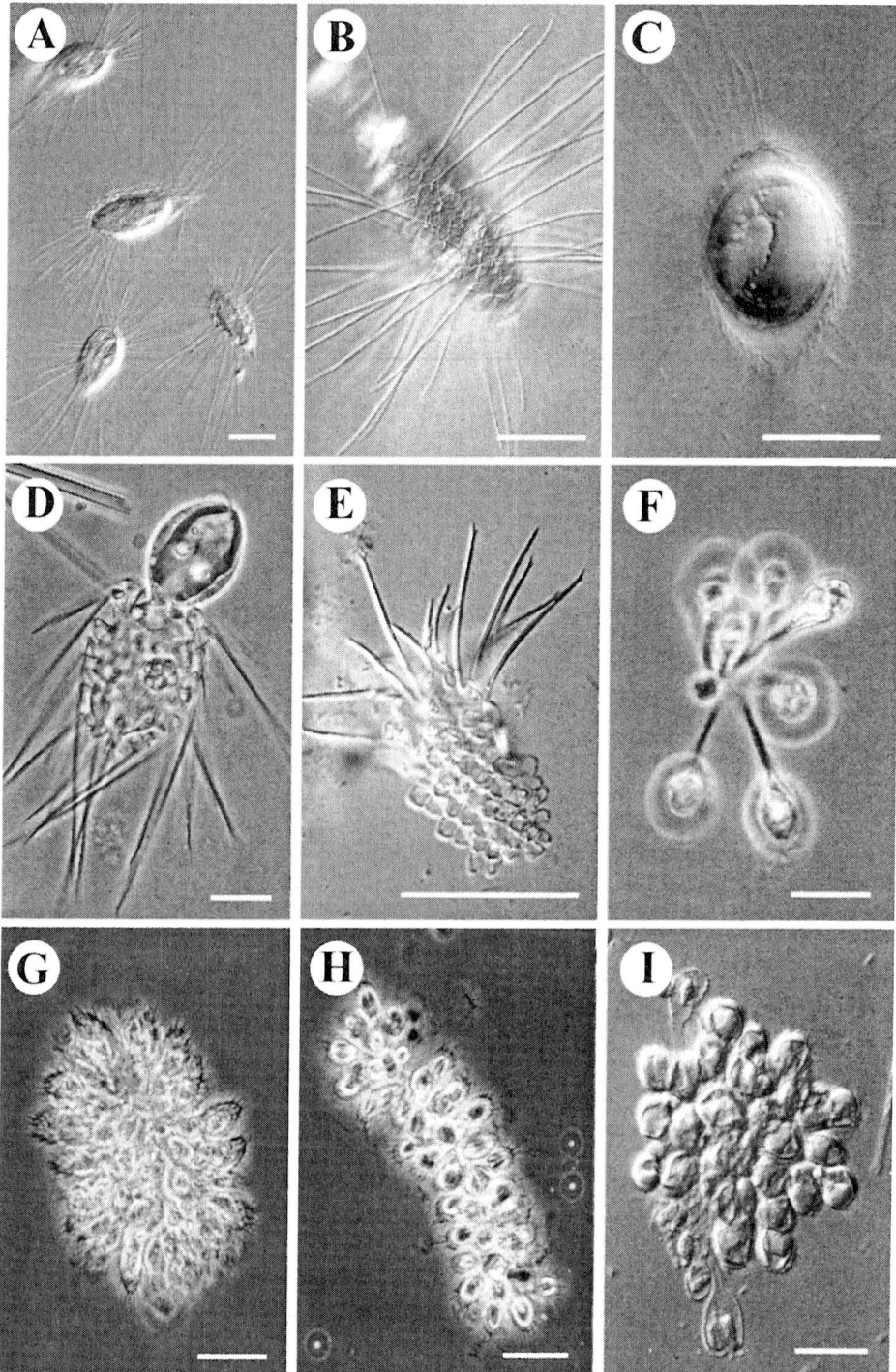

FIGURE 1 Light micrographs of *Mallomonas* cells and *Synura* colonies. A. Four cells of *Mallomonas caudata*. Scale bar = 20 µm. B. Close-up of a cell of *Mallomonas caudata* focused on the outer surface. Note the overlapping oval scales and the proximal ends of the bristles. Scale bar = 20 µm. C. Cell of *Mallomonas caudata* beginning to encyst. Note the large central vacuole, chloroplast, and the overlapping nature of the scales. Scale bar = 20 µm. D. Cell of *Mallomonas corymbosa* where the intact protoplast has squeezed out from the scale coat. The two chloroplasts and a small portion of the flagellum can be observed, as well as scales and bristles. Scale bar = 10 µm. E. Remains of the scale coat of *Mallomonas tonsurata*. Note the overlapping scales, the restriction of the bristles to the anterior end of the cell, and the forked tips of the bristles. Scale bar = 10 µm. F. Colony of *Synura petersenii* with only a few club-shaped cells with long caudal tails. Scale bar = 20 µm. G. Elongated colony of *Synura spinosa*. Note the apical scales with long spines. Scale bar = 20 µm. H. *Chorodesmus*-like form of *Synura spinosa*. Scale bar = 20 µm. I. Squashed colony of *Synura petersenii* denoting club-shaped cells. Scale bar = 20 µm.

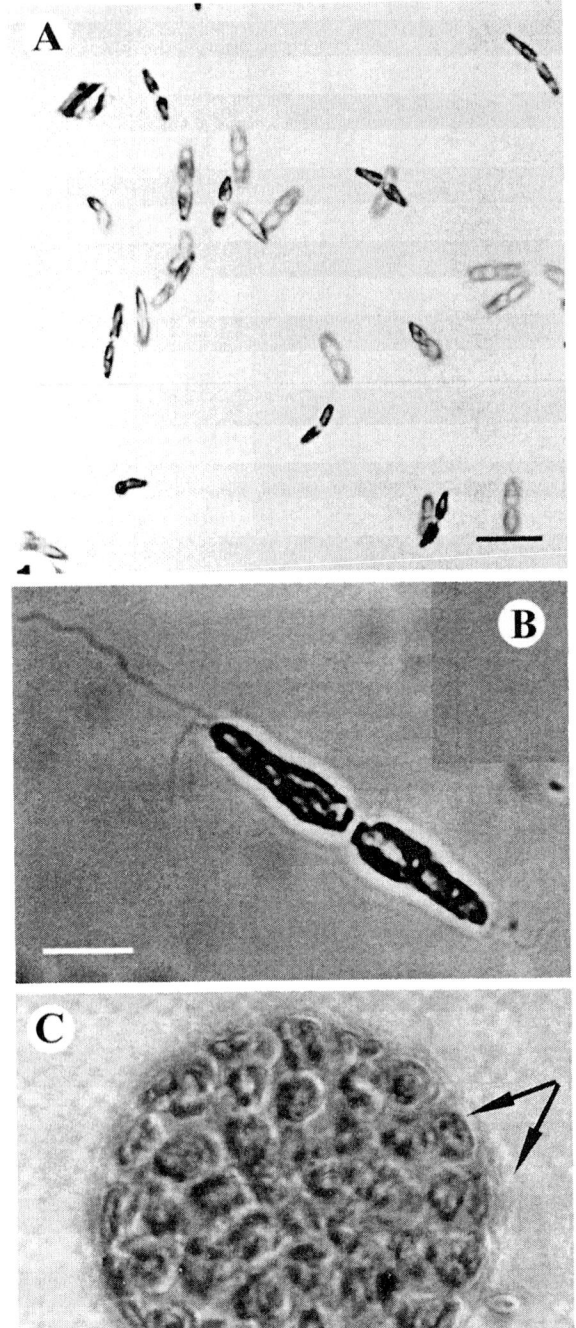

FIGURE 2 Light micrographs of additional synurophycean genera. A and B. Two-celled colonies of *Chrysodidymus*. Scale bar for A = 20 µm and for B = 5 µm. The arrows denote layers of scales around the colony. Scale bar = 50 µm. Reprinted from Graham *et al.* (1993) with permission of The Journal of Phycology. C. Colony of *Tessellaria*. Reprinted from Tyler *et al.* (1989) with permission.

outward at 180° from each other (Fig. 2A and B). The cells of recently formed colonies of *Chrysodidymus* tend to be globose or spherical in shape. As colonies age, the cells become more elongated and vaselike, and the posterior portion is distinctly wider than the flagellated end (Graham *et al.*, 1993). Whereas colonies of *Tessellaria* and *Synura*, including those with two cells, tend to have a rolling or tumbling swimming motion, colonies of *Chrysodidymus* oscillate back and forth along their longitudinal axes (Nicholls and Gerrath, 1985; Graham *et al.*, 1993). Although rarely observed in most collections, *Chrysodidymus* has been found in a number of localities in North America, including Quebec (Puytorac *et al.*, 1972), Ontario (Nicholls and Gerrath, 1985), the midwestern United States (Wujek and Wee, 1983; Wujek and Igoe, 1989; Graham *et al.*, 1993), Florida (Wujek and Bland, 1991), the Adirondacks of New York (Siver, 1988a; Cumming *et al.*, 1992a), Washington (Norris and Munch, 1970), southern New England (Siver and Hamer, 1992), and Louisiana (Wee *et al.*, 1993). *Chrysodidymus* also has been observed in South America (Dürrschmidt, 1982; Wujek and Bicudo, 1993), Greenland (Nygaard, 1978), Germany (Hartmann and Steinberg, 1989), and tropical localities (Prowse, 1962; Hansen, 1995; Cronberg, 1996).

Colonies of *Tessellaria* are primarily spherical in shape, contain a few to several hundred compacted cells, range in size from 25 to 200 µm, and have a spinning swimming motion (Tyler *et al.*, 1989). As previously noted, the entire colony is enclosed in multiple layers of siliceous scales (Fig. 3F) that appear in LM as a "halo" around the colony (Fig. 2C; Tyler *et al.*, 1989). Individual cells lack scales and are elongated with long cytoplasm tails that attach the cells to a ringlike structure in the center of the colony (Tyler *et al.*, 1989). The ringlike structure is best observed with high magnification on squashed specimens. To date, *Tessellaria* is known only from Australia (Tyler *et al.*, 1989; Pipes and Leedale, 1992; Lavau *et al.*, 1997).

Colonies of *Synura* are typically described as being spherical (Fig. 1I) or elongated (Fig. 1G and H) in design and consisting of pyriform or club-shaped cells (Figs. 1F and I, and 3H; Petersen and Hansen, 1956, 1958; Takahashi, 1978; Wee, 1982). There is, however, considerable variation with regard to cell and colony size and shape, rendering such characters of little or no taxonomic value at the species level (Petersen and Hansen, 1956). Within a given species of *Synura*, colonies with a few cells often, but not always, have more spherical-shaped cells (Takahashi, 1978; Calado and Rino, 1994). However, as the number of cells in the colony increases, the cells become distinctly more pyriform or club-shaped (Fig. 3H). As noted by Petersen

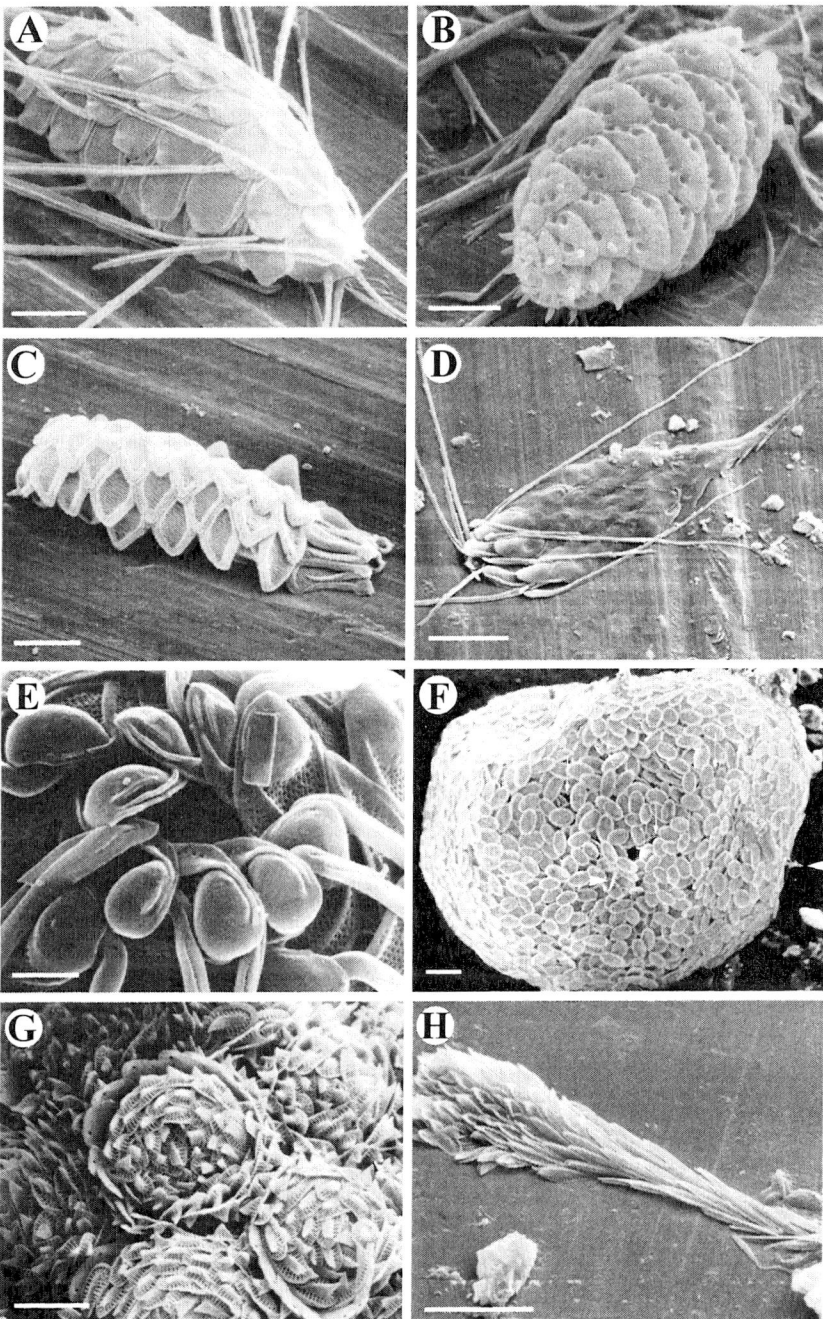

FIGURE 3 Examples of whole cells or colonies as observed with SEM. A. Cell of *Mallomonas transsylvanica* where the bristles are distributed over most of the cell. Note the modified anterior scales from which the flagellum emerges. Scale bar = 5 μm. B. Cell of *Mallomonas lychenensis* viewed from the posterior end. Scale bar = 5 μm. C. Cell of *Mallomonas dickii* with diamond- or rhomboid-shaped body scales and distinctly shaped collar scales that surround the flagellar opening. A very short bristle is associated with each collar scale. Scale bar = 2 μm. Reprinted with permission from P. A. Siver, *The Biology of Mallomonas: Morphology, Taxonomy and Ecology*. Copyright © 1991, Kluwer, Dordrecht. D. Spindle-shaped cell of Mallomonas akrokomos with a long caudal tail. Bristles are restricted to the anterior end. Scale bar = 5 μm. E. Close-up of the anterior end of a cell of *Mallomonas tonsurata* depicting the flagellar opening lined with eight domed scales. Scale bar = 1 μm. Reprinted from Siver and Glew (1990) with permission. F. Colony of *Tessellaria volvocina*. Note that the scales form a layer around the entire colony, not individual cells. Arrows denote flagella and arrowheads denote spined scales. Scale bar = 5 μm. Reprinted from Tyler *et al.* (1989) with permission. G. Close-up of a colony of *Synura petersenii* depicting the spiral rows of scales around individual cells in the colony. Scale bar = 5 μm. H. Club-shaped cell of *Synura petersenii* that had become detached from a colony. Note the long caudal tail. Compare this image with cells in Figure 1I. Scale bar = 10 μm.

and Hansen (1956), and illustrated by Takahashi (1978), cells that become dissociated from intact colonies, which commonly occurs in collections, often change shape and become spherical in nature. Colonies of *Synura petersenii*, *S. spinosa*, *S. sphagnicola*, and *S. echinulata* are commonly 30–40 μm in diameter (Takahashi, 1978; Siver, unpublished data), whereas colonies of *S. uvella* are typically over 200 μm in diameter (Siver, unpublished data) and have been observed as large as 400 μm in diameter (Korshikov, 1929). In addition to forming spherical colonies, *Synura spinosa* often forms elongated colonies (Fig. 1G and H; Takahashi, 1978; Siver, 1987). Extremely elongated colonies of *S. spinosa*, originally described as *Chlorodesmus* (Calado and Rino, 1994), also have been observed in North America (Fig. 1H). Thus, although colony size and shape should not be the sole characters on which to base species determinations, tentative identifications of *S. uvella* and *S. spinosa* often can be made based on the large colony size and formation of distinctly elongated colonies, respectively. *Synura* is a very common, widely observed, and cosmopolitan genus observed in collections throughout North America (e.g., Nicholls and Gerrath, 1985; Siver, 1987).

The distinct golden color is a result of the complement of chloroplast pigments, including chlorophyll-c_1, but not chlorophyll-c_2, and particularly fucoxanthin (Andersen and Mulkey, 1983; Andersen, 1987). Chloroplasts possess thylakoids in groups of three, girdle lamella, and additional external membranes referred to as the chloroplast endoplasmic reticulum (CER) or periplastid endoplasmic reticulum (PER). A large Golgi complex is observed below the flagallar basal bodies (Hibberd, 1978; Beech and Wetherbee, 1990a). A single, rather large nucleus is situated below the Golgi apparatus and between the chloroplasts (Beech and Wetherbee, 1990a). Although evidence suggests that the outer membrane of the nucleus may be confluent with the CER in *Synura* (Mignot and Brugerolle, 1982), it appears to be, at best, weakly developed (Hibberd, 1978; Andersen, 1985). Synurophytes lack eye spots (Moestrup, 1995) and often contain a large chrysolaminarin storage vacuole in the posterior end of the cell; both of these features can be noted with light microscopy.

Synurophytes are heterokont algae that possess a long pleuronematic or tinsel flagellum and a smaller, often highly reduced, acronematic or smooth flagellum (Belcher, 1969; Hibberd, 1976; Moestrup. 1995). In the case of *Mallomonas splendens*, the second flagellum is lacking altogether. The tinsel flagellum bears two rows of tripartite hairs, and both flagella may be covered with organic scales (Moestrup, 1995). The tinsel flagellum moves the cell forward (Belcher, 1969; Wee, 1982) by beating in an S-shaped fashion in a single plane (Jarosch, 1970). In colonial forms the smooth flagellum beats in a more helical motion and provides rotational movement.

As reviewed by Moestrup (1995), the basal bodies of synurophytes range from being parallel to one another to being positioned at an angle of about 20°, and are connected by two or three fibrous bands. A rhizoplast originates at the base of the flagella, extends down along the nuclear membrane, and, at least in *Mallomonas splendens*, becomes splayed into numerous branches that form an inverted cone over the nucleus (Beech and Wetherbee, 1990a). The arrangement of microtubular roots associated with the flagellar basal bodies has been well studied. The R_1 microtubular root, which originates from the rhizoplast and forms a clockwise loop around the basal bodies, is found in all genera (Andersen, 1985, 1987; Beech and Wetherbee, 1990a; Pipes *et al.*, 1991; Graham *et al.*, 1993). The R_3 root appears to be present only in some genera, and the R_2 and R_4 roots are apparently lacking altogether (Andersen *et al.*, 1999).

B. Scale and Bristle Anatomy

All species of synurophytes possess a cell coat or covering composed primarily of siliceous scales. The design of the scale is species-specific and is used to identify each taxon. There are a number of different types of scales including apical, body, caudal, domed, domeless, and spined (Fig. 4). Other scale types, such as the winged scales on *Mallomonas pseudocoronata* (Fig. 9B), are rarer and found on a limited number of species. The cell covering of most species has more than one type of scale, and each type is found in a specific location on the cell coat (e.g., Fig. 3C). Apical, body, and caudal scales are positioned on the anterior, middle, and posterior portions of the cell, respectively. Spines are extensions of the front, or distal, end of the scale. Spined scales are found in all genera, but not on all species, whereas domed and domeless scales are types of scales found exclusively in *Mallomonas*.

In most cases, scales of synurophytes possess a base plate and an upturned rim (Fig. 4). The base plate is typically perforated with minute pores that may be evenly spaced, more dense in specific areas, or lacking altogether in certain regions (e.g., flanges and the dome region). The perforations of *Chrysodidymus* scales are generally larger and often of different diameters on the same scale (Fig. 6A). At least part of the edge of any scale is always turned up and bent over the base plate, forming a rim (Fig. 4A–F). For the majority of taxa, the rim is situated on what is referred to as the proximal

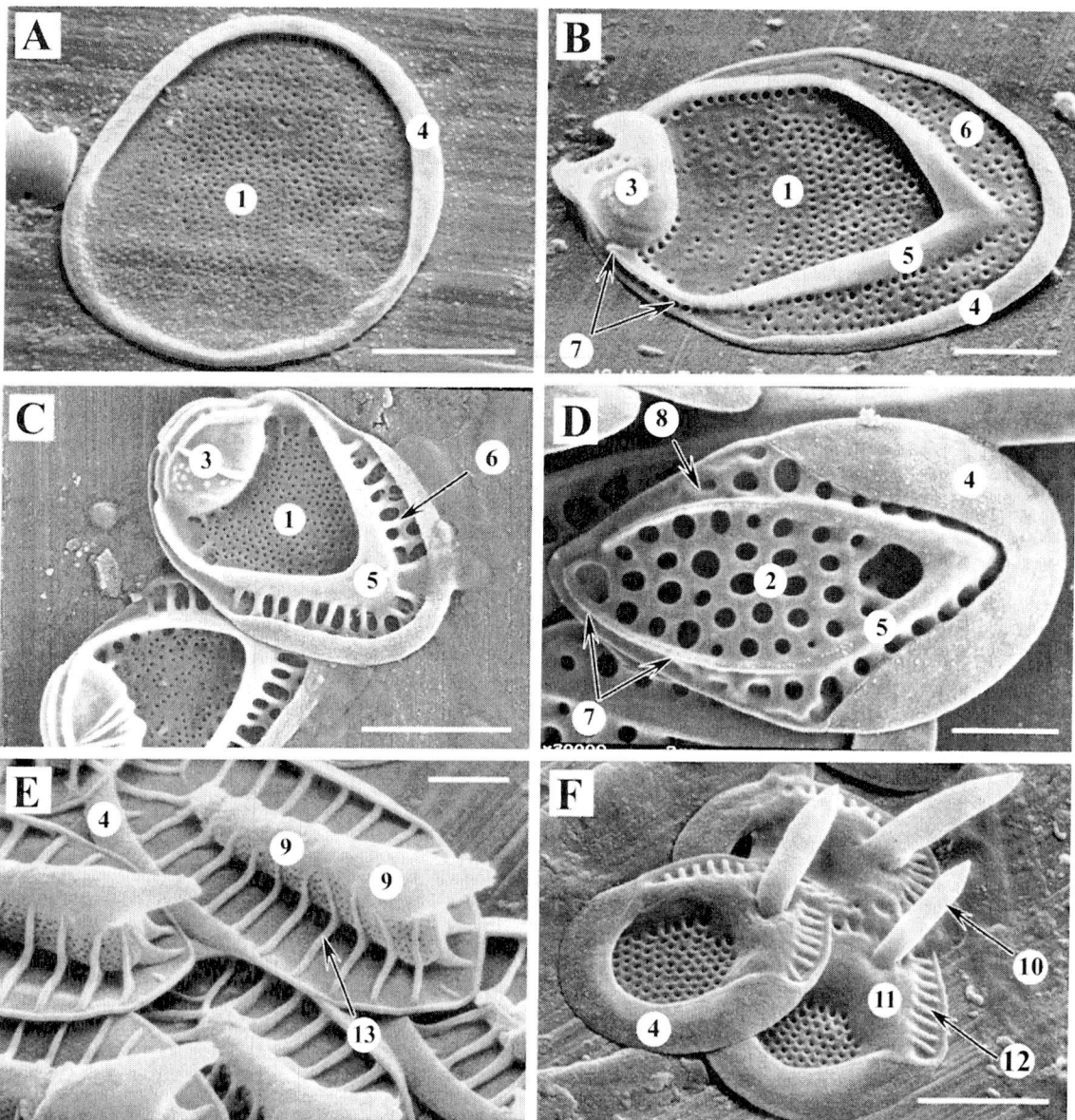

FIGURE 4 The parts of a scale. A. Simple scale of *Mallomonas caudata* with a shield (1) and a posterior rim (4). Scale bar = 2 μm. B. Scale of *Mallomonas elongata* with a shield (1), dome (3), V rib (5), anterior submarginal ribs (7), posterior flange (6), and a posterior rim (4). The arms of the V rib (5) are continuous with those of the anterior submarginal ribs (7). This scale type lacks any secondary structure on the shield or flange regions. Scale bar = 2 μm. Reprinted with permission from P. A. Siver, *The Biology of Mallomonas; Morphology, Taxonomy and Ecology.* Copyright © Kluwer, Dordrecht. C. Scale of *Mallomonas acaroides* var. *muskokana* with a shield (1), dome (3), V rib (5), posterior flange with ribs (6), and a posterior rim (4). Scale bar = 2 μm. D. Domeless scale of *Mallomonas duerrschmidtiae* possessing a shield with a well developed secondary layer of large pores (2), a V rib (5) with arms that extend and become continuous with the anterior submarginal ribs (7), a large posterior rim (4) that covers most of the posterior flange, and a small anterior flange (8). Scale bar = 1 μm. E. Scale of *Synura petersenii* var. *praefracta* with a well developed thorn or keel (9), a series of ribs running from the keel to the perimeter of the scale (13), and a posterior rim (4). Scale bar = 1 μm. F. Three scales of *Synura echinulata* each with a well developed secondary layer on the distal end of the scale (11) where the spine (10) originates, a series of short perpendicular ribs on the distal end (12), and a posterior rim (4). Scale bar = 2 μm.

end of the scale and is referred to as the posterior rim. The region of a scale opposite of the proximal end is referred to as the distal end. The posterior rim generally encircles roughly half of the scale (e.g., Fig. 4B–D), although on scales of some *Synura* species (e.g., *S. sphagnicola* and *S. lapponica*) and *Tessellaria*, the rim may encircle most or all of the scale. The surface of the scale that consists solely of base plate perforations and is in contact with the cell membrane is referred to as the inner or ventral surface. The side of the scale with the rim that faces away from the cell membrane is the outer or dorsal surface.

The majority of synurophytes have scales with additional or secondary layers of silica deposited on the base plate in species-specific designs, as well as other siliceous structures (Fig. 4C and D). In general, a secondary layer in *Synura* is composed of a siliceous layer with pores that are larger than those on the base plate and a series of siliceous ribs or struts. The larger pores of the secondary layer on *Synura* scales are restricted primarily to the distal end of the scale. Scales on the anterior portion of most *Synura* cells possess a spine, an additional diagnostic structure that protrudes from the base plate. For most *Synura* taxa the spine originates on the distal end of the scale and protrudes forward, away from the base plate (Fig. 4F); these taxa are in the Spinosae group. On taxa in the section Peterseniae the spine, more commonly referred to as a keel or thorn (Wee, 1982, 1997), originates along and is fused to the base plate (Fig. 4E). Scales of *S. lapponica* lack spines, but possess a large centrally positioned papillae (Fig. 10G) that is not homologous with the spine of other *Synura* taxa (Wee, 1997). Spines on scales of *Chrysodidymus* also originate on and protrude from the distal end of the scale (Fig. 6A).

Many species of *Mallomonas* possess scales that are considerably more complex than those of other synurophyte genera because of the presence of highly ornamented secondary layers, V ribs, anterior submarginal ribs, and domes (Fig. 4C and D). The V rib is a prominent V-shaped ridge of silica positioned on the scale such that the base of the V lies in the proximal region and the arms of the V rib extend toward the distal end of the scale, often terminating near the midsection of the scale and close to the perimeter (Figs. 4B and C, and 7B). In some cases, the arms of the V rib extend to the base of the dome (e.g., Fig. 9F). Many scales also possess two additional siliceous ribs, known as anterior submarginal ribs (Fig. 4B and D), that originate near the ends of the V-rib arms, run parallel to the margin of the scale, and terminate in the distal region of the scale. On spined scales the anterior submarginal ribs often fuse and become extended to form the spine.

The V rib and the anterior submarginal ribs, collectively referred to as the submarginal rib (Siver, 1991a), serve to divide the scale into distinct regions. The regions bounded on the inside and outside of the submarginal rib are referred to as the shield and the flange, respectively (Fig. 4). The flange is further divided into posterior and anterior flanges. The posterior flange is the region between the V rib and the posterior rim, whereas the anterior flange is that portion of the scale between the anterior submarginal rib and the margin of the scale. The secondary ornamentation of the shield and flange areas often is quite different, and may include pores, ribs, and papillae (Figs. 6–9).

A dome is a raised portion of the distal end of the base plate (Fig. 4B and C) under which the proximal end of a bristle is fitted in a ball and socket fashion. Generally, scales with a dome are associated with a single bristle that emerges from an inverted U-shaped opening along the distal end of the dome. The U-shaped opening is situated to the right of center, imparting a slight asymmetry to the scale. The dome often is ornamented with secondary structures such as ribs and papillae, which are of taxonomic significance (e.g., Figs. 6C and 9C).

Scales of all *Chrysodidymus* and *Synura* species, except *S. lapponica*, have a more or less bilateral symmetry. Although the dome, V rib, and posterior rim may impart slight asymmetries to *Mallomonas* scales, scales in this genus also have essentially a bilateral symmetry. Scales of *S. lapponica* and *Tessellaria* are oval to circular in design with a biradial symmetry.

The siliceous bristles, found only on cells of *Mallomonas*, are produced within the cell independently of scales, but become associated with the latter when the cell covering is constructed. A bristle can be divided into the foot and shaft (Siver, 1991a). The foot consists of the proximal end of the bristle that is slightly bent relative to the long shaft and tucked under the dome. The shaft often is slightly bowed, smooth or ribbed, and serrated; the teeth of the serration may be along most of the shaft or restricted to the distal end. The morphology of the distal ends of bristles is highly variable between species and is of taxonomic importance (Siver, 1991a). Tips are commonly blunt, drawn out into sharp points, bifurcate, or expanded to form a C-shaped or cleftlike opening known as a helmet (e.g., *M. acaroides* v. *muskokana*) or hooked (e.g., *M. heterospina*) bristle. Species commonly possess either one or two types of bristles that are generally found on different positions of the cell. Typically, a smaller type of bristle is associated with the apical scales, and a longer and morphologically distinct type of bristle is allied with body scales.

C. Arrangement of Scales on the Cell Surface

Except for *Synura lapponica* and *Tessellaria*, scales on all synurophytes are arranged on the cell surface in well-ordered, overlapping, spiral rows (Leadbeater, 1986, 1990; Siver and Glew, 1990; Graham *et al.*, 1993). As noted by Siver and Glew (1990), if the scales are traced in a row from the posterior to the anterior of the cell, the rows of scales always appear to be spiraled to the right (e.g., Fig. 3B). In *Mallomonas* and *Synura*, the scales within a given row are overlapped in a posterior to anterior manner; that is, each scale is overlapped by the scale positioned behind it in the same row (Siver and Glew, 1990). The only known exception to this rule is *M. retrorsa*, and most likely *M. fenestrata*, where the orientation of scales on the cell surface is reversed (Siver, 1988b, 1991a).

In all *Mallomonas* taxa except *M. akrokomos*, the spiral rows of scales overlap each other in an anterior to posterior manner (Fig. 3B). However, in *Synura* the rows of scales are overlapped in a posterior to anterior manner (Figs. 13G and H). Thus, for most *Mallomonas* taxa, a given scale is overlapped by the scale positioned behind it in the same row and by scale(s) in the row anterior to it. For *Synura* species, a given scale is overlapped by the scale behind it in the same row and by scale(s) in the row posterior to it. The complexity of the scale coat on *Chrysodidymus* cells has not yet been described.

Within a given row, scales are orientated such that their distal ends point (a) toward the anterior of the cell (parallel arrangement), (b) at 90° to the right (perpendicular arrangement), or (c) at an intermediate position (oblique arrangement; Leadbeater 1986; Siver and Glew, 1990). *Mallomonas retrosa* is the only known exception to this arrangement and orientation of scales. The number of spiral rows of scales on a cell is equal to the number of scales that surround the flagellar pore, and the number of scales in any row on a given cell is approximately equal (Siver and Glew, 1990).

Each spiral row of scales on a cell of *Mallomonas* or *Synura* begins with an apical scale that is part of the ring of scales that encircles the emergent flagella(um), grades into body scales, and terminates with posterior or caudal scales. The change in scale type within a spiral row may be gradual, as in *Synura* (Wee, 1997) and many species of *Mallomonas* (Asmund and Kristiansen, 1986; Siver, 1991a), or very abrupt, as in some sections of *Mallomonas* (e.g., section Torquatae). On *Synura* cells, the spines or keels on scales gradually diminish in length or height from the anterior to the posterior of the cell, and are either minute or lacking on the posterior-most portions of the cells. On cells of *Chrysodidymus*, although the length of scales also diminishes toward the posterior end of the cell, all scales possess spines (Nicholls and Gerrath, 1985). The lack of spines on posterior portions of *Synura* cells may be related to the fact that cells in the colony are in direct contact with each other, a condition not found in two-celled *Chrysodidymus* colonies.

The formation and subsequent deployment of scales and bristles onto the cell surface in *Mallomonas* and *Synura* have been summarized by Leadbeater and Barker (1995), Wetherbee *et al.* (1995), and Wee (1997). A brief synopsis is provided here. Individual scales and bristles form within specialized vesicles known as silica deposition vesicles (SDVs) that are most likely derived from the Golgi complex (McGrory and Leadbeater, 1981; Mignot and Brugerolle, 1982). Initially, the SDV becomes positioned along the outer and anterior surface of the PER of one of the chloroplasts (Belcher, 1969; Mignot and Brugerolle, 1982). Microtubules (Mignot and Brugerolle, 1982; Leadbeater, 1986) and actin-like microfilaments (Brugerolle and Bricheux, 1984; Leadbeater and Barker, 1995) become associated with the SDV and may aid to shape and to move the SDV in a helical pathway down along the PER (Mignot and Brugerolle, 1982). As the SDV is transported along the outer surface of the chloroplast, it is flattened and molded into the shape of a scale. Once the mold of the scale is completed, silica is deposited to produce the finished product (Wujek and Kristiansen, 1978; Mignot and Brugerolle, 1982; Leadbeater, 1990).

Bristles are formed in a similar fashion to scales, but in separate SDV vesicles and apparently at different times (Wujek and Kristiansen, 1978; Mignot and Brugerolle, 1982; Beech *et al.*, 1990). Once a scale or bristle is formed, the SDV moves toward and fuses with the cell membrane, resulting in release of the siliceous component onto the outer surface of the cell. Although the precise mechanism(s) by which the scales and bristles are maneuvered into place remains unclear, hypotheses have been proposed for cells with existing scale coats (Leadbeater 1990; Beech *et al.*, 1990) and for naked cells (Siver and Glew, 1990). Scales and bristles are held in place by an adhesive material (Leadbeater, 1986; Wetherbee *et al.*, 1995).

D. Cysts, Life Cycle, and Cell Division

All synurophytes are believed to form a siliceous resting stage known as a cyst, a stomatocyst, or a statospore (Fig. 5). Cysts can be produced as a result of asexual or sexual reproduction (Skuja, 1950; Cronberg, 1986; Sandgren and Flanagin, 1986; Sandgren, 1988, 1991), and their formation may be triggered by sudden changes in environmental conditions (Cronberg, 1980, 1986; Sandgren, 1981) or population density

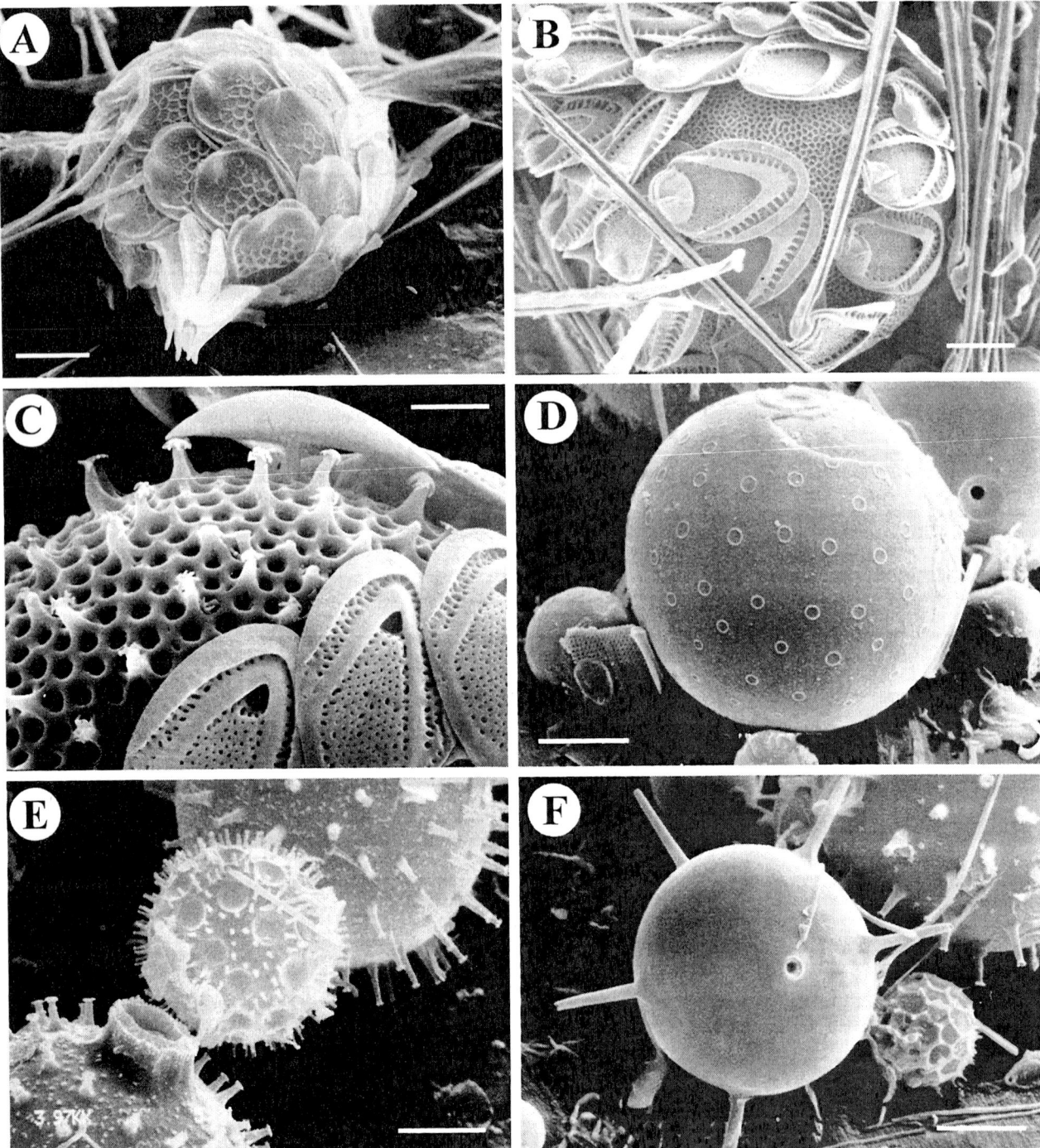

FIGURE 5 Features of cysts. A. Intact cell of *Mallomonas punctifera* that is encysted as viewed from the anterior end. Scale bar = 2 μm. B. Cell of *Mallomonas acaroides* var. *muskokana* containing a partially formed cyst with a reticulated surface. Scale bar = 2 μm. C. Close-up of part of a cell of *Mallomonas tonsurata* with a fully formed cyst. Note spines with splayed tips. Scale bar = 1 μm. D. An immature cyst where the bases of spines (circular structures) have just begun to form. Scale bar = 5 μm. E. Remains of cysts of various sizes and designs. All three cysts possess spines. The two larger cysts also have scabrae (small bumps) and the smaller cyst depressions or tangential circuli. Scale bar = 5 μm. F. A cyst with a smooth surface, long spines, and a small pore surrounded by a collar. Note the small cyst with ridges that form a regular reticulum. Scale bar = 5 μm.

(Sandgren, 1988). Even though hundreds of cyst morphotypes have been observed, only a small percentage have been linked to actual species (Duff *et al.*, 1995).

Cysts are hollow structures that generally are globose in shape, have a single germination pore, and are formed endogenously within a silica SDV (Preisig and Hibberd, 1982a, b; Sandgren, 1991). The SDV encloses the nucleus, chloroplasts, Golgi apparatus, storage material, and a large volume of the cytoplasm (Cronberg, 1986; Sandgren, 1989). The formation of the wall of the cyst is a continuous process that takes place within the SDV in what is believed to be a two-step process (Skogstad, 1984; Sandgren, 1989). The inner, or primary, wall forms first in a rather rapid fashion, and results in a wall that is usually unornamented and morphologically similar in many species (Duff *et al.*, 1995). For many species, the primary wall is also distinctly different in appearance from the mature cysts (e.g., Fig. 5B; Skogstad, 1984; Siver, 1991c). Additional cyst wall layers, including the wall ornamentation and the collar, form in a slower and more controlled manner on the outer surface of the primary wall (Fig. 5D). As the cyst wall is being completed, the pore becomes plugged, usually with an organic material. It is unknown if an immature cyst, in the sense of lacking a fully ornamented wall, is viable.

Duff *et al.* (1995) provided an excellent summary of the terminology associated with the description of cysts. Cysts range in size from ca. 2 to >30 µm (Sandgren and Carney, 1983; Duff *et al.*, 1995), range in shape from spherical to oval to a flat, pancake form, and vary in design from smooth (e.g., Fig. 5F) to highly ornamented (e.g., Fig. 5E). The body of a cyst is divided into anterior and posterior hemispheres, and the pore is located in the anterior hemisphere. The pore is circular and may or may not be surrounded by a thick rim of silica called the collar (Fig. 5E and F). Collars may be simple or complex, the latter consisting of two or more separate collars that surround the pore in a concentric fashion. The morphology of the outer wall and the pore–collar complex is of taxonomic importance (Sandgren and Carney, 1983; Skogstad, 1984; Sandgren, 1989). A multitude of ornamentations is associated with the mature cyst wall, including nodules, spines, ridges, circular ridges (circuli), depressions, and various types of reticulations (Fig. 5).

Vegetative cells and cysts formed through asexual processes are believed to be haploid. Cysts formed by sexual reproduction have two separate and presumably haploid nuclei after cellular fusion; nuclear fusion and meiosis occur upon germination of the cyst (Sandgren, 1991). Sexual reproduction is isogamous and fusion of gametes in several species of *Mallomonas* has been observed to be the result of contact by either the caudal or apical ends of cells (reviewed in Wee, 1982, and Asmund and Kristiansen, 1986). Cytokinesis in *Mallomonas* occurs along the longitudinal axis, beginning from the anterior end, and is completed within minutes (Harris, 1953; Wawrik, 1979; Beech and Wetherbee, 1990b; Beech *et al.*, 1990).

Colonies of *Synura* and *Tessellaria* also divide by binary fission. Colonies of *Synura*, but not *Tessellaria*, may elongate prior to division, and in both genera the resultant daughter colonies are not necessarily composed of equal numbers of cells. In *Tessellaria*, Tyler *et al.* (1989) noted the formation of a furrow channel around the colony that eventually grew deeper and yielded the two daughter colonies.

E. Taxonomic and Phylogenetic Considerations

Species within the genera *Synura* and *Mallomonas* are grouped into sections and series based primarily on ultrastructual features of the scale covering (Momeu and Péterfi, 1979; Asmund and Kristiansen, 1986; Siver, 1991a; Wee, 1997). Petersen and Hansen (1956, 1958) originally proposed three sections for the genus *Synura*: Lapponica, Peterseniae, and Synura. The section Lapponica consists of only one species, *S. lapponica* (Fig. 10G), whereas those taxa with scales that possess either a keel or a spine are positioned within section Peterseniae or section Synura, respectively. Later, Péterfi and Momeu (1977) divided taxa within the section Synura into two series, Synura and Splendidae, based on the presence or absence, respectively, of secondary structures on the scales. Wee (1997) presented a slight modification of this taxonomic grouping. There are at least 27 described species and subspecific taxa within the genus *Synura* (Nicholls and Gerrath, 1985; Siver, 1987, 1988c; Cronberg, 1989), roughly half of which commonly are observed in North America (Nicholls and Gerrath, 1985; Siver, 1987).

Harris and Bradley (1957, 1960) originally divided *Mallomonas* into four series, Tripartitae, Planae, Quadratae, and Torquatae, based primarily on the ultrastructure of scales as observed with both light and electron microscopy. Momeu and Péterfi (1979) later raised the four series to the rank of section. Since that time, the taxonomy of *Mallomonas* has relied progressively more on structures observed with electron microscopy, and there are now over 116 species (Moestrup, 1995) in 17 sections (Asmund and Kristiansen, 1986; Siver, 1988b). Most recently, Péterfi and Momeu (1996) used cluster analysis based on phenetic ultrastructural features of scales and bristles to further modify the genus *Mallomonas*.

The anatomy of the siliceous scales and their orientation on the cell surface clearly indicate that

Chrysodidymus belongs in the Synurophyceae. Graham *et al.* (1993) further demonstrated that the genus shared many cellular and ultrastructural characters with other synurophytes, supporting its inclusion in the Synurophyceae. However, the validity of the genus *Chrysodidymus*, as well as the described species, have been questioned. Prowse (1962) originally described two species of *Chrysodidymus*, *C. synuroideus* and *C. gracilis*, that were differentiated on the basis of cell size and shape. Later, Wujek and Wee (1983) suggested, and Graham *et al.* (1993) concluded that the two species described by Prowse (1962) should be combined into one, *C. synuroideus*. The more important taxonomic question concerned the validity of the genus. Even though the scales of *Chrysodidymus* are clearly *Synura*-like, the genus *Chrysodidymus* has been retained on the basis of the two-celled nature of the colony. The work of Graham *et al.* (1993) clearly demonstrated that *Chrysodidymus* consists solely of two-celled colonies and has several unique ultrastructural characters, supporting separation from the genus *Synura*.

Much work is needed on phylogenetic relationships of the Synurophyceae relative to other groups of heterokonts, especially the Chrysophyceae (Andersen *et al.*, 1999). In their monograph on *Mallomonas*, Asmund and Kristiansen (1986) considered species with scales that lacked domes and secondary structures, that had smooth and bifurcate bristles, and that had cells with two emergent flagella as ancestral. Thus, although their classification system is perhaps somewhat artificial, their treatment of taxa attempted to provide phylogenetic relationships between a diverse array of organisms. Péterfi and Momeu (1996) also provided some indication of the relationships between taxa within the genus *Mallomonas*, but recognized that their use of numerical taxonomy based on phenetic characters may not yield true phylogenetic relationships, a point discussed by Wee (1997).

Despite the need for additional work, five observations on phylogenetic relationships can be made based on the works of Wee (1997) and Lavau *et al.* (1997).

1. The Synurophyceae appears to be monophyletic (Lavau *et al.*, 1997).
2. Both Wee (1997) and Lavau *et al.* (1997) concluded that *Tessellaria volvocina* is at the base of the Synurophyceae lineage, and Wee (1997) further pointed out that *S. lapponica* is closely allied with *Tessellaria volvocina* (Wee, 1997), a hypothesis also supported by Tyler *et al.* (1989). In fact, Petersen and Hansen (1958) suggested that *S. lapponica* be removed from the genus *Synura* because of the biradial symmetry of its scales and the fact that the scales were most likely arranged in multiple layers surrounding the entire colony and not on individual cells.
3. The study by Lavau *et al.* (1997), which utilized both scale and molecular data, weakly supported both *Synura* and *Mallomonas* as monophyletic groups. However, the authors pointed out that when used separately, neither the scale ultrastructure data nor the molecular data were able to resolve *Synura* as a monophyletic unit, and likewise the molecular data alone were not able to resolve *Mallomonas* as monophyletic. In addition, the phylogeny produced by Wee (1997) had *M. caudata* and *S. petersenii* emerging from the same clade within the section Synura, and he concluded that more work was needed to determine if *Mallomonas* was a monophyletic genus.
4. *Chrysodidymus synuroideus* appears to be ancestral to the series Synura, those taxa with scales that have emergent spines and secondary ornamentation (Wee, 1997). On the other hand, Wee (1997) observed that *S. sphagnicola* was removed from other *Synura* taxa with spines, including *S. splendida*, and was less allied with *Chrysodidymus* than suggested by Graham *et al.* (1993).
5. As proposed by Asmund and Kristiansen (1986), both Lavau *et al.* (1997) and Wee (1997) concluded that morphological characters of scales are important and very useful for resolving relationships at the genus, species, and subspecific levels.

III. ECOLOGY AND DISTRIBUTION

A. General Comments

The Synurophyceae are euplanktonic in nature and occur almost exclusively in freshwater habitats. Although a large number of species in the Chrysophyceae are known mixotrophs (Andersen *et al.*, 1999), members of the Synurophyceae are believed to be strict photoautotrophs (Salonen and Jokinen, 1988; Holen and Boraas, 1995). They are most commonly encountered in plankton from ponds and lakes, although substantial numbers also can be found in large slowly flowing rivers, as well as in pools in smaller streams (e.g., Dürrschmidt, 1980; Siver and Vigna, 1997). Generally, taxa of Synurophyceae can be collected in most samples from lacustrine locations during any season, and it is not uncommon to find 10 or more different species from a single collection (Jacobsen, 1985; Siver and Hamer, 1989; Eloranta,

1989; Siver, 1991a; Wee et al., 1993). Over 50 species of Synurophyceae were observed over an annual cycle in a single water body (Siver and Hamer, 1992).

Synurophytes are often observed as components of deep-water metalimnetic algal peaks that develop during stratified periods (e.g., Pick et al., 1984; Pick and Cuhel, 1986; Hoffman and Wille, 1992). Taxa of *Synura* and *Mallomonas*, as well as other colonial scaled chrysophytes, can dominate metalimnetic regions of stratified lakes, may account for over 90% of the phytoplanktonic biomass (Hoffman and Wille, 1992), and form densities well above those found in surface waters (Barbiero and McNair, 1996). Such deep-water layers can form by *in situ* growth, passive accumulation from the epilimnion, or active migration from surface waters, and can be overlooked in routine sampling efforts. Generally, a decrease in the amount of light reaching the metalimnion, depletion of a nutrient, or vernal mixing results in the disruption of such peaks (Barbiero and McNair, 1996).

Species of *Mallomonas* and *Synura*, including *M. crassisquama*, *M. caudata*, *S. uvella*, and *S. petersenii*, are known to form blooms (Clasen and Bernhardt, 1982; Nicholls and Gerrath, 1985; Hoffman and Wille, 1992), and blooms of *Synura* are often associated with taste and odor problems (Wee et al., 1994; Nicholls, 1995). Water tainted with a fishlike odor caused by synurophyte blooms is reported from both soft- and hard-water lakes, and often results in the need to treat the water before consumption by humans (Nicholls, 1995). Although, historically, many taste and odor problems have been attributed to *S. uvella*, most episodes are probably caused by *S. petersenii* (Nicholls and Gerrath, 1985; Wee et al., 1994).

Although members of the Synurophyceae are well documented from many temperate, subtropical, and tropical localities (Nicholls and Gerrath, 1985; Kristiansen, 1986; Cronberg, 1989; Wee et al., 1993), there is some evidence that they diminish in diversity and biomass in subarctic and arctic regions of both the northern (Kristiansen, 1992) and southern (Croome and Tyler, 1988) hemispheres. In fact, Croome and Tyler (1988) did not find a single taxon after examination of many samples from subarctic and arctic islands from the southern hemisphere, and scales are often very scarce in sediment samples from northern arctic ponds (J. Smol, Queens University, personal commmunication). Most in-depth surveys of the Synurophyceae from North America indicate a well developed, temperate flora, but a tropical element has been described in works from Florida (e.g., Siver and Wujek, 1993) and Louisiana (Wee et al., 1993).

Most, but not all, taxa found in North America are cosmopolitan in nature and have been reported on other continents. *Synura petersenii* (Siver, 1987), *Mallomonas caudata* (Siver, 1991a), and *M. crassisquama* (Siver and Skogstad, 1988, and references therein) are among the most common synurophytes found in North America and around the world. A few taxa, however, are known to have restricted geographic ranges. One taxon in particular, *Mallomonas pseudocoronata*, is a common taxon found almost exclusively in North American waters (Siver, 1991a). *Mallomonas duerrschmidtiae*, *M. acaroides* var. *muskokana*, and, to a lesser extent, *M. galeiformis* are other commonly encountered species in the northeastern part of North America, especially in dilute soft-water lakes, but rarely are reported in other regions of the world (Siver, 1988d, 1991a; Siver et al., 1990; Nicholls, 1987, 1988a, b).

Because of their limited distributions along various environmental gradients, many species of synurophytes are excellent bioindicators (Kristiansen, 1986; Siver, 1995; Smol, 1995). Some species are known primarily from warm or cold water, acidic or alkaline conditions, and oligotrophic or eutrophic habitats. Many taxa are limited by the specific conductance of the water (Siver, 1993), and others appear to be sensitive to high concentrations of metals (Gibson et al., 1987; Dixit et al., 1989). As a result of the differential distributions of species along various gradients and the fact that the species-specific siliceous scales or cysts become preserved in lake sediments, the Synurophyceae have become a very valuable organismal group for reconstructing past lake-water conditions (Smol, 1995; Siver et al., 1999).

B. Primary Habitat Requirements

Ecological studies of the Synurophyceae often make observations and draw conclusions that include members of the Chrysophyceae (Chap. 12) that have siliceous scales (i.e., scaled chrysophytes). Therefore, it is unavoidable that much of the following discussion pertains to the scaled chrysophytes. Reviews of the ecological distributions of synurophytes along environmental gradients and their occurrences in blooms have been compiled by Sandgren (1988), Siver (1991a, 1995), Nicholls (1995), and Sandgren and Walton (1995).

As a group, as well as for individual species, synurophytes exhibit very different tolerances along environmental gradients (e.g., Siver, 1989, 1991a, 1995; Siver and Marsicano, 1996; Cumming et al., 1992a). In a broad sense, Siver (1995) summarized the habitats that support the richest floras of synurophytes as those that are slightly acidic (Siver and Hamer, 1989; Siver, 1992; Wee and Gabel, 1989), are low in specific conductance, alkalinity, and nutrient content (Cronberg and Kristiansen, 1980; Roijackers and Kessels, 1986; Siver, 1991a, 1995), and have moderate amounts of humic substances (Cronberg and Kristiansen, 1980; Dürrschmidt, 1980, 1982; Wee and Gabel, 1989; Siver, 1991a; Eloranta, 1995; Siver and Vigna, 1997). Humic-colored lakes and ponds are typically small and shallow, located in forested areas, and often situated in watersheds that drain marsh or wetland areas (see also Chap. 2). A preference for humic-stained localities may explain the common observation that scaled chrysophytes are typically more abundant in smaller ponds (Dürrschmidt, 1980; Kristiansen, 1981; Cronberg, 1989).

The chrysophytes, including the synurophytes, long were considered to be a group that was primarily restricted to cold, oligotrophic habitats and occurred primarily in the spring (Siver, 1995, and references therein). However, more recent studies from warm regions of the world dispute this hypothesis (e.g., Takahashi and Hayakawa, 1979; Kristiansen, 1980, 1981, 1986; Kristiansen and Takahashi, 1982; Dürrschmidt and Croome, 1985; Cronberg, 1989; Saha and Wujek 1990; Siver and Wujek 1993, Siver and Vigna 1997). In North America, rich and diverse floras of synurophytes have been documented from subtropical localities in Florida (Wujek, 1984; Wujek and Bland, 1991; Siver and Wujek, 1993) and Louisiana (Wee et al., 1993). Similarly, even though scaled chrysophytes comprise a larger percentage of the annual phytoplanktonic biomass of oligotrophic and mesotrophic lakes (see subsequent discussion), ample studies demonstrate that large numbers of species also can be found in more eutrophic sites (Kristiansen, 1985, 1988b; Hickel and Maass, 1989; Gutowski, 1989, 1997; Saha and Wujek, 1990). For example, Kristiansen and Tong (1989) and Kristiansen (1985) recorded 40 and 33 taxa of scaled chrysophytes in highly eutrophic ponds in China and Denmark, respectively. Gutowski (1989) and Hickel and Maass (1989) observed over 23 taxa of scaled chrysophytes in nutrient-enriched sites in Germany. In North American localities, Siver and Hamer (1989) found no significant difference in the number of taxa sampled from along a total phosphorus gradient, further supporting the hypothesis that large numbers of synurophyte taxa are capable of growing in nutrient-enriched waters. In addition, although synurophytes are commonly known to comprise a significant portion of vernal blooms, they also can be important components of the flora in summer, autumn, and winter periods (Dürrschmidt, 1980; Siver and Chock, 1986; Kristiansen, 1986; Siver, 1991a; Siver and Hamer, 1989; Sandgren, 1988; Nicholls, 1995).

C. Distribution along a pH Gradient

Lakewater pH repeatedly has been demonstrated to be a primary factor that controls the distribution of synurophytes in freshwater localities (Siver, 1995; Smol, 1995). The importance of pH has been concluded from numerous studies utilizing ordination techniques (e.g., Siver and Hamer, 1989), as well as from more standard floristic surveys (e.g., Takahashi, 1978). In North American-based studies, pH has been demonstrated to be an important variable influence on the composition of scaled chrysophytes in southern New England (Siver and Hamer, 1989), northern New England (Dixit et al., 1990), the Adirondacks (Smol et al., 1984; Cumming et al., 1992a, b; Duff and Smol, 1995), Florida (Siver and Wujek, 1999), Iowa (Wee and Gabel, 1989), and Ontario (Dixit et al., 1988).

The most diverse floras of scaled chrysophytes are typically found at slightly acidic conditions (Siver and Hamer, 1989; Siver and Smol, 1993). In a diverse array of habitats from Connecticut, Siver and Hamer (1989) found the largest number of species between pH 5.5 and 6.5, and significantly fewer taxa below pH 5 and above pH 8. Many species tend to disappear and members of the group as a whole becomes less abundant as the pH drops below 5 (Siver, 1989, 1991a; Hartmann and Steinberg, 1989). Based on a thorough literature search, Siver and Smol (1993) observed a similar pattern for the number of synurophyte species along a pH gradient, which suggests that the effect of pH is similar in different regions of the world. Although not common, large numbers of species have been reported from more alkaline, high-pH habitats (e.g., Gutowski, 1989; Hickel and Maass, 1989), especially if the humic content is high (Siver and Wujek, 1993).

Many synurophyte species have definitive and well-documented distributions along a pH gradient, and many taxa have been characterized by the pH categories defined by Hustedt (1939; see Takahashi, 1978; Siver, 1989, 1991a, 1995; Smol, 1995). Based on a review of the literature, Siver (1989) concluded that many species have similar distributions along a pH continuum and similar weighted mean pH values in widely separated geographic regions. Further, based on the review by Siver (1989), Siver and Smol (1993) identified four groups of scaled chrysophytes that had similar distributions relative to pH gradient. A low-pH group of taxa included *Mallomonas canina*, *M. hindonii*, *M. paludosa*, *M. pugio*, *M. hamata*, *M. acaroides* var. *muskokana*, *Synura sphagnicola*, and *S. echinulata*. Taxa in this group are consistently observed below pH 6, have abundance weighted mean pH (AWMpH) values of less than 6, and are primarily acidobiontic organisms (see Siver, 1995). *Chrysodidymus synuroideus* is another well-documented acidobiontic species, often present in bogs (Graham *et al.*, 1993), that is reported to have an AWMpH of 4.6 (Dixit *et al.*, 1988) and 5.5 (Charles and Smol, 1988).

The mid-pH group consists of species that occur most often below pH 7, but generally above pH 5, have AWMpH values near or above 6, and are best classified as acidophilic (see Siver, 1995, and references therein). Synurophyte taxa in this group included *Mallomonas galeiformis*, *M. heterospina*, *M. duerrschmidtiae*, *M. punctifera*, *M. transsylvanica*, *M. dickii*, *M. doignonii*, *M. torquata*, and *Synura spinosa*. The third group consists of pH indifferent species that have their center of distribution and AWMpH values around pH 7. However, many of the pH indifferent species also can be found at relatively high or low pH. The fourth or high-pH group consists of alkaliphilic taxa that are distributed primarily above pH 7 and have AWMpH values greater than 7. The high-pH group includes *Mallomonas acaroides* var. *acaroides*, *M. tonsurata*, *M. corymbosa*, *M. pseudocoronata*, *M. elongata*, *M. alpina*, and *M. portae-ferreae*. As a result of the rather narrow distributions of many synurophytes along a pH gradient, assemblages have been successfully utilized to infer historical lake-water pH conditions (Siver, 1995; Smol, 1995; Siver *et al.*, 1999).

D. Distribution along a Dissolved Salt Gradient

There is now ample evidence that the concentration of dissolved salts is another important factor that controls the distributions of scaled chrysophytes. As a group, the Synurophyceae are observed more frequently and have significantly higher species diversities in localities low in specific conductance (Sandgren, 1988; Siver and Hamer, 1989), and are virtually absent in saline lakes (Zeeb and Smol, 1995). Based on data from a large and diverse array of lake types, Siver and Hamer (1989) demonstrated that specific conductance was as important as pH in explaining the occurrences of scaled chrysophytes as well as the number of species found per collection. The maximum number of species per collection was observed in water bodies with a specific conductance of only ca. 40 $\mu S\ cm^{-1}$, and dropped significantly as the level reached 200 $\mu S\ cm^{-1}$.

Since the work of Siver and Hamer (1989), a number of additional studies also have demonstrated the importance of dissolved salt concentrations on synurophyte assemblages. Zeeb and Smol (1991) used scaled-chrysophyte remains in a sediment core to trace the effects of road de-icing salts on a small lake in Michigan. The distributions of scale remains (Cumming *et al.*, 1992a) and cysts (Duff and Smol, 1995) in surface sediments of lakes in the Adirondack Mountains of New York also have been reported to be influenced by dissolved salts. Siver (1993) developed an inference model that was used to trace the effects of deforestation and residential development on lake-water specific conductance levels (Lott *et al.*, 1994; Siver *et al.*, 1999). The physiological mechanism(s) that control the responses of specific synurophytes to dissolved salt concentrations, or to specific cations or anions, remains unknown.

E. Distribution along a Trophic Gradient

Scaled chrysophytes, and most likely the Synurophyceae, often account for larger percentages of phytoplankton biomass in oligotrophic and early mesotrophic lakes than in more eutrophic localities (Kristiansen, 1986; Sandgren, 1988; Nicholls, 1995; Siver, 1995; Sandgren and Walton, 1995). Such a trend has been reported from many regions of the world, including temperate and subtropical regions of North America (Kling and Holmgren, 1972; Siver and Chock, 1986; Siver, 1991; Siver and Wujek, 1993), Scandinavia (Eloranta, 1989), Greenland (Jacobsen, 1985), and Australia (Croome and Tyler, 1985, 1988). Based on an in-depth survey of the literature, Sandgren (1988) concluded that chrysophytes (including the Synurophyceae) often account for between 10 and 75% of the biomass of oligotrophic lakes regardless of differences in size, mixing pattern, and geographic location. Sandgren (1988) further noted that the group often comprised <20% of the biomass in more mesotrophic and eutrophic lakes, and generally <5% in very eutrophic sites. Indeed, the presence of scaled chrysophytes usually indicates that the locality is not heavily polluted (Kristiansen, 1981, 1986).

Schindler *et al.* (1973), Findlay (1978), and DeNoyelles and O'Brien (1978) all noted significant declines in chrysophyte biomass following additions of nutrients to experimental lakes and ponds. Conversely, the relative abundance of chrysophytes often increases following management efforts to decrease the nutrient load to eutrophic water bodies (Cronberg *et al.*, 1975; Cronberg, 1982; Nicholls, 1995). Sandgren and Walton (1995) advanced the hypothesis that the inverse relationship between the percentage of phytoplankton biomass composed of chrysophytes and trophic status was partly regulated by zooplankton predation, especially from large-bodied *Daphnia*. Zooplankton feeding patterns may also partially explain why chrysophytes often bloom during discrete periods of the year (Sandgren and Walton, 1995). Despite the generalization that scaled chrysophytes decline with increasing eutrophy, some recent surveys from warm regions indicate that the biomass of scaled chrysophytes can be high, and often greatest, in eutrophic sites (Siver and Wujek, 1993; Wujek and Bicudo, 1993; Yin-Xin and Kristiansen, 1994; Cronberg, 1996).

At the species level, many synurophytes are differentially distributed along a trophic gradient (Siver, 1995; Siver and Marsicano, 1996). Based on literature records, Siver (1991a) recognized a group of *Mallomonas* species common in North American waters, including *M. duerrschmidtiae*, *M. acaroides* var. *muskokana*, *M. galeiformis*, *M. hamata*, *M. asmundiae*, *M. pugio*, *M. paludosa*, and *M. torquata*, that is primarily reported from oligotrophic or mesotrophic habitats. Siver (1991a) further noted that another group of species commonly reported from North America, consisting of *M. tonsurata*, *M. alpina*, *M. corymbosa*, *M. portae-ferreae*, *M. acaroides* var. *acaroides*, *M. lychenensis*, and, to a lesser extent, *M. elongata* and *M. pseudocoronata*, is more common in eutrophic localities. *Mallomonas heterospina* is often reported from dung- or waterfowl-contaminated ponds (Harris and Bradley, 1957; Kristiansen, 1986; Siver, 1991a), and there appears to be ample evidence that *M. matvienkoae* is also common in highly eutrophic habitats (Saha and Wujek, 1990; Siver and Vigna, 1997). *Synura sphagnicola* (Kristiansen, 1986; Siver, 1988a) and *S. spinosa* forma *longispina* (Dürrschmidt, 1982; Siver, 1987) are other oligotrophic indicators, whereas *S. curtispina* (Gutowski, 1989; Santos and Leedale, 1993; Siver and Marsicano, 1996) is often observed to tolerate eutrophic conditions. Because species of scaled chrysophytes are indicative of trophic conditions, they have been useful in paleolimnological studies (Smol, 1995; Siver and Marsicano, 1996; Siver et al., 1999).

F. Distribution along a Temperature Gradient

Although both Roijackers and Kessels (1986) and Siver and Hamer (1989) concluded that pH and related factors were the most important variables that control the development of specific species in a given lake, water temperature is instrumental in determining the time period and degree to which a population developed. Gutowski (1989) and Siver and Hamer (1992) also concluded that temperature plays a key role in the development and subsequent seasonal succession of species in individual lakes, and Siver and Hamer (1992) further showed that the scaled-chrysophyte community could accurately infer water temperature.

Chrysophytes, including synurophytes, often dominate phytoplankton biomass during spring (and, to a lesser extent, autumn) mixing when the water temperature is low (Wetzel, 1983; Sandgren, 1988; Gutowski, 1996). In a survey of 141 water bodies, Sandgren (1988) found that the greatest biomass of chrysophytes occurred between 10 and 20°C, and that the biomass significantly declined above 20°C. Other studies have illustrated maximal concentrations of synurophytes at temperatures below 12°C (Kristiansen, 1975, 1986; Jacobsen, 1985; Siver and Chock, 1986; Hickel and Maass, 1989; Siver 1991a). Despite the fact that the biomass of scaled chrysophytes is often found to be maximal at low temperatures during spring mixing, Siver (1995) and Sandgren and Walton (1995) pointed out that factors other than temperature (e.g., increased nutrients and light, turbulence, and grazing) may play pivotal roles in explaining this phenomenon.

Whether water temperature is involved in regulating species diversity among synurophytes is unclear. Siver and Hamer (1989) found no significant difference between the number of species of *Mallomonas* observed in a lake at a given time and water temperature, but a relationship was observed for species of *Synura*. Other works support the fact that large numbers of synurophyte species can be found at both low (e.g., Jacobsen, 1985; Siver, 1991a) and high (Cronberg, 1989; Saha and Wujek, 1990) water temperatures.

Regardless of the relationship between species diversity and water temperature, many species and subspecific taxa are indeed differently distributed along a temperature gradient (Siver, 1995). Siver (1991a) divided *Mallomonas* taxa into five groups based on abundance along a temperature gradient. The warm-water group consists of species rarely found below 12°C and with abundance weighted mean (AWM) temperature values above 15°C. Cool-water taxa are species observed primarily between 12 and 15°C, and are less often encountered outside of this range. The cool–cold-water species had AWM temperatures below 12°C, but often are found above 15°C. The cold-water class consists of taxa rarely observed above 15°C and that have AWM temperatures below 12°C. The fifth group, temperature-indifferent species, consists of taxa observed along the entire temperature gradient. Species of *Synura* and *Chrysodidymus synuroideus* also are known to be distributed differently along a temperature gradient (Siver, 1995). In particular, *Synura lapponica* (Siver and Hamer 1992), *S. echinulata* (Asmund, 1968; Kristiansen, 1975; Siver and Hamer, 1992), and *S. spinosa* (Kristiansen, 1975; Kies and Berndt, 1984; Roijackers and Kessels, 1986) are all primarily cold water taxa, whereas *S. sphagnicola* (Asmund, 1968; Kristiansen, 1975; Siver and Hamer, 1992) and *C. synuroideus* (Dürrschmidt and Croome, 1985) are most often observed during the warmer summer months.

Only a few observations to relate temperature to changes in scale and bristle structure have been undertaken. Siver and Skogstad (1988) observed different types of bristles on cells of *M. crassisquama* at different temperatures and suggested that bristle type may be related to buoyancy. Gutowski (1996) demonstrated that scales and bristles of *M. tonsurata* decreased in length as the water temperature increased. The role that temperature may play in the formation of cysts by either asexual or sexual processes, as well as in their germination, was discussed in detail by Sandgren and Flanagin (1986) and Sandgren (1988).

IV. COLLECTION AND PREPARATION FOR IDENTIFICATION

A. Collection of Samples

The methods used to collect and concentrate synurophytes depend, in part, on the goals of the investigation. Some projects require collection of samples from near-surface waters, whereas others focus on samples that integrate communities throughout the photic zone. A Van Dorn collecting bottle or some other standard device can be used to collect water from discrete depths, whereas tubing is often used to make an integrated sample of the water column. For quantitative work, standard methods for concentrating and estimating phytoplankton communities should be used; typically, these include either centrifugation or settling techniques. Centrifugation of fixed samples that include synurophytes is best done using low speeds of 2000 rpm (Siver and Hamer, 1990) or less (Wee, 1983). Taylor *et al.* (1986) presented a slight modification of the standard Utermöhl settling method that is specifically designed for scaled chrysophytes.

Wee (1983) reviewed some of the common fixatives used to preserve and to study cells and colonies of scaled chrysophytes. Acidified Lugol's solution is one of the best preservatives for use with synurophytes, because it often, but not always, results in intact cells or colonies that maintain their flagella and it allows for proper concentration of samples. Samples preserved with acidified Lugol's fixative can be maintained for extended periods of time, although samples, especially those in clear glass vials, may require the addition of more preservative. Samples fixed with formalin, formaldehyde, or glutaraldehyde can result in the loss of flagella and cell shape, and aggregation of the material over time will hinder concentration efforts (Takahashi, 1978; Wee, 1983; Graham *et al.*, 1993). However, formaldehyde- and glutaraldehyde-based fixatives added to a sample just prior to observation can yield cells that maintain cell shape and internal structure. Other more specialized methods for observing whole cells and siliceous remains were discussed in detail by Nicholls (1978) and Wee (1983). Jensen's stain (Jensen, 1962) has been used successfully for observation of scales with LM when only bright field optics are available (Nicholls, 1978). Wee (1983) reviewed the use of Nissenbaum's solution for fixation and simultaneous attachment of cells to microscope slides, as well as the burnt-mount method for LM observation of siliceous remains.

Other projects may focus on quantifying the remains of synurophytes in sediment samples. Because the siliceous remains of synurophytes become archived in the sediments, surface sediments can be used to estimate the abundances of all species that grew in a given water body over the last few years. In addition, sediment cores can be used to study historical changes in synurophytes over time periods of tens to hundreds of years. Methods for the collection and preparation of sediment samples for observation of synurophytes are well established (Smol, 1995).

For studies that do not include quantification of cells or for those that require critical identification of taxa, plankton net tows are very useful. Whereas many synurophytes are small, nets with a mesh size of 10 μm or less yield the best results. The net can be lowered to the base of the photic zone and hauled vertically or dragged horizontally through the surface waters. The advantage of plankton net collections is that they can be used directly to prepare samples for observation with LM or electron microscopy (EM) that allow for observation of many cells or their siliceous remains, thus greatly reducing the time needed to study a sample. Such concentrated samples aid finding and identifying taxa, especially rare species, and therefore provide a useful complement to studies that require concentration of cells prior to enumeration. Although net collections are not quantitative, the relative abundances of taxa are often maintained.

I often prepare net samples as soon after collection as possible by pipeting aliquots of unfixed samples onto both glass cover slips and aluminum foil, and immediately drying them at room temperature or under very low heat. This simple technique often results in intact cells. Collections that undergo large changes in temperature, are heated at a high temperature, or are allowed to sit for long periods of time prior to processing most often result in disarticulation of the silica scale coats. Standard critical point drying methods can be used for work that requires observation of intact cells with scanning electron microscopy (SEM).

B. Preparing Samples for Observation with Light Microscopy

The most in-depth study of the preparation and observation of synurophyte scales and bristles with LM was done by Wee (1983). Because the scales and bristles of synurophytes are siliceous in nature, methods to enhance contrast are needed to maximize the detail that can be observed with LM and to make taxonomic determinations. Increased contrast is achieved through either use of interference optics or use of different types of mounting media, or both. Typically, phase contrast or differential interference contrast (DIC) methods are used to view siliceous remains in preparations that have a mounting medium of high refractive index (RI) relative to silica. High magnification oil immersion objective lenses with numerical apertures between 1.3 and 1.4 yield the most information. I find that scales are slightly less likely to be overlooked when phase contrast optics is used. Because the majority of scales are less than 5 µm and because many of the details are close to or below the limits of resolution with LM, it is critical that the light path of the microscope be optimized, including the alignment of phase rings for phase contrast, or the polarizer and analyzer for work with DIC.

Wee (1983) examined six different mounting media for use in observing synurophyte remains, including, Hyrax®, Naphrax, Pleurax, and air. The key to maximizing the contrast is to imbed the remains in a medium with a refractive index that differs significantly from that of silica (RI = ~1.5). The refractive indices of Hyrax®, Naphrax and Pleurax all range between 1.71 and 1.75, and provide for excellent contrast of siliceous remains. However, presently, only Naphrax is commercially available. Mounts of scales and bristles in air (RI = 1.0) provide for the largest difference in RI between the specimen and mounting medium, and therefore the maximum degree of contrast. Air mounts can be made easily by drying the siliceous remains from a sample onto a cover glass and mounting the cover glass onto a glass slide with a thin layer of nail polish along the edges. The nail polish can be applied to either the cover glass or the slide in several layers to provide enough thickness such that a layer of air exists between them in the finished product. Normally, the initial layers of nail polish are allowed to dry and the cover glass is attached to the slide while the final layer of polish it is still sticky.

There are advantages and disadvantages to preparations made using a solid mounting medium versus air mounts. Both preparations can be stored easily and archived for long periods of time. Preparations using solid mounting media can be scribed with a diamond marker and are less likely to be broken, especially when immersion oil is being cleaned off the cover glass. Air mounts provide for the maximum amount of contrast. Both types of preparations are relatively easy to make, but air mounts do not require the purchase of a mounting medium. As pointed out by Wee (1983), images of a silica specimen appear quite different when observed in a solid mounting medium of high RI versus air. The parts of each scale or bristle that have the thickest deposition of silica appear bright in a high RI medium and dark when mounted in air. Thus, structures such as thick ribs appear bright in the high RI medium and dark in air, whereas large pores appear dark and light, respectively.

Air mounts of samples can be viewed easily with an inverted microscope by directly using the cover-glass preparation. The cover glass is positioned onto the stage with the sample side facing up toward the condenser. If the inverted microscope is outfitted with the same objective lenses and condenser system (e.g., one with a short working distance and a high numerical aperature) as an upright microscope, an image of similar quality to a standard upright microscope can be attained. This method has the obvious advantage that the cover glass need not be mounted onto a glass slide and, therefore, it can later be coated and viewed with SEM.

C. Preparing Samples for Observation with Electron Microscopy

A large number of species of synurophytes can be positively identified only after observation with EM. For most routine identification work and to obtain quality micrographs, preparations for either scanning or transmission electron microscopy (TEM) can be made easily and with little expense. In many cases an aliquot from a sample concentrated with a plankton net can be used directly to prepare SEM stubs and TEM grids.

For observation with SEM, samples simply can be dried onto a glass cover glass, a piece of aluminum foil, or directly onto an aluminum stub. The cover-glass or aluminum foil sample is attached to the aluminum stub using a nonconductive wax or double-sided tape, coated with gold, palladium, or a mixture of both, and viewed directly. My personal preference is to use aluminum foil preparations rather than glass cover slips, because they greatly reduce the degree of charging, and the aluminum foil can be trimmed easily to fit multiple samples onto one stub. I have also found that if a field emission SEM is used to view samples, quality images can be attained without coating the samples.

For observation with TEM, aliquots of samples can be dried onto copper grids coated with Formvar or some other material and coated with carbon to increase the stability of the grids. Grids with a mesh size of 200 suffice for most samples.

D. Storage of Samples and Preparations of Scales and Bristles

Glass microscope slide preparations of siliceous remains made with solid media or air mounts can be stored indefinitely. It is best to store such slides in a slotted box such that the slides remain in a horizontal position with the sample side facing upward. Preparations made with a mounting medium that does not totally solidify at room temperature should be avoided for long-term storage of samples.

Aluminum SEM stubs also can serve as a long-term method for storing prepared samples. Although Formvar grids, with or without a carbon coating, used in TEM observation also can be used to store samples, they may break apart with time. In addition to glass slide and aluminum stub preparations, samples from important collections can be dried onto cover glasses and aluminum foil, and stored in small Petri dishes. This method allows for future preparation of SEM or LM preparations using newly devised techniques or mounting media.

V. KEYS TO GENERA AND COMMON SPECIES FROM NORTH AMERICA

A. Introduction to Keys (Figs. 6–11)

The keys that follow can be used to identify genera within the Synurophyceae, as well as the most common species of *Mallomonas* and *Synura* found in North America. The keys are based primarily on characteristics of body scales for *Mallomonas* and apical-spined scales for *Synura* as observed with EM, but are also designed to work for many, but not all, species using high-resolution LM in conjunction with appropriate preparations of scales. Electron microscopy is stressed because observation of many features is required to separate species. Because of the ease of preparation and observation of samples, and the high level of information that can be derived for both scales and cells, the keys and associated micrographs are based on SEM. The reader is urged to make observations based on many scales to more appropriately work through the keys. In addition, the reader should consult the references listed in Section VI if a definitive identification cannot be made and to gain more knowledge of each species.

B. Key to Genera

Organisms are planktonic, unattached, unicellular or colonial in nature, motile, and golden brown in color. Cells generally have two plastids aligned with the long axis of the cell, lack an eye spot, and often possess a chrysolaminarin vesicle in the posterior of the cell.

C. Descriptions of Genera

Chrysodidymus Prowse (Figs. 2A and B and 6A)

This genus comprises motile two-celled colonies that swim in a more or less oscillating motion. The cells have an elongated vase-shaped structure that has slightly wider posterior ends and two emergent flagella. The cells are attached by their posterior ends and face outward; their longitudinal axes become aligned at 180°. Each cell is covered with siliceous spine-bearing scales that rsembles those of *Synura sphagnicola*. The spines tend to be longer on scales positioned on the anterior end of the cell.

A rare but geographically widely known genus reported from a variety of lakes in Quebec (Puytorac *et al.*, 1972), Ontario (Nicholls and Gerrath, 1985), the Adirondacks (Siver 1988a; Cumming *et al.*, 1992a), Florida (Wujek and Bland, 1991), Michigan (Wujek and Igoe, 1989), Washington (Norris and Munch, 1970), southern New England (Siver and Hamer, 1992), and Louisiana (Wee *et al.*, 1993). For ecological details, see Sections IIIA–F.

1a.	Cells unicellular (Figs. 1A–C and 3A–E)..	*Mallomonas*
1b.	Cells in a colony..	2
2a.	Colonies consist of only two basally attached cells that are linearly aligned and face 180° away from each other (Fig. 2A and B)	*Chrysodidymus*
2b.	Colonies of more than two, and usually many cells..	3
3a.	Each cell of the colony individually surrounded by overlapping scales[1]. Colonies most often spherical, but may also be elongated. Commonly observed in North America (Figs. 1F–I and 3G and H)..	*Synura*
3b.	Multiple layers of scales surround the entire colony, not individual cells. To date, this organism has been reported only from Australia (Figs. 2C and 3F)..	*Tessellaria*

[1]*S. lapponica* is an exception (see text for details).

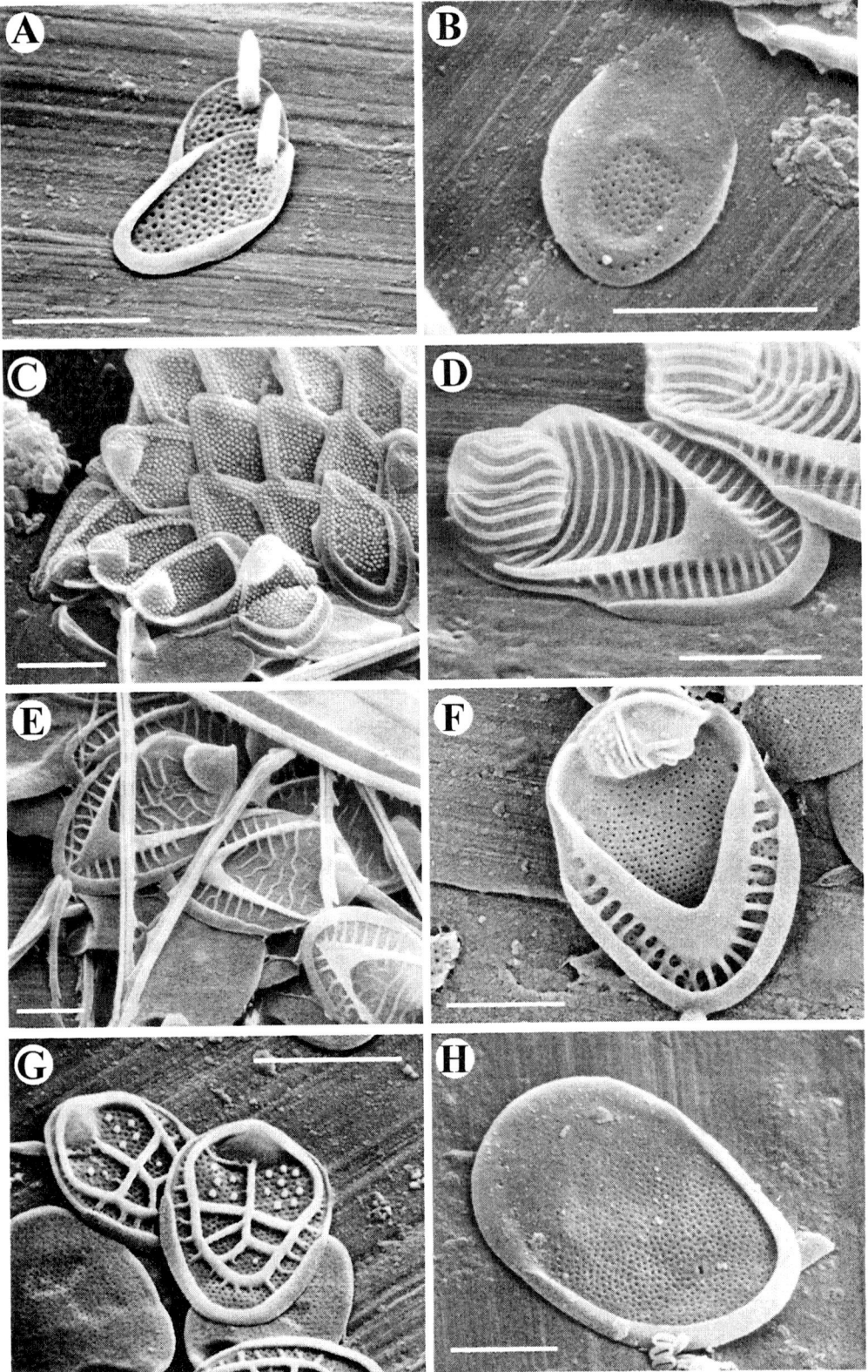

FIGURE 6 Scales of taxa associated with the Keys. A. Scales of *Chrysodidymus synuroideus*. B. Body scale of *Mallomonas akrokomos*. C. Domed and domeless scales of *Mallomonas annulata*. D. Body scale of *Mallomonas asmundiae*. E. Body scales of *Mallomonas acaroides* var. *acaroides*. F. Domed scale of *Mallomonas acaroides* var. *muskokana*. G. Scales of *Mallomonas canina*. H. Scale of *Mallomonas caudata*. All scale bars = 2 μm.

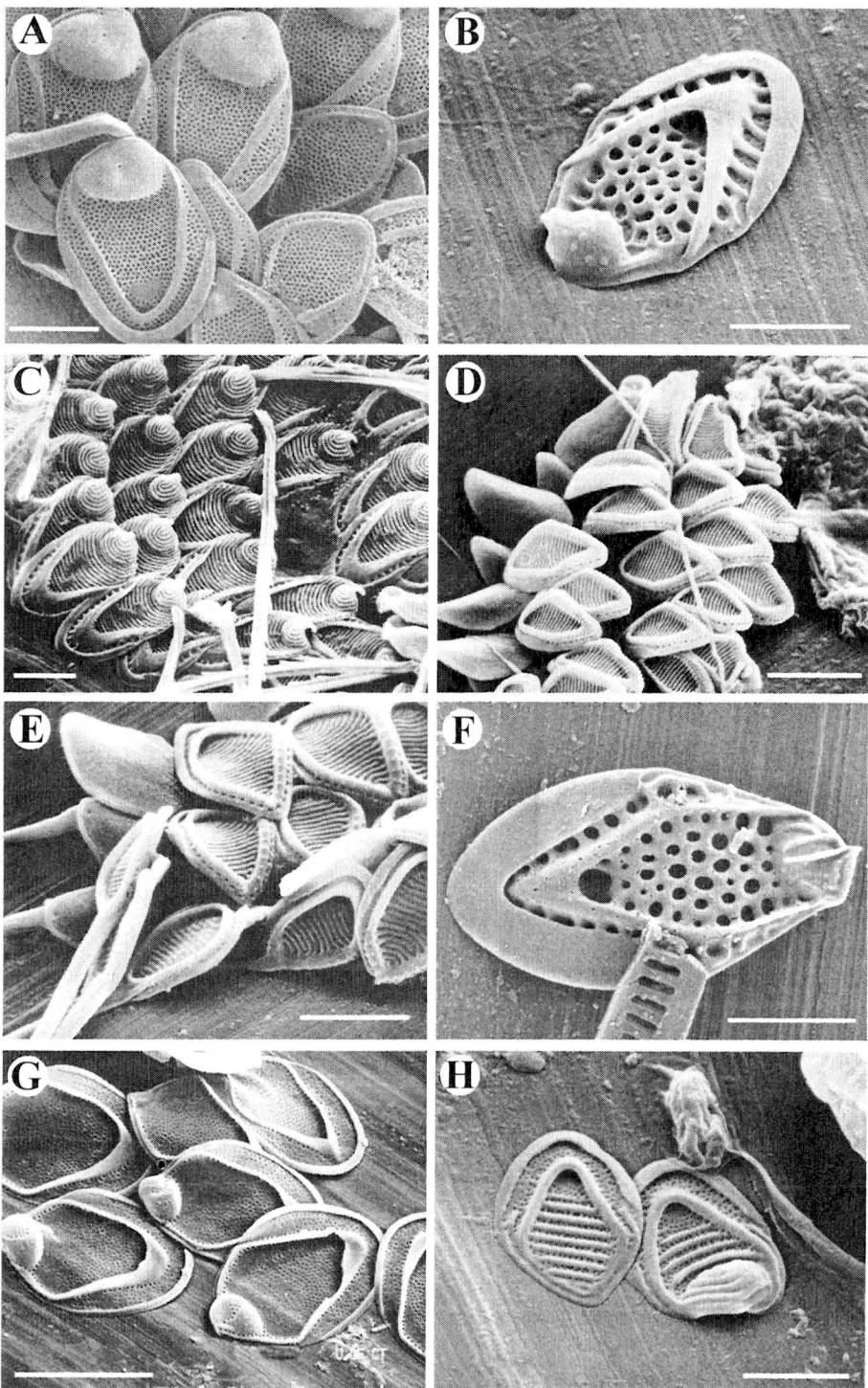

FIGURE 7 Scales of taxa associated with the Keys. A. Domed and domeless scales of *Mallomonas corymbosa*. Scale bar = 2 μm. B. Domed scale of *Mallomonas crassisquama*. Scale bar = 2 μm. C. Scales of *Mallomonas cratis*. Scale bar = 2 μm. Reprinted with permission from P. A. Siver, *The Biology of Mallomonas: Morphology, Taxonomy and Ecology*. Copyright © 1991, Kluwer, Dordrecht. D. Body and collar scales of *Mallomonas dickii*. Scale bar = 2 μm. Reprinted with permission from P. A. Siver, *The Biology of Mallomonas: Morphology, Taxonomy and Ecology*. Copyright © 1991, Kluwer, Dordrecht. E. Body and spined scales of *Mallomonas doignonii* var. *tenuicostis*. Scale bar = 2 μm. F. Domed scale of *Mallomonas duerrschmidtiae*. Scale bar = 2 μm. G. Domed scales of *Mallomonas elongata*. Scale bar = 5 μm. Reprinted with permission from P. A. Siver, *The Biology of Mallomonas: Morphology, Taxonomy and Ecology*. Copyright © 1991, Kluwer, Dordrecht. H. A domed and domeless scale of *Mallomonas galeiformis*. Scale bar = 2 μm.

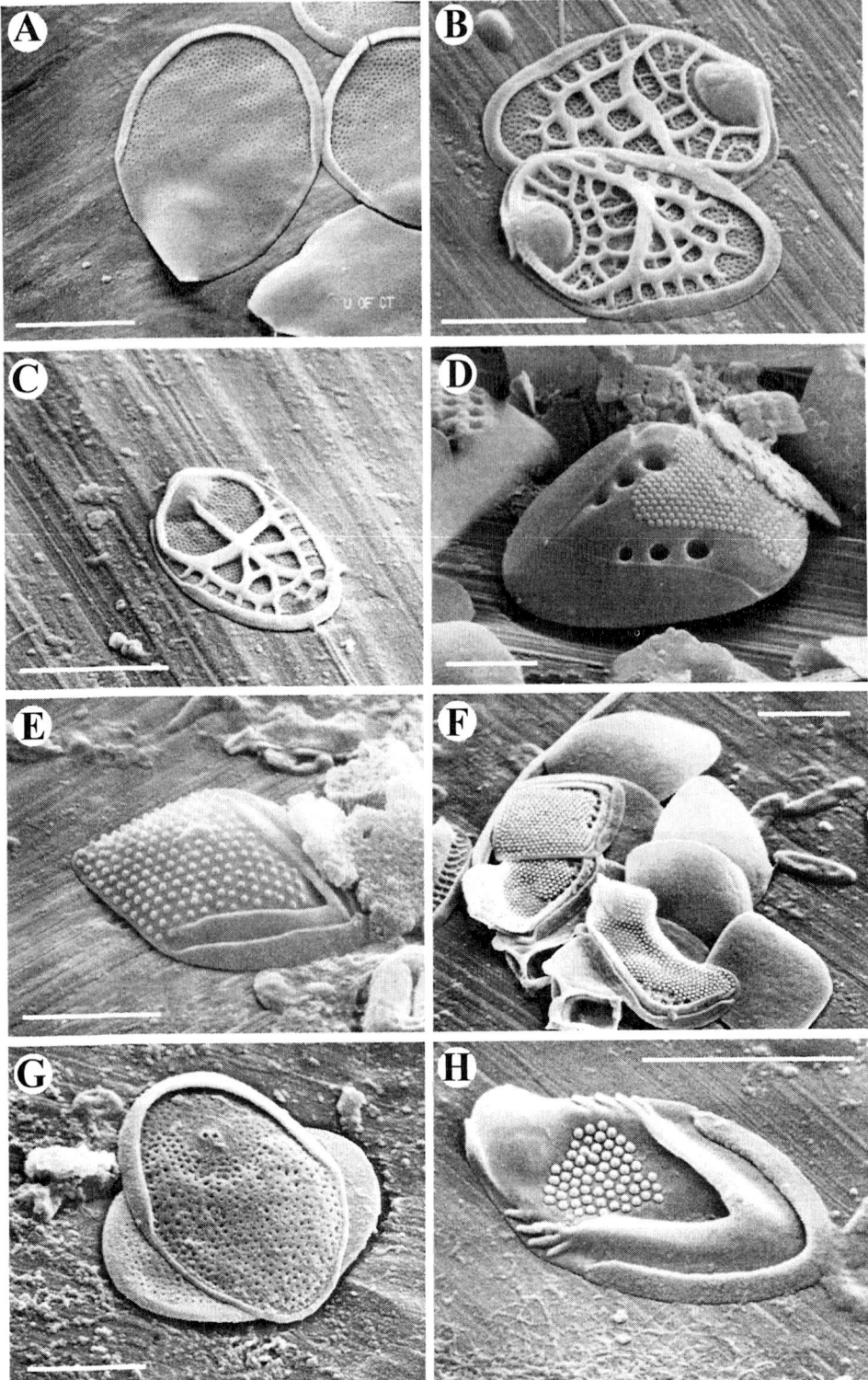

FIGURE 8 Scales of taxa associated with the Keys. A. Scale of *Mallomonas hamata*. Reprinted with permission from P. A. Siver, *The Biology of Mallomonas: Morphology, Taxonomy and Ecology*. Copyright © 1991, Kluwer, Dordrecht. B. Two scales of *Mallomonas heterospina*. C. Scale of *Mallomonas hindonii*. Reprinted with permission from P. A. Siver, *The Biology of Mallomonas: Morphology, Taxonomy and Ecology*. Copyright © 1991, Kluwer, Dordrecht. D. Body scale of *Mallomonas lychenensis*. E. Body scale of *Mallomonas mangofera*. F. Body and collar scales of *Mallomonas mangofera* var. *foveata*. G. Body scale of *Mallomonas matvienkoae*. H. Scale of *Mallomonas papillosa*. Reprinted with permission from P. A. Siver, *The Biology of Mallomonas: Morphology, Taxonomy and Ecology*. Copyright © 1991, Kluwer, Dordrecht. All scale bars = 2 μm.

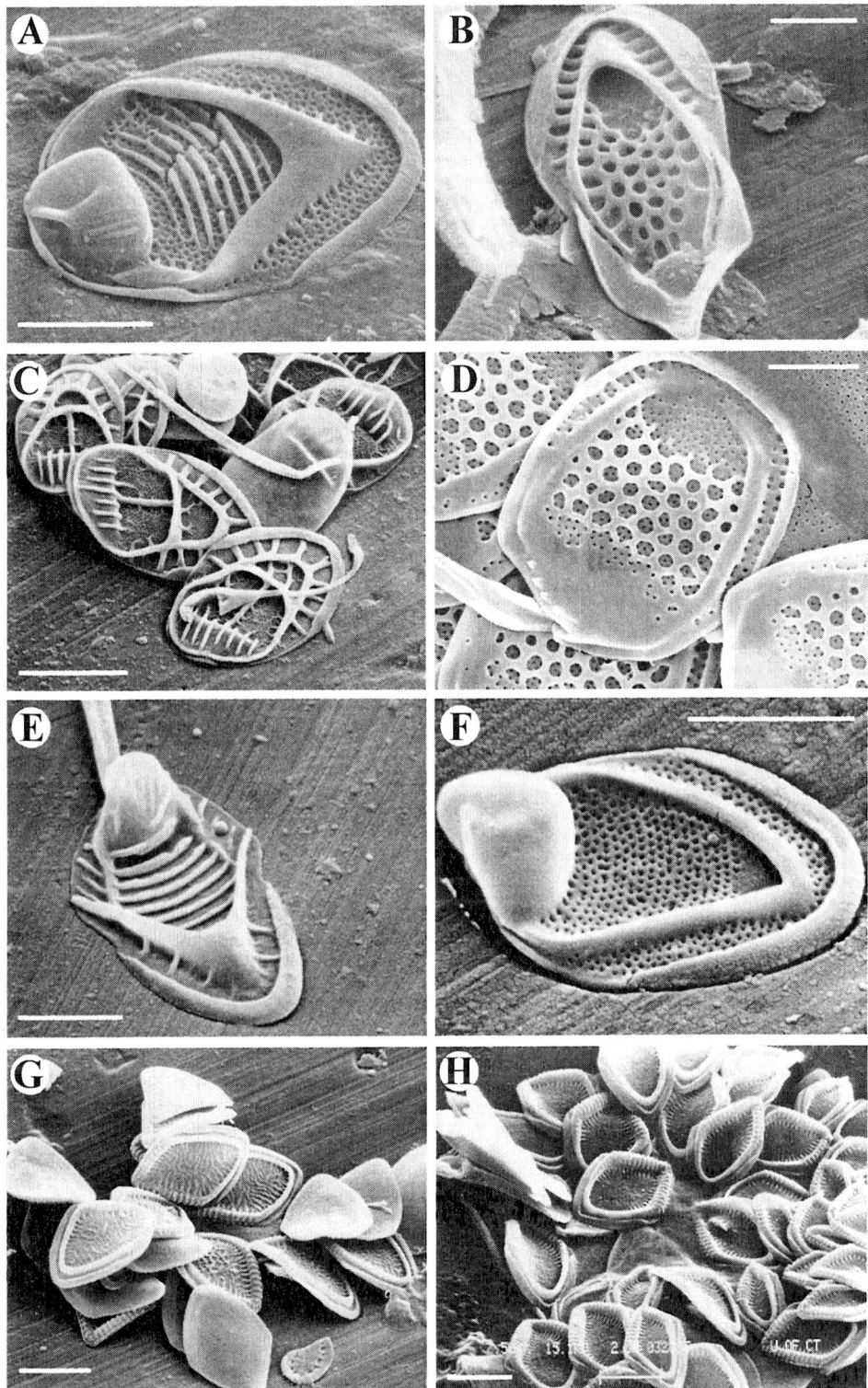

FIGURE 9 Scales of taxa associated with the Keys. A. Domed scale of *Mallomonas portae-ferreae* var. *portae-ferreae*. Scale bar = 2 μm. Reprinted with permission from P. A. Siver and M. S. Vigna, *Nova Hedwigia* 64:421–453. Copyright © 1997, Gebrüder Borntraeger, Stuttgart. B. Winged scale of *Mallomonas pseudocoronata*. Scale bar = 2 μm. C. Scales of *Mallomonas pugio*. Scale bar = 2 μm. D. Scale of *Mallomonas punctifera*. Scale bar = 1 μm. E. Scale of *Mallomonas striata*. Scale bar = 1 μm. Reprinted with permission from P. A. Siver, *The Biology of Mallomonas: Morphology, Taxonomy and Ecology*. Copyright © 1991, Kluwer, Dordrecht. F. Domed scale of *Mallomonas tonsurata*. Scale bar = 2 μm. G. Body scales of *Mallomonas torquata*. Scale bar = 2 μm. H. Body scales of *Mallomonas torquata* forma *simplex*. Scale bar = 2 μm.

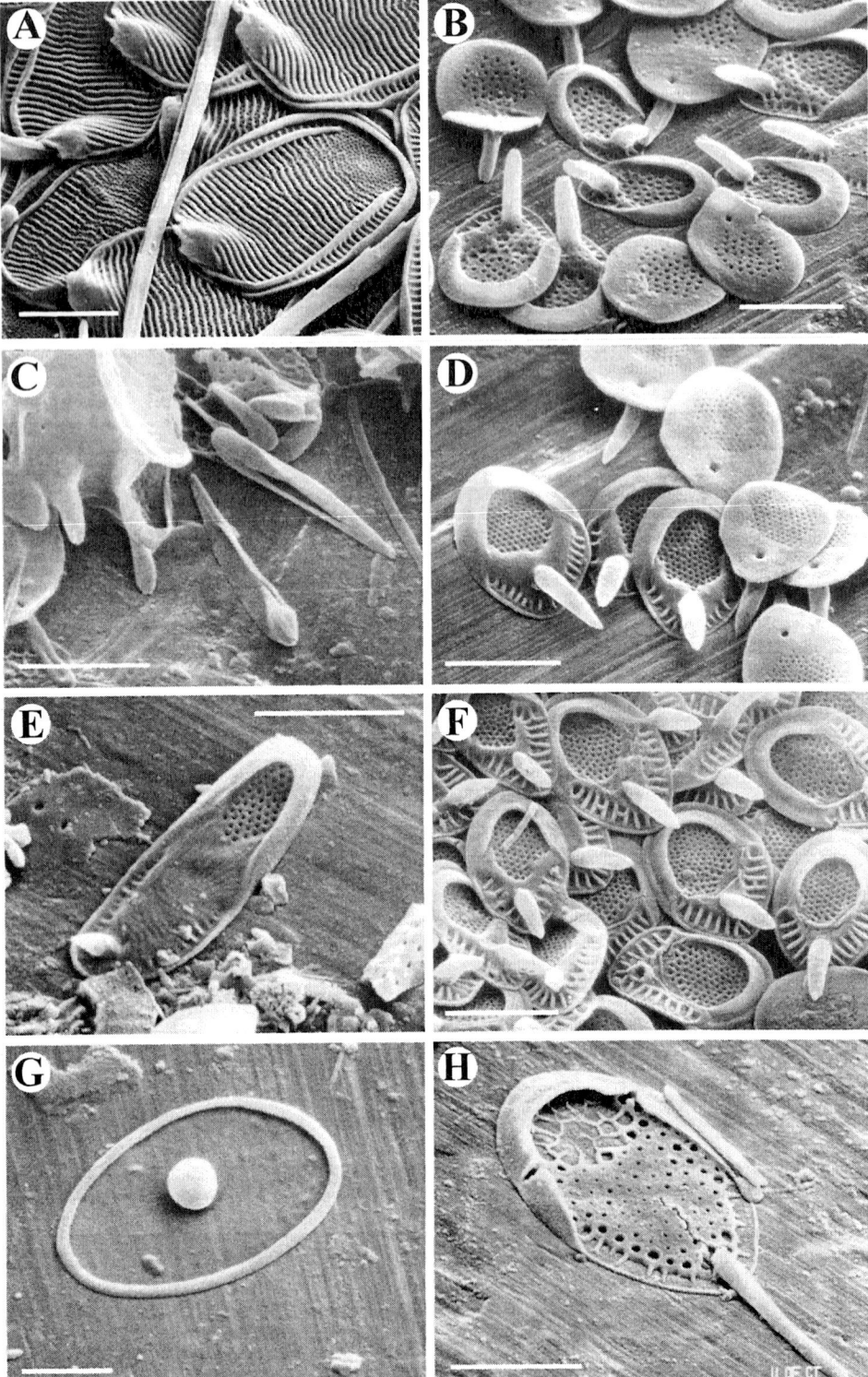

FIGURE 10 Scales of taxa associated with the Keys. A. Body scales of *Mallomonas transsylvanica*. Reprinted with permission from P. A. Siver, *The Biology of Mallomonas: Morphology, Taxonomy and Ecology*. Copyright © 1991, Kluwer, Dordrecht. B. Spined scales of *Synura curtispina*. C. Caudal scales of *Synura curtispina*. Reprinted with permission from P. A. Siver and M. S. Vigna, *Nova Hedwigia* 64:421–453. Copyright © 1997, Gebrüder Borntraeger, Stuttgart. D. Spined scales of *Synura echinulata*. E. Posterior scale of *Synura echinulata*. F. Spined scales of *Synura echinulata* forma *leptorrhabda*. Reprinted with permission from P. A. Siver and M. S. Vigna, *Nova Hedwigia* 64:421–453. Copyright © 1997, Gebrüder Borntraeger, Stuttgart. G. Scale of *Synura lapponica*. H. Spined scale of *Synura mollispina*. All scale bars = 2 μm.

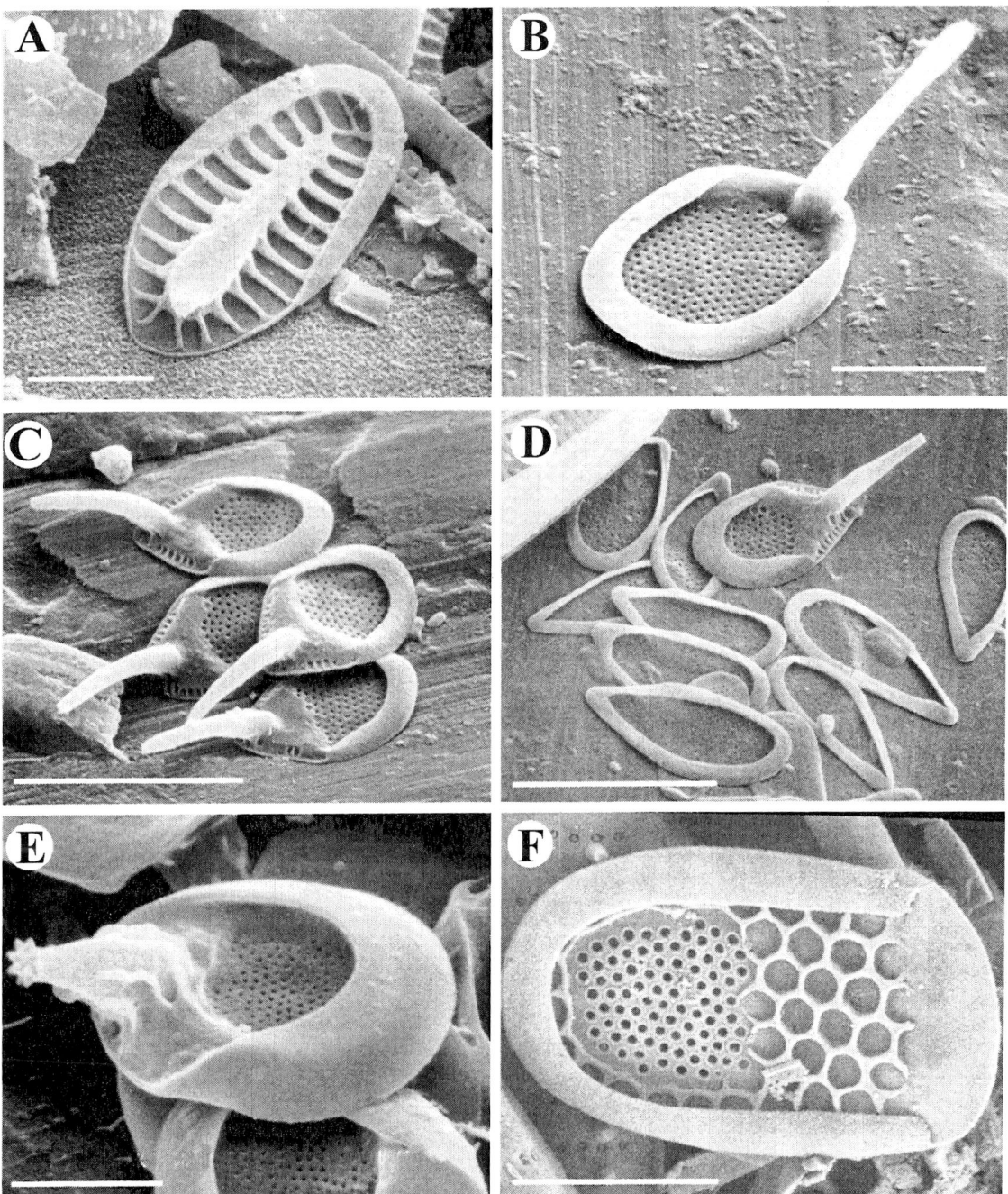

FIGURE 11 Scales of taxa associated with the Keys. A. Scale of *Synura petersenii*. Scale bar = 2 µm. B. Spined scale of *Synura sphagnicola*. Scale bar = 2 µm. C. Spined scales of *Synura spinosa*. Scale bar = 5 µm. D. Spineless scales of *Synura spinosa*. Scale bar = 5 µm. E. Spined scale of *Synura uvella*. Scale bar = 2 µm. F. Spineless scale of *Synura uvella*. Scale bar = 2 µm.

Mallomonas Perty (Figs. 1A–E, 3A–E, and 4A–D)

This genus comprises motile unicellular organisms that range in size from 10 to 70 μm long and 5 to 20 μm wide. Cells are spherical, ovid, or ellipse-shaped, golden in color, possess either a single or two apically emergent flagella, have one deeply lobed or two elongated chloroplasts, and have a posterior-positioned chrysolaminarin vacuole. Cells are covered with a highly organized layer of siliceous scales and bristles. Scales differ in morphology depending on their location on the cell surface. Scales with spines are found only to the posterior end of the cell. Bristles are thin, elongated structures that are tucked under the distal ends of scales and radiate out from the cell. The arrangement of bristles on the cell surface varies between species, but they typically streamlined and oriented toward the posterior of the cell during active swimming.

A cosmopolitan genus that consists of many genera (see Sect. VD), they are planktonic and common in a variety of standing freshwaters across North America and worldwide (for ecological details, see Sects. IIIA–F).

Synura Ehrenberg (Figs. 1F–I, 3G and H, and 4E and F)

A cosmopolitan genus that consists of spherical to elongate-shaped motile colonies that swim in a tumbling or rolling motion. Cells are most often pyriform or club-shaped and have wider anterior ends and slender more elongated posterior tails. The posterior ends of cells come in contact in the center of the colony or, in the case of elongated forms, along a linear axis. The number of cells per colony is highly variable, resulting in an equally variable colony diameter ranging from 30 to over 200 μm. Cells possess two unequal and apically emergent flagella, two golden-colored chloroplasts positioned along the longitudinal axis of the cell, and are covered with a highly organized layer of overlapping siliceous scales (except for *S lapponica*). Spines are typically longer on the anteror ends of cells that face the outside of the colny.

Cosmopolitan and consisting of many genera (see Sect. VE), they are planktonic and common in a variety of standing freshwaters across North America and worldwide (for ecological details, see Sects. IIIA–F).

Tessellaria Playfair (Figs. 2C and 3F)

Tessellaria comprises a rare, spherical-shaped motile colony that consists of from a few to several hundred cells. Colonies range in diameter from 25 to about 200 μm and swim in a rolling motion. Several layers of radially symmetrical scales surround the entire colony, not individual cells, and form in a light microscope what looks like a halo around the colony. Individual cells have two flagella, are golden in color, and are elongated; they have a wider anteror end and a long thin cytoplasmic tail that attaches to a ringlike structure in the center of the colony.

Rare; thus far reported (*T. volvocina*) only from Australia (Tyler *et al.*, 1989; Pipes and Leedale, 1992; Lavau *et al.*, 1997). Planktonic in freshwater, dystrophic, or slightly brackish standing waters, lagoons, and billabongs (Tyler *et al.*, 1989).

D. Key to Common Species of *Mallomonas*

1a.	Scales lack a V rib or with two lateral ribs that are not connected in the proximal region of the scale...	2
1b.	Scales with a distinct V rib or a continuous submarginal rib that is uninterrupted in the distal region...	8
2a.	Body scales with a dome..	3
2b.	Body scales lack a dome	5
3a.	Scales ovate with a shallow dome positioned to one side of the distal end; lack distinct secondary structures, including submarginal ribs; base plate pores are larger in the proximal region, becoming smaller and less distinct or lacking on the distal end. With LM the proximal portion of the scale along the rim appears more transparent and the dome can be observed (Fig. 8A).................	*M. hamata*
3b.	Scales are quadrate in shape with a rounded proximal end and a somewhat squared distal end with a more or less centrally positioned dome. Scales with two, roughly parallel, submarginal ribs that originate at the base of the dome and terminate, but do not connect, in the proximal region of the scale; submarginal ribs clearly seen with LM; shield with additional secondary ornamentation...	4
4a.	Distal two-thirds of the shield with a well developed reticulum of ribs forming large pores; reticulum easily observed with LM (Fig. 9D)..	*M. punctifera*
4b.	Distal two-thirds of shield ornamented with closely spaced, and more or less parallel transverse ribs; ribbing on shield observed with EM, but not LM (Fig. 10A)..	*M. transsylvanica*
5a.	Scales thick, large, heavily silicified, with two rows of large pits or pores aligned parallel to the long axis of the scale and clearly observed with LM; shield surface with prominent, closely spaced papillae not observed with LM (Fig. 14.8D)............	*M. lychenensis*
5b.	Scales lack the two rows of large prominent pits..	6

6a.		Scales small (1.7–3.8 μm × 0.6–2.0 μm), with a rounded U-shaped proximal end and a pointed inverted V-shaped distal end; possess a distinctive patch of base plate pores in the proximal half of the scale and a serrated distal margin. The patch of pores is more clearly observed from the underside with SEM and will appear as a more transparent zone with LM (Fig. 6B)............*M. akrokomos*
6b.		Scales much larger, usually larger than 4 μm, and typically up to 6 μm in length; distal and proximal ends more or less rounded, U-shaped, not pointed; lack distinctive patch of pores....................7
7a.		Scales oval to elliptical, symmetrical, with a secondary layer consisting mostly of small pores covering the distal one-half to two-thirds of the scale. In LM the distal portion of the scale appears opaque due to the secondary thickening (Fig. 8G) ....*M. matvienkoae*
7b.		Scales more rounded, but often slightly asymmetrical; lack a secondary layer or any secondary structures; appear unornamented in both EM and LM (Fig. 6H)............*M. caudata*
8a.		All body scales lack domes; V rib lies close to the rim; arms of the V rib are connected to thick well developed anterior submarginal ribs such as to form a rhomboidal or diamond shape. Bristles restricted to anterior collar scales; posterior scales often have long spines. (organisms within the section *Torquatae*[2])............9
8b.		At least some, if not all, body scales possess domes. Scales not rhomboidal in shape and cells not as described above............13
9a.		Surface of scales covered with closely spaced papillae (Fig. 8E and F)............*M. mangofera*
9b.		Surface of scales with little or no secondary structure, or consisting of a series of ribs, not papillae............10
10a.		Ribs on the scale consist of a series of very short struts radiating from the submarginal rib a short distance onto the shield; otherwise the shield is devoid of secondary structure (Fig. 9H)............*M. torquata* f. *simplex*
10b.		Ribs extend across entire area of the shield............11
11a.		Ribs form a reticulum on the shield (Fig. 19G)............*M. torquata* f. *torquata*
11b.		Ribs are closely spaced, more or less parallel, and each rib traverses the shield............12
12a.		Scales very small with a mean length of only 2.3 μm; cells also small with a mean length of only 11 μm (Fig. 17D)............*M. dickii*
12b.		Scales and cells significantly larger, with mean values of 3.8 and 27 μm, respectively (Fig. 7E)............*M. doignonii* (several varieties)
13a.		Base of the V rib very broad, U-shaped with forward-projecting arms that encircle and fuse at the distal-most edge of the scale, forming a continuous submarginal rib. A number of prominent, thick ribs on the shield, including a single large rib—the transverse rib, that transects the shield perpendicular to the longitudinal axis of the cell............14
13b.		Base of the V rib not broadly U-shaped; the arms of the V rib or anterior submarginal ribs do not completely encircle and fuse in front of the dome; scales lack a single large rib transversing the shield............17
14a.		Body scales widest near the distal end with a very broad dome that has from five to eight evenly spaced parallel ribs (Fig. 9C) ....*M. pugio*
14b.		Domes of body scales smaller and lack a series of parallel ribs............15
15a.		A relatively large number of ribs originate from each side of the transverse rib and radiate onto the shield, each, in turn, dividing or splitting to form a dense reticulum of ribs on the shield; some scales may have ribs on the dome, but they are not aligned in a parallel fashion (Fig. 8B)............*M. heterospina*
15b.		Scales with a single large rib radiating from the center of the transverse rib and terminating near the base of the dome. A similar rib radiates from the transverse rib toward the proximal end of the scale and commonly divides once or twice. The result is that the area of the shield proximal to the transverse rib is divided more than the distal portion of the shield............16
16a.		Areas of the shield between the ribs ornamented with papillae. Needle bristles lack subapical tooth (Fig. 6G)............*M. canina*
16b.		Shield lacks papillae. Needle bristles with subapical tooth (Fig. 8C)............*M. hindonii*
17a.		Secondary structures lacking on the shield and flanges; scales large with well developed V ribs and anterior submarginal ribs. The arms of the V rib bend and become continuous with the anterior submarginal ribs (Fig. 7G)............*M. elongata*
17b.		Scales more complex in nature; possess secondary structures on the shield and/or a rim............18
18a.		Shields with papillae............19
18b.		Shields lacking papillae............20
19a.		All scales with small domes and a series of three to eight evenly spaced parallel ribs on each anterior flange. Papillae cover the shield and often the dome, but are lacking on the flanges (Fig. 8H)............*M. papillosa*
19b.		Each cell has body scales with and without domes. Papillae cover the shield, the dome, and the anterior flanges; the latter lacking parallel ribs. Except for base plate pores, the posterior flange is unornamented (Fig. 6C)............*M. annulata*

20a.	Secondary layer on the shield consists of parallel transverse ribs.	21
20b.	Secondary layer on the shield essentially lacking, or consists of a network of circular holes or pores, or an irregular reticulation of ribs.	25
21a.	Transverse ribs arranged in a parallel fashion and connected by numerous ribs that run perpendicular to the transverse ribs; flange areas lack ribs, but the posterior flange may be covered with a reticulation of pores; scales large, mostly between 4 and 7.5 µm; dome and domeless scales present; domes are large and prominent (Fig. 9A)	*M. portae-ferreae* var. *portae-ferreae*
21b.	Transverse ribs on the shield are prominent, arranged in a parallel fashion, and may be connected by much smaller and less prominent ribs. Flange areas and domes may also possess parallel ribs.	22
22a.	Posterior and anterior flanges lack prominent ribs. The distal-most transverse ribs on the shield, especially on domed scales, are often larger and more prominent; the apical-most domed scales have one or a few transverse shield ribs (Fig. 7H)	*M. galeiformis*
22b.	Posterior flange with prominent ribs.	23
23a.	The posterior flange has a few widely and irregularly spaced ribs. The arms of the V rib extend to the margin of the scale and do not bend, and become continuous with the anterior submarginal ribs. Ribs on the anterior flange prominent, arranged parallel, and generally spaced equally with those on the shield (Fig. 9E)	*M. striata*
23b.	The posterior flange has many evenly spaced ribs. The arms of the V rib extend to the scale margin or bend and are continuous with the anterior submarginal ribs. The anterior flange either lacks ribs or has ribs that are continuous with those on the shield.	24
24a.	Ribs on the posterior flange spaced slightly wider than those on the shield. The anterior submarginal ribs are lacking or at best poorly developed; ribs on the shield are continuous onto the anterior flanges; dome with strongly U-shaped ribs (Fig. 7C)	*M. cratis*
24b.	Ribs on the posterior flange are usually spaced similarly to those on the shield. The arms of the V rib bend and become continuous with the anterior submarginal ribs; ribs are often lacking on the anterior flange; ribs on the dome are spaced similarly to those on the shield and aligned parallel or at a slight angle with the longitudinal axis of the cell (Fig. 6D)	*M. asmundiae*
25a.	Secondary layer on the shield is composed of small to medium-sized pores, close in diameter to those of the base plate; there is a well developed window at the base of the V rib; posterior flange may also contain a secondary layer of pores, but is not traversed with parallel ribs.	26
25b.	Secondary layer on the shield is lacking or composed of a reticulation of ribs or large pores; posterior flange is crossed or traversed with a series of parallel ribs.	27
26a.	Scales are small, most range between 2.7 and 4.4 µm in length; both domed and domeless scales present; on domed scales the arms of the V rib extend to or close to the dome, resulting in anterior submarginal ribs that are very short or lacking altogether (Fig. 9F)	*M. tonsurata*
26b.	Scales often much larger, most ranging in length from 4.2 to 5.2 µm; scales with distinct anterior submarginal ribs (Fig. 7A)	*M. corymbosa*
27a.	Secondary layer on the shield is largely lacking (Fig. 6F)	*M. acaroides* var. *muskokana*
27b.	Secondary layer on the shield consists of large pores or a reticulation of ribs.	28
28a.	Secondary layer on the shield consists of somewhat wavy ribs, most of which originate from the V rib and extend onto the shield. The arms of the V rib extend to the margin of the scale and do not bend, and become continuous with the anterior submarginal ribs (Fig. 6E)	*M. acaroides* var. *acaroides*
28b.	Secondary layer on the shield generally well developed and consisting of a series of large (relative to the base plate pores) pores.	29
29a.	Scales are very large, commonly between 6.4 and 14.1 µm in length, and possess a large, forward-projecting wing (Fig. 9B)	*M. pseudocoronata*
29b.	Scales may be large, but lack a forward-projecting wing.	30
30a.	Domed scales are large, ranging in length from 4.7 to 8.6 µm, with relatively small domes; domes usually marked with parallel ribs; arms of the V rib bend and become continuous with the anterior submarginal ribs (Fig. 7F)	*M. duerrschmidtiae*
30b.	Scales usually smaller, ranging in length from 3.8 to 6.7 µm, with relatively large domes often marked with papillae; arms of V rib extend to the margin of the scale and are not continuous with the anterior submarginal ribs (Fig. 7B)	*M. crassisquama*

[2]Only a few of the many taxa described in the section *Torquatae* are included in the key. Some of the taxa are difficult to positively identify and the reader is urged to consult the references listed in Table I for more details.

E. Key to Common Species of *Synura*

1a. Scales biradially symmetrical with a central papillae (Fig. 10G)..*S. lapponica*

1b. Scales bilaterally symmetrical and lack a central papillae..2

2a. Anterior scales have a keel, a raised hollow cylinder centrally positioned on the base plate; usually with well developed ribs or struts that run perpendicular from the keel to the scale border (Fig. 11A)[3]..................................section Peterseniae including, *S. petersenii*

2b. Anterior scales lack a keel but possess a distally attached and forward-projecting spine...3

3a. Scales with evenly spaced ribs under the rim; spine scales with a short, stout, conical spine with multiple teeth at the distal tip. Ribs beneath the rim are more easily observed with LM or TEM (Fig. 11E and F)...*S. uvella*

3b. Scales lack ribs under the rim; spines not short, stout and conical in shape..4

4a. Base plate consists of rather large, evenly spaced perforations; lack additional secondary structures; rim encircles at least four-fifths of the perimeter on spined scales and the entire perimeter on spineless scales (Fig. 11B)..........................*S. sphagnicola*

4b. Proximal portion of anterior scale with perforations as above, but the distal portion with secondary siliceous features.....................5

5a. Tip of spine a sharp acute point; distal portion of scale with a raised thickened region beneath that lies as either a series of vermiform ribs or a series of closely spaced and linearly arranged papillae (seen only with EM); a series of short ribs run perpendicular to the distal margin; posterior scales elongated, often with a very short spine, and with a more extensive thickened region than found on anterior scales...6

5b. Tip of spine blunt and may possess teeth; secondary ribs form a honeycomb reticulation on at least the distal portion of the scale; a thin siliceous membrane may cover at least part of the secondary honeycomb; a series of short ribs run perpendicular to the distal margin; posterior scales lack spines...8

6a. Distal end of scale composed of closely spaced papillae arranged in linear fashion (observed best with TEM).............*S. mammillosa*

6b. Distal end of scale composed of vermiform ribs, not papillae..7

7a. Series of vermiform ribs rather extensive, covering one-third to one-half of the surface area on anterior scales, and often reaches two-thirds of the surface area on the posterior scales; anterior scales commonly 2 to 3.5 μm long (Fig. 10D and E) ....*S. echinulata*

7b. Vermiform ribbing reduced to a small narrow region; pores on the shield rather large; scales are small, usually around 2 μm long (Fig. 10F)...*S. echinulata* f. *leptorrhabda*

8a. Honeycomb reticulum covers most, if not all, of the scale; a thin membrane may cover ca. two-thirds to three-quarters of the honeycomb layer; posterior scales rectangular or long, narrow, and somewhat triangular, with an elongated distal end (Fig. 10H) ....*S. mollispina*

8b. Honeycomb reticulum covers the distal ca. one-third of the scale; a thin membrane may cover the honeycomb reticulum.................9

9a. Spines generally small, ~0.5–2.0 μm long; posterior-most scales elongated, slipper-shaped, generally lack secondary structure, and are encircled by a thick rim that imparts a diamond shape to the scale (Fig. 10B and C).................................*S. curtispina*

9b. Spines longer, ~2.8–3.5 μm long; posterior scales teardrop-shaped, lack ribbing on the shield, and completely encircled with a rim that does not form a diamond pattern (Fig. 11C and D)..*S. spinosa*

[3] Scales of *S. australiensis* have a morphology similar to those of *S. petersenii*, but are much longer, often over 8 μm in length.

VI. GUIDE TO LITERATURE FOR SPECIES IDENTIFICATION

The monographs by Asmund and Kristiansen (1986) and Siver (1991a) provide excellent general references for *Mallomonas* (Table I). The work by Asmund and Kristiansen (1986) is largely based on TEM images and provides well-written descriptions as well as previous records for all *Mallomonas* taxa described at the time of publication. Siver (1991a) provides many SEM micrographs of whole cells, scales and bristles, morphological analyses, and summaries of ecological preferences for many species of *Mallomonas*.

Nicholls and Gerrath (1985) and Siver (1987) are two of the more useful publications for identifying species of *Synura* in North American waters, and together provide TEM and SEM images of most of the common species. Wujek and Wee (1983) and Graham et al. (1993), and Tyler et al. (1989) provide information on *Chrysodidymus* and *Tessellaria*, respectively.

Three other references are especially valuable for identification of synurophytes as well as other scaled chrysophytes. Although the work of Takahashi (1978) is based largely on water bodies in Japan, it provides much useful information and TEM images of many species commonly found in North America. Wee

TABLE I A List of Useful Publications for the Identification of Species and Subspecific Taxa of Synurophyceae and Other Scaled-Chrysophytes with an Emphasis on References from North American Localities

Reference	Region	Comments
Asmund and Hilliard (1961)	Alaska	Work on *Mallomonas*
Amund and Takahashi (1969)	Alaska	Work on *Mallomonas*
Asmund and Kristiansen (1986)	Monograph	thesis on *Mallomonas*
Cronberg (1989)	Tropics	Review of work from tropical localities
Gretz *et al.* (1979, 1983)	Arizona	TEM
Kling and Kristiansen (1983)	Canada	Central and northern areas
Kristiansen (1975)	Western Canada	Collections from Alberta and British Columbia
Kristiansen (1992)	Greenland	Arctic sites
McKenzie and Kling (1989)	Northwest territories	Arctic regon, Mackenzie Delta
Nicholls (1982)	Ontario	Work on *Mallomonas*
Nicholls (1988a)	Ontario	Provides checklist of *Mallomonas* from NA
Nicholls and Gerrath (1985)	Ontario	Work on *Synura*; discusses taste and odor
Petersen and Hansen (1956, 1958)	General	Works on *Synura*
Siver (1987)	Connecticut	Work on *Synura*; SEM
Siver (1988b)	Adirondacks, New York	Soft-water localities
Siver (1991a)	Mostly Connecticut	Monograph on *Mallomonas*
Siver and Wujek (1993)	Florida	Eutrophic collections; SEM; subtropics
Siver and Wujek (1998)	Florida	Ocala National Forest; acidic habitats; subtropics
Takahashi (1978)	Japan	Thesis on scaled chrysophytes
Tyler *et al.* (1989)	Australia	Work on *Tessellaria*
Wawrzyniak and Andersen (1985)	Northern Boreal	Soft-water sites; scaled chrysophytes
Wee (1982)	Iowa	Thesis on scaled chrysophytes; alkaline sites
Wee *et al.* (1993)	Louisiana	Sothern Atlantic Coastal Plain; subtropics
Wujek (1984)	Florda	Scaled chrysophytes; TEM
Wujek and Hamilton (1972, 1973)	Michigan	Scaled chrysophytes; TEM
Wujek *et al.* (1975, 1977)	Michigan	Scaled chrysophytes; TEM
Wujek and Wee (1983)	General	Work on *Chrysodidymus*
Wujek and Weis (1984)	Kansas	Scaled chrysophytes; TEM

(1982) is one of the few in-depth studies of primarily alkaline habitats from the midwestern United States, and also provides many LM images of whole cells and scales, and a taxonomic key. Cronberg (1989) provides a review of taxa found in tropical localities. References of studies of synurophytes and scaled chrysophytes from specific regions of North American are listed in Table I.

ACKNOWLEDGMENTS

The writing of this chapter was funded, in part, by grants from the National Science Foundation (DEB-9306587 and DEB-9615062). I would like to thank Anne Lott for help in assembling the references, and Ken Nicholls and Jim Wee for helpful comments.

LITERATURE CITED

Andersen, R. A. 1985. The flagellar apparatus of the golden alga *Synura uvella*: Four absolute orientations. Protoplasma 128:289–294.

Andersen, R. A. 1987. Synurophyceae classic nov., a new class of algae. American Journal of Botany 74:337–353.

Andersen, R. A., Van de Peer, Y., Potter, D., Sexton, J. P., Kawachi, M., LaJeunesse, T 1999. Phylogenetic analysis of the ssu rRNA from members of the Chrysophyceae. Protist 150:71–184.

Andersen, R. A., Mulkey, T. J. 1983. The occurrence of chlorophylls c_1 and c_2 in the Chrysophyceae. Journal of Phycology 19:289–294.

Asmund, B. 1968. Studies on Chrysophyceae from some ponds and lakes in Alaska. VI. Occurrence of *Synura* species. Hydrobiologia 31:497–515.

Asmund, B, Hilliard, D. K. 1961. Studies on Chrysophyceae from some ponds and lakes in Alaska. I – *Mallomonas* species examined with the electron microscope. Hydrobiologia 34:305-321.

Asmund, B., Kristiansen, J. 1986. The genus *Mallomonas* (Chrysophyceae). Opera Botanica 85:1–128.

Asmund, B., Takahashi, E. 1969. Studies on Chrysophyceae from some ponds and lakes in Alaska VIII: *Mallomonas* species examined with the electron microscope II. Hydrobiologia 34:305–321.

Barbiero, R. P., McNair, C. M. 1996. The dynamics of vertical chlorophyll distribution in an oligomesotrophic lake. Journal of Plankton Research 18:225–237.

Beech, P. L., Wetherbee, R. 1990a. The flagellar apparatus of *Mallomonas splendens* (Synurophyceae) at interphase and its development during the cell cycle. Journal of Phycology 26:95–111.

Beech, P. L. 1990b. Direct observations on flagellar transformation in

Mallomonas splendens (Synurophyceae). Journal of Phycology 26:90–95.
Beech, P. L., Wetherbee, R., Pickett-Heaps, J. D. 1990. Secretion and deployment of bristles in *Mallomonas splendens* (Synurophyceae). Journal of Phycology 26:112–122.
Belcher, J. H. 1969. Some remarks upon *Mallomonas papillosa* Harris and Bradley and *M. caleolus* Bradley. Nova Hedwigia 18:257–270.
Bourrelly, P. 1981. Les algues d'eau douce. II. Les algues jaunes et brunes. Societe Nouvelle des Editions Boubee, Paris.
Brugerolle, G., Bricheux, G. 1984. Actin microfilaments are involved in scale formation of the chrysomonad cell *Synura*. Protoplasma 123:203–212.
Calado, A. J., Rino, J. A. 1994. *Chlorodesmos hispidus*, a morphological expression of *Synura spinosa* (Synurophyceae). Nordic Journal of Botany 14:235–239.
Cavalier-Smith, T. 1986. The kingdon Chromista: Origin and systematics. Progress in Phycological Research 4:309–347.
Charles, D. F., Smol, J. P. l988. New methods for using diatoms and chrysophytes to infer past pH of low-alkalinity lakes. Limnology and Oceanography 33:1451–162.
Clasen, J., Bernhardt, H. 1982. A bloom of the Chrysophyceae *Synura uvella* in the Wahnbach reservoir as indicator for the release of phosphates from the sediment. Archiv für Hydrobiologie Beiheft 18 61–86.
Cronberg, G. 1980. Cyst development in different species of *Mallomonas* (Chrysophyceae) studied by scanning electron microscopy. Archiv für Hydrobiologica Scandinavica 56:421–434.
Cronberg, G. 1982. Phytoplankton changes in Lake Trummen induced by restoration. Folia Limnologica Scandinavica 18:11–119.
Cronberg, G. 1986. Chrysophycean cysts and scales in lake sediments: A review, *in*: Kristiansen, J., Andersen, R. A., Eds., Chrysophytes: Aspects and problems. Cambridge University Press, pp 281–315.
Cronberg, G. 1989. Stomatocysts of *Mallomonas hamata* and *M. heterospina* (Mallomonadaceae, Synurophyceae) from South Swedish lakes. Nordic Journal of Botany 8:683–692.
Cronberg, G. 1996. Scaled chrysophytes from the Okavango Delta, Botswana, Africa, *in*: Kristiansen, J., cronberg, G. Eds., Chrysophytes: Progress and new horizons. Beiheft zur Nova Hedwigia, Vol. 114. Cramer, Berlin, pp. 99–109.
Cronberg, G., Kristiansen, J. 1980. Synuraceae and other Chrysophyceae from central Småland, Sweden. Botaniska Notiser 133:595–618.
Cronberg, G., Gelin, C., Larsson, K. 1975. Lake Trummen restoration project. II. Bacteria, phytoplankton and phytoplankton productivity. Internationale Vereinigung für Theoretische und Angewandte Limnologie Verhandlungen 19:1907–1996.
Croome, R. L., Tyler, P. A. 1985. Distribution of silica-scaled Chrysophyceae (Paraphysomonadaceae and Mallomonadaceae) in Australian inland waters. Australian Journal of Marine and Freshwater Research 36:839–853.
Croome, R. L. 1988. Phytoflagellates and their ecology in Tasmanian polyhumic lakes. Hydrobiologia 161:245–253.
Cumming, B. F., Smol, J. P., Birks, H. J. B. 1992a. Scaled chrysophytes (Chrysophyceae and Synurophyceae) from Adirondack drainage lakes and their relationship to environmental variables. Journal of Phycology 28:162–178.
Cumming, B. F., Smol, J. P., Kingston, J. C., Charles, D. F., Birks, H. J. B., Camburn, K. E., Dixit, S. S., Uutala, A. J., Selle, A. R. 1992b. How much acidification has occurred in Adirondack region lakes (New York, USA) since preindustrial times? Canadian Journal of Fisheries and Aquatic Science 49:128–141.
DeNoyelles, F., O'Brien, W. J. 1978. Phytoplankton succession in nutrient enriched experimental ponds as related to changing carbon, nitrogen and phosphorus concentrations. Archiv für Hydrobiologie 81:137–165.
Dixit, S. S., Dixit, A. S., Evans, R. D. 1988. Scaled chrysophytes (Chrysophyceae) as indicators of pH in Sudbury, Ontario lakes. Canadian Journal of Fisheries and Aquatic Science 45:1411–1421
Dixit, S. S., Dixit, A. S., Smol, J. P. 1989. Relationship between chrysophyte assemblages and environmental variables in 72 Sudbury lakes as examined by canonical correspondence analysis (CCA). Canadian Journal of Fisheries and Aquatic Science 46:1667–1676.
Dixit, S. S., Smol, J. P, Anderson, D. S., Davis, R. B. 1990. Utility of scaled chrysophytes for inferring lakewater pH in northern New England lakes. Journal of Paleolimnology 3:269–287.
Duff, K. E., Smol, J. P. 1995. Chrysophycean cyst assemblages and their relationship to water chemistry in 71 Adirondack Park (N.Y., U.S.A.) lakes. Archiv für Hydrobiologie 134:307–336.
Duff, K. E., Zeeb, B. A., Smol, J. P. 1995. Atlas of Chrysophycean cysts, Kluwer, Boston, 189 p.
Dürrschmidt, M. 1980. Studies on the Chrysophyceae from Río Cruces, Prov. Valdivia, south Chile by scanning and transmission microscopy. Nova Hedwigia 33:353–388.
Dürrschmidt, M. 1982. Studies on the Chrysophyceae from southern Chilean inland waters by means of scanning and transmission electron microscopy, II. Archiv für Hydrobiologie Supplement 63:121–163.
Dürrschmidt, M., Croome, R. 1985. Mallomonadaceae (Chrysophyceae) from Malaysia and Australia. Nordic Journal of Botany 5:285–298.
Eloranta, P. 1989. Scaled chrysophytes (Chrysophyceae and Synurophyceae) from national park lakes in southern and central Finland. Nordic Journal of Botany 8:671–681.
Eloranta, P. 1995. Biogeography of chrysophytes in Finnish lakes, *in*: Sandgren, C. D., Smol, J. P., Kristiansen, J., eds., Chrysophyte algae: Ecology, phylogeny and development. Cambridge University Press, pp. 214–231.
Findlay, D. L. 1978. Seasonal succession of phytoplankton in seven lake basins in the Experimental Lakes Area, northwestern Ontario, following artificial eutrophication. Data from 1974–1976. Marine Sciences Report 466, Canadian Fisheries and Marine Services, 41 p.
Gibson, K. N., Smol, J. P., Ford, J. 1987. Chrysophycean microfossils provide new insight into the recent history of a naturally acidic lake (Cond Pond, New Hampshire). Canadian Journal of fisheries and aquatic science 44:1584–1588.
Graham, L. E., Graham, J. M., Wujek, D. E. 1993. Ultrastructure of *Chrysodidymus synuroideus* (Synurophyceae). Journal of Phycology 29:330–341.
Gretz, M. R., Sommerfeld, M. R., Wujek, D. E. 1979. Scaled Chrysophyceae of Arizona: A preliminary survey. Journal of the Arizona–Nevada Academy of Science 14:5–80.
Gretz, M. R., Wujek, D. E., Sommerfeld, M. R. 1983. Scaled Chrysophyceae of Arizona: Further additions to the aquatic flora. Journal of the Arizona–Nevada Academy of Science 18:17–21.
Gutowski, A. 1989. Seasonal succession of scaled chrysophytes in a small lake in Berlin, *in*: Kristiansen, J., Cronberg, G., Geissler, U., Eds. Chrysophytes: Developments and perspectives. Beiheft Zur Nova Hedwigia, Vol. 95. 159–77. Cramer, Berlin, pp. 159–177.
Gutowski, A. 1996. Temperature dependent variability of scales and bristles of *Mallomonas tonsurata* Teiling emend Krieger (Synurophyceae), *in*: Kristiansen, J., Cronberg, G. Eds., Chrysophytes: Progress and new horizons. Beiheft Zur Nova Hedwigia Vol. 114. Cramer, Berlin, pp. 125–146.

Gutowski, A. 1997. *Mallomonas* species (Synurophyceae) in eutrophic waters of Berlin (Germany). Nova Hedwigia 65:299–335.

Hansen, P. 1995. Preliminary studies on the silica–scaled Chrysophytes and Synurophytes of Madagascar. M.S. thesis, Department of Phycology, Botanical Institute, University of Copenhagen, 75 pp.

Harris, K. 1953. A contribution to our knowledge of *Mallomonas*. Botany Journal of the Linnean Society 55:88–102.

Harris, K., Bradley, D. E. 1957. An examination of the scales and bristles of *Mallomonas* in the electron microscope using carbon replicas. Proceedings of the Royal Microscopy Society 76:37–46.

Harris, K., Bradley, D. E. 1960. A taxonomic study of Mallomonas. Journal of General Microbiology 22:750–777.

Hartmann, H., Steinberg, C. 1989. The occurrence of silica-scaled chrysophytes in some central European lakes and their relation to pH, in: Kristiansen, J., Cronberg, G., Geissler, U., Eds. Chrysophytes: Developments and perspectives. Beiheft Zur Nova Hedwigia Vol. 95. Cramer, Berlin, pp, 131–158.

Hibberd, D. J. 1976. The ultrastructure and taxonomy of the Chrysophyceae and Prymnesiophyceae (Haptophyceae): A survey with some new observations on the ultrastructure of the Chrysophyceae. Botany Journal of the Linnean Society 72:55–80.

Hibberd, D. J. 1978. The fine structure of *Synura sphagnicola* (Korsh.) Korsh. (Chrysophyceae). British Phycological Journal 13:403–412.

Hickel, B., Maass, I. 1989. Scaled chrysophytes, including heterotrophic nanoflagellates from the lake district in Holstein, northern Germany. Beihefte zur Nova Hedwigia 95:233–257.

Hoffmann, L., Wille, E. 1992. Occurrence of a metalimnetic summer peak of *Mallomonas caudata* (Synurophyceae). Nordic Journal of Botany 12:465–469.

Holen, D. A., Boraas, M. E. 1995. Mixotrophy in chrysophytes, in: Sandgren, C. D., Smol, J. P., Kristiansen, J., Eds., Chrysophyte algae: Ecology, phylogeny and development. Cambridge University Press, pp. 119–141.

Hustedt, F. 1939. Systematische und ökologische Untersuchungen über die Diatomeen-Flora von Java, Bali und Sumatra nach dem Material der Deutschen Limnologischen Sunda-Expedition. III. Die ökologischen Faktoren und ihr Einfluss auf die Diatomeenflora. Archiv für Hydrobiologie Supplement 16: 1–394.

Jacobsen, B. A. 1985. Scale–bearing Chrysophyceae (Mallomonadaceae and Paraphysomonadaceae) from west Greenland. Nordic Journal of Botany: 381–398.

Jarosch, R. 1970. On the flagellar waves of *Synura bioreti* and the mechanics of the uniplanar waves. Protoplasm 69:201–214.

Jensen, W. A. 1962. Botanical histochemistry. Freeman, New York, 523 pp.

Kies, L. Berndt, M. 1984. Die Synura—Arten (Chrysophyceae) Hamburgs und seiner nordostlichen Umgebung. Mitteilung Institut Allgemein Botanie Hamburg 19:99–122.

Kling, H., Holmgren, S. 1972. Species composition and seasonal distribution of phytoplankton in the Experimental Lakes Area, northwestern Ontario technical Report 337, Canadian Fisheries and Marine Services.

Kling, H., Kristiansen, J. 1983. Scale–bearing Chrysophyceae (Mallomondaceae) from central and northern Canada. Nordic Journal of Botany 3:269–290.

Korshikov, A. A. 1929. Studies on Chrysomonads. I. Archiv für Protistenkunde 67:253–290.

Kristiansen, J. 1975. On the occurrence of the species of *Synura* (Chrysophyceae). Internationale vereinigung für Theoretische und Angewandte Limnologie Verhandlungen 19: 2709–2715.

Kristiansen, J. 1980. Chrysophyceae from some Greek lakes. Nova Hedwigia 33:167–194.

Kristiansen, J. 1981. Distribution problems in the Synuraceae (Chrysophyceae). Internationale vereinigung für Theoretische und Angewandte Limnologie Verhandlungen 21:1444–1448.

Kristiansen, J. 1985. Occurrence of scale-bearing Chrysophyceae in a eutrophic Danish lake. Internationale Vereinigung für Theoretische und Angewandte Limnologie Verhandlungen 22: 2826–2829.

Kristiansen, J. 1986. Silica–scale bearing chrysophytes as environmental indicators. British Phycological Journal 21: 425–436.

Kristiansen, J. 1988a. The problem of enigmatic chrysophytes. Arch für Protistenkunde 135:9–15.

Kristiansen, J. 1988b. Seasonal occurrence of silica-scaled chrysophytes under eutrophic conditions. Hydrobiologia 161:171–184.

Kristiansen, J. 1992. Silica–scaled chrysophytes from West Greenland: Disko Island and the Sondre Stromfjord region. Nordic Journal of Botany 12:525–536.

Kristiansen, J., Takahashi, E. 1982. Chrysophyceae: introduction and bibliography, in: Rosowski, J., Parker, B., Eds., Selected papers in phycology, Vol. II. Phycological Society of America, Lawrence, KS, pp. 698–704.

Kristiansen, J., Tong, D. 1989. *Chrysosphaerella annulata* sp. nov., a new scale-bearing chrysophyte. Nordic Journal of Botany 9:329–332.

Lavau, S., Saunders, G. W., Wetherbee, R. 1997. A phylogenetic analysis of the Synurophyceae using molecular data and scale case morphology. Journal of Phycology 33:135–151.

Leadbeater, B. S. C. 1986. Scale–case construction in *Synura petersenii* Korsh. (Chrysophyceae), in: Kristiansen, J., Andersen, R. A., Eds., Chrysophytes: Aspects and problems. Cambridge University Press, pp. 121–131.

Leadbeater, B. S. C. 1990. Ultrastructure and assembly of the scale case in *Synura* (Synurophyceae Andersen). British Phycological Journal 25:117–132.

Leadbeater, B. S. C., Barker, D. A. N. 1995. Biomineralization and scale production in the Chrysophyta, in: Sandgren, C. D., Smol, J. P., Kristiansen, J., Eds., Chrysophyte algae: Ecology, phylogeny, and development. Cambridge University Press, pp. 141–164.

Lott, A. M., Siver, P. A., Marsicano, L. J., Kodama, K. P., Moeller, R. E. 1994. The paleolimnology of a small waterbody in the Pocono Mountains of Pennsylvania, USA: Reconstructing 19th–20th century specific conductivity trends in relation to changing land use. Journal of Paleolimnology 12: 75–86.

Matvienko, A. M. 1941. A contribution to the taxonomy of the genus *Mallomonas*. Proceedings of the Institute of Botany Kharkov 4:41–47.

McGrory, C. B., Leadbeater, B. S. C. 1981. Ultrastructure and deposition of silica in the Chrysophyceae, in: Simpson, T. L., Volcani, B. E., Eds., Silicon and siliceous structures in biological systems. Springer-Verlag, New York, pp. 201–230.

McKenzie, C., Kling, H. 1989. Scale-bearing Chrysophyceae (Mallomonadaceae and Paraphysomonadaceae) from Mackenzie Delta area lakes, Northwest Territories, Canada. Nordic Journal of Botany 9: 103–112.

Mignot, J. P., Brugerolle, G. 1982. Scale formation in chrysomonad flagellates. Journal of Ultrastructure Research 81:13–26.

Moestrup, O. 1995. Current status of chrysophyte 'splinter groups': Synurophytes, pedinellids, silicoflagellates, in: Sandgren, C. D., Smol, J. P., Kristiansen, J., eds. Chrysophyte algae: Ecology, phylogeny and development. Cambridge University Press, pp, 75–91.

Momeu, L., Péterfi, L. S. 1979. Taxonomy of Mallomonas based on the fine structure of scales and bristles. Contributions to Botany Cluj-Napoca 1979:13–20.

Nicholls, K. H. 1978. Jensen staining, a neglected tool in phycology. Transactions of the American Microscopy Society 97:129–132.

Nicholls, K. H. 1982. *Mallomonas* species (Chrysophyceae) from Ontario, Canada including descriptions of two new species. Nova Hedwigia. 36:89–124.

Nicholls, K. H. 1987. The distinction between *Mallomonas acaroides* var. *acaroides* and *Mallomonas acaroides* var. *muskokana* var. nova (Chrysophyceae). Canadian Journal of Botany 65:1779–1784.

Nicholls, K. H. 1988a. Additions to the *Mallomonas* (Chrysophyceae) flora of Ontario, Canada, and a checklist of North American *Mallomonas* species. Canadian Journal of Botany 66:349–360.

Nicholls, K. H. 1988b. Descriptions of three new species of *Mallomonas* (Chrysophyceae): *M. hexagonis* sp. nov., *M. liturata* sp. nov., and *M. galeiformis* sp. nov. British Phycological Journal 23:159–166.

Nicholls, K. H. 1995. Chrysophyte blooms in the plankton and neuston of marine and freshwater systems, *in*: Sandgren, C.D., Smol, J. P., Kristiansen, J., Eds., Chrysophyte algae: Ecology, phylogeny and development. Cambridge University Press, pp. 181–213.

Nicholls, K. H., Gerrath, J. F. 1985. The taxonomy of *Synura* (Chrysophyceae) in Ontario with special reference to taste and odour in water supplies. Canadian Journal of Botany 63:1482–1493.

Norris, R. E., Munch, C. S. 1970. Studies on *Chrysodidymus*, a very rare chrysophyte. Journal of Phycology (Supplement) 6:4.

Nygaard, G. 1978. Freshwater phytoplankton from Narssaq area, South Greenland. Botaniske Tidsskrifter 73:191–238.

Pascher, A. 1925. Neue oder wenig bekannte Flagellaten. Archiv für Protistenkunde 52:565–584.

Péterfi, L. S., Momeu, L. 1977. Remarks on the taxonomy of some *Synura* species based on the fine structure of scales. Stud. Comun. St. Nat. 21:15–23.

Péterfi, L. S. 1996. A revised approach to the numerical taxonomy of *Mallomonas* (Synurophyceae), *in*: Kristiansen, J., Cronberg, G., Eds., Chrysophytes: Progress and new horizons. Beiheft Zur Nova Hedwigia, Vol. 114. Cramer, Berlin, pp. 57–69.

Petersen, J. B., Hansen, J. B. 1956. On the scales of some *Synura* species. I. Biologiske Meddelelser udgivet of det Kongelige danske Videnskabernes Selskab. 23(2):1–27.

Petersen, J. B., Hansen, J. B. 1958. On the scales of some *Synura* species. II. Biologiske Meddelelser udgivet of det Kongelige Danske Videnskabernes Selskab. 23(7):1–14.

Pick, F. R., Cuhel, R. L. 1986. Light quality effects on carbon and sulfur uptake of a metalimnetic population of the colonial chrysophyte *Chrysosphaerella longispina*, *in*: Kristiansen, J., Andersen, R. A., Eds., Chrysophytes: Aspects and problems. Cambridge University Press, pp. 197–206.

Pick, F. R., Nalewajko, C., Lean, D. R. S. 1984. The origin of a metalimnetic peak. Limnology and Oceanography 29:125–134.

Pipes, L. D., Leedale, G. F. 1992. Scale formation in *Tessellaria volvocina* (Synurophyceae). British Phycology Journal 27:1–19.

Pipes, L. D., Leedale, G. F., Tyler, P. A. 1991. Ultrastructure of *Tessellaria volvocina* (Synurophyceae). British Phycological Journal 26:259–278.

Preisig, H. R., Hibberd, D. J. 1982a. Ultrastructure and taxonomy of *Paraphysomonas* (Chrysophyceae) and related genera. I. Nordic Journal of Botany 2:397–420.

Preisig, H. R., Hibberd, D. J. 1982b. Ultrastructure and taxonomy of *Paraphysomonas* (Chrysophyceae) and related genera. II. Nordic Journal of Botany 2:601–638.

Preisig, H. R., Hibberd, D. J. 1983. Ultrastructure and taxonomy of *Paraphysomonas* (Chrysophyceae) and related genera III. Nordic Journal of Botany 3:695–723.

Preisig, H. R., Hibberd, D. J. 1986. Classification of four genera of Chrysophyceae bearing silica scales in a family separate from *Mallomonas* and *Synura*, *in*: Kristiansen, J., Andersen, R.A., Eds., Chrysophytes: Aspects and problems. Cambridge University Press, pp. 71–74.

Prowse, G. A. 1962. Further Malayan freshwater flagellata. Garden Bulletin (Singapore) 19:105–140.

Puytorac, P., Mignot, J. P., Grain, J., Groliere, C. A., Bonnet, L., Couillard, P. 1972. Premier releve de certains de protozoaires libres sure le territoire de la station de biologie de l'universite de Montreal (Saint-Hippolyte, Comte de Terrebonne, Quebec). Naturaliste Canadien (Quebec) 99:417–440.

Roijackers, R. M. M., Kessels, H. 1986. Ecological characteristics of scale-bearing Chrysophyceae from the Netherlands. Nordic Journal of Botany 6:373–383.

Saha, L. C., Wujek, D. E. 1990. Scale-bearing chrysophytes from tropical Northeast India. Nordic Journal of Botany 10:343–354.

Salonen, K., Jokinen, S. 1988. Flagellate grazing on bacteria in a small dystrophic lake. Hydrobiologia 161:203–209.

Sandgren, C. D. 1981. Characteristics of sexual and asexual resting cyst (statospore) formation in *Dinobryon cylindricum* Imhof. Journal of Phycology 17:199–210.

Sandgren, C. D. 1988. The ecology of chrysophyte flagellates: Their growth and perennation strategies as freshwater phytoplankton, *in*: Sandgren, C. D., Ed., Growth and reproductive strategies of freshwater phytoplankton. Cambridge University Press, pp. 9–104.

Sandgren, C. D. 1989. SEM investigations of statospore (stomatocyst) development in diverse members of the Chrysophyceae and Synurophyceae. Beihefte zur Nova Hedwigia 95:45–69.

Sandgren, C. D. 1991. Chrysophyte reproduction and resting cysts: A paleolimnologist's primer. Journal of Paleolimnology 5:1–9.

Sandgren, C. D., Carney, H. J. 1983. A flora of fossil Chrysophycean cysts from the recent sediments of Frains Lake, Michigan, U.S.A. Nova Hedwigia 38:129–163.

Sandgren, C. D., Flanagin, J. 1986. Heterothalic sexuality and density dependent encystment in the Chrysophycean alga *Synura petersenii* Korsh. Journal of Phycology 22:206–216.

Sandgren, C. D., Walton, W. E. 1995. The influence of zooplankton herbivory on the biogeography of chrysophyte algae, *in*: Sandgren, C. D., Smol, J. P., Kristiansen, J., Eds., Chrysophyte algae: Ecology, phylogeny and development. Cambridge University Press, pp. 269–303.

Santos, L.M.A., Leedale, G.F. 1993. Silica-scaled Chrysophytes from Portugal. Nordic Journal of Botny 13:707–716.

Schindler, D. W., Kling, H., Schmidt, R. V., Prokopowich, J., Frost, V. E., Reid, R. A., Capel, M. 1973. Eutrophication of Lake 227 by addition of phosphate and nitrate: The second, third and fourth years of enrichment, 1970, 1971, and 1972. Journal of the Fisheries Research Board of Canada 30:1415–1440.

Siver, P. A. 1987. The distribution and variation of *Synura* species (Chrysophyceae) in Connecticut, USA. Nordic Journal of Botany 7:107–116.

Siver, P. A. 1988a. Distribution of scaled chrysophytes in 17 Adirondack (New York) lakes with special reference to pH. Canadian Journal of Botany 66:1391–1403.

Siver, P. A. 1988b. *Mallomonas retrorsa*, a new species of silica-scaled Chrysophyceae with backwards orientated scales. Nordic Journal of Botany 8:319–323.

Siver, P. A. 1988c. A new form of the common chrysophycean alga *Synura petersenii*. Transactions of the American Microscopy Society 107:380–385.

Siver, P. A. 1988d. Morphology and ecology of *Mallomonas galeiformis* (Chrysophyceae), a potentially useful paleolimnological indicator. Transactions of the American Microscopy Society 107:152–161.

Siver, P. A. 989. The distribution of scaled chrysophytes along a pH gradient. Canadian Journal of Botany 67:2120–2130.

Siver, P. A. 1991a. The biology of Mallomonas: Morphology, taxonomy and ecology. Kluwer, Dordrecht, 230 p.

Siver, P. A. 1991b. Improving paleolimnological inference models utilizing scale-bearing siliceous algae: Transforming scale counts to cell counts. Journal of Paleolimnology 5:219–225.

Siver, P. A. 1991c. The stomatocyst of *Mallomonas acaroides* v. *muskokana* (Chrysophyceae). Journal of Paleolimnology 5:11–17.

Siver, P. A. 1992. A critical literature review on the usefulness of chrysophytes to detect lake chronic or episodic acidification and/or recovery in the context of a long term monitoring program. U.S. Environmental Protection Agency, Corvallis, OR.

Siver, P. A. 1993. Inferring the specific conductivity of lake water with scaled chrysophytes. Limnology and Oceanograpy 38:1480–1492.

Siver, P. A. 1994. *Mallomonas wujekii*, a new species of Synurophyceae from Florida, USA. Nordic Journal of Botany 14:467–471.

Siver, P. A. 1995. The distribution of chrysophytes along environmental gradients: Their use as biological indicators, *in*: Sandgren, C. D., Smol, J. P., Kristiansen, J., Eds., Chrysophyte algae: Ecology, phylogeny and development. Cambridge University Press, pp. 232–268.

Siver, P. A., Chock, J. S. 1986. Phytoplankton dynamics in a chrysophycean lake, *in*: Kristiansen, J., Andersen, R. A., Eds., Chrysophytes: Aspects and problems. Cambridge University Press, pp. 165–183.

Siver, P. A., Glew, J. R. 1990. The arrangement of scales and bristles on *Mallomonas* (Chrysophyceae): A proposed mechanism for the formation of the cell covering. Canadian Journal of Botany 68:374–380.

Siver, P. A., Hamer, J. S. 1989. Multivariate statistical analysis of the factors controlling the distribution of scaled chrysophytes. Limnology and Oceanography 34:368–381.

Siver, P. A., Hamer, J. S. 1990. Use of extant populations of scaled chrysophytes for the inference of lakewater pH. Canadian Journal of Fisheries and Aquatic Science 47:1339–1347.

Siver, P. A., Hamer, J. S. 1992. Seasonal periodicity of Chrysophyceae and Synurophyceae in a small New England lake: Implications for paleolimnological research. Journal of Phycology 28:186–198.

Siver, P. A., Marsicano, L. J. 1993. *Mallomonas connensis sp. nov.*, a new species of Synurophyceae from a small New England lake, U.S.A. Nordic Journal of Botany 13:337–342.

Siver, P. A., Marsicano, L. J. 1996. Inferring lake trophic status using scaled chrysophytes, *in*: Kristiansen, J., Cronberg, G., Eds., Chrysophytes: Progress and new horizons. Beihefte zur Nova Hedwigia Vol. 1J4. Cramer, Berlin, pp. 233–236.

Siver, P. A., Skogstad, A. 1988. Morphological variation and ecology of *Mallomonas crassisquama* (Chrysophyceae). Nordic Journal of Botany 8:99–107.

Siver, P. A., Smol, J. P. 1993. The use of scaled chrysophytes in long term monitoring programs for the detection of changes in lakewater acidity. Water, Air, and Soil Pollution 71:357–376.

Siver, P. A., Vigna, M. S. 1997. The distribution of scaled chrysophytes in the delta region of the Paraná River, Argentina. Nova Hedwigia 64:421–453.

Siver, P. A., Wujek, D. E. 1993. Scaled Chrysophyceae and Synurophyceae from Florida: IV. The flora of Lower Lake Myakka and Lake Tarpon. Florida Science 56:109–117.

Siver, P. A., Wujek, D. E. 1999. Scaled Chrysophyceae and Synurophyceae from Florida, U.S.A.: VI. Observations on the flora from waterbodies in the Ocala National Forest. Nova Hedwigia 68:75–92.

Siver, P. A., Hamer, J. S., Kling, H. 1990. Separation of *Mallomonas duerrschmidtiae* sp. nov. from *M. crassisquama* and *M. pseudocoronata*: Implications for paleolimnological research. Journal of Phycology 26:728–740.

Siver, P. A., Lott, A. M., Cash, E., Moss, J., Marsicano, L. J. 1999. Century changes in Connecticut, U.S.A., lakes as inferred from siliceous algal remains and their relationships to land-use change. Limnology and Oceanography 44:1928–1935.

Skogstad, A. 1984. Vegetative cells and cysts of *Mallomonas intermedia* (Mallomonadaceae, Chrysophyceae). Nordic Journal of Botany 4:275–278.

Skuja, H. 1950. Körperbau und reproduktion bei *Dinobryon borgei* Lemm. Svensk Botanisk Tidskrift 44:96–107.

Skuja, H. 1956. Taxonomische und biologische Studien über das Phytoplankton schwedischer Binnengewässer. Nova Acta Regiae Societatis Scientiarum Upsaliensis Ser. 4 16(3).

Smol, J. P. 1995. Application of chrysophytes to problems in paleoecology, *in*: Sandgren, C. D., Smol, J. P., Kristiansen, J., Eds., Chrysophyte algae: Ecology, phylogeny and development. Cambridge University Press, pp. 232–250.

Smol, J. P., Charles, D. F., Whitehead, D. R. 1984. Mallomonadacean (Chrysophyceae) assemblages and their relationships with limnological characteristics in 38 Adirondack (New York) lakes. Canadian Journal of Botany 62:911–923.

Starmach, K. 1985. Chrysophyceae und Haptophyceae, *in*: Ettl, H., Gerloff, J., Heynig, H., Mollenhauer, D., Eds., Susswasserflora von Mitteleuropa 1. Fischer, Jena, 515 p.

Takahashi, E. 1978. Electron microscopical studies of the Synuraceae (Chrysophyceae) in Japan. Taxonomy, and ecology. Tokai University Press, Tokyo, 194 pp.

Takahashi, E., Hayakawa, T. 1979. The Synuraceae (Chrysophyceae) in Bangladesh. Phykos 18:129–147.

Taylor, W. D., Wee, J. L, Wetzel, R. G. 1986. A modification of the Ütermohl sedimentation technique for improved identification and cell enumeration of diatoms and silica-scaled Chrysophyceae. Transactions of the American Microscopy Society 105:68–72.

Tyler, P. A., Pipes, L. D., Croome, R. L., Leedale, G. F. 1989. *Tessellaria volvocina* rediscovered. British Phycological Journal 24: 329–337.

Wawrik, F. 1979. Eisscluss- und Eisbruchvegetationen in den Teichen des nördlichen Waldiertels 1977/1978. Archiv für Protistenkunde 122:247–266.

Wawrzyniak, L. A., Andersen, R. A. 1985. Silica-scaled Chrysophyceae from North America boreal forest regions in northern Michigan, U.S.A. and Newfoundland, Canada. Nova Hedwigia 5:399–401.

Wee, J. L. 1982. Studies on the Synuraceae (Chrysophyceae) of Iowa. Bibliotheca Phycologica 62:1–183.

Wee, J. L. 1983. Specimen collection and preparation for critical light microscope examination of Synuraceae (Chrysophyceae). Transactions of the American Microscopy Society 102:68–676.

Wee, J. L. 1997. Scale biogenesis in Synurophycean protists: Phylogenetic implications. Critical Reviews in Plant Science 16:497–534.

Wee, J. L., Gabel, M. 1989. Occurrences of silica-scaled chromophyte algae in predominantly alkaline lakes and ponds in Iowa. American Midland Naturalist 121:32–40.

Wee, J. L., Booth, D. J., Bossier, M. A. 1993. Synurophyceae from the Southern Atlantic Coastal Plain of North America: A preliminary survey in Louisiana, USA. Nordic Journal of Botany 13:95–106.

Wee, J. L., Harris, S. A., Smith, J. P., Dionigi, C. P., Millie, D. F.

1994. Production of the taste/odor-causing compound, *trans-2, cis-6* nonadienal, within the Synurophyceae. Journal of Applied Phycology 6:365–369.

Wetherbee, R., Ludwig, M., Koutoulis, A. 1995. Immunological and ultrastructural studies of scale development and deployment in *Mallomonas and Apedinella*, *in*: Sandgren, C. D., Smol, J. P., Kristiansen, J., Eds., Chrysophyte algae: Ecology, phylogeny and development. Cambridge University Press, pp 165–178.

Wetzel, R. G. 1983. Limnology, 2nd ed. Saunders, New York, 765 p.

Wujek, D. E. 1984. Chrysophyceae (Mallomonadaceae) from Florida. Florida Science 47:161–170.

Wujek, D. E., Bicudo, C. E. 1993. Scale-bearing chrysophytes from the state of Sao Paulo, Brazil. Nova Hedwigia 56:247–257.

Wujek, D. E., Bland, R. G. 1991. Chrysophyceae (Mallomonadaceae and Paraphysomonadaceae) from Florida III. Additions to the flora. Florida Science 54:41–48.

Wujek, D. E., Hamilton, R. 1972. Studies on Michigan Chrysophyceae. I. Michigan Botanist 11:51–59.

Wujek, D. E., Hamilton, R. 1973. Studies on Michigan Chrysophyceae. II. Michigan Botanist 12:118–122.

Wujek, D. E., Igoe, M. J. 1989. Studies on Michigan Chrysophyceae. VII. Beihefte zur Nova Hedwigia 95:69–280.

Wujek, D. E., Kristiansen, J. 1978. Observations on bristle and scale production in *Mallomonas caudata*. Archiv für Protistenkunde 120:213–221.

Wujek, D. E., Timpano, P. 1984. The genus *Mallomonopsis* in the United States. Transactions of the Kansas Academy of Science 87:73–82.

Wujek, D. E., Wee, J. L. 1983. *Chrysodidymus* in the United States. Transactions of the American Microscopy Society 102:77–80.

Wujek, D. E., Weis, M. M. 1984. Mallomonadaceae from Kansas. Transactions of the Kansas Academy of Science 87:26–31.

Wujek, D. E., Hamilton, R., Wee, J. 1975. Studies on Michigan Chrysophyceae. III. Michigan Botanist 14:91–94.

Wujek, D. E., Gretz, M., Wujek, M.G. 1977. Studies on Michigan Chrysophyceae. IV. Michigan Botanist 16:191–194.

Yin-Xin, W., J. Kristiansen. 1994. Occurrence and distribution of silica-scaled chrysophytes in Zhejiany, Jiangsu, Hubei, Yunnan and Shandong Provinces, China. Archiv für Protistenkunde 144: 433–449.

Zeeb, B. A., Smol, J. P. 1991. Paleolimnological investigation of the effects of road salt seepage on scaled chrysophytes in Fonda Lake, Michigan. Journal of Paleolimnology 5:263–266.

Zeeb, B. A., Smol, J. P. 1995. A weighted-averaging regression and calibration model for inferring lakewater salinity using chrysophycean stomatocysts from lakes in western Canada. International Journal of Salt Lake Research 4:1–23.

15

CENTRIC DIATOMS

Eugene F. Stoermer
Michigan Herbarium
University of Michigan
Ann Arbor, Michigan 48109

Matthew L. Julius
Department of Biological Sciences
St. Cloud State University
St. Cloud, Minnesota 56301

I. General Introduction to the Diatoms
    A. Introduction
    B. Diversity and Morphology
    C. Collection and Preparation for Identification
II. Introduction to Centric Diatoms
III. Classification
IV. Morphology and Physiology
    A. Frustular Morphology
    B. Cytoplasmic Features
    C. Environmental Physiology
V. Ecology and Evolution
    A. Diversity and Distribution
    B. Reproduction and Life Histories
    C. Ecological Interactions
    D. Evolutionary Relationships
VI. Collection and Study Methods
VII. Key and Descriptions of Genera
    A. Key
    B. Descriptions of Genera
VIII. Guide to Literature for Species Identification
Literature Cited

I. GENERAL INTRODUCTION TO THE DIATOMS
by J. P. Kociolek* and S. A. Spaulding*

A. Introduction

Diatoms are unicellular algae that possess an unusual, and unique, feature of a cell wall composed of silicon dioxide (SiO_2). The opaline silica cell wall, or frustule, is made up of two parts. These parts are called valves, and one valve fits inside the other valve, just as two halves of a pill box or Petri dish might fit together to form a whole container. "Diatom" is a word meaning to split into two, and this term was first applied by early microscopists who were intrigued by these organisms living in nearly every body of water, within microscopic glass boxes.

The earliest known fossil diatoms are from the Cretaceous period, and they are thought to have evolved from a chrysophyte ancestor in silica-rich oceans (Round *et al.*, 1990; Harwood and Nikolaev, 1995). Two major evolutionary lineages are generally recognized: the centric diatoms (described in this chapter) and and the pennate diatoms (described in Chapters 16–19). Centric diatoms are symmetrical in the valve view in more than two planes and many are radial in symmetry; they also have oogamous sexual reproduction. Pennate diatoms are symmetrical about a line, termed "bilateral symmetry," and produce ameboid gametes.

Like many other algal divisions, the life cycles, chloroplast type and pigments, and cytoplasmic features of diatoms are used to develop a classification that reflects evolutionary relationships. Nearly all diatoms are diploid in the vegetative portion of their life cycle. However, great differences in the sexual portion of their life cycle are found among taxa. Variation among taxa is found in meiosis, gamete number, gamete type, gamete behavior, and zygote development (Round *et al.*, 1990). Futhermore, several types of autogamy, or self-fertilization, occur within the diatoms.

The silica cell wall is rigid, and once it has been deposited by a cell undergoing vegetative division, it cannot change in shape or size. Such an inflexible cell wall is unique within living organisms and has novel implications for diatom life history (Kociolek and

---

*Diatom Collection; California Academy of Sciences; Golden Gate Park; San Francisco, California 94118.

Williams, 1987). Mean cell size decreases in a population through the vegetative life cycle, because each valve of a frustule produces a smaller complementary valve. This decrease in cell size, or size diminution series, is accompanied by changes in valve outline and, less frequently, ornamentation (Stoermer and Ladewski, 1982; Stoermer et al., 1986). In most taxa, cell size decreases with each successive vegetative division, until it is within a range where environmental parameters may induce sexual reproduction (Edlund and Stoemer, 1997). Gametes are produced and join to form a zygote, which greatly expands in size to form an auxospore. The auxospore is able to expand because the cell wall is only lightly silicified with bands, or scales. The first valve with a normal morphology, or "Erstlingzelle," formed by the division of the auxospore, is the largest valve in the cell cycle. Hence, in diatoms, sexual reproduction not only increases genetic variability but also results in the enlargement of cells to a maximum size. Most diatomists consider the size diminution of diatoms a consequence of having rigid cell walls, but it has also been interpreted as a timing mechanism or adaptation for sexual reproduction (Lewis, 1984). For thorough reviews of diatom sexual reproduction, see Geitler (1973), Round et al. (1990), Mann (1993), and Edlund and Stoermer (1997).

Diatom chloroplasts are characterized by possession of chlorophyll-*a* and -*c* and the primary accessory pigments, β-carotene and fucoxanthin (van den Hoek et al., 1995). Chloroplast number and arrangement differ among taxa but are consistent within most taxa (Cox, 1996). Most centric diatoms and some araphid pennate diatoms possess numerous small, disc-shaped chloroplasts. In contrast, the chloroplasts of the majority of pennate taxa are large and few per cell. Nevertheless, chloroplasts number and shape may vary at different stages of the cell cycle. Cox (1996) provides a key to freshwater diatoms based on chloroplasts and other cytoplasmic features visible in the light microscope.

The rigid, silica cell walls of diatoms are often highly ornamented, and much of diatom taxonomy has been based on their morphological features. Critical examination of diatom valves with light and electron microscopy has led to a proliferation of terms associated with the minutiae of valve morphology. Given the excellent treatments of Round *et al.* (1990) and Krammer and Lange-Bertalot (1986, 1988, 1991a, b) with regard to diatom ultrastructure, details will not be repeated here. Terminology necessary to utilize the keys and descriptions of genera is presented in the glossary.

In most cases, vegetative cell division results in a high degree of fidelity in valve morphology. Nevertheless, variability in morphology occurs within and between cell lines, as well as within and between populations. This variability has important implications for the identification of diatoms at the species and subspecific levels. Diatom taxonomists usually distinguish taxa based on a lack of intermediates between ranges of variability (Theriot and Stoermer, 1984b; Kociolek and Stoermer, 1988a). In rare cases, heterovalvy occurs within individual cells. In such individuals, the two valves of a frustule may appear to possess the ornamentation of two taxa, even though no intermediate morphologies occur (Stoermer, 1967). In such example, the two "taxa" must be considered as one (Kociolek, 1997). Moreover, differences in environmental conditions (i.e., salt and silica concentrations) may result in alteration of morphology within a taxon (Tuchman et al., 1984). Therefore, size diminution in valve length and width, associated changes in valve outline and morphology with size diminution, morphological variation within populations, and the influence of environment conditions must all be considered when species-level identifications of diatoms are made.

Diatoms live in most fresh waters, in a wide range of aquatic habitats. As a group, they are found across a wide gradient of pH values, within wide concentrations of solutes, nutrients, and organic and inorganic contaminants, and across a range of water temperatures. Individual species are often restricted to specific ecological conditions. Although species, rather than genera, show the strongest relation to limnological parameters, a broad view of microhabitat preferences will be discussed in the following chapters. In addition to ecological parameters, the distribution of diatoms is also related to their historical distribution within geographical regions. Within regions, the assemblages of diatom species may be characteristic of physical, chemical and biological conditions. For example, characteristic taxa are found in acid waters of eastern North America (Battarbee et al., 1999) and in saline lakes of western Canada (Cummins et al., 1995). Consideration of both ecological preferences and historical distributions is likely to provide the most accurate understanding of diatom occurrences.

Because diatom species are specific to the habitats in which they grow, they are valuble as environmental indicators (Stoermer and Smol, 1999, Chap. 23). The use of diatoms as indicators of physical, chemical, and biotic conditions has a long tradition, dating back to the early part of the 20th century (Kolkwitz and Marson, 1908). Diatoms continue to be important components of water quality assessment programs, such as the National Water Quality Assessment Program of the U.S. Geological Survey. They are also powerful tools in the field of paleolimnology, as records in the reconstruction of limnological, watershed, and climatic history. Diatoms can be used to interpret past conditions, because the silica cell walls are often well preserved in lake sediments (Stoermer *et al.*, 1985c; Bradbury, 1988; Smol, 1990).

Fossil diatoms are widespread in lacustrine deposits dating to the late Eocene/earliest Oligocene ages (Lohman and Andrews, 1968; McKnight *et al.*, 1995). Extensive fossil deposits are termed "diatomaceous earth" and are mined for numerous commercial uses (Harwood, 1999). Whereas the largest deposits of diatomaceous earth are of marine origin (e.g., those of Lompoc, CA), large deposits of freshwater fossils are found across North America. Post-Glacial age deposits can be found in eastern Canada (Nova Scotia, New Brunswick, Quebec, and Ontario) (Eardly-Wilmot, 1928; Boyer, 1926), whereas older deposits (post-Glacial and Tertiary age) are found across western North America (Eardley-Wilmot, 1928; Moore, 1937) and Mexico (Ehrenberg, 1843, 1854, 1870).

B. Diversity and Morphology

To catalogue the biodiversity of freshwater diatoms in North America would require an exhaustive effort, an accomplishment that will not be realized in the near future. To date, the literature indicates that the number of taxa is great and the flora is diverse. Estimates of the number of diatom taxa from smaller geographic regions have been studied in some detail, and the floristic diversity of those regions is impressive. For example, nearly 900 subgeneric taxa are reported from fresh waters of the Arctic and Antarctic (Hamilton *et al.*, 1994; Håkansson and Jones, 1994). The diatom flora of central Europe was estimated to be approximately 1600 species (Krammer and Lange-Bertalot, 1986, 1988, 1991a, b), but that estimate may be quite conservative. Recently, Lange-Bertalot and Metzeltin (1996) identified over 800 taxa from just three lakes in northern Europe and nearly 20% of the species were previously unknown.

Estimates of the taxonomic breadth and diversity of the freshwater diatoms in the North American flora are hampered not only by a lack of information (and a toxonomic impediment that is perhaps equal to that in other algal groups) but also by the current state of flux that defines many groups. A testament to this is the large number of revisions and reinterpretations of several major groups within the centrics (e.g., Håkansson and Kling, 1990, 1994), araphids (Williams and Round, 1986, 1987; Krammer and Lange-Bertalot, 1991a), monoraphids (Round, 1996), naviculoids (Lange-Bertalot, 1993; Round *et al.*, 1990), symbelloids (Kammer, 1997a, b), and nitzschioids (Round *et al.*, 1990). Justification for each approach to the nomenclature herein is presented in the individual chapters of this work. Overall we consider 115 genera to be part of the recent freshwater diatom flora of North America.

Our knowledge of the freshwater diatoms of North America is scanty at best. Overarching treatments include those of Wolle (1890) who reported just under 500 freshwater taxa, Boyer (1927a, b) who reported about 525 freshwater forms, and Patrick and Reimer (1966, 1975) whose treatment of taxa within a narrow group of diatoms included 829 taxa (and a list of 63 and 18 additional taxa reported after 1960 and 1971, respectively). Regional compilations include studies from the Laurentian Great Lakes [Stoermer and Kreis (1978), who reported 1530 taxa], Kentucky [Camburn (1982), 514 taxa], Montana [Prescott and Dillard (1979), 542 taxa], Illinois [Dodd (1987), 424 taxa], Alaska [Patrick and Freese (1961), 499 taxa], northern Canada and Alaska (Foged, 1953, 1955, 1973, 1981), and Nebraska [Elmore (1922), 234 taxa]. Ecological characterizations of diatom taxa have been compiled in a few works (Lowe, 1974; Beaver, 1981). Regional ecologically focused studies include work on inland saline lakes of British Columbia (Cummins *et al.*, 1995), Patrick's reviews of several U.S. rivers [Patrick (1961), 280 taxa from eight rivers; Patrick and Roberts (1979), 236 taxa from five rivers], work on three rivers in eastern North America [Hohn and Hellerman (1963), 476 taxa], and the PIRLA study [Camburn *et al.* (1984–1986), 451 taxa] of diatoms in sediments from acid lakes in eastern and midwestern United States. Although compilation of these cited studies and the numerous other individual floristic studies has been published, Patrick's investigations suggest that the distribution of many taxa is limited. This list of works is impressive; however, much of North America has not been surveyed critically for freshwater diatoms. This includes, but is not limited to, the southeastern United States, the Rocky Mountain region, the mountains of the Pacific Northwest, most of California (including Lake Tahoe), much of Alaska, northern Canada, the southwestern United States, and almost all of Mexico.

J. P. Kociolek and C. W. Reimer (unpublished observations) suggest that over 30% of the species present are undescribed in areas where the diatoms have not been treated previously (Sierra Nevada mountains of California, Ozark Plateau, and rivers in the southeastern United States). This estimate is on par with those given for more exotic regions of the world, including the East African Rift Valley lakes (Ross, 1983). Given huge number of taxa already reported from mostly U.S. sites, the land areas almost unexplored, and the significant number of taxa potentially awaiting description, it is not unreasonable to estimate that the freshwater diatom flora of North America may easily exceed 7500 taxa.

Although the diversity of forms is reflected in a complex and variable morphology and specialized jargon has developed to describe the morphology of each of the major groups of freshwater diatoms, there is some terminology that may be applied in a general way to all diatoms. These include terms associated with the overall organization of the siliceous frustules and axes and planes of symmetry for the cells.

C. Collection and Preparation for Identification

Although Cox (1996) offers keys and features to use in the identification of diatoms with live and preserved material, it is the authors' experience that reliable identification of diatoms at the level of species and below is only accomplished by cleaning the samples and preparing permanently mounted slides. Many works have discussed the techniques to collect and prepare diatoms for identification. The overview provided in Patrick and Reimer (1966) has proven useful to many individuals and laboratories. With the recent closure of Custom Research and Development, commercial makers of Hyrax (which was the mounting medium of choice, because of its ease of use and its having reliably stood the test of time for slides over 50 years old), most laboratories have settled on use of Naphrax (Northern Biological Supply, Ipswich, UK) to make permanent slides. It has characteristics that make it less desirable than Hyrax (discoloration and greater britleness), and its "staying power" in terms of holding up over time in collections remains to be tested.

For the identification of diatom genera and species, two excellent taxonomic works have been mainstays in the libraries of diatomists on this continent. Patrick and Reimer's monographs, *The Diatoms of the United States* (1966, 1975), treat many araphid and raphid diatom genera. These works do not, however, cover the centrics or most of the problematic keel-bearing taxa (e.g., *Nitzschia*), and they have been complemented on many bookshelves by the classic work by Hustedt (1930). This latter work has been updated in a four-volume set by Krammer and Lange-Bertalot (1986, 1988, 1991a, b). Other works of utility to freshwater ecologists include those of Cleve-Euler (1951, 1952, 1953a, b, 1955), Hustedt (1927–1930; 1931–1959), Simonsen (1987), and Van Heurck (1880, 1881, 1882, 1883, 1884, 1885).

The present chapters on diatoms deviate somewhat from past taxonomic treaments, particularly those included in large treatises of feshwater algae in North America (e.g., Smith, 1950; Prescott, 1962). These treatises relied heavily on older treatments of diatoms (e.g., Boyer, 1927a, b) or gave diatoms only peripheral consideration (e.g., Prescott's work). The taxonomy of diatoms has undergone a revolution in the last 15 years, with many new taxa (particularly at the genus level) being proposed, based on ultrastructural characteristics. Many of these have gained acceptance by diatom taxonomists, but have not yet found their way into general usage in the phycological literature. The work by Round *et al.* (1990) has led to the recognition of many new taxa, several of which are followed in the present work (e.g., *Luticola* and *Sellaphora*, taxa considered part of *Navicula* in more classical treatments). And the recent work by Lange-Bertalot and colleagues (e.g., Lange-Bertalot and Metzeltin, 1996) continues this approach.

Permanent slides and original material should ultimately be deposited in museums for their long-term care and accessibility to the scientific community. In North America, active repositories include the Academy of Natural Science (Philadelphia), California Academy of Sciences (San Francisco), Canadian Museum of Nature (Ottawa), and Farlow Herbarium of Harvard University (Cambridge). Other active collections include the Natural History Museum (London), Museum National d'Histoire Naturelle (Paris), the Hustedt Collection and Alfred Wegener Institute (Bremerhaven), the Botanical Gardens-Dahlem (Berlin), the Botanical Museum and Library (Copenhagen), and the National Museum (Tokyo).

II. INTRODUCTION TO CENTRIC DIATOMS

Freshwater centric diatoms are a major component of many freshwater systems and, despite large gaps in information concerning their basic biology, strong inferences can be drawn concerning the ecological requirements of some species based on their distribution and occurrence relative to measurable ecological parameters. Centric diatoms have a very rapidly expanding and unstable taxonomy, and information concerning their physiology and life histories is fragmentary at best. In the following account, we attempt to synthesize current understanding of freshwater centric diatoms, with the understanding that rapid developments in diatom studies at all levels may render some of our conclusions less than complete or completely off the mark.

Centric diatoms are most common in marine waters. The majority of genera have no freshwater representatives and relatively few freshwater genera have species represented in strictly marine waters. As understanding of diatom systematics improves, the demarcation between fresh and salt water becomes even sharper (Round and Sims, 1981). On the basis of available evidence it appears that the major genera occurring in fresh water are ultimately derived from multiple invasions from the marine realm. Some have apparently evolved extensively in fresh water. Others retain very close morphological similarities to their marine ancestors.

III. CLASSIFICATION

As we currently understand them, freshwater centric diatoms are a polyphyletic group, composed of at least two major clades (Medlin et al., 1996). A number of characters, usually considered primitive to all diatoms, unite this classification of convenience:

- Organization of valve structure about a point, rather than about a line.
- Lack of significant motility.
- Oogamous sexual reproduction (Fig. 1).

Major groupings contained within freshwater centric diatoms, as presently understood, are shown in Table I. Here we mostly follow the classification of Round et al., (1990) but include some of our conclusions concerning the placement of genera either erected or that have become commonly recognized after 1990.

The classification of centric diatoms is, and probably will remain for some time, very unstable. The causes for this instability are several. We are emerging from a period when diatom classification was dominated by very summary and compressed classification schemes. New technologies, such as digital computers and genetic analyses, and the general availability of electron microscopes have shown that classification systems in common use as late as the 1980s are substantially untenable. This has led to the resurrection of many genera originally described during the grand period of diatom systematics (ca. 1840–1900) but ignored or synonymized in subsequent summary treatises (e.g., Hustedt, 1930) and the description of many new genera (e.g., Round et al., 1990). Species-level classifications have undergone equally large changes. In almost all cases in which common and widely distributed species have been closely inspected, complexes of related species have been identified. This trend is visible in what is probably the most widely used general treatise on diatom classification (Krammer and Lange-Bertalot, 1986, 1988, 1991a, b). The earliest volume of this monograph is mostly in the older tradition and the final volume and certainly these authors' later publications more closely approach modern systematic thought.

This trend has led to a particularly chaotic state in

TABLE 1 A Current Classification Scheme for Freshwater Centric Diatoms

Coscinodiscophyceae
Thalassiosirophycidae
  Thalassiosirales
    Thalassiosiraceae
      Thalassiosira
      Stephanocostis
      Stephanocyclus
      Thalassiocyclus
    Skeletonemataceae
      Cyclotubicoalitus
      Skeletonema
    Stephanodiscaceae
      Crateriportula
      Cyclotella
      Cyclostephanos
      Mesodictyon
      Pelagodictyon
      Pliocaenicus
      Stephanodiscus
Coscinodiscophycidae
  Melosirales
    Melosiraceae
      Melosira
  Paraliales
    Paraliaceae
      Ellerbeckia
  Aulacoseirales
    Aulacoseiraceae
      Aulacoseira
  Orthoseirales
    Orthoseiraceae
      Orthoseira
  Coscinodiscales
    Hemidiscaceae
      Actinocyclus
Biddulphiophycidae
  Triceratiales
    Triceratiaceae
      Pleurosira
  Biddulphiales
    Biddulphiaceae
      Hydrosera
      Terpsinoë
Rhizosoleniophycidae
  Rhizosoleniales
    Rhizosoleniaceae
      Urosolenia
    Acanthoceratacaee
      Acanthoceras
Chaetocerotophycidae
  Chaetocerotales
    Chaetoceraceae
      Chaetoceros

diatom classification in general and in centric diatom classification in particular at the present time. Unfortunately, a large number of researchers are not able to devote themselves to monographic studies of the group, so this condition is liable to persist for some

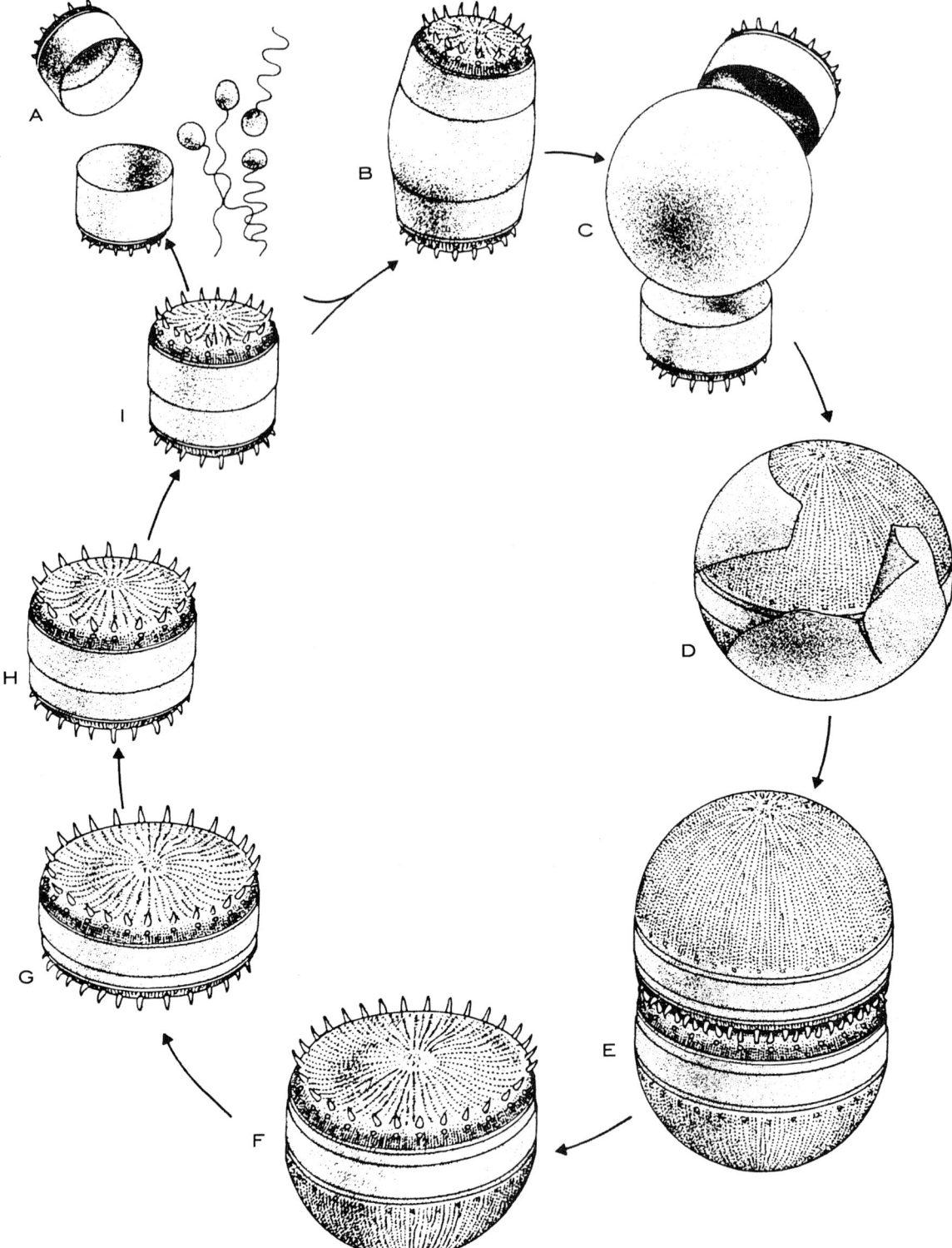

FIGURE 1 Diagram of the life cycle of *Stephanodiscus*. (A) Separated valves of "male" gametangial cell and male gametes. (B) Expanding "female" gametangial cell containing a female gamete. (C) Expanding zygote with shed gametangial cell valves. (D) Expanded auxospore with perizonium. (E–F) Initial cells resulting from first mitotic divisions. (G–I) Size diminution series during asexual reproduction. (I) Cell in size range where sexual reproduction is inducible. (Slightly modified from an original illustration by Dr. D. G. Mann from Round, F. E., Craniford, R. M., Mann, D. G. 1990. The diatoms—Biology and morphology of the genera. Cambridge University Press, Cambridge, UK. (Used with permission.)

time into the future. Students and researchers are well advised to consult recent primary literature, as exemplified by journals found in the Literature Cited of this chapter, for newly described species and new theories of relationships shown in generic and higher classifications. Because centric diatoms are of interest to ecologists and geologists as well as to phycologists, publication of new species and taxonomic revisions is scattered through many literature sources, but some important source publications are *Algologia, Bibliotheca Diatomologica, Diatom (Tokyo), Diatom Research, Journal of Phycology, Nova Hedwigia,* and *Phycologia*. Representative papers published after Krammer and Lange-Bertalot's major work was essentially completed include those of Håkansson and Kling (1989, 1990, 1994) and Prasad *et al.* (1990).

IV. MORPHOLOGY AND PHYSIOLOGY

A. Frustular Morphology

Frustules of centric diatoms are developed in a silica deposition vesicle (Drum and Pankratz, 1964), apparently derived from the Golgi bodies, similar to other diatoms. There are numerous variations in details of valve development in different groups. However, silica deposition vesicle development and subsequent silicification always proceed from a central point. In most freshwater diatoms, silicification appears to proceed from a basal siliceous layer, although this has not been verified in many taxa. Most centric diatoms that have been investigated thoroughly have at least one labiate process (Figs. 2A and 4E). The function or functions of this apparently near-universal morphological feature are not known for certain, although it may serve as an anchor point for valve morphogenesis and may possibly be involved in "recovery" of organic structures left outside newly formed valves (Schmid, 1994).

Several quite different trends in morphological specialization are present within the freshwater centric diatoms. Some (the Biddulphiophycidae in particular) are apparently descended from marine multipolar centrics that have adapted to growth attached to solid substrata. Their cells are attached to substrata and sometimes to daughter cells, forming extended colonies, by organic exudates from elevations of the valve surface bearing differentiated pore fields, usually called "ocelli" (Round *et al.*, 1990). This group is also characterized by lack of strutted processes and central displacement of labiate processes. Although commonly entrained into the plankton because of their colonial growth habit, most species are primarily benthic organisms and display other adaptations to this growth habit, such as thick and apparently strong frustules reinforced by ribs and septae.

Other centric diatoms have morphological adaptations that allow them to grow on sediment surfaces or at least survive impingement on such surfaces. Among freshwater centric taxa, most are contained in the Coscinodiscophycidae. All these taxa lack structures associated with attachment to substrata, but most have highly developed marginal spines (*Aulacoseira, Ellerbeckia,* and *Orthoseira*) and/or structures associated with secretion on the valve surface (*Melosira*), which bind the cells together in colonies (Fig. 6A). In *Aulacoseira*, cells are held together by linking spines, but some species have longer spines, on so-called separation valves (Siver and Kling, 1997), which apparently serve to limit colony length (Fig. 6E). Most species are tychoplanktonic, growing mostly in the plankton, but having the capacity to survive burial in sediments. Colony size is apparently an important factor controlling sinking rates in *Aulacoseira* (Davey, 1986), and variations in frustular morphology which affect colony size may be important in allowing the successful planktonic growth habit of this genus.

Colony formation is also exhibited in some species of euplanktonic (capable of completing their entire life cycle suspended in the water column) genera. Most species of *Chaetoceros* and *Skeletonema* are colonial, the cells in a colony being bound together by more or less elongated spines (Fig. 5H). In *Skeletonema* the degree of spine development appears to be related to salinity level (Hasle and Evensen, 1975, 1976; Paasche *et al.*, 1975). In *Chaetoceros* well-developed spines are found in all species. Within other euplanktonic genera, colonial growth occurs only in certain species and within a given species may take place only under certain conditions. Marine species of *Thalassiosira* are usually bound together into "string of beads" colonies by chitin fibrils secreted by the central strutted processes (Fig. 2A). This is less commonly observed in freshwater species, but it may occur. Within the Stephanodiscaceae, members of some genera may form colonies, but the colonial condition usually occurs only under certain environmental conditions. In *Cyclotella*, some species grow in linear colonies, with cells bound together valve face to valve face (Fig. 3C), apparently by organic secretions (e.g., *C. melosiroides*). Other species of *Cyclotella* form colonies in which the cells are randomly arranged in an organic matrix (e.g., *C. glomerata*). Some species of *Stephanodiscus* are regularly colonial (e.g., *S. binderanus*), whereas others may occur either singly or in colonies (e.g., *S. hantzschii*). In both instances, colonial growth habit is effected by overgrowth and branching of the marginal spines and development of fine projections on these branched

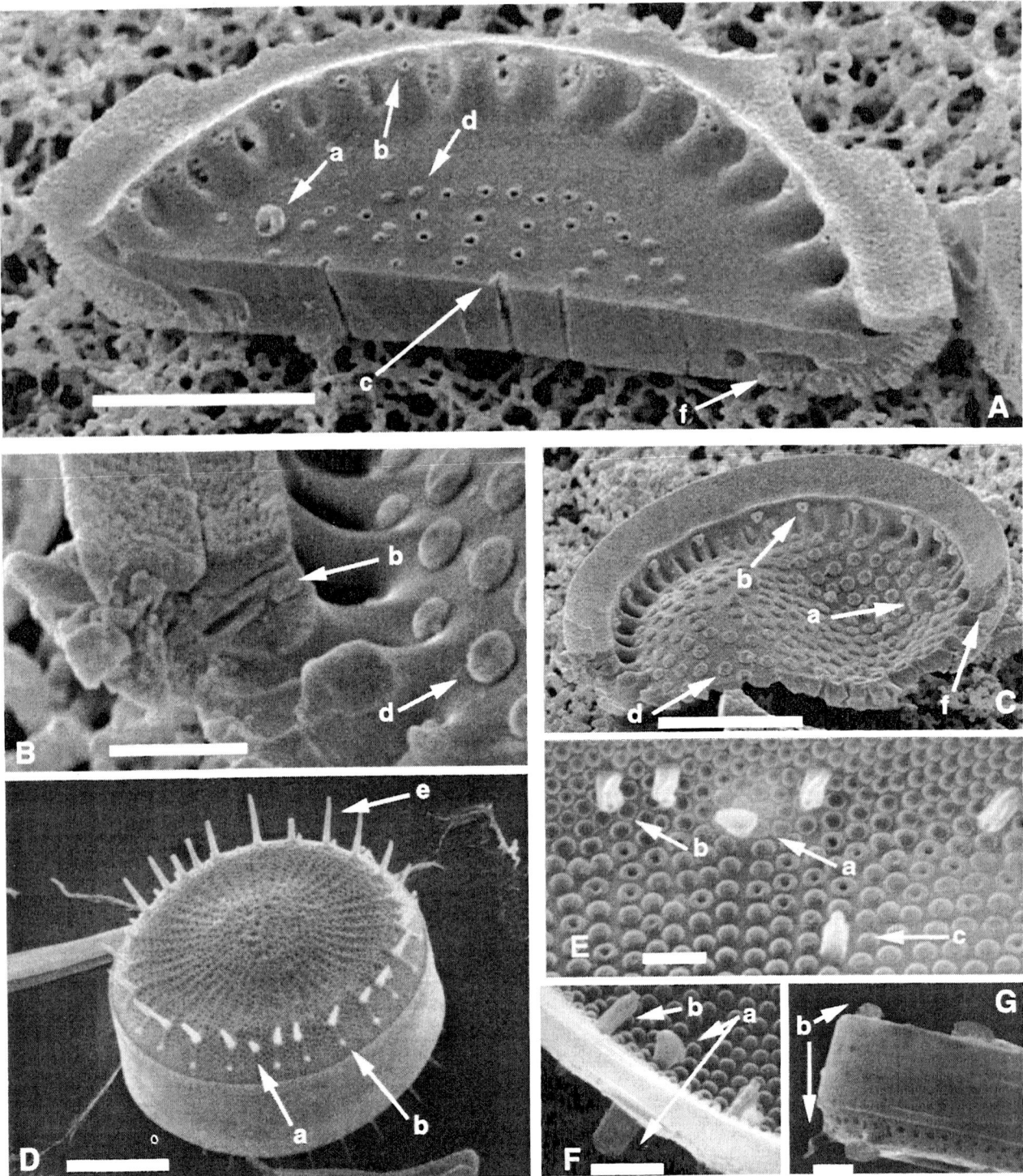

FIGURE 2 Major morphological structures in centric diatoms. (A) *Cyclotella bodanica*, internal scanning eletron micrograph (SEM) of valve in cross section, (B) *Pliocaenicus* sp., internal SEM of valve margin, (C) *Pliocaenicus* sp., internal SEM of valve in cross section, (D) *Stephanodiscus niagarae*, external SEM of frustule, (E) *Stephanodiscus* sp., internal SEM of valve face–mantle junction, (F) *Stephanodiscus* sp., SEM of margin showing internal and external morphologies of valve processes, (G) *Cyclotella pseudostelligera*, external SEM of frustule. Arrows indicate specific morphological features of interest. (a) rimoportula or labiate process; (b) marginal fultoportule or marginal strutted process; (c) valve face fultoportule or central strutted process, (d) valve face cribrum; (e) spine; (f) marginal chamber. Scale bar equal 10 μm for A, C, and D. Scale bar equal 1 μm for B, E, F, and G.

spines which interlock with punctae on the valve mantles of sister valves (Fig. 10B). *Stephanodiscus* species, which may or may not form colonies, also commonly produce valves of strikingly different morphology under different growth conditions (Håkansson and Stoermer, 1984; Kling, 1992). Variable valve morphology is observed in several species of *Stephanodiscus* (Theriot and Stoermer, 1986; Theriot *et al.*, 1988), but it is most striking in species that regularly adopt colonial growth habit (Stoermer *et al.*, 1979).

A morphological adaptation to planktonic growth in many marine species, the formation of "balloon" cells with very lightly silicified frustules and very large vacuoles is found only within the Rhizosoleniaceae in fresh waters. The probable reason is that it is very difficult in fresh water to achieve relative buoyancy by selectively sequestering monovalent versus divalent cations in the vacuolar solution (Gross and Zeuthen, 1948; Moore and Villareal, 1996). Although it might seem intuitively that a balloon cell morphology would confer selective advantage on freshwater diatoms, this does not appear to be the case. In general, freshwater diatom species have more heavily silicified cells than marine species (Conley *et al.*, 1989, 1993).

Perhaps the most successful morphological modification for planktonic centric diatoms, both freshwater and marine, is the strutted process system, characteristic of the Thalassiosirales. The presence of the so-called marginal strutted process or fultoportulae and associated cytoplasmic organelles allows secretion of external organic structures, which appear to reduce sinking velocity (Walsby and Xypolyta, 1977; Herth and Barthlott, 1979). These structures include marginal "skirts" of organic material in *Cyclotella* (Fig. 3D), "bristles" composed of β-chitin secreted from the marginal strutted processes in *Stephanodiscus* and strands connecting cells in "string of bead" colonies in *Thalassiosira* (Hoagland *et al.*, 1993).

A few freshwater centric diatoms form spores [e.g., Fig. 5E, H; (Edlund and Stoermer, 1993; Edlund *et al.*, 1996)], but this does not occur nearly as widely as in neritic marine species (French and Hargraves, 1980). Spore formation appears to be a primitive condition, which is not particularly successful in fresh water.

B. Cytoplasmic Features

The most common cytoplasmic arrangement in centric diatoms consists of a donut-shaped vacuole with a central column of cytoplasm containing the nucleus and peripheral cytoplasm containing chloroplasts and other cytoplasmic organelles. As in all diatoms, the nucleus and chloroplasts are contained in a common membrane system, and the Golgi bodies are arranged around the nucleus in interphase cells. Several chloroplasts are present, and each usually contains a single pyrenoid (van den Hoek *et al.*, 1995).

Cytoplasmic structures are present, which apparently support some of the morphological modifications discussed above. Unfortunately, the ultrastructure of freshwater centric diatoms has not been thoroughly investigated. For example, the precise mechanisms that allow formation of specific structures on diatom valves are known only in modest detail and not at all for many common freshwater genera (Schmid, 1994). The ultrastructure of strutted processes and the cytological structures associated with them have been extensively investigated (reviewed in Hoagland *et al.*, 1993).

C. Envionmental Physiology

There are no particular physiological characteristics that separate centric diatoms, as a group, from other diatoms. The planktonic mode of existence adopted by most centric diatoms does, however, emphasize some of these physiological characteristics. They share the pigments, which allow diatoms to grow under low-light conditions. Accessory pigments are often especially apparent in planktonic centric diatoms, particularly those that inhabit the deep chlorophyll maximum of oligotrophic lakes (Moll and Stoermer, 1982), such as *Cyclotella* spp. (Fahnenstiel and Glime, 1983) or are found in benthic communities of such lakes, such as *Melosira undulata*. (McIntire *et al.*, 1994). Most diatoms accumulate a significant portion of their carbohydrate reserves as lipids, but this tendency is particularly well developed in taxa that can survive prolonged sediment entrainment and burial, such as *Aulacoseira* spp. (Sicko-Goad *et al.*, 1989). Indeed, the ability to survive periodic return to sediment surfaces seems to be one of the major evolutionary adaptations of freshwater plankton diatoms.

Perhaps the major characteristic that separates diatoms from most other physiological groups of algae is their absolute requirement for silicon for cell division and frustule formation. This is a particularly stringent requirement for euplanktonic species because they spend most of their time removed from concentrated sources of silicon, such as bottom sediments. In lakes with long residence times, silicon can be depleted by diatom growth and subsequent sinking of frustules to sediments, resulting in selective advantage for algal groups that do not require silicon for growth (Schelske and Stoermer, 1971, 1972). Silicon limitation also profoundly affects seasonal succession of diatom species, favoring those that grow during cold season circulation in temperate and boreal lakes and reducing or eliminating those adapted to summer growth in the epilimnion or subthermocline maximum (Stoermer et al., 1985a). Because silicon is so often the nutrient limiting plankton diatom growth, many plankton species have adapted remarkable morphological plasticity. Some species produce grossly oversilicified frustules when silicon is available in abundance and greatly reduced wall structure when this nutrient becomes limiting (Stoermer et al., 1985b; Genkal and Håkansson, 1990; Kling, 1992; Genkal and Kiss, 1998). In general, species that are able to survive in either naturally or artificially eutrophic habitats show a high degree of morphological plasticity. Those that grow in oligotrophic habitats are usually less morphologically plastic and more highly silicified (Sicko-Goad et al., 1984).

V. ECOLOGY AND EVOLUTION

A. Diversity and Distribution

The present crude state of diatom taxonomy probably hides the true patterns of diversity and distribution in freshwater centric diatoms. It was formerly thought that most diatom species had "highly variable" morphology and worldwide distribution. This was partially a function of the relatively small number of researchers active in diatom systematics and partially a function of the fact that the few summary identification manuals available were produced in Europe. This situation is now changing rapidly. The majority of generic names now in common usage were either unknown or differently construed in 1985. Although the process of raising our understanding of diatom taxonomy to the levels assumed for most eukaryotic organisms is still ongoing, some rather clear patterns are emerging. At this point, it is abundantly clear that freshwater centric diatoms are grossly underdescribed. One example of the results of modern approaches to taxonomy is the work of Theriot and colleagues on the genus *Stephanodiscus* (Theriot 1987, 1992; Theriot and Stoermer, 1984a, b, 1986; Julius et al., 1997a). By quantifying the effects of environment on morphology, Theriot and co-workers were able to show that what was considered to be a variable (and wildly misnamed) species was actually a species complex, consisting of several taxa with distinct ecological requirements and distribution patterns.

Although centric diatom distribution is still incompletely understood, at this juncture it appears that Hutchinson's "paradox of the plankton" is resolved in the ancient tectonic lakes of the world, which contain very depauperate and almost exclusively endemic floras (Stoermer and Edlund, 1998). Such lakes are now absent from North America, although they were once a conspicuous feature of the intermountain west (Krebs et al., 1987). In glaciated regions of north temperate and boreal zones, at least in Europe and Eastern North America, there are a large number of common species, although there is also evidence of rapid evolution into characteristic regional floras (see Theriot citations above). Recent results from Lake Baikal (Julius et al., 1997b; Edlund, 1998) show a floristic revolution associated with every episode of the latest Pleistocene glaciations. Describing species distribution in nonglaciated regions of North America is more problematic. The centric diatom flora of these regions is not particularly well explored, and common northern species are most commonly reported. However, there are also reports of populations usually associated with South America (Hickel and Håkansson, 1991) and Africa (Stoermer et al., 1992) in the southern United States flora.

B. Reproducion and Life Histories

Diatom life cycles have recently been reviewed by Edlund and Stoermer (1997). As is the case with all diatoms, the rigid frustule of centric species places a finite limit on the number of vegetative divisions most species of freshwater diatoms can undergo between sexual episodes (Round, 1972; Round et al., 1990). (Section IA earlier). There is considerable disparity of opinion about how long a time interval may be involved (Nipkow, 1927; Jewson, 1992a, b). Lewis (1984) has advanced the interesting theory that diatom life histories may constitute a supra-annual clock mechanism. Edlund and Stoermer (1997) postulate that aspects of the sexual life history phases of centric diatoms closely reflect evolutionary relationships.

C. Ecological Interactions

Although the recent ecological literature concerning freshwater phytoplankton, including diatoms, has been dominated by consideration of competitive interactions, particularly for nutrients (Titman, 1976; Tilman et al., 1982; Sommer, 1996), it is evident that life cycle adaptations (Margalef, 1978; Jewson, 1992a) play an important part in ecological success. Many species of freshwater diatoms are able to essentially escape the direct effects of nutrient competition through mechanisms such as subthermocline growth (Fahnenstiel and Glime, 1983) or prolonged vegetative survival in sediments (Sicko-Goad et al., 1989).

Environmental effects on the siliceous portions of diatoms is one of the most common concerns for diatom systematists and ecologists. Virtually all taxonomic identification in this group is based on valve morphology. These structures, at least in theory, are presumed to be stable and consistent, providing a reliable means of consistent identification. Some studies have demonstrated examples of environmental valve modifications so severe that species appear as two different taxa under varying conditions. The extent of this morphological plasticity has to be understood before an accurate understanding of diatom species and their number can be achieved (Schmid, 1994).

It is commonly suggested (Round et al., 1990) that researchers look for "janus valves" [cells in which the frustule has two distinctly different valve types in species for which the two valves normally appear identical (McBride and Edgar, 1998)] if they are concerned about differing environmental forms. Stoermer (1967) identified this phenomenon in field collections of the pennate diatom Mastogloia. Modifications in levels of total dissolved solids apparently caused two distinct morphotypes to appear. Tuchman et al. (1984) noted similar modifications in the valve structure of the centric diatom Cyclotella meneghiniana (Stephanocyclus meneghiniana) with increased salinity under experimental conditions. Other researchers have observed similar taxonomic forms occurring over an environmental gradient and suggested that the forms represent a single highly plastic species. Most notable for centric diatoms is the conclusion that Cyclotella meneghiniana and C. cryptica are conspecific (Schultz and Trainor, 1968; Shoeman and Archibald, 1980). These researchers appear to suggest that our current taxonomic system overestimates the number of species, with many of the current taxonomic entities representing ecomorphotypes (ecological variants) of a single species.

Other researchers arrived at different conclusions. Bourne (1992) used chloroplast DNA to analyze specimens in culture from a number of geographic localities. All of the cultured organisms were identified as C. meneghiniana and exhibited the appropriate morphological features used in identification. The study identified several distinct genetic groups, which could easily be interpreted as individual species. This work suggests that many undescribed species exist and that the number of conspecific ecomorphotypes may be overstated.

The correct interpretation of ecomorphotypes and species may be much more complicated than that presented above. The only way to identify actual species and the extent of environmental modification on these species is through character analysis of individual populations. The taxonomic system for diatoms may be flawed, as ecomorphotype advocates have suggested, but the problems with the system may have nothing to do with the species diversity of the group. Past methods of interpreting species may have simply used features inappropriate for identification. The "defining" characteristics for many of our taxa may be based upon features that vary with the environment rather than consistent homologous characteristics (Julius et al., 1997a). Schmid (1994) suggests that internal valve features used in cell division may be more reliable than external valve surface features prone to variation, although which characteristics to exclude and include remains open to interpretation. Researchers must be able to properly interpret all types of variation in characteristics before a formal systematic analysis can be performed, which will eventually provide a clear classification system for centric diatoms.

Variability in frustule structure is observed widely in centric diatoms. For example, species of Aulacoseira may exist in several different forms, either as a result of growth processes (Müller, 1903, 1906; Cleve-Euler, 1911a, b; Bethge, 1925) or silica limitation (Stoermer et al., 1985b). Although less commonly recognized, similar valve modifications occur in Stephanodiscus and Skeletonema, also as a result of levels of modified dissolved solids (Geissler, 1982; Paasche et al., 1975) or other factors (Stoermer et al., 1979; Håkansson and Stoermer, 1984; Kling, 1992).

The most comprehensive investigations of valve variability for systematic interpretation are with members of the Stephanodiscus niagarae complex (Theriot and Stoermer, 1984a; Theriot, 1987). Three species (S. superiorensis, S. reimerii, and S. yellowstonensis) were distinguished from S. niagarae through principal component analysis of several morphological features of the valve face. Environmentally variable features were identified (Theriot and Stoermer, 1987; Edlund, 1992) along with consistent features potentially representing stable homologous features (Theriot and Stoermer, 1984a). Julius et al., (1997a) noted that the species in this group were significantly different from one another in a statistical analysis, satisfying the requirements for the phylogenetic species definition (Cracraft, 1989). Environmentally stable features were eventually incorporated into a phylogenetic study examining the evolutionary relationships of the taxa (Theriot, 1992).

The taxonomic literature is littered with many varieties and forms of species, often attributing morphological differences to environmental, geographic, or some other factor. What these nomenclatural designations actually mean has never been standardized, and the names generally have little consistency from researcher to researcher. Eventually, the designation of varieties and forms could have some significance for ecotypes. Useful application of a nomenclatural scheme of this type in diatoms is, however, not possible at present. Detailed investigations such as the one above must be performed before a highly organized and meaningful classification scheme can be implemented.

D. Evolutionary Relationships

The freshwater centric diatoms are a polyphyletic group, representing numerous marine invasions. Although explicitly stated by Medlin *et al.* (1993) in a molecular phylogenetic hypothesis, Simonsen (1979) postulated a similar development for the group. Character loss is a exceedingly difficult problem in determining the evolutionary relationships within the group. Early in the fossil record the centric diatoms possessed many complex morphological features and with time these features were simplified or completely lost. A group of Cretaceous marine taxa exemplifies this trend. Gersonde and Harwood (1990), Harwood and Gersonde (1990), and Harwood and Nikolaev (1995) described numerous new marine genera and species possessing an extraordinary number of morphologically complex features. More recent (Tertiary) centric diatom taxa lack most of these complex morphological features, making even the simplest statements of evolutionary relationships between the two floras a difficult task.

This pattern has occurred in freshwater centric diatoms to a lesser, but equally confusing, extent. The genus *Coscinodiscus* is an excellent example. *Coscinodiscus* was traditionally categorized by the absence of features found in other diatom genera. This resulted in a number of freshwater taxa that lack obvious distinguishing features being classified in the genus. In all cases that have been critically examined these supposed relationships have proven false. *Coscinodiscus* almost certainly contains members possessing highly derived taxa with many character losses and ancestral taxa that do not possess any modified morphological features. These two subgroups within *Coscinodiscus* are almost certainly very distantly related to one another and should not be considered members of the same genus, but this possibility has not been considered in work with these taxa.

Despite the importance of production of phylogenetic classifications, few diatom taxonomists have pursued phylogenetic investigations. Williams (1985), Williams and Round (1988), and Kociolek and Stoermer (1988) have dealt with the morphology of pennate diatoms in phylogenetic reconstruction. Medlin *et al.* (1993, 1996, 1997) dealt with the evolutionary relationships of major diatom lineages using rDNA data in a phylogenetic investigation. Within centric diatoms, most phylogenetic investigations involved members of the Thalassiosirales. These represent the most successful centric diatoms in terms of evolution and differentiation in fresh waters. The majority of diverse and ecologically important freshwater genera are contained here. Evidence suggests that the Thalassiosirales is a monophyletic group, but the relationships of taxonomic groups within this order are uncertain (Theriot and Kociolek, 1986; Theriot *et al.*, 1987). One shared feature, the presence of fultoportulae (Fig. 2B, C), exists within this group, demonstrating a common ancestry between species of this order. Freshwater genera are primarily members of the family Stephanodiscaceae. No formal study proposing a phylogeny for this family has been completed. Theriot and Kociolek (1986) suggested that this problem is caused by the lack of primary descriptive work for this group and that only *Stephanodiscus* is a well-defined group. Medlin *et al.* (1991) investigated the relationships of *Skeletonema* clones using rDNA data with phylogenetic systematic techniques. Theriot (1992) produced a phylogenetic hypothesis for some *Stephanodiscus* species, and Theriot (1987), Theriot and Bradbury (1987), Theriot (1990), and Theriot and Serieyssol (1994) have used the principles of phylogenetic systematics to address specific issues with the Thalassiosirales.

The only other freshwater group that contains a large number of species is the Aulacoseirales and the genus *Aulacoseira* in particular. Evolutionary relationships within this order have never been investigated in a formal study, and, like the other major groups of genera and species within these genera, phylogenies are only now beginning to be worked out.

VI. COLLECTION AND STUDY METHODS

To study the taxonomy, cytology, and paleontology of centric diatoms essentially the same methods are used as those for the same aspects of other diatoms. Successful quantitative study of plankton populations presents some unique challenges. The most commonly used technique, fixation with some variant of Lugol's solution followed by settling and observation with an inverted microscope, suffers from poor preservation, the inherent limitation of resolving morphological structures necessary for identification in a low refractive index medium, and the lack of permanent reference mounts. Although Crumpton and Wetzel (1981) and Crumpton (1987) have developed superior methods of phytoplankton preservation and enumeration, these methods still are not widely used. In general, the lack of appropriate and well-established methods of addressing planktonic algal assemblages, which contain many centric diatoms, is a serious hindrance to understanding modern centric diatom population structure and distribution. In fact, it now appears that the most reliable and generally useful analyses of centric diatom occurrence and distribution are derived from paleolimnological techniques. Surficial sediments of most lakes provide a temporal and spatial integration of diatom populations that occur in the lake during the course of one or more years. It would require heroic sampling efforts to duplicate such records by other sampling methods. Core samples of sediments also provide a sequence of estimates of assemblage structure over time, which has proven useful in determining the effects of global environmental problems such as lake acidification, climate change, and eutrophication (Stoermer and Smol, 1999).

VII. KEY AND DESCRIPTIONS OF GENERA

A. Key (Figs. 3–12)

1a.	Valves usually circular in outline; if not circular, lacking specialized structures at poles (Fig. 8A–G)	7
1b.	Valves bipolar or multipolar, with specialized structures at poles (Fig. 7A–G)	2
2a.	Frustules very thin-walled, valves much reduced, most of frustule composed of imbricate girdle bands	3
2b.	Frustules robust, valves conspicuous and usually thick walled, with pore fields or elongate, curved spines at poles	4
3a.	Reduced valve bearing one long, conspicuous spine (Fig. 5A–C)	*Urosolenia*
3b.	Reduced valve bearing two long, conspicuous spines (Fig. 5D and E)	*Acanthoceras*
4a.	Valves with long, curved spines at poles (Fig. 5H)	*Chaetoceros*
4b.	Valves with pore fields at poles	5
5a.	Valves bipolar	6
5b.	Valves tripolar (Fig. 7A–D)	*Hydrosera*
6a.	Valves subcircular, lacking costae (Fig. 7E–H)	*Pleurosira*
6b.	Valves elongate, with strongly developed costae extending into lumen of cell (Fig. 5F and G)	*Terpsinoë*
7a.	Circular valves lacking spines or external process extensions, with an inconspicuous pseudonodulus (Fig. 4A–F)	*Actinocyclus*
7b.	Circular valves, usually with spines or external process extensions, lacking an ocellus	8
8a.	Valves with elongated valve mantle regions, usually seen in girdle view, forming more or less elongate colonies	9
8b.	Valves with short girdle regions, rarely forming colonies	13
9a.	Striation of valve mantle fine, often not resolvable with a light microscope	10
9b.	Striation of valve mantle striate resolvable with light microscopy	12
10a.	Frustules joined into colonies by tips of more or less elongate, spinelike processes arising from margin of valve; valve faces separated (Fig. 12G and H)	*Skeletonema*
10b.	Frustules not joined into colonies by tips of processes, valve faces appressed	11
11a.	Frustules very thick-walled, occurring in colonies bound together by ribs on surface of valve. Ribs differ on sister valves, forming cameo and intaglio valves (Fig. 6J and K)	*Ellerbeckia*
11b.	Frustules less thickened, punctae of valve mantle fine, not resolvable with a light microscope (Fig. 6A–D)	*Melosira*
12a.	Valve mantles with straight or curved rows of punctae, spines simple, valve punctae of the same type as found on valve mantle or lacking (Figs. 3A, B; 6E–H)	*Aulacoseira*
12b.	Valve mantles with anastomosing rows of striae, spines large and bladelike, valve face with one or more large processes contained in a hyaline area at center (Fig. 6I)	*Orthoseira*
13a.	Striation on valve surface separated by ribs or hyaline areas	15
13b.	Striation on valve surface not separated by ribs or hyaline areas	14
14a.	Striae randomly arranged or in anastomosing rows; valve margin with spines or occluded processes; strutted processes present at valve margin and center, external tube of labiate process usually conspicuous (Fig. 9A–C)	*Thalassiosira*
14b.	Striae radial, valve margin with spines fused to marginal strutted processes, labiate process very inconspicuous (Fig. 12I and J)	*Cyclotubicoalitus*

15a.	Valve surface finely punctate, weakly or not arranged into striae, divided into wide sectors	16
15b.	Punctae on valve surface arranged in striae, often fasciculate	18
16a.	Punctae on valve surface divided into sectors by more or less developed external ribs	17
16b.	Punctae on valve surface divided into sectors by internal ribs; structure of central and marginal valve usually appears distinctly different (Figs. 9D–J)	*Stephanocyclus*
17a.	Ribs strongly developed, valve surface plicate; usually found in strictly fresh water (Fig. 12D–F)	*Stephanocostis*
17b.	Ribs wide, but less strongly developed; usually found in estuarine habitats (Fig. 9K)	*Thalassiocyclus*
18a.	Marginal strutted processes grouped, internal openings found in depressions of the inner valve surface; known species are all fossil (Fig. 12A–C)	*Mesodictyon*
18b.	Marginal strutted processes not grouped	19
19a.	Valves with marginal chambers or branched radial ribs near margin and on valve mantle; labiate process either on valve surface or on valve mantle, not at level of spines	20
19b.	Lacking marginal chambers or branched radial ribs; conspicuous spines usually present, emplaced on radial ribs; labiate process at level of spines (Fig. 10A–H)	*Stephanodiscus*
20a.	Valves eccentrically undulate (Fig. 11A–C)	*Pliocaenicus*
20b.	Valves flat or concentrically undulate	21
21a.	Labiate process or processes on valve surface (Fig. 8A–I)	*Cyclotella*
21b.	Labiate process or processes on valve mantle (Fig. 11D–I)	*Cyclostephanos*

B. Descriptions of Genera

Acanthoceras Honigmann (Fig. 5D and E)

Acanthoceras and *Urosolenia* are the primary freshwater examples of very lightly silicified "balloon cells" commonly found in diatoms whose primary habitat is oceanic plankton. The valves are much reduced. Most of the frustule consists of imbricate (overlapping) intercalary bands. The protoplast, with four chloroplasts and a central nucleus, is concentrated in the central region of the cell, and most of the cell volume is occupied by vacuoles. Although we are not aware of any formal estimates of growth rate, *Acanthoceras* can apparently grow very rapidly under favorable conditions. Similar to its marine relatives and *Urosolenia*, it forms resting spores (Edlund and Stoermer, 1993). Edlund and Stoermer (1993) also discuss recent nomenclatural changes concerning *Acanthoceras*.

Acanthoceras is an apparently rare genus but is probably under-reported (Huber-Pestalozzi, 1942). Based on available information, it forms ephemeral summer blooms in shallow, eutrophic lakes and ponds in Europe and North America (Beaver, 1981).

Actinocyclus Ehrenberg (Fig. 4A–F)

The taxonomy of the most common freshwater species, *A. normanii* fo. *subsalsa*, is discussed by Hasle (1977). *Actinocyclus* is distinguished from other centric diatoms by the lack of processes other than rimoportulae and the absence of ribs or areolae regularly arranged into striae on the valve face. The pseudonodulus, a structural character of many marine genera that is present in *Actinocyclus*, is barely visible in freshwater species.

Despite its relatively heavy silicification, *Actinocyclus* has its primary growth period during summer (Hohn, 1969). It is most abundant in regions of lakes shallow enough to be wind-mixed as well as in turbulent rivers and estuaries. It can survive burial in bottom sediments (Sicko-Goad *et al.*, 1989) and is apparently truly tychoplanktonic, spending part of its life suspended in plankton but returning regularly to the nutrient-rich sediment–water interface.

Actinocyclus is a primarily estuarine genus, but it also occurs in the plankton of some salinified or eutrophic inland waters. It became a biomass dominant in Lake Erie during the 1960s (Hohn, 1969) and is also found in polluted nearshore regions of the other Great Lakes and saline streams in the Ohio River drainage.

Aulacoseira Thwaites (Figs. 3A and B; and 6E–H)

Species considered under *Aulacoseira* were previously classified as *Melosira*, a genus now restricted primarily to marine species (Round *et al.*, 1990), based mostly on colonial growth habit. The restricted circumscription of *Aulacoseira* includes species, mostly occurring in fresh water, which grow in linear colonies, joined together by linking spines. Because of this growth habit and the fact that most species have rela-

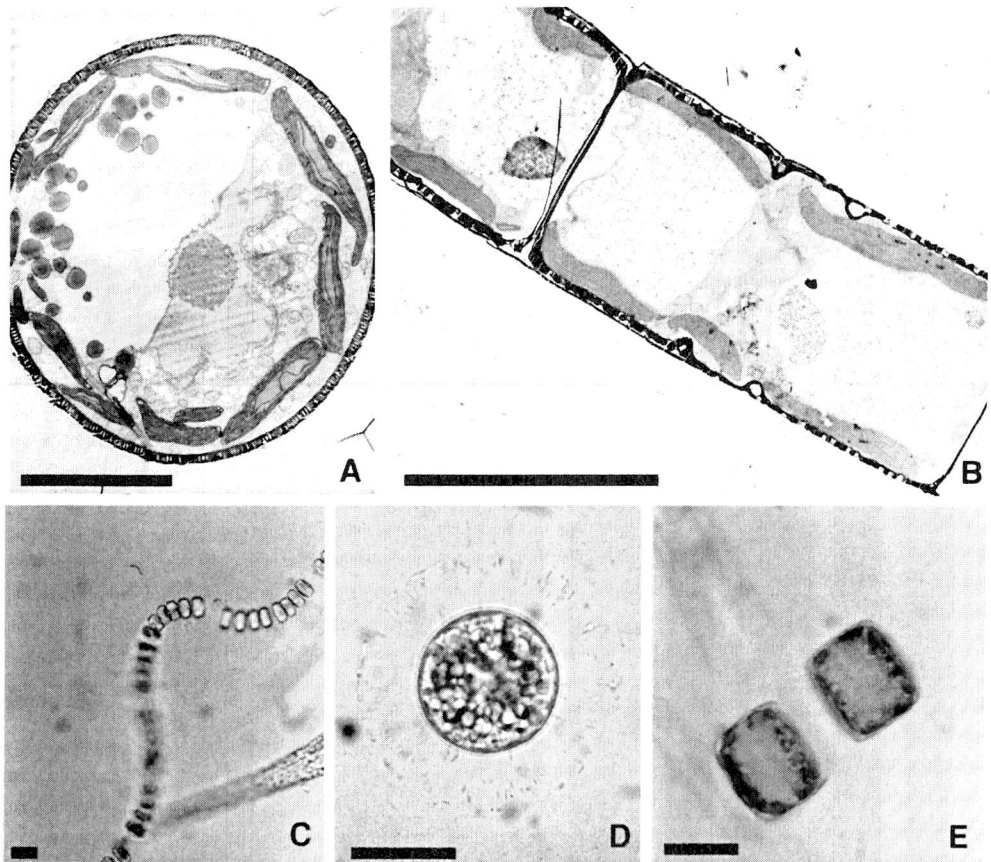

FIGURE 3 (A) *Aulacoseira granulata*, transmission electron micrograph (TEM) longitudinal section showing fine structure of frustule and arrangements of cytoplasmic components, (B) *A. granulata*, TEM cross section through region of central cytoplasmic bridge, (C) *Cyclotella* sp., LM girdle view of linear colony. (D) *Cyclotella* sp. LM valve view of living cell showing extracellular material, (E) *Stephanodiscus* sp. light micrograph (LM) girdle view of two recently divided living cells showing arrangements of chloroplasts. All scale bars equal 10 μm.

tively high valve mantles, cells are usually seen in girdle view. Frustules are usually heavily silicified and are polymorphic in many species. In the older literature (e.g., Hustedt, 1930), colonies composed of heavily silicified, coarsely structured valves were designated as α status, those with relatively thin, finely structured valves in which designated as β status, and colonies in which both types of valve structure occurred were designated as γ status. It is now commonly understood that valve structure responds to silica availability and growth rate. Status designations may convey ecological meaning (Stoermer *et al.*, 1985b) and the degree of plasticity in species needs to be understood to reach reasonable taxonomic decisions.

Aulacoseira is a very large, complex, and widely distributed genus. Some representatives are found in most inland waters, although it appears best adapted to life in lakes, ponds, and larger rivers. At least some species of *Aulacoseira* can survive sediment burial for extraordinary periods of time (Nipkow, 1950; Stockner and Lund, 1970; Sicko-Goad *et al.*, 1989). Most commonly this is accomplished without change in morphology of the frustule (Sicko-Goad *et al.*, 1986), but some species form spores (Edlund *et al.*, 1996). These mechanisms allow some species to thrive in deep lakes during winter circulation or during summer in water bodies shallow enough to be mixed to the bottom by wind stress. Some species also apparently thrive on the bottom sediments of base-poor, highly transparent lakes (Camburn and Kingston, 1986).

Whatever the mechanisms of its success, *Aulacoseira* is one of the most successful, in terms of distribution in time and space, of all freshwater centric diatoms. According to Van Landingham (1964), *Aulacoseira* species are the dominant taxa in most nonmarine diatomites from western North America and occurrences extend back to the Paleocene period. *Aulacoseira* is found in modern lakes from all parts of

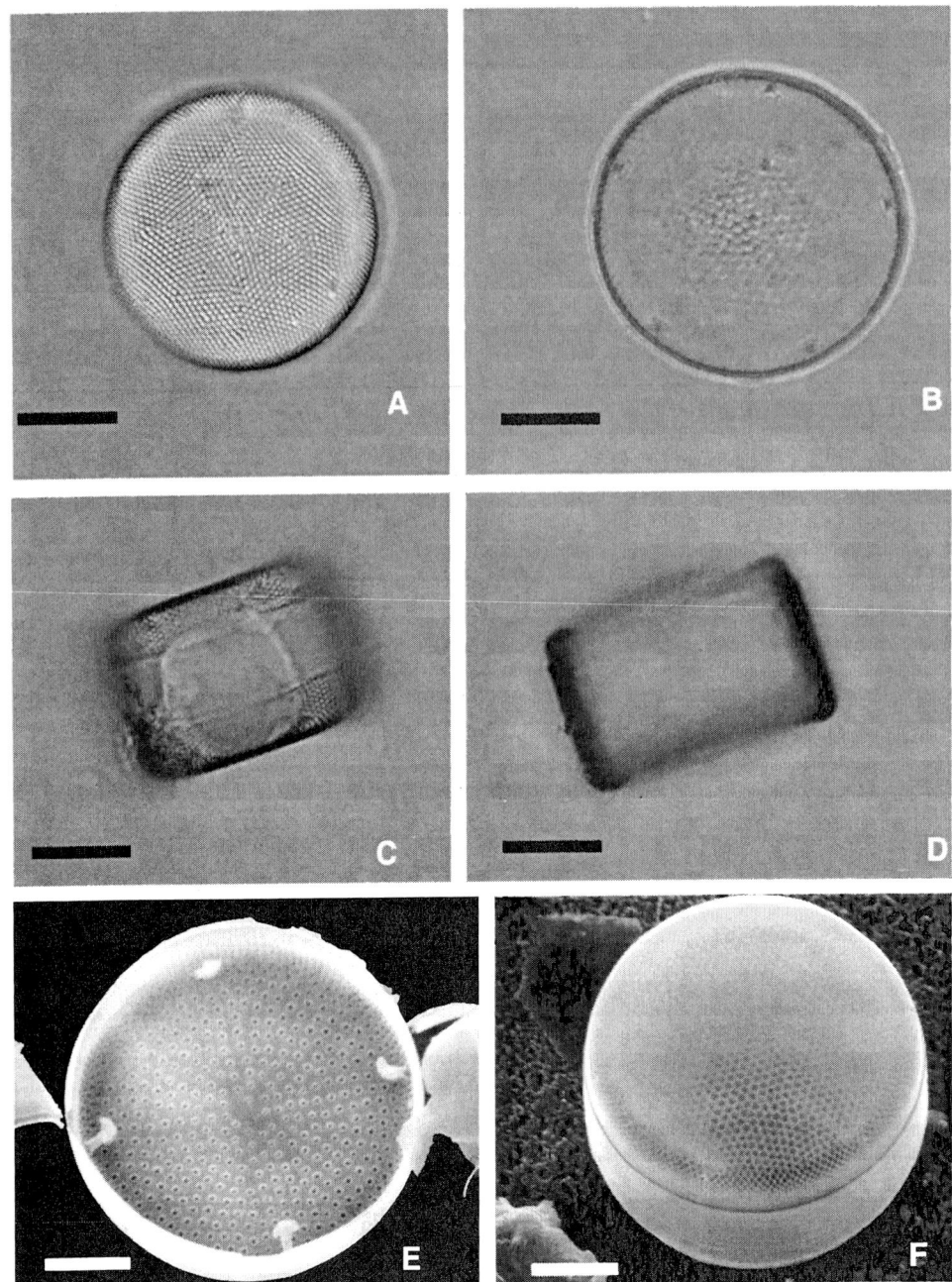

FIGURE 4 *Actinocyclus normanii* fo. *subsalsa*, (A) LM high focus of valve view, showing arrangement of areolae, (B) LM low focus of valve view, showing labiate processes, (C) LM high focus of girdle view, (D) LM mediam focus of girdle view. (E) SEM internal view fo valve, showing labiate processes and arrangement of areolae, (F) SEM external view of complete frustule. All scale bars equal 10 μm.

the world, and ancient tectonic lakes each appear to have their swarms of indigenous species (Edlund *et al.*, 1996).

Chaetoceros Ehrenberg (Fig. 5H)

Chaetoceros species are euplanktonic and their characteristic linear colonial growth habit (most species) and very long, overlapping, spines are thought to minimize sinking rates. Many of those occurring in inland waters have high growth rates and develop high lipid contents (Johansen *et al.*, 1990). Because of this, they may be important primary producers in saline inland waters and have been considered candidates for synthetic fuel production.

Chaetoceros is a primarily marine genus, with only a few representatives in inland waters of North

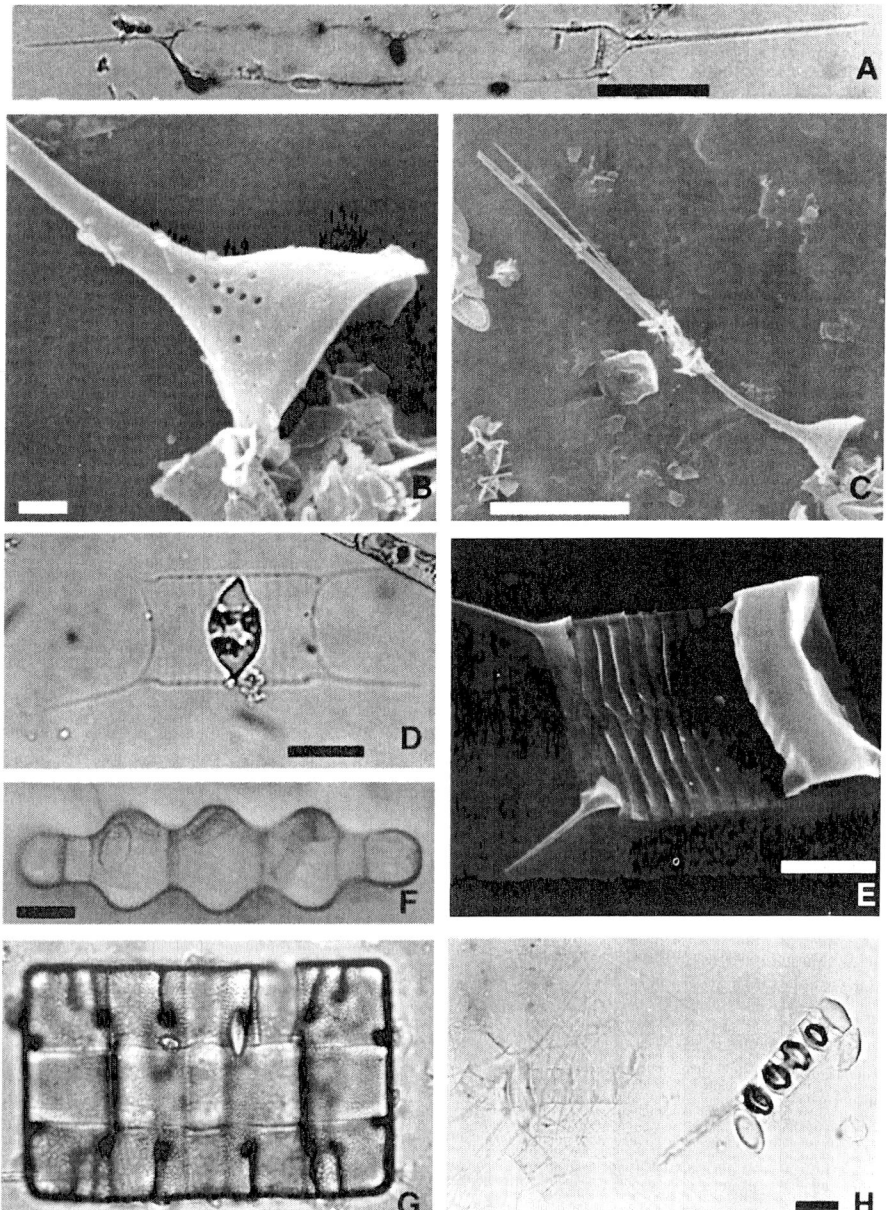

FIGURE 5 (A) LM girdle view of *Urosolenia* sp., (B) higher magnification SEM, showing punctae on valve, (C) SEM valve and spine of *Urosolenia* sp., (D) LM of living cell *Acanthoceras magdeburgense*, showing central cytoplasmic mass with chloroplasts, (E) SEM girdle view of a half frustule of *A. magdeburgense*, showing valve, imbricate girdle bands, and spore, (F) LM valve view of *Terpsinoë musica*, (G) LM girdle view of *T. musica*; (H) LM girdle view of *Chaetoceros elmorei* showing vegetative frustules and spore-forming frustules. Scale bars equal 10 μm for A, C, D, E, F, G, and H. Scale bar equals 1 μm for B.

America (Rushforth and Johansen, 1986). Most occurrences of *Chaetoceros* are from endorheic regions (Johansen and Rushforth, 1985), although some are from either naturally saline or brine-contaminated rivers (Wujek and Graebner, 1980).

Coscinodiscus Ehrenberg (Fig. 9M)

Although there are numerous reports of *Coscinodiscus* from inland localities, those which have been thoroughly investigated turn out to be species of *Actinocyclus* (Hasle, 1977) or *Thalassiosira* (Round et al., 1990). Similar to those of *Actinocyclus*, species of *Coscinodiscus* lack processes other than labiate processes and spines. Partially due to its simple morphology, *Coscinodiscus* has been treated as a classic "not-a group": species which have no obvious distinguishing characteristics are placed in *Coscinodiscus* by default.

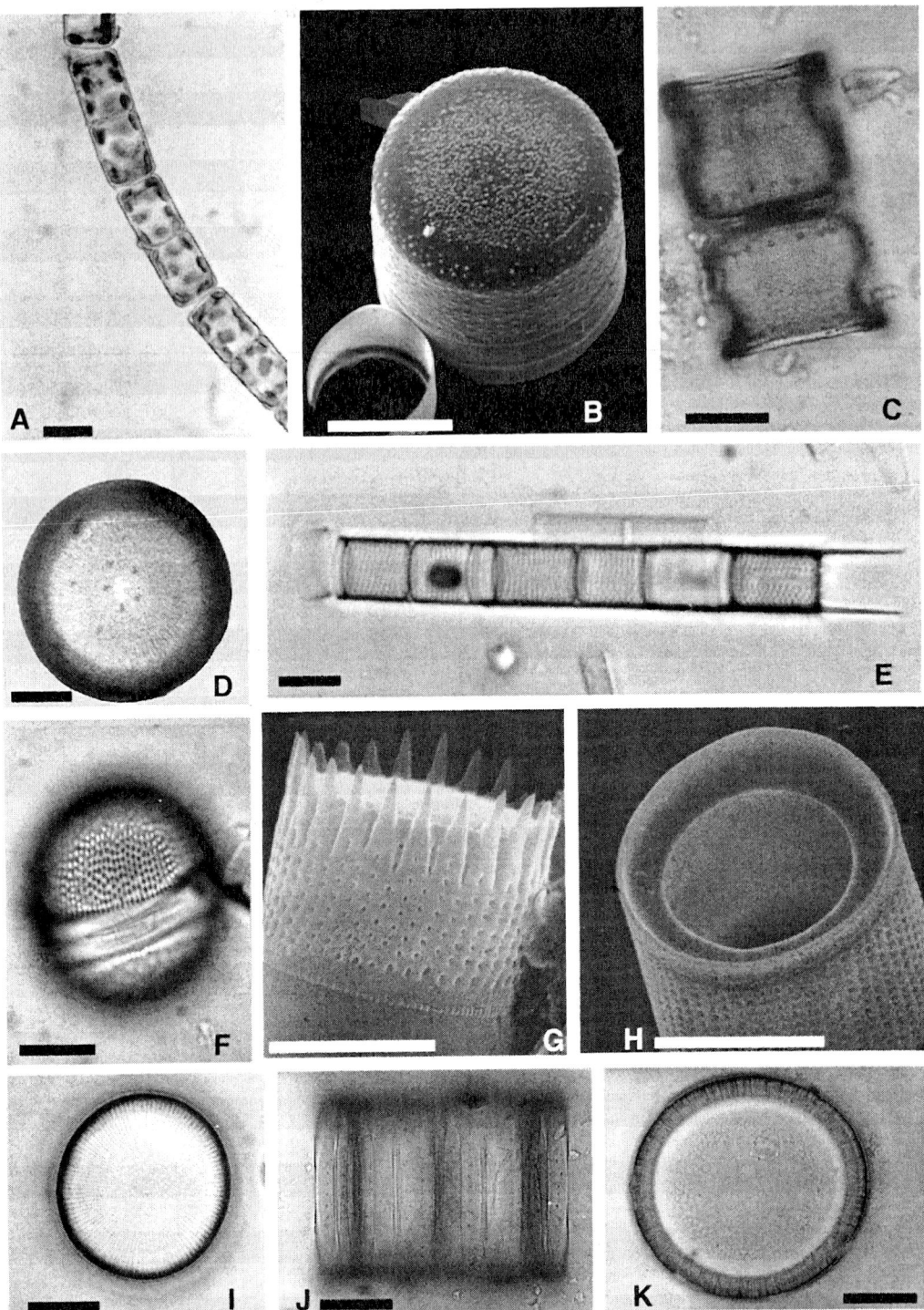

FIGURE 6 (A) LM girdle view of living *Melosira varians* showing cells with numerous chloroplasts, (B) SEM external valve view of *M. varians*, (C) LM girdle view of *M. undulata*, (D) LM valve view of *Melosira undulata*, (E) LM girdle view of *Aulacoseira granulata*, showing variations in silicification and presence of linking spines, (F) LM girdle view of *A. granulata* initial valve showing hemispherical valves; (G) SEM of *Aulacoseira italica* girdle and valve surface, (H) SEM of *A. islandica* girdle and interior, (I) LM valve view of *Orthoseira dendroteres*, (J) LM girdle view of *Ellerbeckia arenaria*, (K) LM valve view of *E. arenaria*. All scale bars equal 10 μm.

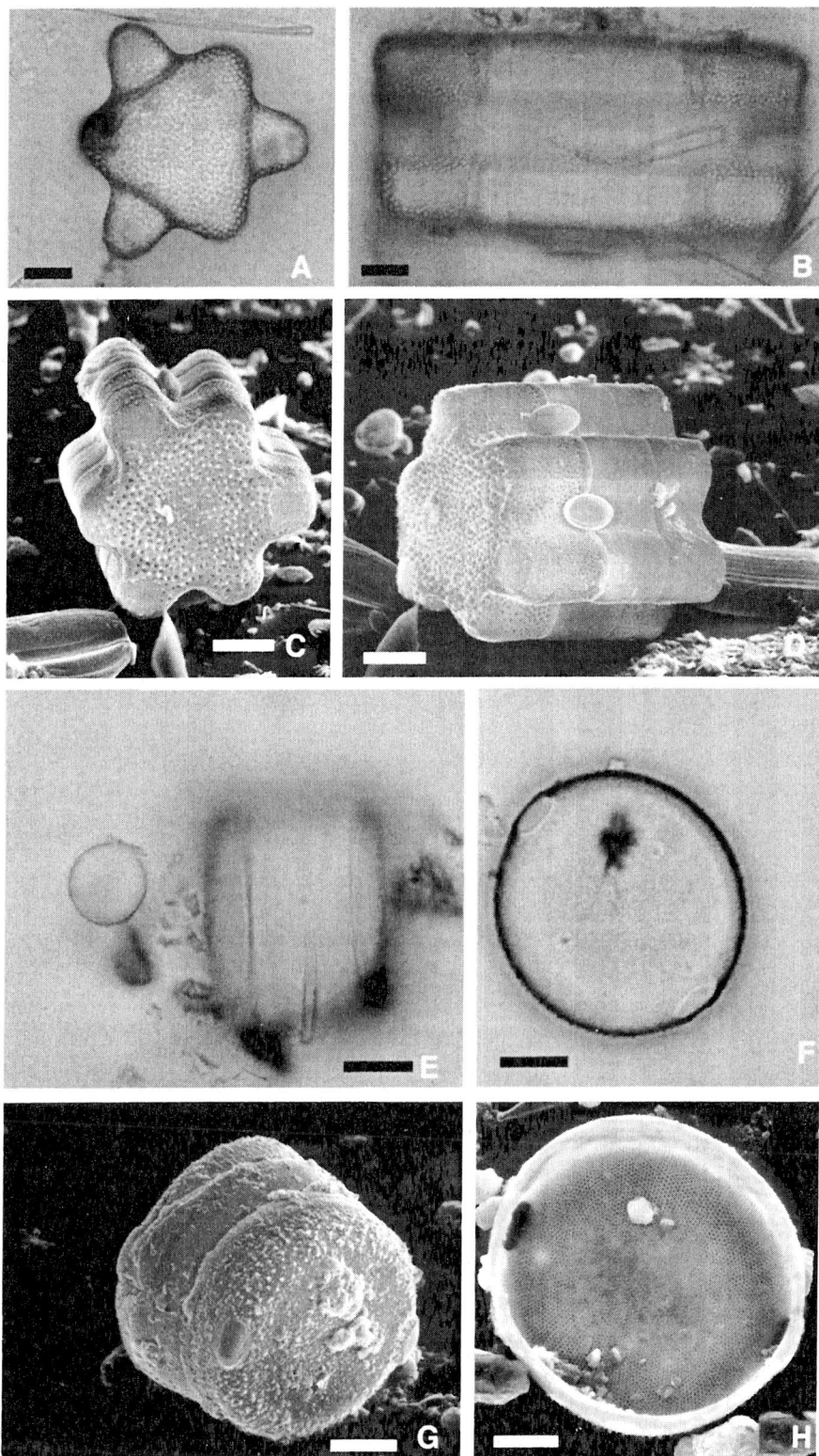

FIGURE 7 (A) LM valve view of *Hydrosera whampoensis*, (B) LM girdle view of *H. whampoensis*, (C) SEM oblique view of *H. whampoensis*, (D) SEM girdle view of *H. whampoensis*, (E) LM girdle view of *Pleurosira laevis*, (F) LM valve view of *P. laevis*, (G) SEM exterior view of *P. laevis*, (H) SEM interior valve view of *P. laevis*. All scale bars equal 10 µm.

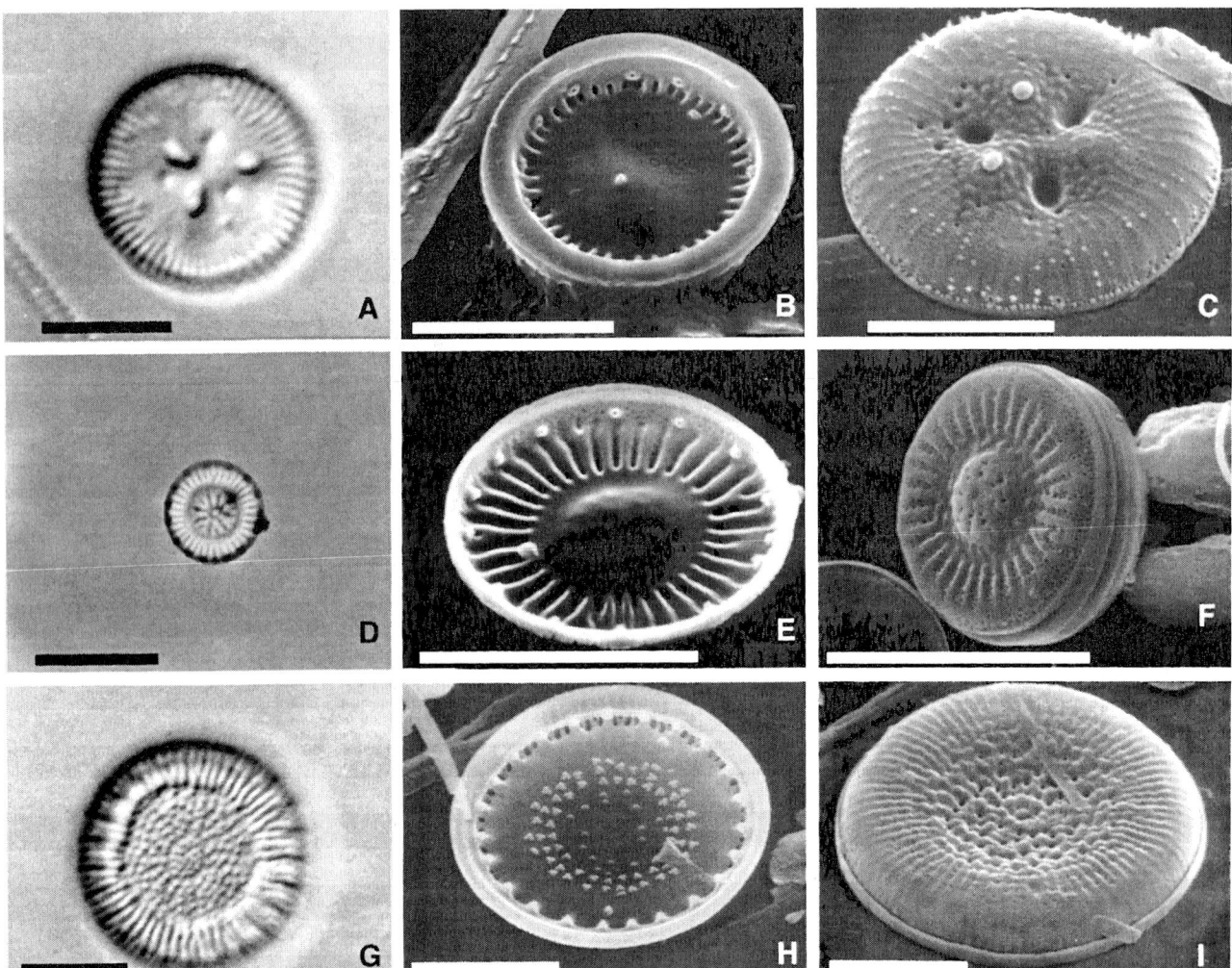

FIGURE 8 (A) LM valve view of *Cyclotella ocellata*, (B) SEM interior valve view of *C. ocellata*, (C) SEM exterior valve view of *C. ocellata*, (D) LM valve view of *C. pseudostelligera*, (E) SEM interior valve view of *C. stelligeroides*, (F) SEM exterior valve view of *C. stelligeroides*, (G) LM valve view of *C. radiosa*, (H) SEM interior valve view of *C. radiosa*, (I) SEM exterior valve view of *C. radiosa*. All scale bars equal 10 μm.

Because of the present state of taxonomic confusion, little can be said about the occurrence and distribution of *Coscinodiscus* species. However, reports from freshwater localities should be treated with caution.

Crateriportula Flower et Håkansson (Fig. 9L)

Crateriportula resembles some species of *Stephanodiscus* in general appearance, but is distinguished by the peculiar crater-like external openings of the strutted processes and the labiate process, only clearly discernible with scanning electron microscopy (SEM). For the original description, a fuller account of morphology, and a discussion of relationships to *Stephanodiscus* and *Cyclostephanos*, see Flower and Håkansson (1994).

Crateriportula is presently known only from Lake Baikal, but we note it here because the lack of current reports may result from the fact that it is easily mistaken for *Stephanodiscus*, and populations may occur in northern North America.

Cyclostephanos Round (Fig. 11D–I)

Cyclostephanos is a relatively recently described genus (Round, 1981; Stoermer et al., 1987; Theriot et al., 1987). As the name implies, it shares characteritics with both *Cyclotella* and *Stephanodiscus*. It is distinguished from the former by the presence of fasciculate striae which extend onto the valve mantle, interfascicular ribs which extend onto the valve mantle and branch below the level of the spines, and occurrence of the labiate process (Fig. 2A) on the valve mantle. It is distinguished from *Stephanodiscus* by the presence of branched ribs on the valve mantle, flattened cribra

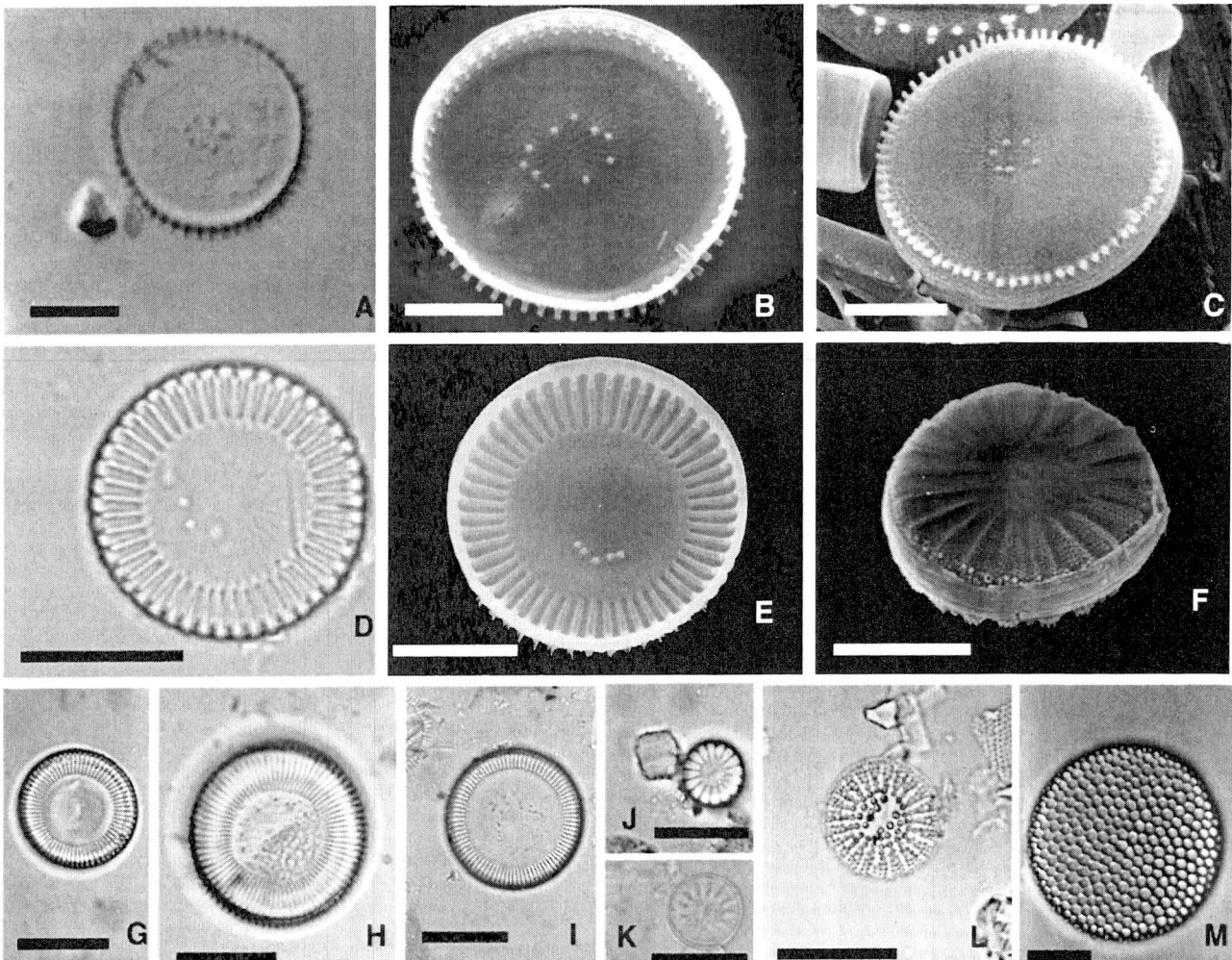

FIGURE 9 (A) LM valve view of *Thalassiosira weissflogii*, (B) SEM interior valve view of *T. weissflogii*, (C) SEM exterior valve view of *T. weissflogii*, (D) LM valve view of *Stephanocyclus meneghiniana*, (E) SEM interior valve view of *S. meneghiniana*, (F) SEM exterior valve view of *S. meneghiniana*, (G) LM valve view of *St gamma*, (H) LM valve view of *S. striata*, (I) LM valve view of *Stephanocyclus* sp., (J) LM valve view of *Stephanocyclus cryptica*, (K) LM valve view of *Thalassiocyclus lucens* (lower specimen), (L) LM valve view of *Crateriportula inconspicuus*; (M) LM valve view of *Coscinodiscus* sp. All scale bars equal 10 μm.

(Fig. 2D) on the valve mantle, and placement of the labiate process below the level of the spines on the valve mantle. Some of these characteristics can only be reliably determined with SEM, particularly in the smaller species.

Although the genus is found in a variety of habitats, many species of *Cyclostephanos* are indicators of eutrophy (Krammer and Lange-Bertalot, 1991a). In North America, smaller species of *Cyclostephanos* are often spring and summer dominants in shallow, eutrophic lakes and large nutrient-rich rivers. Most species in the genus are tolerant of elevated levels of total dissolved solids and are present in highly calcareous or salinified waters. *Cyclostephanos* is apparently very widely distributed, but it is probably under-reported because of taxonomic uncertainty. At least two distinct series, both morphologically and ecologically, occur within the genus as presently conceived. The original concept of the genus was based on species that have strongly developed internal marginal ribs and have a gross appearance similar to those of *Cyclotella*. The identification of *Cyclostephanos* has been expanded to include species that have branched, but less highly developed, marginal ribs. Included are a number of species that closely resemble *Stephanodiscus* and that can only be certainly distinguished with SEM.

An example of the first morphological group is *C. dubius*, a species widely found in Europe and occasion-

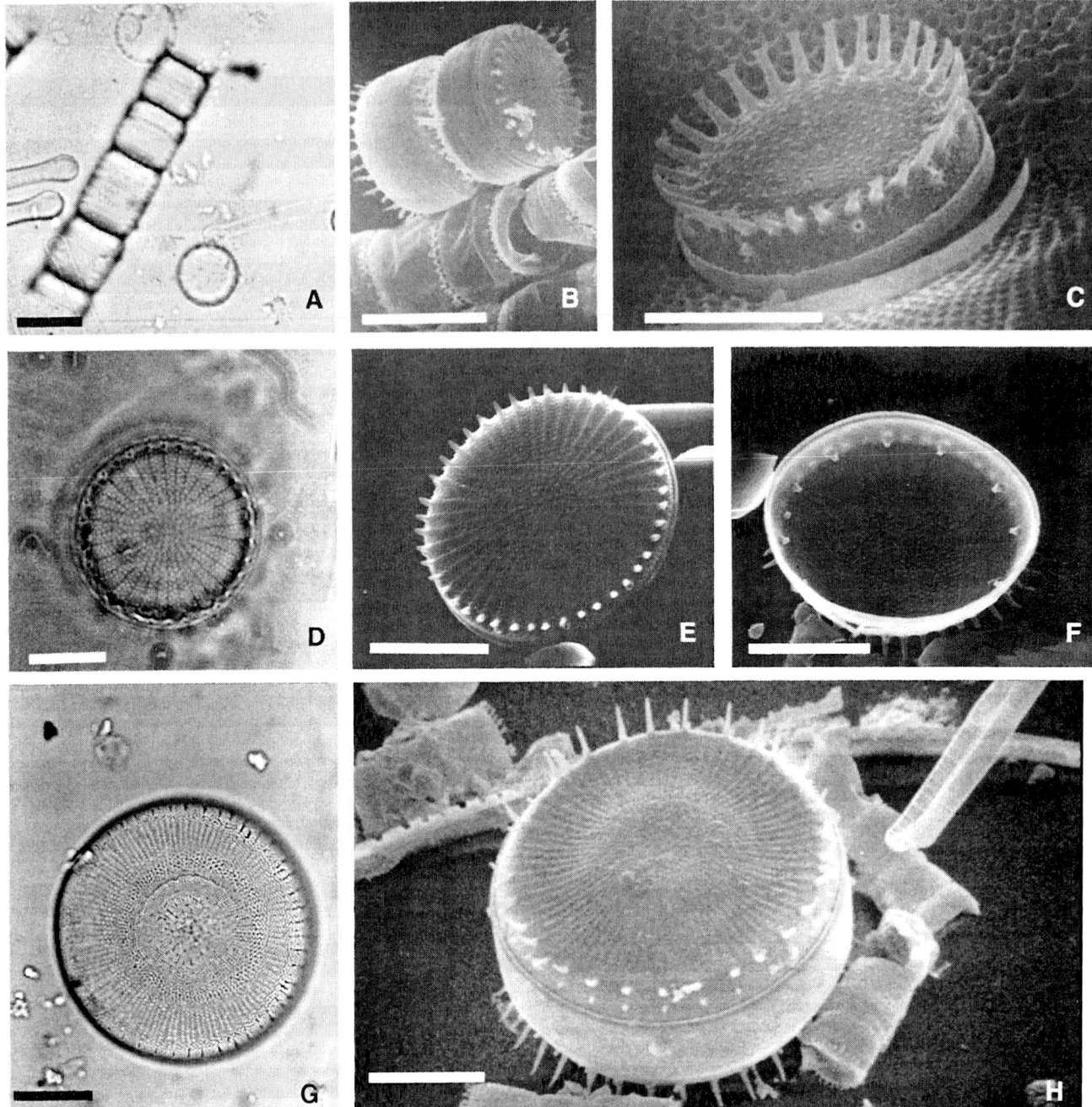

FIGURE 10 (A) LM girdle and valve views of *Stephanodiscus binderanus*, (B) SEM girdle view of *S. hantzschii* (upper) and *S. binderanus* (lower), (C) SEM oblique view of a separation valve of *S. binderanus* var. *oestrupi*, (D) LM valve view of *S. hantzschii*, (E) SEM external valve view of *S. hantzschii* fo. *tenuis*; (F) SEM internal valve view of *S. hantzschii* fo. *tenuis*, (G) LM valve view of *S. niagarae*, (H) SEM valve view of *S. niagarae* (largest specimen). All scale bars equal 10 μm.

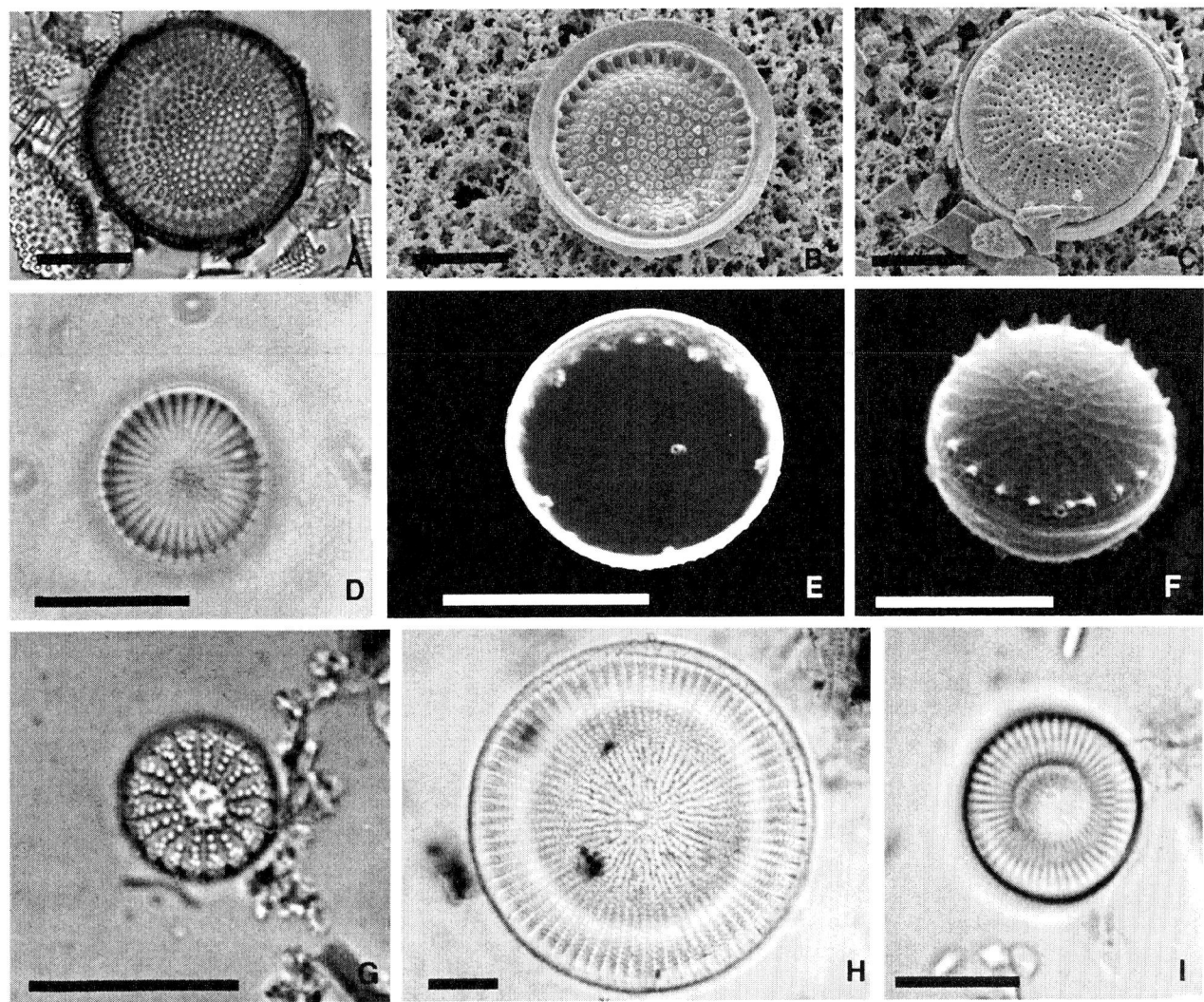

FIGURE 11 (A) LM valve view of *Pliocaenicus* sp. (B) SEM interior valve view of *Pliocaenicus* sp. (C) SEM exterior valve view of *Pliocaenicus* sp. (D) LM valve view of *Cyclostephanos invisitatus*. (E) SEM interior valve view of *C. invisitatus*, (F) SEM exterior valve views of *C. invisitatus*, (G) LM valve view of *Cyclostephanos* sp. (H) LM valve view of *C. damasii*, (I) LM valve view of *C. tholiformis*. All scale bars equal 10 μm.

ally reported from North America. Species that exhibit this morphology are most abundant during spring and fall in temperate regions. Most species that exhibit the second type of morphology occur either in tropical regions, such as *C. damasii* (Stoermer and Håkansson, 1983), a large species common in the African Rift Lakes, or during the summer in temperate regions. Included in this latter group are a number of species (e.g., *C. invisitatus*, and *C. costatilimbus*, and *C. tholiformis*) abundant in North America (Stoermer *et al.*, 1987; Håkansson and Kling, 1990). These differences indicate that further revision of this genus may be necessary.

Cyclotella (Küttzing) Brébisson (Figs. 3D and E; and 8A–I)

Traditionally, the presence of differing central and marginal areas was used to identify members of this group. Under the light microscope (Fig. 8A, D, and G), detailed morphology of individual structures within these regions is poorly resolved. SEM investigation of the taxa shows that these features are greatly dissimilar (Fig. 8C, F, and I). Both strutted and labiate processes (Fig. 2A, B, and C) are found in *Cyclotella* species, but the location and morphology differ in individual species (Fig. 8B, E, and H). These differences indicate that *Cyclotella* is most likely an unnatural group. A

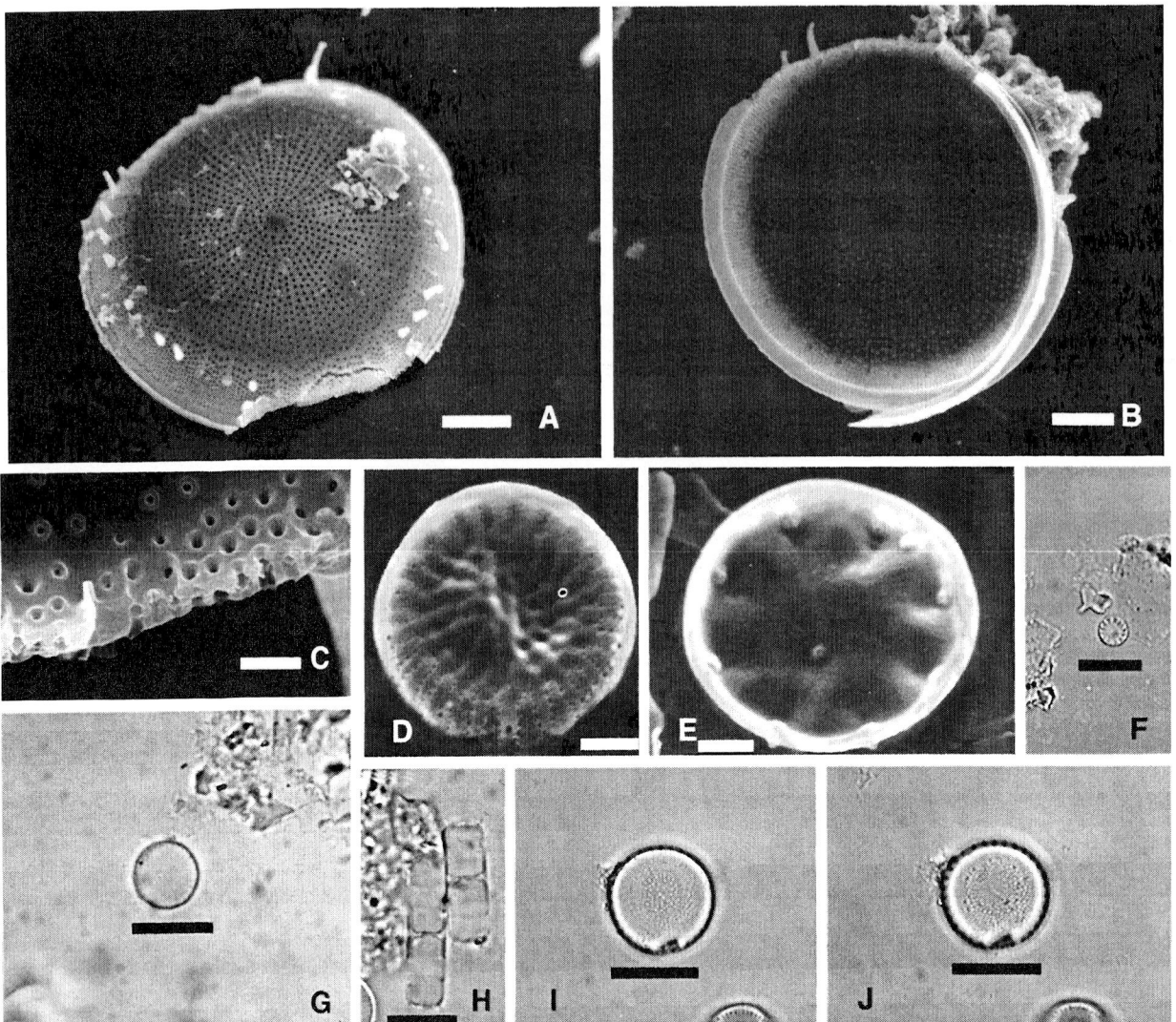

FIGURE 12 (A) SEM exterior view of *Mesodictyon* sp. (B) SEM interior view of *Mesodictyon* sp. (C) SEM view of broken valve margin of *Mesodictyon* sp., showing characteristic median placement of cribra in puncta, (D) SEM exterior view of *Stephanocostis chantaicus*, (E) SEM interior view of *Stephanocostis chantaicus*, (F) LM valve view of *Stephanocostis chantaicus*, (G) LM valve view of *Skeletonema potomos*, (H) LM girdle view of *Skeletonema potomos*, (I) LM valve view of *Cyclotubicoalitus undatus* focused on valve surface, (J) LM valve view of *Cyclotubicoalitus undatus* focused on valve mantle, showing conspicuous processes. Scale bars equal 10 µm for A, B, D, E, F, G, H, I, and J. Scale bar equals 1 µm for C.

number of researchers have broken the genus up into smaller subgenera or species complexes. Lowe (1975) detailed three morphological groups, based upon transmision and scanning electron microscopy observations of strutted and labiate processes (Fig. 2A, B, and C). Serieyssol (1981) further subdivided the genus into six subgroups, using additional information from SEM investigation of the alveolar morphology. Loginova (1988) revisited these infrageneric classifications and described 12 morphological groups from fossil and recent localities.

Other genera and species share characteristics similar to those in individual *Cyclotella* species. The failure of the current classification system to adequately categorize the species and the disparity among researchers regarding the number of subgroups suggest the need to evaluate this group's evolutionary history in the context of other species in the family. Without phylogenetic studies of *Cyclotella* and Thalassiosirales in general, the validity or taxonomic position of the above groups is dubious at best. We are currently confident only in separation of two groups—traditional *Cyclotella* species and *Stephanocyclus* (see below)—as distinct. Many other genera may be separated from *Cyclotella* in the future, but this is best done only after understanding of evolutionary history is more complete.

Cyclotella is portrayed in the literature as an environmentally important, yet poorly understood, group of diatoms. Detailed limnological studies concerning the genus date back to the turn of the century (Brünnthaler *et al.*, 1901). Taxa within the group are recognized as important environmental indicators in a broad range of environments. Hutchinson (1967) refers to the "oligotrophic *Cyclotella* flora" in his classic treatise but species occur under ecological conditions ranging from oligotrophic [e.g., *C. bodanica* (Stoermer and Yang, 1969; Willen *et al.*, 1990)] to hypereutrophic [e.g., *C. pseudostelligera* (Stoermer and Ladewski, 1976)]. Morphological variants appear to respond to more subtle environmental conditions within these general environmental niches (Belcher *et al.*, 1966; Haworth and Hurley, 1986; Wunsam *et al.*, 1995). Other species have been identified as important stratigraphic markers, identifying specific time periods in lake sediments (Stoermer *et al.*, 1996; Julius *et al.*, 1997b).

Cyclotubicoalitus Stoermer, Kociolek, *et* Cody
(Fig. 12I and J)

Under the light microscope *Cyclotubicoalitus* resembles a *Stephanodiscus* with poorly organized fascicles. It is distinguished from other centric diatoms by having unusual marginal structures that appear to be spines fused together with marginal strutted processes. A more complete description of the genus can be found in Stoermer *et al.* (1990).

Cyclotubicoalitus is apparently a very rare genus, at present known only from its type locality, L Lake, Barnwell County, South Carolina, an artificial impoundment formed as a cooling pond for a nuclear reactor on the Savannah River Nuclear Reservation, and the Primorsk reservoir in Russia (Genkal *et al.*, 1998).

Ellerbeckia Crawford (Fig. 6J and K)

Ellerbeckia has been recently described (Crawford, 1988). Most species currently included in this genus were formerly placed in either *Melosira* (*sensu lato*) or *Paralia*. *Ellerbeckia* is distinguished from other freshwater centric diatoms by a unique type of tubular process in the valve mantle. The valve mantle is very thick and is formed of numerous chambers, which are connected to the cell lumen. The valve surface apparently lacks pores or processes, except for a marginal ring in some species. Sibling valves consist of a "cameo" valve bearing a system of ridges and an "intaglio" valve with a complementary system of grooves. This structure is in some ways similar to that of the marine genus *Paralia*, and there is some difference of opinion concerning the separation of the two genera (Round *et al.*, 1990).

The most common species is *E. arenaria*, which is commonly found on sandy sediments in large, oligotrophic to mesotrophic, north temperate to boreal lakes. Unlike most centric diatoms, it is primarily benthic, and colonies are found in the plankton only after strong mixing events. Its cells are relatively large and are connected by their valve faces into short colonies of rarely more than 30 cells. Species of *Ellerbeckia* are also common in many freshwater fossil deposits, and it appears that the genus may have been more widely distributed in the past than it is in the modern flora.

Hydrosera Wallich (Fig. 7A–D)

Hydrosera is one of a few multipolar (Medlin *et al.*, 1996) centric diatoms represented in the freshwater flora. Its valves are very thick and appear as two superimposed triangles in valve view. The protuberances that form one of the triangles have well-developed pore fields at their apices. Living cells are attached in zig-zag colonies by pads of organic material, which are secreted from these pore fields. The terminal cell of a colony is usually attached to a solid substratum by the same mucilaginous secretion (Round *et al.*, 1990). In girdle view the cells are quite elongate and appear grooved due to the shape of the valves.

Hydrosera has restricted distribution in North America. Most reports are from southern coastal areas and the Hawaiian Islands. Other than its primarily subtropical distribution, little is known about the ecology of *Hydrosera* species.

Melosira Agardh (Fig. 6A–D)

Melosira is distinguished from *Aulacoseira* and other freshwater diatoms with similar colonial growth habits by uniformly structured valve walls, without costae or septae, and lack of spines visible under the light microscope. As presently conceived, it has a limited number of freshwater species. The genus was previously construed to include nearly all centric diatoms that grow in tightly bound colonies with their valve faces appressed. The most common freshwater species, according to modern classification schemes, is *M. varians*, that occurs in considerable abundance in naturally eutrophic and polluted streams and lakes throughout North America (personal observation). It is generally benthic in growth habit, but it is also commonly entrained into the plankton. Some authorities (e.g., Cholnoky, 1968) consider it an indicator of organic pollution.

Another less commonly observed species are *M. undulata*. Its ecological requirements and distribution are essentially opposite to those of *M. varians*. It is most commonly found in deep water collections from very large lakes (Hustedt, 1930; Mahood *et al.*, 1984;

McIntire *et al.*, 1994) or from smaller lakes and ponds in the Arctic. It is also a common component of freshwater diatomites, and reworked specimens are sometimes found in collections from lakes and streams, particularly in western North America.

Mesodictyon Theriot *et* Bradbury (Fig. 12A–C)

So far as is presently known, all *Mesodictyon* species lived only during the Miocene period (Theriot and Bradbury, 1987, Serieyssol *et al.*, 1996). The genus is characterized by the placement of the marginal strutted processes in pitlike depressions near the valve margin and by the fact that cribra (finely poroid siliceous membranes) of the punctae are localized in center of each punctum lumen, rather than nearer the internal or external surface of the valve. These characters are only clearly visible at SEM magnifications.

It is included here because reworked *Mesodictyon* species are sometimes found in collections from western rivers and may be confused with *Stephanodiscus* spp.

Orthoseira Crawford (Fig. 6I)

Orthoseira species have morphologies grossly similar to those of *Aulacoseira* and were previously included in that genus. *Orthoseira* is distinguished from *Aulacoseira* by the presence of a unique tube process near the center of the valve and bladelike, rather than simple, spines on the valve margin. Marginal pore fields are present near the valve margins, but these are only clearly visible with SEM. *Orthoseira* species usually occur in short colonies, but the spines do not appear to serve as the mechanism for binding cells together. In living colonies, the tube processes are aligned, and it appears that colonies are bound together by organic material secreted from these central tube processes (Round *et al.*, 1990). In girdle view thickened girdle bands, similar in some respects to those found in *Ellerbeckia*, are visible at median focus.

Orthoseira species are found almost exclusively in subaerial habitats, particularly in bryophyte communities growing on alkaline substrata and are rarely found in lakes or high-order streams.

Pelagodictyon Clarke (not illustrated)

According to its original description (Clarke, 1994), *Pelagodictyon* is distinguished from morphologically similar species of *Cyclostephanos* and *Stephanodiscus* by its very delicately formed valves, which are composed of radial ribs joined by cross ribs (frets) to form polygonal areolae. The areolae are closed by cribra, which lie near the valve surface. Other structures of its valves are very similar to those of *Cyclostephanos* and *Stephanodiscus*, in that marginal and central strutted processes are present and the marginal strutted processes extrude stiff rods (presumably of β-chitin) as in *Thalassiosira* and its allies. Although not well shown in his illustrations, Clarke (1994) indicates that a single labiate process is found at the level of the juncture of the valve surface and the valve mantle. Genkal and Kiss (1998) questioned the validity of this genus, believing that the species described are only partially formed valves of *Stephanodiscus* or *Cyclostephanos*. Although we have not specifically investigated this problem, experience lends credibility to the arguments presented by Genkal and Kiss (1998). Very thin, incompletely formed, valves of several species of *Cyclostephanos* and *Stephanodiscus* are commonly found in lakes marginal to the Laurentian Great Lakes, which are subject to periodic silica depletion brought about by heavy phosphorus loading (Schelske and Stoermer, 1971, 1972). Any reports of the occurrence of *Pelagodictyon* should be treated with skepticism until its biology is further investigated.

Pleurosira (Ehrenberg) Compère (Fig. 7E–H)

Pleurosira is another member of the multipolar centric diatoms (Medlin *et al.*, 1996) present in the freshwater flora. Freshwater species were formerly included in *Biddulphia*, but Compère (1982) formally separated them from marine taxa. Valves are semicircular, with two prominent ocelli opposed to one another on the periphery. Labiate processes are present (Fig. 2A) and are located near the center of the valve. In *P. laevis*, the most common freshwater species, two groups of two to four labiate processes are located in unornamented areas lateral to the axis between the ocelli. The valves are finely poroid, and the poroids are arranged in somewhat sinuous rows, and are closed by external cribra (Johnson and Rosowski, 1992). The frustules are elongate in girdle view and live specimens are joined together in zig-zag colonies by secretions of organic material from the ocelli. Girdle bands are variable in structure and are ornamented with rows of fine poroids (Johnson and Rosowski, 1992).

Pleurosira is widely distributed in naturally saline (e.g., Czarnecki and Blinn, 1978) or anthropogenically salinized (e.g., Kociolek *et al.*, 1983) inland waters in North America and southern Eurasia (Aboal *et al.*, 1996; Khan, 1995). Hustedt (1930) indicates that it is very widely distributed, particularly in the estuarine reaches of coastal rivers, although all reports may not refer to the same entity.

Pliocaenicus Round *et* Håkansson (Fig. 11A–C)

Pliocaenicus is a rare genus, originally described from a fossil locality in Germany (Round and Håkansson, 1992). *Pliocaenicus* shares a number of characteristics with *Cyclotella*, *Cyclostephanos*, and *Stephanodiscus* and the distinction between these genera is not entirely clear. It is identified primarily by a combination of a transversely undulate valve surface, with areolae radiating from the valve center but not fasciculate on the valve face, and the absence of areolae below the marginal strutted processes (Fig. 2B).

Although it has not yet been reported from North America, it is apparently present in freshwater diatomites from western North America (J. P. Kociolek and G. K. Khursevich, personal communication). It is also present in recent material from very oligotrophic lakes in Siberia and Pleistocene deposits in Alaska (Flower *et al.*, 1998). Extant populations may be present in oligotrophic habitats in Alaska and Northern Canada.

Skeletonema Greville (Figs. 12G and H)

Skeletonema is a primarily marine and estuarine genus which has a few representatives in fresh water. The circumscription of freshwater species is difficult. Known species are very lightly silicified, and siliceous structures present approach the limits of resolution of the light microscope. SEM observation is also rendered difficult because the lightly silicified valves collapse when dried. This being the case, it is not surprising that species now understood to belong to *Skeletonema* have previously been reported under *Stephanodiscus*, *Thalassiosira*, and probably *Cyclotella* (Hasle and Evensen, 1975, 1976). The genus is colonial. Cells are connected by more or less elongate, tubular processes. A marginal strutted process occurs at the base of each connecting process and a single labiate process apparently occurs at the same level. Because of its very light silicification, freshwater species of *Skeletonema* may be destroyed by usual cleaning procedures.

Its known distribution appears to be primarily in either naturally (Baltic Sea) or anthropogenically (Lake Erie, Ohio) salinified and eutrophic inland lakes and rivers (Hasle and Evensen, 1975, 1976). In such situations, *Skeletonema* may form massive blooms (Weber, 1970).

Stephanocostis Genkal *et* Kuzmina (Figs. 12D–F)

As presently understood, *Stephanocostis* has a single species, *S. chantaicus*. It superficially resembles *Stephanodiscus*, but lacks an external tube from the labiate process and has markedly raised ribs on the external valve surface which separate irregular fascicles of punctae that are difficult to resolve with a light microscope. This configuration gives the valve surface a folded appearance.

Stephanocostis is apparently a very rare genus, thus far reported from Siberia, fall and winter plankton from Lake Strechlin in Europe, and lakes in Canada (Krammer and Lange-Bertalot, 1991a). Occasional specimens of *S. chantaicus* have been also observed in sediment samples from Lake Superior (E. F. Stoermer unpublished observations).

Stephanocyclus Skabitschevsky (Figs. 9D–J)

As construed here, the genus *Stephanocyclus* includes species that have traditionally been classified under *Cyclotella meneghiniana* and its allies. These taxa have the structurally different marginal and central areas that are characteristic of *Cyclotella*, but lack the two-layered wall characteristic of *Cyclotella sensu stricto*, and marginal chambers, characteristic of *Cyclostephanos*. A study of the chloroplast genome (Bourne, 1992) indicates that *Stephanocyclus* shares its most recent common ancestry with marine and estuarine taxa that have been classified under *Thalassiosira*. This is in accord with its known distribution. *Stephanocyclus* species are most abundant in high conductance waters of inland areas, including other halophilic species such as *S. meneghiniana* (Fritz, 1990) and those found only in highly saline lakes, such as *S. caspia* and *S. quillensis* (Fritz and Battarbee, 1986; Fritz *et al.*, 1991). Confusion between *Cyclotella* and *Stephanocyclus* has had an unfortunate consequence in the ecological literature, in that much of the experimental data (e.g., Titman, 1976) on "*Cyclotella*" refers to *S. meneghiniana*, which has ecological tendencies diametrically opposed to most species of *Cyclotella sensu stricto*.

Stephanocyclus species are very widely distributed in hard water and saline habitats, particularly large rivers and eutrophic lakes and ponds throughout North America (Williams, 1964, 1972; Stoermer, 1978).

Stephanodiscus Ehrenberg (Figs. 10A–H)

Stephanodiscus is one of the most diverse and widespread genera of planktonic diatoms in North America (Håkansson and Kling, 1990) and the northern hemisphere in general (Krammer and Lange-Bertalot, 1991a). It is characterized by fasciculate radial striae separated by more or less strongly developed internal ribs, marginal spines (Fig. 2D), some or all of which are subtended by marginal strutted processes (Fig. 2D), internally domed cribra (Fig. 2F) of the valve and valve mantle punctae, and one or more labiate processes (Fig. 2F), which lie at the level of the spines. The valve surface is flat or concentrically undulate in most modern populations, although strongly eccentrically undulate valves are found in fossil populations included in *Stephanodiscus*, such as *S. rhombus* (Mahood, 1981) and *S. excentricus* (Håkansson and Stoermer, 1987). Central strutted processes (Fig. 2C) are present in most, but not all, species. Several species of *Stephanodiscus* are regularly colonial, and others may also adapt a colonial growth habit under certain conditions. In these taxa, the marginal spines become branched and closely applied to sister valves, binding the cells together in long colonies (Fig. 10A–C). Many species are also highly polymorphic with respect to other aspects of frustular morphology (Stoermer *et al.*, 1979; Håkansson and Stoermer, 1984). Much of this variability depends on ecological conditions (Theriot and Stoermer, 1984a, 1986; Theriot *et al.*, 1988), and it is possible for some species to exist either singly or colonially and have several different types of valve morphology within each condition. Because of this, past trophic conditions can be inferred in paleoecological studies from the morphology of specimens preserved in sediments (Stoermer *et al.*, 1989; Julius *et al.*, 1998). The ability to modify valve structure appears to be an adaptation to differing levels of available dissolved silica (Theriot, 1987) and is present in several diatom genera. Among species of *Stephanodiscus*, those that can modify structure according to silica availability seem best able to survive under severe silica limitation resulting from eutrophication. Those that exhibit constant morphology tend to be more common in oligotrophic habitats and are extirpated when these environments are modified by human activities (Stoermer *et al.*, 1985a).

Stephanodiscus is one of the most widespread and common freshwater planktonic diatom genera. Some species may be found in nearly all lakes, ponds, and large rivers in North America. Despite occasional literature statements to the contrary, there appears to be no consistent ecological preference through the genus. This and the presence of significantly divergent morphologies suggest that *Stephanodiscus* may be underclassified. There is some evidence of very rapid speciation within the genus (Theriot and Stoermer, 1984b; Theriot, 1992). These factors, combined with more than a century without significant taxonomic revision (Spamer and Theriot, 1997), lead to a great deal of uncertainty in *Stephanodiscus* classification.

Terpsinoë Ehrenberg (Fig. 5F and G)

Terpsinoë belongs to the multipolar group of centric diatoms. Its valves are elongate, with undulate margins. Valve surface ornamentation consists of somewhat irregularly spaced punctae that radiate around the center of the valve. The central region also contains a conspicuous labiate process, slightly offset from the center of the valve. Apical pore fields occupy the ends to the valve, and several deep costae transect the long dimension of the valves. These are bent to the side on their internal ends and have the appearance of musical notes in girdle view (Round *et al.*, 1990). This is reflected in the name of the type species, *T. musica* Ehrenberg. Living cells of *Terpsinoë* are attached in zig-zag colonies by organic pads secreted from the terminal apical pore fields. It most often is attached to solid substrata such as aquatic plants, but is also commonly entrained into the plankton and apparently adapts well to planktonic existence (O'Farrell, 1994).

Although not widely reported, *Terpsinoë* is common in fresh water and slightly brackish water habitats in the warmer regions of North America and apparently worldwide. According to published reports, it is most common in the tropics (Sterrenburg, 1994) and is quite rare in Europe (Hustedt, 1930). It appears to be most abundant in rivers and lakes that are connected to the sea, but it has been noted in other habitats. Wujek and Welling (1981) found it together with *Pleurosira laevis* in samples from Lake Michigan, but neither has been noted from this locality either before or since.

Thalassiocyclus Håkansson et Mahood (Fig. 9K)

Thalassiocyclus is, so far as is presently known, a monotypic genus of uncertain affinities. It was originally described by Hustedt (1957) as *Stephanodiscus lucens*. Krammer and Lange-Bertalot (1991a) treat it under that epithet as a taxon of uncertain affinities, but Håkansson and Mahood (1993) later erected *Thalassiocyclus* to accommodate it. It has the same basic valve structure as *Stephanodiscus*, but it lacks marginal spines and a tube-like extension of the labiate process. The ribs of the valve surface are also irregular and the striae are not clearly fasciculate. The valves are also eccentrically undulate, a feature common in *Cyclotella* but rare in modern populations of *Stephanodiscus*. Some fossil populations classified in *Stephanodiscus*, such as *S. rhombus* (Mahood, 1981) and *S. excentricus*. (Håkansson and Stoermer, 1987), have this morphology. The raised ribs on the valve face are similar to those found in *Stephanocostis*, but not nearly so highly developed.

Thalassiocyclus lucens is found in coastal rivers (Hustedt,1957), estuaries (Håkansson and Mahood, 1993), and some salinified inland rivers (E. F. Stoermer, personal observation).

Thalassiosira Cleve (Figs. 9A–C)

Thalassiosira is a very large and complex genus, which, as presently construed, contains species with several divergent morphological trends. Although *Thalassiosira* has not been widely reported as a major component of truly freshwater systems, a number of the morphological types found in the genus are represented in the freshwater flora. Because of the range of morphological diversity present, it is difficult to succinctly characterize *Thalassiosira* and further research may result in additional changes in classification (Hasle, 1978; Hasle and Lange, 1989). As presently understood, freshwater species of *Thalassiosira* share the following characteristics. Its valves are circular, with either flat or undulate surfaces. The valve mantle is usually short and is ornamented differently than the valve surface. The primary surface ornamentation of the valve consists of radial striae, which are not arranged in fascicles. There are one to several strutted processes near the center of the valve (Fig. 2c), often in a circular array. The number present depends on the diameter of valves (Johansen and Theriot, 1987). The margin of the valve has a ring of spines or occluded processes, and there is a ring of marginal strutted processes below them (Fig. 2b), on the valve mantle. The strutted processes have three or four struts. In living cells, the strutted processes exude β-chitin, which may serve to increase form resistance to sinking (Walsby and Xypolyta, 1977). In some marine species, the cells are bound together in "string of bead" colonies by β-chitin strands exuded from the central strutted processes, but this colony type is rare in freshwater. Other than gross position, there is no necessary correspondence between the placement of spines or occluded processes and the placement of marginal strutted processes. One or more labiate processes (Fig. 2a) occur on the valve face, usually just internal to the spines or occluded processes. In freshwater species the external expression of the labiate process is usually a long tube, clearly visible under the light microscope. Many marine species form resting spores (McQuoid and Hobson, 1995), but this is rarely observed in fresh water.

Thalassiosira species are often reported from freshwater fossil localities (Lupikina and Khursevich, 1992; Khursevich and VanLandingham, 1993; Alcantara, 1997; Serieyssol *et al.*, 1998), leading to speculation that they may be ancestral to the eucentric freshwater diatoms. Although perhaps under-reported, modern species are common in estuaries (Mahood *et al.*, 1986; Muylaert and Sabbe, 1996) as well as in high conductance inland waters (Lowe and Busch, 1975; Hasle, 1978; Hasle and Lange, 1989). Most reports come from rivers (Belcher and Swale, 1977; Kiss, 1984; Descy and Willems, 1991), but *Thalassiosira* has also been reported from polluted ponds (Cassie and Dempsey, 1980) and large lakes that have undergone significant anthropogenic modification (Hasle, 1978; Stoermer, 1978; Edlund *et al.*, 2000). Because salinification is often associated with human disturbance, *Thalassiosira* species are considered invasive in many freshwater habitats (Mills *et al.*, 1993, Edlund *et al.*, 2000).

Urosolenia Round *et* Crawford (Fig. 5A–C)

Urosolenia was previously combined with the marine genus *Rhizosolenia* but, as pointed out by Round *et al.* (1990), the resemblances are mostly superficial. The genus substantially resembles *Acanthoceras* (see previous), the main difference being that the reduced valves of *Acanthoceras* bear two elongated spines, whereas those of *Urosolenia* have only one. The valves are small and unipolar. Most of the cell volume is enclosed in very lightly silicified, imbricate, scale-like intercalary bands. Because of their fragile construction, frustules are destroyed by normal diatom cleaning procedures. Valves, with their characteristic spines, are preserved in the sediments of some lakes, but the intercalary bands are usually lost. *Urosolenia* regularly forms spores (Schulz, 1929; Edlund and Stoermer, 1993), usually at the end of a bloom. The ecological affinities of *Urosolenia* are difficult to discern, possibly because of underclassification. Populations reported as the same species are found in lakes of widely divergent trophic types (Poulin *et al.*, 1995; Diaz *et al.*, 1998).

Urosolenia is most commonly reported as being abundant in large, deep lakes (Jackson *et al.*, 1990; Makarewicz and Bertram, 1991; Makarewicz, 1993,), particularly those undergoing ecological change, such as the Laurentian Great Lakes (Deniseger *et al.*, 1986; Yang *et al.*, 1996; Morabito and Curradi, 1997). Part of its ecological preference may have to do with its ability to exist at depth (Jackson *et al.*, 1990; Agbeti *et al.*, 1997), but the genus can also thrive in shallow, hypereutrophic lakes (Edlund and Stoermer, 1993) and shows the same response to disturbance in smaller water bodies (Rhodes and Davis, 1995).

VIII. GUIDE TO LITERATURE FOR SPECIES IDENTIFICATION

Unfortunately, a comprehensive guide to the species of freshwater centric diatoms in North America has never been completed. Guides to species identification largely deal with European floras in either German (Huber-Pestalozzi, 1941; Hustedt, 1927–1930; Krammer and Lange-Bertalot, 1991) or French (Van Huerck, 1885). Although helpful, these works are not complete and do not reflect much of the taxonomic activity occurring in the last half of the 20th centruy.

ACKNOWLEDGMENTS

Although it is patently impossible to properly acknowledge everyone who has contributed to the material discussed here, the senior author would like to particularly thank Dr. H. Håkansson and Dr. C. L. Schelske for long-term research collaboration. Among numerous technical collaborators, T. B. and B. Ladewski and J. J. Yang were particularly instrumental in developing occurrence and distribution data discussed. Several students with special interests in centric diatoms made contributions to understanding the complexity of their taxonomic relationships, including Drs. N. A. Andresen, M. B. Edlund, J. C. Kingston, R. G. Kreis, Jr., E. C. Theriot, M. L. Tuchman, and J. A. Wolin. We thank the authors of *The Diatoms— Biology and Morphology of the Genera* and Cambridge University Press for permission to reproduce Figure 1. Figure 5E was contributed by Dr. M. B. Edlund, Figure 5B and C by Dr. J. C. Kingston, and Figure 12A–C by Dr. E. C. Theriot. Material for *Stephanocostis chantaicus* was provided by the California Academy of Sciences Diatom Herbarium, and specimens of *Cyclotubicoalitus undatus* were provided by the Academy of Natural Sciences Philadelphia Diatom Herbarium.

Research in our laboratory was supported by various Grants from the U.S. Environmental Protection Agency, the U.S. National Science Foundation, and the Natural Sciences Research Council of Sweden. The manuscript was prepared with partial support of National Science Foundation Grant DEB 9521882. This is Contribution 0607 of the Center for Great Lakes and Aquatic Sciences, University of Michigan.

LITERATURE CITED

Aboal, M., Puig, M. A., Soler, G. 1996. Diatom assemblages in some Mediterranean temporary streams in southeastern Spain. Archiv für Hydrobiologie 136:509–527.

Agbeti, M. D., Kingston, J. C., Smol, J. P., Watters, C. 1997. Comparison of phytoplankton succession in two lakes of different mixing regimes. Archiv für Hydrobiologie 140:37–69.

Alcantara, I. I. 1997. Neogene diatoms of Cuitzeo Lake, central sector of the trans-Mexican volcanic belt and their relationship with the volcano-tectonic evolution. Quaternary International 43:137–143.

Barber, H. G., Harworth, E. Y. 1981. A guide to the morphology of the diatom frustule. Freshwater Biological Association, Scientific Publication 44, 112 p.

Battarbee, R. W., Charles, D. F., Dixit, S. S., Renberg, I. 1999. Diatoms as indicators of surface water acidity, in: Stoermer, E. F., Smol, J. P., Eds., The diatoms: Applications for the environmental and earth sciences. Cambridge Univ. Press, Cambridge, UK, pp. 85–127.

Beaver, J. 1981. Apparent ecological characteristics of some common freshwater diatoms. Plankton Taxonomy Unit, Ontario Ministry of the Environment, Rexdale, Ontario, 517 p.

Belcher J. H., Swale, E. M. F. 1977. Species of *Thalassiosira* (diatoms, Bacillariophyceae) in the plankton of English rivers. British Phycological Journal 12:291–297.

Belcher, J. H., Swale, E. M. F., Heron, J. 1966. Ecological and morphological observations on a population of *Cyclotella pseudostelligera* Hustedt. Journal of Ecology 54:335–340.

Bethge, H. 1925. *Melosira* und ihre planktonbegleiter. Pflanzeforschung 3:1–78.

Bourne, C. E. M. 1992. Chloroplast DNA structure, variation and phylogeny in closely related species of *Cyclotella*. Doctoral Dissertation, University of Michigan, School of Natural Resources, 74 p.

Boyer, C. S. 1926. List of quaternary and tertiary Diatomaceae from deposits of southern Canada. Canadian Department of Mines, Victoria Memorial Museum, Museum Bulletin 45:1–26.

Boyer, C. S. 1927a. Synopsis of the North American Diatomaceae. Proceedings of the Academy of Natural Sciences of Philadelphia 78:1–228.

Boyer, C. S. 1927b. Synopsis of the North American Diatomaceae. Proceedings of the Academy of Natural Sciences of Philadelphia 79:229–583.

Bradbury, J. P. 1988. A climatic-limnologic model of diatom succession for paleolimnological interpretation of varved sediments at Elk Lake, Minnesota. Journal of Paleolimnology 1:115–131.

Brünnthaler, J., Prowazek S., von Wettstein, R. 1901. Vorlaufige Mitteilungen uber das Plankton des Attersees in Oberösterreich. Österreichische Botanische Zeitschrift 51:74–82.

Camburn, K. E. 1982. The diatoms (Bacillariophyceae) of Kentucky: A checklist of previously reported taxa. Transactions of the Kentucky Academy of Sciences 43:1–20.

Camburn, K. E., Kingston, J. C. 1986. The genus *Melosira* from soft-water lakes with special reference of northern Michigan, Wisconsin and Minnesota, in: Smol, J. P., Battarbee, R. W., Davis, R. B., Meriläinen, J., Eds., Diatoms and lake acidity. Junk, Dordrecht, pp. 17–34.

Camburn, K. E., Kingston, J. C., Charles, D. F., Eds. 1984–1986. PIRLA Diatom Iconograph. PIRLA Unpublished Report Series, Report 3, Electric Power Research Institute, 53 plates, 1059 figures.

Cassie, V., Dempsey, G. P. 1980. A new freshwater species of *Thalassiosira* from some small oxidation ponds in New Zealand, and its ultrastructure. Bacillaria 3:273–292.

Cholnoky, B. J. 1968. Die Ökologie der Diatomeen in Binnengewässern. Verlag von J. Cramer, Lehre, 699 p.

Clarke, K. B. 1994. *Pelagodictyon*—A new genus of centric diatom from the Norfolk Broads. Diatom Research 9:17–26.

Cleve-Euler, A. 1911a. Das Bacillariaceen-Plankton in Gewaessern bei Stockholm II. Zur Morphologie und Biologie einer pleomorphen *Melosira*. Archiv für Hydrobiologie 7:119–139.

Cleve-Euler, A. 1911b. Das Bacillariaceen-Plankton in Gewaessern bei Stockholm II. Zur Morphologie und Biologie einer pleomorphen *Melosira*. Archiv für Hydrobiologie 7:230–260.

Cleve-Euler, A. 1951. Die Diatomeen von Schweden und Finnland. Kunglica Svenska Vetenkapsakademiens Handlingar 4(2/1):1–163.

Cleve-Euler, A. 1952. Die Diatomeen von Schweden und Finnland. Kunglica Svenska Vetenkapsakademiens Handlingar 4(3/3):1–153.

Cleve-Euler, A. 1953a. Die Diatomeen von Schweden und Finnland. Kunglica Svenska Vetenkapsakademiens Handlingar 4(4/1):1–158.

Cleve-Euler, A. 1953b. Die Diatomeen von Schweden und Finnland. Kunglica Svenska Vetenkapsakademiens Handlingar 4(4/5):1–225.

Cleve-Euler, A. 1955. Die Diatomeen von Schweden und Finnland. Kunglica Svenska Vetenkapsakademiens Handlingar 4(5/4):1–232.

Compère, P. 1982. Taxonomic revision of the diatom genus *Pleurosira* (Eupodiscaceae). Bacillaria 5:165–190.

Conley, D. J., Kilham, S. S., Theriot, E. C. 1989. Differences in silica content between marine and freshwater diatoms. Limnology and Oceanography 34:205–213.

Conley, D. J., Schelske, C. L., Stoermer, E. F. 1993. Modification of the biogeochemical cycle of silica with eutrophication. Marine Ecology Progress Series 101:179–192.

Cox, E. J. 1996. Identification of freshwater diatoms from live material. Chapman & Hall, London, 158 p.

Cracraft, J. 1989. Speciation and its ontology: The empirical consequences of alternative species concepts for understanding patterns and processes of differentiation, in: Otte, D., Endler, J. A., Eds., Speciation and its consequences. Sinauer, Sunderland, Ma, pp.28–29.

Crawford, R. M. 1988. A reconsideration of Melosira arenaria and M. teres resulting in a proposed new genus Ellerbeckia, in: Round, F. E., Ed., Algae and the aquatic environment. Biopress, Bristol, UK, pp.413–433.

Crumpton, W. G. 1987. A simple and reliable method for making permanent mounts of phytoplankton for light and fluorescence microscopy. Limnology and Oceanography 32:1134–1159.

Crumpton, W. G., Wetzel, R. G. 1981. A method of preparing phytoplankton for critical microscopy and counting. Limnology and Oceanography 26:976–980.

Cummins, B. F., Wilson, S. E., Hall, R. I., Smol, J. P. 1995. Diatoms from British Columbia (Canada) lakes and their relationship to salinity, nutrients, and other limnological variables. Koeltz Scientific Books, Stuttgart, 207 p.

Czarnecki, D. B., Blinn, D. W. 1978. Diatoms of the Colorado River in Grand Canyon National Park and Vicinity. Bibliotheca Phycologica 38:1–181.

Davey, M. C. 1987. Seasonal variation in the filament morphology of the freshwater diatom Melosira granulata (Ehrenb.) Ralfs. Freshwater Biology 18:5–16.

Deniseger, J., Austin, A. Roch, M., Clark, M. J. R. 1986. A persistent bloom of the diatom Rhizosolenia eriensis (Smith) and other changes associated with decreases in heavy metal contamination in an oligotrophic lake, Vancouver Island. Environmental and Experimental Botany 26:217–226.

Descy, J.-P., Willems, C. 1991. Contribution to the knowledge of the River Moselle phytoplankton. Cryptogamie Algologie 12:87–100.

Diaz, M. M., Pedrozo, F. L., Temporetti, P. F. 1998. Phytoplankton of two Araucanian lakes of differing trophic status (Argentina). Hydrobiologia 370:45–57.

Dodd, J. J. 1987. Diatoms. The illustrated flora of Illinois. Southern Illinoid Univ. Press, Carbondale, 478 p.

Drum, R. W., Pankratz, H. S. 1964. Post mitotic fine structure of Gomphonema parvulum. Journal of of Ultrastructural Research 10:217–223.

Eardley-Wilmot, V. L. 1928. Diatomite. Its occurrence, preparation and uses. Canadian Department of Mines, Mines Branch 691:1–182.

Edlund, M. B. 1992. Silica related morphological variation in natural and cultured populations of Stephanodiscus niagarae (Bacillariophyta). Master's thesis, University of Michigan, School of Natural Resources and Environment, 102 p.

Edlund, M. B. 1998. Paleoecological evidence of climate change and historical patterns of planktonic diatom diversity inferred from the Lake Baikal (Russia) sediment record. Doctoral dissertation, University of Michigan, School of Natural Resources and Environment, 166 p.

Edlund, M. B., Stoermer, E. F. 1993. Resting spores of the freshwater diatoms Acanthoceras and Urosolenia. Journal of Paleolimnology 9:55–61.

Edlund, M. B., Stoermer, E. F. 1997. Ecological, evolutionary, and systematic significance of diatom life histories. Journal of Phycology 33:897–918.

Edlund, M. B., Stoermer, E. F., Taylor, C. M. 1996. Aulacoseira skvortzowii sp. nov. (Bacillariophyta), a poorly known diatom from Lake Baikal, Russia. Journal of Phycology 32:165–175.

Edlund, M. B., Taylor, C. M., Schelske C. L., Stoermer, E. F. 2000. Thalassiosira baltica (Bacillariophyta), a new exotic species in the Great Lakes. Canadian Journal of Fisheries and Aquatic Sciences 54:610–615.

Ehrenberg, C. G. 1843. Einen Nachtrag zu dem Vortrage über die Verbreitung und Einfluss des mikroskopischen Lebens in Süd- und Nord-Amerika. Abhandlundlungen Königliches Preussicher Akadamie der Wissenschaften zu Berlin 1841:202–209.

Ehrenberg, C. G. 1854, Mikrogeologie. Das Erden und Felsen schaffende Wirken des unsichtbar kleinen selbstständigen Lebens auf der Erde. Leopold Voss, Leipzig.

Ehrenberg, C. G. 1870. Über mächtie Gebirgs-Schichten vorherrschend aus mikroskopischen Bacillarien unter und bei der Stadt Mexiko. Abhandlundlungen Königliches Preussicher Akadamie der Wissenschaften zu Berlin 1869:1–66.

Elmore, C. J. 1922. The diatoms (Bacillarioideae) of Nebraska. Nebraska Geological Survey 8:1–216.

Fahnenstiel, G. L., Glime, J. M. 1983. Subsurface chlorophyll maximum and associated Cyclotella pulse in Lake Superior. Int. Rev. Ges. Hydrobiol. 68:605–616.

Flower, R. J., Håkansson, H. 1994. Crateriportula gen. nov., a new genus with close affinities to the genus Stephanodiscus. Diatom Research 9:249–258.

Flower, R. J., Ozornina, S. P., Kuzmina, A., Round, F. E. 1998. Pliocaenicus taxa in modern and fossil material mainly from eastern Russia. Diatom Research 13:39–62.

Foged, N. 1953. Diatoms from West Greenland. Meddelelser om Grønland 147:1–86.

Foged, N. 1955. Diatoms from Pearyland, North Greenland. Meddelelser om Grønland 128:1–90.

Foged, N. 1973. Diatoms from Southwest Greenland. Meddelelser om Grønland 194:1–84.

Foged, N. 1981. Diatoms in Alaska. Bibliotheca phycologica 53:1–137.

French, F. W., Hargraves, P. E. 1980. Physiological characteristics of plankton diatom resting spores. Marine Biology Letters 1:185–195.

Fritz S. C. 1990. Twentieth-century salinity and water level fluctuations in Devils Lake, North Dakota: Test of a diatom based transfer function. Limnology and Oceanography 35:1771–1781.

Fritz, S. C., Battarbee, R. W. 1986. Sedimentary diatom assemblages in freshwater and saline lakes of the Northern Great Plains, North America: Preliminary results, in: Round, F. E. Ed., Proceedings of the 9th International Diatom Symposium, Biopress, Bristol, UK, pp. 265–271.

Fritz S. C. Juggins, S., Battarbee R. W., Engstrom, D. R. 1991. Reconstruction of past changes in salinity and climate using a diatom based transfer function. Nature 352:706–798.

Geissler U. 1982. Experimentelle Untersuchungen zur Variabilitat der Schalenmerkmale bei einigen zentrischen Süsswasser-Diatomeen. I. Der Einfluss unterschiedlicher Salzkonzentrationen auf den Valva-Durchmesser von Stephanodiscus hantzschii Grunow. Nova Hedwigia 73:211–247.

Geitler, L. 1973. Auxosporenbildung und Systematik bei pennaten Diatomeen und die Cytologie von Cocconeis-Sippen. Österreichische Botanische Zeitschrift 122:299–321.

Genkal, S. I., Håkansson, H. 1990. The problem of distinguishing the newly described diatom genus Pseudostephanodiscus. Diatom Research 5:15–23.

Genkal, S. I., Kiss, K. T. 1998. The taxonomical place of the new centric diatom *Pelagiodictyon* Clarke, *in*: John, J. Ed., Abstracts of the 15th International Diatom Symposium Curtin University, Perth, Australia, pp.147–148.

Genkal, S. I., Makarova, A., Goncharov, A. A. 1998. Centric diatom species (Centrophyceae, Bacillariophyta) new for the waterbodies of Russia. Botanicheskii Zhurnal (St. Petersberg) 83:121–123.

Gersonde, R., Harwood, D. M. 1990. Lower Cretaceous diatoms from ODP leg 113 site 693 (Weddell Sea). Part 1: Vegetative cells, *in*: Barker, P. F., Kennett, J. P. *et al*., Eds., Proceedings of the Ocean Drilling Program, scientific results, Vol. 113, pp.365–401.

Gross, F., Zeuthen, E. 1948. The buoyancy of planktonic diatoms: A problem of cell physiology. Proceedings of the Royal Society of London Series B 135:382–389.

Håkansson, H., Jones, V. J. 1994. The compiled freshwater diatom taxa list for the Maritime Antactic region of the South Shetland and South Orkney Islands. Canadian Technical Reports on Fisheries and Aquatic Sciences 1957:77–83.

Håkansson, H., Kling, H. 1989. A light and electron microscope study of previously described and new *Stephanodiscus* species (Bacillariophyceae) from central and northern Canadian lakes, with ecological notes on the species. Diatom Research 4:269–288.

Håkansson, H., Kling, H. 1990. The current status of some very small freshwater diatoms of the genera *Stephanodiscus* and *Cyclostephanos*. Diatom Research 5:273–287.

Håkansson, H., Kling, H. 1994. *Cyclotella agassizensis* nov. sp. and its relationship to *C. quillensis* Bailey and other prairie *Cyclotella* species. Diatom Research 9:289–301.

Håkansson H., Mahood, A. 1993. *Thalassiocyclus* gen. nov.: A new genus in the Bacillariophyceae with comparison to closely related genera. Beihefte zur Nova Hedwigia 106:197–202.

Håkansson, H., Stoermer, E. F. 1984. Observations on the type material of *Stephanodiscus hantzschii* Grunow in Cleve and Grunow. Nova Hedwigia 39:477–495.

Håkansson, H., Stoermer, E. F. 1987. An investigation of the morphology and taxonomy of *Stephanodiscus excentricus* Hustedt (Bacillariophyta). Archiv für Protistenkunde 134:1–15.

Hamilton, P. B., Douglas, M. S. V., Fritz, S. C., Pienitz, R., Smol, J. P., Wolfe, A. P. 1994. A compiled freshwater diatom taxa list of the Arctic and Subarctic regions of North America. Canadian Technical Reports on Fisheries and Aquatic Sciences 1957:85–102.

Harwood, D. M. 1999. Diatomite, *in*: Stoermer, E. F., Smol, J. P., Eds., The diatoms: Applications for the environmental and earth sciences. Cambridge Univ. Press, Cambridge, UK, pp.436–443.

Harwood, D. M., Gersonde, R. 1990. Lower Cretaceous diatoms from ODP leg 113 site 693 (Weddell Sea). Part 2: Resting spores, chrysophycean cysts, an endoskeletal dinoflagellate, and notes on the origin of diatoms, *in*: Barker, P. F., Kennett, J. P. *et al*., Eds., Proceedings of the Ocean Drilling Program, scientific results, Vol. 113, pp. 403–415.

Harwood, D. M., Nikolaev, V. A. 1995. Cretaceous diatoms; morphology, taxonomy, biostratigraphy, *in*: Blome, C. D. *et al*., Convenors, Siliceous microfossils. Paleontological Society short courses in paleontology, 8, pp. 81–106.

Hasle, G. R. 1977. Morphology and taxonomy of *Actinocyclus normanii* f. *subsalsa* (Bacillariophyceae). Phycologia 16:321–328.

Hasle, G. R. 1978. Some freshwater and brackish water species of the diatom genus *Thalassiosira* Cleve. Phycologia 17:263–292.

Hasle, G. R., Evensen, D. L. 1975. Brackish-water and fresh-water species of the diatom genus *Skeletonema*. Grev. I. *Skeletonema subsalsum* (A. Cleve) Bethge. Phycologia 14:283–297.

Hasle, G. R., Evensen, D. L. 1976. Brackish water and freshwater species of the diatom genus *Skeletonema*. II. *Skeletonema potamos* comb. nov. Journal of Phycology 12:73–82.

Hasle, G. R., Lange, C. B. 1989. Fresh-water and brackish water *Thalassiosira* (Bacillariophyceae)—Taxa with tangentially undulated valves. Phycologia 28:120–135.

Haworth, E. Y., Hurley, M. A. 1986. Comparison of the stelligeroid taxa of the centric diatom genus *Cyclotella*, *in*: Ricard, M., Ed., Proceedings of the 8th International Diatom Symposium. O. Koeltz, Koenigstein, pp. 43–58.

Herth, W., Barthlott, W. 1979. The site of β-chitin fibril formation in centric diatoms. I. Pores and fibril formation. Journal of Ultrastruct Research 68:6–15.

Hickel, B., Håkansson, H. 1991. The freshwater diatom *Aulacoseira herzogii*. Diatom Research 6:299–305.

Hoagland, K. D., Rosowski, J. R., Gretz, M. R., Roemer, S. C. 1993. Diatom extracellular polymeric substances: Function, fine structure, chemistry, and physiology. Journal of Phycology 29:537–566.

Hohn, M. H. 1969. Qualitative and quantitative analysis of plankton diatoms, Bass Island area, Lake Erie, 1938–1965, including synoptic surveys of 1960–1963. Bulletin Ohio Biological Survey New Series 3:1–211+v–xv.

Hohn, M. H., Hellerman, J. 1963. The taxonomy and structure of diatom populations from three eastern North American rivers using three sampling methods. Transactions of the American Microscopic Society 82:250–329.

Huber-Pestalozzi, G. 1942. Das phytoplankton des Süsswasser: Systematik und Biologie, *in*: Thienemann, A. Ed., *Die Binnengewässer*, Band 16, Teil 2, Hälfte 2. E. Schweizerbart'sche Verlagsbuchhandlung, Stuttgart.

Hustedt, F. 1927–1930. Die Kieselalgen Deutschlands, Österreichs und der Schweiz, *in*: Rabenhorst, L., Ed., Kryptogamen-Flora von Deutschland, Österreich und der Schweiz 7(1):1–920.

Hustedt, F. 1930. Dr. L. Rabenhorsts Kryptogamen-Flora von Deutschland, Österreichs und der Schweiz. Band VII. Die Kieselalgen Deutschlands, Österreichs und der Schweiz unter Berücksichtigung der übringen Länder Europas sowie der angrensenden Meeresgebiete. Autorisierter Neudruck (1962), Johnson Reprint Corp. New York.

Hustedt, F. 1930a. Heft 10: Bacillariophyta (Diatomeae), *in*: Pascher, A., Ed. Süsswasser-Flora Mitteleuropas, zweite Auflage. Gustav Fisher Verlag, Jena.

Hustedt, F. 1931–1959. Die Kieselalgen Deutschlands, Österreichs und der Schweiz, *in*: Rabenhorst, L., Ed., Kryptogamen-Flora von Deutschland, Österreich und der Schweiz 7(2):1–845.

Hustedt, F. 1957. Die Diatomeenflora des Flussystems der Weser im Gebiet der Hansestadt Bremen. Abhandlungen Naturwissenschaftlicher der Verein zu Breman 34:181–440.

Hustedt, F. 1961–1966. Die Kieselalgen Deutschlands, Österreichs und de Schweiz, *in*: Rabenhorst, L. Ed., Kryptogamen-Flora von Deutschland, Österreich und der Schweiz, Vol. 7, No. 3, pp. 1–816.

Hutchinson, G. E. 1967. A treatise on limnology. Vol. 2. Introduction to lake biology and the limnoplankton. Wiley, New York, 1115 p.

Jackson L. J., Stockner, J. G., Harrison, P. J. 1990. Contribution of *Rhizosolenia eriensis* and *Cyclotella* spp. to the deep chlorophyll maximum of Sproat Lake, British Columbia. Canadian Journal of Fisheries and Aquatic Sciences 47:128–135.

Jewson, D. H. 1992a. Size reduction, reproductive strategy and the life cycle of a centric diatom. Philosophical Transactions of the Royal Society of London B Biological Sciences 336:191–213.

Jewson, D. H. 1992b. Life cycle of a *Stephanodiscus* sp. (Bacillariophyta). Journal of Phycology 28:856–866.

Johansen, J. R., Rushforth, S. R. 1985. A contribution to the taxonomy of *Chaetoceros muelleri* Lemmermann (Bacillariophyceae) and related taxa. Phycologia 24:437–447.

Johansen, J. R., Theriot, E. C. 1987. The relationship between valve diameter and number of central fultoportulae in *Thalassiosira weissflogii* (Bacillariophyceae). Journal of Phycology 23:663–665.

Johansen, J. R., Barclay, W. R., Nagle, N. 1990. Physiological variability within ten strains of *Chaetoceros muelleri* (Bacillariophyceae). Journal of Phycology 26:271–278.

Johnson, L. M., Rosowski, J. R. 1992. Valve and band morphology of some fresh-water diatoms. 5. Variations in the cingulum of *Pleurosira laevis* (Bacillariophyceae). Journal of Phycology 28:247–259.

Julius, M. L., Estabrook, G. F., Edlund, M. B., Stoermer, E. F. 1997a. Recognition of taxonomically significant clusters near the species level, using computationally intense methods, with examples from the *Stephanodiscus niagarae* complex. Journal of Phycology 33:1049–1054.

Julius, M. L., Stoermer, E. F., Colman, S. M., Moore, T. C. 1997b. Paleoclimatic implications of siliceous microfossil succession in Late Quaternary sediments in Lake Baikal, Siberia. Journal of Paleolimnology 18:187–204.

Julius, M. L., Stoermer, E. F., Taylor, C. M., Schelske, C. L. 1998. Local extirpation of *Stephanodiscus niagarae* Ehrenb. (Bacillariophyta) in the recent limnological record of Lake Ontario. Journal of Phycology 34:766–771.

Khan, M. A. 1995. Floristic studies of diatom communities and habitat characteristics of some inland waters of Libya. Arabandlungen Gulf. Journal of Scientic Research 13:133–149.

Khursevich, G. K., VanLandingham, S. L. 1993. Frustular morphology of some centric diatom species from Miocene fresh-water sedimentary-rocks of western USA and Canada. Nova Hedwigia 56:389–400.

Kiss, K. T. 1984. Occurrence of *Thalassiosira pseudonana* Hasle et Heimdal (Bacillariophyceae) in some rivers of Hungary. Acta Botanica Hungarica 30:277–287.

Kling, H. J. 1992. Valve development in *Stephanodiscus hantzschii* Grunow (Bacillariophyceae) and its implications on species identification. Diatom Research 7:241–257.

Kociolek, J. P. 1997. Historical constraints, species concepts and the search for a natural classification of diatoms. Diatom 13:3–8.

Kociolek, J. P., Stoermer, E. F. 1988a. Taxonomic and systematic position of the *Gomphoneis quadripunctata* species complex. Diatom Research 3:95–108.

Kociolek J. P., Stoermer E. F. 1988. A preliminary investigation of the phylogenetic relationships among the freshwater, apical pore field-bearing cymbelloid and gomphonemoid diatoms (Bacillariophyta). Journal of Phycology 24:377–385.

Kociolek, J. P., Williams, D. M. 1987. Unicell ontogeny and phylogeny: Examples from the diatoms. Cladistics 3:274–284.

Kociolek, J. P., Lamb, M. A., Lowe, R. L. 1983. Notes on the growth and ultrastructure of *Biddulphia laevis* Ehr. (Bacillariophyceae) in the Maumee River, Ohio. Ohio Journal of Science 83:125–130.

Kolkwitz, R, Marson, M. 1908. Ökologie der pflanzlichen Saprobien. Berichte der Deutschen Botanischen Gesellschaft 26a:505–519.

Krammer, K. 1997a. Die cymbelloiden Diatomeen. Eine Monographie der weltweit bekannten Taxa. Teil 1. Allgemeines und *Encyonema* Part. Bibliotheca Diatomologica 36:1–382.

Krammer, K. 1997b. Die cymbelloiden Diatomeen. Eine Monographie der weltweit bekannten Taxa. Teil 2. *Encyonema* Part., *Encyonopsis* und *Cymbellopsis*. Bibliotheca Diatomologica 37:1–469.

Krammer, K., Lange-Bertalot, H. 1986. Bacillariophyceae, 1 Teil, Naviculaceae, *in*: Ettl, H., Gerloff, J., Heynig, H., Mollenhauer, D., Eds., Süsswasserflora von Mitteleuropa, Band 2. Gustav Fischer, Stuttgart New York. 876 p.

Krammer, K., Lange-Bertalot, H. 1988. Bacillariophyceae, 2 Teil, Bacillariaceae, Epithemiaceae, Surirellaceae, *in*: Ettl, H., Gerloff, J., Heynig, H., Mollenhauer, D., Eds., Süsswasserflora von Mitteleuropa, Band 2. Gustav Fischer, Jena, 596 p.

Krammer, K., Lange-Bertalot, H. 1991a. Bacillariophyceae, 3 Teil, Centrales, Fragilariaceae, Eunotiaceae, *in*: Ettl, H., Gerloff, J., Heynig, H., Mollenhauer, D., Eds., Süsswasserflora von Mitteleuropa, Band 2. Gustav Fischer, Stuttgart/Jena, 576 p.

Krammer, K., Lange-Bertalot, H. 1991b. Bacillariophyceae, 4 Teil, Achnanthaceae, Kritische Erganzungen zu Navicula (Lineolatae) und Gomphonema, *in*: Ettl, H., Gerloff, J., Heynig, H., Mollenhauer, D., Eds., Süsswasserflora von Mitteleuropa, Band 2. Gustav Fischer, Stuttgart/Jena. 437 p.

Krebs, W. N., Bradbury, J. P., Theriot, E. C. 1987. Neogene and Quaternary lacustrine diatom biochronology, western USA. Palaios 2:505–513.

Lange-Bertalot, H. 1993. 85 Neue Taxa und über 100 weitere neu definierte Taxa ergänzend zur Süßwasserflora von Mitteleuropa Vol. 2/1–4. Biblotheca Diatomologica 27:1–454.

Lange-Bertalot, H., Metzeltin, D. 1996 Ecology-diversity-taxonomy. Indicators of oligotrophy, Iconographia Diatomologica, Vol. 2. Koeltz Scientific Books, Koenigstein, 390 p.

Lewis, W. M. 1984. The diatom sex clock and its evolutionary significance. America Naturalist 121:825–833.

Loginova, E. I. 1988. Classification of the diatom genus *Cyclotella*, *in*: Simola, H., Ed., Proceeding of the 10th International Symposium on Recent and Fossil Diatoms. O. Koeltz, Koenigstein, pp. 37–46.

Lohman, K. E., Andrews, G. W. 1968. Late Eocene nonmarine diatoms from the Beaver Divide Area, Fremont County, Wyoming. U.S. Geological Survey Professional Papers 593E:1–24.

Lowe, R. L. 1974. Environment requirements and pollution tolerances of freshwater diatoms. EPA-670/4-74-005, Cincinnati, OH, 344 p.

Lowe, R. L. 1975. Comparative ultrastructure of the valves in some *Cyclotella* species (Bacillariophyceae). Journal of Phycology 11:415–424.

Lowe, R. L., Busch, D. E. 1975. Morphological observations on two species of the diatom genus *Thalassiosira* from fresh-water habitats in Ohio. Transactions of the American Microscopical Society 94:118–123.

Lupikina, E. G., Khursevich, G. K. 1992. A new fresh-water species of *Thalassiosira* (Bacillariophyta) from Miocene deposits of Kamchatka. Journal of Paleontology 1:138.

Mahood, A. D. 1981. *Stephanodiscus rhombus*, a new diatom species from Pliocene deposits at Chiloquin, Oregon. Micropaleontology (NY) 27:379–383.

Mahood, A. D., Thomas, R. D., Goldman, C. R. 1984. Centric diatoms of Lake Tahoe. Great Basin Naturalist 44:83–98.

Mahood A. D., Fryxell, G. A., McMillan, M. 1986. The diatom genus *Thalassiosira*: Species from the San Francisco Bay system. Proceedings of the California Academy of Sciences 44:127–156.

Makarewicz, J. C. 1993. Phytoplankton biomass and species composition in Lake Erie, 1970 to 1987. Journal of Great Lakes Research 19:258–274.

Makarewicz, J. C., Bertram, P. 1991. A lakewide comparison study of phytoplankton biomass and its species composition in Lake Huron, 1971 to 1985. Journal of Great Lakes Research 17:553–564.

Mann, D. G. 1993. Patterns of sexual reproduction in diatoms. Hydrobiologia 269/270:11–20.

Margalef, R. 1978. Life forms of phytoplankton as survival alternatives in an unstable environment. Oceanologica Acta 1:493–509.

McBride, S. A., Edgar, R. K. 1998. Janus valves unveiled: Frustular morphometric variability in *Gomphonema angustatum*. Diatom Research 13:293–310.

McIntire, C. D., Phinney, H. K., Larson, G. L., and Buktenica, M. 1994. Vertical-distribution of a deep-water moss and associated epiphytes in Crater Lake. Oregon Northwest Science 68:11–21.

McKnight, B. K., Niem, A. R., Kociolek, J. P., Ranne, P. 1995. Origin of an Oligocene freshwater diatom-rich pyroclastic-debris flow in a shallow-marine forearc basin, NW Oregon, Journal of Sedimentary Research A65:505–512.

McQuoid M. R., Hobson, L. A. 1995. Importance of resting stages in diatom seasonal succession. Journal of Phycology 31:44–50.

Medlin L. K., Elwood, D. J., Stickel, S., Sogin, M. L. 1991. Morphological and genetic variation within the diatom *Skeletonema costatum* (Bacillariophyta): Evidence from a new species, *Skeletonema psuedocostatum*. Journal of Phycology 27:514–524.

Medlin, L. K., William, D. M., Sims, P. A. 1993. The evolution of diatoms (Bacillariophyta). I. Origin of the group and assessment of the monophyly of its major divisions. European Journal of Phycology 28:261–275.

Medlin, L. K., Kooistra, W. H. C. F., Gersonde, R., Wellbock, U. 1996. Evolution of the diatoms (Bacillariophyta) II. Nuclear-encoded small-subunit rRNA sequence comparisons confirm a paraphyletic origin for the centric diatoms. Molecular Biology and Evolution 13:67–75.

Medlin L. K., Kooistra, W. H. C. F., Gersonde, R., Sims, P. A., Wellbrock, U. 1997. Is the origin of the diatoms related to the end-Permian mass extinction? Nova Hedwigia 65:1–11.

Mills, E. L., Leach, J. H., Carlton, J. T., Secor, C. L. 1993. Exotic species in the Great Lakes—A history of biotic crises and anthropogenic introductions. Journal of Great Lakes Research 19:1–54.

Moll, R. A., Stoermer, E. F. 1982. A hypothesis relating trophic status and subsurface chlorophyll maxima of lakes. Archiv für Hydrobiologie 94:425–440.

Moore, B. N. 1937. Non-metallic mineral resources of eastern Oregon. U.S. Geological Survey Bulletin 875:1–180.

Moore, J. K., Villareal, T. A. 1996. Size-ascent rate relationships in positively buoyant marine diatoms. Limnology and Oceanography 41:1514–1520.

Morabito, G., Curradi, M. 1997. Phytoplankton community structure of a deep subalpine Italian lake (Lake Orta, N. Italy)—Four years after the recovery from acidification by liming. Inernational Revue der Gesamte Hydrobiologie 82:487–506.

Muylaert, K., Sabbe, K. 1996. The diatom genus *Thalassiosira* (Bacillariophyta) in the estuaries of the Schelde (Belgium, The Netherlands) and the Elbe (Germany). Botanica Marina 39:103–15.

Müller, O. 1903. Sprungweise Mutation bei Melosireen. Berichte der Deutschen Botanischen Gesellschaft 21:326–333.

Müller, O. 1906. Pleomorphismus, Auxosporen und Dauersporen bei *Melosira*-Arten. Jahrbuch für Wissenschaft Botanica 43:49–88.

Nipkow, F. 1927. Über das Verhalten der Skelette Planktonischer Kieselalgen in geschichteten Tiefschlamm das Zürich-und Baldeggersees. Schweizerische Zeitschrift für Hydrobiologie 4:71–120.

Nipkow, F. 1950. Ruheformen planktischer Kieselalgen in geschichteten Schlamm des Zürichsees. Schwiez. Z. Hydrol. 12:263–270.

O'Farrell, I. 1994. Comparative-analysis of the phytoplankton of 15 lowland fluvial systems of the River Plate basin (Argentina). Hydrobiologia 289:109–117.

Paasche, E., Johansson, S., Evensen, D. L. 1975. An effect of osmotic pressure on the valve morphology of the diatom *Skeletonema subsalsum* (A. Cleve) Bethge. Phycologia 14:205–211.

Patrick, R. 1961. A study of the numbers and kinds of species found in the rivers in eastern United States. Proceedings Academy of Natural Sciences of Philadelphia 113:215–258.

Patrick, R., Freese, L. 1961. Diatoms (Bacillariophyceae) from northern Alaska. Proceedings Academy of Natural Sciences of Philadelphia 112:129–293.

Patrick, R., Roberts, N. A. 1979. Diatom Communities in the Middle Atlantic States. Some factors that are important to their structure. Beihefte zur Nova Hedwigia 54:265–283.

Patrick, R. M., Reimer, C. W. 1966. The diatoms of the United States, Monograph 13. Academy of Natural Sciences of Philadelphia, 688 p.

Patrick, R. R., Reimer, C. W. 1975. The Diatoms of the United States, Vol. 2, Part 1, Monograph 13. Academy of Natural Sciences of Philadelphia, 213p.

Poulin M., Hamilton, P. B., Proulx, M. 1995. Catalog of the freshwater algae of Quebec (Canada). Canadian Field-Naturalist 109:27–110.

Prasad, A. K. S. K., Nienow, J. A., Livingston, R. J. 1990. The genus *Cyclotella* (Bacillariophyta) in Choctawhatchee Bay, Florida, with special reference to *C. striata* and *C. choctawhatcheeana* sp. nov. Phycologia 29:418–436.

Prescott, G. W. 1962. Algae of the western Great Lakes region, revised ed. Wm. C. Brown, Dubuque, IA, 977p.

Prescott, G. W., Dillard, G. E. 1979. A checklist of algal species reported from Montana 1891 to 1977. Monograph No. 1, Proceedings of the Montance Academy of Science (Suppl.) 36:1–102.

Rhodes, T. E., Davis, R. B. 1995. Effects of late Holocene forest disturbance and vegetation change on acidic Mud Pond, Maine, USA. Ecology 76:734–746.

Ross, R. 1983. Endemism and cosmopolitanism in the diatom flora of the East African Great Lakes. Systematics Association (London) 23:157–177.

Ross, R., Cox, E. J., Karayeva, N. I., Mann, D. G., Paddock, T. B. B., Simnsen, R., Sims, P. A. 1979. An amended terminology for the siliceous components of the diatom cell. Beihefte zur Nova Hedwigia 64:513–533.

Round, F. E. 1972. The problems of reduction of cell size during diatom cell division. Nova Hedwigia 23:291–303.

Round, F. E. 1981. *Cyclostephanos*—A new genus within the Sceletonemaceae. Archiv für Protistnenkunde 125:323–329.

Round, F. E., Bukhtiyarova, L. 1996, Four new genera based on *Achnanthes* (*Achnanthidium*) together with a re-definition of *Achnanthidium*. Diatom Research 11:345–361.

Round, F. E., Håkansson, H. 1992. Cyclotelloid species from a diatomite in the Harz Mountains, Germany, including *Pliocaenicus* gen. nov. Diatom Research 7:109–125.

Round F. E., Sims, P. A. 1981. The distribution of diatom genera in marine and freshwater environments and some evolutionary considerations, in: Ross, R., Ed., Proceeedings of the Sixth Symposium on Recent and Fossil Diatoms, Budapest. Otto Koeltz, Koenigstein, pp. 301–320.

Round, F. E., Crawford, R. M., Mann, D. G. 1990. The diatoms—biology and morphology of the genera. Cambridge University Press, Cambridge, 747 p.

Rushforth, S. R., Johansen, J. R. 1986. The inland *Chaetoceros* (Bacillariophyceae) species of North America. Journal of Phycology 22:441–448.

Schelske, C. L., Stoermer, E. F. 1971. Eutrophication, silica depletion and predicted changes in algal quality in Lake Michigan. Science 173:423–424.

Schelske, C. L., Stoermer, E. F. 1972. Phosphorus, silica and eutrophication of Lake Michigan, *in*: Likens, G. E. Ed., Nutrients and eutrophication, Sepecial Sympesiumn, Society Limnology and Oceanography, Vol. I, pp.157–171.

Schmid, A. M. 1994. Aspects of morphogenesis and function of diatom cell walls with implications for taxonomy. Protoplasma 181:43–60.

Schultz M. E., Trainor, F. R. 1968. Production of male gametes and auxospores in the centric diatoms *Cyclotella meneghiniana* and *C. cryptica*. Journal of Phycology 4:85–88.

Schulz V. P. 1929. Über Zellteilung und Dauersporenbildung der Diatomeengattungen *Attheya* und *Rhizosolenia*. Bototanisches Archiv 24:505–523.

Serieyssol, K. K. 1981. *Cyclotella* species of late Miocene age from St. Baulize, France, *in*: Proceedings of the 7th International Diatom Symposium. O. Koeltz, Koenigstein, pp. 27–42.

Serieyssol K. K., Theriot, E. C., Gasse, F. 1996. *Stephanodiscus radiatus* and *Mesodictyon gasseae*, two new Upper Miocene species of the Thalassiosiraceae from Ardeche, France. Nova Hedwigia 62:221–231.

Serieyssol, K. K., Garduno, I. I., Gasse, F. 1998. *Thalassiosira dispar* comb. nov. and *T. cuitzeonensis* spec. nov. (Bacillariophyceae) found in Miocene sediments from France and Mexico. Nova Hedwigia 64:177–186.

Shoeman, F. R., Archibal, R. E. M. 1980. The diatom flora of south Africa. National Report 6, National Institute for Water Research, Council of Science and Industrial Research, 34 p.

Sicko-Goad, L., Schelske, C. L., Stoermer, E. F. 1984. Estimation of carbon and silica content of diatoms from natural assemblages using morphometric techniques. Limnology and Oceanography 29:1170–1178.

Sicko-Goad, L., Stoermer, E. F., Fahnenstiel, G. 1986. Rejuvenation of *Melosira granulata* resting cells from the anoxic sediments of Douglas Lake, Michigan. I. Light microscopy and ^{14}C uptake. Journal of Phycology 22:22–28.

Sicko-Goad L., Stoermer, E. F., Kociolek, J. P. 1989. Diatom resting cell rejuvenation and formation: Time course, species records and distribution. Journal of Plankton Research 11:375–389.

Simonsen, R. 1979. The diatom system: Ideas on phylogeny. Bacillaria 2:9–71.

Simonsen, R. 1987. Atlas and catalogue of the diatom types of Friedrich Hustedt, 3 Vols. Journal of Cramer, Berlin, 525 p.

Siver, P. A., Kling, H. 1997. Morphological observations of *Aulacoseira* using scanning electron microscopy. Canadian Journal of Botany 75:1807–1835.

Smith, G. M. 1950. The freshwater algae of the United States, 2nd ed. McGraw-Hill, New York, 719 p.

Smol, J. P. 1990. Paleolimnology: Recent advances and future challenges. Memorie dell'Istituto di Idrobiologia 47:253–276.

Sommer, U. 1996. Plankton ecology: The past two decades of progress. Naturwissenschaften 83:293–301.

Spamer, E. E., Theriot, E. C. 1997. "*Stephanodiscus minutulus*," "*S. minutus*," and similar epithets in taxonomic, ecological, and evolutionary studies of modern and fossil diatoms (Bacillariophyceae: Thalassiosiraceae)—A century and a half of uncertain taxonomy and nomenclatural hearsay. Proceedings of the Academy Natural Science Philadelphia 148:231–272.

Sterrenburg, F. A. S. 1994. *Terpsinoe musica* Ehrenberg (Bacillariophyceae, Centrales), with emphasis on protoplast and cell division. Netherland Journal of Aquatic Ecology 28:63–69.

Stockner, J. G., Lund, J. W. G. 1970. Live algae in postglacial lake deposits. Limnology and Oceanography 15:41–58.

Stoermer, E. F. 1967. Polymorphism in *Mastogloia*. Journal of Phycology 3:73–77.

Stoermer, E. F. 1978. Phytoplankton as indicators of water quality in the Laurentian Great Lakes. Transaction of the American Microscope Society 97:2–16.

Stoermer, E. F., Yang, J. J. 1969. Plankton diatom assemblages in Lake Michigan. Great Lakes Research Division Special Report Number 47, University of Michigan, 168 p.

Stoermer, E. F., Edlund, M. B. 1998. No paradox in the plankton? Diatom communities in large lakes, *in*: Mayama, S., Idei, M., Koizumi, I., Eds., Proceedings of the 14th International Diatom Symposium, Tokyo, Koeltz Scientific Books, Koenigstein, pp. 51–58.

Stoermer, E. F., Håkansson, H. 1983. An investigation of the morphological structure and taxonomic relationships of *Stephanodiscus damasii* Hust. Bacillaria 6:245–255.

Stoermer, E. F. Kreis, R. G., Jr. 1978. Preliminary checklist of diatoms (Bacillariophyta) from the Laurentian Great Lakes. Journal of Great Lakes Research 4:149–169.

Stoermer, E. F., Ladewski, T. B. 1976. Apparent optimal temperatures for the occurrence of some common phytoplankton species in southern Lake Michigan. Great Lakes Research Division, Special Report 18, University of Michigan, 49 p.

Stoermer, E. F., Ladewski, T. B. 1982. Quantitative analysis of shape variation in type and modern populations of *Gomphoneis herculeana*. Beihefte zur Nova Hedwigia 73:347–386.

Stoermer, E. F., Kingston, J. C., Sicko-Goad, L. 1979. The morphology and taxonomic relationships of *Stephanodiscus binderanus* var. *oestrupi* (A. Cl.) A. Cl. Beihefte zur Nova Hedwigia 64:65–78.

Stoermer, E. F., Smol, J. P. 1999. The diatoms: Applications for the environmental and earth sciences. Cambridge University Press, London. 456 p.

Stoermer, E. F., Wolin, J. A., Schelske, C. L., Conley, D. J. 1985a. An assessment of ecological changes during the recent history of Lake Ontario based on siliceous microfossils preserved in the sediments. Journal of Phycology 21:257–276.

Stoermer, E. F., Wolin, J. A., Schelske, C. L., Conley, D. J. 1985b. Variations in *Melosira islandica* valve morphology in Lake Ontario sediments related to eutrophication and silica depletion. Limnology and Oceanography 30:414–418.

Stoermer, E. F., Kociolek, J. P., Schelske, C. L., Conley, D. J. 1985c. Siliceous microfossil successtion in the recent history of Lake Superior. Proceedings of the Academy of Natural Sciences of Philadelphia 137:106–118.

Stoermer, E. F., Ladewski, T. B., Kociolek, J. P. 1986. Further observations on *Gomphoneis*, *in*: Ricard, H., Ed., Proceedings of the 8th Internatinal Diatom Symposium, O. Koeltz, Koenigstein, pp.205–213.

Stoermer, E. F., Håkansson, H., Theriot, E. C. 1987. *Cyclostephanos* species new to North America: *C. tholiformis* sp. nov. and *C. costatilimbus* comb. nov. British Phycological Journal 22:349–358.

Stoermer, E. F., Emmert, G., Schelske, C. L. 1989. Morphological variation of *Stephanodiscus niagarae* (Bacillariophyta) in a Lake Ontario sediment core. Journal of Paleolimnology 2:227–236.

Stoermer, E. F., Kociolek, J. P., Cody, W. 1990. *Cyclotubicoalitus undatus*, genus et species nova. Diatom Research 5:171–177.

Stoermer, E. F., Andresen, N. A., Schelske, C. L. 1992. Diatom succession in the recent sediments of Lake Okeechobee, Florida, USA. Diatom Research 7:365–386.

Stoermer, E. F., Emmert, G., Julius, M. L., Schelske, C. L. 1996. Paleolimnologic evidence of rapid recent change in Lake Erie's trophic status. Canadian Technical Reports on Fisheries and Aquatic Sciences 53:1451–1458.

Theriot, E. C. 1987. Principal component analysis and taxonomic interpretation of environmentally related variation in silicification in *Stephanodiscus* (Bacillariophyceae). British Phycological Journal 22:359–373.

Theriot, E. C. 1990. Four new species of *Mesodictyon* (Bacillariophyta: Thalassiosiraceae) in the late Miocene lacustrine deposits of the Snake River Basin, Idaho. Proceedings of the Academy of Natural Sciences of Philadephia 142:1–19.

Theriot, E. C. 1992. Clusters, species concepts, and morphological evolution of diatoms. Systematic Biology 41:141–157.

Theriot, E., Bradbury, J. P. 1987. *Mesodictyon*, a new fossil genus of the centric diatom family Thalassiosiraceae from the Miocene Chalk Hills Formation, Western Snake River Plain, Idaho. Micropaleontology (NY) 33:356–367.

Theriot, E. C., Kociolek, J. P. 1986. Two new species of *Cyclostephanos* (Bacillariophyceae) with comments on the classification of the freshwater Thalassiosiraceae. Journal of Phycology 17:64–72.

Theriot, E. C., Serieyssol, K. 1994. Phylogenetic systematics as a guide to understanding features and potential morphological characters of the centric diatom family Thalassiosiraceae. Diatom Research 9:429–450.

Theriot, E. C., Stoermer, E. F. 1984a. Principal components analysis of character variation in *Stephanodiscus niagarae* Ehrenb.: Morphological variation related to lake trophic status, *in*: Mann, D. G. Ed. Proceedings of the VIIth International Diatom Symposium, Philadelphia, O. Koeltz, Koenigstein, pp. 97–111.

Theriot, E. C., Stoermer, E. F. 1984b. Principal component analysis of *Stephanodiscus*: Observations on two new species from the *Stephanodiscus niagarae* complex. Bacillaria 7:37–58.

Theriot, E. C., Stoermer, E. F. 1986. Principal components analysis of *Stephanodiscus*: Field evidence for two varieties of *S. niagarae*, *in*: Ricard, M., Ed., Proceedings of the VIIIth International Diatom Symposium, Paris, Koeltz Scientific Books, Koenigstein, pp. 385–394.

Theriot, E. C., Håkansson, H., Kociolek, J. P., Round, F. E., Stoermer, E. F. 1987 Validation of the centric diatom genus *Cyclostephanos*. British Phycological Journal 22:345–347.

Theriot, E. C., Håkansson, H., Stoermer, E. F. 1988. Morphometric analysis of *Stephanodiscus alpinus* (Bacillariophyceae) and its morphology as an indicator of lake trophic status. Phycologia 27:485–493.

Tilman, D. Kilham, S. S., Kilham, P. 1982. Phytoplankton community ecology: The role of limiting nutrients. Annual Review of Ecology and Systematics 13:349–472.

Titman, D. 1976. Ecological competition between algae: Experimental confirmation of resource-based competition theory. Science 192:463–465.

Tuchman, M. L., Theriot, E. C., Stoermer, E. F. 1984. Effects of low level salinity concentrations on the growth of *Cyclotella meneghiniana* Kütz. Archiv für Protistenkunde 128:319–326.

van den Hoek, C., Mann, D. G., Jahns, H. M. 1995. Algae: An introduction to phycology. Cambridge University Press, Cambridge, U K, 623 p.

Van Heurck, H. 1880 Synopsis des diatomées de Belgique. Atlas, plates 1–30. Anvers, Belgium.

Van Heurck, H. 1881 Synopsis des diatomées de Belgique. Atlas, plates 31–77. Anvers, Belgium.

Van Heurck, H. 1882. Synopsis des diatomées de Belgique. Atlas, plates 78–103. Anvers, Belgium.

Van Heurck, H. 1883 Synopsis des diatomées de Belgique. Atlas, plates 101–132. Anvers, Belgium.

Van Heurck, H. 1884. Synopsis des diatomées de Belgique. Table alphabetique. Anvers, Belgium.

Van Heurck, H. 1885. Synopsis des diatomées de Belgique. Altas, texte, plate A, B, and C. Anvers, Belgium.

Van Landingham, S. L. 1964. Miocene non-marine diatoms from the Yakima region in south central Washington. Nova Hedwigia Beiheft 14:1–78.

Walsby, A. E., Xypolyta, A. 1977. The form resistance of chitin fibres attached to the cells of *Thalassiosira fluviatilis* Hustedt. British Phycological Journal 12:215–223.

Weber, C. I. 1970. A new freshwater centric diatom *Microsiphona potamos* gen. et sp. nov. Journal of Phycology 6:149–153.

Willen, E., Hajdu, S., Pejler, Y. 1990. Summer phytoplankton in 73 nutrient-poor Swedish lakes. Classification, ordination and choice of long-term monitoring objects. Limnologica 21:217–227.

Williams, D. M. 1985. Morphology, taxonomy and inter-relationships of the ribbed diatoms from the genera *Diatoma* and *Meridion* (Diatomaceae: Bacillariophyta). Bibliotheca Diatomologica 8:1–238.

Williams, D. M., Round, F. E. 1986. Revision of the genus *Synedra* Ehrenberg. Diatom Research 1:313–393.

Williams, D. M., Round, F. E. 1987. Revision of the genus *Fragilaria*. Diatom Research 2:267–288.

Williams, D. M., Round, F. E. 1988. Phylogenetic systematics of *Synedra*, *in*: Round, F. E., Ed., Proceedings of the 9th International Symposium on Recent and Fossil Diatoms. O. Koeltz, Koenigstein, pp. 303–315.

Williams, L. G. 1964. Possible relationships between plankton-diatom species numbers and water-quality relationships. Ecology 45:809–823.

Williams, L. G. 1972. Plankton diatom species biomass and the quality of American rivers and the Great Lakes. Ecology 53:1038–1050.

Wolle, F. 1890 *Diatomaceae of North America*. Comenius Press, Bethelem, PA, 42 p.

Wujek, D. I., Graebner, M. 1980. A new freshwater species of *Chaetoceros* from the Great Lakes region. Journal of Great Lakes Research 6:260–262.

Wujek, D. E., Welling, M. L. 1981. The occurrence of two centric diatoms new to the Great Lakes. Journal of Great Lakes Research 7:55–56.

Wunsam, S. Schmidt, R., Klee, R. 1995. *Cyclotella* taxa (Bacillariophyceae) in lakes of the Alpine region and their relationship to environmental variables. Aquatic Science 57:360–386.

Yang J. R., Pick, F. R., Hamilton, P. B. 1996. Changes in the planktonic diatom flora of a large mountain lake in response to fertilization. Journal of Phycology 32:232–243.

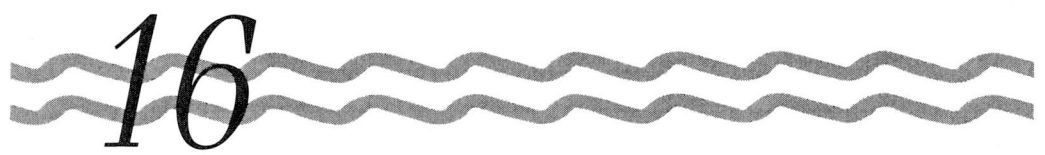

ARAPHID AND MONORAPHID DIATOMS

John C. Kingston

Center for Water and the Environment
Natural Resources Research Institute
University of Minnesota Duluth
Ely, Minnesota 55731

I. Introduction
II. Diversity and Morphology
    A. Araphid Diatoms: Class Fragilariophyceae
    B. Monoraphid Diatoms: Order Achnanthales
III. Ecology and Distribution
IV. Collection and Preparation for Identification
V. Key and Descriptions of Genera
    A. Key
    B. Descriptions of Genera
VI. Guide to Literature for Species Identification
Literature Cited

"Diatoms have proven to be extremely powerful tools with which to explore and interpret many ecological and practical problems. The continuing flood of new information will, without doubt, make the available tools of applied ecology even sharper. It is also apparent that the maturation of this area of science will provide additional challenges. Gone are the comfortable days when it was possible to learn the characteristics of most freshwater genera in a few days and become familiar with the available literature in a few months. Although we might sometimes wish for the return of simpler days, it is clear that this field of study is rapidly expanding, and it is our conjecture that we are on the threshold of even larger changes." (Stoermer and Smol, 1999, p. 8)

I. INTRODUCTION

General discussion of diatom biology and classification is similar in all the diatom chapters and is emphasized in Chapter 15, Section I.

In traditional diatom classifications, four major groups were recognized: centric, araphid, monoraphid, and biraphid. Today systematists believe that the monoraphid condition was multiply derived from a primitive biraphid condition (Kociolek and Williams, 1987; Kociolek and Rhode, 1998). Hence, the monoraphid diatoms are classified within the Class Bacillariophycideae with biraphid diatoms, but the traditional groupings are often considered in discussions of taxonomy, morphology, or ecology.

Freshwater araphid diatoms include many genera with species living in the periphyton (Fig. 1A, B) and a few genera (*Synedra, Fragilaria, Asterionella, Tabellaria*) with species that thrive as phytoplankton in lakes and reservoirs (Fig. 1C–F) (Stoermer and Yang, 1969, 1970; Reynolds, 1984). The common planktonic species have been the subject of much study regarding species succession, nutrients, and food webs (Chap. 2). These diatoms were first reported from the Americas approximately 160 years ago (Ehrenberg, 1841). Freshwater monoraphid diatoms are benthic in habit, living attached to substrata in shallow to mid-depth habitats of lakes and rivers (Fig. 2) (Stoermer, 1980). Many are specialized for life as epiphytes, and others frequently occur attached to sand and rocks.

Several new combinations were made to facilitate discussions of taxonomy in this chapter (Kingston, 2000). Two new combinations are required here:

Karayevia clevei (Grunow) Kingston nov. comb.—Basionym, *Achnanthes clevei* Grunow in Cleve and Grunow (1880, p. 21); Round (1998) validated the generic name but failed to validate the name of the type species.

Karayevia clevei var. *rostrata* (Hustedt) Kingston nov. comb.—Basionym, *Achnanthes clevei* var. *rostrata* Hustedt in Hustedt (1930, p. 204, Fig. 295); this combination was invalid in Kingston (2000) because the species name had not been validated.

II. DIVERSITY AND MORPHOLOGY

From a practical perspective, it is not simple to agree on nomenclatural systems for the araphid and monoraphid diatoms. There are at least three taxonomic systems currently applied to freshwater members of the class Fragilariophyceae (Table I) and the order Achnanthales (Table II):

1. a system deriving from Hustedt's (1930) compression of classical literature and applied by Patrick and Reimer (1966) in their diatom flora of the United States;
2. a system arising from ecological experience in Europe but applied globally, propounded mainly by Lange-Bertalot, Krammer, and others (Lange-Bertalot and Krammer, 1989; Krammer and Lange-Bertalot, 1991a, b);
3. a system arising from a concern for natural classification, which has split many new genera from classical genera, propounded by Williams, Round, Mann, and others (Williams and Round, 1987, 1988b; Round *et al.*, 1990; Williams, 1990a, b, c).

The major problems in choosing between the latter two more modern systems are:

- lack of supporting data, including genetic and morphological data;
- lack of, or incompleteness of, modern systematic analysis and botanical revision of the genera, families, and other categories;
- inconsistency within the systems themselves—system 2 has moved toward system 3, but genera that were considered earlier by the system 2 authors remain large and less natural; for example, Lange-Bertalot and Krammer have

TABLE I Three Common Classification Systems for Genera of Araphid Freshwater Diatoms, Class Fragilariophyceae

System 1 For example (Hustedt, 1959; Patrick and Reimer, 1966)	System 2 For example (Krammer and Lange-Bertalot, 1991a)	System 3 For example (Williams and Round, 1987; Round et al., 1990)
Fragilariaceae Hustedt in Pascher 1. *Tetracyclus* Ralfs 2. *Diatoma* Bory 3. *Meridion* Agardh 4. *Asterionella* Hassall 5. *Tabellaria* Ehrenberg 6. *Synedra* Ehrenberg 7. *Fragilaria* Lyngbye	Fragilariaceae Hustedt 1. *Tetracyclus* Ralfs 2. *Diatoma* Bory 3. *Meridion* Agardh 4. *Asterionella* Hassall 5. *Tabellaria* Ehrenberg 6. *Synedra* Ehrenberg 7. *Fragilaria* Lyngbye Subgenera: • *Fragilaria* Lange-Bertalot • *Alterasynedra* Lange-Bertalot • *Ctenophora* (Grunow) Lange-Bertalot • *Tabularia* (Khtzing) Lange-Bertalot • *Staurosira* (Ehrenberg) Lange-Bertalot	Fragilariaceae Greville 1. *Tetracyclus* Ralfs, different family, Tabellariaceae 2. *Diatoma* Bory 3. *Meridion* Agardh 4. *Asterionella* (freshwater), *Asterionellopsis* (marine) 5. *Tabellaria* Ehrenberg, different family, Tabellariaceae 6. *Oxyneis* Round, different family, Tabellariaceae 7. *Synedra* Ehrenberg 8. *Catacombus* Williams & Round (marine) 9. *Hyalosynedra* Williams & Round (marine) 10. *Tabularia* (Kützing) Williams & Round (mainly marine) 11. *Ctenophora* (Grunow) Williams & Round 12. *Neosynedra* Williams & Round (mainly marine.
8. *Opephora* Petit 9. *Hannaea* Patrick 10. *Centronella* Voigt	8. *Opephora* Petit, part of *Fragilaria* 9. *Hannaea* Patrick 10. *Centronella* Voigt 11. *Frankophila* Lange-Bertalot (biraphid or monoraphid relatives of *Punctastriata*, *Staurosira*, or *Pseudostaurosira*.	13. *Fragilaria* Lyngbye 14. *Staurosirella* Williams & Round 15. *Fragilariforma* (Ralfs) Williams & Round 16. *Punctastriata* Williams & Round 17. *Pseudostaurosira* Williams & Round 18. *Staurosira* (Ehrenberg) Williams & Round 19. *Martyana* Round 20. *Opephora* Petit (marine) 21. *Hannaea* Patrick 22. *Centronella* Voigt (shown to be a persistent form of *Fragilaria crotonensis*, not a genus (Schmid 1997))

TABLE II Three Common Classification Systems for Genera of Monoraphid Freshwater Diatoms, Order Achnanthales.

System 1 For example: (Hustedt, 1959; Patrick and Reimer, 1966)	System 2 For example (Krammer and Lange-Bertalot, 1991b; Lange-Bertalot 1997b)	System 3 For example (Round et al., 1990; Bukhtiyarova and Round, 1996; Round and Bukhtiyarova 1996a)

Family Achnanthoideae Schütt or
Family Achnanthaceae Kützing
1. *Achnanthes* Bory
2. *Cocconeis* Ehrenberg
3. *Rhoicosphenia* Grunow
	Family Achnanthaceae Kützing
	
	1. *Achnanthes* Bory
	  Subgenera:
	  Σ *Achnanthes*
	  Σ *Achnanthidium*
	  "Key group":
	  A
	  B
	  C
	  D
	  E
	  F
	  G
	  H (later, *Achnantheiopsis* nom. illegit.)
	2. *Cocconeis* Ehrenberg
	
	3. *Rhoicosphenia* Grunow
		Family Achnanthaceae Kützing
		
		1. *Achnanthes* Bory
		
		Family Achnanthidiaceae Mann
		2. *Achnanthidium* Kützing
		
		3. *Karayevia* Round & Bukhtiyarova, *Psammothidium*
		   Bukhtiyarova & Round (part)
		4. *Planothidium* Round & Bukhtiyarova (part)
		5. *Kolbesia* Round & Bukhtiyarova
		6. *Eucocconeis* Cleve, *Psammothidium* (part)
		7. *Lemnicola* Round & Basson
		8. *Psammothidium* Bukhtiyarova & Round (part)
		   Achnanthidium (part)
		9. *Rossithidium* Round & Bukhtiyarova
		10. *Kolbesia* Round & Bukhtiyarova
		    Planothidium (part)
		Family Cocconeidaceae Kützing
		11. *Cocconeis* Ehrenberg
		    Rhoicosphenia Grunow treated in family
		    Rhoicospheniaceae

not split out many genera in the araphid diatoms or the keeled diatoms compared to what they have done in biraphid genera such as *Cymbella* and *Navicula*; they have compressed the taxonomy of the genus *Fragilaria* and sunk several other well-recognized genera into their concept;

- The splitting off of genera in system 3 is unevenly distributed among the Bacillariophyta, and it is based on a conglomeration of modern (cladistic) systematics (Kociolek *et al.*, 1989; Williams, 1990a) and somewhat older and less defensible (phenetic) scenarios (Round *et al.*, 1990).

A useful bibliography of publications detailing the fine structure of diatom cell walls has been compiled (Gaul *et al.*, 1993), using the older taxonomy in system 1 above. Diploid chromosome numbers for the few species of araphid and monoraphid diatoms reported are between 8 and 25 (Kociolek and Stoermer, 1989).

In this chapter, I emphasize North American literature and apply system 3, and part of system 2, at the generic level, both for the Fragilariophyceae and for the Achnanthales. Although newer genera often have not been defined rigorously, more modern approaches have more to offer for understanding hypotheses of evolution by promoting discussion and new research. Applying these concepts will provide better ecological information through consideration of differences among the newer, smaller genera and improve data on their distribution and biogeography that might be missed with the fewer categories of the older, compressed system. Round has argued that species within a natural genus may retain similar ecological optima that can be used to help define genera (Round and Bukhtiyarova, 1996a, b; Round, 1997). For example, *Psammothidium* species mainly thrive in soft water, and *Lemnicola* is specialized as an epiphyte on a single higher plant family. Diverse physiological tolerances are observed within some other genera, such as *Asterionella* and *Tabellaria*, which include species possessing divergent pH optima. The problems mentioned above in differentiating the correctness or utility of different classification systems persist, and one hopes that this situation will improve through more rigorous application of modern systematic hypotheses to the understanding of species, genera, and other taxonomic categories (Kociolek, 1998).

Another difficulty in the more recent taxonomic literature is that the upper levels of taxonomic hierarchy, including the family, are often ignored at the time each genus is defined. We have taxonomic systems in transition. Even when the need for a new genus is widely accepted [e.g., *Planothidium* (Round and

Bukhtiyarova, 1996a) and *Achnantheiopsis* (Lange-Bertalot, 1997b) were proposed almost simultaneously to separate the relatives of *Achnanthes lanceolata* to a smaller, more natural genus], phycologists may find themselves working in an older system of systematic hierarchy by default, due to incompleteness of the systematic proposals at levels above genus. Keep in mind that new names do not have to be accepted just because they are proposed, and the older nomenclature does not become automatically invalid. The taxonomy presented in this chapter is selected because it is more natural, not only because it is newer. The acceptance or rejection of hypotheses underlying newly proposed genera will come after more complete systematic analysis (Kociolek, 1998).

Summary tables of selected morphological characteristics and key references are presented for the genera of freshwater araphid diatoms (Table III) and monoraphid diatoms (Table IV) (Patrick and Reimer, 1966; Williams, 1985, 1986, 1989, 1990b; Williams and Round, 1987, 1988a, b; Round *et al.*, 1990; Bukhtiyarova and Round, 1996; Flower *et al.*, 1996; Round and Bukhtiyarova, 1996a; Lange-Bertalot, 1997a, b, 1999; Round and Basson, 1997). The reader will notice that the character matrix is incomplete and information available for each genus is far from uniform, but these tables represent the current state of knowledge. More complete information is in the primary literature cited, and new information is discovered each year. One intrinsic difficulty in obtaining detailed information about diatom genera and species is that diatomists most often do detailed taxonomic work on cell walls cleaned of organic matter, and it is very difficult to relate observations of the cells in living and cleaned condition from mixed assemblages of hundreds of taxa.

A. Araphid Diatoms: Class Fragilariophyceae

The araphid diatoms (considering marine and freshwater genera together) are an unnatural group consisting of diatoms that do not fit within the centric or the raphid diatoms (Williams, 1985). However, the freshwater araphid diatoms are a small subgroup within the class and seem to be closely related, according to cladistic analysis (Williams, 1990a). The freshwater araphid diatoms are represented by members of the orders Fragilariales (containing one family, Fragilariaceae) and the Tabellariales (containing one family, Tabellariaceae). Taxa occur as unicells or in colonies: ribbon-like, stellate, zigzag, or attached to a mucilage pad at one end of the cells. Taxa live in the periphyton, epipsammon, epiphyton, epipelon, epidendron, haptobenthon, and plankton. Disturbance and nutrients can influence their habitat preference (Kuhn *et al.*, 1981). Certain planktonic populations in large lakes can selectively develop either offshore (Stoermer and Yang, 1970) or nearshore (Stoermer, 1968). The silica content and biovolume of many freshwater araphids have been compiled in a paper demonstrating that freshwater diatoms have higher silica content than marine diatoms (Conley *et al.*, 1989).

North American members of the Fragilariophyceae number at least 200 and make up about 10% of the diatom taxa at the specific level and below (species, varieties, and forms) encountered in rivers (Kingston, 1997), 12.3% of the taxa recorded from the Laurentian Great Lakes (Stoermer *et al.*, 1999), and 10.5% of the taxa encountered in softwater lakes (Camburn *et al.*, 1984–1986; P.R. Sweets, pers. comm.). Another biodiversity estimate is that perhaps 300 taxa from this class have been reported from the United States (J. P. Kociolek, pers. comm.). Densities of a single species in riverine collections from the National Water-Quality Assessment Program were as high as 4,000,000 cells cm^{-2} on hard substrata and 27,000,000 cells cm^{-2} in depositional zones (Kingston, 1997). The family Fragilariaceae contains most of the taxa, and the family Tabellariaceae contains a smaller proportion.

B. Monoraphid Diatoms: Order Achnanthales

The order Achnanthales is diverse in fresh waters and represented by members of the families Achnanthidiaceae, Cocconeidaceae, and more rarely by the (mainly marine) Achnanthaceae. Taxa live primarily as unicells, either on short stalks (*Achnanthidium*, *Achnanthes*) or adnate to the substratum (*Cocconeis*, *Psammothidium*, *Planothidium*, *Rossithidium*, *Karayevia*, *Kolbesia*, *Lemnicola*, *Eucocconeis*). In the latter habit, the raphe valve is against the substratum, potentially allowing the cell to move to a new location.

North American members of the freshwater Achnanthales number over 100 and account for about 10% of the diatom taxa at the specific level and below (species, varieties, and forms) encountered in rivers (S. Turner, pers. comm.), 6.5% of the taxa recorded from the Laurentian Great Lakes (Stoermer *et al.*, 1999), and 8.2% of the taxa encountered in softwater lakes (Camburn *et al.*, 1984–1986).

Friedrich Hustedt was the most influential diatomist of the first six decades of the 20th century. Under his influence, the genus *Achnanthes* was used in a very broad sense to include distantly related groups of marine (subgenera *Achnanthidium* and *Achnanthes*) and freshwater (subgenus *Microneis*) organisms. Today, the former subgenera are considered to be the families

TABLE III Selected Morphological and Cytological Characteristics of Araphid Freshwater Diatom Genera and Selected Taxonomic References [Diatom Cell Wall Morphology is Described in Detail by Barber and Haworth (1981)]

Genus	Shape	Striae	Areolae	Pore Field	Labiate process	Spines/plaques	Plastids	Cingulum	Selected references
Fragilaria	Linear, linear-lanceolate	Areolate, regular	External vela	Present, on mantle, ocellulimbus	1, near pole	Yes	2 plates, mainly against the valve faces	A few simple, open, ligulate copulae; single row of poroids	(Patrick and Reimer (1966), Camburn et al. (1984–1986), Williams and Round (1987), Lange-Bertalot (1989), Krammer and Lange-Bertalot (1991a)
Staurosira	Elliptical or cruciform	Areolate, regular, small, round	External vela	Present but reduced	None observed	Yes	2 large parietal plates adjacent to girdle	6–8 open, ligulate, plain copulae	Williams and Round (1987)
Staurosirella	Elliptical, linear, or cruciform	Lineolate	Linear areolae, finely branched closing plates	Large, valve face and mantle	None observed	Yes, often branched	2 large parietal plates adjacent to girdle	8–10 open, plain, ligulate copulae	Williams and Round (1987)
Pseudostaurosira	Elliptical, linear, or cruciform	Up to 4 elliptical areolae/stria, sternum large, sparse marginal areolae	Vela internal, meshlike	If present, reduced	Absent	Spathulate or spathulate-branched	Probably parietal and platelike	Several open, plain, ligulate copulae	Williams and Round (1987)
Punctastriata	Linear-elliptical	At or below valve surface	Closing plates with regular net pore field of transapical and apical bars	1 small apical net pore field	Absent	Small, at junction of mantle and face	Not known	Several open, plain, ligulate copulae	Rosen and Lowe (1981), Williams and Round (1987)
Fragilariforma	Elliptical, lanceolate, or linear; often undulate	Areolate, regular, often continuous, sternum very small	Simple vela	Simple, on face and mantle	1, polar	Yes, simple or spathulate; located on interstria	Many, discoid	4–6 copulae, incomplete, ligulate with 1 row of areolae	Cleve-Euler (1953), Renberg (1976, 1977), Hein (1981), Williams and Round (1987, 1988a), Williams (1990c), Kingston et al. (2001)
Stauroforma	Elliptical to lanceolate, sometimes sub-rostrate ends	Areolate, regular, often continuous, sternum very reduced	Simple vela	Reduced, different in 2 species	Absent	1 taxon with, 1 without	Not known	4–6 copulae	Flower et al. (1996)
Frankophila	Elliptical, linear-lanceolate, or lanceolate	Areolate, bi- to triseriate striae. Short, simple raphe branches without helictoglossae	Not known	None	None	Long spines with bifurcating curved ends locking into adjoining valve punctae	Not known	3 copulae, the second broader, open and not perforated	Lange-Bertalot (1997a)

(Continues)

TABLE III (Continued)

Genus	Shape	Striae	Areolae	Pore Field	Labiate process	Spines/plaques	Plastids	Cingulum	Selected references
					Features				
Martyana	Ovate-elliptical, depression at head-pole	Areolae slit-like, striae sunken between transapical ridges	Not known	Small	Absent	None/none	Not known	Up to 5, or more; valvocopula very broad	Round et al. (1990)
Synedra	Linear, lanceolate	Areolate, uni- or biseriate	Simple vela	Ocellulimbus, plate-like apical pore field on the mantle	2 per valve, a single process near each pole	Spine pair at ends, some taxs with spines, + plaques	2 long plates	All bands closed, single row of poroids	Williams (1986), Williams and Round (1988b), Bound and Maidana (2001)
Ctenophora	Linear to linear-lanceolate, thickened fascia	Elliptical to rectangular areolae	Complex cribra	Ocellulimbus	2 per valve, near ends	None	2, platelike	Partly known, valvocopula non-areolate and open	Round et al. (1990)
Hannaea	Linear and arcuate	Striae uniseriate and sunken, small poroidal areolae	Not known	Fine rows form porefield on mantle	Usually 2 per valve near poles	Small spines	2 along valves with lobes on dorsal side	Split with scalloped advalvar edge	Patrick and Reimer (1966), Round et al. (1990), Bixby (2001)
Diatoma	Elliptical-linear or lanceolate	Striae in groups, separated by costae	Not known	Large, indistinct, on valve and mantle	1 per valve	Some taxa with spines	Small, platelike or discoid	Bands open, 2 rows of pores, Y-shaped ligulae	Patrick and Reimer (1966), Williams (1985, 1990a), Krammer and Lange-Bertalot (1991a)
Meridion	Heteropolar and cuneate or isopolar and linear	Striae in groups, separated by costae	Not known	Indistinct, fine rows mainly on mantle "foot"	1 per valve, near "head"	Spinules and plaques	Numerous, irregular, discoid, along valve face	Split copulae, wider at head-pole	Williams (1985, 1990a)
Asterionella	Linear-lanceolate with capitate ends	Striae unevenly spaced, sternum narrow	Areolae uniseriate, vela indistinct	Both poles, mainly confined to the mantle, simple porelli in apically arranged rows	2 per valve, transversely oriented and within a stria, near each pole	Spines along valve face edge and around pore fields	Chromatophores are numerous small plates	Copulae several, split, with 1 or 2 rows of pores	Körner (1971), Flower and Battarbee (1985b), Pappas and Stoermer (2001a, b)
Distrionella	Lanceolate to linear-lanceolate with capitate ends, sometimes heteropolar; rectangular girdle view	Striae unevenly spaced, scattered at poles, secondary transapical ribs in many valves, no sternum	Not known	Both poles, confined to the mantle, simple porelli	1 per valve, near end, usually central and lying within a stria in a polar rib	No spines	No observations on living cells (but to note their solitary habit)	Girdle with 3 open copulae with poroid rows, valvocopula with slight ligula, single row of poroids, double at ligula	Williams (1990b)

	Shape	Striae	Areolae		Spines				Selected references
Tetracyclus	Elongate to elliptical, with center wide and or constricted, margin undulate	Uniseriate, fine	Simple	None	0, 1, 2 per valve, near center, arranged transversely	No spines	Not known	Copulae numerous, incomplete, ligulate, septate. External rows of pores join internally	Williams (1985, 1989, 1993, 1994, 1996, 1997)
Tabellaria	Elongate, capitate, wider at center	Uniseriate, irregularly spaced	Simple	Large apical pore field on face and mantle near poles	1, near center	Spines usually present	Strip-like chromatophores lie between the septa	Copulae complete or incomplete, septate	Koppen (1975, 1978), Flower and Battarbee (1985a)
Oxyneis	Elliptical or panduriform	Uniseriate, irregularly spaced	Simple	Apical pore field of fine striae rows	1, in a striia next to the apical pore field	Present at the face/mantle boundary, not at the ends; flat, bifurcate	Not known	Numerous open copulae, areolate, septate	Flower (1989), Round et al. (1990)

TABLE IV Selected Morphological Characteristics of Monoraphid Freshwater Diatom Genera and Selected Taxonomic References [Diatom Cell Wall Morphology Is Described in Detail by Barber and Haworth (1981)]

Genus	Shape	Striae	Areolae	Raphe	Plastids	Cingulum/other	Habit	Selected references
Achnanthes	Linear to lanceolate, flexion along transapical axis, raphe valve concave	Uni-, bi-, or triseriate	Poroids with complex cribra bearing volae	Raphe valve usually with a fascia; sternum on both valves, often off-center on rapheless valve	Multiple discoid or 2 H-shaped	3–7 open areolate bands	Haptobenthic on short stalks; solitary or forming short chains	Patrick and Reimer (1966), Camburn et al. (1984-1986), Lange-Bertalot and Krammer (1989), Round et al. (1990), Krammer and Lange-Bertalot (1991b)
Cocconeis	Elliptical or almost circular, flexion along apical axis, raphe valve concave	Uni- to multiseriate, uniseriate in freshwater taxa; many taxa with valves of dissimilar structure, raphe valve with finer structure; area of different structure around edge of raphe valve	Simple to loculate	Fairly simple, some internal deflection of the proximal ends	1, C-shaped, simple or lobed	A few, narrow, plain bands; valvocopula often complete	Haptobenthic, adnate, raphe valve against substratum	Patrick and Reimer (1966), Krammer (1990), Krammer and Lange-Bertalot (1991b)

(Continues)

TABLE IV (Continued)

Genus	Shape	Striae	Areolae	Raphe	Plastids	Cingulum/other	Habit	Selected references
Achnanthidium	Narrow, linear-lanceolate with rounded to rostrate to capitate ends; noticeably curved with concave raphe valve; often 3 to 6 times longer than wide	Striae usually near 30 in 10 µm, finer toward the apices; mantle pores of narrower dimension than the striae; valves of similar structure; raphe valves may have central interruption in striae; areolae aligned within internal depression	Simple round to transapically elongate areolae with internal hymens	Raphe not in channel; species currently treated here have a variety of proximal and distal raphe ending patterns; narrow sternum widened near the center	Single, with pyrenoid	Plain bands; anisogamous fusion noted	Haptobenthic on short stalks, common on a wider range of substrata including plants and rocks	Lange-Bertalot and Ruppel (1980), Round and Bukhtiyarova (1996a), Lange-Bertalot)1999)
Psammothidium	Oval elliptic, sometimes capitate; apices rounded; convex to nearly flat raphe valve, concave rapheless valve; often 2 to 4 times longer than wide	Striae reach the sternum; areolar arrangement similar on both valves; central area larger and more rectangular on the raphe valve; single row of mantle pores	Areolae foramina structure diverse, useful in species identification	Raphe fissures lying in a channel, proximal ends straight externally, curved opposite internally; distal ends with various designs; sternum on both valves	Not known	Plain bands; hoof-like area near center of rapheless valves in a few taxa, similar to features of Planothidium and Lemnicola	Adnate, most abundant on sand, prefers acidic water	Bukhtiyarova and Round (1996), Lange-Bertalot)1999)
Planothidium	Linear-elliptical to elliptical-lanceolate, ends rounded to rostrate; slightly curved in girdle view	Striae of multiseriate rows of areolae	Areolae small, round	Raphe with expanded proximal ends; distal ends deflected to the same side	Single, or 2 in initial cells	A simple closed band; many taxa with horse-hoof-like shape near center, usually on the rapheless valve; isogamous fusion noted	Adnate on sand and pebbles, more common in alkaline water	Round and Bukhtiyarova (1996), Lange-Bertalot (1997b, 1999)
Rossithidium	Narrow linear valves with rounded apices	Striae parallel, raphe valves may have central interruption in striae, valves of similar structure	Not known	Straight raphe, sternum narrow and linear	Not known	Produces 2 auxospores from each cell pair.	Not known	Round and Bukhtiyarova (1996a)

Karayevia	Linear-lanceolate to lanceolate valves, ends elliptic to rostrate, rounded	Valves of dissimilar structure; striae more radiate on raphe valve, parallel to slightly radiate on rapheless valve; striae on raphe valve with oblong areolae along the raphe and the valve margin	Areolae closed internally by cribra on raphe valve, externally by pegged cribra on the rapheless valve	Raphe filiform, in a channel; sternum present	Not known	Plain bands	Adnate on sand, known taxa prefer alkaline water	Round and Bukhtiyarova (1996a)
Kolbesia	Valves of dissimilar structure	Single elongate areolae in the raphe valve striae, 2 or 3 elongate areolae in the rapheless valve striae	Areolae elongate, separated by longitudinal plain regions on the rapheless valve	Raphe not in channel, sternum present; proximal ends expanded externally, distal ends hooked	Not known	Not known	Adnate on sand; described species prefer alkaline water	Round and Bukhtiyarova (1996a)
Lemnicola	Linear to linear-elliptic, with rounded, subrostrate apices	Slightly radial striae, composed of biseriate areolae rows separated by internal transverse ridges; asymmetrical stauros on the raphe valve	Circular areolae	Raphe thin, surrounded by a narrow groove	Single	Similar life history to *Achnanthidium*, *Rossithidium* and *Planothidium*	Epiphytic on *Lemna* and *Wolfia*	Round and Basson (1997)
Eucocconeis	Linear-lanceolate to linear-elliptical with rounded to angular ends; bent a long the median trans-apical axis, with concave raphe valve, but twisted about the sigmoid apical axis	Fine, radiate, uniseriate; valve differentiated from a deep mantle by an unornamented region at the juncture	Round, closed internally by hymens	Raphe in a narrow channel; with a sigmoid sternum centrally expanded on both valves	Single plate-like plastid	Narrow, plain bands	Found in oligotrophic fresh water, common in lacustrine littoral zones, epipsammic to epipelic	Round *et al.* (1990)

Achnanthaceae and Achnanthidiaceae (Round et al., 1990). This change is a good demonstration of the recent consensus that earlier taxonomic hierarchies were often organized at least one taxonomic level lower than reality. After all, species are the ultimate biological reality, and the modern trend is to divide older taxonomic categories above the specific level to create more natural groups.

A more recent cladistic analysis (Kociolek and Stoermer, 1986) of the monoraphid diatoms, as treated by Hustedt (1930, 1927), indicates a polyphyletic origin of the group. The genus *Rhoicosphenia*, considered a close relative to *Cocconeis* and *Achnanthes* by Hustedt (1959), is more related to the biraphid genus *Gomphonema* than it is to the other monoraphid genera treated here. *Cocconeis*, which was considered monophyletic with *Achnanthes* by Hustedt and by Mann (1984), may be more closely related to the biraphid genus *Mastogloia* than it is to other monoraphids (Kociolek and Stoermer, 1986), again indicating the polyphyletic nature of the monoraphid diatom group, but revisionary work along those lines is yet to be attempted.

III. ECOLOGY AND DISTRIBUTION

Much information on the ecology of benthic diatoms can be gleaned from two recent books on algal ecology (Stevenson et al., 1996; Stoermer and Smol, 1999), and the ecology of common phytoplankton diatoms also has been well reviewed (Tilman et al., 1982; Reynolds, 1984; Stoermer and Smol, 1999). Excellent habit photomicrographs of several araphids and *Achnanthidium* have been presented in a book (Canter-Lund and Lund, 1995). Floras and checklists exist for certain localities, states, or regions of North America (Patrick, 1945; Sovereign, 1958; Woodhead and Tweed, 1960; Patrick and Freese, 1961; Hohn and Hellerman, 1963; Bright, 1968; Stoermer and Yang, 1969; Foged, 1971; Sreenivasa and Duthie, 1973; Clark and Rushforth, 1977; Collins and Kalinsky, 1977; Stoermer et al., 1999; Czarnecki and Blinn, 1978; Camburn et al., 1978; Prescott and Dillard, 1979; Johansen and Rushforth, 1981; Kaczmarska and Rushforth, 1983; Bahls et al., 1984; Bateman and Rushforth, 1984; Dodd, 1987; Poulin, 1990; Douglas and Smol, 1993), and hundreds of shorter papers document North American diatom species. There is at least one recent instance of a well-documented North American exiccata slide set being distributed to key museums (Hamilton et al., 1992). Because diatoms are a diverse group with short life cycles, exhibit quick response time to environmental condition and have characteristic cell walls that are often well preserved in lake, pool, and wetland sediments, they are a major tool in paleolimnology, the multidisciplinary science of reconstructing past aquatic environmental conditions (Dixit et al., 1992; Stoermer and Smol, 1999, Chapt. 23).

Species of the Fragilariophyceae and the Achnanthales occur throughout the world in fresh to slightly saline water. The Fragilariophyceae are part of the spring periphyton community of rivers that can respond, with rapid growth, to small additions of organic and inorganic nutrients (Perrin et al., 1987). Along with the other diatoms (Dixit et al., 1992), araphid and monoraphid diatoms can be used to infer environmental conditions at many spatial and temporal scales. They are particularly useful as a diverse group of indicator organisms in multivariate environmental gradient analysis (Dixit et al., 1992), including salinity (Bradbury, 1987; Wilson et al., 1994; Cumming et al., 1995; Laird et al., 1998), nutrients (Bradbury, 1975; Christie and Smol, 1993; Fritz et al., 1993; Reavie et al., 1995; Dixit et al., 1999; Siver, 1999), lake acidity (Koppen, 1978; Charles, 1985; Cumming et al., 1992; Kingston et al., 1992), dissolved organic carbon (Kingston and Birks, 1990), toxic metals (Kingston et al., 1992), temperature (Kingston et al., 1983; Pienitz and Smol, 1992), and lake depth (Yang and Duthie, 1995; Brugam et al., 1998). Many diatom ecological studies use computer programs for multivariate community analysis (ter Braak, 1988; Smilauer, 1992; Birks et al., 1998) or for *ad hoc* environmental index calculation (Lecointe et al., 1993). Further details are given in Chapter 23.

The small, benthic members of the Fragilariaceae are known to occur in great abundance in late-glacial (Florin, 1970) and postglacial sediments (Haworth, 1972; Brugam, 1980; Rawlence, 1988; Hickman and White, 1989; Pienitz et al., 1991) (Fig. 3), and in shallow lakes (Kingston, 1984) or shallow littoral zones (Bradbury and Winter, 1976). One hypothesis links this postglacial high-abundance phenomenon in cold-temperate and Arctic lakes to persistent ice, with marginal summer motes supporting growth of benthic Fragilariaceae, and these finding their way into offshore depositional basins that are typically cored by paleolimnological researchers (Smol, 1988).

Many studies have demonstrated that both araphid and monoraphid diatoms are important components of aquatic communities on pebbles and sand (Kingston and Lowe, 1979; Keithan and Lowe, 1985; Krejci and Lowe, 1987a, b; Miller et al., 1987; Round and Bukhtiyarova, 1996b). The Laurentian Great Lakes contain a very diverse benthic diatom flora with a mixture of widespread taxa and others that reach the southern extension of their boreal distribution in these cold dimictic lakes (Stoermer, 1980). Araphid and monoraphid species were more abundant in mid-depth portions of the Great Lakes prior to massive disturbance by human activities (Stoermer *et al.*, 1991). The significant contribution of diatoms to benthic production of lakes has been documented (Foerster and Schlichting, 1965). Both araphid and monoraphid groups are represented in unusual habitats, including wet cliffs (Lowe and Collins, 1973), caves (St. Clair *et al.*, 1981), warm springs (Kaczmarska and Rushforth, 1983), on tufa deposits (Davis *et al.*, 1989), artesian fens (Reimer, 1990), soil crusts (Ashley *et al.*, 1985), and ombrotrophic bogs (Kingston, 1982). Depending on the species, some thrive in warm water of thermal effluents (Squires *et al.*, 1979) and some in cold water of Arctic lakes (Pienitz and Smol, 1992).

Different species of araphid and monoraphid diatoms have been identified as common in major habitat types in the Great Lakes (Stoermer, 1975). Those common taxa of *Achnanthidium* and *Cocconeis* are typical members of a high-nutrient guild as shown by enrichment experiments (Fairchild *et al.*, 1985; Carrick *et al.*, 1988). Araphid and monoraphid diatoms are often early colonizers of hard substrata and very abundant in rivers (Sherman and Phinney, 1971; Patrick, 1976; Korte and Blinn, 1983; Lamb and Lowe, 1987). Extreme flow events can remove even adnate taxa that attach tenaciously, and the effects of such physical disturbance in rivers are well described (Peterson, 1996). In modern treatments of systems 2 and 3, the family Cocconeidaceae contains marine and freshwater members, much as in system 1 (Table II). Smaller taxa are adnate as part of the prostrate assemblage attached to sand and other hard substrata, and some larger taxa are specialized to attach to macroalgae, such as *Cladophora* (Lowe *et al.*, 1982). As with adnate members of the Achnanthidiaceae, *Cocconeis* species grow with the raphid valve against the substratum, which potentially allows them to move over the surface.

Diatoms often can be the dominant algal group in the phytoplankton of lakes, in terms of abundance and biomass, and the araphid species can be very well represented in many types of North American lakes (Stockner and Benson, 1967; Stoermer and Yang, 1969, 1970; Moore, 1980). The abundance of planktonic araphid and centric diatoms declined during P-loading of Lake Michigan in the mid-20th century and diatoms were replaced by other algal groups (Danforth and Ginsburg, 1980). In smaller lakes, production of *Asterionella formosa* and *Fragilaria crotonensis* has often increased dramatically in response to land development and agriculture in North America (Bradbury and Waddington, 1978; Munch, 1980; Engstrom *et al.*, 1985; Stoermer *et al.*, 1991).

Biogeographical studies of freshwater araphid genera show some trans-Pacific affinities (Williams, 1996), and the probable existence of groups of related species in Europe, North America, and South America (D. M. Williams, pers. comm.). The western (Sovereign, 1963) and southeastern (Hein, 1981) United States have many distinct taxa not occurring in the more-studied midwestern and northeastern regions, as is true for other groups of diatoms (Kociolek and Kingston, 1999).

IV. COLLECTION AND PREPARATION FOR IDENTIFICATION

Collection and preparation are essentially the same for all the diatom groups covered in chapters of this volume (see Chap. 15, Sect. V) and have been summarized in many publications (Patrick and Reimer, 1966; Battarbee, 1986; Kingston, 1986; Dodd, 1987). Diatoms are identified primarily from morphological characteristics of their glass cell walls. Therefore diatom samples should be cleaned of organic matter, oxidizing cytoplasm, extracellular secretions, and nondiatom organic matter, prior to examination of their walls. The cleaned diatoms are mounted in high-refractive-index mounting media such as Hyrax, Pleurax, or Naphrax. Such mountants allow for high-resolution microscopy using research-grade optics (N.A. = 1.3 to 1.4) and create permanent microscope slides that can be used as reference collections for biodiversity and ecological studies (see Chapt. 23).

V. KEY AND DESCRIPTIONS OF GENERA

A. Key

Refer to Figs. 1–19.

1a.	Frustules lacking a raphe system (Fig. 5G)	Araphid diatoms 2
1b.	Frustules with a raphe system on one valve only (Fig. 14J, K)	Monoraphid diatoms 19

2a.	Transapical ribs present (Fig. 5F)	3
2b.	Transapical ribs absent (Fig. 6)	6
3a.	Septa present	4
3b.	Septa absent (Fig. 5D)	5
4a.	Septa large	*Tetracyclus*
4b.	Septa small, not noticeable in light microscopy	*Meridion*
5a.	Raised sternum (Fig. 5G), pore fields on mantle and face, girdle structure more complex	*Diatoma*
5b.	Without sternum, pore fields on mantle only, girdle structure simpler	*Distrionella*
6a.	Septa present (Fig. 13B, C)	7
6b.	Septa absent (Fig. 4E)	8
7a.	Cells elongate, capitate, wider at the center, large polar pore field on face and mantle (Figs. 12H, 13E)	*Tabellaria*
7b.	Cells elliptical or panduriform, apical pore field of fine striae rows (Fig. 12F, G)	*Oxyneis*
8a.	Long, narrow cells (Figs. 4A, 5A, 7A, 12A–E); two labiate processes per valve; solitary, stellate, pincushion, or band-shaped colonial habit	9
8b.	Shorter cells (Figs. 6E, F, 7E, 9, 10, 11A–F); labiate process either one per valve or lacking; solitary, zig-zag or band-shaped colonies	12
9a.	Heteropolar; slender, capitate cells; broader at ends than at the cell center (Figs. 1D, 4A, E); planktonic; stellate colonies	*Asterionella*
9b.	Isopolar, broader at the center than at the poles, benthic or planktonic	10
10a.	Cells arcuate, with central thickening on the ventral side (Figs. 7A, B)	*Hannaea*
10b.	Cells linear to linear-lanceolate	11
11a.	Two terminal spines at each pole in many species, areolae with simple vela	*Synedra*
11b.	No spines, areolae with complex cribra, thickened fascia (Fig. 5A, B), brackish waters	*Ctenophora*
12a.	Broad striae composed of linear areolae (Figs. 7C–E, 11A–F) or composed of multiseriate rows of areolae (Fig. 9E, F)	13
12b.	Narrow striae composed of uniseriate rows of round to transapically elongate areolae	15
13a.	Striae composed of multiseriate rows of areolae	*Punctastriata*
13b.	Striae composed of linear areolae	14
14a.	Cells ovate-elliptical, valve depressed at head-pole (Fig. 7C, D), lacking spines	*Martyana*
14b.	Cells elliptical, linear, or cruciform; possessing spines (Fig. 11A)	*Staurosirella*
15a.	Cells longer, linear to linear-lanceolate, valve margins not undulate (Fig. 6A–E)	*Fragilaria*
15b.	Cells shorter; linear, elliptical, or cruciform	16
16a.	Cells elliptical, lanceolate, or linear; striae very finely punctate; sternum very narrow	17
16b.	Striae composed of larger areolae, sternum wider	18
17a.	Cell margins often undulate (Fig. 6G); one labiate process per valve	*Fragilariforma*
17b.	Cells sometimes with subrostrate ends (Fig. 10C–E); no labiate processes	*Stauroforma*
18a.	Striae made of regularly spaced, round areolae (Fig. 10H, I)	*Staurosira*
18b.	Striae made of few transapically elongate areolae (Fig. 9A–D)	*Pseudostaurosira*
19a.	Valve flexion along apical axis, raphe valve with a rim of ornamentation different from the striae (Fig. 14C–K)	*Cocconeis*
19b.	Valve flexion along transapical axis (Fig. 15L, M) or valves not noticeably flexed (Fig. 16G)	20
20a.	Valves with a sigmoid axial area (Fig. 16A)	*Eucocconeis*
20b.	Valves without a sigmoid axial area	21
21a.	Valves very flexed to a recurved shape in girdle-view, raphe valve concave (Fig. 2F)	22

21b.	Raphe valve not very concave and recurved	23
22a.	Heterovalvar, uniseriate to multiseriate striae, coarse structure, areolae covered by complex cribra, saline or subaerial habitats (Fig. 14A, B)	*Achnanthes*
22b.	Valves of similar structure, narrow linear-lanceolate cells, uniseriate striae, striae usually near 30 per 10 μm, without complex cribra, freshwater habitats (Fig. 15)	*Achnanthidium*
23a.	Striae biseriate to multiseriate	24
23b.	Striae uniseriate	25
24a.	Biseriate striae, asymmetric stauros on the raphe valve (Fig. 17)	*Lemnicola*
24b.	Multiseriate striae, horse-hoof-shaped structure near center in many taxa, usually on the rapheless valve (Fig. 18)	*Planothidium*
25a.	Raphe valve convex to flat; valves oval-elliptic, sometimes capitate; raphe fissures in a channel; adnate, usually on sand; most taxa acidophilic (Fig. 19A–J)	*Psammothidium*
25b.	Raphe valve concave to flat	26
26a.	Valves linear, of similar structure; striae parallel (Fig. 19K–P)	*Rossithidium*
26b.	Valves of dissimilar structure	27
27a.	Raphe valve with radiate striae and a central area with striae of alternating length, rapheless valve with parallel striae (Fig. 16D–G)	*Karayevia*
27b.	Both valves with radiate striae, raphe valve with striae of single elongate areolae, rapheless valve with striae of two or three elongate areolae (Fig. 16H–K)	*Kolbesia*

B. Descriptions of Genera

ARAPHID DIATOMS, CLASS FRAGILARIOPHYCEAE

Fragilariales

Fragilariaceae

Asterionella A. H. Hassall (Figs. 1C, D, 4A–H) Cells attach by basal ends or footpoles to form characteristic stellate colonies. A common component of spring, summer, and fall blooms in lake phytoplankton, *Asterionella formosa* has one of the slowest sinking rates of freshwater planktonic diatoms (Reynolds, 1984). Few species have been described (Körner, 1971) and the freshwater and marine taxa that were formerly treated together have been split into different genera [*Asterionellopsis* Round, in Round *et al.*, (1990), and *Asteroplanus* Gardner and Crawford, in Crawford and Gardner (1997) for marine taxa].

The freshwater taxa are often major components in lake plankton succession (Lund, 1949, 1950), nutrient competition experiments, and paleoecological studies (Duthie, 1989). *Asterionella ralfsii* var. *americana* is common in acidic lakes and is sensitive to toxic metals (Pillsbury and Kingston, 1990; Gensemer, 1991a, b). Colony size in *Asterionella formosa* is affected by P and Si limitation (Tilman *et al.*, 1976). However, *Asterionella* may out-compete *Fragilaria crotonensis* when phosphorus is limiting (Kilham and Kilham, 1978), probably through exceptional ability to store phosphorus (luxury consumption) (Holm and Armstrong, 1981). It remains to be seen, given its worldwide distribution, whether *Asterionella formosa* is a single species or several taxa that are difficult to separate. There may be at least five species in the Laurentian Great Lakes.

Ctenophora (A. Grunow) D. M. Williams et F. E. Round (Figs. 5A, B) A central thickened fascia with ghost striae is characteristic of this linear-lanceolate genus. Two labiate processes per valve, close to the axial area, oriented diagonally within a radial stria near each pole make this genus similar to *Synedra*.

Cells are attached to the substratum by mucilage pads in tuftlike colonies. The genus is common in inland brackish waters, including salinified sites (Tuchman *et al.*, 1984).

Diatoma J. B. M. Bory de Saint-Vincent (Figs. 1B, 5C–G) Valves symmetrical to the longitudinal and transverse axes, with striae and transverse costae, without septa. This genus has been investigated using modern systematic methods (Williams, 1985), and two subgenera are differentiated:

- subgenus *Diatoma*—Striae composed of uniseriate rows; internal transapical ribs are prominent, raised, extending from the sternum to both mantles; tip of the valve mantle interconnected by an overlapping internal rim. Scattered spines near the polar pore fields,

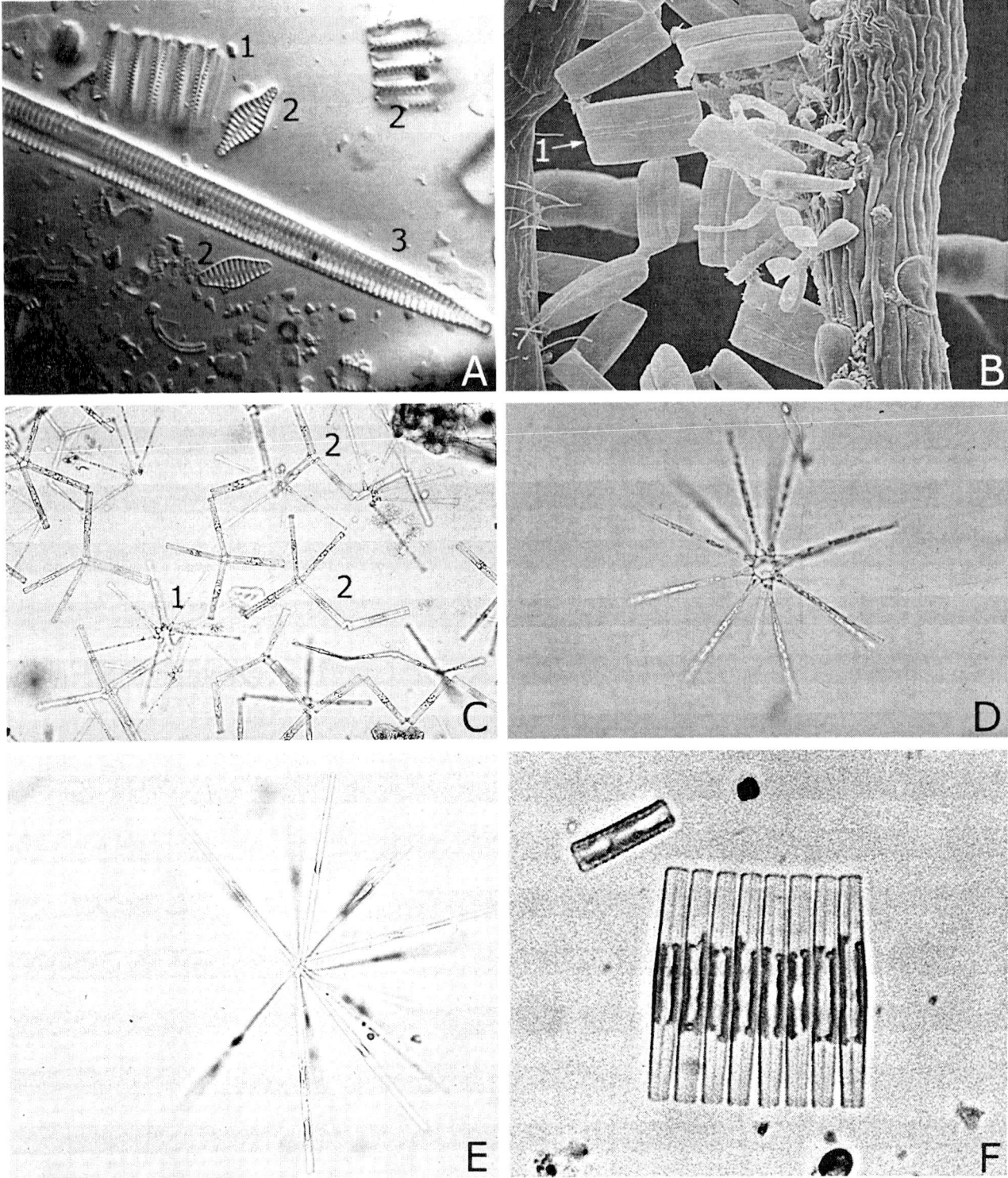

FIGURE 1 Freshwater araphid diatoms, mainly habit images: (A) light micrograph (LM) of cleaned diatom frustules from a riverine benthic assemblage containing *Pseudostaurosira* (1), *Staurosirella* (2), and *Synedra* (3); (B) scanning electron micrograph (SEM) of epiphytes on the green alga *Cladophora glomerata*, including *Diatoma* (1); note the attachment of cells by mucilage pads; (C) LM at low magnification of a mixed collection of preserved phytoplankton including stellate *Asterionella* (1) and zigzag *Tabellaria* (2) colonies; (D) LM at low magnification of a stellate colony of *Asterionella formosa*; (E) LM at low magnification of a "pin-cushion" colony of *Synedra*; (F) LM at low magnification of a ribbon-shaped colony of *Fragilaria crotonensis*, in which the cells are held together by interlocking siliceous spines; note chloroplasts in the LMs of preserved diatoms.

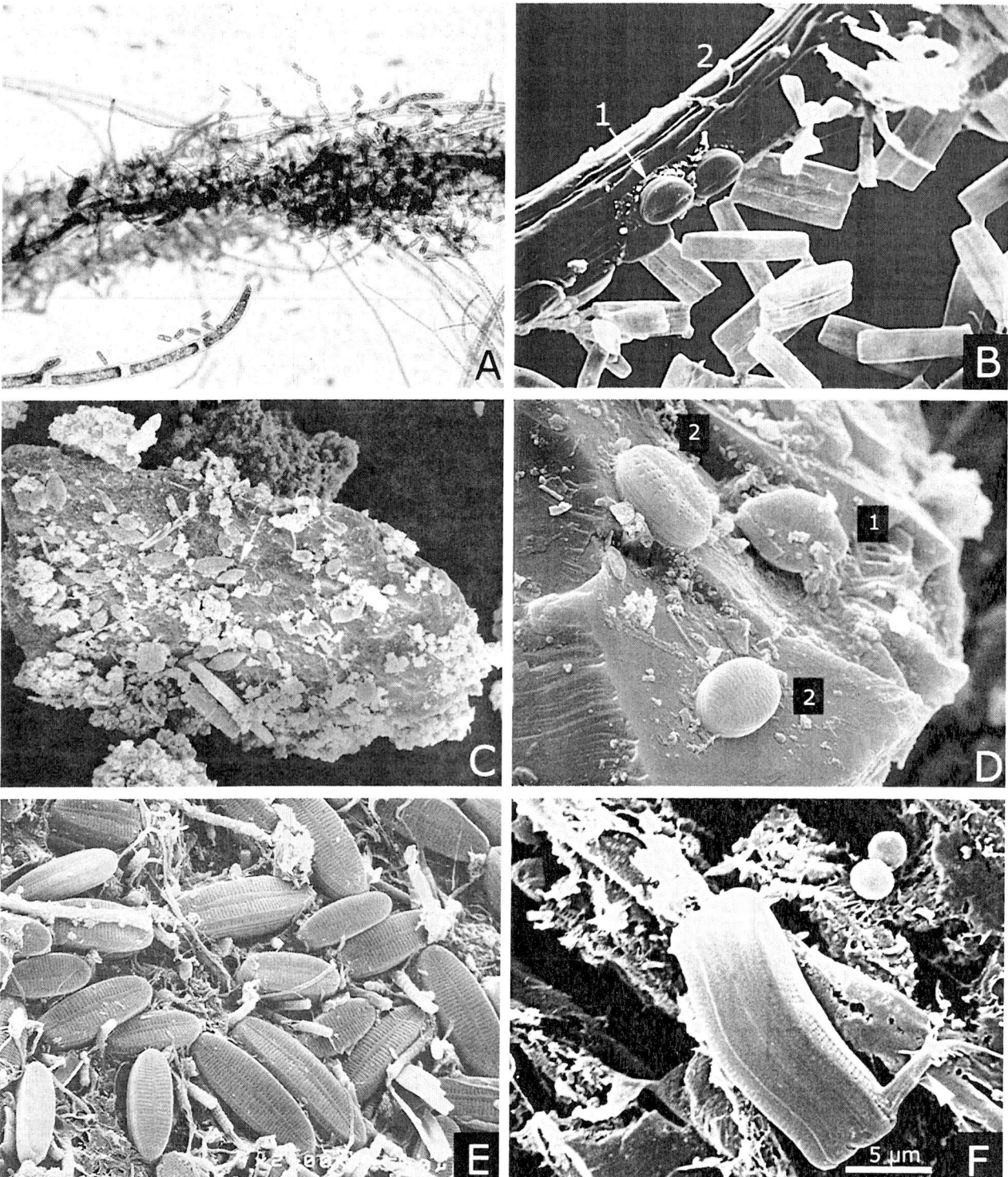

FIGURE 2 Freshwater monoraphid diatoms, habit images: (A) LM at low magnification of epiphytes on *Cladophora glomerata*; (B) SEM of epiphytes on *Cladophora* including *Cocconeis pediculus* (1); note broken frustule (2) with raphe valve still attached to *Cladophora*; (C) SEM of epipsammic algal assemblage including *Karayevia clevei* varieties (arrow); (D) SEM of epipsammic algal assemblage including adnate *Karayevia clevei* (1) and two small species of *Cocconeis* (2); (E) SEM of a dense assemblage of *Rossithidium linearis* and *Achnanthidium* on a pebble; (F) SEM of *Achnanthidium affine* on wood; note polysaccharide stalk attaching the cell to the substratum (arrow).

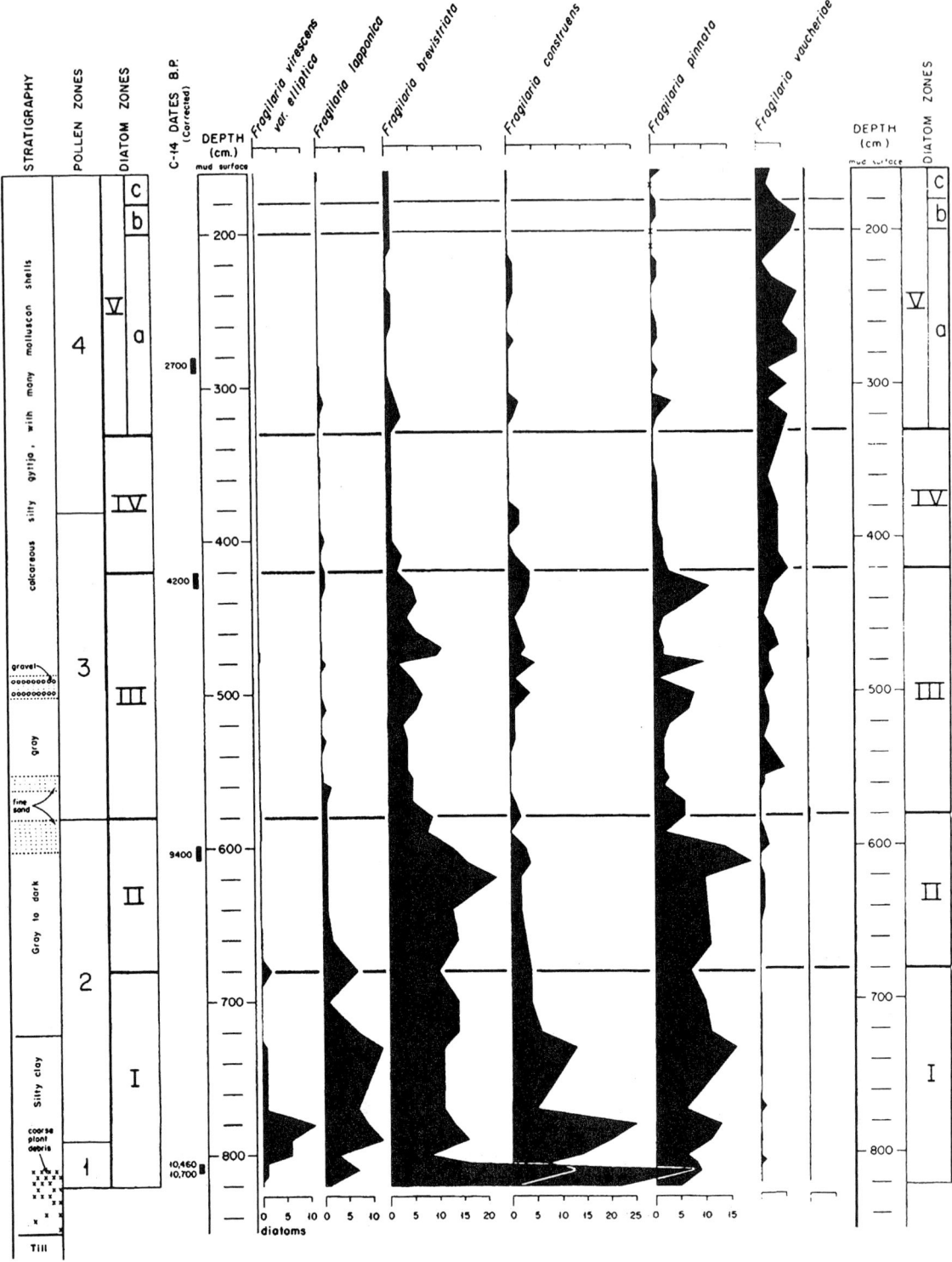

FIGURE 3 Selected parts of a paleoecological (sediment-core) diagram from Pickerel Lake, South Dakota, showing the relative abundances of various Fragilariaceae taxa during the Holocene period, more than 10,000 years at this site. See the original paper for detailed explanations of the inferred climatic, terrestrial, and ecological trends (Haworth, 1972). This is a traditional presentation of paleoecological data, with older samples toward the bottom of the diagram. Depth in lake sediment, geological, pollen, and diatom zones are defined at the sides of the figure. Relative abundance is plotted for each taxon. Note that small benthic Fragilariaceae are very abundant in the early postglacial period. The recent generic designations for these taxa, which were all treated in one genus originally, are from left to right: *Staurosira* (E. Y. Haworth, pers. comm.), *Staurosirella*, *Pseudostaurosira*, *Staurosira*, *Staurosirella*, and *Fragilaria*.

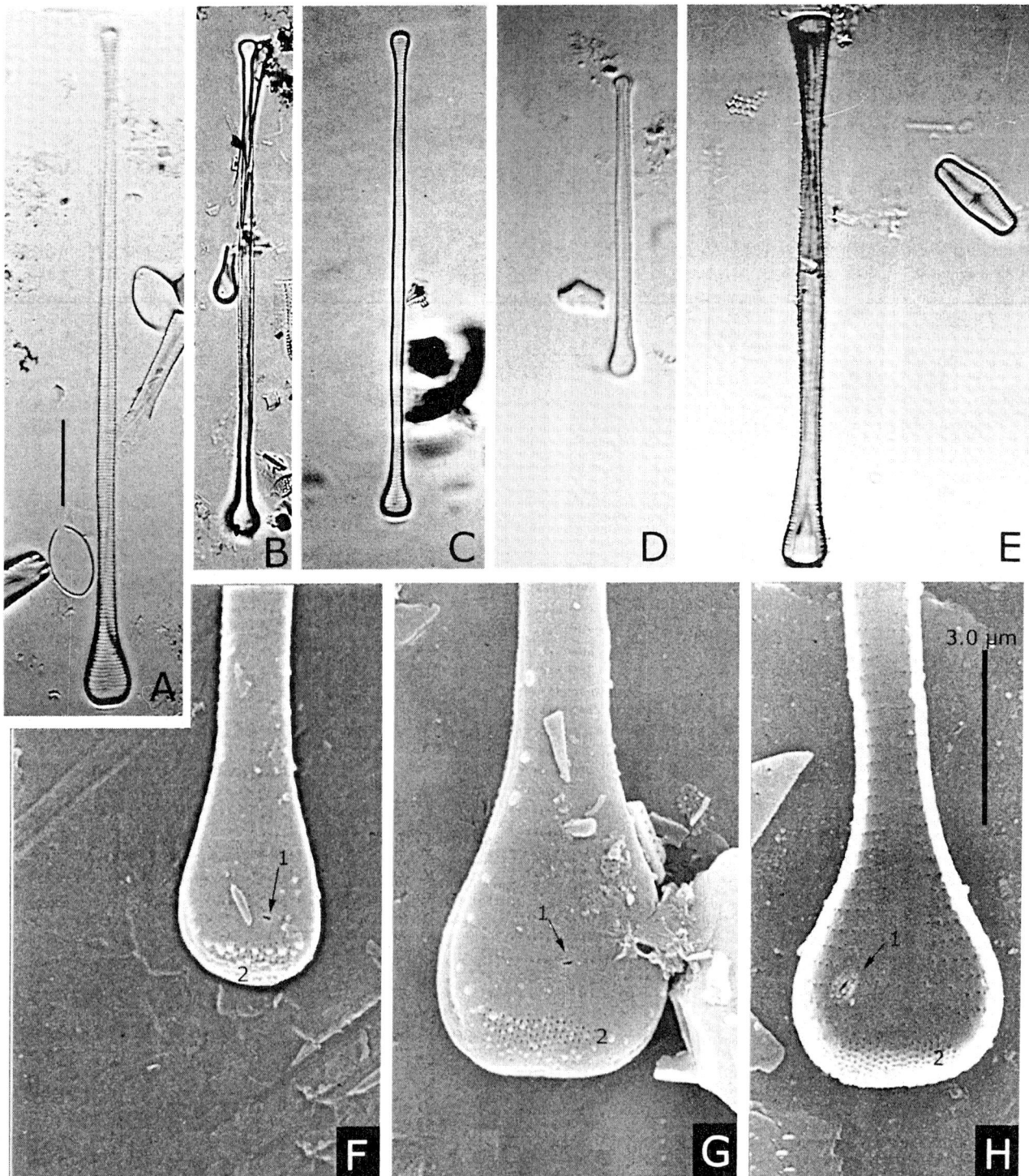

FIGURE 4 *Asterionella*: (A)–(E) LMs of *Asterionella formosa*; (A)–(D) valve views; (E) girdle view of a single cell showing the marginal spines; (F)–(H) SEMs of *Asterionella ralfsii* var. *americana*, same scale bar for these specimens; (F) and (G) external view of foot pole showing labiate process (1) and apical pore field (2); (H) internal view of foot pole showing labiate process (1) and apical pore field (2); scale bar for LMs = 10 μm.

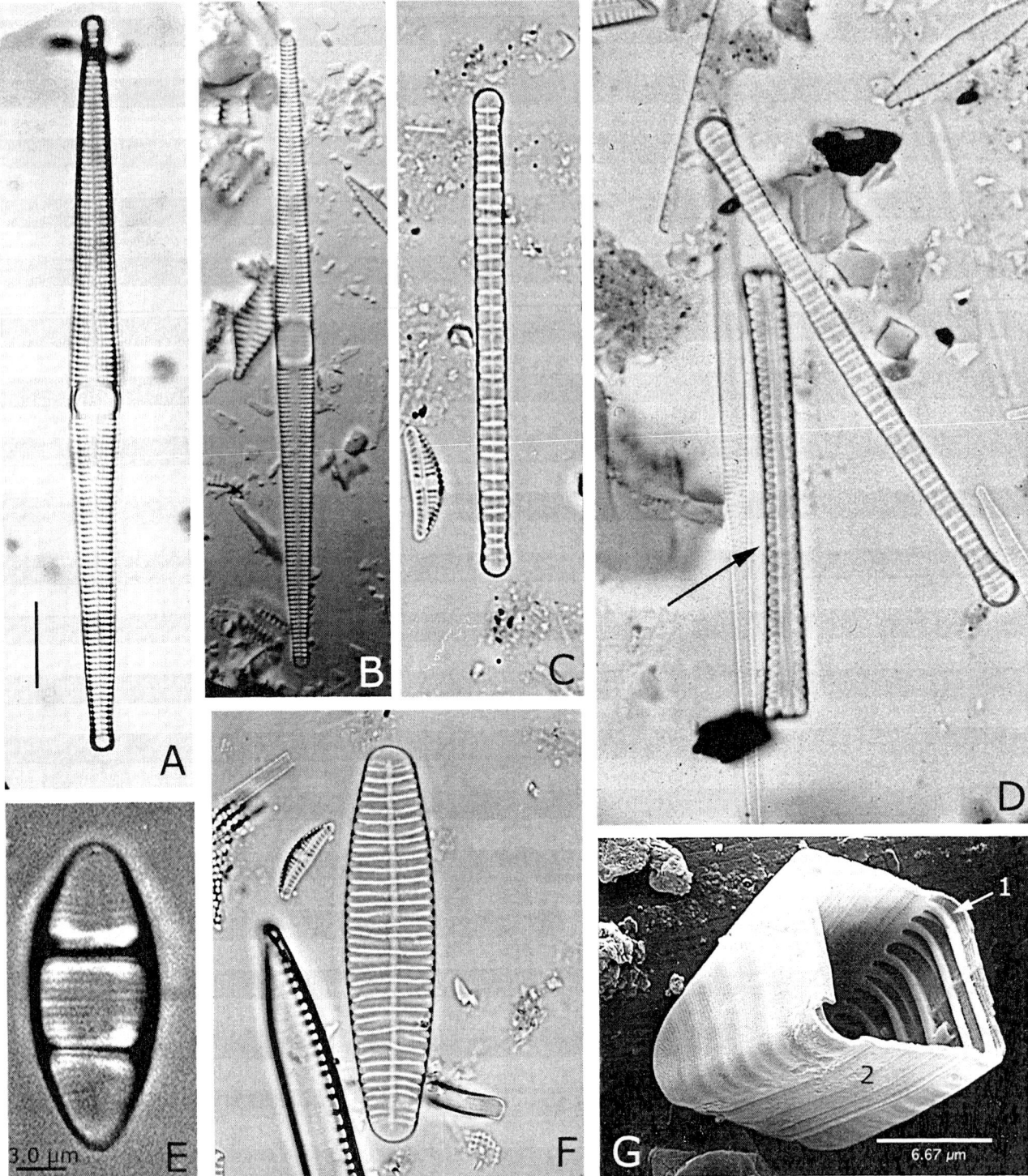

FIGURE 5 Ctenophora and Diatoma: (A) and (B) LM valve views of *Ctenophora pulchella*; (C) and (D) LM of *Diatoma tenue* var. *elongatum* (subgenus *Diatoma*); (C) valve view; (D) valve and girdle view (arrow); (E) LM of *Diatoma mesodon* (subgenus *Odontidium*) (3000×), different LM scale bar for this figure only; (F) LM of *Diatoma vulgare* Bory in valve view (subgenus *Diatoma*); (G) SEM oblique view of a fractured frustule of *Diatoma vulgare* showing external and internal features including striae, internally thickened transapical ribs (1), raised sternum along the axial axis of the valve, and the cingulum [girdle bands (2)]; scale bar for LMs = 10 μm.

absent elsewhere; multiple discoid or platelike chromatophores; examples are *Diatoma vulgare* and *Diatoma tenue*;
- subgenus *Odontidium*—Wide, diffuse sternum; interlocking but non-functional spines; differentiation of the cingulum into three components; copulae with porose ligulae; multiple extra-elliptical lobed chromatophores; examples are *Diatoma mesodon* and *Diatoma hiemale*.

Distrionella D. M. Williams   This genus has two species and is similar in appearance to *Asterionella* and *Diatoma* (Williams, 1990b); it is distinguished by a lighter silicification and weaker transapical ribs than *Diatoma*. It is not yet reported from North America (D. M. Williams, pers. comm.).

Fragilaria H. C. Lyngbye (Figs. 1F, 6A–E) Frustules are rectangular to lanceolate in girdle view, usually forming ribbon- or band-shaped colonies of cells joined by interlocking spines. A single labiate process is present, usually at a pole, and this is different from the two-per-valve condition in the genus *Synedra*. There is currently a problem in the typification of this widely accepted genus, which may be solved by a conservation proposal (Fourtanier and Kociolek, 1999).

Centronella M. Voigt, a triradiate taxon with narrow arms, was considered a separate genus but has been shown to be a persistent growth form of *Fragilaria crotonensis* (Schmid, 1997). Some other members of the family have triradiate forms. A very broad concept of the genus *Fragilaria* has been propounded by Krammer and Lange-Bertalot (Table II) (Lange-Bertalot, 1980; Krammer and Lange-Bertalot, 1991a). The treatment in this chapter follows the narrower circumscription of the genus as presented by Williams and Round (1987).

Fragilaria taxa (*F. crotonensis*, *F. capucina*, *F. vaucheriae*) are often abundant in eutrophic reservoirs (Hoagland and Peterson, 1990). *Fragilaria crotonensis* responds dramatically to elevated phosphorus (Stoermer *et al.*, 1978b), is regarded as a better competitor for silica than *Asterionella formosa* (Kilham and Kilham, 1978), is a dominant summer phytoplankter in eutrophic lakes, and can occur in significant abundance under ice cover (Agbeti and Smol, 1995). *Fragilaria crotonensis* is another putative species with worldwide distribution.

Fragilariforma D. M. Williams *et* F. E. Round (Figs. 6F–H)   Frustules are rectangular in girdle-view, in mainly linear or rarely zig-zag colonies. Valves are elliptical, lanceolate, or linear, with rostrate to capitate apices; valves are often undulate with swelling or constriction at the center. Striae are uniseriate and usually fine; sternum is small or absent. A single, polar labiate process is aligned with a stria. Plastids are numerous, small, discoid.

Fragilariforma has many representatives in boreal bog and fen environments (Cleve-Euler, 1953; Woodhead and Tweed, 1960; Renberg, 1976, 1977; Kingston *et al*, 2001), others in semitropical North American environments (Hein, 1981; Williams, 1990c), others in prairie potholes.

Frankophila H. Lange-Bertalot   Lange-Bertalot (1997a) erected this new genus, containing biraphid (or monoraphid) relatives of the Fragilariaceae (*Punctastriata*, *Staurosirella*, or *Pseudostaurosira*). It is unclear whether this group can stand as an independent genus and how it will be placed into an evolutionary hierarchy. *Frankophila* is very difficult to identify with light microscopy because the simple raphe system is not noticeable. This is reminiscent of, and intermingled with, the problem of identifying *Punctastriata* using light microscopy, because the striae structure at this scale appears identical to that of *Staurosirella*. This genus has not been reported in North America, perhaps due to the lack of SEM studies that would be necessary to observe it.

Hannaea R. Patrick in R. Patrick *et* C. W. Reimer (Figs. 7A, B)   Cells are rectangular in girdle view, linear and arcuate in valve view, with capitate poles. The genus is distinguished by a ventral external swelling, internally depressed, with ghost striae instead of areolae. It has prominent labiate processes, usually one at each end as in *Synedra*.

This benthic genus is commonly found in mountain streams, in Lake Superior, and in Lake Baikal (Bixby, 2001). Frustular ultrastructure of specimens of *Hannaea arcus* from the Canadian arctic have been published (Lichti-Federovich, 1979).

Martyana F. E. Round, in F. E. Round, R. M. Crawford *et* D. G. Mann (Fig. 7C–E)   Frustules are rectangular to wedge-shaped in girdle view. Valves are ovate-elliptical, becoming elliptical in small cells. There is a distinct depression at the broad end of each valve. Striae have regular slitlike areolae and go without a break from the valve face to the undifferentiated valve mantle. Striae are present on the step at the broad end but absent on the narrow end. No spines or labiate process are present. Apical pore field is at the smaller end of the valve; cells are attached to substrata (sand) by mucilage pads at the narrow valve end.

The unnatural grouping of marine and freshwater heteropolar araphid taxa into *Opephora* has been recognized for some time (Sullivan, 1979). *Martyana*

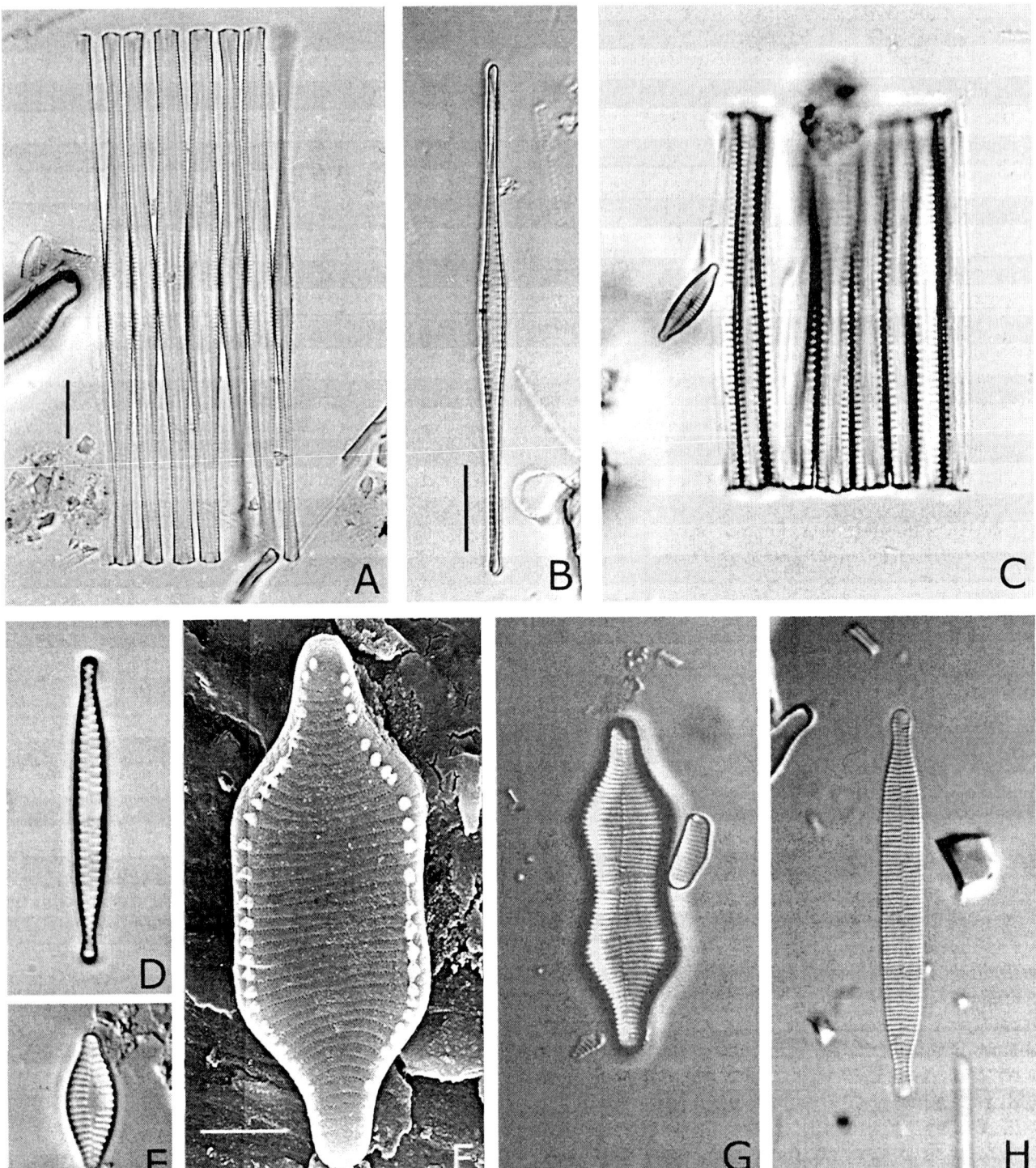

FIGURE 6 *Fragilaria* and *Fragilariforma*: (A) LM girdle view of a colony of *Fragilaria crotonensis* (1000×), different LM scale bar for this figure only, the valves are joined near the centers by interlocking spines; (B) LM valve view of *Fragilaria crotonensis* var. *oregona*; (C) LM girdle view of *Fragilaria intermedia*; (D) LM valve view of *Fragilaria intermedia*; (E) LM valve view of *Fragilaria vaucheriae*; (F) SEM valve view of *Fragilariforma constricta*; (G) LM valve view of *Fragilariforma constricta*; (H) LM valve view of *Fragilariforma virescens*; scale bar for LMs = 10 μm.

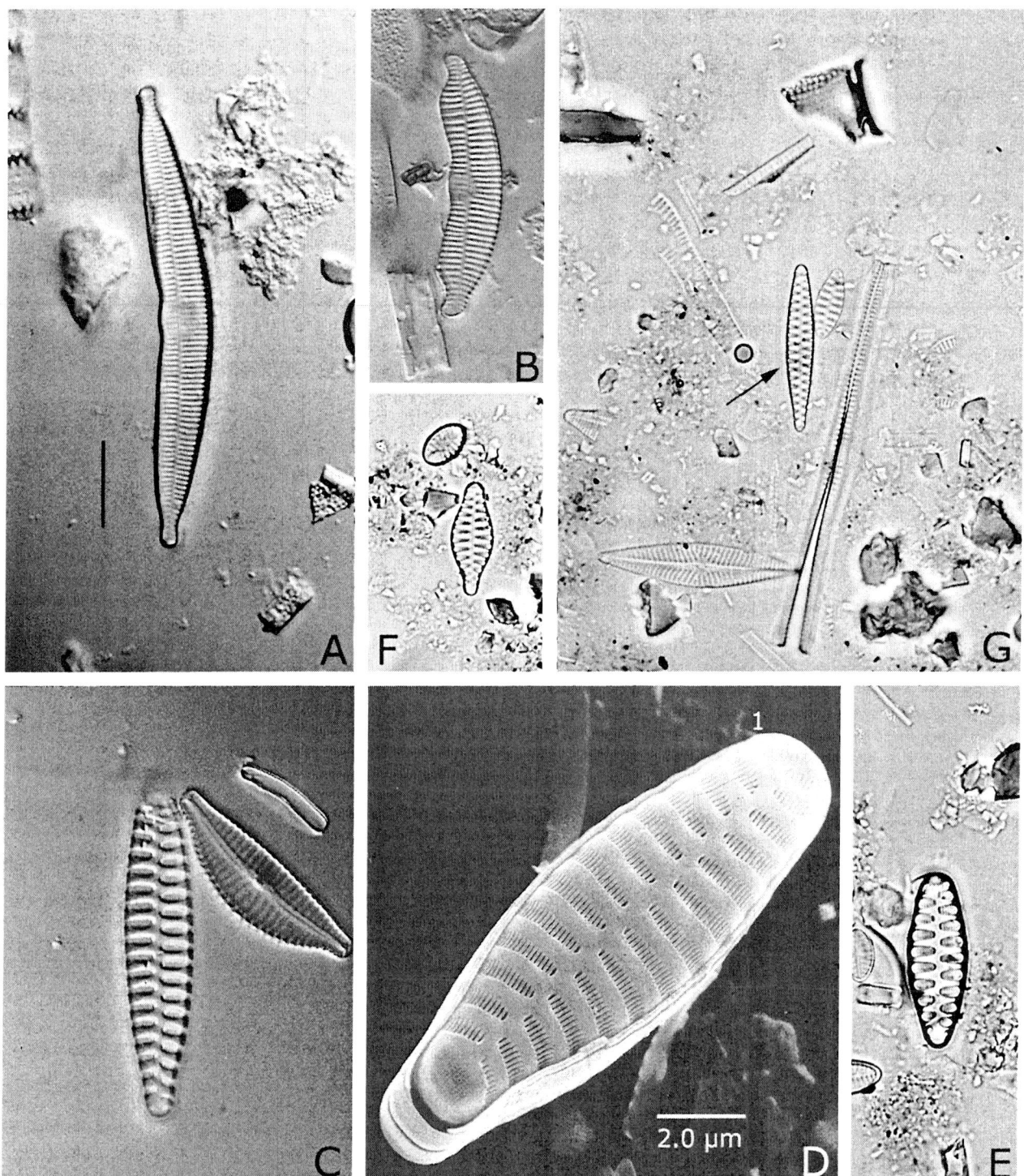

FIGURE 7 *Hannaea*, *Martyana*, and *Staurosirella*: (A) LM valve view of *Hannaea arcus*; (B) LM valve view of *Hannaea arcus* var. *amphioxys*; (C) LM valve view of *Martyana martyi*; note depression at head pole is below focal plane; (D) SEM of *Martyana martyi*; note lack of marginal spines and depression at the head pole (1); (E) LM valve view of *Martyana martyi*; (F) and (G) LMs of heteropolar *Staurosirella* specimens that would have been placed into *Opephora* in system 1, but that do not meet criteria to be placed into *Martyana* (specimen identified by arrow in G); scale bar for LMs = 10 μm.

was erected to include taxa formerly treated under *Opephora martyi*, but not other species described as freshwater *Opephora*, such as *Opephora ansata* (Hohn and Hellerman, 1963). As described, the genus cannot include many small heteropolar species with spines and lacking the unipolar depression (Figs. 7F, G and 11F). Such taxa are classified within the genus *Staurosirella*. Further examination of type specimens could help to determine the validity of the generic concepts.

Meridion C. A. Agardh (Fig. 8A–F) *Meridion* is similar in appearance in LM to *Diatoma*. Earlier, the genus was defined as including only heteropolar taxa with striae and costae, but consideration of character distribution including the presence of septa on girdle band elements caused a reinterpretation of this genus to include some species formerly treated within *Diatoma*. Multiple platelike chromatophores have been observed in species that have been studied in a living condition (Williams, 1985). Examples are *Meridion circulare* and *Meridion anceps*.

Pseudostaurosira D. M. Williams *et* F. E. Round (Figs. 1A, 9A–D) Frustules are rectangular, forming chains. Striae are uniseriate, consisting of a few large, elliptical areolae and sometimes small, round areolae; there are rarely more than four areolae per stria. Spines are along the valve edge, sometimes branched. Plastids are not known.

This genus includes taxa related to *Fragilaria brevistriata* and distinguished by their common striae structure. They are very common in shallow water of streams and lakes.

Punctastriata D. M. Williams *et* F. E. Round (Fig. 9E–F) This genus is rarely reported, as it has been indistinguishable from *Staurosirella* using LM. The presence of two very different striae structures had been known for some time prior to erection of this genus (Rosen and Lowe, 1981). The micrographs presented here show species different than those illustrated in the original paper (Williams and Round, 1987); spines on the Michigan specimens are on the interstriae.

Stauroforma R. Flower, V. Jones, *et* F. E. Round (Fig. 10A–E) This genus was described as containing two species formerly treated as small varieties of *Fragilaria virescens* Ralfs, notably *Fragilaria virescens* var. *exigua* Grun. They are small, linear valves, never centrally expanded, with no labiate processes. The latter feature seems to be the most important in separating this genus from *Fragilariforma*.

Stauroforma exiguiformis is common in dystrophic and northern temperate to boreal ecoregions.

Staurosira C. G. Ehrenberg (Fig. 10F–I) Frustules are rectangular in girdle view, frequently in chains, with interlinking spines. Valves are elliptical or cruciform. Striae are narrow and composed of small rounded areolae.

This taxon was given generic rank and includes relatives of *Fragilaria construens* and *Fragilaria elliptica*. It is a common component of shallow water floras in rivers and lakes. It can be differentiated in LM from *Staurosirella* based on striae structure observable both in valve and girdle views.

Staurosirella D. M. Williams *et* F. E. Round (Figs. 1A, 7F–G, 11A–F) Frustules are rectangular, attached and in chains, linked by the corners or the valve faces. Valves are elliptical to linear, or cruciform. Striae are uniseriate with long cross-members delimiting linear areolae.

This genus was erected to include *Fragilaria lapponica*, *Fragilaria pinnata*, and *Fragilaria leptostauron*. It is a common element of shallow water floras of standing and flowing waters. Cells of *Staurosirella leptostauron* attach by the apical pore field, with short stalks, and the cells form "sprays" fanning out in all directions (B. H. Rosen, pers. comm.). Detailed objections have been raised to the splitting of so many genera from a broader generic concept of *Fragilaria* (Lange-Bertalot, 1989).

Synedra C. G. Ehrenberg (Figs. 1A, E, 11G–I, 12A–E) Frustules are rectangular in girdle view. Cells are linear or lanceolate. There are two labiate processes per valve, one near each end. Many taxa have a prominent spine pair at each end, one on either side of the pore field. Cells often form colonies with cells attaching at a single point, either on substrata or in the plankton.

Krammer and Lange-Bertalot (1991a) argue that each freshwater species of the classical genus *Synedra* has a related counterpart in a broadly conceived genus *Fragilaria*. Based on rules of nomenclatural priority and the types for each genus, they treat nearly all *Synedra* species as members of the genus *Fragilaria*.

Planktonic *Synedra* taxa can be very abundant in oligotrophic, mesotrophic, and eutrophic lakes during different seasons (Stoermer and Yang, 1970; Agbeti and Smol, 1995; Agbeti *et al.*, 1997). Benthic *Synedra* populations can be major components of river communities (Main, 1988).

Tabellariales

Tabellariaceae

Oxyneis F. E. Round, in F. E. Round, R. M. Crawford *et* D. G. Mann (Fig. 12F, G) Cells are

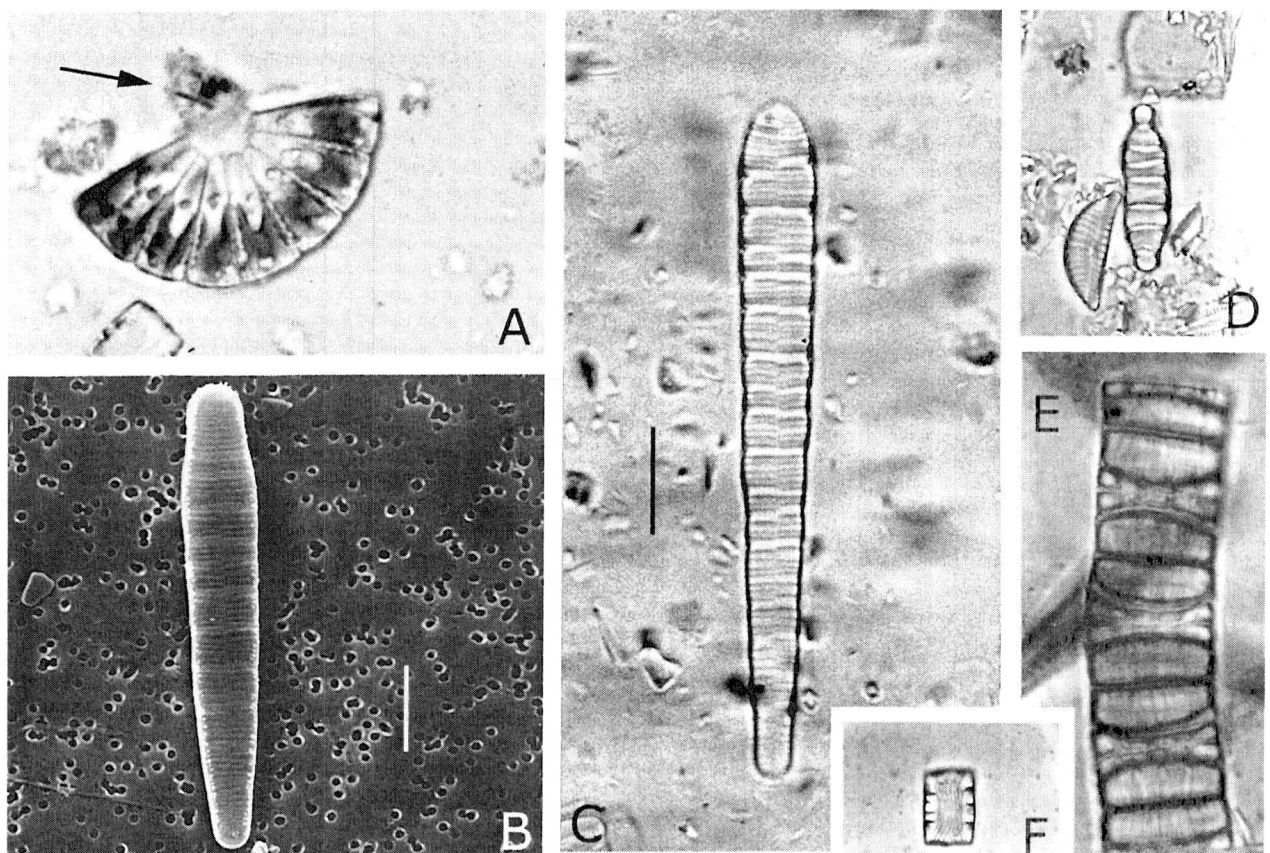

FIGURE 8 Meridion: (A) live colony of *Meridion circulare*; note face-to-face arrangement of heteropolar cells, chloroplasts, and mucilage pad attachment site that has been severed from the substratum (arrow); (B) SEM valve view of *Meridion circulare* var. *constrictum*, bar = 10 µm; (C) LM valve view of *Meridion circulare*; (D) LM valve view of the isopolar *Meridion anceps*; (E) LM girdle view of a chainlike colony of *Meridion anceps*; (F) LM girdle view of a single cell of *Meridion anceps*; scale bar for LMs = 10 µm.

oblong in girdle view, elliptical or panduriform in valve view. Septae are present on the copulae. The cells form short chains or zigzag colonies. This genus was split from *Tabellaria*, with which it shares many characteristics; it is distinguished by generally smaller size, elliptical shape, and band structure.

This genus, the two species of which used to be treated as varieties of *Tabellaria binalis*, is characteristic of acidic lakes and peat pools (Walker and Paterson, 1986; Flower, 1989; Cook *et al.*, 1990). *Oxyneis* typically has very low environmental optima for pH and alkalinity in lake acidification studies (Davis *et al.*, 1994).

Tabellaria C. G. Ehrenberg *ex* F. T. Kützing (Figs. 1C, 12H, 13A–E) Cells are rectangular in girdle view. Valves are elongate, capitate, and generally wider at the center than at the ends. Spines usually are present along the valve face–mantle border, including at the apical pore fields. Copulae are either complete or incomplete, and septate. Number and type of copulae have been used to differentiate species (Koppen, 1975). Short striplike chromatophores lie between the septa. Cells are joined in long zigzag colonies by mucilage pads originating from the apical pore fields; colonies also attach this way to substrata.

The ecology of species and varieties in North America has been described (Koppen, 1978; Davis *et al.*, 1994). The various species have environmental optima ranging from moderately alkaline to very acidic. Planktonic blooms of *Tabellaria* can clog sand filters in surface water treatment facilities in humic waters in Canada (J. C. Kingston, pers. obs.).

Tetracyclus J. Ralfs (Fig. 13F) Valves are elongate to elliptical, often capitate, and often centrally expanded and/or constricted. Valve face is flat; mantles are distinct and tall. Striae are uniseriate not interrupted

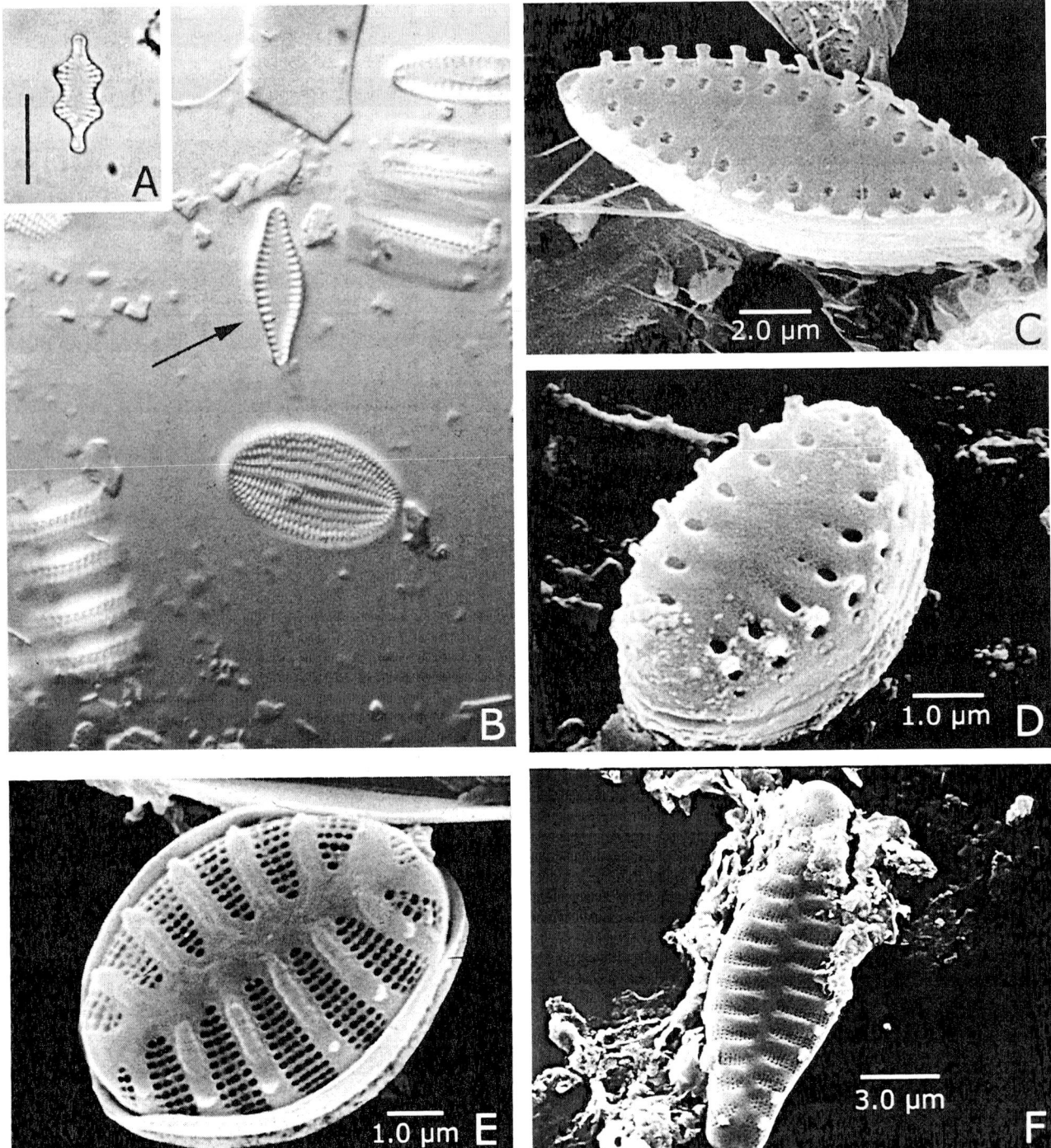

FIGURE 9 *Pseudostaurosira* and *Punctastriata*: (A) LM valve view of *Pseudostaurosira robusta*; (B) LM valve view of *Pseudostaurosira brevistriata* var. *inflata*; (C) SEM external view of *Pseudostaurosira* sp.; (D) SEM external view of *Pseudostaurosira brevistriata* var. *elliptica*; (E) and (F) *Punctastriata* specimens from Lake Kathryn, Michigan (B. H. Rosen, unpublished) [(E) SEM external view of *Punctastriata* sp.; (F) SEM external view of a different, undescribed *Punctastriata* sp.]; scale bar for LMs = 10 μm.

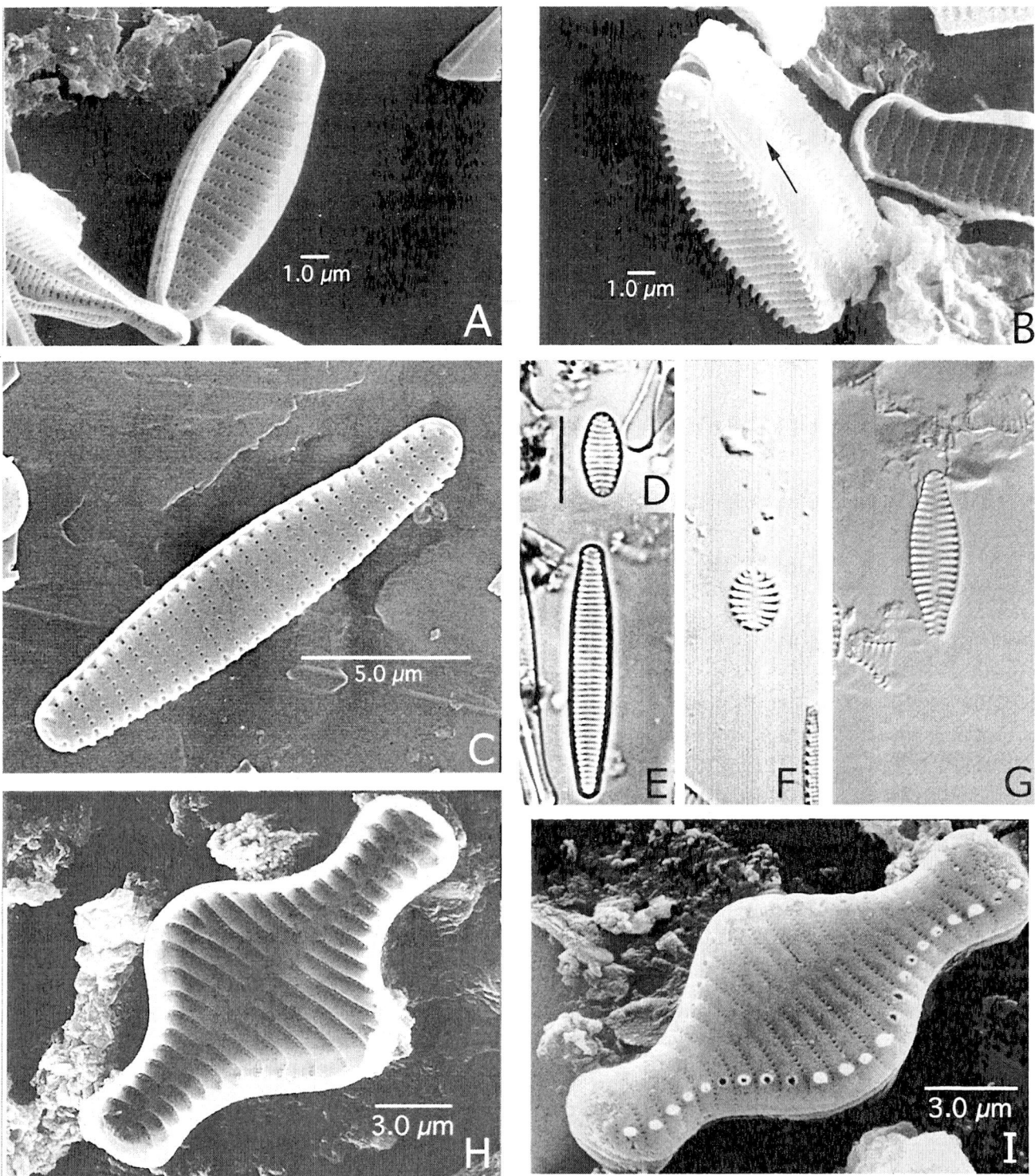

FIGURE 10 *Stauroforma* and *Staurosira*: (A) SEM two valves, internal view of *Stauroforma exiguiformis*; note the absence of labiate processes; (B) SEM oblique external view showing valve features and the girdle bands (arrow) of *Stauroforma exiguiformis*; (C) SEM external view of *Stauroforma exiguiformis*; (D) and (E) LM valve views of *Stauroforma exiguiformis*; (F) LM valve view of *Staurosira construens* var. *venter*; (G) LM valve view of *Staurosira construens* var. *pumila*; (H) and (I) SEM of *Staurosira construens* [(H) internal view; note cribra covering punctae; (I) external view; note slightly eroded, hollow marginal spines]; scale bar for LMs = 10 μm.

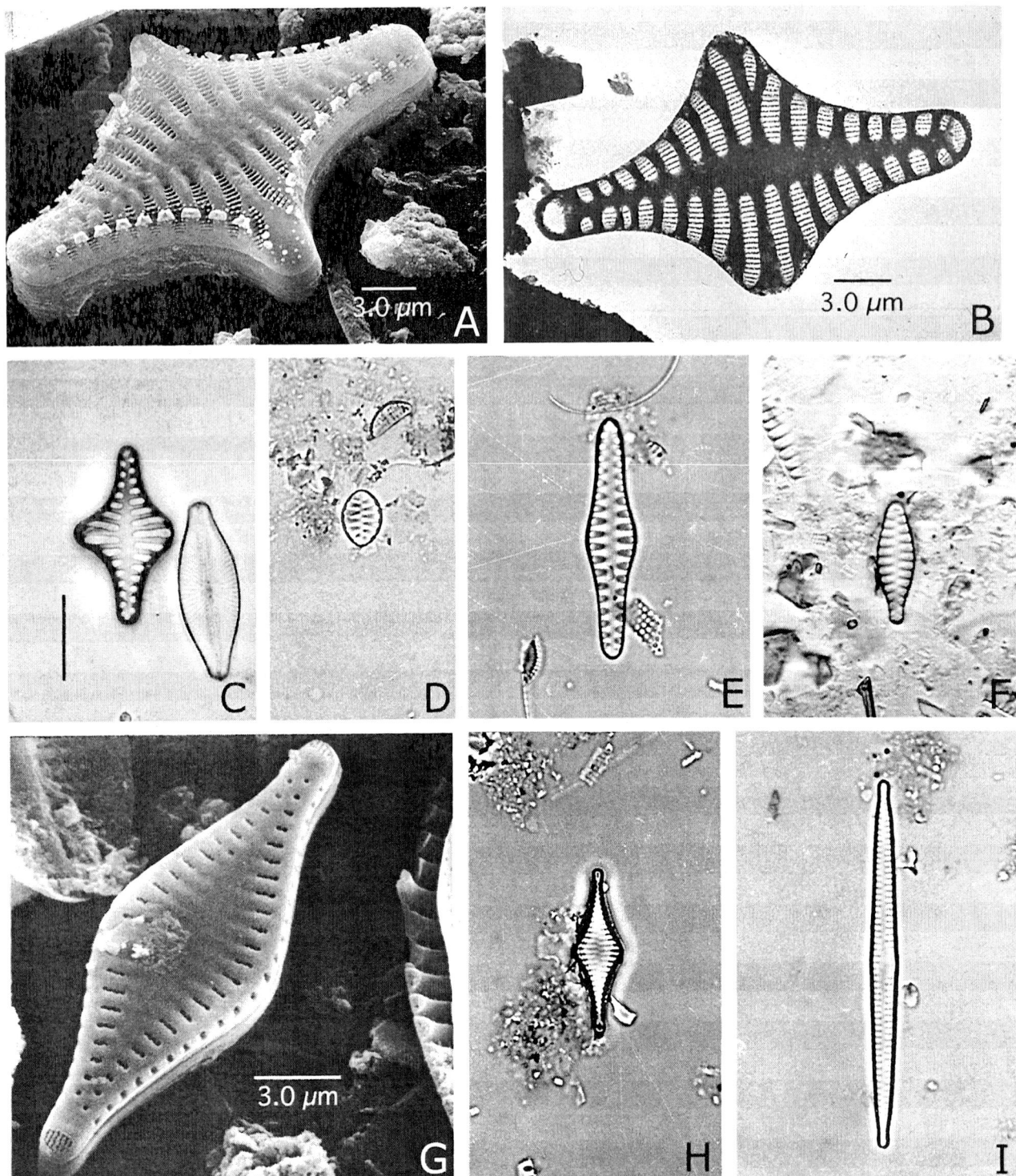

FIGURE 11 *Staurosirella* and *Synedra*: (A)–(C) *Staurosirella leptostauron* [(A) SEM external view; note the bifurcate marginal spines; (B) transmission electron micrograph (TEM) valve view; (C) LM valve view]; (D) LM valve view of *Staurosirella pinnata*; (E) LM valve view of *Staurosirella spinosa*; (F) LM valve view of *Staurosirella ansata*, holotype specimen (G) and (H) *Synedra parasitica* [(G) SEM external view; note pore fields at each end; (H) LM valve view]; (I) LM valve view of *Synedra* sp.; scale bar for LMs = 10 µm.

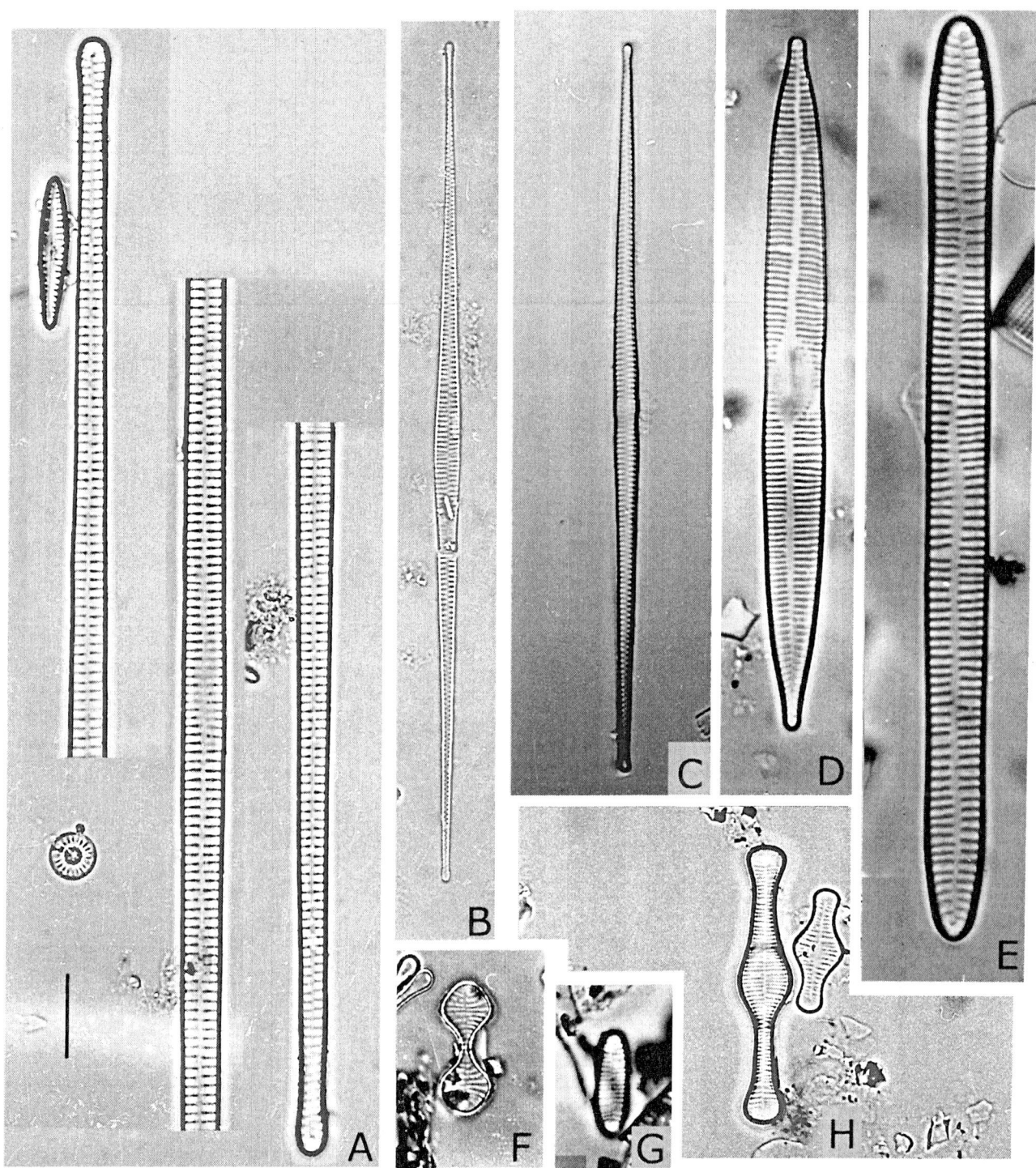

FIGURE 12 *Synedra*, *Oxyneis*, and *Tabellaria*: (A) LM valve view of *Synedra ulna* var. *danica*; image split into three parts to maintain the same magnification as the next taxa; (B) LM valve view of *Synedra ostenfeldii*; (C) LM valve view of *Synedra rumpens* var. *fusa*, holotype specimen; (D) LM valve view of *Synedra ulna*; (E) LM valve view of *Synedra ulna* var. *aequalis*; (F) LM valve view of *Oxyneis binalis*; (G) LM valve view of *Oxyneis binalis* var. *elliptica*; (H) LM valve views of two specimens of *Tabellaria flocculosa* (group IV; Koppen, 1975); scale bar for LMs = 10 μm.

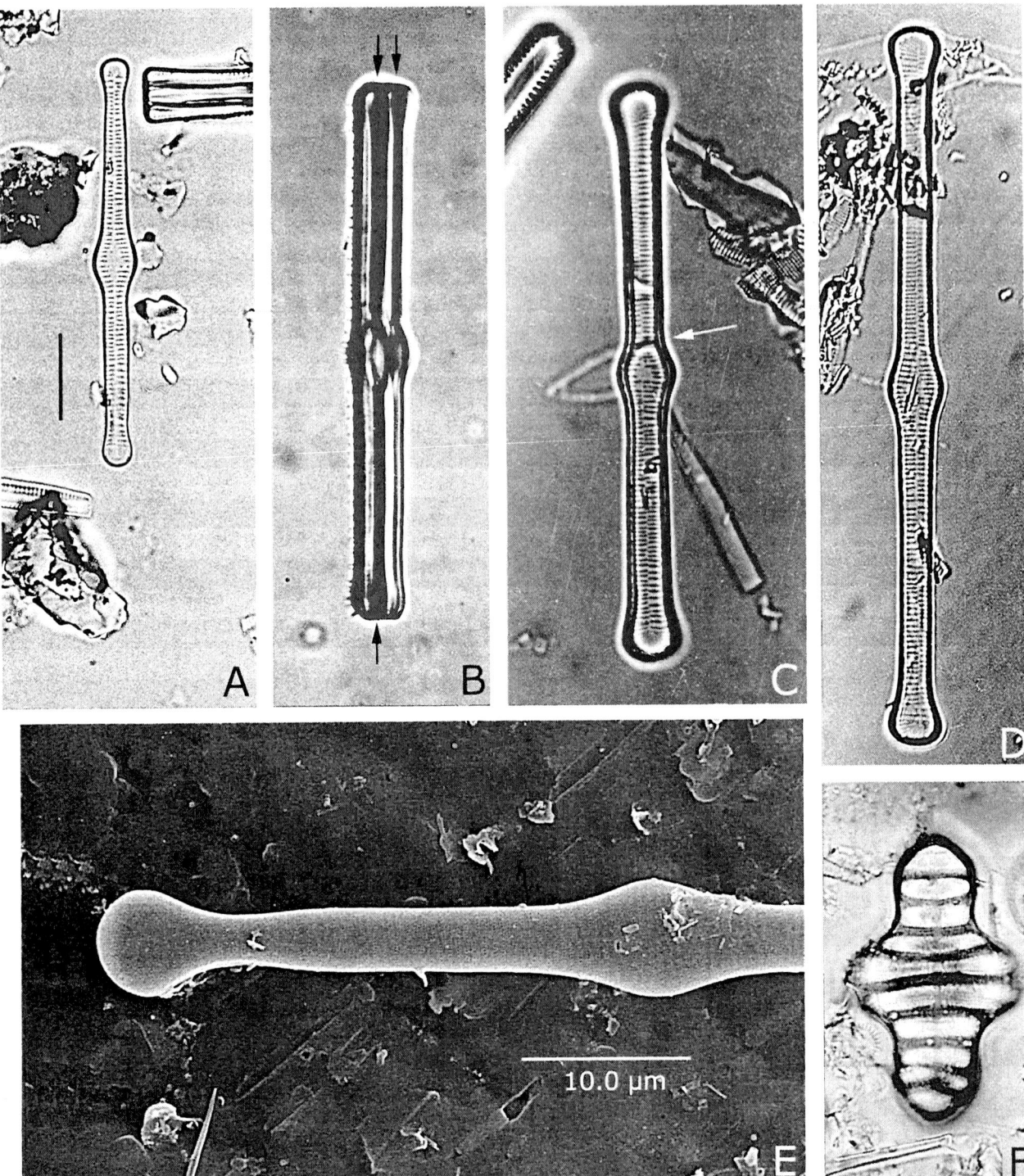

FIGURE 13 Tabellaria and Tetracyclus: (A) LM valve view of *Tabellaria flocculosa* (group III; Koppen, 1975); (B)–(D) LM of *Tabellaria quadriseptata*; note prominent marginal spines [(B) girdle view of one valve and three intercalary bands with septa (arrows); (C) valve view showing the shadow from the edge of a single septum (arrow) on an intercalary band; (D) valve view with no intercalary bands and therefore no line from a septum]; (E) SEM valve view of *Tabellaria fenestrata*, a species usually having no marginal spines, and preferring fairly alkaline water; (F) LM valve view of *Tetracyclus glans*; scale bar for LMs = 10 μm.

externally by underlying costae. There are massive, internal, transapical costae. Valves may have 0–2 labiate processes, situated near the center, arranged transapically along the sternum. Cells form zigzag colonies held together by mucilage pads at the ends.

Tetracyclus is probably most closely related to *Tabellaria* (Williams, 1994, 1996). Better represented in the fossil record, there are only a few species found in modern collections.

MONORAPHID DIATOMS, CLASS BACILLARIOPHYCIDEAE

Achnanthales

Achnanthaceae

Achnanthes J. B. M. Bory de Saint-Vincent (Fig. 14A, B) Cells are heterovalvar (raphid or rapheless). The mantle is more differentiated from the valve face on the rapheless valve. Striae are uniseriate, biseriate, or triseriate, with areolae covered by complex cribra (sieve plates). Raphe valve usually has a fascia or stauros and a central raphe sternum. Rapheless valve is without fascia; narrower rapheless sternum usually is off center. Cells are solitary or form short chains, usually attached to substrata by a mucilage stalk extending through one end of the raphe valve. Plastids are either numerous and discoid (*Achnanthes longipes*) or two and H-shaped (other species).

The genus is predominately marine, but some taxa occur inland; *Achnanthes coarctata* is a common member of the aerophilic assemblage growing associated with moss and lichen (Dodd and Stoermer, 1962). Many common species previously classified in this genus are now recognized as species of *Achnanthidium*, *Psammothidium*, *Planothidium*, or other genera in this chapter.

Cocconeidaceae

Cocconeis C. G. Ehrenberg (Figs. 2B, D, 14C–L) Valves are elliptical or almost circular. The genus is heterovalvar (raphid or rapheless) with the flexion (curvature) along the apical axis. Marine and freshwater, though some species have been removed to new genera [e.g., marine taxa living on cetaceans (Holmes, 1985)]. Striae often are uniseriate, but multiseriate striae and loculate areolae are present in some taxa. Raphe valve has a differentiated region near the edge of the valve, and a more defined mantle than the rapheless valve. Raphe sternum and rapheless sternum are similar in structure. Valvocopula can be complete, with projections ornamented like the valve.

Cells are solitary, adnate, adapted to living on aquatic plants, other algae, and hard substrata such as rock, sand, or shell (Kingston, 1980).

Cocconeis has shown the ability to sequester excess nutrients in polyphosphate bodies (Stevenson and Stoermer, 1982). Krammer has renamed and revised the taxonomy of many small freshwater taxa of *Cocconeis* including *Cocconeis neothumensis*, partly motivated by a lack of type material available from earlier studies (Krammer, 1990; Krammer and Lange-Bertalot, 1991b).

Achnanthidiaceae

Achnanthidium F. T. Kützing (Figs. 2F, 15A–L) Valves are significantly flexed across the transapical axis, forming a shallow V-shape in girdle view. Small, narrow, elongate, linear-lanceolate cells have capitate to rostrate ends. Central area is common on the raphid valve. Distal raphe endings can vary from simple to highly deflected in species considered to be in this genus (Round and Bukhtiyarova, 1996a; Kobayasi, 1997). An isolated row of narrow areolae is on the mantle, separated from those on the valve face. The growth habit is attachment to substrata by a mucilaginous stalk. Sexual reproduction of some taxa has been discussed (Geitler, 1977, 1979, 1980) and contrasted to that in *Planothidium* and *Lemnicola*.

Members of this genus previously classified as *Achnanthes* (including *A. minutissimum*, *A. affine*, *A. microcephalum*) are distinguished by striae, shape, and habit characters. *Achnanthidium*, like many of the species of marine *Achnanthes*, usually have a stipitate habit. They also thrive in moving water and in rapids (Peterson and Hoagland, 1990; Peterson and Stevenson, 1992) and in the wave zones of lakes (Brown, 1973; Kingston, 1980). Cholnoky (1968) even considered some taxa to be "oxygen loving" because they were found in turbulent, well-oxygenated water. *Achnanthidium* taxa are held above the dense prostrate masses of other taxa on stalks, where they can take advantage of more rapid replenishment of the host of chemical constituents flowing past. Small cells like *Achnanthidium minutissimum* (= *Achnanthes minutissima*) are physiologically more active than larger diatom cells, due partly to their large surface to volume ratios (Allen, 1977).

Eucocconeis P. T. Cleve *ex* F. Meister (Fig. 16A–C) Frustules are bent about the median transapical plane as in *Achnanthidium*, with concave raphe valve, but twisted about the apical axis with a sigmoid sternum on both valves.

The genus is often reported from the littoral zone

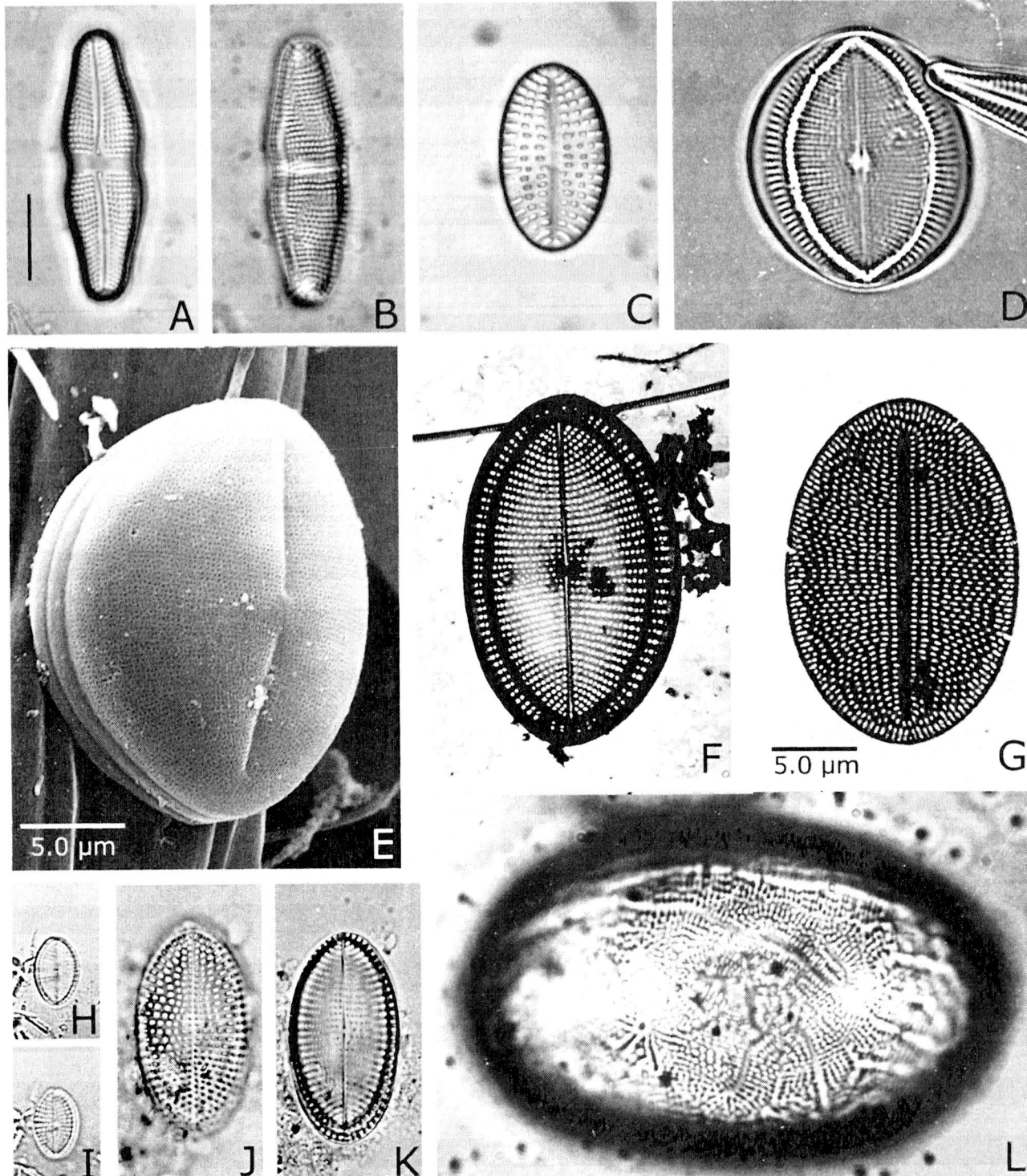

FIGURE 14 *Achnanthes* and *Cocconeis*: (A) and (B) LM valve views of one cell of *Achnanthes coarctata* [(A) raphe valve with the central area expanded into a rectangular fascia; (B) rapheless valve with an off-center axial area]; (C) LM rapheless valve of *Cocconeis disculus*; (D) LM valve view of a raphe valve of *Cocconeis placentula* var. *rouxii*; (E) SEM habit image of *Cocconeis pediculus* growing as an epiphyte on *Cladophora glomerata*; this taxon is very concave along the longitudinal axis of the raphe valve, and the external view of the rapheless valve is observed here; (F) and (G) TEM of valves of *Cocconeis placentula* var. *lineata* [(F) raphe valve showing the typical ornamented rim separated from the striae; (G) rapheless valve]; (H) and (I) LM valve view of a single cell of *Cocconeis neothumensis* [(H) raphe valve; (I) rapheless valve]; (J) and (K) LM valve view of a single cell of *Cocconeis fluviatilis*, holotype specimen [(J) rapheless valve; (K) raphe valve]; (L) LM valve view, postauxospore valve of *Cocconeis pediculus*; note the typical disorganized striae in cells resulting from the joining of haploid gametes in the diatom sexual cycle; scale bar for LMs = 10 µm.

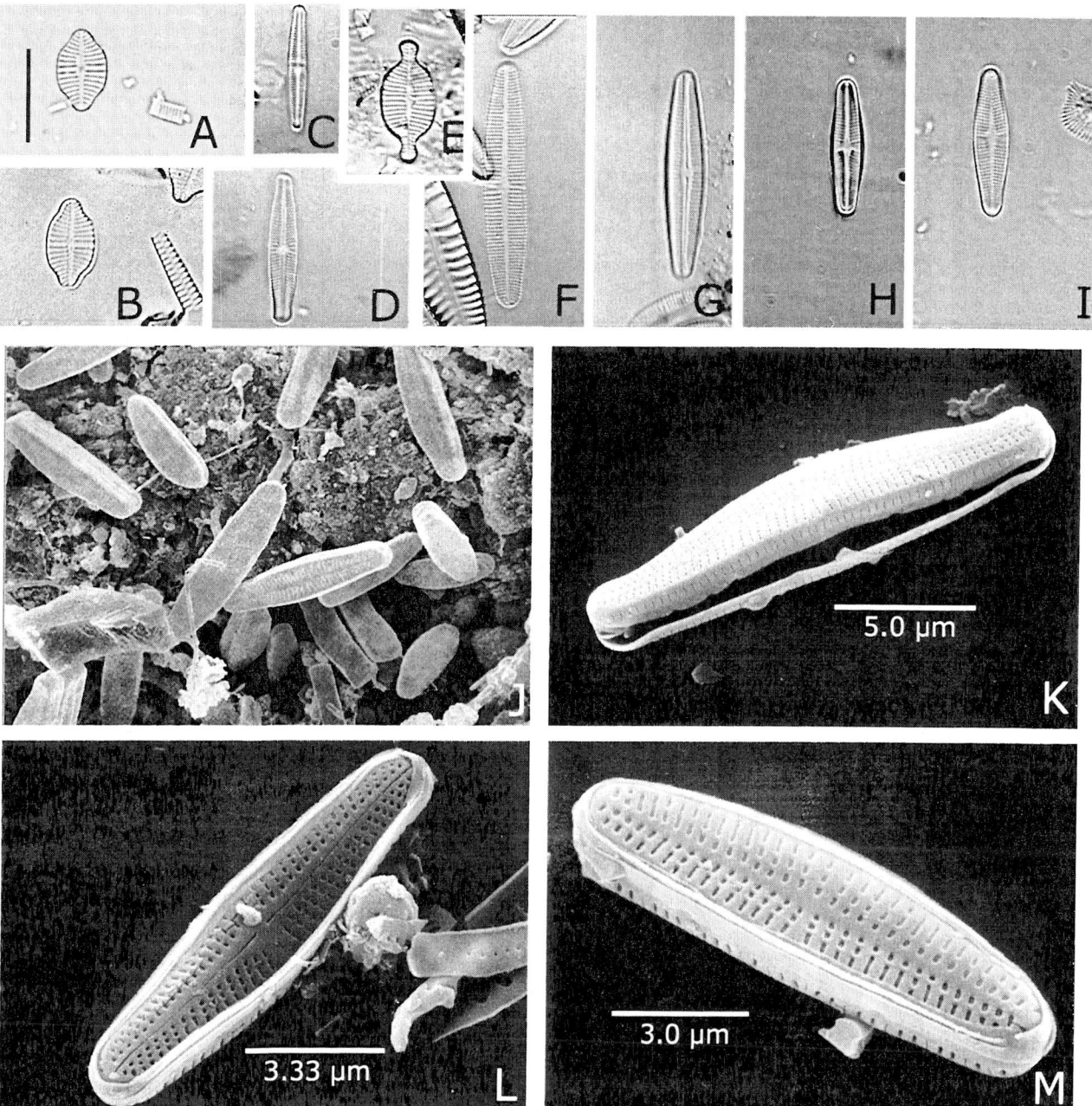

FIGURE 15 Achnanthidium: (A)–(I) LM valve views; (A) and (B) Achnanthidium exiguum [(A) rapheless valve; (B) raphe valve]; (C) and (D) Achnanthidium minutissimum raphe valves; (E) Achnanthidium exiguum var. heterovalvum rapheless valve; (F) and (G) Achnanthidium deflexum, from the holotype slide [(F) rapheless valve; (G) raphe valve]; (H) and (I) Achnanthidium affine [(H) raphe valve; (I) rapheless valve]; (J)–(L) SEM of Achnanthidium minutissimum [(J) habit image of many cells attached to a pebble; (K) oblique external view of a cell showing the rapheless valve; (L) external view of a raphe valve]; (M) SEM oblique view showing the rapheless valve, girdle bands, and the mantle of the raphe valve; a specimen demonstrating the difficulty in distinguishing two genera, Achnanthidium or Rossithidium?; scale bar for LMs = 10 μm.

FIGURE 16 *Eucocconeis*, *Karayevia*, and *Kolbesia*: (A) LM valve view of the raphe valve of *Eucocconeis flexella*; (B) and (C) LM valve view of a single cell of *Eucocconeis flexella* var. *alpestris* (Brun) Hust. [(B) raphe valve; (C) rapheless valve]; (D) LM valve view of the raphe valve of *Karayevia clevei* var. *rostrata*; (E) and (F) LM valve view of a single cell of *Karayevia laterostrata* [(E) raphe valve; (F) rapheless valve]; (G) SEM habit image of two preserved specimens of *Karayevia clevei* var. *rostrata* growing on a sand grain; (H) and (I) LM valve views of *Kolbesia ploenensis* [(H) raphe valve; (I) rapheless valve]; (J) and (K) SEM internal valve views of *Kolbesia kolbei* [(J) raphe valve (arrow); (K) rapheless valve (arrow)]; (J) and (K) are reproduced with permission from Round and Bukhtiyarova [(1996a); Diatom Research 11:345–361, BioPress Ltd.]; scale bar for LMs = 10 μm.

of oligotrophic lakes (especially *E. flexella*), living in the epipelon, epipsammon, or epilithon. Excellent scanning electron micrographs of *Eucocconeis flexella* have been published from arctic material (Lichti-Federovich, 1979).

Karayevia F. E. Round *et* L. Bukhtiyarova *ex* F. E. Round (Figs. 2C, D, 16D–G) Elliptic to lanceolate valves have rounded rostrate to capitate ends. The genus is heterovalvar. Radial striae have variably elongate areolae on the raphid valve, longer areolae toward raphe and the margin. Nearly parallel striae have circular areolae on the rapheless valve. Confusion was introduced in the original description of the genus because the type was not designated, but later corrected (Round, 1998; Fourtanier and Kociolek, 1999).

This genus contains taxa related to *Achnanthes clevei*. Growth habit is adnate on sand grains. Known taxa prefer alkaline water.

Kolbesia F. E. Round *et* L. Bukhtiyarova *ex* F. E. Round (Fig. 16H–K) Valves are elliptic to elliptic-lanceolate with rounded to rostrate apices. Valve ornamentation is dissimilar on raphe and rapheless valves. Striae consist of single extended areole on the raphid valve, and two or three elongate areolae on the rapheless valve face. Typification of the genus occurred after the original publication (Round, 1998; Fourtanier and Kociolek, 1999).

This genus was erected to contain two species formerly treated as *Achnanthes*, *Achnanthes kolbei* and *Achnanthes ploenensis*. The genus is common on sand. *Kolbesia ploenensis* has been recorded in Lake Michigan (Kreis and Stoermer, 1979; Stoermer, 1980) and in the St. Louis River estuary of Lake Superior (J. C. Kingston, pers. obs.). Other taxa from North America include *Kolbesia amoena* and *Kolbesia suchlandtii* (Kreis and Stoermer, 1979; Kingston, 2000).

Lemnicola F. E. Round *et* P. W. Basson (Fig. 17A–D) Valves are linear to linear-elliptic, narrowed to a rounded, subrostrate apex. Cells are solitary. Slightly radial striae are composed of biseriate rows of circular areolae, separated by transverse ridges internally. Sternum is prominent, expanded into an asymmetric stauros on the raphe valve.

This genus was erected for a single species (*Lemnicola hungarica*) previously treated under *Achnanthidium* or *Achnanthes*, and the authors compare and contrast it to *Planothidium* (Round and Basson, 1997). Geitler (1980) pointed out the general similarity of life history for eight members of the Achnanthidiaceae including *Lemnicola*.

Lemnicola is primarily found as an epiphyte on floating angiosperms *Lemna* and *Wolfia*.

Planothidium F. E. Round *et* L. Bukhtiyarova (Fig. 18A–L) (*Achnantheiopsis* Lange-Bertalot) Nearly flat valve is slightly curved about the transapical axis. Shape is elliptic to lanceolate, with rounded or rostrate or capitate apices. Multiseriate striae have multiple areolae rows externally and ribs separating the striae internally. Striae radiate on the raphe valve and less so on the rapheless valve, often 10–15 per 10 μm. There is a horse-hoof-like structure near the center in many taxa, usually on the rapheless valve; a cavum has an internal silica extension over this area, a sinus does not, and this distinction is useful for discrimination of certain species (Moss and Carter, 1982). The raphe is prominent and turned to the same side at the distal ends.

Geitler (1977) described details of sexual reproduction in *Planothidium lanceolatum* (=*Achnanthes lanceolata*). Growth habit is adnate and attached by the raphe valve face. Two prominent research groups independently split this group from *Achnanthidium* almost simultaneously (Round and Bukhtiyarova, 1996a; Lange-Bertalot, 1997b). Other papers have used frustular measurements, including cavum size (structure of the horseshoe shape), to describe population variability in *Planothidium* species (Straub, 1985, 1990).

Psammothidium L. Bukhtiyarova *et* F. E. Round (Fig. 19A–K) Frustule is curved to the apical and transapical axes; raphe valve is convex; rapheless valve, concave (opposite to most monoraphid genera). Valves are small, elliptical, lanceolate-elliptical, or linear-elliptical. Usually twice as long as wide, up to four times longer than wide. Apices are always rounded. Areolae arrangement is similar on both valves. Striae radiate, near 30 in 10 μm. Central area is present on the raphe valve, more rectangular than the expanded axial area on the rapheless valve of some species.

This is a large genus (over 20 taxa) containing species formerly treated as *Achnanthes marginulata* and *Achnanthes levanderi*, with most taxa restricted to acidic waters (Flower and Jones, 1989; Bukhtiyarova and Round, 1996).

Rossithidium F. E. Round *et* L. Bukhtiyarova (Figs. 2E, 15M, 19L–Q) Valves are linear to linear-lanceolate with rounded ends. Striae are parallel often 20–30 in 10 μm. Sternum is narrow and linear on both valves. There may be a central area on the raphe valve. Straight raphe has swollen proximal raphe ends and distal raphe ends turned to one side. Mantle areolae are present. During the sexual cycle, *Rossithidium linearis* produces two auxospores from one cell pair by allogamic fusion of migratory and stationary gametes (Geitler, 1979).

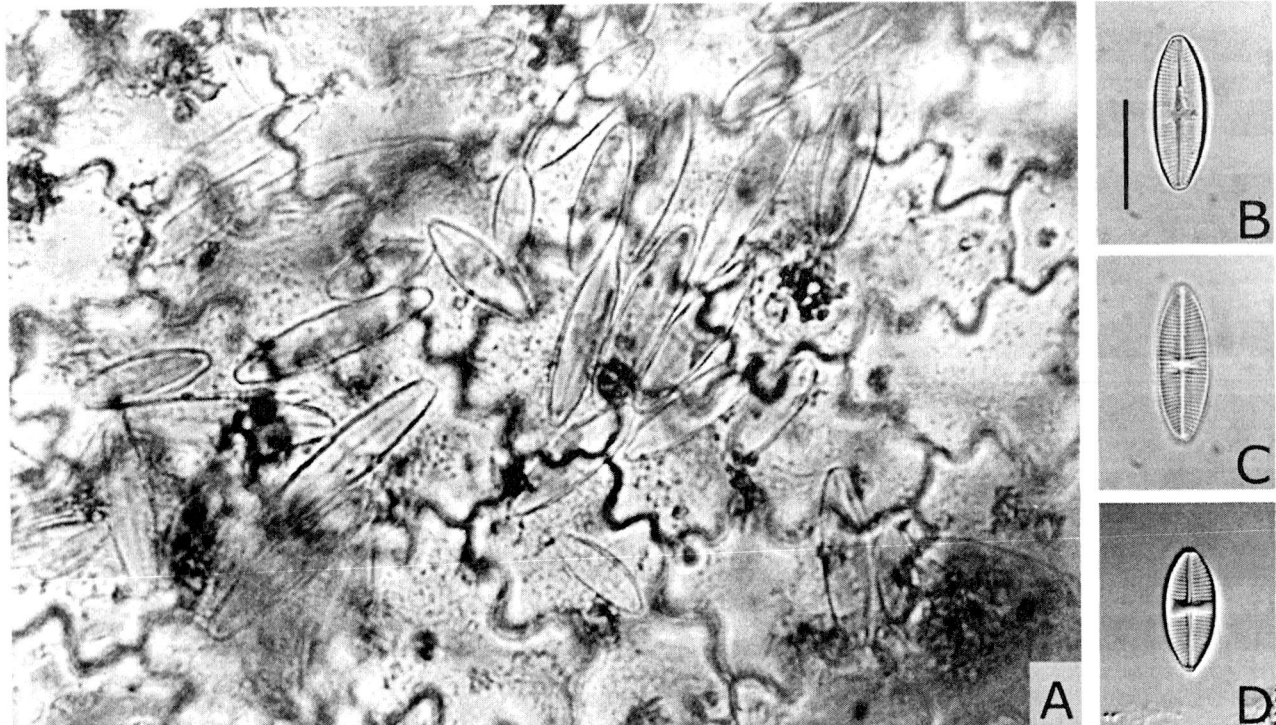

FIGURE 17 *Lemnicola*: (A) habit image of many cells of *Lemnicola* growing on the angiosperm *Lemna minor* L; (B)–(D) LM valve views of *Lemnicola hungarica*; (B) and (C) one cell [(B) raphe valve; (C) rapheless valve]; (D) separate raphe valve; scale bar for LMs = 10 μm.

Rossithidium was erected to include three linear species of *Achnanthidium* (Round and Bukhtiyarova, 1996a) including *Achnanthes linearis* W. Smith. Many other species may be transferred to it. Members of this genus are frequently reported occurring in environments with *Achnanthidium minutissimum*, although polysaccharide stalks are less evident or absent in *Rossithidium*. Many ecological studies do not discriminate small species in the two genera, and there are relatively few published electron micrographs of *Rossithidium* species (Gaul et al., 1993). Considerable work remains to convincingly justify the separation of *Rossithidium* from *Achnanthidium*, and the images presented here are somewhat speculative.

VI. GUIDE TO LITERATURE FOR SPECIES IDENTIFICATION

Most of the same publications useful for identification of diatom genera contain the most up-to-date information about identification of species, varieties, and forms (Tables III and IV). The major European references (Hustedt, 1930, 1959; Cleve-Euler, 1953; Simonsen, 1979, 1987; Germain, 1981; Williams and Round, 1987; Krammer and Lange-Bertalot, 1991a, b) and Asian references (Hustedt, 1937–1939; Skvortzow, 1937; Kobayasi, 1997) may also be consulted. The European references used most in North American ecological studies contain perhaps 40% of the taxa observed in North America. The other 60% must be found in the fragmented literature for North America, South America, and Asia, or are new.

For North American freshwater diatoms, particularly useful publications include Patrick and Reimer (1966) based largely on rivers in the eastern United States, Camburn et al. and coworkers for dilute lakes (Camburn et al., 1984–1986), Rushforth and coworkers in Utah (Johansen and Rushforth, 1981; Kaczmarska and Rushforth, 1983), Czarnecki's flora from Arizona (Czarnecki and Blinn, 1978), Sovereign's floras (Sovereign, 1958, 1963) for western habitats, Stoermer and coworkers for the Laurentian Great Lakes (Stoermer and Yang, 1969; Kreis and Stoermer, 1979; Stoermer, 1980), Duthie and coworkers for Ontario (Sreenivasa, 1971; Duthie and Sreenivasa, 1972; Sreenivasa and Duthie, 1973), and a few floras from Alaska (Patrick and Freese, 1961; Foged, 1971). Some

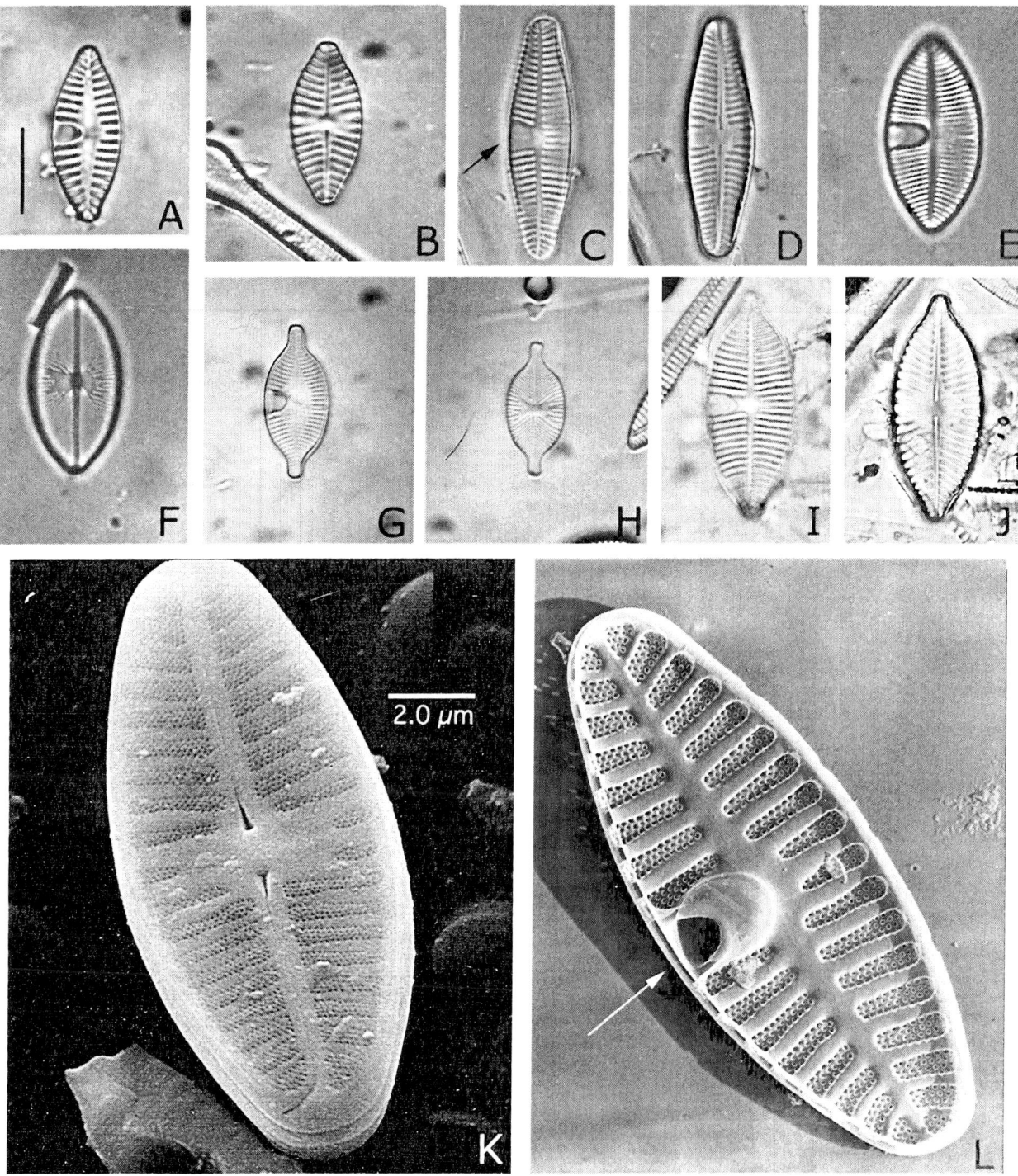

FIGURE 18 *Planothidium*: (A)–(J) LM valve views; (A) and (B) *Planothidium pseudotanense*, holotype specimen of the synonymous *Achnanthes lanceolata* var. *abbreviata* [(A) rapheless valve, with cavum; (B) raphe valve]; (C) and (D) *Planothidium lanceolatum* [(C) rapheless valve, with sinus (arrow); (D) raphe valve]; (E) and (F) *Planothidium oestrupii* [(E) rapheless valve; (F) raphe valve]; (G) and (H) *Planothidium peragalli* [(G) rapheless valve; (H) raphe valve]; (I) and (J) single cell, *Planothidium apiculatum*, holotype specimen [(I) rapheless valve; (J) raphe valve]; (K) SEM external view of raphe valve of *Planothidium frequentissimum*; (L) TEM, metal-shadowed carbon replica, internal view of *Planothidium frequentissimum*, with cavum (arrow); scale bar for LMs = 10 µm.

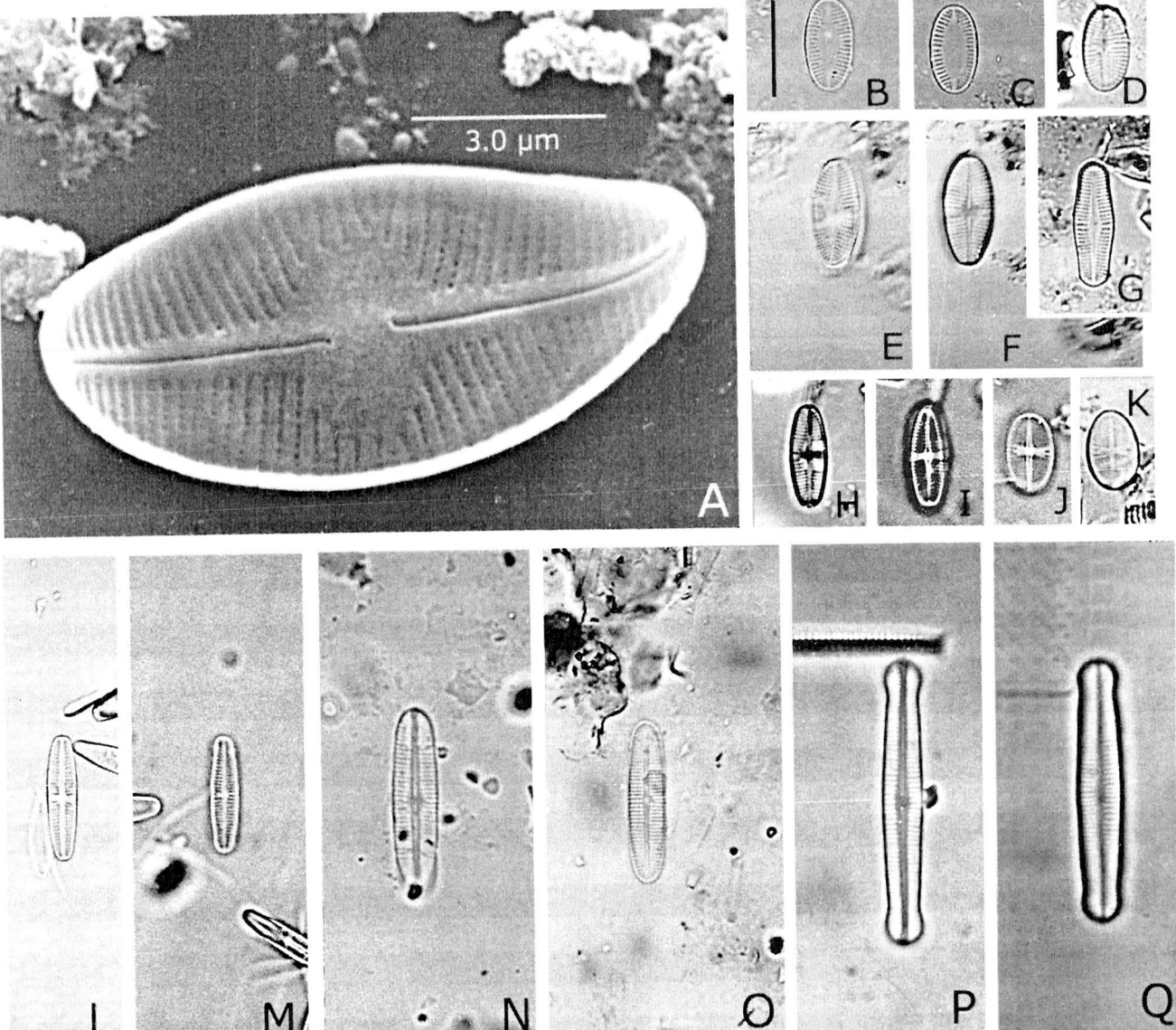

FIGURE 19 *Psammothidium* and *Rossithidium*: (A) SEM internal view, raphe valve of *Psammothidium altaicum*; (B)–(J) *Psammothidium* LM valve views; (B)–(D) *Psammothidium levanderi* [(B) and (C) rapheless valves; (D) raphe valve]; (E)–(F), (H) and (I) *Psammothidium lauenburgianum*, two intact frustules [(E) and (H) rapheless valves; (F) and (I) raphe valves]; (G) *Psammothidium rosenstockii*, rapheless valve; (J) and (K) *Psammothidium subatomoides* [(J) raphe valve; (K) rapheless valve]; (L)–(Q) *Rossithidium* LM valve views; (L) and (M) *Rossithidium linearis* [(L) raphe valve; (M) rapheless valve]; (N) and (O) *Rossithidium pusillum* [(N) raphe valve; (O) rapheless valve]; (P) and (Q) *Rossithidium duthiei* [(P) raphe valve; (Q) rapheless valve]; scale bar for LMs = 10 μm.

ecological works document the diatoms observed, though they are not primarily systematics studies (Cumming *et al.*, 1995).

Specialized information on diatom taxonomy can be obtained at several North American herbaria (California Academy of Sciences; Academy of Natural Sciences of Philadelphia; Farlow Herbarium, Harvard University; Center for Great Lakes and Aquatic Sciences, University of Michigan; Canadian Museum of Nature, Ottawa) and at the biennial North American Diatom Symposium and other workshops emphasizing the North American flora.

ACKNOWLEDGMENTS

I have been fortunate to communicate about diatom systematics with many of the authors cited here, including J. P. Kociolek, E. F. Stoermer, C. W. Reimer, D. M. Williams, F. E. Round, D. G. Mann, H. Lange-Bertalot, K. Krammer, L. Bukhtiyarova, E. Fourtanier, J. R. Carter, and many, many others. I hope that they will forgive my shortcomings in this chapter, and that they will know that I have appreciated the communication. I thank the authors of the other diatom chapters—J. P. Kociolek, S. A. Spaulding, E. F. Stoermer, M. L. Julius, and R. L. Lowe—for their significant help. I could not have completed this chapter without the help and support of past and present colleagues in the U.S. Geological Survey, especially A. R. Brigham, J. P. Lusby, V. A. S. Andrle, L. Marr, S. Turner, and J. Raese. Special thanks to V. A. S. Andrle and M. B. Edlund for editorial comments and to J. P. Lusby for creating the digital figures and plates. P. R. Sweets provided information about taxa found in lake acidification studies. R. G. Kreis, Jr. provided information on Great Lakes diatom taxa and photographs. R. L. Lowe, B. H. Rosen, K. M. Rhode, M. L. Julius, and J. Koppen (nice carbon replica!) provided figures. P. A. Siver provided advice on electronic plates. P. C. Silva helped with botanical nomenclature. Constructive review comments were received from two reviewers and both editors. This chapter is dedicated to the Iowa State University professor, John D. Dodd, who first communicated an excitement for the diatoms to me. This is contribution number 284 of the Center for Water and the Environment.

Since this chapter was written, several new papers have expanded our knowledge of the genera discussed here, but these results could not be fully integrated into the chapter. *Asterionella* (Pappas and Stoermer, 2001a, b), *Fragilariforma* (Kingston et al., 2001; Williams, 2001), and *Hannaea* (Bixby, 2001) have been studied further. Three new genera (*Ulnaria* Compère, *Belonastrum* Round and Maidana, and *Synedrella* Round and Maidana) have been separated from *Synedra* (Lange-Bertalot and Compère, 2001; Compère, 2001; Round and Maidana, 2001). A promising new light microscopy technique (interference reflection contrast) has been developed that allows better differentiation of striae structure in small Fragilariophyceae and other diatoms (Siver and Hinsch, 2000).

LITERATURE CITED

Agbeti, M. D., Smol, J. P. 1995. Winter limnology: A comparison of physical, chemical and biological characteristics in two temperate lakes during ice cover. Hydrobiologia 304:221–234.

Agbeti, M. D., Kingston, J. C., Smol, J. P., Watters, C. 1997. Comparison of phytoplankton succession in two lakes of different mixing regimes. Archiv für Hydrobiologie 140:37–69.

Allen, T. F. H. 1977. Scale in microscopic algal ecology: A neglected dimension. Phycologia 16:253–257.

Ashley, J., Rushforth, S. R., Johansen, J. R. 1985. Soil algae of cryptogramic crusts from the Uintah Basin, Utah, U.S.A. Great Basin Naturalist 45:432–442.

Bahls, L. L., Weber, E. E., Jarvie, J. O. 1984. Ecology and distribution of major diatom ecotypes in the southern Fort Union coal region of Montana. U.S. Geological Survey, Washington, DC.

Barber, H. G., Haworth, E. Y. 1981. A guide to the morphology of the diatom frustule with a key to the British freshwater genera. Freshwater Biological Association, Amblesdie, Cumbria, Scientific Publications No. 44, UK, pp. 1–112.

Bateman, L., Rushforth, S. R. 1984. Diatom floras of selected Uinta Mountain lakes, Utah, U.S.A. Bibliotheca Diatomologica 4:1–99.

Battarbee, R. W. 1986. Diatom analysis, in: Berglund, B. E., Ed., Handbook of Holocene palaeoecology and palaeohydrology. Wiley, New York, Chap. 26 pp. 527–570.

Birks, H. J. B., Frey, D. G., Deevey, E. S. 1998. Review #1. Numerical tools in paleolimnology—progress, potentialities, and problems. Journal of Paleolimnology 20:307–332.

Bixby, R. J. 2001. Morphology, phytogeography, and systematics of the diatom genus *Hannaea* (Bacillariophyceae). Doctoral Dissertation, University of Michigan, Ann Arbor, 150 p.

Bradbury, J. P. 1975. Diatom stratigraphy and human settlement in Minnesota. Special Paper 171, Geological Society of America, Boulder, CO, pp. 1–74.

Bradbury, J. P. 1987. Late holocene diatom paleolimnology of Walker Lake, Nevada. Archiv für Hydrobiologie (Supplement) 79:1–27.

Bradbury, J. P., Waddington, J. C. B. 1978. A paleolimnological comparison of Burntside and Shagawa lakes, northeastern Minnesota. Env. Ecol. Res. Ser. EPA-600/3-78-004, U.S. Environmental Protection Agency, Corvallis, OR, pp. i–51.

Bradbury, J. P., Winter, T. C. 1976. Areal distribution and stratigraphy of diatoms in the sediments of Lake Sallie, Minnesota. Ecology 57:1005–1014.

Bright, R. C. 1968. Surface-water chemistry of some Minnesota Lakes, with preliminary notes on diatoms. Limnological Research Center, University of Minnesota, Minneapolis.

Brown, S.-D. 1973. Species diversity of periphyton communities in the littoral of a temperate lake. Internationale Revue der Gesamten Hydrobiologie 58:787–800.

Brugam, R. B. 1980. Postglacial diatom stratigraphy of Kirchner Marsh, Minnesota. Quaternary Research 13:133–146.

Brugam, R. B., McKeever, K., Kolesa, L. 1998. A diatom-inferred water depth reconstruction for an Upper Peninsula, Michigan, lake. Journal of Paleolimnology 20:267–276.

Bukhtiyarova, L. N. 1995. New taxonomic combinations of diatoms. Algologia (Kiev) 5:417–424.

Bukhtiyarova, L., Round, F. E. 1996. Revision of the genus *Achnanthes sensu lato*. *Psammothidium*, a new genus based on *A. marginulatum*. Diatom Research 11:1–30.

Camburn, K. E., Lowe, R. L., Stoneburner, D. L. 1978. The haptobenthic diatom flora of Long Branch Creek, South Carolina. Nova Hedwigia 30:149–279.

Camburn, K. E., Kingston, J. C., Charles, D. F. 1984–1986. Paleoecological investigation of Recent lake acidification (PIRLA) diatom iconograph. Indiana University, Bloomington.

Canter-Lund, H., Lund, J. W. G. 1995. Freshwater algae. Their microscopic world explored. Biopress, Bristol, 360 p.

Carrick, H. J., Lowe, R. L., Rotenberry, J. T. 1988. Guilds of benthic algae along nutrient gradients: Relationships to algal community diversity. Journal of the North American Benthological Society 7:117–128.

Charles, D. F. 1985. Relationships between surface sediment diatom assemblages and lakewater characteristics in Adirondack lakes. Ecology 66:994–1011.

Cholnoky, B. J. 1968. Die Ökologie der Diatomeen in Binnengewässern. Cramer, Lehre, Stuttgart, 699 p.

Christie, C. E., Smol, J. P. 1993. Diatom assemblages as indicators of lake trophic status in southeastern Ontario lakes. Journal of Phycology 29:575–586.

Clark, R. L., Rushforth, S. L. 1977. Diatom studies of the headwaters of Henrys Fork on the Snake River, Island Park, Idaho, U.S.A. Bibliotheca Phycologica 33:1–204.

Cleve-Euler, A. 1953. Die Diatomeen von Schweden und Finnland II. Araphideae, Brachyraphideae. Kungliga Svenska Vetenskapsakademiens Handlingar 4:1–158 + 483 figs.

Cleve, P. T., Grunow, A. 1880. Beiträge zur Kenntniss der Arctischen Diatomeen. Kongelige Svenska Vetenskaps-Akademiens Handlingar 17 (2):21.

Collins, G. B., Kalinsky, R. G. 1977. Studies on Ohio diatoms I. Diatoms of the Scioto River Basin II. Referenced checklist of the diatoms of Ohio exclusive of Lake Erie and the Ohio River. Ohio Biological Survey Bulletin 5:1–76.

Compère, P. 2001. *Ulnaria* (Kützing) Compère, a new genus name for *Fragilaria* subgen. *Alterasynedra* Lange-Bertalot with comments on the typification of *Synedra* Ehrenberg, in: Jahn, R., Kociolek, J. P., Witkowski, A., Compère, P., Eds. Lange-Bertalot-Festschrift. A. R. G. Gantner Verlag K. G., Koeltz, Koenigstein. pp. 97–101.

Conley, D. J., Kilham, S. S., Theriot, E. 1989. Differences in silica content between marine and freshwater diatoms. Limnology and Oceanography 34:205–213.

Cook, R. B., Kreis, R. G., Jr., Kingston, J. C., Camburn, K. E., Norton, S. A., Mitchell, M. J., Fry, B., Shane, L. C. K. 1990. Paleolimnology of McNearney Lake: An acidic lake in northern Michigan. Journal of Paleolimnology 3:13–34.

Crawford, R.M., Gardner, C. 1997. The transfer of *Asterionellopsis kariana* to the new genus *Asteroplanus* (Bacillariophyceae), with reference to the fine structure. Nova Hedwigia 65:47–57.

Cumming, B. F., Smol, J. P., Kingston, J. C., Charles, D. F., Birks, H. J. B., Camburn, K. E., Dixit, S. S., Uutala, A. J., Selle, A. R. 1992. How much acidification has occurred in Adirondack region lakes (New York, USA) since preindustrial times? Canadian Journal of Fisheries and Aquatic Sciences 49:128–141.

Cumming, B. F., Wilson, S. E., Hall, R. I., Smol, J. P. 1995. Diatoms from British Columbia (Canada) lakes and their relationship to salinity, nutrients and other limnological variables. Cramer, Stuttgart, 207 p.

Czarnecki, D. B., Blinn, D. W. 1978. Diatoms of the Colorado River in Grand Canyon National Park and vicinity (Diatoms of southwestern USA II). Bibliotheca Phycologica 38:1–181.

Danforth, W. F., Ginsburg, W. 1980. Recent changes in the phytoplankton of Lake Michigan near Chicago. Journal of Great Lakes Research 6:307–314.

Davis, J. S., Rands, D. G., Hein, M. K. 1989. Biota of the tufa deposit of Falling Springs, Illinois, U.S.A. Transactions of the American Microscopical Society 108:403–409.

Davis, R. B., Anderson, D. S., Norton, S. A., Ford, J., Sweets, P. R., Kahl, J. S. 1994. Sedimented diatoms in northern New England lakes and their use as pH and alkalinity indicators. Canadian Journal of Fisheries and Aquatic Sciences 51:1855–1876.

Dixit, S. S., Smol, J. P., Kingston, J. C., Charles, D. F. 1992. Diatoms: Powerful indicators of environmental change. Environmental Science and Technology 26:22–33.

Dixit, S. S., Smol, J. P., Charles, D. F., Hughes, R. M., Paulsen, S. G., Collins, G. B. 1999. Assessing water quality changes in the lakes of the northeastern United States using sediment diatoms. Canadian Journal of Fisheries and Aquatic Sciences 56:131–152.

Dodd, J. D., Stoermer, E. F. 1962. Notes on Iowa Diatoms I. An interesting collection from a moss-lichen habitat. Proceedings of the Iowa Academy of Sciences 69:83–87.

Dodd, J. J. 1987. Diatoms, in: Mohlenbrach, R. H., Ed., The illustrated flora of Illinois. Southern Illinois Univ. Press, Carbondale, 478 p.

Douglas, M. S. V., Smol, J. P. 1993. Freshwater diatoms from high arctic ponds (Cape Herschel, Ellesmere Island, N.W.T.). Nova Hedwigia 57:511–552.

Duthie, H. C. 1989. Diatom-inferred pH history of Kejimkujik Lake, Nova Scotia: A reinterpretation. Water, Air, and Soil Pollution 46:317–322.

Duthie, H. C., Sreenivasa, M. R. 1972. The distribution of diatoms on the superficial sediments of Lake Ontario, in: Proceedings of the 15th conference on Great Lakes research, International Association for Great Lakes Research, pp. 45–52.

Ehrenberg, C. G. 1841. Über Verbreitung und Einfluss des mikroskopischen Lebens in Süd- und Nordamerika. Akademie der Wissenschaften zu Berlin, Bericht über die zur Bekanntmachung geeigneten Verhandlungen der Königlichen Preuss. pp 139–144.

Engstrom, D. R., Swain, E. B., Kingston, J. C. 1985. A paleolimnological record of human disturbance from Harvey's Lake, Vermont: Geochemistry, pigments and diatoms. Freshwater Biology 15:261–288.

Fairchild, G. W., Lowe, R. L., Richardson, W. B. 1985. Algal periphyton growth on nutrient-diffusing substrates: An in situ bioassay. Ecology 66:465–472.

Florin, M.-B. 1970. Late-glacial diatoms of Kirchner Marsh, southeastern Minnesota. Beihefte zur Nova Hedwigia 31:667–756.

Flower, R. J. 1989. A new variety of *Tabellaria binalis* (Ehrenb.) Grun. from several acid lakes in the U.K. Diatom Research 4:21–23.

Flower, R. J., Battarbee, R. W. 1985a. The morphology and biostratigraphy of *Tabellaria quadriseptata* (Bacillariophyceae) in acid waters and lake sediments in Galloway, southwest Scotland. British Phycological Journal 20:69–79.

Flower, R. J., Battarbee, R. W. 1985b. Surface Water Acidification Programme workshop, notes on *Asterionella ralfsii* W. Smith. University College, London.

Flower, R. J., Jones, V. J. 1989. Taxonomic descriptions and occurrences of new *Achnanthes* taxa in acid lakes in the U.K. Diatom Research 4:227–239.

Flower, R. J., Jones, V. J., Round, F. E. 1996. The distribution and classification of the problematic *Fragilaria* (*virescens* v.) *exigua* Grun./*Fragilaria exiguiformis* (Grun.) Lange-Bertalot: a new species or a new genus? Diatom Research 11:41–57.

Foerster, J. W., Schlichting, H. E. J. 1965. Phyco-periphyton in an oligotrophic lake. Transactions of the American Microscopical Society 84:485–502.

Foged, N. 1971. Diatoms found in a bottom sediment sample from a small deep lake on the northern slope, Alaska. Nova Hedwigia 21:923–1035.

Fourtanier, E., Kociolek, J. P. 1999. Catalogue of the diatom genera. Diatom Research 14:1–190.

Fritz, S. C., Kingston, J. C., Engstrom, D. R. 1993. Quantitative trophic reconstruction from sedimentary diatom assemblages: A cautionary tale. Freshwater Biology 30:1–23.

Gaul, U., Geissler, U., Henderson, M., Mahoney, R., Reimer, C. W. 1993. Bibliography on the fine-structure of diatom frustules (Bacillariophyceae). Proceedings of the Academy of Natural Sciences of Philadelphia 144:69–238.

Geitler, L. 1977. Entwicklungsgeschichtliche Eigentümlichkeiten einiger *Achnanthes*-Arten (Diatomeae). Plant Systematics and Evolution 126:377–392.

Geitler, L. 1979. Zur Lebensgeschichte der Diatomee *Achnanthes linearis* und Bemerkungen über andere *Achnanthes*-Arten. Plant Systematics and Evolution 132:231–238.

Geitler, L. 1980. Beitrage zur Entwicklungsgeschichte und Taxonomie einiger *Achnanthes*-Arten, Subgenus *Microneis* (Bacillariophyceae). Plant Systematics and Evolution 134:1–10.

Gensemer, R. W. 1991a. The effects of aluminum on phosphorus and silica-limited growth in *Asterionella ralfsii* var. *americana*. Internationale Vereinigung für Theoretische und Angewandte Limnologie Verhandlung 24:2635–2639.

Gensemer, R. W. 1991b. The effects of pH and aluminum on the growth of the acidophilic diatom *Asterionella ralfsii* var. *americana*. Limnology and Oceanography 36:123–131.

Germain, H. 1981. Flore des diatomées (Diatomophycées) eaux douces et saumâtres du Massif Armoricain et des contrées voisines d'Europe occidentale. Boubée, Paris, 444 p.

Hamilton, P. B., Poulin, M., Charles, D. F., Angell, M. 1992. Americanarum diatomarum exsiccata: CANA, voucher slides from eight acidic lakes in northeastern North America. Diatom Research 7:25–36.

Haworth, E. Y. 1972. Diatom succession in a core from Pickerel Lake, northeastern South Dakota. Geological Society of America Bulletin 83:157–172.

Hein, M. K. 1981. Variability in the diatom *Fragilaria floridana* Hanna. Proceedings of the Iowa Academy of Sciences 88:79–81.

Héribaud, J. 1903. Les diatomées d'Auvergne. Libraire des Sciences Naturelles, Vol. 2; Paris, pp. 1–55, Pls. 9–12.

Hickman, M., White, J. M. 1989. Late Quaternary palaeoenvironment of Spring Lake, Alberta, Canada. Journal of Paleolimnology 2:305–317.

Hoagland, K. D., Peterson, C. G. 1990. Effects of light and wave disturbance on vertical zonation of attached microalgae in a large reservoir. Journal of Phycology 26:450–457.

Hohn, M. H., Hellerman, J. 1963. The taxonomy and structure of diatom populations from three eastern North American rivers using three sampling methods. Transactions of the American Microscopical Society 82:250–329.

Holm, N. P., Armstrong, D. E. 1981. Effects of Si:P concentration ratios and nutrient limitation of the cellular composition and morphology of *Astrionella formosa* (Bacillariophyceae). Journal of Phycology 17:420–424.

Holmes R. W. 1985. The morphology of diatoms epizoic on cetaceans and their transfer from *Cocconeis* to two new genera, *Bennettella* and *Epipellis*. British Phycological Journal 20:43–57.

Hustedt, F. 1927. Die Kieselalgen Deutschlands, Österreichs und der Schweiz unter Berüchsichtigung der übrigen Länder Europas sowie der angrenzenden Meeresgebiete, in: Rabenhorst, L. Ed. Kryptogamen-Flora von Deutschlands, Österreichs und der Schweiz 7:1–272.

Hustedt, F. 1930. Bacillariophyta (Diatomeae), A. Pascher, Ed., Die Süsswasser-Flora Mitteleuropas, Vol. 10. Fischer, Jena, 466 p.

Hustedt, F. 1937–1939. Systematische und ökologische Untersuchungen über die Diatomeen-Flora von Java, Bali und Sumatra nach dem Material der Deutschen limnologischen Sunda-Expedition. Archiv für Hydrobiologie (Supplement) 15:131–506, 638–790; 16: 1–394.

Hustedt, F. 1952. Neue und wenig bekannte Diatomeen. IV. Botaniska Notiser 4:366–410, 132 figs.

Hustedt, F. 1959. Die Kieselalgen Deutschlands, Österreichs und der Schweiz, Part 2. Akademische Verlagsgesellschaft Geest und Portig K.-G., Leipzig, 845 p.

Johansen, J. R., Rushforth, S. R. 1981. Diatoms of surface waters and soils of selected oil shale lease areas of eastern Utah. Nova Hedwigia 34:333–390.

Kaczmarska, I., Rushforth, S. R. 1983. The diatom flora of Blue Lake Warm Spring, Utah, U.S.A. Bibliotheca Diatomologica 2:1–123.

Keithan, E. D., Lowe, R. L. 1985. Primary productivity and spatial structure of phytolithic growth in streams in the Great Smoky Mountains National Park, Tennessee. Hydrobiologia 123:59–67.

Kilham, S. S., Kilham, P. 1978. Natural community bioassays: Predictions of results based on nutrient physiology and competition. Internationale Vereinigung für Theoretische und Angewandte Limnologie Verhandlung 20:68–74.

Kingston, J. C. 1980. Characterization of benthic diatom communities in Grand Traverse Bay, Lake Michigan. Ph.D. dissertation, Bowling Green State University, 411 pp.

Kingston, J. C. 1982. Association and distribution of common diatoms in surface samples from northern Minnesota peatlands. Beihefte zur Nova Hedwigia 73:333–346.

Kingston, J. C. 1984. Paleolimnology of a lake and adjacent fen in Southeastern Labrador: Evidence from diatom assemblages. Proceedings of the International Diatom Symposium 7:443–453.

Kingston, J. C. 1986. Diatom analysis—basic protocol, in: Charles, D. F., Whitehead, D. R., Eds., Paleoecological reconstruction of Recent lake acidification (PIRLA), methods and project description. Electric Power Research Institute, Palo Alto, Chap. 6, 11 p.

Kingston, J. C. 1997. Taxonomy and distribution of Fragilariaceae (Bacillariophyta) species from the National Water-Quality Assessment Program (Abstract). Bulletin of the North American Benthological Society 14: 64.

Kingston, J. C. 2000. New combinations in the freshwater Fragilariaceae and Achnanthidiaceae. Diatom Research 15:409–411.

Kingston, J. C., Birks, H. J. B. 1990. Dissolved organic carbon reconstructions from diatom assemblages in PIRLA project lakes, North America. Philosophical Transactions of the Royal Society of London Series B Biological Sciences 327:279–288.

Kingston, J. C., Lowe, R. L. 1979. Attached winter floral assemblages on sand from Grand Traverse Bay, Lake Michigan. Micron 10:203–204.

Kingston, J. C., Lowe, R. L., Stoermer, E. F., Ladewski, T. B. 1983. Spatial and temporal distribution of benthic diatoms in northern Lake Michigan. Ecology 64:1566–1580.

Kingston, J. C., Birks, H. J. B., Uutala, A. J., Cumming, B. F., Smol, J. P. 1992. Assessing trends in fishery resources and lake water aluminum from paleolimnological analysis of siliceous microfossils. Canadian Journal of Fisheries and Aquatic Sciences 49:116–127.

Kingston, J. C., Sherwood, A. R., Bengtsson, R. 2001. Morphology and taxonomy of several *Fragilariforma* taxa from Fennoscandia and North America, in: A. Economou-Amilli, Ed. 16th International Diatom Symposium, Biology, University of Athens, Greece. pp. 73–88.

Kobayasi, H. 1997. Comparative studies among four linear-lanceolate *Achnanthidium* species (Bacillariophyceae) with curved terminal raphe endings. Nova Hedwigia 65:147–163.

Kociolek, J. P. 1998. Does each genus of diatoms have at least one unique feature?—a reply to Round. Diatom Research 13:177–179.

Kociolek, J. P., Kingston, J. C. 1999. Taxonomy, ultrastructure, and distribution of some gomphonemoid diatoms (Bacillariophyceae: Gomphonemataceae) from rivers in the United States. Canadian Journal of Botany 77:686–705.

Kociolek, J. P., Rhode, K. 1998. Raphe vestiges in "*Asterionella*" species from Madagascar: Evidence for a polyphyletic origin of the araphid diatoms? Cryptogamie Algologie 19:57-74.

Kociolek, J. P., Stoermer, E. F. 1986. Phylogenetic relationships and classification of monoraphid diatoms based on phenetic and cladistic methodologies. Phycologia 25:297-303.

Kociolek, J. P., Stoermer, E. F. 1989. Chromosome numbers in diatoms: A review. Diatom Research 4:47-54.

Kociolek, J. P., Williams, D. M. 1987. Unicell ontogeny and phylogeny: Examples from the diatoms. Cladistics 3:274-284.

Kociolek, J. P., Theriot, E. C., Williams, D. M. 1989. Inferring diatom phylogeny: A cladistic perspective. Diatom Research 4:289-300.

Koppen, J. D. 1975. A morphological and taxonomic consideration of *Tabellaria* (Bacillariophyceae) from the northcentral United States. Journal of Phycology 11:236-244.

Koppen, J. D. 1978. Distribution and aspects of the ecology of the genus *Tabellaria* Ehr. (Bacillariophyceae) in the northcentral United States. American Midland Naturalist 99:383-397.

Körner, H. 1971. Morphologie und Taxonomie der Diatomeengattung *Asterionella*. Nova Hedwigia 20:557-724.

Korte, V. L., Blinn, D. W. 1983. Diatom colonization on artificial substrata in pool and riffle zones studied by light and electron microscopy. Journal of Phycology 19:332-341.

Krammer, K. 1990. Zur Identität von *Cocconeis diminuta* Pantocsek and *Cocconeis thumensis* A. Mayer, in: Ricard M., Coste M., Eds., Ouvrage dédié à H. Germain. Koeltz, Koenigstein, pp. 145-156.

Krammer, K., Lange-Bertalot, H. 1991a. Bacillariophyceae: Centrales, Fragilariaceae, Eunotiaceae. Fischer, Stuttgart, 576 p.

Krammer, K., Lange-Bertalot, H. 1991b. Bacillariophyceae: Achnanthaceae, Kritische Ergänzungen zu Navicula (Lineolatae) und Gomphonema. Fischer, Stuttgart, 437 pp.

Kreis, R. G., Jr., Stoermer, E. F. 1979. Diatoms of the Laurentian Great Lakes, III. Rare and poorly known species of *Achnanthes* Bory and *Cocconeis* Ehr. (Bacillariophyta). Journal of Great Lakes Research 5:276-291.

Krejci, M. E., Lowe, R. L. 1987a. The seasonal occurrence of macroscopic colonies of *Meridion circulare* (Bacillariophyceae) in a spring-fed brook. Transactions of the American Microscopical Society 106:173-178.

Krejci, M. E., Lowe, R. L. 1987b. Spatial and temporal variation of epipsammic diatoms in a spring-fed brook. Journal of Phycology 23:585-590.

Kuhn, D. L., Plafkin, J. L., Cairns, J. J., Lowe, R. L. 1981. Qualitative characterization of aquatic environments using diatom life-form strategies. Transactions of the American Microscopical Society 100:165-182.

Laird, K., Fritz, S. C., Cumming, B. F. 1998. A diatom-based reconstruction of drought intensity, duration, and frequency from Moon Lake, North Dakota: A sub-decadal record of the last 2300 years. Journal of Paleolimnology 19:161-179.

Lamb, M. A., Lowe, R. L. 1987. Effects of current velocity on the physical structuring of diatom (Bacillariophyceae) communities. Ohio Journal of Science 87:72-78.

Lange-Bertalot, H. 1980. Zur systematischen Bewertung der bandformigen Kolonien bei *Navicula* und *Fragilaria*. Kriterien für die Vereinigung von *Synedra* (subgen. *Synedra*) Ehrenberg mit *Fragilaria* Lyngbye. Nova Hedwigia 33:723-787.

Lange-Bertalot, H. 1989. Können *Staurosirella*, *Punctastriata* und weitere Taxa sensu Williams, Round als Gattungen der Fragilariaceae kritischer Prüfung standhalten? Nova Hedwigia 49:79-106.

Lange-Bertalot, H. 1997a. *Frankophila*, *Mayamaea*, und *Fistulifera*: Drei neue Gattungen der Klasse Bacillariophyceae. Archiv für Protistenkunde 148:65-76.

Lange-Bertalot, H. 1997b. Revision of the genus *Achnanthes* sensu lato (Bacillariophyceae): *Achnantheiopsis*, a new genus with the type species *A. lanceolata*. Archiv für Protistenkunde 148:199-208.

Lange-Bertalot, H. 1999. Neue Kombinationen von Taxa aus *Achnanthes* Bory (sensu lato). Iconographia Diatomologica 6:276-289.

Lange-Bertalot, H., Compère, P. 2001. *Fragilaria* subgen. *Ulnaria* comb. nov., the correct name of the subgenus including *Synedra ulna*, when treated in *Fragilaria*. Diatom Research 16:103-104.

Lange-Bertalot, H., Krammer, K. 1989. *Achnanthes*, eine Monographie der Gattung mit Definition der Gattung *Cocconeis* und Nachträgen zu den Naviculaceae. Cramer, Stuttgart, 393 p.

Lange-Bertalot, V. H., Ruppel, M. 1980. Zur Revision taxonomisch problematischer, ökologisch jedoch wichtiger Sippen der Gattung *Achnanthes* Bory. Archiv für Hydrobiologie (Supplement) 60:1-31.

Lecointe, C., Coste, M., Prygiel, J. 1993. "OMNIDIA" software for taxonomy, calculation of diatom indices and inventories management. Hydrobiologia 269/270:509-513.

Lichti-Federovich, S. 1979. Contributions to the diatom flora of Arctic Canada: Report I. Scanning electron micrographs of some freshwater species from Ellesmere Island. Current Research, Geological Survey of Canada, Paper 79-01B, pp. 71-82.

Lowe, R. L., Collins, G. B. 1973. An aerophilous diatom community from Hocking County, Ohio. Transactions of the American Microsopical Society 92:492-496.

Lowe, R. L., Rosen, B. H., Kingston, J. C. 1982. A comparison of epiphytes on *Bangia atropurpurea* (Rhodophyta) and *Cladophora glomerata* (Chlorophyta) from northern Lake Michigan. Journal of Great Lakes Research 8:164-168.

Lund, J. W. G. 1949. Studies on *Asterionella*. I. The origin and nature of the cells producing seasonal maxima. Journal of Ecology 37:389-419.

Lund, J. W. G. 1950. Studies on *Asterionella formosa* Hass. II. Nutrient depletion and the spring maximum. Journal of Ecology 38:1-35.

Main, S. P. 1988. Seasonal composition of benthic diatom associations in the Cedar River basin (Iowa). Journal of the Iowa Academy of Sciences 95:85-105.

Mann, D. G. 1984. Structure, life history and systematics of *Rhoicosphenia* (Bacillariophyta). V. Initial cell and size reduction in *Rh. curvata* and a description of the Rhoicospheniaceae Fam Nov. Journal of Phycology 20:544-555.

Miller, A. R., Lowe, R. L., Rotenberry, J. T. 1987. Succession of diatom communities on sand grains. Journal of Ecology 75:693-709.

Moore, J. W. 1980. Seasonal distribution of phytoplankton in Yellowknife Bay, Great Slave Lake. Internationale Revue der Gesamten Hydrobiologie 65:283-293.

Moss, M. O., Carter, J. R. 1982. The resurrection of *Achnanthes rostrata* Østrup. Bacillaria 5:157-164.

Munch, C. S. 1980. Fossil diatoms and scales of Chrysophyceae in the recent history of Hall Lake, Washington. Freshwater Biology 10:61-66.

Pappas, J. L., Stoermer, E. F. 2001a. Fourier shape analysis and fuzzy measure shape group differentiation of Great Lakes *Asterionella* Hassall (Heterokontophyta, Bacillariophyceae), in: A. Economou-Amilli, Ed. 16th International Diatom Symposium, Biology, University of Athens, Greece. pp. 485-501.

Pappas, J. L., Stoermer, E. F. 2001b. *Asterionella* Hassall (Heterokontophyta, Bacillariophyceae), taxonomic history and quantitative methods as an aid to valve shape differentiation. Diatom 17:47-58.

Patrick, R. 1945. A taxonomic and ecological study of some diatoms from the Pocono Plateau and adjacent regions. Farlowia 2:143–221.

Patrick, R. 1976. The formation and maintenance of benthic diatom communities. Proceedings of the American Philosophical Society 120:475–484.

Patrick, R., Freese, L. R. 1961. Diatoms (Bacillariophyceae) from northern Alaska. Proceedings of the Academy of Natural Sciences of Philadelphia 112:129–293.

Patrick, R., Reimer, C. W. 1966. The diatoms of the United States. Academy of Natural Sciences of Philadelphia, Philadelphia, 688 p.

Perrin, C. J., Bothwell, M. L., Slaney, P. A. 1987. Experimental enrichment of a coastal stream in British Columbia: Effects of organic and inorganic additions on autotrophic periphyton production. Canadian Journal of Fisheries and Aquatic Sciences 44:1247–1256.

Peterson, C. G. 1996. Response of benthic algal communities to natural physical disturbance, in: Stevenson, R. J., Bothwell, M. L., Lowe, R. L., Eds., Algal ecology, freshwater benthic ecosystems. Academic Press, San Diego, pp. 375–402.

Peterson, C. G., Hoagland, K. D. 1990. Effects of wind-induced turbulence and algal mat development on epilithic diatom succession in a large reservoir. Archiv für Hydrobiologie 118:47–68.

Peterson, C. G., Stevenson, R. J. 1992. Resistance and resilience of lotic algal communities: importance of disturbance timing and current. Ecology 73:1445–1461.

Pienitz, R., Smol, J. P. 1992. Diatom assemblages and their relationship to environmental variables in lakes from the boreal forest-tundra ecotone near Yellowknife, Northwest Territories, Canada. Proceedings of the International Diatom Symposium 12:391–404.

Pienitz, R., Lortie, G., Allard, M. 1991. Isolation of lacustrine basins and marine regression in the Kuujjuaq area, northern Québec, as inferred from diatom analysis. Géographie Physique et Quaternaire 45:155–174.

Pillsbury, R. W., Kingston, J. C. 1990. The pH-independent effect of aluminum on cultures of phytoplankton from an acidic Wisconsin lake. Hydrobiologia 194:225–233.

Poulin, M. E., Ed. 1990. Proceedings of the third polar diatom colloquium, Ottawa, Ontario, Canada, August 5–10, 1990. Canadian Museum of Nature, Ottawa.

Prescott, G. W., Dillard, G. E. 1979. A checklist of algal species reported from Montana 1891 to 1977. Proceedings of the Montana Academy of Sciences (Supplement) 38:1–102.

Rawlence, D. J. 1988. The post-glacial diatom history of Splan Lake, New Brunswick. Journal of Paleolimnology 1:51–60.

Reavie, E. D., Smol, J. P., Carmichael, N. B. 1995. Postsettlement eutrophication histories of six British Columbia (Canada) lakes. Canadian Journal of Fisheries and Aquatic Sciences 52:2388–2401.

Reimer, C. W. 1966. Consideration of fifteen diatom taxa (Bacillariophyta) from the Savannah River, including seven described as new. Notulae Naturae of the Academy of Natural Sciences of Philadelphia 397:1–15.

Reimer, C. W. 1990. Diatoms (Bacillariophyceae) from the Excelsior Fen-complex, Dickinson Co., Iowa, with the description of two new taxa. Journal of the Iowa Academy of Sciences 97:146–152.

Renberg, I. 1976. Paleolimnological investigations in Lake Prästsjön. Early Norrland 9:113–159.

Renberg, I. 1977. *Fragilaria lata*, a new diatom species. Botaniska Notiser 130:315–318.

Reynolds, C. S. 1984. The ecology of freshwater phytoplankton. Cambridge Univ. Press, Cambridge, UK, 384 pp.

Rosen, B. H., Lowe, R. L. 1981. Valve ultrastructure of some confusing Fragilariaceae. Micron 22:293–294.

Round, F. E. 1997. Does each diatom genus have a unique feature? Diatom Research 12:341–345.

Round, F. E. 1998. Validation of some previously published "achnanthoid" genera. Diatom Research 13:181.

Round, F. E., Basson, P. W. 1997. A new monoraphid diatom genus (*Pogoneis*) from Bahrain and the transfer of previously described species A. *hungarica* and A. *taeniata* to new genera. Diatom Research 12:71–81.

Round, F. E., Bukhtiyarova, L. 1996a. Four new genera based on *Achnanthes* (*Achnanthidium*) together with a redefinition of *Achnanthidium*. Diatom Research 11:345–361.

Round, F. E., Bukhtiyarova, L. 1996b. Epipsammic diatoms—communities of British rivers. Diatom Research 11:363–372.

Round, F. E., Crawford, R. M., Mann, D. G. 1990. The diatoms. Cambridge Univ. Press, Cambridge, UK, 747 p.

Round, F. E., Maidana, N. I. 2001. Two problematic freshwater araphid taxa re-classified in new genera. Diatom 17:21–28.

Schmid, A.-M. M. 1997. Intraclonal variation of the tripolar pennate diatom "*Centronella reicheltii*" in culture: Strategies of reversion to the bipolar *Fragilaria*-form. Nova Hedwigia 65:27–45.

Sherman, B. J., Phinney, H. K. 1971. Benthic algal communities of the Metolius River. Journal of Phycology 7:269–273.

Simonsen, R. 1979. The diatom system: Ideas on phylogeny. Bacillaria 2:9–71.

Simonsen, R. 1987. Atlas and catalogue of the diatom types of Friedrich Hustedt. Cramer, Borntraeger, Berlin, 525, 597, 619 p.

Siver, P. A. 1999. Development of paleolimnological inference models for pH, total nitrogen and specific conductivity based on planktonic diatoms. Journal of Paleolimnology 21:45–59.

Siver, P. A., Hinsch, J. 2000. The use of interference reflection contrast in the examination of diatom valves. Journal of Phycology 36:616–620.

Skvortzow, B. W. 1937. Bottom diatoms from Olhon Gate of Baikal Lake, Siberia. Phillippine Journal of Sciences 62:293–377.

Smilauer, P. 1992. CanoDraw. Environmental Change Research Centre, University College, London.

Smol, J. P. 1988. Paleoclimate proxy data from freshwater Arctic diatoms. Internationale Vereinigung für Theoretische und Angewandte Limnologie Verhandlung 23:837–844.

Sovereign, H. E. 1958. The diatoms of Crater Lake, Oregon. Transactions of the American Microscopical Society 77:96–134.

Sovereign, H. E. 1963. New and rare diatoms from Oregon and Washington. California Academy of Sciences Fourth Series 31:349–368.

Squires, L. E., Rushforth, S. R., Brotherson, J. D. 1979. Algal response to a thermal effluent: Study of a power station on the Provo River, Utah, USA. Hydrobiologia 63:17–32.

Sreenivasa, M. R. 1971. New fossil diatoms from Ontario. II. *Achnanthes duthii* sp. nov., *Achnanthes undulatus* sp. nov., and *Neidium distincte-punctatum* Husted var. *major* var. nov. Phycologia 10:79–82.

Sreenivasa, M. R., Duthie, H. C. 1973. Diatom flora of the Grand River, Ontario, Canada. Hydrobiologia 42:161–224.

St. Clair, L. L., Rushforth, S. R., Allen, J. V. 1981. Diatoms of Oregon Caves National Monument, Oregon. Great Basin Naturalist 41:317–332.

Stevenson, R. J., Stoermer, E. F. 1982. Luxury consumption of phosphorus by five *Cladophora* epiphytes in Lake Huron. Transactions of the American Microscopical Society 101:151–161.

Stevenson, R. J., Bothwell, M. L., Lowe, R. L., Eds. 1996. Algal ecology, freshwater benthic ecosystems. Academic Press, San Diego, 753 p.

Stockner, J. G., Benson, W. W. 1967. The succession of diatom assemblages in the recent sediments of Lake Washington. Limnology and Oceanography 12:513–532.

Stoermer, E. F. 1968. Nearshore phytoplankton populations in the Grand Haven, Michigan vicinity during thermal bar conditions, in: Proceedings of the 11th Conference on Great Lakes Research pp. 137–150.

Stoermer, E. F. 1975. Comparison of benthic diatom communities in Lake Michigan and Lake Superior. Internationale Vereinigung für Theoretische und Angewandte Limnologie Verhandlung 19:932–938.

Stoermer, E. F. 1980. Characteristics of benthic algal communities in the upper Great Lakes. Environmental Research Laboratory, U.S. Environmental Protection Agency, Duluth, MN, pp. 1–73.

Stoermer, E. F., Smol, J. P., Eds. 1999. The diatoms: Applications for the environmental and earth sciences. Cambridge Univ. Press, Cambridge, UK, 469 p.

Stoermer, E. F., Yang, J. J. 1969. Plankton diatom assemblages in Lake Michigan, Special Report 47, Great Lakes Research Division, University of Michigan, Ann Arbor.

Stoermer, E. F., Yang, J. J. 1970. Distribution and relative abundance of dominant plankton diatoms in Lake Michigan. Great Lakes Research Division Publication 16, University of Michigan, Ann Arbor, 64 p.

Stoermer, E. F., Kreis, R. G., Jr., Andresen, N. A. 1978a. Checklist of diatoms from the Laurentian Great Lakes. II. Journal of Great Lakes Research 25:515–566.

Stoermer, E. F., Kreis, R. G., Jr., Andresen, N. A. 1999. Checklist of diatoms from the Laurentian Great Lakes. II. Journal of Great Lakes Research 25:515–566.

Stoermer, E. F., Ladewski, B. G., Schelske, C. L. 1978b. Population responses of Lake Michigan phytoplankton to nitrogen and phosphorus enrichment. Hydrobiologia 57:249–265.

Stoermer, E. F., Kociolek, J. P., Schelske, C. L., Andresen, N. A. 1991. Siliceous microfossil succession in the recent history of Green Bay, Lake Michigan. Journal of Paleolimnology 6:123–140.

Straub, F. 1985. Variabilité comparée d'Achnanthes lanceolata (Breb.) Grun. et d'Achnanthes rostrata Oestrup (Bacillariophyceae) dans huit populations naturelles du Jur suisse. I. Approche morphologique. Bulletin Society Neuchâteloise Sciences Naturalist 108:135–150.

Straub, F. 1990. Compared variability of Achnanthes lanceolata (Bréb) Grunow. 2. Biometrical approach of several races of the sub-species frequentissima Lange-Bertalot, in: Ricard, M., Coste, M., Eds., Ouvrage dédié à la mémoire du Professeur Henry Germain. Koeltz, Koenigstein, pp. 243–250.

Sullivan, M. J. 1979. Taxonomic notes on epiphytic diatoms of Mississippi Sound, USA. Beihefte zur Nova Hedwigia 64:241–249.

ter Braak, C. J. F. 1988. CANOCO a FORTRAN program for canonical community ordination by [partial] [detrended] [canonical] correspondence analysis and redundancy analysis. Agricultural Mathematics Group, Wageningen, Netherlands.

Tilman, D., Kilham, S. S., Kilham, P. 1976. Morphometric changes in Asterionella formosa colonies under phosphate and silicate limitation. Limnology and Oceanography 21:883–886.

Tilman, D., Kilham, S. S., Kilham, P. 1982. Phytoplankton community ecology: the role of limiting nutrients. Annual Review of Ecology and Systematics 13:349–372.

Tuchman, M. L., Stoermer, E. F., Carney, H. J. 1984. Effects of increased salinity on the diatom assemblage in Fonda Lake, Michigan. Hydrobiologia 109:179–188.

Van Heurck, H. 1881. Synopsis des diatomées de Belgique. Atlas. Ducaju, Anvers, Belgium, Pls. 31–77.

Walker, I. R., Paterson, C. G. 1986. Associations of diatoms in the surficial sediments of lakes and peat pools in Atlantic Canada. Hydrobiologia 134:265–272.

Williams, D. M. 1985. Morphology, taxonomy and inter-relationships of the ribbed araphid diatoms from the genera Diatoma and Meridion (Diatomaceae: Bacillariophyta). Cramer, Hirschberg, Germany, 255 pp.

Williams, D. M. 1986. Comparative morphology of some species of Synedra Ehrenb. with a new defintion of the genus. Diatom Research 1:131–152.

Williams, D. M. 1989. Observations on the genus Tetracyclus Ralfs (Bacillariophyta) II. Morphology and taxonomy of some fossil species previously classified in Stylobiblium Ehrenberg. British Phycological Journal 24:317–327.

Williams, D. M. 1990a. Cladistic analysis of some freshwater araphid diatoms (Bacillariophyta) with particular reference to Diatoma and Meridion. Plant Systematics and Evolution 171:89–97.

Williams, D. M. 1990b. Distrionella D. M. Williams, nov. gen., a new araphid diatom (Bacillariophyta) genus closely related to Diatoma Bory. Archiv für Protistenkunde 138:171–177.

Williams, D. M. 1990c. Fragilaria floridana Hanna: Ultrastructure of the valve and girdle and its transference to Fragilariforma Williams, Round, in: Ricard, M., Coste, M., Eds., Ouvrage dédié à la mémoire du Professeur Henry Germain. Koeltz, Koenigstein, pp. 259–265.

Williams, D. M. 1993. Hustedt's study of the diatom species Tetracyclus ellipticus: Why history is not just a chronicle of events. Proceedings of the International Diatom Symposium 12:21–30.

Williams, D. M. 1994. Ontogeny and phylogeny in the genus Tetracyclus. Proceedings of the International Diatom Symposium 11:247–256.

Williams, D. M. 1996. Fossil species of the diatom genus Tetracyclus (Bacillariophyta, 'ellipticus' species group): Morphology, interrelationships and the relevance of ontogeny. Philosophical Transactions of the Royal Society Series B, Biological Sciences 351:1759–1782.

Williams, D. M. 1997. Notes on the diatom species Tetracyclus castellum (Ehrenb.) Grunow with a description of Tetracyclus pseudocastellum nov. sp. Bulletin of the Natural History Museum of London (Botany) 27:1–5.

Williams, D. M. 2001. Comments on the structure of "post-auxospore" valves of Fragilariforma virescens, in: Jahn, R., Kociolek, J. P., Witkowski, A., Compère, P., Eds. Lange-Bertalot-Festschrift. A. R. G. Gantner Verlag K. G., Koeltz, Koenigstein. pp. 103–117.

Williams, D. M., Round, F. E. 1987. Revision of the genus Fragilaria. Diatom Research 2:267–288.

Williams, D. M., Round, F. E. 1988a. Fragilariforma, nom. nov., a new generic name for Neofragilaria Williams, Round. Diatom Research 3:265–267.

Williams, D. M., Round, F. E. 1988b. Phylogenetic systematics of Synedra. Proceedings of the Diatom Symposium 9:303–315.

Wilson, S. E., Cumming, B. F., Smol, J. P. 1994. Diatom–salinity relationships in 111 lakes from the Interior Plateau of British Columbia, Canada: The development of diatom-based models for paleosalinity reconstructions. Journal of Paleolimnology 12:197–221.

Woodhead, N., Tweed, R. D. 1960. Additions to the algal flora of Newfoundland. Hydrobiologia 15:309–362.

Yang, J.-R., Duthie, H. C. 1995. Regression weighted averaging models relating surficial sedimentary diatom assemblages to water depth in Lake Ontario. Journal of Great Lakes Research 21:84–94.

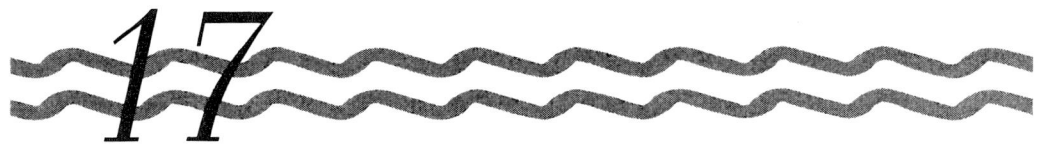

SYMMETRICAL NAVICULOID DIATOMS

J. P. Kociolek, S. A. Spaulding

Diatom Collection
California Academy of Sciences
Golden Gate Park
San Francisco, California 94118

I. Introduction
II. Ecology and Distribution
III. Key and Descriptions of Genera
    A. Key
    B. Descriptions of Genera
IV. Guide to Literature for Species Identification
    Literature Cited

I. INTRODUCTION

The diversity of species, range in cell morphology, number of types of sexual reproduction, and extent of ecological breadth is remarkable within freshwater naviculoid diatoms. Until the 1990s, symmetrical naviculoid diatoms had been considered to be a natural group; they were treated as if they shared a common evolutionary lineage. However, as morphological, cytological, and genetic evidence continues to demonstrate, most diatoms classified in the family Naviculaceae are a variety of forms that are not necessarily closely related to one another. They were included in Cleve's (1894, 1895) monumental work and have received various treatments by Hustedt (1961–1966), Patrick and Reimer (1966), and Krammer and Lange-Bertalot (1986). A more natural classification for the symmetrical naviculoid diatoms has been attempted (Round *et al.*, 1990), especially at higher taxonomic categories, but this approach to classification lacks a formal basis.

The genus *Navicula*, with over 10,000 species, varieties, and forms, is the largest genus governed by the International Code of Botanical Nomenclature. To a dramatic degree, *Navicula* perhaps best exemplifies the current state of taxonomy within the naviculoid diatoms. The group is largely a repository for any symmetrical bi-raphid diatom, which may or may not share common descent. However, many new genera have been dissected out of *Navicula* (Round *et al.*, 1990). Although nontaxonomists may be dismayed at the prospect of learning yet another diatom genus, current differences in diatom genera might be closer to differences at the family level, or higher, of terrestrial plants. Thus, partitioning *Navicula* into separate genera serves to bring the higher classification of diatoms closer in accord to other organisms. Although some of these genera had been recognized previously and then subsumed within *Navicula* (such as *Sellaphora* and *Diadesmis*), others were established as new genera (such as *Luticola* and *Geissleria*). Relationships between these resurrected or newly created genera are still not well understood, and the separation of new genera from *Navicula* at least is an acknowledgment of the problem. Most of the new genera have been diagnosed on the basis of features discernable with the scanning electron microscope (e.g., *Nupela*), or with corresponding cytoplasmic characteristics (such as *Sellaphora*; Mann, 1984a, 1985). Nevertheless, future work is likely to distinguish finer degrees of separation. The Lineolate group (*Navicula sensu stricto*) of *Navicula* species was thought to be a relatively homogeneous group, yet variability in ultrastructure suggests that new genera may still be established from within it (e.g., Witkowski *et al.*, 1998; Kociolek *et al.*, 1998).

Other symmetrical naviculoid genera contain several hundred to thousands of described species, (*Pinnularia, Stauroneis*) whereas others are represented by a few species [*Amphipleura, Muelleria* (Frenguelli) Frenguelli, *Geissleria*], including a few that have but a single species reported in the North American freshwater flora (*Diatomella, Capartogramma*).

Our understanding of the North American flora has benefitted from the work of Patrick and Reimer (1966), which is based in part on their previous taxonomic work (Patrick, 1959; Reimer, 1959, 1961). In addition, the genera *Muelleria* (Spaulding and Stoermer, 1997; Spaulding *et al.*, 1999) *Brachysira* (Lange-Bertalot and Moser, 1994), and *Neidium* (Reimer, 1959; Hamilton *et al.*, 1990, 1994, 1995, 1996) have been revised in part by considering specimens from North America. Stoermer and coworkers have helped document taxa from the Great Lakes (Stoermer *et al.*, 1999). Investigation of diatoms in northern European lakes indicates that we have severely underestimated species-level diversity (Lange-Bertalot and Metzeltin, 1996); thus we have a tremendous amount of work to describe and document the diversity of diatoms in North America. In this chapter, we summarize the large number of symmetrcal naviculoid diatom genera (37).

The features of symmetrical naviculoid diatoms are myriad in form. Morphological features of the valve include a variety of structures and their expression, including pseudosepta, longitudinal canals, stigmata (isolated perforations at the center of the valve that have rounded external openings), conopea (a thick siliceous covering found in the genus *Fallacia*), unique striae and areolae (holes that perforate the valve) construction, distinct raphe structure, as well as variable features of the cingula (also called girdle bands, two to many cingula join the two halves of the diatom together to form the frustule) (Round *et al.*, 1990; Gaul *et al.*, 1993). Number, position, and pattern of division of the plastids are variable in this group and are useful in taxonomy (Mann, 1984a; Cox, 1996). The mode of sexual and asexual reproduction is known for only a small fraction of naviculoid diatoms, yet a great diversity is evident in patterns of reduction division, behavior of gametes, and number and orientation of resulting auxospores (Geitler, 1973; Mann, 1984b, 1989; Edlund and Stoermer, 1997).

II. ECOLOGY AND DISTRIBUTION

Given the tremendous taxonomic breadth represented in the group considered here, it should be no surprise that generalities regarding ecological preferences and geographic distribution at the generic level are difficult at best. Species-level identification is necessary to make relevant correlations to ecology and distribution. Moreover, ecological habitats and geographic regions may or may not overlap with one another. For example, it may be possible to identify similar ecological habitats within eutrophic lakes across North America. However, arid regions with saline lakes are found in midwestern and western arid regions. Therefore, in characterizing ecological preferences, we may easily misinterpret the influence of ecology (Cholnoky, 1965) or region. *Diatomella balfouriana* is a species found in oligotrophic, high-elevation lakes. It is conjecture whether *D. balfouriana* is distributed based on the chemistry of the water, has a better ability than other diatoms to obtain sparse nutrients, or is present as a relict of past distributions. Although *Diatomella* has a narrow environmental niche, more typically the genera show a wide range of ecological and geographical ranges. Yet, even at the species level, it is not clear why some taxa are eurytrophic and other stenotrophic. For North America, summaries of the ecological preferences for more common taxa can be found in Beaver (1981), Lowe (1974), and Cumming *et al.* (1995).

Nevertheless, a few generalities can be made. All representatives of this group possess a raphe system, indicating they are capable of movement, and they may be attached to a variety of substratum types. In contrast to the asymmetric naviculoid diatoms (Chap. 18), only one genus, *Brachysira*, is known to form stalks. Indeed, some stalk-forming *Brachysira* are asymmetric, reflecting that asymmetry may be an adaptation to a stalked habit. Colonies of *Amphipleura* grow within mucilaginous tubes, whereas *Mastogloia* forms mucilagious enclosures around individual cells. Several genera are found in unique aerophilic habitats of soils, growing in association with mosses, and in splash zones of small streams [*Adlafia, Cavinula, Chamaepinnularia, Cosmioneis, Diadesmis, Luticola, Microcostatus,* and *Muelleria* (Frenguelli) Frenguelli]. Many of the remaining genera are considered to be epipelic.

Most of the genera are restricted to fresh waters, but some have both freshwater and marine species (*Caloneis, Diploneis, Fallacia, Gyrosigma* and *Pleurosigma*). The brackish or highly saline waters of inland lakes and estuaries may contain species of *Anomoeoneis, Craticula, Gyrosigma, Plagiotropis, Pleurosigma,* and *Scoliopleura* Patrick and Reimer, 1975; Cumming *et al.*, 1995, (Chapter 2, Section VI). Genera that may be found in oligotrophic waters include *Cavinula, Diadesmis,* and *Oestrupia*. Lakes and bogs that are low in pH characteristically are habitats for *Brachysira*, *Frustulia*, *Kobayasia*, *Neidium*, and *Pinnularia*. Alkaline waters often contain members of *Cosmioneis*, *Mastogloia*, and *Stauroneis*, whereas eutrophic waters may have locally abundant populations of *Fistulifera*.

III. KEY AND DESCRIPTIONS OF GENERA

A. Key

This key represents a traditional approach to keying out freshwater diatom genera in the Naviculaceae *sensu lato*, reflected in the keys of Hustedt (1930), Patrick and Reimer (1966, 1975), Patrick (in Edmonson, 1959), and Krammer and Lange-Bertalot (1986, 1991). This approach uses features discernable with light microscopy to identify taxa. More recently, however, circumscription of some taxa has been proposed using features only detected with SEM. When necessary, taxa are distinguished with SEM features. (See Figs. 1–5.)

1a.	Frustules with septa	2
1b.	Frustules lacking septa	3
2a.	Septa with partecta (chambered structure unique to the genus *Mastogloia*, which functions to secrete mucilage from the cell) (Fig. 4H, I)	*Mastogloia*
2b.	Septa without partecta (Fig. 5H, I)	*Diatomella*
3a.	Valves and/or raphe sigmoid in shape	4
3b.	Valve and raphe not sigmoid in shape	6
4a.	Valve and raphe both sigmoid	5
4b.	Raphe slightly sigmoid, valve torsionally twisted, boat shaped (Fig. 1I)	*Scoliopleura*
5a.	Puncta and striae form lines in apical and transapical (longitudinal) directions (Fig. 1N, O)	*Gyrosigma*
5b.	Puncta and striae form patterns transapically and in two diagonal directions (Fig. 1M)	*Pleurosigma*
6a.	Striae without distinct punctation; valves generally > 15 μm long	7
6b.	Striae composed of distinctly punctate striae	9
7a.	Striae appear broad, riblike	8
7b.	Striae not appearing broad, more like thin lines (Fig. 4A–C)	*Caloneis*
8a.	Striae interrupted, specimens very rare (Fig. 4J)	*Oestrupia*
8b.	Striae continuous, specimens more common (Fig. 4D–F)	*Pinnularia*
9a.	Valves with images of longitudinal lines	10
9b.	Longitudinal lines not present	13
10a.	Longitudinal lines near axial area; raphe in a thickened rib (Fig. 2L–N)	*Diploneis*
10b.	Raphe not in thickened rib	11
11a.	Longitudinal lines near margin; proximal raphe ends deflected in opposite directions or not deflected at all	12
11b.	Longitudinal lines near axial area; proximal raphe ends deflected in same direction (Fig. 1J–K)	*Muelleria*
12a.	Valve outline variable, distal raphe ends divergent (Fig. 2D–F)	*Neidium*
12b.	Valve outline linear and narrow, distal raphe ends unilaterally deflected (Fig. 2G)	*Neidiopsis*
13a.	Raphe elevated above face of valve (Fig. 1L)	*Plagiotropis*
13b.	Raphe not elevated above face of valve	14
14a.	Raphe contained between thick ribs	15
14b.	Raphe not contained between thickened rib	16

15a.	Raphe running full length of valve; most species with porto crayon (a thickened rib at the distal valve unique to the genus *Frustulia*) (Fig. 5A–C)	*Frustulia*
15b.	Raphe very short; no porto crayon striae very fine, easily distinguished; (Fig. 5D)	*Amphipleura*
16a.	Central area expanded into X- or H-shaped stauros	17
16b.	Central area not with stauros or not shaped thusly	20
17a.	Central area X-shaped (Fig. 4G)	*Capartogramma*
17b.	Central area H-shaped or with a staurose	18
18a.	H-shaped central area present	19
18b.	Central area with stauros (Fig. 2A–C)	*Stauroneis*
19a.	H-shaped central area in larger, more coarsely ornamented cells (>25 μm) (Fig. 5E)	*Fallacia*
19b.	H-shaped central area in small, more finely ornamented cells (Fig. 5M)	*Microcostatus*
20a.	Central area with a stigma (Fig. 4L, M)	*Luticola*
20b.	Central area without a stigma	21
21a.	Striae interrupted at ends of valve; isolated punctum may be present in central area (Fig. 4D)	*Geissleria*
21b.	Striae otherwise	22
22a.	Striae composed of more widely spaced, radial and elongated puncta around central area (Fig. 5F)	*Cavinula*
22b.	Striae otherwise	23
23a.	Longitudinal hyaline area bordering axial area, with unornamented T-shape siliceous thickening near terminus (Fig. 3C–E)	*Sellaphora*
23b.	Valves otherwise	24
24a.	Row of puncta bordering axial area; this row of puncta separated from rest of puncta by hyaline area (Fig. 2K)	*Anomoeoneis*
24b.	Valves otherwise	25
25a.	Striae lineate, areolae oriented with long axes parallel to apical axis of valve (Fig. 1A–D) (also *Placoneis*, compare genus descriptions)	*Navicula*
25b.	Striae not lineate	26
26a.	Valve with longitudinal ribs or lines running parallel to the apical axis of the valve	27
26b.	Valves without long apically oriented ribs	28
27a.	Valves large, with puncta arranged in parallel rows, ribs across the valve face, forming internal valves with thick costae (craticulular stage), more common in alkaline environments (Fig. 3A, B)	*Craticula*
27b.	Valve smaller, with elevated ribs around the margin of the valve, valve face dotted with siliceous spinules, more common in soft or acid waters (Fig. 2I, J)	*Brachysira*
28a.	Raphe undulate or sinuous, complex (Fig. 3F)	*Aneumastus*
28b.	Raphe otherwise	29
29a.	Cells small, mostly in chains, maybe without raphe system, external proximal raphe ends simple or T-shaped, external striae appear as open foraminae (an elongate opening) (Fig. 5J–L)	*Diadesmis*
29b.	Cells usually occurring as individuals (not in chains)	30
30a.	Striae not visible, except may be visible around central area	31
30b.	Striae visible along the entire length of valve	32
31a.	Cells small, ovoid, weakly silicified, striae invisible in LM (Fig. 3N)	*Fistulifera*
31b.	Cells elongate, usually capitate, more strongly silicified, with striae visible around the central area in some species (Fig. 5N)	*Kobayasia*
32a.	Striae appearing interrupted near valve terminus	33
32b.	Valves otherwise	34
33a.	Valves heavily silicified, pseudoseptum forming a thickened area at terminus (Fig. 3L, M)	*Hippodonta*

33b.	Terminal nodules strongly angular, and unilaterally bent (Fig. 3K)	Adlafia
34a.	Valves small	35
34b.	Valves larger, striae composed of round-to-oval puncta, striae strongly radiate (Fig. 5G)	Cosmioneis
35a.	Valves small, internal proximal raphe ends recurved in a single direction (Fig. 3G, H)	Chamaepinnularia
35b.	Valves small, internal proximal raphe ends forming a "T" (Fig. 3I, J)	Nupela

B. Descriptions of Genera

Adlafia Moser et al. 1998 (Fig. 3K)

Valve outline is linear to linear-lanceoate with ends abruptly terminated to rostrate or subcapitate. Valves are under 25 µm in length and grow singly. The terminal nodules are unilaterally bent and strongly angular. Areolae are formed in single rows and form radial striae.

Adlafia comprises species formerly classified in *Navicula*, including *A. muscora* and *A. bryophila*. It is distinguished from other genera by small size and the features of the terminal nodules. Members of *Adlafia* are widespread across North America, but often rare within a given assemblage. They are characteristic of aerophilic habitats, especially around mosses. Some species are found in oligotrophic lakes.

Amphipleura Kützing 1844 (Fig. 5D)

Valve outline is linear or spindle-shaped. The striae are composed of punctae that are extremely fine (0.25 µm), so that striae are difficult to resolve except under conditions of optimal resolution in the light microscope. A simple, narrow, median rib (central sternum) is evident on the internal valve face, except near the poles. At the poles, the medium rib is split into two, forming apparent "needle eyes," which are parallel to the raphe. The raphe is short compared to other naviculoid genera. Live cells contain one central H-shaped plastid with a central pyrenoid.

Amphipleura is widely distributed in the epipelon of standing or slowly moving waters. Cells occur individually, or may be enclosed in diffuse, gelatinous tubes. This genus occurs predominantly in alkaline waters.

Aneumastus Mann et Stickle 1990 (Fig. 3F)

Valve outline is lanceolate and valve ends are rostrate or capitate. The valve mantle is narrow, and the areola are complex in structure. Until the 1990s, species, now placed in *Aneumastus* had been included in *Navicula* (*Navicula tuscula* group). *Aneumastus* differs from *Navicula* in possessing plastids composed of two plates, which appear H-shaped in girdle view. This genus is thought to be closely related to *Mastogloia*, but lacks partecta.

This genus contains only a few species, broadly distributed across North America, but locally uncommon in occurrence. It grows in epipelic and epipsammic habitats of alkaline waters.

Anomoeoneis Pfitzer 1871 (Fig. 2J)

Shape is lanceolate to elliptical-lanceolate with broadly rounded to capitate apices. Striae are distinctly punctate, interrupted by hyaline areas on either side of the axial area. The central area may be symmetrical (appearing lyrelike) or asymmetrical, extending unilaterally to the valve margin in some specimens. The distal raphe ends are deflected and usually distinct. The concept of *Anomoeoneis* employed here is restricted, considering that smaller species containing external siliceous ridges and spinules are currently recognized to belong to *Brachysira* (Round and Mann, 1981; Round et al., 1990). An irregular pattern of striae, and hyaline areas on both sides of the axial area separate this genus from *Brachysira* and *Navicula*. A single multilobed plastid with a large spherical pyrenoid is present.

Anomoeoneis is epipelic in habitats of high conductance and brackish waters. This genus is especially common in saline inland waters of the midwestern United States and western Canada. It is also found in the brackish waters of estuaries.

Brachysira Kützing 1836 (Fig. 2H, I)

Valve outline is linear to linear-lanceolate, and the valve apices are rounded to protracted. The valves are symmetrical to the apical axis, and some specimens have a slight to strong asymmetry about the transapical axis. An elevated siliceous ridge separating the valve face and mantle on the external valve surface is evident in the light microscope for most species in this genus. The striae are finely punctate, and longitudinal undulations of the striae are apparent. The raphe is straight, and the axial area is narrow. Cells possess a single plastid. *Brachysira* may grow singly and unattached, or grow at the ends of narrow mucilaginous stalks. A monograph of the genus has been published by Lange-Bertalot and Moser (1994).

Brachysira is broadly distributed across North America over a range of trophic status; it is most common in oligotrophic to dystrophic waters. It is often abundant in waters with low conductance and

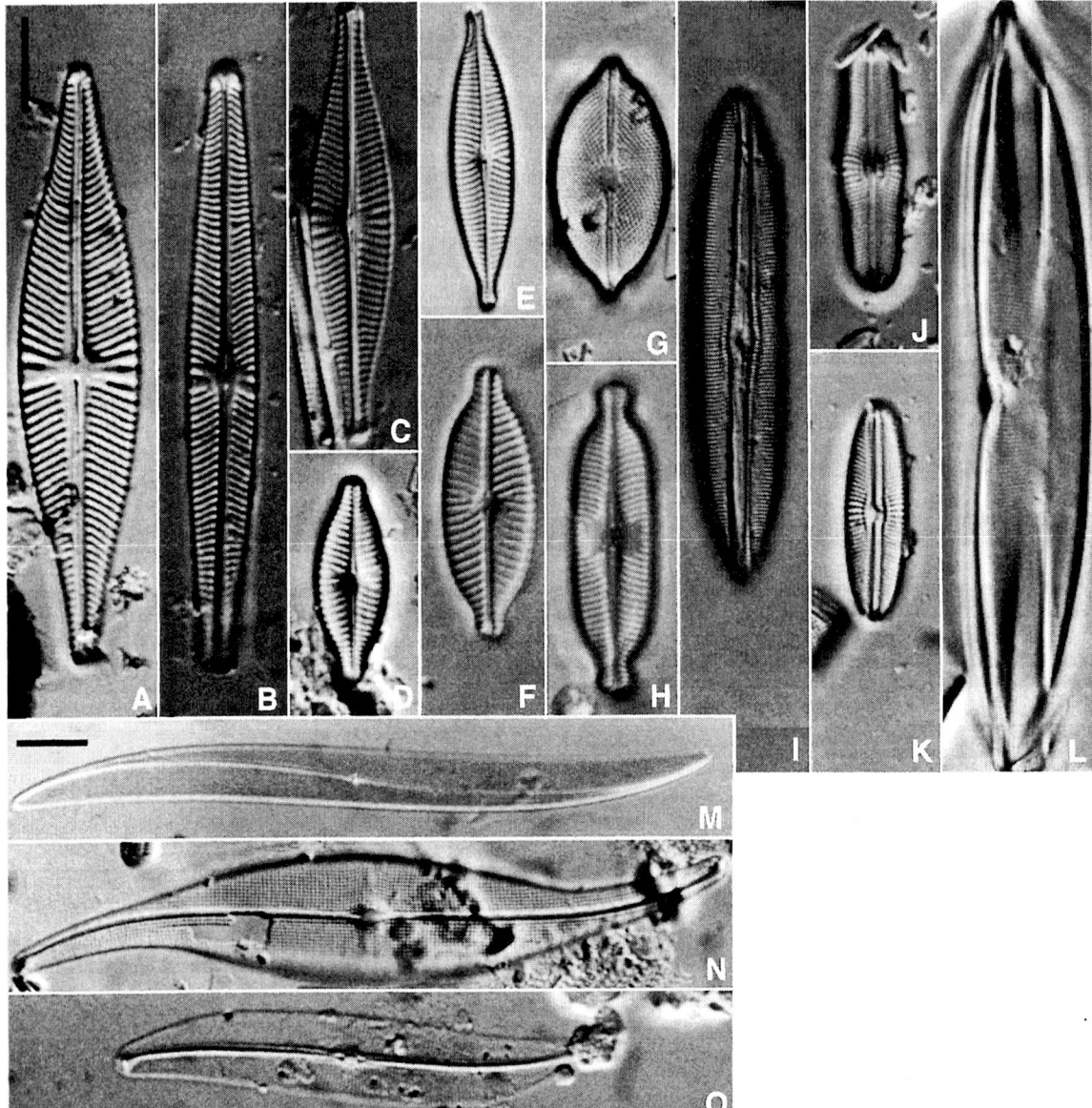

FIGURE 1 (A) *Navicula hasta*, striae are composed of areolae that are lineate to the apical axis; (B) *Navicula* sp.; (C) *Navicula rhynchocephala*; (D) *Placoneis* sp., striae are uniserate and composed of circular puncta; (E) *Navicula* sp.; (F) *Placoneis* sp., striae are uniserate and composed of circular puncta; (G) *Navicula placenta*, this taxon will soon be transferred out of *Navicula*; with an oval central area, and puncta forming diagonal striae, it is not a member of *Navicula sensu stricto*; (H) *Placoneis abiskoensis*, striae are uniserate and composed of circular puncta; (I) *Scoliopleura peisonis*, longitudinal canal is present, bordering the raphe; valves have a torsional twist; (J) and (K) *Muelleria gibbula*, longitudinal canals are present, and proximal raphe ends hook unilaterally; (L) *Plagiotropis* sp., raphe is located in a raised wing on the valve face; (M) *Pleurosigma elongatum*, valve outline is sigmoid, and striae form diagonal rows. N. *Gyrosigma parkeri*. Valve outline is sigmoid, and striae form axial and transverse rows; (O) *Gyrosigma scalproides*, valve outline is sigmoid, and striae form axial and transverse rows. Scale bars equal 10 μm. Scale bar in (A) applies to (A)–(L); scale bar in (M) applies to (M)–(O).

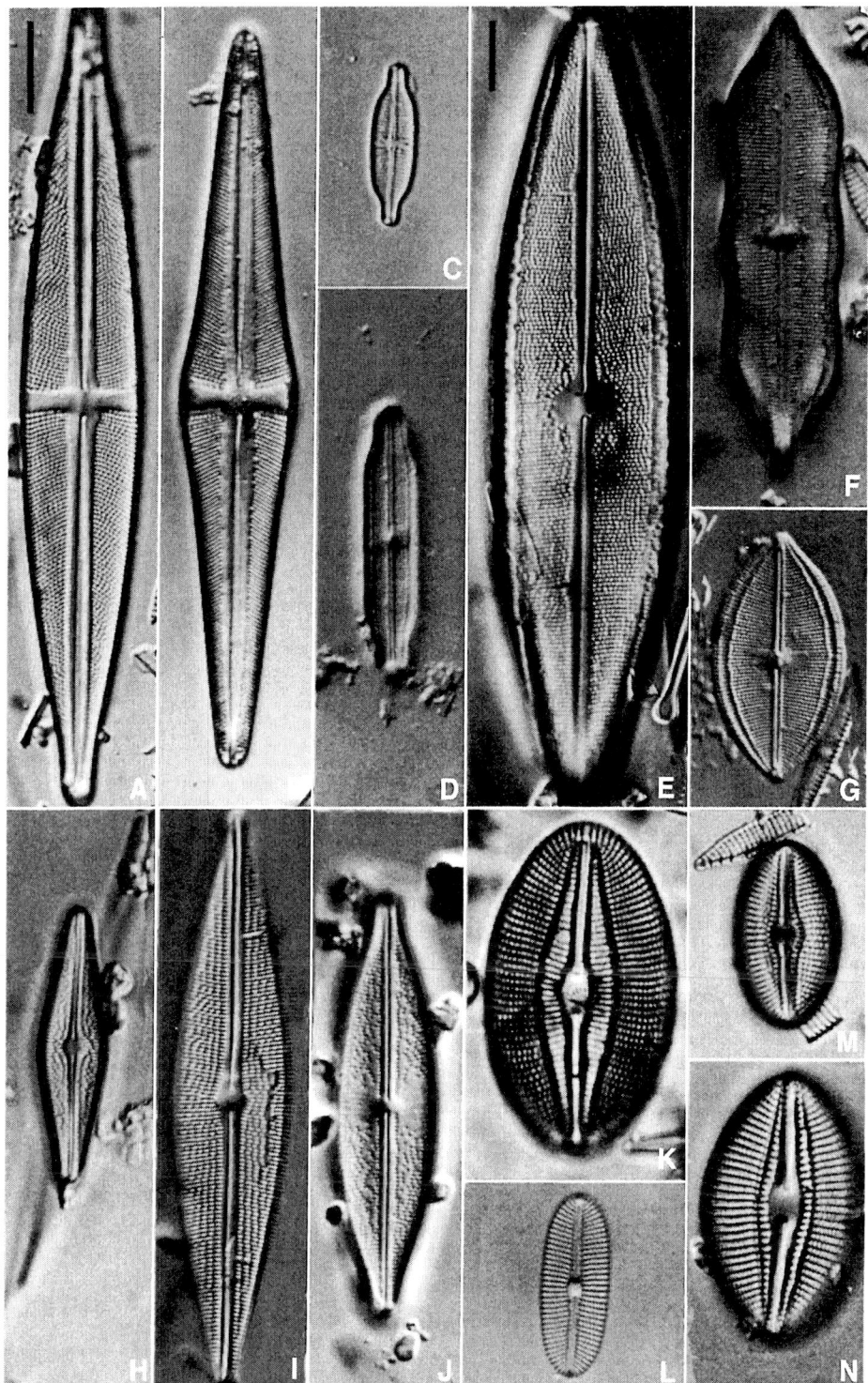

FIGURE 2 (A) *Stauroneis phoenicenteron*, a prominent stauros is visible, extending toward the valve margins from the central area; (B) *Stauroneis acuta*, a prominent stauros is visible, extending toward the valve margins from the central area; (C) *Stauroneis kriegeri*, stauros is small, and the thickening of silica is seen by optical dissection in the microscope; (D) *Neidiopsis levanderi*, valves are linear, and a longitudinal line is present near the valve margins; (E) *Neidium* cf. *iridis*, proximal raphe ends hook bilaterally; (F) *Neidium hitchcockii*, proximal raphe ends hook bilaterally; (G) *Neidium densestriatum*, proximal raphe ends are straight; (H) *Brachysira styriaca*, striae appear to form longitudinal undulations; (I) *Brachysira serians*, striae appear to form longitudinal undulations; (J) *Anomoeoneis sphaerophora*, striae are punctate and interrupted by hyaline areas; (K) *Diploneis finnica*, large, thickened longitudinal canal is present, bordering the raphe; (L) *Diploneis oblongella*, longitudinal canal is narrow and borders the raphe; (M) *Diploneis elliptica*; (N) *Diploneis smithii* var. *dilatata*. Scale bars equal 10 μm. Scale bar in (A) applies to (A)–(N); except where marked in (E).

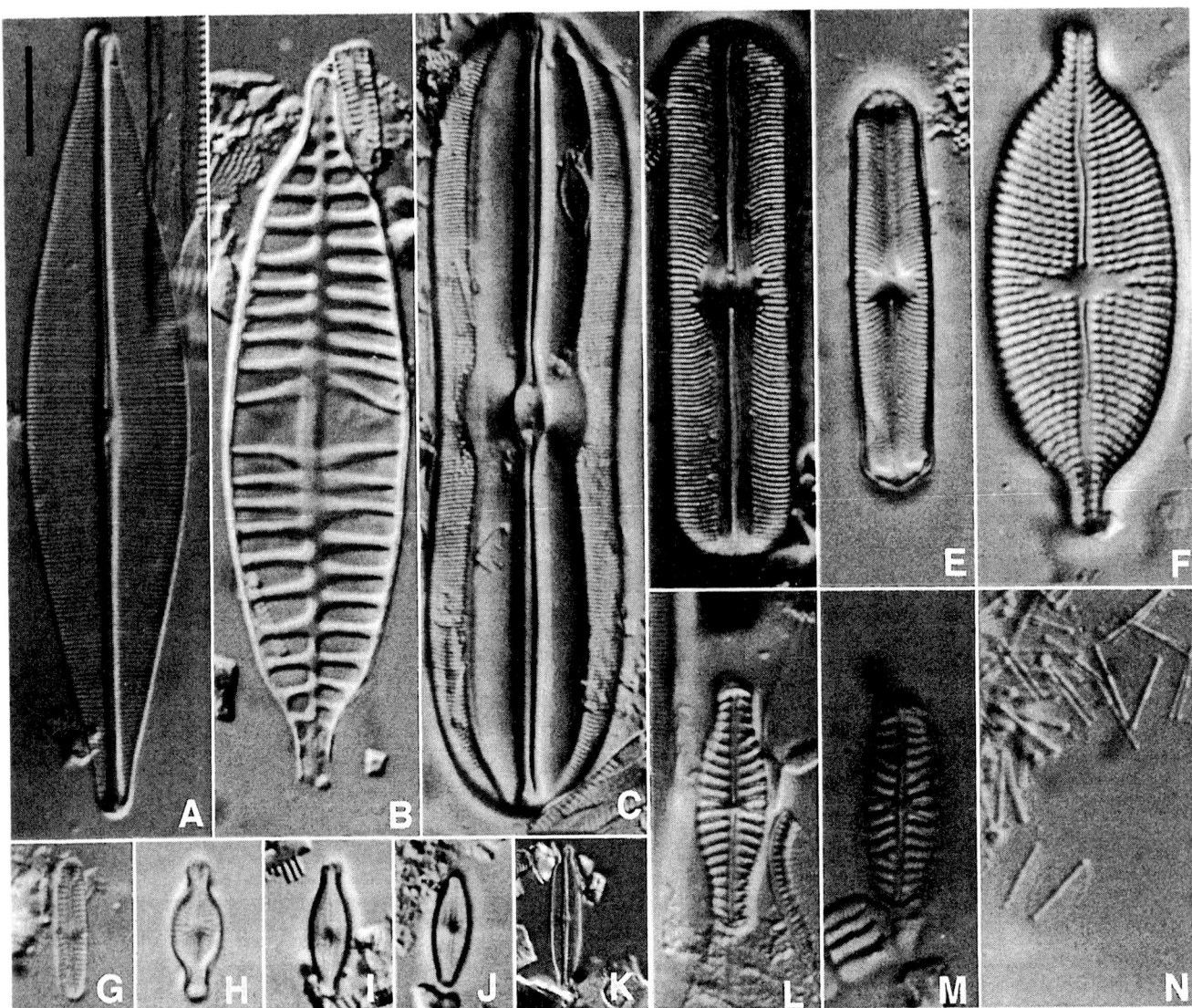

FIGURE 3 (A) *Craticula cuspidata*, striae are parallel or nearly parallel; (B) *Craticula cuspidata*, craticula is a characteristic, internal valve with thickened costae; (C) *Sellaphora americana*, axial area is expanded in this species; (D) *Sellaphora* sp; (E) *Sellaphora pupula*, silicious thickenings are present at the apices; (F) *Aneumastus tuscula*, raphe is undulate; (G) *Chamaepinnularia mediocris*, striae are uniserate, composed of simple chambers; (H) *Chamaepinnularia* sp.; (I) and (J) *Nupela* spp., striae form narrow longitudinal bands; (K) *Adlafia* sp., terminal nodules are strongly angular; (L) *Hippodonta capitata*, terminus of the raphe is thickened with silica; (M) *Hippodonta hungarica*; (N) *Fistulifera saprophila*, numerous valves are present, valves are lightly silicified, and the median rib is distinct. Scale bars equal 10 μm. Scale bar in (A) applies to (A)–(N).

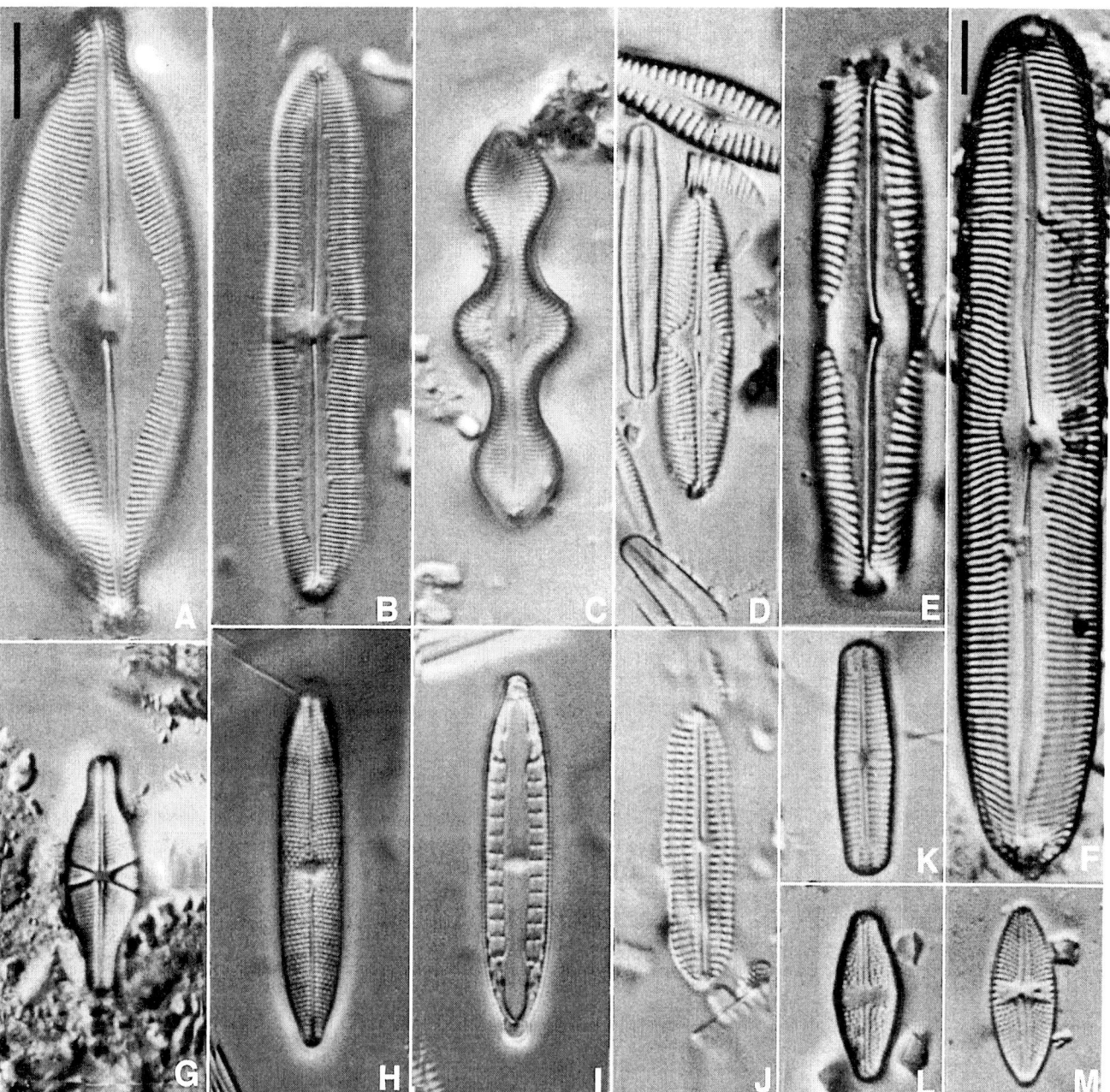

FIGURE 4 (A) *Caloneis amphisbaena*, individual puncta are not resolved, and striae are crossed by a longitudinal line; (B) *Caloneis silicula*, longitudinal lines are indistinct; (C) *Caloneis schumanniana*; (D) *Pinnularia* sp., striae are alveolate, appearing as chambers; (E) *Pinnularia mesolepta*; (F) *Pinnularia viridis*; (G) *Capartogramma crucicula*, the internal valve surface is marked by a distinctive X-shaped hyaline region of thickened silica; (H) *Mastogloia smithii* var. *lacustris*, high focus, of the external valve surface; (I) *Mastogloia smithii* var. *lacustris*, low focus, showing the valvocopula with partecta; the partecta has chambers (locules) from which mucilage is secreted; (J) *Oestrupia zachariasii*, striae are composed of large, coarse puncta; (K) *Geissleria ignota* var. *palustris*, near the terminal raphe (at the poles) the striae are interrupted, forming an annulus; (L) *Luticola mutica*, striae are distinctly punctate; the distal raphe ends curve unilaterally; (M) *Luticola goeppertiana* striae are distinctly punctate, and a single stigma is present in the central area. Scale bars equal 10 μm. Scale bar in (A) applies to (A)–(M); except where marked in (F).

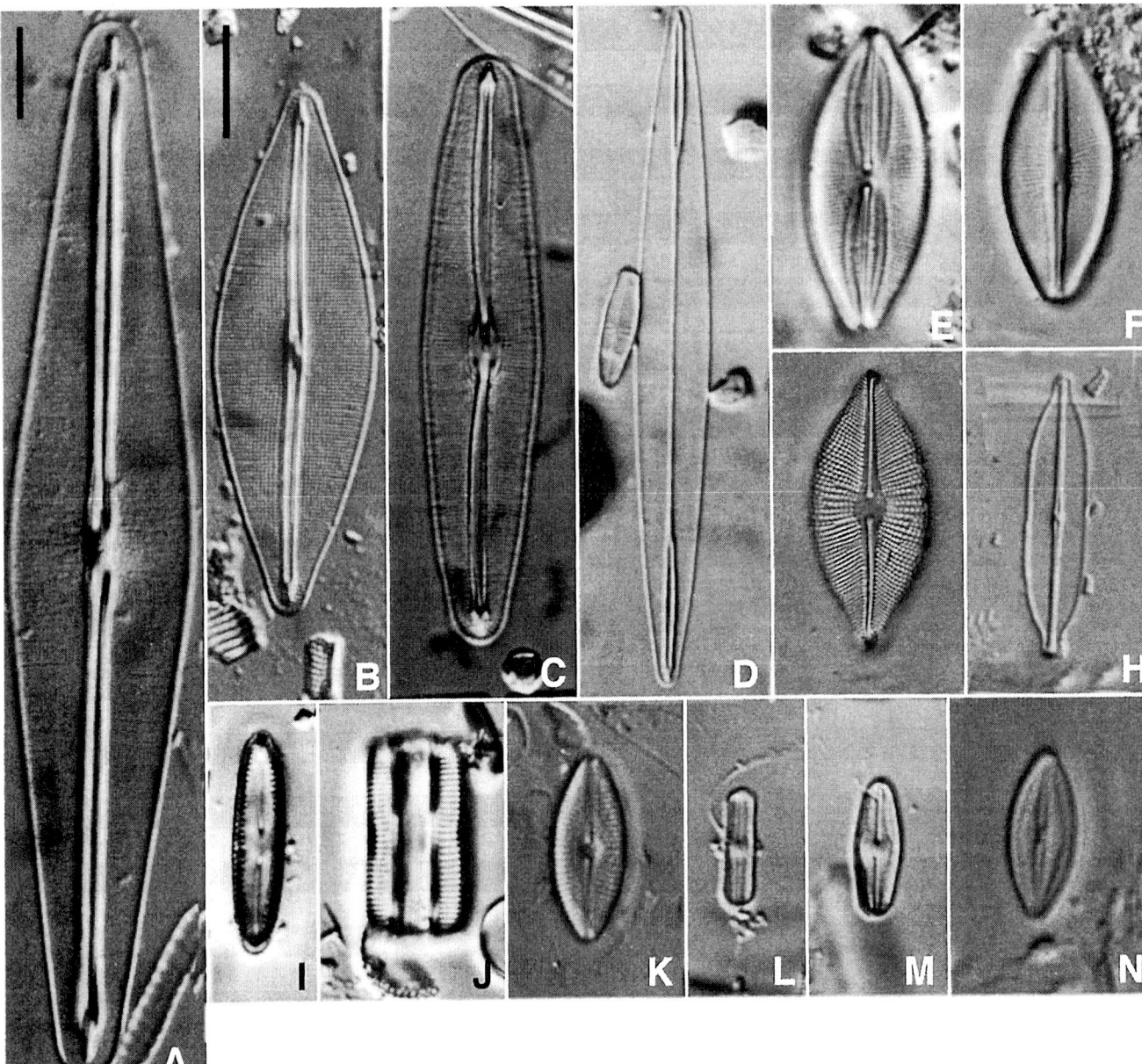

FIGURE 5 (A) *Frustulia rhomboides*, the raphe is contained between longitudinal ribs that extend most of the valve length; distally, the ribs join to form a "pencil tip," or porto crayon; (B) *Frustulia rhomboides* var. *saxonica*, the raphe is contained between longitudinal ribs that extend most of the valve length; distally, the ribs join to form a "pencil tip," or porto crayon;. (C) *Frustulia vulgaris*, the raphe is contained between longitudinal ribs that extend most of the valve length; distally, the ribs join to form a "pencil tip," or porto crayon; (D) *Amphipleura pellucida*, a simple, narrow median rib (central sternum) is evident on the valve face, except near the poles; the raphe is short and located between the ribs at the valve ends; (E) *Fallacia pygmaea*, the striae are interrupted by a lateral hyaline area, or sternum, which resembles a lyre; (F) *Cavinula cocconeiformis*, the central sternum is distinctly thickened, and the striae are radial; (G) *Cosmioneis* sp., striae are composed of round to oval puncta that are irregular in length near the central area; striae are strongly radiate; (H) *Kobayasia subtilissima*, axial area is narrow, and the raphe is striaght with dialated proximal raphe ends; striae are very fine and may not be visible in the light microscope; (I) *Diatomella balfouriana*, high level of focus of the valve face; striae are short and do not reach the central sternum; the raphe is filiform, with dialated proximal ends; (J) *Diatomella balfouriana*, girdle view, showing the septa as distinct thickenings; (K) *Diadesmis confervacea*, puncta are characteristically elongate; (L) *Diadesmis contenta*, puncta are characteristically elongate; the raphe has secondarily filled with silica and is not apparent; (M) *Diadesmis perpusilla*; (N) *Microcostatus krasskei*, the axial area has depressions on either side of a prominent central sternum. Scale bars equal 10 μm. Scale bar in (A) applies to (A); scale bar in (B) applies to (B)–(N).

low pH values, such as bogs. It is primarily benthic, but may become entrained in the water column of lakes.

Caloneis Cleve 1894 (Fig. 4A–C)

Valves are symmetrical about the apical and transapical axes. Individual puncta are not resolved, and striae are composed of alveoli crossed by one or two longitudinal lines. These longitudinal lines may be indistinct in smaller specimens. The central area may have thickened areas that are lunate or irregular in shape. Some diatomists (e.g., Patrick and Reimer, 1966; Round et al., 1990) suggest that *Caloneis* is a synonym of the closely related genus *Pinnularia*. Cells may possess one or two plastids.

Several species are commonly found thoughout North America in alkaline habitats. *Caloneis* species are also found in brackish and marine waters.

Capartogramma Kufferath 1956 (Fig. 4G)

Valves are lanceolate in shape. The internal valve surface is marked by a distinctive X-shaped hyaline region of thickened silica. Pseudosepta are present internally at the valve apices. The axial area is narrow axial area and the striae are punctate.

Capartogramma is monotypic in North America (represented by *C. crucicula*). This species is found in flowing waters of varying conductance in the eastern and southern United States and in the Laurentian Great Lakes.

Cavinula Mann et Stickle 1990 (Fig. 5F)

Valves are symmetrical to the apical and transapical axes. The valves are linear-lanceolate to nearly elliptical in outline. Striae are composed of round to elongate puncta that form uniseriate, radial striae. Internally, the central sternum is thickened. The external proximal raphe is expanded, whereas the internal proximal raphe is straight. *Cavinula* posssesses one or two chloroplasts, which are H-shaped in girdle view.

This genus is known from oligotrophic lakes and moist subaerial habitats across the United States. It includes the formerly recognized *Navicula cocconeiformis* and its allies. *Cavinula* differs from *Navicula* in that it possesses uniserate, radial striae and the terminal raphe ends curve to opposite sides of the valve.

Chamaepinnularia Lange-Bertalot et Krammer 1996 (Fig. 3 G, H)

Valve margins are more or less linear, and sometimes may be undulate. The valves are usually not more than 25 µm long and 4 µm broad. The raphe system may be simple or complex, as in *Pinnularia* or *Navicula*. External distal raphe fissure are hooked and terminate internally with small helictoglossa. The external proximal raphe ends are inconspicuous, and internally the proximal ends are unilaterally hooked. The striae are uniseriate and composed of simple pores, with external openings with covered by vela. The internal portion of the alveolar openings are divided with internal plates of silica. Cells occur singly.

Many of the species are aerophilic and commonly collected from splash zones of streams and associated with moist moss and lichen habitats. The genus includes a number of smaller species formerly part of *Navicula*, including *N. soehrensis*.

Cosmioneis Mann et Stickle 1990 (Fig. 5G)

Frustules have lanceolate to elliptical valve outline and capitate or rostrate poles. Valve mantles are high. Axial area is linear and central area is elliptical. The striae are composed of round to oval puncta, which are irregular in length near the central area. Striae are strongly radiate. Terminal raphe ends curve to the same side of the valve. Cells possess two H-shaped plastids.

Cosmioneis is a small genus found in alkaline, aerophilic habitats of North America. It includes the species *C. pusilla*, formerly classified in *Navicula*.

Craticula Grunow 1867 (Fig. 3A, B)

Valves are lanceolate with narrowly rostrate or capitate poles. The striae are parallel or nearly parallel. This genus is characteristically polymorphic, with distinct internal valves (craticula) that are produced in response to osmotic stress. These internal, craticular valves consist of a raphe–sternum and robust transverse bars. Cells contain two plastids, which are elongate and simple.

Craticula occurs in epipelic habitats across North America, in a wide range of fresh to brackish waters. In some species, the striae appear in both the longitudinal and the transverse axis. For a recent review see Mann and Stickle (1991).

Diadesmis Kützing 1844 (Fig. 5K–M)

Valves are small (usually < 20 µm in length), and linear to linear-lanceolate in valve outline. Although *Diadesmis* is biraphid, the raphe may be secondarily filled with silica and may not be visible in the light microscope. The characteristic elongate puncta of many species may also be difficult to resolve. The cells form bandlike colonies, which are often linked by spines. The plastids are single and slightly lobed.

The genus is characteristically aerophilic and tends to grow in association with mosses and attached to damp rocks. It occurs across North America, often in waters of low conductance and slightly acid waters.

Diatomella Greville 1855 (Fig. 5I, J)

Valves are symmetrical to the apical and transapical axes. The valve outline is linear-elliptical. The striae are short, and the raphe is filiform with dilated proximal ends. The frustules may form filaments. Distinct septa are present, with three openings, or holes. The presence of a true septum (in contrast to a pseudoseptum) in naviculoid diatoms is known only in *Diatomella*. The structure of the plastids is uncertain.

A single species, *D. balfouriana*, occurs in high-elevation alpine regions of Colorado, Utah, and Wyoming that are characteristically low in nutrient concentration. Although this taxon is relatively rare across given regions, it may be locally abundant in some habitats.

Diploneis Ehrenberg *ex* Cleve 1894 (Fig. 2K–N)

Valves are elliptical to panduriform (valve shape in which the margin of the central part of the valve is narrow, expanding toward the valve ends) in outline. Well-developed, thickened, longitudinal canals are present on either side of the raphe. Striae are composed of alveola that are complex in structure. Puncta may also be present on the longitudinal canal as single pores or elongate puncta. Two plastids, each with a pyrenoid, are adnate to the cinculum on either side of the frustule.

Diploneis is a predominantly marine genus, but is also widely distributed within freshwater epipelic environments.

Fallacia Stickle *et* Mann 1990 (Fig. 5E)

Valves are linear-lanceolate to elliptical in outline with bluntly rounded poles. The striae are interrupted by a lateral hyaline area (sterna), which resembles a lyre. A siliceous canopea covers the striae on the external valve surface. Although this conopea is not resolved in the light microscope, linear openings of the conopea (slits) on either side of the raphe are sometimes visible at the valve apices. In living cells, the conopea is filled with nitrogen-fixing cyanobacteria. Cells possess an H-shaped plastid consisting of two plates that lie against the girdle. The two plates of the plastid are connected by a narrow isthmus that lies against the epivalve.

Fallacia includes the formerly recognized *Navicula pygmaea* and its allies. Many species are found in marine tropical waters, but a few species occur in North American epipelic habitats of high conductance.

Fistulifera Lange-Bertalot 1997 (Fig. 3N)

Valves range in outline from linear to elliptical and they are very lightly silicified, and small. The striae are not visible with light microscopy. A median rib along the central sternum and distal ends of the raphe are both distinct.

Fistulifera includes the former *Navicula pelliculosa* and *N. saprophila*, species that occur in widely in fresh waters, and can reach great abundance in eutrophic and polluted waters.

Frustulia Rabenhorst 1853 (Fig. 5A–C)

Valve are rhomboid to linear-lanceolate in shape. Valve margins vary from straight to undulate. The raphe is contained in distinct median, longitidinal ribs that extend most of the length of the valve. At the apices, the ribs join to form a single tip (termed porto crayon). The proximal and distal raphe ends are not clearly observed with light microscopy. The striae are composed of fine puncta, arranged so as to produce a pattern of apical and transapical rows. Cells possess one H-shaped plastid.

Frustulia occurs in benthic habitats singly or in mucilaginous tubes. Species are found across North America, and they are often abundant in slightly acid waters, high in dissolved organic carbon, and low in conductance.

Geissleria Lange-Bertalot *et* Metzeltin 1996 (Fig. 4K)

Valves are elliptical to linear-elliptical, becoming obtuse to broadly rounded. The raphe is straight and filiform. Central endings of the raphe are straight, and inconspicuous to more or less bent, and the terminal raphe ends are hooked. The internal aspect of the raphe sternum with internal fissure is simple as compared to *Navicula sensu stricto*. Lineolae are fine, 50–80 per 10 μm. An isolated puncta may be present in the central area. A distinct feature of *Geissleria* is the presence of one to four transapical striae that are interrupted near the poles, forming an annulus. Thus, this group was formerly included in the "Annulatae" section of *Navicula*, including the former *Navicula paludosa* and *N. similis*.

Geissleria is broadly distributed across North America, but usually locally rare in occurrence. Some species of *Geissleria* are restricted to oligotrophic waters, whereas others are more common in eutrophic waters.

Gyrosigma Hassall 1845 (Fig. 1N, O)

Valves are symmetrical to the apical and transapical axes. Valve outline is sigmoid and the raphe is sigmoid with external proximal raphe ends recurved in opposite directions from one another. The axial area is narrow; central area, round to elliptical. The striae are composed of punctate striae that form rows perpendicular and parallel to the axial area. Two distinct, platelike or lobed plastids are present and positioned adnate to the cingulum.

Gyrosigma is broadly distributed in North America in epipelic and endopelic habitats. Species within this genus are often found in dense algal mats growing on the bottoms of lakes and reservoirs. These mats may become dislodged and float up from the bottom, bringing cells into the plankton. *Gyrosigma* is mainly epipelic, with brackish and, rarely, marine species.

Hippodonta Lange-Bertalot, Witkowski, *et* Metzeltin 1996 (Fig. 3L, M)

Valve outline is lanceolate and variously shaped. Valves of *Hippodonta* are characteristially heavily silicified. The striae are distinct, broad, and composed of double rows of puncta (visible under the scanning electron microscope). The raphe is straight, with dilated external, proximal ends. The terminus of the valve has an apparent thickened area, caused by the presence of a pseudoseptum. The genus includes the formerly recognized *Navicula capitata, N. hungarica,* and their relatives.

Members of *Hippodonta* are found in a wide range of benthic habitats, some tolerating high salinity in closed-basin lakes in the western United States.

Kobayasia Lange-Bertalot 1996 (Fig. 5H)

Valve outline is linear to linear-lanceolate, with swollen to capitate apices. The axial area is narrow, and the raphe is straight with dilated proximal raphe ends. Striae are very fine, not always visible in the light microscope. The striae are radiate and may be crossed by longitudinal lines.

Kobayasia is broadly distributed in acid waters, including *Sphagnum* bogs. This genus contains the formerly recognized *Navicula subtilissima* and its allies.

Luticola Mann in Round *et al.* 1990 (Fig. 4L, M)

Valves symmetrical to the apical and transapical axes. The valve apices are protracted. *Luticola* is marked by punctate striae and an expanded central area with a distinct stigma. The raphe is straight, with the proximal raphe ends slightly recurved in the same direction, and the distal ends hooked. The plastid is single, with two lobes and a central pyrenoid.

Luticola is found across North America and is characteristic of soil, aerophilic, and moss habitats. The genus includes the formerly recognized *Navicula mutica* and its allies.

Mastogloia Thwaites *ex* W. Smith 1856 (Fig. 4H, I)

Valves symmetrical to the apical and transapical axes. The valve margin is elliptical to lanceolate, with apices rounded to capitate. The valvocopulae possesses partecta, a distinguishing character of the genus. The raphe is straight, and the striae are distinctly punctate. The axial area is narrow, with slightly expanded central area. Cells contain two plastids, connected by a large, central pyrenoid.

In fresh waters across North America, *Mastogloia* is often found in calcareous, benthic habitats. Only a few species are present in fresh water, whereas several hundred species are found in marine habitats. Some taxa may resemble *Aneumastus* when valvocopulae are missing.

Microcostatus Johansen *et* Sray 1998 (Fig. 5N)

Valves are symmetrical to the apical and transapical axes. The axial area has depressions on either side of a prominent central sternum. The axial depressions are composed of small costae ("microcostae") and form the shape of a very small lyre in the central area. Striae are not visible with the light microscope, but are composed of single rows of areolae. Two plastids are present and adpressed to the cingulum.

The genus is monotypic, and contains *M. krasskei*, a small (up to 15 μm long) species thus far found in aerophilous habitats of Ohio.

Muelleria (Frenguelli) Frenguelli 1945 (Fig. 1J, K)

Outline of valves is variable, ranging from linear to linear-elliptical. The raphe is straight and associated with two longitudial canals parallel to the raphe, one on each side. The raphe has hooked external proximal raphe ends extending unilaterally. Distal raphe ends arc divergent. Puncta are loculate. Cells contain four plastids.

In North America, *Muelleria* includes *M. gibbula* and *M. terrestris*; *M. gibbula* appears to be widespread in aerophilic habitats, in association with mosses and in ephemeral pools of the midwestern and western United States. Other species of the genus are geographically restricted to the Arctic, Antarctic, or South America. Revisions of the genus have been published by Spaulding and Stoermer (1997) and Spaulding *et al.* (1999).

Navicula Bory 1822 (Fig. 1A–C, E, G)

Valves are symmetrical to the apical and transapical axes and are elliptical to broadly lanceolate in outline, with ends capitate, acute, rounded, or not produced. Striation is variable and may be not resolved in the light microscope to grossly punctate. Pseudosepta may or may not be present. The raphe is straight, filiform, or lateral. An isolated puncta may be present or absent. The central area may be expanded, but is not thickened into a stauros.

Navicula has traditionally been a genus where species that do not fit into other genera have been placed, making it a large collection of taxa that are probably not closely related. Taxa within *Navicula* that have structural similarities have traditionally been placed in subgenera of *Navicula* (e.g., Patrick and Reimer, 1966; Krammer and Lange-Bertalot, 1986); many of these have recently been erected to genus status (Round *et al.*, 1990). Although we have followed Round *et al.* in many of their newly proposed genera, *Navicula*, as followed here, is still a diverse (probably unnatural) group.

As it now stands, *Navicula* is broadly distributed, in nearly every freshwater habitat across North America.

Neidiopsis Lange-Bertalot *et* Metzeltin 1999 (Fig. 2D)

Valve outline is linear-lanceolate to linear-elliptical, with rostrate ends. Longitudinal lines are present on both sides of the axial area. The raphe is simple and straight, with proximal ends that are simple or slightly unilaterally deflected. The distal raphe ends are deflected.

Neidiopsis contains the formerly recognized *Navicula levanderi* and may be a close relative of *Neidium*. It is reported infrequently from North America, but it is known to occur in clearwater lakes of western Oregon and Washington and in the Laurentian Great Lakes region.

Neidium Pfitzer 1871 (Fig. 2E–G)

Valves have one or more longitudinal lines (formed by internal canals) on both sides of the axial area. The striae are distinctly punctate. Proximal raphe ends may recurve in opposite directions or terminate as straight ends. The distal raphe ends appear to bifurcate, forming an apical flap, in many taxa. Discontinuities (called Voigt faults) in the striae are common in this genus, being found on one side of the axial area along each raphe branch. Two plastids are present, each with lobes positioned adnate to the cingulum.

There are a large number of species within *Neidium* and they are found across North America. They tend to occur in neutral to slightly acid waters and are rarely abundant. Our understanding of the genus has been developed by the work of Reimer (1959), Stoermer (1963), and Hamilton *et al.* (1990, 1994, 1995, 1996). Mann (1984c) has documented sexual reproduction and development in the genus.

Nupela Vyverman *et* Compère 1991 (Fig. 3I, J)

Valves are slightly asymmetric to the apical axis. The central area is rectangular and extends nearly to the valve margin. Striae appear as narrow logitudinal bands under the light microscope. With SEM examination, areolae are found to be elliptical and arranged in rows of variable length. The number of longitudinal rows, or bands, varies from two to four and may vary between different sides of the raphe. The external proximal raphe is slightly expanded, whereas the internal proximal raphe is somewhat curved, and terminates in a "T".

This genus was described from high-elevation ponds of Papua New Guinea, in low-conductance, circumneutral-pH waters. Species within *Nupela* also have been reported from Europe, South America, and North America.

Oestrupia Heiden *ex* Hustedt 1935 (Fig. 4J)

Valves are symmetrical to the apical and transapical axes. The raphe is simple and straight. Large, coarse punctate striae are interrupted by apically-oriented ribs running the length of the valve. *Oestrupia* resembles *Pinnularia*, but differs by having apically oriented ribs.

This genus is primarily marine, but *O. zachariasii* occurs in cold, oligotrophic lakes, including those in Alaska (Foged, 1971, 1981).

Pinnularia Ehrenberg 1843 (Fig. 4D–F)

Valves are symmetrical to the apical and transapical axes. Valves are small to large (>250 μm). Striae are alveolate. In many species, the internal opening of the alveolus forms the appearance of longitudinal lines on either side of the axial area. In other species, such longitudinal lines are absent. The raphe system may be straight or complex. Usually, the external proximal raphe ends are dilated and bent slightly in the same direction, whereas the distal raphe ends are deflected. The central area may be expanded to one or both sides, and often valves of the same frustule differ in the shape or size of the central area. Two plastids, with many finger-like projections, extend from apex to apex along the valve mantle.

Pinnularia occurs across North America and is often abundant in low conductance, slightly acidic freshwaters. Krammer (1992a, b) has reviewed this genus.

Placoneis Mereschkowsky 1903 (Fig. 1D, F, H)

Valve outline is linear to lanceolate, sometimes with rostrate or capitate ends. Striae are uniserate and composed of loculate areola that are internally occluded. The axial area is narrow. The proximal raphe ends are straight and slightly expanded. The plastid is large, distinct, and divided into two X-shaped plates that are apprised against each valve.

Placoneis is a small, primarily freshwater genus that occurs in epipelic habitats of lakes and streams. This taxon includes the former *Navicula gastrum* and was reviewed by Cox (1987).

Plagiotropis Pfitzer 1871 (Fig. 1L)

Valve margins are lanceolate with narrow poles. The raphe is elevated from the valve to form a keel. The valve face is folded, with folds appearing as lines on either side of the raphe. The areolae are loculate, forming striae that are parallel and distinct. The axial area is narrow, and the central area is variable in shape. In girdle view, frustules appear constricted (depressed central area) at the center on both valves.

Plagiotropis comprises epipelic and brackish water species. At least two freshwater species are collected in alkaline, saline waters (Patrick and Reimer, 1975; Czarnecki and Blinn, 1978). It has been reported from western and southwestern states of North America (Czarnecki and Reinke, 1981; Bahls, 1982).

Pleurosigma Wm. Smith 1852 (Fig. 1M)

Valves are elongate and sigmoid in outline. The raphe is similar in sigmoid shape, following the curvature of the valve face. Areolae are loculate with slits or small pores opening to the external surface. The puncta are regularly spaced and form striae that give the appearance of transverse rows, as well as two diagonal rows. Two to four plastids extend from apex to apex. See also *Gyrosigma* to compare differences in striae.

Pleurosigma is primarily a marine genus, with several brackish-water and high-conductance inland or freshwater species. This genus grows in epipelic habitats of eastern North America.

Scoliopleura Grunow 1860 (Fig. 1I)

Valves have a torsional twist, so that the raphe is slightly sigmoid. Valve margins are linear-lanceolate in outline. Longitudinal canals border the raphe. The proximal raphe ends are deflected in opposite directions. Distal raphe ends are divergent. Striae are formed of distinct, loculate puncta.

A single species, *S. peisonis* is known from western North America, occurring in waters of high mineral content, such as in Great Salt Lake, Utah.

Sellaphora Mereschkowsky 1902 (Fig. 3C–E)

Valve outline ranges from linear, lanceolate, to elliptical with bluntly rounded poles. The striae are distinctly punctate. The axial area is distinct and may be expanded along the apical axis to form a conopeum. The external proximal raphe ends are dilated, whereas the distal raphe ends are deflected. Transapical thickenings occur interally, near the apices, in some species.

Sellaphora is widespread in North America in alkaline fresh to brackish waters. It occurs in habitats of circumneutral pH. This genus contains the formerly recognized *Navicula pupula* group, a complex of closely related species resurrected by Mann (1989).

Stauroneis Ehrenberg 1843 (Fig. 2A–C)

Valve outline is elliptical, and in some smaller specimens may be lanceolate. Central area is composed of the thickened central nodule being expanded to, or nearly to, the valve margin. Pseudosepta may be present in some species, where they are evident at the valve apices. The striae are distinctly or indistinctly punctate, and the axial area is usually narrow. Cells have two distinct plastids, which are closely associated with the cingulum.

Stauroneis is a diverse freshwater genus found across North America. It occurs in both benthic and planktonic habitats of lakes and streams. Some species of *Navicula* have expanded central areas or pseudosepta and therefore may be mistaken for *Stauroneis*. *Stauroneis* possesses a distinctly thickened central nodule, rather than an expanded, but not thickened central area of some *Navicula*.

IV. GUIDE TO LITERATURE FOR SPECIES IDENTIFICATION

The following is a list of key references that deal with species found in North America. Each contains citations to older literature and resources for other continents. The great flux in genus names for the group of diatoms considered in this chapter makes it difficult to list references for species identifications in the taxonomic scheme presented here. General references: Hohn and Hellerman (1963); Hustedt (1930): Hustedt (1931–1959); Krammer and Lange-Bertalot (1986, 1991b); Patrick and Reimer (1966).

Taxon-specific references:
1. *Navicula*—Krammer and Lange-Bertalot (1986, 1991b)
2. *Neidium*—Reimer (1959); Stoermer (1963); Hamilton et al. (1990, 1994, 1995, 1996)
3. *Microcostatus*—Johansen and Sray (1998)
4. *Muelleria*—Spaulding and Stoermer (1997); Spaulding et al. (1999)
5. *Pinnularia*—Krammer (1992a, b).

ACKNOWLEDGMENTS

We thank P. Hamilton, R. Lowe, R. Sheath, and E. F. Stoermer for comments and suggestions that greatly improved this chapter.

LITERATURE CITED

Bahls, L. L. 1982. Eight new diatom genus records for Montana. Proceedings of the Montana Academy of Sciences 41:79–86.

Barber, H. G., Haworth, E. Y. 1981. A guide to the morphology of the diatom frustule. Freshwater Biological Association, Scientific Publication 44, 112 p.

Beaver, J. 1981. Apparent ecological characteristics of some freshwater diatoms. Water Quality Branch, Ministry of Environment, Toronto, Ontario, 249 p.

Cleve, P. T. 1894. Synopsis of the naviculoid diatoms, Part 1. Kongelige Svenska Vetenskaps-Akademiens Handlingar 26:1–194.

Cleve, P.T. 1895. Synopsis of the naviculoid diatoms, Part 2. Kongelige Svenska Vetenskaps-Akademiens Handlingar 27:1–219.

Cox, E. J. 1987. *Placoneis* Mereschkowsky: The re-evaluation of a diatom genus originally characterized by its chloroplast type. Diatom Research 2:145–157.

Cox, E. J. 1996. Identification of freshwater diatoms from live material. Chapman & Hall, London, 158 p.

Cumming, B. F., Wilson, S. E., Hall, R. I., Smol, J. P. 1995. Diatoms from British Columbia (Canada) lakes and their relationship to salinity, nutrients, and other limnological variables. Koeltz Scientific Books, Stuttgart, 207 p.

Czarnecki, D. Reinke, D. C. 1981. Diatoms new to Kansas and nomenclatural notes on previous reports. Technical Publications of the State Biological Survey of Kansas 10:20–31.

Czarnecki, D. B., Blinn, D. W. 1978. Diatoms of the Colorado River in Grand Canyon National Park and Vicinity. Bibliotheca Phycologica 38:1–181.

Edlund, M., Stoermer, E. F. 1997. Review: Ecological, evolutionary, and systematic significance of diatom life histories. Journal of Phycology 33:897–918.

Edmondson, W. T., Ed. 1959. Freshwater biology. Wiley, New York, 1248 p.

Ehrenberg, C. G. 1843. Einen Nachtrag zu dem Vortrage über die Verbreitung und Einfluss des mikroskopischen Lebens in Süd- und Nord-Amerika. Abhandlungen der Königlichen Akademie der Wissenschaften Berlin 1841:202–209.

Foged, N. 1971. Diatoms found in bottom sediment from a small deep lake on the northern slope, Alaska. REF 4052.

Foged, N. 1981. Diatoms in Alaska. Bibliotheca Phycologica 53:1–137.

Gaul, U., Geissler, U., Henderson, M., Mahoney, R., Reimer, C. W. 1993. Bibliography on the fine structure of diatom frustules (Bacillarophyceae). Proceedings of the Academy of Natural Sciences of Philadelphia 144:69–238.

Geitler, L. 1973. Auxosporenbildung und Systematik bei pennaten Diatomeen und die Cytologie von *Cocconeis*-Sippen. Österreichische Botanische Zeitschrift 122:299–321.

Hamilton, P. B., Poulin, M. 1993. A taxonomic and morphological study of an acidobiontic diatom, Neidium holstii (Cleve) Krammer from North America and Greenland. Beihefte zur Nova Hedwigia 106:109–119.

Hamilton, P. B., Poulin, M., Taylor, M. C. 1990. *Neidium alpinum* var. *quadripunctatum* (Hustedt) comb. nov. an important acidobiontic taxon from northeastern North America. Diatom Reséarch 5:289–299.

Hamilton, P. B., Douglas, M. S. V., Fritz, S. C., Pienitz, R., Smol, J. P. Wolfe, A. P. 1994 A compiled freshwater diatom taxa list for the Arctic and Subarctic regions of North America. Canadian Technical Reports—Fisheries and Aquatic Sciences 1957:85–102.

Hamilton, P. B., Poulin, M., Walker, D. 1995. *Neidium hitchkockii* (Ehrenberg) Cleve, a morphologically complex taxon within the genus Neidium (Naviculales, Bacillariophyta), in: Kociolek, J. P., Sullivan, M. J. Eds., A century of diatom research in North America: A tribute to the distinguished careers of Charles W. Reimer and Ruth Patrick. Koeltz Scientific Books, Champaign, IL, pp. 61–77.

Hamilton, P. B., McNeely, R., Poulin, M. 1996. The morphology and distribution of *Neidium distinctepunctatum* Hustedt and its systematic position within the genus. Diatom Research 11:59–71.

Hustedt, F. 1930. Bacillariophyta, in: Pascher, A., Ed., Süßwasserflora von Mitteleuropa, Vol. 10, Fischer, Jena, pp. 1–466.

Hustedt, F. 1931–1959. Die Kieselalgen Deutschlands, Österreichs und der Schweiz, in: Rabenhorst, L., Ed., Kryptogamen-Flora von Deutschland, Österreich und der Schweiz 7(2):1–845.

Hustedt, F. 1961–1966. Die Kieselalgen Deutschlands, Österreichs und der Schweiz, in: Rabenhorst, L., Ed., Kryptogamen-Flora von Deutschland, Österreich und der Schweiz 7(3):1–816.

Johansen, J. R., Sray, J. C. 1998. *Microcostatus* gen. nov., a new aerophilic diatom genus based on Navicula krasskei Hustedt. Diatom Res. 13:93–101.

Kociolek, J. P., Spaulding, S. A., Kingston, J. C. 1998. Valve morphology and sytematic position of *Navicula walkeri* (Bacillariophyceae), a diatom endemic to Oregon and California. Nova Hedwigia 67:235–245.

Krammer, K. 1992a. *Pinnularia*. Die Gattung Pinnularia in Bayern. Hoppea 52:1–308.

Krammer, K. 1992b. *Pinnularia*. Eine Monographie der europäischen Taxa. Bibliotheca Diatomologica 26: 1–353.

Krammer, K. 1997a. Die cymbelloiden Diatomeen. Eine Monographie der weltweit bekannten Taxa. Teil 1. Allgemeines und Encyonema Part. Bibliotheca Diatomologica 36:1–382.

Krammer, K. 1997b. Die cymbelloiden Diatomeen. Eine Monographie der weltweit bekannten Taxa. Teil 2. Encyonema Part., Encyonopsis und Cymbellopsis. Bibliotheca Diatomologica 37:1–469.

Krammer, K, Lange-Bertalot, H. 1986. Naviculaceae, in: Ettl, H., Gerloff, J., Heynig, H., Mollenhaer, D., Eds. Süßwasserflora von Mitteleuropa, Vol. 2. Bacillariophyceae, Part 1. Gustav Fischer, Stuttgart, 876 p.

Krammer, K, Lange-Bertalot, H. 1991. Centrales, Fragilariaceae, Eunotiaceae, in: Ettl, H., Gerloff, J., Heynig, H., Mollenhaer, D., Eds., Süßwasserflora von Mitteleuropa, Vol. 2. Bacillariophyceae, Part 3. Fischer, Stuttgart, 576 p.

Lange-Bertalot, H., Metzeltin, D. 1996. Ecology—Diversity—Taxonomy. Indicators of Oligotrophy. Iconographia Diatomologica, Vol. 2. Koeltz Scientific Books, Koenigstein, 390 p.

Lange-Bertalot, H., Moser, G. 1994. *Brachysira*. Monographie der Gattung. Bibliotheca Diatomologica 29:1–212.

Lowe, R. L. 1974. Environmental requirements and pollution tolerances of freshwater diatoms. EPA-670/4–74–005, Cincinnati, OH, 344 p.

Mann, D. G. 1984a. Protoplast rotation, cell division and frustule symmetry in the diatom *Navicula bacillum*. Annals of Botany (London) 53: 295–302.

Mann, D. G. 1984b. Observation on copulation in *Navicula pupula* and *Amphora ovalis* in relation to the nature of diatom species. Annals of Botany (London) 54:429–38.

Mann, D. G. 1984c. Auxospore formation and development in *Neidium* (Bacillariophyta). British Phycological Journal 19:319–331.

Mann, D. G. 1985. In vivo observations of plastid and cell division in raphid diatoms and their relevance to diatom systematics. Annals of Botany (London) 55:95–108.

Mann, D. G. 1989. The diatom genus *Sellaphora*: Separation from Navicula. British Phycological Journal 24:1–20.

Mann, D. G. 1993. Patterns of sexual reproduction in diatoms. Hydrobiologia 269/270:11–20.

Mann, D. G., Stickle, A. J. 1991. The genus *Craticula*. Diatom Research 6:79–107.

Moser, G., Lange-Bertalot, H., Metzeltin, D. 1998. Insel der Endemiten. Geobotanisches Phänomen Neukaledonien. Bibliotheca Diatomologica 38:1–464.

Patrick, R. M. 1959. New subgenera and two new species of the genus *Navicula* (Bacillariophyceae). Notulae Naturae of the Academy of Natural Sciences of Philadelphia 324:1-11.

Patrick, R. M., Reimer, C. W. 1966. The diatoms of the United States. Monograph 13. Academy of Natural Sciences of Philadelphia, 688 p.

Patrick, R. M., Reimer, C. W. 1975. The Diatoms of the United States. Vol. 2, Part 1. Monograph 13. Academy of Natural Sciences of Philadelphia, 213 p.

Reimer, C. W. 1959. The diatom genus *Nedium*. I. New species, new records and taxonomic revision. Proceedings of the Academy of Natural Sciences of Philadelphia 111:1–35.

Reimer, C. W. 1961. New and variable taxa of the diatom genera *Anomoeoneis* Pfitz. and *Stauroneis* Ehr. (Bacillariophyta) from the United States. Proceedings of the Academy of Natural Sciences of Philadelphia 113:187–214.

Round, F. E., Mann, D. G. 1981. The diatom genus *Brachysira*. I. Typification and separation from Anomoeoneis. Archiv fur Protistenkunde 124:221–231.

Round, F. E., Crawford, R. M., Mann, D. G. 1990. The diatoms. Biology and morphology of the genera. Cambridge Univ. Press, Cambridge, UK, 747 p.

Spaulding, S. A., Stoermer, E. F. 1997. Taxonomy and distribution of the genus *Muelleria* Frenguelli. Diatom Research 12:95–113.

Spaulding, S. A., Kociolek, J. P., Wong, D. 1999. The genus *Muelleria* Frenguelli: A systematic revision, taxonomy and biogeography. Phycologia 38:314–341.

Stoermer, E. F. 1963. New taxa and new United States records of the diatom genus Neidium from West Lake Okoboji. Notulae Naturae of the Academy of Natural Sciences of Philadelphia 358:1–9.

Stoermer, E. F., Kreis, R. G., Jr., Andresen, N. A. 1999. Checklist of diatoms from the Laurentian Great Lakes. II. Journal of Great Lakes Research 25:515–566.

Vyverman, W., Compere, P. 1991. *Nupela giluwensis* gen. + spec. nov. a new genus of naviculoid diatoms. Diatom Research 6:175–179.

Witkowski, A., Lange-Bertalot, H., Stachura, K. 1998. New and confused species in the genus *Navicula* (Bacillariophyceae) and the consequences of restrictive generic circumscription. Cryptogamie Algologie 19:83–108.

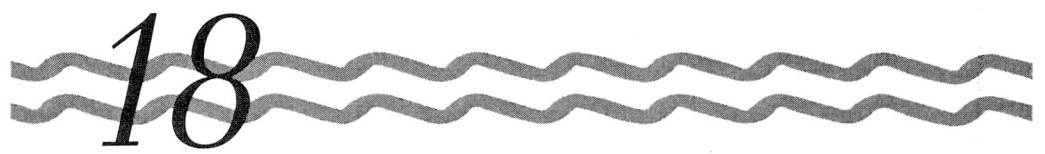

EUNOTIOID AND ASYMMETRICAL NAVICULOID DIATOMS

J. P. Kociolek, S. A. Spaulding

Diatom Collection
California Academy of Sciences
Golden Gate Park
San Francisco, California 94118

I. Introduction
II. Diversity and Morphology
    A. Eunotiaceae
    B. Catenulaceae
    C. Cymbellaceae
    D. Gomphonemataceae
    E. Rhoicospheniaceae
III. Ecology and Distribution
    A. Physical Factors
    B. Chemical Factors
    C. Biotic Factors
IV. Key and Descriptions of North American Genera
    A. Key
    B. Descriptions of Genera
V. Guide to Literature for Species Identification
Literature Cited

I. INTRODUCTION

In this chapter, we discuss the diverse assemblage of asymmetric diatoms that lack a keel. Grouping these taxa is a matter of convenience, because asymmetry about the apical or transapical axis has evolved in various lineages (Kociolek and Stoermer, 1988c, 1993a). Asymmety has arisen several times over evolutionary history as an adaptation to a stalked, or sessile, existence. Only recently have diatomists made explicit hypotheses about convergent evolution of characters and the extent to which members of these groups are phylogenetically distant taxa. Symmetry is generally unreliable for diagnosing natural groups, yet the older taxonomic literature relied heavily on symmetry features (e.g., Bessey, 1899; Karsten, 1928; Hustedt, 1930; Patrick and Reimer, 1966, 1975), and features of asymmetry may still be helpful in identification.

Five families and 14 genera are considered here. Taxonomic differentiation is based on the number and organization of the plastids, features of the nucleus and pyrenoids (Geitler, 1932; Round, *et al.*, 1990, Cox 1996), methods of sexual reproduction (Geitler, 1932, 1973a), valve morphology, growth habits, and ecological breadth. Due to their widespread distribution and dominance in many habitats, eunotioid and asymmetrical naviculoid diatoms have been the focus of valve ultrastructural studies (Hasle, 1973; Dawson, 1972, 1973a–c, 1974; Cox, 1996; Gaul *et al.*, 1993). Still, less than 5% of most diatom taxa have been studied from an ultrastructural point of view.

Members of these groups have been useful in environmental applications, such as paleolimnological reconstructions and as indicators of water quality (Patrick *et al.*, 1954; Smol, 1990). However, this group also includes forms acknowledged to pose some of the most vexing taxonomic problems, such as the species identification of many *Eunotia* and *Gomphonema*.

Patrick and Reimer (1966, 1975) have provided a foundation of North American taxa on which to document and understand diversity of these groups. Later monographs or concentrated taxonomic work on the following genera have been accomplished using wholly, or based in part on, North American specimens: *Actinella* (Kociolek *et al.*, 1997), *Gomphonema* (e.g., Kociolek and Stoermer, 1987b, 1991; Kociolek and Kingston, 1999; Reichardt, 1999), *Gomphoneis* (Kociolek and Rosen, 1984; Kociolek and Stoermer,

1986, 1988a, b), *Cymbella* (Krammer, 1982, 1997a, b), and *Reimeria* (Kociolek and Stoermer, 1987a).

The following classification is proposed for the genera considered herein:

Bacillariophyceae
  Naviculales
    Eunotiaceae
      Amphicampa (Ehrenberg) Ralfs in Pritchard
      Actinella Lewis
      Eunotia Ehrenberg
      Peronia Brébisson *et* Arnott *ex* Kitton
      Semiorbis Patrick in Patrick *et* Reimer
    Catenulaceae
      Amphora Ehrenberg *ex* Kützing
    Cymbellaceae
      Cymbella C. A. Agardh
      Didymosphenia M. Schmidt
      Encyonema Kützing
      Reimeria Kociolek *et* Stoermer
    Gomphonemataceae
      Gomphoneis Cleve
      Gomphonema Ehrenberg
    Rhoicospheniaceae
      Gomphosphenia Lange-Bertalot
      Rhoicosphenia Grunow

II. DIVERSITY AND MORPHOLOGY

A. Eunotiaceae

The Eunotiaceae is an unusual and morphologically heterogeneous group. In earlier literature, this family was referred to as rhaphidioid diatoms (Hustedt, 1926, 1952), due to the peculiar orientation of the raphe system. The proximal raphe ends and most of the length of the raphe are found on the valve mantle; only the distal raphe ends may extend onto the valve face. An exception to this arrangement occurs in *Peronia*, (Fig. 1E), in which the raphe system is positioned eccentrically on the valve face.

Of the raphe-bearing diatoms, the Eunotiaceae is the only group with one or more rimoportules. These structures are slitlike openings in the valve, with internally thickened edges. In much of the older literature, rimoportules are termed labiate processes, or jelly pores. Rimoportules are also present in centric and araphid diatoms and are thought to be a primitive feature. Rimoportules have also been suggested to be homologous to the raphe system itself (Hasle, 1973). Because of the possession of both rimoportules and a raphe, eunotioid diatoms are thought to constitute an evolutionary intermediate, between the primitive araphids and more advanced raphe-bearing groups (Hustedt, 1926, 1952).

Cells may grow singly, in filaments, or at the ends of short stalks, attached to a variety of substratum types. Several members of the group have prominent spines around the periphery of the valve, whereas others (e.g., *Semiorbis*) have deep ridges running transversely across the valve. Within the family, the most diverse genus by far in terms of number of species is *Eunotia* (Van Landingham, 1969). *Eunotia* and its close allies (e.g., *Actinella, Amphicampa*) characteristically have a valve face traversed by striae and interrupted close to the ventral margin (e.g., Fig. 1A–C). All other genera are relatively small, from the monotypic *Semiorbis* to genera with only several taxa in the North American flora. Only one species of *Actinella* is reported from North America. *Peronia*, a relatively rare genus, possesses a raphe on the valve face, which is displaced to one side (Fig. 1E). The valve of *Peronia* also includes a central sternum (an unornamented area or rib of silica running the length of the valve). Vyverman *et al.*, (1998) have described a unique genus within the Eunotiaceae, with valve and symmetry features similar to *Amphora*. This genus, named *Eunophora*, is an endemic of Tasmania and has no representatives from North America (Vyverman *et al.*, 1998).

B. Catenulaceae

Catenulaceae is represented by the genus *Amphora*, a widely distributed genus encompassing a broad spectrum of morphology. Members of *Amphora* share an eccentrically placed raphe system, which is located toward the ventral margin of the valve (Fig. 2G–I). The ventral mantle is distinctly more shallow than the dorsal mantle. The position of the raphe and height of the ventral and dorsal mantles form a frustule that is highly adapted to grow on the curved surfaces of aquatic macrophytes. *Amphora* is represented by fewer than 25 known species in the North American flora (Stoermer and Yang, 1971; Patrick and Reimer, 1975). Kingston *et al.*, (1980) discussed differentiation of the genus from *Cymbella*.

C. Cymbellaceae

Members of the Cymbellaceae are asymmetrical about the apical axis and herein include *Cymbella, Encyonema, Reimeria,* and *Didymosphenia*. Several additional genera have been subdivided out of *Cymbella* and *Encyonema* (Krammer, 1997a, b). *Reimeria, Cymbella,* and *Didymosphenia* bear apical pore fields, (specialized pores, differentiated in size or position from areolae) (Fig. 2). Apical pore fields of these genera are not bisected by the external distal raphe ends.

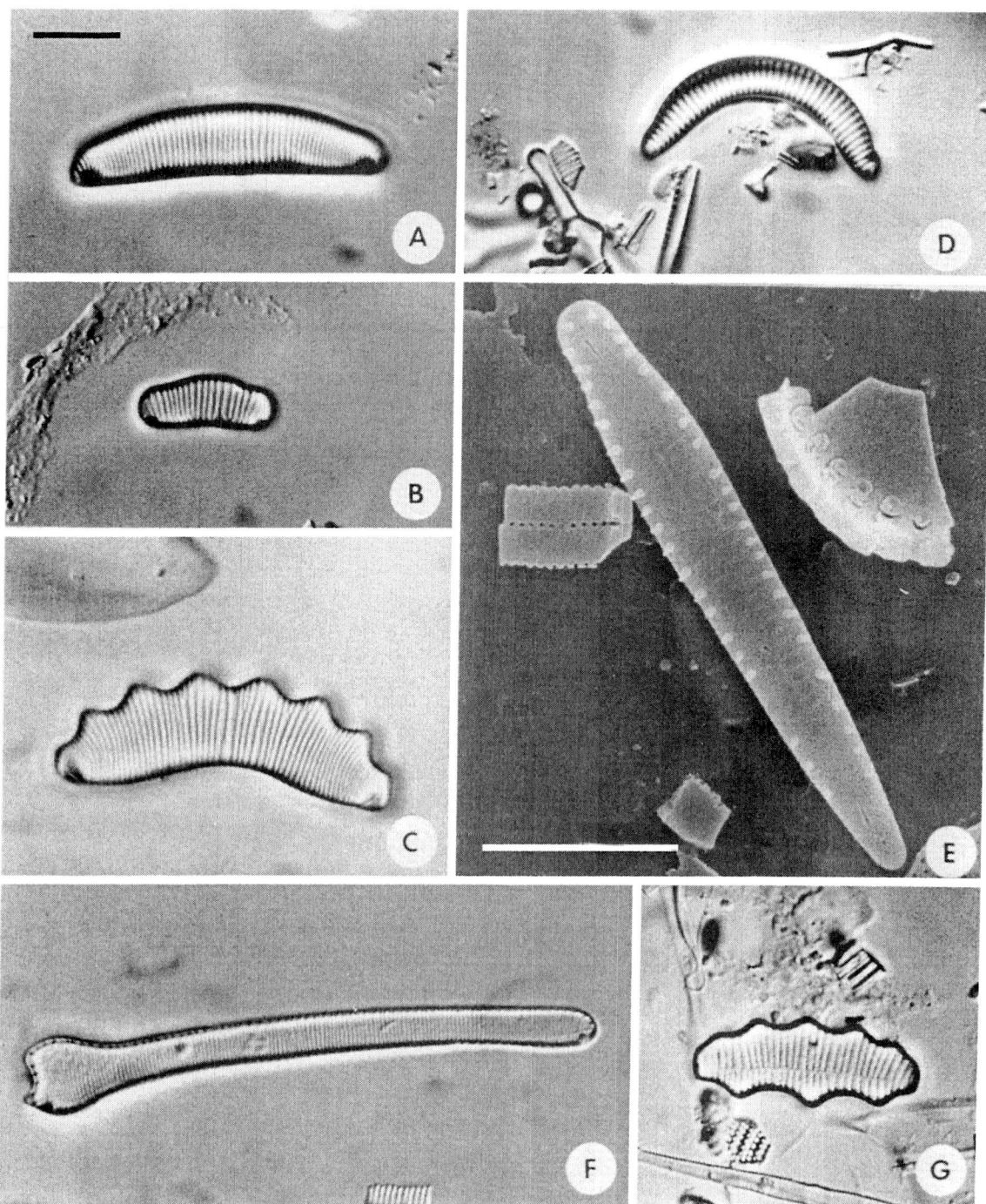

FIGURE 1 (A)–(C) *Eunotia* spp., the apical axis is asymmetrical and the transapical axis is symmetrical (A) *Eunotia faba*; (B) *Eunotia* sp.; (C) *Eunotia serra*; (D) *Semiorbis hemicyclus*, valves are highly arched and lack undulations; (E) *Peronia fibula*; on one valve the raphe is short and straight and a rimoportule is present near the apex; the other valve lacks a raphe, or it is only rudimentary; (F) *Actinella punctata*, valves are distinctly symmetrical to the apical and transapical axes; (G) *Amphicampa eruca*, both dorsal and ventral margins are distinctly undulate. Scale bars equal 10 μm. Scale bar in (A) applies to (A)–(G); except where marked in (E).

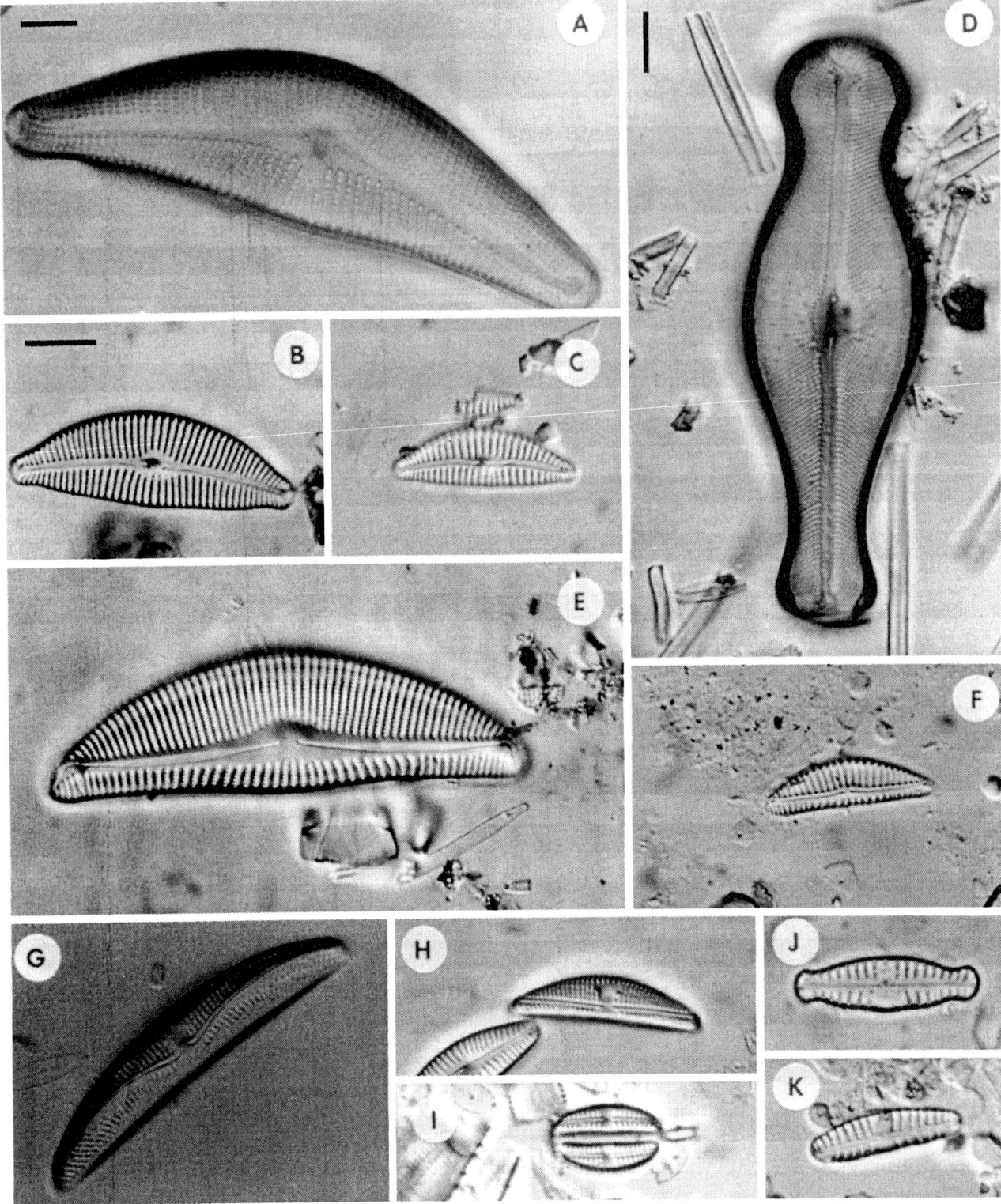

FIGURE 2 (A)–(C) *Cymbella* spp., valves are slightly to strongly asymmetrical to the apical axis and symmetrical to the transapical axis (A) *Cymbella proxima*; (B) *Cymbella turgidula*; (C) *Cymbella affinis*; (D) *Didymosphenia geminata*, valves are markedly large and robust, and the apical pore field is not bisected by the raphe; (E), (F) *Encyonema* spp., the distal raphe ends are ventrally deflected (E) *Encyonema muelleri*; (F) *Encyonema minuta*; (G)–(I) *Amphora* spp., the frustules are markedly wedge-shaped (G) *Amphora* sp.; (H) *Amphora pediculus*; (I) *Amphora perpusilla*; (J), (K) *Reimeria* taxa, the dorsal margin is slightly arcuate, whereas the ventral margin is straight to slightly concave, and tumid at the central area (J) *Reimeria sinuata* f. *antiqua*; (K) *Reimeria sinuata*. Scale bars equal 10 μm. Scale bar in (B) applies to (B)–(K); except where marked in (A) and (D).

Discovery of evolutionary convergence in asymmetry has led to changes in classification of some genera (Kociolek and Stoermer, 1988c; Kociolek, 1998). Both Cymbellaceae and Gomphonemataceae possess asymmetry about the transapical axis, as well as similar overall valve morphology. Also, both families contain genera with apical pore fields. However, within the Gomphonemataceae, apical pore fields are bisected by the distal raphe. Cladistic analysis of cytologic and other morphologic characters showed evolutionary parallelism between *Didymosphenia* and the Gomphonemataceae (Kociolek and Stoermer, 1988c). Thus, *Didymosphenia* is now considered to be a member of the cymbelloid diatoms, rather than its previous inclusion with gomphonemoid taxa.

Stigmata (isolated perforations at the center of the valve that have rounded external openings but slitlike internal openings) are also common in all four genera. These structures may be variable with respect to size, placement, and internal structure. In both Cymbellaceae and Gomphonemataceae, a phylogenetic relationship exists between the position of stigmata, external distal raphe ends, nucleus, and pyrenoid (Geitler, 1981; Kociolek and Stoermer, 1988c). In taxa where the stigma is dorsal, the nucleus is also dorsally placed. These taxa also possess a ventral pyrenoid and ventral deflection of the distal raphe ends (as in *Encyonema*; dorsal and ventral being determined by the curvature of the valve outline). In other taxa (e.g., *Cymbella sensu stricto*, *Didymosphenia*) stigmata are ventral, the nucleus is ventral, the pyrenoid is dorsal, and the external distal raphe ends are deflected dorsally.

Cymbella is the most diverse genus within the Cymbellaceae in terms of number of species (Krammer, 1982, 1997a, b). Diversity within *Reimeria* and *Didymosphenia* has been masked more by a lack of detailed taxonomic work than by a lack of variation (Skvortzow and Meyer, 1928; Sala *et al.*, 1993; Metzeltin and Lange-Bertalot, 1995).

D. Gomphonemataceae

The Gomphonemataceae appears to be a natural group, diagnosed by several shared, derived features (Kociolek and Stoermer, 1988c, 1993a). Most species produce stalks of mucopolysaccharides, which are secreted through the apical pore field. Stalks form an attachment to benthic substrata, and dense aggregations of vegetative clones are often visible to the naked eye.

Both *Gomphoneis* and *Gomphonema* are well-represented genera in the North American flora. A remarkable aspect of the North American flora is the diversity and geographic distribution of *Gomphoneis*. Within the *G. herculeana* species complex (Kociolek and Stoermer, 1988a), several large taxa may be abundant in freshwater diatom communities in western North America, the Laurentian Great Lakes, and the southeast United States. *Gomphoneis* as originally construed (i.e. the doubly punctate species with longitudinal lines) is not a monophyletic group (Cleve, 1894; Kociolek and Rosen, 1984; Kociolek and Stoermer, 1988c). The *Gomphoneis herculeana* species complex, with true stigmata and bilobed apical pore fields, is more closely related to *Gomphonema* than to the other species complex in *Gomphoneis* (Kociolek and Stoermer, 1993a). This second species complex, the *Gomphoneis elegans* group, has stigmoids instead of stigmata and condensed striae instead of apical pore fields at the footpole. Some astigmate species may also occur in the *G. elegans* group.

Gomphonema is a large genus, with over 400 taxa (Van Landingham, 1978) and worldwide distribution. Within North America, it is common and diverse. *Gomphonema* species may be astigmate or bear stigmata. Morphology has been well documented with the scanning electron microscope for some species (e.g., Dawson, 1972, 1973c), but the diversity of the genus is just now being revealed. Heterogeneity is becoming more widely known with respect to the presence or absence of stigmata. Furthermore, the structure of stigmata, type of areolar occlusions, and structure and position of apical pore fields are variable in form. Taxonomic and morphological diversity in North America is becoming more apparent (Lowe and Kociolek, 1984; Kociolek and Stoermer, 1991; Kociolek *et al.*, 1995; Kociolek and Kingston, 1999; Reichardt 1999).

E. Rhoicospheniaceae

The Rhoicospheniaceae is a group of freshwater and marine taxa, with only two genera in fresh waters of North America. *Rhoicosphenia* has many distinctive features, including flexed valves and one valve with very small raphe branche (Fig. 3I–K). The genus also shares some features with members of the Cymbellaceae and Gomphonemataceae, including type of sexual reproduction, organization of cytoplasmic organelles, stalked growth form, and, with the Gomphonemataceae, presence of pseudosepta. A single extant species is found in fresh waters in North America, *R. abbreviata* (known until relatively recently as *R. curvata*). The nomenclature (Lange-Bertalot, 1980) and morphology (Mann, 1982a, b) of the genus have been documented.

Gomphosphenia lacks stigmata, apical pore fields, septa, and pseudosepta while having characteristic anchor-shaped internal proximal raphe ends (Fig. 3L). Besides its asymmetry about the transapical axis, it

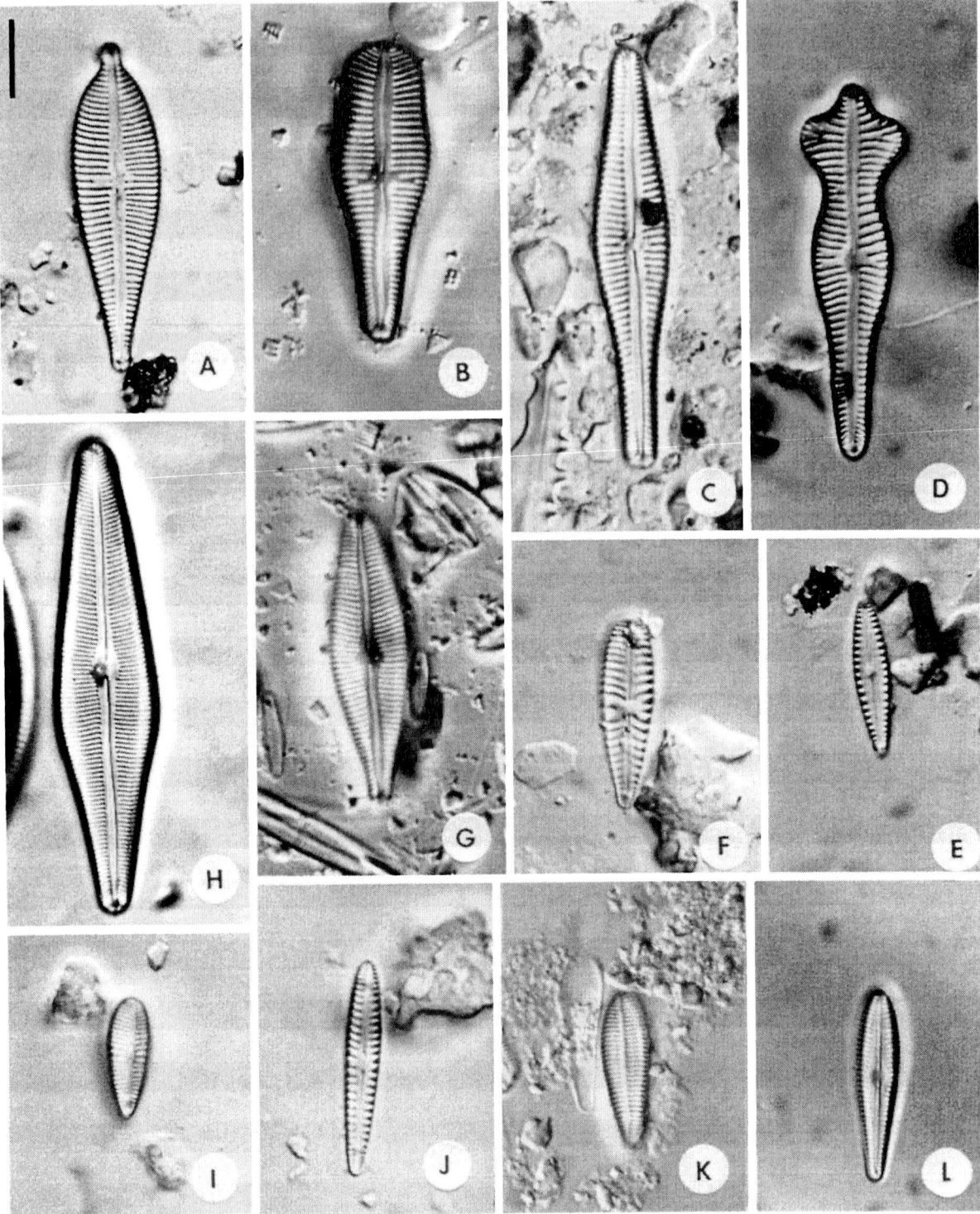

FIGURE 3 (A)–(E) *Gomphonema* spp., a single stigma is usually present on one side of the central area and longitudinal lines are absent (A) *Gomphonema sphaerophorum*; (B) *Gomphonema truncatum*; (C) *Gomphonema* sp.; (D) *Gomphonema acuminatum*; (E) *Gomphonema apuncto*; (F)–(H) *Gomphoneis* (F) *Gomphoneis olivacea*, stigmoids number 0–4, longitudinal lines are absent, and striae are composed of indistinct puncta; (G) *Gomphoneis eriense* var. *variabilis*, valves possess stigmata, longitudinal lines, and doubly punctate striae; (H) *Gomphoneis minuta* var. *lowei*.; (I)–(K) *Rhoicosphenia curvata*, the concave valve possesses a raphe that extends almost the length of the valve; the convex valve possesses short raphe branches near apices; (L) *Gomphosphenia lingulataeforme*. Scale bar equal 10 μm, applies to (A)–(L).

appears to be distantly related to *Gomphonema* and *Gomphoneis*. *Gomphosphenia* includes a taxon cited in older literature as "*Gomphonema brasiliense*" (e.g., Wallace, 1960; Patrick and Reimer, 1975) and *G. grovei* (Kociolek et al., 1988). The former, now known as *Gomphosphenia lingulataeforme*, has been reported almost exclusively from eastern North America (Kociolek and Kingston, 1999). *Gomphosphenia grovei* is found mostly as a fossil in western North America and occurs elsewhere only rarely.

III. ECOLOGY AND DISTRIBUTION

Given the diversity of phylogenetic lineages and forms found in the group of diatoms dealt with in this chapter, it is difficult to generalize about ecological distributions, especially at the level of genus. These forms all possess a raphe system, assume a variety of growth forms, and occur on a wide range of substratum types. Like the symmetrical naviculoid diatoms, they occur in lentic, lotic, and aerophilic habitats. Much of the literature on the distribution of freshwater diatoms in North America is now dated (Patrick and Reimer, 1966). Cholnoky (1965) covered a wide range of topics concerning the interplay of ecology and diatom distributions. Lowe (1974) provides a summary of ecological preferences and tolerances for selected freshwater taxa, based on literature reports. Krammer and Lange-Bertalot (1986, 1991) provide a wide range of summary information for specific taxa considered in their floristic study of the diatoms of central Europe.

A. Physical Factors

Members of the Eunotiaceae may be cosmopolitan (e.g., *Eunotia praerupta*) or narrowly restricted in geographic distribution. For example, *E. papilio* is considered circumboreal (Krammer and Lange-Bertalot, 1991), and *Actinella punctata* is found across eastern North America and Scandinavia (Kociolek et al., 1997). Other species from eastern North America are known only from their type localities (Patrick, 1958). Some *Eunotia* species have been reported from specific types of physical habitat types, including some from strictly flowing or still waters (*E. formica*) (Foged, 1948). Other *Eunotia* species have also been characterized as being aerophilic (*E. pectinalis* var. *minor*) (Manguin, 1952).

Within the Catenulaceae, species are usually benthic and are found in swamps, wetlands, lakes, and large rivers with slow-moving currents. Some taxa are limited to alpine or colder regions (Krammer and Lange-Bertalot, 1986), including the upper Great Lakes (Stoermer and Yang, 1971). For example, *Amphora calumetica* is restricted to the upper Great Lakes (Patrick and Reimer, 1975; Edlund and Stoermer, 1999).

Members of the Cymbellaceae are widely distributed in freshwater ecosystems, including both lakes and rivers. A few taxa have been reported exclusively from either lentic (e.g., *Cymbella lata*, *C. cymbiformis* Agardh, *C. proxima*, *C. cistula*) habitats or lotic (e.g., *C. turgidula*) habitats (Cholnoky, 1965; Patrick and Reimer, 1975). The group is found across a broad temperature spectrum, but some species seem restricted to cool-water environments, such as *Didymosphenia geminata* and *Encyonema norvegica* (Patrick and Reimer, 1975).

Members of the Gomphonemataceae are characterized by their eurytolerance of flow conditions. Some forms (e.g., *Gomphonema gracile*) have been characterized as being lake forms (Hustedt, 1930; Foged, 1954) whereas others are considered to be rheophils (e.g., *G. parvulum*) (Foged, 1948, 1954; Patrick and Reimer, 1975). At times, some taxa are found in the plankton of lakes. Cells may grow individually or form massive benthic colonies that blanket any available substrate.

Although many species are considered to be cosmopolitan in distribution (e.g., *G. parvulum* and *G. angustatum*, among others) some are restricted to specific regions. For example, *G. hebridense* is confined to the northern hemisphere (Krammer and Lange-Bertalot, 1986). *Gomphonema apuncto* and *G. manubrium* are known exclusively from the east coast of North America (Patrick and Reimer, 1975; Kociolek and Kingston, 1999), whereas other species are known from single localities in the United States (Patrick and Reimer, 1975; Reichardt, 1999).

Rhoicosphenia is common in lotic ecosystems or along the shores of lakes. It is found across a wide range of nutrient and other chemical regimes.

B. Chemical Factors

The Eunotiaceae grow across the spectrum of hydrogen ion concentration (pH), whether the habitats are bogs rich in humic acids, drainages of the Atlantic coastal plain, or clear lakes acidified by natural or anthropogenic sources. Certain species, such as *E. exigua* have been reported from habitats with pH values as low as 2–3. Others are found in circumneutral or even slightly alkaline waters, whereas some may be indifferent to pH (Lowe, 1974). Within the genus *Eunotia*, species show a range of tolerance of nutrient concentrations and organic contaminants. Some species are restricted to oligotrophic waters (Patrick and Reimer, 1966; Krammer and Lange-Bertalot,

1991), and others, such as *E. ruzickae*, are characterized as preferring eutrophic waters. *Eunotia septentrionalis* is known to occur in habitats with high sulfate levels (Krammer and Lange-Bertalot, 1991).

Amphora, of the Catenulaceae, is most commonly a benthic, marine form. In fresh water, some species of *Amphora* are found in circumneutral to alkaline waters, often in habitats with high ionic concentrations (Krammer and Lange-Bertalot, 1986; Round et al., 1990). Other species are found only in very dilute, oligotrophic conditions. Cholnoky (1965) reports that *A. inflata* is found in waters with high oxygen concentrations.

The Cymbellaceae are often characterized as occurring in circumneutral fresh waters across a wide spectrum of nutrient concentrations and water chemistries. Yet, at the species level, habitat preferences become more apparent. *Encyonema helvetica*, *C. aspera* and *D. geminata* grow in oligotrophic waters. *Reimeria sinuata* was reported as having a pH optimum of 8 (Cholnoky, 1965), and *C. cesatii* occurs in acid waters (Patrick and Reimer, 1975). *Cymbella amphicephala* and *E. microcephala* are reported from high oxygen habitats (Krammer and Lange-Bertalot, 1986). *Cymbella pusilla* Grunow is tolerant of relatively high levels of specific conductance and may grow in abundance in estuaries (Patrick and Reimer, 1975).

In the Gomphonemataceae, most species are found across a wide range of pH values, but the group might be best characterized as being circumneutral. Several species have been reported to be alkaliphilous, including *Gomphonema acuminatum* var. *brebissonii*, *G. angustatum*, and *G. abbreviatum* (Lowe, 1974). Many species are tolerant of highly eutrophic waters, including *G. augur*, *G. sphaerophorum*, *G. minutum*, *G. parvulum*, and *Gomphoneis olivaceum* (Patrick and Reimer, 1975; Krammer and Lange-Bertalot, 1986). Other species, including *Gomphonema subtile* Ehrenberg, are indicative of oligotrophic waters (Krammer and Lange-Bertalot, 1986). Krammer and Lange-Bertalot (1986) suggest *G. truncatum* is found in waters high in electrolytes. Local extinction of *Gomphoneis* species in the lower Laurentian Great Lakes was attributed to sensitivity to eutrophication and low concentrations of silica (Kociolek and Stoermer, 1988a).

C. Biotic Factors

Sphagnum bogs and other organic-rich wetland areas commonly contain a rich diversity of Eunotiaceae, although the biotic interactions, if any, have not been detailed. *Eunotia* species are epiphytic on a wide range of bryophytes and aquatic vascular plants (Krammer and Lange-Bertalot, 1991). *Eunotia bilunaris* var. *mucophila* is found in the mucilage or other algae, including the red alga *Batrachospermum* (Krammer and Lange-Bertalot, 1986).

Czarnecki (1995) reports several species of *Cymbella*, *Encyonema*, and *Gomphonema* found within the mucilage of the protozoan *Ophrydium versatile*. Geitler's (1975) suggestion that *C. cesatii* is an obligate inhabitant of the mucilage of this protozoan has not been confirmed (Czarnecki, 1995). *Amphora* species, many of the Cymbellaceae, and the Gomphonemataceae are common epiphytes on a variety of aquatic algae, bryophytes, and vascular plants.

IV. KEY AND DESCRIPTIONS OF NORTH AMERICAN GENERA

A. Key

1a.	Valve outline asymmetrical about apical axis (Figs. 1A, 2A)	2
1b.	Valve outline symmetrical about apical axis	9
2a.	In valve view, raphe restricted to ends of valve; in girdle view, most of raphe branch on valve mantle or not evident (Fig. 1C)	3
2b.	In valve view, raphe clearly in axial area on valve face; in girdle view, only tip or short section of raphe on mantle (Fig. 3C)	6
3a.	Valves asymmetrical about transapical axis (Fig. 1F)	*Actinella*
3b.	Valves symmetrical about transapical axis	4
4a.	Raphe branches indistinct, with striae uninterrupted across valve face (Fig. 1D)	*Semiorbis*
4b.	Raphe branches visible, striae commonly interrupted near ventral margin of valve	5
5a.	Dorsal and ventral margins undulate (Fig. 1G)	*Amphicampa*
5b.	Ventral margin smooth; dorsal margin undulate or smooth	*Eunotia*
6a.	Valves with distal raphe ends deflected dorsally, or not as below (Fig. 2A)	*Cymbella* (sensu lato)
6b.	Valves with distal raphe ends deflected ventrally (Fig. 2E)	7
7a.	Stigma present, centrally positioned, with relatively small, swollen, unornamented area on ventral margin (Fig. 2J, K)	*Reimeria*

7b.	Stigma dorsal or absent; ventral margin otherwise	8
8a.	Valve mantle height dissimilar between dorsal and ventral margin; dorsal mantle height higher than ventral; interruption in striation along dorsal portion of raphe (Fig. 2H, I)	*Amphora*
8b.	Valve mantle height similar on dorsal and ventral margins (Fig. 2E, F)	*Encyonema*
9a.	Valves large, robust, with multiple stigmata and single, unlobed apical pore field	*Didymosphenia*
9b.	Valves not as above	10
10a.	Raphe of reduced length on one or both valves (Figs. 1E, 3K)	11
10b.	Raphe long, of similar length on both valves	12
11a.	Septa and pseudosepta present; one valve convex, the other concave (Fig. 3.I–K)	*Rhoicosphenia*
11b.	Septa and pseudosepta absent, valves not highly arched (Fig. 1E)	*Peronia*
12a.	Proximal raphe fissure with anchor-shaped ends	*Gomphosphenia*
12b.	Proximal raphe fissure not anchor shaped	13
13a.	Valves with longitudinal lines on either side of the axial area *or* with 4 or more stigmoids around central area (one exception, see genus description)	*Gomphoneis*
13b.	Valves without longitudinal lines and with 0, 1 or (sometimes) 2 stigmata on one side of the central area (some exceptions, see genus description)	*Gomphonema*

B. Descriptions of Genera

Catenulaceae

Amphora Ehrenberg *ex* Kützing 1844 (Fig. 2G–I)

Valves are asymmetrical to the apical axis and symmetrical to transapical axis. On the dorsal margin, the valve mantle is deeper, or higher, than that on ventral margin. As a result, the frustule is wedge-shaped. The wedge shape prevents complete focus in one focal plane. The raphe is in a position of moderate to strong eccentricity. Furthermore, the raphe may be straight, arched, or slightly sigmoid. Striae on the dorsal margin are usually crossed by hyaline areas, whereas striae on the ventral margin are short. Depending on the position of the valve, the striae on the ventral margin may difficult to see. Stigmata are lacking, and terminal nodules are indistinct. The ultrastructure of *Amphora* has been considered by Archibald and Schoeman (1984), Archibald and Barlow (1983), Karayeva *et al.*, (1984), and Lee and Round (1987, 1988). Stoermer and Yang (1969) described several new species of *Amphora* from the Laurentian Great Lakes. Mann (1984) has documented sexual reproduction in *Amphora*.

Although primarily a marine genus, some species (i.e., *A. ovalis*) of *Amphora* are widely distributed in fresh waters across North America. Other species (i.e., *A. calumetica*), are restricted in distribution to the Laurentian Great Lakes (Edlund and Stoermer, 1999). The extreme asysmmetry of *Amphora* makes it a benthic specialist, where it may attach firmly to substrata, especially other algae and aquatic plants.

Cymbellaceae

Cymbella C. A. Agardh 1830 (Fig. 2A–C)

Valves are slightly to stongly asymmetrical to the apical axis, and symmetrical to the transapical axis. Species with slight asymmetry may be very close to naviculoid symmetry. The raphe is centrally or eccentrically positioned. Stigmata are absent, or if present, located on the ventral side of the central area. The distal raphe ends are straight or deflected dorsally. Cells grow in benthic habitats, often surrounded by mucilage or producing mucilaginous stalks. Apical pore fields may be absent, or present at both poles. Taxa with apical pore fields are likely to be found growing on stalks.

The broad sense of *Cymbella* embraced here (Krammer, 1982; Krammer and Lange-Bertalot, 1986; Round *et al.*, 1990) is probably not a natural group of taxa. As such, it contains a heterogeneous assemblage of taxa that are not referred to *Encyonema* or *Reimeria*. A natural group within *Cymbella* contains species with ventrally placed stigmata, dorsally deflected external distal raphe ends, and apical pore fields. Krammer's work has led to a broad understanding of the valve (Krammer, 1982), raphe (Krammer, 1979), and girdle components of cymbelloid frustule (Krammer, 1981). Krammer (1997a, b) has begun to revise the genus and identify clusters of morphologically similar species. Krammer and Lange-Bertalot (1986) offer an excellent account of common members of the genus. *Cymbella* is a large genus and is often common in benthic habitats across North America. As a genus, it varies greatly in size and degree of asymmetry, with some species with only slight asysmmetry to the apical axis. North American species have been considered by Hufford and Collins (1972a, b), Kingston (1978), and Selva (1981). Geitler (1927, 1956, 1967) and Geitler and Mack (1953) have documented sexual reproduction.

Didymosphenia M. Schmidt 1899 (Fig. 2D)

Valves are markedly large and robust. Symmetry relations are asymmetry to the transapical axis and symmetry to the apical axis (although some populations are asymmetrical to the apical axis). Both apices are capitate, and valves are wedge-shaped in girdle view. A marginal ridge of silica is present on both sides of the valves, terminating near the headpole as spines. Septa and pseudosepta are absent. Two to several stigmata occur on one side of the central area. The apical pore field is not bisected by the raphe. Kociolek and Stoermer (1988c) showed that this taxon is more closely allied to the cymbelloid diatoms than to the gomphonemoid groups to which it had previously been referred. Dawson (1973a, b) studied the ultrastructure of Asian specimens; Stoermer et al., (1986) quantified and compared shape and shape change across several populations. Meyer (1929) documented sexual reproduction in the genus.

A single species, *D. geminata*, occurs in western North America. It is locally abundant in some lakes and streams, at times producing high concentrations of biomass. The sheer mass of mucilaginous stalks of *D. geminata* cover surfaces and may foul water intake pipes, reaching nuisance porportions.

Encyonema Kützing 1833 (Fig. 2E, F)

Valves are asymmetrical to the apical axis and symmetrical to the transapical axis. The dorsal margin is highly arched, whereas the ventral margin is straight or nearly so. Stigmata are lacking or, if they are present, they are located on the dorsal side of the central area. The distal raphe ends are ventrally deflected. Apical pore fields are absent. Cells occur singly and free, produce mucilaginous sheaths, or form colonies within mucilaginous tubes. Krammer (1982) showed that the ultrastructure of the species in *Encyonema* consistently differs from other species included in *Cymbella*. Kociolek et al., (1994) documented the cytoplasmic fine structure of two *Encyonema* species with transmission electron microscopy. Taxonomy and ultrastructure of the genus are considered in Krammer (1997a, b).

Encyonema is broadly distributed across North America, primarily in benthic habitats. In the classical literature many of the species in this genus are included in *Cymbella*.

Reimeria Kociolek *et* Stoermer 1987 (Fig. 2J, K)

Valves are symmetrical to the transapical axis and asymmetrical to the apical axis. The dorsal margin is slightly arcuate, whereas the ventral margin is straight to slightly concave, and tumid at the central area. The axial area is narrow, and the central area is unilaterally expanded toward the ventral margin. A single stigma is located just ventral to the dilated proximal raphe ends. Apical pore fields are visible at both poles on the ventral margin. The distal raphe ends are deflected toward the ventral margin. Kociolek and Stoermer (1987a) proposed this genus for the *Cymbella sinuata* species complex.

Reimeria is limited to a small number of species and is broadly distributed across North America, primarily in benthic habitats.

Eunotiaceae

Actinella Lewis 1864 (Fig. 1F)

Valves are distinctly symmetrical to the apical and transapical axes. In valve view, the raphe is not evident. The raphe is highly reduced and restricted to the ventral margin. Terminal nodules are visible. Striae are clearly visible, with small spines evident along the outline of the valve. The cells grow singly or in small colonies that are joined by mucilage secreted at the narrow ends of the cells.

Almost all reports of this genus are from soft or acidic waters in eastern and southeastern North America. *Actinella* is differentiated from *Eunotia* by asymmetry about the transapical axis. A single species, *A. punctata*, is extant. Morphology and distribution records of *A. punctata* are documented in Kociolek et al. (1997).

Amphicampa (Ehrenberg) Ralfs in Pritchard 1861 (Fig. 1G)

Valves are asymmetrical to apical axis, and symmetrical to the transapical axis. Both dorsal and ventral margins are distinctly undulate. The raphe is short and restricted to the distal ends.

A single species, *A. eruca*, is rare to uncommon in circumneutral to alkaline waters in the midwestern and western United States (Kociolek, 2000). The report by Patrick and Reimer (1966) of *A. mirabilis* in North America cannot be confirmed.

Eunotia Ehrenberg 1837 (Fig. 1A–C)

Apical axis is asymmetrical, and transapical axis is symmetrical. Dorsal margins of the valve are convex and may be smooth or undulate. Ventral margins of the valve are straight or concave. The raphe is restricted to the ends and lies along the valve mantle. Distal ends of the raphe curve slightly or strongly onto the valve face at the apices. Usually, the terminal nodules are conspicuous. In girdle view, the frustules are rectangular, or boxlike. Also, in girdle view the raphe branches are evident. Ultrastructure of the genus has been documented by Alles *et al.*, (1991), Mayama and Kobayashi (1990, 1991), and Kobayashi *et al.*, (1981). The treatments by Krammer and Lange-Bertalot (1991) and Lange-Bertalot and Metzeltin (1996) are very helpful for species-level identifications. Geitler (1951a, b, 1973c) documented sexual reproduction.

Cells occur singly, free or attached by mucilaginous stalks, or in long ribbon-like colonies. The genus is large and widespread across North America, but most diverse in softwater or acid-water habitats (Patrick and Reimer, 1966; Krammer and Lange-Bertalot, 1991).

Peronia Brébisson *et* Arnott *ex* Kitton 1868 (Fig. 1 E)

Valve outline is clavate, with a narrow central sternum. On one valve, the raphe is short and straight, and a rimoportule is present near the apex. The other valve lacks a raphe, or it is only rudimentary. This interesting genus is similar to the Naviculaceae, in that the position of the raphe is on the valve face. It also shares the feature of a rimoportule with the Eunotiaceae (Hasle, 1973; Round *et al.*, 1990). The morphology of a South American species has been documented (Kociolek, 2000).

Peronia occurs rarely in North America, with scattered reports from the midwestern, southern, and eastern United States, and occurs only in acidic waters.

Semiorbis Patrick in Patrick *et* Reimer 1966 (Fig. 1D)

Valves are highly arched and lack undulations. They are asymmetrical about the apical axis and symmetrical about the transapical axis. The raphe is inconspicuous, and the terminal nodules are not evident. Cells occur singly or in short filaments. Krammer and Lange-Bertalot (1991) consider *Semiorbis* within the genus *Eunotia*. Moss *et al.*, (1978) documented the ultrastructure of *S. hemicyclus* with the scanning electron microscope.

Patrick and Reimer (1966) report *Semiorbis* from waters of low mineral content and humic rich waters, especially bogs, of the eastern United States. They included this genus with the araphid diatoms despite Kolbe's (1956) report of a short raphe system.

Gomphonemataceae

Gomphoneis Cleve 1894 (Fig. 3F–H)

Valves are wedge-shaped: asymmetrical to the transapical axis, and symmetrical to the apical axis. Both septa and pseudosepta are evident. Species possess either stigmata, longitudinal lines, and doubly punctate striae (*G. herculeana, G. eriense*), or 0–4 stigmoids, no longitudinal lines, and striae composed of indistinct puncta (*G. olivacea, G. quadripunctata*). Taxonomy, ultrastructure, distribution, and phylogeny of the genus have been documented (Kociolek and Rosen, 1984; Kociolek and Stoermer, 1986, 1988a, b, 1989).

Gomphoneis is found predominantly in midwestern and western North America. Like *Gomphonema*, they produce long, mucilaginous stalks, with the cells growing upward from the point of attachment of the stalk. The colonies form thick, mucilaginous masses attached to surfaces along the shores of lakes, rivers, and streams.

Gomphonema Ehrenberg 1832 (Fig. 3A–E)

Cells are asymmetrical to the transapical axis and asymmetrical to the apical axis, forming a wedge shape. A single stigma is usually present on one side of the central area. Longitudinal lines are absent. North American species possess striae that are usually punctate and not costalike. Species commonly occur individually or growing at the ends of mucilaginous stalks. Some taxa form stellate colonies or occur in mucilaginous masses. Our understanding of the taxonomy of the genus has benefitted from contributions by Reichardt and Lange-Bertalot (1991), Ueyama and Kobayashi (1988), and Lange-Bertalot (1993), who have helped to provide a better understanding of valve ultrastructure. North American specimens have been treated by Wallace and Patrick (1950), Wallace (1960), Kalinsky (1984), Czarnecki and Blinn (1979), Hohn (1959), Van Landingham (1967), Kociolek and Stoermer (1987b, 1991), Kociolek *et al.* (1995), Kociolek and Kingston (1999), and Lowe and Kociolek (1984). Geitler (1951c, 1973b) has documented sexual reproduction in *Gomphonema* species.

An estimated 150 *Gomphonema* taxa have been reported from the across the United States (Kociolek, unpublished). The genus is found in nearly every habitat type within circumneutral lakes and streams.

Rhoicospheniaceae

Gomphosphenia Lange-Bertalot 1995 (Fig. 3L)

Valves are cuneate, with straight or undulate sides, never bent or flexed. Raphe is straight, filiform with external terminal fissures (both proximal and distal) straight. Internally, distal raphe fissure ends in a helictoglossa; the proximal raphe fissure has characteristic anchor-shaped endings. Stigmata are lacking. Striae are open foramina, sometimes obscured externally to give the appearance of "ghost" striae. Apical pore fields, septa, and pseudosepta are all wanting. Kociolek *et al.* (1988), Lange-Bertalot (1995), and Kociolek and Kingston (1999) document the ultrastructure of this genus.

Gomphosphenia is a small genus, relatively recently separated from *Gomphonema* (Lange-Bertalot, 1995).

Two species, *G. grovei* and *G. lingulataeforme*, have been collected from widely distributed benthic habitats of fresh waters of North America. Wallace (1960) documented variation in a taxon called *Gomphonema brasiliense*, which is now considered a species of *Gomphosphenia*.

Rhoicosphenia Grunow 1860 (Fig. 3I–K)

Valves are heterovalvate and flexed about the transapical axis, in girdle view. The concave valve possesses a raphe that extends almost the length of the valve. The convex valve possesses short raphe branches near apices only, which may be difficult to discern. Asymmetrical to the transapical axis and symmetrical to apical axis in valve view. The raphe is straight, and the proximal ends are dilated. Cells may be free or attached to surfaces by mucilage stalks.

A single species, *R. abbreviata*, is extant in the freshwater flora. It is widely distributed and common across North America. An extinct species swarm is found in diatomite deposits in western North America (Schmidt, 1899).

V. GUIDE TO LITERATURE FOR SPECIES IDENTIFICATION

The following is a list of key references that deal with species found in North America. Each contains citations to older literature and resources for other continents. General references: Hustedt (1930); Hustedt (1927–1930); Krammer and Lange-Bertalot (1986, 1991b); Lange-Bertalot and Metzeltin (1996); Patrick and Reimer (1966, 1975).

Taxon-specific references:
1. *Actinella*—Kociolek *et al.* (1997)
2. *Amphora*—Stoermer and Yang (1971)
3. *Encyonema*—Krammer (1997a, b)
4. *Cymbella*—Krammer (1997a, b)
5. *Reimeria*—Kociolek and Stoermer (1987)
6. *Gomphoneis*—Kociolek and Rosen (1984); Kociolek and Stoermer (1986, 1988a, b, 1989)
7. *Gomphonema*—Wallace and Patrick (1950); Wallace (1960); Kalinsky (1984); Czarnecki and Blinn (1979); Hohn (1959); Kociolek and Stoermer (1987b, 1991); Kociolek and Lowe (1984); Kociolek and Kingston (1999)
8. *Gomphosphenia*—Wallace (1960); Kociolek *et al.* (1988); Kociolek and Kingston (1999)
9. *Rhoicosphenia*—Lange-Bertalot (1980).

ACKNOWLEDGMENTS

We thank R. Sheath and E. F. Stoermer for comments and suggestions that greatly improved this chapter.

LITERATURE CITED

Alles, E., Nörpel-Schempp, M., Lange-Bertalot, H. 1991. Taxonomy and ecology of characteristic *Eunotia* species in headwaters with low electric conductivity. Beihefte zur Nova Hedwigia 53:171–213.

Archibald, R. E. M., Barlow, D. J. 1983. On the raphe ledge in the genus *Amphora* (Bacillariophyta). Bacillaria 6:257–266.

Archibald, R. E. M., Schoeman, F. R. 1984. *Amphora coffeaeformis* (Agardh) Kützing: A revision of the species under light and electron microscopy. South African Journal of Botany 3:83–102.

Bessey, C. E. 1899. The modern conception of the structure and classification of diatoms. Transactions of the American Microscopical Society 21:61–85.

Cholnoky, B. J. 1965. The relationship between algae and the chemistry of natural waters. Council for Scientific and Industrial Research, Special Report 129:215–225.

Cleve, P. T. 1894. Synopsis of the naviculoid diatoms, Part 1. Kongliga Svenska Vetenskaps-Akademiens Handlingar 26:1–194.

Cox, E. J. 1996. Identification of freshwater diatoms from live material. Chapman & Hall, London, 158 p.

Czarnecki, D. B. 1995. Additions and confirmations to the algal flora of Lake Itasca (MN) State Park. III. The intramucilaginous diatom flora of the colonial peritrich ciliate, *Ophrydium versatile* (Ophrydiidae), in: Kociolek, J. P., Sullivan, M. J. Eds., A century of diatom research in North America. A tribute to the distinguished careers of Charles W. Reimer and Ruth Patrick. Koeltz Scientific Books, Champaign, pp. 183–194.

Czarnecki, D. B., Blinn, D. W. 1979. Observations on southwestern diatoms. II. *Caloneis latiuscula* var. *reimeri*, *Cyclotella pseudostelligera* f. *parva* and *Gomphonema montezumense* n. sp., new taxa from Montezuma Well National Monument. Transactions of the American Microscopical Society 98:110–114.

Dawson, P. A. 1972. Observations on the structure of some forms of *Gomphonema parvulum* Kütz. I. Morphology based on light microscopy, transmission and scanning electron microscopy. British Phycological Journal 7:255–271.

Dawson, P. A. 1973a. The morphology of the siliceous components of *Didymosphenia geminata*. British Phycological Journal 8:65–78.

Dawson, P. A. 1973b. Further observations on the genus *Didymosphenia*. British Phycological Journal 8:197–210.

Dawson, P. A. 1973c. Observations on some species of the diatom genus *Gomphonema* C. A. Agardh. British Phycological Journal 8:413–423.

Dawson, P. A. 1974. Observations on diatom species transferred from *Gomphonema* C. A. Agardh to *Gomphoneis* Cleve. British Phycological Journal 9:75–82.

Edlund, M. B., Stoermer, E. F. 1999. Taxonomy and morphology of *Amphora calumetica* (B.W. Thomas ex Wolle) Perag., an epipsammic diatom from post-Pleistocene large lakes, in: Mayama, S., Idei, M., Koizumi, I., Eds., Proceedings of the XIVth international diatom symposium. Koeltz Scientific Books, Koenigstein, pp. 65–76.

Gaul, U., Geissler, U., Henderson, M., Mahoney, R., Reimer, C. W. 1993. Bibliography on the fine-structure of diatom frustules (Bacillariophyceae). Proceedings of the Academy of the Natural Sciences Philadelphia 144:69–238.

Geitler, L. 1927. Die Reduktionsteilung und Kopulation von *Cymbella lanceolata*. Archiv für Protistenkunde 58:465–507.

Geitler, L. 1932. Der Formwechsel der pennaten Diatomeen (Kieselalgen). Archiv für Protistenkunde 78:1–226.

Geitler, L. 1951a. Zelldifferenzierung bei der Gametenbildung und Ablauf der Kopulation von *Eunotia*. Biologisches Zentralblatt 70:385–390.

Geitler, L. 1951b. Kopulation und Formwechsel von *Eunotia arcus*. Österreichische Botanische Zeitschrift 98:293–337.

Geitler, L. 1951c. Die Stellungen der Kopulationpartner und der Auxosporen bei *Gomphonema*-Arten. Abhandlungen der Mathematisch-Naturwissenschaftlichen Klasse, Akademie der Wissenschaften Lit. Mainz 6:217–225.

Geitler, L. 1956. Autogamie, Geschlechtbestimmung und Pyknose von Gonenkernen bei *Cymbella aspera*. Planta 47:393–396.

Geitler, L. 1967. Paarung und Auxosporenbildung bei *Cymbella*. Österreichische Botanische Zeitschrift 114:484–489.

Geitler, L. 1973a. Auxosporenbildung und Systematik bei pennaten Diatomeen und die Cytologie von *Cocconeis*-Sippen. Österreichische Botanische Zeitschrift 122:299–321.

Geitler, L. 1973b. Zur lebensgeschichte und Morphologie pennater Diatomeen. I. Allogamie bei *Gomphonema constrictum* var. *capitatum* (Ehr.) Cleve. Österreichische Botanische Zeitschrift 122:35–49.

Geitler, L. 1973c. Bewegungs- und Teilungverhalten der Chromatophoren von *Eunotia pectinalis* var. *polyplastidica* und anderer *Eunotia*-Arten bei der Zellteilung. Österreichische Botanische Zeitschrift 122:185–194.

Geitler, L. 1975. Über bau Algenflora der Gallertkolonien des Ciliaten *Ophridium versatile*. Archiv für Hydrobiologie 76:24–32.

Geitler, L. 1981. Die Lage des Chromatophors in Beziehung zur Systematik von *Cymbella*-Arten (Bacillarophyceae). Plant Systematics and Evolution 138:153–156.

Geitler, L., Mack, B. 1953. Die Achsenlagen der Auxosporen und erstlingzellen bei der Diatomee *Cymbella*. Österreichische Botanische Zeitschrift 100:261.

Hasle, G. R. 1973. The "mucilage pore" of pennate diatoms. Beihefte zur Nova Hedwigia 45:167–186.

Hohn, M. H. 1959. Variability in three species of *Gomphonema* undergoing auxospore formation. Proceedings of the Academy of Natural Sciences of Philadelphia 316:1–7.

Hufford, T. L., Collins, G. B. 1972a. Some morphological variations in the diatom *Cymbella cistula*. Journal of Phycology 8:192–195.

Hufford, T. L., Collins, G. B. 1972b. The stalk of the diatom *Cymbella cistula*. Journal of Phycology 8:208–210.

Hustedt, F. 1926. Untersuchungen über den Bau der Diatomeen. I. Raphe und Gallertporen der Eunotioideae. Berichte der Deutschen Botanischen Gesellschaft 44:142–150.

Hustedt, F. 1930. Bacillariophyta, *in*: Pascher, A. Ed., Süsswasserflora von Mitteleuropa. Vol. 10. Fischer, Jena, 466 p.

Hustedt, F. 1952. Neue und wenig bekannte Diatomeen. III. Phylogenetische Variationen bei den rhaphidioiden Diatomeen. Berichte der Deutschen Botanischen Gesellschaft 65:133–144.

Kalinsky, R. G. 1984. Notes on Louisiana diatoms. III. Some new, rare and interesting diatoms from northwestern Louisiana, *in*: Proceedings of the 7th International diatom symposium. Koeltz, Koenigstein, pp. 299–306.

Karayeva, N. I., Maggerramova, N. R., Rhazeva, S. G. 1984. Morphology of the diatom frustule of the genus *Amphora* based on the data of electron microscopy. Botanicheskii Zhurnal 69:492–497.

Karsten, G. 1928. Bacillariophyta (Diatomeae), *in*: Engler, A., Prantl, K., Eds., Die naturlichen Pflanzenfamilien. Vol. 2. Englemann, Leipzig, pp. 105–303.

Kingston, J. C. 1978. Morphological variation of *Cymbella delicatula* and *C. hustedtii* from northern Lake Michigan. Transactions of the American Microscopical Society 97:311–319.

Kingston, J. C., Lowe, R. L., Stoermer, E. F. 1980. The frustular morphology of *Amphora thumensis* (Mayer) A. Cl. from northern Lake Michigan and consideration of its systematic position. Transactions of the American Microscopical Society 99:276–283.

Kobayashi, H., Ando, K., Nagumo, T. 1981. On some endemic species of the genus *Eunotia* in Japan, *in*: Proceedings of the 6th International diatom symposium. Koeltz, Koenigstein, pp. 93–114.

Kociolek, J. P. 1998. Does each genus of diatoms have at least one unique feature?—A reply to Round. Diatom Research 13:177–179.

Kociolek, J. P. 2000. Valve ultrastructure of some Eunotiaceae (Bacillariophyceae), with comments on the evolution of the raphe system. Proceedings of the California Academy of Sciences Occasional Proceedings, 4th Series 52:11–21.

Kociolek, J. P., Kingston, J. C. 1999. Taxonomy, ultrastructure and distribution of gomphonemoid diatoms (Bacillariophyceae: Gomphonemataceae) from rivers of the United States. Canadian Journal of Botany 77:686–705.

Kociolek, J. P., Rosen, B. H. 1984. Observations on North American *Gomphoneis* (Bacillariophyceae). I. Valve ultrastructure of *G. mammilla* with comment on the taxonomic status of the genus. Journal of Phycology 20:361–368.

Kociolek, J. P., Stoermer, E. F. 1986. Observations on North American *Gomphoneis* (Bacillariophyceae). II. Descriptions and ultrastructure of two new species. Transactions of the American Microscopical Society 105:141–151.

Kociolek, J. P., Stoermer, E. F. 1987a. Ultrastructure of *Cymbella sinuata* (Bacillariophyceae) and its allies, and their transfer to *Reimeria*, gen. nov. Systematic Botany 12:451–459.

Kociolek, J. P., Stoermer, E. F. 1987b. Geographic range and variability of the diatom (Bacillariophyceae) *Gomphonema ventricosum* Gregory. Beihefte zur Nova Hedwigia 45:223–236.

Kociolek, J. P., Stoermer, E. F. 1988a. Taxonomy, ultrastructure, and distribution of *Gomphoneis herculeana*, *G. eriense* and closely related species. Proceedings of the Academy of Natural Sciences of Philadelphia 140: 24–97.

Kociolek, J. P., Stoermer, E. F. 1988b. Observations on North American *Gomphoneis* (Bacillariophyceae). IV. Ultrastructure and distribution of *Gomphoneis elegans*. Transactions of the American Microscopical Society 107:386–396.

Kociolek, J. P., Stoermer, E. F. 1988c. A preliminary investigation of the phylogenetic relationships of the freshwater, apical pore field–bearing cymbelloid and gomphonemoid diatoms (Bacillariophyceae). Journal of Phycology 24:377–385.

Kociolek, J. P., Stoermer, E. F. 1989. Phylogenetic relationships and evolutionary history of the diatom genus *Gomphoneis*. Phycologia 28:438–454.

Kociolek, J. P., Stoermer, E. F. 1991. Taxonomy and ultrastructure of some *Gomphonema* Ehrenberg and *Gomphoneis* Cleve taxa from the upper Laurentian Great Lakes. Canadian Journal of Botany 69: 1557–1576.

Kociolek, J. P., Stoermer, E. F. 1993a. Freshwater gomphonemoid diatom phylogeny: Preliminary results. Hydrobiologia 269/270: 31–38.

Kociolek, J. P., Stoermer, E. F., Sicko-Goad, L. 1994. Cytoplasmic fine structure of two *Encyonema* species. Memoirs of the California Academy of Sciences 17:235–246.

Kociolek, J. P., Stoermer, E. F., Edlund, M. A. 1995. Two new freshwater diatom species, *in*: Kociolek, J. P., Sullivan, M., Eds., A century of progress in diatom research in North America. A tribute to the distinguished careers of C. W. Reimer and R. M. Patrick. Koeltz Scientific Books, Champaign, pp. 9–20.

Kociolek, J. P., Rhode, K., Williams, D. M. 1997. Taxonomy, ultrastructure and biogeography of the *Actinella punctata* (Bacillariophyta: Eunotiaceae) species complex. Beihefte zur Nova Hedwigia 65:177–193

Kolbe, R. W. 1956. Zur Phylogenie des Raphe-Organs der Diatomeen: *Eunotia (Amphicampa) eruca* Ehr. Botanische Noten 109:91–97.

Krammer, K. 1979. Zur Morphologie der Raphe bei der Gattung *Cymbella*. Beihefte zur Nova Hedwigia 21:993–1029.

Krammer, K. 1981. Morphologic investigations of valve and girdle of the diatom genus *Cymbella* Agardh. Bacillaria 4:125–146.

Krammer, K. 1982. Valve morphology in the genus *Cymbella*, in: Helmcke, J. G., Krieger, W., Krammer, K. Eds., Micromorphology of Diatom Valves 11: 1–300.

Krammer, K. 1997a. Die cymbelloiden Diatomeen. Eine Monographie der weltweit bekannten Taxa. Teil 1. Allgemeines und *Encyonema* Part. Bibliotheca Diatomologica 36:1–382.

Krammer, K. 1997b. Die cymbelloiden Diatomeen. Eine Monographie der weltweit bekannten Taxa. Teil 2. Encyonema Part., *Encyonopsis* und *Cymbellopsis*. Bibliotheca Diatomologica 37:1–469.

Krammer, K., Lange-Bertalot, H. 1986. Bacillariophyceae. Part 1. Naviculaceae, in: Ettl, H., Gerloff, J., Heynig, H., Mollenhauer, D., Eds., Süsswasserflora von Mitteleuropa, Vol. 2. Fischer, Stuttgart, 876 p.

Krammer, K., Lange-Bertalot, H. 1991. Vol. 2. Bacillariophyceae. Part 3. Centrales, Fragilariaceae, Eunotiaceae, in: Ettl, H., Gerloff, J., Heynig, H., Mollenhauer, D., Eds. Süsswasserflora von Mitteleuropa, Vol. 2. Fischer, Stuttgart, 576 p.

Lange-Bertalot, H. 1980. Ein Beitrag zur Revision der Gattungen *Rhoicosphenia* Grun., *Gomphonema* C. Ag., *Gomphoneis* Cl. Botanische Noten 133:585–594.

Lange-Bertalot, H. 1993. 85 Neue Taxa und über 100 weitere neu definierte Taxa ergänzend zur Süsswasserflora von Mitteleuropa Vo. 2/1–4. Bibliotheca Diatomologica 27:1–454.

Lange-Bertalot, H. 1995. *Gomphosphenia paradoxa* nov. sp. et nov. gen. und Vorschlag zur Lösung taxonomischer Probleme infolge eines veränderten Gattungskonzepts von *Gomphonema* (Bacillariophyceae). Beihefte zur Nova Hedwigia 60:241–252.

Lange-Bertalot, H, Metzeltin, D. 1996. Ecology–diversity–taxonomy. Indicators of oligotrophy. Iconographia Diatomologica, Vol. 2. Koeltz Scientific Books, Koenigstein, 390 p.

Lee, K., Round, F. E. 1987. Studies on freshwater *Amphora* species. I. Amphora ovalis. Diatom Research 2:193–203.

Lee, K., Round, F. E. 1988. Studies on freshwater *Amphora* species. II. *Amphora copulata* (Kütz.) Schoeman and Archibald. Diatom Research 3:217–225.

Lowe, R. L. 1974. Environmental Requirements and Pollution Tolerance of Freshwater Diatoms. EPA-670/4-74-005. Environmental Protection Agency, Cinncinnati, 334 p.

Lowe, R. L., Kociolek, J. P. 1984. New and rare diatoms from Great Smoky Mountains National Park. Beihefte zur Nova Hedwigia 39:465–476.

Mann, D. G. 1982a. Structure, life history and systematics of *Rhoicosphenia*. I. The vegetative cell of *Rhoicosphenia curvata*. Journal of Phycology 18:162–176.

Mann, D. G. 1982b. Structure, life history and systematics of *Rhoicosphenia*. II. Auxospore formation and perizonium structure of *Rhoicosphenia curvata*. Journal of Phycology 18:264–274.

Mann, D. G. 1984. Observation on copulation in *Navicula pupula* and *Amphora ovalis* in relation to the nature of diatom species. Annals of Botany (London) 54:429–438.

Mayama, S., Kobayashi, H. 1990. Studies on *Eunotia* species in the classical "Degernäs Material" housed in the Swedish Museum of Natural History. Diatom Research 5:351–366.

Mayama, S., Kobayashi, H. 1991. Observations on *Eunotia arcus* Ehr., type species of the genus *Eunotia*. Japanese Journal of Phycology 39:131–141.

Metzeltin, D., Lange-Bertalot, H. 1995. Kritische Wertung der Taxa in *Didymosphenia* (Bacillariophyceae). Beihefte zur Nova Hedwigia 60:381–406.

Meyer, K. 1929. Über die Auxosporenbildung von *Gomphonema geminatum*. Archiv für Protistenkunde 66:421–435.

Moss, M. O., Gibbs, G., Gray, V., Ross, R. 1978. The presence of a raphe on *Semiorbis hemicyclus* (Ehr.) R. Patr. Bacillaria 1:137–150.

Patrick, R., Hohn, M., Wallace, J. H. 1954. A new method for determining the pattern of the diatom flora. Notulae Naturae of the Academy of Natural Sciences of Philadelphia 259:1–12.

Patrick, R. M., Reimer, C. W. 1966. The diatoms of the United States. Monograph 13. Academy of Natural Sciences of Philadelphia, 688 p.

Patrick, R. M., Reimer, C. W. 1975. The diatoms of the United States. Vol. 2, Part 1. Monograph 13. Academy of Natural Sciences of Philadelphia, 213 p.

Reichardt, E. 1999. Zur Revision der Gattung *Gomphonema*. Die Arten um *G. affine/insigne*, *G. angustatum/micropus*, *G. acuminatum* sowie gomphonemoide Diatomeen aus dem Oberoligozän in Böhmen. Iconographia Diatomologica 8:1–203.

Reichardt, E., Lange-Bertalot, H. 1991. Taxonomische Revision des Artenkomplexes um *Gomphonema angustatum*–*G. dichotomum*–*G. intricatum*–*G. vibrio* und ähnlicher Taxa (Bacillariophyceae). Beihefte zur Nova Hedwigia 53:519–544.

Round, F. E., Crawford, R.M., Mann, D. G. 1990. The diatoms. Biology and morphology of the genera. Cambridge Univ. Press, Cambridge, UK, 747 p.

Sala, S. E., Guerrero, J. M., Ferrario, M. E. 1993. Redefinition of *Reimeria sinuata* (Gregory) Kociolek and Stoermer and recognition of *Reimeria uniseriata* nov. sp. Diatom Research 8:439–446.

Schmidt, M. 1899. Tafel 215, in: A. Schmidt et al., Eds., Atlas der Diatomaceenkunde. Reisland, Leipzig.

Selva, J. 1981. Tertiary freshwater diatoms from the Ogallala of western Kansas. Proceedings of the Iowa Academy of Sciences 88:85–90.

Skvortzow, B.W., Meyer, K. 1928. A contribution to the diatoms of Baikal Lake. Proceedings of the Sungaree River Biological Station 1:1–55.

Smol, J. P. 1990. Paleolimnology: recent advances and future challenges. Memorie Istituto Italiano di Idrobiologia 47:253–276.

Stoermer, E. F., Yang, J. J. 1971. Contributions to the diatom flora of the Laurentian Great Lakes. I. New and little-known species of *Amphora* (Bacillariophyta, Pennatibacillariophyceae). Phycologia 10:397–409.

Stoermer, E. F., Qi, Y., Ladewski, T. B. 1986. A quantitative investigation of shape variation in *Didymosphenia* (Lyngbye) M. Schmidt (Bacillariophyta). Phycologia 25: 494–502.

Ueyama, S., Kobayashi, H. 1988. Two *Gomphonema* species with strongly capitate apices: *G. sphaerophorum* and *G. pseudosphaerophorum* sp. nov., in: Proceedings of the 9th international diatom symposium Koeltz, Koenigstein, pp. 449–458.

Van Landingham, S. L. 1967. A new species of *Gomphonema* from Mammoth Cave, Kentucky. International Journal of Speleology 2:405–406.

Van Landingham, S. L. 1969. Catalogue of the fossil and Recent genera and species of diatoms and their synonyms. Part III. *Coscinophaena* through *Fibula*. Cramer, Weinheim, Germany, pp. 1087–1756.

Van Landingham, S. L. 1978. Catalogue of the fossil and Recent genera and species of diatoms and their synonyms. Part IV. *Fragilaria* through *Naunema*. Cramer, Weinheim, Germany, pp. 1757–2385.

Vyverman, W., Sabbe, K., Mann, D. G., Kilroy, C., Vyverman, R., Vanhutte, K., Hodgson, D. 1998. *Eunophora* gen. nov. (Bacillariophyta) from Tasmania and New Zealand: Description and comparison with *Eunotia* and amphoroid diatoms. European Journal of Phycology 33:95–111.

Wallace, J. H. 1960. New and variable diatoms. Notulae Nature 331:1–8.

Wallace, J. H., Patrick, R. M. 1950. A consideration of *Gomphonema parvulum* Kütz. Butler University Botanical Studies 9:227–234.

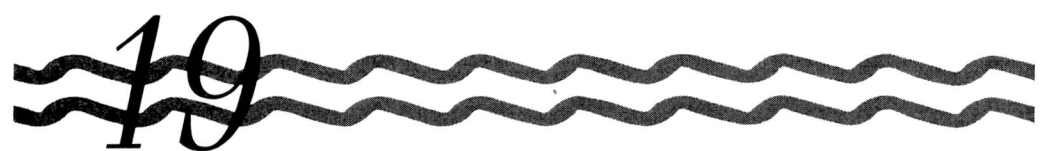

KEELED AND CANALLED RAPHID DIATOMS

Rex L. Lowe
Biological Sciences
Bowling Green State University
Bowling Green, Ohio 43403

and
University of Michigan Biological Station
Pellston, Michigan 49769

I. Introduction
II. Diversity and Morphology
III. Ecology and Distribution
IV. Collection and Preparation for Identification
V. Keys and Descriptions of Genera
   A. Key to Orders
   B. Key to North American Genera of Bacillariales
   C. Key to North American Genera of Rhopalodiales
   D. Key to North American Genera of Surirellales
   E. Descriptions of Genera: Bacillariales
   F. Descriptions of Genera: Rhopalodiales
   G. Descriptions of Genera: Surirellales
VI. Guide to Literature for Species Identification
Literature Cited

I. INTRODUCTION

The diatoms presented in this chapter belong to the class Bacillariophyceae and in one of three orders, Bacillariales, Rhopalodiales, and Surirellales. All of the species may be motile and possess relatively complex and advanced raphe systems that occupy specialized siliceous structures within the cell wall. In both the Bacillariales and Surirellales the raphe occupies a keel that is elevated from the cell surface (Fig. 1). The presence of a "keeled" raphe in these diatom genera effectively elevates this locomotory structure from the valve surface allowing more intimate contact between the raphe and substrata that may have fine-scale irregularities, such as unconsolidated fine sediments like silt (Round, 1981). Thus, these diatoms are often found in the epipelon. The keeled raphe enables these organisms to move efficiently through epipelic habitats, and species of this group often reach their maximum abundance in the epipelon (see Chapter 2 for details on habitats). Epipelic habitats are most abundant in slow-moving streams or in lentic environments such as lakes, ponds, and wetlands (Burkholder, 1996; Goldsborough and Robinson, 1996). Epipelic communities, containing populations of diatoms from these two families, can also be found in quiet pools of swift streams where slower current velocities allow fine sediments to fall from suspension.

The Bacillariales and Surirellales include some of the most speciose genera (*Nitzschia* with > 600 described species; van Landingham, 1978), with some of the most difficult species to identify. Although hundreds of species of the genus *Nitzschia* have been described thus far, many appear to be synonyms, as new names are attributed to previously described taxa. This genus has received the attention of several taxonomists (Lange-Bertalot and Krammer, 1993; Krammer and Lange-Bertalot, 1988), but remains a difficult genus to work with at the species level.

Genera belonging to the order Rhopalodiales, *Epithemia* and *Rhopalodia*, are asymmetric to the apical axis in valve view and possess a raphe that occupies a canal connecting with the interior of the cell via round or oval holes (portulae) (Round *et al.*, 1990). In addition, the raphe of some species of *Rhopalodia* is also slightly elevated from the valve surface into a keel

(Round *et al.*, 1990). Cells are usually relatively large and heavily silicified. Most, and perhaps all, species in this order contain endocellular symbiotic cyanobacteria (blue green algae) (Drum and Pankratz, 1965; Geitler, 1977). Diatom cells containing endosymbiotic cyanobacteria have been shown to be capable of fixing nitrogen (Floener and Bothe, 1980). DeYoe *et al.* (1992) suggested that the endosymbionts have been modified into nitrogen-fixing "organelles" and their intracellular density is partially a function of the diatom's nitrogen needs.

II. DIVERSITY AND MORPHOLOGY

Diatom orders and genera considered in this chapter are listed in Table I. Diatoms in the orders Bacillariales, Surirellales, and Rhopalodiales share common pigments (chlorophylls and carotenoids) (Stoermer and Julius, 2002) with other diatoms (Bacillariophyta). Most taxa in this group are benthic in habitat and may inhabit dimly lighted microhabitats in deep lakes or turbid rivers and frequently maintain increased quantities of accessory pigments in such situations. Deep-living members of the Surirellaceae in particular are often deep chocolate brown in color (Round *et al.*, 1990).

TABLE I Keeled or Canal-Bearing Freshwater Diatoms of North America

Class	Order	Family	Genus
Bacillariophycidae			
	Bacillariales		
		Bacillariaceae	
			Bacillaria Gmelin
			Cylindrotheca Rabenhorst
			Cymbellonitzschia Hustedt
			Denticula Ehrenberg
			Hantzschia Grunow
			Nitzschia Hassall
			Tryblionella W. Smith
	Surirellales		
		Entomoneidaceae	
			Entomoneis Ehrenberg
		Surirellaceae	
			Campylodiscus Ehrenberg *ex* Kützing
			Cymatopleura W. Smith
			Stenopterobia Brébisson *ex* Van Heurck
			Surirella Turpin
	Rhopalodiales		
		Rhopalodiaceae	
			Epithemia Brébisson
			Rhopalodia O. Müller

Taxa within these three orders vary greatly in size. For example, the smallest members of the genus *Nitzschia* may be as small as 5 μm in length (*Nitzschia inconspicua*) whereas some species may be over two orders of magnitude larger (*Nitzschia scalaris*). Some species of *Surirella* and *Cymatopleura* (Surirellaceae) are among the most massive of freshwater diatoms. For example, *Cymatopleura solea* may constitute more than 50 times the biovolume of some co-occurring species (Lowe and Pan, 1996). It may be important to bear these size differences in mind when analyzing data on algal community structure, because a taxon comprising just a few percent of the community cell numbers may comprise a majority of the community biomass.

All diatoms within these three orders possess a raphe and most species are highly motile. Genera within Bacillariales (*Bacillaria*, *Cylindrotheca*, *Cymbellonitzschia*, *Denticula*, *Hantzschia*, *Nitzschia*, and *Tryblionella*) have a raphe on a keel that is elevated above the surface of the valve face (Fig. 1). Heavy siliceous ribs (fibulae) are regularly spaced along the keeled raphe where it opens into the cell interior (Round *et al.*, 1990). These fibulae vary in length among species and have been referred to as keel punctae or carinal punctae in older literature (Fig. 2) (Patrick and Reimer, 1966; Anonymous, 1975). The external raphe opening may be continuous from one pole of the diatom to the opposite pole or the raphe may be interrupted in the center of the valve. Genera within Surirellaceae (Surirellales), which include *Campylodiscus*, *Cymatopleura*, *Stenopterobia*, and *Surirella*, are generally heavily silicified and all have raphe systems that occupy both margins of each valve (Fig. 3). Thus, the raphe runs around the entire perimeter of each valve rendering them highly motile with substratum contact in almost any position. Fibulae associated with the raphes of species in the Surirellaceae are usually much broader than those in members of the Bacillariales. *Entomoneis*, in the family Entomoneidaceae (Surirellales) has a keeled raphe that is in two arches that are elevated a great distance from the valve surface, with each arch beginning at either end of the valve and terminating near the center of the valve. In addition, the entire frustule is often twisted about the apical axis, resulting in a relatively complex morphology and assuring its contact with substrata in the epipelic community (see Sec. V, Fig. 22).

Much splitting and revision of taxonomic groups within the Bacillariales and Surirellales has occurred, particularly at the generic level (Round *et al.*, 1990). Some of the proposed changes are a result of re-erecting older names that had been synonymized (e.g., *Tryblionella*), whereas other genera have been split from the

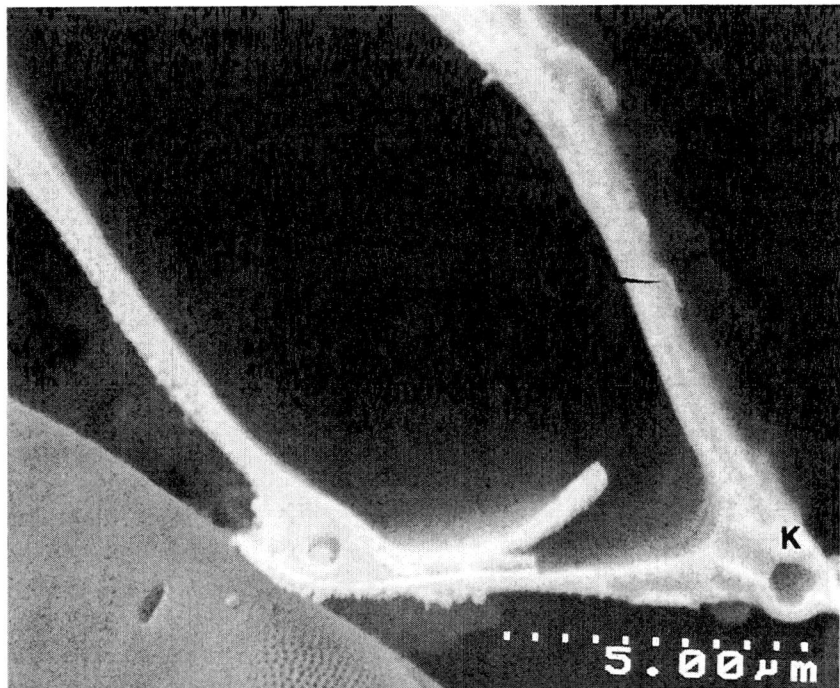

FIGURE 1 Scanning electron micrograph of a cross section of a frustule of *Nitzschia* sp. Note the keel (K) projecting from the margin of the frustule.

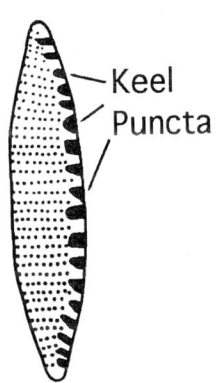

FIGURE 2 Keel puncta of *Nitzschia amphibia*, 2000×.

large genus *Nitzschia* based on details revealed largely by electron microscopy or by consideration of cytoplasmic characters (*Psammodictyon*, a marine genus). Genera within Rhopalodiales (*Epithemia* and *Rhopalodia*) have the raphe located in a canal that opens into the interior of the cell via round or oval pores (portulae). *Epithemia* and *Rhopalodia* also have internal fibulae, which usually cross the entire valve and appear as heavy lines when viewed under the light microscope (Fig. 4). *Rhopalodia* may have its canalled raphe slightly elevated into a keel (Round et al., 1990). Cells of diatoms belonging to the Rhopalodiales are usually heavily silicified and robust and although they may be highly motile they often adnate on aquatic vegetation (epiphytic) (Marks and Lowe, 1993; Lowe, 1996).

III. ECOLOGY AND DISTRIBUTION

Species in the orders Bacillariales, Surirellales and Rhopalodiales are found worldwide in a variety of freshwater systems, including lakes, rivers, and wetlands (Stevenson et al., 1996). The Bacillariales and Surirellales also have estuarine and marine littoral species. The microhabitats of genera in the Bacillariales and Surirellales are primarily the epipelon and endopelon where the presence of the raphe in an elevated keel allows cells to move more efficiently over and through unconsolidated sediment. Substantial populations of *Nitzschia* can be found several hundred micrometers beneath the sediment (Fig. 5). In undisturbed sediments, populations of *Nitzschia* can accumulate to great numbers (Fig. 6), resulting in a golden brown color on the sediment surface. Many species of the Surirellales can also be found at considerable depth associated with lake sediments, especially species of *Campylodiscus*, *Surirella*, *Entomoneis*, *Tryblionella*, and *Cymatopleura*. Some of the benthic species of *Surirella* and *Nitzschia* are large enough to support

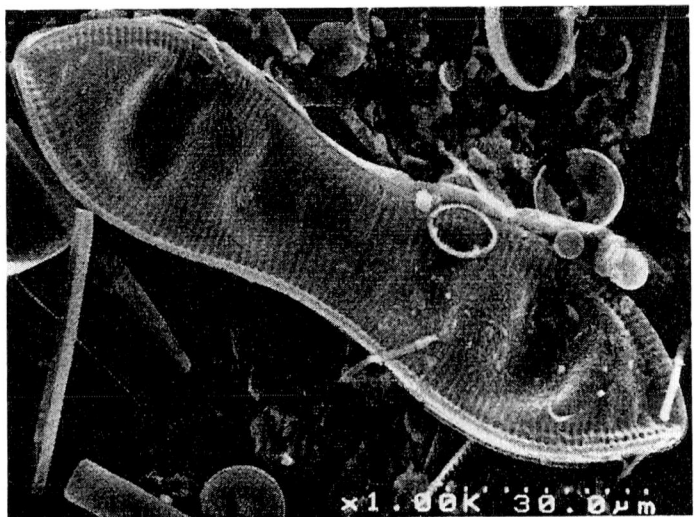

FIGURE 3 Scanning electron micrograph of *Cymatopleura solea* with raphe in a wing around the entire valve margin.

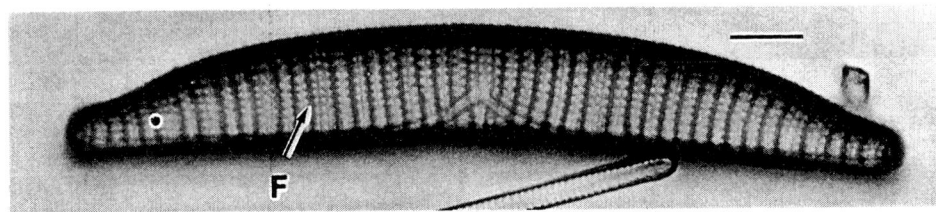

FIGURE 4 Light micrograph of *Epithemia turgida* illustrating the fibulae (F) characteristic of genera in the Rhopalodiales. Scale the bar = 10 μm.

other epiphytic diatoms such as *Synedra parasitica* and *Amphora perpusilla* (Belanger *et al.*, 1985).

Several species of Bacillariales, Surirellales, and Rhopalodiales also occupy microhabitats other than those in the periphyton. Some species in the genus *Nitzschia*, particularly *N. acicularis* and *N. holsatica*, are lightly silicified and may be planktonic (Hustedt, 1942). Other species of keeled diatoms may be encountered in the seston of lakes and rivers (resuspended) but are rarely truly planktonic and are more properly tychoplanktonic, such as *Cymatopleura solea* and *Entomoneis ornata*. One genus, *Hantzschia*, contains a species (*H. amphioxys*), that is universally present in soils around the world (Round *et al.*, 1990) and has been reported as part of the soil flora on every continent.

Stenopterobia, a genus in the Surirellales, is almost entirely restricted to waters of low pH. Although a relatively small genus, its presence in diatom assemblages is an indication of acidic environments (Round *et al.*, 1990). *Stenopterobia sigmatella*, a narrow sigmoid species is common in bogs and fens in North America (Stokes and Yung, 1986).

Many species of the genus *Nitzschia* are recognized as indicators of organic enrichment or pollution of the water in which they are found (Lowe, 1974). Cholnoky (1968) has attributed this pollution-tolerant behavior to the fact that many *Nitzschia* species are nitrogen heterotrophs and capitalize on organic nitrogen molecules in the water. Thus, they are found in largest numbers where organic nitrogen is abundant, areas of organically polluted water, and in this sense might be thought of as "pollution dependent" rather than "pollution tolerant." *Nitzschia palea* is one of the most common and pollution-dependent species in this genus (Palmer, 1969). Tuchman (1996) reported that *N. palea* was able to utilize 21 different organic substrates.

Members of the Rhopalodiales, *Epithemia* and *Rhopalodia*, contain endosymbiotic cyanobacteria first described by Geitler (1977) as "Spharoidkörper." These were shown to be coccoid cyanobacteria (Drum and Pankratz, 1965) that are capable of nitrogen fixation within the diatom host (Floener and Bothe, 1980). Fairchild and Lowe (1984) and DeYoe *et al.*, (1992) observed that these endosymbiotic cyanobacteria

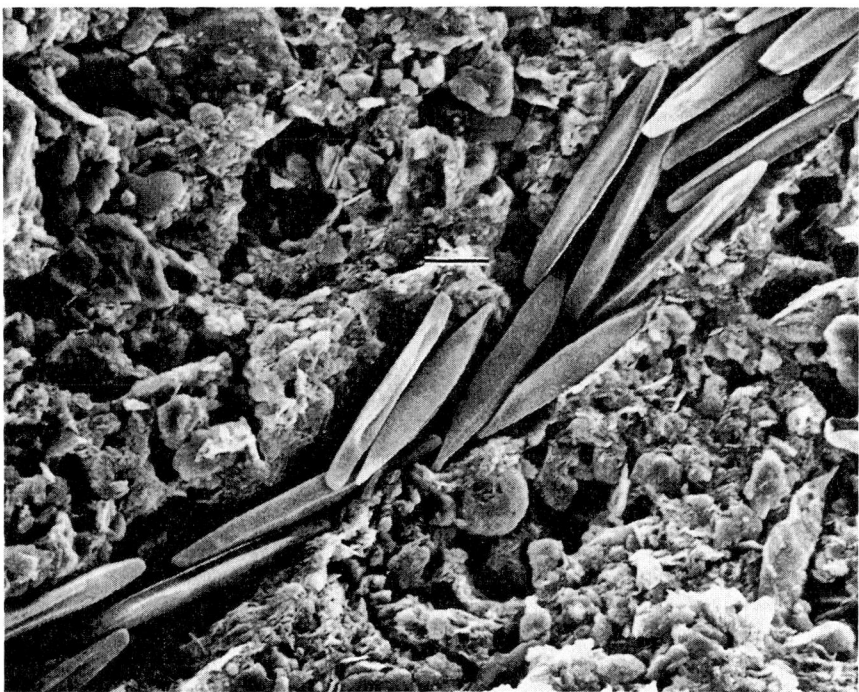

FIGURE 5 Scanning electron micrograph of freeze-fractured stream sediments with a colony of *Nitzschia filiformis* in a mucilaginous tube several 100 μm below the sediment surface. Scale bar = 10 μm.

FIGURE 6 Scanning electron micrograph of freeze-fractured stream sediments with a colony of *Nitzschia* spp. in a layer on the sediment surface. Scale bar = 100 μm.

confer a competitive advantage on species of *Epithemia* and *Rhopalodia* in microhabitats with inadequate nitrogen supplies and it has been suggested that these highly modified endosymbionts may be on the evolutionary pathway toward a nitrogen-fixing organelle in this family of diatoms (DeYoe *et al.*, 1992). Species in the genera *Epithemia* and *Rhopalodia* are uncommon among the diatoms in their ability to fix nitrogen with endosymbiotic blue green algae (cyanobacteria). *Epithemia* and *Rhopalodia* are most common in alkaline water, while occupying microhabitats that are relatively poor in quantities of fixed nitrogen (NO_3, NH_4) (Fairchild *et al.*, 1985). Such microhabitats are often in nitrogen-poor lakes and streams (Fairchild and Lowe, 1984; Bahls and Weber, 1988; Peterson and Grimm, 1992), usually on submerged plants (macrophyton) that provide colonizable surface area for diatoms and other microalgae (epiphyton) (Power, 1990) (Fig. 7). These host plants may be increasing PO_4 availability to epiphytes (Burkholder *et al.*, 1990), thus decreasing the local N:P ratio and conferring a competitive advantage on nitrogen-fixing epiphytes, such as heterocyst-bearing blue green algae and diatoms belonging to the Rhopalodiales (DeYoe *et al.*, 1992). DeYoe *et al.* (1992) have shown that the number of endosymbionts per cell is a function of nitrogen availability in the microenvironment of the diatom cell.

IV. COLLECTION AND PREPARATION FOR IDENTIFICATION

The reader is referred to Chapter 15 of this book for collection and preparation methodology useful for all diatoms. The keeled diatoms, however, warrant some special attention due to their predominance in the epipelon. If care is taken in the collection of epipelic algae, the investigator can be rewarded with collections rich in diatoms with little sediment or detritus. The tools of choice for collecting are an epipelic periphyton sampler (turkey baster) or micropipette with a rubber bulb for sucking up specimens. The epipelic periphyton sampler is a relatively coarse tool given the size of diatoms but if the sampler is deftly operated at the sediment surface and not thrust deeply into the sediment rich collections can be obtained. The micropipette is a better tool for fine-scale sampling where one is interested in different microhabitats and perhaps different clones of epipelic diatoms on the sediment. Careful observation in the field often reveals subtle patterns of algal distribution through slight changes in color, texture, and oxygen bubble distribution.

Another habitat that should not be ignored is the endopelon. This community is best collected with a coring device and should be studied with scanning electron microscopy to observe patterns of distribution (Fig. 5) (Greenwood *et al.*, 1999).

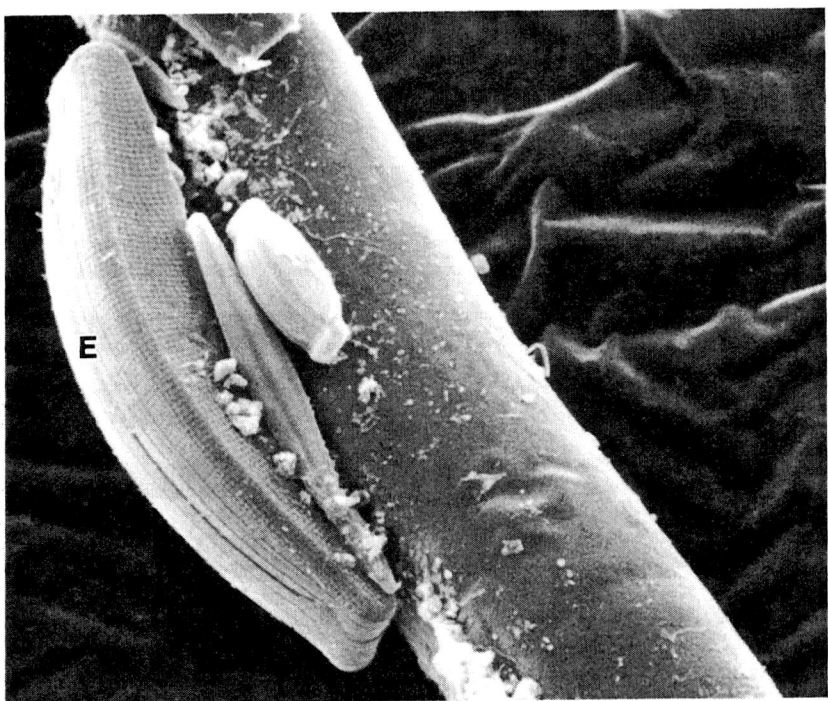

FIGURE 7 Scanning electron micrograph of *Epithemia* sp. (E) epiphytic on the filamentous green alga *Dichotomosiphon*, 1000×.

V. KEYS AND DESCRIPTIONS OF GENERA

See Figures 8–37.

A. Key to Orders

1a.	Two keels with raphes on each valve, located at the valve margin (Fig. 3)	Surirellales
1b.	One raphe per valve, located either in a keel or in a canal	2
2a.	Cells with raphe in a keel, most taxa symmetric to the apical axis, endosymbiotic cyanobacteria lacking	Bacillariales
2b.	Cells with raphe in a canal, always asymmetric to the apical axis and bearing endosymbiotic cyanobacteria	Rhopalodiales

B. Key to North American Genera of Bacillariales

1a.	Cells gregarious in large raftlike colonies; colonies dynamic with cells sliding back and forth—first stretching out in an elongated linear colony, then sliding back into a tabular colony of cells touching along the entire cell length (Figs. 9, 10)	Bacillaria
1b.	Cells usually singular; if colonial, colony not dynamic as above	2
2a.	Cells weakly or strongly lunate, asymmetric to the apical axis	3
2b.	Cells linear, sygmoid, or some other shape	4
3a.	Cells elongate and slightly lunate; dorsal margin slightly convex, ventral margin slightly concave; raphe on concave margin on both valves (Fig. 8)	Hantzschia
3b.	Cell strongly lunate with a strongly convex dorsal margin and a flat ventral margin; raphe on same margin of each valve (Fig. 12)	Cymbellonitzschia
4a.	The two valves of the frustule twisted around each other, resulting in the raphe systems wrapping around the cell in a spiral pattern; in epipelic habitats, moving in a screwlike fashion through sediments (Fig. 11)	Cylindrotheca
4b.	Frustule not twisted	5
5a.	Cells with fibulae extending entirely across valve face appearing like thick internal costa; raphe submarginal (Fig. 13)	Denticula
5b.	Fibulae projecting, but not entirely, across valve face	6
6a.	Cells normally narrow, linear and straight or sygmoid, one plastid in each end of the cell; striae not interrupted by sterna (Figs. 15–21)	Nitzschia
6b.	Cells usually robust and relatively broad; may be elliptical, linear, or panduriform; valve surface with an undulation along the apical axis; one plastid in each end of the cell; usually epipelic (Figs. 14, 34)	Tryblionella

C. Key to North American Genera of Rhopalodiales

1a.	Cells strongly lunate but not canoe-shaped; raphe at distal ends ventral becoming more dorsal toward the proximal ends often forming a peak in the center of the valve (Fig. 28)	Epithemia
1b.	Valves of frustule lunate but frustule shaped like a canoe; thus, cells normally seen in girdle view; raphe normally along the dorsal margin of the valve (Figs. 27, 37)	Rhopalodia

D. Key to North American Genera of Surirellales

1a.	Frustule with one raphe on each valve; cells with a high narrow keel that occupies an arch on each end of each valve; frustules often twisted about the apical axis, resulting in a cell that appears bilobate (Fig. 22)	Entomoneis
1b.	Frustule with two raphes on each valve; raphes marginal and cells usually relatively large and robust; growing most often on sediments	2
2a.	Valves saddle-shaped (Figs. 29, 30)	Campylodiscus
2b.	Valves may be twisted, but not saddle-shaped	3

3a.	Valves with undulations or waves along valve face; frustule usually robust and heavily silicified (Figs. 3, 31, 32)..........*Cymatopleura*
3b.	Valves lacking undulations along valve face..4
4a.	Valves S-shaped or straight and narrow; specimens usually restricted to acidic habitats (Fig. 33)...........................*Stenopterobia*
4b.	Valves robust, broad, and heavily silicified; may be twisted, isopolar, or heteropolar, but not S-shaped (Figs. 23, 26, 34) ....*Surirella*

E. Descriptions of Genera: Bacillariales

Bacillaria Gmelin (Figs. 9, 10)

Cells of *Bacillaria* are elongate and occur in unique colonies that dynamically change their morphology. The cells are held together along their margins by interlocking ridges and grooves (Fig. 9). Cells in the colony slide back and forth from a side-by-side position to a pole-to-pole position in an oscillating fashion. Individual frustules of *Bacillaria* possess a keeled raphe positioned near the center of each cell (Fig. 10). There are usually two plastids in each cell positioned toward the poles.

Bacillaria paradoxa is a common North American species and is normally found in slightly brackish water and water high in dissolved solids. Colonies are normally epipelic but often become tychoplanktonic.

Cylindrotheca Rabenhorst (Fig. 11)

Frustules of *Cylindrotheca* are relatively long and narrow with attenuated apices. Each frustule is twisted about the apical axis so that the valve and girdle areas are spiraled around the length of the cell. The raphe is fibulate and marginal on each of the twisted valves. Cells have two to several platelike or disk-shaped plastids.

Cylindrotheca is a relatively small genus (four species) found mostly in brackish water or marine habitats. The most widely distributed freshwater species is *C. gracilis*; it occurs primarily in the epipelon of streams of high conductance (Christensen and Reimer, 1968). *Cylindrotheca* can usually be identified under the microscope in living material by its screwlike movement across the substratum. The frustules are lightly silicified and are susceptible to destruction during acid or peroxide cleaning of samples.

Cymbellonitzschia Hustedt (Fig. 12)

Cells of *Cymbellonitzschia* are usually solitary or may form short chains in benthic areas of still water. Each cell has two small platelike plastids positioned on opposite poles of the cell. The valves are asymmetrical to the apical axis and thus have dorsal and ventral margins (Fig. 12). The fibulate raphe may be on the ventral margin of each valve (similar to *Hantzschia*), such as in the most common species (*Cymbellonitzschia diluviana*) but may be on the dorsal margin in other species. At first glance this taxon might be confused with *Cymbella* or *Amphora* due to its characteristic shape, but can quickly be identified as *Cymbellonitzschia* by the presence of the fibulate raphe.

Denticula Ehrenberg (Fig. 13)

Cells of *Denticula* are relatively small and usually solitary, but they may occur in short chains. Each cell contains two plastids on either side of the transapical plane near the center of the cell. Valves are symmetrical to both the apical and transapical axes and are linear to lanceolate with a fibulate raphe on each valve that is slightly eccentric on the valve surface (Fig. 13). Raphes on opposite valves display nitzschioid symmetry.

Denticula is a relatively small genus that is uncommon, but occurs most abundantly in benthic hardwater habitats. This genus had formerly been included in the Epithemiaceae (Patrick and Reimer, 1975). More recently, *Denticula* has been suggested to be more closely related to *Nitzschia* and its allies by virtue of the "nitzschioid symmetry" of its raphe systems (Round *et al.*, 1990). In addition, *Denticula* species in North America have not been recorded with cyanobacterial endosymbionts that appear universally in the Rhopalodiaceae, although Geitler (1977) indicated European species of the genus possess endosymbionts.

Hantzschia Grunow (Fig. 8)

Cells of *Hantzschia* are curved and thus are asymmetric to the apical axis in valve view (Fig. 8). The raphe occupies a keel on the concave (ventral) margin of each valve and its presence is best detected in the light microscope by the appearance of the associated fibulae. The convex (dorsal) margin of each valve lacks a raphe. Plastids may be rounded or lobed, are usually two in number, and are typically ventral.

This is a relatively small genus but one species, *Hantzschia amphioxys*, is present in soil flora worldwide. This species is also a common component of atmospheric algal particulates.

Nitzschia Hassall (Figs. 15–20)

Cells of *Nitzschia* are usually long, straight, and narrow but may be ovoid or even slightly sygmoid.

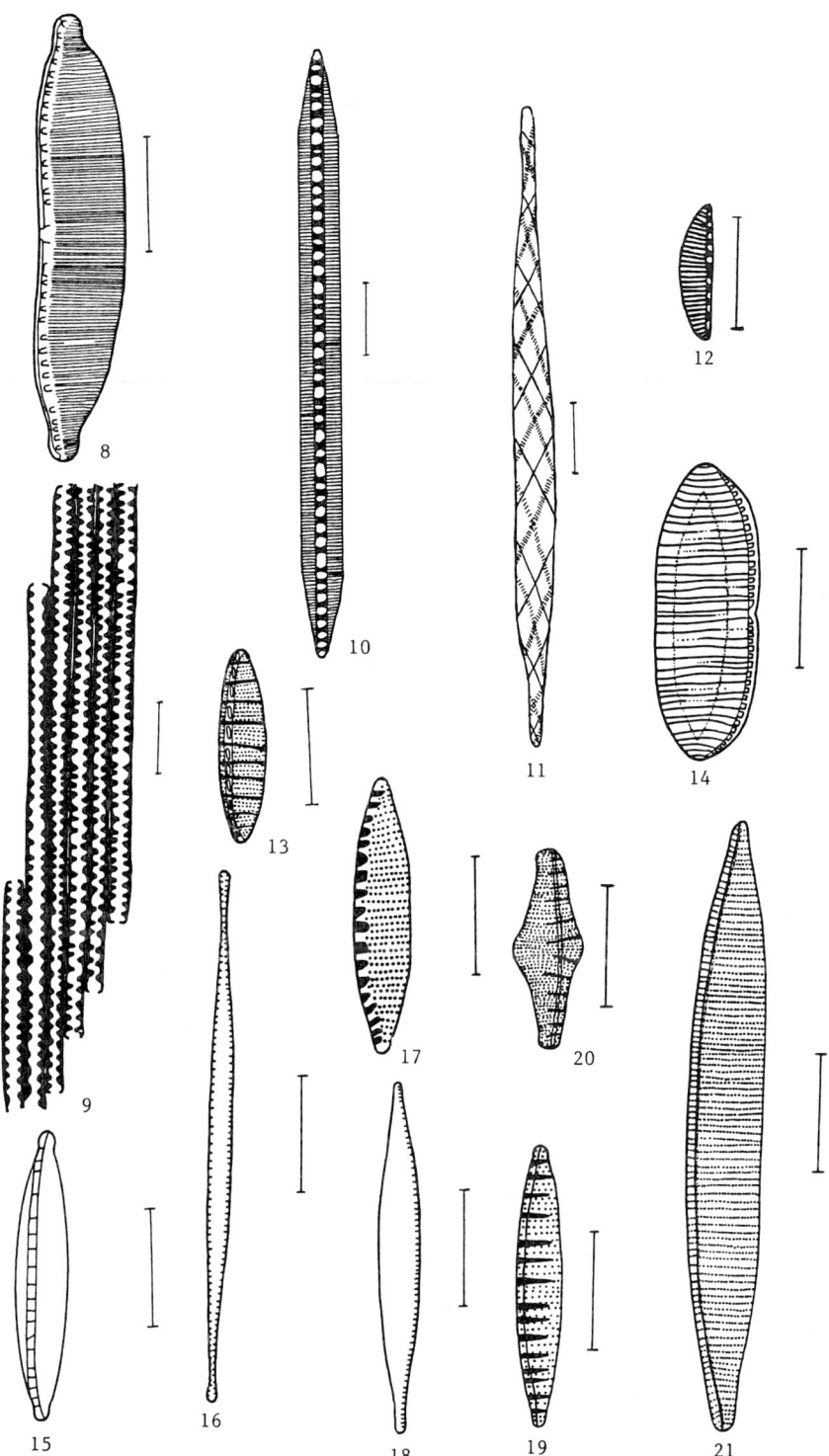

FIGURE 8 *Hantzschia amphioxys*, valve view. Scale bar = 10 µm. FIGURE 9 Colony of *Bacillaria paradoxa*. Scale bar = 10 µm. FIGURE 10 *Bacillaria paradoxa*, valve view. Scale bar = 10 µm. FIGURE 11 *Cylindrotheca gracilis*, valve view. Scale bar = 10 µm. FIGURE 12 *Cymbellonitzschia diluviana*, valve view. Scale bar = 10 µm. FIGURE 13 *Denticula tenuis*, valve view. Scale bar = 10 µm. FIGURE 14 *Tryblionella* sp., valve view. Scale bar = 10 µm. FIGURE 15 *Nitzschia dissipata*, valve view. Scale bar = 10 µm. FIGURE 16 *Nitzschia acicularis*, valve view. Scale bar = 10 µm. FIGURE 17 *Nitzschia amphibia*, valve view. Scale bar = 10 µm. FIGURE 18 *Nitzschia palea*, valve view. Scale bar = 10 µm. FIGURE 19 *Nitzschia denticula*, valve view. Scale bar = 10 µm. FIGURE 20 *Nitzschia sinuata* v. *tabellaria*, valve view. Scale bar = 10 µm. FIGURE 21 *Nitzschia angustata*, valve view. Scale bar = 10 µm.

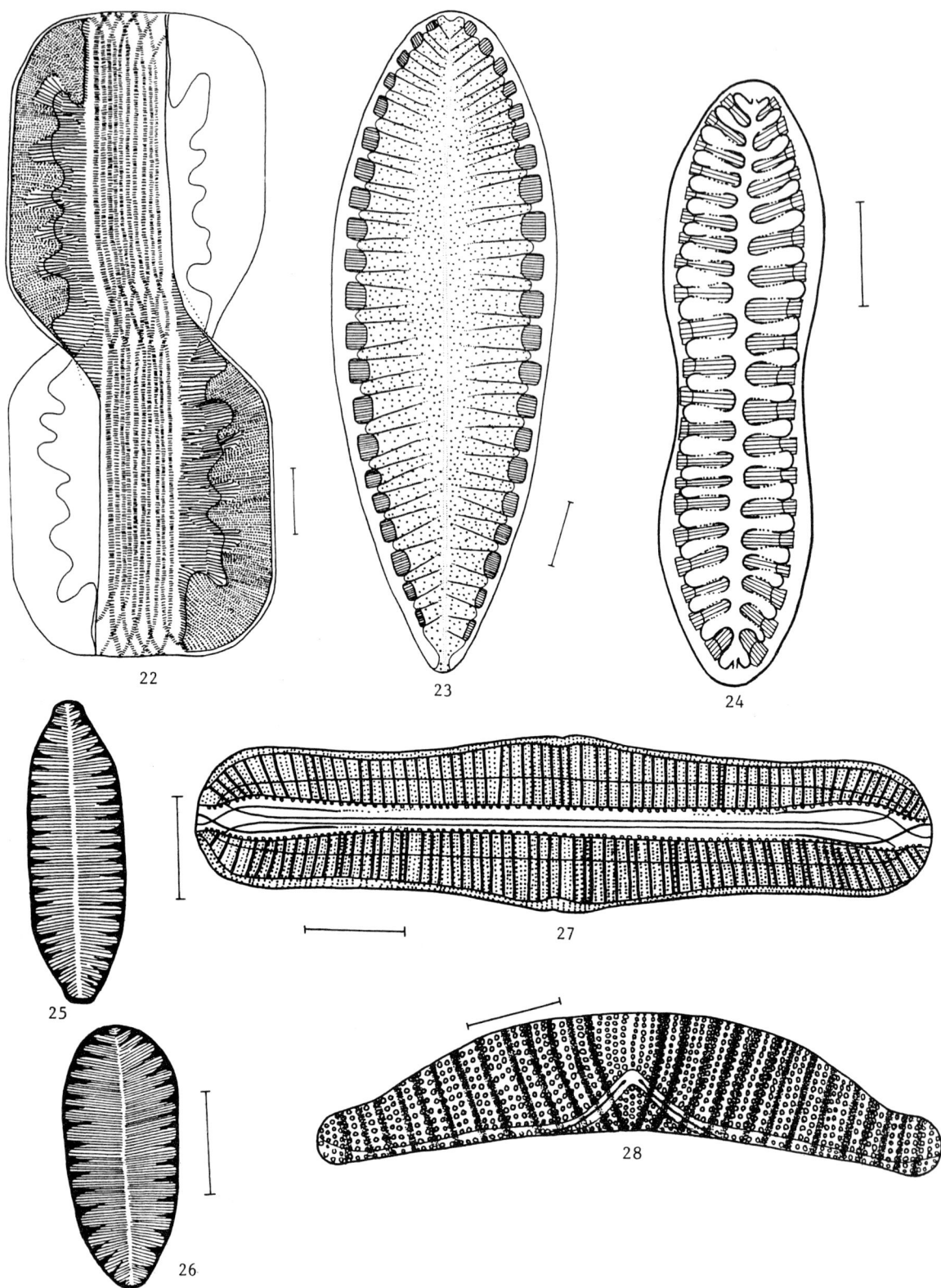

FIGURE 22 *Entomoneis ornata*, valve view. Scale bar = 10 μm. FIGURE 23 *Surirella tenera*, valve view. Scale bar = 10 μm. FIGURE 24 *Surirella linearis* v. *constricta*, valve view. Scale bar = 10 μm. FIGURE 25 *Surirella angustata*, valve view. Scale bar = 10 μm. FIGURE 26 *Surirella ovata*, valve view. Scale bar = 10 μm. FIGURE 27 *Rhopalodia gibba*, valve view. Scale bar = 10 μm. FIGURE 28 *Epithemia sp.*, valve view. Scale bar = 10 μm.

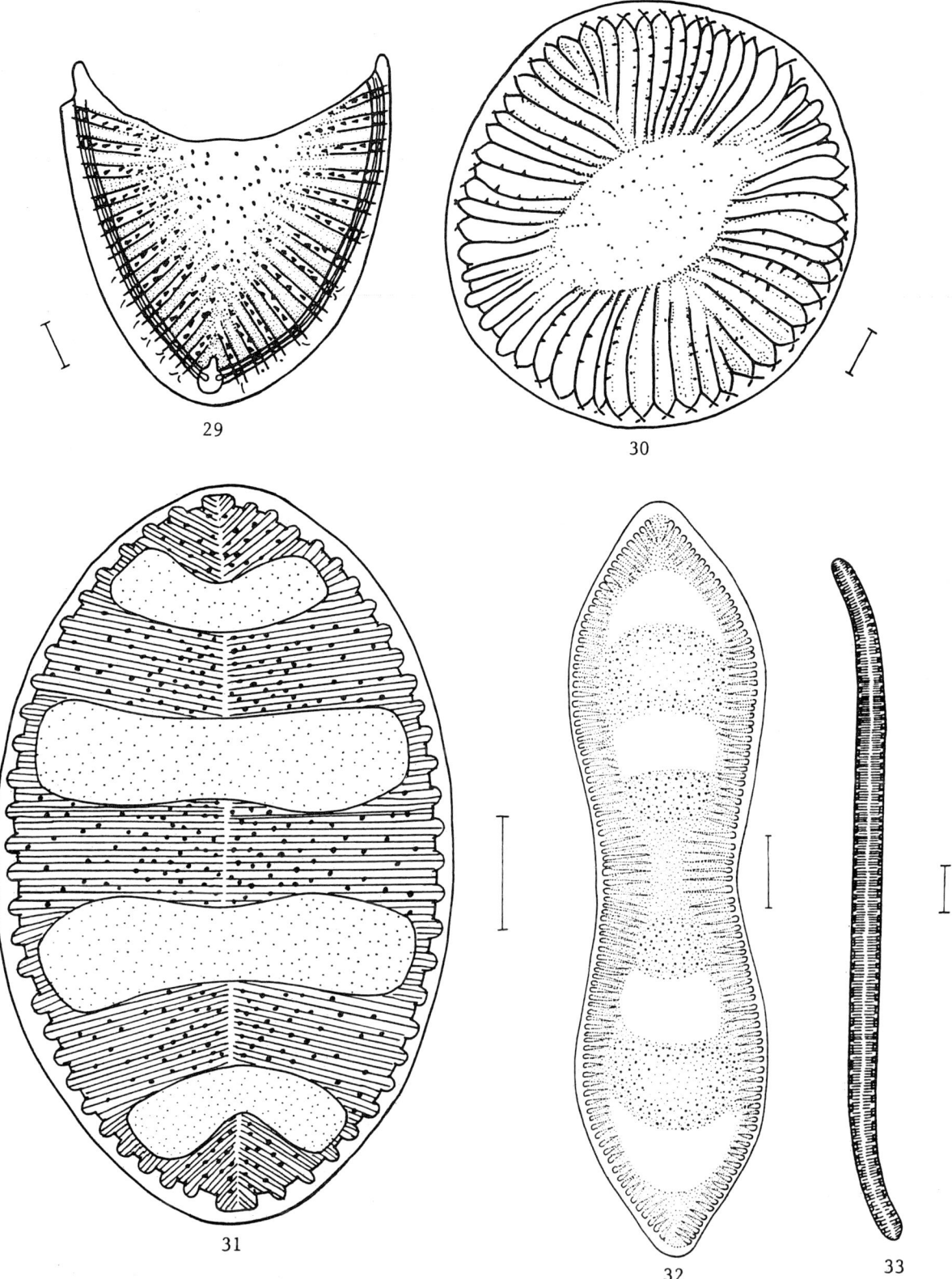

FIGURE 29 *Campylodiscus noricus*, valve view of saddle-shaped valve from the side. Scale bar = 10 µm.
FIGURE 30 *Campylodiscus noricus*, valve view of saddle-shaped valve from the top. Scale bar = 10 µm.
FIGURE 31 *Cymatopleura elliptica*, valve view. Scale bar = 10 µm. FIGURE 32 *Cymatopleura solea*, valve view. Scale bar = 10 µm. FIGURE 33 *Stenopterobia sigmatella*, valve view. Scale bar = 10 µm.

FIGURE 34 Scanning electron micrograph of *Tryblionella* (foreground) and *Surirella* (background). Note the longitudinal undulation of *Tryblionella* and the wing along the margin of *Surirella* with raphe in a wing around the entire valve margin.

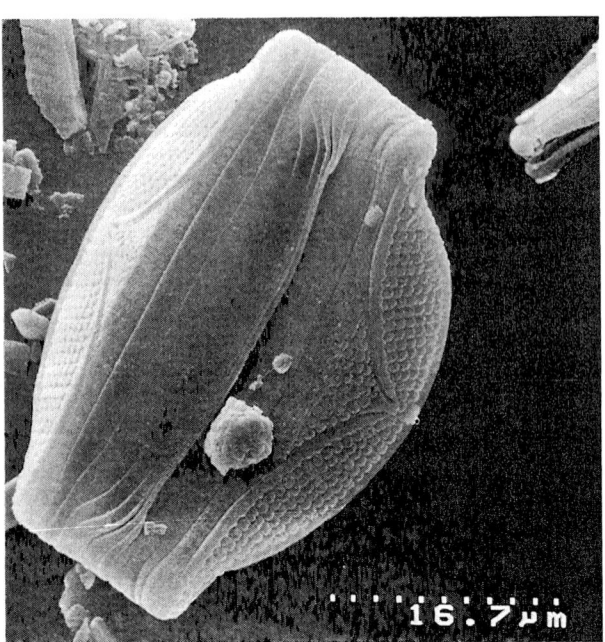

FIGURE 35 Scanning electron micrograph of *Epithemia* frustule in girdle view illustrating the slightly wedge-shaped nature of the cell.

FIGURE 36 Scanning electron micrograph of an *Epithemia* valve.

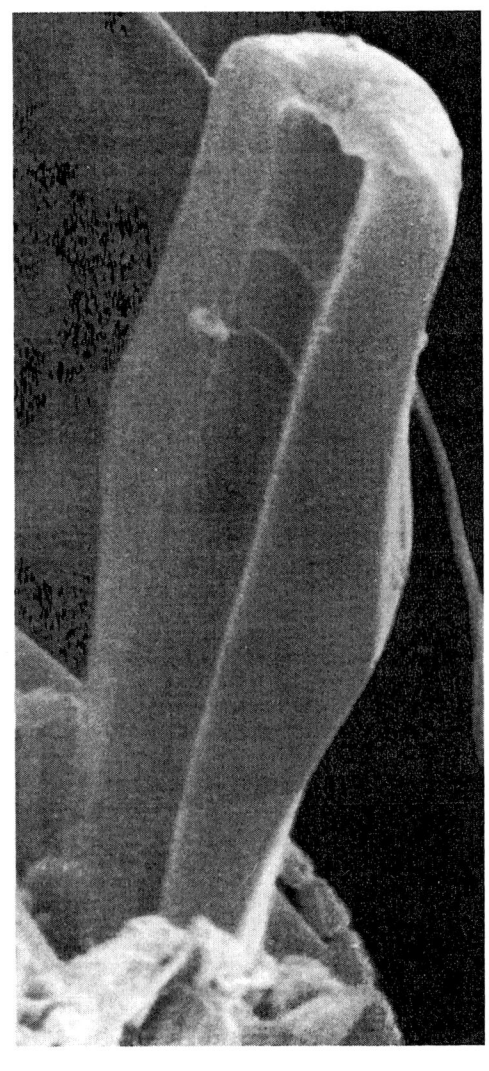

FIGURE 37 Scanning electron micrograph of *Rhopalodia* frustule illustrating the wedge-shaped nature of the cell, 2500×.

They usually occur singly but may form stellate colonies or live in mucilage tubes (Fig. 6). Cells usually contain two plastids that are toward each pole of the cell. The raphe system in *Nitzschia* is fibulate and is normally on or near the margin of the valve surface. The raphe is on opposite margins of the two valves of a frustule (nitzschioid symmetry), in contrast to the raphe position of *Hantzschia*.

Nitzschia is a relatively large genus with hundreds of freshwater and marine species. Most species are epipelic in microhabitat but *Nitzschia* also contains planktonic, epilithic, and epiphytic species. *Nitzschia* contains many pollution-tolerant species (Lowe, 1974) that have been used as indicators of deteriorated water quality (Whitton et al., 1991; Whitton and Rott, 1996).

Tryblionella W. Smith (Figs. 14, 21, 34)

Frustules of *Tryblionella* are symmetric to both the apical and transapical axes in valve view (Fig. 14). Each valve possesses a marginal raphe elevated in a keel. Raphes of the two valves are on opposite margins of the frustule. Each valve is undulated about the apical axis (Round et al., 1990) (Fig. 34). Each cell contains two plastids located on opposite ends of the cell.

Tryblionella is common but seldom abundant in the epipelon of a variety of hard-water habitats and can be distinguished from *Nitzschia* by its apically undulating valves.

F. Descriptions of Genera: Rhopalodiales

Epithemia Brébisson (Figs. 28, 35, 36)

Cells of *Epithemia* are lunate in valve view with a dorsal and ventral margin and usually occur singly (Fig. 28). The two valves are often oriented at a slight angle to each other resulting in wedge-shaped frustule (Fig. 35). A single large lobed plastid is located near the ventral margin of the cell. The raphe on each valve is ventral toward the ends of the valve and arcs toward the middle of the valve where it is more dorsal (Fig. 36). The raphe opens internally into a canal, which is connected to the cell interior by circular or subcircular holes or portulae. In addition, frustules are normally heavily silicified and the valves have thickened internal costae that appear as heavy lines in the light microscope. All species examined thus far contain endosymbiotic cyanobacteria-like cells that function in nitrogen fixation (Floener and Bothe, 1980).

Epithemia is to be found in benthic hard-water habitats reaching maximum abundance in microhabitats where phosphorus is relatively more available (low N/P microhabitats) such as the surface of submerged aquatic plants.

Rhopalodia O. Müller (Fig. 27)

Cells of *Rhopalodia* are solitary and are dorsoventral (lunate) in valve view like *Epithemia* (Fig. 28). However, the cell is normally seen in girdle view because the entire cell is strongly wedge-shaped similar to a canoe (Fig. 37). This is a result of the cingulum being much wider on one side of the cell than the other. Cells have a single ventral platelike plastid similar to *Epithemia*. The raphe is normally along the dorsal margin of the valve and may be difficult to see in valve view. It opens into a canal similar to the structure displayed by *Epithemia*. Transapical costae are thickened internally appearing as fibulae. Endosymbiotic cyanophytes are present in all species of this genus studied and function in nitrogen fixation (Floener and Bothe, 1980).

This genus is found in hard-water nitrogen-poor benthic habitats like those characteristic of *Epithemia*.

G. Descriptions of Genera: Surirellales

Entomoneidaceae

Entomoneis Ehrenberg (Fig. 22)

Cells of *Entomoneis* are solitary and highly motile. The frustule is twisted about the apical axis (Fig. 22) and thus may lie in a variety of positions on the microscope slide. Frustules often appear in girdle view as hourglass-shaped or panduriform. They would appear slightly sigmoid in valve view but are rarely seen in this view. Each valve has a raphe in a keel that arches above the valve surface. Species of *Entomoneis* have one or two large platelike plastids.

This genus is rarely abundant, but broadly distributed in epipelic habitats in water of high conductance. *Entomoneis* may occasionally become entrained in the water column as tychoplankton. Approximately six species have been reported from North America (Patrick and Reimer, 1975).

Surirellaceae

Campylodiscus Ehrenberg *ex* Kützing (Figs. 29, 30)

Cells of *Campylodiscus* are relatively large, solitary, and saddle-shaped (Fig. 29) and thus appear subcircular or crescent-shaped under the microscope, depending on their orientation. Each valve possesses two raphes that lie on the valve margins and are elevated on a wing supported by radially arranged ribs. Multiseriate striae alternate with the ribs. The two valves of the frustule have their apical axes at right angles to each other (Round et al., 1990). The single large plastid is divided into two plates appressed against each of the valve surfaces. It is useful to examine specimens of this large saddle-shaped diatom without a coverglass on the slide because the three-dimensional cells are easily broken under the weight of a cover glass.

There is just one freshwater species in North America, *Campylodiscus noricus*, normally restricted to epipelic habitats in lentic ecosystems.

Cymatopleura W. Smith (Figs. 3, 31, 32)

This benthic diatom genus is characterized by having relatively large solitary cells with a raphe elevated on a wing along each margin of each valve. The valve surfaces display regular transapical undulations (Fig. 3). Valves are isopolar or heteropolar and may be elliptical, linear, or panduriform in shape (Round *et al.*, 1990) and some species are slightly twisted about the apical axis (Hustedt, 1930). Each cell has a single plastid that is divided into two large plates appressed against each valve.

This genus is quite common in epipelic habitats of lakes, rivers, and wetlands. The most common North American species is *Cymatopleura solea*, which is easily identified by its characteristic size and shape (Fig. 32).

Stenopterobia Brébisson *ex* Van Heurck (Fig. 33)

Cells of *Stenopterobia* are relatively long and narrow with a keel along both margins of each valve that is elevated from the valve surface (Fig. 33). Each cell has a single plastid divided into two plates lying against opposite valves. The cells may be either straight or sygmoid in outline.

Few species have been reported from North America. Of these, *Stenopterobia sigmatella* is probably the most common (Stokes and Yung, 1986). This genus is closely related to *Surirella*, from which it can be differentiated by its narrow linear to sygmoid shape and its near restriction to acidic habitats. Species of *Stenopterobia* are normally benthic in acidic habitats.

Surirella Turpin (Fig. 23–26, 34)

Cells of *Surirella* are solitary and may be either isopolar or heteropolar and are slightly wedge-shaped in girdle view. Some species are twisted about the apical axis (Lowe, 1974; Krammer and Lange-Bertalot, 1988). Each cell contains one large platelike plastid divided into two halves with each half flattened against opposite valves of the frustule. Each valve has two raphes that are located on both margins of the valve. Many species are relatively large, heavily silicified, and may be ornamented with siliceous spines and protuberances.

All species are benthic and are found most often on epipelic habitats but may also occupy the epilithon and epiphyton.

VI. GUIDE TO LITERATURE FOR SPECIES IDENTIFICATION

The following is a list of key references for species identification. An attempt has been made to list references from North America. However, many of the references listed here are from references outside North America. This is an unfortunate outcome resulting from our incomplete knowledge of the North American diatom flora.

Bacillaria—Dodd (1987), Hustedt (1930), Krammer and Lange-Bertalot (1988)
Campylodiscus—Hustedt (1930), Krammer and Lange-Bertalot (1988)
Cylindrotheca—Christiansen and Reimer (1968)
Cymatopleura—Hustedt (1930), Krammer and Lange-Bertalot (1988)
Cymbellonitzschia—Hustedt (1930), Krammer and Lange-Bertalot (1988)
Denticula—Johansen *et al.* (1990), Patrick and Reimer (1975)
Entomoneis—Patrick and Reimer (1975)
Epithemia—Patrick and Reimer (1975)
Hantzschia—Mann (1977, 1980a, b 1981), Hustedt (1930), Krammer and Lange-Bertalot (1988)
Nitzschia—Archibald (1970), Reimer, (1954), Mann (1986), Lange-Bertalot (1976, 1980), Krammer and Lange-Bertalot (1988), Lobban and Mann (1987)
Rhopalodia—Patrick and Reimer (1975)
Stenopterobia—Hustedt (1930), Krammer and Lange-Bertalot (1988)
Surirella—Hustedt (1930), Krammer and Lange-Bertalot (1988)
Tryblionella—Krammer and Lange-Bertalot (1988)

ACKNOWLEDGMENTS

I thank the following individuals for assistance with the figures: Todd Clason, Jennifer Greenwood, David Johnson, Katherine Jones, Gina LaLiberte, and Susan Makosky.

LITERATURE CITED

Anonymous. 1975. Proposals for a standardization of diatom terminology and diagnoses. Beihefte zur Nova Hedwigia, 53:323–354.

Archibald, R. E. M. 1970. Key to the genus *Nitzschia*. Cont. News Lett. Limnological Society of South Africa 19:37–55.

Bahls, L., Weber, C. I. 1988. Ecology and distribution in Montana of *Epithemia* sorex Kütz., a common nitrogen fixing diatom. Proceedings of the Montana Academy of Sciences 48:15–20.

Belanger, S. E., Lowe, R. L., Rosen, B. H. 1985. Epiphytism of *Synedra parasitica* on *Surirella robusta*: Observations of populations and associations in a Virginia pond. Transactions of the American Microscopical Society 104:378–386.

Burkholder, J. M. 1996. Interactions of benthic algae with their substrata, in: Stevenson, R. J., Bothwell, M. L., Lowe, R. L., Eds., Algal ecology. Academic Press, San Diego, pp. 253–297.

Burkholder, J. M., Wetzel, R. G., Klomparens, K. L. 1990. Direct comparison of phosphate uptake by adnate and loosely attached microalgae within an intact biofilm matrix. Applied and Environmental Microbiology 56:2882-2890.

Cholnoky, B. J. 1968. Die Ökologie der Diatomeen in Binnengewasser. Cramer, Lehre, 699 p.

Christensen, C. L., Reimer, C. W. 1968. Notes on the diatom *Cylindrotheca gracilis* (Breb. ex Kütz) Grun: Its ecology and distribution. Journal of the Iowa Academy of Science 75:36-41.

DeYoe, H. R., Lowe, R. L., Marks, J. C. 1992. The effect of nitrogen and phosphorus on the endosymbiont load of *Rhopalodia gibba* and *Epithemia turgida* (Bacillariophyceae). Journal of Phycology 28:773-777.

Dodd, J. J. 1987. *Diatoms* (The illustrated flora of Illinois). Southern Illinois Univ. Press, Carbondale, 477 p.

Drum, R. W., Pankratz, S. 1965. Fine structure of an unusual cytoplasmic inclusion in the diatom genus *Rhopalodia*. Protoplasma 60:141-9.

Fairchild, G. W., Lowe, R. L. 1984. Artificial substrates which release nutrients: Effects upon periphyton and invertebrate succession. Hydrobiologia 184:29-37.

Fairchild, G. W., Lowe, R.L., Richardson, W. B. 1985. Nutrient-diffusing substrates as an in situ bioassay using periphyton: Algal growth responses to combinations of N and P. Ecology 66: 465-472.

Floener, L., Bothe, H. 1980. Nitrogen fixation in *Rhopalodia gibba*, a diatom containing blue-greenish inclusions symbiotically, in: Schwemmler, W., Shenk, H. E. A., Eds., Endocytobiology: Endosymbiosis and cell biology, a synthesis of recent research, Vol. 1. de Gruyter, Berlin, pp. 541-552.

Geitler, L. 1977. Zur Entwicklungsgeschichte der Epithemiaceen *Epithemia*, *Rhopalodia* and *Denticula* (Diatomophyceae) und ihre vermutlich symbiotischen Spharoidkörper. Plant Systematics and Evolution 128:259-275.

Goldsborough, L. G., Robinson, G. G. C. 1996. Pattern in wetlands, in: Stevenson, R. J., Bothwell, M. L., Lowe, R. L., Eds., Algal ecology. Academic Press, San Diego, pp. 77-117.

Greenwood, J., Clason, T., Lowe, R. L., Belanger, S. E. 1999. Examination of endopelic and epilithic algal community structure employing scanning electron microscopy. Freshwater Biology 41:821-828.

Hustedt, F. 1930. Dr. L. Rabenhorsts Kryptogamen-Flora von Deutschland, Österreichs und der Schweiz, Vol. VII. Die Kieselalgen Deutschlands, Österreichs und der Schweiz unter Berücksichtigung der übrigen Länder Europas sowie der angrenzenden Meeresgebiete. Autorisierter Neudruck (1962), Johnson Reprint Corporation, New York.

Hustedt, F. 1942. Das Phytoplankton des Süsswassers. Teil 2, Hälfte 2, Die Binnengewässer. Schweizerbart'sche Verlagsbuchhandlung, Stuttgart, 549 p.

Johansen, J., Cognata, S. L., Kociolek, J. P. 1990. Examination of type material of *Denticula rainierensis* Sovereign. Memoirs of the California Academy of Sciences 17:211-219.

Krammer, K., Lange-Bertalot, H. 1988. Süsswasserflora von Mitteleuropa. Bacillariophyceae, Part 2. Teil: Bacillariaceae, Epithemiaceae, Surirellaceae. Fischer, Stuttgart, 596 p.

Lange-Bertalot, H. 1976. Eine Revision zur Taxonomie der Nitzschiae lanceolatae Grunow. Beihefte zur Nova Hedwigia 28:253-307.

Lange-Bertalot, H. 1980. New species, combinations and synonyms in the genus *Nitzschia*. Bacillaria 3:41-77.

Lange-Bertalot, H., Krammer, K. 1993. Observations on Simonsenia and some small species of *Denticula* and *Nitzschia*. Beihefte zur Nova Hedwigia 106:93-99.

Lobban, C. S., Mann, D. G. 1987. The systematics of the tube-dwelling diatom *Nitzschia martiana* and *Nitzschia* section Spathulatae. Canadian Journal of Botany 65:2396-2402.

Lowe, R. L. 1974. Environmental requirements and pollution tolerance of freshwater diatoms. EPA-670/4-74-007, 340 p.

Lowe, R. L. 1996. Periphyton patterns in lakes, in: Stevenson, R. J., Bothwell, M. L., Lowe, R. L., Eds., Benthic algal ecology in freshwater ecosystems. Academic Press, San Diego, pp. 57-76.

Lowe, R. L., Pan, Y. 1996. Use of benthic algae in water quality monitoring, in: Stevenson, R. J., Bothwell, M. L., Lowe, R. L., Eds., Benthic algal ecology in freshwater ecosystems. Academic Press, San Diego, pp. 705-739.

Mann, D. G. 1977. The diatom genus *Hantzschia* Grünow, an appraisal. Beihefte zur Nova Hedwigia 54:323-354.

Mann, D. G. 1980a. *Hantzschia* fenestrata Hust., *Hantzschia* or *Nitzschia*. British Phycological Journal 15:249-260.

Mann, D. G. 1980b. Studies on the diatom genus *Hantzschia*, II. *H. distinctepunctata*. Beihefte zur Nova Hedwigia 33:341-352.

Mann, D. G. 1981. Studies on the diatom genus *Hantzschia*, 3. Interspecific variation in *H. virgata*. Annals of Botany 47:377-395.

Mann, D. G. 1986. *Nitzschia* subgenus *Nitzschia* (notes for a monograph of the Bacillariaceae, 2), in: Ricard, M., Ed., Proceedings of the 8th International Diatom Symposium. Koeltz, Koenigstein, pp. 215-226.

Marks, J. C., Lowe, R. L. 1993. Interactive effects of nutrient availability and light levels on the periphyton composition of a large oligotrophic lake. Canadian Journal of Fisheries and Aquatic Sciences 50:1270-1278.

Palmer, C. M. 1969. A composite rating of algae tolerating organic pollution. Journal of Phycology 5:78-82.

Patrick, R., Reimer, C. W. 1966. The diatoms of the United States exclusive of Alaska and Hawaii. Monograph 13. Academy of Natural Sciences of Philadelphia, 672 p.

Patrick, R., Reimer, C. W. 1975. The diatoms of the United States exclusive of Alaska and Hawaii. Monograph 13. Academy of Natural Sciences of Philadelphia, 672 p.

Peterson, C. G., Grimm, N. B. 1992. Temporal variation in enrichment effects during periphyton succession in a nitrogen-limited stream ecosystem. Journal of the North American Benthological Society 11:20-36.

Power, M. E. 1990. Effects of fish in river food webs. Science 250:811-814.

Reimer, C. W. 1954. Re-evaluation of the diatom species *Nitzschia frustulum* (Kütz.) Grun. Butler University Botanical Studies 11:178-191.

Round, F. E. 1981. The ecology of algae. Cambridge Univ. Press, New York.

Round, F. E., Crawford, R. M., Mann, D. G. 1990. The diatoms. Biology and morphology of the genera. Cambridge Univ. Press, Cambridge, UK, 747 p.

Stevenson, R. J., Bothwell, M. L. Lowe, R. L. Eds., Benthic algal ecology in freshwater ecosystems. Academic Press, San Diego, 753 p.

Stoermer, E. F., Julius, M. L. 2002. Centric diatoms, in: Wehr, J. D., Sheath, R. G., Eds., Freshwater algae of North America. Academic Press, San Diego, pp. 559-594.

Stokes, P. M., Yung, Y. K. 1986. Phytoplankton in selected LaCloche (Ontario) lakes, pH 4.2-7.0, with special reference to algae as indicators of chemical characteristics, in: Smol, J. P., Battarbee, R. W., Davis, R. B., Merilainen, J., Eds., Diatoms and lake acidity. Vol. 4. Junk, pp. Dordrecht, 4:57-72.

Tuchman, N. C. 1996. The role of heterotrophy in algae, in: Stevenson, R. J., Bothwell, M. L., Lowe, R. L. Eds., Benthic algal ecology in freshwater ecosystems. Academic Press, San Diego, pp. 299-319.

Van Landingham, J.L. 1978. catalogue of the fossil and Recent genera and species of diatoms and their synonyms. Part VI. *Neidium through Rhoicosphenia*. Cramer, Weinheim, Germany, pp. 2964–3605.

Whitton, B. A., Rott, E. 1996. Use of algae for monitoring rivers, II. Institut für Botanik, AG Hydrobotanik, Universität Innsbruck, Austria, 196 p.

Whitton, B. A., Rott, E., Friedrich, G. 1991. Use of algae for monitoring rivers. Institut für Botanik, AG Hydrobotanik, Universität Innsbruck, Austria, 193 p.

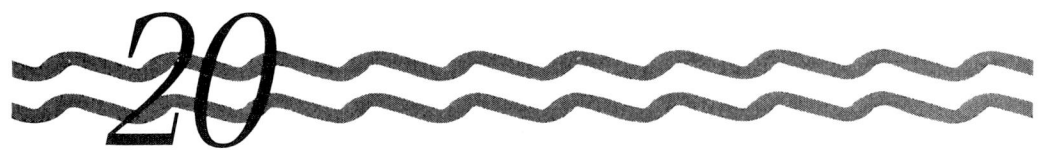

DINOFLAGELLATES

Susan Carty

Department of Biology
Heidelberg College
Tiffin, Ohio 44883

I. Introduction
II. Morphology and Diversity
   A. Morphology
   B. Life Cycle
   C. Classification
III. Ecology and Distribution
   A. Dinoflagellate Blooms
   B. Tropic States
   C. Specificity of Habitat
   D. Geographical Distribution
IV. Collection and Preparation for Identification
   A. Collection
   B. Fixation
   C. Preparation for Identification
V. Key and Descriptions of Genera
   A. Key
   B. Descriptions of Genera
VI. Guide to Literature for Species Identification
   A. Compendia
   B. Local Floras
Literature Cited

I. INTRODUCTION

Dinoflagellates (Division or Phylum Pyrrhophyta) are a group of primarily unicellular organisms united by a suite of unique characteristics, including flagellar insertion, pigmentation, organelles, and features of the nucleus, that distinguishes them from other groups. The name dinoflagellate comes from *dinos* (Greek), "whirling," which describes their distinctive swimming pattern, and *flagellum* (Latin), "a whip." Pyrrhophyta comes from the Greek *pyrrh* "flame colored," "reddish." Freshwater dinoflagellates, including one of the most recognizable unicells of all the plankton, *Ceratium hirundinella*, can be dominant members of the summer phytoplankton and may be responsible for taste and odor problems in drinking water. Currently there are about 250–300 species of freshwater dinoflagellates known worldwide, and about 150 have been reported from North America.

Dinoflagellates, however, are best known to the public as the source of marine red tides leading to various types of human illness caused by their toxins: paralytic shellfish poisoning (PSP), neurotoxic shellfish poisoning, diarrhetic shellfish poisoning, and ciguatera (Hallegraeff *et al.*, 1995; Burkholder, 1998). These red tide species are marine or estuarine species, as is the ambush predator, the "cell from hell," *Pfiesteria piscicida* (Burkholder *et al.*, 1992; Steidinger *et al.*, 1996a). While *Pfiesteria* is known for its fish-killing and neurological effects (Levin *et al.*, 1997), its remarkable multi-stage life cycle has caught the attention of scientists from many disciplines who now ponder the placement of this extraordinary group of protists in the evolution of eukaryotes. Freshwater dinoflagellate taxa exhibit much of the variability found within their marine counterparts, including autotrophy, heterotrophy, and fish parasitism—apparently lacking only toxin production. It is the aim of this chapter to provide an introduction to the freshwater dinoflagellates, and most of the text will refer to the commonly encountered taxa.

TABLE I Unique Features of Dinoflagellates

Nuclear
   Permanently condensed chromosomes
   Dinomitosis (nuclear membrane intact, spindle external, chromosomes attached to nuclear envelope, nucleoli remain)
   Lack histones and nucleosomes
   Unusual base: 5-hydroxymethyl uracil
Cellular
   Arrangement of flagella in cingulum and sulcus
   Pusule
   Trichocysts
Chemical
   Dinosterol
   Peridinin
   Toxins

The characteristics that unite the dinoflagellates and make them unique (Table I) also make determining their phylogenetic affinities difficult (Table II). Dinoflagellates have a nucleus with permanently condensed chromosomes and unique bases, which lacks histones and nucleosomes (Soyer-Gobillard, 1996), although stages in some complex life histories may have a more typical eukaryotic nucleus (Steidinger *et al.*, 1996a; Buckland-Nicks *et al.*, 1997). The terms *mesokaryotic* (Dodge, 1965, 1966) and *dinokaryotic* have been used for this unique nucleus. Many autotrophic dinoflagellates have the unique carotenoid pigment peridinin associated with a unique peridinin-chlorophyll protein (Larkum, 1996) and an atypical photosynthesis with an unusual form of ribulose 1,5 bisphosphate carboxylase (RuBisCo), an enzyme critical to the initiation of the Calvin cycle (reviewed in Palmer, 1995, 1996). The distinctive whirling swimming pattern while being propelled forward is due to the presence of a tinsel flagellum in a transverse groove (cingulum) and a whiplash flagellum in a longitudinal groove (sulcus), again, a unique organization (Fig. 1A). Internally distinctive organelles such as trichocysts (ejectile) and the pusule (nutrient uptake) are found.

Plantlike features of dinoflagellates include cellulose walls in some taxa and the synthesis of starch. Distantly related photosynthetic protists may include the Euglenophyta (some nuclear similarities) and Cryptophyta. The animal-like features of some dinoflagellates (e.g., heterotrophy, trichocysts, eyespots), together with some strong evidence that dinoflagellates have come by photosynthesis through symbiosis (Tomas and Cox, 1973; Wilcox and Wedemayer, 1984; Fields and Rhodes, 1991; Chesnick *et al.*, 1996), lead some researchers to look to the protozoa for nearest relatives.

Many of the characters that make dinoflagellates unique are visible in the light microscope. Swimming cells show the pattern of whirling while being propelled forward and, if photosynthetic, are a (yellow-) golden (-brown) color. When cells are not moving, the cingulum is usually visible as a cinched-in waist, and the nucleus can be seen. The nucleus confirms the identification; it is large, filling 25-35% of the cell, and often centrally located, and the permanently condensed chromosomes appear as a fingerprint pattern of swirls. There is a broad range of cell size: small cells (the aptly named *Peridinium inconspicuum*) may be 10 μm wide by 12 μm long, and large *Ceratium* may reach 400 μm in length.

This chapter deals with freshwater dinoflagellates, but there is also an extensive literature on marine dinoflagellates. In addition to their role as red-tide organisms, marine dinoflagellates are important member of oceanic phytoplankton, and as symbionts with reef-building corals (e.g., zooxanthellae such as *Symbio-*

TABLE II Comparison of Dinoflagellate Features to Other Taxa

Dinoflagellate feature	Most like	Reference
(N base) hydroxymethyluracil	Bacteriophage	Herzog *et al.* (1984)
17S rRNA gene	Archaebacteria	Herzog and Maroteaux (1986)
No histones	Prokaryotes	
Arch-shaped nuclear fibrils	Bacteria	Herzog *et al.* (1984)
Chromosomes attached to nuclear membrane	Bacteria	Herzog *et al.* (1984)
RuBisCo	Bacteria	Palmer (1996)
18S rRNA gene	Apicomplexa	Wright and Lynn (1997a,b)
Trichocysts	*Paramecium*	Hausmann (1978)
Condensed chromosomes	Euglenoids	
Chlorophyll *c*	Chromophytes	
Store starch, cellulose walls	Green algae/plants	
2Fe 2S ferredoxin	Chlorophytes	Yoshikawa *et al.* (1997)
17S rRNA gene	Plants	Herzog and Maroteaux (1986)

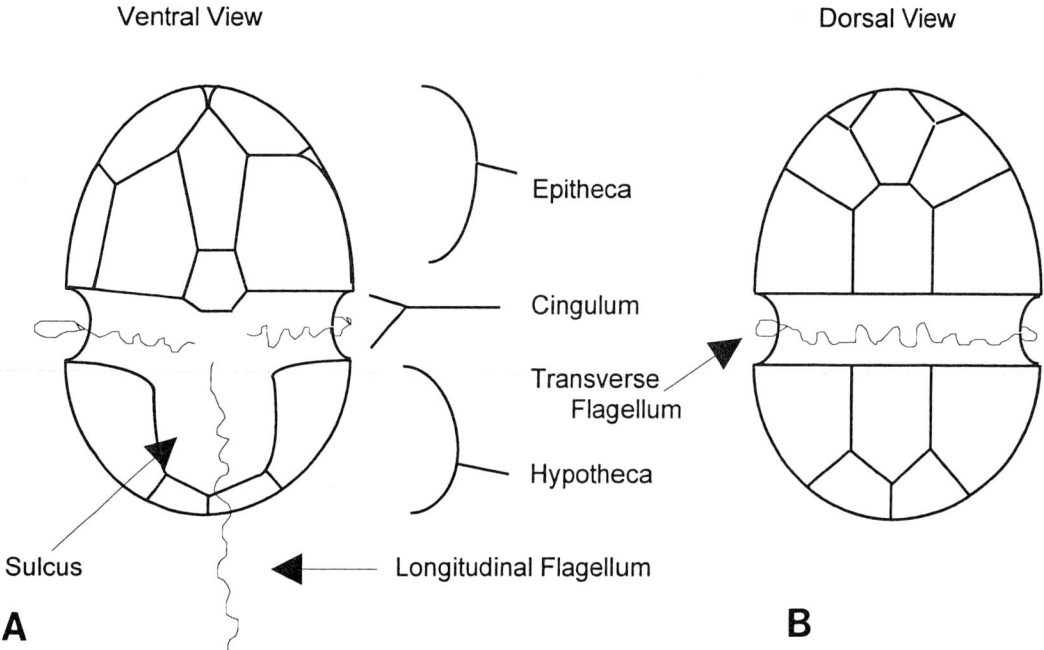

FIGURE 1 Typical thecate motile dinoflagellate cell. (A). Ventral view. (B). Dorsal view.

dinium; Battey, 1992; Blank, 1992) they are vital to maintaining coral reef systems; bleaching occurs when dinoflagellates abandon their coral host. Dinoflagellates form symbiotic relationships with other organisms, including some jellyfish (Trench and Thinh, 1995). Some dinoflagellates are bioluminescent and have been widely studied to understand the circadian rhythm of the flashes (Knaust et al., 1998) and their function (Mensinger and Case, 1992).

Another extensive literature exists on fossil dinoflagellate cysts. Dinoflagellates have a long (from the Silurian; Sarjeant, 1978), widespread, and extensive fossil record. Initially cysts were important stratigraphic markers used by oil companies (Lentin and Williams, 1989) and are increasingly used to understand global environmental conditions (MacRae et al., 1996). Work linking cysts to motile cells has clarified the identity of many fossil cysts (Wall and Dale, 1968). Currently, geologists are at the forefront of much of the work on dinoflagellate systematics (Fensome et al., 1993) and evolution (MacRae et al., 1996; Fensome et al., 1996).

Books devoted to the biology of dinoflagellates include Spector (1984), Taylor (1987), (though many examples and topics relate to marine taxa), and Evitt (1985) (emphasis on cysts, but excellent diagrams and explanations of thecal morphology).

II. MORPHOLOGY AND DIVERSITY

A. Morphology

Freshwater dinoflagellates occur as single cells either in the plankton or attached to substrates such as fish, algal filaments, etc. Historically they have been described as thecate (armored with cellulose plates) or naked (lacking plates). In either case, a typical motile cell has the following characteristic features (Fig. 1A, B). A transverse groove or girdle, the cingulum, encircles the cell and divides it into an epitheca (anterior portion) and hypotheca (posterior portion), or an epicone and hypocone in taxa without plates. The cingulum houses a ribbon-like tinsel flagellum that is responsible for the whirling motion during swimming. The cingulum usually divides the cell into two approximately equal halves, but may be found higher, dividing cells into 1/3 epitheca, 2/3 hypotheca (*Amphidinium* Fig. 2C) or lower, dividing the cell into 2/3 epitheca, 1/3 hypotheca (*Katodinium* Fig. 2G). The ends of the cingulum may face each other or may be displaced to various degrees (very offset in *Gyrodinium* (Fig. 2F) and *Gonyaulax* (Fig. 3B). Displacement is measured in girdle widths and is usually left-handed (descending). A longitudinal groove, the sulcus, defines the ventral face

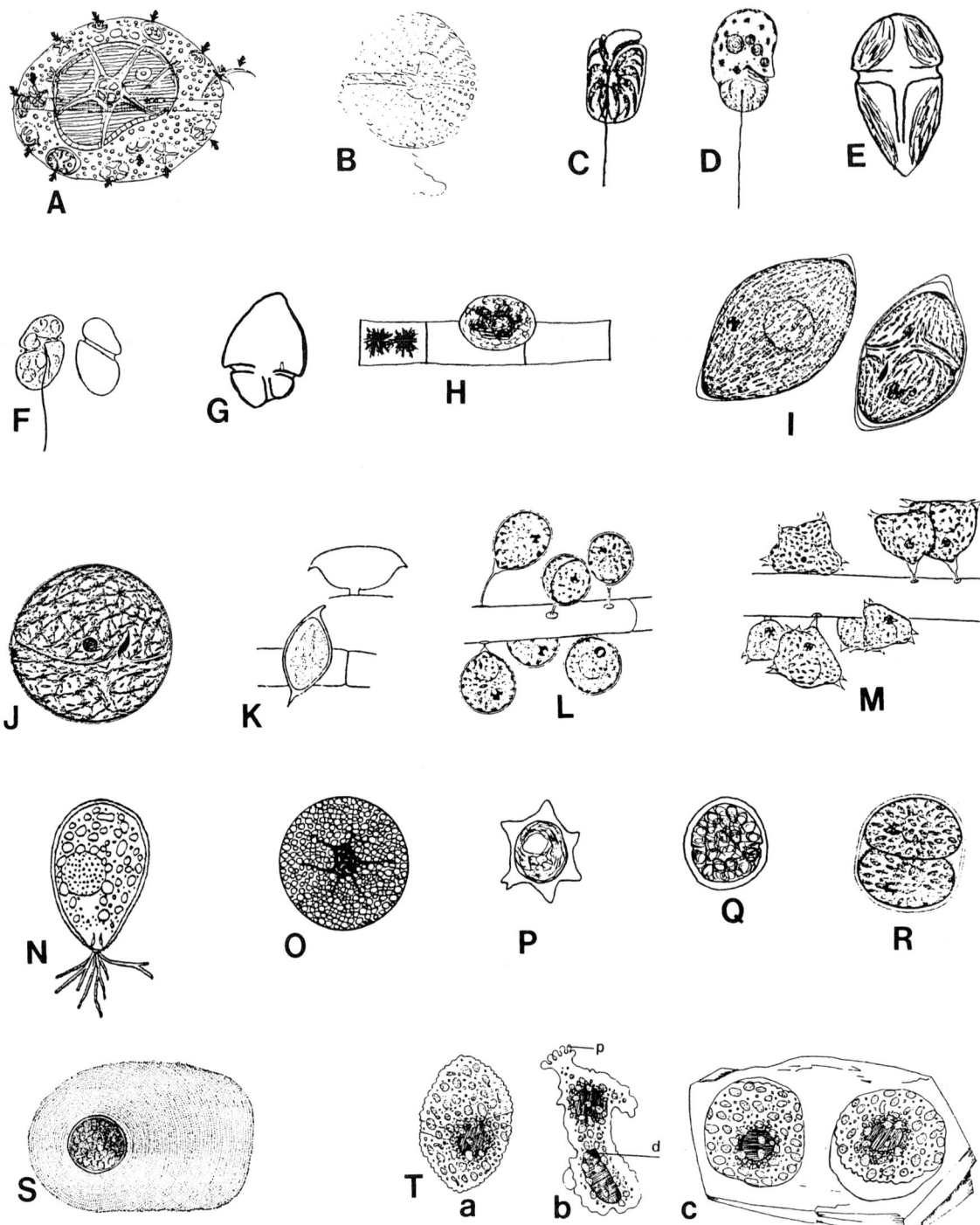

FIGURE 2 Line drawings of unarmored dinoflagellates. No line scale, refer to text for cell size. (A) *Actiniscus pentasterias* v. *arcticus*. Arrows indicate rudimentary pentasters. (B) *Pseudoactiniscus apentasterias*. (C) *Amphidinium klebsii*. (D) *Bernardinium bernardinense*. (E) *Gymnodinium acidotum*. (F) *Gyrodinium pusillum*. (G) *Katodinium spiroidinoides*. (H) *Cystodinedria inermis*. (I) *Cystodinium bataviense*. (J) *Hypnodinium sphaericum*. (K) *Dinococcus* as *Raciborskia bicornis*. (L) *Stylodinium globosum*. (M) *Tetradinium javanicum*. (N) *Oodinium limneticum*. (O) *Haidadinium ichthyophilum*. (P) *Dinastridium sexangulare*. (Q) *Phytodinium simplex*. (R) *Hemidinium* as cyst form *Gloeodinium montanum*. (S) *Rufusiella insignis*. (T) *Dinamoebidium coloradense*, gymnodinioid cell (a), amoeboid cell (b), cysts on sand grain (c). p = "papilli-forming pseudopodia", d = was not labeled, possibly a phagocytosed diatom. A and B are from Bursa (1969) with permission. C and D are from Thompson (1950) with permission. F is from Thompson (1947) with permission. I, J, L, M, R are from Thompson (1949) with permission. K is from Prescott (1951) with permission. N is from Jacobs (1946) with permission. O is from Buckland-Nicks *et al.* (1997) with permission. S is from Smith (1950) with permission. T is from Bursa (1970) and is reproduced from Arctic and Alpine Research with permission of the Regents of the University of Colorado.

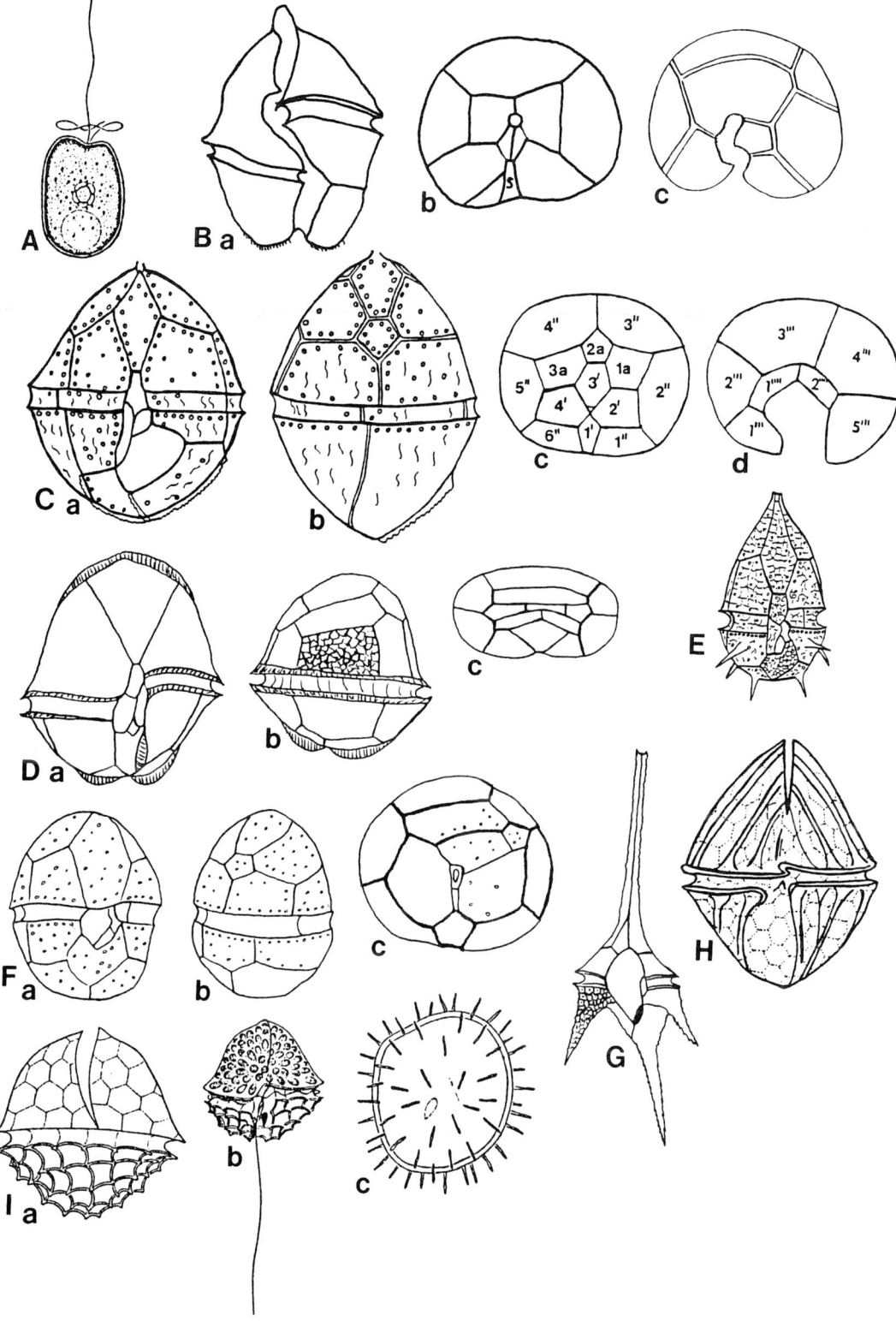

FIGURE 3 Line drawings of armored dinoflagellates. Where there is no line scale, refer to the text for cell size. (A) *Exuviella compressa*. (B) *Gonyaulax spinifera*. a, ventral; b, epitheca (s is a sulcal plate); c, hypotheca. (C) *Thompsodinium intermedium*. a, ventral (note trichocyst pores on plates); b, dorsal; c, epitheca; d, hypotheca. (D) *Peridinium willei*. a, ventral; b, dorsal; c, epitheca. (E) *Peridiniopsis quadridens* ventral. (F) *Durinskia baltica*. a, ventral; b, dorsal; c, epitheca. (G) *Ceratium hirundinella*, ventral H. *Lophodinium polylophum*. ventral. (I) *Woloszynskia reticulata*. a, dorsal; b, ventral; c, cyst. A, I are from Thompson (1950) with permission. C, E are from Carty (1989) with permission.

of the cell, extends some distance into the hypotheca, may extend into the epitheca, and houses the whiplash flagellum, which propels the cell forward (Leadbeater and Dodge, 1967).

Most dinoflagellates have a motile stage with a cingulum-sulcal arrangement at some time in their life cycle. The motile stage is frequently the assimilative stage during which nutrients are absorbed (photosynthetic or heterotrophic) but may be reproductive and/or dispersive. Nonmotile assimilative forms include the free-floating forms (*Cystodinium* (Fig. 2I), *Hypnodinium* (Fig. 2J), those with stalks attached to a substrate (*Tetradinium* (Fig. 2M), *Stylodinium* (Fig. 2L), and those attached directly to the substrate (*Cystodinedria* (Fig. 2H), *Oodonium* (Fig. 2N)). Nonmotile cysts may occur in the plankton or in sediments and may be more distinctive than the motile stage.

Dinoflagellates are eukaryotic cells with membrane-bound organelles. The nucleus is usually large, variously shaped (round, C-shaped, curved), variously located (centrally, in epitheca or hypotheca), and with large, permanently condensed chromosomes. Mitochondria are tubular. Some species have a red stigma or eyespot in the sulcal region (Dodge, 1969), which may be part of a chloroplast or may be free (Dodge, 1969). Unique organelles include trichocysts and the pusule. Trichocysts are ejectile organelles whose construction and chemical composition are somewhat similar to those of the spindle trichocysts of *Paramecium* (Hausmann, 1978). The pusule is a tubular or vesicular organelle formed by invagination of the plasma membrane and is covered by additional membrane (Klut *et al.*, 1987) located in the sulcus near the base of the flagella. Dodge (1972) identified seven types of pusule. Several functions have been hypothesized for it, including fluid intake for nutrients, a buoyancy device, waste expulsion, and osmoregulation (summarized in Spector, 1984). Experimental evidence of the uptake of molecular markers (Klut *et al.*, 1987) would support nutrient uptake (Kofoid, 1909; Kofoid and Swezy, 1921).

Most photosynthetic dinoflagellates have discoid or lobed chloroplasts located in a peripheral position. Chloroplasts are typically surrounded by three membranes, lack chloroplast endoplasmic reticulum (CER), and have thylakoid membranes in stacks of three and some type of pyrenoid (Dodge, 1975). Photosynthetic pigments are chlorophylls *a* and c_2, with the carotenoid peridinin contributing the usually golden color, although cells may appear yellowish or almost brown (as in *Peridinium gatunense*). Other carotenoids include diadinoxanthin, dinoxanthin, and beta carotene (Jeffrey *et al.*, 1975). Dinoflagellate photosynthetic systems are unique in having a water-soluble Chl *a*-peridinin system and a membrane-bound Chl *a*-Chl c_2-peridinin system (summarized in Iglesias-Prieto, 1996). Evidence of other chlorophylls (c_1), carotenoids (fucoxanthin), phycoerythrin, and alloxanthin in some dinoflagellates suggests previous engulfment of diatoms, prymnesiophytes, and cryptomonads (Meyer-Harms and Pollehne, 1998). Kleptoplastidy, the usage of prey chloroplasts, has been demonstrated for several dinoflagellates (Fields and Rhodes, 1991; Lewitus *et al.*, 1999). Heterotrophic cells may lack chloroplasts but have a pink protoplasm (*Gymnodinium helveticum*, *Entzia acuta*), food vacuoles, and brightly colored (red, orange, yellow) accumulation bodies. Accumulation bodies may also be found in photosynthetic species. The storage material in most dinoflagellates is starch localized outside the chloroplast, although red oil droplets may be seen in the cytoplasm, especially near the end of the growing season.

Dinoflagellates have a multilayered cell covering, the amphiesma or theca. The outermost layer is a unit membrane (considered by many the plasma membrane), beneath which are located vesicles that may contain plate material (Loeblich, 1969; Dodge and Crawford, 1970), then a membrane bounding the cell contents. Some dinoflagellates have a pellicle beneath the thecal layer (Morrill and Loeblich, 1981). Thecate dinoflagellates have cellulose plates in the vesicles beneath the outer membrane. Plate boundaries may have ridges. The taxonomy of thecate forms is based on the number and arrangement of plates. Kofoid (1909) proposed the system that is now widely used. Plates are arranged in concentric rings in relation to the cingulum, with prime (′) designations indicating which ring, and numbered from the most sulcal in a counter-clockwise manner (Fig. 4A–D). Apical plates (′) are followed by precingular (″), postcingular (‴), and antapical (⁗). Plates between apical and precingular are anterior intercalary (a), and plates between postcingular and antapical, not in contact with the sulcus, are posterior intercalary (p). Cingular and sulcal plates have also been designated (C for cingular, T for a transition plate between cingulum and sulcus, and S for sulcal: Sa, Sd, Ss, Sm, Sp, Spa) (Fig. 4E) (Balech, 1974, 1980). The plate pattern of *Peridinium cinctum* is 4′, 3a, 7″, 5C, 5‴, 2⁗. Some species seem more prone to plate shifting, fusion, and splitting than others (Lefèvre 1932). There have been other systems devised to designate plates, notably that of Eaton (1980), who organized plates to reflect hypothesized evolutionary changes.

Apical pores occur on some thecate species. The pore may be surrounded by a pore plate (Po), pore canal plates, and other small plates (Dodge and Hermes, 1981; Toriumi and Dodge, 1993). The pres-

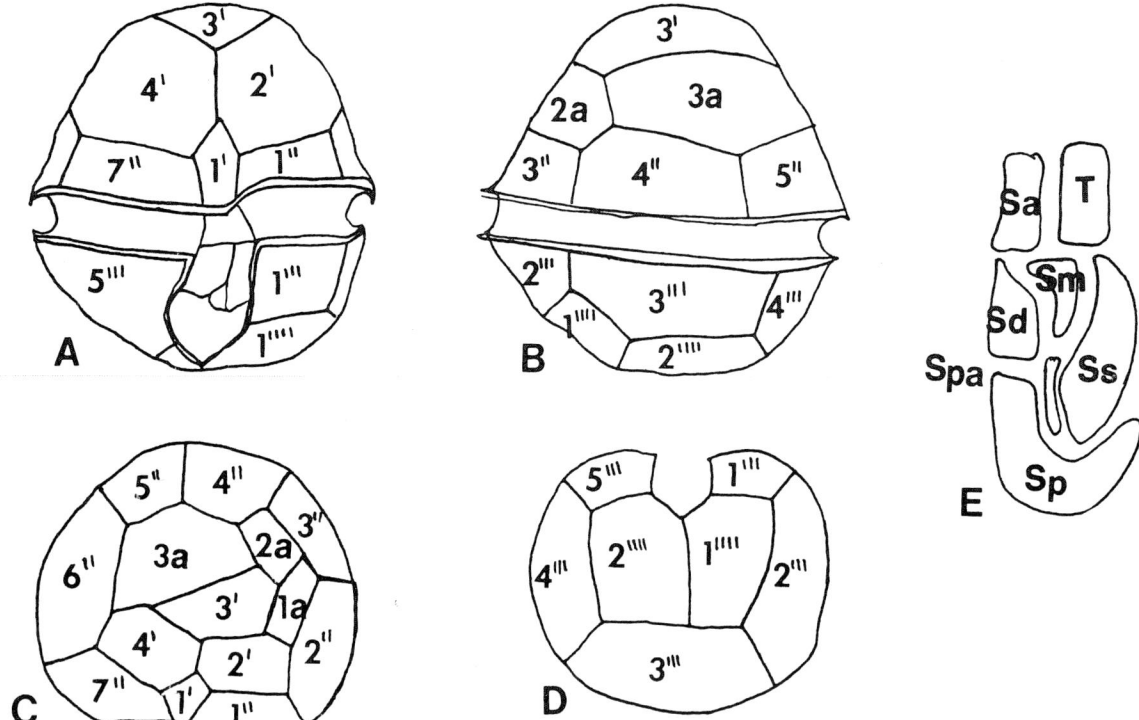

FIGURE 4 Typical thecate motile cell (*Peridinium gatunense*). Plates are numbered according to the system of Kofoid (1909). (A) Ventral view. (B) Dorsal view. (C) Apical view. (D) Antapical view. (E) Sulcal plates, generalized diagram to show relative positions of plates. T, transition plate between cingular and sulcal,; Sa, sulcal anterior; Sp, sulcal posterior; Ss, sulcal sinister (left); Sd, sulcal dexter (right); Sm, sulcal medial; Spa, sulcal plate anterior to Sp.

ence of an apical pore alters the outline of the cell from smooth in those without a pore, to slight ridges, to a distinct chimney. Apical slits occur in *Lophodinium* and some species of *Woloszynskia* and may not be evident with the light microscope unless the cell ecdyses (sheds the theca).

Some species have thick cellulose plates, some thinner, and some too thin to be resolved without electron microscopy. Plates have trichocyst pores and may have various types of ornamentation. Older cells show striated bands (Fig. 5J) where plates expand during growth, forming distinct overlap patterns. Cells with thicker plates may have plate extensions, either spines or lists (winglike flanges), along plate margins or on plate interiors (lists on *Peridinium willei*, Figs. 3D and 5I; *Sphaerodinium fimbriatum*, Figs. 6C and 7A; spines on *Peridiniopsis quadridens*, Fig. 3E). Plates may be involved in forming cell extensions called horns (*Ceratium*; Fig. 3G and 8D) or lobes (*Peridinium cinctum* f. *tuberosum*).

Cell shape varies from spherical to oval/ovate, with alternative shapes appreciated by the taxonomist (i.e., *Ceratium*). There are degrees of dorsoventral compression from none (*Peridinium gatunense*) to extreme concavity (*Peridiniopsis polonicum*). Lateral compression is rare in freshwater dinoflagellates (*Amphidiniopsis* being the exception). Apical–antapical depression is uncommon (some in *Peridinium gatunense*). Morphological variability within a species is slight, except for *Ceratium hirundinella* (Pearsall, 1929; Hutchinson, 1967).

Athecate cells may be "gymnodinioid" or not. Gymnodinioid genera (*Amphidinium*, *Bernardinium*, *Gymnodinium*, *Gyrodinium*, *Katodinium*) are motile with a cingulum and sulcus. Their taxonomy is based on cell shape, location and completeness of cingulum, penetration of the sulcus into the epicone and hypocone, presence and color of chloroplasts, location of nucleus, color of cytoplasm, size, eyespot, and presence of accumulation bodies. Athecate cells that are nonmotile (nongymnodinioid) may have typical dinoflagellate coloring (*Cystodinium*, *Hypnodinium*, *Tetradinium*). Nonmotile genera are free floating or are attached by long (*Stylodinium*; Fig. 2L), short (*Tetradinium*, Fig. 2M), or rhizoidal (*Oodinium*, Fig. 2N) stalks or attachment discs (*Dinococcus*; Fig. 2K) or directly (*Cystodinedria*; Fig. 2H) to various substrata.

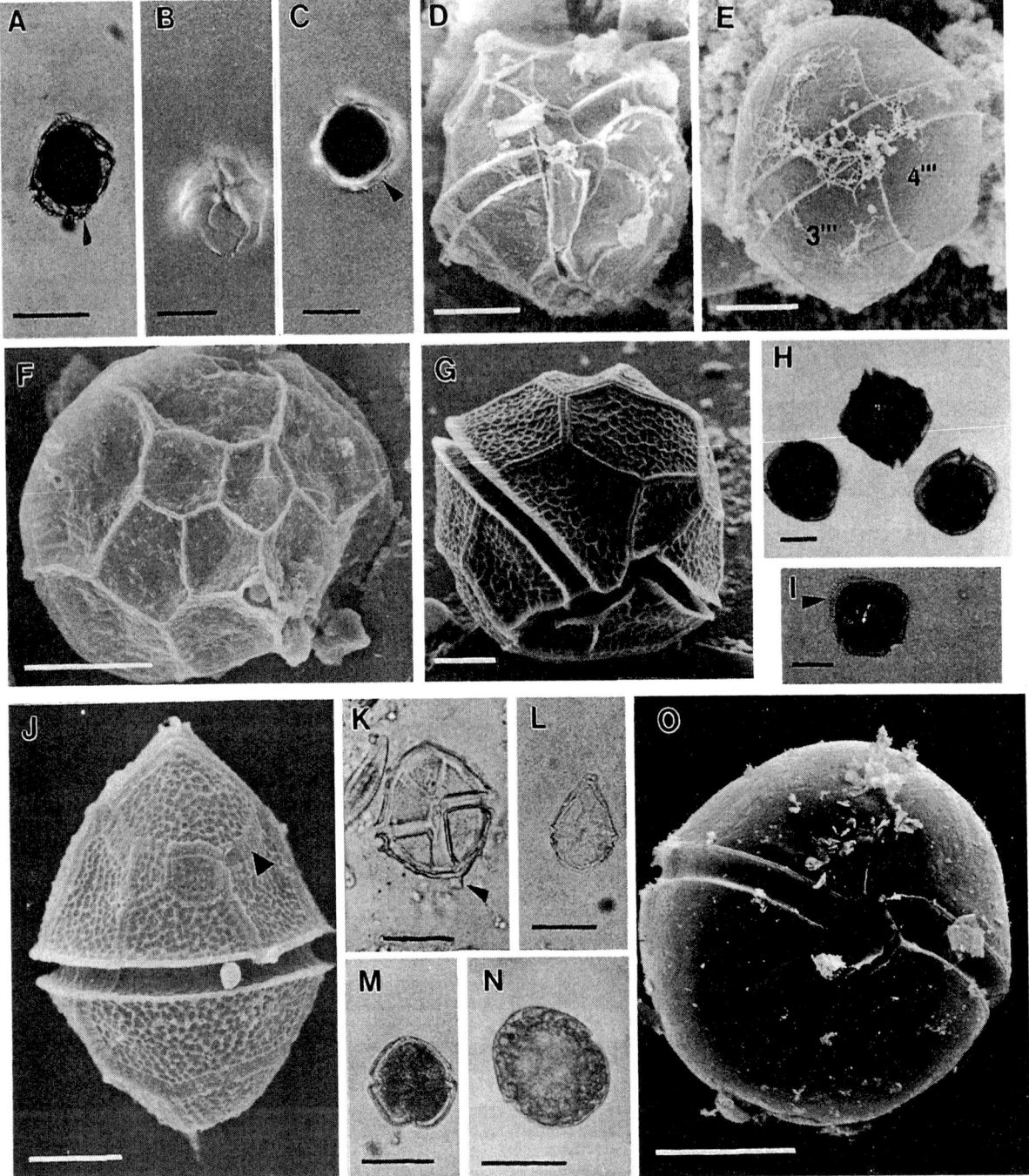

FIGURE 5 Light (LM) and SEM micrographs of armored dinoflagellates, location included (county, state). Line scale on SEM = 10 μm, line scale on LM = 20 μm. (A) *Gonyaulax spinifera*, LM, note offset cingulum, apical pore, hypothecal spines (arrowhead) (Suffolk, NY). (B) *Thompsodinium intermedium*, LM; ventral view, empty cell (Suffolk, NY). (C) *Thompsodinium intermedium*, LM, note hypothecal fringe (arrowhead) (Suffolk, NY). (D) *Thompsodinium intermedium*, SEM, ventral view (Brazos, TX). (E) *Thompsodinium intermedium*, SEM, dorsal view; note large postcingular plates (Brazos, TX). (F) *Thompsodinium intermedium*, SEM, epithecal plate pattern; note apical pore, star pattern of apical and apical intercalary plates (Brazos, TX). (G) *Peridinium gatunense*, SEM, ventral view; note reticulate ornamentation (Burleson,TX). (H) *Peridinium gatunense*, LM, three cells; note round cross sections, "lumpy" appearance of cell. (I.) *Peridinium willei*, LM; note apical flange (arrowhead), posterior flanges (Seneca, OH). (J) *Peridiniopsis polonicum*, SEM, dorsal view; note two apical intercalary plates, striated growth bands (arrowhead), single posterior spine (Hamilton, OH). (K) *Peridiniopsis polonicum* LM, ventral view; note large 1' plate, single posterior spine (Brazos, TX). (L) *Peridiniopsis quadridens*, LM, empty cell (Huron, OH). (M) *Peridiniopsis penardiforme*, LM (Huron, OH). (N) *Durinskia baltica*. (O) *Durinskia Daltica*, SEM, ventral view (Sandusky, OH).

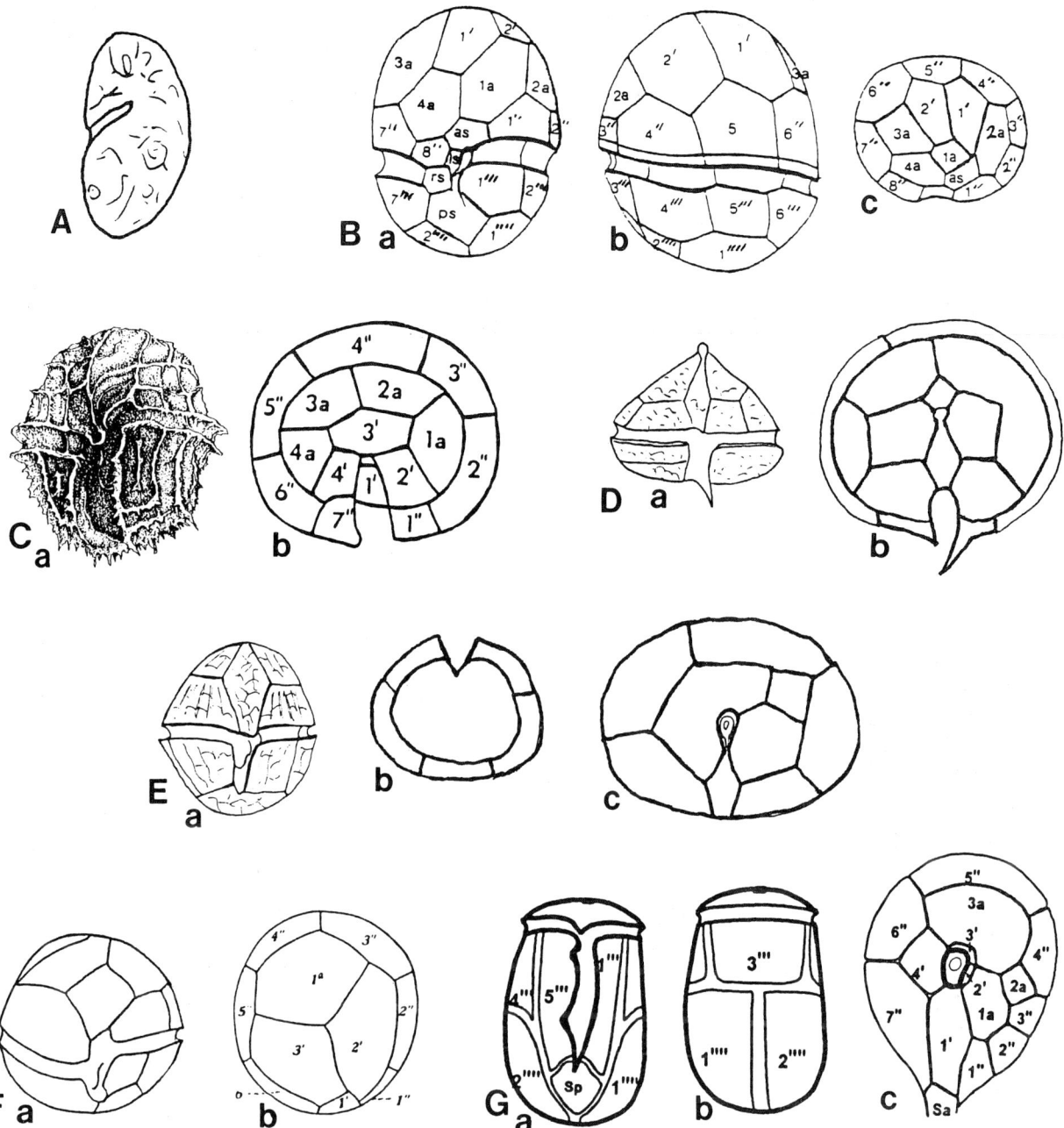

FIGURE 6 Line drawings of armored dinoflagellates. Where there is no line scale, refer to the text for cell size. (A) *Hemidinium nasutum*. side. (B) *Glenodiniopsis steinii*. a, ventral; b, dorsal; c, epitheca. (C) *Sphaerodinium fimbriatum*. a, ventral; b, epitheca. (D) *Entzia acuta*. a, ventral; b, epitheca. (E) *Kansodinium ambiguum*; a, ventral; b, hypotheca; c, epitheca. (F) *Dinosphaera palustris*. a, ventral; b. epitheca. (G) *Amphidiniopsis sibbaldii*. a, ventral; b, dorsal; c, epitheca. B is from Highfill and Pfiester (1992b) with permission. Cb is from Carty (1986) Fa is from Prescott (1951) with permission. Fb is from Kofoid and Michner (1912) G is from Nicholls (1998) with permission.

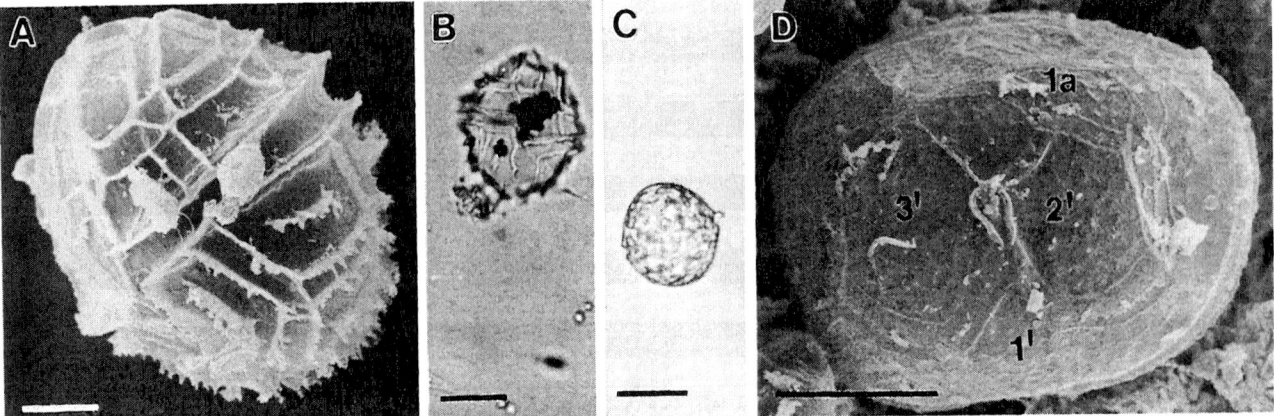

FIGURE 7 Light (LM) and SEM micrographs of armored dinoflagellates, location included (county, state). Line scale on SEM = 10 μm, line scale on LM = 20 μm. (A) *Sphaerodinium fimbriatum* SEM, side view. Note frimbriae on plates (Williams, OH). (B) *Sphaerodinium fimbriatum*, LM, empty cell (Williams, OH). (C) *Kansodinium ambiguum*, LM (Hill, TX). (D) *Kansodinium ambiguum*, SEM, epitheca. Note apical pore, three apical plates (Hill, TX).

B. Life Cycle

1. Life Cycles of Motile Taxa

Dinoflagellate life cycles include a growing, dividing, assimilative stage and a resting (cyst) stage (Fig. 9). About half of the genera have an assimilative stage that is motile, with a distinct cingulum, and are planktonic. Cells are typically haploid and divide mitotically to produce other assimilative cells or gametes. Cells undergoing mitosis may remain motile or form temporary nonmotile cells. *Gymnodinium* undergoes binary fission (Pfiester and Anderson, 1987). Some taxa ecdyse from the parental cell before division, and some may divide within the parental theca (Fig. 8G), which is then shed (Pfiester and Anderson, 1987). *Ceratium* donates part of the parental theca to each daughter cell (Fig. 8A).

The sexual life cycles of some dinoflagellates have been determined (Table III). Assimilative, haploid cells undergo mitosis to produce cells that function as gametes. In culture, producton of gametes has been induced by nitrogen deficiency in some taxa (Pfiester, 1975, 1976, 1977; Chapman and Pfiester, 1995). Gametes may be the same size and shape as the parental cell (hologametes) or smaller. Gametes may be the same (isogamy) or different (anisogamy) sizes, clones may be monoecious and able to produce zygotes, or dioecious and require different clones to produce zygotes (Pfiester and Skvarla, 1979). Gametes fuse in the sulcal region, and, in *Peridinium cinctum*, the nuclei meet in a fertilization tube between the gametes (Pfiester, 1984). The resulting planozygote may have two trailing flagella and remain motile for a period of growth. Subsequently, it becomes a nonmotile hypnozygote and settles to the sediment. Encystment occurs in north temperate areas during the autumn to overwinter. In subtropical Lake Kinneret, Israel, cysts are instead formed to oversummer extreme conditions (Pollingher *et al.*, 1993). There is a period of dormancy following cyst formation before excystment can occur. The dormancy period may require cold and dark (*Peridiniopsis cunningtonii*, Sako *et al.*, 1985; *Peridinium bipes*, Park and Hayashi, 1992) followed by light (*Peridinium bipes*, Park and Hayashi, 1992).

Studies of excystment include long-term observations of *Ceratium hirundinella* in Esthwaite Water,

TABLE III Sexually Reproducing Freshwater Dinoflagellates

Taxon	Reference
Ceratium furcoides	Hickel (1988)
Ceratium cornutum	Stosch (1965)
Cystodinium bataviense	Pfiester and Lynch (1980)
Durinskia baltica	Chesnick and Cox (1987, 1989)
Glenodiniopsis steinii	Highfill and Pfiester (1992a)
Gloeodinium montanum	Kelley and Pfiester (1990)
Gymnodinium paradoxum	Stosch (1972)
Peridiniopsis cunningtonii	Sako *et al.* (1985)
Peridiniopsis lubieniensiforme	Diwald (1938)
Peridiniopsis penardii	Sako *et al.* (1987)
Peridinium bipes	Park and Hayashi (1993)
Peridinium cinctum	Pfiester (1975)
Peridinium gatunense	Pfiester (1977)
Peridinium inconspicuum	Pfiester *et al.* (1984)
Peridinium limbatum	Pfiester and Skvarla (1980)
Peridinium volzii	Pfiester and Skvarla (1979)
Peridinium willei	Pfiester (1976)
Woloszynskia apiculata	Stosch (1973)
Woloszynskia pseudopalustre	Stosch (1973)

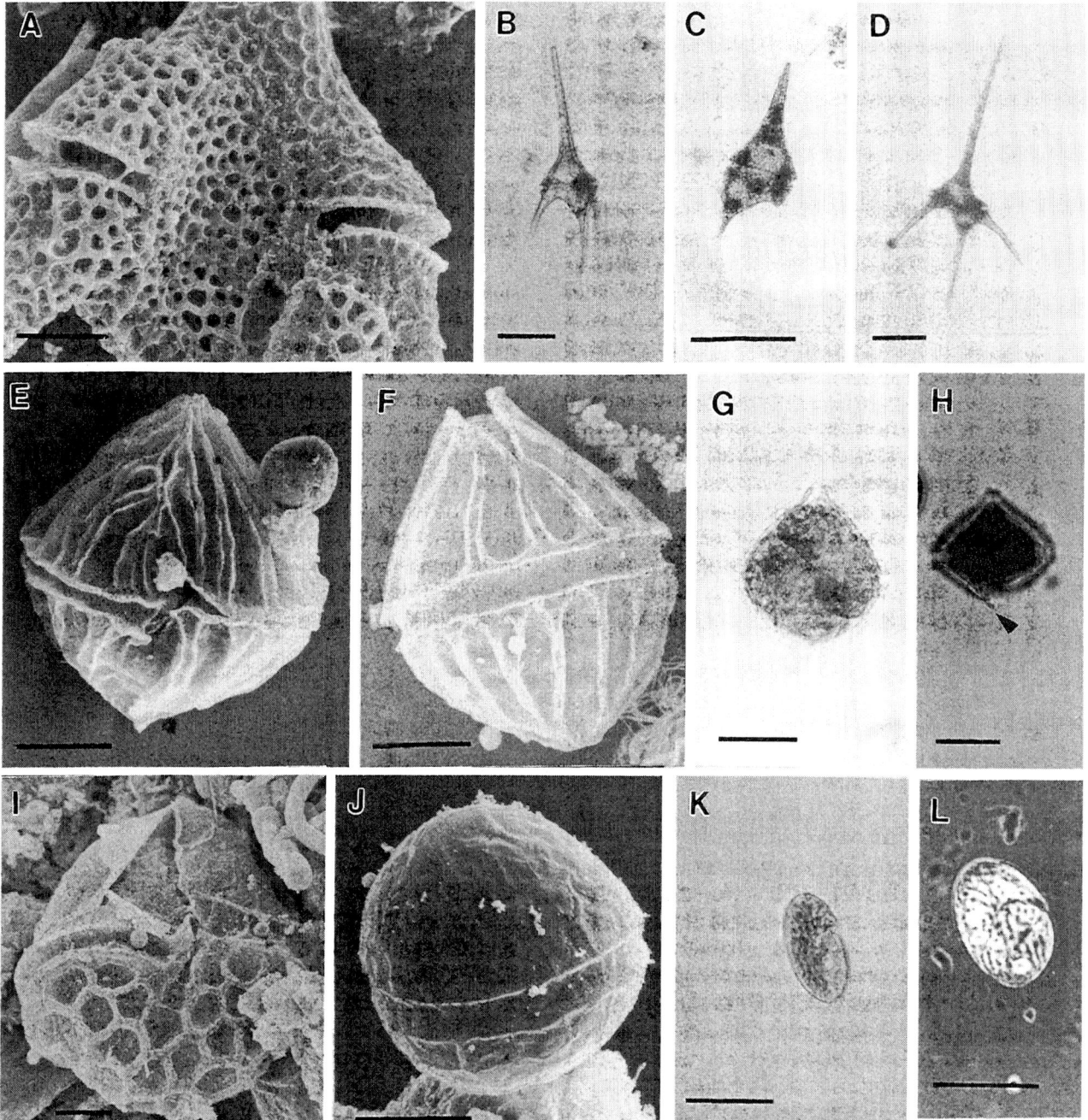

FIGURE 8 Light (LM) and SEM micrographs of armored dinoflagellates, location included (county, state). Line scale on SEM = 10 μm, line scale on LM = 20 μm except as noted (LMs of *Ceratium*). (A) *Ceratium hirundinella* SEM. Note that reticulate ornamentation is heavier on older, lower (left in micrograph) sections and lighter on upper (regenerating) section; trichocyst pores are evident in upper section (Washington, TX). (B) *Ceratium hirundinella* f. *hirundinella* LM, line scale = 50 μm (Seneca, OH). (C) *Ceratium brachyceros*, LM, line scale = 50 μm (Huron, OH). (D) *Ceratium hirundinella* f *piburgense*, LM, line scale = 50 μm (Huron, OH). (E) *Lophodinium polylophum* SEM ventral view; note ridges (Brazos, TX). (F) *Lophodinium polylophum*, SEM, dorsal view (Brazos, TX). (G) *Lophodinium polylophum*, LM, dividing cell (Brazos, TX). (H) *Entzia acuta*, LM. Note posterior sulcal extension (arrowhead), apical pore (Stark, OH). (I) *Woloszynskia reticulata* SEM. Note heavy hypothecal sutures. (J) *Woloszynskia* SEM. Note thin, polygonal plates, (Fulton, OH). K. *Hemidinium nasutum* LM, (Brazos, TX). L. *Hemidinium nasutum*, LM. Note slashed appearance of cingulum (Brazos, TX). E, F, G are from Carty and Cox (1985) with permission.

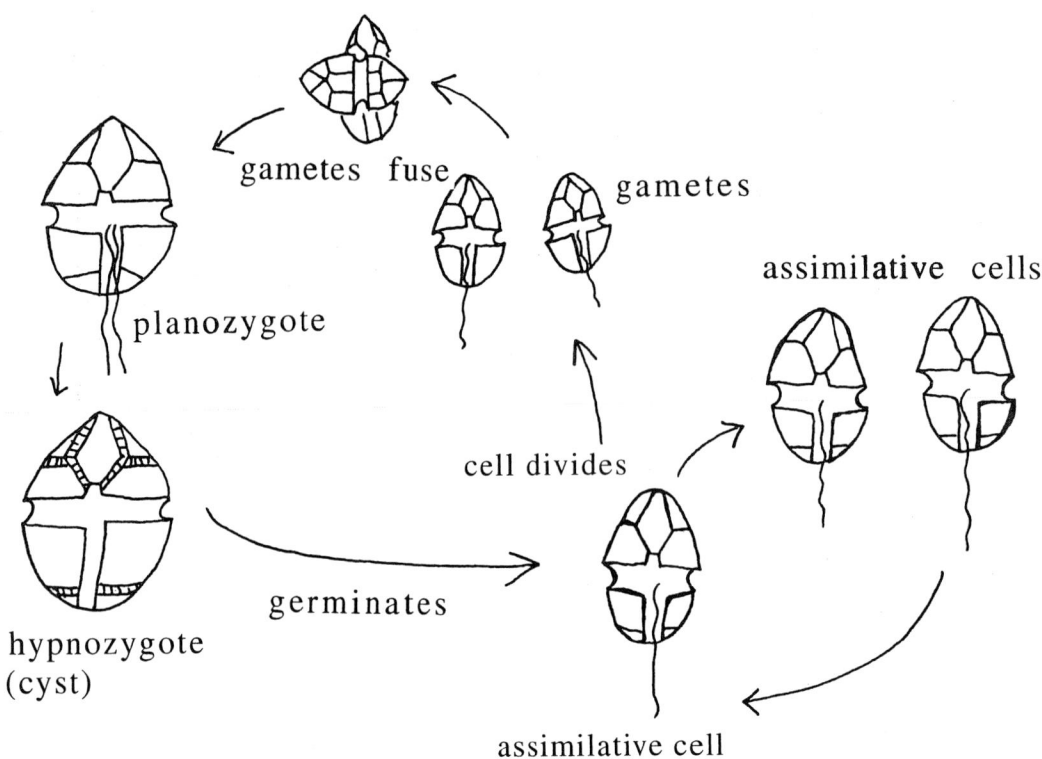

FIGURE 9 Generalized life cycle of a motile, photosynthetic, thecate dinoflagellate.

English Lake District, where a rise in temperature from 3°C to 5°C at the end of cold winter temperatures corresponded to a large increase in the number of vegetative cells in the plankton and empty cysts in the sediments (Heaney et al., 1983). In Japan, cysts collected in April from bottom sediments at 10–20°C showed maximum excystment in the lab between 20°C and 25°C, with none at 5°C (Kawabata and Banba, 1993). Work with other taxa found the optimum temperature for excystment of *Peridiniopsis cunningtonii* to be 22°C (Sako et al., 1985) and 15–25°C for *Peridinium bipes* (Park and Hayashi, 1992). The seemingly conflicting reports concerning the effects of temperature and light on excystment may indicate that beyond a required dormancy period, excystment is controlled by an internal biological clock (Perez et al., 1998; Rengefors and Anderson, 1998). Excystment ceases in temperate climates in midspring, possibly because of anoxia of sediments (Heaney et al., 1983).

2. Life Cycles of Nonmotile Taxa

The order Phytodiniales (Dinococcales) includes cells that are nonmotile in the assimilative form and frequently include a parasitic or amoeboid stage. This group has complex, heteromorphic life histories, including parasitic or photosynthetic assimilative stages, gymnodinioid or amoeboid swarmers (gametes?), and cysts. In the past, different generic names have been assigned to different stages of the same life cycle, such as *Hemidinium* (photosynthetic, motile swarmer) and *Gloeodinium* (cyst), *Cystodinedria* (cyst?), and *Vampyrella* (motile cell). It is also possible that different genus designations have been assigned to slight differences in morphology of the same entity, such as *Cystodinium*, *Hypnodinium*, and *Dinococcus*.

Examples of complex life cycles include *Stylodinium*, which is an attached, round to oval cell on a stalk that produces amoebae that parasitize the filamentous chlorophyte *Oedogonium*. The amoebae then swell into the *Stylodinium* shape, which may release amoebae or gymnodinioid cells. Some gymnodinioid cells, with yellow-brown chromatophores and stigmas, behave like gametes (Pfiester and Popovsky, 1979). *Cystodinedria inermis* attached to *Oedogonium* was observed to release amoebae that later fed on filaments of *Spirogyra*, after which they became immobile, rounded up, and eventually took on the brownish appearance typical of *Cystodinedria* (Pfiester and Popovsky, 1979). *Cystodinium bataviense* reproduces both motile gymnodinioid zoospores and parasitic amoeboid stages (Pfiester and Lynch, 1980).

The fish parasite *Haidadinium* also has a complex life history. The photosynthetic vegetative cyst, with dinokaryon, induces hyperplasia in stickleback fish, although a trophont (feeding) stage is rarely observed (Buckland-Nicks and Reimchen, 1995). This cyst stage repeatedly divides, producing dinospores, rhizopodial amoebae, or lobose amoebae. Motile dinospores contain chloroplasts and a dinokaryon and may be the infective stage. Rhizopodial amoebae are heterotrophic, ingest bacteria, and eventually produce yellow resting cysts. Lobose amoebae, with a eukaryotic nucleus, may generate spheroid amoebae (Buckland-Nicks and Reimchen, 1995). Many of the stages contain symbiotic bacteria. This organism is placed in the Phytodiniales, as its nutrition does not depend on fish, and it has features in common with other members of the order.

Another fish parasite, *Oodinium*, has a parasitic trophont stage that feeds on the fish. A gymnodinioid swarmer is the infective form, settling onto the fish, extending a feeding tube, and then increasing greatly in size. This is followed by encystment and multiple divisions of the cyst, eventually producing the swarmer stage (Jacobs, 1946).

C. Classification

A classification scheme is sought that reflects the phylogenetic relationship of dinoflagellates to other protists and relationships within the dinoflagellates. Dinoflagellates appear to be monophyletic, sharing a large suite of characters unique to themselves. This uniqueness is the root of the difficulty in determining evolutionary directions. Dinoflagellates lack an outgroup for comparison, and attempts have been made to organize the distinctive morphological groups without clear agreement on what consititutes ancestral characteristics; these include a plate-increase model and a plate-decrease model. In the former, few-plated taxa (*Prorocentrum*) become increasingly fragmented, leading through *Peridinium*-like taxa to many plated forms (*Woloszynskia*) and ultimately gymnodinioids (Loeblich, 1976; Taylor, 1980). The plate-decrease model places gymnodinioids in the ancestral position and ends with few plated *Prorocentrum* (Eaton, 1980; Dodge, 1983). There is also a plate fragmentation model that attempts to reconcile the variation in living cells with the fossil record (Bujak and Williams 1981). Molecular sequencing data are beginning to resolve some relationships and confirm monophyly (Saunders *et al.*, 1997; Taylor, 1999). This monophyletic group has been historically claimed and named under both the Code of Botanical Nomenclature and Code of Zoological Nomenclature. Fensome *et al.*, (1993) use botanical nomenclature, but call the division Dinophyta rather than Pyrrhophyta (used by phycologists) or a division name based on a genus. The present account recognizes dinoflagellates as the division Pyrrhophyta.

Classification within the dinoflagellates (Tables IV and V) is complicated by species with complex life cycles. Since previous classification was based in part on morphology, there has been synonymizing as different parts of one species life cycle are united (Pfiester and Highfill, 1993). This chapter includes "cyst" names because (a) the cell may the assimilative form; (b) the cyst form may not yet be correlated with

TABLE IV Classification of Freshwater Dinoflagellates[a]

Kingdom: Protista
Division: Pyrrhophyta
Class: Dinophyceae
Order Blastodiniales
   Family: Oodiniaceae
      Genus: *Oodinium*
Order Dinamoebales
   Family Dinamoebaceae
      Genus: *Dinamoebidium*
Order Gymnodiniales
   Family: Actiniscaceae
      Genera: *Actiniscus, Pseudoactiniscus*
   Family: Gymnodiniaceae
      Genera: *Amphidinium, Bernardinium, Gymnodinium, Gyrodinium, Katodinium*
Order: Peridiniales
   Family: Gonyaulacaceae
      Genera: *Gonyaulax, Thompsodinium*
   Family: Peridiniaceae
      Genera: *Peridinium, Peridiniopsis, Durinskia*
   Family: Ceratiaceae
      Genus: *Ceratium*
   Family: Lophodiniaceae
      Genera: *Lophodinium, Woloszynskia*
   Family: Hemidiniaceae
      Genus: *Hemidinium*
   Family: Glenodiniopsidaceae
      Genera: *Glenodiniopsis, Sphaerodinium*
   Family: Dinosphaeraceae
      Genera: *Dinosphaera, Entzia, Kansodinium*
   Family Thecadiniaceae
      Genus: *Amphidiniopsis*
Order: Phytodiniales (Dinococcales)
   Family: Phytodiniaceae
      Genera: *Cystodinedria, Cystodinium, Dinastridium, Dinococcus, Haidadinium, Hypnodinium, Phytodinium, Rufusiella, Stylodinium, Tetradinium*
Order Prorocentrales
   Family Prorocentraceae
      Genus: *Exuviaella*

[a] Modified from Loeblich (1982), with new genera included and marine taxa excluded. Only genera reported from North America are included.

TABLE V Number of Species Reported from North American Literature[a]

Genus (#sp + forms)	U.S.	Canada	Mexico	Carib.+ C.A.
Oodinium (1)	1		1	
Dinamoebidium (1)	1			
Actiniscus (2)		2		
Pseudoactiniscus (1)		1		
Amphidinium (6)	6	2		
Bernardinium (1)	1	1		
Gymnodinium (29)	23	18	1	
Gyrodinium (1)	1			
Katodinium (7)	5	3		
Gonyaulax (3)	2			
Thompsodinium (1)	1			1
Peridinium (31+9)	22	23	7	6
Peridiniopsis (17)	13	11	2	4
Durinskia (1)	1	1		
Ceratium (5+8)	9	8	5	1
Lophodinium (1)	1		1	
Woloszynskia (5)	4	2		
Hemidinium (2)	2	1		1
Glenodiniopsis (1)	1	1		
Sphaerodinium (3)	3	1	1	
Dinosphaera (1)		1	1	
Entzia (1)	1			
Kansodinium (1)	1			
Amphidiniopsis (1)		1		
Cystodinedria (1)	1			
Cystodinium (6)	6	4		
Dinastridium (1)	1			
Dinococcus (2)	2			
Haidadinium (1)		1		
Hypnodinium (1)	1	1		
Phytodinium (1)	1			
Rufusiella (1)	1			
Stylodinium (3)	3	1		
Tetradinium (4)	3	1		
Exuviaella (1)	1			

[a] From Buckland-Nicks et al. (1997), Bursa (1969), Bursa (1970), Carty (1986), Carty (1993), Duthie and Socha (1976), Forest (1954), Haberyan et al. (1995), Meyer and Brook (1969), Nicholls (1998), Ortega (1984), Popovsky (1970), Poulin et al. (1995), Stein and Borden (1979), Taft and Taft (1971), Thompson (1947), Thompson (1949), Thompson (1950), Whitford and Schumacher (1984).

though *Spiniferites* had precedent, the name of the motile cell would be used. By this convention, all species of *Cystodinium* would become species of *Gymnodinium*. Some freshwater dinoflagellate genera have a nonmotile assimilative stage, are capable of reproducing similar cells via autospores, and have a brief, motile (gametic?) stage (e.g., *Cystodinium*). It makes more sense that the assimilative form have precedence.

Another complication for classification has been the division of species into thecate (with cellulose plates) or naked (lacking plates). While heavily thecate taxa are obvious in the light microscope, taxa with thin theca or apparently no theca are problematic. Steidinger et al., (1996b) have demonstrated that gymnodinioid (naked) cells may show evidence of plates when examined with the scanning electron microscope (SEM). Confounding the problems of life cycle and cell covering is the unknown degree of variability inherent in the arrangement of thecal plates. Taxa may be synonymized as they are recognized as nutritional or developmental variants, or speciated if differences are found to be stable.

Some generic distinctions are unclear, such as those between *Peridinium* and *Peridiniopsis*. *Peridinium* consists of at least two genera, those with larger cells and three intercalary plates (*Peridinium sensu stricto* according to Boltovskoy in Bujak and Davies, 1983) and those with smaller cells and two intercalary plates (the Umbonatum group). The species within the genus *Peridinium* need careful evaluation. Popovsky and Pfiester (1990) synonymized many species in *Peridinium*, which makes keying them easier, but this is less accurate than consulting a reference that includes all species. *Peridiniopsis* has clearly defined species, but it is questionable that they belong in the same genus. The variety of plate patterns (Table VI), coupled with currrent understanding of plate variation due to plate shifting, loss, and fusion, should enable us to determine which are basal patterns and which are derived.

TABLE VI Apical Plate Patterns of *Peridiniopsis*

Apical plates	Apical intercalary	Precingular	Example
3	1	6	P. borgei
3	1	7	P. lindemannii
4	0	6	P. penardii
4	0	7	P. elpatiewskyi
4	1	7	P. lubiensiforme
4	1	6	P. cunningtonii
5	0	6	P. cunningtonii
5	0	7	P. thompsonii
5	1	7	P. quadridens

an assimilative stage, or one cyst form may be correlated with more than one assimilative genus (as with the cyst genera *Hypnodinium* and *Dinastridium*); (c) many times only one form may be reported (either *Hemidinium* or *Gloeodinium*); and (d) the name of the motile cell rather than the cyst form may be used, according to the taxonomic convention. Linking cysts to motile cells from the marine environment allowed a better understanding of the taxonomic affinities of fossil cysts, many of which looked similar to modern cysts. The cyst genus *Spiniferites* was linked to the motile cell *Gonyaulax*, and it is proposed that even

Separation of species into new genera can give us better clues to the evolution of the thecate taxa.

Defining and distinguishing a species has always been a difficult task. A practical consideration should be if the taxon is morphologically distinctive enough to separate it from similar taxa, and whether this separation adds to our understanding of their ecological roles and evolutionary position. For example, forms of *Ceratium hirundinella*, hypothesized to be seasonal variations, may later be recognized as individual species based on distinctive cysts and stable morphology.

III. ECOLOGY AND DISTRIBUTION

A. Dinoflagellate Blooms

Generally, freshwater dinoflagellates are minor members of the summer phytoplankton maxima. However, a few species are capable of blooms. Most freshwater blooms are benign, almost unialgal assemblages (Tables VII and VIII). Harmful algal blooms (HABs) and red tides are generally not found in freshwater, although better definitions are needed (Smayda, 1997). The best known freshwater bloom former is *Ceratium hirundinella*, which has been studied in many parts of the United States, Canada, Europe, Africa, and Japan (Table VIII).

Densities of dinoflagellates in a body of water are related to both bottom-up (factors promoting growth) and top-down (factors causing loss) processes. Factors promoting growth include inoculum, inorganic nutrients, vitamins, light, oxygen, temperature, pH, and lake morphometry (depth, stratification, ratio of epilimnion to hypolimnion) (Pollingher, 1987). Factors causing loss include predation, disease, life history characteristics (cyst production), and outflow.

TABLE VII Freshwater Dinoflagellate Blooms

Organism	Location	Intensity	Comment	Reference
Peridinium bipes	Japan	Max 192 cells/ml		Park and Hayashi (1993)
	Japan	99%		Yoshikawa *et al.* (1997)
	Taiwan	95.7% biomass	Reservoir	Wu and Chou (1998)
Peridinium gatunense	Israel	95% biomass	Water odoriferous and brown	Lindström (1991)
	Israel	1672 cells/ml		Pollingher and Hickel (1991)
Peridinium inconspicuum	Finland	"Virtual monoculture"	Acidic lake, July–Aug	Holopainen (1992)
	OH, USA	88% composition, 690 cells/ml		Koryak (1978)
Peridinium limbatum	Canada	70%	Acid lake	Yan and Stokes (1978)
Peridinium lomnickii	England	>95% sample	Eutrophic lake	Cranwell *et al.* (1985)
Peridinium pusillum	GA, USA	88%, 1868 cells/L	July, dystrophic lake	Stoneburner and Smock (1980)
Peridinium willei	USA	1.5×10^5 cells/L	Shallow, sheltered areas	Stewart and Blinn (1976)
Peridiniopsis cunningtonii	Japan	4×10^4 cells/ml	Reservoir, summer >17°C	Sako *et al.* (1984)
Peridiniopsis penardii	Japan		winter–spring	Sako *et al.* (1987)
Woloszynskia reticulata	OK, USA	"Bloom condition"	August	Pfiester *et al.* (1980)

TABLE VIII *Ceratium hirundinella* Blooms

Location	Intensity	Comment
Utah L., USA (1)	89–100% total standing crop	Shallow, TDS 795–1650, high silt, PO_4, NO_3, pH 8.5+
Eau Galle Res. WI, USA (2)	36–94% total biomass	Eutrophic, moderately alkaline, shallow
Kam, NWT, Canada (4)	55×10^3 cells/m^3	Eutrophic, Aug, pH 6.9–7.9
Properus, NWT, Canada (4)	50×10^3 cells/m^3	Oligotrophic, Aug, DO > 80%, pH 7–7.4, low TP, low NO_3
Prelude, NWT, Canada (4)	48×10^3 cells/m^3	Oligotrophic, Aug, DO > 80%, pH 7–7.4, low TP, low NO_3
Grace, NWT, Canada (4)	92×10^3 cells/m^3	Mesotrophic, June/Sept, pH 6.8–7.7
Long, NWT, Canada (4)	26×10^3 cells/m^3	Mesotrophic, Aug/mSept, pH 6.8–7.7
Madeline, NWT, Canada (4)	100×10^3 cells/m^3	Mesotrophic, Aug, pH 6.8–7.7
Ishitigawa Res. Japan (3)	1300 cells/ml	
Esthwaite Water, GB (5)	10^3 cells/ml	
L. Balaton, Hungary (6)	4.85×10^4 cells/ml	Shallow
Goczalkowice Res. Poland (7)	4.3×10^5 cells/dm^3	
L. Sempach, Switz (8)	380 cells/ml	

(1) Whiting *et al.* (1978); (2) James *et al.* (1992); (3) Kawabata and Kagawa (1988); (4) Moore (1981); (5) Heaney and Talling (1980); (6) Padisák (1985); (7) Bucka and Zurek (1992); (8) Pollingher *et al.* (1993).

Bottom-up factors include cysts that may eventually serve as inoculum and which form during a previous bloom (see also Section II.B on the Life Cycle). *Ceratium hirundinella* has a high percentage of conversion between assimilative cells and cysts but low cyst viability (1–6%) (Pollingher et al., 1993). *Peridinium willei* has a high rate of survival of cysts (50–81%), though fewer cysts are produced from the planktonic population (Pollingher et al., 1993). *Peridinium gatunense* also has a low (1%) proportion of the population forming cysts, but this is enough to produce blooms in Lake Kinneret (Pollingher, 1987). The actual timing, intensity, and duration of a bloom involve factors such as turbulence, which resuspends the cysts (Pollingher and Hickel, 1991), and water depth. Shallow lakes have greater emergence rates (*Peridinium limbatum*, Sanderson and Frost, 1996).

Inorganic nutrients, particularly high levels of nitrates and phosphates, are often cited as factors necessary to trigger blooms (Whiting et al., 1978). *Ceratium hirundinella* was collected in higher numbers near a river inflow with elevated NO_3 and PO_4 than the reservoir (Kawabata and Kagawa, 1988). Uptake of PO_4 by *Ceratium hirundinella* may be favored in June by downward migration at night (up to four meters) to the upper hypolimnion, but in July anoxia restricts migration and phosphate uptake, and a decrease in *Ceratium hirundinella* biomass is noted (James, et al., 1992). The mechanism explaining the effect of nitrates on blooms may be a low affinity for nitrate by nitrogen reductase, which allows the dinoflagellate to outcompete other algae (Witt et al., 1999, for *Peridinium gatunense*). *Peridinium* abundance may also correlate with phosphorus levels, but not with nitrogen (Wu and Chou, 1998). Other studies seem to indicate that above a minimal level nutrients are not limiting (Padisák, 1985; Sanderson and Frost, 1996). In addition to inorganic nutrients, many dinoflagellates seem to require vitamins such as B_{12} (Bruno and McLaughlin, 1977; Holt and Pfiester, 1981).

Light is an important factor in the vertical distribution of dinoflagellates. Over 95% of cell counts were from subsurface waters (3–15 m) in Feitsui Reservoir, Taiwan (Wu and Chou, 1998). Heaney and Talling (1980) similarly found that *C. hirundinella* avoided surface waters of Esthwaite Water, U.K. In culture, higher light intensity yielded higher cell densities for this species (Bruno and McLaughlin, 1977), but in nature the compromise between surface light and nutrient supply at depth may explain why *C. hirundinella* can be situated at depths corresponding to about 10% of summer surface irradiance values (Heaney and Furnass, 1980), or even a deep chlorophyll maximum at 1–3% surface incident light (Gálvez et al., 1988; Echevarria and Rodriguez, 1994).

Most dinoflagellates are long day or warm temperature organisms with maximum growth during the summer. *Ceratium hirundinella* has been observed in plankton in March–April, with exponential growth during June and July, followed by a stationary period in July–September and a decline in October–November (Padisák, 1985). Growth occurs in subarctic lakes from June to September, with maximum abundance when water is between 4°C and 18°C, suggesting that temperature may not be a controlling factor. Blooms occur as days shorten and when there are strong thermoclines (Moore, 1981).

In culture, an axenic strain of *Ceratium hirundinella* from Calder Lake, NY, grew at all tested pH values from 5.0 to 8.5, and grew best at pH 7.0–7.5 (Bruno and McLaughlin, 1977). This same strain also grew best at total dissolved solids ranging from 48 to 960 mg/L. (Bruno and McLaughlin, 1977), but in the field this species has been found blooming between 795 and 1650 mg/L (Utah Lake; Whiting et al., 1978).

Interwoven in many of the resource factors is the influence of stratification. Dinoflagellates, being motile, can remain in the light, oxygenated, warmer surface waters and avoid sedimentation into the dark, anoxic hypolimnion (Heaney and Talling, 1980). Experiments with *Ceratium* in unmixed columns support these observations (Klemer and Barko, 1991). It has been suggested that in eutrophic lakes, dinoflagellates are absent only from system with high flushing rates or unstratified water columns (Klemer and Barko, 1991, citing Sommer et al., 1986).

Top-down factors may be important for freshwater dinoflagellates, although little documentation exists, and some of that is conflicting. This is particularly true for *Ceratium*, which has been described as a "large, relatively ungrazed species" (Smayda, 1997). There even is some evidence that it reduces the number of zooplankters during a bloom (Bucka and Zurek, 1992). Another study in a fishless pond found that a crash in the population of *Daphnia* (filter feeding cladoceran) corresponded with a bloom of *Ceratium hirundinella* that displaced populations of smaller phytoplankton species that *Daphnia* fed upon. Lacking *Daphnia* as a food source, larger dipteran predators ate *Ceratium* (Xie et al., 1998). Other large dinoflagellates such as *Peridinium limbatum* may also be unaffected by grazing pressure, although effects may also depend on zooplankton size (Sanderson and Frost, 1996).

Dinoflagellate losses can also occur through lake outflows, especially if wind direction has concentrated cells at the outflow end. Loss of 880 cells mL^{-1} × a discharge of 87×10^3 m^3 d^{-1} produced a loss of 7.65×10^{13} cells d^{-1} (Heaney and Talling, 1980). Changing the site of withdrawal from the hypolimnion to the surface may also drastically reduce biomass (James et al., 1992).

The number of dinoflagellates occurring in a body of water is also determined by biological factors such as motility and physical features of the lake. Motility allows for vertical migration related to a circadian rhythm. *Ceratium hirundinella* in particular may be localized at the surface during early hours and deeper by the afternoon (Heaney and Furnass, 1980; Padisák, 1985). In a deep, subarctic Canadian lake, densities of *C. hirundinella* were $> 4 \times 10^4$ cells m^{-3} at 60 m and $> 2 \times 10^3$ cells m^{-3} at 70-m depths (Moore, 1981). *Ceratium* also apparently position themselves between the foam rows caused by Langmuir circulation in larger lakes (Squires et al., 1979) and show variable horizontal distributions within reservoirs from inflow to dam (Padisak, 1985; Kawabata and Kagawa, 1988). High densities of *Peridinium* have also been found near the inflow (Wu and Chou, 1998).

B. Trophic States

Dinoflagellates are highly variable in trophic status, including pigmented autotrophs, auxotrophs (require exogenous vitamins), mixotrophs (combine autotrophy with phagotrophy), and organotrophs (strict heterotrophs lacking chloroplasts) (Holt and Pfiester, 1981; Gaines and Elbrächter, 1987; Stoecker, 1998). Mixotrophic phagotrophy is exhibited in *Ceratium hirundinella* through production of a feeding veil (pallium), extracellular digestion, and a pseudopod that draws prey into the cell. Dinoflagellates utilizing a pallium begin the sequence with a pre-capture swimming pattern in which a tow filament rapidly connects with prey and a pseudopod is extended over the prey. Following digestion of the prey cell contents, the pallium is retracted into the theca (Jacobson and Anderson, 1986). Details of the feeding behavior of *Peridiniopsis berolinensie* indicate a chemosensory attraction to injured prey, attachment via a capture filament, extension of a feeding tube (form of peduncle), and suction of the prey contents (Calado and Moestrup, 1997). A feeding rate of 0.6 cells hr^{-1} has been measured with cryptophyte prey in the lab (Weisse and Kirchhoff, 1997). Organotrophy includes osmotrophy (absorbing dissolved organics), phagotrophy (particle ingestion, dinoflagellates ingest other dinoflagellates, diatoms, cyanobacteria, ciliates, metazoans), myzocytosis (cell contents sucked out via peduncle, seen in *Katodinium fungiforme*, *Cystodinedria*, and *Stylodinium*; Frey and Stoermer, 1980; Spero, 1985), and ectoparasitism of fish (*Oodinium*, Jacobs, 1946; *Haidadinium*, Buckland-Nicks et al., 1997).

C. Specificity of Habitat

Dinoflagellates occur most often in lentic habitats. Some are more common in large bodies of water, such as reservoirs and lakes (*Ceratium hirundinella*, *Peridiniopsis polonicum*), and others in ponds (Dinococcales found in duckweed ponds). Some species are more frequently encountered in soft water (acid to neutral), such as *Peridinium limbatum*, *Gymnodinium caudatum*, and *Ceratium carolinianum*), while others are more common in hard water (alkaline), such as *Ceratium hirundinella*. Some species may be more prevalent in eutrophic systems (*Ceratium hirundinella*) and others in brackish systems (*Gonyaulax*, *Exuviaella*).

Unusual habitats may also harbor dinoflagellates. *Rufusiella* was reported "on the under surface of a dripping sandstone ledge" (Richards, 1962), and *Dinamoebidium* was attached to sand grains in an alpine stream (Bursa, 1970). Sand dwelling dinoflagellates are known from marine sands (Saunders and Dodge, 1984) and have been collected from a freshwater sandy beach (Nicholls, 1998). Fish parasites are well known, especially by aquarium hobbyists (Ling et al., 1993; Buckland-Nicks and Reimchen, 1995).

D. Geographical Distribution

Too little is known about the geographic distribution of dinoflagellate species to be able to determine if patterns exist. However, of 11 armored species and 13 unarmored species first reported from North America (Table IX), only three (*Peridinium gatunense*, *P. limbatum*, *Ceratium carolinianum*) have been reported elsewhere. Some species, such as *Ceratium hirundinella* and *Peridinium gatunense*, are cosmopolitan, but the lack of reports for other species should not be considered evidence of rarity. Species may be common but unnoticed, and some species have patchy distributions. Three examples from my own experience may provide some insight.

Lophodinium polylophum is probably a rare species. It was first collected in Paraguay (Daday, 1905) and has since been collected once from Mexico (Osorio-Tafall, 1942), and from one small pond in College Station, TX (Carty and Cox, 1985). It has not been reported in the literature since, although it is quite distinctive (Figs. 3H and 8E and F).

A case of a common species absent from a state record is *Peridiniopsis polonicum*, another distinctive phytoplanker (Fig. 5J and K). It was not previously reported from Ohio (Taft and Taft, 1971) but has since been found in 24 countries (Carty, 1993; Carty and Fazio, 1997). It is generally widespread and commonly reported from North America and was probably absent

TABLE IX Armored and Unarmored Freshwater Taxa First Identified in North America

Name	Year	Locality
Ceratium carolinianum	1850	South Carolina
Peridinium limbatum	1888	New Jersey
Peridinium gatunense	1925	Panama
Peridinium wisconsinense	1930	Wisconsin
Gymnodinium caudatum	1944	Wisconsin
Oodinium limneticum	1946	Minnesota
Peridiniopsis thompsonii	1947	Kansas
Gymnodinium marylandicum	1947	Maryland
Tetradinium simplex	1949	Michigan
Stylodinium longipes	1949	Maryland
Rufusiella insignis	1949	Kansas
Gymnodinium cruciatum	1950	Kansas
Kansodinium ambiguum	1950	Kansas
Thompsodinium intermedium	1950	Kansas
Sphaerodinium fimbriatum	1950	Kansas
Woloszynskia reticulata	1950	Kansas
Woloszynskia cestocoetes	1950	Kansas
Actiniscus canadensis	1969	Northwest Territories
Pseudoactiniscus apentasterias	1969	Northwest Territories
Dinamoebidium coloradense	1970	Colorado
Katodinium auratum	1970	Colorado
Amphidinium cryophilum	1982	Wisconsin
Haidadinium ichthyophilum	1997	British Columbia
Amphidiniopsis sibbaldii	1998	Ontario

because of the limited geographical area covered by the authors.

Thompsodinium intermedium was first reported by Thompson in 1950 from Kansas. Later it was found in Cuba (Popovsky, 1970), Texas (Carty, 1986, 1989), and one location in Ohio (Carty, 1993). I have since found large numbers in one deep kettle lake on Long Island, NY, and in large numbers in one tiny, brown-water pond on a peninsula in Belize, (Carty, unpublished observations). This may be a widespread species that is not recognized because of difficulties in identification.

IV. COLLECTION AND PREPARATION FOR IDENTIFICATION

A. Collection

Dinoflagellates are predominently planktonic and may be collected in whole-water samples or concentrated with a plankton net (10-μm mesh). It is advantageous to sample from different depths, in either discrete aliquots (e.g., Van Dorn sampler) or a tow, as the motile cells can travel throughout the water column and may not be at the surface. Squeezings from submerged vegetation (algal and macrophyte) can also be collected (Thompson, 1947). Whole-water samples may be centifuged to concentrate organisms or placed in tapered-bottomed containers and the sediment examined. It is also useful to scrape fish. Any freshwater source may be examined, including reservoirs, lakes, ponds, marshes, stock tanks, permanent ditches, fish hatcheries, rivers, and creeks (see also Section III.C). While I have not sampled swimming pools, I have found dinoflagellates in waters treated with commercial shading-type chemicals. In reviewing 330 samples from Ohio, USA, collected in 1997, one-third (107) had no dinoflagellates, another third (114) had one species (frequently *Ceratium hirundinella*), and a third (107) had more than one, and up to nine, dinoflagellate species (Carty, unpublished observations).

Dinoflagellates are in greatest abundance during the summer months, but autumn, spring, and winter collections should also be made. Some species seem more prevalent in cooler months, and some are exclusively cold-weather species. Thompson (1947) has collected cells from beneath ice.

B. Fixation

Samples are usually fixed with Lugol's iodine (to which glycerin is added), which is added until a tea color is reached. Lugol's is the most widely used preservative, as it is known to maintain delicate flagella and its pH is not detrimental to cell coverings. Fixing a subsample of all collections while in the field will prevent further predation of algal cells by zooplankton and may highlight the presence of scarce dinoflagellates as iodine turns their starch black. After the live sample has been examined and found to contain useful material, the remainder of the sample may be preserved with Lugol's or glutaraldehyde for later work with the SEM.

Long-term storage of samples may be in Lugol's, although the iodine bleaches with time and fungi can grow. Permanent slides may be prepared with syrup medium (Taft, 1978).

C. Preparation for Identification

Examination of living cells gives details of shape, coloring, size, and swimming pattern crucial for identification. Most naked dinoflagellates lose their shape at death and become spherical. Some species seem to congregate in the water outside of the coverslip, and this area should be examined. While most features can be seen with light microscopy, both phase and interference contrast optics improve the visualization of features.

Positive identification of thecate dinoflagellates requires reconstruction of the plate pattern and/or recognition of certain defining characteristics. Many cells will retract from their outer wall, a useful feature for identifying the presence of a theca in those taxa with an obscure or thin theca such as *Woloszynskia*, especially in reaction to heat stress from the microscope light. Live samples may be left overnight, and the sedimented material may be examined the next day before too many cells have ecdysed. Both living cells and empty thecas may be rolled in the viewing field by gently blowing near the coverslip. Thecae may be dissociated and better seen after the addition of 5% sodium hypochlorite under the coverslip; slight to moderate pressure on the coverslip may be necessary for some species (Boltovskoy, 1975, 1976, 1989). Various stains may be used to aid in identification or to gain additional information about the cell. Examples of stains include Sudan stains for lipids, acetocarmine, feulgen, or propionic-lactic-orcein stain for the nucleus, and hydroiodic acid, trypan blue, or chlor-zinc iodine for cellulose (Pearsall, 1929; Sournia, 1978).

Scanning electron microscopy (SEM) provides definitive information of plate shapes, numbers, and relationships and may be the easiest method for the reconstruction of sulcal plates. Samples preserved with 2% glutaraldehyde or other fixatives should be critical-point dried before sputter coating with gold-paladium. It may be necessary to remove the outer cell membrane to see thin plates (Steidinger *et al.*, 1996b). Hexadimethyldisilazane (HMDS) and air drying may be satisfactory for heavily thecate species.

V. KEY AND DESCRIPTIONS OF GENERA

A. Key

The frequently encountered cell-stage in the life cycle of freshwater dinoflagellates.

1a.	Cell attached (or with attachment stalk and disc)	2
1b.	Cell free	9
2a.	No cingulum evident	3
2b.	Attached to sand, moss, slight cingulum (Fig. 2T)	*Dinamoebidium*
3a.	Spherical cell on distinctive stalk, attached to filamentous algae (Figs. 2L and 10G)	*Stylodinium*
3b.	Cell on short stalk, attachment disc or directly attached	4
4a.	Cell contours rounded, sessile, attached to filamentous algae (Figs. 2 and 10D)	*Cystodinedria*
4b.	Cell otherwise	5
5a.	Cell ovoid, with short stalk, attached to fish	6
5b.	Cell with angles, drawn out into spines	7
6a.	Cell ovoid, with short stalk (Fig. 2N)	*Oodinium*
6b.	Cell round with translucent, fenestrated matrix (Fig. 2O)	*Haidadinium*
7a.	Oval cell with spines at both ends (Fig. 2K)	*Dinococcus*
7b.	Angular cell	8
8a.	Tetragonal cell with spines at corners (Figs. 2 and 10H)	*Tetradinium*
8b.	Irregularly polygonal cell, corners with blunt spines (Figs. 2 and 10I)	*Dinastridium*
9a.	No cingulum evident	10
9b.	Cingulum present	15
10a.	Cell surrounded by sheath	11
10b.	Thecate cell lacking sheath (Fig. 3A)	*Exuviaella*
11a.	Cells in thick mucilaginous sheaths	12
11b.	Cells in firm, discrete envelope	13
12a.	Cell in several-layered envelope (like *Gloeocystis*; Figs. 2R and 10J)	*Hemidinium* (cyst)
12b.	Cell in many-layered, tubelike envelope (like *Hormotilia*) (Fig. 2S)	*Rufusiella*

13a.	Cell oval, brown, may have terminal spine(s) (Fig. 2J)	*Cystodinium*
13b.	Cell round	14
14a.	Cell with chloroplasts in roseate clusters, large vacuoles (Fig. 2J)	*Hypnodinium*
14b.	Cell not as above (Figs. 2Q and 10F)	*Phytodinium*
15a.	Cingulum incomplete	16
15b.	Cingulum encircles cell	17
16a.	Cell photosynthetic (with chloroplasts; Fig. 6A)	*Hemidinium*
16b.	Cell heterotrophic (lacking chloroplasts; Fig. 2D)	*Bernardinium*
17a.	Cingulum not medial	18
17b.	Cingulum medial	21
18a.	Cingulum divides cell into ≤1/3 epitheca, ≥2/3 hypotheca	19
18b.	Cingulum divides cell into 2/3 epitheca, 1/3 hypotheca (Fig. 2G)	*Katodinium*
19a.	Cell thecate	20
19b.	Cell nonthecate (Fig. 2C)	*Amphidinium*
20a.	Cell with two large plates, often near marine environment (Fig. 3A)	*Exuviaella*
20b.	Cell with more plates, laterally compressed (Fig. 6G)	*Amphidiniopsis*
21a.	Cingulum ends offset more than 1.5 cingulum widths	22
21b.	Cingulum ends offset none to less than 1.5 cingulum widths	23
22a.	Cell athecate (Fig. 2F)	*Gyrodinium*
22b.	Cell heavily thecate, often near marine environments (Fig. 3B)	*Gonyaulax*
23a.	Cell with distinctive plates	24
23b.	Cell athecate, with very thin plates, or uncertain	31
24a.	Cell with one apical and 2–3 hypothecal horns (Figs. 3G and 8A–C)	*Ceratium*
24b.	Cell without extended horns	25
25a.	Cell heterotrophic, may have pink cytoplasm, 4′, 2a, 7″, 5‴, 1″″ (Figs. 6D and 8H)	*Entzia*
25b.	Cell yellow-golden brown	26
26a.	Cell has spine(s) on hypothecal plates	27
26b.	Cell lacks spines, may have lists or fimbrae on plates	28
27a.	Plate tabulation 4′, 2–3a, 7″, 5‴, 2″″ (Figs. 4A–D)	*Peridinium* (in part)
27b.	Plate tabulation otherwise 3–5′, 0–1a, 6–7″, 5‴, 2″″ (Fig. 3E)	*Peridiniopsis* (in part)
28a.	Cell with definite thick plate sutures only on hypotheca (Fig. 3I)	*Woloszynskia reticulata*
28b.	Sutures uniform on cell	29
29a.	Single, fringed list on 1″″ plate, 4′, 3a, 6″, 5‴, 2″″ (Fig. 3C)	*Thompsodinium*
29b.	Plate tabulation otherwise	30
30a.	Plate tabulation 4′, 2–3a, 7″, 5‴, 2″″ (Figs. 4A–D)	*Peridinium* (in part)
30b.	Plate tabulation otherwise, 3–5′, 0–1a, 6–7″, 5‴, 2″″ (Fig. 3E)	*Peridiniopsis* (in part)
31a.	Athecate with cell membrane or pellicle	32
31b.	Theca thin (can discern cell contents separate from outer covering), plate sutures not or barely visible	34
32a.	Cell from high Arctic lake, may have pentasters within	33
32b.	Cell from more temperate climate, athecate (Figs. 2E and 10B)	*Gymnodinium*
33a.	Cell with internal siliceous pentasters, rare (Fig. 2A)	*Actiniscus*

33b.	Cell lacking pentasters, rare (Fig. 2B)	Pseudoactiniscus
34a.	Cell with many, thin, polygonal plates	35
34b.	Cell with definite plate pattern	36
35a.	Cell spindle shaped, with vertical ridges (Figs. 3H and 8E–G)	Lophodinium
35b.	Cell round (Figs. 3I and 8I–J)	Woloszynskia
36a.	Three apical plates, one antapical plate	37
36b.	Four apical plates, two antapical plates	38
37a.	Five precingular plates, apical pore, 3′, 1a, 5″, 5‴, 1″″ (Fig. 6E)	Kansodinium
37b.	Six precingular plates, no apical pore, 3′, 1a, 6″, 5‴, 1″″ (Fig. 6F)	Dinosphaera
38a.	Two anterior intercalary plates, 4′, 2a, 6″, 5‴, 2″″ (Fig. 3F)	Durinskia
38b.	Four anterior intercalary plates	39
39a.	Seven precingular plates, 4′, 4a, 7″, 6‴, 2″″ (Figs. 6C and 7A–B)	Sphaerodinium
39b.	Eight precingular plates, 4′, 4a, 8″, 6–8‴, 2″″ (Fig. 6B)	Glenodiniopsis

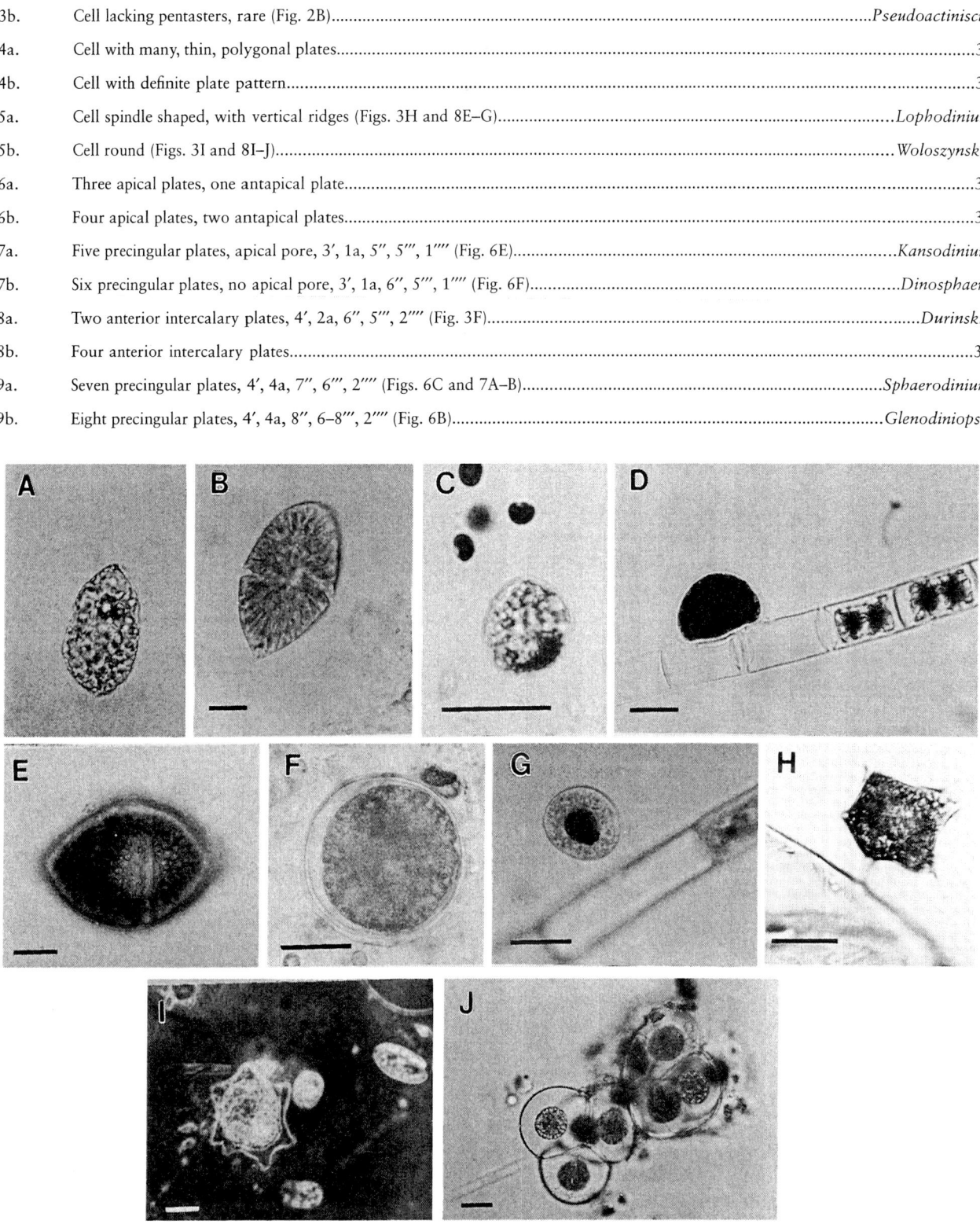

FIGURE 10 Light micrographs of unarmored dinoflagellates, location included (county, state). Line scale = 20 μm. (A) *Bernardinium bernardinense* (Brazos, TX). (B) *Gymnodinium fuscum* (Burleson, TX). (C) *Katodinium spiroidinoides* (Marion, OH). (D) *Cystodinedria inermis* (Williams, OH). (E) *Cystodinium bataviense* (Montgomery, TX). (F) *Phytodinium simplex* (Brazos, TX). (G) *Stylodinium globosum* (Kenedy, TX). (H) *Tetradinium javanicum* (Brazos, TX). (I) *Dinastridium sexangulare* (Brazos, TX). (J) *Hemidinium* cyst stage (Brazos, TX).

B. Descriptions of Genera

This section includes the genus name, authority, figure references, the number of species reported from North America, name of species if just one, a brief description, plate pattern, sizes (based on North American specimens where possible), and topical reference if applicable. Several included taxa are from single (worldwide) reports; they are included to call attention to how little is known about dinoflagellates and how much remains to be done. This listing follows the order of genera in Table IV.

Order Blastodiniales
Family Oodiniaceae

Oodinium Chatton *O. limneticum.* (Fig. 2N) Epizooic, life cycle includes parasitic, nonmotile cell in the assimilative stage, a cyst, and gymnodinioid swarmers that infect fish. The parasitic cell has light-green to olive chromoplasts, starch, no eyespot, and grows from about 28 μm to 60 μm. It is attached by tentacle-like rhizoids and is enclosed in a thin cellulose wall. Spherical cysts are about 71 μm in diameter. Cysts undergo several divisions, the final forming gymnodinioid swarmers about 15 μm long. Swarmers have a deep cingulum dividing the cell into a larger epicone and smaller hypocone, and a sulcus that extends to the antapex, no eyespot, and numerous yellow-green chromoplasts (Jacobs, 1946). Lom (1981) differentiated parasitic dinoflagellates based primarily on host and features of the trophont stage (especially the mode of attachment). *O. limneticum* was transferred to *Piscinoodinium*, a genus erected for freshwater ectoparasites (Lom, 1981). *O. limneticum* does not attach to hosts like the type species *P. pillulare*, so further study is required. In addition to the original report, there has been one report from Mexico.

Order Dinamoebales
Family Dinamoebaceae

Dinamoebidium Pascher (Fig. 2T) Athecate, gymnodinioid/amœboid cell with irregularly scalloped periphery, nonmotile, no eyespot, with golden chromatophores, cells 19.8–38.5 μm long, 13–23 μm wide, amoebae 28–46.5 μm. Found in cold, alpine stream attached to sand grains and moss. Dinoflagellate features include a large nucleus and an "equatorial girdle-like groove" in the gymnodinioid cells. Amoeboid cells ingest algae. There has been one report from CO (USA) (Bursa, 1970).

Order Gymnodiniales
Family Actiniscaceae Kützing

Actiniscus Ehrenberg (Fig. 2A) Athecate with internal siliceous starlike element (pentaster), best known from fossils and marine species, found in Great Bear Lake, Canada. There has been one report (Bursa, 1969). The two freshwater species were not validly described, but marine species are known; additional work needs to be done. While no measurements are given in the text, based on the magnification of the figure, cells are about 59 μm long × 72 μm in diameter.

Pseudoactiniscus Bursa *P. apentasterias* (Fig. 2B) Similar to *Actiniscus* but lacking pentasters, found in Canada, one report (Bursa, 1969). The genus and species have not been validly described, and additional work needs to be done. While no measurements are given in the text, based on the magnification of the figure, cells are about 70 μm long × 69 μm diameter.

Family Gymnodiniaceae (Bergh) Schütt

Amphidinium Claparède *et* Lachmann (Fig. 2C) Athecate, cingulum divides cell into ≤1/3 epitheca, ≥2/3 hypotheca, cells 18.6–22 μm long, 10–14 μm wide, 7–10 μm thick. Four species; reported infrequently from the United States and Canada.

Bernardinium Chodat (Figs. 2D and 10A) Like *Hemidinium* with incomplete cingulum, but nonphotosynthetic, athecate. Cytoplasm may contain red or orange inclusions (food vacuoles?); cells 13–16 μm wide × 20–21 μm long, with eyespot. There have been several reports from the United States.

Gymnodinium Stein (Figs. 2E and 10B) Classic "naked" dinoflagellate, some photosynthetic with peridinin, two bluegreen, others heterotrophic. Cingulum bisects cell. Must be examined alive for species determination. Cells small (*G. triceratium* 16 μm × 13 μm) to large (*G. fuscum* 55–60 μm × 80–100 μm). Genus reported (29 species) throughout the United States and Canada.

Gyrodinium Kofoid *et* Swezy (Fig. 2F) Cingulum offset 1.5 times, athecate. One report from MD (USA) (Thompson, 1947); cells dorsoventrally compressed, with eyespot, chromatophores present, 25–32 μm long, 18–20 μm wide, 14–15 μm thick, collected in January.

Katodinium Fott (Figs. 2G and 10C) Athecate, cingulum divides cell into 2/3 epitheca, 1/3 hypotheca, many heterotrophic; most are small, 4–16 μm × 6–12 μm. *Massartia* is a synonym. Reported infrequently from the United States and Canada. Must be viewed alive and swimming for genus and species identification (Christen, 1961). Species (7 known) are determined by size, shape, +/− stigma, +/− plastids, sulcal features.

Order Peridiniales Schütt

Family Gonyaulacaceae Lindemann

Gonyaulax Diesing (Figs. 3B and 5A) Heavily thecate, mainly found in brackish water, most species marine; 2–4′, 0–3a, 5–6″, 1p, 5–6‴, 1–2⁗, cingulum very displaced, cells 28–62 μm × 29–58 μm, infrequently reported (3 species).

Thompsodinium Bourrelly (Figs. 3C and 5B–F) Photosynthetic golden chloroplasts, with eyespot, fringed list near antapex; 2‴, 3‴, and 4‴ plate, 4′, 3a, 6″, C6, 5‴, 2⁗, 4 sulcal plates, the 2a plate is variable in shape. Cells 28–43 μm long × 26–40 μm in diameter with slight dorsoventral compression; cingulum without displacement. Collected in the United States, Cuba, Belize (see IIID).

Family Peridiniaceae Ehrenberg

Peridinium Ehrenberg (Figs. 4A–D, 3D, 5G–I) Thecal tabulation 4′, 2–3a, 7″, 5‴, 2⁗; heavily thecate, plates ornamented, most photosynthetic, either with an apical pore (*Poroperidinium Lefèvre*) or without (*Cleistoperidinium Lefèvre*). While most species (ca. 31 described) are round to oval, some have lobes (*P. cinctum* f. *tuberosum*, *P. willei*, *P. limbatum*), and some have distinctive lists (*P. willei*). Most of the subgroups within *Peridinium* produce large, robust cells 60–65 μm long and wide, the Umbonatum group includes smaller cells 13–30 μm long × 11–27 μm in diameter. (Lefèvre, 1932). Widely reported (especially *P. inconspicuum* and *P. willei*) from Canada, Caribbean, Mexico, and the United States.

Peridiniopsis Lemmermann (Figs. 3E, 5J–M) Thecal tabulation 3–5′, 0–1a, 6–7″, 5‴, 2⁗, slight to moderately thecate, some with plate ornamentation, some photosynthetic, some heterotrophic, 32–47 μm long × 21–39 μm in diameter. The wide range of plate patterns requires plate determination for identification (Table VI). Species in the genus *Glenodinium* with known plate patterns were transferred to this genus by Bourrelly (1968), although the genus *Glenodinium* persists in the literature. Widely reported (17 species) from Canada, Caribbean, Mexico, and the United States.

Durinskia Carty et Cox (Figs. 3F, 5N–O) Thecal tabulation 4′, 2a, 6″, 5‴, 2⁗, photosynthetic, with eyespot, cell round to slightly oval, 26–33 μm long × 26–32 μm in diameter, with an apical pore, thin theca, no ornamentation. Collected from freshwater and saline environments. Infrequently reported, found in TX and OH (USA), and, as *Peridinium dybowski*, reported from Canada and MN (USA). As a *Peridinium balticum* isolate from the Salton Sea, CA (USA), it has been intensely studied because of its binucleate status (Tomas and Cox, 1973; Chesnick and Cox, 1987, 1989).

Family Ceratiaceae (Schütt) Lindemann

Ceratium Schrank (Figs. 3G and 8A–D) Thecal tabulation 4′, 5″, 5‴, 2⁗; only genus with 1–2 horns formed from postcingular plates, with an apical horn (apical plates) which may have an apical pore and antapical horn (anatapical plates), heavily thecate, plates ornamented, mixotrophic, pale yellow to golden, no eyespot. Stubby *C. brachyceros* may be 33–40 μm wide × 65–80 μm long; *C. hirundinella* may be over 400 μm long. *Ceratium hirundinella* (and many of its forms) are widely reported.

Family Lophodiniaceae Lemmermann

Lophodinium Lemmermann (Figs. 3H and 8E–G) Thin theca of many hexagonal plates arranged into vertical ridges, photosynthetic with numerous oval golden chloroplasts, eyespot in sulcus, 42–44 (–62 in dividing cells) μm long × 31–41(–54) μm in diameter (Osorio-Tafall (1942) reports cells 70–80 μm long × 63–67 μm in diameter). Apical slit bounded by carina (ridge) not continuous with sulcus. Rarely encountered, it may have a brief planktonic stage. Reported from Mexico and TX (USA) (Carty and Cox, 1985).

Woloszynskia Thompson (Figs. 3I and 8I–J) Cells round with numerous polygonal thin plates which are difficult to see, photosynthetic, infrequently reported. Size: 20–52 μm long × 14–46 μm wide, may have an apical slit. Distinctive cysts may verify the presence of the genus in the plankton. Reported (five species) from OK, TX, KS, OH, MD (USA).

Family Hemidiniaceae Bourrelly

Hemidinium Stein (Figs. 6A and 8K–L; cyst Figs. 2R and 10J) Motile cell with thin plates (very difficult to see), incomplete cingulum gives cell a "slashed" appearance, photosynthetic with golden brown chloroplasts, 24–29 μm long × 11–20 μm in diameter. *Gloeodinium*, a genus of immobile round cells in a thick gelatinous matrix was found to be a stage in the life cycle of *Hemidinium* (Pfiester and Highfill, 1993). The cyst (*Gloeodinium*) stage has 2–4 nonmotile, round cells in mucilage. Cells have numerous brown chromatophores, are 18–28 μm in diameter; colony is 69–74 μm in diameter. Cysts are epiphytic or free in the plankton.

Family Glenodiniopsidaceae Schiller

Glenodiniopsis Woloszynska (Fig. 6B) Thecal tabulation 4′, 4a, 8″, 6–8‴, 2⁗; no apical pore, photosynthetic, thin asymmetrically arranged plates may require SEM for verification; 26–50 μm long × 26–33 μm in diameter (Highfill and Pfiester 1992a,b). Three reports (OK, MN, BC).

Sphaerodinium Woloszynska (Figs. 6C and 7A–B) Thecal tabulation 4′, 4a, 7″, 6‴, 2⁗; photosynthetic with golden chloroplasts, with an eyespot. Plate extensions give a distinctive cell outline for *S. fimbriatum*. Size: 42–53 μm long × 32–46 μm in diameter, infrequently reported. Genus is considered a synonym of *Glenodinium*, based on the description of *Glenodinium cinctum* with a horseshoe-shaped eyespot, which was also found in *Sphaerodinium polonicum* (Loeblich, 1980). There were no plates originally figured for *Glenodinium*, but there is a known pattern for *Sphaerodinium*. It is useful to maintain *Sphaerodinium* until *Glenodinium* is defined.

Family Dinosphaeraceae Lindemann

Dinosphaera Kofoid *et* Michener (Fig. 6F) Thecal tabulation 3′, 1a, 6″, 5‴, 1⁗; no apical pore, photosynthetic, cells 25–30 μm diameter, 27–34 μm long. Reported, as *Glenodinium palustre*, from soft water lakes and bogs in WI (USA) (Prescott, 1951) and as *Gonyaulax palustris* from IA, IL, MA, MN.

Entzia Lebour (Figs. 6D and 8H) Thecal tabulation 4′, 2a, 7″, 5‴, 1⁗; heavily thecate with lightly reticulate ornamentation, related to marine genus *Diplopsalis*, heterotrophic, with apical pore, may have pink cytoplasm, distinctive feature a prominent sulcal list that extends past the antapex. Cell may ecdyse a pink, motile, gymnodinioid/katodinioid cell. Size: 30–38 μm long × 26–38 μm in diameter. Reported from OH (USA).

Kansodinium Carty *et* Cox (Figs. 6E and 7C–D) Thecate, 3′, 1a, 5″, 5‴, 1⁗, apical pore surrounded by apical collar, photosynthetic, with eyespot, 32–42 μm long × 27–39 μm in diameter × 27–31 μm thick, rounded cell with thin plates which require empty cells or SEM for verification. Reported from KS, TX (USA) (Carty and Cox, 1986).

Family Thecadiniaceae

Amphidiniopsis Woloszynska (Fig. 6G) Thecate, laterally compressed, P_o, 4′, 3a, 7″, 5c, 4(?)s, 5‴, 2⁗, apical pore, lacking chloroplasts, cells brownish-gray. Cingulum divides cell into about 1/5 epitheca, 4/5 hypotheca, cells 33–45 μm long, 19–28 μm lateral width, 24–34 μm dorsoventral width. Collected from sand, Ontario (Nicholls, 1998).

Order Phytodiniales (Dinococcales)

Family Phytodiniaceae Klebs

Cystodinedria Pascher (Figs. 2H and 10D) Oval cell, athecate, 38–48 μm long × 22–32 μm in diameter, epiphytic and parasitic on filaments (like *Oedogonium*, *Zygnema*, *Spirogyra*), nonmotile in assimilative stage. Assimilative cell gives rise to amoebae that parasitize green algal filaments and then form the ovoid shape. Formation of gymnodiniod cells is uncertain (Pfiester and Popovsky, 1979). May not be a dinoflagellate at all, but a digestive cyst of the protozoan *Vampyrella* (M. Elbrächter, personal communication, based on Röpstorf *et al.*, 1994), or *Vampyrella* may be a stage in the dinoflagellate life history (Pfiester and Popovsky, 1979). The golden-brown color of the cell may be from the parasitized cell. This may not be a valid genus but rather part of the life cycle of another organism (Popovsky, 1982). Reported from OH (USA).

Cystodinium Klebs (Figs. 2I and 10E) Athecate, *Gymnodinium*-like cell inside casing, photosynthetic, nonmotile in assimilative stage, planktonic, 65–118 μm long × 52–66 μm in diameter. Most commonly sampled from small ponds and marshes (*Lemna* ponds), several reports (six species) from the United States and Canada. Includes a parasitic amoeboid stage (Pfiester and Lynch, 1980).

Dinastridium Pascher (Figs. 2P and 10I) Irregularly-shaped cells, usually with six sides (five to seven species), each angle with a single or double short spine, discoid, parietal plastids. Reproduce via gymnodinioid zoopores or autospores (Bourrelly, 1970). Considered by Popovsky and Pfiester (1990) to be hypnospores of other taxa, but recognizable in this form, described as follows: "color and chromatophores are typical for most dinoflagellates"; cell 28–40 μm in diameter (Forest, 1954). From TN (USA).

Dinococcus Fott (Fig. 2K) Athecate, epiphytic with short stalk with one spine at either end of the elliptical cell, nonmotile in assimilative stage. Collected in WI (USA) by Prescott (1951) at 11 m depth. Cells 25–35 μm long including spines, 9–12 μm in diameter. *Raciborskia* Woloszynska is a synonym; may not be a valid genus but rather part of the life cycle of *Cystodinium* (Pfiester and Lynch, 1980).

Haidadinium Buckland-Nicks *et al.*, (Fig. 2O) Ectoparasite on stickleback fish, with complex life history including vegetative cysts (typical dinoflagellate nucleus and chloroplast), lobose and rhizopodial amoeboid stages, and swarmer stage. There has been one report from British Columbia, from small lakes in *Sphagnum* bogs with low cation concentrations (Buckland-Nicks et al., 1997).

Hypnodinium Klebs (Fig. 2J) Spherical planktonic cell, photosynthetic with chloroplasts arranged in roseate clusters, athecate, nonmotile in assimilative stage, 64–66 μm in diameter. There have been a few reports from MD, MN, NC, OH, Quebec.

Phytodinium Klebs (Figs. 2Q and 10F) Athecate, nonmotile, round cells enclosed by thick wall, photosynthetic with oval chloroplasts, similar to *Cystodinium*, reproduce via autospores There has been one report (Meyer and Brook, 1969, from MN, USA) from a dystrophic pond.

Rufusiella Loeblich (Fig. 2S) Athecate, cells embedded in an asymetrically layered mucilagenous envelope. Noted to produce gymnodinioid swarmers (Thompson in Smith, 1950), but later work found thecate *Hemidinium*-like motile cells. Cell single or two to four, protoplast brownish, may contain red oil globules. Cells 30–83 μm in diameter, 41–100 μm long including sheath. Cells collected from scrapings or grown on solid media have typical eccentrically layered sheath; cells grown in liquid culture are more "Gloeodinium"-like in appearance. Collected in KS (USA) as scrapings from wet areas (Richards, 1962).

Stylodinium Klebs (Figs. 2L and 10G) Athecate, epiphytic, nonmotile in assimilative stage, may contain colored globules, round cell atop distinct stalk, cell 20 μm long × 17.5 μm in diameter; stalk 12.5 μm long. Parasitic on *Oedogonium* and *Fragilaria*, produces gymnodiniod zoospores or amoebae (Pfiester and Popovsky, 1979). Three species known; river habitats, few reports from BC, MD, MN, NY, OH.

Tetradinium Klebs (Figs. 2M and 10H) Athecate, epiphytic, photosynthetic with many golden-brown discoid chloroplasts, nonmotile in assimilative stage. Name derived from tetragonal shape of cell; has two spines at each of the four corners. Cell about 20–73 μm across, stalk length variable, 13–23 μm. Reported (four species) from the United States and Canada, attached to filamentous algae (Thompson, 1949).

Order Prorocentrales Bourrelly

Family Prorocentraceae Engler

Exuviaella Cienkowski (Fig. 3A) Thecate, usually a marine genus, two large heavy plates with smaller plates at the apex. One report from MD (USA), with two parietal chromatophores, a large posterior nucleus, no eyespot, 22–26 μm long, 15–18 μm in diameter, 11–12 μm thick (Thompson, 1950). Two species placed in *Prorocentrum* have been found in fresh waters of Tasmania, Australia (Croome and Tyler, 1987). Heavily thecate species with two large plates are placed in *Prorocentrum* or *Exuviaella*. Dodge and Bibby (1973) merged the two genera into *Prorocentrum*; McLachlan et al., (1997) reinstated *Exuviaella*. Grzebyk et al., (1998), using 18S rDNA sequences and morphological differences, allowed that there was sufficient heterogeneity to support separation of *Prorocentrum* species at the genus level.

VI. GUIDE TO LITERATURE FOR SPECIES IDENTIFICATION

There are a few keys to species within genera; Lefèvre (1932) for *Peridinium* and Christen (1961) for *Katodinium* are two. Compendia should be consulted for the dinoflagellate species within each genus. Local floras include illustrations and may contain keys to species found in that area. Where appropriate I have included citations to pertinent literature under the genus description.

A. Compendia

Lefèvre, M. 1932. Monographie des espèces d'eau douce du genre *Peridinium*. *Archivio Botanico* 2:1–208. Mém. No. 5. Need to check for current genus, as many species of *Peridinium* have been transferred. Good illustrations and comprehensive.

Popovsky, J., Pfiester, L. A. 1990. Süßwasserflora von Mitteleuropa, Band 6: Dinophyceae (Dinoflagellida). Gustav Fischer Verlag, Jena, 272 pp. Many lumped species that are probably valid. Check other sources.

Schiller, J. 1933/37. Dinoflagellatae (Peridineae), in: Kolkwitz, R., Ed., Rabenhorst's KryptogamenFlora von Deutschland, Österreich und der Schweiz, 2. Aufl., 10(3):1–2.

Starmach, K. 1974. Cryptophyceae, Dinophyceae, Raphidophyceae. *Flora Slodkowodna Polski* 4:1–520. Panstwowe Wydawnictwo Naukowe. Warszwa, Kraków. Need to check for current genera names, since some taxa have been transferred. Good illustrations and comprehensive.

B. Local Floras

Forest, H. S. 1954. *Handbook of algae with special reference to Tennessee and the southeastern United States*. The University of Tennessee Press, Knoxville, TN.

Prescott, G.W. 1951. *Algae of the western Great Lakes area exclusive of desmids and diatoms*. Cranbrook Institute of Science Bulletin 31:1–946.

Taft, C. E., Taft, C. W. 1971. *The algae of western Lake Erie*. Bulletin of the Ohio Biological Survey 4:1–189.

Tiffany, L. H., Britton, M. E. 1952. *The algae of Illinois*. University of Chicago Press, Chicago, 407 pp.

Wailes, G. H. 1934. Freshwater dinoflagellates of North America. *Museum and Art Notes, Vancouver City Museum* 7, (Supp. II):1–10, 4 plates.

Whitford, L. A., Schumacher G. J. 1984. *A manual of fresh-water algae*. Sparks Press, Raleigh, NC.

ACKNOWLEDGMENTS

I would like to acknowledge my dinoflagellate mentors, Elenor Cox and Lois Pfiester, friends and family who have been encouraging, Thomas Bermudez for the original drawing 6Ca, Victor W. Fazio, III, for the original drawing 3Ba, the reviewers and editors of the manuscript, the EM center and Dan Schwab at Bowling Green State University (OH), and the librarians at Heidelberg College for cheerful, extensive processing of interlibrary loan requests. Special thanks to Vic Fazio and Dale Ritter for patient reading of the manuscript.

LITERATURE CITED

Balech, E. 1974. El genero *Protoperidinium* Bergh, 1881 (*Peridinium* Ehrenberg, 1831, partim). *Revista del Museo Argentino de Ciencias Naturales "Bernardino Rivadavia"* (Buenos Aires) IV, No. 1.

Balech, E. 1980. On thecal morphology of dinoflagellates with special emphasis on circular and sulcal plates. Anales del Instituto Ciencias del Mar y Limnologia, Universidad Nacional Autonoma de Mexico 7:57–68.

Battey, J. F. 1992. Carbon metabolism in zooxanthellae-coelenterate symbioses, *in*: Reisser, W., Ed., *Algae and symbioses*. Biopress Limited, Bristol, England, pp. 153–187.

Blank, R. J. 1992. Taxonomy of *Symbiodinium*—the microalgae most frequently found in symbiosis with marine invertebrates, *in*: Reisser, W., Ed., Algae and symbioses. Biopress Limited, Bristol, England, pp. 189–197.

Boltovskoy, A. 1975. Estructura y estereoultraestructura tecal de dinoflagelados. II. *Peridinium cinctum* (Müller) Ehrenberg. Physis Seccion B Los Aguas Continentales y Sus Organismos 34:73–84.

Boltovskoy, A. 1976. Estructura y estereoultraestructura tecal de dinoflagelados. III. *Peridinium bipes* Stein forma *apoda*, n.f. Physis Seccion B Los Aguas Continentales y Sus Organismos 35:147–155.

Boltovskoy, A. 1989. Thecal morphology of the dinoflagellate *Peridinium gutwinskii*. Nova Hedwigia 49:369–380.

Bourrelly, P. 1968. Notes sur les Péridiniens d'eau douce. Protistologica 4:5–16.

Bourrelly, P. 1970. *Les Algues d'Eau Douce*. Initiation à la Systématique. Tome III, Les algues bleues et rouges: Les Eugléniens, Peridiniens et Cryptomonadines. Éditions N. Boubée & Cie, Paris, 512pp.

Bruno, S. F., McLaughlin, J. J. A. 1977. The nutrition of the freshwater dinoflagellate *Ceratium hirundinella*. Journal of Protozoology 24:548–553.

Bucka, H., Zurek, R. 1992. Trophic relations between phyto- and zooplankton in a field experiment in the aspect of the formation and decline of water blooms. Acta Hydrobiologica 34:139–155.

Buckland-Nicks, J., Reimchen, T. E. 1995. A novel association between an endemic stickleback and a parasitic dinoflagellate. 3. Details of the life cycle. Archiv für Protistenkunde 145:165–175.

Buckland-Nicks, J., Reimchen, T. E., Garbary, D. J. 1997. *Haidadinium ichthyophilum* gen. nov. et sp.nov. (Phytodiniales, Dinophyceae), a freshwater ectoparasite on stickleback (*Gasterosteus aculeatus*) from the Queen Charlotte Islands, Canada. Canadian Journal of Botany 75:1936–1940.

Bujak, J. P., Davies, E. H. 1983. Modern and fossil Peridiniineae. AASP Contribution Series Number 13. American Association of Stratigraphic Palynologists, Dallas, 203pp.

Bujak, J. P., Williams, G. L. 1981. The evolution of dinoflagellates. Canadian Journal of Botany 59:2077–2087.

Burkholder, J. M. 1998. Implications of harmful microalgae and heterotrophic dinoflagellates in management of sustainable marine fisheries. Ecological Applications 8(Suppl.): S37–S62.

Burkholder, J. M., Noga, E. J., Hobbs, C. H., Glasgow, H. B. 1992. New "phantom" dinoflagellate is the causative agent of major estuarine fish kills. Nature 358:407–410.

Bursa, A. S. 1969. *Actiniscus canadensis* n. sp., *A. pentasterias* Ehrenberg v. *arcticus* n. var., *Pseudoactiniscus apentasterias* n. gen., n. sp., marine relicts in Canadian arctic lakes. Journal of Protozoology 16:411–418.

Bursa, A. S. 1970. *Dinamoebidium coloradense* spec. nov. and *Katodinium auratum* spec. nov. in Como Creek, Boulder County, Colorado. Arctic and Alpine Research 2:145–151.

Calado, A. J., Moestrup, Ø. 1997. Feeding in *Peridiniopsis berolinensis* (Dinophyceae): New observations on tube feeding by an omnivorous, heterotrophic dinoflagellate. Phycologia 36:47–59.

Carty, S. 1986. The taxonomy and systematics of freshwater armored dinoflagellates. Ph.D. dissertation, Texas A&M University, 286pp.

Carty, S. 1989. *Thompsodinium* and two species of *Peridiniopsis* (Dinophyceae): taxonomic notes based on scanning electron micrographs. Transactions of the American Microscopical Society 108:64–73.

Carty, S. 1993. Contribution to the dinoflagellate flora of Ohio. Ohio Journal of Science 93:140–6.

Carty, S., Cox, E. R. 1985. Observations on *Lophodinium polylophum* (Dinophyceae). Journal of Phycology 21:396–401.

Carty, S., Cox, E. R. 1986. *Kansodinium* gen. nov. and *Durinskia* gen. nov.: Two genera of freshwater dinoflagellates (Pyrrhophyta). Phycologia 25:197–204.

Carty, S., Fazio, V. W., III. 1997. //aves.net/algaeweb/ppolncum.htm.

Chapman, A. D., Pfiester, L. A. 1995. The effects of temperature, irradiaance, and nitrogen on the encystment and growth of the freshwater dinoflagellates *Peridinium cinctum* and *P. willei* in culture (Dinophyceae). Journal of Phycology 31:355–359.

Chesnick, J. M., Cox, E. R. 1987. Synchronized sexuality of an algal symbiont and its dinoflagellate host, *Peridinium balticum* (Levander) Lemmermann. BioSystems 21:69–78.

Chesnick, J. M., Cox, E. R. 1989. Fertilization and zygote development in the binucleate dinoflagellate *Peridinium balticum* (Pyrrhophyta). American Journal of Botany 76:1060–1072.

Chesnick, J. M., Morden, C. W., Schmieg, A. M. 1996. Identity of the endosymbiont of *Peridinium foliaceum* (Pyrrhophyta): Analysis of the rbcLS operon. Journal of Phycology 32:850–857.

Christen, Von H. R. 1961. Über die Gattung *Katodinium* Fott (= *Massartia* Conrad). Schweizerische Zeitschrift für Hydrologie 23:309–341.

Cranwell, P. A., Robinson, N., Eglinton, G. 1985. Esterified lipids of the freshwater dinoflagellate *Peridinium lomnickii*. Lipids 20:645–651.

Croome, R. L., Tyler, P. A. 1987. *Prorocentrum playfairi* and *Prorocentrum foveolata*, two new dinoflagellates from Australian freshwaters. British Phycological Journal 22:67–75.

Daday, E. von 1905. Untersuchungen über die Süsswasser-Mikrofauna Paraguays. Zoologica (Stuttgart) 18:1– 374.

Diwald, K. 1938. Die ungeschlechtliche und geschlechtlich Fortpflanzung von *Glenodinium lubieniensiforme* sp. nov. Flora (Jena) 32:174–192.

Dodge, J. D. 1965. Chromosome structure in the dinoflagellates and the problem of the mesocaryotic cell, *in*: Abstracts of the Second International Conference on Protozoology, London.

Dodge, J. D. 1966. The Dinophyceae, *in*: M.B.E. Godward, Ed., The chromosomes of the algae. Arnold, London, pp 96–115.

Dodge, J. D. 1969. A review of the fine structure of algal eyespots. British Phycological Journal 4:199–210.

Dodge, J. D. 1972. The ultrastructure of the dinoflagellate pusule: A unique osmo-regulatory organelle. Protoplasma 75:285–302.

Dodge, J. D. 1975. A survey of chloroplast ultrastructure in the Dinophyceae. Phycologia 14:253–263.

Dodge, J. D. 1983. Dinoflagellates: Investigation and phylogenetic speculation. British Phycological Journal 18:335–356.

Dodge, J. D., Bibby, B. T. 1973. The Prorocentrales (Dinophyceae) I. A comparative account of fine structure in the genera *Prorocentrum* and *Exuviaella*. Botanical Journal of the Linnean Society 67:175–187 + 7 plates.

Dodge, J. D., Crawford, R. M. 1970. A survey of thecal fine structure in the Dinophyceae. Botanical Journal of the Linnean Society 63:53–67, + 7 plates.

Dodge, J. D., Hermes, H. B. 1981. A scanning electron microscopical study of the apical pores of marine dinoflagellates (Dinophyceae). Phycologia 20:424–430.

Duthie, H. C., Socha, R. 1976. A checklist of the freshwater algae of Ontario, exclusive of the Great Lakes. Naturaliste Canadien (Quebec) 103:83–109.

Eaton, G. L. 1980. Nomenclature and homology in peridinialean dinoflagellate plate patterns. Palaeontology 23:667–688.

Echevarria, F., Rodriguez, J. 1994. The size structure of plankton during a deep bloom in a stratified reservoir. Hydrobiologia 284:113–124.

Evitt, W. R. 1985. Sporopollenin Dinoflagellate cysts, their morphology and interpretation. American Association of Stratigraphic Palynologists, Dallas, 333p.

Fensome, R. A., MacRae, R. A., Moldowan, J. M., Taylor, F. J. R., Williams, G. L. 1996. The early Mesozoic radiation of dinoflagellates. Paleobiology 22: 329–338.

Fensome, R. A., Taylor, F. J. R., Norris, G., Sarjeant, W. A. S. Wharton, D. I., Williams, G. L. 1993. A classification of living and fossil dinoflagellates. Micropaleontology Special Publication 7:1–351.

Fields, S. D., Rhodes, R. G. 1991. Ingestion and retention of *Chroomonas* spp. (Cryptophyceae) by *Gymnodinium acidotum* (Dinophyceae). Journal of Phycology 27:525–529

Forest, H. S. 1954. Handbook of algae with special reference to Tennessee and the southeastern United States. The University of Tennessee Press, Knoxville, TN.

Frey, L. C., Stoermer, E. F. 1980. Dinoflagellate phagotrophy in the upper Great Lakes. Transactions of the American Microscopical Society 99:439–444.

Gaines, G., Elbrächter, M. 1987. Heterotrophic nutrition, *in*: Taylor, F. J. R., Ed., The Biology of Dinoflagellates. Blackwell Science, Malden, MA, pp. 224–268.

Gálvez, J. A. Niell, F. X., Lucena, J. 1988. Description and mechanism of formation of a deep chlorophyll maximum due to *Ceratium hirundinella* (O.F. Müller) Bergh. Archiv für Hydrobiologie 112:143–155.

Grzebyk, D. Sako, Y., Berland, B. 1998. Phylogenetic analysis of nine species of *Prorocentrum* (Dinophyceae) inferred from 18S ribosomal DNA sequences, morphological comparisons, and description of *Prorocentrum panamensis*, sp. nov. Journal of Phycology 34:1055–1068.

Haberyan, K. A., Umaña, G., Collado, C., Horn, S. P. 1995. Observations on the plankton of some Costa Rican lakes. Hydrobiologia 312:75–85.

Hallegraeff, G. M., Anderson, D. M., Cembella, A. D., Eds. 1995. Manual on harmful marine microalgae. United Nations Educational, Scientific and Cultural Organization, Parts, 551p.

Hausmann, K. 1978. Extrusive organelles in protists. International Review of Cytology 43:197–276.

Heaney, S. I., Chapman, D. V., Morison, H. R. 1983. The role of the cyst stage in the seasonal growth of the dinoflagellate *Ceratium hirundinella* within a small productive lake. British Phycological Journal 18:47–59.

Heaney, S. I., Furnass, T. I. 1980. Laboratory models of diel vertical migration in the dinoflagellate *Ceratium hirundinella*. Freshwater Biology 10:163–170.

Heaney, S. I., Talling, J. F. 1980. Dynamic aspects of dinoflagellate distribution patterns in a small productive pake. Journal of Ecology 68:75–94.

Herzog, M., Boletzky, S. von, Soyer, M–O. 1984. Ultrastructure and biochemical nuclear aspects of eukaryotic classification: Independent evolution of the dinoflagellates as a sister group of the actual eukaryotes? Origins of Life 13:205–215.

Herzog, M., Maroteaux, L. 1986. Dinoflagellate 17S rRNA sequence inferred from the gene sequence: Evolutionary implications. Proceedings of the National Academy of Sciences of the United States of America 83:8644–8648.

Hickel, B. 1988. Sexual reproduction and life cycle of *Ceratium furcoides* (Dinophyceae) *in situ* in the lake Plußsee (F.R.G.). Hydrobiologia 161:41–48.

Highfill, J. F., Pfiester, L. A. 1992a. The sexual and asexual life cycle of *Glenodiniopsis steinii* (Dinophyceae). American Journal of Botany 79:899–903.

Highfill, J. F., Pfiester, L. A. 1992b. The ultrastructure of *Glenodiniopsis steinii* (Dinophyceae). American Journal of Botany 79:1162–1170.

Holopainen, I. J. 1992. The effects of low pH on planktonic communities. Case history of a small forest pond in eastern Finland. Annales Zoologici Fennici 28:95–103.

Holt, J. R., Pfiester, L. A. 1981. A survey of auxotrophy in five freshwater dinoflagellates (Pyrrhophyta). Journal of Phycology 17:415–416.

Hutchinson, G. E. 1967. A treatise on Limnology, Vol. II. Introduction to lake biology and the limnoplankton. Wiley, New York.

Iglesias-Prieto, R. 1996. Biochemical and spectroscopic properties of the light-harvesting apparatus of dinoflagellates, *in*: Chaudhary, B. R., Agrawal, S. B., Eds., Cytology, Genetics and Molecular Biology of Algae. SPB Academic Publishing, Amsterdam, The Netherlands, pp. 301–322.

Jacobs, D. L. 1946. A new parasitic dinoflagellate from fresh-water fish. Transactions of the American Microscopical Society 65:1–17.

Jacobson, D. M., Anderson, D. M. 1986. Thecate heterotrophic dinoflagellates: Feeding behavior and mechanisms. Journal of Phycology 22:249–258.

James, W. F., Taylor, W. D., Barko, J. W. 1992. Production and vertical migration of *Ceratium hirundinella* in relation to phosphorus availability in Eau Galle Reservoir, Wisconsin. Canadian Journal of Fisheries and Aquatic Sciences 49:694–700.

Jeffrey, S. W., Sielicki, M., Haxo, F. T. 1975. Chloroplast pigment patterns in dinoflagellates. Journal of Phycology 11:374–384.

Kawabata, Z., Banba, D. 1993. Effect of water temperature on the excystment of the dinoflagellate *Ceratium hirundinella* (O.F. Müller) Bergh. Hydrobiologia 257:17–20.

Kawabata, Z., Kagawa, H. 1988. Distribution pattern of the dinoflagellate *Ceratium hirundinella* (O. F. Müller) Bergh in a reservoir. Hydrobiologia 169:319–325.

Kelley, I., Pfiester, L. A. 1990. Sexual reproduction in the freshwater dinoflagellate *Gloeodinium montanum*. Journal of Phycology 26:167–173.

Klemer, A., Barko, J. 1991. Effects of mixing and silica enrichment on phytoplankton seasonal succession. Hydrobiologia 210:171–181.

Klut, M. E., Bisalputra, T., Antia, N. J. 1987. Some observations on the structure and function of the dinoflagellate pusule. Canadian Journal of Botany 65:736–744.

Knaust, R., Urbig, T., Li, L., Taylor, W., Hastings, J. W. 1998. The circadian rhythm of bioluminescence in *Pyrocystis* is not due to differences in the amount of luciferase: A comparative study of three bioluminescent marine dinoflagellates. Journal of Phycology 34:167–172.

Kofoid, C. A. 1909. On *Peridinium steini* Jörgensen, with a note on the nomenclature of the skeleton of the Peridinidae. Archiv für Protistenkunde 16:25–47.

Kofoid, C. A., Michener, J. R. 1912. On the structure and relationships of *Dinosphaera palustris* (Lemm.). University of California Publications in Zoology 11:21–28.

Kofoid, C. A., Swezy, O. 1921. The free-living unarmoured dinoflagellates. Memoires of the University of California 5:1–562.

Koryak, M. 1978. The occurrence of *Peridinium inconspicuum* Lemmermann (Dinophyceae) in minerally acid waters of the upper Ohio River basin. Proceedings of the Academy of Natural Sciences of Philadephia 130:21–25.

Larkum, T. 1996. How dinoflagellates make light work with peridinin. Trends in Plant Science 1:247–248.

Leadbeater, B., Dodge, J. D. 1967. An electron microscope study of dinoflagellate flagella. Journal of General Microbiology 46:305–314, + 4 plates.

Lefèvre, M. 1932. Monographie des espèces d'eau douce du genre *Peridinium*. Archivio Botanico Mem. Caen 2:1–208.

Lentin, J. K., Williams, G. L. 1989. Fossil dinoflagellates: Index to genera and species, 1989 edition. American Association of Stratigraphic Palynologists, Dallas, Contribution Series Number 20. AASP, 473p.

Levin, E. D., Schmechel, D. E., Burkholder, J. M., Glasgow, H. B. Jr., Deamer-Melia, N. J., Moser, V. C., Harry, G. J. 1997. Persisting learning deficits in rats after exposure to *Pfiesteria piscicida*. Environmental Health Perspectives 105:1320–1325.

Lewitus, A. J., Glasgow, H. B., Jr., Burkholder, J. M. 1999. Kleptoplastidy in the toxic dinoflagellate *Pfiesteria piscicida* (Dinophyceae). Journal of Phycology 35:303–312.

Lindström, K. 1991. Nutrient requirements of the dinoflagellate *Peridinium gatunense*. Journal of Phycology 27:207–219.

Ling, K. H., Sin, Y. M., Lam, T. J. 1993. Protection of goldfish against some common ectoparasitic protozoans using *Ichthyophthirius multifiliis* and *Tetrahymena pyriformis* for vaccination. Aquaculture 116:303–314.

Loeblich, A. R., III. 1969. The amphiesma or dinoflagellate cell covering, in: Proceedings of the North American Paleontology Convention Part G, Geological Society, Berkeley, CA. pp. 867–929.

Loeblich, A. R., III. 1976. Dinoflagellate evolution: Speculation and evidence. Journal of Protozoology 23:13–28.

Loeblich, A. R., III. 1980. Dinoflagellate nomenclature. Taxon 29:321–323.

Loeblich, A. R., III. 1982. Dinophyceae, in: Parker, S. P., Ed., Synopsis and classification of living organisms. McGraw-Hill, New York, pp. 101–115.

Lom, J. 1981. Fish invading dinoflagellates: A synopsis of existing and newly proposed genera. Folia Parasitologica (Praha) 28:3–11.

MacRae, R. A., Fensome, R. A., Williams, G. L. 1996. Fossil dinoflagellate diversity, originations, and extinctions and their significance. Canadian Journal of Botany 74:1687–1694.

McLachlan, J. L., Boalch, G. T., Jahn, R. 1997. Reinstatement of the genus *Exuviaella* (Dinophyceae, Prorocentrophycidae) and an assessment of *Prorocentrum lima*. Phycologia 36:38–46.

Mensinger, A. F., Case, J. F. 1992. Dinoflagellate luminescence increases susceptibility of zooplankton to teleost predation. Marine Biology 112:207–210.

Meyer, R. L., Brook, A. J. 1969. Freshwater algae from the Itasca State Park, Minnesota. III. Pyrrhophyta and Euglenophyta. Nova Hedwigia 18:369–382.

Meyer-Harms, B., Pollehne, F. 1998. Alloxanthin in *Dinophysis norvegica* (Dinophysiales, Dinophyceae) from the Baltic Sea. Journal of Phycology 34:280–285.

Moore, J. W. 1981. Seasonal abundance of *Ceratium hirundinella* (O. F. Müller) Schrank in lakes of different trophy. Archiv für Hydrobiologie 92:535–546.

Morrill, L. C., Loeblich, A. R., III. 1981. The dinoflagellate pellicular wall layer and its occurrence in the division Pyrrhophyta. Journal of Phycology 17:315–323.

Nicholls, K. H. 1998. *Amphidiniopsis sibbaldii* sp. nov. (Thecadiniaceae, Dinophyceae), a new freshwater sand-dwelling dinoflagellate. Phycologia 37:334–339.

Ortega, M. M. 1984. *Catálogo de algas continentales recientes de México*. Univ. Nac. Autón. de Méx. México. 296pp.

Osorio-Tafall, B. F. 1942. Estudios sobre el plancton de Mexico 1. El genero *Lophodinium* Lemm. (Dinophyceae Peridiniales). Ciencia; Revista Hispano-Americana de Ciencias Puras y Aplicadas 3:111–119.

Padisák, J. 1985. Population dynamics of the freshwater dinoflagellate *Ceratium hirundinella* in the largest shallow lake of Central Europe, Lake Balaton, Hungary. Freshwater Biology 15:43–52.

Palmer, J. D. 1995. Rubisco rules fall; gene transfer triumphs. BioEssays 17:1005–1008.

Palmer, J. D. 1996. Rubisco surprises in dinoflagellates. Plant Cell 8:343–345.

Park, H.-D., Hayashi, H. 1992. Life cycle of *Peridinium bipes* f. *occulatum* (Dinophyceae) isolated from Lake Kizaki. Journal of the Faculty of Science Shinshu University 27:87–104.

Park, H.-D., Hayashi, H. 1993. Role of encystment and excystment of *Peridinium bipes* f. *occulatum* (Dinophyceae) in freshwater red tides in Lake Kizaki, Japan. Journal of Phycology 29:435–441.

Pearsall, W. H. 1929. Form variation in *Ceratium Hirundinella* O.F.M. Proceedings of the Leeds Philosophical and Literary Society Scientific Section 1:432–439.

Perez, C. C., Roy, S., Levasseur, M., Anderson, D. M. 1998. Control of germination of *Alexandrium tamarense* (Dinophyceae) cysts from the lower St. Lawrence estuary (Canada). Journal of Phycology 34:242–249.

Pfiester, L. A. 1975. Sexual reproduction of *Peridinium cinctum* f. *ovoplanum* (Dinophyceae). Journal of Phycology 11:259–265.

Pfiester, L. A. 1976. Sexual reproduction of *Peridinium willei* (Dinophyceae). Journal of Phycology 12:234–238.

Pfiester, L. A. 1977. Sexual reproduction of *Peridinium gatunense* (Dinophyceae). Journal of Phycology 13:92–95.

Pfiester, L. A. 1984. Sexual reproduction, in: Spector, D.L., Ed., *Dinoflagellates*. Academic Press, New York, pp. 181–199.

Pfiester, L. A., Anderson, D. M. 1987. Dinoflagellate reproduction, *in*: Taylor, F.J.R., Ed., The biology of Dinoflagellates, Botanical Monographs, Vol. 21. Blackwell Science, Malden, MA.

Pfiester, L. A., Highfill, J. F. 1993. Sexual reproduction of *Hemidinium nasutum* alias *Gloeodinium montanum*. Transactions of the American Microscopical Society 112:69–74.

Pfiester, L. A., Lynch, R. A. 1980. Amoeboid stages and sexual reproduction of *Cystodinium bataviense* and its similarity to *Dinococcus* (Dinophyceae). Phycologia 19:178–183.

Pfiester, L. A., Lynch, R. A., Skvarla, J. J. 1980. Occurrence, growth, and SEM portrait of *Woloszynskia reticulata* Thompson (Dinophyceae). Transactions of the American Microscopical Society 99:213–217.

Pfiester, L. A., Popovsky, J. 1979. Parasitic, amoeboid dinoflagellates. Nature 279:421–424.

Pfiester, L. A., Skvarla, J. J. 1979. Heterothallism and thecal development in the sexual life history of *Peridinium volzii* (Dinophyceae). Phycologia 18:13–18.

Pfiester, L. A., Skvarla, J. J. 1980. Comparative ultrastructure of vegetative and sexual thecae of *Peridinium limbatum* and *Peridinium cinctum* (Dinophyceae). American Journal of Botany 67:955–958.

Pfiester, L. A., Timpano, P, Skvarla, J. J., Holt, J. R. 1984. Sexual reproduction and meiosis in *Peridinium inconspicuum* Lemmermann (Dinophyceae). American Journal of Botany 71:1121–1127.

Pollingher, U. 1987. Freshwater ecosystems, *in*: Taylor, F. J. R., Ed., The biology of Dinoflagellates. Blackwell Science, Malden, MA, pp. 502–529.

Pollingher, U., Burgi, H. R., Ambühl, H. 1993. The cysts of *Ceratium hirundinella*: Their dynamics and role within a eutrophic (Lake Semach, Switzerland). Aquatic Sciences 55:10–18.

Pollingher, U., Hickel, B. 1991. Dinoflagellate associations in a subtropical lake (Lake Kinneret, Israel). Archiv für Hydrobiologie 120:267–285.

Popovsky, J. 1970. Some thecate dinoflagellates from Cuba. Archiv für Protistenkunde 112:252–258.

Popovsky, J. 1982. Another case of phagotrophy by *Gymnodinium helveticum* Penard f. *achroum* Skuja. Archiv für Protistenkunde 125:73–78.

Popovsky, J., Pfiester, L. A. 1990. *Süßwasserflora von Mitteleuropa*, Band 6: Dinophyceae (Dinoflagellida). Gustav Fischer Verlag, Jena, 272 pp.

Poulin, M., Hamilton, P. B., Proulx, M. 1995. Catalogue des algues d'eau douce du Québec, Canada. Canadian Field-Naturalist 109:27–110.

Prescott, G. W. 1951. Algae of the western Great Lakes area exclusive of desmids and diatoms. Cranbrook Institute Science Bulletin 31:1–946.

Rengefors, K., Anderson, D. M. 1998. Environmental and endogenous regulation of cyst germination in two freshwater dinoflagellates. Journal of Phycology 34:568–577.

Richards, B. C. 1962. The morphology, cytology and life history of *Urococcus insignis* (Hass.) Kutz. M.S. thesis, University of Kansas, 55 pp. + plates.

Röpstorf, P., Hülsmann, N., Hausmann, K. 1994. Comparative fine structural investigations of interphase and mitotic nuclei of vampyrellid filose amoebae. The Journal of Eukaryotic Microbiology 4:18–30.

Sako, Y., Ishida, Y., Kadota, H., Hata, Y. 1984. Sexual reproduction and cyst formation in the freshwater dinoflagellate *Peridinium cunningtonii*. Bulletin of the Japanese Society of Scientific Fisheries 50:743–750.

Sako, Y., Ishida, Y., Kadota, H., Hata, Y. 1985. Excystment in the freshwater dinoflagellate *Peridinium cunningtonii*. Bulletin of the Japanese Society of Scientific Fisheries 51:267–272.

Sako, Y., Ishida, Y., Nishijima, T., Hata, Y. 1987. Sexual reproduction and cyst formation in the freshwater dinoflagellate *Peridinium penardii*. Nippon Suisan Gakkaishi 53:473–478.

Sanderson, B. L., Frost, T. M. 1996. Regulation of dinoflagellate populations: Relative importance of grazing, resource limitation, and recruitment from sediments. Canadian Journal of Fisheries and Aquatic Sciences 53:1409–1417.

Sarjeant, W. A. S. 1978. *Arpylorus antiquus* Calandra, emend., a dinoflagellate cyst from the Upper Silurian. Palynology 2:167–179.

Saunders, R. D., Dodge, J. D. 1984. An SEM study and taxonomic revision of some armoured sand-dwelling marine dinoflagellates. Protistologica 20:271–283.

Saunders, G. W., Hill, D. R. A., Sexton, J. P., Andersen, R. A. 1997. Small subunit ribosomal RNA sequences from selected dinoflagellates: Testing classical evolutionary hypotheses with molecular systematic methods. Plant Systematics and Evolution (Suppl.) 11:237–259.

Smayda, T. J. 1997. What is a bloom? A commentary. Limnol. Oceanogr. 42:1132–1136.

Smith, G. M. 1950. The fresh-water algae of the United States, 2nd Edn., McGraw-Hill Book Company, New York, 719 pp.

Sommer, U., Gliwicz, Z. M., Lampert, W., Duncan, A. 1986. The PEG model of seasonal succession of planktonic events in freshwaters. Archiv für Hydrobiologie 106:433–471.

Sournia, A., Ed. 1978. *Phytoplankton Manual*. UNESCO, Paris, 337pp.

Soyer-Gobillard, M.-O. 1996. The genome of the primitive eukaryote dinoflagellates: Organization and functioning. Zoology Studies 35:78–84.

Spector, D. L. Ed. 1984. *Dinoflagellates*. Academic Press, New York.

Spero, H. J. 1985. Chemosemsory capabilities in the phagotrophic dinoflagellate *Gymnodinium fungiforme*. Journal of Phycology 21:181–184.

Squires, L. E., Whiting, M. C., Brotherson, J. D., Rushforth, S. R. 1979. Competitive displacement as a factor influencing phytoplankton distribution in Utah Lake, Utah. Great Basin Naturalist 39:245–252.

Steidinger, K. A., Burkholder, J. M., Glasgow, H. B., Hobbs, C. H., Garrett, J. K., Truby, E. W., Noga, E. J., Smith, S. A. 1996a. *Pfiesteria piscicida* gen. et sp. nov. (Pfiesteriaceae fam. nov.), a new toxic dinoflagellate with a complex life cycle and behavior. Journal of Phycology 32:157–164.

Steidinger, K. A., Landsberg, J. H., Truby, E. W., Blakesley, B. A. 1996b. The use of scanning electron microscopy in identifying small "gymnodiniod" dinoflagellates. Nova Hedwigia 112:415–422.

Stein, J. R., Borden C. A. 1979. Checklist of freshwater algae of British Columbia. Syesis 12:3–39.

Stewart, A. J., Blinn, D. W. 1976. Studies on Lake Powell, USA: Environmental factors influencing phytoplankton success in a high desert warm monomictic lake. Archiv für Hydrobiologie 78:139–164.

Stoecker, D. K. 1998. Conceptual models of mixotrophy in planktonic protists and some ecological and evolutionary implications. European Journal of Protistology 34:281–290.

Stoneburner, D. L., Smock, L. A. 1980. Plankton communities of an acidic, polymictic, brown-water lake. Hydrobiologia 69:131–137.

Stosch, H. A. von 1965. Sexualität bei *Ceratium cornutum* (Dinophyta). Naturwissenschaften 52:112–113.

Stosch, H. A. von. 1972. La signification cytologique de la "cyclose nucléaire" dans le cycle de vie des Dinoflagellés. Mémoires Société Botanique de France (1949–1973):201–212.

Stosch, H. A. von. 1973. Observations on vegetative reproduction and sexual life cycles of two freshwater dinoflagellates, *Gymnodinium pseudopalustre* Schiller and *Woloszynskia apiculata* sp. nov. British Phycological Journal 8:105–134.

Taft C. E. 1978. A mounting medium for fresh-water plankton. Transactions of the American Microscopical Society 97:263–264.

Taft, C. E., Taft, C. W. 1971. The algae of western Lake Erie. Bulletin of Ohio Bio. Survey 4:1–189.

Taylor, F. J. R. 1980. On dinoflagellate evolution. BioSystems 13:65–108.

Taylor, F. J. R. Ed. 1987. *The biology of dinoflagellates*, Botanical Monographs, Vol. 21. Blackwell Science, Malden, MA.

Taylor, F. J. R. 1999. Morphology (tabulation) and molecular evidence for dinoflagellate phylogeny reinforce each other. Journal of Phycology 35:1–3.

Thompson, R. H. 1947. Fresh-water dinoflagellates of Maryland. State of Maryland Board of Natural Resources No. 67. 3–24. Chesapeake Biological Laboratory, Solomons, MD.

Thompson, R. H. 1949. Immobile Dinophyceae. I. New records and a new species. American Journal of Botany 36:301–308.

Thompson, R.H. 1950. A new genus and new records of fresh water Pyrrophyta in the Desmokontae and Dinophyceae. Lloydia 13:277–299.

Tomas, R. N., Cox, E. R. 1973. Observations on the symbiosis of *Peridinium balticum* and its intracellular alga. I. Ultrastructure. Journal of Phycology 9:304–323.

Toriumi, S., Dodge, J. D. 1993. Thecal apex structure in the Peridiniaceae (Dinophyceae). European Journal of Phycology 28:39–45.

Trench, R. K., Thinh, L. V. 1995. *Gymnodinium linucheae* sp. nov.: The dinoflagellate symbiont of the jellyfish *Linuche unguiculata*. European Journal of Phycology 30:149–154.

Wall, D., Dale, B. 1968. Modern dinoflagellate cysts and evolution of the Peridiniales. Micropaleontology 14:265–304, plates 1–4.

Weisse, T., Kirchhoff, 1997. Feeding of the heterotrophic freshwater dinoflagellate *Peridiniopsis berolinense* on cryptophytes: analysis by flow cytometry and electronic particle counting. Aquatic Microbial Ecology 12:153–164.

Whitford, L. A., Schumacher, G. J. 1984. A manual of fresh-water algae. Sparks Press, Raleigh, NC, 324pp.

Whiting, M. C., Brotherson, J. D., Rushforth, S.R. 1978. Environmental interaction in summer algal communities of Utah Lake. Great Basin Naturalist 38:31–41.

Wilcox, L. W., Wedemayer, G. J. 1984. *Gymnodinium acidotum* Nygaard (Pyrrhophyta), a dinoflagellate with an endosymbiotic cryptomonad. Journal of Phycology 20:236–242.

Witt, F. G., Stöhr, C., Ullrich, W. R. 1999. Soluble and membrane-associated nitrate reductases in the dinoflagellate *Peridinium gatunense*. New Phytology 142:27–34.

Wright, A.-D. G., Lynn, D. H. 1997a. Phylogenetic analysis of the rumen ciliate family Ophryoscolecidae based on the 18S ribosomal RNA sequences, with new sequences from *Diplodinium*, *Eudiplodinium*, and *Ophryoscolex*. Canadian Journal of Zoology 75:963–970.

Wright, A.-D. G., Lynn, D. H. 1997b. Monophyly of the trichostome ciliates (Phylum Ciliophora: Class Litostomatea) tested using new 18S rRNA sequences from the vestibuliferids, *Isotricha intestinalis* and *Dasytricha ruminantium*, and the haptorian, *Didinium nasutum*. European Journal of Protistology 33:305–315.

Wu, J.-T., Chou, J.-W. 1998. Dinoflagellate associations in Feitsui Reservoir, Taiwan. Botanical Bulletin of Academia Sinica (Taipei) 39:137–145.

Xie, P., Iwakuma T., Fujii, K. 1998. Changes in the structure of a zooplankton community during a *Ceratium* (dinoflagellate) bloom in a eutrophic fishless pond. Journal of Plankton Research 20:1663–1678

Yan, N. D., Stokes, P. 1978. Phytoplankton of an acidic lake, and its responses to experimental alterations of pH. Environmental Conservation 5:93–100.

Yoshikawa, T., Takishita, K., Ishida, Y., Uchida, A. 1997. Molecular cloning and nucleotide sequence analysis of the gene coding for chloroplast-type ferredoxin from the dinoflagellates *Peridinium bipes* and *Alexandrium tamarense*. Fisheries Science (Tokyo) 63:692–700.

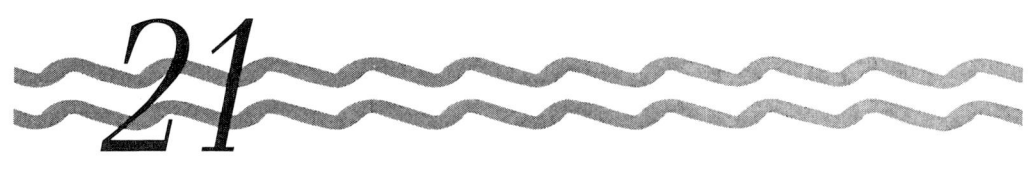

CRYPTOMONADS

Paul Kugrens
Department of Biology
Colorado State University
Fort Collins, Colorado 80523

Brec L. Clay*
UTEX Culture Collection
University of Texas
Austin, Texas 78713

I. Introduction
II. Unique Features of Cryptomonads
    A. External Cell Architecture
    B. Periplast Structure
    C. Flagella and Flagellar Apparatus
    D. Ejectisomes
    E. Ejectisome Digestion Vesicles
    F. Mitochondria and Chloroplasts
    G. Nucleomorphs
    H. Reproduction
    I. Nucleus and Mitosis
    J. Contractile Vavuoles
    K. Starch
III. Origin of Cryptomonads
IV. Ecology
    A. Abiotic Factors
    B. Biological Factors
    C. Cryptomonad Endosymbiotins and Pathogens
    E. Types of Nutrition—Carbon Sources
V. Collection, Preparation for Isolation, and Culturing
VI. Classification, Key, and Descriptions
    A. Introduction
    B. General Features Useful in Determining Genera and Species
    C. Classification of the Phylum Cryptophyta
    D. Key
    E. Freshwater Cryptomonad Genera and Species
    F. Guide to Literature for Species Identification
VII. Availability of Cryptomonads
VIII. Family Kathablepharidaceae
    A. Ecology
    B. Cell Structure
    C. Classification of Kathablepharidaceae
    D. Isolation and Culturing Techniques for Kathablepharids
Literature Cited

I. INTRODUCTION

Cryptomonads, cryptoprotists, or cryptophytes, as these algae are commonly called, are unicellular, biflagellate protists. They are variously classified as belonging to the phylum (division) Cryptophyta, class Cryptophyceae, order Cryptomonadales, or phylum Cryptista *sensu* Cavalier-Smith (1986). Cryptomonads are important primary producers in freshwater and marine habitats (Gillott, 1990; Klaveness, 1988a, b), and many are cosmopolitan in their distribution, although they appear to be more common in cooler water. Many are collected in phytoplankton samples, but their cells are extremely delicate and rupture when fixatives are added or when temperatures are elevated. As a consequence, their numbers generally are low in preserved samples and therefore are assumed to be a small and obscure taxonomic group of protists. To the contrary, they often assume dominant phytoplankton status in temperate lakes and reservoirs, where they may dominate the under-ice, early spring and late fall populations. In fact, the variations in cell structure discovered with specialized electron microscopic techniques (Hill, 1991a, b; Hill and Wetherbee, 1986, 1988, 1989;

*Present address: CH Diagnostic and Consulting Service, Loveland, Colorado 80538.

Kugrens and Lee, 1986, 1991; Kugrens *et al.*, 1986, 1987; Lee and Kugrens, 1986; Clay and Kugrens, 1999a–c; Clay *et al.*, 1999) strongly indicate that there are numerous unrecognized freshwater genera and species (Andersen, 1992).

While cryptomonads represent a well-circumscribed group of algae, there is another group of colorless flagellates that historically has been included in the Cryptophyta. This group, the katablepharids, includes the genera *Leucocryptos* and *Kathablepharis*, which are common in both freshwater and marine habitats. Except for the presence of ejectisomes and placement of flagella, their other cellular features do not support the placement of this group within the cryptomonads (Lee and Kugrens, 1991b; Lee *et al.*, 1991; Vørs, 1992a, b; Clay and Kugrens, 1999a, b). Currently their systematic/phylogenetic position remains undetermined, but, for convenience, we have retained this group in the Cryptophyta, realizing that they probably are not cryptomonads. For clarity in presentation, the discussion and description of cryptomonad characteristics are presented in the first portion of this chapter (Figs. 1–16), and a separate description of kathablepharids (Figs. 17 and 18) follows the cryptomonad description.

II. UNIQUE FEATURES OF CRYPTOMONADS

A. External Cell Architecture

The external cell architecture is influenced primarily by a furrow/gullet complex (Figs. 1, 7A–C, 8A, 11A–C, 12A–C, 16A, C) which tends to impart an asymmetrical shape to the cells. In addition, cell shapes may be oval, compressed, lunulate, caudate, acute, elongate, sigmoid, or otherwise contorted. All cells possess an anterior, outwardly facing depression called a *vestibulum*, from which the flagella originate on the right side. In freshwater crytpomonads the contractile vacuole is located in the anterior end of the cell and usually discharges through a predetermined site in the dorsal portion of the vestibulum (Kugrens *et al.*, 1986). In some genera, such as *Campylomonas*, a vestibular ligule, which is a small, flat extension of the cell, covers the discharge site of the contractile vacuole (Hill, 1991c; Kugrens and Lee, 1991; Kugrens *et al.*, 1986).

A gullet, some type of furrow, or a combination of a furrow-gullet (Fig. 7A–C) is one of the primary diagnostic features for the genera. A furrow is a ventral groove, of variable length, that begins in the vestibular region of the cell and extends posteriorly, terminating somewhere in the anterior half of the cell (Hill and Wetherbee, 1986, 1988, 1989; Klaveness, 1985;

Kugrens *et al.*, 1986; Munawar and Bistricki, 1979). A tubular invagination called the gullet may extend posteriorly from the vestibulum or from the end of the furrow (Munawar and Bistricki, 1979; Hill and Wetherbee, 1986, 1988, 1989; Kugrens *et al.*, 1986).

Several types of furrows have been described in cryptomonads (Munawar and Bistricki, 1979; Klaveness, 1985; Kugrens *et al.*, 1986) and these variations have become major features in cryptomonad systematics (Hill and Wetherbee, 1986, 1988, 1989; Hill, 1990, 1991b; Clay *et al.*, 1999). There are at least five variations in the furrow/gullet complex (Kugrens *et al.*, 1986). Furthermore, a gullet may have evolved from the fusion of a furrow (Kugrens and Lee, 1991; Clay *et al.*, 1999). Consequently, many genera have a combination of a furrow and gullet, and these may represent intermediate stages in the evolution of a gullet. Cryptomonads may have only a gullet (Figs. 11A, C, 12A–C, 13A, B), a simple furrow only (Fig. 16A, B, 8A), a simple furrow and gullet, or a complex furrow structure with or without a gullet (Fig. 16A). The gullet and furrow are lined with large ejectisomes, and ejectisome discharge in this area would not destroy cell organelles by this action.

Specialized furrow plates may be associated with each type of furrow, thereby providing additional variations to the furrow structure. Furrow plates are of two types, scalariform (Fig. 1) and fibrillar. A scalariform furrow plate is in the form of a ladder that has sides connected by lateral, crystalline "rungs." Fibrillar furrow plates are made up of microfibrils that are oriented parallel to each other and occur as a thin plate along one side of the furrow.

Surprisingly little is known about the actual cell architecture in cryptomonads. Prior to Hill's studies, as well as our own, we relied on light microscopic descriptions for this aspect (Skuja, 1948; Huber-Pestalozzi, 1950; Butcher, 1967; Bourelly, 1970) because most electron microscopic preparatory procedures tend to distort the cell shapes. Parducz's (1967) fixation preserves cell shape well; however, freeze drying is the best method for examining the external features and determining whether a gullet and/or furrow is/are present. In fact, the morphology of cryptomonad cells has been drastically revised as a result of studies using cryofixation techniques. The existence of tubular gullets, furrows, and furrow/gullet combinations and their variations has proved to be a significant delineator of genera and should serve to enhance light microscopic identifications. Unfortunately, facilities that utilize electron microscopic cryotechniques are fewer, and it is becoming increasingly difficult to find a facility that still employs these techniques.

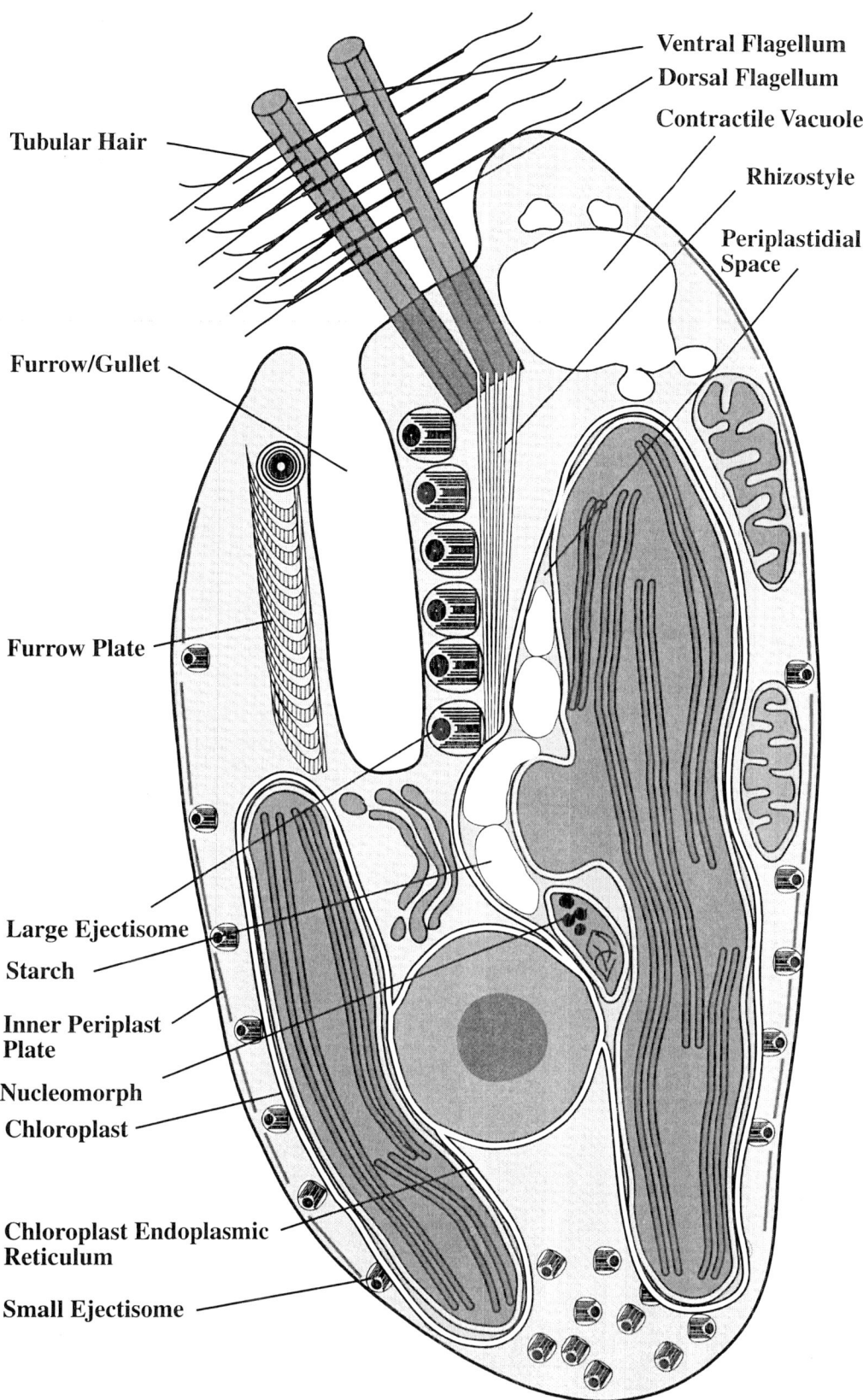

FIGURE 1 Diagram of a generalized cryptomonad cell, showing the cellular details described in the text.

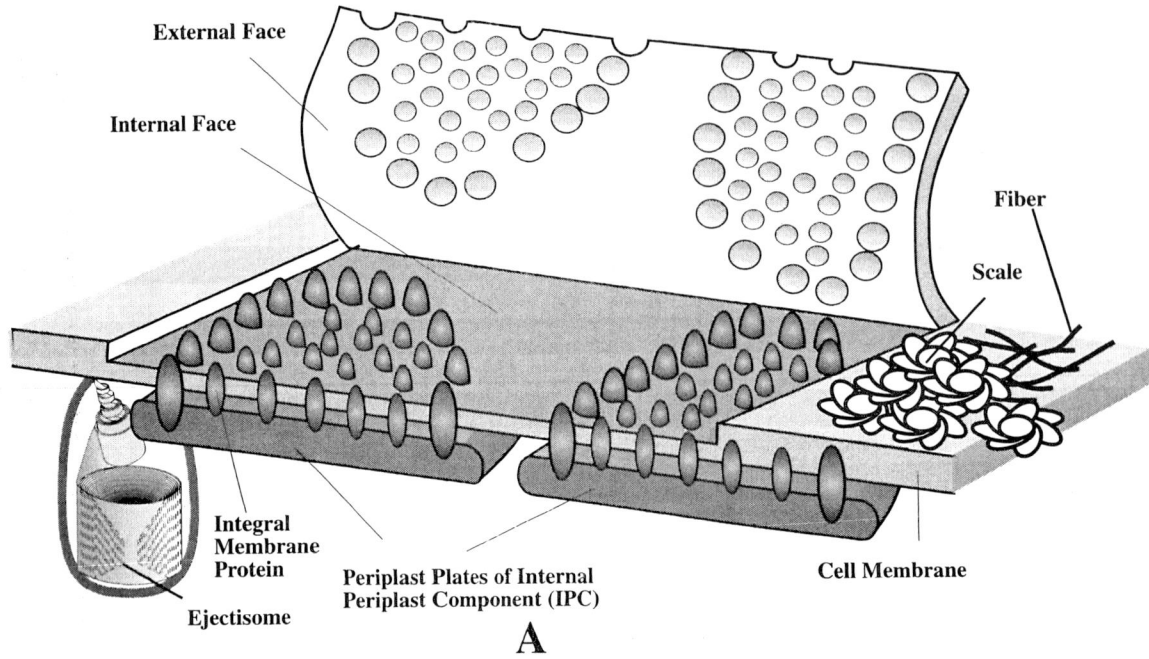

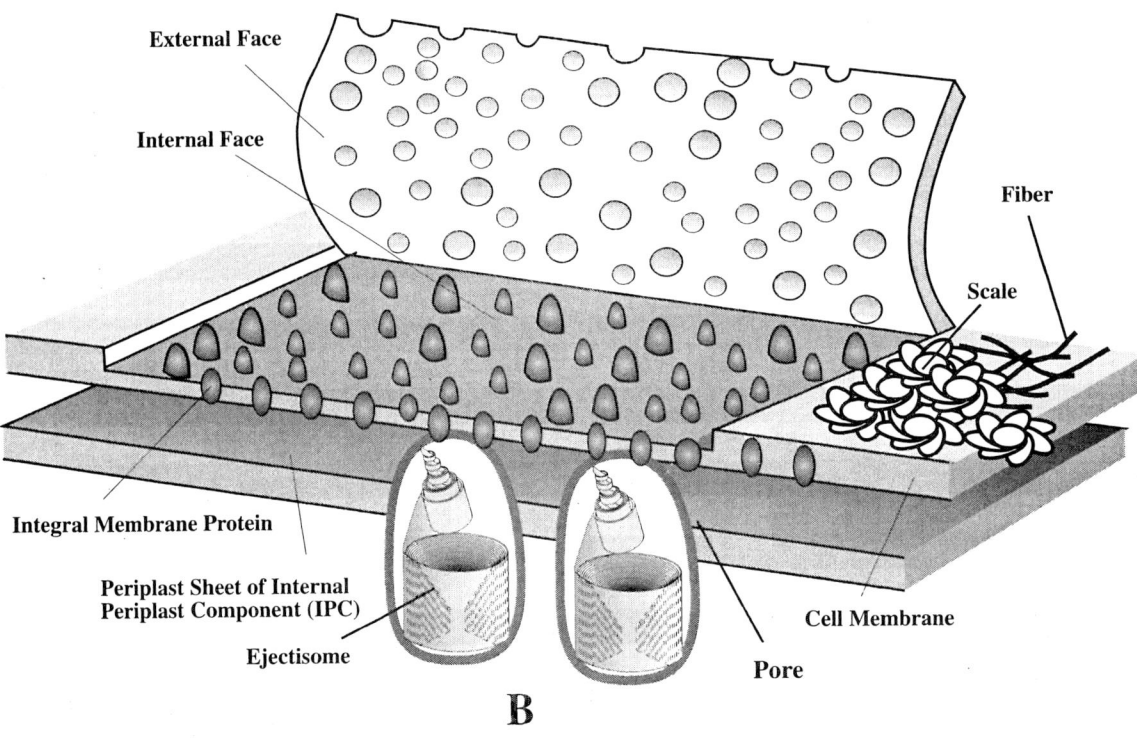

FIGURE 2 Diagrams of two major periplast types and their association with the plasma membrane and ejectisomes. (A) Periplast structure in cells where the inner periplast component is composed of plates. The plates are attached to the plasma membrane by transmembrane particles. (B) Periplast structure in cells having an inner periplast component consisting of a single sheet. The sheet is not associated with the plasma membrane.

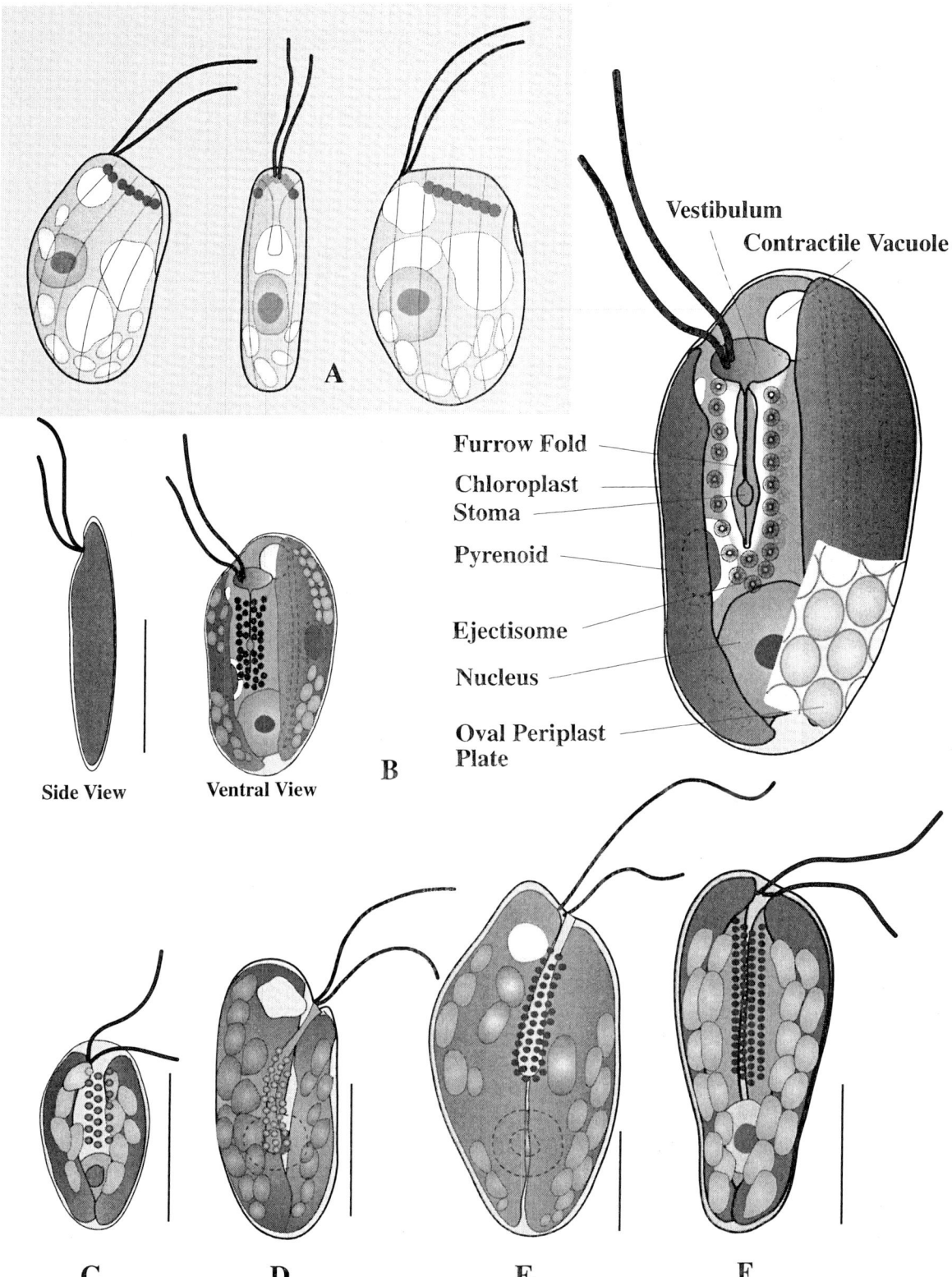

FIGURE 3 Light micrographic illustrations of *Goniomonas* and *Cryptomonas* spp. (A) *Goniomonas truncata*, showing the flattened nature of cells, the dorsal nucleus, and the dorsally inserted flagella. Vertical striations shown in the diagrams are visible with a light microscope. (B) *Cryptomonas ovata* from various views. (C) *Cryptomonas obovata*. (D) *Cryptomonas erosa*. (E) *Cryptomonas ovolinii*. Scale bars = 10 μm.

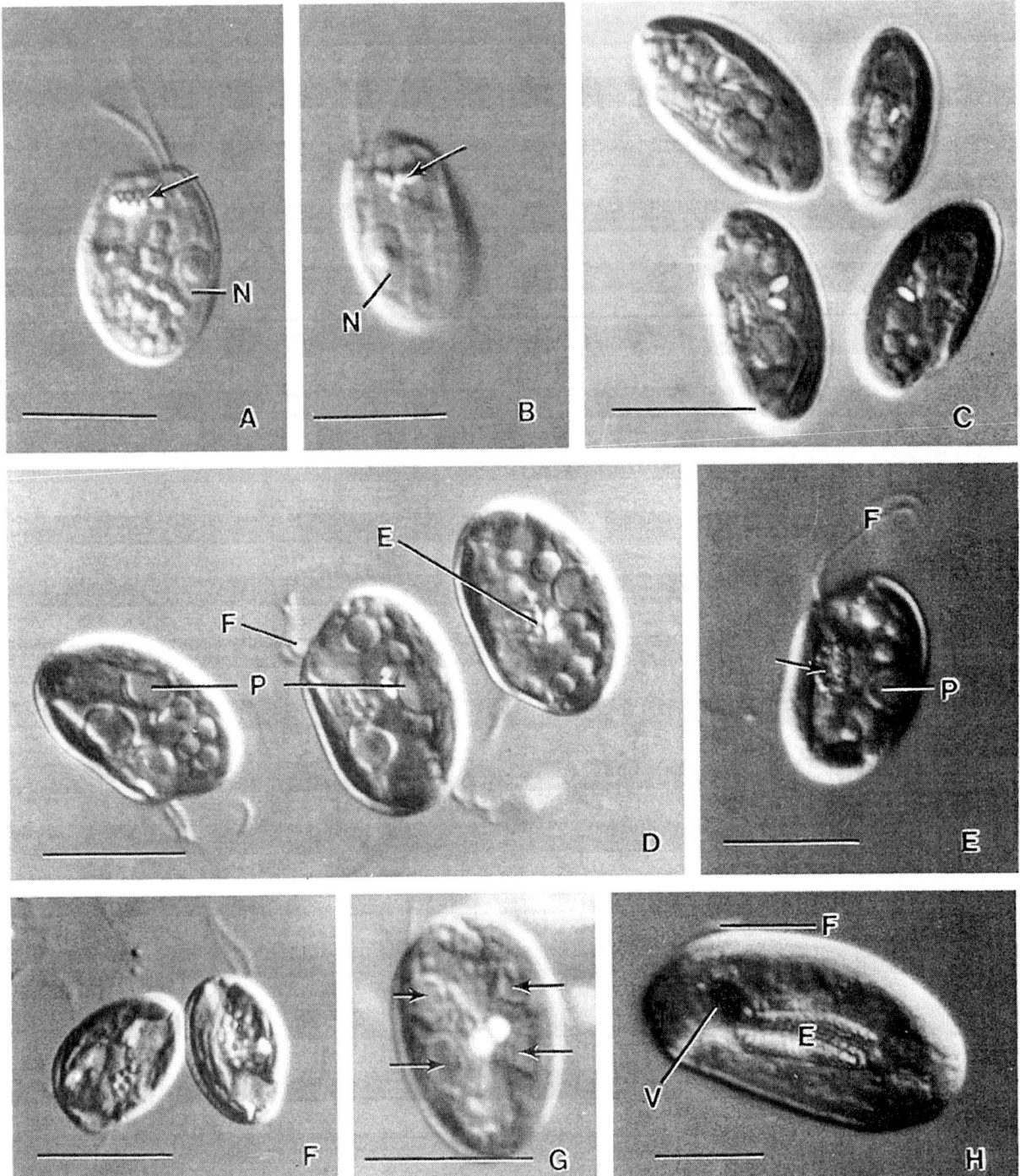

FIGURE 4 Light micrographs of *Goniomonas* and *Cryptomonas* species. (A) *Goniomonas truncata* cell, showing the location of the nucleus (N) and ejectisomes (E). (B) *G. truncata* cell, showing surface striations (arrows). (C) *Cryptomonas obovata* cells, with a pyrenoid visible in some cells. (D) *Cryptomonas ovata* cells, with two pyrenoids (arrows) in each cell. Flagella (F) and ejectisomes (E) are also evident. (E) *Cryptomonas ovata* cell, showing ejectisomes (E) around furrow, with flagella and a nucleus (N) also visible. (F) *Cryptomonas phaseolus* cell. (G) *Cryptomonas tetrapyrenoidosa*, showing four pyrenoids (arrows). (H) *Cryptomonas erosa* with the furrow flanked by ejectisomes (E), a portion of a flagellum (F), and the location of the vestibulum. Scale bars = 10 μm.

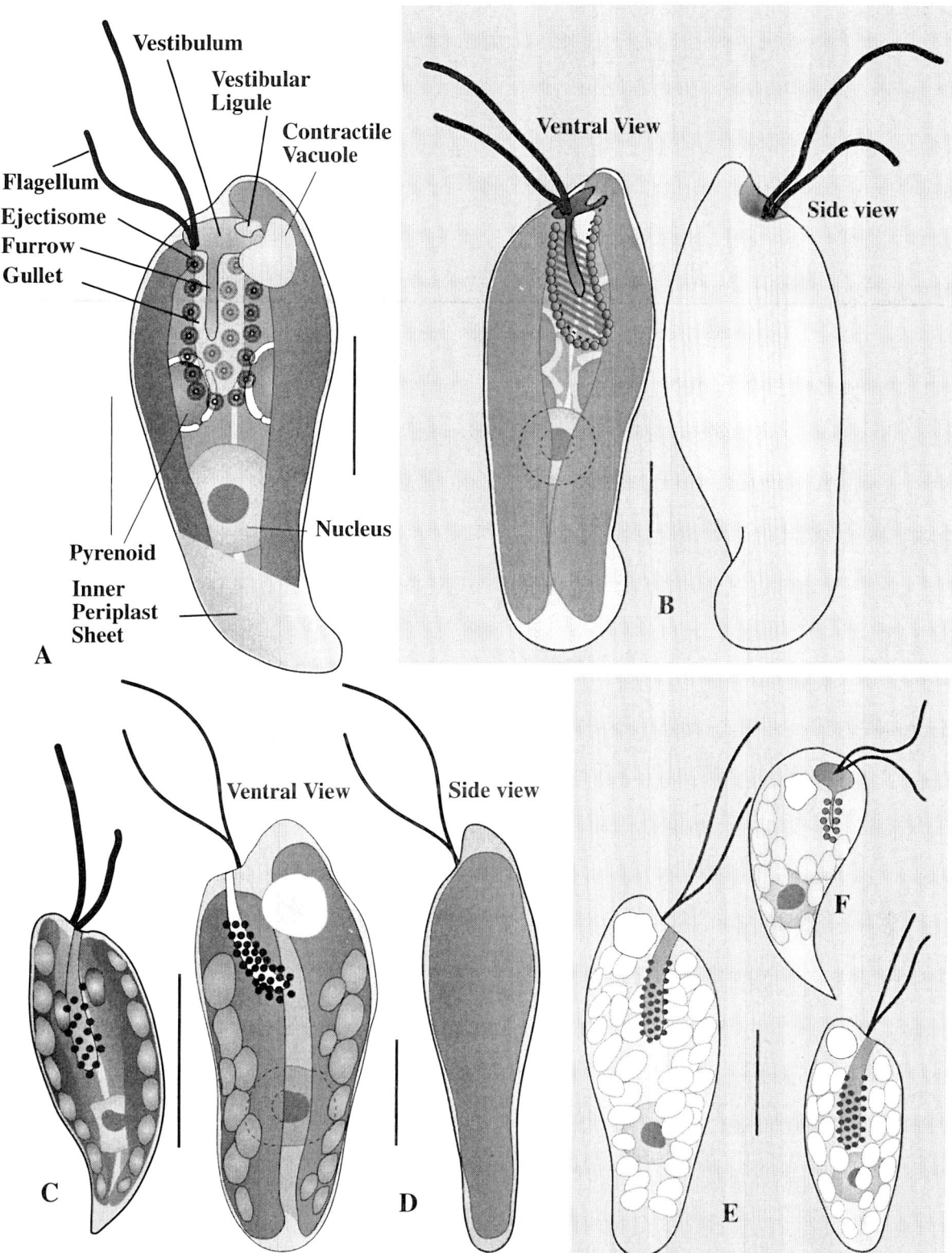

FIGURE 5 Diagrams of genera belonging to the family Campylomonadaceae. (A) *Campylomonas reflexa*. (B) *Cryptomonas (Campylomonas) rostratiformis*. (C) *Cryptomonas (Campylomonas) marssonii*. (D) *Cryptomonas (Campylomonas) platyuris*. (E) *Chilomonas paramecium*. (F) *Chilomonas acuta*. Scale bars = 10 μm.

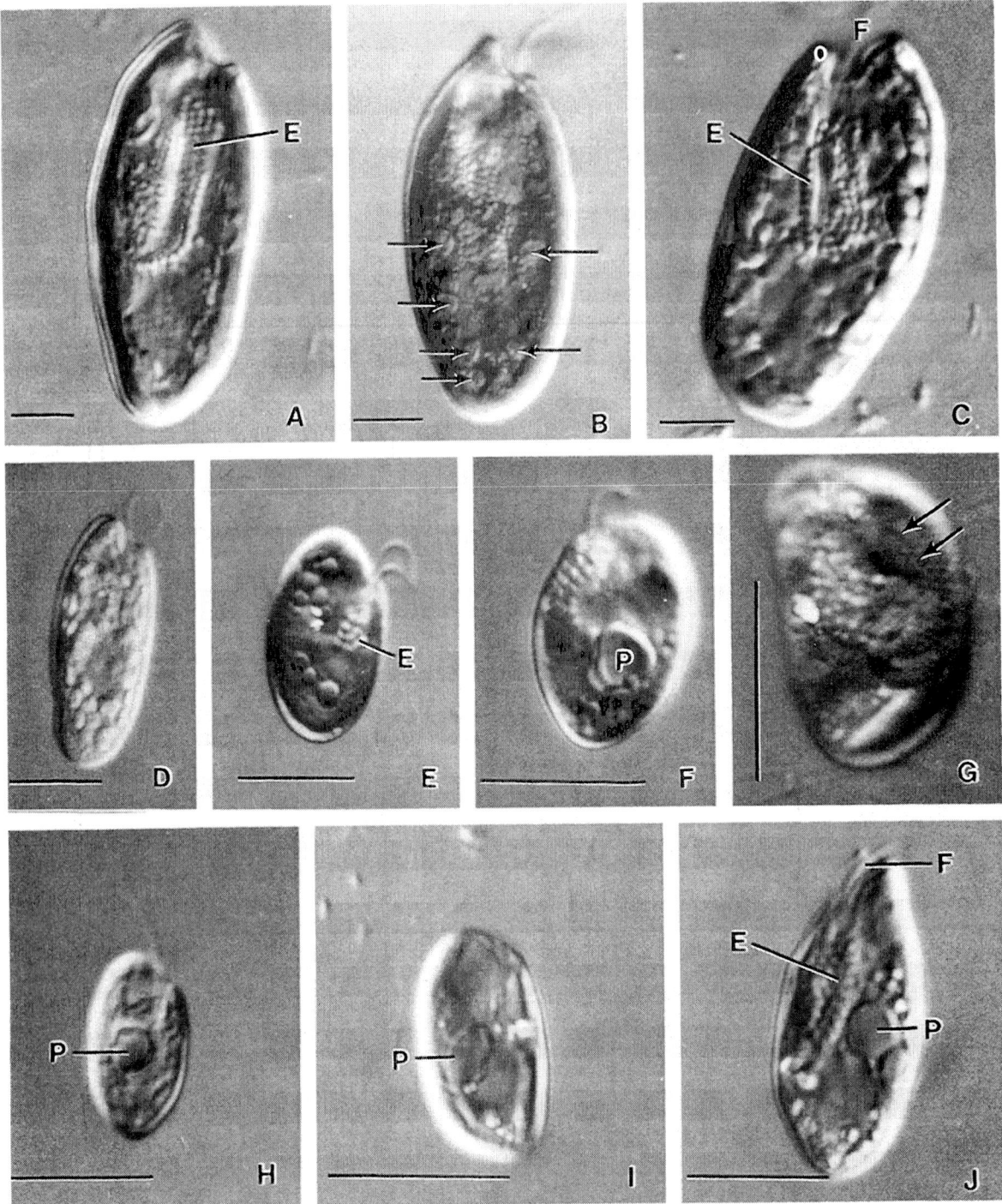

FIGURE 6 Light micrographs (differential interference contrast) of several species from the Campylomonadaceae and Pyrenomonadaceae. (A) *Campylomonas rostratiformis* cell, with its characteristic rhinote anterior and numerous ejectisomes (E) lining the furrow. (B) Somewhat flattened cell of *Cryptomonas (Campylomonas) rostratiformis*, showing multiple pyrenoids (arrows). (C) *Cryptomonas (Campylomonas) platyuris* in ventral view, showing the broad shape of the cell in this view and the ejectisomes (E) lining the furrow. The cell is filled with considerable starch. (D) *Chilomonas paramecium* cell, with starch filling most of the cell. (E) *Pyrenomonas ovalis* cell with ejectisomes (E). (F) *Pyrenomonas ovalis* with the characteristic prominent pyrenoid (P). (G) Higher magnification of a *P. ovalis* cell, showing periplast plates (arrows). (H) *Storeatula rhinosa* cell with a prominent pyrenoid (P). (I) *S. rhinosa* cell as viewed obliquely, showing the asymmetrical cell shape and a prominent pyrenoid (P). (J) *Storeatula* sp. cell with an extensive ejectisome region and a prominent pyrenoid (P). Portions of the flagella are visible and are adhering to the cell. Scale bars = 10 μm.

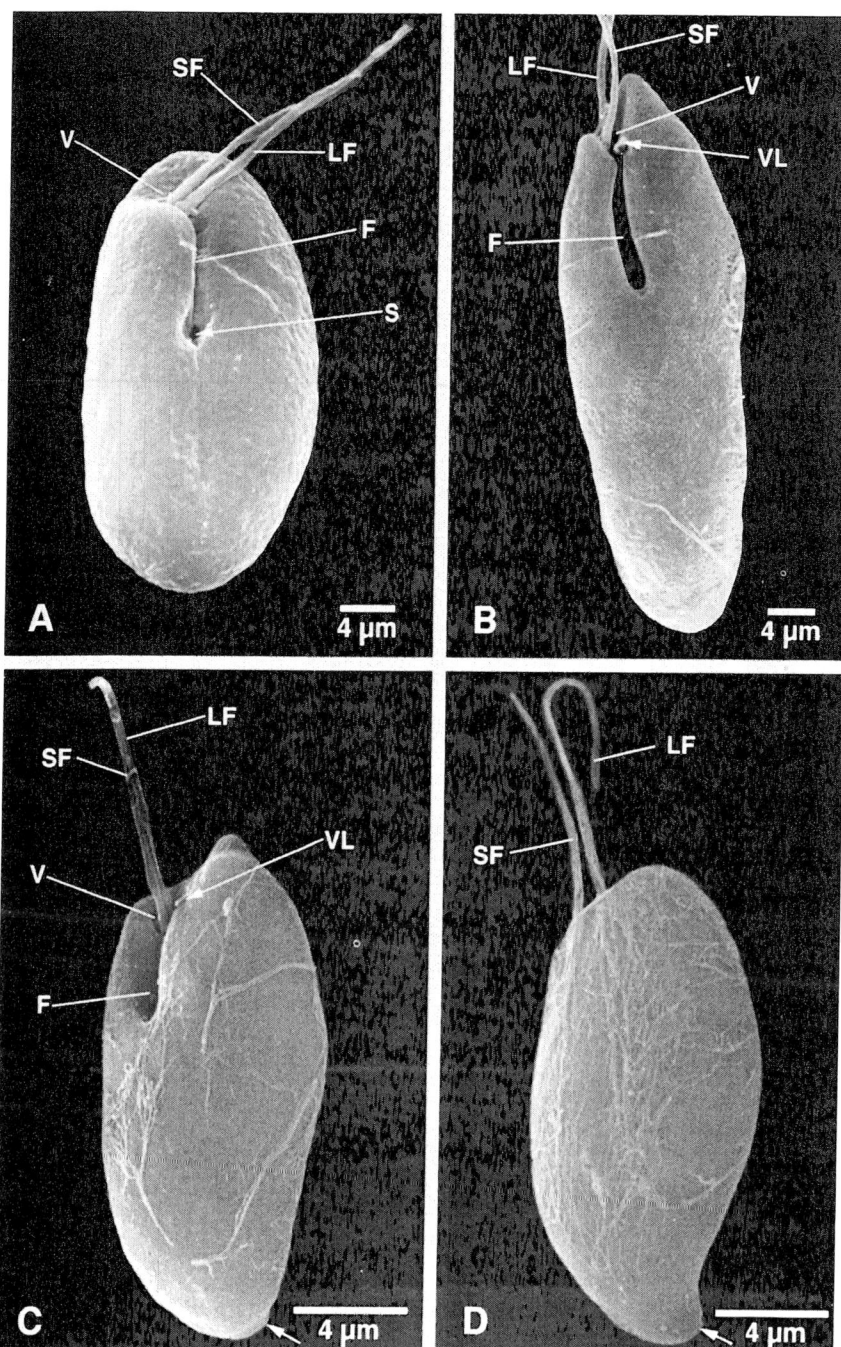

FIGURE 7 Scanning electron micrographs of *Cryptomonas* and *Campylomonas* species. (A) Cell of *Cryptomonas tetrapyrenoidosa* with long (LF) and short (SF) flagella inserted on the right side of the vestibulum (V). A long furrow extends from the vestibulum, and a stoma (S) is present. (B) Cell of *Campylomonas rostratiformis* with long (LF) and short (SF) flagella inserted on the right side of the vestibulum. A vestibular ligule (vl) is seen attached to the dorsal wall of the vestibulum. A slightly curved, oblique furrow (F) runs for approximately one-third of the cell length. Note the rostrate anterior of the cell. (C) Oblique view of a cell of *Campylomonas reflexa*, showing the long (LF) and short (SF) flagella inserted on the right side of the vestibulum. A vestibular ligule (vl) attaches to the dorsal wall of the vestibulum. A furrow (F) extends posteriorly for a third of the cell length. Note the reflexed tail (arrow). (D) Lateral view of *C. reflexa*, showing the reflexed shape of the cell. A and B are from Kugrens *et al.* (1986), with permission.

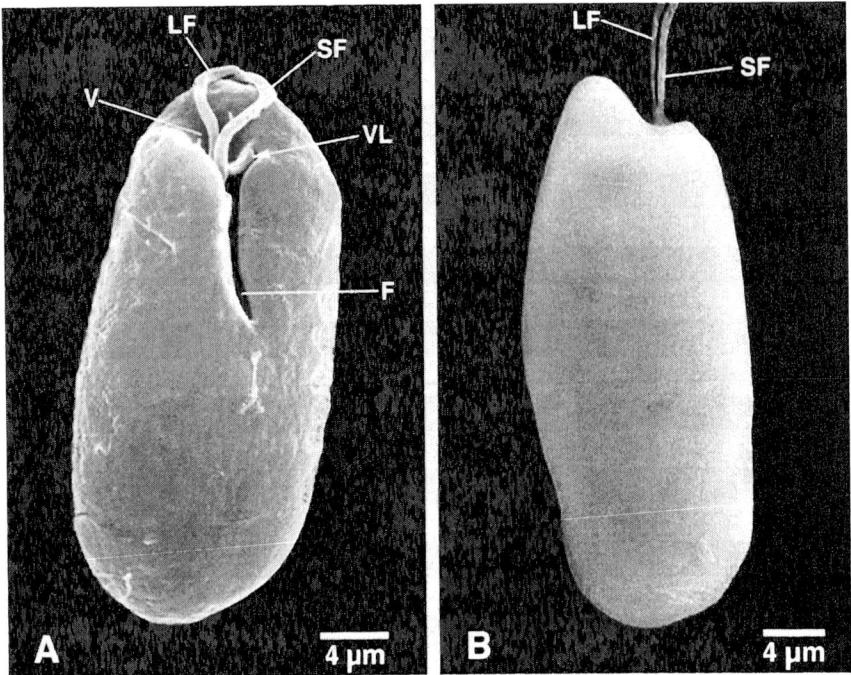

FIGURE 8 Scanning electron micrographs *Campylomonas platyuris*. (A) Ventral view with long (LF) and short (SF) flagella inserted in the vestibulum (V). A vestibular ligule (vl) occurs on the dorsal side of the vestibulum. An oblique furrow (F) runs for almost one-half of the cell length from the vestibulum. (B) Dorsal view of *C. platyuris*. Note the slightly reflexed cell shape. A is from Kugrens *et al.* (1986) with permission.

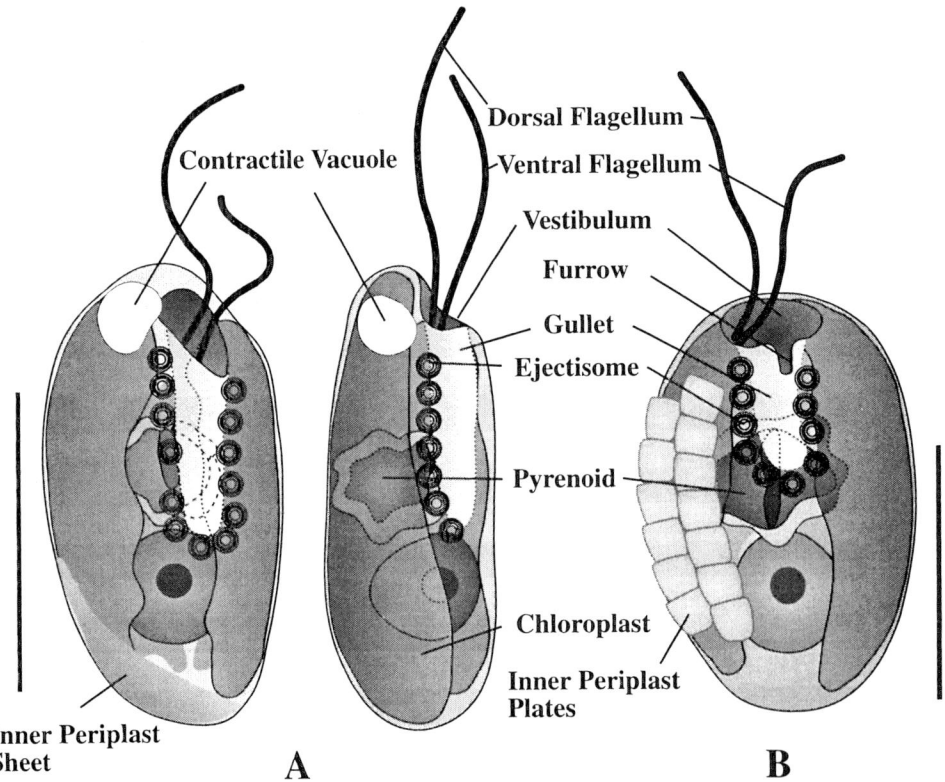

FIGURE 9 Light microscopic illustrations of red-colored cryptomonad species. (A) *Pyrenomonas ovalis*. (B) *Storeatula rhinosa*. Scale bars = 10 μm.

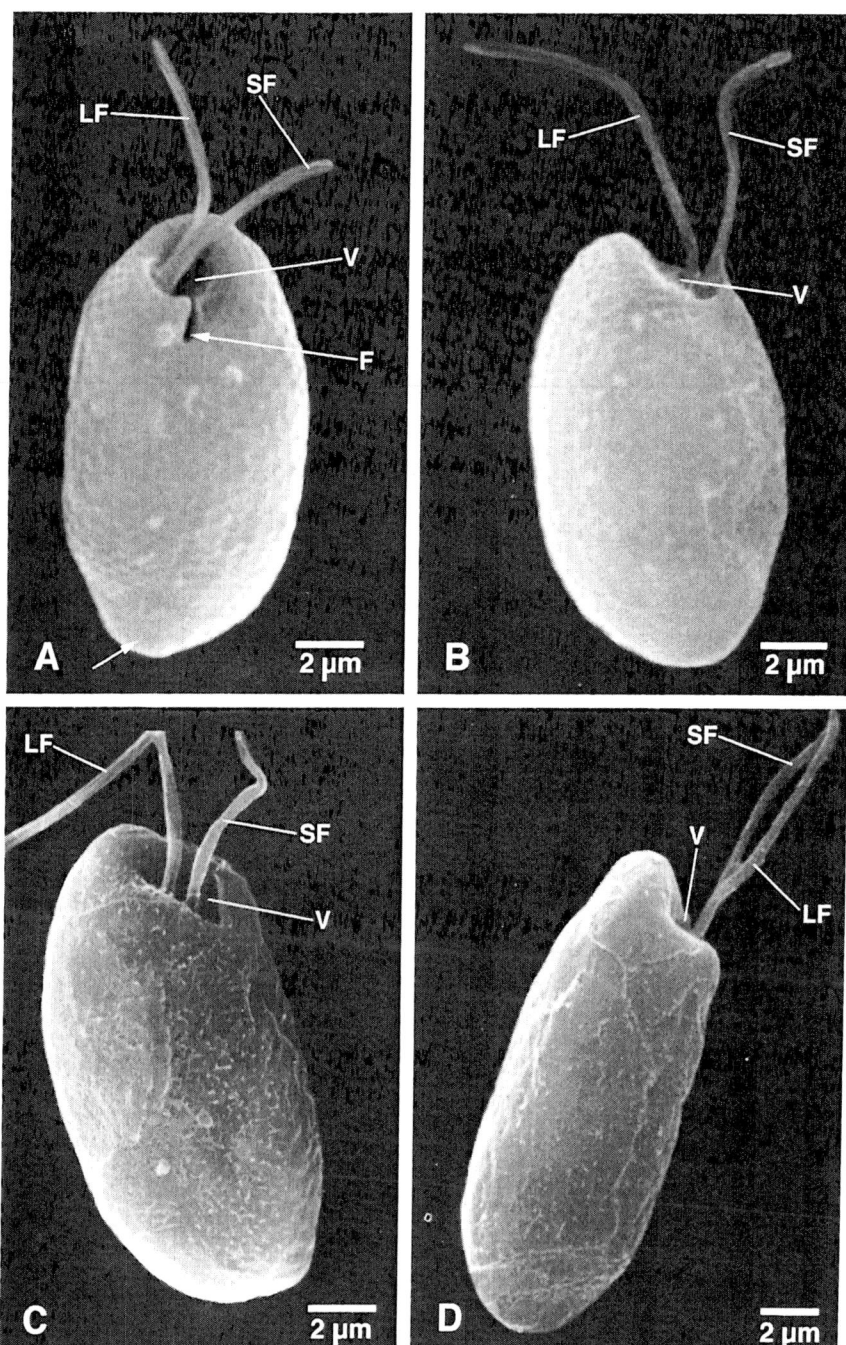

FIGURE 10 Scanning electron micrographs of red-colored cryptomonad species. (A) Ventral view of *Rhodomonas ovalis*, showing long (LF) and short (SF) flagella inserted on the right side of the vestibulum (V). A short furrow (F) and vestibulum are shown near the anterior, ventral surface. Note the absence of distinct plates at the cell posterior (arrow). (B) Lateral view of *R. ovalis*. (C) Oblique view of a cell of *Storeatula rhinosa*, showing long (LF) and short (SF) flagella inserted on the right side of the vestibulum (V). (D) Lateral view of *S. rhinosa*, showing the narrower elongate shape of the cell. C and D from Kugrens *et al.* (1999), with permission.

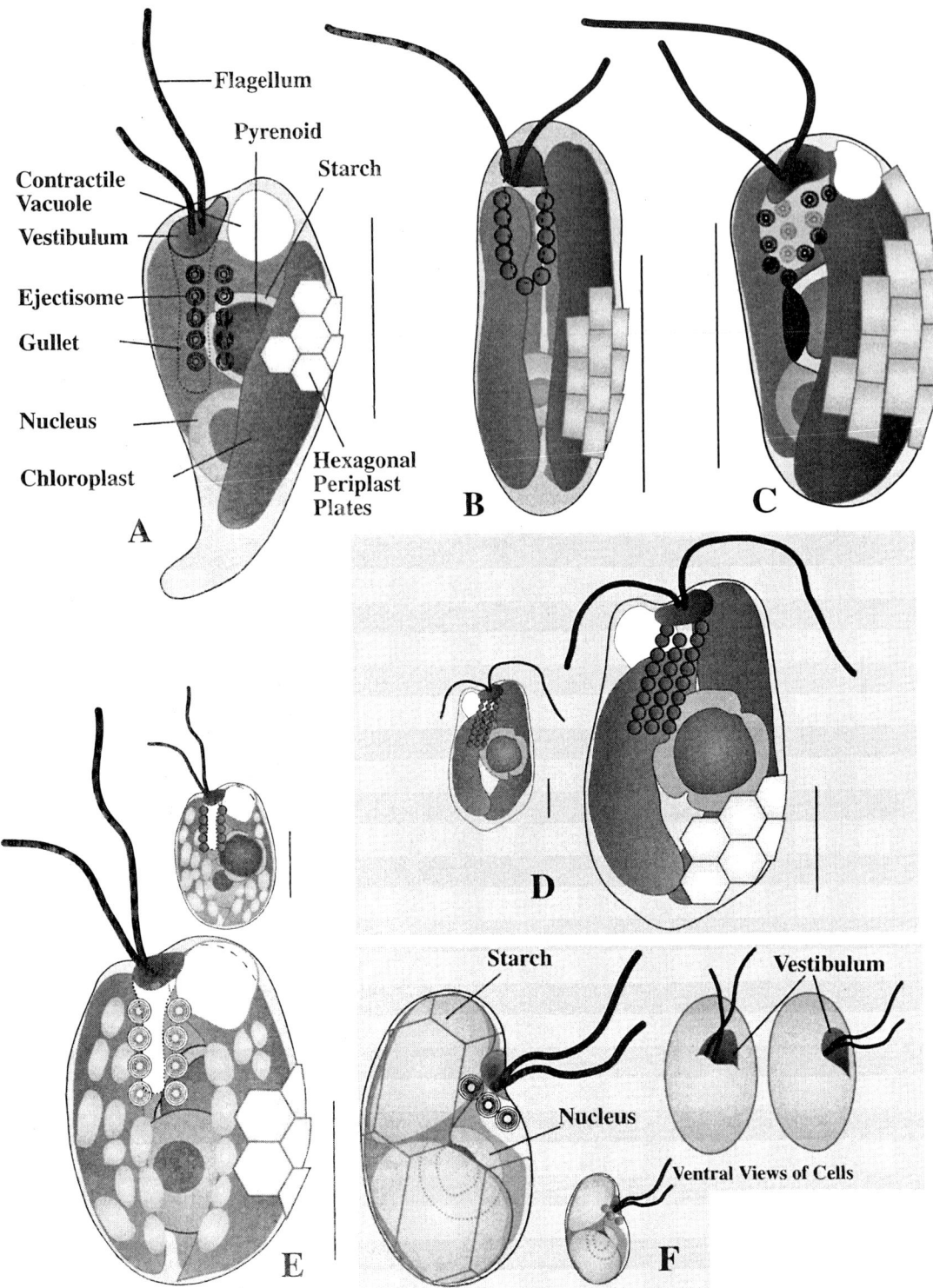

FIGURE 11 Light microscopic illustrations of blue-green cryptomonad species. (A) *Komma caudata*. (B) *Chroomonas oblonga*. (C) *Chroomonas coerulea*. (D) *Chroomonas nordstedtii*. (E) *Chroomonas pochmanni*. (F) *Hemiselmis amylosa*. Scale bars = 10 μm. Scanning electron micrographs of blue-green cryptomonad species. (A) Cell of *Chroomonas coerulea*, showing a short flagellum (SF) and a long flagellum (LF) inserted subapically in the right side of the vestibulum (V). (B) Cell of *Chroomonas oblonga*, showing a short flagellum (SF) and a long flagellum (LF) inserted subapically in the right side of the vestibulum (V). (C) Ventral view of a cell of *Chroomonas* sp., showing a short flagellum (SF) and a long flagellum (LF) inserted subapically in the right side of the vestibulum (V). Scale bars = 10 μm.

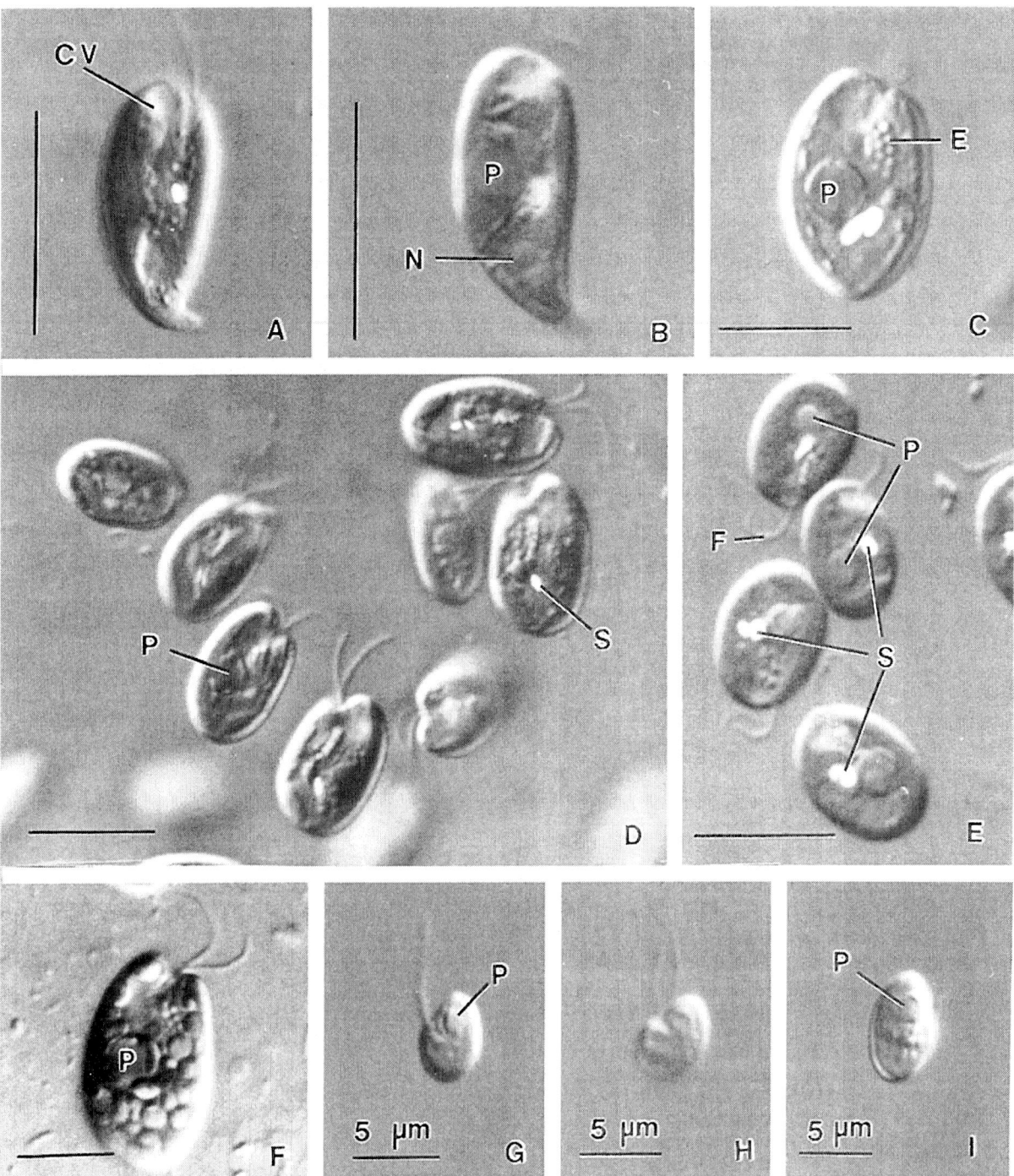

FIGURE 12 Light micrographs of blue-green cryptomonads species. (A) *Komma caudata* cell showing the typical shape. (B) *Komma caudata* cell with a dorsal pyrenoid (P) and a nucleolus located in the nucleus (N). (C) *Chroomonas pochmanni* cell, displaying the typical shape and a pyrenoid (P) and ejectisomes (E). D) *Chroomonas coerulea* cells with some stigmas (S) and pyrenoids (P) visible. (E) Slightly flattened *C. coerulea* cells, showing the stigma (S) associated with the pyrenoid (P). (F) *Chroomonas nordstedtii* cell displaying the typical cell shape. A pyrenoid (P) and starch grains are present in the cell. (G) *Hemiselmis amylosa* illustrating the typical bean-shaped cell in lateral view. A flagellum (F) and pyrenoid (P) are evident. (H) *Hemislemis amylosa* cell showing two flagella arising from a slight depression near the middle of the cell. (I) Slightly larger *Hemiselmis amylosa* cell with a pyrenoid (P). Scale bars = 10 μm, unless otherwise indicated.

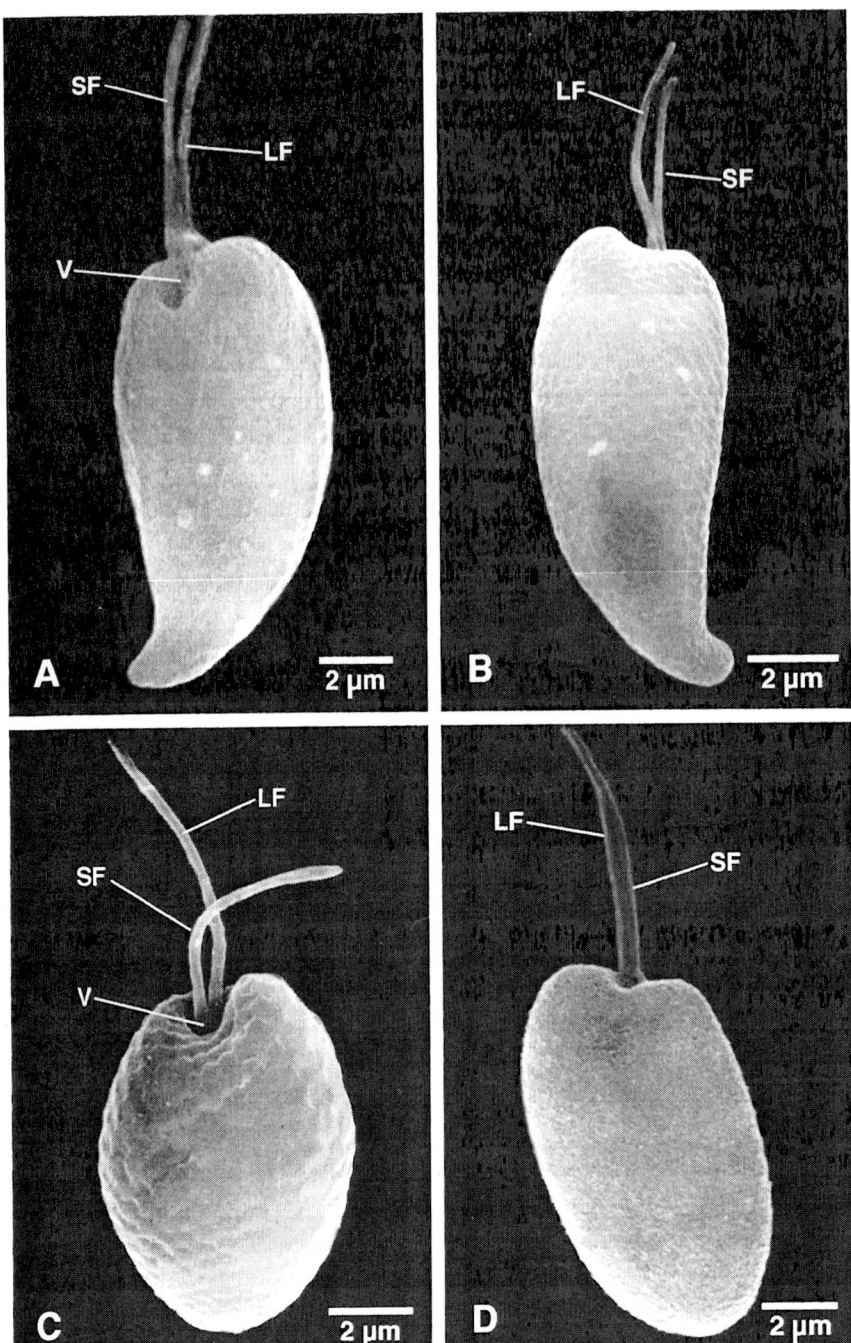

FIGURE 13 Scanning electron micrographs of blue-green cryptomonad. (A) Ventral view of a cell of *Komma caudata*, showing a short flagellum (SF) and a long flagellum (LF) inserted subapically in the right side of the vestibulum (V). Note the acuminate tail. (B) Lateral view of *Komma caudata*. Plate elevations are visible. C. Ventral view of a cell of *K. pochmanni*, showing a short flagellum (SF) and a long flagellum (LF) inserted subapically in the right side of the vestibulum (V). (D) Lateral view of *K. pochmanni*. C is from Kugrens and Lee (1991), with permission.

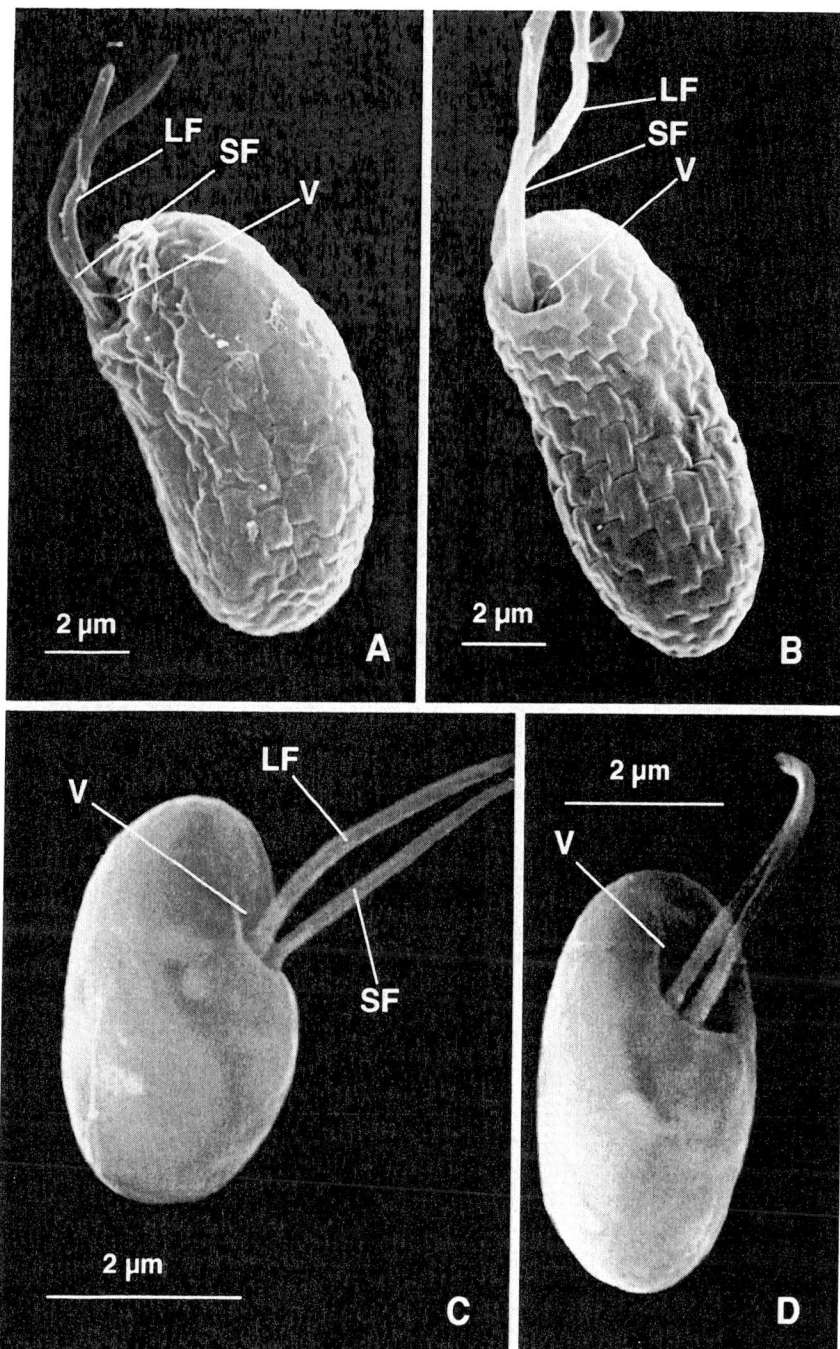

FIGURE 14 Scanning electron micrographs of blue-green cryptomonads. (A) Cell of *Chroomonas coerulea*, showing a short flagellum (SF) and a long flagellum (LF) inserted subapically in the right side of the vestibulum (V). Rectangular surface plates are visible. (B) Cell of *Chroomonas oblonga*, showing rectangular surface palted, and the short (SF) and long (LF) flagellar insertion. (C) Lateral view of *Hemiselmis amylosa* cell with long flagella (LF) and short flagella (SF) inserted in the vestibulum (V), approximately one-third of the distance down from the cell apex. (D) Ventral view of *P. amylosa*. showing the location of the vestibulum (V). A and B after Kugrens *et al.* (1986), and C and D are from Clay and Kugrens (1999b), with permission.

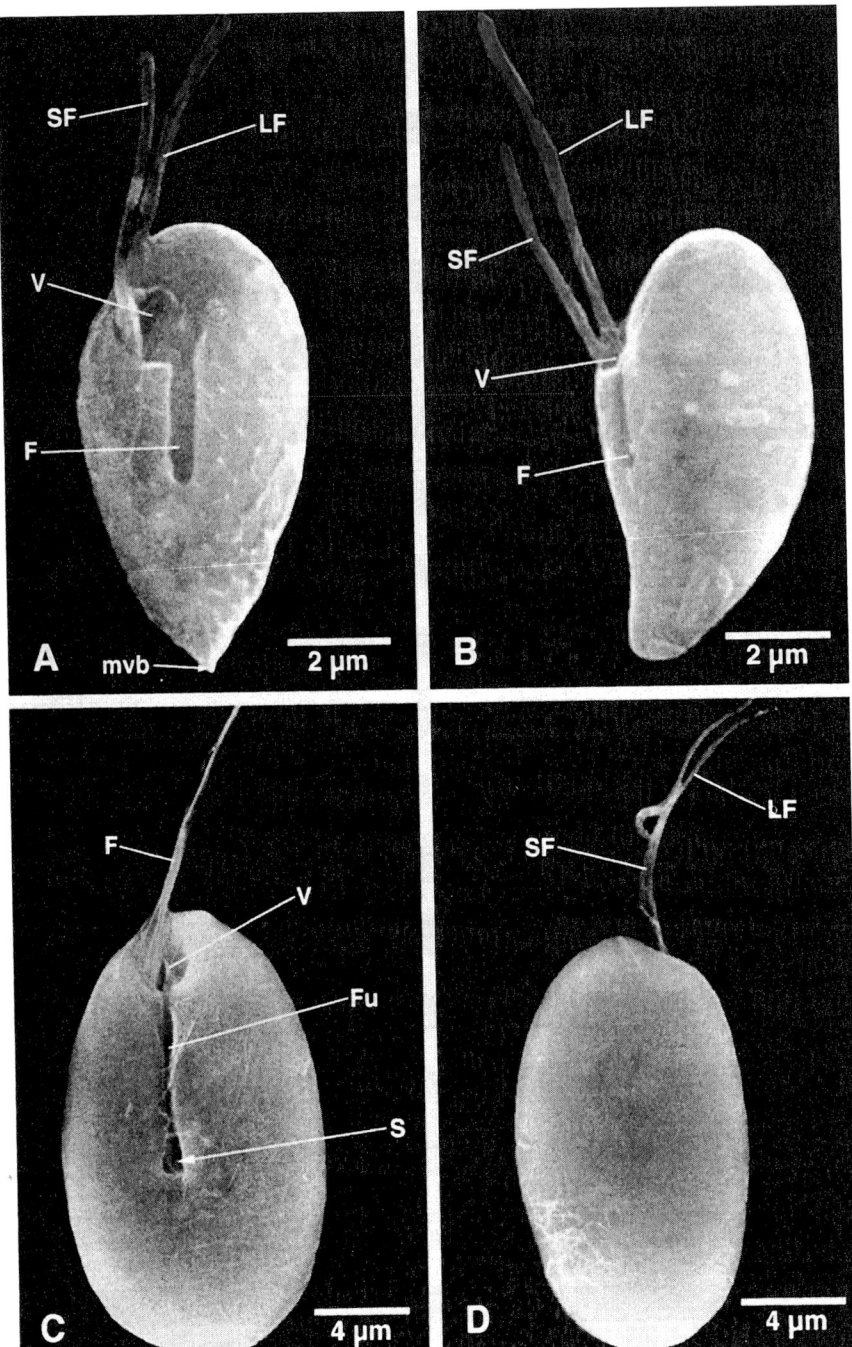

FIGURE 15 Scanning electron micrographs of *Plagioselmis* and *Cryptomonas ovata*. (A) Cell of *Plagioselmis nanoplanctica*, showing the long (LF) and short (SF) flagella inserted on the right side of the vestibulum (V). A furrow (F) extends from the vestibulum for approximately half the length of the cell. Note the midventral band (mvb) on the cell posterior. (B) Slightly lateral view of *P. nanoplanctica* to show the cell shape. (C) Cell of *Cryptomonas ovata* with long (LF) and short (SF) flagella inserted on the right side of the vestibulum (V). A long furrow extends from the vestibulum. Note the stoma (S). (D) Dorsal view of *Cryptomonas ovata*.

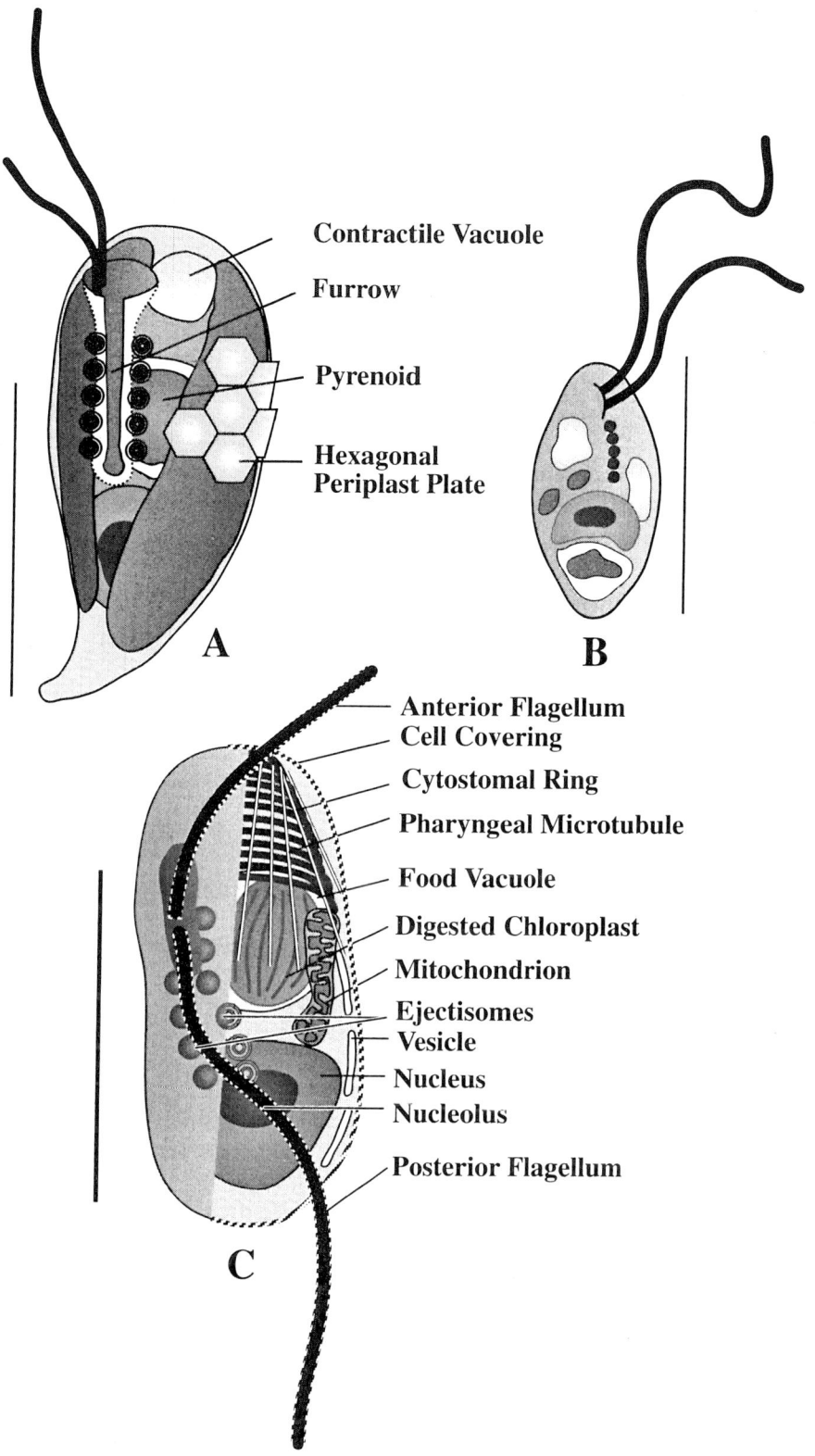

FIGURE 16 Light microscopic illustrations of *Plagioselmis* and *Kathablepharis* species. (A) Diagram of *Plagioselmis nanoplanctica*. (B) *Kathablepharis ovalis*, showing its general features. (C) Diagram of a *Kathablepharis phoenikoston* cell, as interpreted from electron microscopic data.

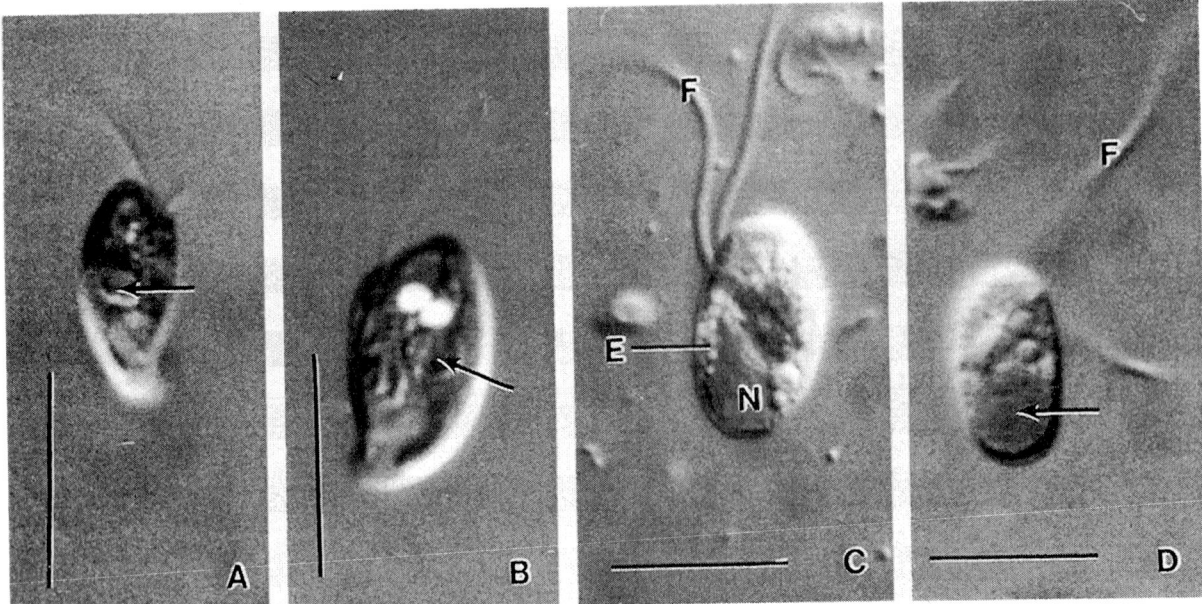

FIGURE 17 Light micrographs of *Plagioselmis* and *Kathablepharis*. (A) Lateral view of *Plagioselmis nanoplanctica*, showing the typical comma shape of the cell. The pyrenoid is dorsal (arrow). (B) Ventral view of *Plagioselmis nanoplanctica*, showing that this cell is broad in this view. The pyrenoid is on the dorsal side of the cell (arrow). (C) *Kathablepharis ovalis* with ingested food in an enlarged food vacuole. A row of ejectisomes (E) extends posteriorly from the site of flagellar insertion. The nucleus (N) is in the posterior of the cell. (D) *Kathablepharis phoenikoston* cell with the nucleus in the posterior. Globular contents probably represent ingested food. Scale bars = 10 µm.

B. Periplast Structure (Fig. 2)

Periplasts are special cell coverings found only in cryptomonads, although the same term is applied to euglenoids, which have a different arrangement. In cryptomonads periplasts consist of inner and surface components (Hibberd et al., 1971; Hill and Wetherbee, 1986, 1988, 1989; Kugrens et al., 1987; Kugrens and Lee, 1991; Wetherbee et al., 1986, 1987; Hill, 1990, 1991b; Clay and Kugrens, 1999a–c) with the plasma membrane located between the two components. Both components are variable in their structure, depending on the genus.

The inner periplast component (IPC) represents the protein component beneath the plasma membrane and is of two general types. One type consists of multiple plates of various shapes (Hibberd et al., 1971; Hill and Wetherbee, 1986, 1988, 1989; Kugrens and Lee, 1986; Hill, 1990, 1991b), and a second type consists of a single sheet (Grim and Staehelin, 1984; Kugrens and Lee, 1986; Hill, 1991b). The plates of the inner periplast component in the first type are connected to the cell membrane by intramembrane particles or proteins (Brett and Wetherbee, 1986; Hill and Wetherbee, 1986, 1988, 1989; Kugrens and Lee, 1986, 1991; Wetherbee et al., 1986, 1987; Clay et al., 1999). The arrangement of these IMP domains conforms to the plate shapes (Brett and Wetherbee, 1986; Kugrens and Lee, 1986, 1991; Wetherbee et al., 1987; Clay et al., 1999). Variations in the shapes of these plates are major features in establishing genera. Multiple-plated periplasts may have hexagonal, square, oval/round, rectangular, or irregularly shaped plates making up the inner and/or surface components.

In genera with a periplast sheet, which is a continuous internal protein cover, there are numerous closely spaced pores through which the ejectisome membranes penetrate to dock with the cell membrane (Grim and Staehelin, 1984; Kugrens et al., 1994). The periplast sheet itself is not connected to the plasma membrane. The surface periplast component (SPC) does not have a sheetlike variant, but consists of plates, heptagonal scales, mucilage, or a combination of any of these. It appears that both types of inner periplast components, and some of the surface plates, are composed of protein (Gantt, 1971; Faust, 1974).

Since plate shapes are used for generic designations, it is critical that the plate shapes be determined accurately. Oakley and Santore (1982), Gantt (1971), and Faust (1974) suggested that periplast shapes may undergo conformational changes. However, plates do not undergo shape changes unless they are subjected

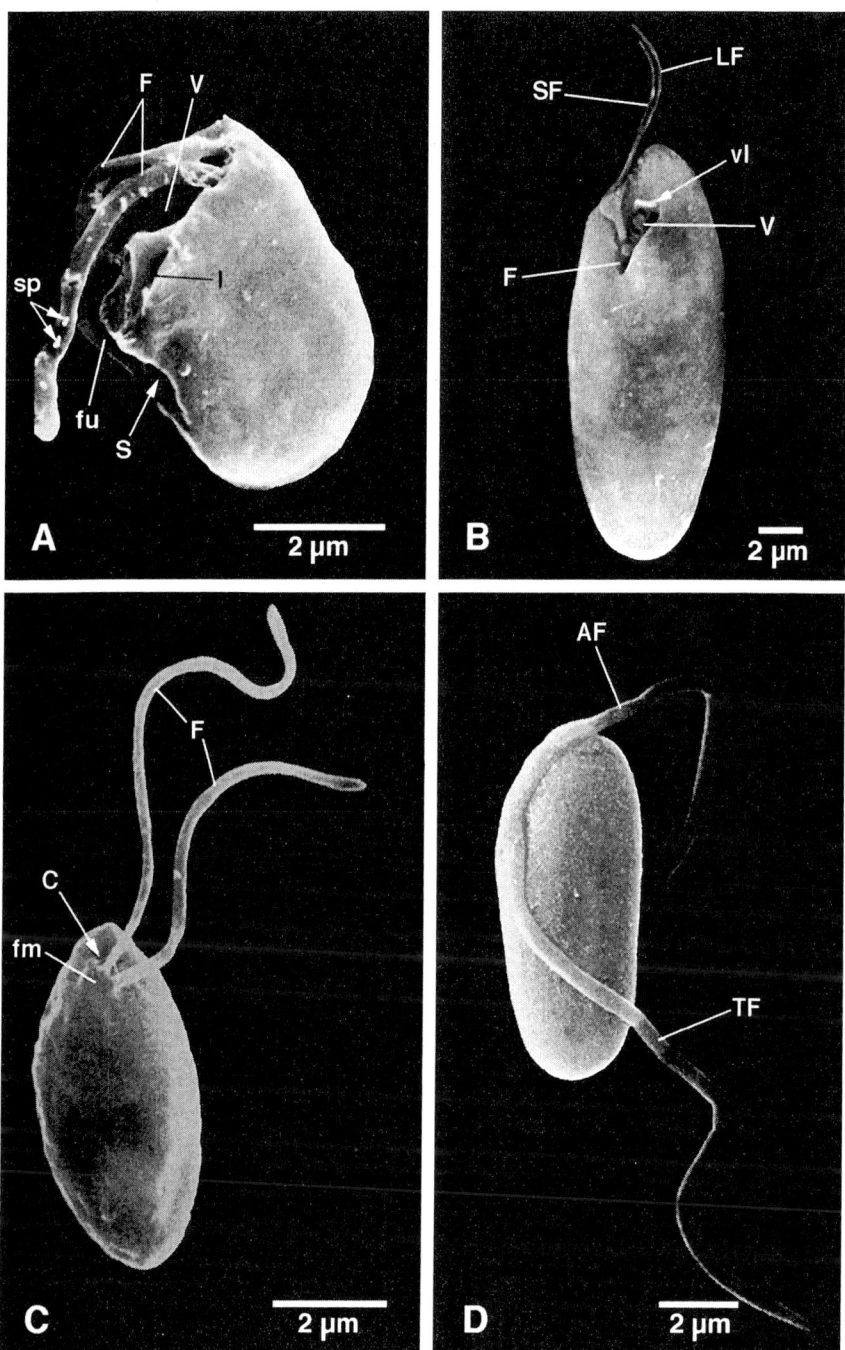

FIGURE 18 Scanning electron micrographs of colorless cryptomonad and kathablepharid species. (A) Cell of *Goniomonas truncata* with both flagella (F) inserted on the dorsal side of the vestibulum (V). Note some unilateral spikes (sp) remaining on the left flagellum. A ventral furrow (F) continues from the vestibulum and features a persistent opening termed the stoma (S). A second tubular invagination termed the infundibulum (I) is present on the left side of the cell. (B) Cell of *Chilomonas paramecium* showing two subapically inserted flagella (F) on the right side of the vestibulum (V). Note the vestibular ligule (vl) attached to the dorsal side of the vestibulum. A short furrow (F) extends from the vestibulum. (C) Cell of *Kathablepharis ovalis*, showing two subapically inserted flagella (F) arising from a flagellar mound (fm). A cytostome (C) is located between the flagellar mound and the cell apex. (D) Cell of *Kathablepharis phoenikoston*, showing an anteriorly directed flagellum (AF) and a trailing flagellum (TF), both inserted subapically. A and B are from Kugrens and Lee (1991), C is from Lee and Kugrens (1991), and D is from Clay and Kugrens (1999a), with permission.

to drastic treatments, such as desiccation, fixation, or excessive centrifugation. Any technique other than quick freezing may create artifacts, particularly in those periplasts having circular or oval plates, the shapes of which may be modified by pressures from adjacent plates. For example, in instances where the plates are approximately the same size, each plate is surrounded by six others (1 by 6 arrangement). When the plates are forced against each other, a hexagonal pattern may result (Kugrens et al., 1987).

Some studies attempted to use either light (Novarino, 1993a,b) or scanning electron microscopy (Santore, 1977; Novarino, 1991a,b; Novarino and Lucas, 1993; Novarino et al., 1994) to determine plate shapes, but with a few exceptions (Munawar and Bistricki, 1979; Klaveness, 1985; Kugrens et al., 1986; Hill, 1990), scanning electron microscopy is inadequate for studying subsurface components. If possible, quick-freezing freeze-fracture procedures should be used to study periplast plate shapes. With this technique cells are quick-frozen ("slammed") without pretreatment or fixation (Boyne, 1979; Chandler, 1984; Phillips and Boyne, 1984). Consequently, the periplast shapes in cryptomonads and the intimate association that exists between the plates and the plasma membrane can be examined accurately.

C. Flagella and Flagellar Apparatus (Fig. 1)

With the exception of *Goniomonas*, where the flagella are inserted on the dorsal side of the vestibulum, the flagella of cryptomonad cells are inserted subapically on the right side of the cell. The two flagella are subequal in length and consist of a dorsal and ventral insertion. Bipartite tubular hairs occur on at least one of the flagella (Kugrens et al., 1987; Clay et al., 1999), and there appear to be at least five variations in the arrangement of tubular and nontubular hairs on the flagella (Kugrens et al., 1987). Most of these features can be seen only with EM.

The most common arrangement of hairs on the flagella is one in which the longer or dorsal flagellum bears two laterally opposed rows of tubular hairs and the shorter or ventral flagellum bears a single row of hairs. Tubular hairs on the dorsal flagellum have one solid extension called a terminal filament, whereas the tubular hairs of the ventral flagellum have two unequal terminal filaments. In addition, flagella may bear heptagonal scales (Pennick, 1981; Lee and Kugrens, 1986). Instead of tubular hairs, *Goniomonas* has a unilateral row of curved spikes on one of its flagella and fine, nontubular hairs on both flagella (Kugrens et al., 1987; Kugrens and Lee, 1991).

The flagellar transition region is unique and consists of a doublet system of septa in all cryptomonads (Grain et al., 1988; Kugrens and Lee, 1991). A rhizostyle is an integral component of the flagellar apparatus in most cryptomonads, and it consists of microtubules that originate near the basal bodies and then extend posteriorly into the cell. One type of rhizostyle, found in *Chilomonas* (Roberts et al., 1981; Kugrens and Lee, 1991), *Hanusia phi* (Gillot and Gibbs, 1983; Gillott, 1990; Deane et al., 1998) *Teleaulax* (Hill, 1991b), *Storeatula* (Hill, 1991c), *Geminigera* (Hill, 1991c) and *Proteomonas* (Hill and Wetherbee, 1986), passes close to the nucleus and terminates near the posterior end of the cell. Each microtubule has a wing-like extension (lamella), and the lengths of these lamellae may vary with the respective microtubule (Gillot and Gibbs, 1983). A second type of rhizostyle, reported for *Cryptomonas ovata* (Roberts, 1984; Hill, 1990) and *Cryptomonas theta* (= *Guillardia theta*) (Gillot and Gibbs, 1983), lacks wings on the microtubules. Usually, this rhizostyle terminates anterior to the nucleus (Roberts, 1984).

D. Ejectisomes (Fig. 2)

The cryptomonad ejectisomes (formerly called *trichocysts*) were the first to be described (Anderson, 1962), but different types have been discovered in other organisms, and these are noted later. Ejectisomes are the extrusive organelles of all cryptomonads, and they appear to be identical in all genera that have been investigated (Kugrens et al., 1994). Cryptomonad cells contain two sizes of ejectisomes (Schuster, 1970; Kugrens et al., 1994; Clay et al., 1999). Large ejectisomes are located near the gullet/furrow complex, whereas small ejectisomes occur elsewhere in the peripheral cytoplasm. Both types of ejectisomes consist of two unequal sized components that are joined together and enclosed by a membrane (Kugrens et al., 1994). Each of these components has a tightly wound, tapered ribbon. The widest part of the tape is toward the outside of the ribbon. The smaller ribbon generally faces toward the outside of the larger ribbon. The tape has a crystalline substructure (Morrall and Greenwood, 1980; Grim and Staehelin, 1984; Kugrens et al., 1994). The ejectisomes discharge when the organism is irritated. Discharged ejectisomes form a long tube, with the short portion oriented at a slight angle to the long tube. The tube is formed because of a spiral rolling of the ribbon (Kugrens et al., 1994).

The ejectisome ribbons are formed in Golgi-derived vesicles, where they are tightly wound. Initially, the tape consists of a few turns, but the number increases as material is added. As the ejectisomes mature and enlarge, they are transported to the periphery of the cell. The ejectisome membranes contain intramembrane particles (IMPs) arranged in a rosette configuration (Grim and Staehelin, 1984), and these attach the ejectisome membrane to the cell membrane in a process called *docking*. In those species with a single periplast sheet, such as *Chilomonas*, there are numerous pores within the sheet. The ejectisomes are anchored to the sides of these pores (Grim and Staehelin, 1984).

Some *Pyramimonas* species (chlorophytes) and the colorless *Kathablepharis* (Lee and Kugrens, 1991; Kugrens *et al.*, 1994; Clay and Kugrens, 1999a,b) and *Leucocryptos* (Vørs, 1992a, b) also possess ejectisomes; however, they differ in structure from cryptomonad ejectisomes. In all three genera, only the large ribbon makes up the ejectisome. After release all form a tube, just as in cryptomonads, but the small component is lacking. Discharge of ejectisomes is forceful and rapid, propelling the organism in the direction opposite the discharge. Thus the ejectisome could function as an escape mechanism to avoid predators, or it could be a defense mechanism to inflict damage on a potential predator (Kugrens *et al.*, 1994).

E. Ejectisome Digestion Vesicles

Ultrastructural evidence indicates that some vesicles in cryptomonads are specialized for ejectisome autolysis (Kugrens *et al.*, 1994). These vesicles originate from the fusion of several ejectisome chambers and continue to enlarge by the fusion of additional ejectisome membranes. Individual ejectisomes disaggregate within these vesicles. In older vesicles, components of expanded ejectisomes make up most of the contents. In later stages most of the tubular, expanded components of ejectisomes are no longer recognizable, and the contents appear fibrillar or granular. The vesicle sizes are larger in cells from older cultures, and there may be several vesicles per cell. Golgi vesicles have been observed to fuse with the existing vesicles, perhaps to add lytic enzymes. The vesicles apparently represent specific repositories for defective or surplus ejectisomes; thus they represent another unique component of cryptomonad cells. These may be the refractive vesicles that are frequently seen with the light microscope and formerly may have been referred to as the *Corps de Maupas* (Lucas, 1970b).

F. Mitochondria and Chloroplasts (Fig. 1)

A single reticulate mitochondrion with flattened cristae (Santore and Greenwood, 1977; Roberts *et al.*, 1981; Kugrens and Lee, 1991) apparently occurs in cells of all cryptomonads.

Chloroplasts may be olive green, brownish, blue-green, or red, depending on the pigments present. Pigments consist of chlorophylls a and c_2, alpha and beta carotene, alloxanthin, diadinoxanthin, and several forms of blue and red phycobiliproteins called Cr-phycocyanin and Cr-phycoerythrin, to differentiate them from cyanobacterial and rhodophyte phycobiliproteins (Glazer and Appel, 1977; Hill and Rowan, 1989). With the exception of a marine endosymbiont (Hibberd, 1977), only one or two chloroplasts occur in the cells of pigmented genera (Santore, 1984, 1987; Hill, 1991a,b,c). Either Cr-phycocyanin or Cr-phycoerythrin (Hill and Rowan, 1989) is located in the intrathylakoidal lumens of the photosynthetic lamellae (Gantt *et al.*, 1971; Faust and Gantt, 1973; Gantt, 1979, 1980; Ludwig and Gibbs, 1989).

Chloroplasts are surrounded by a double membrane called the periplastidial envelope, periplastidial compartment, periplastidal complex, or chloroplast endoplasmic reticulum (CER), which originates from an evagination of the outer membrane of the nuclear envelope and surrounds the chloroplast, starch granules, and a reduced nucleus known as the *nucleomorph* (Gillott and Gibbs, 1980; Santore, 1982c; Ludwig and Gibbs, 1985a). Starch granules are formed within the periplastidial compartment, not in the chloroplast, and they generally are associated with a pyrenoid if present. The number of thylakoids penetrating the pyrenoid has been suggested as a possible taxonomic character (Santore, 1984). Thylakoids in the chloroplasts usually are arranged in pairs (Gantt *et al.*, 1971; Dwarte and Vesk, 1982, 1983; Santore, 1984), sometimes in groups of three (Klaveness, 1981; Hill, 1991b), or in stacks of variable number (Hill, 1991b). *Chilomonas* has a reduced chloroplast which lacks pigments and is called a leucoplast (Sespenwol, 1973; Heywood, 1988; Kugrens and Lee, 1991). *Goniomonas* lacks plastids and a nucleomorph, consequently it also lacks the periplastidial compartment.

G. Nucleomorphs (Fig. 1)

Nucleomorphs are highly reduced endosymbiont nuclei located in the periplastidial compartment (Gillott and Gibbs, 1980; McKerracher and Gibbs, 1982; Morrall and Greenwood, 1982; Santore, 1982c, 1984, 1987; Ludwig and Gibbs, 1985a; Kugrens and Lee, 1989, 1991). They represent a vestigial nucleus, which remains from an ancestral endosymbiont, possibly a red alga (Douglas *et al.*, 1991; McFadden *et al.*, 1997). The nucleomorph is small, it is limited by a double membrane and it contains DNA (Ludwig and Gibbs, 1985a; Douglas *et al.*, 1991; McFadden, 1993; McFadden *et al.*, 1997). In addition, the nucleomorph contains a fibrillo-granular region and dense bodies. Its location within the compartment may have systematic applications (Santore, 1984; Hill and Wetherbee, 1989), specifically in *Rhodomonas* and *Storeatula*, where the nucleomorph is located in the pyrenoidal bridge (Hill and Wetherbee, 1989; Novarino, 1991a,b).

H. Reproduction

Reproduction usually occurs by mitotic divisions, although sexual cycles have been documented for *Proteomonas* (Hill and Wetherbee, 1986) and *Chroomonas* (Kugrens and Lee, 1988). Resistant spore production is rare, but some species produce cysts or palmelloid stages to withstand adverse conditions (Santore, 1978). It has been suggested that palmelloid cells, surrounded by extensive mucilage, might be an adaptation to deter grazing (Klaveness, 1988a).

I. Nucleus and Mitosis

The nucleus is located in the cell posterior. In interphase it contains dispersed chromatin and a prominent nucleolus. The outer membrane of the nuclear envelope expands to form the chloroplast endoplasmic reticulum around the chloroplast, nucleomorph, and starch. The Golgi apparatus, the gullet/furrow, and contractile vacuole in freshwater species are situated anterior to the nucleus.

Mitosis has been studied in *Chroomonas salina* (Oakley and Dodge, 1973, 1976; Meyer and Pienaar, 1981, 1984b), *Cryptomonas* sp. (Oakley and Bisalputra, 1977; Oakley and Heath, 1978), *Cryptomonas theta* (McKerracher and Gibbs, 1982), and *Chroomonas africana* (Meyer and Pienaar, 1981, 1984b). Generally, microtubules proliferate near the flagellar bases at the onset of mitosis. The nuclear envelope disaggregates as the microtubules move away from the basal bodies to form a spindle. At metaphase the chromosomes appear as a solid mass. Chromosomal microtubules terminate in the chromosomal mass (Oakley and Dodge, 1973, 1976; McKerracher and Gibbs, 1982), but kinetochores, which are special sites of microtubular attachment to chromosomes, have not been observed. At anaphase the solid chromosomal mass splits and the two masses move to the poles. A cytokinetic ring is formed at metaphase. The ring constricts to cleave the cells into two daughter cells. Cytokinesis and cell separation follow a pole reversal when daughter cells are formed (Perasso et al., 1993).

J. Contractile Vavuoles (Fig. 1)

Most freshwater cryptomonads possess a pulsating vacuole, known as a contractile vacuole, that functions in osmoregulation. The contractile vacuole expels excess water and waste metabolites from the cell (Patterson, 1981). Since freshwater cryptomonads exist in a hypo-osmotic medium, there is a net influx of water into the cell, and the contractile vacuole is the organelle that actively expels water from the cell to prevent a rupture of the cell, since a cell wall is lacking. Contractile vacuoles generally are located in the anterior of the cell.

K. Starch (Figs. 1, 3, 6C, D, 9)

The principal storage product is starch, and it is found as granules in the periplastidial space and not in the chloroplast. If a pyrenoid is present some starch accumulates around the pyrenoid in large plates. Cryptomonad starch is an α-1-4-glucan composed of 30% amylose and amylopectin and is similar to starch found in green algae and dinoflagellates (Antia et al., 1979), it stains purple with iodine.

III. ORIGIN OF CRYPTOMONADS

Cryptomonads consist of both colorless and pigmented cells, and one of the colorless forms lacks any vestiges of plastids. With the exception of *Goniomonas*, cryptomonads have one of the most complex cells known, consisting of four genomes—the host genome, the mitochondrial genome, the chloroplast genome, and the endosymbiont nuclear genome. These associations apparently originated from three distinct symbiotic events: two prokaryote–eukaryote events and a eukaryote–eukaryote endosymbiotic association (McKerracher and Gibbs, 1982; Ludwig and Gibbs, 1985a,b, 1989; Gillott, 1990; Douglas et al., 1991; McFadden et al., 1994, 1997). Therefore, except for *Goniomonas* (McFadden et al., 1994), cryptomonads consist of a eukaryotic host cell, two prokaryotic endosymbionts (the mitochondrion and chloroplast), and a eukaryotic endosymbiont.

The endosymbiotic events leading to a pigmented cryptomonad cell presumably are due to a secondary endosymbiosis whereby a colorless phagocytic host cell ingested a red algal cell (McFadden, 1993; McFadden et al., 1997). Once the red algal cell was ingested, it became surrounded by two concentric membranes, one of which is continuous with the outer membrane of the host nuclear envelope. Therefore, the inner membrane probably represents the plasma membrane of the red algal cell, whereas the outermost of these membranes may have originated from the food vacuole membrane fusing with the outer membrane of the nuclear envelope, which bore 80S ribosomes, thereby creating one continuous outer membrane. This outer membrane subsequently became studded with 80S ribosomes. Presumably, this occurred by lateral diffusion, through the lipid bilayer, of pre-existing ribosome receptors (i.e., ribophorins) derived from the outer nuclear membrane, resulting in an outer membrane covered with 80S ribosomes. These 80S ribosomes occurring on the outer membrane are distinct from the 80S ribosomes found in the cytoplasm of the red algal endosymbiont. Over time, numerous genes of the endosymbiont were translocated to the host nucleus. This event necessitated the evolution of a complex protein import mechanism whereby chloroplast gene products transcribed on host cytosolic ribosomes could be targeted through the two topogenically unique membranes of the chloroplast endoplasmic reticulum, back into the chloroplast. Two putatively novel N-terminal signal sequences would serve as the import mechanism. Together, these complex events resulted in the chloroplast and the endosymbiont cytoplasm being surrounded by two membranes. This complex, including the smooth inner membrane, is termed the chloroplast endoplasmic reticulum (CER), and the former endosymbiont cytoplasm, with its contents that surround the chloroplast(s), is termed the periplastidial space or compartment.

The endosymbiont subsequently transferred many of its genes to the host nucleus. These evolutionary events created a situation in which the endosymbiont cell became fully dependent on the host for its survival. Molecular evidence indicates that the nucleomorph is of red algal origin, and the presence of starch in the periplastidial cytoplasm is visual evidence of this origin. Moreover, cryptomonad chloroplasts contain phycobiliproteins similar to those found in red algae (Glazer and Appell, 1977) and possess the type I purple form of Rubisco, which, among eukaryotes, occurs only in red algae and cryptomonads (Martin *et al.*, 1992). The nucleomorph contains three linear chromosomes bearing multiple ribosomal RNA genes (McFadden *et al.*, 1997). Finally, it is now widely accepted that extant photosynthetic cryptomonads evolved monophyletically from the permanent fusion of a phagocytic protozoan host and a red algal unicell which originally possessed both phycoerythrin and phycocyanin phycobiliproteins. Differential loss of phycoerythrin or phycocyanin produced photosynthetic cryptomonads that are either blue-green or red to reddish brown in coloration.

The most comprehensive molecular data regarding cryptomonad phylogeny (Marin *et al.*, 1998) indicate that a switch in coloration occurred from a red to a blue-green cryptomonad, and perhaps back to a red cryptomonad again. Such an event is biochemically feasible (Wemmer *et al.*, 1993; Glazer and Wedemeyer, 1995), and in this context, it is proposed that the photosynthetic ancestor lost phycocyanin and allophycocyanin. Thus, phycoerythrin was the ancestral accessory pigment in these cryptomonads. and blue-green cryptomonads were derived from phycoerythrin-containing types. Blue-green cryptomonads actually possess true phycoerythrins, but they appear blue-green because linear phycoerythrobilin chromophores have been replaced with phycocyanobilin chromophores (Apt *et al.*, 1995; Glazer and Wedemeyer, 1995).

Recent cryptomonad molecular phylogenies (Marin *et al.*, 1998; Clay and Kugrens, 1999; Clay *et al.*, 1999) are largely consistent with the most recently proposed cryptomonad classification system (Clay *et al.*, 1999). In this scheme the division Cryptophyta consists of two classes with the following three orders: the Goniomonadales, which contains a single family; the Cryptomonadales, which contains two families; and the Pyrenomonadales, which contains five families. Six of the eight families appear to be natural groups (i.e., monophyletic) and are supported by molecular phylogenies. The remaining two families represent well-circumscribed taxa, but each also contains a monotypic clade.

Based on early sequence data, the host nuclear-encoded rRNA appeared to be related to the protozoan *Acanthameba* and the green algal lineage. However, considerably more sequence data have accumulated in recent years, and this proposed relationship is no longer tenable. At present, the large-scale phylogenetic affinities of the host component of cryptomonads continues to be one of the major conundrums in evolutionary protistology. Ultrastructurally, they appear to be related to the "stramenopiles" *sensu* Patterson (Heterokonta *sensu* Cavalier-Smith, 1986) by virtue of possessing tubular hairs and chloroplasts not free in the cytosol, but rather delimited by the two extra membranes of the chloroplast endoplasmic reticulum. Cryptomonads, however, have flat cristae, which demonstrates their affinities with other platycristate eukaryotes. Sequence analyses have not provided evidence of the nature of the host that engulfed the endosymbionts. It has been suggested that Glaucophytes are the sister group to the cryptomonads (Cavalier-Smith *et al.*, 1996), and this is consistent with the analyses by Bhattacharya *et al.* (1995). It should also be realized that in all published trees to date, cryptomonad host affinities are contravening, and none are considered reliable. In contrast, the origin of the cryptomonad nucleomorph is well understood and is corroborated by several lines of evidence. In particular, many independent molecular analyses support the hypothesis that nucleomorph encoded ribosomal rRNA genes are related to red algae (Douglas *et al.*, 1991; Cavalier-Smith *et al.*, 1996; McFadden *et al.*, 1997).

IV. ECOLOGY

Cryptomonads are ubiquitous and have been reported from nearly all types of water throughout the world, including arctic, temperate, and tropical oceans; streams, lakes, and reservoirs; and environments of variable salinity (Klaveness, 1988a,b). However, it is the lakes in temperate regions of the world where cryptomonads display their largest diversity, and where they are found under a wide variety of conditions (Taylor et al., 1979). Several have even exploited intracellular environments, serving as functional endosymbiotic chloroplasts for some ciliates (Hibberd, 1977; Klaveness, 1988a) and dinoflagellates (Lewitus et al., 1999). However, the number of cryptomonad species generally is underestimated in phytoplankton inventories, largely because of the preservation of samples with destructive fixatives such as formaldehyde (Klaveness, 1988a).

A. Abiotic Factors

In lentic, estuarine, and marine habitats, cryptomonads are permanent residents of the phytoplankton community. In fact, when other populations are diminishing, cryptomonads increase in numbers (Rott, 1983; Klaveness, 1988a, b). In many lakes there appears to be a population peak during autumn destratification when the epilimnion and hypolimnion mix (Pollingher, 1981). In an extensive survey conducted by Taylor et al., (1979) of lakes in the eastern and southeastern United States, cryptomonads seemed to prefer colder waters, an observation that we have also made in the Rocky Mountain region. In small temperate lakes, they display a variety of seasonal strategies, including stable stratified populations, diel vertical migration, and formation of resting stages during some parts of the year. In most lakes, cryptomonads exhibit maximal population densities far below the surface (Reynolds, 1980, 1984; Rott, 1983). Optimal depths have been reported from 15 to 25 m, with the deepest occurring in late spring and early summer and the shallowest in late autumn and early winter. This pattern occurs primarily in more productive, buffered lakes with low turbulence and, therefore, reduced but presumably adequate light. In low buffered, eutrophic lakes, the decrease in pH due to photosynthesis favors productivity of cyanobacteria and a decrease in cryptomonads.

Several cryptomonad species are able to thrive under low light conditions through the phenomenon of chromatic acclimation. Moreover, some are adept at surviving prolonged periods of darkness. One particular strain was shown to survive a dark period for more than 24 weeks, whereas two species, *Hemiselmis virescens* and *Rhodomonas lens*, survived for only a 4-week maximum. Survival during winter under lake ice in near-dark conditions has been shown, in part, to be an effect of low temperatures resulting in low respiration, a change in lipid composition (Henderson and Mackinlay, 1989), and low grazing pressures (Morgan and Kalff, 1979). These phenomena, coupled with chromatic acclimation to efficiently harvest available light, appear to be sufficient for winter survival.

Day length and its relation to water depth have been shown to be important factors shaping phytoplankton communities (Arvola et al., 1991). Because of the attenuating effects of water, daybreak occurs later and nighttime occurs earlier with increasing depth. At extremes, low light intensity translates into low productivity, and extreme light intensity has proved adverse to some phytoplanktonic algae. Consequently, day length acts as a selective force that favors those planktonic algae most capable of responding to the day length–light intensity variable. Cryptomonads are capable of sensing light intensity and avoid the extremes through diel, vertical migration behavior (Watanabe et al., 1976; Watanabe and Furuya, 1982a, b; Arvola et al., 1991).

Variation in pH among lakes causes differences to appear in cryptomonad populations and even among strains of the same species isolated from different sources (Pringsheim, 1968; Klaveness, 1988a). However, it is often difficult to determine what is ultimately responsible for these differences, since other environmental factors are coupled with pH under natural conditions. Nevertheless, laboratory studies indicate that strains of a given species isolated from different geographic regions exhibit growth at restricted pH ranges that correlate with those from the locality where they were isolated (Pringsheim, 1968). Another laboratory study conducted on a strain of *Rhodomonas lacustris* showed good growth between pH 6 and pH 8.5, although pH 10 was tolerated in the light cycle in unbuffered media, but growth was reduced (Klaveness, 1977). As with other environmental parameters, cryptomonads occur over a wide range of pH, but generally the range of pH tolerance is narrow. Based on our own observations, cryptomonads in Colorado and Wyoming favor alkaline conditions and are most prominent in lakes with pH values above 7.5.

It has been proposed that algae with a chloroplast endoplasmic reticulum, such as the cryptomonads, might have an advantage in high-pH environments (Lee and Kugrens, 1998, 2000). Dissolved inorganic carbon (DIC) is present mainly as bicarbonate in waters that have a high pH. The carbon-fixing enzyme Rubisco can only utilize DIC in the form of CO_2. Therefore, if the space within the chloroplast endoplasmic reticulum is acidic, then it could be a reservoir of DIC in the form of CO_2 that algae without chloroplast endoplasmic reticulum would not have. The availability of CO_2 in this space would impart a competitive advantage to these algae in waters high in pH and low in CO_2.

A putative cryptomonad, *Cyanomonas*, has been reported to cause massive catfish deaths in Texas ponds (Pfiester and Holt, 1978). Neither the organism nor the toxin was isolated; therefore, the identity of *Cyanomonas* remains in doubt. Furthermore, the light micrographs presented in this publication lacked the resolution needed to determine whether it indeed was a *Cyanomonas* or even a member of the cryptomonads. *Rhodomonas* sp. also has been implicated in exotoxin production (Stemberger and Gilbert, 1985), although it was not demonstrated that the observed inhibition of rotifer growth was specifically due to *Rhodomonas*. Bacteria in the cultures may have been the causative agents.

B. Biological Factors

Cryptomonads are optimal or near-optimal food organisms for zooplankton (Guillard, 1975; Klaveness, 1984; Stemberger and Gilbert, 1985; Sarnelle, 1993). Additionally, they are routinely ingested by various colorless dinoflagellates, ciliates, and *Kathablepharis* sp. (Stemberger and Gilbert, 1985; Clay and Kugrens, 1999a,b; Lewitus *et al.*, 1999). Observations of *Daphnia hyalina* and *Diaptomus gracilis* showed that *Rhodomonas* sp. was present most of the time (Ferguson *et al.*, 1982). Moreover, various studies suggest that when given a choice among flagellate food items, rotifers appear to select cryptomonads preferentially (Stemberger and Gilbert, 1985), and the presence of cryptomonads enhances the reproduction of planktonic rotifers (Edmondson, 1965). Pejler (1977) observed that various rotifers prefer *Rhodomonas* to *Chrysochromulina*. Given that many zooplankton selectively choose cryptomonads as prey, grazing probably plays a significant role in regulating planktonic cryptomonad population dynamics (Sarnelle, 1993). Cryptomonad populations generally reach a maximum following periods of moderate turbulence, when they are disseminated throughout the water column and mixed with higher nutrient waters (Reynolds, 1984), and grazing is reduced. When turbulence decreases and nutrients are depleted, grazing once again reduces numbers, thereby regulating the populations.

Perhaps one of the most interesting aspects of cryptomonad ecology is the phenomenon known, as kleptoplastidy, which is a process in which the chloroplast of an ingested photoautotroph is retained, and the chloroplast remains functional for a period of time (Schnepf *et al.*, 1989; Lewitus *et al.*, 1999). Kleptoplastidy of cryptomonad chloroplasts occurs in ciliates (Stoecker and Silver, 1990) and dinoflagellates (Skovgaard, 1988; Schnepf *et al.*, 1989; Putt, 1990), and these chloroplasts photosynthesize and produce starch, which may be an available carbon source for the host (Putt, 1990; Schnepf and Elbrächter, 1992), or it may fulfill metabolic requirements under limited food availability (Lewitus *et al.*, 1999).

Kleptoplastidity is particularly important in the survival of the ichthyotoxic dinoflagellate *Pfiesteria piscicida*, where cryptomonad chloroplasts are selectively ingested by nontoxic zoospores. Cryptomonad chloroplasts are retained for approximately 9 days, and they remain functional during that time, as determined by the uptake of C^{14}-bicarbonate (Lewitus *et al.*, 1999). This retention of chloroplasts promotes the survival of this intermediate stage in the life cycle of *Pfiesteria*.

C. Cryptomonad Endosymbiotins and Pathogens

Cryptomonad cells are susceptible to prokaryotic or eukaryotic infections; however, pathogenicity in the cryptomonads has not been the object of detailed studies. Although mixotrophy is rare in cryptomonads (see Section IV.D), bacteria can enter the cell and become endosymbionts, as demonstrated by Schnepf and Melkonian (1990). These bacteria apparently do not have any adverse effects on cells and might represent a mutualistic association. In addition, these bacteria harbor bacteriophages (viruses which infect the bacteria). However, other bacteria and viruses (Pienaar, 1976) may adversely affect cryptomonads (Klaveness, 1982). Klaveness (1982) has shown that *Cualobacter* can attach to cells externally, causing malformations of the cells.

Canter (1968) reported that certain cryptomonads are vulnerable to chytrid parasites. Specifically, *Rhizophydium fugax* has been observed on species of *Cryptomonas* that are resting in palmelloid colonies. The chytrid may initially be attracted to the polysaccharide mucilage that envelopes the cells of the palmelloid colony. Upon reaching a cell, the chytrid apparently situates itself in the furrow, and the rhizoidal system invades the cell through this depression. Chytrids have also been reported parasitizing *Chilomonas striata* (Caljon, 1983). In addition to chytrid parasites, intracellular parasites of undetermined taxonomic status have been observed in *Cryptomonas* (= *Campylomonas*) *rostratiformis* (Ettl and Moestrup, 1980).

D. Types of Nutrition—Carbon Sources

Cryptomonads are photoautotrophic, heterotrophic, or mixotrophic. Photoautotrophs synthesize organic molecules from CO_2 by photosynthesis. Heterotrophic cryptomonads require some organic materials for their carbon source, via osmotrophy or phagotrophy. Osmotrophs utilize dissolved organic matter, whereas phagotrophs ingest particulate matter, including other organisms. Both types are restricted to colorless cryptomonads and kathablepharids. For instance, *Chilomonas* is colorless; thus, it is unable to manufacture its own basic organic molecules. It is strictly osmotrophic and obtains organic compounds by incorporating dissolved organic molecules into its cell for metabolism (unpublished observations). It does not ingest particulate materials, the type of nutrition that this alga was assumed to have. In fact, the cell covering is a major obstacle to phagotrophy. *Goniomonas*, on the other hand, is phagotrophic and routinely ingests bacteria (Mignot, 1965), presumably through a specialized structure in the cell known as the infundibulum (Mignot, 1965; Kugrens and Lee, 1991). Whether it is also osmotrophic is unknown.

The vast majority of cryptomonads, however, are strictly photoautotrophic and do not require dissolved organic matter (DOM) in their metabolism. Several studies indicated that dissolved organic matter does not enhance the growth of photosynthetic cryptomonads (Lewitus and Caron, 1991; Arvola and Tulonen, 1998). Reported increased growth when DOM was added probably was due to bacterial respiration, where the bacteria oxidized the organic molecules and released CO_2. Cryptomonad growth increased because of increased CO_2 for photosynthesis (Arvola and Tulonen, 1998) and not through a utilization of dissolved organic matter.

Mixotrophy is an ecologically important type of nutrition in many flagellates (Boraas *et al.*, 1988) where a photosynthetic organism can also ingest particulate matter, primarily other cells, both prokaryotes and eukaryotes. This type of nutrition is common in chrysophytes, but it also has been reported in a few cryptomonads (Tranvik *et al.*, 1989; Kugrens and Lee, 1991). For example, *Cryptomonas* was studied with respect to bacterial ingestion (Tranvik *et al.*, 1989), but electron microscopic examinations were not conducted. One species of *Chroomonas* may be mixotrophic, as determined by ultrastructural studies (Kugrens and Lee, 1991). This study revealed a specialized bacterial incorporation vesicle and bacteria in various stages of digestion; it is the only genus of cryptomonad in which mixotrophy has been documented with electron microscopy. As was the case with phagotrophy in *Chilomonas*, mixotrophy in other cryptomonads actually is precluded because of the presence of the periplast and furrow plates, which would impede phagocytosis.

V. COLLECTION, PREPARATION FOR ISOLATION, AND CULTURING

Cryptomonads can be collected with phytoplankton nets from lakes or other bodies of standing water, or by grab samples. Attached cryptomonads in mucilage can be scraped off various substrata with a putty knife and placed in a collecting bottle with water, where they become motile. Samples must be kept cold during transport. Fixation of cells is impractical since cells either rupture or distort drastically. Lugol's fixative, however, provides the least distortion, but because iodine is a component of the fixative, the cells often appear purple because of stained starch, and cells are not their original color. For proper identification living cells must be examined with a microscope. Photomicrography is difficult since usually the cells are actively swimming, and a flash attachment for photomicrography is helpful. Phase contrast or differential interference contrast (DIC) is also helpful in identifying features of cells.

Before isolations are attempted, field samples should be enriched with growth media to establish populations that are capable of growing in a given medium. Then individual cells are isolated in the medium with confidence that they will grow. Isolations of cryptomonads are most successful with the serial dilution pipetting technique (Hoshaw and Rosowski, 1973), with the use of either a dissecting microscope or inverted microscope, depending on the dexterity of the individual. All freshwater cryptomonads studied thus far grow profusely in sterilized lake water with added Bold's Basal Medium (Nichols, 1973) or Alga-Gro (Carolina Biological Supply Company) concentrate at 40 mL per L of lake water. Cultures should be grown in media with a pH of 7.8 or higher and maintained at approximately 18°C (the optimum temperature range is 16°–20°C) in 16:8 h light:dark regimes. In addition, a sterilized wheat seed must be added to colorless cultures such as *Chilomonas* and *Goniomonas*. *Kathablepharis* can only be maintained in mineral medium that also contains its food organism. For instance, *K. ovalis* requires *Chrysochromulina parva*, and *K. phoenikoston* requires *Chroomonas*.

VI. CLASSIFICATION, KEY, AND DESCRIPTIONS

A. Introduction

By virtue of their unicellularity and small size,

proper cryptomonad identification generally cannot be accomplished by light microscopy. Therefore, current and future classification schemes must rely on electron microscopic techniques to reveal many of the unique features of cryptomonad cells (Brett and Wetherbee, 1986; Dodge, 1969; Dwarte and Vesk, 1983; Faust, 1974; Grim and Staehelin, 1984; Hibberd *et al.*, 1971; Gantt, 1971, 1980; Greenwood and Griffiths, 1971; Hill, 1991a, b; Hill and Wetherbee, 1986, 1988, 1989; Klaveness, 1985; Kugrens and Lee, 1987, 1991; Kugrens *et al.*, 1986, 1987; Lucas, 1970a, b, 1982; Munawar and Bistricki, 1979; Santore, 1977, 1982a, b, 1983, 1984, 1987; Sespenwol, 1973; Wetherbee *et al.*, 1986). The pertinent features were described in detail earlier in this chapter. Furthermore, with the use of information from the structures discussed it has been possible to delineate 18 genera, 11 of which occur in freshwater. *Cryptomonas*, *Campylomonas*, and *Komma* are strictly freshwater genera.

It is significant that the number of genera has increased since Santore's 1984 and 1987 review articles, in which he recognized only five genera. Since that time, an expansion and revision of genera has occurred, as well as the elimination of the genus *Rhodomonas* (Erata and Chihara, 1989; Novarino, 1991; Novarino and Lucas, 1993), which was erected by Santore (1984, 1987), although there were compelling arguments to retain the genus (Hill and Wetherbee, 1989; Hill, 1991a).

More recent ultrastructural investigations also point out the need for reexamining the structural characters that were proposed by Santore (1984, 1987) and Novarino (1991, 1994) as generic characters. More importantly, a correct interpretation of the actual structures must be made since the key characters, such as cell architecture and periplast, easily produce artifacts. Therefore, specialized techniques for scanning electron microscopy and freeze-fracture must be employed to observe the unaltered variations in these features (Munawar and Bistricki, 1979; Grim and Staehelin, 1984; Klaveness, 1985; Brett and Wetherbee, 1986; Wetherbee *et al.*, 1986, 1987; Hill and Wetherbee, 1986, 1988, 1989; Hill 1990, 1991b; Kugrens *et al.*, 1986, 1987; Wetherbee *et al.*, 1986; Kugrens and Lee, 1987, 1991). For information on earlier classification schemes based on light microscopy, the publications by Bourrelly (1970), Huber-Pestalozzi (1950), and Skuja (1939, 1948) should be consulted for freshwater cryptomonads, whereas the extensive treatise by Butcher (1967) is the major reference for marine cryptomonads.

Since cryptomonads and kathablepharids are combined in this chapter, the unifying characteristics of this grouping would be the presence of ejectisomes and the subapical insertion of flagella. The following section, however, involves only those that are within the cryptomonads with periplasts. Kathablepharids are described separately at the end of the chapter.

B. General Features Useful in Determining Genera and Species

Initially, strains can be separated artificially into two broad groups based on the presence or absence of pigments. Then a second separation can be based upon the presence of either phycocyanin (blue) or phycoerythrin (red) as the major accessory pigment (Hill and Rowans, 1989). The blue-green and non-blue-green colored genera are separated easily into distinct groups when live cells are observed. The third separation could be based on life histories. For instance, the marine *Proteomonas* apparently is the only genus studied that displays an alternation of generations. The remainder reproduce asexually or are haplobionts (Kugrens and Lee, 1988).

The fourth and most specific method for separating genera is based on a combination of characters involving the furrow/gullet complex (Figs. 1, 7, 8) and the type of periplast component (Fig. 9), primarily the inner component. Plate types, sizes, shapes, and plate arrangements vary among genera. The first periplast separation is based on laminate (single sheet) plate forms vs. multiple plated forms. Genera in each group could further be separated even when pigmentation is lacking. For instance, *Chilomonas* has a single inner periplast sheet (Grim and Staehelin, 1984; Kugrens and Lee, 1991), and *Goniomonas* has rectangular plates (Kugrens and Lee, 1991). In the multiple-plated forms, plate shapes, the size of plates, and their arrangement could further delineate genera, in conjunction with the furrow/gullet type and cell shape.

The following characteristics are most useful in delineating species (based mainly on SEM features); a brief discussion of these features follows.

1. Flagellar Hair Arrangement and Scale Morphology

The arrangement of tubular and/or nontubular hairs, scales, and other structures on flagellar surfaces could circumscribe species. Variations were reported by Kugrens *et al.* (1986) and Lee and Kugrens (1986).

2. Variations in the Structure of the Flagellar Apparatus

The reconstruction of the entire flagellar apparatus for each genus, particularly, the complexity of the rhizostyle, can be examined for comparative purposes in delineating species, as well as for possible future phylogenetic considerations.

3. The Presence and Structural Variations of the Furrow Plate

There are currently two types associated with cryptomonad cells. These appear to correlate with the type of periplast for a given species. Other variations are expected to be found and could serve as an additional character when morphological cladistic analyses are conducted.

4. The Number and Location of the Nucleomorph(s) in the Periplastidal Compartment

Variations have been summarized by Santore (1982c); and these variations were used in delineating *Pyrenomonas/Rhodomonas* and *Storeatula* (Hill and Wetherbee, 1989; Kugrens *et al.*, 1999), and *Teleaulax*, *Geminigera*, and *Campylomonas* (Hill, 1991b). The genera that have a nucleomorph embedded within a pyrenoid that bridges two lobes of a chloroplast constitute the family Pyrenomonadaceae (Clay *et al.*, 1999). In fact, these genera form a phylogenetic clade based on ssu rDNA sequence data (Marin *et al.*, 1998; Clay *et al.*, 1999).

5. Presence and Location of Eyespots

Eyespots have been confirmed ultrastructurally only for *Chroomonas* spp. (Santore, 1987; Hill, 1991a) Many cryptomonads have orange bodies that do not represent eyespots.

6. Types of Thylakoid Arrangements within the Chloroplasts

Several arrangements have been reported, with doublet thylakoids being the most common arrangement. However, caution must be exercised since environmental conditions may influence this feature (Klaveness, 1981).

7. Type of Scales Comprising the Outer Periplast Component

Only one type has been found to date, but the sample size has been limited and other types are expected. Only heptagonal scales have been reported (Pennick, 1981; Hill, 1990, 1991b; Lee and Kugrens, 1986).

8. The Number, Location, and Types of Pyrenoids

Pyrenoids may be absent, or there may be one, two, or several pyrenoids per chloroplast (Huber-Pestalozzi, 1950). Thylakoids may or may not penetrate the pyrenoid, and this character may be a diagnostic feature for species determination (Clay and Kugrens, 1999; Clay *et al.*, 1999).

9. The Number of Chloroplasts per Cell

This variation may be species specific or may even be useful in determining genera (Hill, 1991b).

C. Classification of the Phylum Cryptophyta

The following classification scheme for cryptomonads conforms to the rules and regulations set forth by the International Code of Botanical Nomenclature (ICBN). This scheme is based on the most current and reliable ultrastructural and phylogenetic information and was proposed by Clay *et al.* (1999a). Phycobiliprotein pigment types are important features in this classification scheme.

Phylum Cryptophyta (syn. Cryptista) Cavalier-Smith (1986)

Plastidial complex with nucleomorphs may be present or absent; chloroplasts (when present) contain chlorophylls *a* and c_2, and phycobiliproteins are located in the lumen of the thylakoids; bipartite tubular hairs on flagella occur in members possessing the plastidial complex; cell covering comprises inner and superficial periplast components (IPC and SPC); ejectisomes are present. Two classes are recognized:

Class Goniomonadophyceae Cavalier-Smith (1993)

Plastids and nucleomorphs are absent. Bipartite tubular hairs on flagella are lacking. Spikes occur on one flagellum. Cells possess an infundibulum. One order:

Order Goniomonadales (Goniomonadida) Novarino and Lucas (1993)   Diagnosis identical to the class.

Family Goniomonadaceae Hill (1991a)   Synonymous with *Cyathomonadaceae* Pringsheim (1944). Characters as for order.

Goniomonas Stein.

Class Cryptophyceae

Plastidial complex with nucleomorphs present. Chloroplasts possess either the phycobiliprotein Cr-phycoerythrin or Cr-phycocyanin in the intrathylakoidal space. Leucoplast is present in some. Bipartite tubular hairs appear on at least one flagellum. Two orders:

Order Cryptomonadales   Not equivalent to Cryptomonadales *sensu* Novarino and Lucas (1993). Chloroplasts possess the phycobiliprotein Cr-phycoerythrin 566 (PE III). Leucoplast is present in some. Two families:

Family Cryptomonadaceae Furrow and gullet complex with a stoma present. The IPC comprises multiple plates. Nucleomorphs are positioned between the pyrenoid and nucleus. Possess a short rhizostyle without wings (lamellae). A fibrous furrow plate is present. One genus:

Cryptomonas Ehrenberg

Family Campylomonadaceae Furrow and gullet present; IPC composed of a sheet; nucleomorphs positioned between pyrenoid and nucleus or similar position if a leucoplast is present; possesses a long, keeled flagellar rhizostyle with wings (lamellae); scalariform furrow plate present; vestibular ligule present. Two genera:

Campylomonas Hill
Chilomonas Ehrenberg

Order Pyrenomonadales Not equivalent to Pyrenomonadales *sensu* Novarino *et* Lucas (1993, 1995).

Chloroplasts possess the phycobiliprotein Cr-phycoerythrin 545 (PE I) or Cr-phycoerythrin 555 (PE II), never Cr-phycoerythrin 566 (PE 566); or possess Cr-phycocyanin. Four families:

Family Pyrenomonadaceae Novarino *et* Lucas (1993) Chloroplasts possess Cr-phycoerythrin 545 (PE I); nucleomorphs positioned within pyrenoid. Two genera:

Pyrenomonas Santore, Synonym: *Rhodomonas* Karsten
Storeatula Hill
Rhinomonas Hill *et* Wetherbee

Family Geminigeraceae Clay, Kugrens *et* Lee Chloroplasts possess Cr-phycoerythrin 545 (PE I); IPC comprises a sheet or a sheet and multiple plates if diplomorphic; nucleomorphs never positioned in the pyrenoid; possesses a long, keeled rhizostyle with wings (lamellae); scalariform furrow plate present. Five genera, and all are marine:

Geminigera Hill
Teleaulax Hill
Hanusia Deane, Hill, Brett, *et* McFadden
Guillardia Hill *et* Wetherbee
Proteomonas Hill *et* Wetherbee

Family Chroomonadaceae Clay, Kugrens, *et* Lee (1999) Chloroplasts possess Cr-phycocyanin 630 (PC III), 645 (PC IV), or Cr-phycocyanin 569; rhizostyle absent. Three genera:

Chroomonas Hansgirg
Falcomonas Hill
Komma Hill

Family Hemiselmidaceae Butcher (1967) Chloroplasts possess Cr-phycocyanin 615 (PC II) or Cr-phycoerythrin 555 (PE II), never possess the other three types of phycocyanins or other two types of phycoerythrins; gullet only; nucleomorphs positioned anterior to pyrenoid; rhizostyle absent; thylakoids penetrate pyrenoid; flagella inserted laterally. One genus:

Hemiselmis Parke

D. Key

1a.	Cells colorless..	2
1b.	Cells pigmented..	4
2a.	Cells with leucoplast..	*Chilomonas*
2b.	Cells without leucoplasts and chloroplast endoplasmic reticulum...	3
3a.	Cells with a furrow/gullet complex..	*Goniomonas*
3b.	Cells lacking a furrow/gullet but have a distinct feeding apparatus...	*Kathablepharis*
4a.	Cells blue-green in color, because of presence of phycocyanin...	5
4b.	Cells olive, brown, or red in color because of presence of phycoerythrin...	7
5a.	Flagella inserted approximately one-third of the cell length behind anterior..	*Hemiselmis*
5b.	Subapically inserted flagella, near anterior end...	6
6a.	Cells with hexagonal inner and surface periplast plates..	*Komma*
6b.	Cells with inner and surface rectangular periplast plates...	*Chroomonas*
7a.	Cells with multiplated periplast..	8

7b.	Single sheetlike inner periplast component, cells may appear somewhat contorted	10
8a.	Cells with square inner periplast plates with beveled corners	*Pyrenomonas*
8b.	Cells with periplast plates with hexagonal or oval shapes	9
9a.	Inner periplast plates oval; cells with a complex furrow	*Cryptomonas*
9b.	Inner periplast plates hexagonal; cells with a simple furrow	*Plagioselmis*
10a.	Furrow absent, gullet only	*Storeatula*
10b.	Simple furrow and gullet present	*Campylomonas*

E. Freshwater Cryptomonad Genera and Species

The following descriptions provide characteristics of genera and some species currently recognized. Scanning electron micrographs (SEMs) and light micrographs and diagrams are provided so that cellular features can be interpreted properly. Furthermore, SEM is becoming more common in phytoplankton identification, and the SEMs provided should serve as baseline information for identification. In addition to descriptions for genera, some of the most commonly found species for each genus are described and depicted.

Class Cryptophyceae

Order Goniomonadales

   Family Goniomonadaceae

Goniomonas Stein (*Syn. Cyathomonas* (Figs. 3A, 4A, B, 19A)

This genus represents a cell that is related to the ancestral type and is the least complex cryptomonad. It lacks any vestige of a plastid and nucleomorph, thus a periplastidial compartment is lacking. However, *Goniomonas* definitely is a cryptomonad based on the presence of ejectisomes, the structure of the flagellar transition region, and the presence of a periplast. Large ejectisomes are arranged in a ring around the anterior of the cell (Mignot, 1965; Schuster, 1968; Kugrens and Lee, 1991; Kugrens, 1998), and small ejectisomes occur at the corners of the periplast plates, just beneath the periplast, but these are not evident with the light microscope. Cells are laterally compressed, colorless, and phagocytic on bacteria. The nucleus is situated in the dorsal portion of the cell. Flagella are inserted on the dorsal side of the vestibulum, with one flagellum bearing recurved spines, while the other flagellum has fine fibrillar hairs (Kugrens and Lee, 1991). The vestibulum connects to a ventral furrow, which connects to a furrow with posterior stoma; a gullet is absent. An opening, the infundibulum, is located on the left side of the cell and presumably is the ingestion site for particulates. The periplast has inner and outer rectangular plates that are not offset. *Goniomonas* has freshwater and marine representatives. Refer to Mignot (1965), Schuster (1968), Hill (1991a), Kugrens and Lee (1991), and Kugrens (1998) for additional information.

Goniomonas truncata (Figs. 6B, 7A)

Cells 5–12 μm long, 3–5 μm wide, and 4–10 μm deep. This species generally possesses the features listed for the genus.

Order Cryptomonadales

   Family Cryptomonadaceae

Cryptomonas Ehrenberg (Figs. 3B–E, 4C–H, 6A–C, 7, 8, 15C, D)

Cells often form palmelloid colonies with cells embedded in extensive mucilage. Motile cells possess two flagella that originate from the right side of the vestibulum, and flagella have the most common arrangement of tubular hairs. Cells have a complex type of vestibular-furrow-gullet complex, with the furrow consisting of furrow ridges, furrow folds, and a persistent oval opening called the stoma that is located at the posterior of the furrow. The furrow appears to have the ability to open and close. The periplast consists of an inner component of round to oval-shaped plates and a surface component of a thin layer of fibrils. The periplastidial compartment contains two chloroplasts with two pyrenoids not traversed by thylakoids, and two nucleomorphs, each located between the nucleus and the pyrenoids. Chloroplasts possess Cr-phycoerythrin with maximum absorption at 566 nm. This genus is ubiquitous in temperate lakes, reservoirs, and streams and occurs only in freshwater habitats. Light microscopic observations generally are unreliable with respect to species characteristics. For instance, the type species, *C. ovata*, was described as lacking pyrenoids, but pyrenoids were found (Roberts, 1984; Hill, 1991c; Fig. 4D,E). Based on this character, it would meet the characteristics for *C. pyrenoidosa* Harvey. For further information on the structure of this genus refer to Santore (1977, 1984), Munawar and

Bistricki (1979), Roberts (1984), Brett et al. (1986), Kugrens et al. (1986), Kugrens and Lee (1987), and Hill (1991b).

Cryptomonas ovata (Figs. 3B, 4D, E, 15C, D)

Cells are ellipsoid to oval, and they may be slightly curved. Cells measure 20–80 μm in length, 6–20 μm in width, and 5–18 μm in depth and appear somewhat flattened. The furrow is complex, and cells have a short gullet. There are two chloroplasts per cell, each with a pyrenoid, that are olive green to dark brown in color.

Cryptomonas obovata (Fig. 3C)

Cells are 24–46 μm long and 13–24 μm in diameter and slightly curved. The vestibulum is below the cell apex, imparting a lobed appearance to the cell apex. Cells have two olive or brown chloroplasts without pyrenoids. Usually many starch grains are present in the cells.

Cryptomonas phaseolus (Figs. 3D, 4F)

This is the smallest *Cryptomonas* species, measuring 8–13 μm in length and 5–8 μm in diameter. It is ellipsoid in lateral view and oval in cross section, with rounded ends. The anterior end has a rounded protrusion anterior to the site of flagellar insertion. The posterior of the cell is slightly narrower than the anterior end. Each cell has two brownish chloroplasts without pyrenoids.

Cryptomonas tetrapyrenoidosa (Fig. 4G)

Cells measure 20–60 μm in length, 10–27 μm in width, and 5–17 μm in depth. There are two chloroplasts per cell, each with two pyrenoids; thus there are four total pyrenoids. The periplast type has not been investigated.

Cryptomonas erosa (Figs. 3E, 4H)

Cells are oval or slightly elliptical, flat and slightly contorted, ranging in size from 13 to 45 μm in length, and from 6 to 26 μm in width. There are two chloroplasts per cell which are olive, and pyrenoids are absent. The periplast type has not been investigated.

Cryptomonas ozolini Skuja (Fig. 3F)

Cells are slightly egg shaped and compressed. The anterior end is the widest portion of the cell. Cells measure 17–29 μm in length, 9–13 μm in width, and 6–9 μm in depth. Cells contain two olive-green chloroplasts, each with a pyrenoid. The ultrastructure of this species indicates that it should be a new genus (unpublished observations).

Family Campylomonadaceae

Campylomonas Hill (Figs. 5A–D, 6A–C, 7B–D, 8B)

Members of this genus were formerly placed in the genus *Cryptomonas* because of a lack of ultrastructural evidence that demonstrated that the two genera are different (Hill, 1991b). A basic distinction at the light microscopic level is that *Campylomonas* spp. are slightly contorted, sigmoid-shaped cells with a characteristic recurved posterior, which imparts a sigmoid shape to the cells in lateral view. At the electron microscopic level, the main distinguishing features are the presence of a periplast sheet, a vestibular ligule, and a simple furrow with a gullet of variable length extending posteriorly from the furrow. A surface periplast component may be lacking, or it may consist of fibrillar material or heptagonal scales. Cells contain two chloroplasts, which may or may not have pyrenoids. If pyrenoids are present, each chloroplast has its own pyrenoid and it is not traversed by thylakoids. Two nucleomorphs are located either near and posterior to each pyrenoid when present, or close to the nucleus. Chloroplasts possess Cr-phycoerythrin with a maximum absorption at 566 nm, but not in quantities that impart a reddish coloration. The genus is strictly freshwater. For additional information refer to Munawar and Bistricki (1979), Klaveness (1985), Kugrens et al. (1986), Kugrens and Lee (1987), Hill (1991b). Note that in all publications, except for *Campylomonas reflexa* Hill (1991b), *Campylomonas* is still incorrectly identified as *Cryptomonas*.

Campylomonas reflexa (Figs. 5A, 7C, D)

This is the type species, and it has the characteristics described for the genus. Two pyrenoids are present. Nucleomorphs are located posterior to the pyrenoids and anterior to the nucleus. Cells are highly variable in size and can range from 15 to 60 μm in length and from 10 to 30 μm in width.

Campylomonas rostratiformis (= *Cryptomonas rostratiformis*) (Figs. 5B, 7B).

This species is the largest cryptomonad, ranging in size from 45 to 80 μm in length, 16 to 40 μm in width, and from 14 to 24 μm in depth. It has not been officially transferred to *Campylomonas*, although it has been shown that it does not belong in *Cryptomonas* (Kugrens and Lee, 1986; Kugrens et al., 1987). Cells are slightly recurved at the posterior, and they have a rostrate anterior. The furrow is curved slightly toward the left, and the vestibular ligule is pointed and attached to the left side of the vestibulum. The cells have two chloroplasts, each with numerous pyrenoids.

Starch is often present in large amounts throughout the chloroplast, obscuring the pyrenoids.

Campylomonas platyuris (Figs. 5D, 6C, 8)

As was the case with *C. rostratiformis*, this species also has not been transferred officially to *Campylomonas*, even though it possesses the periplast and furrow/gullet complex of *Campylomonas* (Kugrens and Lee, 1986; Kugrens *et al.*, 1987). Cells range in size from 30 to 55 μm in length, from 15 to 28 μm in width, and from 9 to 16 μm in depth, and are characterized by a flattened posterior portion or tail when viewed from the side. There are two chloroplasts per cell, but pyrenoids are lacking. Considerable starch usually is present in the cells.

Campylomonas marssoni (Fig. 5C)

Cells range in size from 16 to 38 μm in length and from 8 to 14 μm in width. Cells are somewhat fusiform and slightly sigmoid in shape, with a pointed posterior end. Each cell contains two chloroplasts without pyrenoids, which differs from *C. reflexa*.

Chilomonas Ehrenberg (Figs. 5E, 6D, 18B)

Three freshwater species have been described. The cells are colorless, but not phagocytic. The furrow/gullet complex in *C. paramecium* consists of a vestibulum, a short furrow, and a long tubular gullet (Kugrens and Lee, 1991). A vestibular ligule covers the area of contractile vacuole discharge. Both flagella have a unilateral row of tubular hairs (Kugrens and Lee, 1991). The inner periplast component consists of an inner sheet with numerous ejectisome pores, and the sheet is not connected to the plasma membrane (Grim and Staehelin, 1984; Kugrens *et al.*, 1986). The surface periplast component consists primarily of fibrils. Ejectisomes are attached to the pore edges in the periplast sheet. The periplastidial compartment contains leucoplasts that lack thylakoids, numerous large starch grains, and two nucleomorphs located in the periplastidial compartment anterior to the nucleus. Refer to Anderson (1962), Schuster (1970), Sespenwol (1973), Roberts *et al.* (1981), Grim and Staehelin (1984), Kugrens *et al.* (1986), Kugrens and Lee (1987, 1991), Heywood (1988), and Kugrens (1998) for additional information.

Chilomonas paramecium (Figs. 5E, 6D, 19B)

Cells are 20–40 μm long and 10–20 μm in diameter, with a rhinote anterior and a blunt, reflexed posterior, imparting a sigmoid shape to the cells. Two leucoplasts and nucleomorphs are located in the periplastidial compartment. Evolutionarily this species appears to be derived from *Campylomonas* (Clay *et al.*, 1999).

Chilomonas acuta (Fig. 5F)

This species has been described from freshwater, but it has not been cultured. It is possible that this genus might be *Leucocryptos acuta* in Bourrelly (1970). Its ultrastructural features are unknown. This species was observed in an enrichment culture from a reservoir near Severance, Colorado, but isolation was unsuccessful.

Order Pyrenomonadales

Family Pyrenomonadaceae

Pyrenomonas Santore (= *Rhodomonas* Karsten) (Figs. 6E–G, 9B, 10A, B)

Pyrenomonas is the most accepted name, but some authors continue to use *Rhodomonas*; the generic names are synonyms. Cells may be red, brown or golden brown, in coloration. However, in freshwater collections we have encountered only red-colored forms. Cells have a short furrow and a deep gullet. The periplast consists of inner more or less square plates with beveled corners. The plates taper slightly toward the posterior. The surface periplast component consists of intertwining fibrils. The periplastidial compartment usually has a single, bilobed chloroplast with a pyrenoid situated between the two lobes of the chloroplast. Thylakoids do not traverse the pyrenoid, and a nucleomorph is located in an invagination of the pyrenoid. Chloroplasts contain a preponderance of phycoerythrin. For additional information refer to Santore (1984), Erata and Chihara (1989), Hill and Wetherbee (1989), Novarino (1993), and Kugrens *et al.* (1999).

Pyrenomonas ovalis (Figs. 6E–G, 9B, 10A, B)

Cells are oval to ellipsoid, 14–15.5 μm long and 7–8 μm wide and have a single red chloroplast with two lobes. The pyrenoid is attached to both lobes, forming a bridge between the two lobes, making the chloroplast appear H-shaped. The nucleomorph is embedded within the pyrenoid. Cells have a short furrow and an anterior tubular, deep gullet. Currently this is the only species described from freshwater, and it was collected from Great Western reservoir near Broomfield, Colorado (Kugrens *et al.*, 1999).

Storeatula Hill (Figs. 6H–J, 9A, 10C, D)

Cells are ellipsoid with a slightly rhinote anterior. A furrow is lacking, and the tubular gullet extends to approximately the middle of the cell and is lined with several rows of ejectisomes. The periplast has an inner sheet and an outer component of coarse fibrils. The periplastidial compartment contains a single, bilobed chloroplast, with a pyrenoid that connects the two lobes of the chloroplast. A nucleomorph is located in an anterior groove or depression in the pyrenoid. Chloroplasts contain the biliprotein Cr-phycoerythrin. Only one freshwater species has been identified thus far. For additional information refer to Hill (1991b) and Kugrens et al. (1999).

Storeatula rhinosa (Figs. 6H–J, 9A, 10C, D)

Cells are 16–20 µm long, 7–8 µm wide, and 8–10 µm deep and ellipsoid with a slightly pointed anterior; a furrow is lacking, with a tubular gullet; single chloroplast with pyrenoid. This species has been collected from Hanratty's Ditch near Beulah, Colorado, and Sheldon Lake and North Shields Pond in Fort Collins, Colorado.

Family Chroomonadceae

Komma Hill (Figs. 11A, 12A, B, 13A, B)

Based on one isolate, this genus originally was described as being comma-shaped or acuminate with a rounded anterior end, tapering to a pointed or acutely rounded posterior and the absence of a furrow. A tubular gullet extends posteriorly from the base of the vestibulum. The periplast consists of relatively small internal and surface hexagonal plates, with the surface plates being crystalline in composition, and, occasionally, rosulate, heptagonal scales lie on the surface of these external plates. The periplastidial compartment contains a single blue-green chloroplast with a central pyrenoid lacking traversing thylakoids and projects from the chloroplast. The chloroplast occupies a dorsocentral position in the cell and contains C-phycocyanin with a maximum absorption at 645 nm. The nucleomorph is situated at the level of the pyrenoid. Refer to Hill (1990) for more specific descriptions. Other blue-green cryptomonads also have hexagonal plates (Kugrens and Lee, 1991), but they are not comma shaped. The genus is strictly freshwater, and only one species has been described.

Komma caudata (Figs. 11A, 12A, B, 13A, B)

Any blue-green acuminate cryptomonads probably represent *Komma*. Cells are comma shaped or acuminate with an acute posterior end and a rounded anterior end. Cells measure 8–12 µm in length and 4–6 µm in width. Cells contain a single, dorsal, blue-green chloroplast with a single pyrenoid projecting from the center of the chloroplast. Cells lack a furrow and possess a gullet only.

Chroomonas Hansgirg (Figs. 11B–E, 12C–F, 13C, D, 14B)

Cells are subobovate and chloroplasts are blue-green in color. Cells lack a furrow, but a tubular gullet extends posteriorly from the vestibulum. Inner and outer components of the periplast consist of offset rectangular plates (Hill, 1991a), with the anterior of the plate edges raised, because of rows of intramembrane particles in the cell membrane attaching the plates tightly at the posterior end of each plate. Scales or fibrils may be present, in addition to the surface plates in some species. The periplastidial compartment contains one or two chloroplasts that may have a pyrenoid. The nucleomorph usually is located near the pyrenoid. Chloroplasts contain Cr-phycocyanin, imparting a blue-green color to the cells. A stigma may be present in chloroplasts of some species. Refer to Dodge (1969), Gantt (1971), Antia et al. (1973), Meyer and Pienaar (1984), Kugrens et al. (1986), Kugrens and Lee (1987), and Hill (1990) for additional features and descriptions of *Chroomonas* spp.

Chroomonas oblonga (Fig. 11B)

Cells are ellipsoid, measuring 15 µm in length and 6 µm in width. Two chloroplasts are present per cell, each with a pyrenoid. A stigma may be present. The periplast comprises small, rectangular plates in the inner and surface components. Isolated from Fossil Creek south of Fort Collins, Colorado.

Chroomonas coerulea (Figs. 11C, 12D, E, 14A)

Cells are ellipsoid and sometimes slightly concave dorsiventrally, measuring 8–12 µm in length and 4–6 µm in width. The cell posterior is rounded. Inner and surface periplast components consist of small rectangular periplast plates. A single blue-green chloroplast is present, with a single pyrenoid and a prominent stigma, which is associated laterally with the pyrenoid. Only a vestibulum and gullet are present. Flagella are shorter than the cell length. Cells often form extensive mucilage in which groups of cells remain embedded and attached to a substrate. Ubiquitous in Rocky Mountain lakes, reservoirs, and streams.

Chroomonas pochmanni (Figs. 11E, 12C, 13C, D)

Cells are barrel shaped to ovoid, 10–18 µm long and 8–13 µm wide, and blue-green in color. A single chloroplast and a massive pyrenoid are present in the cell. This species may be mixotrophic (Kugrens and

Lee, 1990, 1991). Cells have a gullet but lack a furrow. One massive blue-green chloroplast with a large pyrenoid is present per cell. A large, prominent contractile vacuole is located in the anterior of the cell. In addition a bacterial ingestion vacuole also may be visible with a light microscope. Numerous large starch grains usually are present in the cell. This species does not conform to the characters for *Chroomonas* or any other blue-green cryptomonad and therefore will need to be a new genus.

Chroomonas nordstedtii (Figs. 11D, 12F)

Cells are slightly elongated and egg-shaped, with the posterior larger in diameter than the anterior, with the anterior end obliquely truncate. Cells are slightly curved on the dorsal side and range in size from 10 to 30 μm in length and 7 to 15 μm in diameter, which is larger than described by Huber-Pestalozzi (1950). A blue-green parietal chloroplast with a prominent pyrenoid and a small stigma fill the posterior three-fourths of the cell. Common in Wyoming lakes.

Family Hemiselmidaceae

Hemiselmis Parke (Figs. 11F, 12H–I, 14C, D)

Cells are rounded anteriorly and posteriorly, slightly flattened dorso-ventrally, and appear somewhat bean-shaped in the lateral view, becausae of the vestibulum and gullet located approximately one-third of the cell length from the anterior. A furrow is absent. The surface periplast component comprises large hexagonal plates, and the inner component probably consists of a periplast sheet. The periplastidial compartment contains a single dorsal, boat-shaped chloroplast with a centrally situated, stalked pyrenoid that is traversed by a single thylakoid. The nucleomorph is located anterior to the pyrenoid. Chloroplasts are blue-green in color because of the presence of Cr-phycocyanin. This genus is ubiquitous in Colorado and Wyoming lakes, specifically Lake John and Cowdrey Lake in Colorado and Diamond and Twin Buttes Lakes in southern Wyoming. Currently there is only one freshwater species described, and it is the smallest cryptomonad described from fresh water.

Hemiselmis amylifera (Figs. 11F, 12H–I, 14C, D)

Cells generally are suspended throughout the water column and usually do not swim unless disturbed. Cells range in size from 4 to 5.5 μm in length and from 2.5 to 3 μm in width and are 3 μm in depth. They are slightly compressed laterally, and they are ovate or bean-shaped in the lateral view. The vestibulum and gullet are oval and shallow, located one-third the distance from the anterior. Cells have few ejectisomes and a single parietal chloroplast with a prominent dorsal pyrenoid in the anterior portion of the cell. Cells are blue-green in color.

Cryptomonads of Uncertain Taxonomic Status

Cyanomonas Oltmanns (Figures Not Available)

It must be pointed out that the existence of this genus is in doubt. Cells contain several blue-green chloroplasts. It has never been cultured and might represent a species of *Chroomonas*, in which large starch grains have been mistaken for chloroplasts (Hill, 1990). The cell shape is similar to that of some *Chroomonas* spp. Since this genus has not been investigated with the electron microscope, and it has not been cultured, its status as a legitimate genus remains in doubt. There have been suggestions that numerous large starch grains that have a blue-green refraction might have been mistaken for chloroplasts. Nevertheless, a description is provided. One species, *C. americana*, has been described, and there has been one collection reported in North America, that from the Charles River near Boston, Massachusetts.

Plagioselmis Butcher (Figs. 15A, B, 16A, 17A, B)

This genus originally was described as being from the marine environment by Butcher (1967), with *P. prolonga* representing the type species, until Novarino *et al.* (1994) transferred the freshwater *Rhodomonas minuta* into the genus *Plagioselmis*, as *P. nanoplanctica*. Cells are pink or red in color because of the presence of phycoerythrin, and they are comma-shaped, with an acute tail that lacks plates. There is a ventral furrow (= sulcus, a deep furrow rather than a gullet). The periplast consists of an internal component (IPC) of hexagonal plates, whereas rosette scales make up the surface periplast component (SPC). The acute tail has a continuous periplast sheet rather than plates. Thylakoids usually occur in groups of three (Klaveness, 1981). Novarino *et al.* (1994) described several of these features from three isolates of *Plagioselmis*, but one major characteristic differed among the isolates. This was used to delineate taxa at the species level. For instance, they noted that some strains of *P. prolonga* and *P. nanoplanctica* lacked a furrow (= sulcus). Cells have a depression on the ventral side, but they attributed these depressions to folds that were induced by cell shrinkage. If the absence of a furrow indeed is correct, then either the characters for *Plagioselmis* must be expanded or the nonfurrow strains should be described as a new genus; an investigation is needed to determine whether the folds are artifacts or whether the collapsed folds were obscure a furrow. However, it is clear from the figures provided in this chapter that *Plagioselmis* has a furrow.

P. nanoplanctica (Fig. 15A, B, 16A, 17A, B)

This is the only recognized freshwater species. Cells are comma-shaped with an acute posterior end, and chloroplasts are red or pink colored. Cells range in size from 12 to 18 μm in length and from 8 to 10 μm in diameter at the widest portion of the cell. This is the only recognized freshwater species at this time.

F. Guide to Literature for Species Identification

1. *Campylomonas*—Hill (1991c), Kugrens et al. (1986)
2. *Chilomonas*—Hill (1991), Kugrens and Lee (1991)
3. *Chroomonas*—Hill (1991a)
4. *Cryptomonas*—Kugrens et al. (1986)
5. *Goniomonas*—Hill (1991), Kugrens and Lee (1991)
6. *Hemislemis*—Clay and Kugrens (1999b)
7. *Kathablepharis*—Lee and Kugrens (1991), Lee et al. (1991), Clay and Kugrens (1999a, c)
8. *Komma*—Hill (1991a)
9. *Plagioselmis*—Novarino et al. (1994)
10. *Rhodomonas/Pyrenomonas*—Kugrens et al. (1999)
11. *Storeatula*—Kugrens et al. (1999)

VII. AVAILABILITY OF CRYPTOMONADS

Cryptomonad cultures are available from a variety of sources. At Colorado State University 172 strains of cryptomonads are maintained, and most are not duplicate strains. In addition, Dr. Michael Melkonian's laboratory at the University of Cologne and Dr. Dag Klaveness's laboratory at the University of Oslo, Norway, have numerous cryptomonads in culture. Each of these culture collections contains over 50 isolates. Other culture collections with a considerable number of cryptomonad cultures are the University of Texas Culture Collection, the Japanese NIES Collection, the Culture Collection of Algae and Protozoa, and Provasoli-Guilliard Culture Collection of Marine Phytoplankton (CCMP) at the Bigelow Laboratories, Boothbay Harbor, Maine.

VIII. FAMILY KATHABLEPHARIDACEAE

The flagellates belonging to this family include the genera *Kathablepharis* and *Leucocryptos*, but the distinction between the two is rather tenuous (Vørs, 1992a, b). Higher taxonomic affiliations are not possible at this time, and the family is considered *incertae sedis* (Clay and Kugrens, 1999). As mentioned earlier, these probably should be placed in their own phylum, but currently they are still classified as cryptomonads. These organisms represent one of the most enigmatic groups of protists. Some of their cellular structures are unusual and are not found in any other groups, yet other cellular structures transcend several groups. Their classification is uncertain, but traditionally the katablepharids have been placed in the cryptomonads because of the presence of ejectisomes and the site of flagellar insertion. These features were based on earlier light microscopic observations, but electron microscopic examination has revealed distinct differences. For instance, the ejectisomes are not similar to either the cryptomonad ejectisomes or the *Pyramimonas parkeii* ejectisomes. *Kathablepharis* also has a complex feeding apparatus similar to that of suctorian ciliates or the apical complex found in apicomplexans (Kugrens et al., 1994), based on the similarity of the conoid rings and associated microtubules in the feeding apparatus, resembling an apical complex.

In his treatment of the kingdom Protozoa, Cavalier-Smith (1993) places *Kathablepharis* in the phylum Opalozoa, a position somewhat removed from that of the rest of the cryptomonads, because it differs in many ultrastructural aspects, including the presence of tubular mitochondrial cristae and the purported absence of the subsidiary scroll that is present in the ejectisomes of all cryptomonads. In addition, a periplast is absent in kathablepharids, and they possess the distinct feeding apparatus described earlier (Lee and Kugrens, 1991b).

A. Ecology

Based on collecting data from the Rocky Mountain region, kathablepharids appear to be tolerant of a variety of temperatures, pH, salinities, and nutrient conditions, a fact also noted by Vørs (1992) in her autecological studies. The lack of more global information regarding these flagellates probably is due to the fact that they may be overlooked in plankton samples or misidentified, especially in fixed material.

Kathablepharids are voracious predators, attacking their prey in groups, and the size of their groups ranges up to several hundred cells. They feed on both bacteria and various eukaryotes, but each freshwater species that has been studied in detail appears to prefer a specific food organism (Lee and Kugrens, 1991; Lee et al., 1991). For instance, *Kathablepharis ovalis* feeds on the chrysophyte *Chrysochromulina parva*, and *K. phoenikoston* prefers to feed on *Chroomonas*.

B. Cell Structure (Fig. 8C)

1. Cell Covering

A distinctive cell covering surrounds the cell, including the flagella (Fig. 8C). The cell covering is composed of outer and inner components. The outer component of the cell covering occurs in distinct rows encircling the body of the cell. The inner component of the cell covering occurs between the outer compartment of the wall and the plasma membrane. The inner component is composed of randomly arranged fibrils and covers the whole cell. The outer compartment of the cell covering is absent over the area of the cytostome, the area where the flagella are inserted into the cell and the area posterior to the cytostome, where the rows of ejectisomes occur under the plasma membrane. The outer cell covering appears to be made up of rows of hexagonal subunits, perhaps scales (Lee and Kugrens, 1991), which overlap each other in the rows. The rows of subunits are oriented approximately 45° to the cell surface. The subunits and thus the rows are approximately 25 nm wide.

2. Flagella

Flagella are inserted subapically, and their orientation is variable among species. The flagella appear thick when viewed with the light microscope because of a scale covering.

3. Feeding Apparatus

All kathablepharids have a distinctive, complex feeding apparatus that is similar to that of suctorian ciliates or apicomplexans (Kugrens et al., 1994; Lee et al., 1991). Depending on the species, the feeding apparatus consists of a stack of 2 to 10 rings that are located just posterior to the depression at the anterior end known as the stoma or mouth. Microtubular bundles are attached to these rings, and the microtubules extend toward the posterior of the cell. Small vesicles occur inside along the cytopharyngeal microtubules. The successive rings increase in diameter when they are farther from the stoma, forming a cone or "conoid" of rings. The microtubules collectively are known as the cytopharyngeal skeleton (Vørs, 1992), and there are two of these circular arrays. One array is associated with the conoid rings, and the others occur just beneath the cell cover and probably function as a cytoskeleton. This second array of microtubules is known as the pellicular skeleton (Vørs, 1992).

4. Nucleus and Mitosis

The interphase nucleus is typical of a eukaryotic nucleus, with chromatin attached to the inner membrane of the nuclear envelope. A single nucleolus occurs in the nucleus. When cells divide, the nucleolus disperses, the nuclear envelope detaches from the chromatin and is converted into rough endoplasmic reticulum, and the chromatin condenses into a single disc-shaped mass where individual chromosomes cannot be discerned. Microtubules penetrate the chromosome mass, but kinetochores were not observed. Spindle microtubules end in a number of minipoles in the cytoplasm. The chromosome mass separates at anaphase, and each mass migrates to the poles. Then the nuclear envelope attaches to the chromatin and the nucleolus reappears. Cytokinesis is longitudinal, forming two cell products.

5. Mitochondria

Mitochondria have tubular cristae, and they usually are found between the outer and inner arrays of microtubules of the pellicular and cytopharyngeal skeletons. Serial sections of the cell have not been made; thus there is the possibility that only one mitochondrion occurs per cell.

6. Ejectisomes

Ejectisomes consist of only one component rather than two as described for cryptomonads. Large ejectisomes occur in the area of the cell posterior to, and to the right of, the flagella. The large ejectisomes occur in one or two rows oriented parallel to the long axis of the cell. Smaller ejectisomes occur under the plasma membrane in the medial and posterior areas of the cell. Both large and small ejectisomes consist of a single ribbon wound into a spiral that is contained within a membrane. On discharge into the surrounding medium, the ejectisomes consist of a long, straight ribbon. After discharge of the ejectisome, the edges of the ribbon roll inward, creating a tubular structure. Near the tip of the discharged ejectisome, the ribbon tapers rapidly to a spatula-like point.

7. Food Vacuoles

Food vacuoles are located in the posterior portion of the cell, and food particles are transferred from the cytosome to these food vacuoles. Both bacteria and chloroplasts from the food organisms are commonly found in these food vacuoles.

8. Alveolate-like Structures

Alveoli are flattened vesicles that occur beneath the plasma membrane and are characteristic of dinoflagellates and ciliates. Kathablepharids have similar vesicles that are derived from endoplasmic reticulum. These differ from dinoflagellates, however, because they have ribosomes attached to them.

C. Classification of Kathablepharidaceae

The colorless flagellate *Kathablepharis* Skuja comprises eight freshwater species based on light microscopic studies. Two of the most common freshwater representatives, *K. ovalis* and *K. phoenikoston*, are described in this chapter.

Kathablepharis Skuja (Figs. 16B, C, 17C, D, 18C, D)

Cells of *Kathablepharis* vary in size and morphology, but all have two flagella that are subapically inserted on the ventral side, a continuous covering of fused scales outside of the plasma membrane, and ejectisomes. Cells may be oval to cylindrical in shape. Chloroplasts are absent, although when chloroplasts are present in food vacuoles, there could be the impression that cells possess chloroplasts. Two species of freshwater *Kathablepharis* are discussed.

Kathablepharis ovalis (Figs. 16B, 17C. 18C)

K. ovalis is a common flagellate in freshwater habitats in the Rocky Mountain region, occurring in a variety of lentic and lotic habitats. It is small and colorless and often contains one to several cells of *Chrysochromulina parva*, making it appear as though it contains chloroplasts. Cells range in size from 8 to 15 µm in length, but the size is dependent on the number of ingested cells. Cells have two subapical flagella, and the flagella emerge laterally from a subapical mound. The anterior flagellum is approximately 15 µm long, and the posterior flagellum is approximately 12 µm long. The flagella are encased in the same cell covering that encloses the entire cell. The cells have a large central nucleus. One to several large food vacuoles occupy the posterior portion of the cell. Two arrays of microtubules, one inside the other, begin at the anterior end of the cell and continue into the posterior region of the cell. A Golgi apparatus is just anterior to the nucleus and inside the inner array of microtubules. Six large ejectisomes occur in two rows posterior to and just to the right of the flagella, while smaller ejectisomes occur under the plasma membrane in the posterior and medial portion of the cell. At the light microscopic level, cells of *K. ovalis* are ovate to subovate, with both flagella directed anteriorly during swimming. At the ultrastructural level, the feeding apparatus in *K. ovalis* has two cytopharyngeal rings.

K. ovalis has been collected from ponds in the Department of Energy's Rocky Flats Nuclear Weapons Plant, Jefferson County; in Horsetooth Reservoir and North Shields Pond, Larimer County, and in South Delaney Buttes Lake and Lake John, Jackson County, Colorado, USA. This organism, however, is easily overlooked because it is small and colorless and has no striking features when examined in the light microscope. In addition to dispersed cells, swarms of *Kathablepharis* are common, consisting of aggregations of 20–100 cells that attack the prey cell.

Kathablepharis phoenikoston Skuja (Figs. 16C, 17D, 18D)

Cells of *K. phoenikoston* are cylindrical and have one anteriorly directed flagellum and one trailing flagellum when swimming. *K. phoenikoston* possesses 9 to 10 conoid-like rings that are associated with the feeding apparatus. The cell covering and other features are similar to those described for *K. ovalis*.

D. Isolation and Culturing Techniques for Kathablepharids

Isolation involves the same techniques as described for cryptomonads. In addition to the usual mineral media, these organisms require the presence of their specific food organism in the culture since they do not survive on bacteria alone. The mineral medium promotes growth of the photosynthetic prey organism so that kathablepharids can actively feed and survive by ingesting the selected alga.

LITERATURE CITED

Andersen, R. A. 1992. Diversity of eukaryotic algae. Biodiversity and Conservation 1:267–292.

Anderson, E. 1962. A cytological study of *Chilomonas paramaecium* with particular reference to the so-called trichocysts. Journal of Protozoology 9: 380–395.

Antia, N. J., Cheng, J. Y., Foyle, R. A., Percival, E. 1979. Marine cryptomonad starch from autolysis of glycerol-grown *Chroomonas salina*. Journal of Phycology 15:57–62.

Antia, N. J., Kalley, J. P., McDonald, T., Bisalputra, T. 1973. Ultrastructure of the marine cryptomonad *Chroomonas salina* cultured under conditions of photoautotrophy and glyceroheterotrophy. Journal of Protozoology 20:377–385.

Apt, K. E., Collier, J. L., Grossman, A. R. 1995. Evolution of the phycobiliproteins. Journal of Molecular Biology 248:79–96.

Arvola, L., Ojala, A., Barbosa, F., Heaney, S. I. 1991. Migration behaviour of three cryptophytes in relation to environmental gradients: An experimental approach. Phycologia 26:361–373.

Arvola, L., Tulonen, T. 1998. Effects of allochthonous dissolved organic matter and inorganics on the growth of bacteria and algae from a highly humic lake. Environment International 24:509–520.

Bhattacharya, D., Helmchen, T., Bibeau, C., Melkonian, M. 1995. Comparisons of nuclear-encoded small-subunit ribosomal RNAs reveal the evolutionary position of the Glaucocystophyta. Molecular Biology and Evolution 12:415–420.

Boraas, M. E., Estep, K. W., Johnson, P. W., Sieburth, J. McN. 1988. Phagotrophic phototrophs: The ecological significance of mixotrophy. Journal of Protozoology 35:249–252.

Bourelly, P. 1970. Les algues d'eau douce. Tome III: Les agues bleues et rouges. Les eugleniens, peridiniens et cryptomonadines. Editions N. Boubee, Cie, Paris, 512 p.

Boyne, A. F. 1979. A gentle, bounce free assembly for quick freezing tissues for electron microscopy: Application to isolated *Torpedine* ray electrode stacks. Journal of Neuroscience Methods 1: 353–364.

Brett, S. J., Wetherbee, R. 1986. A comparative study of periplast structure in *Cryptomonas cryophila* and *C. ovata* (Cryptophyceae). Protoplasma 131: 23–31.

Butcher, R. W. 1967. An introductory account of the smaller algae of British coastal waters. IV. Cryptophyceae. Fishery Invest., Ser. 4, London, pp. 1–54.

Caljon, A. 1983. Brackish-water phytoplankton of the Flemish lowland. Developments in Hydrobiology 18:1–272.

Canter, H. M. 1968. Studies on British chytrids. XXVII. *Rhizophydium fugax* sp. nov., a parasite of planktonic cryptomonads with additional notes and records of planktonic fungi. Transactions of the British Mycological Society 51:699–705.

Cavalier-Smith T. 1986. The Kingdom Chromista: Origin and systematics, in: Round, F. E., Chapman, D. J., Eds, Progress in phycological research, Vol. 4. Biopress Ltd., Bristol, 309–347.

Cavalier-Smith, T. 1993. Kingdom protozoa and its 18 phyla. Microbiological Reviews Dec: 953–994.

Cavalier-Smith, T., Couch, J. A., Thorsteinsen, K. E., Gilson, P., Deane, J. A., Hill, D. R. A., McFadden, G. I. 1996. Cryptomonad nuclear and nucleomorph 18S rRNA phylogeny. European Journal of Phycology 31:315–328.

Chandler, D. E. 1984. Comparison of quick-frozen and chemically fixed sea urchin eggs: Structural evidence that cortical granule exocytosis is preceded by a local increase in membrane mobility. Journal of Cell Science 72:23–36.

Clay, B. L., Kugrens, P. 1999a. Systematics of the enigmatic kathablepharids, including EM characterization of the type species, *Kathablepharis phoenikoston*, and new observations on *K. remigera* comb. nov. Protistology 150:43–59.

Clay, B. L., Kugrens, P. 1999b. Characterization of *Hemiselmis amylosa* sp. nov. and phylogenetic placement of the blue-green cryptomonads *H. amylosa* and *Falcomonas daucoides*. Protistology 150:297–310.

Clay, B. L., Kugrens, P. 1999c. Description and ultrastructure of *Kathablepharis tenuis* sp. nov. and *K. obesa* sp. nov.—two new freshwater kathablepharids (Kathablepharididae) from Colorado and Wyoming. European Journal of Protistology 35:435–447.

Clay, B. L., Kugrens, P., Lee, R. E. 1999. A revised classification of Cryptophyta. Botanical Journal of the Linnean Society 131:131–151.

Deane, J. A., Hill, D. R. A., McFadden, G. I. 1998. *Hanusia phi* gen. et sp. nov. (Cryptophyceae): Characterization of *Cryptomonas* sp. φ. European Journal of Phycology 33:149–154.

Dodge, J. D. 1969. The ultrastructure of *Chroomonas mesostigmatica* Butcher (Cryptophyceae). Archiv für Mikrobiologie 69:266–280.

Douglas, S. E., Murphy, C. A., Spencer, D. F., Gray, M. W. 1991. Cryptomonad algae are evolutionary chimeras of two phylogenetically distinct unicellular eukaryotes. Nature 350:148–151.

Dwarte, D., Vesk, M. 1982. Freeze-fracture thylakoid ultrastructure of representative members of chlorophyll *c* algae. Micron 13:325–326.

Dwarte, D., Vesk, M. 1983. A freeze-fracture study of cryptomonad thylakoids. Protoplasma 117:130–141.

Edmondson, W. T. 1965. Reproductive rate of planktonic rotifers as related to food and temperature in nature. Ecological Monographs 35:61–111.

Erata, M., Chihara, M. 1989. Re-examination of *Pyrenomonas* and *Rhodomonas* (class Cryptophyceae) through ultrastructural survey of red pigmented cryptomonads. Botanical Magazine of Tokyo 102:429–442.

Eschbach, S., Wolters, J., Sitte, P. 1991. Primary and secondary structure of the nuclear small subunit ribosomal RNA of the cryptomonad *Pyrenomonas salina* as inferred from the gene sequence: Evolutionary implications. Journal of Molecular Evolution 32:247–252.

Ettl, H., Moestrup, O. 1980. Uber einen intrazellularen Parasiten be *Cryptomonas* (Cryptophyceae), I. Plant Systematics and Evolution 135:211–226.

Faust, M. A. 1974. Structure of the periplast of *Cryptomonas ovata* var. *palustris*. Journal of Phycology 10:121–124.

Faust, M. A., Gantt, E. 1973. Effect of light intensity and glycerol on the growth, pigment composition and ultrastructure of *Chroomonas* sp. Journal of Phycology 9:489–495.

Ferguson, A. J. D., Thompson, J. M., Reynolds, C. S. 1982. Structure and dynamics of zooplankton communities maintained in closed systems, with special reference to the algal food supply. Journal of Plankton Research 4:523–543.

Francke, J. A., Coesel, P. F. M. 1985. Isozyme variation within and between Dutch populations of *Closterium ehrenbergii* and *C. moniliferum* (Chlorophyta, Conjugatophyceae). British Phycological Journal 20:201–209.

Gantt, E. 1971. Micromorphology of the periplast of *Chroomonas* sp. (Cryptophyceae). Journal of Phycology 7:177–184.

Gantt, E. 1979. Phycobiliproteins of Cryptophyceae, in: Levandowsky, M., Hunter, S. H., Eds., Biochemistry and physiology of protozoa, Vol. 1. Academic Press, New York and London, pp. 121–138.

Gantt, E. R. 1980. Photosynthetic cryptophytes, in: Cox, E. R., Ed., Phytoflagellates, developments in marine biology, Vol. 2. Elsevier, North-Holland, Amsterdam, pp. 381–405.

Gantt, E., Edwards, M. R., Provasoli L. 1971. Chloroplast structure of the Cryptophyceae, evidence for Phycobiliproteins within intrathylakoidal spaces. Journal of Cell Biology 48:280–290.

Gillot, M. 1990. Phylum Cryptophyta (Cryptomonads), in: Margulis, L., Corliss, J. O., Melkonian, M., Chapman, D. J., Eds., Handbook of protoctista, Jones, Bartlett Publishers, Boston, pp. 139–151.

Gillott, M. A., Gibbs, S. P. 1980. The cryptomonad nucleomorph: Its ultrastructure and evolutionary significance. Journal of Phycology 16:558–568.

Gillott, M. A., Gibbs, S. P. 1983. Comparison of the flagellar rootlets and periplast in two marine cryptomonads. Canadian Journal of Botany 61:1964–1980.

Glazer, A. N., Appell, G. S. 1977. A common evolutionary origin for the biliproteins of cyanobacteria, Rhodophyta, and Cryptophyta. Federation of European Microbiological Societies Microbiology Letters 1: 113–116.

Glazer, A. N., Wedemeyer, G. J. 1995. Cryptomonad biliproteins—an evolutionary perspective. Photosynthesis Research 46:93–105.

Grain, J., Mignot, J. P., Puytorac, P. 1988. Ultrastructures and evolutionary modalities of flagellar and ciliary systems in protists. Biology of the Cell 63:219–237.

Greenwood, A. D., Griffiths, H.B., Santore, U.S. 1977. Chloroplast and cell compartments in Cryptophyceae. British Phycological of Journal 12:112–119.

Grim, J. N., Staehelin, L. A. 1984. The ejectisomes of the flagellate *Chilomonas paramecium*: Visualization by freeze-fracture and isolation techniques. Journal of Protozoology 3:259–267.

Guillard, R. R. L. 1975. Culture of phytoplankton for feeding marine invertebrates, in: Smith, W. L., Chanley, M. H., Eds., Culture of marine invertebrate animals. Plenum Press, New York, pp. 29–60.

Hausmann, K., Walz, B. 1979. Periplaststruktur und Organisation der Plasmamembran von *Rhodomonas* sp. (Cryptophyceae). Protoplasma 101:349–354.

Henderson, R. J., Mackinlay, E. E. 1989. Effect of temperature on lipid composition of the marine cryptomonad *Chroomonas salina*. Phytochemistry 28:2943–2948.

Heywood, P. 1988. Ultrastructure of *Chilomonas paramecium* and the phylogeny of cryptomonads. BioSystems 21:293–298.

Hibberd, D. J. 1977. Observations on the ultrastructure of the cryptomonad endosymbiont of the red-water ciliate *Mesodinium rubrum*. Journal of the Marine Biological Association of the United Kingdom 57:45–61.

Hibberd, D. J., Greenwood, A. D., Griffiths, H. B. 1971. Observations on the ultrastructure of the flagella and periplast in the Cryptophyceae. British Phycological Journal 6:61–72.

Hill, D. R. A. 1990. *Chroomonas* and other blue-green cryptomonads. Journal of Phycology 27: 133–145.

Hill, D. R. A. 1991a. *Chroomonas* and other blue-green cryptomonads. Journal of Phycology 26: 133–145.

Hill, D. R. A. 1991b. Diversity of heterotrophic cryptomonads, in: Patterson, D.J., Larsen, J., Eds., The biology of free-living heterotrophic flagellates, Systematics Association Special Volume 45, pp. 235–240.

Hill, D. R. A. 1991c. A revised circumscription of *Cryptomonas* (Cryptophyceae) based on examination of Australian strains. Phycologia 30:170–188.

Hill, D. R. A., Rowan K. S. 1989. Biliproteins of the Cryptophyceae. Phycologia 28:455–463.

Hill, D. R. A., Wetherbee, R. 1986. *Proteomonas sulcata* gen. et sp. nov. (Cryptophyceae) a cryptomonad with two morphologically distinct and alternating forms. Phycologia 27:521–543.

Hill, D. R. A., Wetherbee, R. 1988. The structure and taxonomy of *Rhinomonas pauca* gen. et sp. nov. (Cryptophyceae). Phycologia 27:355–365.

Hill, D. R. A., R. Wetherbee. 1989. A reappraisal of the genus Rhodomonas (Cryptophyceae). Phycologia 28:143–158.

Hofmann, C. J. B., Rensing, S. A., Haeuber, M. M., Martin, W. F., Mueller, S. B., Couch, J., McFadden, G. I., Igloi, G. L., Maier, U.G. 1994. The smallest known eukaryotic genomes encode a protein gene: Towards an understanding of nucleomorph functions. Molecular and General Genetics 243:600–604.

Hoshaw, R. W., Rosowski, J. R. 1973. Isolation and purification. 3: Methods for microscopic algae, in: Stein J. R., Ed., Handbook of phycological methods: Culture methods and growth measurements. Cambridge University Press, New York, pp. 53–67.

Huber-Pestalozzi, G. 1950. *Das Phytoplankton das Süsswassers, Teil 3. Cryptophyceen, Chloromonadinen, Peridineen*, in: Thienemann, A., Ed., Die Binnengewasser. Stuttgart, pp. 2–78.

Klaveness, D. 1977. Morphology, distribution and significance of the manganese-accumulating microorganism *Metallogenium* in lakes. Hydrobiologia 56:25–33.

Klaveness, D. 1981. *Rhodomonas lacustris* (Pascher, Ruttner) Javornicky (Cryptomonadida): Ultrastructure of the vegetative cell. Journal of Protozoology 28:83–90.

Klaveness, D. 1982. The *Cryptomonas-Caulobacter* consortium: Facultative ectocommensalism with possible taxonomic consequences? Nordic Journal of Botany 2:183–188.

Klaveness, D. 1984. Studies on the morphology, food selection and growth of two planktonic freshwater strains of *Coleps* sp. Protistologica 20:335–349.

Klaveness, D. 1985. Classical and modern criteria for determining species of Cryptophyceae. Bulletin of Plankton Society of Japan 32:111–128.

Klaveness, D. 1988a. Ecology of the Cryptomonadida: A first review, in: Sandgren, C. D., Ed., Growth and reproductive strategies of freshwater phytoplankton. Cambridge University Press, New York, pp. 105–133.

Klaveness, D. 1988b. Biology and ecology of the Cryptophyceae: Status and challenges. Biological Oceanography 6:257–270.

Kugrens, P. 1999. Cryptomonad systematics—an algal enigma? in: Seckbach, J., Ed., Enigmatic algae. Kluwer Academic Publishers, Dordrecht, the Netherlands, pp. 127–138.

Kugrens, P., Clay, B. L., Lee, R. E. 1999. Ultrastructure and systematics of two new freshwater cryptomonads, *Pyrenomonas ovalis* and *Storeatula rhinosa* sp. nov. Journal of Phycology 35:1079–1089.

Kugrens P., Lee, R. E. 1986. An ultrastructural survey of cryptomonad periplasts using quick-freezing freeze-fracture techniques. Journal of Phycology 23:365–376.

Kugrens, P., Lee, R. E.. 1988. Ultrastructure of fertilization in a cryptomonad. Journal of Phycology 24:510–518.

Kugrens, P., Lee, R. E. 1990. Ultrastructural evidence for bacterial incorporation and mixotrophy in the photosynthetic cryptomonad *Chroomonas pochmanni* Huber-Pestalozzi (Cryptomonadida). Journal of Protozoology 37:263–267.

Kugrens, P., Lee, R. E. 1991. Organization of cryptomonads, in: Patterson, D. J., Larsen, J., Eds., The biology of free-living heterotrophic flagellates. The Systematics Association Special Volume 45, pp. 219–233.

Kugrens, P., Lee, R. E., Andersen, R. E. 1986. Cell form and surface patterns in *Chroomonas* and *Cryptomonas* cells (Cryptophyta) as revealed by scanning electron microscopy. Journal of Phycology 22:512–522.

Kugrens, P., Lee, R. E., Andersen, R. E. 1987. Ultrastructural variations in cryptomonad flagella. Journal of Phycology 23:511–518.

Kugrens, P., Lee, R. E., Corliss, J. O. 1994. Ultrastructure, function and biogenesis of extrusive organelles in selected non-ciliate protists. Protoplasma 181:164–190.

Lee, R. E., Kugrens, P. 1986. The occurrence and structure of flagellar scales in some freshwater cryptophytes. Journal of Phycology 22:549–552.

Lee, R. E., Kugrens, P. 1991. *Katablepharis ovalis*, a colorless flagellate with interesting cytological characteristics. Journal of Phycology 27:505–513.

Lee, R. E., Kugrens, P. 1998. Hypothesis: The ecological advantage of chloroplast endoplasmic reticulum—the ability to outcompete at low dissolved CO_2 concentrations. Protistology 149:341–345.

Lee, R. E., Kugrens, P. 2000. Ancient atmospheric CO_2 and the timing of evolution of secondary endosymbioses. Phycologia 39:167–172.

Lee, R. E., Kugrens, P., Mylnikov, A. P. 1991. Feeding apparatus of the colorless flagellate *Katablepharis* (Cryptophyceae). Journal of Phycology 27:725–733.

Lee, R. E., Miller-Hughes, C., Kugrens, P. 1993. Ultrastructure of mitosis and cytokinesis in the colorless flagellate *Katablepharis ovalis* Skuja. Journal of Eukaryotic Microbiology 40:377–383.

Lewitus, A. J., Caron, D. A. 1991. Physiological responses of phytoflagellates to dissolved organic substrate additions. 2. Dominant role of autotrophic nutrition in *Pyrenomonas salina* (Cryptophyceae). Plant and Cell Physiology 32:791–801.

Lewitus, A. J., Glasgow, H. B., Burkholder, J. M. 1999. Kleptoplastidy in the toxic dinoflagellate *Pfiesteria piscicida* (Dinophyceae). Journal of Phycology 35:303–312.

Lucas, I. A. N. 1970a. Observations on the ultrastructure of representatives of the genera *Hemiselmis* and *Chroomonas* (Cryptophyceae). British Phycological Journal 5:29–37.

Lucas, I. A. N. 1970b. Observation on the fine structure of the Cryptophyceae. I. The genus *Cryptomonas*. Journal of Phycology 6:30–38.

Lucas, I. A. N. 1982. Observation on the fine structure of the

Cryptophyceae. II. The eyespot. British Phycological Journal 17:113–119.

Ludwig, M., Gibbs, S. P. 1985a. DNA is present in the nucleomorph of cryptomonads: Further evidence that the chloroplast evolved from a eukaryotic endosymbiont. Protoplasma 127:9–20.

Ludwig, M., Gibbs, S. P. 1985b. DNA is present in the nucleomorph of cryptomonads: Further evidence that the chloroplast evolved from a eukaryotic endosymbiont. Protoplasma 127:9–20.

Ludwig, M., Gibbs, S. P. 1989. Localization of phycoerythrin at the lumenal surface of the thylakoid membrane in *Rhodomonas lens*. Journal of Cell Biology 108:875–884.

Lund, J. W. G. 1962. A rarely recorded but very common British alga, *Rhodomonas minuta* Skuja. British Phycological Bulletin 2:133–139.

Marin, B., Klingberg, M., Melkonian, M. 1998. Phylogenetic relationships among the Cryptophyta: Analysis of nuclear-encoded SSU rRNA sequences support the monophyly of extant plastid-containing lineages. Protistology 149:265–276.

Martin W., Sommerville C. C., Loiseaux-de Goer S. 1992. Molecular phylogenies of plastid origins and algal evolution. Journal of Molecular Evolution 35: 385–404.

McFadden, G. I. 1993. Second-hand chloroplasts: Evolution of cryptomonad algae. Advances in Botanical Research 19:189–230.

McFadden, G. I., Gilson, P. R., Douglas, S. E., Cavalier-Smith, T., Hofmann, C. J. B., Maier, U-G. 1997. Bonsai genomics: Sequencing the smallest eukaryotic genomes. Trends in Genetics 13:46–49.

McFadden, G. I., Gilson, P. R., Hill, D. R. A. 1994. *Goniomonas*: rRNA sequences indicate that this phagotrophic flagellate is a close relative of the host component of cryptomonads. European Journal of Phycology 29:29–32.

McKerracher, L., Gibbs, S.P. 1982. Cell and nucleomorph division in the alga *Cryptomonas*. Canadian Journal of Botany 60:2440–2452.

Meyer, S. R., Pienaar, R. N. 1981. The ultrastructure of mitosis and cytokinesis in a new species of *Chroomonas* (Cryptophyceae). Electron Microscopy Society of Southern Africa Proceedings 11:163–164.

Meyer, S. R., Pienaar, R. N. 1984a. The microanatomy of *Chroomonas africana* sp. nov. (Cryptophyceae). South African Journal of Botany 3:306–319.

Meyer, S. R., Pienaar, R. N. 1984b. Mitosis and cytokinesis in *Chroomonas africana* Meyer, Pienaar (Cryptophyceae). South African Journal of Botany 3:320–330.

Mignot, J. P. 1965. Etude ultrastructurale de *Cyathomonas truncata* From. (Flagelle Cryptomonadine). Journal de Microscopie (Paris) 4:239–252.

Morgan, K., Kalff, J. 1975. The winter dark survival of an algal flagellate *Cryptomonas erosa* (Skuja). Internationale Vereinigung für Theoretische und Angewandte Limnologie Verhandlungen 19:2735–2740.

Morrall, S., Greenwood, A. D. 1980. A comparison of the periodic substructure of the trichocysts of the Cryptophyceae and Prasinophyceae. Biosystems 12:71–83.

Morrall, S., Greenwood, A. D. 1982. Ultrastructure of nucleomorph division in species of Cryptophyceae and its evolutionary implications. Journal of Cell Science 54:311–328.

Munawar, M., Bistricki T. 1979. Scanning electron microscopy of some nanoplankton cryptomonads. Scanning Electron Microscopy 3:247–252.

Nichols, H. W. 1973. Growth media—freshwater, *in*: Stein, J. R., Ed., Handbook of Phycological Methods. Culture Methods and Growth Measurements. Cambridge University Press, New York, pp. 7–24

Novarino, G. 1991a. Observations on *Rhinomonas reticulata* comb. nov. and *R. reticulata* var. *eleniana* var. nov. (Cryptophyceae), with comments on the genera *Pyrenomonas* and *Rhodomonas*. Nordic Journal of Botany 11:2453–2252.

Novarino, G. 1991b. Observations on some new and interesting Cryptophyceae. Nordic Journal of Botany 11:599–611.

Novarino, G. 1993a. A comparison of some morphological characters in *Chroomonas ligulata* sp. nov. and *C. placoides* sp. nov. (Cryptophyceae). Nordic Journal of Botany 13:583–589.

Novarino, G. 1993b. Possible detection of the periplast areas and the nucleomorph of cryptomonads by light microscopy: Some early observations by Künstler, Skuja and Hoolande. Quekett Journal of Microscopie (Paris) 37:45–51.

Novarino, G., Lucas, I. A. N. 1993. Some proposals for a new classification system of the Cryptophyceae. Botanical Journal of the Linnean Society, London 111:3–21.

Novarino, G., Lucas, I. A. N., Morrall, S. 1994. Observations on the genus *Plagioselmis* (Cryptophyceae). Cryptogamie, Algologie 15:87–96.

Oakley, B. R., Bisalputra, T. 1977. Mitosis and cell division in *Cryptomonas* (Cryptophyceae). Canadian Journal of Botany 55:2789–2800.

Oakley, B. R., Dodge, J. D. 1973. Mitosis in the Cryptophyceae. Nature 244:521–522.

Oakley, B. R., Dodge, J. D. 1976. The ultrastructure of mitosis in *Chroomonas salina* (Cryptophyceae). Protoplasma 88:241–254.

Oakley, B. R., Heath, I. B. 1978. The arrangements of microtubules in serially sectioned spindles of the alga *Cryptomonas*. Journal of Cell Science 31:53–70.

Oakley, B. R., Santore, U. J. 1982. Cryptophyceae: Introduction and bibliography, *in*: Rosowski, J. R., Parker, B. C., Eds., Selected papers of phycology. Allen Press, Lawrence, KS, pp. 682–686.

Parducz, B. 1967. Ciliary movement and coordination in ciliates. International Review of Cytology 21:91–128.

Patterson, D. J. 1981. The behaviour of contractile vacuole complexes of cryptophycean flagellates. British Phycological Journal 16:429–439.

Pennick, D. L. 1981. Flagellar scales in *Hemiselmis brunnescens* Butcher and *H. virescens* Droop (Cryptophyceae). Archiv für Protistenkunde 124:267–270.

Pejler, B. 1977. *Experience with rotifer cultures based on Rhodomonas*. Archiv für Hydrobiologie Ergebnisse der Limnologie 8:264–266.

Perasso, L., Brett, S. J., Wetherbee, R. 1993. Pole reversal and the development of cell asymmetry during division in cryptomonad flagellates. Protoplasma 174: 19–24.

Pfiester, L. A., Holt, J. R. 1978. A freshwater "red tide" in Texas. Southwestern Naturalist 23:103–110.

Phillips, D., Boyne, A. F. 1984. Liquid nitrogen-based quick freezing: Experiences with bounce-free delivery of cholenergic nerve terminals to a metal surface. Journal of Electron Microscopy Technique 1:9–29.

Pienaar. R. N. 1976. Virus-like particles in three species of phytoplankton from San Juan Island, Washington. Phycologia 15:185–190.

Pollinger, U. 1981. The structure and dynamics of the phytoplankton assemblages in lake Kinneret, Israel. Journal of Plankton Research 3:93–105.

Pringsheim, E. G. 1968. Zur kenntnis der Cryptomonaden des Süsswassers. Nova Hedwigia 16:367–401.

Putt, M. 1990. Metabolism of photosynthate in the chloroplast-retaining ciliate *Loboea strobila*. Marine Ecology Progress Series 60:271–282.

Reynolds, C. S. 1980. Phytoplankton assemblages and their periodicity in stratifying lake systems. Holarctic Ecology 3:141–159.

Reynolds, C. S. 1982. Phytoplankton periodicity: Its motivation, mechanisms and manipulation. Freshwater Biological Association Annual Report 3:141–159.

Reynolds, C. S. 1984. Phytoplankton periodicity: The interactions of form, function and environmental variability. Freshwater Biology 14:111–142.

Roberts, K. R. 1984. Structure and significance of the cryptomonad flagellar apparatus. I. *Cryptomonas ovata* (Cryptophyta). Journal of Phycology 20:159–167.

Roberts, K. R., Stewart, K. D., Mattox, K. R. 1981. The flagellar apparatus of *Chilomonas paramecium* (Cryptophyceae) and its comparison with certain zooflagellates. Journal of Phycology 17:159–167.

Rott, E. 1983. Sind die Veränderung im Phytoplanktonbild dem Pilburger Sees Auswirkungen der Tiefenwasserableitung? Archiv für Hydrobiologie Supplement band 67:29–80.

Santore, U. J. 1977. Scanning electron microscopy and comparative micromorphology of the periplast of *Hemiselmis rufescens*, *Chroomonas* sp., *Chroomonas salina* and members of the genus *Cryptomonas* (Cryptophyceae). British Phycological Journal 12:255–270.

Santore, U. J. 1978. Light- and electron-microscopic observations of the palmelloid phase in members of the genus *Cryptomonas* (Cryptophyceae). Archiv für Protistenkunde 120:420–435.

Santore, U. J. 1982a. Comparative ultrastructure of two members of the Cryptophyceae assigned to the genus *Chroomonas*—with comments on their taxonomy. Archiv für Protistenkunde 125:5–29.

Santore, U. J. 1982b. The ultrastructure of *Hemiselmis brunnescens* and *Hemiselmis virescens* with additional observations on *Hemiselmis rufescens* and comments about the Hemiselmidaceae as a natural group of the Cryptophyceae. British Phycological Journal 17:81–89.

Santore, U. J. 1982c. The distribution of the nucleomorph in the Cryptophyceae. Cell Biology International Reports 6:1055–1063.

Santore, U. J. 1983. Flagellar and body scales in the Cryptophyceae. British Phycological Journal 18:239–248.

Santore, U. J. 1984. Some aspects of taxonomy in the Cryptophyceae. New Phytologist 98:627–646.

Santore, U. J. 1987. A cytological survey of the genus *Chroomonas*—with comments on the taxonomy of this natural group of the Cryptophyceae. Archiv für Protistenkunde 134:83–114.

Santore, U. J., Greenwood, A. D. 1977. The mitochondrial complex in Cryptophyceae. Archives of Microbiology 112:207–218.

Sarnelle, O. 1993. Herbivore effects on phytoplankton succession in a eutrophic lake. Ecological Monographs 63:129–149.

Schnepf, E., Elbrächter, M. 1992. Nutritional strategies in dinoflagellates: a review with emphasis on cell biological aspects. European Journal of Protistology 28:3–24.

Schnepf, E., Melkonian, M. 1990. Bacteriophage-like particles in endocytic bacteria of Cryptomonas (Cryptophyceae). Phycologia 29:338–343.

Schnepf, E., Winter, S., Mollenhauer, D. 1989. *Gymnodinium aeruginosum* (Dinophyta): A blue-green dinoflagellate with a vestigial, anucleate, cryptophycean endosymbiont. Plant Systematics and Evolution 164:75–91.

Schuster, F. L. 1968. The gullet and trichocysts of *Cyathomonas truncata*. Experimental Cell Research 49:277–284.

Schuster, F. L. 1970. The trichocysts of *Chilomonas paramecium*. Journal of Protozoology 17:521–526.

Sespenwol, S. 1973. Leucoplast of the cryptomonad *Chilomonas paramecium*; evidence for the presence of a true plastid in a colorless flagellate. Experimental Cell Research 76:395–409.

Skovgaard, A. 1998. Role of chloroplast retention in a marine dinoflagellate. Aquatic and Microbiology Ecology 15:293–301.

Skuja, H. 1939. Beitrag zur Algenflora Lettlands. II. Acta Horti Botanici Universitatis Latviensis. 11/12:41–168.

Skuja, J. 1948. Taxonomie des Phytoplanktons einiger Seen in Uppland, Schweden. Symboiae Botanicae Upsaliensis 9:1–399.

Sommer, U. 1982. Vertical niche separation between two closely related planktonic flagella species (*Rhodomonas lens* and *Rhodomonas minuta* v. *nanoplanctica*). Journal of Plankton Research 4:137–142.

Stemberger, R. S., Gilbert, J. J. 1985. Body size, food concentration, and population growth in planktonic rotifers. Ecology 66:1151–1159.

Stoecker, D. K., Michaels, A. E., Davis, L. H. 1987. Large proportion of marine planktonic ciliates found to contain functional chloroplasts. Nature 3216:790–792.

Stoecker, D. K., Silver, M. W. 1990. Replacement and aging of chloroplasts in *Strombidium capitatum* (Ciliophora: Oligotrichida). Marine Biology (Berlin) 107:491–502.

Stoecker, D. K., Silver, M. W., Michaels, A. E., Davis, L. H. 1988/1989. Enslavement of algal chloroplasts by four *Strombidium* spp. (Ciliophora, Oligotrichida). Marine Microbial Food Webs 3:79–100.

Taylor, W. D., Hern, S. C., William, L. R., Lambou, V. W., Morris, M. K., Morris, F. A. 1979. Phytoplankton Water Quality Relationships in U.S. Lakes. Part 6. The Common Phytoplankton Genera from Eastern and Southeastern Lakes. U.S. Environmental Protection Agency, Environmental Monitoring and Support Laboratory, Working Paper No. 710.

Tranvik, L. J., Porter, K. G., Sieburth, J. 1989. Occurrence of bacterivory in *Cryptomonas*, a common freshwater phytoplankter. Oecologia 78:473–476.

Vørs, N. 1992a. Heterotrophic amoebae, flagellates and Heliozoa from the Tvarminne area, Gulf Of Finland, in 1988–1990. Ophelia 36:1–109.

Vørs, N. 1992b. Ultrastructure and autecology of the marine, heterotrophic flagellate *Leucocryptos marina* (Braarud) Butcher 1967 (Katablepharidaceae/Kathablepharidae), with a discussion of the genera *Leucocryptos* and *Katablepharis/Kathablepharis*. European Journal of Protistology 28: 369–389.

Watanabe, M., Furuya, M. 1982a. Phototactic behaviour of cells of *Cryptomonas* sp. in response to continuous and intermittent light stimuli. Photochemistry and Photobiology 35:559–563.

Watanabe, M., Furuya, M. 1982b. Effects of viscosity on phototactic movement and period of cell rotation in *Cryptomonas* sp. Physiologie Plantarum 56:194–198.

Watanabe, M., Miyoshi, M.& Furuya, M. 1976. Phototaxis in *Cryptomonas* sp. under conditions suppressing photosynthesis. Plant and Cell Physiology 17:683–690.

Wemmer, D. E. Wedemeyer, G. J., Glazer, A. N. 1993. Phycobilins of cryptophycean algae. Novel linkage of dihydrobiliverdin in a phycoerythrin 555 and a phycocyanin 645. Journal of Biological Chemistry 268:1658–1669.

Wetherbee, R., Hill, D. R. A., Brett, S. J. 1987. The structure of the periplast components and their association with the plasma membrane in a cryptomonad flagellate. Canadian Journal of Botany 65:1019–1026.

Wetherbee, R., Hill, D. R. A., McFadden, G. I. 1986. Periplast structure of the cryptomonad flagellate *Hemiselmis brunnescens*. Protoplasma 131:11–22.

Willen, E., Oke, M., Gonzalez, F. 1980. *Rhodomonas minuta* and *Rhodomonas lens* (Cryptophyceae)—aspects of form-variation and ecology in lakes Mälaren and Vättern, central Sweden. Acta Phytogeographica Suecica 68:163–172.

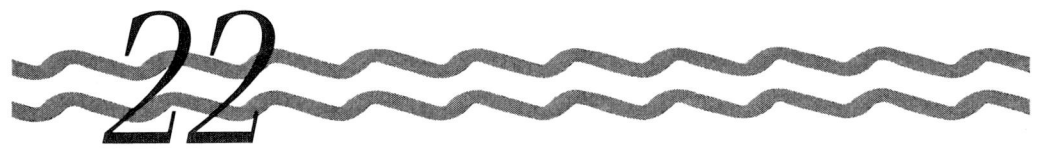

BROWN ALGAE

John D. Wehr

*Louis Calder Center—Biological Station
and Department of Biological Science
Fordham University,
Armonk, New York 10504*

I. Introduction
II. Diversity and Morphology
    A. Diversity and Classification
    B. Morphology and Reproduction
III. Ecology and Distribution
    A. Ecological Factors
    B. Geographical Distribution
IV. Methods for Collection and Identification
V. Key and Descriptions of Genera
    A. Key
    B. Descriptions of Genera
VI. Guide to Literature for Species
    Identification
    Literature Cited

I. INTRODUCTION

Brown algae, the Phaeophyceae (or Fucophyceae; Christensen, 1978), are a class (or division, Phaeophyta; Papenfuss, 1951) of algae consisting mainly of complex, macroscopic seaweeds whose brown color comes from a carotenoid pigment, fucoxanthin, and in some species, various phaeophycean tannins. Of perhaps 2000 species (in 265 genera) of brown algae (Van den Hoek *et al.*, 1995), less than 1% are known from freshwater habitats, although some marine species may colonize brackish waters (Wilce, 1966; Dop, 1979; West and Kraft, 1996). Various authors cite between 3 and 7 genera, and up to 12 species of freshwater brown algae worldwide (see Sect. 21.II.A). Members of the group have many features in common with chrysophytes, synurophytes (Chrysophyta), and diatoms (Bacillariophyta), including chloroplast structure (thylakoids in stacks of three, girdle lamella, chloroplast endoplasmic reticulum), heterokont motile stage (unequal flagella), major pigments (chlorophylls *a*, c_1, and c_2, β-carotene, violaxanthin, diatoxanthin, and large amounts of fucoxanthin), as well as the storage reserve laminarin (Craigie, 1974; Goodwin, 1974; Pueschel and Stein, 1983; Lee, 1989). However, no members of the Phaeophyceae are unicellular or colonial in the vegetative phase—the predominant morphology in other golden-brown groups. Brown algae have cell walls composed of cellulose, which is often supplemented with the mucopolysaccharide alginic acid. In seaweeds, this material is produced in sufficient quantities in some species to be harvested for commercial purposes, but in freshwater species, alginates appear to be less prevalent. Further descriptions of the group can be found in reviews by Papenfuss (1951), Van den Hoek *et al.* (1995), and Graham and Wilcox (2000).

Freshwater species of brown algae have been known for more than 100 years [*Pleurocladia* was described by Braun (1855), *Heribaudiella* by Gomont (1896); Bourrelly (1981)], but today most are still known only from scattered locations. Freshwater phaeophytes occur in a variety of streams and rivers, as well as in the littoral zone of lakes, but their biology has largely remained obscure for most phycologists and freshwater ecologists. This is unfortunate, because some species, most notably *Heribaudiella fluviatilis*, can at times be one of the dominant species of benthic algae in smaller rivers (Kann, 1978a). One reason for their obscurity may be that most species form crusts or brown colonies that may be mistaken for members of other algal groups or lichens. The most recent monograph of freshwater algae in North America (Smith, 1950) listed only one species (*H. fluviatilis*), which was recorded from only one location [considered doubtful by Smith (1950), but see Sect. V.B], and even its identity has been questioned (Pueschel and Stein, 1983). Since that time, other genera and many more locations have been identified on this continent, suggesting that at least some species are less rare than previously thought, although their distribution is still far from well known. Whitford (1977) proposed that few species of freshwater algae may actually be rare, but instead are simply under-reported. Several species long thought to be uncommon have turned out to be fairly cosmopolitan, such as several species of freshwater red algae (Sheath and Hambrook, 1990). For these reasons, Whitford challenged biologists to describe the habitats of newly discovered algae more carefully. Because of the paucity of information on freshwater Phaeophyta, the present chapter describes all known taxa, although at present only five species (in four genera) have thus far been confirmed from sites in North America.

II. DIVERSITY AND MORPHOLOGY

A. Diversity and Classification

Freshwater brown algae are undoubtedly the least diverse of all groups of freshwater algae. Although some species can at times form substantial populations, no habitats are known that have several species of freshwater brown algae within a single location. Kann (1993) has observed filamentous *Pleurocladia* and encrusting *Heribaudiella* occurring together on stones in the littoral zone of Lake Erken, Sweden, and Kusel-Fetzmann (1996) noted that the *Pleurocladia* may grow as an epiphyte on *Heribaudiella* in some Austrian streams. More than 60 years ago, Israelsson (1938) demonstrated that different species of freshwater brown algae exhibit different geographic patterns, which appear to be the result of different ecological requirements. Given the small number of species overall, it is not surprising that their local diversity is low.

Accounts of the number of genera and species of phaeophytes from freshwater vary among authors, largely due to lack of study. Their classification is also unsettled, mainly because of uncertainty regarding the reliability of certain morphological features (e.g., branching pattern, colony shape, presence of hairs) as taxonomic attributes, the possible synonymy of several taxa, and whether historical collections were described and identified accurately (Waern, 1952; Müller and Geller, 1978; Dop, 1979; Bourrelly, 1981; Pueschel and Stein, 1983; Wehr and Stein, 1985). Further work with field populations and cultured material to better describe their reproduction and genetic relationships will undoubtedly reveal new groupings and perhaps new (or fewer) species within the group. The present account recognizes six freshwater genera and seven species worldwide within the division (Table I). Of these, five species have thus far been reported from sites in North America. Following the general schemes of Bold and Wynne (1985) and Van den Hoek et al. (1995), all fresh water phaeophytes are classified as members of the Ectocarpales (five genera) or Sphacelariales (one genus, two species).

B. Morphology and Reproduction

The morphologies of all freshwater phaeophyte species are based upon a relatively simple filamentous structure and do not form parenchymatous (tissue-like) thalli, characteristic of more complex brown seaweeds. Their size range is also substantially smaller than those that colonize marine habitats. Crustose forms, although visually conspicuous, may only be 10–30 cells tall (1–2 mm), and form colonies of perhaps 0.2–50 cm^2 in area (Wehr and Stein, 1985; Kusel-Fetzmann, 1996). Filamentous forms can form macroscopic tufts 2–10 mm in size. Several colonies may coalesce to form larger expanses on rocks, but these dimensions are in great contrast to species of intertidal brown algae, which reach sizes of several meters, or subtidal kelp forests that may be as tall as 20–60 m (Bold and Wynne, 1985).

Among freshwater forms, three basic morphologies are seen (Fig. 1): (1) most consist of uniseriate (single axis), branched filaments (members of the Ectocarpales), which develop to form either (Fig. 1A) spreading or cushion-like tufts [*Bodanella, Ectocarpus, Pleurocladia, Porterinema* (syn. = *Pseudobodanella*)]; (2) others consists of prostrate filaments, which produce an upright series of densely packed, vertical filaments, forming a crustose morphology (*Heribaudiella*; Fig. 1B); and (3) in *Sphacelaria* cells are arranged in multiseriate (multiaxial), branched filaments, and also form spreading cushions on submerged substrata (Fig. 1C). Thalli in most species seem capable of forming hyaline, multicellular filaments or hairs (Waern, 1952; Wilce, 1966; Dop, 1979; Schloesser and Blum, 1980; Yoshizaki et al., 1984; Kusel-Fetzmann, 1996; Wujek et al., 1996). It seems doubtful that these hairs can be used as diagnostic features, as studies suggest they may be produced in response to reduced Cl$^-$ or long photoperiods (Dop, 1979), or P-limitation (Fig. 2A). Such patterns are similar to those observed in filamentous species of green algae (Gibson and Whitton, 1987)

TABLE I Species of Brown Algae Reported from Freshwater Environments, with Morphology (UF = Uniseriate Filaments; CR = Crustose, MF = Multiseriate Filaments), Habitats, and Localities

Taxon	Morphology	Habitat	Localities
Ectocarpales			
Bodanella lauterborni	UF	Lake	North America: unknown
			Other: Lake Constance,[a] Europe
Ectocarpus siliculosus[b]	UF	Stream, estuary	North America: unknown
			Other: Hopkins River, Australia
Pleurocladia lacustris	UF	Stream, lake	North America: Green River (UT, CO), Devon Island (NWT)
			Other: Austria, Germany, Poland, Scandinavia, England
Heribaudiella fluviatilis[c]	CR	Stream, lake	North America: at least 30 sites
			Other: many locations in Europe, also Japan, China
Porterinema fluviatile	UF	Lake	North America: unknown[d]
			Other: Germany, Netherlands, United Kingdom
Sphacelariales			
Sphacelaria fluviatilis	MF	Stream, lake	North America: Gull Lake, MI
			Other: China
S. lacustris	MF	Lake	North America: Lake Michigan
			Other: unknown

[a] Known locally as Bodensee.
[b] Ectocarpus confervoides has been collected from the River Werra, Germany, polluted by potassium mines (Geißler, 1983).
[c] Previously reported as Lithoderma arvernensis, L. fluviatile, and L. fontanum; L. zonatum (Jao, 1941) is retained by some authors.
[d] Freshwater and euryhaline (= Pseudobodanella peterfii) reported from North America from marine and estuarine sites only.

and cyanobacteria (Sinclair and Whitton, 1977). A few species (e.g., Pleurocladia) become encrusted with $CaCO_3$, which may cause the thallus to appear pale brown or gray macroscopically; microscopically carbonates may even cloak filaments in a crystalline tube (Kirkby et al., 1972; Kusel-Fetzmann, 1996). All known freshwater species have a diplohaplontic life history (both diploid and haploid vegetative phases) and most are isomorphic (diploid and haploid stages identical or very similar). For details on the alternation of generations in members of this division, see Papenfuss (1951) and Van den Hoek et al. (1995).

Cellular features of freshwater brown algae are much like that of the division (Schloesser, 1977; Pueschel and Stein, 1983; West and Kraft, 1996); differences among genera are discussed later (Sect. V.B). Cells contain one to several golden-brown chloroplasts, which may be discoid, ribbon-like, or irregular-shaped, and usually parietal; pyrenoids are present in some species. The ultrastructure of freshwater species investigated (Heribaudiella, Sphacelaria) suggest typical phaeophyte features: thylakoids in triplets, chloroplast envelope consisting of four membranes, and plasmodesmata traversing crosswalls (Schloesser, 1977; Schloesser and Blum, 1980; Timpano, 1980; Pueschel and Stein, 1983). Most species possess numerous refractive bodies, including physodes, darkly pigmented bodies that may store phaeophycean tannins (fucosan), other polyphenolics, and terpenes (Chadefaud, 1950; Graham and Wilcox, 2000; for cytological methods, see Sect. IV). When thalli are exposed to the air these tannins become oxidized and darken (Lee, 1989), giving dried or exposed colonies a dark brown or black color.

Reproductive structures are distinctive for this group, but life cycles are incompletely known among the freshwater species. In general, species within the Ectocarpales and Sphacelariales produce two types of terminal sporangia: unilocular (large, single chamber) and plurilocular (multichambered), although in some species, only one of the two structures has been observed (Papenfuss, 1951; Hamel, 1931–1939; Bourrelly, 1981). Unilocular sporangia are typically produced on sporophytes (diploid), appear as large, ovate or clavate structures (Figs. 1A, C, 3F), and are the usual sites of meiosis. Initially unilocular sporangia (often arising from elongated filaments; Svedelius, 1930; Kumano and Hirose, 1959) contain several brown chloroplasts, which later condense. Following meiosis the sporangium produces (usually eight) biflagellate zoospores (or zooids), which are thought to serve as gametes (Kumano and Hirose, 1959; Müller and Geller, 1978). Plurilocular sporangia (Fig. 2F) are produced on either gametophyte (1n) or sporophyte (2n) plants, which divide repeatedly from erect narrow threadlike cells (e.g., Ectocarpus, Heribaudiella) or

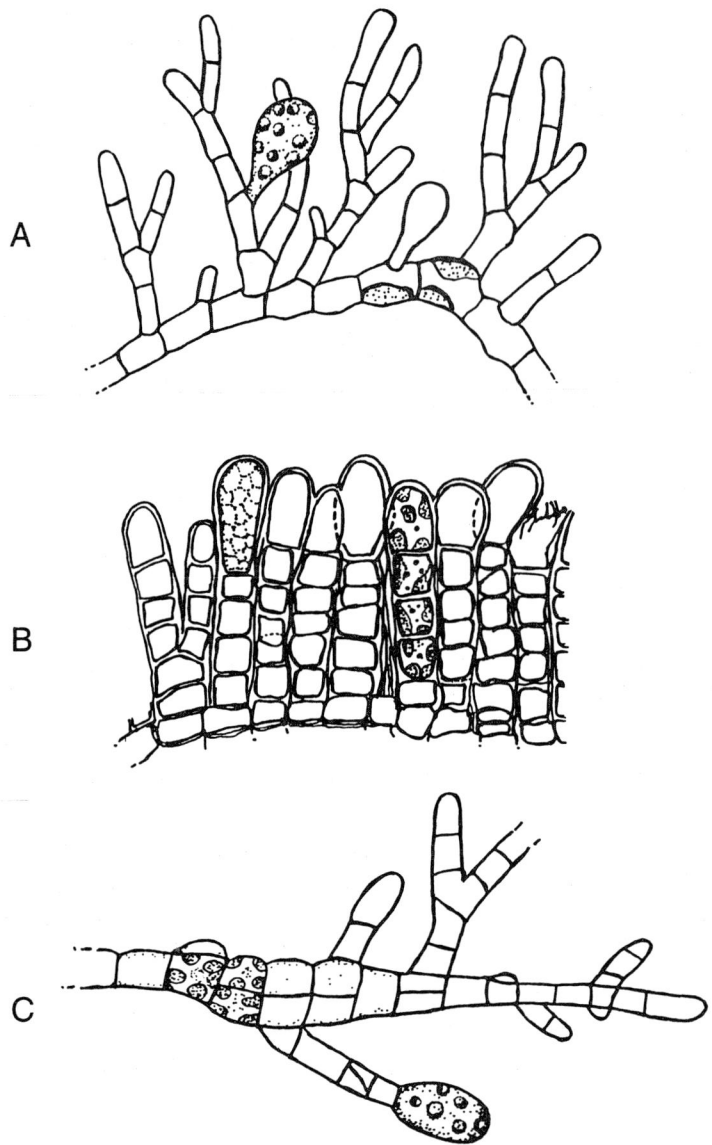

FIGURE 1 Three general morphologies exhibited by freshwater brown algae: (A) uniseriate, branched filaments forming cushion-like tufts (e.g., *Pleurocladia*); (B) branched prostrate filaments giving rise to tightly packed, vertical filaments, forming crustose thalli (e.g., *Heribaudiella*); (C) multiseriate branched filaments (quasicorticated), forming spreading cushions (*Sphacelaria*).

terminal branches (e.g., *Bodanella*) to form multicellular structures that produce asexual zoospores or zoospores that later settle and germinate to produce new filaments.

In *Heribaudiella* and *Porterinema*, biflagellate zoospores are pear-shaped with two laterally inserted flagella, and possess a single parietal chloroplast and an apical stigma (Kumano and Hirose, 1959; Dop, 1979). Dispersal of freshwater phaeophytes is likely favored by zoospores released from unilocular sporangia. After release they attach to available substrata and form germination tubes that later form filaments that develop the typical prostrate or disclike basal system (Yoshizaki *et al.*, 1984). More recent success in isolating some freshwater phaeophytes into pure culture [*Bodanella*; Müller and Geller (1978); *Pleurocladia*; Kusel-Fetzmann and Schagerl (1992) and Kusel-Fetzmann (1996); and *Ectocarpus*, West and Kraft (1996)] offers promise in addressing questions of relatedness among genera and species, dominant reproductive or ploidy phases, morphological plasticity, and mechanisms of their reproduction and spread. For

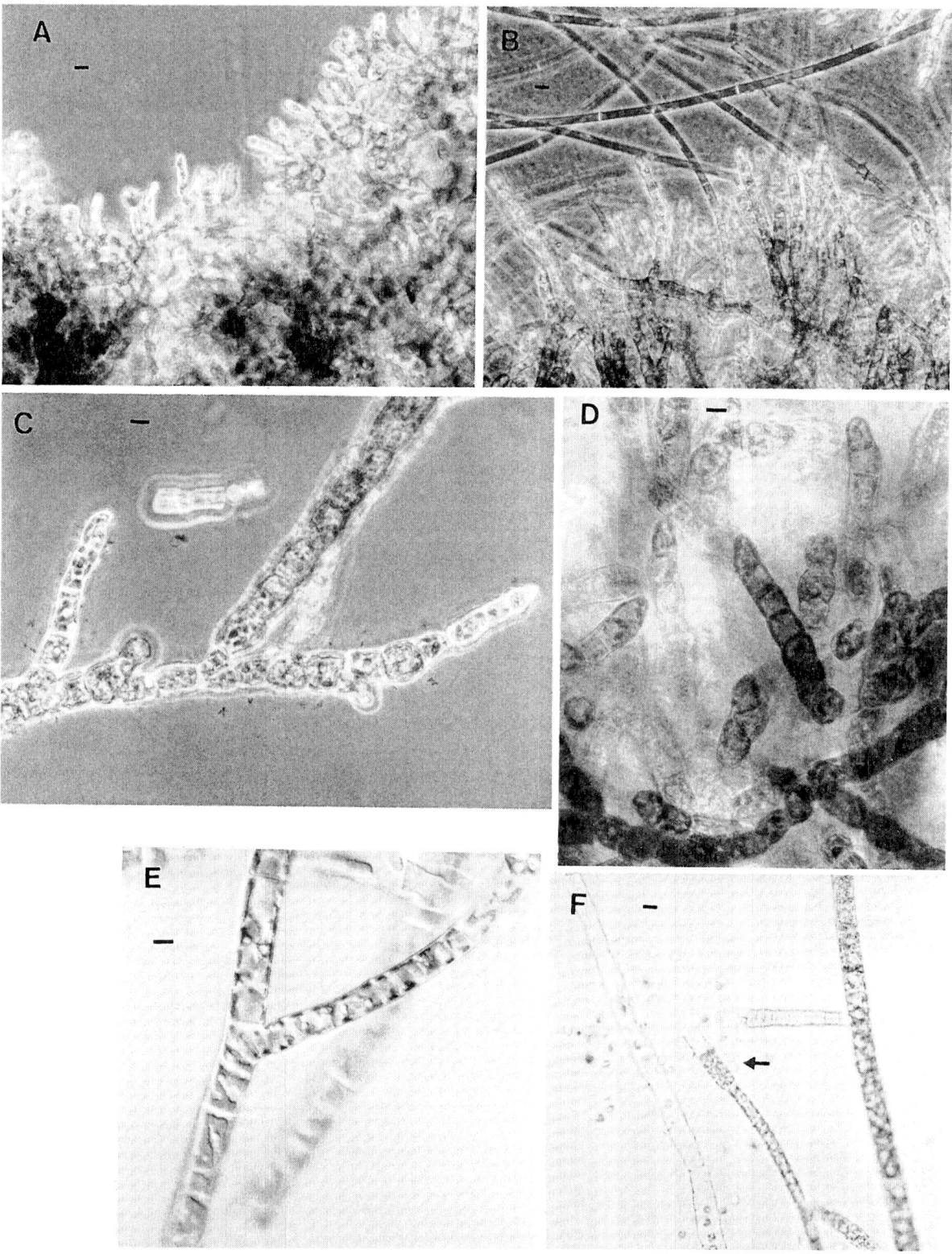

FIGURE 2 Freshwater phaeophytes, *Pleurocladia*, *Bodanella*, *Ectocarpus* (scale bars = 10 μm): (A) and (B) differences in hair formation by *Pleurocladia lacustris* in response to P-rich (A, +22 μM NaH$_2$PO$_4$) and P-limited (B, no P added) conditions in culture; hairs (B, upper) are terminal, hyaline, multicellular filaments 100–300 μm long; (C) and (D) *Bodanella lauterborni*, showing detail of cells and branching pattern (C) and general morphology (D); (E) and (F) *Ectocarpus siliculosus*, showing sparse branching pattern, ribbon-like chloroplasts (E, photo by Jason Sonneman, with permission), and plurilocular sporangium (arrow) (F, photo by Jason Sonneman, with permission).

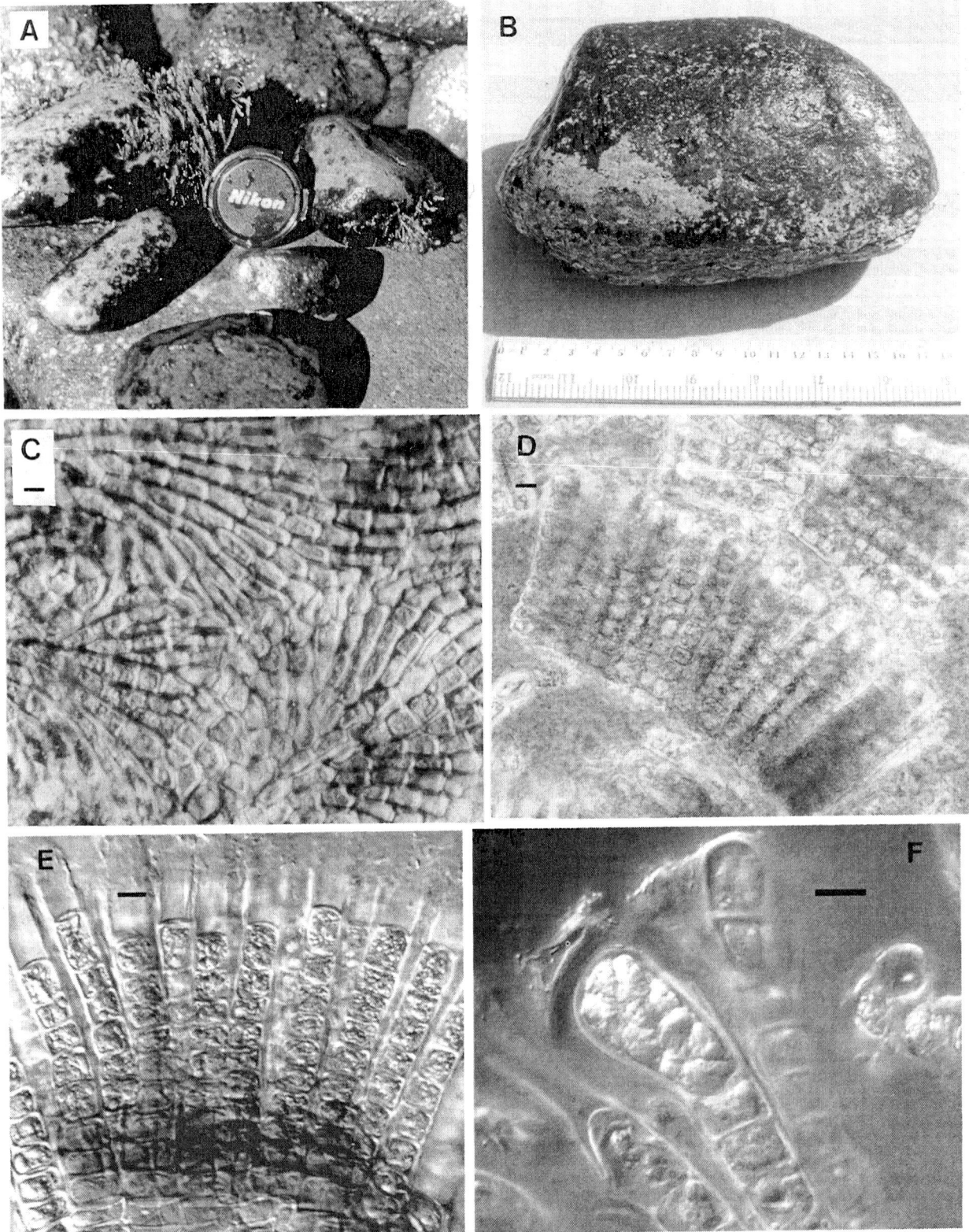

FIGURE 3 Freshwater phaeophyte, *Heribaudiella* (scale bars = 10 μm, except where indicated): (A) and (B) macroscopic appearance of individual colonies (A, McKenzie River, Oregon) and coalescing colonies (B, Bonaparte River, British Columbia) on rocks; (C) prostrate filaments; (D) columns of vertical filaments following removal from rocks and pressure applied to the coverslip; (E) detail of cells and chloroplasts in vertical system; (F) unilocular sporangium.

example, it is possible that in some freshwater species, portions of the typical phaeophyte life cycle may not occur. Further work on the reproductive ecology of freshwater brown algae based on field populations (as shown for the red alga *Batrachospermum*; Hambrook and Sheath, 1991), in conjunction with purified cultures, is clearly needed for members of this group.

III. ECOLOGY AND DISTRIBUTION

A. Ecological Factors

Due to the relatively sparse literature on freshwater brown algae, knowledge of the ecological factors governing their abundance and distribution is correspondingly limited. In some accounts, precise or even approximate locations of populations are difficult to discern (see methods, Sect. IV). However, despite these gaps in our knowledge, some generalizations can be made.

All freshwater species are benthic and most are epilithic in habit, particularly *Heribaudiella fluviatilis*, which almost exclusively colonizes stones in streams or lakes. With this encrusting species, there appears to be a preference for more resistant rocks, such as basalt, quartz, schist, and gneiss (Allorge and Manguin, 1941; Wehr and Stein, 1985), although Kusel-Fetzmann (1996) also reported this alga colonizing bricks in one Austrian stream. Other taxa, such as *Pleurocladia* and *Porterinema fluviatile*, are fairly nonspecific with regard to substratum. Some are epiphytes on larger algae (e.g., *Cladophora*, *Rhizoclonium*) and macrophytes (*Phragmites*, *Typha*) or may colonize artificial substrata (Israelsson, 1938; Waern, 1952; Kirkby *et al.*, 1972; Dop, 1979). *Pleurocladia lacustris* has been found attached to stones and boulders in the Green River (Utah–Colorado; Ekenstam *et al.*, 1996) and in several streams in southern Austria (Kusel-Fetzmann, 1996). It may also be attached to reeds in the littoral zone of lakes, such as in Lake Wigry, Poland (Szymanska and Zakrys, 1990), Brasside Ponds, United Kingdom (Kirkby *et al.*, 1972), and several Swedish lakes (Israelsson, 1938). *Pleurocladia* was also observed on glass slides that were placed in Lake Erken, Sweden (Kann, 1993). *Porterinema fluviatile* similarly colonizes reed stems, as well as submerged glass slides in eutrophic lakes (Dop and Vroman, 1976). With further study, the ecological breadth of this alga may prove to be quite broad, as earlier studies have reported it from brackish sites (0–8‰); and growing as an endophyte in *Enteromorpha* and *Cladophora* (Waern, 1952). As mentioned earlier, some of those with encrusting or entangled growth forms may also be complexed with $CaCO_3$, particularly *Sphacelaria* spp. and *P. lacustris* (Israelsson, 1938; Waern, 1952; Kirkby *et al.*, 1972; Schloesser and Blum, 1980). Although several species are found in both lakes and rivers, certain species, especially *Heribaudiella fluviatilis*, *Pleurocladia lacustris*, and *Sphacelaria fluviatilis*, appear to be best developed in flowing waters (Jao, 1943; Kann, 1978a; Wehr and Stein, 1985; Kusel-Fetzmann, 1996).

Among those species that occur in running waters, most reports mention that freshwater brown algae occur in rocky, clear-water streams and are largely absent from turbid or muddy habitats (Budde, 1927; Fritsch, 1929; Allorge and Manguin, 1941; Jao, 1941; Chadefaud, 1950; Holmes and Whitton, 1975; Kann, 1966, 1978a, b; Starmach, 1977; Wehr and Stein, 1985; Yoshizaki and Iura, 1991; Kusel-Fetzmann, 1996). The most complete (although still fragmentary) ecological data for any freshwater phaeophyte is for *Heribaudiella fluviatilis*. In summarizing sites throughout Europe at the time, Israelsson (1938) noted that the species colonized a wide range of streams and lakes spanning oligotrophic to eutrophic conditions. More recent data suggest that this species most often occurs in stony streams, with moderately alkaline water (most ≥ pH 7.0), but fairly broad Ca, P, and N concentrations (Table II). Kann (1966, 1978a, b) reported that *Heribaudiella* is typical of "calcium-poor, summer warm streams," although her data indicate that the species is rarely observed in either extremely softwater or hardwater systems, or at very low nutrient levels. Nearly all populations occupy clear-water habitats, although a few may tolerate moderate levels of humic materials (Kann, 1978a; Wehr and Stein, 1985), but not low pH, humus-rich waters (Israelsson, 1938). Although this alga is most commonly a lotic species, it has also been reported from rocky-shore habitats in some European lakes (Kann, 1945, 1993). Many studies from Europe and Japan (e.g., Budde, 1927; Fritsch, 1929; Geitler, 1932; Allorge and Manguin, 1941; Yoneda, 1949; Holmes and Whitton, 1975, 1977a; Kusel-Fetzmann, 1996) report that *Heribaudiella* often co-occurs with the encrusting red alga *Hildenbrandia* (particularly in shaded reaches), although this pattern is far from consistent. Holmes and Whitton (1977a, b, c) report both species in the rivers Tees, Swale, and Wear (United Kingdom), but each alga has also been recorded from sites where the other is apparently absent. Jao (1944) described *Heribaudiella* (as *Lithoderma zonatum*; see Sect. V.B.) as an indicator species for rapidly flowing, stony streams in China, along with the encrusting cyanobacterium *Schizothrix* (also *Homoeothrix*, *Lemanea*, *Bangia*, and several diatom species), but without *Hildenbrandia*. Kann (1978a, b) suggests that *Heribaudiella* and *Hildenbrandia* often occur separately because the latter tends to colonize more Ca-rich streams. Surveys of North American

TABLE II Summary of Selected Ecological Conditions of Streams and Rivers from Three Regions in Which *Heribaudiella fluviatilis* Has Been Collected [N/A = Data Not Available; Based on Kann (1978a), Holmes and Whitton (1977a, b, 1981), Wehr and Stein (1985), Kusel-Fetzmann (1996), and Wehr (unpublished data)]

Variable	Northern UK	Austria	North America
Light	Open to partial shade	Shaded	Open to partial shade
Substratum	Boulders, cobbles	Boulders, cobbles, pebbles, bricks	Boulders, cobbles
Geology	Basalt, sandstone	Quartz, basalt, marble	Basalt, quartz, granite
Width (m)	5–25	1–10	1–60
Depth (cm)	N/A	10–100	10–100
Current velocity (cm s^{-1})	Up to 1700	N/A	200–1600
Temperature (°C)	3–20	5–21	5–25
Conductance (μS cm^{-1})^a	150–600	110–640	50–450
pH	7.5–8.0	7.4–8.3	7.0–8.7
PO$_4$-P (μg P L^{-1})	20–1300	17–62	< 2.5–25
NO$_3$-N (μg N L^{-1})	200–1050	300–3600	2–200
NH$_4$-N (μg N L^{-1})	20–300	20–280	< 2–100
Ca (μg L^{-1})	20–50	30–45	10–70

^{a}Corrected to 25°C.

streams thus far have encountered *Heribaudiella* only without *Hildenbrandia* (Wehr and Stein, 1985; Wehr, unpublished), but with other macroalgal species (e.g., *Nostoc parmelioides, N. verrucosum, Cladophora glomerata*). Kann (1945, 1978a) has also pointed out that several encrusting forms (e.g., *Chamaesiphon, Gongrosira, Heribaudiella, Homoeothrix*) may co-occur in turbulent streams, simply because they are well adapted to these habitats. Taken together, these community-level data suggest that so-called associations of algae described in earlier publications (e.g., Geitler, 1932; Waern, 1938; Luther, 1954) may not be consistent in different parts of the world.

Some forms are equally common in lakes and rivers, and *Pleurocladia lacustris* is the most widely reported freshwater species from lentic systems (although it is also present in streams). This species colonizes a wide range of substrata, including stones, wood, many aquatic plant species, and artificial substrata, yet in freshwaters, it occurs in a fairly narrow range of chemical conditions. Early distribution data for Europe suggest that *Pleurocladia* prefers nutrient-rich, strongly calcareous waters (Israelsson, 1938). Subsequent studies support this suggestion; nearly all reports of freshwater population sites from Europe and North America mention eutrophic conditions and the presence of CaCO$_3$ precipitates associated with older colonies of *Pleurocladia* (Waern, 1952; Kirkby et al., 1972; Szymanska and Zakrys, 1990; Kann, 1993; Ekenstam et al., 1996; Kusel-Fetzmann, 1996; D. Ekenstam, pers. comm.). Limited chemical data report specific conductance levels > 600 μS cm^{-1} (Kirkby et al., 1972; Kusel-Fetzmann, 1996) and its absence in nearby sites with conductance $\approx$ 450 μS cm^{-1} (Kusel-Fetzmann, 1996). Algal species that often co-occur with *Pleurocladia*, including *Gloeotrichia pisum, Rivularia* spp., and *Chaetophora incrassata*, are also typical of nutrient-rich, hard waters (Kann, 1993). *Pleurocladia* has been observed in Hell Kettles, a series of very hard-water ponds in northern England (B. A. Whitton, pers. comm.), also colonized abundantly by *Chara hispida* and known for their especially rich marsh flora (Wheeler and Whitton, 1971). *Pleurocladia* may also occur in some sites with *Heribaudiella*, especially those with higher dissolved calcium (Kann, 1993; Kusel-Fetzmann, 1996; D. Ekenstam, pers. comm.). Occurrences of *P. lacustris* in some brackish or intermittently marine habitats (Waern, 1952), including arctic sites in North America (Wilce, 1966), suggest that the alga may also be a euryhaline species. However, Waern (1952) noted that, despite a large number of freshwater *Pleurocladia* populations in northern Europe, no nearby marine (or even brackish) populations have been identified (for further discussion of disjunct distributions, see Sect. III.B). Incomplete knowledge of the ecological factors affecting *Pleurocladia lacustris* is undoubtedly due to a paucity of collections. Kusel-Fetzmann (1996) pointed out that the alga was first discovered in Austria after observing "curious pale hairs" when examining crusts of *Heribaudiella*; it was an epiphyte on the other alga and produced small cushion-like colonies on rocks. Kirkby et al. (1972) comment that *Pleurocladia* was observed only on rotting (not recently dead) *Typha* leaves in the littoral zone of a eutrophic pond in England, whereas Waern (1952) observed that *Pleurocladia* may also grow as an endophyte in aquatic plants and macroalgae, which may lead to its under-reporting. Future studies require careful attention to the observations made by earlier researchers, as well as more complete collection of ecological data at these sites.

The light requirements of freshwater phaeophytes have not been studied in detail, although scattered reports suggest that *Heribaudiella* may more often colonize shaded reaches of streams (Kusel-Fetzmann, 1996). Surveys of other rivers in North America and the United Kingdom have encountered this alga in habitats spanning a broad range of light environments, from small shaded streams (< 1 m wide) to wide rivers (> 50 m) with little or no shade (Holmes and Whitton, 1977a, b, c; Wehr and Stein, 1985). In contrast, *Bodanella lauterborni* has been collected only from deep epilithic habitats (> 15 m) on limestone rocks in lakes (Geitler, 1928; Müller and Geller, 1978; Kann, 1982). Similarly, the only known populations of *Sphacelaria lacustris* have been collected from deep (5–15 m), poorly illuminated areas of the sublittoral region of western Lake Michigan, and not in shallower habitats (Schloesser and Blum, 1980). Culture studies with this alga found that optimum growth and reproduction (gemmae-like propagules and unilocular sporangia) was achieved under reduced (screened) light levels (reported as 1000 lux) and short-day (8L : 16D) conditions (Schloesser, 1977; Schloesser and Blum, 1980).

Very little is known of the competitive abilities or community importance of brown algae in freshwater habitats. Many populations of *Heribaudiella fluviatilis* colonize and may completely encrust certain rocks, with few other species present (Holmes and Whitton, 1975; Wehr, unpublished data). *Heribaudiella* may overgrow crusts of *Hildenbrandia rivularis* in streams where they co-occur (Fritsch, 1929; Geitler, 1932; Kusel-Fetzmann, 1996). In running waters, *Sphacelaria fluviatilis* apparently also grows in nearly monospecific stands (Jao, 1944). Thalli of *Heribaudiella* are usually free of epiphytes, but may occasionally serve as a host for diatoms, small cyanobacterial species (*Chamaesiphon incrustans*, *Homoeothrix varians*), chantransia stages of red algae, and *Pleurocladia lacustris* (Svedelius, 1930; Pueschel and Stein, 1983; Kusel-Fetzmann, 1996). The influence of herbivores on any freshwater species is unknown, although studies of marine species suggest that the large quantities of polyphenolics produced by many phaeophytes (> 2% of dry mass) may inhibit herbivore activity (Targett and Arnold, 1998). Field and lab herbivory experiments are clearly needed for freshwater species.

B. Geographical Distribution

Despite several hundred documented populations of freshwater phaeophytes recognized world wide, knowledge of their distribution and biogeography is fragmentary. The paucity of information on these algae makes each discovery of a new locality still worthy of publication, and renews speculation as to their origins (e.g., West, 1990; Kusel-Fetzmann, 1996; Wujek et al., 1996). Some may still be safely regarded as rare (or at least very poorly known), whereas others are world wide in their distribution. Many species distributions are regarded as disjunct, such as *Sphacelaria fluviatilis*, an epilithic species with only two known locations, a stream in south-central China and a small lake in Michigan (Jao, 1943; Thompson, 1975; Wujek et al., 1996). *Sphacelaria lacustris* is thus far known only from western Lake Michigan (Schloesser and Blum, 1980). *Bodanella lauterborni* is apparently known from only three locations, all in western Europe (Bourrelly, 1981).

Heribaudiella fluviatilis may be the most widespread of all freshwater phaeophytes, occurring in many locations in Europe, western North America, Japan, and China (Wehr and Stein, 1985; Kusel-Fetzmann, 1996), although there are currently no records from Africa, South America, Australia, or New Zealand, despite many phycological studies in these areas. More recent surveys of more than 250 river reaches for *H. fluviatilis* in North America have located 30 populations, all within western coniferous or boreal forests, and none have been located in any biomes east of the Mississippi River or south of Oregon (Wehr, unpublished). The discovery of *Heribaudiella* from a stream near Yellowknife, Northwest Territories (Sheath and Cole, 1992), extends its distribution more than 1000 km north (11°N latitude). No populations are known from Mexico or Central America, although given the diversity of freshwater habitats in these regions (see Chap. 2), it is reasonable to expect this alga in rocky streams from these regions as well. Comments more than 70 years ago by Budde (1927) and Fritsch (1929) that this species is easily missed are still true.

Because more than 99% of all known species within the Phaeophyceae occupy marine habitats, questions often focus on the possible dispersal and adaptation of taxa from marine to freshwater habitats. A review of the published ecological information for all freshwater phaeophytes suggests little evidence of (at least recent) marine invasions by most species. One reason is that at least four of the seven recognized species from fresh waters have no counterparts in marine environments. Two exceptions, *Ectocarpus siliculosus* and *Porterinema fluviatile*, seem to be true euryhaline species that colonize a broad range of salinities, including fresh waters (Dop, 1979; West and Kraft, 1996). *Pleurocladia lacustris* has been almost entirely reported from freshwater locations distant from the ocean, but at least one North American location is intermittently saline, and morphological data strongly indicate that these populations are the same species (Wilce, 1966). Current biogeographic data, however, show no patterns that suggest a marine invasion. Analyses of *Pleurocladia* distribution patterns in northern Europe (Israelsson, 1938; Waern, 1952) indicate a complete lack of marine populations in a region where freshwater populations are most common. With additional freshwater populations in Austria, France, and the Ukraine, data suggest that the freshwater history of this species may be quite old, perhaps pre-glacial (Waern, 1952). The discovery of *Pleurocladia lacustris* in the Green River (Utah, Colorado) more than 1000 km from any marine water (Ekenstam *et al.*, 1996), is in agreement with reports from European sites (Szymanska and Zakrys, 1990; Kusel-Fetzmann, 1996).

Similarly, the North American distribution of *Heribaudiella fluviatilis* in the Northwest Territories, British Columbia, Washington, Montana, Oregon, and Utah shows a near absence of the alga from river sites near the coast and an abundance of populations in interior and upland regions (Wehr and Stein, 1985; West, 1990; D. Ekenstam, pers. comm.). In addition, none of the inland populations of *Heribaudiella*, *Pleurocladia*, or *Sphacelaria* in North America or Europe are reported to be influenced by elevated salinity. Finally, there are no studies that have yet identified marine species of Phaeophyta from inland saline lakes, which are scattered across most continents. An exception may be found in *Ectocarpus siliculosus*, which was discovered in a waterfall of the Hopkins River (Australia), roughly 40m above sea level, with a specific conductance of 3.0 mS cm^{-1}; an isolate of this population has been shown to tolerate a wide range of salinities in culture (West and Kraft, 1996). As this is the first documented case of a population of *Ectocarpus* from any freshwater site, it is too early to speculate on the causes for its distribution. An intriguing but obscure species, *Porterinema fluviatile*, has been sampled from many brackish and freshwater sites in Europe (Waern, 1952), whereas Wilce *et al.* (1970) described a North American population from a freshwater site adjacent to a salt marsh in Massachusetts. It has since been sampled from freshwater sites in Netherlands and isolated into culture, using Wood's Hole freshwater medium (Dop, 1979). The ecological and distributional history of this species complex clearly requires further attention. Molecular analyses (e.g., 18S rRNA and rbcL genes) in conjunction with biogeographic studies of most freshwater phaeophytes are sorely needed to elucidate distribution patterns and genetic relationships among these apparently disjunct populations.

IV. METHODS FOR COLLECTION AND IDENTIFICATION

Because all known freshwater brown algae are benthic, methods used for sampling or removing various substrata are needed (e.g., Weitzel, 1979; Stevenson, 1996). Also, most species are macroalgal, that is, colonies or filaments which are recognizable (if not easily identified) with the naked eye. Nonetheless, freshwater phaeophytes are still cryptic and difficult to find. A survey of >1000 stream segments in North America (Sheath and Cole, 1992) located only one additional population of *Heribaudiella fluviatilis*, from a stream in the Northwest Territories. In most investigations, new populations are discovered by researchers who have encountered the species nearby or elsewhere (Wehr and Stein, 1985; Kusel-Fetzmann, 1996). Because colonies or thalli of benthic algae may be inconspicuous in the field, Sheath (Chap. 5) recommends the use of a plastic view box, which provides a clear view through calm water and allows the investigator to distinguish different growth forms, pigmentation, and microhabitats much more easily. The present author has found this device reduces search time in both streams and shallow littoral areas of lakes. If quantitative samples are needed, transects (along tape measures sampled at regular intervals) or quadrats may be required. With macroalgae some authors may use visual estimates of cover for visually distinct physiognomies (e.g., Holmes and Whitton, 1977a, c; Wehr and Stein, 1985; Sheath and Cole, 1992) and later assign species names based on microscopic examination. This chapter thus provides descriptions of genera (Sect. V.B.) based on macroscopic and microscopic appearances, which may be helpful in field sampling.

Our understanding of the distribution and ecology of freshwater brown algae requires many more thorough surveys. With the development of inexpensive global positioning systems (GPS), relatively precise geographic information (latitude, longitude, altitude) should be relatively easy to collect in all future studies. Whenever possible, ecological data, particularly type of substratum and size, current velocity, irradiance, temperature, conductance, turbidity, and water chemistry (especially pH, N, P, Ca), should be measured in all collections and surveys of freshwater phaeophytes. In addition, simple relative scales of abundance or cover estimates (Wehr and Stein, 1985; Sheath and Cole, 1992) will greatly aid future syntheses of the ecology of this group of algae. In rivers, surveys have also been conducted by wading specific lengths (10-m to 0.5-km reaches) and recording presence or absence, relative abundance, or percentage of cover estimates of each macroalga from that area (Holmes and Whitton, 1977a, b). A field microscope (e.g., Swift Instruments, San Jose, CA) is helpful in distinguishing colonies or crusts that appear similar in the field (see also Holmes and Whitton, 1975).

Methods for removal of material from substrata depend on the growth form and type of substratum. Encrusting or firmly attached epilithic species (mainly *Heribaudiella*, *Sphacelaria*) are best sampled by removing entire rocks when possible, then scraping material (using a razor blade) into vials for later identification, whereas smaller stones may be transported intact. It is important to note the macroscopic appearance (shape, margin, size, thickness) and color of the colonies in the field during sampling, as material may change even during brief storage times. Most epiphytic forms (e.g., *Pleurocladia*) in fresh water are usually firmly attached to plant hosts, permitting removal of plants or plant fragments with dip nets or collected using SCUBA. However, more delicate or gelatinous forms may be best sampled using forceps or à modification of the half-bottle sampler (Douglas, 1958), which can isolate water plus algal material and permit removal of thalli without loss. Kann (1976, 1978a) has also detected some species of brown algae (*Pleurocladia*) from lakes by observing settled zoospores on artificial (glass, plastic) substrates, although their selectivity for or against freshwater phaeophytes is unknown.

Material is best examined live and soon after collection for recognition of pigmentation, chloroplast form, and presence of hairs. Algal samples can be kept alive for several days if stored cool (5–10°C) and wet or moist. Some authors (Waern, 1952; Kusel-Fetzmann, 1996; West and Kraft, 1996) have reported that filamentous forms retain their normal growth form in sample water for several weeks or months, and some species (as discussed earlier) can be brought into culture using standard media (Müller and Geller, 1978; Schloesser and Blum, 1980; Kusel-Fetzmann and Schagerl, 1992). Identification of some species frequently requires reproductive structures (Sect. V). Their position, number, and arrangement are diagnostic, but there is some doubt whether shape of sporangia may be used (Waern, 1952; West and Kraft, 1996). Motile stages are readily produced with field material or in culture if maintained under moderate (10–20°C) temperatures (Kumano and Hirose, 1959; Müller and Geller, 1978; Wehr, unpublished). Long photoperiods (16:8) apparently favor formation of plurilocular sporangia (and zooids) in *Porterinema* (Dop, 1979), whereas short-day conditions (8:16) may induce the production of unilocular sporangia in *Sphacelaria lacustris* (Schloesser and Blum, 1980).

If samples are to be stored for long periods, excised specimens are best preserved using 2–3% glutaraldehyde or 2% paraformaldehyde, and stored cool and in the dark. Alternatively, encrusting species may be "preserved" on rocks (and suitable for herbaria) by simply air-drying the entire rock, although cells and plastids tend to become distorted upon drying. Isolates of *Sphacelaria lacustris* from Lake Michigan that were grown on agar plates, later were air dried to produce thin, dry specimens that were stored on herbarium sheets (Scholesser and Blum, 1980). Inspection of this material (holotype in U.S. National Herbarium, Algal Collection) more than 20 years later found that most of the morphological and cellular features were retained (Wehr, unpublished).

When preparing thalli for light microscopy, specimens of crustose species may require vigorous chopping (with razor blade) and/or crushing (with cover slip) to separate densely packed filaments. The morphology of filamentous forms may also be more clearly revealed in squash preparations. Species that have become calcified with $CaCO_3$ (especially *Pleurocladia*, *Porterinema*) may require treatment with dilute acid (1–2% HCl) before observation. Stains may be used to emphasize important structures, particularly cresyl blue (Chadefaud, 1950; Schloesser and Blum, 1980), which may be used to demonstrate physodes (which store phaeophycean tannins). Vanillin–HCl may also be used, which stains physodes red (Chadefaud, 1950: Lee, 1989).

V. KEY AND DESCRIPTIONS OF GENERA

A. Key

Taxa not reported from North America in fresh waters are marked with an asterisk.

1a.	Thalli small cushion-like tufts or expanses of spreading filaments (Fig. 1A, C)..2
1b.	Thalli not cushion-like (Figs. 1B, 3C); creeping or crustose...3

2a.	Filaments uniseriate and multiseriate; with numerous disc-shaped chloroplasts (Figs. 1C, 4E, F)	Sphacelaria
2b.	Filaments uniseriate only; single (rarely two) chloroplasts; basal filaments curved or arching; erect system spreading (comblike) (Fig. 4A, B)	Pleurocladia
3a.	Thalli crustose, forming dark brown patches on stones; branched basal filaments with short, densely packed upright filaments (Figs. 1B, 3)	Heribaudiella
3b.	Thalli not crustose, spreading or simple, variously branched	4
4a.	Filaments sparingly branched, vegetative cells narrow, cylindrical, chloroplasts several, ribbon-like; plurilocular sporangia (if present) narrow-elongate (Fig. 2E)	Ectocarpus*
4b.	Filaments frequently or irregularly branched, cells inflated or quadrate, chloroplasts few to several, plurilocular sporangia unknown or (if present) broad or inflated in shape	5
5a.	Branched filaments with prostrate and erect forms	6
5b.	Branched filaments prostrate only, creeping along substrata; may have rhizoid-like branches, chloroplasts many Fig. 2C, D)	Bodanella*
6a.	Basal filaments often curved or arching, erect filaments spreading; cells with single (rarely two) chloroplast, unilocular sporangia ovoid	Pleurocladia
6b.	Basal and erect filaments irregularly arranged, with two parietal, lobed chloroplasts; plurilocular sporangia in crown-shaped clusters of four or more (Fig. 4C, D)	Porterinema

B. Descriptions of Genera

Ectocarpales

Bodanella Zimmermann (Fig. 2C, D)

Thalli basal or creeping filaments form on rocky substrata, without erect filaments. Filaments are uniseriate, frequently but irregularly branched, composed of irregularly shaped cells; inflated, quadrate, angular, ovoid, or "wavy" in shape; vegetative cells are 10–16 µm wide, 10–25 µm long. General form may be confused with *Sphacelaria lacustris*, but the only latter genus possesses multiseriate axes; filaments of *Bodanella* are uniseriate. Terminal, short, narrow hairs (6–10 µm diameter), or basal rhizoid-like filaments may also be present. Parietal chloroplasts are small, numerous (10–15 per cell), and discoid. Unilocular sporangia are ovoid or globose; 15–20 µm wide by 25–30 µm long. Zoospores are pyriform (10–12 µm by 5–6 µm), with laterally inserted flagella. Plurilocular sporangia are unknown.

A monotypic genus, *Bodanella lauterborni* was named for its original location, Bodensee (Lake Constance, Austria–Germany), where it colonizes limestone deep (15–35 m) in lakes (with *Hildenbrandia* and *Cladophora*). It is not known from North America; worldwide distribution consists of three European lacustrine populations, in Lake Constance (Zimmermann, 1928; Müller and Geller, 1978), Lunzer Untersee (Austria; Geitler, 1928), and Traunsee (Austria; Kann, 1982).

Ectocarpus Lyngbye (Fig. 2E, F)

The freshwater form is sparingly or irregularly branched; thalli are mostly erect filaments consisting of cylindrical (isodiametric) cells 15–40 µm diameter, up to four times as long as broad. Narrow, hair-like filaments also are present (8–12 µm diameter). Chloroplasts are several (2 to 4?), ribbon-like, lobed, and parietal, with pyrenoids, arrangement variable (netlike, spiral, etc.). Plurilocular sporangia are terminal, narrowly ellipsoid, conical, or linear, with numerous divisions; length is 70–200 (500) µm by 15–35 µm diameter. Unilocular sporangia are unknown from freshwater populations, but well established in marine population (Lee, 1989).

Ectocarpus is an ecologically and geographically widespread marine and estuarine genus (Müller, 1979), recently discovered (*E. siliculosus* Dillw.) in a freshwater waterfall of the Hopkins River (Australia) with other freshwater taxa (*Mougeotia*, *Cladophora*; West and Kraft, 1996). This is the only record of any freshwater phaeophyte in the southern hemisphere, according to Entwisle et al. (1997). The alga is unknown from freshwater sites in North America, but is common in coastal brackish sites on this continent. It was also observed (*E. confervoides*) with freshwater taxa (e.g. *Anabaena*, *Cyclotella*, *Scenedesmus*) in the River Werra (Germany) in sites polluted by potassium mine waste (Geißler 1983).

Heribaudiella Gomont (Fig. 3A–F)

Thalli are olive-brown to dark brown crusts on rocks in streams and lakes; colonies are 1–5 cm (up to ≈ 20 cm) diameter with rounded or irregular outline (Fig. 3A), but with distinct margins (*Chamaesiphon* spp. also forms brown crusts, but margins are indis-

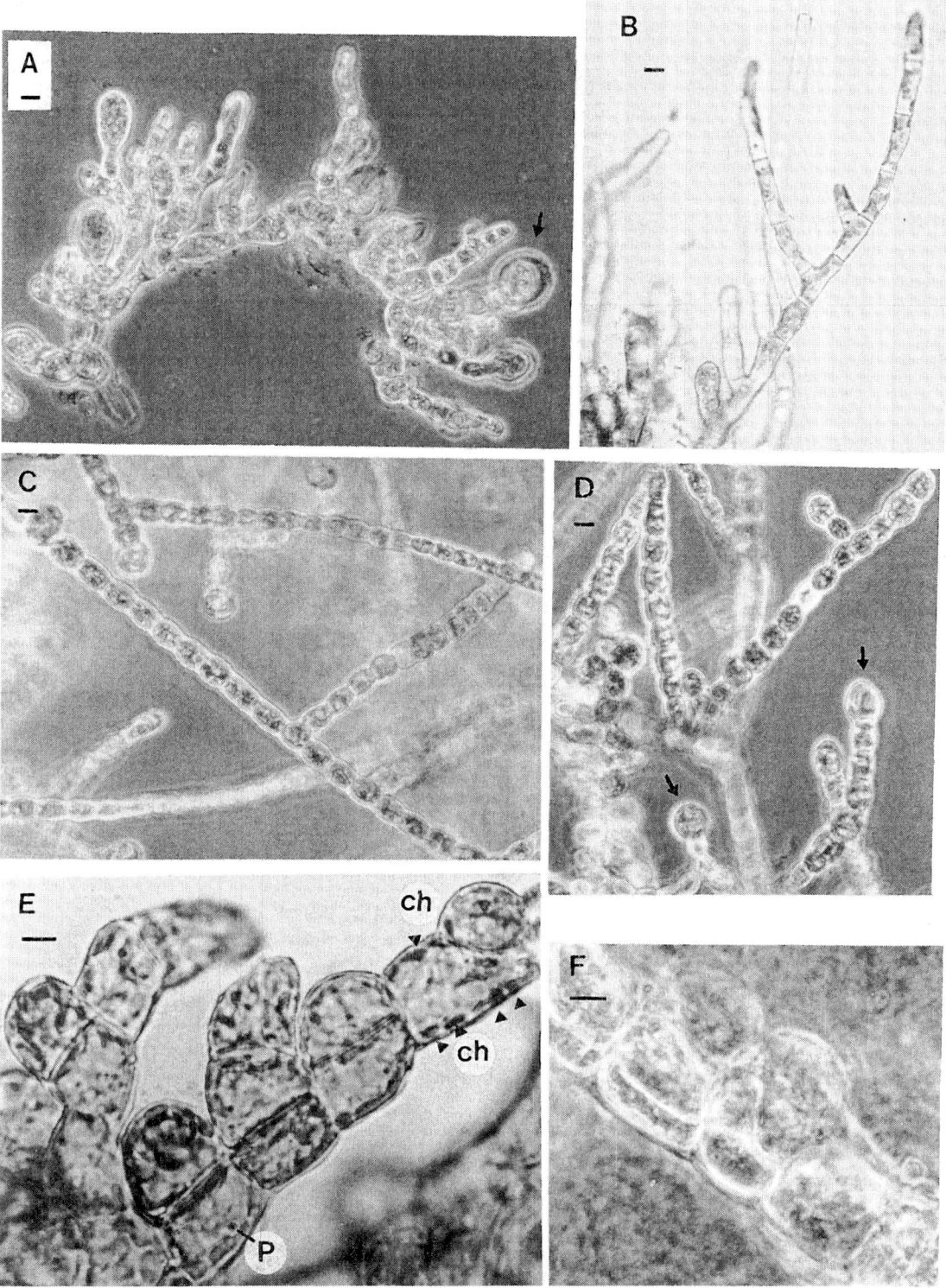

FIGURE 4 Freshwater phaeophytes, *Pleurocladia*, *Porterinema*, *Sphacelaria* (scale bars = 10 μm, except where indicated): (A) and (B) *Pleurocladia lacustris*, detail of filament and chloroplasts, showing centrifugal or arched growth patterns, with unilocular sporangium (arrow) (A); and erect, branched filaments, showing one or two parietal chloroplasts per cell (B, photo by E. L. Kusel-Fetzmann, with permission); (C) and (D) *Porterinema fluviatile*, showing irregularly branched filaments with short erect filaments (C) and clusters of filaments with apical, unilocular sporangia (arrows); (E) and (F) *Sphacelaria lacustris* (holotype, from Lake Michigan) [(E) filament showing complex, multiply branched, multiseriate growth form, (ch) many small chloroplasts and (P) physodes (photo reproduced with permission of the Journal of Phycology from Schloesser and Blum (1980) 16:201–207, Fig. 5); (F) detail of multiseriate primary axis; note cell divisions in two planes (from herbarium specimen)].

tinct smudges or flecks; Holmes and Whitton, 1975; Wehr and Stein, 1985). Multiple colonies may coalesce to cover entire rocks or boulders (Fig. 3B). When scraped off carefully, colonies appear as a series of vertical columns at low magnification (Fig. 3D). Filaments in basal system are repeatedly branched (Fig. 3C); erect filaments are sparingly and dichotomously branched. *Heribaudiella* forms an erect system of appressed vertical filaments that do not easily separate under pressure; cells are mostly quadrate, 8–15 μm diameter, 5–15 cells long. Chloroplasts are oval or discoid, numerous (4–10 per cell; Fig. 3E); physodes are present. Multicellular hyaline hairs may be present (up to 1 mm long). Unilocular sporangia are terminal, inflated, ovoid or clavate (Fig. 3F); 10–25 μm wide, 15–35 μm long [development described by Svedelius (1930), Fig. 4]. Biflagellate zoospores are pyriform or irregular shape (≈ 6–8 μm). Plurilocular sporangia are obscure (reported as rare), produced terminally in narrow-celled columns four (rarely eight) cells tall; immature plurilocular sporangia are difficult to distinguish from smaller vegetative filaments. Svedelius (1930) united several taxa (e.g., *Lithoderma fluviatile*, *L. fontanum*) under this name. *Lithoderma zonatum* was described from fresh waters by Jao (1941), based on frequent plurilocular sporangia (reported occasionally in *H. fluviatilis*) and several layers or zones of erect filaments; this taxon was still reported by Bourrelly (1981). Jao (1941) was apparently unaware of Svedelius' study, because this layered morphology is exactly shown in his earlier study (Svedelius, 1930, Fig. 13) and was suggested to perhaps represent annual layers. For these reasons, the Chinese species (*L. zonatum* Jao) should be regarded as synonymous with *Heribaudiella fluviatilis* (Aresch.) Sved. No freshwater taxa are now assigned to the genus *Lithoderma*.

A monotypic genus, *Heribaudiella fluviatilis* is the most widely observed freshwater phaeophyte worldwide, reported from at least 30 locations in western North America, but no extant populations are known east of the Mississippi or south of Oregon. A very early record (Collins *et al.*, 1898; as *Lithoderma fluviatile*) from Island Brook, Connecticut, was included in the *Phycotheca Boreali-Americana*, although its identity has been questioned due to its possible marine habitat (Smith, 1950) and appearance of some samples (Pueschel and Stein, 1983). Further examination by the present author of the 1888 material deposited in the New York Botanical Garden, University of Michigan, and U.S. National (Smithsonian) herbaria suggests this population may in fact be *Heribaudiella fluviatilis*, based on morphology (vertical series of erect filaments) and co-occurring diatoms (>99% were freshwater species; Wehr, unpublished data). However, surveys of Island Brook in 1998 and adjacent streams failed to find an extant population in the area (Wehr, unpublished). In North American streams it often co-occurs with *Audouinella hermannii*, *Chamaesiphon* spp., *Nostoc parmelioides*, *N. verrucosum*, and *Cladophora glomerata*; in Europe, *Hildenbrandia rivularis* may colonize the same rock.

Pleurocladia A. Braun (Figs. 2A, B, 4A, B)

Thalli are small, brown to pale brown or tan (depending on calcification) hemispherical tufts or cushions (up to 3 mm diameter by 100–300 μm tall) on rocks or attached to plants (angiosperms, mosses, macroalgae), or endophytic, present in both streams and lakes. Waern (1952) reports that colonies appear macroscopically like a brown-colored *Gloeotrichia*. In less calcareous regions, colonies may be gelatinous. At low magnification (100×), colonies appear as a dense network of radiating, branched filaments (Fig. 2B). Two filament systems are evident: (1) creeping or basal system, with (infrequently) branched filaments usually consisting of rounded or inflated cells 8–16 μm diameter (occasionally elongate), often exhibiting centrifugal or arched growth patterns (Fig. 4A), which give rise to (2) upright, irregularly (alternate or opposite) branched long filaments (Fig. 4B), which are usually narrower (6–12 μm), elongate (cells 12–35 μm long), and more nearly isodiametric. Vegetative cells contain one (rarely two) large golden-brown parietal chloroplast (with pyrenoids); darker granules (physodes?) and refractive (lipid?) bodies may be common. Unilocular sporangia are common, single, clavate or globose [15–30 μm in diameter by 25–60 (–80) μm long], and borne laterally or terminally (Fig. 1A). Plurilocular sporangia are uncommon in freshwater populations (but see Waern, 1952), linear-elongate, narrow. Long (100–300 μm) multicellular hairs (5–7 μm diameter) are common in field populations (environmentally induced; Fig. 2A, B), arising from upright filaments, giving the colony a fuzzy appearance when viewed macroscopically.

Pleurocladia lacustris is currently known from a few freshwater sites in North America [Wyoming, Colorado (Green River), and Devon Island, Northwest Territories]; suitable freshwater habitats undoubtedly occur in other locations on this continent. Many more populations are known from Europe. The species has also been recorded from marine and brackish habitats. The relationship between *P. lacustris* and related species and genera are discussed by Waern (1952), Wilce (1966), and Bourrelly (1981).

Porterinema Waern (Fig. 4C, D)

Thalli are monostromatic, brown disc-shaped plates of loosely arranged filaments. The genus occurs as an epiphyte on or endophyte in other algae (e.g., *Rhizoclonium, Enteromorpha*) or macrophytes (e.g., *Elodea*); it may also colonize stones and artificial substrata (e.g., glass slides). Thalli are creeping, composed of irregularly branched filaments, with short (a few cells) erect filaments produced infrequently (Fig. 4C). Basal cells are barrel-shaped, or occasionally enlarged on proximal ends (6–12 μm diameter by 6–12 μm long); erect cells are few but more elongate (up to 40–50 μm long). Vegetative cells occur with one to three lobed, golden-brown, parietal chloroplasts. Terminal, multicellular hairs (3–8 μm diameter; up to 200 μm long) are common and may be sheathed at their base. Unilocular sporangia are rarely reported, on basal or erect filaments, pear- or club-shaped (15–30 μm wide; up to 80 mm long). Plurilocular sporangia (6–8 μm diameter) are common, intercalary (occasionally terminal), typically four-celled clusters (or "crowns") on pedicels (short filaments) or sessile; sometimes they are produced in clusters of up to 32 sporangia.

One species, *Porterinema fluviatile*, is distributed mainly among brackish sites in Europe and North America, but several truly freshwater sites are known in Europe (Waern, 1952; Dop, 1979), and one site in North America: a stream draining into a salt marsh near Ipswich, Massachusetts (Wilce *et al.*, 1970). As such it should be regarded as part of the North American freshwater algal flora, but requires further study. The report of a new genus, *Pseudobodanella peterfii* in Europe (Gerloff, 1967), appears to be identical to *Porterinema fluviatile*, lacking only hairs, and is very likely synonymous (Bourrelly, 1981; D. M. Müller, pers. comm). Bourrelly (1981) and Dop (1979) suggest that past records of *Apistonema pyrenigerum* (previously classified in Chrysophyceae), such as from a small pond in the United Kingdom (Belcher, 1959), are also *Porterinema*. Other synonymous or related taxa (e.g., *Apistonema expansum, Porterinema marina*) have been similarly considered by Wilce *et al.* (1970) and Dop (1979).

Sphacelariales

Sphacelaria Lyngbye (Fig. 4E, F)

Freshwater thalli are small (1–2 mm) brown tufts or cushions on rocks in streams or lakes; they may be calcified. Vegetative growth is the result of basal (creeping) and erect filaments; rhizoidal cells form where basal filaments contact substrata. The genus is distinguished by axes variably multiaxial (biseriate or multiseriate) and uniaxial (uniseriate); branches are hemiblastic (primordial cells arising from upper position), resulting in apical growth pattern. Branching pattern is irregular (*S. lacustris*) or opposite (*S. fluviatilis*); it may become pseudoparenchymatous. Cells comprising the main axis are rectangular or inflated (12–25 μm diameter); they are broader prior to lateral cell division, cylindrical on erect filaments. Cells contain numerous (10–20), small (3–8 μm), peripheral, disc-shaped chloroplasts and physodes (especially meristematic regions); pyrenoids are lacking. Multicellular hairs (> 500 μm length) develop from basal and erect filaments in some plants. Unilocular sporangia are known in one species (*S. lacustris*; see Schloesser and Blum, 1980, Fig. 3). Plurilocular sporangia are unknown in fresh waters for either species. Clusters of vegetative, gemmae-like propagules common; sessile or borne on short (1- or 2-celled) branches.

Two freshwater species are known: *S. lacustris*, reported on rocks in western Lake Michigan at depths of 5–15 m (Schloesser and Blum, 1980); *S. fluviatilis*, reported in rapidly flowing water in the Kialing River, China (Jao, 1943, 1944), and in the shallow (≤1 m) littoral zone of Gull Lake, Michigan (Thompson, 1975; Timpano, 1978, 1980; Wujek *et al.*, 1996). The two species were separated largely on the basis of branching pattern (alternate or irregular in *S. lacustris*, opposite in *S. fluviatilis*), lateral cell divisions (infrequent and irregular in *S. lacustris*, two to four regular divisions in main axes of *S. fluviatilis*), and hairs (lacking in *S. fluviatilis*?) (Schloesser and Blum, 1980). Further studies are needed on the ecological requirements, geographic distribution, and genetic differences of the two species. No populations are yet known from Europe, Central America, or South America.

VI. GUIDE TO LITERATURE FOR SPECIES IDENTIFICATION

All but one freshwater phaeophyte genus (*Sphacelaria*) are monotypic; however, the primary literature is still helpful for identification. A few general keys are useful for most freshwater species (Starmach, 1977; Bourrelly, 1981), but the reader should be aware that recent combinations (e.g., *Pseudobodanella* = *Porterinema*) are not included. Many of the most complete descriptions may be older literature, come from other continents, or are in German or Japanese.

1. *Bodanella*—Zimmermann (1928), Müller and Geller (1978)
2. *Ectocarpus*—West and Kraft (1996)
3. *Heribaudiella*—Holmes and Whitton (1975), Yoshizaki *et al.* (1984), Kusel-Fetzmann (1996)
4. *Pleurocladia*—Wilce (1966), Kirkby *et al.* (1972), Kusel-Fetzmann (1996)
5. *Porterinema*—Waern (1952), Dop (1979)
6. *Sphacelaria*—Jao (1941), Schloesser and Blum (1980), Wujek *et al.* (1996)

ACKNOWLEDGMENTS

I express my thanks to Dean Blinn, Janet Stein, and Brian Whitton, who helped instill in me an enthusiasm for stream algae. Advice, samples, records, and/or photographs from Robert Wilce, Dieter Müller, John West, Elsa Kusel-Fetzmann, Devon Ekenstam, Robert Sims, and Willem Prud'homme van Reine were especially helpful in the preparation of this manuscript. Loans of herbarium specimens from the National Herbarium (Smithsonian), University of Michigan, New York Botanical Garden, and Trinity College, Connecticut, were extremely useful. Thanks also to many people who have helped in my searches for freshwater phaeophytes, Deb Donaldson, Syd and Dick Cannings, Bob Sheath, Greg Flannery, and Alissa Perrone.

LITERATURE CITED

Allorge, P., Manguin, E. 1941. Algues d'eau douce des Pyrénées basques. Bulletin de la Société Botanique de France 88:159–191.
Belcher, J. H. 1959. Some uncommon Chlorophyceae from the Lee Valley. British Phycological Bulletin 1:73–74.
Bold, H. C., Wyne, M. J. 1985. Introduction to the algae. 2nd ed. Prentice-Hall, Englewood Cliffs, NJ, 720 pp.
Bourrelly, P. 1981. Les algues d'eau douce, Vol. II. Les algues jaunes et braunes, Chrysophycées, Phaeophycées, Xanthoophycées, et Diatomées, rev. ed. Soc. Nouvelle des Éditions Boubée, Paris.
Braun, A. 1855. Decade XLV + XLVI, in: Rabenhorst, L., Ed. (1848–1860) Die Algen Sachsens, Respective Mittel-Europa's. Dresden, Germany.
Budde, H. 1927. Die Rot- und Braunalgen des Westfälischen Sauerlandes. Berichte der Deutschen Botanischen Gesellschaft 45:143–150.
Chadefaud, M. 1950. Observations cytologiques sur la Phéophycée d'eau douce: Heribaudiella fluviatilis (Aresch.) Sved. Bulletin Société Botanique de France 97:198–199.
Christensen, T. 1978. Annotations to a textbook of phycology. Botanisk Tidsskrift 73:65–70.
Collins, F. S., Holden, I., Setchell, W. A. 1898. Phycotheca boreali-Americana, Vol. XI. Malden, MA, 536 p.
Craigie, J. S. 1974. Storage products, in: Stewart, W. D. P., Ed., Algal physiology and biochemistry. Univ. California Press, Berkeley, pp. 206–235.
Dop, A. J. 1979. Porterinema fluviatile (Porter) Waern (Phaeophyceae) in the Netherlands. Acta Botanica Neerlandica 28:449–458.
Dop, A. J., Vroman, M. 1976. Observations on some interesting freshwater algae from the Netherlands. Acta Botanica Neerlandica 25:321–328
Douglas, B. 1958. The ecology of the attached diatoms and other algae in a small stony stream. Journal of Ecology 45:295–322.
Ekenstam, D., Bozniak, E. G., Sommerfeld, M. R. 1996. Freshwater Pleurocladia (Phaeophyta) in North America. Journal of Phycology (Supplement) 32:15.
Entwisle, T. J., Sonneman, J., Lewis, S. H. 1997. Freshwater algae in Australia. A guide to conspicuous genera. Sainty Associates, Potts Point, NSW, Australia.
Fritsch, F. E. 1929. The encrusting algal communities of certain fast-flowing streams. New Phytologist 28:165–196.
Geibler, U. 1983. Die salzbelastete Flußstrecke der Werra—ein Binnenlandstandort für Ectocarpus confervoides (Roth) Kjellman. Nova Hedwigia 37:193–217.
Geitler, L. 1928. Über die Tiefenflora an Felsen im Lunzer Untersee. Archiv für Protistenkunde 62:96–104.
Geitler, L. 1932. Notizen über Hildenbrandia rivularis und Heribaudiella fluviatilis. Archiv für Protistenkunde 76:581–588.
Gerloff, J. 1967. Eine neue Phaeophyceae aus dem Süsswasser: Pseudobodanella peterfii nov. gen. et nov. spec. Revue Roumaine de Biologie—Botanique 12:27–35.
Gibson, M. T., Whitton, B. A. 1987. Hairs, phosphatase activity and environmental chemistry in Stigeoclonium, Chaetophora and Draparnaldia (Chaetophorales). British Phycological Journal 22:11–22.
Gomont, M. 1896. Contribution à la flore algologique de la Haut-Auvergne. Bulletin de la Société Botanique de France 43:373–393.
Goodwin 1974. Carotenoids and biliproteins, in: Stewart, W. D. P., Ed., Algal physiology and biochemistry. Univ. California Press, Berkeley, pp. 176–205.
Graham, L. E., Wilcox, L. W. 2000. Algae. Prentice-Hall, Upper Saddle River, NJ.
Hambrook, J. A., Sheath, R. G. 1991. Reproductive ecology of the freshwater red alga Batrachospermum boryanum Sirodot in a temperate headwater stream. Hydrobiologia 218:233–246.
Hamel, G. 1931–1939. Phaéophycées de France. Trait de Botanique 47. Paris, 432 p. + 10 pl.
Holmes, N. T. H., Whitton, B. A. 1975. Notes on some macroscopic algae new or seldom recorded for Britain: Nostoc parmelioides, Heribaudiella fluviatilis, Cladophora aegagropila, Monostroma bullosum, Rhodoplax schinzii. Vasculum 60:47–55.
Holmes, N. T. H., Whitton, B. A. 1977a. The macrophytic vegetation of the River Tees in 1975: Observed and predicted changes. Freshwater Biology 7:43–60.
Holmes, N. T. H., Whitton, B. A. 1977b. The macrophytic vegetation of the River Swale, Yorkshire. Freshwater Biology 7:545–558.
Holmes, N. T. H., Whitton, B. A. 1977c. Macrophytes of the River Wear: 1966–1976. Naturalist (Hull) 102:53–73.
Holmes, N. T. H., Whitton, B. A. 1981. Phytobenthos of the River Tees and its tributaries. Freshwater Biology 11:139–163.
Israelsson, G, 1938. Über die Süsswasserphaeophycéen Schwedens. Botanische Notiser 1938:113–128.
Jao, C.-C. 1941. Studies on the freshwater algae of China. VII. Lithoderma zonatum, a new freshwater member of the Phaeophyceae. Sinensia 12:239–44.
Jao, C.-C. 1943. Studies on the freshwater algae of China. XI. Sphacelaria fluviatilis, a new freshwater brown alga. Sinensia 14:151–154.
Jao, C.-C. 1944. Studies on the freshwater algae of China. XII. The attached algal communities of the Kialing River. Sinensia 15:61–73.
Kann, E. 1945. Zur Ökologie der Litoralalgen in ostholsteinischen Waldseen. Archiv für Hydrobiologie 41:14–42.
Kann, E. 1966. Der Algenaufwuchs in einigen Bächen Österreichs. Verhandlungen—Internationale Vereinigung für Theoretische und Angewandte Limnologie 16:646–654.
Kann, E. 1976. Algenaufwuchs unter natürlichen Bedingungen auf Kunststoffen. Chemie Kunststoffe Aktuell 2:63–71.
Kann, E. 1978a. Systematik und Ökologie der Algen österreichischer Bergbäche. Archiv für Hydrobiologie (Supplement) 53:405–643.
Kann, E. 1978b. Typification of Austrian streams concerning algae. Verhandlungen—Internationale Vereinigung für Theoretische und Angewandte Limnologie 20:1523–1526.
Kann, E. 1982. Qualitative Veränderungen der litoralen Algenbiocönose österreichischer Seen (Lunzer Untersee, Traunsee, Attersee) im Laufe der letzen Jahrzehnte. Archiv für Hydrobiologie Supplement 62:440–490.

Kann, E. 1993. Der litorale Algenaufwuchs im See Erken und in seinem Abfluß (Uppland, Schweden). Algological Studies 69:91–112.

Kirkby, S. M., Hibberd, D. J., Whitton, B. A. 1972. *Pleurocladia lacustris* A. Braun (Phaeophyta)—a new British record. Vasculum 57:51–56.

Kumano, S., Hirose, H. 1959. On the swarmers and reproductive organs of a phaeophyceous fresh-water alga of Japan, *Heribaudiella fluviatilis* (Areschoug) Svedelius. Bulletin of the Japanese Society of Phycology 7:45–51 (in Japanese).

Kusel-Fetzmann, E. L. 1996. New records of freshwater Phaeophyceae from lower Austria. Nova Hedwigia 62:79–89.

Kusel-Fetzmann, E., Schagerl, M. 1992. Verzeichnis der Sammlung von Algen-Kulturen an der Abteilung für Hydrobotanik am Institut für Pflanzenphysiologie der Universität Wien. Phyton 32:209–234.

Lee, R. E. 1989. Phycology, 2nd ed. Cambridge Univ. Press, Cambridge, UK.

Luther, H. 1954. Über Krustenbewuchs an Steinen fliessender Gewässer, speziell in Südfinnland. Acta Botanica Fennica 55:1–66.

Müller, D. G. 1979. Genetic affinity of *Ectocarpus siliculosus* (Dillw.) Lyngb. from the Mediterranean, North Atlantic and Australia. Phycologia 18:312–318.

Müller, D. G., Geller, W. 1978. Einige Beobachtungen an Kulturen der Süsswasser-Braunalge *Bodanella lauterborni* Zimmermann. Nova Hedwigia 29:735–741.

Papenfuss, G. F. 1951. Phaeophyta, in: Smith, G. M., Ed., Manual of phycology. Chronica Botanica, Waltham, MA, pp. 119–158.

Pueschel, C. M., Stein, J. R. 1983. Ultrastructure of a freshwater brown alga from western Canada. Journal of Phycology 19:209–215.

Schloesser, R. E. 1977. The identification of a new freshwater brown alga from the Lake Michigan sublittoral zone. M.Sc. thesis, University of Wisconsin, Milwaukee, 80 p.

Schloesser, R. E., Blum, J. L. 1980. *Sphacelaria lacustris* sp. nov., a freshwater brown alga from Lake Michigan. Journal of Phycology 16:201–207.

Sheath, R. G., Cole, K. M. 1992. Biogeography of stream macroalgae in North America. Journal of Phycology 28:448–460.

Sheath, R. G., Hambrook, J. A. 1990. Freshwater ecology, in: Cole, K. M., Sheath, R. G., Eds., Biology of the red algae. Cambridge Univ. Press, Cambridge, UK, pp. 423–453.

Sinclair, C., Whitton, B. A. 1977. Influence of nutrient deficiency on hair formation in the Rivulariaceae. British Phycological Journal 12:297–313.

Smith, G. M. 1950. The fresh-water algae of the United States, 2nd ed. McGraw-Hill, New York.

Starmach, K. 1977. Phaeophyta—brunatnice. Rhodophyta—krasnorosty. Flora słodkowodna Polski, Vol. 14. Panstwowe Wydawnictwo Naukowe, Warsaw/Krakow.

Stevenson, R. J., 1996. An introduction to algal ecology in freshwater benthic habitats, in: Stevenson, R. J., Bothwell, M. L., Lowe, R. L., Eds., Algal ecology: Freshwater benthic ecosystems. Academic Press, San Diego, pp. 3–30.

Svedelius, N. 1930. Über die sogenannten Süsswasser-Lithodermen. Zeitschrift für Botanik 23:892–918.

Szymanska, H., Zakrys, B. 1990. New phycological records from Poland. Archiv für Hydrobiologie Supplement 87:25–32.

Targett, N. M., Arnold, T. M. 1998. Predicting the effects of brown algal phlorotannins on marine herbivores in tropical and temperate oceans. Journal of Phycology 34:195–205.

Thompson, R. H. 1975. The freshwater brown alga *Sphacelaria fluviatilis*. Journal of Phycology (Supplement) 11:5.

Timpano, P. 1978. A preliminary report of the fresh-water phaeophyte, *Sphacelaria fluviatilis*. Journal of Phycology (Supplement) 14:34.

Timpano, P. 1980. The ultrastructure of the fresh-water phaeophyte, *Sphacelaria fluviatilis*. Journal of Phycology (Supplement) 16:44.

Van den Hoek, C., Mann, D. G., Jahns, H. M. 1995. Algae. An introduction to phycology. Cambridge Univ. Press, Cambridge, U.K.

Waern, M. 1938. Om *Cladophora aegagropila*, *Nostoc pruniforme*, och andra alger i Lilla Ullevifjärden, Mälaren. Botanische Noten 1938:129–142.

Waern, M. 1952. Rocky-shore algae in the Öregund Archipelago. Acta Phytogeographica Suecica 30:1–298.

Wehr, J. D., Stein, J. R. 1985. Studies on the biogeography and ecology of the freshwater phaeophycean alga *Heribaudiella fluviatilis*. Journal of Phycology 21:81–93.

Weitzel, R. L. [Ed.] 1979. Methods and measurements of periphyton communities: A review. ASTM special technical publication 690. American Society for Testing and Materials, Philadelphia.

West, J. A. 1990. Noteworthy collections. Washington. *Heribaudiella fluviatilis* (Areschoug) Svedelius. Madroño 37:144.

West, J. A., Kraft, G. T. 1996. *Ectocarpus siliculosus* (Dillwyn) Lyngb. from Hopkins River Falls, Victoria—the first record of a freshwater brown alga in Australia. Muelleria 9:29–33.

Wheeler, B. D., Whitton, B. A. 1971. Ecology of Hell Kettles. 1. Terrestrial and sub-aquatic vegetation. Vasculum 55:25–37.

Whitford, L. A. 1977. Are there any rare freshwater algae? Journal of Phycology 13:73.

Wilce, R. T. 1966. *Pleurocladia lacustris* in Arctic America. Journal of Phycology 2:57–66.

Wilce, R.T., Webber, E. E., Sears, J. R. 1970. *Petroderma* and *Porterinema* in the New World. Marine Biology 5:119–135.

Wujek, D. E., Thompson, R. H., Timpano, P. 1996. The occurrence of the freshwater brown alga *Sphacelaria fluviatilis* Jao from Michigan. Michigan Botanist 35:111–114.

Yoneda, Y. 1949. Notes on the freshwater algae of Kikusui-sen, a rheocrene at Yoro-mura in Province Mino. Journal of Japanese Botany 24:169–175.

Yoshizaki, M., Iura, K. 1991. Notes on *Heribaudiella fluviatilis* from Chiba Prefecture and Ibaraki Prefecture. Chiba Seibutu-si 40:37–39 (in Japanese).

Yoshizaki, M., Miyaji, K,, Kasaki, H. 1984. A morphological study of *Heribaudiella fluviatilis* (Areschoug) Svedelius (Phaeophyceae) from Central Japan. Nankiseibutu 26:19–23 (in Japanese).

Zimmermann, W. 1928. Über Algenbestände aus der Tiefenzone des Bodensees. Zur Ökologie und Soziologie der Tiefseepflanzen. Zeitschrift für Botanik 20:1–28 + 2 pl.

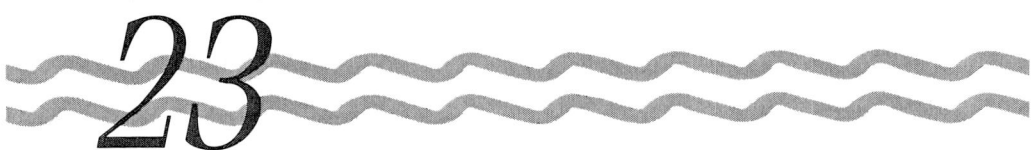

USE OF ALGAE IN ENVIRONMENTAL ASSESSMENTS

R. Jan Stevenson
Department of Zoology
Michigan State University
East Lansing, Michigan 48824

John P. Smol
Department of Biology
Paleoecological Environmental Assessment
  and Research Laboratory (PEARL)
Queen's University
Kingston, Ontario
Canada K7L 3N6

I. Introduction
II. Goals of Environmental Assessment with Algae
III. Sampling and Assessing Algal Assemblages for Environmental Assessment
    A. Sampling Algae in Freshwater Habitats
    B. Attributes of Algal Assemblages for Environmental Assessment
IV. Developing Metrics for Hazard Assessment
    A. Relating Goals to Ecological Attributes
    B. Testing Metrics
    C. Multimetric Indices
    D. Multivariate Statistics and Hazard Assessment
V. Exopsure Assessment: What Are Environmental Conditions?
VI. Stressor–Response Relations
VII. Risk Characterization and Management Decisions
VIII. Conclusions
Literature Cited

I. INTRODUCTION

Algae have long been used to assess environmental conditions in aquatic habitats throughout the world. During the early part of the twentieth century, algae were exploxed as indicators of organic pollution in European streams and rivers (Kolkwitz and Marsson, 1908). Between 20 and 50 years ago, use of algal indicators of environmental conditions flourished based on the environmental sensitivities and tolerances of individual taxa and species composition of assemblages (e.g., Butcher, 1947; Fjerdingstad, 1950; Zelinka and Marvan, 1961; Slàdecek, 1973; Lowe, 1974; Lange-Bertalot, 1979). Nutrient stimulation of algal growth made algae part of the problem in the eutrophication of lakes such that trophic status of lakes was also characterized by the amount of algae (Vollenweider, 1976; Carlson, 1977). In North America, Ruth Patrick and C. Mervin Palmer were pioneers in the development of large monitoring programs to assess the ecological health of rivers and nuisance algal growths (Patrick, 1949; Patrick et al., 1954; Palmer, 1969). More recently, the sensitivity of many algal taxa to pH, combined with preservation of certain algal cell wall components (e.g., diatom frustules and chrysophyte scales) in sediments, has been employed to assess problems with acid deposition and to determine if rates of lake acidification have been enhanced by human contributions to acid deposition (Smol, 1995; Battarbee et al., 1999). Government agencies throughout the world now use algae to monitor and assess ecological conditions in many types of aquatic ecosystems (e.g., Weber, 1973; Dixit and Smol, 1991; Dixit et al., 1992, 1999; Bahls, 1993; Kentucky Division of Water, 1993; Whitton and Rott, 1996; Biggs et al., 1998; Kelly et al., 1998; Stevenson and Bahls, 1999). Thus, characterization of algal assemblages has been important in environmental assessment, both in indicating changes in environmental conditions that impair or threaten ecosystem health and in determining if algae themselves are causing problems.

Algae are particularly valuable in environmental assessments. Algae are the base of most aquatic food chains, are important in biogeochemical cycling, and serve as habitat for many organisms in aquatic ecosystems (e.g., Minshall, 1978; Wetzel, 1983; Power, 1990; Carpenter and Kitchell, 1993; Vymazal, 1994; Bott, 1996; Lamberti, 1996; Mulholland, 1996; Wetzel, 1996). Thus, a natural balance of species and assemblage functions is important for ecosystem health (Angermeier and Karr, 1994). Increases in algal biomass and shifts in species composition can cause problems with many ecosystem services by causing taste and odor problems in water supplies (Sigworth, 1957; Palmer, 1962; Arruda and Fromm, 1989), toxic algal blooms (Bowling and Baker, 1996; Burkholder and Glasgow, 1997), and low dissolved oxygen levels (Lasenby, 1975).

In many aquatic habitats, algae are the most diverse assemblage of organisms that can be easily sampled and readily identified to species (particularly diatoms and desmids). The great species-specific sensitivity of algae to environmental conditions and their high diversity in habitats provide the potential for very precise and accurate assessments of the physical, chemical, and biological conditions that may be causing problems. Moreover, algae and paleolimnological techniques can be used to infer historical conditions in lakes, wetlands, and even reservoirs and rivers (Fritz, 1990; Smol, 1992). Algae occur in all aquatic habitats, so they could be very valuable for comparison among ecosystems with the same group of organisms. From a logistical perspective, algae are relatively easy to sample, and analysis is relatively inexpensive compared with bioassessment with other groups of organisms. In addition, many characteristics of algal assemblages can be measured and used as multiple lines of evidence for whether ecological integrity has been altered and the causes of those alterations. Algal bioassessment complements physical and chemical data by providing corroborative evidence for environmental change.

Both structural and functional characteristics of algae can be used to assess environmental conditions in aquatic habitats. Algal biomass (measured as chlorophyll *a*, cell numbers, and/or algal biovolume; Stevenson, 1996) can be used to indicate the presence of toxic pollutants as well as trophic status and nuisance algal growths (Carlson, 1977; Dodds *et al.*, 1998). Taxonomic composition and diversity of algal assemblages are used to assess ecological health of habitats and to infer probable environmental causes of ecological impairment (e.g., Patrick *et al.*, 1954; Smol, 1992; Stevenson and Pan, 1999). Ratios of chemicals in algal samples can be used to indicate algal health (phaeophytin:chlorophyll *a*) and nutrient limitation (N:P) (Weber, 1973; Hecky and Kilham, 1988; Biggs, 1995). Photosynthesis, respiration, and phosphatase activity are examples of algal metabolism that can be used to assess the amount of algae in habitats, physiological impairment, and phosphorus limitation (Blanck, 1985; Hill *et al.*, 1997; Newman *et al.*, 1994).

In this chapter, the abundant and diverse methods of using algae to assess environmental conditions in all aquatic habitats are organized in a risk assessment framework (U.S. Environmental Protection Agency, 1992, 1996, 1998). Many reviews of algal methods for environmental assessment have been published in recent years (Stevenson and Lowe, 1986; Round, 1991; Coste *et al.*, 1991; Smol, 1992; Whitton and Kelly, 1995; Rosen, 1995; Reid *et al.*, 1995; Lowe and Pan, 1996; Stevenson, 1998; McCormick and Stevenson, 1998; Wehr and Descy, 1998; Kelly and Whitton, 1998; Kelly *et al.*, 1998; Ibelings *et al.*, 1998; Prygiel *et al.*, 1999a; Stevenson and Pan, 1999; Stevenson and Bahls, 1999; see many chapters in Whitton *et al.*, 1991; Whitton and Rott, 1996; Stoermer and Smol, 1999; Prygiel *et al.*, 1999b). We take this abundance of recent reviews as an indication of the growing importance of algae in environmental assessment. In our chapter, we emphasize understanding the goals of environmental programs, developing and testing hypotheses that address program goals, and selecting the simplest and most direct methods for achieving program goals. We present the characteristics of algae that can be used in environmental assessments and then elaborate on how these characteristics can be related by using them in the five steps of ecological risk assessment. Although algae have been used for such assessments in habitats throughout the world, great potential exists for developing indices that more directly meet the needs of specific environmental assessment programs. Thus, in this chapter, we present the many approaches for developing algal methods for environmental assessment, and then we describe the application of algal methods for assessment.

II. GOALS OF ENVIRONMENTAL ASSESSMENT WITH ALGAE

The goals of environmental assessment programs can be established by legislation, by government officials and policy decision makers, by scientists, or by the general public. In most cases, scientists play an important role in translating the official goals of an environmental program into hypotheses that can be tested and in developing a practical study plan that can be implemented within the budget allocated for the project. The United States Environmental Protection Agency (U.S. EPA) risk assessment and risk management framework (U.S. EPA, 1992, 1996, 1998) (Fig. 1) is valuable for translating the many goals of 'environmental problem solving' into a series of testable hypotheses and for providing a sound scientific approach for solving problems.

The overall goals of environmental assessment, with algae or other organisms, are to characterize the effects or potential effects of human activities and to implement management strategies that reduce the risk of ecological impairment and restore valued ecological conditions. In addition to the actual state of the ecosystem, factors such as economic, social, and legal issues may affect decisions about how to protect or restore valued ecological characteristics (Fig. 1). Because of the complexity of many environmental issues, clearly stated goals permit the development of testable hypotheses, sampling and statistical design, and choice of the best methods. The ecological risk assessment (ERA) framework helps to organize and relate the many issues associated with environmental problems and form these hypotheses. The ERA helps distinguish between the ecological conditions that the public wants to protect, the stressors that threaten those conditions, the human activities causing those stressors, and the many other factors involved in decisions on how to solve the problem.

In general, most environmental assessments involve one or more steps in the ERA (Fig. 1). A full risk assessment involves five steps of ERA: problem formulation, hazard (response) assessment, exposure (stressor) assessment, evaluating the stressor–response relationships, and then characterizing the risk associated with each stressor and responses of interest. Problem formulation is the identification of the ecological attributes and stressors with which the public are most concerned. We may start from the perspective that a valued ecological attribute, such as water clarity or biotic integrity (*sensu* Karr and Dudley, 1981), is threatened or impaired and that we need to determine the cause of the problem. Are aesthetics or taste and odor impaired by nuisance growths of algae? Are toxic algal blooms occurring? Alternatively, we may be concerned about how a stressor, such as acid deposition or nutrients, could be affecting ecosystems. During problem formulation, identifying and distinguishing valued ecological attributes and stressors is very important, so that cause–effect relationships can be identified and targeted in an ERA.

Our adaptation of the ERA for algal bioassessments has broad applications in the integration of the diversity of information that can be obtained in ecological assessments with algae and applying them to hazard assessment and exposure assessment. Hazard assessment is determining whether the ecological conditions in a habitat are impaired and is an assessment of the dependent variable (valued ecological attribute or an indicator of that attribute, a response variable) in the problem (Fig. 2). For example, have algae accumulated to nuisance levels or have sensitive species been lost from the habitat. Exposure assessment is an evaluation of the intensity, frequency, and duration of altered habitat conditions or contamination. For example, what is the pH, total phosphorus concentration, or organic load in the habitat, and how long does it last? Exposure assessment is a measurement of the stressor, which is the independent variable in the stressor–response relationship.

The stressor–response relationship permits determination of the stressor that is likely to be most threatening or causing impairment of ecological conditions (Fig. 2). Stressor–response relationships may be found in previously published literature or in studies that accompany the ERA. Ecological risk associated with each stressor should then be characterized by comparing assessed response (hazard assessment) and stressor levels (exposure assessment) with the stressor–response relationship (Fig. 2). One or more stressors should be identified as being the likely stressors that most threaten impairment or cause impairment (see Stevenson, 2001, for more discussion). Most stressors should be too low to cause the observed impairment. However, at least one stressor, or an interaction of multiple stressors, should be high enough to cause the observed response.

Algae can be used in ERA to determine whether a problem exists, to infer levels of specific stressors in a

FIGURE 1 Elements of the risk assessment and risk management framework. Modified from U.S. Environmental Protection Agency (1996).

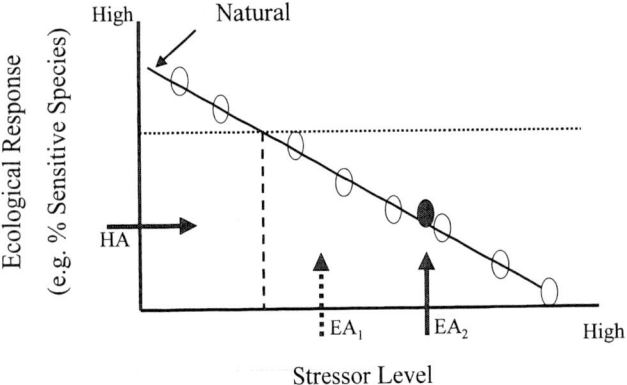

FIGURE 2 Response–stressor relationship between hypothetical ecological response and stressor with hazard assessment and exposure assessment indicated. Acceptable levels of the ecological response are indicated by the horizontal dotted line. The stressor level that produces that level of response is indicated by the vertical dashed line. The observed ecological response (indicated by horizontal solid arrow marked HA on the y axis) is below acceptable levels. Two stressors were measured or inferred based on algal indices (indicated by the vertical arrows marked EA (exposure assessment) on the x axis). One stressor (indicated by EA_1) is too low to cause the observed response, whereas the other has a high certainty of causing the observed ecological condition.

habitat, and to characterize stressor–response relationships. Algal indices are used to assess both stressors as well as valued ecological attributes. Probably more than any other group of organisms, algae have been used to infer physical and chemical conditions (potential stressors) in a habitat through the determination of species composition of assemblages and species' ecological preferences. Potential for inferring stressor conditions exists for other organisms, but algae are used much more often than fish, macroinvertebrates, or insects in terrestrial habitats to infer levels of pH, conductivity, trophic status, and sewage contamination. Therefore, distinguishing between whether algal indices are indicating valued ecological attributes or stressors is important for relating results of algal bioassessments to goals of the ERA. This opportunity to use algae to determine whether problems exist, potentially even forecast problems, and to diagnose causes of problems should be emphasized in algal bioassessments.

III. SAMPLING AND ASSESSING ALGAL ASSEMBLAGES FOR ENVIRONMENTAL ASSESSMENT

A. Sampling Algae in Freshwater Habitats

Sampling techniques and design may vary with the objectives of the assessment, probable factors affecting algae and anticipated problems, water body type, targeted habitat within the water body, and budget. A complete discussion of the most appropriate sampling methods and design is beyond the scope of this chapter, but we will review some of the important issues related to sampling for bioassessment. In general, the same basic approaches are useful for solving most problems in any water body or habitat. Algal indicators may, however, be more precise if they are refined with regional datasets for specific water body types, but this problem will be discussed later with development of indicators of exposure assessment (weighted averaging inference models).

1. Sampling Design

Objectives of an environmental assessment should be defined as clearly as possible for the formulation of hypotheses and a sampling design (e.g., Green, 1979). Testable hypotheses should be formed so that results provide answers that address the objectives of the assessment and that have a defined error or uncertainty. Testing hypotheses requires a sampling design that includes replicate sampling or some form of assessment of error variation. Estimates of ecological health in small, local studies are usually based on replicate sampling at all sites and provide means and estimates of variation at each site to test the hypothesis that conditions at a tested site are significantly different from conditions at a reference site or from a criterion. However, replicate sampling at all sites in large surveys is often not affordable; then replicate sampling at a random subset of sites (often 10% or more) can be used to characterize variation in estimates of ecological health. The error variation associated with the random subset of sites in large surveys is a measure of the precision (standard deviation or standard error) of assessments at all sites.

Before sampling, investigators should define the extent of each sample site so that it can be repeated at all locations. The open-water region of lakes usually defines the horizontal extent of plankton habitats sampled at a lake site, but the vertical extent (depth of sampling) may vary with goals of the project. In many river programs, for example, sampling one riffle has been considered sufficient to characterize the ecological health of a stream with benthic algae (Bahls, 1993), but in other programs the extent of the habitat is defined as a stream reach with a length that is 40 times the width of the stream (Klemm and Lazorchak, 1994; Stevenson and Bahls, 1999). Similarly, a representative portion of wetland should be chosen to sample (e.g., Stevenson et al., 1999).

Most algal sampling strategies focus on a specific habitat within the water body (Wehr and Sheath, Chap. 2, this volume), such as plankton, algae on rocks (epilithic) in riffles, or algae on plants (epiphytic). Alternatively, objectives of a project may call for characterizing the diversity of algae in a habitat and consequently sampling all suitable habitats within a water body (defined site) (Porter *et al.*, 1993). The presumed advantage of sampling targeted habitats is that algal indicators are more sensitive and can more precisely detect changes in environmental conditions if interhabitat variability is reduced (Rosen, 1995). In a recent review, Kelly *et al.* (1998) make strong arguments for sampling rocks or other hard substrata, if they are present. In some cases, however, sampling the same targeted habitat in all water bodies of a project is impractical. Although finding plankton in lakes is usually not a problem, finding cobble riffles in all streams or open water in all wetlands can be a problem. Multihabitat sampling is one solution to the problem of habitat diversity among sites. One advantage to multihabitat sampling is a more complete assessment of all taxa at a site, which potentially is a better characterization of biodiversity and biointegrity than assemblages from targeted habitats. Another solution is to classify streams and wetlands by size and hydrogeomorphology (Vannote *et al.*, 1980; Biggs and Close, 1989; Rosgen, 1994; Biggs, 1995; Brinson, 1993; Goldsborough and Robinson, 1996; Biggs *et al.*, 1998), so they have similar habitats, and then to develop sampling strategies and indicators for specific hydrogeomorphic classes of streams and wetlands.

Estimates of algal biomass, which are particularly important in characterizing trophic state, require quantitative sampling of habitats. Habitats are quantitatively sampled by measuring the volume of water collected or area of substratum sampled and accounting for the proportion of sample assayed. Algal attributes (e.g., biomass or productivity) can be expressed on a volume-specific or area-specific basis by correcting measurements for volume or area sampled and proportion of sample assayed (Wetzel and Likens, 1991; APHA, 1998). The main disadvantage of quantitative sampling is the time required and the practicality of precisely characterizing the area sampled. Measuring sample volume, in the field or in the laboratory, requires relatively little extra time. The benefits of quantitative sampling are also reduced when habitat conditions are spatially or temporally variable such that biomass is affected. Thus, quantitative sampling is particularly problematic in structurally diverse and hydrologically variable streams and wetlands.

Qualitative algal sampling is recommended in habitats that have great spatial and temporal variation and when sufficient time is not available to measure the substratum area. One goal of qualitative sampling could be sampling all species at a site (Porter *et al.*, 1993), which would call for sampling all habitats, water column, rocks, plants, and sediment in different physical settings (e.g., light, depth, current velocity, etc.). Another goal could be to determine the dominant algae at a site, which would require estimating the relative areas of different habitats and proportional sampling of those habitats.

Variation in quantitative or qualitative estimates of algal attributes, due to spatial variation in habitat conditions, can be reduced by composite sampling. Reducing spatial variation calls for subsampling of many areas throughout the defined extent of the study site and putting all subsamples into a composite sample. Variation in estimates of algal attributes due to temporal variation in habitats is more difficult to reduce because it requires sampling throughout a study period and return visits to a site, which may not be practical. To reduce effects of temporal variation on quantitative attributes in highly variable ecosystems like streams, it is best to sample after an extended (1–2-week) period of stable habitat conditions so that algal assemblages have reached peak or sustainable biomass and a relatively predictable state (Stevenson, 1990; Peterson and Stevenson, 1992; Biggs, 1996; Stevenson, 1996).

Semiquantitative approaches for assessing algal biomass and percentage cover of different algal groups have been used to reduce field and lab time and to increase the spatial and temporal extent of algal assessments in streams. Secchi disc transparency is a semiquantitative approach for assessing plankton biomass in lakes (Brezonik, 1978; Davies-Colley and Vant, 1988; Wetzel and Likens, 1991). In streams, visual characterization of algal type, percentage cover, filament length, and periphyton mat thickness along multiple transects have been used in many situations (Holmes and Whitton, 1981; Sheath and Burkholder, 1985; Rout and Gaur, 1990; Stevenson and Bahls, 1999). These techniques can easily be employed in all sampling programs because they require little time and can provide biomass assessments over large areas. They may be excellent quantitative tools for volunteer programs because they require little taxonomic expertise.

2. Sampling Techniques

a. Sampling Present-Day Assemblages Numerous algal habitats can be sampled within water bodies. Phytoplankton can be sampled at specific depths with Van Dorn, Kemmerer, or similar discrete-depth samplers (APHA, 1998). Depth-integrated samples can be collected with devices (e.g., peristaltic pumps) that allow water to slowly enter a sampling chamber or by compositing samples collected from specific depths.

Phytoplankton biomass is almost always expressed per unit volume (e.g., Wetzel and Likens, 1991). Qualitative samples can be collected with plankton nets; however, we recommend collecting whole water samples with known volumes whenever possible so that small algae are not missed and samples can be assayed quantitatively. Algae in whole water samples can be concentrated by filtering or settling (e.g., Wetzel and Likens, 1991; APHA, 1998).

Metaphyton are macroalgal and microalgal masses suspended in the water column and entangled among macrophytes or along shorelines, typically in slow or still water (Hillebrand, 1983; Goldsborough and Robinson, 1996). Quantitative sampling of metaphyton requires collecting algae from a vertical column through the assemblage. Coring tubes can be used to isolate and collect a column of metaphyton. Scissors are useful for cutting horizontal filaments that block the insertion of the tube through the metaphyton assemblage. The depth of the core should not extend to the substratum surface. The diameter of the core depends upon the spatial variability of the metaphyton, on necessary sample size, and on the ability to isolate the core of algae from surrounding metaphyton (Stevenson, personal observation). Metaphyton in the form of unconsolidated green clouds requires wider cores (ca. 10 cm) because filaments are difficult to isolate in narrow cores. Narrower cores (ca. 3 cm) can be used to sample consolidated microalgal mats. Metaphyton biomass should be expressed on an areal basis (e.g., m^{-2}). Qualitative samples of metaphyton can be gathered with grabs, forceps, strainers, spoons, or cooking basters.

Benthic algae are sampled by scraping hard or firm substrata, such as rocks, plants, and tree branches, usually after they have been removed from the water (Stevenson and Hashim, 1989; Aloi, 1990; Porter et al., 1993). Cores of algae should be collected on soft or unconsolidated substrata, such as sediments and sand (Stevenson and Stoermer, 1981; Stevenson and Hashim, 1989). Area of substrata sampled should be recorded to quantify samples. Some substrata, such as bedrock and logs, cannot be removed from the water. In those cases, vertical tubes can be used to isolate an area of substratum. After algae are scraped from the substratum in the tube, algae and water in the tube can be removed with a suction device.

Artificial substrata are also used to assess benthic algal assemblages (Patrick et al., 19547; Tuchman and Stevenson, 1980; Aloi, 1990). They are typically uniform substrata (e.g., glass or acrylic slides, clay tiles, acrylic or wooden dowels) that can be used across many water body types (streams, rivers, wetlands, lakes). If placed in similar light and current environments in all habitats, differences in assemblages among sites should be highly sensitive to water chemistry. However, placement and sampling of artificial substrata require more than one trip to sample sites. Samples are often lost because they are subject to vandalism. Finally, assemblages on artificial subtrates may not reflect historical changes in habitats or changes in physical habitat structure as well as assemblages on natural substrata (see also Section II). Most large national and state programs have chosen to sample natural substrata (Porter et al., 1993; Klemm and Lazorchak, 1994), but artificial substrata can be valuable in smaller-scale programs where travel time is limited or in ecosystems with great habitat diversity.

b. Sampling Historic Assemblages Sediment sampling in lakes, streams, rivers, and wetlands with deposited sediments can include an algal assemblage that has accumulated for months, years, or centuries, depending upon the depth and disturbance of sediments. A large variety of coring apparatuses are available to retrieve sediment cores (several are illustrated in Smol and Glew, 1992), with the choice of equipment largely dependent on the type of system being studied and the temporal resolution required. The resolution is also dependent on the type of sectioning techniques and equipment one uses. Close-interval sectioning equipment and techniques (e.g., Glew, 1988) are available that can provide lake managers with a high degree of temporal resolution.

The overall paleolimnological approach is summarized in the accompanying schematic (Fig. 3). Once the study site is chosen, a sediment core is removed, usually from near the center of the lake. In general, the central, flat portion of a basin integrates indicators from across the lake, and so a more holistic record of past environmental change is archived.

Once the core is retrieved and sectioned, the "depth-time" profile must be established. This requires dating a sufficient number of sediment layers to attain a reliable chronology. For most paleolimnological studies dealing with recent environmental assessments, ^{210}Pb dating is most often used (Oldfield and Appleby, 1984), as the half-life (22.26 years) of this naturally occurring isotope enables one to date, with reasonable certainty, approximately the last century or so of sediment accumulation. In some lake systems, close to annual (and sometimes subannual) resolution is possible.

Analyzing a sediment core at close intervals (e.g., every cm or every 0.5 cm) is time consuming and may not be practical for some large-scale, regional environmental assessments. For these cases, paleolimnologists have sometimes used a "snapshot" approach, which

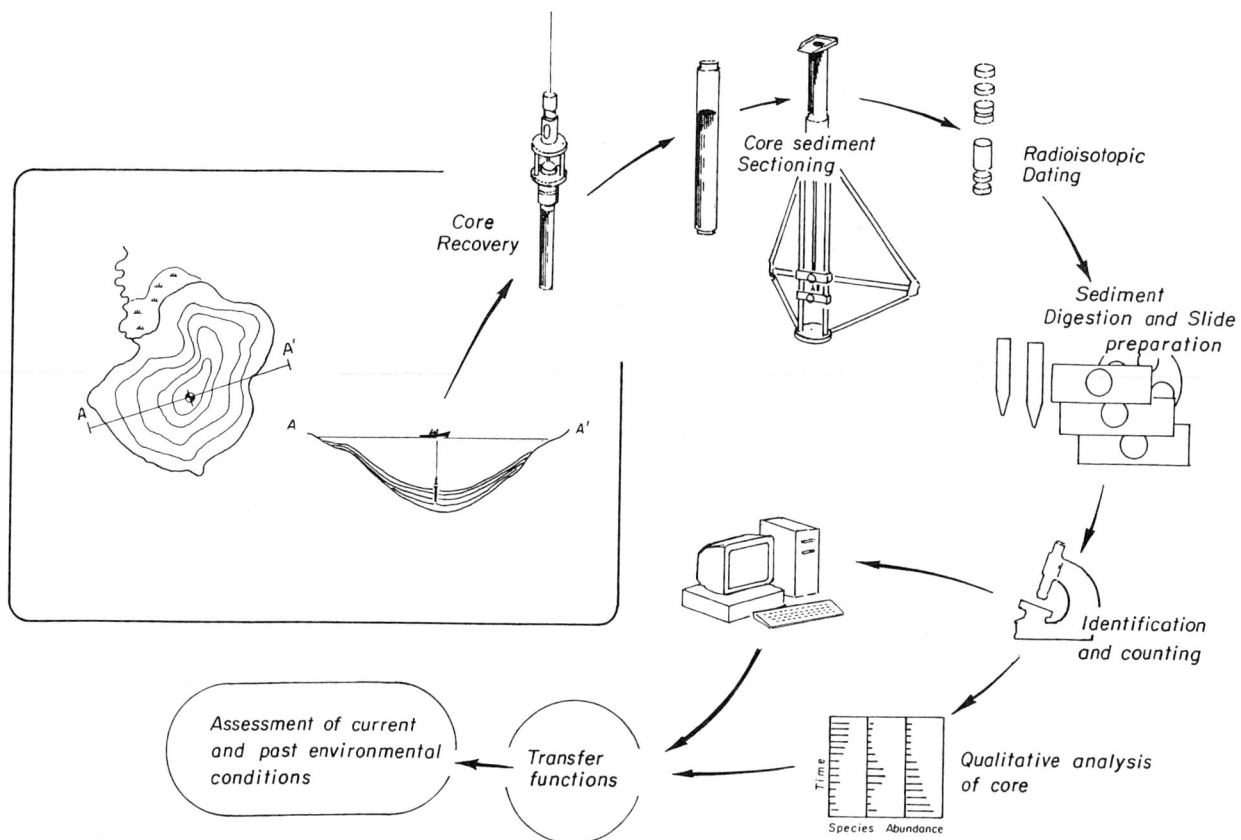

FIGURE 3 Schematic diagram showing the major steps involved in a paleolimnological assessment. Modified from Dixit et al., (1992a).

attempts to estimate what conditions were like before anthropogenic impacts and how much degradation has occurred. This so-called top/bottom approach is a very simple but effective tool for obtaining regional assessments of environmental change. Paleolimnologists remove surface sediment cores as they would in a detailed paleoenvironmental assessment; but instead of sectioning and analyzing the entire core, they simply analyze, for example, diatom valves and/or chrysophyte scales in the top 1 cm of sediment (= present-day conditions) and from a sediment level known to have been deposited before anthropogenic impact (i.e., the bottom sediment section). This before-and-after approach has been effectively used to infer environmental change, such as acidification (Cumming et al., 1992a; Battarbee et al., 1999; Dixit et al., 1999) and eutrophication (e.g., Dixit and Smol, 1994; Hall and Smol, 1999) that has occurred on regional scales. In addition to providing some estimate of degradation, these paleoenvironmental data also provide important information on the natural background conditions of a system and therefore provide important mitigation targets for environmental remediation efforts (Smol,

1992). To determine the rates and trajectories of past changes, more detailed paleoenvironmental assessments are required.

The next step is to recover any paleoenvironmental information archived in dated sediment cores. Our focus here is on the paleophycological data, but many other types of proxy data are available. For example, past changes in terrestrial vegetation can be inferred from the analyses of fossil pollen grains (the field of palynology); paleomagnetic measurements and other techniques can be used to estimate past erosion rates (e.g., Dearing et al., 1987); and isotope and geochemical analyses of metals and other contaminants (e.g., PCBs, DDT, etc.) can be analyzed from the sedimentary profiles (Autenrieth et al., 1991). Despite these other powerful approaches, the mainstay of many paleolimnological assessments is algal data. Virtually every algal group leaves some sort of morphological or chemical fossil in the sedimentary record, but the indicators that are most often used are diatom valves (Dixit et al., 1992), chrysophyte scales and cysts (Smol, 1995), and fossil pigments (Leavitt, 1996). As shown in examples given later in this chapter, these indicators can be used to reconstruct past limnological characteristics (such as pH, eutrophication variables, and salinity). Paleolimnological approaches have been subjected to a large amount of quality assurance and quality control considerations; if undertaken carefully and correctly, the paleolimnological approach is robust, reproducible, and powerful.

B. Attributes of Algal Assemblages for Environmental Assessment

Many attributes of algal assemblages can be used to assess environmental conditions in a habitat or site (Table I). Structural attributes (e.g., species composition) and functional attributes (e.g., productivity) can be measured in the field or the laboratory. The diversity of these attributes and the pros and cons of their uses have been discussed in recent studies (Stevenson, 1996; Stoermer and Smol, 1999; Stevenson and Pan, 1999; Stevenson and Bahls, 1999). In this treatment, we will focus more on the value of these attributes in detecting effects of humans on ecological systems.

For this review, we will use a set of terms recommended by Karr and Chu (1999) to clarify discussions of using biological data for bioassessment. An *attribute* is any characteristic of an assemblage that can be measured, such as chlorophyll *a* (chl *a*), number of species, or net primary productivity. A *metric* is an attribute that responds to human disturbances of habitats. Some attributes, such as the number of diatom genera in a 200-valve count, may not reliably respond to human impacts. Karr and Chu (1999) also recommend the use of the term *index* for statistical and other mathematical summaries of many metrics and indices. Multimetric indices of biotic integrity, such as the diatom bioassessment index (Kentucky Division of Water, 1993), are examples of multiple metrics being averaged or summed to compose a single, summary index. Many of the water quality indices that are commonly used in Europe (e.g., pollution tolerance index (Lange-Bertlalot 1979), generic diatom index (Coste and Ayphassorho, 1991), and trophic diatom index (Kelly and Whitton, 1995)) should probably be characterized as metrics, according to the method of Karr and Chu (1999), but we do not recommend renaming these indicators and being encumbered by semantics.

Algal assemblages can be characterized through the use of two basic kinds of attributes, structural and functional (Table I). Structural attributes are instantaneous characterizations of assemblages, such as biomass per unit area or volume of habitat, taxonomic and chemical characterizations of community composition, and diversity of community taxa (e.g., species richness). Functional attributes are measures or indicators of assemblage metabolism, such as photosynthetic rate (gross primary production), respiration rate, net primary productivity, nutrient cycling, phosphatase activity, and population growth rates. Functional assessments usually require more time in the field or multiple trips, so are used less in routine environmental assessments. However, they can be important for understanding impairment of algal and microbial activity.

1. Biomass

Biomass of algae usually increases with resource availability and decreases with many stressors caused by humans (Vollenweider, 1976; Dodds *et al.*, 1998). Removal of riparian (stream-side) canopies along streams and nutrient loading in all water bodies increase light, temperature, and nutrient availability, which can limit algal growth rates and biomass accrual (see reviews in Biggs, 1996; Hill, 1996; DeNicola, 1996; Borchardt, 1996). Sediments, toxic substances, and removal of benthic habitat can limit algal growth and accrual (Genter, 1996; Hoagland *et al.*, 1996). Because biomass and the potential for nuisance algal growths vary temporally with season and weather (Whitton, 1970; Wong *et al.*, 1978; Lembi *et al.*, 1988), timing of sampling is important. In most habi-

TABLE I Basic Attributes of Algal Assemblages That Can Be Measured and Potentially Be Used to Assess Environmental Conditions

Structural attributes
  Biomass
    chl *a*
    Ash free dry mass
    Cell density
    Cell biovolume
  Taxonomic composition
    Species relative abundances
    Species relative biovolume
    Functional group biovolume
  Diversity
    Species richness
    Genus richness
    Evenness
  Chemical composition
    chl *a*:(Phaeophytin+Chl *a*) ratio
    chl *a*:ash free dry mass ratio
    P or N/ash free dry mass
    N:P ratio of algal assemblages
Functional attributes
  Photosynthesis rates
  Respiration rates
  Net primary productivity
  Growth rates
  Nutrient uptake rates

For a more detailed list with literature citations, see McCormick and Cairns (1994).

tats, peak biomass occurs after periods of undisturbed habitat conditions (e.g., post-flood) when algal biomass has had an opportunity to accrue. Spring and summer blooms of phytoplankton are common in lakes with spring turnover and warm summer temperatures (Wetzel, 1983; Harper, 1992). Nuisance algal growths may occur, with filamentous benthic algal accrual during seasonal optima (Whitton, 1970; Biggs and Price, 1987; Dodds and Gudder, 1992; Lembi et al., Chap. 24, this volume), or in the water column during summer low flow, when water residence time is sufficient for algal accrual in the water column (Bowling and Baker, 1996). Biomass is an important attribute in environmental assessments because it is related to productivity and nuisance problems.

Biomass of algal assemblages can be estimated with laboratory assays of chl a, dry mass, ash-free dry mass, algal cell density, biovolume, or chemical mass of samples. All of these measurements have pros and cons (see Stevenson, 1996, for a review), because none directly measure all constituents of algal biomass or only algal biomass. However, all are reasonable estimates of algal biomass in different situations. Chl a must be extracted from cells in organic solvents, such as acetone or methanol, and then assayed by spectrophotometry, fluorometry, or high-performance liquid chromatography (HPLC) (Lorenzen, 1967; Mantoura and Llewellyn, 1983; Wetzel and Likens, 1991; APHA, 1998; Van Heukelem et al., 1992; Millie et al., 1993). Spectrophotometric and flourometric chl a assays should be corrected for phaeophytin. Dry mass and ash-free dry mass are measured by drying and combusting samples (APHA, 1998). Cell density is measured after cells are counted microscopically (Lund et al., 1958; APHA, 1998; Stevenson and Bahls, 1999). Algal biovolume can be measured by distinguishing sizes of cells during microscopic counts, multiplying biovolume by cell size for all size categories, and finally summing biovolumes for all size categories in the sample (Stevenson et al., 1985; Wetzel and Likens, 1991; APHA, 1998; Hillebrand et al., 1999). Because of variation in vacuole size and cell walls (e.g., Sicko-Goad et al., 1977), cell surface area may be a valuable indicator of biomass because much cytoplasm is within 1–2 μm of the cell membrane.

Biomass can also be estimated rapidly with field assays, such as secchi depth in the water column and percentage cover and thickness of algal assemblages on substrata (Wetzel and Likens, 1991). Secchi depth characterizes light attenuation in the water column; assessments of algal biomass by this method are confounded by suspended inorganic materials, dissolved substances, and other factors (Preisendorfer, 1986). The strengths of assessing benthic algal biomass from percentage cover and thickness of algal assemblages is that biomass throughout a stream reach can be readily characterized (Holmes and Whitton, 1981; Sheath and Burkholder, 1985; Stevenson and Bahls, 1999). Remote sensing of algal biomass also shows promise for assessing spatially and temporally variable growths (Cullen et al., 1997). These rapid assessment techniques may permit more thorough spatial and temporal assessments, which may improve the notoriously variable relationships between biomass and nutrient concentrations or loading.

2. Taxonomic Composition

Taxonomic composition of algae is a powerful tool for assessing biotic integrity and diagnosing the direct and indirect causes of environmental problems (Stevenson, 1998). Differences in taxonomic composition of assemblages between an assessed site and a reference (desired) site can indicate impairment of biotic integrity and environmental conditions, if natural variation in assemblage composition is well documented (e.g., McCormick and O'Dell, 1996; McCormick and Stevenson, 1998). When natural seasonal or interhabitat variation in composition is not well known, changes in taxonomic composition can be related to human activities by comparing shifts in taxonomic composition to environmental change with autecological characteristics of species and relating inferred environmental changes to human activities (e.g., Kwandrans et al., 1998). Shifts in functional groups of algae (defined as different growth forms and divisions of algae; e.g., Pan et al., 2000) can also indicate an important change in food quality and in habitat structure for benthic invertebrates. For example, the food quality and accessibility of diatoms are usually greater than cyanobacteria and filamentous green algae for many herbivores (Porter, 1977; Lamberti, 1996). In addition, the habitat structure for benthic invertebrates differs greatly with changes from microalgae (e.g., diatoms) to macroalgae (e.g., *Cladophora*) (Holomuzki and Short, 1988; Power, 1990).

Taxonomic composition of algal assemblages can provide a highly precise and accurate characterization of biotic integrity and environmental conditions (Stoermer and Smol, 1999). Taxonomic composition of assemblages develops over periods of time, ranging from weeks to years, and should reflect environmental changes during that period. Even though taxonomic composition varies spatially and temporally in a water body, autecological characterizations of environmental conditions based on taxonomic composition should consistently reflect the physical and chemical changes caused by humans. For example, if trophic status of a habitat is being assessed, only low-nutrient indicator taxa should occur in low-nutrient habitats, even though temperature and shading and stage of community development may change with time and local habitat structure. Species presence and success in assemblages are fundamentally constrained by environmental conditions and interactions (e.g., competition) with other species in the habitat (Stevenson, 1997). Thus, attributes of assemblages based on percentage taxonomic similarity of assemblages at a test site and a reference site (Raschke, 1993; Stevenson, 1984) and percentage sensitive species should be good metrics because typically they sensitively, precisely, and monotonically change along gradients of human disturbance.

Taxonomic composition of algal assemblages usually requires microscopic assessments of samples, but some basic information about growth form and class can be obtained with rapid field assessments (e.g., Sheath and Burkholder, 1985; Stevenson and Bahls, 1999). The methods used for microscopic identification and counting of algae depend upon the objectives of data analysis and type of sample. A two-step process has been adopted by the national stream assessment programs in the United States. The first step is to count all algae and identify only nondiatom algae in a wet mount at either 400× (e.g., Palmer cell) or 1000×, if many small algae occur in samples. Algae can be counted in wet mounts at 1000× with an inverted microscope (Lund et al., 1958) or with a regular microscope by drying samples onto a coverglass, inverting the sample onto a microscope slide in 20 µL of water, and sealing the sample by ringing the coverglass with fingernail polish or varnish (Stevenson, unpublished method). The second step is to count diatoms after oxidizing organic material out of diatoms and mounting them in a highly refractive mounting medium (Stevenson and Bahls, 1999). This technique provides the most complete taxonomic assessment of an algal assemblage. Using counts of 300 algal cells, colonies, or filaments and about 500 diatom valves is a standard approach of some U.S. national programs (Porter et al., 1993; Pan et al., 1996) and usually provides relatively precise estimates of the relative abundances of the dominant taxa in a sample. Alternatively, counting rules have been defined so that cells of all algae are identified and counted until at least 10 cells (or natural counting units = cells, colonies, or filaments) of the 10 dominant taxa are counted (Stevenson, unpublished data). This type rule, rather than a fixed total number of cells, ensures precision in estimates of a specified number of taxa. Some assessment programs primarily use diatoms (Bahls, 1993; Kentucky Division of Water, 1993; Kelly et al., 1998; Kwandrans et al., 1998), because the number of species in diatom assemblages is usually sufficient to show a response.

Taxonomic composition can be recorded as presence/absence, percentage or proportional relative abundances, percentage or proportional relative biovolumes, or absolute densities and biovolumes of taxa (cells or μm^3 cm^{-2} or mL^{-1}). Although there is no published comparison of these forms of data, they represent scales of biological resolution and probably reflect a gradient from least sensitive to most sensitive and least variable to most variable. Presence/absence records of species should be based on observations of thousands of cells and conceptually should reflect long-term changes in habitat conditions if immigration and colonization of habitats are a selective barrier. Relative abundances and biovolumes of taxa probably reflect recent habitat conditions more than long-term conditions because of recent species responses to environment. Densities and biovolumes of taxa change daily, so absolute densities and biovolumes may be too sensitive to detect more long-term environmental changes. Relative abundance of cells is more commonly used than relative biovolumes (because of ease of use), but the latter is particularly valuable when cell sizes vary greatly among taxa within samples.

3. Diversity

Richness and evenness of taxa abundances are two basic elements of diversity (Shannon, 1948; Simpson, 1949; Hurlbert, 1971) of biological assemblages. Richness and evenness are hypothesized to decrease with increasing human disturbance of habitats; however, evenness of species abundances may increase if toxic stresses retard the growth of dominant taxa more than rare taxa (e.g., Patrick, 1973). Two problems develop with use of diversity measures in environmental assessment: standard counting procedures may not accurately assess diversity (Patrick et al., 1954; Stevenson and Lowe, 1986), and diversity may not change monotonically across the gradient of human disturbance (Stevenson, 1984; Jüttner et al., 1996; Stevenson and Pan, 1999). Species diversity and evenness are highly correlated with standard 300–600 cell counts (Archibald, 1972). In these counts many species have usually not been identified, so richness is more a function of evenness than evenness is a function of richness (Patrick et al., 1954; Stevenson and Lowe, 1986). As shown by standard counting procedures, nonmonotonic (showing both positive and negative changes as the independent variable increases) responses of algal diversity to some environmental gradients seem to be related to maximum evenness of tolerant and sensitive taxa at midpoints along environmental optima, to fewer species being adapted to environmental extremes at both ends of environmental gradients, and to subsidy-stress perturbation gradients

(Odum et al., 1979). Despite these difficulties, species richness and evenness may respond monotonically (having only positive or negative changes, but not necessarily linear changes, as the independent variable increases), sensitively, and precisely to gradients of human disturbance in some settings and should be tested for use as metrics.

4. Chemical Composition

The chemical composition of algal assemblages can be used to assess the trophic status of water bodies (e.g., Carlson, 1977), such as total phosphorus (TP) and nitrogen (TN) concentrations of water and periphyton (Dodds et al., 1998; Biggs, 1995). TN:TP ratios are widely used to infer which nutrient regulates algal growth (Healey and Hendzel, 1980; Hecky and Kilham, 1988; Biggs, 1995). In many of these assessments, most of the total P and N are particulate, and much of the particulate matter is algae. Thus, measurements of TP or TN per unit volume or area of habitat largely reflect the amount of algae in the habitat. Of course, the most widespread use of trophic assessments with TP and TN is phytoplankton in lakes (Carlson, 1977), but use has also been proposed for streams, rivers, and wetlands (Dodds et al., 1998; McCormick and Stevenson, 1998). TP and TN per unit biomass in benthic algae have also been positively correlated to benthic algal biomass in streams; however, negative density-dependent effects may reduce biomass-specific concentrations of benthic algal TP and TN and confound estimates of P and N availability to cells (Humphrey and Stevenson, 1992). Volume-specific, area-specific, and biomass-specific estimates of TP and TN do increase monotonically with most gradients of human disturbance and may be good metrics for trophic status in streams, rivers, and wetlands as well as lakes.

Chemical assessments are also valuable for monitoring heavy metal contamination in rivers, lakes, and estuaries (Briand et al., 1978; Whitton et al., 1989; Say et al., 1990). Many algae accumulate heavy metals when exposed to them in natural environments (Whitton, 1984). While toxicity of heavy metals to algae is one reason for monitoring heavy metals in algae, other reasons include bioaccumulation and metal removal from waste streams and movement of heavy metals into the food web (Whitton and Shehata, 1982; Vymazal, 1984; Radwin et al., 1990).

5. Functional Attributes

Metabolism of algal assemblages is highly sensitive to environmental conditions and is important to the assessment of ecosystem function and many ecosystem services. Estimates of photosynthesis (gross primary productivity), respiration, net primary productivity, nutrient uptake and cycling, and phosphatase activity are common functions measured in ecological studies (Bott et al., 1978; Healey and Hendzel, 1979; Wetzel and Likens, 1991; Marzolf et al., 1994; Hill et al., 1997; Whitton et al., 1998). These techniques are rarely incorporated into routine monitoring and survey work because they require more field time than typical water, phytoplankton, and periphyton sampling. However, they can be valuable additions to bioassessment projects. Biggs (1990) describes using algal growth rates to assess stream enrichment. Metabolism can be based on an area-specific, volume-specific, or biomass-specific basis. The most direct measurement of cellular performance is biomass-specific rates of metabolism; however area-specific and volume-specific measurements directly relate to community performance and ecosystem services. Caution in the accurate use of these attributes must be exercised. Area- and volume-specific measurements of productivity and respiration increase with biomass in the habitat, irrespective of human influence of biomass-specific rates. However, for periphyton, biomass-specific rates of productivity and nutrient uptake decrease substantially with increasing biomass in the habitat (e.g., Hill and Boston, 1989), presumably because of shading and impairment of nutrient mixing through the microbial matrix (Stevenson and Glover, 1993).

Gross and net productivity and respiration can be measured in the field with light and dark chambers and changes in oxygen concentration (Bott et al., 1978; Wetzel and Likens, 1991). Alternatively, productivity can be estimated with changes in oxygen concentration in the water during the sampling period or at two locations in a stream, if diffusion of oxygen from the water column is properly accounted for (Kelly et al., 1974; Marzolf et al., 1994). Furthermore, algae secrete an enzyme called phosphatase when in low-P environments. The phosphatase enzyme cleaves PO_4 from organic molecules and makes it biologically available. Phosphatase is measured with water samples in the laboratory (Healey and Hendzel, 1979).

6. Bioassay

In this discussion we define bioassays as in-lab culture of organisms in waters from the study site. A valuable, field-based use of this technique is the *Selenastrum* bottle assay, in which known quantities of this highly culturable green alga are added to water from the study site and growth is monitored over a predefined period (Cain and Trainor, 1973; U.S. Environmental Protection Agency, 1971; Trainor and Shubert, 1973; Greene et al., 1976; Ghosh and Gaur, 1990; McCormick et al., 1996). Alternatively, planktonic or benthic assemblages from reference or test sites could be cultured in bioassays with waters from those habitats or different dilution levels of effluents entering those regions. Twist *et al.* (1997) introduced the novel approach of embedding test organisms in alginate and culturing them *in situ*. Nutrient-diffusing substrata, microcosms, and mesocosms are also valuable bioassay techniques in field settings (e.g., Cotê, 1983; Fairchild *et al.*, 1985; Gensemer, 1991; Hoagland *et al.*, 1993; Lamberti and Steinman, 1993; Thorp *et al.*, 1996). Response of organisms to bioassays can provide another valuable line of evidence for identifying causes of environmental stress. Bioassay results with specific chemicals or effluents added can be used to confirm cause–effect relations between parameters for which only observational correlations can be obtained in field surveys (e.g., McCormick and O'Dell, 1996; Pan *et al.*, 2000).

IV. DEVELOPING METRICS FOR HAZARD ASSESSMENT

A. Relating Goals to Ecological Attributes

Hazard assessments are the determination of the intensity, spatial extent, frequency, and duration of environmental problems or the threat of environmental problems in which ecological conditions do not meet designated use (U.S. EPA, 1996). Designated use describes the goals for environmental protection in U.S. management plans, such as preserving biotic integrity and biodiversity, maintaining fishable and swimmable conditions, minimizing taste and odor problems or risks to human health in water supplies, optimizing sustainable fisheries production, and protecting human health. Designated use emphasizes valued ecological attributes that the public wants to protect. Many algal attributes can be related to these designated uses. We recommend selecting as many metrics as possible to develop multiple lines of evidence to help assess ecological conditions, which can be altered in many ways. Thus, hazard assessment requires identifying the goals of environmental assessment, selecting algal metrics that represent qualities of ecosystems that are related to designated uses, and then measuring those algal metrics to determine if goals are being met.

One of the most fundamental goals of environmental assessment is to determine if the natural balance of flora and fauna has been altered in a habitat. The concept of biotic integrity and natural balance of flora and fauna is a legislated goal in the United States' Clean Water Act (Karr and Dudley, 1981; Adler, 1995), is fundamental to ecosystem protection and sustaining biodiversity (Angermeier and Karr, 1994), and is broadly applied in U.S. monitoring programs in which indices of biotic integrity are used to identify ecological problems (Plafkin *et al.*, 1989; Barbour *et al.*, 1999; Karr and Chu, 1999). Biotic integrity or ecosystem health can be defined as the similarity between assemblages in an evaluated habitat and assemblages in a set of reference habitats (*sensu* Hughes, 1995). That assessment of similarity can be based on structural and functional characteristics. In most cases, assessment of biotic integrity has been based on changes in diversity, species composition, functional groups (such as proportions of diatoms versus green algae and cyanobacteria), and changes in ecological conditions inferred by species composition and species autecological characteristics (Karr, 1981; Smol, 1992; Kerans and Karr, 1994; Stevenson and Bahls, 1999).

Alternatively, more specific assessments of algal nuisance can be the goals of projects. Such nuisances may cause reduced water clarity, hypolimnetic deoxygenation, taste and odor problems, habitat alteration, or toxic effects on other organisms, including humans (Carmichael, 1994, Chap. 24). In these cases, algal biomass or specific problem taxa may be important attributes for assessment.

Attributes that respond to gradients of human disturbance are classified as metrics. A good metric for hazard assessment is unambiguous, sensitive, precise, reliable, and transferable among regions and perhaps water body types (Murtaugh, 1996; Karr and Chu, 1999). McCormick and Cairns (1994) list other ideal qualities of indicators that should also be considered, such as relevance, redundancy, and cost-effectiveness. An unambiguous attribute responds monotonically to increasing levels of human disturbance (Fig. 4a and b). An attribute that can be equal at low and high levels of human disturbance (Fig. 4c) is ambiguous and should not be used as a metric. A good attribute may respond nonlinearly to human disturbance, but most change substantially (i.e., be sensitive) over the range of human disturbance being assessed. Precision (low variability in repeated measures) of metrics is important for detecting responses. In addition, metrics will be much more effective if they respond to human disturbance at many times of year and if they respond in many regions.

Two categories of attributes can be identified that characterize or indicate valued ecological attributes, and these categories are distinguished by the degree to which reference conditions need to be characterized. All attributes in the first category can be directly related to gradients of human disturbance by regression analysis to determine their use as metrics without rigorous identification and characterization of reference conditions and assemblages at reference sites. The first category includes attributes such as biomass,

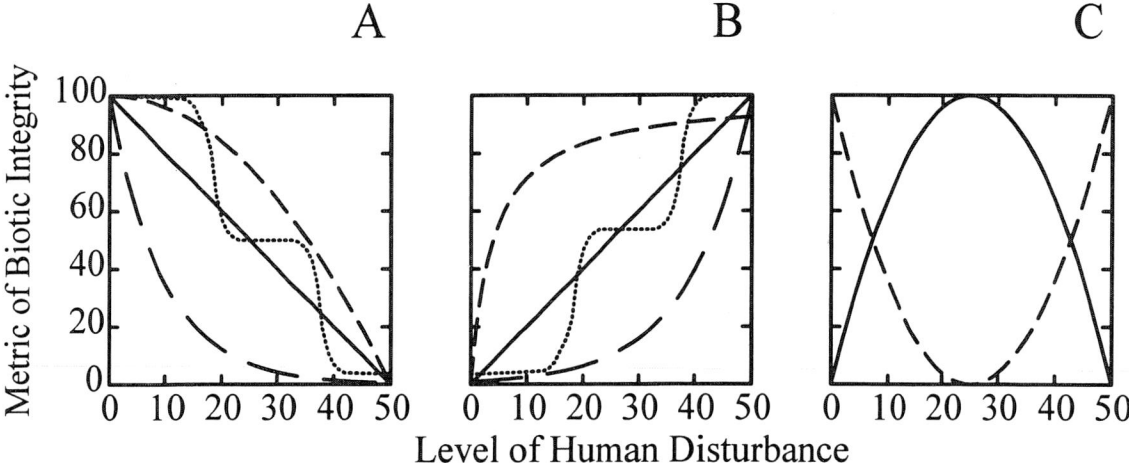

FIGURE 4 Examples of attribute responses that provide useful metrics. (A) Negative attribute responses along a gradient of human disturbance. (B) Positive responses along a gradient of human disturbance. (C) Ambiguous responses along a gradient of human disturbance. The different lines in each figure (A–C) represent different patterns that would fit the categories represented in the figures (negative, positive, and ambiguous, respectively).

species richness, Shannon (1948) diversity, relative abundances of specific nuisance taxa or functional groups of taxa, relative abundances of pollution-sensitive and pollution-tolerant taxa, and function of assemblages (Tables I and II; Stevenson and Bahls, 1999).

A second category of metrics requires comparison of differences between reference and test sites as the response variable to the gradient of human disturbance (Table II). Similarity of species composition between test sites and reference sites is one potential metric. Similarity of species composition between reference and test sites should decrease with increasing human disturbance at test sites. Many different formula can be used to calculate percentage similarity. One compares

TABLE II Examples of Metrics That Are Based on Taxonomic Composition of Assemblages and Autecological Characteristics of Species

Class		Stressor	Representative references
SC	BI	Pollution in general (based on species)	Lange-Bertalot (1979), Descy (1979), Coste (1982), Bahls (1993)
SC	BI	Pollution in general (based on genera)	Rumeau and Coste (1988), Coste and Ayphassorho (1991)
SC	D	pH	Whitmore (1989)
SC	D	Trophic status	Whitmore (1989), Kelly and Whitton (1995)
SC	D	Organic wastes	Zelinka and Marvan (1961), Palmer (1969), Sládeček (1973, 1986), Watanabe et al. (1986)
SC	D	Salinity	Zeimann (1991)
WA	D	pH	Charles and Smol (1988), ter Braak and van Dam (1989), Sweets (1992), Cumming et al. (1992a, b)
WA	D	Salinity	Fritz (1990), Cumming and Smol (1992)
WA	D	TP	Anderson et al. (1993), Reavie et al. (1995), Pan et al. (1996), Pan and Stevenson (1996)
WA	D	Fish presence	Kingston et al. (1992)

Metrics are classified based on whether they are based on simple categorical autecological characterizations (SC) or accurate weighted-average (WA) autecological characterizations and on whether they could be used to infer biotic integrity (BI) of sites or are highly diagnostic (D) of stressors that may threaten or impair biotic integrity.

relative abundances of species in two samples and bases similarity upon the sum of the lower relative abundances of each taxon in the two communities. The percentage community similarity (PS_C; Whittaker, 1952) is a straightforward example, where

$$PS_C = \Sigma_{i=1,S} \min(a_i, b_i)$$

Here a_i is the percentage of the ith species in sample a, and b_i is the percentage of the same ith species in sample b.

A second kind of similarity measurement is based on a distance measurement, which is a dissimilarity measurement rather than a similarity measurement, because the index increases with the greater dissimilarity (Pielou, 1984; Stevenson, 1984). Euclidean distance (ED) is a standard, where

$$ED = \sqrt{(\Sigma_{i=1,S}(a_i - b_i)^2)}$$

When using these dissimilarity indices, we recommend log-transforming relative abundances to reduce the importance and variability of common taxa.

A third category of similarity indices only compares the observed taxa at an assessed site with the taxa that are expected at that site to determine the proportion of species that have been lost from assessed sites. This technique has been used widely with macroinvertebrates (Moss et al., 1999) and was recently employed with diatom genera (Chessman et al., 1999). This approach has the potential for increasing the precision of algal metrics by distinguishing species in three categories: species that should be there and are still there, species that have been lost, and species that have invaded. First, loss of species is an important impairment of biodiversity. Second, diagnosis of causes of impairment may be substantially enhanced by linking autecological information to whether species have resisted disturbance, not resisted disturbance, or invaded to exploit disturbance.

In our application of the ERA framework, we emphasize the distinction between algal indicators that characterize designated use and ecological values in which the public are most interested and algal indicators that diagnose the stressors that may threaten or cause impairment of designated use. Some protocols recommend that metrics indicating the status of designated use be called "response" or "condition" indicators, and metrics that diagnose the physical, chemical, or biological factors that could be impairing designated use be called "stressor" or "causal" indicators (Paulsen et al., 1991; U.S. Environmental Protection Agency, 1998). This distinction of types of metrics emphasizes the diversity of information that can be obtained with algal assessments and how to apply that information in environmental problem solving. Therefore, algal indicators that use environmental preferences of species, such as weighted average indices of pH or TP, will be described later under exposure assessment, rather that here under hazard assessment.

B. Testing Metrics

Metrics can be tested with measurements of attributes at multiple sites with varying levels of human disturbance and either parametric or nonparametric statistical methods (see Sokal and Rohlf, 1998). Sites with different levels of human impact should be chosen and sampled to assess the ecological response to human disturbance. The level of human disturbance at sites can be characterized with multiple lines of evidence. When point sources of pollution occur, environmental gradients are relatively simple to establish with a reference condition upstream from the point source and a decreasing gradient of disturbance at increasing distances downstream from the point source. When non-point-source pollution is a concern, human disturbance can be estimated by land use type, intensity, and proximity to a habitat or by concentrations of contaminants. The multivariate nature of complex non-point-source contamination can be simplified with the use of ordination techniques and axis scores as a ranking scale of human disturbance.

Reference sites help define expected conditions in a habitat if it had not been affected by human activities (Hughes, 1995); these are typically the least impacted ecosystems in the region. Reference sites can be sites upstream from a point source of pollution in a stream, whereas test sites can be downstream. Alternatively, reference sites for a specific climatic and hydrogeomorphic class of habitats can be defined as a set of sites with lowest human disturbance or greatest riparian buffer within their watersheds. Reference sites may have the lowest level of a specific stressor in them, such as low phosphorus and other specific indicators of human disturbance.

Algae are particularly useful in establishing reference conditions in lakes and wetlands, where sediments are continuously deposited because algal remains in sediments from times of low human disturbance can be used to infer historical conditions in those habitats. Paleolimnological approaches provide direct measurements of long-term environmental trends at a specific site, which increases certainty about how fast and the extent to which a system is deteriorating. To propose realistic mitigation procedures, paleolimnological reconstruction of past conditions can provide a realistic target for restoration. Long-term data can also show critical loads of pollutants or stressors that a system can handle before negative effects are manifested

(Smol, 1990, 1992, 1995; Anderson and Battarbee, 1994; and papers in Stoermer and Smol, 1999).

Paleolimnological approaches are based on fairly straightforward principles. Under ideal conditions, sediments slowly accumulate at the bottom of lakes, without disruptions. Certainly, in some cases, problems may occur (e.g., excessive bioturbation), but these problems can usually be recognized and assessed. Over time, therefore, the history of the lake and its watershed is archived in the depth/time profile of the sediments. Incorporated in these sediments is a surprisingly large library of information on the conditions present in the lake (from autochthonous indicators), as well as environmental conditions that existed outside the lake (from allochthonous indicators). Physical, chemical, and biological information is archived in sediments; however, for the purpose of this chapter, we will primarily focus on algal data. Paleolimnology is now widely recognized as a robust environmental management tool. We mainly discuss lake paleoenvironmental studies in this chapter, as most of the research has centered on these systems. However, many paleo approaches can easily be transferred to other aquatic systems such as ponds (Douglas et al., 1994), rivers (e.g., Reavie and Smol, 1987, 1998; Amoros and Van Urk, 1989; Reavie et al., 1998), wetlands (Bunting et al., 1997), estuaries (Cooper, 1999), and marine systems (Anderson and Vos, 1992).

After algal attributes from habitats with different levels of human disturbance have been assessed, their response to human disturbance can be characterized. Log-transformation of biomass-related variables (such as chl *a*, cell density, and biovolume) is recommended to meet the equal variances assumption of regression analysis and sometimes to make patterns more linear. Other data transformations may be necessary to meet assumptions of statistical tests (e.g., Green, 1979; Sokal and Rohlf, 1998) or to manage the sensitivity of metrics. For example, arc-sine transformations of proportional data and log or square-root transformations of relative abundances should increase the normality of the data (Sokal and Rohlf, 1998) and can reduce the importance of highly variable abundant taxa, which increases the precision of some metrics.

Both nonparametric and parametric statistical techniques can be used to determine whether algal attributes respond to gradients of human disturbance. The simplest and most direct method is to compare attributes to the gradient of human disturbance by regression or correlation (e.g., Hill et al., 2000), if human disturbance can be quantified on a continuous scale. Alternatively, human disturbance can be categorized as low and high, and ANOVA or Mann–Whitney U tests can be used to test for differences in metrics with differences in human disturbance (Green, 1979; Barbour et al., 1992; Sokal and Rohlf, 1998; Barbour et al., 1999).

C. Multimetric Indices

Summarizing data in the form of multimetric indices has been a valuable method for communicating results of complex analyses that often involve multiple lines of evidence. This method has been used commonly with fish and invertebrate assemblages as multimetric indices of biotic integrity (IBI) (Karr, 1981; Kerans and Karr, 1992), but it has also been used with periphyton (Kentucky Division of Water, 1993; Hill et al., 2000). Development of a multimetric index calls for selecting 6–10 metrics that describe a diversity of responses of assemblages that will be sensitive to all probable environmental stressors. For example, species richness, percentage of diatoms, percentage similarity to reference assemblages, number of taxa sensitive to pollution, percentage motile diatoms, percentage aberrant diatoms, inferred trophic status, inferred salinity (conductivity), inferred saprobity, and inferred pH could be used in a multimetric index, if they all responded to gradients of human disturbance (i.e., performed as good metrics). The range of each metric should be normalized, for example, to a range of 0–10, so that each metric has equal weight (see Hill et al., 2000). Then values of each metric for a sample can be summed. In an example with 10 metrics and each ranging from 0 to 10 in scale, the multimetric index would then range from 0 to 100.

The value of multimetric indices is that they provide a single number as a summary of multiple lines of evidence. Such a summary statistic is highly valuable for communicating information to a lay audience, especially when compared with interpretations of ordination analyses (Karr and Chu, 1999). The disadvantage is that they may mask effects on one or two metrics; however, they can provide a hierarchically decomposable system of metrics for assessing ecological risk and even diagnosing causes or threats to impairment (Stevenson and Pan, 1999).

D. Multivariate Statistics and Hazard Assessment

Multivariate statistics are powerful and informative statistical tools for determining the major patterns of change in species composition and relating them to physical, chemical, or other biological characteristics of the habitats studied. We regard a multimetric index of biotic integrity (IBI) and multivariate statistics as complementary tools.

Cluster analysis and ordination provide multivariate methods for grouping stations by similarity in assemblage structure, exploring patterns in data, and illustrating those patterns (Hill, 1979; Pielou, 1984; Jongman *et al.*, 1995). Cluster analyses can be bottom-up, such as UPGMA, or top-down, like TWINSPAN. A recent evaluation comparing these approaches showed that UPGMA, compared with TWINSPAN, grouped artificial assemblages better (developed based on a selected set of assumptions and assignment of species' relative abundances, based on a probabilistic distribution) (Belbin and McDonald, 1993). However, many researchers use TWINSPAN and find results with actual data to be highly interpretable (e.g., Pan *et al.*, 2000).

Ordination (e.g., correspondence analysis and principal components analysis) condenses patterns in assemblage characteristics to axes that explain the covariation in assemblage characteristics among samples. Similarity in species composition among sites and species responses to environmental conditions at sites can be compared by plotting sample scores and species loadings in ordination space (Fig. 5). Environmental factors can then be incorporated into correspondence analyses (CAs) to relate variation in species distributions and sampled sites to variation in environmental conditions among sites. Many canonical ordination techniques can be applied to relating species and environmental variance among sites. If great differences occur in species composition among sites, a U-shaped pattern in ordination scores of sites often occurs; this artifact can be reduced by using detrending techniques (e.g., DCA) (Jongman *et al.*, 1995).

Multivariate statistics are valuable for the early stages of environmental programs when initial relationships between changes in assemblages and environmental conditions are being explored to develop metrics and multimetric IBI (e.g., Pan *et al.*, 1996). They can be an important part of any program when it is necessary to reduce the complexity of multivariate data. Even though cluster analysis and ordination group assemblages in classes that are often biologically interpretable, they do not test hypotheses that are directly related to questions of whether a site or a group of test sites is impaired or not. Testing these hypotheses calls for multivariate analysis of variance or discriminate function analysis (Pan *et al.*, 2000). These latter approaches may be valuable for finding thresholds along gradients and being able to establish a probability that sites with specific characteristics were exposed to unacceptable levels of human disturbance and were impaired. Another weakness in multivariate statistics as an endpoint in ERA is that results are often not easily and repeatedly interpretable by audiences that are not trained in the use of multivariate statistics. Reviews of multivariate statistics and their use can be found in Green (1979), Pielou (1984), Jongman *et al.* (1995), and Birks (1995, 1998).

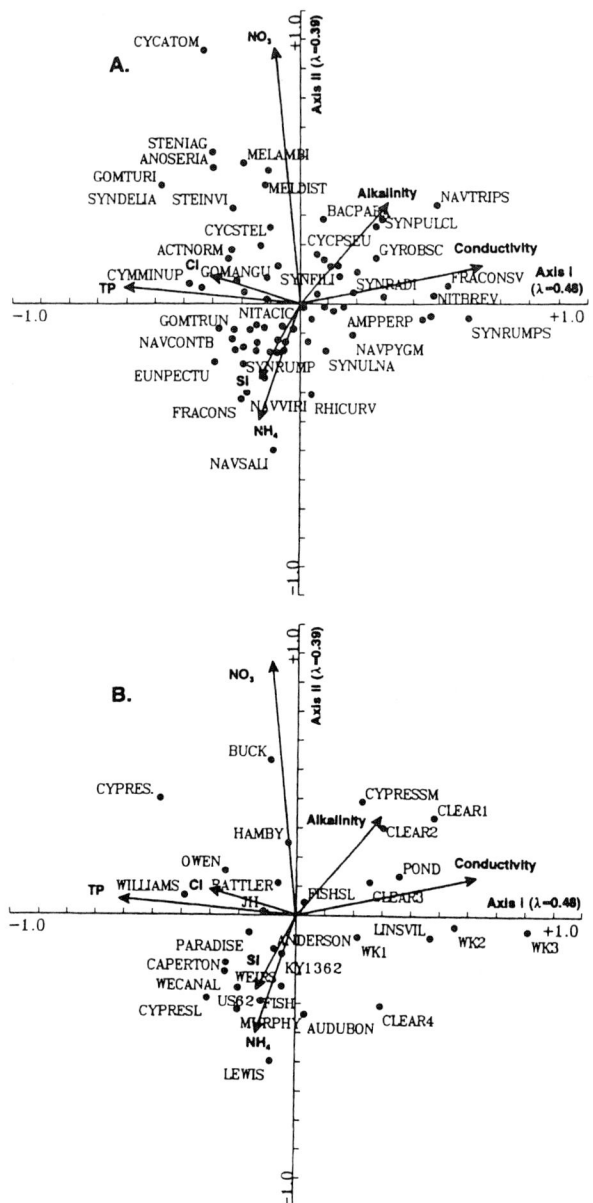

FIGURE 5 Plots of sample sites and species along ordination axes to indicate the relationship between similarity in species composition among sites and the species that are most important in defining that similarity (from Pan and Stevenson, 1996). (A) Plot of species and environmental variables (arrows) along ordination axes to show which species are most important in defining similarity among assemblages. (B) Plot of sample sites that shows sites with similar species composition located in similar locations along ordination axes with related environmental variables (arrows).

V. EXPOSURE ASSESSMENT: WHAT ARE ENVIRONMENTAL CONDITIONS?

Exposure assessment may be as simple as measuring stressors, such as pH or phosphorus concentration, directly, or it may call for using biological indicators of exposure. Exposure assessment is important for precisely characterizing the level or intensity of environmental conditions that may be affecting valued ecosystem components or services. Often, environmental conditions in a habitat cannot be measured accurately or precisely, particularly in shallow-water habitats like streams and wetlands, where environmental conditions change diurnally and seasonally with biological activity and from day to day with weather.

Many algal taxa have long been recognized to be, with varying degrees of specificity, restricted to certain aquatic environments (Kolkwitz and Marsson, 1908); therefore, they can potentially be used as bioindicators of environmental conditions. The simplest of the quantitative stressor or causal indicators are simply the sum of relative abundances of organisms that are either tolerant or sensitive to a specific environmental stressor, such as the relative abundance of motile diatoms or aberrant diatoms, which indicate silt and heavy metal pollution, respectively (Bahls, 1993; McFarland et al., 1997). Alternatively, the relative abundance of organisms adapted to environmental extremes could be used to diagnose stressors, such as high organic contamination, high salinity, low dissolved oxygen, or low pH (Stevenson and Bahls, 1999). More complex quantitative approaches use species composition of algal assemblages and categorical rankings of species environmental preferences, with either weighted average equations (Zelinka and Marvan, 1961) or regression equations (Renberg and Hellberg, 1982) to infer the stressor level. Recently, new accessibilities to personal computers and new statistical techniques (weighted average assessment of species preferences) have enabled the development of more accurate characterizations of species environmental preferences and more accurate and precise biological indicators of stressors in ecosystems (see Birks, 1995, 1998, for reviews). Thus, stressor levels in a habitat can be inferred with weighted average equations with species autecologies that were developed based on a categorical ranking of species environmental preferences or with autecologies determined with weighted average techniques.

Simple autecological ranks have been assigned to characterize environmental preferences for many taxa and many environmental characteristics (see van Dam et al., 1994, for a review). These autecological characteristics of taxa have been compiled in several reviews: Lowe (1974), Beaver (1981), Denys (1991), Hofmann (1994), and van Dam et al. (1994). Using a weighted average formula, stressor levels in habitats can be inferred based on the categorical autecological ranks of taxa (often eight or fewer categories) and relative abundances of taxa in samples. For example, a simple autecological index (SAI) for trophic status can be developed based on autecological ranks (Θ_i) of 0-7 for taxa observed to be most abundant in waters classified as ultraoligotrophic, oligotrophic, oligo- to mesotrophic, mesotrophic, meso- to eutrophic, eutrophic, and hypereutrophic, respectively (see van Dam et al., 1994). Then the trophic index can be calculated as

$$SAI_{TI} = \Sigma_{i=1,S} p_i \Theta_i$$

where p_i is the proportion of the ith species and Θ_i is the ecological condition in which the highest relative abundances of the ith species are collected. If autecological information is not known for all taxa, valuable information can still be obtained by correcting the index for the proportion of taxa with autecological characterizations. Next, one can redefine the community as the subset of taxa for which autecological characteristics are known by dividing the SAI by the sum of the proportional abundances of taxa with known autecological information.

This weighted average approach with simple autecological characterizations of taxa has been used extensively in stream assessments with algae, particularly in Europe (Table II) (see reviews in Whitton et al., 1991; Whitton and Rott, 1996; Prygiel et al., 1999). Software and databases of diatom autecological characteristics (e.g., OMNIDIA; Lecointe et al., 1993) have been developed that can be used to calculate these indices. In tests of these indices, some perform better than others when used in regions other than those for which they were originally developed (e.g., Kwandrans et al., 1998). Regional calibration of these indices may be required to improve performance by reassessment of algal autecological characteristics.

Stressor indicators based on accurate weighted-average assessments of species' environmental optima are more precise than indicators based on categorical characterizations of species' autecologies (e.g., ter Braak and van Dam, 1989; Agbeti, 1992). However, acquiring accurate descriptions of species' autecologies may be more difficult than using categorical characterizations. Weighted-average inference (WAI) models have been widely used to precisely infer environmental characteristics (Birks, 1998). The general principle for characterizing species autecological preferences is that, under most circumstances, the distributions of most algal taxa will exhibit a unimodal, Gaussian response curve (Fig. 6), if the gradient is long enough. The optimum (m) is estimated by the position along the environmental gradient where the taxon is most common, and the tolerance (t) can be estimated by the standard deviation of the curve. Because different taxa will have different optima and tolerances to environmental variables, these data can be used to make quantitative inferences of these variables (Fig. 7).

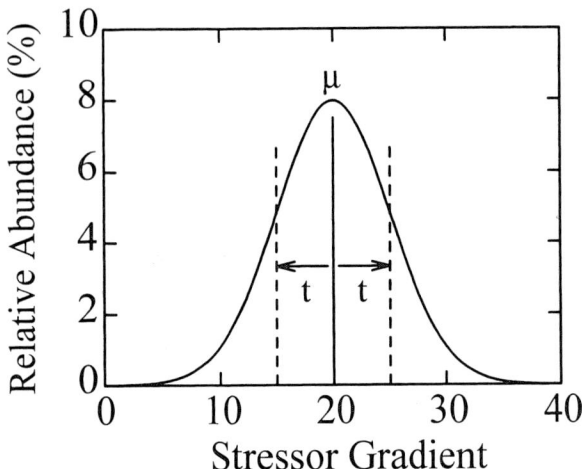

FIGURE 6 Idealized pattern in relative abundances of a single algal species along an environmental gradient showing a unimodal curve with optima (μ) and tolerance (t) to environmental gradients.

Some of the largest and most robust ecological calibration sets, or training sets, have been developed by paleolimnologists using diatoms and chrysophytes preserved in the surface (recent) sediments from a set of calibration lakes. Briefly, a calibration set is constructed by choosing a suite of sites that have been well studied (i.e., limnological characteristics are well defined) and span the gradient of interest (e.g., pH, trophic status, etc.). For example, one of the most pressing environmental questions in North America in the 1980s was, "Have lakes acidified because of acid precipitation?" Very little long-term monitoring data were available, so paleolimnological approaches were used to infer past pH and related limnological variables.

These approaches, which have been used extensively in both paleo- and neo-environmental assessments, are statistically robust and ecologically sound (reviewed in Charles and Smol, 1994; Birks, 1995, 1998). In general, these quantitative inferences are based on transfer functions that have been derived from paleolimnological studies, typically with the use of surface sediment calibration sets (reviewed by Charles and Smol, 1994) or with the use of present-day algal assemblages from a large suite of sites (e.g., Siver, 1995; Reavie and Smol, 1997, 1998b; Stevenson et al., 1999; Pan et al., 1996).

A major research effort was focused on the lakes in Adirondack Park (New York) (Charles et al., 1990). To develop transfer functions to infer past lakewater acidity levels in a suite of Adirondack lakes, a calibration set of 71 lakes was chosen that ranged in present-day pH from 4.4 to 7.8 (Dixit et al., 1993). From each of these lakes, the surface sediments (e.g., top 1 cm of sediment accumulation, representing the last few years of sediment deposition) were removed with a gravity corer (Glew, 1991). The indicators preserved in these sediments, in this case diatom valves (Dixit et al., 1993) and chrysophyte scales (Cumming et al., 1992a), were analyzed (identified to the species level, counted, and expressed as relative frequencies) from the surface sediments of the calibration lakes. This provides one of the matrices (i.e., the 71 lakes and the percentages of the taxa found in the recent sediments of lakes)

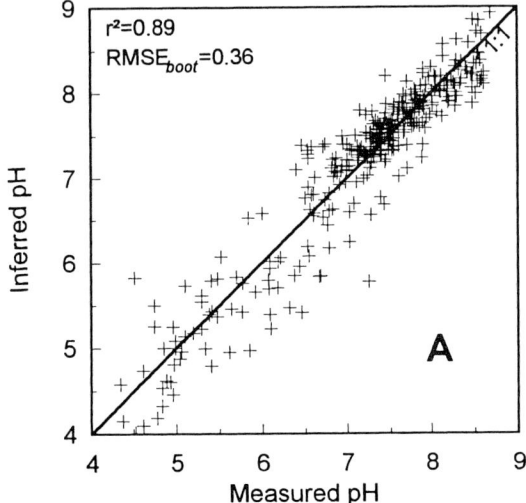
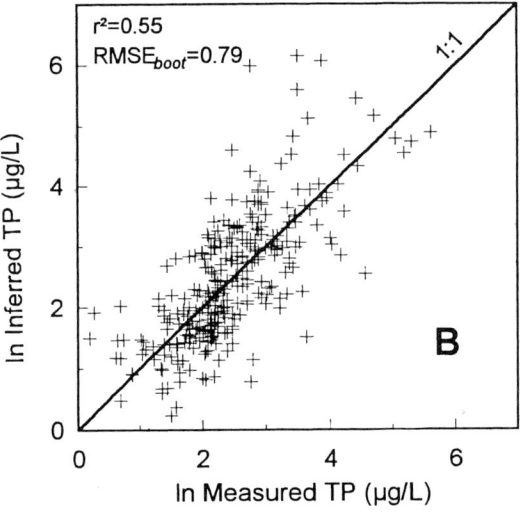

FIGURE 7 Weighted averaging calibration models for inferring lakewater pH (A) and lakewater total phosphorus (B) from diatom assemblages preserved in the surface sediments of 309 lakes in the northeastern United States. Modified from Dixit et al., (1999). The error is estimated by the bootstrapped root mean squared error.

required for the calibration (i.e., the species matrix). The second matrix was composed of the present-day environmental data collected for the 71 calibration lakes (in the above example, 21 limnological variables, such as lakewater pH, monomeric aluminum, dissolved organic carbon, nutrient levels, and depth were recorded). Once these two matrices are constructed, a variety of direct gradient analysis techniques (see reviews in Charles and Smol, 1994; Birks, 1995, 1998), such as Canonical Correspondence Analysis (CCA), can then be used to determine which environmental variables are most closely related to species distributions. Thereafter, weighted averaging calibration and regression (e.g., WACALIB; Line *et al.*, 1994) were used to construct robust inference equations to infer lakewater characteristics (with known errors) from the diatoms or chrysophyte assemblages recorded in the sediments (e.g., Cumming *et al.*, 1992b, 1994). Such approaches (e.g., Fig. 8) have been used in a variety of management issues (see Smol, 1992, 1995; Anderson and Battarbee, 1994).

The above calibration approaches can also be applied, in slightly modified forms, in settings where sediment accrual is not as regular as in lakes (e.g., some rivers and wetlands). For example, Reavie and Smol (1997, 1998) developed inference models from diatom assemblages attached to a variety of substrata in the St. Lawrence River and then used these transfer functions to infer past river conditions from sediment cores taken from fluvial lakes in the river system (Reavie *et al.*, 1998).

In situations where environmental conditions are highly variable, weighted averaging inference models have been shown to be better indicators of environmental conditions than one-time sampling and measurement of physical and chemical conditions (Stevenson, 1998). Field travel and sampling is an expensive part of program budgets, so habitats are often only sampled once. Water chemistry (TP concentration, for example) could be estimated based on a single water chemistry sample from a habitat or inferred based on algal species composition and autecological characteristics of those algae in a habitat. The precision of the estimate of mean TP concentration in a stream with a single water sample is the standard deviation of that assessment. The precision of an estimate of mean TP concentration in a stream with a single algal sample is the root mean square error of a weighted average regression model. Hence, 66% of estimates of mean TP concentration in a stream should be within one standard deviation of the measured TP concentration in a water sample and within one root mean square error of an inferred TP concentration from a weighted averaging inference model. Based on estimates of temporal variability in measured concentrations of TP along a wetland P gradient and among streams in a regional assessment, the standard deviation of the measured TP concentration was greater than the root mean square error of inferred TP with a weighted average model (Fig. 9; Stevenson, unpublished data).

Although intuitively one might think that it would be imperative to use ecological calibration data taken only from the region of study, and that autecological data are not readily transferable from region to region, experience suggests that this is not strictly the case. What is most critical in ecological calibration is to capture the range of environmental conditions that one will need to infer from the biological indicators. Although regional calibration data would be more likely to contain analogues, several studies have shown that data from geographically distant regions can also be used effectively. For example, as part of the SWAP acidification program in Europe, diatom calibration

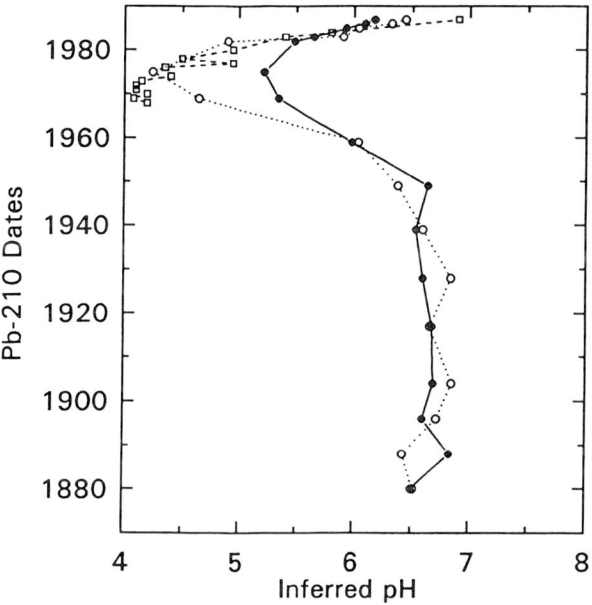

FIGURE 8 Paleolimnological assessment of acidification and recovery in Baby Lake, Sudbury, Ontario (modified from Dixit *et al.*, 1992b). The core has been dated with ^{210}Pb chronology. Closed circles represent inferred pH values from diatom assemblages, open circles represent values from scaled chrysophyte assemblages, and squares represent measured pH data (from Hutchinson and Havas, 1986) collected for the lake from various sources over the last three decades. These paleolimnological data clearly show a marked acidification of the lake in the middle part of this century, as a result of the emissions from the Sudbury smelters. Following the closure of the Coniston smelter in 1972, the fossil algal assemblages track a recovery pattern, which is matched by the measured pH data for this period. As is often the case, chrysophytes typically record more extreme acidification sequences, perhaps because they primarily bloom during spring, when the effects of acidification may be most severe.

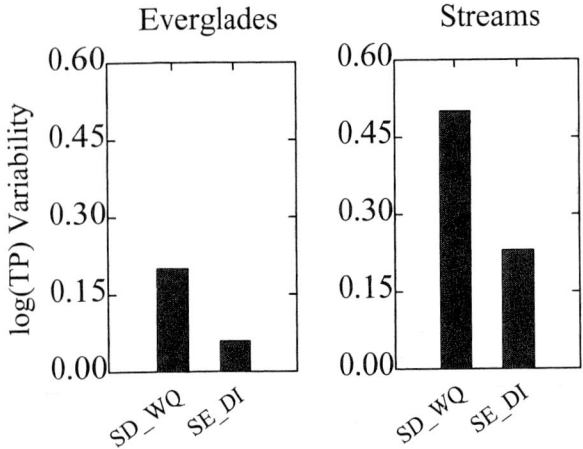

FIGURE 9 A comparison of total phosphorus (TP) variability estimated by one-time sampling and assay of water chemistry (SD_WQ) and by one-time sampling of diatom assemblages and inference with weighted average models (SE_DI). Variation in TP assessed in water chemistry (SD_WQ) is based on the standard deviation in assessed TP in the Everglades at one location over 2.5 years (provided by Paul McCormick, South Florida Water Management District, West Palm Beach, FL) and in a stream near Louisville, KY (Stevenson, unpublished data). The TP variability is based on the RMSE of diatom inference models from Everglades data (Slate, 1998) and mid-Atlantic Highlands streams (Pan et al., 1996).

data were effectively pooled from England, Norway, Scotland, Sweden, and Wales (Birks et al., 1990). Bennion et al. (1996) have combined regional diatom calibration sets from England, Wales, Northern Ireland, Denmark, and Sweden to develop robust transfer functions to infer lakewater epilimnetic phosphorus concentrations. As part of the CASPIA project, diatom calibration sets from North America, Africa, Australia, and Europe were pooled (Juggins et al., 1994). Ongoing work in our labs is also showing that calibration data are not as regional as may have once been thought, but that careful attention must be paid to taxonomic consistency and to the development of calibration data sets that effectively capture the necessary range of environmental variables.

VI. STRESSOR–RESPONSE RELATIONS

Effects of specific environmental changes on assemblage characteristics can be determined at many temporal and spatial scales and with both observational and experimental approaches. The largest scale employs surveys of large regions (e.g., ecoregions) and correlates changes in environmental conditions and assemblages, which provides observations of potential stressor–response relationships. Correlations between environmental factors and assemblage responses from surveys may show great changes and precision, and cause–effect relations may be biologically reasonable. However, experimental manipulation of environmental factors and measurement of assemblage response is important for more reliable confirmation of cause–effect relations. Experimental confirmation of stressor–response relations is particularly valuable in large-scale projects where expensive restoration efforts are planned and identification of the principal stressor is critical (e.g., McCormick and O'Dell, 1996; McCormick and Stevenson, 1998; Pan et al., 2000). Experimental approaches are also valuable in small-scale projects, where surveys of a large number of habitats are not practical because of budget or availability of habitats, and in toxicological studies of chemicals when chemicals are not yet widespread in the environment (Hoagland et al., 1993; Belanger et al., 1994). Observation of stressor–response correlations in large-scale surveys is useful because the relative importance of multiple environmental factors can be compared. In addition, observation of stressor–response relations in surveys shows that responses occur in the natural setting and that the relation can be expected to hold over the range of conditions studied in the survey.

Distinction between stressors and human activities that cause stressors is important in assessing stressor–response relations and in developing management strategies; this is the distinction between direct and indirect relations (U.S. Environmental Protection Agency, 1993; Yoder and Rankin, 1995; Kentucky Natural Resources and Environmental Protection Cabinet, 1997). Ecological responses to human disturbance may be caused directly by changes in physical, chemical, or biological conditions in a habitat (stressors) and indirectly by the human activities that cause those stressors. Distinguishing which stressors are most responsible for undesirable ecological responses and which human activities can be regulated to control those stressors is important for developing a plan to protect or restore environmental conditions. Many human activities (e.g., farming, logging, urbanization, sewage treatment plants) may be the source of a specific stressor (e.g., P enrichment). Many stressors (e.g., N and P enrichment, siltation, organic enrichment, and flow and light regimes) may be altered by a single human activity (e.g., farming).

The magnitude and linearity of stressor–response relationships may vary with the attribute tested. Many results indicate that metrics based on higher levels of biological organization (ecosystem/community level: e.g., biomass and productivity) are less sensitive to environmental change than metrics based on lower levels of biological organization (community/population: e.g., species composition) (Schindler, 1990; Leland, 1995). Because of high dispersal rates and high species numbers in microbial assemblages, species adapted to altered environmental conditions are probably able to invade and populate a habitat relatively quickly. Thus, impaired populations may be replaced by populations that are adapted to the altered conditions and thereby maintain ecosystem function (Stevenson, 1997). Therefore, biomass and many functional attributes of algal assemblages are often less sensitive to environmental change than changes in species composition (Schindler, 1990).

Relating the stressor–response relations between algal species composition and environmental factors to stressor–response relations for ecosystem attributes may help in the understanding of ecosystem dynamics and establishing criteria to protect ecological integrity. Recently, results from the Everglades show punctuated (sudden, discrete) changes in algal species composition along a phosphorus gradient (Pan et al., 2000) (Fig. 10). Along the same gradient, higher-level biological attributes, such as biomass and productivity, change linearly or asymptotically. Thus, punctuated changes in species composition may result in multiple stable states (May, 1974) along an environmental gradient, which result from changes in the factor or factors that are the most important constraints on species composition. Punctuated changes in biomass and productivity may be blurred by species replacement along the phosphorus gradient, and spatial and temporal variability in other factors that have more short-term effects on biomass. Criteria for protecting the ecological integrity of a habitat may be established at thresholds along

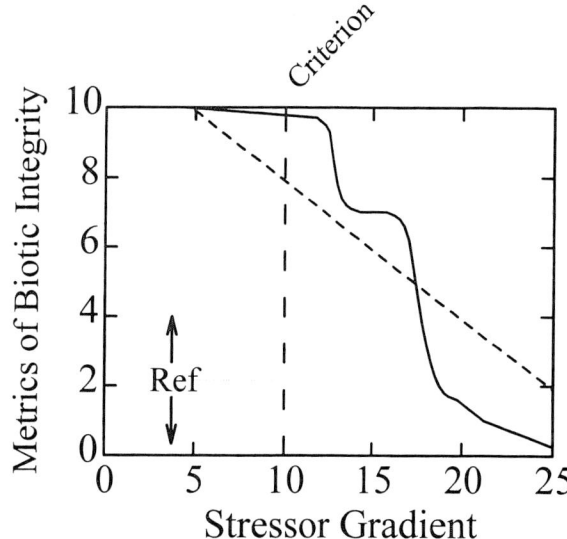

FIGURE 11 Linear (diagonal dashed line) and nonlinear (solid line) ecological responses along an environmental gradient (e.g., µg TP/L). Setting criteria for protection and remediation of ecological integrity may be facilitated by nonlinear ecological responses. Reference conditions (Ref) are indicated by the vertical arrow. The vertical line indicates a criterion.

gradients where punctuated changes in species composition occur that correspond to undesirable changes in more than one ecological attribute (Fig. 11).

VII. RISK CHARACTERIZATION AND MANAGEMENT DECISIONS

Risk characterization relates assessments of exposures to stressor–response relationships to evaluate the level of threat to an unimpaired system or the probable stressors and intensity of stress of impaired systems (Fig. 2) (U.S. Environmental Protection Agency, 1992). Thus, based on a set of metrics and stressor–response relationships, we can predict ecological effects of exposures to specific stressors, assess the likelihood that effects will occur, or assess the likelihood that effects were caused by specific stressors or interactions among multiple stressors. Standard risk characterizations also increase risk ratings with increasing uncertainty in information. Therefore, we are concerned with both underprotecting as well as overprotecting our resources.

Quantitatively, risk characterization on a metric-by-metric and stressor-by-stressor basis can be conducted to assess the sustainability of ecological conditions and the level of impairment or restorability of impaired conditions (Stevenson, 1998) (Fig. 12). Sustainability can be defined quantitatively as the difference between

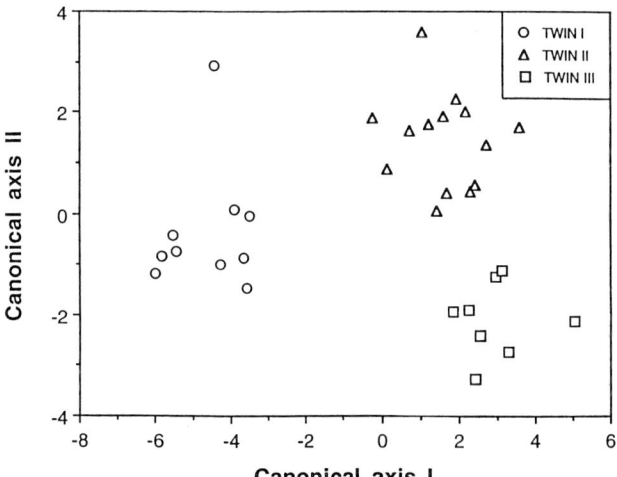

FIGURE 10 Ordination of sites based on species composition of algal assemblages at sites along a phosphorus gradient in the Everglades (Pan et al., 2000).

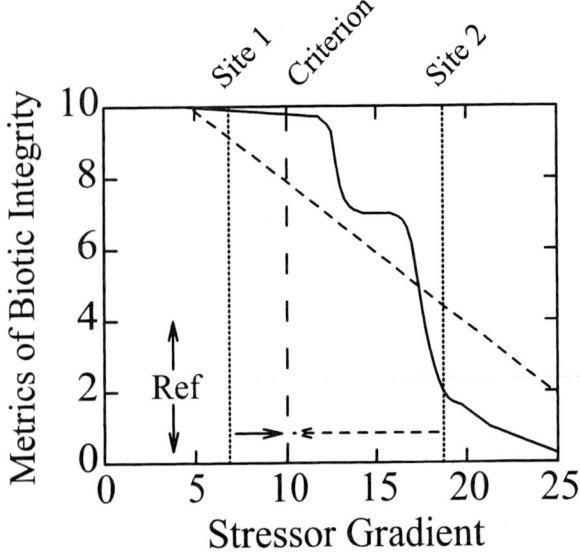

FIGURE 12 Relating ecological response to the sustainability of ecological conditions in unimpaired ecosystems and to restorability of conditions in impaired ecosystems. Reference conditions (Ref) are indicated by the vertical arrow. The vertical lines indicate a criterion and assessed conditions at two sites. Assessed conditions at Sites 1 and 2 are unimpaired and impaired, respectively. Site 1 illustrates that the greater the difference between assessed conditions and the criterion for unimpaired sites, the greater is the sustainability of valued ecological attributes at a site. Site 2 indicates that the greater the difference between assessed conditions and the criterion for impaired sites, the less restorable a habitat is.

measured exposure (stressor) levels and the exposure criteria for protecting ecological integrity (or designated use of the ecosystem). Restorability is assessed for impaired ecosystems and is equal to 1/(difference between measured exposure levels and the criteria for protecting ecological integrity). These two measures of risk characterization can provide quantitative assessments of probable effects on ecological systems—the greater the sustainability, the lower the probability of stressor effects on ecological integrity, and the lower the restorability, the greater the probability of stressor effects on ecological integrity.

Management decisions can be linked to risk assessment through risk characterizations and the stressor linkage to human activities and management options (Fig. 1). After risk characterizations are completed, management options for protecting or remediating environmental stressors will be considered with the many other factors that affect risk management decisions (e.g., political, economic, and regulatory factors). Management options depend upon the human activities that produce stressors that are causing or threatening environmental impairment. Biological assessments can actually be used to infer the human activities that cause the main ecological stressors (Yoder and Rankin, 1995; Stevenson, 1998), but detailing that approach is beyond the scope of this chapter. Thus, management options for controlling nutrients and deoxygenation in streams include reducing point sources and nonpoint sources of nutrients and BOD to streams. Point sources (e.g., municipal waste treatment plants) are easier to regulate and significantly reduce nutrient loading and related problems in many lakes and streams. However, further reduction in nutrient problems will require addressing nonpoint sources of nutrients. Thus, distinguishing the direct and indirect effects of stressors and human activities provides for a more focused approach to evaluating management options and developing a risk management decision.

VIII. CONCLUSIONS

Algae have been used successfully for environmental assessment of many streams, larger rivers, lakes, and wetlands around the world. Many different approaches can be used, which vary in the level of technical expertise required and in the amount of time required per sample. In this chapter we have tried not to make specific recommendations for which methods should be used, because different programs may call for different methods. In addition, we would not want to constrain the great promise for further development and linkage of algal methods for environmental assessment by making specific recommendations of methods. However, we have started to develop a framework for relating the different methods of environmental assessment and to explore the specific situations in which different methods should be used. We have linked the algal framework for environmental assessment to a standardized risk assessment framework so that assessments with algae can be better related to assessments with other organisms and other approaches (such as laboratory-based toxicology).

Future developments in algal methods for environmental assessment should be directed to the ultimate goal of solving environmental problems. More rigorous hypothesis testing of the precision and sensitivity of algal indicators will be important in documenting the performance of algal indicators and encouraging their application. Transferability of algal indicators among regions should be rigorously evaluated because this allows development of consistent approaches across regions (and perhaps habitat type) and saves costs of developing regional indicators. Current efforts to develop web-based access to autecological characteristics of taxa will be valuable for making this information available to a broader audience, but this approach also requires cautious evaluation of the quality of information on the web. Great challenges also exist for our more consistent application and integration of multiple lines of evidence, which may be facilitated by using a framework, such as the risk assessment framework. In addition, we must learn to communicate the results of our research more effectively and clearly to a broad audience, some of whom have little experience with interpreting mathematically complex information.

Algae can cause ecological problems, but can also perform valuable ecological services. In a recent report by the National Research Council of the US (CEIMATE, 2000), total and native species diversity, productivity, trophic status, and nutrient-use efficiency were listed as fundamental ecological indicators for the United States. Algal attributes related to these indicators, as well as other valued ecological attributes of algae, should provide one part of a rationale for assessing algal properties of ecosystems. The second part of that rationale is the extraordinary sensitivity and diagnostic power of algae to detect environmental problems and identify their causes.

ACKNOWLEDGMENTS

We acknowledge the contributions of anonymous reviewers and comments by the editors of this book, John Wehr and Bob Sheath. In addition, we are grateful for the openness of interactions among colleagues and students, which has moved our discipline forward so successfully. Stevenson's efforts were supported by the U.S. Environmental Protection Agency. Smol's efforts were supported by the Natural Sciences and Engineering Research Council of Canada.

LITERATURE CITED

Adler, R. W. 1995. Filling the gaps in water quality standards: Legal perspectives on biocriteria, *in*: Davis, W. S., Simon, T. P., Eds., Biological assessment and criteria: Tools for water resource planning and decision making. Lewis Publishers, Boca Raton, FL, pp. 31–47.

Agbeti, M. D. 1992. The relationship between diatom assemblages and trophic variables: a comparison of old and new approaches. Canadian Journal of Fisheries and Aquatic Sciences 49:1171–1175.

Aloi, J. E. 1990. A critical review of recent freshwater periphyton methods. Canadian Journal of Fisheries and Aquatic Sciences 47:656–670.

Amoros, C., Van Urk, G. 1989. Palaeoecological analyses of large rivers: Some principles and methods, *in*: Petts, G. E., Möller, H., Roux, A. L., Eds., Historical change of large alluvial rivers: Western Europe. Wiley, Chicester, U.K., pp. 143–165.

Anderson, N. J., Battarbee, R. W. 1994. Aquatic community persistence and variability: A paleolimnological perspective, *in*: Giller, P. S., Hildrew, A. G., Raffaelli D. G., Eds., Aquatic ecology: Scale, pattern and process. Blackwell Scientific, Oxford, U.K., pp. 233–259.

Anderson, N. J., Ripply, B., Gibson, C. E. 1993. A comparison of sedimentary and diatom-inferred phosphorus profiles: Implications for defining pre-disturbance nutrient conditions. Hydrobiologia 253:171–213.

Anderson, N. J., Vos, P. 1992. Learning from the past: Diatoms as palaeoecological indicators of change in marine environments. Netherlands Journal of Aquatic Ecology 26:19–30.

Angermeier, P. L., Karr, J. R. 1994. Biological integrity versus biological diversity as policy directives. BioScience 44:690–697.

APHA. 1998. Standard methods for the evaluation of water and wastewater, 20th ed. American Public Health Association, Washington, DC.

Archibald, R. E. M. 1972. Diversity in some South African diatom assemblages and its relation to water quality. Water Research 6:1229–1238.

Arruda, J. A., Fromm, C. H. 1989. The relationship between taste and odor problems and lake enrichment from Kansas lakes in agricultural watersheds. Lake and Reservoir Management 5:45–52.

Autenrieth, R., Bonner, J., Schreiber, L. 1991. Aquatic sediments. Research Journal of Water Pollution Control Federation 63:709–725.

Bahls, L. L. 1993. Periphyton bioassessment methods for Montana streams. Water Quality Bureau, Department of Health and Environmental Sciences, Helena, MT.

Barbour, M. T., Gerritsen, J., Snyder, B. D., Stribling J. B. 1999. Rapid bioassessment protocols for use in wadeable streams and rivers: Periphyton, benthic macroinvertebrates, and fish, 2nd Ed., EPA 841-D-97-002. United States Environmental Protection Agency, Washington, DC.

Barbour, M. T., Plafkin, J. L., Bradley, B. P., Graves, C. G., Wisseman, R. W. 1992. Evaluation of EPA's rapid bioassessment benthic metrics: Metric redundancy and variability among reference stream sites. Environmental Toxicology and Chemistry 11:437–449.

Battarbee, R. W., Charles, D. F., Dixit, S. S., Renberg, I. 1999. Diatoms as indicators of surface water acidity, *in*: Stoermer, E. F., Smol, J. P., Eds., The diatoms: Applications for the environmental and earth sciences. Cambridge University Press, Cambridge, U.K., pp. 85–127.

Beaver, J. 1981. Apparent ecological characteristics of some common freshwater diatoms. Ontario Ministry of the Environment, Technical Support Section, Central Region, Don Mills, Ontario, Canada.

Belanger, S. E., Barnum, J. B., Woltering, D. M., Wowling, J. W., Ventullo, R. M., Schermerhorn, S. D., Lowe, R. L. 1994. Algal periphyton structure and function in response to consumer chemicals in stream mesocosms, *in*: Graney, R. L., Kennedy, J. H., Rogers, J. H., Eds., Aquatic mesocosm studies in ecological risk assessment, SETAC Special Publications Series, Lewis Publishers, Ann Arbor, MI.

Belbin, L., McDonald, C. 1993. Comparing three classification strategies for use in ecology. Journal of Vegetation Science 4:341–348.

Bennion, H., Juggins, S., Anderson, N. J. 1996. Predicting epilimnetic phosphorus concentrations using an improved diatom-based transfer function and its application to lake eutrophication management. Environmental Science & Technology 30:2004–2007.

Biggs, B. J. F. 1990. Use of relative specific growth rates of periphytic diatoms to assess enrichment of a stream. New Zealand Journal of Marine and Freshwater Research 24:9–18.

Biggs, B. J. F. 1995. The contribution of flood disturbance, catchment geology and land use to the habitat template of periphyton in stream ecosystems. Freshwater Biology 33:419–438.

Biggs, B. J. F. 1996. Patterns of benthic algae in streams, in: Stevenson, R. J., Bothwell, M., Lowe, R. L., Eds., Algal ecology: Freshwater benthic ecosystems. Academic Press, San Diego, pp. 31–55.

Biggs, B. J. F., Close, M. E. 1989. Periphyton biomass dynamics in gravel bed rivers: The relative effects of flow and nutrients. Freshwater Biology 22:209–231.

Biggs, B. J. F., Kilroy, C., Mulcock, C. M. 1998. New Zealand stream health monitoring and assessment kit. Stream monitoring manual, Version 1. NIWA Technical Report, Christchurch, New Zealand, 150 pp.

Biggs, B. J. F., Price, G. M. 1987. A survey of filamentous algal proliferations in New Zealand rivers. New Zealand Journal of Marine and Freshwater Research 21:175–191.

Birks, H. J. B. 1995. Quantitative palaeoenvironmental reconstructions, in: Maddy, D., Brew, J. S., Eds., Statistical modelling of quaternary science data. Technical Guide 5. QUATERNARY Research Association, London, U.K., pp. 161–254.

Birks, H. J. B. 1998. Numerical tools in palaeolimnology: Progress, potentialities, and problems. Journal of Paleolimnology 20:307–332.

Birks, H. J. B., Line, J. M., Juggins, S., Stevenson, A. C., ter Braak, C. J. F. 1990. Diatoms and pH reconstruction. Philosophical Transactions of the Royal Society (London) B327:263–278.

Blanck, H. 1985. A simple, community level, ecotoxicological test system using samples of periphyton. Hydrobiologia 124:251–261.

Borchardt, M. A. 1996. Nutrients, in: Stevenson, R. J., Bothwell, M. L., Lowe, R. L., Eds., Algal ecology: Freshwater benthic ecosystems. Academic Press, San Diego, pp. 184–228.

Bott, T. L. 1996. Algae in microscopic food webs, in: Stevenson, R. J., Bothwell, M. L., Lowe, R. L., Eds., Algal ecology: Freshwater benthic ecosystems. Academic Press, San Diego, pp. 574–608.

Bott, T. L., Brock, J. T., Cushing, C. E., Gregory, S. V., King, D., Petersen, R. C. 1978. A comparison of methods for measuring primary productivity and community respiration in streams. Hydrobiologia 60:3–12.

Bowling, L. C., Baker, P. D. 1996. Major cyanobacterial bloom in the Barwon-Darling River, Australia, in 1991, and underlying limnological conditions. Marine and Freshwater Research 47:643–657.

Brezonik, P. L. 1978. Effect of organic color and turbidity of Secchi disk transparency. Journal of the Fisheries Research Board of Canada 35:1410–1416.

Briand, F., Trucco, R., Ramamoorthy, S. 1978. Correlations between specific algae and heavy metal binding in lakes. Journal of the Fisheries Research Board of Canada 35:1482–1485.

Brinson, M. M. 1993. A hydrogeomorphic classification for wetlands. Wetlands Research Program Technical Report WRP-DE-4. U. S. Army Corps of Engineers Waterways Experimental Station, Vicksburg, MS.

Bunting, M. J., Duthie, H., Campbell, D., Warner, B., Turner, L. 1997. A paleoecological record of recent environmental change at Big Creek Marsh, Long Point, Lake Erie. Journal of Great Lakes Research 23:349–368.

Burkholder, J. M., Glasgow, H. B. Jr. 1997. *Pfiesteria piscicida* and other *Pfiesteria*-like dinoflagellates: Behavior, impacts, and environmental controls. Limnology and Oceanography 42:1052–1075.

Butcher, R. W. 1947. Studies in the ecology of rivers. IV. The algae of organically enriched water. Journal of Ecology 35:186–191.

Cain, J. R., Trainor, F. R. 1973. A bioassay compromise. Phycologia 12:227–232.

Carlson, R. E. 1977. A trophic state index for lakes. Limnology and Oceanography 22:361–69.

Carmichael, W. W. 1994. The toxins of cyanobacteria. Scientific American 270:78–86.

Carpenter, S. R., Kitchell, J. F., Eds. 1993. The trophic cascade in lakes. Cambridge University Press, Cambridge, UK. 283 pp.

CEIMATE (Committee to Evaluate Indicators for Monitoring Aquatic and Terrestrial Environments). 2000. Ecological indicators for the nation. National Academy Press, Washington, DC, 180 pp.

Charles, D. F., Binford, M. W., Furlong, E. T., Hites, R. A., Mitchell, M. J., Norton, S. A., Oldfield, F., Paterson, M. J., Smol, J. P., Uutala, A. J., White, J. R., Whitehead, D. R., Wise, R. J. 1990. Paleoecological investigation of recent lake acidification in the Adirondack Mountains, New York. Journal of Paleolimnology 3:195–141.

Charles, D. F., Smol, J. P. 1988. New methods for using diatoms and chrysophytes to infer past pH of low-alkalinity lakes. Limnology and Oceanography 33:1451–1462.

Charles, D. F., Smol, J. P. 1994. Long-term chemical changes in lakes: Quantitative inferences using biotic remains in the sediment record, in: Baker, L, Ed., Environmental chemistry of lakes and reservoirs, Advances in Chemistry Series 237. American Chemical Society, Washington, DC, pp 3–31.

Chessman, B., Growns, I., Currey, J., Plunkett-Cole, N. 1999. Predicting diatom communities at the genus level for the rapid biological assessment of rivers. Freshwater Biology 41:317–331.

Chilton, E. W., Lowe, R. L., Schurr, K. M. 1986. Invertebrate communities associated with *Bangia atropurpurea* and *Cladophora glomerata* in western Lake Erie. Journal of Great Lakes Research 12:149–153.

Cooper, S. R. 1999. Estuarine paleoenvironmental reconstructions using diatoms, in: Stoermer, E. F., Smol, J. P., Eds., The diatoms: Applications for the environmental and earth sciences. Cambridge University Press, Cambridge, U.K., pp. 352–373.

Coste, M., Bosca, C., Dauta, A. 1991. Use of algae for monitoring rivers in France, in: Whitton, B. A., Rott, E., Friedrich, G., Eds., Use of algae for monitoring rivers. Institut für Botanik, Universität Innsbruck, Innsbruck, Austria, pp. 75–88.

Coté, R. 1983. Aspects toxiques du cuivre sur la biomasse et la productivité du phytoplankton de la rivière du Saguenay, Québec. Hydrobiologia 98:85–95.

Cullen, J. J., Ciotiti, A. M., Lewis, M. R. 1997. Optical detection and assessment of algal blooms. Limnology and Oceanography 42:1223–1239.

Cumming, B. F., Davey, K., Smol, J. P., Birks, H. J. 1994. When did Adirondack Mountain lakes begin to acidify and are they still acidifying? Canadian Journal of Fisheries and Aquatic Sciences 51:1550–1568.

Cumming, B. F., Smol, J. P., Birks, H. J. B. 1992a. Scaled chrysophytes (Chrysophyceae and Synurophyceae) from Adirondack (N.Y., USA) drainage lakes and their relationship to measured environmental variables, with special reference to lakewater pH and labile monomeric aluminum. Journal of Phycology 28:162–178.

Cumming, B. F., Smol, J. P., Kingston, J. C., Charles, D. F., Birks, H. J. B., Camburn, K. E., Dixit, S. S., Uutala, A. J., Selle, A. R. 1992b. How much acidification has occurred in Adirondack region (New York, USA) lakes since pre-industrial times? Canadian Journal of Fisheries and Aquatic Sciences 49:128–141.

Davies-Colley, R. J., Vant, W. N. 1988. Estimation of optical properties of water from Secchi disk depths. Water Research Bulletin 24:1329–1335.

Dearing, J. A, Håkansson, H., Liedberg-Jönsson, B., Persson, A., Skansjö, S., Windholm, D., El-Dahousy, F. 1987. Lake sediments used to quantify the erosional response to land use changes in southern Sweden. Oikos 50:60–78.

Denicola, D. M. 1986. Periphyton responses to temperature at different ecological levels, in: Stevenson, R. J., Bothwell, M. L., Lowe, R. L., Eds., Algal ecology: Freshwater benthic ecosystems. Academic Press, San Diego, pp. 150–183.

Denys, L. 1991. A check-list of the diatoms in the Holocene deposits of the western Belgian coastal plain with a survey of their apparent ecological requirements. I. Introduction, ecological code, and complete list. Professional paper. Ministère des Affaires Economiques, Service Géologique de Belgique. Brussels, Belgium, 41 p.

Descy, J. P. 1979. A new approach to water quality estimation using diatoms. Nova Hedwigia 64:305–323.

Dixit, S. S., Smol, J. P. 1994. Diatoms as environmental indicators in the Environmental Monitoring and Assessment—Surface Waters (EMAP-SW) program. Environmental Monitoring and Assessment 31:275–206.

Dixit, A. S., Dixit, S. S., Smol, J. P. 1992a. Algal microfossils provide high temporal resolution of environmental trends. Water, Air, and Soil Pollution 62:75–87.

Dixit, S. S., Smol, J. P., Kingston, J. C., Charles, D. F. 1992a. Diatoms: Powerful indicators of environmental change. Environmental Science & Technology 26:22–33.

Dixit, S. S., Cumming, B. F., Kingston, J. C., Smol, J. P., Birks, H. J. B., Uutala, A. J., Charles, D. F., Camburn, K. 1993. Diatom assemblages from Adirondack lakes (N.Y., USA) and the development of inference models for retrospective environmental assessment. Journal of Paleolimnology 8:27–47.

Dixit, S. S., Smol, J. P., Charles, D. F., Hughes, R. M., Paulsen, S. G., Collins, G. B. 1999. Assessing water quality changes in the lakes of the northeastern United States using sediment diatoms. Canadian Journal of Fisheries and Aquatic Science 56:131–152.

Dodd, W. K., Gudder, D. A. 1992. The ecology of *Cladophora*. Journal of Phycology 28:415–427.

Dodds, W. K., Jones, J. R., Welch, E. B. 1998. Suggested criteria for stream trophic state: Distributions of temperate stream types by chlorophyll, total nitrogen and phosphorus. Water Research 32:1455–1462.

Douglas, M. S. V., Smol, J. P., Blake, W., Jr. 1994. Marked post-18th century environmental change in high Arctic ecosystems. Science 266:416–419.

Fairchild, G. W., Lowe, R. L., Richardson, W. B. 1985. Algal periphyton growth on nutrient-diffusing substrates: An in situ bioassay. Ecology 66:465–472.

Fjerdingstad, E. 1950. The microflora of the River Molleaa with special reference to the relation of benthic algae to pollution. Folia Limnologica Scandanavica 5:1–123.

Fritz, S. C. 1990. Twentieth-century salinity and water-level fluctuations in Devils Lake, North Dakota: test of a diatom-based transfer function. Limnology and Oceanography 35:1771–1781.

Fritz, S. C., Juggins, S., Battarbee, R. W., Engstrom, D. R. 1991. Reconstruction of past changes in salinity and climate using a diatom-based transfer function. Nature 352:706–708.

Gensemer, R. W. 1991. The effects of pH and aluminum on the growth of the acidophilic diatom *Asterionella ralfsii* var. *americana*. Limnology and Oceanography 36:123–131.

Genter, R. B. 1996. Ecotoxicology of inorganic chemical stress to algae, in: Stevenson, R. J., Bothwell, M., Lowe, R. L., Eds., Algal ecology: Freshwater benthic ecosystems. Academic Press, San Diego, pp. 403–468.

Ghosh, M., Gaur, J. P. 1990. Application of algal assay for defining nutrient limitation in two streams at Shillong. Proceedings of Indian Academy of Sciences 100:361–368.

Glew, J. R. 1991. Miniature gravity corer for recovering short sediment cores. Journal of Paleolimnology 5:285–287.

Glew, J. R. 1988. A portable extruding device for close interval sectioning of unconsolidated core samples. Journal of Paleolimnology 1:235–239.

Green, R. H. 1979. Sampling design and statistical methods for environmental biologists. Wiley, New York, 257 pp.

Greene, J. C., Miller, W. E., Shiroyama, T., Soltero, R. A., Putnam, K. 1976. Use of laboratory cultures of *Selenastrum*, *Anabaena*, and the indigenous isolate *Sphaerocystis* to predict effects of nutrient and zinc interactions upon phytoplankton growth in Long Lake, Washington. Internationale Vereinigung für Theoretische und Angewandte Limnologie Mitteilungen 21:372–384.

Hall, R. I., Smol, J. P. 1992. A weighted-averaging regression and calibration model for inferring total phosphorus from diatoms from British Columbia (Canada) lakes. Freshwater Biology 27:417–437.

Hall, R. I., Smol, J. P. 1996. Paleolimnological assessment of long-term water quality changes in south-central Ontario lakes affected by cottage development and acidification. Canadian Journal of Fisheries and Aquatic Sciences 53:1–17.

Hallegraeff, G. M. 1993. A review of harmful algal blooms and their apparent global increase. Phycologia 32:79–99.

Hanson, J. M., Leggett, W. C. 1982. Empirical prediction of fish biomass and yield. Canadian Journal of Fisheries and Aquatic Sciences 39:257–263.

Harper, D. 1992. Eutrophication of freshwaters: Principles, problems, and restoration. Chapman and Hall, New York. 327 p.

Harvey, R. S., Patrick, R. 1968. Concentration of ^{137}Cs, ^{65}Zn, and ^{85}Sr by fresh-water algae. Biotechnology and Bioengineering 9:449–456.

Healey, F. P., Hendzel, L. L. 1979. Fluorometric measurement of alkaline phosphatase activity in algae. Freshwater Biology 9:429–439.

Healey, F. P., Hendzel, L. L. 1980. Physiological indicators of nutrient deficiency in lake phytoplankton. Canadian Journal of Fisheries and Aquatic Sciences 37:442–453.

Hecky, R. E., Kilham, P. 1988. Nutrient limitation of phytoplankton in freshwater and marine environments: A review of recent evidence on the effects of enrichment. Limnology and Oceanography 33:796–822.

Hill, B. H., Herlihy, A. T., Kaufmann, P. R., Stevenson, R. J., McCormick, F. H., Johnson, C. B. 2000. The use of periphyton assemblage data as an index of biotic integrity. Journal of the North American Benthological Society 19:50–67.

Hill, B. H., Lazorchak, J. M., McCormick, F. H., Willingham, W. T. 1997. The effects of elevated metals on benthic community metabolism in a Rocky Mountain stream. Environmental Pollution 95:183–190.

Hill, M. O. 1979. TWINSPAN—A FORTRAN program for detrended correspondence analysis and reciprocal averaging. Cornell University, Ithaca, NY.

Hill, W. R., Boston, H. L. 1991. Community development alters photosynthesis-irradiance relations in stream periphyton. Limnology and Oceanography 36:1375–1389.

Hillebrand, H. 1983. Development and dynamics of floating clusters of filamentous algae, in: Wetzel, R. G., Ed., Periphyton of freshwater ecosystems. Dr. W. Junk Publishers, The Hague, pp. 31–39.

Hillebrand, H., Dürlsen, C. D., Kirschtel, D., Pollingher, U., Zohary,

T. 1999. Biovolume calculation for pelagic and benthic microalgae. Journal of Phycology 403–424.

Hoagland, K. D., Carder, J. P., Spawn, R. L. 1996. Effects of organic toxic substances, *in*: Stevenson, R. J., Bothwell, M., Lowe, R. L., Eds., Algal ecology: Freshwater benthic ecosystems. Academic Press, San Diego, pp. 469–497.

Hoagland, K. D., Drenner, R. W., Smith, J. D., Cross. D. R. 1993. Freshwater community responses to mixtures of agricultural pesticides: Effects of atrazine and bifenthrin. Environmental Toxicology and Chemical 12:627–637.

Hofmann, G. 1994. Aufwuchs Diatomeen in Seen und ihre Eignung als Indikatoren der Trophie. Bibliotheca Diatomologica 30:1–24.

Holmes, N. T. H., Whitton, B. A. 1981. Phytobenthos of the River Tees and its tributaries. Freshwater Biology 11:139–163.

Holomuzki, J. R., Short, T. M. 1988. Habitat use and fish avoidance behaviors by the stream-dwelling isopod, *Lirceus fontinalis*. Oikos 52:79–86.

Hughes, R. M. 1995. Defining acceptable biological status by comparing with reference conditions, *in*: Davis, W. S., Simon, T. P., Eds., Biological assessment and criteria: Tools for water resource planning and decision making. Lewis Publishers, Boca Raton, FL, pp. 31–47.

Humphrey, K. P., Stevenson, R. J. 1992. Responses of benthic algae to pulses in current and nutrients during simulations of subscouring spates. Journal of the North American Benthological Society 11:37–48.

Hurlbert, S. H. 1971. The nonconcept of species diversity: A critique and alternative parameters. Ecology 52:577–586.

Hutchinson, T., Havas, M. 1986. Recovery of previously acidified lakes near Coniston, Canada, following reductions in atmospheric sulphur and metal emissions. Water, Air, and Soil Pollution 29:319–333.

Ibelings, B., Admiraal, W., Bijker, R., Letswaart, T., Prins, H. 1998. Monitoring of algae in Dutch rivers: Does it meet its goals? Journal of Applied Phycology 10:171–181.

Jongman, R. H. G., ter Braak, C. J. F., Van Tongeren, O. F. R. 1995. Data analysis in community and landscape ecology. Cambridge University Press, Cambridge, U.K., 299 p.

Juggins, S., Battarbee, R., Fritz, S., Gasse, F. 1994. The CASPIA project: Diatoms, salt lakes, and environmental change. Journal of Paleolimnology 12:1–196.

Juggins, S., ter Braak, C. J. F. 1992. CALIBRATE—A program for species-environment calibration by [weighted averaging] partial least squares regression. Environmental Change Research Center, University College, London, 20 p.

Jüttner, I., Rothfritz, H., Omerod, S. J. 1996. Diatoms as indicators of river water quality in the Nepalese Middle Hills with consideration of the effects of habitat-specific sampling. Freshwater Biology 36:475–486.

Karr, J. R. 1981. Assessment of biotic integrity using fish communities. Fisheries 6:21–27.

Karr, J. R., Chu, E. W. 1999. Restoring life in running waters. Island Press, Washington, DC, 206 p.

Karr, J. R., Dudley, D. R. 1981. Ecological perspective on water quality goals. Environmental Management 5:55–68.

Kelly, M. G., Cazaubon, A., Coring, E., Dell'Uomo, A., Ector, L., Goldsmith, B., Guasch, H., Hürlimann, J., Jarlman, A., Kawecka, B., Kwandrans, J., Laugaste, R., Lindstrrm, E.-A., Leitao, M., Marvan, P., Padisák, J., Pipp, E., Pyrgiel, J., Rott, E., Sabater, S., van Dam, H., Viznet, J. 1998. Recommendations for the routine sampling of diatoms for water quality assessments in Europe. Journal of Applied Phycology 10:215–224.

Kelly, M. G., Hornberger, G. M., Cosby, B. J. 1974. Continuous automated measurement of rates of photosynthesis and respiration in an undisturbed river community. Limnology and Oceanography 19:305–312.

Kelly, M. G., Penny, C. J., Whitton, B. A. 1995. Comparative performance of benthic diatom indices used to assess river water quality. Hydrobiologia 302:179–188.

Kelly, M. G., Whitton, B. A. 1989. Interspecific differences in Zn, Cd and Pb accumulation by freshwater algae and bryophytes. Hydrobiologia 175:1–11.

Kelly, M. G., Whitton, B. A. 1995. The trophic diatom index: A new index for monitoring eutrophication in rivers. Journal of Applied Phycology 7:433–444.

Kentucky Division of Water. 1993. Methods for assessing biological integrity of surface waters. Kentucky Natural Resources and Environmental Protection Cabinet, Frankfort, Kentucky, 139 p.

Kentucky Natural Resources and Environmental Protection Cabinet. 1997. Kentucky outlook 2000: A strategy for Kentucky's third century. Executive summary and guide to the technical committee reports. Frankfort, Kentucky.

Kerans, B. L., Karr, J. R. 1994. A benthic index of biotic integrity (B-IBI) for rivers of the Tennessee Valley. Ecological Applications 4:768–785.

Kingston, J. C., Birks, H. J. B., Uutala, A. J., Cumming, B. F., Smol, J. P. 1992. Assessing trends in fishery resources and lake water aluminum from paleolimnological analyses of siliceous algae. Canadian Journal of Fisheries and Aquatic Science 49:127–138.

Klemm, D.J., Lazorchak, J. M. [Eds.] 1994. Pilot field operation and methods manual for streams, EPA/620/R-94/004. Environmental Monitoring Systems Laboratory Office of Research and Development, U.S. Environmental Protection Agency, Cincinnati, OH.

Kolkwitz, R., Marsson, M. 1908. Ökologie der pflanzliche Saprobien. Berichte der Deutschen Botanischen Gessellschaft 26:505–519.

Kwandrans, J., Eloranta, P., Kawecka, B., Wojtan, K. 1998. Use of benthic diatom communities to evaluate water quality in rivers of southern Poland. Journal of Applied Phycology 10:193–201.

Lamberti, G. A. 1996. The role of periphyton in benthic food webs, *in*: Stevenson, R. J., Bothwell, M., Lowe, R. L., Eds., Algal ecology: Freshwater benthic ecosystems. Academic Press, San Diego, pp. 533–572.

Lamberti, G. A., Steinman, A. D., Eds. 1993. Research in artificial streams: Applications, uses, and abuses. Journal of the North American Benthological Society 12:313–384.

Lange-Bertalot, H. 1979. Pollution tolerance of diatoms as a criterion for water quality estimation. Nova Hedwigia 64:285–304.

Lasenby, D. C. 1975. Development of oxygen deficits in 14 southern Ontario lakes. Limnology and Oceanography 20:993–999.

Leavitt, P. R. 1993. A review of factors that regulate carotenoid and chlorophyll deposition and fossil pigment abundance. Journal of Paleolimnology 9:109–127.

Lecointe, C., Coste, M., Prygiel, J. 1993. "Omnidia": Software for taxonomy, calculations of diatom indices and inventories management. Hydrobiologia 269/270:509–513.

Leland, H. V. 1995. Distribution of phytobenthos in the Yakima River basin, Washington, in relation to geology, land use and other environmental factors. Canadian Journal of Fisheries and Aquatic Sciences 52:1108–1129.

Lembi, C. A., O'Neil, S. W., Spencer, D. F. 1988. Algae as weeds: economic impact, ecology, and management alternatives, *in*: Lembi, C. A., Waaland, J. R., Eds., Algae and human affairs. Cambridge University Press, Cambridge, U.K., pp. 455–481.

Line, J. M., ter Braak, C. J. F, Birks, H. J. B. 1994. WACALIB version 3.3—a computer program to reconstruct environmental variables from fossil assemblages by weighted averaging and to derive sample-specific errors of prediction. Journal of Paleolimnology 10:147–152.

Lorenzen, C. J. 1967. Determinations of chlorophyll and pheopigments: spectrophotometric equations. Limnology and Oceanography 12:343–346.

Lowe, R. L. 1974. Environmental requirements and pollution tolerance of freshwater diatoms, U.S. Environmental Protection Agency, EPA-670/4-74-005. Cincinnati, OH, 334 pp.

Lowe, R. L., Pan, Y. 1996. Benthic algal communities and biological monitors, in: Stevenson, R. J., Bothwell, M., Lowe, R. L., Eds., Algal ecology: Freshwater benthic ecosystems. Academic Press, San Diego, pp. 705–739.

Lund, J. W. G., Kipling, C., LeCren, E. D. 1958. The inverted microscope method of estimating algal numbers and the statistical basis of estimations by counting. Hydrobiologia 11:143–170.

Mantoura, R. F. C., Llewellyn, C. A. 1983. The rapid determination of algal chlorophyll and carotenoid pigments and their breakdown products in natural waters by reverse-phase high-performance liquid chromatography. Analytica Chimica Acta (Amsterdam) 151:297–314.

Marzolf, E. R., Mulholland, P. J., Steinman, A. D. 1994. Improvements to the diurnal upstream-downstream dissolved oxygen change technique for determining whole-stream metabolism in small streams. Canadian Journal of Fisheries and Aquatic Sciences 51:1591–1599.

May, R. M. 1974. Biological populations with nonoverlapping generations: Stable points, stable cycles, and chaos. Science 186:273–275.

McCormick, P. V., Cairns, J. Jr. 1994. Algae as indicators of environmental change. Journal of Applied Phycology 6:509–526.

McCormick, P. V., O'Dell, M. B. 1996. Quantifying periphyton responses of phosphorus in the Florida Everglades: A synoptic-experimental approach. Journal of the North American Benthological Society 15:450–468.

McCormick, P. V., Rawlik, P. S., Lurding, K., Smith, E. P., Sklar, F. H. 1996. Periphyton-water quality relationships along a nutrient gradient in the northern Florida Everglades. Journal of North American Benthological Society 14:433–449.

McCormick, P. V., Stevenson, R. J. 1998. Periphyton as a tool for ecological assessment and management in the Florida Everglades. Journal of Phycology 34:726–733.

McFarland, B. H., Hill, B. H., Willingham, W. T. 1997. Abnormal *Fragilaria* spp. (Bacillariophyceae) in streams impacted by mine drainage. Journal of Freshwater Ecology 12:141–149.

Millie, D. F., Pearl, H. W., Hurley, J. P. 1993. Microalgal pigment assessments using high-performance liquid chromatography: A synopsis of organismal and ecological applications. Canadian Journal of Fisheries and Aquatic Sciences 50:2513–2527.

Minshall, G. W. 1978. Autotrophy in stream ecosystems. BioScience 28:767–771.

Moss, D., Wright, J. F., Furse, M. T., Clarke, R. T. 1999. A comparison of alternative techniques for prediction of the fauna of running water sites in Great Britain. Freshwater Biology 41:167–181.

Mulholland, P. J. 1996. Role of nutrient cycling in streams, in: Stevenson, R. J., Bothwell, M., Lowe, R. L., Eds., Algal ecology: Freshwater benthic ecosystems. Academic Press, San Diego, pp. 609–639.

Murtaugh, P. A. 1996. The statistical evaluation of ecological indicators. Ecological Application 6:132–139.

Neill, C., Cronwell, J. C. 1992. Stable carbon, nitrogen, and sulfur isotopes in a prairie marsh food web. Wetlands 12:217–214.

Newman, S., Aldridge, F. J., Phlips, E. J., Reddy, K. R. 1994. Assessment of phosphorus availability for natural phytoplankton populations from a hypereutrophic lake. Archiv für Hydrobiologie 130:409–427.

Niederlehner, B. R., Cairns, J. C., Jr. 1994. Consistency and sensitivity of community level endpoints in microcosm tests. Journal of Aquatic Ecosystem Health 3:93–99.

Odum, E. P., Finn, J. T., Franz, E. H. 1979. Perturbation theory and the subsidy-stress gradient. BioScience 29:349–352.

Oldfield, F., Appleby, P. G. 1984. Empirical testing of ^{210}Pb-dating models for lake sediments, in: Haworth, E. Y., Lund J. W. G., Eds., Lake sediments and environmental history. Leicester University Press, Leicester, U.K., pp. 93–124.

Palmer, C. M. 1962. Algae in water supplies. U.S. Department of Health, Education and Welfare, Washington, D.C., 88 p.

Palmer, C. M. 1969. A composite rating of algae tolerating organic pollution. Journal of Phycology 5:78–82.

Pan, Y., Stevenson, R. J. 1996. Gradient analysis of diatom assemblages in western Kentucky wetlands. Journal of Phycology 32:222–232.

Pan, Y., Stevenson, R. J., Hill, B. H., Herlihy, A. T., Collins, G. B. 1996. Using diatoms as indicators of ecological conditions in lotic systems: A regional assessment. Journal of the North American Benthological Society 15:481–495.

Pan, Y., Stevenson, R. J., Vaithiyanathan, P., Slate, J., Richardson, C. J. 2000. Changes in algal assemblages along observed and experimental phosphorus gradients in a subtropical wetland, U.S.A. Freshwater Biology 43:1–15.

Patrick, R. 1949. A proposed biological measure of stream conditions based on a survey of the Conestoga Basin, Lancaster County, Pennsylvania. Proceedings of the Academy of Natural Sciences of Philadelphia 101:277–341.

Patrick, R. 1973. Use of algae, especially diatoms, in the assessment of water quality, in: Biological methods for the assessment of water quality, ASTM STP 528, American Society for Testing and Materials, Philadelphia, pp. 76–95.

Patrick, R., Hohn, M. H., Wallace, J. H. 1954. A new method for determining the pattern of the diatom flora. Notulae Naturae (Philadelphia) No. 259, 12 p.

Paulsen, S. G., Larsen, D. P., Kaufmann, P. R., Whittier, T. R., Baker, J. R., Peck, D. V, McGue, J., Hughes, R. M., McMullen, D., Stevens, D., Stoddard, J. L., Larzorchak, J., Kinney, W., Selle, A. R., Hjort, R. 1991. Environmental monitoring and assessment program (EMAP) surface waters monitoring and research strategy—fiscal year 1991. Environmental Research Laboratory, Office of Research and Development, U.S. Environmental Protection Agency, Corvallis, OR.

Peterson, C. G., Stevenson, R. J. 1992. Resistance and recovery of lotic algal communities: Importance of disturbance timing, disturbance history, and current. Ecology 73:1445–1461.

Pielou, E. C. 1984. The interpretation of ecological data: A primer on classification and ordination. Wiley, New York, 263 pp.

Plafkin, J. L., Barbour, M. T., Porter, K. D., Gross, S. K., Hughes, R. M. 1989. Rapid bioassessment protocols for use in streams and rivers: Benthic macroinvertebrates and fish, U.S. EPA Office of Water, Washington, D.C. EPA/444/4–89–001.

Porter, K. G. 1977. The plant-animal interaction in freshwater ecosystems. American Scientist 65:159–170.

Porter, S. D., Cuffney, T. F., Gurtz, M. E., Meador, M. R. 1993. Methods for collecting algal samples as part of the national water-quality assessment program. Report 93–409. U.S. Geological Survey, Raleigh, NC.

Power, M. E. 1990. Benthic turfs versus floating mats of algae in river food webs. Oikos 58:67–79.

Preisendorfer, R. W. 1996. Secchi disk science: Visual optics of natural waters. Limnology and Oceanography 31:909–926.

Pringle, C. M., Bowers, J. A. 1984. An in situ substratum fertilization technique: Diatom colonization on nutrient-enriched, sand substrata. Canadian Journal of Fisheries and Aquatic Sciences 41:1247–1251.

Prygiel, J., Coste, M., Bukowska, J. 1999a. Review of the major diatom-based techniques for the quality assessments of rivers—state of the art in Europe, *in*: Prygiel, J., Whitton, B. A., Bukowska, J., Eds., Use of algae for monitoring rivers III. Agence de l'Eau Artois-Picardi, Douai, France, 271 p.

Prygiel, J., Whitton, B. A., Bukowska, J. 1999b. Use of algae for monitoring rivers III. Agence de l'Eau Artois-Picardi, Douai, France, 271 pp.

Radwan, S., Kowalik, W., Kowalczyk, C. 1990. Occurrence of heavy metals in water, phytoplankton and zooplankton of a mesotrophic lake in eastern Poland. Science of the Total Environment 96:115–120.

Raschke, R. L. 1993. Diatom (Bacillariophyta) community response to phosphorus in the Everglades National Park, USA. Phycologia 32:48–58.

Reavie, E. D., Smol, J. P. 1997. Diatom-based model to infer past littoral habitat characteristics in the St. Lawrence River. International Journal of Great Lakes Research 23:339–348.

Reavie, E. D., Smol, J. P. 1998. Epilithic diatoms of the St. Lawrence River and their relationships to water quality. Canadian Journal of Botany 76:251–257.

Reavie, E. D., Smol, J. P., Carignan, R., Lorrain, S. 1998. Diatom paleolimnology of two fluvial lakes in the St. Lawrence River: A reconstruction of environmental changes during the last century. Journal of Phycology 34:446–456.

Reid, M. A., Tibby, J. C., Penny, D., Gell, P. A. 1995. The use of diatoms to assess past and present water quality. Australian Journal of Ecology 20:57–64.

Renberg, I., Hellberg, T. 1982. The pH history of lakes in Southwestern Sweden, as calculated from the subfossil diatom flora of the sediments. Ambio 11:30–33.

Rosen, B. H. 1995. Use of periphyton in the development of biocriteria, *in*: Davis, W. S., Simon, T. P., Eds., Biological assessment and criteria: Tools for water resource planning and decision making. Lewis Publishers, Boca Raton, FL, pp. 209–215.

Rosgen, D. L. 1994. A classification of natural rivers. Amsterdam, 50 pp.

Round, F. E. 1991. Diatoms in river water-monitoring studies. Journal of Applied Phycology 3:129–145.

Rout, J., Gaur, J. P. 1990. Comparative asessment of line transect and point intercept methods for stream periphyton. Archiv für Hydrobiologie 119:293–298.

Rumeau, A., Coste, M. 1988. Initiation à la systématique des diatomées d'eau douce pour l'utilisation pratique d'un indice diatomique générique. Bulletin Francais de la Pêche et de la Pisciculture 309:1–69.

Say, P. J., Burrows, L. G., Whitton, B. A. 1990. *Enteromorpha* as a monitor of heavy metals in estuaries. Hydrobiologia 195:119–126.

Schindler, D. W. 1990. Experimental perturbations of whole lakes as tests of hypotheses concerning ecosystem structure and function. Oikos 57:25–41.

Shannon, C. F. 1948. A mathematical theory of communication. Bell Systems Technical Journal 27:37–42.

Sheath, R. G., Burkholder, J. M. 1985. Characteristics of soft water streams in Rhode Island. II. Composition and seasonal dynamics of macroalgal communities. Hydrobiologia 128:109–118.

Sicko-Goad, L., Stoermer, E. F., Ladewski, B. G. 1977. A morphometric method for correcting phytoplankton cell volume estimates. Protoplasma 93:147–163.

Sigworth, E. A. 1957. Control of odor and taste in water supplies. Journal of American Water Works Association 49:1507–1521.

Simpson, E. H. 1949. Measurement of diversity. Nature 163:688.

Siver, P. A. 1995. The distribution of chrysophytes along environmental gradients: Their use as biological indicators, *in*: Sandgren, C., Smol, J. P., Kristiansen, J., Eds., Chrysophyte algae: Ecology, phylogeny and development. Cambridge University Press, Cambridge, U.K., pp. 232–268.

Siver, P. A., Smol, J. P. 1993. The use of scaled chrysophytes in long term monitoring programs for the detection of changes in lakewater acidity. Water, Air, and Soil Pollution 73:357–376.

Sládeček, V. 1973. System of water quality from the biological point of view. Archiv für Hydrobiologie und Ergebnisse Limnologie 7:1–218.

Sládeček, V. 1986. Diatoms as indicators of organic pollution. Acta Hydrochimica et Hydrobiologica 14:555–566.

Slate, J. E. 1998. Inference of present and historical environmental conditions in the Everglades with diatoms and other siliceous microfossils. Ph.D. dissertation, University of Louisville, Louisville, Kentucky.

Smol, J. P. 1990. Paleolimnology—recent advances and future challenges, *in*: De Bernardi, R., Giussani, G., Barbanti, L., Eds., Scientific Perspectives in Theoretical and Applied Limnology. CNDR, Pallanza. Memorie dell' Istituto Italiano di Idrobiologia Dott Marco de Marchi 47:253–276.

Smol, J. P. 1992. Paleolimnology: An important tool for effective ecosystem management. Journal of Aquatic Ecosystem Health 1:49–58.

Smol, J. P. 1995. Application of chrysophytes to problems in paleoecology, *in*: Sandgren, C., Smol, J. P., Kristiansen, J., Eds., Chrysophyte algae: Ecology, phylogeny and development. Cambridge University Press, Cambridge, U.K., pp. 303–329.

Smol, J. P., Glew, J. R. 1992. Paleolimnology, *in*: Nierenberg, W. A., Ed., Encyclopedia of earth system science, Vol. 3. Academic Press, San Diego, pp. 551–564.

Sokal, R. R., Rohlf, F. J. 1998. Biometry: The principles and practice of statistics in biological research. W. H. Freeman, New York, 887 p.

Stevenson, R. J. 1984. Epilithic and epipelic diatoms in the Sandusky River, with emphasis on species diversity and water quality. Hydrobiologia 114:161–175.

Stevenson, R. J. 1990. Benthic algal community dynamics in a stream during and after a spate. Journal of the North American Benthological Society 9:277–288.

Stevenson, R. J. 1996. An introduction to algal ecology in freshwater benthic habitats, *in*: Stevenson, R. J., Bothwell, M., Lowe, R. L., Eds., Algal ecology: Freshwater benthic ecosystems. Academic Press, San Diego, pp. 3–30.

Stevenson, R. J. 1997. Scale-dependent causal frameworks and the consequences of benthic algal heterogeneity. Journal of the North American Benthological Society 16:248–262.

Stevenson, R. J. 1998. Diatom indicators of stream and wetland stressors in a risk management framework. Environmental Monitoring and Assessment 51:107–118.

Stevenson, R. J. 2001. Using algae to assess wetlands with multivariate statistics, multimetric indices, and an ecological risk assessment framework, *in*: Rader, R., Batzger, D., Wissinger, S., Eds., Biomonitoring and management of North American freshwater wetlands. Wiley, New York.

Stevenson, R. J., Bahls, L. L. 1999. Periphyton protocols, *in*: Barbour, M. T., Gerritsen, J., Snyder, B. D., Eds., Rapid bioassessment protocols for use in wadeable streams and rivers: Periphyton, benthic macroinvertebrates, and fish, 2nd Ed., EPA 841-B-99-002. U.S. Environmental Protection Agency, Washington, DC, pp 6–1–6–22.

Stevenson, R. J., Glover, R. 1993. Effects of algal density and current on ion transport through periphyton communities. Limnology and Oceanography 38:1276–1281.

Stevenson, R. J., Hashim, S. 1989. Variation in diatom

(Bacillariophyceae) community structure among microhabitats in sandy streams. Journal of Phycology 25:678–686.
Stevenson, R. J., Lowe, R. L. 1986. Sampling and interpretation of algal patterns for water quality assessment, in: Isom, B. G., Ed., Rationale for sampling and interpretation of ecological data in the assessment of freshwater ecosystems, ASTM STP 894. American Society for Testing and Materials Publication, Philadelphia, pp. 118–149.
Stevenson, R. J., Pan, Y. 1999. Assessing ecological conditions in rivers and streams with diatoms, in: Stoermer, E. F., Smol, J. P., Eds., The diatoms: Applications to the environmental and earth sciences. Cambridge University Press, Cambridge, U.K., pp. 11–40.
Stevenson, R. J., Stoermer, E. F. 1981. Quantitative differences between benthic algal communities along a depth gradient in Lake Michigan. Journal of Phycology 17:29–36.
Stevenson, R. J., Singer, R., Roberts, D. A., Boylen, C. W. 1985. Patterns of benthic algal abundance with depth, trophic status, and acidity in poorly buffered New Hampshire lakes. Canadian Journal of Fisheries and Aquatic Sciences 42:1501–1512.
Stevenson, R. J., Sweets, P. R., Pan, Y., Schultz, R. E. 1999. Algal community patterns in wetlands and their use as indicators of ecological conditions, in: McComb, A. J., Davis, J. A., Eds., Proceedings of INTECOL's Vth international wetland conference, Gleneagles Press, Adelaide, Australia, pp. 517–527.
Stoermer, E. F., Smol, J. P. 1999. The diatoms: Applications for the environmental and earth sciences. Cambridge University Press, Cambridge, U.K., 484 pp.
Sweets, P. R. 1992. Diatom paleolimnological evidence for lake acidification in the Trial Ridge region of Florida. Water, Air and Soil Pollution 65: 43–57.
ter Braak, C. J. F., van Dam, H. 1989. Inferring pH from diatoms: A comparison of old and new calibration methods. Hydrobiologia 178:209–223.
Tett, P., Gallegos, C., Kelly, M. G., Hornberger, G. M., Cosby, B. J. 1978. Relationships among substrate, flow, and benthic microalgal pigment density in the Mechums River, Virginia. Limnology and Oceanography 23:785–797.
Thorp, J. H., Black, A. R., Jack, J. D., Casper, A. F. 1996. Pelagic enclosures—modification and use for experimental study of riverine plankton. Archiv für Hydrobiologie (Supplement) 113:583–589.
Trainor, F. R., Shubert, L. E. 1973. Growth of *Dictyosphaerium*, *Selenastrum*, and *Scenedesmus* (Chlorophyceae) in a dilute algal medium. Phycologia 12:35–39.
Tuchman, M., Stevenson, R. J. 1980. Comparison of clay tile, sterilized rock, and natural substrate diatom communities in a small stream in southeastern Michigan, U.S.A. Hydrobiologia 75:73–79.
Twist, H., Edwards, A. C., Codd. G. A. 1997. A novel in-situ biomonitor using alginate immobilized algae (*Scenedesmus* subspicatus) for the assessment of eutrophication in flowing surface waters. Water Research 31:2066–2072.
U.S. Environmental Protection Agency. 1978. The *Selenastrum capricornutum* Printz algal assay bottle test, EPA-600/9-78-018. Office of Research and Development, U.S. Environmental Protection Agency, Corvallis, OR, 126 p.
U.S. Environmental Protection Agency. 1992. Framework for ecological risk assessment, EPA 630/R–92/001. Washington, DC.
U.S. Environmental Protection Agency. 1993. A guidebook to comparing risks and setting environmental priorities, EPA 230-B-93-003. Washington, DC.
U.S. Environmental Protection Agency. 1996. Strategic plan for the Office of Research and Development, EPA/600/R-96/059. Washington, DC.
U.S. Environmental Protection Agency. 1998. Guidelines for ecological risk assessment. EPA/630/R-95/002F. Washington, D.C. 114 p.
Van Dam, H., Mertenes, A., Sinkeldam, J. 1994. A coded checklist and ecological indicator values of freshwater diatoms from the Netherlands. Netherlands Journal of Aquatic Ecology 28:117–133.
Van Heukelem, L., Lewitus, A. J., Kana, T. M. 1992. High-performance liquid chromatography of phytoplankton pigments using a polymeric reverse-phase C_{18} column. Journal of Phycology 28:867–872.
Vannote, R. L., Minshall, G. W., Cummins, K. W., Sedell, J. R., Cushing, C. E. 1980. The river continuum concept. Canadian Journal of Fisheries and Aquatic Sciences 37:130–137.
Vinebrooke, R. D., Graham, M. D. 1997. Periphyton assemblages as indicators of recovery in acidified Canadian Shield lakes. Canadian Journal of Fisheries and Aquatic Sciences 54:1557–1568.
Vollenweider, R. A. 1976. Advances in defining critical loading levels for phosphorus in lake eutrophication. Memorie dell'Istituto Italiano di Idrobiologia Dott Marco de Marchi 33:53–83.
Vymazal, J. 1984. Short-term uptake of heavy metals by periphyton algae. Hydrobiologia 119:171–179.
Vymazal, J. 1994. Algae and element cycling in wetlands. Lewis Publishers, Boca Raton, FL, 689 pp.
Watanabe, T., Asai, K., Houki, A., Tanaka, S., Hizuka, T. 1986. Saprophilous and eurysaprobic diatom taxa to organic water pollution and diatom assemblage index (DAIpo). Diatom Research 2:23–73.
Weber, C. I. 1973. Recent developments in the measurement of the response of plankton and periphyton to changes in their environment, in: Glass, G., Ed., Bioassay techniques and environmental chemistry. Ann Arbor Science Publishers, Ann Arbor, MI, pp. 119–38.
Wehr, J. D., Descy, J.-P. 1998. Use of phytoplankton in large river management. Journal of Phycology 34:741–749.
Wetzel, R. G. 1983. Limnology, 2nd Ed. Saunders College Publishing, Philadelphia, 767 pp.
Wetzel, R. G. 1996. Benthic algae and nutrient cycling in lenthic freshwater ecosystem, in: Stevenson, R. J., Bothwell, M., Lowe, R. L., Eds., Algae ecology: Freswater benthic ecosystems. Academic Press, San Diego, pp. 641–667.
Wetzel, R. G., Likens, G. E. 1991. Limnological analyses. 2nd Ed. Springer-Verlag, New York, 391 p.
Whitehead, D. R., Charles, D. F., Goldstein, R. A. 1990. The PIRLA project (Paleoecological Investigation of Recent Lake Acidification): An introduction to the synthesis of the project. Journal of Paleolimnology 3:187–194.
Whitmore, T. J. 1989. Florida diatom assemblages as indicators of trophic state and pH. Limnology and Oceanography 34:882–895.
Whitton, B. A. 1970. Biology of *Cladophora* in freshwaters. Water Research 4:457–476.
Whitton, B. A. 1984. Algae as monitors of heavy metals in freshwaters, in: Shubert, L. E., Ed., Algae as ecological indicators. Academic Press, London, pp. 257–280.
Whitton, B. A., Shehata, F. H. A. 1982. Influence of cobalt, nickel, copper and cadmium on the blue-green alga *Anacystis nidulans*. Environmental Pollution 27:275–281.
Whitton, B. A., Kelly, M. G. 1995. Use of algae and other plants for monitoring rivers. Australian Journal of Ecology 20:45–56.
Whitton, B. A., Rott, E., Eds. 1996. Use of algae for monitoring rivers II. Universität Innsbruck, Innsbruck, Austria, 196 pp.
Whitton, B. A., Burrows, I. G., Kelly, M. G. 1989. Use of *Cladophora glomerata* to monitor heavy metals in rivers. Journal of Applied Phycology 1:293–299.

Whitton, B. A., Rott, E., Friedrich, G., Eds. 1991. Use of algae for monitoring rivers. Institut für Botanik, Universität Innsbruck, Innsbruck, Austria, 193 pp.

Whitton, B. A., Yelloly, J. M., Christmas, M., Hernández, I. 1998. Surface phosphatase activity of benthic algae in a stream with highly variable ambient phosphate concentrations. Internationale Vereinigung für Theoretische und Angewandte Limnologie Verhandlungen 26:967–972.

Wong, S. L., Clark, B. Kirby, M., Kosciuw, R. F. 1978. Water temperature fluctuations and seasonal periodicity of *Cladophora* and *Potamogeton* in shallow rivers. Journal of the Fisheries Research Board of Canada 35:866–870.

Yoder, C. O., Rankin, E. T. 1995. Biological response signatures and the area of degradation value: new tools for interpreting multimetric data, *in*: Davis, W. S., Simon, T. P., Eds., Biological assessment and criteria: Tools for water resource planning and decision making. Lewis Publishers, Boca Raton, FL, pp. 263–286.

Zeimann, H. 1991. Veränderungen der Diatomeenflora der Werra unter dem Einfluß des Salzgehaltes. Acta Hydrochimica et Hydrobiologica 19:159–174.

Zelinka, M., Marvan, P. 1961. Zur Prazisierung der biologischen Klassifikation des Reinheit fliessender Gewässer. Archiv für Hydrobiologie 57:389–407.

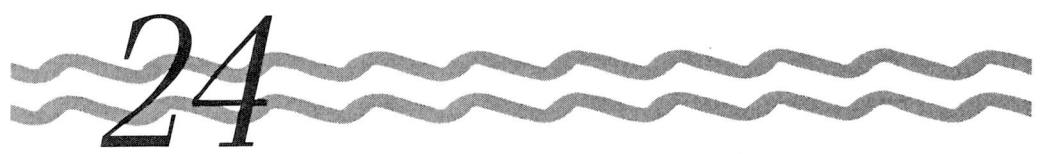

CONTROL OF NUISANCE ALGAE

Carole A. Lembi

*Department of Botany and
Plant Pathology
Purdue University
West Lafayette, Indiana 47907*

I. Introduction
II. Problems Associated with Algae
    A. Microscopic Algae
    B. Macrophytic Filamentous Algae
    C. *Chara* and *Nitella*

III. Control Methods for Nuisance Algae
    A. Nutrient Manipulation
    B. Direct Control Methods
Literature Cited

I. INTRODUCTION

Algae play many important and beneficial roles in freshwater environments. They produce oxygen and consume carbon dioxide, act as the base for the aquatic food chain, remove nutrients and pollutants from water, and stabilize sediments. Excessive algal growths, however, can cause detrimental effects on aquatic systems, endangering the organisms that live in or depend on these systems and hampering or preventing human uses of the infested waterways.

When we refer to the kinds of problems that algae cause, it is helpful to divide algae into three groups according to their growth habits: microscopic algae (primarily phytoplanktonic), filamentous mat-forming algae, and the *Chara/Nitella* group. Each group poses its own unique problems to aquatic systems. This chapter describes the problems caused by each of these three groups and then covers the control methods that typically are used for these algae.

II. PROBLEMS ASSOCIATED WITH ALGAE

Many of the problems that the public associates with algae occur in more or less static bodies of water (i.e., ponds, lakes, and reservoirs) with long residence times. Algae also produce excessive or unwanted growths in flowing waters such as streams, rivers, and water delivery systems. Control of algae in these sites is more typically achieved with watershed management techniques that reduce nutrient inputs than with direct control methods. Nuisance algae found in irrigation canals and drainage systems can be and are managed using direct control techniques, but the options are more limited than those used in static systems. Although the algae of flowing waters are discussed, the emphasis in this chapter is on the problems caused by the algae of lakes and ponds.

A. Microscopic Algae

The term "bloom" is typically reserved for excessive growths of microscopic, planktonic algae. Any discussion of bloom-forming algae starts with the cyanobacteria (also referred to as blue-green algae) *Microcystis*, *Anabaena*, and *Aphanizomenon* (see Chap. 3) Blooms of these prokaryotic organisms give a characteristic green or yellow-green color to water. Under static conditions, they rise to the surface to form very distinctive films and windrows of greenish scum (Fig. 1) [for a review of mechanisms regulating buoyancy, see Oliver (1994)]. Their notoriety is well deserved. In addition to being indicators of nutrient (particularly

FIGURE 1 A surface scum formed by a cyanobacterial bloom of *Microcystis*.

phosphorus)-enriched waters, the presence of cyanobacterial blooms is a very visible symptom of deteriorated bodies of water. Population crashes (death) and the microbial decomposition of cyanobacterial cells result in the depletion of dissolved oxygen. Anoxic conditions can cause fish kills in bodies of water as small as prairie potholes (Barica, 1975, 1978) or as large as Lake Okeechobee, Florida (Jones, 1987). The largest cyanobacterial bloom recorded on Lake Okeechobee, which occurred on June 30, 1986, covered 337 km², almost 20% of the lake surface (Lamon, 1995).

The most infamous example of the adverse impact of algal blooms on a large body of water was the deterioration of the western basin of Lake Erie in the 1960s (Beeton, 1969; Rosa and Burns, 1987). Increases in the quantity of algal biomass (Davis, 1964) and shifts in species composition from diatoms to cyanobacterial blooms (Ogawa and Carr, 1969; Munawar and Munawar, 1976; Nicholls et al., 1980; Stoermer, 1988) exacerbated an already deteriorating water quality situation. The resulting oxygen-depleted conditions hastened the demise of native fish species and their replacement by invasive species such as alewife and lamprey.

Crashes of cyanobacterial blooms also have an adverse effect on the aquaculture industry. A fish kill in an 8.9 ha aquaculture pond (Boyd et al., 1975) caused by the die-off of a cyanobacterial bloom resulted in oxygen depletion that killed 6800 kg of catfish. At today's market value, the loss would have been worth between U.S. $11,000 and $19,000.

Many other cyanobacteria, as well as planktonic chlorophytes, euglenoids, diatoms, synurophytes, and dinoflagellates can bloom in nutrient-enriched waters. An important phenomenon is the occurrence of red water and red surface scums, caused by blooms of *Oscillatoria* (*Planktothrix*) *rubescens*, generally in large lakes in early spring (Jaag, 1972; Konopka, 1982a, b), and by species of *Euglena* and *Trachelomonas* in static waters in mid-to-late summer (Lackey, 1968). These red water-causing organisms should not be confused with red tides, which are marine, are composed primarily of dinoflagellates, and often produce toxic compounds. Other than being symptoms of highly nutrient-enriched waters, the red color-producing euglenoids of freshwaters are not toxic. The *O. rubescens/agardhii* complex has been reported to produce hepatotoxins (Carpenter and Carmichael, 1995) and may be responsible for dermatitis or skin irritation when people come in contact with contaminated water (Gorham and Carmichael, 1988), but documented reports of incidents are rare (W. W. Carmichael, personal communication) in comparison to reports of those caused by other cyanobacteria (discussed below).

The presence of blooms of microscopic algae has long been associated with eutrophication (Schelske and Stoermer, 1971; Schindler, 1975, 1977; Reavie et al., 1995). As a result, most trophic classification systems (e.g., Carlson, 1977; Wetzel, 1983; EPA, 1990) are based on some measure of algal biomass (e.g., chlorophyll or cell volumes), Secchi disk readings (a measure of transparency that can be affected by algal biomass), and types of algae present. Cooke et al. (1993b) summarized the effects of eutrophication as follows: "Symptoms of eutrophication, such as algal blooms (including surface scums), low transparency, rapid loss of volume in reservoirs, noxious odors, tainted fish flesh, impaired potable water supplies, dissolved oxygen depletions, fish kills, and the development of nuisance or exotic animal populations (e.g., common carp) can bring about economic losses in the forms of decreased property values, high cost treatments of raw drinking water, illness, depressed recreation industries, expenditures for management and restoration, and the need to build new reservoirs." While these authors describe algal blooms as one of several symptoms of eutrophication, the presence of the algal blooms themselves can lead to low transparency, noxious odors, tainted fish flesh, impaired potable water, dissolved oxygen depletions, and fish kills.

From an economic standpoint, the most important problem caused by algal blooms is production of taste and odor in surface water supplies (Raman, 1985; Hawkins and Griffiths, 1987). The potential for significant taste and odor problems exists in the United States because more water for human uses is obtained from surface water than from groundwater. Approximately 65% of the 399 billion gallons of freshwater withdrawn for all purposes in the United States in 1985 was obtained from surface water sources (Solley *et al.*, 1988). The cyanobacteria *Microcystis*, *Anabaena*, *Aphanizomenon*, and *Pseudanabaena* and the golden flagellates *Synura*, *Mallomonas*, and *Dinobryon* are common causes of taste and odor in water supplies, but diatoms, dinoflagellates, and even some green algae also cause problems (Palmer, 1962; Nicholls and Gerrath, 1985; American Public Health Association, 1992). Colonial species of the diatom *Stephanodiscus* were reported to cause undesirable odors and clog filter runs at municipal water plants on the Great Lakes (Stoermer, 1988). The fishy tastes and odors produced by *Synura* spp. are frequently reported in softwater lakes in Ontario, and the haptophyte *Chrysochromulina breviturrita* has been cited as a producer of particularly offensive odors in that area (Nicholls *et al.*, 1982). The tastes and odors produced by cyanobacteria such as *Anabaena* and *Oscillatoria* are caused by two compounds: 2-methylisoborneol (2-MIB) at concentrations greater than 12 ng L^{-1} and geosmin at concentrations greater than 7 ng L^{-1} (Simpson and MacLeod, 1991).

The cost of treating water for taste and odor problems is high. Copper sulfate ($CuSO_4$) is widely used to control or eliminate potential taste- and odor-causing algae in water supply reservoirs. Water treatment with activated carbon is then used to remove undesirable tastes and odors that do occur. Chlorine has no effect on the removal of musty/earthy aromas nor does treatment with ozone or aeration (Maga, 1987). The cost to treat Lake Manatee (Florida) Reservoir water with activated carbon where blooms have occurred has exceeded U.S. $14,000 a day (Clarke *et al.*, 1997). Additional expenses are incurred to replace the activated carbon filters, which frequently become clogged with humic substances, and to treat the lake water with copper sulfate.

Taste and odor problems are typical not only of surface waters where blooms occur but also of deeper waters. In a study of six Kansas lakes, Arruda and Fromm (1989) noted that the mostly fishy and grassy tastes in the surface waters were caused by algae and correlated with the trophic status of the lake. The foul taste of the bottom water (musty, sulfurous, and rotten egg-like) was typical of highly organic, anoxic sediments and overlying water. Even this problem was related to algae because it was probably caused by the deposition and slow decomposition of organic matter contributed over time by dying algal blooms. Having to draw from deep waters to avoid infested surface water can lead to other problems such as the deposition of iron and manganese in pipes and on clothing in washing machines.

The other major taste and odor problem caused by cyanobacterial blooms is off-flavors in the flesh of aquaculture-produced fish (particularly catfish) and other animals (Jüttner *et al.*, 1986; Maga, 1987; Martin *et al.*, 1991; Schrader and Blevins, 1993). Some of the causative organisms are species of *Lyngbya*, *Oscillatoria*, *Aphanizomenon*, *Anabaena*, and *Phormidium*. Geosmin (Brown and Boyd, 1982; Lovell *et al.*, 1986) and 2-MIB (Martin *et al.*, 1988; van der Ploeg *et al.*, 1995) are the primary chemicals that produce off-flavors. In general, once fish have been tainted with these compounds, they must be moved to clean water for several weeks so the off-flavor can dissipate before they are marketed. The harvest and transport of the fish to a new site for cleansing are expensive and laborious. Off-flavor problems have been estimated to add $50 million each year to the cost of producing catfish in the United States (Schrader *et al.*, 1997).

Various cyanobacteria produce toxins that are harmful to humans and animals [for review, see Carmichael (1997)]. The major genera are *Anabaena*, *Aphanizomenon*, *Microcystis*, *Cylindrospermopsis*, *Nodularia*, and *Oscillatoria*. There are many reported instances of livestock, pets, wild animals, and birds that have died after drinking tainted water, but, in general, the adverse effects of cyanobacterial blooms to humans have been limited to forms of dermatitis and irritation of the mucous membranes. Some evidence that human gastrointestinal disorders were associated with consumption of water from reservoirs with blooms of cyanobacteria can be found (Carmichael *et al.*, 1985; Carmichael and Falconer, 1993), and cyanobacterial toxins were implicated in the deaths of 26 people in Brazil when contaminated water was used for hemodialysis (Jochimsen *et al.*, 1998). An association between toxins in water supplies and primary liver cancer in China has been suggested (Carmichael, 1994). Fortunately, the instances of human poisoning are rare because the unattractiveness and foul odors of water in which an algal bloom occurs usually deter people from using or drinking the water.

Algal blooms can have adverse impacts on the health of organisms other than fish and humans. High nutrient concentrations in the water column trigger algal blooms, which reduce light penetration. Light reduction can severely limit the growth of submersed vascular plants (Spence, 1976; Jupp and Spence, 1977; Jones *et al.*, 1983), thus decreasing habitat and shelter available for fish and fish food organisms. As a result, the most eutrophic lakes are those that are dominated by bloom-forming algae, with little or no submersed vascular plant production (Wetzel, 2001).

In addition to loss of habitat, cyanobacterial blooms may cause a loss of system-level productivity. Some cyanobacterial species are allelopathic to other algae that are considered to be food sources for zooplankton (Keating, 1976, 1977, 1978), and they themselves are not significantly grazed by zooplankton (which is one reason why they can dominate aquatic systems). Cladoceran populations (e.g., *Daphnia*) decline or disappear when cyanobacteria, particularly the filamentous forms, predominate (Burns, 1968; Keating, 1976; Infante and Abella, 1985; Rothhaupt, 1991). The major reason for lack of predation appears to be a mechanical interference with feeding when the filamentous forms accumulate in the filtering apparatus. The loss in available energy suppresses zooplankton reproduction. Webster and Peters (1978) showed that as densities of filaments of *Anabaena* spp., *Aphanizomenon flos-aquae*, *Oscillatoria tenuis*, or *Lyngbya* spp. increased, the larger-sized cladocerans filtered at lower rates, increased rejection rates, and decreased brood sizes. The increased energy expenditure of trying to obtain sufficient food appears to increase the respiration rates of cladocerans (Porter and McDonough, 1984), further reducing assimilation efficiency, growth, and reproduction. As few as 50 cyanobacterial filaments per mL of water have an adverse effect on zooplankton feeding rates (Infante and Abella, 1985). The presence of filamentous cyanobacteria can cause a shift from large-bodied to small-bodied zooplankters, which feed on other materials such as small algae, bacteria, and organic debris. Some evidence suggests that the adverse effect of cyanobacteria on cladocerans also is due to toxin production (Infante and Abella, 1985; Fulton and Paerl, 1987; DeMott *et al.*, 1991).

B. Macrophytic Filamentous Algae

The problems caused by macrophytic filamentous algae in aquatic systems are primarily due to their ability to form large mats of vegetation (Fig. 2). These algae are typically found in shallow water where they may be free-floating (e.g., *Pithophora*, *Rhizoclonium*, *Spirogyra*, and *Hydrodictyon*) or attached (e.g., *Cladophora*, *Ulothrix*, *Stigeoclonium*, and *Oedogonium*) to substrata, either living (plants and other algae) or nonliving (rocks, cement linings, and sediments). The free-floating forms are generally restricted to static waters such as ponds and the sheltered littoral zones of lakes. Attached forms occupy a much wider range of habitats. They are found in both static and flowing systems, including the wave-scoured edges of lakes (e.g., *Cladophora* in the Laurentian Great Lakes), fast-flowing streams, and the extensive irrigation systems and aqueducts of the western United States (Fig. 3).

The genera listed above are all green algae. Filamentous cyanobacteria also can form free-floating mats. The cells of filaments of *Lyngbya wollei* (Speziale

FIGURE 2 Filamentous algal mats (*Spirogyra*) causing obvious aesthetic problems in a lake.

FIGURE 3 Filamentous algal mats (*Cladophora*) in an irrigation canal that have broken away from the sides and are floating downstream. These mats can clog irrigation intakes and pumps. Photo courtesy of Lars Anderson, USDA-Agricultural Research Service.

and Dyck, 1992) are quite large (cell diameter: 25–64 µm, length: 2–11 µm) for a cyanobacterium, and the dimensions, coarseness, and even color (dark green) of the filaments may cause the untrained observer to think that they are handling a filamentous green alga. The mats are dense and can completely cover ponds and shallow areas of lakes. Many species of *Oscillatoria* form benthic mats that break free from the bottom and float to the surface (Fig. 4A) when gas bubble accumulation dislodges the mats (Halfen and McCann, 1975). These growths are typically dark blue-green to black in color and are quite slimy. The mats are often coated with sediments that were deposited on them while they were still associated with the bottom

FIGURE 4 Mats of macrophytic *Oscillatoria*. (A) Infestation of free-floating mats that have broken loose from the sediments in the shallow cove of a lake. (B) Close-up photo of the free-floating mats. The light color is due to the sediment deposited on the surface of the mats. Some of the mats have been turned upside down and show the natural dark color of the organism. Photos courtesy of Neil Gerber, Aquatic Management, Bluffton, Indiana.

substratum (Fig. 4B). Several species of *Phormidium* form the "black algae" growths that attach to the cement linings of swimming pools (Fitzgerald, 1959; Adamson and Sommerfeld, 1978).

The occurrence of filamentous algae is widespread. The Florida Department of Environmental Protection, Bureau of Aquatic Plant Management, recently surveyed 451 public bodies of water in Florida (Schardt, 1994). Filamentous algae were the only submersed "plant" grouping observed in more than 50% of the water bodies (62%) and were the dominant group in 16% of the waters. They represented one of only five plant groupings that showed an increase over the previous 12 years (increasing by 22%, more than any of the other plants). The two genera that were singled out as being predominant were *Pithophora* spp. and *L. wollei*. In an unpublished 1995 survey (J. Schardt, personal communication), *L. wollei* was collected in more than 218 lakes in Florida. An indication of the severity (or perceived severity) of algal problems in the state is the fact that approximately 90% of the calls from individuals or lake associations to aquatic plant management services in Florida seek information on algae control (J. Williams, The Lake Doctors, Winter Springs, FL; personal communication).

Extensive survey data are less available for other parts of North America, but filamentous algae are a widespread management problem. In the midwestern United States, approximately 60–70% of the total volume of aquatic plant control chemicals applied is for the control of (primarily) filamentous algae (R. Johnson, Aquatic Control, Seymour, IN; personal communication). *Cladophora glomerata*, *Pithophora* spp., and *Rhizoclonium* spp. are listed as causing problems in the western United States and Canada, and *Cladophora* and *Chara* are on the list of the 13 most consistently problematic aquatic weed species in these regions (Anderson, 1993).

Excessive growths of mat-forming algae, either alone or in combination with aquatic vascular plants, impair recreational activities such as swimming, fishing, and boating. Swimming beaches fouled with algal mats are not only unappealing but also hazardous when ladders, rocks, and submerged concrete are coated by slime-producing species such as *Spirogyra* (Bennett, 1971) and cyanobacteria. *Cladophora* growths in the Great Lakes were noted as posing a potential danger to young and inexperienced swimmers who might become entangled in the mats and drown (Herbst, 1969). When large odiferous masses are washed up on the shore they decay and are aesthetically unpleasant, are a barrier to recreation, and are implicated in taste and odor events in drinking water supplies (Brownlee *et al.*, 1984; Painter and Kamaitis, 1987; American Public Health Association, 1992). The loss of recreational and aesthetic values can have a significant economic impact on waterfront properties. Ormerod (1970) reported that the value of real estate on Lake Erie fronted with *Cladophora* mats averaged 80–85% of the value of clean frontage.

There is some evidence that mat-forming cyanobacteria (e.g., *L. wollei*, *Oscillatoria* spp.) also produce toxins similar to those produced by the phytoplanktonic cyanobacteria (Gunn *et al.*, 1992; Carmichael *et al.*, 1997; Onodera *et al.*, 1997). Although there are no reports to date of animals being killed by the toxins produced by *L. wollei*, dogs were reportedly killed by the toxins from benthic mats of *Oscillatoria* (Gunn *et al.*, 1992).

Macrophytic algae restrict and greatly reduce the efficiency of culture and harvest activities in fish culture ponds (Tucker et al., 1983). They may compete with phytoplankton, thus reducing the base of the aquatic food chain (Boyd, 1982). A 50% reduction in fish production in farm ponds was attributed to heavy *Pithophora* growth and concomitant loss of phytoplankton (Lawrence, 1954). Although reports implicating macrophytic algae as direct causes of fish kills are few (e.g., Robinson and Hawkes, 1986), excessive algal growth must add to the oxygen deficits that result from respiration of submersed plant growth and/or phytoplankton at night, during periods of cloudy weather, or under snow-covered ice. Fish kills in ponds that are dominated by filamentous algae are not uncommon (personal observation). Oxygen deficits can be stressful to fish in other ways by causing declines in food consumption and growth and by making them more prone to bacterial infections (Boyd, 1982).

Overpopulation, stunted or reduced growth, and a decline in the capture efficiency of forage species by predatory fish have been attributed to dense vascular plant growth (Colle and Shireman, 1980; Savino and Stein, 1982; Shireman et al., 1983), and algal growth is assumed to contribute to this problem. However, the need for some macrophytic growth to provide protection for young fish and habitat for fish food organisms is also recognized (Barnett and Schneider, 1974; Wiley et al., 1984). The proper balance of macrophyte coverage and density for optimal fisheries is at this point somewhat controversial (Bettoli et al., 1992, 1993; Hoyer and Canfield, 1996a, b; Maceina, 1996), in part because of the variability in study sites and interpretations. The possible beneficial or detrimental roles of macrophytic algae as fish habitat clearly require further study (Hinkle, 1986).

Cladophora, *Stigeoclonium*, *Oedogonium*, and *Ulothrix* have been cited as presenting serious problems in irrigation canals because they attach to concrete canal linings, thus reducing both flow rate and capacity. *Cladophora* can become associated with beds of pondweeds (*Elodea* spp.) and coontails (*Ceratophyllum* spp.), which increases resistance to the flow of water (Mitchell et al., 1989). Mats that break away from the linings float downstream where they foul pump inlets, irrigation siphons, trashracks, and sprinkler heads (Hansen et al., 1984). *C. glomerata* in particular is a major problem in water delivery systems in the western states. For example, in the 620 km length of the Salt River Project Canal, which supplies water and electricity for the cities of Phoenix and Tempe, Arizona, the control of aquatic weeds is a major operation/maintenance task (Corbus, 1982). The annual budget in the late 1980s for aquatic macrophyte control (primarily *Cladophora*) in this system was approximately U.S. $1.5 million.

The clogging of rivers, canals, and drainage ditches by aquatic plants and algae can prevent adequate drainage so that water backs up, even to the point of causing flooding. Another problem associated with water conveyance or storage systems, particularly in arid parts of the world, is the potential to lose water through evaporation from floating or emergent plant surfaces. Although considerable data are available to show that substantial loss does occur with coverage of vascular plants such as water hyacinth (Brezny et al., 1973), nothing is known of the potential of surface-floating algae to add to the problem.

The impacts of filamentous algae on the dynamics of food webs in natural systems have not been well documented. Most studies that have been conducted have focused on *Cladophora*. For example, *Cladophora* is not a major food source for the invertebrates or fish that live in lakes (see reviews by Lembi et al., 1988; Dodds and Gudder, 1992), although it is grazed by fish in river systems (Power, 1990). *Cladophora* provides an extensive surface area for colonization by periphyton and invertebrates, and the limited grazing that does occur may have more to do with ingesting these associated organisms than with the filamentous alga itself (Dodds and Gudder, 1992). On the other hand, dense growths of *Cladophora* were reported to reduce invertebrate diversity and to have disrupted shoal spawning by walleye, whitefish, and lake trout in the Great Lakes (Neil, 1975). Filamentous algal mats compete with submersed vascular plants for space and light. Examples include the replacement of angiosperms, such as *Najas marina*, by *Spirogyra* (Phillips et al., 1978), diverse macrophytes by *C. glomerata* (Bolas and Lund, 1974), *Elodea* by *Cladophora* and *Spirogyra* (Simpson and Eaton, 1986), and *Potamogeton pectinatus* by *Cladophora* (Ozimek et al., 1991). Phillips et al. (1978) suggested that replacement of submersed vascular plants in lakes undergoing eutrophication may be due more to the shading by epiphytic and filamentous algae than to phytoplankton.

The diversity among filamentous algae, for example, the sliminess of *Spirogyra* (which appears to prevent colonization by periphyton) in contrast to the thick cell walls of *Cladophora* (which provide an excellent substratum for periphyton), the summer domination by *Pithophora* in contrast to the spring/fall distribution of *Cladophora*, and the net-like habit of *Hydrodictyon* or the unbranched habit of *Spirogyra* in contrast to the branched habit of *Cladophora*, suggests that much remains to be learned about the micro-niches that these algae make available to invertebrate and fish communities and their impacts on food webs of static freshwater systems.

C. Chara and Nitella

Charophytes are usually viewed as being beneficial components of aquatic systems, and their reestablishment is an important factor in lake restoration (van den Berg et al., 1998b). Chara and Nitella are considered excellent habitats for littoral invertebrates (Rosine, 1955; Quade, 1969; Allanson, 1973; Hargeby et al., 1994) and fish (Fassett, 1957; Schardt, 1994), and they are a major food source for herbivorous waterbirds (Hargeby et al., 1994; van den Berg et al., 1998b). Their ability to form low-growing meadows of vegetation reduces the resuspension of sediments (van den Berg et al., 1998b). These macroalgae, however, can cause problems in shallow water when their growths reach the surface of the water, thereby preventing successful angling, swimming, and boating (Fig. 5). Chara and Nitella are typically named when weedy submersed species (mostly vascular plants, such as Ceratophyllum, Myriophyllum, and Elodea) are listed. For example, Steward (1993) listed Chara and Nitella as among the plant groups causing weed problems in the eastern United States, and Anderson (1993) cited these genera for the western United States also.

Charophytes also produce repellent (allelopathic) materials that exclude certain limnetic species of invertebrates (Pennak, 1966, 1973) and phytoplankton (Gibbs, 1973; Anthoni et al., 1980; Wium-Andersen et al., 1982). The latter finding may provide a partial explanation for the lack of epiphytes and clear water conditions frequently associated with some charophyte species [Crawford, 1979; Wium-Andersen et al., 1982 (but see Chap. 2, Sect. II.F.4)].

Chara is common in regions with hard water (e.g., areas of the Midwest with a limestone bedrock), and Nitella is more characteristic of soft waters (e.g., granitic regions of the Northeast). In a survey of 451 water bodies in Florida (which has regions of both soft and hard waters), Schardt (1994) collected Nitella in 64 bodies, and found that it was dominant in 28 of them. Chara was found in 52 bodies and was dominant in 15 of them.

An interesting characteristic of Chara is that it tends to colonize sites in which vascular plants have been controlled (Nichols, 1984). Drawdown, the draining and exposure of shallow or shoreline areas to desiccation, eliminated water shield (Brasenia schreberi), restricted the spread of parrot feather (Myriophyllum brasiliense) and water lily (Nymphaea odorata), but enhanced the infestation of Chara vulgaris in a Louisiana reservoir (Lantz et al., 1964). In a survey of the effects of drawdowns, C. vulgaris increased in 33 cases, decreased in 15 cases, and stayed the same in 44 cases (Cooke et al., 1993b). Invasion or expansion by Chara has also been documented after dredging (Born et al., 1973; Nichols, 1984), mechanical harvesting (Anonymous, 1990), and the application of herbicides for the control of vascular plants (C. A. Lembi, personal observations).

Opening of sites disturbed by weed control activities to light is the major reason cited for the invasion by Chara (Born et al., 1973), and recent studies seem to confirm that irradiance is a major factor regulating charoid distribution (Steinman et al., 1997). Although some evidence suggests that as eutrophication proceeds, charophyte populations may be reduced because of their sensitivity to "toxic" levels of phosphorus (P) (Forsberg, 1965), other studies show that increased P levels do not have an adverse effect on charoid growth (Blindow, 1988). Melzer et al. (1977) suggested that increased P concentrations play an indirect role in the disappearance of Chara, primarily by causing an increase in phytoplankton growth and turbidity, which in turn shades out charoid growths. The restoration of

FIGURE 5 A solid stand of Chara infests this pond. Although the vegetation has not formed a surface canopy, the underwater growth limits swimming and fishing success.

a *Chara* community in one system was achieved by reducing P concentrations, which resulted in higher water transparencies (Simons *et al.*, 1994). However, light may not be the only factor critical to *Chara* establishment, particularly in mixed plant communities. van den Berg *et al.* (1998a) experimentally observed that *Chara* was negatively impacted by shading from sago pondweed (*Potamogeton pectinatus*), but the fact that *Chara* dominates sago pondweed in some clear water lakes suggests that light is probably not a key factor in that domination. The more efficient use by *Chara* of carbon (HCO^{3-}) at low concentrations, which are typical within *Chara* meadows, has been suggested as a possible reason for its dominance (van den Berg *et al.*, 1998b).

III. CONTROL METHODS FOR NUISANCE ALGAE

Management practices for nuisance algae are divided into two major categories: nutrient manipulation and direct control techniques. Nutrient manipulation, particularly reduction of nutrient inputs, should be viewed as the best approach for long-term control of algal problems. There are situations for which significant nutrient reduction is impractical or ineffective; under these conditions, direct control of the algal biomass may be the only alternative available. Direct control methods should only be viewed as temporary solutions and should be coupled with longer-term strategies for reducing nutrient inputs.

A. Nutrient Manipulation

It has long been known that inputs of nutrients, particularly P, stimulate algal growth. Many studies have shown a strong correlation between total phosphorus (TP) and planktonic algal biomass (Dillon and Rigler, 1974; Jones and Bachmann, 1976; Carlson, 1977; Schindler, 1978; Prepas and Trew, 1983). The positive relationship between chlorophyll-*a* concentrations (or shallower Secchi disk transparencies) and TP is a commonly used tool to predict water quality and trophic status (Vollenweider, 1969; Dillon and Rigler, 1974; Dillon *et al.*, 1988).

Some lakes are nitrogen (N)-limited. For example, a number of Florida lakes are surrounded by rich phosphate-containing deposits and soils and therefore may be N-limited. In a study of 223 Florida lakes, 27% were considered N-limited (Canfield, 1983). Also, studies of lakes in the semiarid and mountainous regions in the western United States indicate that the importance of N may be equal to or greater than that of P in limiting phytoplankton growth (Elser *et al.*, 1990; Reuter *et al.*, 1993). N limitation, as well as P limitation, has been implicated in the regulation of filamentous algal growth (Spencer and Lembi, 1981; O'Neal *et al.*, 1985; Dodds and Gudder, 1992).

In addition to the amount of phytoplankton biomass that is produced with P or N additions, another consideration is species composition, particularly in relation to nutrient ratios. When sufficient silicon (Si) and N are available in relation to P (high Si:P and N:P ratios), diatom growth appears to be favored (Tilman and Kiesling, 1984). These conditions are typical of spring periods in temperate lakes following turnover or when sediment deposition occurs with spring rains. In late spring or early summer, green algae may dominate over diatoms as Si concentrations decrease [lower Si:P ratios (Sommer, 1983)].

From a management standpoint, the most critical nutrient ratios are low N:P or Si:P. These generally occur under conditions of excessive P loading, and it is under these circumstances that N-fixing cyanobacteria (*Anabaena*, *Aphanizomenon*, and others), which fix atmospheric N_2 when the water becomes N-depleted, become dominant (Schindler, 1977; Smith, 1983), particularly during the summer months. For example, the shift from high Si:P and N:P ratios to low ratios was concomitant with the shift in dominance from diatoms to cyanobacteria in the Great Lakes in the 1960s (Schelske and Stoermer, 1971; Schelske, 1975). For this reason the emphasis on nutrient removal for generally improving water quality has been placed on P rather than on N.

Another approach for maintaining a high N:P ratio is to increase N rather than to decrease P. In fact, several researchers (Leonardson and Ripl, 1980; Smith, 1983) suggested that N removal could be counterproductive and that N addition might actually be helpful in increasing populations of green algae or diatoms in relation to cyanobacteria. Barica *et al.* (1980) added N to ponds with low N:P ratios to see if it could reduce the incidence of cyanobacterial blooms, particularly blooms of *Aphanizomenon* that were causing fish kills when they crashed. The initial N:P ratios in these ponds were around 4 to 5 [Schindler (1977) found that cyanobacteria dominate when N:P ratios dropped from 15 to 5]. When low amounts of N were added ($0.1 \text{ g N m}^{-3} \text{ d}^{-1}$) prior to bloom formation (but not during the bloom) or when high amounts ($1 \text{ g N m}^{-3} \text{ d}^{-1}$) were added during the bloom, a shift from *Aphanizomenon* to green algae and cryptomonads occurred. The technique worked, but it was not considered to be a realistic approach because N would have to be added over several to many weeks. Although similar results were reported by Stockner and Shortreed (1988), the general consensus is that, when possible, it is much better to reduce P concentrations than to elevate N concentrations. In fact, increasing the N:P ratio stimulated the growth of non-nitrogen fixing cyanobacteria such as *Lyngbya, Oscillatoria,* and *Chroococcus* in mesocosms placed in Lake Okeechobee (Havens and East, 1997).

In an analysis of the literature, Elser *et al.* (1990) suggested that both P and N potentially limited algal growth. They found little support for P alone as a causative factor. However, they recommended that efforts should concentrate on P reduction because it is easier to achieve from a technical standpoint than N reduction. Clearly where N fixation by planktonic cyanobacteria is a response to N reduction, P should be the more reliable means to lower algal biomass. The same general approach is used for the control of filamentous algae and *Chara,* although the situation regarding specific nutrient ratios and target amounts is far from clear.

There are three general approaches for achieving P reduction: decrease external P loading, suppress internal P loading, and increase P output from the system. External inputs of P can be decreased with diversion and advanced wastewater treatment, with detention basins and wetlands, and by the initiation of other watershed management techniques. Internal P loading can be suppressed with alum applications, dredging, and aeration. P-laden waters from the site can be released with hypolimnetic withdrawal.

1. Diversion and Advanced Wastewater Treatment

These two techniques (used together) are the most frequently used methods to reduce external loading. Diversion is achieved primarily through sewage collection systems, and the water is then subjected to tertiary treatment in which P is removed by alum (aluminum sulfate), lime (calcium carbonate), or iron (ferric chloride). There have been a number of successes (for case histories, see Cooke *et al.,* 1993b). Probably the best example is Lake Washington in Seattle, Washington (Edmundson and Lehman, 1981; Edmundson, 1994), in which 88% of the lake's external loading was diverted from 1964 to 1967. TP declined from a mean annual concentration of 64 μg L^{-1} prior to diversion to 21 μg L^{-1} 5 years after diversion. Chlorophyll-*a* decreased from 36 to 7 μg L^{-1} by 1969. Secchi disk depth increased from 1 to 3.1 m. Further reductions in algal biomass were attributed to increased populations of *Daphnia* [following a decline in planktonic *Oscillatoria,* which negatively impacts *Daphnia* feeding (Infante and Abella, 1985)] and a decrease in a planktivorus crustacean (*Neomysis mercedis*) population. The condition of the lake in the late 1970s was 17 μg L^{-1} TP, 3 μg L^{-1} chlorophyll-*a*, and 7 m Secchi disk depth, and it had clearly shifted from a eutrophic to a meso- or oligomesotrophic state.

During the 1970s, significant reductions in P loading to Lake Erie also were achieved through legislation that upgraded sewage treatment to include chemical precipitation of P and reduced the allowable levels of phosphates in laundry detergents (Phosphorus Management Strategies Task Force, 1980). Declines in phytoplankton biomass averaged about 5% per year over the period from 1970 to 1985 (Nicholls and Hopkins, 1993) and were correlated with the resultant reduction in P loading (Nicholls *et al.,* 1977, 1980). Interestingly, phytoplankton populations continue to decline due to removal by zebra mussels (Nicholls and Hopkins, 1993). Filamentous algal growths also respond to nutrient diversion. For example, *Cladophora* biomass and tissue P concentrations at seven sites in Lake Ontario steadily decreased from 1972 to 1983 in response to P control programs introduced in the early 1970s (Painter and Kamaitis, 1987).

Diversion and treatment work best where there are distinct point sources of nutrient inputs. They are less successful at sites impacted by nonpoint sources or in which significant concentrations of nutrients have been stockpiled in the sediments and are a major source of internal loading.

2. Detention Basins and Wetlands

Discharge of domestic wastewater and urban runoff into detention basins (also called retention ponds) or natural or constructed wetlands is often recommended for improving water quality before release into a river or lake (Mitsch and Gosselink, 1993; Olson, 1993; Etnier and Guterstam, 1997). Many local and some state ordinances now mandate the construction of retention ponds in new housing developments, industrial parks, and similar sites. These ponds mostly serve as settling basins for sediments and associated nutrients and other pollutants (Walker, 1987; Robbins *et al.,* 1991). Of course, these sites themselves become ideal environments for the development of algal blooms and mats and are in large part the cause for the substantial increase in the number of companies offering aquatic plant and algal management services in recent years (C. A. Lembi, personal observation). Although algae and other aquatic plants in retention ponds serve as a filtration system for nutrients, urban residents frequently complain about having to look at scummy water!

Highly vegetated wetland areas also act as settling basins; in addition, they provide biological filtering, uptake and storage, and transformation (e.g., denitrification) of nutrients (Mitsch and Gosselink, 1993). Although the P storage capability can be lost in temperate areas in winter when the plants are no longer taking up P and nutrients are released, wetlands do tend to store considerable P in the summer, which is the critical time for algal blooms to occur in downstream sites (Cooke et al., 1993b).

3. Watershed Management

The importance of a broad watershed management program to reduce both point and nonpoint sources of fertilizers and other pollutants is gaining increased recognition at local, state, and federal levels. In agricultural areas, the promotion of best management practices (BMPs) has resulted in widespread acceptance of practices that reduce erosion of nutrient-laden soils (Scholze, 1994; EPA, 1998). Such practices include no-til and conservation tillage, vegetated filter strips and grass waterways, lowering of fertilizer application rates, and proper handling of animal manures. The adoption of BMPs in the United States is voluntary although cost-sharing programs are available through federal agencies such as the Farm Services Agency and the Natural Resources Conservation Service. Section 303(d) of the Clean Water Act calls for the implementation of total maximum daily loads into streams and lakes that have low water quality, and the Clean Lakes Program provided assistance in watershed management and improving water quality in lakes prior to 1995. Clearly, there is general recognition of the importance of watershed management in improving water quality, and the erosion of sediments into waterways has been considerably lessened. There are, however, still areas where the implementation of programs has been slow. An example is the discharge of animal wastes into rivers in North Carolina and Maryland watersheds, which appears to have resulted in fish-killing blooms of the estuarine dinoflagellate *Pfiesteria piscida* (Burkholder et al., 1997).

4. Alum

In many situations, reduction of external P loading does reduce algal growth. This is particularly true in water where most of the P loading to the photic zone is from external sources and in deep stratified lakes where P released from the anaerobic bottom sediments and hypolimnion does not reach the photic zone. In shallow lakes, on the other hand, significant quantities of P can be released from the sediments and reach the photic zone (Wetzel, 1990; Cooke et al., 1993b). Therefore, reduction of external P loading may not have much of a short-term impact on phytoplankton growth in these sites. Resuspension of sediments is considered a potential source of nutrients for phytoplankton production in many shallow lakes (Carper and Bachmann, 1984; Riley and Prepas, 1984; Hansen et al., 1997; Havens and James, 1997). Stauffer and Lee (1973) calculated that all of the summer algal blooms in Lake Mendota, Wisconsin, could be accounted for by internal loading of P from the lower waters and sediments to the photic zone. Therefore, steps to reduce internal P cycling in many of these lakes may be more effective in reducing algal growth than the reduction of external P inputs. The methods used to reduce internal P loading include chemical treatment with alum, the removal of sediments by dredging, and aeration.

Alum ($Al_2[SO_4]_3$) is used to lower P availability through P precipitation and to retard P release from the lake sediments (P inactivation). When added to water, alum and P form aluminum phosphate and a colloidal aluminum hydroxide floc to which certain P fractions are bound (Cooke et al., 1993b). The floc settles to the sediment and continues to sorb and retain P within the lattice of the molecule, thereby preventing further release of P. Sodium aluminate ($AlNaO_2$), which is a good buffering material, is added to alum treatments to maintain pH values between 6 and 8 (Kortmann and Rich, 1994) because a severe shift in pH can be detrimental to fish populations. In addition, alum is not recommended for use in waters with an acidic pH or low alkalinity because of the potential for aluminum toxicity to fish at pH values below 5.5. Iron salts also can be used to inactivate P (Kortmann and Rich, 1994), and treatments with calcium salts ($Ca(OH)_2$ and $CaCO_3$) have successfully reduced P loading from bottom sediments in Canadian lakes (Prepas et al., 1990; Babin et al., 1994).

There are numerous examples of success with alum treatments in shallow lake systems, and many treatments last from 2 to 15 years (Welch et al., 1988; Smeltzer, 1990; Cooke et al., 1993a; Jacoby et al., 1994; Welch and Cooke, 1995). In some lakes, internal loading has been significantly reduced for up to 20 years (Welch and Cooke, 1999). Holz and Hoagland (1999) reported improved water clarity, decreased chlorophyll-*a* concentrations, reduced cyanobacterial biomass and abundance, increased *Daphnia* biomass and abundance, and increased usable fish habitat in a shallow (mean depth = 4 m), alum-treated lake in Nebraska.

To ensure success and long-lasting effects, reduction of internal P cycling must be accompanied by a reduction in external P loading. Factors that can lead to failure of an alum treatment include continued high external P loading (Welch et al., 1988; Barko et al., 1990), redistribution of alum floc to the lake center by wind mixing (Garrison and Knauer, 1984), and P recycling from senescing rooted macrophytes or from macrophytes that expand their range due to improved water clarity (Welch et al., 1988; Welch and Cooke, 1999). There also is evidence that cyanobacteria newly recruited from the sediments can transport P into the water column, even in alum-treated lakes (Perakis et al., 1996).

5. Dredging

Dredging offers a more permanent solution to internal P loading in shallow lakes than alum treament because sediments, the actual source of the P loading, are removed from the system. Dredging, however, is much more expensive than alum. According to Cooke et al. (1993b), dredging costs nearly 30 times more than alum initially although over the long term (repeat alum treatments every 10 years), the cost differential is only 5 times greater if totaled over 50 years. The cost of alum treatment averages about U.S. $700 per ha and the cost of dredging is about U.S. $20,000 per ha (Cooke et al., 1993b). As with all nutrient reduction approaches, external loading must be reduced or eliminated to achieve long-term results.

6. Aeration

P is released from sediments under anoxic conditions. The function of aerators (other than to improve habitat for fish) is to oxygenate the water column, or portions of the water column, and the upper layers of the sediments, thereby preventing the occurrence of low-O_2 conditions. In theory, oxidized forms of P are not released into the photic zone to encourage phytoplankton blooms.

There are two major methods for aerating pond or lake water (Cooke et al., 1993b; Kortmann and Rich, 1994). The first is artificial circulation. This method oxygenates the whole water column. Air is pumped from a compressor on shore through a tube to a weighted diffuser unit that is placed on the bottom. Air bubbles pass from the diffuser into the water and are often visible as a surface "boil" (Fig. 6A). This method destroys or prevents thermal stratification; therefore, it is not feasible in sites where deep cold water is necessary to maintain coldwater fish populations. It is, however, a good solution to potential oxygen depletion problems for warmwater fish species.

The second method is termed "hypolimnetic" aeration. This method maintains stratification because the water is removed from the hypolimnion, oxygenated at the surface, and then returned to the bottom. Hypolimnetic aeration is used in deep lakes to overcome anoxia, improve coldwater fisheries habitat, and control sediment P release.

The impacts of either type of aeration method on algal blooms have been difficult to document. The sediment–water interface in many shallow lakes may already be oxygenated, in which case aeration will not have an impact. Cooke et al. (1993b) summarized data

FIGURE 6 Aeration. (A) The surface boil from an underwater circulating aerator. (B) A fountain has been attached to the aerator to improve aesthetics. Photos courtesy of Neil Gerber, Aquatic Management, Bluffton, Indiana.

from a number of aerated lakes and observed that phytoplankton content decreased in less than half of the lakes examined. Cyanobacterial blooms, however, decreased and green algal populations increased in the majority of cases. This shift was attributed to several factors. For example, aeration may increase the carbon dioxide concentration in the water, thus lowering the pH and favoring green algal development. The turbulence created may disrupt the ability of the cyanobacteria to form surface scums, which normally shade out other, potentially competitive, algae. In those cases in which cyanobacterial blooms were not affected, water mixing and aeration may have been incomplete.

There is presently no evidence to suggest that aeration has an impact on filamentous algae or *Chara*. Frodge et al. (1991) reported that *Pithophora* mats growing among vascular plant canopies were associated with high concentrations of P in the surface water. This trend was thought to be the result of the conversion of iron-bound P to OH^--bound P at high pH values (>10), so that the P was not precipitated even at high dissolved oxygen levels. Thus, high pH values, often associated with high photosynthetic rates of dense vegetation, can potentially offset aeration effects.

The presence of a fountain in a body of water does not mean that the water is being aerated. Aerators are specifically designed pieces of equipment that move air into the water column, not spray water into the air. A fountain can be attached to an aerator for aesthetic purposes (Fig. 6B). Although fountains may cause some surface circulation and aeration, they do little to prevent nutrient cycling or fish kills.

7. Hypolimnetic Withdrawal

The principle behind hypolimnetic withdrawal is the pumping or siphoning of bottom waters that have a high P content into receiving waters. The technique has not been used frequently. Although evidence suggests that it can be successful in reducing the P content of lake water (Cooke et al., 1993b), few studies provide convincing data that algal blooms are reduced. Replacement of water to maintain depth must come from a source with a low P content. Another problem with this technique is the potential for damage downstream caused by releasing anoxic, nutrient-laden, polluted waters.

8. Summary of Nutrient Manipulation Methods

Nutrient manipulation techniques, particularly those that regulate P inputs or internal cycling, can successfully reduce the incidences and severity of algal blooms. In some instances, a single technique is not sufficient. Water supply lakes for the city of St. Paul, Minnestoa, were infested with blooms of *Anabaena* and *Aphanizomenon* (Walker et al., 1989). Chemicals (powdered carbon and potassium permanganate) were added at the water treatment plant to reduce taste and odors, and copper sulfate was applied every week during the growing season. These approaches were unsuccessful. It was only when the lakes were subjected to a multimethod approach that included the reduction of external and internal P concentrations by using iron chloride to inactivate P, the construction of detention ponds to reduce P loadings from runoff from urban watersheds, and hypolimnetic aeration that some success was achieved.

The St. Paul example illustrates the complexities involved in nutrient manipulation procedures. Probably the greatest impediment to initiation of a nutrient removal plan is the watershed analysis and water quality testing (and financial outlay). This analysis is necessary to determine which approach or combination of approaches is most likely to succeed. It is important for water management agencies and property owner associations to accumulate and allocate sufficient resources for a thorough lake and watershed monitoring program before implementing nutrient manipulation techniques. Once the management approach has been chosen, however, the science is now to the point where successes far outnumber failures.

One outcome of nutrient control techniques to reduce algal populations can be increased colonization by submersed vascular plants. Submersed vascular plants obtain the majority of their nutrients from the sediments rather than from the water (Carignan and Kalff, 1980; Barko and Smart, 1980, 1981). Therefore, the techniques described above have little impact on submersed vascular plant growth. When shading by phytoplankton or filamentous algal mats is removed, these plants can become established or reestablished. The presence of submersed plants is advantageous in many instances, but sometimes it has serious consequences. For example, the improved clarity of Lake Washington has led to invasion by Eurasian watermilfoil (*Myriophyllum spicatum*) in shallow areas, and there are other examples of similar shifts in populations (Spencer and King, 1984; van Donk et al., 1990). Madsen (1996) recorded that Secchi disk values doubled (1.5 to 3 m) in Lake St. Clair, Michigan, between 1967 and 1995, as a result of the introduction of zebra mussels. Macrophytic plant range expanded from 60 to 95% of the lake over this period, and Eurasian watermilfoil range expanded from 20 to 44% of the lake. Hence, management plans may need to consider which is the better alternative: a eutrophic, algal-dominated lake vs. relatively clean water with abundant vascular plant growth. The latter problem is somewhat ameliorated when native vascular species and *Chara*, which tend to have shorter growth habits and provide valuable fish habitat, colonize an area. It is exacerbated, however, when the colonizer is an invasive species such as Eurasian watermilfoil or hydrilla (*Hydrilla verticillata*). The stems of these species grow up through the water column to form a canopy of vegetation at the surface (Barko et al., 1986; Smith and Barko, 1990). This growth form not only prevents use of the water but can shade out stands of native vegetation. Such shifts among populations further illustrate the complexity of dealing with algal and aquatic plant management issues.

B. Direct Control Methods

The goal of direct control methods is to remove the algal biomass as quickly, efficiently, and cost effectively as possible. Although choices have to be made about which technique to use, the extensive watershed and in-lake monitoring that should precede nutrient manipulation does not have to be conducted. On the other hand, the efficacy of direct control methods often causes the user to overlook the need to initiate a long-term, nutrient management program. Certainly direct control techniques have their place in an overall management plan, but they should not be viewed as the only approach to solving noxious algal blooms or extensive filamentous algal mat growth.

The major methods of direct control of algal biomass are harvesting, biomanipulation, biological controls, allelochemicals, and algicides.

1. Harvesting

Harvesting methods can range from hand-pulling or raking to use of large mechanized harvesting equipment (Fig. 7). The vegetation is gathered and preferably moved away from the site so that it cannot wash back into the water. This is obviously not a technique that will remove phytoplankton, but it can be used with some success for the removal of floating filamentous algal mats and charophytes.

Hand harvesting or raking of filamentous algal mats and *Chara* is considered difficult because these growths fragment very easily. The tremendous amounts of biomass (and associated water) make hand labor exhausting and time-consuming. However, probably more hand harvesting occurs than one would expect. Pond and lake property owners, for example, frequently clean off beach and dock areas with rakes and other handheld devices. The beaches of Lake Ontario have at times been hand raked of *Cladophora* growth washed up on shore by municipal workers. Hand harvesting sometimes is encouraged prior to algicide treatment. This is particularly true in late summer when large amounts of mat material have accumulated. The death of an excessive amount of biomass (which leads to decomposition and associated bacterial growth, which uses the oxygen in the water) can lead to oxygen depletion and fish kills. Removing at least some of the biomass prior to treatment can help prevent severe oxygen depletion situations.

Most mechanical harvesting activities are directed at the removal of rooted submersed vascular plants. There are, however, reports of mechanical harvesting used successfully for *Chara* control (Conyers and Cooke, 1982; Cooke *et al.*, 1993b), particularly in shallow water where the harvesting blades can cut at the sediment–water interface. *Chara* is almost invariably collected along with submersed vascular vegetation when the two grow intermixed.

Some evidence suggests that populations of phytoplankton, filamentous mat-forming algae, and *Chara* can increase after intensive mechanical harvesting of submersed vascular plants (Neel *et al.*, 1973; Nichols 1973; Cooke and Kennedy, 1989; Anonymous, 1990). Although the increases have not been clearly associated with harvesting, the opening up of areas to light and the potential increase in nutrients after harvest may make algal growth more likely to occur.

Another method of harvesting has been used in irrigation systems in the western United States. Racks are inserted at intervals along the canal to collect the algal mats that slough off the sides of the canal and float downstream. The racks are removed periodically, cleaned of algae and other debris, and returned to the canal (Fig. 8).

A major consideration in harvesting is to ensure that the collected vegetation does not wash back into

FIGURE 7 A mechanical harvester. Photo courtesy of United Marine International, Div. of Liquid Waste Technology, Inc., Somerset, Wisconsin.

FIGURE 8 A rack to collect floating material taken from an irrigation canal in California. The majority of vegetation is *Cladophora*. Photo courtesy of Lars Anderson, USDA-Agricultural Research Service.

the body of water. Even though it may appear that the algal mats have dried out once they are exposed to the air and sun, the underlying portions of the mats may still be viable. Akinetes found in *Pithophora* mats that were exposed to the drying effects of the sun still showed 80% viability 136 days after initial stranding (Lembi *et al.*, 1980).

Little use has been made of harvested aquatic vegetation. It generally has little value as food for livestock or humans, and the energy costs to dry and pellet the vegetation can be prohibitive (National Academy of Sciences, 1976; Joyce, 1993). Some research has been conducted on the potential use of filamentous algae such as *Pithophora* and *Cladophora* to make paper, and the protein content of *Hydrodictyon*, *Spirogyra*, and *Pithophora* was reported to be 18–26%, which is comparable to the protein content in some cyanobacteria and vegetables (Khan *et al.*, 1996); however, the major (but sporadic) use of harvested algal mats at present is as mulch and fertilizer for gardens.

2. Biomanipulation

The observation that the relationship between algal growth and P concentrations is not perfect [in fact, in a study of 66 lakes by Schindler (1978), the regression statistic explained only 48% of the variance in algal productivity] led in part to the theories behind biomanipulation. Much research in the past (reviewed by Cooke *et al.*, 1993b) has indicated that the type of zooplankton present in a body of water can have an effect on phytoplankton populations. The type of zooplankton is affected in turn by the types of fish that are present. The potential for zooplankton (and fish) to have an effect on phytoplankton populations, irrespective of P content in the water, may explain why, in some instances, phytoplankton populations are lower or higher than those predicted by the P content of the lake.

The term "biomanipulation" was coined by Shapiro *et al.* (1975). It is also referred to as top-down feeding and involves manipulating the components of the trophic cascade (Paine, 1980; Carpenter *et al.*, 1985). The premise of biomanipulation, as elucidated by Shapiro (1980) and Carpenter *et al.* (1985, 1987), is that top predators, such as piscivorous fish, can influence the abundance of planktivorous fish, which in turn can determine the abundance, size structure, and productivity of zooplankton and phytoplankton (Fig. 9). For example, planktivorous fish tend to feed on large-bodied zooplankton, which results in domination by small-bodied zooplankton. Because it is the large-bodied zooplankton (some species of *Daphnia*) that feed most effectively on algae, their reduction is typically accompanied by a relatively high phytoplankton

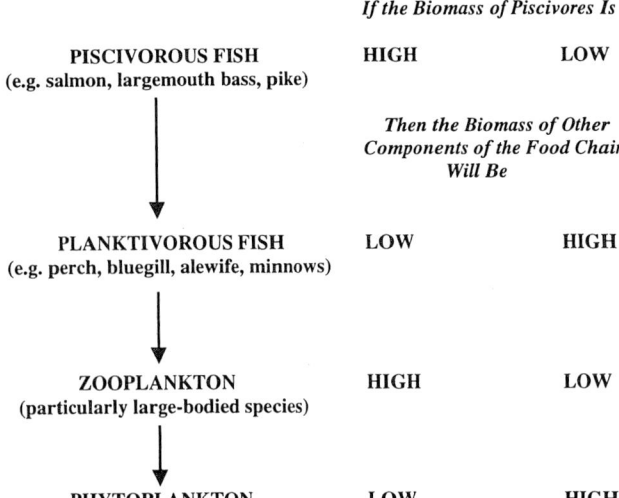

FIGURE 9 A flowchart illustrating the connections between food-chain levels as determined by the biomass of piscivores. Biomanipulation attempts to decrease the mass of planktivores in order to increase the numbers of large-bodied zooplankton, which, in turn, graze on phytoplankton.

biomass. The key to success in biomanipulation is the addition of piscivorous fish, which theoretically should reduce the numbers of planktivorous fish, which in turn enhances the development of populations of large-bodied *Daphnia* species and a decline in phytoplankton populations. The actual manipulation involves the introduction, where necessary, of piscivorous fish and/or the removal of planktivorous fish. An example would be the removal of planktivorous fish from a site by rotenone treatment and the stocking of piscivorous fish to eliminate any planktivorous fish that might be introduced later. Observations of fish–zooplankton–phytoplankton relationships and the successful manipulation of all or part of the trophic cascade in enclosures, ponds, and lakes (Spencer and King, 1984; Carpenter *et al.*, 1985, 1987; Elser and MacKay, 1989; Gulati, 1990; Mazumder *et al.*, 1990; Quirós, 1995) have provided evidence that the principle is essentially valid.

However, the practice of biomanipulation has been marked by inconsistencies that suggest that aquatic systems are much more difficult to manipulate than originally anticipated. Analyses of data in which biomanipulation did not provide the expected results include those cited by McQueen *et al.* (1989), McQueen (1990), Vanni and Findlay (1990), Badgery *et al.* (1994), and Noonan (1998). McQueen *et al.* (1989) suggested that the various trophic levels depend on nutrients and energy flow, which is essentially a bottom-up process rather than top-down. In less productive oligotrophic lakes, a top-down cascade may extend all the way to phytoplankton, but in nutrient enriched eutrophic systems the results are less clear because bottom-up forces are large relative to top-down forces. In other words, more phytoplankton populations can be supported by high P concentrations than can be effectively grazed by zooplankton. Benndorf (1989) concluded that the long-term success of top-down manipulation requires a reduction of external P loading. Another complication is that even if large-bodied zooplankton populations increase, the gelatinous (Porter, 1973) or large-celled ungrazable or undigestible algae can still proliferate.

The conditions under which biomanipulation will work are still unclear. DeMelo et al. (1992) indicated that a careful analysis of the data does not support the eutrophic/oligotrophic differences proposed by McQueen (1990) and others. Differences in the N and P requirements for the growth and reproduction of *Daphnia* further complicate the situation. For example, *Daphnia* growth and reproduction are strongly suppressed when they are fed P-limited algae having a high C:P ratio (Sommer, 1992; Urabe et al., 1997; MacKay and Elser, 1998). When zooplankton that have high P needs ingest phytoplankton with high N:P ratios, there is a disproportionate release of unused N to the system, which in turn can affect algal community structure because of changing nutrient ratios in the water (Urabe, 1993; Steinman, 1996).

Biomanipulation can have effects on components of the aquatic ecosystem that are not directly involved in the trophic cascade. For example, Spencer and King (1984) reported that zooplankton successfully reduced phytoplankton densities in ponds with no fish or with dense populations of largemouth bass (a piscivore), but the resulting clear water stimulated dense growths of *Cladophora* spp. and the submersed vascular plants *Elodea canadensis* and *Potamogeton* spp. This kind of a shift is frequently the outcome of any control or nutrient manipulation technique that reduces phytoplankton growth. Biomanipulation as a tool is still experimental and should not be recommended without considerable analysis of the composition of the various trophic levels and regulating environmental factors.

3. Biological Controls

The use of one organism to control unwanted organisms has been widely studied in aquatic plant management. Most of the research has focused on the control of aquatic vascular plants, but there are studies that suggest potentially useful agents for algae control.

Reports that cyanophages (viruses) lyse cyanobacterial cells date to the 1960s (Safferman and Morris, 1963; Safferman et al., 1969). The first described cyanophage was named LPP-1 for its ability to lyse cells of *Lyngbya*, *Phormidium*, and *Plectonema* (Safferman and Morris, 1963). Since then, other cyanophages have been isolated and evaluated as possible biocontrol agents (Padan et al., 1971; Stewart and Daft, 1977; Martin and Benson, 1988; Phlips et al., 1990; Monegue and Phlips, 1991). Although most studies have been conducted in the laboratory, a few have shown successful results in experimental field enclosures and ponds (Martin et al., 1978; Desjardins and Olsen, 1983). It is generally agreed that cyanophage application is most effective when it is applied before the host populations are well established (Desjardins and Olsen, 1983; Monegue and Phlips, 1991), but its impact on established cyanobacterial populations is poorly known. Eukaryotic algae, including the green alga *Chlorella*, are also susceptible to viruses (Van Etten et al., 1991), but their biocontrol potential is unknown.

Although cyanobacterial and eukaryotic algal populations clearly are affected by phages in nature and although lysis can be induced in the laboratory and in small-scale tests, no virus has been developed for control purposes. Extensive field testing under a variety of environmental conditions has not been conducted, and the information needed on amount of inoculum, application technique, and the environmental factors conducive for replication and lysis is not available. The selectivity of viruses to one or a few species further limits the broad application of the technique. Other microorganisms that have shown activity on planktonic algae include bacteria (Shilo, 1967; Burnham and Fraleigh, 1983; Walker and Higginbotham, 2000) and fungi (Redhead and Wright, 1978; Canter and Jaworski, 1979; Kudoh and Takahashi, 1990). Clearly, all of these organisms play a role in natural successional patterns in lakes, but their potential as biological controls remains unexplored. Even so, this approach shows promise for the future.

A wide variety of insect and other invertebrate grazers, including snails, caddisfly larvae, mayfly larvae, chironomid larvae, and shrimp have reduced benthic algal growths (Fulton, 1988; Steinman, 1996), but their effect on prolific growths of filamentous algae (outside of river and stream systems) is apparently minimal, and none has been investigated as a potential biological control agent. The crayfish *Oronectes immunis* significantly reduced stands of *Chara* (and submersed vascular plants) in a New York lake (Letson and Makarewicz, 1994), but the treatment was more expensive and a higher stocking density was required compared with treatments with grass carp (discussed below). When the vegetation was removed, the crayfish themselves became subject to predation. Therefore, it was difficult to maintain sufficient crayfish densities to consume vegetation regrowth without additional stocking.

A number of fish species have been investigated for their potential to control algae and aquatic vascular plants. The silver carp (*Hypophthalmichthys molitrix*) and bighead carp (*Aristichthys nobilis*) consume phytoplankton and zooplankton (Dimitrov, 1984; van der Zweerde, 1993). Many tilapia (*Tilapia* spp.) are herbivorous (Hauser *et al.*, 1976; Smith, 1985). Some are filter-feeders that consume phytoplankton; others feed on macrophytes including filamentous algae and *Chara*. The distribution of tilapia is restricted by temperature (they are native to India, Africa, and South America); they do not survive in waters colder than 10°C. Therefore, in the United States their use has been confined to the South and to sites that receive heated discharge (Crutchfield *et al.*, 1992). In addition, their use has many disadvantages, such as the ability to switch to animal food when they have eliminated plant and algal growth, a high reproductive potential, and interference with native fish species.

The most successful and widely used biological control agent for aquatic vascular plants and some algae has been the grass carp (*Ctenopharyngodon idella*; Fig. 10) (see reviews by Cooke *et al.*, 1993b; van der Zweerde, 1993). This fish is native to northern China and was introduced into the United States in the 1960s. It was originally introduced into Arkansas but is now used in at least 35 states for aquatic weed control (Sanders *et al.*, 1991). Because there are no predators in its native range, grass carp do not show typical avoidance behaviors and are extremely susceptible to predation. Therefore, fish must be at least 20 to 25 cm long (~450 g) when stocked to avoid predation by largemouth bass and other native predators. Under ideal conditions grass carp can grow to a weight of at least 23 kg within 5–10 years. Recommended stocking densities vary from region to region. Recommendations from the Indiana Department of Natural Resources, for example, suggest 37 grass carp vegetated ha^{-1} if maintenance of some vegetation is desired and 74 fish ha^{-1} if elimination of vegetation is desired. The grass carp survives in cold water and begins to feed regularly at about 14°C. Feeding peaks at about 20–26°C and decreases when the water temperature reaches about 33°C.

The concern about the potential for grass carp to reproduce and crowd out native species led to the development of sterile, triploid grass carp. Even with this precaution, grass carp must not be introduced in areas in which the elimination of vegetation in wetland areas would destroy valuable habitat for waterfowl and other animal life.

Young grass carp up to 50 mm (about 2 in) in length feed mostly on zooplankton. After that they shift to a diet of filamentous algae, duckweed, and submersed vascular plants. There has never been any evidence of the fish shifting to an animal diet once they exceed a length of about 100 mm. The fish clearly has preferences for the kinds of plants it eats. Numerous lists of preferred plant species have been published (Fowler and Robson, 1978; Cassani and Caton, 1983; Shireman *et al.*, 1983a; Pine and Anderson, 1991; Sanders *et al.*, 1991; Cooke *et al.*, 1993b), and in almost every case, *Chara* and *Nitella* are listed as preferred or highly preferred plants. Bauer and Willis (1990) reported that grass carp introduced at 49 fish ha^{-1} almost totally eliminated *Chara* (the dominant macrophyte) in 2 years in two small South Dakota lakes.

The effectiveness of grass carp for the control of filamentous algal mats is less clear. Some of the references listed above claim good control of relatively coarse species such as *Cladophora* and *Pithophora*; others indicate no or only weak control of filamentous algae. In some cases, only high stocking densities of more than 123 fish ha^{-1} have succeeded in controlling filamentous algae. *Spirogyra* seems to be least preferred, probably because its slimy nature prevents effective ingestion; however, it too can be eaten if no other vegetation is available (Lembi *et al.*, 1978). In mixed populations of vegetation, grass carp will clearly consume soft-bodied vascular plants such as pondweeds, elodea, and naiads, and *Chara* (even though *Chara* can be coated with a hard coat of calcium carbonate, fish appear to be attracted to it) in preference to filamentous algae. When filamentous algae are the only plant material present, grass carp will feed on it, probably to avoid starvation.

Problems with grass carp include lack of consistency and predictability. Use of grass carp does not work under all circumstances, and visible control may not be

FIGURE 10 The grass carp (*Ctenopharyngodon idella*), a widely used biological control agent for certain macrophytic algae and submersed vascular plants.

achieved until several years after introduction, particularly in heavily infested areas. Increased turbidity due to increased phytoplankton populations has been noted in some situations (Shireman et al., 1985; Maceina et al., 1992), and overstocking can result in the total elimination of all vegetation, a situation that is not considered beneficial for fish and other animal life in natural bodies of water. In fact, considerable controversy in the sportfishing industry has erupted over the potential for elimination of vegetation cover by grass carp, particularly because the effect of removing weed beds on angling success is still being debated (Bain, 1993; Bettoli et al., 1993; Killgore et al., 1998). Other control methods (e.g., mechanical harvesting or the use of chemicals) can be used for weed beds in certain areas of a lake so that some vegetation can be selectively retained. Unfortunately, there is little ability, once grass carp have been introduced, to dictate where they will graze and how much vegetation they will consume over time. Stocking rates can be reduced to avoid elimination of plant populations, but because the grass carp work so slowly and because it is so difficult to predict their impact on the vegetation, herbicides/algicides or other methods may have to be used if the fish cannot provide adequate control. Finally, grass carp may consume desirable native species (including *Chara* and *Nitella*) and leave less desirable species, such as the invasive weed Eurasian watermilfoil (Fowler and Robson, 1978).

Another method of biological control is the use of waterfowl, specifically geese or swans. Filamentous algae are consumed by waterfowl, and charophytes are a favored food of herbivorous ducks, coots, and swans (Martin et al., 1961; Hargeby et al., 1994). A pair of swans reportedly will keep a 0.4 ha pond free of submersed vegetation, as will 7–20 geese or ducks ha^{-1} (Holm and Yeo, 1981). Unfortunately, there are many problems associated with the presence of waterfowl. Birds that are introduced for aquatic weed control are usually rendered flightless. Therefore, their diet of aquatic vegetation must be supplemented to provide adequate nutrition, they must be protected from predators, and lake managers must be willing to tolerate their aggressiveness during the breeding season. As with free-living waterfowl, their waste materials can litter the banks and stimulate phytoplankton blooms.

An unusual form of biological control is the use of competitive plants. Doyle and Smart (1998) showed that established plantings of the flowering plants pickerelweed (*Pontederia cordata*) and American pondweed (*Potamogeton nodosus*) reduced the biomass of *L. wollei* by 50% and prevented the formation of floating mats. The effect was attributed to shading and possibly to competition for nutrients in the sediments. The main drawback of this method is that it only works in very shallow areas where the flowering plants can root. In a larger sense, the water user must be willing to accept the shift from an algal infestation to dense stands of flowering plants. This fact alone negates the potential benefits of shading by free-floating flowering plant species such as water hyacinth (*Eichhornia crassipes*) or duckweed (*Lemna* spp.), both of which tend to be weedy. On the other hand, the premise of using competitive submersed plants, possibly those that have been selected or genetically modified to produce shortened stems with minimal canopy formation, should be explored further.

The use of biological agents shows great promise for the control of weedy algae and plants, but it has drawbacks. More research is necessary, particularly on those organisms that might be relatively selective, such as phages. The introduction of nonselective agents, such as grass carp, has been used successfully in some situations, but it also has the potential to cause adverse ecological impacts in others.

4. Allelochemicals

Allelopathic chemicals are chemicals produced by plants that have either an adverse or beneficial effect (usually adverse) on other plants (Rice, 1984). Although this area has not received much attention in controlling nuisance algae, there is evidence that allelochemicals may be useful.

A number of cyanobacteria have been investigated for their potential to produce allelochemicals that inhibit the growth of other cyanobacteria or algae (Mason et al., 1982; Flores and Wolk, 1986). Allelochemicals from some fungi (Redhead and Wright, 1978) and terrestrial vascular plants (mostly phenylpropanoids) (Della Greca et al., 1992) have been shown to have algicidal activity. Studies have been conducted on allelochemicals produced by aquatic vascular plants, but most of the bioassays have been conducted using vascular plant species, such as duckweed or lettuce seedlings, rather than algae (Elakovich and Wooten, 1989; Sutton and Porter, 1989; Wooten and Elakovich, 1991). One exception is the study by Gross et al. (1996) in which extracts from Eurasian watermilfoil inhibited cyanobacteria. Although some potential algicidal allelochemicals have been identified, the major constraints to further development is the expense of culturing the organisms and extracting sufficient amounts of the allelochemicals for application. An alternative is to synthesize the active chemical, but this is also costly, particularly in view of the relative cheapness and availability of copper sulfate. Thus, the financial incentive for industry to develop these compounds for the aquatic market is lacking.

A cheaper method of allelopathic control may be the "bale of hay" technique. For years, farmers have applied straw or hay to ponds to reduce algal growth. This method has been tested and substantiated in a series of experimental studies in England (Welch et al., 1990; Gibson et al., 1990; Pillinger et al., 1992, 1994; Newman and Barrett, 1993). In the laboratory, rotting barley (Hordeum) straw inhibited the growth of several planktonic (including Microcystis) and filamentous algae. Additions to a canal over a 3-year period resulted in reduced biomass of C. glomerata after 2 years. The apparent mechanism is through the release of quinone compounds (Pillinger et al., 1994).

Attempts to replicate these effects in North American waters have had mixed results. Nicholls (1996) showed decreases in chlorophyll a in ponds treated with barley in Ontario, but tests using barley, wheat (Triticum), and rye (Secale) in ponds with Pithophora in North Carolina were unsuccessful (Kay, 1997). Discrepancies among results may be due to differences in straw concentrations, target species, environmental conditions, and length of exposure. Because barley is not grown in many parts of the United States and thus is not widely available, other forages should be tested. Laboratory and small-scale field testing suggests that alfalfa (Medicago) hay may be effective for filamentous algae control (Marencik and Lembi, 1998), but additional research is needed to determine if its rapid breakdown in water could cause oxygen depletion problems. Further research in this area is warranted.

5. Algicides

A common method of controlling algal infestations is the use of chemicals (algicides). Of the algicides, the most commonly used compounds are copper-containing products. Other products, such as diquat and the mono (N,N-dimethylalkylamine) salt of endothall, are registered for algae control, but their use is relatively minor compared to that of the copper-containing products.

Of the copper products, copper sulfate ($CuSO_4$) is the most widely used algicide for controlling algal populations in water supplies, recreational lakes, and reservoirs (Elder and Horne, 1978; Effler et al., 1980; Raman, 1985). It has been used since at least 1905 and probably was used earlier (Moore and Kellerman, 1905; Murphy and Barrett, 1993). In the late 1960s and early 1970s, more than 9 million kg of $CuSO_4$ were applied annually in waters in the United States (Fitzgerald, 1971). Approximately 68% of all water area in the United States treated with a chemical product in 1992 was treated with an algicide (unpublished industry data). $CuSO_4$ was applied to 70% of this area, and the remainder was treated mostly with other copper compounds (chelated copper compounds). Even though algicides were used on 68% of the area treated, they accounted for just 20% of the total sales ($32.5 million) of all aquatic chemicals (herbicides and algicides) due to their relatively low cost.

Copper is effective on a wide range of algae (Maloney and Palmer, 1956). The toxic agent is free cupric ion (Cu^{2+}), and toxic cupric ion activities range from greater than 10^{-6} to 10^{-11} M for species of diatoms, dinoflagellates, microscopic green algae, and cyanobacteria (McKnight et al., 1983). The fact that cyanobacteria are more sensitive to copper than some of the eukaryotic algae (Whitton, 1973; Swain et al., 1986) accounts for its widespread success and acceptance. Diatoms are probably next in sensitivity followed by the green algae (Swain et al., 1986; Havens, 1994). A copper concentration of 25–40 µg L^{-1} effectively controlled A. flos-aquae in shallow eutrophic lakes in Manitoba (Whitaker et al., 1978), a dose that is considerably lower than the typical doses of 125–250 µg L^{-1}. Suppression of nitrogen fixation by Anabaena and Aphanizomenon was observed after copper additions of only 5–10 µg L^{-1} (Horne and Goldman, 1974), leading to the suggestion that maintaining low doses to inhibit N_2 fixation would be an alternative to a single large dose. This approach has not been practical because the short residence time of copper in the water column mandates almost continual copper application.

Certain microscopic green [e.g., Oocystis (Meador et al., 1993; personal observations)] and euglenoid planktonic (Hawkins and Griffiths, 1987) algae are relatively tolerant to copper treatments. For example, the microscopic green algae Ankistrodesmus, Scenedesmus, and Pandorina may require copper concentrations as high as 500 µg L^{-1} for control (Copper Sulfate Fine Crystals product label).

Among mat-forming green algae, Spirogyra and Oedogonium are very susceptible to copper (Whitton, 1970; Francke and Hillebrand, 1980), whereas Pithophora and Hydrodictyon are considerably more tolerant (Table I). In addition to inherent tolerance or susceptibility, mat structure also may dictate relative tolerance to copper (or other exogenously applied materials). Mat structure appears to be governed in part by branching pattern. Pithophora filaments, which are branched, produce intertwined, tighter mats than filaments of Spirogyra and Oedogonium, which are unbranched (Table I). It may be more difficult for copper to penetrate the extremely dense, massive mats that are produced by Pithophora (Lembi et al., 1984) than the loose, less tangled mats formed by Spirogyra and Oedogonium. Of all mat-forming species, the cyanobacteria are the most tolerant to copper, which

TABLE I Taxa, Filament Morphologies, Mat Structures, and Susceptibilities to Copper[a] of Filamentous Mat-Forming Algae

Taxon	Morphology	Mat structure	EC_{50}[b]
Chlorophyta			
Spirogyra	Unbranched	Loose	1
Oedogonium	Unbranched	Loose	3
Hydrodictyon	Net-like	Moderate	48
Pithophora	Branched	Dense	46
Cyanobacteria			
Oscillatoria	Unbranched	Dense, slime + associated sediment	290
Lyngbya	Unbranched	Dense	1630

[a] Lembi, 2000; data on *Lyngbya* from Hallingse and Phlips (1996).
[b] EC_{50} = concentration of Cu^{2+} in µg L^{-1} required to reduce biomass (dry weight) of alga by 50% under laboratory (not field) conditions.

seems unusual given the susceptibility of planktonic cyanobacteria. Both an inherent tolerance (at least 6-fold greater than that of *Pithophora*; Table I) plus the presence of thick slime and a coating of sediment make *Oscillatoria* mats extremely difficult to control. Likewise, *Lyngbya* (which also produces sheaths) produces thick, dense mats which probably add to its tolerance. Unfortunately, it is likely (although not well documented) that the elimination of susceptible species (both microscopic and mat-forming) has led to their replacement by tolerant species.

The mechanisms by which copper affects algae appear to vary. The list of reported copper effects [taken from Gledhill *et al.* (1997); see references therein] indicates that it inhibits photosynthesis (see also Kallqvist and Meadows, 1978), disrupts electron transport in photosystem II (see also Cedeno-Maldondo and Swader, 1974), reduces pigment concentrations, affects the permeability of the plasma membrane and induces losses in cations, inhibits nitrate uptake, restricts growth, affects cell motility, and affects the distribution of proteins, lipids, sterols, sterol esters, and free fatty acids in the cell. Although this list was developed primarily for marine algae, there is no reason to think that the same effects should not be expected in freshwater algae. In addition, copper has been reported to inhibit P uptake (Peterson *et al.*, 1984) and to precipitate proteins in the cell (Murphy and Barrett, 1993). All of these presumed modes of action suggest that copper is a general algal cell toxicant. When applied at the recommended dosage (250 µg L^{-1} copper), copper acts rapidly, usually within a period of hours.

A number of mechanisms have been proposed to account for the differential tolerance among algae and include (Lage *et al.*, 1996) intracellular accumulation of copper in polyphosphate bodies, storage of copper in membrane-bound vesicles, excretion into the medium of organic compounds that bind copper, intracellular chelation of copper by organic compounds like phytochelatins, and efflux of the copper. In addition, copper accumulation in the cell walls of some algae and higher plants has been reported (Pearlmutter and Lembi, 1986; Allan and Jarrell, 1989). As noted above, mat structure also should be considered a factor in species tolerance.

Water chemistry, particularly pH and alkalinity (a measure of bicarbonates, carbonates, and hydroxides), plays an important role in copper toxicity (McKnight *et al.*, 1983). Below neutral pH, Cu^{2+} is the major copper species; above neutral pH the major forms of copper are the copper carbonate complexes and malachite and tenorite. The various complexes and precipitants formed above neutral pH values effectively prevent Cu^{2+} from being taken up by target organisms. At high pH and alkalinity, the concentration of soluble Cu^{2+} in the water is extremely low and possibly ineffective for algal control (Button *et al.*, 1977). In low alkalinity, acidic waters, the recommended dose (250 µg L^{-1} Cu^{2+}) of $CuSO_4$ will kill algae, but given the relatively high amounts of soluble Cu^{2+}, particularly sensitive fish species, such as trout, can also be killed. This phenomenon is even true for the chelated copper products (discussed below), and all currently registered copper products have a statement similar to the one on the Cutrine-Plus label: "Do not use in water containing trout if the carbonate hardness of the water does not exceed 50 mg L^{-1}."

As total alkalinity increases, so must the $CuSO_4$ dosage to overcome the precipitation problem. Toxicity to fish is essentially nonexistent at high alkalinities because of the low concentrations of Cu^{2+} in the water [as little as 0.5% of the total dissolved copper has been calculated to be present as free cupric ion (Wagemann and Barica, 1979)]. For example, at the low alkalinity of 18.7 mg L^{-1} (as $CaCO_3$ + HCO_3), the LC_{50} (concentration that will kill 50% of the population) for bluegill (*Lepomis*) is 884 µg L^{-1} copper (3.5 mg L^{-1} $CuSO_4$) (Herbicide Handbook, 1994). At the moderate to high alkalinity of 166 mg L^{-1}, the LC_{50} is 7300 µg L^{-1} copper (29.2 mg L^{-1} $CuSO_4$). The upper legal limit for copper use in water is 1000 µg L^{-1} copper (4 mg L^{-1} $CuSO_4$); this concentration (which is seldom recommended) could kill fish in low alkalinity waters. However, there is almost a 30-fold safety factor between the LC_{50} and the typical recommended dosage [250 µg L^{-1} copper (1 mg L^{-1} $CuSO_4$)] in high alkalinity waters.

Although direct copper toxicity is seldom a problem to fish, the depletion of oxygen during algal death and decomposition can cause fish kills. Treatments must never be made to bodies of water that have heavy algal infestations.

Chelation with organic compounds stabilizes soluble copper and theoretically retards its precipitation and adsorption (McKnight et al., 1983). This principle is the basis for the formulation of the chelated forms of copper as substitutes for copper sulfate. Most commercial chelated formulations are variations of ethanolamine complexes. One of the presumed advantages of using a copper chelate is that a longer persistence of active copper in water should increase algal contact and control. In a comparative study of copper chelates and copper sulfate used at the same copper dose, Masuda and Boyd (1993) found that chelates slowed the loss of total copper (from an initial level of 500 to 100 µg L^{-1}) from the water from 4.3 to 6.3 days, but they concluded that this advantage did not compensate for the greater cost of the chelated formulation. On the other hand, the chelated copper products provide flexibility in application because they are formulated either as granules or as liquids, whereas $CuSO_4$ is packaged as a solid material only. The combination of liquid formulations with other liquid aquatic herbicides is useful for commercial applicators to control a broad spectrum of species that include both algae and vascular plants.

Phytoplanktonic blooms are usually treated by pumping the copper compounds (dissolved in water in a tank mounted in the boat or airboat) through a boom and trailing hoses into the water (Fig. 11A). The hoses can be adjusted to deliver the compound to different depths in the water column. For example, hoses that disperse the copper in the upper meter of water (or surface treatments with a spray) are effective for the control of cyanobacteria that have formed surface scums. The hoses can be set lower in the water column to deliver copper to filamentous algal mats that are still lying on the bottom and have not floated to the surface. Once the mats have floated to the surface, spot treatments directly on the mats with a spray can ensure that contact with the cells is maximal (Fig. 11B). This ability to place the treatment directly on or near the target organism allows the applicator to selectively treat some areas and not others, thus reducing both the amount of copper needed and the volume of water that comes into contact with the chemical. Other methods of application include slow dispersal through a burlap sack or drip or single high dose applications in flowing water systems such as irrigation and drainage canals,.

The persistence time of copper in water is relatively short. Button et al. (1977) found that 95% of the

FIGURE 11 Algicide applications. (A) A unit set up to deliver chemical through trailing hoses. (B) A unit set up to deliver a spray directly to the algal mats. B courtesy of Neil Gerber, Aquatic Management, Bluffton, Indiana.

copper sulfate distributed over a lake surface dissolved in the upper 1.8 m of the water column, but the total copper concentration was at pretreatment levels within 24 h. Tucker and Boyd (1978) made 10 applications of 0.84 kg ha^{-1} $CuSO_4$ at 2-week intervals to ponds without causing an appreciable increase in total copper concentration. Wagemann and Barica (1979) reported a half-life of total dissolved copper from 1 to 7 days. Anderson and Dechoretz (1984) reported a half-life of about 1 day with all of the copper gone from the water column at 14–28 days. In other bodies of water, copper has persisted for up to 30 days after treatment (Elder and Horne, 1978; Whitaker et al., 1978; McKnight, 1981; Hawkins and Griffiths, 1987).

The implications of a short residence time in water are several. First of all, copper effects, whether on target or nontarget organisms, are temporary. In most cases, algal populations rebound, although not necessarily at the same densities or with the same species. Although this is a problem from the standpoint that repeated treatments may be needed during a single season to provide adequate control, the short persistence reduces the exposure time of nontarget organisms, including humans.

One of the major concerns with the use of copper is its ultimate fate. The copper complexes, as well as decaying copper-containing algae, fall to the bottom where the copper is readily adsorbed onto sediments. Copper, as a heavy metal, persists in the sediment for prolonged periods of time (Frank, 1972; Brown, 1978). The key issue is that copper will accumulate in sediments and be toxic to benthic organisms and then serve as a source of copper to the water after treatment is discontinued. The evidence is somewhat conflicting. Sanchez and Lee (1978) noted that copper-enriched sediments in Lake Monona, Wisconsin (treated over 50 years to 1950), were not interacting with the more recent sediments deposited or with overlying waters. The copper content of the water was no different from that of local hardwater lakes which had not been treated, and they concluded that there were no long-term adverse effects resulting from the copper treatments. Ankley et al. (1993) studied sediment and pore water from Steilacoom Lake, Washington, which had been "grossly contaminated" by copper because of copper sulfate treatments. Extracted copper concentrations in sediments ranged from 0.6 to 3.0 μmol g^{-1} dry weight, but pore water and overlying water concentrations were less than the analytical detection limit of 7 μg L^{-1}. Toxicity tests showed no effects of the water on the amphipod *Hyalella azteca*.

Probably the most negative report of copper treatment effects on sediments is from a string of interconnected lakes in southern Minnesota that were treated over a period of 58 years (Hanson and Stefan, 1984). Effects included elevated concentrations of copper in the sediments, fish kills due to oxygen depletion or possibly copper toxicity, increased internal P cycling, rapid recovery of algal populations within 7–21 days, shifts of game fish to rough fish, disappearance of macrophytes, and reductions of benthic macroinvertebrates. Some conditions improved when the sediments had been dredged and the use of copper discontinued. It is difficult to evaluate the results of this study because factors other than copper applications may have caused some of these effects. Fortunately, such extreme effects have not been observed in most copper-treated lakes. If they did occur in large numbers, copper should have been banned long ago. In fact, Sanchez and Lee (1978) concluded that 50 years of copper treatments in Lake Monona had not resulted in the loss of the excellent sport fisheries supported by that lake. Nevertheless, the Minnesota study should serve as a warning that a program that includes long-term, whole-lake treatments with copper must be scrutinized and continually monitored for potential deleterious effects.

An additional concern is the sensitivity of zooplankton to copper. The LC$_{50}$ for planktonic crustaceans is 60–90 μg L^{-1} copper; for rotifers it is 1100–1700 μg L^{-1} (Demayo et al., 1982). Other studies on invertebrates indicate LC$_{50}$ values ranging from 10 to 130 μg L^{-1} copper (McIntosh and Kevern, 1974; Winner, 1985; Meador et al., 1993). Concentrations as low as 8 μg L^{-1} copper have caused significant effects on cladocerans in life cycle toxicity tests (Belanger et al., 1989), and Hedtke (1984) found large reductions in zooplankton biomass (along with that of snails, total macroinvertebrates, and midges) when laboratory microcosms were treated with 30–270 μg L^{-1} copper. Most of these concentrations are well within the range of normal use dosages. Studies using lake mesocosms also showed reductions in zooplankton populations (Moore and Winner, 1989; Havens, 1994) as well as species shifts (Moore and Winner, 1989; Winner et al., 1990; Havens, 1994). Both phytoplankton and zooplankton communities were more sensitive to copper in the spring than in the summer or fall (Winner et al., 1990), an observation that was attributed to differences in levels of copper-complexing compounds in the water during the various seasons.

Effects of copper on zooplankton communities in lakes have been somewhat mixed. Effler et al. (1980) found no effects of a low-level CuSO$_4$ treatment on several zooplankton populations in a lake; however, McKnight (1981) found decreases of populations of *Bosmina*, *Tetramastix*, and *Keratella* following treatment of Mill Pond, Massachusetts. Long-term effects of copper treatments on zooplankton in lakes have not been adequately studied, although the short-term persistence of copper within the water column should allow most zooplankton populations to recover. The loss of zooplankton populations, even if temporary, may explain why algal populations can recover, sometimes to levels higher than original levels. The loss of grazing impact on phytoplankton due to copper effects on zooplankton has been suggested by the work of McKnight (1981), Taub et al. (1989), and Havens (1994). In addition, bacterial biomass has been reported to rapidly recover or even increase following copper treatment (Effler et al., 1980; Havens, 1994; Dionigi and Champagne, 1995), and in some cases this may be due to loss of grazing pressure from zooplankton (Havens, 1994).

Copper is a trace element that is required for the survival of many plant and animal species, including humans. The low mammalian toxicity of copper when diluted in water, its short persistence time in water, and its lack of bioaccumulation in fatty tissues are the reasons that it is the only one of the algicides/herbicides registered in the United States for which the use of water following treatment at normal doses is not restricted. This includes use of the water for drinking, swimming, fishing, livestock watering, and irrigation, although a 24-hour waiting period after treatment is desirable just to be cautious. Copper-containing compounds have not been reported to induce cancers in humans or experimental animals (Sunderman, 1978). The only warning on EPA-approved products is the potential toxicity to trout under softwater conditions.

6. Summary of Direct Control Methods

Methods for the direct control of algae are available, but none of them ensure that algae problems will be solved other than in the short term. Even grass carp, which live for up to 16 years, become less efficient feeders after about 5 years and restocking is required for long-term control.

The adverse effect of copper on food webs must be considered, and appropriate long-term ecosystem studies must be undertaken. These concerns are offset in part by the overall safety of copper to humans, by the short persistence of copper in the water column, and by our ability to place copper directly on or in the vicinity of the target algae. However, lakes and reservoirs will be better protected if repeated whole-lake copper treatments can be avoided. A program of watershed and water quality management through nutrient manipulation is the best alternative, particularly for phytoplanktonic blooms.

There will still be a role for the use of copper sulfate, at least in the United States. Copper treatments may be needed periodically or at certain sites where filamentous mat-forming algae or *Chara* are problems in lakes and reservoirs despite nutrient reduction efforts. Residents on small lakes and ponds that receive large inputs of nutrients from nonpoint sources will continue to require direct control techniques, particularly because they may have few financial or political resources to minimize these inputs. The ability to maintain irrigation systems in the western states free of algae and provide maximal water delivery rates to urban and rural users will still require copper applications for the foreseeable future.

Regulatory agencies in several regions of the United States have indicated an interest in eliminating the use of copper for algae control. Copper is not used to any great extent in Canada because of that country's very cautious approach to water quality (H. Vandermeulen, Fisheries & Oceans Canada, personal communication). However, the immediate benefits and favorable economics of copper sulfate usage suggest that this compound will continue to be used in the United States. Unfortunately, the ready availability and efficacy of the copper compounds reduce incentives needed for development of alternative control methods, such as organically based algicides and allelochemicals or viral and bacterial biological control organisms.

The short-term nature of direct control techniques plus the difficulties inherent in regulating nutrient inputs in every situation dictate that algae will pose water quality problems for many years to come. These problems will continue to pose a challenge to all of us who have an interest in maintaining or restoring healthy aquatic ecosystems.

ACKNOWLEDGMENTS

I thank Erik C. Brockman for his tremendous assistance in the literature search for this chapter. I also acknowledge the contributions and insights of numerous associates, both academic and those who deal with algal management on a day-to-day basis. The constructive comments of reviewers Alan D. Steinman and Herb Vandermeulen are greatly appreciated.

LITERATURE CITED

Adamson, R. P., Sommerfeld, M. R. 1978. Survey of swimming pool algae of the Phoenix, Arizona, metropolitan area. Journal of Phycology 14:519–521.

Allan, D. L., Jarrell, W. M. 1989. Proton and copper adsorption to maize and soybean root cell walls. Plant Physiology 89:823–832.

Allanson, B. R. 1973. The fine structure of the periphyton of *Chara* sp. and *Potamogeton natans* from Wytham Pond, Oxford, and its significance to the macrophyte–periphyton metabolic model of R. G. Wetzel and H. L. Allen. Freshwater Biology 3:535–542.

American Public Health Association. 1992. Standard methods for the examination of water and wastewater, 18th ed. Greenberg, A. E., Clesceri, L. S., Eaton, A. D., Eds. American Public Health Association, Washington, DC.

Anderson, L. W. J. 1993. Aquatic weed problems and management in the western United States and Canada, in: Pieterse, A. H., Murphy, K. J., Eds., Aquatic weeds, the ecology and management of nuisance aquatic vegetation. Oxford University Press, Oxford, UK, pp. 371–391.

Anderson, L. W. J., Dechoretz, N. 1984. Laboratory and field investigations of a potential selective algicide, PH4062. Journal of Aquatic Plant Management 22:67–75.

Ankley, G. T., Mattson, V. R., Leonard, E. N., West, C. W., Bennett, J. L. 1993. Predicting the acute toxicity of copper in freshwater sediments: Evaluation of the role of acid-volatile sulfide. Environmental Toxicology and Chemistry 12:315–320.

Anonymous. 1990. Environmental assessment, aquatic plant management (NR 107) program, 3rd ed. Wisconsin Department of Natural Resources, Madison.

Anthoni, U., Christophersen, C., Madsen, J. O., Wium-Andersen, S.,

Jacobsen, N. 1980. Biological active sulfur compounds from the green alga *Chara globularis*. Phytochemistry 19:1228–1229.

Arruda, J. A., Fromm, C. H. 1989. The relationship between taste and odor problems and lake enrichment from Kansas lakes in agricultural watersheds. Lake and Reservoir Management 5:45–52.

Babin, J., Prepas, E. E., Murphy, T. P., Serediak, M., Curtis, P. J., Zhang, Y., Chambers, P. A. 1994. Impact of lime on sediment phosphorus release in hardwater lakes: The case for hypereutrophic Halfmoon Lake, Alberta. Lake and Reservoir Management 8:131–142.

Badgery, J. E., McQueen, D. J, Nicholls, K. H., Schaap, P. R. 1994. Biomanipulation at Rice Lake, Ontario, Canada. Lake and Reservoir Management 10:163–173.

Bain, M. B. 1993. Assessing impacts of introduced aquatic species: Grass carp in large systems. Environmental Management 17:211–224.

Barica, J. 1975. Summerkill risk in prairie ponds and possibilities of its prediction. Journal of the Fisheries Research Board of Canada 32:1283–1288.

Barica, J. 1978. Collapses of *Aphanizomenon flos-aquae* blooms resulting in massive fish kills in eutrophic lakes: Effect of weather. International Vereinigung für Theoretische und Angewandte Limnologie Verhandlungen 20:208–213.

Barica, J., Kling, H., Gibson, J. 1980. Experimental manipulation of algal bloom composition by nitrogen addition. Canadian Journal of Fisheries and Aquatic Sciences 37:1175–1183.

Barko, J. W., Smart, R. M. 1980. Mobilization of sediment phosphorus by submersed freshwater macrophytes. Freshwater Biology 10:229–238.

Barko, J. W., Smart, R. M. 1981. Sediment-based nutrition of submersed macrophytes. Aquatic Botany 10:339–352.

Barko, J. W., Adams, M. S., Clesceri, N. L. 1986. Environmental factors and their consideration in the management of submersed aquatic vegetation: A review. Journal of Aquatic Plant Management 24:1–10.

Barko, J. W., James, W. F., Taylor, W. D., McFarland, D. G. 1990. Effects of alum treatment on phosphorus and phytoplankton dynamics in Eau Galle Reservoir: A synopsis. Lake and Reservoir Management 6:1–8.

Barnett, B. S., Schneider, R. W. 1974. Fish populations in dense submerged aquatic plant communities. Hyacinth Control Journal 12:12–14.

Bauer, D. L., Willis, D. W. 1990. Effects of triploid grass carp on aquatic vegetation in two South Dakota lakes. Lake and Reservoir Management 6:175–180.

Beeton, A. M. 1969. Changes in the environment and biota of the Great Lakes, in: Eutrophication, causes, consequences, correctives. National Academy of Sciences, Washington, DC, pp. 150–187.

Belanger, S. E., Farris, J. L., Cherry, D. S. 1989. Effects of diet, water hardness, and population source on acute and chronic copper toxicity to *Ceriodaphnia dubia*. Archives of Environmental Contamination and Toxicology 18:601–611.

Benndorf, J. 1989. Food-web manipulation as a tool in water-quality management. Aqua (JWSRT) 38:296–304.

Bennett, G. W. 1971. Management of lakes and ponds. Van Nostrand Reinhold, New York.

Bettoli, P. W., Maceina, M. J., Noble, R. L., Betsill, R. K. 1992. Piscivory in largemouth bass as a function of aquatic vegetation abundance. North American Journal of Fisheries Management 12:509–516.

Bettoli, P. W., Maceina, M. J., Noble, R. L., Betsill, R. K.1993. Response of a reservoir fish community to aquatic vegetation removal. North American Journal of Fisheries Management 13:110–124.

Blindow, I. 1988. Phosphorus toxicity in *Chara*. Aquatic Botany 32:393–395.

Bolas, P. M., Lund, J. W. G. 1974. Some factors affecting the growth of *Cladophora glomerata* in the Kentish Stour. Water Treatment and Examination 23:25–51.

Born, S. M., Wirth, T. L., Brick, E., Peterson, J. O. 1973. Restoring the recreation potential of small impoundments: The Marion Millpond experience. Bulletin No. 71, Wisconsin Department of Natural Resource Technology, Madison, 20 p.

Boyd, C. E. 1982. Water quality management for pond fish culture. Elsevier, Amsterdam.

Boyd, C. E., Prather, E. E., Parks, R. W. 1975. Sudden mortality of a massive phytoplankton bloom. Weed Science 23:61–66.

Brezny, D., Mehta, I., Sharma, R. K. 1973. Studies of evapotranspiration of some aquatic weeds. Weed Science 21:197–204.

Brown, A. W. A. 1978. Herbicides in water, in: Brown, A. W. A., Ed., Ecology of pesticides. John Wiley, Sons, New York, pp. 426–431.

Brown, S. W., Boyd, C. E. 1982. Off-flavor in channel catfish from commerical ponds. Transactions of the American Fisheries Society 111:379–283.

Brownlee, B. G., Painter, D.S., Boone, R. J. 1984. Identification of taste and odour compounds from western Lake Ontario. Water Pollution Research Journal of Canada 19:111–118.

Burkholder, J. M., Mallin, M. A., Glasgow, H. B., Larsen, L. M., McIver, M. R., Shank, G. C., Deamer-Melia, N., Briley, D. S., Springer, J., Touchette, B. W., Hannon, E. K. 1997. Impacts to a coastal river and estuary from rupture of a large swine waste holding lagoon. Journal of Environmental Quality 26: 1451–1466.

Burnham, J. C., Fraleigh, P. D. 1983. Predatory myxobacteria: Lytic mechanisms and prospects as biological control agents for cyanobacteria (blue-green algae), in: Lake restoration, protection and management. EPA-440/5-83-001, pp. 249–256.

Burns, C. W. 1968. The relationship between body size of filter-feeding Cladocera and maximum size of particle ingested. Limnology and Oceanography 13:675–678.

Button, K. S., Hostetter, H. P., Mair, D. M. 1977. Copper dispersal in a water supply reservoir. Water Research 11:539–544.

Canfield, D. E. 1983. Prediction of chlorophyll *a* concentration in Florida lakes: The importance of P and N. Water Research Bulletin 19:255–262.

Canter, H. M., Jaworski, G. H. M. 1979. The occurrence of a hypersensitive relation in the planktonic diatom *Asterionella formosa* Hassal parasitized by the chytrid *Rhizophydium planktonicum* Canter emend., in culture. New Phytology 82:187–202.

Carignan, R., Kalff, J. 1980. Phosphorus sources for aquatic weeds: Water or sediments? Science 207:987–989.

Carlson, R. E. 1977. A trophic state index for lakes. Limnology and Oceanography 22:363–369.

Carmichael, W. W. 1994. The toxins of cyanobacteria. Scientific American 270:78–86.

Carmichael, W. W. 1997. The cyanotoxins. Advances in Botany Research 27:211–256.

Carmichael, W. W., Evans, W. R., Yin, Q. Q., Bell, P., Moczydlowski, E. 1997. Evidence for paralytic shellfish poisons in the freshwater cyanobacterium *Lyngbya wollei* (Farlow ex Gomont) comb. nov. Applied and Environmental Microbiology 63:3104–3110.

Carmichael, W. W., Falconer, I. R. 1993. Diseases related to freshwater blue-green algal toxins, and control measures, in: Falconer, I. R., Ed., Algal toxins in seafood and drinking water. Academic Press, San Diego.

Carmichael, W. W., Jones, C. L. A., Mahmood, N. A., Thiess, W. C. 1985. Algal toxins and waterbased diseases. CRC Critical Reviews in Environmental Science and Technology 15:275–313.

Carpenter, E. J., Carmichael, W. W. 1995. Taxonomy of cyanobacteria, in: Hallegraeff, G. M., Anderson, D. M., Cembella, A. D., Eds., Manual on harmful marine microalgae. IOC Manuals and Guides No. 33, UNESCO 1995k pp. 373–380.

Carpenter, S. R., Kitchell, J. F., Hodgson, J. R. 1985. Cascading trophic interactions and lake ecosystem productivity. BioScience 35:635–639.

Carpenter, S. R., Kitchell, J. F., Hodgson, J. R., Cochran, P. A., Elser, J. J., Elser, M. M., Lodge, D. M., Kretchmer, D., He, X., von Ende, C. 1987. Regulation of lake primary productivity by food web structure. Ecology 68:1863–1876.

Carper, G. L, Bachmann, R. W. 1984. Wind resuspension of sediments in a prairie lake. Canadian Journal of Fisheries and Aquatic Sciences 41:1763–1767.

Cassani, J. R., Caton, W. E. 1983. Feeding behavior of yearling and older hybrid grass carp. Journal of Fish Biology 22:35–41.

Cedeno-Maldonodo, A., Swader, J. A. 1974. Studies on the mechanism of copper toxicity in *Chlorella*. Weed Science 5:433–449.

Clarke, R. A., Stanley, C. D., MacLeod, B. W., McNeal, B. L. 1997. Relationship of seasonal water quality to chlorophyll a concentration in Lake Manatee, Florida. Lake and Reservoir Management 13:253–258.

Colle, D. E., Shireman, J. V. 1980. Coefficients of condition for largemouth bass, bluegill, and redear sunfish in *Hydrilla*-infested lakes. Transactions of the American Fisheries Society 109:521–531.

Conyers, D. L., Cooke, G. D. 1982. A comparison of chemical and mechanical methods for macrophyte control. Lake Line 2:8.

Cooke, G. D., Kennedy, R. H. 1989. Water quality management for reservoirs and tailwaters. Report 1, in: Reservoir water quality management techniques. Techical Report E-89-1, U. S. Army Corps of Engineers, Washington, DC.

Cooke, G. D., Welch, E. B., Martin, A. B., Fulmer, D. G., Hyde, J. B., Schrieve, G. D. 1993a. Effectiveness of Al, Ca, and Fe salts for control of internal phophorus loading in shallow and deep lakes. Hydrobiologia 253:323–335.

Cooke, G. D., Welch E. B., Peterson, S. A., Newroth, P. R. 1993b. Management and restoration of lakes and reservoirs. Lewis, Boca Raton, FL.

Corbus, F. G. 1982. Aquatic weed control with endothall in a Salt River Project Canal. Journal of Aquatic Plant Management 20:1–3.

Crawford, S. A. 1979. Farm pond restoration using *Chara vulgaris* vegetation. Hydrobiologia 62:17–31.

Crutchfield, J. U., Jr., Schiller, D. H., Herlong, D. D., Mallen, M. A. 1992. Establishment and impact of redbelly tilapia in a vegetated cooling reservoir. Journal of Aquatic Plant Management 30:28–35.

Davis, C. C. 1964. Evidence for the eutrophication of Lake Erie from phytoplankton records. Limnology and Oceanography 9:275–283.

Della Greca, M., Monaco, P., Pollio, A., Previtera, L. 1992. Structure–activity relationships of phenylpropanoids as growth inhibitors of the green alga *Selenastrum capricornutum*. Phytochemistry 31:4119–4123.

Demayo, A., Taylor, M. C., Taylor, K. W. 1982. Effects of copper on humans, laboratory and farm animals, terrestrial plants, and aquatic life. CRC Critical Reviews in Environmental Science and Technology 12:183–255.

DeMelo, R., France, R., McQueen, D. J. 1992. Biomanipulation: Hit or myth? Limnology and Oceanography 37:192–207.

DeMott, W. R., Zhang, Q-X., Carmichael, W. W. 1991. Effects of toxic cyanobacteria and purified toxins on the survival and feeding of a copepod and three species of *Daphnia*. Limnology and Oceanography 36:1346–1357.

Desjardins, P. R., Olsen, G. B. 1983. Viral control of nuisance cyanobacteria (blue–green algae). California Water Resources Center Contribution No. 185, pp. 1–35.

Dillon, P. J., Nicholls, K. H., Locke, B. A., de Grosbois, E., Yan, N. D. 1988. Phosphorus–phytoplankton relationships in nutrient-poor soft-water lakes in Canada. International Vereinigung für Theoretische und Angewandte Limnologie Verhandlungen 23:258–264.

Dillon, P. J., Rigler, F. H. 1974. The phosphorus–chlorophyll relationship in lakes. Limnology and Oceanography 19:767–773.

Dimitrov, M. 1984. Intensive polyculture of common carp (*Cyprinus carpio*) and herbivorous fish (silver carp: *Hypophthalmichthys molitrix* and grass carp: *Ctenopharyngodon idella*). Aquaculture 38:241–253.

Dionigi, C. P., Champagne, E. T. 1995. Copper-containing aquatic herbicides increase geosmin biosynthesis by *Streptomyces tendae* and *Penicillium expansum*. Weed Science 43:196–200.

Dodds, W. K., Gudder, D. A. 1992. The ecology of *Cladophora*. Journal of Phycology 28:415–427.

Doyle, R. D., Smart, R. M. 1998. Competitive reduction of noxious *Lyngbya wollei* mats by rooted aquatic plants. Aquatic Botany 61:17–32.

Edmondson, W. T. 1994. Sixty years of Lake Washington: A curriculum vitae. Lake and Reservoir Management 10:75–84.

Edmondson, W. T., Lehman, J. T. 1981. The effect of changes in the nutrient income on the condition of Lake Washington. Limnology and Oceanography 26:1–29.

Effler, S. W., Litten, S., Field, S. D., Tong-Ngork, T., Hale, F., Meyer, M., Quirk, M. 1980. Whole lake response to low level copper sulfate treatment. Water Research 14:1489–1499.

Elakovich, S. D., Wooten, J. W. 1989. Allelopathic potential of sixteen aquatic and wetland plants. Journal of Aquatic Plant Management 27:78–84.

Elder, J. F., Horne, A. J. 1978. Copper cycles and copper sulphate algicidal capacity in two California lakes. Environmental Management 2:17–30.

Elser, J. J., MacKay, N. A. 1989. Experimental evaluations of effects of zooplankton biomass and size distribution on algal biomass and productivity in 3 nutrient-limited lakes. Archiv fur Hydrobiologie 114:481–496.

Elser, J. J., Marzolf, E. R., Goldman, C. R. 1990. Phosphorus and nitrogen limitation of phytoplankton growth in freshwaters of North America: A review and critique of experimental enrichments. Canadian Journal of Fisheries and Aquatic Sciences 47:1468–1477.

EPA. 1990. Lake and reservoir restoration guidance manual, 2nd ed. Office of Water, WH-553, EPA-440/4-90-006, U.S. Environmental Protection Agency, Washington, DC.

EPA. 1998. Clean water action plan: restoring and protecting America's waters. U.S. Environmental Protection Agency, Washington, DC.

Etnier, C., Guterstam, B., Eds. 1997. Ecological engineering for wastewater treatment. CRC Press, Boca Raton, FL.

Fassett, N. C. 1957. A manual of aquatic plants. University of Wisconsin Press, Madison.

Fitzgerald, G. P. 1959. Bacterial and algicidal properties of some algicides for swimming pools. Applied Microbiology 7:205–211.

Fitzgerald, G. P. 1971. Algicides. University of Wisconsin Water Resources Research Center, Eutrophication Information Program, Literature No. 2, Madison, WI.

Flores, E., Wolk, C. P. 1986. Production, by filamentous, nitrogen-fixing cyanobacteria, of a bacteriocin and of other antibiotics that kill related strains. Archives of Microbiology 145:215–219.

Forsberg, C. 1965. Nutritional studies of *Chara* in axenic cultures. Physiologia Plantarum 18:275–290.

Fowler, M. C., Robson, T. O. 1978. The effects of the food preferences and stocking rates of grass carp (*Ctenopharyngodon idella* Val.) on mixed plant communities. Aquatic Botany 5:261–276.
Francke, J. A., Hillebrand, H. 1980. Effects of copper on some filamentous Chlorophyta. Aquatic Botany 8:285–289.
Frank, P. A. 1972. Herbicidal residues in aquatic environments, *in*: Fate of organic pesticides in the aquatic environment. Advances in Chemistry Series, Oxford University Press, Oxford, UK, Vol. 111, pp. 135–148.
Frodge, J. D., Thomas, G. L., Pauley, G. B. 1991. Sediment phosphorus loading beneath dense canopies of aquatic macrophytes. Lake and Reservoir Management 7:61–71.
Fulton, R. S., III. 1988. Grazing on filamentous algae by herbivorous zooplankton. Freshwater Biology 20:263–271.
Fulton, R. S., Paerl, H. W. 1987. Toxic and inhibitory effects of the blue-green alga *Microcystis aeruginosa* on herbivorous zooplankton. Journal of Plankton Research 9:837–855.
Garrison, P. J., Knauer, D. R. 1984. Long term evaluation of three alum treated lakes, *in*: Lake and reservoir management. EPA 440/5-84-001, pp. 513–517.
Gibbs, G. W. 1973. Cycles of macrophytes and phytoplankton in Pukepuke lagoon following a severe drought. Proceedings of the New Zealand Ecological Society 20:13–20.
Gibson, M. T., Welch, I. M., Barrett, P. R. F., Ridge, I. 1990. Barley straw as an inhibitor of algal growth II: Laboratory studies. Journal of Applied Phycology 2:241–248.
Gledhill, M., Nimmo, M., Hill, S. J., Brown, M. T. 1997. The toxicity of copper(II) species to marine algae with particular reference to macroalgae. Journal of Phycology 33:2–11.
Gorham, P. R., Carmichael, W. W. 1988. Hazards of freshwater blue–green algae (cyanobacteria), *in*: Lembi, C. A., Waaland, J. R., Eds., Algae and human affairs. Cambridge University Press, Cambridge, UK, pp. 403–431.
Gross, E. M., Meyer, H., Schilling, G. 1996. Release and ecological impact of algicidal hydrolysable polyphenols in *Myriophyllum spicatum*. Phytochemistry 41:133–138.
Gulati, R. D. 1990. Structural and grazing response of zooplankton community to biomanipulation of some Dutch water bodies. Hydrobiologia 200/201:99–118.
Gunn, G. J., Raferty, A. G., Rafferty, G. C., Cockburn, N., Edwards, C., Beattie, K. A., Codd, G. A. 1992. Fatal canine neurotoxicosis attributed to blue-green algae (cyanobacteria). Veterinary Record 130:301–302.
Halfen, L. N., McCann, M. T. 1975. Behavioral aspects of benthic communities of filamentous blue–green algae in lentic habitats. Michigan Botanist 14:49–56.
Hallingse, M. W., Phlips, E. J. 1996. Effects of Cutrine-Plus and Cide-Kick II on the growth of algae and cyanobacteria. Journal of Aquatic Plant Management 34:39–40.
Hansen, G. W., Oliver, F. E., Otto, N. E. 1984. Herbicide manual. U.S. Department of the Interior, Bureau of Reclamation, Denver, CO.
Hansen, P. S., Phlips, E. J., Aldridge, F. J. 1997. The effects of sediment resuspension on phosphorus available for algal growth in a shallow subtropical lake, Lake Okeechobee. Lake and Reservoir Management 13:154–159.
Hanson, M. J., Stefan, H. G. 1984. Side effects of copper sulfate treatment of the Fairmont Lakes, Minnesota. Water Research Bulletin 20:889–900.
Hargeby, A., Andersson, G., Blindow, I., Johansson, S. 1994. Trophic web structure in a shallow eutrophic lake during a dominance shift from phytoplankton to submerged macrophytes. Hydrobiologia 280:83–90.
Hauser, W. J., Legner, E. F., Medved, R. A., Platt, S. 1976. *Tilapia*—A management tool. Fisheries 1:24.

Havens, K. E. 1994. Structural and functional responses of a freshwater plankton community to acute copper stress. Environmental Pollution 86:259–266.
Havens, K. E., East, T. 1997. In situ responses of Lake Okeechobee (Florida, USA) phytoplankton to nitrogen, phosphorus, and Everglades agricultural area canal water. Lake and Reservoir Management 13:26–37.
Havens, K. E., James, R. T. 1997. A critical evaluation of phosphorus management goals for Lake Okeechobee, Florida, USA. Lake and Reservoir Management 13:292–301.
Hawkins, P. R., Griffiths, D. J. 1987. Copper as an algicide in a tropical reservoir. Water Research 21:475–480.
Hedtke, S. F. 1984. Structure and function of copper-stressed aquatic microcosms. Aquatic Toxicology 5:227–244.
Herbicide Handbook, 7th ed. 1994. Weed Science Society of America, Lawrence, KS.
Herbst, R. P. 1969. Ecological factors and the distribution of *Cladophora glomerata* in the Great Lakes. American Midland Naturalist 82:90–98.
Hinkle, J. 1986. A preliminary literature review on vegetation and fisheries with emphasis on the largemouth bass, bluegill, and hydrilla. Aquatics 8:9–14.
Holm, L. G., Yeo, R. 1981. The biology, control, and utilization of aquatic weeds. Part III. Weeds Today 12:7–10.
Holz, J. C., Hoagland, K. D. 1999. Effects of phosphorus reduction on water quality: Comparison of alum-treated and untreated portions of a hypereutrophic lake. Lake and Reservoir Management 15:70–82.
Horne, A. J., Goldman, C. R. 1974. Suppression of nitrogen fixation by blue-green algae in a eutrophic lake with trace additions of copper. Science 183:409–411.
Hoyer, M. V., Canfield, D. E., Jr. 1996a. Lake size, aquatic macrophytes, and largemouth bass abundance in Florida lakes: A reply. Journal of Aquatic Plant Management 34:48–50.
Hoyer, M. V., Canfield, D. E., Jr. 1996b. Largemouth bass abundance and aquatic vegetation in Florida lakes: An empirical analysis. Journal of Aquatic Plant Management 34:23–32.
Infante, A., Abella, S. E. B. 1985. Inhibition of *Daphnia* by *Oscillatoria* in Lake Washington. Limnology and Oceanography 30:1046–1052.
Jaag, O. 1972. *Oscillatoria rubescens* D. C, *in*: Desikachary, T. V., Ed., Taxonomy and biology of blue–green algae. University of Madras, India, pp. 296–299.
Jacoby, J. M., Gibbons, H. L., Stoops, K. B., Bouchard, D. D. 1994. Response of a shallow, polymictic lake to buffered alum treatment. Lake and Reservoir Management 10:103–112.
Jochimsen, E. M., Carmichael, W. W., 10 other authors. 1998. Liver failure and death after exposure to microcystins at a hemodialysis center in Brazil. New England Journal of Medicine 338:873–878.
Jones, B. L. 1987. Lake Okeechobee eutrophication research and management. Aquatics 9:21–26.
Jones, J. R., Bachmann, R. W. 1976. Prediction of phosphorus and chlorophyll levels in lakes. Journal of Water Pollution Control Federation 48:2176–2182.
Jones, R. C., Walti, K., Adams, M. S. 1983. Phytoplankton as a factor in the decline of the submersed macrophyte *Myriophyllum spicatum* L. in Lake Wingra, Wisconsin. Hydrobiologia 107:213–219.
Joyce, J. C. 1993. Practical uses of aquatic weeds, *in*: Pieterse, A. H., Murphy, K. J., Eds., Aquatic weeds, the ecology and management of nuisance aquatic vegetation. Oxford University Press, Oxford, UK, pp. 274–291.
Jupp, B. P., Spence, D. H. N. 1977. Limitations on macrophytes in a eutrophic lake, Lock Leven. I. Effects of phytoplankton. Journal of Ecology 65:175–186.

Jüttner, F., Höfflacher, B., Wurster, K. 1986. Seasonal analysis of volatile organic biogenic substances (VOBS) in freshwater plankton populations dominated by *Dinobryon*, *Microcystis*, and *Aphanizomenon*. Journal of Phycology 22:169–175.

Kallqvist, T., Meadows, B. S. 1978. The toxic effects of copper on algae and rotifers from a soda lake (Lake Nakaru, East Africa). Water Research 12:771–775.

Kay, S. H. 1997. Barley straw for algae control: A North Carolina experience. Aquatics, Spring, pp. 9–10.

Keating, K. I. 1976. Algal metabolite influence on bloom sequence in eutrophied freshwater ponds. U.S. EPA Ecological Research Service EPA-600/3-76-081.

Keating, K. I. 1977. Allelopathic influence on blue–green bloom sequence in a eutrophic lake. Science 196:885–887.

Keating, K. I. 1978. Blue-green algal inhibition of diatom growth: Transition from mesotrophic to eutrophic community structure. Science 199:971–973.

Khan, M. A. R., Begum, Z. N. T., Rahim, A. T. M. A., Salamatullah, Q. 1996. A comparative study on proximate composition and mineral content of three fresh water green algae. Bangladesh Journal of Botany 25:189–196.

Killgore, K. J., Kirk, J. P., Foltz, J. W. 1998. Response of littoral fishes in upper Lake Marion, South Carolina following hydrilla control by triploid grass carp. Journal of Aquatic Plant Management 36:82–7.

Konopka, A. 1982a. Buoyancy regulation and vertical migration of *Oscillatoria rubescens* in Crooked Lake, Indiana. British Phycological Journal 17:427–442.

Konopka, A. 1982b. Physiological ecology of a metalimnetic *Oscillatoria rubescens* population. Limnology and Oceanography 27:1154–1161.

Kortmann, R. W., Rich, P. H. 1994. Lake ecosystem energetics: The missing management link. Lake and Reservoir Management 8:77–97.

Kudoh, S., Takahashi, M. 1990. Fungal control of population changes of the planktonic diatom *Asterionella formosa* in a shallow eutrophic lake. Journal of Phycology 26:239–244.

Lackey, J. B. 1968. Ecology of *Euglena*, in: Buetow, D. E., Ed., The biology of *Euglena*, Vol. I, General biology and ultrastructure. Academic Press, New York, pp. 27–44.

Lage, O. M., Parente, A. M., Vasconcelos, M. T. S. D., Gomes, C. A. R., Salema, R. 1996. Potential tolerance mechanisms of *Prorocentrum micans* (Dinophyceae) to sublethal levels of copper. Journal of Phycology 32:416–423.

Lamon, E. C., III. 1995. A regression model for the prediction of chlorophyll *a* in Lake Okeechobee, Florida. Lake and Reservoir Management 11:283–290.

Lantz, K. E., Davis, J. T., Hughes, J. S., Shafer, H. E. 1964. Water level fluctuation—Its effect on vegetation control and fish population management. Proceedings of the Southeast Associated Game and Fish Commissons 18:483–494.

Lawrence, J. M. 1954. Control of a branched alga, *Pithophora*, in farm fish ponds. Progress in Fisheries Culture 16:83–86.

Lembi, C. A. 2000. Relative tolerance of mat-forming algae to copper. Journal of Aquatic Plant Management. 38:68–70.

Lembi, C. A., O'Neal, S. W., Spencer, D. F. 1988. Algae as weeds: economic impact, ecology, and management alternatives, in: Lembi, C. A., Waaland, J. R., Eds., Algae and human affairs. Cambridge University Press, Cambridge, UK, pp. 455–481.

Lembi, C. A., Pearlmutter, N. L., Spencer, D. L. 1980. Life cycle, ecology and management considerations of the green filamentous alga, *Pithophora*. Technical Report No. 130, Water Resources Research Center, Purdue University.

Lembi, C. A., Ritenour, B. G., Iverson, E. M., Forss, E. C. 1978. The effect of vegetation removal by grass carp on water chemistry and phytoplankton in Indiana ponds. Transactions of the American Fisheries Society 107:161–171.

Lembi, C. A., Spencer, D. F., O'Neal, S. W. 1984. Evaluation of nonpoint pollution control practices and copper treatments on control of filamentous algae. Technical Report No. 170, Purdue University Water Resources Research Center, U. S. Geological Survey, Reston, VA.

Leonardson, L., Ripl, W. 1980. Control of undesirable algae and induction of algal successions in hypertrophic lake ecostystems, in: Barica, J., Mur, L. R., Eds., Hypertrophic ecosystems. Junk, The Hague, The Netherlands, pp. 57–65.

Letson, M. A., Makarewicz, J. C. 1994. An experimental test of the crayfish (*Orconectes immunis*) as a control mechanism for submersed aquatic macrophytes. Lake and Reservoir Management 10:127–132.

Lovell, R. T., Lelana, I. Y., Boyd, C. E., Armstrong, M. S. 1986. Geosmin and musty–muddy off-flavors in pond-raised channel catfish. Transactions of the American Fisheries Society 115:485–489.

Maceina, M. J. 1996. Largemouth bass abundance and aquatic vegetation in Florida lakes: An alternative interpretation. Journal of Aquatic Plant Management 34:43–47.

Maceina, M. J., Cichra, M. F., Betsill, R. K., Bettoli, P. W. 1992. Limnological changes in a large reservoir following vegetation removal by grass carp. Journal of Freshwater Ecology 7:81–95.

MacKay, N. A., Elser, J. A. 1998. Factors potentially preventing trophic cascades: Food quality, invertebrate predation, and their interaction. Limnology and Oceanography 43:339–347.

Madsen, J. D. 1996. Aquatic plant management evaluation and environmental assessment, in: Luff, C., Ed., Lake St. Clair, Michigan aquatic plant management investigation. U.S. Army Corps of Engineers, Detroit District, April 1996.

Maga, J. A. 1987. Musty/earthy aromas. Food Reviews International 3:269–284.

Maloney, T. E., Palmer, C. M. 1956. Toxicity of six chemical compounds to thirty cultures of algae. Water and Sewage Works 103:509–513.

Marencik, J., Lembi, C. A. 1998. Bioalgicidal potential of reed canary grass and fresh cut alfalfa. Journal of Phycology 34(suppl.):38.

Martin, A. C., Zim, H. S., Nelson, A. L. 1961. American wildlife and plants. A guide to wildlife food habits. Dover, New York.

Martin, E., Benson, R. 1988. Phages of cyanobacteria, in: Calendar, R., Ed., Bacteriophages, Vol. 2. Plenum, New York, pp. 607–645.

Martin, E. L., Leach, J. E., Kuo, K. J. 1978. The biological regulation of bloom-causing blue–green algae, in: Louitit, M. W., Miles, J. A. R., Eds., Microbial ecology. Springer-Verlag, Berlin, pp. 62–67.

Martin, J. F., Izaguirre, G., Waterstrat, P. 1991. A planktonic *Oscillatoria* species from Mississippi catfish ponds that produces the off-flavor compound 2-methyliso-borneol. Water Research 12:1447–1451.

Martin, J. F., McCoy, C. P., Greenleaf, W., Bennett, L. 1988. Analysis of 2- methylisoborneol in water, mud, and channel catfish (*Ictalurus punctatus*) from commercial culture ponds in Mississippi. Canadian Journal of Fisheries and Aquatic Sciences 44:909–912.

Mason, C. P., Edwards, K. R., Carlson, R. E., Pignatello, J., Gleason, F. K., Wood, J. M. 1982. Isolation of chlorine-containing antibiotic from the freshwater cyanobacterium *Scytonema hofmanni*. Science 215:400–402.

Masuda, K., Boyd, C. E. 1993. Comparative evaluation of the solubility and algal toxicity of copper sulfate and chelated copper. Aquaculture 117:287–302.

Mazumder, A., Taylor, W. D., McQueen, D. J., Lean, D. R. S. 1990. Effects of fish and plankton on lake temperature and mixing depth. Science 247:312–315.

McIntosh, A. W., Kevern, N. R. 1974. Toxicity of copper to zooplankton. Journal of Environmental Quality 3:166–171.

McKnight, D. 1981. Chemical and biological processes controlling the response of a freshwater ecosystem to copper stress: A field study of the $CuSO_4$ treatment of Mill Pond Reservoir, Burlington, Massachusetts. Limnology and Oceanography 26:518–531.

McKnight, D. M., Chisholm, S. W., Harleman, D. R. F. 1983. $CuSO_4$ treatment of nuisance algal blooms in drinking water reservoirs. Environmental Management 7:311–320.

McQueen, D. J. 1990. Manipulating lake community structure: Where do we go from here? Freshwater Biology 23:613–620.

McQueen, D. J., Johannes, M. R. S., Post, J. R., Stewart, T. J., Lean, D. R. S. 1989. Bottom-up and top-down impacts on freshwater pelagic community structure. Ecological Monographs 59:289–309.

Meador, J. P., Taub, F. B., Sibley, T. H. 1993. Copper dynamics and the mechanism of ecosystem level recovery in a standardized aquatic microcosm. Ecological Applications 3:139–155.

Melzer, A., Haber, W., Kohler, A. 1977. Foristisch-ökologische Charakterisierung und Gliederung der Ostersseen (Oberbayern) mit Hilfe von submersen Makrophyten. Mitteilungen der Floristisch-Soziologischen Arbeitsgemeinschaft. 19/20:139–151.

Mitchell, D. S., Pieterse, A. H., Murphy, K. J. 1989. Aquatic weed problems and management in Africa, in: Pieterse, A. H., Murphy, K. J., Eds., Aquatic weeds. The ecology and management of nuisance aquatic vegetation. Oxford Science Publications, pp. 341–354.

Mitsch, W. J., Gosselink, J. G. 1993. Wetlands, 2nd ed. Van Nostrand Reinhold, New York.

Monegue, R. L., Phlips, E. J. 1991. The effect of cyanophages on the growth and survival of *Lyngbya wollei*, *Anabaena flos-aquae*, and *Anabaena circinalis*. Journal of Aquatic Plant Management 29:88–93.

Moore, G. T., Kellerman, K. F. 1905. Copper as an algicide and disinfectant in water supplies. Bulletin of the Bureau of Plant Industry USDA. 76:19–55.

Moore, M. V., Winner, R. W. 1989. Relative sensitives of *Ceriodaphnia dubia* laboratory tests and pond communities of zooplankton and benthos to chronic copper stress. Aquatic Toxicology 15:311–330.

Munawar, M., Munawar, I. F. 1976. A lakewide study of phytoplankton biomass and its species composition in Lake Erie, April–December, 1970. Journal of the Fisheries Research Board of Canada 33:581–600.

Murphy, K. J., Barrett, P. R. F. 1993. Chemical control of aquatic weeds, in: Pieterse, A. H., Murphy, K. J., Eds., Aquatic weeds. The ecology and management of nuisance aquatic vegetation. Oxford Science, Oxford, UK, pp. 136–173.

National Academy of Sciences. 1976. Making aquatic weeds useful: Some perspectives for developing countries. Washington, DC.

Neel, J. K., Peterson, S. A., Smith, W. L. 1973. Weed harvest and lake nutrient dynamics. Ecological Research Series, EPA-660/3-73-001, U. S. Environmental Protection Agency, Washington, DC.

Neil, J. H. 1975. Ecology of the *Cladophora* niche, in: Shear, H., Konasewich, D. E., Eds., *Cladophora* in the Great Lakes. International Joint Commission Regional Office, Windsor, Ontario, Canada, pp. 125–127.

Newman, J. R., Barrett, P. R. F. 1993. Control of *Microcystis aeruginosa* by decomposing barley straw. Journal of Aquatic Plant Management 31:203–206.

Nicholls, K. H. 1996. Barley straw for algae control in ponds. 1996. STB Technical Bulletin No. AqSS-1, Ministry of Environment and Energy, Ontario, Canada.

Nicholls, K. H., Beaver, J. L., Estabrook, R. H. 1982. Lakewide odours in Ontario and New Hampshire caused by *Chrysochromulina breviturrita* Nich. (Prymnesiophyceae). Hydrobiologia 96:91–95.

Nicholls, K. H., Gerrath, J. F. 1985. The taxonomy of *Synura* (Chrysophyceae) in Ontario with special reference to taste and odour in water supplies. Canadian Journal of Botany 63:1482–1493.

Nicholls, K. H., Hopkins, G. J. 1993. Recent changes in Lake Erie (North Shore) phytoplankton: Cumulative impacts of phosphorus loading reduction and the zebra mussel introduction. Journal of Great Lakes Research 19:637–647.

Nicholls, K. H., Standen, D. W., Hopkins, G. J. 1980. Recent changes in the near-shore phytoplankton of Lake Erie's western basin at Kingsville, Ontario. Journal of Great Lakes Research 6:146–153.

Nicholls, K. H., Standen, D. W., Hopkins, G. J., Carney, E. C. 1977. Declines in the near-shore phytoplankton of Lake Erie's western basin since 1971. Journal of Great Lakes Research 3:72–78.

Nichols, S. A. 1973. The effects of harvesting aquatic macrophytes on algae. Transactions of the Wisconsin Academy of Sciences, Arts and Letters 61:165–172.

Nichols, S. A. 1984. Macrophyte community dynamics in a dredged Wisconsin lake. Water Research Bulletin 20:573–576.

Noonan, T. A. 1998. Como Lake, Minnesota: The long-term response of a shallow urban lake to biomanipulation. Lake and Reservoir Management 14:92–109.

Ogawa, R. E., Carr, J. F. 1969. The influence of nitrogen on heterocyst production in blue-green algae. Limnology and Oceanography 14:342–351.

Oliver, R. L. 1994. Floating and sinking in gas-vacuolate cyanobacteria. Journal of Phycology 30:161–173.

Olson, R. K., Ed. 1993. Created and natural wetlands for controlling nonpoint source pollution. C. K. Smoley, Boca Raton, FL.

O'Neal, S. W., Lembi, C. A., Spencer, D. L. 1985. Productivity of the filamentous alga *Pithophora oedogonia* (Chlorophyta) in Surrey Lake, Indiana. Journal of Phycology 21:562–569.

Onodera, H., Satake, M., Oshima, Y., Yasumoto, T., Carmichael, W. W. 1997. New saxitoxin analogues from the freshwater filamentous cyanobacterium *Lyngbya wollei*. Natural Toxins 5:146–151.

Ormerod, G. K. 1970. The relationship between real estate values, algae, and water levels. Report of Lake Erie Task Force, Department of Public Works, Canada.

Ozimek, T., Pieczynska, E., Hankiewicz, A. 1991. Effects of filamentous algae on submerged macrophyte growth: A laboratory experiment. Aquatic Botany 41:309–315.

Padan, E., Rimon, A., Ginzburg, D., Shilo, M. 1971. A thermosensitive cyanophage (LPP-1G) attacking the blue-green alga *Plectonema boryanum*. Virology 40:773–776.

Paine, R. T. 1980. Food webs: Linkage, interaction strength and community infrastructure. Journal of Animal Ecology 49:667–685.

Painter, D. S., Kamaitis, G. 1987. Reduction of *Cladophora* biomass and tissue phosphorus in Lake Ontario, 1972–1983. Canadian Journal of Fisheries and Aquatic Sciences 44:2212–2215.

Palmer, C. M. 1962. Algae in water supplies. Public Health Service Publication No. 657. Superintendent of Documents, U.S. Government Printing Office, Washington, DC.

Pearlmutter, N. L., Lembi, C. A. 1986. The effect of copper on the green alga *Pithophora oedogonia*. Weed Science 34:842–849.

Pennak, R. W. 1966. Structure of zooplankton populations in the

littoral macrophyte zone of some Colorado lakes. Transactions of the American Microscopical Society 85:329–349.
Pennak, R. W. 1973. Some evidence for aquatic macrophytes as repellents for a limnetic species of *Daphnia*. International Revue der Gesamten Hydrobiologie 58:569–576.
Perakis, S. S., Welch, E. B., Jacoby, J. M. 1996. Sediment-to-water blue–green algal recruitment in response to alum and environmental factors. Hydrobiologia 318:165–177.
Peterson, H. G., Healey, F. P., Wagemann, R. 1984. Metal toxicity to algae: A highly pH dependent phenomenon. Canadian Journal of Fisheries and Aquatic Sciences 41:974–979.
Phillips, G. L., Eminson, D., Moss, B. 1978. A mechanism to account for macrophyte decline in progressively eutrophicated waters. Aquatic Botany 4:103–126.
Phlips, E. J., Monegue, R. L., Aldridge, F. J. 1990. Cyanophages which impact bloom-forming cyanobacteria. Journal of Aquatic Plant Management 28:92–97.
Phosphorus Management Strategies Task Force. 1980. Phosphorus management of the Great Lakes. Final Report to the Great Lakes Water Quality Board and the Great Lakes Science Advisory Board. International Joint Commission, Windsor, Ontario, Canada.
Pillinger, J. M., Cooper, J. A., Ridge, I. 1994. Role of phenolic compounds in the antialgal activity of barley straw. Journal of Chemical Ecology 20:1557–1569.
Pillinger, J. M., Cooper, J. A., Ridge, I., Barrett, P. R. F. 1992. Barley straw as an inhibitor of algal growth. III: The role of fungal decomposition. Journal of Applied Phycology 4:353–355.
Pine, R. T., Anderson, L. W. J. 1991. Plant preferences of triploid grass carp. Journal of Aquatic Plant Management 29:80–82.
Porter, K. G. 1973. Selective grazing and differential digestion of algae by zooplankton. Nature 244:179–180.
Porter, K. G., McDonough, R. 1984. The energetic cost of response to blue-green algal filaments by cladocerans. Limnology and Oceanography 29:365–369.
Power, M. E. 1990. Effects of fish in river food webs. Science 250:811–814.
Prepas, E. E., Murphy, R. P., Crosby, J. M., Walty, D. T., Lim, J. T., Babin, J., Chambers, P. A. 1990. Reduction of phosphorus and chlorophyll a concentrations following $CaCO_3$ and $Ca(OH)_2$ additions to hypereutrophic Figure Eight Lake, Alberta. Environmental Science and Technology 24:1252–1258.
Prepas, E. E., Trew, D. O. 1983. Evaluation of the phosphorus chlorophyll relationship for lakes of the Precambrian Shield in western Canada. Canadian Journal of Fisheries and Aquatic Sciences 40:27–35.
Quade, H. W. 1969. Cladoceran fauna associated with aquatic macrophytes in some lakes in northwestern Minnesota. Ecology 50:170–179.
Quirós, R. 1995. The effects of fish assemblage composition on lake water quality. Lake and Reservoir Management 11:291–298.
Raman, R. K. 1985. Controlling algae in water supply impoundments. Journal of the American Water Works Association 78:41–43.
Reavie, E. D., Smol, J. P., Carmichael, N. B. 1995. Postsettlement eutrophication histories of six British Columbia (Canada) lakes. Canadian Journal of Fisheries and Aquatic Sciences 52:2388–2401.
Redhead, K., Wright, S. J. L. 1978. Isolation and properties of fungi that lyse blue-green algae. Applied and Environmental Microbiology 35: 962–969.
Reuter, J. E., Rhodes, C. L., Lebo, M. E., Kotzman, M., Goldman, C. R. 1993. The importance of nitrogen in Pyramid Lake (Nevada, U.S.A.), a saline, desert lake. Hydrobiologia 167:179–189.

Rice, E. L. 1984. Allelopathy. Academic Press, Orlando, FL.
Riley, E. T., Prepas, E. E. 1984. Role of internal phosphorus loading in two productive lakes in Alberta, Canada. Canadian Journal of Fisheries and Aquatic Sciences 41:845–855.
Robbins, R. W., Glicker, D. M., Bloem, D. M., Niss, B. M. 1991. Effective watershed management for surface water supplies. American Water Works Association Research Foundation, Denver, CO.
Robinson, P. K., Hawkes, H. A. 1986. Studies on the growth of *Cladophora glomerata* in laboratory continuous-flow culture. British Phycological Journal 21:437–444.
Rosa, R., Burns, N. M. 1987. Lake Erie central basin oxygen depletion changes from 1929–1980. Journal of Great Lakes Research 13:684–696.
Rosine, W. N. 1955. The distribution of invertebrates on submerged aquatic plant surfaces in Muskee Lake, Colorado. Ecology 36:308–314.
Rothhaupt, K. O. 1991. The influence of toxic and filamentous blue–green algae on feeding and population growth of the rotifer *Brachionus rubens*. International Revue der Gesamten Hydrobiologie 76:67–72.
Safferman, R. S., Morris, M. E. 1963. Algal virus: Isolation. Science 140:679–680.
Safferman, R. S., Schneider, I. R., Steare, R. L., Morris, M. E., Diener, T. O. 1969. Phycovirus Sm-1: A virus infecting unicellular blue-green algae. Virology 37:386–395.
Sanchez, I., Lee, G. F. 1978. Environmental chemistry of copper in Lake Monona, Wisconsin. Water Research 12:889–903.
Sanders, L., Hoover, J. J., Killgore, K. J. 1991. Triploid grass carp as a biological control of aquatic vegetation. Publication A-91-2, Aquatic Plant Control Research Program, U.S. Army Corps of Engineers, Waterways Experiment Station, Vicksburg, MS.
Savino, J. E., Stein, R. A. 1982. Predator–prey interactions between largemouth bass and bluegills as influenced by simulated, submersed vegetation. Transactions of the American Fisheries Society 111:255–266.
Schardt, J. D. 1994. 1994 Florida aquatic plant survey. Technical Report No. 972-CGA, Distributed by the Bureau of Aquatic Plant Management, Tallahassee, FL.
Schelske, C. L. 1975. Silica and nitrate depletion as related to rate of eutrophication in Lakes Michigan, Huron, and Superior, in: Hasler, A. D., Ed., Coupling of land and water systems. Springer-Verlag, New York, pp. 277–298.
Schelske, C. L., Stoermer, E. F. 1971. Eutrophication, silica depletion, and predicted changes in algal quality in Lake Michigan. Science 173:423–424.
Schindler, D. W. 1975. Whole lake eutrophication experiments with phosphorus, nitrogen, and carbon. Internationale Vereinigung für Theoretische und Angewandte Limnologie Verhandlungen 19: 3221–3231.
Schindler, D. W. 1977. Evolution of phosphorus limitation in lakes. Science 195:260–262.
Schindler, D. W. 1978. Factors regulating phytoplankton production and standing crop in the world's freshwaters. Limnology and Oceanography 23:478–486.
Scholze, R. J. 1994. A summary of best management practices for nonpoint source pollution. U.S. Army Corps of Engineers, USACERL Technical Report EP-93/06, National Technical Information Service, Springfield, VA.
Schrader, K. K., Blevins, W. T. 1993. Geosmin-producing species of *Streptomyces* and *Lyngbya* from aquaculture pools. Canadian Journal of Microbiology 39:834–840.
Schrader, K. K., de Regt, M. Q., Tucker, C. S., Duke, S. O. 1997. A rapid bioassay for selective algicides. Weed Technology 11:767–74.

Shapiro, J. 1980. The importance of trophic-level interactions to the abundance and species composition of algae in lakes, *in*: Barica, J., Mur, L. R., Eds., Hypertrophic ecosystems. Junk, The Hague, The Netherlands, pp. 105–116.

Shapiro, J., LaMarra, V., Lynch, M. 1975. Biomanipulation: An ecosystem approach to lake restoration, *in*: Brezonik, P. L., Fox, J. L., Eds., Water quality management through biological control. Department of Environmental Engineering Sciences, University of Florida, Gainesville, pp. 85–96.

Shilo, M. 1967. Formation and mode of action of algal toxins. Bacteriological Reviews 31:180–193.

Shireman, J. V., Haller, W. T., Colle, D. E., DuRant, D. F. 1983a. Effects of aquatic macrophytes on native sportfish populations in Florida, *in*: Proceedings of the international symposium on aquatic macrophytes, Nijmegen, The Netherlands, pp. 208–214.

Shireman, J. V., Hoyer, M. V., Maceina, M. J., Canfield, D. E. 1985. The water quality and fishing of Lake Baldwin, Florida: 4 years after macrophyte removal by grass carp. Lake and Reservoir Management 1:201–206.

Shireman, J. V., Rottmann, R. W., Aldridge, F. J. 1983b. Consumption and growth of hybrid grass carp fed four vegetation diets and trout chow in circular tanks. Journal of Fish Biology 22:685–693.

Simons, J., Ohm, M., Daalder, R., Boers, P., Rip, W. 1994. Restoration of Botshol (The Netherlands) by reduction of external nutrients load—Recovery of a characean community, dominated by *Chara connivens*. Hydrobiologia 276:243–253.

Simpson, M. R., MacLeod, B. W. 1991. Comparison of various powdered activated carbons for the removal of geosmin and 2-methylisoborneol in selected water conditions, *in*: 1991 Proceedings of the American Water Works Association Annual Conference.

Simpson, P. S., Eaton, J. W. 1986. Comparative studies of the photosynthesis of the submerged macrophyte *Elodea canadensis* and the filamentous algae *Cladophora glomerata* and *Spirogyra* sp. Aquatic Botany 24:1–12.

Smeltzer, E. 1990. A successful alum/aluminate treatment of Lake Morey, Vermont. Lake and Reservoir Management 6:9–19.

Smith, C. S., Barko, J. W. 1990. Ecology of Eurasian watermilfoil. Journal of Aquatic Plant Management 28:55–64.

Smith, D. W. 1985. Biological control of excessive phytoplankton growth and enhancement of aquacultural production. Canadian Journal of Fisheries and Aquatic Sciences 42:1940–1945.

Smith, V. H. 1983. Low nitrogen to phosphorus ratios favor dominance by blue–green algae in lake phytoplankton. Science 221:669–671.

Solley, W. B., Merk, C. F., Pierce, R. R. 1988. Estimated use of water in the U.S. in 1985. U.S. Geological Survey Circular 1004, USGS, Denver, CO.

Sommer, U. 1983. Nutrient competition between phytoplankton species in multispecies chemostat experiments. Archiv für Hydrobiologie 96:399–416.

Sommer, U. 1992. Phosphorus-limited *Daphnia*: Intraspecific facilitation instead of competition. Limnology and Oceanography 37:966–973.

Spence, D. H. N. 1976. Light and plant response in freshwater, *in*: Evans, R. B., Rockham, O., Eds., Light as an ecological factor: II. Blackwell, Oxford, UK, pp. 93–133.

Spencer, C. N., King, D. L. 1984. Role of fish in regulation of plant and animal communities in eutrophic ponds. Canadian Journal of Fisheries and Aquatic Sciences 41:1851–1855.

Spencer, D. L., Lembi, C. A. 1981. Factors regulating the spatial distribution of the filamentous alga *Pithophora oedogonia* (Chlorophyceae) in an Indiana lake. Journal of Phycology 17:168–173.

Speziale, B. J., Dyck, L. A. 1992. *Lyngbya* infestations: Comparative taxonomy of *Lyngbya wollei* comb. nov. (Cyanobacteria). Journal of Phycology 28:693–706.

Stauffer, R. E., Lee, G. F. 1973. The role of themocline migration in regulating algal blooms, *in*: Middlebrooks, E. J., Falkenbort, D. H., Maloney, T. E., Eds., Modeling the eutrophication process. Utah State University, Water Resources Center, Logan, UT, pp. 73–82.

Steinman, A. D. 1996. Effects of grazers on freshwater benthic algae, *in*: Stevenson, R. J., Bothwell, M. L., Lowe, R. L., Eds., Algal ecology. Freshwater benthic systems. Academic Press, San Diego, pp. 341–373.

Steinman, A. D., Meeker, R. H., Rodusky, A. J., Davis, W. P., Hwang, S. J. 1997. Ecological properties of charophytes in a large subtropical lake. Journal of North American Benthological Society 16:781–793.

Steward, K. K. 1993. Aquatic weed problems and management in the eastern United States, *in*: Pieterse, A. H., Murphy, K. J., Eds., Aquatic weeds, the ecology and management of nuisance aquatic vegetation. Oxford University Press, Oxford, UK, pp. 391–405.

Stewart, W. D. P., Daft, M. J. 1977. Microbial pathogens of cyanophycean blooms, *in*: Droop, M. R., Jannasch, H. W., Eds., Advances in aquatic microbiology, Vol. 1. Academic Press, New York, pp. 177–218.

Stockner, J. G., Shortreed, K. S. 1988. Response of *Anabaena* and *Synechococcus* to manipulation of nitrogen: Phosphorus ratios in a lake fertilization experiment. Limnology and Oceanography 33:1348–1361.

Stoermer, E. F. 1988. Algae and the environment: The Great Lakes case, *in*: Lembi, C. A., Waaland, J. R., Eds., Algae and human affairs. Cambridge University Press, Cambridge, UK, pp. 57–83.

Sunderman, F. W. 1978. Carcinogenic effects of metals. Federation Proceedings 37:40–46.

Sutton, D. L., Porter, K. M. 1989. Influence of allelochemicals and aqueous plant extracts on growth of duckweed. Journal of Aquatic Plant Management 27:90–95.

Swain, E. B., Monson, B. A., Pillsbury, R. W. 1986. Use of enclosures to assess the impact of copper sulfate treatments on phytoplankton. Lake and Reservoir Management 11:303–308.

Taub, F. B., Kindig, A. C., Meador, J. P., Swartzman, G. L. 1989. Effects of seasonal succession and grazing on copper toxicity in aquatic mesocosms. International Vereinigung für Theoretische und Angewandte Limnologie Verhandlungen 24:2205–2214.

Tilman, D., Kiesling, R. L. 1984. Freshwater algal ecology taxonomic trade-offs in the temperature dependence of nutrient competitive abilities, *in*: Klug, M. J., Reddy, C. A., Eds., Current perspectives in microbial ecology. American Society for Microbiology, Washington, DC, pp. 314–319.

Tucker, C. S., Boyd, C. E. 1978. Consequences of periodic applications of copper sulfate and simazine for phytoplankton control in catfish ponds. Transactions of the American Fisheries Society 107:316–320.

Tucker, C. S., Busch, R. L., Lloyd, S. W. 1983. Effects of simazine treatment on channel catfish production and water quality in ponds. Journal of Aquatic Plant Management 21:7–11.

Urabe, J. 1993. N and P cycling coupled by grazers' activities: Food quality and nutrient release by zooplankton. Ecology 74:2337–2350.

Urabe, J., Clasen, J., Sterner, R. W. 1997. Phosphorus limitation of *Daphnia* growth: Is it real? Limnology and Oceanography 42:1436–1443.

van den Berg, M. S., Coops, H., Simons, J., De Keizer, A. 1998a. Competition between *Chara aspera* and *Potamogeton pectinatus* as a function of temperature and light. Aquatic Botany 60:241–250.

van den Berg, M. S., Scheffer, M., Coops, H., Simons, J. 1998b. The role of characean algae in the management of eutrophic shallow lakes. Journal of Phycology 34:750–756.

van der Ploeg, M., Dennis, M. E., de Regt, M. Q. 1995. Biology of *Oscillatoria* cf. *chalybea*, a 2-methylisoborneol producing blue-green alga of Mississippi catfish ponds. Water Science and Technology 31:173–180.

van der Zweerde, W. 1993. Biological control of aquatic weeds by means of phytophagous fish, *in*: Pieterse, A. H., Murphy, K. J., Eds., Aquatic weeds. The ecology and management of nuisance aquatic vegetation. Oxford Science, pp. 201–221.

van Donk, E., Grimm, M. P., Gulati, R. D., Kline Breteler, J. P. G. 1990. Whole-lake food-web manipulation as a means to study community interactions in a small ecosystem. Hydrobiologia 200/201:291–301.

Van Etten, J. L., Lane, L. C., Meints, R. H. 1991. Viruses and virus-like particles of eukaryotic algae. Microbiological Reviews 55:586–620.

Vanni, M. J., Findlay, D. L. 1990. Trophic cascades and phytoplankton community structure. Ecology 71:921–937.

Vollenweider, R. A. 1969. A manual on methods for measuring primary production in aquatic environments, IBP Handbook 12. Blackwell Science, Oxford, UK.

Wagemann, R., Barica, J. 1979. Speciation and rate of loss of copper from lakewater with implications to toxicity. Water Research 13:515–523.

Walker, H. L., Higginbotham, L. R. 2000. An aquatic bacterium that lyses cyanobacteria associated with off-flavor of channel catfish (*Ictalurus punctatus*). Biological Control 18:71–78.

Walker, W. W., Jr. 1987. Phosphorus removal by urban runoff detention basins. Lake and Reservoir Management 3:314–326.

Walker, W. W., Jr., Westerberg, C. E., Schuler, D. J., Bode, J. A. 1989. Design and evaluation of eutrophication control measures for the St. Paul water supply. Lake and Reservoir Management 5:71–83.

Webster, K. E., Peters, R. H. 1978. Some size-dependent inhibitions of larger cladoceran filterers in filamentous suspensions. Limnology and Oceanography 23:1238–1244.

Welch, E. B., Cooke, G. D. 1995. Internal phosphorus loading in shallow lakes: Importance and control. Lake and Reservoir Management 11:273–281.

Welch, E. B., Cooke, G. D. 1999. Effectiveness and longevity of phosphorus inactivation with alum. Lake and Reservoir Management 15:5–27.

Welch, E. B., DeGasperi, C. L., Spyridakis, D. E., Belnick, T. J. 1988. Internal phosphorus loading and alum effectiveness in shallow lakes. Lake and Reservoir Management 4:27–33.

Welch, I. M., Barrett, P. R. F., Gibson, M. T., Ridge, I. 1990. Barley straw as an inhibitor of algal growth I: Studies in the Chesterfield Canal. Journal of Applied Phycology 2:231–239.

Wetzel, R. G. 2001. *Limnology*, 3rd ed. Academic Press, San Diego.

Wetzel, R. G. 1990. Land–water interfaces: Metabolic and limnological regulators. International Vereinigung für Theoretische und Angewandte Limnologie Verhandlungen 24:6–24.

Whitaker, J., Barica, J., Kling, H., Buckley, M. 1978. Efficacy of copper sulphate in the suppression of *Aphanizomenon flos-aquae* blooms in prairie lakes. Environmental Pollution 15:185–194.

Whitton, B. A. 1970. Toxicity of heavy metals to freshwater algae: A review. Phykos 9:116–125.

Whitton, B. A. 1973. Freshwater plankton, *in*: Carr, N. G., Whitton, B. A., Eds., The biology of blue–green algae. University of California Botanical Monographs, Vol. 1, pp. 353–367.

Wiley, M. J., Gordon, R. W., Waite, S. W., Powless, T. 1984. The relationship between aquatic macrophytes and sport fish production in Illinois ponds: A simple model. North American Journal of Fisheries Management 4:111–119

Winner, R. W. 1985. Bioaccumulation and toxicity of copper as affected by interactions between humic acid and water hardness. Water Research 19:449–455.

Winner, R. W., Owen, H. A., Moore, M. V. 1990. Seasonal variability in the sensitivity of freshwater lentic communities to a chronic copper stress. Aquatic Toxicology 17:75–92.

Wium-Andersen, S., Anthoni, U., Christophersen, C., Houen, G. 1982. Allelopathic effects on phytoplankton by substances isolated from aquatic macrophytes (Charales). Oikos 39:187–190.

Wooten, J. W., Elakovich, S. D. 1991. Comparisons of potential allelopathy of seven freshwater species of spikerushes (*Eleocharis*). Journal of Aquatic Plant Management 29:121–125.

GLOSSARY

accessory pigment pigment capable of capturing radiant energy and transferring it to chlorophyll-*a*.
accumulation body colored (red, orange, or yellow) lipid material in the cytoplasm, but with uncertain function.
acicular needle shaped.
acidophile (acidophilic) species that typically occurs in dilute waters (soft water) with lower pH values.
aerobic oxygenated environment or metabolism requiring oxygen.
aerophytic tendency to colonize subaerial or terrestrial habitats, often under conditions of high humidity.
aerotope in cyanobacteria, clusters of gas vesicles; the older term is gas vacuole.
agarose gelatinous substance (a sulfated polygalactan) produced in the walls of certain species of red algae.
agglomeration loosely arranged mass (of cells); usually colonies with no definite shape or cellular arrangement.

Note: Glossary terms were defined by individual authors in reference to their chapter. Consequently, terms may actually apply to other taxa and have broader definitions than indicated here.

akinete thick-walled cell produced by members of several algal classes; may be released by vegetative cells or attached to filaments; functions as an asexual resting stage and typically is resistant to harsh conditions (e.g., low temperatures).
algal mat thallus composed of tightly interwoven filaments.
algicide various chemicals used to control algal growth in lakes and ponds, such as $CuSO_4$ or Diquat.
alkalophile (alkalophilic) species that is typically localized in high ion waters of neutral to high pH values.
allochthonous in aquatic ecosystems, matter or production formed outside of the water body, such as terrestrial carbon entering a stream from a forest.
allophycocyanin type of blue colored phycobilin pigment produced by members of the cyanobacteria and red algae.
alternation of generations life history with two alternating multicellular phases, in which the haploid (n) phase (= gametophyte) produces gametes, which later fuse to form a zygote and later form a diploid (2n) phase (sporophyte). Spores are produced by the sporophyte

often through meiosis. The two phases may be isomorphic or heteromorphic; also termed diplohaplobiontic.

alum various forms of aluminum sulfate (e.g., $Al_2[SO_4]_3$) used to precipitate and reduce dissolved phosphorus in lakes.

alveolate pitted or with many small cavities or pores.

amoeboid type of cell organization that lacks a cell wall and possesses flexible and frequent changes in shape; present in some members of Chrysophyceae.

amorphous without a definite shape.

amphiesma in dinoflagellates, a cell covering that includes an outer membrane and a layer of vesicles that may contain cellulosic plate material.

ampulla (1) flasklike reservoir that covers the cell of some euglenoids; (2) the reproductive region within the cortical cells of some red algae.

amylopectin component of true starch that is a branched glucose polymer composed of α-1,4 linkages and α-1,6 branches; similar to glycogen.

amylose component of true starch composed of α-1,4 linkages; an unbranched glucose polymer.

anaerobic oxygen-free environment or metabolism that either does not require oxygen or requires an absence of oxygen.

anastomosing combination of, or communication between separate branches, sheaths, or filaments (as in the joining of tributaries with rivers).

androspore zoospore that becomes a dwarf male (in Oedogoniales).

anisogamy (anisogamous) sexual reproduction in which the two types of gametes differ in size or appearance but are both typically motile.

antheridium (antheridia) male gametangium in a wide variety of algal groups.

antherozoid motile male gametes (e.g., *Sphaeroplea*), spermatozoid.

anthropogenic caused by human activities, often referring to disturbances to ecological systems, such as acidic precipitation.

apical cell cell positioned at the end of a filament or thallus; often the site of meristematic growth.

apical pore field group of pores at one or both poles in freshwater cymbelloid and gomphonemoid diatoms; functions in secretion of mucilage and forms stalks that attach cells to surfaces.

apical series plates at the apex of a cell and that may surround an apical pore.

aplanogamete nonmotile gamete produced by a member of an algal group that normally produces flagellated gametes (e.g., flagellated green algae).

aplanospore nonmotile (nonflagellated) spore produced by divisions of parental cell; may have the potential to produce flagella in some algal groups.

araphid diatom a pennate diatom with no raphe system on either valve.

archeopyle in dinoflagellates, the exit pore from which a geminating cell is released from the cyst.

arcuate arched or crescent shaped; in diatoms, pennate cells that are bent along the apical axis.

areola in diatoms, a perforation or pore through the valve that is bounded by an internal or external sieve membrane; a more specific term than "puncta," which is a general term for an opening in the valve wall.

armor cellulose plates that form a covering in some dinoflagellates.

ash free dry mass measure of biomass in algal and limnological studies, based on the difference between dry mass and the ash remaining after combustion in a muffle furnace (generally between 450 and 600°C); used as an estimate of the organic mass of biological materials.

assimilative growth phase in which the cell takes in nutrients (either autotrophic or heterotrophic), not reproductive or dispersive.

astaxanthin orange–red xanthophyll (carotenoid) accessory pigment present (temporary or permanent) in some algae (e.g., *Haematococcus*, *Euglena*), rendering cells a reddish color. Also occurs in the eyespot of several species; termed haematochrome in older literature.

athecate in dinoflagellates, lacking cellulose plates (naked).

autochthonous in aquatic ecosystems, matter or production (e.g., primary production) formed *in situ*, within the water body, such as carbon formed by phytoplankton.

autocolony colony formed asexually as coenobia, usually within parent cells and often a miniature of the parent colony (e.g., *Scendesmus*).

autospore nonflagellated spore similar in appearance to the vegetative cell that produced it.

autotrophic cells capable of synthesizing organic matter; in algae via photosynthesis.

auxospore large cell resulting from sexual reproduction or autogamy. In diatoms, they lack the rigid valve construction of other cells; instead, they are covered by delicate siliceous scales or bands (perizonium).

auxotrophy nutritional requirement for external sources of vitamins.

axial area in diatoms, an unornamented area along the apical axis; includes the central sternum. In the older literature, the axial area is referred to as the pseudoraphe when applied to araphid diatoms.

baeocyte spores produced by certain cyanobacteria or green algae via repeated (often rapid) divisions of vegetative cells, yielding smaller daughter cells.

band-shaped diatom colony type caused by the attachment of cells along the entire valve face; also ribbon-shaped.

basal toward or at the base or point of origin of a thallus or filament.

basal siliceous layer first layer of a diatom frustule deposited during its formation.

basionym original name given to a genus or species, and which is retained if transferred to another position or grouping (also written basonym).

benthic organisms that grow on or are associated with the bottom (e.g., sediments or rocks) zone of a water body.

beta-carotene carotenoid pigment (also β-carotene) in several groups of algae.

biogeography study of the geographical distribution of organisms, their patterns, origins, movements, and sources.

biomass mass of biological material in a specific area (per square meter) or volume (per liter or cubic meter); in algae, expressed as dry mass, ash-free dry mass, chlorophyll-*a*, or carbon.

biovolume apparent cell volume as calculated from external dimensions.

bipartite consisting of or divided into two parts.

biraphid diatom diatom with raphe systems on both valves.

bloom massive or conspicuous growth of algae, typically planktonic and often forming surface scums; often a large percentage of the total cells are one or a few species.

brackish slightly saline, a mixture of fresh and marine water; often used to describe salinity conditions in estuaries.

brittlewort see stonewort.

caespitose clustered, in thick tufts or clumps, forming a turf.

calcification process of depositing calcium carbonate ($CaCO_3$), which occurs on algal cell surfaces or in walls, especially in alkaline habitats.

calcite form of calcium carbonate with rhombohedral crystals.

calyptra thickened or enlarged tip; in filamentous cyanobacteria, occurs at the tip of some trichomes.

cameo raised figure; refers to the raised portion of some diatom valves.

canal raphe general term for a raphe that opens into a channel or canal (e.g., *Epithemia*); internal openings (when present) of the canal raphe are called portules.

canopeum in diatoms, an external siliceous covering slightly to distinctly elevated from the valve face; widths are variable, sometimes extending to the valve margin; visible with scanning electron microscope only. Found to contain nitrogen-fixing cyanobacteria in several taxa.

capitate with a distinct head; knoblike or swollen at the end.

carboxysomes granules or inclusions in cyanobacterial cells (may appear like polyhedrons); contain the enzyme ribulose-1,5-bisphosphate carboxylase/oxygenase.

carina ridge (dinoflagellates).

carotenoids group of lipid-soluble pigments, including carotenes (yellow or orange) and xanthophylls (yellow or golden), that serve as accessory pigments in photosynthesis.

carpogonial branch supports the female gametangium (carpogonium) in the red algae; may or may not be differentiated from surrounding vegetative cells.

carpogonium female gametangium in red algae, usually consisting of the receptive region (trichogyne) for the male gamete (spermatium) and a base.

carposporangium sporangium of many red algae that typically forms diploid spores (carpospores) on tips of the postfertilization stage (carposporophyte). Carpospores germinate into a free-living diploid phase of the life history.

carposporophyte postfertilization stage of many red algae, typically consisting of small diploid filaments (gonimoblast) localized on the haploid gametophyte. Gonimoblast filaments form diploid spores (carpospores) at their apices.

cell wall typically rigid external structure enclosing the cell membrane; in algae, may consist of cellulose, silica, pectin, or other materials.

central area in diatoms, an unornamented area in the central or middle part of the valve face, near or surrounding the proximal raphe ends.

central nodule internally, the area between the proximal raphe ends that is usually thickened.

central sternum in diatoms, a region of solid silica deposited during the initial stages of silica deposition of the valve. This apically oriented region may or may not be perforated by a raphe slit; nonperforated is termed pseudoraphe in older literature.

central strutted process in diatoms, a strutted process that occurs on the valve face, as opposed to the valve margin; see also fultoportule.

centric diatom diatom that has a valve structure symmetrical around a point, also termed radial symmetry.

centroplasm less pigmented, central region of a cell, especially in cyanobacteria; less heavily pigmented than the chromatoplasm.

CER chloroplast endoplasmic reticulum; encircles chloroplasts of some algae.

chantransia stage diploid life-history phase of members of the red algal order Batrachospermales and *Rhododraparnaldia* of the Balbianiales. Forms the haploid gametophyte stage directly attached to it by meiosis of an apical cell.

character as applied here, specific structure or attribute useful in classification.

chlorophyll-*a* primary photosynthetic pigment and light receptor (maximum absorption about 663 nm) in algae and higher plants (photosystem I) that is in the form of a porphyrin ring with a central magnesium atom; also used as a chemical marker for algal biomass in ecological studies; also associated with PS-II.

chlorophyll-*b* secondary photosynthetic pigment present in higher plants, green algae, prochlorophytes, and euglenophytes (absorption maximum about 645 nm).

chlorophyll-*c* secondary class of photosynthetic pigment, which occurs in chrysophytes, synurophytes, diatoms, cryptophytes, tribophytes, dinoflagellates, and brown algae; includes two components (both forms not found in all algal groups) termed c_1 and c_2, each has several different absorption peaks.

chloroplast membrane-bound organelle that contains photosynthetic thylakoids, chlorophyll-*a*, and other pigments (occasionally as "plastid").

chromatic adaptation see photoacclimation.

chromatophore organelle that contains pigments (syn. plastid).

chromatoplasm portion of a cell that contains pigments; in cyanobacteria, typically a peripheral pigmented region.

chrysolaminarin polysaccharide storage product, a β-1,3-linked glucan; occurs in several algal groups, including chrysophytes, synurophytes, haptophytes, tribophytes, and diatoms. Also termed leucosin.

chytrids group of small fungi that parasitize many organisms, including algae; appear as colorless globose cells on various planktonic diatoms, desmids, and other algae.

cingulum in dinoflagellates, a transverse groove that encircles the cell (usually) and holds the transverse flagellum in place. In diatoms, another name for all the girdle bands of a diatom cell (elements of girdle region).

cirque in limnology, a glacially created lake that formed at the head of a glacial valley in mountainous region; amphitheater shaped.

clade grouping of taxa that is recognized to have descended from a common ancestor.

cladoceran group of mostly microscopic, planktonic crustaceans (e.g., *Daphnia*), common in freshwater environments.

clathrate with irregular perforations or openings; lattice-like.

clone cells or organisms with identical genetic complements derived from an single ancestor.

coccoid simple cell type that is spherical, subspherical, or rod-shaped.

coccolith calcite structure (scale), aggregations of which surround the exterior of coccolithophorid haptophyte cells.

coccolithophorid coccolith-bearing haptophyte.

coenobium (coenobia) form of colony in which the number of cells is fixed (genetically determined) at an early develop mental stage; cell number does not change during development.

coenocyst in some green algae, condition where the cytoplasm divides in multinucleate portions and develops into a thick-walled resting stage (e.g., *Protosiphon*).

coenocytic multinucleate condition and lacking cross walls; some coenocytic taxa produce cross walls infrequently or during reproduction (e.g., *Vaucheria*).

colony group of cells that may be connected or held together by cytoplasmic strands, mucilage, or parent cell wall.

confluent growing together or intermingled; sheaths that join together.

conjugation form of sexual reproduction in which nonflagellated gametes join by means of a specialized tube or papillae (e.g., Zygnemataceae).

conspecific belonging to the same species.

contractile vacuole membrane-bound vesicle that expands and contracts to regulate water and/or osmotic conditions within cells; especially in nonwalled forms.

copulae another name for girdle bands that make up the cingulum (in diatoms).

cordate heart shaped; also termed cordiform.

corona ring of cells at the apex of the oogonium of charophyte algae.

cortex peripheral layer of a differentiated thallus that is typically photosynthetic and surrounds the inner colorless medulla

cortical filament filament that is part of a surrounding layer that covers the main axis of the thallus (e.g., red algae and charophytes).

corticated possessing cortical filaments or cells.

costa(e) Latin for rib; as used here, refers to linear thickenings in a diatom valve.

craticula network of siliceous bars or ribs formed internally in the diatom genus *Craticula*, usually under environmental conditions of increased solute concentration.

crenate with an edge or margin that is notched, wavy, or scalloped.

cribra finely poroid siliceous membranes, which occlude puncta.

cricolith heterococcolith (in haptophytes) with two narrow subhorizontal shields connected by a central tube.

cruciform cross-shaped; also termed cruciate.

crust closely adherent, flat thallus composed of compacted tiers of cells.

cryophilic organisms that colonize ice or snow environments.

cuboidal shape similar to a cube.

cuneate wedge shaped.

cyanelle cyanobacterium-like plastid endosymbiotic within certain protist cells (e.g., green algae *Chalarodora* and *Gloeochaete*), although other symbiotic cyanobacteria exist as whole cells within other organisms.

cyanophage virus that infects cyanobacteria.

cyanophycean starch type of starch produced in cyanobacterial cells that is composed of α-1,4 linkages and α-1,6 branches; similar to glycogen and does not react positively to an iodine test.

cyclomorphosis repeated or cyclic change in form, or plasticity in the morphology of an organism, such as elongation of spines (e.g., *Scenedesmus*) or horns (e.g., *Ceratium*).

cyst nonmotile stage that protects the cell during adverse conditions; also termed resting cyst.

daughter cells cells derived from a parent by a mitotic division.

dendroid tree-like morphology.

desmid unicellular and filamentous green algae that give rise to nonflagellated, amoeboid gametes that conjugate.

desmokont in dinoflagellates, a flagellar arrangement where both flagella extend longitudinally from the posterior of the cell.

detritus dead or partially degraded organic matter.

diadinoxanthin carotenoid pigment (euglenophytes, tribophytes, chrysophytes, synurophytes, dinoflagellates, and diatoms); acts as an accessory pigment in photosynthesis.

diatomite type of rock formed primarily from diatom frustules; also known as diatomaceous earth.

diatoxanthin golden carotenoid pigment produced by many algal groups (euglenophytes, tribophytes, chrysophytes, synurophytes, haptophytes, dinoflagellates, and diatoms, brown algae); acts as an accessory pigment in photosynthesis.

dichotomous branch forked branch consisting of two approximately equal branches.

dimictic lake lake that mixes twice per year and stratifies in summer and winter.

dinokaryon unique nucleus found in dinoflagellates with permanently condensed chromosomes.

dinokont in dinoflagellates, a flagellar arrangement with one transverse flagellum that encircles the cell and a second, longitudinal flagellum that extends out at the posterior of the cell.

dinospore motile (flagellated) stage in the life cycle of parasitic forms.

dioecious literally, two households; organisms in which male and female (or + and −) gametes are produced on different individuals.

diplobiontic life cycle in which there are two free-living growth phases.

diplontic life cycle in which there is one multicellular, diploid phase, and the haploid phase consists of only gametes.

disjunct in biogeography, locations or populations that are widely separated and generally isolated from one another.

distal away from the base or center of a thallus or filament or origin (opposite of basal).

distal raphe ends in diatoms, the external terminus of raphe at the poles or ends of valves.

dorsal back of a cell or thallus.

dystrophic waters rich in organic matter (often from allochthonous sources), that have slow rates of organic matter decay and typically are colored yellow to reddish brown; in many, the pH can be low (≤ 4.0).

ecdysis shedding of thecal walls (e.g., dinoflagellates).

ecomorphotype type of morphological modification caused by or related to certain ecological conditions.

ecorticate condition of a main axis or branch that lacks a surrounding layer of smaller filaments (cortical cells).

edaphic referring to the soil.

ejectisome in cryptomonads, projectile-like structure (often in a spiral) that is discharged from the cell; may serve as an escape mechanism or direct defense against other organisms.

emarginate uneven margin or surface; may have notches or concavities.

endemism geographic distribution pattern that is restricted to a particular habitat or locality.

endolithic (endolithan) growing within rocks or intercrystalline spaces.

endophytic (endophyton) growing within plants or algae; may be intra- or intercellular, or within another species' mucilage (endogloeic).

endoplasmic reticulum series of cytoplasmic membranes in cells that function to process and transport materials or cellular components.

endospore in cyanobacteria, specialized spores see baeocytes. In conjugating green algae, the innermost layer of the zygospore wall; see also mesosopore.

endosymbiont organism that lives within another, resulting in a mutually beneficial and intimate relationship between the two (e.g., cyanobacteria in diatoms and algae in lichens).

environmental optimum apparent preference of a species along a controlling environmental gradient.

epicingulum all the girdle elements associated with epivalves.

epicone in dinoflagellates, the upper or anterior portion of an athecate cell.

epidendric (epidendron) growing on wood.

epilimnion upper, well-mixed region of stratified lakes, above the thermocline; typically has greater light availability and is isothermal.

epilithic (epilithon) growing attached to the surface of rock or stone substrata.

epipelic (epipelon) growing on sediments, clays, and silt.

epiphytic (epiphyton) growing on plants and other algae.

epipsammic (epipsammon) growing attached to or living among grains of sand.

epitheca upper or anterior portion of thecate cell; in diatoms and dinoflagellates.

epivalve in diatoms, the larger and therefore older of two valves of a frustule.

epizooic (epizoon) growing attached to animal surfaces (e.g., *Colacium* on copepods and cladocerans).

eucentric centric diatoms that are circular or approximately circular, in valve view.

eukaryotic cell type characterized by the presence of membrane-bound organelles, including a nucleus, mitochondria, endoplasmic reticulum, and (in most algae and plants) chloroplasts.

euphotic zone upper depths of a water body that receive sufficient light to support photosynthesis.

euplanktonic capable of completing an entire life cycle suspended in the water column.

eutrophic literally, well nourished; water bodies that have high levels of dissolved nutrients (esp. N and P) and high levels of organic production; in lakes, often shallower with a broad littoral zone, depleted summer hypolimnetic oxygen, and reduced transparency.

eutrophication process of becoming eutrophic.

evenness in ecological studies, a measure (or index) of the relative proportion of the community (in numbers or biomass) represented by the species present.

eversion process by which a colony in some flagellated green algae fold backward and turns inside out; also termed inversion.

exiccata slide set permanent microscope slides and accompanying data distributed to subscribers or established museums for the purpose of allowing access to the specimens at multiple locations.

exocyte in cyanobacteria, daughter cells that divide off of distinctly polarized cells (e.g., *Chamaesiphon*); produced singly or successively in rows; also as known as exospore.

exospore in conjugating green algae, the outer layer of the zygospore wall; see also mesosopore.

eyespot light sensitive, typically red-colored (carotenoids pigments) spot in many algal cells (typically flagellates or motile reproductive stages of multicellular forms), and typically in the anterior region of the cell.

FAA common fixative 10:7:2:1 parts of 95% ethanol:distilled water:formalin: acetic acid.

facultative heterotroph ability in cells to use photosynthesis and also to obtain external sources of organic matter, such as under low light conditions.

false branching condition where branches arise from a break (or cell death) in the main filament and the continued growth of one or both ends; in cyanobacteria, only the sheath splits while the vegetative trichomes separate.

fascia literally a band. In the pennate diatoms, an unornamented area across the middle portion of a valve and visible in valve view, generally rectangular and band-shaped.

fascicle literally a bundle. In filamentous algae, a cluster of multiple filaments; in some red algae, lateral vegetative branches; in centric diatoms, a group of rows of areolae oriented radially.

fasciculate occurring in bundles, as clusters of filamentous cyanobacteria (e.g., *Aphanizomenon*) or the striae on the diatom *Stephanodiscus*.

fenestra in keel-bearing diatoms, such as *Surirella* and *Cymatopleura*, part of the valve surface and mantle fuse together to form openings (fenestra or windows).

fenestrated window-like; divided into shiny portions.

fibulae internal struts that provide structural support to the raised keel that contains the raphe; extend transapically from valve face to valve mantle (e.g., *Nitzschia*) or under the raphe in the central region (e.g., *Denticula*).

filament common type of thallus in which cells are arranged in a linear series and in which adjacent cells share a common cross wall.

filiform thread-like.

fimbriate fringed appearance or bordered with hairs or hair-like structures.

flagellar transition region zone between the flagellum and its basal body.

flagellate nonphyletic term for protists that possess one or more flagella; also the condition of possessing a flagellum (-a).

flagellum (flagella) long, threadlike organelle that projects out of the cell and functions in motility; in eukaryotic cells, they consist of a 9 doublet + 2 central singlet array of microtubules.

floridean starch carbohydrate storage granules localized in the cytoplasm of red algae; similar in composition to glycogen; consists of glucose polymers with α-1,4 linkages and α-1,6 branches.

footpole in a heteropolar diatom, pole by which the cells attach to each other or to substrata.

fresh water ill-defined term used to describe largely inland, low-salinity (<0.1 g L^{-1}) bodies of water; amounts of dissolved salts vary, but generally much less saline than seawater (ca. 35 g of salts L^{-1}).

frustule literally, a box. As applied here, the entire siliceous covering of a diatom cell; the collective term for the epivalve and hypovalve, and their associated cingulum elements.

fucosan vesicles small, refractive bodies or vesicles present in brown algal cells and that contain fucosan; also termed physodes.

fucoxanthin brown-colored carotenoid pigment produced by members of the golden-pigmented algae (chrysophytes, synurophytes, haptophytes, and diatoms) and brown algae (Phaeophyta); acts as an accessory pigment in photosynthesis.

fultoportule another term for strutted process (in diatoms).

furcate forked or branched, in filaments, spines, or other processes; used to denote characteristic branching number (e.g., 2 = bifurcate; 3 = trifurcate).

fusiform spindle-shaped; any elongate morphology that is widest near the middle and tapering at either end.

gametangium (gametangia) any structure or feature of a thallus that produces gametes

gametophyte multicellular, usually haploid, gamete producing stage in organisms that exhibit alternation of generations.

gas vesicles gas containing spherical or cylindrical structure in cyanobacterial cells, often gathered in clusters termed aerotopes (the older term is gas vacuoles) and visible with light microscopy as red or brownish bodies (due refraction of light, may be mistaken as actual pigmentation). Function to provide buoyancy to planktonic forms (e.g., *Microcystis*, *Anabaena*, and *Aphanizomenon*).

gemmae dense (asexual) aggregations of cells formed in cavities in the thallus (e.g., red alga *Hildenbrandia*); when released, form new thalli.

geniculate having joints like a knee; occurs in some filamentous green algae (e.g., *Klebsormidium*, *Sirogonium*, and *Zygnema*).

girdle bands in diatoms, another term for elements of the cingulum. In desmids, formed from very short wall segments; there may be several per cell and they may be obvious in some species (e.g., *Closterium* and *Penium*).

girdle view in diatoms, a frustule oriented so that the girdle bands and valve mantle are visible, opposed to valve view, where the valve surface and, in some cases, a portion of the valve mantle are visible.

globose in the form of a globe; completely or nearly spherical.

gloeocapsin dark blue pigment present in the sheaths of certain cyanobacteria

glomerate in a compact cluster.

glycoprotein protein with attached carbohydrates, often in wall or external gelatinous coatings.

Golgi body organelle that consists of a series or stack of flattened, membrane-bound sacs, in which materials are assembled or modified and from which are secreted; also termed dictyosome.

gone cell cell derived from a germinating zygote in some green algae.

gonidium large, nonmotile cell in colonies of volvocalean green algae that can generate a new daughter colony.

gonimoblast filament in red algae, a diploid filament that composes the postfertilization stage (carposporophyte) and is localized on the gametophyte. Diploid spores (carpospores) are formed at the filament tips.

guild assemblage of co-occurring taxa with similar ecological requirements or functional roles.

gullet in some euglenoids, chrysophytes, cryptomonads, and other flagellates, a depression in the anterior region of the cell where the flagellum (-a) emerges.

haematochrome see astaxanthin.

hair cell typically colorless, thin, and elongate cell at branch apices that increases the thallus surface area for nutrient and gas exchange; in some filamentous cyanobacteria, green algae and red algae, produced under periods of nutrient deficiency.

haplobiontic life cycle in which there is one haploid vegetative phase, where zygotes are the only diploid phase.

haptobenthon community of organisms living on hard substrata.

haptonema (haptonemata) flagellum-like organelle, the base of which is linked to the basal bodies of flagella; present in haptophyte algae; in longer haptonemata, there is a tendency to form tight coils (e.g., *Chrysochromulina parva*). In contrast to 9 + 2 ultrastructure of a flagellum, the haptonema consists of 6–7 peripheral microtubules enclosed in a cylinder of endoplasmic reticulum (ER), which is confluent with the peripheral ER of the cell.

haptophyte any member of the algal division Haptophyta that possess a well-developed or vestigial haptonema.

haptophyte scale platelike organic structure with microfibrillar basal structure (and another type with spinelike superstructure in some species), produced in Golgi bodies and released to the cell exterior where they form a layer surrounding the cell.

helictoglossa in diatoms, an internal, liplike structure that terminates the distal raphe.

heterococcolith in certain haptophytes, a coccolith with an organic base plate constructed within a specialized Golgi vesicle.

heterocyst in certain cyanobacteria (e.g., *Anabaena*, *Nostoc*, and *Calothrix*), a thick-walled, multilayered (apparently gas-tight, anaerobic), and weakly pigmented cell; contains the nitrogenase enzyme, which enables fixation of gaseous nitrogen (N_2) to ammonium; also termed heterocyte.

heterokont possessing flagella of unequal length or different ornamentation.

heteromorphic life history life history in which the gametophyte and sporophyte have different morphologies (i.e., alternation of generations).

heteropolar asymmetric to the transverse (longitudinal) axis in a filament, diatom valve, cell, or other structure. In cyanobacteria, morphologically differentiated in basal and apical parts.

heterothallic two different strains that are compatible for sexual fertilization; self-incompatible.

heterotrichous filamentous thallus with two forms of organization; typically with prostrate (creeping) and erect components (e.g., *Gongrosira*, *Heribaudiella*, *Hildenbrandia*, and *Trentepohlia*) or otherwise dissimilar forms of branching.

heterotrophy (heterotrophic) type of nutrition in which organic matter is obtained from external sources; may involve osmotrophy (dissolved) or phagotrophy (particulate).

heterovalvar condition when the two valves of a diatom cell have different structure, such as the density, direction, and ultrastructure of the striae.

holdfast unicellular or multicellular structure that attaches a thallus or filament to the substratum

homologous character evolutionarily derived character, as opposed to an analogous character, which may appear similar, but is the product of convergent evolution.

homonym name applied to an organism that has also been used for another different organism (typically the latter name is invalid).

hormogonium (hormogonia) in filamentous cyanobacteria, a means of vegetative reproduction (and dispersal) formed via fragmentation of the trichome, forming distinct segments that are often motile (via gliding).

hyaline transparent or colorless.

hypersaline extremely saline habitats.

hypnospore thick-walled resting cyst.

hypnozygote in green algae, dinoflagellates, and some other flagellated forms, a thick-walled, nonmotile resting zygote (cyst) that is dormant in sediments under adverse conditions; may germinate under favorable conditions (in dinoflagellates called a dinocyst).

hypocingulum girdle elements associated with the hypovalve (in diatoms).

hypocone lower or posterior portion of athecate cell (in dinoflagellates).

hypolimnion deeper region of a stratified lake, below the thermocline; typically colder, more poorly illuminated, and richer in nutrients.

hypotheca lower or posterior portion of thecate cell; in diatoms and dinoflagellates (below the cingulum).

hypovalve in diatoms, the smaller and therefore younger of two valves of a frustule (= hypotheca).

imbricate overlapping or arranged like scales or roof tiles; an overlapping series.

incrassate swollen or thickened in form or appearance.

indeterminant growth pattern or structure (e.g., filament or branch) with no genetically determined form or limit to growth.

infrageneric taxon that is below the level of a genus, including species, sections, varieties, and forms.

intaglio in diatoms, a surface that is impressed; as used here, the opposite of cameo.

intercalary growth growth in the middle of the thallus.

internal valves in diatoms, some valve form that are not associated with cell division; they are morphologically similar to normal valves and produced inside normal valve walls; a cell may form several internal valves.

internode portion of a thallus, filament or axis between two nodes.

inversion see eversion.

involucral lateral branch of the main branch (carpogonial branch) bearing the female gametangium (carpogonium) in red algae.

isodiametric diameters equal; in filamentous forms, the diameter is more or less constant along its length.

isogamy (isogamous) sexual reproduction in which the two types of gametes are morphology indistinguishable, although functionally different.

isokont flagella of equal size and form (as recognized by light or electron microscopy).

isolated puncta puncta set off from others, in striae.

isomorphic life history life history in which the gametophyte and sporophyte have similar (or indistinguishable) vegetative morphologies.

isopolar symmetric to the transverse (longitudinal) axis of a filament, diatom valve, cell, or other structure.

isthmus in desmids, the constriction (narrow region) between semicells; typically the location of the nucleus.

ITS internal transcribed spacers of the ribosomal RNA array; there are two, designated ITS-1 and ITS-2.

Janus valves diatom cells where the frustule has two distinctly different valve types in species where the two valves usually appear to be identical.

keel in diatoms, a raised or elevated ridge that contains the raphe, formed from a folding of the valve wall. Many keel-forming diatoms have fibulae under the raised ridge for structural support.

keritomized netlike in appearance; within cells, sometimes the result of vacuolation or irregular arrangement of materials.

kettle lake glacial lake formed after melting glacier; typically bowl-shaped depressions, also called kettle holes; common across Wisconsin and the prairie provinces of Canada.

kleptoplastidy stealing plastids from prey; plastids usually remain functional in the new host.

K_m Michaelis Menton constant; a measure of the affinity of an enzyme for its substrate.

labiate process tubular process, extending through the valve, with a slitlike inner opening that has thickened margins (giving the appearance of lips); the external opening may be an extended tube or a simple circular pore; also called a rimoportula.

lacuna opening or cavity.

lacustrine referring to lake (standing water) environments.

lamellae stack of thylakoids; also refers to any structures that occur in layers or stacks of plates.

lamellate layered or arranged in layers.

laminarin in brown algae, a food storage polysaccharide composed mainly of β-1,3 linked glucose units that exists in cells outside of chloroplasts, in vacuoles.

laminate platelike.

lanceolate shape that is wider near the center than the ends; dory shaped.

landslide lake lake basin formed when water flows through an existing depression and is blocked by rock or other material (e.g., Spirit Lake, Mount St. Helens, WA, and Mountain Lake, VA).

lateral conjugation in members of the Zygnemataceae, a form of conjugation that occurs between two adjacent cells on the same filament.

lentic referring to standing waters; ponds, lakes, or marshes.

lenticular lenslike or lentil shaped; flattened or thinning toward the ends.

leucoplast colorless plastid with few or no thylakoids; may contain starch (= amyloplast); in some green algae and cryptomonads.

leucosin see Chrysolaminarin.

lichen mutualistic symbiosis between an alga or cyanobacterium and a fungus, in which the two organisms grow together in an intimate organization. The symbiosis typically forms a distinct macroscopic morphology; often occur in terrestrial, epiphytic, xeric, and occasionally aquatic habitats.

ligula silica extension and part of the cingula, in diatoms; in some taxa, projects to intersect with neighboring cingula.

lime encrusted calcified thallus; see calcification.

linear morphologies with generally parallel sides.

linking spine in diatoms, a spine modified to bind daughter valves together, thereby effecting colony formation.

list in dinoflagellates, a winglike plate extension.

littoral region of lakes or large rivers near the shore; often defined as the area from the shore to the maximum depth of rooted macrophytes.

LM light microscope; light microscopy.

loculate valve walls that have more than one layer of silica may have a complex, chambered structure. Areolae may also be considered to be chambered, or loculate (the chamber is termed the loculus).

locules older term used for the chambers and other components associated with the girdle bands in *Mastogloia*; see also partecta.

longitudinal canal tubelike chamber apically oriented, extending most of or the entire length of the valve, and may be interrupted in the central area; a general term that applies to structures that may not have the same evolutionary origin (not homologous) such as the canals of *Neidium* and *Diploneis*.

longitudinal lines lines that run along the apical axis, on either side of the axial area; term is used for lines visible in the light microscope, although they may be formed by different structures.

lorica protective investment or envelope that surrounds the protoplast (naked, nonwalled cell); present in genera from several flagellate groups (e.g., *Dinobryon* and *Trachelomonas*).

lotic running waters springs, streams, and rivers.

Lugol's iodine common fixative; a saturated solution of I_2KI: 2 g potassium iodide + 1 g resublimed iodine +300 mL distilled water; it can be acidified with 10 mL glacial acetic acid or made neutral or slightly basic with 1 g of sodium acetate; sometimes used to preserve algal samples; often used for quantifying phytoplankton by inverted microscope method.

lutein carotenoid pigment present in many algal divisions.

macrandrous sexual reproduction where the male filament or thallus is as large as the female thallus (e.g., species of *Oedogonium* and *Bulbochaete*).

macroalga algae that form macroscopic or plantlike morphologies with a thallus structure that is recognizable with the naked eye; can be important in the benthic communities of streams and lakes.

marginal strutted process tubular process through the valve, with two or more satellite pores on the inner valve surface; external expression may be a tube or a simple pore.

marl $CaCO_3$ deposits on substrata or other surfaces in hard water or alkaline lakes and streams; appears whitish and may be precipitated on the surface of algal cells during photosynthesis (termed lime in some literature).

mastigonemes hairlike appendages on flagella; numbers and arrangement vary among some algal groups.

mastigote motile cell.

medulla cells of the inner layer of a differentiated thallus; often colorless and surrounded by an outer photosynthetic layer (the cortex).

meiospores spores produced from a zygote via meiosis.

meromictic lake lake that mixes in the upper layers (mixolimnion), but deeper waters (monimolimnion) do not; generally the result of much greater density or salt concentration.

mesokaryotic nucleus of dinoflagellates once thought to be intermediate between prokaryotic and eukaryotic.

mesoplankton plankton size category between 200 and 2000 µm; includes only taxa that form very large colonies, macroscopic clumps, or mats in the plankton.

mesospore middle layer of the zygospore wall of conjugating green algae (with exospore and endospore); its structure varies from smooth to ornamented, and is used in identification of genera and species.

mesotrophic water bodies with intermediate levels of plant nutrients and organic production.

metaboly in euglenoids, motility and flexibility in the pellicle (outer cell covering) without the aid of a flagellum.

metaphyton microscopic floating community among the littoral zone plants of lakes and ponds; often loosely associated with aquatic plants.

microplankton plankton size category between 20 and 200 µm; includes the largest unicellular algae and many colonial taxa.

microplasm connecting pore between neighboring cells (at cross walls) in filamentous cyanobacteria.

mitochondrion (mitochondria) in eukaryotes, an organelle with internal and external membranes, the former arranged in folds called cristae; responsible for respiration and contains DNA.

mixotrophy (mixotrophic) cells with both photosynthetic and heterotrophic (typically phagotrophic) nutrition (e.g., *Dinobryon*); occasionally spelled myxotrophy.

moniliform filamentous morphology that looks like a string of beads.

monoecious literally, one household; organisms in which male and female (or + and −) gametes are produced on the same thallus.

monomictic lake lakes that mix once per year, typically in the autumn (warm monomictic).

monophyletic natural taxonomic group that shares a most recent common ancestor.

monoraphid diatom pennate diatom with a raphe system on one valve only.

monosporangium asexual sporangium of some red algae that produces one spore (monospore) that germinates into the same life history phase that produced it.

monostromatic thallus composed of a single layer of cells; also monolayer.

monotypic genus with a single species.

morainal lake lake basin formed through glacial activity wherein the resultant valley is closed by glacial deposits (e.g., Finger Lakes of New York).

morphometry sizes and shapes of lake basins; also used to describe a suite of morphological features of an organism.

morphotype particular recognizable form of a species that has variable morphology.

mucocyst saclike vesicle in certain flagellates that may be released by the cell (= muciferous body).

multiaxial thallus with the main axis composed of several to numerous parallel filaments (= multiseriate); in some definitions, differs from multiseriate in that the latter is not necessarily in parallel.

multinucleate having many nuclei.

multipolar as used here, centric diatom in which the valve outline is not circular.

multiseriate see multiaxial.

myxophycean starch see cyanophycean starch.

myzocytosis in dinoflagellates, particle uptake using the peduncle.

nannandrous in some species of green algae (e.g., within *Oedogonium*), a very small (often unicellular) male filament that attaches to a much larger female filament.

nanoplankton plankton size category between > 2 and 20 µm; includes many taxa of unicellular algae and small colonies.

natural classification taxonomy that takes into account hypotheses of lineage and evolution, leading to monophyletic groups.

necridia in cyanobacteria, dead cells that function to aid separation of filaments or pseudofilaments for vegetative reproduction (e.g., hormogonia) or the formation of false branches.

net plankton older term that refers to planktonic organisms captured by a standard plankton net with a mesh size of 25, 40, or 64 µm.

net primary productivity quantity of primary production in an ecosystem minus respiration costs (gross − respiration); generally the amount available to the next trophic level.

neuston (neustonic) planktonic organisms living at the air–water interface, either above the water (epineuston) or just below the surface (hyponeuston).

niche hypothetical construct, with several definitions, including (1) an organism's role in the biotic environment, in relation to its competitors and enemies; and (2) the set of all ecological conditions and resources that a species can exploit effectively for growth and reproduction, expressed as an n-dimensional hypervolume of resource axes.

nitrogen fixation process by which cyanobacteria (and other bacteria) convert atmospheric (gaseous) nitrogen (N_2) to biologically useful forms (e.g., NH_4^+).

node location along an axis or thallus that produces branches.

Nomarski optics type of high resolution optics in light microscopy; also termed differential interference contrast (DIC).

nomenclature as used here, formal rules for naming and referring to organisms.

nuclear associated organelle (NAO) structure that appears at the poles of the spindle apparatus during the process of mitosis.

nucleomorph highly reduced endosymbiont nuclei present in the cells of some protists (e.g., cryptomonads).

obovoid inversely ovoid, with the broader end upward or outward; also termed obovate.

occluded process tubular process found near the valve surface margin of some centric diatoms, which have no associated internal opening or structure.

ocellulimbus in diatoms, a platelike apical pore field on the mantle.

ocellus (ocelli) literally, eyes. (1) In dinoflagellates, a light-sensitive organelle; (2) in diatoms, a large group of areolae or porelli that are physically separated from areolae by an unornamented rim or hyaline area.

oligotrophic literally, poorly nourished. Water bodies with low levels of nutrients (esp. N and P) and low levels of organic production; often deep and steep sided with high transparency, narrow littoral zone, abundant dissolved oxygen with depth, and larger relative hypolimnion volume.

oogamy (oogamous) sexual reproduction in which there is a fusion of a smaller flagellated male gamete (e.g., spermatozoid) with a larger, nonflagellated female gamete (e.g., egg or oogonium).

oogonium (oogonia) large, single-celled female gametangium that may produce one or more eggs.

organelle membrane-bound intracellular structure with a specific function or functions (e.g., nucleus, chloroplast).

osmotrophy nutrition involving the absorption of dissolved organic molecules.

ovoid egg shaped; rounded with one pole broader than the other.

paedogamous conjugation of gametes within a gametangium.

pallium feeding veil of extruded cytoplasm from heterotrophic and mixotrophic dinoflagellates that engulfs prey and may serve as the site of extracellular digestion.

palmelloid in many algal groups, a nonmotile, colonial stage of indefinite cell number and arrangement; often within a mucilaginous matrix.

papilla protuberance or swelling.

papillate covered with papillae or granulate.

paramylon storage compound in euglenoids and some tribophytes and haptophytes; a polymer of many β-1,3 linked glucans organized in a membrane-bound crystalline structure; appears as distinct rods or disks in the light microscope; does not stain with iodine.

paraphyletic taxonomic group that does not include all of the descendents from a specified parent group or ancestor.

parenchyma(tous) type of thallus with true tissues, formed from cell division in three planes; often differentiation of cells into an outer photosynthetic cortex and an inner colorless medulla.

parietal pertaining to the outer or peripheral surface of a cell or thallus; parietal chloroplasts are localized within plasma membrane near the cell wall.

partecta in some diatoms, a loculate chamber that is associated with the valvocopula in *Mastogloia*.

parthenospore spores formed in association with sexual reproduction in conjugating green algae; usually produced in cells that have failed to pair up with a compatible sexual partner.

peduncle cytoplasmic extension used as a means of attachment. In dinoflagellates, a feeding structure exerted through the flagellar pore in the sulcus; see also pallium.

pelagic open water region of lakes; also termed the limnetic zone.

pellicle in some dinoflagellates, the layer beneath the amphiesmal vesicles, which remains upon ecdysis of outer layers. In Euglenoids, the outer proteinaceous surface layer, often in a helical arrangement.

penicillate in a small tuft; brushlike.

pennate diatom diatom with features symmetrical about a line, termed bilateral symmetry, and produces amoeboid gametes.

PER periplastid (surrounding the plastid) endoplasmic reticulum.

pericentral cell formed immediately adjacent to the main axis.

peridinin golden brown carotenoid pigment produced by dinoflagellates that acts as an accessory pigment in photosynthesis (esp. in dinoflagellates).

periphyton collection of organisms (algae, bacteria, fungi, and protozoa) and detritus attached to surfaces without stipulating substratum type; in common usage refers to algal cells in this habit.

periplast in cryptomonads, the outer cell covering, consisting of a series of overlapping plates plus the plasma membrane.

phagopod in dinoflagellates, a specialized feeding tube that lacks microtubules and is used to obtain nutrients phagotophically from other organisms.

phagotrophy type of heterotrophic nutrition based on ingestion of particulate organic carbon. Particle capture may involve pseudopodia, flagella, or haptonema (haptophytes).

photoacclimation ability to vary the amounts of different photosynthetic pigments in response to changes in the light environment; in cyanobacteria this may result in distinctly different colors of the thallus.

photoautotroph(ic) organism that synthesizes its own organic compounds through photosynthesis.

photoheterotroph pigmented species capable of photosynthesis and metabolizing external dissolved organic molecules to meet its energy needs.

phototaxis movement of an organism toward (positive) or away from (negative) a light source.

phragmoplast form of cell division in which the microtubules lie perpendicular to the plane of division; cytokinesis thus occurs via centrifugal cell plate formation. Occurs in certain groups of green algae and true plants.

phycobiliproteins water soluble pigments of the cyanobacteria, red algae, and cryptomonads; there are three basic types red-colored phycoerythrin, and blue-colored phycocyanin and allophycocyanin. These molecules act as accessory pigments in photosynthesis.

phycobilisomes granules (usually spherical or disk shaped) located on the thylakoids, consisting of the phycobiliprotein accessory pigments; in cyanobacteria and red algae.

phycocyanin blue pigment phycobiliprotein (water soluble) pigment produced by cyanobacteria, red algae, and cryptomonads.

phycoerythrin red pigment phycobiliprotein (water soluble) pigment produced by cyanobacteria, red algae, and cryptomonads.

phycoplast form of cell division in which the array of microtubules is oriented parallel to the plain of division; a new set of microtubules lie perpendicular to them to form a new cell wall; occurs in members of the green algal class Chlorophyceae.

phylogenetic attributes or studies that recognize the evolutionary relationships among organisms; classification systems based on these relationships.

physode see fucosan vesicles.

phytoplankton portion of the microscopic floating (plankton) community represented by algae, including cyanobacteria.

picoplankton plankton size category between 0.2 and 2 μm; includes several genera of cyanobacteria (e.g., *Synechococcus* and *Synechocystis*), a few green (e.g., *Nannochloris*), eustigmatophyte, and chrysophyte algae.

placoderm desmid members in the class Desmidiaceae; cells with distinct semicells (walls of either half are of different ages), in which there is a median constriction (= sinus), a connecting zone (= isthmus), and pores in their walls. Cells may be solitary, or joined in colonies or filaments.

planktonic growing suspended, floating, or drifting in water.

planospores motile cells, asexual zoospores or sexual gametes.

planozygote motile, thick-walled cyst resulting from sexual reproduction; in some flagellates (e.g., *Woloszynskia*), may store lipids and/or starch.

plaque in diatoms, thin external layers of silica along the girdle bands.

plastids organelles bounded by two membranes and typically with thylakoids and photosynthetic pigments (e.g., chloroplasts); also, an alternate term for chloroplast. Others, such as amyloplasts (+ starch) lack pigments.

plicate folded or in furrows, giving the appearance of folds.

plurilocular sporangium in brown algae, a reproductive structure that becomes subdivided into many compartments (locules), from which single flagellated zoospores (swarmers) are released.

pole terminal portions of cells, filaments, pennate diatom valves.

polygonal in the shape of a polygon; many sided; angular.

polymictic lake shallow lake with frequent or continuous mixing, in tropical and equatorial areas.

polymorphic having more than one form.

polyphosphate storage granules of condensed polymers of inorganic phosphate that are visible with light microscopy; in several algal groups.

polyphyletic artificial taxonomic group; not sharing a recent, common ancestor; members may be more closely related to organisms outside the group.

polysiphonous thallus that appears as multiple parallel filaments (or siphons), formed from a series of adjacent filaments; in some genera of red algae (e.g., *Polysiphonia*).

pore field in diatoms, a specialized area of pores unlike the pores of the striae, usually associated with organelles that extrude material from the cell.

porelli small, closely packed perforations through the valve wall of diatoms; perforations in ocelli, pseudocelli, and apical pore fields.

primary production amount of new organic matter synthesized by autotrophic organisms.

prokaryotic cell type that lacks membrane-bound organelles; also lack true (9+2) flagella; also as procaryotic.

protist phyletic group (Kingdom) of eukaryotic organisms not classified as members of the Plant, Fungi, or Animal kingdoms; includes all eukaryotic algae, plus non-photosynthetic protists (protozoa).

protolichen association of subaerial or soil algae growing in close association with fungi, but not forming an intimate symbiosis; see lichen.

protoplast living material inside cells, excluding the organelles.

proximal raphe ends raphe ends on the central nodule (internal valve surface) and near the central portion of the valve (external valve surface).

pseudocelli group of areolae set off from the pattern of the rest of the valve, which decreases in size from areolae on the main part of the valve. Pseudocelli are not physically separated from areolae by unornamented band or ring.

pseudocilia see pseudoflagella.

pseudofilament loose chain of cells held together with a common gelatinous matrix, or linked by fibrils or other connections; cells are typically separated from each other. Unlike true filaments, they do not share a common cross wall; also termed a chain.

pseudoflagellum giving the appearance of flagella, but not functional in motility (e.g., *Tetrapora*).

pseudonodule in diatoms, a differentiated area or structure on the valve that has a variable form; essentially an area that may resemble an ocellus or pseudocellus, but is structurally different. In freshwater diatoms, this structure is found only in *Actinocyclus*.

pseudoparenchyma(tous) thallus that is tissue-like but composed of compacted or interwoven filaments; resembles parenchyma. Often, the main axis consists of a single filament (uniaxial) or a series of parallel filaments (multiaxial).

pseudopyrenoid smaller, naked pyrenoid.

pseudoraphe literally, a false raphe; an unornamented linear region in the axial area of some pennate diatom valves. Usually called the central sternum in current terminology.

pseudosepta plate or lamina of silica projecting internally from the apical portion of the valve.

puncta(e) pore/perforation through the valve when substructure (i.e., sieve membrane) is unknown or lacking.

pusule in dinoflagellates, an organelle composed of a series of tubes and thought to function in nutrient uptake and possibly expulsion of excess water.

pyrenoid distinct, proteinaceous structure (often spherical), embedded in or associated with chloroplasts of algae; in some, contains the enzyme RuBisCO; associated with starch in some green algae.

pyriform pear shaped.

quadrate square or rectangular.

raphe structure in monoraphid and biraphid diatoms that consists of a slit through the valve face, and associated cytoplasmic structures; usually situated along the apical axis or within a marginal keel; composed of (usually) two branches per valve. The raphe enables a diatom cell to move over substrata.

raphe branch single raphe slit, extending from the proximal end to distal ends.

rapheless in monoraphid diatoms, an adjective that describes a valve that lacks a raphe system; also araphid.

raphid in monoraphid diatoms, an adjective that describes a valve that has a raphe system.

***Rbc*L** gene encoding the large subunit of the RuBisCO enzyme; sequences are used in phylogenetic analyses.

refractive index quantity reflecting the degree to which a substance refracts light.

reniform bean or kidney shaped.

replicate folded back.
reticulate netlike or arranged to form a network.
rhizoid downward-growing cell or chain of cells that is typically involved in thallus attachment to substrata.
rhizoplast striated strand that connects flagellar basal bodies with the nucleus.
rhizostyle in cryptomonads, a part of the flagellar apparatus that consists of microtubules that extend from the basal bodies into the cell.
rib in diatoms, a linear siliceous thickening on the diatom valve; see costa. In synurophyte algae, a peripheral thickening on siliceous scales.
rimoportule(e) in diatoms, another term for labiate process.
16S rRNA gene encoding the small subunit (SSU) of the ribosomal RNA array of prokaryotes, mitochondria and chloroplasts; sequences are used in phylogenetic analyses.
18S rRNA gene encoding the small subunit (SSU) of the nuclear ribosomal RNA array of eukaryotes; sequences are used in phylogenetic analyses.
RuBisCO ribulose-1,5-bisphosphate carboxylase/oxygenase; the enzyme in photosynthetic organisms that catalyses ("fixes") the incorporation of CO_2 into carbohydrate.
saccate like a sac or balloon.
saccoderm desmid members in the family Mesotaeniaceae (older term); mostly simple unicells with no distinct semicells (see placoderm desmids) and with no pores in their walls; some become linked to form filaments.
scalariform conjugation literally, ladder-like; in members of the Zygnemataceae, a type of conjugation, where the conjugation tube forms between two parallel filaments.
scrobiculate surface pitted or furrowed; with many small depressions.
scytonemin yellow–brown pigment that is present in the sheaths of certain cyanobacteria; may protect against extremes of UV radiation.
seasonal succession regular seasonal changes in the species composition of phytoplankton over an annual cycle.
SEM scanning electron microscope; scanning electron microscopy.
semicell one-half of a cell in members of the Placoderm desmids.
semierect nonrigid thallus that is capable of bending and may exhibit branch reconfiguration to reduce drag in high current velocities.
Sensu lato Latin, in the broad sense.
Sensu stricto Latin, in the strict sense.
separation valve specialized valve with elongated spines that facilitates separation of daughter valves and serves to limit colony size in some centric diatoms.
septum (septa) in most algal cells, a cross wall. In diatoms, an invagination into the cell lumen in the valve plane, usually arising from the valvocopulae.
seta stiff hair or bristle (e.g., *Coleochaete*); also an elongate, external projection from the cell wall. In some forms it may be a hollow, narrow extension of the wall.
sheath mucilaginous covering over cells, colonies, or filaments (trichomes); may be firm or loose, narrow or broad.

silica deposition vesicle vesicle in which the siliceous diatom cell wall is deposited during cell division; also functions in some chrysophytes and synurophytes to form siliceous spores.
silica scale siliceous coverings produced by members of the Chrysophyceae and Synurophyceae.
siliceous composed primarily of silicon.
sinkhole lake lake formed from the dissolution of (mainly limestone) bedrock by surface and underground waters charged with CO_2.
sinus constricted region in a cell; the median constriction in desmids (class Desmidiaceae).
siphonous type of thallus with large multinucleate cells and few (or no) cross walls, except where reproductive structures occur.
smooth flagellum flagellum lacking hairs; see whiplash flagellum.
species richness ecological attribute of communities that describes the total number of species present in an area (or volume of water).
specific conductance ability of a water sample to conduct electricity; it is the reciprocal of resistance and is measured in units of micromhos per centimeter ($\mu\Omega$ cm^{-1}) or microsiemens per centimeter (μS cm^{-1}). Used as an estimate of total dissolved solids (TSS); the older term is conductivity.
spermatangium in red algae, a male gametangium that is typically a colorless, obovoid cell produced at the tips of vegetative branches; each produces a single male gamete (spermatium) that is released to fertilize the female gametangium (carpogonium). Some red algae form spermatangia on specialized stalk cells.
spermatium colorless male (nonflagellated) gamete in red algae.
spine thin projection, simple or branched, cylindrical or conical, ending in a point or flattened; common in many algal groups. In diatoms, appears as a granule or a tube.
spinule minute spinelike or thornlike protuberances.
sporangium (sporangia) structure that produces spores.
spore specialized (asexual) reproductive structure that germinates without fusion.
sporic meiosis meiosis that occurs during spore production.
sporophyte typically diploid, spore producing stage in organisms that exhibit alternation of generations.
sporopollenin resistant polymer produced in the vegetative walls (e.g., some Chlorococcales) or spore walls (e.g., Charales) of certain green algae that provides added strength and is thought to help prevent desiccation.
starch storage polysaccharide; a polymer composed of glucose units with α-1,4 and α-1,6 linkages. Composed of amylose and amylopectin.
statospore siliceous resting stage produced by several types of algae (e.g., Chrysophyceae and Synurophyceae).
status in diatoms, differing types of frustules regularly formed by some genera, notably *Aulacoseira*.
stauros in diatoms, where the central nodule (more heavily silicified) is expanded to the valve mantle to form a crosslike structure.
stellate literally star shaped; describes several features of algal

cells, such as chloroplasts or cell shape. In diatoms, a colony type caused by attachment of cells at the poles at one end of each cell (e.g., *Asterionella*).

stenotherm organism with a narrow range of temperature tolerance or preference.

sternum silica thickening along the axis of many pennate diatoms along which the valve structure is built during cell division.

stichidia inflated, multichambered structures at the tips of vegetative branches (red algae).

stigma see eyespot.

stigmata in some diatoms, an isolated perforation through the valve face, usually in the central area, where the external opening is rounded (or nearly so) and internal openings may be slitlike or otherwise modified.

stigmoids perforation through valve face the external opening of which is similar to puncta of the valve and the internal opening of which is slightly modified from the other puncta.

stipulode unicellular outgrowth of branchlets (bractlike) present in some members of the Charales.

stomatocyst see statospore.

stonewort common name for charophyte algae; refers to the common occurrence of calcification on the thallus in many species.

stratification property of lakes in which water forms two or more layers that differ in temperature and/or density, with warmer upper waters and cooler, deeper strata.

striae literally a line. In diatoms, an approximately linear array of puncta or areolae.

strutted process in diatoms, a tubular process found in some centric diatoms, usually associated with secretion of β-chitin. Internally a tube surrounded by two or more satellite pores; externally either a tube or a simple pore in the valve wall; also as fultoportula.

subaerial ecological habit exposed to the air, but usually attached to various substrata; also terrestrial.

subarctic regions that are nearly arctic; somewhat south of the Arctic Circle.

sulcus in dinoflagellates, a longitudinal groove in the ventral face of the cell (in hypotheca/hypocone) that may extend into epitheca/epicone; holds the longitudinal, whiplash, flagellum.

suture in dinoflagellates, the edge of a plate in thecate species, sometimes raised.

tabulation system used to classify the plate morphology in thecate dinoflagellates.

taxon (taxa) species or a group at any taxonomic level.

taxonomy formal system of classification.

tectonic lake lake formed by movements of the Earth's crust, usually continental rifting [e.g., Lake Tahoe (USA), Lake Baikal (Siberia), Lake Chapala (Mexico), and African rift lakes].

TEM transmission electron microscope; transmission electron microscopy.

terminal fissure continuation of the raphe on the valve exterior beyond the point where the raphe penetrates to the internal part of the valve.

terminal nodules in raphid pennate diatoms, the terminal nodule is a siliceous thickening distal to the internal raphe terminus.

tetrachotomous dividing into four equal or nearly equal braches or divisions.

tetrasporangium meiotic sporangium of certain red algae that forms four spores (tetraspores) that germinate into the gametophyte.

tetraspore haploid spores are formed by meiosis and germinate into the gametophytic stage in red algae.

tetrasporophyte diploid, typically free-living stage of some red algae. Forms four haploid spores (tetraspores) by meiosis. These spores germinate into the gametophyte stage.

thallus (thalli) general form or body of an alga.

theca in dinoflagellates, cellulose plate covering. In diatoms, another term for a diatom frustule. In flagellate green algae, a rigid wall.

thermocline region in lakes in which there is an abrupt change of at least $1°$ m^{-1}; the depth where the rate of temperature change with depth is greatest.

thermokarst lake lakes formed from the freezing and thawing action in ice and soil; common in the arctic.

thylakoid flattened membranous vesicles (or sacs) that form the photosynthetic membranes and contain photosynthetic pigments; arranged in various patterns in cyanobacterial cells or within plastids in eukaryotic algae.

tinsel flagellum flagellum with dense mastigonemes (bristle-like extensions); in some groups they occur in two rows of tripartite hairs (e.g., members of Chrysophyceae, Synurophyceae, and Phaeophyceae).

tomont in dinoflagellates, a cyst formed following feeding in parasitic forms.

travertine calcium carbonate (concretionary limestone) that is precipitated from alkaline water, often with the participation of algae; common in several species of cyanobacteria.

tremalith heterococcolith composed of pentagonal calcite elements fused into a ringlike or tubelike structure (e.g., in the freshwater species *Hymenomonas roseola*).

trichoblasts elongate hair cells, often terminal; occur in cyanobacteria, red and green algae.

trichocyst in dinoflagellates and raphidophytes, a proteinaceous ejectile organelle.

trichogyne receptive portion of the female gametangium (carpogonium) of many red algae.

trichome in cyanobacteria, a term referring to a filament without its sheath.

trichotomous dividing into three equal or nearly equal braches or divisions.

trophont feeding stage in certain dinoflagellates.

true branching branches formed from lateral divisions of the primary axis; in cyanobacteria, where divisions occur longitudinally and resultant cells grow roughly perpendicular to the original trichome. True branches are physiologically connected to the original trichome, unlike false branches.

tufts short radiating filaments without a common matrix.

turbidity degree of cloudiness or opacity of a water sample as influenced by suspended matter (sediment, silt, organisms), and which causes light to be scattered and absorbed; typically measured via nephelometry in standard units (NTUs) against formazin-based reference suspensions.

tychoplanktonic species that are predominately planktonic, but capable of prolonged survival on or in sediments.

ultrastructure morphological features observable with an electron microscope.

uniaxial (= uniseriate) thallus with the main axis composed of a single chain of cells (filament).

unilocular sporangium in brown algae, a reproductive structure (sporangium) that becomes enlarged (often spherical or club-shaped) but does not subdivide into compartments, and which later releases many spores, typically the result of meiosis.

valve top (epivalve) and bottom (hypovalve) elements of a diatom frustule.

valve face portion of the valve visible in valve view, that is, the valve oriented to the valvar plane.

valve mantle portion of the valve, differentiated by slope, that is apparent in the girdle view (oriented to the apical plane).

valvocopula girdle elements directly attached to valves in diatom frustules. Accessory structures, such as septa, usually arise from valvocopulae.

velum in diatoms, a thin plate or flap of silica that covers the openings of loculate areolae; may be perforated with smaller openings (in diatoms).

ventral front (in dinoflagellates, sulcal) side of a cell or thallus.

verruca(e) short or wartlike irregular projection(s) on the surface of cell walls or spores.

voigt faults irregularities in striae ornamentation bordering one side of the axial area, approximately equidistant from the valve center. These interruptions in the pattern of the striae are the result of valve ontogeny, representing the last areas of silica deposition.

volcanic lake lake basin formed via volcanic activity; include maars (volcanic explosion) and caldera (collapsed crater); examples include Crater Lake in Oregon and Lake Nicaragua.

whiplash flagellum smooth flagellum with no hairs.

whorl type of branching that has several lateral filaments arising at a common position in the main axis.

wing in diatoms, a complex type of keel that results from an extensive fold in the valve wall around the raphe. Extreme folding can result in a siliceous fusion (or partial fusion) of the internal valve surface below the raphe.

zigzag colony type (e.g., diatoms) caused by the attachments at both ends of cells; also a growth patterns in some filamentous green algae (e.g., *Stichococcus*).

zooplankton portion of the plankton community represented by mainly microscopic (up to a few millimeters) animals and protists (in freshwater: cladocerans, copepods, rotifers, heterotrophic flagellates, ciliates, and fish larvae).

zoospore motile (flagellated) spore ("naked swarmer") formed by vegetative cells or in specialized sporangia; typically the same ploidy as the parental cell.

zygospore thick-walled resting cyst formed after fertilization of an oogonium or fusion of gametes in other algal groups.

zygotic meiosis meiosis that occurs after maturation (cell division) or germination of the zygote.

Author Index

A
Abel, R. A., Olson, D. M., Dinerstein, E., Hurley, P. (2000), 12, 45
Aboal, M., Puig, M. A., Soler, G. (1996), 584, 588
Adamson, R. P., Sommerfield, M. R. (1978), 809, 826
Adler, R. W. (1995), 786, 797
Admiraal, D. M. J. (1993), 38, 45
Admiraal, W., Breebaart, L., Tubbing, D. M. J., Van Zanten, B., de Ruitjer van Steveninck, E. D., Bijerk, R. (1994), 38, 46
Admiraal, W., Mylius, S. D., de Ruitjer van Steveninck, E. D., Tubbing, D. M. J. (1993), 32, 38, 46
Agbeti, M. D. (1992), 791, 797
Agbeti, M. D., Kingston, J.-C., Smol, J. P., Watters, C. (1997), 587–588, 616, 631
Agbeti, M. D., Smol, J. P. (1995), 613, 616, 631
Ahlstrom, E. G. (1937), 477, 488, 498, 503
Ahmadjian, V. (1993), 45–46
Al-Dhaheri, R. S., Willey, R. L. (1996), 408, 416
Al-Thukair, A. A., Golubic, S., Rosen, G. (1994), 107, 110
Albertano, P. Capucci, E. (1997), 65, 110
Albertano, P., Kovácik, L. (1994), 135, 191
Albertano, P., Pinto, G., Pollio, A. (1994), 489, 503
Alcantara, I. I. (1997), 587–588
Allanson, B. R. (1973), 811, 826

Allegre, C. F., Jahn, T. L. (1943), 415–416
Allen, D. L., Jarrell, W. M. (1989), 823, 826
Allen, D. M., Northcote, D. H. (1975), 484, 504
Allen, J. D. (1995), 28, 34, 46
Allen, M. M. (1968), 81, 110
Allen, T. F. H. (1977), 623, 631
Allen, W. E. (1921), 36, 46
Alles, E., Nörpel-Schempp, M., Lange-Bertalot, H. (1991), 665, 666
Allorge, P., Manguin, E. (1941), 763, 772
Aloi, J. E. (1990), 24–25, 46, 780, 797
American Public Health Association (1992), 807, 809, 826
American Public Health Association (1998), 779–780, 783, 797
Amoros, C., Van Urk, G. (1989), 789, 797
Anagnostidis, K. (1961), 120–121, 127, 190, 191
Anagnostidis, K. (2001), 150, 192
Anagnostidis, K., Golubić, S. (1966), 132, 192
Anagnostidis, K., Komárek, J. (1988), 110, 117–120, 128, 135, 139, 144, 150, 155, 192
Anagnostidis, K., Komárek, J. (1990), 60, 110, 120, 192
Anagnostidis, K., Komárek, J. (2001), 132, 139, 144, 192
Anagnostidis, K., Pantazidou, A. (1991), 66, 110
Anagnostidis, K., Roussonoustakaki, M. (1985), 144, 145, 192
Andersen, R. A. (1982), 480, 504

Andersen, R. A. (1985), 528, 552
Andersen, R. A. (1986), 481, 501, 504
Andersen, R. A. (1987), 471, 474, 504, 523–524, 528, 552
Andersen, R. A. (1989), 513, 519
Andersen, R. A. (1992), 716, 751
Andersen, R. A. (1996), 410, 416
Andersen, R. A., Barr, D. J. S., Lynn, D. H., Melkonian, M., Moestrup, Ø., Sleigh, M. A. (1991), 402, 416
Andersen, R. A., Brett, R. W., Potter, D., Sexton, J. P. (1998), 424–425, 427, 465–466
Andersen, R. A., Brett, R. W., Potter, D., Sexton, J. P. (1998a), 471, 473, 504
Andersen, R. A., Mulkey, T. J. (1983), 474, 504, 528, 552
Andersen, R. A., Potter, D., Bidigare, R. R., Latasa, M., Rowan, K., O'Kelly, C. J. (1998b), 471, 473, 491, 504
Andersen, R. A., Potter, D., Daugberg, N., Bailey, J. C. (1998c), 503–504
Andersen, R. A., Saunders, G. W., Paskind, M. P., Sexton, J. P. (1993), 471, 504
Andersen, R. A., Van de Peer, Y., Potter, D., Sexton, J. P., Kawachi, M., LaJeunesse, T. (1999), 528, 534, 552
Andersen, R. A., van de Peer, Y., Potter, D., Sexton, J. P., Kawachi, M., LaJeunesse, T. (1999), 473–474, 504
Andersen, R. A., Wetherbee, R. (1992), 20, 46, 473, 504
Anderson, E. (1962), 734, 746, 751
Anderson, L. W. J. (1993), 809, 811, 826
Anderson, L. W. J., Dechoretz, N. (1984), 824, 826
Anderson, N. J., Battarbee, R. W. (1994), 789, 793, 797
Anderson, N. J., Ripply, B., Gibson, C. E. (1993), 787, 797
Anderson, N. J., Vos, P. (1992), 789, 797
Andresen, N. A., Stoermer, E. F. (1978), 666
Andrews, H. T. (1970), 483, 501, 504
Andreyeva, V. M. (1998), 307
Angeler, D. G. (2000), 385, 416
Angermeier, P. L., Katt, J. R. (1994), 776, 786, 797
Anita, N. J., Cheng, N. Y., Foyle, F. A., Percival, E. (1979), 736, 751
Anita, N. J., Kalley, J. P., McDonald, T., Bisalputra, T. (1973), 747, 751
Ankley, G. T., Mattson, V. R., Leonard, E. N., West, C. W., Bennett, J. L. (1993), 825, 826
Ann, S. S., Friedl, T., Hegewald, E. (1999), 256–257, 307
Anonymous (1975), 670, 682
Anonymous (1990), 811, 817, 826
Anthoni, U., Christophersen, C., Madsen, J. O., Wium-Andersen, S., Jacobsen, N. (1980), 811, 826
Apt, K. E., Collier, J. L., Grossman, A. R. (1995), 737, 751
Archibald, R. E. M., Barlow, D. J. (1983), 663, 666
Archibald, R. E. M., Schoeman, F. R. (1984), 663, 666
Archiblad, P. (1973), 291, 307
Archiblad, P., Bold, H. C. (1970), 255, 307
Archiblad, R. E. M. (1970), 682
Archiblad, R. E. M. (1972), 784, 797
Ariztia, E. V., Andersen, R. A., Sogin, M. L. (1991), 424, 466
Arnoldi, V. P. (1916), 247–248
Arruda, J. A., Fromm, C. H. (1989), 776, 797, 807, 827
Arvola, L., Ojala, A., Barbosa, F., Heaney, S. I. (1991), 738, 751
Arvola, L., Tulonen, T. (1998), 740, 751
Asaul, Z. I. (1975), 415–416
Ashley, J., Rushforth, S. R., Johansen, J. R. (1985), 605, 631
Asmund, B. (1968), 538, 552
Asmund, B., Hilliard, D. K. (1961), 552
Asmund, B., Kristiansen, J. (1986), 524, 531, 533–534, 551–552
Asmund, B., Takahashi, E. (1969), 552
Auer, M. T., Graham, J. M., Graham, l. E., Kranzfelder, J. A. (1983), 25–26, 46

Autenrieth, R., Bonner, J., Schreiber, L. (1991), 797
Aziz, A., Whitton, B. A. (1988), 39, 46

B
Babin, J., Prepas, E. E., Murphy, T. P., Serediak, M., Curtis, P. J., Zhang, Y., Chambers, P. A. (1994), 814, 827
Bachmann, H. (1908), 478, 504
Bachmann, H. (1921), 472, 504
Bachmann, M. D., Carlton, R. G., Burkholder, J. M., Wetzel, R. G. (1986), 45–46
Badgery, J. E., McQueen, D. J., Nichols, K. H., Schaap, P. R. (1994), 818, 827
Bahal, M. Talpasayi, E. R. S. (1972), 119, 192
Bahls, L. L. (1982), 650, 651
Bahls, L. L. (1993), 778, 784, 787, 791, 797
Bahls, L. L., Weber, E. E., Jarvie, J. O. (1984), 604, 631
Bahls, L. Weber, C. I. (1988), 674, 682
Bailey, D., Mazurak, A. P., Rosowski, J. R. (1973), 63, 110
Bailey, J. C., Andersen, R. A. (1998), 429, 466
Bailey, J. C., Bidigare, R. R., Christensen, S. J., Andersen, R. A. (1998), 443, 466, 473, 491, 503–504
Bain, M. B. (1993), 821, 827
Baker, A. L., Baker, K. K. (1979), 37–38, 46, 110
Baker, A. L., Baker, K. K. (1981), 65, 110
Baker, E. R., McLaughlin, J. J. A., Hutner, S., DeAngelis, B., Feingold, S., Frank, O., Baker, H. (1981), 406, 416
Balch, W. M., Holligan, P. M., Ackelson, S. G., Voss, K. J. (1991), 511, 519
Balech, E. (1974), 690, 710
Balech, E. (1980), 690, 710
Bando, T. (1988), 379
Barber, H. G., Haworth, E. Y. (1981), 20, 46, 599, 631, 651
Barber, L. (1980), 385, 416
Barbiero, P. R., McNair, C. M. (1996), 535, 552
Barbiero, R. P., Welch, E. B. (1992), 23, 46
Barbiero, R. P., Welch, H. (1992), 64, 110, 166, 192
Barbour, M. T., Plafkin, J. L., Bradley, B. P., Graves, C. G., Wisseman, R. W. (1992), 789, 797
Barbour, M.T., Gerritsen, J., Snyder, B. D., Stribling, J. B. (1999), 786, 789, 797
Barica, J. (1975), 806, 827
Barica, J. (1978), 806, 827
Barica, J., Kling, H., Gibson, J. (1980), 812, 827
Barker, G. L. A., Handley, B. A., Vacharapiyasophon, P., Stevens, J.R., Hayes, P. K. (2000), 177, 192
Barko, J. W., Adams, M. S., Clesceri, N. L. (1986), 816, 827
Barko, J. W., James, W. F., Taylor, W. D., McFarland, D. G. (1990), 814, 827
Barko, J. W., Smart, R. M. (1980), 816, 827
Barko, J. W., Smart, R. M. (1981), 816, 827
Barnett, B. S., Schneider, R. W. (1974), 810, 827
Barr, D. J. S., Hickman, C. J. (1967), 364, 379
Barsanti, L., Passarelli, V., Walne, P. L., Gualtieri, P. (1997), 402, 416
Bartlett, R., Willey, R. (1998), 408, 416
Bateman, L., Rushforth, S. R. (1984), 604, 631
Batko, A. (1970), 235, 248
Batko, A., Zakrys, B. (1995), 389, 402, 412, 415–416
Battarbee, R. W., Charles, D. F., Dixit, S. S., Renberg, I. (1999), 560, 588, 775, 781, 797
Batterbee, R. W. (1986), 605, 631
Battey, J. F. (1992), 687, 710
Bauer, D. L., Willie, D. W. (1990), 820, 827
Beaumont, P. (1975), 28, 46
Beaver, J. (1981), 561, 572, 588, 638, 651, 791, 797
Beech, P. L. (1990b), 533, 552

Beech, P. L., Wetherbee, R., (1990a), 528, 552
Beech, P. L., Wetherbee, R., Pickett-Heaps, J. D. (1990), 531, 533, 553
Beeton, A. M. (1969), 806, 827
Beijernick, M. W. (1980), 254, 307
Belanger, S. E., Barnum, J. B., Woltering, D. M., Wowling, J. W., Ventullo, R. M. Schermerhorn, S. D., Lowe, R. L. (1994), 794, 797
Belanger, S. E., Farris, J. L., Cherry, D. S. (1989), 825, 827
Belanger, S. E., Lowe, R. L., Rosen, B. G. (1985), 672, 682
Belay, A., Kato, T., Ota, Y. (1996), 121, 192
Belbin, L., McDonald, C. (1993), 790, 797
Belcher, J. H. (1959), 771–772
Belcher, J. H. (1966), 481, 484, 499, 504
Belcher, J. H. (1968), 484, 504
Belcher, J. H. (1969), 484, 504, 524, 528, 531, 553
Belcher, J. H., Swale, E. M. F. (1976), 502, 504
Belcher, J. H., Swale, E. M. F. (1997), 587, 588
Belcher, J. H., Swale, E. M. F., Heron, J. (1966), 583, 588
Bell, R. A. (1993), 45, 46
Benke, A. C., Hall, C. A. S., Hawkins, C. P., Lowe-McConnell, R. H., Stanford, J. A., Suberkrop, K., Ward, J. V. (1988), 29, 31–32, 46
Benndorf, J. (1989), 819, 827
Bennett, G. W. (1971), 809, 827
Bennett, H. D. (1969), 41–42, 46
Bennion, H., Juggins, S., Anderson, N. J. (1996), 794, 797
Benson, C. E., Rushforth, S. R. (1975), 454, 459, 461, 463, 466
Berg, C. O. (1963), 13, 16, 42, 46
Bergman, B. (2002), 141, 192
Bergquist, A. M., Carpenter, S. R. (1986), 19, 46
Berlyn, G. P., Micksche, J. P. (1976), 365, 379
Berninger, U. G., Caron, D. A., Sanders, R. W. (1992), 23, 46
Berninger, U. G., Finlay, B. J., Canter, H. M. (1986), 45–46
Bessey, C. E. (1899), 655, 666
Bethge, H. (1925), 569, 588
Bettoli, P. W., Maceina, M. J., Noble, R. L., Betsill, R. K. (1992), 810, 827
Bettoli, P. W., Maceina, M. J., Noble, R. L., Betsill, R. K. (1993), 810, 821, 827
Bhattacharya, D., Helmchen, T., Bibeau, C., Melkonian, M. (1995), 737, 751
Bhattacharya, D., Weber, K., An, S. S., Berning-Koch, W. (1998), 228, 248
Bianchi, T. S., Findlay, S., Dawson, R. (1993), 39, 46
Bicuda, C. (1984), 379
Bicudo, C. E. deM., De-Lamonica-Freire, E. M. (1993), 414, 416
Bicudo, C. E. deM., Wolowski, K. (1998), 388, 416
Bidigare, R. R., Ondrisek, M. E., Kennicutt, M. C., Iturriaga, R., Harvey, H. R., Hohman, R. W., Macko, S. A. (1993), 43–44, 46
Biebel, P. (1968), 335, 349
Biggs, B. F., Goring, D. G., Nikora, V. I. (1998), 35–36, 46
Biggs, B. J. F. (1990), 785, 798
Biggs, B. J. F. (1995), 776, 779, 785, 798
Biggs, B. J. F. (1996), 33–35, 46, 779, 782, 798
Biggs, B. J. F., Close, M. E. (1989), 779, 798
Biggs, B. J. F., Kilroy, C., Mulcock, C. M. (1989), 775, 779, 798
Biggs, B. J. F., Ondrisek, M. E., Kennicutt, M. C., Iturriaga, R., Harvey, H. R., Hohman, R. W., Macko, S. A. (1993), 46
Biggs, B. J. F., Price, G. M. (1987), 783, 798
Billen, G., Garnier, J., Hanset, P. (1994), 38, 46
Bird, D. F., Kalff, J. (1986), 473, 504
Bird, D. F., Kalff, J. (1987), 23, 46, 473, 504
Birks, H. J. B. (1995), 785, 790–793, 798
Birks, H. J. B. (1998), 604, 631, 790–793, 798
Birks, H. J. B., Line, J. M., Juggins, S., Stevenson, A. C., ter Braak, C. J. E. (1990), 794, 798

Bixby, R. J. (2001), 631
Blackburn, S. I., Jones, G. J. (1995), 177, 192
Blackwell, J. R., Cox, E. J., Gilmour, D. J. (1991), 255, 307
Blanck, H. (1985), 776, 798
Blank, R. J. (1992), 687, 710
Bliding, C. (1963), 349–350
Bliding, C. (1968), 349–350
Blindow, I. (1987), 24, 28, 46
Blindow, I. (1988), 811, 827
Blindow, I. (1992), 28, 46
Blinn, D. W. (1971), 42, 46
Blinn, D. W. (1993), 42–43, 46
Blinn, D. W., Fredericksen, A., Korte, V. (1980), 36, 46
Blinn, D. W., Prescott, G. W. (1976), 210, 221
Blinn, D. W., Shannon, J. P., Benenati, P. L., Wilson, K. P. (1998), 34, 46
Blinn, D. W., Stein, J. R. (1970), 333, 348, 350
Blinn, D. W., Truitt, R. T., Pickart, A. (1989), 25, 46
Bloem, J., Bär Gilissen, M.-J. B. (1988), 514, 519
Blum, J. L. (1951), 463, 466
Blum, J. L. (1956), 31–32, 34, 46
Blum, J. L. (1972), 465–466
Bodyl, A. (1996), 385, 416
Bolas, P. M., Lund, J. W. G. (1974), 810, 827
Bold, H. C. (1958), 349–350
Bold, H. C., MacEntee, F. J. (1973), 403–404, 416
Bold, H. C., Starr, R. C. (1953), 236, 248
Bold, H. C., Wynne, M. J. (1978), 255, 257–258, 307
Bold, H. C., Wynne, M. J. (1985), 225, 230, 232, 235, 238, 245, 248, 415–416, 423–424, 426, 429, 432, 466, 758, 772
Boltovskoy, A. (1975), 703, 710
Boltovskoy, A. (1976), 703, 710
Boltovskoy, A. (1989), 703, 710
Boney, A. D. (1980), 354, 363, 379
Boney, A. D. (1981), 353, 363, 379
Boney, A. D. (1982), 354, 363, 379
Booton, G. C., Floyd, G. L., Fuerst, P. A. (1998), 257, 307, 313, 350
Boraas, M. E., Estep, K. W., Johnson, P. W., Sieburth, J. McN. (1998), 740, 751
Borchardt, M. A. (1996), 782, 798
Borchardt, M. A., Hoffmann, J. P., Cook, P. W. (1994), 35, 46
Born, S. M., Wirth, T. L., Brick, E., Peterson, J. O. (1973), 811, 827
Bornet, E., Flahault, C. (1886-1888), 155, 158, 161, 164, 166, 171, 174, 177, 180–181, 184, 187, 189, 192
Boschker, H. T. S., Dekkers, E. M. J., Pel, R., Cappenberg, T. E. (1995), 39, 46
Bothe, H. (1982), 20, 45, 46, 192
Bott, T. L. (1996), 776, 798
Bott, T. L., Brock, J. T., Cushing, C. E., Gregory, S. V., King, D., Petersen, R. C. (1978), 785, 798
Boucher, P., Blinn, D. W., Johnson, D. B. (1984), 16, 46, 274, 308
Bouck, G. B. (1982), 402, 416
Bouck, G. B., Ngô, H. (1996), 399, 401, 416
Bouck, G. B., Rogalski, A., Valaitis, A. (1978), 402, 416
Bourne, C. E. M. (1992), 569, 585, 588
Bourrelly, P. (1957), 497, 503, 504
Bourrelly, P. (1963), 485, 504
Bourrelly, P. (1966), 110, 126, 192, 268–269, 271–273, 275, 278–279, 282–283, 285–287, 289–290, 292–294, 296–297, 299–300, 302, 304–308
Bourrelly, P. (1968), 707, 710
Bourrelly, P. (1968, 1981), 434, 447, 451, 453, 458–459, 465–466
Bourrelly, P. (1970), 119, 160, 192, 387–388, 411, 415, 416, 708, 710, 716, 741, 746, 751

Bourrelly, P. (1981), 503–504, 512, 516, 519, 523, 553, 757–759, 765, 770–772
Bourrelly, P. (1985), 12, 40, 46, 63, 100, 102, 104, 110
Bourrelly, P. (1988), 307–308, 313, 345, 349–350
Bourrelly, P., Manguin, E. (1952), 75, 102, 110, 234, 241–242, 248, 412, 415–416
Bovee, E. C. (1982), 404, 416
Bowling, L. C., Baker, P.D. (1966), 776, 783, 798
Boxhorn, J. E., Holden, D. A., Boraas, M. E. (1998), 501, 504
Boyd, C. E. (1982), 810, 827
Boyd, C. E., Prather, E. E., Parks, R. W. (1975), 806, 827
Boyer, C. S. (1926), 561, 588
Boyer, C. S. (1927a), 561–562, 588
Boyer, C. S. (1927b), 561–562, 588
Boyne, A. F. (1979), 734, 752
Braarud, T. (1954), 516, 519
Bradbury, J. P. (1975), 604, 631
Bradbury, J. P. (1987), 604, 631
Bradbury, J. P., Waddington, J. C. B. (1978), 605, 631
Bradbury, J. P., Winter, T. C. (1976), 604, 631
Bradbury, J. P.(1988), 561, 588
Braun, A. (1855), 757, 772
Brett, S. J., Wetherbee, R. (1986), 732, 741, 745, 752
Brezny, D., Mehta, I., Sharma, R. K. (1973), 810, 827
Brezonik, P. L. (1978), 779, 798
Briand, F., Trucco, R., Ramamoorthy, S. (1978), 785, 798
Bright, R. C. (1968), 604, 631
Brinson, M. M. (1993), 779, 798
Broadwater, S. T., Scott, J. L. (1994), 201, 221
Broady, P. (1984), 121, 189, 192
Broady, P., Given, D., Greenfield, L., Thompson, K. (1987), 189, 192
Brock, M. L., Wiegert, R. G., Brock, T. D. (1969), 41, 46
Brock, T. D. (1967), 41, 47
Brock, T. D. (1973), 40, 42, 47
Brock, T. D. (1985a), 12, 19, 47
Brock, T. D. (1985b), 41, 47
Brock, T. D. (1986), 41, 47
Brocklesby, J. (1851), 384, 416
Brönmark, C., Klosiewski, S. P., Stein, R. A. (1992), 25, 47
Brook, A. J. (1981), 354, 361, 363–364, 379
Brook, A. J. (1997), 371, 379
Brook, A. J. (1998), 371, 380
Brook, A. J., Williamson, D. B. (1988), 40, 47
Brooks, A. E. (1966), 243, 248
Brooks, A. E. (1972), 227, 248
Brooks, J. L., Deevey, E. S. (1963), 19, 40, 47
Browder, J. A., Gleason, P. J., Swift, D. R. (1994), 39, 47, 66, 110
Brown, A. W. A. (1978), 825, 827
Brown, C. W., Yoder, J. A. (1993), 511, 519
Brown, D. L, Weier, T. E. (1968), 200, 221
Brown, L. M., Smith, R. J., Shivers, R. R., Day, A. W. (1986), 518–519
Brown, R. M., Herth, W., Franke, W. W., Romanovicz, D. (1973), 484, 504
Brown, S.-D. (1973), 623, 631
Brown, S. W., Boyd, C. E. (1982), 807, 827
Brownlee, B. G., Painter, D. S., Boone, R. J. (1984), 809, 827
Brugam, R. B. (1980), 604, 631
Brugam, R. B., McKeever, K., Kolesa, L. (1998), 604, 631
Brugerolle, G., Bricheux, G. (1984), 531, 553
Bruno, S. F., McLaughlin, J. J. A. (1977), 700, 710
Brunskill, G. J., Ludlam, S. D. (1969), 16, 47
Brunthaler, J., Prowazek, S., von Wettstein, R. (1901), 583, 588
Bryant, D. A. (1994), 61, 110

Buchheim, M. A., Buccheim, J. A., Chapman, R. L. (1997), 234, 248
Buchheim, M. A., Chapman, R. L. (1997), 234, 249
Buchheim, M. A., Lemieux, C., Otis, C., Gutell, R. R., Chapman, R. L., Turmel, M. (1996), 234, 249
Buchheim, M. A., McAuley, M. A., Zimmer, E. A., Theriot, E. C., Chapman, R. L. (1994), 225, 232, 249
Bucka, H., Zurek, R. (1992), 699–700, 710
Buckland-Nicks, J., Reimchen, T. E. (1995), 697, 701, 710
Buckland-Nicks, J., Reimchen, T. E., Garbary, D. J. (1997), 686, 688, 698, 701, 709–710
Budde, H. (1927), 763, 765, 772
Büdel, B. (1985), 103, 110
Büdel, B., Henssen, A. (1983), 103, 110
Büdel, B., Lüttge, U., Stelzer, R., Huber, O., Medina, E. (1994), 110
Büdel, B., Wessels, D. C. J. (1991), 110
Buell, H. F. (1938), 82–83, 110
Buetow, D. E. (1968), 384, 416
Bujak, J. P., Davies, E. H. (1983), 698, 710
Bujak, J. P., Williams, G. L. (1981), 697, 710
Bukhtiyarova, L. N. (1995), 631
Bukhtiyarova, L., Round, F. E. (1996), 597–598, 602, 627, 631
Bunting, M. J., Duthie, H., Campbell, D., Warner, B., Turner, L (1997), 789, 798
Burkholder, J. M., Wetzel, R. G., Klomparens, K. I. (1990), 674, 683
Burkholder, J. M. (1996), 669, 682
Burkholder, J. M. (1998), 685, 710
Burkholder, J. M., Glasgow, H. B., Jr. (1997), 776, 798
Burkholder, J. M., Malin, M. A., Glasgow, H. B., Larsen, L. M., McIver, M. R., Shank, G. C., Deamer-Melis, N., Briley, D. S., Springer, J., Touchette, B. W., Hannon, E. K. (1997), 814, 827
Burkholder, J. M., Noga, E. J., Hobbs, C. H., Glasgow, H. B. (1992), 685, 710
Burkholder, J. M., Sheath, R. G. (1984), 364, 380
Burkholder, J. M., Wetzel, R. G. (1990), 25, 47
Burnham, J. C., Fraleigh, P. D. (1983), 819, 827
Burns, C. W. (1968), 808, 827
Burns, C. W., Stockner, J. G. (1991), 20, 47, 65, 110
Bursa, A. A. (1969), 688, 698, 706, 710
Bursa, A. A. (1970), 688, 698, 701, 706, 710
Butcher, R. W. (1933), 29, 47
Butcher, R. W. (1947), 775, 798
Butcher, R. W. (1967), 716, 741, 743, 748, 752
Button, K. S., Hostetter, H. P., Mair, D. M. (1977), 823–824, 827

C

Cain, J. R., Trainor, F. R. (1973), 785, 798
Calado, A. J., Moestrup, Ø. (1997), 701, 710
Calado, A. J., Rino, J. A. (1994), 524, 526, 528, 553
Caljon, A. (1983), 739, 752
Calkins, G. N. (1926), 385, 416
Callieri, C., Bertoni, R., Amicucci, E. Pinolini, M. A. (1995), 65, 110
Callow, P. (1973a), 26, 47
Callow, P. (1973b), 26, 47
Camburn, K. E. (1982), 434, 459, 466, 561, 588
Camburn, K. E., Kingston, J. C. (1986), 573, 588
Camburn, K. E., Kingston, J. C., Charles, D. F. (1984–1986), 561, 588, 598–599, 601, 628, 631
Camburn, K. E., Lowe, R. L., Stoneburner, D. L. (1978), 604, 631
Camburn, K. E., Warren, M. L., Jr. (1983), 212, 221
Cameron, R. E. (1963), 63, 110
Cameron, R. E., Morelli, F. A., Blank, G. B. (1965), 63, 110
Cameron, W. A., Larson, G. L. (1993), 15, 47
Campbell, C. E., Prepas, E. E. (1986), 43, 47
Campbell, E. O., Sarafis, V. (1972), 349–350
Campeau, S., Murkin, H. R., Titman, R. D. (1994), 39, 47

Canavan, R. W., Siver, P. A. (1995), 19, 47
Canfield, D. E. (1983), 812, 827
Canter, H. M. (1968), 739, 752
Canter, H. M., Jaworski, G. H. M. (1979), 819, 827
Canter-Lund, H., Lund, J. W. G. (1995), 45, 47, 171, 192, 314, 350, 387, 416, 604, 631
Caraco, N. F., Cole, J. J., Raymond, P. A., Strayer, D. L., Pace, M. L., Findlay, S. E. G., Fischer, D. T. (1997), 37, 47
Carefoot, J. R. (1966), 242, 249
Carignan, R., Kalff, J. (1980), 816, 827
Carlson, R. E. (1977), 775–776, 785, 798, 806, 812, 827
Carmichael, W. W. (1992), 64, 110, 169, 192
Carmichael, W. W. (1994), 786, 798, 807, 827
Carmichael, W. W. (1997), 64, 110, 121, 155, 192, 807, 827
Carmichael, W. W., Evans, W. R., Yin, Q. Q., Bell, P., Moczydlowski, E. (1977), 809, 827
Carmichael, W. W., Ewans, W. R., Yin, Q. Q., Bell, P., Mocauklowski, E. (1997), 121, 155, 192
Carmichael, W. W., Falconer, I. E. (1993), 807, 827
Carmichael, W. W., Jones, C. L. A., Mahmood, N. A., Thiess, W. C. (1985), 807, 818, 827
Carmona-Jiménez, J., Gold-Morgan, M. (1994), 181–182, 192
Caron, D. A., Lim, E. L., Dennett, M. R., Gast, R. J., Kosman, C., DeLong, E. F. (1999), 473, 504
Caron, D. A., Sanders, R. W., Lim, E. L., Marrase, C., Amaral, L. W., Whitney, S., Aoki, R. B., Porter, K. G. (1993), 23, 47
Carpenter, E. J. Carmichael, W. W. (1995), 806, 828
Carpenter, S. R., Kitchell, J. F. (1993), 776, 798
Carpenter, S. R., Kitchell, J. F., Hodgson, J. R. (1985), 818, 828
Carpenter, S. R., Kitchell, J. F., Hodgson, J. R., Cochran, P. A., Elser, J. J., Elser, M. M., Lodge, D. M., Kretchmer, D., He, X., von Ende, C. (1987), 818, 828
Carpenter, S. R., Lathrop, R. C., Munoz-del-Rio, A. (1993), 19, 47
Carper, G, L., Bachmann, R. W. (1984), 814, 828
Carr, N. G., Whitton, B. A. (1973), 59, 63, 68, 110, 120–121, 192
Carr, N. G., Whitton, B. A. (1982), 59, 111, 121, 192
Carrick, H. J., Lowe, R. L., Rotenberry, J. T. (1998), 605, 631
Carson, J. L., Brown, R. M. (1978), 45, 47
Carty, S. (1986), 693, 698, 702, 710
Carty, S. (1989), 689, 702, 710
Carty, S. (1993), 698, 701–702, 710
Carty, S., Cox, E. R. (1985), 695, 701, 707, 710
Carty, S., Cox, E. R. (1986), 708, 710
Carty, S., Fazio, V. W., III (1997), 701, 710
Cassani, J. R., Caton, W. E. (1983), 820, 828
Cassie, V., Dempsey, G. P. (1980), 587, 588
Castenholz, R. W. (1960), 42, 47
Castenholz, R. W. (1969a), 66, 82, 111, 192
Castenholz, R. W. (1969b), 82, 111, 192
Castenholz, R. W. (1970), 82, 111
Castenholz, R. W. (1976), 66, 111, 129, 192
Castenholz, R. W. (1977), 111
Castenholz, R. W. (1982), 24, 47
Castenholz, R. W. (1992), 60, 62, 111, 117–118, 192
Castenholz, R. W. (2001), 59–60, 111, 192
Castenholz, R. W., Waterbury, J. B. (1989), 59, 111
Castenholz, R. W., Wickstrom, C. E. (1975), 41, 47, 66, 111, 189, 192
Catling, P. M., McKay, S. M. (1980), 347, 350
Cattaneo, A. (1983), 25, 47
Cattaneo, A. (1996), 35, 47
Cattaneo, A., Amireault, M. C. (1992), 25, 47
Cattaneo, A., Galanti, G., Gentinetta, S. Romo, S. (1998), 24–25, 47
Cattaneo, A., Kerimian, T., Roberge, M., Marty, J. (1997), 36, 47
Cavalier-Smith, T. (1981), 383, 416
Cavalier-Smith, T. (1986), 523, 553, 715, 737, 742, 752
Cavalier-Smith, T. (1991), 387, 416
Cavalier-Smith, T. (1993), 383, 385, 387, 416, 742, 749, 752
Cavalier-Smith, T. (1998), 383–385, 402, 416
Cavalier-Smith, T. (1999), 385, 387, 402, 416
Cavalier-Smith, T., Chao, E. E. (1996), 423–424, 466
Cavalier-Smith, T., Couch, J. A., Thorssteinsen, K. E., Gilson, P., Deane, J. A., Hill, D. R. A., McFadden, G. I. (1996), 737, 752
Cedeno-Maldondo, A., Swader, J. A. (1974), 823, 828
CEIMATE (2000), 797, 798
Cepak, V. (1993), 61, 111
Chadefaud, M. (1950), 759, 763, 767, 772
Chandler, D. E. (1984), 734, 752
Chapman, A. D., Pfiester, L. A. (1995), 694, 710
Chapman, R. L., Waters, D. A. (1992), 311, 315, 332, 343–344, 348, 350
Charles, D. F. (1985), 604, 631
Charles, D. F., Binford, M. W., Furlong, E. T., Hites, R. A., Mitchell, M. J., Norton, S. A., Oldfield, F., Paterson, M. J., Smol, J. P., Utala, A. J., White, J. R., Whitehead, D. R., Wise, R. J. (1990), 792, 798
Charles, D. F., Smol, J. P. (1988), 537, 553, 787, 798
Charles, D. F., Smol, J. P. (1994), 792–793, 798
Chesnick, J. M., Cox, E. R. (1987), 694, 707, 710
Chesnick, J. M., Cox, E. R. (1989), 694, 707, 710
Chesnick, J. M., Morden, C. W., Schmieg, A. M. (1996), 686, 710
Chessman, B., Growns, I., Currey, J., Plunkett-Cole, N. (1999), 788, 798
Chiavelli, D. A., Mills, E. L., Threlkeld, S. (1993), 408, 416
Chilton, E. W., Lowe, R. L., Schurr, K. M. (1986), 798
Chisholm, S. W., Stross, R. G. (1976a), 406, 417
Chisholm, S. W., Stross, R. G. (1976b), 406, 417
Chodat, R. (1922), 482, 504
Choi, H.-G., Kraft, G. T., Saunders, G. W. (2000), 218, 222
Cholnoky, B. J. (1965), 661–662, 666
Cholnoky, B. J. (1968), 583, 588, 623, 631, 672, 683
Chorus, I., Bartram, J. (1999), 64, 92, 111, 121, 169, 192
Christen. see von Christen
Christensen, C. L., Reimer, C. W. (1968), 676, 682–683
Christensen, T. (1962), 471, 504, 511, 519
Christensen, T. (1964), 471, 504
Christensen, T. (1968), 465–466
Christensen, T. (1969), 465–466
Christensen, T. (1978), 757, 772
Christensen, T. (1987a), 465–466
Christensen, T. (1987b), 465–466
Christensen (1987), 379, 380
Christie, C. E., Smol, J. P. (1993), 604, 631
Christoffersen, K., Riemann, B., Hansen, L. R., Klysner, A., Sorensen, H. B. (1990), 20, 47
Cienkowski, C. (1870), 485, 504
Clark, R. L., Rushforth, S. L. (1977), 604, 631
Clarke, K. B. (1994), 584, 588
Clarke, R. A., Stanley, C. D., MacLeod, B. W., McNeal, B. L. (1997), 807, 828
Clasen, J., Bernhardt, H. (1982), 535, 553
Claus, G. (1962), 44, 47
Clay, B. L., Kurgens, P., Lee, R. E. (1999), 716, 732, 734, 737, 742–743, 746, 752
Clay, B. L., Kurgens, P. (1999a), 716, 732–733, 735, 737, 739, 742, 749, 752
Clay, B. L., Kurgens, P. (1999b), 716, 729, 732, 735, 737, 739, 742, 749, 752
Clay, B. L., Kurgens, P. (1999c), 716, 732, 737, 742, 749, 752
Clayton, C., Häusler, T., Blattner, J. (1995), 386, 417
Clements, F. E., Shantz, H. L. (1909), 86, 111

Cleve, P. T. (1894), 637, 646–647, 651, 659, 665, 666
Cleve, P. T. (1895), 637, 651
Cleve, P. T., Grunow, A. (1880), 596, 631
Cleve-Euler, A. (1911a), 569, 588
Cleve-Euler, A. (1911b), 569, 588
Cleve-Euler, A. (1951), 562, 588
Cleve-Euler, A. (1952), 562, 588
Cleve-Euler, A. (1953), 599, 613, 628, 631
Cleve-Euler, A. (1953a), 562, 588
Cleve-Euler, A. (1953b), 562, 588
Cleve-Euler, A. (1955), 562, 588
Cmiech, H. A., Leedale, G. F., Reynolds, C. S. (1986), 120, 192
Cochran-Stafira, D. L., Andersen, R. A. (1984), 40, 47
Codd, G. A. (1995), 64, 111, 121, 169, 192
Coesel, P. F. M. (1983), 364, 380
Coesel, P. F. M. (1991), 364, 380
Coesel, P. F. M. (1993), 354, 374, 376, 379–380
Coesel, P. F. M. (1996), 364, 380
Coesel, P. F. M., Delfos, A. (1986), 373, 380
Cohen-Bazire, G., Bryant, D. A. (1982), 61, 111
Cole, G. A. (1963), 15–16, 40–41, 47
Cole, G. A. (1994), 12, 15–16, 40, 42, 47
Cole, J. J., Caraco, N. F., Peierls, B. (1992), 32, 37, 47
Coleman, A. W. (1959), 226–227, 241, 249
Coleman, A. W. (1977), 226–227, 241, 249
Coleman, A. W. (1996), 227, 249
Coleman, A. W. (1999), 242, 249
Coleman, A. W., Goff, L. J. (1991), 473, 504
Coleman, A. W., Suarez, A., Goff, L. (1994), 227, 249
Colle, D. E., Shireman, J. V. (1980), 810, 828
Colletti, P. J., Blinn, D. W. Pickart, A., Wagner, V. T. (1987), 35, 47
Collins, F. S. (1905), 463, 466
Collins, F. S., Holden, I., Setchell, W. A. (1898), 770, 772
Collins, G. B., Kalinsky, R. G. (1977), 604, 631
Colt, L. C. (1974), 429, 433, 461, 463, 466
Colt, L. C. (1985), 433, 463, 466
Colt, L. C., Jr., Saumure, R. A., Jr., Baskinger, S. (1995), 12, 47, 339, 350
Comas, A. (1996), 303, 307–308
Compère, P. (1982), 584, 588
Compère, P. (1996), 376, 379–380
Compère, P. (2001), 631–631
Conforti, V. (1991), 405, 417
Conforti, V. (1999), 415, 417
Conforti, V., Joo, G.-J. (1994), 416, 417
Conforti, V., Perez, M. del C. (2000), 405, 417
Conforti, V., Ruiz, L. (2000), 414, 417
Conforti, V., Walne, P. L., Dunlap, J. R. (1994), 414–417
Conley, D. J., Kilham, S. S., Theriot, E. (1989), 598, 632
Conley, D. J., Kilham, S. S., Theriot, E, C, (1989), 567, 588
Conley, D. J., Schelske, C. L., Stoermer, E. F. (1993), 567, 588
Conn, H. W., Edmondson, C. H. (1918), 385, 417
Conrad, W. (1926), 247, 249
Conrad, W. (1928), 516, 519
Conrad, W. (1934), 415, 417
Conrad, W., van Meel, L. (1952), 415–417
Conyers, D. L., Cooke, G. D. (1982), 817, 828
Cook, R. B., Kreis, R. G., Jr., Kingston, J. C., Camburn, K. E., Norton, S. A., Mitchell, M. J., Fry, B., Shane, L. C. K. (1990), 617, 632
Cooke, G. D., Kennedy, R. H. (1989), 817, 828
Cooke, G. D., Welch, E. B., Martin, A. B., Fulmer, D. G., Hyde, J. B., Schrieve, G. D. (1993a), 814, 828
Cooke, G. D., Welch, E. B., Peterson, S. A., Newroth, P. R. (1993b), 806, 811, 813–818, 820, 828

Cooke, M. C. (1883), 235, 249
Cooksey, K. E., Cooksey, B. (1978), 39, 47
Cooper, S. R. (1999), 789, 798
Copeland, J. J. (1936), 60, 66–67, 72–73, 75–82, 88, 94–95, 104–105, 107, 111, 118, 121, 126, 128, 131–132, 134–137, 139–140, 144, 148, 154–155, 158, 186–189, 192
Corbus, F. G. (1982), 810, 828
Corliss, J. O. (1975), 384, 417
Corliss, J. O. (1984), 383, 385, 417
Corliss, J. O. (1989), 384, 391, 417
Corliss, J. O. (1990), 388, 417
Corliss, J. O. (1991a), 388, 417
Corliss, J. O. (1991b), 385, 417
Corliss, J. O. (1995), 385, 388, 417
Corliss, J. O., Esser, S. C. (1974), 404, 417
Corpe, W. A., Jensen, T. E. (1992), 65, 79, 89, 111
Coste, M., Bosca, C., Dauta, A. (1991), 776, 782, 787, 798
Coté, R. (1983), 786, 798
Cottingham, K. L., Carpenter, S. R., St. Amand, A. L. (1998), 487, 504
Couch, J. N. (1932), 432, 466
Couté, A. (1983), 480, 504
Couté, A., Sarthou, C. (1990), 202, 222
Couté, A., Tell, G. (1981), 365, 380
Couté, A., Thérézien, Y. (1994), 416–417
Covach, A. P. (1976), 47
Cowles, R. P., Brambel, C. E. (1936), 428, 463, 465–466
Cox, E. J. (1987), 650, 651
Cox, E. J. (1990), 490, 504
Cox, E. J. (1996), 560, 562, 588, 638, 651, 655, 666
Cox, E. R., Bold, H. C. (1966), 337, 349–350
Cox, E. R., Hightower, J. (1972), 45, 48
Cracraft, J. (1989), 569, 588
Craige, J. S. (1974), 757, 772
Craige, J. S. (1990), 200, 222
Cramer, M., Myers, J. (1952), 406, 417
Cranwell, P. A., Robinson, N., Eglinton, G. (1985), 699, 710
Crawford, R. M. (1988), 583, 588
Crawford, R. M., Gardner, C. (1997), 607, 632
Crawford, S. A. (1979), 811, 828
Croasdale, H. (1973), 60, 80, 111, 433, 459, 461, 466
Croasdale, H. T., Bicudo, C., Prescott, G. W. (1983), 374–380
Cronberg, G. (1980), 531, 553
Cronberg, G. (1982), 537, 553
Cronberg, G. (1986), 531, 533, 553
Cronberg, G. (1989), 485, 504, 533, 535–536, 538, 552–553
Cronberg, G. (1996), 485, 504, 526, 538, 553
Cronberg, G., Gelin, C., Larsson, K. (1975), 537, 553
Cronberg, G., Komárek, J. (1994), 67, 89, 111, 135, 192
Cronberg, G., Komárek, J. (2002), 171, 192
Cronberg, G., Kristiansen, J. (1980), 536, 553
Cronberg, G., Lindmark, G., Bjork, S. (1988), 428, 466
Cronk, J. K., Mitsch, W. J. (1994), 39, 48
Croome, R. L. (1988), 537, 553
Croome, R. L., Tyler, P. A. (1985), 535, 537, 553
Croome, R. L., Tyler, P. A. (1987), 709, 711
Crumpton, W. G. (1987), 570, 588
Crumpton, W. G., Wetzel, R. G. (1981), 570, 588
Crumpton, W. G., Wetzel, R. G. (1982), 48
Crumpton, W. G., Wilson, S. E., Hall, R. I., Smol, J. P. (1995), 588
Crutchfield, J. R., Jr., Schiller, D. H., Herlong, D. D., Mallen, M. A. (1992), 820, 828
Cullen, J. J., Ciotiti, A. M., Lewis, M. R. (1997), 783, 798
Cumming, B. F., Davey, K., Smol, J. P., Birks, H. J. (1994), 793, 798

Cumming, B. F., Smol, J. P., Birks, H. J. B. (1992a), 526, 535–537, 541, 553, 781, 787, 792, 798
Cumming, B. F., Smol, J. P., Kingston, J. C., Charles, D. F., Birks, H. J. B., Camburn, K. E., Dixit, S. S., Uutala, A. J., Selle, A. R. (1992), 604, 632
Cumming, B. F., Smol, J. P., Kingston, J. C., Charles, D. F., BIrks, H. J. B., Camburn, K. E., Dixit, S. S., Uutala, A. J., Selle, A. R. (1992b), 793, 798
Cumming, B. F., Smol, J. P., Kingston, J. C., Charles, D. F., Birks, H. J. B., Camburn, K. E., Dixit, S. S., Uutala, A. J., Selle, A. R. (1992b), 553
Cumming, B. F., Wilson, S. E., Hall, R. I., Smol, J. P. (1995), 604, 630, 632
Cumming, B. F., Wilson, S. E., Hall, R. I., Smol, P. J. (1995), 638–639, 651
Cummins, B. F., Wilson, S. E., Hall, R. I., Smol, J. P. (1995), 560–561, 588
Czarnecki, D. B. (1979), 26, 48
Czarnecki, D. B. (1995), 662, 666
Czarnecki, D. B., Blinn, D. W. (1978), 584, 589, 604, 628, 632, 650, 651
Czarnecki, D. B., Blinn, D. W. (1979), 665, 666
Czarnecki, D. B., Reinke, D. C. (1981), 650, 651

D
Dahm, C. N., Grimm, N. B., Marmonier, P. Vallett, H. M., Vervier, P. (1998), 29, 48
Daily, F. K. (1952), 314, 350
Daily, W. A. (1942), 60, 80, 82, 95, 110–111
Daily, W. A. (1943), 139, 192
Daily, W. A. (1946), 60, 111
Damann, K. E. (1945), 461, 466
Danforth, W. F., Ginsburg, W. (1980), 605, 632
Daugbjerg, N., Andersen, R. A. (1997), 424, 429, 466
Daugbjerg, N., Andersen, R. A. (1997a), 473, 504
Daugbjerg, N., Andersen, R. A. (1997b), 473, 504
Davey, M. C. (1987), 565, 589
Davis, B. M. (1904), 432, 466
Davis, C. C. (1964), 20–21, 48, 806, 828
Davis, J. S., Rands, D. G., Hein, M. K. (1989), 605, 632
Davis, L. W., Hoffmann, J. P. Cook, P. W. (1990a), 44, 48
Davis, L. W., Hoffmann, J. P. Cook, P. W. (1990b), 48
Davis, R. B., Anderson, D. S., Norton, S. A., Ford, J., Sweets, P. R., Kahl, J. S. (1994), 617, 632
Davis-Colley, R. J., Vant, W. N. (1988), 779, 799
Dawson, N. S., Dunlap, J. R., Walne, P. L. (1988), 412, 417
Dawson, N. S., Walne, P. L. (1991), 413, 417
Dawson, N. S., Walne, P. L. (1994), 385, 391, 417
Dawson, P. A. (1972), 655, 659, 666
Dawson, P. A. (1973a), 655, 664, 666
Dawson, P. A. (1973b), 655, 664, 666
Dawson, P. A. (1973c), 655, 659, 666
Dawson, P. A. (1974), 655, 666
Dayner, D. M., Johansen, J. R. (1991), 44, 48
de Noyelles, F., Knoechel, R., Reinke, D., Treanor, D., Altenhoifen, C. (1980), 22, 48
de Ruyter van Steveninck, E. D., Admirall, W., Breebaart, L., Tubbingm G, M. J., van Zanten, B. (1992), 37, 48
De Toni, G. (1936), 80, 111
Deane, J. A., Hill, D. R. A., McFadden, G. I. (1998), 734, 752
Dearing, J. A., Håkansson, H., Liedberg-Jönsson, B., Persson, A., Skansjö, S., Windholm, D., El-Dahousy, F. (1987), 799
Deason, T., Silva, P. C., Watanabe, S., Floyd, G. L. (1991), 256, 308
Deason, T. (1959), 303, 308
Deason, T. R. (1969), 326, 350

Deason, T. R. (1971a), 432, 466
Deason, T. R. (1971b), 432, 466
Deason, T. R., Bold, H. (1960), 326, 349–350
Deflandre, G. (1926), 416–417
Deflandre, G. (1930), 414, 417
Della Greca, M., Monaco, P., Pollio, A., Previtera, L. (1992), 821, 828
DeLuca, P., Gambardella, R., Merola, A. (1979), 206, 208, 222
DeLuca, P., Morretti, A. (1983), 41, 48
Demayo, A., Taylor, M. C., Taylor, K. W. (1982), 825, 828
DeMelo, R., France, R., McQueen, D. J. (1992), 819, 828
DeMott, W. R., Zhang, Q.-X., Carmichael, W. W. (1991), 808, 828
Denicola, D. M. (1986), 782, 799
Deniseger, J., Austen, A., Roch, M., Clark, M. J. R. (1986), 587, 589
DeNoyelles, F., O'Brien, W. J. (1978), 537, 553
Denys, L. (1991), 791, 799
Descy, J.-P. (1979), 787, 799
Descy, J.-P., Gosselain, V. (1994), 32, 48
Descy, J.-P., Gosselain, V., Evrard, F. (1994), 32, 37, 48
Descy, J.-P., Servais, P. Smitz, J. S., Billen, G., Everbecq, E. (1987), 37–38, 48
Descy, J.-P., Willems, C. (1991), 587, 589
Desikachary, T. V. (1959), 59, 64, 92, 111, 119–121, 161, 164, 174, 176, 181, 192
Desjardins, P. R., Olsen, G. B. (1983), 819, 828
deVecchi, L., Grilli-Caiola, M. (1986), 118, 192
DeYoe, H. R., Lowe, R. L., Marks, J. C. (1992), 670, 673–674, 683
Diaz, M. M., Lorenzo, L. E. (1990), 513, 519
Diaz, M. M., Pedrozo, F. L., Temporetti, P. F. (1998), 587, 589
DiCastri, E., Younez, T. (1994), 60, 111
Dillard, G. E., Crider, S. B. (1970), 496, 504
Dillard, G. E. (1967), 454, 463, 466
Dillard, G. E. (1970), 483, 495, 504
Dillard, G. E. (1989), 228, 230–231, 234–236, 238, 241–242, 244–245, 247, 249, 254–255, 270, 272, 280, 284, 288, 291, 298, 307–308, 313, 348, 350
Dillard, G. E. (1999), 253, 258, 307–308, 414, 417, 463, 466
Dillard, G. E. (2000), 388, 400, 410–417
Dillard, G. E., Moore, S. P., Garret, L. S. (1976), 433, 449, 461, 466
Dillon, P. J., Nicholls, K. H., Locke, B. A., de Grosbois, E., Yan, N. D. (1988), 812, 828
Dillon, P. J., Rigler, F. H. (1974), 812, 828
Dimitrov, M. (1984), 820, 828
Dionigi, C. P., Champagne, E. T. (1995), 825, 828
Diwald, K. (1938), 694, 711
Dixit, A. S., Dixit, S. S., Smol, J. P. (1992a), 775, 781, 799
Dixit, S. S., Cumming, B. F., Kingston, J. C., Smol, J. P., Birks, H. J. B., Uutala, A. J., Charles, D. F., Camburn, K. (1993), 792, 799
Dixit, S. S., Dixit, A. S., Evans, R. D. (1988), 536–537, 553
Dixit, S. S., Dixit, A. S., Smol, J. P. (1989), 535, 553
Dixit, S. S., Smol, J. P. (1994), 775, 781, 799
Dixit, S. S., Smol, J. P., Anderson, D. S., Davies, R. B. (1990), 536, 553
Dixit, S. S., Smol, J. P., Charles, D. F., Hughes, R. M., Paulsen, S. G., Collins, G. B. (1999), 604, 632, 775, 781, 792, 799
Dixit, S. S., Smol, J. P., Kingston, J. C., Charles, D. F. (1992), 604, 632
Dixit, S. S., Smol, J. P., Kingston, J. C., Charles, D. F. (1992b), 793, 799
D'Lacoste, V., Ganesan, E. K. (1987), 202, 222
Dodd, J. D., Stoermer, E. F. (1962), 623, 632
Dodd, J. J. (1987), 561, 589, 604–605, 632, 682–683
Dodd, W. K., Gudder, D. A. (1992), 783, 799
Doddema, H., van der Veer, J. (1983), 490, 504

Dodds, W. K. (1989), 36, 48, 180, 192
Dodds, W. K., Gudder, D. A. (1992), 35, 48, 340, 350, 810, 812, 828
Dodds, W. K., Gudder, D. A., Mollenhauer, D. (1995), 27, 48, 180, 192
Dodds, W. K., Jones, J. R., Welch, E. B. (1998), 31, 38, 48, 776, 782, 785, 799
Dodge, J. D. (1965), 686, 711
Dodge, J. D. (1966), 686, 711
Dodge, J. D. (1969), 690, 711, 741, 747, 752
Dodge, J. D. (1972), 690, 711
Dodge, J. D. (1975), 690, 711
Dodge, J. D. (1983), 697, 711
Dodge, J. D., Bibby, B. T. (1973), 709, 711
Dodge, J. D., Crawford, R. M. (1970), 690, 711
Dodge, J. D., Hermes, H. B. (1981), 690, 711
Doemel, W. N., Brock, T. D. (1971), 197, 206, 208, 222
Doers, M. P., Parker, D. L. (1988), 111
Dokkulil, M., Mayer, J. (1996), 174, 192
Domozych, D. S. (1989), 235, 249
Domozych, D. S., Nimmons, T. T. (1992), 235, 249
Dop, A. J. (1978), 481, 496–497, 504
Dop, A. J. (1979), 757–758, 760, 763, 766–767, 771–772
Dop, A. J. (1980), 484, 500–502, 505
Dop, A. J., Vroman, M. (1976), 763, 772
Douglas, B. (1958), 36, 48, 767, 772
Douglas, M. S. V., Smol, J. P. (1993), 604, 632
Douglas, M. S. V., Smol, J. P. (1995), 24, 26, 48, 490, 505
Douglas, M. S. V., Smol, J. P., Blake, W., Jr. (1994), 789, 799
Douglas, S. E., Murphy, C. A., Spencer, D. F., Gray, M. W. (1991), 735–737, 752
Dow, C. S., Swoboda, U. K. (2000), 12, 32, 48, 121, 192
Downes, B. J., Lake, P. S., Schreiber, E. S. G., Glaister, A. (1998), 36, 48
Doyle, R. D., Smart, R. M. (1998), 821, 828
Dragos, N., Péterfo, L. S., Popescu, C. (1997), 400, 417
Drew, K. M. (1935), 201, 222
Drews, G. (1959), 119, 192
Drews, G. (1973), 118, 192
Drews, G., Prauser, H., Uhlmann, O. (1961), 79, 111
Drews, G., Weckesser, J. (1982), 61, 111
Drouet, F. (1933), 463, 466
Drouet, F. (1934), 154–155, 193
Drouet, F. (1936a), 60, 111
Drouet, F. (1936b), 60, 111
Drouet, F. (1937), 131–132, 136, 193
Drouet, F. (1938), 60, 111, 148, 193
Drouet, F. (1942), 60, 78, 82, 94, 111, 131, 140, 145, 155, 193
Drouet, F. (1943a), 144, 149–151, 193
Drouet, F. (1943b), 149, 193
Drouet, F. (1954), 461, 463, 466
Drouet, F. (1968), 118, 193
Drouet, F. (1973), 118, 193
Drouet, F. (1981a), 118, 193
Drouet, F. (1981b), 118, 193
Drouet, F., Cohen, A. (1935), 428, 463, 466
Drouet, F., Daily, W. A. (1948), 111
Drouet, F., Daily, W. A. (1952), 111
Drouet, F., Daily, W. A. (1956), 80–81, 88, 96, 101, 111, 118, 193
Drum, R. W., Pankratz, H. S. (1964), 565, 589
Drum, R. W., Pankratz, S. (1965), 670, 673, 683
Duff, K. E., Smol, J. P. (1995), 536–537, 553
Duff, K. E., Zeeb, B. A., Smol, J. P. (1995), 485, 505, 533, 553
Dujardin, F. (1841), 413, 417
Dunlap, J. R., Walne, P. L. (1985), 417

Dunlap, J. R., Walne, P. L. (1987), 406, 417
Dunlap, J. R., Walne, P. L., Bentley. J. (1983), 414, 417
Dunlap, J. R., Walne, P. L., Kivic, P. A. (1986), 403, 414, 417
Dunlap, J.R., Walne, P. L., Preisig, H. R. (1987), 485, 505
Durrell, L. W., Norton, C. (1960), 433–434, 454, 459, 466
Dürrschmidt, M. (1980), 534, 536, 553
Dürrschmidt, M. (1982), 526, 536, 538, 553
Dürrschmidt, M., Croome, R. (1985), 536, 538, 553
Duthie, H. C. (1989), 607, 632
Duthie, H. C., Ostrofsky, M. L. (1975), 411, 417
Duthie, H. C., Ostrofsky, M. L. (1978), 491, 505
Duthie, H. C., Ostrofsky, M. L., Brown, D. J. (1976), 429, 451, 466
Duthie, H. C., Socha, R. (1976), 72, 80–82. 87–88, 92, 95, 111, 118, 132, 135, 141, 155, 158, 161, 164, 166, 169, 171, 174, 177, 180, 184, 193, 230, 232, 234–235, 238–239, 241–242, 245, 247, 249, 411–413, 415, 417, 425, 428–429, 432, 434, 447, 449, 451, 453–454, 463, 466, 491, 505
Duthie, H. C., Sreenivasa, M. R. (1972), 628, 632
Duthie, H., Socha, R. (1976), 698, 711
Dwarte, D., Vesk, M. (1982), 735, 752
Dwarte, D., Vesk, M. (1983), 735, 741, 752

E
Eardley-Wilmot, V. L. (1928), 561, 589
Eaton, G. L. (1980), 690, 697, 711
Eaton, J. W., Moss, B. (1996), 27, 48
Echevarria, F., Rodriguez, J. (1994), 700, 711
Eddy, S. (1934), 36, 48
Edgren, R. A., Egren, M. K., Tiffany, L. H. (1953), 314, 350
Edlund, M. B. (1992), 569, 589
Edlund, M. B. (1998), 568, 589
Edlund, M. B., Stoermer, E. F. (1993), 567, 572, 587, 589
Edlund, M. B., Stoermer, E. F. (1997), 560, 568, 589, 638, 652
Edlund, M. B., Stoermer, E. F. (1999), 661, 663, 666
Edlund, M. B., Stoermer, E. F., Taylor, C. M. (1996), 567, 573–574, 589
Edlund, M. B., Taylor, C. M. Schelske, C. L., Stoermer, E. F. (2000), 587, 589
Edmondson, W. T. (1959), 639, 652
Edmondson, W. T. (1963), 12, 16, 48
Edmondson, W. T. (1965), 739, 752
Edmondson, W. T. (1969), 405, 417
Edmondson, W. T. (1977), 23, 48
Edmondson, W. T. (1994), 813, 828
Edmondson, W. T., Lehman, J. T. (1981), 23, 48, 813, 828
Edvardsen, B, Eikrem, W., Green, J. C., Andersen, R. A., Moon-Van der Staay, S. Y., Medlin, L. K. (2000), 511, 515, 519
Edwards, P. (1975), 342, 349–350
Effler, S. W., Litten, S., Field, S. D., Tong-Ngork, T., Hale, F., Meyer, M., Quirk, M. (1980), 822, 825, 828
Effler, S. W., Owens, E. M. (1996), 42, 48
Eguchi, M., Oketa, T., Miyamoto, N. Maeda, H., Kawai, A. (1996), 65, 111
Ehara, M., Hayashi-Ishimaru, Y., Inagaki, Y., Ohama, Y. (1997), 447, 459, 466
Ehrenberg, C. G. (1830), 384, 417
Ehrenberg, C. G. (1831), 384, 411, 417
Ehrenberg, C. G. (1833), 384, 417
Ehrenberg, C. G. (1838), 384, 418, 471, 505
Ehrenberg, C. G. (1841), 595, 632
Ehrenberg, C. G. (1843), 561, 589, 649–650, 652
Ehrenberg, C. G. (1854), 561, 589
Ehrenberg, C. G. (1870), 561, 589
Ekenstam, D., Bozniak, E. G., Sommerfeld, M. R. (1996), 763–764, 772

Elakovich, S. D., Wooten, J. W. (1989), 821, 828
Elder, J. F., Horne, A. J. (1978), 822, 824, 828
Elenkrn, A. A. (1936, 1938, 1949), 110, 111
Ellis-Adams, A. C. (1983), 478–479, 505
Elmore, C. J. (1992), 561, 589
Eloranta, P. (1989), 505, 534–535, 537, 553
Eloranta, P. (1995), 536, 553
Elser, J. J., MacKay, N. A. (1989), 818, 828
Elser, J. J., Marzolf, E. R., Goldman, C. R. (1990), 812–813, 828
Elster, J., Svoboda, J., Komárek, J., Marvan, P. (1997), 158, 193
Eminson, D., Moss, B. (1980), 24–25, 48
Engstrom, D. R., Swaitn, E. B., Kingston, J. C. (1985), 605, 632
Entwisle, T. J. (1987), 465–466
Entwisle, T. J. (1988a), 465–466
Entwisle, T. J. (1988b), 465–466
Entwisle, T. J. (1989), 33, 48
Entwisle, T. J., Sonneman, J. A., Lewis, S. H. (1998), 226–227, 249
Entwisle, T. J., Sonneman, J., Lewis, S. J. (1997), 768, 772
Entwistle, T. J., Andersen, R. A. (1990), 505
Environmental Protection Agency (1990), 806, 828
Environmental Protection Agency (1998), 814, 828
Erata, M., Chihara, M. (1989), 741, 746, 752
Erikson, R., Pum, M., Vammen, K. Cruz, A., Ruiz, M., Zamora, H. (1997), 18, 48
Eschbach, S., Wolters, J., Sitte, P. (1991), 752
Eskew, D. L., Ting, I. O. (1978), 63, 111
Esser, S. C., Valkenburg, S. D. (1977), 485, 505
Etnier, C., Guterstam, B. (1997), 813, 828
Ettl, H. (1978), 426, 429, 431–432, 434, 440, 442, 444–445, 448, 452, 454–460, 462, 464–466
Ettl, H. (1980), 235, 249
Ettl, H. (1983), 225–228, 230–232, 234–236, 238–239, 241, 243, 245, 247–249
Ettl, H., Gärtner, G. (1995), 254, 268–269, 271–273, 275, 277–279, 283, 287, 290, 292, 297, 300, 304, 306–308, 322, 348, 350
Ettl, H., Moestrup, O. (1980), 739, 752
Ettl, H., Popovský, J. (1986), 391, 418
Evans, J. C., Arts, M. T., Robarts, R. D. (1996), 42–43, 48
Evans, J. C., Prepas, E. E. (1996), 42–43, 48
Everitt, D. T., Burkholder, J. M. (1991), 203, 216, 222
Evitt, R. W. (1985), 687, 711

F
Faber, Jr., W. W., Preisig, H. R. (1994), 511, 519
Fabry, S., Köhler, A., Coleman, A. W. (1999), 227, 243, 249
Fahnenstiel, G, L., Glime, J. M. (1983), 567–568, 589
Fahnenstiel, G. L., Sicko-Goad, L., Scavia, D., Stoermer, E. F. (1986), 65, 111
Fairchild, E. C., Wilson, D. L. (1967), 63, 111
Fairchild, G. W., Lowe, R. L. (1984), 673–674, 683
Fairchild, G. W., Lowe, R. L., Richardson, W. B. (1985), 605, 632, 674, 683, 786, 799
Fairchild, G. W., Sherman, J. W., Acker, F. W. (1989), 26, 48
Falconer, I. R. (1998), 121, 193
Fallon, R. D., Brock, T. D. (1981), 111
Faridi, M. A. F. (1962), 339, 348, 350
Farmer, J. N. (1980), 386–387, 418
Fassett, N. C. (1957), 811, 828
Faust, M. A. (1974), 732, 741, 752
Faust, M. A., Gantt, E. (1973), 735, 752
Fay, P. (1983), 61, 111, 121, 193
Fay, P., Van Baalen, C. (1987), 63, 111, 117, 121, 193
Feldmann, J. (1958), 111
Felip, M., Sattler, B., Psenner, R., Catalan, J. (1995), 44, 48, 487, 505
Feminella, J. W., Hawkins, C. P. (1995), 35, 48

Feminella, J. W., Power, M. E., Resh, V. H. (1989), 35, 48
Fensome, R. A., MacRae, R. A., Moldowan, J. M., Taylor, F. J. R., Williams. G. J. (1996), 687, 711
Fensome, R. A., Taylor, F. J. R., Norris, G., Sarjeant, W. A. S., Wharton, D. L., Williams. G. L. (1993), 687, 697, 711
Ferguson, A. J. D., Thompson, J. M., Reynolds, C. S. (1982), 739, 752
Fields, S. D., Rhodes, R. G. (1991), 686, 690, 711
Findlay, D. L. (1978), 537, 553
Findlay, D. L., Kasian, S. E. M. (1987), 513, 519
Findlay, D. L., Kasian, S. E. M. (1996), 488, 505
Findlay, D. L., Kling, H. J. (1979), 147, 193
Findlay, S., Likens, G. E., Hedin, L., Fisher, S. G., McDowell, W. H. (1997), 31, 48
Fitzgerald, G. P. (1959), 809, 828
Fitzgerald, G. P. (1971), 822, 828
Fjerdingstad, E. (1950), 775, 799
Flechtner, V. R., Boyer, S. L. Johansen, J. R., DeNoble, M. L. (2002), 191, 193
Fleming, R. F. (1989), 254, 308
Fling, E. M. (1939), 463, 466
Flint, L. H. (1955), 34, 48
Flint, L. J. (1970), 197, 222
Floener, L. Bothe, H. (1980), 670, 673, 681, 683
Flores, E., Wolk, C. P. (1986), 821, 828
Florin, M.-B. (1970), 604, 632
Flower, R. J. (1989), 601, 617, 632
Flower, R. J., Battarbee, R. W. (1985a), 601, 632
Flower, R. J., Battarbee, R. W. (1985b), 600, 632
Flower, R. J., Häkasson, H. (1994), 578, 589
Flower, R. J., Jones, V. J. (1989), 627, 632
Flower, R. J., Jones, V. J., Round, F. E. (1996), 598–599, 632
Flower, R. J., Ozornina, S. P., Kuzmina, A., Round, F. E. (1998), 585, 589
Floyd, G. K., Watanage, S., Deacon, T. R. (1993), 257, 308
Foerster, J. W., Schlichting, H. E. J. (1965), 605, 632
Foged, N. (1953), 561, 589
Foged, N. (1955), 561, 589
Foged, N. (1971), 604, 628, 632, 649, 652
Foged, N. (1973), 561, 589
Foged, N. (1981), 561, 589, 649, 652
Fogg, G. E. (1949), 119, 193
Fogg, G. E. (1986), 65, 111
Fogg, G. E., Stewart, W. D. P., Fay, P., Walsby, A. E. (1973), 59, 63, 111, 117, 120–121, 193
Forest, H. S. (1954), 698, 708, 711
Forest, H. S. (1956), 348, 350
Forest, H. S., Weston, C. R. (1966), 63, 111
Forsberg, C. (1965), 811, 828
Fott, B. (1949), 245, 249
Fott, B. (1959), 485, 505
Fott, B. (1963), 238, 248–249
Fott, B. (1967), 245, 249
Fott, B. (1968), 465–466
Fott, B. (1971), 427, 466
Fourreau, J. (1868), 245, 249
Fourtanier, E., Kociolek, J. P. (1999), 613, 627, 632
Fowler, M. C., Robson, T. O. (1978), 820–821, 829
Francke, J. A., Coesel, P. F. M. (1985), 752
Francke, J. A., Hillebrand, H. (1980), 822, 829
Frank, P. A. (1972), 825, 829
Frémy, P. (1930a), 64, 111, 145, 165, 174, 181–183, 193
Frémy, P. (1930b), 111, 145, 165, 174, 181–183, 193
Frémy, P. (1933), 66, 111
Frémy, P. (1949), 72, 111

French, F. W., Hargraves, P. E. (1980), 567, 589
Fresnel, J. (1994), 512, 519
Frey, D. G. (1963), 12, 48
Frey, L. C., Stoermer, E.F. (1990), 701, 711
Friedl, T. (1995), 257, 308
Friedl, T. (1997), 225, 228, 231, 249, 257, 308
Friedl, T. (1998), 313, 350
Friedl, T., Zeltner, C. (1994), 257, 308
Friedmann, E. I. (1955), 186–187, 193
Friedmann, E. I. (1971), 60, 66, 111
Friedmann, E. I. (1979), 186–187, 193
Friedmann, E. I. (1980), 111
Friedmann, E. I., Ocampo, R. (1985), 66, 112
Friedmann, E. I., Ocampo-Friedmann, R. (1984), 66, 112
Fritsch, F. E. (1922), 45, 49
Fritsch, F. E. (1929), 33, 49, 205, 222, 763, 765, 772
Fritsch, F. E. (1935), 356, 361–364, 380
Fritsch, F. W. (1945), 385, 411, 414–415, 418
Fritz, S. C. (1990), 585, 589, 776, 787, 799
Fritz, S. C., Battarbee, R. W. (1986), 585, 589
Fritz, S. C., Juggins, S., Battarbee, R. W., Engstrom, D. R. (1991), 585, 589, 799
Fritz, S. C., Juggins, S., Batterbee, R. W. (1993), 42–43, 49
Fritz, S. C., Kingston, J. C., Engstrom, D. R. (1993), 604, 632
Frodge, J. D., Thomas, G. L., Pauley, G. B. (1991), 815, 829
Fromentel, E. de (1874), 502, 505
Frost, T. M., Graham, L. E., Elias, J. E., Haase, M. J., Kretchmer, D. W., Kranzfelder, J. A. (1997), 45, 49
Frost, T. M., Williamson, C. E. (1980), 45, 49
Fuller, R. L., Roelofs, J. L., Fry, T. J. (1986), 31, 35, 49
Fulton, A. B. (1978), 226, 249
Fulton, R. S., III (1988), 819, 829
Fulton, R. S., Paerl, H. W. (1987), 64, 112, 808, 829

G
Gabor, T. S., Murkin, H. R., Stainton, M. P., Boughen, J. A., Titman, R. D. (1994), 39, 49
Gaines, G., Elbrachter, M. (1987), 701, 711
Galat, D. L. Verdin, J. P., Sims, L. L. (1990), 177, 193
Gálvez, J. A., Niell, F. X., Lucena, J. (1988), 700, 711
Gantt, E. (1971), 732, 741, 747, 752
Gantt, E. (1979), 735, 752
Gantt, E. (1980), 735, 752
Gantt, E., Edwards, M. R., Provasoli, L. (1971), 735, 752
Gantt, E., Scott, J. Lipschultz, C. (1986), 200, 222
Garbary, D. J., Hansen, G. I., Scagel, R. F. (1980), 201, 222
Garcia-Pichel, F., Belnap, J. (1996), 63, 112
Garcia-Pichel, F., Castenholz, R. W. (1991), 61, 112
Garcia Reina, G. (1997), 68, 112
Gardner, N. L. (1906), 112
Gardner, N. L. (1918), 104–105, 107–108, 110, 112
Gardner, N. L. (1927), 60, 66–67, 72, 75, 78, 80, 82, 89, 92–93, 95–97, 99, 102–104, 106–108, 110, 112, 118, 126, 130, 132, 134, 138–139, 143, 145, 148, 150–153, 157, 159–162, 165, 176, 178, 181, 183, 193
Garric, R. K. (1965), 81, 112
Garrison, P.J., Knauer, D. R. (1984), 814, 829
Gartner, G. (1992), 311, 350
Gartner, G., Ingolic, E. (1989), 315, 348, 350
Garwood, P. E. (1982), 314, 350
Gaufin, A. R., Prescott, G. W., Tibbs, J. F. (1976), 434, 443, 447, 449, 459, 466
Gaul, U., Geissler, U., Henderson, M., Mahoney, R., Reimer, C. W. (1993), 597, 628, 632, 638, 652, 655, 666
Gayral, P., Haas, C., Lepailleur, H. (1972), 485, 505

Geibler, U. (1983), 759, 768, 772
Geissler, U. (1982), 569, 589
Geitler, L. (1925), 60, 77, 91, 101, 112
Geitler, L. (1927), 664, 666
Geitler, L. (1928), 765, 768, 772
Geitler, L. (1932), 59–61, 65, 79, 82, 91, 93, 99–100, 106, 108, 112, 118, 120, 127, 143–145, 147, 151, 154, 158–160, 163, 166, 174–178, 181, 185, 188. 193, 197, 206, 222, 655, 666, 763–765, 772
Geitler, L. (1942), 60–61, 80, 112, 118, 158, 193
Geitler, L. (1951a), 665, 666
Geitler, L. (1951b), 665, 666
Geitler, L. (1951c), 665, 666
Geitler, L. (1956), 664, 666
Geitler, L. (1960), 98, 112, 118, 158, 193
Geitler, L. (1967), 664, 666
Geitler, L. (1973), 560, 589, 638, 652
Geitler, L. (1973a), 655, 666
Geitler, L. (1973b), 665, 666
Geitler, L. (1973c), 665, 666
Geitler, L. (1975), 662, 667
Geitler, L. (1977), 623, 627, 632, 670, 673, 676, 683
Geitler, L. (1979), 623, 627, 632
Geitler, L. (1980), 623, 627, 632
Geitler, L. (1981), 659, 667
Geitler, L. (1982), 118, 193
Geitler, L., Mack, B. (1953), 664, 667
Geitler, L., Ruttner, F. (1935), 67, 82, 104–105, 112
Gektidis, M., Golubic, S. (1996), 107, 112
Gelin, F., Boogers, I., Noordelos, A. A. M., Sinninghe Damsté, J. S., Riegman, R., de Leeuw, J. W. (1997), 254, 308
Gelin, F., Boogers, I., Noordelos, A. A. M., Sinninghe Damsté, J. S., Riegman, R., de Leeuw, J. W. (1997), 312, 350
Genkal, S. I., Håkansson. (1990), 568, 589
Genkal, S. I., Kiss, K. T. (1998), 568, 583–584, 589
Genkal, S. I., Makarova, A., Goncharov, A. A. (1998), 583, 589
Gensemer, R. W. (1991), 786, 799
Gensemer, R. W. (1991a), 607, 632
Gensemer, R. W. (1991b), 607, 632
Genter, R. B. (1996), 782, 799
Gerber, S., Häder, D.-P. (1993), 407, 418
Gerloff, J. (1967), 771–772
Germain, H. (1981), 628, 632
Gerrath, J. F. (1970), 354, 374, 380
Gerrath, J. F. (1993), 363–365, 380
Gersonde, R., Harwood, D. M. (1990), 570, 589
Ghosh, M. Gaur, J. P. (1990), 785, 799
Gibbs, G. W. (1973), 811, 829
Gibbs, S. P. (1978), 385, 402, 418
Gibbs, S. P. (1981), 385, 402, 418, 472, 505
Gibson, C. E. (1975), 129, 193
Gibson, C. E., Smith, R. V. (1982), 64, 112
Gibson, K. N., Smol, J. P., Ford, J. (1987), 535, 553
Gibson, M. T., Welch, I. M., Barrett, P. R. F., Ridge, I. (1990), 822, 829
Gibson, M. T., Whitton, B. A. (1987), 204, 222, 758, 772
Gillot, M. (1990), 715, 734, 736, 752
Gillot, M., Gibbs, S. P. (1980), 734–735, 752
Gillot, M., Gibbs, S. P. (1983), 734–735, 752
Glazer, A. N., Appell, G. S. (1977), 735, 737, 752
Glazer, A. N., Wedemeyer, G. J. (1995), 737, 752
Gledhill, M., Nimmo, M., Hill, S. J., Brown, M. T. (1997), 823, 829
Glew, J. R. (1991), 792, 799
Glew, J. R. (1998), 780, 799
Glime, J. M., Wetzel, R. G., Kennedy, B. J. (1982), 40, 49

Goff, L. J., Stein, J. R. (1978), 45, 49, 291, 308
Gojdics, M. (1953), 412, 415, 418
Gold-Morgan, M., Montejano, G., Komárek, J. (1994), 60, 106–107, 112, 151, 193
Gold-Morgan, M., Montejano, G., Komárek, J. (1996), 98, 100–102, 112
Golden, J. W., Robinson, S. J., Haselkorn, R. (1985), 119, 193
Goldman, C. R. (1988), 19, 49
Goldsborough, L. G., Robinson, G. G. C. (1996), 38–40, 49, 66, 112, 669, 683
Goldstein, A. K., Manzi, J. J. (1976), 425, 427, 433, 449, 463, 466
Goldstein, M. (1964), 226, 239, 241, 247–249
Golecki, J. R., Drews, G. (1974), 119, 193
Golubic, S. (1965), 60, 92, 112
Golubic, S. (1967a), 60, 63, 65, 92, 112, 121, 166, 193
Golubic, S. (1967b), 60, 92, 112, 121, 166, 193
Golubic, S. (1980), 60, 66, 112, 120–121, 193
Golubic, S., Focke, J. W. (1978), 193
Golubic, S., Friedmann, E. I., Schneider, J. (1981), 66–67, 112, 121, 193
Golubic, S., Hernandez-Mariné, M., Hoffmann, L. (1996), 119, 193
Golubic, S., Perkins, R. D., Lukas, K. J. (1975), 67, 107, 112, 121, 193
Golubic, S., Yun, Z., Campbell, S. E. (1985), 62, 112
Gomont, M. (1882), 142–144, 148–149, 193
Gomont, M. (1892), 59, 112
Gomont, M. (1896), 757, 772
Good, R. H., Chapman, R. L. (1978), 312, 350
Goodrich, S. G. (1859), 385, 418
Goodwin, T. W. (1974), 20, 49
Goodwin (1974), 757, 772
Gorham, E., Eisenreich, S. J., Ford, J., Santelmann, M. V. (1985), 40, 49
Gorham, P. R., Carmichael, W. W. (1988), 92, 112, 806, 829
Gorham, P. R., McLachlan, J., Hammer, U. T., Kim, W. K. (1964), 64, 112
Gosse, P. H. (1859), 383, 385, 418
Gosse, P. H. (1896), 385, 418
Gosselain, V., Descy, J.-P., Everbecq, E. (1994), 38, 49
Gosselain, V., Viroux, L., Descy, J.-P. (1998), 38, 49
Gough, S. B., Woelkerling, W. J. (1976), 24, 49
Graham, J. M., Kranzfelder, J. A., Auer, M. T. (1985), 26, 49
Graham, L. E., Graham, J. M., Wujek, D. E. (1993), 526, 528, 531, 533–534, 537, 539, 551, 553
Graham, L. E., Wilcox, L. W. (2000), 1, 5, 9, 254–255, 308, 312–313, 350, 384, 387, 399, 418, 423–424, 427, 432, 465, 467, 757–759, 772
Graham, M. D., Vinebrooke, R. D. (1998), 24, 26, 49
Graham, T. P., McCoy, J. J. (1974), 406, 418
Grain, J., Mignot, J. P., Puytorac, P. (1988), 734, 752
Green, J. (1953), 411, 418
Green, J. C., Leadbeater, B. S. C. (1994), 512, 519
Green, J. C., Piennar, R. N. (1977), 512, 519
Green, R. H. (1979), 778, 789–790, 799
Greenberg, A. E. (1964), 36–37, 49
Greene, J. C., Miller, W. E., Shiroyama, T., Soltero, R. A., Putnam, K. (1976), 785, 799
Greenwood, A. D. (1959), 432, 467
Greenwood, A. D., Griffiths, H. B., Santore, U. S. (1977), 741, 752
Greenwood, J., Clason, T., Lowe, R. L., Belanger, S. E. (1999), 674, 683
Gretz, M. R., Sommerfeld, M. R., Wujek, D. E. (1979), 552–553
Gretz, M. R., Sommerfield, M. R., Athey, P. V. Aronson, J. M. (1991), 200, 222
Gretz, M. R., Wujek, D. E. Sommerfeld, M. R. (1983), 552–553

Grim, J. N., Staehelin, L. A. (1984), 732, 734–735, 741, 746, 752
Gromov, B. V., Mamkayeva, K. A., Bobina, V. D. (1988), 489, 505
Gross, E. M., Meyer, H., Schilling, G. (1996), 821, 829
Gross, F., Zeuthen, E. (1948), 567, 589
Grote, M. (1977), 364, 380
Gruendling, G. K. (1971), 27, 49
Grzebyk, D., Sako, Y., Berland, B. (1998), 709, 711
Gualtieri, P. (1993), 402, 418
Gualtieri, P., Pelosi, P., Passarelli, V., Barsanti, L. (1992), 402, 418
Guglielmi, G., Cohen-Bazire, G. (1982a), 118, 193
Guglielmi, G., Cohen-Bazire, G. (1982b), 118, 193
Guillard, R. R. L. (1975), 739, 752
Guillard, R. R. L., Lorenzen, C. J. (1972), 424, 467
Gulati, R. D. (1990), 818, 829
Gunn, G. J., Raferty, A. G., Rafferty, G. C., Cockburn, N., Edwards, C., Beatie, K. A., Codd, G. A. (1992), 809, 829
Guo, M., Harrison, P. J., Taylor, F. R. J. (1996), 512, 519
Gurnee, J. (1994), 44, 49
Gutowski, A. (1989), 536, 538, 553
Gutowski, A. (1996), 538–539, 553
Gutowski, A. (1997), 536, 553

H
Haberyan, K. A., Umana, G., Collado, C., Horn, S. P. (1995), 13, 49, 234–235, 249, 412, 415, 418, 698, 711
Hadas, O., Malinsky-Rushansky, N., Pinkas, R., Cappenberg, T. E. (1998), 20, 49
Häder, D.-P. (1974), 119, 193
Häder, D.-P. (1987), 407, 418
Haffen, L. M., McCann, M. T. (1975), 808, 829
Håkansson, H., Jones, V. J. (1994), 561, 589
Håkansson, H., Kling, H. (1989), 565, 589
Håkansson, H., Kling, H. (1990), 561, 565, 581, 585, 589
Håkansson, H., Kling, H. (1994), 561, 565, 590
Håkansson, H., Mahood, A. (1993), 586, 590
Håkansson, H., Stoermer, E. F. (1984), 567, 569, 586, 590
Håkansson, H., Stoermer, E. F. (1987), 585–586, 590
Halfen, L. N. (1979), 119, 193
Halingse, M. W., Phlips, E. J. (1996), 823, 829
Hall, R. I., Smol, J. P. (1992), 799
Hall, R. I., Smol, J. P. (1996), 781, 799
Hallegraeff, G. M. (1993), 799
Hallegraeff, G. M., Anderson, D. M., Cembella, A. D. (1995), 685, 711
Hällfors, G., Hällfors, S. (1988), 486, 505
Hällfors, G., Munsterhjelm, R. (1982), 98, 112
Hallick, R. B., Hong, L., Drager, R. G., Favreau, M. R. Monfort, A., Orsat, B., Spielmann, A., Stutz, E. (1993), 387, 418
Hambrook, J. A., Sheath, R. G. (1987), 205, 222
Hambrook, J. A., Sheath, R. G. (1991), 202–204, 222, 763, 772
Hamel, G. (1931–1939), 759, 772
Hamilton, P. B., Douglas, M. S. V., Fritz, S. C., Pienitz, R., Smol, J. P., Wolfe, A. P. (1994), 561, 590, 638, 649, 652
Hamilton, P. B., Edlund, S. A. (1994), 45, 49
Hamilton, P. B., McNeely, R., Poulin, M. (1996), 638, 649, 652
Hamilton, P. B., Poulin, M., Charles, D. F., Angell, M. (1992), 604, 632
Hamilton, P. B., Poulin, M., Taylor, M. C. (1990), 638, 649, 652
Hamilton, P. B., Poulin, M., Walker, D. (1995), 638, 649, 652
Hammer, U. T. (1981), 43, 49
Hammer, U. T. (1986), 42, 49
Hammer, U. T., Shamess, J., Haynes, R. C. (1983), 42, 49
Hanagata, N. (1998), 257, 308
Hann, B. J. (1991), 39, 49
Hanninen, O., Ruuskanen, J., Oksanen, J. (1993), 45, 49

Hansen, G. W., Oliver, F. E., Otto, N. E. (1984), 829
Hansen, L. R., Kristiansen, J., Rasmussen, J. V. (1994), 512–513, 519
Hansen, P. (1995), 526, 554
Hansen, P. (1996), 485, 505
Hansen, P. S., Phlips, E. J., Aldridge, F. J. (1997), 814, 829
Hanson, J. M., Leggett, W. C. (1982), 799
Hanson, M. J., Stefan, H. G. (1984), 810, 825, 829
Hansson, L.-A., Rudstam, L. G., Johnson, T. B., Soranno, P., Allen, Y. (1994), 23, 49
Happey, C., Moss, B. (1967), 490, 505
Happey-Wood, C. M. (1976), 490, 505
Happey-Wood, C. M. (1988), 258, 308, 364, 380
Harding, J. P. C., Whitton, B. A. (1981), 205, 222
Hardwick, G. G., Blinn, D. W., Usher, H. D. (1992), 34, 49
Hargeby, A., Andersson, G., Blindow, I., Johansson, S. (1994), 811, 821, 829
Hargreaves, J. W., Lloyd, E. J. H., Whitton, B. A. (1975), 41–42, 49, 363, 380
Hargreaves, J. W., Whitton, B. A. (1976), 407, 418
Hargreaves, J. W., Whitton, B. A. (1976a), 42, 49
Hargreaves, J. W., Whitton, B. A. (1976b), 42, 49
Harper, D. (1992), 782, 799
Harris, D. O. (1964), 433, 463, 467
Harris, D. O., Starr, R. C. (1969), 227, 241, 248–249
Harris, E. H. (1989), 226, 249
Harris, K. (1953), 533, 554
Harris, K., Bradley, D. E. (1957), 533, 538, 554
Harris, K., Bradley, D. E. (1960), 533, 554
Harrison, S. S. C., Hildrew, A. G. (1998), 26, 49
Hartmann, H., Steinberg, C. (1989), 526, 536, 554
Harvey, R. S., Patrick, R. (1968), 799
Harvey, W. H. (1836), 312, 350
Harwood, D. M. (1999), 561, 590
Harwood, D. M., Gersonde, R. (1990), 570, 590
Harwood, D. M., Nikolaev, V. A. (1995), 559, 570, 590
Haselkorn, R. (1986), 119, 193
Hasle, G. R., Evensen, D. L. (1975), 565, 585, 590
Hasle, G. R., Evensen, D. L. (1976), 565, 585, 590
Hasle, G. R., Lange, C. B. (1989), 587, 590
Hasle, G. R. (1973), 655–656, 665, 667
Hasle, G. R. (1977), 572, 575, 590
Hasle, G. R. (1978), 587, 590
Haüber, M. M., Müller, S. B., Maier, U.-G. (1994), 385, 402, 418
Haughey, A. (1970), 406, 418
Haupt, W. (1972), 354, 380
Haupt, W., Schönbohm, W. (1970), 354, 380
Hauschild, C. A., McMurrer, H. J. G., Pick, F. R. (1991), 65, 112
Hauser, W. J., Legner, E.F., Medved, R. A., Platt, S. (1976), 820, 829
Hausmann, K. (1978), 403, 418, 686, 690, 711
Hausmann, K., Walz, B. (1979), 752
Havens, K. E. (1994), 822, 825, 829
Havens, K. E., Bull, L. A., Warren, G. L., Crisman, T. L., Phlips, E. J., Smith, J. P. (1996), 24, 49
Havens, K. E., East, T. (1997), 813, 829
Havens, K. E., III (1989), 428, 467
Havens, K. E., James, R. T. (1997), 814, 829
Hawes, I., Schwarz, A.-M. (1996), 24, 28, 49
Hawkes, H. A. (1975), 31, 49
Hawkins, P. R., Griffiths, D. J. (1987), 806, 822, 824, 829
Hawley, G. R. W., Whitton, B. A. (1991), 20, 49
Haworth, E. Y. (1972), 604, 610, 632
Haworth, E. Y., Hurley, M. A. (1986), 583, 590
Hayaski, M., Toda, K., Kitaoka, S. (1993), 387, 418
Hayden, A. (1910), 463, 467

Hayes, P. K., Barker, G. L. A., Walsby, A. E. (1997), 177, 193
Hazen, T. E. (1902), 325, 348, 350
Healey, F. P., Hendzel, L. L. (1979), 785, 799
Healey, F. P., Hendzel, L. L. (1980), 785, 799
Healy, F. P. (1982), 61, 112
Heaney, S. I., Chapman, D. V., Morison, H. R. (1983), 696, 711
Heaney, S. I., Furnass, T. I. (1980), 20, 49, 700–701, 711
Heaney, S. I., Lundm J. W. G., Canter, H., Gray, K. (1988), 20, 50
Heaney, S. I., Talling, J. F. (1980), 23, 49, 699–700, 711
Hecky, R. E., Hesslein, R. H. (1995), 26
Hecky, R. E., Kilham, P. (1988), 23, 50, 776, 785, 799
Hedley, S., Patterson, D. J. (1992), 418
Hedtke, S. F. (1984), 825, 829
Hegewald, E. (1976), 434, 447, 449, 459, 467
Hegewald, E., Schmidt, A. (1987), 255, 308
Hegewald, E., Schmidt, A. (1991), 255, 308
Hegewald, E., Silva, P. C. (1988), 254–256, 308
Hegner, R. W. (1922), 412, 415, 418
Hegner, R. W. (1923), 412, 415, 418
Hein, M. K. (1981), 599, 605, 613, 633
Heinonen, P. (1980), 488, 505
Henderson, R. J., Mackinlay, E. E. (1989), 738, 753
Henson, E. B. (1984), 314, 350
Hepperle, D. (1997), 236, 249
Hepperle, D., Krientitz, L. (1996), 238, 249
Hepperle, D., Nozaki, H., Hohenberger, S., Huss, V. A. R., Morita, E., Krienitz, L. (1998), 236, 249
Herbicide Handbook (1994), 823, 829
Herbst, D. B., Blinn, D. W. (1998), 43, 50
Herbst, D. B., Bradley, T. J. (1989), 43, 50
Herbst, D. B., Castenholz, R. W. (1994), 333, 350
Herbst, R. P. (1969), 809, 829
Heribaud, J. (1903), 633
Herth, W., Barthlott, W. (1979), 567, 590
Herth, W., Kuppel, A., Brown, R. M. (1975), 484, 505
Herth, W., Kuppel, A., Schnepf, E. (1977), 501, 505
Herth, W., Zugenmaier, P. (1979), 484, 505
Herzog, M., Maroteaux, L. (1986), 686, 711
Herzog, M., von Boletzky, S., Soyer, M.-O. (1984), 686, 711
Hesse, L. W., Schmulbach, J. C., Carr, J. M., Keenlyne, K. D., Unkenholz, D. G., Robinson, J. W., Mestl, G. E. (1989), 28, 50
Heynig, H. (1963), 513, 519
Heywood, P. (1973), 465, 467
Heywood, P. (1978a), 427, 467
Heywood, P. (1978b), 427, 467
Heywood, P. (1980), 427–428, 467
Heywood, P. (1988), 735, 753
Heywood, P. (1989), 427, 467
Heywood, P. (1990), 427–428, 463, 465, 467
Hibberd, D. J. (1976), 523, 528, 554
Hibberd, D. J. (1976a), 471, 473–474, 505, 511, 519
Hibberd, D. J. (1976b), 498, 505, 512, 519
Hibberd, D. J. (1977), 485, 505, 735, 738, 753
Hibberd, D. J. (1978), 528, 554
Hibberd, D. J. (1980), 429, 432, 465, 467
Hibberd, D. J. (1981), 424, 465, 467
Hibberd, D. J. (1983), 499, 505, 512, 519
Hibberd, D. J. (1985), 512, 519
Hibberd, D. J. (1990), 384, 387, 418
Hibberd, D. J. (1990a), 429, 431–432, 434, 446, 467
Hibberd, D. J. (1990b), 424–425, 427, 443, 465, 467
Hibberd, D. J., Greenwood, A. D., Griffiths, H. B. (1971), 732, 741, 753
Hibberd, D. J., Leedale, G. F. (1970), 424, 467
Hibberd, D. J., Leedale, G. F. (1971a), 424, 467

Hibberd, D. J., Leedale, G. F. (1971b), 424, 432, 467
Hibberd, D. J., Leedale, G. F. (1972), 424, 467
Hickel, B. (1981), 73, 112
Hickel, B. (1988), 694, 711
Hickel, B., Häkansson, H. (1991), 568, 590
Hickel, B., Maass, I. (1989), 536, 538, 554
Hickman, M. (1978), 27, 50
Hickman, M., White, J. M. (1989), 604, 633
Hietala, J., Lauren-Maatta, C., Walls, M. (1997), 64, 112
Highfill, J. F., Pfiester, L. A. (1992a), 693–694, 708, 711
Highfill, J. F., Pfiester, L. A. (1992b), 693–694, 708, 711
Hilenski, L. L., Walne, P. L. (1983), 403, 418
Hill, B. H., Herlihy, A. T., Kaufmann, P. R., Stevenson, R. J., McCormick, F. H., Johnson, C. B. (2000), 789, 799
Hill, B. H., Lazorchak, J. M., McCormick, F. H., Willingham, W. T. (1997), 776, 785, 799
Hill, D. J. (1969), 193
Hill, D. J. (1970a), 174, 180, 194
Hill, D. J. (1970b), 174, 180. 194
Hill, D. J. (1972), 194
Hill, D. J. (1976a), 169, 170, 194
Hill, D. J. (1976b), 169, 170, 194
Hill, D. J. (1976c), 169, 170, 194
Hill, D. R. A. (1990), 716, 732, 734, 741–742, 747, 753
Hill, D. R. A. (1991a), 715, 735, 741–742, 744, 747, 749, 753
Hill, D. R. A. (1991b), 715–716, 732, 734–735, 741–742, 745, 747, 749, 753
Hill, D. R. A. (1991c), 716, 734–735, 744, 749, 753
Hill, D. R. A., Rowan, K. S. (1989), 735, 741, 753
Hill, D. R. A., Wetherbee, R. (1986), 715–716, 732, 734, 736, 741, 746, 753
Hill, D. R. A., Wetherbee, R. (1988), 715–716, 732, 741, 753
Hill, D. R. A., Wetherbee, R. (1989), 715–716, 732, 736, 741–742, 753
Hill, M. O. (1979), 790, 799
Hill, W. (1996), 489, 505
Hill, W. R., Boston, J. L. (1991), 785, 799
Hillebrand, H. (1983), 780, 799
Hillebrand, H., Dürlsen, C. D., Kirschtel, D., Pollingher, U., Zohary, T. (1999), 783, 799
Hilliard, D. K. (1966), 478, 487, 499, 505
Hilliard, D. K. (1967), 487, 499, 501, 505
Hilliard, D. K. (1968), 488, 498, 505
Hilliard, D. K. (1971a), 477, 495, 498, 505
Hilliard, D. K. (1971b), 505
Hilliard, D. K., Asmund, B. (1963), 477, 484, 486, 498, 505
Hindák, F. (1963), 349–350
Hindák, F. (1981), 447, 467
Hindák, F. (1984), 81, 112
Hindák, F. (1985), 135, 194
Hindák, F. (1996), 81, 112, 345, 350
Hindák, F. (2002), 120, 194
Hindák, F. Hindakova, A. (1992), 301, 308
Hindák, F., Moustaka, M. P. (1988), 81, 112
Hinkle, J. (1986), 810, 829
Hirose, H., Hirano, M. (1981), 82, 112
Hirsch, A., Palmer, C. M. (1958), 432, 459, 467
Hoagland, K. D., Carder, J. P., Spawn, R. L. (1996), 782, 800
Hoagland, K. D., Drenner, R. W., Smith, J. D., Cross, D. R. (1993), 786, 794, 800
Hoagland, K. D., Peterson, C. G. (1990), 24–25, 50, 613, 633
Hoagland, K. D., Roemer, S. C., Rosowski, J. R. (1982), 25, 50
Hoagland, K. D., Rosowski, J. R., Gertz, M. R., Roemer, S. C. (1993), 567, 590
Hoek. *see* van den Hoek

Hoffman, L. R. (1967), 349, 350
Hoffman, L. R., Vesk, M., Pickett-Heaps, J. D. (1986), 499, 505
Hoffmann, J. P. (1998), 12, 44, 50
Hoffmann, L. (1989), 197, 206, 222
Hoffmann, L. (1996), 12, 50
Hoffmann, L., Demoulin, V. (1985), 161, 194
Hoffmann, L., Willie, E. (1992), 535, 554
Hofmann, C. J. B., Rensing, S. A., Haeuber, M. M., Martin, W. F., Mueller, S. B., Couch, J., McFadden, G. I., Igloi, G. L., Maier, U. G. (1994), 753
Hofmann, G. (1994), 791, 800
Hoham, R. W. (1973), 315, 350
Hoham, R. W. (1974a), 234, 249
Hoham, R. W. (1974b), 234, 249
Hoham, R. W. (1975), 44, 50
Hoham, R. W. (1980), 226, 249
Hoham, R. W., Blinn, D. W. (1979), 43–44, 50, 407, 418
Hoham, R. W., Mohn, W. W. (1985), 44, 50
Hoham, R. W., Mullet, J. E., Roemer, S. C. (1983), 235, 249
Hoham, Roemer, S. C., Mullet, J. E. (1979), 235, 249
Hohman, R. W., Mullet, J. E. (1977), 43, 50
Hohn, M. H. (1959), 665, 667
Hohn, M. H. (1969), 572, 590
Hohn, M. H., Hellerman, J. (1963), 561, 590, 604, 616, 633, 652
Holcomb, G. E. (1986), 343, 350
Holdway, P. A., Watson, R. A., Moss, B. (1978), 512, 519
Holen, D. A., Boraas, M. E. (1995), 534, 554
Hollenberg, G. J. (1939), 94–95, 107, 109, 112
Holm, L. G., Yeo, R. (1981), 821, 829
Holm, N. P., Armstrong, D. E. (1981), 607, 633
Holmes, N. T. H., Whitton, B. A. (1975), 763, 765, 767, 770–772
Holmes, N. T. H., Whitton, B. A. (1977), 30, 33–34, 50
Holmes, N. T. H., Whitton, B. A. (1977a), 763–766, 772
Holmes, N. T. H., Whitton, B. A. (1977b), 763–766, 772
Holmes, N. T. H., Whitton, B. A. (1977c), 763, 765–766, 772
Holmes, N. T. H., Whitton, B. A. (1981), 34, 50, 764, 772, 779, 783, 800
Holmes, R. W. (1985), 623, 633
Holmgren, S. K. (1984), 513, 519
Holmquist, E., Willén, T. (1993), 512, 519
Holomuzki, J. R., Short, T. M., (1988), 783, 800
Holopainen, I. J. (1992), 699, 711
Holt, J. R., Pfiester, L. A. (1981), 700–701, 711
Holz, J. C., Hoagland, K. D. (1999), 814, 829
Hooper, C. A. (1981), 40, 50
Hooper-Reid, N. M., Robinson, G. G. C. (1978), 39, 50
Hoops, H. J. (1984), 242, 249
Hoops, H. J., Floyd, G. L. (1982), 245, 249
Horecká, M., Komárek, J. (1979), 173–174, 194
Hori, H., Osawa, S. (1987), 387, 418
Horne, A., Goldman, C. R. (1994), 21, 50
Horne, A. J., Goldman, C. R. (1974), 822, 829
Horner, R. R., Welch, E. B., Veensstra, R. B. (1990), 35, 50
Hörtzel, G., Croome, R. (1994), 65, 112
Hoshaw, McCourt, R. M., Wang, J. C. (1990), 353, 362–363, 380
Hoshaw, R. W. (1968), 364–365, 380
Hoshaw, R. W., McCourt, R. M. (1988), 363–365, 380
Hoshaw, R. W., Rosowski, J. R. (1973), 409, 418, 740, 753
Hoshaw, R. W., Wang, J. C., McCourt, R. M., Hull, H. M. (1985), 371, 380
Hoshaw, R. W., Wells, C. V., McCourt, R. M. (1987), 371, 380
Hosiaisluoma, V. (1975), 363, 380
Howard, R. V., Parker, B. C. (1980), 203, 210, 221–222
Howarth, R. W., Cole, J. J. (1985), 23, 50
Howe, M. A. (1924), 89, 112

Howe, M. A. (1932), 81, 112
Howell, E. T., South, G. R. (1981), 363, 380
Hoyer, M. V., Canfield, D. E., Jr. (1996a), 810, 829
Hoyer, M. V., Canfield, D. E., Jr. (1996b), 810, 829
Hua, M. S., Friedmann, E. I., Ocampo-Friedmann, R., Campbell, S. (1989), 1112
Huang, T. C., Grobbelaar, N. (1989), 45, 50
Huber, A. L. (1984), 120, 177, 194
Huber-Pestalozzi, G. (1938), 92, 113
Huber-Pestalozzi, G. (1941), 503, 505
Huber-Pestalozzi, G. (1942), 572, 590
Huber-Pestalozzi, G. (1950), 716, 741–742, 748, 753
Huber-Pestalozzi, G. (1955), 387–388, 401, 415–416, 418
Huber-Pestalozzi, G. (1961), 247, 249
Hudon, C., Paquet, S., Jarry, V. (1996), 37, 50
Hufford, T. L., Collins, G. B. (1972a), 664, 667
Hufford, T. L., Collins, G. B. (1972b), 664, 667
Hufford, T. L., Collins, G. B. (1976), 34, 50
Hughes, E. O. (1948), 434, 441, 454, 459, 461, 467
Hughes, R. M. (1995), 786, 788, 800
Humm, H. J., Wicks, S. R. (1980), 113, 151, 194
Humphrey, K. P., Stevenson, R. J. (1992), 785, 800
Hunt, M. E., Floyd, G. L., Stout, B. B. (1979), 45, 50, 63, 113
Hurlbert, S. H. (1971), 784, 800
Huss, V. A. R., Frank, C., Hartmann, E. C., Hirmer, M., Kloboucek, A., Seidel, B. M., Wenzler, P., Kessler, E. (1999), 253, 257, 308
Huss, V. A. R., Sogin, M. L. (1990), 257, 308
Hustedt, F. (1926), 656, 667
Hustedt, F. (1927), 604, 633
Hustedt, F. (1927–1930), 562, 590
Hustedt, F. (1930), 596, 604, 628, 633, 639, 652, 655, 661, 667, 682–683
Hustedt, F. (1930b), 562, 573, 583–584, 586, 590
Hustedt, F. (1931–1959), 562, 590, 641, 652
Hustedt, F. (1937–1939), 628, 633
Hustedt, F. (1939), 536, 554
Hustedt, F. (1942), 672, 683
Hustedt, F. (1952), 633, 656, 667
Hustedt, F. (1957), 586, 590
Hustedt, F. (1959), 596–597, 604, 628, 633
Hustedt, F. (1961–1966), 637, 652
Hutchinson, G. E. (1957), 12–13, 15–16, 19, 41, 50
Hutchinson, G. E. (1961), 19, 50
Hutchinson, G. E. (1967), 12, 16, 19, 50, 583, 590, 691, 711
Hutchinson, G. E. (1969), 405, 418
Hutchinson, G. E. (1975), 12, 17, 24–25, 27, 50, 314, 339, 350
Hutchinson, T., Havas, M. (1986), 793, 800
Hymes, B. J., Cole, K. M. (1983), 201, 222
Hynes, H. B. N. (1970), 28–32, 34, 50, 203, 222

I

Ibelings, B., Admiraal, W., Bijker, R., Letswaart, T., Prins, H. (1998), 776, 800
Iglesias-Prieto, R. (1996), 690, 711
Ikävalko, J., Kristiansen, J., Thomsen, H. A. (1994), 474, 502, 505
Inagaki, Y., Hayashi-Ishimaru, Y., Ehara, M., Igarashi, I., Ohama, T. (1997), 418
Infante, A., Abella, S. E. B. (1985), 808, 813, 829
Islam, A. K. M. N. (1961), 332, 348, 350
Islam, A. K. M. N. (1963), 349–350
Islam, A. K. M. N., Khondker, M. (1994), 465, 467
Israelson, G. (1938), 758, 763–764, 766, 772
Israelson, G. (1942), 204, 222
Israelson, G. (1949), 35, 50, 363, 380
Ito, H. (1989), 513, 519
Ito, H., Takahashi, E. (1982), 487, 505
Iyengar, M. O. P. (1925), 432, 467
Iyengar, M. O. P., Desikachary, T. V. (1981), 226–228, 230, 232, 235, 238, 247–249

J

Jaag, O. (1941), 113
Jaag, O. (1945), 61–64, 92, 113
Jaag, O. (1972), 806, 829
Jackim, E., Gentile, J. (1968), 121, 194
Jackson, A. E. (1997), 45, 50
Jackson, J. E., Castenhloz, R. W. (1975), 189, 194
Jackson, L. J., Stockner, J. G., Harrison, P. J. (1990), 587, 590
Jacobs, D. L. (1946), 688, 697, 701, 706, 711
Jacobs, J. E. (1968), 434, 441, 454, 467
Jacobs, J. E. (1971), 447, 459, 461, 467, 491, 505
Jacobsen, B. (1985), 486, 506
Jacobsen, B. A. (1985), 537–538, 554
Jacobson, D. M., Anderson, D. M. (1986), 701, 711
Jacoby, J. M., Gibbons, H. L., Stoops, K. B., Bouchard, D. D. (1994), 814, 829
Jagg, O. (1945), 120, 158, 194
Jahn, T. L. (1946), 399, 404, 407, 413, 415, 418
Jahn, T. L. (1951), 404, 412, 418
Jahn, T. L., Bovee, E. C. (1968), 401, 418
Jahn, T. L., Bovee, E. C., Jahn, F. F. (1979), 407, 418
James, T. L., de la Cruz, A. (1989), 512, 519
James, W. F., Taylor, W. D., Barko, J. W. (1992), 699–701, 711
Jao, C.-C. (1941), 759, 763, 770–772
Jao, C.-C. (1943), 763, 765, 771–772
Jao, C.-C. (1944), 763, 765, 771–772
Jarosch, R. (1970), 528, 554
Jasby, A. D., Goldman, C. R., Reuter, J. E. (1995), 26, 50
Jeffrey, S. W. (1989), 474, 506
Jeffrey, S. W., Sielicki, M., Haxo, F. T. (1975), 690, 711
Jensen, T. E. (1984), 61, 114
Jensen, T. E. (1985), 61, 113
Jensen, W. A. (1962), 539, 554
Jewson, D. J. (1992a), 568, 590
Jewson, D. J. (1992b), 568, 590
Jimenez, J. C. (1999), 208, 222
Jochimsen, E. M., Carmichael, W. W. (1998), 807, 829
Johansen, J., Cognata, S. L., Kociolek, J. P. (1990), 682–683
Johansen, J. R. (1993), 45, 50, 113
Johansen, J. R. (1999), 45, 50
Johansen, J. R., Ashley, J., Rayburn, W. R. (1993), 113
Johansen, J. R., Doucette, G. J., Barclay, W. R., Bull, J. D. (1988), 43, 50, 512, 514–515, 519
Johansen, J. R., Rushforth, S. R. (1981), 604, 628, 633
Johansen, J. R., Rushforth, S. R. (1985), 113
Johansen, J. R., Stray, J. C. (1998), 648, 652
Johanson, J. R., Barclay, W. R., Nagle, N. (1990), 574, 590
Johanson, J. R., Rushforth, S. R. (1985), 574, 590
Johanson, J. R., Theriot, E. C. (1987), 587, 590
John, D. M. (1994), 311–312, 350
John, D. M., Johnson, L. R. (1987), 336, 350
John, D. M., Johnson, L. R. (1989), 349, 351
John, D. M., Moore, J. A. (1985), 314, 351
Johnson, L. M. Rosowski, J. R. (1992), 584, 590
Johnson, L. P. (1944), 391, 404–405, 408, 412, 418
Johnson, L. P. (1968), 404, 418
Johnson, L. P., Jahn, T. L. (1942), 412, 418
Johnson, L. R., John, D. M. (1990), 348, 351
Jones, B. L. (1987), 806, 829
Jones, H. L. J., Leadbeater, B. S. C., Green, J. C. (1994), 514, 520

Jones, J. R. Bachmann, R. W. (1976), 812, 829
Jones, R. C., Walti, K., Adams, M. S. (1983), 807, 829
Jones, R. I., Rees, S. (1994), 489, 506
Jongman, R. H. G., ter Braak, C. K. F., Van Tongeren, O. F. R. (1995), 790, 800
Jordan, R. W., Chamberlain, A. H. L. (1997), 512, 520
Jordan, R. W., Green, J. C. (1994), 512, 520
Jordan, R. W., Kleinjne, A., Heimdal, B. R., Green, J. C. (1995), 511–512, 516, 520
Joyce, J. C. (1993), 818, 829
Jügensen, M. F. (1973), 63, 113
Juggins, S., Battarbee, R., Fritz, S., Gasse, F. (1994), 794, 800
Juggins, S., ter Braak, C. J. F. (1992), 800
Julius, M. L. Estabrook, G. F., Edlund, M. B., Stoermer, E. F. (1997a), 568–569, 590
Julius, M. L., Stoermer, E. F., Colman, S. M., Moore, T. C. (1997b), 568, 583, 590
Julius, M. L. Stoermer, E. F., Taylor, C. M., Schelske, C. L. (1998), 586, 590
Junk, W. J., Bayley, P. B., Sparks, R. E. (1989), 32, 50
Jupp, B. P., Spence, D. H. N. (1997), 807, 830
Jürgens, U. J., Weckesser, J. (1985), 118, 194
Jüttner, F. (1987), 64, 113
Jüttner, F., Höfflacher, B., Wurster, K. (1986), 807, 830
Jüttner, I., Rothfritz, H., Omerod, S. J. (1996), 784, 800

K
Kaczmarczyk, D., Sheath, R. G. (1991), 203, 222
Kaczmarczyk, D., Sheath, R. G., Cole, K. M. (1992), 203, 214, 216, 221–222
Kaczmarska, I., Rushforth, S. R. (1983), 604–605, 628, 633
Kadlubowska, J. Z. (1972), 362, 369–372, 379–380
Kalff, J., Knoechel, R. (1978), 19, 50
Kalinsky, R. G. (1984), 665, 667
Kallqvist, T., Meadows, B. S. (1978), 823, 830
Kann, E. (1941), 25, 50
Kann, E. (1945), 763–764, 772
Kann, E. (1959), 26, 50
Kann, E. (1966), 763, 772
Kann, E. (1972), 101, 113
Kann, E. (1973), 65, 101, 113
Kann, E. (1976), 767, 772
Kann, E. (1978), 25, 33, 50
Kann, E. (1978a), 757, 763–764, 767, 772
Kann, E. (1978b), 763, 772
Kann, E. (1982), 765, 768, 772
Kann, E. (1993), 758, 763–764, 772
Kantz, T. S., Theriot, E. C., Zimmer, E. A., Chapman, R. L. (1960), 257, 308
Karayeva, N. I., Maggerramova, N. R., Rhazeva, S. G. (1984), 663, 667
Karim, A. G., Round, F. E. (1967), 484, 506
Karlström, U. (1978), 36, 50
Karol, K. G., McCourt, R. M., Cimino, M. T., Delwiche, C. F. (2001), 228, 249
Karr, J. R. (1981), 786, 800
Karr, J. R., Chu, E. W. (1999), 782, 786, 789, 800
Karr, J. R., Dudley, D. R. (1981), 777, 786, 789, 800
Karsten, G. (1928), 655, 667
Kato, S. (1991), 428, 467
Kato, S. (1994), 415., 418
Kato, T., Watanabe, M. (1993), 63, 113
Kawabata, Z., Banba, D. (1993), 696, 712
Kawabata, Z., Zagawa, H. (1988), 699–701, 712
Kawachi, M., Inouye, I. (1995), 512, 520

Kawai, H., Inouye, I. (1989), 513, 520
Kawecka, B. (1980), 33, 45, 50
Kawecka, B. (1981), 32, 51
Kawecka, B. (1990), 35, 51
Kay, S. H. (1997), 822, 830
Keating, K. I. (1976), 808, 830
Keating, K. I. (1977), 808, 830
Keating, K. I. (1978), 808, 830
Keithan, E. D., Lowe, R. L. (1985), 604, 633
Kelley, I., Pfiester, L. A. (1990), 694, 712
Kelly, M. G., Cazaubon, A., Coring, E., Dell'Uomo, A., Ector, L., Goldsmith, B., Guasch, H., Hürlimann, J., Jarlman, A., Kawecka, B., Kwandrans, J., Laugaste, R., Lindstrrm, E.-A., Leitao, M., Marvan, P., Padisák, J., Pipp, E., Pyrgiel, J., Rott, E., Sabater, S., van Dam, H., Viznet, J. (1998), 775–776, 779, 784, 800
Kelly, M. G., Hornberger, G. M., Cosby, B. J. (1974), 785, 800
Kelly, M. G., Penny, C. J., Whitton, B. A. (1995), 800
Kelly, M. G., Whitton, B. A. (1989), 800
Kelly, M. G., Whitton, B. A. (1995), 776, 782, 787, 800
Kelmer, A., Barko, J. (1991), 700, 712
Kempner, E. S., Miller, J. H. (1972), 407, 418
Kentucky Division of Water (1993), 775, 782, 784, 789, 800
Kentucky Natural Resources and Environmental Protection Cabinet (1997), 794, 800
Keough, J. R., Hagley, C. A., Ruzycki, E., Siersen, M. (1998), 39, 51
Kerans, B. L., Karr, J. R. (1994), 786, 789, 800
Kesler, D. H. (1981), 25, 51
Kessler, E., Huss, V. A. R. (1992), 307–309
Kessler, E., Schäfer, M. Hümmer, C., Kloboucek, A., Huss, V. A. R. (1997), 256, 308
Khan, M. A. (1993), 384, 412, 418
Khan, M. A. (1995), 584, 591
Khan, M. A. R., Begum, Z. N. T., Rahim, A. T. M. A., Salamatullah, Q. (1996), 818, 830
Khursevich, G. K., VanLandingham, S. L. (1993), 587, 591
Kiener, W. (1944), 407, 418
Kies, L., Berndt, M. (1984), 538, 554
Kilham, P., Tilman, D. (1979), 487, 506
Kilham, S. S., Kilham, P. (1978), 22, 51, 487, 506, 607, 613, 633
Killgore, K. G., Kirk, J. P., Foltz, J. W. (1998), 821, 830
Kim, J. T., Boo, S. M. (1998), 406, 418
Kindle, E. M. (1934), 314, 351
King, J. M. (1983), 501, 506
King, J. M. (1984), 497, 506
King, J. M., Ward, C. H. (1977), 45, 51, 63, 113
Kingsley, J. S. (1888), 384, 418
Kingston, J. C. (1978), 664, 667
Kingston, J. C. (1980), 623, 633
Kingston, J. C. (1982), 40, 51, 605, 633
Kingston, J. C. (1984), 604, 633
Kingston, J. C. (1986), 605, 633
Kingston, J. C. (1997), 598, 633
Kingston, J. C. (2000), 596, 613, 627, 633
Kingston, J. C., Birks, H. J. B. (1990), 604, 633
Kingston, J. C., Birks, H. J. B., Uutala, A. J., Cumming, B. F., Smol, J. P. (1992), 604, 633, 787, 800
Kingston, J. C., Lowe, R. L. (1979), 604, 633
Kingston, J. C., Lowe, R. L., Stoermer, E. F. (1980), 656, 667
Kingston, J. C., Lowe, R. L., Stoermer, E. F., Ladewski, T. B. (1983), 27, 51, 604, 633
Kingston, J. C., Sherwood, A. R., Bengtsson, R., (2001), 631, 633
Kirjakov, I. K. (1983), 416, 419
Kirjakov, I. K. (1998), 413, 419
Kirk, D. L. (1998), 226, 242, 250

Kirk, D. L., Birchem, R., King, N. (1986), 226, 250
Kirkby, S. M., Hibberd, D. J., Whitton, B. A. (1972), 759, 763–764, 771, 773
Kiselev, I. A. (1947), 80, 113
Kiss, K. T. (1984), 587, 591
Kiss, K. T. (1994), 38, 51
Kivic, P. A., Vesk, M. (1972), 399, 419
Kivic, P. A., Vesk, M. (1974), 399, 401, 419
Kivic, P. A., Walne, P. L. (1984), 385, 419
Klaveness, D. (1977), 738, 753
Klaveness, D. (1981), 735, 742, 748, 753
Klaveness, D. (1982), 739, 753
Klaveness, D. (1984), 739, 753
Klaveness, D. (1985), 716, 734, 741, 745, 753
Klaveness, D. (1988a), 715, 736, 738, 753
Klaveness, D. (1988b), 715, 738, 753
Klebs, G. (1883), 401, 419
Klebs, G. (1893a), 471, 484, 506
Klebs, G. (1893b), 471, 484, 506
Klemm, D. J., Lazorchak, J. M. (1994), 778, 780, 800
Kling, H. (1975), 171, 194
Kling, H., Findlay, D. L. Komárek, J. (1994), 171–172, 194
Kling, H., Holmgren, S. (1972), 129, 147, 194, 537, 554
Kling, H. J. (1981), 514, 518–520
Kling, H. J. (1992), 567–569, 591
Kling, H. J., Kristiansen, J. (1983), 500, 503, 506
Kling, H., Kristiansen, J. (1983), 552, 554
Klut, M. E., Bisalputta, T., Antia, N. J. (1987), 690, 712
Knaust, R., Urbig, T., Li, L., Taylor, W., Hastings, J. W. (1998), 687, 712
Kobayashi, H. (1997), 623, 628, 633
Kobayashi, H., Ando, K., Nagumo, T. (1981), 665, 667
Koch, W. J. (1951), 432, 467
Kociolek, J. P. (1997), 560, 591
Kociolek, J. P. (1998), 597–598, 633, 659, 667
Kociolek, J. P. (2000), 664–665, 667
Kociolek, J. P., Herbst, D. B. (1992), 43, 51
Kociolek, J. P., Kingston, J. C. (1999), 605, 633, 655, 659, 661, 665, 667
Kociolek, J. P., Lamb, M. A., Lowe, R. L. (1983), 584, 591
Kociolek, J. P., Rhode, K. (1998), 595, 633
Kociolek, J. P., Rhode, K., Williams, D. M. (1997), 655, 661, 664, 667
Kociolek, J. P., Rosen, B. H. (1984), 655, 659, 665, 667
Kociolek, J. P., Spaulding, S. A., Kingston, J. C. (1998), 12, 51, 637, 652
Kociolek, J. P., Stoermer, E. F. (1986), 604, 633, 655, 665, 667
Kociolek, J. P., Stoermer, E. F. (1987a), 656, 664, 667
Kociolek, J. P., Stoermer, E. F. (1987b), 655, 664–665, 667
Kociolek, J. P., Stoermer, E. F. (1988a), 560, 591, 655, 659, 662, 667
Kociolek, J. P., Stoermer, E. F. (1988b), 570, 591, 655, 667
Kociolek, J. P., Stoermer, E. F. (1988c), 655, 659, 664, 667
Kociolek, J. P., Stoermer, E. F. (1989), 597, 633, 665, 667
Kociolek, J. P., Stoermer, E. F. (1991), 655, 659, 665, 667
Kociolek, J. P., Stoermer, E. F. (1993a), 655, 659, 667
Kociolek, J. P., Stoermer, E. F. (1993b), 667
Kociolek, J. P., Stoermer, E. F., Edlund, M. A. (1995), 659, 665, 667
Kociolek, J. P., Stoermer, E. F., Sicko-Goad, L. (1994), 664, 667
Kociolek, J. P., Theriot, E. C., Williams, D. M. (1989), 597, 633
Kociolek, J. P., Williams, D. M. (1987), 559–560, 591
Kociolek, J. P., Williams. D. M. (1987), 595, 633
Kociolek, J. P., Yang, J.-R., Stoermer, E. F. (1988), 661, 665, 667
Kofoid, C. A. (1899), 227, 240–241, 250
Kofoid, C. A. (1903), 36, 51
Kofoid, C. A. (1908), 36, 51
Kofoid, C. A. (1909), 690, 712
Kofoid, C. A., Michener, J. R. (1912), 693, 712
Kofoid, C. A., Swezy, O. (1921), 690, 712
Kohl, J.-G., Dudel, G., Schlangstedt, M., Kuhl, H. (1985), 120, 194
Kohl, J.-G., Nicklisch, A. (1981), 126, 129, 194
Kohl, J.-G., Schlangstedt, M., Dudel, G. (1987), 119, 194
Köhler, J. (1993), 37, 51
Köhler, J. (1995), 38, 51
Kol, E. (1942), 355, 363–364, 369, 380
Kol, E. (1944), 363, 369, 380
Kol, E. (1964), 363, 369, 380
Kol, E. (1968), 43–44, 51
Kolbe, R. W. (1956), 665, 667
Kolkwitz, R., Marson, M. (1908), 560, 591, 775, 790, 800
Komagata, K. (1987), 121, 194
Komárek, J. (1958), 67, 83, 90, 113, 120, 127, 141, 145, 167, 172, 194
Komárek, J. (1969), 68, 76, 113
Komárek, J. (1970), 113
Komárek, J. (1975), 98, 113
Komárek, J. (1976), 60, 68, 113
Komárek, J. (1984), 89, 113, 142, 163–164, 194
Komárek, J. (1985), 104, 107, 109, 113
Komárek, J. (1989), 76, 82–83, 86–87, 99, 102, 113, 131, 146, 150, 152, 170, 173–174, 184–185, 194
Komárek, J. (1991), 410, 419
Komárek, J. (1993), 62, 64, 92–93, 113
Komárek, J. (1994), 60, 93, 95, 113, 126, 163–164, 194
Komárek, J. (1995), 72, 77, 79, 83, 87, 113
Komárek, J. (1999), 60–61, 65, 113, 120, 194
Komárek, J. (2001), 129, 145, 174, 194
Komárek, J., Albertano, P. (1994), 147, 194
Komárek, J., Anagnostidis, K. (1986), 60, 62, 101, 113
Komárek, J., Anagnostidis, K. (1989), 117–121, 161, 171, 177, 180, 194
Komárek, J., Anagnostidis, K. (1995), 113
Komárek, J., Anagnostidis, K. (1998), 60, 62–63, 65, 67, 72, 74, 79, 81–82, 87–89, 91–92, 95–96, 101–102, 104, 107, 110, 113
Komárek, J., Cáslavaská, J. (1991), 118, 194
Komárek, J., Cepak, V. (1998), 77, 79, 113
Komárek, J., Cronberg, G. (2001), 135, 141, 194
Komárek, J., Fott, B. (1983), 253–256, 302, 306–308
Komárek, J., Hindák, F. (1975), 103–105, 113
Komárek, J., Hindák, F. (1988), 87–88, 113
Komárek, J., Hindák, F. (1989), 113
Komárek, J., Hübel, H. _amarda, J. (1993), 120, 177, 194
Komárek, J., Kling, H. (1991), 133, 135, 194
Komárek, J., Komárková-Legnerová, J. (1992), 64–65, 67, 83, 85–87, 89, 113
Komárek, J., Komárková-Legnerová, J. (1993), 113
Komárek, J., Komárková-Legnerová, J. (2002), 60, 80, 87, 113
Komárek, J., Komárková-Legnerová, J., Sant'Anna, C. L., Azevedo, M. T. P., Senna, P. A. C. (2002), 92, 113
Komárek, J., Kopecky, J., Cepák, V. (1999), 79, 113
Komárek, J., Ludvik, J., Pokorny, V. (1975), 61, 113
Komárek, J., Lukavsky, J. (1988), 66–67, 113
Komárek, J., Montejano, G. (1994), 60, 67, 96–97, 113
Komárek, J., Novelo, E. (1994), 94–95, 113
Komárek, J., Watanabe, M. (1990), 158, 160, 194
Komáreková, J. (1998), 174, 194
Komáreková, J. (2001), 120, 194
Komáreková, J., Ladares-Silva, R., Senna, P. A. C. (1999), 174, 194
Komáreková-Legnerová, J., Tavera, R. (1996), 135, 174, 180, 194
Komárková, J. (2001), 65, 114
Komárková-Legnerová, J. (1991), 74, 113

Komárková-Legnerová, J., Cronberg, G. (1994), 74, 113
Komárková-Legnerová, J., Tavera, R. (1996), 82, 113
Kondrateva, N. V. (1961), 118, 194
Kondrateva, N. V. (1968), 63, 113, 133, 138, 146, 149, 154, 159, 163, 165–168, 175, 179, 194
Kondrateva, N. V. (1972), 120, 194
Kondrateva, N. V., Kovalenko, O. V., Prichod'kova, I. P. (1984), 75, 113
Konopka, A. (1982a), 806, 830
Konopka, A. (1982b), 806, 830
Koppen, J. D. (1975), 601, 617, 621–622, 633
Koppen, J. D. (1978), 601, 604, 617, 633
Korch, J. E., Sheath, R. G. (1989), 201, 222
Korner, H. (1971), 600, 607, 633
Korshikov, A. A. (1924), 247, 250
Korshikov, A. A. (1925), 235, 247, 250
Korshikov, A. A. (1928), 245–247, 250
Korshikov, A. A. (1929), 481, 506, 528, 554
Korshikov, A. A. (1942), 479, 506
Korte, V. L., Blinn, D. W. (1983), 605, 633
Korte, V. L., Blinn, D. W. (1993), 36, 51
Kortmann, R. W., Rich, P. H. (1994), 814–815, 830
Koryak, M. (1978), 699, 712
Kosinskaja, E. K. (1948), 93, 100, 114, 151, 195
Kouwets, F. A. C. (1980), 477, 506
Kouwets, F. A. C. (1995), 256, 308
Kouwets, F., Coesel, P. (1984), 373, 380
Kovácik, L. (1988), 89, 114
Krammer, K. (1979), 663, 667
Krammer, K. (1981), 663, 667
Krammer, K. (1982), 656, 659, 663–664, 667
Krammer, K. (1990), 601, 623, 633
Krammer, K. (1992a), 650, 652
Krammer, K. (1992b), 650, 652
Krammer, K. (1997a), 561, 591, 656, 659, 663–664, 667
Krammer, K. (1997b), 561, 591, 656, 659, 663–664, 667
Krammer, K., Lange-Bertalot, H. (1986), 560–563, 591, 637, 639, 646, 652, 661–664, 668
Krammer, K., Lange-Bertalot, H. (1988), 560–563, 591, 669, 682–683
Krammer, K., Lange-Bertalot, H. (1991), 639, 652, 661–662, 665, 668
Krammer, K., Lange-Bertalot, H. (1991a), 560–563, 579, 585–586, 591, 596, 599–600, 613, 616, 628, 633
Krammer, K., Lange-Bertalot, H. (1991b), 560–563, 591, 596–597, 601, 623, 628, 633
Krebs, W. N., Bradbury, J. P., Theriot, E. C. (1987), 568, 591
Kreis, R.G., Jr., Stoermer, E. F. (1979), 627–628, 633
Krejci, M. E., Lowe, R. L. (1987a), 604, 634
Krejci, M. E., Lowe, R. L. (1987b), 604, 634
Kremer, B. (1983), 203, 222
Krieger, W. (1933), 361, 380
Krienitz, L., Hehmann, A., Caspar, S. J. (1997), 488, 506
Krienitz, L., Hepperle, D., Stich, J., Weiler, W. (2000), 425, 427, 467
Krienitz, L., Takeda, H., Hepperle, D. (1999), 257, 308
Kristiansen, J. (1969), 484, 497, 506
Kristiansen, J. (1971), 513, 520
Kristiansen, J. (1972), 484, 497, 506
Kristiansen, J. (1975), 538, 552, 554
Kristiansen, J. (1980), 536, 554
Kristiansen, J. (1981), 536–537, 554
Kristiansen, J. (1985), 536, 554
Kristiansen, J. (1986), 474, 506, 535–538, 554
Kristiansen, J. (1988), 487, 506
Kristiansen, J. (1988a), 524, 554
Kristiansen, J. (1988b), 536, 554
Kristiansen, J. (1990), 471, 506
Kristiansen, J. (1992), 486, 506, 535, 552, 554
Kristiansen, J. (1995a), 471, 506
Kristiansen, J. (1995b), 491, 506
Kristiansen, J., Andersen, R. A. (1983), 503, 506
Kristiansen, J., Cronberg, G. (1996), 503, 506
Kristiansen, J., Cronberg, G., Geissler, U. (1989), 503, 506
Kristiansen, J., Takahashi, E. (1982), 536, 554
Kristiansen, J., Tong, D. (1989), 536, 554
Kristiansen, J., Tong, D. (1995), 486, 506
Kristiansen, J., Vigna, M. S. (1996), 486, 506
Krogmann, D. W., Buttala, R., Sprinkle, J. (1986), 37, 51, 65, 114
Krolikowska, J. (1997), 27, 51
Kudoh, S., Takahashi, M. (1990), 819, 830
Kugrens, P. (1999), 744, 746, 753
Kugrens, P., Clay, B. L., Lee, R. E. (1999), 725, 749, 753
Kugrens, P., Lee, R. E. (1986), 716, 723–724, 729, 732, 745–746, 753
Kugrens, P., Lee, R. E. (1988), 736, 741, 746, 753
Kugrens, P., Lee, R. E. (1990), 748, 753
Kugrens, P., Lee, R. E. (1991), 716, 728, 732–735, 740–741, 744, 746–749, 753
Kugrens, P., Lee, R. E., Andersen, R. E. (1986), 716, 734, 741, 745–747, 749, 753
Kugrens, P., Lee, R. E., Andersen, R. E. (1987), 716, 732, 734, 741, 745–746, 753
Kugrens, P., Lee, R. E., Corliss, J. O. (1994), 732, 734–735, 749–750, 753
Kuhn, D. L., Plafkin, J. L., Cairns, J. J., Lowe, R. L. (1981), 598, 634
Kumano, S. (1978), 202, 222
Kumano, S., Hirose, H. (1959), 759–760, 767, 773
Kumano, S., Necchi, O., Jr. (1987), 202, 222
Kusel-Fetzmann, E. L. (1996), 758–760, 763–767, 771, 773
Kusel-Fetzmann, E. L., Schagerl, M. (1992), 760, 767, 773
Kützing, T. F. (1849), 62, 114
Kuzinicki, L., Mikolajczyk, E., Walne, P. L. (1990), 391–392, 419
Kwandrans, J., Eloranta, P., Kawecka, B., Wojtan, K. (1998), 783–784, 791, 800

L
Lackey, J. B. (1938), 42, 51, 487, 506
Lackey, J. B. (1939), 512, 520
Lackey, J. B. (1942), 428, 434, 459, 467
Lackey, J. B. (1968), 405–407, 419, 806, 830
Laessle, A. M. (1961), 45, 51
Lage, O. M., Parente, A. M., Vasconcelos, M. T. S. D., Gomes, C. A. R., Salema, R. (1996), 823, 830
Lagerheim, G. (1989), 432, 467
Lair, N., Reyes-Marchant, P. (1997), 32, 51
Laird, K., Fritz, S. C., Cumming, B. F. (1998), 604, 634
Lakshminarayana, J. S., Devi, J. S. (1975), 499, 506
Lamb, M. A., Lowe, R. L. (1987), 605, 634
Lamberti, G. A. (1996), 776, 783, 800
Lamberti, G. A., Gregory, S. V., Ashkenas, L. R., Steinman, A. D., McIntire, C. D. (1989), 34–35, 51
Lamberti, G. A., Steinman, A. D. (1993), 786, 800
Lamberti, G. A., Steinman, A. D. (1997), 32, 51
Lamon, E. C., III (1995), 806, 830
Lang, N. J., Fay, P. (1971), 119, 195
Lange, T. R., Rada, R. G. (1993), 38, 51
Lange-Bertalot, H. (1976), 682–683
Lange-Bertalot, H. (1979), 775, 782, 787, 800
Lange-Bertalot, H. (1980), 613, 634, 659, 668, 682–683

Lange-Bertalot, H. (1989), 599, 616, 634
Lange-Bertalot, H. (1993), 561, 591, 665, 668
Lange-Bertalot, H. (1995), 665–666, 668
Lange-Bertalot, H. (1997a), 598–599, 613, 634
Lange-Bertalot, H. (1997b), 597–598, 602, 627, 634
Lange-Bertalot, H. (1999), 598, 602, 634
Lange-Bertalot, H., Compère, P. (2001), 631, 634
Lange-Bertalot, H., Krammer, K. (1989), 596, 601, 634
Lange-Bertalot, H., Krammer, K. (1993), 669, 683
Lange-Bertalot, H., Metzeltin, D. (1996), 561–562, 591, 638, 647, 649, 652, 665, 668
Lange-Bertalot, H., Moser, G. (1994), 638, 646, 652
Lange-Bertalot, H., Ruppel, M. (1980), 602, 634
Lantz, K. E., Davis, J. T., Hughes, J. S., Schafer, H. E. (1964), 811, 830
LaRivers, I. (1978), 429, 433–434, 451, 459, 463, 467
Larkum, T. (1996), 686, 712
Larsen, A., Eikrem, W., Paasche, E. (1993), 512, 520
Larsen, J., Patterson, D. J. (1991), 387, 404, 419
Larson, A., Kirk, M. M., Kirk, D. L. (1992), 242, 250
Larson, G. K. (1989), 15, 51
Larson, G. L., McIntire, C. D., Hurley, M. Buktenica, M. W. (1996), 15, 51
Lasenby, D. C. (1975), 776, 800
Lavau, S., Saunders, G. W., Wetherbee, R. (1997), 524, 526, 534, 548, 554
Lawrence, J. M. (1954), 810, 830
Lawry, N. H., Simon, R. D. (1982), 61, 114
Leadbeater, B., Dodge, J. D. (1967), 690, 712
Leadbeater, B. S. C. (1986), 531, 554
Leadbeater, B. S. C. (1990), 531, 554
Leadbeater, B. S. C., Barker, D. A. N. (1995), 524, 531, 554
League, E. A., Greulach, V. A. (1955), 433, 465, 467
Leander, B. S., Farmer, M, A, (2000), 399, 403, 419
Leavitt, P. R. (1993), 781, 800
LeCampion-Alsumard, T., Golubic, S. (1985), 62, 109, 114
Lecointe, C., Coste, M., Prygiel, J. (1993), 604, 634, 791, 800
Lee, J. J., Capriulo, G. M. (1990), 383, 419
Lee, K., Round, F. F. (1987), 663, 668
Lee, K., Round, F. F. (1988), 663, 668
Lee, R. E. (1989), 424, 429, 432, 467, 757, 759, 767–768, 773
Lee, R. E. (1999), 353, 380, 384, 399, 419
Lee, R. E., Kugrens, P. (1986), 716, 734, 741–742, 745, 753
Lee, R. E., Kugrens, P. (1991), 716, 733, 735, 744, 749–750, 753
Lee, R. E., Kugrens, P. (1998), 474, 506, 738, 753
Lee, R. E., Kugrens, P. (2000), 738, 753
Lee, R. E., Kugrens, P., Mylnikov, A. P. (1991), 716, 749–750, 753
Lee, R. E., Miller-Hughes, C., Kugrens, P. (1993), 753
Lee, Y.-K., Soh, C.-W. (1991), 45, 51
Leedale, G. F. (1964), 386, 399, 403, 419
Leedale, G. F. (1967a), 406, 415, 419
Leedale, G. F. (1967b), 383–384, 387–388, 397, 399, 401–402, 404–405, 407, 410, 413, 415, 419
Leedale, G. F. (1982), 402–403, 419
Leeper, D. A., Porter, K. G. (1995), 501, 506
Lefébure, P., Cheneviére, E. (1938), 245, 250
Lefèvre, M. (1932), 690, 707, 709, 712
Leitch, A. R., John, D. M., Moore, J. A. (1990), 312, 351
Leland, H. V. (1995), 795, 800
Lembi, C. A. (1975), 235, 250
Lembi, C. A. (1980), 226, 234, 250
Lembi, C. A. (2000), 823, 830
Lembi, C. A., O'Neal, S. W., Spencer, D. F. (1988), 782–783, 800, 810, 830
Lembi, C. A., Pearlmutter, N. L., Spencer, D. L. (1980), 818, 830

Lembi, C. A., Ritenour, B. G., Iverson, E. M., Forss, E. C. (1978), 820, 830
Lembi, C. A., Spencer, D. F., O'Neal, S. W. (1984), 822, 830
Lemieux, C., Otis, C., Turmel, M. (2000), 228, 250
Lemmermann, E. (1899), 89, 114
Lemmermann, E. (1905), 91, 114
Lentin, J. K., Williams, G. L. (1989), 687, 712
Leonardson, L., Ripl, W. (1980), 812, 830
Leopold, L. B., Wolman, M. G., Miller, J. P. (1964), 28, 51
Lepisto, L., Antikainen, S., Kivinen, J. (1994), 428, 467
Lepistö, L. Rostenström, U. (1998), 489, 506
Letson, M. A., Makarewicz, J. C. (1994), 819, 830
Leukart, P., Hanelt, D. (1995), 203, 222
Levin, E. D., Schmechel, D. E., Burkholder, J. M., Glasgow, H. B., Jr., Deamer-Melia, N. J., Moser, V. C., Harry, G. J. (1997), 685, 712
Lewandowski, K., Ozimek, T. (1997), 28, 51
Lewin, R., Robinson, X. (1979), 89, 114
Lewis, I. F., Zirkle, C., Patrick, R. (1933), 461, 463, 467
Lewis, L. A. (1997), 256, 308
Lewis, L. A., Wilcox, L. W., Fuerst, P. A., Floyd, G. L. (1992), 256, 308
Lewis, W. M. (1984), 560, 568, 591
Lewitus, A. J., Caron, D. A. (1991), 740, 753
Lewitus, A. J., Glasgow, H. B., Burkholder, J. M. (1999), 690, 712, 738–739, 753
Lhotsky, O., Komárek, J. (1981), 68, 114
Li, R., Watanabe, M. M. (1998), 117, 195
Li, R., Watanabe, M., Watanabe, M. M. (2000), 195
Lichti-Federovich, S. (1979), 613, 627, 634
Likens, G. E. (1985), 19, 51
Likens, G. E., Bormann, F. H., Pierce, R. S., Eaton, J. S., Johnson, N. M. (1997), 28, 51
Lim, E. E., Amaral, L., Caron, D. A., DeLong, E. F. (1993), 473, 506
Lim, E. E., Caron, D. A., DeLong, E. F. (1996), 473, 506
Lim, E. E., Caron, D. A., Dennett, M. R. (1999), 473, 506
Lin, C. K., Blum, J. L. (1977), 26, 51, 205, 222
Lindeman, R. L. (1941a), 40, 51
Lindeman, R. L. (1941b), 40, 51
Lindeman, R. L. (1942), 40, 51
Lindström, K. (1991), 699, 712
Line, J. M., ter Braak, C. J. F., Birks, H. J. B. (1994), 792, 800
Ling, H. U. (1996), 226, 250
Ling, K. H., Sin, Y. M., Lam, T. J. (1993), 701, 712
Linnaeus, C. (1753), 312, 351
Linnaeus, C. (1758), 225, 250
Linne von Berg, K.-H., Kowallik, K. V. (1996), 12, 51
Linton, E. W., Hittner, D., Lewandowski, C., Auld, T., Triemer, R. E. (1999), 385, 387, 391, 419
Linton, E. W., Triemer, R. E. (1999a), 387, 391, 401, 419
Linton, E. W., Triemer, R. E. (1999b), 385, 391, 419
Livingstone, D., Khoja, T. M., Whitton, B. A. (1983), 118, 195
Livingstone, D., Whitton, B. A. (1984), 39, 51
Livingstone, D.A. (1963), 14, 51
Lobban, C. S., Mann, D. G. (1987), 682–683
Lock, M. A., Wallace, R. R., Costerton, J. W., Ventulloa, R. M., Charlton, S. E. (1984), 32, 51
Lock, M. A., Williams, D. D. (1981), 28, 32, 51
Lodge, D. L., Kershner, M. W., Aloi, J. E., Covich, A. P. (1994), 25, 51
Lodge, D. M. (1986), 24–25, 51
Loeb, S. L., Reuter, J. E., Goldman, C. R. (1983), 25, 51
Loeblich, A. R., III (1969), 690, 712
Loeblich, A. R., III (1976), 697, 712
Loeblich, A. R., III (1980), 708, 712
Loeblich, A. R., III (1982), 697, 712

Loginova, E. I. (1988), 582, 591
Lohman, K. E., Andrews, G. W. (1968), 561, 591
Lokhorst, G. M. (1978), 349, 351
Lokhorst, G. M. (1991), 345, 351
Lokhorst, G. M. (1992), 465, 467
Lokhorst, G. M. (1996), 346, 351
Lokhorst, G. M. (1999), 341, 351
Lokhorst, G. M., Rongen, G. P. J. (1994), 335, 351
Lom, J. (1981), 706, 712
Lorenzen, C. J. (1967), 783, 801
Losee, R. F., Wetzel, R. G. (1983), 24, 52
Lott, A. M., Siver, P. A., Marsicano, L. J., Kodama, K. P., Moeller, R. E. (1994), 537, 554
Lovell, R. T., Lelana, I. Y., Boyd, C. E., Armstrong, M. S. (1986), 807, 830
Lowe, C. W. (1927), 432–434, 458, 461, 467
Lowe, R. L. (1974), 561, 591, 638, 652, 661–662, 668, 775, 791, 801
Lowe, R. L. (1975), 582, 591
Lowe, R. L. (1996), 25, 52, 672, 683
Lowe, R. L., Busch, D. E. (1975), 587, 591
Lowe, R. L., Collins, G. B. (1973), 605, 634
Lowe, R. L., Kociolek, J. P. (1984), 659, 665, 668
Lowe, R. L., Pan, Y. (1996), 34, 52, 670, 683, 776, 801
Lowe, R. L., Rosen, B. H., Kingston, J. C. (1982), 26, 52, 605, 634
Lucas, I. A. N. (1970a), 741, 753
Lucas, I. A. N. (1970b), 735, 741, 753
Lucas, I. A. N. (1982), 742, 753
Ludwig, M., Gibbs, S. P. (1985a), 735–736, 754
Ludwig, M., Gibbs, S. P. (1985b), 736, 754
Ludwig, M., Gibbs, S. P. (1989), 735–736, 754
Lukas, K. G. Golubic, S. (1981), 62, 114
Lukas, K. G. Golubic, S. (1983), 60, 107, 114
Lukas, K. G., Hoffman, E. J. (1984), 107, 114
Lukesová, A. (1993), 63, 114
Lund, J. W. G. (1942), 408, 413, 419, 514, 520
Lund, J. W. G. (1949), 607, 634
Lund, J. W. G. (1950), 607, 634
Lund, J. W. G. (1955), 146–147, 195
Lund, J. W. G. (1962), 754
Lund, J. W. G. (1964), 20, 52
Lund, J. W. G., Kipling, C., LeCren, E. D. (1958), 490, 506, 783–784, 801
Lund, J. W. G., Reynolds, C. S. (1982), 23, 52
Lupikina, E. G., Khursevich, G. K. (1992), 587, 591
Luther, A. (1899), 432, 467
Luther, H. (1954), 764, 773

M

Maceina, M. J. (1996), 810, 830
Maceina, M. J., Cichra, M. F., Betsill, R. K. Bettoli, P. W. (1992), 821, 830
MacEntree, F. J., Schreckenberg, S. G., Bold, H. C. (1972), 63, 114
MacFarlane, J. J., Raven, J. A. (1985), 203, 222
MacKay, N. A., Elser, J. A. (1998), 819, 830
MacRae, R. A., Fensome, R. A., Williams, G. L. (1996), 687, 712
Madsen, J. D. (1996), 816, 830
Maga, J. A. (1987), 807, 830
Mahmood, N. A., Carmichael, W. W. (1986), 121, 195
Mahood, A. D. (1981), 585–586, 591
Mahood, A. D., Fryxell, G. A., McMillan, M. (1986), 587, 591
Mahood, A. D., Thomas, R. D., Goldman, C. R. (1984), 583, 591
Main, S. (1988), 616, 634
Makarewicz, J. C. (1993), 21, 52, 587, 591
Makarewicz, J. C., Baybutt, R. I. (1981), 22, 52
Makarewicz, J. C., Bertram, P. (1991), 587, 591

Malin, G., Liss, P. S., Turner, S. M. (1994), 511, 520
Maloney, T. E., Palmer, C. M. (1956), 822, 830
Maltais, M.-J., Vincent, W. F. (1997), 24, 26, 52
Mann, D. G. (1977), 682–683
Mann, D. G. (1980a), 682–683
Mann, D. G. (1980b), 682–683
Mann, D. G. (1981), 682–683
Mann, D. G. (1982a), 659, 668
Mann, D. G. (1982b), 659, 668
Mann, D. G. (1984), 604, 634, 663, 668
Mann, D. G. (1984a), 637–638, 652
Mann, D. G. (1984b), 638, 652
Mann, D. G. (1984c), 649, 652
Mann, D. G. (1985), 637, 652
Mann, D. G. (1986), 682–683
Mann, D. G. (1989), 638, 650, 652
Mann, D. G. (1993), 560, 591, 652
Mann, D. G., Stickle, A. J. (1991), 644, 646–647, 652
Mann, N. H., Carr, N. G. (1992), 63, 114
Manny, B. A., Edsall, T. A., Wujek, D. F. (1991), 212, 222
Manton, I., Peterfi, L. S. (1969), 516, 519–520
Mantoura, R. F. C., Llewellyn, C. A. (1983), 783, 801
Marencik, J., Lembi, C. A. (1998), 822, 830
Margalef, R. (1947), 454. 467
Margalef, R. (1978), 568, 591
Margulis, L. (1974), 385, 419
Margulis, L., Corliss, J. O., Melkonian, M., Chapman, D. J. (1990), 383, 385, 387–388, 419
Marin, B., Klingberg, M., Melkonian, M. (1998), 737, 742, 754
Marino, R., Howarth, R. W., Shamess, J., Prepas, E. (1990), 23, 52
Marks, J. C., Lowe, R. L. (1993), 25–26, 52, 672, 683
Martin, A. C., Zim, H. S., Nelson, A. L. (1961), 821, 830
Martin, E., Benson, R. (1988), 819, 830
Martin, E. L., Leach, J. E., Kuo, K. J. (1978), 819, 830
Martin, J. F., Izaguirre, G., Waterstrat, P. (1991), 807, 830
Martin, J. F., McCoy, C. P., Greenleaf, W., Bennett, L. (1988), 807, 830
Martin, T. C., Wyatt, J. T. (1974), 118–120, 195
Martin, W., Sommerville, C. C., Loiseaux-de Goer, S. (1992), 737, 754
Marzoulf, E. R., Mulholland, P. J., Steinman, A. D. (1994), 785, 801
Mason, C. P., Edwards, K. R., Carlson, R. E., Pignatello, J., Gleason, F. K., Wood, J. M. (1982), 821, 830
Massalski, A., Leedale, G. F. (1969), 432, 467
Masuda, K., Boyd, C. E. (1993), 824, 831
Mataloni, G., Tell, G. (1996), 40, 52
Mattox, K. R., Stewart, K. D. (1984), 225, 250, 257, 308
Mattox, K. R., Stewart, K. D. (1985), 313, 351
Matula, J. (1992), 348, 351
Matvienko, A. M. (1941), 524, 554
Matvienko, A. M. (1954), 480, 503, 506
Maxwell, C. D. (1991), 63, 114
May, R. M. (1974), 795, 801
Mayama, S., Kobayashi, H. (1990), 665, 668
Mayama, S., Kobayashi, H. (1991), 665, 668
Mayer, M. S., Likens, G. E. (1987), 31, 52
Mazumder, A., Taylor, W. D., McQueen, D. J., Lean, D. R. S. (1990), 818, 831
McBride, S. A., Edgar, R. K. (1988), 569, 591
McCormick, P. V., Carins, J., Jr. (1994), 782, 786, 801
McCormick, P. V., O'Dell, M. B. (1996), 783, 786, 794, 801
McCormick, P. V., Rawlik, P. S., Lurding, K., Smith, E. P., Sklar, F. H. (1996), 785, 801
McCormick, P. V., Stevenson, R. J. (1998), 39, 52, 66, 114, 776, 783, 785, 794, 801

McCourt, R. M. (1995), 313, 351
McCourt, R. M., Hoshaw, R. W., Wang, J. C. (1986), 363, 380
McCourt, R. M., Karol, K. G., Kaplan, S., Hoshaw, R. W. (1995), 353, 380
McFadden, G. I. (1993), 735–736, 754
McFadden, G. I., Gilson, P. R., Douglas, S. E., Cavalier-Smith, T., Hofmann, C. J. B., Maier, U.-G. (1997), 735–737, 754
McFadden, G. I., Gilson, P. R., Hill, D. R. A. (1994), 736, 754
McFarland, B. H., Hill, B. H., Willingham, W. T. (1997), 791, 801
McGrory, C. B., Leadbeater, B. S. C. (1981), 531, 554
McInteer, B. B. (1930), 461, 463, 467
McInteer, B. B. (1939), 461, 463, 467
McIntire, C. D., Phinney, H. K., Larson, G. L., Buktenica, M. (1994), 567, 584, 591
McIntire, C. D., Tinsley, I. J., Lowry, R. R. (1969), 35, 52
McIntosh, A. W., Kevern, N. R. (1974), 825, 831
McKay, R. M. L., Gibbs, S. P. (1990), 200, 222
McKenzie, C., Deibel, D., Paranjape, M., Thompson, R. J. (1995), 489, 506
McKenzie, C., Kling, H. (1989), 486, 506, 552, 554
McKerracher, L., Gibbs, S. P. (1982), 735–736, 754
McKnight, B. K., Niem, A., Kociolek, J. P., Ranne, P. (1995), 561, 591
McKnight, D. (1981), 825, 831
McKnight, D. M., Chisholm, S. W., Harleman, D. R. F. (1983), 822–824, 831
McLachlan, J. L., Boalch, G. T., Jahn, R. (1997), 709, 712
McLachlan, J. L., Curtis, J. M., Boutilier, K. Keusgen, M., Seguel, M. R. (1999), 408, 419
McLachlan, J. L., Seguel, M. R., Fritz, L. (1994), 388, 401, 408, 419
McNabb, C. D. (1960), 490, 506
McQueen, D. J. (1990), 818–819, 831
McQueen, D. J., Johannes, M. R. S., Post, J. R., Stewart, T. J., Lean, D. R. S. (1989), 818, 831
McQuoid, M. R., Hobson, L. A. (1995), 587, 591
Meador, J. P., Taub, F. B., Sibley, T. H. (1993), 822, 825, 831
Medlin, L. K., Elwood, D. J., Stickel, S., Sogin, M. L. (1991), 570, 591
Medlin, L. K., Kooistra, W. H. C. F., Gersonde, R., Sims, P. A., Wellbrock, U. (1997), 570, 592
Medlin, L. K., Kooistra, W. H. C. F., Gersonde, R., Wellbock, U. (1996), 563, 570, 583, 584, 592
Medlin, L. K., Kooistra, W. H. C. F., Potter, D., Saunders, G. W., Andersen, R. A. (1997), 473, 506
Medlin, L. K., William, D. M., Sims, P. A. (1993), 570, 592
Meffert, M. E. (1987), 128–129, 195
Meffert, M. E. (1988), 128–129, 195
Meffert, M. E., Krambeck, H. J. (1977), 129, 195
Meijer, M.-L., Hosper, H. (1997), 28, 52
Melkonian, M. (1990), 336, 351
Melkonian, M. (1996), 387, 419
Melkonian, M., Robenek, H., Reize, I. B., Preisig, H. (1987), 473, 506
Melkonian, M., Surek, B. (1995), 257, 308
Melzer, A., Haber, W., Kohler, A. (1977), 811, 831
Mensinger, A. F., Case, J. F. (1992), 687, 712
Mercado, A. (1977), 180, 195
Metzeltin, D., Lange-Bertalot, H. (1995), 659, 668
Meulemans, J. T., Heinis, F. (1983), 39, 52
Meyer, K. (1929), 664, 668
Meyer, R. L. (1969), 433–434, 446–447, 453, 459, 461, 467
Meyer, R. L. (1971), 478, 499, 501, 507
Meyer, R. L. (1995), 443, 467
Meyer, R. L., Brook, A. J. (1969), 412, 419, 491, 507, 512, 520, 698, 709, 712
Meyer, R. L., Wheeler, J. H., Brewer, J. R. (1970), 429, 432–434, 447, 449, 454, 459, 461, 463, 467
Meyer, S. R., Pienaar, R. N. (1981), 736, 754
Meyer, S. R., Pienaar, R. N. (1984a), 747, 754
Meyer, S. R., Pienaar, R. N. (1984b), 736, 747, 754
Meyer-Harms, B., Pollehne, F. (1998), 690, 712
Mickle, A. M., Wetzel, R. R. (1978), 39, 52
Mignot, J. P. (1965), 740, 744, 754
Mignot, J. P. (1967), 427–428, 468
Mignot, J. P. (1976), 427–428, 468
Mignot, J. P., Brugerolle, G. (1982), 528, 531, 554
Mignot, J. P., Brugerolle, G., Bricheux, G. (1987), 404–405, 419
Miller, A. R., Lowe, R. L., Rosenberry, J. T. (1987), 604, 634
Miller, P. E., Scholin, C. A. (1998), 257, 308
Miller, S. R., Wingard, C. E., Castenholz, R. W. (1998), 66, 114
Millie, D. R., Pearl, H. W., Hurley, J. P. (1993), 783, 801
Milliman, J. D., Meade, R. H. (1983), 28, 52
Mills, E. L., Leach, J. H., Carlton, J. T., Secor, C. L. (1993), 587, 592
Minshall, G. W. (1978), 29, 32, 52, 776, 801
Minshall, G. W., Petersen, R. C., Cummins, K. W., Bott, T. L., Sedell, J. R., Cushing, C. E., Vannote, R. L. (1983), 29, 31–32, 52
Mitchell, D.S., Pieterse, A. H., Murphy, K. J. (1989), 810, 831
Mitsch, W. J., Gosselink, J. G. (1993), 38, 52, 813, 831
Moeller, R. E., Burkholder, J. M., Wetzel, R. G. (1988), 25, 39, 52
Moestrup, Ø. (1982), 432, 468
Moestrup, Ø. (1991), 225, 228, 230, 250
Moestrup, Ø. (1994), 512, 514, 520
Moestrup, Ø. (1995), 484, 507
Moestrup, Ø., Andersen, R. A. (1995), 474, 507
Moestrup, Ø., Thomsen, H. A. (1980), 491, 507, 514, 520
Moestrup, Ø., Thomsen, H. A. (1995), 512, 520
Moestrup, Ø (1995), 524, 528, 533, 554
Moewus, F. (1940), 432, 468
Moewus, F. (1959), 235, 250
Moll, R. A., Stoermer, E. F. (1982), 567, 592
Mollenhauer, D. (1970), 180, 195
Mollenhauer, D., Mollenhauer, R. (1996), 180, 195
Mollenhauer, D., Mollenhauer, R., Kluge, M. (1996), 180, 195
Momeu, L., Péterfi, L. S. (1979), 524, 533, 554
Monastersky, R. (1998), 384, 419
Monegue, R. L., Phlips, E. J. (1991), 819, 831
Montegut-Felkner, A. E., Triemer, R. E. (1997), 387, 391, 401, 419
Montejano, G., Gold, M., Komárek, J. (1993), 100, 106–107, 114
Montejano, G., Gold, M., Komárek, J. (1997), 62, 99, 102, 114
Montoya, H.T., Golubic, S. (1991), 66, 114
Moore, B. N.(1937), 561, 592
Moore, G. T., Carter, N. (1923), 82, 114
Moore, G. T., Carter, N. (1926), 434, 451, 468
Moore, G. T., Kellerman, K. F. (1905), 822, 831
Moore, J. K., Villareal, T. A. (1996), 567, 592
Moore, J. W. (1974), 27, 52
Moore, J. W. (1979), 407, 419
Moore, J. W. (1980), 605, 634
Moore, J. W. (1981), 699–701, 712
Moore, M. V., Winner, R. W. (1989), 825, 831
Morabito, G., Curradi, M. (1997), 587, 592
Moraczewski, I. R., Zakrys, B. (1992), 419
Morgan, K., Kalff, J. (1975), 738, 754
Morisawa, M. (1968), 28, 52
Morita, E., Abem T., Tsuzuki, M., Fujiwara, S., Sato, M., Hirata, A., Sonoike, K., Nozaki, H. (1998), 234, 250
Morita, E., Abem T., Tsuzuki, M., Fujiwara, S., Sato, M., Hirata, A., Sonoike, K., Nozaki, H. (1999), 235, 250
Morrall, S., Greenwood, A. D. (1980), 734, 754
Morrall, S., Greenwood, A. D. (1982), 735, 754

Morrill, L. C., Loeblich, A. R., III (1981), 690, 712
Morris, I., (1980), 485, 507
Moser, G., Lange-Bertalot, H., Metzeltin, D. (1998), 643, 652
Moss, B. (1973), 407, 419
Moss, B., Beklioglu, M., Carvalho, L., Klinic, S., McGowan, S., Stephen, D. (1997), 23, 44, 52
Moss, D., Wright, J. F., Furse, M. T., Clarke, R. T. (1999), 788, 801
Moss, M. O., Carter, J. R. (1982), 627, 634
Moss, M. O., Gibbs, G., Gray, V., Ross, R. (1978), 665, 668
Mosser, J. L., Mosser, A. G., Brock, T. D. (1977), 23, 43, 52
Mosto-Cascallar, P. (1987), 270, 308
Mrozinska, T. (1995), 348, 351
Mühling, M., Scott, M., Harris, N., Whitton, B. A. (1997), 132, 195
Mulholland, P. J. (1996), 776, 801
Müller, D. G. (1979), 768, 773
Müller, D. G., Geller, W. (1978), 758–760, 765, 767–768, 771, 773
Müller, K. M., Sheath, R. G., Vis, M. L., Crease, T. J., Cole, K. M. (1998), 12, 34, 52, 199, 206, 210, 221–222
Müller, K. M., Sherwood, N. R., Rueschel, C. M., Gutell, R. R., Sheath, R. G. (2002), 221–222
Müller, K. M., Vis, M. L., Chiasson, W. B., Whitick, A., Sheath, R. G. (1997), 203, 205, 221–222
Müller, O. (1903), 569, 592
Müller, O. (1906), 569, 592
Müller, O. F. (1786), 471, 495, 507
Munawar, M. (1972), 406, 419
Munawar, M., Bistricki, T. (1979), 716, 734, 741, 744, 754
Munawar, M., Munawar, I. F. (1976), 806, 831
Munawar, M., Munawar, I. F. (1978), 454, 461, 468
Munawar, M. Munawar, I. F. (1981), 230, 234, 238, 250
Munawar, M., Munawar, I. F. (1982), 513, 520
Munawar, M., Munawar, I. F. (1996), 19, 21, 52
Munawar, M., Munawar, I. F. (2000), 19, 52
Munawar, M., Talling, J. F. (1986), 19, 52
Munch, C. S. (1980), 605, 634
Mundie, J. R. (1929), 432, 468
Mur, L., Bejsdorf, R. O., (1978), 23, 52
Murkin, H. R., Stainton, M. P. Boughen, J. A., Pollard, J. B., Titman, R. D. (1991), 39, 52
Murphy, K. J., Barrett, P. R. F. (1993), 823, 831
Murtaugh, P. A. (1996), 786, 801
Muylaert, K., Sabbe, K. (1996), 587, 592

N
Nägeli, C. (1849), 62, 114
Nagy, J. P. (1965), 197, 208, 222
Nakamura, H. (1994), 385, 419
Nakayama, T., Watanabe, S. Inouye, I. (1996a), 230, 250
Nakayama, T., Watanabe, S. Inouye, I. (1996b), 226, 250
Nakayama, T., Watanabe, S., Mitsui, K., Uchida, H., Inouye, I. (1996), 257, 308
Nakazawa, A., Krienitz, L., Nozaki, H. (2001), 235, 248, 250
National Academy of Sciences (1976), 818, 831
National Academy of Sciences (1987), 43, 52
Nauwerck, A. (1979), 478, 507
Neale, P. J., Talling, J. F., Heaney, S. I., Reynolds, C. S., Lund, J. W. G. (1991), 20, 52
Necchi, O., Jr. (1993), 202, 204, 222
Necchi, O., Jr. (1995), 202, 222
Necchi, O., Jr. (1997), 203, 222
Necchi, O., Jr., Entwisle, T. J. (1990), 215–216, 223
Necchi, O., Jr., Sheath, R. G., Cole, K. M. (1993a), 201, 203–204, 211–213, 221, 223
Necchi, O., Jr., Sheath, R. G., Cole, K. M. (1993b), 201, 212, 221, 222

Necchi, O., Jr., Sheath, R. G., Cole, K. M. (1993c), 214–215, 221, 222
Neel, J. K., Peterson, S. A., Smith, W. L. (1973), 817, 831
Neely, R. K (1994), 24, 39, 52
Neil, J. H. (1975), 810, 831
Neilan, B. A., Jacobs, D., De Lot, T., Blackall, L. L., Hawkins, P. R., Cox, P. T., Goodman, E. (1997), 63, 114
Neill, C., Cronwell, J. C. (1992), 801
Newbold, J. D., Elwood, J. W., O'Neill, R. V., Van Winkle, W. (1981), 31, 52
Newman, J. R., Barrett, P. R. F. (1993), 822, 831
Newman, S., Aldridge, F. J., Phlips, E. J., Reddy, K. R. (1994), 776, 801
Nicholls, K. H. (1978), 491, 507
Nicholls, K. H. (1979), 490, 507
Nicholls, K. H. (1981a), 489, 507
Nicholls, K. H. (1981b), 478, 495, 497, 500, 507
Nicholls, K. H. (1981c), 503, 507
Nicholls, K. H. (1984a), 484, 500, 503, 507
Nicholls, K. H. (1984b), 500, 503, 507
Nicholls, K. H. (1984c), 497, 503, 507
Nicholls, K. H. (1984d), 507
Nicholls, K. H. (1985), 500, 507
Nicholls, K. H. (1987), 478, 503, 507
Nicholls, K. H. (1988), 500, 507
Nicholls, K. H. (1989), 503, 507
Nicholls, K. H. (1990), 480, 501, 503, 507
Nicholls, K. H. (1995), 487, 490, 507
Nicholls, K. H. (1996), 822, 831
Nicholls, K. H. (2000), 477, 498, 503, 507
Nicholls, K. H., Beaver, J. L., Estabrook, R. H. (1982), 807, 831
Nicholls, K. H., Carney, E. C., Robinson, G. W. (1977), 486, 507
Nicholls, K. H., Gerrath, J. F. (1985), 807, 831
Nicholls, K. H., Hopkins, G. J. (1993), 813, 831
Nicholls, K. H., Nakamoto, L., Keller, W. (1992), 488, 507
Nicholls, K. H., Standen, D. W., Hopkins, G. J. (1980), 806, 813, 831
Nicholls, K. H., Standen, D. W., Hopkins, G. J., Carney, E. C. (1977), 813, 831
Nichols, H. W. (1973), 740, 754
Nichols, H. W., Bold, H. C. (1965), 349, 351
Nichols, H. W., Nichols, M. S., Thomas, C. M., Deacon, J. S., Veith, M. (1991), 301, 309
Nichols, K. H., Beaver, J. L., Estabrook, R. H. (1982), 512–513, 520
Nichols, K. H., Nakamoto, L., Keller, W. (1992), 513, 520
Nichols, K. H. (1978), 518–520, 539, 554
Nichols, K. H. (1979), 512, 520
Nichols, K. H. (1982), 552, 555
Nichols, K. H. (1987), 535, 555
Nichols, K. H. (1988a), 524, 535, 552, 555
Nichols, K. H. (1988b), 535, 555
Nichols, K. H. (1995), 535–537, 555
Nichols, K. H. (1998), 693, 698, 701, 708, 712
Nichols, K. H., Beaver, J. R., Estabrook, R. H. (1982), 23, 52
Nichols, K. H., Gerrath, J. F. (1985), 526, 528, 531, 533, 535, 541, 551–552, 555
Nichols, S. A. (1973), 817, 831
Nichols, S. A. (1984), 811, 831
Niederlehner, B. R., Cairns, J. C., Jr. (1994), 801
Niiyama, Y., Watanabe, M., Umezaki, I. (1993), 141, 147, 195
Nipkow, F. (1927), 568, 592
Nipkow, F. (1950), 573, 592
Nixdorf, B., Mischke, U., Le_mann, D. (1998), 489, 507
Noonan, T. A. (1998), 818, 831

Nordin, R. N., Stein, J. R. (1980), 177, 195
Norris, R. E. (1967), 114
Norris, R. E. (1980), 225, 227, 230, 250
Norris, R. E., Munch, C. S. (1970), 526, 541, 555
Northcote, T. G., Larkin, P. A. (1963), 13, 40, 52
Norton, T. A., Melkonian, M., Andersen, R. A. (1996), 410, 419
Novarino, C. (1991a), 734, 736, 741, 754
Novarino, C. (1991b), 734, 736, 741, 754
Novarino, C. (1993a), 734, 754
Novarino, C. (1993b), 734, 754
Novarino, C., Lucas, I. A. N. (1993), 734, 741–743, 754
Novarino, C., Lucas, I. A. N., Morrall, S. (1994), 734, 741, 748–749, 754
Nozaki, H. (1981), 240–241, 250
Nozaki, H. (1982), 240, 242, 250
Nozaki, H. (1983), 226, 243, 248, 250
Nozaki, H. (1986), 227, 245, 247–248, 250
Nozaki, H. (1988), 227, 241–242, 250
Nozaki, H. (1989a), 238–239, 243, 250
Nozaki, H. (1989b), 247–248, 250
Nozaki, H. (1993), 257, 309
Nozaki, H. (1994), 234, 250
Nozaki, H. (1995), 227, 240, 245, 250
Nozaki, H., Aizawa, K., Watanabe, M. M. (1994a), 233–234, 247–248, 250
Nozaki, H., Aizawa, K., Watanabe, M. M. (1994b), 227, 250
Nozaki, H., Ito, M. (1994), 226, 238, 243, 250
Nozaki, H., Ito, M., Sano, R., Uchida, H., Watanabe, M. M., Kuroiwa, T. (1995a), 226, 239, 242, 250
Nozaki, H., Ito, M., Sano, R., Uchida, H., Watanabe, M. M., Kuroiwa, T. (1996a), 234, 250
Nozaki, H., Ito, M., Sano, R., Uchida, H., Watanabe, M. M., Kuroiwa, T. (1996b), 244, 251
Nozaki, H., Ito, M., Sano, R., Uchida, H., Watanabe, M. M., Kuroiwa, T. (1997a), 226, 239, 241–243, 247, 250
Nozaki, H., Ito, M., Sano, R., Uchida, H., Watanabe, M. M., Kuroiwa, T. (1997b), 226, 239, 250
Nozaki, H., Ito, M., Sano, R., Uchida, H., Watanabe, M. M., Kuroiwa, T. (1997c), 226, 251
Nozaki, H., Krienitz, L. (2001), 251
Nozaki, H., Kuroiwa, H., Kuroiwa, T. (1994b), 227, 251
Nozaki, H., Kuroiwa, H., Mita, T., Kiroiwa, T. (1989), 226, 239, 241, 248, 251
Nozaki, H., Kuroiwa, T. (1990), 242, 248, 251
Nozaki, H., Kuroiwa, T. (1991), 241, 248, 251
Nozaki, H., Kuroiwa, T. (1992), 238–239, 241–242, 248, 251
Nozaki, H., Misawa, K. Kajita, T., Kato, M., Nohara, S., Watanabe, M. M. (2000), 239, 243–244, 251
Nozaki, H., Ohta, N., Morita, E., Watanabe, M. M. (1998b), 227, 233–234, 247, 251
Nozaki, H., Ohta, N., Takano, H., Watanabe, M. M. (1999), 239, 242–244, 251
Nozaki, H., Ohtani, S. (1992), 227, 245, 248, 251
Nozaki, H., Onishi, K., Morita, E. (2002a), 226, 235, 242, 251
Nozaki, H., Song, L.-R., Liu, Y.-D., Hiroki, M., Watanabe, M. M. (1998a), 242, 247, 251
Nozaki, H., Takahara, M., Nakazawa, A., Kita, Y., Yamada, T., Takano, H., Kawano, S., Kato, M. (2000b), 251
Nozaki, H., Watanabe, M. M. Aizawa, K. (1995b), 234, 251
Nudelman, M. A., Lombardo, R., Conforti, V. (1998), 414, 419
Nultsch, W. (1974), 119, 195
Nygaard, G. (1978), 472, 486, 503, 507, 526, 555
Nygaard, G. (1984), 139, 195
Nyström, P. Brönmark, C., Granéli, W. (1996), 26, 53

O
Oakley, B. R., Bisalputra, T. (1977), 736, 754
Oakley, B. R., Dodge, J. D. (1973), 736, 754
Oakley, B. R., Dodge, J. D. (1976), 736, 754
Oakley, B. R., Heath, I. B. (1978), 736, 754
Oakley, B. R., Santore, U. J. (1982), 732, 754
Odum, E. P., Finn, J. T., Franz, E. H. (1979), 784–785, 801
O'Farrell, I. (1994), 586, 592
Ogawa, R. E., Carr, J. F. (1969), 806, 831
O'Grady, K., Brown, L. M. (1989), 513, 520
Ohtani, S. (1986), 363, 380
Ohtani, S. (1990), 379–380
O'Kelly, C. J. (1989), 432, 468
O'Kelly, C. J. Watanabe, S., Floyd, G. I. (1994), 257, 309
Oldfield, F., Appleby, P. G. (1984), 780, 801
Oliver, R. L. (1994), 805, 831
Oliver, R. L., Ganf, G. G. (2000), 12, 53, 120, 195
Olli, K. (1996), 404, 408, 419
Olrik, K. (1998), 489, 507
Olsen, J. L. (1990), 473, 507
Olsen, Y., Vadstein, O., Andersen, T., Jensen, A. (1989), 22, 53
Olson, R. K. (1993), 813, 831
Oltmanns, F. (1895), 432, 468
O'Neal, S. W., Lembi, C. A., Spencer, D. L. (1985), 812, 831
Onodera, H., Satake, M., Oshima, Y., Yasumoto, T., Carmichael, W. W. (1997), 809, 831
Ormerod, G. K. (1970), 809, 831
Ortega, M. M. (1984), 227, 232, 234–235, 241–242, 251, 411–413, 415, 419, 472, 507, 698, 712
Osorio-Tafall, B. F. (1942), 701, 712
Ostrofsky, M. L., Duthie, H. (1975), 488, 507
Ott, D. W., Brown, R. M. (1974), 432, 468
Ott, D. W., Brown, R. M. (1978), 468
Ott, D. W., Hommersand, M. H. (1974), 432, 463, 468
Ott, F. D. (1972), 206, 221, 223
Ott, F. D. (1976), 203, 208, 221, 223
Owens, J. L., Crumpton, W. G. (1995), 37, 53
Owens, K. J., Farmer, M. A., Triemer, R. E. (1988), 385, 401, 411, 415, 419
Ozimek, T., Pieczynska, E., Hankiewicz, A., (1991), 810, 831

P
Paasche, E., Johansoon, S., Evenson, D. L. (1975), 565, 569, 592
Pace, M. L., Findlay, S. E. G., Links, D. (1992), 38, 53
Padan, E., Rimon, A., Ginzburg, D., Shilo, M. (1971), 819, 831
Padisák, J. (1985), 699–701, 712
Padisák, J. (1990), 174, 195
Padisák, J. (1997), 174, 195
Paerl, H. W., Bowles, N. D. (1987), 37, 53
Paerl, H. W., Priscu, J. C., Brawner, D. L. (1989), 20, 53
Paerl, H.W., Bowles, N. D. (1987), 65, 114
Paerl, H.W. (1996), 64, 114
Paerl, H.W., Millie, D. F. (1996), 64, 114
Paine, R. T. (1980), 818, 831
Painter, D. S., Kamaitis, G. (1987), 809, 813, 831
Palamar-Mordvinsteva, G. (1976), 378, 380
Palinska, K. A., Leisack, W., Rhiel, E., Krumbein, W. E. (1996), 66, 114
Palmer, C. M. (1962), 44, 53, 776, 801, 807, 831
Palmer, C. M. (1969), 406, 419, 673, 683, 775, 787, 801
Palmer, C. M. (1980), 405–406, 419
Palmer, J. D. (1995), 686, 712
Palmer, J. D. (1996), 686, 712
Palmer, J. D., Round, F. E. (1965), 23, 53, 405, 407–408, 412, 419
Palmer, T. C. (1902), 415, 420

Pan, Y., Stevenson, R. J. (1996), 39, 53, 787, 801
Pan, Y., Stevenson, R. J., Hill, B. H., Herlihy, A. T., Collins, G. B. (1996), 784, 787, 790, 792, 794, 801
Pan, Y., Stevenson, R. J., Vaithiyanathan, P., Slate, J., Richardson, C. J. (2000), 783, 786, 790, 794–795, 801
Pandian, T. J., Marian, M. P. (1986), 35, 53
Pankow, H. (1976), 166, 195
Papenfuss, G. F. (1951), 757, 759, 773
Pappas, J. L., Stoermer, E. F. (2001a), 600, 631, 634
Pappas, J. L., Stoermer, E. F. (2001b), 600, 631, 634
Parducz, B. (1967), 716, 754
Park, H.-D., Hayashi, H. (1992), 694, 696, 712
Park, H.-D., Hayashi, H. (1993), 694, 699, 712
Park, N. E., Karol, K. G., Hoshaw, R. W., McCourt, R. M. (1996), 353, 380
Parke, M., Lund, J. W. G., Manton, I. (1962), 513, 517, 520
Parke, M., Martin, I., Clarke, S. (1955), 511, 514, 519–520
Parker, B. C., Preston, R. E., Fogg, G. E. (1963), 431, 468
Parker, B. C., Samsel, G. E., Prescott, G. W. (1973), 488–489, 499, 507
Parker, B. C., Wenkert, L. J., Parson, M. J. (1991), 18, 53
Parker, B. C., Wolfe, H. E., Howard, R. V. (1975), 15, 53
Parker, D. L. (1982), 60, 114
Pascher, A. (1910), 480, 507
Pascher, A. (1914), 312, 351, 471, 507
Pascher, A. (1917), 482, 507
Pascher, A. (1925), 481, 483–484, 507, 524, 555
Pascher, A. (1927), 233, 246, 251
Pascher, A. (1929), 483, 507
Pascher, A. (1931), 312, 351, 388, 420, 484, 507
Pascher, A. (1937-1939), 429, 432, 465, 468
Pascher, A. (1939), 503, 507
Pascher, A. (1940a), 478–479, 507
Pascher, A. (1940b), 479, 507
Patrick, R. (1945), 604, 634
Patrick, R. (1949), 775, 801
Patrick, R. (1959), 638, 652
Patrick, R. (1961), 561, 592
Patrick, R. (1973), 784, 801
Patrick, R. (1976), 605, 634
Patrick, R., Freese, L. (1961), 561, 592, 604, 628, 634
Patrick, R., Hohn, M., Wallace, J. H. (1954), 655, 658, 668, 775–776, 780, 784, 801
Patrick, R., Reimer, C. W. (1966), 561–562, 592, 596–601, 605, 628, 634, 637–639, 646, 649, 652, 655, 661–662, 664–665, 668, 670, 676, 683
Patrick, R., Reimer, C. W. (1975), 561–562, 592, 639, 650, 652, 655–656, 661–662, 668, 681–683
Patrick, R., Roberts, N. A. (1979), 561, 592
Patterson, D. J. (1981), 736, 754
Patterson, D. J. (1989), 423, 468
Patterson, D. J., Hedley, S. (1992), 385, 410, 420
Paulsen, S. G., Larsen, D. P., Kaufmann, P. R., Whittier, T. R., Baker, J. R., Peck, D. V., McGue, J., Hughes, R. M., McMullen, D., Stevens, D., Stoddard, J. L., Larzorchak, J., Kinney, W., Selle, A. R., Hjort, R. (1991), 788, 801
Pavoni, M. (1963), 480, 507
Pearlmutter, N., Lembi, C. A. (1986), 823, 832
Pearsall, W. H. (1929), 691, 703, 712
Pejler, B. (1977), 739, 754
Pennak, R. W. (1966), 811, 832
Pennak, R. W. (1973), 811, 832
Pennak, R. W. (1989), 385, 388, 420
Pennick, D. L. (1981), 734, 742, 754
Penno, S., Campbell, L., Hess, W. R. (2000), 384, 420

Pentecost, A. (1982), 44, 53
Pentecost, A. (1991), 364, 380
Pentecost, A., Whitton, B. A. (2000), 44, 53, 119, 121, 195
Perakis, S. S., Welch, E. B., Jacoby, J. M. (1996), 815, 832
Perasso, L., Brett, S. J., Wetherbee, R. (1993), 736, 754
Perez, C. C., Roy, S., Levasseur, M., Anderson, D. M. (1998), 696, 712
Perez, M. C., Bonilla, S., deLe\'f3n, L., \'8amarda, J., Komáek, J. (1999), 177–178, 195
Pernthaler, J., Simek, K., Sattler, B., Schwarzenbacher, A., Bobkova, J. Psenner, R. (1996), 20, 53
Perrin, C. J., Bothwell, M. L., Slaney, P. A. (1987), 604, 634
Perty, M. (1849), 413, 420
Perty, M. (1852), 413, 420
Péterfi, L. S. (1969), 477, 507
Péterfi, L. S. (1996), 555
Péterfi, L. S., Momeu, L. (1977), 533–534, 555
Peters, M. C., Andersen, R. A. (1993), 480, 484, 503, 508
Petersen, J. B., Hansen, J. B. (1956), 526, 533, 552, 555
Petersen, J. B., Hansen, J. B. (1958), 526, 533–534, 555
Peterson, C. G. (1996), 36, 53, 605, 634
Peterson, C. G., Grimm, N. B. (1992), 674, 683
Peterson, C. G., Hoagland, K. D. (1990), 623, 634
Peterson, C. G., Stevenson, R. J. (1989), 37, 53
Peterson, C. G., Stevenson, R. J. (1992), 623, 634, 779, 801
Peterson, H. G., Healey, F. P., Wagemann, R. (1984), 823, 832
Pettersson, K., Herlitz, E., Istvanovics, V. (1993), 23, 53
Petts, G., Calow, P. (1996), 28, 53
Pfiester, L. A. (1975), 694, 712
Pfiester, L. A. (1976), 694, 712
Pfiester, L. A. (1977), 694, 712
Pfiester, L. A. (1984), 694, 712
Pfiester, L. A., Anderson, D. M. (1987), 694, 712
Pfiester, L. A., Highfill, J. F. (1993), 697, 707, 713
Pfiester, L. A., Holt, J. R. (1978), 739, 754
Pfiester, L. A., Lynch, R. A., Skvarla, J. J. (1980), 694, 696, 708, 713
Pfiester, L. A., Popovsky, J. (1979), 696, 708–709, 713
Pfiester, L. A., Skvarla, J. J. (1979), 694, 713
Pfiester, L. A., Skvarla, J. J. (1980), 694, 699, 713
Pfiester, L. A., Timpano, P., Skvarla, J.J., Holt, J. R. (1984), 694, 713
Philipose, M. T. (1982), 384, 402, 404–407, 412–413, 420
Philipose, M. T. (1984), 413, 420
Philipose, M. T. (1988), 416, 420
Phillips, D., Boyne, A. E. (1984), 734, 754
Phillips, G. L., Eminson, D., Moss, B. (1978), 27, 53, 810, 832
Phlips, E. J., Monegue, R. L., Aldridge, F. J. (1990), 819, 832
Phosphorous Management Strategies Task Force (1980), 813, 832
Pick, F. R., Cuhel, R. L. (1986), 535, 555
Pick, F. R., Lean, D. R. S. (1987), 64, 114
Pick, F. R., Nalewajko, C., Lean, D. R. S. (1984), 17, 53, 535, 555
Pickett-Heaps, J. D. (1975), 257, 309, 313, 351
Pickett-Heaps, J. D., Staehelin, L. (1975), 254, 309
Pielou, E. C. (1984), 788, 790, 801
Pienaar, R. N. (1976), 739, 754
Pienitz, R., Lortie, G., Allard, M. (1991), 604, 634
Pienitz, R., Smol, J. P. (1992), 604–605, 634
Pierce, S. (1987), 15, 53
Pillinger, J. M., Cooper, J. A., Ridge, I. (1994), 822, 832
Pillinger, J. M., Cooper, J. A., Ridge, I., Barrett, P. R. F. (1992), 822, 832
Pillsbury, R. W., Kingston, J. C. (1990), 607, 634
Pinder, A. W., Friet, S. C. (1994), 45, 53
Pine, R. T., Anderson, L. W. J. (1991), 820, 832
Pipes, L. D., Leedale, G. F. (1992), 526, 548, 555
Pipes, L. D., Leedale, G. F., Tyler, P. A. (1991), 524, 528, 555

Pipp, E., Rott, E. (1994), 204, 223
Pithart, D., Pechar, L., Mattsson, G. (1997), 428, 468
Pizarro, H. (1995), 270, 309, 465, 468
Plafkin, J. L., Barbour, M. T., Porter, K. D., Gross, S. K., Hughes, R. M. (1989), 786, 801
Platt, T., Li, W. K. W. (1986), 60, 65, 114
Playfair, G. (1921), 415, 420
Playfair, G. I. (1914), 247, 251
Pochmann, A. (1942), 416, 420
Pocock, M. A. (1954), 242–243, 251
Pocock, M. A. (1955), 243–244, 251
Pocock, M. A. (1960), 231–232, 251
Pohlman, J. W., Iliffe, T. M., Cifuentes, L. A. (1997), 44, 53
Pollinger, U. (1981), 738, 754
Pollinger, U. (1986), 513, 520
Pollingher, U. (1987), 20, 53, 699–700, 713
Pollingher, U., Burgi, H. R., Ambühl, H. (1993), 694, 699–700, 713
Pollingher, U., Hickel, B. (1991), 699–700, 713
Polunin, N. (1954), 433, 463, 468
Popovsky, J. (1970), 698, 702, 713
Popovsky, J. (1982), 708, 713
Popovsky, J., Pfiester, L. A. (1990), 698, 708, 713
Porter, K. G., McDonough, R. (1984), 808, 832
Porter, K. G. (1973), 819, 832
Porter, K. G. (1977), 783, 801
Porter, K. G. (1988), 20, 23, 53
Porter, K. G., Cuffney, T. F., Gurtz, M. E., Meador, M. R. (1993), 779–780, 784, 801
Potter, D., Saunders, G. W., Andersen, R. A. (1997), 424, 468
Potter, M. C. (1888), 351
Poulin, M. E. (1990), 604, 634
Poulin, M., Hamilton, P. B., Proulx, M. (1995), 587, 592, 698, 713
Poulton, E. M. (1930), 424–425, 429, 432–433, 441, 459, 461, 463, 468
Power, M. E. (1990), 674, 683, 776, 783, 801, 810, 832
Power, M. E., Stewart, A. J. (1987), 36, 53
Power, M. E., Stout, R. J., Cushing, C. E., Harper, P. P., Hauer, F. R., Matthews, W. J., Moyle, P. B., Statzner, B., Wais De Badgen, I. R. (1988), 32, 53
Prasad, A. K. S. K., Nienow, J. A., Livingston, R. J. (1990), 565, 592
Prát, S. (1929), 121, 141, 144, 195
Preisendorfer, R. W. (1996), 783, 801
Preisig, H. R. (1986), 485, 508
Preisig, H. R. (1989), 432, 468
Preisig, H. R. (1995), 471, 508
Preisig, H. R. (1999), 257, 309
Preisig, H. R., Anderson, O. R., Corliss, J. O., Moestrup, Ø., Powell, M. J., Roberson, R. W., Wetherbee, R. (1994), 399, 420
Preisig, H. R., Hibberd, D. J. (1982a), 503, 508, 533, 555
Preisig, H. R., Hibberd, D. J. (1982b), 503, 508, 533, 555
Preisig, H. R., Hibberd, D. J. (1983), 523, 555
Preisig, H. R., Hibberd, D. J. (1986), 523, 555
Preisig, H. R., Hibberd, D. J. (1987), 503, 508
Preisig, H. R., Melkonian, M. (1984), 231, 247, 251
Preisig, H. R., V\'f8rs, N., Hällfors, G. (1991), 474, 508
Prepas, E. E., Murphy, R. P., Crosby, J. M., Walty, D. T., Lim, J. T., Babin, J., Champers, P. A. (1990), 814, 832
Prepas, E. E., Trew, D. O, (1983), 812, 832
Prescott, G. W. (1931), 412, 415, 420, 429, 432–434, 439, 449, 454, 458, 461, 463, 468
Prescott, G. W. (1942), 244, 251
Prescott, G. W. (1944), 432, 441, 468
Prescott, G. W. (1951), 126, 137, 141, 151, 155, 177, 195, 323, 325, 327–328, 348, 351, 688, 693, 708, 713
Prescott, G. W. (1953), 433, 463, 468

Prescott, G. W. (1955), 239, 251, 391, 404, 412, 420
Prescott, G. W. (1962), 6, 9, 12, 40, 53, 60, 63, 67, 72, 80–82, 87–88, 92, 95, 114, 118, 139, 141, 150, 155, 171, 174, 180–181, 184, 189, 195, 307, 309, 313, 348, 351, 369, 380, 397–399, 405, 413, 415, 420, 429, 442, 457, 468, 496, 499, 503, 508, 562, 592, xvi
Prescott, G. W. (1963), 463, 468
Prescott, G. W. (1978), 253, 258, 268–269, 271, 275, 279, 285–286, 293–294, 297, 299–300, 305, 307, 309, 384, 388, 405, 407–415, 420, 425–429, 432, 434, 439–441, 443, 445, 450–452, 458, 463, 468
Prescott, G. W. (1983), 315, 325–326, 329, 332, 345–346, 348, 351
Prescott, G. W. (1984), 388, 420
Prescott, G. W., Bicudo, C., Vinyard, W. C. (1982), 377–379, 381
Prescott, G. W., Croasdale, H. T. (1937), 86, 114
Prescott, G. W., Croasdale, H. T. Vinyard, W. C. (1972), 353, 363, 369–372, 379–380
Prescott, G. W., Croasdale, H. T. Vinyard, W. C. (1975), 372–373, 375, 378–379–380
Prescott, G. W., Croasdale, H. T. Vinyard, W. C. (1977), 375–377, 379–380
Prescott, G. W., Croasdale, H. T. Vinyard, W. C., Bicudo, C. (1981), 373–374, 376–377, 379–380
Prescott, G. W., Dillard, G. E. (1979), 443, 451, 454, 458, 461, 463, 468, 561, 592, 604, 634
Prescott, G. W., Vinyard, W. C. (1965), 89, 114, 432–433, 443, 446, 453–454, 458–459, 461, 468
Pringle, C. M., Naiman, R. J., Bretschko, G., Karr, J. R., Oswood, M. W., Webster, J. R., Welcomme, R. L., Winterbourn, M. J. (1988), 32, 53
Pringle, C. M., Rowe, G., Triska, F. J., Fernandez, J. F., West, J. (1993), 41–42, 53
Pringle, C. W., Bowers, J. A. (1984), 801
Pringsheim, E. G. (1946), 68, 114
Pringsheim, E. G. (1948), 391, 399, 420
Pringsheim, E. G. (1953a), 391, 414–416, 420
Pringsheim, E. G. (1953b), 391, 414–416, 420
Pringsheim, E. G. (1955), 481, 508
Pringsheim, E. G. (1956), 388, 391, 402–403, 406, 412, 415, 420
Pringsheim, E. G. (1960), 247, 251
Pringsheim, E. G. (1963), 387, 420
Pringsheim, E. G. (1968), 738, 742, 754
Pringsheim, E. G., Wiessner, W. (1960), 227, 251
Prinsep, M. R., Caplan, F. R., Moore, R. E., Patterson, G. M. L., Honkanen, R. E., Boynton, A. L. (1992), 121, 195
Printz, H. (1964), 348, 351
Pröschold, T., Marin, B., Schlösser, U. G., Melkonian, M. (2001), 235, 248, 251
Provasoli, L. (1958), 406, 420
Provasoli, L. (1961), 420
Provasoli, L. (1968), 465, 468
Provasoli, L. (1969), 406, 420
Provasoli, L., Pintner, I. J. (1953), 406, 420
Prowse, G. A. (1959), 24, 53
Prowse, G. A. (1962), 526, 534, 555
Prygiel, J., Coste, M., Bukowska, J. (1999a), 776, 791, 802
Prygiel, J., Whitton, B. A., Bukowska, J. (1999b), 776, 791, 802
Pueschel, C. M. (1990), 41, 53, 200, 223
Pueschel, C. M., Sanders, G. W., West, J. A. (2000), 201, 223
Pueschel, C. M., Stein, J. R. (1983), 757–759, 765, 770, 773
Pueschel, C. M., Sullivan, P. G., Titus, J. E. (1995), 33, 53, 218, 223
Putt, M. (1990), 739, 754
Puytorac, P., Mignot, J. P., Grain, J., Groliere, C. A., Bonner, L., Couillard, P. (1972), 526, 541, 555

Q
Quade, H. W. (1969), 811, 832
Quirós, R. (1995), 818, 832

R
Radwan, S., Kowalik, W., Kowalczyk, C. (1990), 785, 802
Rae, R., Vincent, W. F. (1998), 37–38, 53
Raman, R. K. (1985), 806, 822, 832
Ramanthan, K. R. (1964), 342, 349, 351
Ranch, D. C. (1981), 347, 351
Raschke, R. L. (1993), 784, 802
Rashash, D. M. C., Dietrich, A. M., Hoehn, R. C., Parker, B. C. (1995), 487, 508
Raven, J. (1992), 314, 351
Raven, J. A. (1993), 202, 223
Raven, J. A., Johnston, A. M., Newman, J. R., Scrimgeour, C. M. (1994), 204, 223
Rawlence, D. J. (1988), 604, 634
Rawson, D. S. (1956), 454, 461, 468
Rawson, D. S., Moore, A. J. (1944), 42, 53
Rayburn, W. R., Starr, R. C. (1974), 242, 248, 251
Reavie, E. D., Smol, J. P. (1997), 789, 792–793, 802
Reavie, E. D., Smol, J. P. (1998), 789, 792–793, 802
Reavie, E. D., Smol, J. P., Carignan, R., Lorrain, S. (1998), 787, 789, 793, 802
Reavie, E. D., Smol, J. P., Carmichael, N. B. (1995), 604, 635, 806, 832
Redhead, K., Wright, S. J. L. (1978), 819, 821, 832
Reed, R. H., Warr, S. R. C., Kerby, N. W., Stewart, W. D. P. (1986), 65, 114
Reichardt, E. (1999), 655, 659, 661, 668
Reichardt, E., Lange-Bertalot, H. (1991), 665, 668
Reid, M. A., Tibbey, J. C., Penny, D., Gell, P. A. (1995), 776, 802
Reimer, C. W. (1954), 682–683
Reimer, C. W. (1959), 638, 649, 652
Reimer, C. W. (1961), 638, 652
Reimer, C. W. (1966), 635
Reimer, C. W. (1990), 605, 635
Reinersten, H. (1982), 513, 520
Reinhard, E. G. (1931), 36, 53, 65, 114
Reinke, D. C. (1979), 415, 420
Rejmánková, E. Roberts, D. R., Manguin, S., Pope, K. O., Komárek, J., Post, R. A. (1996), 66, 114
Renberg, I. (1976), 599, 613, 635
Renberg, I. (1977), 599, 613, 635
Renberg, I., Hellberg, T. (1982), 791, 802
Rengefors, K., Anderson, D. M. (1998), 696, 713
Reuter, J. E., Loeb, S. L., Goldman, C. R. (1986), 26, 53
Reuter, J. E., Rhodes, C. L., Lebo, M. E., Klotzman, M., Goldman, C. R. (1993), 43, 53, 812, 832
Reynolds, C. S. (1980), 738, 754
Reynolds, C. S. (1982), 738, 755
Reynolds, C. S. (1984), 258, 309, 595, 604, 607, 635, 739, 755
Reynolds, C. S. (1984a), 19–20, 23, 54
Reynolds, C. S. (1984b), 19–20, 54
Reynolds, C. S. (1988), 36–38, 54
Reynolds, C. S. (1995), 37–38, 54
Reynolds, C. S. (1996), 32, 54
Reynolds, C. S., Descy, J.-P. (1996), 36–38, 54, 65, 114
Reynolds, C. S., Jawarski, G. H. M., Roscoe, J. V., Hewitt, D. P., George, D. G. (1998), 487, 508
Reynolds, C. S., Jaworski, G. H. M., Cmiech, H. A., Leedale, G. F. (1981), 64, 114
Reynolds, C. S., Walsby, A. E. (1975), 59, 64, 92, 114
Rhee, G. Y., Goldman, I. J. (1980), 22, 54

Rhodes, R. G., Stofan, P. E. (1967), 429, 468
Rhodes, R. G., Terzis, A. J. (1970), 461, 468
Rhodes, T. E., Davis, R. B. (1995), 587, 592
Rice, E. L. (1984), 821, 832
Richards, C. B. (1962), 709, 713
Rieth, A. (1980), 429, 431, 465, 468
Riley, E. T., Prepas, E. E. (1984), 814, 832
Rindi, F., Guiry, M. D. Barbiero, R. (1999), 342, 349, 351
Rindi, F., Guiry, M. D., Barbiero, R. P., Cinelli, F. (1991), 45, 54
Rintoul, T. C. Sheath, R. G., Vis, M. L. (1999), 212, 221, 223
Rippka, R., Deruelles, J. B., Waterbury, J. B., Herdman, M., Stanier, R. Y. (1979), 60, 62, 114, 117–118, 120, 195
Robarts, R. D. Evans, M. S., Arts, M. T. (1992), 43, 54
Robbins, R. W., Glicker, D. M., Bloem, D. M., Niss, B. M. (1991), 813, 832
Roberts, D. A., Boylen, C. W. (1988), 27, 54
Roberts, D. A., Boylen, C. W. (1989), 27, 54
Roberts, K. R. (1984), 734, 744–745, 755
Roberts, K. R., Stewart, K. D., Mattox, K. R. (1981), 734–735, 746, 755
Robinson, G. G. C. (1983), 25, 54
Robinson, G. G. C., Gurney, S. E., Goldsborough, L. G. (1997), 38, 54
Robinson, P. K., Hawkes, H. A. (1986), 810, 832
Roijackers, R. M. M. (1983), 487, 508
Roijackers, R. M. M., Kessels, H. (1986), 489, 508, 536, 538, 555
Rojo, C., Cobelas, M. a., Arauzo, M. (1994), 36, 54
Röpstorf, P., Hülsmann, N., Hausmann, K. (1994), 708, 713
Rosa, R., Burns, N. M. (1987), 806, 832
Rosemarin, A. S. (1975), 36, 54
Rosemond, A. D. (1993), 205, 223
Rosen, B. H. (1995), 776, 779, 802
Rosen, B. H., Lowe, R. L. (1981), 599, 616, 635
Rosen, G. (1981), 364, 381
Rosenberg, M. (1941), 508
Rosgen, D. L. (1994), 779, 802
Rosine, W. N. (1955), 811, 832
Rosowski, J. R. (1977), 401–404, 420
Rosowski, J. R., Couté, A. (1996), 415, 420
Rosowski, J. R., Glider, W. V. (1977), 402, 404, 406, 420
Rosowski, J. R., Hoagland, K. D., Roemer, S. C., Lee, K. W. (1981), 414, 420
Rosowski, J. R., Hoshaw, R. M. (1970), 410, 420
Rosowski, J. R., Hoshaw, R. M. (1971), 402, 420
Rosowski, J. R., Kugrens, P. (1973), 393, 403–405, 408, 411, 415, 420
Rosowski, J. R., Langenberg, W. G. (1994), 415, 420
Rosowski, J. R., Lee, K. W. (1978), 384, 396, 399, 411–412, 420
Rosowski, J. R., Vadas, R. L. Kugrens, P. (1975a), 414, 420
Rosowski, J. R., Walne, P. L., West, L. K. (1975b), 414, 420
Rosowski, J. R., Wiley, R. L. (1975), 388, 394, 404, 408, 411, 420
Rosowski, J. R., Wiley, R. L. (1977), 392, 403, 420
Ross, R. (1954), 432–433, 454, 461, 468
Ross, R. (1983), 561, 592
Ross, R., Cox, E. J., Karayeva, N. L., Mann, D. G., Paddock, T. B. B., Simnsen, R., Sims, P. A. (1979), 592
Rother, J. A., Fay, P. (1977), 120, 195
Rothhaupt, K. O. (1991), 808, 832
Rott, E. (1983), 738, 755
Rott, E., Pfister, P. (1988), 29, 54
Rott, E., Pipp, E. (1999), 121, 195
Round, F. E. (1963), 353, 381
Round, F. E. (1971), 21, 54, 353, 381
Round, F. E. (1972), 22, 26, 54, 568, 592
Round, F. E. (1981), 11–12, 19–20, 24, 32, 43–44, 54, 433, 468, 485, 508, 578, 592, 669, 683

Round, F. E. (1984), 253, 309, 351, 407, 420
Round, F. E. (1991), 776, 802
Round, F. E. (1997), 597, 635
Round, F. E. (1998), 596, 627, 635
Round, F. E., Basson, P. W. (1997), 598, 603, 627, 635
Round, F. E., Bukhtiyarova, L. (1996), 561, 592
Round, F. E., Bukhtiyarova, L. (1996a), 597–598, 602–603, 623, 626–628, 635
Round, F. E., Bukhtiyarova, L. (1996b), 597, 602, 604, 635
Round, F. E., Crawford, R. M., Mann, D. G. (1990), 559–565, 568, 572, 575, 583–584, 586–587, 592, 596–598, 600–601, 603–604, 607, 635, 637–638, 645–646, 648–649, 652, 655, 662–663, 665, 668, 669–670, 672, 676, 681–683
Round, F. E., Häkanson, H.. (1992), 584, 592
Round, F. E., Maidana, N. I. (2001), 600, 631, 635
Round, F. E., Mann, D. G. (1981), 645, 652
Round, F. E., Palmer, J. D. (1966), 405, 407–408, 412, 420
Round, F. E., Sims, P. A. (1981), 563, 592
Roussomoustakaki, M., Anagnostidis, K. (1991), 61, 66, 114
Rout, J., Gaur, J. P. (1990), 779, 802
Rowell, H. C. (1996), 42–43, 54
Rudi, K., Skulberg, O. M., Jakobsen, K. S. (1998), 62–63, 115, 117, 195
Rudi, K., Skulberg, O. M., Skulberg, R., Jakobsen, K. S. (2000), 117, 195
Rumeau, A., Coster, M. (1988), 787, 802
Rumpf, R. R., Vernon, d., Schreiber, D., Birky, C. W., Jr. (1996), 235, 251
Ruse, L., Love, A. (1997), 37, 54
Ruse, L. P., Hutchings, A. J. (1996), 37–38, 54
Rushforth, S. R., Johansen, J. R. (1986), 575, 592
Ruzicka, J. (1977), 353, 361, 364, 381

S

Sabater, S., Gregory, S. V., Sedell, J. R. (1998), 36, 54, 314, 351
Safferman, R. S., Morris, M. E. (1963), 819, 832
Safferman, R. S., Schneider, I. R., Steare, R. L., Morris, M. E., Diener, T. O. (1969), 819, 832
Saha, L. C., Wujek, D. E. (1990), 485, 508, 536, 538, 555
Sako, Y., Ishida, Y., Kadota, H., Hata, Y. (1984), 699, 713
Sako, Y., Ishida, Y., Kadota, H., Hata, Y. (1985), 694, 696, 713
Sako, Y., Ishida, Y., Kadota, H., Hata, Y. (1987), 694, 699, 713
Sala, S. E., Guerrero, J. M., Ferrario, M. E. (1993), 659, 668
Salonen, K., Jokinen, S. (1988), 489, 508, 534, 555
Salonen, K., Jones, R., Arvola, L. (1984), 23, 54
Sanchez, I., Lee, G. F. (1978), 825, 832
Sand-Jensen, K., Pedersen, O., Gertz-Hansen, O. (1997), 45, 54
Sand-Jensen, K., Søndergaard, M. (1981), 24, 54
Sanders, L., Hoover, J. J., Killgore, K. J. (1991), 820, 832
Sanders, R. W., Porter, K. G., Caron, D. A. (1990), 473, 508
Sanderson, B. L., Frost, T. M. (1996), 700, 713
Sandgren, C. D. (1980a), 485, 508
Sandgren, C. D. (1980b), 485, 508
Sandgren, C. D. (1981), 485, 508, 531, 555
Sandgren, C. D. (1983), 485, 488, 508
Sandgren, C. D. (1988), 19, 54, 485, 490, 508, 531, 535–539, 555
Sandgren, C. D. (1989), 533, 555
Sandgren, C. D. (1991), 531, 533, 555
Sandgren, C. D., Carney, H. J. (1983), 533, 555
Sandgren, C. D., Flanagin, J. (1986), 485, 508, 531, 539, 555
Sandgren, C. D., Smol, J. P., Kristiansen, J. (1995), 503, 508
Sandgren, C. D., Walton, W. E. (1995), 535, 537–538, 555
Santore, U. J. (1977), 734, 741, 744, 755
Santore, U. J. (1978), 736, 755
Santore, U. J. (1982a), 741, 755
Santore, U. J. (1982b), 741, 755
Santore, U. J. (1982c), 735, 742, 755
Santore, U. J. (1983), 741, 755
Santore, U. J. (1984), 735–736, 741, 744, 746, 755
Santore, U. J. (1987), 735, 741–742, 755
Santore, U. J., Greenwood, A. D. (1977), 735, 755
Santos, L. M. A. (1996), 424, 465, 468
Santos, L. M. A., Leedale, G. F. (1993), 538, 555
Santos, L. M. A., Melkonian, M., Kreimer, G. (1996), 424, 468
Sarjeant, W. A. S. (1978), 687, 713
Sarma, P. (1986), 351
Sarnelle, O. (1993), 739, 755
Sarojini, Y. (1994), 406, 421
Saunders, G. W., Hill, D. R. A., Sexton, J. P., Andersen, R. A. (1997), 697, 713
Saunders, G. W., Potter, D., Paskind, M. P., Andersen, R. A. (1995), 473, 508
Saunders, R. D., Dodge, J. D. (1984), 701, 713
Savino, J. E., Stein, R. A. (1982), 810, 832
Say, P. J., Burrows, J. G., Whitton, B. A. (1990), 785, 802
Say, P. J. Whitton, B. A. (1980), 407, 412, 421
Schagerl, M., Angeler, D. (1998), 244, 251
Schalles, J. F., Shure, D. F. (1989), 17, 40, 54
Schardt, J. D. (1994), 809, 811, 832
Schelske, C. L. (1975), 812, 832
Schelske, C. L., Carrick, H. J., Aldridge, F. J. (1995), 23, 54
Schelske, C. L., Stoermer, E. F. (1971), 567, 584, 592, 806, 812, 832
Schelske, C. L., Stoermer, E. F. (1972), 567, 584, 592
Schiller, J. (1933-1937), 709, 713
Schiller, J. (1956), 66, 93, 95, 115
Schindler, D. W. (1975), 806, 832
Schindler, D. W. (1977), 23, 54, 806, 812, 832
Schindler, D. W. (1978), 23, 54, 812, 818, 832
Schindler, D. W. (1985), 23, 54
Schindler, D. W. (1990), 795, 802
Schindler, D. W., Frost, V. E., Schmidt, R. V. (1973), 26, 54
Schindler, D. W., Kling, H., Schmidt, R. V., Prokopowich, J., Frost, V. E., Reid, R. L., Capel, M. (1973), 486, 508, 537, 555
Schlichting, H. E., Jr. (1960), 407, 421
Schlichting, H. E., Jr. (1964), 407, 421
Schloesser, R. E. (1977), 759, 765, 773
Schloesser, R. E., Blum, J. L. (1980), 758–759, 763, 765, 767, 769, 771, 773
Schmid, A. M. (1994), 565, 567, 569, 592
Schmid, A. M. (1997), 596, 613, 635
Schmidle, W. (1902), 481, 508
Schmidt, A. (1994), 37, 54
Schmidt, M. (1899), 664, 666, 668
Schneider, C. W., Lane, C. E. (2000), 463, 468
Schneider, C. W., Lane, C. E., Norland, A. (1999), 433, 465, 468
Schneider, C. W., MacDonald, L. A., Cahill, J. F., Heminway, S. W. (1993), 433, 465, 468
Schnepf, E., Elbrächer, M. (1992), 739, 755
Schnepf, E., Melkonian, M. (1990), 739, 755
Schnepf, E., Niemann, A., Christian, W. (1996), 427, 449. 468
Schnepf, E., Winter, S., Mollenhauer, D. (1989), 739, 755
Scholz, R. J. (1994), 814, 832
Schönbohm, W. (1972), 354, 381
Schoonoord, M. P., Ellis-Adams, A. C. (1984), 485, 508
Schrader, K. K., Blevins, W. T. (1993), 807, 832
Schrader, K. K., de Regt, M. Q., Tucker, C. S., Duke, S. O. (1997), 807, 833
Schultz, M. E., Trainor, F. R. (1968), 569, 592
Schulz, V. P. (1929), 587, 592

Schulz-Baldes, M., Lewin, R. A. (1976), 115
Schumacher, G. J., Bellis, V. J., Whitford, L. A. (1963), 432–434, 453–454, 457, 459, 468
Schumacher, G. J., Kim, Y. C., Whitford, L. A., Dillard, G. E. (1966), 424–425, 427, 446, 454, 458–461, 468
Schumacher, G. J., Whitford, L. A. (1965), 203, 223
Schumm, S. A. (1977), 29, 54
Schuster, F.L. (1968), 744, 755
Schuster, F.L. (1970), 734, 746, 755
Schwartzbach, S. D. Osafune, T., Löffelhardt, W. (1998), 385, 391, 421
Schwarz, A.-M., Hawes, I. (1997), 27, 54
Scott, J. (1983), 201, 223
Seckbach, J. (1991), 41, 54, 201, 221, 223
Sedell, J. R., Riley, J. E., Swanson, F. J. (1989), 32, 54
Segal, S. (1969), 45, 54
Selva, J. (1981), 664, 668
Senna, P. A. C., Komárek, J. (1998), 147–148, 195
Senna, P. A. C., Peres, A. C., Komárek, J. (1998), 83, 115
Serieyssol, K. K. (1981), 582, 592
Serieyssol, K. K., Garduno, I. I., Gasse, F. (1998), 587, 592
Serieyssol, K. K., Theriot, E. C., Gasse, F. (1996), 584
Serruya, C., Pollingher, U. (1983), 15, 54
Seshadri, C. V., Jeeri-Bai, N. (1992), 121, 195
Sespenwol, S. (1973), 735, 741, 746, 755
Setchell, W. A. (1924), 110, 115
Setchell, W. A., Gardner, N. L. (1905), 115
Setchell, W. A., Gardner, N. L. (1918), 150, 195
Setchell, W. A., Gardner, N. L. (1919), 78, 110, 115
Setchell, W. A., Gardner, N. L. (1924), 96, 115, 150, 195
Setchell, W. A., Gardner, N. L. (1930), 101, 107, 115
Seto, R. (1977), 202, 223
Shannon, C. F. (1948), 784, 787, 802
Shannon, E. E., Brezonik, P. L. (1972), 16, 54
Shapiro, J. (1980), 818, 833
Shapiro, J., LaMarra, V., Lynch, M. (1975), 818, 833
Sheath, R. G. (1984), 4, 9, 197–203, 214–215, 217, 219, 223
Sheath, R. G. (1986), 14, 19, 54
Sheath, R. G. (1987), 204, 206, 223
Sheath, R. G., Burkholder, J. M. (1985), 779, 783–784, 802
Sheath, R. G., Burkholder, J. M., Hambrook, J. A. Hogeland, A. M., Hoy, E., Kane, M. E., Morison, M. O., Steinman, A. D., Van Alstyne, K. L. (1986), 205, 223
Sheath, R. G., Cole, K. A. (1992), 339, 342, 351
Sheath, R. G., Cole, K. M. (1980), 26, 34, 54
Sheath, R. G., Cole, K. M. (1984), 200, 205, 223
Sheath, R. G., Cole, K. M. (1990), 200, 223
Sheath, R. G., Cole, K. M. (1992), 33–34, 45, 55, 141, 195, 206, 223, 765–766, 773
Sheath, R. G., Hambrook, J. A. (1988), 200, 203, 223
Sheath, R. G., Hambrook, J. A. (1990), 4–5, 9, 33, 55, 197–200, 202–206, 215–216, 223, 758, 773
Sheath, R. G., Hambrook, J. A., Nerone, C. A. (1988), 206, 223
Sheath, R. G., Havas, M., Hellebust, J. A., Hutchinson, T. C. (1982), 42, 55
Sheath, R. G., Hellebust, J. A. (1978), 424, 427, 429, 433, 454, 459, 461, 463, 468
Sheath, R. G., Hellebust, J. A., Sawa, T. (1979), 200, 223
Sheath, R. G., Hymes, B. J. (1980), 203, 210, 221, 223
Sheath, R. G., Kaczmarczyk, D., Cole, K. M. (1993a), 203–204, 218, 221, 223
Sheath, R. G., Morrison, M. O. (1982), 25, 55, 205–206, 223
Sheath, R. G., Morrison, M. O., Korch, J. E., Kaczmarczyk, D., Cole, K. M. (1986), 461, 469
Sheath, R. G., Müller, K. M. (1997), 34, 55, 202, 223, 433, 461, 469
Sheath, R. G., Müller, K. M., Colbo, M. H.., Cole, K. M. (1996a), 205, 224
Sheath, R. G., Müller, K. M., Larson, D. J., Cole, K. M. (1995), 205, 224
Sheath, R. G., Müller, K. M., Sherwood, A. R. (2000), 218, 224
Sheath, R. G., Müller, K. M., Vis, M. L., Entwisle, T. K. (1996b), 217, 221, 224
Sheath, R. G., Steinman, A. D. (1981), 434, 454, 459, 461, 463, 469
Sheath, R. G., Steinman, A. D. (1982), 80, 82, 87–88, 92, 94, 101, 115, 118, 132, 135, 137, 144, 150, 155, 158, 161, 164, 166, 169, 171, 174, 177, 180, 184, 195, 230, 232, 234, 239, 241–242, 251, 412, 415, 421
Sheath, R. G., Van Alstyne, K. L., Cole, K. M. (1985), 34, 55
Sheath, R. G., Vis, M. L. (1995), 223
Sheath, R. G., Vis, M. L., Cole, K. M. (1992), 201, 203, 213, 221, 223
Sheath, R. G., Vis, M. L., Cole, K. M. (1993b), 200, 203–204, 217–219, 221, 223
Sheath, R. G., Vis, M. L., Cole, K. M. (1993c), 202, 204, 218–219, 220–221, 223
Sheath, R. G., Vis, M. L., Cole, K. M. (1993d), 221, 223
Sheath, R. G., Vis, M. L., Cole, K. M. (1994), 40, 55
Sheath, R. G., Vis, M. L., Cole, K. M. (1994a), 213, 221, 223
Sheath, R. G., Vis, M. L., Cole, K. M. (1994b), 203, 223
Sheath, R. G., Vis, M. L., Cole, K. M. (1994c), 199, 221, 224
Sheath, R. G., Vis, M. L., Hambrook, J. A., Cole, K. M. (1996), 34, 55, 433, 461, 463, 469
Sheath, R. G., Whittick, A.., Cole, K. M. (1994d), 201–203, 212–213, 221, 224
Sheath, R., Munawar, M. (1974), 486, 508
Sheldon, S. P., Wellnitz, T. A. (1998), 36, 55
Sherman, B. J., Phinney, H. K. (1971), 605, 635
Sherwood, A. R., Garbary, D. J., Sheath, R. G. (2000), 342, 351
Sherwood, A. R., Sheath, R. G.(1999a), 203, 224
Sherwood, A. R., Sheath, R. G.(1999b), 218, 221, 224
Sherwood, A. R., Sheath, R. G.(2000a), 218, 224
Sherwood, A. R., Sheath, R. G.(2000b), 202, 218, 221, 224
Shi, Z. (1995), 388, 416, 421
Shi, Z. (1996a), 388, 407, 421
Shi, Z. (1996b), 421
Shi, Z. (1998), 387, 421
Shi, Z. (1999), 387, 391, 421
Shi, Z., Zao, C. (1998), 416, 421
Shilo, M. (1967), 819, 833
Shimmel, S. M., Darley, W. M. (1995), 63, 115
Shin, W., Boo, S. M. (1999), 384, 421
Shireman, J. V., Haller, W. T., Colle, D. E., DuRant, D. F. (1983a), 810, 820, 833
Shireman, J. V., Hoyer, M. V., Maciena, M. J., Canfield, D. E. (1985), 821, 833
Shireman, J. V., Rottmann, R. W., Aldridge, F. J. (1983b), 833
Shoeman, F. R., Archibal, R. E. M. (1980), 569, 593
Shubert, L. E. (1975), 256, 309
Shubert, L. E. (1998), 258, 309
Shubert, L. E., Trainor, F. R. (1974), 256, 309
Sicko-Good, L., Schelske, C. L., Stoermer, E. F. (1984), 568, 593
Sicko-Good, L., Stoermer, E. F., Fahnenstiel, G. (1986), 573, 593
Sicko-Good, L., Stoermer, E. F., Kociolek, J. P. (1989), 567–568, 572–573, 593
Sicko-Good, L., Stoermer, E. F., Ladewski, B. G. (1977), 783, 802
Sieburth, J. M., Smetacek, V., Lenz, J. (1978), 20, 55
Sigworth, E. A. (1957), 776, 802
Silva, H., Sharp, A. J. (1944), 461, 469
Silva, P. C. (1959), 247, 251
Silva, P. C. (1972), 247, 251

Silva, P. C., Papenfuss, G. F. (1953), 226, 251
Simms, R. W., Freeman, P. Hawksworth, D. L. (1988), 348, 351
Simon, N., Brenner, J., Edvardsen, B., Medlin, L. K. (1997), 515, 520
Simons, J., Ohm, M., Daalder, R., Boers, P., Rip, W. (1994), 812, 833
Simons, J., van Beem, A. P., de Vries, P. J. R. (1986), 334, 337, 351
Simonsen, R. (1979), 570, 593, 628, 635
Simonsen, R. (1987), 562, 593, 628, 635
Simonsen, S., Moestrup, Ø. (1997), 512, 520
Simpson, E. H. (1949), 784, 802
Simpson, M. R., MacLeod, B. W. (1991), 807, 833
Simpson, P. S., Eaton, J. W. (1986), 810, 833
Sinclair, C., Whitton, B. A. (1977), 61, 115, 759, 773
Singh, K. P. (1956), 416, 421
Siver, P. A. (1977), 24, 55
Siver, P. A. (1987), 503, 508, 528, 533, 538, 551–552, 555
Siver, P. A. (1988a), 489, 502, 508, 538, 541, 555
Siver, P. A. (1988b), 503, 508, 531, 533, 552, 555
Siver, P. A. (1988c), 533, 555
Siver, P. A. (1988d), 535, 555
Siver, P. A. (1989), 535–536, 556
Siver, P. A. (1991a), 524, 530–531, 533, 535–536, 538, 551–552, 556
Siver, P. A. (1991b), 556
Siver, P. A. (1991c), 533, 556
Siver, P. A. (1992), 556
Siver, P. A. (1993), 487, 508, 535, 537, 556
Siver, P. A. (1994), 556
Siver, P. A. (1995), 535–538, 556, 792, 802
Siver, P. A. (1999), 604, 635
Siver, P. A., Chock, J. S. (1986), 536–538, 556
Siver, P. A., Glew, J. R. (1990), 524, 527, 531, 556
Siver, P. A., Hamer, J. S. (1989), 534–538, 556
Siver, P. A., Hamer, J. S. (1990), 539, 556
Siver, P. A., Hamer, J. S. (1992), 488, 508, 526, 535, 541, 556
Siver, P. A., Hamer, J. S., Kling, H. (1990), 535, 556
Siver, P. A., Hinsch, J. (2000), 631, 635
Siver, P. A., Kling, H. (1997), 565, 593
Siver, P. A., Lott, A. M., Cash, E., Moss, J., Marsicano, L. J. (1999), 535, 537–538, 556
Siver, P. A., Marsicano, L. J. (1993), 556
Siver, P. A., Marsicano, L. J. (1996), 535, 538, 556
Siver, P. A., Skogstad, A. (1988), 535, 539, 556
Siver, P. A., Smol, J. P. (1993), 536, 556
Siver, P. A., Vigna, M. S. (1997), 534, 536, 538, 545–546, 556
Siver, P. A., Wujek, D. E. (1993), 535–538, 552, 556
Siver, P. A., Wujek, D. E. (1999), 552, 556
Siver, P. A.,Smol, J. P. (1993), 802
Skácelová, O., Komárek, J. (1989), 141, 196
Skogstad, A. (1984), 533, 556
Skogstad, A., Reymond, O. L. (1989), 474, 480, 485, 508
Skovgaard, A. (1998), 739, 755
Skuja, H. (1938), 199, 224
Skuja, H. (1939), 75, 115, 741, 755
Skuja, H. (1948), 73, 81, 115, 479, 502, 508, 716, 741, 755
Skuja, H. (1950), 531, 556
Skuja, H. (1956), 524, 556
Skuja, H. (1964), 80, 115
Skulberg, O. M., Carmichael, W. W., Codd, G. A., Skulberg, R. (1993), 141, 169, 174, 196
Skulberg, O. M., Skulberg, R. (1985), 141, 147, 196
Skvortzow, B. W. (1937), 628, 635
Skvortzow, B. W., Meyer, K. (1928), 659, 668
Sládecek, V. (1973), 775, 787, 802
Sládecek, V. (1986), 787, 802
Sládecková, A. (1962), 24–25, 55
Sládecková, A., Marvan, P., Vymazal (1983), 44, 55
Slate, J. E. (1998), 794, 802
Slobodkin, L. B. (1964), 45, 55
Sluiman, H. J., Guihal, C. (1999), 275, 309
Smajs, D., Samarda, J. (1999), 60, 65, 115
Smarda, J. (1991), 61, 115
Smarda, J., Smajs, D., Komrska, J. (1996), 61, 115
Smarda, J., Smajs, D., Komrska, J., Krzyzanek, V. (2002), 61, 115
Smayda, T. J. (1970), 20, 55
Smayda, T. J. (1997), 699–700, 713
Smeltzer, E. (1990), 814, 833
Smilauer, P. (1992), 604, 635
Smith, A. J. (1982), 61, 115
Smith, C. S., Barko, J. W. (1990), 816, 833
Smith, D. W. (1985), 820, 833
Smith, D. W., Brock, T. D. (1973), 206, 224
Smith, G. M. (1916), 429, 434, 461, 469
Smith, G. M. (1918), 453, 469
Smith, G. M. (1920), 60, 74, 76, 78, 83, 86, 90, 92, 94, 115, 118, 132, 142, 147, 196, 425, 432, 458, 465, 469
Smith, G. M. (1925), 60, 115
Smith, G. M. (1933), 60, 115, 323–330, 333, 344, 348, 351
Smith, G. M. (1944), 227, 241–242, 248, 251
Smith, G. M. (1950), 5–6, 9, 11–12, 33, 55, 60, 63, 74, 78, 80, 83, 85, 90, 96, 99–101, 104, 107, 109–110, 115, 118, 127–129, 132–133, 135, 137–139, 141–147, 149, 154–156, 158–160, 164, 166, 168–169, 171–175, 177–180, 182, 184, 187, 189, 196, 225, 227–228, 230–232, 234–236, 238–239, 241–242, 244, 247, 251, 255, 307, 309, 315, 344, 348, 351, 353–354, 363, 370, 381, 385, 387–388, 399, 410–413, 415, 421, 426, 429, 432, 457, 464, 469, 502, 508, 562, 593, 688, 709, 713, 758, 770, 773, xv, xvi
Smith, G. M. (1955), 239, 251
Smith, T. E., Stevenson, R. J., Caraco, N. F., Cole, J. J. (1998), 37, 55
Smith, V. H. (1983), 812, 833
Smol, J. P. (1988), 604, 635
Smol, J. P. (1990), 561, 593, 655, 668, 789, 802
Smol, J. P. (1992), 776, 781, 786, 789, 793, 802
Smol, J. P. (1995), 535–539, 556, 775, 789, 793, 802
Smol, J. P., Charles, D. F., Whitehead, D. R. (1984), 536, 556
Smol, J. P., Glew, J. R. (1992), 780, 802
Soballe, D. M., Kimmel, B. L. (1987), 37, 55
Sokal, R. R., Rohlf, F. J. (1998), 788–789, 802
Solley, W. B., Merk, C. F., Pierce, R. R. (1988), 807, 833
Solomon, J. A., Walne, P. L., Dawson, N. S., Wiley, R. L. (1991), 385, 401, 421
Sommaruga, R. (1995), 20, 55
Sommer, J. R. (1965), 399, 421
Sommer, U. (1982), 755
Sommer, U. (1983), 812, 833
Sommer, U. (1984), 19, 55
Sommer, U. (1988), 20, 55
Sommer, U. (1992), 819, 833
Sommer, U. (1996), 568, 593
Sommer, U., Gliwicz, Z. M., Lampert, W., Duncan, A. (1986), 700, 713
Sommer, U., Kilham, S. S. (1985), 22, 55
Sournia, A. (1978), 703, 713
Sovereign, H. E. (1958), 604, 628, 635
Sovereign, H. E. (1963), 605, 628, 635
Soyer-Gobillard, M.-O. (1996), 686, 713
Spamer, E. E., Theriot, E. C. (1997), 586, 593
Sparks, R. E. (1995), 28, 55
Spaulding, S. A., Kociolek, J. P., Wong, D. (1999), 638, 649, 652
Spaulding, S. A., Stoermer, E. F. (1997), 638, 649, 652

Spector, D. L. (1984), 687, 690, 713
Spence, D. H. N. (1976), 807, 833
Spencer, C. N., King, D. L. (1984), 816, 818–819, 833
Spencer, D. L., Lembi, C. A. (1981), 812, 833
Spencer, L. B. (1971), 465, 469
Spero, H. J. (1985), 701, 713
Speziale, B. J., Dyck, L. A. (1992), 808, 833
Squires, L. E., Rushforth, S. R., Brotherson, J. D. (1979), 605, 635
Squires, L. E., Rushforth, S. R., Endsley, C. J. (1973), 488, 508
Squires, L. E., Whiting, M.C., Brotherson, J. D., Rushforth, S. R. (1979), 701, 713
Sreenivasa, M. R. (1971), 628, 635
Sreenivasa, M. R., Duthie, H. C. (1973), 604, 628, 635
St. Clair, L. L., Rushforth, S. R., (1976), 44, 55
St. Clair, L. L., Rushforth, S. R., Allen, J. V. (1981), 605, 635
Stal, L. (2000), 119, 121, 196
Stanier, R. Y., Cohen-Bazire, G. (1977), 59, 62, 115
Starks, T. L., Shubert, L. E. (1979), 254, 309
Starks, T. L., Shubert, L. E., Trainor, F. R. (1981), 45, 55, 63, 115, 258, 309
Starling, M. B., Chapman, V. J., Brown, J. M. A. (1974), 27, 55
Starmach, K. (1957), 144, 146, 196
Starmach, K. (1966), 91, 98, 115, 128, 136, 147–148, 176–177, 196
Starmach, K, (1969), 202, 224
Starmach, K. (1972), 347–348, 351
Starmach, K. (1973), 77, 115
Starmach, K. (1974), 709, 713
Starmach, K. (1977), 763, 771, 773
Starmach, K. (1983), 387–388, 399, 415, 421
Starmach, K. (1985), 501, 503, 508, 523, 556
Starr, R. C. (1955), 256, 309
Starr, R. C. (1962), 242, 251
Starr, R. C. (1970), 242, 248, 251
Starr, R. C., Zeikus, J. A. (1993), 234, 247, 251, 409–410, 421
Stauffer, R. E., Lee, G. F. (1973), 814, 833
Steidinger, K. A., Burkholder, J. M., Glasgow, H. B., Hobbs, C. H., Garrett, J. K., Truby, E. W., Noga, E. I., Smith, S. A. (1996a), 685–686, 713
Steidinger, K. A., Landsberg, J. H., Truby, E. W., Blakesley, B. A. (1996b), 698, 703, 713
Steil, W. N. (1944), 349, 351
Stein, F. (1878), 411, 415, 421, 471, 502, 509
Stein, J. R. (1958a), 226, 242–243, 251
Stein, J. R. (1958b), 226, 243, 252
Stein, J. R. (1959), 244–245, 248, 252
Stein, J. R. (1965), 244, 252
Stein, J. R. (1973), 258, 309
Stein, J. R. (1975), 180, 196, 247, 252, 363, 370–371, 373, 377, 381, 425–429, 433–434, 443, 446–447, 451, 454, 458–459, 461, 463, 469, 477, 483, 509
Stein, J. R., Amundsen, C. C. (1967), 43–44, 55
Stein, J. R., Borden, C. A. (1978), 425, 428–429, 434, 443, 446, 449, 451, 453–454, 458–459, 461, 463, 469
Stein, J. R., Borden, C. A. (1979), 72, 80–82, 87–88, 92, 95–96, 101, 115, 118, 126, 132, 135, 137, 141, 147, 150–151, 155, 158, 161, 164, 166, 169, 171, 174, 177, 180, 184, 189, 196, 232, 234–236, 238–239, 241–242, 244–245, 252, 411–413, 415, 421, 491, 509, 698, 713
Stein, J. R., Gerrath, J. F. (1969), 373, 381, 429, 433–434, 451, 453–454, 459, 463, 469
Steinkötter, J., Bhattacharya, D., Semmerlroth, I., Bibeau, C., Melkonian, M. (1994), 230, 252
Steinman, A. D. (1996), 32, 55, 819, 833
Steinman, A. D., McINtire, C. D. (1986), 205, 224
Steinman, A. D., McIntire, C. D., Lowry, R. R. (1988), 35, 55

Steinman, A. D., Meeker, R. H., Rodusky, A. J., Davis, W. P., Hwang, S.-J. (1997), 28, 55, 811, 833
Steinman, A. D., Sheath, R. G. (1984), 33, 55
Steinmüller, K., Kaling, M., Zetsche, K. (1983), 41, 55
Stemberger, R. S., Gilbert, J. J. (1985), 739, 755
Stephens, D. W., Gillespie, D. M. (1976), 43, 55
Sterrenberg, F. A. S. (1994), 586, 593
Stevenson, R. J. (1983), 36, 55
Stevenson, R. J. (1984), 784, 788, 802
Stevenson, R. J. (1990), 779, 802
Stevenson, R. J. (1996), 766, 773, 776, 779, 782–783, 802
Stevenson, R. J. (1996a), 24, 33, 55
Stevenson, R. J. (1996b), 35, 55
Stevenson, R. J. (1997), 34, 55, 784, 795, 802
Stevenson, R. J. (1998), 776, 783, 793, 795–796, 802
Stevenson, R. J. (2001), 777, 802
Stevenson, R. J., Bahls, L. L. (1999), 775–776, 778–779, 782–784, 786–787, 791, 802
Stevenson, R. J., Bothwell, M. L., Lowe, R. L. (1996), 24, 55, 604, 635, 672, 683
Stevenson, R. J., Glover, R. (1993), 785, 802
Stevenson, R. J., Hashim, S. (1989), 780, 802
Stevenson, R. J., Lowe, R. L. (1986), 776, 784, 803
Stevenson, R. J., Pan, Y. (1999), 776, 782, 784, 789, 803
Stevenson, R. J., Singer, R., Roberts, D. A., Boylen, C. W. (1985), 783, 803
Stevenson, R. J., Stoermer, E. F. (1981), 27, 55, 780, 803
Stevenson, R. J., Stoermer, E. F. (1982), 623, 635
Stevenson, R. J., Sweets, P. R., Pan, Y., Schultz, R. E. (1999), 778, 792, 803
Steward, K. K. (1993), 811, 833
Stewart, A. J., Blinn, D. W. (1976), 699, 713
Stewart, K. D., Mattox, K. R. (1975), 313, 351
Stewart, K. D., Mattox, K. R. (1978), 313, 351
Stewart, W. D. P. (1972), 119, 196
Stewart, W. D. P. (1980), 119, 121, 196
Stewart, W. D. P., Daft, M. J. (1977), 819, 833
Stiller, J. W., Hall, B. D. (1997), 387, 421
Stockner, J. G. (1967), 41, 55
Stockner, J. G. (1988), 65, 115
Stockner, J. G., Armstrong, F. A. J. (1971), 26, 55
Stockner, J. G., Benson, W. W. (1967), 605, 635
Stockner, J. G., Callieri, C., Cronberg, G. (2000), 20, 56, 120, 196
Stockner, J. G., Lund, J. W. G. (1970), 573, 593
Stockner, J. G., Shortreed, K. S. (1988), 22, 56, 812, 833
Stockner, J. G., Shortreed, K. S. (1991), 65, 115
Stoecker, D. K. (1998), 490, 509, 701, 713
Stoecker, D. K., Michaels, A. E., Davis, L. H. (1987), 755
Stoecker, D. K., Silver, M. W. (1990), 739, 755
Stoecker, D. K., Silver, M. W., Michaels, A. E., Davis, L. H. (1988/89), 755
Stoermer, E. F. (1963), 650, 652
Stoermer, E. F. (1967), 560, 569, 593
Stoermer, E. F. (1968), 598, 635
Stoermer, E. F. (1975), 605, 635
Stoermer, E. F. (1978), 585, 587, 593
Stoermer, E. F. (1980), 596, 605, 627–628, 635
Stoermer, E. F. (1988), 806–807, 833
Stoermer, E. F., Andersen, N. A., Schelske, C. L. (1992), 568, 593
Stoermer, E. F., Edlund, M. B. (1998), 568, 593
Stoermer, E. F., Emmert, G., Julius, M. L., Schelske, C. L. (1996), 583, 593
Stoermer, E. F., Emmert, G., Schelske, C. L. (1989), 586, 593
Stoermer, E. F., Håkansson, H. (1983), 581, 593
Stoermer, E. F., Håkansson, H., Theriot, E. C. (1987), 578, 581, 593

Stoermer, E. F., Julius, M. L. (2002), 670, 683
Stoermer, E. F., Kingston, J. C., Sicko-Goad, L. (1979), 567, 569, 586, 593
Stoermer, E. F., Kociolek, J. P., Cody, W. (1990), 583, 593
Stoermer, E. F., Kociolek, J. P., Schelske, C. L., Andresen, N. A. (1991), 605, 635
Stoermer, E. F., Kociolek, J. P., Schelske, C. L., Conley, D. J. (1985c), 561, 593
Stoermer, E. F., Kreis, R. G., Jr. (1978), 561, 593
Stoermer, E. F., Kreis, R. G., Jr., Andresen, N. A. (1978a), 635
Stoermer, E. F., Kreis, R. G., Jr., Andresen, N. A. (1999), 598, 604, 635, 638, 653
Stoermer, E. F., Ladewski, B. G., Schelske, C. L. (1978b), 613, 635
Stoermer, E. F., Ladewski, T. B. (1976), 583, 593
Stoermer, E. F., Ladewski, T. B. (1982), 560, 593
Stoermer, E. F., Ladewski, T. B., Kociolek, J. P. (1986), 560, 593
Stoermer, E. F., Qi, Y., Ladewski, T. B. (1986), 664, 668
Stoermer, E. F., Sicko-Goad, L. (1977), 512, 520
Stoermer, E. F., Smol, J. P. (1999), 10, 56, 560, 571, 593, 595, 597, 604, 635, 776, 782–783, 788, 803
Stoermer, E. F., Wolin, J. A., Schelske, C. L., Conley, D. J. (1985a), 568, 586, 593
Stoermer, E. F., Wolin, J. A., Schelske, C. L., Conley, D. J. (1985b), 568–569, 573, 593
Stoermer, E. F., Yang, J. J. (1969), 583, 593, 595, 604–605, 628, 635
Stoermer, E. F., Yang, J. J. (1970), 595, 598, 605, 616, 635
Stoermer, E. F., Yang, J. J. (1971), 656, 661, 663, 668
Stokes, A. C. (1885), 477, 502, 509
Stokes, A. C. (1886), 472, 498, 509
Stokes, P. M. (1986), 26, 39, 56
Stokes, P. M., Yung, Y. K. (1986), 672, 682–683
Stoneburner, D. L., Smock, L. A. (1980), 699, 713
Stosch. *see* von Stosch
Strahler, A. N. (1957), 28, 56
Straub, F. (1985), 627, 635
Straub, F. (1990), 627, 635
Stross, R. G., Sokol, R. C., Schwarz, A.-M., Howard-Williams, C. (1995), 27, 56
Suda, S., Watanabe, M. M., Inouye, I. (1989), 228, 230, 252
Sugawara, H., Miyazaki, S. (1999), 68, 115, 121, 196
Sulli, C., Fang, Z. W., Muchhal, U., Schwartzbach, S. D. (1999), 385, 421
Sullivan, M. J. (1979), 613, 635
Sunderman, F. W. (1978), 826, 833
Surek, B.. Beemelmanns, U., Melkoniam, M., Bhattacharya, D. (1994), 351
Surek, B., Melkonian, M. (1986), 385, 401, 421
Sutherland, J. M., Reaston, J., Stewart, W. D. P., Herdman, M. (1985a), 120, 135, 196
Sutherland, J. M., Reaston, J., Stewart, W. D. P., Herdman, M. (1985b), 120, 196
Suttle, C. A., Harrison, P. J. (1988), 65, 115
Suttle, C. A., Stockner, J. G., Harrison, P. J. (1987), 20, 22, 56
Suttle, C. A., Stockner, Shortreed, K. S., J. G., Harrison, P. J. (1988), 20, 56
Sutton, D. L., Porter, K. M. (1989), 821, 833
Suxena, M. R. (1955), 415, 421
Suykerbuyk, R. E. M., Roijackers, R. M. H., Houtman, S. S. J. (1995), 487, 509
Svedelius, N. (1930), 759, 765, 770, 773
Swain, E. B., Monson, B. A., Pillsbury, R. W. (1986), 822, 833
Sweets, P. R. (1992), 787, 803
Sze, P. (1986), 381
Sze, P. (1998), 1, 5, 9
Sze, P., Kingsbury, J. M. (1972), 42–43, 56

Szymanska, H., Spalik, K. (1993), 332, 348, 351
Szymanska, H., Zakrys, B. (1990), 763–764, 766, 773

T
Taft, C. E. (1964), 347, 349, 352
Taft, C. E. (1978), 702, 714
Taft, C. E., Taft, C. W. (1970), 139, 180, 196
Taft, C. E., Taft, C. W. (1971), 415, 421, 698, 701, 714
Takahashi, E. (1978), 524, 526, 528, 536, 539, 551–552, 556
Takahashi, E., Hayakawa, T. (1979), 536, 556
Tanaka, J., Kamiya, M. (1993), 202, 224
Tani, Y., Tsumura, H. (1989), 406, 421
Tarapchak, S. J. (1972), 424–427, 429, 432–434, 441–453, 456–459, 469
Targett, N. M., Arnold, T. M. (1998), 765, 773
Taub, F. B., Kindig, A. C., Meador, J. P., Swartzman, G. L. (1989), 825, 833
Tavera, R., Komárek, J. (1996), 60, 67, 83, 87, 96–97, 101, 103–104, 107, 115, 134, 177, 196
Taylor, F. J. R. (1980), 697, 714
Taylor, F. J. R. (1987), 687, 714
Taylor, F. J. R. (1999), 697, 714
Taylor, J. F. R. (1990), 388, 421
Taylor, W. D., Hern, S. C., William, L. R., Lambou, V. W., Morris, M. K., Morris, F. A. (1979), 738, 755
Taylor, W. D., Sanders, R. W. (1991), 385, 421
Taylor, W. D., Wee, J. L., Wetzel, R. G. (1986), 539, 556
Taylor, W. R. (1928), 115, 462, 469
Taylor, W. R. (1934), 429, 469
Teiling, E. (1941), 90, 115
Teiling, E. (1948), 379, 381
Teiling, E. (1950), 361, 381
Teiling, E. (1952), 362, 381
Teiling, E. (1957), 376, 381
Teiling, E. (1967), 379, 381
Tell, G., Conforti, V. (1984), 416, 421
Tell, G., Conforti, V. (1986), 388, 414–416, 421
ter Braak, C. J. F. (1988), 604, 635
ter Braak, C. J. F., van Dam, H. (1989), 787, 791, 803
Tett, P., Gallegos, C., Kelly, M. G., Hornerger, G. M., Cosby, B. J. (1978), 803
Thérézien, Y. (1999), 388, 414, 416, 421
Theriot, E. C. (1987), 568–570, 586, 593
Theriot, E. C. (1990), 570, 593
Theriot, E. C. (1992), 568–570, 586, 593
Theriot, E. C., Bradbury, J. P. (1987), 570, 584, 593
Theriot, E. C., Fritz, S. C., Gresswell, R. E. (1997), 41, 56
Theriot, E. C., Håkansson, H., Kociolek, J. P., Round, F. E., Stoermer, E. F. (1987), 570, 578, 593
Theriot, E. C., Håkansson, H., Stoermer, E. F. (1988), 567
Theriot, E. C., Kociolek, J. P. (1986), 570
Theriot, E. C., Serieyssol, K. (1994), 570
Theriot, E. C., Stoermer, E. F. (1984a), 568–569, 586
Theriot, E. C., Stoermer, E. F. (1984b), 560, 568, 586
Theriot, E. C., Stoermer, E. F. (1986), 567–569, 586
Thirb, H. H., Benson-Evans, K. (1982), 203, 224
Thomas, W. H., Duval, B. (1995), 44, 56
Thompson, R. H. (1938), 83, 86, 115, 412, 421, 433, 461, 463, 469
Thompson, R. H. (1947), 688, 698, 702, 706, 714
Thompson, R. H. (1949), 688, 698, 709, 714
Thompson, R. H. (1950), 688–689, 702, 709, 714
Thompson, R. H. (1954), 241–242, 252
Thompson, R. H. (1972a), 315, 352
Thompson, R. H. (1972b), 330, 352
Thompson, R. H. (1975), 765, 771, 773

Thompson, R. H., Halicki, P. J. (1965), 513, 520
Thompson, R. H., Wujek, D. E. (1989), 232, 248, 252
Thompson, R. H., Wujek, D. E. (1996), 345, 349, 352
Thompson, R. H., Wujek, D. E. (1997), 45, 56, 315, 343, 348, 352
Thompson, R. H., Wujek, D. E. (1998a), 483, 503, 509
Thompson, R. H., Wujek, D. E. (1998b), 496–498, 509
Thomsen, H. A., Zimmerman, B., Moestrup, Ø., Kristiansen, J. (1981), 503, 509
Thorp, J. H., Black, A. R., Haag, K. H., Wehr, J. D. (1994), 28, 32, 38, 56
Thorp, J. H., Black, A. R., Jack, J. D., Casper, A. F. (1996), 786, 803
Thorp, J. H., Delong, M. D. (1994), 31–32, 56
Threlkeld, S. T., Chiavelli, D. A., Willey, R. L. (1993), 408, 411, 421
Threlkeld, S. T., Willey, R. L. (1993), 408, 421
Tiffany, L. H. (1930), 38, 341, 352
Tiffany, L. H. (1934), 83, 115
Tiffany, L. H. (1936), 381
Tiffany, L. H. (1937), 325, 348, 352, 429, 433, 461, 469
Tiffany, L. H. (1944), 348, 352
Tiffany, L. H., Britton, M. E. (1944), 433–434, 459, 462–463, 469
Tiffany, L. H., Britton, M. E. (1952), 118, 158, 161, 164, 166, 169, 174, 177, 196, 325, 330, 348, 352, 710, 714
Tiftickjian, J. D., Rayburn, W. R. (1986), 364, 381
Tilden, J. (1910), 60, 91, 96, 115, 117, 132, 135, 137, 139, 141, 144, 147, 150–151, 155, 158, 161, 164, 166, 169, 171, 174, 180–181, 184, 189, 196
Tillett, D., Parker, D. L. Neilan, B. A. (1999), 63, 115
Tilman, D. (1977), 22, 56
Tilman, D. (1982), 22, 56
Tilman, D., Kiesling, R. L. (1984), 812, 833
Tilman, D., Kiesling, R., Sterner, S. S., Johnson, F. A. (1986), 487, 509
Tilman, D., Kilham, S. S. (1976), 487, 509
Tilman, D., Kilham, S. S., Kilham, P. (1976), 607, 636
Tilman, D., Kilham, S. S., Kilham, P. (1982), 487, 509, 568, 604, 636
Timpano, P. (1978), 771, 773
Timpano, P. (1980), 759, 771, 773
Tippet, R. (1970), 25, 56
Titman, D. (1976), 568, 585
Tokuyasu, K., Scherbaum, O. H. (1965), 403, 421
Tomas, R. N., Cox, E. R. (1973), 686, 707, 714
Tomaselli, L., Palandri, M. R., Tredici, M. R. (1996), 132, 196
Toriumi, S., Dodge, J. D. (1993), 690, 714
Trainor, F. R. (1978), 255–256, 309
Trainor, F. R. (1991), 255, 309
Trainor, F. R. (1998), 254–256, 309
Trainor, F. R., Cain, J. R., Shubert, L. E. (1976), 256, 309
Trainor, F. R., Egan, P. (1990a), 255, 309
Trainor, F. R., Egan, P. (1990b), 255, 309
Trainor, F. R., Egan, P. (1990c), 255, 309
Trainor, F. R., Morales, E. A. (1999), 256–257, 309
Trainor, F. R., Shubert, L. E. (1973), 785, 803
Transeau, E. (1913), 424–425, 463, 469
Transeau, E. (1916), 364, 381
Transeau, E. (1917), 429, 459, 469
Transeau, E. (1925), 369, 381
Transeau, E. (1926), 381
Transeau, E. (1933), 372, 381
Transeau, E. (1951), 355, 362–363, 369–372, 379, 381
Tranvik, L. J., Porter, K. G., Sieburth, J. (1989), 740, 755
Trelease (1889), 171, 196
Trench, R. K., Thinh, L. V. (1995), 687, 714
Triemer, R. E. (1992), 405, 421
Triemer, R. E., Farmer, M. A. (1991a), 385, 387, 391, 401, 421

Triemer, R. E., Farmer, M. A. (1991b), 387, 391, 401, 405, 421
Triemer, R. E., Lewandowski, C. L. (1994), 385, 401, 408, 421
Tschermak-Woess, E., Kasel-Fetzman, E. (1992), 509
Tschermak-Woess, W. (1980), 482, 509
Tuchman, M. L., Stevenson, R. J. (1980), 36, 56
Tuchman, M. L., Stoermer, E. F., Carney, J. J. (1984), 607, 636
Tuchman, M. L., Theriot, E. C., Stoermer, E. F. (1984), 560, 569
Tuchman, M., Stevenson, R. J. (1980), 780, 803
Tuchman, N. C. (1996), 673, 683
Tucker, C. S., Boyd, C. E. (1978), 824, 833
Tucker, C. S., Busch, R. L., Lloyd, S. W. (1983), 810, 833
Tupa, D. D. (1974), 322–323, 332, 348, 352
Turner, M. A., Howell, E. T., Robinson, G. G. C., Brewster, J. F., Sigurdson, L. J., Findlay, D. L. (1995), 370, 372, 381
Turner, M. A., Howell, E. T., Robinson, G. G. C., Campbell, P., Hecky, R. E., Schindler, E. U. (1994), 26, 56
Turner, M. A., Schindler, E. U., Findlay, D. L., Jackson, M.B., Robinson, G. G. C. (1995), 26, 39, 56
Turpin, D. H., Harrison, P. J. (1979), 19, 56
Twist, H., Edwards, A. C., Codd, G. A. (1997), 786, 803
Tyler, P. A. (1996), 12, 56
Tyler, P. A., Pipes, L. D., Croome, R. L., Leedale, G. F. (1989), 524, 526–527, 533–534, 548, 551–552, 556

U

U. S. Environmental Protection Agency (1978), 803
U. S. Environmental Protection Agency (1992), 776–777, 795, 803
U. S. Environmental Protection Agency (1993), 794, 803
U. S. Environmental Protection Agency (1996), 776–777, 786, 803
U. S. Environmental Protection Agency (1998), 776–777, 788, 803
Ueyema, S., Kobayashi, H. (1988), 665, 668
Umana-Villalobos, G. (1993), 15, 56
Umezaki, I. (1974), 147, 196
Urabe, J. (1993), 819, 833
Urabe, J., Clasen, J., Sterner, R. W. (1997), 819, 833
Utermöhl, H. (1958), 67, 115

V

Van Baalen, C. (1965), 68, 115
Van Baalen, C. (1987), 121, 196
Van Baalen, C., O'Donnell, R. (1972), 88, 115
Van Dam, H., Mertenes, A., Sinkeldam, J. (1994), 791, 803
van den Berg, M. S., Coops, H., Noordhuis, R., van Schie, J., Simons, J. (1997), 28, 56
van den Berg, M. S., Coops, H., Simons, J. (1998b), 811–812, 834
van den Berg, M. S., Coops, H., Simons, J., DeKeizer, A. (1998a), 812, 834
van den Hoek, C. (1982), 340, 348, 350
van den Hoek, C., Mann, D. G., Jahns, H. M. (1995), 1, 5, 9, 225, 252, 312–313, 322, 340–341, 345–347, 350, 353, 381, 384, 387, 399, 401, 405, 421, 423–425, 427, 429, 432, 469, 474, 509, 560, 567, 757–759, 773
van der Ploeg, M., Dennis, M. E., de Regt, M. Q. (1995), 807, 834
van der Zweerde, W. (1993), 820, 834
van Donk, E., Grimm, M. P., Gulati, R. D., Kline Breteler, J. P. G. (1990), 816, 834
Van Etten, J. L., Lane, L. C., Meints, R. H. (1991), 819, 834
Van Everdingen, R. P. (1970), 41, 56
Van Heukelem, L., Lewitus, A. J., Kana, T. M. (1992), 783, 803
Van Heurck, H. (1880), 562
Van Heurck, H. (1881), 562, 621, 636
Van Heurck, H. (1882), 562
Van Heurck, H. (1883), 562
Van Heurck, H. (1884), 562
Van Heurck, H. (1885), 562

Van Landingham, J. I. (1978), 669, 683
Van Landingham, S. L. (1964)
Van Landingham, S. L. (1969), 656, 665, 668
Van Landingham, S. L (1967), 665, 668
Van Landingham, S. L (1978), 659, 668
van Leeuwenhoek, A. (1700), 225, 252
Van Niewenhuyse, E. E., Jones, J. R. (1996), 38, 56
Vanni, M. J., Findlay, D. L. (1990), 818, 834
Vanni, M., Layne, C. D. (1997), 22, 56
Vannote, R. L., Minshall, G. W., Cummins, K. W., Sedell, J. R., Cushing, C. E. (1980), 31, 56, 779, 803
Vazquez, G., Moreno-Casasola, P., Barrera, O. (1998), 45, 56
Venkataraman, G. S. (1961), 465, 469
Venkataraman, G. S. (1969), 68, 115
Verb, R. G., Vis, M. L., Ott, D. W., Wallace, R. L. (1999), 463, 469
Vesk, M. Hoffman, L. R., Pickett-Heaps, J. D. (1984), 499, 509
Vickerman, K. (1990), 385, 421
Vickerman, K., Brugerolle, G., Mignot, J.-P. (1991), 385, 422
Vigna, M. S., Kristiansen, J. (1996), 486, 509
Villeneuve, V., Vincent, W. F., Komárek, J. (2001), 166, 196
Vincent, W. F. (2000), 121, 196
Vinebrook, R. D. (1996), 27, 56
Vinebrook, R. D., Hall, R. I., Leavitt, P. R., Cumming, B. F. (1998), 43, 56
Vinebrooke, R. D., Graham, M. D. (1997), 803
Vinyard, W. C. (1955), 314, 352
Vinyard, W. C. (1958), 434, 454, 459, 461, 469
Vinyard, W. C. (1966), 458–459, 469
Viroux, L. (1997), 38, 56
Vis, M. L, Carlson, T. A., Sheath, R. G. (1991), 203, 224
Vis, M. L, Entwisle, T. J. (2000), 215, 224
Vis, M. L., Saunders, G. W., Sheath, R. G., Dunse, K., Entwisle, T. J. (1998), 218, 221, 224
Vis, M. L, Sheath, R. G. (1992), 203–204, 216–217, 221, 224
Vis, M. L, Sheath, R. G. (1993), 204, 208, 210, 221, 224
Vis, M. L, Sheath, R. G. (1996), 203, 213, 221, 224
Vis, M. L, Sheath, R. G. (1997), 221, 224
Vis, M. L, Sheath, R. G. (1998), 221, 224
Vis, M. L, Sheath, R. G. (1999), 216, 221, 224
Vis, M. L, Sheath, R. G., Cole, K. M. (1992), 203–204, 212, 221, 224
Vis, M. L, Sheath, R. G., Cole, K. M. (1996a), 203, 213–214, 221, 224
Vis, M. L, Sheath, R. G., Cole, K. M. (1996b), 213, 221, 224
Vis, M. L., Sheath, R. G., Hambrook, J. A., Cole, K. M. (1994), 4, 9
Vogel, S. (1984), 200, 224
Vollenweider, R. A. (1969), 514, 520, 812, 834
Vollenweider, R. A. (1976), 775, 782, 803
von Christen, H. R. (1961), 706, 709–710
von Daday, E. (1905), 711, 714
von Stosch, H. A. (1965), 694, 713
von Stosch, H. A. (1972), 694, 713
von Stosch, H. A. (1973), 694, 713
Vonshak, A. (1997), 121, 196
Vørs, N. (1992a), 716, 735, 749–750, 755
Vørs, N. (1992b), 716, 735, 749–750, 755
Vørs, N. Johansen, B., Havskum, H. (1990), 486–487, 509
Vymazal, J. (1984), 785, 803
Vymazal, J. (1994), 776, 803
Vyverman, W., Compere, P. (1991), 649, 653
Vyverman, W., Sabbe, K., Mann, D. G., Kilroy, C., Vyverman, R., Vanhutte, K., Hodgson, D. (1998), 656, 668

W
Waern, M. (1938), 764, 773

Waern, M. (1952), 758, 763–767, 770–771, 773
Wagemann, R., Barica, J. (1979), 824, 834
Wales, G. H. (1934), 710, 714
Walker, H. L., Higginbotham, L. R. (2000), 819, 834
Walker, I. R., Paterson, C. G. (1986), 617, 636
Walker, W. W., Jr. (1987), 813, 834
Walker, W. W., Jr., Westberg, C. E., Schuler, D. J., Bode, J. A. (1989), 816, 834
Wall, D., Dale, B. (1968), 687, 714
Wallace, J. H. (1960), 661, 665–666, 668
Wallace, J. H., Patrick, R. M. (1950), 665, 668
Walne, P. L. (1971), 391–392, 422
Walne, P. L., Dawson, N. S. (1993), 387, 402, 422
Walne, P. L., Gualtieri, P. (1994), 402, 422
Walne, P. L., Kivic, P. A. (1990), 387, 406, 422
Walne, P. L., Möestrup, P., Norris, R. E., Ettl, H. (1986), 402, 413, 422
Walne, P. L., Passarelli, V., Lenzi, P., Barsanti, L., Gualtieri, P. (1998), 392, 402, 422
Walsby, A. E. (1972), 61, 115
Walsby, A. E. (1978), 61, 115
Walsby, A. E. (1981), 61, 115
Walsby, A. E., Xypolyta, A. (1977), 567, 587
Walters, C. J., Krause, E., Neill, W. E., Northcote, T. G. (1987), 19, 56
Walton, L. B. (1915), 415, 422
Wang, J. C., Hoshaw, R. W., McCourt, R. M. (1986), 371, 381
Ward, A. K., Dahm, C. N., Cummins, K. W. (1985), 36, 56, 180, 196
Ward, D. M., Castenholz, R. W. (2000), 41, 56, 121, 196
Ward, J. V., Stanford, J. A. (1983), 31, 56
Ward, K. A., Willey, R. L. (1981), 411, 422
Warner, R. W. (1971), 42, 56
Watanabe, M. (1971), 120, 196
Watanabe, M. (1999), 60, 115
Watanabe, M., Furuya, M. (1982a), 738, 755
Watanabe, M., Furuya, M. (1982b), 738, 755
Watanabe, M., Komárek, J. (1989), 118, 150, 156, 196
Watanabe, M., Miyoshi, M., Furuya, M. (1976), 738, 755
Watanabe, S., Floyd, G. (1989), 253, 256, 309
Watanabe, T., Asai, K., Houki, A., Tanaka, S., Hizuka, T. (1986), 787, 803
Waterbury, J. B. (1979), 79, 115
Waterbury, J. B. (1989), 115
Waterbury, J. B., Rippka, R. (1989), 63, 116
Waterbury, J. B., Stanier, R. Y. (1977), 62, 116
Waterbury, J. B., Stanier, R. Y. (1978), 60, 62, 79, 99, 110, 116
Wawrik, F. (1979), 533, 556
Wawrzyniak, L. A., Andersen, R. A. (1985), 552, 556
Weber, C. I. (1970), 585
Weber, F. I. (1973), 775–776, 803
Weber van Bosse, A. A. (1925), 110, 116
Webster, K. E., Peters, R. h. (1978), 808, 834
Webster, P. (1989), 401, 422
Wee, J. L. (1982), 523–524, 526, 528, 530, 533, 551–552, 556
Wee, J. L. (1983), 491, 509, 539–540, 556
Wee, J. L. (1996), 473, 509
Wee, J. L. (1997), 530–531, 533–534, 556
Wee, J. L., Booth, D. J., Bossier, M. A. (1993), 526, 535–536, 541, 552, 556
Wee, J. L., Gabel, M. (1989), 535–536, 556
Wee, J. L., Harris, S. A., Smith, J. P., Dionigi, C. P., Millie, D. F. (1994), 535, 556
Wee, J. L., Hinchey, J. M., Nguyen, K. X., Hurley, D. L. (1996), 473, 509

Wehr, J. D. (1981), 34, 56
Wehr, J. D. (1989), 20, 57, 65, 116
Wehr, J. D. (1990), 20, 57, 65, 116
Wehr, J. D. (1991), 20, 57
Wehr, J. D. (1992), 65, 116
Wehr, J. D. (1993), 22, 57
Wehr, J. D., Brown, L. M. (1985), 23, 57, 513, 520
Wehr, J. D., Brown, L. M., O'Grady, K. (1985), 513, 520
Wehr, J. D., Brown, L. M., O'Grady, K. (1987), 23, 57, 513, 520
Wehr, J. D., Descy, J.-P. (1998), 28, 32, 37, 57, 65, 116, 776, 803
Wehr, J. D., Lonergan, S. P., Thorp, J. H. (1997), 38, 57
Wehr, J. D., Stein, J. R. (1985), 33, 57, 342, 352, 758, 763–766, 770, 773
Wehr, J. D., Thorp, J. H. (1997), 32, 37, 57, 65, 116
Wehr, J. D., Whitton, B. A. (1983), 42, 57
Wei, Y. (1996), 513, 520
Weisse, T. (1993), 20, 57, 65, 116
Weisse, T., Kirchhoff (1997), 701, 714
Weitzel, R. L. (1979), 766, 773
Welch, E. B., Cooke, G. D. (1995), 814, 834
Welch, E. B., Cooke, G. D. (1999), 814, 834
Welch, E. B., DeGasperi, C. L., Spyridakis, D. E., Belnick, T. J. (1988), 814, 834
Welch, I. M., Barrett, P. R. F., Gibson, M. T., Ridge, I. (1990), 822, 834
Welcomme, R. L. Ryder, R. A., Sedell, J. A. (1989), 31, 57
Wemmer, D. E., Wedemeyer, G. J., Glazer, A. N. (1993), 737, 755
Wenrich, D. H. (1923), 412, 415, 422
Wenrich, D. H. (1924), 412, 415, 422
West, G. S. (1904), 463, 469
West, J. A. (1990), 765–766, 773
West, J. A., Kraft, G. T. (1996), 757, 759–760, 766–767, 771, 773
West, W., West, G. S. (1904), 379, 381
West, W., West, G. S. (1905), 379, 381
West, W., West, G. S. (1908), 379, 381
West, W., West, G. S. (1912), 379, 381
West, W., West, G. S., Carter (1923), 379, 381
Westbroek, P., Brown, C. W., van Bleijswijk, J., Brownlee, C., Brummer, G., Conte, M., Egge, J., Fernandéz, E., Jordan, R., Knappertsbusch, M., Stefels, M., Velduis, M., van der Wal, P., Young, J. (1993), 511, 520
Wetherbee, R., Andersen, R. A. (1992), 473, 509
Wetherbee, R., Hill, D. R. A., Brett, S. J. (1987), 732, 741, 755
Wetherbee, R., Hill, D. R. A., McFadden, G. I. (1986), 732, 741, 755
Wetherbee, R., Ludwig, M., Koutoulis, A. (1995), 531, 557
Wetherbee, R., Platt, S. J., Beech, P. L., Pickett-Heaps, J. D. (1988), 473, 509
Wettstein, R. (1924), 60, 116
Wetzel, R. G. (1964), 43, 57
Wetzel, R. G. (1975), 32, 57
Wetzel, R. G. (1983), 538, 557, 776, 783, 803
Wetzel, R. G. (1983a), 12, 18–19, 23–24, 40, 42, 57
Wetzel, R. G. (1983b), 24, 57
Wetzel, R. G. (1983c), 24, 57
Wetzel, R. G. (1990), 16, 57, 814, 834
Wetzel, R. G. (1996), 776, 803
Wetzel, R. G. (2001), 806–807, 834
Wetzel, R. G., Likens, G. E. (1991), 514, 520, 779–780, 783, 785, 803
Wetzel, R. G., Ward, A. K. (1992), 32, 57
Whale, G. F., Walsby, A. E. (1984), 119, 196
Whatley, J. M. (1993), 387, 422
Wheeler, B. D., Whitton, B. A. (1971), 764, 773
Whelden, R. M. (1941), 99, 116, 459, 469
Whelden, R. M. (1947), 82, 96, 98, 101, 116, 118, 134, 137, 155, 158, 161, 164, 166, 169, 180, 184, 196
Whitaker, J., Barica, J., Kling, H., Buckley, M. (1978), 822, 824, 834
Whitehead, D. R., Charles, D. F., Goldstein, R. A. (1990), 803
Whitford, L. A. (1960), 35, 57, 203, 224
Whitford, L. A. (1969), 314, 352, 496, 509
Whitford, L. A. (1970), 483, 509
Whitford, L. A. (1977), 758, 773
Whitford, L. A. (1979), 428, 434, 443, 447, 451, 469, 512, 514, 520
Whitford, L. A., Schumacher, C. J. (1984), 363, 370, 381
Whitford, L. A., Schumacher, G. J. (1963), 196
Whitford, L. A., Schumacher, G. J. (1969), 63, 67, 72, 79–80, 82, 88, 92, 95–96, 116, 118, 126, 132, 137, 139, 141, 144, 147, 150–151, 155, 158, 161, 164, 166, 169, 171, 174, 180–181, 184, 189, 196, 315, 335, 348, 352, 413, 415, 422, 424, 428–429, 433–434, 443, 449–453, 458–459, 461, 465, 469, 477–482, 496, 500, 509
Whitford, L. A., Schumacher, G. J. (1973), 503, 509
Whitford, L. A., Schumacher, G. J. (1984), 239, 252, 408, 422, 698, 714, xvi
Whitford, L. A., Shumacher, G. J. (1984), 295, 303, 306, 309
Whiting, M. C., Brotherson, J. D., Rushforth, S. R. (1978), 699–700, 714
Whitmore, T. J. (1989), 787, 803
Whittaker, R. H. (1975), 38, 40, 57
Whitton, B. A. (1970), 340, 352, 782–783, 803, 822, 834
Whitton, B. A. (1973), 822, 834
Whitton, B. A. (1975), 11–12, 28, 32–35, 57, 203–204, 224
Whitton, B. A. (1977), 61, 116
Whitton, B. A. (1984), 28, 57, 121, 196, 785, 803
Whitton, B. A. (1987), 118, 164, 196
Whitton, B. A. (1992), 61, 66–67, 116
Whitton, B. A. (2000), 121, 196
Whitton, B. A., Burrows, I. G., Kelly, M. G. (1989), 785, 803
Whitton, B. A., Carr, N. G. (1982), 116
Whitton, B. A., Kelly, M. G. (1995), 776, 803
Whitton, B. A., Peat, A. (1969), 129, 196
Whitton, B. A., Potts, M. (1982), 66, 116
Whitton, B. A., Potts, M. (2000), 19, 57, 63–64, 116, 117, 120, 196
Whitton, B. A., Rott, E. (1996), 681, 684, 775–776, 791, 803
Whitton, B. A., Rott, E., Friedrich, G. (1991), 681, 684, 776, 791, 804
Whitton, B. A., Shehata, F. H. A. (1982), 785, 803
Whitton, B. A., Yelloly, J. M., Christmas, M., Hernández, I. (1998), 785, 804
Wiedner, C., Nixdorf, B. (1998), 487, 509
Wilce, R. T. (1966), 757–758, 764, 766, 770–771, 773
Wilce, R. T., Weber, E. E., Sears, J. R. (1970), 766, 771, 773
Wilcox, L. W., Lewis, L. A., Fuerst, P. A., Floyd, G. L. (1992), 257, 309
Wilcox, L. W., Wedemayer, G. J. (1984), 686, 714
Wiley, M. J., Gordon, R. W., Waite, S. W., Powless, T. (1984), 810, 834
Wilken, L. R., Kristiansen, J., Jürgensen, T. (1995), 486, 509
Willén, E. (1992), 364, 381
Willen, E., Hajdu, S., Pejler, Y. (1990), 583
Willen, E., Oke, M., Gonzalez, F. (1980), 755
Willey, R. L. (1972), 408, 411, 422
Willey, R. L. (1980), 411, 415, 422
Willey, R. L. (1982), 411, 415, 422
Willey, R. L., Cantrell, P. A. (1990), 408, 422
Willey, R. L., Durbin, E. M., Bowen, W. R. (1973), 404, 422
Willey, R. L., Threlkeld, S. T. (1993), 408, 422
Willey, R. L., Walne, P. L., Kivic, P. (1988), 387, 391, 422
Willey, R. L., Ward, K., Russin, W., Wibel, R. G. (1977), 405, 410, 422

Willey, R. L., Wibel, R. G. (1985a), 385, 401, 422
Willey, R. L., Wibel, R. G. (1985b), 385–386, 391, 401, 422
Willey, R. L., Willey, R. B., Threlkeld, S. T. (1993), 408, 411, 422
Williams, D. M. (1985), 570, 598, 600–601, 605, 616, 636
Williams, D. M. (1986), 598, 600, 636
Williams, D. M. (1989), 598, 601, 636
Williams, D. M. (1990a), 596–598, 600, 636
Williams, D. M. (1990b), 596, 598, 600, 613, 636
Williams, D. M. (1990c), 596, 599, 613, 636
Williams, D. M. (1993), 601, 636
Williams, D. M. (1994), 601, 604, 623, 636
Williams, D. M. (1996), 601, 605, 623, 636
Williams, D. M. (1997), 601, 636
Williams, D. M. (2001), 631, 636
Williams, D. M., Round, F. E. (1986), 561, 570
Williams, D. M., Round, F. E. (1987), 561, 596, 598–599, 613, 616, 628, 636
Williams, D. M., Round, F. E. (1988a), 598–599, 636
Williams, D. M., Round, F. E. (1988b), 596, 598, 600, 636
Williams, L. G. (1964), 585
Williams, L. G. (1972), 585
Williams, W. D. (1996), 42, 57
Wilmotte, A., Stam, W. T. (1984), 116
Wilson, S. E., Cumming, B. F., Smol, J. P. (1994), 604, 636
Wilson, S. E., Cumming, B. F., Smol, J. P. (1996), 42–43, 57
Winkenbach, F., Wolk, C. P. (1973), 119, 196
Winner, J. M. (1975), 38, 57
Winner, R. W. (1985), 825, 834
Winner, R. W., Owen, H. A., Moore, M. V. (1990), 825, 834
Witkowski, A., Lange-Bertalot, H., Stachura, K. (1998), 637, 653
Witt, F. G., Stöhr, C., Ullrich, W. R. (1999), 700, 714
Wium-Andersen, S., Anthoni, U., Christophersen, C., Houen, G. (1982), 811, 834
Woelkerling, W. J. (1976), 40, 57
Woelkerling, W. J. (1990), 197, 224
Wojciechowski, I. (1971), 90, 116
Wolk, C. P. (1973), 119, 121, 196
Wolk, C. P. (1982), 119, 121, 196
Wolken, . J., Palade, G. E. (1952), 501, 509
Wolle, F. (1890), 561
Wolowski, K. (1992), 416, 422
Wolowski, K. (1993), 402, 422
Wolowski, K., Walne, P. L. (1997), 387, 410, 415, 422
Wong, S. L., Clark, B., Kirby, M., Kosciuw, R. F. (1978), 782, 804
Wood, H. C. (1869), 92, 116
Wood, H. C. (1872), 116
Wood, R. D. (1948), 348, 352
Wood, R. D. (1950), 27, 57
Wood, R. D. (1967), 348, 352
Wood, R. D., Imahori, D. (1964), 331, 348, 352
Woodhead, N., Tweed, R. D. (1960), 604, 613, 636
Woodson, B. R. (1962), 433, 469
Woodson, B. R., Afazal, M. (1976), 433, 469
Woodson, B. R., Holoman, V. (1964), 429, 433–434, 461, 463, 469
Woodson, B. R., Wilson, W. (1973), 461, 463, 469
Wooten, J. W., Elakovich, S. D. (1991), 821, 834
Wootton, J. T., Parker, M. S., Power, M. E. (1996), 32, 57
Wright, A.-D. G., Lynn, D. H. (1997a), 686, 714
Wright, A.-D. G., Lynn, D. H. (1997b), 686, 714
Wu, J.-T., Chou, J.-W. (1998), 699–701, 714
Wujek, D. (1971), 335, 349, 352
Wujek, D. E. (1967), 499, 509
Wujek, D. E. (1976), 503, 509
Wujek, D. E. (1983), 500, 509
Wujek, D. E. (1984), 536, 552, 557

Wujek, D. E. (1996), 474, 509
Wujek, D. E. (1999), 474, 509
Wujek, D. E., Bicudo, C. E. (1993), 526, 538, 557
Wujek, D. E., Bland, R. G. (1988), 503, 509
Wujek, D. E., Bland, R. G. (1991), 526, 536, 541, 557
Wujek, D. E., Gardiner, W. E. (1985), 503, 509, 514, 518–519, 521
Wujek, D. E., Graebner, M. (1980), 575
Wujek, D. E., Gretz, M., Wujek, M. G. (1977), 552, 557
Wujek, D. E., Hamilton, R. (1972), 552, 557
Wujek, D. E., Hamilton, R. (1973), 552, 557
Wujek, D. E., Hamilton, R., Wee, J. (1975), 552, 557
Wujek, D. E., Igoe, M. J. (1989), 541, 557
Wujek, D. E., Kristiansen, J. (1978), 531, 557
Wujek, D. E., Saha, L. C. (1991), 513, 521
Wujek, D. E., Saha, L. C. (1995), 485, 509
Wujek, D. E., Saha, L. C. (1996), 485, 509
Wujek, D. E., Thompson, R. H. (2001), 502–503, 509
Wujek, D. E., Thompson, R. H. (2002), 499, 503, 509
Wujek, D. E., Thompson, R. H., Timpano, P. (1996), 758, 765, 771, 773
Wujek, D. E., Timpano, P. (1984), 524, 557
Wujek, D. E., Timpano, P. (1986), 197, 201, 210, 221, 224
Wujek, D. E., Wee, J. L. (1983), 524, 526, 534, 551–552, 557
Wujek, D. E., Wee, J. L. (1984), 453, 458, 469
Wujek, D. E., Weis, M. M. (1984), 552, 557
Wujek, D. E., Welling, M. I. (1981), 586
Wunsam, S., Schmidt, R., Klee, R. (1995), 583
Wurtsbaugh, W. A., Berry, T. S.(1990), 43, 57
Wynn-Williams, D. D. (2000), 121, 196

X
Xavier, M. B., Mainardes-Pinto, C. S. R., Takino, M. (1991), 408, 422
Xie, P., Iwakuma, T., Fujii, K. (1998), 700, 714

Y
Yamagishi, T., Couté, A. (1995), 416, 422
Yan, N. D., Stokes, P. (1978), 699, 714
Yang, J.-R., Duthie, H. C. (1995), 604, 636
Yang, J.-R., Pick, F. R., Hamilton, P. B. (1996), 587
Yin-Xin, W. J., Kristiansen (1994), 538, 557
Yoder, C. O., Rankin, E. T. (1995), 794, 796, 804
Yoneda, Y. (1949), 763, 773
Yong, Y. Y. R., Lee, Y.-K. (1991), 45, 57
Yoshida, T. (1959), 202, 224
Yoshikawa, T., Takishita, K., Ishida, Y., Uchida, A. (1997), 686, 699, 714
Yoshizaki, M., Iura, K. (1991), 763, 773
Yoshizaki, M., Miyaji, K., Kasaki, H. (1984), 758, 760, 771, 773
Youngs, H. L., Gretz, M. R., West, J. A., Sommerfield, M. R. (1998), 200, 224
Yung, Y.-K., Sawa, T., Stokes, P. M. (1986), 349, 352
Yung, Y.-K., Stokes, P., Gorham, E. (1986), 40, 57, 205, 224, 363, 381

Z
Zacharias, O. (1898), 36, 57
Zakrys, B. (1986), 388, 399–400, 402–405, 407, 409, 412, 415, 422
Zakrys, B. (1994), 388, 403, 422
Zakrys, B. (1997a), 388, 407, 422
Zakrys, B. (1997b), 388, 407, 422
Zakrys, B., Kucharski, R., Moraczewski, I. (1997), 407, 422
Zakrys, B., Moraczewski, I., Kucharski, R. (1996), 404, 407, 422
Zakrys, B., Walne, P. L. (1994), 388, 390–391, 394–395, 397, 399, 402–405, 407, 410, 412–416, 422

Zakrys, B., Walne, P. L. (1998a), 402, 422
Zakrys, B., Walne, P. L. (1998b), 403, 422
Zalessky, M. M. (1926), 74, 116
Zeeb, B. A., Smol, J. P. (1991), 537, 557
Zeeb, B. A., Smol, J. P. (1995), 537, 557
Zeimann, H. (1991), 787, 804
Zelinka, M., Marvan, P. (1961), 775, 787, 791, 804
Zimmerman, W. (1928), 768, 771, 773
Zohary, T., Breen, C. M. (1989), 64, 116
Zohary, T., Robarts, R. D. (1990), 64, 116

Subject Index

Accumulation body (-ies), 690–691
Acid (-ic),
   Acid environment (also acidification), 6, 7, 40–42, 66, 81, 87–88, 184, 204, 227, 284, 363–364, 370–379, 407, 424, 428, 487–489, 513, 535–537, 552, 560–561, 602, 604, 627, 640, 647–650, 661–662, 664–665, 676, 682, 699, 701, 792–793, 814, 823
   Acid mine drainage, 41–42,
   Acid precipitation (or deposition), 39, 44, 513, 571, 607, 661, 792–793
   Acid soils, 206, 369
   Acid spring (stream), 41, 80, 208, 498
   Acidic (lower) pH, 7, 40–41, 44, 204, 206, 208, 280–281, 341, 363, 371, 406, 407, 412, 427–428, 433, 488–489, 513, 535–536, 552, 602, 607, 617, 627, 650, 665, 672, 676, 682, 699, 739, 814, 823
Adirondack Mountains, 44, 526, 536–537, 541, 552, 792
Aerotope (also Gas vacuole), 20, 61, 70, 72, 81–82, 87, 89, 92, 122–123, 125–126, 129, 132, 139, 141, 144, 147, 151, 155, 158, 164, 166, 169, 171, 174, 177, 180–181, 184
Akinete – see Cyst, Akinete
Alabama, 210, 303, 345, 459, 463
Alaska, 13, 41, 80, 89, 101, 129, 137, 141, 150, 203, 212, 365, 377, 379, 425, 427–428, 433–434, 439, 441, 443, 446–447, 449, 451, 453–454, 458–459, 461, 463, 487–488, 552, 561, 585, 628, 649
Alberta, 552
Algae,
   Acidophilic, 27, 41–42, 80, 206, 208, 227, 363, 375, 378, 489, 536–537, 607, 617, 673
   Aerophytic (also aerophilic), 64, 67, 277, 284, 288, 623, 638, 643, 646–649, 661
   Alkalophilic, 66, 72, 79, 86, 129, 158, 187, 204, 338, 537, 602
   Auxotrophic, 383, 385, 387, 401, 406, 409, 701
Benthic, 5, 8, 9, 23–28, 30–36, 38–39, 42–43, 65–66, 68, 79–80, 82, 89, 124–126, 129, 132, 139, 147, 150–151, 155, 161, 169, 173–174, 177, 180–181, 199, 205, 210, 316, 363, 371, 489–490, 565, 567, 583, 601–606, 608, 610, 613, 616, 646–648, 650, 661–663, 664, 666, 672, 676, 681–682, 757–758, 762–764, 766, 778, 780, 782–783, 785–786, 808–809, 819, 825–826
Cryophilic, 44
Culture (laboratory), 59–60, 63, 67–68, 81, 99, 121, 126, 131, 141, 216, 131, 141, 216, 227, 229, 235, 240, 243–246, 248, 254–258, 291, 315–316, 333, 335–338, 340–341, 345–346, 365, 374, 388, 393–394, 402–411, 428, 432–433, 461–465, 487–489, 497, 501, 512–514, 569, 605, 694, 699, 700, 709, 735, 739–740, 746, 748–749, 751, 758, 760–761, 763, 765–767, 785–786, 806–807, 810
Edaphic – See Algae, soil,

Algae (continued)
  Endolithic, 6, 66–68, 71–72, 104, 107–109, 314–315
  Endophytic, 45, 72, 95, 107, 180, 259, 274, 295, 298, 314–316, 319–321, 330–332, 342–343, 435, 441, 461, 770
  Epilithic, 24–26, 35–36, 43, 65, 67, 71, 101–102, 107, 158, 164, 169, 174, 180–181, 187, 205, 208, 210, 212, 346, 424, 433, 441, 454, 456, 458, 461, 500, 681, 763, 765, 767, 779
  Epipelgic, 27, 38, 40, 43, 65, 80, 88, 120, 137, 139, 180, 406–408, 433, 453, 463, 603, 638, 644, 646–648, 650, 669–670, 674–676, 681–682
  Epiphytic, 24–28, 34, 38, 39, 40, 45, 65, 67, 71, 76, 79, 81, 87, 96–97, 101–104, 107–109, 120, 122, 124, 132, 150, 158, 161, 164, 166, 169, 205, 208, 210–12, 271–271, 274, 284, 291, 298, 303, 312, 315–316, 319–338, 341–345, 347, 424–435, 432–433, 435–437, 441–444, 446, 450–456, 458–459, 461–463, 482–483, 492, 495–500, 603, 662, 672, 674, 681, 707–709, 767, 779, 810
  Epizooic, 12, 88, 171, 212, 271, 303, 319, 339–340, 384–385, 388, 393–394, 411, 433
  Filamentous, 2–4, 6, 8, 20, 22, 24–28, 3234, 36, 39–43, 61–63, 65–70, 81, 101–102, 107, 117–190, 199–221, 260–261, 264, 267, 270, 272, 288, 298, 311–349, 353–355, 357, 359–360, 362–374, 376–379, 429–433, 438, 441–443, 446, 451, 454, 458–463, 473, 488, 491–494, 496–497, 499, 502, 674, 696, 703, 709, 758, 767, 783, 805, 808–810, 812–824, 826
  Gelatinous, 1–3, 60–61, 69–71, 79–81, 89, 91–92, 95–97, 101–104, 110, 124–125, 141, 144, 151, 166, 177, 180, 184, 187, 189, 198–200, 232–236, 239–249, 259, 262, 264–267, 270, 280, 284, 288, 291–292, 294–295, 297, 301, 304, 321, 332–333, 344, 361, 363, 365, 367, 369, 371, 375–377, 429, 435, 437, 441, 443–444, 453–455, 458, 474, 476, 481–483, 488, 493–503, 643, 707, 770, 819
  Lithogenic, 67, 81, 96, 121
  Mat, 6, 26, 40–41, 44, 60, 63, 65–67, 79, 104, 119–126, 129, 132, 135, 137, 139, 141, 144, 147–151, 155, 158, 161, 164, 166, 169, 171, 174, 177, 180, 184, 189, 198–199, 314–316, 319, 337, 340, 344, 363, 365, 429, 463, 779, 805, 809, 817, 822–823, 816
  Neustonic, 291, 423–433, 441, 490, 493, 496, 501
  Planktonic, 5–9, 19–23, 26, 32, 36–38, 42–43, 59, 61, 64–65, 67–68, 70–76, 79–89, 92–95, 119–125, 129–132, 135, 139–144, 147, 151, 155, 166–167, 169–172, 174–177, 180, 227, 257–260, 267–306, 314–317, 323, 344–346, 363–365, 372, 376, 378, 383, 406–409, 412–413, 424–428, 432, 443, 446–454, 458–459, 461, 473, 478, 484–490, 492–503, 512–518, 541, 548, 567, 570, 585–586, 595, 598, 606–607, 617, 650, 672, 681, 694, 700, 702, 707–709, 738–739, 805–806, 812–813, 819, 822–823, 825
  Snow, 43–44, 81, 130, 226–227, 235–236, 306, 315–316, 318, 346, 364, 367, 369, 407, 433, 810
  Soil, 6, 11, 45, 60, 63, 72, 79, 82, 89, 95, 120–121, 126, 129, 135, 137, 141, 144, 150, 158, 166, 177, 180–181, 184, 198–199, 206, 208, 210, 253, 255, 258, 270, 272, 274, 276, 280, 282, 284, 291, 294–295, 298, 301, 303, 315–322, 332–337, 341–342, 344–347, 363, 369, 372, 404, 407–408, 424–425, 429, 433, 451, 453, 463, 605, 638, 648, 672, 676, 812, 814
  Subaerial, 6, 60, 63–64, 70, 79–80, 82, 92, 104, 119, 121–122, 135, 137, 139, 144, 151, 158, 161, 166, 169, 174, 177, 181–184, 187–189, 258, 267, 272, 274–276, 289–301, 306, 315–316, 319, 321–322, 343, 345–346, 363, 369–374, 490, 584, 607, 646
  Unicellular (or solitary), 1–2, 6, 8, 9, 24, 44, 60–62, 65, 68, 70–84, 92–96, 129, 199–201, 205, 207–209, 225–239, 253–259, 268–270, 274–276, 284–291, 294, 297–298, 317, 322, 354–363, 372, 379, 383, 388, 410–415, 424, 427–429, 436, 439–459, 464–465, 473–475, 477–478, 480, 482, 491–501, 524, 541, 548, 559, 600–601, 606, 623, 627, 676, 681–682, 685, 715, 740, 757
Algicide, 817, 821–822, 824, 826
Alkaline (higher) pH, 41–42, 63, 189, 204, 208, 210, 212, 218, 221, 364, 370, 412, 428, 433, 461, 489, 535–536, 552, 584, 602–603, 617, 622, 627, 639, 640, 643–644, 646, 650, 661
Alkaline environments, 40–41, 66–67, 72, 79, 86–87, 95–96, 104, 129, 161, 164, 166, 171, 177, 184, 187, 204, 212, 218, 220–221, 338, 369–370, 372, 374, 407, 535, 552, 584, 602–603, 617, 622, 627, 639–640, 643–644, 650, 661–662, 674, 699, 701, 738, 763
Alkaline phosphatase, 25
Alkalinity, 7, 485, 488, 536, 617, 814, 823
Allelochemical (-chemistry, -pathy), 817, 821, 826
Allochthonous (organic) matter, 31–32, 789
Allophycocyanin – see Pigment, Allophycocyanin
Alternation of generations, 201, 202, 256, 312, 741
Alum ($Al_2[SO_4]_3$), 813–815
Amoeboid stage (cell), 353, 362, 374, 429–430, 432, 434–436, 439–442, 450–451, 476, 480, 482, 485, 494–496, 501, 688, 696, 706, 708–709
Amylopectin, 5, 254, 736
Amylose, 5, 254, 736
Anaerobic environment, 18, 119, 227, 248, 409, 814
Anisogamy (-ous), 226, 233, 235–236, 240, 254, 312, 347, 366, 432, 485, 602, 694
Anoxic – See Anaerobic environment
Antarctic, 18, 189, 227, 246, 369, 485–486, 561, 649
Antheridium (-a), 331, 338–342, 344, 432, 464–465
Antherozoid, 342
Anthropogenic effect, 584–585, 587, 661, 781
Apical pore field, 586, 600–601, 606, 611, 613, 616–617, 656, 658–659, 663–665, 690–692, 694–695, 705, 707–708
Aplanogamete, 235
Aplanospore – see Spore, aplanospore
Aquaculture, 387, 512, 807
Araphid diatom, 7, 9, 560–562, 595–623, 625, 627, 629
Archeopyle – see Cyst, archeopyle
Arctic, 12–14, 18–19, 26–27, 33–34, 38, 82, 98, 101, 118, 134, 137, 155, 158, 161, 164, 166, 169, 171, 180, 184, 197, 231, 276, 280, 284, 288, 295, 299, 303, 306, 342, 347, 364, 369, 407, 424, 433, 443, 446, 453–454, 458–459, 461, 463, 486, 490, 535, 552, 561, 584, 604–605, 613, 627, 649, 688, 700–701, 704, 738, 764
Areola (-ae), 572, 574, 584, 599, 600–603, 606–607, 613, 616, 623, 627, 638, 640–641, 643–644, 648–650, 656, 659
Arizona, 40, 44, 139, 208, 210, 439, 552, 628, 810
Arkansas, 41, 203, 425–428, 433–434, 441, 443, 446–447, 449, 451, 453–454, 458–459, 461, 463, 478, 820
Artificial substratum (-a), 789
  Clay tile, 26–27, 36
  Glass slide, 25, 313, 315–316, 332, 394, 411, 463, 763, 771
  Plexiglas, 25
  Styrofoam, 25
Ash free dry mass, 782–783
Astaxanthin – see Pigment, Astaxanthin
Athecate dinoflagellate, 691, 704, 706, 708–709
Aufwuchs, 24, 314
Australia, 7, 65, 218, 316, 503, 524, 526, 537, 541, 548, 552, 709, 759, 765–766, 768, 794
Autospore – see Spore, autospore
Autotrophic (growth or metabolism), 20, 37, 63, 66, 120–121, 202, 254, 387, 406, 489, 514, 534, 685–686, 701, 739–740
Auxospore – see Spore, Auxospore
Auxotrophic metabolism – see Algae, Auxotrophic
Axial area, 606–607, 624, 627, 639–640, 643, 645–646, 648–650, 662–664

Bacteria (Eubacteria), 59, 61–62, 81, 384, 406, 668, 697, 739–740, 808, 810, 825–826

Chemoautotrophic, 18, 41
Epiphytic, 415
Heterotrophic, 20, 23–24, 36, 41, 44, 406, 409, 489, 490, 497, 499, 817
Symbiotic (also endosymbiotic), 490, 492, 494, 497, 499, 697, 739, 744, 747, 749–751
Bacterivory (*see also* Phagotrophy, Mixotrophy), 20, 23, 41, 383, 441, 473, 489, 697, 740
Bahamas, 89, 96, 107
Beetle, 41, 205
Belize, 66, 72, 80, 212, 218, 702, 707
Benthic,
Diatom, 7, 26–27, 36, 39, 490, 565, 583, 595, 601–602, 604–606, 608, 610, 613, 616, 646–648, 650, 661–664, 666, 672, 676, 681–682
Habitat, 37, 24–28, 32–37, 80, 120, 139, 161, 181, 565, 659, 663, 567, 648, 659, 664, 670, 782
Invertebrate, 25–26, 31, 34–36, 39, 783, 825
Bermuda, 89, 107, 151, 459, 481
Beta-carotene (also b-carotene), 254, 474, 560, 690, 735, 757
Bicarbonate (HCO_3^-), 41–42, 204, 738–739, 812, 823
Big Soda Lake (Nevada), 15, 42
Bioassay, 22, 39, 68, 785–786, 821
Biodiversity, 6, 12, 19–20, 32–34, 62–63, 118, 199–200, 225–226, 311, 354, 387–388, 424, 427, 429, 473, 512, 561, 598, 779, 786, 788
Biogeography, 12, 19, 32–34, 60, 63, 67, 203–205, 486–487, 597, 605, 765–766
Biomanipulation, 22, 817–819
Biomass, 19–20, 22–27, 31–32, 34–40, 43–44, 63–66, 120, 155, 204–205, 423, 487–488, 513, 535–538, 605, 664, 670, 699–701, 776, 779–780, 782–783, 785–786, 789, 794–795, 806, 812–814, 816–818, 822–823, 825
Biovolume (also Cell volume), 24, 36, 598, 670, 776, 782–784, 789
Biraphid diatom, 9, 595–597, 604, 613, 637–650, 655–666
Bird bath, 45
Bloom, 19, 23, 37, 43, 53, 60, 64–65, 68, 72, 81, 89, 92, 117, 120, 141, 147, 169–174, 177, 227, 254, 274, 315, 335, 364, 370, 372, 405–407, 409, 411, 415, 428, 473, 487, 489, 503, 511–512, 513–514, 535–537, 572, 585, 587, 607, 617, 699–701, 783, 793, 805–807, 812–817, 821, 824, 826
Bog, 14, 19, 38, 40, 66, 88–89, 189, 198, 205, 210, 215, 267, 276, 280–281, 284, 291, 295, 298, 301, 303, 306, 335, 339–341, 344–345, 363, 369–373, 375, 378, 407, 424–429, 433, 441, 443, 446–447, 449, 451, 453–454, 458, 463, 488, 490, 498, 537, 605, 613, 639, 646, 661–662, 665, 672, 709
Boreal, 34, 171, 205, 215–216, 433, 441, 446, 489, 552, 567–568, 583, 605, 613, 616, 661, 765
Brackish, 60, 64, 72, 82, 88, 132, 151, 166, 171, 177, 199, 202, 205, 208, 220–221, 311, 338, 347, 413, 432, 463, 485, 512, 514, 515, 548, 586, 607, 638, 646–648, 650, 676, 701, 707, 757, 763–764, 768, 770–771
Brazil, 84, 138, 140–141, 202, 204, 807
British Columbia, 13, 17, 42–43, 95, 101, 132, 137, 147, 174, 181, 184, 189, 233, 235–237, 239, 243–244, 267, 270, 276, 280, 284–285, 298, 301, 339, 363, 370, 412–413, 415, 425, 428–429, 443, 446–447, 449, 451, 453–454, 458–459, 461, 463, 552, 561, 702, 709, 762, 766
Bryophyte, 4, 45, 205, 432, 441, 443, 461, 584, 662
Buoyancy, 20, 23, 488, 539, 567, 690, 805

C:N ratio, 39
C:P ratio, 819
Caddisfly (Trichoptera), 26, 105, 819
Calcification, 44, 137, 147, 152, 169, 239, 366, 513–515, 767, 770–771
Calcium (Ca^{2+}), 239, 363, 364, 366, 374, 376–377, 497, 763–764, 766, 814
Calcium carbonate ($CaCO_3$), 27, 39, 44, 59, 66, 69, 79–80, 89, 119, 137, 147, 158, 166, 169, 187, 206, 237, 311, 316, 319–320, 332, 334–335, 338–339, 759, 763–764, 767, 813–814, 819–820, 823
California, 42, 78, 80–81, 89, 92, 94–95, 100, 102, 104–105, 108–110, 139–140, 149, 201, 216, 246, 364, 377, 434, 459, 463, 561–562, 630, 707, 767, 817
Canadian Shield, 171, 363, 488, 513
Canopeum (-a), 647
Carbon (source), 31–32, 39, 44, 204, 406, 604, 647, 738–740, 793, 807, 812
Carboxysome, 61, 200
Caribbean, 66, 72, 80, 87, 95, 104, 150, 174, 212, 218, 220, 267, 270, 274, 276, 280, 284–285, 289, 291, 295, 298, 302–303, 306, 332, 707
Carotenoid – *see* Pigment, Carotenoid
Carp, 806, 814, 819–821, 826
Carpogonial branch, 202–213, 215–216, 218
Carpogonium, 201–202, 205, 210–216, 218–219
Carposporangium, 211–216, 218
Carpospore, 199, 201–202, 212, 215–218
Carposporophyte, 201–201, 207–208, 211–218
Catfish, 739, 806–807
Cave, 6, 11, 44, 125, 187, 197, 208, 605
Cedar Bog Lake (Minnesota), 14, 40
Cell wall, 1–9, 26, 59, 61–62, 118–120, 125–126, 129, 132, 135, 137, 139, 144, 147, 151, 158, 161, 174, 177, 180–181, 200, 226, 228–229, 231–236, 242, 246, 248, 255, 257, 260–261, 263–267, 270, 274, 276–277, 280–282, 284, 288–289, 291, 295, 298–299, 301, 303, 306, 311–312, 318, 322, 353–354, 357, 362, 365–369, 372–376, 378, 388, 425, 427, 430–432, 434–438, 441, 332, 445–447, 449, 451, 453–454, 457–458, 461, 494–495, 500, 503, 559–561, 597–599, 601, 604–605, 669, 736, 757, 775, 783, 823
Cellulose, 2, 5–6, 200, 226, 255, 311–312, 343, 431, 484, 498, 686–687, 690–691, 698, 703, 706, 757
Central America, 60, 67, 81, 90, 94, 100, 102, 147, 164, 171, 267, 270, 765
Central Canada, 73, 80, 85, 89, 96, 156, 164, 171, 267, 270
Centroplasm, 61, 135
Chantransia stage, 201–202, 212–216, 218, 765
China, 387, 513, 536, 759, 763, 765, 771, 807, 820
Chironomid (Midge), 205, 319, 819, 825
Chitin, 255, 411, 473, 501, 565, 567, 584, 587
Chloride (Cl^-), 514
Chlorophyll-a – *see* Pigment, Chlorophyll-a
Chlorophyll-b – *see* Pigment, Chlorophyll-b
Chlorophyll-c – *see* Pigment, Chlorophyll-c
Chloroplast (*see also* Plastid), 5–8, 41, 197–200, 202, 207–210, 212, 215–216, 218, 225–226, 229, 231–237, 239–244, 246, 248, 254, 256–257, 259–267, 270–272, 274, 276–277, 280–282, 284–285, 288–289, 291–292, 295, 298–299, 301–303, 306, 311–312, 317–322, 330, 332–347, 353–355, 357, 361–362, 365–379, 384–391, 394–397, 401–404, 411–415, 472, 474, 491, 493–503, 512, 515–518, 524–525, 528, 531, 533, 548, 559–560, 567, 569, 572–573, 575, 576, 585, 608, 617, 646, 690–691, 697, 701, 704, 707–709, 717, 719, 724, 726, 731, 735–739, 742–748, 750–751, 757, 759–760, 761–762, 767, 768–771
Chromatic adaptation – *see* Photoacclimation
Chromatophore – *see* Chloroplast
Chromatoplasm, 61, 69, 72, 79, 81, 87–88, 101, 121, 135, 147
Chrysolaminarin – *see* Storage product, chrysolaminarin
Chytrid, 364, 379, 739
Ciliate, 20, 45, 272, 385, 701, 738–739, 749–750
Cilium (-ia), 312, 385, 512
Cingulum, 599–603, 612–623, 648–650, 681, 686–687, 690–692, 694–695, 703–704, 706–707
Cladistic, 239, 244, 388, 597–598, 604, 659, 742
Cladocera (-an), 16, 20, 38–39, 408, 411, 700, 808, 825
Classification
Brown algae, 757–759
Chrysophytes, 471–473, 484–485
Cryptomonads, 715–716, 740–743, 749
Cyanobacteria, 62–63

Classification (*continued*)
  Diatoms, 559–561, 563–565, 595–604, 637–638, 655–661, 669–670, 675
  Dinoflagellate, 685–686, 697–699
  Euglenoids, 383–385, 387
  Eustigmatophyte, 423–424
  Green algae, 228–229, 232–233, 236–237, 239–240, 243–246, 253–254, 312–313, 353–354
  Haptophyte, 511–513
  Major groups, 5–9
  Raphidophyte, 427–428
  Red algae, 197–199
  Synurophyte, 523–524, 533–534
  Tribophyte, 423–424, 429–432
CO_2 (carbon dioxide), 15, 39, 59, 204, 365, 474, 805
Coastal (habitat), 14, 16, 18, 40, 64, 66–77, 72, 78, 80–82, 88, 95–96, 101–102, 107–108, 110, 150, 155, 158, 177–178, 203, 258, 315, 332, 335, 338, 363, 463, 489, 512, 552, 583–584, 586, 661, 768
Coccolith, 511–516, 518–519
Coccolithophorid, 43, 511–513
Coenobium (-ia), 237–240, 246, 261–263, 267, 274, 276, 284, 288, 295
Coenocyte (-cytic), 7, 254, 266–267, 276–277, 298, 311, 319, 344, 429, 431–434, 463–464
Colony (-ial), 1–3, 3, 6, 8–9, 20, 27, 37, 60, 69–72, 79–110, 119, 122–125, 132, 135–139, 144, 158, 161–169, 171–173, 177, 179, 180–181, 207–208, 225–229, 232–233, 239–248, 253–306, 315–316, 325, 332, 335, 344, 354, 356, 360, 368, 374–376, 388, 405, 410–411, 424–425, 429, 434–437, 443–435, 450–464, 473–485, 489–503, 524–528, 531, 533–535, 539, 541, 548, 565, 567, 571–574, 583–587, 598, 605–608, 613–614, 616–617, 622–623, 638, 647, 661, 664–665, 673–677, 681, 707, 739, 744, 757–759, 762–768, 770, 777, 784, 807, 810–811, 816
  Amorphous (irregular), 60, 69, 70–72, 79, 81, 88–89, 92, 95–96, 102, 104, 107, 119, 122, 141, 147, 155, 177, 180, 187, 208, 260, 263–264, 274, 276, 284, 288, 298, 295, 301, 303, 320, 322, 332, 334, 336–337, 340, 343, 354, 443, 452–453, 455, 498, 768
  Autocolony, 226, 239, 246
  Chain (chainlike), 81, 199, 282, 341, 431, 437, 452–453, 494, 496, 501, 601, 616–617, 623, 640, 676
  Coenobium, 239–240, 246, 261–263, 267, 274, 276, 284, 288, 295
  Cube-like, 89, 264, 295
  Daughter colony, 226, 241, 243–245
  Hemispherical, 72, 96, 107, 124, 137, 151, 166, 330, 332, 334, 499, 770
  Moniliform, 102, 121–122, 125, 139, 177, 184, 187–189
  Plate-like (flat), 69, 87, 280, 317
  Radial (radiating), 69–71, 81–82, 87, 89, 96, 104, 107, 124, 135, 144, 166, 169, 199, 240, 243–244, 264–265, 267, 270, 284, 298, 301, 314, 337, 340, 488, 496, 770
  Saccate (sacklike), 125, 207, 209–210, 246, 259, 303
  Spherical, 69–70, 72, 79, 81–82, 87, 88–89, 92, 95–96, 119, 123–125, 147, 180, 187, 233, 242–243, 258, 284, 288, 291–292, 295, 301, 303, 306, 320, 330, 340, 435, 437, 443, 452–454, 493, 496, 526, 528, 541, 548
  Zigzag, 3, 318
Colorado (CO), 17, 86, 454, 459, 647, 702, 738, 746, 748–749, 751, 763, 766, 770
Colorado River, 34
Condensed chromosomes, 7, 686, 690
Conjugation – *see* Reproduction, conjugation
Connecticut (CT), 80, 158–159, 425, 427, 459, 461, 463, 489, 563, 552, 770, 772
Contractile vacuole, 226, 229, 231–237, 239–240, 242–244, 246, 248, 254, 272, 274, 292, 295, 301, 332, 392, 401, 409, 427–430, 434, 439, 441, 443, 472, 485, 494–503, 514–518, 716–717, 719, 724, 731, 736, 747
Copepod, 20, 39, 212, 411
Copper sulfate ($CuSO_4$), 807, 816, 821–826
Core sample, 571
Cortex, 4, 210, 212, 216, 218, 220, 338
Cortical filament (or cells), 3–4, 207–208, 210–212, 215–217, 220
Costa Rica, 13, 40–42, 210, 212, 218, 221, 235, 343, 415
Costae, 357, 367, 372, 571–572, 583, 586, 600, 602, 616, 623, 640, 643, 648, 681
Crater Lake (Oregon), 12–15
Cricolith, 511, 515, 518
Crust (algal), 4, 24–25, 33, 36, 63, 80, 107, 110, 136–137, 181, 199–200, 204, 206–207, 215, 218, 314, 316, 320, 328, 340, 364–366, 376, 458, 488, 605, 757–760, 763–765, 767–768, 813
Crustacean (micro-), 260, 271, 393, 404, 410–411, 454, 490, 813
Cryophilic algae – *see* Algae, cryophilic
Cuba, 66, 72, 76–66, 79–84, 86, 93, 99, 101, 103–105, 107, 109, 131, 142, 145, 150, 156, 163, 175, 184–185, 189, 291, 303, 307, 454, 702, 707
Culture (studies) – *see* Algae, culture
Current velocity, 28–30, 34–37, 200, 203–204, 206, 210, 212, 215–216, 218, 433, 473, 764, 766, 779
Cyanelle, 263, 266, 270, 284, 385
Cyanophage, 819
Cyanophycean starch – *see* Storage product, cyanophycean starch
Cyclomorphosis, 255, 488
Cyprinoid fish, 205, 212
Cyst (*see also* Spore)
  Akinete, 19, 119–121, 124–125, 139, 151, 155, 158, 161, 164, 166, 169, 171, 174, 176–177, 180–181, 184, 189, 210, 246, 295, 312, 320–321, 325, 327–328, 330, 332–341, 343–347, 355, 362, 372, 432, 442, 453, 462–463, 818
  Coenocyst, 344
  Hypnospore, 43–44, 245–246, 463
  Hypnozygote, 226, 240, 243–244, 694
  Planozygote, 248, 694
  Tomont, 1
  Zygospore, 19, 330, 355–356, 362–363, 369–379

Dam, 16–17, 28, 31, 34
Darling River, 65
Daughter cell – *see* Reproduction, daughter cell
Delta Marsh (Manitoba), 38
Denmark, 536, 794
Desert, 6, 29, 32, 40, 42, 60, 66, 121, 141, 203
Desiccation, 26, 39, 45, 123, 206, 255, 363, 734
Desmid, 2, 6, 8, 354, 356–379, 458, 776
Desmokont – *see* Flagellum, desmokont
Detritus, 24, 26, 31, 35, 38–39, 132, 169, 205, 378, 409, 674
Diadinoxanthin – *see* Pigment, Diadinoxanthin
Diatomite, 573, 584–585, 666
Diatoxanthin – *see* Pigment, Diatoxanthin
Dictyosome – *see* Golgi body
Dinokaryon, 697
Dinokont – *see* Flagellum, dinokont
Dinospore, 697
Disjunct, 12, 276, 433, 764–766
Dispersal, 33, 70, 202, 407, 486, 760, 765, 795, 824
Diversity (*see also* Biodiversity), 12, 19–20, 24–25, 32–36, 39–44, 62–67, 118, 197–199, 225–226, 254, 311, 354, 387–388, 424, 427–431, 473–474, 512–513
  Evenness, 782, 784
  Shannon index, 784, 787
  Species richness, 40, 776, 779, 782, 784
Drinking water, 6, 16, 685, 806, 809
Dystrophic – *see* Lake, dystrophic

Eastern Canada, 96, 126, 270, 280, 284–285, 288, 291, 295, 298, 302–303, 306, 365, 377, 485, 561
Eastern North America, 12, 40, 42, 137, 169, 203, 205, 210, 216, 280–281, 332, 377, 424, 459, 488, 513, 560–561, 650, 661
Eastern United States, 92, 189, 210, 270, 274, 276, 280–281, 284–285, 288–289, 291, 295, 298, 301–303, 306, 315, 332, 340, 344, 363–364, 377, 459, 461, 485, 561, 665, 792, 811
Ecdysis, 691, 694, 703, 708
Ecological risk assessment (ERA), 776–777
Ecomorphotype (also Morphotype), 67, 89, 104, 135, 189, 406, 410, 461, 485, 569
Ecorticate, 317, 338–339

Ejectisome, 715–722, 724, 726–727, 731–732, 734–735, 741–742, 744, 746, 748–751
El Salvador, 15, 41, 206
Electron microscopy (EM) – *see* Microscopy, electron
Elevation, 15, 204, 565, 638, 647, 649, 728
Ellesmere Island, 144, 166, 203, 303
Encrusting algae – *see* Algae, encrusting
Endemic species, 12, 16, 66, 72, 80, 82, 87, 89, 101, 227, 365, 377, 568, 656
Endogloeic algae – *see* Algae, endogloeic
Endolithic algae – *see* Algae, endolithic
Endophytic algae – *see* Algae, endophytic
Endosymbiont (-biosis, also Symbiont), 11, 36, 45, 63, 180, 226, 258, 263, 271, 284, 306, 311, 315, 385, 387–388, 402, 490, 492, 497, 499, 670, 673–676, 681, 686–687, 697, 735–737, 739
English Lake District, 21, 487, 513, 696
Environmental optima (also Ecological optima), 23, 26, 44, 206, 487, 513, 597, 617, 784, 791–792
Environmental Protection Agency (EPA), 776–777, 786, 788, 826
Epilimnion, 17–18, 23, 258, 535, 567, 699, 738, 794
Epilithic algae – *see* Algae, epilithic
Epipelic algae – *see* Algae, epipelic
Epiphytic algae – *see* Algae, epiphytic
Epipsammic algae – *see* Algae, epipsammic
Epitheca, 687, 689, 690, 693–694, 704, 706, 708
Epizooic algae – *see* Algae, epizooic
Eubacteria – *see* Bacteria, Eubacteria
Euphotic zone (also Photic zone), 18, 20, 258, 539, 814–815
Eutrophic environment (*see* Lake, eutrophic)
Eutrophication, 19, 21, 23, 27, 64, 129, 571, 586, 662, 775, 781–782, 806, 810–811
Everglades (Florida), 39, 66, 72, 131, 133–134, 140, 154, 158, 162, 173, 184, 434, 794–795
Eversion (also Inversion), 228, 239, 242, 244
Evolutionary (relationships), 41, 60–62, 226, 257, 313, 362, 384–385, 387, 391, 401–402, 405, 559, 567–570, 582, 656–657, 659, 674, 690, 697, 699, 737
Exocyte – *see* Reproduction, exocyte
Exospore – *see* Spore, exospore
Experimental Lakes Area (ELA), 26, 129, 434, 486, 488, 537
Eyespot, 226, 229–233, 235–237, 239–240, 242–244, 246, 248, 301, 332, 383–385, 388, 391–392, 397, 399, 401–402, 411–4388, 391–392, 397, 399, 401–402, 411–413, 439, 472, 494–503, 686, 690–691, 706–709, 742

FAA – *see* Fixative, FAA
Facultative heterotroph, 254, 406
False branching, 3, 119, 123–126, 132, 141, 153, 155, 158–161, 164–165, 168, 184, 189, 207–209

Fatty acid, 35, 147, 387, 408, 823
Fibulae, 670, 672, 675–676, 681
Filament (true; *see* Algae, filamentous)
Finger Lakes (New York), 12–14, 18, 42
Finland, 486, 489, 699
Fish, 6, 22–23, 31, 38, 43, 171, 205, 313–314, 335, 387, 408, 514, 685, 687, 697, 701–703, 708, 778, 786–787, 789, 806–807, 810–811, 814–816, 818, 820–821, 824–825
Fish kill, 685, 806, 810, 812, 814, 816–817, 823, 825
Fixative
 FAA, 259, 316, 365, 491
 Formaldehyde (formalin), 67–68, 206, 227, 259, 316, 365, 491, 514, 539, 738, 767
 Glutaraldehyde, 67–68, 206, 227, 259, 409, 463, 465, 491, 539, 702, 703, 767
 Lugol's iodine, 67, 206, 259, 409, 463–465, 491, 514
 M^3, 259
 Nissenbaum's solution, 539
 Osmic acid, 514
 Transeau's solution, 491
Flagellate, 2, 5–8, 40, 45, 225–248, 255, 257, 383–416, 423–428, 439–454, 471, 490–503, 511–519, 524–528, 685–709, 715–751, 806–807, 814, 822
Flagellum (-a), 1–2, 5–6, 8, 20, 23, 197, 227, 229–230, 235, 237, 254, 316, 339, 341, 387, 393, 401–402, 405–406, 408, 410, 412, 424, 427–429, 441–443, 472–473, 484, 490–491, 501, 512, 514, 518, 524–525, 527–528, 531, 534, 539, 541, 548, 685–690, 694, 696, 702, 716–717, 719–724, 726–734, 741–744, 746–747, 749–751, 759–760, 768, 770
 Heterokont (unequal), 1, 229–231, 334, 388, 408, 413, 426430, 434, 439, 492–493, 495, 498–499, 501–503, 512, 515, 548, 719, 721, 724, 726, 728–731, 757
 Isokont (equal), 1, 226, 229–244, 246, 248, 256, 322, 332, 388, 410, 432, 493, 513, 515, 519
 Subapical, 7, 392, 401, 410–414, 428–429, 435, 439–440, 733–734, 751
 Tinsel, 392, 402, 423, 472–473, 491, 495, 515, 528, 686–687, 717, 734, 742, 744, 746
 Whiplash (smooth), 225–226, 254, 423–424, 427, 432, 472, 495, 500, 512, 528, 686, 690
Florida, 15–16, 23, 41, 45, 66, 72, 86, 88, 99, 102, 104, 110, 128, 131, 134, 139–141, 151, 154, 156, 158, 162, 173–174, 184, 187, 216, 218, 221, 371, 377, 433–434, 459, 463, 514, 526, 535–536, 541, 552, 794, 806–807, 809, 811–812
Floridean starch – *see* Storage product, Floridean starch
Food web, 31–32, 34–35, 39, 488, 595, 810, 826

Fossil
 Algae, 387, 687, 706, 793
 Cyst, 687, 698
 Diatom, 559, 561, 572, 582–587, 661
 Pigment, 43, 781
 Pollen, 781
 Record, 570, 623, 687, 697, 781
Frustule, 5, 7–9, 559–560, 562, 565–569, 571–575, 584, 587, 605, 608–609, 612–613, 616, 623, 627, 630, 638–639, 646–647, 650, 656, 658, 663, 665, 670–671, 675–676, 680–682, 775
Fucosan vesicle (also Physode), 759, 767–771
Fucoxanthin – *see* Pigment, Fucoxanthin
Fungus (-i), 24, 44–45, 180, 236, 258, 315, 364, 514, 702, 819, 821

Gametangium (-ia) – *see* Reproduction, gametangium
Gametophyte, 201–202, 204–205, 212–216, 218, 759
Garibaldi Lake, 13–14
Gas vacuole – *see* Aerotope
Gas vesicle, 23, 61, 64–65, 70, 79–82, 87, 89, 92, 95–96, 101–102, 120, 122, 125, 129, 135, 141, 147, 166, 169, 177
Gelatinous envelope (or matrix; *see also* Mucilage), 1–3, 60–61, 69–71, 79–82, 89, 92, 95–96, 101, 124–125, 141, 151, 166, 177, 180, 184, 187, 189, 199–200, 206, 208, 210, 212, 215, 226, 228–229, 232–236, 239–244, 246, 248, 259, 262–267, 270, 284–285, 288, 291, 295, 321, 332–333, 344, 363, 365, 367, 369, 371, 375–376, 429, 437, 453–454, 458, 474, 476, 488, 493, 495–502
Gemmae – *see* Reproduction, gemmae
Geology (-ical), 28–29, 34, 40–42, 565, 610, 764
 Basalt, 29, 763–764
 Granite (*see also* Canadian Shield), 29, 184, 764
 Limestone, 14–16, 29, 36, 44, 63–65, 72, 89, 92, 102, 104, 107, 121, 164, 169, 208, 314, 316, 319, 334, 765, 811
 Sandstone, 36, 65, 701, 764
Georgia, 41, 203, 463
Georgian Bay, 206, 513
Germany, 37, 488–489, 514, 526, 536, 584, 759, 768
Girdle band, 361, 372–373, 571, 573, 575–576, 584, 612, 616, 619, 625, 638, 647
Girdle lamella (-ae), 423–424, 432, 474, 528
Glass microscope slide, 25, 316, 410, 490, 514, 539–541, 562, 604–605, 681, 702, 763, 771, 780, 784
Gloeocapsin – *see* Pigment, Gloeocapsin
Golgi body (apparatus), 385, 387, 404, 427, 472, 511–512, 516–517, 528, 531, 533, 565, 567, 734–736, 751
Gone cell, 226, 240, 242–244, 248
Gonidium – *see* Reproduction, gonidium
Grass carp, 819–821, 826

Grazer, 23, 25–27, 31, 34–35, 38–39, 43, 205, 819
Great Lakes (Laurentian), 12–14, 18, 25–26, 40, 95, 101, 118, 150, 177, 180–181, 184, 205, 208, 212, 231, 235, 239, 267, 270, 276, 284, 292, 295, 298, 301–302, 306–307, 314, 338–340, 347, 415, 461, 499, 513, 561, 584, 587, 598, 604–605, 607, 628, 638, 659, 661, 663, 807–810, 812
Great Salt Lake, 12, 650
Great Slave Lake, 12, 14, 301
Greater Antilles, 88, 102, 151, 184, 189
Greenland, 18, 139, 144, 184, 369, 472, 486, 526, 537, 552
Guadeloupe, 75, 101, 102, 184, 235–236, 242–243, 412, 415, 446, 459
Guatemala, 14–15
Guild, 605
Gullet (also Flagellar canal, Furrow), 384, 401, 428, 716–717, 721, 724, 726, 734, 736, 741, 743–748

Haematochrome, *see* Pigment, Astaxanthin
Hair cells (*see also* Seta, Trichoblast), 70, 101, 118, 122–124, 137, 151, 157–158, 164–169, 204, 208, 212–213, 320–322, 324, 330, 333–334, 337–338, 334, 758, 761, 764, 767–768, 770–771
Halophilic species, 66–657, 89, 95, 101, 585
Haplobiontic, 312, 585
Haptonema, 5, 7, 9, 498–499, 511–512, 514–519
Haptophyte scale, 511–519
Hard water (*see also* calcification; marl), 16, 101, 204–205, 208, 210, 338–339, 369, 376, 535, 585, 676, 681, 701, 764, 763, 811, 825
Harvest (algae or weeds), 757, 811, 817–818, 821
Hawaii, 17, 89, 96, 158, 184, 583
Heavy metal – see Pollution, heavy metal
Hepatotoxin (*see also* Toxin), 64, 806
Herbarium, 68, 81, 206, 316, 465, 562, 588, 630, 767, 769, 772
Herbivory, 19, 22, 25, 32, 35, 64, 205, 408, 765, 783
Heterocyst – *see* Heterocyte
Heterocyte, 20, 116, 119–125, 139, 154–155, 157–184
Heterokont flagella – *see* Flagellum, heterokont
Heteromorphic life history, 201, 512, 696
Heteropolar morphology, 60, 62, 68, 96–102, 106–107, 121–125, 134–136, 151, 158–169, 174–175, 474, 600, 606, 613, 615–617, 676, 682
Heterotrichous morphology, 3, 313
Heterotrophic nutrition (photoheterotrophic), 27, 37, 41, 226, 254, 383, 385, 387, 401, 404, 406, 441, 473, 486, 690, 697, 714, 706–708, 740
Holdfast, 24, 26, 35, 212, 320, 393
Homologous character, 530, 569

Homonym, 89, 246, 248
Hormogonium (-ia), 78, 80, 118–120, 123, 126–127, 129, 132, 135–139, 141, 144, 147, 150–151, 153, 155–158, 160–161, 163–164, 166, 169, 174, 177, 180–181, 183–184, 187, 189
Hot spring (thermal water), 6, 40–41, 59, 66–69, 72–73, 75, 77–78–79, 82, 88, 92, 100, 102, 104, 107, 121, 125–126, 129–130, 132, 135, 139, 141, 144, 147, 158, 184, 187, 189, 197–199, 206–208, 363, 605
Hudson River, 29, 32–33, 37–39, 218
Humic material (substances), 7, 40, 485, 536, 617, 661, 665, 763, 807
Hypocone, 687, 706
Hypolimnion, 18–19, 699–701, 738, 786, 813–816
Hypotheca, 687, 689–690, 692–693, 695, 704, 706
Hyrax, 540, 562, 605

Ice, 11–14, 18, 23, 26, 43–44, 226–227, 258, 346, 363, 367, 369, 409, 433, 463, 486–487, 604, 613, 702, 715, 738, 810
Ichthyotoxin, 512, 739
Illinois, 36, 39, 118, 164, 169, 174, 463, 561
Indeterminant growth, 208, 214–215
Indiana, 15, 41, 80, 243, 246, 340, 408, 809, 815, 820, 824
Indicator species, 7, 364, 405–406, 489–490, 513, 535, 538, 560, 579, 583, 604, 655, 763, 775, 777–782, 786, 788–793, 796, 805–806
Inorganic carbon, 26, 204, 738–739
Intaglio, 571
Intercalary growth (development), 119, 124–125, 187, 210, 321, 333, 339–340, 343, 572, 587, 622, 771
Internal valve, 569, 580, 626, 643–646
Internode, 338
Invasive species (or Invader), 26, 34, 199, 205, 563, 570, 587, 765–766, 795, 811, 816
Inversion – *see* Eversion
Iowa, 80, 337, 404, 412, 449, 454, 458, 461, 463, 536, 552
Irrigation, 180, 805, 808, 810, 817, 826
Isokont flagella – *see* Flagellum, isokont
Isomorphic (stage, life history), 61, 72, 79, 81, 125, 759
Isthmus, 354, 361–362, 368, 373–379, 647

Jamaica, 139, 147, 221, 434, 447, 449, 459
Japan, 68, 82, 428, 513, 551, 696, 699, 759, 763, 765

Kansas, 83, 86, 339, 345, 347, 461, 463, 497, 552, 702, 807
Keel, 7, 9, 413, 529–531, 533, 551, 562, 597, 650, 655, 669–677, 681–692
Kentucky, 561, 775, 782

Keritomy (-ized), 79, 102, 122, 147
Kleptoplastidy, 690, 739

Labiate process, 565–566, 571–572, 574–575, 578–579, 581–582, 584–587, 599–601, 606–607, 611, 613, 616, 619, 623, 656
Labrador, 411, 451, 454
Lake Baikal, 12, 14, 568, 578, 613
Lake Catemaco, 82, 174, 181
Lake Erie, 20, 84, 180, 206, 210, 513, 572, 585, 806, 809, 813
Lake Huron, 26, 206, 210, 212, 314
Lake Mendota, 19, 814
Lake Michigan, 22, 27, 206, 210, 314, 586, 605, 627, 759, 765, 767, 769, 771
Lake Okeechobee, 14–15, 806, 813
Lake Ontario, 206, 210, 340, 813, 817
Lake Simcoe, 206, 210
Lake Superior, 39, 206, 454, 585, 613, 627
Lake Tahoe, 14–15, 18–19, 26, 561
Lake Washington, 23, 813, 825
Lake
  Arctic, 12–14, 18–19, 26–27, 33, 82, 98, 118, 134, 137, 155, 158, 166, 169, 171, 180, 184, 231, 276, 280, 284, 288, 299, 303, 306, 347, 407, 424, 433, 446, 453–454, 458–459, 461, 490, 535, 552, 561, 584, 604–605, 613, 704
  Caldera, 13–16
  Cirque, 13
  Deep, 12–16, 18–19, 27–28, 42, 87, 107, 573, 583, 587, 670, 701–702, 765, 768, 814–815
  Dimictic, 18, 605
  Dystrophic, 40, 79, 376, 424–425, 427–429, 433, 439, 441, 443, 446–447, 449, 451, 453–454, 458–459, 461, 489, 616, 646, 699, 709
  Eutrophic, 18–21, 23, 25–28, 31, 64–66, 81, 87, 89, 92, 120, 129, 166, 171, 174, 226, 248, 258, 274, 322, 337, 340, 364, 372, 405, 409, 428, 433, 446–447, 449, 451, 453–454, 458, 461, 473, 487, 489, 513, 535–538, 552, 568, 572, 579, 583, 585, 613, 616, 638, 647–648, 662, 699–701, 738, 763, 765, 807, 813, 816, 819, 822
  Graben, 14–15
  Kettle, 13–14, 702
  Landslide, 14–15
  Maar, 14–15
  Meromictic, 16, 18–19, 42,
  Mesotrophic, 20, 25, 65–66, 72, 79–82, 87–89, 120, 126, 129, 135, 166, 171, 258, 274, 363, 424, 427, 433, 439, 441, 443, 446–447, 449, 451, 453–454, 458–459, 461, 499, 536–538, 583, 616, 699, 791, 813
  Monomictic, 18, 42
  Morainal, 14
  Oligotrophic, 18–20, 22–23, 25–28, 31, 39, 64–65, 79–81, 126, 129, 139, 155, 258, 363–364, 369–370, 372–379, 451,

453–454, 458, 461, 473, 487, 489, 499, 513, 535–538, 567–568, 585–586, 603, 616, 627, 638–639, 643, 646, 648, 661–662, 699, 791, 818–819
Polymictic, 18
Shallow, 13, 15–17, 19, 23–24, 27, 42, 126, 150, 155, 258, 314, 335, 337, 413, 433, 495, 512, 536, 572–573, 579, 587, 595, 604, 616, 699–700, 766, 771, 808–809, 811, 814–815, 817, 822
Sinkhole, 15–16
Stratification, 17–22, 28, 258, 699–700, 738, 815
Subtropical, 15, 18, 23, 72, 79–80, 82, 87, 95, 135, 139, 161, 166, 169, 171, 174, 184, 187, 215, 315, 332, 340, 343, 377, 535, 537, 694
Tectonic, 14–15, 568, 574
Thermokarst, 13
Tropical, 13, 15, 18, 26, 34, 40, 60, 64–65, 67, 72, 79, 81–82, 129, 135, 139, 164, 171, 174, 189, 340, 365, 377, 485, 535, 552, 581, 647
Volcanic, 16
Laminarin – see Storage product, laminarin
Latitude, 32, 42, 204, 486, 765
Laurentian Great Lakes – see Great Lakes, Laurentian
Lentic environment – see Lake
Leucoplast, 735, 742–743, 746
Leucosin – see Pigment, chrysolaminarin
Lichen, 206, 306, 315, 623, 646, 758
Light
High light, 407, 513
Irradiance, 17, 23, 26, 35, 38, 64, 489, 700, 766, 811
Low light, 23, 27, 44, 65, 407, 489, 567, 738
Shading, 26, 31–32, 35, 45, 203, 206, 215, 258, 258, 333, 343, 412, 702, 763, 765, 784–785, 819–812, 815–816, 821
Ultraviolet (UV), 44, 66
Limpet, 25–26
Line-encrusted, see CaCO$_3$
Linking spine, 565, 572
Lipid (see also Fatty acid), 35, 312, 474, 574, 738
Littoral (habitat or zone), 6, 16–17, 19, 24–28, 38–40, 66–67, 82, 104, 120, 137, 150, 164, 166, 169, 177, 180–181, 189, 315, 340, 370, 372, 409, 427, 495, 603–604, 623, 757–758, 763–764, 766, 771, 808, 811
Liverwort, 17, 184, 314, 332, 341, 369
Longitudinal canal, 641–642, 647
Lorica, 1–2, 3, 5, 224, 228–229, 236–239, 388, 397–400, 406, 410–411, 414–415, 430, 434–434, 441–442, 458, 473–478, 480, 484–485, 487–488, 491–493, 495–499, 501–502
Lotic habitat – see River
Louisiana, 80, 101, 181, 198, 203, 218, 244, 343, 463, 526, 535, 541, 552, 811
Lugol's iodine – see Fixative, Lugol's iodine
Lutein – see Pigment, Lutein

Macroalga (-ae), 17, 27, 33–34, 205–206, 313, 332, 336, 339, 605, 764–767, 770, 780, 783, 809–811, 814, 819–820
Macrophyte, 6, 24, 31, 39–40, 139, 428, 656, 674, 702, 771, 780, 808, 814, 816, 820
Emergent, 17, 24, 38–39, 315, 810
Submersed, 24–28, 39, 126, 137, 141, 158, 161, 166, 205, 210, 298, 314, 332, 335, 443, 807, 809–811, 816–817, 819, 821
Magnesium (Mg^{2+}), 364
Maine, 203, 463, 749
Malaysia, 202
Management (environmental), 38, 314, 350, 537, 777, 786, 789, 793–796, 805–806, 809, 812–814, 816, 817, 819, 824
Manitoba, 38, 85, 139, 163, 822
Marine (habitats or ecosystems), 5–6, 8, 11–12, 15, 20, 34, 43, 59–60, 64, 66–67, 72, 79, 82, 87–88, 93–94, 96, 101–102, 104–105, 107–109, 129, 132, 135, 137, 144, 147, 150–151, 155, 164, 166, 169, 197, 199–200, 205, 218, 221, 232, 253, 258, 311, 313, 333–337, 339, 342, 346–347, 387–388, 401, 406, 408, 413, 427–428, 485, 511–512, 514, 519, 562–563, 646–648, 676, 698, 701, 704, 716, 738, 748, 758–759, 764–765, 770, 789
Marl, 27–28, 141, 150, 314
Maryland, 497, 702, 814
Massachusetts, 497, 41, 80, 131–132, 144, 169, 181, 237, 243, 340, 425, 428, 451, 461, 463, 748, 766, 771, 825
Mastigoneme, 427
Mayfly (Ephemeroptera), 35, 205, 411, 819
Medulla, 217–219
Mesocosm, 43, 786, 813, 825
Mesokaryotic, 686
Mesospore, 362
Mesotrophic environment – see Lake, mesotrophic
Metaboly (-ic), 388, 390, 393–394, 397, 400–401, 410–414, 428–429, 435, 439–440, 485, 495, 499,
Metalimnion, 17, 535
Metaphyton (-ic), 38, 59, 64, 70, 72, 79–82, 87–89, 95–96, 119, 122, 124–126, 129, 132, 135, 139, 141, 147, 151, 155, 158, 171, 174, 177, 181, 189, 267, 270, 276, 280–281, 284, 288, 291, 295, 298, 301–303, 306, 424–425, 427, 432–433, 439, 441, 443, 446–447, 4498, 451, 453–454, 459, 461, 463, 780
Mexico, 14–15, 42, 66–67, 72, 80, 82–83, 87, 92–94, 96–104, 106–107, 127, 129, 134, 137, 139, 141, 144, 150–151, 153, 174, 180–182, 184, 203, 206, 208, 215–216, 227, 233, 235–236, 242–244, 411–415, 486, 561, 698, 701, 706–707, 765
Michigan, 337, 363, 370, 411, 424–425, 459, 463, 497, 537, 541, 552, 586, 605, 616, 618, 627, 630, 702, 759, 765, 767, 770, 772, 816,

Microplankton, 20
Microscopy, 33, 384
Brightfield – see Microscopy, light
Differential interference contrast (DIC), 257, 259, 431, 491, 409, 514, 540, 738–740
Dissection, 206
Electron, 61, 67, 70, 219, 234–236, 242, 246, 253, 259, 335, 365, 388, 392, 396, 400, 402–406, 411, 414–415, 432, 463, 465, 472, 475, 485–486, 491, 500–501, 511–512, 523, 533, 539–541, 561, 672, 691, 740
Inverted, 409
Light, 60–61, 68, 132, 206, 225, 229, 235, 242, 253–254, 258–259, 312–313, 316, 365, 385, 388, 394, 399–400, 402–404, 410–411, 413–414, 427, 471, 484, 491–492, 497, 499–500, 514, 518, 523–524, 528, 533, 539, 561, 571, 605–606, 639, 647, 702, 741, 767
Normarski – see Microscopy, differential interference contrast
Oil immersion, 259
Phase contrast, 70, 81, 257, 259, 394, 431, 497, 500–501, 514
Scanning electron (SEM), 234–235, 365, 402, 406, 414, 511, 527, 539–541, 549, 551–552, 566, 574–580, 582, 584–585, 608, 612–615, 617–620, 622, 624–625, 630, 639, 649, 674, 692–693, 695, 703, 708, 734, 741
Transmission electron TEM, 236, 253, 257, 405, 411, 499, 514, 540–541, 551–553, 573, 620, 624, 629, 664
Midge (Diptera), 36, 180
Midwestern United States, 118, 171, 267, 270, 274, 363, 526, 552, 561, 605, 638, 646, 649, 665, 809
Mississippi, 101, 288, 377, 770
Mississippi River, 28–29, 32, 36, 38, 65, 765
Missouri, 152, 451, 463
Missouri River, 28
Mitochondrion (-ia), 387, 459, 690, 731, 735–736, 749–750
Mixotrophy (-ic), 20, 23, 489–490, 514, 534, 701, 739–740, 747
Moniliform morphology – see Colony, moniliform
Monitoring, 7, 23, 487, 775, 785–786, 792, 816
Mono Lake (California), 42–43
Monophyletic taxa, 218, 226, 235–236, 239, 244, 245, 387, 534, 570, 604, 659, 697, 737
Monoraphid diatom, 7, 9, 595–598, 601, 604–605, 609, 613, 623, 627
Monosporangium (-ia), 199, 201–202, 207, 210–213, 215, 217–219
Monostromatic (morphology), 207, 210, 218, 220, 317, 329, 347, 494, 503, 770
Montana, 411, 425, 427–428, 443, 447, 449, 451, 453–454, 458–459, 461, 463, 561, 766
Montezuma Well (Arizona), 14, 16, 26

Morphotype – *see* Ecomorphotype
Moss, 17, 23–25, 44–45, 63–65, 144, 180, 184, 187, 258, 260, 274, 298, 313, 330, 335–336, 341, 363, 369–370, 378, 407, 490, 703, 706
Mountain (montane) regions, 12–13, 29, 44, 64–65, 68, 81–82, 89, 92, 96, 101, 107, 137, 151, 153, 158, 181, 204, 212, 342, 346, 363, 369, 488, 561, 631, 738, 747, 749, 751
Mucilage – *see* Algae, gelatinous
Mucocyst, 403, 412, 427
Mud, 141, 144, 147, 150–151, 155, 180, 298, 314, 405–407, 429, 433, 463
Multiaxial (filament or thallus), 3–4, 8, 107, 118, 125, 181, 184, 200, 207–210, 217, 219, 295, 317–318, 335, 337–338, 340, 342, 344–345, 434, 459, 758, 771
Multiseriate: *See* Multiaxial
Myxophycean starch – *see* Cyanophycean starch
Myzocytosis, 701

N:P ratio, 22–23, 674, 776, 782, 812–813, 819
Nanoplankton, 20, 64–65, 67, 80–82, 87–89
Naphrax, 540, 562, 605
Nebraska, 16, 407, 561, 814
Necridia (necridic cell), 80, 118–120, 132, 135, 139, 141, 147, 150–151, 155, 158, 164, 166, 168, 189
Neustonic – *see* Algae, neustonic
Nevada, 15, 42, 177, 451, 459, 463, 561
New Brunswick, 332, 454
New England, 19, 29, 40, 181, 434, 526
New Hampshire, 184, 210
New Jersey, 16, 181, 463, 498, 702
New Mexico, 212, 512, 514–515
New York, 13, 16, 19, 42, 79, 89, 216, 218, 340, 347, 463, 526, 537, 552, 770, 792, 817, 819
Newfoundland, 141, 205, 215–216, 339, 489
Nicaragua, 15–16, 18
Nitrate (NO_3^-) – *see* Nutrient, nitrate
Nitrogen (N_2) fixation, 23, 36, 63, 119, 121, 681, 822
North Carolina, 65, 79, 88, 92, 95, 96, 118, 132, 137, 144, 147, 150, 158, 164, 169, 174, 180–181, 184, 216, 218, 288, 295, 303, 338, 363, 370, 413, 415, 425, 427–429, 434, 443, 446–447, 449, 451, 453–454, 458–459, 814, 822
North Dakota, 82, 181
Northwest Territories, 14, 42, 95, 144, 169, 203, 270, 302, 412, 427, 449, 552, 702, 765–766
Nova Scotia, 363, 369, 441, 454, 561
Nunavut, 231, 233, 270, 274, 412, 414–415
Nutrient
  Ammonium (NH_4^+), 23, 119, 406, 513, 674, 764
  Competition, 19–20, 22, 65, 406, 485, 487, 489, 568, 607, 821
  Iron (Fe), 23, 37, 41, 76, 235–237, 301, 406–407, 412, 414–415, 484–485, 495, 497, 501–502, 807, 813–816
  Nitrate (NO_3^-), 7, 23, 119, 208, 257, 364, 406, 513, 674, 699, 700, 764, 823
  Nitrogen (N), 45, 117, 119, 121, 124, 177, 254–255, 257, 342, 365, 406, 487, 489, 513, 647, 670, 673–674, 681, 694, 700, 766, 782, 785, 794, 812–813, 819, 822
  Nutrient limited (deficiency), 7, 26–27, 37, 39, 118, 164, 204, 254, 257, 512–513, 567–569, 586, 607, 694, 758, 776, 812
  Nutrient rich (surplus), 17, 19–20, 39, 43, 45, 280, 295, 299, 342, 406, 572, 579
  Phosphorus (P; also phosphate, PO_4^{3-}), 7, 22–23, 25–27, 32, 36–39, 42–44, 61, 65, 118, 203–204, 256, 258, 406, 485–487, 489, 512–513, 520, 536, 584, 607, 613, 623, 682, 674, 686, 699–700, 764, 776–777, 785, 788, 790, 792, 794–795, 806, 811–814, 823
  Requirement, 6, 22–23, 26–27, 44–45, 257, 406, 433, 513, 567
  Selenium (Se), 23, 513
  Silica (Si), 8, 20–23, 38, 41, 414, 432, 474, 524, 530–531, 533, 539–540, 559–561, 565, 569, 573, 584, 586, 598, 607, 613, 627, 642–647, 656, 662, 664, 812
  Trace element, 15, 23, 825

Odor (water quality problem), 38, 64, 338, 473, 487–488, 499, 503, 513, 535, 685, 776–777, 786, 806–807, 809
Ohio, 36, 41, 246, 411, 426, 428, 459, 461, 463, 487, 586, 648, 701–702
Ohio River, 28, 28–29, 32, 37, 572
Oklahoma, 314, 454, 458–459, 461, 463
Oligotrophic environment (*see* Lake, oligotrophic)
Ontario, 81, 95, 129, 132, 139, 169, 171, 180, 208, 210, 231, 233, 239, 242, 248, 267, 270, 377, 411–413, 415, 425, 428, 434, 447, 449, 451, 453–454, 486, 488–489, 513–514, 526, 536, 541, 552, 561, 702, 708, 793, 807, 813, 817, 822
Oogamy – *see* Reproduction, oogamy
Oogonium, 312, 330–332, 338–342, 344, 432, 464–465
Oregon, 12–13, 42, 189, 201, 212, 235, 243, 463, 614, 762, 765, 766, 770
Organic carbon, 23, 31, 204, 406, 604, 647, 793
Organic scale, 480, 484, 493, 499, 512, 514–516, 518–519
Organic-rich (environment), 204, 236, 243–244, 248, 342, 405, 447, 449, 662
Oxygen (O_2), 15, 17–19, 27–28, 38–39, 41, 45, 59, 63, 66, 155, 189, 204, 409, 433, 623, 662, 674, 699, 776, 785, 791, 805–806, 810, 816–817, 822, 824–825
Oxygen, depletion, 18–19, 30, 189, 409, 806, 815, 817, 822, 824–825

Pacific Northwest, 40, 561,
Paleolimnology (paleoecology), 43, 538, 560, 570, 585, 604, 607, 610, 655, 776, 780–782, 788–789, 792–793
Pallium, 701
Palmelloid (morphology; also Palmella), 257, 260–261, 271, 294–295, 336, 346, 388, 394, 403–404, 406, 409–412, 427, 429–430, 432, 434, 473, 484, 492, 497–498, 500, 502, 736, 739, 744
Panama, 181, 210, 240, 242, 702
Papilla, 231, 235, 248, 266, 301, 319–320, 343, 370, 398, 414, 495, 530, 548–551
Parallelism, 429, 659
Paramylon – *see* Storage product, paramylon
Parasitic (species or stage), 45, 206, 298, 311, 315, 319, 442, 685, 696–697, 701, 706, 708–709, 739, 751
Parenchymatous (morphology), 4–5, 110, 311, 313, 317, 320, 322, 332, 337, 342, 347–348, 494, 500, 758, 771
Parthenospore – *see* Spore, parthenospore
Peat, 38, 40, 66, 88–89, 210, 335, 363, 372, 617
Peduncle, 701
Pelagic (habitat), 17, 23, 39, 66, 147
Pellicle, 5–6, 8, 383–384, 390–392, 395, 399–401, 403–404, 410–413, 690, 704
Pennsylvania, 26, 42, 147, 150, 218, 463
Peridinin – *see* Pigment, Peridinin
Periphyton, 24, 66, 126, 135, 139, 144, 147, 151, 158, 166, 205, 314, 372, 374–376, 378, 595, 598, 604, 672, 674, 779, 785, 789, 810
pH, 11, 15, 23, 26, 43–45, 61, 215–216, 363–364, 485, 487, 490
pH effect, 26, 42, 44, 92, 204, 363–364, 407, 488–489, 513
pH, Inferred pH, 789, 793
Phagotrophy, 383, 385, 387, 401, 403, 408, 473, 489–490, 512, 514, 701, 740
Phenology, 205
Phosphatase (enzyme), 25, 776, 782, 785
Phosphorus – *see* Nutrient, phosphorus
Photoacclimation (also Chromatic adaptation), 45, 61, 65, 738
Photosynthesis (-thetic), 17–18, 20, 23, 38, 41, 43–44, 59, 61, 120–121, 200, 202–204, 217–218, 226, 257, 311, 383–385, 387, 401–402, 406, 415, 423, 473, 485, 489, 686, 690, 696–697, 704, 706–709, 735, 737–740, 751, 776, 782, 785, 816, 823
Phototaxis (-tactic), 119, 465
Phycobiliprotein – *see* Pigment, Phycobiliprotein
Phycobilisomes, 61, 200, 208
Phycocyanin – *see* Pigment, Phycocyanin
Phycoerythrin – *see* Pigment, Phycoerythrin
Phylogenetic (relationships), 1, 68, 225, 227, 229, 235–236, 239–240, 242, 244, 254, 257, 312, 424, 524, 533–534, 569–570, 582, 655, 659, 661, 686, 697, 716, 737, 741–742

Phylogeny, 225, 231, 388, 473, 534, 570, 665, 737
Physode – *see* Fucosan vesicles
Phytoplankton – *see* Algae, planktonic
Phytoplankton bloom – *see* Bloom
Picoplankton, 20, 22, 60, 64–65, 67–68, 89
Pigment
    Accessory, 20, 23, 200, 254, 423–423, 429, 474, 560, 567, 670, 737, 741
    Allophycocyanin, 5, 61, 737
    Astaxanthin, 43, 312, 342, 392
    Carotenoid, 43–45, 61, 126, 141–147, 254, 257, 312, 315, 343, 392, 474, 560, 670, 686, 690, 733, 757
    Chlorophyll-*a*, 6–8, 19, 23, 36–37, 43, 59, 61, 254, 311, 423, 560, 812–814
    Chlorophyll-*b*, 6, 8, 225–226, 254, 311, 384, 423, 451
    Chlorophyll-*c*, 6–8, 423–424, 474, 512–513, 560
    Diadinoxanthin, 427, 429, 474, 690, 735
    Diatoxanthin, 429, 474, 757
    Fucoxanthin, 5, 7–8, 424, 443, 473–474, 528, 560, 690, 757
    Gloeocapsin, 61
    Heteroxanthin, 427, 429, 443
    Loroxanthin, 245
    Lutein, 254
    Oscillaxanthin, 141
    Peridinin, 5, 7–8, 686, 690, 706
    Phaeophytin (or Pheophytin), 776, 782–783
    Phycobiliprotein (also Phycobilin), 6, 59, 61, 147, 197, 384, 735, 737, 742–743
    Phycocyanin, 5, 7–8, 61, 126, 200, 203, 208
    Phycoerythrin, 5, 7–8, 20, 61, 126, 147, 200, 203
    Scytonemin, 61, 158
    Tracer, 39
    Vaucheriaxanthin, 427, 429
    Xanthophyll, 141, 245, 474
Plankton net, 67, 227, 258, 365, 408, 463, 490, 539, 702
Planozygote – *see* Cyst, planozygote
Plastid (*see also* Chloroplast), 208, 254–256, 261, 270, 276, 280, 291, 295, 298, 301, 423–425, 427, 432, 435, 438–439, 441, 443, 451, 461, 465, 474, 485, 541, 599–602, 613, 661, 623, 638, 643–650, 655, 675–676, 681–682, 690, 704, 708, 735–737, 742, 744, 767
Plurilocular sporangium, 761
Pollution, 11, 28, 34, 42, 126, 132, 150, 333, 336, 405–406, 409
    Acid rain (acidic precipitation), 39, 41, 44, 363–364, 372, 513
    Fertilizer, 19, 487, 513, 814, 818
    Heavy metals, 41, 205, 340, 345, 414, 785, 781, 825
    Nutrient enrichment – *see* Eutrophication
    Sewage, 19, 23, 44, 336–337, 406, 514, 778, 794, 813
Polyphosphate (body [-ies]), 61, 623, 823
Polyphyletic taxa, 225, 242, 257, 563, 570, 604

Polysaccharide, 61, 118, 200, 304, 474, 609, 628, 659, 739, 757
Pond, 6–8, 12–16, 19, 24, 26–27, 40, 42, 44–45, 66,72, 80, 82, 87–88, 129, 132, 141, 155, 164, 166, 169, 171, 174, 180–181, 197–199, 205, 226, 257, 267, 270, 272, 274, 276, 280–281, 284, 288–289, 291, 295, 298–299, 301–303, 306, 314–315, 330, 333, 335, 337–341, 344, 346–347, 363, 365, 369–370, 372–379, 385, 405–409, 412–413, 424–425, 427–429, 433, 439, 441, 443, 446–447, 449, 451, 453–454, 458–459, 461, 463, 485–490, 493, 495–496, 499, 501, 512–515, 534–538, 573, 583–587, 649, 668, 700–702, 708–709, 739, 747, 751, 763–765, 771, 805–806, 808, 810–813, 815–819, 821–822, 824–825
Population dynamics, 20–21, 27, 32, 37, 258, 739
Potomac River, 65
Precambrian Shield – *see* Canadian Shield
Preservative – *see* Fixative
Primary production, 19–20, 31–32, 34, 38–39, 41, 43, 63, 782
Protist, 20, 383–385, 388, 403, 405, 423, 473, 490, 511–512, 685–686, 697, 710, 715, 749
Protolichen, 315
Pseudocilia – *see* Pseudoflagella
Pseudocyst, 480, 484, 495–496
Pseudofilament, 1, 3, 62, 69–72, 80–81, 96, 102, 107, 110, 129, 199–201, 208–210, 276, 317, 336, 344, 346, 434, 459–460, 515
Pseudoflagella, 254, 259–261, 263, 267, 270, 284, 295, 298
Pseudoparenchya, 4–5, 96, 181, 199–200, 203, 207, 210, 216, 260, 264, 270, 298, 320, 322, 332, 334–338, 340, 342–343, 434–435, 459–461, 473
Pseudopyrenoid, 274
Puerto Rico, 75, 78, 80, 82, 92–93, 95–97, 99, 101, 104, 106–108, 110, 118, 126–127, 132, 134, 138–139, 145–146, 148, 150–151, 154, 156–157, 160, 162, 176, 178, 183, 212, 220, 363, 370, 433–433, 459, 463
Puncta (-ae, -ate), 333, 438, 567, 571–572, 575, 582, 584–586, 599, 606, 639–641, 643–650, 659–660, 665, 670–671
Pusule, 686, 690
Pyrenoid, 199–200, 207–208, 210, 215, 226–227, 231–233, 235–237, 239–244, 246, 248, 254, 256–257, 260, 263–267, 270–272, 274, 276–277, 280–282, 284–285, 288–289, 291–292, 295, 298–299, 301–303, 306, 312, 317–322, 330, 332–347, 354, 362, 366–367, 369–379, 384, 389–390, 394–395, 397, 402–403, 410–415, 424, 427, 432, 437, 439, 441, 443, 446–447, 449, 451, 453–454, 458–459, 461, 463, 496–498, 500, 515–516, 518, 720–724, 726–727,

731–732, 735–736, 742–748, 759, 768, 770–771

Quebec, 14, 16, 26, 35, 38, 203, 215, 377, 379, 434, 458, 526, 541, 561, 709

Raphe, 9, 24, 36, 599, 601–603, 605, 607, 613, 623, 627, 638–650, 656, 658–659, 661–666, 669–670, 672, 675–676, 680–682
    Canal (-ed) raphe, 7, 9, 669–670, 672, 675, 681
    Keeled raphe, 669–670, 676
    Raphe branch, 648, 659, 662, 664–666
    Raphe ending, 602, 623, 627, 639–642, 644, 646–650, 656, 658–659, 662–664
    Raphe sternum, 603, 623, 647
Raphid valve, 598–599, 601–603, 605, 607, 623–625, 627–630, 641
Reproduction
    Asexual, 201, 210, 226, 231–237, 239, 241–243, 245–246, 254–255, 257, 288, 312, 322, 332–333, 340, 342–344, 362, 424, 432, 485, 531, 533, 539, 564, 638
    Conjugation, 6, 8, 226, 235, 311, 353–356, 362–376
    Daughter cell, 61–62, 70–71, 79, 81, 87, 89, 92, 102, 126, 132, 226, 346, 354, 356, 374, 485, 496, 565, 693, 736
    Exocyte, 62, 70–71, 101–102, 98–99
    Gametangium (-ia), 201–202, 212, 319–320, 322, 338, 343–344, 355–356, 362, 364–366, 369–376, 564
    Gemmae, 199, 202, 218–219, 771, 765
    Gonimoblast, 207–208, 211–215
    Isogamous, 226, 229, 231–233, 235–237, 239–240, 243–244, 246, 248, 254–255, 312, 330, 333–225, 337, 339, 341, 343–344, 347, 365–366, 432, 461. 463, 485, 533, 602, 694
    Oogamous, 226, 233, 235–236, 240, 312–313, 318, 322, 330, 332, 338, 340–342, 344, 346, 432, 463, 559, 563
    Sexual, 8, 62, 201, 207–208, 210, 212, 218, 221, 226–227, 229, 231–237, 239–246, 248, 254–255, 257, 312, 322, 330–335, 337, 340–347, 353, 362–365, 369–377, 388, 405, 424, 432, 461–463, 485, 531, 539, 559–560, 563–564, 568, 622, 624, 627, 637–638, 649, 655, 659, 664–665, 694, 736
Reservoir (water body), 14, 16–17, 28, 64–65, 72, 81–82, 129, 135, 139, 171, 174, 177, 227, 405, 454, 458, 461, 583, 699–700, 739, 746, 751, 807, 811
Restoration, 788, 974, 806, 811
Rhine River, 37–38
Rhizoid, 24, 27, 35, 200, 207, 209–212, 215, 219–221, 260, 298, 317–320, 322, 332–334, 337–339, 340–342, 344, 347, 363, 370–371, 463–464, 768
Rhizoplast, 528
Rhizostyle, 427, 717, 734, 741, 743

Rhode Island, 201, 203, 216, 364, 463
Rice field (also crop, paddy), 45, 66, 79, 121, 174, 177, 180–181, 226
Riffle, 19–30, 35–36, 778–779
Risk assessment, 776–777, 796–797
River / stream, 6, 8, 11–12, 14–17, 28–41, 44–45, 65, 68, 82, 89, 96, 101, 104, 107, 110, 126, 137, 139, 141, 144, 147, 150–151, 155, 158, 161, 164, 166, 177, 180–181, 184, 187, 197–210, 212, 215–218, 220–221, 226, 257, 280, 284, 288–289, 295, 298–299, 303, 314, 330, 332–333, 335–342, 344–347, 363–364, 366, 370, 372, 376, 405, 407, 412, 424, 427–429, 433, 446, 449, 453–454, 459, 461, 463, 487–488, 490, 498, 500, 512–514, 534, 561, 572–573, 575, 579, 584–587, 596, 598, 601, 605, 616, 628, 661, 665, 670, 672, 682, 702, 709, 757, 759, 763, 764–765, 767, 775
   Discharge (see also Current Velocity), 28, 32, 37–38
   Flood, 16, 28–29, 32–36, 40, 205, 342, 810
   Headwater stream (see also Spring–stream), 28, 31, 203, 514
   Large river, 28, 32, 37–38, 65, 586, 661
   River continuum concept (RCC), 30–31
   River order, 28–29, 31–32, 38, 206
   Spring-stream, 16, 28, 40–42, 80, 104, 141, 150–151, 155, 161, 181, 198–199, 203, 208, 210, 218, 303, 330, 333, 335, 337–338, 433
River Thames (U.K.), 37, 81, 206
Rocky Mountains, 81
Rotifer, 16, 32, 38, 41, 43, 129, 383, 411, 463, 739, 825
Rubisco, 34, 61, 200, 206, 686, 737–738

Saccoderm desmid, 353
Saline (environments), 6, 11, 23, 42–43, 59, 64, 66, 81–82, 87–88, 96, 107, 123, 132, 135, 137, 135, 137, 139, 150–151, 169, 171, 177, 258, 321, 333, 512, 514–515, 537, 560–561, 572, 574–575, 584–585, 604, 606–607, 638, 646, 650, 707, 766
Salinity, 11–12, 34, 42, 65–67, 80, 177, 189, 514, 565, 569, 604, 648, 738, 766, 782, 787, 789, 791
Sampling method, 25, 206, 315, 365, 486, 490–491, 514, 778
Saskatchewan, 490–491, 514
Scandinavia, 129, 485–486, 513, 537, 661, 759
SCUBA, 315, 767
Scytonemin – see Pigment, Scytonemin
Seasonal succession, 19–22, 25–27, 44–45, 203–205, 258, 312, 345, 364, 433, 487–488, 538, 567
Secchi (depth or disk), 15, 19, 513, 779, 783, 806, 812–813, 816
Sediment core, 537, 539, 610, 780–781
Semicell, 354, 356–357, 361–362, 365–369, 372–379

Seta (see also Hair cell), 262, 284, 294, 373
Sheath, 1–3, 20, 60–62, 67, 70–72, 74–80, 82–86, 89–97, 101–102, 104, 107, 110, 118–127, 132, 135–138, 141, 143–151, 154–169, 171, 174, 177, 181–187, 189–190, 208, 210, 239, 242, 244–248, 254, 257–258, 260–263, 265, 267, 274, 276–277, 280, 282, 288–289, 291, 301, 312, 318322, 332, 335, 340, 344–346, 354, 371, 376, 378, 409, 725, 435, 437, 453–455, 482–483, 501, 703, 709
Si:P ratio, 20, 22, 812
Siberia, 12, 585
Sierra Nevada Mountains, 44, 561
Silica (Si) – see Nutrient, silica (Si)
Silica scale (also Siliceous scale), 5, 7, 9, 493, 497, 499–502, 512–513, 474–475, 486, 491
Silicon dioxide (SiO_2), 559
Sinus, 627, 629
Siphon (also Siphonous morphology), 4, 297–298, 312–313, 344, 462, 465, 810
Sloth, 6, 45, 197–199, 207, 210, 315, 319, 338
Snail, 25–26, 39, 205, 210, 215, 339, 819, 825
Snow algae – see Algae, snow
Sodium (Na^+), 41–42, 333
Softwater (habitat), 26, 39, 126, 205, 210, 215, 340, 344, 364, 372–373, 412, 485–486, 489, 495, 499, 513, 598, 665, 763, 807, 826
Soil, 6, 11, 38, 40, 44–45, 60, 63, 72, 79, 82, 89, 95, 121–121, 126, 129, 135, 137, 141, 141, 150, 158, 166, 177, 180–181, 184, 197–199, 206, 208, 210, 253, 255, 258, 270, 272, 274, 276, 280, 284, 288, 291, 295, 298, 301, 303, 311, 315–322, 332–337, 341–342, 344–347, 363, 369, 372, 393, 404–409, 424–425, 429, 433, 451, 453, 463, 490, 605, 648, 672, 676
South America, 202, 338, 513, 526, 568, 605, 628, 649, 765, 771
South Carolina, 425, 427, 434, 441, 449, 454, 459, 461, 463, 495, 583, 702
South Dakota, 610, 820
Southeastern United States, 203, 216, 235, 263, 267, 270–271, 274, 276, 280–281, 284, 288, 291–292, 295, 298, 301–303, 306–307, 348, 363, 387, 407, 413–415, 428, 561, 664, 709
Southwestern United States, 42, 104, 203, 216, 334, 407
Species richness, 40, 782, 785, 787, 789
Specific conductance (also Conductivity), 6–7, 40, 88, 171, 204, 206, 208, 210, 212, 215–216, 218, 220–221, 363, 428, 485, 487–488, 513–514, 535–537, 540, 585, 587, 646, 647, 649–650, 662, 676, 681, 738, 740, 742, 764, 766–767, 778, 789, 795, 810
Spermatangium (-ia), 40, 88, 171, 204, 206, 210, 212, 215–216, 218, 220–221, 363, 433, 485, 487–488, 513–514
Spermatium (-ia), 201–202, 210–213

Sponge, 45, 258, 272
Spore
   Spore, aplanospore, 212, 235, 245, 255, 265, 288, 291, 312, 320, 322, 332–336, 340–342, 344–345, 347, 355, 362, 370–372, 375–376, 432, 447, 451, 454, 457–459, 461, 463, 496
   Spore, autospore, 288,–289, 298, 322, 341, 424–427, 432, 443–447, 449, 451–456, 458–459, 461, 463, 485, 497, 500, 502. 698, 708
   Spore, auxospore, 560, 564, 602, 624, 627, 638
   Spore, baeocyte, 62, 70–72, 101–104, 107, 110
   Spore, endospore, 199, 201, 207–208, 210, 362
   Spore, exospore (see also Exocyte), 62, 70–71, 101–102
   Spore, hypnospore, 43–44, 245–246, 463, 708
   Spore, parthenospore, 362, 371–372
   Spore, planospore, 212, 255–256
   Spore, statospore, 432, 439–440, 474, 480, 485
   Spore, zoospore, 226, 231–232, 235–237, 239, 254–256, 265–267, 270, 274, 276–277, 280, 282, 284–285, 288, 291, 295, 298, 301–303, 306, 311–312, 320, 322, 326, 330, 332–347, 423–428, 432, 434, 439, 441–451, 453–456, 458–464, 473, 483, 485, 495–502, 696, 709, 739, 759–760, 767–768, 770
   Spore, zygospore, 19, 330, 355–356, 362–363, 369–379
Sporophyte, 201–202, 207–208, 211–218, 759
Sporopollenin, 245–255, 312
Spring-fed streams – see River, spring-stream
St. Lawrence River, 34, 37, 210, 793
Stagnant water, 6, 89, 204, 322, 336, 340, 347, 363, 371, 413
Stain (microscopy), 70, 259, 403, 411, 491, 498, 500
   Alcian Blue, 198, 410
   Iodine (test for starch), 5, 226, 259, 403, 409, 454, 463, 491, 514, 702–703, 736, 740
   Jensen's stain, 491, 497, 500
   Methylene blue, 295, 403, 476, 491–492, 498, 502
   Toluidine Blue O (TBO), 211, 217, 219
Starch – see Storage product, starch
Statistical methods, 569, 777, 788–792
Statospore – see Spore, statospore
Stauros, 603, 607, 623, 627, 640, 642–643
Stenotherm, 80, 141, 147, 488
Sternum, 599–606, 612–613, 623, 627, 643, 645–648, 656, 665
Stichidia, 199, 202, 208, 219–220
Stickleback (fish), 697, 708
Stigma (-ata) – see Eyespot
Stomatocyst, 414, 476, 480, 485–486, 496, 499–503, 531
Stoneflies (Plecoptera), 205

Stonewort, 311, 314, 332
Storage product
  Chrysolaminarin, 5–7, 423, 472, 474, 528, 541, 548
  Cyanophycean starch, 6
  Floridean starch, 6, 197
  Laminarin, 5–6, 8, 757
  Paramylon, 5–6, 384, 388–390, 395–397, 402, 407, 409, 411–415
  Polyphosphate, 61, 623, 823
  Starch (true starch), 5–8, 225–226, 231, 254, 259, 265, 270, 274, 276, 280, 282, 291, 298, 301, 303, 311–312, 318, 335–336, 340–341, 384, 423, 427, 686, 690, 702, 706, 717, 722, 726–727, 735–737, 739–740, 745–748
Stratification – see Lake, stratification
Stream – see River / stream (habitats)
Stria (-ae), 357, 365–368, 372–373, 403, 437–438, 518, 546, 572, 578, 599–607, 612–613, 616–618, 623–624, 627, 638–640
Stromatolite, 59, 63, 67, 121
Strutted process, 565–567, 571–572, 583–585, 587
Subaerial – see Algae, subaerial
Subarctic (habitat or biome), 12, 26, 38, 40, 60, 80, 88, 101, 141, 158, 171, 289, 513, 535, 700–701
Substratum (-a) (surface), 4, 7, 24–30, 32–37, 44, 63–68, 70, 72, 80, 87, 89, 92, 95–96, 101–102, 104, 107, 110, 119, 121–122, 125–126, 132, 135, 137, 139, 141, 144, 147, 150–151, 155, 158, 161, 164, 166, 169, 177, 180–181, 184, 187, 189, 205–206, 220, 260, 270, 272, 280, 288, 291, 295, 303, 316, 354, 363, 371, 376, 378, 388, 393, 405, 408, 410–411, 436, 441, 458, 463, 474, 488, 492, 494–497, 499–503, 565, 583–584, 586, 595, 598, 601–602, 605, 607, 609, 613, 616–617, 623, 638, 656, 659, 661, 663, 669–670, 673, 677, 687, 690–691, 740, 747, 758, 760, 763–764, 766–768, 771, 779, 780, 783, 786, 808–810
Succession – see Seasonal succession
Sulcus, 388, 395–396, 410–411, 686–687, 690–691, 706–707, 788
Sulfate (SO_4^{2-}), 23, 41–42, 66, 372, 377, 406, 514, 662, 807, 813, 816, 821–822, 824–826
Surface bloom – see Bloom
Suture, 357, 361, 372–373, 695, 704
Sweden, 204, 513, 588, 758, 763, 794
Synonym (-ous), 96, 158, 164, 212, 229, 232, 267, 344, 376, 379, 388, 425, 427, 458–459, 512, 524, 563, 629, 646, 669–670, 697–698, 706, 708, 742–743, 746, 758, 770–771

Tadpole, 26, 410–412
Taste (water quality problem), 38, 64, 467, 535, 552, 685, 776–777, 786, 806, 809, 816

Temperate environment (region), 8, 12, 17–20, 26, 31, 33, 37, 64–65, 80–82, 87–89, 95, 101–102, 126, 129, 132, 135, 139, 147, 158, 161, 164, 169, 171, 174, 177, 187–189, 202–204, 216, 227, 246, 258, 284–289, 291, 295, 298, 301, 303, 306, 334, 339–340, 345, 364, 485–486, 535, 537, 568, 581, 583, 616, 694, 696, 704, 715, 738, 812–813
Temperature
  Cold, 12, 17, 40, 59, 66, 80, 88, 95, 101, 107, 129, 141, 147, 288, 303, 342, 488, 499, 535–536, 538, 567, 605, 649, 694, 696, 702, 740, 815, 820
  Hot – see Hot spring
  Warm, 18, 23, 33, 37, 43–44, 65, 80, 88, 101–102, 107, 150, 174, 203–204, 212, 218, 220–221, 314, 407, 409, 434, 463, 488, 513, 535–537, 605, 700, 763, 783
Temporary pond (pool), 19, 44, 66, 257, 364–365, 371, 433, 501, 512
Tennessee, 15, 428, 434, 459, 461, 463
Terrestrial (habitat), 11, 25, 45, 59, 63, 68, 120–121, 144, 150–151, 174, 180, 187, 260, 267, 276–277, 284–285, 288–289, 306, 312, 315, 317–319, 321–322, 334, 341–343, 424, 433, 463, 610, 637, 778, 781, 821
Tetrasporangium (-ia), 201–202, 212–213, 220
Tetrasporophyte, 201–202
Texas, 16–17, 40, 150, 208, 212, 218, 243, 270, 291, 332, 345–346, 425, 427, 512, 702, 739, 749
Theca, 2, 5, 7, 229, 231–232, 690–691, 694, 698, 701, 703–704, 707
Thermal spring – see Hot spring
Thermocline, 17, 23
Thylakoid, 5–8, 44, 61, 69, 73, 75–77, 79, 81, 87–88, 101–102, 104, 118–119, 121–122, 126–127, 129–130, 132, 135, 137, 139, 141–142, 144, 147, 149–150, 153, 177, 187, 197, 200, 208, 257, 397, 402, 423, 474, 690, 742, 748
Tierra del Fuego, 486
Total nitrogen (TN), 489, 785
Total phosphorus (TP), 36, 43, 486, 489, 513, 536, 699, 777, 787, 785, 792–793, 794–795, 812–813
Toxin (also Toxic), 6, 63–64, 92, 139, 141, 147, 169, 512, 514, 604, 607, 685–686, 739, 776–777, 782–783, 785–786, 806–808
Travertine, 16, 59, 67, 69, 81, 119, 121, 141, 158, 169
Tree (also Tree bark), 45, 64, 89, 121, 144, 161, 181, 184, 203, 206, 258, 267, 315, 330, 344, 346, 780
Trichoblast, 208, 220–221, 689, 691, 695
Trichocyst, 390, 403, 409, 412, 426–429
Trichogyne, 201–202, 211–216, 218–219
Trichome, 118–132, 134–144, 147–152, 155–158, 161, 164–166, 168–177, 179–181, 184, 187, 189
Trichoptera (see Caddisfly)

Trinidad, 463
Trophont stage, 696, 706
True branch, 3–4, 119, 121, 123, 125, 181–190, 200–202, 207–208, 210–221, 259–260, 263–264, 267, 271, 284–285, 298, 313–315, 317–325, 327, 330–344, 347, 354, 364–366, 376, 434–436
Tundra (habitat), 19, 33–34, 121, 158, 180, 184, 197, 203, 215, 422, 446
Turbid (-ity), 23, 27, 31–21, 37–38, 202, 204, 206, 258, 407, 670, 763, 766, 811, 820
Turtle, 12, 313–314, 316, 319, 321, 334, 339–340
Tychoplankton (-ic), 88, 146–147, 408, 424–425, 427, 432–433, 453–454, 458–459, 461, 463, 565, 572, 672, 676, 681

Ultraviolet radiation (UV_R), 44–45, 66
Uniaxial (also Uniseriate), 3, 69, 80, 102, 118, 125, 135, 158, 161, 169, 171, 177, 180–181, 184, 187, 189, 200, 207–208, 210–216, 218–219, 264, 317, 322, 326, 330, 332–342, 344–345, 354, 365, 369–377, 434, 459, 461, 600–602, 606, 613, 616–617, 623, 646, 758–760, 768, 771
Unilocular sporangium, 759–760, 762, 765, 767–771
Uniseriate – see Uniaxial
United Kingdom (U.K.), 42, 206, 338, 700, 759, 763, 765, 771
Urban (-ization), 19, 39, 45, 342, 794, 813, 816, 826
Utah, 12, 72, 191, 454, 459, 461, 463, 628, 647, 650, 699–700, 763, 766
Utermöhl technique (sedimentation), 67, 514, 539

Valve face, 565–566, 569, 571–572, 583, 585, 587, 600, 613, 616–617, 623, 627, 640–641, 643, 645, 650, 656, 660, 662, 665, 670, 675–676,
Valve mantle, 567, 571–573, 578–579, 582–583, 585, 607, 644, 646, 650, 656, 662–663, 665
Valvocopula, 600–601, 623, 644, 648
Van Dorn sampler, 539, 702, 779
Vascular plant, 40, 65, 180, 315, 3198, 321, 363, 365, 377, 405, 424–425, 432, 439, 441, 443, 454, 458–459, 461, 463, 490, 662, 807–808, 810–811, 815–817, 820–821, 824
Venezuela, 93, 95, 202
Vermont, 189
Virginia, 41, 212, 434, 461, 463
Virus (-es), 384, 739, 819
Vitamin, 387, 406, 409, 490, 699–701

Washington (state), 13, 15, 19, 23, 42, 44, 203, 235, 346, 526, 541, 649, 766

Water bird (waterfowl), 333, 407, 486, 538, 811, 820–821
Water quality, 13, 39, 59, 406, 489, 560, 598, 655, 782
Watershed (also Catchment), 28–29, 32, 34, 36, 204, 536, 560, 788, 789, 805, 813–814, 816, 826
Wave action (also Scour), 24–27, 137, 347, 453, 808
Weed (or Weedy species), 174, 821, 817, 809–812, 820–821
West Virginia, 42, 459, 463
Western Canada (*see also* specific provinces), 33, 126, 267, 274, 276, 282, 284, 288–289, 292, 295, 302–303, 306, 552, 646
Western United States (*see also* specific states), 42, 274, 282, 284, 288–289, 291–292, 295, 298, 301–303, 306, 334, 459, 461, 526, 552, 561, 648, 808–809, 811, 817
Wetland (also Swamp, Marsh), 11, 28–29, 37–40, 66–67, 72, 79–83, 86–89, 95, 102, 104, 137, 147, 151, 155, 158, 151, 164, 166, 174, 177, 180–181, 184, 189, 257, 341, 345, 372, 375–378, 413, 425, 428–429, 433, 458–459, 463, 536, 604, 661–662, 669, 672, 682, 702, 776, 778–780, 785, 788–789, 791, 793, 796
Whorl (branching pattern), 3–4, 207–208, 213–215, 217, 317, 321, 333–334, 337–339, 367
Wisconsin, 12, 19, 22, 74, 76, 78, 83–87, 90, 94, 118, 143–144, 174, 242, 339, 363, 370, 425, 453, 459, 461, 487, 702, 814, 825
Yellowstone National Park, 15, 41, 66–67, 72–73, 75–82, 94–95, 104–105, 107, 118, 126, 128–129, 130–137, 139–140, 144, 147–148, 151–152, 155, 157–158, 186–187, 189–190, 206, 363
Yucatan Peninsula, 15–16, 66
Zebra mussel, 28, 37–38, 813, 816
Zooplankton (*see also* specific groups), 6, 20, 38, 43, 64, 271, 303, 488, 537, 700, 702
Zoospore, 226, 231, 232–233, 235–237, 239, 254–255, 265–267, 270, 274, 276–277, 280, 284–285, 288, 291, 298, 301–303, 306, 311–312, 320, 322, 330, 332–347, 423–428, 432, 434, 439, 441–451, 453–456, 458–464, 473, 483–485, 495–502, 698, 709, 739, 759

Taxonomic Index

Acanthameba, 737
Acanthoceras, 563, 571–572, 587
   magdeburgense, 575
Acanthoica, 512, 515
   schilleri, 515
Acanthosphaera, 262, 267
   zachariasii, 268
Achnanthaceae, 597–598, 604, 623
Achnanthales, 595–598, 623
Achnantheiopsis, 598, 627
Achnanthes, 34, 41, 45, 597, 601, 604, 607
   affine, 623
   clevei, 627
   clevei var. *rostrata*, 596
   coarctata, 623–624
   kolbei, 627
   lanceolata, 598, 627
   lancolata var. *abbreviata*, 629
   levanderi, 627
   linearis, 628
   longipes, 623
   marginulata, 627

   microcephalum, 623
   minutissima, 623
   ploenensis, 627
   pusilla, 43
Achnanthidiaceae, 597, 604–605, 623, 627
Achnanthidium, 597–598, 602, 604–605, 607, 623, 627
   affine, 609, 625
   deflexum, 625
   exiguum, 625
   exiguum var. *heterovallvum*, 625
   lanceolata, 36
   minutissima, 36
   minutissimum, 25–26, 33, 623, 625, 628
Achnanthoideae, 597
Acrochaetiales, 198, 201, 202, 204, 206, 211–212, 213
Actidesmium, 265, 267
   hookeria, 268
Actinastrum, 254, 262, 267
   hantzchii, 268
Actinella, 655–656, 662

   punctata, 657, 661, 664
Actiniscaceae, 697, 706
Actiniscus, 697, 698, 704, 706
   canadensis, 702
   pentasterias v. *arcticus*, 688
Actinochloris, 266–267
   sphaerica, 268
Actinocyclus, 563, 571, 575
   normanii, 572
   normanii f. *subsalsa*, 574
   subsalsa, 572
Actinotaenium, 368, 373
   diplosporum, 373
   perminutum, 373
   rufescens, 357
Adlafia, 638, 641, 643
   bryophila, 643
   muscora, 643
Aeronemum, 431, 435, 459
   polymorphum, 460
Agmenellum
   quadruplicatum, 88

Akanthochloris, 430, 434, 438, 443, 445
  bacillifera, 445
  brevispinosa, 445
  scherffelii, 445
Albrightia, 125, 184
  tortuosa, 186, 187
Alterasynedra, 596
Amblystoma, 45, 291
Amblystomatis, 45
Ambrosia, 298
Ammatoidea, 123, 151–152
  normannii, 151–152
  yellowstonensis, 151–152
Ammatoideoideae, 151
Amphicampa, 656, 662
  eruca, 657, 664
  mirabilis, 664
Amphichrysis, 493, 495
  compressa, 481
Amphidiniopsis, 691, 697–698, 704, 708
  sibbaldii, 693, 702
Amphidinium, 687, 691, 697–698, 704, 706
  cryophilum, 702
  klebsii, 688
Amphipleura, 638, 640, 643
  pellucida, 645
Amphithrix, 137
Amphora, 39, 43, 656, 663, 676
  calumetica, 661, 663
  coffeiformis, 43
  inflata, 662
  ovalis, 663
  pediculus, 658
  perpusilla, 658, 672
Anabaena, 20, 23, 37, 43, 45, 64–65, 68, 79, 119–120, 125, 169–171, 176, 180, 303, 768, 805, 807, 808, 812, 816
  azollae, 45, 180
  circinalis, 22
  cycadearum, 180
  fertilissima, 176
  flos-aquae, 20
  lutea, 176
  mendotae, 169
  oblonga, 170
  perturbata, 169–170
  subtropica, 176
  variabilis, 176
  viguieri, 170
Anabaenoideae, 120, 169
Anabaenopsis, 119–120, 124, 171–172
  elenkina, 172
Anabaenopsis (Cylindrospermopsis)
  raciborskii, 174
Anacanthoica, 512, 515
  ornata, 515
Anacystis, 74, 104
  nidulans, 81
  nigroviolacea, 93
  rupestris, 74
Anathece (subgenus of *Aphanothece*), 72
Ancylonema, 363, 367, 369
  nordenskioeldii, 355, 364, 369
Ancylus
  fluviatilis, 26

Aneumastus, 640, 644
  tuscula, 643
Animalcula, 471
Ankistrodesmus, 254–255, 257, 262, 267, 289, 822
  falcatus, 269
Ankylonoton, 430, 432, 435, 439
  pyreniger, 439–440
  salinis, 439
Anomoeoneis, 638, 640, 645–646
  brachysira, 40
  sphaerophora, 43, 642
Anthophysa, 492, 495
  vegetans, 471, 477, 495
Apatococcus, 45, 258, 264, 267, 322, 333, 346
  lobatus, 269, 323
Aphanizomenon, 20, 23, 37, 43, 64–65, 119–120, 124, 171–172, 805, 807, 812, 816
  aphanizomenoides, 171
  flos-aquae, 21, 43, 171–172, 808, 822
  gracile, 171
  issatschenkoi, 171
  schindleri, 171–172
  skujae, 171
Aphanocapsa, 40–41, 44, 64–65, 70, 82
  arctica, 82
  botryoides, 82
  conferta, 82
  delicatissima, 82
  farlowiana, 78, 82
  grevillei, 78
  holsatica, 82
  incerta, 78, 82
  intertexta, 82
  marina, 82
  montana, 89
  muscicola, 82
  planctonica, 82
  protea, 82
  pulchra, 92
  saxicola, 37
  thermalis, 82
  tolliana, 82
Aphanochaete, 314, 320, 322
  repens, 324
Aphanothece, 62–66, 68–69, 72, 74
  bachmannii, 72
  bacilloidea, 67, 72
  bullosa, 72
  castagnei, 72, 74
  clathrata, 72, 74
  conglomerata, 72
  cylindracea, 72
  halophytica, 43
  karukerae, 72
  microscopica, 72
  minutissima, 72
  nidulans, 72
  opalescens, 72
  pallida, 72
  saxicola, 72
  smithii, 72, 74
  stagnina, 64, 72, 74
  thermalis, 72

  uliginosa, 72
  utahensis, 72
Apiocystis, 258, 260, 267
  brauniana, 269
Apistonema
  expansum, 771
  pyrenigerum, 771
Araceae, 319, 344
Arachnochloris, 430, 433–434, 438, 443, 446–447
  major, 445
Arisaema, 315
  triphyllum, 328, 344
Aristata, 215
Aristichthys
  nobilis, 820
Artemia, 43
  salina, 512
Arthrodesmus, 376, 378–379
Arthronema
  africanum, 67
Arthrospira, 66, 122, 139–141
  fusiformis, 139
  gomontiana, 141
  jenneri, 139–141
  khanmnae, 140
  maxima, 139–140
  platensis, 139–140
  skujae, 140
Ascoglena, 388, 410, 414–415
  vaginicola, 411
Askenasyella, 265, 270
  chlamydopus, 271
Astasia, 385
Asterionella, 3, 20–22, 595–597, 600, 606, 613, 631
  formosa, 20, 21–23, 605, 607–608, 611, 613
  ralfsii var. *americana*, 607, 611
Asterionellopsis, 596, 607
Asterocapsa, 62, 64, 71, 92–93
  divina, 92–93
  magnifica, 92
  pulchra, 92
Asterococcus, 266, 270
  superbus, 271
Asterogloea, 431, 437, 453
  gelatinosa, 455
Asteroplanus, 607
Astrephomene, 226, 238, 242, 248
  eugenea, 212
  gubernaculifera, 243
  hermannii, 212
  perforata, 243
  pygmaea, 212
Audouinella, 34, 202, 205, 207, 211–212
  eugenea, 198
  helminthosum, 203
  hermannii, 36, 198, 201, 204–205, 211, 213, 770
  macrospora, 198, 201
  pygmaea, 198
  tenella, 198, 201, 204
Aulacoseira, 20, 563, 565, 567, 569–572, 583–584
  granulata, 573, 576

islandica, 576
italica, 576
Aulacoseiraceae, 563
Aulacoseirales, 563, 570
Aulakochloris, 430, 438, 443
   striata, 445
Aulosira, 125, 173–174
   fertilissima, 174
   implexa, 173
   laxa, 173
Autosira, 171
Azolla, 45, 180, 338

Bacillaria, 670, 675
   paradoxa, 676–677
Bacillariaceae, 670
Bacillariales, 669, 670, 671, 672, 675, 676, 681
Bacillariophyceae, 7, 201, 656, 669
Bacillariophycidae, 670
Bacillosiphon, 77, 79
   gracilis, 79
   induratus, 77
Bacteria (see Subject Index)
Bacularia, 67, 69, 77
   gracilis, 77
   indurata, 67, 72, 77
Baetis
   tricaudatus, 35
Balbianiales, 198, 212, 213
Ballia, 202, 207, 218
   prieurii, 199, 218–219
Bambusina, 366, 373, 375
   brebissonii, 357
Bangia, 3, 26, 34, 200, 204, 206, 210, 763
   atropurpurea, 25, 198–201, 205, 209
Bangiales, 198, 210
Bangiophycidae, 197, 198, 200, 209, 211
Basichlamys, 244–245, 248
   sacculifera, 239, 244–245
Basicladia, 314, 321, 339–340
   chelonum, 12, 327
Batrachospermaceae, 198, 202, 206, 212–213, 215–216
Batrachospermales, 198, 200–202, 204–205, 212, 214–216, 218
Batrachospermum, 4, 17, 35, 197–198, 200–203, 205, 207, 212–216, 322, 662, 763
   ambiguum, 198, 213
   anatinum, 198, 201, 213
   androinvolucrum, 198
   arcuatum, 198
   atrum, 198
   boryanum, 198, 203
   carpocontorium, 198, 203
   carpoinvolucrum, 198, 203
   confusum, 198
   elegans, 198
   gelatinosum, 198, 200, 203, 204, 213–215
   gelatinosum forma *spermatoinvolucrum*, 198, 203, 213
   globosporum, 198, 214
   helminthosum, 198, 202–203, 205, 213

   heterocorticum, 198
   intortum, 198, 201, 213
   involutum, 198, 202–203
   keratophytum, 40
   louisianae, 198, 213
   macrosporum, 198
   procarpum, 198
   pulchrum, 198
   skujae, 199
   trichocontortum, 199
   trichofurcatum, 199
   turfosum, 40, 198–199, 203, 205, 215
   virgato-decaisneanum, 198, 215
Belonastrum, 631
Bernardinium, 691, 697–698, 704, 706
   bernardinense, 688, 705
Bicosoeca, 477, 491, 495
   borealis, 477
   kenaiensis, 477
Bicosoecaeae, 484–485
Biddulphia, 584
Biddulphiaceae, 563
Biddulphiales, 563
Biddulphiophycidae, 563, 565
Binuclearia, 318, 344
   tatrana, 326
Bitrichia, 478, 492, 495
   chodati, 478, 495
   longispina, 495
   ollula, 478, 495
Blastodiniales, 697, 706
Blennothrix, 119, 123, 150–151, 153
   cantharidosma, 151, 153
   coerulea, 151
   comoides, 151
   ganeshii, 151, 153
   glutinosa, 151
   groesbeckiana, 151
   heterotricha, 151
   majus, 151
   mirifica, 151
   ravenelii, 151
Bodanella, 758, 760, 768, 771
   lauterbornii, 759, 761, 765, 768
Bodo, 387
   sultans, 386
Boldia, 4, 203, 207, 210
   erythrosiphon, 198, 209, 211
Borzia, 122, 138–139
   trilocularis, 138–139
Borziaceae, 122, 139
Borzinemataceae, 125, 184
Bosmina, 38, 825
Bostrychia, 202, 205, 208, 218, 220
   moritziana, 199, 200, 202, 219–220
   radicans, 199
   tenella, 199, 220
Botrydiaceae, 431
Botrydiales 429, 431, 433–434, 463–464
Botrydiopsidaceae, 430, 436, 450–451
Botryochloridaceae, 430, 436, 451–452
Botrydiopsis, 430, 432, 436, 451, 463
   arhiza, 450
Botrydium, 429, 431–434, 463
   granulatum, 464

Botryochloris, 430, 437, 451, 453
   cumulata, 452
Botryococcus, 43, 264, 270
   braunii, 43, 271
Bourrellia, 494–495
   skuja, 483, 495
Bracchiogonium, 430, 438, 446
   ophiaster, 445
Brachionus, 411
Brachysira, 638–640, 645–646
   serians, 642
   styriaca, 642
Brachytrichia, 119
Bracteococcus, 254, 256, 266, 270
   minor, 272
Bradypus, 210, 315
Brasenia
   schreberi, 811
Bulbochaete, 39, 320, 341
   minor, 330
Bumilleria, 431–434, 459, 462
   klebsiana, 462
   sicula, 462
Bumilleriopsis, 431–432, 436, 457–458
   biverruca, 457
   closterioides, 457

Calliglena, 412
Caloglossa, 4, 208
   leprieurii, 199, 202, 208
   ogasawaerensis, 199, 202
Caloneis, 638–639, 646
   amphisbaena, 644
   schumanniana, 644
   silicula, 644
Calothrix, 26, 39, 44–45, 65, 118–119, 164–165
   ascendens, 164
   braunii, 164
   contarenii, 164
   donnellii, 164
   elenkinii, 164
   epiphytica, 164
   fusca, 164–165
   juliana, 164
   kawraiskii, 164
   parietina, 164
   pulvinata, 164
   rivularis, 164
   scytonemicola, 164
   simplex, 165
   stagnalis, 164
   stellaris, 164
   tenella, 165
Campylodiscus, 43, 670, 672, 675, 681
   noricus, 679
Campylomonadaceae, 721–722, 743, 745–746
Campylomonas, 716, 723, 741–745, 748
   marssoni, 746
   platyuris, 724, 745
   reflexa, 721, 723, 745–746
   rostratiformis, 722–723, 745
Capartogramma, 638, 640
   crucicula, 644, 646

Capitulariella, 431, 435, 459
　radians, 460
Capsosira, 182
　brebissonii, 181–182
Capsosiraceae, 181
Carex, 439, 458
Carteria, 227, 232, 234, 247–248
　eugametos, 233–234
　nivale, 44
Castor
　canadensis, 16
Catacombus, 596
Catenochrysis, 523
Catenulaceae, 655–656, 661–662, 663
Cavinula, 638–640, 646
　cocconeiformis, 645
Celloniella, 494–495, 499
　palensis, 483, 499
Centritractaceae, 431, 436, 457–458
Centritractus, 431, 434, 436, 457–459
　belenophorus, 457
　brunneus, 457
　ellipsoideus, 457
　globulosus, 457
Centronella, 596, 613
Cephaleuros, 45, 315, 319, 328, 342
　virescens, 315, 328
Cephalomonas, 236, 248
　granulata, 236–237
Ceramiales, 199, 202–204, 218–219, 220–221
Cerasterias, 254, 262, 270
　irregularis, 272
Ceratiaceae, 697, 707
Ceratium, 2, 23, 691, 697–698, 704
　brachyceros, 695, 707
　carolinianum, 701–702
　cornutum, 694
　furcoides, 694
　hirundinella, 20, 685, 689, 691, 694–695, 699–702, 707
　hirundinella f. *hirundinella*, 695
　hirundinella f. *piburgense*, 695
Ceratophyllum, 17, 314, 459, 810–811
　demersum, 24, 314
Chadefaudiothrix, 431, 434, 459
　gallica, 460
Chaetoceraceae, 563
Chaetocerophycidae, 563
Chaetoceros, 563, 565, 571, 574
　elmorei, 575
　muelleri, 43
Chaetocerotales, 563
Chaetomorpha, 318, 339
Chaetonema, 321–322
　irregulare, 324
　ornatum, 322
Chaetonemopsis, 330
Chaetopedia, 431, 435, 459
　stigeoclonioides, 460
Chaetopeltis, 260, 270
　orbicularis, 272
Chaetophora, 314, 320, 322, 330, 497
　elegans, 325
　incrassata, 764
Chaetophoraceae, 314–315, 348

Chaetophorales, 267, 313, 319, 321–322, 330, 332–338, 340, 347–348, 454
Chaetosphaeridium, 322, 330
　globosum, 324
Chalarodora, 266, 270
　azurea, 272
Chalkopyxis, 495
　tetrasporoides, 484, 496
Chamaecalyx, 62, 65, 70, 96, 100–101
　calyculatus, 100–101
　clavatus, 100
　suffultus, 100
　swirenkoi, 100–101
Chamaepinnularia, 638, 641, 646
　mediocris, 643
Chamaesiphon, 24–25, 33, 62, 64–65, 68, 70–71, 99, 101, 764, 768, 770
　amethystinus, 99
　britannicus, 101
　confervicolus, 101
　geitleri, 101
　incrustans, 99, 101, 765
　polonicus, 99, 101
　regularis, 101
　rostafinskii, 101
　subglobosus, 99
　willei, 99
Chamaesiphonaceae, 60, 62, 68, 71, 96
Chamaetrichon, 320, 332
　capsulatum, 323, 332
Chara, 17, 24, 27–28, 39, 311–312, 317, 322, 335, 338–339, 809, 811, 815–817, 819–821, 826
　aculeolata, 27
　canescens, 331
　hispida, 764
　tomentosa, 27
　vulgaris, 811
Characidiopsidaceae, 430
Characidiopsis, 430, 432, 435, 441, 444
　acuta, 444
　ellipsoidea, 444
　elongata, 444
Characiochloris, 260, 270
　characiodes, 272
Characiopsis, 260, 270, 429, 431, 433, 436, 454, 456, 465
　acuta, 456
　minuta, 272, 427
　ovalis, 427
　pyriformis, 456
Characium, 254, 258, 260, 270, 441, 454
　minutum, 427
　sieboldii, 272
Chara/Nitella, 805
Charales, 311–313, 317, 338–339
Charophyceae, 229, 313, 353, 384
Charophyta, 12
Chilomonas, 734–735, 740–741, 743, 748
　acuta, 721, 746
　paramecium, 721–722, 733, 746
　striata, 739
Chlainomonas, 227, 232, 234
　kollii, 234
　rubra, 234

Chlamydobotrys, 247
Chlamydomonadaceae 225, 228, 231, 233, 235–236
Chlamydomonadales, 225
Chlamydomonas, 23, 45, 225–228, 232, 234–235, 238, 245, 248, 253, 257, 291–292, 301, 429
　acidophila, 42
　bohemica, 245
　sonowiae, 233–234
　tetragama, 234
Chlamydophyceae, 313
Chlamydomyxa, 430, 434–435, 441
　labyrinthuloides, 442
Chlorakys, 446
Chlorallanthus, 446
Chlorallantus, 430, 431, 433, 446
　oblongus, 445
Chloramoeba, 430, 432, 435, 439
Chloramoebaceae, 430
Chloramoebales, 429–430, 432, 434, 439–440
Chlorangiella, 260, 271
　pygmaea, 273
Chlorangium, 271
Chlorarkys, 430, 438, 447
　reticulata, 445
Chlorcorona, 246, 248
　bohemica, 246
Chlorella, 11, 44–45, 254, 255, 257–259, 265, 271, 307, 446
　miniata, 44
　vulgaris, 273
Chlorellidiopsis, 430, 437, 451
　separabilis, 452
Chlorellidium, 430, 437, 451
　tetrabotrys, 452
Chloremys, 261, 272
　sessilis, 273
Chloridella, 429–430, 437, 445–446
　ferruginea, 445
　neglecta, 445
Chlorobotryaceae, 424
Chlorobotrys, 424–425, 431, 434, 437, 453–454, 463
　regularis, 424–426, 453
　simplex, 455
　stellata, 425, 427, 453
Chlorobrachis
　gracilima, 247
Chlorochytrium, 260, 274, 316, 321, 332
　lemnae, 273, 314, 323
Chlorocloster, 430, 432, 438, 446
　angulus, 445
Chlorococcopsis, 256
Chlorococcum, 45, 254–256, 266, 274, 303, 333
　regulare, 425
　submarinum, 255, 258
　wimmeri, 273
Chlorodesmus, 523, 528
　hispidus, 524
Chlorogibba, 430, 438, 446
　trochisciaeformis, 445

Chlorogloea, 62, 67, 71, 96, 97
　cuauhtemocii, 96
　epiphytica, 96–97
　lithogenes, 67, 96, 97
　regularis, 96
　tuberculosa, 96
Chlorogonium, 226, 232, 233, 247–248
　capillatum, 233, 234
　elongatum, 233
　euchlorum, 233
　tetragamum, 234
Chlorokardion, 430, 435, 439
　pleurochloron, 439, 440
Chlorokoryne, 436, 454
　petrovae, 456
Chlorokybus, 264, 274
　atmophyticus, 273
Chloromeson, 423, 430, 432, 435, 439
　agile, 439, 440
　parvum, 439
　viridis, 439
Chloromonas, 227, 232, 234, 235, 248
　anglica, 235
　brevispina, 44, 235
　clathrata, 235
　depauperata, 235
　granulosa, 235
　minima, 233, 235
　nivalis, 44, 235
　pinchiae, 44, 235
　platystigma, 235
　polypyera, 235
Chloromonas (previously *Chlamydomonas*)
　nivalis, 43
Chloropedia, 431, 436, 458
　plana, 457
Chloropediaceae, 431, 436, 457, 458
Chlorophyceae, 12, 35, 225, 231, 313, 384, 427, 429, 431, 447, 454
Chlorophysema, 261, 274
　contractum, 275
Chlorophyta, 6, 12, 45, 253, 259, 311, 313, 353, 384, 441, 446–447, 453, 458, 465, 823
Chlorosaccus, 437
　fluidus, 453, 455
Chlorosarcina, 265, 274
　brevispinosa, 275
Chlorosarcinopsis, 254, 258
Chlorothecium, 431, 436, 456, 458
　capitatum, 456
　crassiapex, 456
Chlorotylium, 319, 332
　cataractum, 325
Chodatella, 288
Choleochaete
　pulvinata, 324
　scutata, 324
Choloepus, 210
Chondrocystis, 70, 89, 91
　bracei, 89
　dermochroa, 89, 91
　schauinslandii, 89, 91
Chorogloea, 65
Chromista, 384

Chromophyton, 493, 496
　rosanofii, 480
Chromulina, 473, 485–486, 489, 493, 495–498, 524
　chionophila, 44
　globulifera, 497
　palensis, 495
　stellata, 481
Chromulinaceae, 484
Chroococcaceae, 62, 71, 92
Chroococcidiopsis, 64, 66, 71, 104
　cubana, 104
　cyanosphaera, 103
　thermalis, 104
Chroococcidium, 67, 71, 103–104
　gelatinosum, 103–104
Chroococcopsis, 72, 106
　fluviatilis, 104, 106
Chroococcus, 40, 41, 43–44, 60, 62, 64–66, 71, 94–95, 813
　aeruginosus, 95
　cubicus, 95
　deltoides, 95
　distans, 95
　endophyticus, 95
　heanogloios, 95
　limneticus, 95
　limneticus var. *subsalsus*, 94
　microscopicus, 95
　minimus, 95
　minutus, 95
　mipitanensis, 95
　multicoloratus, 67, 95
　pallidus, 95
　polyedriformis, 95
　prescottii, 95
　refractus, 95
　rufescens, 95
　schizodermaticus, 95
　sonorensis, 95
　submarinus, 95
　tenacoides, 95
　thermalis, 95
　turgidus, 95
　varius, 95
　yellowstonensis, 95
Chroococcus (*Linmococcus*)
　dispersus, 94
　limneticus, 94
　sonorensis, 94
Chroodactylon, 3, 204, 207–208
　ornatum, 25, 198–199, 201, 205
　ramosum, 25
Chroomonadaceae, 743, 747–748
Chroomonas, 736, 740, 742–743, 747–749
　africana, 736
　americana, 748
　coerulea, 726, 727, 729, 747
　nordstedtii, 726–727, 747
　oblonga, 726, 729, 747
　pochmanni, 726–727, 747
　salina, 736
Chroothece, 207–210
　mobilis, 198, 203

Chrysamoeba, 485, 494–496, 501
　mikrokonta, 476
　radians, 482
Chrysapion, 503
Chrysapsis, 480, 493, 496
　agilis, 480
　fenestrata, 480
Chrysarachnion, 494, 496
　insidians, 482
Chrysidiastrum, 494
　catenatum, 476, 496
　epiphyticum, 482
Chrysoamphipyxis, 492, 496, 499, 503
　canadensis, 478
Chrysoamphitrema, 492, 496
　nygaardii, 478
Chrysobotrys, 497
Chrysocapsa, 491, 495–496
　planktonica, 484
Chrysocapsella, 496
Chrysocapsopsis, 483, 494, 503
　rupicola, 483, 496
Chrysochaete, 494, 496–497, 500
　britannica, 482
Chrysochromulina, 23, 484, 499, 512–513, 515, 518, 739
　breviturrita, 7, 23, 512–515, 518–519, 807
　inornata, 513, 515, 518
　laurentiana, 513, 515, 518–519
　onornata, 519
　parva, 512–515, 517, 519, 740, 749, 751
　strobilus, 512
Chrysoclonium, 473, 501, 503
Chrysococcus, 484, 486, 492, 497, 499
　minutus, 478
　rufescens, 487
Chrysocrinus, 484–485
Chrysodictyon, 503
Chrysodidymus, 523–526, 528, 530–531, 533–534, 541–542, 551–552
　gracilis, 534
　synuroideus, 534, 536, 538, 542
Chrysoikos, 497
Chrysolepidomonas, 484, 493, 497, 503
　dendrolepidota, 480, 497
Chrysolepidomonadaceae, 484
Chrysolykos, 478, 486–487, 492, 497, 501
　planktonicus, 478
　skujae, 478
Chrysomallus, 443
Chrysomonadaceae, 471
Chrysophaerella, 523
Chrysophyceae, 19, 73, 313, 414, 429, 440, 443, 471–472, 484–486, 488, 491, 497, 500, 503, 511–513, 523–524, 534–535, 771
Chrysophyta, 5, 9, 24, 259, 757
Chrysopyxis, 492, 497
　canadensis, 496
　stenostoma, 478
Chrysosaccus, 491, 495, 497
　incompletus, 484
Chrysosphaera, 495, 497

Chrysosphaerella, 473–474, 486–488, 491, 493, 497
  brevispina, 475, 479, 486, 487–488, 491
  longispina, 475, 486–488, 491
Chrysostephanosphaera, 494, 497–499
  globulifera, 476
  globulosa, 482
Chrysoxys, 493, 498
  maior, 479
Chytridiochloris, 431, 436, 458
  acus, 456
  viridis, 458
Cladocera, 16, 20, 38–39, 408, 411, 700, 808, 825
Cladonia, 258
Cladophora, 4, 24, 26–27, 34–35, 39, 101, 107, 208, 288, 314, 316, 318, 321–322, 334–337, 339–340, 605, 763, 768, 783, 808–810, 813, 817, 818–820
  amethystinus, 101
  fallax, 101
  glomerata, 25–26, 34, 43, 314, 327, 608–609, 624, 764, 809–810, 822
  halophilus, 101
  minutus, 101
  portoricensis, 101
Cladophorales, 311–313, 317, 319, 339–340, 454
Cladophorophyceae, 313
Clastidium, 70, 98, 101
  cylindricum, 98, 101
  setigerum, 98, 101
Cloniophora, 321, 332
  spicata, 325
Closteriaceae, 372–373
Closteriopsis, 263, 274
  longissima, 275
Closterium, 40, 361–362, 364, 367–368, 372, 377
  aciculare, 364, 372
  actum, 356
  acutum, 372
  angustatum, 357
  archerianum, 357
  closteroides, 357
  dianae, 357
  gracile, 357
  navicula, 357
Coccolithales, 511
Coccomonas, 236, 248
  orbicularis, 236–237
Coccomyxa, 258, 265, 274
  dispar, 275
Cocconeidaceae, 597–598, 605, 623
Cocconeis, 24, 597–598, 601, 604–606
  disculus, 624
  fluviatilis, 624
  globularis, 28
  neothumensis, 623–624
  pediculus, 34, 609, 624
  placentula, 25, 28, 33, 43
  placentula var. *lineata*, 624
  placentula var. *rouxii*, 624
  tomentosa, 28

Codioliales, 313, 347
Codiolum, 347
Coelastrum, 257–258, 264, 274, 453
  reticulum, 275
Coelomoron, 65, 70, 82, 84
  microcystoides, 84
  minimum, 82
  pusillium, 82
  regulare, 67, 82
  tropicalis, 84
Coelosphaerium, 2, 23, 64–65, 70, 82, 84, 87, 303
  aerugineum, 84, 87
  collinsii, 82
  confertum, 87
  dubium, 87
  kuetzingianum, 84, 87
  microcystoides, 82
  minimum, 82
  minutissimum, 87
  naegelianum, 85, 89
  subarcticum, 84, 87
  vestitum, 82
Colacium, 12, 384, 385–386, 388, 404, 405–406, 408, 409–412, 415
  calvum, 408, 411
  gojdicsae, 408, 411
  libellae, 388, 394, 404, 408, 411
  oblonga, 404
  oblongata, 404
  pisciformis, 404
  proxima, 404
  rubra, 404
  sanguinea, 404
  schmitzii, 404
  sociabilis, 404
  splendens, 404
  vesiculosum, 393, 403–404, 406, 408, 411
Coleochaetales, 308, 312–313, 319, 330, 332–333
Coleochaete, 24, 25, 320, 332, 353
  scutata, 26
Coleodesmium, 119, 124, 158–159
  floccosum, 158–159
  wrangelii, 158–159
Colteronema, 125
  funebre, 186–187
Compsopogon, 4, 200, 205, 207, 210, 212
  coeruleus, 198, 200, 204, 211–212
  prolificus, 198
Compsopogonales, 198, 201, 210, 212
Compsopogonopsis, 205, 207, 211–212
  leptocladus, 198
Conferva, 312
Conjugatophyceae, 353, 744
Conjugatophyta, 353
Conochaete, 322, 332
Conradiella, 523–524
Contorta (section of *Batrachospermum*), 215
Corona, 245
Coronastrum, 263, 274
  aestivale, 275
Corvomeyenia
  everetti, 45

Coscinodiscales, 563
Coscinodiscophycidae, 563, 565
Coscinodiscus, 570, 575, 578–579
Cosmarium, 354, 356, 364, 368, 374–376
  contractum, 359
  eloiseanum, 365
  margaritatum, 359
  montrealense, 359
  pseudoconnatum, 359
  quadrifarium f. *hexastichum*, 359
Cosmioneis, 638–639, 641, 645
  pusilla, 646
Cosmoastrum, 378
Cosmocladium, 354, 368, 374, 376
  constrictum, 374
  pulchellum, 374
  pusillum, 376
  saxonicum, 356, 374
  tuberculatum, 376
  tumidium, 376
Crateriportula, 563, 578
  inconspicuus, 579
Craticula, 638, 640, 646, 647
  cuspidata, 643
Crucigenia, 3, 43, 264, 276
  quadrata, 275
Cryptista, 715, 742
Cryptoglena, 384, 385, 388, 401–402, 410, 411, 415
  pigra, 396, 400, 412
Cryptomonadaceae, 743–745
Cryptomonadales, 715, 742, 744
Cryptomonas, 23–24, 719–720, 723, 736, 740–741, 743–745, 749
  erosa, 719–720, 745
  obovata, 719–720, 745
  ovata, 719–720, 730, 734, 745
  ozolini, 719, 745
  phaseolus, 720, 745
  pyrenoidosa, 744
  rostratiformis, 745
  tetrapyrenoidosa, 720, 723, 745
Cryptomonas (*Campylomonas*)
  marssonii, 721
  platyuris, 721–722
  rostratiformis, 721–722, 739
Cryptophyceae, 19, 715, 742, 744
Cryptophyta, 7, 24, 686, 715–716, 737, 742
Ctenocladales, 333, 335
Ctenocladus, 43, 321, 333
  circinnatus, 43, 327, 333
Ctenopharyngodon
  idella, 820
Ctenophora, 596, 600, 606–607
  pulchella, 612
Cualobacter, 739
Cyanidioschyzon
  merolae, 41
Cyanidium, 200, 206, 208
  caldarium, 41, 198, 201, 206, 207
Cyanobacteria(-um), 1–3, 5–6, 8, 12, 19–20, 22–25, 27, 33–37, 39–45, 59–68, 117–121, 132, 135, 137, 139, 144, 187, 189, 205, 315, 320, 332, 364,

372, 384, 487, 489, 647, 670,
    674–675, 681, 701, 735, 738, 759,
    763, 765, 805–809, 812–815, 819,
    821–824
Cyanobacterium, 66, 69, 79
    cedrorum, 75, 79
    diachloros, 75
    minervae, 66, 75, 79
Cyanobium, 20, 65, 69, 73, 79, 120
    amethystinum, 79
    eximium, 73, 79
    gracile, 73
    roseum, 73, 79
Cyanocystis, 65, 71, 100, 102
    hemisphaerica, 102
    mexicana, 100, 102
    olivacea, 102
    pacifica, 100, 102
    pseudoxenococcoides, 102
    sphaeroidea, 102
    valiae-allorgei, 100, 102
    violacea, 102
Cyanoderma
    bradypodis, 315
Cyanodictyon, 68–69, 73, 79
    filiforme, 79
    planctonicum, 73, 79
    reticulatum, 79
    tubiforme, 79
Cyanokybus, 71, 93, 95
    venezuelae, 93, 95
Cyanomonas, 739, 748
Cyanosaccus, 67
Cyanosarcina, 62, 71, 95
Cyanosarcina (Chroococcus)
    minutus, 94
    minutus var. thermalis, 94
    mipitanensis, 94
    polymorphus, 94
    turgidus, 94
    yellowstonensis, 94
Cyanotetras, 70, 83, 87
    aerotopa, 87
    crucigenielloides, 83, 87
Cyanothece, 20, 66, 69, 77, 79
    aeruginosa, 77, 80
    lineata, 80
    major, 77, 80
Cyanothrix, 78, 80
    primaria, 78
    willei, 78
Cyathobodo, 498
Cyathomonas, 744
Cyclonexis, 493
    annularis, 479, 498
Cyclops, 393, 411
Cyclostephanos, 563, 572, 578–579, 584–585
    costatilimbus, 581
    damasii, 581
    dubius, 579
    invisitatus, 581
    tholiformis, 581
Cyclotella, 37, 43, 563, 565, 567, 572–573,
    579, 581–582, 585, 768
    bodanica, 566, 583

cryptica, 569
glomerata, 565
melosiroides, 565
meneghiniana, 569, 585
ocellata, 578
pseudostelligera, 566, 578, 583
radiosa, 578
Cyclotella (Stephanocyclus)
    meneghiniana, 22
Cyclotubicoalitus, 563, 571, 582–583
    undatus, 588
Cylindriastrum, 378
Cylindrocapsa, 318, 340
    geminella, 328
Cylindrocapsales, 317, 340–341
Cylindrocystis, 363, 367, 369
    brebissonii, 358, 369
    crassa, 369
Cylindrospermopsis, 64, 120, 124, 174–175,
    807
    catemaco, 174
    raciborskii, 174–175
Cylindrospermum, 65, 119, 124, 174–175
    catenatum, 174
    longisporum, 175
    minutissimum, 175
    stagnale, 175
Cylindrotheca, 670, 675
    gracilis, 676–677
Cymatopleura, 676
    elliptica, 679
    solea, 670, 672, 679
Cymbella, 24, 33, 39, 597, 656, 663–664, 676
    affinis, 658
    amphicephala, 662
    aspera, 662
    cesatii, 662
    cistula, 661
    cymbiformis, 661
    lata, 661
    minuta, 36
    proxima, 658, 661
    pusilla, 662
    sinuata, 664
    turgidula, 658, 661
Cymbellaceae, 655–656, 659, 661–664
Cymbellonitzschia, 670, 675
    diluviana, 676–677
Cystodinedria, 690–691, 697, 701, 703, 708
    inermis, 688, 696, 705
Cystodinium, 690–691, 697–698, 704,
    708–709
    bataviense, 688, 694, 696, 705

Dactylococcus, 288
Dacytlococcopis, 40
    raphidioides, 76
Daphnia, 411, 700, 808, 813–814, 819
    hyalina, 739
    laevis, 408
    pulex, 408
Dasygloea, 123, 147–148
    amorpha, 147
    brasiliense, 148

calcicola, 147
lamyi, 147–148
yellowstonensis, 147–148
Deasonia, 266–276
    granata, 277
Debarya, 366, 369–370
    glyptosperma, 355
    smithii, 355
Dendromonas, 491, 498
    cryptostylis, 477
Denticula, 670, 675–676
    tenuis, 677
Derepyxis, 492
    anomala, 498
    dispar, 478
    ollula, 475
Dermatochrysis, 2, 494, 498, 503
    reticulata, 482
Dermatophyton, 314, 319, 340
    radians, 327
Dermocarpa
    pacifica, 100
Dermocarpella, 71, 100, 102
    hemisphaerica, 100
    prasina, 102
    protea, 102
Dermocarpellaceae, 60, 70–71, 102
Desmatractum, 266, 276
    bipyramidatum, 277
Desmidiaceae, 373–379, 441
Desmidiales, 313, 353–354, 361–365,
    372–379
Desmidium, 366, 374
    baileyi, 357
    grevillii, 357
Desmococcus, 258, 264, 276, 315, 322, 333,
    346
    olivaceus, 277, 323
Desmodesmus, 254, 256–258, 261, 276,
    288, 299
    armatus, 255
    protuberans, 278
Desmonema
    wrangelii, 158–159
Diacanthos, 261, 276
    belenophorus, 278
Diachros, 430–432, 437, 446
    simplex, 445
Diacronema, 512, 515
Diadesmis, 637–640, 647
    confervacea, 645
    contenta, 645
    perpusilla, 645
Diaptomus
    gracilis, 739
Diatoma, 596, 600, 606–608, 616
    hiemale, 613
    mesodon, 612–613
    tenue, 36, 613
    tenue var. elongatum, 612
    vulgare, 34, 36, 612–613
Diatomella, 638–639
    balfouriana, 638, 645, 647
Dicellula, 261, 276
    planctonica, 278

Dichothrix, 124, 164–166
　baueriana, 166
　calcarea, 166
　compacta, 166
　gypsophila, 166
　hosfordii, 166
　inyoensis, 166
　meneghiniana, 166
　orsiniana, 165
　rupicola, 166
　spiralis, 166
　willei, 165
Dichotomococcus, 431, 437, 453
　elongatus, 452
　lunatus, 451
Dichotomosiphon, 319, 344
　tubersosus, 328
Dicranochaete, 321, 333
　reniformis, 324
Dictyochlorella, 264, 276
　reniformis, 278
Dictyochloris, 266, 276
　fragrans, 278
Dictyochloropsis, 276
　splendida, 279
Dictyococcus, 266, 277
　varians, 279
Dictyosphaerium, 254, 257–258, 263, 280
　pulchellum, 279
Didymogenes, 262, 280
　palatina, 279
Didymosphenia, 656, 659, 663–664
　geminata, 33, 658, 661–662
Difflugia, 458, 461
Dilabifilum, 315, 320, 333
　printzi, 325
Dimorphococcus, 263, 280
　lunatus, 279
Dinamoebaceae, 697, 706
Dinamoebales, 697, 706
Dinamoebidium, 697–698, 701, 703, 706
　coloradense, 688, 702
Dinastridium, 697–698, 703, 708
　sexangulare, 688, 705
Dinobryaceae, 484
Dinobryon, 3, 23, 471, 473, 477, 480, 484–489, 492, 497–498, 501–503
　acuminatum, 489
　attenuatum, 498
　balticum, 489
　bavaricum, 488–489
　borgei, 488–489, 498
　cylindricum, 485, 487–489
　dilatatum, 498
　dillonii, 477
　divergens, 477, 487, 489
　lorica, 480
　pediforme, 488–89
　sertularia, 22, 487–490
　suecicum, 475, 488, 498
　tubaeforme, 498
Dinobryopsis, 498
Dinococcales, 696–697, 701, 708
Dinococcus, 688, 696–698, 703, 708

Dinosphaera, 697–698, 705, 708
　palustris, 693
Dinophyceae, 697
Dinophyta, 697
Dinosphaeraceae, 697, 708
Dioxys, 431, 436, 456, 458
　inermis, 456
　tricornuta, 456
Diplocolon
　heppii, 158
Diploneis, 638–639, 647
　elliptica, 642
　finnica, 642
　oblongella, 642
　smithii var. *dilatata*, 642
Diplonema, 387, 405
Diplopsalis, 708
Diptera, 700
Discoglena, 402, 412
Dispora, 264, 280
　crucigenoides, 279
Distigma, 385
Distrionella, 600, 606, 613
Docidium, 367, 375
　baculum, 357
Dolichospermum, 169
Draparnaldia, 4, 35, 314, 321–322, 333–334, 483, 498
　glomerata, 325
　ravenelii, 325
Draparnaldiopsis, 321, 334
　alpinis, 325
Dreissena
　polymorpha, 37
Ducelliera, 431, 437, 453
　chodati, 452
Dunaliella
　salina, 43
　viridus, 43
Durinskia, 697–698, 705, 707
　baltica, 689, 692, 694
Dysmorphococcus, 236, 248
　globosus, 236
　variabilis, 236–237

Echinosphaerella, 262, 280
Ectocarpales, 758–759, 768
Ectocarpus, 758, 760, 768
　confervoides, 759, 768
　siliculosus, 759, 761, 765–766, 768
Ectogeron, 261, 280
Eichhornia
　crassipes, 821
Eirmodesmus, 494, 498
　phaeotilus, 483, 498
Elakatothrix, 263, 280, 317, 344
　americana, 345
　gelatinosa, 326
　viridis, 326
Ellerbeckia, 563, 565, 571, 584
　arenaria, 576, 583
Ellipsoidion, 424–425, 430, 438, 446
　acuminatum, 425–427, 446
　stellatum, 445

Elodea, 314, 771, 810–811
　canadensis, 819
Enallagma
　civile, 408
Encephalartos, 45
Encyonema, 656, 659, 663–664
　helvetica, 662
　norvegica, 661
Endochloridion, 430, 432, 438, 446
　polychloron, 445
Endoclonium
　rivulare, 338
Endospora
　rubra, 95
Enteromorpha, 27, 34, 317, 347, 763, 771
　flexuosa, 347
　intestinalis, 43, 329, 347
　prolifera, 347
Entocladia, 321, 334, 338
　polymorpha, 323
Entomoneidaceae, 670, 681
Entomoneis, 670, 675, 681–682
　ornata, 672, 678
Entophysalidaceae, 62, 71, 96
Entophysalis, 62, 67, 71, 96–97
　atrata, 96
　cornuana, 96
　lemaniae, 96
　lithophila, 96, 97
　rivularis, 96
　willei, 96, 97
Entransia, 363, 366, 369
　fimbriata, 355
Entzia, 697–698, 704, 708
　acuta, 690, 693, 695
Epigloeosphaera, 69, 74, 80
　glebulenta, 74, 80
Epihydra, 41
Epipyxis, 484, 486–487, 491–492, 498, 501
　pulchra, 473
　ramosa, 476–477
　tabellariae, 475
Epithemia, 25, 34, 39, 43, 669–670, 673, 675, 678, 680–681
　turgida, 672
Epithemiaceae, 676
Eremosphaera, 265, 280
　viridis, 282
Eremotyl, 280
Erkenia, 493, 498–499
　subaequiciliata, 480
Errerella, 288
Euastrum, 361, 368, 375–376
　boldtii, 359
　divaricatum, 359
　humerosum, 359
　pseudoboldtii, 359
　verrucosum, 359
Eucapsis, 66, 70, 86, 89
　alpina, 86, 89
　alpina var. *maior*, 89
　minor, 89
Eucocconeis, 597, 598, 603, 606, 623
　flexella, 626, 627
　flexella var. *alpestris*, 626

Eucyonema
 minuta, 658
 muelleri, 658
Eudinobryon, 498
Eudorina, 24, 239, 241, 242, 247–248
 conradii, 239
 cylindrica, 239
 elegans, 23, 239, 240
 illinoisensis, 239
 interconnexa, 239
 unicocca, 239
Euglena, 2, 23, 24, 383, 385–386, 388, 389–390, 401–406, 407, 408–409, 410, 412, 414–415, 806
 acus, 390, 406
 adhaerens, 390, 403
 agilis, 388
 caudata, 391
 chadefaudii, 390, 402–403
 clavata, 390
 deses, 399, 404, 406
 geniculata, 402, 406
 gracilis, 387, 392, 399, 404–407
 granulata, 403
 jirovecii, 391
 mutabilis, 42, 390, 404, 406–407
 myxocylindracea, 388, 403–404, 406
 oblonga, 390
 obtusa, 406, 407–408
 orientalis, 404
 oxyuris, 390, 403, 406
 oxyuris var. *charkowiensis*, 406
 pisciformis, 404, 406–407
 polymorpha, 403, 405
 proxima, 406
 repulsans, 390
 rubra, 405, 407, 412
 sanguinea, 407, 412
 schmitzii, 405
 sima, 400
 sociabilis, 391, 406
 spirogyra, 390, 401, 406, 412
 splendens, 403, 405
 stellata, 403–404
 texta, 390
 tripteris, 390, 392, 404, 406
 tristella, 403, 406
 truncata var. *baculifera*, 391
 tuba, 391, 404–405, 412
 velata, 403
 viridis, 384, 390, 403–406
 walnei, 403
Euglenamorpha, 388, 410, 412, 415
 hegneri, 397
Euglenida, 383, 385
Euglenophyceae, 383
Euglenophyta, 6, 383, 686, 806
Euglenozoa, 383, 385, 387
Eunophora, 656
Eunotia, 33, 39, 655–656, 664–665
 bilunaris var. *mucophila*, 662
 elegans, 40
 exigua, 40, 661
 faba, 657
 formica, 661

 microcephala, 662
 papilio, 661
 pectinalis, 24
 pectinalis var. *minor*, 661
 praerupta, 661
 ruzickae, 662
 septentrionalis, 662
 serra, 657
 tenella, 42
Eunotiaceae, 655–656, 661, 664–665, 681
Euplotes, 45
Eusphaerella, 493, 499
 turfosa, 479, 499
Eustigmatales, 424
Eustigmatophyceae, 6, 8, 423–424, 426–427, 446–447, 449, 465
Eustigmatos, 424–425, 447, 463
 magnus, 425–426
 vischeri, 425
Eustropsis, 263, 280, 282
Eutreptia, 385, 388, 401, 404, 410, 413, 415
 globulifera, 394, 413
 pertyi, 413
 viridis, 413
Eutreptiella, 388, 402, 413
 eupharyngea, 413
 gymnastica, 404
Euvolvox (section of *Volvox*), 242
Exanthemachrysis, 512, 515
 noctivaga, 515
Excentrochloris, 430, 436, 451
 gigas, 450
Excentrosphaera, 265
 viridis, 282
Exuviaella, 697–698, 701, 703–704, 709
 compressa, 689

Falcomonas, 743
Fallacia, 638, 640, 647
 pygmaea, 645
Fasciculochloris, 265
 boldii, 283
Ferrissia
 fragilis, 25
Filoprotococcus, 315, 318, 345
 polymorphum, 329
Fischerella, 118, 120, 125, 184–185
 ambigua, 184–185
 letestui, 184
 major, 184
 thermalis, 184–185
Fischerellaceae, 184
Fistulifera, 639–640, 647
 saprophila, 643
Flintiella, 208
 sanguinaria, 198, 200, 203
Florideophycidae, 197–198, 200–201
Fortiea, 124, 161, 163–164
 bossei, 164
 monilispora, 163–164
 salinicola, 163–164
Fottea, 282
 cylindrica, 283

Fottiella, 263, 284
 quadrangularis, 283
Fragilaria, 20, 33, 39, 43, 595–596, 599, 606, 610, 709
 brevistriata, 616
 capucina, 613
 construens, 616
 crotonensis, 20–21, 605, 607–608, 613–614
 crotonensis var. *oregona*, 614
 elliptica, 616
 intermedia, 614
 lapponica, 616
 leptostauron, 616
 pinnata, 616
 vaucheriae, 25, 613–614
 virescens, 616
 virescens var. *exigua*, 616
Fragilariaceae, 596, 598, 604, 607, 610, 613
Fragilariales, 598, 607
Fragilariforma, 596, 599, 606, 613, 616, 631
 constricta, 614
 virescens, 614
Fragilariophyceae, 595–598, 604, 607
Franceia, 255–256, 262, 284
 droescheri, 283
Frankophila, 599, 613
Fremya, 431, 435, 461
 sphagni, 460
Fremyella, 164
 diplosiphon, 164
 robusta, 164
 tenera, 164
Fridaea, 320, 334
 torrenticola, 325
Fritschiella, 319, 334
 tuberosa, 315, 325
Frustulia, 24, 39, 639, 640, 647
 rhomboides, 40, 645
 rhomboides var. *saxonica*, 645
 saxonica, 40
 vulgaris, 645

Galdiera
 sulphuraria, 41
Gamophyta, 353
Geissleria, 637–638, 640, 647–648
 ignota var. *palustris*, 644
Geitleria, 125
 calcarea, 186–187
 floridana, 187
Geitleribactron, 70, 98, 101–102
 crassum, 98, 102
 periphyticum, 98
Geitlerinema, 119, 122, 126–127
 amphibium, 126
 claricentrosa, 127
 claricentrosum, 126
 earlei, 126
 lemmermannii, 127
 splendidum, 126–127
 unigranulatum, 127
Geminella, 319, 345
 interrupta, 326
 minor, 326

Geminigera, 734, 742–743
Geminigeraceae, 743
Genicularia, 367–368, 373
  elegans, 358
Geosiphon, 180
Glaucospira, 122, 129, 131–132
  laxissima, 131
  yellowstonensis, 132
Glenodiniopsidaceae, 697, 708
Glenodiniopsis, 697–698, 705, 708
  steinii, 693–694
Glenodinium, 707
  cinctum, 708
  palustre, 708
Gleosphaeridium, 431
Gloechloris, 435
Gloeoactinium, 264, 284
  limneticum, 285
Gloeobotrydaceae, 431, 436, 453, 455
Gloeobotrys, 431, 433–434, 437, 454
  limneticus, 455
Gloeocapsa, 2, 20, 26, 44, 60, 62–64, 70, 89, 91, 95
  acervata, 92
  alpicola, 92
  alpina, 91, 92
  arenaria, 92
  atrata, 92
  calcicola, 92
  caldariorum, 92
  conglomerata, 91, 92
  crepidinum, 93
  decorticans, 92
  dermochroa, 91
  fusco-lutea, 92
  gelatinosa, 91, 92
  granosa, 92
  kuetzingiana, 92
  magma, 95
  nigrescens, 92
  rupestris, 92
  sanguinea, 91, 92
  sparsa, 92
  sphaerica, 92
  thermophila, 92
Gloeocapsopsis, 62, 71, 93, 95
  crepidinum, 93
  magma, 93
Gloeochaete, 263, 284
  wittrockiana, 285
Gloeochloris, 430, 443
  planctonica, 444
  smithiana, 443
Gloeococcus, 257, 259, 284
  pyriformis, 285
Gloeocystis, 258, 265, 284
  bacillus, 285
Gloeocystopsis, 291
Gloeodendron, 259, 284
  catenatum, 286
Gloeodinium, 696, 707
  montanum, 688, 694
Gloeomonas, 232, 248
  kupfferi, 235
  ovalis, 235

Gloeopodiaceae, 431, 436, 454, 457
Gloeopodium, 431, 436, 454
  rivulare, 457
Gloeoskene, 431, 437, 454
  turfosa, 455
Gloeosphaeridium, 437, 454
  firmum, 455
Gloeotaenium, 263, 284
  loitlebergerianum, 286
Gloeothece, 64, 69, 75, 80
  confluens, 80
  distans, 80
  endochromatica, 80
  fusco-lutea, 80
  heufleri, 75
  interspersa, 75, 80
  linearis, 80
  linearis var. *composita*, 76
  membranacea, 80
  opalothecata, 80
  palea, 80
  prototypa, 80
  rupestris, 75, 80
  transsylvanica, 81
Gloeotila, 319, 345
  contorta, 326
Gloeotilopsis, 318, 345
  sterile, 326
Gloeotrichia, 25, 39, 119, 124, 166–167, 770
  echinulata, 23, 166–167
  natans, 166
  pilgeri, 166
  pisum, 166–167, 764
Godlewskia, 65, 71, 101
Golenkinia, 257, 262, 284
  radiata, 286
Golenkiniopsis, 262, 284
  solitaria, 286
Gomontia, 319, 334
  holdenii, 323
  perforans, 314
Gomphoneis, 655–56, 659, 661, 663
  elegans, 659
  eriense, 665
  eriense var. *variabilis*, 660
  herculeana, 33, 35, 659, 665
  olivacea, 660, 665
  olivaceum, 662
  quadripunctata, 665
Gomphonema, 24–25, 39, 604, 655–656, 659, 663, 665
  abbreviatum, 662
  acuminatum, 660
  acuminatum var. *brebissonii*, 662
  angustatum, 661–662
  apuncto, 660–661
  augur, 662
  brasiliense, 661, 666
  gracile, 24, 661
  grovei, 661, 666
  hebridense, 661
  lingulataeforme, 666
  manubrium, 661
  minutum, 662

  olivaceum, 36
  parvulum, 661–662
  sphaerophorum, 660, 662
  subtile, 662
  truncatum, 660, 662
Gomphonemataceae, 655–656, 659, 661–662, 665
Gomphosphaeria, 60, 65, 70, 86–87
  aponina, 64, 86–87
  cordiformis, 87
  irieuxii, 86
  lacustris, 85
  multiplex, 64, 87
  natans, 64, 67, 86–87
  salina, 87
  semen-vitis, 67, 86–87
  virieuxii, 87
  wichurae, 89
Gomphosphenia, 663, 665–666
  grovei, 661
  lingulataeforme, 661
Gonatozygon, 366–368, 373
  brebissonii, 358
  monotaenium, 359
Gongrosira, 33, 319–320, 335, 764
  debaryana, 314, 325, 335
  incrustans, 25
  pseudoprostrata, 335
  scourfieldia, 335
Goniaceae, 225, 228, 239, 243–246
Goniochloris, 430–431, 433, 438, 446
  fallax, 465
  sculpta, 445
Goniomonadaceae, 742, 744
Goniomonadales, 742, 744
Goniomonadophyceae, 742
Goniomonas, 719,–20, 734–736, 740–744, 749
  truncata, 719–720, 733, 744
Gonium, 242–243, 248
  discoideum, 244
  formosum, 244
  multicoccum, 243–244
  octonarium, 244
  pectorale, 227, 243–244
  sacculifera, 245
  sociale, 227, 238
Gonyaulacaceae, 697, 707
Gonyaulax, 687, 697–698, 701, 704, 707
  palustris, 708
  spinifera, 689, 692
Gonyostomum, 426,–428, 463, 465
  depressum, 428
  latum, 428
  semen, 428
Granulochloris, 236, 248
  spinifera, 237–238
Groenbladia, 366, 375
  neglecta, 375
  undulata
Groenlandiella, 443
Guillardia, 743
  theta, 734
Gumaga
  nigricula, 35

Gymnodiniaceae, 697, 706
Gymnodiniales, 697, 706
Gymnodinium, 691, 697–698, 704, 708
   acidotum, 688
   caudatum, 702
   cruciatum, 702
   fuscum, 705–706
   helveticum, 690
   marylandicum, 702
   paradoxum, 694
   triceratium, 706
Gyrodinium, 687, 691, 697–698, 704, 706
   pesillium, 688
Gyromitus
   disomatus, 512
Gyrosigma, 638–639, 648, 650
   parkeri, 641
   scalproides, 641

Haematococcaceae, 225, 228, 232–233
Haematococcus, 231–232, 248
   carocellus, 232
   lacustris, 45, 232
   pluvialis, 231–232
Haidadinium, 697–698, 701, 703, 708
   ichthyophilum, 688, 702
Hammatoidea, 151
Hannaea, 596, 600, 606, 631
   arcus, 33, 36, 613, 615
   arcus var. *amphioxys*, 615
Hantzschia, 45, 670, 675, 681
   amphioxys, 672, 676–677
Hanusia, 743
Hapalosiphon, 24, 40, 44, 65, 119, 125, 188–189
   aureus, 189
   brasiliensis, 189
   confervaceus, 189
   delicatulus, 189
   flexuosus, 189
   fontinalis, 189
   hibernicus, 188–189
   intricatus, 189
   pumilus, 27, 189
   tenuis, 189
   welwitschii, 188–189
Haplotaenium, 362, 367, 375, 377, 379
   minitum, 357, 375
   sceptrum, 375
Haptophyceae, 7, 471, 511
Haptophyta, 1, 5, 7, 9, 807
Harpochytrium, 458
Hassallia, 119, 124, 158, 160
   byssoidea, 160
   discoidea, 160
   granulata, 160
Hazenia, 315, 320, 335
   mirabilis, 323
Heimansia, 354, 368, 374–376, 379
   pusilla, 360
Heliapsis, 492, 499, 502
   mutabilis, 479
Helicodictyon, 315, 320, 335
   planktonicum, 323

Helminthogloea, 430, 435, 443
   ramosa, 444
Hemidiniaceae, 697, 707
Hemidinium, 688, 696–698, 703–707, 709
   nasutum, 693, 695
Hemidiscaceae, 563
Hemiselmidaceae, 743, 748
Hemiselmis, 743, 748–749
   amylifera, 748
   amylosa, 726–727, 729
   virescens, 738
Hemisphaerella, 431, 436, 458
   operculata, 456
Heribaudiella, 4, 760, 762, 768, 771
   fluviatilis, 25, 33, 757–759, 763–766, 770
Heterochloris, 429, 430, 435, 439
   mutabilis, 439
   viridis, 439–440
Heterochromonas, 502
Heterococcus, 431, 435, 459, 461, 465
   ramosissimum, 460
Heterodendraceae, 431
Heterodendron, 431, 434, 461
   pascheri, 460
Heterodesmus, 431, 437, 453
   bichloris, 452
Heterogloea, 430, 434, 436, 443–444
   endochloris, 444
   minor, 444
Heterogloeaceae, 430
Heterogloeales, 429–430, 432, 434, 441, 444
Heterohormogonium, 80
   schizodichotomum, 78
Heteroleibleinia, 65, 122, 132, 134–135
   kuetzingii, 134
   minor, 134
   profunda, 134
   pusilla, 134–135
   versicolor, 135
Heteroleibleinioideae, 135
Heteromastix, 230
   angulata, 230
Heteropediaceae, 431
Heteropedia, 431, 435, 459, 461
   polychloris, 460
Heterothrix, 431, 434, 461
   exilis, 462
Hildenbrandia, 4, 34, 202, 204, 207, 218, 763–764, 768
   angolensis, 199, 218–219
   rivularis, 25, 33, 204, 218, 765, 770
Hildenbrandiales, 199, 201–203, 218–219
Hippodonta, 640, 648
   capitata, 643
   hungarica, 643
Homoeothrix, 65, 122, 135–137, 763–764
   crustacea, 136–137
   janthina, 136–137
   stagnalis, 137
   varians, 136–137
   violacea, 137
Hordeum, 822
Hormathonema, 96

Hormidiopsis, 319, 345
   crenulatum, 345
   ellipsoideum, 326
Hormidium, 313, 346
Hormotila, 285
   blennista, 287
Hormotilopsis, 255, 285
Hyalella
   azteca, 825
Hyalobryon, 498
Hyalosynedra, 596
Hyalotheca, 364, 366, 376, 441, 458
   dissiliens, 358, 376
   mucosa, 376
Hybrida (section of *Batrachospermum*), 215
Hydra, 45, 272
   viridis, 258
Hydrilla
   verticillata, 816
Hydrococcaceae, 71, 96
Hydrococcus, 62, 65, 71, 96, 98
   cesatii, 96, 98
   rivularis, 96
Hydrocoleum, 123, 147, 149–151
   groesbeckianum, 149
   homoeotrichum, 149–150
Hydrocoleus, 151
Hydrodictyon, 24, 255, 257, 259, 285, 808, 810, 818, 822–823
   reticulatum, 287
Hydrosera, 563, 571, 583
   whampoensis, 577
Hydrurus, 494, 500
   foetidus, 12, 35, 476, 483, 488, 499
Hyella, 60, 62, 67, 72, 107, 109
   balani, 67, 107, 109
   caespitosa, 107, 109
   fontana, 107, 109
   gigas, 107
   kalligrammos, 107
   linearis, 107
   littorinae, 107
   pyxis, 107
   seriata, 109
   tenuior, 67, 107
   vacans, 107
Hyellaceae, 60, 62, 71, 107
Hymenomonas, 512–513, 515, 519
   coccolithophora, 516
   danubiensis, 516
   prenanti, 512
   roseola, 511–512, 514, 516, 519
   scherffeli, 516
Hypheothrix (*Symplocastrum*)
   parciramosa, 148
Hypnodinium, 690–691, 696–698, 704, 709
   sphaericum, 688
Hypophthalmichthys
   molitrix, 820

Ilsteria, 431, 437, 453
   quadrijuncta, 452
Imantonia, 512
Inactis, 119, 137

Infusoria, 384–385, 471
Ischnura
  verticalis, 394, 408
Isochrysidales, 511, 513
Isocystis, 124, 174, 176–177
  infusionum, 176
  planctonica, 176
Isthmochloron, 430, 439, 445, 447
  lobulatum, 445
  trispinatum, 445

Jaaginema, 119, 122, 126, 128
  filiforme, 126
  neglecta, 128
  subtilissima, 128
Johannesbaptistia, 69, 78, 80
  pellucida, 78, 80
  primaria, 78
  schizodichotoma, 78, 80

Kansodinium, 697–698, 708
  ambiguum, 693, 694, 702
Karayevia, 596–598, 603, 607, 627
  clevei, 596, 609
  clevei var. *rostrata*, 596, 626
  laterostrata, 626
Kathablepharidaceae, 715–716, 749–751
Kathablepharis, 716, 732, 735, 739–740, 743, 749, 750–751
  ovalis, 731–732, 733, 740, 750–751
  phoenikoston, 731–732, 733, 740, 749–751
Katodinium, 687, 691, 697–698, 704, 706, 709
  auratum, 702
  fungiforme, 701
  spiroidinoides, 688, 705
Kentrosphaera, 266, 288
  facciolae, 287
Kephyrion, 486–487, 492, 497, 499, 503
  obliquum, 478
Kephyriopsis, 501
Keratella, 393, 411, 825
Keratococcus, 263, 288
  bicaudatus, 287
Keriochlamys, 265, 288
  styriaca, 287
Keriosphaera, 430, 438, 447
  gemma, 448
Khawkinea, 384–385
Kinetoplastida, 385–387, 402
Kirchneriella, 262, 288
  obesa, 289
Klebsiella, 388, 411
Klebsormidiales, 313, 345–347
Klebsormidiophyceae, 313
Klebsormidium, 45, 313, 318, 344–345
  flaccidum, 315
  klebsii, 326
  rivulare, 42
Kobayasia, 639, 648
  subtilissima, 645
Kolbesia, 597–598, 603, 607
  amoena, 627

  kolbei, 626
  ploenensis, 626–627
  suchlandtii, 627
Koliella, 318, 326, 346
Komma, 741, 743, 749
  caudata, 726–728, 747
  pochmanni, 728
Komvophoron, 122, 139–140
  groenlandicum, 139
  jovis, 139–140
  minutum, 140
  schmidlei, 140
Kybotion, 492, 499
  eremita, 479
Kyliniella, 207, 210
  latvica, 198, 209

Lagerheimia, 255–256, 262, 288–289
Lagynion, 492, 499
  macrotrachelum var. *oedotrachelum*, 478
Lamprothamnium, 27, 317, 338
  buckellii, 338
  longifolium, 338
Larix, 429
Lauterborniella, 261, 288
  elegantissima, 289
Leibleinia, 122, 132–133, 135
  calotrichicola, 133
Lemanea, 33, 203–205, 208, 211, 216–217, 763
  borealis, 199, 203
  fluviatilis, 33, 35, 199, 203–205, 214, 216–217
  fucina var. *parva*, 199
  mamillosa, 202, 204
Lemaneaceae, 199, 203–206, 216–217, 219
Lemmermanniella, 68
Lemna, 274, 314, 316, 332, 429, 443, 459, 627, 708
  minor, 628
  wollei, 821
Lemnicola, 597–598, 603, 607, 623
  hungarica, 627–628
Lepidochrysis, 474, 502
Lepochromulina, 492, 496, 498–499
  bursa, 478
Lepocinclis, 388, 394, 401, 406–407, 410, 413, 415
  capito, 394
  fusiformis, 394
  marssonii, 394
  ovum, 394, 413
  playfairiana, 413
  salina, 413
Leptochaete, 137
Leptolyngbya, 63, 65–66, 119, 122, 132, 134–135
  cartilaginea, 134
  foveolarum, 134
  nostocorum, 134
Leptolyngbyoideae, 132
Leptosira, 315, 320, 335–336
  mediciana, 323
Leptosiropsis, 335

Leucocryptos, 716, 735, 749
  acuta, 746
Leuvenia
  natans, 451
Limbata, 450
Limnococcus, 95
Limnothrix, 126, 128–129, 155
  guttulata, 129
  redekei, 128–129
  vacuolifera, 129
Linnothrix, 122
Lithoderma
  arvenensis, 759
  fluviatile, 759, 770
  fontanum, 759, 770
  zonatum, 759, 763, 770
Lithomyxa, 69, 80–81
  calcigena, 81
Lobococcus, 289
Lobocystis, 264, 288
  dichotoma var. *mucosa*, 289
Lobomonas, 232, 248
  rostrata, 233
Lophodiniaceae, 697, 707
Lophodinium, 691, 697–698, 705, 707
  polylophum, 689, 695, 701
Loriellaceae, 125, 184
Lucas, 743
Lutherella, 431, 433, 436, 456, 458
  adhaerens, 456
  globulosa, 456
Luticola, 637–638, 640, 648
  goeppertiana, 644
  mutica, 644
Lyngbya, 3, 24–25, 39–41, 43, 45, 66, 123, 132, 135, 151, 154–155, 807–809, 813, 819, 823
  aeruginea, 132
  aestuarii, 151
  angustissima, 135
  bijahensis, 135
  birgei, 151, 154
  calitrichicola, 132
  cartilaginea, 135
  confervoides, 151
  contorta, 135, 155
  giuseppei, 155
  hahatonkensis, 155
  intermedia, 154
  lagerheimii, 135
  magnifica, 154
  maior, 151
  martensiana, 151
  meneghiniana, 151
  patrickiana, 155
  rubra, 135
  salina, 151
  spirulinoides, 151
  splendens, 154
  subterranea, 135
  subtilis, 135
  tenuis, 135
  vesiculosa, 135
  wollei, 808–809
  yellowstonensis, 135

Lyngbyopsis, 123, 145, 150
  willei, 145, 150

Macrozamia, 45
Magnolia, 328, 343
Malleochloris, 260, 288
  sessilis, 290
Malleodendraceae, 430
Malleodendron, 430, 435, 443
  caespitosum, 443
  gloeopus, 443–444
Mallomonadaceae, 523–524
Mallomonas, 7, 523–552, 807
  acaroides, 532
  acaroides var. *acaroides*, 537–538, 542, 550
  acaroides var. *muskokana*, 529–530, 535–536, 538, 542, 550
  adamas, 524
  akrokomos, 531, 542, 549
  alpina, 537
  annulata, 542, 549
  asmundiae, 538, 542, 550
  canina, 536, 542, 549
  caudata, 525, 529, 535, 542, 549
  corymbosa, 525, 537–538, 543, 550
  crassisquama, 535, 539, 543, 550
  cratis, 543, 550
  dickii, 524, 527, 537, 543, 549
  doignonii, 537
  doignonii var. *tenuicostis*, 543
  duerrschmidtiae, 529, 535, 537–538, 543, 550
  elongata, 529, 537–538, 543, 549
  fenestrata, 531
  galeiformis, 535, 537–538, 543, 550
  hamata, 524, 536, 538, 544, 548
  heterospina, 530, 537, 538, 544, 549
  hindonii, 536, 544, 549
  lychenensis, 527, 538, 544, 548
  mangofera, 544, 549
  mangofera var. *foveata*, 544
  matvienkoae, 538, 544, 549
  paludosa, 536, 538
  papillosa, 544, 549
  portae-ferreae, 537–538, 545, 550
  pseudocoronata, 528, 537–538, 545, 550
  pugio, 536, 538, 545, 549
  punctifera, 532, 537, 545, 548
  retorsa, 531
  splendens, 528
  striata, 545, 550
  tonsurata, 524–525, 527, 532, 537, 539, 545, 550
  torquata, 524, 537–538, 545
  torquata f. *simplex*, 545, 549
  torquata f. *torquata*, 549
  transsylvanica, 524, 537, 546, 548
Mallomonopsis, 523–524
Mantellum, 70, 83, 87
  rubrum, 83, 87
Martyana, 596, 600, 606, 613
  martyi, 615
Massartia, 706

Mastigocladaceae, 125, 189
Mastigocladus, 41, 66, 125
  laminosus, 66, 189–190
Mastigophora, 385
Mastogloia, 39, 43, 569, 604, 638–639, 648
  smithii var. *lacustris*, 644
Medicago, 822
Melastomaceae, 343
Melosira, 20, 33–34, 563, 565, 571–572, 583
  undulata, 567, 576, 583
  varians, 33, 576, 583
Melosiraceae, 563
Melosirales, 563
Meridion, 596, 600, 606
  anceps, 616–617
  circulare, 616–617
  circulare var. *constrictum*, 617
Meringosphaera, 430, 438, 447
  tenerrima, 448
Merismogloea, 431, 437, 454
  polychloris, 455
Merismopedia, 43, 65–66, 69, 83, 87–88
  angularis, 83, 88
  convoluta, 88
  elegans, 83, 88
  elegans var. *maior*, 83
  gardner, 88
  glauca, 88
  major, 88
  punctata, 83, 88
  smithii, 67, 83, 88
  tenuissima, 88
Merismopediaceae, 69, 82
Merotrichia, 427–428
  capitata, 426, 428
Mesodictyon, 563, 582, 584
Mesostigma, 228, 248
  grande, 228–229
  viride, 228
Mesotaeniaceae, 353
Mesotaenium, 354, 363, 367, 369
  berggrenii, 369
  degreyi, 369
  endlicherianum, 358
  kramstai, 364
Micractinium, 261, 288
  pusillum, 290
Micrasterias, 2, 40, 361, 364, 368, 375–376
  foliacea, 359, 366, 376
  johnsonii var. *ranoides*, 360
  muricata, 365
  pinnatifida, 360
Microchaetaceae, 123, 158, 161
Microchaete, 124, 163–164
  robinsonii, 163–164
  tenera, 163
Microchaetoideae, 161
Microcoleoideae, 147
Microcoleus, 20, 24, 39, 66, 123, 139, 149–150
  chthonoplastes, 149
  lacustris, 150
  purpureus, 139
  vaginatus, 149

Microcostatus, 638, 640
  krasskei, 645, 648
Microcrocis, 70, 84, 88
  gigas, 88
  irregulare, 84
  irregularis, 88
  obvoluta, 84, 88
  pulchella, 84, 88
Microcystaceae, 62, 70, 89
Microcystis, 23, 37, 43, 60–61, 63–65, 68, 70, 90, 92, 805, 807, 822
  aeruginosa, 20, 43, 90, 92
  comperei, 90, 92
  flos-aquae, 92
  glauca, 92
  ichthyoblabe, 92
  natans, 92
  pulchra, 90
  smithii, 92
  splendens, 94–95
  viridis, 92
  wesenbergii, 67, 90, 92
Microglena, 493, 499, 523–524
  butcheri, 481, 499
  punctifera, 499
Micromonadophyceae, 229
Microneis, 604
Microspora, 318, 341, 431
  willeana, 326
Microsporales, 317, 341
Microthamniales, 336
Microthamnion, 321, 335
  kuetzingianum, 325, 336
  strictissimum, 336
Mimosaceae, 343
Mischococcaceae, 431, 436, 457
Mischococcales, 270, 430, 432, 434, 443, 445, 448, 450–459
Mischococcus, 429, 431, 433, 436, 454
  confervicola, 457
Moina, 411
Monadodendron, 498
Monallantus, 430, 432, 438, 447
  pyreniger, 448
Monimiaceae, 343
Monocilia, 459
Monochrysis, 487, 493, 499–500
  vesiculifera, 481
Monodopsidaceae, 424
Monodopsis, 424–425, 427
  subterranea, 426, 447
Monodus, 430, 439, 447
  chodatii, 448
  ovalis, 427, 447
  subterraneus, 447
Monoraphidium, 263, 289
  nanum, 447
  pusillum, 290
Monostroma, 317, 347
  latissimum, 329
Mougeotia, 24–26, 39, 354–355, 363, 366, 369–370, 370, 372, 393, 768
Mougeotiopsis, 363, 366, 370
  calospora, 355, 370

Muelleria, 638–639, 648
　gibbula, 641, 649
　terrestris, 649
Myriophyllum, 314–315, 322, 335, 811
　brasiliense, 811
　spicatum, 24, 816
Myrmecia, 254, 266, 276, 289
　pyriformis, 290
Myxochloridaceae, 430
Myxochloris, 430, 435, 441
　sphagnicola, 441–442
Myxosarcina, 71, 104
　amethystina, 104
　gloeocapsoides, 104
　rubra, 104

Naegeliella, 494, 500
　flagellifera, 483
Najas, 314–315, 322, 335
　marina, 24, 810
Nannochloris, 20, 24, 265, 291
　bacillaris, 290
　occulata, 427
Nannochloropsis, 424–425, 427, 465
　occulata, 427
　oculata, 426
Nautococcus, 266, 291
　piriformis, 292
Navicula, 24, 34, 39, 43, 45, 597, 637, 640–641, 643, 645, 647
　avenacea, 35
　capitata, 648
　cocconeiformis, 646
　contenta, 44
　cryptocephala, 27
　gastrum, 650
　gregaria, 36
　hasta, 641
　hungarica, 27, 648
　levanderi, 649
　mutica, 648
　paludosa, 648
　pelliculosa, 647
　placenta, 641
　pupula, 650
　pygmaea, 647
　rhychocephala, 641
　saprophila, 647
　similis, 648
　soehrensis, 646
　subinflatoides, 43
　subtilissima, 40, 648
　tantula, 44
　tenuicephala, 27
　(*tuscula* group), 644
Naviculaceae, 637, 639, 665
Naviculales, 656
Neidiopsis, 639, 649
　levanderi, 642
Neidium, 638–639, 649
　densestriatum, 642
　hitchcockii, 642
　iridis, 642
Nemalionopsis, 207, 218

　tortuosa, 199, 217
Neochloris, 256, 266, 291
　aquatica, 292
Neomysis
　mercedis, 813
Neonema, 431, 434, 461
　quadratum, 462
Neonemataceae, 431
Neospongiococcum, 266, 291, 303
　gelatinosum, 292
Neosynedra, 596
Nephrochloris, 430, 435, 439
　incerta, 439–441
　salina, 439
Nephrocytium, 262, 291
　agardhianum, 292
Nephrodiella, 430, 438, 447
　brevis, 447
　nana, 447
　phaseolus, 448
Nephroselmis, 228, 230, 247
　olivacea, 229–230
Netrium, 354, 363, 367, 370, 379
　digitus, 358
　minus, 370
　oblongum, 358
Nitella, 17, 27, 317, 332, 339, 348, 811, 820, 821
　flexilis, 331, 339
　hookeri, 27
Nitellopsis, 27–28, 338
　obtusa, 28
Nitzschia, 22, 24, 39, 43, 669, 672, 675–676, 681
　acicularis, 672, 677
　amphibia, 671, 677
　angustata, 677
　communis, 43
　denticula, 677
　filiformis, 673
　frustulum, 43
　holsatica, 672
　inconspicua, 670
　monoensis, 43
　palea, 43–44, 673, 677
　scalaris, 670
　sinuata var. *tabellaria*, 677
Noctiluca, 388
Nodularia, 64, 66, 120, 125, 177–178, 807
　baltica, 178
　harveyana, 67, 177–178
　litorea, 178
　sphaerocarpa, 177
　spumigena, 43, 177–178
　spumigena var. *minor*, 178
　willei, 177–178
Nostoc, 11, 26–27, 39–40, 45, 63, 68, 119–120, 125, 158, 177, 179–180
　aureum, 180
　commune, 179–180
　edaphicum, 179
　linckia, 180
　minutum, 180
　paludosum, 179–180

　parmelioides, 36, 180, 764, 770
　pruniforme, 180
　sphaericum, 180
　sphaeroides, 180
　verrucosum, 33, 764, 770
Nostocaceae, 118, 123, 169
Nostocales, 117–120, 123, 155, 158, 161, 164, 169
Nostochopsaceae, 125, 187
Nostochopsis, 125, 187–188
　lobata, 188–189
Nostocoideae, 120
Notosolenus, 407
Novarino, 743
Nupela, 637, 641, 643, 649
Nuphar, 40, 427, 459, 461
Nymphaea, 427, 459, 461
　odorata, 811

Ochromonas, 2, 429, 472–473, 485, 487, 493, 495–496, 500–502, 523–524
　monicis, 490
　sphaerocystis, 476, 480
　tuberculata, 485
　vulcania, 489
Octacanthium, 376, 379
　octocorne, 360, 376
Octogoniella, 261, 291
　sphagnicola, 293
Odontidium, 612
Oedocladium, 319, 341
　hazenii, 330
Oedogoniales, 311–313, 315, 317, 319, 341, 348
Oedogonium, 24, 35, 39, 40, 44, 314, 316, 318, 322, 340, 441, 458, 497, 696, 708, 709, 808, 810, 822–823
　croasdaleae, 330
Oestrupia, 639
　zachariasii, 644, 650
Oligochaetophora, 322, 336
　simplex, 324
Oncobyrsa, 96
Onychonema, 366, 376
　filiforme, 360
Oocardium, 354, 364, 366, 376
　stratum, 356
Oocystidium, 264, 291
　ovale, 293
Oocystis, 43, 255, 264, 291, 301, 822
　lacustris, 293
Oodesmus
　oederleinii, 481
Oodiniaceae, 697, 706
Oodinium, 690, 691, 697–698, 701, 703
　limneticum, 688, 702, 706
Oophila, 260, 291
　amblystomalis, 293
Opalozoa, 749
Opephora, 596, 613, 615
　ansata, 616
　martyi, 616
Ophiocytiaceae, 431, 456, 459

Ophiocytium, 429, 431–433, 436, 457, 459
　arbusculum, 457
　capitatum, 457
　longipes, 457
　majus, 459
　mucronatum, 457
　parvulum, 457
Ophrydium, 45
　versatile, 662
Oronectes
　immunis, 819
Orthoseira, 563, 565, 571, 584
　dendroteres, 576
Orthoseiraceae, 563
Orthoseriales, 563
Oscillatoria, 20, 23–24, 27, 39–41, 43–45, 65, 123, 126–127, 132, 141, 155–156, 807, 809, 813, 823
　amphibia, 126
　amphigranulata, 129
　anguina, 155
　claricentrosa, 127
　curviceps, 155
　depauperata, 155
　earlei, 127
　filiforme, 126
　funiformis, 155
　jenensis, 156
　lacustris, 147
　limnetica, 129
　limosa, 155–156
　margaritifera, 155
　obtusa, 156
　ornata, 155
　princeps, 155–156
　proboscidea, 155
　redekei, 129
　refringens, 156
　rhamphoidea, 155
　rubescens/agardhii, 806
　sancta, 155–156
　splendida, 126
　tenuis, 808
Oscillatoria (Planktothrix)
　rubescens, 806
Oscillatoriaceae, 118, 120, 122, 150–151
Oscillatoriales, 117–119, 121, 126, 129, 132, 135, 137, 139, 147, 151
Ourococcus, 288
Oxyneis, 596, 601, 606, 616–617, 621
　binalis, 621
　binalis var. *elliptica*, 621

Pachycladella, 261, 292
　umbrinus, 293
Pachycladon, 292
Palmella, 260, 292, 295
　miniata var. *aequalis*, 294
Palmellochaete, 261, 295
　tenerrima, 294
Palmellococcus, 271
Palmellocystis, 301
Palmellopsis, 260, 295
　gelatinosa, 294

Palmodactylon, 295
Palmodictyon, 255, 264, 295
　varium, 294
Pandorina, 226, 239, 241, 248, 822
　charkowiensis, 241
　colemaniae, 241
　morum, 226–227, 240–241
　unicocca, 241–242
Pannus
　spumosus, 65
Parabasalia, 387
Paracoenia, 41
Paradoxia, 261, 295
　multiseta, 294
Paralemanea, 203, 205, 208, 216–217
　annulata, 199–200
　catenata, 199
　mexicana, 199, 217
Paralia, 583
Paraliaceae, 563
Paraliales, 563
Paramecium, 686, 690
　bursaria, 258
Paraphysomonadaceae 474, 513, 523
Paraphysomonas, 474, 486, 488, 491, 500, 523
　butcheri, 485
　campanulata, 486
　foraminifera, 486
　imperforata, 485–486
　sediculosa, 486
　siderophora, 486
　sigillifera, 486
　vestita, 477, 486–487, 491
Parietochloris, 256
Pascheriella, 247
Pascherina, 245, 248
　tetras, 246, 247
Paulschulzia, 263, 295
　pseudovolvox, 296
Pavlova, 512, 515
Pavlovales, 511, 513
Pectodictyon, 264, 295
　cubicum, 296
Pediastrum, 43, 254–255, 257–258, 263, 295, 301
　duplex var. *typicum*, 296
Pedimellophyceae, 471
Pedinomonas, 228, 230, 247
　maior, 230
　minor, 229–230
　noctilucae, 388
　rotunda, 230
Pedinopera, 236, 238, 248
　granulosa, 237–238
　rugulosa, 238
Pedinophyceae, 225, 229, 231
Pelagodictyon, 563, 584
Pelagophyceae, 471
Peniaceae, 373
Penium, 361–362, 367–368, 373
　didymocarpum, 373
　exiguum, 357
　phymatosporum, 373
　silvae-nigrae, 373

　spinospermum, 373
　spirostriolatum, 357
Peranema
　trichophorum, 403
Percolozoa, 387
Peridiniaceae, 697, 707
Peridiniales, 697, 707
Peridiniopsis, 697–698, 704, 707
　berolinensie, 701
　cunningtonii, 694, 696, 699
　lubieniensiforme, 694
　penardiforme, 692
　penardii, 694, 699
　polonicum, 691–692, 701
　quadridens, 689, 691–692
　thompsonii, 702
Peridinium, 697–698, 704, 709
　balticum, 707
　bipes, 694, 696, 699
　cinctum, 690, 694
　cinctum f. *tuberosum*, 691, 707
　dybowski, 707
　gatunense, 690–692, 694, 699–702
　inconspicuum, 23, 686, 694, 699, 707
　limbatum, 694, 699–702, 707
　lomnickii, 699
　pusillum, 699
　volzii, 694
　willei, 691–692, 694, 699, 707
　wisconsinense, 702
Perone, 430, 436, 451
　dimorpha, 450
Peronia, 656, 663, 665
　fibula, 657
Peroniella, 431, 434, 436, 458
　hyalothecae, 456, 458
　planctonica, 458
Petalomonas, 387
Petalonema, 124, 160–161
　alatum, 160–161
　byssoidea, 161
　involvens, 160
Pfiesteria
　piscicida, 685, 739, 814
Phacotaceae, 225, 228, 236–237, 239
Phacotus, 236, 238, 248
　angustus, 238
　glaber, 238
　lenticularis, 237–238
　subglobosus, 238
Phacus, 388, 395, 399–402, 404, 406–407, 410, 413, 416
　agilis, 395
　chloroplastes, 413
　chloroplastes f. *incisa*, 413
　curvicauda f. *anomalus*, 395
　elegans, 395
　longicauda, 395
　longicauda var. *insecta*, 395
　longicauda var. *tortus*, 395
　mariana, 395
　monilata, 395
　orbicularis var. *longicauda*, 395
　pleuronectes, 395
　polytrophos, 395

Phacus (continued)
  trimarginatus, 395
  triqueter, 395
Phaeaster, 500
Phaeobotrys, 503
Phaeocystales, 511
Phaeocystis, 512
Phaeodermatium, 488, 494, 500
  rivulare, 481, 500
Phaeogloea, 491, 495, 500, 503
  mucosa, 483, 500
Phaeophyceae, 8, 757, 765
Phaeophyta, 12, 19, 24, 758, 766
Phaeoplaca, 473, 494, 497, 500
  thallosa, 481
Phaeoschizochlamys, 473, 495, 500, 503
  mucosa, 484
Phaeosphaera, 491, 495, 500
  gelatinosa, 481
Phaeothamnion, 473, 494, 500–503
  confervicola, 481, 502
Phaeothamniophyceae, 443, 471, 473, 491, 503
Phalansterium, 499
Phillipsiella, 523
Phormidium, 26, 34–35, 41, 63, 65–66, 123, 132, 135, 141–142, 807, 809, 819
  autumnale, 141–142
  favosum, 141
  fonticolum, 141
  formosum, 142
  geysericola, 128
  inundatum, 141
  minnesotense, 141
  retzii, 141
  richardsii, 142
  uncinatum, 141
Phormidiaceae, 120, 122, 132, 139, 147, 150–151
Phormidioideae, 139
Phragmites, 39
Phycopeltis, 315, 319, 343
  arundinacea, 328
Phyllobium, 260, 295
  sphagnicola, 296
Phyllogloea, 259, 295
  fimbriatum, 296
Phyllosiphon, 315, 319, 328, 344
  arisari, 328
Phymatodocis, 364, 366, 377
  alternans, 377
  nordstedtiana, 358, 377
Physolinum, 344
Physomonas, 500
Phytodinales, 696–697, 708
Phytodiniaceae, 697, 708
Phytodinium, 697–698, 704, 709
  simplex, 688, 705
Pinnularia, 24, 39, 41, 638–639, 646, 649–650
  braunii, 42
  mesolepta, 644
  microstauron, 42
  viridis, 40, 644

Piscinoodinium
  pillulare, 706
Pithophora, 321, 334, 340, 808–810, 815, 818, 820, 823
  oedogonia, 327
Placoma, 96
Placoneis, 650
  abiskoensis, 641
Plagioselmis, 731–732, 744, 749
  nanoplanctica, 730–732, 748
  ovalis, 732
  prolonga, 748
Plagioselmis (as Rhodomonas)
  minuta, 43
Plagiotropis, 638–639, 641, 650
Planktolyngbya, 122, 133, 135
  bipunctata, 135
  capillaris, 135
  contorta, 133, 135, 155
  limnetica, 133, 135
  regularis, 135
  tallingii, 133, 135
Planktosphaeria, 265, 270, 295
  gelatinosa, 297
Planktothrix, 37, 65, 120, 123, 141, 143, 147, 155
  agardhii, 141, 143
  cryptovaginata, 141
  mougeotii, 141, 143
  prolifica, 141
  rubescens, 141, 143
Planktothrix (Oscillatoria)
  rubescens, 23
Planorbis
  contortus, 26
Planothidium, 597–598, 602, 607, 623, 627
  apiculatum, 629
  frequentissimum, 629
  lanceolatum, 627, 629
  oestrupii, 629
  peragalli, 629
  pseudotanense, 629
Plantae, 384
Platydorina, 227, 239, 241, 248
  caudata, 240–241
Platymonas, 231
  subcordiformis, 231
Plectonema, 20, 45, 119, 123, 144, 155, 819
  batrachospermi, 144, 146
  edaphicum, 146
  flexuosum, 146
  murale, 146
  tenue, 144
  tomasinianum, 152, 155
  wollei, 155
Pleodorina, 239, 241, 248
  californica, 240–241
  indica, 241
  japonica, 241
Pleurastromsarcina, 274
Pleurastrophyceae, 313
Pleurastrum, 313, 322, 335–336
  insigne, 323
Pleurocapsales, 60

Pleurocapsa, 62, 72, 107–108, 110
  crepidinum, 108
  epiphytica, 108
  fluviatilis, 104
  minor, 44, 108, 110
  minuta, 108
  varia, 110
Pleuroceridae, 210
Pleurochloridaceae, 430, 443, 445, 448, 450
Pleurochloridella, 430, 435, 443, 473
  vacuolata, 444
Pleurochloridellaceae, 430
Pleurochloris, 430, 437, 446–447
  commutata, 44, 425, 447
  magna, 425, 447
  polyphem, 425, 447
  pyrenoidosa, 448
Pleurochrysis, 484, 515
  carterae, 43
  carterae var. *dentata*, 512, 514–515, 518
Pleurocladia, 757–758, 760, 767–768, 771
  lacustris, 759, 761, 763–765, 766, 769–770
Pleurococcus, 267, 276, 333
Pleurodiscus, 363, 366, 370
  borniquinae, 355, 370
Pleurogaster, 430, 438, 447
  lunaris, 448
Pleurosigma, 638–639, 650
  elongatum, 641
Pleurosira, 563, 571, 584
  laevis, 577, 584, 586
Pleurotaenium, 361–362, 366–367, 375, 377
  coronatum, 358
Pliocaenicus, 563, 566, 572, 581, 584
Polyblepharaceae, 225, 228, 229, 231–232
Polyblepharides, 228, 230, 247
  fragariiformis, 230
  singularis, 229–230
Polychaetophora, 321–322, 336
  lamellosa, 324
Polycystis, 92
  firma, 92
  incerta, 78
  marginata, 92
  montana f. *minor*, 82
  pulverea, 92
Polyedriella, 430, 433, 439, 447–448
  aculeata, 448
  helvetica, 427, 447–448
Polyedriopsis, 262, 298
  spinulosa, 297
Polygoniochloris, 430, 431, 433, 437, 448–449
  regularis, 448
  tetragona, 448
Polykyrtos, 430, 435, 441
  vitreus, 440
Polylepidomonas
  vaculata, 489
  vestita, 489
Polysiphonia, 202, 208, 221
  subtilissima, 199, 220–221
Polytaenia, 371
  trabeculata, 355

Polytoma, 226, 232, 235, 248
  granuliferum, 235
  uvella, 233, 235
Polytomella, 228, 230, 248
  agilis, 230
  citrii, 229–230
Pontederia
  cordata, 821
Pontosphaera
  stagnicola, 516
Porochloris, 261, 298
  filamentarum, 297
Porphyridiales, 198, 208–209
Porphyridium, 3, 200, 206–208
  aerugineum, 200
  purpureum, 198, 209
  sordidum, 198
Porphyrosiphon, 123, 144–145
  fuscus, 144–145
  notarisii, 144–145
  robustus, 145
  versicolor, 144–145
Porterinema, 758, 760, 767–768, 770
  fluviatile, 759, 763, 765–766, 769, 771
  marina, 771
Potamogeton, 17, 315, 335, 459, 461, 819
  nodosus, 821
  pectinatus, 28, 810, 812
Poterioochromonas, 23, 473, 492, 501
  malhamensis, 477
Prasinophyceae, 225, 229, 231–232, 313, 384
Prasiola, 11, 34, 45, 317, 342
  fluviatilis, 35
  mexicana, 33, 329, 342
Prasiolales, 312, 317, 342
Printzina, 319, 343
  ampla, 343
Prismatella, 430, 439, 449
  hexagona, 448–449
Proales
  werneckii, 463
Prorocentrum, 697, 709
Proteomonas, 734, 736, 741, 743
Protista, 383–385, 473, 511–512, 685–686, 697, 715
Protococcus, 267, 276
Protococcus-Pleurococcus, 315, 322
Protoctista, 385
Protoderma, 260, 298, 314, 320, 332, 336
  involvens, 319
  sarcinodeum, 297
  viridae, 323
Protoeuglena, 388
Protosiphon, 256, 258, 260, 298, 322, 344
  botryoides, 297, 315, 328
Protozoa, 24, 44, 171, 260, 272, 383–385, 391, 404, 686, 737, 749
Prymnesiales, 511, 513
Prymnesiophyceae, 471, 511, 516, 690
Prymnesium, 511, 515
  parvum, 512, 514–515
  saltans, 514
Psammodictyon, 672
Psammothidium, 597–598, 602, 607, 623, 627
  altaicum, 630

  lauenburgianum, 630
  levanderi, 630
  rosenstockii, 630
  subatomoides, 630
Pseudanabaena, 65, 122, 129–130, 139, 155, 807
  catenata, 44, 129
  galeata, 130
  limnetica, 129–130
  lonchoides, 130
  mucicola, 129
  thermalis, 129
Pseudanabaenaceae, 118, 120–121, 126, 129, 132
Pseudanabaenoideae, 126
Pseudendoclonium, 320, 336
  basiliense, 323
  basiliense var. *brandii*, 337
  prostratum, 336
  submarinum, 337
Pseudoactiniscus, 697–698, 705
  apentasterias, 688, 702, 706
Pseudobodanella, 758
  peterfii, 759, 771
Pseudocharaciopsidaceae, 424
Pseudobohlinia, 261, 298
  americana, 297
Pseudobumilleriopsis, 459
Pseudochaete, 321, 337
  gracilis, 324, 337
Pseudocharaciopsis, 424–425, 441
  minuta, 426–427
  ovalis, 425, 427, 446–447
  texensis, 427
Pseudocharacium, 270, 441
Pseudodendromonas, 498
Pseudokephyrion, 486–487, 492, 498–499, 501
  alaskanum, 478
  auroreum, 475
Pseudophormidium, 63, 119, 123, 144, 146
  batrachospermi, 144
  flexuosum, 146
Pseudopolyedriopsis, 430, 438, 449
  skujae, 448
Pseudoschizomeris, 315, 317, 346
  caudata, 326
Pseudostaurastrum, 424–425, 430, 439, 447–449, 465
  enorme, 448, 465
  hastatum, 448, 465
  limneticum, 426–427, 449, 465
Pseudostaurosira, 596, 599, 606, 608, 610, 613, 616
  brevistriata var. *elliptica*, 618
  brevistriata var. *inflata*, 618
  robusta, 618
Pseudotetraedron, 430–431, 439, 447, 449
  neglectum, 450
Pseudulvella, 320, 337
  americana, 323, 337
Pteromonas, 236, 238
  aculeata, 237–238
  angulosa, 238
  cordiformis, 238

  cruciata, 238
  sinuosa, 238
Pulchrasphaera, 289
Punctastriata, 596, 599, 606, 613, 616, 618
Pyramimonas, 735
  parkeii, 749
Pyrenomonadaceae, 722, 743, 746–747
Pyrenomonadales, 737, 743, 746–748
Pyrenomonas, 2, 742–744, 746, 749
  ovalis, 722, 724, 746
Pyrobotrys, 226–227, 245, 247, 248
  casinoensis, 246–247
  stellata, 247
Pyrrhophyta, 7, 12, 124, 687, 697

Quadricoccus, 263, 298
  verrucosus, 299
Quadrigula, 262, 298
  closteroides, 299

Raciborskia, 708
  bicornis, 688
Radaisia, 72, 108, 110
  confluens, 110
  epiphytica, 108
  gardneri, 108, 110
  willei, 110
Radiococcus, 264, 298
  nimbatus, 299
Radiocystis, 69, 73, 81
  elongata, 73, 81
  fernandoi, 81
  geminata, 73, 81
Radiofilum, 318, 346
  conjunctivum, 326
Radiosphaera, 267
Rana, 45
  clamitans, 412
  pipiens, 412
Raphidiastrum, 378
Raphidiella, 431, 437, 453
  fascicularis, 452
Raphidiopsis, 124, 173, 180
  curvata, 173, 180
  mediterranea, 174, 180
Raphidium, 280
Raphidomonadales, 427
Raphidonema, 315, 318, 346
  nivale, 44, 326
Raphidonemopsis, 318, 346
  sessilis, 326
Raphidophyceae, 40, 423, 427, 465
Rayssiella, 263, 298
  hemisphaerica, 299
Reimeria, 656, 659, 663–664
  sinuata, 658, 662
  sinuata f. *antiqua*, 658
Rhabdoderma, 65–66, 69, 76, 81
  compositum, 76, 81
  curtum, 81
  lineare, 76, 81
  zygnemicolum, 76

Rhabdogloea, 69, 76, 81
    hungarica, 81
    smithii, 76, 81
    subtropica, 81
Rhinomonas, 743
Rhizochloridaceae, 430
Rhizochloridales, 430, 432, 434, 441–442
Rhizochloris, 430, 435, 441
    mirabilis, 441–442
    stigmatica, 441
Rhizochrysis, 494, 496, 501
    scherffelii, 482
Rhizoclonium, 208, 318, 321–322, 334, 340, 763, 771, 808–809
    hieroglyphicum, 314, 327, 340
Rhizolekane, 430, 435, 441
    sessilis, 441–442
Rhizoochromonas, 485, 493, 501, 503
    endoloricata, 480, 501
Rhizophydium
    fugax, 739
    sphaerocarpum, 364
Rhizosoleniaceae, 563, 567
Rhizosoleniales, 563
Rhizosoleniophycidae, 563
Rhodochytrium, 260, 298
    spilanthidis, 300
Rhododendron, 343
Rhododraparnaldia, 201–202, 207, 212
    oregonica, 198, 201, 212–213
Rhodomonas, 736, 739, 741–743, 746, 749
    lacustris, 738
    lens, 738
    minuta, 748
    ovalis, 725
Rhodophyta, 6, 12, 19, 34–35, 41, 197–198, 200, 202–203, 204–205, 735
Rhoicosphenia, 597, 604, 656, 661, 663
    abbreviata, 659, 666
    curvata, 34, 659
Rhoicospheniaceae, 597, 655–656, 659, 661, 665–666
Rhomboidella, 430, 439, 449
    oblique, 450
Rhopalodia, 43, 669, 670, 672–675, 680–681
    gibba, 678
Rhopalodiaceae, 670, 676
Rhopalodiales, 669–672, 674–675, 681
Richteriella, 288
Rigida, 412
Rivularia, 65, 118–119, 124, 166, 168–169, 764
    aquatica, 168
    biasolettiana, 169
    compacta, 169
    dura, 168, 169
    globiceps, 169
    haematites, 169
    minutula, 169
    planctonica, 166
Rivulariaceae, 118, 124, 164
Romeria, 121, 129–130
    alascense, 129
    elegans, 129
    elegans var. *nivicola*, 129
    heterocellularis, 129
    hieroglyphica, 129
    leopoliensis, 129–130
    mexicana, 129
    nivicola, 130
Rosenvingiella, 342
Rossithidium, 597–598, 602, 607, 625, 628
    duthiei, 630
    linearis, 609, 627, 630
    pusillium, 630
Rotifera, 383
Roya, 367, 370
Rufusia, 45, 199, 207, 210
    pilicola, 198, 201
Rufusiella, 697–698, 701, 703, 709
    insignis, 688, 702

Saccochrysis, 493, 501
    piriformis, 481, 501
Sacconema, 124, 167, 169
    rupestris, 167, 169
Sagittaria, 322, 332
Salicornia, 333
Sarcomastigophora, 383
Scenedesmus, 44, 254, 256–258, 262, 298, 301, 453, 768
    acutus, 300
    armatus, 255, 276, 299
    f. *quadricauda*, 256
    quadricauda, 44
    trainorii, 256
Scherffelia, 228, 230, 248
    phacus, 229–230
Schizochlamydella, 259, 263, 301
    gelatinosa, 300
Schizochlamys, 301
Schizodictyon, 284
Schizomeris, 317, 337
    leibleinii, 329
Schizothrix, 34, 39, 44, 63, 65–66, 119, 122, 136–137, 148, 150, 763
    acuminata, 150
    aikenensis, 137
    californica, 150
    chalybea, 150
    constricta, 136–137, 150
    friesii, 150
    giuseppei, 150
    hancockii, 150
    mexicana, 150
    muelleri, 150
    parciramosa, 148, 150
    penicillatum, 150
    purpurascens, 148, 150
    richardsii, 150
    rivularis, 150
    sauterianum, 150
    stricklandii, 150
    taylori, 150
    telephoroides, 150
    thermophila, 135
    violacea, 136
    wollei, 148

Schizotrichaceae, 121, 137, 150
Schmidleinema, 125, 184–185
    cubanum, 184–185
    indicum, 184
    roberti-lamyi, 184
Schroederia, 261, 301
    setigera, 300
Scirpus, 17, 439, 458–459, 461
Scoliopleura, 638–639
    peisonis, 641, 650
Scotiella, 266, 301
    cryophila, 44
    turberculata, 300
Scotiellopsis, 266, 301
    rubescens, 300
Scourfieldia, 228, 230, 247
    cordiformis, 229–230
Scytonema, 3, 39, 66, 119, 123, 155, 157–158, 161
    arcangelii, 158
    capitatum, 157
    cincinnatum, 158
    crispum, 158
    crustaceum, 158
    dubium, 158
    endolithicum, 67
    longiarticulatum, 157
    myochrous, 158
    ocellatum, 158
    tolypothrichoides, 158
Scytonemataceae, 123, 155
Scytonematopsis, 118, 123, 157–158
    fuliginosa, 157–158
    hydnoides, 157–158
Secale, 822
Selenastrum, 262, 301, 785
    capricornatum, 302
Selenophaea, 491, 495, 501, 503
    granulosa, 482, 501
Sellaphora, 637, 640, 650
    americana, 643
    pupula, 643
Semiorbis, 656, 662
    hemicyclus, 657, 665
Setacea (section of *Batrachospermum*), 215
Siderocelis, 265, 301
    minutissimus, 302
Simulium, 35
Siphonales, 312, 319, 321, 344
Sirodotia, 200, 205, 208, 215
    delicatula, 203
    huillensis, 199, 216
    suecica, 199, 214, 215
Sirogonium, 354, 363, 366, 370
    illinoiense, 355
Skeletonema, 563, 565, 569–571, 585
    potomos, 582
Skeletonemataceae, 563
Sklerochlamys, 430, 438, 449
    pachyderma, 450
Smithsonimonas, 227
Snowella, 64–65, 70, 85, 88
    fennica, 85, 88
    lacustris, 88
    litoralis, 85, 88

Snowella (continued)
  rosea, 67
  septentrionalis, 88
Sorastrum, 261, 301
  spinulosum, 302
Sphacelariales, 758–759, 771
Sphaeropleales, 312, 317, 342
Spermatozopsis, 228, 230–231, 247
  exsultans, 229, 231
Sphacelaria, 758, 760, 767–768
  fluviatilis, 759, 763, 765, 771
  lacustris, 759, 765, 767–769, 771
Sphaerellocystis, 265, 301
  aplanospora, 301
  ellipsoidea, 302
Sphaerellopsis, 236
Sphaeridiothrix, 473, 494, 501, 503
  compressa, 483
Sphaerocystis, 43, 258, 265, 301
  schroeteri, 302
Sphaerodinium, 697–698, 705
  fimbriatum, 691, 693–694, 702, 708
  polonicum, 708
Sphaeroplea, 318, 342
  annulina, 328
Sphaerosorus, 431, 437, 453
  coelastroides, 452
Sphaerozosma, 366, 376–377, 458
  vertebratum var. *latus*, 358
Sphagnum, 40, 291, 295, 298, 363, 370–371, 373, 375, 378, 427–428, 432, 439, 441–443, 451, 458, 461, 463, 490, 662, 709
Sphinoclosterium, 364–365
Spiniferites, 698
Spiniferomonas, 473, 485–488, 491, 493, 497, 501, 523
  abei, 475, 486, 489
  bilacunosa, 487, 489
  bourrellyi, 474, 480, 487–489
  cornuta, 475, 489
  coronacircumspina, 489, 491
  minuta, 491
  serrata, 488, 491
  silverensis, 475, 486
  takahashii, 491
  trioralis, 486–489
Spinoclosterium, 367, 372
  cuspidatum, 356, 372–373
Spinocosmarium, 361, 368, 377
  laconiense, 377
  quadridens, 359, 365, 377
Spirirestis, 191
  rafaelensis, 191
Spirogyra, 24, 39, 314, 340, 353–354, 362–365, 370–372, 708, 808–810, 818, 820, 822–823
  floridana, 371
  juergensii, 355
  majuscula, 364
  wrightiana, 355
Spirotaenia, 280, 354, 363, 367, 371
  condensata, 357
  densata, 371
Spirulina, 122, 131–132

  caldaria var. *magnifica*, 131
  gigantea, 132
  labyrinthiformis, 132
  laxa, 132
  major, 131–132
  meneghiniana, 131
  nordstedtii, 132
  platensis, 139
  princeps, 132
  stagnicola, 131–132
  subalsa, 131
  subtilissima, 132
  weissii, 131–132
Spirulinoideae, 129
Splendidae, 533
Spondylomoraceae, 225, 228, 246, 248
Spondylomorum, 245, 248
  quaternarium, 246–247
Spondylosium, 366, 377
  pulchellum, 377
  pulchrum, 358
  rectangulare, 358
Spongilla, 272
  lacustris, 45, 258
Spongiochloris, 256, 266, 302
  spongiosa, 302
Spongiococcum, 291, 303
  alabamense, 302
Spongomonas, 499
Spumella, 491, 502
  sociabilis, 477
Stanieria, 66, 71, 102–104
  cyanosphaera, 67, 103–104
  sphaerica, 104
Staurastrum, 354, 361, 363–364, 368–369, 377–379, 447, 456, 458
  anatinum f. *curtum*, 360
  bioculatum, 360
  claviferum, 360
  dejectum, 23
  longipes, 364
  tribedrale, 360
  turgescens, 360
Staurodesmus, 363, 368–368, 376, 378–379
  crassus, 364
  cuspidatus, 360, 364
  extensus var. *joshuae*, 364
  sellatus, 364
  subtriangularis, 360
  triangularis var. *limneticus*, 364
Stauroforma, 599, 606
  exiguiformis, 616, 619
Stauromatonema, 125, 181
  viride, 181–182
Stauroneis, 24, 40, 638–640, 650
  acuta, 642
  anceps, 27
  kriegeri, 642
  phoenicenteron, 642
Staurosira, 596, 599, 606, 610, 616
  construens var. *pumila*, 619
  construens var. *venter*, 619
Staurosirella, 596, 599, 606, 608, 610, 613, 615–616
  ansata, 620

  leptostauron, 616, 620
  pinnata, 620
  spinosa, 620
Stenopterobia, 670, 676
  sigmatella, 672, 679
Stephanocostis, 563, 572
  chantaicus, 582, 585, 588
Stephanocyclus, 563, 572, 585
  caspia, 585
  cryptica, 579
  gamma, 579
  meneghiniana, 37, 569, 579, 585
  quillensis, 585
  striata, 579
Stephanocyclus-(Cyclotella)
  meneghiniana, 22
Stephanodiscaceae, 563, 570
Stephanodiscus, 20–22, 37, 563–565, 567–568, 570, 572–573, 578–579, 583–586, 807
  binderanus, 565, 580
  binderanus var. *oestrupi*, 580
  excentricus, 585–586
  hantzschii, 38, 565, 580
  hantzschii f. *tenuis*, 580
  lucens, 586
  niagarae, 566, 569, 580
  reimerii, 569
  rhombus, 585–586
  superiorensis, 569
  yellowstonensis, 569
Stephanoporos, 492, 502
  sphagnicola, 478
Stephanosphaera, 231–232, 248
  pluvialis, 231–232
Stichochrysis, 493, 502
  immobilis, 481, 502
Stichococcus, 44, 318, 345–346
  bacillaris, 42
  subtilis, 326
Stichogloea, 473, 491, 495, 502–503
  doederleinii, 481
  olivacea, 487
Stichosiphon, 62, 65, 70, 99, 102
  exiguus, 99, 102
  filamentosus, 102
  gardneri, 102
  regularis, 102
  sansibaricus, 99, 102
  willei, 99, 102
Stigeoclonium, 39, 44, 314, 320–321, 334, 337, 808, 810
  farctum, 338
  lubricum, 325
  tenue, 44
Stigonema, 35, 40, 119, 125, 181, 183–184
  congestum, 183
  elegans, 183
  hormoides, 184
  informe, 181, 184
  mamillosum, 183–184
  mesentericum, 184
  minutum, 181, 184
  minutum var. *saxicola*, 184
  mirabile, 184

Stigonema (continued)
  ocellatum, 184
  panniforme, 184
  thermale, 184
  turfaceum, 184
Stigonematales, 117–120, 123, 181, 184, 187, 189
Stipitococcaceae, 430
Stipitococcus, 430, 433, 435, 441–442
  apiculata, 442
  crassistipitatus, 442
  vas, 442
  vasiformis, 442
Stipitoporos, 430, 435
  polychloris, 442
Stomatochroon, 319, 343
  lagerheimii, 328
Storeatula, 734, 736, 742–744, 749
  rhinosa, 722, 724–725, 746
Streptophyta, 229
Stromatochroon, 315
Strombomonas, 2, 388, 403, 410, 413–414, 416
  acuminata, 400
  conspersa, 414
  costata, 414
  deflandrei, 400
  fluviatilis, 400
  giardinana, 400
  gibberosa, 400
  lackeyi, 400
  longicauda, 400
  ovalis, 400, 414
  rotunda, 400
  schauinslandii, 400
  taiwanensis var. *bigeonii*, 413–414
  tambowika, 400
  urceolata, 397, 400
  verricosa, 400
  verricosa var. *zmiewika*, 400
  volgensis, 400
Stylobryon, 491, 502
  abbotti, 472, 477, 502
Stylochrysalis, 492, 502
  aurea, 478
  parasitica, 502
Stylococcus, 502
Stylodinium, 690–691, 696–698, 701, 703, 709
  globosum, 688, 705
  longipes, 702
Stylonema, 208
Stylosphaeridium, 260, 303
  stipitatum, 304
Surirella, 39, 670, 672, 680
  angustata, 678
    linearis var. *constricta*, 678
  ovata, 678
  tenera, 678
Surirellaceae, 670, 681–682
Surirellales, 669–672, 675, 681–682
Symbiodinium, 686–687
Symploca, 119, 123, 142, 144
  borealis, 144
  cartilaginea, 144
  cavernarum, 144
  ciliata, 144
  dubia, 144
  hydnoides, 142
  kieneri, 144
  muralis, 144
  muscorum, 142, 144
  nemecii, 144
  thermalis, 144
Symplocastrum, 119, 123, 137, 148, 150
Syncrypta, 493
  volvox, 479
Synechococcaceae, 62, 68
Synechococcus, 20, 22, 24, 41, 63, 65–66, 69, 73, 76–77, 79, 81–82
  aeruginosus, 77, 80
  bigranulatus, 82
  cedrorum, 79
  koidzumii, 82
  lividus, 66, 76, 82
  maior, 77
  minervae, 79
  nidulans, 76, 81, 88
  sigmoideus, 82
  vulcanus, 66, 82
Synechocystis, 20, 65, 69, 78, 88–89
  aquatilis, 78, 88
  fuscopigmentosa, 88
  minuscula, 88
  primigenia, 89
  salina, 88
  thermalis, 78, 88
  willei, 78, 89
Synedra, 22, 595–596, 600, 606–608, 613, 616, 620, 631
  ostenfeldii, 621
  parasitica, 620, 672
  radians, 22
  rumpens var. *fusa*, 621
  ulna, 25, 36, 621
  ulna var. *aequalis*, 621
  ulna var. *danica*, 621
Synedra-Fragilaria, 25
Synedrella, 631
Synochromonas, 502
Synura, 7, 471, 502, 523–526, 528–531, 533–536, 538, 541, 546–548, 551–552
  australiensis, 551
  curtispina, 538, 546, 551
  echinulata, 528, 529, 536, 538, 546, 551
  echinulata f. *leptorrhabda*, 551
  echinulata f. *leptorrhabda*, 546
  lapponica, 524, 530, 533–534, 538, 541, 546, 548, 551
  mammillosa, 551
  mollispina, 546, 551
  petersenii, 525, 527–528, 534–535, 547, 551
  petersenii var. *praefracta*, 529
  sphagnicola, 528, 530, 534, 536, 538, 541, 547, 551
  spinosa, 524–525, 528, 537, 547, 551
  spinosa f. *longispina*, 538
  splendida, 534
  uvella, 22, 528, 535, 547, 551
Synuraceae, 523
Synurophyceae, 7, 471, 474, 494, 513, 523–526, 528–530, 533–537, 541, 552
Synuropsis, 493, 502–503
  gracilis, 479

Tabellaria, 3, 22, 24, 39, 595–597, 601, 606, 608, 621, 623
  binalis, 617
  fenestrata, 20, 622
  flocculosa, 21, 621, 622
  quadriseptata, 622
Tabellariaceae, 596, 598, 616
Tabellariales, 598, 616
Teilingia, 366, 378
  granulata, 359
Teleaulax, 734, 742–743
Termemorus
  laevis, 357
Terpsinoë, 563, 571
  musica, 575, 586
Tessellaria, 524, 526, 530, 533, 541, 548, 551, 552
  volvocina, 527, 534, 548
Tetmemorus, 367–368, 378
Tetrabaena, 244, 248
  socialis, 227, 238, 244–245
Tetrabaenaceae, 225, 228, 239, 244–246
Tetracanthium, 368
Tetrachrysis, 473, 494, 496, 500, 502–503
  dendroides, 476, 484, 502
Tetracyclus, 596, 601, 606, 617, 623
  glans, 622
Tetracystis, 254, 258, 264, 303
  texensis, 304
Tetradesmus, 262, 303
  wisconsinensis, 304
Tetradinium, 690, 691, 697–698, 703, 709
  javanicum, 688, 705
  simplex, 702
Tetraedriella, 430, 438, 449–450
  enorme, 465
  hastatum, 465
  limneticum, 465
  regularis, 450, 465
  spinigera, 450
  trigonum, 465
Tetraedron, 254, 262, 303, 465
  regulare, 465
  victoriae, 304
Tetragoniella
  gigas, 449
  regularis, 449
Tetraktis, 431, 437, 453
  aktinastroides, 452
Tetrallantos, 262, 303
  lagerheimii, 304
Tetramastix, 825
Tetrapion, 503
Tetraplektron, 430, 433, 439, 449–450
  torsum, 450
  tribulus, 450

Tetraselmis, 228, 230–231, 248
   cordiformis, 229, 231
   subcordiformis, 231
Tetraspora, 255, 259, 303, 322, 498
   gelatinosa, 304
Tetrasporopsis, 473, 482, 494, 498, 503
   fuscescens, 482
   perforata, 482
Tetrastrum, 261, 303
   heterocanthum, 305
Tetratomococcus, 298
Tetreutreptia, 385, 388, 408
   pomquetensis, 401
Thalassiocyclus, 563, 572, 586
   lucens, 579, 586
Thalassiosira, 563, 565, 567, 571, 575, 584–587
   pseudonana, 43
   weissflogii, 579
Thalassiosiraceae, 563
Thalassiosirophycidae, 563
Thalpophila, 125
   cossyrensis, 189
   imperialis, 189–190
Thamniochaete, 320, 337–338
   huberi, 324
Thecadiniaceae, 697, 708
Thompsodinium, 697–698, 704, 707
   intermedium, 689, 692, 702
Thoracomonas, 236, 238, 248
   feldmanii, 238
   phacotoides, 237–238
Thorea, 207, 218–219
   hispida, 199, 219
   violacea, 33, 199, 218–219
Thoreales, 199–200, 203, 216–217, 219
Tilapia, 820
Tolypella, 27, 317, 339, 345
   nidifica, 331
Tolypothrix, 26, 34, 119, 124, 161–162, 184
   amoena, 162
   bouteillei, 161
   distorta, 25, 161
   lanata, 161
   papyracea, 162
   penicillata, 161–62
   robusta, 162
   setchellii, 161
   tenuis, 161–162
   tenuis f. *minor*, 161
   willei, 162
Tolypotrichoideae, 158
Tomaculum, 264, 306
   catenatum, 305
Torquatae (section of *Mallomonas*), 550
Tortitaenia, 371, 379
   obscura, 371
Torytaemia, 367
Trachelomonas, 23, 384, 388, 403–404, 406, 410, 414–416, 806
   abrupta, 398
   acanthostoma, 398
   aculeata f. *brevispinosa*, 399
   armata, 398
   armata f. *inevoluta*, 398
   armata var. *longispina*, 398
   armata var. *steinii*, 398
   bulla, 399
   charkowiensis, 399
   cylindrica, 397–398
   dubia, 397–398
   dybowskii, 398
   erecta, 397
   girardiana, 399
   grandis, 397, 407
   hexangulata, 399
   hispida, 398, 406
   hispida var. *coronata*, 398
   hispida var. *cremulatocollis* f. *recta*, 398
   hispida var. *papillata*, 398
   hispida var. *punctata*, 398
   horrida, 398
   intermedia, 397
   kelloggii, 397
   lacustris, 397
   lefevrei, 406
   mammillosa, 399
   playfairii, 399
   pulcherrima, 398, 407
   pulcherrima var. *minor*, 398
   robusta, 398
   rotunda, 397
   scabra var. *longicollis*, 399
   similis, 399
   speciosa, 399
   spectabilis, 399
   spirillifera, 415
   superba, 398
   superba var. *duplex*, 398
   superba var. *spinosa*, 398
   superba var. *swirenkiana*, 398
   sydneyensis, 398
   triangularis, 397
   volvocina, 397, 407
   volvocina var. *compressa*, 397
   volvocina var. *punctata*, 397
Trachychloron, 430, 438, 449–450
   depauperatum, 450
   fusiforme, 450
Trachycystis, 430, 438, 447, 451
   subsolitaria, 450–451
Trachydiscus, 430, 439, 450–451
   ellipsoideus, 450
   lenticularis, 450
   sexangulatus, 450
Trapa
   natans, 24
Trebouxia, 254, 258, 266, 306
   parmeliae, 305
Trebouxiophyceae, 313
Trentepohlia, 45, 315, 320, 343–344
   aurea, 328
Trentepohliales, 312–313, 342–344
Trentepohliophyceae, 309
Treubaria, 261, 306
   triappendiculata, 305
Tribonema, 44, 341, 429, 431–432, 433–434, 461–463
   aequale, 462
   affine, 461
   regulare, 462
   viride, 461–462
Tribonemataceae, 431
Tribonematales, 429, 431, 433–434, 459–460, 462
Tribophyceae, 270, 423–424, 427, 429–430, 432, 441, 447, 465, 473
Triceratiaceae, 563
Trichocoleus, 122, 137–139, 150
   acutissimus, 138–139
   erectiusculus, 138
   minor, 139
   purpureus, 138
   sociatus, 139
Trichodesmium, 64, 119, 123, 143–144, 147
   iwanoffianum, 143
   lacustre, 143, 147
Trichodiscus, 338
   elegans, 324, 338
Trichophilus, 319, 338
   welcheri, 315, 323
Trichoptera, 35
Trichormus, 65, 120, 125, 171, 176, 180
   anomalus, 180
   doliolum, 180
   fertilissimus, 176, 180
   luteus, 176
   subtropicus, 176
   variabilis, 176, 180
Trichosarcina, 315
   polymorphum, 345
Triploceras, 362–363, 365, 367, 378
   gracile, 357
Triticum, 822
Trochiscia, 262, 306
   hystrix, 305
Tryblionella, 670, 672, 675, 677, 680–681
Tuomeya, 203, 205, 208, 216
   americana, 199, 204–205, 214
Turfosa (section of *Batrachospermum*), 215
Tychonema, 122, 146–147
   bornetii, 147
   bourrellyi, 146–147
Typha, 17, 332, 427–428, 443, 458–459, 461, 764

Ulnaria, 631
Ulothrix, 34, 313, 318, 345, 347, 370, 808, 810
   zonata, 25–26, 35, 314, 326
Ulotrichaceae, 348
Ulotrichales, 40, 313, 317, 337, 340, 344–347
Ulva, 253, 312
Ulvales, 311, 313, 317, 336–337, 347
Ulvella, 336
Ulvophyceae, 313, 384
Uroglena, 23, 485, 487–488, 493, 502–503
   americana, 489
   volvox, 488
Uroglenopsis, 473, 476, 485, 487–488, 491, 493, 499, 502–503
   articulatus, 489

Uronema, 313, 318, 344, 347
  elongatum, 326
Urosolenia, 563, 571–572, 575, 587
Utricularia, 363

Vacuolaria, 427–428
  virescens, 426
Vallisneria, 17
Vampyrella, 696, 708
Vaucheria, 4, 11, 322, 341, 423, 429, 431–434, 463–465
  aversa, 464
  bursata, 433, 464
  dillwynii, 433, 464
  fontinalis, 433, 464
  geminata, 433, 464
Vaucheriaceae, 431
Vaucheriales, 429, 431, 433–434, 463–464
Virescentia (section of *Batrachospermum*), 215
Vischeria, 424–425, 427
  helvetica, 427, 447
  punctata, 427
  stellata, 426–427
Vitreochlamys, 232, 248
  fluviatilis, 233
Volvocaceae, 225–228, 233, 239–240, 242–243, 246
Volvocales, 37, 225–227, 235–236, 245
Volvochrysis, 502
Volvox, 225–227, 239, 241–242, 248, 312
  aureus, 240, 242
  carteri, 242
  carteri f. *weismannia*, 242
  dissipatrix, 242
  globator, 242
  perglobator, 242
  powersii, 242

  prolificus, 242
  rousseletii, 242
  spermatosphaera, 242
  tertius, 242
  vegetans, 471, 495
  weismannia, 242
Volvulina, 239, 242, 248
  pringsheimii, 242
  steinii, 240–241, 243
Vorticella, 171

Westella, 264, 306
  botryoides, 305
Wislouchiella, 236, 238, 248
  planctonica, 237–238
Wolfia, 627
Wollea, 125, 180–81
  bharadwajae, 181
  saccata, 173, 181
Woloszynskia, 691, 697–698, 703, 705, 707
  apiculata, 694
  cestocoetes, 702
  pseudopalustre, 694
  reticulata, 689, 695, 699, 702, 704
Woronichinia, 60, 61, 64, 68, 70, 85, 87, 89
  elorantae, 89
  fremyi, 89
  karelica, 89
  klingae, 67, 85, 89
  naegeliana, 67, 85, 89

Xanthidium, 363, 368, 376, 378–379
  controversum, 379
  hastiferum, 360
Xanthophyceae, 429, 440, 473
Xanthophyta, 24

Xenococcaceae, 60, 71, 104
Xenococcus, 65, 72, 106–107
  angulatus, 107
  bicudoi, 106–107
  candelariae, 107
  chaetomorphae, 107
  cladophorae, 107
  deformans, 107
  gilkeyae, 107
  lamellosus, 107
  pallidus, 107
  pyriformis, 107
  schousboei, 107
  willei, 106–107
  yellowstonensis, 107
Xenotholos, 72, 106–107
  huastecanus, 106
  kerneri, 106–107

Yamagishiella, 239, 241, 242, 248
  unicocca, 240, 242

Zoochlorella, 271
Zygnema, 3, 24, 33, 36, 39, 354, 363, 366, 371–372, 708
  conspicuum, 355
  frigidum, 355
Zygnemaphyceae, 353
Zygnemaphyta, 353
Zygnematales, 12, 19, 39–40, 311, 313, 353–354, 362–365, 369–372, 379
Zygnematophyceae, 313, 353
Zygnemopsis, 366, 371–372
  decussata, 355
Zygogonium, 39, 45, 363, 366, 372, 497
  ericetorum, 355

QK 570.5 .F74 2003

Freshwater algae of North America